McGraw-Hill
Encyclopedia of Environmental Science

McGraw-Hill

Encyclopedia of Environmental Science

McGRAW-HILL BOOK COMPANY

NEW YORK
SAN FRANCISCO
ST. LOUIS

DUSSELDORF
JOHANNESBURG
KUALA LUMPUR
LONDON
MEXICO
MONTREAL
NEW DELHI

PANAMA
PARIS
SAO PAULO
SINGAPORE
SYDNEY
TOKYO
TORONTO

Library of Congress Cataloging in Publication Data

McGraw-Hill encyclopedia of environmental science.

   Includes bibliographies.
   1. Ecology—Dictionaries. 2. Man—Influence on nature—Dictionaries. 3. Environmental protection—Dictionaries. I. Title: Encyclopedia of environmental science.
QH540.4.M3   301.31'03   74-13065
ISBN 0-07-045260-1

EDITORIAL STAFF

Consulting Editors

Preface

The ecosystem has existed from the time that life in its simplest form appeared on Earth. As evolution occurred and new species emerged, each species found its place in the ecosystem enlarging or shrinking, even to the point of extinction, as the ecosystem changed. The human race evolved, and it too found its niche in the ecosystem, increasing in number when food, climate, and other environmental factors were favorable and decreasing when conditions were unfavorable.

As civilization developed, human beings changed from participants to exploiters and masters of nature. Over the centuries humans acquired knowledge of science and technology. The knowledge accumulated at a slow pace, then in the middle of the 19th century grew exponentially. As the human race spread over the surface of the Earth in increasing numbers, their growing technology enabled them to exploit the environment at a faster and faster rate for the production of goods. As a result, natural resources, especially those which are nonrenewable, such as fossil fuels and metallic ores, but also renewable ones such as arable lands, forests, and ocean fisheries, have been consumed at an ever-increasing rate, with the prospect that the essential nonrenewable natural resources will be exhausted in the not too distant future.

The effects on the environment of the increasing number of humans and their personal and industrial activities have at times been disastrous. Some of the damage to the environment can be seen in the streams polluted with industrial waste and sewage, in air polluted with noxious gases, in algae-choked lakes, in the oceans that are losing their productivity from overfishing or from pollution by dumped sewage sludge or from oil spills.

There has been a growing concern by all segments of society about the effects upon the environment of the human race's activities. The *McGraw-Hill Encyclopedia of Environmental Science* focuses on the study of the external conditions that affect and influence life. In readable, authoritative articles, it covers the present state of knowledge of meteorology and climatology, conservation, ecology, oceanography, soil science, pollution, sewage treatment, nuclear engineering, agriculture, mining, and petroleum engineering. The encyclopedia is written for the nonspecialist, but of course every specialist is a nonspecialist in a field other than his or her own.

The encyclopedia contains 500,000 words in the more than 300 alphabetically arranged articles. The articles, some drawn from the *McGraw-Hill Encyclopedia of Science and Technology* (3d edition, 1971) and its Yearbooks (1971–1974), and some written especially for this volume, were selected or suggested by a board of consultants. All articles are signed, and the authors and their affiliations are provided in the List of Contributors beginning on page 705. Almost every article opens with a definition of the subject and ends with a bibliography for further reading. The cross-references to other articles and the analytic index beginning on page 713 interrelate the articles.

DANIEL N. LAPEDES
Editor in Chief

McGraw-Hill

Encyclopedia of Environmental Science

Aerobiology

Zooplankton

Aerobiology

The word aerobiology came into use in the 1930s as a collective term for studies of the "air spora" — airborne fungus spores, pollen grains, and microorganisms. Since the start of the International Biological Program (IBP) in 1964, the term has been extended to include investigations of airborne materials of biological significance. In addition to studies of air spora dispersed by the atmosphere, as before, aerobiology now embraces algae and protozoans, minute insects such as aphids, and in some instances even pollution gases and particles that exert specific biological effects. Thus the discipline of aerobiology is held together by common principles of atmospheric dispersion, some common biological qualities ("viability," "source strength"), and adaptability of these atmospheric transport events to treatment as similar ecological systems.

The 10-year term of the IBP affords a wholly new set of opportunities for multinational and intercontinental studies that had been hampered previously by lack of direct communication between scientists and by absence of cooperation between governments. Aerobiology is emerging as a distinct interdisciplinary field, with new capabilities, such as calculating the probabilities for spread of airborne plant or animal diseases in various directions, or determining the influences of the environment on allergenic particles in the air, or tracing aerial pathways of minute organisms for various scientific and practical purposes.

Long-distance aerial transport. An example of the intercontinental scope of aerobiology is research on long-range dispersal of minute organisms and spores. The northward spread of black stem rust of wheat every year from Texas and Mexico to the northern states and Canada was discovered by E. C. Stakman and colleagues in the 1920s. Similar great seasonal wind-driven infections of grain crops occur also in the plains of eastern Europe, the Soviet Union, and India. Recently much new evidence of such long-range dispersals has come to light. U. Hafsten in 1960 reported finding pollen grains in sediments on Tristan da Cunha and Gough Island — grains that could have been transported only from continental areas to those isolated tiny islands in the South Atlantic Ocean. In 1964 L. Maher proved that pollen of the desert shrub *Ephedra*, whose growth is restricted in North America to the Great Basin and Southwest deserts, is blown as far east as the central Great Lakes region. Still more recently N. Moar identified pollen grains of the Australian beefwood tree, *Casuarina*, in snow samples at 2100 m altitude on the Tasman Glacier in the Mount Cook region of the South Island of New Zealand and from two other localities which also could receive wind drift only from the west. At one of these localities the pollen was associated with fine brick-red dust typical of Australian dust storms.

The improved quality and availability of meteorological data are making it possible to reconstruct the atmospheric trajectories that bring exotic biological matter to an area. A new virus disease

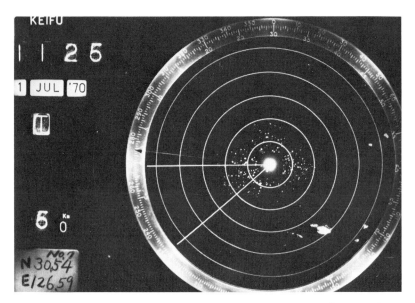

Fig. 1.   Radar screen indicates swarms of plant hoppers as bright "angels" above the sea south of Kyushu, Japan, July 1, 1970. (*Courtesy of Eiichi Inoue, National Institute of Agricultural Sciences, Tokyo*)

drop splash to dislodge and launch the spores. it is true: but once airborne. the spores possibly can be transported long distances in viable condition. Here. as in most questions of this kind. knowledge of air-transport processes and of temperature. humidity. and solar irradiation of the air mass is necessary. Such information is being supplied by cooperating agricultural meteorologists and biometeorologists.

Many organisms get rides on dust particles. For example. H. E. Schlichting. Jr.. showed that the number of algae and protozoans in the air over north-central Texas usually ranges from 0 to 283 organisms per cubic meter. whereas during a dust storm the air contained 3000 organisms per cubic meter. Dusts from dry grasslands and deserts are capable of carrying very significant numbers of minute biotas. but the actual organisms have not yet been assessed. *See* PLANT DISEASE.

Predicting disease outbreaks. Aerial transport of disease spores and microorganisms and of insects is becoming predictable within certain limits. Studies of airborne fungus diseases of crop plants led the way. In Minnesota spore trapping has been used for 20 years for forecasting the advent of wheat stem rust in the northern Plains states. In South Africa spore trapping is an aid in predicting outbreaks of citrus black spot. The blister blight of tea. caused by the fungus *Exobasidium vexans*. has been known in Assam since 1868. but it suddenly appeared in southern India and Ceylon in 1949. The discovery and use of copper fungicides managed to check the blister blight. and now spore trapping is used to warn against attacks and to signal spraying of the fungicides.

Japanese entomologists. phytopathologists. and meteorologists are engaged in vigorous research on insect pests of crops. especially the sucking insects which transmit diseases from one plant to another and from one field to another. These scientists are amassing considerable evidence in support of the view that pheromones. volatile substances given off by organisms to the air. are used to guide these insects to the crop plants. Although these insects are weak fliers and tiny. not more than about 2 mm long. and must depend upon wind for travel over any appreciable distance. they can in some manner control where they fall out of the wind stream. Plant hoppers blown from mainland China across the East China Sea in early summer bring rice crop diseases that then spread northward from Kyushu to the other islands of the Japanese Archipelago. Rainstorms over the sea do not entirely stop these migrations—plant hoppers can become airborne from the sea surface. The economic importance of this insect vector of rice diseases had led Eiichi Inoue and a team of other meteorologists to plan "Operation Plant Hopper" in 1973 as part of a synoptic study of the changing characteristics of continental air masses from the Asiatic mainland as they come over the East China Sea. The meteorologists will use shipborne and aircraft radar to locate and track swarms of plant hoppers (Fig. 1). They will trap the insects from shipboard. with kite-suspended nets. and from aircraft. with towed nets (Fig. 2.).

The Anti-Locust Control Center in London is an example of effective union between basic research and immediate application of findings. For several

in grain crops on the Canterbury Plain of the South Island of New Zealand was identified within several days of its first appearance and related to a species of aphid not previously known from New Zealand but common to southeastern Australia. Subsequently R. C. Close worked with government meteorologists in Wellington on reconstructing air-mass trajectories in the several days preceding the disease outbreak. They found that almost precisely in the time interval suggested by the biological facts. winds over the Tasman Sea were such that the insects (which have a life-span of 4 days) could have been transported from Australia and around the south end of South Island in favorable temperatures at 700–1000 m altitude in slightly more than 2 days.

Rothamsted Experimental Station in England has been the site of many fundamental investigations of fungus spore production and dispersal in the atmosphere. especially through the work of P. H. Gregory and J. M. Hirst. In recent years. scientists there have examined the diurnal surges of spore release by different species of saprophytic and parasitic fungi associated with crops and grassland. Some of the daily surges of identifiable spores were traced downwind. eastward. onto the Continent by sampling from airplanes. Although they gradually disintegrate. these spore clouds could still be detected at 1000 m altitude after crossing the North Sea provided they had not been washed out by rain.

Similar investigations may soon yield an answer to a question that has spawned much controversy among phytopathologists and others concerned with air spora: how the African coffee rust reached Brazil several years ago. Some believe that long-range intercontinental aerial transport brought the disease to South America. Others deny this and point to a number of other possible avenues of introduction (on bootlegged breeding plants. on packing material. or by many other conceivable ways). The local spread is dependent upon rain-

decades this center has directed a program of research and forecasting by field workers in locust-plagued areas across Africa between the Sahara and the equatorial forest and through the Middle East into southwestern Asia. Every workday in the London laboratories the staff gathers to review situation reports and to draft forecasts, much as an army general staff would gather at headquarters for intelligence briefings and battle directives. This organization has saved untold tonnages of crops, and thereby many human lives, through knowledge of the way locusts behave in response to different weather and crop conditions. Now the scientists are employing radar and other electronic remote-sensing devices to track swarms and to monitor positions of weather fronts, which strongly influence locust swarm movements. The center's methods will undoubtedly be emulated in approaches to problems of other diseases and pests.

Aeroallergens. Pollinosis (hay fever) affects an estimated 10% of the population of the United States, and it is probably responsible for more discomfort and disabling illness than the direct effects of all forms of air pollution. The pollen of ragweed (*Ambrosia*) is a leading cause of allergic responses. The biology and ecology of ragweed are relatively well known, so that it is now possible to eliminate the plants by land management practices. New assessments of weather and land-use conditions that affect atmospheric concentrations of aeroallergens are in progress. Such surveys, notably those by E. C. Ogden, G. Raynor, and J. V. Hayes in New York State and A. M. Solomon in New Jersey and Arizona, will help victims of pollinosis to plan for avoidance or protection.

Certain green algae borne by the air have been found by I. L. Bernstein and R. S. Safferman to cause both skin sensitivity and respiratory allergy. They found viable algae, notably the genera *Chlorella* and *Chlorococcum*, in many samples of house dust from Los Angeles, Milwaukee, and Caracas. *Chlorella* has been found to be common in trickling filter sewage disposal systems in Ohio, and H. E. Schlichting in 1969 reported collecting varying concentrations of microorganisms in the atmosphere near such systems. Various workers have recently demonstrated that bursting bubbles at water surfaces are important means of launching microorganisms from water into the atmosphere.

Although much has been learned about the air spora of atmospheric masses, little has been done in studying the air of indoor environments. (Exceptions are microbiological studies of hospitals and other health-care spaces such as those by R. L. Dimmick and others at the Naval Biological Laboratory in California.) House dusts have been found to contain plant and animal debris, pollen, spores, and algae. Some air-conditioning and humidifying systems of buildings harbor sources of dust and mold spores which may be causes of allergic reactions in some people. Since indoor environments reflect many of the air-pollution characteristics of the ambient outdoor air, there are possibilities of synergistic, or reinforcing, effects of air pollutants with indoor allergens. Although investigations of the great variety of indoor environments have only begun, these studies promise to bring considerable benefits in human comfort and health.

New approaches. The Aerobiology Theme of the IBP, and its associated national activities such as the United States Aerobiology Program, has helped to promulgate new approaches and methods in aerobiology. Better sampling equipment and reporting techniques are being developed so that disease outbreaks or risk of pollinosis can be predicted earlier and with more certainty. Japan, for example, has an elaborate system of stations for observing crop disease. The stations are connected by efficient communication links to government evaluation offices and to radio and newspaper outlets for advising farmers. However, detection and assessment of most human- and animal-disease microbes collected in air samples are far more difficult. These microbes cannot be identified by simple microscopic examination but must be cultured in special media and often under strict conditions of temperature and moisture. It is likely, therefore, that the aerobiology of airborne diseases of humans will be largely confined to indoor environments for several more years.

The ecological systems approach is being used extensively in the IBP efforts. This approach views a process, in this case some aspect of aerial transport, as a machine in which various parts function together, with certain feedback controls that regulate the speed of internal reactions and the kind and magnitude of responses to the external environment. This approach has been aided by computer programs that can predict atmospheric trajectories and changes in state, such as precipitation. Paul E. Waggoner of the Connecticut Agricultural Experiment Station brought out in 1968 a mathematical simulation model of a generalized plant disease of the kind that reaches epidemic proportions on row crops. He called this computer model EPIDEM and demonstrated how it could be used to predict the progress of a disease if appropriate values are entered for source strength of inoculum, meteorological variables such as wind speed and direction, and age and condition of the exposed crop. The excellent promise of simulation modeling, especially for prediction, has convinced many biologists that it will be a valuable management tool in the future.

The systems approach has led scientists in the

Fig. 2. Japanese aircraft tows a net to catch airborne plant hoppers over the sea south of Kyushu. (*Courtesy of Eiichi Inoue, National Institute of Agricultural Sciences, Tokyo*)

Fig. 3. General view of the Mauna Loa Observatory of the National Oceanic and Atmospheric Administration (NOAA), a unique location for making observations of the chemistry and particulate matter of the atmosphere. (*Environmental Research Laboratories of NOAA*)

United States Aerobiology Program into attempts to improve methods for estimating crop losses. A group led by L. Calpouzos is endeavoring to develop more reliable methods for estimating losses by taking into account different environments and geographical areas. The United Nations Food and Agriculture Organization is concerned with world reports on crop losses but is in serious need of improved information. Certainly, as human consumption in various parts of the world approaches the limits of crop yields, accurate prediction of crop losses from disease will become increasingly important.

Global and regional monitoring. Great interest has developed in the 1970s in monitoring and warning systems for biological and environmental changes. Local and regional monitoring of the atmosphere for gaseous pollutants from industry and internal combustion engines, for example, is expanding at remarkable rates in Europe and North America. The impetus for this monitoring comes from such startling evidence as dying conifer forests on mountainsides in coastal California, friezes dissolving on centuries-old buildings under acidified rains, and eye-stinging smog over urban areas. More subtle, but nonetheless significant, effects of gaseous pollution are being discovered, such as the killing of some airborne fungal spores by sulfur dioxide at normal pollution levels in humid air. These effects raise serious questions. What does pollution do to the airborne pollen necessary for seed set in flowering plants? Where do the pollutants come to rest? How much can the world system stand before some irreversible change occurs?

International organizations, with the International Council of Scientific Unions (ICSU) taking the scientific lead and United Nations agencies doing much of the work, are moving toward regional and global networks of stations to monitor such things as the ozone and carbon dioxide content of the atmosphere, atmospheric turbidity (an index to the

concentration of microscopic particles and droplets in the air), dissolved salts in the seas, and the amount of green plant material grown on a given land area in a year. A few of these stations are already in operation. One of these is the meteorological observatory on Mauna Loa, Hawaii, where measurements of ozone, carbon dioxide, turbidity, and other factors have been in progress longer than at any other station (Fig. 3). At Mauna Loa, tests are being made of a suction-type spore trap and an insect trap for sampling biological materials in the atmosphere. Scientists of the IBP and the United States Aerobiology Program will set out simple sedimentation-type traps at many stations around the world to gather base-line information on the air spora. They will install suction traps and more sophisticated samplers, operated for daily or hourly records, at selected stations. Testing for viability of spores and microbes may be feasible within a few years, but routine culturing by microbiological techniques may be many years away.

Meanwhile, since the data of aerobiology will be accumulating at a much more rapid rate, provisions are being made for keeping these data in order and available for analysis. Aerobiologists are beginning to use the Environmental Protection Agency's Storage and Retrieval of Aerometric Data (SAROAD) system and the National Aerometric Data Bank. Both of these are designed to make data more readily available for surveys and comparative analyses.

These new data on dispersal of minute organisms, spores, and pollen grains in the atmosphere, over both short- and long-range trajectories, will serve many useful purposes. Biogeographers, for example, will be able to improve the probability calculations for gene flow in populations or to assess quantitatively the migration ability of species in which aerial transport is a part of the life cycle. Through such knowledge, aerobiological monitoring may become an important tool in the protection and management of endangered genetic stocks, as well as in the protection of special biotic communities against invading species. *See* SYSTEMS ECOLOGY.

[WILLIAM S. BENNINGHOFF]

Bibliography: W. S. Benninghoff and R. L. Edmonds (eds.), *Ecological Systems Approaches to Aerobiology, I: Identification of Component Elements and Their Functional Relationships*, U.S. Aerobiol. Program Handb. no. 2, 1972; W. S. Benninghoff and R. L. Edmonds (eds.), *Proceedings of a Conference on Aerobiology Objectives in Atmospheric Monitoring*, U.S. Aerobiol. Program Handb. no. 1, 1971; Commission on Monitoring of the Scientific Committee on Problems of the Environment (SCOPE) of the International Council of Scientific Unions (ICSU), *Global Environmental Monitoring: A Report Submitted to the United Nations Conference on the Human Environment*, SCOPE 1, 1971; R. L. Dimmick and A. B. Akers (eds.), *An Introduction to Experimental Aerobiology*, 1969; R. B. Felch and G. L. Barger, *U.S. Weather Bur. Weekly Weather Crop Bull.*, 58(43):13–17, 1971; R. W. Romig and L. Calpouzos, *Phytopathology*, 60:1801–1805, 1970; I. H. Silver (ed.), *Aerobiology: Proceedings of the 3d International Symposium Held at the University of Sussex, England, September 1969*, 1970; P. E. Waggoner and J. G. Horsfall, *Conn. Agr. Exp. Sta. Bull.*, no. 698, 1969.

Agricultural chemistry

The science of chemical compositions and changes involved in the production, protection, and use of crops and livestock. As a basic science, it embraces in addition to test-tube chemistry all the life processes through which man obtains food and fiber for himself and feed for his animals. As an applied science or technology, it is directed toward control of those processes to increase yields, improve quality, and reduce costs. One important branch of it, chemurgy, is concerned chiefly with utilization of agricultural products as chemical raw materials.

Scope of field. The goals of agricultural chemistry are to expand man's understanding of the causes and effects of biochemical reactions related to plant and animal growth, to reveal opportunities for controlling those reactions, and to develop chemical products that will provide the desired assistance or control. So rapid has progress been that chemicalization of agriculture has come to be regarded as a 20th-century revolution. Augmenting the benefits of mechanization (a revolution begun at mid-19th century), the chemical revolution has advanced farming much further in its transition from art to science.

Every scientific discipline that contributes to agricultural progress depends in some way on chemistry. Hence agricultural chemistry is not a distinct discipline, but a common thread that ties together genetics, physiology, microbiology, entomology, and numerous other sciences that impinge on agriculture. Chemical techniques help the geneticist to evolve hardier and more productive plant and animal strains; they enable the plant physiologist and animal nutritionist to determine the kinds and amounts of nutrients needed for optimum growth; they permit the soil scientist to determine a soil's ability to provide essential nutrients for the support of crops or livestock, and to prescribe chemical amendments where deficiencies exist. *See* FERTILIZER; SOIL.

Chemical materials developed to assist in the production of food, feed, and fiber include scores of herbicides, insecticides, fungicides, and other pesticides, plant growth regulators, fertilizers, and animal feed supplements. Chief among these groups from the commercial point of view are manufactured fertilizers, synthetic pesticides (including herbicides), and supplements for feeds. The latter include both nutritional supplements (for example, minerals) and medicinal compounds for the prevention or control of disease.

To supply chemicals for agriculture, sizable industries have developed. Total sales of manufactured fertilizers in the United States topped $1,000,000,000 in 1952; they have continued to climb, and passed the $2,000,000,000 mark in 1967–1968. For their expenditure of $2,161,000,000 for fertilizer and lime in 1967, farmers received 38,300,000 tons of fertilizer, analyzing on the average as 38.1% primary nutrients (nitrogen; phosphorus, calculated as phosphorus pentoxide; and potassium, calculated as potassium oxide).

Sales of pesticides (including herbicides), which totaled about $285,000,000 at the manufacturer's level in 1960, had more than doubled to $584,-000,000 by 1966 and approximated $950,000,000 in 1968. *See* HERBICIDE.

Important chemicals. Chemical supplements for animal feeds may be added in amounts as small as a few grams or less per ton of feed, but the tremendous tonnage of processed feeds sold, coupled with the high unit value of some of the chemical supplements, makes this a large market. Precise statistics are not available, but total sales of animal health and nutrition chemicals in the United States were reliably estimated at $340,000,000 for 1967. Such sales have been growing at about 6% per year.

Of increasing importance since their commercial introduction have been chemical regulators of plant growth. Besides herbicides (some of which kill plants through overstimulation rather than direct chemical necrosis), the plant growth regulators include chemicals used to thin fruit blossoms, to assist in fruit set, to defoliate plants as an aid to mechanical harvesting, to speed root development on plant cuttings, and to prevent unwanted growth, such as sprouting of potatoes in storage. *See* DEFOLIANT AND DESICCANT.

Striking effects on growth have been observed in plants treated with gibberellins. These compounds, virtually ignored for two decades after they were first isolated from diseased rice in Japan, attracted widespread research attention in the United States in 1956; first significant commercial use began in 1958. The gibberellins are produced commercially by fermentation in a process similar to that used to manufacture penicillin.

In the perennial battle with insect pests, chemicals that attract or repel insects are increasingly important weapons. Attractants (usually associated with the insect's sexual drive) may be used along with insecticides, attracting pests to poisoned bait to improve pesticidal effectiveness. Often highly specific, they are also useful in insect surveys; they attract specimens to strategically located traps, permitting reliable estimates of the extent and intensity of insect infestations.

Repellents have proved valuable, especially in the dairy industry. Milk production is increased when cows are protected from the annoyance of biting flies. Repellents also show promise as aids to weight gain in meat animals and as deterrents to the spread of insect-borne disease. If sufficiently selective, they may protect desirable insect species (bees, for instance) by repelling them from insecticide-treated orchards or fields.

Agricultural chemistry as a whole is constantly changing. It becomes more effective as the total store of knowledge is expanded. Synthetic chemicals alone, however, are not likely to solve all the problems man faces in satisfying his food and fiber needs. Indeed, many experts are coming to the view that the greatest hope for achieving maximum production and protection of crops and livestock lies in combining the best features of chemical, biological, and cultural approaches to farming. *See* AGRICULTURAL SCIENCE (ANIMAL); AGRICULTURAL SCIENCE (PLANT); AGRICULTURE.

[RODNEY N. HADER]

Agricultural meteorology

The study and application of relationships between meteorology and agriculture. It involves simple concepts such as timing the planting of crops to avoid damage from freezing temperatures, and more complicated problems such as the combined effects of temperature and humidity in producing an outbreak of a disease such as potato blight.

Here, meteorology is used in its broadest sense to include observing, reporting, and forecasting day-to-day variations in weather, as well as the study and use of past climatological data. Agriculture includes all farming, ranching, orchard, nursery, and forestry operations concerned with the production, harvesting, processing, and shipping of foods, fibers, flowers, leather, and lumber. Protection from and control of plant and animal diseases and insect pests are also included.

Major participating agencies. The U.S. Department of Agriculture and the state agricultural colleges and experiment stations are active in research in this field. The U.S. Weather Bureau assists in many of these projects, operates various service programs, and conducts research through cooperative agreements.

International cooperation and exchange of information are carried out through the Commission for Agricultural Meteorology (CAgM) of the World Meteorological Organization (WMO). This commission promotes meteorological development and standardizes methods, procedures, and techniques in the application of meteorology to problems in agriculture.

Microfocus of investigation. Microclimatology is of major importance in the study of agricultural meteorology. The interrelationships of climate and soil and the many important phenomena involved in the interchange of heat and moisture at the air-soil interface are of critical importance to vegetation and animal life in the biosphere. *See* CLIMATOLOGY; MICROMETEOROLOGY.

Water and moisture problems. Water is often the most critical limiting factor in food production. More than 25,000,000 acres of land is irrigated in the United States. Much of this is in arid regions of the West and Southwest but a surprisingly large and increasing amount is in the more humid East.

Moisture is withdrawn from the soil by direct evaporation from the soil surface and by transpiration through the plants. The first process is capable of quickly drying a shallow surface layer of soil, and in hot summer weather the water from brief showers can be removed by evaporation before it can enter the root zone. Transpiration removes moisture from the soil layers penetrated by roots. *See* EVAPOTRANSPIRATION.

The rate of loss by both processes (evapotranspiration) is closely related to the energy available for evaporation. Sunshine and temperature are used as indicators. Wind and humidity are also critical. Evapotranspiration rates of 0.25–0.30 in./day are not uncommon during warm summer seasons. Several methods have been suggested for computation of these rates from meteorological parameters. None has been completely accepted but several give approximations useful in determining irrigation requirements.

Precipitation data for agricultural planning include more information than simple monthly and annual average amounts. The range and the frequency distribution of various amounts about the mean are necessary in planning land use and crop risks. The duration and frequency of drought periods are needed in planning potential irrigation requirements.

Temperature factors. Temperature, both of the air and the soil, is an important factor in agricultural meteorology. The length of growing season and the average temperature during the season often determine the choice of crop species or variety. Sugarcane requires temperatures high enough to permit rapid growth for at least 8 months. Rice requires a warm moist environment with mean temperatures of 70°F or higher for 4–6 months. Cotton requires warm summer months (75–80°F), a long (180–200 days) growing season, and warm sunny days during harvest. Corn is considered to be of tropical origin but varieties have been developed for widely differing climates. However, the principal corn belt lies in a region having warm summer months (70–80°F) and a growing season of 140–160 days. Wheat production is divided into winter-wheat and spring-wheat areas largely on the basis of the severity of the winter.

The frequency and duration of freezing temperatures at certain seasons are of critical importance in many areas. The U.S. Weather Bureau operates localized frost-warning services in winter citrus- and truck-producing areas of Florida and California, in apple and other fruit regions of Washington and Oregon, and in summer in cranberry areas of Wisconsin and Massachusetts.

The temperature range of 60–80°F is optimum for milk production in dairy cattle. At 85°F production is reduced by as much as 25%, and at 95°F it is reduced 50%. Hogs gain little or no weight at 90°F. At 95°F or higher, fattened animals lose weight. When high temperatures are expected, hog shipments are postponed or arrangements are made for artificial cooling en route.

Diverse responses to weather. It is difficult to generalize in agricultural meteorology. Species and varieties of crops and breeds of animals all have their own peculiar responses to weather factors. Effects are often cumulative. Many special problems are isolated and studied separately.

Maleic hydrazide (MH-30) acts as a growth regulator on some plants. Used as a spray, it controls the growth and development of suckers on tobacco, runners on strawberries, branches and shoots on trees, and new growth in grasses. Turgidity of plants and the weather (moisture, temperature, sunshine) during a period following application are important in determining final effectiveness. Much experimentation is being concentrated on this and similar problems to determine optimum spraying weather conditions.

Rapid increases in the use of aircraft for agricultural purposes have brought new meteorological problems. Weather reports and forecasts are needed to determine optimum times of application for control of diseases, insects, and weeds and also to determine conditions of wind, visibility, and temperature for safe, effective operation of aircraft.

Potato blight, a fungus disease caused by *Phytophthora infestans*, occurs in serious to epidemic outbreaks in many areas. It can be controlled by spraying above-ground plant parts with a fungicide, but economical and effective protection depends upon application at the right time. If too early, it wastes the spray and permits new unprotected growth to be exposed to later infections. If too late, the crop may be lost. Timing is based on certain critical combinations of temperatures and moisture which favor spore germination. Special warning services are developed to assist farmers in timing their spraying programs.

In the forests there are many special problems. Forecast services are established to warn of low humidity, high temperatures, and winds which favor the outbreak and rapid spread of fires. Forest insect pests are favored by certain weather conditions. For example, the spruce budworm emerges in largest numbers after a snowy cold winter which is followed by a warming period with rain in the spring. The tent caterpillar is favored by a warm sunny spring followed by warm humid weather.

Principal technical literature. The sources of literature references and reports on recent work in agricultural meteorology in the United States are varied and numerous. Examples are publications of the American Meteorological Society and the U.S. Weather Bureau, *Agronomy Journal, Soil Science Society of America Proceedings, Agricultural Engineering, American Geophysical Union Transactions, Journal of Agricultural Research,* and *Agricultural Meteorology.*

[MILTON L. BLANC; WOODROW C. JACOBS]
Bibliography: *Climate and Man,* USDA Yearb. Agr., 1941; R. Geiger, *The Climate near the Ground,* 1950; J. W. Smith, *Agricultural Meteorology: The Effect of Weather on Crops,* 1920; Soil Science Society of America, American Agronomy Society, *Plant Environment and Efficient Water Use,* 1967; A. Stefferud (ed.), *Water,* USDA Yearb. Agr., 1955; A. C. True, *A History of Agricultural Experimentation and Research in the United States 1607–1925,* USDA Misc. Publ. no. 251, 1937; R. O. White, *Crop Production and Environment,* 1960; Jen-Yu Wang, *Agricultural Meteorology,* San Jose, Calif., Agricultural Weather Information Service, 1967; Jen-Yu Wang and G. L. Barger, *Bibliography on Agricultural Meteorology,* 1962.

Agricultural science (animal)

The science which deals with the selection, breeding, nutrition, and management of domestic animals for economical production of meat, milk, eggs, wool, hides, and other animal products. Horses for draft and pleasure and bees for honey production may also be included in this group.

When primitive man first domesticated animals, they were kept as means of meeting his immediate needs for food, transportation, and clothing. Sheep probably were the first and most useful animals to be domesticated, furnishing milk and meat for food, and hides and wool for clothing.

As chemistry, physiology, anatomy, genetics, nutrition, parasitology, pathology, and other sciences developed, their principles were applied to the field of animal science. Since the beginning of the 20th century, great strides have been made in livestock production. Today, farm animals fill a highly important place in the life of man. They convert raw materials, such as pasture grasses which are of little use to man as food, into animal products having nutritional values not directly available in plant products. In the United States the animal industry produced about 44% of the annual cash farm income in 1967. Of this 44%, beef cattle and calves contributed 22%, dairy products 12%, hogs 9%, sheep and lambs 0.7%, and poultry and eggs 9%.

Ruminant animals (those with four compartments or stomachs in the fore portion of their digestive tract, such as cattle and sheep) have the ability to consume large quantities of roughages because of their particular type of digestive system. They also consume large tonnages of grains, as well as mill feeds, oil seed meals, and other materials not suitable for human food. Estimates show that in the United States on Jan. 1, 1968, there were 108,800,000 cattle, of which 86,600,000 were beef cattle and 22,200,000 were dairy cattle. As of Jan. 1, 1967, there were 51,000,000 hogs, 23,700,000 sheep and lambs, 427,600,000 chickens, and 7,340,000 turkeys. Not included are over 2,000,000,000 commercial broiler chickens and 100,000,000 turkeys raised for the fresh and frozen turkey market. The total value of all cattle, hogs, sheep, chickens, and turkeys on farms was estimated at $18,889,000,000 as of Jan. 1, 1967. There were 4,000,000 horses and mules estimated in 1963. The number of pleasure horses continues to grow rapidly, with an estimated 6- to 7,000,000 horses and mules in the United States in 1968. An estimated 142,400,000 tons of grains, mill feeds, and high-protein feeds were consumed by these livestock in 1963.

Products of the animal industry furnish raw materials for many important processing industries, such as meat packing, dairy manufacturing, poultry processing, textile production, and tanning. Many services are based on the needs of the animal industry, including livestock marketing, milk deliveries, poultry and egg marketing, poultry hatcheries, artificial insemination services, and veterinary services. Thus, animal science involves the application of scientific principles to all phases of animal production, furnishing animal products efficiently and abundantly to consumers.

Livestock breeding. The breeding of animals began thousands of years ago. During the last half of the 19th century, livestock breeders made increasing progress in producing animals better suited to the needs of man by simply mating the best to the best. However, in the 20th century animal breeders began to apply the scientific principles of genetics and reproductive physiology. Much of the progress made in the improvement of farm animals has resulted from selected matings based on knowledge of body type or conformation. Many breeders of dairy cattle, poultry, beef cattle, sheep, and swine make use of production records or records of performance. Some of their breeding plans are based on milk fat production or egg production, as well as on body type or conformation. The keeping of poultry and dairy cow production records began in a very limited way late in the 19th century. The first Cow-Testing Association in the United States was organized in Michigan in 1906. Now over 1,500,000 cows are tested regularly in the United States. *See* BREEDING (ANIMAL).

Many states now have production testing for beef cattle, sheep, and swine, in which records of rate of gain, efficiency of feed utilization, incidence of twinning, and other characteristics of production are maintained on part or all of the herd or flock. These records serve as valuable information in the selection of animals for breeding or sale.

Breeding terminology. A breed is a group of animals that has a common origin and possesses characteristics that are not common to other individuals of the same species.

A purebred breed is a group that possesses certain fixed characteristics, such as color or markings, which are transmitted to the offspring. A record, or pedigree, is kept which describes their ancestry for five generations. Associations have been formed by breeders primarily to keep records, or registry books, of individual animals of the various breeds.

A purebred is one that has a pedigree recorded in a breed association or is eligible for registry by such an association. A grade is an individual having one parent, usually the sire, a purebred and the other parent a grade or scrub. A scrub is an inferior animal of nondescript breeding. A hybrid is one produced by crossing parents that are genetically pure for different specific characteristics. The mule is an example of a hybrid animal produced by crossing two different species, the American jack, *Equus asinus*, with a mare, *E. caballus*.

Systems of breeding. The modern animal breeder has genetic tools which he may apply, such as selection and breeding, and inbreeding and outbreeding. Selection involves directly the retaining or rejecting of a particular animal for breeding purposes, being based largely on qualitative characteristics. Inbreeding is a system of breeding related animals. Outbreeding is a system of breeding unrelated animals. When these unrelated animals are of different breeds, the term crossbreeding is usually applied. Crossbreeding is in common use by commercial swine producers. About 80–90% of the hogs produced in the Corn Belt states are now crossbred. Crossbreeding is also used extensively by commercial sheep producers.

Grading-up is the process of breeding purebred sires of a given breed to grade females and their female offspring for generation after generation. Grading-up offers the possibility of transforming a nondescript population into one resembling the purebred sires used in the process. It is an expedient and economical way of improving large numbers of animals.

Formation of new breeds. New breeds of farm animals have been developed from crossbred foundation animals. Montadale, Columbia, and Targhee are examples of sheep breeds developed from crossbred foundations. The Santa Gertrudis breed of beef cattle was produced by crossing Brahman and Shorthorn breeds on the King Ranch in Texas.

Artificial insemination. In this process spermatozoa are collected from the male and deposited in the female genitalia by instruments rather than by natural service. In the United States this practice was first used for breeding horses. Artificial insemination in dairy cattle was first begun on a large scale in New Jersey in 1938. In 1958 over 6,000,000 cows were bred artificially in the United States. Freezing techniques for preserving and storing spermatozoa have been applied with great success to bull semen, and it is now possible for outstanding bulls to sire calves years after the bulls have died. The use of artificial insemination for beef cattle and poultry (turkeys) is becoming more common since 1965. Drugs are being developed which stimulate beef cow herds to come into heat (ovulate) at approximately the same time. This will permit the insemination of large cow herds without the individual handling and inspection which is used with dairy cattle.

Livestock feeding. Scientific livestock feeding involves the systematic application of the principles of animal nutrition to the feeding of farm animals. The science of animal nutrition has advanced rapidly since 1930, and the discoveries are being utilized by most of those concerned with the feeding of livestock. The nutritional needs and responses of the different farm animals vary according to the functions they perform and to differences in the anatomy and physiology of their digestive systems. Likewise, feedstuffs vary in usefulness depending upon the time and method of harvesting the crop, the methods employed in drying, preserving, or processing them, and the forms in which they are offered to the animals consuming them.

Chemical composition of feedstuffs. The various chemical compounds that are contained in animal feeds have been divided into groups called nutrients. These include proteins, fats, carbohydrates, vitamins, and mineral matter.

Proteins are made up of amino acids. Twelve amino acids are essential for all nonruminant animals and must be supplied in their diets.

Fats and carbohydrates provide mainly energy. In most cases they are interchangeable as energy sources for farm animals. Fats furnish 2.25 times as much energy per pound as do carbohydrates because of their higher proportion of carbon and hydrogen to oxygen. Thus the energy concentration in poultry and swine diets can be increased by inclusion of considerable portions of fat. Ruminants cannot tolerate large quantities of fat in their diets, however.

Vitamins essential for health and growth include fat-soluble A, D, E, and K, and water-soluble vitamins thiamine, riboflavin, niacin, pyrodoxine, pantothenic acid, and cobalamin. *See* VITAMIN.

Mineral salts that supply calcium, phosphorus, sodium, chlorine, and iron are often needed as supplements, and those containing iodine and cobalt may be required in certain deficient areas. Zinc may also be needed in some swine rations. Many conditions of mineral deficiency have been noted in recent years by using rations that were not necessarily deficient in a particular mineral but in which the mineral was unavailable to the animal because of other factors in the ration or imbalances with other minerals. For example, copper deficiency can be caused by excess molybdenum in the diet.

By a system known as the "proximate analysis," feeds have long been divided into six fractions including moisture, ether extract, crude fiber, crude protein, ash, and nitrogen-free extract. The first five fractions are determined in the laboratory. The nitrogen-free extract is what remains after the percentage sum of these five has been subtracted from 100%. Although proximate analysis serves as a guide in the classification, evaluation, and use of feeds, it gives very little specific information about particular chemical compounds in the feed.

The ether extract fraction includes true fats and certain plant pigments, many of which are of little nutritional value.

The crude fiber fraction is made up of celluloses and lignin. This fraction, together with the nitrogen-free extract, makes up the total carbohydrate content of a feed.

The crude protein is estimated by multiplying the total Kjeldahl nitrogen content of the feed by the factor 6.25. This nitrogen includes many forms of nonprotein as well as protein nitrogen.

The ash, or mineral matter fraction, is determined by burning a sample and weighing the residue. In addition to calcium and other essential mineral elements, it includes silicon and other nonessential elements.

The nitrogen-free extract (NFE) includes the more soluble and the more digestible carbohydrates, such as sugars, starches, and hemicelluloses. Unfortunately, most of the lignin, which is not digestible, is included in this fraction.

A much better system of analysis has been developed for the crude fiber fraction of feedstuffs. This system separates more specifically the highly digestible portion and the less digestible fibrous portion of plant feedstuffs. This system of non-nutritive residue analysis was developed at the U.S. Department of Agriculture laboratories in Maryland.

Digestibility of feeds. In addition to their chemical composition or nutrient content, the nutritionist and livestock feeder should know the availability or digestibility of the different nutrients in feeds. The digestibility of a feed is measured by determining the quantities of nutrients eaten by an animal over a peroid of time and those recoverable in the fecal matter. By assigning appropriate energy values to the nutrients, total digestible nutrients (TDN) may be calculated. These values have been determined and recorded for a large number of feeds.

Formulation of animal feeds. The nutritionist and livestock feeder finds TDN values of great use in the formulation of animal feeds. The TDN requirements for various classes of livestock have been calculated for maintenance and for various productive capacities. However, systems have been developed for expressing energy requirements of animals or energy values of feeds in units which are more closely related to the body process being supported (such as maintenance, growth, and milk or egg production). New tables of feeding standards are being published using the units of metabolizable energy and net energy, which are measurements of energy available for essential body processes. Recommended allowances of nutrients for all species of livestock and some small animals (rabbit, dogs, and mink) are published by the National Academy of Sciences – National Research Council. These are assembled by experts in the field of animal science and are available for distribution through the Superintendent of Documents, Washington, D.C.

Nutritional needs of different animals. The nutritional requirements of different classes of animals are partially dependent on the anatomy and physiology of their digestive systems. Ruminants can digest large amounts of roughages, whereas horses, hogs, and poultry, with simple stomachs, can digest only limited amounts and require more concentrated feeds, such as cereal grains. In simple-stomached animals the complex carbohydrate starch is broken down to simple sugars which are absorbed into the blood and utilized by the body for energy. Microorganisms in the rumen of ruminant animals break down not only starch but the fibrous carbohydrates of roughages, namely, cellulose and hemicellulose, to organic acids which are absorbed into the blood and utilized as energy. Animals with simple stomachs require high-quality proteins in their diets to meet their protein requirement. On the other hand, the microorganisms in ruminants can utilize considerable amounts of simple forms of nitrogen to synthesize high-quality protein which is, in turn, utilized to meet the ruminant's requirement for protein. Thus, many ruminant feeds now contain varying portions of urea, an economical simple form of nitrogen, which is synthesized commercially from nonfeed sources. Simple-stomached animals require most of the vitamins in the diet. The microorganisms in the rumen synthesize adequate quantities of the water-soluble vitamins to supply the requirement for the ruminant animal. The fat-soluble vitamins A, D, and E must be supplied as needed to all farm animals. Horses and mules have simple stomachs but they also have an enlargement of the cecum (part of the large intestine), in which bacterial action takes place similar to that in the rumen of ruminants.

Livestock judging. The evaluation, or judging, of livestock is important to both the purebred and the commercial producer.

Show-ring judging. The purebred producer usually is much more interested in show-ring judging, or placings, than is the commercial producer. Because of the short time they are in the show-ring, the animals must be placed on the basis of type or appearance by the judge who evaluates them. The show-ring has been an important influence in the improvement of livestock by keeping the breeders aware of what judges consider to be desirable types. The shows have also brought breeders together for exchange of ideas and breeding stock and have helped to advertise breeds of livestock and the livestock industry. The demand for better meat-animal carcasses has brought about more shows in which beef cattle and swine are judged, both on foot and in the carcass. This trend helps to promote development of meat animals of greater carcass value and has a desirable influence upon show-ring standards for meat animals.

Selection of animals for breeding. The evaluation or selection of animals for breeding purposes is of importance to the commercial as well as to the purebred breeder. In selecting animals for breeding, desirable conformation or body type is given careful attention. The animals are also examined carefully for visible physical defects, such as blindness, crooked legs, jaw distortions, and abnormal udders. Animals known to be carriers of genes for heritable defects, such as dwarfism in cattle, should be discriminated against.

When they are available, records of performance or production should be considered in the selection of breeding animals. Some purebred livestock record associations now record production performance of individual animals on their pedigrees.

Grading of market animals. The grading on foot of hogs or cattle for market purposes requires special skill. In many modern livestock markets, hogs are graded as no. 1, 2, or 3 according to the estimated values of the carcasses. Those hogs grading no. 3 are used to establish the base price, and sellers are paid a premium for better animals. In some cases the grade is used also to place a value of the finished market product. For example, the primal cuts from beef cattle are labeled prime, choice, or good on the basis of the grade which the rail carcass received.

Livestock pest and disease control. The innumerable diseases of farm livestock require expert diagnosis and treatment by qualified veterinarians. The emphasis on intensive animal production has increased stresses on animals and generally increased the need for close surveillance of herds or flocks for disease outbreaks. Both external and internal parasites are common afflictions of livestock but can be controlled by proper management of the animals. Sanitation is of utmost importance in the control of these pests, but under most circumstances sanitation must be supplemented with effective insecticides, ascaricides, and fungicides. *See* FUNGISTAT AND FUNGICIDE; INSECTICIDE.

Internal parasites. Internal parasites, such as stomach and intestinal worms in sheep, cannot be controlled by sanitation alone under most farm conditions. For many years the classic treatment for sheep was drenching with 25 g of phenothiazine per adult sheep (1/2 dose for 25- to 50-lb lambs) and continuous free choice feeding of one part phenothiazine mixed with nine parts of salt. A new drug, thiabendazole, has been developed which breaks the life cycle of the parasitic worm. The dose recommended for sheep is 1/2 g of thiabendazole for animals under 50 lb weight and 1 to 1 1/2 g for adults, depending on the size of the animal and severity of parasitism. This can be given as a bolus or drench.

Control of gastrointestinal parasites in cattle can be accomplished in many areas by sanitation and the rotational use of pastures. In areas of intensive grazing, animals, especially the young ones, may become infected. They can be treated with a therapeutic dose of 20 g phenothiazine per 100 lb of body weight followed by 2 g of phenothiazine per day for a month.

The use of drugs to control gastrointestinal parasites and also certain enteric bacteria in hogs is commonplace. Control is also dependent on good sanitation and rotational use of nonsurfaced lots and pastures. Similarly, because of intensive housing systems, the opportunities for infection and spread of both parasitism and disease in poultry flocks are enhanced by poor management conditions. The producer has a large choice of materials to choose from in preventing or treating these conditions, including sodium fluoride, piperazine salts, nitrofurans, and arsenicals and antibiotics such as penicillin, tetracyclines, hygromycin, and tylosin.

External parasites. Control of horn flies, horseflies, stable flies, lice, mange mites, ticks, and fleas on farm animals is in the process of rapid change with the introduction of many new insecticides. Such compounds as DDT, methoxychlor, toxaphene, lindane, and malathion are very effective materials for the control of external parasites. However, the use of these materials is restricted to certain conditions and classes of animals by the provisions of Public Law 518, which is the Miller Amendment to the Federal Food, Drug, and Cosmetic Act. For example, use of DDT is not permitted on dairy animals. Reliable information should be obtained before using these materials for the control of external parasites.

Control of cattle grubs, or the larvae of the heel fly, may be accomplished by dusting the backs of the animals with a 1.5% rotenone powder or by spraying under high pressure with 7.5 lb of 5% rotenone in 100 gal of water. Three applications of spray should be made at 30-day intervals. Trolene and Reulene, systemic insecticides for grub control, have been given approval if used according to the manufacturer's recommendation.

Fungus infections. Actinomycosis is a fungus disease commonly affecting cattle, swine, and horses. In cattle this infection is commonly known as lumpy jaw. The lumpy jaw lesion may be treated with tincture of iodine or by local injection of streptomycin in persistent cases. Most fungus infections, or mycoses, develop slowly and follow a prolonged course. A veterinarian should be consulted for diagnosis and treatment.

[RONALD R. JOHNSON]

Bibliography: D. Acker, *Animal Science and Industry*, 1963; H. M. Briggs, *Modern Breeds of Livestock*, rev. ed., 1958; L. E. Card, *Poultry Production*, 1953; W. E. Carroll, J. L. Krider, and F. N. Andrews, *Swine Production*, 3d ed., 1962; H. H. Cole, *Introduction to Livestock Production*, 1962; E. W. Crampton, *Applied Animal Nutrition*, 1956; R. R. Dykstra, *Animal Sanitation and Disease Control*, 6th ed., 1964; C. H. Eckles and E. L. Anthony, *Dairy Cattle and Milk Production*, 5th ed., 1956; M. E. Ensminger, *Sheep and Wool Science*, 3d ed., 1964; W. G. Kammlade, Sr., and W. G. Kammlade, Jr., *Sheep Science*, 1955; D. J. Kays, *The Horse*, 1953; L. A. Maynard and J. K. Loosli, *Animal Nutrition*, 5th ed., 1962; F. B. Morrison, *Feeds and Feeding*, 22d ed., 1959; O. H. Siegmund, *The Merck Veterinary Manual*, 2d ed., Merck and Co., Inc., 1961; R. R. Snapp and A. L. Neumann, *Beef Cattle*, 5th ed., 1960.

Agricultural science (plant)

The plant sciences, both pure and applied, have contributed immeasurably to the development of agriculture. The greatest advancements have occurred in those countries where practical application has been made of the ever-increasing knowledge of plant physiology, ecology, morphology and anatomy, taxonomy, pathology, cytology, genetics, plant breeding and reproduction, agronomy, horticulture, and forestry. Increased production per man-hour of labor and other significant changes have been due to a number of factors individually and in combination. Some of these factors are covered in this article under the subheadings mechanization; fertilizers and plant nutrition; insecticides, fungicides, and nematocides; herbicides; viruses; growth regulators; and photoperiodism. Detailed descriptions of these and of the many important roles of the plant sciences in agriculture are discussed in separate articles. *See* BREEDING (PLANT).

Mechanization. About 1800 the grain cradle came into general use in the United States and some other areas. A good cradler could cut 2–2.5 acres per day, with another worker to rake and bind the grain. Then the reaper was invented. A test of an early model in 1852 demonstrated that 9 men with a reaper could do the work of 14 men with cradles, a gain of 36% in efficiency. Thirty years later the first combines appeared. In 1930, 16-ft combines had a daily capacity of 20–25 acres, and they not only harvested but also threshed the grain, almost a 75-fold gain over the cradle and flail methods of a century earlier. The mechanical cotton picker harvests a 500-lb bale in 75 min, 40–50 times as much as the average hand picker. The peanut harvester turns out about 300 lb of shelled peanuts per hour, a 300-hour job if done by hand labor. In the Orient, where over 90% of the world's rice crop is raised, it takes 100–125 hours of coolie labor to plant an acre of paddy rice; in California one person seeds 50 acres per hour by airplane. Hand-setting 7500 celery plants is a day's hard labor for one man; a modern transplanting machine with two people operating it readily sets 40,000, reducing labor costs by two-thirds.

Fertilizers and plant nutrition. No one knows when or where the practice originated of burying a fish beneath the spot where a few seeds of corn were to be planted, but it was common among North American Indians when Columbus discovered America and is evidence that the value of fertilizers was known to primitive peoples. Farm manures have been utilized almost from the time animals were first domesticated and crops grown. It was not until centuries later, however, that these animal fertilizers were supplemented by mineral forms of lime, phosphate, and potash. Rational use of the latter substances had its beginnings in the mid-19th century as an outgrowth of soil and plant analyses made by such pioneers in agricultural chemistry as Justus von Liebig of Germany. Usage has become more widespread with time; in 1967 approximately 37,000,000 tons of mineral fertilizers was applied to crop and pasture lands in the United States.

An early point of view was that crop production and fertilizer application are mostly problems in addition and subtraction, that a crop removes so many pounds of the nutrient elements from the soil and therefore the same number of pounds must be replaced if the original fertility is to be maintained. Furthermore, it was thought that the soil's supplies of some of the nutrient elements, for example, magnesium and sulfur, are more or less unlimited and fully adequate to meet requirements indefinitely. As time went on, however, it was realized that some of these elements, such as iron, are sometimes unavailable to plants even though present in the soil in relatively large amounts. Not only may much of the soil's supply of iron be unavailable to plant roots, but once within the plant it may, under some conditions, accumulate in certain tissues and be immobilized for use at points where it is needed for growth. Simultaneously, evidence was accumulating which indicated that plant nutrients could be used to influence growth processes, accelerating some and retarding others—that they could have more than just mass effects. A major breakthrough came when it was discovered that some kinds of nutrients, particularly those containing nitrogen, could be used to regulate type of growth, for example, to cause the plant to make further vegetative growth, or to channel its resources into reproductive activities. This put into the plantsman's hands a tool for controlling the behavior of plants, making them develop so as to serve his needs best.

During the latter part of the 19th and early part of the 20th centuries, it was observed that crop plants often failed to develop normally even though they were growing on soils known to contain sufficient amounts of the 10 elements considered essential for their proper nutrition, and despite the fact that other conditions, such as light, temperature, and moisture supply, were favorable. Since no pathogens were present to cause poor growth, the general appearance of the plants indicated a nutrient deficiency. In 1900 a Japanese investigator, M. Nagaoka, discovered that manganese had a marked stimulating effect on the growth of rice, and a decade later a German, E. Hazelhoff, found that boron similarly influenced the growth of a number of plants. The next four decades witnessed the addition of copper, zinc, and molybdenum to the list of essential mineral nutrients. These 5 elements are required in only very small amounts as compared to the first 10 and hence have been classified as trace or micronutrients. From a quantitative standpoint they are truly minor, but in reality they are just as important as any of the others, for without them the plant cannot survive. Many heretofore puzzling plant disorders are now known to be due either to insufficient supplies of micronutrients in the soil, or to their presence in forms unavailable to plants. Thus, dwarfing, cupping, and wrinkling of pecan leaves, often called mouse ear, is a symptom of manganese deficiency. Tomatoes maturing on copper-deficient plants are prone to be few-seeded or seedless. Zinc deficiency results in shortened internodes, leaf dwarfing, and a rosetted appearance in many deciduous fruits, in dwarfing and sickle leaf in cacao, and in stunting and poor filling of pods in beans. Boron deficiency leads to brown heart in the roots of beets, turnips, and swede (rutabagas); it causes blossom blast in the pear, poor setting of cotton bolls, and gum deposits in citrus fruits. Whiptail of cauliflower is associated with molybdenum deficiency. With some elements, for example, zinc, deficiency symptoms tend to follow a rather definite pattern; in others, such as boron, the symptoms vary greatly with species. *See* PLANT DISEASE.

Soil deficiencies of trace elements are usually not so widespread as those of major ones. However, manganese deficiencies have been reported in half of the United States and in such widely separated areas as Japan, New Zealand, and Sicily. Copper deficiencies have been reported in 16 states and are especially common in organic soils in this country, Ireland, and South Africa.

Light applications of trace elements in fertilizers are generally adequate. Sometimes a pound to the acre is sufficient. Heavy or frequent applications may result in their accumulation in toxic amounts. For instance, in Florida many soils used for citrus culture are deficient in copper, and light applications of this element are desirable. However, three or four such applications will last for one to three decades and heavier or more frequent applications are likely to prove harmful. Overliming acid soils renders their iron unavailable to many kinds of plants; overliming of some Florida soils has resulted in increased absorption of molybdenum by plants to the point where herbage growing on these soils becomes toxic to animals feeding on it. An important result of the discoveries relating to the absorption and utilization of trace elements is that they have served to emphasize the complexity of soil fertility and fertilizer problems. Far from being problems of simple addition and subtraction, they involve questions of interaction and balance within the soil and plant.

Insecticides, fungicides, and nematocides. Total destruction of crops by swarms of locusts and subsequent starvation of many people have occurred throughout the world. The pioneers in the Plains states suffered disastrous crop losses from hordes of grasshoppers and marching army worms. An epidemic of potato blight brought hunger to much of western Europe and famine to Ireland in 1846–1847. The perfection and use of new insecticides and fungicides in the more technologically advanced countries have done much to prevent such calamities. Various mixtures, really nothing more than nostrums (unscientific concoctions), were in use centuries ago, but the first really trustworthy insect control measure appeared in the United States in the mid-1860s, when Paris green was first used to halt the eastern spread of the Colorado potato beetle. This was followed by other arsenical poisons in the next three decades, culminating in lead arsenate in the last decade of the 19th century. By 1950 approximately 100,-000,000 lb of arsenical poison was being used annually in the United States. Various oil emulsions and tobacco extracts came into use for sucking insects in the 1880–1900 period. A major development occurred during World War II, when the value of DDT (dichlorodiphenyltrichloroethane) for control of many insects was discovered. Though this material was known to the chemist decades earlier, it was not until 1942 that its value as an insecticide was definitely established and a new chapter written in the continual contest between man and insects. Three or four applications of DDT give better control of many pests at lower cost than was afforded by a dozen materials used formerly. For instance, the number of gallons of spray used per acre in the Pacific Northwest decreased from approximately 2000 in 1944 to 600 in 1967 and labor and material costs decreased from 23 to 14 cents per bushel of fruit. Furthermore, DDT affords control of some kinds of pests that were practically immune to materials formerly available. In the meantime other materials, such as chlordane, parathion, and aldrin, have been developed that are even more effective for specific pests. *See* PESTICIDES, PERSISTENCE OF.

Sulfur-containing dusts and solutions have long been used to control mildew on foliage, but the first really effective fungicide was discovered accidentally in the early 1880s, when to discourage theft a combination of copper sulfate and lime was used near Bordeaux, France, to give grape vines a poisoned appearance. Bordeaux mixture has remained the standard remedy for many fungus diseases. However, other materials have gradually been replacing Bordeaux mixture, which has harmful side effects, such as accelerated transpiration, causing skin russeting of apples and premature defoliation of peach trees. Today a new insecticide or fungicide is evaluated not only for its ability to repel or destroy a certain pest but also for its influence on the physiological processes of the plant, such as transpiration, respiration, photosynthesis, fruit setting, or duration of dormancy.

Another important group of plant pests is the nematodes, minute roundworms that harbor principally in soil and roots. Their importance was not fully realized until studies conducted during the 1930s and 1940s revealed that nematodes are carriers of some of the most important virus diseases. Methods of nematode control were limited for many years to such measures as steaming the soil, usually an impractical operation except in greenhouses and small plant beds. Development of such fumigants as ethylene dibromide has greatly extended the conditions under which nematodes can be effectively controlled.

Herbicides. Because they can be seen, weeds were recognized as crop competitors long before microscopic bacteria, fungi, and viruses. The time-honored methods of controlling them have been to pull, dig, hoe, and cultivate them out, or smother them with mulches. These methods are still effective and practicable in many instances. However, under other conditions some kinds of weeds can be poisoned more cheaply. Thus the use of such materials as sodium arsenite in courtyards and other areas where no vegetation of any kind is wanted became fairly common. In the early part of the 20th century it was learned that ammonium sulfamate, sodium chlorate, and somewhat later, sodium trichloroacetate (TCA) could be used for the eradication of vegetation of all kinds in courtyards, parking lots, driveways, railroad beds, and along fences. All of these herbicides destroy plants by oxidizing or "burning" the tissues, by plasmolyzing the cells of the shoot, or by similar systemic toxic action after being absorbed by the roots. With all of them comparatively large dosages per unit area are required, for example, 50 lb or more of sodium trichloroacetate per acre and even more of sodium arsenite.

In the early 1940s new types of herbicides made their appearance. The first of these was 2,4-D (2,4-dichlorophenoxyacetic acid), followed shortly by 2,4,5-T (2,4,5-trichlorophenoxyacetic acid) and others, many of which are now used in a number of formulations. This class of materials kills by upsetting normal growth processes, causing hypertrophies (abnormal increase in size), distortions, and various cancerlike abnormalities that lead to the death of the plant. As a group these herbicides are much less toxic to parallel-veined, narrow-leaved plants (monocotyledons) than to even netted-veined, broad-leaved species (dicotyledons). Individually, they are highly selective. For instance, 2,4-D is more efficient for use against herbaceous weeds and 2,4,5-T for woody or brushy species. Consequently, mixtures of two or more are often preferred to a single kind. For some crops, pre-emergence applications are preferred. Thus 3–5 lb of an 80% sodium salt of 2,4-D in 20–40 gal of water per acre is effective in preventing emergence of weeds in sugarcane plantations, and does not at all impede sprouting and growth of the cane cuttings. Similarly 2 lb of CMU, 3-(p-chlorophenyl)-1,1-dimethylurea, per acre is effective in vineyards. For brush control in pastures or weed control in lawns, application is made directly to the foliage. With virtually all of these hormone-type herbicides much smaller amounts are required than with the caustic types. Concentrations of 1000 ppm or less are generally effective. However, some crops and ornamental species, such as cotton, papaya, and poinsettia, are very sensitive to them and much care is necessary in making applications in nearby areas. Wind-carried particles of 2,4-D dusts have been known to cause injury to cotton located 2–4 mi distant from where airplane applications were made to rice fields. The place that herbicides, particularly of the hormone type, have come to occupy in agriculture is indicated by the fact that farmlands treated in the United States alone increased from a few thousand acres in 1940 to over 150,-000,000 acres in 1967, not including large areas of swamp and overflow lands treated for aquatic plant control, and thousands of miles of treated highways, railroad tracks, and drainage and irrigation ditches.

Viruses. Agricultural journals of the 19th and early 20th centuries carried many reports of the "running out" of varieties and the more or less wholesale deterioration of entire populations of crop plants over extensive areas. Some of these maladies assumed the proportions of epidemics. As no organisms could be implicated, the outbreaks were attributed to unfavorable soil conditions, overbearing, drought, and other supposed causes. As the nature of virus diseases came to be more fully understood, suspicion grew that many of these epidemics were due to viruses. Therefore, in the mid-20th century, when the "quick decline" disease struck in a number of important citrus-producing areas of the world, there was less trouble in determining its cause. Studies of other virus diseases of both plants and animals formed the basis for undertaking a scientific approach to its control. Thus subjecting strawberry propagating stock to a temperature of 150°F for a very short period will destroy one of its virus diseases without injury to the plants. In the majority of cases, however, it is still more practicable to destroy infected plants and thus prevent spread to others.

Growth regulators. Many of the plantsman's century-old practices may be classified as methods of regulating growth. Thus, hedges are pruned to produce barriers of vegetation of desired height, width, and appearance. Vines have been trained to cover arbors, to provide shade, or to please the eye. Crops have been irrigated and fertilized to increase growth and yield. Use of specific substances to influence particular plant functions, however, has been a more recent development, though these modern uses, and even some of the substances used, had their antecedents in century-old practices in certain parts of the world. For example, in Japan old sake casks have long been used for temporary storage of persimmons to remove astringency and render the fruits sweet and edible. In the Near East people have known for a long time that putting a drop of olive oil in the "eye" of a fully grown but still green fig would cause the fig to ripen a week or two earlier. Those who employed these practices did not know how or why they brought about the changes. These were successful practices that had been handed down from generation to generation, and that was all they cared to know.

In the late 19th and the early 20th century, however, people began to wonder why some of these practices operated as they did. In the case of the persimmons, it was determined that small amounts of ethylene gas given off from the sake-soaked casks hastened ripening. Some trials were made on other kinds of still-green fruits. In some cases it worked. Exposure to ethylene fumes in ripening rooms in the northern part of the United States became standard procedure in ripening southern-grown tomatoes. Then it was learned that exposing potato tubers to the vapor of CIPC, isopropyl N-(3-chlorophenyl) carbamate, inhibits sprouting in storage and that such sprouting of both potatoes and onions can be long delayed by spraying the plants a few weeks before harvesting with a 1000–2500 ppm solution of maleic hydrazide. Since about 1930 many uses have been discovered for a considerable number of organic compounds having growth-regulating influences. For instance, several of them applied as sprays with a few days to several weeks before normal harvest will prevent or markedly delay dropping of such fruits as apples and oranges. Somewhat higher concentrations can be used to pre-thin fruits of the same and other kinds at the blossoming stage. Benzothiazoyl-2-oxyacetic acid and some of its derivatives can cause unfertilized Smyrna fig ovules to develop into "seeds," and thus are a substitute for an artificial method of pollinating cultivated figs known as caprification. β-Naphthoxypropionic acid has much the same effect on seedless Black Corinth and Thompson Seedless grapes. Runner development in strawberries can be inhibited by sprays of maleic hydrazide, and sucker formation in tobacco may be prevented with indoleacetic acid and certain bland mineral oils. Japanese workers report striking effects from gibberellins and fumaric acid, the first greatly increasing vegetative growth and the latter causing dwarfing. In practice, growth-inhibiting or growth-retarding agents are finding wider use than growth-stimulating ones. In higher concentration many growth-retarding agents become inhibiting agents.

Formulations of N-dimethylamine succinamic acid (known in the trade as B 9) are used to retard growth in length of shoots and spurs of a number of fruit-producing and ornamental species, thus serving as dwarfing agents; in many instances these same materials promote flower bud initiation and induce heavier production of flowers and fruits. Various formulations of methyl caprate kill rapidly expanding tissue, for example, the tips of lengthening shoots, and cause branching from subterminal lateral buds; thus they act essentially as a substitute for mechanical pinching or pruning. There are marked differences between plant species and even varieties in their response to most plant growth regulators. Many, if not most, growth regulators are highly selective; a concentration of even 100 times that effective for one species or variety is necessary to produce the same response in another. Furthermore, the specific formulation of the substances, for example, the kind or amount of wetting agent used with them, is important in determining their effectiveness. In brief, growth regulators are essentially new products, though there are century-old instances of the empirical use of a few of them. Some of them influence growth rate, some direction of growth, others plant structure, anatomy or morphology. With the discovery of new ones the indications are that it is only a matter of time before almost every feature of plant growth and development may be directly or indirectly controlled by them to a marked degree. Applications of these substances in agriculture are unfolding rapidly, and their use is one of the many factors making farming more of a science and less of an art.

Photoperiodism. The bases of most scientific discoveries are observations that require explanation. Attempts to find the reasons often lead further than was expected. This has been especially true of some of the seasonal peculiarities of plants. Much of their seasonal behavior is obviously a response to temperature, growth starting as it becomes warmer in spring and ceasing with the advent of cooler weather in autumn. Many plant functions, however, proceed more or less independently of temperature. A new flush of growth occurs or the plant flowers or matures fruits and seeds at a definite time of the year, provided latitude is the same. So definite are these adjustments to calendar date in many cases that it is almost possible to tell the time of year from the stage of development of certain plants. One important reason for this correlation of plant activities to time of year was first reported in 1920, the result of experiments in which growing plants of different kinds were exposed to alternating light and dark periods of definite lengths. Numerous plants behave in certain definite ways when the length of day is 14 hours and the period of darkness is shorter. If the lengths of the light and dark periods are reversed, these plants behave in an entirely different manner. In other words the plant is sensitive to length of day, or photoperiod. An interesting thing about this function of light is that intensities far below those necessary for photosynthesis are fully effective photoperiodically. Thus, while the optimum intensity for photosynthesis in the aster is approximately 3000 footcandles, 0.3 foot-candle is fully effective in influencing its photoperiodic function, a 10,000:1 difference.

One of the most striking influences of photoperiod is the flower bud initiation. For example, *Kalanchoe blossfieldiana*, a so-called short-day plant, produces only vegetative growth when day and night are of about equal length. However, six to eight consecutive exposures to 9-hr days will serve to initiate flower buds, followed rather promptly by flowering, if the temperature is favorable. Peppermint, however, will initiate flower buds only under the influence of very long days (16–18 hr). If this plant is to flower in time to mature seeds before the frosts of autumn, it must be grown in a latitude like that of southern England or New Zealand. Varieties of corn adapted to the North American corn belt become premature-flowering dwarfs at the Equator and semidwarfs in the latitude of Mexico City. As a consequence, Mexico City is about as far south as they can be grown, and growth is not very successful there. Varieties adapted to the shorter-day summers of northern Mexico have a prolonged vegetative period in the North American corn belt and flower so late in the summer that they fail to mature their seeds before the frosts of autumn. Photoperiod similarly exerts a controlling influence on many other plant functions and structures. Thus formation of potato tubers, onion bulbs, and strawberry runners depends on relative lengths of day and night, as does stooling (tillering) in many grains and grasses. Many other species, such as carnation, cacao, and banana, are "day neutral" and little, if at all, influenced by photoperiod.

In general, the plantsman has little direct control over the naturally occurring photoperiod of his location. However, he sometimes has a choice between several planting dates. Thus, he can bring his plants to a predetermined ideal flowering or stooling size at a date coinciding with a particular photoperiod. With some kinds of plants grown under glass structures he may want to lengthen the photoperiod directly by artificial lighting or shorten it by shading and thus induce flowering and fruit production at will, for example, to ready poinsettias for the Christmas holiday trade or asters for Easter. Natural photoperiods can also be capitalized on through plant breeding, that is, by developing varieties adjusted to certain localities. Through knowledge of photoperiods and the responses of different plants to them, the plantsman is able to make many crop plants behave for him as he wants them to behave. He can make some more vegetative, others less so; he can induce dormancy or continued growth; he can bring on the maturing of flowers, fruits, or seeds at a particular time or perhaps prevent it; he can determine whether or not there will be stooling or bulbing and whether the flowers produced will be pistillate, staminate, or both. In short, the plantsman uses his knowledge of photoperiod as a kind of plant regulator. Indeed it may be postulated that photoperiod exerts its effects on plants through its influence on the natural development of growth regulators. Though at present these are undetermined, eventually they may be identified chemically and synthesized.

[VICTOR R. GARDNER]

Bibliography: F. C. Bawden, *Plant Viruses and Virus Diseases*, 1964; F. Bear, *Soil in Relation to Crop Growth*, 1965; V. R. Gardner, *Principles of Horticultural Production*, 1966; B. Hocking, *Six-legged Science*, 1967; L. Kavaler, *Wonders of Fungi*, 1964; H. B. Tukey, *Plant Regulators in Agriculture*, 1954; J. C. Walker, *Plant Disease Handbook*, 1957; F. W. Went, *Experimental Control of Plant Growth*, 1957.

Air mass

A term applied in meteorology to an extensive body of the atmosphere which approximates horizontal homogeneity in its weather characteristics. An air mass may be followed on the weather map as an entity in its day-to-day movement in the general circulation of the atmosphere. The expressions air mass analysis and frontal analysis are applied to the analysis of weather maps in terms of the prevailing air masses and of the zones of transition and interaction (fronts) which separate them. The relative horizontal homogeneity of an air mass stands in contrast to the sharp horizontal changes in a frontal zone. The horizontal extent of important air masses is reckoned in millions of square miles. In the vertical dimension an air mass extends at most to the top of the troposphere, and frequently is restricted to the lower half or less

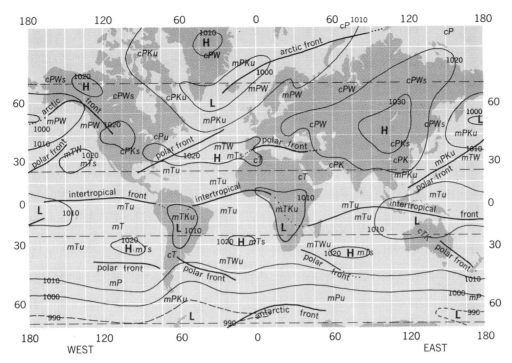

Fig. 1. Air-mass source regions, January. High- and low-atmospheric-pressure centers are designated *H* and *L* within average pressure lines numbered in millibars (such as 1010). Major frontal zones are labeled along heavy lines. (*From H. C. Willett and F. Sanders, Descriptive Meteorology, 2d ed., Academic Press, 1959*)

of the troposphere. The frontal zones between air masses usually slope in such a manner that the colder air mass underlies the warmer as a wedge. In the vertical direction the properties of an air mass, specifically its content of heat and moisture, may vary between a high degree of stratification and one of homogeneity produced by vertical mixing. *See* FRONT; METEOROLOGY; WEATHER MAP.

Development of concept. Practical application of the concept to the air mass and frontal analysis of daily weather maps for prognostic purposes was a product of World War I. A contribution of the Norwegian school of meteorology headed by V. Bjerknes, this development originated in the substitution of close scrutiny of weather map data from a dense local network of observing stations for the usual far-flung internation network. The advantage of air-mass analysis for practical forecasting became so evident that during the three decades following World War I the technique was applied in more or less modified form by nearly every progressive weather service in the world. However, the rapid increase of observational weather data from higher levels of the atmosphere during and since World War II has resulted in a progressive tendency to drop the careful application of air-mass analysis techniques in favor of those involving the kinematical or dynamic analysis of upper-level air-flow patterns.

Origin. The occurrence of air masses as they appear on the daily weather maps depends upon two facts, the existence of air-mass source regions, and the large-scale character of the branches or elements of exchange of the general circulation. Air-mass source regions consist of extensive areas of the Earth's surface which are sufficiently uniform so that the overlying atmosphere acquires similar characteristics throughout the region; that is, it approximates horizontal homogeneity. The designation of an area of the Earth's surface as a source region assumes that the overlying atmosphere in that area normally remains there long enough to approximate thermodynamic equilibrium with respect to the underlying surface, or in other words, to acquire the weather characteristics that typify that particular source region.

The large-scale character of the elements by which the general circulation is accomplished is observed on the daily weather maps in the major atmospheric currents of polar or tropical origin, whose southward or northward progress can be traced from day to day. These major currents, together with the associated polar and tropical air masses, are the means by which surplus tropical heat is effectively transported to polar latitudes. *See* ATMOSPHERIC GENERAL CIRCULATION.

Weather significance. The thermodynamic properties of air mass determine not only the general character of the weather in the extensive area covered by the air mass but also, to some extent, the severity of the weather activity in the frontal zone of interaction between air masses. Those properties which determine the primary weather characteristics of an air mass are defined by the vertical distribution of the two elements, water vapor and heat (temperature). On the vertical distribution of water vapor depend the presence or absence of condensation forms and, if present, the elevation and thickness of fog or cloud layers. On the vertical distribution of temperature depend the relative warmth or coldness of the air mass and, more importantly, the vertical gradient of temperature, known as the lapse rate. The lapse rate determines the stability or instability of the air

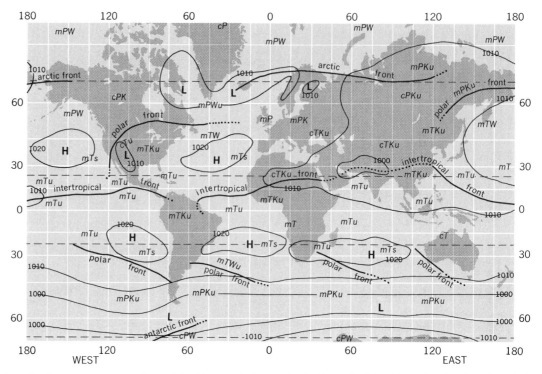

Fig. 2. Air-mass source regions, July. The symbols which are used in this figure are the same as those for

Fig. 1. (*From H. C. Willett and F. Sanders, Descriptive Meteorology, 2d ed., Academic Press, 1959*)

mass for thermal convection and, consequently, the stratiform or convective cellular structure of the cloud forms and precipitation. The most unstable moist air mass, in which the vertical lapse rate may approach 1°C/100 m, is characterized by severe turbulence and heavy showers or thundershowers. In the most stable air mass there is observed an actual increase (inversion) of temperature with increase of height at low elevations. With this condition there is little turbulence, and if the air is moist there is fog or low stratus cloudiness and possible drizzle, but if the air is dry there will be low dust or industrial smoke haze. *See* TEMPERATURE INVERSION.

Classification. A wide variety of systems of classification and designation of air masses was developed by different weather services around the world. The usefulness of a system with type designators to be applied in the analysis of weather maps is directly proportional to its effectiveness in accurately expressing the thermodynamic properties which determine the weather characteristics of the air mass. These properties are imparted to the air masses primarily by the particular source region of its origin, and secondarily by the modifying influences to which it is subjected after leaving the source region. Consequently, most systems of air-mass classification are based on a designation of the character of the source region and the subsequent modifying influences to which the air mass is exposed. Probably the most effective and widely applied system of classification is a modification of the original Norwegian system that is based on the following four designations.

Polar versus tropical origin. All primary air-mass source regions lie in polar (P in Figs. 1 and 2) or in tropical (T) latitudes. In middle latitudes

there occur the modification and interaction of air masses initially of polar or tropical origin. This difference of origin establishes the air mass as cold or warm in character.

Maritime versus continental origin. To be homogeneous, an air-mass source region must be exclusively maritime or exclusively continental in character. On this difference depends the presence or absence of the moisture necessary for extensive condensation forms. However, a long trajectory over open sea transforms a continental to a maritime air mass, just as a long land trajectory, particularly across major mountain barriers, transforms a maritime to a continental air mass. On Figs. 1 and 2, m and c are used with P and T (mP, cP, mT, and cT) to indicate maritime and continental character, respectively.

Heating versus cooling by ground. This influence determines whether the air mass is vertically unstable or stable in its lower strata. In a moist air mass it makes the difference between convective cumulus clouds with good visibility on the one hand and fog or low stratus clouds on the other. Symbols W (warm) and K (cold) are used on maps—thus, mPK or mPW.

Convergence versus divergence. Horizontal convergence at low levels is associated with lifting and horizontal divergence at low levels with sinking. Which condition prevails is dependent in a complex manner upon the large-scale flow pattern of the air mass. Horizontal covergence produces vertical instability of the air mass in its upper strata (u on maps), and horizontal divergence produces vertical stability (s on maps). On this difference depends the possibility or impossibility of occurrence of heavy air-mass showers or thundershowers or of heavy frontal precipitation. Examples of the designation of these tendencies and the intermediate conditions for maritime polar air masses are *mPWs, mPW, mPWu, mPs, mPu, mPKs, mPK, and MPKu.*   [HURD C. WILLETT]

Bibliography: W. L. Donn, *Meteorology*, 3d ed., 1965; H. C. Willett and F. Sanders, *Descriptive Meteorology*, 2d ed., 1959.

Air pressure

The force per unit area that the air exerts on any surface in contact with it, arising from the collisions of the air molecules with the surface. It is equal and opposite to the pressure of the surface against the air, which for atmospheric air in normal motion approximately balances the weight of the atmosphere above, about 15 psi at sea level. It is the same in all directions and is the force that balances the weight of the column of mercury in the Torricellian barometer, commonly used for its measurement.

Units. The basic units of pressure, which are the ones mainly used in meteorology, are based on the "bar," which is defined as equal to 1,000,000 dynes/cm². One bar equals 1000 millibars (mb) equals 100 centibars (cb).

Also widely used in practice are units based on the height of the mercury barometer under standard conditions, expressed commonly in millimeters or in inches. The standard atmosphere (760 mm Hg) is also used as a unit, mainly in engineering where large pressures are encountered. The following equivalents show the conversions between the commonly used units of pressure, where (mm Hg)$_n$ and (in. Hg)$_n$ denote the millimeter and inch of mercury, respectively, under standard (normal) conditions, and where (kg)$_n$ and (lb)$_n$ denote the weight of a standard kilogram and pound mass, respectively, under standard gravity.

$$1 \text{ mb} = 1000 \text{ dynes/cm}^2 = 0.750062 \text{ (mm Hg)}_n$$
$$= 0.0295300 \text{ (in. Hg)}_n$$
$$1 \text{ atm} = 1013.250 \text{ mb} = 760 \text{ (mm Hg)}_n$$
$$= 29.9213 \text{ (in. Hg)}_n = 14.6959 \text{ (lb)}_n/\text{in.}^2$$
$$= 1.03323 \text{ (kg)}_n/\text{cm}^2$$
$$1 \text{ (mm Hg)}_n = 1 \text{ torr} = 1.333224 \text{ mb}$$
$$= 0.03937008 \text{ (in. Hg)}$$
$$1 \text{ (in. Hg)}_n = 33.8639 \text{ mb} = 25.4 \text{ (mm Hg)}_n$$

Variation with height. Because of the almost exact balancing of the weight of the overlying atmosphere by the air pressure, the latter must decrease with height, according to the hydrostatic equation, Eq. (1), where P is air pressure, ρ is air

$$dP = -g\rho \, dZ \qquad (1)$$

density, g is acceleration of gravity, Z is altitude above mean sea level, dZ is infinitesimal vertical thickness of horizontal air layers, dP is pressure change which corresponds to altitude change dZ, P_1 is pressure at altitude Z_1, and P_2 is pressure at altitude Z_2. The expressions on the right-hand side of Eq. (2) represent the weight of the column of air between the two levels Z_1 and Z_2.

$$P_1 - P_2 = \int_{Z_1}^{Z_2} \rho g \, dZ \qquad (2)$$

In the special case in which Z_2 refers to a level above the atmosphere where the air pressure is nil, one has $P_2 = 0$, and Eq. (2) yields an expression for

air pressure P_1 at a given altitude Z_1 for an atmosphere in hydrostatic equilibrium.

By substituting in Eq. (1) the expression for air density based on the well-known perfect gas law and by integrating, one obtains the hypsometric equation for dry air under the assumption of hydrostatic equilibrium, Eq. (3), valid below about 90

$$\log_e \left(\frac{P_1}{P_2}\right) = \frac{M}{R} \int_{Z_1}^{Z_2} \frac{g}{T} \, dZ \qquad (3a)$$

$$Z_2 - Z_1 = \frac{R}{M} \int_{P_2}^{P_1} \frac{T}{g} \frac{dP}{P} \qquad (3b)$$

km, where g is the gravitational acceleration; M is the gram-molecular weight, 28.97 for dry air; R is the gas constant for 1 mole of ideal gas, or 8.31470×10^7 erg/(mole)(°K); and T is the air temperature in °K.

Equation (3) may be used for the real moist atmosphere if the effect of the small amount of water vapor on the density of the air is allowed for by replacing T by T_v, the virtual temperature given by Eq. (4), in which e is partial pressure of water va-

$$T_v = T\left[1 - \left(1 - \frac{M_w}{M}\right)\frac{e}{P}\right]^{-1} \qquad (4)$$

por in the air, M_w is gram-molecular weight of water vapor (18.0160 g/mole), and $(1 - M_w/M) = 0.37803$.

Equation (3a) is used in practice to calculate the vertical distribution of pressure with height above sea level. The temperature distribution in a standard atmosphere, based on mean values in middle latitudes, has been defined by international agreement. The use of the standard atmosphere permits the evaluation of the integrals of Eqs. (3a) and (3b) to give a definite relation between pressure and height. This relation is used in all altimeters which are basically barometers of the aneroid type. The difference between the height estimated from the pressure and the actual height is often considerable; but since the same standard relationship is used in all altimeters, the difference is the same for all altimeters at the same location, and so causes no difficulty in determining the relative position of aircraft. Mountains, however, have a fixed height, and accidents have been caused by the difference between the actual and standard atmospheres.

Horizontal and time variations. In addition to the large variation with height discussed in the previous paragraph, atmospheric pressure varies in the horizontal and with time. The variations of air pressure at sea level, estimated in the case of observations over land by correcting for the height of the ground surface, are routinely plotted on a map and analyzed, resulting in the familiar "weather map" representation with its isobars showing highs and lows. The movement of the main features of the sea-level pressure distribution, typically from west to east, produces characteristic fluctuations of the pressure at a fixed point, varying by a few percent within a few days. Smaller-scale variations of sea-level pressure, too small to appear on the ordinary weather map, are also present. These are associated with various forms of atmospheric motion, such as small-scale wave motion and turbulence. Relatively large variations are found in and near thunderstorms, the most intense being the low-pressure region in a tornado. The pressure drop within a tornado can be a large fraction of an atmosphere, and is the principal cause of the explosion of buildings over which a tornado passes. *See* WEATHER MAP.

It is a general rule that in middle latitudes at localities below 1000 m (3280 ft) in height above sea level, the air pressure on the continents tends to be slightly higher in winter than in spring, summer, and autumn; whereas at considerably greater heights on the continents and on the ocean surface, the reverse is true.

Various maps of climatic averages indicate certain regions where systems of high and low pressure predominate. Over the oceans there tend to be areas or bands of relatively high pressure, most marked during the summer, in zones centered near latitude 30°N and 30°S. The Asiatic landmass is dominated by a great high-pressure system in winter and a low-pressure system in summer. Deep low-pressure areas prevail during the winter over the Aleutian, the Icelandic-Greenland, and Antarctic regions. These and other centers of action produce offshoots which may travel for great distances before dissipating.

Thus during the winter, spring, and autumn in middle latitudes over the land areas, it is fairly common to experience the passage of a cycle of low- and high-pressure systems in alternating fashion over a period of about 6–9 days in the average, but sometimes in as little as 3–4 days, covering a pressure amplitude which ranges on the average from roughly 15–25 mb less than normal in the low-pressure center to roughly 15–20 mb more than normal in the high-pressure center. During the summer in middle latitudes the period of the

pressure changes is generally greater, and the amplitudes are less than in the cooler seasons (see table).

Within the tropics where there are comparatively few passages of major high- and low-pressure systems during a season, the most notable feature revealed by the recording barometer (barograph) is the characteristic diurnal pressure variation. In this daily cycle of pressure at the ground there are, as a rule though with some exceptions, two maxima, at approximately 10 A.M. and 10 P.M., and two minima, at approximately 4 A.M. and 4 P.M., local time.

The total range of the diurnal pressure variation is a function of latitude as indicated by the following approximate averages (latitude N and range in millibars): 0°, 3 mb; 30°, 2.5 mb; 35°, 1.7 mb; 45°, 1.2 mb; 50°, 0.9 mb; 60°, 0.4 mb. These results are based on the statistical analysis of thousands of barograph records for many land stations. Local peculiarities appear in the diurnal variation because of the influences of physiographic features and climatic factors. Mountains, valleys, oceans, elevations, ground cover, temperature variation, and season exert local influences; while current atmospheric conditions also affect it, such as amount of cloudiness, precipitation, and sunshine. Mountainous regions in western United States may have only a single maximum at about 8–10 A.M. and a single minimum at about 5–7 P.M., local time, but with a larger range than elsewhere at the same latitudes, especially during the warmer months (for instance, about 4 mb difference between the daily maximum and minimum).

At higher levels in the atmosphere the variations of pressure are closely related to the variations of temperature, according to Eq. (3a). Because of the lower temperatures in higher latitudes in the lower 10 km, the pressures at higher levels tend to decrease toward the poles. The figure shows a typical pattern at approximately 10 km above sea level. As is customary in representing pressure patterns at upper levels, the variation of the height of a surface of constant pressure, in this case 300 mb, is shown, rather than the variation of pressure over a horizontal surface.

Besides the latitudinal variation, the figure also shows the characteristic wave pattern in the pressure field, and the midlatitude maximum in the wind field known as the jet stream, with its "waves in the westerlies." In the stratosphere the temperature variations are such as to reduce the pressure variations at higher levels, up to about 80 km, except that in winter at high latitudes there are relatively large variations above 10 km. At altitudes above 80 km the relative variability of the pressure increases again. Although the pressure and density at these very high levels are small, they are important for rocket and satellite flights, so that their variability at high altitudes is likewise important.

Relations to wind and weather. The practical importance of air pressure lies in its relation to the wind and weather. It is because of these relationships that pressure is a basic parameter in weather forecasting, as is evident from its appearance on the ordinary weather map.

Horizontal variations of pressure imply a pressure force on the air, just as the vertical pressure variation implies a vertical force that supports the

Mean atmospheric pressure and temperature in middle latitudes, for specified heights above sea level*

Altitude above sea level			
Standard geopotential meters, m'	m at latitude 45°32'40"	Air pressure, mb	Assumed temperature,°K
0	0	1.01325×10^3	288.15
11,000	11,019	2.2632×10^2	216.65
20,000	20,063	5.4747×10^1	216.65
32,000	32,162	8.6798×10^0	228.65
47,000	47,350	1.1090×10^0	270.65
52,000	52,429	5.8997×10^{-1}	270.65
61,000	61,591	1.8209×10^{-1}	252.65
79,000	79,994	1.0376×10^{-2}	180.65
88,743	90,000	1.6437×10^{-3}	180.65†

*Approximate annual mean values based on radiosonde observations at Northern Hemisphere stations between latitudes 40 and 49°N for heights below 32,000 m and on observations made from rockets and instruments released from rockets. Some density data derived from searchlight observations were considered. Values shown above 32,000 m were calculated largely on the basis of observed distribution of air density with altitude. In correlating columns 1 and 2, G is 98,066.5 cm²/sec² per standard geopotential meter (m'). Data on first three lines are used in calibration of aircraft altimeters.

†Above 90,000 m there occurs an increase of temperature with altitude and a variation of composition of the air with height, resulting in a gradual decrease in molecular weight of air with altitude.

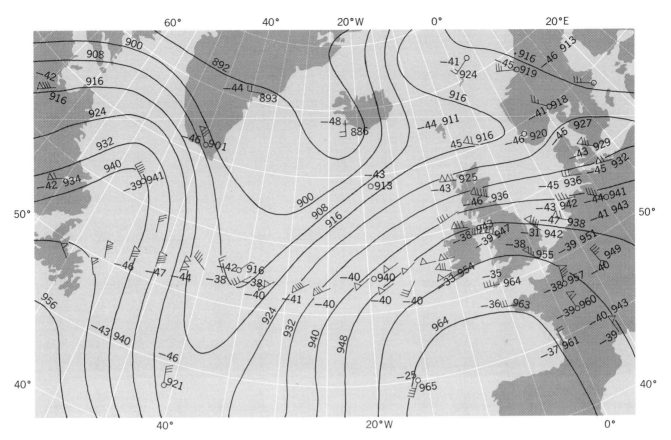

Contours of 300-mb surface, in tens of meters, with temperature in °C, and measured winds at the same level, on June 16, 1960. Winds are plotted with arrow pointing in direction of the wind, with each bar of the tail representing 10 m/sec. Triangle represents 50 m/sec.

weight of the air, according to Eq. (1). This force, if unopposed, accelerates the air, causing the wind to blow from high to low pressure. The sea breeze is an example of such a wind. However, if the pressure variations are on a large scale and are changing relatively slowly with time, the rotation of the Earth gives rise to geostrophic or gradient balance such that the wind blows along the isobars. This situation occurs when the pressure variations are due to the slow-moving lows and highs that appear on the ordinary weather map, and to the upper air waves shown in the figure, in which the relationship is well illustrated.

The wind near the ground, in the lowest few hundred meters of the atmosphere, is retarded by friction with the surface to a degree that depends on the smoothness or roughness of the surface. This upsets the balance mentioned in the previous paragraph, so that the wind blows somewhat across the isobars from high to low pressure.

The large-scale variations of pressure at sea level shown on a weather map are associated with characteristic patterns of vertical motion of the air, which in turn affect the weather. Descent of air in a high heats the air and dries it by adiabatic compression, giving clear skies, while the ascent of air in a low cools it and causes it to condense and produce cloudy and rainy weather. These processes at low levels, accompanied by others at higher levels, usually combine to justify the clear-cloudy-rainy marking on the household barometer.

[RAYMOND J. DELAND]

Bibliography: H. R. Byers, *General Meteorology*, 1959; R. G. Fleagle and J. A. Businger, *An Introduction to Atmospheric Physics*, 1963; A. Miller, *Meteorology*, 1966; H. Riehl, *Introduction to the Atmosphere*, 1965; O. G. Sutton, *Understanding the Weather*, 1960.

Air-pollution control

Air pollution, according to the definition developed by the Engineers Joint Council, means the presence in the outdoor atmosphere of one or more contaminants, such as dust, fumes, gas, mist, odor, smoke, or vapor, in quantities, of characteristics, and of duration such as to be injurious to human, plant, or animal life or to property, or to interfere unreasonably with the comfortable enjoyment of life and property. The sources of airborne wastes are many. They may be roughly divided into natural, industrial, transportation, agricultural activity, commercial and domestic heat and power, municipal activities, and fallout. *See* ATMOSPHERIC POLLUTION.

Sources of pollution. Natural sources include the pollen from weeds, water droplet or spray evaporation residues, wind storm dusts, meteoritic dusts, and surface detritus. Industrial sources include ventilation products from local exhaust systems, process waste discharges, and heat, power, and waste disposal by combustion processes. Transportation sources include motor vehicles, rail-mounted vehicles, airplanes, and vessels. Agricultural activity sources include insecticidal and

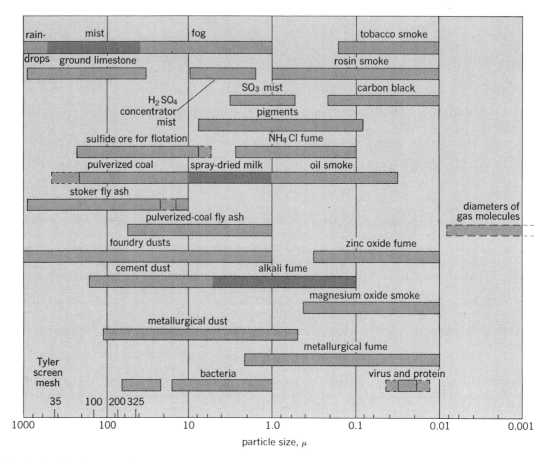

Fig. 1. Particle size ranges for aerosols, dusts, and fumes. (*From W. L. Faith, Air Pollution Control, Wiley, 1959*)

pesticidal dusting and spraying, and burning of vegetation. Commercial heat and domestic heat and power sources include gas-, oil-, and coal-fired furnaces used to produce heat or power for individual dwellings, multiple dwellings, commercial establishments, utilities, and industry. Municipal activity sources include refuse disposal, liquid waste disposal, road and street plant operations, and fuel-fired combustion operations. Fallout is a term applied to radioactive pollutants in mass atmosphere resulting from thermonuclear explosion.

The sources are so varied that pollution of the atmosphere is a matter of degree. Pollution from natural sources is in effect a base line of pollution. The major problems of pollution are associated with community activity as opposed to rural activity, because community air is generally more grossly polluted and may contain harmful and dangerous substances affecting property, plant life, and, on occasion, health. Environment is made less desirable by the polluting influence, and there is ample reason to conserve the air resource in many ways parallel to the need for conservation of the water resource. In actuality, the engineer is concerned with engineering management of the air resource, a broader concept than the control of air pollution.

Control. Air-pollution control suggests in its simplest form a background of knowledge concerning ideal atmospheres and criteria of clean air, the existence of specific standards setting limits on the allowable degree of pollution, means of precise measurement of pollutants, and practical means of treating polluting sources to maintain the desired degree of air cleanliness. There are many areas in the above listing that are under research at the present time. University research foundations, Federal, state, and municipal air-pollution control agencies, and all of the professional engineering societies are actively engaged in the development of criteria, standards, design factors, and equipment for the control of air pollution.

Reduced visibility has been a focal point of air pollution for over 700 years. The burning of soft coal in England combined with the fog of the atmosphere forms a particularly opaque mixture which may at times reduce visibility to zero. The word smog has been coined to describe this mixture.

Microscopic water droplets condense about nucleating substances in the air to form aerosols. An aerosol is a liquid or solid submicron particle dispersed in a gaseous medium. An atmosphere having an aerosol concentration of about 1 mg/m³ has been estimated to limit visibility to 1600 ft. The mass would contain perhaps 16,000 particles/ml. Restriction in visibility is the result of light scattering by these particles. Chemical condensation of reaction products in the air may also nucleate and grow to size that will bring about light scattering. Sulfur dioxide is also a nucleating substance as it oxidizes and hydrolyzes to form sulfuric acid mist.

Elimination of sources of pollution has been one of the favored means of controlling pollution.

There are many means of accomplishing the reduction of pollution, but complete elimination is not always practicable. Sulfur dioxide release can be reduced by choosing low sulfur-bearing fuel. An industrial process with a gaseous effluent can be changed to eliminate the gaseous waste. Gases and particulates can be removed from a gas stream by air-cleaning equipment.

Air-cleaning devices. Air-cleaning devices to remove particulates are selected to remove particles and aerosols on the basis of their size (Fig. 1). Screens will remove coarse solids. Settling chambers are containers which by expanded cross section reduce velocity below 10 feet per second (fps) and thereby allow particles to settle. Particles down to 10 μ in size may be recovered with such chambers. Cyclone separators operate by injecting a gas stream tangentially at the top of a cylindrical chamber. A high-velocity spiral motion is created. Particles are centrifuged out of the gas stream, hit the side wall, and fall to a conical bottom out of the airflow, which turns up through the core or vortex beginning at the bottom and flows to the top through a pipe inserted into the core and extending into the body of the cyclone. Particles from 10 to 200 μ are removed with 50–90% efficiency. Filters are made of cloth, fiber, or glass. Air velocities are low and efficiency is about 50% for dry fiber filters. Efficiency is increased by using a low volatile oil viscous coating. Cloth filters are usually tubular and a number of bags are enclosed in a large chamber. Particles are trapped as air passes through the cloth from inside to outside. Dust is knocked down by shaking and falls to a hopper. Bag filters remove 99% of particles above 10-μ size. Wet collectors, or scrubbers, operate by passing and contacting the gas with a liquid. Water is sprayed, atomized, or distributed over a geometric shape. Deflectors may be added to provide an impinging surface. Scrubbers are efficient on 1- to 5-μ size particles (Figs. 2 and 3). Electrostatic precipitators operate by charging or ionizing particles as the gas flow passes through the unit (Fig. 4). Opposite-pole high-voltage plates, or electrodes, are provided to trap particles. Precipitators operate at 80–99% efficiency of ionizable aerosols down to 0.1-μ size.

Scrubbers may also remove water-soluble gases. Chemicals may be added to the liquid to provide improved absorption. Filters packed with activated charcoal are used to adsorb gases.

Packed towers, plate towers, and spray towers are also used to absorb gaseous pollutants from a gas stream. These devices provide for mixing a gas stream under treatment with water or a chemical solution, so that gases are taken into solution and possibly converted chemically as well.

Atmospheric dilution. This provides another means of reducing air pollution. Meteorology of a region, local topography, and building configuration are critical factors in determining suitability of atmosphere as a dispersal, diffusion, and dilution medium. Basic meteorological conditions of atmosphere that must be considered include wind speed and direction, gustiness of wind, and vertical temperature distribution. Humidity is also important under certain circumstances of emission.

In general, diffusion theories predict that the ground concentration of a gas or a fine particle

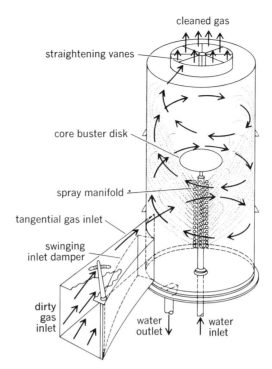

Fig. 2.   Typical cyclonic spray scrubber. (*From W. L. Faith, Air Pollution Control, Wiley, 1959*)

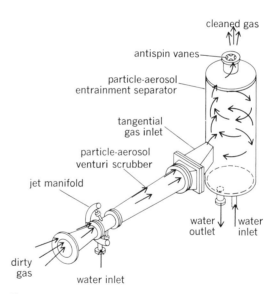

Fig. 3.   Typical venturi scrubber. (*From W. L. Faith, Air Pollution Control, Wiley, 1959*)

effluent with very low subsidence velocity is inversely proportional to the mean wind speed. Vertical temperature distribution is an important factor, determining the distance from stack of known height at which maximum ground concentration occurs. Temperature of the stack gas has the effect of increasing stack height, as does stack gas velocity. Gas does not normally come to the ground under inversion conditions, but may accumulate aloft under calm or near calm conditions and be brought down to the surface as the Sun heats the ground in the early morning. Effect of building

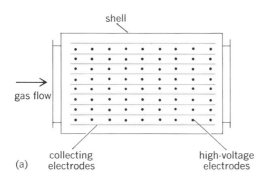

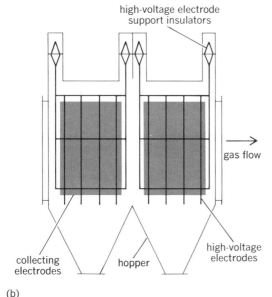

Fig. 4. Diagram of horizontal-flow electrostatic precipitator. (a) Plan. (b) Elevation. (*From W. L. Faith, Air Pollution Control, Wiley, 1959*)

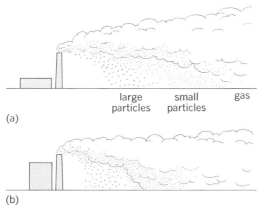

Fig. 5. Effect of building configuration on dispersal of gas plume. (a) Favorable configuration. (b) Unfavorable configuration. (*Research Division, New York University School of Engineering and Science*)

configuration is shown in Fig. 5. The turbulence introduced by buildings and topography is so complex that it is difficult to make theoretical calculations of effect. Model studies in wind tunnels have been used successfully to make predictions based on measurements of gas concentration and visible pattern of smoke (Fig. 6).

Nonventilating conditions may be present over an area for several days as a result of certain meteorological phenomena. During such periods the pollution emitted from various sources, such as fuel-fired combustion and automobile exhaust, continues to increase in concentration until ventilation sufficient to dilute the accumulated gases and particulates takes place. Figure 7 illustrates the record of sulfur dioxide–concentration measurement in the atmosphere over New York City during one such period of poor ventilation lasting for several days.

Incineration. The need for municipalities to find a means of disposing of refuse when land values are high and little land is available for sanitary landfill has resulted in increased use of incineration for refuse disposal. Incineration introduces problems of air pollution that are quite different from those of fuel-fired combustion. The material is not homogeneous, and has a wide variation in fuel value ranging from 600 to 6500 Btu/lb of refuse as fired. Volatiles are driven off by destructive distillation and ignite from heat of the combustion chamber. Gases pass through a series of oxidation changes in which time-temperature relationship is important. The gases must be heated above 1200°F to destroy odors. End products of refuse combustion pass out of the stack at 800°F or less after passing through expansion chambers, fly ash collectors, wet scrubbers, and in some instances electrostatic precipitators. The end products include carbon dioxide; carbon monoxide; water; oxides of nitrogen; aldehydes; unoxidized or unburned hydrocarbons; particulate matter comprising unburned carbon, mineral oxides, and unburned refuse; and unused or excess air. Particulates are reduced in quantity. Normally only micron-size and submicron particles should escape with the flue gases. Care in operation is required to hold down particulate loading. Dust emissions in stacks may be in the range of 2–3 lb/ton/hr of refuse charged at a well-operated unit equipped with air-cleaning devices.

Incinerator design. There are several types of incinerator design promoted by manufacturers of incinerator equipment. Kiln shape may be round, rectangular, or rotary. The hearth may be horizontal fixed with grates, traveling with grates, multiple, step movement, or barrel-type rotary (Fig. 8). Drying hearths are provided on some types. Feed into the incinerator may be continuous, stoker, gravity, or batch.

It is necessary to know or estimate water content, percentage combustible material and inert material, Btu content, and weight of refuse to complete a rational design of incinerators. Heat balance can be calculated from several estimates based on averages. Available heat from the refuse must be balanced against the heat losses due to radiation, as well as from moisture, excess air, flue gas, and ash. Each type design has recommended sizings suggested by the manufacturer. There is fair agreement on the need for over 100% excess air. An allowance of 20,000 Btu/ft³ has been suggested for approximating chamber volume, and an allowance of 300,000 Btu/ft² for grate area. Incinerator loading rates of 40–70 lb/(ft² grate area) (hr) have been used. Small incinerators for apartment houses and institutions are loaded at much lower rates. The Incinerator Institute of America in its

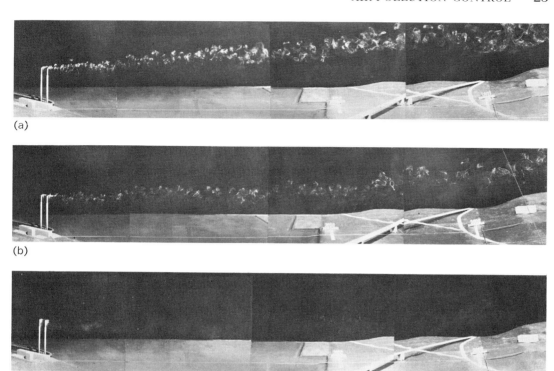

(a)

(b)

(c)

Fig. 6. Wind tunnel demonstration of dispersal patterns at specified wind speeds. (a) 20 mph. (b) 25 mph. (c) 30 mph.

standards has suggested loading rates for household or domestic-type refuse from 20 lb/(ft²) (hr) in 100 lb/hr burning units up to 30 lb/(ft²) (hr) in 1000 lb/hr units.

The Building Research Advisory Board (National Academy of Sciences, National Research Council) suggests that apartment house single-chamber incinerators should be sized on the basis of 0.375 ft³ capacity per person, 0.075 ft² grate area per person, and heat release rate of not more than 18,000 Btu/ft³ of capacity, where the burning period is 10 hr or less.

Air-monitoring instruments. Air-sampling methods may be classified generally as those for sampling particulates or gases or both concurrent-

ly. The samples may be analyzed for specific pollutants or for general pollution levels. Sampling devices have been constructed with many variants. Generally, however, they follow reasonably well-defined principles which include gravity and suction-type collection, with passage through thermal and electrostatic precipitators; impingers and impactors; cyclones; absorption and adsorption trains; scrubbing apparatus; filters of various materials, such as paper, glass, plastic, membrane, and wool; glass plates; and impregnated papers. Combination instruments that measure wind direction and velocity and direct air samples into multiple sample units, each of which represents a wind sector, are used for general sampling

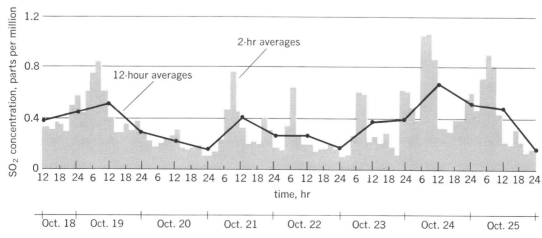

Fig. 7. Air-pollution episode in New York City, Oct. 18–25, 1963. The SO₂ values were at Christodora Station, 189 ft above ground. (*Research Division, New York University School of Engineering and Science*)

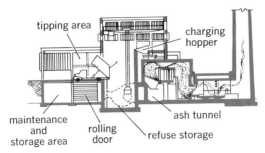

Fig. 8. Diagram of incinerator with rectangular grate. (*American Society of Civil Engineers*)

and locating of emission sources. Samples may be taken as single samples, or as a composite over a predetermined time period, or as a continuing monitoring operation. Some instruments are designed to extract a sample from the air, analyze it automatically, and record the result on a chart. Others take a sample which must be examined in a laboratory.

Many instruments have been developed during the 1950s and 1960s that are mechanized, automatic, and recording, so that they can be used with a minimum of attendance and manipulation. Such instruments require careful initial calibration with standard test substances that are to be measured, and a continuing field check with recalibration at frequent intervals to maintain accuracy.

Particulate samples may be analyzed for weight, size range, effect on visibility, chemical character, shape, and other specific information required. Gas samples may be analyzed to determine the presence of a specific gas or of a group of gases of the same chemical family; that is, nitrogen oxide may be determined or the concentration of all oxides of nitrogen may be established by analysis, or total hydrocarbons may be measured, or by more specific analysis the fractions of several specific hydrocarbons making up the total may be found.

Several types of units with air pumps drawing air through paper tapes mounted on a spool have been developed. The tape is moved automatically so that successive samples on fresh paper are taken at timed intervals.

High-volume samplers are used at many sampling network stations in the United States. The electron microscope has been employed for the examination of aerosols and fine particles. Spectrographic instruments are used for analyzing hydrocarbons and oxides of nitrogen and carbon. Automatically operated units take a sample and then pump chemicals into it at the appropriate time to produce a succession of chemical reactions; they are used to obtain a continuous record of concentration fluctuation of gaseous pollutants, such as sulfur dioxide and others where a wet-chemical analytical method is appropriate. Other instruments use the principle of conductance for measurement of a gas dissolved in a liquid medium. The sample is passed into the liquid medium and a change in electrical energy is measured and recorded. Such instruments measure the effect of any substance that is ionizable in the medium. Orsat analyses are made on flue gases. Photoelectric cells are used to control alarm systems connected to stacks.

Analytical instruments that use several principles of measurement have made it possible to take more data at less cost and manpower. Systems are being developed whereby data from a number of monitoring stations can be transmitted to a central point, transferred to computer program operations, and become statistical information concerning air-pollution concentration. Other systems under development provide for measurement of certain index pollutants while in motion by utilizing automatic instruments mounted in vehicles.

Methods of sampling and methods of analysis have yet to achieve widespread agreement or standardization. The American Society for Testing and Materials (ASTM) has published some 17 standard methods of tests applicable to atmospheric analysis. ASTM has also published definitions of terms relating to air sampling and analysis. Numerous industrial associations and professional organizations are in the process of bringing together in published form the multitude of sampling and testing methods in use.

Air-quality control. This is predicated on standards or guides that take form in official regulations and laws according to three control approaches: restriction of all sources of pollution so that pollution levels in community air are not in excess of certain levels chosen as a safe standard of air quality; limitation on the amount of a specific pollutant that may be present in the exhaust gas from a duct or a stack; or limitation on the amount of impurity in raw materials whose residues reach the community air. These approaches are frequently combined in an attempt to achieve maximum control. The guides or standards may consist of one or more of the following: ambient air-quality standards, emission standards, and material-quality standards.

Official control agencies at local, state, and Federal levels are now in the process of establishing ambient air-quality levels for such pollutants as sulfur dioxide and particulates. Many municipal control agencies have specified the limits of pollutants such as sulfur dioxide, particulates, and solvents that may be emitted from a single source. Many control agencies at municipal and state levels have adopted standards limiting the amount of sulfur in fuel. The U.S. Department of Health, Education, and Welfare publishes from time to time a digest of state air-pollution control laws. All major cities of the United States have adopted laws and regulations based on one or more of the quality control approaches mentioned.

Standards of ambient air quality may vary. In the state of New York, for example, there is a recognized difference in the kind and quantity of pollutants that may be emitted in rural areas, as opposed to highly urbanized areas. Within any region, land use may vary as to its industrial, commercial, residential, or rural components. Subregions based on the predominant land use and on air-quality objectives to be obtained may be established, each having its own air-quality guides.

[WILLIAM T. INGRAM]

Bibliography: *Air Conservation*, Amer. Ass. Advan. Sci. Publ. no. 80, 1965; Air Quality Committee of the Manufacturing Chemists Association, *Source Materials for Air Pollution Control Laws*, 1968; American Industrial Hygiene Association, Air Pollution Committee, *Air Pollution Man-*

ual, pt. 1: Evaluation, 1960, pt. 2: Control equipment, 1968; American Society for Testing and Materials Standards, *ASTM Standards on Methods of Atmospheric Sampling and Analysis*, pt. 23, 1967; *Apartment House Incinerators*, Nat. Acad. Sci.–Nat. Res. Counc. Publ. no. 1280, 1965; W. T. Ingram et al., Adaption of Technicon Auto-Analyzer for continuous measurement while in motion, *Technicon Symposia, 1967*, vol. 1: *Automation in Analytical Chemistry*, 1968; W. T. Ingram and L. C. McCabe, *The Effects of Air Pollution on Airport Visibility*, Amer. Soc. Chem. Eng., J. Sanit. Eng. Div., Pap. no. 1543, 1958; W. T. Ingram, C. Simon, and J. McCarroll, *Air Research Monitoring Station System*, J. Sanit. Eng. Div., Proc. Amer. Soc. Chem. Eng. 93, no. SA2, 1967; Interbranch Chemical Advisory Committee, U.S. Department of Health, Education, and Welfare, *Selected Methods for the Measurement of Air Pollutants*, Environmental Health Series, 1965; A. J. Johnson and G. H. Auth, *Fuels and Combustion Handbook*, 1951; W. C. McCrone, R. G. Draftz, and J. G. Gustav, *The Particle Atlas*, 1967; Metropolitan Engineers Council on Air Resources, *Incineration of Solid Wastes*, Proc. MECAR Symp., Mar. 21, 1967; A. C. Stern (ed.), *Air Pollution*, vols. 1 and 2, 2d ed., 1968; C. D. Yaffe et al. (eds.), *Air Sampling Instruments for Evaluation of Atmospheric Contaminants*, 2d ed., 1962.

Animal community

Theoretically an aggregation of animal species characteristically associated with one another. Animals of a community are usually held together in such aggregations by ties to the physical environment, to the vegetation with which they are associated, or to other animals of the community. *See* PLANT COMMUNITY.

With few exceptions, mainly in water, vegetation dominates the physiographic landscape and animals usually live among the plants or under their influence. Because of the intimate interrelationships between the physical environment or habitat and the plants and animals, an animal community must be treated as a part of a biological or biotic unit. In this unit, certain materials (chemical substances) of the Earth and cosmic energy (mainly solar radiation) are organized into organic substances and incorporated into living plant bodies. When eaten, these in turn are transformed into components of animal bodies and eventually returned to the physical environment after use by the animals. *See* ECOSYSTEM; NITROGEN CYCLE.

Habitat. That part of the physical environment in which particular animals live is called their habitat. The ecological position that a species occupies in a particular ecosystem is called a microhabitat or niche. Animals may spread into or enter many habitats, but instinctive traits may largely determine the ones in which they live. For example, birds in migration may pass over many habitats, but at the end of their journey they select one to which they are adapted.

Community animals are bound to their physical environment by requirements for life, such as their need for oxygen, water, heat, light, or other necessities that they obtain from the medium in which

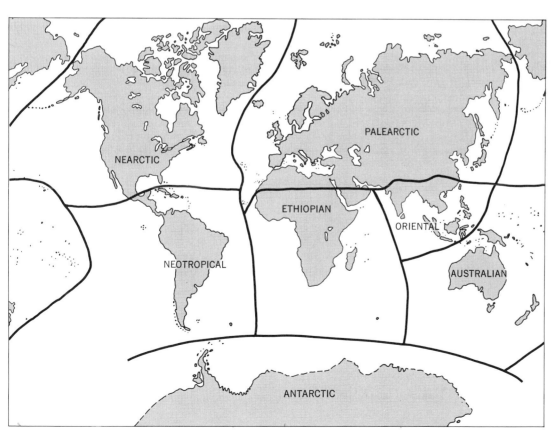

Fig. 1. Terrestrial faunal regions. (*After A. R. Wallace, Geographical Distribution of Animals, Macmillan, 1876*)

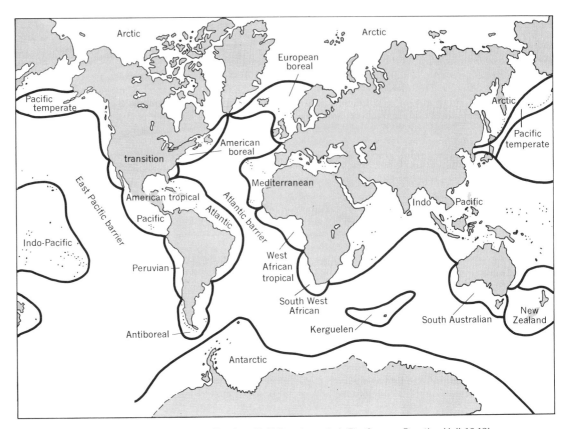

Fig. 2. Marine littoral faunal regions. (*Modified from H. U. Sverdrup, et al., The Oceans, Prentice-Hall, 1942*)

they live (air, water, or soil), and to plants for their food supplies, their shelter or protection, and in some cases for home, reproductive sites, nesting materials, perches, shade, and other detailed needs.

Community divisions. In a broad sense, there is only one complete, discrete animal community, the entire fauna of the Earth. So far as is known, the biosphere, which includes all living organisms of the Earth, is completely separated from any other biological community anywhere else in the universe. For practical reasons, the world community of animals, the zoosphere, must be subdivided for more detailed study into subordinate animal communities, but there are few, if any, boundaries between them absolutely separating one community from another. Most have zones of intergradation between then, known as ecotones. *See* BIOGEOGRAPHY; COMMUNITY.

Because of this intergradation, the boundaries between communities are drawn arbitrarily within the ecotone as near to a natural division as knowledge about the animals permits. The community units are usually selected to contain a relatively consistent composition of animal species. The boundaries of these units are generally placed at lines where the community composition changes sufficiently to warrant recognition of a different unit.

Distribution of animal communities. The existing animal communities occur in those situations where in the march of history the evolution of their species and the interactions of these with their environments have left them. The zoosphere of the Earth has been divided into marine and ter-

restrial faunal regions on the basis of great geographic divisions separated by important barriers. These regions are indicated on the maps in Figs. 1 and 2. *See* ZOOGEOGRAPHIC REGION; ZOOGEOGRAPHY.

Criteria for communities. Animal communities to be suitable for classification purposes should portray the interrelationships of the animals in nature. Because animals are elusive, mobile, and difficult to observe, the more obvious features of landscape and vegetation are often used as criteria for field recognition of communities that have been established, for example, animal communities of tall-grass prairie or vegetated sand dunes. Careful study over large areas is needed to identify units of relatively uniform species composition and to locate their ecotones with units of different composition. *See* COMMUNITY CLASSIFICATION.

A community to be recognized as a unit must usually meet the following criteria: (1) correlation of ecologic dispersion of the animals with particular features of the habitat and vegetation; (2) an associated group of species that interact with one another and may have in part evolved together; (3) occupation of a geographic area that may be either continuous or discontinuous, as on different mountaintops; (4) boundaries that may be relatively abrupt or may intergrade gradually with adjacent communities; and (5) capability of recognition in the field by observable characteristics of landscape, vegetation, and conspicuous animals.

Community operation. Because animals in general depend upon plants for food, any animal community must be associated with plants or else have special means of obtaining food. Animals such as birds or carnivores living in cliffs make periodic trips to adjacent areas in search of food, but spiders with food-catching webs in the same cliffs wait for insects nurtured upon plant foods to come to them. Sedentary aquatic animals, such as sponges and bivalve mollusks, pump water through their bodies and trap microscopic plankton that enter with the water.

Plants in general are capable of making more food than they need for their own operation and are adapted to having the surplus removed by animals. The amount of this surplus generally limits the number of animals that can be supported in a community. As a rule, a huge mass of vegetation is required in order to support a relatively small mass of animal flesh. *See* BIOLOGICAL PRODUC-TIVITY.

Among the animals in a community, some are adapted to take food from plants and transform it for use in their own bodies. These vegetarian animals in turn fall prey to carnivorous animals, which are adapted to take, digest, and assimilate animal flesh. Also, parasites are adapted to utilize food from green plants, vegetarian animals, and carnivores. Parasites in general are smaller than the hosts. After death, if not used in other ways, the bodies of both plants and animals will be decomposed and returned to inorganic matter by microorganisms, collectively known as saprophytes. This pyramidal food relationship is illustrated in Fig. 3.

Community organization. Community animals are ecologically interrelated in ways that are adapted to the physical habitat and the associated plants. On land, animal communities are located at the bottom of the atmosphere, where gravity holds them, but aquatic animal communities may be suspended in the water as well as located on the bottom. Communities which are suspended in the ocean may be floating or swimming groups; those animals on the bottom may attach to solid support, burrow in the mud, crawl on the bottom, or swim.

Terrestrial communities also have great variety in organization, caused largely by their complex interrelationships with habitat, physiography, and associated plants. In forested areas, some animals live on the ground, some burrow in the soil, some climb the plants and may be restricted to particular levels of the vegetation, forming layer communities. In each situation a number of species are intimately associated under the dominating influence of shade, shelter, and food supplies furnished by plants.

In shrubby plant communities, the animals are restricted to fewer layers. In short-grass prairie, low desert shrubs, or moss-lichen tundra, they are restricted to a single plant layer in which only small animals can obtain shade or shelter; and larger animals, such as reindeer of the Arctic tundra and American bison formerly of the prairies, are exposed to the rigors of the climate with little or no amelioration by plants.

Consortism. Some animals individually consort in a mutually helpful way with individuals of a different species. Thus crabs may be bedecked with hydroids and sponges that find a base for attachment and in return help to camouflage the crab.

Some species associate on a symbiotic basis, in which one member derives benefit without damage to the other, for example, spiders which spin their webs in burrows made by rodents. Others may associate as parasites, in which case one species benefits at the expense of the host, as lice and fleas do in the fur of mammals. Still others may form mixed bands which are composed of several to many species. *See* ECOLOGICAL IN-TERACTIONS.

Social relations. Environmental forces may drive individual animals of the same species to collect in an aggregation. Thus a playa lake of a desert diminishing in size by evaporation may concentrate tadpoles in the lake into close contact with each other. If, on the contrary, individuals are bound together by mutual attractions that result in social behavior, the group is termed a society. There may be many social groups in a single community, exemplified by ant colonies, flocks of birds, and herds of elk. *See* SPECIES POPULATION.

There are many ways in which plants and animals of a community aid each other. Many flowers, for example, are pollinated by insects or hummingbirds, which obtain nectar or pollen for food while providing this service to the plants. Consortive conduct, both individual and collective, helps to knit the animals into a community organization. These behavioral relationships are distinctive of animal communities and are not found in plant communities.

[ANGUS M. WOODBURY/FRANCIS C. EVANS]

Bibliography: H. G. Andrewartha, *Introduction to the Study of Animal Populations*, 1961; E. D. Le Cren (ed.), *The Exploitation of Natural Animal Populations: A Symposium of the British Ecological Society, Durham, 28th-31st March 1960*, 1962; V. E. Shelford, *Animal Communities in Temperate America*, 2d ed., 1937; A. M. Woodbury, *Principles of General Ecology*, 1954.

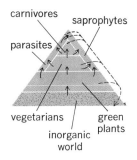

ANIMAL COMMUNITY

Fig. 3. Diagram of the food pyramid showing the general trends in food circulation.

Antarctica

A snow- and ice-covered continent roughly centered on the South Pole and surrounded by an ocean consisting of the southern parts of the Atlantic, Pacific, and Indian oceans. It is the fifth largest continent, having an area of 5,400,000 mi^2, which is $1\frac{1}{2}$ times the size of the United States (Fig. 1). Antarctica is the coldest and highest of all the con-

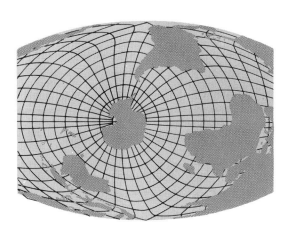

Fig. 1. An equal-area projection which shows Antarctica's size in relation to that of the nearest continents. (*U.S. Antarctic Projects Officer*)

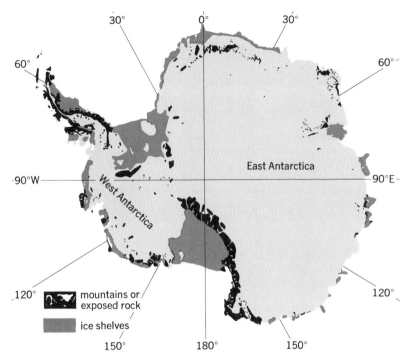

Fig. 2. Ice shelves and regions of mountains or exposed rock. (*After American Geographical Society's Antarctic Map Folio Series*)

tinents and there are no permanent inhabitants. For millions of years snow and ice have built up on the land, burying all but 1% of it (Fig. 2).

Ice sheet. The continent holds 5,750,000 mi³ of ice, which is 90% of all the ice on Earth. The mean height of the ice surface is 6700 ft above sea level, giving Antarctica almost three times the mean height of the other continents. The South Pole is 9500 ft above sea level. About 10% of the area is made up of ice shelves, which are ice sheets floating on the sea and commonly 1000 ft thick, but in places over 4000 ft thick. The remainder of the ice sheet rests on rock. Average ice thickness over this vast area is 6000 ft, though measurements are not yet sufficient to give figures to a better accuracy than ±20%. If all this ice were to melt, an unlikely event which would in any case take thousands of years, world sea levels would rise some 200 ft.

As in glaciers, the ice of this huge sheet is moving slowly downhill at speeds varying from a few feet to about 3000 ft per year. When it reaches the coast, it breaks off to create the tabular icebergs which are characteristic of the Antarctic waters. *See* ANTARCTIC OCEAN.

Geology. Antarctica is divided into two parts by one of the world's great mountain chains, the Transantarctic Mountains (Figs. 3 and 4). Radioactive dating indicates that parts of East Antarctica are at least 1,800,000,000 years old. Similarities of rock types and sequences in other continents suggest that at one time it may have been part of a single great landmass known as Gondwanaland, including what is now the Indian peninsula, Australia, South America, and Africa. One of the crucial lines of evidence for this former supercontinent arises from the discovery in recent years of Permo-Carboniferous tillites (about 280,000,000 years old)—consolidated ground moraines from a former widespread and broadly synchronous glaciation of the southern continents. The Antarctic coal measures, perhaps forming the world's largest coalfields, correspond in age with the main coal horizons of South Africa and Australia. According to the Gondwanaland hypothesis, the superconti-

Fig. 3. Aerial view of Transantarctic Mountains. (*U.S. Navy photograph*)

nent broke up around 150,000,000 years ago, and East Antarctica began to drift south toward its present position.

Most of the geological history of West Antarctica postdates the disruption of Gondwanaland. West Antarctica represents a southerly continuation of the South American Andes; it is a relatively young mountain belt welded on to the continental shield of East Antarctica. If the present ice sheet were removed, West Antarctica would consist of a group of islands, whereas much of East Antarctica would be above sea level. About 60% of the present ice sheet rests on rock which is above sea level. The weight of the ice has depressed the land surface beneath it by as much as 2000 ft, so that if the ice load were removed, isostatic recovery would far exceed the rise in sea level caused by the melting of the ice. The rock surface would rise to a mean elevation of about 3000 ft above sea level, leaving Antarctica still as high as Asia, now the second-highest continent. The highest peak in Antarctica is 16,860-ft Vinson Massif in the Sentinel Range; it was climbed by an American party in 1966. There are very many unclimbed peaks in Antarctica.

Climate. Antarctica is the coldest region on Earth, the interior being colder than Siberia in winter and colder than the North Pole. The world record low temperature, −126.9°F, was recorded at the Soviet Vostok Station on Aug. 24, 1960. The

Fig. 4. Motor toboggan hauling the sledges of an exploratory party in the Transantarctic Mountains.

mean annual temperature at Vostok, the coldest inhabited place, is −71°F. At the South Pole there are 6 months (Mar. 21 to Sept. 21) during which the Sun remains below the horizon. The length of the period of continuous darkness in winter and continuous daylight in summer decreases with distance from the pole, so that many coastal points have only a few weeks of each. There are considerable differences of climate between coast and interior: The average temperature at the South Pole is −60°F, while at Anvers Island off the Antarctic Peninsula it is 26°F. There is very little precipita-

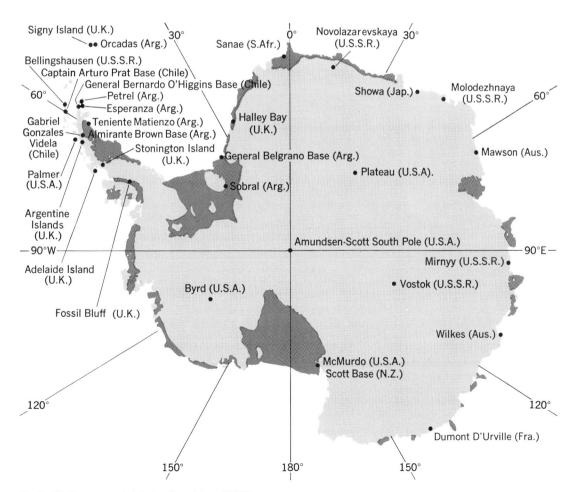

Fig. 5. Stations occupied during the winter of 1968.

tion and what there is falls as snow; the high plateau that covers most of Antarctica is the world's largest desert. Precipitation at the South Pole is equivalent to 2 in. of rainfall per year, though increasing amounts toward the coasts give figures up to about 25 in. The windiest part of Antarctica is Adélie Coast south of Australia, where one temporary research station had a mean annual wind speed of 37 knots, with many gusts over 100 knots. However, winds on the ice sheet in the interior are not exceptional; Plateau Station has a mean wind speed of only 9 knots.

Biology. Algae, mosses, and lichens are the characteristic plants of Antarctica; only two genera of flowering plants are known. Lichens cling to rock outcrops on all the main mountain ranges, but they are often inconspicuous. There are no vertebrate land animals; the largest creature living entirely on land is an insect less than 1/4 in. long. Four species of seal and four species of flightless bird, the penguins, breed on or near the shores of Antarctica, though they probably spend more than half their lives at sea. The most widespread of the penguins is the Adélie, which nests on land in summer. The emperor penguin nests in winter on sea ice attached to the continent, and it can endure extremely low temperatures. The gentoo penguin and the chinstrap, or ringed, penguin do not extend further south than the Antarctic Peninsula. Weddell seals are the commonest seals on the coast of Antarctica, followed by the crabeater seal. Leopard seals and Ross seals are much less common. Apart from the penguins, only 10 species of birds breed on the coast of Antarctica at higher latitudes than the Antarctic Peninsula. Evidently because of the lack of food, few birds stray inland, though skuas and snow petrels have been seen far inland during the summer months.

Exploration and research. Antarctica was first sighted in 1820 by American and British sealing vessels and by a Russian expedition. But it was not known to be a continent until the discoveries of Dumont d'Urville of France, Charles Wilkes of the United States, and Sir James Clark Ross of Great Britain in the period 1838–1843. The first extensive land explorations were made by British expeditions under Captain R. F. Scott and Sir Ernest Shackleton during 1901–1912. The South Pole was first reached by a Norwegian party under Roald Amundsen in 1911 and a few weeks later by a British party under Scott. Scott's party perished on the return journey from the pole. The first extensive aerial explorations were made by American expeditions under Admiral Richard E. Byrd (1928–1930 and 1933–1935). Lincoln Ellsworth made the first flight across Antarctica in 1935.

Since World War II voyages of discovery and great journeys have given way to scientific research on a semipermanent footing. The International Geophysical Year 1957–1958 provided a great stimulus to research in meteorology, glaciology, auroral physics, geomagnetism, ionospherics, seismology, and gravimetry. Twelve nations manned a total of 44 research stations south of latitude 60°S. The wintering population of Antarctica, which consists entirely of the men at scientific stations, has remained at about 800 since 1957 (Fig. 5). In summer, however, logistic support operations may temporarily raise the population to about 5000. The Antarctic Treaty of 1959, ratified by 16 countries, encourages cooperation and ensures freedom of scientific investigation in any part of the continent. It guarantees nonmilitarization by allowing mutual inspection and requiring the exchange of information on current activities. Technical developments include the regular use of satellite photography in weather forecasting and in the navigation of ships through pack ice. Airborne radar is used to measure the thickness of the ice sheet, which in places has been found to be 14,000 ft thick.

[CHARLES W. SWITHINBANK]

Bibliography: H. S. Francis, Jr., and P. M. Smith, *Defrosting Antarctic Secrets*, 1962; T. Hatherton (ed.), *Antarctica*, 1966; R. S. Lewis, *A Continent for Science*, 1965; L. B. Quartermain, *Down to the Ice*, 1966; U.S. Naval Support Force, Antarctica, *Introduction to Antarctica*, 1967.

Aquifer

A subsurface zone that yields economically important amounts of water to wells. The word aquifer is synonymous with the term water-bearing formation.

An aquifer may be porous rock, unconsolidated gravel, fractured rock, or cavernous limestone. Economically important amounts of water may vary from less than a gallon per minute for cattle water in the desert to thousands of gallons per minute for industrial, irrigation, or municipal use. *See* ARTESIAN SYSTEMS; GROUNDWATER.

Aquifers differ widely in shape, area, and thickness. Among the most productive are the sand and gravel formations of the Atlantic and Gulf Coastal plains of the southeastern United States. These layers commonly extend for hundreds of miles and may be several hundred feet thick. Also highly productive are some of the deposits of sand and gravel washed out from the continental glaciers in the northern United States; the outwash gravel deposits from the western mountain ranges; certain cavernous limestones such as the Edwards limestone of Texas and the Ocala limestone of Florida, Georgia, and South Carolina; and some of the volcanic rocks of the Snake River Plain in Idaho and the Columbia Plateau.

[ALBERT N. SAYRE/RAY K. LINSLEY]

Arctic and subarctic islands

Defined primarily by climatic rather than latitudinal criteria, arctic islands are those in the Northern Hemisphere where the mean temperature of the warmest month does not exceed 50°F and that of the coldest is not above 32°F. Subarctic islands are those in the Northern Hemisphere where the mean temperature of the warmest month is over 50°F for less than 4 months and that of the coldest is less than 32°F.

Such islands generally are in high latitudes. Distribution of land and sea masses, ocean currents, and atmospheric circulation greatly modify the effect of latitude so that it is often misleading to use location relative to the Arctic Circle as a significant criterion for designation of arctic or subarctic. The largest proportion by area of the islands lies in the Western Hemisphere, located primarily in Greenland and in the Canadian Arctic Archipelago.

Diversity of land surfaces. Physiographically, the islands include all the varied major landforms found elsewhere in the world, from rugged mountains over 10,000 ft high, through plateaus and hills, to level plains only recently emerged from the sea. All have been glaciated except Sakhalin and some of the islands in the Bering Sea sector. R. F. Flint reported that of the 5,800,000 mi² (10%) of land area of the world still ice covered, over 5,000,000 mi² is in Antarctica, and over 700,000 mi² lies in the arctic islands with over 600,000 mi² of this in Greenland. Removal of the weight of ice sheets and the resultant crustal rebound has exposed prominent marine beaches and wave-cut cliffs on many of the islands. These now commonly occur at elevations of over 300 ft above sea level.

Climate. The general climatic pattern of these islands is set by their location relative to the two semipermanent centers of low pressure over the Aleutian Islands and over Iceland (the Aleutian Low and the Icelandic Low). Especially during the winters these low-pressure centers affect the areas from Kamchatka to southeastern Alaska and from Newfoundland to Novaya Zemlya, respectively. The intervening areas, including the islands of the northwestern Canadian Arctic Archipelago and those of the central and eastern Russian Arctic, are much less subject to cyclonic storms; rather they are dominated by stable, dry air masses often linked across the polar basin from either continent. During the summer this pattern decreases in intensity. Most of the precipitation is cyclonic in origin. Because they are marine areas, the islands receive more precipitation than they otherwise would, yet even so this is very light for most of the arctic islands removed from the zone of cyclonic activity. Also, because they are marine areas, the islands, regions of low temperatures by definition, are not regions of extreme low temperatures. Much lower temperatures are reported from continental land areas farther south than from the most northern arctic islands. In general, the larger the island and the closer its proximity to a continental landmass, the higher are the summer temperatures and the lower its winter temperature.

Vegetation and soils. The climatic differences between arctic and subarctic islands are reflected in their natural vegetation. The arctic islands are treeless. Natural vegetation consists of the tundra—mosses, sedges, lichens, grasses, and creeping shrubs. Luxuriance and continuity of ground cover vary with such factors as moisture, insolation, and soil nutrient conditions. Bare ground is often exposed and in places plant growth may be lacking completely except for a few rock-encrusting lichens. In such places the ground surface may consist of frost-shattered rock fragments, tidal mud flats, boulder-strewn fell fields, or snow patches and ice. Permafrost (permanently frozen ground) occurs throughout the Arctic (and in parts of the subarctic) and is reflected in impeded drainage and patterned ground.

The natural vegetation of subarctic islands characteristically is the boreal forest or taiga, composed predominantly of conifers such as spruce, fir, pine, and larch with deciduous trees such as birch, aspen, and willow; the latter are especially common in regrowth of clearings in the forest.

Impeded drainage because of permafrost or glaciation gives rise to numerous ponds and muskeg areas. A transitional type of vegetation, the forest-tundra, is recognized on some subarctic islands in sectors where smaller trees are widely spaced and abundant mosses cover the ground.

The typical soils of the subarctic islands are podzols—the surface soil grayish-white beneath the raw humus layer and highly acidic in nature. The tundra soils of the Arctic islands really consist only of a dark-brown peaty surface layer over poorly defined thin horizons, and much of the ground cannot properly be termed soil.

Character of major islands. Within this general description, individual islands vary considerably (see table and illustration). A brief summary of some of the larger islands and archipelagos in the Western Hemisphere is given in the following sections.

Aleutian Islands. Extending southwest for more than 1000 mi, from the Alaska Peninsula at 163°W to Attu Island at 175°W, the Aleutians separate the Bering Sea to the north from the Pacific Ocean to the south. Their rugged, mountainous surface consists of the drowned continuation of the Alaska and Aleutian ranges. Several active volcanoes are included among them. Most of the islands were glaciated by local mountain glaciers, a few of which persist in reduced size on some of the eastern islands (Unimak, Unalaska, and Umnak). There are about 150 islands and innumerable reefs. They are grouped from east to west as the Fox, Andreanof, Rat, and Near islands. Unimak, the easternmost island, is the largest, about 65 mi long and 25 mi wide.

Size and elevation of larger arctic and subarctic islands*

Name	Area, mi²	Highest point, elevation, ft
Aleutian Is.		
Unimak I.	15,500	Shishaldin Volcano, 9978
Unalaska I.	10,800	Makushin Volcano, 6680
St. Lawrence I.	18,200	Kookooligit Mts., 2207
Nunivak I.	16,000	Roberts Mt., 1675
Kodiak I.	37,400	Grayback Mt., 3317
Canadian Arctic		
Archipelago	500,000	
Baffin I.	183,810	Penny Highlands, 8500
Ellesmere I.	82,119	British Empire Range and United States Range, 10,000
Victoria	81,930	Shaler Mts., 2000
Banks	23,230	Durham Heights, 2500
Devon	20,861	Treuter Mts., 6190
Melville	16,141	Raglan Range, 3500
Axel Heiberg	15,779	Name not available, 8400
Southampton	15,700	Porsild Mts., 1750
Prince of Wales	12,830	Name not available, 500
Newfoundland	42,734	Gros Morn, 2666
Greenland	840,000	Mt. Forel, 11,286
Iceland	39,961	Oraefajökull, 6955
Svalbard (archipelago)	24,100	
Vest-Spitsbergen	15,250	Newtontopp, 5630
Franz Josef Land (archipelago)	7,000	Name not available, 3000
Novaya Zemlya (archipelago)	36,000	
Severny I.	21,000	Name not available, 3510
Yughny I.	15,000	
Severnaya Zemlya (archipelago)	14,000	Name not available, 1500
New Siberian Is.	12,000	Name not available, 1033
Wrangel I.	2,000	Peak Berry, 3510
Sakhalin I.	27,000	Nevelskoi Mt., 6600
Kurile Is.	6,000	Alaid Volcano, 7674

*Approximate only in some cases because of incomplete mapping.

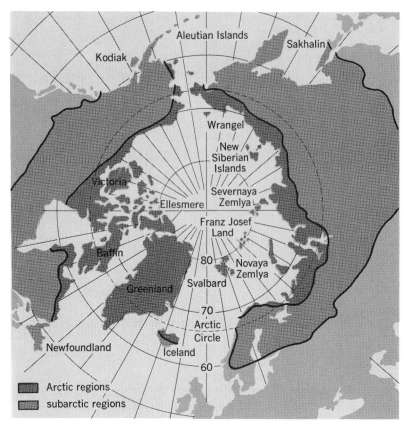

Islands of the arctic and the subarctic regions.

Arctic regions

subarctic regions

maximum elevation is about 3000 ft in Melville Island but is commonly less than 1000 ft. The northwestern part of the Queen Elizabeth Islands is a recently emerged coastal plain, in which salt domes occur at places. Most of the islands in the western part of the southern archipelago are low and comparatively flat.

Most of the islands were glaciated, although there still is not sufficient information to establish conditions in the northwestern islands. Relic ice caps remain upon Ellesmere, Axel Heiberg, Devon, Bylot, and Baffin islands, because of their higher elevations and greater precipitation.

The Archipelago has long, cold winters. For 3–4 months mean monthly temperatures range from −20 to −30°F over most of the area. February is usually the coldest month; temperatures of −35 to −40°F are commonly recorded then. Rarely do temperatures drop below −50°F, although a minimum of −63°F has been recorded at three stations. The southeastern and eastern parts of the archipelago have milder winters, because of the proximity to open water in Davis Strait and the passage of cyclonic storms. During the cool summer, temperatures are more uniform over the archipelago, with the July mean ranging only from 40 to 50°F. Maximum temperatures exceed 60°F at most but not all stations.

Precipitation is light. Most of the archipelago receives less than 10 in. a year, with only three stations (in the southeast) recording more than 15 in. One station (Eureka) reported less than 2 in. a year over a 2-year period. Snow may fall in any month, but over most of the archipelago rain accounts for about half the precipitation. Snowfall averages from 12.5 to 100 in. (in the southeast). Natural vegetation is of the tundra type.

Sea ice closes archipelago waters to shipping for 9–10 months in the year. Breakup usually occurs between late June and mid-July with freezeup between mid-September and late October, depending on location. The waters around the northwestern islands are rarely clear of ice. Icebergs are confined to the eastern parts, chiefly in the waters separating the archipelago from Greenland.

Newfoundland. Generally a plateau of rolling surface, this subarctic island is tilted west to east from the Long Range Mountains (over 2000 ft) to the Avalon Peninsula (700 ft). The northeast-southwest grain of the island, the result of folding and faulting, is reflected in the coastal configuration and physiography. The island was completely glaciated and most of it shows the results of intense ice scour. The indented seacoasts are commonly cliffed and rugged, with marked fiords in the high sectors of the Northern Peninsula. The only significant lowland area is a coastal plain in the west.

From the standpoint of climate and natural vegetation the island is predominantly subarctic, yet it includes aspects of the arctic and of the more continental regions. The moderating marine influence on climate is lessened by the island's east (that is, lee) location relative to the continental landmass, and by the cold Labrador Current, which flows south along the east coast and swings around to affect the south and parts of the west coasts.

Winters are cold, particularly in the interior (January mean, 15–20°F), but are somewhat mild-

The Aleutian Islands experience extremely variable weather because of the frequent cyclonic storms, complicated by the mountainous terrain. An associated gusty wind is known as the williwaw. Swift sea currents run among the islands and there is much fog. At Dutch Harbor (Unalaska Island) the mean temperatures are January, 32°F; July, 51°F; while at Atka they are 33°F and 50°F, respectively. Mean annual precipitation at the two stations is 56.7 and 70.2 in., respectively. The islands are treeless but generally support a luxuriant growth of grass, willow, and alder.

Canadian Arctic Archipelago. All those islands lying north of the continental mainland and west of Greenland to 141°W are included in this group. In rough outline the archipelago resembles a triangle, from a rather irregular base at about 61°N in the east and 67°N in the west, to its apex at the northernmost tip of Ellesmere Island in latitude 83°30′N. Those lying north of 74°N are known as the Queen Elizabeth Islands.

The easternmost islands of the archipelago are mountainous, whereas those to the west and northwest are plateaulike or plains. The mountains in Baffin Island, eastern Devon Island, and southeastern Ellesmere Island are composed of Precambrian rock and average 5000–7000 ft elevation. Continuing the mountain line northward through Ellesmere Island and Axel Heiberg Island, summit elevations exceed 10,000 ft (partly reflecting a different fold axis, the Innuitian). In contrast the more southerly of the Queen Elizabeth Islands are more typically plateaulike, even though they contain more of the Innuitian fold structure: here the

er on the coast, especially in the southeast (St. John's January mean, 24°F). Conversely, the coasts remain cooler in the summer (about 55°F in July), whereas the western lowlands exceed 60° in mean daily July temperatures. The entire island has abundant precipitation, well distributed through the year, with heaviest falls (over 55 in.) in the south. Snowfall is abundant (over 100 in.) everywhere except along the south coasts. Sea ice seals off all coasts at its maximum extent, except the south. Fog is fairly frequent both on the coast and inland. It is most common in summer and on the southeast coasts. The juxtaposition of the cold Labrador Current and the warm Gulf Stream in the offshore ocean is largely responsible.

Boreal forest covers less than half of the island, with the best stands in the western lowlands and the north central valleys. Poor drainage, resulting from recent glaciation, and elevation restrict its extent. Where the altitude exceeds 1200 ft, the forest gives way to barrens—extensive areas of tundra—in the west and southwest.

Greenland. The world's largest island, Greenland extends over 1600 mi from Cape Farewell (59°46′N) at the south to Cape Morris Jesup (83°39′N) at the north, the latter being the nearest land to the North Pole. Its greatest width (77°N) is just under 700 mi. Five-sixths of the surface is buried beneath the largest remaining land ice sheet in the Northern Hemisphere, and numerous smaller ice caps and glaciers occur as separate bodies around its margin. In profile the ice sheet is similar to a flat shield rising gently to form three broad domes, the highest of which exceeds 10,000 ft. A maximum thickness of 7000 ft of ice has been estimated, but more work has to be done on this as well as on the suggestion that the underlying surface may consist of several separate islands rather than just one. Occasional peaks (nunatak) project through the ice sheet near its margin. Tongues of ice from the main glacier descend to the sea at many points. Most of the icebergs in the North Atlantic originate from such glaciers in southwest Greenland.

The ice-free margins are widest in the southwest and on the northeast coasts, although access to the latter is impeded by the Arctic pack ice (storis) in the southward-moving East Greenland Current. All the land has been intensely glaciated, except the northernmost (Peary Land), which probably received insufficient precipitation. The margins include alpine mountains, plateaus (particularly in areas of basalt bedrock in the central areas on both east and west coasts), and lowlands. The skaergaard, a swarm of low islands and reefs, is very prominent along the southwest coast. The shore is much indented by bays and fiords.

The full significance of the ice cap in the climatic pattern is still not known. Temperatures there usually range from 27 to −49°F through the year. One associated phenomenon is the outward movement of strong katabatic winds from the ice margins which often produce a foehn or chinook effect on the valleys through which they are channeled. Climatic conditions in the ice-free margins are extremely variable, and the local complex topography has a great influence: for example, the inner parts of the fiords usually are warmer in summer than the outer, and colder in winter. Mean winter temperatures range rather uniformly on the west coast, from a February mean of 18°F at Ivigtut in the southwest, to −20°F at Smith Sound in the northwest. During the summer there is much less contrast, with July mean temperatures of 50°F in the south and 35–40°F in the north. Precipitation, mainly as snow, decreases rapidly from 46 in. at Ivigtut in the south to less than 10 in. north of 69°N. Notable climatic change has occurred in Greenland and its adjacent seas within historical times. Fog is common through the summers, especially near the broken sea ice. The island becomes ice locked in winter except for part of the southwest coast.

The natural vegetation of the ice-free areas is essentially tundra. Plant growth reaches its maximum development in the inner parts of the fiords, particularly in southwest Greenland. Five varieties of arboreal growth occur in the latter area, and in the Julianehaab district some birches grow to the height of 20 ft. Copses up to 7 ft high occur in favored areas as far north as Disko Bay but the grasses, mosses, and stunted growth of the true tundra are more typical and commonly make up the only vegetation over much of the island.

[WILLIAM C. WONDERS]

Bibliography: P. D. Baird, *The Polar World,* 1964; R. F. Flint, *Glacial Geology and the Pleistocene Epoch,* 1947; G. H. T. Kimble and D. Good (eds.), *Geography of the Northlands,* Amer. Geog. Soc. Spec. Publ. no. 32, 1955.

Arctic biology

The Arctic, with its small land areas surrounding the frozen arctic seas, has been exploited from the south for its yield of oil and furs of animals. Until the recent prospect of arctic petroleum and minerals, only traders and explorers from temperate lands came as transients among the scant population of indigenous arctic people. Terence Armstrong described how the recent development of arctic natural resources has multiplied invasion of the Asian Arctic by strangers from the south. Many new residents are now moving northward to exploit American arctic petroleum and minerals. This new and still unsettled movement of people into the Arctic encounters life that is strange to them in the long cold and darkness of arctic winters. Interesting biological, social, and economic conditions, as well as problems of adjustment to the Arctic, face the newcomers.

Life on land. Only about 20 species of mammals of the more than 3000 species in the world live on the treeless arctic tundra, where production of plants is so sparse and specialized that few kinds of animals can find a living. The only large arctic herbivores are musk-ox and caribou or reindeer, but many small herbivorous mice and lemmings live in obscure ways under winter snow. These small herbivores sustain carnivorous bears, wolves, foxes, weasels, and a surprising variety of small shrews.

The sparse vegetation on arctic lands must accomplish the annual production on which all arctic life depends during a short cool summer in which periodic freezing requires that all plants must retain resistance to frost and adjust reproduction opportunistically to brief spells of sufficient warmth. Spiders, insects, and many cold-blooded animals meet requirements of the Arctic by devices that could not be projected by imagination based on southern experience. Their success demonstrates the surprising adaptability of some species. Their few kinds show the rigorous selection by the arctic seasonal regimes.

No vertebrate animal seems to endure freezing, as do so many plants and invertebrates. Arctic flowers protruding through snow in early spring freeze to brittle hardness in repeated episodes of cold with only brief interruptions of the progress of forming seeds. John Baust finds that a small boreal beetle that winters in stumps above snow contains over 20% of glycerol in its fluids. The attractive view that the substance serves as an antifreeze is spoiled by the observed freezing of the beetle in arctic temperatures. In warm summer the beetle contains no glycerol, and is killed at the temperature at which water freezes. It is possible that glycerol protects against the destructive consequences of freezing as glycerol and some other cryoprotective agents allow long frozen storage of bull's sperm for artificial insemination and tissues like cornea for surgical repair of damaged eyes. Natural protection from injury by freezing occurs by diverse ways in the Arctic. Results of research into natural adaptation to arctic temperatures may lead to cryoprotective ways for preservation of cells and tissues.

Life on seas. Peter Freuchen and Finn Salomonsen narrated how the arctic seas support the wandering polar bear and the small arctic fox that often accompanies the bear far out on winter ice-covered seas. These ice-going mammals nevertheless return to shore to breed. The main food of the bear is the small arctic ringed seal, but it also takes eagerly to carrion remains of sea mammals. On sea ice the arctic foxes take remnants of the prey of bears or any flesh. On land in summer their prey is mice and birds.

The large bowhead whale drew whalers into the margins of Atlantic and Pacific ice with such success that the Atlantic bowhead is very nearly extinct. Fur seals and sea lions do not enter arctic life, but large numbers of hair seals breed at the margins of arctic ice. In fact, about 90% of the world's hair seals bear their pups on antarctic or arctic ice, attesting the productivity of polar seas and the capability of warm-blooded seals for living in icy waters.

Preservation of warmth. The preservation of warmth in arctic winter is a matter of real concern. Early explorers recorded that some arctic birds and mammals were as warm as those of warmer regions. L. Irving and J. Krog found that arctic and tropical birds and mammals are similarly warm in their interiors. Exposed experimentally to a temperature of −40° an arctic fox is comfortable in its thick fur. It generates no more heat in cold and so conserves heat by its insulation. This model of adaptation to arctic cold through insulation indicates an economy of heat that sustains a large arctic mammal at an expenditure of metabolism no greater than that required for mammals in a milder climate. The large arctic mammal is thus not handicapped by cold.

Arctic people. Arctic man, with meager natural insulation, wore fur clothing skillfully made from arctic animals, built shelters, and utilized fire.

From Greenland across North America to the edge of Siberia, Eskimos lived with a similar language and a culture distinct from those of northern American Indians.

Across arctic Eurasia, people of different cultures and languages reflected the fact that they were derived from peoples of steppes and forests who had migrated northward along great rivers and coastal routes. Indigenous Eurasian people are now far exceeded in numbers by fresh migrants who perform technical tasks for exploitation of minerals. Migration of strangers into arctic America seems to have just begun.

Migration of life into Arctic. Throughout the last Pleistocene glaciation, arctic plants and animals were confined to limited refuge areas free from ice. David Hopkins showed that 10,000 years ago melting of the ice sheets allowed spread of plants and animals from arctic refuges and invasion from the south by plants and animals that had retained versatility to adapt for arctic life. The distribution and many forms of life are modern characteristics of the recently changing Arctic.

Spectacular in the arctic spring are migrations of birds coming to nest for a few summer months on arctic lands and coasts after wintering on temperate lands and seas and even over South America and Africa. These great migrations visibly demonstrate the inclinations of birds to occupy opportunistically suitable parts of the world and their capability for organized flights and long navigation. In the seas fishes, whales, and seals annually migrate northward to breed and harvest the production of the arctic year. On land a residue of American caribou still migrate in summer into the Arctic. As arctic migrants return south for winter, they export from the Arctic the annual increment in their populations, an exploitation of arctic production.

It is an interesting speculation that these visible annual migrations in some ways recapitulate the postglacial resettlement of the Arctic. Plant, animal, and human life seem to have been ever pressing their capability for life in the Arctic and developing adaptability for the changing conditions of the arctic seasons. *See* BIOCLIMATOLOGY; BIOGEOGRAPHY; VEGETATION ZONES, WORLD.

[LAURENCE IRVING]

Bibliography: T. E. Armstrong, *Russian Settlement in the North*, 1965; J. Baust, Seasonal variations in the glycerol content and its influence on cold hardiness in the Alaskan carabid beetle, *Pterostichus brevicornis*, *J. Insect Physiol.*, in press; P. Freuchen and F. Salomonsen, *The Arctic Year*, 1958; D. M. Hopkins (ed.), *The Bering Land Bridge*, 1967; L. Irving, Adaptations to cold, *Sci. Amer.*, 214:94-101, 1966.

Artesian systems

Groundwater conditions formed by water-bearing rocks (aquifers) in which the water is confined above and below by impermeable beds. These systems are named after the province of Artois in France, where artesian wells were first observed. *See* GROUNDWATER.

Because the water table in the intake area of an artesian system is higher than the top of the aquifer in its artesian portion, the water is under sufficient head to cause it to rise in a well above the top of the aquifer. Many of the systems have sufficient head to cause the water to overflow at the surface, at least where the land surface is relatively low. Flowing artesian wells were extremely important during the early days of the development of groundwater from drilled wells, because there was no need for pumping. Their importance has diminished with the decline of head that has occurred in many artesian systems and with the development of efficient pumps and cheap power with which to operate the pumps. When they were first tapped, many artesian aquifers contained water that was under sufficient pressure to rise 100 ft or more above the land surface. Besides furnishing water supplies, many of the wells were used to generate electric power. With the increasing development of the artesian aquifers through the drilling of additional wells, the head in most of them has decreased and it is now from a few feet to several hundred feet below the land surface in many areas of former artesian flow. A majority of artesian wells are now equipped with pumps.

Perhaps the best-known artesian aquifer in the United States is the Dakota sandstone, of Cretaceous age, which underlies most of North Dakota, South Dakota, and Nebraska, much of Kansas, and parts of Minnesota and Iowa at depths ranging from 0 to 2000 ft. The water is highly mineralized, as a general rule, but during the latter part of the 19th century, when these areas were being settled, the Dakota sandstone provided a valuable source of water supply under high pressure. Few wells in this aquifer flow more than a trickle of water today. The St. Peter sandstone and deeper-lying sandstones of early lower Paleozoic age, which underlie parts of Minnesota, Wisconsin, Iowa, Illinois, and Indiana, form another well-known artesian system. Formerly, wells on low ground flowed abundantly, but now wells have to be pumped throughout most of the area. Some of the water is highly mineralized, but in many places it is of good quality. In New Mexico, in the Roswell artesian basin, cavernous limestone of Permian age provides water to irrigate thousands of acres of cotton and other farm crops. Although the head has been steadily declining, many wells still have large flows and others yield copious supplies by pumping. Among the most productive artesian systems are the Cretaceous and Tertiary aquifers of the Atlantic and Gulf Coastal plains. These provide large quantities of water for irrigation and industrial use and supply large cities, such as Savannah, Ga., Memphis, Tenn., and Houston and San Antonio, Tex. Numerous artesian basins are found in intermontane valleys of the West. Some of the best known are in the Central Valley, Calif., where confined aquifers provide water to irrigate millions of acres of farmland, and the San Luis Valley, Colo. Numerous other lesser artesian systems are found in all parts of the United States.

[ALBERT N. SAYRE/RAY K. LINSLEY]

Atmosphere

The gaseous envelope surrounding a celestial body. The terrestrial atmosphere, by its composition, control of temperature, and shielding effect from harmful wavelengths of solar radiation, makes possible life as known on Earth. The atmosphere, which is retained on the Earth by gravitational attraction and in a large measure rotates with it, is a system whose chemical and physical properties and fields of motion constitute the subject matter of meteorology. The changing atmospheric conditions which affect man's environment, particularly temperature, wind, humidity, cloudiness, and precipitation, constitute weather, and the synthesis of these conditions over a period determines the climate at any place. *See* CLIMATOLOGY; METEOROLOGY; WEATHER.

The average atmospheric pressure at the Earth's surface is about 1013 mb and the density 1.2 kg m^{-3}, and these vary by only a few percent over the globe. They both decrease rapidly and roughly exponentially with height, and at several earth radii the density can be said to have fallen to that of interplanetary space.

Composition. The atmosphere is thought to have developed as the result of chemical and photochemical processes combined with differential escape rates from the Earth's gravitational field. Chemical abundances in the atmosphere, therefore, are not directly related to cosmic abundances; in particular, the atmosphere is highly oxidized and contains very little hydrogen.

The atmosphere, apart from its highly variable water-vapor content in the troposphere, its solid matter in suspension, and its variable ozone content in the stratosphere, is well mixed and constant in composition up to about 100 km. This region is termed the homosphere. At higher levels where there is little mixing, diffusive separation tends to take place, with the lighter elements becoming progressively more dominant with height. Moreover, in this region oxygen and the minor constituents, such as carbon dioxide and water vapor, are dissociated by solar ultraviolet radiation. At about 300 km atomic oxygen probably becomes the most important constituent, until about 800 km where helium and hydrogen in turn predominate. This

Table 1. Composition of the atmosphere*

Molecule	Fraction by volume near surface	Vertical distribution
Major constituents		
N_2	7.8084×10^{-1}	Mixed in homosphere; photochemical dissociation high in thermosphere
O_2	2.0946×10^{-1}	Mixed in homosphere; photochemically dissociated in thermosphere, with some dissociation in mesosphere and stratosphere
A	9.34×10^{-3}	Mixed in homosphere with diffusive separation increasing above
Important radiative constituents		
CO_2	3.3×10^{-4}	Mixed in homosphere; photochemical dissociation in thermosphere
H_2O	highly variable	Forms clouds in troposphere; little in stratosphere; photochemical dissociation above mesosphere
O_3	variable	Small amounts, 10^{-8}, in troposphere; important layer, 10^{-6}–10^{-5}, in stratosphere; dissociated above
Other constituents		
Ne	1.1818×10^{-5}	Mixed in homosphere with diffusive separation increasing above
He	5.24×10^{-6}	
Kr	1.14×10^{-6}	
CH_4	1.6×10^{-6}	Mixed in troposphere; dissociated in upper stratosphere and above
H_2	5×10^{-7}	Mixed in homosphere; product of H_2O photochemical reactions in lower thermosphere, and dissociated above
NO	$\sim 10^{-8}$	Photochemically produced in mesosphere

*Other gases, for example, CO, N_2O, and NO_2, of less importance also exist in small amounts.

Table 2. Atmospheric structure, a selection of mean midlatitude values*

Height, km	Pressure, mb	Temperature, °K	Density, kg m⁻³	Mean molecular weight	Layer
0	1.01×10^3	288	1.23×10^0	28.96	Troposphere
5	5.40×10^2	256	7.36×10^{-1}	28.96	
10	2.65×10^2	223	4.14×10^{-1}	28.96	
20	5.53×10^1	217	8.89×10^{-2}	28.96	Stratosphere
40	2.87×10^0	250	4.00×10^{-3}	28.96	
60	2.25×10^{-1}	256	3.06×10^{-4}	28.96	Mesosphere
80	1.04×10^{-2}	181	2.00×10^{-5}	28.96	
100	3.01×10^{-4}	210	4.97×10^{-7}	28.88	Thermosphere
150	5.06×10^{-6}	893	1.84×10^{-9}	26.92	
200	1.33×10^{-6}	1236	3.32×10^{-10}	25.56	
300	1.88×10^{-7}	1432	3.59×10^{-11}	22.66	
400	4.03×10^{-8}	1487	6.50×10^{-12}	19.94	
500	1.10×10^{-8}	1499	1.58×10^{-12}	17.94	
600	3.45×10^{-9}	1506	4.64×10^{-13}	16.84	

*Based on United States Standard Atmosphere, 1962.

region of highly variable composition is termed the heteorosphere. Ionization of the various constituents also due to absorption of ultraviolet solar radiation becomes a major factor above about 60 km, the base of the ionosphere, which is of major importance to radio communications. At levels above 600–800 km collisions between atmospheric particles become so infrequent that some traveling outward may escape from the atmosphere. This region is termed the exosphere. *See* ATMOSPHERIC CHEMISTRY; ATMOSPHERIC POLLUTION.

Table 1 gives a summary of the composition of the atmosphere. The water vapor is a very variable constituent having a mass mixing ratio of 10^{-3} to 10^{-2} gram per gram of dry air (gg⁻¹) in the lower troposphere and 2.10^{-6} gg⁻¹ in the stratosphere. The ozone layer in the stratosphere has its maximum concentration of about 10^{-5} gram per gram of air at 30–35 km, and is formed by molecular-atomic collisions following dissociation of molecular oxygen by solar radiation wavelengths below about 2400 A. The minor constituents water vapor, carbon dioxide, and ozone play a vital role in the atmosphere's primary biological, heating, and shielding functions.

Thermal structure and circulation. A convenient division of the atmosphere is by spherical shells, "spheres," each characterized by the way its temperature varies in the vertical, and with tops denoted by "pauses," as in the figure. Table 2 lists characteristic values for temperature, pressure, density, and mean molecular weight at a selection of levels within these shells.

Troposphere. The average temperature in the troposphere decreases with height from the surface (~300°K in low latitudes and 260°K in high latitudes) to its upper level, the tropopause, which is around 16 km (temperature 180°K) in the tropics and 10 km (230°K) near the poles. The troposphere includes the layer in which man lives and is the seat of all the important weather phenomena affecting the environment. Its thermal structure is primarily due to the heating of the Earth's surface by solar radiation, followed by upward heat transfer by turbulent mixing and convection. Heat is also transferred poleward by atmospheric motions from the more strongly heated equatorial regions.

The processes involved include evaporation of water from the surface and condensation with release of latent heat, leading to clouds and precipitation. The general circulation of the troposphere includes wind systems on all scales—prevailing winds, monsoons, long waves, anticyclones and depressions, fronts, hurricanes, thunderstorms, and shower clouds—each system being associated with characteristic weather patterns. The application of physical and dynamical principles to prognosis of atmospheric motions and weather is one of the main tasks of modern meteorology. *See* CLOUD PHYSICS; WEATHER FORECASTING AND PREDICTION.

Stratosphere. The stratosphere extends from

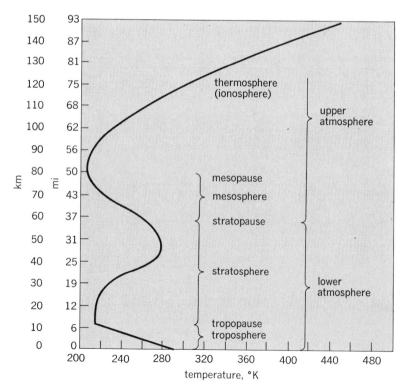

The thermal structure of the atmosphere, showing major divisions.

about 10–16 km to about 55 km. Its thermal structure is mainly determined by its radiation balance, in contrast to the approach to convective equilibrium of the troposphere. It is generally very stable with low humidity and no weather in the popular sense. The only clouds in this region are the "mother-of-pearl" clouds seen infrequently at 20–30 km in high latitudes in winter. The ozone layer strongly absorbs solar radiation of 2000–3000 A in the mesosphere and upper stratosphere, which results in the high temperature (250–290°K) region around the stratopause. The resultant energy source driving the circulation at these levels is the excess of absorbed energy over infrared energy emitted by the atmosphere (mainly by carbon dioxide and ozone) at the summer pole, compared with the deficit or sink at the winter pole. This produces an atmospheric circulation system in the upper stratosphere and mesosphere which is separate from that of the lower stratosphere, whose circulation is broadly driven by that of the troposphere below.

Mesosphere. There is a decrease of temperature with height throughout the mesosphere from 55 km to about 80 km, the mesopause, where there is a temperature minimum. In summer its value is about 150°K and occasionally noctilucent clouds are formed, while in winter it has a higher value, around 220°K. This distribution is probably mainly due to dynamic effects, as there is comparatively little direct absorption of solar ultraviolet radiation at this level. The lowest ionized region, the D layer, with $10–10^3$ electrons cm^{-3}, is from about 60 to 90 km.

Thermosphere. The thermosphere is a very high temperature region and extends from 80 km to the outer edge of the atmosphere. It receives its energy by the direct absorption of the solar beam below about 2000 A. The response to change in solar radiation, for example, from night to day, with solar activity and with sunspot cycle, is very marked, and this region is under direct solar control. The atmospheric motions seem to be mainly thermally induced solar tides, but the dynamics of this region are not well understood. The physical phenomena also are complicated and include excitation, dissociation, and ionization of the constituents and effects, such as the aurora following corpuscular radiation from the Sun.

The increasing importance of ionization with height also means that electrical and magnetic forces have important effects on the atmospheric motions and these, in turn, produce measurable geomagnetic effects at the Earth's surface. The principal ionized layers or wedges are the E layer at 90–160 km, the F1 layer at 160–200 km, and the F2 layer above 200 km with $10^4–10^5$ electrons cm^{-3}. The effects of viscosity increasing with height also become manifest, and above 100–110 km, the turbopause, atmospheric flow no longer appears to be turbulent.   [R. J. MURGATROYD]

Bibliography: R. A. Craig, *The Upper Atmosphere: Meteorology and Physics*, 1965; R. M. Goody, *Atmospheric Radiation*, 1964; R. M. Goody, *The Physics of the Stratosphere*, 1954; F. K. Hare, *The Restless Atmosphere*, 1966; E. N. Lorenz, *The Nature and Theory of the General Circulation of the Atmosphere*, World Meteorol. Organ. Tech. Publ. no. 115, 1967; T. F. Malone (ed.), *Compendium of Meteorology*, 1951; J. A. Ratcliffe (ed.), *Physics of the Upper Atmosphere*, 1960; K. Rawer (ed.), *Winds and Turbulence in Stratosphere, Mesosphere and Ionosphere*, 1968; A. N. Strahler, *The Earth Sciences*, 1963.

Atmospheric chemistry

Discipline of meteorology concerned with production, transport, modification, and removal of atmospheric constituents in the troposphere and stratosphere. Most of these are present as trace substances with concentrations of less than 100 μg/m^3 of air. The number of constituents known to be present increases as analytical techniques improve.

All atmospheric constituents cycle through the ocean, soil, or biosphere, or a combination of these. The cycle may, for example, consist of simple exchange with dissolved fractions in ocean water or may be accompanied by chemical transformations. In general, the variability of constituents decreases with increasing lifetime. Water vapor has the shortest-known lifetime, about 10 days. Trace substances comprise gases, aerosols, and naturally and artificially radioactive material. The isotopic compositions of constituents and their variations due to fractionation processes can provide important information about the cycle of these constituents.

Gases. The vertical distribution of water vapor in the troposphere is determined by the temperature distribution. The stratosphere is very dry and most likely shows a constant mixing ratio above 20 km. The O^{18}/O^{16} and deuterium-hydrogen (D/H) ratios of H_2O vary with latitude and season and are used for polar ice stratigraphy. The tritium content of H_2O is now dominated by production in atomic tests.

Carbon dioxide cycles through the ocean and the biosphere. The CO_2 content of the oceans is about 60 times that of the atmosphere and is controlled by the temperature and the pH value of sea water. Release of CO_2 into the atmosphere over tropical oceans and uptake by polar oceans results in residence times of about 5 years. The cycle through the land plants by assimilation and decay of organic matter has a turnover time of a few decades and is associated with global seasonal variations of the atmospheric CO_2 content, primarily in the Northern Hemisphere. Release of CO_2 by volcanic activity through the ages was compensated by the formation of lime sediments. Incorporation of cosmic ray–produced C^{14} into living plants forms the basis of the radiocarbon dating method. *See* SEA WATER.

The small amount of about 0.3 cm of ozone under normal pressure is of vital importance for life, by absorbing lethal solar ultraviolet radiation below 3000 A. This absorption results in a warm upper stratosphere, with profound effects on the general atmospheric circulation. Ozone is produced photochemically via atomic oxygen in the upper stratosphere. Below 25 km, O_3 is a quasi-conservative atmospheric constituent and is destroyed primarily by contact with the Earth's surface. Locally, it can be produced in polluted areas. *See* AIR POLLUTION CONTROL.

Knowledge about the cycle of most other gaseous constituents is very scanty (Table 1). Nitrous oxide, methane, molecular hydrogen, ammonia, and hydrogen sulfide are known to be produced in soil or in swamps. Sulfur dioxide probably originates from H_2S oxidation, and partially from human activity. Some aerosol constituents such as SO_4^{--}, NH_4^+, and NO_3^- can be the product of trace gas reactions.

Aerosols. Natural aerosols cover a wide size spectrum, from radii of about 3×10^{-3} to about 10^2 μ. The chart indicates the importance of different parts of the spectrum to various meteorological disciplines and gives a survey of size distribution and chemical composition. Below 0.5 μ, concentrations over the ocean are lower than over land, suggesting the continental origin of these particles. Over the ocean particles larger than 0.5 μ consist almost entirely of sea salt. Over land the composition is complex with some of the common compounds indicated. The total number of particles is controlled by those smaller than 0.1 μ, the total mass by those larger than 0.1 μ: the lower and upper end of the spectrum are determined by co-

agulation and sedimentation, respectively. For particles larger than 0.1 μ, the distribution can be approximated by a power law.

Water-soluble matter results in a growth of aerosol particles with relative humidity. Radioactive nuclides produced by cosmic rays or by radon decay become attached to aerosols. Tropospheric aerosols have an average lifetime of a few weeks. Stratospheric aerosols have a maximum concentration around 20 km and contain large fractions of sulfate. They are responsible for the purple light after sunset. Except for magnetite spherules no extraterrestrial dust has been indentified beyond doubt in the troposphere and stratosphere.

Radioactivity. Radioactive nuclides in the atmosphere are produced by decay of radon and by cosmic radiation. Radon isotopes are produced by decay of U^{238}, U^{235}, and Th^{232} in soil and decay in the atmosphere. Pb^{210} (radium D) with a half-life of 22 years is always found in air even in the stratosphere. Cosmic rays produce a number of radioactive isotopes primarily in the lower stratosphere

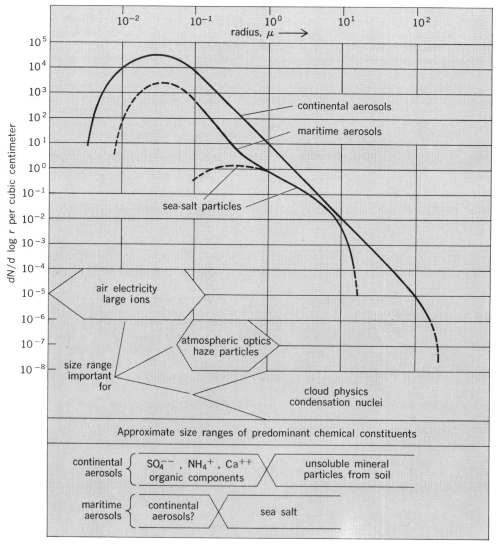

Chart of the average size distributions and the predominant chemical constituents of some natural aerosols. The size ranges which are important for the various fields of meteorology are shown.

Table 1. Atmospheric gases except oxygen and nitrogen

Gas	Low-troposphere concentration, $\mu g/m^3(STP)$*	Residence time
Argon	1.6×10^7	—
Neon	1.6×10^4	—
Krypton	4100	—
Helium	920	$\sim 10^6$ years
Xenon	500	—
Water vapor	$(3-3000) \times 10^4$	10 days
Carbon dioxide	$(4-8) \times 10^5$	~ 5 years
Methane	$(7-14) \times 10^2$	Few years
Nitrous oxide	$(5-12) \times 10^2$	Few years
Carbon monoxide	50–500	0.5 years (?)
Ozone	0–100	0.5 years (?)
Hydrogen	30–90	—
Sulfur dioxide	0–50	About 1 week
Ammonia	0–15	—
Nitrogen dioxide	0–6	—

*Standard temperature and pressure.

(see Table 2). Those with short lifetimes have stratospheric concentrations which correspond to their decay equilibria. Longer-lived nuclides are diluted by mixing with the troposphere. Natural radioactivity is useful as tracer for atmospheric motions.

Atomic tests have injected large quantities of fission products into the stratosphere. The most dangerous is Sr^{90} because it enters human bones and has a long half-life of 27.7 years. The fission products remain on the average about 1 year in the lower stratosphere, before removal into the troposphere and deposition over the entire surface of the Earth as radioactive fallout. Studies of fission-product distribution in the stratosphere have provided important information about the general circulation. *See* RADIOACTIVE FALLOUT.

Precipitation. Removal of aerosols and of some gas traces by precipitation is of great importance for cleaning the troposphere. In precipitation trace substances are concentrated by a factor of about 10^6 compared with air. The efficiency of concentration varies so that the composition of precipitation reflects only quantitatively that of air. Rain-analysis networks have provided information on the geographical distribution of atmospheric trace substances. Sea salt is the dominant compound over the ocean but decreases rapidly in concentration inland. There it is replaced by SO_4^{--}, CA^{++},

Table 2. Cosmic ray–produced nuclides in the atmosphere

Nuclide	Half-life	Produced from
$H^3(T)$	12.3 years	
Be^7	53 days	N^{14}, O^{16}
Be^{10}	2.7×10^6 years	
C^{14}	5568 years	N^{14}
Na^{22}	2.6 years	
Si^{32}	710 years	
P^{32}	14 days	
P^{33}	25 days	Ar
S^{35}	87 days	
Cl^{36}	3×10^5 years	
Cl^{39}	55 min	

NH_4^+, and other ions. In some areas the material deposited by rain can be important for soil composition and plant nutrition, particularly SO_4^{--} and NO_3^-. [C. E. JUNGE]

Bibliography: International Symposium on Atmospheric Chemistry, Circulation, and Aerosols, *Tellus*, vol. 18, nos. 2 and 3, 1966; International Symposium on Trace Gases and Natural and Artificial Radioactivity in the Atmosphere, *J. Geophys. Res.*, vol. 68, 1963; H. Israel and A. Krebs, *Nuclear Radiation in Geophysics*, 1962; C. E. Junge, *Air Chemistry and Radioactivity*, 1963.

Atmospheric electricity

The electrical processes constantly taking place in the lower atmosphere. This activity is of two kinds, the intense local electrification accompanying storms, and the much weaker fair-weather electrical activity over the entire globe, which is produced by the many electrified storms continuously in progress over the Earth.

The mechanisms by which storms generate electric charge are presently unknown, and the role of atmospheric electricity in meteorology has not been determined. Some scientists believe that electrical processes may be of importance in precipitation formation and in cloud dynamics.

Disturbed-weather phenomena. Almost all precipitation-producing storms throughout the year are accompanied by energetic electrical activity. The most intense of these are the thunderstorms, in which the electrification attains values sufficient to produce lightning. Electrical measurements show that most other storms, even though they do not give lightning, are also quite strongly electrified.

Thunderstorms begin as little fair-weather clouds that usually form and disappear without producing rain or electrical effects. When the air is sufficiently thermally unstable, a few of these clouds undergo a rapid and dramatic change. Suddenly, for no obvious reason, one will begin a very rapid growth, increasing in height as fast as 1 km/min. When this happens, other significant changes occur. Often in only a few minutes, strong electric fields and rain appear; shortly after, if the cloud is sufficiently vigorous, lightning appears.

The usual height of a thunderstorm is about 10 km; however, they can be as low as 3 km or as high as 20 km. A common feature of these storms is their strong updrafts and downdrafts, which often have speeds in excess of 30 m/sec.

The electric fields are most intense within the cloud, where they reach values as high as 3000 volts/cm. Above the top of the cloud, fields in excess of 1000 volts/cm have been observed. On the ground beneath the storm, the fields are usually much smaller and seldom exceed 100 volts/cm. Although the distribution of the electric fields in and about the thunderstorm is complex and variable, most storms approximate a vertical dipole with positive charge above and negative charge below. A few storms appear to have just the reverse polarity.

The origin and nature of the charged regions responsible for the electric fields of thunderstorms are not understood. A variety of explanations has been proposed, most of which are based on the idea that electrification is caused by the falling of charged precipitation particles. The charging of the precipitation is variously ascribed to processes such as selective ion capture, contact electrification, freezing electrification, and break-up of raindrops. According to other suggestions, electrification occurs independently of precipitation and is brought about by charge transported in updrafts and downdrafts. There is no general agreement on the mechanism, because no theory offers a satisfactory, coherent, and quantitative explanation of the observed facts.

The electric fields of thunderstorms cause three currents to flow, each of a few amperes: lightning, point discharge from the ground beneath, and conduction in the surrounding air. Because the external field and conductivity are greatest over the top of the cloud, most of the conduction current flows to the ionosphere, the upper, highly conductive layer of the atmosphere.

Fair-weather field. Fair-weather measurements, irrespective of place and time, show the invariable presence of a weak negative electric field caused by the estimated several thousand electrified storms continually in progress. Together these storms cause a 2000-amp current from the earth to the ionosphere that raises the ionosphere to a positive potential of about 400,000 volts with respect to the earth. This potential difference is sufficient to cause a return flow of positive charge to the earth by conduction through the intervening lower atmosphere equal and opposite to the thunderstorm supply current. The fair-weather field is simply the voltage drop produced by the flow of this current through the atmosphere. Because the electrical resistance of the atmosphere decreases with altitude, the field is greatest near the Earth's surface and gradually decreases with altitude until it vanishes at the ionosphere.

The fair-weather field at the Earth's surface is observed to fluctuate somewhat with space and time, largely as a result of local variations in atmospheric conductivity. However, in undisturbed locations far at sea or over the polar regions, the field is observed to have a diurnal cycle independent of position or local time and quite similar to the diurnal variation of thunderstorm activity over the globe. No importance is presently attached to fair-weather atmospheric electricity except that according to some theories it is responsible for the initiation of the thunderstorm electrification process. *See* CLOUD PHYSICS; STORM DETECTION; THUNDERSTORM.

[BERNARD VONNEGUT]

Bibliography: J. A. Chalmers, *Atmospheric Electricity*, 2d ed., 1967; S. C. Coroniti (ed.), *Problems of Space and Atmospheric Electricity*, 1964; Q. J. Malan, *Physics of Lightning*, 1963; B. F. J. Schonland, *Atmospheric Electricity*, 1953; B. F. J. Schonland, *The Flight of Thunderbolts*, 1950; P. E. Viemeister, *The Lightning Book*, 1961.

Atmospheric evaporation

The exchange of water between the liquid and vapor state. The transfer from the solid state to the vapor state is technically known as sublimation, but in discussing the atmospheric process, exchange from snow and ice is commonly included with evaporation. The primary source of atmospheric water vapor is the oceans, which cover the major part of the Earth. Evaporation from lakes, rivers, ice, snow, and soil also contributes large amounts of water vapor to the atmosphere, although these sources are small in comparison with evaporation from the oceans. Annual evaporation from the oceans is estimated to be about 118,000 mi^3 of water.

Heat of vaporization. The process of evaporation requires large amounts of heat. The heat of vaporization of water at 70°F is 584 cal/g or 1047 Btu/lb. Because of this heat requirement, water vapor in air masses transports large quantities of heat from warmer to cooler regions of the world. As the vapor is condensed to form clouds, the heat of vaporization is released. Atmospheric water vapor also is essential for the formation of precipitation (rain, snow, hail, and sleet). When an air mass containing vapor is cooled, the vapor condenses to form clouds of small water droplets. If the cloud droplets coalesce into drops large enough to fall to the ground, precipitation results. Under some conditions the released heat of vaporization may be important in increasing the bouyancy of the air, which causes the air to rise and cool, leading to more condensation.

Evaporation cools a water surface because the escaping water carries with it the heat of vaporization. As the water cools, its vapor pressure decreases and the evaporation rate drops. Unless additional heat is provided from some source, evaporation will cease. In a deep water body the cooled surface water may sink and be replaced by warmer water from below.

Estimating evaporation. There are several methods for estimating evaporation: the difference in vapor pressure between air and water surface, the energy balance of the system, and the use of evaporimeters.

Vapor-pressure difference. Many factors influence evaporation from a water surface, and in nature, where none of the factors can be controlled, the process is complex. J. Dalton in 1802 first pointed out that evaporation is proportional to the difference in vapor pressure between the water surface and the air above it. On the basis of Dalton's law evaporation is possible only when the dew-point temperature of the air is less than the temperature of the water surface. The greater this temperature difference, the greater is the evaporation rate. In absolutely still air a vapor blanket forms at the water surface, and evaporation decreases rapidly to rates limited by the diffusion of vapor from the blanket. With wind the vapor is carried away and replaced by dryer air and evaporation continues.

Energy balance. Because solar radiation is the primary source of heat for the evaporation from any water body, another basis for estimating evaporation is the energy-balance method. If the total radiation input to the water body less the heat loss by long-wave radiation and convection can be determined, the excess of heat input over heat outgo by other means divided by the latent heat of vaporization indicates the volume of evaporation.

Evaporimeters. Estimates of evaporation also are made by use of evaporimeters. The most effective evaporimeters are pans of water 3–4 ft in diameter from which evaporation can be measured. Because of heat transfers through the sides and bottoms of such pans, evaporation from pans is always higher than from lakes or oceans. One of the most commonly used pans is the U.S. Weather Bureau class A pan. This pan is 4 ft in diameter and 10 in. deep. It is supported a few inches above the ground on a timber grid. Water depth in the pan is maintained at about 8 in. Evaporation from the class A pan must be reduced about 30% to estimate lake evaporation. Other evaporimeters use paper or porous ceramic as the evaporating surface, but it has not proved possible to relate evaporation loss from such surfaces to evaporation from large water bodies.

Salt, ice, and snow factors. Addition of salts to water reduces the vapor pressure at the water surface. Thus evaporation per unit area from the oceans and salt lakes is slightly less than from a fresh-water body. Roughly, evaporation is reduced about 1% for each percent of dissolved salts, so that evaporation from the oceans is about $3\frac{1}{2}\%$ less than from an equivalent fresh-water surface. Evaporation from ice and snow is also less than from water at the same temperature because the vapor pressure over ice is somewhat less than over water. Evaporation from snow and ice is also low, because vapor pressure (and vapor-pressure differences) are very low at temperatures below freezing.

Based largely on evaporimeter data, it can be said that evaporation varies from 15 to 20 ft/year of water in tropical deserts to near zero in polar regions.   [RAY K. LINSLEY]

Bibliography: T. A. Blair and R. C. Fite, *Weather Elements*, 4th ed., 1957; A. A. Gordon, *Elements of Dynamic Meteorology*, 1962; S. L. Hess, *Introduction to Theoretical Meteorology*, 1959; R. K. Linsley, M. A. Kohler, and J. L. H. Paulhus, *Hydrology for Engineers*, 1958; O. G. Sutton, *Micrometeorology*, 1953; R. C. Ward, *Principles of Hydrology*, 1967.

Atmospheric general circulation

In a broad sense the statistical mean global flow pattern of the atmosphere. By superposing the motion averaged over time in every meridional plane onto a mean meridional cross section, the longitudinal variations of the flow pattern are eliminated. In such a cross section the mean zonal, meridional, and vertical components of the air motion appear as functions of latitude and height only. In this restricted sense the general atmospheric circulation is discussed in this article.

Practically all motions in the atmosphere are ultimately caused by differential radiative heating or cooling. In low latitudes the Earth-atmosphere system is, on the average, heated and in higher latitudes cooled by radiation processes. The Earth's surface receives generally more radiative heat than it emits, whereas the opposite is true for the atmosphere. Because the atmosphere and the underlying surface of the Earth maintain a nearly constant temperature over sufficiently long periods of time, heat has to be transferred from regions of radiative heat surplus to regions of heat deficit. At the Earth-atmosphere interface this occurs in form of turbulent flux of sensible heat and through evapotranspiration (flux of latent heat). In the atmosphere the latent heat is released in connection with condensation of water vapor. *See* CLIMATOLOGY; HEAT BALANCE, TERRESTRIAL ATMOSPHERIC.

Radiation and heat exchange with the Earth's surface heat the atmosphere in low latitudes and cool it in higher latitudes. Heat must therfore be transferred both upward and poleward from the principal heat sources at low levels and low latitudes toward the heat sinks at upper levels and higher latitudes. This transport is provided by the complex circulation pattern of the atmosphere. Only the meridional and vertical wind components transfer heat and other properties in the plane of a mean meridional cross section, but the transfer mechanism cannot be studied without taking into account the longitudinal and time variations of the wind components and the properties considered.

Mean circulation. In a mean meridional cross section the zonal wind component is almost everywhere dominant and in quasi-geostrophic balance with the mean meridional pressure gradient. *See* METEOROLOGY.

Only in the lowest friction layer and in the vicinity of the Equator, where the geostrophic balance is weak, the mean meridional wind component may be of the same magnitude. Because of the nature of the atmosphere as a shallow layer, the mean vertical wind component is everywhere very weak and can therefore not be observed directly. Whereas the magnitude of the mean zonal wind varies between 0 and 40 m/sec, and the corresponding magnitude of the mean meridional wind between 0 and 3 m/sec, the mean vertical wind component nowhere exceeds a magnitude of 1 cm/sec.

The pattern of the tropospheric zonal circulation is essentially the same in winter and summer with easterlies in low latitudes and westerlies in higher latitudes, except in small regions of low-level easterlies around the poles (Fig. 1). The strongest west winds, about 40 m/sec, are in the winter hemispheres observed around the horse latitudes at the 200-mb level (roughly 12 km). These wind maxima represent the subtropical jet streams of both hemispheres. In the summer hemispheres the corresponding west-wind maxima are considerably weaker and are located farther poleward. In the stratosphere the westerly circulation generally decreases with height. The boundaries between the low-latitude easterlies and the higher-latitude westerlies show an equatorward slope up to about 200 mb, and a poleward slope above that level. In the upper stratosphere easterly circulation prevails in the summer hemispheres and westerly circulation in the winter hemispheres outside the tropical region. *See* JET STREAM.

The much weaker meridional circulation consists essentially of four separate cells, two in low latitudes and two in higher latitudes. The former two cells are mainly called the Hadley cells, the latter two the Ferrel cells. In the Hadley cells the air moves toward the thermal equator at low levels and in the opposite direction in the upper troposphere. In the Ferrel cells the circulation is reversed. The mean annual mass circulation in the northern Hadley cell may be estimated at about 110×10^{12} g/sec and in the southern Hadley cell at about 130×10^{12} g/sec. The mass circulation in the Ferrel cells is much weaker; it has been estimated at about 20×10^{12} g/sec and 30×10^{12} g/sec, respectively. Figure 2 shows meridional circulation.

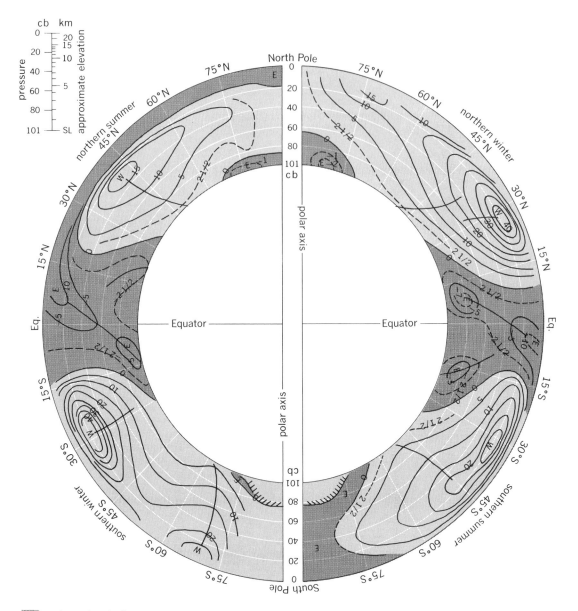

regions of easterlies

Fig. 1. Pattern of mean zonal (east-west) wind speed averaged over all longitudes as a function of latitude, height, and season (*after Y. Mintz*). Height, greatly enlarged relative to earth radius, is shown on linear pressure (α mass) scale with geometrical equivalent given at upper left. Mean zonal wind (westerly positive, easterly negative) has same value along any one line and is shown in meters per second (1 m/sec $\cong$ 2 knots) on the line. Note subtropical westerly jet in high troposphere at about 30° latitude (or more in northern summer). This jet is not to be confused with polar front jet of higher latitudes; latter is migratory and does not appear in mean.

The zonal circulation has the character of a quasi-horizontal motion nearly balanced by the meridional pressure gradient. In a coordinate system with the horizontal axes along a latitude circle and a meridian and with pressure as vertical axis, this balance is expressed by Eqs. (1), (2), and (3).

$$\overline{u}\,(2\Omega a \sin \varphi + \overline{u} \tan \varphi) = -g\frac{\partial \overline{Z}}{\partial \varphi} \tag{1}$$

$$p\,\frac{\partial \overline{u}}{\partial p} = \frac{R}{2(\Omega a \sin \varphi + \overline{u} \tan \varphi)}\frac{\partial \overline{T}}{\partial \varphi} \tag{2}$$

$$\frac{1}{p}\frac{\partial p}{\partial Z} = -\frac{g}{R\overline{T}} \tag{3}$$

The notation is as follows: u denotes zonal wind component; Ω, angular velocity of the Earth; a, radius of the Earth; φ, latitude; g, acceleration of gravity; Z, height of an arbitrary isobaric surface; p, pressure; R, the gas constant for air; and T, absolute temperature. The bars denote values averaged over longitude and time. Equation (3) represents the transformation between pressure and height as vertical coordinates.

Since the zonal pressure gradient vanishes if averaged over longitudes, no corresponding relationships exist for the mean meridional wind component $\overline{v}$. This component must therefore be computed from wind observations. If $\overline{v}$ is known in a mean meridional plane, the corresponding vertical component in pressure coordinates, $\omega = dp/dt$, can be evaluated from the continuity equation in the

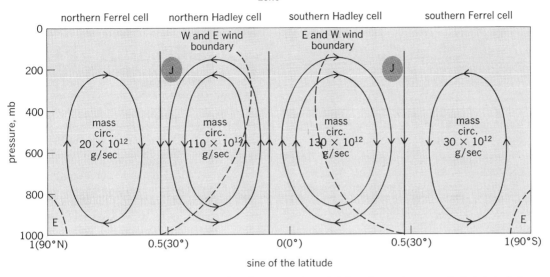

Fig. 2. Mean mass circulation and position of Hadley cells and Ferrel cells of both hemispheres. Dashed lines indicate boundaries between easterlies and westerlies, and symbol J marks maximum west winds in both hemispheres (subtropical jet streams).

form of Eq. (4). Here p_0 denotes the pressure at the

$$\overline{\omega} = \frac{1}{a} \int_p^{p_0} \left(\frac{\partial \overline{v}}{\partial \varphi} - \overline{v} \tan \varphi \right) dp \qquad (4)$$

ground, where $\overline{\omega}$ is assumed to vanish. The relationship between vertical velocity in height coordinates $\overline{w}$ and $\overline{\omega}$ is given by Eq. 5.

$$\overline{w} = -\frac{R\overline{T}}{gp} \overline{\omega} \qquad (5)$$

The maintenance of the zonal circulation presupposes conservation of the meridional temperature and pressure distribution, according to Eqs. (1) and (2). Hence the heat changes caused by heat sources and sinks in the meridional cross section must be compensated by meridional and vertical energy fluxes. This principle permits the computation of the fluxes.

Angular momentum flux. There is, however, another aspect to be considered. The surface easterlies in low latitudes and the surface westerlies in higher latitudes exert a frictional torque on the Earth tending to speed up the westerly circulation in the regions of surface easterlies and to slow down the westerly circulation in regions of surface westerlies. This is equivalent with an exchange of angular momentum between the Earth and the atmosphere. Per unit area the torque is expressed by $\tau_x a \cos \varphi$, where τ_x denotes the frictional stress, positive for east wind and negative for west wind. The source of angular momentum in a zone limited by two latitude circles φ_1 and φ_2 is given by notation (6). If the integration is performed over the

$$2\pi a^3 \int_{\varphi_1}^{\varphi_2} \overline{\tau}_x \cos^2 \varphi d\varphi \qquad (6)$$

whole globe, the net torque must vanish. However, from the principle of conservation of the zonal circulation, it also follows that angular momentum has to be transferred from the belts of surface eas-

terlies to the zones of westerlies. Investigations have shown that this meridional flux of angular momentum primarily occurs in the upper troposphere. Hence angular momentum has to be brought upward from the surface layer in low latitudes, transferred poleward, primarily in higher levels, and ultimately be brought down to the Earth in the belts of westerlies.

The observed general circulation hence represents a self-consistent system, in which the meridional and vertical fluxes of energy and angular momentum must satisfy balance requirements determined by the distribution of the corresponding sources and sinks. At the same time the general circulation with all its disturbances must act as a heat engine which converts a part of the incoming solar heat into kinetic energy. If the ratio between the rate of energy conversion and the incoming solar radiation defines the efficiency of the atmospheric heat engine, the observed efficiency is low, amounting to about $1-2\%$.

Balance requirements. Let X denote a certain property per unit area of the Earth's surface. Then the northward flux of the same property F_φ across an entire latitude circle is given by Eq. (7).

$$F_\varphi = -2\pi a^2 \int_\varphi^{\pi/2} \frac{dX}{dt} \cos \varphi d\varphi \qquad (7)$$

The following sources will be considered: R_a, the difference between the radiative heat absorbed and emitted by the atmosphere; R_e, the same difference for the Earth's surface; Q_s, the turbulent flux of sensible heat from the surface to the atmosphere; LE, the flux of latent heat at the surface-atmosphere interface (E denoting the rate of evapotranspiration and L the heat of vaporization); LP, the release of latent heat in the atmosphere estimated from the observed rate of precipitation P; and $\tau_x a \cos \varphi$, the surface frictional torque. In Table 1 the different sources are summarized.

Table 1. Sources of atmospheric properties

Different sources	$\dfrac{\overline{dX}}{dt}$
Atmospheric heat	$\overline{R}_a + \overline{Q}_s + L\overline{P}$
Latent heat	$L(\overline{E} - \overline{P})$
Heat of Earth's surface	$\overline{R}_e - \overline{Q}_s - L\overline{E}$
Heat of atmosphere and Earth	$\overline{R}_a + \overline{R}_e$
Surface torque	$\overline{\tau}_x a \cos\varphi$

By inserting the above values of $\overline{dX}/dt$ in Eq. (7), one can compute the northward fluxes of energy and latent heat in the atmosphere, the heat flux in the Earth's surface layer (mainly in the oceans), the total energy flux in the atmosphere-ocean system, and the required transfer of angular momentum in the atmosphere.

Figure 3 shows the average annual fluxes of atmospheric energy (sensible heat and potential energy), latent heat, oceanic heat, and the sum of all these flux components. All fluxes are given in calories per minute and in kilojoules per second. For the computations empirical values of $\overline{R}_a$, $\overline{R}_e$, $\overline{Q}_s$, $\overline{E}$, and $\overline{P}$ are used; these are, of course, subject to further improvements.

Of special interest are the curves showing the fluxes of atmospheric energy and latent heat in the tropics. Both fluxes approach zero around latitude 5°N, which represents the mean annual position of the intertropical convergence zone (thermal equator). Toward this zone latent heat is carried from both hemispheres, and from the same zone released heat again is transferred away. In the tropics the lower branches of the Hadley cells (Fig. 2)

carry water vapor into the convergence zone, where latent heat is released in connection with abundant rain. This zone appears therefore as the most prominent heat source in the global atmosphere. Another interesting feature in Fig. 3 is the remarkable role the oceans play in transporting heat poleward in lower latitudes; around the horse latitudes the heat flux associated with ocean currents amounts to about 30–40% of the total energy flux required. *See* OCEAN-ATMOSPHERE RELATIONS.

In Fig. 4 the mean northward flux of angular momentum and the corresponding frictional torques for zones of latitude are presented. Because the surface stresses are not very well known, the computation is limited to the Northern Hemisphere, where it can be supported by atmospheric flux computations.

Mechanism of transfer processes. Transfer processes may be better understood if the fluxes are computed directly from aerological data. If x denotes a given property per unit mass of air, the northward flux through an arbitrary latitude circle φ is given by Eq. (8), where the integration is

$$F_\varphi = \frac{2\pi a \cos\varphi}{g} \int_0^{p_0} (\overline{x}\,\overline{v} + \overline{x'v'})\,dp \qquad (8)$$

extended from the bottom ($p = p_0$) to the top of the atmosphere ($p = 0$). Similarly, the upward flux between the latitude circles φ_1 and φ_2 is given by Eq. (9). Here $\overline{x}$, $\overline{v}$, and $\overline{\omega}$ denote mean values over

$$F_z = -\frac{2\pi a^2}{g} \int_{\varphi_1}^{\varphi_2} (\overline{x}\,\overline{\omega} + \overline{x'\omega'})\cos\varphi\,d\varphi \qquad (9)$$

time and longitude, and x', v', and ω' the corresponding deviations from these mean values.

The first terms of the integrands represent the

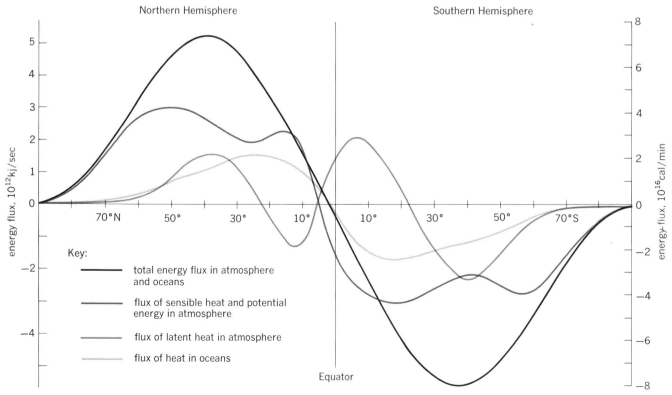

Fig. 3. Average annual northward energy fluxes in the atmosphere and in the oceans.

meridional and vertical fluxes associated with the mean meridional mass circulation, whereas the second terms give the fluxes resulting from the fluctuations x', v', and ω'. The former contributions to the total fluxes may be called circulation fluxes, and the latter parts eddy fluxes. By inserting for x the expressions in Table 2, the two different modes of the fluxes can be computed. In the table c_p denotes specific heat at constant pressure, gZ potential energy, and q specific humidity.

Table 3 shows, as an example, the partition of the flux of atmospheric energy into circulation and eddy fluxes. In middle and high latitudes the eddy fluxes dominate and in low latitudes the circulation fluxes. Similar results concerning the flux of water vapor (latent heat) have been found. *See* HYDRO-METEOROLOGY.

Figure 5 gives a comparison between the total mean annual flux of atmospheric energy, as computed separately from Eqs. (7) and (8). The results agree fairly well, but the differences indicate that considerable improvement of the data is needed.

The eddy fluxes comprise a whole spectrum of processes of very different scale. Roughly, they can be divided into two groups: large-scale (synoptic) disturbances and small-scale eddies associated either with mechanical turbulence or with thermal convection. Only the large-scale disturbances play a significant role for the meridional eddy fluxes of properties. Concerning the vertical flux of properties, however, the entire spectrum of eddies has to be considered. This principal difference depends on the shallowness of the atmosphere and the variability of its vertical stability. Generally it may be stated that the mechanical turbulence tends to transport heat and angular momentum downward but moisture upward, whereas the convective turbulence transports heat and moisture upward and angular momentum downward. Especially in the tropics the convective turbulence is very active in this respect. In the ex-

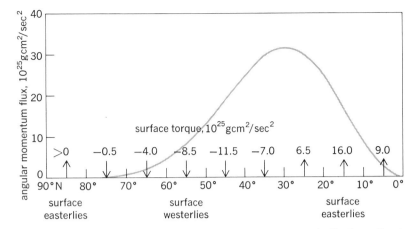

Fig. 4. Average annual northward flux of angular momentum in Northern Hemisphere and corresponding surface torques in belts of latitude.

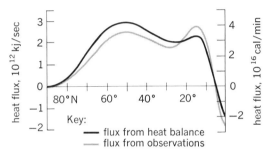

Fig. 5. Comparison between the average northward flux of heat, computed separately from the balance requirement and aerological data.

tratropical regions, where the convective phenomena are less prominent, especially in winter, the large-scale eddies, associated with polar-front cyclones, are very active in transporting heat and moisture upward. *See* FRONT; STORM.

A scheme of the different fluxes associated with the maintenance of the general atmospheric circulation is presented in Table 4. Here the symbols P, E, U, and D denote poleward, equatorward, upward, and downward fluxes, respectively. The dominating part is underlined. It should also be stressed that P and E are referred to the thermal equator around latitude 5°N (Fig. 1). In the table the circulation fluxes are marked by $\overline{F}_\varphi$ and $\overline{F}_z$, the eddy-fluxes by F'_φ and F'_z, and the total fluxes by F_φ and F_z, respectively. The scheme is valid for the troposphere, or roughly for about 9/10 of the total mass of the atmosphere.

Seasonal changes. Because of the limited scope of this article, it has not been possible to consider the seasonal changes of the atmospheric circulation. Because of the strong seasonal variations in the solar radiation and related phenomena, corresponding seasonal changes in the circulation pattern and fluxes of different type must occur. In the winter hemisphere both the zonal and meridional circulation are considerably stronger than they are in the summer hemisphere. The boundaries between the Hadley and Ferrel cells and the intertropical convergence zone are displaced toward the summer hemisphere. Especially in the

Table 2. Atmospheric properties used for flux computations

Property	x (per unit mass)
Atmospheric energy	$c_p T + gZ$
Latent heat	Lq
Angular momentum	$\Omega a^2 \cos^2\varphi + ua \cos\varphi$

Table 3. Circulation and eddy northward fluxes of atmospheric energy (unit, 10^{16} cal/min)

Latitude	70°N	60°	50°	40°	30°	20°	10°	0°
Circulation flux	0.0	−0.4	−0.7	−0.5	1.0	3.0	3.2	−2.4
Eddy flux	1.7	3.4	4.2	3.7	1.5	0.3	0.1	0.0
Total flux	1.7	3.0	3.5	3.2	2.5	3.3	3.3	−2.4

Table 4. Main direction of different types of fluxes

Latitude zones:	Hadley cells						Ferrel cells					
Fluxes:	$\overline{F}_\varphi$	F'_φ	F_φ	$\overline{F}_z$	F'_z	F_z	$\overline{F}_\varphi$	F'_φ	F_φ	$\overline{F}_z$	F'_z	F_z
Heat	<u>P</u>	P	P	U	<u>U</u>	U	E	<u>P</u>	P	D	<u>U</u>	U
Moisture	<u>E</u>	P	E	U	<u>U</u>	U	P	<u>P</u>	P	–	<u>U</u>	U
Angular momentum	P	<u>P</u>	P	U	<u>D</u>	U	E	<u>P</u>	P	<u>D</u>	D	D

Northern Hemisphere with its much stronger continental influence, the seasonal changes are very prominent. On the whole, however, the circulation pattern and the fluxes of heat, water vapor, and angular momentum show the same characteristic features in all seasons. [E. PALMEN]

Bibliography: E. N. Lorenz, *The Nature and Theory of the General Circulation of the Atmosphere*, World Meteorological Organization, 1967; W. D. Sellers, *Physical Climatology*, 1965.

Atmospheric pollution

All airborne particulate matter, liquid and solid, and gases—except water in its several phases—which exist in the atmosphere in variable amounts. Typical natural contaminants are salt particles from the oceans or dust and gases from active volcanoes; typical artificial contaminants are waste smokes and gases formed by industrial, municipal, household, and automotive combustion processes. Pollens, spores, rusts, and smuts are natural aerosols augmented artificially by man's land-use practices. *See* ATMOSPHERIC CHEMISTRY.

Sources. Air contaminants are produced in many ways and come from many sources. It is difficult to identify all the various producers. For example, it is estimated that in the United States 60% of the air pollution comes from motor vehicles and 14% from plants generating electricity. Industry produces about 17% and space heating and incineration the remaining 9%. Other sources, such as pesticides and man's earth-moving and agricultural practices, lead to vastly increased atmospheric burdens of fine soil particles and of pollens, spores, rusts, and smuts; the latter are referred to as aeroallergens because many of them induce allergic responses in sensitive persons.

Sources may be characterized in a number of ways. A frequent classification is in terms of stationary and moving sources. Examples of stationary sources are power plants, incinerators, industrial operations, and space heating. Examples of moving sources are motor vehicles, ships, aircraft, and rockets. Another classification describes sources as point (a single stack), line (a line of stacks), or area (a city). The types of pollution released sources are described in the next section.

Types. Different types of pollution are conveniently specified in various ways: gaseous, such as carbon monoxide, or particulate, such as smoke; inorganic, such as hydrogen fluoride, or organic, such as the mercaptans; oxidizing substances, such as ozone, or reducing substances, such as the sulfur oxides; radioactive substances, such as iodine-131, or inert substances, such as pollen or fly ash; or thermal pollution, such as the heat produced by nuclear power plants.

The annual emission over the United States of many contaminants is very great (Fig. 1). Motor vehicles contribute about 60% of total pollution: nearly all the carbon monoxide, two-thirds of the hydrocarbons, one-half of the nitrogen oxides, and much smaller fractions in other categories.

Individual cities and their environs, representing large area sources, are heavy contributors to atmospheric pollution. As shown in Fig. 2, Los Angeles County releases daily thousands of tons of contaminants, mainly carbon monoxide, nitrogen oxides, and hydrocarbons from motor vehicles;

stringent controls have greatly limited emissions of the sulfur oxides and of particulates.

Dispersion. Dispersion of pollution is dependent on atmospheric conditions. Winds transport and diffuse contaminants; rain may wash them to the surface; and under cloudless skies, solar radiation may induce important photochemical reactions.

Primary meteorological influences. Wind direction, speed, and turbulence influence atmospheric pollution. Wind direction determines the area into which the pollution is carried. Dilution of contaminants from a source is directly proportional, other factors being constant, to wind speed, which also determines the intensity of mechanical turbulence produced as the wind flows over and around surface objects, such as trees and buildings.

Eddy diffusion by wind turbulence is the primary mixing agency in the atmosphere; molecular diffusion is negligible in comparison. In addition to mechanical turbulence, there is thermal turbulence which occurs in an unstable layer of air. Thermal turbulence and associated intense mixing develop in an unsaturated layer in which the temperature decreases with height at a rate greater than 1°C/100 m, the dry adiabatic rate of cooling. When the temperature decreases at a lower rate, the air is stable and turbulence and mixing, now primarily mechanical, are less intense. If the temperature increases with height—a condition known as an inversion—the air is very stable and horizontal turbulence and mixing are still appreciable, but vertical turbulence and mixing are almost completely suppressed.

Secondary meteorological influences. Precipitation, fog, and solar radiation exert secondary meteorological influences. Falling raindrops may collect particles with radii greater than 1 μ or may entrain gases and smaller particles in their wakes and carry them to the ground. Gas reactions with aerosols also occur; neutralizing cations in fog

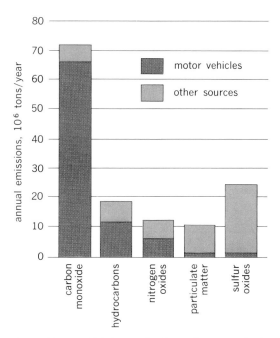

Fig. 1. Motor vehicle's contribution to five atmospheric contaminants in the United States.

droplets or traces of ammonia, NH₃, in the air act as catalysts to accelerate reaction rates leading to rapid oxidation of sulfur dioxide, SO₂, in fog droplets. For highly polluted city air it is estimated that, in the presence of NH₃, the oxidation of the SO₂ present in a fog to ammonium sulfate, (NH₄)₂SO₄, is completed in 1 hr for fog droplets 10 μ in radius. Photochemical oxidation of hydrocarbons in sunlight is frequent. Most hydrocarbons do not have appropriate absorption bands for a direct photochemical reaction; nitrogen dioxide, NO₂, when present acts as an oxidation catalyst by absorbing solar radiation strongly and subsequently transferring the light energy to the hydrocarbon and thereby oxidizing it.

Natural ventilation. Natural ventilation in the atmosphere is best when the winds are strong and turbulent so that mixing is good, and when the volume in which mixing occurs is large so that dilution of pollution is rapid. As cities have grown in size, air pollution has become more widespread and acute, and it has become necessary to think of whole urban complexes as large area sources of pollution. The rate of natural ventilation of such an urban area is dependent on two quantities: the wind speed and the mixing volume over the city. Active mixing upward is often limited by a stable layer, perhaps even a very stable inversion layer, aloft. The upward extent of this region of active mixing, known as the mixing height, determines the magnitude of the mixing volume over the city. The number of air changes per unit time in this mixing volume specifies the rate of natural ventilation of the urban area. The problems of air pollution become highly complex, however, because the mixing height is rarely constant for long. Some of the factors causing it to vary are described below.

At night when the sky is clear and the wind light, Earth's surface loses heat by long-wave radiation to space. As a result, the ground cools and a surface radiation inversion is formed, which by 3 A.M. may appear as shown in Fig. 3. The inversion inhibits mixing, and the mixing height is effectively zero, so that pollution accumulates. Solar heating of the ground causes a reversal of the lapse rate, which may exceed the dry adiabatic rate of cooling by 9 A.M.; active mixing in this unstable layer results in an appreciable mixing height H, as shown.

The mixing may bring pollution down from aloft, causing a temporary peak in the surface concentrations, a process known as an inversion breakup fumigation. By 3 P.M. H has reached its maximum value for the day, and surface concentrations tend to be low as the natural ventilation improves. As the Sun goes down, the lapse rate becomes stable again, being shown as isothermal at 9 P.M., and accumulation of contaminants recommences as the cycle repeats, if clear skies and light winds still prevail.

The accumulation of pollution for longer periods of time is especially likely to occur if a persistent inversion aloft exists. Such an inversion aloft is the subsidence inversion formed by the sinking and vertical convergence of air in an anticyclone, as illustrated in Fig. 4. A layer of air at high levels descends, diverging horizontally and hence converging in the vertical, and warms at the dry adiabatic rate of heating of 1°C/100 m. Figure 4a shows how a low-level inversion may result from

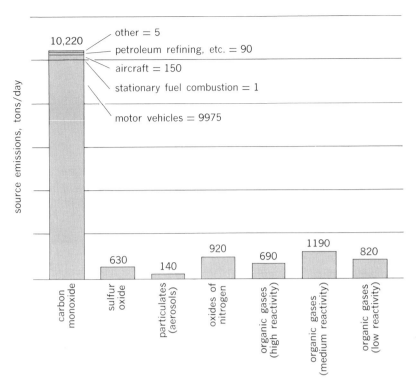

Fig. 2. Comparative emissions of air pollutants in Los Angeles County, January, 1967. Mean daily rate based on annual emissions.

this process, while Fig. 4b depicts how the mixing height H is limited in vertical extent by the subsidence inversion aloft, so that pollution accumulates within and just above the city. It is the presence of such a subsidence inversion aloft associated with the Pacific subtropical anticyclone which is the primary cause of Los Angeles and California smogs; these are made even worse by local mountain and valley sides which prevent horizontal dispersion.

The worst pollution occurs when, in addition to subsidence inversions accompanying slowly moving or stationary anticyclones, fog also develops. All the major air pollution disasters, such as those listed in a later section, took place when fog persisted during protracted stagnant anticyclonic conditions. The reasons for the adverse influence of fog are shown in Fig. 5. When there is no fog (Fig. 5a), solar radiation heats the ground, which in turn causes a lapse rate equal to, or greater than, the dry adiabatic rate of cooling, with good mixing and hence a substantial mixing height H. On the other hand, with a fog layer (Fig. 5b), up to 70% of the solar radiation incident at the top of the fog is reflected back to space, with relatively little left to heat the fog and ground below. With the cloudless skies characteristic of anticyclonic weather, there is a continuous loss of heat to outer space from the upper surface of the fog bank, which acts radiatively as an elevated ground surface. More heat is lost to space than is gained from the Sun, and an inversion develops above the fog and persists night and day until the anticyclone dissipates or moves away. If the air is polluted, the fog particles may become acids and salts in solution; the saturation vapor pressure over such particles may decrease

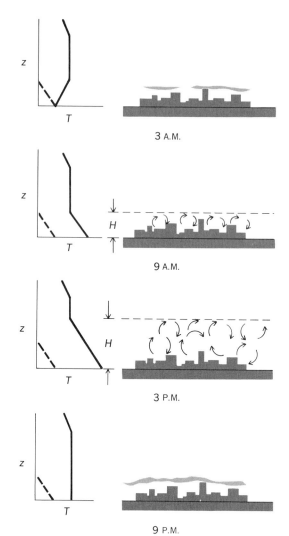

Fig. 3. Diurnal variation of mixing height H over a city with light winds and clear skies. Full curves show temperature T plotted against height z; dashed lines represent the dry adiabatic rate of cooling.

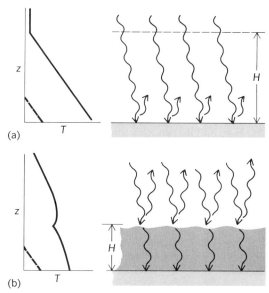

Fig. 5. The influence of fog in reducing mixing height H. (a) Without fog. (b) With fog. The dashed lines represent the dry adiabatic rate of cooling. Arrows represent solar and reflected radiation.

ATMOSPHERIC POLLUTION

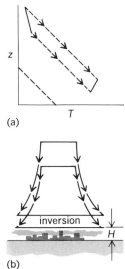

Fig. 4. Subsidence inversion. (a) Solid lines show temperature T and height z before and after dry adiabatic descent of air; dashed lines represent dry adiabatic rate of heating. (b) Inversion limits mixing height H over city.

to 90 or 95%, so the smog becomes even more persistent than if it remained as a pure water fog. Disastrous concentrations of contaminants may, and sometimes do, accumulate during prolonged foggy conditions of this kind.

Another significant inversion aloft is associated with a slowly moving warm frontal surface. Consider two cities, one lying to the southwest and the other lying to the northeast of a warm front extending from the southeast to the northwest, as illustrated in Fig. 6a. City B lies in the cool air, with the warm frontal surface above it. With characteristic southeast winds in the cool air ahead of the warm front, the pollution from City B is trapped below the warm frontal inversion, as shown in the vertical cross section of Fig. 6b, and may travel for many miles with large surface concentrations. On the other hand, the prevailing southwest winds in the warm sector will carry pollution from City A up and above the warm frontal inversion, which effectively prevents its diffusion downward to the surface. This figure brings out an important point:

An inversion layer may be advantageous, not disadvantageous, if it inhibits diffusion down to the ground.

Stack dispersion. Dispersion from an elevated point source, such as a stack, is conveniently expressed by the equation below, where χ is ground-level concentration of contaminant in mass per unit volume; Q is the source strength in mass per unit time; $\bar{u}$ is mean wind speed; x is the horizontal direction coinciding with that of the mean wind $\bar{u}$;

$$\chi = \frac{Q}{\pi \sigma_y \sigma_z \bar{u}} \exp\left[-\frac{1}{2}\left(\frac{y^2}{\sigma_y^2} + \frac{h^2}{\sigma_z^2}\right)\right]$$

y is the horizontal direction perpendicular to the mean wind $\bar{u}$; σ_y and σ_z are diffusion coefficients expressed in length units in the y and z directions, respectively, the z direction being vertical; and h is the height of the source above ground.

This diffusion equation should be used only under the simplest conditions, for example, in flat, uniform terrain and well away from hills, slopes, valleys, and shorelines. Table 1 lists various meteorological categories as A, B, C, D, E, and F, and Table 2 gives values of the diffusion coefficients appropriate for each category. It should be noted that the values to be used depend on distance from the source at which concentrations are to be calculated. A variety of other forms is available for more complex conditions of terrain and meteorology.

Natural cleansing processes. Pollution is removed from the atmosphere in such ways as washout, rain-out, gravitational settling, and turbulent impaction. Washout is the process by which contaminants are washed out of the atmosphere by raindrops as they fall through the contaminants; in rain-out the contaminants unite with cloud droplets which may later grow into precipitation.

Gravitational settling is significant mainly for large particles, those having a diameter greater

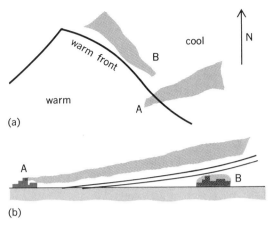

Fig. 6. Influence of slowly moving warm frontal inversion aloft. (*a*) Map of cities A and B. (*b*) Side view.

Table 2. Values of diffusion coefficients

Distance from source, m		Diffusion coefficient, m^2					
		Meteorological categories					
		A	B	C	D	E	F
10^2	σ_y	22	16	12	8	6	4
10^3		210	150	105	75	52	36
10^4		1700	1300	900	600	420	360
10^5		11000	8500	6300	4100	2800	2000
10^2	σ_z	14	11	7.6	4.8	3.6	2.2
10^3		500	120	70	32	24	14
10^4		–	–	420	140	90	46
10^5		–	–	2100	440	170	92

than 20 μ. Agglomeration of finer particles may result in larger ones which settle out by gravitation. Fine particles may also impact on surfaces by centrifugal action in very small turbulent eddies. Gases may be converted to particulates, as by photochemical action of sunlight in Los Angeles, Denver, and Mexico City. These particulates may then be removed by settling or impaction.

The rate of natural cleansing may be slower than the rate of injection of pollutants into the atmosphere, in which case pollution may increase on a global scale. There is evidence that the concentration of atmospheric carbon dioxide has been increasing slowly since the beginning of the century because of combustion of fossil fuels. The burden of very small particles may also be increasing. What appreciable effects, if any, these increases might have on the climate is subject only to conjecture at this time. Worldwide pollution by radioactive debris produced by fission- and fusion-bomb testing is removed slowly, mainly by rainfall originating in the high troposphere.

Effects. A number of the many effects of atmospheric pollution are described briefly below.

Man. The effects of many pollutants on human health under most ordinary circumstances of living, rural or urban, are difficult to specify with confidence. In the United States a number of animal experiments under controlled conditions and of epidemiological studies have been made, but the results are difficult to interpret in terms of human health. A study on the lower East Side of New York City indicated that, in children less than 8 years old, the occurrence of respiratory symptoms was associated with the levels of particulate matter and of carbon monoxide in the atmosphere. With heavy smokers, however, eye irritation and headache were directly related to increasing concentrations of carbon monoxide.

Air pollution is suspected as a causative agent in the occurrence of chronic bronchitis, emphysema, and lung cancer, but the evidence is not clear cut. On the other hand, from mid-August to late September the potent aeroallergen, ragweed pollen, is a substantial cause of allergic rhinitis and bronchial asthma for over 10,000,000 persons living east of the Rockies. In Los Angeles County the effects of smog on health are becoming better understood. For example, people living in less smoggy areas of the county survive heart attacks more readily than others: In 1958 in high-pollution areas the mortality rate per 100 hospital admissions for heart attacks averaged 27.3 in comparison with 19.1 for low-smog areas. Other studies show a small but significant relation between motor vehicle accidents and oxidant levels. Medical authorities in Los Angeles are becoming concerned about the long-term influences of various pollutants, including photochemical smog, despite the lack of comprehensive knowledge of the nature of such effects.

In Great Britain there is a similar lack of precise

Table 1. Meteorological categories*

Surface wind speed, m/sec	Daytime insolation			Thin overcast or $\geq$ 4/8 cloudiness†	$\leq$ 3/8 cloudiness
	Strong	Moderate	Slight		
<2	A	A–B	B		
2	A–B	B	C	E	F
4	B	B–C	C	D	E
6	C	C–D	D	D	D
>6	C	D	D	D	D

*A = Extremely unstable conditions.
B = Moderately unstable conditions.
C = Slightly unstable conditions.
D = Neutral conditions (applicable to heavy overcast, day or night)
E = Slightly stable conditions.
F = Moderately stable conditions.

†The degree of cloudiness is defined as that fraction of the sky above the local apparent horizon which is covered by clouds.

knowledge. There are indications that emphysema, bronchitis, and other respiratory diseases are not caused primarily by increased atmospheric pollution, but because more people are living longer. Despite a substantial reduction in the concentrations of atmospheric particulates since the Clean Air Act was passed in 1956, the respiratory disease rate continues to rise. These facts do not prove that air pollution is not a factor, but that its influence may be synergistic and therefore difficult to identify precisely. For example, it is known that the combined effect of sulfur oxides and particulates is substantially greater than the sum of the two separate effects, and many other such synergistic effects doubtless occur.

In the Netherlands special efforts have been made to relate SO₂, both concentration values and exposure times, to effects on humans and on vegetation. The results of these studies, based on investigations in England, the United States, West Germany, Italy, and the Netherlands, are illustrated in Fig. 7. Influences on humans are shown in the lower family of curves: A first-degree effect is a small increase in functional disturbances, symptoms, illnesses, diseases, and deaths; a second-degree effect is a more prevalent or more pronounced effect of the same kind; and a third-degree effect is a substantial increase in the number of deaths. It should be emphasized that the exposures were, in general, to SO₂ in dusty and sooty atmospheres.

Under extreme circumstances when stagnant atmospheric conditions with persistent low wind and fog exist, major disasters involving many deaths occurred, as in and around London in 1873, 1880, 1891, 1948, 1952, 1956, and 1962. Similar disasters occurred in the Meuse Valley of Belgium in 1930 and at Donora, Pa., in 1948.

Atmospheric pollution has a substantial influence on the social aspects of human life and activity. For example, the distribution of urban populations is being increasingly affected by such pollution, and recreational patterns are similarly influenced. The atmospheric burden of pollution is thus becoming more and more important as a determinant in social decision making.

Animals. Studies of the response of laboratory animals to specified concentrations of pollutants have been conducted for many years, but the interpretation of the results in terms of corresponding human response is most difficult. Assessment of the effects of certain contaminants on livestock is relatively straightforward, however. Thus contamination of forage by airborne fluorides and arsenicals from certain industrial operations has led to the loss of large numbers of cattle in the areas adjacent to such chemical industries.

Plants. Damage to vegetation by air pollution is of many kinds. Sulfur dioxide may damage such field crops as alfalfa, and trees such as pines, especially during the growing season; some general relations are presented in Fig. 7. Both hydrogen fluoride and nitrogen dioxide in high concentrations have been shown to be harmful to citrus trees and ornamental plants which are of economic importance in central Florida. Ozone and ethylene are other contaminants which cause damage to certain kinds of vegetation.

Materials. Corrosion of materials by atmospheric pollution is a major problem. Damage occurs to ferrous metals; to nonferrous metals, such as aluminum, copper, silver, nickel, and zinc; to building materials; and to paint, leather, paper, textiles, dyes, rubber, and ceramics.

Weather. Atmospheric pollution may affect weather in a number of ways. Heavy precipitation at Laporte, Ind., is attributed to a substantial source of air pollution there, and similar but less pronounced effects have been observed elsewhere. Industrial smoke reduces visibility and also ultraviolet radiation from the Sun, and polluted fogs are more dense and more persistent than natural fogs occurring under similar conditions. Possible major effects of air pollution on Earth's climate have been mentioned earlier.

Costs. It is extremely difficult to estimate accurately the economic costs of air pollution. Dollar values are not readily established for losses due to illness caused by exposure to air pollution. At the present time a reasonable estimate appears to be that air pollution costs the United States about $10,000,000,000 a year. For Great Britain the corresponding figure—probably a conservative one—is £400,000,000 a year.

Controls. Four main methods of air-pollution control are indicated below.

Prevention. This method has, until recently, been applied mainly to reduce pollution from combustion processes. Improved equipment design and smokeless fuels have reduced pollution both from industrial and motor vehicle sources. Industry is giving serious consideration to designing new plants in such a way that atmospheric contaminants will not be produced at all, or in much smaller amounts than at present. Such radical re-

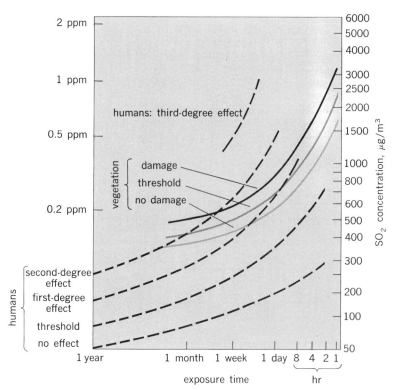

Fig. 7. Effects of sulfur dioxide on humans and vegetation. (*L. J. Brasser et al.*)

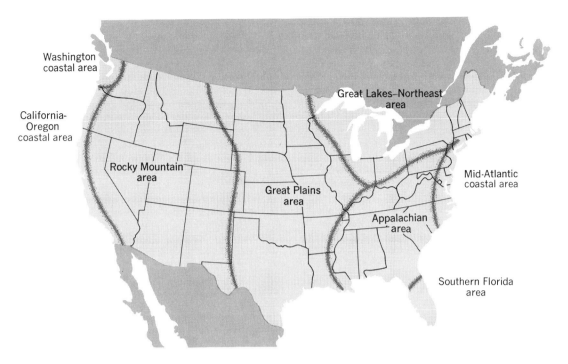

Fig. 8. Atmospheric areas of the United States, specified in accordance with the Air Quality Act of 1967.

thinking of plant processes holds great hope for substantial reduction in air pollution per unit output of product.

Collection. Collection of contaminants at the source has been one of the important methods of control. Many types of collectors have been employed successfully, such as settling chambers, cyclone units employing centrifugal action, bag filters, liquid scrubbers, gas-solid adsorbers, ultrasonic agglomerators, and electrostatic precipitators. The optimum choice for a given industrial process depends on many factors. A major problem is disposal of the collected materials. Sometimes they can be used in by-product manufacture on a profitable or a break-even basis. Neither disposal in water nor outdoor piling or ponding near the plant is acceptable. It is sometimes necessary to haul waste materials to remote locations — an expensive procedure. Collected solid wastes have sometimes been hauled to low-lying land, leveled, and then given a surface treatment which is capable of sustaining the growth of vegetation.

Containment. This method is useful for pollutants whose noxious characteristics may decrease with time, such as radioactive contaminants from nuclear power plants. For contaminants with a short half-life, containment may allow the radioactivity to decay to a level which permits their release to the atmosphere. Containment, with destruction or conversion of the offending substances, often malodorous or toxic, is used in certain chemical, oil refining, and metallurgical processes and in liquid scrubbing.

Dispersion. Atmospheric dispersion as a control method has a number of advantages, especially for industrial processes which can be varied to take advantage of the periods when dispersion conditions are so good that contaminants may be distributed very widely in such small concentrations that they inconvenience no one. Some coal-burn-

ing electrical power stations are building high stacks, up to 1000 ft, to lift the SO_2-bearing stack gases well above the ground. Some plants store low-sulfur anthracite coal for use when atmospheric dispersion is poor.

The use of cooling towers by plants which would otherwise cause thermal pollution of rivers and lakes may lead to air-pollution problems. The larger amounts of water vapor injected into the atmosphere may increase the frequency and duration of fogs in the plant vicinity, perhaps with serious consequences. *See* AIR-POLLUTION CONTROL.

Laws. Many new laws designed to limit air pollution have been enacted. Major steps forward were taken by the Netherlands in 1952, by Great Britian in 1956, by Germany in 1959 and 1962, by France in 1961, by Norway in 1962, by the United States in 1963 and 1967, and by Belgium in 1964.

Efforts to control air pollution by legal means commenced many years ago in Great Britain. In 1906 the Alkali Act consolidated and extended previous similar acts, the first of which was passed in 1863. This calls for the annual registration of scheduled industrial processes, and requires that the escape of contaminants to the atmosphere from scheduled processes must be prevented by the "best practicable means." The Alkali Act functions by interpretation and not by statutory requirement, the Alkali Inspector being the sole judge of the "best practicable means." The Clean Air Act of 1956 provided more effective ways of limiting air pollution by domestic smoke, industrial particulates, gases and fumes from the processes registrable under the Alkali Act, and smoke from diesel engines. This legislative program has had considerable success in alleviating air-pollution problems in Great Britain.

In the United States, air pollution control had been considered to be a matter of local concern only. By 1963 only one-third of the states had air-pollution control programs, most of which were relatively ineffective. Only in California were local programs, at the city and county level, supported adequately. The Clean Air Act of 1963 brought the Federal government into a regulatory position of increased scope by granting the Secretary of Health, Education, and Welfare specific abatement powers under certain circumstances. It also established a Federal program of financial assistance to local control agencies and recommended more vigorous action to combat pollution by motor vehicle exhausts and by smoke from incinerators.

The Air Quality Act of 1967 brought the Federal government into a more substantial regulatory role. One of its important effects has been to change the emphasis in legislation from standards based on emissions from sources, such as stacks, to standards based on concentrations of contaminants in the ambient air which result from such emissions. The Air Quality Act of 1967 consists of three main portions, listed below.

Title I: Air-Pollution Prevention and Control. The first section of the Air Quality Act amends the Clean Air Act to encourage cooperative activities by states and local governments for the prevention and control of air pollution and the enactment of uniform state and local laws; to establish new and more effective programs of research, investigation, training, and related activities; to give special emphasis to research related to fuel and vehicles; to make grants to agencies to support their programs; to provide strong financial support for interstate air-quality agencies and commissions; to define atmospheric areas—as done subsequently in January, 1968, and shown in Fig. 8—and to assist in establishing air quality control regions, criteria, and control techniques; to provide for abatement of pollution of the air in any state or states which endangers the health or welfare of any persons and to establish the necessary procedures; to establish the President's Air Quality Advisory Board and Advisory Committees; and to provide for control of pollution from Federal facilities.

Title II: National Emission Standards Act. This section is concerned mainly with pollution from motor vehicles which accounts for some 60% of the total for the United States. The act covers such matters related to motor vehicle emissions as the following: establishment of effective emission standards and of procedures to ensure compliance by means of prohibitions, injunction procedures, penalties, and programs of certification of new motor vehicles or motor vehicle engines and registration of fuel additives. The act also calls for a comprehensive report on the need for, and the effect of, national emission standards for stationary sources.

Title III. The final section is general and covers matters such as comprehensive economic cost studies, definitions, reports, and appropriations.

There is no doubt that this far-reaching legislative program, stimulating new approaches at the local, state, and Federal levels, will play a major role in controlling air pollution within the United States. The other industrial nations of the world are preparing to meet their growing air-pollution problems by initiatives appropriate to their own particular circumstances. [E. WENDELL HEWSON]

Bibliography: L. J. Brasser et al., *Sulphur Dioxide: To What Level is it Acceptable?*, Research Institute of Public Health Engineering, Delft, Netherlands, Rep. no. G300, 1967; P. A. Leighton, Geographical aspect of air pollution, *Geogr. Rev.*, 56:151, 1966; P. A. Leighton, *Photochemistry of Air Pollution*, 1961; A. R. Meetham, *Atmospheric Pollution*, 1964; R. Scorer, *Air Pollution*, 1968; A. R. Smith, *Air Pollution*, 1966; A. C. Stern (ed.), *Air Pollution*, 3 vols., 1968; H. Wolozin (ed.), *The Economics of Air Pollution*, 1966.

Avalanche

A mass of snow or ice moving rapidly down a mountain slope or cliff. Avalanches, or snowslides, range from small movements on established avalanche tracks to large, sporadic, very rapid movements capable of taking a heavy toll of life. Avalanches are infrequent on slopes of less than 25°, especially numerous on those exceeding 35°, and most commonly start on convex slopes. Both old and new snow may avalanche, but serious avalanches are always possible when 12 in. or more of new snow is present. Movement may be set off by temperature, vibration, shearing, or other slope disturbance.

Dry snow avalanches usually occur during, or within several days after, snowfall. They may affect whole slopes, even if wooded, and may exceed 100 mph. Wet snow avalanches are formed during thaws or rainy weather. Their movement is less rapid but may be destructive. Slab avalanches of wind-packed snow are broader and deeper and move rapidly. This type of avalanche is an extreme hazard to life and property.

Avalanche prevention and prediction has become an important service. Reforestation, snow-sheds, and avalanche breakers and barriers help to prevent or control movement. Careful coordination in planning land use, avalanche forecasting, and safety patrolling, with artificial release of slides and temporary closing of hazardous zones, help make mountain areas safe for human use.

[C. F. STEWART SHARPE]

Bioclimatology

A study of the effects of the natural environment on living organisms. These effects may be direct, such as the influence of ambient temperature on body heat, or indirect, such as the influences on composition of food. Only direct effects are discussed in this article. Natural and artificial elements cannot be sharply distinguished. For example, smoke from a lightning-induced forest fire is natural, but smoke from a chimney is artificial.

Bioclimatology encompasses biometeorology, climatophysiology, climatopathology, air pollution, and other fields. The interplay between disciplines, such as meteorology and physiology, is emphasized. Concerning the time scale, the study of events enduring for hours to days is called biometeorology; for years to centuries, bioclimatology; for millennia and more, paleobioclimatology. For intrinsic reasons bioclimatology should serve as an overall term. Bioclimatology falls naturally into the two areas of plants and of animals and man.

In plant bioclimatology solution of any problem involves study of the natural climatic values, such as air temperature, precipitation, and wind speed and its variations; the transfer mechanism, such as the eddy diffusivity; and the effect of these agents on the plant. The most important mechanisms are (1) photochemical effects, such as photosynthesis and (blue) phototaxis; photosynthesis needs blue and red light, water from the roots (and possibly from the air), and CO_2 from the air; all these components vary with weather and climate; (2) evapotranspiration through integument and stomata of plants, a process depending highly on availability of water, transfer of liquids from root to leaf, temperature of the leaf, water vapor of the air, and ventilation; (3) picking up compounds of N, K, P, Ca, and others from the ground; and (4) avoidance of destructive conditions, such as freezing, drying out of leaves (if water supply is smaller than evaporative demand), and overheating. *See* AGRICULTURAL METEOROLOGY.

Since there has been little systematic work done on animal bioclimatology, the following discussion concerns man.

Photochemical bioclimatology. Essentially this is the investigation of the effects of light from the Sun and sky. Sunburn of the skin and cornea of the eye is initiated by denaturization of nucleic acid and skin proteins, causing a local histaminelike action. In nature the effect is restricted and is induced by ultraviolet radiation with wavelengths around $0.3-0.31 \mu$. Sunburn is delayed or prevented by three screening agents in the skin: pigment, horny layer, and urocanic acid. The pigment of permanently brown or black races, as well as the radiation-induced pigment in variably colored races (the so-called white race), acts as a protective filter. The horny layer of the skin reportedly grows in thickness by ultraviolet exposure. It is still debated how much protection is offered this way. The third screen is urocanic acid, a substance derived by enzymatic action from the amino acid L-histidine in the sweat. A 1-mm sweat layer absorbs 50% of the natural ultraviolet at 300 mm and nearly 100% of the mercury lamp ultraviolet at 250 mm. Artificial ultraviolet sources, such as the cold mercury lamp, act very differently from these sources. Solar erythema can be seen within minutes of exposure. A simple rule to avoid solar-overexposure is this: As soon as the dividing line between exposed and shielded (clothed) areas can be discerned, stop sun bathing.

There are two kinds of solar pigmentation, late and direct. The late seems more prevalent in fair races; it occurs in conjunction with sunburns days later. The direct type is found more in southern Europeans and Japanese. It occurs, without sunburn, during a 30-min exposure, and is caused by long-ultraviolet $(0.32-0.4 \mu)$ and possibly visible light.

Frequent exposure over years leads to skin elastosis (sailor's skin) and finally skin cancer. Skin carcinomas occur more frequently on facial skin exposed to sun. Proof that exposure to sun is carcinogenic comes from statistical evidence and from the results of animal experiments, in which rodents were exposed to very strong artificial ultraviolet radiation.

Although bacteria are easily killed with artificial ultraviolet radiation of short wavelength, natural sunlight probably has very little bactericidal action because it lacks the short wavelengths.

Vitamin D is produced from natural sterols in plant and animal foodstuffs and in human skin by short-wavelength solar ultraviolet radiation (0.3μ).

It has been claimed that solar ultraviolet radiation favorably affects blood circulation and general health, but sunbather faddism has not been conducive to serious research efforts. It is difficult to separate the effects of ultraviolet radiation, thermohygric exposure, and mental stimuli on a sunbather.

Most sunlight and sky light received by the eye is that reflected by clouds and surfaces. The intensity of the incoming light, its angle of incidence, and the amount and kind (specular or diffuse) of albedo control the amount of light received by the eye. The specular reflections of water, ice, metals, snow, clouds, and white sand are bright enough to irritate the eye. Most sunglasses dampen the whole visible spectrum uniformly and eliminate ultraviolet and infrared rays.

All the important ultraviolet effects mentioned above occur at about 0.3μ, near the end of the solar spectrum. This ultraviolet radiation is controlled by absorption in stratospheric ozone and scattering by air molecules, clouds, and smog. Dependence of ultraviolet radiation on solar and geometric altitude is pronounced. Usually, the scattered sky ultraviolet radiation exceeds that of the Sun. Snow reflects both ultraviolet rays and visible light to cause snow blindness and sunburn below the chin. No other natural substance reflects more than a few percent of natural ultraviolet; however, metals reflect highly.

Natural penetrating ionizing radiation is composed of β- and γ-rays from radioactive minerals, including their emanations and secondary rays from cosmic radiation, mainly mesons and neutrons. Ionization by both effects at altitudes below 18 km is less than 40 milliroentgens per day, probably an insignificant figure.

Air bioclimatology. Gaseous air constitutents, such as oxygen and water vapor, influence body chemistry and heat balance. The oxygen partial pressure (pO_2) of inhaled air is vital for blood oxygenation and depends mainly on altitude. At or above an altitude of about 3000 m, the pO_2 is low enough to cause air sickness, or mountain sickness. Residence at an altitude of $1-2$ km is supposed to benefit circulation, but may be harmful for some heart patients.

Low water vapor pressure may cause drying of the skin and of the mucous membranes of upper respiratory organs. Water vapor pressure, even indoors, is especially low when outdoor air temperatures are low. Many skin and respiratory complaints in winter are caused more by dryness than by cold. Water vapor influence on heat balance is discussed below.

Carbon dioxide, CO_2, water, H_2O, and organic vapors emitted by man contribute to the unpleasantness of crowded and ill-ventilated rooms, which may also be overheated.

Man's industries, volcanoes, fires, dust storms, and other more or less violent events spew large amounts of matter into the air known as fallout, smog, air pollution, and so forth. No bioclimatic problem in this field can be reported as solved unless source, transfer mode, change in transfer, concentration near the sink, the mode of intake by the sink, and the biological and chemical action of the sink are clear.

Ozone, O_3, may serve as an example. It is produced by solar ultraviolet radiation in the higher stratosphere, by lightning, and by sunlight falling on smog (Los Angeles). Stratospheric ozone is brought down by turbulence, and might reach toxic levels for crews of airplanes at about 15-km altitude. Ozone is blamed for some smog-induced injuries, for example, ocular pain in Los Angeles. A very important sink action is the inhalation into the respiratory organs, where especially the lung's surface is changed by oxidation. This causes coughing at first and tiredness later. Amounts of O_3 less than 0.5 ppm are toxic. See OZONE.

Carbon monoxide, unburned hydrocarbons, NO, NO_2, SO_2, and other by-products are prevalent where there is incomplete combustion. Many by-products are also highly hygroscopic, and serve as nuclei of condensation to cause the low-humidity type of smog, such as that occurring in Los Angeles. This smog condition is further complicated by the occurrence of light-induced reactions between hydrocarbons and NO_2 and other pollutants. The high-moisture type of smog seems typical for London. See SMOG.

Very few gases, in fact very little of any gas, pass the skin, with the exception of more easily permeated areas such as chapped lips, scrotum, and labia.

Aerosol bioclimatology. Solid or liquid suspensions in the air affect breathing organs or skin. Widely discussed elements of the atmospheric aerosols are the ions, which are particles containing one or more electrical charges. For a long time beneficial or dangerous effects of an abundance of particles of one charge type have been claimed. These electrical space charges reportedly have influenced results of physiological tests, growing of living cells, and severity of hay fever attacks. As a rule, negative space charge of the order of 1000–10,000 ions/cm³ is reportedly beneficial. These results are expected to explain some observations of statistical bioclimatology. See ATMOSPHERIC CHEMISTRY: ATMOSPHERIC POLLUTION.

Thermal bioclimatology. This subdiscipline concerns the heat balance of man as controlled by his environment. The basic relation is shown by Eq. (1), where M is metabolic heat production; C is

$$M - C d\theta_b/dt + H_h = H_r + H_a + H_e + R \quad (1)$$

body heat capacity; H is heat exchanged through skin; R is respiratory heat exchange; θ is temperature (°C); b refers to total body; h refers to solar radiation; r refers to infrared radiation; s refers to skin; a refers to air (convection); e refers to evaporation; and t is time. The H terms are defined in Eq. (2), where S is visible and near-infrared radiant

$$H_h = S \cdot \epsilon_h \cdot A_h \qquad H_r = h_r \cdot A_r (\theta_s - \theta_r)$$
$$H_a = h_a A_a (\theta_s - \theta_a) \qquad H_e = h_e \cdot A_e (e_s - e_a) \quad (2)$$

heat flow from the Sun, sky, and environment; and ϵ_h is absorptivity of skin (0.6 for white and 0.9 for black skin). The A factors, that is, the respective surface areas, depend on body posture and on the process involved, such as radiation or convection; these areas are always smaller than the geometric surface. The terms e_s and e_a are the vapor pressures of skin and air, respectively.

The h factors are heat conductances at absolute temperature T, as shown by Eq. (3), where ϵ_r is

$$h_r = \epsilon_r \sigma (T_s{}^4 - T_r{}^4)/(\theta_s - \theta_r) \quad (3)$$

0.98 (infrared absorptivity of skin), and σ is Stefan's constant, 4.9 kcal·m²/(hr)(deg⁴). From experiments with men, and using kilocalorie units (kcal), $h_a = 6.3\sqrt{v}$ kcal/(m²)(hr)(deg) for wind velocity v m/sec; further, $h_a = 3.3$ in calm. Equation (4)

$$h_e = 1.63 h_a \text{ kcal}/(m^2)(hr)(mb)(deg) \quad (4)$$

follows from h_a. These data are valid for supine adults at sea level, and refer to the heat- or vapor-exchanging area.

Equation (5) applies for the clothed body, if h_c is

$$\theta_s - \theta_a = M/A_a h_c + (M + H_h)/A_a (h_a + h_r) \quad (5)$$

conductance of clothes, and if $\theta_r = \theta_a$, $A_r = A_a$, $H_c = 0$, and $d\theta_b/dt = 0$. The absorptivity of the clothing is inserted for ϵ_h. If $h_a \gg h_e$ (strong wind, thick clothes), solar heating H_h becomes unimportant.

H_e can be measured directly with a good balance. The relative skin humidity $r_s = e_s/e^*_s$, where $e_s{}^*$ is saturation vapor pressure at skin temperature, can be measured with a hair hygrometer or can be derived from H_e, h_e, e_a, and e_s. During sweating r_s is 100%; otherwise, it is usually below 60%. If skin relative humidity is below 10%, the skin may crack.

Skin water transfer is accomplished by sweating and diffusion. The latter corresponds usually to 1 kcal/(m²)(hr); it may reverse, so that water or vapor is transferred into the skin. Sweat is a powerful emergency measure capable of producing $H_e = 600$ kcal/(m²)(hr) or more. Bioclimatic sweat control works two ways, via body temperatures and via water coverage. The prime control for the amount secreted seems to be the temperature of a section of the hypothalamus, skin temperature acting as a moderator. These temperatures are of course bioclimatologically controlled. Skin totally covered with sweat or bath water lowers its sweat water loss strongly, about 4:1. The effect is absent in saline or with sweat highly concentrated by evaporation.

The respiratory heat and vapor loss is small except during hyperventilation at reduced pressure. Ordinarily, this loss varies little because the exhaled temperature drops as θ_a and e_a fall.

Of the climatic elements, S, h_a (wind), and precipitation are easiest to control. High values of θ_a, θ_r, and e_a are much harder to influence than low values. To ensure a constant θ_b, the body alters the peripheral blood flow and thus alters θ_s. In low temperatures M is raised by shivering or work; in a hot environment H_e is raised by sweat. These emergency regulations are effective only for certain periods. Short-time limits are set by variations of θ_b and finally θ_s (skin burn or freezing). All body controls vary with age, sex, health, exercise, and adaptation. Reportedly, frequent limited exposures to adverse thermal conditions, particularly cold, invigorate many body functions, especially if exposure is combined with exercise.

The most important climatic element is local air temperature; in the cold, wind increases the rate of body cooling. For a well-insulated man or house, Eq. (5), air temperature alone is the deciding factor.

The house heating bill is proportional to the number of degree days, that is, time multiplied by $(\alpha - \theta_a)$, where α is the preferred room temperature, about 22°C. No usable formula for the cooling effect of open air on average individuals can be derived because clothing styles change.

At temperatures greater than 25°C sweating starts. Its evaporation is restricted by high values of e_a. Hence, the combination of θ_a and e_a describes livability in the heat. Tests on sensation experienced by young men exposed to different pairs of θ_a and e_a led to the effective temperature (ET), which approximately equals $(\theta_a + \text{dew-point temperature})/2$. The area of high ET or extreme summer discomfort in the United States is the Gulf Coast, especially the lower Rio Grande Valley (average ET, 26–27°C). In the Indus Valley ET is 28.5°C; and the world maximum is in Zeila, Somali Coast, where ET is 29.7°C. Figure 1 shows a United States map of the sum of heating and cooling degrees days. Since cooling also involves drying, the cooling degree days are multiplied by two. Large numbers of the sum mean climatic hardship for the outdoors man and high bills for heating fuel and cooling wattage. See SENSIBLE TEMPERATURE.

The microclimate can enhance discomfort. The values of θ_a and θ_r are much below normal on calm clear nights, especially when the terrain is concave. The Sun can raise the surface temperatures of such features as sand, walls, and vehicles to as much as 40° above the air temperature.

Extreme climates and microclimates. Certain climatic conditions can be tolerated only for a limited time, since they either overrun the physiological defense mechanism or injure and destroy body parts directly. The first mechanism usually leads to a breakdown of temperature regulation and of the body circulation, and the latter usually involves skin injuries.

The safe survival time depends primarily on air and wall temperature, if there is no artificial ventilation and no radiation other than the walls. Other conditions may suitably be converted into equivalent temperature data. Figure 2 contains such data, listing safe times between seconds (in a fire) to life-span. Some case studies are listed below.

1. The hot desert: Temperatures are in the fifties, vapor pressure is 10–15 mb, and there is sunshine and wind. Over a period of a few hours no equilibrium of body data is reached, and especially heart rate and core temperatures rise. The skin is dry and salt covered. The skin water loss is much below the level needed to compensate by evaporation for the large heat input by solar and infrared radiation and convection.

2. The jungle: Normal jungle is the habitat of a large segment of people. However, extreme jungle conditions are around 35°C and 40 mb vapor pressure. The skin is totally wet, but evaporation becomes insufficient due to external moisture. The effective temperature is a good measure for this danger signal. Worst in this respect are locally wet areas surrounded by a hot desert.

3. Fire: A forest or house fire has flame and glowing fuel temperatures of 600–900°C. Injury comes from skin burns via contact, convection, or radiation. The latter can be warded off by aluminum-covered suits. Near a large fire radiant

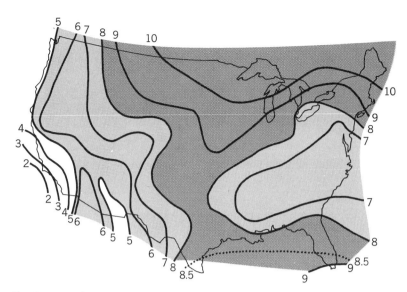

Fig. 1. Sum of heating degree days plus two times cooling degree days. The larger the sum, the larger is the integrated year-round outdoors discomfort, and also total fuel and power bill for heating and air conditioning. Units 1000°F times days. (*From K. J. K. Buettner, Human aspects of bioclimatological classification, in S. W. Tromp, ed., Biometeorology, Pergamon Press, 1962*)

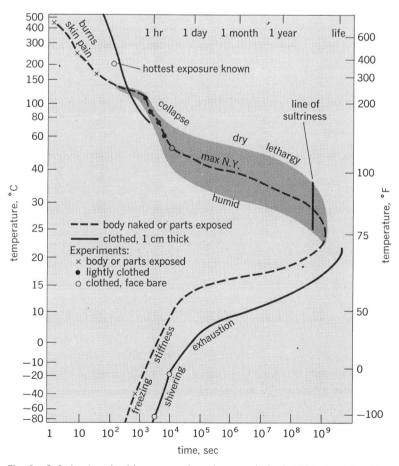

Fig. 2. Safe heat and cold exposure times in seconds for healthy, normal men at rest, with body wholly or partly exposed. Room is free from artificial ventilation and radiation; air and walls have identical temperature. Humidity influence is shown by shading. Max N.Y.= highest data of temperature and humidity in New York City. (*From K. J. K. Buettner, Space medicine of the next decade as viewed by an environmental physicist, U.S. Armed Forces Med. J., 10:416, 1959*)

heat of $40,000$kcal m^{-2} hr^{-1} may cause skin pain in 1 sec and burns in a few seconds. This heat transfer is superior to flame contact. Inhaling of hot air and lack of O_2 are frequently minor problems compared to poisoning by CO.

4. Frostbite and freezing fast: Strong convective cooling by wind and low temperatures may overrun the local defenses, causing defective circulation in the skin and breakdown of small vessels and blood particles. Developing of ice crystals may puncture cells. Skin touching metals below $-30°C$ freezes on contact, causing mechanical loss of skin layers upon removal.

5. Weather accident and disaster: Yearly average of fatalities by weather catastrophes for the United States during 1901–1960 are as follows: hurricanes, 100; tornadoes, 150; lightning, 175; floods, 80; and snowstorms, blizzards, and ice storms, 200. There are probably deaths of similar numbers from heat waves and smog periods. Weather-caused farming disasters have produced many mass-killing famines. These figures fortunately have declined, as have death figures from hurricanes and tornadoes and, probably, lightnings. All others mentioned here have risen.

Statistical bioclimatology. This aspect of bioclimatology investigates correlation between weather and climate phenomena and their effects on man. The most important work concerns clinical data correlated with daily weather. Two weather types seem to be involved: Fronts, central low, and ascending air coincide with frequent attacks of angina pectoris and embolism; incidence of foehn and cyclogenesis to the west are found to correlate with increasing circulatory disorders, mental troubles, increase of accidents, and kidney colic. It has been claimed that there is a correlation between increased solar activity and all deaths in big cities, especially deaths from mental diseases. *See* CLIMATOLOGY.

Climatotherapy. This health treatment is more venerated than exact. Going to a resort is intimately connected with nonclimatic changes of housing, mode of living, food, exercise or rest, and phychological stimuli. The vacation effect itself may be quite helpful.

A health climate location should avoid extremes of temperature, especially high effective temperature felt as sultriness. Air should be free of smoke, smog, allergens, ozone, car exhausts, and industrial and volcanic effluents.

Solar ultraviolet and natural radioactive substances and rays are generally harmful. Beneficial outdoor living usually involves exposure of parts of the skin to solar ultraviolet, and the ensuing pigment is then taken as proof of health, a rather doubtful conclusion. A certain moderate thermal stimulation by wind, waves, and temperature changes may be helpful to most people. Adaptation to the new time zone may severely interfere with adaptation of intercontinental East-West travelers.

If a physician prescribes a very balmy climate for a weak elderly patient, the patient should be aware of the difference between the dry summer heat of for example, El Paso, Texas, and the moist heat of the Lower Rio Grande.

Paleobioclimatology. This deals with possible environmental influences on the development of species, especially of man. The local microclimate has to be considered. A cave, for example, has a very constant temperature and humidity and no sunlight. Fire, as an adjunct, was introduced in China and Hungary more than 500,000 years ago. In a tropical jungle one is not exposed to solar rays. The reduction of fur in man seems correlated with the perfection of eccrine sweating as part of a well-developed cranial temperature control. It might also permit man to utilize his skin temperature sensors to function as infrared "eyes" in the dark of a cave.

In most animals skin and fur color seem to promote camouflage or its opposite. Should this be true for man, one would find dark skin in the tropical jungle; while in a temperate maritime clime white in winter and brown in summer seems adequate. Thermally, a black skin is less suited for sunny hot climate than is a lighter one. Fair-skinned but unprotected man at a latitude of $40-50°$ may seasonally adapt himself to the varying solar ultraviolet. Sunburn is unlikely if daily exposure is routine. During the ultraviolet-rich summer season a sufficient deposit of vitamin D can be produced in the body to last for the darker seasons. There might have been sufficient vitamin D in the food of Stone Age man. So neither sunburn nor vitamin D seems to have affected evolution.

Ideal bioclimate. There is no ideal climate for all men. One's physiological and psychological preference seems to be set in childhood. History starts, long before effective room heating, in such subtropical climates as Egypt, Mesopotamia, and southern China, or in tropical lowlands, for example, Yucatan, or in tropical highlands, as Peru. It moves in the Old World to higher latitudes, where Romans developed heating. Later, civilized power centers around the $45°$ latitude, where the house microclimate was adequate the year around, even before air cooling.

Statements that culture best develops in a particular climate are of little value.

In the United States people entirely free to settle where they like the climate seem to prefer the Mediterranean climate of Southern California and the desert of Arizona. A large immigration proves this preference.　　[KONRAD J. K. BUETTNER]

Bibliography: F. Becker et al., *A Survey of Human Biometeorology*, World Meteorol. Organ. Tech. Note no. 65, 1964; K. Buettner, *Physikalische Bioklimatologie*, 1938; K. Buettner et al., Biometeorology today and tomorrow, *Bull. Amer. Meteorol. Soc.*, 48:378–393, 1967; F. Daniels, Man and radiant energy: Solar radiation, in D. B. Dill (ed.) *Handbook of Physiology*, sect. 4, 1964; G. E. Folk, *Introduction to Environmental Physiology*, 1966; H. E. Landsberg, Bioclimatic work in the Weather Bureau, *Bull. Amer. Meteorol. Soc.*, 41: 184–187, 1960; S. Licht (ed.), *Medical Climatology*, 1964; F. Sargent and R. G. Stone (eds.), Recent studies in bioclimatology, *Meteorol. Monogr.*, vol. 2, no. 8, 1954.

Biogeochemical balance

Chemical elements that are essential components of living organisms may be influenced in their abundance and distribution by basic biological processes, such as photosynthesis, assimilation, respiration, and excretion. At the same time the abundance and distribution of species and communities of organisms may be influenced by the availability of the elements that are essential to them. Elements of particular concern to ecologists are carbon, oxygen, nitrogen, phosphorus, and sulfur (Fig. 1), but others are involved in biogeochemical processes. Recent studies have shown that the cycles of elements essential to living organisms are complex and varied and that much interaction exists between living and nonliving processes. The influence of technology on biogeochemical cycles, particularly in the form of pollution by waste products, has become a subject of both scientific investigation and public concern. The overfertilization of bodies of water—and its result, known as eutrophication—is caused by an increased flow of essential elements brought about by increasing population and increasing industrialization. Concern also has been expressed about technological effects on the cycles of the atmospheric gases, especially oxygen and carbon dioxide. *See* FERTILIZER; WATER POLLUTION.

Eutrophication. Much attention has focused on the biologically essential elements as limiting factors in ecosystems. Liebig's law of the minimum states that the growth of a population will be limited by that element which is present in least amount, relative to the requirement for it. This law was extended to whole ecosystems by ecologists. However, it is now understood that the effect of limiting elements on ecosystems is to shape them by controlling the mixture of species present. A change in the concentration of essential elements will not bring about a change in density of existing populations but a change in the species that are dominant in the system. Changes of this kind may occur naturally or they may be brought about by pollution from sewage, agricultural wastes and fertilizers, and some organic industrial wastes. For example, organic pollution of a lake often produces a change in populations from slow-growing diatoms to fast-growing blue-green algae. This causes changes in the populations that eat the algae and so on through the entire food web.

Systems with high concentrations of essential elements are called eutrophic whether they are naturally enriched or polluted. Naturally eutrophic systems are highly productive. The regions of the ocean where nutrient-rich subsurface water is upwelled, such as the region off Peru, support the world's largest fisheries. Systems which have become eutrophic through pollution may become less productive of desirable organisms because they are unstable and lack a varied set of organisms at terminal consumer stages of the food web, such as fishes. This may make them less satisfactory from a human viewpoint.

Eutrophication of water bodies is a reversible process. In the case of Lake Washington, on which Seattle is located, it was shown that a eutrophic system returned to its former condition in a few years when the source of pollution was eliminated completely. However, in most cases waste-water input cannot be eliminated, and wastes must be treated. In ordinary two-stage sewage treatment a large fraction of the nutrients in it find their way through the treatment plant and into the waste water. Either the inflow of waste water must be eliminated, as was the case in Lake Washington,

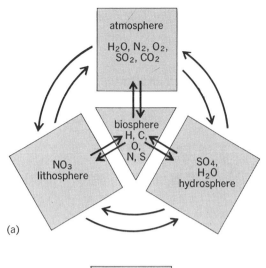

(a)

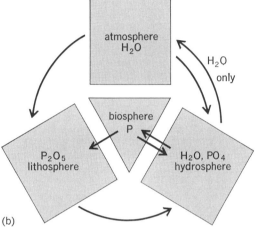

(b)

Fig. 1. Compartmental models of biogeochemical cycling. (*a*) Cycles of carbon, oxygen, nitrogen, and sulfur involve atmospheric processes. (*b*) Cycle of phosphorus does not become directly involved in atmospheric processes, although it is dependent on the cycling of water through the atmosphere to produce rainfall. (*From E. S. Deevey, Biogeochemistry of lakes: Major substances, in G. E. Likens (ed.), Nutrients and Eutrophication, Amer. Soc. Limnol. Oceanogr. Spec. Symp., 1:14–20, 1972*)

or there must be additional treatment to remove such essential elements as nitrogen and phosphorus. *See* SOLID WASTE DISPOSAL.

Phosphorus has been suggested as most likely to be the limiting element in natural waters, and therefore the most troublesome pollutant, because it is present in the biosphere in least amount relative to the requirements of aquatic plants. However, any of several elements may in fact be limiting, and often two or more elements approach depletion at the same time. Therefore the best generalized approach to pollution is to eliminate excess sources of several essential elements, especially nitrogen and phosphorus.

Aquatic systems, except some shallow ones such as ponds and estuaries, have little reserve of essential elements. Most of the supply of nitrogen, phosphorus, and certain other elements is incorporated in living tissue at any instant. Therefore new growth depends on the release of the elements,

nutrient inputs

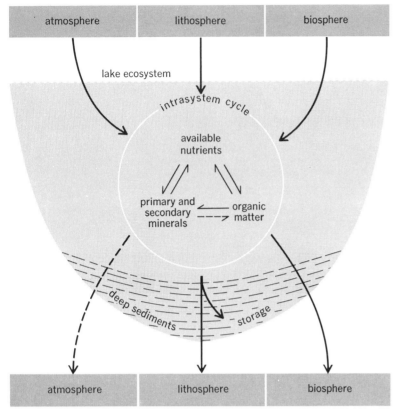

Fig. 2. Nutrient flow in a lake ecosystem. Essential elements move into the system from the surrounding atmosphere, lithosphere, and biosphere, and are lost to the sediments. Short-term biological and chemical cycling goes on within the system. (*From G. E. Likens, Eutrophication and aquatic systems, in G. E. Likens (ed.), Nutrients and Eutrophication, Amer. Soc. Limnol. Oceanogr. Spec. Symp., 1:3–13, 1972*)

primarily through digestion and excretion of wastes by the organisms that make up the system. Most life in natural waters is supported by rapid recycling of elements within the system (Fig. 2). This is why external sources of nutrients tend to produce changes in the direction of eutrophication.

Toxins. There is concern that toxic elements, such as heavy metals, and synthetic organic toxins, such as DDT and polychlorinated biphenyls (PCBs), will have deleterious effects on biogeochemical cycling by poisoning life processes that make the cycles go. The detection of DDT and PCBs in marine organisms throughout the oceans and heavy metals such as lead in polar ice demonstrates the capacity of modern technology to distribute toxins globally. Just what the impact of them will be on living systems is not yet clear, but thorough study of the problem is justified by the known facts. *See* POLYCHLORINATED BIPHENYLS.

Oxygen and carbon dioxide. L. Van Valen has shown that the maintenance of the oxygen content of the atmosphere is not primarily controlled by photosynthesis and respiration. At the present time these processes are nearly in balance on a global basis, the total result of all living processes being a trivial net change. Although the mecha-

nisms are sketchily known, oxygen appears to be regulated primarily by nonliving processes of the Earth and its atmosphere. Oxygen may be introduced into the atmosphere by the photolysis of water vapor in the upper atmosphere and subsequent loss of hydrogen to outer space, as well as by reduction of sulfates and carbon in anaerobic basins, such as the Black Sea, by biological processes. Oxygen is lost from the atmosphere by the oxidation of volcanic gases, by the oxidation of reduced forms of iron, sulfur, and manganese, and by the formation of water by reaction with hydrogen that is introduced into the atmosphere by the solar wind.

Although it is still difficult to understand in a quantitative way the processes that control the amount of oxygen in the atmosphere, it is easy to calculate that neither biological nor technological processes can have a very large effect. Total oxidation of all living organisms would produce a change of 1% in atmospheric oxygen, and oxidation of the known reserves of all fossil fuels would produce a change of another 1%.

The carbon cycle has substantial biological components. The most important probably is the deposition of calcium carbonate by marine organisms. Carbon dioxide in the atmosphere is in equilibrium with carbon dioxide and bicarbonate in solution in the ocean. If carbon dioxide is added to the atmosphere from the burning of fossil fuels, it is absorbed by the ocean in polar regions because gases are more soluble in cold water. In the tropics, where the water is warm, some carbon dioxide is returned to the atmosphere from the sea and some is used by marine organisms to produce shells or skeletons of calcium carbonate. These shells become limestone beds.

As fossil fuels are burned, the carbon dioxide content of the atmosphere is increasing very slightly. This increasing partial pressure of carbon dioxide permits the ocean to absorb more of it, and ultimately more calcium carbonate will be deposited on the ocean bottom. In the long run, the ocean will control the carbon dioxide content of the atmosphere, but over the next few decades atmospheric carbon dioxide can be expected to continue to increase because the rate of utilization of fossil fuels is still rising and there is a substantial delay in the oceanic regulatory processes.

Analysis of cycles. Biogeochemical cycles are difficult to comprehend because they involve processes operating over a broad range of scales of space and time, from seconds to millions of years and from living cells to the cosmos. Moreover, it is often necessary to consider interactions between two or more chemical elements. Strategies to cope with such complexity now include automated analytical methods and computer simulation of cycles that are reduced to mathematical models. With computer techniques it is possible to simulate rapidly events of great complexity which involve long time periods in the real world. These approaches are still in their formative stages, and their full utilization may be some years away. *See* BIOSPHERE, GEOCHEMISTRY OF; EUTROPHICATION, CULTURAL.

[LAWRENCE R. POMEROY]

Bibliography: Committee on Oceanography, *Chlorinated Hydrocarbons in the Marine Environ-*

ment, National Academy of Sciences, 1971; G. E. Likens (ed.), *Nutrients and Eutrophication*, Amer. Soc. Limnol. Oceanogr. Spec. Symp., 1:1–328, 1972; L. R. Pomeroy, *Ann. Rev. Ecol. System.*, 1: 171–190, 1970; L. Van Valen, *Science*, 171:439–443, 1971.

Biogeography

The science concerned with distribution of life on the Earth. Plants and animals are irregularly distributed, both on the continents and in the oceans. Some areas have a great abundance and variety of life forms, whereas others are relatively sterile. Biogeographic studies are concerned with learning the manner in which living organisms are arranged on the Earth and the causative factors of this arrangement. All animals are dependent in the final analysis on plants to supply food. Therefore, the distribution of plant life is the basic component of biogeography. Animal life is the secondary or dependent component.

Plant geography. Many factors regulate the growth of plants, but of particular importance are climatic conditions and the supply of nutrient substances on which plants feed. Driving across the western United States, during the course of a few miles, one may pass from a barren desert through wooded foothills, then up the slopes of a heavily timbered mountain to the crest which is above timberline and sparsely covered with low vegetation. This stratification of vegetation is controlled largely by conditions of temperature and moisture, which vary from the flat lowlands to the mountaintop. The nature of the underlying rock, from which plant nutrients are derived by weathering, may strongly influence the growth of plants, but the effect is less striking than that of climate. *See* TERRESTRIAL ECOSYSTEM.

Animal life in turn is stratified in conformance with the zonation of plants. There are typical animals of the desert, the foothills, the forest zone, and the alpine crest. Each of these major communities of plants and animals may be called a life zone, a vegetation zone, a biome, or a climax formation according to various systems of classification. All of these communities are part of the temperate region of the world. To the north is the boreal region, which extends generally from the North Pole to the latitude of southern Canada, and south of the temperate region is the tropical region, extending from the Equator roughly to central Mexico. The Southern Hemisphere may be

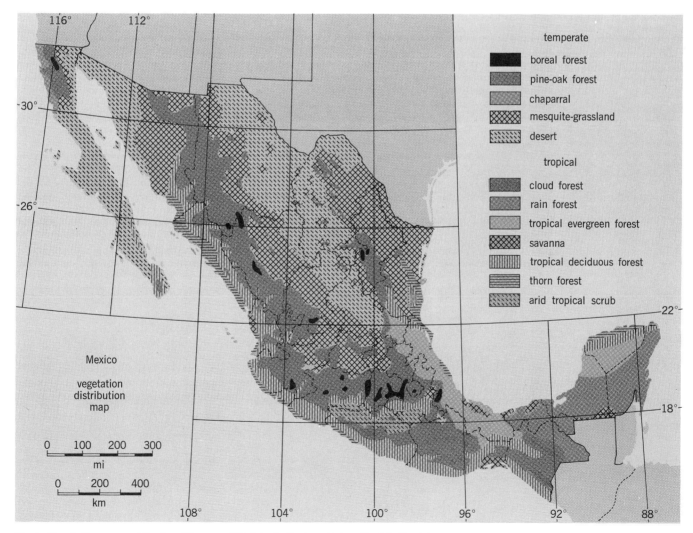

Fig. 1. Vegetation zones of Mexico. (*Museum of Vertebrate Zoology, University of California*)

similarly divided into biotic regions, and these in turn into vegetation zones or communities. *See* COMMUNITY; PLANT GEOGRAPHY.

As an example of the subdivision of a landscape into component types, Fig. 1 shows the distribution of vegetation zones in Mexico. This country spans the border of the temperate and tropical regions; the vegetation zones are arranged correspondingly in two series. *See* PLANT FORMATIONS, CLIMAX.

Tropical region. The greatest variety of plant life exists in the tropics, especially in wet forested areas. In southeastern Mexico, for example, there are tall, dense rainforests made up of many species of trees stratified in layers. The trees are heavily laden with climbing vines and with epiphytic plants, such as orchids and bromeliads. The ground beneath has a complex flora of shrubs, herbs, and lesser plants. On a few acres there may be hundreds of different kinds of plants. In general, as one proceeds north or south from the tropics, the variety of plant life decreases and fewer kinds dominate the vegetation. In polar latitudes, as on islands of the Arctic or Antarctic seas, floras are very simple.

Temperate regions. Within temperate zones, although the variety of plants is less than in the tropics, the total mass of vegetation may be as great, at least in wet climates. The coniferous forests of coastal Washington or the original hardwood forests of Ohio or central Europe were as massive as the great tropical rainforests of Mexico or the Philippines.

Arid zones. Arid zones in any region of the world support relatively little vegetation, for the obvious reason that lack of water limits plant growth. The sparse development of plants in polar regions may be due as much to frozen ground as to the cold itself.

Changes in distribution. The distribution of plants on the surface of the Earth is never static, but is constantly changing with shifts in climate, with emergence and submergence of land masses, and with such surface disturbances as volcanic activity or geologic erosion. As environmental conditions change, floras tend to migrate and in so doing some species become extinct and new species evolve. Certain floristic communities have moved long distances in geologic time. During the Cretaceous Period, redwood trees dominated a rich forest in the American Arctic. Redwood fossils are abundant along the Colville River in Alaska. Cold and drought of subsequent periods pushed the redwood flora southward, both in the Americas and in Asia, so that now redwoods are found in California and in China but not in their original northern homeland. *See* POPULATION DISPERSAL.

Thus the study of plant geography must consider the historic background of plant communities as well as problems of ecologic adaptation to existing conditions.

Animal geography. As stated in the discussion of plant geography, terrestrial animal populations are distributed largely in accordance with vegetation patterns. Every kind of animal is to some degree specific in the type of habitat it requires. The distribution of individual species of animals often conforms rather closely with the distribution of particular vegetation zones. As an example, Fig. 2 shows the range of the Montezuma quail (*Cyrtonyx montezumae*) in Mexico. Reference to the vegetation map (Fig. 1) will show that this quail occupies chiefly the pine-oak zone. This association is so close that the Montezuma quail may be taken as an indicator of pine-oak vegetation.

Not all animals are as specific as the Montezuma quail in their habitat requirements. Most species are adaptable enough to occupy two or more vegetation zones, although they may achieve highest density in one favored plant association. But in all the world there is no known animal that is completely nonspecific in its environmental needs. Requirements for particular types of food, shelter or cover, and amounts of water are the main components of habitat that limit the range, and locally the density, of every kind of animal.

Some animals make seasonal use of different habitats by migrating. Thus many species of migratory birds nest in northern boreal zones in summer but return southward to temperate or even tropical habitats in winter. On a more local scale, some species of both birds and mammals migrate vertically in mountainous terrain, that is, upward in summer and downward in winter. The device of seasonal migration permits utilization of environments habitable only part of the year.

Distribution in time. In geologic time, whole faunas have moved up or down the continental land masses in response to major climatic changes. As is true of the plants, many cases of discontinuous distribution of families or orders of animals are best explained on the basis of past movements. Thus the marsupials are found in Australia and nearby lands and in the Americas, but nowhere else. The most plausible explanation, strongly supported by the fossil record, is that marsupials originated in the north and moved down the continents to the present areas of proliferation when the northern climate became too severe for their existence. The ranges of other animal groups have been similarly altered over geologic time.

Oceanic distribution. Animal life in the oceans is distributed as irregularly as on land. The temperature of the water and the concentration of mineral nutrients varies with air temperature, ocean currents, and especially with points of upwelling of enriched waters from the ocean depths. In favorable waters, floating microscopic plants and ani-

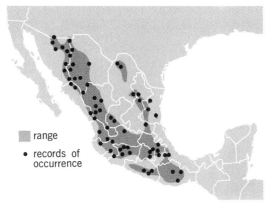

■ range

• records of occurrence

Fig. 2.   Range of the Montezuma quail in Mexico.

mals, collectively called plankton, are abundant and serve as rich pastures for larger organisms which in turn feed the fishes and even large mammals such as baleen whales. Fish-eating birds, seals, and porpoises thrive in turn. Areas rich in oceanic life are situated as a rule in cold currents and along continental shelves. Thus the great oceanic fisheries of the world and the centers of whaling are mostly in northern or southern waters and not in equatorial latitudes. *See* MARINE ECOSYSTEM.

Human biogeography. The distribution of human populations on the Earth bears a close relationship to the distribution of natural plant and animal life. Throughout history, man has thrived best in areas that are highly productive of foodstuffs. By and large the most favorable habitats for man are situated in the temperate regions of the world, where climates are moderate and soils are rich. The highest development of agricultural and pastoral industries today is in the regions of temperate climate and moderate rainfall. Various human societies, however, have become adapted to many specialized environments. Eskimos and Lapps, for example, live successfully in the icy northern wastelands, Incas and Sherpas in towering mountains, and Mayans and Pygmies in wet tropical lowlands.

Unlike most other animals, man has to a considerable degree altered his environment in supplying his needs. Thus, forests have been cleared for crops, deserts irrigated with water brought from distant mountains, marshes filled and cities built thereon. Foodstuffs, ores, and fuels are transported all over the world. The system which man calls civilization has become so complex that many people lose track of the fact that it still depends upon the function of natural resources, just as it always did. Urban dwellers in particular come to think that industry supports us, forgetting what supports industry.

Through blind destruction of agricultural soils and exploitation of other resources, many past civilizations have become impoverished or have even destroyed themselves. Since Biblical times, the rich lands of the Mediterranean region have largely been reduced to semideserts. The glory that was the Mayan Empire persists only as ruins overgrown with jungle. The great experiments of civilization have not been unqualified successes. In the last analysis, man is still an animal dependent upon many natural features of the environment for his welfare. He can modify and use these features but he cannot with impunity destroy them. *See* ECOLOGY, APPLIED; ECOLOGY, HUMAN.

Perhaps the outstanding aspect of human biogeography today is the explosive increase of population since 1900. Modern medicine has reduced the normal death rate among people and technological improvements have intensified the processes of resource extraction, permitting, for the moment at least, more people than ever before to live upon the Earth. How long the resources of the Earth will support such spiraling demands is conjectural. *See* ZOOGEOGRAPHIC REGION; ZOOGEOGRAPHY.   [A. STARKER LEOPOLD]

Bibliography: H. G. Andrewartha and L. C. Birch, *The Distribution and Abundance of Animals*, 1954; *Cold Spring Harbor Symposia on Quantitative Biology*, vol. 22, 1957; P. M. Dansereau, *Biogeography*, 1957; P. J. Darlington, *Zoogeography: The Geographic Distribution of Animals*, 1957; W. George, *Animal Geography*, 1962; R. Hesse, W. C. Allee, and K. P. Schmidt, *Ecological Animal Geography*, 2d ed., 1951; C. L. Hubbs (ed.), *Zoogeography*, AAAS Symposium, vol. 51, 1959; N. V. Polunin, *Introduction to Plant Geography and Some Related Sciences*, 1960.

Biological productivity

The quantity of organic matter (in the form of living matter, stored food, waste products, and material taken by consumers) or its equivalent in dry matter, carbon, or energy content which is accumulated during a given time period. This quantity may apply to a single organism, a single species population, or a group of organisms. Following discussions within the International Biological Program, the term biological productivity is now applied in more general and nonquantitative contexts, while biological production has been defined as "increase in biomass (weight of living matter in a population); this includes any reproductive products released during the period concerned." In practice, studies of production are always made in the context of a particular time period and are thus expressed as production rate. They are also related to a defined population or a defined area (or volume) of the environment. *See* BIOMASS.

Productivity. This may be a useful parameter of a single population, a group of organisms belonging to the same trophic level, or to a whole ecosystem. In all cases it refers to rate of gain of material, as distinct from the amount actually present (measured as standing crop of biomass) or the turnover rate (the proportion of biomass incorporated in unit time). *See* ECOSYSTEM; FOOD CHAIN.

The general term may be qualified by giving the further definitions: (1) Production is used by fisheries biologists to exclude released reproductive products. (2) Gross production (confined to plants) is the total product of photosynthesis, much of which will be oxidized by plant respiration. (3) Primary production is the production by plants (autotrophic organisms). (4) Secondary production is production by animals (or other heterotrophic organisms). (5) Gross primary production is the total primary production by photosynthesis. (6) Net primary production is gross production less losses within the plant due to respiration. (7) Potential production is the maximum possible production within certain clearly defined environmental circumstances.

Related concepts, all of which relate to amounts of matter (or equivalent energy or carbon content) and which have sometimes been confused with production include (1) yield, which is that proportion of production that is exploited by one particular predator of the organism in question (often by man); (2) consumption, represented by C, the amount which is eaten by a heterotrophic organism (this may be much less than that killed because many animals are wasteful feeders and many also cause additional damage to their food organisms); and (3) assimilation, or A, the amount actually digested or absorbed. This equals consumption minus production of feces, or F (egesta), and other rejected material (rejecta), such as feeding pellets. It follows that assimilation is the sum of production P, (of food P_f, gametes P_g, sloughed skins P_s, and other products) and of respiration, represented by R, (loss of assimilate due to the liberation of energy from organic material and of inorganic materials including carbon dioxide and nitrogenous compounds). These relationships can be expressed as $C = A - F$, $A = P + R$, and $P = P_f + P_g + P_s$.

Matter and energy in ecosystem. All foodstuffs are organic compounds which on combustion yield large amounts of energy (around 5000 cal/g). The energy and matter remain linked until separated by respiration. In the context of the whole ecosystem, energy and matter are incorporated together in photosynthesis and separated whenever respiration occurs. In a system which is isolated and in balance, the losses of energy, whenever oxidation occurs, equate with the gain at photosynthesis. The magnitude of the energy flow (in calories per unit time and unit area) is a precise measure of the activity of the system and can be used for quantitative comparisons between whole systems and between their parts.

Matter, whether as carbon or inorganic elements, is associated with the energy flow but recycles to an indeterminate extent. Thus the cycles of carbon, sulfur, phosphorus, or nitrogen cannot be considered a universal measure of activity in the same way that energy can.

Measurement of production. Practical methods for the measurement of primary production depend greatly on the type of organism involved. In the case of plankton, for instance, production is estimated by comparing the effect of the algae on dissolved oxygen and carbon dioxide concentrations in closed containers. One container is held under conditions of respiration only (darkened) and another under conditions of respiration plus photosynthesis (exposed to sunlight). Labeled C^{14} can be used also to measure carbon uptake. Such experiments take but a few hours and must be repeated throughout the year.

Production in forests, on the other hand, is measured by adding together gain in biomass, loss due to stem, leaf, and litter fall, and loss due to predators. This may involve repeated seasonal observations, but use can also be made of girth increment, revealed by growth rings, and of regression analysis, which is used to link different measurements of height, weight, and leaf area. In nearly all terrestrial plants reliable data are lacking for root production.

Secondary production can be estimated from measurements of growth rate and reproductive rate. It can also be derived from measurements of mortality because these equate with growth in a balanced situation. Finally, estimates can be derived from the difference between respiration and assimilation. A classical study is that of K. R. Allen, who was the first to show that annual production of a fish population may greatly exceed the biomass of food organisms available to it at any one time. This is because the invertebrate food animals grow quickly and constantly renew their numbers.

In many young, and most mature, animals the greater part of assimilation contributes not to growth but to respiration. It follows that, from the

point of view of whole ecosystem studies (and since assimilation and respiration are nearly equal and metabolism is usually easier to measure than production or assimilation), respiration rates provide a more practicable index for determining the main paths of energy flow. In all cases, however, it may be very difficult to extrapolate from measurements of confined animals in the laboratory to populations in the fields. Many of the quantities (assimilation, consumption, and production) and their ratios vary greatly between species, age groups, and sexes. They also vary seasonally with food quality, with climate, with the type of food eaten, and with many other factors.

Efficiency. The ratios between quantities or calorific contents of foodstuffs at different stages of treatment by the organism are termed efficiencies. These are best designated in terms of the two quantities being compared. Important examples are: (1) Production-consumption efficiency; this is of major interest to the farmer and

Table 1. Some representative figures of biological productivity

Type of productivity	Rate, kcal/(m²)(yr)
Marine	
Solar energy reaching surface of North Sea (55°N lat.)	884,000
Gross productivity, organic matter	
North Sea	360
Sargasso Sea	800
Long Island Sound	4,700
Coral reefs	26,600
Fish yield	
North Sea fisheries	2.3
Tropical seas, maximum	192
Fresh water	
Gross productivity	
Temperate lakes, oligotrophic	1,020
Temperate lakes, eutrophic	3,060
Florida stream, Silver Springs	25,600
Fish yield	
German rivers	1.9
German carp ponds, maximum	3.8
Chinese fish ponds, maximum	27
Land	
Deserts, gross productivity, maximum	700
Temperate grassland, gross productivity, maximum	14,000
Yield of temperate crops (edible part only)	
Wheat, Europe	890
Potatoes, Europe	2,280
Sugar beet, Europe	3,240
Timber, Europe	1,000
Alfalfa, United States	372
Carrot (very productive), Europe	2,250
Asparagus (low productivity), Europe	140
Yield of tropical crops	
Sugarcane	16,100
Banana	24,000
Cassava	34,800
Meat	
Beef, United States*	29.8
Beef, Germany†	10.8
Pig, Germany†	16.8
Human metabolism, Pennsylvania, United States (0.3 men/acre)	67.5

*E. P. Odum, *Fundamentals of Ecology*, 2d ed., Saunders, 1959.

†K. Kalle, *Deut. Hydrograph. Z.*, vol. 1, no. 1, 1948.

Table 2. Calorie content of foodstuffs and organic substances

Type	Calorie content, kcal/g
Pure substances	
Fats	8.4–9.4
Carbohydrates	4.5–5.7
Proteins	3.9–4.2
Plant products	
Wood	4.8
Leaf litter	4.8
Cereals	3.6–4.3
Vegetables, root	3.4–3.8
Vegetables, leguminous	3.2–3.8
Vegetables, green	2.6–3.1
Fruits	3.3–3.8
Animal products	
Meats	6.4–7.4
Poultry	5.9–7.0
Fish	4.0–4.4
Milk, cow's	5.2
Cheese	6.3

other predators. In nature this is usually of the order of 10% or less, but special breeding of domestic animals has raised it to 30% or more. (2) Food chain efficiency; this is the ratio of food available to two successive trophic levels. (3) Growth-assimilation efficiency, or growth efficiency; in this case assimilation includes contributions from the mother to the embryo via the placenta and to the young animal in milk. *See* BREEDING (ANIMAL).

The photosynthetic process itself is not very efficient; in ideal laboratory conditions the yield of carbohydrate from light of the correct wavelength may reach 30%, but the efficiency of plants under field conditions is seldom above 1%.

Units of measurement. Productivity figures are usually quoted in terms of calories per square meter per annum where calorific contents are known or can reasonably be derived (Tables 1 and 2). In some cases dry weights are substituted, and it may be more appropriate to use an area of a hectare. Anglo-Saxon measurements of pounds and acres are obsolescent and the SNU (standard nutritional unit) of 10^9 calories or 10^6 kcal has been suggested, but not widely adopted. This has considerable merit, being about one human being's annual dietary requirement and of the order of the yield per acre of many plant crops.

[AMYAN MACFADYEN]

Bibliography: K. R. Allen, *The Horokiwi Stream*, N.Z. Mar. Dep. Fish. Bull. no. 10, 1951; A. N. Duckham, *Agricultural Synthesis: The Farming Year*, 1963; P. M. Jonasson and J. Kristiansen, Primary and secondary productivity in Lake Esrom, *Int. Rev. ges. Hydrobiol.*, 52:163–217, 1967; A. Macfadyen, *Animal Ecology: Aims and Methods*, 2d ed., 1963; E. F. Menhinick, Structure, stability and energy flow in a *Sericea lespedeza* stand, *Ecol. Monogr.*, 37:255–272, 1967; P. J. Newbould, *Methods for Estimating the Primary Production of Forests*, I.B.P. Handb. no. 2, International Biological Program, London, 1967; E. P. Odum, C. E. Connell, and L. B. Davenport, Population energy flow of three primary consumer components of old-field ecosystems, *Ecology*, 43:88–96, 1962; E. P. Odum and H. T. Odum, *Fundamentals of Ecology*, 1959; K. Petrusewicz, *Secondary Productivity of Terrestrial Ecosystems: Principles and Methods*, International Biological Program and Institute of Ecology, Warsaw, 1967; J. Phillipson, *Ecological Energetics*, 1967; W. E. Ricker (ed.), *Methods for Assessment of Fish Production in Fresh Waters*, I.B.P. Handb. no. 3, International Biological Program, London, 1968.

Biomass

The dry weight of living matter, including stored food, present in a species population and expressed in terms of a given area or volume of the habitat. Biomass is an expression used chiefly in relation to food chains. For example, a comparison of herbivores feeding on grass should include cows as well as leaf hoppers, and weight of flesh is a fairer basis for comparison than numbers of individuals. *See* FOOD CHAIN.

Measurement. The conversion of census figures to biomass strictly requires data on age distribution of weight and mortality, but these vary with biotic and physical factors and with season and have been computed only for some fish populations. The synthesis of biomass figures for whole communities is usually based on adult animals only.

Metabolic rate. The ultimate criterion of biological productivity is metabolic rate, and this is not simply related to biomass. Smaller species have higher metabolic rates per unit mass according to a 2/3 power law in general, but there are many exceptions to this law. Also, some forms, such as the mollusks, contain nonliving minerals (an objection which can be overcome by measuring total body nitrogen and converting results to weight of protoplasm). The biomass of a species at a given time (which is a measure of standing crop) is a poor index of productivity. As an example, the biomass of marine animals in lower in tropical than artic plankton of equal productivity, but the rate of synthesis and breakdown of organic matter is higher. Again, metabolism is faster at some seasons than at others, and restriction of nutrients may result in accumulation of biomass but reduction in productivity. *See* BIOLOGICAL PRODUCTIVITY.

Biomass pyramids. Biomass pyramids can be constructed on the same lines as pyramids of numbers to summarize the biomass structure of communities. Excellent examples are given by E. Odum, who shows that the shape of the pyramid depends upon whether the community is self-contained (as regards primary production of food substances), or imports or exports matter to or from the different trophic levels of the food chain.

[AMYAN MACFADYEN]

Bibliography: E. P. Odum, *Ecology*, 1963; E. P. Odum, *Fundamentals of Ecology*, 2d ed., 1959.

Biome

A complex biotic community covering a large geographic area and characterized by the distinctive life forms of important climax species of plants and animals. A life form is the common morphological features that characterize a group of organisms. On the land, a biome is identified by the life form of the dominant climax plants, as well as by the distinctive types of vegetation and landscape in which they grow. In the ocean, the life forms of the predominant animals serve as the criterion. The biome incorporates all seral (successional) as well as climax stages of the community, and its distribution is controlled by climate or, in the ocean, by other physical factors of the environment. Principal terrestrial biomes are the temperate deciduous forests, coniferous forest, woodland, chaparral, tundra, grassland, desert, tropical savanna, and tropical broad-leaved forest. Principal marine biomes are the pelagic, the barnacle–gastropod–brown algae, the sea urchin–large snail, the bivalve-annelid, the coral reef, and the abyssal-benthic. *See* DESERT VEGETATION; GRASSLAND; SAVANNA; TUNDRA.

Terrestrial biomes are divisible into plant associations, which are distinguished by distinctive combinations of climax dominant species. For instance, the temperate deciduous forest biome includes the beech-maple, oak-hickory, and other associations. Both terrestrial and marine biomes are divisible into animal biociations. A biociation is distinguished by the distinctiveness of the predominant animal species. Some of the biociations within the temperate deciduous forest biome are the North American deciduous forest, with gray squirrel and eastern chipmunk among its predominant animals; the European deciduous forest, with the red deer; and the Eurasian deciduous forest, with the Manchurian tiger, musk deer, and true wolf. *See* COMMUNITY; ECOLOGY.

[S. CHARLES KENDEIGH]

Bibliography: P. Dansereau, *Biogeography*, 1957; L. R. Dice, *Natural Communities*, 1952; S. C. Kendeigh, *Animal Ecology*, 1961.

Biosphere

The life zone of the Earth. The biosphere, out of analogy with the atmosphere, hydrosphere, and lithosphere, consists of the outer sphere of the Earth that is inhabited by living organisms. It includes the lower part of the atmosphere, the entire hydrosphere down to its maximum depth (10,863 m), soils, and the lithosphere to a depth of about 2 km, where some bacteria occur in conjunction with petroleum deposits. The term has also been used to designate the total quantity of living matter on the Earth, but such usage should be discontinued. The concept of the biosphere was introduced by J. B. Lamarck and later defined by E. Suess in 1875. Because of the great geological activities of *Homo sapiens*, the modern biosphere is sometimes referred to as the anthroposphere or noosphere. *See* ATMOSPHERE; BIOSPHERE, GEOCHEMISTRY OF; HYDROSPHERE.

[JOHN R. VALLENTYNE]

Biosphere, geochemistry of

The term biosphere indicates the totality of organisms which are alive on Earth at a given time. Biogeochemistry is the description and the reconstruction of the genetical history of the biosphere. This article first discusses briefly the problem of the primitive genesis of the biosphere, and the possible evolutionary changes in the chemistry of the biosphere throughout the geological ages since the emergence of the protoorganism. Then the chemical properties and other characteristics of the biosphere at the present time are described. The article terminates with a brief description of the processes of degradation of deeply buried biological debris.

Chemical interaction exists between the atmosphere, hydrosphere, lithosphere, and biosphere. The biosphere is particularly closely interconnected with nonliving organic matter within the Earth's crust, which may be referred to by the term carbosphere. The carbosphere is the subject matter of organic geochemistry. Most of the interactions between the biosphere and its surroundings are, as a first approximation, of a reversible-fluxing character. The chemical elements enter the organism, are later ejected from the biosphere through the products of metabolism and death, and are thereafter incorporated into another organism, and so forth. These processes are termed the biogeochemical cycles. A relatively smaller portion of organic matter within certain localities, however, becomes so deeply buried within the Earth's crust that it would not be available for further biological recycling.

The importance of living organisms in geochemical processes was first appreciated by J. B. Lamarck, J. Liebig, E. Suess, and other investigators of the 19th century, but it was not until the beginning of the 20th century that biogeochemical data were systematically accumulated into a coherent body of information. The first comprehensive studies were undertaken by V. I. Verdansky (1863–1945) of the Soviet Union, who is regarded as the modern founder of the subject. Other leaders in the field have been A. P. Vinogradov (Soviet Union), G. Bertrand and D. Bertrand (France), and F. W. Clarke and G. E. Hutchinson (United States).

Genesis of biosphere. The spectral data relating to stars, interstellar space, and smaller celestial bodies indicate that simple gases such as CH_4, CO, CO_2, NH_3, and H_2O are abundant throughout the cosmos. Numerous experiments clearly demonstrate that the above mentioned gases can condense to complex organic molecules, including amino acids and so on, on irradiation with ionizing rays or on being subjected to electrical discharges; these conditions are presumed to occur in the course of the condensation of a given celestial body. Therefore, prebiogenic carbonaceous complexes would be expected to have preceded the emergence of the organism on any condensing celestial body, and the cosmochemical and geological evidences for such carbonaceous complexes are briefly outlined below. *See* LIFE, ORIGIN OF.

There are some 30 carbonaceous meteorites, which contain 0.3–4.0% carbon, corresponding to approximately twice as much organic complex. According to the generally accepted theory, the meteorites originate from asteroid-sized celestial bodies during the course of condensation, of which the conditions have been adequate for the genesis of a prebiogenical carbonaceous complex, but too variable and most likely too dry for the emergence of the organism. The abiogenic nature of the carbonaceous complex is partially supported by the following evidence: (1) There is absence or virtual absence of optical rotation, which is characteristic to all biological products which contain asymmetrical molecules; (2) paraffins and other hydrocarbons do not show a marked trend toward dominance of carbon chains of odd numbers, as is the case in most biogenic products; (3) C_{12}/C_{13} ratio is well within the range of juvenile carbons; and (4) the carbonaceous complex contains up to 5% organic chlorine, and chlorinated compounds are extremely rare within the organism of the terrestrial type of biochemistry.

Of the terrestrial geological settings, there are several candidates for organic substances which may have been preserved since times prior to the emergence of life, or they are condensation products from the interior of the Earth. Thus, small quantities of amino acids have been detected in impermeable chert from the 3,200,000,000-year-old Fig-tree formation of Transvaal, South Africa. It is possible that these sediments deposited prior to the genesis of life. Bitumenlike substances with a high percentage of organic acids, and ashes rich in uranium and thorium, termed thucholites, are found in numerous pegmatite veins, which traverse exclusively igneous rocks. It appears, therefore, that the hot vapors or solutions which produced such veins must have transported the organic matter from the deeper zones of the Earth. Distillation from surrounding organic sediments is unlikely, because they were not available within these areas. The same considerations may apply to the small quantities ($1-2$ ppm) of amino acids, which have been detected within some active fumaroles from the Carribean, Hawaii, and Iceland.

The above-mentioned evidence does not establish the presence of prebiogenic carbonaceous complex either on the parent bodies of meteorites or on the Earth, but all the same are strongly indicative of its existence.

Evolution of biosphere. Little is known about possible chemical changes which may have occurred in the course of evolution from the protoorganism to the present-day biosphere. At the early Precambrian the presence of exclusively heterotrophic organisms in a reducing atmosphere is postulated. The presence of some three valent iron in the form of magnetite ($FeO \cdot Fe_2O_3$), however, does not seem to corroborate the hypothesis of a markedly reducing atmosphere even at that early age. The evolution of photosynthetizing organisms in the sea, possibly in later Precambrian, may have quite fundamentally changed the cycles of the diverse elements, resulting in the liberation of considerable quantities of oxygen into the atmosphere. This trend may have received yet another impetus on the spreading of photosynthetizing plants to the land surface during the Paleozoic and particularly during the Carboniferous age. It is likely that since the Paleozoic, the chemical composition and the geochemical cycles of the biosphere did not change in any fundamental manner.

Living matter. Living matter occurs in the environment wherever water exists in the liquid state in conjunction with a source of energy for metabolism. Some extreme conditions under which certain species maintain permanent populations are listed in Table 1.

The fundamental distinction between organisms with regard to geochemical activities and importance results in two classes: (1) those organisms that contain chlorophyll pigments and utilize the energy of visible light for the synthesis of energy-rich organic compounds (pigmented algae, photosynthetic bacteria, and higher plants); and (2) those lacking the photosynthetic apparatus (viruses, nonphotosynthetic bacteria, fungi, and animals). Those without chlorophyll derive their energy for metabolism from the respiratory oxidation of food materials in the form of organic and in some cases inorganic compounds.

In separate parts of the biosphere, fairly distinct communities of organisms can be recognized, for example, the plankton communities of natural waters, coral reefs, sphagnum bogs, spruce-fir forests, and steppes. Because there is an exchange of energy and matter between the living community and the nonliving environment, the complex is often studied as a unit, called an ecosystem. The physical dimensions of an ecosystem may range from a puddle of water to a maximum which is the entire biosphere. The passage of energy through an ecosystem can be conceived as proceeding through a series of trophic (feeding) levels. These trophic levels are recognized: photosynthetic plants, herbivorous animals and microorganisms of decay, primary carnivores, and higher carnivorous levels. Under steady-state conditions, the rate of energy production by one trophic level must necessarily exceed the rate of energy utilization by a succeeding trophic level. This is in accordance with the second law of thermodynamics as applied to steady-state systems. It follows that, for the biosphere as a whole, green plants tend to be of greater geochemical importance than consumer organisms, in terms of both biomass and metabolism. However, photosynthetic plants are unable to maintain permanent populations at depths greater than 100 m in the hydrosphere or greater than a few meters (in the case of plant roots) in soils, because of the lack of light. In such environments, all biochemical activity is the result of animals and nonphotosynthetic microorganisms. This situation exists in the more peripheral parts of the biosphere. *See* FOOD CHAIN.

The linear dimensions attained by living organisms lie between the limits of 10^{-6} cm in the case of viruses and 10^4 cm for sequoia trees. Within a given taxonomic group, the rate of respiration per unit weight of tissue tends to increase as the two-thirds power of the weight. There is, however, much variation from one taxonomic group to another, and many exceptions to this general rule are known. The overall result is that, per unit weight,

Table 1. Conditions under which living organisms are known to reproduce in nature

Factor	Minimum	Maximum
Temperature	$-7°C$ (*Pyramidomonas* in saline lakes)	$95°C$ (blue-green algae in hot springs)
Hydrostatic pressure	Not determined	1070 atm (deep-sea forms)
pH	1.7 (the diatom *Pinnularia*)	12.0 (blue-green algae in Kenya lakes)
Oxidation potential, E_h	-0.35 volt (marine sedimentary bacteria)	0.83 volt (aerobic organisms)
Salinity	Very low as in rainwater	220 parts per thousand (halophytic bacteria and *Artemia salina*)

smaller individuals tend to respire and also to increase in mass at faster rates than do larger individuals. Thus, under maximal growth conditions a virus population will double its weight in a matter of minutes, bacteria in less than 1 hr, small animals in days or weeks, and larger forms in months or years. As a consequence, small individuals tend to be more active geochemically than their biomass might suggest.

Although the total mass of living matter in the biosphere has never been accurately estimated, its probable value is $n \times 10^{17}$ g live weight (where n is any small integer), which corresponds to $20n$ mg per cm² of Earth surface. The amount of carbon dioxide annually fixed in photosynthesis in the entire biosphere is probably equivalent to 7×10^{16} g of carbon, with the true value not less than one-half nor greater than twice this amount. This rate approximately corresponds to a 0.1% efficiency in the utilization of the visible light energy that actually reaches the Earth's surface. Photosynthesis by marine plants accounts for at least two-thirds of the total photosynthesis in the biosphere. The greatest amount of photosynthesis on land occurs in forests, followed in decreasing order by cultivated land, steppes, and deserts. *See* BIOLOGICAL PRODUCTIVITY; ECOLOGY; ECOSYSTEM.

Influence of living matter. As a result of enzymatic action, many chemical reactions occur at low temperatures within the bodies of living organisms that would otherwise proceed only at slow rates in the biosphere. Prime examples are photosynthesis and the formation of nitrates from molecular nitrogen by the combined action of nitrogen-fixing and nitrifying microorganisms. In spite of the small mass of living matter in the biosphere relative to that of the atmosphere, hydrosphere, and lithosphere, living organisms do have an appreciable geochemical influence. Terrestrial plants are known to regulate soil erosion rates, to transport certain elements from the lower soil layers to the surface, and to accelerate the decomposition of some minerals and rocks by the excretion of carbon dioxide and polyvalent organic compounds that act as ligands (chelating agents). In addition, many of the physical and chemical characteristics of soils and aquatic sediments are determined by the activities of microorganisms, such as pH, oxidation potential E_h, and texture.

The concentrations of certain elements in plants and animals may exceed those in the media of growth by factors up to several hundred thousand. One of the more marked cases is the accumulation of vanadium by some ascidians. The concentration of vanadium in some species of *Ascidia* may reach 0.01% of the live weight, whereas sea water contains 0.0000002% vanadium. The extensive deposition of foraminiferan and radiolarian oozes on the sea floor emphasizes the importance of these forms in the deposition of $CaCO_3$ and SiO_2, respectively. Evidence of the former activity or organisms in the biosphere is provided by acaustobioliths (inorganic bioherms) such as reef limestones, and by caustobioliths (organic accumulations) such as oil shales, and deposits of coal and petroleum.

Considerable indirect geochemical evidence, as well as more direct indications from the occurrence of fossil microorganisms in Precambrian deposits, indicates that the biosphere has been in existence for some 2,000,000,000 years, and possibly for twice that length of time. As a result of evolutionary processes, both the numbers and kinds of species have been changing from time to time. New environments have been occupied and even biologically created in the process. The quantitative importance of living organisms in the biosphere is quickly appreciated when it is realized that a weight of water equivalent to that present in all the oceans (1.4×10^{24} g) is decomposed by photosynthetic plants once every 10,000,000 years. The supply of oxygen in the Earth's atmosphere, a characteristic that serves to differentiate the Earth from all other planets, with the possible exception of Mars, has been derived mostly from photosynthesis resulting from a lack of balance in the biospheric carbon cycle. The total amount of carbon photosynthetically incorporated into green plants during the history of Earth is approximately 10^{26} g, an amount which is equivalent to one-fiftieth of the weight of the globe. It is estimated that about 10^{21} g of this carbon still persists as fossil organic compounds in coal, petroleum, and organic shales.

Following the appearance of *Homo sapiens* in the biosphere, both direct and indirect geochemical changes have arisen as a result of land clearing, water conservation practices, road construction, combustion of fossil fuels, mining operations, and extinction and redistribution of species. The term anthroposphere (or noosphere) has been used to designate the modern biosphere in which human changes play a significant part. As a result of atomic fission processes, both in weapons testing and in nuclear reactors, new biogeochemical problems have arisen. The radioactive nuclides present in atomic fallout and in low-level reactor wastes are accumulated by organisms; the nuclide-enrichment factors sometimes amount to as much as 100,000 times the environmental concentrations. In general, the radioactive nuclides most concentrated by organisms are those of elements that occur in the environment in physiologically limiting concentrations. This is the case, for example, in the accumulation of P^{32} by phytoplankton and some vertebrates in the Columbia River (United States) below the level of the Hanford atomic energy plant.

Chemical composition of living matter. A living organism can perhaps be best described in general terms as a complex colloidal system, enclosed and separated from the environment by a semipermeable membrane, with enzyme-controlled feedback mechanisms that serve to maintain and reproduce the entire system. A characteristic which serves to distinguish the living or once living from the nonliving is the molecular asymmetry of biological matter; certain enantiomorphs (optically active isomers) predominate. Thus D-glucose is the only known biological enantiomorph of glucose, and in proteins all the amino acids appear to be of the L series. In chemical syntheses, a racemate (50:50 mixture of D- and L-isomers) is usually produced. *See* AMINO ACIDS; GLUCOSE.

Elements of living matter. Excluding technetium, promethium, astatine, francium, and the transuranium elements, all of which are either very rare or nonexistent in nature, there are 88 elements that could conceivably enter into the composition of living matter. Of this number, 71 have been identified, but the data for three of these (niobium, zirconium, and tantalum) are questionable and need confirmation. Among the elements still undetected in living matter are four of the noble gases (helium, neon, krypton, and xenon) that undoubtedly occur there as atmospheric contaminants, all the platinum metals (ruthenium, rhodium, palladium, osmium, iridium, and platinum), and seven other elements of atomic numbers in the range 49–91 (indium, tellurium, hafnium, rhenium, polonium, actinium, and protactinium). It is improbable that any of the unidentified elements is completely excluded from protoplasm.

The major biological elements, hydrogen (H), carbon (C), nitrogen (N), and oxygen (O), compose 96–99% of the live weight of nonskeletal tissues. These elements enter into the composition of the four main groups of protoplasmic molecules, namely, water (89.4% O and 10.6% H by weight), proteins (51.3% C, 22.4% O, 17.8% N, and 6.9% H, by weight), carbohydrates (49.4% O, 44.4% C, and 6.2% H, by weight), and lipids (69.0% C, 17.9% O, and 10.0% H, by weight). In terms of the number of atoms, the abundance of elements in living matter decreases in the order hydrogen > oxygen > carbon > nitrogen. Skeletal or supporting structures may be composed of organic compounds such as cellulose, lignin, chitin, and sclero-proteins; or inorganic compounds such as calcium carbonate, silicon dioxide, and calcium phosphate.

Concentrations of elements. The average percentage composition of terrestrial vegetation is listed in Table 2 for those elements that have been studied in sufficient detail. The concentration of a minor element in living matter provides no indication of its physiological importance or necessity. Concentrations of essential elements may vary by a factor of 100,000 or higher. Some elements of unknown function occur in very small amounts. The cow liver cell, for example, contains only some 23 atoms of radium, probably in nonessential sites. The protistan *Euglena gracilis*, on the other hand, requires only 5000 molecules per cell of the essential cobalt-containing vitamin cyanocobalamin (vitamin B$_{12}$).

Several attempts have been made to construct a

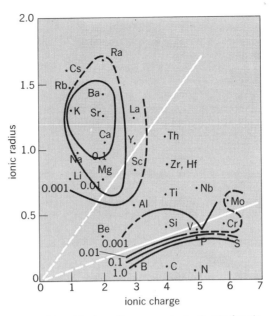

Fig. 1. Plot of ionic radius versus ionic charge for elements of constant valence. Curved lines show ratios of average concentration in terrestrial plants to concentrations in accessible lithosphere. *(From Goldschmidt, in G. E. Hutchinson, The biogeochemistry of aluminum and of certain related elements, Quart. Rev. Biol., 18(4):337, 1943)*

biological classification of the elements, but no single satisfactory tabulation has yet been discovered. The biological elements are generally those of low atomic weight; however, iodine, with an atomic weight of 127, is a notable exception. Partly as a result of the high percentage of water in living tissues, there is actually more resemblance between the cosmic abundance of elements and the abundance of elements in living matter than there is between that of living matter and the lithosphere. Perhaps the best-documented relationship is that discovered by G. E. Hutchinson for elements of constant valence, as shown in Fig. 1. The elements enriched by plants relative to the upper part of the lithosphere are those with either a high or a low ionic potential (ratio of ionic charge to radius). The basis of the rule is water solubility. The elements in the upper left group of Fig. 1 form soluble cations in water; those in the lower right group form anionic complexes with oxygen that are soluble in water, whereas the elements in the central group tend to form insoluble oxides.

The chemical composition of single-species populations cannot be taken as a completely invariable characteristic, although it is approximately so. Chemical composition changes with genetic constitution, growth, and environmental conditions. The availability of many soil elements for plant growth depends upon the pH of the soil, the state of the element, concentration, and many other factors. The ability of plants to reflect subsoil concentrations of certain elements has been used as a basis of biogeochemical prospecting for subsurface ore deposits. In the nickel-rich areas of the southern Ural Mountains, some abnormal forms of *Anemone patens* may contain up to 8 times more cobalt and 50 times more nickel than the same species growing on normal soils. Biogeochemical

Table 2. The average percentage by weight of elements in terrestrial vegetation*

Element	Percentage	Element	Percentage	Element	Percentage
O	70	Fe	2×10^{-2}	Br	1×10^{-4}
C	18	Mn	7×10^{-3}	Mo	5×10^{-5}
H	10.5	F	3×10^{-3}	Y†	4×10^{-5}
Ca	0.5	Ba	3×10^{-3}	Ni	2×10^{-5}
N	0.3	Al	2×10^{-3}	V	2×10^{-5}
K	0.3	Sr	2×10^{-3}	Pb	2×10^{-5}
Si	0.15	B	1×10^{-3}	Li	1×10^{-5}
Mg	7×10^{-2}	Zn	3×10^{-4}	U	1×10^{-5}
P	7×10^{-2}	Rb	2×10^{-4}	Ga	3×10^{-6}
S	5×10^{-2}	Cs	2×10^{-4}	Co	2×10^{-6}
Cl	4×10^{-2}	Ti	1×10^{-4}	I	1×10^{-6}
Na	2×10^{-2}	Cu	1×10^{-4}	Ra	2×10^{-12}

*Modified slightly from a tabulation by G. E. Hutchinson.

†Data are given for yttrium and the rare earths.

prospecting has been similarly used for copper, lead, selenium, and uranium deposits.

Trace elements and isotopes. Many of the trace metals are incorporated into protoplasm by forming complexes with ligands such as proteins, porphyrins, amino acids, mucopolysaccharides, and other polyvalent compounds. Among metals of the first transition series, there exists a stability sequence for the metal-ligand complexes that is largely independent of the nature of the ligand: $Mn^{++} < Fe^{++} < Co^{++} < Ni^{++} < Cu^{++} > Zn^{++}$.

It has occasionally been suggested that specific isotopes of some elements may be essential for living matter, but there is no evidence for such a concept. A fractionation of isotopes does occur in most biological processes, with a preferential accumulation of the lighter members in the pairs $C^{12} - C^{13}$ and $S^{32} - S^{34}$. Isotope ratios of natural materials provide useful data in determining the mode of origin of certain deposits, and in the case of $O^{16} - O^{18}$ in carbonate shells, in determining paleotemperatures.

Biogeochemical cycles of elements. Elements go through characteristic processes or cycles of entering an organism, returning to the organism's surroundings through the products of metabolism and death, entering another organism, and so forth.

Primary elements. Each of the primary elements of living matter (hydrogen, carbon, nitrogen, oxygen, and phosphorus) constitutes more than 0.5% of the weight of the living body.

Hydrogen occurs in nearly all biological compounds. Molecular hydrogen is both produced and oxidized by different kinds of bacteria. Hydrogen sulfide and methane are also metabolized under anaerobic conditions. The basis of many enzymatic oxidation reactions is a dehydrogenation.

Carbon is unique in that it forms more compounds than all the other elements combined. The cycle of carbon in the biosphere is shown in Fig. 2. Photosynthetic plants reduce carbon dioxide during photosynthesis to organic carbon compounds that are ultimately converted back to carbon

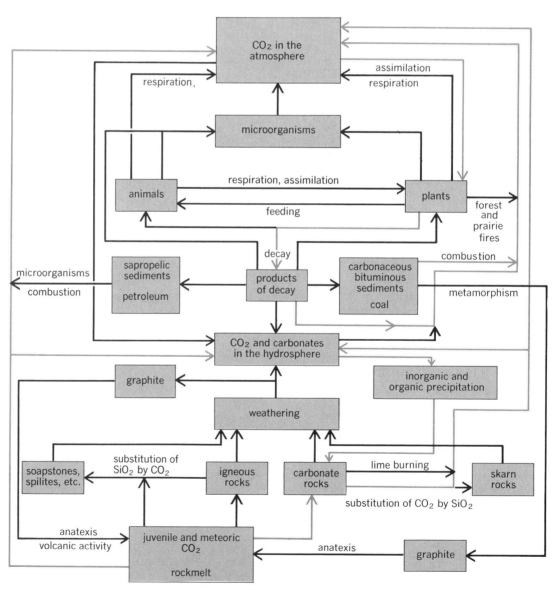

Fig. 2.   Cycle of carbon. (*From K. Rankama and T. G. Sahama, Geochemistry, University of Chicago Press, 1950*)

dioxide in plant and animal respiration. Because some carbon compounds are not readily attacked by animals and microorganisms, there is a slight but steady loss of carbon to sediments, a process that has led to the accumulation of coal and petroleum deposits.

Much of the carbonate deposition in the oceans is mediated by calcareous organisms such as corals, mollusks, brachiopods, and some bacteria and algae. Carbonates may be indirectly precipitated from natural waters by a rise in pH following periods of intense phytoplankton photosynthesis. About 1 mg of carbon dioxide per cm² of Earth surface is added annually to the atmosphere as a result of the combustion of fossil fuels. This appears to have led to a 9% increase in atmospheric carbon dioxide in the period 1900–1960.

Oxygen is an essential element for all organisms; however, certain microorganisms are adapted to an anaerobic existence which in some cases is facultative and in others obligate. The oxygen produced by green plants during photosynthesis is used in plant and animal respiration, the overall reaction (proceeding to the right) being a reversal of photosynthesis, as shown in the equation below. The oxygen produced during photosynthesis arises from the decomposition of water.

$$(CH_2O)_n + nO_2 \rightleftharpoons nCO_2 + nH_2O + 106n \text{ kcal}$$

The geochemical cycle of nitrogen is outlined in Fig. 3. In most plant tissues, nitrogen averages 2–6% of the dry weight, whereas in animals and some bacteria the corresponding value is 6–13% for ash-free weights. *Azotobacter*, some blue-green algae, and *Rhizobium* in association with plant roots convert molecular nitrogen into organic compounds. Hutchinson estimates that the total annual biological fixation of nitrogen amounts to $n \times 10^{12}$ g of nitrogen, or approximately $n/5 \mu$g of nitrogen per cm² of Earth surface. The return of molecular nitrogen to the atmosphere is accomplished by a number of facultative anaerobic bacteria that convert nitrates and nitrites into molecular nitrogen. The most important biological forms of nitrogen are proteins and nucleic acids. The predominant excreted forms of nitrogen in terrestrial animals are uric acid and urea, whereas in aquatic forms the nitrogen is usually excreted as ammonia.

Phosphorus usually forms 0.2–1.0% of the dry weight of plant and animal tissues. Many organic phosphorus compounds occur biologically (for example, phosphoproteins, phospholipids, carbohydrate-phosphates, and nucleic acids). In all these compounds, the phosphorus occurs as phosphate esters. Phosphate is taken up from natural waters by algae and bacteria when the concentrations are only a few parts per 1,000,000,000 parts of water. The concentrations of phosphate in both waters and soils are low enough to be limiting to plant growth. Because of erosion and agricultural practices, about 10^{13} g of phosphorus is annually delivered to the sea. This is largely a unidirectional process because only $n \times 10^{10}$ g of phosphorus is annually removed from the sea, mostly in the form of guano deposited by marine birds.

Secondary elements. The other essential elements for living organisms in approximate order of

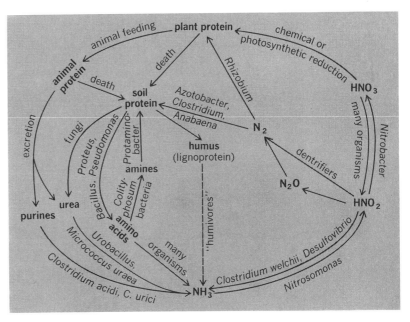

Fig. 3. Outline of the major processes involved in the nitrogen cycle in the biosphere. (*From K. V. Thimann, The Life of Bacteria, Macmillan, 1950*)

decreasing concentration by weight are sodium, calcium, magnesium, sulfur, chlorine, potassium, iron, silicon, vanadium, zinc, boron, fluorine, manganese, cobalt, copper, and iodine.

Among the alkali-metal cations, sodium (Na) occurs primarily in circulating fluids, whereas potassium (K) is located intracellularly. The biological K/Na ratio is usually higher than in the growth medium. Lithium is a minor and variable constituent of living matter. Rubidium, on the other hand, is a normal constituent, behaving essentially like potassium; however, potassium cannot be completely replaced by rubidium. Chlorine occurs as chloride in biological fluids. The antibiotic chloromycetin is one of the few known organic chloride compounds. Iodine is concentrated from sea water by brown and red algae where it may occur in concentrations of 1% of the live weight. In sponges, corals, and the thyroid glands of vertebrates, iodine occurs in the form of iodinated amino acids. Bromine occurs biologically as bromide and also in the form of organic compounds. Fluorine enters the apatite lattice of bones and teeth.

Calcium is biologically the most important member of the alkaline-earth family, occurring as skeletal matter in the form of carbonates (aragonite and calcite) and as phosphate (hydroxyapatite). Calcite skeletons are generally rich (2–28%) in $MgCO_3$, whereas aragonite skeletons usually have less than 1%. Corals and some red and brown algae accumulate strontium (Sr) preferentially to calcium (Ca) from sea water, but most plants and animals are characterized by higher Sr/Ca ratios in their diets than in their bodies. Sr^{90} is the most serious of the radioactive fallout nuclides because it has a long biological residence time as well as a long physical half-life. Sr^{90} fallout is apt to be most serious in areas of calcium-deficient soils. Magnesium is an integral part of the chlorophyll molecule and is also required as a cofactor in enzymatic phosphate reac-

Table 3. Comparison of Recent and Oligocene marine mud*

Content	Recent	Oligocene
Carbonate carbon, %	0.71	1.49
Organic carbon, %	0.53	0.27
Organic nitrogen, %	0.044	0.032
Amino acids, $\mu M/g$	3.0	0.51

*From J. G. Erdman, E. M. Marlett, and W. E. Hanson, Survival of amino acids in marine sediments, *Science*, 124:1026, 1956.

tions. Radium is accumulated by some plants. Marine animals (especially fish) appear to contain less radium than marine plants.

Boron in the form of borates is essential for the growth of higher plants, but its function is unknown. Silicon is primarily a skeletal element. It is accumulated by diatoms, radiolarians, some sponges, and some terrestrial plants. Copper is accumulated from sea water by certain mollusks and crustaceans where it occurs in the form of the respiratory pigment hemocyanin. Copper is also known to participate in some enzyme reactions, but ionic copper is decidedly toxic to most forms of life when it is present in appreciable concentrations.

The geochemical cycle of sulfur is markedly affected by biological reactions. The photosynthetic sulfur bacteria have a photosynthesis based on the metabolism of sulfur compounds. Hydrogen sulfide is generated during the anaerobic decomposition of plankton. Selenium is selectively accumulated by some land plants such as *Astragalus*, but it is otherwise toxic to plants and animals.

Aluminum is the third most abundant element in the lithosphere, but it is scant in most living tissues. One exception is the club moss *Lycopodium alpinum*, which may contain up to 33% Al_2O_3 in its ash.

Vanadium, manganese, iron, and cobalt appear to be indispensable but only for some species in the cases of the ferrides vanadium and cobalt. Iron occurs biologically in a number of organic compounds, the most important of which are hemoglobin and the respiratory enzymes (cytochromes). The geochemical cycle of iron is influenced directly by the deposition of hydroxides by the iron bacteria and indirectly through changes in pH and E_h.

Table 4. Principal amino acids in Recent and Oligocene sediments*

Recent[†]	Oligocene[†]
Valine	Alanine
Leucines	Glutamic acid
Alanine	Glycine
Glutamic acid	Proline
Aspartic acid	Leucines
Glycine	Aspartic acid
Proline	
Tyrosine	
Phenylalanine	

*From J. G. Erdman, E. M. Marlett, and W. E. Hanson, Survival of amino acids in marine sediments, *Science*, 124:1026, 1956.

[†]Arranged in order of decreasing abundance.

conditions. Manganese is a cofactor for some enzymes, and vanadium appears to play some role in photosynthesis. Vanadium also occurs in the form of vanadyl porphyrins in petroleum.

The other elements of the periodic table are, so far as is known, neither essential for organisms nor present therein in appreciable amounts. The geochemistry of such elements may be secondarily influenced by biological events, as in the precipitation of heavy metals by organic sediments rich in hydrogen sulfide.

Degradation of biogenic molecules. Under average conditions of burial and subsequent low-temperature maturing of biogenic sediments, the main chemical trend appears to be dehydration. As a result, all the molecules which have an H/O ratio close to that of water have a relatively low geological time-scale stability. Such types of molecules are lignin, cellulose, proteins, and nucleic acids.

The higher H/O ratios of the lipids render these molecules of greater geological time-scale stability. The same applies to fossil resins, which can become preserved as ambers from the Mesozoic age onward, particularly in specimens from East Prussia; and occasionally insects, mosses, and even a small lizard, which lived some 150,000,000 years ago, are present in a state of perfect preservation. Some pigments of high H/O ratios are also well preserved. This applies in particular to the V, Ni, and Ca complexes of porphyrins, which can be detected in Paleozoic sediments. Hydroxyquinoline dyes have been identified from fossil sea lilies of Mesozoic age, which are closely comparable to those which exist in Recent species.

It appears from works on members of the coal series that diamino acids are preserved only in peats, lignite, and subbituminous coals, whereas the geologically more stable monoamino acids can be detected also in bituminous coals and anthracites. Table 3 indicates that the percentage of carbonate increases with the aging of marine muds and that amino acids are depleted at a considerably higher rate than either organic carbon or nitrogen. According to the data in Table 4, alanine is the amino acid of major geological stability, and it is interesting to note that the results of thermal decomposition experiments also demonstrated that alanine is the stablest of all the amino acids.

Little is presently known regarding the stabilities of the individual sugar and other carbohydrate molecules on a geological time scale. Analyses of samples from a test bore hole from the bottom of the Pacific, collected in the course of the Mohole project, demonstrated the reduction in the total organic C of sugars from 3.8 to 0.2% at a depth of 138 m. Galactose showed in this case at least a relatively lower stability than glucose. Depolymerization of cellulose in Miocene lignites has also been proven.   [GEORGE MUELLER]

Bibliography: I. A. Breger (ed.), *Organic Geochemistry*, 1963; F. W. Clarke, *The Data of Geochemistry*, 5th ed., USGS Bull. no. 770, 1924; U. Colombo and G. D. Hobson (eds.), *Advances in Organic Geochemistry*, 1964; S. W. Fox (ed.), *The Origins of Prebiological Systems*, 1965; G. E. Hutchinson, *A Treatise on Limnology*, vol. 1, 1957; K. Rankama and T. G. Sahama, *Geochemistry*, 1950; H. U. Sverdrup, M. W. Johnson, and R. H.

Fleming, *The Oceans*, 1946; W. Vernadsky, *La Biosphere*, 1929; A. P. Vinogradov, The elementary chemical composition of marine organisms, *J. Mar. Res.*, Sears Foundation for Marine Research, 1953.

Biotic isolation

The occurrence of organisms in isolation from others of their species. Many organisms live in aggregations or social groups. Among many others, however, there appears an opposite tendency toward relative isolation, with each individual or pair living by itself, separated by at least a small distance from others of its kind. Thus in a desert, shrubs may grow widely scattered, each with its roots occupying an area around it larger than that of its foliage. A shrub seedling which begins to grow close to an established larger shrub encounters more intense root competition for water and nutrients, and in some cases stronger unfavorable chemical effects of the larger shrub, than does a seedling which grows at a point more distant from other shrubs. Seedlings most likely to survive are those near the center of spaces between other shrubs. The shrubs consequently grow not in clumps but in relative isolation from each other, with a spacing between individuals which may be more even than in a random distribution. They may show a degree of negative contagion in their distribution. In many animals a breeding pair may isolate themselves, occupying a definite territory from which others of the species are excluded. Such biotic isolation provides each individual or pair with a living space from which it can obtain food, or water and nutrients, more or less free from competition with others of its species. *See* ECOLOGICAL INTERACTIONS; POPULATION DISPERSION.

[ROBERT H. WHITTAKER]

Bibliography: E. P. Odum, *Ecology*, 1963; E. P. Odum, *Fundamentals of Ecology*, 1959.

Breeding (animal)

The application of genetic principles to farm animals to improve heredity for economically important characteristics, such as milk production in dairy cattle, meatiness in pigs, and feed requirements in beef cattle. Experience has demonstrated that domestic animals, despite a 6000–8000-year history of domestication, still respond readily to selection. Many specialized breeds and strains of animals, adapted to widely different environmental conditions, have been developed to meet local needs for meat, milk, fiber, and draft.

Methods of selection. Selection is the primary tool for generating directed hereditary changes in animals. Selection may be concentrated on one characteristic, may be directed independently toward several traits, or may be directed toward an index on total score which includes information on several important traits. The first method is most effective in generating improvement in the one chosen trait, but generates only correlated response in other traits. The second method, independent culling levels, generates greater overall improvement in total merit than the first, but is less effective for the chosen trait. The total score or index can generate maximum overall improvement, especially when emphasis given the traits is weighed by their economic importance and herita-

bilities, but has the disadvantage of delaying culling until all of the pertinent information has been collected.

In general, the second and third methods are preferable to the first when several equally important and equally heritable traits need attention. The relative difference between the latter two methods is great only when little culling, and consequently little improvement, can be achieved. In practice, selection is likely to be a mixture of the latter two methods, depending upon the age at which selection has to be practiced and what information is available at that time. Practical considerations, such as availability of feed and space, market prices, or age of animals, play an important part in decisions as to which animals are retained.

Accuracy of selection. Those characteristics whose variation is determined largely by genetic differences are termed highly heritable, or are said to have high heritability, whereas those influenced strongly by environmental fluctuations are lowly heritable. With the former, selection can be accurate and effective in the sense that phenotypic differences are closely related to genetic differences or breeding values. Genes may also have effects which differ according to the other genes present. This tends to make selection inaccurate and hence ineffective, since the ultimate effect is similar to that of random environmental variations.

The improvement achieved by selection is directly related to the accuracy with which the breeding values of the subjects can be recognized. Accuracy in turn depends upon the heritabilities of the traits and upon whether they can be measured directly upon the subjects for selection (mass selection), upon their parents (pedigree selection), upon their brothers and sisters (family selection), or upon their progeny (progeny testing). Mass selection is effective for traits with high heritability which are expressed before breeding age. Pedigree selection is most useful for traits which are expressed relatively late in life and for traits which are limited in expression to only one sex. Pedigree selection can never be highly accurate as compared with other kinds of selection, because a subject can have only two parents, a maximum of four different grandparents, and so on, whereas the number of sibs or progeny can be large. For traits of medium heritability, the following sources of information are approximately equally accurate for predicting breeding values of subjects: (1) one record measured on the subject himself; (2) one record on each ancestor for three previous generations; (3) one record each on five brothers or sisters where there is no environmental correlation between family members; and (4) one record each on five progeny having no environmental correlations, each from a different mate.

Artificial insemination is an effective tool for increasing the intensity of selection, especially in males. It has been most widely applied in dairy cattle, where carefully planned young sire sampling programs are being combined with thoroughly conducted progeny testing schemes to identify genetically superior sires. The use of artificial insemination with beef cattle, sheep, and pigs has been limited to date but appears to be growing. Performance testing and family selection have played more important roles in the notable improvement in meatiness in pigs than has progeny testing.

Purebred breeds. Propagation of improved animal seed stocks is achieved primarily with purebred strains descended from imported or locally developed groups of animals which have been selected and interbred for a long enough period to be reasonably uniform for certain trademark characteristics, such as coat color. Each breed is promoted and sponsored by its farmer breeders organized as a purebred society. Because the number of breeding animals is finite and because breeders tend to prefer certain bloodlines and sires, some inbreeding occurs with the pure breeds. This inbreeding rate in those breeds that have been studied has been about 0.4% per generation.

As a rule, each breed is characterized by certain easily identifiable characteristics, such as coat color, size and shape of ear, and horns or polledness. The genes controlling such traits are almost homozygous in breeds that have been subjected to intense selection for many generations. The frequency of genes influencing functional traits of economic importance is likely to be much lower for several reasons. First, functional characteristics, such as growth rate or milk production, are influenced by many genes, including those concerned with appetite, ability to obtain and digest food, temperament, disease resistance, and energy metabolism. Second, the effects of individual genes upon functional characteristics are likely to be small relative to the total variation. Hence selection for improved performance changes the frequency of any one desired gene only slightly, although the cumulative improvement over many generations may be large. Third, selection has not always been consistently for the same goal, frequently changing intensity or direction as a consequence of changing styles or economic conditions.

Crossbreeding. Most of the commercial pigs and sheep in the United States are produced by some form of systematic rotational crossbreeding, wherein the crossbred males are marketed and enough of the crossbred females are retained to replace the female breeding stock. These crossbred females are then mated to purebred males of a different breed. After a cycle of males involving three or four breeds has been used, the usual practice is to return to the original breed of male and repeat the cycle.

Genetic fingerprinting. The discovery of methods of analyzing body tissues for genetic differences offers some opportunity of increasing rate of genetic change in animal improvement over that which can be realized by the usual methods of selecting for phenotypic differences. Studies of blood antigens, serum proteins, and metabolic enzymes indicate clearly that these are inherited in most cases according to the basic hereditary mechanism. Identification of chemical entities in animal tissues thus promises to provide more accurate clues to the genes contained in the tissues. Application of such "genetic fingerprinting" would have to be carried out on an extensive basis to have noticeable genetic impact, but this may eventually be achieved. These procedures would be particularly useful for males used extensively in artificial insemination. *See* AGRICULTURAL SCIENCE (ANIMAL).

[L. N. HAZEL]

Bibliography: D. S. Falconer, *Introduction to Quantitative Genetics*, 1960; J. F. Lasley, *Genetics of Livestock Improvement*, 1963.

Breeding (plant)

The application of genetic principles to the improvement of cultivated plants, with heavy dependence upon the related sciences of statistics, pathology, physiology, and biochemistry. The aim of plant breeding is to produce new and improved types of farm crops or decorative plants, to better serve the needs of the farmer, the processor, and the ultimate consumer. New varieties of cultivated plants can result only from genetic reorganization that gives rise to improvements over the existing varieties in particular characteristics or in combinations of characteristics. In consequence, plant breeding can be regarded as a branch of applied genetics, but it also makes use of the knowledge and techniques of many aspects of plant science, especially physiology and pathology. Related disciplines, like biochemistry and entomology, are also important, and the application of mathematic statistics in the design and analysis of experiments is essential.

Plant breeding has made major contributions to increasing the yields of crops and to diminishing their susceptibility to hazards that limit their productivity. It has been estimated that the annual production of corn in the United States has been increased by 750,000,000 bu by plant breeding methods, especially by the exploitation of hybrid corn. In western Europe the yields of wheat and barley have been increased by approximately 1% per annum since the late 1940s. Perhaps the most dramatic impact of plant breeding occurred during the 1960s in Mexico where, as the result of the stimulus from a Rockefeller Foundation program, a wheat-importing country was changed into a wheat-exporting country because of a surplus of wheat.

Scientific method. The cornerstone of all plant breeding is selection. By selection the plant breeder means the picking out of plants with the best combinations of agricultural and quality characteristics from populations of plants with a variety of genetic constitutions. Seeds from the selected plants are used to produce the next generation, from which a further cycle of selection may be carried out if there are still differences. Much of the early development of the oldest crop plants from their wild relatives resulted from unconscious selection by the first farmers. Subsequent conscious acts of selection slowly molded crops into the forms of today. Finally, since the early years of the 20th century, plant breeders have been able to rationalize their activities in the light of a rapidly expanding understanding of genetics and of the detailed biology of the species studied.

Plant breeding can be divided into three main categories on the basis of ways in which the species are propagated. Species that reproduce sexually and that are normally propagated by seeds occupy two of these categories. First come the species that set seeds by self-pollination, that is, fertilization usually follows the germination of pollen on the stigmas of the same plant on which it was produced. The second category of species sets seeds by cross-pollination, that is, fertilization usually follows the germination of pollen on the stigmas of different plants from those on which it was produced. The third category comprises the species that are asexually propagated, that is, the commercial crop results from planting vegetative parts or by grafting. Consequently, vast areas can be occupied by genetically identical plants of a single clone that have, so to speak, been budded off from one superior individual. The procedures used in breeding differ according to the pattern of propagation of the species.

Self-pollinating species. The essential attribute of self-pollinating crop species, such as wheat, barley, oats, and many edible legumes, is that, once they are genetically pure, varieties can be maintained without change for many generations. When improvement of an existing variety is desired, it is necessary to produce genetic variation among which selection can be practiced. This is achieved by artificially hybridizing between parental varieties that may contrast with each other in possessing different desirable attributes. All members of the first hybrid (F_1) generation will be genetically identical, but plants in the second (F_2) generation and in subsequent generations will differ from each other because of the rearrangement and reassortment of the different genetic attributes of the parents. During this segregation period the breeder can exercise selection, favoring for further propagation those plants that most nearly match the ideal he has set himself and discarding the remainder. In this way the genetic structure is remolded so that some generations later, given skill and good fortune, when genetic segregation ceases and the products of the cross are again true-breeding, a new and superior variety of the crop will have been produced.

This system is known as pedigree breeding, and while it is the method most commonly employed, it can be varied in several ways. For example, instead of selecting from the F_2 generation onward, a bulk population of derivatives of the F_2 may be maintained for several generations. Subsequently, when all the derivatives are essentially true-breeding, the population will consist of a mixture of forms. Selection can then be practiced, and it is assumed, given a large scale of operation, that no useful segregant will have been overlooked. By whatever method they are selected, the new potential varieties must be subjected to replicated field trials at a number of locations and over several years before they can be accepted as suitable for commercial use.

Another form of breeding that is often employed with self-pollinating species involves a procedure known as backcrossing. This is used when an existing variety is broadly satisfactory but lacks one useful and simply inherited trait that is to be found in some other variety. Hybrids are made between the two varieties, and the F_1 is crossed, or backcrossed, with the broadly satisfactory variety which is known as the recurrent parent. Among the members of the resulting first backcross (B_1) generation, selection is practiced in favor of those showing the useful trait of the nonrecurrent parent and these are again crossed with the recurrent parent. A series of six or more backcrosses will be necessary to restore the structure of the recurrent parent, which ideally should be modified only by the incorporation of the single useful attribute sought from the nonrecurrent parent. Backcross-

ing has been exceedingly useful in practice and has been extensively employed in adding resistance to diseases, such as rust, smut, or mildew, to established and acceptable varieties of oats, wheat, and barley. See PLANT DISEASE.

Cross-pollinating species. Natural populations of cross-pollinating species are characterized by extreme genetic diversity. No seed parent is true-breeding, first because it was itself derived from a fertilization in which genetically different parents participated, and second because of the genetic diversity of the pollen it will have received. In dealing with cultivated plants with this breeding structure, the essential concern in seed production is to employ systems in which hybrid vigor is exploited, the range of variation in the crop is diminished, and only parents likely to give rise to superior offspring are retained.

Inbred lines. Here plant breeders have made use either of inbreeding followed by hybridization (see illustration) or of some form of recurrent selection. During inbreeding programs normally cross-pollinated species, such as corn, are compelled to self-pollinate by artificial means. Inbreeding is continued for a number of generations until genetically pure, true-breeding, and uniform inbred lines are produced. During the production of the inbred lines rigorous selection is practiced for general vigor and yield and disease resistance, as well as for other important characteristics. In this way desirable attributes can be maintained in the inbred lines, which are nevertheless usually of poor vigor and somewhat infertile. Their usefulness lies in the vigor, high yield, uniformity, and agronomic merit of the hybrids produced by crossing different inbreds. Unfortunately, it is not possible from a mere inspection of inbred lines to predict the usefulness of the hybrids to which they can give rise. To estimate their value as the parents of hybrids, it is necessary to make tests of their combining ability. The test that is used depends upon the crop and on the ease with which controlled cross-pollination can be effected.

Tests may involve top crosses (inbred × variety), single crosses (inbred × inbred), or three-way crosses (inbred × single cross). Seeds produced from crosses of this kind must then be grown in carefully controlled field experiments designed to permit the statistical evaluation of the yields of a range of combinations in a range of agronomic environments like those normally encountered by the crop in agricultural use. From these tests it is possible to recognize which inbred lines are likely to be successful as parents in the development of seed stocks for commercial growing. The principal advantages from the exploitation of hybrids in crop production derive from the high yields produced by hybrid vigor, or heterosis, in certain species when particular parents are combined.

Economic considerations. The way in which inbred lines are used in seed production is dictated by the costs involved. Where the cost of producing F_1 hybrid seeds is high, as with many forage crops or with sugar beet, superior inbreds are combined into a synthetic strain which is propagated under conditions of open pollination. The commercial crop then contains a high frequency of superior hybrids in a population that has a similar level of variability to that of an open-pollinated variety. However, because of the selection prac-

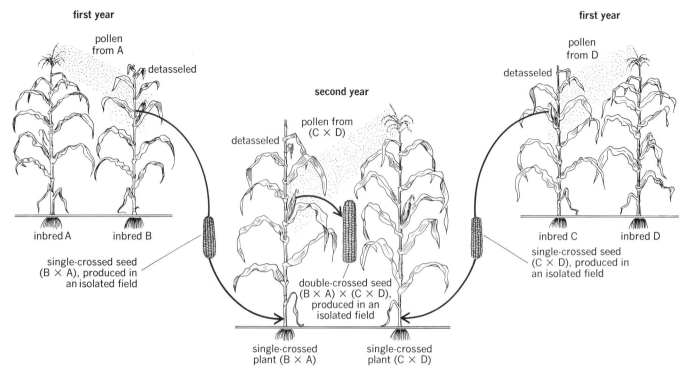

Sequence of steps in crossing inbred plants and using the resulting single-crossed seed to produce double-crossed hybrid seed. (*Crops Research Division, Agricultural Research Service, USDA*).

ticed in the isolation and testing of the inbreds, the level of yield is higher because of the elimination of the less productive variants.

When the cost of seed is not of major significance relative to the value of the crop produced, and where uniformity is important, F_1 hybrids from a single cross between two inbred lines are grown. Cucumbers and sweet corn are handled in this way. By contrast, when the cost of the seeds is of greater significance relative to the value of the crop, the use of single-cross hybrids is too expensive and then double-cross hybrids (single cross A × single cross B) are used, as in field corn. As an alternative to this, triple-cross hybrids can be grown, as in marrow stem kale, in the production of which six different inbred lines are used. The commercial crop is grown from seeds resulting from hybridization between two different three-way crosses.

Recurrent selection. Breeding procedures designated as recurrent selection are coming into limited use with open-pollinated species. In theory, this method visualizes a controlled approach to homozygosity, with selection and evaluation in each cycle to permit the desired stepwise changes in gene frequency. Experimental evaluation of the procedure indicates that it has real possibilities. Four types of recurrent selection have been suggested: on the basis of phenotype, for general combining ability, for specific combining ability, and reciprocal selection. The methods are similar in the procedures involved, but vary in the type of tester parent chosen, and therefore in the efficiency with which different types of gene action (additive and nonadditive) are measured. A brief description is given for the reciprocal recurrent selection.

Two open-pollinated varieties or synthetics are chosen as source material, for example, A and B. Individual selected plants in source A are self-pollinated and at the same time outcrossed to a sample of B plants. The same procedure is repeated in source B, using A as the tester parent. The two series of testcrosses are evaluated in yield trials. The following year inbred seed of those A plants demonstrated to be superior on the basis of testcross performance are intercrossed to form a new composite, which might be designated A_1. Then B_1 population would be formed in a similar manner. The intercrossing of selected strains to produce A_1 approximately restores the original level of variability or heterozygosity, but permits the fixation of certain desirable gene combinations. The process, in theory, may be continued as long as genetic variability exists. In practice, the hybrid $A_n \times B_n$ may be used commercially at any stage of the process if it is equal to, or superior to, existing commercial hybrids.

Asexually propagated crops. A very few asexually propagated crop species are sexually sterile, like the banana, but the majority have some sexual fertility. The cultivated forms of such species are usually of widely mixed parentage, and when propagated by seed, following sexual reproduction, the offspring are very variable and rarely retain the beneficial combination of characters that contributes to the success of their parents. This applies to such species as the potato, to fruit trees like apples and pears that are propagated by grafting, and to raspberries, grapes, and pineapples.

Varieties of asexually propagated crops consist of large assemblages of genetically identical plants, and there are only two ways of introducing new and improved varieties. The first is by sexual reproduction and the second is by the isolation of sports or somatic mutations. The latter method has often been used successfully with decorative plants, such as chrysanthemum, and new forms of potato have occasionally arisen in this way. When sexual reproduction is used, hybrids are produced on a large scale between existing varieties with different desirable attributes in the hope of obtaining a derivative possessing the valuable characters of both parents. In some potato-breeding programs many thousands of hybrid seedlings are examined each year. The small number that have useful arrays of characters are propagated vegetatively until sufficient numbers can be planted to allow the agronomic evaluation of the potential new variety.

Special techniques. So far attention has been concentrated on the general principles of plant breeding and on methods of general applicability. However, there are a few special procedures that should be mentioned.

Cytoplasmic male sterility. In several crop species variants have been found that do not produce fertile pollen. However, pollen fertility can be restored in hybrids that result when the male sterile lines are pollinated by other, so-called restorer, lines. Using crosses between male sterile lines and restorer lines, F_1 hybrid seed supplies can easily be obtained on a large scale. Systems of this kind are used in the commercial production of hybrids in corn, sorghum, sugar beet, and onions. In addition, the potentialities of male sterile breeding systems are being explored in wheat and field beans.

Polyploids. Many crop species are naturally polyploid, and polyploidy can be induced artificially by colchicine treatment and in other ways. Polyploids are often characterized by a more sturdy growth habit and by larger roots, leaves, and flowers than the related diploids. Artificial polyploids are grown commercially in clover, watercress, sugar beet, and forage grasses. In watermelons crosses between normal diploids and artificial polyploids are used to produce hybrids that are sterile and so produce fruit without seeds—obviously a desirable attribute.

Multilinear varieties. Varieties of self-pollinating species like wheat generally consist of genetically identical plants. Multilinear varieties are also made up of genetically similar plants but contain several lines that differ in having genetically different forms of disease resistance. Each component line is produced by backcrossing different forms of resistance into a common recurrent parental variety. The advantage of the multilinear constitution is that, if the resistance of one of the constituent lines breaks down, production will be maintained by the remaining resistant lines.   [RALPH RILEY]

Bibliography: R. W. Allard, *Principles of Plant Breeding*, 1960; F. N. Briggs and P. F. Knowles, *Introduction to Plant Breeding*, 1967; A. Müntzing, *Genetics: Basic and Applied*, 1967; W. Williams, *Genetical Principles and Plant Breeding*, 1964.

Climate, man's impact on

Twenty thousand years ago the places where many of the world's great cities now stand were deeply covered by ice, as were regions where the food for millions of people is now grown. The reasons why the ice receded are not known, but it did so with extraordinary speed, in a length of time compara-

ble with that for which there exists written record of the doings of civilized man. Some of the ice remains. In Greenland and in the Antarctic there are land-borne masses of ice several thousand feet thick. If the Greenland ice alone were to melt with a speed comparable with that reached during the last retreat of the northern continental ice, the sea level would be raised by about 25 ft and many major seaports and the spaces where millions of people now live would be inundated. There is much more ice in the Antarctic, carrying the potential for sea-level changes of several hundred feet.

There is no reason to suggest that the human race would not survive either the return of the Quaternary ice caps or the melting of the ice caps existing today, but current social and political organizations would not survive. A change of climate could cause either catastrophe, and in recent years each has been put forward as a possible long-term consequence of man's modification of the atmosphere and the Earth's surface. There are, as shall be discussed later, other consequences of climatic change which, although less universal and less spectacular, are nonetheless undesirable.

"Climate" is not readily defined, except loosely as "weather" averaged over a period of time. Statistics of temperature and wind, rain, snow, cloud, sunshine, and humidity are the usual measures of the climate of an area. On the scale of the whole planet, the concept of climate is closely tied to the idea of the "general circulation of the atmosphere," a concept equally difficult to define but embodying the regularities and variations of the motion and temperature of the whole atmosphere. *See* CLIMATOLOGY.

CHANGE OF CLIMATE

One undoubted attribute of climate, local or global, is that it changes. This is evident from the records of measuring instruments, from historical

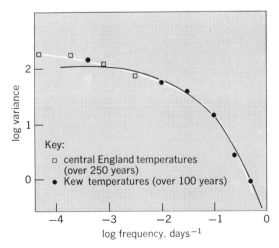

Fig. 1. Spectrum of temperature variations. Light line shows the relative variance associated with periodicities of 2 days to 40 years in the temperatures of southern and central England; due to J. M. Craddock. Dark line indicates the variance in a stationary series having the same statistics as the temperatures in the range of 2 days to 2 years. Diurnal and annual periodicities are removed.

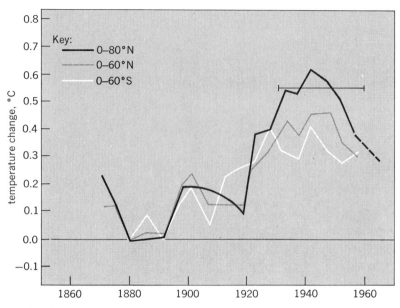

Fig. 2. Mean annual temperature for various latitudes for 1870–1967. Horizontal bar shows mean value for 0 to 80°N for 1931–1960. (*From Inadvertent Climate Modification, M.I.T. Press, 1971*)

texts, from prehistoric artifacts, and from geological evidence.

The longest instrumental record which has been appropriately analyzed is that of temperature in southern and central England. Change is clearly evident in this record, after it has been processed

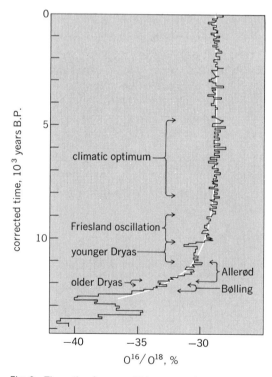

Fig. 3. The ratio of oxygen-16 to oxygen-18 as a function of depth in an ice core sample from the Greenland ice sheet. An increase in this ratio accompanies an increase in temperature at middle and high latitudes in the Northern Hemisphere. (*From Inadvertent Climate Modification, M.I.T. Press, 1971*)

mathematically to remove diurnal and annual periodic variations. Figure 1 shows variance calculations for two series of monthly and daily mean temperature measurements (one series is for central England over 250 years, and the other is for Kew over 100 years). The two series are not stationary (a stationary curve is included for comparison), and the variance increases with observation time. Figure 2 is more direct evidence of change, from temperatures reported for 1870–1967 over extensive areas of the globe; data are due to J. M. Mitchell and R. Scherhag.

An example of historical evidence of climatic change (and one that clearly shows the effect of such change on human society) is the Norse settlement of southern Greenland about 1000 years ago. A pastoral community appears to have been well established, but it vanished with little trace as the land ice readvanced over the pastures and the sea ice cut communications with Iceland and Europe. It is estimated that when the settlement was established, temperatures in the area were 2–4°C higher than now.

Modern techniques of isotope analysis have produced a remarkable confirmation and extension of fragmentary records from various other sources. (When water evaporates or condenses there is temperature-dependent fractionation of the isotopic constituents. Thus the isotopic make-up of an ice sample is a clue to the temperature that prevailed when it was formed.) W. Dansgaard and others have examined the annual layers in the Greenland ice cap. Figure 3 shows their findings concerning the O^{16}/O^{18} ratio, which in turn relates to surface temperatures over the seas of the Northern Hemisphere. The long sequence of warm years ending about A.D. 1000 and the "little ice age" that followed show clearly. There is some indication of comparatively large recent fluctuations, which will be discussed when the possibility of human intervention is considered here.

Only one aspect of the fossil evidence of climatic change need be mentioned here. Interpretation of the fossil record is complicated by geographical changes accompanying the drift of the land masses, but it is accepted that for the great bulk of its lifetime the Earth has been free of land ice. The present anomalous situation appears to have been initiated about 5,000,000 years ago. This conclusion is the more remarkable since the Sun is regarded as a star in the main sequence whose output of radiation should have been appreciably lower than it is at present at epochs when geologists assert the planet to have been free of ice.

THEORIES AND MODELS OF CLIMATE

The atmosphere is often compared to a heat engine. Solar energy falling on the planet is partially reflected to space; the remainder is absorbed, to some extent in the atmosphere but mainly at the irradiated surface. Air in contact with the ground or sea is heated and water is evaporated, and the heated moist buoyant air rises. Water condenses in clouds which, together with the Earth's surface and certain trace gases, notably water vapor and carbon dioxide, radiate energy to space. There is a net absorption of solar energy at low levels, low latitudes, and high temperatures and a net radiation to space at high levels, high latitudes, and low temperatures. Heat is transported between source and sink, the "boiler" and "condenser" of the atmospheric engine, by the moving air. Pressure gradients and winds are established. The Earth's rotation provides deflecting forces. Instabilities at various scales develop in the global patterns of motion—the traveling storms of the weather map. The work generated in the thermodynamic cycle is dissipated by friction in the working fluid of the engine.

Mathematical expression. Individually and in principle, the major processes of weather and climate are well understood and in this sense there exists a "theory" of climate. This theory is expressed mathematically in terms of the continuity equations for momentum, mass, and energy, the integral equation of radiative transfer, the equations governing the phase transitions of water substance, and the equation of state of dry air. There are as many equations as variables, so that the system appears soluble in principle.

Much effort has been devoted to such calculations in the past 20 years, the incentive being the prediction of weather more than the understanding of climate. Systems ("models") have been developed for the numerical solution of equations for the time change of weather elements (wind, temperature, pressure, and other factors) at points covering a hemisphere and as little as 100 mi apart. Such models are now the major tool of the weather forecaster. However, they treat empirically—parameterize, in the jargon of the trade—processes which are important on small space scales and on long time scales. The long-time-scale processes are important in the investigation of climate, and in practice the integration of these processes into the models has proved so complex that a satisfactory theory or even numerical model of global climate is still far from being formalized. One cannot yet assess analytically or computationally the effect of any changes which man might impose on the atmosphere or the Earth's surface. It may be possible to start the chain of reasoning, and identify the initial reactions of the atmospheric system, but it is not possible as yet to work analytically through the feedback loops to the final result.

Model types. Currently feasible quantitative investigation of climatic change is of two types. One type employs what have been termed "global average models," which are of varying degrees of complexity but all of which neglect the horizontal components of motion of the atmosphere. It is generally conceded that the global average models can do little more than indicate the direction of the initial reaction of certain variables (for example, surface temperature) to imposed change in others (such as carbon dioxide content of the atmosphere). The second type of investigation employs empirical models of the whole atmosphere, adding to the global average models certain empirically adjusted terms in an attempt to simulate the effects of atmospheric and oceanic motions. For example, some empirical relation must be assumed between the temperature difference between two latitudes and the heat transport between them, and between the temperature of a locality and the solar radiation absorbed there (the local albedo).

The pronouncements now current concerning man's impact on the global climate are based in the main on these empirical and global average models. But although they are based on empiricism and intuitive judgment to a degree unusual in the physical sciences, they are not just idle speculation. The problem they imply deserves and is receiving serious consideration; it is as difficult as any problem which has ever required the attention of applied scientists.

It is clear that it is not going to be a simple matter to detect man-made changes in a system which changes significantly and, so far as we now understand it, capriciously. Climate has changed and will change, with no assistance from man, and it may be that man will influence its course and never know that he has done so.

CLIMATE AND ATMOSPHERIC POLLUTION

One must consider possible mechanisms by which man's activities might influence the climate. He is changing the composition of the atmosphere and the nature of the Earth's surface. These changes affect the magnitude and distribution of the sources and sinks of heat and moisture in the thermodynamic system whose workings produce the climate.

Carbon dioxide pollution. There is no doubt that the carbon dioxide, CO_2, content of the atmosphere has increased during the 20th century and continues to increase at a rate which correlates with the consumption of fossil fuels. Figure 4 shows the increase in the mean annual concentration of CO_2 measured at three land stations in remote areas and from aircraft during transpolar flights. The measurements are not completely in-

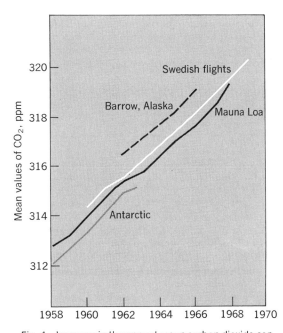

Fig. 4. Increase in the annual mean carbon dioxide content of air measured at three stations in remote areas and on commercial flights in northern polar regions at altitudes of 30,000 to 40,000 ft, according to data of C. D. Keeling and B. Bolin and W. Bischof. (*From G. D. Robinson, Gobal environmental monitoring, Technol. Rev., May 1971*)

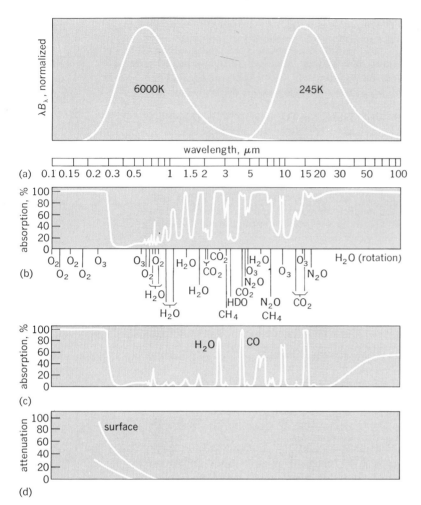

Fig. 5. Radiative properties of atmospheric gases. (a) Blackbody emission for 6000 and 245K, being approximate emission spectra of the Sun and Earth, respectively (since inward and outward radiation must balance, the curves have been drawn with equal areas—though in fact 30% of solar radiation is reflected unchanged). (b) Atmospheric absorption spectrum for a solar beam reaching the ground. (c) Atmospheric absorption spectrum for a beam reaching the tropopause in temperate latitudes. (d) Attenuation of the solar beam by Rayleigh scattering, at the ground and at the temperature tropopause. (From R. M. Goody and G. D. Robinson, Quart. J. Roy. Meteorol. Soc., vol. 77, 1951)

changes. One must therefore expect the atmospheric CO_2 concentration to continue to increase at about its present rate relative to the consumption of fossil fuel, and seek to estimate its climatic impact.

Temperature change. Figure 5 shows diagrammatically the radiative properties of the atmospheric gases. Carbon dioxide can absorb solar radiation in the near infrared, but its absorption bands overlap the wavelengths at which water absorbs strongly. A change in the CO_2 content of the atmosphere therefore has only a secondary effect on absorption of solar radiation, and consequently, on the supply of energy to the planet and its mean temperature. CO_2 is an efficient radiator at terrestrial temperatures because of the strong band centered around 15 μm—a region where other atmospheric gases are relatively transparent.

CO_2 tends to distribute itself fairly uniformly throughout the atmosphere, and therefore an increase in the amount of CO_2 tends to increase the proportion of CO_2 to other gases in the upper atmosphere, thereby increasing the radiation from the upper atmosphere. Since the total radiative loss is not changed (because the heat supply is not changed), there will be a corresponding decrease in the loss of heat to space from the surface and lower atmosphere. Increase in atmospheric carbon dioxide content, with no other change in atmospheric composition, would therefore be expected to lead to a decrease of temperature high in the atmosphere and an increase near the surface—the "greenhouse effect." See GREENHOUSE EFFECT, TERRESTRIAL.

More than 30 years ago G. S. Callendar made the first reasonably accurate quantitative estimates of the surface temperature increases to be expected from an increase of CO_2 content in an otherwise unchanged atmosphere. (He also pointed out the systematic increase in the measurements of CO_2 content and the increased consumption of fossil fuel, which was sufficient to account for it.) At that time, the late 1930s, there was also an obvious upward trend in surface temperatures, at least over much of the Northern Hemisphere and particularly the polar sea, with a marked recession of the sea ice. It was natural to associate the trends as cause and effect and to conclude that man was polluting the air and changing the climate. It was natural that the proviso that no other change should accompany the increase of CO_2 should go relatively unnoticed by many interpreters of Callendar's work.

Water vapor content. There is no question that "if nothing else changed," an increase in CO_2 concentration of the magnitude observed would cause a general, climatically significant increase in near-surface temperatures. But there is also no question that something else would change. At this time the nature and magnitude of this change cannot be decided, but there are certain obvious possibilities. The water content of the atmosphere would increase because of increased evaporation from the oceans. If this were a change in water vapor content alone, two things would happen. First, more solar radiation would be absorbed before it reached the ground, tending to reduce the buildup in near-surface temperatures. Second, the greenhouse effect would be enhanced, the increased

dependent, being referred to a common standard gas mixture, but there is sufficient other evidence to give them credence.

CO_2 reservoirs. The annual increase in total atmospheric content of CO_2 varies between about one-third and two-thirds of the yearly output from the burning of fossil fuels. The fate of the remainder of man's output of CO_2 is not understood in detail. There is a potential reservoir in the biosphere since the total amount of living and decaying matter may be increasing. However, the average time of temporary biomass storage is uncertain. The current "permanent" removal by fossilization is clearly negligible compared to destruction of the fossil store. Solution of CO_2 in the oceans is another reservoir, depending in the short term on near-surface temperatures and not readily estimated. The long-term fixation of CO_2 in the ocean as carbonate rocks by biological processes and by reaction with silicate minerals is very slow, taking as long a time as major glacial and climatic

H₂O content behaving in the same way as the increased CO_2 content. This would tend to augment the increase in near-surface temperatures. Examination of the net effect of these opposing trends has proved a tractable problem: There can be no doubt that the net effect of an increase of water vapor content (alone) would be to reinforce the tendency for an increase in surface temperature. This is a destabilizing condition, a positive feedback. But it is unlikely that the water vapor content of the atmosphere would increase without a corresponding increase in cloud, either in amount or depth, since an increase in evaporation implies a corresponding increase in precipitation. Either form of cloud increase would lead to an increase in reflection of solar radiation—an increase in the global albedo. Reduced absorption of solar radiation means a reduced mean temperature of the planet, and almost certainly reduced near-surface temperatures. This chain of events forms a negative feedback, a climate-stabilizing reaction.

With computational resources that are available at present, detailed investigation of the outcome is not a tractable problem. Since it is not known quantitatively what change in the amount, nature, and global distribution of cloud would accompany a global increase in atmospheric CO_2 content, one cannot specify the ultimate climatic effect of such an increase. One can be confident, however, that the initial change would be in the direction of increased surface temperature.

Atmospheric particles. As Fig. 2 shows, the steady rise in temperature which was a feature of world weather early this century appears to have ended by about 1940. The arrangements for collecting and processing of world weather records at that time were slow and were disrupted by World War II, so that it was not until the mid-1950s that the reversal of the temperature trend was confirmed.

Those who had attributed the earlier temperature increase to increasing atmospheric CO_2 content sought a reason, and found one possibility in a postulated increase in the particle content of the atmosphere associated with the rapid increase in human populations. The argument was that atmospheric particles scatter solar radiation to space: More particles would scatter more radiation, less solar energy would be absorbed by the planet (the planetary albedo would increase), and the mean planetary temperature would fall. A probable, but not inevitable, consequence would be a drop in surface temperature. These speculations called for three lines of investigation. Is the particle load of the atmosphere increasing? Are the optical properties of the particulate products of combustion such that an increase in their number would produce an increase in the Earth's albedo? How would an increase in albedo affect surface temperatures? *See* ATMOSPHERIC POLLUTION.

Particle load. The first question has proved surprisingly difficult to answer. There is no doubt that, at least until a few years ago, particle concentration in the air near major cities was increasing as population and the energy conversion per head of population increased. Air pollution control technology and legislation appear to have abated the rate of increase, but reversals of the trend of particle production can be expected only in areas pre-

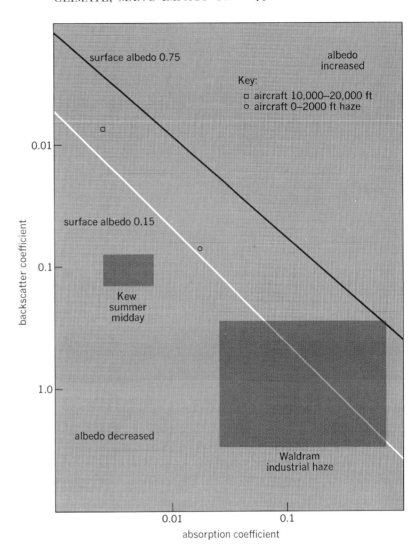

Fig. 6. Effect of atmospheric particles on the global albedo. Particles are characterized by their absorption and backscattering coefficients. They reduce or increase the global albedo according to the position of their characteristic point relative to the appropriate surface albedo line. Data are due to M. Atwater and G. D. Robinson. (*From G. D. Robinson, Particles in the atmosphere and the global heat budget, in Air and Water Pollution: Proceedings of the Summer Workshop, University of Colorado, Aug. 3– 15, 1970, Colorado Associated University Press, 1972*)

viously subject to intensive industrial pollution. Generally, increased consumption will negate the effects of improved control. However, there is no clear evidence of increase on a continental, let alone a global, scale of the number of particles in the atmosphere.

Evidence of long-term increase in the scattering of solar radiation downward to the Earth's surface was found at the two sites in the world where the consistency of standardization of records over a long period could not be questioned, but one site, Washington, D.C., is at the heart of an unrepresentative increase in energy conversion and the other, Davos, Switzerland, is within 100– 200 mi of several such sites. In surface air over the North Atlantic there is some evidence of an increase in the number of certain very small particles produced, among other ways, by combustion. But similar measurements show no such increase over the Central Pacific.

Furthermore, researchers have gradually realized that the proportion of the atmospheric particle load attributable to man's activities is small. The extreme limits of published estimates of this proportion are 5 and 45%; most estimates are less than 10%. It is also clear that industrially produced particles are only a fraction of the total of man-made particles. Agricultural malpractice appears to be responsible for many more particles than industry. In fact, much of the uncertainty in estimating man's contribution to the particle load stems from the near-impossibility of distinguishing between natural and man-induced forest fires and natural and man-induced wind erosion. The table summarizes what we know about the source of the particle load of the atmosphere.

Optical properties. Whether or not the present level of human activity has an appreciable impact on the particle load, it is conceivable that future developments may have an impact. The optical properties of particles, therefore, should be investigated. For example, the argument that since particles scatter sunlight, they must increase the global albedo is not necessarily true. The surface below the particles plays a key role. White particles over a black surface clearly increase the albedo, but black particles over a white surface decrease it. The effects of brown over brown, or gray over gray, or brown over green—the real-life cases—are not so easy to predict. Similarly, the exact solution for real particles of mixed composition and and irregular shape over real inhomogeneous surfaces with topographical irregularities is not possible.

Various approximations are possible, however. Figure 6 shows some of the results of an investigation using a relatively simple approach. In this figure, a particle is characterized by its absorption and backscatter coefficients (a numerical measure of the proportion of incident radiation scattered upward to space). If the characteristic point for a particle is above and to the right of the surface

Estimates of particles smaller than 20-μm radius emitted into, or formed in, the atmosphere

Source	Amount, 10^6 metric tons/year
Natural	
Soil and rock debris*	100–500
Forest fires and slash-burning debris*	3–150
Sea salt	300
Volcanic debris	25–150
Gaseous emissions	
Sulfate from H_2S	130–200
Ammonium salts from NH_3	80–270
Nitrate from NO_2	60–430
Hydrocarbons from plant exudations	75–200
Subtotal	773–2200
Man-made	
Direct emissions	10–90
Gaseous emissions	
Sulfate from SO_2	130–200
Nitrate from NO_2	30–35
Hydrocarbons	15–90
Subtotal	185–415
Total	958–2615

*Includes unknown amounts of indirect man-made contributions.
SOURCE: Inadvertent Climate Modification, M.I.T. Press, 1971.

albedo line, the system albedo is increased. The surface albedo lines in Fig. 6 correspond to total solar radiation and new snow or thick cloud (0.75) and thickly settled areas with mixed agriculture and industry (0.15). The characteristic points plotted refer to measurements made from aircraft using total solar or visible radiation. The diagram contains most of the observational material available in 1970 and illustrates both the paucity of data (though the amount should increase rapidly within the very near future) and the probability that the atmospheric particle load has little direct effect on the Earth's albedo. It is just possible that hygroscopic particles originating in combustion may have an indirect effect on albedo by modifying the nature of cloud. A cloud forming in air polluted by combustion nuclei tends to have a relatively large number of relatively small particles, whereas a cloud forming in cleaner air but in otherwise similar circumstances concentrates the same amount of water on fewer nuclei. The "polluted" cloud is the more efficient light scatterer with the higher albedo. Considering all possibilities and the few available measurements, it seems likely that an increase in atmospheric particle load would lead to an increased albedo, but that the effect might not be large on a global scale. Confirmation by measurement is certainly required.

CLIMATE AND SURFACE CHANGES

There is evidence that, within the present century, man's industrial activities have begun detectably to affect significant properties of the atmosphere. The effect of man's search for, and production of, food is much more evident, and has been clear throughout his history. The significant properties are the albedo, the ability to retain and release water, and the associated ability to store heat. Together these properties dictate the amount of energy returned to the atmosphere and the partition of energy between latent heat, which can be transported and released elsewhere, and sensible heat. Major differences in albedo exist between vegetated, bare, and ice- or snow-covered surfaces. The type of vegetation, except when associated with snow cover, has a relatively small effect on albedo, but a greater effect on water retention, particularly in the case of cultivated and harvested crops. Man has changed the type of vegetation over considerable areas of the globe, clearly without globally catastrophic consequences (though he has been responsible for local catastrophes). But there is now in sight an increase of population in the tropics of such magnitude that climate changes following surface changes connected with tropical agriculture seem at least as likely to occur as any associated with industrial development.

Agriculture. Man is continually changing the nature of the Earth's surface, and there is little doubt that he has already changed the climate of certain regions—for the worse from his own point of view. These known deteriorations have been the consequence of destruction of forests, mainly to create pastures and cultivate food crops but often to exploit the timber crop. The surface that replaces the forest has a higher albedo and less capacity for retaining water and returning it locally to the atmosphere by evaporation.

Overuse of cleared land makes it unsuitable for further agricultural use and accelerates the spread of cleared areas. Particularly in low-rainfall areas, "dust-bowl" conditions can develop. The effect of the dust is to stabilize the lower atmosphere and reduce the incidence of convective rainfall. This process is believed to have increased the aridity of much of the area from the southern Mediterranean shores to Afghanistan. It appears to be operative today in semiarid areas of India and Pakistan. There, however, it has been demonstrated that, at least locally, the desert-producing process is not yet irreversible and that considerably increased vegetative cover quickly follows the exclusion of grazing animals.

The situation is somewhat different in areas where the tropical rainforest is under attack by man. Here the principal destructive process is that of "slash and burn" agriculture—relatively small areas of jungle are cut, the vegetation burnt, and the clearing exploited for rather primitive agriculture. When in a few years the soil becomes uncultivable and infertile, the exploiters move on. The forest does not recover, since the soil has become as unsuitable for forest growth as for agriculture. The surface albedo has changed and the water-retention properties have changed. Figures 7 and 8, pictures relayed from the geosynchronous Advanced Technology Satellite *ATS 3*, show the intense diurnal activity, which accounts for a considerable portion of the atmosphere's turnover of energy, over the Amazon basin. Substantial change in the albedo and evaporation from the tropical land areas must be expected to affect the world's climate. As a further complication, the fires of the "slash and burn" cycle may be a major source of the man-made particle load of the atmosphere, comparable to the direct emissions from industrial sources.

Sea ice. Arctic sea ice has assumed a prominent role in much recent speculation on climatic change because most of the simplified empirical models of climate suggest that the extent of sea ice is extremely sensitive to changes in global temperature. The underlying cause of this sensitivity is the high albedo of ice. In the formulation of the models, an increase in sea-ice cover increases the global albedo, with a consequent further reduction in global temperature and increase in ice. Decrease in sea ice decreases the albedo, increases the solar energy absorbed, and accelerates the decrease in ice cover. There is a "positive feedback" instability with no constraint in the model formulation to counteract it. However, the historical record suggests that some constraints which are not modeled do exist—there have been quite considerable trends in the area of the ice, all ultimately reversed—but the balance of instability and constraint may be delicate. Man may disturb this balance by initiating changes of unprecedented magnitude or speed.

Apart from the possible global temperature changes resulting from unintentional modification of the atmosphere, there have been some projects proposed which could directly or indirectly change the area of the ice in a calculated effort to modify the climate and navigability of arctic waters. The potential dangers of such schemes are now realized.

Fig. 7. Satellite photo taken from the Advanced Technology Satellite *ATS 3* on Dec. 30, 1971, at 13:04 hr. (*NASA photo; Space Science and Engineering Center, The University of Wisconsin*)

Fig. 8. Satellite photo taken from the Advanced Technology Satellite *ATS 3* on Dec. 30, 1971, at 18:01 hr, 5 hr later than Fig. 7. (*NASA photo; Space Science and Engineering Center, The University of Wisconsin*)

It is generally conceded that implementation of any such projects must await a better understanding of their less immediate consequences. However, there is still occasional discussion of projects whose less immediate consequences could be a major disturbance of the arctic sea ice. One such is diversion of the waters of rivers draining into the Arctic Basin—the Ob, Yenisei, and Mackenzie. Such diversion would change the salinity of arctic waters, a change which would in turn affect the thickness and extent of the sea ice and set in train a sequence of climatic reactions perhaps trivial, perhaps catastrophic, but certainly—with present resources and knowledge—incalculable. Whatever the value to the human race of availability of the Mackenzie basin waters in Southern California, and it is debatable, the theory of climatic change in its present state suggests that the interests of more than the inhabitants of the two watersheds might be affected.

Heat release. Direct release of heat by man's activities is very small indeed compared with that supplied by the Sun. The United Nations estimate of world power consumption in 1967 was about 5×10^9 kW, against an average solar supply (absorbed power) of more than 10^{14} kW. There is little prospect of man's changing the global climate by directly heating the environment until his energy consumption attains more than 100 times its present level. There are those who believe that this level will be reached in the future, but at least we have a little time to consider the consequences.

Nevertheless, this direct heating—often accomplished by release of hot water or water vapor—is clearly changing the local climate of some cities. It is conceivable that the future growth and aggregation of cities in certain areas might produce a heat source strategically placed to trigger major climatic change. One such location is the northeastern coast of the United States, where in winter there is a strong land-to-sea temperature gradient which breeds and steers traveling storms and which might be modified by intense energy conversion in the coastal strip.

SUMMARY

Man has changed the climate of his cities. They are warmer than their surroundings, often by up to 10°C at night and in calm winter weather, and 1 or 2°C at other times. The amount of solar radiation reaching the ground in cities is lower than that in surrounding areas, sometimes by as much as 10%. There is a little, not yet conclusive, evidence of an increase, or at least a redistribution, of summer rainfall near a few large cities. Subarctic cities have their own near-permanent blanket of ice fog in winter.

While no evidence of man-induced change in global climate has been found, scientists are unable to be sure that it has not occurred, and have reason to believe that it could occur in the future. Man's energy conversion is at present negligible in comparison with the supply of solar energy, and any influence he may have on global climate must be by way of the triggering of instabilities. Two such "positive-feedback" mechanisms have been identified—one the greenhouse effect, a surface temperature increase, initiated by added carbon dioxide and accelerated by water vapor added as the surface temperature increases; the second, a change in ice or snow cover tending to change the amount of solar radiation absorbed by the Earth in the direction calculated to augment the rate of change of ice cover. During the first 30–40 years of this century, the CO_2 content of the atmosphere increased. The temperature also increased and the ice cover diminished. These tendencies have been reversed, in spite of the continuing increase in CO_2.

Scientists, therefore, seek a constraining or negative-feedback mechanism. The possible effect of atmospheric particles added by man has not yet been confirmed—neither the fact of a significant increase nor the direction of the temperature changes which would follow it has been estab-

lished. There is a possible "natural" constraint in the increase of condensed water expected in the atmosphere if surface temperatures rise and evaporation increases. This extra cloud would increase the Earth's albedo, with a corresponding tendency to reduced temperatures. Man's addition of particles to the atmosphere might affect the nature of cloud in a direction reinforcing this constraint.

There is no doubt that in the natural state of the Earth-biosphere-atmosphere system, constraints exist as well as instabilities. The Earth has never been completely covered by ice; instead ice has formed and spread within limits on a previously ice-free globe. The composition of the atmosphere has varied considerably and enormous amounts of CO_2 have passed through it, to be fixed temporarily in fossil fuels and permanently in carbonate rocks; yet surface conditions have always permitted life, once it appeared.

But although scientists may conclude that there are constraints on climatic instabilities, little is known about the time scale on which they may operate. It is possible that a man-induced perturbation may operate so rapidly that the unknown constraint is unable to counteract it. Man should not be content with the argument that since life appears to have been cared for, or to have cared for itself, on Earth for 1,000,000,000 years, it is likely to continue to do so. The life of the future may not be the kind of life existing now. The situation should be watched carefully, and until there is a better understanding of the workings of the environment, excessive interference with it by man should be avoided. At the same time it should be realized that the environment cannot be preserved entirely without change. No living thing has ever done that.

[G. D. ROBINSON]

Climatology

That branch of meteorology concerned with climate, that is, with the mean physical state of the atmosphere together with its statistical variations in both space and time, as reflected in the totality of weather behavior over a period of many years. Climatology encompasses not only the description of climate but also the physical origins and the wide-ranging practical consequences of climate and of climatic change. Thus it impinges on a wide range of other sciences, including solar system astronomy, oceanography, geography, geology and geophysics, biology and medicine, agriculture, engineering, economics, social and political science, and mathematical statistics.

Like meteorology, climatology is conveniently resolved into subdisciplines in a way that recognizes a hierarchy of geographical scales of climatic phenomena and their governing physics. Macroclimatology refers to the largest (planetary) scale of regimes and phenomena; regional climatology to the scale of continents and subcontinental areas; mesoclimatology to the scale of individual physiographic features such as a mountain, lake, or urban area; and microclimatology to the smallest scale, for example, a house lot or the habitat of an insect. Climatology is also resolved in another way that distinguishes between the theoretical, descriptive, and applied aspects of the science: phys-

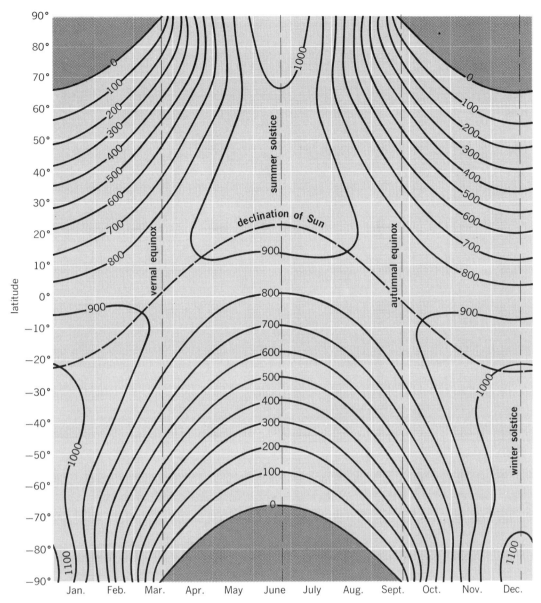

Fig. 1. Total daily solar radiation for Earth without atmosphere. Solid curves show total radiation per day, in g-cal/cm², on horizontal surface. (*After J. B. Leighly, as given by R. J. List, Smithsonian Meteorological Tables*)

ical and dynamic climatology, concerned with the governing physical laws; descriptive and synoptic climatology, concerned with comparisons of climatic norms and anomalies, respectively, as concurrently observed in different places; and applied climatology, concerned with the practical utilization of climatological data in engineering design, operations strategy, and activity planning. Other important subdisciplines are climatography, concerned with the comprehensive documentation of climate by means of data summaries, maps, and atlases; and statistical climatology, concerned mostly with the estimation of climatological expectancies, risks of extreme events, and probability distributions of climatic variables (including joint distributions of combinations of variables) as needed to solve problems in applied climatology.

Origins and pattern of global climate. The climate of the Earth as a whole, including its geographical and seasonal variations, is fundamentally prescribed by the disposition of solar radiant energy that is intercepted by the atmosphere. This disposition depends in part on astronomical factors and in part on terrestrial factors, such as atmospheric composition, the distribution of land and sea, and the physiographic nature of the land. At the mean Earth-Sun distance, solar energy arrives in the vicinity of the Earth at the rate of approximately 1.95 g-cal/(cm²)(min), as measured outside the atmosphere perpendicular to the incident rays. This value, known as the solar constant, has been measured to an accuracy of about ±1% and is believed to vary by not more than a fraction of 1% from year to year. When reckoned per unit horizontal area, however, incoming solar radiation (or insolation) is intercepted at different rates over

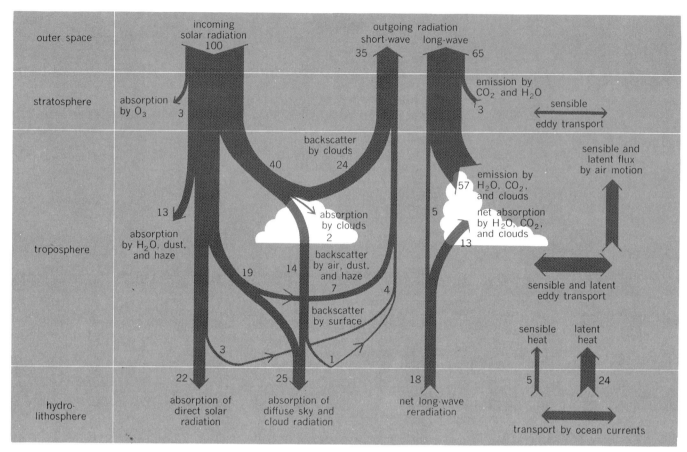

Fig. 2. Mean annual heat balance of whole Earth, showing magnitudes of principal items in percent of total solar radiation arriving at top of atmosphere. (*Based mostly on estimates for Northern Hemisphere by M. I.* *Budyko and K. Y. Kondratiev, from J. M. Mitchell, Jr., Theoretical Paleoclimatology, in H. E. Wright, Jr., and David G. Frey, eds., The Quarternary of the United States, Princeton University Press, 1965*)

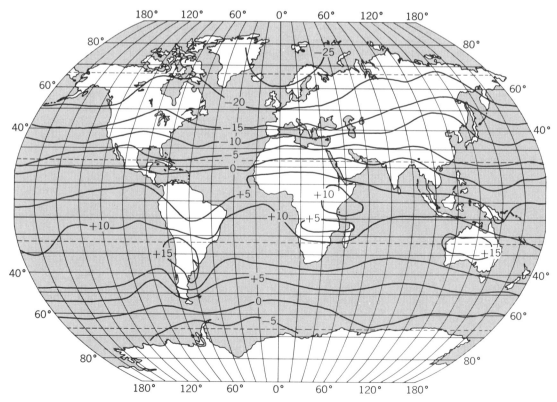

Fig. 3. Net radiation over Earth, January. Values in kg-cal/(cm²) (month). Regions of net surplus and deficit of radiation energy are shown. (*Modified after G. C. Simpson; base map copyright Denoyer-Geppert Co., Chicago*)

different parts of the Earth. The rotation of the Earth causes this rate to change locally with the time of day. In addition, the shape of the Earth, together with the seasonal changes in orbital distance and axial tilt of the Earth relative to the Sun, causes solar radiation (per unit horizontal area and per unit time) to vary with latitude and with the time of year (Fig. 1).

Terrestrial heat balance. The disposition of solar energy after it penetrates the atmosphere is a complex process that locally depends on the solar elevation angle and on various properties of both the atmosphere and its underlying surface. Averaged over the whole Earth and over all seasons of the year, the disposition is as shown in Fig. 2. The figure represents the planetary average heat balance. This term is appropriate because in the observed absence of large changes of atmospheric temperature from year to year the income and outgo of heat energy through the Earth's surface, within the atmosphere, and through the top of the atmosphere must balance closely. In the annual mean for the whole planet, only about half of the solar energy incident on the top of the atmosphere penetrates as far as the Earth's surface. Of this amount, more than 90% is absorbed in the surface, which is used in about equal shares to heat the surface material and to evaporate water, primarily from the oceans. Approximately 35% of the incident solar energy is returned unused to space, mostly by backscatter from clouds; this corresponds to the planetary albedo. The remainder of the incident solar energy (18%) is absorbed on the way down through the atmosphere by water vapor, ozone, dust, haze, and clouds. This provides a direct source of heat to the atmosphere.

More than twice as much heat is added indirectly to the atmosphere by way of the surface. Nearly half of this indirect heating is supplied by a net upward transfer of thermal (infrared) radiation energy that is constantly being exchanged between the surface and atmospheric water vapor, carbon dioxide, and clouds. The remainder consists mainly of the latent heat content of water evaporated from the oceans, which is made available to the atmosphere at the point of condensation as clouds and precipitation. Finally, balance is achieved by a flow of thermal (infrared) radiation energy back to space (65% of the incident solar energy), nearly all of it by emission from water vapor, carbon dioxide, and the tops of clouds in the atmosphere. Less than one-tenth of this thermal radiation loss to space originates at the Earth's surface because the atmosphere is almost completely opaque to infrared radiation except in relatively narrow-wavelength bands (notably between 8 and 12 μ).

The fact that the atmosphere intercepts most of the thermal radiation from the Earth's surface, which in a transparent atmosphere would flow unimpeded into space, is very important to climate because it leads to the maintenance of much higher surface temperatures than those otherwise possible. This warming (the greenhouse effect) is largely attributable to the presence of water vapor and carbon dioxide in air. The magnitude of the warming effect importantly depends on the variable concentration of water vapor, and thus it changes with latitude, time of year, and atmospheric conditions generally. *See* GREENHOUSE EFFECT, TERRESTRIAL.

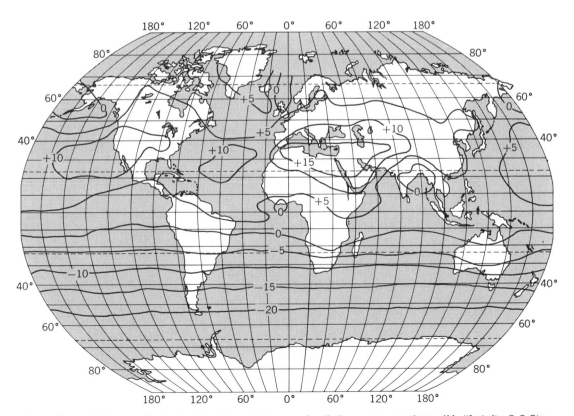

Fig. 4. Net radiation over the Earth, July. Values in kg-cal/(cm²) (month). Regions of net surplus and net deficit of radiation energy are shown. (*Modified after G. C. Simpson; base map copyright Denoyer-Geppert Co., Chicago*)

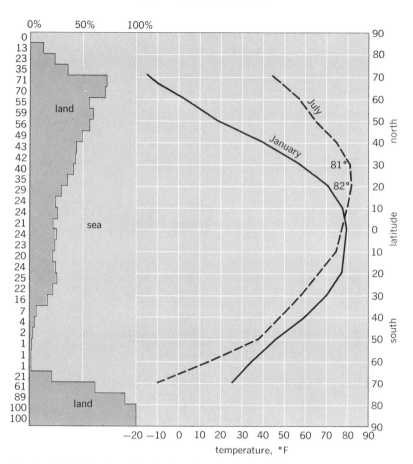

Fig. 5.   The distribution of land and sea by 5° latitude zones and the mean temperature by latitude circles between 70°N and 70°S for January and July. (*Land-sea percentages after E. Kossinna; temperatures after J. Hann*)

Role of atmospheric circulation. In regions of intense solar radiation, primarily in the tropics, there is a net surplus of radiation energy. Similarly in regions of relatively little solar radiation, primarily in the polar regions, there is a net deficit of radiation energy (Figs. 3 and 4). These radiative imbalances lead to differential heating and cooling of the atmosphere, and thereby generate a large-scale circulation of air that transports heat and moisture from areas of surplus to areas of deficiency, generally into higher latitudes. In this way global air currents (along with ocean currents) importantly modify the zonation of world climate otherwise dictated by the astronomical factors. The characteristic unsteadiness of these air currents causes the familiar changeability of weather whose statistics are another important aspect of climate. The pattern of atmospheric circulation, as reflected at the Earth's surface by differences of atmospheric pressure and their associated winds, is discussed in fuller detail below. *See* ATMOSPHERIC GENERAL CIRCULATION.

Land- and sea-surface influences. The climatic significance of the distribution of land and sea lies in the contrast between these two major kinds of surfaces regarding heating and cooling. Except in regions of snow cover or inland water bodies, the lands are heated to higher temperatures than are the seas, both diurnally during the daytime and seasonally during the summer. Obversely, during the nighttime and during the winter the lands are cooled to lower temperatures than are the seas. As a direct result, both diurnal and annual temperature ranges are greater over the land than over the sea. *See* OCEAN-ATMOSPHERE RELATIONS.

The distribution of land and sea by 5° latitude

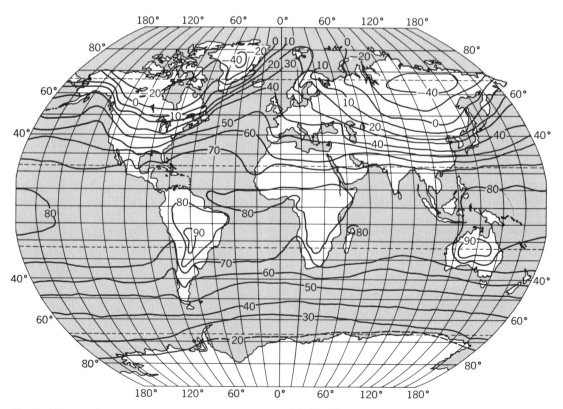

Fig. 6.   Mean sea-level temperatures, °F, January. Note location of cold poles in interior of great landmasses.

(*Modified from Sir Napier Shaw et al.; base map copyright Denoyer-Geppert Co., Chicago*)

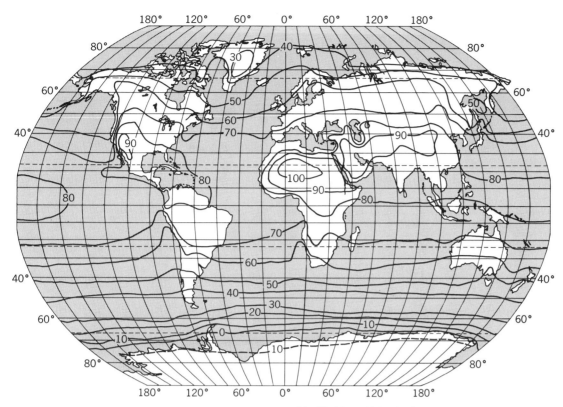

Fig. 7. Mean sea-level temperatures, °F, July. Note displacement of warmest zone northward from Equator. (*Modified from Sir Napier Shaw et al.; base map copyright Denoyer-Geppert Co., Chicago*)

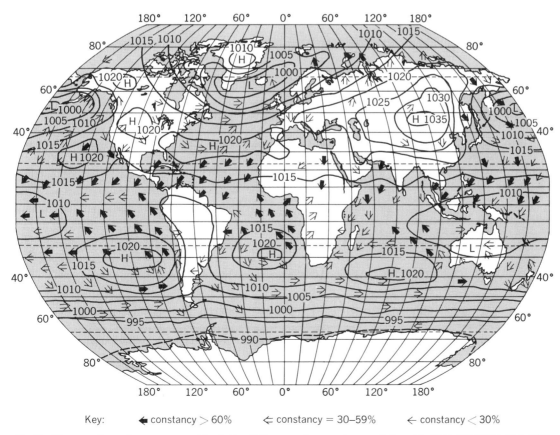

Key: ◄ constancy > 60%   ⇐ constancy = 30–59%   ← constancy < 30%

Fig. 8. Mean surface air pressure and surface winds, January. Pressures are in millibars; wind constancy is indicated by type of arrow. Wind directions are shown to nearest eight compass points; E-W directions true with reference to parallels, and other directions true with reference to meridians. (*Modified from McDonald et al.; base map copyright Denoyer-Geppert Co., Chicago*)

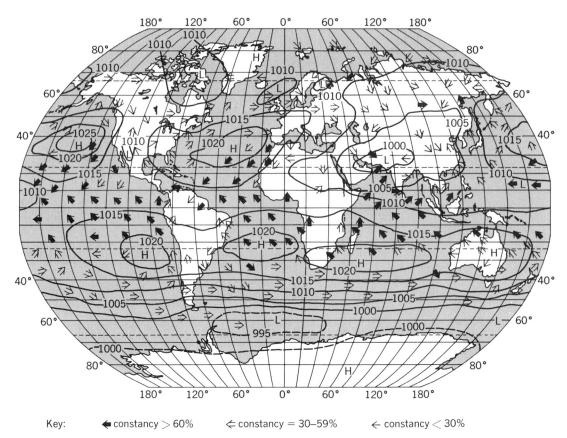

Key: ◄ constancy > 60%   ⇐ constancy = 30–59%   ← constancy < 30%

Fig. 9. Mean surface air pressure and surface winds, July. Pressures are in millibars; wind constancy indicated by type of arrow. Wind directions are shown to nearest eight compass points; E-W directions true with reference to parallels, and other directions true with reference to meridians. (*Modified from McDonald et al.; base map copyright Denoyer-Geppert Co., Chicago*)

zones is shown in Fig. 5. The importance of this factor is evident from the companion curves, showing mean temperatures by latitude circles between 70°N and 70°S for January and for July. Note, for example, that the January temperature is 19°F at 50°N, where land occupies more than half the latitude circle; and it is 38° in the corresponding month of July at 50°S, where there is no land.

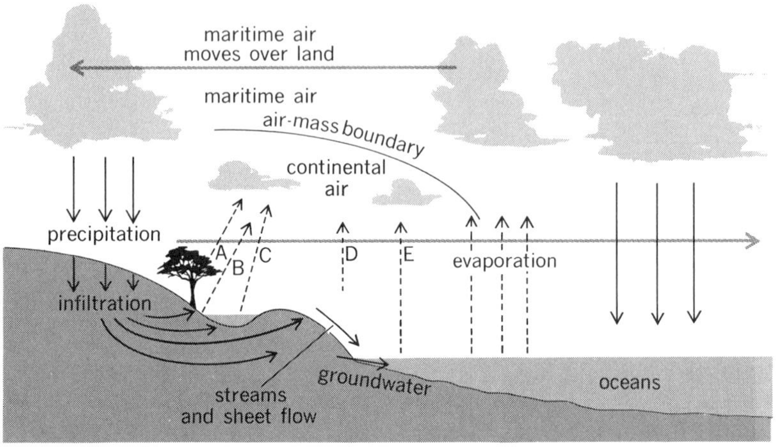

Land-sea distribution and the distribution of radiation are two of the three major factors that determine the broad features of the distribution of temperature. The third factor is the circulation of the atmosphere and of the oceans. In the mean, both circulations transfer energy from lower to higher latitudes. Of the transport, 65% is by the atmospheric circulation, which carries energy both in the form of sensible heat and latent heat (as water vapor that will give up heat upon condensing). The high-latitude regions of great net radiation loss (Figs. 3 and 4) are in large part fed by energy derived from latent heat. Of the transport, another 35% is accomplished through the circulation of ocean waters, with warm water moving poleward along the western sides of the oceans and cool water moving equatorward along the eastern sides. This circulation is reflected in the mean temperature maps for January and July (Figs. 6 and 7), as in the northward bulge of the isotherms in the Iceland-Spitsbergen-Norway area. Other notable features of these maps are the location of the cold poles in the interior of the great landmasses and the displacement of the warmest zone northward from the Equator, a displacement associated with the great landmass of Africa in latitudes 10–30°N. *See* OCEAN CURRENTS; WIND.

Relations with air pressure and wind. The global distribution of temperature is related to the distribution of surface air pressure and of wind (Figs. 8

Fig. 10. Diagram of hydrologic cycle. Water returns chiefly to dry continental air through transpiration (A), evaporation from soil (B), lakes and ponds (C), and streams (D). Continental air moves over ocean to become more moist (E) with conversion to maritime air with precipitation over the oceans. (*Adapted from B. Holzman*)

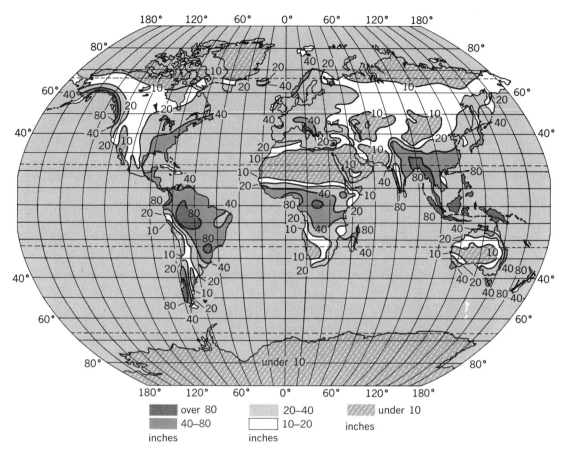

Fig. 11. Mean annual precipitation on major land areas. (*Base map copyright Denoyer-Geppert Co., Chicago*)

and 9). These relationships are neither simple nor direct, but there tends to be an inverse relationship between mean temperature and mean surface pressure over the continents, with the hot desert areas sustaining thermal lows and the cold polar areas sustaining thermal highs. The mean low- and high-pressure areas over the oceans are quite different. In the first instance they represent the summation of moving, dynamic lows (extratropical cyclones), which are particularly common synoptic features in the Aleutian and Icelandic areas, as well as in the waters north of Antarctica. In the second instance they represent the summation of subtropical high cells, which are persistent synoptic circulation features over the subtropical oceans, even though they vary continuously in size, intensity, and location. *See* AIR PRESSURE: METEOROLOGY.

The mean pressure patterns are approximately matched by the patterns of dominant windflow. The most constant winds are those of the trades, which lie on the equatorward side of the subtropical oceanic highs, and those of the monsoon circulation in the Asiatic-Australian area. In this great monsoon circulation, winds blow outward from Asia and inward onto Australia during the Northern Hemisphere winter, with a reversal of the circulation during the summer. Other major wind flow regions are those of the westerlies, to poleward of the oceanic highs, and of the polar easterlies, off Antarctica and Greenland. Neither the westerlies nor the polar easterlies are as constant as the trades or the major monsoon winds because both occur in areas that are frequently the scene

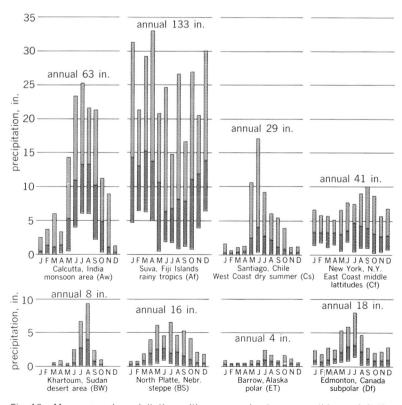

Fig. 12. Mean annual precipitation, with mean and extreme monthly precipitation bar graphs, at selected stations. Top of each bar shows extreme high value; bottom of each bar shows extreme low value; mean indicated by horizontal line within each bar. Letter symbols according to Koeppen type of climate. (*Based on 20-year data for 1921–1940, World Weather Records, Smithsonian Miscellaneous Collections*)

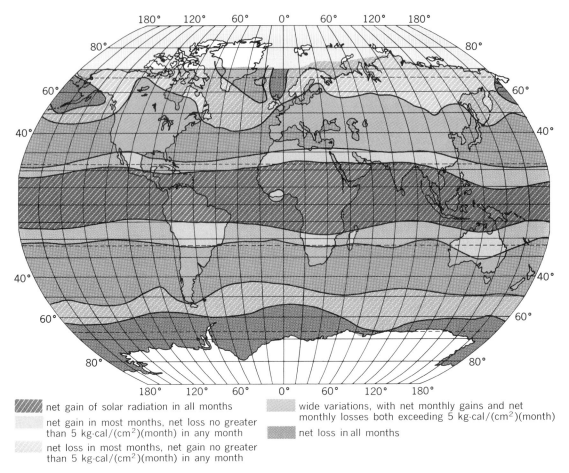

net gain of solar radiation in all months

net gain in most months, net loss no greater than 5 kg·cal/(cm²)(month) in any month

net loss in most months, net gain no greater than 5 kg·cal/(cm²)(month) in any month

wide variations, with net monthly gains and net monthly losses both exceeding 5 kg·cal/(cm²)(month)

net loss in all months

Fig. 13. Major net radiation regions. (*Modified from G. C. Simpson; base map copyright Denoyer-Geppert Co., Chicago*)

of moving cyclones, which may bring winds from any direction.

Marine atmospheric influences. Air that moves long distances across the oceans acquires great quantities of moisture through evaporation from the ocean surface. Because evaporation is greatest where cold dry air moves across much warmer ocean water, the most rapid flux of moisture from

Principal quantitative definitions of major climatic regions*

Symbol	Name of region	Mean temperature, °F	Precipitation Main season	Precipitation Amount, formula for inches†
Af	Tropical rain forest	>64.4‡	All	Driest month >2.4 Annual >40
Aw	Tropical savanna	>64.4‡	Summer	Driest month <2.4 Annual >40
BW	Desert		Winter Summer Even	Annual <0.22t − 7 Annual <0.22t − 1.5 Annual <0.22t − 4.25
BS	Steppe		Winter Summer Even	Annual <0.44t − 14 Annual <0.44t − 3 Annual <0.44t − 8.5
Cf	Humid mesothermal	<64.4 >26.6‡	Even	Annual >0.44t − 8.5
Cw	Humid mesothermal, winter dry	<64.4 >26.6‡	Summer	Annual >0.44t − 3
Cs	Humid mesothermal, summer dry	<64.4 >26.6‡	Winter	Annual >0.44t − 14
Df	Humid microthermal	<26.6‡ >50§	Even	Annual >0.44t − 8.5
Dw	Humid microthermal, winter dry	<26.6‡ >50§	Summer	Annual >0.44t − 3
ET	Tundra	<50 >32§		
EF	Perpetual frost	<32§		

*After W. Koeppen, as outlined in G. T. Trewartha, *An Introduction to Climate*, McGraw-Hill, 3d ed., 1954. Arrangement modified from one by H. E. Landsberg.

†All temperature values are in °F (including t, which is mean value). Precipitation values are in inches.

‡Temperature of coldest month.   §Temperature of warmest month.

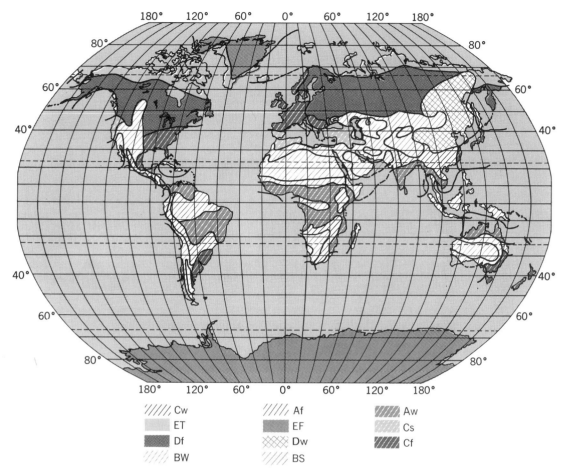

Cw
ET
Df
BW

Af
EF
Dw
BS

Aw
Cs
Cf

Fig. 14. Major climatic regions of the world according to Koeppen classification. Except in the Tibetan area, high mountain effects (including ET or EF) are not shown.   Types are defined in the table. (*Base map copyright Denoyer-Geppert Co.,* Chicago)

sea to air occurs on the western sides of the northern oceans in middle latitudes during the winter season, and over the waters off Antarctica during winter. For example, at 35°N in the western Atlantic an average of about 600 g-cal/(cm²) (day) of thermal energy is expended for evaporation of sea water. In contrast, at latitudes 0–20°N in the eastern Atlantic, the value does not exceed 200 g-cal during any season.

Whatever the variations in detail, the oceans everywhere make a large net contribution of water to the air (except when there is ice). Subsequently, this moisture is carried onto the land by maritime air masses and there it is partly precipitated as rain, snow, hail, dew, or frost. Thereafter, the water returns to the oceans in streams, as sheet wash along the margins of the lands, as underground water, or in moving air (chiefly dry continental air) that acquires water from the land through evaporation and transpiration. This worldwide water cycle is diagrammed in Fig. 10. *See* HYDROLOGY.

Continental precipitation patterns. Figure 11 shows the mean annual precipitation over the continents. Appreciable precipitation occurs only when moist air is forced to rise and this takes place largely through convection, through orographic lifting (the forced lifting of air upslope, as up the flanks of a mountain), or through convergence and the forced ascent of air within an eddy, particular-

ly within an extratropical or tropical cyclone. Hence the precipitation patterns of Fig. 11 can be viewed in terms of the relative frequency with which moist maritime air is present and the frequency with which the air is forced to rise to appreciable heights. The vertical structure of the air with reference to temperature and moisture is also important, because the structure may be stable and so resist vertical movement, or it may be conditionally unstable, so that when forced lifting has produced condensation, the release of latent heat causes the air to rise to still greater heights.

Annual precipitation is very high on the west coasts of the continents in high middle latitudes, in the major monsoon areas, and in equatorial areas. On the west coasts of the continents, the high totals are associated with frequent and prolonged cyclonic precipitation and with the orographic lifting of maritime air. In these areas most of the precipitation is in the winter half-year, when cyclonic storms are most common. In contrast, the monsoon areas have their maximum rains during summer, with an influx of very moist, conditionally unstable air. Here, as in the wet equatorial areas, the rainfall mechanisms are convection, orographic lifting, and convergence in minor eddy systems. However, in the equatorial areas the rainfall is fairly well distributed throughout the year.

The extremely dry areas are characterized by

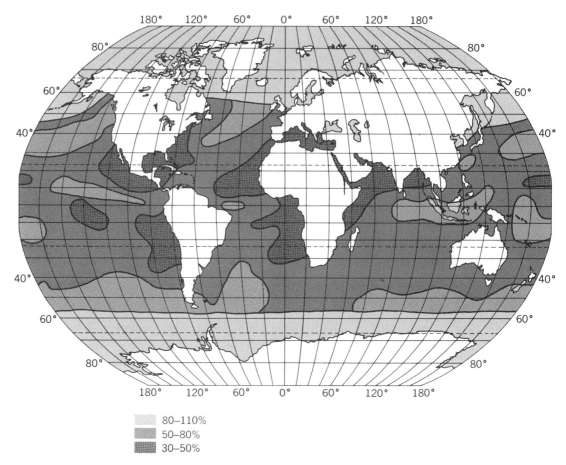

80–110%
50–80%
30–50%

Fig. 15. Climatic regions for fresh-water supply on the oceans, December-February, according to scheme for problem solving. How well climatic conditions meet requirements of rainfall and sunshine, as related to use of solar stills, is indicated in percentages. (*After W. C. Jacobs; base map copyright Denoyer-Geppert Co., Chicago*)

infrequent invasions of moist unstable air and relatively little cyclonic activity. These dry areas are the west coast deserts, such as the Sahara; the desert basins that are shielded from fresh maritime air by high mountains, such as the Tarim Basin; and the polar lands.

In absolute terms, the widest swings in annual precipitation from year to year occur in the wettest areas of the monsoon and the equatorial regions. In these regions it is not uncommon during a 20-year period for the annual rainfall to range from a minimum of 60–80 in. to a maximum of 200–300 in. In percentage terms, the greatest variability is in the driest areas—the deserts and dry polar regions. Here the minimum is less than 2 in. in many localities with maximum annual amounts reaching 15 in. or more in so-called wet years.

The water circumstance of a locality is not only a function of the annual precipitation and its variability but even more of the distribution of precipitation throughout the year. Figure 12 shows the seasonal precipitation regimes at eight selected localities. Also shown are the extreme monthly precipitation amounts during the 20-year period covered by the graphs. The general form of these graphs is in each instance broadly representative of the precipitation regime throughout the regions specified in the diagrams. The variability as shown by the extreme values is typical.

Climatic regions. A wide variety of schemes is available for the definition of climatic regions. These schemes fall into two classes, those intended to bring out one or another aspect of relationships and processes within the atmosphere, and those intended to bring out relationships between areal variations in climatic conditions and corresponding variations in phenomena related to climate.

The first class of schemes is represented by Fig. 13, which shows the major radiation regions of the Earth. The distinction is among regions in which the monthly radiation balance is always positive (in the mean), always negative, never strongly positive, never strongly negative, or highly variable and ranging from strongly positive to strongly negative.

The second class of schemes is represented by Figs. 14 and 15. Figure 14 shows the major climatic regions of the Earth according to W. Koeppen's classification (contrast with Fig. 13). This Koeppen classification was intended to bring out the general coincidence between the distribution of natural vegetative formations and climatic conditions. It also serves, however, as a useful classification for purposes of general description, in terms of quantitative definitions of precipitation and temperature conditions. The major aspects of Koeppen's classification scheme are summarized in the table, whose arrangement is taken from H. E. Landsberg. Koeppen and his followers revised his classification twice and carried it beyond the broad scheme given here. For example, cold and hot deserts were distinguished, isothermal regions (mean annual temperature range less than 9°F) were identified, and subregions where fog was common were designated. Clearly a climatic classification scheme such as this could be extended in any desired way to show areal variations in greater and greater detail.

Quite different from Koeppen's scheme is a class of schemes illustrated by Fig. 15. This is a classification developed to solve a particular problem in applied climatology, that of designing life rafts for use at sea. The regions were defined primarily through consideration of the rainfall probabilities and the effective sunshine, related to the efficiency of portable solar stills. In applied climatology, special classifications of these kinds abound, each designed to serve a practical purpose.

It is feasible and instructive to distinguish minor climatic regions, not on a global scale, but within much smaller areas. Where such regions are defined upon the lands, the topographic factor becomes very important. This is especially true where the area is of the order of a few square miles or less. *See* MICROMETEOROLOGY.

[J. MURRAY MITCHELL, JR.]

Bibliography: H. J. Critchfield, *General Climatology*, 2d ed., 1966; R. Geiger, *The Climate near the Ground*, 2d rev. ed., 1966; B. Haurwitz and J. M. Austin, *Climatology*, 1944; W. G. Kendrew, *The Climates of the Continents*, 4th ed., 1953; H. Landsberg, *Physical Climatology*, 1958; W. D. Sellers, *Physical Climatology*, 1965; H. C. Willett and F. Sanders, *Descriptive Meteorology*, 2d ed., 1959.

Climax community

A particular grouping of natural populations of plants and animals. In any local region, the biota (group of living organisms) becomes sorted into a variety of combinations or communities according to the local variations in environmental factors. The relatively stable combination community which occurs on the most mesophytic sites is termed the climax community of the region. It commonly contains the most shade-tolerant plants of the area and those with the fewest adaptations to extreme conditions of moisture, temperature, and nutrient supply. From a dynamic standpoint, the climax community is in a biological steady state, such that time variations in its average composition occur about a mean. The capture and utilization of energy is presumed to be at a higher level than in any other community of the region in the sense that more kinds of organisms, both plant and animal, participate in the gradual degradation of the original stored material. The climax community is the end product or terminus of a series of successional or developmental changes that occurs on mesic sites. Its practical importance largely stems from this fact. Land managers and others interested in natural landscapes can be guided in their management techniques by a knowledge of potential end points and the conditions that obtain within them. *See* COMMUNITY CLASSIFICATION; SUCCESSION, ECOLOGICAL.

Kinds of climax. The major controlling factor of the climax community just described is the climate, since all other environmental factors are assumed to be medium for the region. This climatic climax is the only one recognized by some investigators, who are thus adherents of the monoclimax hypothesis. Other workers extend the term climax to any community of the region which is reasonably stable, in apparent equilibrium with the environment, and which shows no evidence portending future changes in composition. They are adherents of the polyclimax hypothesis, which holds that soils, topography, fire, animal interactions, or other environmental factors in addition to climate may exert primary control over a biotic community, and that the stable end products of such control are fully worthy of recognition as climaxes. In most cases, the climatic climax is an important factor in elucidating the active controlling agents of nonclimatically determined communities.

Monoclimax. In studying the climatic climax, the greatest amount of information about the nature of the climax community has been gathered in the summer-green temperate forests of the world. The concept may have its greatest meaning in that region, although it probably is important also in the forested tropics. The attributes of climax are less apparent in the major grasslands, savannas, and subarctic conifer forests of the world, and become hazy or nonexistent in deserts and arctic tundras where the idea of medium or mesic environment is difficult to apply.

In the summer-green temperate forest, a rather definite set of microenvironmental conditions exists within the climax stands, for which the herbaceous plants and the seedlings of the trees possess distinct adaptations. Perhaps the most important microclimatic factor is the low internal light intensity during the summer. Light values near the forest floor are usually less than 1% of the intensity of full sunlight, because of the dense leaf cover of the canopy of the dominant trees. Few plants are able to endure such low light conditions, but the few species with the ability (which include the seedlings of the climax trees) necessarily have a very low compensation point. They have also a high chlorophyll content (8–10 μg/mg of dry weight) relative to that in more open communities (2–4 μg/mg of dry weight), and this chlorophyll is present in a chemical complex that makes it sensitive to photodestruction. The environmental supply of moisture within the forest in the summer season is adequate for plant growth and, above all, is uniform, because of reduced air turbulence, reduced evapotranspirational loss, and good internal soil drainage. The plants tend to have a low osmotic pressure in their cell sap (usually less than 10 atm). They exist in a fully distended or turgid condition, but the loss of as little as 3% of their original water content causes them to wilt. *See* PHOTOSYNTHESIS.

Many of the herbaceous species in the climatic climax community are unable to grow under the low light conditions prevailing during the summer. They have developed adaptations which enable them to complete most or all of their life cycle in early spring before the leaves appear on the overhead trees. They do this by very rapid growth supported by extensive food reserves in undergound bulbs, corms, tubers, or fleshy rhizomes, and by the simple, rapid elongation and enlargement of flower buds which were formed during the preceding summer. Their leaves commonly die down and disappear shortly after the tree canopy is complete, and therefore they are called spring ephemerals.

Few of the plants of the climatic climax forest have seeds which are dispersed by mammals, but many species have adaptations for transport of seeds by ants. Plants with wind-pollinated flowers are rare, but insect-pollinated flowers are common. There are few burrowing mammals and hence little mass mechanical disturbance of the soil, but the earthworm populations are high. The latter, plus many other types of soil fauna and large numbers of fleshy fungi, lead to a rapid decay of leaf litter and an extensive incorporation of organic matter into the top layer of the mineral soil, forming a rich mull humus. This incorporation produces excellent water-retaining capacities in the soil and also leads to a relatively high level of available nutrients.

Polyclimax. All of the plant communities of a region must be able to tolerate the full range of climate affecting that region, but they may be influenced by many factors other than climate. In some cases, these other influences may be of a temporary nature, subject to change by activities of the community itself, for example, the improvement of soil through incorporation of organic matter, the increase in atmospheric moisture and decrease in light through the development of a shielding canopy in forests, or the decrease in depth of water in lakes and swamps through biotically influenced sedimentation or increased transpiration. In such instances, the communities present give internal evidence of the changes by obvious instability in species composition. In other cases, however, the nonclimatic agents may be so extreme as to resist biotic change or at least to slow it to a nonsignificant rate. The soil or the topography may be of such a nature that only geomorphological changes associated with base leveling or stream cutting are capable of inducing mesic or medium conditions. These changes are so slow as to confer a high degree of permanence or stability on the communities which are in equilibrium with the extreme conditions. Such communities then represent the polyclimax view and are best discussed under the major factors modifying the environment.

Edaphic modifiers. Soils may deviate markedly from those found under climatic climax communities. Such deviations result from unusual parent materials, as serpentine rocks; from faulty drainage, as in topographic depressions; from nonloamy textures, as in pure sands or clays; or from a number of other causes. Regardless of the reasons for the unusual soils, the plant communities found on them commonly differ greatly from the climatic normal, usually by an increase in the dominance of species adapted to the special conditions. Sometimes these changes may affect the entire gross appearance of the community, as in the replacement of forest by savanna or grassland. In tropical regions with a pronounced wet and dry season, soils developed on geologically mature flatlands commonly exhibit impeded internal drainage due to a subsurface layer of impervious claylike materials or concretions (laterite). These sites are wet, almost inundated, during the rainy season but become depleted of nearly all water in the layer above the impervious pan during the dry season. Such a severe seasonal alternation in moisture conditions is unfavorable for the growth of most trees but is tolerated well by a number of grasses and grassland plants. The community which develops there is a grassland with scattered trees called a savanna, which is an edaphic climax in a climate fully capable of supporting closed tropical forest on normal soils with good drainage. *See* SAVANNA.

The reverse effect is more frequently seen in temperate zones where excessively good drainage, resulting from high sand or gravel content in the soil, leads to xeric or drought modifications in the climax community. If the sands are siliceous, the soils tend to have a reduced nutrient content as well as a reduced water supply. As a result, total production is low and little organic matter exists. Hence the improvement of the soil through nutrient pumping and incorporation of water-holding colloids is slow. Given sufficient time, this soil improvement can eventually make the area suitable for the climatic climax, but in many regions the inherent dryness of the soil leads directly to an increase in severe fires. The more or less permanent community which develops is in equilibrium with both the soil conditions and the fires and is made up of species of xerophytic tendencies which also are fire-resistant. The sand plains and pine barrens of New Jersey, the Great Lakes states, and other regions are typical examples of such edaphic communities. They have a high stability in the presence of repeated fires but show rather quick changes in composition if artificially protected from fire.

Topographic modifiers. The influence on vegetation of such major topographic features as mountains is usually difficult to detect because of accompanying differences in major climatic factors. At a given elevation under a uniform macroclimate, however, minor topographical features like hills, cliffs, rivers, valleys, or undrained depressions may exert controlling influence over the vegetation. Large hills have a significantly different microclimate on their south slopes from that on their north slopes. In the Northern Hemisphere, the south slopes tend to be both hotter and drier than the north slopes. In regions near the boundary of two major biocoenoses, as along the prairie-forest border in central North America, these directional differences often result in grassland communities just below the crest on the south slopes, while the north slopes have forest communities of the prevailing climatic climax. Near the center of the area of any one biocoenosis, topography does not result in such a severe change in physiognomy but only in a change in composition of species. The two slopes can develop to convergence only following geological baseleveling of the hill.

Another major effect of topography is seen in lowland sites, as along streams and lakes or other depressions. Such sites have excessive moisture due to runoff from adjacent uplands, and the biotic communities developing there are composed of species with special adaptations to the higher water supply. Ordinarily, the community is able to do little to change the conditions toward mesic moisture levels; such changes are dependent upon stream entrenchment, alluviation, or sedimentation, all of which are very slow processes. The communities that come into equilibrium with the high moisture conditions, therefore, are very stable, some having persisted for 10,000 years or more, as in the case of certain conifer swamps in northern glaciated regions. Such stable communities are topographic climaxes, which can develop to the climatic climax only if the controlling factor of excess moisture is removed by some external action.

Disclimax. The climatic climax and the edaphic and topographic modifications of it are presumed to represent natural conditions in which the biota have come into equilibrium with the physical environment unaffected by other outside agencies. When disturbances are caused by man, the communities may readjust their composition in response. If a new stability is achieved, as when the disturbance is uniformly and continually applied, the community is called a disturbance climax or a disclimax. Sometimes the same term is applied to a plant community which is clearly influenced by one of its natural animal components, such as the bison on the western shortgrass plains, but such usage is often restricted to the effects or abnormal population levels of the animal, particularly where these levels are themselves the result of man's activities.

Overgrazing. Populations of native herbivorous animals exert an effect on the plant composition of the world's grasslands, but these effects are of a minor and nondisrupting nature. When domesticated animals, especially cattle, sheep, and goats, are introduced to the same grasslands, pronounced changes occur in the community. The reasons are that the animals are confined within an enclosure and hence must use and reuse the same plants, and that the populations per unit area are usually higher than those of the native animals. The plant species change from those with erect stems and aerial growing points to recumbent, prostrate, or rosette plants with their leaves and buds close to the soil surface. There is also an increase in species with spines and thorns or other devices which reduce their palatability. Reductions in stature and growth form also occur in the understory of forests which are pastured. When the size of the animal herd is carefully adjusted to the carrying capacity of the land, as in some of the grazing regions of Europe, a stable or pasture disclimax may result. In many regions, however, the size of the herd is governed mostly by economic reasons, such that animal-plant interactions are varying and often on the side of continual degradation of the vegetation. Under these conditions, the requisite stability of a climax is not present. Such combinations are called disturbance communities, rather than disclimaxes.

More or less similar changes in community composition sometimes occur when a native animal population is allowed to become excessively large, as is the case of deer and rabbits in regions where natural predators have been eliminated.

Lumbering. Another man-made disturbance which is prevalent in the forests of the world is that brought about by lumbering. This varies from total destruction of the original community over large areas by clear-cutting, as in many commercial conifer forests, to the long-term removal of a few trees each year, as by selective harvest of private woodlots. While these activities may permanently change the nature of the community which can later develop on the lumbered areas, they rarely result in a truly stable community and hence are considered to produce disturbance communities rather than disclimaxes.

Fire. Fires of natural origin, especially those set by dry lightning, are a common feature of many large regions of the world and as such are as much a part of the natural environment as rainfall, sunlight, or wind. However, such fires have the same effect as those deliberately or accidentally set by man, and both are commonly considered disturbance factors. In any case, there have been fire-using cultures in all major regions of the world for at least 30,000 years, and hence many biotic communities have had time to come into equilibrium with fire, whether natural or artificial.

The effect of fire varies greatly with the type of vegetation and the general climate concerned. Large areas of the tropical rainforests cannot burn unless they are first cut down. In temperate zones, the climatic climax hardwood forests are similarly fire-resistant and may burn only under unusual conditions of drought. Northern conifer forests also burn primarily during drought, but their greater inflammability permits wider destruction than in the hardwood region. Most of the permanent pine forest areas of the world are fire disclimaxes, in that they are perpetually maintained at the pine stage of succession through recurrent destruction and regeneration. The chaparral of California is another major fire disclimax of similar behavior.

Fire can create a differential competition; for instance, it burns grass more readily than trees. This is seen along the boundary between grassland and forest and in certain other places where fires are exceptionally effective, as on the flatlands of the American Coastal Plain. The forests are reduced to a scattered stand of the most fire-resistant tree species, with an understory of grasses and other fire-resistant herbaceous plants. Nearly all temperate zone savannas and many tropical savannas as well are clear examples of fire disclimax.

In the grasslands proper, fire brings about no important change in biotic composition. On the more humid edges of the grassland, however, fire is necessary to maintain the community in a healthy condition and to prevent it from smothering itself with undecomposed mulch, and the same may be true of all grasslands. Thus prairies may be thought of as climatic climaxes and fire a natural part of the environment. Where there is artificial fire protection, the resulting community is a disclimax.

Fire is of infrequent occurrence on the major deserts and is similarly rare on the arctic tundra, although there have been few detailed studies of the effects in these regions and little can be said of the nature of possible disclimaxes.

Wind, salt spray, and fog. Several other environmental factors usually associated with restricted geographical locations may cause deviations from a climatic climax. In high mountains, unidirectional winds of considerable strength may produce a gnarled and stunted forest at timberline which contains tree species not found elsewhere in the region. These trees are well adapted to resist the shearing and twisting action of the strong winds and may remain alive for long periods with only a few green branches. Along exposed seacoasts, prevailing onshore winds create a similar effect, aided by the salty droplets picked up from the waves. This salt spray climax is widespread along the Atlantic and Gulf coasts and is composed of specialized plants most able to tolerate the harmful effects of the salt.

Persistent occurrence of a fog blanket along coastal areas or on abrupt mountain promontories may induce a fog climax. The redwood forests along the Pacific Coast and the elfin woodland variant of the tropical cloud forests are examples of such control. See ECOLOGY; ENVIRONMENT; PLANT COMMUNITY. [JOHN T. CURTIS/JERRY S. OLSON]

Bibliography: P. Dansereau, Biogeography, 1957; H. P. Hanson and E. D. Churchill, The Plant Community, 1961; H. J. Oosting, Study of Plant Communities, 1956.

Cloud

Suspensions of minute droplets or ice crystals produced by the condensation of water vapor (the ordinary atmospheric cloud). Other clouds, less commonly seen, are composed of smokes or dusts. See ATMOSPHERIC POLLUTION; DUST STORM.

This article presents an introductory outline of cloud formation upon which to base an understanding of cloud classifications. For a more technical consideration of the physical character of atmospheric clouds, including the condensation and precipitation of water vapor see CLOUD PHYSICS.

Rudiments of cloud formation. A grasp of a few physical and meteorological relationships aids in an understanding of clouds and skies. First, if water vapor is cooled sufficiently, it becomes saturated, that is, in equilibrium with a plane surface of liquid water (or ice) at the same temperature. Further cooling in the presence of such a surface causes condensation upon it; in the absence of any surfaces no condensation occurs until a substan-

Fig. 2. Small cumulus. (U.S. Weather Bureau)

tial further cooling provokes condensation upon random large aggregates of water molecules. In the atmosphere, even in the apparent absence of any surfaces, they are in fact always provided by invisible motes upon which the condensation proceeds at barely appreciable cooling beyond the state of saturation. Consequently, when atmospheric water vapor is chilled sufficiently, such motes, or condensation nuclei, swell into minute water droplets and form a visible cloud. The total concentration of liquid in the cloud is controlled by its temperature and the degree of chilling beyond the state in which saturation occurred, and in most clouds approximates to 1 g in 1 m³ of air. The concentration of droplets is controlled by the concentrations and properties of the motes and the speed of the chilling at the beginning of the condensation. In the atmosphere these are such that there are usually about 100,000,000 droplets/m³. Because the cloud water is at first fairly evenly shared among them, these droplets are necessarily of microscopic size, and an important part of the study of clouds is concerned with the ways in which they often become aggregated into drops large enough to fall as rain.

The chilling which produces clouds is almost always associated with the upward movements of air which carry heat from the Earth's surface and restore to the atmosphere that heat lost by radiation into space. These movements are most pronounced in storms, which are accompanied by thick, dense clouds, but also take place on a smaller scale in fair weather, producing scattered clouds or dappled skies. See STORM.

Fig. 1. Cirrus, with trails of slowly falling ice crystals at a high level. (F. Ellerman, U.S. Weather Bureau)

Fig. 3. An overcast of stratus, with some fragments below the hilltops. (U.S. Weather Bureau)

Fig. 4.   View from Mount Wilson, Calif. High above is a veil of cirrostratus, and below is the top of a low-level layer cloud. (*F. Ellerman, U.S. Weather Bureau*)

Fig. 6.   Altocumulus, which occurs at intermediate levels. (*G. A. Lott, U.S. Weather Bureau*)

Rising air cools by several degrees Celsius for each kilometer of ascent, so that even over equatorial regions temperatures below 0°C are encountered a few kilometers above the ground, and clouds of frozen particles prevail at higher levels. Of the abundant motes which facilitate droplet condensation, very few cause direct condensation into ice crystals or stimulate the freezing of droplets, and especially at temperatures near 0°C their numbers may be vanishingly small. Consequently, at these temperatures, clouds of unfrozen droplets are not infrequently encountered (supercooled clouds). In general, however, ice crystals occur in very much smaller concentrations than the droplets of liquid clouds, and may by condensation alone become large enough to fall from their parent cloud. Even small high clouds may produce or become trails of snow crystals, whereas droplet clouds are characteristically compact in appearance with well-defined edges, and produce rain only when dense and well-developed vertically (2 km or more thick).

Classification of clouds. The contrast in cloud forms mentioned above was recognized in the first widely accepted classification, as well as in several succeeding classifications. The first was that of L. Howard (London, 1803), recognizing three fundamental types: the stratiform (layer), cumuliform (heap), and cirriform (fibrous). The first two are indeed fundamental, representing clouds formed respectively in stable and in convectively unstable atmospheres, whereas the clouds of the third type are the ice clouds which are in general higher and more tenuous and less clearly reveal the kind of air

motion which led to their formation. Succeeding classifications continued to be based upon the visual appearance or form of the clouds, differentiating relatively minor features, but later in the 19th century increasing importance was attached to cloud height, because direct measurements of winds above the ground were then very difficult, and it was hoped to obtain wind data on a great scale by combining observations of apparent cloud motion with reasonably accurate estimates of cloud height, based solely on their form.

WMO cloud classification. The World Meteorological Organization (WMO) uses a classification which, with minor modifications, dates from 1894 and represents a choice made at that time from a number of competing classifications. It divides clouds into low-level (base below about 2 km), middle-level (about 2 to 7 km), and high-level (between roughly 7 and 14 km) forms within the middle latitudes. The names of the three basic forms of clouds are used in combination to define 10 main characteristic forms, or "genera."

1. Cirrus are high white clouds with a silken or fibrous appearance (Fig. 1).

2. Cumulus are detached dense clouds which rise in domes or towers from a level low base (Fig. 2).

Fig. 5.   Cirrocumulus, high clouds with a delicate pattern. (*A. A. Lothman, U.S. Weather Bureau*)

Fig. 7.   Altostratus, a middle-level layer cloud. Thick layers of such cloud, with bases extending down to low levels, produce prolonged rain or snow, and are then called nimbostratus. (*C. F. Brooks, U.S. Weather Bureau*)

Fig. 8. Cumulonimbus clouds photographed over the upland adjoining the upper Colorado River valley. Note the rain showers which appear under some of the clouds. (*Lt. B. H. Wyatt, U.S.N., U.S. Weather Bureau*)

3. Stratus are extensive layers or flat patches of low clouds without detail (Fig. 3).

4. Cirrostratus is cirrus so abundant as to fuse into a layer (Fig. 4).

5. Cirrocumulus is formed of high clouds broken into a delicate wavy or dappled pattern (Fig. 5).

6. Stratocumulus is a low-level layer cloud having a dappled, lumpy, or wavy structure. See the foreground of Fig. 4.

7. Altocumulus is similar to stratocumulus but lies at intermediate levels (Fig. 6).

8. Altostratus is a thick, extensive, layer cloud at intermediate levels (Fig. 7).

9. Nimbostratus is a dark, widespread cloud with a low base from which prolonged rain or snow falls.

10. Cumulonimbus is a large cumulus which

Cloud classification based on air motion and associated physical characteristics

Kind of motion	Typical vertical speeds, cm/sec	Kind of cloud	Name	Characteristic dimensions, km		Characteristic precipitation
				Horizontal	Vertical	
Widespread slow ascent, associated with cyclones (stable atmosphere)	10	Thick layers	Cirrus, later becoming: cirrostratus altostratus altocumulus	10^3	1–2	Snow trails
			nimbostratus	10^3	10	Prolonged moderate rain or snow
Convection, due to passage over warm surface (unstable atmosphere)	10^2	Small heap cloud	Cumulus	1	1	None
	10^3	Shower- and thunder-cloud	Cumulo-nimbus	10	10	Intense showers of rain or hail
Irregular stirring causing chilling during passage over cold surface (stable atmosphere)	10	Shallow low layer clouds, fogs	Stratus Stratocumulus	10^2 $<10^3$	<1	None, or slight drizzle or snow

produces a rain or snow shower (Fig. 8).

Classification by air motion. Modern detailed studies of clouds were stimulated by the discovery of cheap methods of seeding supercooled clouds with artificial motes, promoting ice-crystal formation, aimed at stimulating or increasing snowfall. These studies show that the external form of clouds gives only indirect and incomplete clues to the physical properties which determine their evolution. Throughout this evolution the most important properties appear to be the air motion and the size-distribution spectrum of all the cloud particles, including the condensation nuclei. These properties vary significantly with time and position within the cloud, so that cloud studies demand the intensive examination of individual clouds with expensive facilities such as aircraft and radar. The interplay of physical processes in atmospheric clouds is very complicated, and until it is better understood, no satisfactory physical classification will be possible. From a general meteorological point of view a classification can be based upon the kind of air motion associated with the cloud, as shown in the table.   [FRANK H. LUDLAM]

Bibliography: H. R. Byers, *General Meteorology*, 3d ed., 1959; C. E. Koeppe and G. C. DeLong, *Weather and Climate*, 1958; F. H. Ludlam and R. S. Scorer, *Cloud Study*, 1957; R. S. Scorer and H. Wexler, *A Colour Guide to Clouds*, 1964.

Cloud physics

The study of the physical and dynamical processes governing the structure and development of clouds and the release from them of snow, rain, and hail (collectively known as precipitation).

The factors of prime importance are the motion of the air, its water-vapor content, and the numbers and properties of the particles in the air which act as centers of condensation and freezing. Because of the complexity of atmospheric motions and the enormous variability in vapor and particle content of the air, it seems impossible to construct a detailed, general theory of the manner in which clouds and precipitation develop. However, calculations based on the present conception of laws governing the growth and aggregation of cloud particles and on simple models of air motion provide reasonable explanations for the observed formation of precipitation in different kinds of clouds.

Cloud formation. Clouds are formed by the lifting of damp air which cools by expansion under continuously falling pressure. The relative humidity increases until the air approaches saturation. Then condensation occurs (Fig. 1) on some of the wide variety of aerosol particles present; these exist in concentrations ranging from less than 100 particles per cubic centimeter (cm^3) in clean, maritime air to perhaps $10^6/cm^3$ in the highly polluted air of an industrial city. A portion of these particles are hygroscopic and promote condensation at relative humidities below 100%; but for continued condensation leading to the formation of cloud droplets, the air must be slightly supersaturated. Among the highly efficient condensation nuclei are the salt particles produced by the evaporation of sea spray, but it now appears that particles produced by man-made fires and by natural combustion (for example, forest fires) also make a major contribution. Condensation onto the nuclei contin-

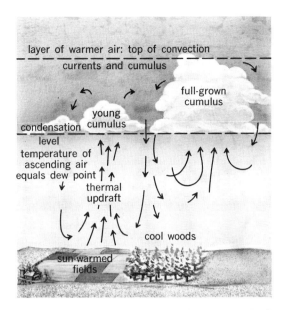

Fig. 1. Conditions leading to birth of a cumulus cloud. (*Based on photodisplay in Willetts Memorial Weather Exhibit, Hayden Planetarium, New York City*)

ues as rapidly as the water vapor is made available by cooling of the air and gives rise to droplets of the order of 0.01 millimeter (mm) in diameter. These droplets, usually present in concentrations of a few hundreds per cubic centimeter, constitute a nonprecipitating water cloud.

Mechanisms of precipitation release. Growing clouds are sustained by upward air currents, which may vary in strength from a few centimeters per second (cm/sec) to several meters (m) per second. Considerable growth of the cloud droplets (with falling speeds of only about 1 cm/sec) is therefore necessary if they are to fall through the cloud, survive evaporation in the unsaturated air beneath, and reach the ground as drizzle or rain. Drizzle drops have radii exceeding 0.1 mm, while the largest raindrops are about 6 mm across and fall at nearly 10 m/sec. The production of a relatively few large particles from a large population of much smaller ones may be achieved in one of two ways.

Coalescence process. Cloud droplets are seldom of uniform size for several reasons. Droplets arise on nuclei of various sizes and grow under slightly different conditions of temperature and supersaturation in different parts of the cloud. Some small drops may remain inside the cloud for longer than others before being carried into the drier air outside.

A droplet appreciably larger than average will fall faster than the smaller ones, and so will collide and fuse (coalesce) with some of those which it overtakes (Fig. 2). Calculations show that, in a deep cloud containing strong upward air currents and high concentrations of liquid water, such a droplet will have a sufficiently long journey among its smaller neighbors to grow to raindrop size. This coalescence mechanism is responsible for the showers that fall in tropical and subtropical regions from clouds whose tops do not reach the 0°C level and therefore cannot contain ice crystals

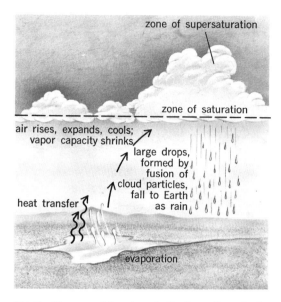

Fig. 2. Diagram of the steps in the formation of rain. (*Based on photodisplay in Willetts Memorial Weather Exhibit, Hayden Planetarium, New York City*)

which are responsible for most precipitation. Radar evidence also suggests that showers in temperate latitudes may sometimes be initiated by the coalescence of waterdrops, although the clouds may later reach to heights at which ice crystals may form in their upper parts.

Initiation of the coalescence mechanism requires the presence of some droplets exceeding 20 microns (μ) in diameter (1 μ = 0.0001 cm). Over the oceans and in adjacent land areas they may well be supplied as droplets of sea spray, but in the interiors of continents, where so-called giant salt particles of marine origin are probably scarce, it may be harder for the coalescence mechanism to begin.

Ice crystal process. The second method of releasing precipitation can operate only if the cloud top reaches elevations where temperatures are below 0°C and the droplets in the upper cloud regions become supercooled. At temperatures below −40°C the droplets freeze automatically or spontaneously; at higher temperatures they can freeze only if they are infected with special, minute particles called ice nuclei. As the temperature falls below 0°C, more and more ice nuclei become active, and ice crystals appear in increasing numbers among the supercooled droplets. But such a mixture of supercooled droplets and ice crystals is unstable. The cloudy air, being usually only slightly supersaturated with water vapor as far as the droplets are concerned, is strongly oversaturated for the ice crystals, which therefore grow more rapidly than the droplets. After several minutes the growing crystals will acquire definite falling speeds, and several of them may become joined together to form a snowflake. In falling into the warmer regions of the cloud, however, this may melt and reach the ground as a raindrop.

Precipitation from layer-cloud systems. The deep, extensive, multilayer-cloud systems, from which precipitation of a usually widespread, persistent character falls, are generally formed in cyclonic depressions (lows) and near fronts. Such cloud systems are associated with feeble upcurrents of only a few centimeters per second, which last for at least several hours. Although the structure of these great raincloud systems, which are being explored by aircraft and radar, is not yet well understood, it appears that they rarely produce rain, as distinct from drizzle, unless their tops are colder than about −12°C. This suggests that ice crystals may be responsible. Such a view is supported by the fact that the radar signals from these clouds usually take a characteristic form which has been clearly identified with the melting of snowflakes.

Production of showers. Precipitation from shower clouds and thunderstorms, whether in the form of raindrops, pellets of soft hail, or true hailstones, is generally of greater intensity and shorter duration than that from layer clouds and is usually composed of larger particles. The clouds themselves are characterized by their large vertical depth, strong vertical air currents, and high concentrations of liquid water, all these factors favoring the rapid growth of precipitation elements by accretion.

In a cloud composed wholly of liquid water, raindrops may grow by coalescence with small droplets. For example, a droplet being carried up from the cloud base would grow as it ascends by sweeping up smaller droplets. When it becomes too heavy to be supported by the vertical upcurrents, the droplet will then fall, continuing to grow by the same process on its downward journey. Finally, if the cloud is sufficiently deep, the droplet will emerge from its base as a raindrop.

In a dense, vigorous cloud several kilometers deep, the drop may attain its limiting stable diameter (about 5 mm) before reaching the cloud base and thus will break up into several large fragments. Each of these may continue to grow and attain breakup size. The number of raindrops may increase so rapidly in this manner that after a few minutes the accumulated mass of water can no longer be supported by the upcurrents and falls out as a heavy shower. The conditions which favor this rapid multiplication of raindrops occur more readily in tropical regions.

In temperate regions, where the 0°C level is much lower in elevation, conditions are more favorable for the ice-crystal mechanism. However, many showers may be initiated by coalescence of waterdrops.

The ice crystals grow initially by sublimation of vapor in much the same way as in layer clouds, but when their diameters exceed about 0.1 mm, growth by collision with supercooled droplets will usually predominate. At low temperatures the impacting droplets tend to freeze individually and quickly to produce pellets of soft hail. The air spaces between the frozen droplets give the ice a relatively low density; the frozen droplets contain large numbers of tiny air bubbles, which give the pellets an opaque, white appearance. However, when the growing pellet traverses a region of relatively high air temperature or high concentration of liquid water or both, the transfer of latent heat of fusion from the hailstone to the air cannot occur sufficiently rapidly to allow all of the deposited water to freeze immediately. There then forms a wet coating of slushy ice, which may later freeze to form a layer of compact, relatively transparent ice. Alternate layers of opaque and clear ice are characteristic of large hailstones, but their formation and detailed structure are determined by many factors such as the number concentration, size and impact velocity of the supercooled cloud droplets, the temperature of the air and hailstone surface, and the size, shape, and aerodynamic behavior of the hailstone. Giant hailstones, up to 10 cm in diameter, which cause enormous damage to crops, buildings, and livestock, most frequently fall not from the large tropical thunderstorms, but from storms in the continental interiors of temperate latitudes. An example is the Nebraska-Wyoming area of the United States, where the organization of larger-scale wind patterns is particularly favorable for the growth of severe storms.

The development of precipitation in convective clouds is accompanied by electrical effects culminating in lightning. The mechanism by which the electric charge dissipated in lightning flashes is generated and separated within the thunderstorm has been debated for more than 200 years, but there is still no universally accepted theory. However, the majority opinion holds that lightning is closely associated with the appearance of the ice phase, and the most promising theory suggests that the charge is produced by the momentary collision between ice crystals and pellets of soft hail or by the ejection of small charged splinters of ice during the freezing of cloud droplets collected by these small hailstones, or by both of these processes.

Basic aspects of cloud physics. The various stages of the precipitation mechanisms raise a number of interesting and fundamental problems in classical physics. Worthy of mention are the supercooling and freezing of water; the nature, origin, and mode of action of the ice nuclei; and the mechanism of ice-crystal growth which produces the various snow crystal forms.

It has now been established how the maximum degree to which a sample of water may be supercooled depends on its purity, volume, and rate of cooling. The freezing temperatures of waterdrops containing foreign particles vary linearly as the logarithm of the droplet volumes for a constant rate of cooling. This relationship, which has been established for drops varying between 10 μ and 1 cm in diameter, characterizes the heterogeneous nucleation of waterdrops and is probably a consequence of the fact that the ice-nucleating ability of atmospheric aerosol increases logarithmically with decreasing temperature.

When extreme precautions are taken to purify the water and to exclude all solid particles, small droplets, about 1 μ in diameter, may be supercooled to −40°C and drops of 1 mm diameter to −35°C. Under these conditions freezing occurs spontaneously without the aid of foreign nuclei.

The nature and origin of the ice nuclei, which are necessary to induce freezing of cloud droplets at temperatures above −40°C, are still not clear. Measurements made with large cloud chambers on aircraft indicate that the most efficient nuclei, active at temperatures above −10°C, are present in concentrations of only about 10 in a cubic meter of air, but as the temperature is lowered, the num-

bers of ice crystals increase logarithmically to reach concentrations of about 1 per liter at −20°C and 100 per liter at −30°C. Since these measured concentrations of nuclei are less than one-hundredth of the numbers that apparently are consumed in the production of snow, it seems that there must exist processes by which the original number of ice crystals are rapidly multiplied. Laboratory experiments suggest the fragmentation of the delicate snow crystals and the ejection of ice splinters from freezing droplets as probable mechanisms.

The most likely source of atmospheric ice nuclei is provided by the soil and mineral-dust particles carried aloft by the wind. Laboratory tests have shown that, although most common minerals are relatively inactive, a number of silicate minerals of the clay family produce ice crystals in a supercooled cloud at temperatures above −18°C. A major constituent of some clays, kaolinite, which is active below −9°C, is probably the main source of highly efficient nuclei.

The fact that there may often be a deficiency of efficient ice nuclei in the atmosphere has led to a search for artificial nuclei which might be introduced into supercooled clouds in large numbers. Silver iodide is a most effective substance, being active at −4°C, while lead iodide and cupric sulfide have threshold temperatures of −6°C for freezing nuclei.

In general, the most effective ice-nucleating substances, both natural and artificial, are hexagonal crystals in which spacings between adjacent rows of atoms differ from those of ice by less than 16%. The detailed surface structure of the nucleus, which is determined only in part by the crystal geometry, is of even greater importance. This is strongly indicated by the discovery that several complex organic substances, notably steroid compounds, which have apparently little structural resemblance to ice, may act as nucleators for ice at temperatures as high as −1°C.

The collection of snow crystals from clouds at different temperatures has revealed their great variety of shape and form. By growing the ice crystals on a fine fiber in a cloud chamber, it has been possible to reproduce all the naturally occurring forms and to show how these are correlated with the temperature and supersaturation of the environment. With the air temperature along the length of a fiber ranging from 0 to −25°C, the following clear-cut changes of crystal habit are observed (Fig. 3):

Hexagonal plates — needles — hollow prisms — plates — stellar dendrites — plates — prisms

This multiple change of habit over such a small temperature range is remarkable and is thought to be associated with the fact that water molecules apparently migrate between neighboring faces on an ice crystal in a manner which is very sensitive to the temperature. Certainly the temperature rather than the supersaturation of the environment is primarily responsible for determining the basic shape of the crystal, though the supersaturation governs the growth rates of the crystals, the ratio of their linear dimensions, and the development of dendritic forms.

CLOUD PHYSICS

(a) thin hexagonal plate

(b)   needles

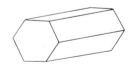

(c) hexagonal prismatic column

(d) dendritic star-shaped crystal

Fig. 3. Ice crystal types formed in various temperature ranges (°C); (a) 0 to −3°, −8 to −12°, −16 to −25°; (b) −3 to −5°; (c) −5 to −8°, below −25°; (d) −12 to −16°.

Artificial stimulation of rain. The presence of either ice crystals or some comparatively large waterdroplets (to initiate the coalescence mechanism) appears essential to the natural release of precipitation. Rainmaking experiments are conducted on the assumption that some clouds precipitate inefficiently, or not at all, because they are deficient in natural nuclei; and that this deficiency can be remedied by "seeding" the clouds artificially with dry ice or silver iodide to produce ice crystals, or by introducing waterdroplets or large hygroscopic nuclei. In the dry-ice method, pellets of about 1-cm diameter are dropped from an aircraft into the top of a supercooled cloud. Each pellet chills a thin sheath of air near its surface to well below $-40°C$ and produces perhaps 10^{12} minute ice crystals, which subsequently spread through the cloud, grow, and aggregate into snowflakes. Only a few pounds of dry ice are required to seed a large cumulus cloud. Some hundreds of experiments, carried out mainly in Australia, Canada, South Africa, and the United States, have shown that cumulus clouds in a suitable state of development may be induced to rain by seeding them with dry ice on occasions when neighboring clouds, untreated, do not precipitate. However, the amounts of rain produced have usually been rather small.

For large-scale trials designed to modify the rainfall from widespread cloud systems over large areas, the cost of aircraft is usually prohibitive. The technique in this case is to release a silver iodide smoke from the ground and rely on the air currents to carry it up into the supercooled regions of the cloud. In this method, with no control over the subsequent transport of the smoke, it is not possible to make a reliable estimate of the concentrations of ice nuclei reaching cloud level, nor is it known for how long silver iodide retains its nucleating ability in the atmosphere. It is usually these unknown factors which, together with the impossibility of estimating accurately what would have been the natural rainfall in the absence of seeding activities, make the design and evaluation of a large-scale operation so difficult.

In published data, no convincing evidence can be found that large increases in rainfall have been produced consistently over large areas. Indeed, in temperate latitudes most rain falls from deep layer-cloud systems whose tops usually reach to levels at which there are abundant natural ice nuclei and in which the natural precipitation processes have plenty of time to operate. It is therefore not obvious that seeding of these clouds would produce a significant increase in rainfall, although it is possible that by forestalling natural processes some redistribution might be effected.

Perhaps more promising as additional sources of rain or snow are the persistent supercooled clouds produced by the ascent of damp air over large mountain barriers. The continuous generation of an appropriate concentration of ice crystals near the windward edge might well produce a persistent light snowfall to the leeward, since water vapor is continually being made available for crystal growth by lifting of the air. The condensed water, once converted into snow crystals, has a much greater opportunity of reaching the mountain surface without evaporating, and might accumulate in appreciable amount if seeding were maintained for many hours.

The results of trials carried out in favorable locations in the United States and Australia suggest that in some cases seeding has been followed by seasonal precipitation increases of about 10%, but rarely have the effects been reproduced from one season to the next, and overall the evidence for consistent and statistically significant increases of rainfall is not impressive. Experiments of improved design will probably have to be continued for several more years before the effects of large-scale cloud seeding can be realistically assessed.

Attempts to stimulate the coalescence process by spraying waterdroplets or dispersing salt crystals into the bases of incipient shower clouds have been made in places as far apart as Australia, the Caribbean, eastern Africa, and Pakistan. The results of these experiments, though encouraging, are not yet sufficient to allow definite conclusions.

On a more optimistic note, it is relatively easy to clear quite large areas of the supercooled cloud and fog by seeding and converting it into ice crystals, which grow at the expense of the waterdrops and fall out to leave large holes.

The science and technology of cloud modification is, however, still in its infancy. It may well be that further knowledge of cloud behavior and of natural precipitation mechanisms will suggest new possibilities and improved techniques which will lead to developments far beyond those which now seem likely. *See* WEATHER MODIFICATION.

[BASIL J. MASON]

Bibliography: N. H. Fletcher, *The Physics of Rainclouds*, 1962; B. J. Mason, *Clouds, Rain and Rainmaking*, 1962; B. J. Mason, *The Physics of Clouds*, 1957.

Coal gasification

The conversion of coal, char, or coke to a gaseous product by reaction with air, oxygen, steam, or carbon dioxide or mixtures of these. Research and development efforts for technically feasible and economically sound processes to produce pipeline-quality gas (over 950 Btu/ft³) have been greatly accelerated since 1971.

The American Gas Association (A.G.A.) and the Office of Coal Research (OCR) of the U.S. Department of the Interior have initiated a $30,000,000/year 4-year program to investigate the feasibility of several existing processes in large pilot plants (1 to 3 tons of coal/hr). These processes include Consolidation Coal Company's CO_2 Acceptor process, Bituminous Coal Research's BI-GAS process, and three versions of the Institute of Gas Technology's HYGAS process. Several other processes will be added to the accelerated program in the near future. In addition, the U.S. Bureau of Mines will pilot their Synthane process. The gas industry is also investigating the process owned by the Lurgi Gesellschaft fur Warme und Chemotechnik M.G.H. of Frankfurt (Main), West Germany, which is the only currently available commercial process applicable to pipeline-quality gas. These processes will be briefly discussed below. From this work a process, perhaps including elements of each of the existing processes, will be developed which can be tested on a large scale (80,000,000 ft³/day) beginning in 1975 or 1976.

Lurgi gasifier. Many commercial Lurgi plants, comprising high-pressure, nonslagging, steam-oxygen gasifiers have been built and most are still in operation. Each gasifier is capable of producing up to 25,000,000 ft³/day of 400–450 Btu/ft³ heating-value gas at pressures of up to 450 psig. The relatively low-heating-value gas produced in Lurgi gasifiers would be upgraded by catalytic methanation to remove carbon monoxide to less than 0.1% and raise the Btu value to 950 or higher to be usable as pipeline gas. This vital process step which is common to all the processes has not yet been proved commercially, although several versions have been demonstrated on a small pilot plant scale (on the order of 1000 ft³/hr output). Up to the present, Lurgi gasifiers were only designed for noncaking coals (lignite, brown coal, and many subbituminous coals). Most United States bituminous coals would not appear to be usable in these gasifiers without prior carbonization because of their caking properties. Some low-rank coals must also be pretreated and briquetted to form a suitable feed. However, recently a modification of the conventional Lurgi generator, containing rotating rabble arms to break up caking coals, has been offered for commercial use with United States bituminous coals. The Lurgi generators also cannot handle fines. Any fines produced in mining in excess of power requirements would have to be briquetted.

The Lurgi gasifier receives coal through a lock hopper system to introduce it into the high-pressure environment. The coal flows downward as a fixed bed; consequently the coal size must be plus 1/4 in. to avoid excessive pressure drop. Even then, the gas flow rate has to be low and the production rate per gasifier is low. Production is only about 8,000,000–10,000,000 ft³/day of high-Btu gas per gasifier compared to 80,000,000 for the other processes. Lurgi is the only fixed-bed reactor system being considered for coal gasification.

Heat is supplied by burning the residual carbon at the bottom of the reactor with oxygen and steam. This produces a mixture of hydrogen and carbon monoxide as well as heat. As the hydrogen-rich gas flows upward through the coal bed, methane is formed and the coal is devolatilized to produce a raw gas of the typical volumetric composition shown in Table 1.

The sulfur present in the raw coal appears primarily as H_2S in this gas. It would be removed in a Claus plant as elemental sulfur. The products of gasification also contain light oil and impurities such as ammonia which have to be removed.

New synthetic pipeline gas from coal. The new synthetic pipeline gas from coal processes under development are intended to improve upon Lurgi.

All of the new gasification processes incorporate an initial hydrogasification stage to a greater or lesser degree and all have the following common features that distinguish them from earlier technology: continuous fluid-bed, falling-bed, or en-

trained-bed gasification at a pressure of about 250 psig and above; contacting of raw or lightly pretreated coal with hot, hydrogen-rich gas at these high pressures to form one-third or more of the methane directly by hydrogasification of the coal, substantially more than in Lurgi; and the use of less oxygen, or of no oxygen at all, in the production of the hydrogen-rich (synthesis) gas from the residual char remaining after the primary gasification step.

Because of the relatively low thermal efficiency of steam-oxygen gasification and catalytic methanation, the best practical thermal efficiency of processes employing the older technology tends to be low. The newer processes have efficiencies up to 75%.

HYGAS gasifier. The most advanced version of the various hydrogasification processes is the HYGAS process of the Institute of Gas Technology (IGT). The program for development of this process has been funded jointly by the OCR and A.G.A. since 1964. A.G.A. sponsored IGT's earlier synthetic pipeline gas from coal research beginning in 1946.

Construction of a HYGAS process pilot plant capable of producing 1,000,000–1,500,000 ft³/day of synthetic pipeline gas from 75 tons/day of coal was completed by Procon Inc. in 1971. With the exception of the coal-based hydrogen-rich gas-generation system, the plant is now in operation. In this process, pulverized coal is reacted in a first stage at 1200–1800°F. The coal is reacted with hot, raw, hydrogen-rich gas containing a substantial amount of steam at a nominal pressure of 1000 psig. Pretreatment is currently necessary for most bituminous coals because they tend to agglomerate during hydrogasification. However, the HYGAS reactor has provisions to potentially avoid pretreatment. Pretreatment consists of mild surface oxidation of the pulverized coal with air in a fluid-bed reactor at about 800°F and atmospheric pressure. Although under these conditions most of the valuable reactive portion ("volatile matter") of the coal is preserved, the pretreatment step still generates a substantial volume of a low-heating-value (on the order of 40–50 Btu/std ft³) fuel gas, containing as much as 25% of the sulfur content of the coal in the form of sulfur dioxide. Small amounts of tar and light oil are also formed. Lignite does not require pretreatment.

The hydrogasification reactor of the HYGAS process pilot plant has three separate stages: a fluidized preheat stage in which the unreacted residual char gives up its heat to the incoming reaction gas, a fluidized-bed gasification stage in which a portion of the less reactive coal constituents are converted to methane and carbon oxides by hydrogen and steam reactions, and a dilute-phase stage in which the most reactive coal constituents are converted to methane by hydrogenolysis reactions. About half of the coal is converted in the hydrogasification reactor, resulting in the formation of about two-thirds of the final methane product. Final upgrading to pipeline quality is accomplished in a fixed-bed catalytic methanation step after purification of the gas. About one-third of the methane is formed in this step.

The three versions of HYGAS differ from one another in the method by which the hydrogen-rich gas is produced.

Table 1. Raw gas composition in Lurgi gasifier

Component	Volume, %
CO	9.2
CO_2	14.7
H_2	20.1
H_2O	50.2
CH_4	4.7
C_2H_6	0.5
Other	0.6

Table 2.   Raw gas composition in electrothermal HYGAS gasifier

Component	Volume, %
CO	21.3
CO_2	14.4
H_2	24.2
H_2O	17.1
CH_4	19.9
C_2H_6	0.8
Other	2.3

Table 4.   Raw gas composition in steam-iron-hydrogen HYGAS gasifier

Component	Volume, %
CO	7.4
CO_2	7.1
H_2	22.5
H_2O	32.9
CH_4	26.2
C_2H_6	1.0
Other	2.9

Electrothermal version. In the electrothermal version the residual char from the hydrogasifier contacts steam in a fluidized bed. The bed is heated to about 1900°F by electric resistance heating. At this temperature, the highly endothermic steam-carbon reaction proceeds at a rapid rate to produce carbon monoxide and hydrogen plus some methane. This gas then is the source of hydrogen for the hydrogasifier. The spent char from this reactor is burned with air to produce the electric power. Thus combustion in air is the heat source and the use of oxygen is eliminated. The typical composition of the raw gas made in the HYGAS electrothermal process is shown in Table 2.

Oxygen version. In the oxygen version the char from the hydrogasifier is burned with a mixture of oxygen and steam at about 1800°F. A deficiency of oxygen is used so that a hydrogen-rich gas containing carbon monoxide and some methane is produced for use in the hydrogasifier. Typical composition of the raw gas produced in the HYGAS oxygen process is shown in Table 3.

The oxygen version will be tested in the pilot plant by modifying the electrothermal reactor. As with the Lurgi and two of the other processes, oxygen is the source of heat for the process. The oxygen is produced from power derived from spent char or waste-heat streams. Although several of the processes use oxygen and steam to generate the hydrogen-rich gas, there are distinct differences between them.

Steam-iron-hydrogen version. The steam-iron-hydrogen version of the HYGAS process is quite complex compared to the other versions. Potentially, however, it is far superior in efficiency and in the cost of the gas produced.

The spent char from the hydrogasifier is burned with a deficiency of air and steam to produce a gas capable of reducing iron oxide to iron. This reaction is carried on at 1000 psi in a fluid bed of iron oxide particles or in a raining bed. The partly reduced iron oxide is transferred to a second vessel, in which it is reoxidized by steam. This results in producing a hydrogen-steam mixture containing

about 50% hydrogen at 1000 psi and 1200°F, an ideal gas to feed into the hydrogasifier. The resultant raw gas produced in the hydrogasifier using steam-iron-hydrogen has the typical composition shown in Table 4.

The steam-iron-hydrogen process is expected to be tested at the large pilot scale in the next year as part of the A.G.A-OCR accelerated development program. There is no plan to integrate this pilot plant with the HYGAS hydrogasifier except that the hydrogasifier char will be used. The present initial HYGAS operation is being conducted with a steam-hydrogen mixture (produced from natural gas) identical to the mixture that would be produced in the steam-iron-hydrogen reactor. The heat input in this version of the HYGAS process is from air combustion, as is true of the HYGAS electrothermal process and the CO_2 Acceptor process.

CO_2 Acceptor process. Another process that is nearing large pilot plant operation is the Consolidation Coal Company's CO_2 Acceptor process. It is similar to the electrothermal and steam-iron versions of the HYGAS process in that it eliminates the need for oxygen in the production of hydrogen-rich synthesis gas by a novel technique and provides for contacting of raw coal and hydrogen-rich synthesis gas to form a substantial portion of the methane directly. Ground was broken near Rapid City, S. Dak., in 1969 for a pilot plant with a nominal capacity of 1 ton/hr (moisture- and ash-free basis) of Dakota lignite; construction was completed in the early part of 1971. Startup operation is now under way. This development program was supported by the OCR but is now part of the A.G.A.-OCR program. Stearns-Roger Corp. is Consolidation Coal Company's subcontractor for construction and operation of the pilot plant.

In this process the heat source for the endothermic steam-char reaction is hot, calcined dolomite, or limestone (the acceptor). The acceptor also releases additional heat when the calcium oxide forms calcium carbonate by combining with the carbon dioxide evolved in the steam-carbon and carbon monoxide shift reactions taking place in

Table 3.   Raw gas composition in oxygen HYGAS gasifier

Component	Volume, %
CO	18.0
CO_2	18.5
H_2	22.8
H_2O	24.4
CH_4	14.1
C_2H_6	0.5
Other	1.7

Table 5.   Raw gas composition in CO_2 Acceptor process

Component	Volume, %
CO	14.1
CO_2	5.5
H_2	44.6
H_2O	17.1
CH_4	17.3
C_2H_6	0.3
Other	1.1

Table 6. Raw gas composition in BI-GAS process

Component	Volume, %
CO	22.9
CO_2	7.3
H_2	12.7
H_2O	47.9
CH_4	8.1
C_2H_6	0.0
Other	1.1

the gasifier. The spent acceptor is continuously regenerated by reconverting the calcium carbonate to calcium oxide and carbon dioxide, utilizing residual carbon from the gasification step to supply the necessary heat by combustion with air, thus eliminating the need for oxygen. The driving force for regeneration is provided by the higher temperature and lower carbon dioxide partial pressure in the regenerator. Dolomite is preferred to limestone as an acceptor because of its superior ability to withstand physical decrepitation and chemical deactivation in cycling through the process.

The CO_2 Acceptor process is designed primarily for lignite, which is much more reactive than bituminous coal and, in the United States, costs approximately one-third less per unit of heat energy content. It may also be applicable to the highly reactive subbituminous coals that are in abundant supply in the western parts of the United States. The typical composition of the raw gas produced by this process is shown in Table 5.

BI-GAS process. The BI-GAS process is under development by Bituminous Coal Research and is also part of the accelerated program. A contractor is currently being selected for the design, construction, and operation of a large-scale pilot plant. Spent char is burned with oxygen and steam in a slagging gasifier (about 3000°F). A deficiency of oxygen is used so that the gas produced is composed of hydrogen, carbon oxides, and steam. This gas flows upward and entrains fresh, ground coal. The sensible heat in the gas heats the coal and produces a certain amount of hydrogasification. The unreacted char is collected in a cyclone and recycled to the slagging gasifier. The typical composition of gas produced is shown in Table 6.

Synthane process. The Bureau of Mines is developing the Synthane process. Superficially the process is much like the HYGAS-oxygen process. There are, however, significant differences. In the Synthane process bituminous coal is pretreated in free-fall with oxygen and steam at the top of the reactor rather than externally, as in the HYGAS process. The gas produced is, therefore, part of the

Table 7. Raw gas composition in Synthane process

Component	Volume, %
CO	10.5
CO_2	18.2
H_2	17.5
H_2O	37.1
CH_4	15.4
C_2H_6	0.5
Other	0.8

product. The reactor design is simpler, being one compartment in which the oxygen-steam combustion, the hydrogasification, and the pretreatment are carried out. The raw gas produced is reported to have the typical composition shown in Table 7.

Raw gas purification. In all of the processes for production of synthetic pipeline gas from coal, the raw gas contains not only carbon dioxide as an undesirable constituent but also substantial amounts of hydrogen sulfide, ammonia, and benzene and lesser amounts of organic sulfur compounds such as carbon oxysulfide and carbon disulfide. The sulfur compounds must be reduced to extremely low concentrations prior to methanation because they poison the very active nickel catalysts used in this step. It is also convenient to remove carbon dioxide prior to methanation since most commercial gas purification processes remove the "acid gas," that is, carbon dioxide and hydrogen sulfide, simultaneously. A typical purification scheme would comprise (1) hot potassium carbonate scrubbing for carbon dioxide and hydrogen sulfide removal, followed by (2) passage through fixed beds of hot alkalized iron or hot zinc oxide to remove any remaining hydrogen sulfide and a portion of the organic sulfur compounds, (3) water scrubbing for removal of ammonia and, finally, (4) passage through activated carbon to remove any remaining traces of organic sulfur and benzene. Elemental sulfur can be recovered from the acid gas stream by the Claus process.

[G. J. TANKERSLEY]

Coastal engineering

A branch of civil engineering pertaining to the study of the action of the seas on shorelines and to the design of structures to protect against this action. The design hypothesis is essentially empirical, based on lengthy observations of the immediate area, on measurements of existing structures, and on large- and small-scale model tests. The two basic problems are the destructive power of waves and their ability to transport materials in the nearshore area. Of the types of protective construction, groins, bulkheads, and sea walls or revetments are primarily for beach and shore protection; breakwaters and jetties for protection of maritime properties and activities.

BEACH EROSION AND ACCRETION

When subjected to wave action, beaches and shorelines continuously change in configuration through the processes of erosion and accretion, the latter being the gradual buildup of the beach area. Material transported from erosion zones to accretion or deposition zones is referred to as littoral drift, and its movement under the influence of waves and currents as littoral transport. A beach is stabilized when erosion and accretion have been arrested. The use of groins is one of the principal methods of stabilizing a beach.

Waves. Waves in deep water are usually of the oscillatory type, in which the water particles making up the wave oscillate in a circular orbit about a mean position. In an area of water over which the wind is blowing and generating waves, known as the fetch or fetch area, the growth of waves is governed by at least three factors: the force of the wind, the length of time it has been blowing, and the length of the fetch. In relatively shallow water areas, in addition to these factors, the depth of water in the fetch limits the growth of waves. Wave velocity varies with the depth of water in which the wave is moving; as the depth decreases, the velocity decreases. At a certain point in its advance toward shore a wave becomes unstable, peaks up, and breaks. The water motion changes from a circular orbital motion to turbulent white water conditions. The determination of the point of breaking and of wave heights at breaking is of major importance in the design of coastal protective installations. *See* SEA STATE.

Tides. The effect of the tide on water level results in a constantly changing breaking point of the waves. Along the Atlantic Coast of the United States the two tides each day are of nearly the same height. On the Gulf Coast the two tides are low but in some instances have a pronounced inequality. Pacific Coast tides compare in height with those on the Atlantic Coast, but the two daily tides have a decided inequality. In addition to the tidal effects, changes in winds and barometric conditions cause variation in sea level from day to day. There are also seasonal variations in sea level, but at coastal stations these are usually less than half a foot.

Currents. In most coastal locations the wind may induce a surface current in the general direction of the wind movement, thus causing an increase or decrease in water level above or below that due to tidal action. This results from tangential stresses at the water surface between wind and water. The wind-induced surface current produces a piling up of water at the leeward side of a water body and a lowering of water level at the windward side with a return flow along the bottom. This deviation from still-water level caused by wind-driven currents is called wind setup, wind tide, or storm tide. This effect may be significantly higher than normal tidal action in coastal areas: for example, in Fort Meyer, Fla., where the normal tidal range is about 3 ft, hurricane Donna caused a water level up to 11 ft above normal in September, 1960. Daily, seasonal, and extreme water levels, which determine the elevation at which the wave energy will act on the shore, must be considered when planning shore protective measures. *See* STORM SURGE.

Littoral processes. A stable shoreline is one in which the supply of material to the area under consideration is approximately equal to the loss of material from the area. On an accreting shoreline the supply of material exceeds that lost, and the reverse is true of an eroding shoreline. Accordingly, the need for protective works and choice of type are dependent on the net balance between supply and loss of material. The main natural sources of material to any beach segment are the material moving into the area by natural littoral transport from adjacent beach areas, contributions by streams, and contribution through erosion of coastal formations other than beaches.

Waves and currents supply the necessary forces to move the littoral materials. The mechanics of littoral transport are not exactly known, but it may be generally stated that littoral material is moved by one of three basic modes of transport. Material known as beach drift is moved along the foreshore in a more or less scalloped path due to the uprush and backwash of obliquely approaching waves. Material that is principally in suspension in the surf zone is moved by littoral currents and the turbulence of breaking waves. Within and seaward of the surf zone, material close to the bottom, known as bed load, is moved by sliding, rolling, and saltation caused by the oscillating currents of passing waves. Significant bottom movement has been observed in depths exceeding 100 ft on exposed seacoasts. Regardless of the mode of transport, the direction and rate of littoral drift depend primarily upon the direction and energy of waves approaching the shore. Exceptions exist on short reaches of shore adjoining tidal inlets where the pattern of the tidal currents may be dominant.

In studying the effect of littoral drift, the existence of previously placed structures provides the most reliable means of determining littoral transport characteristics and will ordinarily outweigh all other evidence.

SHORE-PROTECTIVE STRUCTURES

In determining the size, type, and location of protective works, the objective should be an engineering design which will accomplish the desired results most economically, with full consideration of its effects on adjacent shorelines.

Bulkheads, sea walls, and revetments. These structures are placed parallel or nearly parallel to the shorelines, separating a land area from a water area. The primary purpose of a bulkhead is to retain or prevent sliding of the land; its secondary purpose is to protect the fill behind against damage by wave action. A sea wall or revetment is used to protect the land and upland structures from damage by wave forces, with incidental functions as a retaining wall or bulkhead. There is really no sharp distinction between the three structures, and the same type of structure may be called by different

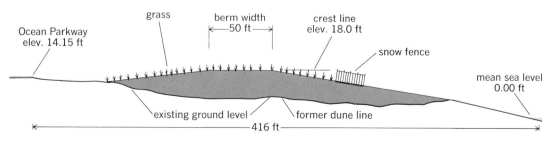

Fig. 1. Dune and beach restoration, Gilgo State Park, Long Island, N.Y. (*New York State Department of Public Works*)

names in different localities. All these structures, however, have one feature in common: They separate land and water areas. Consequently, they are generally used where it is necessary to maintain the shore in an advanced position relative to that of adjacent shores, where there is scant supply of littoral material in the area and little or no protective beach, or where it is desired to maintain a depth of water along the shore, as for a wharf. These structures protect only the land behind them, not adjacent areas up and down the coast. When built on a receding shoreline, the recession will continue on the downdrift side. In addition, any tendency for loss of beach material in front of such a structure may be intensified. Where it is desired to maintain a beach in the immediate vicinity, companion works may be necessary.

Artificially formed beaches. Since beaches, when maintained to adequate dimensions, are the most effective means of dissipating wave energy, they are the most effective protection for adjoining upland. In an erosion problem, in addition to the use of remedial structures such as groins, the use of sand fill for artificially nourishing and maintaining an adequate beach should be considered. Where fill is available for beach nourishment, this treatment directly remedies the basic cause of most erosion problems, that is, a deficiency in natural sand supply, and thereby benefits rather than damages the shore beyond the immediate problem area. However, beach nourishment without a confining structure such as a groin may prove only temporary unless done periodically.

Artificially formed dunes. Sand dunes perform the function of beach fills in preventing waves from reaching the upland but are located landward of the beach itself. A belt of sand dunes will effectively protect upland property as long as the tops of the dunes remain above the limit of wave uprush. At locations that have an adequate natural supply of sand and are subject to inundation by storm tide and high seas, a belt of artificially formed and nourished sand dunes may provide more effective protection at a lower cost than either a bulkhead or sea wall. The surface of a dune may be protected from the wind by planting beach grass. A snow fence placed on the dune will retard the drifting of the sand as it does with snow.

Figure 1 shows the design of artificial beach and dune by the New York State Department of Public

Fig. 3. Groin system of protection at Atlantic Beach (top) and Rockaway Beach (bottom), separated by East Rockaway Inlet. (*Aerial mosaic by Lockwood, Kessler, and Bartlett, Inc., Syosset, N.Y.*)

Works for Gilgo State Park, Long Island. Erosion of the dune is retarded by sprigging with American beach grass (*Ammophila breviligulata*) and by a snow fence set in a zigzag pattern. This design in association with groins can be expected to give reasonable stability to the beach. *See* DUNE; DUNE VEGETATION.

Groins. These structures are used to widen a protective beach by trapping littoral drift and to prevent further loss by erosive processes. They are usually approximately perpendicular to the shore, extending from a point landward of possible shoreline recession into the water a distance sufficient to stabilize the shore of the desired location. A series of groins acting together to protect a long section of shoreline is a groin field or groin system. The groin acts as a partial barrier intercepting a portion of the normal drift. As material accumulates on the updrift side, supply to the downdrift shore is correspondingly reduced and the latter shore recedes. The accretion of material on the updrift side of the groin changes the alignment of that portion of the shore and produces a stable

Fig. 2. Groin at East Atlantic Beach. (*New York State Department of Public Works*)

alignment normal to the resultant of wave attack. Groins must be used with caution, for if the natural supply of littoral material is used to restore or widen an area of beach, a deficiency in supply is likely to be created in adjoining areas, with resulting expansion of the problem area. Groins may be used in connection with artificial fill or beach nourishment mentioned above. Groins differ from jetties in that jetties are used primarily at river mouths and channel entrances to direct and confine the stream or tidal flow and to prevent shoaling in the channel. In some areas groins are commonly referred to as jetties.

Figure 2 shows one of a field of groins at East Atlantic Beach, Long Island, N.Y. The height of the groin is designed so that it is topped by waves near its inshore end, permitting littoral transport. Accretion has taken place on the left or updrift side of the groin, and some deposition of sand can be seen on the downdrift side.

Figure 3 shows groin fields protecting the beaches at Rockaway and Atlantic beaches, Long Island, N.Y.

Breakwaters. The usual requirement of a breakwater is to provide a calm anchorage at a port or harbor. An offshore breakwater, usually parallel to, but unconnected with, the shore and some distance out, has quite a different function: to exert a direct effect upon the rate of littoral transport. The offshore breakwater, by damping wave action and thus reducing the forces responsible for transport, causes sand to be deposited in its lee. The accretion area thus formed tends to widen the beach inshore of the breakwater. If the breakwater is of sufficient length in relation to the distance from the shore to act as a complete littoral barrier, the beach widening may continue until it reaches the breakwater. Such is the case with an offshore breakwater at Venice, Calif.

[EDWARD J. QUIRIN]

Bibliography: P. Brunn, Tidal inlets and littoral drift, in *Stability of Coastal Inlets*, vol. 2, 1966; Council on Wave Research, *Proceedings of the Conference on Coastal Engineering*, vol. 1, 1961; C. King, *Beaches and Coasts*, 1960; W. C. Krumbein, Geological aspects of beach engineering, in *Application of Geology to Engineering Practice*, Geological Society of America, reprint, 1958; A. D. Quinn, *Design and Construction of Ports and Marine Structures*, 1961; R. B. Thorn, *Design of Sea Defense Works*, 1960; U.S. Beach Erosion Board, Corps of Engineers, *Shore Protection Planning and Design*, Tech. Rep. no. 4, 1954.

Community

An aggregation of organisms characterized by a distinctive combination of species, such as deciduous forest, grassland, pond, or mud flat. An ecological community is sometimes called a biocenose. Each community occupies a particular habitat and together with its habitat constitutes an ecosystem. A community may be composed of any distinctive combination of plant or animal species. When both are included, the term biotic community is commonly used.

Major communities together with their habitats are self-sustaining units. Minor communities, sometimes called societies, are secondary combinations of species within major communities. *See* ECOSYSTEM.

Dominants. Within most but not all major communities, certain organisms, called dominants, exert a commanding role in determining the character and composition of the community. On land, plants are commonly dominants because they receive the full impact of the climate and environment. By reactions on the habitat plants establish the conditions of light, moisture, air movement, space, and other factors under which all other organisms must live. The plant dominants are ordinarily the most prominent species and serve as the major source of food, substrate, and shelter for the animals that are present. Thus trees are the dominant species in a forest community, while grasses dominate in a prairie. Animals sometimes exert dominance, especially in aquatic habitats. For instance, carp and suckers in ponds may root out all submerged vegetation and make the water so turbid by stirring up the bottom mud that some other species of fish cannot exist. Dominance may thus be affected through coactions (interaction of one organism on another) as well as by reactions, although this is less common. *See* ECOLOGICAL INTERACTIONS.

Structure. The structure of a community is evident in various ways in addition to the relation between the dominants and other constituent species. Terrestrial vegetation is commonly stratified into subterranean, ground, herb, shrub, and one or more tree layer societies, although tree and shrub strata are absent from some communities, for instance, the grassland. The microclimate of each stratum is somewhat different and this, together with the variation in substrate, divides the animal distribution into two principal layer societies: subterranean-ground and herb-shrub-tree.

Community changes. Except for tropical rainforests, seasonal changes of climate during the year bring important changes in the kinds of plants that are active and in the activities of animals. In humid north temperate climates, different aspects may be recognized: hiemal, from mid-November to mid-March; vernal, to late May; estival, to mid-August; and autumnal. In addition, the community changes between day and night, especially in the predominant animals that are active. Species are commonly divided into diurnal, crepuscular (evening, dawn), nocturnal, and arhythmic, or those exhibiting no periodicity.

Species composition. Species may be classified in respect to their fidelity to a particular community into exclusive, if confined to one community; characteristic, if significantly more abundant in one community than in any other; or ubiquitous in several communities. In respect to abundance, they may be predominant and member constituents. Two aggregations of species in different localities or habitats are not considered as belonging to distinct communities unless they are more different than they are alike; that is, more than 50% of the predominant species in each one are characteristic or exclusive to it alone. Communities are commonly named after two or three dominant or predominant species or the type of vegetation or habitat. *See* BIOME; PLANT FORMATIONS, CLIMAX; PLANT GEOGRAPHY; VEGETATION ZONES, ALTI-

TUDINAL; VEGETATION ZONES, WORLD.

Community balance. All organisms fit into food chains, which may have only three links, or trophic levels—such as grass, antelope, coyote—or as many as five, for instance, bacteria, protozoan, rotifer, small fish, large fish. In each community the food chains anastomose in various ways to form a characteristic food web. See ECOLOGY; FOOD CHAIN.

The number, biomass or number times weight, and rate of reproduction of organisms decrease progressively at higher trophic levels. There is an inverse relation between number and size of organisms, the pyramid of numbers. Since each trophic level depends on the surplus production of food by lower trophic levels, a balance of nature becomes established within the community. Thus, the abundance of each species tends to become stabilized or to fluctuate mildly around some definite level. A temporary excess or diminution in numbers of any species may reverberate throughout the community until a new balance becomes established. See BIOMASS.

Succession. Because of fixation of nitrogen by certain bacteria or influx of nutrients from outside, the habitat may increase in fertility, and new species may invade it. These species may modify the physical characteristics of the habitat by decreasing light intensity or monopolizing space or by other factors, such as eliminating species already present. This may be sufficiently extensive in the course of time, decades or centuries, to bring a biotic succession of different communities replacing each other until a climax is reached which becomes stabilized under the particular climate. Communities do not remain permanently constant because each is dependent upon climate, physiography, and organic evolution, factors which are themselves continuously changing. See CLIMAX COMMUNITY; SUCCESSION, ECOLOGICAL.

Concepts of structure. The organismic concept of the community is that it is a complex organic entity representing the highest stage in the organization of living matter, that is, cell, tissue, organ, system, organism, community. It behaves as a unit in its relation to other communities, response to climate, local and geographic distribution, and in its evolution. It possesses such emergent characteristics as density, population dynamics, trophic balance, dominance, and succession beyond those of the individual organisms of which the community is composed. On the other hand, the individualistic concept considers the community as a chance combination of species that occur together because of similarity in their environmental relations. There is no intrinsic interrelation between the species in respect to distribution or evolution. The population density of each species is distributed in a form of a more or less normal curve when plotted along the gradient of a given environmental factor, and the curves of different species overlap in a heterogeneous or random manner without exact agreement between any two species. The community is simply an arbitrary portion of a continuum of constantly changing composition of plant and animal life along environmental gradients that extend from one extreme to another. Adherents of both concepts agree, however, that

there exist functional relations between species that occur in the same place, and, regardless of one's point of view, the community is a useful unit for detailed ecological investigation. See ANIMAL COMMUNITY; COMMUNITY CLASSIFICATION; PLANT COMMUNITY; POPULATION DYNAMICS.

[S. CHARLES KENDEIGH]

Community classification

The arrangement of communities in any of several systems. A community is an assemblage of two or more ecologically related species. Organisms of a single species would not be regarded as a community, but as a population. The term ecosystem means a community and its physical environment. Communities may be classified in respect to their complexity and extent, or magnitude, their stage of ecological succession, or their primary production. Classification provides an orderly basis for describing the distribution of organisms in relation to their physical and biotic environments, as well as providing a framework in which community ecology may be studied. A community may be thought of as being more than the sum of its component organisms in that it has a definite organization which results in additional attributes. At a functional level, separate plant and animal communities do not exist because of interactions of the component organisms.

The hierarchy based on extent and complexity has at one extreme the biosphere, a community consisting of all living organisms on the face of the Earth. Any organism could theoretically affect the conditions of existence of any other organism in this huge community, but on the average the effects would be relatively very minute and would involve a considerable time lag. Within the biosphere, land surfaces, fresh water, and salt water provide contrasting conditions for existence. See ECOSYSTEM.

Terrestrial biocycle. Within the land biomes, there are major biotic realms which owe their existence to the distribution of land surfaces in a pattern of major continents and to past histories of dispersal of land-living organisms. A formal classification of the land masses into zoological realms has long been in use, and these realms are also biotic ones. Each biotic realm has subdivisions in which similarity of climate has resulted in the evolution of similar life forms and generally similar adaptations, for example, grasslands or deciduous forests. Various classifications have been developed for these subdivisions. The biome, or plant-animal formation, is a community that corresponds to the ecological climax and its successional stages. The biome is not continuously distributed but recurs wherever the climax recurs. In another classification, the biotic province is a community occupying an area where similarity of climate, physiography, and soils leads to the recurrence of similar combinations of organisms. The biotic province is continuously distributed. Biotic provinces are divisible into biotic districts on the basis of lesser differences than those separating provinces. See BIOME; LIFE ZONES.

Ecological associations. These are the complex of communities which have developed in accord with variations of physiography, soil, and successional history within the major subdivisions of a biotic realm, whether the subdivision is called a biome, biotic province, or life zone. Thus an elm-hackberry association may occupy the deep alluvial soils of a river floodplain, while an oak-hickory association may occupy the adjacent uplands. In the concrete sense, an association may be referred to as a stand, or all of the organisms occurring together at a specific geographic place. As the dominant oaks and hickories and their associated plants and animals at a particular spot comprise a stand, recurrent combinations of approximately similar constitution throughout the biotic province belong to an abstraction, or association type, the oak-hickory association.

As may be seen in the above example, ecological associations are usually given names indicative of a few of their dominant plants, since these are usually the most obvious and stable organisms in the community. It is possible to recognize subdivisions below the level of the association. The organisms living in a rotting log, for example, would comprise a microstand, and in the aggregate, these microstands would comprise a microassociation. The number of microassociations within an association is potentially great. In the hierarchy of classification of communities, the degree of interaction of different organisms in the community increases from the top down. The greatest interrelationship would be found in the microassociation or association, and the least would be found among members of the biosphere.

Stage of ecological succession. This aspect provides another basis for the classification of communities. Succession involves invasion of an area by a pioneer community, and modification of the environment by the component organisms through a series of stages. *See* CLIMAX COMMUNITY; SUCCESSION, ECOLOGICAL; TERRESTRIAL ECOSYSTEM; VEGETATION MANAGEMENT.

Aquatic systems. Classification in respect to primary production is useful and has been used principally for aquatic communities. A eutrophic community has relatively high primary productivity, or rate at which producer organisms incorporate energy in chemical compounds usable as food; an oligotrophic community is one with relatively low primary productivity.

Marine and fresh-water biocycles may be recognized by the same general kinds of criteria used for the terrestrial biocycle, and there is a similar hierarchy of communities of decreasing magnitude and increasing interaction. The aquatic biocycles differ from the terrestrial one in the much greater importance of the vertical dimension, since water is a sufficiently dense medium to support communities independently of the substrate, and since the penetrance of light changes greatly with depth.

Marine biocycle. This biocycle is classically divided into a benthal division (sea bottom) and pelagial division (open sea). The benthal is divided into a littoral zone, which is a lighted zone corresponding generally to the continental shelf and extending out to depths of about 200 meters, and an abyssal or lightless zone. The pelagial division is divided into a neritic region, or lighted zone above the littoral, and an oceanic region which is above the abyssal. The upper level of the oceanic

region down to the limits of effective light penetrance for photosynthesis is the euphotic zone. Stands and microstands of organisms, comparable in the aggregate to the associations and microassociations of the terrestrial biocycle, occur in these subdivisions of the marine biocycle. Proponents of the biotic province and biome concepts hold that such divisions can be recognized in the marine biocycle. *See* MARINE ECOSYSTEM.

Fresh-water biocycle. It is difficult to classify the fresh-water biocycle, because of its discontinuous distribution in fragments of various sizes and characteristics. The communities of large bodies of water, such as the Great Lakes, may be relatively stable and permanent, while those of small bodies may exist more briefly as seral stages of a terrestrial climax.

Waters may be categorized into lotic or flowing and into lentic or standing, although one type grades into the other. Lotic waters may be roughly divided into pools and rapids. Stands and microstands of organisms, comprising associations and microassociations in the aggregate, are distributed in relation to these divisions of lotic waters, and also in relation to such factors as type of stream bottom, depth, and physical and chemical characteristics of the water. Lentic waters show vertical and horizontal zonation comparable to that of marine waters. The littoral zone of ponds and lakes is composed of shallow water in which light penetrates to the bottom. The limnetic zone includes the upper levels of open water down to the level of effective light penetration. The profundal zone includes the deep water and bottom below the level of effective light penetration. Deep lakes in summer develop a characteristic zonation. There is a metalimnion (thermocline), or zone of rapid temperature decrease with increased depth. Above this there is an epilimnion, or zone of relatively warm water, in which mixing occurs as the result of wind action and convection currents. Below the thermocline there is a hypolimnion of cold, noncirculating water. Stands and microstands, comprising associations and microassociations, occur in the various zones of lakes just as they do in other environmental subdivisions. *See* COMMUNITY; FRESH-WATER ECOSYSTEM; LIMNOLOGY.

[W. FRANK BLAIR]

Bibliography: F. E. Clements and V. E. Shelford, *Bio-ecology*, 1939; L. R. Dice, *Natural Communities*, 1952; C. S. Elton, *The Pattern of Animal Communities*, 1966; R. Hesse, W. C. Allee, and K. P. Schmidt, *Ecological Animal Geography*, 2d ed., 1951; E. P. Odum, *Fundamentals of Ecology*, 2d ed., 1959.

Conservation of resources

Conservation is concerned with the utilization of resources—the rate, purpose, and efficiency of use. This article emphasizes integrated conservation trends and policies. For treatment of individual resources *see* FISHERIES CONSERVATION; FOREST MANAGEMENT AND ORGANIZATION; LAND-USE PLANNING; MINERAL RESOURCES CONSERVATION; RANGELAND CONSERVATION; SOIL CONSERVATION; WATER CONSERVATION; WILDLIFE CONSERVATION.

Nature of resources. Universal natural resources are the land and soil, water, forests, grassland and other vegetation, fish and wildlife, rocks and minerals, and solar and other forms of energy. Some natural resources, such as metallic ores, coal, petroleum, and stone, are called fund or stock resources. They usually are referred to as nonrenewable natural resources because extraction from the stock depletes the usable quantity remaining and, even if some is being formed, the rate of formation is too slow for practical meaning. Other natural resources, such as living organisms and their products and solar and atomic radiation, are called flow resources. They usually are referred to as renewable natural resources because they involve organic growth and reproduction or because they are relatively quickly recycled or renewed in nature, as in the case of water in the hydrologic cycle and certain atmospheric phenomena. Some natural resources are difficult to fit into such a simple system. Soil, for example, is commonly thought of as renewable, as erosion and nutrient depletion can in some cases be rather quickly corrected, but if the upper layers of the soil are removed or bedrock is exposed, renewal may take thousands of years. Water also is commonly renewable, but rapid extraction of water by wells from deep aquifers may be equivalent to mining minerals. *See* HYDROLOGY.

Human resources are of two types: the people themselves (their numbers, qualities, knowledge, and skills) and their culture (the tools and institutions of society). The three broad classes of resources correspond to the economic factors of production: land (natural resources), labor (personal human resources), and capital (cultural tools and institutions). The natural resources have meaning only as there is human ability to make use of them.

Nature of conservation. Conservation has received many definitions because it has many aspects, concerns issues arising between individuals and groups, and involves private and public enterprise. Conservation receives impetus from the social conscience aware of an obligation to future generations and is viewed differently according to one's social and economic philosophy. To some extent, the meaning of conservation changes with the time and place. It is understood differently when approached from the natural sciences and technologies than when it is approached from the social sciences. Conservation for the petroleum engineer is largely the avoidance of waste from incomplete extraction; for the forester it may be sustained yield of products; and for the economist it is a change in the intertemporal distribution of use toward the future. In all cases, conservation deals with the judicious development and manner of use of natural resources of all kinds.

No definition of conservation exists that is satisfactory to all elements of the public and applicable to all resources. In its absence, an operational or functional definition can be arrived at by considering a series of conservation measures.

1. Preservation is the protection of nature from commercial exploitation to prolong its use for recreation, watershed protection, and scientific study. It is familiar in the establishment and protection of parks and reserves of many kinds.

2. Restoration, another widely familiar conservation measure, is essentially the correction of past willful and inadvertent abuses that have impaired the productivity of the resources base. This measure is familiar in modern soil and water conservation practices applied to agricultural land.

3. Beneficiation is the upgrading of the usefulness or quality of something, for instance, the utilization of ores that were formerly of uneconomic grade. Modern technology has provided many examples of this type of conservation.

4. Maximization includes all measures to avoid waste and increase the quantity and quality of production from resources.

5. Reutilization, in industry commonly called recycling, is the reuse of waste materials, as in the use of scrap iron in steel manufacture or of industrial water after it has been cleaned and cooled.

6. Substitution, an important conservation measure, has two aspects: the use of a common resource instead of a rare one when it serves the same end and the use of renewable rather than nonrenewable resources when conditions permit.

7. Allocation concerns the strategy of use—the best use of a resource. For many resources and products from them, the market price, as determined by supply and demand, establishes to what use a resource is put, but under certain circumstances the general welfare may dictate usage and resources may be controlled by government through the use of quotas, rationing, or outright ownership.

8. Integration in resources management is a conservation measure because it maximizes over a period of time the sum of goods and services that can be had from a resource or a resource complex, such as a river valley; this is preferable to maximizing certain benefits from a single resource at the expense of other benefits or other resources. This is one of the meanings of multiple use, and integration is a central objective of planning.

A generalized definition that fits many but not all meanings of conservation is "the maximization over time of the net social benefits in goods and services from resources." Although it is technologically based, conservation cannot escape socially determined values. There is an ethic involved in all aspects of conservation. Certain values are accepted in conservation, but they are the creation of society, not of conservation.

Conservation trends. There has been an important trend in conservation from an almost exclusive interest in production from individual natural resources to a balanced interest in that need and in the human resources and social goals for which resources are managed. The conservation movement originated with the realization that the economic doctrine of laissez-faire and quick profits —whether from forest, farm, or oil field—was resulting in tremendous waste that was socially harmful, even if it seemed to be good business. The beginning of the conservation movement stemmed from revulsion against destructive and wasteful lumbering. In time, the movement spread to farm and grazing lands, water, wildlife, and oil and gas. It was gradually learned that conservation management was good business in the long run.

The second trend in conservation was toward integrated management of resources. Students and administrators of resources—in colleges, business, and government— were discovering that the way one resource was handled affected the usability of others. Forestry broadened its interest to include forest influences on the watershed and the relations of the forest to wildlife and human uses for recreation. For many industries working directly with natural resources or their products, as in paper and chemical manufacturing and in coal mining, it was discovered that waste products produced costly and dangerous pollution of streams and air. Engineering on great river systems moved on from problems of flood hazard abatement and hydropower development to the design of structures with regard to fisheries and recreational values, and it was slowly realized that the way the land of a valley was managed affected erosion, siltation, water retention, and flooding.

Conservation in its third phase extends the ecological or integrated approach to resources management to include a more complete acceptance of the force of societal factors (such as economics, government, and social conditions) in determining resource management. There also is a closer examination of social costs and benefits and of human goals for which resources are employed.

Conservation of human resources. Many students of conservation prefer for practical reasons to limit conservation to the management of natural resources and to leave problems of human resources to other disciplines and fields of action. However, because of the inevitable interplay of natural resources and the resourcefulness of man in utilizing them, others emphasize the role of man himself as a resource. This leads to the application of conservation measures to man and his institutions. The many measures that tend to preserve, rehabilitate, renew, maximize, allocate, and integrate human abilities are coming to be referred to as conservation measures. These measures tend to be organized and institutionalized so that the institutions themselves become means to an end within the society and thus are considered to be resources as truly as water, petroleum, or the labor force. *See* ECOLOGY, HUMAN.

Conservation of recreation resources. More than 30,000,000 acres of rural public land in the United States are managed for recreational use or to preserve scenic, historical, or scientific and natural history values, and additional space is allocated for more intensive recreational activities by units of local government. There has been in recent decades a tremendous growth in the use of such lands for all forms of recreation, and the forecast is that outdoor recreation will increase more rapidly than population growth because of the increase in urban living and personal incomes, more and better roads, the shorter workweek, and paid vacations. Rapidly growing new trends in recreation require special uses of space, such as bow and arrow shooting, skiing, and motorboating. With increasing numbers of persons camping, picnicking, and swimming, the provision of adequate facilities and physical maintenance of crowded sites have become problems.

Except for the more intensive recreational uses, rural and wildland areas serve multiple purposes. An abundance of game can be raised in agricultural regions. Watershed protection and the maintenance of scenic values go naturally together. Stream impoundments for multiple purposes create new recreational facilities. Yet some land uses are incompatible: municipal, manufacturing, and mining pollution destroys water recreation values; commercial developments destroy wilderness values.

Many trends indicate that there are not enough acres dedicated to recreation, especially smaller areas near strongly urbanized regions. There is a growing need for research on trends in wildland usage by people and for better use of space in large parks and forests so as to avoid wearing out the sites by the persistent concentration of people, as at campsites. It is clear that certain human facilities must be provided where many people use wildlands and that inappropriate ones, such as amusement facilities, must be kept separate from activities requiring special terrain. Some growing recreational demands are being met by private enterprise, such as in ski resorts and privately owned public hunting grounds. Although game is owned by the public, it lives and breeds largely on private land. In time more farm and ranch owners will be paid for hunting privileges. Further development in recreation is coming from increased knowledge of the biology of fish and wildlife and the consequent improvement of management arising from it and from a better understanding by the sportsman of the management problems.

Conservation policies. Individuals, corporations, and governmental entities at all levels have policies pertaining to resources. There could be a single national policy concerning a natural resource, such as water, only if the central government had complete authority over all governmental agencies and private enterprises. In the United States, however, many Federal and state departments, bureaus, agencies, and commissions have some authority over natural resources. There usually are several policies concerning the use and management of soil, water, forests, rangelands, fish, wildlife, minerals, and space, and they are not necessarily uniform as to objectives or program. One exception is the Atomic Energy Commission, which has centralized authority in the creation and execution of policy regarding radioactive resources. This authority is granted by Congress and can be modified by congressional action.

Because each resource is capable of being utilized for a variety of goods and services, and because individuals, enterprises, and regions tend to value certain uses more than others, conflicts arise in the allocation of a resource among competing uses when the supply is inadequate for all desirable uses. The demand for water, for example, may be for rural domestic needs, urban and industrial needs, irrigation, power development, and recreation in its many aspects.

Because situations such as this exist or are potential with respect to every resource, two outstanding needs arise in the conservation of natural resources. Detailed information is needed concerning the location, quantity, and quality of each resource and of the interrelations among the resources. As a result, each agency of government needs to strengthen its own fact-finding, analysis, and programming machinery. The second need is for coordinating machinery that will improve the efficiency of allocation of resources so as to maximize the net private and public benefits from them. This is the goal of conservation policy, and it must, in a democratic society, be approached as far as possible through the voluntary cooperation of government and private enterprise. However, as pressures on society increase, whether because of actual depletion of resources or because of an increase in critical demand, as in war, authority to allocate resources tends to be delegated by the citizenry to the central government.

Federal conservation legislation. National resources policy in the United States is framed by acts of Congress and in the states by acts of legislatures. The language of specific acts, however, usually permits some freedom for administrative decisions and also permits different interpretations that must be settled in the courts of law. As the country has developed and conditions have changed, a sequence of laws has been passed to deal with the exigencies of natural resource conditions. Also, there has occurred some evolution of political philosophy to meet the changing conditions. In addition to legislative and court actions, some natural resources are subject to treaties and other international agreements.

Water. In 1824 Congress assigned to the Army Corps of Engineers responsibility for improving rivers and harbors for navigation, and this assignment has become its principal civilian activity. The Inland Waterways Commission in 1907 and the National Waterways Commission in 1909 were created by Congress with broad responsibilities for planning the development and conservation of water and related resources. The Flood Control Act of 1927 provided for Federal surveys, and the act of 1936 provided for Federal construction of projects. Subsequent amendments led to the 1954 Watershed Protection and Flood Prevention Act, which clearly recognized the relation of upland management practices to erosion, siltation, streamflow, and other water-related resource problems. Interest in irrigation has resulted in a series of acts from 1866 onward (the Desert Land Act of 1877 being significant), while water facilities, drainage, hydroelectric power, and water pollution also have received repeated attention. The Water Power Act of 1920 set up the Federal Power Commission and aimed to safeguard the rights of government and hydroelectric companies. Water pollution control is largely state responsibility, but the Federal Water Pollution Control Act of 1948, amended in 1956, enables financial and other assistance to the states.

Soil. Although soil surveys had been carried out for a century, it was the Soil Conservation Act of 1935 that created the Soil Conservation Service and the autonomous Soil Conservation Districts. Soil conservation payments were started under the Agricultural Adjustment Act of 1933. Research, education, cooperation, and financial inducements have been stressed more than regulatory measures for soil and water conservation on the land.

Minerals. In 1866 Congress provided for the sale of mineral lands, but a general minerals policy did not exist until later. The Minerals Leasing Act of 1920 provided that nonmetalliferous minerals on public lands could be utilized only under lease. In 1953 Congress confirmed jurisdiction of the states over minerals under navigable inland waters and on the continental shelf to state boundaries.

Forests. A long series of acts has been concerned with the disposal of the public lands and the exploitation of timber and forage. An act in 1891 empowered the President to set aside forest reserves. In 1901 a Forestry Division was created in the General Land Office in the Department of Interior, and in 1905 forestry activities were transferred to the Department of Agriculture. An act in 1907 changed the reserves to the national forests. The Forest Service soon became recognized as one of the most efficient bureaus in Washington as it dealt, on a decentralized basis, with the conservational use of all resources, not just timber. The Weeks Act in 1911 inaugurated cooperation with the states and is best known because it started the policy of land acquisition. Today management of the nation's forest lands has become a balanced cooperation between government and private enterprise on a mosaic of ownership.

Wildlife. The first Federal recognition of responsibility for fisheries and wildlife was in regard to research, with the establishment in 1905 of the Bureau of Biological Survey. An international convention for protection of fur seals was signed in 1911, with protection of other marine animals coming later. The Migratory Bird Act was passed in 1913. Federal refuges were started in 1903, and sizable funds for refuges became available with the Migratory Bird Conservation Act of 1929 and at an accelerated rate from 1933 to 1953. State activities in the fields of wildlife and fishery management have been greatly facilitated by Federal aid under the Pittman-Robertson Act of 1937 and the Dingell-Johnson Act of 1950.   [STANLEY A. CAIN]

Bibliography: S. W. Allen and J. W. Leonard, *Conserving Natural Resources: Principles and Practice in a Democracy*, 2d ed., 1966; G. Borgstrom, *The Hungry Planet: The Modern World at the Edge of Famine*, 1965; C. H. Callison (ed.), *America's Natural Rsources*, 1957; S. V. Ciriacy-Wantrup, *Resource Conservation: Economics and Policies*, 1952; D. C. Coyle, *Conservation: An American Story of Conflict and Accomplishment*, 1957; S. T. Dana, *Forest and Range Policy*, 1956; R. F. Dasman, *Environmental Conservation*, 1959; E. S. Helfman, *Rivers and Watersheds in America's Future*, 1965; G.-H. Smith, *Conservation of Natural Resources*, 3d ed., 1965; E. W. Zimmermann, *World Resources and Industries*, rev. ed., 1951.

Continental shelf and slope

The continental shelf is the zone around the continent, extending from the low-water line to the depth at which there is a marked increase in slope to greater depth. The continental slope is the declivity from the edge of the shelf extending down to great ocean depths. The shelf and slope comprise the continental terrace, which is the submerged fringe of the continent, connecting the shoreline with the 2½-mi-deep (4-km) abyssal ocean floor (see illustration).

Continental shelf. This comparatively featureless plain, with an average width of 45 mi (72 km), slopes gently seaward at about 10 ft/mi (1.9m/km). At a depth of about 70 fathoms (128 m) there generally is an abrupt increase in declivity called the shelf break, or the shelf edge. This break marks the limit of the shelf, the top of the continental

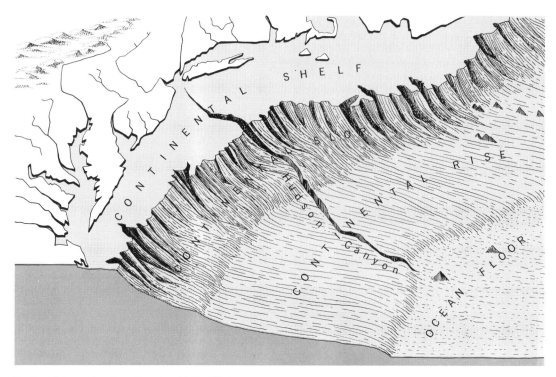

Continental margin off northeastern United States.

slope, and the brink of the deep sea. However, some shelves are as deep as 200–300 fathoms (180–550 m), especially in past or presently glaciated regions. For some purposes, especially legal, the 100-fathom line or 200-m line is conventionally taken as the limit of the shelf. Characteristically, the shelves are thinly veneered with clastic sands, silts, and silty muds, which are patchily distributed but show a slight tendency to become finer offshore. Geologically, the shelf is an extension of, and in unity with, the adjacent coastal plain. The position of the shoreline is geologically ephemeral, being subject to constant prograding and retrograding, so that its precise position at any particular time is not important. Genetically, the origin of the shelf seems to be primarily related to shallow wave cutting (waves cut effectively as breakers and surf only down to about 5 fathoms (9 m), the depth of vigorous abrasion), shoreline deposition, and oscillations of sea level, which have been especially strong during the Pleistocene and Recent.

Although worldwide in distribution and comprising 5% of the area of the Earth, shelves differ considerably in width. Off the east coast of the United States the shelf is about 75 mi wide (120 km), while off the west coast it is about 20 mi wide (32 km). Especially broad shelves fringe northern Australia, Argentina, and the Arctic Ocean. In the Barents Sea the shelf is 750 mi across (1200 km). As along the eastern United States, continental shelves commonly acquire a prism of sediments as the continental margin downflexes. Such capping prisms appear to be nascent miogeosynclines, common in the geologic record.

Continental slope. The drowned edges of the low-density "granitic" or sialic continental masses are the continental slopes. They are the walls of

the deep sea basins. The continental plateaus float like icebergs in the Earth's mantle with the slopes marking the transition between the low-density continents and the heavier oceanic segments of the Earth's crust. Averaging $2\frac{1}{2}$ mi high (4 km) and in some places attaining 6 mi (10 km), the continental slopes are the most imposing escarpments on the Earth. The slope is comparatively steep with an average declivity of 4.25° for the upper 1000 fathoms (1.8 km). Most slopes resemble a straight mountain front but are highly irregular in detail; in places they are deeply incised by submarine canyons, some of which cut deeply into the shelf. Usually the slope does not connect directly with the sea floor; instead there is a transitional area, the continental rise, or apron, built by the shedding of sediments from the continental block.

[ROBERT S. DIETZ]

Bibliography: F. P. Shepard, *Submarine Geology*, 2d ed., 1963.

Crops, irrigation of

The artificial application of water to the soil to produce plant growth. Irrigation also cools the soil and atmosphere, making the environment favorable for plant growth as well as removing undesirable salts when present. The use of some form of irrigation is well documented throughout the history of mankind.

Use of water by plants. All growing plants use water continuously. Growth of crops under irrigation is stimulated by optimum moisture but retarded by excessive or deficient amounts. Conditions influencing the rate of water use by plants include the type of plant and stage of growth, temperature, wind velocity, humidity, available water supply, and soil structure. Plants use the least amount of water upon emergence from the soil and at the end

Table 1. Consumption of water by various crops, in inches

Crop	April	May	June	July	Aug.	Sept.	Oct.	Seasonal total
Alfalfa	3.3	6.7	5.4	7.8	4.2	5.6	4.4	37.4
Beets		1.9	3.3	5.3	6.9	5.8	1.1	24.3
Cotton	1.1	2.0	4.1	5.8	8.6	6.7	2.7	31.0
Peaches	1.0	3.4	6.7	8.4	6.4	3.1	1.1	30.0
Potatoes			0.7	3.4	5.8	4.4		14.0

of the growing period. Irrigation and other management practices should be coordinated with the various stages of growth. A vast amount of research has been done on the use of water by plants, and results are available for crops under varying conditions.

Consumptive use. Consumptive use is an important factor in determining the amount of water required in planning new irrigation projects or in supplying supplemental water to older ones. It has become an important consideration in the arbitration of controversies over water rights in major stream systems.

Consumptive use, or evapotranspiration, is defined as water entering plant roots to build plant tissues, water retained by the plant, and water transpired by leaves into the atmosphere. It also includes water evaporated from adjacent soil surfaces and from free water surfaces such as those in irrigation ditches. Included also is evaporated water that has fallen on plant leaves from sprinklers or rainfall. Consumptive use of water by various crops under varying conditions has been determined by soil moisture studies and computed by other well-established methods for many regions of the United States and other countries. Factors which have been shown to influence consumptive use are precipitation; air temperature; humidity; wind movement; the growing season; and latitude, which influences hours of daylight. Table 1 shows how consumptive use varies by months during the growing season.

The data presented here are for various locations or different years, and cannot be applied specifically. When consumptive use data are presented for an extensive area, such as an irrigation project, the results may be given in acre-feet per acre for a given month of the growing season or for the entire irrigation period. An acre-foot is the amount of water required to cover 1 acre 1 ft deep.

Soil, plant, and water relationships. Soil of root zone depth is the storage reservoir from which plants obtain moisture to sustain growth. Plants take from the soil not only water but dissolved minerals necessary to build plant cells. How often this reservoir must be filled by irrigation is determined by the storage capacity of the soil, depth of the root zone, and water use by the crop. Table 2 shows the approximate amounts of water held by soils of various textures available to plants.

Water enters coarse, sandy soils and silt loams quite readily, but in heavy clays the entry rate may be as slow as 0.2 in./hr or less.

Soil conditions, position of the water table, length of growing season, and other factors exert strong influence on root zone depth. Table 3 shows typical root zone depths in well-drained, uniform soils. The depth of rooting increases during the

entire growing period, given a favorable, unrestricted root zone. Plants in deep, uniform soils usually consume water more slowly from the lower root zone area than from the upper. Thus the upper portion is the first to be exhausted of moisture. However, shallow, frequent irrigations should not be used except on shallow-rooted crops. For most crops the entire root zone should be supplied with moisture when needed.

Maximum production can usually be obtained with most irrigated crops if not more than 50% of the available water in the root zone is exhausted during the maximum stages of growth. Soil and crop conditions may alter this irrigation scheduling somewhat. Application of irrigation water should not be delayed until plants signal a need for moisture; wilting in the hot part of the day may reduce crop yields considerably. Determination of the amount of water in the root zone can be done by laboratory methods which are slow and costly. In modern irrigation practice, tensiometers and resistance blocks are used to make rapid determinations directly with enough accuracy for practical use. These instruments placed in selected field locations will permit an operator to schedule periods of water application for best results. The irrigation system should be designed to supply sufficient water to care for periods of most rapid evapotranspiration of the crop season. The rate of evapotranspiration may vary from 0.12 to 0.25 in. or more per day.

Water quality. Saline and alkaline conditions may affect the productivity and value of one-fourth of the irrigated land of the United States and lesser amounts of nonirrigated acreage. Salt-affected

Table 2. Approximate amounts of water in soils available to plants

Soil texture	Water capacity, in inches for each foot of depth
Coarse sandy soil	$\frac{1}{2}-\frac{3}{4}$
Sandy loam	$1\frac{1}{4}-1\frac{1}{2}$
Silt loam	$1\frac{3}{4}-2\frac{1}{4}$
Heavy clay	$1\frac{3}{4}-2$ or more

Table 3. Approximate root zone depths for various crops

Crop	Root zone depth, ft
Alfalfa	6–10
Corn	5–6
Cotton	4–6
Potatoes	2–3
Grasses	1–2
Trees	5–10

Fig. 1. Furrow method of irrigation. (a) Water from a pump or elevated source is furnished to irrigated furrows through a pipe line. (b) The flow to each furrow through the small pipe is controlled by a small gate valve. The supply pipes 6–12 in. in diameter are made of metal or butyl rubber.

soils are found mostly in arid and semiarid regions. In humid regions soluble salts are usually carried downward by precipitation to the groundwater and ultimately in streams to the sea. Salt-affected soils may be found under natural conditions but the main problem in agriculture arises when a once-productive soil becomes saline because of the application of irrigation water carrying saline constituents. High levels of salt accumulation seriously affect plant growth by limiting the entry of water to the roots; also, some salt constituents are toxic to some crops.

Saline accumulations can often be alleviated by overapplication of irrigation water during and after the crop season. Water is sometimes ponded on fields and allowed to percolate downward below the crop root zone, carrying the saline constituents with it to natural or artificial drains. Under some conditions sprinkling or natural rainfall will produce the desired results.

Soils which have come in contact with high sodium irrigation or groundwater usually contain adsorbed sodium, calcium, and magnesium. The adsorbed sodium has a detrimental effect on soils, making tillage difficult, and often leaving the soil poorly drained and unproductive.

Chemical amendments added to the sodium soils have the effect of replacing sodium with calcium, resulting in improvement of the cultivated land, better drainage, and increased productive capacity. Gypsum and sulfur are the amendments most commonly used to improve sodium soils; others are limestone, calcium chloride, sulfuric acid, lime sulfur, and iron sulfate. Barnyard manure will often improve cultivation, drainage and productivity of the soil. Some soils which show evidence of poor physical condition and drainage may have little or no sodium but have a high clay content and may lack of organic matter.

Visual observations of soils and the crops growing on them are seldom sufficient to diagnose salinity or alkali problems properly. Without proper information, the wrong additives might be used or salt-sensitive crops planted where salt-tolerant ones were indicated. Crops vary greatly in their ability to tolerate saline soils.

Federal agencies and many state universities have facilities for soil testing and are prepared to give advice regarding corrective measures and to indicate the proper crops to grow where soil problems exist.

Methods. Water is applied to crops by surface furrows, flooding, subirrigation, or sprinkling.

Furrow method. Irrigation by furrows is a common method of application to row crops (Fig. 1). The flow, carried in furrows between the rows of plants, percolates into the soil and replenishes the soil reservoir. However, water carried in furrows down slopes may result in serious erosion. The safe furrow stream may be approximately computed by the formula $Q = 10/S$, in which Q is the safe furrow stream in gal/min and S is the slope of the furrow in percentage. Maximum furrow length depends on the size of the furrow stream and the rate that water is absorbed by the soil. Flow to furrows from the supply is controlled by temporary ditch openings, siphon tubes, spiles, or gated surface pipe. Corrugations are small, closely spaced furrows used in irrigation of close-growing crops. They are often employed on soils which crust badly when flooded.

Flood method. This is a method of surface application which may be done with border strips, contour or bench borders, flooding from contour ditches, or basins.

Border strip irrigation is accomplished by advancing a sheet of water down a long narrow area between low ridges known as borders. Moisture enters the soil as the sheet advances. The strips may vary from 20 to 100 ft in width, depending on available water supply and other factors. The border strip must be well leveled and the grade uniform to prevent ponding. The ridges are usually low and well rounded so that crops can be grown and harvested on them. Where saline conditions exist the borders are usually made narrow and are not crossed by machinery. Saline constituents have a tendency to collect in the ridges since they are not carried downward by irrigation water. The border strip method is well adapted to irrigation of close-growing crops and pasture, but it can be used as well for row crops where land is relatively level. Hay and grain can often be irrigated by this method or by flooding from contour ditches on slopes up to 3% and up to 6% on well-established pastures. Slopes below 1% are less liable to erosion damage and the water application is more uniform than on steeper slopes. Border strip irrigation is economical of labor and water if designed and constructed well.

Bench border irrigation is well adapted to uniform, moderately gentle slopes with fairly deep soils. The border strips, instead of running down the slope, are constructed across it. Since each strip must be level in width, considerable earth-moving may be necessary. The width of each strip should remain the same throughout its length and be arranged in multiples of usual row-crop machinery to avoid point rows. The irrigation process is the same as for border strip or basin irrigation. The advantage of bench borders is that they allow efficient, economical use of water on slopes where ordinary methods might require more labor and might result in serious erosion and loss of water.

Flooding the land surface from contour ditches is a common method for irrigation of close-growing crops on land with uneven topography which is not adapted to other methods. Field ditches are usually closely spaced to follow the contour of the slope. Water is diverted from them to flow over the intervening cropped area. This method is being replaced by sprinklers, which are more efficient.

Basin irrigation is well adapted to flatlands. It is done by flooding a diked area, using electrically driven pumps (Fig. 2), to a predetermined depth and allowing the water to enter the soil uniformly throughout the root zone. It may be used for all types of crops, including orchards, where soil and topographic conditions permit.

Subirrigation. This method is accomplished by raising the water table to the root zone of the crop or by carrying moisture to the root zone by perforated underground pipe. Either method requires special soil conditions for successful operation. The pipe system has long been in use for irrigation of baseball diamonds and other small areas. It may have much wider application as it is developed by research since it is economical of water and labor once established.

Sprinkling systems. A sprinkler system consists of pipelines which carry water under pressure from a pump or elevated source to lateral lines along which sprinkler heads are spaced at appropriate intervals. By mechanization, modern sprinkler systems have overcome the objection of high labor costs for moving lateral lines from irrigated to nonirrigated areas. Long lateral lines are now towed to new positions by tractors. One or more systems use a long lateral line pivoted at a central water-supply point. The lateral, carrying the sprinkler heads, is motivated by hydraulic or electrical

Fig. 2. An electrically driven irrigation pumping plant. Natural underground basins furnish water for millions of irrigated acres. Such pumps often supply water at rates of more than 1000 gal/min.

power to move in a circular course irrigating an area containing up to 140 acres. Other systems use the lateral, carrying the sprinkler heads, as an axle for wheels which roll it into position by power or by a small amount of man labor (Fig. 3).

Sprinkler irrigation has the advantage of being adaptable to soils too porous for use with other systems. It can be used on land where soil or topographic conditions cannot be prepared for surface methods. It can be used on steep slopes and operates efficiently with a small water supply. Some disadvantages are high initial cost, distortion of the sprinkler pattern by wind, and some wasteful evaporation losses in hot, dry, windy climates.

Automated systems. Research is in progress on many types of automated irrigation systems and many are now in use. Some simple ones have time clocks which, at predetermined times, close or

Fig. 3. A side-roll sprinkler system which uses the main supply line (often more than 1000 ft long) to carry the sprinkler heads and as the axle for wheels.

open check dams in irrigation ditches, directing water to areas to be watered. Some automated sprinkler systems have laterals carrying the sprinkler heads in place, allowing an entire field to be irrigated with no moving of pipe; watering is started and stopped electrically by connections to tensiometers buried in the field soil.

Some entirely automated systems are in use and others are in the development stage. Irrigation in these systems is begun when electrical resistance blocks or tensiometers placed in the soil of the field signal that the soil is low on moisture by sending an electrical signal; this starts the pump, which is electrically programmed to serve various field areas. When a field is sufficiently watered, the flow is stopped and directed to an unwatered area. Waste water that is usually lost at the end of the field or furrow is channeled back to a pump for reuse in the system.

Fertilizing operations. Liquid fertilizer may be applied to crops along with the irrigation water. This has the advantage of reducing labor costs and producing yields equal to those of mechanical fertilizing methods. When the furrow method of irrigation is used, the liquid fertilizer is mixed with the water; turbulence and thorough mixing is accomplished by placing an obstruction in the supply ditch. With sprinkler irrigation, measured amounts of liquid fertilizer are forced into the supply line by pumping. Careful computations must be made regarding the amount of water to be applied per unit if land area and the amount of fertilizer required are to give desired results. Water carrying the fertilizer must be uniformly distributed over the crop area.

Irrigating humid and arid regions. The acreage of irrigated land is increasing in humid regions and arid and semiarid areas where the limiting factor is lack of water, not land. However, irrigation programs are often more satisfactory when the farmer does not depend on rainfall for crop growth. Good yields are obtained by well-timed irrigation, maintenance of high fertility, well-cultivated land, and the use of superior crop varieties.

There is little difference in the principles of crop production under irrigation in humid and arid regions. The programming of water application is more difficult in humid areas because natural precipitation cannot be accurately predicted. Sprinkler systems are better adapted to humid areas.

To be successful, any irrigation system in any location must have careful planning with regard to soil conditions, topography, climate, cropping practices, water quality, and supply, as well as engineering requirements.

Outlook. It has been estimated that about 36,-000,000 acres are irrigated in the United States; in 1960 a Senate committee estimated that 17,-000,000 acres of potentially new irrigation land existed in the West. The limit of using simple diversions of natural rivers and streams has been reached with few exceptions, and future development of large acreages of irrigated land must come through extensive reservoir storage; high-lift pumping projects, characteristic of Federal programs; transportation of surplus water sources to water-poor areas; and better conservation of water supplies.

Development of full irrigation potential in the western United States will be costly and will depend upon many factors, the nature and extent of which are not fully known. Major diversions of stream flow to regions outside the watershed will involve many complicated interstate problems and compacts. As the population increases, there will be greater competition for water by industry, for municipalities, power development, recreational facilities, and wildlife reserves. Some underground water supplies that irrigate large areas by pumping are being depleted, and must at some time be replenished by transported water if the irrigated area is to remain under cultivation.

Better water conservation could assist in expanding the irrigated acreage. It is estimated that phreatophites (water-loving plants) along streams and irrigation canals transpire 25,000,000 acre-feet of water to the atmosphere; research centers on ways to save at least one-fourth of this. Evaporation from reservoirs accounts for the loss of millions of additional acre-feet, part of which may be saved by monomolecular films spread on the water surface. Seepage from canals, the use of small, shallow reservoirs with large exposed surfaces, and inferior irrigation practices all waste water which might be put to good use. *See* LAND DRAINAGE (AGRICULTURE); WATER CONSERVATION.

[IVAN D. WOOD]

Bibliography: O. W. Israelsen and V. E. Hansen, *Irrigation Principles and Practices*, 3d ed., 1962; W. B. Langbein and W. G. Hoyt, *Water Facts for the Nation's Future*, 1959; USDA, *Yearbook*, 1955.

Dam

A structure that bars or detains the flow of water in an open channel or watercourse. Dams are constructed for several principal purposes. Diversion dams divert water from a stream; navigation dams raise the level of a stream to increase the depth for navigation purposes; power dams raise the level of a stream to create or concentrate hydrostatic head for power purposes; and storage dams store water for municipal and industrial use, irrigation, flood control, river regulation, recreation, or power production. A dam serving two or more purposes is called a multiple-purpose dam. Dams are commonly classified by the material from which they are constructed, such as masonry, concrete, earth, rock, timber, and steel. Most dams now are built either of concrete or of earth and rock.

Concrete dams. Concrete dams may be typed as gravity, arch, or buttress type. Gravity dams depend on weight for stability against overturning and for resistance to sliding on their foundations (Figs. 1 and 2). An arch dam may have a near-vertical face or, more usually, one that curves upstream (Figs. 3 and 4). The dam acts as an arch to transmit most of the horizontal thrust from the water pressure against the upstream face of the dam to the abutments of the dam. The buttress type of concrete dam includes the slab-and-buttress, or Ambursen, type; round- or diamond-head buttress type; multiple-arch type; and multiple-dome type. Buttress dams depend on the weight of the structure and of the water on the dam to resist overturning and sliding.

Fig. 1. Green Peter Dam, concrete gravity type on Middle Santiam River, Willamette River Basin, Ore. Gate controlled overflow-type spillway through crest of dam; powerhouse is at downstream toe of dam. (*U.S. Army Corps of Engineers*)

Forces acting on concrete dams. Principal forces acting on a concrete dam are (1) vertical forces from weight of the structure and vertical component of water pressure against the upstream and downstream faces of the dam, (2) uplift pressures under the base of the structure, (3) horizontal forces from the horizontal component of the water pressure against the upstream and downstream faces of the dam, (4) forces from earthquake accelerations in regions subject to earthquakes, (5) temperature stresses, (6) pressures from silt deposits and earth fills against the structure, and (7) ice pressures.

The uplift pressure under the base of a dam varies with the effectiveness of the foundation drainage system and the perviousness of the foundation. A grout curtain to reduce underseepage is usually provided beneath a dam and parallel to its axis.

Horizontal forces caused by earthquakes often are assumed to equal 0.05–0.10 times the force of gravity, with the vertical force being somewhat less; these forces act on the mass of the dam. In addition, hydrodynamic forces are produced against the dam by water in the reservoir and by tailwater downstream of the dam. The water adjacent to the dam resists the movement of the dam and increases the forces acting on it.

Stresses resulting from temperature changes must be considered in analyzing arch dams. These stresses are usually disregarded in the design of concrete gravity dams, but must be controlled to acceptable limits by concreting and curing methods, discussed below.

Pressure from silt deposited in the reservoir against the dam is considered only after sedimentation studies indicate that it may be a significant factor. Backfill pressures are important where a concrete gravity dam ties into an embankment.

Ice pressure, applied at the maximum elevation at which the ice will occur in project operations, is considered when conditions indicate that it would be significant. The pressure, commonly assumed to be 10,000–20,000 lb per linear foot, results from the thermal expansion of the ice sheet and varies with the rate and magnitude of temperature rise and thickness of the ice.

Stability and allowable stresses. Stability of a concrete gravity dam is evaluated by analyzing the available resistance to overturning and sliding. To satisfy the former, the resultant of forces is required to fall within the middle third of the base under normal load conditions. Sliding stability is assured by requiring available shear and friction resistance to be greater by a designated safety factor than the forces tending to produce sliding. The strengths used in computing resistance to sliding are based on investigation and tests of the foundation. Bearing strength of the foundation for a gravity dam is a controlling factor only for weak foundations or for high dams. Because an arch dam depends on the competency of the abutments, the rock bearing strength must be sufficient to provide an adequate safety factor for the compressive stresses, and the resistance to sliding along any weak surface must be great enough to provide an adequate safety factor.

Concrete stresses control the design of arch dams, but ordinarily not gravity dams. The stresses adopted for arch dams are conservative because

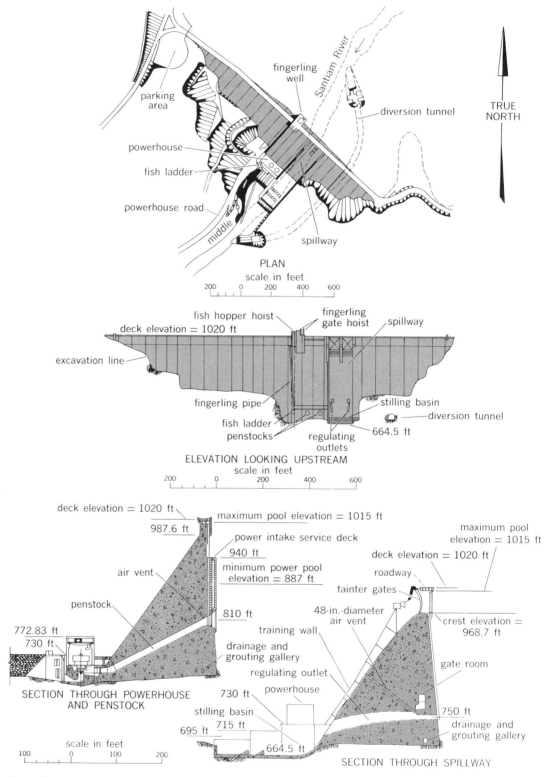

Fig. 2. Plan and sections of Green Peter Dam. (*U.S. Army Corps of Engineers*)

this type of structure must have a high factor of safety. A safety factor of 4 on concrete compressive strength is commonly used for normal load conditions.

Concrete temperature control. Volume changes accompanying temperature changes in a concrete dam tend to cause the development of tensile stresses. A major factor in development of temper-

ature changes within a concrete mass is the heat developed by chemical changes in the concrete after placement. Uncontrolled temperature changes can cause cracking which may endanger the stability of a dam, cause leakage, and reduce durability. Temperatures are controlled by using cementing materials having low heat of hydration, and by artificial cooling by precooling the concrete

Fig. 3. East Canyon Dam, a thin-arch concrete structure on the East Canyon River, Utah. Note uncontrolled overflow-type spillway through crest of dam at right center of photograph. (*U.S. Bureau of Reclamation*)

mix or circulating cold water through pipes embedded in the concrete or both.

Concrete dams are constructed in blocks, with the joints between the blocks serving as contraction joints (Fig. 5). In arch dams the contraction joints are filled with cement grout after maximum shrinkage has occurred to assure continuous bearing surfaces normal to the compressional forces set up in the arch when the water load is applied to the dam.

Quality control. During construction, continuing testing and inspection are performed to ensure that the concrete will be of required quality. Tests are also made on materials used in manufacture of the concrete, and concrete batching, mixing, transporting, placing, curing, and protection are continuously inspected.

Earth dams. Earth dams have been used for water storage since early civilizations. Improvements in earth-materials techniques, particularly the development of modern earth-handling equipment, have brought about a wider use of this type of dam, and today as in primitive times the earth embankment is the most common dam (Figs. 6 and 7). Earth dams may be built of loose rock, gravel, sand, silt, or clay in various combinations.

Most earth dams are constructed with an inner impervious core, with upstream and downstream zones of more pervious materials, sometimes including rock zones at the downstream toe. Earth dams limit the flow of water through or under the dam by use of fine-grained soils. Where possible, these soils are formed into a relatively impervious core connected to bedrock by a cutoff trench backfilled with compacted soil. If such cutoffs are not economically feasible because of the great depth of pervious foundation soils, then the central impervious core is connected to a long horizontal upstream impervious blanket that increases the length of the seepage path. The impervious core is often encased in pervious zones of sand, gravel, or rock fill for stability. When there is a large difference in the particle sizes of the core and pervious zones, transition zones are required to prevent the core material from being transported into the pervious zones by seeping water. In some cases where pervious soils are scarce, the entire dam may be a homogeneous fill of relatively impervious soil. Downstream pervious drainage blankets are provided to collect seepage passing through, under, and around the abutments of the dam.

Materials can be obtained from required excavations for the dam and appurtenances or from borrow areas. Rock fill is generally used when large quantities of rock excavation are needed or when soil borrow is scarce.

Earth-fill embankment is placed in layers and compacted by sheepsfoot rollers or heavy pneumatic rollers. Moisture content of silt and clay soils is carefully controlled to facilitate optimum compaction. Sand and gravel fills are compacted in

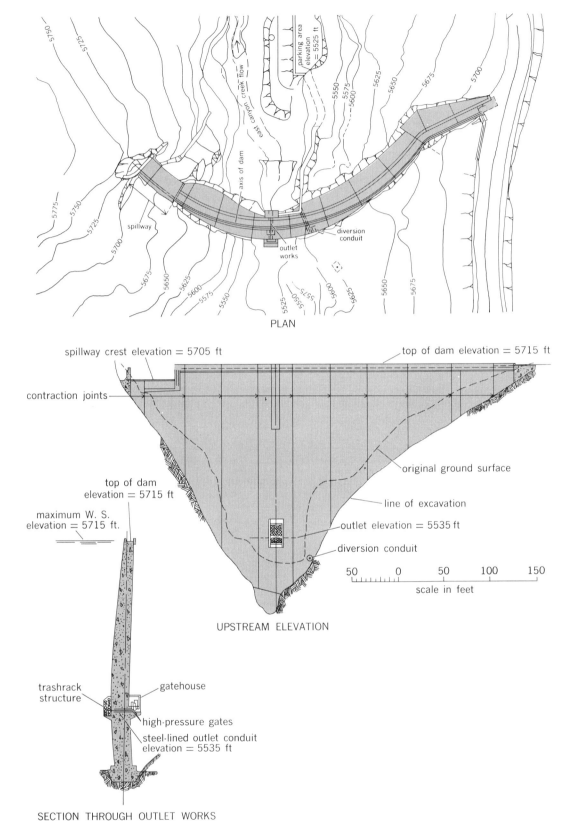

PLAN

UPSTREAM ELEVATION

spillway crest elevation = 5705 ft

top of dam elevation = 5715 ft

contraction joints

original ground surface

line of excavation

outlet elevation = 5535 ft

diversion conduit

50   0   50   100   150

scale in feet

top of dam
elevation = 5715 ft

maximum W. S.
elevation = 5715 ft.

trashrack
structure

gatehouse

high-pressure gates

steel-lined outlet conduit
elevation = 5535 ft

SECTION THROUGH OUTLET WORKS

Fig. 4. Plan and sections of East Canyon Dam. (*U.S. Bureau of Reclamation*)

slightly thicker layers by pneumatic rollers, vibrating steel drum rollers, or placement equipment. The placement moisture content of pervious fills is less critical than for silts and clays. Rock fill usually is placed in layers 1–3 ft deep and is com-

pacted by placement equipment and vibrating steel drum rollers.

Spillways. A spillway releases water in excess of storage capacity so that the dam and its foundation are protected against erosion and possible failure.

Fig. 5. Block method of construction on a typical concrete gravity dam. (*U.S. Army Corps of Engineers*)

Fig. 6. Aerial view of North Fork Dam, a combination earth and rock embankment on the North Fork of Pound River, Va. Channel-type spillway (left center) has simple overflow weir. (*U.S. Army Corps of Engineers*)

All dams must have a spillway, except small ones where the runoff can be safely stored in the reservoir without danger of overtopping the dam. Ample spillway capacity is of particular importance for large earth dams, which would be destroyed or severely damaged by being overtopped.

Types. Spillways are of two general types: the overflow type, constructed as an integral part of the dam; or the channel type, located as an independent structure discharging through an open chute or tunnel. Either type may be equipped with gates to control the discharge. Various control structures have been used for channel spillways, including the simple overflow weir, side-channel overflow weir, and drop or morning-glory inlet where the water flows over a circular weir crest and drops directly into a tunnel.

Unless the discharge end of a spillway is remote from the toe of the dam or erosion-resistant bedrock exists at shallow depths, some form of energy dissipator must be provided to protect the toe of the dam and the foundation from spillway

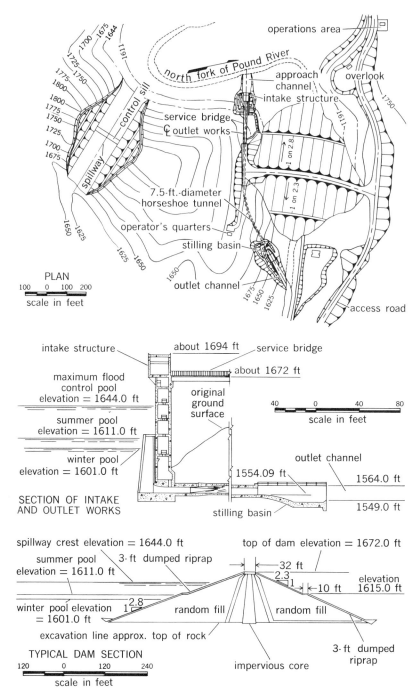

Fig. 7. Plan and sections of North Fork of Pound Dam. (*U.S. Army Corps of Engineers*)

discharges. For an overflow spillway the energy dissipator may be a stilling basin, a sloping apron downstream from the dam, or a submerged bucket. When a channel spillway terminates near the dam, it usually has a stilling basin. A flip bucket is used for both overflow and channel spillways when the flow can be deflected far enough downstream, usually onto rock, to prevent erosion at the toe of the dam or end of the spillway.

Gates. Several types of gates may be used to regulate and control the discharge of spillways (Fig. 8). Tainter gates are comparatively low in cost and require only a small amount of power for operation, being hydraulically balanced and of low friction. Drum gates, which are operated by reservoir

pressure, are costly but afford a wide, unobstructed opening for passage of drift and ice over the gates. Vertical-lift gates of the fixed-wheel or roller type are sometimes used for spillway regulation, but are more difficult to operate than the others. Floating ring gates control the discharge of morning-glory spillways. Like the drum gate, this type offers a minimum of interference to the passage of ice or drift over the gate and requires no external power for operation.

Reservoir outlet works. These are used to regulate the release of water from the reservoir; they consist essentially of an intake and an outlet connected by a water passage, and are usually provided with gates. Outlet works usually have trash-racks at the intake end to prevent clogging by debris. Bulkheads or stop logs are commonly provided to close the intakes so that the passages may be unwatered for inspection and maintenance. A stilling basin or other type of energy dissipator is usually provided at the outlet end.

Locations. Outlets may be sluices through concrete dams with control valves located in chambers in the dam or on the downstream end of the sluices, tunnels through the abutments of the dam, or cut-and-cover conduits extending along the foundation through an earth-fill dam. In the last case special precautions must be taken to prevent leakage of water along the outside of the conduit.

Outlet control gates. Various gates and valves are used for regulating the release of water from reservoirs, including high-pressure slide gates, tractor gates (roller or wheel), and radial or tainter gates (Fig. 9); also needle valves of various kinds, butterfly valves, fixed cone dispersion valves, and cylinder or sleeve valves. They must be capable of operating, without excessive vibration and cavitation, at any opening and at any head up to the maximum to which they may be subjected. They also must be capable of opening and closing under the maximum operating head. Emergency gates generally are used upstream of the operating gates, where stored water is valuable, so that closure can be made if the service gate should fail to function.

The slide gate, which consists of a movable leaf that slides on a stationary seat, is the most commonly used control gate. The high-pressure slide gate is of rugged design, having corrosion-resisting metal seats on both the movable rectangular leaf and the fixed frame. This gate has been used for regulating discharges under heads of over 600 ft.

Provision of low-level outlet. The usual storage reservoir has low-level outlets near the elevation of the stream bed to enable release of all the stored water. Some power and multiple-purpose dams have relatively high-level dead storage pools and do not require low-level outlets for ordinary operation. In such a dam, provision of a capability for emptying the reservoir in case of an emergency must be weighed against the additional cost.

Penstocks. A penstock is a pipe that conveys water from a forebay, reservoir, or other source to a turbine in a hydroelectric plant. It is usually made of steel, but reinforced concrete and wood-stave pipe have also been used. Pressure rise and speed regulation must be considered in the design of a penstock.

Pressure rise, or water hammer, is the pressure change that occurs when the rate of flow in a pipe or conduit is changed rapidly. The intensity of this pressure change is proportional to the rate at which the velocity of the flow is accelerated or decelerated. Accurate determination of the pressure changes that occur in a penstock involves consideration of all operating conditions. For example, one important consideration is the pressure rise that occurs in a penstock when the turbine wicket gates are closed after the loss of load.

Selection of dam site. This depends upon such factors as hydrologic, topographic, and geologic conditions; storage capacity of reservoir; accessibility; cost of lands and necessary relocations of prior occupants or uses; and proximity of sources of suitable construction materials. For a storage dam the objective is to select the site where the

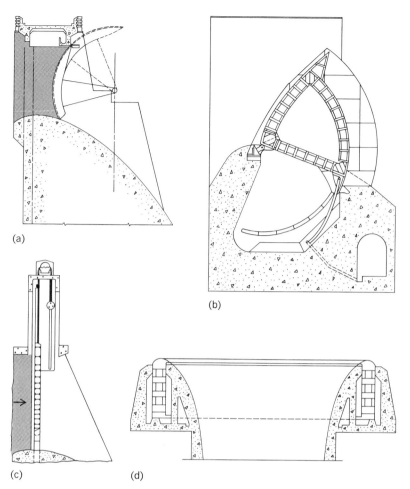

Fig. 8. Spillway gates. (*a*) Tainter gate. (*b*) Drum gate. (*c*) Vertical lift gate. (*d*) Ring gate. (*U.S. Army Corps of Engineers and U.S. Bureau of Reclamation*)

desired amount of storage can be most economically developed. Power dams must be located to develop the desired head and storage. For a diversion dam the site must be considered in conjunction with the location and elevation of the outlet canal or conduit. Site selection for navigation dams involves special factors such as desired navigable depth and channel width, slope of river channel, natural river flow, amount of bank protection, amount of channel dredging, approach and exit conditions for tows, and locations of other dams in the system.

Unless topographic and geologic conditions for a proposed storage, power, or diversion dam site are satisfactory, hydrological features may need to be subordinated. Important topographic characteristics include width of the floodplain, shape and height of valley walls, existence of nearby saddles for spillways, and adequacy of reservoir rim to retain impounded water. Controlling geologic conditions include the depth, classification, and engineering properties of soils and bedrock at the dam site, and the occurrence of sinks, faults, and major landslides at the site or in the reservoir area. The elevation of the groundwater table is also significant because it will influence the construction operations and suitability of borrow materials. The beneficial effect of reservoir water on groundwater recharge may become an important consideration, as well as the adverse effects on existing

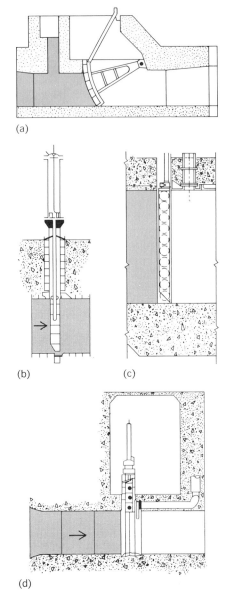

(a)

(b)   (c)

(d)

Fig. 9. Outlet gates. (a) Tainter gate. (b) High-pressure slide gate. (c) Tractor gate. (d) Jet flow gate. (*U.S. Army Corps of Engineers and U.S. Bureau of Reclamation*)

or potential mineral resources and developments that would be destroyed or require relocation at the site or within the reservoir.

Selection of type of dam. This is made on the basis of the estimated costs of various types. The most important factors are topography, foundation conditions, and the accessibility of construction materials. In general, a hard-rock foundation is suitable for any type of dam, provided the rock has no unfavorable jointing, there is no danger of movement in existing faults, and foundation underseepage can be controlled at reasonable cost. Rock foundations of high quality are essential for arch dams because the abutments receive the full thrust of the water pressure against the face of the dam. Rock foundations are necessary for all medium and high concrete dams. An earth dam may be built on almost any kind of foundation if properly designed and constructed.

The chance of an embankment dam being most economical is improved if large spillway and outlet capacities are required and topography and foundation are favorable. In a wide valley a combination of an earth embankment dam and a concrete dam section containing the spillway and outlets often is economical. Availability of suitable construction materials frequently determines the most economical type of dam. A concrete dam requires adequate quantities of suitable concrete aggregate and reasonable availability of cement, while an earth dam requires sufficient quantities of both pervious and impervious earth materials. If quantities of earth materials are limited and enough rock is available, a rock-fill dam with an impervious earth core may be the most economical.

Determination of dam height. The dam must be high enough to (1) store water to the normal full-pool elevation required to meet intended functions of the project, (2) provide for temporary storage of floods exceeding the normal full-pool elevation, and (3) provide sufficient freeboard height above the maximum surcharge elevation to assure an acceptable degree of safety against possible overtopping.

Physical characteristics of the dam and reservoir site or existing developments within the reservoir area may impose upper limits in selecting the normal full-pool level. In other circumstances economic considerations govern.

With the normal full-pool elevation established, flood flows of unusual magnitude may be passed by providing spillways and outlets large enough to discharge the maximum probable flood without raising the reservoir above the normal full elevation or, if it is more economical, by raising the height of the dam and obtaining additional lands to permit the reservoir to temporarily attain surcharge elevations above the normal pool level during extreme floods. Use of temporary surcharge storage capacity also serves to reduce the peak rates of spillway discharge.

Freeboard height is the distance between the maximum reservoir level and the top of the dam. Usually 3 ft or more of freeboard is provided to avoid overtopping the dam by wind-generated waves. Additional freeboard may be provided for possible effects of surges induced by earthquakes, landslides, or other unpredictable events.

Diversion of stream. During construction the dam site must be unwatered so that the foundation may be prepared properly and materials in the structure may be moved easily into position. The stream may be diverted around the site through tunnels, passed through or around the construction area by flumes, passed through openings in the dam, or passed over low sections of a partially completed concrete dam. Diversion may be conducted in one or more stages, with a different method used for each stage.

Foundation treatment. The foundation of a dam must support the structure under all operating conditions. For concrete dams, following removal of unsatisfactory materials to a sound foundation surface, imperfections such as adversely oriented rock joints, open bedding planes, localized soft seams, and faults lying on or beneath the foundation surface receive special treatment. Necessary foundation treatment prior to dam construction may include "dental excavation" of surface weaknesses, or shafting and mining to remove deeper localized weaknesses, followed by backfilling with concrete or grout. Such work is sometimes supplemented by pattern grouting of foundation zones after construction of the dam. Foundation features such as rock joints, bedding planes, or faults that do not require preconstruction treatment are consolidated by curtain grouting from a line of deep grout holes located near the upstream heel and extending the full length of the dam. Although a grout curtain controls seepage at depth, and thus also reduces hydrostatic pressures from the reservoir acting on the base of the dam, the effectiveness of the grouting or its permanency cannot be relied upon alone. As a result, drain holes are drilled into the foundation just downstream of the grout curtain to intercept seepage passing through it. Occasionally chemical solutions such as acrylamide, sodium silicate, chrome-lignin, and polyester and epoxy resins are used for consolidating soils or rocks with fine openings.

The foundation of an earth dam must safely support the weight of the dam, limit seepage of stored water, and prevent transportation of dam or foundation material away or into open joints or seams in the rock by seepage. Earth-dam foundation treatment may include removal of excessively weak surface soils to prevent both potential sliding and excessive settlement of the dam, excavation of a cutoff trench to rock, and grouting of joints and seams in the bottom and downstream side of the cutoff trench. The cutoff trenches and grouting extend up the abutments, which are first stripped of weak surface materials.

When weak soils in the foundation of an earth dam cannot be removed economically, the slopes of the embankment must be flattened to reduce shear stress in the foundation to a value less than the soil strength. Relief wells are installed in pervious foundations to control seepage uplift pressures and to reduce the danger of piping when the depth of the pervious material is such as to preclude an economical cutoff.

Instrumentation. Instruments are installed at dams to observe structural behavior and physical conditions during construction and after filling, to check safety, and to provide information for design improvement.

In concrete dams instruments are used to measure stresses either directly or to measure strains from which stresses may be computed. Plumb lines are used to measure bending, and clinometers to measure tilting. Contraction joint openings are measured by joint meters spanning between two adjacent blocks of a dam. Temperatures are measured either by embedded electrical resistance thermometers or by adapting strain, stress, and joint measuring instruments. Water pressure on the base of a concrete dam at the contact with the foundation rock is measured by uplift pressure cells. Interior pressures in a concrete dam are measured by embedded pressure cells. Measurements are also made to determine horizontal and vertical movements; strong-motion accelerometers are being installed on and near dams in earthquake regions to record seismic data.

Instruments installed in earth-dam embankments and foundations are piezometers to determine pore water pressure in the soil or bedrock during construction and seepage after reservoir impoundments; settlement gages to determine settlements of the foundation of the dam under dead load; vertical and horizontal markers to determine movements, especially during construction; and inclinometers to determine horizontal movements along a vertical line.

Inspection of dams. Because failure of a dam may result in loss of life or property damage in the downstream area, it is essential that dams be inspected systematically both during construction and after completion. The design of dams should be reviewed to assure competency of the structure and its site, and inspections should be made during construction to ensure that the requirements of the design and specifications are incorporated in the structure.

After completion and filling, inspections may vary from cursory surveillance during day-to-day operation of the project to regularly scheduled comprehensive inspections. The objective of such inspections is to detect symptoms of possible distress in the dam at the earliest time. These symptoms include significant sloughs or slides in embankments; evidence of piping or boils near embankments; abnormal changes in flow from drains; unusual increases in seepage quantities; unexpected changes in pore water pressures or uplift pressures; unusual movement or cracking of embankments or abutments; significant cracking of concrete structures; appearance of sinkholes or localized subsidence near foundations; excessive deflection, displacement, erosion, or vibration of concrete structures; erratic movement or excessive deflection or vibration of outlet or spillway gates or valves; or any other unusual conditions in the structure or surrounding terrain.

Detection of any such symptoms of distress should be followed by an investigation of the causes, probable effects, and remedial measures required. Inspection of a dam and reservoir is particularly important following significant seismic events in the locality. Systematic monitoring of the instrumentation installed in dams is essential to the inspection program.

[U.S. ARMY CORPS OF ENGINEERS
AND U.S. BUREAU OF RECLAMATION]

Bibliography: American Concrete Institute, *Symposium on Mass Concrete*, Spec. Publ. SP-6, 1963; W. P. Creager, J. D. Justin, and J. Hinds, *Engineering for Dams*, 1945; C. V. Davis, *Handbook of Applied Hydraulics*, 1952; J. L. Sherard et al., *Earth and Earth-Rock Dams*, 1963; G. B. Sowers and G. F. Sowers, *Introductory Soil Mechanics and Foundations*, 1961; U.S. Bureau of Reclamation, *Design of Small Dams*, 1960; U.S. Bureau of Reclamation, *Trial Load Method of Analyzing Arch Dams*, Boulder Canyon Proj. Final Rep., pt. 5, Bull. no. 1, 1938; H. M. Westergaard, Water pressures on dams during earthquakes, *Trans. ASCE*, 98:418–472, 1933.

Defoliant and desiccant

Defoliants are chemicals that cause leaves to drop from plants; defoliation facilitates harvesting. Desiccants are chemicals that kill leaves of plants; the leaves may either drop off or remain attached; in the harvesting process the leaves are usually shattered and blown away from the harvested material. Defoliants are desirable for use on cotton plants because dry leaves are difficult to remove from the cotton fibers. Desiccants are used on many seed crops to hasten harvest; the leaves are cleaned from the seed in harvesting.

True defoliation results from the formation of an abscission layer at the base of the petiole of the leaf. Most of the chemicals bring about this type of defoliation, and the leaves abscise and drop from the plant. Certain fortified oil-emulsion contact herbicides will kill plant leaves at a low application cost, but usually such killing does not result in abscission.

The most common agency of defoliation in nature is frost; frosted leaves dry up and fall off after a few days.

Many years ago it was found that a dust of calcium cyanamide applied to cotton plants would result in defoliation. For success, however, it was necessary that the plants be wet with dew and that the humidity remain high for a period of several days and nights.

Because of the advantages of defoliation, much effort has gone into the search for other chemical defoliants. Improved formulations containing calcium cyanamide have been produced; borate-chlorate combinations have proved to be good defoliants; and 3,6-endoxohexahydrophthalic acid (endothall), aminotriazole (ATA), ammonium thiocyanate, magnesium chlorate, arsenic acid, diquat, paraquat, cacodylic acid, tributylphosphorotrithioate, and hexachloroacetone in oil have been used. A new compound, sodium *cis*-3-chloroacrylate, defoliates at 0.5–2.0 lb/acre; desiccation requires up to 5.0 lb/acre.

Alfalfa, clovers, soybeans, field beans, and a good many flower and nursery crops are now defoliated. Other crops such as rice and grain sorghums are sprayed with contact herbicides to prepare them for timely harvest. The term preharvest desiccation is used to describe this process. Since the Pacific campaigns of World War II, work has been going on to discover defoliants for use in jungle warfare. In the war in Vietnam, chemical defoliation of trees became an essential feature of the military effort to prevent ambush and to combat guerrilla methods. Since 1962, millions of pounds of chemical compounds, principally esters of 2,4-D and 2,4,5-T, cacodylic acid, and picloram, have been used in this way. Twin-engined C-123 cargo planes, fitted with 1000-gal tanks and spray booms, carried out the defoliation assignments. Strips bordering highways and waterways were defoliated, as well as borders around military installations, airfields, ammunition dumps, and villages. Defoliants and desiccants were also used to destroy crops. It was in an effort to find such chemical agents that 2,4-D was discovered in 1942 by E. J. Kraus of the department of botany, University of Chicago. More than 20 years later 2,4-D and 2,4,5-T were the principal compounds used for crop destruction in Vietnam.

[ALDEN S. CRAFTS]

Degree-day

A unit used in estimating energy requirements for building heating and, to a lesser extent, for building cooling. It is applied to all fuels, district heating, and electric heating. Origin of the degree-day was based on studies of residential gas heating systems. These studies indicated that there existed a straight-line relation between gas used and the extent to which the daily mean outside temperature fell below 65°F.

The number of degree-days to be recorded on any given day is obtained by averaging the daily maximum and minimum outside temperatures to obtain the daily mean temperature. This procedure was found to be adequate and less time-consuming than the averaging of 24 hourly temperature readings. The daily mean so obtained is subtracted from 65°F and tabulated. Monthly and seasonal totals of degree-days obtained in this way are available from local weather bureaus. They currently use the 30-year period 1931–1960 as their basis.

The base temperature of 65°F was established by observation of the outside temperatures at which the majority of homeowners started their gas heating systems in the early fall. Interpreted in another way, the internal heat gains in residences, which appear to be about 5000 Btu/hr, are sufficient to provide heat down to a balance temperature of 65°F.

Provided that the efficiency of the heating equipment remains constant, the fuel usage of a given house will be proportional to the number of degree-days. For purposes of comparing one heating season with another, constant efficiency is a reasonable assumption. However, efficiency of the heating equipment decreases markedly in the warmer portions of the heating season. Therefore, the fuel use per degree-day will be higher in spring and fall than in winter. The same considerations apply to use of degree-day data in warmer climates compared to cooler regions.

A frequent use of degree-days for a specific building is to determine before fuel storage tanks run dry when fuel oil deliveries should be made.

Number of Btu which the heating plant must furnish to a building in a given period of time is

$$\text{Btu required} = \text{Heat rate of building} \times 24 \times \text{degree-days}$$

where "Btu required" is the heat supplied by the heating system to maintain the desired inside temperature. "Heat rate of building" is the hourly building heat loss divided by the difference between inside and outside design temperatures.

To determine fuel requirement from this relationship, two decisions are needed. First, the correct seasonal efficiency of the heating system must by obtained or estimated; second, appropriate base temperature for the degree-day data must be calculated. Seasonal efficiency is generally less than the test efficiency of the heating unit although the load efficiency curves are surprisingly level down to quite low loads. Base temperature for degree-day values may be chosen from consideration of internal heat gains of the building. (As already discussed, these steps need not be taken for residences, which tend to be supplied with relatively small yet similar internal gains.) When the estimating procedure is applied to buildings with high levels of internal heat gains, as in a well-lighted office building, then degree-day data on other than a 65°F basis are required. Balance temperatures for both day operation and night operation may also need to be taken into consideration. Then it would be proper to replace the term degree-days in the equation with the new term adjusted degree-days. *See* CLIMATOLOGY; HEATING, COMFORT.

[C. GEORGE SEGELER]

Desert vegetation

No desert on Earth is so extreme as to be totally lacking in plant life. However, some deserts, such as the Atacama in Chile or Rub' al Khali in Arabia, have such long periods of drought (often years) that vegetation is sparse and ephemeral and the land appears to be devoid of plants. There are places in all deserts that have no visible plants: saline flats or playas, areas of bare rock, and active sand dunes. Deep, stable, nonsaline soils in most deserts, however, do have a vegetation of small shrubs and herbaceous plants. In warmer deserts, this vegetation may be interspersed with spectacular large cacti, euphorbias, yuccas, or small trees.

The density of desert vegetation depends upon the amount of water available for plant growth. Since water is available more often and in greater amounts in some deserts than in others, a greater diversity and abundance of vegetation will be found in such deserts. The density gradient of desert vegetation parallels the moisture gradient, whether this be on a continental or very local scale, from arroyo bottom to ridgetop. Temperature plays a secondary but important role; diversity in numbers of species and kinds of life-forms is far greater in warm deserts than in deserts that have a cold winter.

Life-form adaptations. There is a great variety of ways by which plants can grow and survive in deserts. These modifications are combinations of structural and physiological adaptations. All are concerned with an ability to get, utilize, and conserve water. Most desert vegetation is a mixture of more than one of these life-forms, which belong to several species. The warmer, moister, and more diverse the topography, the more kinds of desert-plant life-forms will occur together. A description of the principal kinds follows.

Ephemerals. These are annual plants that live their whole life cycle from germinating seed to flower to seed again in one short moist season. They are usually very small plants, but most of them have the ability to grow larger and produce more seeds in wetter years. They are not really drought-resisters but drought-escapers and spend the dry season, no matter how long, in the form of a hard, dormant seed. The more severe the desert (greater length of drought), the greater the percentage of flora with the ephemeral life-form.

Phreatophytes. The root systems of these plants, termed well-plants, tap the liquid groundwater that underlies desert basins and dry watercourses. Their roots—for example, those of the mesquite (*Prosopis juliflora*)—may penetrate more than 50 ft, making them relatively independent of temporary soil moisture near the surface. Because of their dependable and easily obtained source of water, they too are not so much drought-resistant as drought-escaping.

Succulents. These are the fleshy water-storing plants that most people associate with deserts and comprise cacti, euphorbs, and thick-leaved mem-

bers of the lily family, such as *Aloe*. They not only store water in stems and leaves but conserve water by opening their stomates, mainly at night. Carbon dioxide enters the plant through the stomates at night, and the carbon is stored as malic acid until utilized in photosynthesis the following day (crassulacean acid metabolism). Succulents are drought-resisters because they store and conserve water. With few exceptions, they are confined to warm deserts.

Deciduous xerophytes. These are woody plants (shrubs or small trees) that put out small leaves very quickly after a rain—but just as quickly drop them when drought returns in a few days or weeks. Examples are the ocotillo (*Fouquieria splendens*) of the warm deserts of the Southwest and the shrubby *Oxalis* of the Atacama Desert (*O. gigantea*). Leaves and flowers appear on plants of the latter species so quickly after a rain as to be almost "overnight." In a way, such plants are drought-escapers in that their leaves do not withstand drought stress. However, the main plant body is extremely drought-resistant and for this reason they are classified as xerophytes (drought plants).

True xerophytes. Plants that retain their leaves during long periods of drought are true xerophytes; their leaves carry on some metabolism even during drought stress. They do this by a combination of a number of traits. All of these plants can develop very low water potentials in their leaves. Many of them apparently can carry on photosynthesis and respiration at such low potentials. Almost all have small or narrow leaves, often light-colored, which reflect a high proportion of solar radiation. These small, light-colored leaves have a favorable heat balance in spite of a low transpiration (evaporative heat loss) rate. Because of their low leaf water potentials and extensive root systems, true xerophytes can extract much of the soil moisture and utilize it efficiently in metabolism even through rather extended droughts. Most of them can drop their leaves and survive through droughts of several years. Many true xerophytes are small spiny shrubs, but others are grasses and small perennial herbaceous plants. Large or thickened root systems are common. *See* ECOLOGY, PHYSIOLOGICAL (PLANT).

Halophytes. Plants that can tolerate salty soils are termed halophytes. They are common in the lower parts of desert basins and may exist in any of the life forms mentioned above.

Effects of physical environment. Unlike the vegetation of moist regions, desert vegetation, because of its open and sparse nature, allows an almost complete control of the community by the physical environment. The sun heats the rocks; the dry wind blows between the small gray shrubs carrying sand, which abrades and which piles up under the branches. At night, outward radiation results in low temperatures at the soil surface; a daily swing of 100°F between maximum and minimum temperatures is common at this level. In such an environment scarce water becomes even scarcer, for the heat of the soil drives it into the dry air.

The local pattern of plant community types is determined by the substratum with the most available soil moisture. The amount of soil moisture depends upon those topographic and substratum

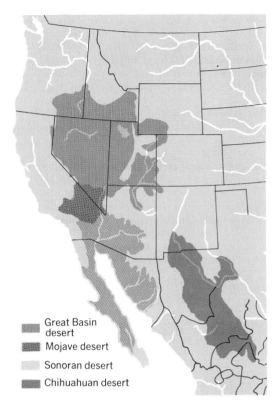

Great Basin
desert

Mojave desert

Sonoran desert

Chihuahuan desert

Fig. 1. The four desert regions of western North America. (From F. Shreve, The desert vegetation of North America, Bot. Rev., 8 195–246, 1942)

properties affecting absorption, infiltration, and evaporation of water. Thus, in any desert, the local vegetational mosaic is largely coincidental with the pattern produced by the interaction of geologic substratum and topography. Limestone supports quite a different vegetation than does sandstone or granite. Stabilized sand dunes have vegetational types distinct from those of nearby areas of saline clay or rocky alluvial fans. Except along the dry arroyos, where phreatophytes are common, the most productive areas in the desert are stabilized sand dunes. There are many reasons for increased production in these areas: Precipitation sinks easily into sand rather than running off; the surface sand dries into a mulch which prevents further evaporation from the soil; sand does not hold water tightly, as do silts and clays, so that soil moisture is readily available to deep plant roots; and sands generally have been leached of inhibiting salts. Salty soils are usually the least productive areas because of the strong osmotic concentrations (low water potentials) of the soil moisture. See DUNE VEGETATION.

Western North America. Even though the deserts of Southwestern North America are relatively small compared to those of Africa and Asia, they are quite diverse and their vegetations are representative of the kinds found elsewhere. As illustrated in Fig. 1, F. Shreve has divided the North American desert into four large regions, each with a characteristic climate and a unique array of vegetational types. There are, however, no really sharp lines between the regions of their climates and vegetations. Because of their lower lati-

tudes, winters are milder and summers are hotter in the Mojave, Sonoran, and Chihuahuan deserts. These can therefore be termed warm deserts. The Great Basin Desert is a cold winter desert because it is farther north and at higher elevations. Snow and very low temperatures (often far below 0°F), make its environment unsuitable for many of the kinds of plants that live in the three warm deserts.

Great Basin Desert. The vegetation is a monotonous open shrubland with almost no cacti or large plants. It occupies the valleys, foothills, and lower mountain ranges of most of Nevada and Utah and parts of adjacent states. The northern and upper edges, which have more precipitation, are characterized by sagebrush (*Artemisia tridentata*); the drier parts (sometimes with dry saline soils) are covered with a very open vegetation (Fig. 2) of small spiny shrubs such as shadscale (*Atriplex confertifolia*), bud sage (*Artemisia spinescens*), and blackbrush (*Coleogyne ramosissima*). Saline flats usually have very open stands of greasewood (*Sarcobatus vermiculatus*) with species of *Atriplex*, or they may be bare. Indian rice grass (*Oryzopsis hymenoides*), indigo bush (*Dalea polyadenia*), fourwing saltbush (*Atriplex canescens*), and a number of other shrubs and herbaceous plants are also to be found on the Great Basin Desert.

Mojave Desert. The characteristic xerophytic shrub of all the American warm deserts is the creosote bush (*Larrea divaricata*). Its bronze-green color contrasts in the Mojave with the gray color of the Great Basin shrubs. The Mojave is also the home of the Joshua tree (*Yucca brevifolia*), which in places stands out prominently above the shrubs. There are more cacti (but no really large ones) here than in the Great Basin; also more spring ephemerals because of the mild winters with some rain.

Sonoran Desert. This is the most subtropical of the North American deserts and extends south well into Mexico. In addition to the creosote bush and its associates, there are many cacti, some of which are quite large. The most prominent cactus is the saguaro (*Cereus giganteus*), typical of the desert around Tucson, Ariz. Barrel cacti (*Echinocactus* spp.), hedgehog cacti (*Echinocereus* spp.), and pricklypears (*Opuntia* spp.) are also common. Small trees of the legume family are prominent in this desert. Examples are palo verde (*Cercidium floridum* and *C. microphyllum*) and

Fig. 2. Shadscale (*Atriplex confertifolia*) vegetation in Nevada, typical of deserts that have cold winters.

ironwood (*Olneya tesota*). Elephant tree (*Bursera microphylla*) is typical of the southern two-thirds of this desert. Ocotillo (*Fouquieria splendens*) and boojum tree (*Idria columnaris*) are related odd-looking plants that are restricted to this desert. Ephemerals are of two kinds, spring and summer, because of the winter and summer bimodal rainfall characteristics of this desert, in contrast to the Mojave and Great Basin deserts.

Chihuahuan Desert. This easternmost of the North American deserts has cooler winters than do the other two warm deserts. The larger cacti do not occur here, and even ocotillo is smaller than in the other warm deserts. Small cacti, however, occur in great diversity among the creosote bushes. The most conspicuous of the larger plants are various species of *Yucca*, *Agave*, and *Dasylirion*. Mesquite (*Prosopis*) is common, as it is in the other warm deserts, but it, too, is smaller in the Chihuahuan Desert. Summer rainfall makes up most of the annual total precipitation in this desert. The vegetation is at its best in the warm season, and there are no winter and early spring ephemerals, as in the other deserts.

[W. D. BILLINGS]

Bibliography: M. Kassas, Plant life in deserts, in E. S. Hills (ed.), *Arid Lands: A Geographical Appraisal*, 1966; D. T. MacDougal, *Botanical Features of North American Deserts*, Carnegie Inst. Wash. Publ. no. 99, 1908; W. G. McGinnies, *Inventory of Research on Vegetation of Desert Environments*, Arid Lands Studies, University of Arizona, 1967; F. Shreve, The desert vegetation of North America, *Bot. Rev.*, 8:195–246, 1942; F. Shreve, *Vegetation of the Sonoran Desert*, Carnegie Inst. Wash. Publ. no. 591, 1951.

Dew point

The temperature at which air becomes saturated when cooled without addition of moisture or change of pressure. Any further cooling causes condensation; fog and dew are formed in this way.

Frost point is the corresponding temperature of saturation with respect to ice. At temperatures below freezing, both frost point and dew point may be defined because water is often liquid (especially in clouds) at temperatures well below freezing; at freezing (more exactly, at the triple point, +.01°C) they are the same, but below freezing the frost point is higher. For example, if the dew point is −9°C, the frost point is −8°C. Both dew point and frost point are single-valued functions of vapor pressure.

Determination of dew point (or frost point) can be made directly by cooling a flat polished metal surface until it becomes clouded with a film of water or ice; the dew point is the temperature at which the film appears. In practice, the dew point is usually computed from simultaneous readings of wet- and dry-bulb thermometers. [J. R. FULKS]

Bibliography: R. J. List (ed.), *Smithsonian Meteorological Tables*, 6th ed. rev., 1951.

Disease

An alteration of the dynamic interaction between an individual and his environment sufficient to be deleterious to the well-being of the individual. The cause of a disease may be environmental, or an altered reactivity of the individual to his environment, or a combination of both. The environmental change may be a physical, chemical, or biological agent.

A disease state is life under altered conditions; thus it represents a deviation from a norm and is a summation of the characteristics of the deviation from normal structure and function. The norm, called homeostasis, is the constellation of dynamic bodily activities that serve to maintain the functional integrity of an individual. These activities range from the biochemical subcellular level of organization to the cellular, tissue, organ, and organismal levels. The dynamic equilibrium of homeostasis for one person is not necessarily the same for another, varying with age, sex, race, and environment. The individual, as a biological system, is an open system, which means that homeostasis is a steady state in a dynamic system. Open systems have two principal characteristics: They are inseparable from, and constantly interact with, their external environment and they do not maintain unique stationary levels. Therefore there is an infinite variety of normal homeostatic systems, all of which have a common feature—the constancy of change. *See* HOMEOSTASIS.

Recognition. The terms health or disease are relative, and the condition of perfect health probably does not exist. Consequently, the recognition of a disease is a subjective concept, the limits of which cannot be precisely defined. The term disease is used in many ways and is often employed synonymously with illness, sickness, or a specific condition. It should be recognized that there is no sharp division between extremes of physiological response to stimuli and the ill-defined beginnings of a disease process. Consideration must be given to the individual—sex, age, habits, health pattern, genetic background, and many other pertinent factors.

The existence of disease in a person is arbitrarily defined, based on certain recognizable objective (signs) and subjective (symptoms) changes. Examples of the signs of disease are abnormal changes in temperature, pulse rate, and respiratory rate and laboratory measurements of blood and other body fluids. Symptoms such as pain, lassitude, restlessness, or emotional upset cannot be recorded or quantitated, though they may be no less real to the ill individual. Methods for observing and recording objective signs are based on the application of the principles of biochemistry, biophysics, and morphology. To compare the results of a test with a norm, a physician must note the environmental agent, as well as the mechanisms by which the individual adapts to its influence. In some instances the response of the individual may be so characteristic for a particular agent that by observing the response the cause of the disease may be reliably inferred. The goal of the medical scientist is to recognize all fundamental factors related to the initiating agent of disease (cause or etiology) and the mechanisms of adaptation to the agent (pathogenesis).

Death occurs when the rate or degree of interaction is so altered that the individual can no longer maintain homeostasis and irreversible

Common exogenous and endogenous causes of disease

Causative agent	Disease	Causative agent	Disease
Exogenous factor		*Biological (cont.)*	
Physical		Bacteria	Abscess, scarlet fever, pneumonia, meningitis, typhoid, gonorrhea, food poisoning, cholera, whooping cough, undulant fever, plague, tuberculosis, leprosy, diphtheria, gas gangrene, botulism, anthrax
Mechanical injury	Abrasion, laceration, fracture		
Nonionizing energy	Thermal burns, electric shock, frostbite, sunburn		
Ionizing radiation	Radiation syndrome		
Chemical			
Metallic poisons	Intoxication from methanol, ethanol, glycol		
Nonmetallic inorganic poisons	Intoxication from phosphorus, borate, nitrogen dioxide	Spirochetes	Syphilis, yaws, relapsing fever, rat bite fever
Alcohols	Intoxication from methanol, ethanol, glycol	Virus	Warts, measles, German measles, smallpox, chickenpox, herpes, roseola, influenza, psittacosis, mumps, viral hepatitis, poliomyelitis, rabies, encephalitis, trachoma
Asphyxiants	Intoxication from carbon monoxide, cyanide		
Corrosives	Burns from acids, alkalies, phenols		
Pesticides	Poisoning		
Medicinals	Barbiturism, salicylism	Rickettsia	Rocky Mountain spotted fever, typhus
Warfare agents	Burns from phosgene, mustard gas	Fungus	Ringworm, thrush, actinomycosis, histoplasmosis, coccidiomycosis
Hydrocarbons (some)	Cancer		
Nutritional deficiency		Parasites (animal)	
Metals (iron, copper, zinc)	Some anemias	Protozoa	Amebic dysentery, malaria, toxoplasmosis, trichomonas vaginitis
Nonmetals (iodine, fluorine)	Goiter, dental caries	Helminths (worms)	Hookworm, trichinosis, tapeworm, filariasis, ascariasis
Protein	Kwashiorkor		
Vitamins:		**Endogenous factor**	
A	Epithelial metaplasia		
D	Rickets, osteomalacia	*Hereditary*	Phenylketonuria, alcaptonuria, glycogen storage disease, Down's syndrome (trisomy 21), Turner's syndrome, Klinefelter's syndrome, diabetes, familial polyposis
K	Hemorrhage		
Thiamine	Beriberi		
Niacin	Pellagra		
Folic acid	Macrocytic anemia		
B_{12}	Pernicious anemia		
Ascorbic acid	Scurvy		
Biological		*Developmental*	Many congenital anomalies
Plants (mushroom, fava beans, marijuana, poison ivy, tobacco, opium)	Contact dermatitis, systemic toxins, cancer, hemorrhage	*Hypersensitivity*	Asthma, serum sickness, eczema, drug idiosyncrasy

changes ensue. The individual enters a steady state of equilibrium with his environment. Without dynamic change, life ceases.

In common usage the term disease indicates a constellation of specific signs and symptoms which can be attributed to altered reactions in the individual (form and function) produced by certain agents that affect the body or its parts. Alterations of body structure or function are often called lesions or pathologic lesions, which represent the reactions of cells, tissues, organs, or the organism to injury. The physician observes signs and symptoms in an individual, evaluates their relative significance, and compares them with known patterns of signs and symptoms. Therapeutic measures either protect the individual from the etiologic agent or alter the body's response to it.

Etiology. The causative agents of disease may be classified as exogenous (environmental) and endogenous (altered capacity of the individual to adapt to environmental changes). Whether environmental agents can produce disease in an individual depends on the amount, rate, and duration of exposure to the agent, as well as the individual's capacity to react (resistance) in an appropriate manner. Exogenous agents include virtually every feature of the environment. A partial list of exogenous and endogenous causes of disease is given in the table.

Frequently the interaction of multiple endogenous and exogenous factors is involved in the genesis of a disease. For example, excessive exposure to ultraviolet rays (physical agent) of 2800–3100-A wavelengths produces a disease called erythema solare (sunburn). The intensity of the radiation per minute varies with the latitude north or south of the Equator, season of the year, altitude, and atmospheric conditions (for example, dust and moisture in the air may act as a filter). Individuals vary greatly in their reaction and sensitivity to the ultraviolet rays (hereditary factors). Blondes and the rufous Celtic types are more susceptible than darker or black-skinned individuals or those previously exposed to small doses. Men react more than women, and infants more than adults (age factor). Women have increased sensitivity during menses. Coexistence of diseases such as hyperthyroidism or tuberculosis increases sensitivity, as do some medicines. Therefore, a given exposure to sunshine may produce the disease in one individual and not in another.

Environmental factors. Though some diseases may have an endogenous basis, the expression of the defect may depend on the environment of the individual. For example, nutritional diseases are a general class of exogenous diseases. If the individual fails to ingest sufficient quantities of ascorbic acid (vitamin C), a chemical substance found in abundance in citrus fruits and tomatoes and many other raw vegetables, the disease scurvy develops. Ascorbic acid is essential to the metabolism of certain amino acids (components of proteins), especially in connective tissue fibroblasts.

For some animals, such as birds, exogenous vitamin C is not a requirement. The body tissue of the animal can synthesize its own ascorbic acid and thus the animal is not subject to scurvy. Other animals, such as guinea pigs and humans, are unable to synthesize ascorbic acid (the endogenous defect) and must ingest it from the environment. In those environments where vitamin C is plentiful, scurvy does not exist. One may hypothesize that during the evolution of man the capacity of the cells to synthesize this substance was lost. However, the loss was not detrimental to man's survival because of the plentifulness of vitamin C in many environments. *See* MALNUTRITION.

Whether a particular disease is harmful or not may also depend on the environment. Sickle-cell trait is an inherited feature of hemoglobin in some individuals, usually those of African Negro descent. Under conditions of decreased blood oxygen the red blood cells assume a sickle shape rather than the usual round shape; these cells clump and may occlude (thrombose) blood vessels, producing death of the adjacent tissues; red blood cell destruction results in severe anemia. If this trait occurs in an individual in North America, it is obviously deleterious to his health and well-being. However, the same trait is protective in the tropical zones, for individuals with the sickle-cell trait have increased resistance to malaria. Thus the trait may be regarded as a disease in one environment and an improved adaptation to maintain homeostasis in another.

Cytological response. Regardless of the etiologic agent and the functional changes that ensue, cells are the basic units of body structure that are altered by disease. Cells can react to injury in a limited number of ways. They can undergo degeneration and death (necrosis) or increase in size (hypertrophy) or number (hyperplasia). Some cells can change their functional specialization (metaplasia), or acquire new growth properties (neoplasia). Furthermore, cells may decrease in size (atrophy) or number (hypoplasia).

[N. KARLE MOTTET]

Bibliography: R. P. Morehead, *Human Pathology*, 1965; R. Perez-Tamayo, *Mechanisms of Disease*, 1961.

Drought

A general term implying a deficiency of precipitation of sufficient magnitude to interfere with some phase of the economy. Agricultural drought, occurring when crops are threatened by lack of rain, is the most common. Hydrologic drought, when reservoirs are depleted, is another common form. The Palmer index has become popular among agriculturalists to express the intensity of drought as a function of rainfall and hydrologic variables.

The meteorological causes of drought are usually associated with slow, prevailing, subsiding motions of air masses from continental source regions. These descending air motions, of the order of 200 or 300 m/day, result in compressional warming of the air and therefore reduction in the relative humidity. Since the air usually starts out dry, and the relative humidity declines as the air descends, cloud formation is inhibited—or if clouds are formed, they are soon dissipated. The area over which such subsidence prevails may involve several states, as in the 1962–1966 Northeast drought or in the dust bowl drought of the 1930s over the Central Plains.

The atmospheric circulations which lead to this subsidence are the so-called centers of action, like the Bermuda High, which are linked to the planetary waves of the upper-level westerlies. If these centers are displaced from their normal positions or are abnormally developed, they frequently introduce anomalously moist or dry air masses into regions of the temperate latitudes. More important, these long waves interact with the cyclones along the Polar Front in such a way as to form and steer their course into or away from certain areas. In the areas relatively invulnerable to the cyclones, the air descends, and if this process repeats time after time, a deficiency of rainfall and drought may occur. In other areas where moist air is frequently forced to ascend, heavy rains occur. Therefore, drought in one area, say the northeastern United States, is usually associated with abundant precipitation elsewhere, like over the Central Plains.

After drought has been established in an area, there seems to be a tendency for it to persist and expand into adjacent areas. Although little is known about the physical mechanisms involved in this expansion and persistence, some circumstantial evidence suggests that numerous "feedback" processes are set in motion which aggravate the situation. Among these are large-scale interactions between ocean and atmosphere in which variations in ocean-surface temperature are produced by abnormal wind systems, and these in turn encourage further development of the same type of abnormal circulation. Then again, if an area, such as the Central Plains, is subject to dryness and heat in spring, the parched soil appears to influence subsequent air circulations and rainfall in a drought-extending sense.

Finally, it should be pointed out that some of the most extensive droughts, like those of the 1930s dust bowl era, require compatibly placed centers of action over both the Atlantic and the Pacific.

In view of the immense scale and complexity of drought-producing systems, it will be difficult for man to devise methods of eliminating or ameliorating them. [JEROME NAMIAS]

Bibliography: J. Namias, *Factors in the Initiation, Perpetuation and Termination of Drought*, Int. Union Geod. Geophys. Ass. Sci. Hydrol. Publ. no. 51, 1960; W. C. Palmer, *Meteorological Drought*, U.S. Weather Bur. Res. Pap. no. 45, 1965.

Dune

A mobile mound, hill, or ridge of windblown material, usually sand. Dunes develop where the velocity of sand-driving winds is checked or reduced, often as a result of topographic barriers, vegetation, or manmade obstructions. Dunes require not only a suitable site for development but an adequate sand source as well. Consequently, most dunes are found downwind from sandy lake or ocean beaches; dry stream beds, particularly in the desert; and barren, sandy, glacial outwash plains. A few dunes appear to be derived from the disintegration of exposed sandstone or other bedrock.

Dune surfaces become rippled, apparently because of slight initial surface irregularities. The portions of irregularities facing the wind receive more sand grains than those parts facing away from the wind. Because sand grains tend to splash or rebound when they hit a sand surface, more splashes or rebounds occur from the sides of the hollows facing the wind. The grains, being nearly equal in size, tend to land at about the same place, and thus a system of nearly parallel ridges or ripples develops. Coarser grains commonly collect at the crests of the ripples, because as ripples are built upward and exposed to increasingly stronger wind action, the finer particles are removed.

Dunes typically possess on the leeward side slip faces with slopes of 32–34°. The slip faces develop as sand piles up on the windward side of a dune until it becomes unstable and slides or slips downward. As a slip face increases in height, sand continues to be added at the top either as grains rolled over the crest by the wind or dropped from above.

Types. Dunes may occur in various forms as listed below.

1. Barchan dunes are crescent-shaped and develop on nearly flat, barren desert surfaces, where wind direction is nearly constant and sand supply somewhat limited. Barchans migrate, horns first, across the desert floor (Fig. 1). The smaller the barchan, the more rapid is the movement, wind conditions being comparable. Some California barchans with slip faces 30 ft high advance about 50 ft a year.

2. Climbing dunes develop on the windward sides of mountains or hills.

3. Falling dunes occur in sheltered places on the leeward sides of mountains over which sand has been blown (Fig. 2). Usually, climbing dunes lie on the opposite or windward side of the mountain to provide the source of sand.

4. Seif dunes or linear ridges are sometimes hundreds of miles long, as in the Sahara and Australian deserts. These are produced by winds blowing roughly parallel to the dunes, but varying through an angle of 15–30°. Barchan chains pass gradually into seif ridges as winds change from unidirectional to multidirectional; and as sand-driving winds come from increasingly divergent directions, the seif form is lost altogether (Fig. 3).

5. Transverse dunes are most important in

Fig. 1.   Barchans, or crescent-shaped dunes, near Biggs, Ore. These riverbed dunes move from left to right with steep side away from the wind. (*USGS*)

coastal localities where vegetation growing on the sand affects its movement. Typically, several parallel ridges are found, with the smallest, barest ridge near the beach and the highest, most extensively vegetated ridge farthest inland. These dunes are oriented more or less at right angles to the direction of the sand-driving winds.

6. Parabolic or blowout dunes are hairpin or spoon-shaped with horns pointing toward the wind, opposite to those of the barchan. These usually develop when vegetation on a dune surface is destroyed, so that the wind is allowed to remove the underlying sand.

7. Complex dunes develop in areas where sand-driving winds come from widely varying directions.

Geographic distribution. All the desert areas of the world have dunes. The most extensive area and greatest variety of dune forms occur in the great Saharan and Arabian deserts, where approximately 30% of the surface is covered with sand. Dune sand covers about 1% of the surface of the North American desert. The central Australian desert is characterized by the linear sand ridge, which is practically the only dune form there and covers thousands of square miles. The desert coast of Peru is famous for its perfectly formed barchans.

Coastal dunes are found along many sea and lake shores of the world. Especially prominent and

Fig. 2.   A falling dune, the "Cronese Cat," Cronese Valley, San Bernardino County, Calif.

Fig. 3. Seif dune or linear ridge, located in the Algodones dunes, Imperial County, Calif.

well known are the dunes along the coasts of the Low Countries and France. In North America extensive coastal dunes are found along both the Atlantic and Pacific coasts and along some Great Lakes shores.

Materials. Most dunes are composed of quartz and feldspar grains, minerals both durable and extremely abundant and the principal constituents of beach and stream sands. The peculiar gypsum dunes at White Sands, N.Mex., are derived from exposed ribs or low ridges of Permian bedrock containing gypsum. Some dunes along tropic shores are derived chiefly from limy or coral beach sands. In Australia windblown fine silt and clay form dunelike features called lunettes. In polar regions dunelike features are sometimes produced from wind-driven snow.

Size. Some complex dunes in southern Iran rise 700 ft above their bases. The Algodones dunes, a chain of large overlapping barchans about 40 mi long in southeastern California, rise as much as 250 ft above the desert floor. The great sand ridges of the Egyptian desert are as much as 150 ft high and 180 mi long. The Australian sand ridges are 30–100 ft high, and single ridges are frequently 100 mi or more in length. Barchans are commonly 30–50 ft high and 500–800 ft long. The giant Pur-Pur dune of the Virú Valley of Peru is more than 1 mi long and about 170 ft high.

Fossil dunes. Ancient desert dunes are preserved in many parts of the world and are recognized by the long, curved, intersecting layers characteristic of modern dunes; the polished, rounded, and frosted sand grains; the red iron stain; and the association with gypsum, salt, and other desert lake deposits. Examples are the Navajo sandstone of Jurassic age found in the Colorado Plateau of the United States, some of the red Triassic sandstones of the Connecticut Valley and the Maritime Provinces of Canada, and the New Red Sandstone of Great Britain.   [ROBERT M. NORRIS]

Bibliography: R. A. Bagnold, *Physics of Blown Sand and Desert Dunes*, reprint, 1954; W. S. Cooper, *Coastal Sand Dunes of Washington and Oregon*, 1958; R. M. Norris, Barchan dunes of Imperial Valley, Calif., *J. Geol.*, 74:292–306, 1966; R. P. Sharp, Kelso dunes, Mojave Desert, Calif., *Bull. Geol. Soc. Amer.*, 77:1045–1074, 1966; H. T. U. Smith, Geological studies in southwestern Kansas, *Kans. Geol. Surv. Bull.*, no. 34, 1940; A. N. Strahler, *A Geologist's View of Cape Cod*, 1966.

Dune vegetation

The plants which occupy dunes and the swales and flats between dunes. Along many shores these plants form distinct zones of vegetation. Plants also grow over desert dunes during short periods of precipitation but seldom form any distinct zones of vegetation except around oasis basins. *See* DUNE.

Succession and zonation. The vegetation on coastal dunes usually becomes denser and the plants become larger and more persistent as the dunes become stabilized—a process known as plant succession. Generally there are three main zones of this succession: (1) a pioneer zone composed mostly of herbaceous plants on the foredunes near the beaches; the plants in this zone are very instrumental in forming and stabilizing the succeeding dunes further inland; (2) a scrub zone of woody shrubs and vines and dwarfed trees; this zone continues stabilization of the dunes and adds humus to the sands; and (3) a forest zone, dominated by trees, which is known as the mesophytic climax of the dune succession. The plants of the pioneer and scrub zones are usually adapted to dry conditions and are xerophytes. Those plants near salt waters, the halophytes, are also resistant to saline conditions. Some types of vegetation in a dune succession are illustrated in Figs. 1, 2, and 3. *See* SUCCESSION, ECOLOGICAL.

The vegetation of dunes along salt waters and along fresh waters is essentially similar. Some plants of desert dunes are similar to those of coastal dunes. Nearly all withstand periods of dryness, and many withstand shifting sands. Those plants growing near salt water withstand salt spray which, together with winds, often deforms them, especially the trees and shrubs. This complex of plants, their forms and adaptations, the successional stages, and their zonation have been described for many coastal strand areas. The following is a brief summary of North American dune vegetation.

Atlantic and Gulf coasts. Along the Atlantic Ocean and Gulf of Mexico there is a gradual change in dune plants and features of the vegetation from the areas of temperate climate to those of tropical climate. In the north, the herbs of the pioneer zone are mostly annuals, and severe winters present the development of dense scrub and forest zones. Southward the variety and number of species increase from an average of less than 60 to more than 150 (on subtropical dunes in Florida). This abundance and variety of plants on subtropical dunes have the effect of creating more dune ridges of greater stability. Many evergreen shrubs and trees, fibrous and succulent plants, and palms increase toward the tropics. A fringe of palm forest, especially coconut trees, is common in the tropics on the upper beach and foredunes. The following plants are common along coasts of middle latitudes, from Maryland to Palm Beach, Fla.

Pioneer zone. Plants commonly found in this zone include beach grass (*Ammophilia*), sea rocket (*Cakile*), sea oats (*Uniola paniculata*), panic grass (*Panicum amarum*), salt wort (*Salsola kali*), orach (*Atriplex arenaria*), and in subtropical areas the railroad vine (*Ipomoea pes-caprae*) and sea purslane (*Sesuvium portulacastrum*).

Scrub zone. Plants that are typical of this zone include march elder (*Iva imbricata*), myrtles

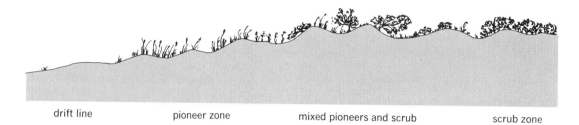

drift line   pioneer zone   mixed pioneers and scrub   scrub zone

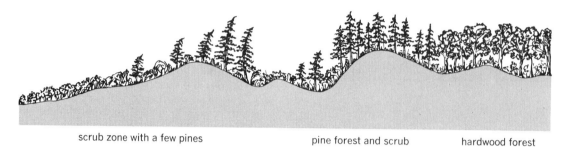

scrub zone with a few pines   pine forest and scrub   hardwood forest

Fig. 1. Diagrammatic cross section showing types of vegetation over broad coast. (*Adapted from J. H. Davis, Dune Formation and Stabilization by Vegetation and Plant-ings, Beach Erosion Board, U.S. Army Corps of Engineers, Tech. Memo. no. 101, 1957*)

Fig. 2. Pioneer zone composed of sea oats. These plants assist in formation of the foredunes.

Fig. 3. Scrub zone with bare sand of a blowout area in foreground. Dwarfed evergreen magnolias are a part of this zone, near Panama City, Fla.

(*Myrica*), scrub forms of oaks (*Quercus*), cedars (*Juniperus*), holly (*Ilex opaca*), groundsel bush (*Baccharis*), buckthorns (*Bumelia*), and alders (*Alnus*). Northward, sand myrtle (*Leiophyllum buxifolium*), birches (*Betula*), false heather (*Hudsonia tomentosa*), and some true heaths (*Erica*) are locally abundant. Southward, the saw palmetto (*Serenoa repens*) is very abundant and is important as a dune stabilizer. In Florida the sand pine (*Pinus clausa*) is typical of the scrub vegetation. *Agave* and species of cactus (*Opuntia*) are common along dry coasts, and the sea grape (*Coccoloba uvifera*) becomes very frequent in subtropical areas. Many of these plants, especially the oaks, are densely branched shrubs and dwarfed, wind-formed trees that often form thickets (Fig. 4). Between the thickets are bare migrating sands or partly stabilized sands populated by the pioneer plants.

Forest zone. Many different trees form the dune forests. Along northern Atlantic coasts, poplars and aspens (*Populus*), pines (*Pinus*), birches (*Betula*), cherries (*Prunus*), and oaks (*Quercus*) are common. Southward, the pines and oaks, particularly the evergreen live oak (*Quercus virginiana*), are common. Hickories (*Carya*) are frequent, and the palms begin to increase toward the tropics. The cabbage palm (*Sabal palmetto*) is very frequent from North Carolina to the tip of Florida and along much of the Gulf of Mexico coast. The coconut palm (*Cocos nucifera*) is now naturalized and abundant along American tropical coasts.

Pacific Coast. The dune vegetation along the Pacific Coast varies from a dry semidesert type in Mexico and southern California to a humid-temperature-climate type along coastal areas north of central California. Introduced European and American beach or marram grass (*Ammophila arenaria* and *A. breviligulata*) have been used to stabilize foredune ridges and parts of the very large blowout areas of bare sand, and are spreading from areas planted in the 1930s more vigorously than native pioneer species. Trees resistant to salt spray include lodgepole pine (*Pinus contorta*), which persists or regenerates on dry, leached soils after burning, and sitka spruce (*Picea sitchensis*).

Southern dry type. Along the dry parts of the coast are *Yucca gloriosa, Opuntia serpentina*, sand ambrosia (*Franseria chamissonis*), sagebrush (*Ar-*

Fig. 4. Typical deformed trees, live oak, caused by winds and salt spray, on edge of dune forest.

temesia pychocephala), mariposa (*Calochortus venustus*), and *Abronia latifolia*, with a few seasonal grasses and other herbs over the young forming dunes. The distinct pioneer and scrub zones are usually present with *Haplopappus erocoides* and lupine (*Lupinus chamissonis*). The shore shrubs and dwarfed trees include Monterey pine (*Pinus radiata*), a live oak (*Quercus agrifolia*), leather oak (*Q. durata*), coffeeberry (*Rhamnus californica*), *Yucca*, *Opuntia*, and *Ceanothus dentatus*. These plants seldom form a dense, distinct forest zone.

Northern moist type. In contrast to the southern dry-climate coast, the northern moist-climate coastal areas have a forest zone on stabilized dunes, and scrub and pioneer zones where there are active dunes. A few plants are the same as, or similar to, those of the northern Atlantic Coast. Redwoods (*Sequoia*), pines, spruces (*Picea*), aspens, and manzanitas (*Arctostaphylus*) are common.

Great Lakes coasts. Dunes of the Great Lakes shores became famous as examples of successional change shown by progressively older dunes. The dune's external environment did not have a gradient of salt spray confounded with gradients of age since surface stabilization. Both beaches and blowout sands have annuals: mostly sea rocket (*Cakile*), bugseed (*Corispermum*), winged pigweed (*Cycloloma*), and beach pea (*Lathyrus*). Dune-building grasses include marram (*Ammophila breviligulata*) and sand reed (*Calamovilfa longifolia*), in order of tolerance to rapid sand burial on new dune ridges along the beach and on lee slopes and crests of the U- or V-shaped blowout ridges. The latter open toward the lake or the dominant westerly winds. Little bluestem bunchgrass typically invades within a few years after deposition stops or becomes reduced to a few centimeters per year. Sand cherries (*Prunus pumila*), several willows (*Salix*), and poplars (*Populus deltoides* or *P. balsamifera* and hybrids) germinate in hollows of young dunes but tolerate burial well, as do grape vines (*Vitis*) and basswood (*Tilia americana*) on lee slopes. The latter is joined by red oak (*Quercus rubra*), maple (*Acer sabcharum* and *A. rubrum*), and in Michigan by beech (*Fagus grandifolia*) and hemlock (*Tsuga canadensis*) within a few hundred

or thousand years after stabilization on protected slopes and hollows that are relatively favorable for rapid succession. However, the majority of old Indiana, Wisconsin, and even Michigan and Ontario dunes follow alternative successions through pines (*Pinus banksiana* and *P. strobus*, in addition to *P. resinosa* northward) to oak (*Quercus velutina*, *Q. alba*, and northward *Q. rubra*). Evidence of replacement of the oak-blueberry vegetation, on acid soils, by the mesophytic forests dominated by the trees mentioned earlier is rare or conjectural, if not absent, even on old dunes (stabilized 8000–12,000 years ago). *See* CLIMAX COMMUNITY.

[JOHN H. DAVIS]

Bibliography: S. G. Boyce, The salty spray community, *Ecol. Monogr.*, 24:29-67, 1954; H. C. Cowles, The ecological relations of the vegetation on sand dunes of Lake Michigan, *Bot. Gaz.*, 27: 95–117, 167–202, 281–308, 361–391, 1899; J. H. Davis, *Dune Formation and Stabilization by Vegetation and Plantings*, Beach Erosion Board, U.S. Corps of Engineers, Tech. Memo. no. 101, 1957; J. S. Olson, Rates of succession and soil changes on southern Lake Michigan sand dunes, *Bot. Gaz.*, 119:125–170, 1968; H. J. Oosting, Ecological processes and vegetation of the maritime strand in the southeastern United States, *Bot. Rev.*, 20(4): 226–262, 1954; E. J. Salisbury, *Downs and Dunes: Their Plant Life and Environment*, 1952; L. A. Starker and *Life* (eds.), *The Desert*, 1961.

Dust and mist collection

The physical separation and removal of particles, either solid or liquid, from a gas in which they are suspended. Such separation is required for one or more of the following purposes: (1) to collect a product which has been processed or handled in gas suspension, as in spray-drying or pneumatic conveying; (2) to recover a valuable product inadvertently mixed with processing gases, as in kiln or smelter exhausts; (3) to eliminate a nuisance, as a fly-ash removal; (4) to reduce equipment maintenance, as in engine intake air filters; (5) to eliminate a health, fire, explosion, or safety hazard, as in bagging operations or nuclear separations plant ventilation air; and (6) to improve product quality, as in cleaning of air used in processing pharmaceutical or photographic products. Achievement of these objectives involves primarily gas-handling equipment, but the design must be concerned with the properties and relative amounts of the suspended particles as well as with those of the gas being handled.

All particle collection systems depend upon subjecting the suspended particles to some force which will drive them mechanically to a collecting surface. The known mechanisms by which such deposition can occur may be classed as gravitational, inertial, physical or barrier, electrostatic, molecular or diffusional, and thermal or radiant. There are also mechanisms which can be used to modify the properties of the particles or the gas to increase the effectiveness of the deposition mechanisms. For example, the effective size of particles may be increased by condensing water vapor upon them or by flocculating particles through the action of a sonic vibration. Usually, larger particles simplify the control problem. To function successfully, any collection device must have an adequate means for continuously or periodically removing

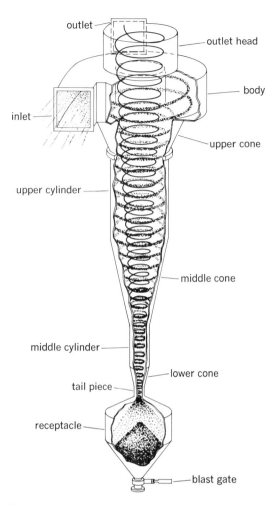

Fig. 1. Cyclone dust separator, an inertial device. (*American Standard Industrial Division*)

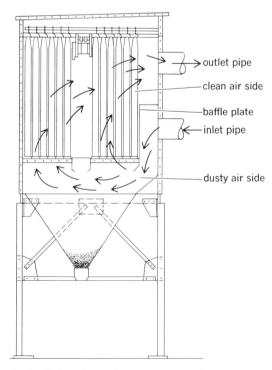

Fig. 2. Cloth collector. (*Wheelabrator Corp.*)

collected material from the equipment.

Devices for control of particulate material may be considered, by structural or application similarities, in eight categories as follow.

Gravity settling chamber. In this, the simplest type of device but not necessarily the least expensive, the velocity of the gas is reduced to permit particles to settle out under the action of gravity. Normally, settling chambers are useful for removing particles larger than 50 microns (μ) in diameter, although with special configurations they may be used to remove particles as small as 10 μ.

Inertial device. The basis of this type of device is that the particles have greater inertia than the gas. The cyclone separator, typical of this type of equipment, is one of the most widely used and least expensive types of dust collector. In a cyclone, the gas usually enters a conical or cylindrical chamber tangentially and leaves axially. Because of the change of direction, the particles are flung to the outer wall, from which they slide into a receiving bin or into a conveyor, while the gases whirl around to the central exit port (Fig. 1). A large variety of configurations is available. For large air-handling capacities, an arrangement of multiple small-diameter units in parallel is often used to attain high collection efficiencies and to permit lower headroom requirements than a single unit would.

Mechanical inertial units are similar to cyclones except that the rotational motion of the gas is induced by the action of a rotating member. Some such units are designed to act as fans in addition to their dust-collecting function. There are also a wide variety of other units; many are called impingement separators. Most separators used to remove entrained liquids from steam or compressed air fall into this category.

Packed bed. A particle-laden gas stream may be cleaned by passing it through a bed or layer of packing composed of granular materials such as sand, coke, gravel, and ceramic rings, or fibrous materials such as glass wool, steel wool, and textile staples. Depending on the application, the bed depth may range from a fraction of an inch to several feet. Coarse packings, which are used at relatively high throughput rates (1–15 ft/sec superficial velocity) to remove large particles, rely primarily on the inertial mechanism for their separating action. Fine packings, operated at lower throughput rates (1–50 ft/min superficial velocity) to remove relatively small suspended particles, usually depend on a variety of deposition mechanisms for their separating effect. Packed beds, because of a gradual plugging caused by particle accumulation, are usually limited in use to collecting particles present in the gas at low concentration, unless some provision is made for removing the dust—for example, by periodic or continuous withdrawal of part of the packing for cleaning. Depending on the application and design, the collection efficiencies of packed beds range widely (50–99.999%).

Cloth collector. In such a collector, also known as a bag filter, the dust-laden gas is passed through a woven or felted fabric upon which the gradual deposition of dust forms a precoat, which then serves as a filter for the subsequent dust. These units are analogous to those used in liquid filtration and represent a special type of packed bed. Because the dust accumulates continuously, the resistance to gas flow gradually increases. The cloth must, therefore, be vibrated or flexed periodically to dislodge accumulated dust (Fig. 2). A wide variety of filter media is available. Cotton or wool sateen or felts are usually used for temperatures below 212°F. Some of the synthetic fibers may be used at temperatures up to 300°F. Glass and asbestos or combinations thereof have been used for temperatures up to 650°F. For special high-temperature applications, metallic screens and porous ceramics or stainless steel have been employed. Collection efficiencies of over 98% are attained readily with cloth collectors, even with very fine dusts.

Scrubber. A scrubber uses a liquid, usually water, to assist in the particulate collection process. An extremely wide variety of equipment is available, ranging from simple modifications of corresponding dry units to permit liquid addition, to devices specifically designed for wet operation only (Fig. 3). When properly designed for a given application, scrubbers can give very high collection efficiency, although the mere addition of water to a gas stream is not necessarily effective. For a given application, collection efficiency is primarily a function of the amount of power supplied to the gas stream, in the form of gas pressure drop, water pressure, or mechanical energy. With scrubbers it is important that proper attention be given to liquid entrainment separation in order to avoid a spray nuisance. Consideration must also be given to the liquid-sludge disposal problem.

Electrostatic precipitator. Particles may be charged electrically by a corona discharge and caused to migrate to a collecting surface. The single-stage unit, which is commonly known as a Cottrell precipitator, and in which the charging and collecting proceed simultaneously, is the type generally used for industrial or process applications. These units normally employ direct current at voltages ranging from 30,000 to 100,000 volts. The two-stage unit, in which charging and collection are carried out successively, is commonly used for air conditioning applications. These units also employ direct current, ranging from 5,000 to 13,000 volts, and involve close internal clearances (0.25–0.5 in.). Electrical precipitators are capable of high collection efficiency of fine particles. The reentrainment of collected material as flocs in the exhaust gas, a phenomenon known as snowing, must be avoided to prevent a possible accentuated nuisance or vegetation damage problem.

Air filter. This is a unit used to eliminate very small quantities of dust from large quantities of air, as in air conditioning applications. Although units in this class actually fall into one of the previous classes, they are given a special category because of wide usage and common special features. In this category are viscous-coated fiber-mat filters and dry filters. These are actually a form of packed bed and are frequently known as unit filters; they are available as standard packaged units from a large number of manufacturers. The domestic furnace filter is an example of a viscous-coated unit filter. Automatic filters provided with continuous and automatic cleaning arrangements are available in both the viscous-coated and dry

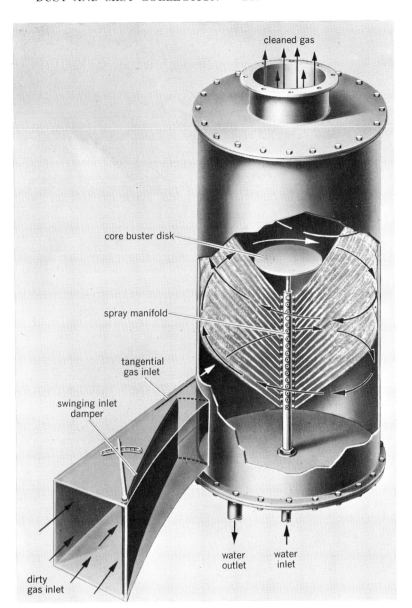

Fig. 3. Cyclonic liquid scrubber. (*Chemical Construction Corp.*)

forms, as well as with electrostatic provisions.

Miscellaneous equipment. Acoustic or sonic vibrations imparted to a gas stream cause particulates to collide and flocculate, forming larger particles that are more readily collected in conventional apparatus. This principle has been employed, but has had extremely limited application because of economic and other practical considerations. In thermal precipitation, suspended particles are caused to migrate toward a cold surface or away from a heated surface by the action of a temperature gradient in the gas stream. This principle has found extensive use in atmospheric sampling work. *See* AIR-POLLUTION CONTROL: ATMOSPHERIC POLLUTION.

[CHARLES E. LAPPLE]

Bibliography: R. E. Kirk and D. F. Othmer (eds.), *Encyclopedia of Chemical Technology*, vol. 7, 1951; C. E. Lapple, Elements of dust and mist collection, *Chem. Eng. Progr.*, 50(6):283–287, 1954; K. E. Lunde and C. E. Lapple, Dust and mist collection, *Chem. Eng. Progr.*, 53(8):385–391, 1957; G. Nonhebel, *Gas Purification Processes*, 1964; R. H. Perry (ed.), *Chemical Engineers' Handbook*, 4th ed., 1963; W. Strauss, *Industrial Gas Cleaning*, 1966; U.S. Atomic Energy Commission, *Handbook on Air Cleaning*, 1952.

Dust storm

A strong, turbulent wind carrying large clouds of dust. In a large storm, clouds of fine dust may be raised to heights well over 10,000 ft and carried for hundreds or thousands of miles.

Sandstorms differ by the larger mass, more rapid settling speeds of the particles involved, and the stronger transporting winds required. The sand cloud seldom rises above 1–2 m and is not carried far from the place where it was raised.

Dust storms cause enormous erosion of the soil, as in the dust bowl disasters of 1933–1937 in the Great Plains of the United States. Besides causing acute physical discomfort, they present a severe hazard to transportation by reducing the visibility to very low ranges. Conditions required are an ample supply of fine dust or loose soil, surface winds strong enough to stir up the dust, and sufficient atmospheric instability for marked vertical turbulence to occur.

Mechanics of dust raising. Dust (or sand) is initially raised when particles become dislodged by aerodynamic stresses of the strong winds upon exposed grains. The larger particles fall obliquely after attaining considerable horizontal speed, bombarding other particles on the surface which in turn become dislodged and further the process.

R. A. Bagnold classifies as dust the particles having diameters of $10^{-3}–10^{-2}$ mm, with free-fall speeds ranging $10^{-2}–2$ cm/sec; sand particles, 0.1–1 mm in diameter, have fall speeds 40–600 cm/sec. Dust can be readily carried upward by ordinary turbulent eddies in an unstable air mass, but these are generally too feeble to sustain large sand particles, which attain only small heights by bouncing. In any dust storm, particles of various sizes are raised, the smallest being carried to greatest heights from which they may take days or weeks to settle while remaining as a dust haze.

Soil factors. Soil condition is the most decisive criterion for development of dust storms. This depends on vegetative cover and upon binding of soil by moisture, both factors being dependent upon prior rain- or snowfall. Dust storms are most frequent in spring, when in semiarid regions the Earth is least covered by vegetation. Loosening of soil and overturning of humus by spring plowing and overgrazing of grasslands are prime contributors to setting up soil conditions favorable for dust storms.

Meteorological factors. Surface wind speeds required vary according to soil characteristics. In some desert regions sandstorms occur with winds of 15–20 mph. Extensive dust storms in North America usually require winds of 25–30 mph or more. Such winds are present over large areas in the circulations of many well-developed cyclones, which may raise dense dust clouds several hundred miles across if soil conditions are right.

A further requisite is thermodynamic instability (strong decrease of temperature with height), necessary for development of the vertical eddies required to transport dust aloft from surface layers. Most dust storms occur in daytime, particularly in the afternoon when the air is warmest at the ground, hence most unstable just above. Major dust storms in the United States are almost exclusively confined to maritime polar air masses from the Pacific Ocean, which are characteristically unstable to great heights. *See* AIR MASS.

Dust storms associated with large-scale wind systems are also common in the Sahara and Gobi deserts. More local but often severe dust storms resulting from thundersqualls are common in all the desert regions.

Optical and electrical effects. Small dust particles increase scattering of light, mainly in short (blue) wavelengths. The Sun often appears a deep orange or red when seen through a dust cloud; however, optical effects are variable. Large particles are effective reflectors, and an observer in an aircraft above a dust storm may see a solid sheet with an apparent dust horizon.

Because of friction with air or ground, dust particles acquire appreciable electrostatic charges, and on striking radio antennas may cause severe static. Visible electrical discharges sometimes occur within the dust cloud.   [CHESTER W. NEWTON]

Bibliography: R. A. Bagnold, *The Physics of Blown Sand and Desert Dunes*, 1965; H. R. Byers, *Synoptic and Aeronautical Meteorology*, 1937.

Earth, resource patterns of

The physical character and distribution of natural resources at the face of the Earth. No section of the Earth is exactly like any other in its resource endowment. Nevertheless, latitudinal differences in insolation, the great difference between land and ocean, and geological composition of the Earth's crust together provide the basis for distinguishing definite geographical patterns of resource availability over the world.

Delineation of the Earth's resource pattern begins with differentiation between land and marine resources. Although marine resources have been used by men since earliest times, the 3,500,000,000 people on the Earth are highly dependent upon the resources of the land for their continued existence.

Five principal resources associated with land are: fresh water, agricultural soils, mineral deposits, forest lands, and grasslands. Such other resources as the native animal life and genetic stocks of plants and animals are very mobile and currently cannot be considered significant in differentiation of resource patterns. On the other hand, the five principal resources do have unique associations which differ from one broad area to another.

The underlying causes for distinctive features in the pattern of resource endowment are the regime of energy receipts from the Sun, the effects of planetary atmospheric circulation on the distribution of moisture and heat, the geological composition of the continents and islands, and the structural history of different sections of the Earth's surface.

Ready keys for understanding the resource pattern of a section of the Earth's surface are given in the character of climate, Earth surface configuration, and rock composition. A basic pattern is outlined in regional climates with associated characteristic types of agricultural, forest, and

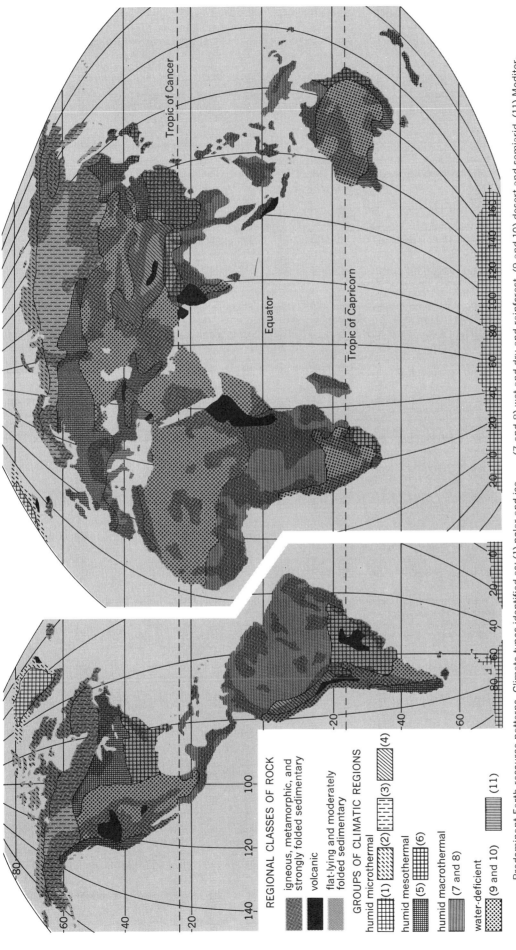

REGIONAL CLASSES OF ROCK

igneous, metamorphic, and strongly folded sedimentary

volcanic

flat-lying and moderately folded sedimentary

GROUPS OF CLIMATIC REGIONS

humid microthermal
(1) (2) (3) (4)

humid mesothermal
(5) (6)

humid macrothermal
(7 and 8)

water-deficient
(9 and 10) (11)

Predominant Earth resource patterns. Climate types identified as: (1) polar and ice cap, (2) tundra, (3) taiga, (4) puna, (5) upper midlatitude, (6) humid subtropical, (7 and 8) wet-and-dry and rainforest, (9 and 10) desert and semiarid, (11) Mediterranean. The map is a flat polar quartic equal-area projection.

grazing lands and water availability. On this is placed an overlay of differences in rock composition which alters the pattern within the climatic regions. A second overlay of differences in surface configuration produces further alteration on the previously shown patterns.

Eleven regional climatic types (numbered consecutively on map) in four groups are recognized in describing the Earth's resource pattern. This number of types is fewer than that normally employed to describe regional climates, but is considered adequate to outline the basic resource pattern. Distinction is made between the so-called humid climates and the water-deficient climates with subtypes as follows:

Humid microthermal
    (1) Polar and ice cap
    (2) Tundra
    (3) Taiga
    (4) Puna
Humid mesothermal
    (5) Upper midlatitude
    (6) Humid subtropical
Humid macrothermal
    (7) Wet-and-dry
    (8) Rainforest
Water-deficient
    (9) Desert
    (10) Semiarid
    (11) Mediterranean

Humid microthermal regions. These areas of predominantly low temperatures are characterized by unfavorability to crop and natural vegetative growth and, under present techniques, by a relatively low carrying capacity for people.

(1) In polar and ice-cap areas available resources are dominantly marine and land animal life, on which the sparse human settlement is almost wholly dependent. Despite the enormous ice-cap area of the Antarctic, settled sections are exclusively in the Arctic. Polar regions may be of future importance for mineral deposits, but their geology is as yet relatively unknown.

(2) Tundra, except for minor alpine locations, is entirely within the Northern Hemisphere. The principal resources are lichens and the native animal life, such as reindeer and caribou, which can use these as food. Parts of the tundra may be considered a grazing land, as managed herds of reindeer are pastured nomadically. Potential mineral deposits are imperfectly known, although a few commercial workings are now located within tundra regions. See TUNDRA.

(3) Millions of acres of taiga located in Siberia and northern Soviet Union, Scandinavia, Canada, and Alaska are capable of producing boreal coniferous forests, the most valuable known resource of the microthermal regions. Varieties of spruce, fir, and larch are of particular significance to the pulping industries. Some taiga may be considered potential agricultural land, but the possible incidence of frost in every month and the short growing season (80 days or less) do not presage major agricultural development under present techniques. Deposits of uranium, copper, iron, and other minerals are exploited in the taiga. It is also a commercial source of furs. See TAIGA.

(4) The puna type of climate is found in elevations high above sea level and is characteristically cold. Certain plateaus, such as the high plateaus of Tibet and limited sections of the South American upland, belong in this group. In these treeless places the principal resource is low-productivity grazing land. Cultivated lands are limited to hardy small grains and root crops. South American puna areas contain some significant metal deposits, particularly copper, tin, lead, and zinc.

Humid mesothermal regions. On the average, temperatures are intermediate in these parts. Considered in the light of present-day technology, the heart of the world's natural resource base is in the mesothermal regions. They contain a large share of the crop and pasture lands, and some productive coniferous and broadleaf forests. Several distinct climates lie within these regions, but only the two divisions are distinguished here.

(5) Upper midlatitude climate contains most of the lands adapted to the raising of wheat, barley, rye, and oats. Certain areas, such as the North American Corn Belt and the Danubian Valley, also produce maize. Extensive forage cropping supports dairying and meat-animal raising. Central and east central North America, western and central Europe, the nonarid part of southern Soviet Union, and Argentina account for most of the area within this climate.

(6) Humid subtropical areas have an ample water supply and relatively long growing season (200 days or longer), making these rice and cotton lands highly productive. Soils in these areas, unless well fertilized, are less capable of sustaining cultivation over long periods than those of poleward—mostly northern—parts of the middle latitudes.

Humid macrothermal regions. Predominantly winterless regions of warm to hot temperatures, such areas are divided according to the regime of rainfall: (7) wet-and-dry, with a pronounced dry season, and (8) rainforest, with year-round precipitation. Macrothermal regions have year-round growing seasons and lateritic soils. Problems induced by fungal growth, bacterial disease, and insect abundance handicap the agricultural land resources. Within the great alluvial valleys, extensive flooding also may be disrupting. Agricultural land resources are developed only in spots. Particularly within the rainforest regions, extensive and rapid-growing forests occur but are mostly unexploited commercially in the 20th-century economy. Sites of enormous potential hydroelectric generation are largely undeveloped. See RAINFOREST.

Water-deficient regions. Receipts of moisture are scant or lacking during much of the year. Except where exotic water supplies are available for irrigation, the agricultural lands inherently are less productive than those of any humid region with similar land surface. There are three major subdivisions.

(9) Deserts differ strikingly in their form and in temperature conditions, but everywhere present meager resources for agriculture. Where water is available, desert oases blossom, but vast areas contain only scrub growth, ephemerals, or virtually no vegetation. Some grazing resources are available, but carrying capacity of deserts is very low. Sparsity of vegetation has made mineral prospecting and exploration somewhat easier than in

vegetation-covered areas, and in this century desert occupance often started with mineral discoveries. *See* DESERT VEGETATION.

(10) The semiarid regions are basically grasslands, which have grazing as their characteristic resource. Because of cyclic rainfall variability, men have converted the inherently fertile soils, as in China and the United States, to cereal growing during periods of higher rainfall. Rainfall fluctuations, however, make the land resource unstable under cultivation. For this reason, these regions have suffered consistently from the accelerated and destructive erosion resulting from unsuited cultivation and overgrazing. Semiarid lands are responsive to and most productive under irrigation. Irrigation is not developed to its full potential because its value depends heavily upon temperature conditions and location with reference to markets. *See* GRASSLAND.

(11) Regions of Mediterranean climate, because of their winter rainfall, generally are classed as humid lands. However, the greater part of the growing season is water deficient, and the most productive agricultural lands are dependent on irrigation. Agricultural lands are the major resource, since water deficiency is pronounced enough to discourage forest productivity. Major mineral deposits may complement agricultural lands in a few areas.

Rock composition and surface configuration. Imposed on the basic land pattern induced by climatic differences are variations in rock composition and surface configuration which cause intraregional differences within the pattern. Although not exactly the same in their effects, the variations caused by these two geographical elements are often concomitant and may be treated together as follows:

1. Rock composition and structure
   a. Flat-lying and moderately folded sedimentary rocks
   b. Igneous, metamorphic, strongly folded sedimentary rocks
   c. Volcanic rocks
2. Surface configuration
   a. Flood plains and other flat or gently sloped surface
   b. Mountains and maturely dissected hill lands, plateau faces, or faces of cuestas

These elements of crustal variation produce the following six geographical differences in resource endowment:

First, all major agricultural lands are on flat-lying or moderately folded sediments and have gentle slopes, well exemplified in such alluvial valleys as the Mississippi, Nile, Huang, and Ganges-Brahmaputra.

Second, productive secondary agricultural land resources are located on volcanic areas where soils have been formed through weathering or wind action. The Deccan section of India, for example, is one of the largest areas of this kind.

Third, agricultural lands are extremely limited on igneous rock areas, no matter what the surface configuration, particularly in regions north of 40°N latitude, as illustrated in the occupance of the Laurentian Shield area (Canada) and the Fenno-Scandian Shield (Europe). Exceptions are found in the humid tropical and subtropical climates where weathering has proceeded long enough to produce a substantial soil mantle, as on the Piedmont of southeastern United States.

Fourth, forest lands are not limited in their extent (although limited in productivity) by either crustal rock composition or by the surface configuration.

Fifth, mountainous areas are important catchment areas and sources of fresh water and the services which may be derived from water. Most hydroelectric generation, or generation potential, is associated with mountains. In arid and semiarid regions mountains are sources of water for irrigation, domestic and industrial water supply, wood products, and warm-season grazing lands.

Sixth, mineral resources have definite patterns which are associated with rock structure and composition. Major deposits of coal, petroleum, natural gas, and lignite are, with few exceptions, found in flat-lying or gently folded and faulted sedimentary rocks, as in Texas oil fields and the coal fields of the Allegheny Plateau. Sedimentary nonmetallics (the phosphates, potash, sulfur, nitrates, and limestone) as well as bauxite and uranium (carnotite) also are associated with sedimentary rocks.

Associated with the igneous and metamorphic rock areas are most metals, for example, iron (usually), lead, copper, tin, the ferroalloys, gold, and silver. Most gems and some nonmetallic minerals (mica and asbestos) are found in the same association. Uranium (pitchblende) occurs in these rocks.

While these associations are well recognized, the mineral deposits have an erratic geographical occurrence.

Employed and potential resources. Resources have meaning insofar as they are placed in use or are available for future exploitation. Distinction must be made between the employed and the potential resources. In practice, this distinction is complex, but here only the simple geographical distinction will be noted. Employed resources are those which are significant to the present support of mankind, at least locally. In general, the denser the population and the more advanced the technical arts of an area, the greater the need for production from resources, and employed resources become more nearly synonymous with all known resources. Thus, the recognized resources of the European peninsula and of northeastern United States are mainly employed resources. On the other hand, the natural resources of the Amazon basin or of Alaska and western Siberia are still largely potential resources.

Marine resources. Although the physical and biotic geography of the oceans is much less fully explored than that of the continents, enough is known to indicate that both living and mineral resources extend far beyond those presently exploited. The employed resources are rather sharply localized. The principal exploitation of marine animals and vegetation is (1) over the continental shelves; (2) in the vicinity of the mixing of warm and cold currents; (3) near large upwellings which characteristically occur in lower middle latitudes; and (4) adjacent to densely settled countries. Thus, the North Atlantic near Europe and from New

England to Newfoundland contains heavily exploited fishing grounds, as do the seas near Japan, Korea, and southern California, and also waters of the Gulf of Mexico. Minerals of the sea are employed resources in few localities; salt is the principal mineral extracted, mainly in dry, low-latitude areas.

There seem to be large potential marine resources which include (1) the population of life forms now exploited in some parts of the world, but not in others; (2) animal and vegetative species currently unused; (3) fresh water from desalted sea water; (4) the minerals which are in solution, are precipitated to the ocean bottom, or lie within rock below the water. One of the interesting speculative resources appears in the large quantities of so-called manganese nodules that cover some sections of sea bottom at intermediate depths. Several of the minerals not commonly found in land deposits, like manganese and nickel, occur in these deposits. *See* MARINE RESOURCES.

Resources of the continents. The resource pattern of the Earth may be summarized in a brief description of that for each continent and its neighboring waters.

Eurasian continent. As the largest landmass in the world, the Eurasian continent has the largest area of agricultural land in use, a very extensive total forest land area, and a wide variety of mineral deposits. Great differences mark the several sections of the continent. The most productive agricultural areas are generally near the edge of the landmass, in western and central Europe, European Soviet Union, India, and mainland China. Much of interior Siberia, central Asia, and Asia Minor, shut off from productive agriculture by cold and drought, contain forest resources of major potential significance and mineral reserves, including petroleum, coal, and the metals. The southeastern and eastern borders of the heartland, moreover, have some of the great, but still undeveloped, hydrogeneration sites of the world. Off the coasts of western Europe and north China and south Siberia are the two most productively employed fisheries of the world.

Africa and Australia. Africa is handicapped by heat and drought. A major section of the continent must be classed as desert or semiarid, with few exotic water sources. Much of the remainder has wet-and-dry or rainforest climates, with the attendant handicaps to forest or agricultural exploitation. The east African highlands from Ethiopia southward, the Cape area of South Africa, and the Nile Valley are the only noteworthy exceptions. Except in the Nile Valley, there are still potential agricultural land resources, but they are comparatively minor. Africa's chief resources over the long term may prove to be mineral, including fuels and other energy resources. Although still far from fully explored, major metal deposits include copper, bauxite, and iron; and diamonds, gold, and uranium are well known. Large potential but undeveloped hydroelectrical resources also exist.

Similar general remarks may be made about Australia, whose smaller area is covered mostly by desert, semiarid, and tropical wet-and-dry environments. Most productivity and population are peripheral, especially in the southeast.

South America. The land resource is dominated by the unbroken extent of rainforest and wet-and-dry land stretching east of the Andes from Colombia to northern Argentina, and by a substantial area of water-deficient territory on the west, south, and northeast ot the continent. Some highly fertile flat subtropical lands about the Parana and La Plata valleys are of minor extent by comparison to the whole. South America's resources must be classed largely as potential, although the rapidly growing population may bring changes within the 20th century. Like Africa, South America has some important mineral deposits, chiefly the metals and petroleum.

North America. Large sections of North American lands benefit from the advantages which characterize middle latitude humid-land resources under present technology. Disadvantages of desert and semiarid environments on this continent are tempered somewhat by the interspersal of mountain ranges throughout these drier regions. Taiga and other northern climatic environments are coincident with the igneous rock in the Laurentian Shield; and tropical environments are of small extent. In sum, this continent may be considered to have one of the best balanced sets of resources, considering its substantial endowment in minerals of many different kinds, extensive forest lands, and great and varied agricultural lands, and the productive fisheries off both Atlantic and Pacific shores. North America has the highest ratio of employed resources to land area of all continents. In addition, it still contains significant potential resources.

Summary comment. The Earth's resource pattern has certain general characteristics. (1) Minerals usable under present technology are found in every environment, although mineral types differ according to location in sedimentary or igneous and metamorphic rock areas. Mineral exploration will continue indefinitely in all land areas, but the mineral resource possibilities of North America and the European part of the Eurasian continent have been examined in greater detail than those of any other large area. Ocean basins are the least-known part of the world as to mineral possibilities. (2) Agricultural lands and forest lands usable under present technology are dominated by those lying in middle latitudes. Currently, sections of the taiga are becoming more important as forest resources. (3) The great potential agricultural and forest resources, if some technological improvement is assumed, lie within the tropical environments, possibly in northern South America. *See* CLIMATOLOGY; LAKE; SOIL, ZONALITY OF; VEGETATION ZONES, WORLD.

[EDWARD A. ACKERMAN; DONALD J. PATTON]

Bibliography: M. Clawson (ed.), *Natural Resources and International Development*, 1964; J. L. Fisher and N. Potter, *World Prospects for Natural Resources*, 1964; M. R. Huberty, *Natural Resorces*, 1960; D. Patton, *The United States and World Resources*, 1968.

Earth sciences

Sciences primarily concerned with the atmosphere, the oceans, and the solid Earth. They deal with the history, chemical composition, physical characteristics, and dynamic behavior of solid Earth, fluid streams and oceans, and gaseous atmosphere. Because of the three-phase nature of the Earth system, Earth scientists generally have to consider the interaction of all three phases — solid, liquid, and gaseous — in most problems that they investigate.

The geosciences (geology, geochemistry, and geophysics) are concerned with the solid part of the Earth system. Geology is largely a study of the nature of Earth materials and processes, and how these have interacted through time to leave a record of past events in existing Earth features and materials. Hence, geologists study minerals, rocks, ore deposits, mineral fuels and fossils, and the long-term effects of terrestrial and oceanic waters and of the atmosphere. They also investigate present processes in order to explain past events.

Geochemistry involves the composition of the Earth system and the way that matter has interacted in the system through time. For example, by studying the behavior of radioactive substances it is possible to determine how old the substances are and how much energy has been released through time as a result of decay of radioactive compounds.

Geophysics deals with the physical characteristics and dynamic behavior of the Earth system and thus concerns itself with a great diversity of complex problems involving natural phenomena. For example, earthquakes, vulcanism, and mountain building throw light on the structure and constitution of the Earth's interior and lead to consideration of the Earth as a great heat engine. Study of the magnetic field involves considering the Earth as a self-sustaining dynamo.

The atmospheric sciences, commonly grouped together as meteorology, are concerned with all chemical, physical, and biological aspects of the Earth's atmosphere. Although the study of weather used to be the chief occupation of meteorologists, man's entry into the space age calls for a vast increase in knowledge of the environment through which vehicles and ultimately living things will go and return. Consequently, many aspects of the Earth's atmosphere are now being studied intensively for the first time. As an example, great planetary currents in the atmosphere, and also in the oceans, are now being investigated not only for the light they may shed on a better understanding of weather but also as a basis for understanding more fully the motion of the entire atmosphere and oceans. *See* METEOROLOGY.

Oceanography encompasses the study of all aspects of the oceans — their history, composition, physical behavior, and life content. Before World War II little was known about the oceans of the world. During that war many important characteristics of the ocean were discovered, and since then, with instruments and facilities developed during the war, oceanographic research has been going on at a quickened pace. *See* OCEANOGRAPHY.

[ROBERT R. SHROCK]

Bibliography: *Investigating the Earth*, Earth Science Curriculum Project of the American Geological Institute, 1967; K. Krauskopf, *Introduction to Geochemistry*, 1967; H. Takeuchi, S. Uyeda, and H. Kanamori, *Debate About the Earth*, 1967.

Ecological interactions

Relations between species that live together in the same community. The interaction refers to the effect that an individual of one species may exert

upon an individual of another species, and this effect may be physiological, such as excretion of toxins, or behavioral, such as fighting. Interactions may occur between two species or more than two, between animals, between plants, or between protists, or in any combination of these groups. These interactions may be harmful, neutral, or beneficial to the individual and can be described in a spectrum from positive, or attraction, to negative, or repulsion. The importance for man may be direct if human groups are interacting with each other or their environment, or indirect if interactions among other organisms or environments have a subtle or indirect effect on man. *See* ECOLOGY, APPLIED; ECOLOGY, HUMAN; ECOSYSTEM.

Symbiosis. Although the term symbiosis is often restricted to an interaction that is beneficial to both species, the original meaning refers to living together and thus includes interactions that are positive or negative. This broad meaning has the approval of the American Society of Parasitologists. The usage of "beneficial" and "harmful" presents various problems — principally, the fact that an interaction that is harmful to an individual may benefit the species or social group. A number of parts of the spectrum of symbiotic interactions may be distinguished.

An aggregation is a temporary group of animals that are attracted to some area for a specific purpose. An aggregation is an important means for utilizing the resources of the habitat, since the activity and call notes attract individuals to the location of environmental necessities. Some aggregations may utilize physical aspects, for example, snakes sunning on a rock, or may utilize food, as birds feeding on fruit trees. In many cases specific call notes or activities serve to notify other members of the species that some habitat requirement is available. For example, the persistent circling of vultures attracts others from many miles.

Commensalism. Commensal interaction refers to the joint utilization of food. Literally, the term means "at the same table" but the relationship is rarely equal. Generally, one member provides the food and the other consumes some part of it. The relationship is not harmful to either party, at least directly. Sometimes the term is extended to include other habitat necessities, such as shelter or transportation.

A spectacular example of commensalism occurs in the association of certain fish with sea anemones. These little fish, *Amphiprion percula*, are able to enter the anemone (*Discosoma*) and to avoid harm from the poison tentacles. Subsequently, the fish eat the debris of other fish that have been captured by the anemone. Another example is the relation of commensal rats (*Rattus*) to humans. These mammals derive their food and shelter directly and, although they do not normally eat "at the same table," they use man's food in urban areas. However, they can survive very well in some wild habitats, without human assistance. The rats rarely harm humans directly but may soil food and, more important, serve as a reservoir for diseases.

In many cases the relationship is facultative, or unessential for survival of one member, but often it is obligate, or essential. For example, certain oligochaete worms, the Branchiobdellidae, attach to crayfish in an almost ectoparasitic manner and subsist on refuse from the crayfishes' meals. Obligate commensalism grades imperceptibly into parasitism and, indeed, the term host is used for one partner. Usually, the guest lives on the outside of the host, but it may live in the respiratory or digestive tract.

The commensals may be about the same size or may differ greatly. As the disparity in size develops, there is a trend toward parasitism by the smaller. This relation occurs whether the guest obtains food or shelter or transport from its host. Naturally, a very frequent relation is the use of the host simply as a place to live. The chain of commensals may become quite long, as in the case of the barnacles that perch on other barnacles that are attached to whales.

Commensalism does not suggest a taxonomic relationship. Indeed, plants commonly are commensal with animals, and closely related species rarely are commensal. Presumably through the centuries the competition among closely related animals forced a taxonomic separation and precluded a commensal relation.

Frequently, the commensals develop anatomical specialization or behavioral peculiarities in relation to their host. Means of attachment, processes for collection of food, and senses for detecting the host are some characters that may evolve. These peculiarities may develop so elaborately that the taxonomic relations of the species are obscured or the species is utterly incapable of living without provision of food by its host.

Mutualism. The term mutualism is now replacing the term symbiosis to refer to relations that are beneficial to both species. Such relations are vitally important to animals because the survival of the individual, or of the species, may depend upon the success of a particular mutual interaction.

Some of the mutual interactions are spectacular and intricate, whereas others are rather simple. Perhaps the least complicated relation is the simple exchange of carbohydrate and oxygen from a plant for nitrogeneous wastes and carbon dioxide from an herbivore. Usually, if the interaction is remote and at the level of the ecosystem rather than the species, it may be considered to be mutual; in other cases the relation is still closer, as in fungal mycorhizae, commonly associated with plant roots. Mycorhizae have specialized processes that enter the root structure. In a few cases the interaction is intimate, for example, the similar exhange in a lichen, which has formalized the interaction into a single structure that appears to be an individual. *See* ECOLOGY.

A somewhat different situation occurs in the interaction of certain animals and their domesticated animals and plants. Some beetles, ants, and termites grow a number of fungi, and man grows animals and plants that can no longer survive unaided in nature. Although the individual domesticated animal or plant is destroyed, the species is maintained and thus a mutual relation occurs. In many cases the relation is very specific; one kind of beetle grows only one kind of fungus, whereas another species grows another kind. The transmission of the fungus to subsequent generations may require elaborate behavioral mechanisms.

The mutual dependence of two species is extended more elaborately in the examples of digestion of an animal's food by the bacteria or protozoa in the gut. For example, cockroaches cannot exist without their protozoa which break down cellulose so that the cockroach can utilize it. Many such situations exist. Indeed, it seems likely that most species, including man, are partially dependent upon bacteria and protozoa in the gut for digestion. Some insects have elaborate structures for storing or transmitting the organisms. This mutual relation occurs in plants as well. The nitrogen-fixing bacteria on the roots of legumes are a well-known case. *See* NITROGEN CYCLE.

Most of the cases of mutualism are examples of a common effort to provide food either directly or indirectly. However, another process necessary for survival of the species, namely, fertilization of gametes, is frequently arranged by a mutual interaction. Many flowers have conspicuous devices that attract birds or insects and encourage cross-pollination. These mutualistic adaptations occur in some species of plants that are distantly related and thus must have developed independently. Some examples are simple, such as arrangement of anthers so that the pollen falls on the insect as it feeds on the nectar. Other examples are more complex, such as the wasp that fertilizes figs. The larvae of the wasp develop within the flower, and thus the plant sacrifices some of its reproductive potential of the rest to be ensured.

The variation in detail of the examples of mutualism may distract from the general trend toward interdependence. At one extreme, closely related forms may have some minor beneficial exchange, while at the other extreme, completely unrelated forms may be so dependent that they cannot exist alone. This trend led to the concepts of a "superorganism," which is a combination of two or more species into a functional whole and ecosystem. Examples of all stages in this evolutionary story exist.

Neutralism. Neutral interactions are frequent at certain times or stages of life history. Meadowlarks and sparrows may exist together in an area with no direct interaction except possibly the mutual use of an abundant food supply. Foxes may prey upon mice during the winter but ignore them during the summer in favor of grasshoppers and berries. All members of an ecologic community, however, have some indirect interactions. While the term neutral interaction is somewhat self-contradictory, it is helpful to express the fact that some potential interactions result in no evident effect on either species at any time or only at certain times. Furthermore, the interaction can change from neutral to positive (aggregation, commensal, symbiotic) or to negative (avoidance, competition, predation, parasitism) under various conditions.

A negative interaction is harmful to some extent to an individual of the combination. Probably the simplest negative type is avoidance or repulsion. Small animals, such as crows, surrounding a carcass may learn to avoid larger species, like vultures, which drive them away; thus harmful results are prevented. The avoidance may be direct, as in the previous example, or may be based on signs, such as trails or feces. The avoidance of a potentially harmful interaction obviously improves survival of the individual.

Competition. This interaction results when several individuals share an environmental necessity. Indirect competition occurs when there is a difference in time so that one individual uses the resource before the other. This is illustrated when an elk eats an item of food in the summer that the deer would eat in the winter. Direct competition occurs when an elk drives a deer away from an item of food. It is at once apparent that competition stretches from avoidance on the one hand to actual fights, and even death, on the other hand.

Competition may be either intraspecific, that is, among individuals of the same species, or interspecific, that is, among individuals of different species. In many cases the results may be similar. It matters little to a deer whether another deer or an elk ate the food, for the food is gone.

The importance of competition lies in the survival of the individuals that compete. Generally, the competition is relatively passive, as one individual soon leaves the area. However, if he is repeatedly driven from place to place, he will either starve or fail to reproduce. Thus, while competition rarely leads to a fight and death, it may frequently result in the elimination of individuals from the population.

Behavioral devices may mitigate competition. A simple means, naturally, is flight and return, as when an individual flees but returns at a later time to obtain the desired item. Animals of the same species, or even different species, may formalize this device into a social rank or hierarchy that regulates the order of precedence. Under this arrangement a low-ranking individual simply waits his turn at the food supply. If the supply is inadequate, the low-ranking individuals do without. If shortage becomes more acute, the low-ranking individuals flee or eventually die. The use of social organization to systematize the effects of competition occurs in many variations in many species, but is most highly developed in vertebrates. While the result may be disastrous to the individual, the consequences for the population may be beneficial. If there exists only enough food for two roosters, it is better for survival of the population to let the two have enough to survive than to divide the food equally among the four, resulting in the death of all four. This illustrates the earlier point that while competition may be negative for the individual, it may be positive, or beneficial, for the species.

Another behavioral device, territorialism, may mitigate the effects of competition. Individuals or pairs defend an area against intruders of the same species or of different species. Although most cases of territorial interactions occur among individuals of the same species, in some cases the interactions occur among members of various species. In the winter redheaded woodpeckers defend an area in which they store acorns. The individual drives out other redheaded woodpeckers and also titmice and chickadees. Under these circumstances the interaction is both intra- and interspecific.

Examples of competitive interactions are surprisingly difficult to document quantitatively, even in the laboratory. The behavioral devices of flight, rank, and territorialism mitigate the effects in natural populations to such an extent that only gradual changes occur. Flickers have decreased in numbers in many areas since the European starling arrived, yet direct combat for nest holes is rarely seen and it is difficult to measure the interaction. Laboratory experiments on the competition of beetles, mice, viruses, and a few other forms show results that resemble the interactions in nature. The precise effect of an interaction is difficult to predict, even in the laboratory, because it depends upon many variables.

Predation. This act is the killing and eating of an individual of one species by an individual of another species. A conventional example is the killing of a mouse by a cat. However, the term refers also to the killing of an individual by a group, such as an attack of a pack of wolves upon a deer.

Predation also refers to a process in a population. It is the statistical effect on the population by the various predators. Thus predation is said to reduce a population or to alter the age composition. In these cases the mass effect of some or all of the predators on the population is being considered.

Examples of acts of predation are commonplace. Kingfishers and herons catch fish, dragonflies catch insects, ant lions build a trap for ants, and starfish capture clams. The importance of predation on individuals exists primarily in terms of natural selection. It was known, even before Darwin, that predators may select certain types of individuals from among a population, but only recently have quantitative data become available. In England predators captured light-colored moths at a higher rate than dark-colored individuals. Thus during many generations selection will favor the dark forms.

The evolution of devices that improve ability to capture and kill the prey has produced some unusual structures. Simple examples are claws and teeth, while more complex examples are the pitfalls of ant lions, and the traps of several plants, like the Venus' flytrap. Obviously, the guns, nets, and traps created by humans are merely extensions of devices for predation. It is, of course, apparent that some devices used for predation can also be used in competitive interactions.

The result of the interspecies interaction of predation when individuals are concerned is relatively simple. The prey is killed or escapes, and the predator gets food or searches elsewhere. However, a number of compensatory processes complicate the result on the population of the prey. Indeed, a continuous spectrum of results on the population can be described from inadequate predation on the one hand to excessive predation on the other hand. The terms inadequate and excessive refer to the welfare of the prey species. Examples of inadequate predation occur in many fish populations. Predators, such as man or herons, may be unable to remove enough fish to prevent a population increase that reduces the food supply, with the result that the individual fish are stunted and may not reproduce. An increase in predation may reduce the population numerically but result in an increase in size and in the rate of reproduction.

In other cases predation has no measurable effect on the population other than the removal of particular individuals, since predation does not affect the number of prey, except momentarily, if the total mortality rate remains constant. Increased hunting did not affect pheasant population in Michigan, partly because the pheasants learned to avoid hunters, and partly because of a compensatory decrease in mortality from other causes.

However, in some cases predation can be clearly

associated with a decline in population, such as the reduction in the number of mosquitoes in an area by the mosquito fish *Gambusia*, or the reduction of rabbits in Australia by the virus disease myxomytosis. Humans, of course have reduced many species to low levels.

The extreme effect of the interaction is the extinction of a prey species by a predator. Man has exterminated many vertebrates in recent years. However, the extinction of one vertebrate by another vertebrate, other than man, is rare. The only examples adequately proven occur when rats or mongooses exterminate some birds on islands. Disease provides some spectacular cases of extinction, such as the extermination of the American chestnut by blight, a virus-caused disease.

These examples demonstrate the spectrum of interspecific interactions that occur under the general term of predation. While they may harm the individual, in many cases the survival of the species is improved.

Parasitism. In this interaction a small species lives in or on another species and usually derives food or shelter from it. As might be expected, relations that are mutual or commensal grade imperceptibly into parasitic. Indeed, the same two species may have mutual or commensal relations at one time and parasitic at another. The distinction among the three types of symbiosis is based upon the relation to the survival of an individual or to its welfare. If one individual is harmed, the relation is called parasitic. However, usually only certain of the parasitized individuals are harmed or, in the case of ectoparasites, the harm is trivial.

Two types of parasites are distinguished: ectoparasites, which are external, and endoparasites, which are internal. The parasite may live only a part of its life cycle on or in the host. In most examples the parasite spends only a fraction of its time in any one host, but may inhabit two or more hosts during its life.

Some general relations among parasites may be mentioned. Parasites may be facultative or obligate. The former are able to exist independently but may be parasitic on certain occasions, whereas the latter cannot survive at certain stages without the host. Parasites may be specific, living only in a certain host species, or nonspecific, living in many hosts.

Parasitic interrelations may exist between practically any group of plants and animals. Plants may parasitize plants or animals and animals may parasitize animals or plants. The taxonomic relations may be close in some cases or distant in others. Indeed, every conceivable combination occurs, even the example of one sex parasitizing the other, as in the angler fish.

A rather aberrant type of parasitism called social parasitism occurs in some birds. The female lays her eggs in the nests of other species and permits the foster parents to raise the young. This habit has appeared independently in five families.

While the interaction of parasitism may be harmful to the individual, it is not necessarily harmful to the species. As for predation, the effect of parasitism may prevent overpopulation and stimulate the evolution of new structures or adaptations. A value judgment about its effect on the species is precarious. *See* POPULATION DISPERSAL.

[DAVID E. DAVIS]

Bibliography: S. M. Henry (ed.), *Symbiosis: Associations of Microorganisms, Plants, and Marine Organisms*, 1966; C. B. Knight, *Basic Concepts of Ecology*, 1965; R. L. Smith, *Ecology and Field Biology*, 1966.

Ecological pyramids

The concept of a pyramid of numbers was introduced by C. S. Elton to express the idea of using feeding relationships as a basis for the quantitative analysis of ecological systems in nature.

If the series of organisms involved in a food chain are allocated to feeding groups (subsequently called trophic levels) such as plants, herbivores, and carnivores, and if these groups are conventionally defined as low to high (in that order), Elton showed that numbers are usually greatest in the low levels and least in the high. This result may be expressed in a pyramid-shaped diagram, the shape being due to a tendency for herbivores to be smaller, more numerous, and faster breeding than the predators which feed on them. In practice the diagram based on numbers is not always pyramid-shaped because these conditions are seldom all met. (Lions are smaller than antelopes; aphids are more numerous and breed faster than many food plants.)

As a criterion of biological "importance," biomass (total quantity of living tissue in a population of one species) is preferable to numbers of individuals, and population metabolic activity (respiration rate) is better still. "Biomass" and "metabolic" or "energy pyramids" are usually pyramid-shaped, having a broad base (plants) and a narrow apex (carnivores).

However, two additional factors should be considered in applying the pyramid concept: (1) It applies strictly only to "complete" ecosystems that are isolated from other systems in the sense that little material is gained or lost between them. For example, caves and the sea bottom gain food from elsewhere and lack photosynthetic levels. (2) The attribution of a particular species to a distinct trophic level is not always possible. *See* BIOMASS; ECOSYSTEM; FOOD CHAIN.

[AMYAN MACFADYEN]

Ecology

A study of the relation of organisms to their environment, or in more simple terms, environmental biology. Ecology is concerned especially with the biology of groups of organisms and with functional processes on the lands, in the oceans, and in fresh waters. Ecology is the study of the structure and function of nature (mankind being considered part of nature).

Ecology is one of the basic divisions of biology which are concerned with principles or fundmentals common to all life. Since biology may also be divided into taxonomic divisions which deal with specific kinds of organisms, ecology is not only a basic division of biology but also an integral part of any and all of the taxonomic divisions. Both approaches are profitable. If it often productive to restrict certain studies to taxonomic groups, insects, for example, because different kinds of organisms require different methods of study. (Some groups of organisms are of greater interest to man than others.) It is also important to seek and to test unifying principles which may be applicable to nature as a whole. *See* ANIMAL COMMUNITY; PLANT COMMUNITY.

Approaches to ecology. From the standpoint of general ecology, an instructive subdivision is in terms of the concept of level of organization. For convenience, a biological spectrum can be visualized as follows: protoplasm, cells, tissues, organs, organ systems, organisms, populations, communities, and ecosystems. Ecology is concerned largely with the levels beyond that of the individual organism. In ecology the term of population, originally used to denote a group of people, is broadened to include groups of individuals of any one species of organism. Community in the ecological sense, sometimes designated as biotic community, includes all of the species populations of a given situation. The community or individual and the nonliving environment function together as an ecological system or ecosystem. The Earth as an ecosystem is conveniently designated as the biosphere. The term autecology is often used to refer to the study of environmental relations of individuals or species, whereas the term synecology refers to the study of groups of organisms such as communities. Mathematical concepts have permeated the basic foundations of ecology and have provided new means of exploring qualitative and quantitative relations at all levels. This new brand of ecology is known as systems ecology. *See* COMMUNITY; POPULATION DYNAMICS; SYSTEMS ECOLOGY.

For convenience, subdivisions of ecology may also be based on the kind of environment or habitat to be considered or studied. Marine ecology, fresh-water ecology, and terrestrial ecology are the three broad divisions from this point of view; estuarine ecology, stream ecology, or grassland ecology represent more restricted interests. *See* FRESH-WATER ECOSYSTEM; MARINE ECOSYSTEM; TERRESTRIAL ECOSYSTEM.

Some attributes, obviously, become more complex and variable during the procession from cells to ecosystems; however, it is often overlooked that other attributes may become less complex and less variable from the small to the large unit. The reason is that homeostatic mechanisms, which may be defined as checks and balances or forces and califorces, operate all along the line to produce a certain amount of integration, and smaller units function within larger units. For example, the rate of photosynthesis of a whole forest or a whole cornfield may be less variable than that of the individual trees or corn plants within the communities, because when one individual or species slows down, another may speed up in a compensatory manner. *See* HOMEOSTASIS.

Furthermore, it is important that findings at any one level aid in the study of another level, but never completely explain the phenomena occurring at that level. The old saying that the forest is more than a collection of trees will illustrate very well what is meant here. For a full understanding of the forest it is necessary to study both the trees as separate units and also the forest as a whole. During recent years ecology has made the most progress, not from intensive study at one level, but by coordinated, simultaneous attack at all levels.

Coral reef study. The importance of studying both the part and the whole may be illustrated by the following example. A coral reef represents one of the most beautiful and best-adapted ecosystems. Corals are small animals that have tentacles adapted to seize small animals called zooplankton which are in the water. Embedded in the tissues of the coral animal and in the calcareous skeleton are numerous small plants or algae.

Some years ago C. M. Yonge carried out a carefully planned series of experiments with isolated coral colonies in tanks in an effort to clarify the relationship between corals and their contained algae. He found that when the corals were supplied with abundant zooplankton they thrived and grew normally, if all of the algae were killed by keeping the colonies in the dark. On the basis of these experiments Yonge concluded that the algae do not contribute to the well-being of the coral, and therefore are of no importance in the reef-building activities of corals. Some years later H. T. Odum and E. P. Odum were able to measure the metabolism of an intact coral reef and thus determine the amounts of food which corals require. It was soon evident that there were not enough zooplankton in the surrounding infertile oceans to account for the large population and rapid growth of the coral reef. It was suggested, therefore, that the corals must indeed obtain some of their food from the algae. Other investigators became interested in the problem, and L. Muscatine and C. Hand (1958) showed by the use of carbon-14 isotopic tracer that food does indeed diffuse from the algae, which manufacture food by photosynthesis, to the coelenterate. Still other investigators have shown that activities of the symbiotic algae are important in skeleton formation. Thus it was proved that what was true of an isolated colony in a tank was not entirely true for the coral living in its natural ecosystem.

Functional ecology. Returning to the definition of ecology as the study of structure and function of nature, it should be noted that until fairly recently ecologists often described how nature looked. Now equal emphasis is being placed on what nature does, because the changing face of nature can never be understood unless her metabolism is also studied. Consideration of function brings the small organisms, which may be inconspicuous but very active, into true perspective with the large organisms, which may be conspicuous but relatively inactive. As long as a purely descriptive viewpoint was maintained, there seemed little in common between such structurally diverse organisms as trees, birds, and bacteria. In real life, however, all these are intimately linked functionally in ecological systems according to well-defined laws. Likewise from a descriptive standpoint, a forest and a pond seem to have little in common, yet both function according to exactly the same principles. Thus general ecology is essentially an ecology of function.

THE ECOSYSTEM

Any area or volume of nature that includes living organisms and nonliving (abiotic) substances interacting so that a flow of energy leads to characteristic trophic structure and material cycles. *See* ECOSYSTEM.

Biotic components. From a functional standpoint, an ecosystem has two biotic components: an autotrophic component able to fix light energy and manufacture food from simple inorganic sub-

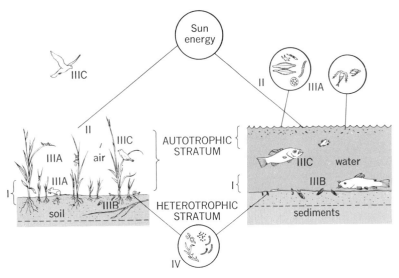

Fig. 1. Comparison of trophic structure of a simple terrestrial ecosystem (grassland) with an open-water aquatic ecosystem (either fresh water or marine). Basic units of the ecosystem are I, abiotic substances; II, producers; III, macroconsumers (A= direct herbivores, B= detritus eaters or saprovores, C= carnivores); and IV, decomposers (bacteria and fungi of decay).

The concept of the ecosystem is and should be a broad one, its main function in ecological thought being to emphasize obligatory relationships, interdependences, and casual relationships. Ecosystems may be conceived and studied in various sizes. A small pond, a large lake, a tract of forest, or even a small aquarium may provide a convenient unit of study. As long as the major components are present and operate together the entity may be considered an ecosystem. Throughout the entire biosphere, ecosystems have a very similar functional makeup, though they may differ markedly in structural features and degree of stability.

Very frequently the basic functions and the organisms responsible for the basic processes in an ecosystem are partially separated in space and also in time. The autotrophic and heterotrophic components are often stratified one above the other, and there is often a considerable delay in the heterotrophic utilization of the products of the autotrophic organisms. For example, photosynthesis predominates in the aboveground portion of a forest ecosystem. Only a part, often very small, of the food manufactured in the upper layers of the forest is immediately and directly used by the plants and by herbivores and parasites which feed on foliage and new wood. Consumption of much of the synthesized food material in the form of leaves, wood, and stored food in seeds and roots is delayed until it eventually falls to the ground and becomes part of the litter and soil, which constitute a well-defined heterotrophic stratum. A similar functional stratification and time lag may be observed in the pond ecosystem (Fig. 1); the phytoplankton comprise the well-defined autotrophic layer, whereas the sediments constitute the heterotrophic system.

The ecological categories which have been discussed above are ones of function rather than of species. There are no hard and fast lines between such categories as producers, consumers, and decomposers, because some species of organisms occupy intermediate positions in the series and others are able to shift their mode of nutrition according to environmental circumstances. For example, certain species of algae and bacteria are able to function either as autotrophs under certain conditions, or at least partly as heterotrophs under other conditions. The separation of heterotrophs into large consumers and small decomposers is arbitrary. Many decomposers (bacteria and fungi) are relatively immobile (usually embedded in the medium being decomposed) and are very small with high rates of metabolism and turnover. They obtain their energy by heterotrophic absorption of decomposition products. Their specialization is more evident biochemically than morphologically; consequently, their role in the ecosystem cannot be determined by such direct methods as looking at them or counting their numbers. Rather, their actual functions must be measured. Macroconsumers obtain their energy by heterotrophic ingestion of particulate organic matter. These are largely the animals in the broad sense. In contrast to the decomposers the macroconsumers are larger, have slower rates of metabolism, and are more readily studied by direct means. They tend to be morphologically adapted for active food seeking or food gathering, with the development in higher forms of complex sensory and neuromotor as well as diges-

stances, and a heterotrophic component which utilizes, rearranges, and decomposes the complex materials synthesized by the autotrophs. From a structural standpoint it is convenient to recognize four constituents composing the ecosystem, as shown in Fig. 1: (I) abiotic substances, basic elements and compounds of the environment; (II) producers, the autotrophic organisms, largely the green plants such as phytoplankton algae or terrestrial vegetation; (III) the large consumers or macroconsumers (also called biophages), heterotrophic organisms, chiefly animals, which ingest other organisms or particulate organic matter; and (IV) the decomposers or microconsumers (also called saprophages or saprophytes), heterotrophic organisms, including the bacteria and fungi, which break down the complex compounds of dead protoplasm, absorb some of the decomposition products, and release simple substances usable by the producers. Microorganisms also release growth-promoting and growth-inhibiting substances that are important in regulating the metabolism of the ecosystem as a whole. As shown in Fig. 1, a pond is a good example of an ecosystem which exhibits a recognizable unity both in regard to function and structure.

The stratification of autotrophic and heterotrophic functions and basic trophic levels is the same for land and water systems, but the kinds and relative sizes of the organisms are different. On land, the individual producer organisms tend to be relatively large in relation to consumers, whereas in water the producers are often microscopic in size. Although the biomass of the land vegetation per unit of area may be much greater than that of the biomass of aquatic phytoplankton, photosynthesis of the latter can be as great, if light and nutrients are equivalent. Marshes, ponds, and shallow margins of lakes and the oceans have a trophic structure intermediate between these two systems, because both large and small producers may be present.

tive, respiratory, and circulatory systems. Much may be inferred about the functioning of the macroconsumer groups by observing them and counting their numbers. Even in this case it is necessary to devise means of assaying rate of functions to fully understand their role in the ecosystem.

A sizable ecosystem may contain only producers and microorganism decomposers (pioneer aquatic communities, for example). Almost everywhere on Earth, however, the macroconsumers or animals invade sooner or later and play a prominent role in the functioning of the ecosystem. It is convenient to consider soil, fallen logs, or deep-sea basins as ecosystems (because they show consistent structural and functional characteristics), provided it is recognized that these systems consist only of the heterotrophic components and are therefore ecologically incomplete, as subsystems of a larger system.

Biotic influence on environment. The physical environment controls the organisms, and organisms also influence and control the abiotic environment in many ways. Organisms return new compounds and isotopes to the nonliving environment; for example, the chemical content of sea water and of air is largely determined by the action of organisms. Plants growing on a sand dune build up a soil radically different from the original substrate. Thus the biosphere is important not only as a place in which living organisms can exist, but also as a region in which the incoming radiation energy of the Sun brings about fundamental chemical and physical changes in the inert material of the Earth, chiefly through the functioning of various ecosystems.

Homeostatic mechanisms. Some equilibriums between organisms and environment may be maintained by a balance of nature which tends to resist change in the ecosystem as a whole. Much has been written about this, but the fundamentals involved are not yet clearly understood. These homeostatic mechanisms include those which regulate the storage and release of nutrients, as well as those which regulate the growth of organisms and the production and decomposition of organic substances. Many organisms, particularly decomposer and producer groups, release organic substances into the environment during their growth processes. These substances often have a profound influence on other organisms and on the regulation of function of the whole ecosystem. Some substances are antibiotic in that they inhibit the growth of other organisms, whereas other substances may be stimulatory, as, for example, various vitamins and other growth-promoting substances. Such external metabolites may be considered to be ectocrines or environmental hormones, in that there is growing evidence that many such substances influence and control the functioning of the ecosystem in the same general manner that endocrines control metabolic rates within individual organisms. Much needs to be learned about the specific action of these substances.

As a result of the evolution of the central nervous system and brain, man has gradually become a most powerful organism, as far as the ability to modify the ecosystem is concerned. Man's power to change and control, unfortunately, seems to be increasing faster than his realization and understanding of the profound changes of which he is capable. Although nature has remarkable resilience, the limits of homeostatic mechanisms can easily be exceeded by the actions of man. When treated sewage is introduced into a stream at a moderate rate, for example, the system is able to purify itself and return to the previous condition within a comparatively few miles downstream. However, if the pollution is great or if toxic substances for which no natural homeostatic mechanisms have been evolved are included, the stream may be permanently altered or even destroyed as far as usefulness to man is concerned. Consequently, the concept of the ecosystem and the realization that mankind is, and must always be, part of an ecosystem and that he has increasing power to modify these systems are concepts basic to modern ecology. These also are points of view of extreme importance to human affairs generally. Conservation of natural resources, one of the most important practical applications of ecology, must be built around these points of view. Thus if understanding of ecological systems and moral responsibility among mankind can keep pace with man's power to effect changes, the old practice of unlimited exploitation of resources will give way to unlimited ingenuity in perpetuating a cyclic abundance of resources. *See* CONSERVATION OF RESOURCES; ECOLOGY, APPLIED; WATER POLLUTION.

ENERGY FLOW

Everyone is familiar with the fact that the kinds of organisms to be found in any particular part of the world depend on the local conditions of existence and on the geography, because each major region of the Earth, especially if isolated from other regions, tends to have its own special flora and fauna. Often, however, ecologically similar or ecologically equivalent species have evolved in different parts of the globe where the basic environment is similar. The species of grasses in the temperate, semiarid part of Australia are largely different from those of a similar climatic region in North America, but they perform the same basic function as producers in the ecosystem. Likewise, the grazing kangaroos of the Australian grasslands are ecological equivalents of the grazing bison, or the cattle which have replaced them, on North American grasslands. The kangaroo and bison, although not closely related taxonomically, occupy the same ecological niche in the sense that they have a similar functional position in the ecosystem in a similar type of habitat. It is also true that the same species may function differently: that is, they may occupy different niches, in different habitats. The point to emphasize is that a list of species to be found in an area is not sufficient information in itself to determine how the biotic community works. For a full understanding of nature, rate functions must also be investigated. *See* BIOGEOGRAPHY; BIOTIC ISOLATION; GRASSLAND; ISLAND FAUNAS AND FLORAS; PLANT FORMATIONS, CLIMAX; POPULATION DISPERSAL; VEGETATION ZONES, WORLD.

Energy and materials. In any ecosystem the number of organisms and the rate at which they live depend on the rate at which energy flows through the system, and the rate at which materials circulate within the system and are exchanged with adajcent systems, or both. Materials circulate, but energy does not. Nitrogen, carbon, water,

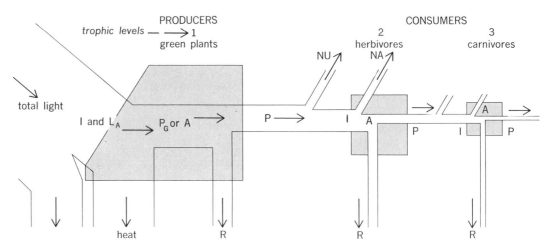

Fig. 2. A simplified energy-flow diagram of an ecosystem. L_A, absorbed light; P_G, total photosynthesis (gross production); P, net primary production; I, energy intake; R, respiration; A, assimilation; NA, ingested but not assimilated; and NU, not used by trophic level shown in diagram. (*After E. P. Odum, Fundamentals of Ecology, 2d ed., Saunders, 1959*)

and other materials of which living organisms are composed may circulate many times between living and nonliving entities; that is to say, the atoms may be used over and over again. On the other hand, free energy is used once by a given organism or population and then is converted into heat and lost from the ecosystem. Life is kept going by the continuous inflow of solar energy from the outside. The interaction of energy and materials in the ecosystem is of primary concern to ecologists. The one-way flow of energy and the circulation of materials are the two great principles or laws of general ecology, because these principles are equally applicable to any type of environment and any type of organism, including man. *See* ECOLOGY, HUMAN.

Energy flow diagrams. A simplified energy-flow diagram which might, in principle, be applied to any ecosystem is shown in Fig. 2. The "boxes" represent the population mass or biomass; the "pipes" depict the flow of energy between the living units.

Only about one-half of the average sunlight impinging upon green plants (producers) is absorbed by the photosynthetic machinery, and only a small portion of absorbed energy, about 1–5% in productive vegetation, is converted into food energy. The assimilation rate of producers in an ecosystem is designated as primary production or primary productivity. Gross primary production (P_G in Fig. 2) is the total amount of organic matter fixed, including that used up by plant respiration during the measurement period; net primary production is organic matter stored in plant tissues in excess of plant respiration during the period of measurement. Net production represents food potentially available to heterotrophs. In Fig. 2, net primary production is represented by the flow P which leaves the producer component. When plants are growing rapidly under favorable light and temperature conditions, plant respiration may account for as little as 10% of gross production. However, under most conditions in nature net production is much less than 90% of gross. *See* BIOLOGICAL PRODUCTIVITY; BIOMASS.

Energy transfer. The transfer of food energy from the source in plants through a series of organisms with repeated eating and being eaten is known as the food chain or web. In complex natural communities, organisms whose food is obtained from plants by the same number of steps are said to belong to the same trophic level. Thus green plants occupy the first trophic level, that is, the producer level; ideally, plant-eaters (herbivores), the second level; carnivores which eat the herbivores, the third level; and secondary carnivores, a fourth level. This trophic classification is one of function and not of species as such; a given species population may occupy one trophic level or more, according to the source of energy actually assimilated, and this may change with time. *See* FOOD CHAIN.

Energy degradation. At each transfer of energy from one organism to another, or from one trophic level to another, a large part of the energy is degraded into heat as required by the second law of thermodynamics. The shorter the food chain or the nearer the organism to the beginning of the food chain, the greater will be the available food energy. As shown in Fig. 2, the energy flows are greatly reduced with each successive trophic level, whether considering the total flow or the production P or respiration R components. The reduction with each link in the food chain is about one order of magnitude. Thus, for every 100 cal of net plant production in a stable community about 10 kcal would probably be reconstituted into primary consumers, the P/P_{T-1} efficiency ratio shown in Fig. 2, which then becomes available to carnivores, which in turn might convert 1 kcal of the 10 into new organic materials. Although a small amount of the primary food energy might be involved in a number of transfers, it is evident that the amount of food remaining after two or three successive transfers is so small that few organisms could be supported if they had to depend entirely on food available at the end of a long food chain. For all practical purposes, then, the food webs are limited to three or four links along most pathways.

Ecological efficiencies. Ecological efficiencies

may be expressed by formulas which indicate the ratios between and within trophic levels:

$$E_I = \frac{I_T}{I_{T-1}} \quad \text{or} \quad \frac{I_2}{I_1} \qquad E_E = \frac{A_T}{A_{T-1}}$$

$$E_P = \frac{P_T}{P_{T-1}} \qquad E_U = \frac{I_T}{P_{T-1}} \quad \text{and} \quad \frac{A_T}{P_{T-1}}$$

Intake, growth, production, and utilization efficiencies between trophic levels are expressed by the above equations. The Lindemann efficiency or the efficiency of trophic level intake E_I is expressed as the ratio between two successive trophic levels; the subscripts indicate the trophic levels being compared. Growth efficiency E_E is the ratio between the assimilation at one trophic level with that of the preceding trophic level. Trophic level production E_P is the ratio of production rates at different trophic levels. The efficiency of utilization E_U is the ratio of intake at one level to the production rate at the preceding level, or it may be expressed as the assimilation rate A_T to the production rate P_{T-1}. Ecological efficiencies within trophic levels are measured as tissue-growth and assimilation efficiencies. Tissue-growth efficiency E_G is the ratio between the production and assimilation rates at the same trophic level or between production and intake rates. Assimilation efficiency E_A is expressed as a ratio of the rate of assimilation to the energy intake or consumption:

$$E_G = \frac{P_T}{A_T} \quad \text{or} \quad \frac{P_T}{I_T} \qquad E_A = \frac{A_T}{I_T}$$

Energy-flow models. In the simplified diagram of Fig. 2, bacteria and fungi which decompose plant tissues and stored plant food would be placed in the primary consumer "box" along with herbivorous animals; likewise, microorganisms decomposing animal remains would go along with the carnivores. However, because there is usually a considerable time lag between direct consumption of living plants and animals and the ultimate utilization of dead organic matter, not to mention the metabolic differences between animals and microorganisms, a more realistic energy-flow model is obtained if the decomposers are placed in a separate box connected by appropriate energy flows to the other components. For example, the *NU* (not utilized) and *NA* (not assimilated) flow components, if not exported from the system, would ultimately be utilized by decay organisms, which in turn might supply at least part of the food used by such macroconsumers as detritus-feeding or filter-feeding animals. Organic excretions, sugars, amino acids, and other organic materials which leak out of organisms into the environment may be considered to be a part of production, because these materials are ultimately consumed by microorganisms aggregated in bubbles or particles and possibly returned through consumer food chains.

Standing crop and energy flow. The boxes in Fig. 2 represent the energy in the dry biomass of organisms functioning at the trophic level indicated. The number of organisms per unit area at any one time, or the average quantity over a period of time, may be informally designated as the standing crop. The relationship between the "boxes" and the "pipes," that is, between standing-crop energy and the energy flows P, A, or I, is of great interest and importance. The energy flow must always decrease with each successive trophic level. Likewise, in many situations the energy in biomass of standing crop also decreases. However, standing-crop biomass is much influenced by the size of the individual organisms making up the trophic group in question. In general, the smaller the organism, the greater will be the rate of metabolism per gram of weight (inverse size–metabolism law). Thus, 1 g of bacteria may have an energy flow equal to many grams of cow, or 1 g of small algae may be equal in metabolism to many grams of tree leaves. Consequently, if the producers of an ecosystem are composed largely of very small organisms and the consumers are large, then the standing-crop biomass of consumers may be greater than that of the producers; of course, the energy flow through the latter must average greater, assuming that food used by consumers is not being imported from another ecosystem. Such a situation often exists in marine environments of moderate depths because bottom invertebrate consumers (clams, crustacea, and echinoderms) and fish often outweigh the phytoplankton (microscopic floating algae) on which they depend. By harvesting at frequent intervals, man, as well as the clam, may obtain as much food (net production) from mass cultures of small algae as he obtains from a grain crop which is harvested after a long interval of time. However, the standing crop of algae at any one time would be much less than that of a mature grain crop. To summarize, standing-crop biomass is usually expressed in terms of grams of organic matter, grams of carbon, or kilocalories per unit of area or volume. Productivity is expressed as grams or calories per unit of time. As indicated by the above examples, these two quantities should not be confused: the relationship between the two depends on the kind of organisms involved.

Gross production and respiration. The relationship between gross production P_G and total community respiration (the sum of all the Rs in Fig. 2) is important in the understanding of the total function of the ecosystem and in predicting future events. One kind of ecological climax or steady state exists if the annual production of organic matter equals total consumption, that is, $P/R = 1$, and if exports and imports of organic matter are either nonexistent (as in a self-sufficient climax) or equal. In a mature tropical rainforest the balance may be almost a day-by-day affair, whereas in mature temperate forests an autotrophic regime in spring and summer is balanced by a heterotrophic regime in autumn and winter. Another type of steady state exists if gross production plus imports equals total respiration, as in some types of stream ecosystems, or if gross production equals respiration plus exports, as in stable agriculture.

Species, individuals, and energy flow. An important and little-known area of ecology deals with the relationship between the number of species, the number of individuals, and the energy flow. Most natural communities contain a few species in each trophic group which are abundant and account for most of the energy flow. Usually, however, the few common species are associated with a large number of rare species. The number of species relative to the number of individuals

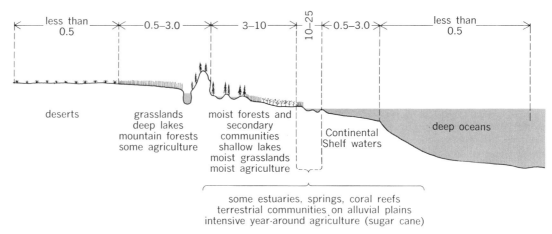

Fig. 3. World distribution of primary production showing gross production rate (grams of dry matter per square meter per day) expected. (*After E. P. Odum, Fundamentals of Ecology, 2d ed., Saunders, 1959*)

(species diversity) often increases with the maturity of stability of the ecosystem. Increasing diversity is not necessarily accompanied by increasing total productivity (in fact, the reverse may be the case). One advantage of diversity, for the survival value to the community, lies in increased protection against perturbations in physical environment (drought stress, for example). The more species present, the greater are the possibilities for adaptation of the given type of ecosystem to changing conditions, whether these be short-term or long-term changes in climate or other factors.

Primary production in the world. The world distribution of primary production is shown schematically in Fig. 3. Values represent the average gross production rate, in grams of dry organic matter per square meter per day, to be expected over an annual cycle. For an estimate of total annual production, multiply by 365. To visualize these values in terms of approximate kilocalories of potential food, multiply by 5. Much less than 90% of gross production may be available to heterotrophs. Man or any other single species cannot assimilate all energy fixed as dry matter by plants. For example, corn stalks and wheat stubble and roots would be included in the total production of these crops, but only the grain is currently consumed by man. Figure 3 shows about three orders of magnitude in potential biological fertility of the world: (1) some parts of the open oceans and land deserts, which range around 0.1 g/(m²) (day) or less; (2) semiarid grasslands, coastal seas, shallow lakes, and ordinary agriculture, which range between 1 and 10; (3) certain shallow-water systems such as estuaries, coral reefs, and mineral springs, together with moist forests, intensive agriculture (such as year-round culture of sugarcane or cropping on irrigated deserts), and natural communities on alluvial plains, which may range from 10 to 20 g/(m²) (day). Production rates higher than 20 have been reported for experimental crops, polluted waters, and limited natural communities, but these values are based on short-term measurements; values higher than 25 have not been obtained for extensive areas over long periods of time.

Productivity controls. Two tentative generalizations may be made from the data at hand. First, basic primary productivity is not necessarily a function of the kind of producer organism or of the kind of medium (whether air, fresh water, or salt water), but is controlled by local supply of raw materials and solar energy, and the ability of local communities as a whole (including man) to utilize and regenerate materials for continuous reuse. Terrestrial systems are not inherently different from aquatic situations if light, water, and nutrient conditions are similar. However, large bodies of water are at a disadvantage because a large portion of light energy may be absorbed by the water before it reaches the site of photosynthesis. Second, a very large portion of the Earth's surface is open ocean or arid and semiarid land and thus is in the low-production category, because of lack of nutrients in the former and lack of water in the latter. Many deserts can be irrigated successfully and it is theoretically possible, and perhaps feasible in the future, to bring up nutrients from the bottom of the sea and thus greatly increase production. Such an upwelling occurs naturally in some coastal areas which have a productivity many times that of the average ocean.

Efficiency limits. It now seems clear that there is a rather definite upper limit to the efficiency with which light may be converted into organic matter on any large scale: this maximum has apparently been achieved by some natural communities (coral reefs, for example) as well as by the most efficient agriculture. In the former, of course, production is consumed by a large variety of organisms, whereas in the latter a large portion of the net is temporarily stored and then harvested by man. Average agriculture is far below the maximum: world average grain production, for example, is 2 g/(m²) (day). The best immediate possibilities for increasing food production for man lie in measures which reduce physical limiting factors and increase the season of growth so that sunlight is utilized for as large a part of the annual cycle as possible.

BIOGEOCHEMICAL CYCLES

The more or less circular paths of chemical elements passing back and forth between organisms and environment are known as biogeochemical cycles. The rate at which vital elements become available to biological components of the ecosystem is more important in determining primary and secondary productivity than flow of solar energy. If an essential element or compound is in short supply in terms of potential growth, the substance may be said to be a limiting factor. The productivity of an entire ecosystem is sometimes limited by one material available in least amount in terms of need, according to the principle of Liebig's law of the minimum. Thus water limits the desert ecosystem, and nitrogen or phosphorus often limits ocean ecosystems. However, nature has considerable powers of adaptation and compensation. In many environments, species and varieties have evolved which have low requirements for scarce materials. Also, the amount of one substance which, in itself, may not be limiting often greatly affects the requirement for another substance which is approaching a critical minimum. Consequently, it is usually necessary to consider the interaction of the essential materials if the limiting factors operating in a given situation are to be determined.

Nutrients. Dissolved salts essential to life may be conveniently termed biogenic salts or nutrients. They may be divided into two groups, the macronutrients and the micronutrients. The macronutrients include elements and their compounds needed in relatively large quantities, for example, carbon, hydrogen, oxygen, nitrogen, potassium, calcium, phosphorus, and magnesium. The micronutrients include those elements and their compounds necessary for the operation of living systems but which are required only in minute amounts. At least 10 micronutrients are known to be necessary for primary production: iron, manganese, copper, zinc, boron, sodium, molybdenum, chlorine, vanadium, and cobalt. Several others such as iodine are essential for certain heterotrophs. Because minute requirements are often associated with an equal or even greater minuteness in environmental occurrence, the micronutrients deserve equal consideration along with the macronutrients as possible limiting factors.

Types of cycle. From the standpoint of the biosphere as a whole, biogeochemical cycles fall into two groups: the gaseous-type cycles, as illustrated by the nitrogen cycle in Fig. 4; and the sedimentary-type cycles illustrated in Fig. 5. Cycles of oxygen, carbon, and water resemble the cycle of nitrogen, in that a large gaseous pool is important in the continuous flow between inorganic and organic states. As shown in Fig. 4, the nitro-

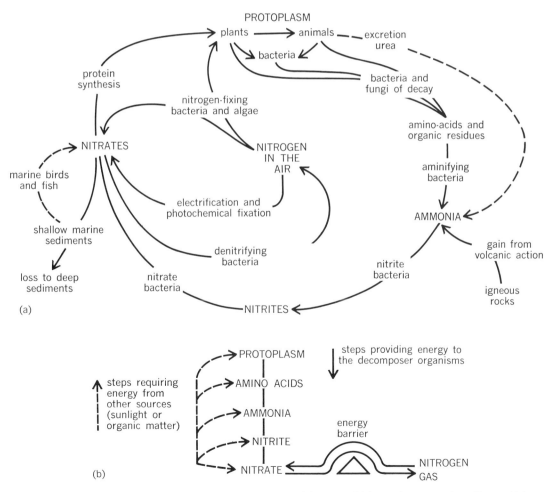

Fig. 4. Nitrogen biogeochemical cycle. (a) Circulation of nitrogen between organisms and environment, with microorganisms responsible for key steps. (b) Same basic steps, with the high-energy forms on top to distinguish steps which require energy from those which release energy. (*After E. P. Odum. Fundamentals of Ecology, 2d ed., Saunders, 1959*).

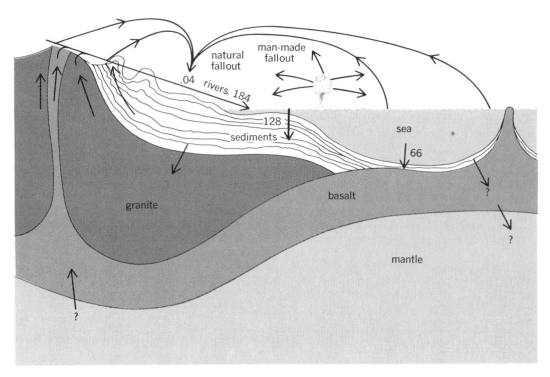

Fig. 5.   Diagram of the sedimentary cycle involving movement of the more earthbound elements. Where estimates are possible, the amounts of material are estimated in geograms/10^6 years (1 geogram = 10^{20} g).

gen of protoplasm is broken down from organic to inorganic form by a series of decomposer bacteria, each specialized for a particular part of the job. The nitrogen ends up as nitrate or other form usable by green plants in the synthesis of new organic matter. The air is the great reservoir and safety valve of the system. Nitrogen is continually entering the air by action of denitrifying bacteria and continually returning to the cycle through the action of nitrogen-fixing organisms and also through nonbiological fixation. Only certain bacteria and algae (which, however, are abundant in water and on land) can fix nitrogen. No so-called higher plant or animal has this ability: legumes fix nitrogen only because of the symbiotic bacteria which live in their roots. The steps from protoplasm down to nitrates provide energy for the decomposer organisms, whereas the return steps require energy from other sources, such as from organic matter or sunlight. Likewise, nitrogen fixers must use up some of their carbohydrate or other energy stores in order to transform atmospheric nitrogen into nitrates.

Feedback mechanisms. The self-regulating feedback mechanisms, as shown in a simplified manner in Fig. 4, make the nitrogen and the other gaseous-type cycles relatively perfect when large areas of the biosphere are considered. Thus increased movement of materials along one path is quickly compensated for by adjustments along other paths. Nitrogen often becomes a limiting factor locally, either because regeneration is too slow, or because loss from the local system is too rapid.

Most biogenic substances are more earthbound than nitrogen, and their cycles follow the pattern of erosion, sedimentation, mountain building, and volcanic activity, as shown in Fig. 5. Biological activity on land and in the upper layers of water re-

sults in local cycles from which there is usually a continual loss downhill and replacement from uphill runoff and from solid matter moving in the air as dust, that is, natural fallout. Man often disrupts the sedimentary cycles because he tends to increase the downhill movement. The phosphorus cycle is a good example. Phosphorus is relatively rare in the surface materials of the Earth in terms of biological demand. At present more phosphorus is apparently escaping to the deep ocean sediments (where it is unavailable to producers) than is being replaced by natural processes. For the time being, man is able to mine the considerable reserves of underground phosphate rock and make up some of this loss, but eventually a means may have to be found to recover phosphorus from the sea.

Nonessential and radioactive elements. Biogeochemical cycles involve elements essential to life. The nonessential elements pass back and forth between organisms and environment and many of them are involved in the general sedimentary cycle, as pictured in Fig. 5. Although they have no known value to the organism, many of these elements become concentrated in tissues, apparently because of similarity to specific vital elements. The ecologist would have little interest in most of the nonessential elements were it not for the fact that atomic bombs and nuclear power operations produce radioactive isotopes of some of these elements, which then find their way into the environment and into food chains. Even a rare element, in the form of a radioactive isotope, can be of biological concern because a very small amount of material from a geochemical standpoint can have marked biological effects. Thus the cycling of such things as strontium, cesium, cerium, ruthenium, and many others may be of great concern in coming years. Strontium is receiving special attention; it behaves like calcium and follows it into biological systems. None of the strontium which is naturally involved in the calcium cycle is radioactive, but radioactive strontium is becoming widespread in the biosphere as a result of fallout from atomic weapons tests. Small amounts of radiostrontium have now followed calcium from soil and water into vegetation, animals, milk, and other human food, and into human bones in almost all parts of the world. In 1958 several hundred micromicrocuries (1 micromicrocurie $= 10^{-12}$ curie) of radiostrontium were present for every gram of calcium in some soils. The bones of children in North America and Europe averaged 1.2 micromicrocuries/g of calcium. There is considerable controversy as to whether these small amounts are detrimental, but most scientists agree that it would be desirable to keep as much radioactivity out of the food chain as possible.

Tracers. There is a bright side to the atomic-age production of radioactive isotopes. Tiny amounts of such isotopes provide convenient tracers whereby the movement of materials in ecosystems can be accurately measured. Much has already been learned about the cycling and turnover rates of phosphorus through the use of the isotope P^{32}. The radioactive isotope of carbon, C^{14}, has proved invaluable in measuring the rate of primary production in the ocean. It is certain that intelligent use of radioisotopes as tools can help solve problems of the atomic age. Radioisotopes not only offer unprecedented means of studying flow in ecological networks, but are used as sources of high-energy radiation for experimental manipulation. *See* BIOSPHERE, GEOCHEMISTRY OF; RADIOECOLOGY.

ECOSYSTEM DEVELOPMENT

Ecosystems undergo orderly development as do individual organisms because the biotic components are capable of modifying and controlling the physical environment to varying degrees. This process of change is commonly called ecological succession or ecosystem development. The table summarizes the major changes that occur in the development of an ecosystem, irrespective of the physiographic site or the climatic region. Trends are emphasized by contrasting the situation in early development with that at maturity. The table, in essence, contrasts "young nature" with "mature nature" and lays the foundation for an ecological solution to the growing conflicts between man and nature.

As shown, youthful types of ecosystems, such as agricultural crops or the early stages of forest development, have characteristics that contrast to those of adult systems, such as mature forests or climax prairies, for example:

1. High net productivity that is available for harvest or storage.

2. Relatively small standing crop at any one time with a consequent low ratio between biomass and productivity.

3. Unbalanced metabolism with an excess of production over utilization.

4. Linear grazing food chains (that is, grass-cow-man) in contrast to complex weblike energy flows involving animal-microbial detritus utilization that characterize older systems.

5. Low biotic diversity (for example, few species but large numbers of individuals).

6. Rapid one-way flux of inorganic nutrients in contrast to recycling and retention within the organic structure.

7. Selection for species with high birth rates, rapid growth rates, and simple life histories.

These attributes combine to produce a general lack of stability in that young systems are more easily disrupted by drought, storms, disease, or other perturbations.

These contrasts underline the basic conflict between the strategy of man and nature. For example, the goal of agriculture or intensive forestry, as now generally practiced, is to achieve high rates of production of readily harvestable products with little standing crop left to accumulate on the landscape or, in other words, a high production-biomass efficiency. Nature's strategy, on the other hand, as seen in the outcome of the ecosystem development process, is directed toward the reverse efficiency, namely, a high biomass-production ratio. The natural strategy of maximum protection (that is, optimizing for the support of a complex living structure that buffers the physical environment) conflicts with man's goal of maximum production (that is, maximizing for high efficiency in production). Since the environment is man's living space as well as a supply depot

Trends in the ecological development of landscapes.

Ecosystem attribute	Developmental stages	Mature stages
Gross primary productivity (total photosynthesis)	Increasing	Stabilized at moderate level
Net primary community productivity (yield)	High	Low
Standing crop (biomass)	Low	High
Ratio growth to maintenance (production-respiration)	Unbalanced	Balanced
Ratio biomass to energy flow (growth + maintenance)	Low	High
Utilization of primary production by heterotrophy (animals and man)	Predominantly via linear grazing food chains	Predominantly weblike detritus food chains
Diversity	Low	High
Nutrients	Inorganic (extrabiotic)	Organic (intrabiotic)
Mineral cycles	Open	Closed
Selection pressure	For rapidly growing species adapted to low density	For slow growing species adapted to equilibral density
Stability (resistance to outside perturbations)	Low	High

for food and fiber, its gas exchange, waste purification, climate control, and recreational and esthetic capabilities are as vitally important in the long run as its capacity to yield consumable products. Many "protective" or "living space" functions are best provided by the more stable or mature-type ecosystems. It is clear, then, that there must be both "productive" and "protective" landscapes in reasonable balance to safeguard the gaseous, water, and mineral cycles on which life depends.

As the human population of the world increases and more and more changes in the face of the Earth are contemplated, increasing attention must be given to the total, and not just the immediate, effect of the changes. It is necessary to maintain a moderate amount of diversity in nature, and to preserve a maximum amount of "open space" if the quality of man's environment is to be protected. Attention must be directed toward conservation of the ecosystem, and not just conservation of individual organisms which are in demand at the moment.   [EUGENE P. ODUM]

Bibliography: W. C. Allee et al., *Principles of Animal Ecology*, 1949; C. Elton, *The Pattern of Animal Communities*, 1966; P. Farb, *Ecology*, 1963; K. A. Kershaw, *Quantitative and Dynamic Ecology*, 1965; E. J. Kormondy (ed.), *Readings in Ecology*, 1965; R. H. MacArthur and J. H. Connell, *The Biology of Populations*, 1966; R. F. Morgan, *Environmental Biology*, vol. 1, 1963; E. P. Odum, *Fundamentals of Ecology*, 2d ed., 1959; E. P. Odum, *Ecology*, 1963; H. J. Oosting, *The Study of Plant Communities*, 2d ed., 1956; V. E. Shelford, *The Ecology of North America*, 1964; R. L. Smith, *Ecology and Field Biology*, 1966.

Ecology, applied

Mankind is changing the face of the Earth to such an extent that many of the present plant and animal communities, and even the land surfaces themselves, bear the marks of his interference. Applied ecology deals with man's activities in managing natural resources, a management based upon a knowledge of basic ecology and upon how man's efforts can change actions, reactions, and interactions with the communities of plants and animals. An ecological approach is implicit in all agriculture, forestry, range management, wildlife management, fisheries management, and other aspects of natural resource management. The term applied ecology is usually restricted, however, to management practices specifically based upon ecological science.

A natural-resource ecosystem is an integrated system, one element of which is a product of direct or indirect use to man. The product may be biological, as in the case of forests, ranges, agricultural products, fish, and wildlife; physical, as in the case of water, air, and soil; or both. The distinguishing facet of a natural-resource ecosystem is that man has a direct involvement in the complex set of ecological interactions, and exerts a more or less direct manipulation of the system. Beneficial management involves goals to maximize the returns to man, while exploitation is sometimes meant to imply reduction of the productivity of the ecosystem to mankind over a period of time. The ecological principles of natural resource ecosystems are generally applicable regardless of the particular natural commodity, as are the tools of manage-

ment and the basic rules governing their application. The principles of ecosystem management apply equally to wilderness and to the urban environment, but they are most clearly understood today with regard to the wild-land resources of forest, range, wildlife, and the like. *See* ECOSYSTEM.

Successful applications of ecology require a knowledge of the life histories of the plants and animals to be managed, the effect of environments upon these life histories, the interactions of the plants and animals making up the community, and the means at the disposal of man to change these relationships. These applications are illustrated for major fields of natural resource management.

Forestry. The management of forests for man's own use through silviculture provides a prime example of applied ecology. Forest communities are usually much closer to natural communities than agricultural crop communities, so that the relationship is clear between natural ecological processes and these processes as affected by man's activities. *See* FOREST MANAGEMENT AND ORGANIZATION.

Classically, forestry is oriented toward the growing of sawtimber as a source of lumber, plywood, and veneers. In the 20th century, however, the rapid development of the paper and plastics industries has resulted in increasing emphasis upon growing trees as a source of fibers and raw materials for chemical products, especially cellulose. During the same period, the concept of multiple use of forests has gained increasing acceptance so that management for compatible uses, including wood production, wildlife production, grazing, recreation, and watershed protection, is correspondingly emphasized. *See* FOREST RESOURCES.

Occasionally, it may be to man's interest to perpetuate climax communities that develop over a long period of time in the absence of severe disturbances. The hemlock–beech–sugar maple forest in the northeastern United States and the Norway spruce–silver fir–beech forest in central Europe are examples of economically valuable climax forest associations. More often than not, however, the valuable forest types will be replaced in time by less valuable forest types unless man intervenes. Thus, the white and red pine forests of New England and the Lake States will be replaced in the absence of disturbance by shade-tolerant hemlock and hardwoods that develop as an understory under the aging pine. Similarly, the pines of the southeastern United States are subclimax to mixed hardwoods on most sites, and Douglas fir in the Pacific Northwest is usually subclimax to western hemlock, western red cedar, and other shade-tolerant species. *See* ANIMAL COMMUNITY; CLIMAX COMMUNITY; PLANT COMMUNITY.

To encourage the desired tree species and to obtain the desired stand structure, the silviculturist employs cutting, planting, fire, chemicals, and mechanical treatment. By logging the existing stand in a patterned partial cutting or by clear-cutting, environmental conditions may be created that will favor the desired species (see illustration). If natural seeding is not successful, the desired species may be planted. Unwanted species may be killed or discouraged by prescribed burning or by chemicals. Mechanical scarification of the site combines brush control with the preparation of the soil for the establishment of seedlings.

Cutover areas, Olympic National Forests, Wash. (*U.S. Forest Service*)

Once the forest is established, it must be kept vigorous and healthy. A knowledge of growing space requirements for individual trees is used as a basis for systematic weedings and thinnings. Forest insects and diseases must be held in check largely through cultural and biological control by the establishment of conditions unfavorable to pests or favorable to their parasites and predators. Direct control of forest pests is usually uneconomic because forests are relatively low in value per unit of area.

The success of a silvicultural treatment depends largely upon how well the silviculturist understands the ecology of the particular forest community being managed and upon the timing and intensity of his treatment. *See* FOREST ECOLOGY; PLANT FORMATIONS, CLIMAX.

Agriculture. The artificial or man-made nature of most agricultural-crop and domestic-animal communities tends to obscure the ecological nature of agriculture. The requirements of a given economic plant (such as corn, cotton, or coconut) or animal (such as cattle, sheep, or chickens) as to climate, soil, and other growing conditions have been determined over the ages by trial and error and are fairly well known. The present-day production of many new hybrids and other products of plant and animal breeding, however, constantly poses the problem of what the environmental requirements of these new organisms are and where and how they can be introduced into existing plant and animal communities.

The effect of environment upon plant and animal growth is being studied increasingly in controlled environmental chambers where plants and animals can be subjected to specified day lengths, temperatures, humidities, and other aspects of the environment.

Of critical importance to continued agriculture is the prevention of soil erosion and of chemical depletion of the soil upon which crops are grown or upon which farm animals are reared. The soil constitutes an ecological system formed of a complex of minerals, organic residues, and living plants and animals. Only through the application of ecological principles can this complex be maintained in such a condition as to permit the indefinite use of land for the maximum production of agricultural products. The effects of chemical removals through crops, leaching, and erosion and of chemical additions through fertilizers, cover crops, and rainfall can be assessed only if the soil ecosystem as a whole is clearly comprehended.

Range management. As an aspect of applied ecology, range management represents more intensive effort on the part of man than most forestry activities but less effort than most cropland practices. The open grasslands and savanna woodlands of the American West and many other parts of the world can support large numbers of cattle, sheep, and other grazing animals only if the food and water supply of the range is maintained. Grasslands may readily be overgrazed with resultant development of less palatable communities, soil erosion, and even destruction of the site. The maintenance of the proper number and kind of stock at the various times of the year is the basis of modern range management. Uniform use of the available range can be approached through the development of water supplies, the placing of salt licks, and the forced movement of the stock. *See* GRASSLAND; RANGELAND CONSERVATION; SAVANNA.

Wildlife management. The scientific production of deer, grouse, quail, ducks, and other desired wildlife is a rapidly growing aspect of applied ecology. The near extermination of wolf, puma, and other natural predators; the heavy pressures of hunting by man that have greatly reduced populations of bison, elk, caribou, and other game animals; and the change in ecological balances produced by the introduction of exotic animals and plants such as the pheasant in the northern United States and the eucalyptus in California, have all changed the numbers and distribution of game animals, both mammals and birds. Maximum healthy populations of desired animals can be maintained by creating optimum environments, mainly through the development of increased food supplies and increased shelter. Also, through controlled hunting and through the management of natural predators, populations can be regulated at a level that will permit the appropriate number of animals to weather unfavorable seasons and at the same time minimize losses through starvation and disease. *See* WILDLIFE CONSERVATION.

Fisheries management. The management of fisheries for food or sport provides another example of applied ecology. In contained bodies such as ponds, lakes, reservoirs, and rivers, a fair degree of population and environmental control may be possible. Bodies of water may be drained, undesirable fish may be trapped, poisoned, or otherwise removed, desirable fish may be introduced, food plants and animals may be added, and even fertilizers may be spread. Here again, the maintenance of the wanted fish at the highest levels is the desired end, and this can be achieved only through understanding the community as a whole and managing it as a whole. In the oceans and other large bodies of water, management of the whole community is seldom practicable, but certain aspects of the life cycle of certain commercial fishes can be regulated through controlled fishing and management of spawning areas. The development of favorable environmental conditions for the spawning of salmon in rivers and streams and the creation of unfavorable conditions for the spawning of the lamprey in the Great Lakes are examples of applied ecology in commercial fisheries management. *See* FISHERIES CONSERVATION.

Water-resource management. Production of clear, unpolluted water for human use is rapidly becoming a critical problem in many areas of the United States and has long been a controlling factor in determining the distribution of human populations. In the hydrologic cycle, much of the water that falls as precipitation is returned to the atmosphere through evaporation and through transpiration by plants. Through changing the nature and distribution of the plant cover—changes which lead to changes in the soil structure—the amount of water that enters the ground can be increased and the amount lost through evapotranspiration can be regulated. The hydrologic management of streams through such devices as channel improvement, dams, and canals can bring larger quantities of clearer water to points of human use, whether for agriculture, industry, or human consumption. Control of stream pollution through regulation of the source of impurities, filtration, and chemical treatment represents applications of ecology that increase available water supply, both to man and to other animals. *See* ECOLOGY, HUMAN: WATER CONSERVATION. [STEPHEN H. SPURR]

Environmental pollution. Rising public concern about the quality of water, air, and many other aspects of man's environment has led to the use of new methods for evaluating the movement and ultimate impact of contaminants through various parts of that environment, and their possible effects on man or other parts of the ecosystems on which human life depends.

The development of nuclear energy added a new pollutant; its potential impact, if uncontrolled, could be even more dramatic than that of chemicals being released at alarming and accelerating rates. However, recognition of the potential danger led to unprecedented intensity of research and controls on release of radioactive waste and fallout. Essential for this research was the use of isotopes and other tracers in the environment, and

development of mathematical approaches for dealing with kinetics of movement and possible effects.

A second effect of nuclear energy was thus to stimulate the analytical data and the theoretical foundations of basic ecology. These in turn provide approaches and tools for research on other problems of environmental disruption that have been recognized more gradually than the threats posed by nuclear contamination itself. *See* SYSTEMS ECOLOGY.

[JERRY S. OLSON]

Bibliography: P. Farb, *Ecology*, 1963; A. Leopold, *Game Management*, 1933; E. P. Odum, *Fundamentals of Ecology*, 2d ed., 1959; H. J. Oosting, *The Study of Plant Communities*, 2d ed., 1956; J. D. Ovington, *Woodlands*, 1965; R. L. Smith, *Ecology and Field Biology*, 1966; S. H. Spurr, *Forest Ecology*, 1964; Symposium on applied ecology, *Ecology*, 38:46–64, 1957.

Ecology, human

The branch of ecology that considers the relations of individual men and of human communities with their particular environments.

Each human being, like every other kind of animal, possesses numerous regulatory mechanisms that maintain a dynamic balance between his internal physiology and the constantly fluctuating conditions in his local habitat. Energy is acquired from food and is lost in radiated heat, muscular work, metabolism, and growth. Man is very adaptable in his food and can exist on vegetable or animal foods alone or on various combinations of these. By acclimation, each person can to some degree adjust his physiology to fluctuations in weather or season or to new climates that he may encounter by migration. Man differs from other animals, however, by being able to modify his personal environment by the use of clothing, houses, and heating and cooling devices of various kinds. Man is consequently able to thrive in many habitats that otherwise would be inhospitable to him. *See* HOMEOSTASIS.

Human community. A human ecologic community is composed of human beings and also of members of numerous associated species of animals and of plants. Man is dependent upon certain of these species for his food. From others he secures materials for making clothing and tools, for constructing buildings, and for numerous other uses. Wild herbivores living in the same community with man may eat or damage some of the useful plants. These herbivores may be preyed upon by carnivores, which thereby benefit man indirectly. Bacteria, viruses, or larger parasites may injure or destroy certain of the herbivores, carnivores, other parasites, or even man himself. Scavengers and saprophytes convert dead organic matter into substances that can be reused by the plants. The numbers of the several species that together compose such a community are controlled by various ecologic regulatory mechanisms that maintain a working balance within the community.

The kinds of plants and animals that can exist in each geographic area are limited by the local climate, physiography, and soils. The plants, animals, and to some degree man, however, may modify the soil and the microclimate and thereby may make the habitat more suited to the community.

Each human community together with its local habitat thus constitutes an interacting system that may be called a human ecosystem. *See* ECOSYSTEM.

Cultural evolution. By the evolution of language and through social cooperation man has developed various types of culture. He thereby is able to modify his immediate environment to a considerable degree and to expand greatly his resources of food and other useful materials. Each of the cultures that have evolved in the past in diverse parts of the world may be presumed to have been adapted, at least tolerably, to the climate, minerals, and other physical features of the local habitat, to the wild and domestic plants and animals of the region, and to the local diseases to which the people and their domestic crops and herds were subject. The most widespread existing cultures are based upon the domestication of plants and animals, exploitation of biologic and mineral resources, division of labor, accumulation of capital, invention of specialized technical processes, and the operation of numerous kinds of social, economic, and political institutions.

Regulatory mechanisms. Many of the natural ecosystems of the world have been greatly altered by the activities of man. Forests, for example, have been cut down or burned, swamps have been drained, pasturelands have been overgrazed, fertile soils have become eroded, fields have been planted to cultivated crops, rivers and lakes have become polluted by sewage and industrial wastes, native species of plants and animals have been reduced in numbers or locally extirpated, foreign species have been introduced, and extensive areas of productive land are covered by highways and cities. Numerous new types of ecologic communities also have developed in cultivated fields, in managed forests, about farmsteads, along roadsides, and within cities. The natural regulatory mechanisms usually are rendered ineffective in communities that are much modified by man's activities. In consequence, certain native or introduced species may become destructive pests. Man, therefore, either must establish effective new regulatory mechanisms in these disturbed communities or must himself control the pests.

Regulatory mechanisms that maintain a working balance among the several species that compose a natural community mostly have become ineffective for controlling the density of human populations. Predators dangerous to man largely have been extirpated from the vicinity of human habitations. Communicable disease is becoming increasingly eliminated by the adoption of improved sanitary and medical practices. Human starvation in local areas usually is prevented by food storage and by the transportation of food from areas of abundance. With the natural checks to his multiplication thus greatly removed, man has increased in numbers until he now is one of the most abundant species of mammals.

Special regulatory mechanisms have been evolved by every human society to control the activities of its members and to keep the society adjusted to the fluctuations in its resources and in the conditions in its habitat. Among these special cultural mechanisms are public opinion, punishments, rewards, supply and demand, differential taxation, and cooperative procedures of various kinds. Every social group has customs, ideals, and beliefs that govern the behavior of its members toward each other and toward strangers and regulate the utilization of the resources of the group.

War among tribes and nations in the past often has resulted from competition for land or other resources. Fortunately, other less destructive mechanisms for adjusting nations to their resources and to each other are being developed.

Scope. Human ecology is involved in many branches of natural and social science. Physiology, nutrition, hygiene, psychology, and climatology are basic to an understanding of the ecology of individuals. The ecology of human communities involves anthropology, botany, demography, economics, geography, geology, history, political science, sociology, and zoology. Also involved are the several branches of applied ecology, including agriculture, animal husbandry, conservation, forestry, game and fish management, medicine, parasitology, public health, and range management. Most of the concepts of human ecology are those of the individual disciplines concerned. The concept that each person and each human community operates as an ecologic unit that must possess effective regulatory mechanisms to maintain stability in its fluctuating habitat, however, serves to unify all branches of inquiry concerned in any way with the relations between man and his physical, biotic, and social environment.   [LEE R. DICE]

Bibliography: M. Bates, *The Forest and the Sea: A Look at the Economy of Nature and the Ecology of Man*, 1960; E. Borgstrom, *The Hungry Planet: The Modern World on the Edge of Famine*, 1965; J. B. Bresler (ed.), *Human Ecology: Collected Readings*, 1966; L. R. Dice, *Man's Nature and Nature's Man: The Ecology of Human Communities*, 1955; E. H. Graham, *Natural Principles of Land Use*, 1944; A. H. Hawley, *Human Ecology: A Theory of Community Structure*, 1950; E. S. Rogers, *Human Ecology and Health*, 1960; G. A. Theodorson, *Studies in Human Ecology*, 1961.

Ecology, physiological (plant)

The study of biological processes and growth under natural or simulated environments. This article deals only with the physiological ecology of plants the principal life-cycle processes are germination, growth, photosynthesis, respiration, absorption and loss of water and nutrients, and reproduction. Each of these is a complex series of physical and chemical reactions in cells and tissues of the plant. *See* ENVIRONMENT.

Scope. Physiological ecology attempts to explain why a certain species of organism is able to grow or survive in a specific environment. This question has not been answered completely for any species. An answer requires a knowledge of the genetic structure of the species and the interactions between the deoxyribonucleic acid (DNA) of the genes and physiological processes in an individual as they operate under the impact of the total environment.

The potential geographical range of a species, both natural and cultivated, depends upon the nature and degree of genetic variation among its individuals, the range of physiological and morphological adaptability within each individual (phenotypic plasticity), and the frequency and extent of occurrence of suitable environments. These three complex variables, operating together, determine the rates of physiological processes and thus the degree of success in growth and reproduction of individuals of a species in a particular environment.

Almost all widespread species are genetically diverse. Within such species, ecological races (ecotypes) have evolved in response to environmental selection of genetic material. Sometimes these races merge gradually with each other to form a genetic gradient (ecocline) of adaptability to an environmental gradient. The races and clines consist of local populations, usually with enough genetic diversity to allow some survival within certain limits of environmental change. Studies of process rates within local populations, ecoclines, and ecotypes provide a measure of the environmental tolerance limits of a species. Such comparative physiological ecology is a relatively new interdisciplinary field. The data which it is providing are improving the understanding of the adaptability of plants and of the evolution of physiological processes.

Such data may be obtained in natural environments under field conditions or in controlled simulated environments. Field measurements are difficult because of logistics and the vagaries of uncontrolled environments but have the advantage of providing realistic values. Measurements under controlled laboratory environments are more precise, but the technical difficulties of reproduction of natural environments are great, costly, and often insurmountable. A research strategy combining both is the future goal.

Field measurements. Genetically determined ecotypes or ecoclines are found by transplanting individuals of a species from different environments to uniform gardens and comparing their growth rates and appearance. Controlled-environment rooms or chambers may be used in place of the uniform garden. Plants may also be started from field-collected seed rather than by transplanting older plants. If inherent morphological and physiological differences exist, they show up rather quickly in the uniform and controlled environments of the garden or growth room.

Physiological rates can be measured in the field if portable instrumentation is available. This can be difficult logistically because of weight of equipment, distances, and environmental severity. Miniaturization of circuits and the use of solid-state components is helping to solve this problem. Mobile laboratories in enclosed trucks or trailers make it easier to get laboratory precision in the natural environment of the plant. The processes most often studied in the field or laboratory are germination, growth, energy exchange and water loss of the leaf, photosynthesis, respiration, and flowering.

Germination and growth. Germination percentages are easy to measure in the field by planting known numbers of seed. Germination is studied in the laboratory in seed-germination chambers or on temperature-gradient bars, usually run under constant light and temperature conditions. However, alternating temperatures result in better germination in many species. Growth may be measured in height, in terms of fresh or dry weight, or in caloric values.

Energy and water relations. The temperature of a leaf is a result of the balance between heat income and heat loss. The measurement of these components of the energy budget of the leaf involves radiation receipts and losses, evaporative losses, convection, net radiation, and metabolism. Detailed techniques for measuring energy budget will not be described here. However, leaf temperatures may be measured by thermocouples, thermistors, or by noncontact infrared thermometers (bolometers). Leaf temperature on a sunny day is largely under the control of transpiration (evaporative heat loss); the effect is greater in a large leaf than in a small one. In turn, leaf temperature affects transpiration rate itself and the important metabolic processes of photosynthesis and respiration. All these processes are influenced by mean wind speed, turbulent variations, and thermal stability of the air. *See* METEOROLOGY.

Transpiration rates of nonwoody plants or young trees or shrubs may be measured by loss of weight of systems consisting of a plant having its roots in a sealed container of moist soil. A plant may be enclosed in a transparent airstream chamber and its water-vapor losses measured by an infrared gas analyzer adapted for the absorption spectrum of water vapor or by the use of electrical humidity-sensing devices, including wet and dry thermocouples. Transpiration rates of large woody plants are difficult to measure directly. Immersing the cut end of a twig in water in a buretlike potometer measures transpiration indirectly through absorption. Another method involves weighing a freshly cut twig with leaves and reweighing after a few minutes; transpiration is equal to loss in weight per unit time. Transpiration rates must be related to leaf surface area, leaf fresh weight, or to soil-moisture tensions.

The ability of a plant to absorb water from the soil is best measured by the water potential in the leaves. Water potential is the difference in free energy or chemical potential (per unit molal volume) between pure water and water in cells and solutions. The potential of pure water is set at zero: the potential of water in cells, solutions, and soil therefore is less than zero, or negative. In rapidly transpiring plants there is a gradient of decreasing water potential from the soil through the roots, stems, and leaves to the air. This causes water to move from the soil through the plant to the atmosphere. In order to absorb water, plants must maintain a water potential lower (more negative) than that of the soil. Leaf water potential can be measured in the field by the Schardakov dye method or with a Scholander pressure bomb. More precise measurements are possible in the laboratory with a thermocouple psychrometer apparatus.

Photosynthesis. Net carbon assimilation may be measured by enclosing a plant or leaf in a sealed transparent chamber, passing air through the chamber, and measuring decrease in carbon dioxide (CO_2) by absorbing CO_2 in potassium hydroxide solution and titrating. A better method employs an infrared gas analyzer adapted for the CO_2 absorption spectrum. The decreased CO_2

Portable infrared gas analyzer being used to measure photosynthesis of alpine plants at 11,000 ft in Wyoming. The plant is in the small chamber in the right background and is connected by tubing to the pumping and flow-meter apparatus at the left, then to the analyzer, amplifier, and recorder on the right.

content of the air after passing through the chamber can be continuously recorded in either a closed or open system (see illustration). Overheating of the air and leaves within the chamber by radiation is a difficult problem to correct, particularly under remote field conditions. Refrigeration and a relatively high air-flow rate help to keep chamber and leaf temperatures close to that of the outside air. Leaf temperatures should be measured with thermocouples or thermistors. Photosynthesis may also be measured by using radiocarbon as $^{14}CO_2$ in a closed-chamber system for a given amount of time and making subsequent counts of the amount fixed in the leaf or stem tissues. Another field method utilizes measurements of the CO_2 flux within and above wholly unconfined foliage of natural or cultivated vegetation. The rates of photosynthesis and respiration should be related to the leaf area, the fresh and dry weight, or the chlorophyll content.

Respiration. Respiration of leaves, branches, or small individuals may be measured by darkening the chamber by either of the two gas-flow methods and measuring CO_2 production by the plant, but it is not certain whether such results compare with respiration in the light. It may be measured at night in the open-vegetation flux method, or by CO_2 accumulation under strong inversions (chilling of air near the ground). Soil and other respiration factors are then included, and these are difficult to measure independently. The temperature dependence of respiration may be studied further by measuring the temperature relationships of the oxidative rates of the mitochondria.

Productivity. Since photosynthesis is the basic process introducing energy into the food chain, measurements of photosynthesis may be used to estimate net primary productivity in ecosystems. Such productivity may also be measured by periodic harvesting of the vegetation and determining the dry weight and caloric value per unit of land surface per unit time.

Flowering and fruiting. These phenomena may be observed in the field as phenological events and correlated with trends in environmental conditions through time. They may also be studied experi-

mentally both in the field and growth chamber by manipulation and control of temperature and photoperiod.

Measurements in controlled environments. All processes may be measured with greater ease, precision, and environmental variety in controlled-environment facilities. Disadvantages of such facilities are the difficulties of exactly reproducing natural environments, even in small space, and their cost.

Two principal types of controlled environment facilities are in use: (1) relatively small reach-in or walk-in chambers in which thermoperiod, photoperiod, light intensity and quality, and sometimes humidity are controlled, and (2) phytotrons, relatively large air-conditioned buildings in which a variety of controlled environments are available. There are several phytotrons. Among them are one at the Earhart Laboratory at California Institute of Technology (the original "phytotron"), one at Gif-sur-Yvette near Paris, the C.S.I.R.O. phytotron at Canberra, Australia, the North Carolina phytotron at Duke University and North Carolina State University, one at the Royal College of Forestry in Stockholm, and the biotron at the University of Wisconsin. Controlled growth chambers are manufactured by several companies and are available in a variety of models and sizes.

Chambers and phytotrons are valuable tools in physiological ecology because their environments can be controlled and programmed and are reproducible so that an experiment may be repeated. The principal problems in environmental control are in obtaining light intensities and leaf energy budgets approaching those in full sunlight. Humidity control is also difficult.

One important contribution of controlled-environment research is evidence for better germination and growth in many species under cycling temperatures rather than at the constant temperatures of most laboratory experiments. This is in agreement with growth and temperature cycles under natural conditions.

Dependent upon the data of physiological ecology is knowledge of the effects of changing environments on metabolic rates, growth, and reproduction of native and economic plants. This science is also basic to an understanding of primary productivity and efficiency in ecosystems. Physiological ecology, in turn, is dependent upon development of portable and precise field instruments and mobile laboratories, more space in phytotrons, and the training of ecologists capable of adapting physiological methods to field conditions.

[W. D. BILLINGS]

Bibliography: W. D. Billings, Physiological ecology, *Annu. Rev. Plant Physiol.*, 8:375–392, 1957; W. D. Billings, *Plants and the Ecosystem*, 1964; D. M. Gates, Energy exchange and ecology, *BioScience*, 18(2):90–95, 1968; W. M. Hiesey, The physiology of ecological races, *Annu. Rev. Plant Physiol.*, 16:203–216, 1965; P. J. Kramer and T. T. Kozlowski, *Forest Tree Physiology*, 1960; H. A. Mooney and W. D. Billings, Comparative physiological ecology of arctic and alpine populations of *Oxyria digyna*, *Ecol. Monogr.*, 31:1–29, 1961; F. W. Went, *The Experimental Control of Plant Growth*, 1957.

Ecosystem

A functional system which includes the organisms of a natural community together with their environment. The term, a contraction of ecological system, was coined by the British ecologist A. G. Tansley in 1935. The concept has been of increasing importance in the development of ecology and related fields. All natural communities, or assemblages of organisms which live together and interact with one another, are closely related to their environments. It is consequently appropriate to conceive of community and environment as forming a single, complex whole—the ecosystem. *See* ANIMAL COMMUNITY; ECOLOGY; ECOLOGY, APPLIED; ENVIRONMENT; PLANT COMMUNITY.

The term ecosystem can be applied to any size or level of environment chosen for study. Communities of small and distinctive environments, such as the organisms inhabiting an animal's intestine, a decaying log, or the foliage of a particular plant, are sometimes referred to as microcommunities, and their environments, as microcosms or microenvironments. These are generally parts of larger ecosystems, such as a forest and its environment or a lake and its organisms, for which the terms macrocommunity and macroenvironment are appropriate. At the opposite extreme from microcommunities, all the organisms of the world together may be regarded as a world community which, with its environment, forms a single world ecosystem. The living part of this world ecosystem is often termed the biosphere. The transfer and concentration of materials in the world ecosystem by both organisms and physical processes are the subject of biogeochemistry. Since man himself is part of the world ecosystem, man's place and effects on it are part of the study of human ecology. *See* BIOSPHERE, GEOCHEMISTRY OF; ECOLOGY, HUMAN.

Any organism lives in a natural context which includes an environment and a community of other organisms. The natural context of the life of an organism is thus an ecosystem. Two aspects of an organism's relation to the ecosystem in which it lives are often distinguished, though they cannot be sharply separated. Its location, as described in terms of physical and chemical factors, and often, kind of community, constitute its functional habitat. Its place within the community, including relations to other organisms, is its niche.

Within an ecosystem certain major fractions, or divisions, may be recognized. A terrestrial ecosystem is often thought of as having four or five such subdivisions: (1) the physical environment, such as climate; (2) the soil, if distinguished from physical environment; (3) the vegetation or plant community; (4) the animal community; and (5) the saprobe community, which consists of bacteria and fungi. These same divisions may be recognized in a marine or fresh-water ecosystem, with the substitution of substrate or bottom material for soil. Aquatic ecosystems, however, are often divided in a different way. These divisions are the communities and environments of the open water, those of the shores, and those of the deeper bottom. *See* FRESH-WATER ECOSYSTEM; MARINE ECOSYSTEM; TERRESTRIAL ECOSYSTEM.

Some major features of ecosystems important in ecological understanding are discussed here.

Complexity and interrelations. The scientist studying an ecosystem is confronted by a bewildering variety of interrelations. The environmental factors which can be specified and measured are variously interrelated in their effects on one another, and in their significance to the physiology of the organisms. Thus, the effect of humidity cannot simply be measured by relative humidity, for the effect on any given organism is conditioned also by the existing temperature and wind velocity, by the position in the community and characteristics of the organism concerned, and by the results of water loss on the physiology of that organism. The environment affecting the organisms in a community is often conceived as the environmental complex, that is, the sum of all distinguishable environmental factors together with the interrelations among these. The factors may, in fact, be regarded as isolates from the whole environmental complex, as a subset of those particular features abstracted from the whole and chosen for measurement and study. Different organisms in the community are variously affected by different factors and combinations of factors. Organisms are also most variously affected by major kinds of interrelations, such as food relations, shelter, competition for space and other resources of environment, shading, chemical influences, and others. These interrelations among species in the community, which so link together the species that almost any one may directly or indirectly affect almost any other, are referred to as the web of life. The species' position in the web of life, in relation to other species and the community as a whole, is one way of conceiving of the species' niche. Such is the complexity of an ecosystem that it is generally not possible to study all its features at once or to seek complete understanding of all its interrelations. *See* ECOLOGICAL INTERACTIONS.

Environmental dependence and change. Environment and community are always intimately interrelated. The community is necessarily dependent on environment and cannot be understood apart from it. The relation between environment and community is not one of simple cause and effect, however, for environmental factors are, in various ways and to varying degrees, modified by the community. Some factors, such as salinity of sea water, temperature of water of a pond, and others, may be scarcely affected by the presence of organisms; others, such as humidity and wind velocity inside a forest and concentration of phosphates in the water, are strongly modified by organisms, or are determined by the changing activities of organisms. Still others, for example, organic matter in water and soil, exist only because of the activities of organisms. Some communities, such as a highly developed, stable forest, modify the local environments in which their component organisms live more than do other communities, such as those of deserts and the first stages of successions. In general, however, there is a pervasive reciprocity between environment and community in determining the characteristics of both environment and community.

Community adaptation. Characteristics of environment are reflected in characteristics of the community. The community may be said to express the environment surrounding the local ecosystem. Certain broad correlations between kinds of environments and kinds of communities around the world can be observed.

Among terrestrial communities climate is expressed in the overall structure and composition in terms of major kinds of plants, the physiognomy of vegetation. Thus, wherever tropical climates with high rainfall throughout the year occur, these climates support the kind of community known as tropical rainforest. Wherever temperate continental climates with sufficient summer rainfall occur, these support temperate deciduous forests. One may thus recognize the general biological phenomenon of adaptation to environment on two levels: first, the organism, which must have structure and function suited to survival in its niche and the environmental complex of its ecosystem; and second, the community, which must have structure and function appropriate to utilization of the resources of, and persistence in, its environment. *See* BIOME: PLANT FORMATIONS, CLIMAX: RAINFOREST.

Spatial patterning. Many environmental factors form gradients in space, as they are followed from one point to another on the Earth's surface. The many factors of the environment change in different directions through space, some in correlation with one another, others in partial or complete independence. At each point in this environmental pattern a natural community develops in dependence upon the environment of that point. To the pattern of environments in space there consequently corresponds a pattern of communities, and the populations and other characteristics of communities form complex patterns of gradients in space, as do the factors of environment. Often this patterning in space has a self-repeating character, as in an area of low mountains in New York, where one may find a pattern of hemlock forests in ravines and on north-facing slopes, oak forests on most other slopes, and grassy oak openings on dry south slopes, with this pattern repeating itself across a series of valleys and ridges. Environments and communities, in general, grade continuously into one another along spatial gradients of environment. Unlike organisms, ecosystems and communities consequently are not distinct, clearly bounded, and separate from each other. There are exceptions to this. A lake may be regarded as an ecosystem with a fairly clear boundary. Relative discontinuities between communities occur because of disturbance, and where environment is relatively discontinuous, as at the shores of lakes and seas. When such relative discontinuities, or ecotones, are studied, they are often found to represent not merely the meeting place of two communities but distinctive communities in their own rights.

Rhythms. Environments vary also in time, in part irregularly, but more generally and more significantly in regular, periodic rhythms. Only the abyssal communities of the ocean depths live in a constant environment. Other communities are subject to annual and diurnal cycles, and communities of the seashore, to more complex rhythms of tides. Organisms of the community respond in various ways in their cycles of activity to these rhythms of environment. Consequently the environmental rhythms impose self-repeating patterns

of change in time on the community. Similar activities may be pursued by different species at different times in these cycles. Thus, flycatchers feeding in the daytime on diurnal insects may be replaced at dusk by nighthawks feeding on nocturnal insects; spring flowers, and the insects visiting them, may be replaced in fall by quite different groups of flowers and insect visitors. Daily and yearly rhythms thereby permit specialization by time of function among the species of a community, and differentiation in time of the community as a whole. *See* PERIODICITY IN ORGANISMS.

Community differentiation. A natural community comprises a large number of species, but seldom are two species exact competitors. The species of the community are specialized to fill different, though often overlapping, niches. Species with similar niche requirements may be active at different times or in different places in the community. The specialization of species and differentiation of the community are illustrated in the stratification of terrestrial communities. A forest may include five or more strata: the upper and lower tree layers, and shrub, herb, and moss layers, with each stratum having its characteristic plant and animal species. Parallel activities may be pursued by different species in different strata. For example, one bird species may nest and feed on insects in the tree stratum, another on the ground. Such vertical differentiation in space is a basis of one of the major characteristics of communities—their species diversity, or relative richness in numbers of species. One may generalize that species diversity, differentiation into strata, and productivity of terrestrial communities are maximal in most favorable environments, such as that of the tropical rainforest, and decrease toward environments which are unfavorable, unstable, or extreme. There are, however, various limitations to this statement, and species diversity and productivity do not simply parallel one another. *See* SPECIATION.

Productivity relations. Productivity is, even more than species diversity, a fundamental characteristic of communities and ecosystems. It is perhaps the most important single feature of the community. Productivity may be defined as the rate at which energy is bound or matter combined into organic compounds by organisms, per unit time per unit of the Earth's surface. Energy is dissipated back to environment and organic matter reduced to inorganic forms by the respiration of organisms in the community. Productivity is thus counterbalanced by various loss rates. Although particular environmental factors may effectively limit productivity, it is in general a complex resultant of all factors of the environment, an expression in biological activity of the environmental resources and of other factors affecting utilization of these resources by the community. Major factors affecting productivity include those of temperature, nutrient availability, and water availability. Certain broad correlations of productivity with environments may be observed, such as the decline along the temperature gradient from tropical rainforest to the treeless arctic tundra, and along the moisture gradient from tropical rainforest to desert. Productivity thus underlies such other characteristics of communities as their structure or

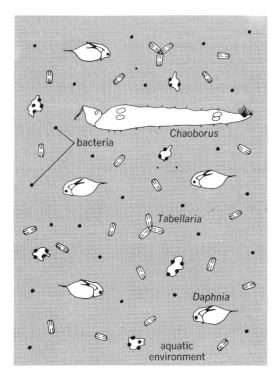

Fig. 1. Simplified ecosystem consisting of water and plankton organisms. Components of this ecosystem are environment, or water and its dissolved material; green plants, or producers (*Tabellaria*, diatom); herbivorous animals, or primary consumers (*Daphnia*, water flea); carnivorous animals, or secondary consumers (*Chaoborus*, fly larva); and bacteria, or decomposers. The various organisms are not drawn in correct size relations.

physiognomy and stratification. *See* BIOGEOGRAPHY; BIOLOGICAL PRODUCTIVITY; COMMUNITY.

Food chains and trophic levels. Productivity of a community is based wholly on the activities of green plants, except for certain autotrophic bacteria, and the communities of such lightless environments as the depths of oceans and lakes, which are dependent on other communities for their food. Green plants use the energy of sunlight to produce organic substances from carbon dioxide and water. These plants are then eaten by animals, and these animals by other animals. In the community illustrated in Fig. 1, organic matter and food energy might be passed along from the diatoms, to the water fleas, to the *Chaoborus* larva, and from this to a small fish, and to a heron. Such a sequence of organisms is referred to as a food chain. Since many species feed on a number of other species, food chains, as they are variously interrelated, form an important part of the web of life in a com-

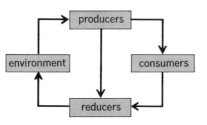

Fig. 2. Generalized cycle of materials in an ecosystem.

munity. Certain major steps (trophic levels) in food webs are the green plants, or producers; the herbivorous animals; the carnivorous animals; and carnivorous animals feeding on carnivorous animals. At each level much of the energy available to organisms is expended in the life activities of those organisms. Only a fraction of this energy can be harvested and used by the next trophic level. There is, consequently, a stepwise decrease of productivity through the sequence of consumer trophic levels. This relation is known as the pyramid of life. *See* FOOD CHAIN.

Cycling and functional kingdoms. Food chains and pyramids are part of a broader phenomenon, the cycling of materials between organisms and environment in the ecosystem. A given substance, such as phosphate in the soil, may be taken up by the roots of green plants and used in organic syntheses and then passed along food chains through animals until, with the death and decomposition of these organisms, the phosphate is released and returned to the soil, where it is again available to green plants. Decomposition of organic material results largely from the activities of certain organisms, such as bacteria and fungi. One may thus recognize three major nutritional groupings and evolutionary directions as the functional kingdoms in communities. These are the true or green plants, or producers, characterized by photosynthesis as their mode of nutrition; the animals, or consumers, characterized by consumption or ingestion of organic food; and the bacteria and fungi, the saprobes, decomposers, or reducers, characterized by absorption of organic food and by their contribution to the breakdown of dead organic matter. The generalized relation among these in cycling of materials through the ecosystem is illustrated in Fig. 2. When the movement of particular substances is studied in detail, a great complexity of routes of movement in the ecosystem may be revealed. This is best known for the nitrogen cycle. Concentrations of many substances in the environment are determined by the cycling of the material through the community. An experiment was conducted by R. H. Whittaker, who used a radioactive phosphorous (P^{32}) tracer to determine the pattern of phosphate movement in a pond (Fig. 3). The amount of P^{32} in water at a given time depends on the various rate values, and the temporary steady state of P^{32} distribution between water and organisms. Productivity relations also are determined, not simply by the amounts of nutrients, and in terrestrial communities by the amount of water in the environment, but by the rate and manner of cycling of these materials through the ecosystem.

Energy flow and steady state. Energy also flows in a cycle of uptake from the environment, movement through the community, and dissipation back to the environment. The community and the ecosystem are thus, like organisms, open energy systems. In a stable ecosystem, the community possesses a stable pool of available free energy of organic compounds, and the flow of energy into this pool by photosynthesis is balanced by the outward flow by respiration, which dissipates energy into environment. With regard to both energy and matter, the community is in steady state, a dynamic equilibrium of apparent constancy underlying which there is a continuing flow of energy and mat-

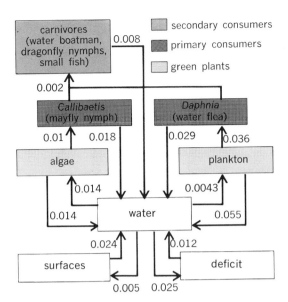

Fig. 3. Simplified pattern of phosphate movement in pond, based on an experiment by R. H. Whittaker with radioactive phosphorus (P³²) tracer. Numbers are transfer rates (that fraction of P³² in box at tail of arrow moving in direction indicated per hour). Surface is film of bacteria and other microorganisms on rock surfaces; deficit refers to P³², mainly in depths of rocky substrate, not otherwise accounted for.

ter. As an open energy system in steady state, the community, like the organism, resists running down to maximum entropy, according to the second law of thermodynamics, by its continuing intake of energy and passing of this energy through the trophic levels as the negative entropy of food. Rates of uptake of nutrients and other substances from environment and rates of return of these substances to environment are in balance. The biomass, or mass of organic materials making up the community, and the chemical composition of this biomass are consequently in steady state. Not only the community as a whole, but also the species populations of a stable community may be in steady state. The steady state for the species involves a balance of individual births and deaths, while the population itself remains essentially stable. *See* POPULATION DYNAMICS.

Succession. Not all communities are fully stabilized in this sense, and many become so only approximately, after an extended process of community development, or succession. If a bare area of the Earth's surface is exposed, as by a landslide, this bare area may be occupied first by such simple plants as lichens and mosses. These plants contribute, along with physical processes, to the breakdown of the rock and formation of soil until higher plants, such as grasses, may occupy the environment. The grasses may in turn make it possible for shrubs to grow, and the shrubs make growth possible for trees, until finally a stable forest may result. Through the course of succession there is usually increasing productivity, species diversity, and stability of the community, with increasing development of the soil, increasing stocks of material in circulation, and increasing modification of the environment by the community. There

are, however, exceptions to these trends in particular successions. *See* SUCCESSION, ECOLOGICAL; VEGETATION MANAGEMENT.

Climax. The stable ecosystem which ends the succession is termed climax. Climax communities are characterized by self-maintenance and a considerable degree of permanence when free from external disturbance. That is, the populations in the climax stage reproduce and maintain themselves, as they do not during the successional stages. The climax thus represents a steady state of energy flow, materials circulation, and population reproduction. It may represent a maximum of sustained productivity and biomass for the ecosystem. (Or it can be a system of degraded productivity, after irreversible losses of nutrients, for example.) The climax has a self-stabilizing character, and tends to return to normal after disturbance of its equilibria. For instance, a heavy phosphate fertilization of a lake produces only a temporary increase in productivity, and phosphate in the water rapidly declines back to the original level. Thus a temporary overpopulation of a given species is counteracted by increased mortality from predation, disease, or other factors, until the population returns to its normal level. It appears that these stability mechanisms, in general, depend not so much on any single factor or interaction as on the reestablishment of a steady-state balance in relation to various and complexly interrelated factors. A pattern of climax communities may correspond to the pattern of environments in any large area. Within a given area, however, many environments may be similarly affected by the general climate of the area. Many of the successions of a given area consequently converge toward similar climax communities which occupy the largest part of the landscape surface. In the low mountains referred to previously, the oak forests would occupy the greatest part of the mountain slopes, give the vegetation its predominant character, and express the climate. Such a community may be termed a climatic climax or prevailing climax. *See* CLIMAX COMMUNITY.   [ROBERT H. WHITTAKER]

Bibliography: W. C. Allee et al., *Principles of Animal Ecology*, 1949; G. L. Clarke, *Elements of Ecology*, 1954; L. R. Dice, *Natural Communities*, 1952; S. C. Kendeigh, *Animal Ecology*, 1961; E. P. Odum, *Ecology*, 1963; E. P. Odum, *Fundamentals of Ecology*, 1959; R. H. Whittaker, *Communities and Ecosystems*, 1969.

Electric power systems

A complex assemblage of equipment and circuits for generating, transmitting, transforming, and distributing electric energy. Principal elements of a typical power system are shown in Fig. 1. They are described briefly here; many of them are discussed more fully in cross-referenced articles.

Generation. Electricity in the large quantities required to supply electric power systems is produced in generating stations, commonly called power plants. Such generating stations should be considered as conversion units converting the heat energy of fuel (coal, gas, oil, or uranium) or the hydraulic energy of falling water to electricity.

A little over 80% of the electric power used in the United States is obtained from generators driven by steam turbines. The largest turbine in serv-

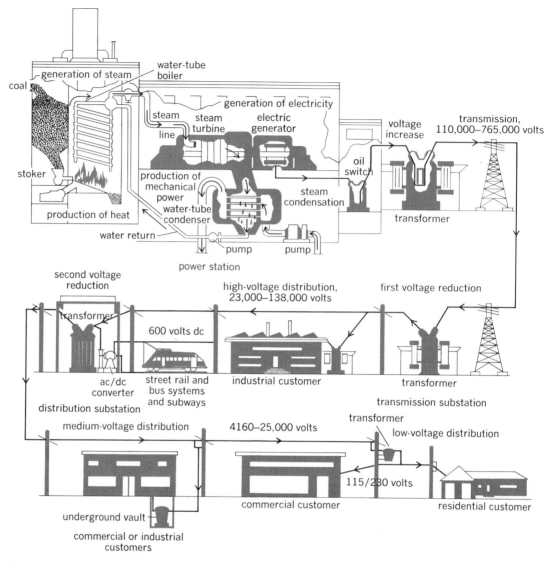

Fig. 1. Major steps in the generation, transmission, and distribution of electricity.

ice at this time (1968) is rated 1,130,000 kilowatts (kw), equivalent to about 1,500,000 horsepower (hp). Larger machines, up to 1,300,000 kw, are on order, and many units of 500,000 to 1,000,000 kw are in service or under construction.

Coal is the fuel for more than half of the steam-turbine generation. Natural gas is used extensively in the southern part of the United States, and heavy fuel oil is burned in a number of power plants situated where they can take delivery from oceangoing tankers.

Uranium, despite the rapid development of nuclear power plants, is the energy source for only about 2 or 3% of the steam-turbine generation in the United States; this low figure for nuclear power is a direct reflection of the lengthy Atomic Energy Commission reviews and public hearings deemed necessary to ensure safe use of this concentrated form of energy. As the numerous nuclear plants under construction or planned go into commercial operation, however, they will cause a sharp rise in the percentage of steam-turbine generation fueled by uranium. The largest nuclear unit in service

(1968) is capable of generating 873,000 kw of electrical output; others projected for the early 1970s are expected to generate 1,000,000 kw or more.

Waterpower is still an important energy source in the United States, although in 1970 it supplied only about 17% of the electric power consumed. This is because most of the attractive sites, where sufficient water drops far enough in a reasonable distance to drive reasonable-sized hydraulic turbines, have been developed.

Waterpower offers one distinct advantage over steam power plants: it has greater flexibility in adapting to changes in power demands. Hydrogenerators can even be shut down to store water for later load periods; in such operation the idle generators can be restarted and put into service in minutes. Many existing hydroplants have been expanded by installing additional generating units to increase their effectiveness for such intermittent operation.

Some plants, totaling several million kilowatts of installed capacity, actually draw power from other generating facilities during light system-load periods to pump water from a river or lake into an artificial reservoir at a higher elevation from which it can be drawn through a hydroelectric station when the power system needs additional generation. Such installations, known as pumped storage, consume about 50% more energy than they return to the power system and, accordingly, cannot be considered energy sources. Their use is justified, however, by their ability to convert surplus power during low-demand periods into prime power to serve system needs during peak-demand intervals—a need that otherwise would require building more generating stations for operation during the relatively few hours of high system demand.

Gas-turbine-driven generators have gained wide acceptance as an economical source of additional power for heavy load periods. Typical unit ratings in the United States are 10,000–20,000 kw, but many installations have two, three, or four such units and one has 16 units for a total of 256,000 kw. Gas-turbine units offer extremely flexible operation, being capable of frequent start-ups and loading to full rating within 5–10 min. Thus they are extremely useful as emergency power sources, as well as for operating during the few hours of daily load peaks. Gas turbines total about 3% of the total installed generating capacity in the United States but supply less than 1% of the total energy generated.

Internal combustion engines of the diesel type drive generators in many small power plants. In addition, they offer the ability to start quickly for operation during load peaks or emergencies. But their small size, commonly about 2000 kw per unit although a few approach 10,000 kw, has limited their use in recent years. Such installations account for less than 1% of the total power-system generating capacity in the United States, and an even smaller contribution to the total electric energy consumed.

Because of their simplicity and efficient use of conductors, three-phase, 60-Hz alternating-current (ac) systems are used almost exclusively in the United States. Consequently power-system generators are wound for three-phase output at a voltage usually limited by design features to a range from about 11,000 volts for small units to 30,000 volts for large ones. The output of modern generating stations is usually stepped up by transformers to the voltage level of transmission circuits used to deliver power to distant load areas.

Transmission. The transmission system carries electric power efficiently and in large amounts from the generating stations to areas where it is consumed. Such transmission also is used to interconnect adjacent power systems for mutual assistance in case of emergency and to gain for the interconnected power systems the economies possible in regional operation. Interconnections have expanded to the point where most of the generation east of the Rocky Mountains regularly operates in parellel, and over 90% of all generation in the United States, exclusive of Alaska and Hawaii, can be linked.

Transmission circuits are designed to operate at voltages up to 765,000, depending on the amount of power to be carried and the distance.

Table 1. Power capability of typical three-phase open-wire transmission lines

Line-to-line voltage	Capability, kva
110,000 ac	50,000
138,000 ac	80,000
230,000 ac	225,000
345,000 ac	500,000
500,000 ac	1,100,000
765,000 ac	2,500,000
800,000 dc	1,440,000

The permissible power loading of a circuit depends upon many factors, such as the thermal limit of the conductors and their clearances to ground, the voltage drop between the sending and receiving end, the degree to which system service reliability depends on it, and how much the circuit is needed to hold various generating stations in synchronism.

An approximation to the voltage appropriate for a transmission circuit, widely followed, is that the permissible load-carrying ability varies as the square of the voltage. Typical examples of ratings determined in this manner are listed in Table 1.

Transmission as a distinct function began about 1886 with a 17-mi 2000-volt line in Italy. Transmission began at about the same time in the United States, and by 1891 a 10,000-volt line was operat-

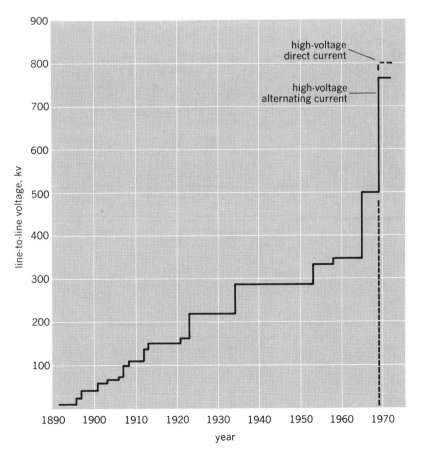

Fig. 2. Growth of transmission voltages from 1890.

ing. Subsequent lines have been at successively higher voltages, reaching 500,000 volts in 1965 and 765,000 volts in 1969 (Fig. 2). Transmission engineers are anticipating even higher voltages, 1,100,000 or perhaps 1,500,000 volts, but they are fully aware that this objective may prove too costly in spacings and funds to gain wide acceptance. Experience already gained at 500,000 volts and, to a limited extent, at 765,000 volts verifies that the prime requirement no longer is insulating the lines to withstand lightning discharges, but to tolerate continuous operation at rated voltage and occasional voltage surges caused by the operation of circuit breakers. The power capability of transmission circuits depends upon many factors, for example, conductor size, length of circuit, and characteristics of the entire power system of which the circuit is a part.

Experience has shown that, within about 10 years after the introduction of a new voltage level for overhead lines, it becomes necessary to begin connecting underground cable. This has already occurred for 345,000 volts; the first overhead line was completed about 1958, and by 1967 about 100 mi of pipe-type cable had been installed to take power received at this voltage level into metropolitan load areas. This experience suggests that 500,000-volt cable will be needed by 1975 and 765,000-volt cable a few years later. In both cases, however, growing governmental and public pressure in favor of underground transmission circuits may speed the use of cable.

In anticipation of transmission circuits of still higher load capability, an extensive research program is in progress, spread among several large and elaborately equipped research centers. Among these are Project EHV for extra-high-voltage overhead lines operated by General Electric Co. near Pittsfield, Mass., for the Electric Research Council (ERC); the Waltz Mills Cable Test Center operated by Westinghouse Electric Co. near Pittsburgh, Pa., also for ERC; and the Frank B. Black Research Center built and operated by Ohio Brass Co. near Mansfield, Ohio. All three include equipment for testing full-scale line or cable components at well over 1,000,000 volts. In addition, many utilities and specialty manufacturers have test facilities related to their fields of operation.

A new approach to high-voltage long-distance transmission has been developed recently. It is high-voltage direct current, which offers advantages of less costly lines, lower transmission losses, and insensitivity to many system problems that restrict ac systems. Its greatest disadvantage is the need for costly conversion equipment for stepping up the sending-end power and converting it to direct current, and for converting the receiving-end power to alternating current for distribution to consumers. Starting in the late 1950s with a 65-mi 100,000-volt system in Sweden, high-voltage direct current has been applied successfully to a series of special cases around the world, each one for a higher voltage and greater power capability. The first such installation in the United States, built by a group of utilities and the Federal government, is operating at 800,000 volts line-to-line to interchange 1,440,000 kw between the Pacific Northwest and the Pacific Southwest. A second circuit of similar capability is scheduled for service in the early 1970s, and system engineers today look upon this system as an attractive alternative to extra-high-voltage alternating current for some applications.

In addition to these high-capacity circuits, every large utility has many miles of lower-voltage transmission, usually operating at 110,000–345,000 volts, to carry bulk power to the numerous cities, towns, and large industrial plants that they serve. These circuits often include extensive lengths of underground cable where they pass through densely populated areas. Their design, construction, and operation are based upon research done some years ago, augmented by extensive experience.

Substations. Power delivered by transmission circuits must be stepped down in facilities called substations to voltages more suitable for use in industrial and residential areas. On transmission systems these facilities are commonly called bulk-power substations; on or near factories or mines they are termed industrial substations; and where they supply residential and commercial areas, distribution substations.

Basic equipment in a substation includes circuit breakers, switches, transformers, lightning arresters and other protective devices, instrumentation, control devices, and other apparatus related to specific functions in the power system.

Distribution. That part of the electric power system that takes power from a bulk-power substation to customers' switches, commonly about 40% of the total plant investment, is called distribution. This category includes distribution substations, subtransmission circuits that feed them, primary circuits that extend from distribution substations to every street and alley, distribution transformers, secondary lines, house service drops or loops, street and highway lighting, metering equipment, and a wide variety of associated devices.

Primary distribution circuits usually operate at 4160–34,500 volts line-to-line and may be overhead open wire on poles, overhead or aerial cable, or underground cable. These circuits supply large commercial, institutional, and some industrial customers directly. Smaller customers are supplied through numerous distribution transformers. There is a growing movement today to place most of these facilities underground.

At conveniently located distribution transformers in residential and commercial areas, the voltage is stepped down again to 120 and 240 volts for secondary lines, from which service drops or loops extend to every customer's lights and appliances. These low voltages, which include 125/216 in commercial areas and sometimes 480 in industrial areas, are known as utilization voltages.

Electric utility industry. In the United States, which has the third highest per capita use of electricity in the world and more electric power capacity than any other nation, the electric utility systems are deemed by some criteria the largest industry. The total plant investment as of Dec. 31, 1969, was about $110,000,000,000. The electric utility industry spends nearly $8,000,000,000 a year for plant additions to supply the growing load, and collects about $19,500,000,000 per year from 70,100,000 customers. This industry comprised

Table 2. Electric utility industry statistics, 1963–1969*

Item	1963	1964	1965	1966	1967	1968	1969
Generating capacity installed, $kw \times 10^3$†	222,000	228,500	240,900	252,700	267,075	292,000	312,000
Electric energy output, $kwhr \times 10^6$	921,800	986,800	1,060,100	1,158,200	1,223,200	1,305,900	1,392,500
Energy sales, $kwhr \times 10^6$							
Total	830,811	890,356	953,414	1,038,982	1,106,000	1,179,200	1,256,400
Residential	241,692	262,010	280,970	306,572	330,900	351,700	377,800
Small light and power	166,516	183,539	202,112	225,878	242,100	264,900	286,100
Large light and power	388,399	409,356	433,365	465,077	486,100	513,200	540,100
Other	34,204	35,451	36,967	41,455	46,900	49,400	52,400
Customers $\times 10^3$	62,857	64,149	65,558	66,910	68,159	68,950	70,100
Revenue $\times \$10^6$	13,697	14,408	15,158	16,196	17,185	18,300	19,500
Average residential use, kwhr/year	4,440	4,703	4,933	5,263	5,555	5,853	6,186
Average residential bill rate, cents/kwhr	2.37	2.30	2.25	2.19	2.15	2.10	2.06
Coal consumption, lb/kwhr	0.856	0.857	0.858	0.869	0.871	0.872	0.872

*From *Elec. World*, p. 121, Sept. 18, 1967; p. 81, Jan. 22, 1968; p. 69, Feb. 19, 1968   †Fuel and hydro.

3139 public and investor-owned systems producing electricity in nearly 4000 generating plants with a combined operating capability of 312,000,000 kw at the close of 1969.

The 3139 systems include 298 investor-owned companies operating 78% of the generating capacity and serving 79% of the ultimate customers. The remaining 21% are served by 1790 municipal systems, 932 rural electric cooperatives, 67 public power districts, 8 irrigation districts, 26 United States government systems, 12 state-owned authorities, 1 county authority, and 5 mutual systems.

The industry's annual output reached 1,-392,500,000,000 kwhr in 1969 and its sales to ultimate consumers, 1,256,400,000,000 kwhr at an average of 1.55 cents/kwhr. Of this, about half was consumed by industrial and other large power users, the remainder going mostly to residential and commercial customers. Residential usage has climbed steadily for many years, being 6186 kwhr/year in 1969, with an average bill of 2.06 cents/kwhr (Table 2).   [LEONARD M. OLMSTED]

Bibliography: J. Bleiweis, Today's overhead systems reflect the new look, *Elec. World*, 168:99–118, July 10, 1967; Cable engineers explore path to 750-kv application, *Elec. World*, 164:60–62, July 26, 1965; W. H. Dickinson, Muddy Run pumped storage: Big capacity at low cost, *Elec. World*, 166(1):31–33, 1966; R. H. Dunham, C. D. Durfee, and R. R. Lewis, Design concept of TVA's Paradise steam plant 1150-Mw unit, *Proc. Amer. Power Conf.*, 28:318, 1966; *EHV Transmission Line Reference Book*, Edison Electric Institute, 1968; Electric Research Council, *Electric Transmission Structures*, Edison Electric Institute, 1968; *Electrical World Directory of Electric Utilities*, 1969; I. E. Grant et al., Bull Run coal-fired unit goes automatic, *Elec. World*, 165(2):44–47, 1966; D. F. Heyburn, The steam generator for Paradise unit no. 3, *Proc. Amer. Power Conf.*, 28:336, 1966; 750-KV dc intertie rescheduled for April 1969, *Elec. World*, 164(8):98–101, 1965; F. J. Kovalcik and W. C. Hayes, New ideas in underground distribution, *Elec. World*, 169(11):97–116, 1968; E. B. Kurtz, *The Lineman's and Cableman's Handbook*, 5th ed., in press; L. M. Olmsted, 11th steam station design survey, *Elec. World*, 190(17):83–102, 1968; L. M. Olmsted, 15th steam station cost survey, *Elec. World*, 158(5):99–114, 1967; L. M. Olmsted and A. J. Stegeman, EHV: Today's designs, tomorrow's plans, *Elec. World*, 154(20):95–118, 1965; J. A. Rawls, VEPCo evolves construction methods for 500-kv project, *Elec. World*, p. 58, 1965; F. S. Sanford and C. W. Anthony, SCEC expands 500-kv transmission, adds $67 million to grid's cost, *Elec. World*, pp. 23–24, Jan. 20, 1964; K. Smalling, URD gains momentum in Long Island, *Elec. World*, 166(12):169, 1966; R. C. Spencer and Y. S. Hargett, The 1,130,000-kw turbine generator for the Pardise station of TVA, *Proc. Amer. Power Conf.*, 28:349, 1966; P. Sporn et al., The Cardinal plant story, *Elec. World*, 167(12):88–109, 1967; Substation trends, *Elec. World*, 170(2):89–104, 1968; E. Vennard, *Government in the Power Business*, 1968.

Engineering

Most simply, the art of directing the great sources of power in nature for the use and the convenience of man. In its modern form engineering involves men, money, materials, machines, and energy. It is differentiated from science because it is primarily concerned with how to direct to useful and economical ends the natural phenomena which scientists discover and formulate into acceptable theories. Engineering therefore requires above all the creative imagination to innovate useful applications of natural phenomena. It is always dissatisfied with present methods and equipment. It seeks newer, cheaper, better means of using natural sources of energy and materials to improve man's standard of living and to diminish toil.

Types of engineering. Traditionally there were two divisions or disciplines, military engineering and civil engineering. As man's knowledge of natural phenomena grew and the potential civil applications became more complex, the civil engineering discipline tended to become more and more specialized. The practicing engineer began to restrict his operations to narrower channels. For instance, civil engineering came to be concerned primarily with static structures, such as dams, bridges, and buildings, whereas mechanical engineering split off to concentrate on dynamic structures, such as machinery and engines. Similarly, mining engineering became concerned with the discovery of, and removal from, geological structures of metalliferous ore bodies, whereas metallurgical engineering involved extraction and refinement of the metals from the ores. From the practical applications of electricity and chemistry, electrical and chemical engineering arose.

This splintering process continued as narrower specialization became more prevalent. Civil engineers had more specialized training as structural engineers, dam engineers, water-power engineers, bridge engineers; mechanical engineers as machine-design engineers, industrial engineers, motive-power engineers; electrical engineers as power and communication engineers (and the latter divided eventually into telegraph, telephone, radio, television, and radar engineers, whereas the power engineers divided into fossil-fuel and nuclear engineers); mining engineers as metallic-ore mining engineers and fossil-fuel mining engineers (the latter divided into coal and petroleum engineers).

As a result of this ever-increasing utilization of technology, mankind and his environment have been affected in various ways—some good, some bad. Sanitary engineering has been expanded from treating the waste products of humans to also treating the effluents from technological processes. The increasing complexity of specialized machines and their integrated utilization in automated processes has resulted in physical and mental problems for the operating personnel. This has led to the development of bioengineering, concerned with the physical effects upon man, and management engineering, concerned with the mental effects.

Integrating influences. While the specialization was taking place, there were also integrating influences in the engineering field. The growing complexity of modern technology called for many specialists to cooperate in the design of industrial processes and even in the design of individual machines. Interdisciplinary activity then developed to coordinate the specialists. For instance, the design of a modern structure involves not only the static structural members but a vast complex including moving parts (elevators, for example); electrical machinery and power distribution; communication systems; heating, ventilating, and air conditioning; and fire protection. Even the structural members must be designed not only for static loading but for dynamic loadings, such as for wind pressures and earthquakes. Because men and money are as much involved in engineering as materials, machines, and energy sources, the management engineer arose as another integrating factor.

The typical modern engineer goes through several phases of activity during his career. His formal education must be broad and deep in the sciences and humanities underlying his field. Then comes an increasing degree of specialization in the intricacies of his discipline, also involving continued postscholastic education. Normal promotion thus brings interdisciplinary activity as he supervises the specialists under his charge. Finally, he enters into the management function as he interweaves men, money, materials, machines, and energy sources into completed processes for the use of man.

[JOSEPH W. BARKER]

Engineering, social implications of

The rapid development of man's ability to bring about drastic alterations of his environment has added a new element to the responsibilities of the engineer. Traditionally, the ingredients for sound engineering have been sound science and sound economics. Today, sound sociology must be added if engineering efforts are to meet the challenge of continuing to improve man's way of life.

Engineering practices and achievements, even when they are valid so far as science and economics are concerned, may cause new problems to emerge: the exhaust gases from millions of automobiles; the stack gases from fuel-burning power plants, incinerators, and process furnaces; the effluent from sewage systems and from waste disposal operations; the strip mining of coals and ores; the noise of trucks, trains, riveting hammers, factory operations, and impatient automobile drivers; the residual sulfites from paper mills; the chemical and particulate effusions from metallurgical and manufacturing establishments; and the dust storms and soil erosion accompanying intensified and mechanized farming. All are examples of the contamination which has been the handmaid of industrialization. See ECOLOGY, APPLIED.

The pattern of progress often causes shifts in the type of problem generated. For example, the substitution of nuclear fuels for fossil fuels in large power-plant operations will go far to relieve the burden of atmospheric pollution by stack gases. But other types of pollution accompany nuclear fuels. One type is "thermal" pollution. The present thermal efficiency of nuclear plants, being lower than that of fossil-fuel-fired power plants, increases by at least 50% the thermal pollution of the bodies of water used for condensing purposes. The question is whether this increase will be acceptable in the long view. Other factors must also be considered. There is evidence that the incidence of cancer among uranium miners may be much higher than among other hard-rock miners. Not only is this a serious medical problem for the miners themselves, but the radioactivity carried with them may constitute a hazard to life in their surroundings. Will the solution of such a problem require that uranium miners be prohibited from living in the densely populated megalopolis areas?

The sociological consequences of technological changes must be faced by the engineer. He must incorporate another ingredient in the definition of sound engineering so that it includes three significant factors: sound science, sound economics, and sound sociology. See ENGINEERING.

[THEODORE BAUMEISTER]

Entomology, ecological

Ecological entomology is the study of insects' relation to their environment. By number of species alone, insects outclass all other species of plants and animals combined. In total living weight, insects probably outweigh any other animal group, including humans, as well as play an important role in keeping the life system functioning. Insects also exert a significant impact on the quality of man's food, homes, and health. In the United States insects cause a greater loss of crops than any other pest group. On a worldwide basis, they are vectors of some of the most serious diseases infecting man.

Role in cycle of matter. Insects help concentrate and convert plant and animal material into food for other animals. Insects are the prime source of food for many species of fish, including trout and salmon. Toads, frogs, and other amphibians consume insects. The basic food items of many birds, such as the bluebird and house wren, consist primarily of insects. Moles, shrews, skunks, and other small mammals feed almost exclusively on insects. See FOOD CHAIN.

Several groups of insects feed on dead and decaying animals and plants and thereby help to degrade these wastes. Because of this role and the tremendous living mass of insects, they play a dominant role in recycling the vital elements of life for reuse in the life system.

Pollination role. Insects not only keep the life system functioning by eating and being eaten, but also they serve a most important role in plant reproduction—pollination. Despite man's many technological advances, substitutes for general insect pollination of plants, cultivated and wild, have not been found. The production of fruits, vegetables, and forages depends upon the activity of bees and other insects which pollinate or fertilize the blossoms.

The role of honeybees alone in pollination is impressive. For example, a single honeybee may visit and pollinate 1000 blossoms in a single day (10 trips with 100 blossoms visited per trip). In New York State, with about 3,000,000 beehives and each with 10,000 worker bees, honeybees could visit 30 trillion blossoms in a day. Wild bees pollinate a number of blossoms equal to or more than that pollinated by honeybees. In fact, on a bright, sunny day, more than 60 trillion blossoms may be pollinated—a herculean task carried out by the "busy bees."

Honey and wax production. Insects make several products of value to humans, of which honey and wax are the two most valuable. About 200 million lb (1 lb = 0.45 kg) of honey and 4 million lb of wax are produced in the United States annually, with a value of $23 million and $2 million respectively. To produce 1 lb of honey, 1500 bees work 10 hr, and to make 1 lb of beeswax (an ingredient of most lipsticks), the honeybee must consume about 8 lb of its honey.

Pest control. In addition to pollinating plants, some insects, because of their tremendous numbers, are useful in controlling and preventing outbreaks of pest populations of insects and plants. The impact of insect feeding pressure for weed control is illustrated throughout nature. For example, the plant pest known as Klamath weed was introduced into California about 1900, and by 1944 the weed had become a dominant plant in about 2 million acres (1 acre = 4047 m²) of pasture and range land. In 1945 a small beetle (*Chrysolina quadrigemina*) which fed on the weed was introduced to the area. This beetle population has increased, dispersed, and eliminated 99% of the weed. The previous forage vegetation has regrown, with the pasture and range land once again suitable for cattle production.

In United States agriculture about 95% of control of pest insects depends upon various natural controls, including insect predators and parasites. Surprisingly, only 5% of total agricultural acres in the United States is treated with insecticides. Therefore, the prime means of pest insect control is biological control. See INSECT CONTROL, BIOLOGICAL.

Insecticide use. What about the pest insects and insecticides used to control them? Nearly a half billion pounds of insecticides is used annually in the United States for insect control. Slightly more than half of this insecticide is used directly in agriculture, with the remainder used by government agencies, industries, and homeowners.

Insecticides used in agriculture are not evenly distributed over all crops. For example, 50% of all insecticide applied in agriculture is used on the nonfood crops of cotton and tobacco. Of the food crops, corn, fruit, and vegetables receive the largest amounts of insecticide (see table).

Although cotton receives nearly 30% of the total insecticide used in agriculture, about half (46%) of the total cotton acreage receives no insecticide treatments at all (see table). The largest percentage (79%) of the acres treated is in the Southeast and Delta states, whereas the smallest percentage (37%) treated is in the Southern plains. Of the food crops, only citrus, apples, and potatoes have more than 85% of their acreages treated with insecticides. Many acres of small grains and pastures receive little or no insecticide treatment. This is the reason that only 5% of the crop acres receives any insecticide treatment at all.

Crop loss due to insects. Despite a 20-fold increase in the use of insecticides during the past 25 years, crop losses due to insects have nearly doubled; they have increased from about a 7% loss in the 1940s to about a 13% loss currently. Although crop losses due to insects have generally increased despite the significant use of insecticides, important advances have been made in reducing crop losses from certain pests. For example, losses from potato insects have declined from about 22% previous to 1935 to about 14% today. In contrast, corn losses due to insects have been increasing. In the 1940s corn losses due to insects averaged about 3.5%; however, this loss has increased significantly and now averages about 12%. Factors contributing to increased corn losses due to insects include the continuous culture of corn on the same land and the planting of insect-susceptible types instead of resistant types.

If the 5% of crop acres now treated with insecticides was no longer treated, it is estimated that crop losses would increase from the current total of 13% to about 16.3%. This would amount to a loss of about $1 billion annually. If no insecticides were used, the increased losses would probably increase retail food costs about 5.7%. Based on these estimates, the nation would not starve if in-

Some examples of percentages of crop acres treated, of pesticide amounts used on crops, and of acres planted to this crop

Crops	Insecticides		Herbicides		Fungicides		% of total crop acres
	% acres	% amount	% acres	% amount	% acres	% amount	
NONFOOD	1	50	0.5	NA*	<0.5	NA	1.26
Cotton	54	47	52	6	2	1	1.15
Tobacco	81	3	2	NA	7	NA	0.11
FOOD	4	NA	11.5	NA	<0.5	NA	98.74
Field crops	NA	NA	NA	NA	NA	19	NA
Corn	33	17	57	41	2	NA	7.43
Peanuts	70	NA	63	3	35	4	0.16
Rice	10	NA	52	2	0	NA	0.22
Wheat	2	NA	28	7	0.5	NA	6.11
Soybeans	4	2	37	9	0.5	NA	4.19
Pasture hay + range	0.5	3	1	9	0	NA	68.40
Vegetables	NA	8	NA	5	NA	25	NA
Potatoes	89	NA	59	NA	24	12	0.16
Fruit	NA	13	NA	NA	NA	NA	NA
Apples	92	6	16	NA	72	28	0.07
Citrus	97	2	29	NA	73	13	0.08
All Crops	5	54	12	36	0.5	10	NA

*Not available.

secticides were not used; supplies of food for the nation would be ample, but quantities of certain fruits and vegetables such as apples, peaches, plums, oranges, onions, potatoes, and cabbages would be significantly reduced. Because of this, substitutes for some fruits and vegetables that people normally like to eat would have to be found.

Actually, the loss in some fruits and vegetables could be reduced if "cosmetic standards" were modified. Although safe and nutritionally sound, some fruits and vegetables with skin blemishes are not sold because consumers generally rate them unacceptable. Small blemishes do not adversely affect these product's overall palatability and nutrition.

Home and health hazards. Insects not only attack crops, but they damage homes and more importantly are health hazards. As nuisances, they bite and sting man, but of greater significance on a worldwide basis is the fact that insects are vectors of some of the most serious diseases of mankind.

In much of the world the ancient afflictions of mankind—malaria, filariasis, yellow fever, and dengue—continue to limit progress. In terms of numbers affected, mosquito-transmitted malaria continues to be the number one disease of man. About 400 million people live in areas where malaria is highly endemic; it is estimated that at least 100 million cases occur annually and result in 1 million deaths. *See* DISEASE.

At present, mosquito control is in a state of crisis because man has become heavily dependent upon insecticides. The very properties of the insecticides that made them so useful (persistent residues and high toxicity) have caused serious environmental problems.

Environmental impact. Despite the fact that only a small percentage (5%) of the crop acres in the United States is treated, serious insecticide pollution problems have occurred. Insecticides have destroyed natural enemies of other pests, resulting in other pest outbreaks and thereby requiring additional pesticide sprays. This has oc-

curred in several crops, but especially in fruit and cotton pest control.

Some insecticides have caused fish kills and sometimes have stopped reproduction in a species, such as lake trout in the Northeast. In addition, several bird species, including eagles, falcons, and pelicans, have declined in part due to insecticides in some habitats in the United States.

Insecticide pollutants are now widespread and occur in water, soil, air, and most living organisms, including humans. In the United States alone, nearly 40,000 lb of DDT is estimated to be present in humans. However, no harmful effects are believed to have resulted from this dosage.

Even the amount of insecticide drifting in the atmosphere is disturbing. Insecticides can enter the atmosphere either during application by volatilization and codistillation, or they can be picked up from the soil and plants by the wind. Pesticide application via aircraft spraying is of especial concern in polluting the environment. Various studies indicate that as little as 25% of the insecticide applied by aircraft reaches crop level; the other 75% drifts away in the atmosphere. With 60% of all insecticides used in agriculture being applied by aircraft, the problem of polluting the atmosphere and the environment is significant.

Systems approach. In addition to reducing insecticide applications by aircraft, there is need for an ecosystem approach to pest management technology. This is referred to as a systems approach and includes three parts: (1) source reduction of pests by minor environmental manipulations; (2) integrated controls or combinations of species-specific controls; (3) analysis of the costs and benefits of the pest management options to humans and environment.   [DAVID PIMENTEL]

Bibliography: D. J. Borror and D. M. DeLong, *An Introduction to the Study of Insects*, 1964; D. Pimentel, *Ecological Effects of Pesticides on Nontarget Species*, 1971; D. Pimentel, Extent of pesticide use, food supply, and pollution, *J. N.Y. Entomol. Soc.*, 1973.

Environment

Ecologically, the environment is the sum of all external conditions and influences affecting the life and development of organisms. Various ecological principles and concepts have been developed in regard to the environment. Two main aspects of the environment are usually considered, the abiotic and the biotic. These divisions are artificial in the sense that neither can be separated when organisms are studied. All environmental aspects and their influences on living organisms must be considered together.

Nature of the environment. Holocoenosis, the nature of the action of the environment, pertains to those factors which exist as a vast complex and therefore do not act separately and independently, but as a whole. This principle, which lies at the core of all ecological thinking, is shown in the illustration.

The particular environment occupied by an or-ganism or by groups and communities of organisms is the habitat. Environmental conditions generally exist in an area independently of the occurrence of any species of plant or animal that may depend on them, and area is used to express this concept, for example, in contrasting humid versus dry climates. The whole system of interaction between a particular organism and its physical and biotic environment is the niche of that organism. The term indicates not only the habitat, or microenvironment, but also the activity of the organism. The environment of small areas in contrast to that of larger ones, or of particular organisms in contrast to generalized environments of communities, is recognized as the microenvironment. Other terms commonly applied to this concept, but much more restricted in scope, are microclimate and bioclimate. Being relative, the term microenvironment may signify the environment of a pine stand or, equally well, a lichen which is located within that stand.

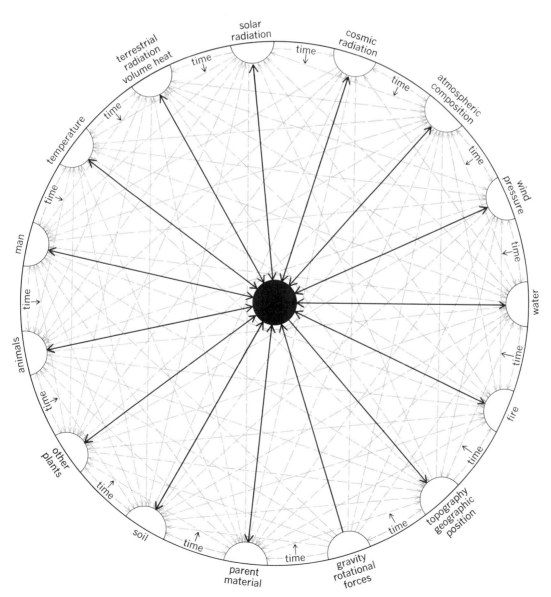

A holocoenotic environmental complex. (*From W. D. Billings*)

Dynamic concept. Variations in space and time are responsible for a dynamic concept of the environment. In space, there is continual change from the Equator to the poles, from sea level to outer space or to ocean depths, and from the forest floor to the treetops. Variations in time may be geological, as climates, carbon dioxide content of the air, and salt content of the ocean, or be concerned with present-day cycles and periodicities. The more familiar include day and night, ocean tides, Moon phases, and seasonal cycles.

The change or rate of change in an environmental variable is the gradient. This is pertinent, through all units of space and time, to an understanding of the continuous adjustment life must make to its environment.

Action systems provide a convenient way of referring to the interaction between organisms and their complex environment. Ecological action is the effect of the physical environment on organisms; reaction is the reciprocal effect of organisms on the physical environment; and coaction is the interaction of organisms on each other.

Limiting factors. The concept of limiting factors recognizes that any factor of the environmental complex which approaches or exceeds the limits of tolerance of any organism or group of organisms acts as a control on them. This concept is derived from two laws. The law of the minimum, originally given by Justus Liebig in 1840, has been restated by W. P. Taylor as follows: "The growth and functioning of an organism is dependent upon the amount of the essential environmental factor presented to it in minimal quantity during the most critical season of the year, or during the most critical years of a climatic cycle." In 1913, by adding the concept of an ecological maximum, V. Shelford showed, through his law of tolerance, that all organisms or groups of organisms must live in a range of conditions between the maximum and minimum, which then represent the limits of tolerance. A series of terms has come into general use attaching the prefixes "steno-" meaning narrow, and "eury-" meaning wide, to indicate the relative degree of tolerance, for example, stenothermal, which indicates a narrow tolerance to temperature.

Other factors. Those factors which upset the delicate balance of an ecosystem are termed trigger factors. They cause a chain reaction which may not end until drastic changes have occurred. For example, the permanent addition of water to a desert environment may ultimately change the whole ecosystem.

The substitution of elevation for latitude is an example of a compensating factor which allows plants of northern distribution to grow far southward on high mountain ranges. Some factors such as temperature may substitute for one another without apparent effect on the organism, but factors such as day length are different.

Environmental substances such as nutrients are necessary for the growth and development, or nutrition, of organisms. Thus, inorganic substances provide for the nutrition of green plants, which in turn provide the organic nutrients of nongreen plants and animals. The characteristic pathways in which these substances circulate from the environ-ment to organisms and back again are known as biogeochemical cycles. Oxygen, carbon, and nitrogen are the principal gases involved, while more than 40 minerals are obtained from the crust of the Earth, among them phosphorus and potassium. The synthesis of protoplasm by living organisms and their decomposition after death from the energy basis for these cycles and provide for a classification of organisms into producers, consumers, and reducers. *See* BIOSPHERE, GEOCHEMISTRY OF.

ABIOTIC ENVIRONMENT

The physical or abiotic environment includes all those physical and nonliving chemical aspects which exert an influence on living organisms. Among these factors are soils, water, and the atmosphere, as well as the influence of energy from various sources.

Various forms of energy exert an influence on, or modify, the environment in which organisms are found. With minor exceptions, the ecologist is concerned with radiant energy received directly from the Sun, infrared radiation or heat, visible radiation, and ionizing radiation. These energies, coming from the same source, are subject to the same modifying factors, the primary one being daily and seasonal cycles which in turn are modified by latitude, altitude, slope, exposure, atmospheric conditions, and the color, texture, and cover of the ground surface.

Temperature. The intensity aspect of heat energy is temperature, and, with moisture and light, it is one of the most familiar of environmental conditions. In nature, temperature ranges from a low of $-70°C$ ($-94°F$) recorded in Siberia, to highs of $60°C$ ($140°F$) in deserts and almost $100°C$ ($212°F$) in hot springs. With few exceptions, however, almost all life exists within the relatively narrow range of $-17.8°C$ ($0°F$) to $45°C$ ($113°F$). The specific responses developed by organisms to temperature include metabolic activity, behavior, abundance, and distribution. With the exception of birds and mammals, the body temperature of plant and animal life is determined primarily by the external environment. *See* HOMEOSTASIS.

Light. The source of light may be the Sun, Moon, stars, or luminescence, but usually only the first two are of sufficient intensity to influence life. Light varies according to wavelength, intensity, and duration. Sunlight may vary at noon in the open from a few hundred footcandles (ftc) on overcast days to more than 14,000 ftc on clear days in southern latitudes. In woods, it may vary from a few to 1000–2000 ftc under the same conditions. Maximum moonlight is about 0.05 ftc. Light penetration in relatively clear water is about 30% at 5 m and 10% at 10 m. In addition to the well-known uses of light for photosynthesis in plants and vision in more advanced animals, it also affects growth, development, and survival, and through many types of photoperiodicities, affects the behavior, life cycle, and distribution of many plants and animals.

Ionizing radiation. Ultraviolet, nuclear, and cosmic radiations have sufficient energy to damage protoplasm on relatively short exposures. The potential significance of ionizing radiation as an environmental factor was not recognized until the mid-1950s, when worldwide attention was focused on ionizing radiation received from fallout. The atmosphere is opaque to all ultraviolet light shorter than 2900 A, so that little or no injury results at moderate altitudes. In the stratosphere and above, the lethal action of ultraviolet light is important, the bacterial effects beginning at 3650 A. Nuclear radiation, composed of alpha and beta particles and gamma rays, comes from radioactive substances naturally occurring in the crust of the Earth, and from living and nonliving materials which have these incorporated in them. The intensity of this background radiation varies on an average from 0.1 to 0.5 roentgen unit (r) per year, although many areas give much higher readings. Because of fallout, this natural radiation is being added to each year. Cosmic rays come from space and vary appreciably with altitude. Some geneticists estimate that 10% of the mutations in geological time are attributable to these ionizing radiations. Very little is yet known about their direct effects at natural levels on organisms.

Water. Water and air are the fundamental media in which life exists. As such, they provide the basis for the division of the world into two major environments, aquatic and terrestrial. Water not only covers 70% of the Earth's surface, but also provides for the existence of life throughout its depths. Water has universal solubility, low chemical activity, high surface tension, high ionization, high freezing point, latent heat of fusion, latent heat of evaporation, and high specific heat. As the chief components of protoplasm, water and air have many common physicochemical characteristics.

The distribution of water through the atmosphere and over the terrestrial environment is controlled by the hydrological cycle. Principal marine environments are tidal zones (neritic), and oceanic zones (euphotic, pelagic, and benthic). Fresh-water environments are standing water (lakes, ponds, swamps, and bogs) and running water (springs, streams, and rivers). On land, the term xeric refers to relatively dry, hydric to relatively wet, and mesic to moist environments. See HYDROLOGY.

Atmospheric moisture may exist in three states, solid (hail, sleet, or snow), liquid (rain), or gas (water vapor). Water vapor content of the atmosphere varies with temperature and is referred to as humidity. The ratio of precipitation to evaporation is the most important factor in the distribution of vegetation zones, such as desert, grassland, woodland, and forest.

Soil moisture varies with the structure and texture of the soil, as well as with precipitation-evaporation factors. Soil is saturated when all pore spaces are filled. The pore spaces make up 40–60% of the volume of soil. Water which subsequently drains away is gravitational water, while that retained by capillary forces is capillary water. In very dry soil, the extremely thin film which cannot move as a liquid is hygroscopic water, and that chemically bound with soil materials is combined water. All gravitational and most capillary water is available to plants, while bound water and hygroscopic water cannot be utilized. Plants which can no longer obtain moisture from the soil reach a physiological state at which wilting occurs. This varies among different plant species and is known as the wilting coefficient of the soil.

Atmosphere. The atmosphere is composed of 78% nitrogen, 21% oxygen, 0.03% carbon dioxide, several other gases in very small quantities, and varying amounts of water vapor. All exist in a simple physical mixture. In addition to the physical characteristics of the atmosphere as a medium, its oxygen and carbon dioxide chemically affect all forms of life through photosynthesis and respiration, that is, through their reciprocal relationship of synthesis and decomposition. Concentration of these gases varies with organic and industrial activity, with altitude in the atmosphere, with organic activity, and with their solubility in fresh and salt water. Carbon dioxide also influences the hydrogen ion concentration of soil and water.

Wind is a significant atmospheric phenomenon acting through its physical force. It exerts an influence on the rate of transpiration in plants and evaporation of moisture. It modifies weather and climate, and is a disseminating agent for pollen, spores, and microscopic organisms.

Substratum. This is any solid on whose surface an organism rests or moves, or within which it lives, thus providing anchorage, shelter, and food. Aquatic substrata are rock, sand, and mud, while terrestrial substrata are rock, sand, and soil, soil being the most complex.

Soil is the link between the mineral core of the Earth and life on its surface. It consists of decomposed mantle rock and organic matter, with spaces for water and air, and is organized into distinct layers or horizons. The A horizon, or surface soil, is the zone of most abundant life and also of leaching. The B horizon, as subsoil, is a zone of deposition from the layer above, while the C horizon is composed of the parent materials. The major factors in soil formation are climate, topography, living things, and time. The differential nature of these factors is responsible for the great soil groups of the world. See SOIL.

Other abiotic factors. These include such conditions as gravity, pressure, sound, and fire. Gravity affects the environment through its action in isostasy, earthquakes, and stratification in air and water. Its effects on organisms are through polarity, orientation, structure, distribution, and behavior. Pressure acts mechanically, and environmental pressures range from a half atmosphere pressure at 5800 m elevation to 1000 atm at a depth of 10,000 m in the ocean. Since pressure changes much more rapidly in water than in air, it becomes of paramount importance in the ocean through its effects on the structure, physiology, and distribution of organisms.

Sounds are produced and conveyed by mechanical vibrations, substratal vibrations, and mechanical shock, which are interconnected phenomena. While aerial, and usually audible, vibrations are important to man and many animals, particularly birds and mammals, the substratal vibrations such as earthquakes, especially when sudden and intense, are the ones of prime importance in ecology.

Fire has always been an important factor in the terrestrial environment. Classified as crown, surface, and ground types, fires exert direct effects through injury and destruction, and indirect effects through often drastic alterations of environmental conditions and changes in community relationships. The beneficial values of controlled burning in forestry and game management are well established, and fire is an important tool in shifting agriculture in the humid tropics.

BIOTIC ENVIRONMENT

The biotic environment consists of living organisms, which both interact with each other and are inseparably interrelated with their abiotic environment.

Interactions between organisms. Within a population, interactions include such aspects as density, birth rate, death rate, age distribution, dispersion, and growth forms. *See* ECOLOGICAL INTERACTIONS.

Interactions between organisms at the interspecies population level include competition, which arises from utilization of the same thing when it is in short supply. Predation and parasitism are negative interactions which result in harm or destruction of a prey or host population by the predator or parasite population. Positive interactions also occur, such as commensalism, in which one population is benefited; cooperation, in which both populations are benefited; and mutualism, in which both populations are benefited and are completely dependent on each other. At the community levels, interactions include dominance, where one or several species control the habitat. In succession, there is an orderly process of community change. Rhythmic change in community activity is termed periodicity.

Interactions with abiotic environment. These interactions include (1) the effects of organisms on the microenvironment through temperature, water, wind, and light, (2) their modification of the substrate, as through soil building, and (3) their modification of the medium, as in aquatic habitats. *See* ECOSYSTEM.   [ROBERT B. PLATT]

Bibliography: A. H. Benton and W. E. Wagner, Jr., *Field Biology and Ecology*, 1966; R. F. Daubenmire, *Plants and Environment*, 2d ed., 1962; P. Farb, *Ecology*, 1963; R. Geiger, *The Climate near the Ground*, 4th ed., 1965; J. P. Huxley, *World of Life*, 1958; E. P. Odum, *Fundamentals of Ecology*, 2d ed., 1964; R. B. Platt and J. Griffiths, *Environmental Measurement and Interpretation*, 1964; R. L. Smith, *Ecology and Field Biology*, 1966.

Environmental protection

The protection of man and equipment against stresses of climate and other elements of the environment. Research in this field aims to improve the effectiveness of human performance under all environmental conditions through interrelated studies of environment, man, and materials, including clothing, shelter, food, machines, and equipment. Systematic research in this field was begun in 1942 to meet the need of the Quartermaster Corps to provide adequate clothing for military personnel overseas in World War II. The principles evolved are equally applicable to civilians under conditions of field or outdoor life. For discussion of protection of the environment against man *see* AIR-POLLUTION CONTROL; ECOLOGY, HUMAN; WATER TREATMENT.

Physical geographic environment. In order to know the environmental stresses from which men and materiel must be protected, the scientist must study all stressful aspects of the geographic environment, especially climate. Both the essential character and the regional and seasonal distribution of environments require attention. Heat and cold have received primary emphasis, and ambient temperatures have been mapped for all parts of the world. Since 1943 U.S. Army clothing authorization has been based upon clothing zones developed from mean monthly temperature maps. Clothing almanacs show the month-by-month clothing needs. Refinement of the basis of mapping clothing- or equipment-use zones proceeds as data are processed for maximum and minimum temperatures and, most realistic of all, for frequency and duration of critical temperatures. Direct solar radiation and wind also modify the effects of heat and cold induced by basic ambient temperatures. The term windchill is used to express the combined cooling effect of wind and temperature upon the body.

Factors of moisture, including rainfall, snow, glaze, hail, humidity, dew, and fog, introduce other stresses, only in part thermal. Wind is a potent mechanical factor that must be considered in the design of tents and other structures, and airborne sand and dust are irritants and abrasives. For work with combined climatic stresses, individual elements are grouped into climatic types. Army clothing, for example, is classified for research purpos-

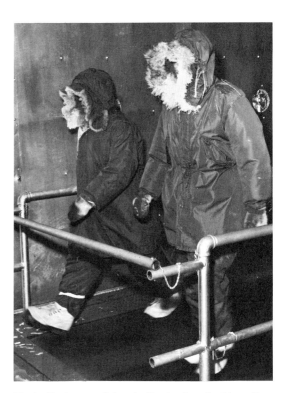

Fig. 1. Environmental protection testing of cold-weather clothing at −40°F with gentle breeze in arctic wind tunnel, Quartermaster Environmental Research Center. Skin temperatures of the men on the treadmill are measured by thermocouples that are attached to various parts of the body. (*U.S. Army photograph*)

es into four moisture-temperature types: cold-dry, cold-wet, hot-dry, and hot-wet.

Microclimates are studied to gain an understanding of the conditions experienced by man and materiel at various levels at, above, and below the ground, and under diverse terrain or vegetation situations. The designer of shoes, for example, needs to know not merely that the "standard" air temperature in a desert may reach 125°F but also that the ground temperature may reach 160°F.

The terrain, both physical and biological, introduces additional stresses. The composition of the ground surface, such as sand, rough rock, or clay (hard when dry, mud when wet), affects movement of vehicles and adequacy and longevity of clothing. Surface configuration, especially slope steepness, is of paramount importance for vehicular movement, and ways of making quantitative regional generalizations must be devised. Water, in the form of streams, lakes, swamps, or ice, adds other problems of personnel protection. For example, the invention of crevasse detectors has greatly increased the safety of travel on glaciers.

Biological elements of the environment cannot be overlooked. Natural vegetation such as taiga, tropical forest, brush, grassland, desert, and tundra, and cultivated vegetation such as cane fields, rice paddies, and orchards affect durability of clothing, trafficability of vehicles, food and fuel for survival, and concealment against hostile eyes. The camouflager needs to know the seasonal changes of color of vegetation, ground, and snow. Noxious insects are usually the most serious animal pests. Character, seasonal occurrence, and distribution must be studied to permit development of insect control or individual protection with netting or repellents. Microorganisms and air conditions favoring them are part of the problem of minimizing deterioration of food and materiel in storage. *See* BIOCLIMATOLOGY.

Man. In the environmental protection complex, man himself is studied from several viewpoints. Physiological studies are made under controlled conditions of environment in climatic chambers (Fig. 1), or outdoors where all environmental factors converge upon the test subject in the complexity of their interactions (Fig. 2). Attention is paid to effects on body and skin temperatures of exposure to heat or cold for varying periods of time and types of activity in which the subject wears various types of clothing. Sweating rates and water requirements, work capabilities, metabolism, and food requirements in extreme environments are observed and measured. To anthropologists falls the task of making body measurements essential for the design of clothing that fits properly to give maximum protection and comfort. Psychologists design questionnaires to elicit statistically meaningful reactions to clothing in varying environments, determine problems of human factors in the use of materiel, study effects of environmental extremes on the psychophysiological capabilities of man, check equipment to assure operability by men clothed protectively, and determine acceptability of protective items to be sure they will be used by the men who need them. *See* HOMEOSTASIS.

Research on materials. The third element of the environmental complex, materials, is studied

Fig. 2.   Metabolism testing of men in desert. (*U.S. Army photograph*)

to determine its effectiveness in interposing a protective barrier between man and the environment.

Clothing. Studies must deal with the characteristics of textiles such as wind perviousness, insulating value, wicking capability, durability, and comfort of individual garments and of clothing ensembles, through the activity of the biophysicist and the clothing specialist. The widely used "layer system" takes advantage of the insulating value of air layers between cloth layers. Even in arctic clothing, provisions must be made for ventilation to prevent overheating and moisture accumulation. Disposable paper garments, which obviate the necessity for laundering, have been developed. Laundering has its own problems, especially in desert areas where water is limited, and in cold areas in which much fuel must be expended to melt ice. Hand wear must retain dexterity and provide insulation; headgear must be designed to permit sight, hearing, and breathing while providing insulation or shade. Trench foot, the dreaded scourge of cold, wet weather, has been nearly eliminated by the development of the insulated boot, with a layer of felt or sponge rubber sandwiched between layers of waterproof rubber—an item invented in the Quartermaster Corps but since applied widely to civilian use. Sleeping gear involves problems associated with the ground surface as well as with the ambient air.

Shelter. Shelter must be designed to protect against extremes of cold, heat, and rain. Unlike clothing, the shelter must be heated mechanically by means of a stove in cold climates, unless it is to be used for the storage of nonperishable articles, thus adding to the great bulk of petroleum fuel products that must be supplied any large modern force of men in the field. In summer in hot climates, ways must be found to reduce the heat load occasioned by intense solar radiation and air temperature. The careful study of geographical sites is an added environmental requirement in the establishment of shelter, especially for permanent and semipermanent buildings.

Relations with materiel testing. The term environmental protection, originally applied to the protection of man against the elements, has been extended to include protection of materials and the development of environment-resistant materials. The whole field of testing paint, vehicles, or electronic equipment against environmental stresses in the field or in the laboratory (including climatic chambers large enough to contain an assembled aircraft) is in a sense a part of environmental protection research. Materiel testing requires that quantitative environmental criteria for design must be formulated in advance, and that the role of man as operator in the system must be considered. In modern materiel testing, much attention is paid to vibration, shock, and acceleration, which are considered to be environmental factors, though they are usually man-made rather than natural factors.

Special environmental complexes. Man-made regional features and man's frontiers in space add complexities to environment. The environments resulting from military operations give rise to studies of such matters as the use of body armor against projectiles, physiological and psychological effects of explosions, minimization of radiation effects, and the psychological effects of limited rations. Large solar furnaces have been constructed to obtain radiation intensities that are comparable to those obtained by the use of nuclear reactors.

Recently human environments have been included among the factors to be considered in environmental protection. The social, legal, political, and economic practices and attitudes of local populations have a profound bearing upon the degree of success likely to be achieved by visitors or invaders. Certainly the material portions of man-made environments, the settlements, roads, and forms of land use, may have as direct an influence as the natural terrain and vegetation.

Space environments pose special problems of environmental protection, mostly intensified forms of the ordinary stresses of radiation, vibration, acceleration, and high-altitude conditions, though weightlessness is a space factor without parallel on Earth. To the increasing number of all-weather climatic testing chambers, there are now being added chambers to simulate the rarefied conditions and intense radiation of 400 mi and more above the Earth. Research on selenography, or lunar "geography," is essential in the planning of environmental protection of man on the Moon. *See* CLIMATOLOGY.   [PEVERIL MEIGS]

Bibliography: D. H. K. Lee and H. Lemons, Clothing for global man, *Geogr. Rev.*, 39:181–213, 1949; L. H. Newburg (ed.), *Physiology of Heat Regulation and the Science of Clothing*, 1949; S. W. Tromp, *Medical Biometeorology*, 1963; U.S.

Army, *Operation of Material under Extreme Conditions of Environment*, AR 705–15, 1957; U.S. Department of Defense, *Climatic Extremes for Military Equipment*, MIL-STD-Z10A, 1957.

Epidemiology

The study of the mass aspects of disease. The word epidemic literally means "upon the people," and was coined to describe the way an infectious disease, in spreading through a group of people, gives the impression of an affliction that has been placed upon the community as a whole. An observer reflecting on such an epidemic would ask questions about the circumstances which allowed the disease to develop and flourish, what permitted some members to escape, and what brought the epidemic to an end. These are questions about the disease as it affects groups of people, rather than individuals in the group, and they are the typical problems of epidemiology.

In recent years diseases other than infectious diseases have been included in the subject matter of epidemiology. It is quite natural, therefore, that these massive noninfectious diseases should be studied as group phenomena, and that one should speak of the epidemiology of accidents, cancer, heart disease, or mental disease. For both infectious and noninfectious diseases there are similar general problems such as identifying cases in the population at risk, and finding which circumstances lead to an increase and which to a decrease in the disease. In spite of the differences in detail arising in a study of Asian influenza and a study of suicide in a particular population, both are fundamentally investigations of disease rates and a comparison of the way these rates vary from one subgroup to another; for example, is there a difference in the rates for the two sexes, the various age groups, or different socioeconomic classes? There are traditionally a number of rates that have proved useful in the study of epidemiology. Some of the most important are the mortality rate, morbidity rate, disease rates (prevalence ratio and incidence rate), and case fatality ratio.

Mortality rate. The mortality rate for a given year is the number of deaths occurring in the year per 1000 total midyear population. In the United States in 1946 there were 1,395,600 deaths and the population as of July 1 was 139,893,000. Hence the mortality rate was

$$\frac{1,395,600 \times 1000}{139,893,000} = 10.0$$

This rate is sometimes referred to as the crude death rate because no account is taken of such factors as the age and sex of the members of the population. Since France has a higher percentage of old people than the United States and Israel a lower percentage, it is not surprising to find that the crude death rate in France is considerably higher than that of the United States and in Israel considerably lower. These objections to the crude death rate can be met by computing separate rates for particular age groups in each sex, for example, the rate for males aged 15–24, another rate for males 25–34, and so on. The process can be carried further by computing separate death rates for individual diseases such as tuberculosis, or groups of diseases such as those associated with child bearing.

The rates for particular age groups, particular causes of death, or specified sex are known as specific rates. They are often essential for the thorough study of an epidemic. However, it is true in general that the more specific a rate, the more difficult it is to collect accurately the information necessary for the computation. Thus to compute the death rate from accidents in 1959 among male children aged 5–14 years, it is necessary to know both the number of deaths from accidents among the group in 1959 and also what is called the population at risk, that is, the number of boys in the given age group at the time and place under consideration. Specific rates are more difficult to obtain than crude rates, and it is for this reason that crude rates are still used and relied upon in the methodology of epidemiology. An example of the effect of a major epidemic on this rate is shown by the death rate for the United States during the influenza epidemic of 1918. The figure for that year was 18.1 deaths per 1000 population; for the previous year it was 14.0, and for the succeeding year 12.9.

Morbidity rates. There are certain diseases which cannot be studied by a means of mortality rates of any kind. Mental diseases, arthritis, and the common cold are examples of conditions the epidemiology of which is reflected inadequately by death rates. A more useful approach is to consider morbidity rates which give the number of cases of disease per 100,000 population, either as cases existing at a particular point in time or as cases occurring in a particular period of time. Knowledge of these rates for the population in general is inadequate. There are many countries where the reporting of death is virtually 100%; but there is no comparable record by which one can ascertain how many people suffer from disease. It is true that there is compulsory reporting in many countries of all discovered cases of certain infectious diseases, such as smallpox and leprosy; that schools, hospitals, factories, and health insurance groups have records giving useful information about the diseases found in special segments of the population; and that morbidity surveys such as the United States National Health Survey, based on household inquiry by skilled interviewers, have proved useful. But, in general, whenever the information about morbidity covers the whole population or a representative sample of it, the data are limited in detail. Thus the course of the epidemic of Asian influenza in the United States in 1957–1958 can be traced accurately, as to the number of cases involved, from the returns of the National Health Survey, but there is no information on the finer details about the disease, such as can be studied when samples of blood are available for special examinations.

Disease rates. In the investigation of a particular epidemic, ascertainment of disease rates is of great importance. It is necessary to distinguish two main kinds of rates, the one based on the incidence of the disease and the other on its prevalence.

Incidence rate. The incidence rate is concerned with the cases of disease that develop or, more strictly, that come under diagnosis in a given period of time. The incidence rate per 100,000 population for a given year is as shown in the equation here. This rate measures the risk of developing the disease in the given time period; for convenience, it is expressed in terms of so many cases per 100,000 population at risk.

$$\text{Incidence rate} = \frac{\text{number of new cases occurring during year}}{\text{midyear population}} \times 100,000$$

Prevalence. The prevalence ratio is concerned with the cases of disease that exist at a particular point in time. The concept is clearest when the point of time is thought of as being instantaneous; for example, one might ask how many cases of tuberculosis per 100,000 population existed at a specified moment, and this would then be a prevalence ratio for the disease. Prevalence ratios are also computed, especially for chronic diseases, over periods as long as a year. The essential characteristic is still maintained, however, that the count is based on cases existing at some part of the given period, and not necessarily developing within the given period. The prevalence ratio takes in cases that developed prior to the period under consideration, provided the disease persisted until the prevalence count was made.

In the case of diseases which last for several years, such as tuberculosis and diabetes, the prevalence may be much greater than the incidence. In fact, there is some evidence that the modern effective treatment of diabetes, which holds the disease in check without curing it in a fundamental sense, has actually increased the prevalence of the disease. A similar point arose during World War II in connection with the epidemiology of tuberculosis. Since it is well known that in Tennessee the death rate from tuberculosis is higher among the Negroes than the whites, it was expected that in the Selective Service examinations a higher prevalence of the disease would be found in Negroes than in whites. This was not so, presumably because for various reasons the disease ran a longer but less fatal course in the whites.

The distinction between incidence and prevalence is crucial for much epidemiological work. In particular, if one wishes to decide whether a factor is associated with a particular disease, for example, baldness with disease of the coronary arteries, one should make the correlation on cases that arise in a given period of time. The cases of disease that exist at some point in time represent a biased sample of the total, since they include only those who have survived for a period of time, and if survival were better for bald men than for the rest, one would be in danger of concluding that baldness predisposed to coronary artery disease even if, in fact, this were not true.

During the period an incidence rate is being measured, a person may have more than one attack or may suffer from more than one disease. Consequently it is necessary to state whether the unit of measurement is a person, an episode of illness, or a particular disease; the count will vary according to the definition. There is the further obstinate problem of defining when a person is sick. For some purposes this may be done by the subject himself, in terms of the number of days of absence from work. In other instances, especially when long-lasting diseases such as diabetes are involved, periodic examinations may be made by a physician. The problem of the careful definition of cases is important in epidemiological research and this has led to a close cooperation in modern work among physicians, who identify cases by clinical means; laboratory workers, who identify cases by microbiological, biochemical, or other laboratory examinations; and statisticians, who make the comparisons of the disease rates in various subgroups of the population at risk.

Case fatality ratio. In addition to studying the presence or absence of a disease, the epidemiologist often uses some measure of the severity of the disease. One index of severity is the case fatality ratio, which is defined as the number of deaths expressed as a percentage of the number of cases.

Experimental epidemiology. The observational methods so far discussed do not exhaust the possibilities of studying the factors influencing epidemics. The experimental approach has also been used and experimental epidemiology has great potential for improving knowledge of the subject. An epidemic is started by introducing into an animal colony a certain number of infected animals. The course of the subsequent epidemic can then be followed and a study made of the influence of factors such as diet, the percentage of susceptible animals, and the intimacy of the contact between the animals. It is easy to show by such methods, for example, that the introduction of fresh susceptibles into the population from time to time has a marked effect in sustaining an epidemic.

[COLIN WHITE]

Estuarine oceanography

The study of the chemical, physical, biological, and geological properties of estuaries. An estuary is a semienclosed coastal body of water having a free connection with the open sea and within which sea water is measurably diluted with fresh water. Because of similarities in the controlling physical and chemical processes and in the techniques employed in their study, the field of estuarine oceanography is often considered to include investigations of certain coastal water bodies which are not, by strict definition, estuaries. Bays and lagoons in which evaporation is equal to or exceeds freshwater inflow, so that the salt content equals or exceeds that of the adjacent open sea, are generally included in the field of study.

Classification. Estuaries may be classified according to several different criteria; no one system is universally used. A broad classification separates true estuaries, called positive estuaries, in which fresh-water inflow exceeds evaporation, from those embayments, called inverse estuaries, in which evaporation exceeds fresh-water inflow. In this broad classification neutral estuaries are those embayments in which neither evaporation nor fresh-water inflow dominates.

Estuaries are also classified on the basis of geomorphological structure. Coastal-plain estuaries are formed by the drowning of river mouths and the lower reaches of river valleys, either from subsidence or from a rise in sea level. This type of estuary is usually an elongate indenture of the coastline with a river flowing into the upper end. The resulting body of water is typically relatively shallow, sometimes with an irregular (dendritic) shore-line. The eastern coast of North America is characterized by estuaries of this class, such as Delaware Bay, the lower Hudson River, the lower Savannah River, and Chesapeake Bay with its tributary estuaries. These systems are true estuaries; however, in some cases, either nature or man has diverted the inflowing river to a new outlet to the sea, leaving an embayment of the coastal-plain type in which evaporation exceeds fresh-water inflow.

Glacially cut fiords of the Norwegian and Canadian Pacific coasts are representatives of a second geomorphological class of estuaries, called deep-basin estuaries. These elongated indentures of the coastline contain a relatively deep basin with a shallow sill at the mouth.

A third group of estuaries results from the development of an offshore bar on a shoreline of low relief. These bar-built estuaries have a rather narrow connection with the open sea. They are elongated parallel to the coast, rather than being indentures into the coastline. Many bodies of water formed by offshore bars are not true estuaries, as they frequently exhibit an excess of evaporation over fresh-water inflow.

An essential feature of an estuary is the intermixing of sea water with fresh water derived from land drainage. This intermixing produces a variation, both vertically and horizontally, in the salt content of estuarine waters. Oceanographers use the term salinity to designate the concentration of dissolved solids in oceanic and estuarine waters. Salinity is generally expressed in terms of grams of dissolved material per thousand grams of sea water, that is, in parts per thousand ($^{\circ}/_{\circ\circ}$). Salinities in estuaries range from near zero at the head to approximately 30 $^{\circ}/_{\circ\circ}$ at the mouth. There is also generally an increase in salinity with depth. *See* SEA WATER.

Physical structure and circulation. Estuarine circulation refers to the gross features of the pattern of estuarine water movements. Water movements in estuaries result primarily from the interaction of three processes: tides, river flow, and wind. The first two of these are usually dominant, and control the distribution of salinities, though in bar-built estuaries wind-induced motion is often the most important. The physical structure of estuarine waters refers to the vertical and horizontal distribution of density. Since variations in density are related primarily to variations in salt content, any treatment of estuarine circulation logically includes consideration of the physical structure.

Coastal-plain estuaries. Estuaries of this type have a physical structure and circulation pattern controlled primarily by the relative magnitude of oscillatory tidal movements and motion induced by river inflow. The oscillatory tidal velocities moving over the shallow bottom of the estuary produce turbulent mixing of the river inflow and the salt waters from the sea. The lower density of fresh water as compared to that of sea water produces a stable vertical stratification which resists vertical mixing. The relative magnitudes of the tidal flow and the river inflow are thus an important control on the physical structure of the estuary.

Salt-wedge estuary. In coastal-plain estuaries in which the river flow is large compared to the tidal flow, the sea water enters as a salt-water wedge along the bottom. If no frictional drag existed be-

Fig. 1. Freighter stirring up clear Gulf water in South-west Pass of Mississippi River. Muddy river water flows seaward over the denser, upstream flow of Gulf water. (*U.S. Corps of Engineers*)

tween the denser salt-water wedge and the lighter river water, the wedge would extend upstream to the point where the river bottom intersected mean sea level. The river inflow would move seaward over the stationary wedge as a thin fresh-water layer.

In the real fluid, however, some frictional drag always exists. The extent of intrusion of the wedge up the estuary depends upon the magnitude of this frictional drag, which in turn depends upon the relative velocities in the upper, seaward-flowing layer and in the wedge. Thus the volume rate of flow of the river water which moves seaward on top of the salt-water wedge controls the position of the wedge. Under conditions of high river flow, the wedge extends only a short distance into the estuary, while for low river flows the wedge extends many miles upstream. The Mississippi River is an example of this salt-wedge estuary (Fig. 1). When the flow in the Mississippi is low, the undiluted salt wedge extends upstream for over 100 miles. Under conditions of high river flow the salt wedge extends only a mile or so above the mouth of the river.

Comparatively little mixing occurs at the interface between the seaward-flowing upper layer and the salt wedge; hence the salinity throughout the wedge is nearly that of full sea water. At the upper boundary of the wedge, unstable interfacial waves form and break into the upper layers, producing a slowly increasing salt content in these layers as they move seaward. Even so, a very sharp vertical salinity gradient exists between the upper layer of relatively low salinity and the salt wedge. For a given river flow the horizontal position of the wedge and its vertical dimension remain stationary. The loss of salt water from the wedge to the upper layer must be compensated for by flow of water into the wedge from the sea. Since this loss of salt water to the upper layer takes place all along the upper boundary of the wedge, there must be, in order to maintain the position and shape of the wedge, a flow directed toward the upstream tip of the wedge at all positions within the wedge.

Thus the circulation pattern in this salt-wedge estuary involves a seaward-flowing upper layer riding over the landward-directed flow in the salt wedge. The compensating flow in the wedge is small compared to the seaward flow of the low-salinity surface layers.

Partially mixed estuaries. In coastal-plain estuaries in which tidal movements are large compared to the river inflow, vertical mixing is sufficiently strong to destroy the sharp boundaries separating the salt wedge from the upper layer, and the wedge ceases to exist as a readily identifiable feature. There still exists a transition layer of relatively rapid increase in salinity with depth, called the halocline, which separates the low-salinity surface layers from the higher-salinity deeper water. The salinity in both the surface layers and the bottom layers decreases steadily from the mouth to the head of the estuary. Most of the estuaries along the eastern coast of the United States fall into this class of partially mixed estuaries.

The oscillatory tidal currents produce the most obvious motion in these partially mixed estuaries. Superimposed on the tidal currents there is a net circulation pattern, with a net seaward flow in the surface layers, and a net flow directed from the mouth toward the head of the estuary in the deeper layers. There is also a small net vertical motion directed from the deeper layers to the surface layers. The volume of water flowing toward the head of the estuary per unit time decreases as one proceeds from the mouth to the head of the estuary, since water is being transferred through vertical motion from these deeper layers to the surface layers. Hence, the volume rate of seaward flow in the surface layers increases as one proceeds from the head toward the mouth of the estuary.

The flows involved in this net circulation pattern in partially mixed estuaries are many times larger than the fresh-water inflow. This results from the potential energy gained through the tidal mixing. The center of gravity of a column of water, the upper half of which is fresh and the lower half salt water, will be raised when the two halves are mixed together. Thus kinetic energy of the oscillatory tidal motion is transferred through the disorganized motion of turbulence to increased potential energy of the water column, which in turn provides for increased kinetic energy of the net circulation pattern.

Vertically homogeneous estuaries. In this third group of coastal-plain estuaries the tidal movements are very large compared to the motion induced by the inflowing fresh water. The mixing induced by the tidal motion completely overcomes the stability resulting from the fresh-water inflow, producing uniform salinity from surface to bottom. The salinity decreases from the mouth to the head of the estuary. Also, in relatively wide estuaries the salinity on the right side (looking toward the mouth) will be lower than the salinity on the left side, as a result of the effects of the Earth's rotation (Fig. 2).

The circulation pattern in a vertically homogeneous estuary shows no variation in water movement with depth. In relatively wide estuaries, net seaward flow occurs along the right side (looking toward the mouth of the estuary), and a net motion directed toward the head of the estuary occurs on

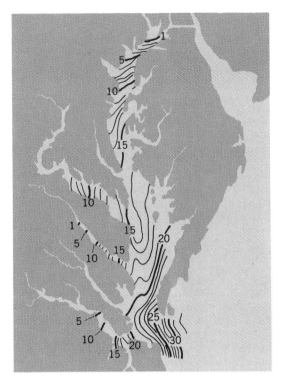

Fig. 2. Typical surface salinity distribution in Chesapeake Bay. (*From D. W. Pritchard, Estuarine Hydrography, in H. E. Landsberg, ed., Advances in Geophysics, vol. 1, Academic Press, 1952*)

the left side. A laterally directed flow carries water from the left side of the estuary to the right, and large-scale horizontal mixing occurs between the counterflows on the two sides of the estuary.

In narrow, vertically homogeneous estuaries tidal mixing may be sufficient also to destroy the lateral salinity gradient. Such an estuary is said to be sectionally homogeneous. The only variation in salinity is the decrease from the mouth to the head of the estuary. The net circulation pattern is quite simple, being a slow seaward movement at all depths.

Deep-basin estuary. The deep-basin, or fiord, type of estuary has a physical structure and circulation pattern similar to those which would be expected if the solid bottom in a partially mixed coastal-plain estuary were replaced by a deep basin filled with sea water. The significant circulation pattern and variations in salinity are restricted to the upper 20 m or so. Below this depth the salinities are nearly constant and approximately the same as the salinity of the corresponding depths, above the depth of the sill, in the adjacent open sea.

Many fiords have rivers flowing into the upper ends. In such estuaries the salinity of the surface layers is low at the head of the fiord and increases steadily toward the mouth. This surface layer is several meters deep as a result of tidal and wind mixing. Below the surface layer a halocline marks the transition layer, in which the salinity increases with depth from the low value found in the surface layer to the high salinity of the deep basin.

Superimposed on the oscillatory tidal motion there exists a net circulation pattern in which the water in the transition zone flows toward the head of the fiord. There is also a slow net vertical movement from the transition zone to the surface layers, such that the volume rate of flow seaward in the surface layers increases in the seaward direction, while the volume rate of flow toward the head of the fiord in the transition layer decreases along the direction of flow. Partial mixing takes place between the surface layer and the transition zone. There is very little motion in the deep waters of the basin.

Bar-built estuary. This class of estuary exhibits no well-established pattern of salinity distribution and circulation. Except at the narrow inlets connecting the sound (as this type of estuary is frequently called) with the open sea, these estuaries are usually so shallow that wind mixing produces vertical homogeneity. Tidal-current velocities through the inlets are relatively large. Because of the narrowness of the inlet, however, the total volume of water which flows in and out with the tide is relatively small, and the tidal rise and fall, as well as the tidal currents, are greatly reduced within the estuary. In the inlets proper there frequently occur a salinity structure and circulation pattern characteristic of a partially mixed estuary. Within the sound the wind produces the most significant water movements.

There must occur a net movement through and out of the estuary just sufficient to remove the fresh water added to the estuary by river inflow, runoff, and direct precipitation. Because of the relatively large cross-sectional area, this net flow is usually not directly measurable within the bar-built estuary, though long-time averages of the flow through the inlets do reveal this net discharge superimposed on the more evident oscillatory tidal motion.

Flushing of estuaries. Estuarine flushing denotes the composite processes whereby the water within the estuary is renewed, both through fresh-water inflow and through exchange with the waters of the adjacent coastal area.

Renewal and flushing time. The renewal time is the time required to replace a stated percentage of the water within the estuary with new water. Thus the 99% renewal time is the time required to replace 99% of the water in the estuary with new water, both through fresh-water inflow and through exchange with sea water.

Theoretically, an infinite time is required to renew all the water in the estuary, the mathematical expression for the process having an exponential form. Thus, the average retention time is not the time required to replace 50% of the water in the estuary, but rather the time required to replace all but $1/e$ of the total volume, or approximately 63% of the water in the estuary. This average retention time has also been called the flushing time of an estuary.

The 50% renewal time, or the time interval required to replace 50% of the water present in the estuary at any time with new water, is called the half-life of the estuary, by analogy with the processes of radioactive decay.

The concepts of renewal time and flushing time can be applied to segments of the estuary as well as to its total volume. When used in this sense, the source of new water for the exchange with any given segment is the next-most-seaward segment of the estuary, rather than the sea water at the mouth of the estuary. Theoretical treatment of estuarine flushing has been successful only for the relatively simple, sectionally homogeneous estuary.

Interest in estuarine flushing has been stimulated by the problems associated with the introduction of pollutants into the estuarine environment, both as a result of man's activity and through natural processes. The overall problem involves the dilution and dispersal of the contaminant within the estuary, as well as its ultimate discharge from the estuary through processes of exchange. Thus estuarine flushing as discussed above includes only part of the problem. There has been a recent tendency to include within the area of study of estuarine flushing the processes of dilution and dispersion of a contaminant within the estuary.

Distribution of pollutants. When a contaminant, such as an industrial or municipal waste, is introduced into an estuary, an initial dilution occurs which is related by the manner of introduction and to the physical properties of the contaminant. The resulting contaminated volume then participates in the oscillatory tidal motion and in the net circulation pattern of the estuary, and is dispersed through the process of turbulent diffusion. Finally, the exchange of waters within the estuary with the open sea leads to the transport of the contaminant out of the estuary.

If the contaminant is in the form of a suspension of particulate material, then the added processes of settling and resuspension must be included. Waste material initially in solution may be adsorbed on the natural suspended silt, or may be taken up by the flora and fauna of the estuary, introducing added processes which must be considered.

The initial density of the introduced contaminant, relative to that of the receiving estuarine waters, greatly influences the initial dilution. The most efficient initial dilution results when the densities of the contaminant and of the receiving estuarine waters are equal. A contaminant with a lower density than the surface waters of the estuary is initially confined to a thin surface lens, and the initial dilution as well as the subsequent diffusion is low. Likewise, poor dilution and slow diffusion result from the introduction of a contaminant with a much higher density than the bottom water, since the waters are initially confined to a thin bottom layer.

Maximum initial dilution of wastes which are less dense than the receiving estuarine waters is achieved when the wastes are introduced near the bottom, since considerable entrainment of surrounding water occurs as the plume of contaminant ascends to the surface. Likewise, wastes that are denser than the receiving estuarine waters are subject to maximum initial dilution if introduced near the surface.

If a contaminant with an initial density equal to, or greater than, that of the receiving waters is introduced into the deep layers of a salt-wedge estuary, the contaminated volume disperses only very slowly within the wedge, since the currents there are weak. The waste is carried by the slow upstream-directed flow to the very tip of the wedge. The process of upward exchange of water from the wedge slowly transfers contaminated water to the seaward-flowing upper layer. Once in the upper layer, the waste is carried seaward with the surface-water layers and transported out of the estuary. The time required to flush the wastes from the estuary in this case is relatively long, since the rate of transfer of the waste from the salt wedge to the surface layers is slow.

A contaminant introduced into the surface layers of a salt-wedge estuary, with an initial density equal to, or less than, that of the receiving waters, is rapidly flushed from the estuary with the seaward-flowing surface waters. None of the contaminated water enters the salt wedge, since exchange across the upper interface of the wedge is essentially one-way, directed from the wedge to the fresher surface layers.

A contaminant initially introduced into the bottom layers in a partially mixed coastal-plain estuary, in addition to participating in the oscillatory movement of the tidal currents, is carried in the net motion toward the head of the estuary. At the same time, turbulent mixing leads to horizontal dispersion in both the longitudinal and lateral directions, and to vertical dispersion into the surface layers. The wastes which become mixed with the surface layers are carried in the net flow toward the mouth. Seaward from the point of introduction, the contaminant being carried toward the ocean in the surface layers is partially mixed downward into the deeper layers, and reintroduced into layers moving toward the head of the estuary.

A contaminant introduced into the surface layers is initially carried in the net flow toward the mouth of the estuary. Turbulent mixing leads to both horizontal and vertical dispersion, and the wastes are thus also added to the deeper layers having a net flow directed towards the head of the estuary.

In the region of the estuary headward from the point of introduction, the concentrations of the contaminant will always be greater in the deeper layers than in the surface layers, while seaward from the point of introduction the converse will be true. These conditions prevail regardless of whether the wastes are initially introduced into the surface layers or into the deeper layers.

The contaminant is ultimately flushed from the estuary in the seaward-directed flow of the surface layers.

Estuarine environments. Estuarine ecological environments are complex and highly variable compared to other marine environments. They are richly productive. A number of commercially important marine forms are indigenous to the estuary, and this environment serves as a spawning or nursery area for other prominent species.

River inflow and land drainage provide the primary source of nutrients required for the production of plant life in the estuary. Relative shallowness coupled with tidal mixing provides for the return of nutrients from decayed organic matter on the bottom to the productive surface layers.

Many estuaries are quite turbid as a result of suspended silt carried into the environment with the river inflow, as well as from shore erosion. The depth to which the solar energy necessary for photosynthesis can penetrate is thus limited to a rela-

Fig. 3. Aerial photograph of tidal flats showing the areas of pans, marshes, and vegetation between the channels, Scott Head Island, England. (*Photograph by J. K. St. Joseph, Crown copyright reserved*)

tively thin surface layer. However, a producing phytoplankton population may be maintained over a much thicker layer as a result of the stirring induced by the tidal motion and by the wind. This stirring keeps the phytoplankton in the layers of adequate light intensity for sufficient time to produce organic matter in excess of utilization. For a more complete treatment of the ecology of estuarine environments from a biological viewpoint *see* MARINE ECOSYSTEM.

The close relationship between the circulation pattern and certain faunal distributions should be pointed out. The net headward flow in the deeper layers of a partially mixed coastal-plain estuary is important in transporting certain larval forms up the estuary. Thus fingerling fish, notably the croaker *Micropogon undulatus*, which are spawned in the coastal waters off the mouth of coastal-plain estuaries along the eastern coast of the United States, are carried into the estuary, where a suitable nursery environment exists. The flow in the bottom layers is also important in transporting oyster larvae from brood-stock areas located in the lower reaches of some estuaries to oyster seed beds further up the estuary.   [D. W. PRITCHARD]

Estuarine sediments. The inorganic sediments of estuaries are derived from inflowing rivers, bordering sea cliffs, the sea floor outside the estuary, and the reworked deposits of tidal flats and marshes along the shores. Regardless of the source, much reworking of sediment occurs within estuaries and lagoons. Erosion, too, is evidenced by the migration of tidal channels and the muddy color of the water when no river inflow is taking place. Some estuaries have entrances narrow enough so that tidal currents scour the bottom locally, leaving rocky or gravelly bottoms. The prevailing condition, however, must be one of deposition, and the average rate of deposition is greater than that of the open sea.

Distribution of grain sizes. The coarsest sediment in most estuaries is on the barrier or bay-mouth bar, and consists of sand and cobbles. Generally, this material is too coarse to have been transported across the tidal flats, but is derived from erosion of a sea cliff, then transported and deposited by longshore currents and waves. The excellent sorting and absence of much silt and clay may result from the turbulence of the waves.

The flat portions of the floors of estuaries that are deeper than about 18 ft are usually covered by sediment which becomes progressively finer with depth of water. A smooth concentric pattern of sediments may occur, ranging from sand along the shore to fine mud at depth. Such a distribution occurs only where the bottom is relatively flat and wave and current conditions are mild. In estuaries where the deeper areas are extremely irregular, mud occurs only in depressions, and coarse sediments characterize the shallower bottoms.

The sediment distribution in shallow areas, mostly the tidal flats, is more complex but usually follows a systematic pattern. Most of the flow of water is confined to well-defined channels which slowly migrate over the tidal flat (as shown by remapping at intervals of several years). The velocity of the water is such that the finer grains are swept out, leaving the coarse sediment in the channel. The areas between the channels consist of poorly sorted mud, which becomes finer with distance from the tidal channels. Probably most, but not all, of the reworking of sediment in estuaries takes place on the shallow flats where the ebbing and flooding currents erode and redeposit the sediment (Fig. 3).

Organic constituents. The sediments contain the remains of all phyla of animals and much plant debris. Even though the remains become scattered by scavenging, decomposition, and diagenesis, the organisms still have enriched the sediments in organic matter, calcium carbonate, silica, nutrients, and other constituents.

Sediments in estuaries located in areas where precipitation exceeds evaporation have organic nitrogen contents from 0.2 to 0.6%, and sediments in hypersaline areas, below 0.2%. The percentage used to differentiate between these areas, or 0.2% organic nitrogen, corresponds to about 1.7% organic carbon, or 2.9% total organic matter. Phosphorus is also abundant in sediments of normal environments, ranging from 0.1 to 0.4%.

Calcium carbonate is variable because of the presence or absence of shells and because of solution induced by acidic conditions. In coastal bays in temperate and arctic regions, calcium carbonate ranges between 0 and 6%, whereas in bays of tropical regions it is 10 to 47%.

Manner of deposition. In the tidal section of a fresh-water river a transition takes place, and the distribution of sediments may be quite variable and confused. When the estuary proper is reached there is some admixture of sea salts, and where the net upstream flow in lower layers occurs, there is a distinct change in sediment distribution. Finer sediments tend to be deposited in the channel (the reverse of conditions commonly found in river channels). In most streams, the bulk of suspended material probably is silt which is deposited directly out of suspension. Clay sizes, however, may be deposited through flocculation. The clays then fall to the deeper floors of the estuaries. Sediment may also travel down rivers at or near the surface in large floating floccules containing organic debris. When these settle to the bottom or are stranded by lowering water level, they are held by capillary action. Near the mouth of the estuary coarser sediments are again found in the channel as a result of wave action and because much of the silt load has already been deposited in the channel farther upstream.

[ROBERT E. STEVENSON]

Bibliography: A. B. Arons and H. Stommel, A mixing-length theory of tidal flushing, *Trans. Amer. Geophys. Union*, 32(3):419–421, 1951; K. O. Emery and R. E. Stevenson, *Estuaries and Lagoons*, Geol. Soc. Amer. Mem. no. 67, 1:673–750, 1957; M. N. Hill (ed.), *The Sea, Ideas and Observations on Progress in the Study of the Seas*, vol. 2, 1963; H. E. Landsberg (ed.), *Advances in Geophysics*, vol. 1, 1952; G. H. Lauff (ed.), *Estuaries*, Amer. Ass. Advan. Sci. Publ. no. 83, 1967; G. L. Pickard, *Descriptive Physical Oceanography*, 1964; D. W. Pritchard, Estuarine circulation patterns, *ASCE Proc.*, 81(717), 1955; G. K. Reid, *Ecology of Inland Waters and Estuaries*, 1961.

Eutrophication, cultural

The deterioration of the esthetic and life-supporting qualities of natural and man-made lakes and estuaries, caused by excessive fertilization from effluents high in phosphorus, nitrogen, and organic growth substances. Algae and aquatic plants become excessive, and, when they decompose, a sequence of objectionable features arise. Water for consumption from such lakes must be filtered and treated. Diversion of sewage, better utilization of manure, erosion control, improved sewage treatment and harvesting of the surplus aquatic crops alleviate the symptoms. Prompt public action is essential. *See* WATER CONSERVATION.

Extent of problem. In inland lakes this problem is due in large part to excessive but inadvertent introduction of domestic and industrial wastes, runoff from fertilized agricultural and urban areas, precipitation, and groundwaters. The interaction of the natural process with the artificial disturbance caused by the activities of man complicates the overall problem and leads to an accelerated rate of deterioration in lakes. Since population increase necessitates an expanded utilization of lakes and streams, cultural eutrophication has become one of the major water resource problems in the United States and throughout the world. A more thorough understanding must be obtained of the processes involved. Without this understanding and the subsequent development of methods of control, the possibility of losing many of the desirable qualities and beneficial properties of lakes and streams is great.

Cultural eutrophication is reflected in changes in species composition, population sizes, and productivity in groups of organisms throughout the aquatic ecosystem. Thus the biological changes caused by excessive fertilization are of considerable interest from both the practical and academic viewpoints. *See* FRESH-WATER ECOSYSTEM; LAKE.

Phytoplankton. One of the primary responses to eutrophication is apparent in the phytoplankton, or suspended algae, in lakes. The nature of this response can be examined by comparing communities in disturbed and undisturbed lakes or by following changes in the community over a period of years during which nutrient input is increased.

The former approach was utilized in studies at the University of Wisconsin, in which the overall structure of the phytoplankton communities of the eutrophic Lake Mendota, at Madison, and the oligotrophic Trout Lake, in northern Wisconsin, was analyzed. These investigations showed that in the eutrophic lake the population of species is slightly lower, although the average size of organisms is considerably larger, indicating higher levels of production than in the oligotrophic lake. When compared in terms of an index of species diversity, the community of the eutrophic lake displayed values lower than those observed in the oligotrophic lake. Seasonal changes and bathymetric differences in the index of diversity were also more apparent in the eutrophic lake.

Often the low species diversity of the phytoplankton in eutrophic lakes is a result of high populations of blue-green algae, such as *Aphanizomenon flosaquae* and *Anabaena spiroides* in Lake Mendota. Frequently, however, species of diatoms such as *Fragillaria crotonensis* and *Stephanodiscus astrae* also attain high degrees of dominance in the community. W. T. Edmondson reported that dense populations of the blue-green algae *Oscillatoria rubescens* were indicative of deteriorating conditions in Lake Washington. The same species has been observed in several European lakes that have undergone varying degrees of cultural eutrophication. The relatively high nutrient concentrations in eutrophic waters appear to be capitalized on by one or two species that outcompete other species and periodically develop extremely high population levels. Because of the formation of gas vacuoles during metabolism, senescent forms of the blue-green algae rise to the surface of the lake, causing nuisance blooms.

In addition to nuisance scums in the pelagial, or open-water, regions, the rooted aquatic plants and the attached algae of the littoral, or shoreward, region often prove to be equally troublesome in eutrophic lakes. Species of macrophytes, such as *Myriophyllum* and *Ceratophyllum*, and algal forms, such as *Cladophora*, frequently form dense mats of vegetation, making such areas unsuited for both practical and recreational uses.

Bottom fauna. Often in eutrophic lakes the bottom fauna display characteristics similar to those observed in the algal community. Changing environmental conditions appear to allow one or two species to attain high degrees of dominance in the community. Generally higher levels of production are associated with the change in structure of the community—the result being nuisance populations of organisms. In Lake Winnebago, in Wisconsin, for example, the lake fly or midge *Chironomus plumosus* develops extremely high populations, which as adults create an esthetic as well as an economic problem in nearby cities.

Great Lakes. It was originally thought that eutrophication would not be a major problem in large lakes because of the vast diluting effect of their size. However, evidence is accumulating that indicates eutrophication is occurring in the lower Great Lakes. Furthermore, the undesirable changes in the biota appear to have been initiated in relatively recent years. Charles C. Davis, utilizing long-term records from Lake Erie, has observed both qualitative and quantitative changes in the phytoplankton of that large body of water owing to cultural eutrophication. Total numbers of phytoplankton have increased more than threefold since 1920, while the dominant genera have changed from *Asterionella* and *Synedra* to *Melosira, Fragillaria,* and *Stephanodiscus.*

Other biological changes usually associated with the eutrophication process in small lakes have also been observed in the Great Lakes. Alfred M. Beeton has summarized the literature pertaining to the trophic status of the Great Lakes in terms of their biological and physicochemical characteristics and indicated that, of the five lakes, Lake Erie has undergone the most noticeable changes due to eutrophication. In terms of annual harvests, commercially valuable species of fish, such as the lake herring or cisco, sauger, walleye, and blue pike, have been replaced by less desirable species, such as the freshwater drum or sheepshead, carp, and smelt. Similarly, in the organisms living in the bottom sediments of Lake Erie, drastic changes in species composition have been observed. Where formerly the mayfly nymph *Hexagenia* was abundant to the extent of 500 organisms per square meter, it presently occurs at levels of 5 and less per square meter. Chironomid midges and tubificid worms now are dominant members in this community.

Oxygen demand. It is apparent that the increase in organic matter production by the algae and plants in a lake undergoing eutrophication has ramifications throughout the aquatic ecosystem. Greater demand is placed on the dissolved oxygen in the water as the organic matter decomposes at the termination of life cycles. Because of this process, the deeper waters in the lake may become entirely depleted of oxygen, thereby destroying fish habitats and leading to the elimination of desirable species. The settling of particulate organic matter from the upper, productive layers changes the character of the bottom muds, also leading to the replacement of certain species by less desirable organisms. Of great importance is the fact that nutrients inadvertently introduced to a lake are for the most part trapped there and recycled in accelerated biological processes. Consequently, the damage done to a lake in a relatively short time requires a many-fold increase in time for recovery of the lake.

Action programs and future studies. Lake eutrophication represents a complex interaction of biological, physical, and chemical processes. The problem, therefore, necessitates basic research in a wide variety of scientific disciplines. Moreover, to be profitable, such research must be coordinated into well-integrated team-research efforts requiring extensive monetary support.

Studies are needed in such areas as the identification of those nutrients that reach critical levels in lake waters and lead to the development of nuisance growths of plants and algae. Nitrogen and phosphorus are undoubtedly important; however, biologically active substances such as vitamin B_{12} and other organic growth factors may also play an important role.

In Lake Mendota sewage effluent is the major contributor of nitrogen and phosphorus to lakes,

followed by runoff from manured and fertilized land. Considering nitrogen alone, rain adds more than any single source. Its nitrogen content comes from combustion engines and smokestacks.

Where sewage effluent has been diverted from lakes (as at Lakes Monona, Waubesa, and Kegonsa in Wisconsin, Lake Washington near Seattle, and two lakes in Germany), an improvement in nuisance conditions occurs. Hence this treatment is the first step in alleviation.

The conditions observed in lakes and streams reflect not only the processes operating within the body of water but also the metabolism and dynamics of the entire watershed or drainage basin. After precise identification of the critical nutrient compounds, it is necessary to determine the nutrient budget of the whole drainage basin before action can be taken to alleviate the undesirable fertilization of a lake or stream.

New methods of treating sewage plant effluent are being explored, and further support for these efforts is justified. For example, the complete evaporation of effluent to a powder would alleviate to a great extent the present problem of dealing with large volumes of these materials. Such methods will be expensive, but this may be inevitable for some pollution problems. *See* SEWAGE TREATMENT.

It is known that aquatic plants concentrate nutrients from the lake waters in their tissues. The removal or harvesting of aquatic plants in eutrophic lakes, consequently, is a good potential method for reducing nutrient levels in these lakes. Similarly, significant amounts of nitrogen and phosphorus are concentrated in fish flesh. Efficient methods of harvesting these organisms are important in impoverishing a well-fertilized lake.

Utilization or land disposal of farm manure is a major problem. Animal manures are largely unsewered, yet in the Midwest it is equivalent to the sewage of 350,000,000 people.

In addition to improvements in waste disposal, more research and development are needed on the profitable utilization of surplus algae, aquatic plants, fish, manure, and sewage. The over-fertilized lake needs to be impoverished of its nutrients as well as protected from inflowing sources. Chemicals have been used to poison the plants and algae, but this is not a good conservation practice because the plants and algae rot and provide more nutrients. Moreover, eventual harm to other species has not been assessed.

It would seem desirable to set aside certain lake areas for research purposes. More information is needed to decide upon the best plans for allowing the domestic development of these areas with the least disturbance to the water resources. Steps will be necessary to devise optimum zoning laws and multiple-use programs in light of the intense economic and recreational uses made of water resources. Legislation and law enforcement in relation to public interactions undoubtedly will be a complex problem to overcome in this respect.

Cultural eutrophication is a paradoxical condition, since it is in large part due to the economic and recreational activities of man and at the same time eventually conflicts with these same activities of society in general.

[ARTHUR D. HASLER]

Bibliography: D. G. Frey and F. E. J. Fry (eds.), *Fundamentals of Limnology*, 1966; A. D. Hasler, Eutrophication of lakes by domestic drainage, *Ecology*, 28(4):383–395, 1947; A. D. Hasler and B. Ingersoll, Dwindling lakes, *Natural History*, Nov. 1968; A. D. Hasler and M. E. Swenson, Eutrophication, *Science*, 158(3798):278–282, 1967; G. E. Hutchinson, *A Treatise on Limnology*, vol. 1: *Geography, Physics, and Chemistry*, 1957, and vol. 2: *Introduction to Lake Biology and the Limnoplankton*, 1967; G. A. Rohlich and K. M. Stewart, *Eutrophication: A Review*, Water Qual. Contr. Board Calif. Publ. no. 34, 1967.

Evapotranspiration

A term applied to the discharge of water from land surfaces to the atmosphere by evaporation from lakes, streams, and soil surfaces and by transpiration from plants. The term is applied both to the process and to the quantity of water discharged. On the average, two-thirds of the precipitation is returned to the atmosphere by evapotranspiration, but this ratio varies from nearly 100% in deserts to one-third or less where precipitation is high and evaporation relatively small.

Evaporation and transpiration are considered together because of the great difficulty of measuring or estimating them separately. Transpiration is the process by which water is taken in by plant roots, moved up through the stem or trunk, and released as vapor through the leaves. The maximum possible evapotranspiration, termed potential evapotranspiration, is governed by the available heat energy and is taken to equal the evaporation from a large water surface. Actual evapotranspiration is determined by available moisture and is generally much less than the potential evapotranspiration. Actual evapotranspiration is never greater than precipitation except on irrigated land. For a river system actual evapotranspiration can be estimated approximately as the precipitation over the tributary area less the streamflow leaving the basin. This approximation ignores retention of water in the groundwater or in reservoirs and diversions from the basin. *See* ATMOSPHERIC EVAPORATION.

Local estimates of actual evapotranspiration are made with lysimeters, large tanks of soil with growing plants, for which the gain or loss of water can be determined by weighing or accounting for water supplied to the tank.

Phreatophytes. Evaporation can remove water from only a relatively thin surface layer of soil, but transpiration removes water from the entire root depth of plants. Phreatophytes obtain water from the groundwater or the capillary fringe above the water table and have root systems extending to great depths, for example, mesquite, 60 ft, and alfalfa, 100 ft. In addition to mesquite and alfalfa, typical phreatophytes include salt grass, salt cedar (tamarisk), willow, cottonwood, and greasewood. It has been estimated that 25,000,000 acre-ft (1 acre-ft = 325,400 gal) of water is transpired annually by nonbeneficial phreatophytes in the 17 western states of the United States.

Xerophytes and mesophytes. Xerophytes are plants of desert regions and are adapted to survive for long periods without moisture. However, when moisture is available, they transpire at the same rates as mesophytes, plants of humid regions, that require nearly continuous water supply. *See* PLANTS, LIFE FORMS OF.   [RAY K. LINSLEY]

Bibliography: V. T. Chow, *Handbook of Applied Hydrology*, 1964; R. K. Linsley, M. A. Kohler, and J. L. H. Paulhus, *Hydrology for Engineers*, 1958; R. C. Ward, *Principles of Hydrology*, 1967.

Faunal extinction

The death and disappearance throughout the world of diverse groups of organisms under circumstances that suggest common or related causes. This phenomenon is also known as mass extinction. The time span involved in mass extinctions varies greatly, but commonly it is short compared with evolutionary history. It may, however, seem to be long as compared with human records, and there is no scientific evidence that such mass extinctions were cataclysmic or involved extraordinary processes.

One of the greatest of all mass extinctions started a few hundred years ago and is still in progress. Approximately 500 species of animals and many plants have disappeared in that time or may soon become extinct as the result of ecological disturbances caused by man. This revolution in the organic world is expected to become climactic within a few decades because of the rapid spread of human population.

Evidence from fossils. Mass extinctions of past eras are recorded in hundreds of successive assemblage zones of fossil animals, many of which contain widespread and abundant species and genera that first appear together at one level and disappear abruptly at a higher stratigraphic level without any, or only a few, known descendants.

The most critical episodes of faunal extinction are recorded at the close of the Cambrian (52% of all families of the time), Devonian (30% of families), Permian (50% of families), Triassic (35% of families), and Cretaceous (26% of families) (Fig. 1). These, and many lesser episodes of mass extinction, are used by biostratigraphers as a basis for delimiting many major and minor stratigraphic divisions.

Patterns. Mass extinctions are of two sorts. They may take place as a result of competition between native and immigrant faunas, or they may stem from external physical changes in environment in which competition is not a major factor. In the former case, the immediate cause of extinction stems from biological stresses between better- and poorer-adapted species that are using the same environmental facilities. Portions of an invading fauna replace the native animals in a sort of relay.

Within historical times there have been many well-documented examples of such competition and replacement of indigenous species by aggressive elements from other regions. For example, the terrestrial biotas of Australia, Madagascar, the Antilles, New Zealand, innumerable smaller islands, and even the major continents have been greatly modified because of competition with species introduced by man and consequent extinctions of native species.

Natural migrations of animals in the past have also resulted in mass extinctions. The Isthmus of Panama rose out of the sea in the Pliocene Epoch, some 6,000,000 or 7,000,000 years ago, forming a land connection between the Americas and an intermingling of faunal elements of the previously separate continents. Many species, especially the marsupial mammals of South America, were decimated by unequal competition.

In a second pattern of mass extinction there is little or no intermingling of old and new faunas. In this case, many niches may be vacated for a time until new animals eventually replace those that had died out before. Evidently, invasion of a region by a new fauna is not the immediate cause of the elimination of the native species. By implication, physical factors of the environment are responsible.

The Paleocene Epoch (of some 11,000,000 years' duration) elapsed before many of the niches vacated by the vanished dinosaurs were reoccupied by the arising mammals. The pattern of such episodes is rapid extinction followed by an interval of recovery and subsequent replacement by new communities (Fig. 1) that were not in intimate contact with the vanished forms.

Predation. The disappearance of many kinds of large herbivorous mammals in the Americas (Fig. 2) about 8000–14,000 years ago, shortly after the advent of man in the Western world, has been attributed to early human hunters.

This hypothesis stresses the fact that man frequently kills animals for motives other than hunger—overkill. Several authorities, however, question the premise that human populations were sufficiently dense and widely distributed or techni-

cally so advanced 10,000 years ago as to threaten the existence of widespread abundant animals such as the horses, which then ranged in great numbers over most of the Americas. These persons point to the recently increasing harshness and drying of climate, especially in intermediate latitudes, as a more likely cause of this extinction.

Mechanics. In simple terms, extinction results from an excess of deaths over births over much of the geographic range of a species. The equation $dn/dt = an - bn$ indicates the rate of change in population density through time, where n represents the number of individuals per unit area (that is, population density); t, time interval; a, birth rate; b, death rate. The values of a and b commonly depend on both population density and environmental conditions. When n falls below a critical value, population variability may be significantly reduced, breeding patterns disturbed, and a threshold passed below which the species cannot longer cope with the ordinary vicissitudes of the environment, and extinction follows. On the other hand, if a population irrupts and the density exceeds a critical level, irreversible damage to the environment may occur, and intraspecific competition increases greatly. These changes may be followed by rapid reduction in population numbers, and extinction. Thus, there seems to be an optimal population density, or climax, which varies according to the carrying capacity of the environment. *See* CLIMAX COMMUNITY.

Causes. Many ingenious hypotheses, ranging

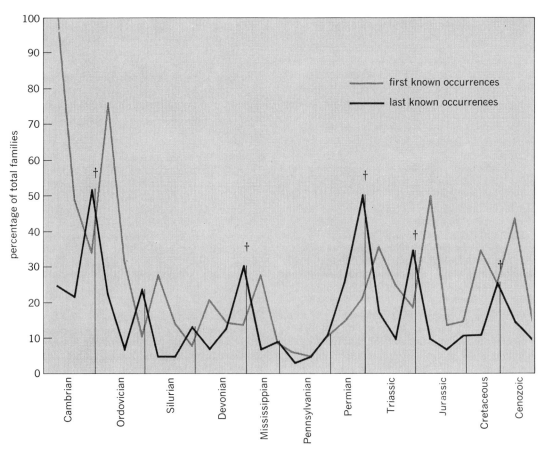

Fig. 1.   Episodic fossil record of animals. (*N. D. Newell, Geol. Soc. Amer. Spec. Pap., no. 89, pp. 63–91, 1967*)

from unsupported speculation to sophisticated scientific arguments, have been advanced to explain past episodes of mass extinction. Commonly, these are specially designed for particular examples, and most are unsatisfactory because they do not suggest significant tests or corollaries needed to assess probabilities for a particular hypothesis of extinction.

Published hypotheses include meteorite storms, diseases, shifts of the Earth's axis or crust, variations in composition of the atmosphere, variations in supply of metallic trace elements, reduction of salinity in the ocean, racial old age, insufficient nutrient salts in the sea, excessive juvenile mortality, and climatic changes. All of these hypotheses have been advanced as causes of mass extinctions, but mostly they are not supported by compelling evidence.

Ionizing radiation. An idea currently popular and repeated in variant forms since the 1930s is that mass extinctions and subsequent accelerated evolution of faunas might be attributed to bursts of ionizing radiation from the Sun or supernovae. K. Terry and W. Tucker estimated the frequency of such explosions to be about once in 60,000,000 years.

R. Uffen suggested, in 1963, that each of many reversals of polarity of the Earth's magnetic field recorded in remanent magnetism of stratified rocks must have been accompanied by temporary dissipation of the protective Van Allen belt, thus allowing floods of lethal and mutagenic radiation to reach the surface of the Earth and cause mass extinctions of the most sensitive animals. Most of the proponents of this hypothesis have postulated accelerated genetic mutations among the survivors to explain the appearance of mainly new faunas after the major episodes of extinction.

Apparent confirmation of this argument came with announcement, in 1966, by N. D. Opdyke, B. Glass, J. D. Hays, and J. Foster that magnetic reversals in Antarctic deep-sea cores of Pleistocene sediments lie at approximately the same stratigraphic levels as extinctions of fossil plankton.

This interesting hypothesis needs critical reappraisal. It now appears that the evidence of faunal extinctions generally does not correspond as closely to times of magnetic reversal as originally thought. Furthermore, although aquatic animals would have been relatively free from deleterious effects of excessive radiation, fossil marine faunas show about the same extinction effects as terrestrial animals. Since evolutionary rate is controlled by natural selection and is not likely to be much influenced by mutation rate, it is not considered a reliable clue to episodes of excessive radiation.

C. Waddington argues that the shielding effect of the Earth's atmosphere is so great that a radiation dose at sea level during a magnetic reversal would be negligible.

Geologic factors. Historical geology has demonstrated that physical environments have been in flux since the origin of the Earth. Of these changes, no single set of factors has influenced the later history of life more than the continuous modifications of topographic relief and distribution of continents and ocean basins and the attendant climatic changes. The resulting disturbances to ecosystems

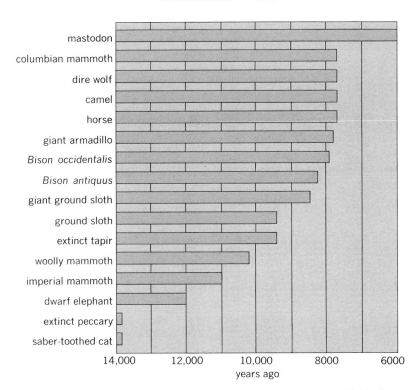

Fig. 2. Recent mass extinction of large mammals in North America. (*K. MacGowan and J. A. Hester, Early Man in the New World, Doubleday, 1962*)

at times must have been very great. During intervals of very low crustal relief, the changes probably were also rapid as compared with organic evolution. The stratigraphic levels marked by mass extinctions commonly coincide with episodes of sweeping geographic changes and argue for a causal, if complex, relationship between the two. Scientific investigations of these relationships in the future may provide definite conclusions about the causes of mass extinctions.

[NORMAN D. NEWELL]

Bibliography: P. S. Martin and H. E. Wright, Jr., *Pleistocene Extinctions*, 1968; N. D. Newell, *Geol. Soc. Amer., Spec. Pap.*, no. 89, pp. 63–91, 1967; N. D. Opdyke et al., Paleomagnetic study of Antarctic deep-sea cores, *Science*, 154:349–357, 1966; K. D. Terry and W. H. Tucker, Biological effects of supernovae, *Science*, 159:421–423, 1968; C. J. Waddington, Paleomagnetic field reversals and cosmic radiation, *Science*, 158:913–915, 1967.

Fertilizer

Materials added to the soil to supply elements needed for plant nutrition. This article discusses fertilizer types and the effect that overuse and misuse of farm fertilizers has on surface and underground water quality.

Fertilizers may be products manufactured for the purpose, by-products from other chemical manufacturing operations, or by-product natural materials. By-products, particularly natural organic materials, were important nutrient sources in the early days of the industry. The growing need for fertilizers, however, has outstripped the supply of by-products. Today manufactured materials are the major type by far and by-products have only

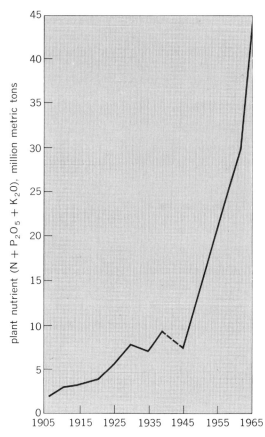

World fertilizer consumption for fiscal years 1906–1965. (*Food and Agriculture Organization*)

minor significance.

The fertilizer industry once was mainly a simple materials-handling and mixing technology. However, it has become a major segment of the chemical industry, with giant plants embodying advanced developments in chemical engineering producing very large quantities of fertilizer. Several factors have contributed to its change in status: a maturing in agricultural practice in developed countries, rapidly increasing use of fertilizer in countries that had used little before, and a growing realization that massive production of fertilizer is the first line of defense against the problems of growing populations. The increase in world fertilizer consumption is indicated in the illustration.

Fertilizer types. The nutrient elements cannot be supplied to plants as such; they must be combined with other elements in the form of suitable compounds. Phosphorus, for example, is toxic to plants in the elemental form but is a good fertilizer when combined with oxygen and ammonia to form ammonium phosphate. For each of the elements there are several compounds that can be used. The choice between them, in most cases, is based on economic considerations. The material is used that gives the lowest cost per unit of nutrient applied to the soil.

Nitrogen, phosphorus, and potassium are called the macronutrients because of the relatively large amounts needed. The usual requirement range per acre per cropping season is 50–200 lb of nitrogen, 10–40 lb of phosphorus, and 30–150 lb of potassium.

Nitrogen is supplied in the ammoniacal (NH_4^+) or nitrate (NO_3^-) form or as urea, $(NH_2)_2CO$. Except in unusual situations, one form is as good as the other because in the soil the ammoniacal and urea forms are rapidly converted to nitrate. The principal fertilizer nitrogen materials are ammonium nitrate, urea, ammonia, ammonium sulfate, and ammonium phosphate.

Phosphorus is supplied mainly as calcium phosphates and ammonium phosphates. Calcium phosphate is the older type and is still the leader. Ammonium phosphates have certain advantages, however, that have given them a major place in the industry. The main differences between the two types are in nutrient concentration and solubility; ammonium phosphates generally are higher in both categories, which favors them in regard to handling and shipping cost and, in some instances, in agronomic value.

Potassium is supplied almost entirely as potassium chloride, a mineral widely available in various parts of the world and usable without any further treatment.

Calcium and magnesium are essential elements, but neither is considered part as a fertilizer industry product because ample quantities are supplied incidentally in macronutrient fertilizers, for example, calcium in calcium phosphates, or from soil minerals. Sulfur was once in this category also but is becoming a fertilizer material because in some areas deficiencies have developed that are not met by incidental or natural means. These three elements are called secondary nutrients because of the intermediate quantities needed per acre per cropping season: 5–25 lb of calcium, 3–30 lb of magnesium, and 5–50 lb of sulfur.

The remaining essential elements are called micronutrients; the amounts required are very small, from a trace to 1 lb/acre. They are usually supplied as cations or anions of salts such as borates or sulfates.

The nutrient amounts given refer to the quantity taken up from the soil by various crops. This does not mean necessarily that such an amount added to the soil will supply the plant adequately; there are many ways in which nutrient can be lost before the plant can take it up. Moreover, some nutrient is obtained from materials already present in the soil. Therefore, the amount of fertilizer actually applied depends on several factors that vary widely. Beyond the factor of plant need, a major consideration is the financial ability and technological status of the farmer; these are major factors in the very wide variation in fertilizer application per acre in various parts of the world. Highly developed and crowded countries use the most; in Japan, for example, average macronutrient application is over 180 lb/acre. In contrast, less-developed areas, such as India, use as little as 2 lb/acre.

Nitrogen. This nutrient is required by plants in the largest quantity. Its use is expanding more rapidly than for any of the other nutrients. It has been found that, if adequate amounts of phosphorus, potash, and other nutrients are applied, very large amounts of nitrogen, coupled with high plant population per acre, give yields much higher than those attainable under previous fertilization practice. In 1956 world consumption was higher for phosphate and potash than for nitrogen (tonnage is on the

Table 1. Fertilizer nitrogen supplied by various materials

Material	Nitrogen content, %	Supply, 1000 metric tons of nitrogen*	
		U.S.	World
Ammonium nitrate	33.5	610	3878
Ammonium sulfate	20.5	326	2590
Complex fertilizers	—†	373	1519
Urea	45	297	1453
Anhydrous ammonia (used as such)	82	1148	—
Ammoniated superphosphate	—†	790	—
Nitrogen solutions (used as such)	—†	668	—
Calcium nitrate	15.5	5	430
Sodium nitrate	16	44	231

*In 1964–1965.   †Varies.

oxide basis, P_2O_5 and K_2O, the common reporting method in the industry). But by 1961 nitrogen had passed both and continues to climb rapidly; in 1966–1967 the estimated world use of fertilizer nitrogen was 21,800,000 metric tons.

The fertilizer nitrogen supply is divided between several materials; the approximate division is shown in Table 1.

Ammonia. The basic nitrogen fertilizer material is ammonia, used both as a primary fertilizer and as the starting material for all the other leading nitrogen fertilizers. Ammonia is also an important general chemical, but its principal use (almost 85% of the total on a worldwide basis) is in the fertilizer industry. In 1967 there were about 98 plants in the United States, with a production capacity of more than 13,000,000 tons per year.

In past decades the fertilizer industry used large quantities of natural nitrogen materials, such as animal by-products, ammonia recovered from coal during coke production, and sodium nitrate from localized natural deposits in Chile. Today these sources are inadequate or too expensive; elemental nitrogen, present in inexhaustible quantities in the atmosphere, is the major source by far, accounting for more than 95% of the fertilizer nitrogen used in the United States.

Ammonia production technology changed rapidly in the 1960s. Natural gas became firmly established as the major feedstock, particularly after discovery and development of gas bodies in Europe and Japan. The use of elevated pressure in reforming, that is, reaction of gas with steam to give synthesis gas (hydrogen plus carbon monoxide), added considerable economy to the process. Also, low-pressure shift conversion, that is, reaction of synthesis gas with steam to eliminate carbon monoxide and produce more hydrogen, was widely adopted.

After purification the hydrogen is combined with nitrogen (introduced earlier in the process as air) under pressure (2000 psig or more) and high temperature (425–650°C) in the presence of promoted iron catalyst to form ammonia.

The synthesis step in ammonia production has undergone major changes. Centrifugal compressors, used to compress the nitrogen-hydrogen mixture to synthesis pressure, have reduced plant cost. Single converters produce up to 1500 tons of ammonia per day, and converter design has been simplified to make the large size feasible.

Use of ammonia directly as a fertilizer requires use of pressure equipment since ammonia is a gas at atmospheric pressure. The material is carried in pressure tanks on applicators and injected into the soil through injector knives. Care is necessary to avoid loss by volatilization, and with deep placement, the ammonia is quickly sorbed by the clay and other constituents of the soil.

In 1958 ammonia became the leading nitrogen fertilizer in the United States. Use in other parts of the world, although growing, is relatively small.

Ammonium nitrate. The most important nitrogen fertilizer, from the standpoint of world consumption, is ammonium nitrate. Large tonnages are used directly as a fertilizer, and considerable amounts go into mixtures with other materials. The material plays the somewhat remarkable double role of fertilizer and explosive; in fact, several plants built to make ammonium nitrate for munitions use in World War II now produce the material as fertilizer. The explosive tendency has been somewhat of a problem in handling and shipping for fertilizer use. In modern practice, however, the hazard has been adequately controlled.

Ammonia and nitric acid are the raw materials for ammonium nitrate production. Since nitric acid is made by oxidizing ammonia with air, the only basic raw material needed is ammonia. This gives ammonium nitrate some advantage because other major nitrogen fertilizers require some raw material in addition to ammonia.

Nitric acid plants differ mainly in the pressure used at various stages of the process. Ammonia is oxidized catalytically with air by passing the mixture through a platinum catalyst screen at high temperature, after which the product gases are cooled and then passed through a scrubber, where water absorbs the nitrogen oxides to form nitric acid. The absorption and sometimes the oxidation are carried out at pressures up to about 100 psig.

Ammonium nitrate is made by neutralizing nitric acid with ammonia, usually in pressure-type neutralizers so that steam can be recovered from the highly exothermic reaction. The neutralized solution is then concentrated to a melt of low moisture content in an evaporator. Practice differs in the method used for converting the melt to the final solid product. The various types of crystallizers used formerly have been replaced by prilling towers, rotary drums, and pan granulators. Prilling is the most popular method, but the others show promise. In all, the objective is to make agglomerates of fairly large size, on the order of 1/16 to 1/8 in.

Finally, ammonium nitrate is usually conditioned to reduce moisture absorption and caking. The main practice is coating with up to about 2% of a parting agent, such as clay.

Ammonium sulfate. Until 1959, when ammonium nitrate moved ahead, ammonium sulfate was the leading nitrogen fertilizer in the world. One of the major reasons for its drop in popularity is low nutrient concentration, only 20–21% nitrogen as compared with 33.5% for ammonium nitrate.

Much of the ammonium sulfate used is by-product material from coke manufacture (by scrubbing ammonia out of the oven gases with sulfuric acid), or from production of caprolactam. Coke ovens, for example, supply about 750,000 tons of ammonium sulfate annually in the United States. A considerable tonnage is made also by reacting ammonia with sulfuric acid, which often is a by-product from petroleum refining or organic syntheses.

Manufacture of ammonium sulfate is relatively

simple. Solution, either by-product or made by neutralizing sulfuric acid, is concentrated and then crystallized in one of several crystallization processes.

Urea. In third place among world nitrogen fertilizers and growing rapidly in consumption, urea is the highest in concentration (45% nitrogen) of any of the major types. With growing costs for handling and shipping, the high concentration is assuming more and more importance.

Some urea is used in chemical industry, mainly in plastics production and in animal feeds, but the major proportion by far goes into fertilizers, either for direct application or as a constituent in mixes.

Urea is made by reacting ammonia and carbon dioxide under pressure and at elevated temperature but without a catalyst. The carbon dioxide is readily available because it is a by-product, usually wasted, in manufacture of ammonia from carbonaceous materials. Thus there is little or no cost for the carbon dioxide, but the urea plant must be built close to an ammonia plant. *See* UREA.

The urea reaction does not go to completion; the product from the reactor is a solution of urea and ammonium carbamate, an unstable compound of ammonia, water, and carbon dioxide. The solution must be heated to decompose the carbamate and strip out the resulting gases, which are then recycled back to the reactor. The urea solution is finished in much the same way as ammonium nitrate, usually by prilling.

Processes differ mainly in the method for recycling the unreacted gases. There has been a shift in practice from methods that involve separation of the gases and recycling separately to absorption in water and recycling the resulting solution.

Corrosion is a major problem in urea manufacture. Reactors lined with stainless steel are widely used, and some have been lined with highly resistant but expensive materials, such as titanium. Another problem is formation of biuret, $(NH_2CO)_2$-NH, during manufacture. Formation can be kept at a minimum by reducing the time during which the urea solution is exposed to elevated temperature. Biuret is harmful to some types of plants.

The use of urea presents the added problem that surface application may result in nitrogen loss because of urea decomposition (by hydrolysis) back to the gases from which it was made. The decomposition is accelerated by high soil temperature, low pH, and an enzyme (urease) found in the soil. By proper attention to conditions of application, losses can be kept to an acceptable level.

Urea is less hygroscopic than ammonium nitrate and gives less trouble with caking. It is also nonexplosive.

Nitrogen solutions. In the United States liquid forms of fertilizer are quite popular because they can be handled and applied with a minimum amount of labor. The use of anhydrous ammonia, already described, is an example of this. Ammonia, however, must be handled under high pressure and must be applied carefully to prevent loss. To avoid these problems a class of products called nitrogen solutions has been developed. These are of various types, but in general they can be classified as water solutions of nitrogen salts (plus ammonia in some), all characterized by low vapor pressure (above atmospheric) or, in some instances, no gage pressure at all. The nonpressure solutions can be handled in ordinary tanks without special precautions and can be sprayed on the soil surface without loss. The pressure type requires some care in handling and must be injected into the soil, but the pressure is so low, only a few pounds, that requirements are much less rigorous than for anhydrous ammonia.

Nitrogen solutions are popular in the United States but are little used in other parts of the world. In 1966–1967 consumption in the United States was more than 2,500,000 tons, or about 750,000 tons of nitrogen—second only to anhydrous ammonia.

The leading nitrogen solution is a combination of urea and ammonium nitrate. These two salts are soluble separately only to the extent of about 20% nitrogen, but they have a mutual solubility effect on each other that produces a 32% nitrogen content when the two are dissolved together. About 80 lb of the combined salts can be dissolved by 20 lb of water.

Pressure-type solutions that contain ammonia and urea or ammonium nitrate, or in some products all three together, are also popular. The advantage of the ammonia is that it increases the total nitrogen content of the solution. Several of the pressure solutions contain over 40% nitrogen, with vapor pressures on the order of 10–15 psig.

Aqua ammonia, a solution of ammonia in water, also has the advantage of easier handling than anhydrous ammonia, but nitrogen content must be reduced to about 20% to get a satisfactorily low vapor pressure. Nevertheless, large quantities of aqua ammonia are used (about 800,000 tons in 1966–1967).

Other compounds. Numerous other nitrogen compounds are used as fertilizers, but the amounts are much smaller than for the leading nitrogen fertilizers described.

Sodium nitrate was once a major fertilizer but in 1968 supplied only about 3% of world fertilizer nitrogen. The relative low nitrogen content (16%) is a major disadvantage in view of rising labor and shipping costs. About one-third of the world tonnage is consumed in the southeastern part of the United States, where the use is to some extent traditional.

Calcium nitrate is used mainly in Europe and in 1968 supplied about 4% of world nitrogen. Almost half of the production is in Norway, where the material was made in the past as a method of using nitric acid made by the now-obsolete arc process for nitrogen fixation; production still continues from nitric acid made by ammonia oxidation.

Calcium cyanamide, made from calcium carbide and atmospheric nitrogen, in 1968 made up about 2% of the nitrogen supply. Most of the production is in Europe and Japan.

Organic nitrogen materials once were major nitrogen fertilizers but in 1968 supplied only about 1% of the nitrogen. Organics may regain a significant place, however, because of the growing problem in disposing of urban organic waste. When the problem becomes severe enough in a particular situation to warrant waste processing, fertilizer usually is the most appropriate product to make.

Controlled-release nitrogen materials are not used in large quantities but may become sig-

Table 2. Fertilizer phosphate supplied by various materials

Material	Phosphate content, %*	Supply, 1000 metric tons of phosphate*†	
		U.S.	World
Ordinary superphosphate	16–22	1010	6210
Concentrated superphosphate	44–47	1065	1932
Complex fertilizers	–‡	–	2090
Basic slag	17.5	6	1932
Ground phosphate rock	–‡	130	1650
Ammonium phosphate	20–46	860	–

*As P_2O_5.   †In 1964–1965.   ‡Varies.

nificant in the future. By using a slowly soluble nitrogen compound or depositing an impermeable coating on a soluble one, dissolution and release of nitrogen in the soil is delayed, thus reducing leaching and other losses that can be incurred with soluble nitrogen materials.

Multinutrient fertilizers supply the remainder of the fertilizer nitrogen, about 15 and 25%, respectively, of world and United States consumption. These are classed with phosphate or mixed fertilizers rather than with nitrogen products.

Phosphate. World consumption of fertilizer phosphate is not as large as that of nitrogen. In 1966–1967 the estimated world usage was 15,800,000 metric tons, with about 25% of the total used in the United States. The principal phosphatic fertilizers and their consumption and phosphate content are listed in Table 2.

Phosphoric acid. Modern high-analysis phosphatic fertilizers, such as ammonium phosphate and concentrated superphosphate, require phosphoric acid in their manufacture. The acid in turn is made from phosphate rock, an ore found mainly in the Soviet Union, North Africa, and the United States (principally in Florida).

The leading method for phosphoric acid manufacture involves treatment of phosphate rock with sulfuric acid. The insoluble calcium phosphate in the ore dissolves in the acid and crystals of calcium sulfate (gypsum) form. After separation of the gypsum by filtration, the acid is concentrated to the level required in making the various fertilizer phosphates. The gypsum is discarded in storage piles and becomes a problem because of the large amount produced; the tonnage is larger than that of the phosphate rock used.

A growing problem in phosphoric acid production is the supply of sulfur needed for making the sulfuric acid. The rapid increase in production of fertilizer phosphates, coupled with declining reserves of minable sulfur, has produced a severe sulfur shortage and a high price. New sources are being sought, and ways of reducing sulfur requirement in phosphatic fertilizer manufacture are being investigated.

One method of eliminating sulfur requirement is the electric furnace process for phosphoric acid. Phosphate rock, mixed with coke and silica, is heated to a high temperature in the furnace to reduce the phosphate to elemental phosphorus. The phosphorus is then burned with air and the resulting oxide absorbed in water to give phosphoric acid. The furnace method normally is not competitive with the sulfuric acid process, but rising sulfur cost and prospects for low-cost electric power by

the nuclear route have improved its economic potential.

A major development is polyphosphoric acid (or superphosphoric), made up of various chain-type acid species and containing about 50% more phosphate (as P_2O_5) than ordinary phosphoric acid. Superphosphoric acid is made either by concentrating wet-process acid (made with sulfuric acid) to a much higher concentration than usual (70–72% P_2O_5 versus 54%), or by using less water in the furnace method. The acid has the advantages of higher concentration, lower suspended solids content (in the wet-process type), and certain chemical properties that make it superior in production of several phosphatic fertilizers.

Superphosphate. The oldest of the phosphatic fertilizers, and still the most popular on a world basis, is normal superphosphate. The manufacturing process is quite simple: Phosphate rock is mixed with sulfuric acid (a smaller amount than in phosphoric acid production), the resulting slurry is held in a container for a few minutes until it solidifies, and the material is removed to a storage pile, where it cures for 3 weeks or so to complete the reaction between rock and acid.

The cured superphosphate is used to some extent as a fertilizer without further treatment, but most of it serves as one of the starting materials in making other fertilizers. The status of the material is declining, mainly because of its low nutrient concentration. In the period 1957–1965 the percentage of world phosphate supplied by normal superphosphate declined from 57 to 45%.

Concentrated superphosphate (usually called triple superphosphate) is in a much better position because of its higher concentration, over twice that of normal superphosphate. Its popularity has been growing in the United States for several years, and in 1964 it passed normal superphosphate, the leader since the beginning of fertilizer history. On a world basis, however, triple superphosphate has not made as much headway; in 1965 it supplied 14% of the phosphate and the normal type supplied 45%.

Triple superphosphate is made by treating phosphate rock with phosphoric acid rather than with sulfuric acid. Therefore there is no sulfate present to dilute the product; instead, the acid supplies more nutrient phosphate than does the phosphate rock, and a very high nutrient content is realized.

Triple superphosphate is manufactured in much the same manner as the normal type. The rock and acid are mixed, the slurry held in a "den" until it solidifies, and the moist mass transferred to storage for curing. Unlike normal superphosphate, much of the triple is granulated; that is, finely divided material cut from the curing pile is agglomerated to particles of fairly large size, on the order of $\frac{1}{16}$–$\frac{1}{8}$ in. in diameter. In this form it is easier to handle in direct application and in mixing with other granular materials to make a multinutrient, nondusty fertilizer.

Ammonium phosphate. The highest nutrient concentration yet attained is in ammonium phosphates, fertilizers made by neutralizing phosphoric acid with ammonia, agglomerating the resulting slurry, and drying the granules. No curing is needed because the reaction goes to completion rapidly. Both the cation and anion contain nutrients — in contrast to the superphosphates, which have calcium as the cation—and therefore the nutrient concentration is quite high. A typical product, diammonium phosphate, contains 18% nitrogen and 46% phosphate (P_2O_5).

Several processes have been developed for making ammonium phosphates. In one of the more popular ones, the acid is treated with part of the ammonia in a tank-type vessel, the partially ammoniated slurry flows onto a rolling bed of solids (solidified ammonium phosphate) in a rotary granulating drum, the rest of the ammonia is injected under the bed of solids, the granules are dried, and the finished product of the desired particle size is screened out. The fine material passing through the screen is recycled to the drum to provide the bed of solids.

Ammonium phosphate consumption grew rapidly in the 1960–1970 decade, mainly because of the high nutrient content. Growth has been particularly rapid in the United States, where numerous giant ammonium phosphate plants were built in the late 1960s. Production is also large in the United Kingdom, but there are relatively few plants elsewhere.

The newest product of the ammonium phosphate type is ammonium polyphosphate, made by reacting ammonia with superphosphoric acid. The product has high nutrient content, typically 15% nitrogen and 60% phosphate, and possesses certain properties, for example, high solubility, that make it promising for the future.

Nitric phosphate. A major problem in making ammonium phosphate or the superphosphates is that sulfuric acid is required. The short supply and high price of sulfur has created interest in acidulation with nitric acid rather than with sulfuric. The product is called nitric phosphate.

Substitution of nitric acid for sulfuric or phosphoric acids, however, causes some special problems. Reactions with phosphate rock gives phosphoric acid plus calcium nitrate, rather than the phosphoric acid–calcium sulfate combination formed when sulfuric acid is used in making phosphoric acid. Calcium sulfate precipitates and can be separated, but calcium nitrate remains dissolved in the acid. If allowed to remain through further processing, it has a diluting effect and, more seriously, makes the product hygroscopic.

Various methods have been developed for coping with the calcium nitrate problem. In Europe it is crystallized out (by cooling the solution) and used directly as a fertilizer. Hygroscopicity is a problem, but with moistureproof bags the poor physical properties can be tolerated. Other acids that give an insoluble calcium salt can be used along with the nitric to convert the calcium nitrate, for example, to calcium sulfate or calcium phosphate.

In a typical process, phosphate rock is treated with 20 moles of nitric acid and 4 moles of phosphoric per mole of calcium phosphate in the rock. The acidulate slurry is then neutralized with ammonia and granulated. The phosphoric acid and ammonia convert the calcium nitrate to dicalcium phosphate and ammonium nitrate, both acceptable products.

In the crystallization version, enough calcium nitrate is crystallized out so that, when the solution is separated and ammoniated, the remaining calcium precipitates as dicalcium phosphate and the nitrate is converted to ammonium nitrate.

One disadvantage to nitric phosphate is reduced nutrient content because part or all of the calcium is left in the product. A typical product contains 20% nitrogen and 20% phosphate; in comparison, without the calcium the nutrient content would be about 25% higher.

Nitric phosphates are popular in continental Europe, but production in other major fertilizer-producing areas, the United Kingdom, the United States, and Japan, is relatively low. The relation of sulfur in the economy, however, may alter this situation.

Other compounds. A few other phosphates contribute to the fertilizer supply. The most important of these is basic slag, third in world phosphate supply but losing ground to the newer highly concentrated materials. Basic slag, which contains only about 17% P_2O_5, is popular because it is a by-product of steel production in Europe and therefore sells at a relatively low price. Over 1,500,000 tons of P_2O_5 was supplied by basic slag in Europe in 1965, but practically none was made in other parts of the world.

Phosphate rock itself is also a leading phosphate fertilizer. Although the ore is so insoluble that acidulation or furnace treatment normally is considered necessary to make it usable, there are special soil-crop situations in which finely ground phosphate rock gives enough crop response to make its use justifiable, even though phosphate utilization is much lower than for processed phosphates. Elimination of the processing cost makes it economical even though only part of the phosphate is used.

Phosphate rock consumption as a direct fertilizer is quite large, about 1,650,000 metric tons of P_2O_5 in 1965 (world usage). The full amount is not counted when ranking phosphate rock with other phosphate fertilizers, however, because much of the P_2O_5 in the rock is inert and unused.

Thermal phosphates are used to a limited extent, mainly in Germany and Japan. Phosphate rock is heated, usually to fusion, with some solid material that reacts with the rock at high temperature. In Japan, for example, the reactant is a magnesium silicate ore. In Germany soda ash plus silica has been used. The product has low concentration (19–24% P_2O_5) and is declining in popularity.

Bones and other organic sources of phosphate, once the leading type, have little significance in phosphate supply.

Potash. Tonnage of potash is the lowest of any of the macronutrients, estimated at 13,100,000 metric tons (of K_2O) in 1966–1967 as compared with 15,800,000 for P_2O_5 and 21,800,000 for nitrogen. The United States leads in consumption, with France, Germany, and the Soviet Union also major consumers.

The potash industry is quite simple in comparison to the nitrogen and phosphate industries. Atmospheric nitrogen and phosphate rock must be subjected to expensive processing to make them usable as fertilizers, but potash ore is soluble and can be used directly as mined without any treatment other than removing impurities. Hence, potash supply is more of a mining industry than a chemical processing.

The principal potash ore, supplying more than 90% of the world total, is potassium chloride. The natural deposits are tremendous; a single potash bed in Canada is estimated to contain 6,400,000,000 tons of recoverable K_2O.

The United States leads in potash production, followed by West Germany, East Germany, France, and the Soviet Union. Most of the potash mined, over 90% (94% in the United States), is in the form of potassium chloride. Potassium sulfate, also a soluble, readily available material, is the only other potash ore of significance.

Potassium chloride is a high-grade material with more than 50% potassium after impurities are removed. Potassium sulfate is somewhat lower in nutrient content, about 45% potassium. Both are normally used in fertilizer mixtures without any intended reaction with other constituents, although reaction does occur incidentally in some cases, as in mixtures with ammonium nitrate, in which reaction with potassium chloride produces ammonium chloride and potassium nitrate in the mixture.

One of the very few instances in which a potash ore is purposely reacted with another material is in the production of potassium nitrate by a process (only one plant in operation in 1968) in which potassium chloride is treated with nitric acid to form potassium nitrate and chlorine.

Mixed fertilizers. Most fertilizers are of the mixed type; that is, they contain more than one nutrient. This is mainly a convenience for the farmer since his soils need nutrients in certain proportions. Although he could buy single-nutrient materials and apply them in the desired ratios, it is more convenient and usually less expensive to buy them already mixed.

A mixed fertilizer may be simply a mechanical mixture, or the constituent materials may be reacted to form a "chemical" mixture. In the early days of the industry, the mechanical mixture was the prevalent type; today chemically reacted products occupy an important place in the industry.

One of the most important reaction-type combinations, particularly in the United States, is ammonia plus superphosphate. Superphosphates, both the normal and concentrated types, are acidic in nature and will take up a considerable amount of ammonia in a reaction easily carried out in simple equipment. Either ammonia or ammoniating solutions may be used. The solutions have the advantage that they supply ammonium nitrate or urea as well as ammonia, and therefore give a higher content of nitrogen in the mix. Such a combination is quite economical and plays an important part in the mixed fertilizer industry. Through the period bounded roughly by 1935 and 1955, it was unchallenged as the primary supplier of nutrients in mixed fertilizers. The rapid growth in use of ammonium phosphates, however, has provided another major source of nitrogen and phosphate.

Much of the mixed fertilizer is produced in granular form; the powdery, dusty mixtures of the past have met with increasing resistance on the part of farmers. Granulation, usually carried out in a rotary drum or a paddle mixer, is accomplished by moistening the mix with water or with solution un-

til the dry solids agglomerate. The modern practice is to carefully adjust the amount and proportion of soluble materials used (urea, ammonium nitrate, or acids), so that, at the elevated reaction temperature reached in the granulator, the proper proportion between liquid and solid phases will be reached and granulation accomplished at low water content. This method reduces the expense of drying, which is relatively high when water alone is used for granulation.

Liquid mixed fertilizers are also an important fertilizer type. Low handling cost has popularized the liquid type to the extent that 5–10% of all mixed fertilizers in the United States are supplied in liquid form. In other parts of the world, however, liquids are not used to any significant extent.

The basic step in liquid fertilizer production is reaction of phosphoric acid to make an ammonium phosphate solution. Other materials, such as urea–ammonium nitrate solution and potassium chloride, are then added to give the nutrient ratio desired.

Although liquid fertilizers are simple to handle with pumps and through pipelines, they have the drawback that the water required to keep the constituents in solution dilutes the product. Development of polyphosphates has improved this situation; for example, a solution containing 11% nitrogen and 37% phosphate can be made from superphosphoric acid, but with the usual type of acid the contents are 8% nitrogen and 24% phosphate.

However, if the liquid mix contains a large proportion of potash, the polyphosphate is not nearly so effective in increasing concentration. For such products, suspension fertilizers, a relatively new fertilizer type, give a very high nutrient content. Suspensions are made by restricting the amount of water and carrying the resulting crystallized salts in suspension by use of 1% or so of clay as a suspending agent.

Bulk blending. The trend to granulation has brought about a revival of the mechanical-mixing practice prevalent in the early days of the industry, but with the difference that granular rather than powdered materials are used and the product is handled mainly in bulk rather than in bags. The main advantage is that a very simple mixing plant is adequate and the resulting low investment makes small, community-type plants feasible. Materials brought into such a plant do not have to be shipped very far after they are mixed, in contrast to "chemical-mixing" plants, which must be relatively large because of higher investment. Such plants must distribute the product over a larger area, and therefore raw materials may be hauled back over part of the route they traveled initially.

The principal materials used in bulk blending are ammonium nitrate, ammonium sulfate, urea, superphosphate, ammonium phosphate, and potassium chloride—all granular. They should be of uniform size to minimize segregation in handling and shipping.

The favorable economics of bulk blending have caused the practice to grow rapidly. It is estimated that more than one-third of the mixed fertilizer consumed in the United States is in the bulk blend form.

Micronutrients. Supplying micronutrients is not yet a very significant activity in the fertilizer industry. Although such nutrients are as essential as the macronutrients, natural supplies in the soil are adequate in most instances. The number of identified deficient areas is growing, however, and use of micronutrient materials is increasing.

From the standpoint of amount used, the principal micronutrient appears to be zinc, followed by boron, iron, manganese, and copper. Reliable figures on the comparative amounts consumed are not available. The tonnage is low, however, probably not more than 30,000 tons of total material annually in the United States.

Practice is split between "shotgun" and prescription application. Most agronomists and technical people prefer the prescription method, in which soil analysis and crop response are used as the basis for prescribing application in a particular situation. The "shotgun" approach involves mixing small amounts of micronutrient material into standard mixed fertilizers, on the basis that the micronutrient is needed generally as insurance against the development of deficiency.

Micronutrient salts can be incorporated in liquid fertilizers if polyphosphate is a constituent. The polyphosphate sequesters the micronutrient metals (zinc, iron, copper, and manganese), holding them in solution when they would otherwise precipitate.

There are often problems in incorporating micronutrients into solid fertilizers of the bulk blend type. The micronutrient must be finely divided because such a small amount is required; if granules were used they would be too far apart when applied to the soil. Mixing the fine material with granular mixed fertilizer, however, is a problem because the different sizes segregate in handling. Methods have been developed to cause the micronutrient to adhere to the surface of the granules.

[A. V. SLACK]

Bibliography: R. Noyes, *Ammonia and Synthesis Gas*, 1967; R. Noyes, *Potash and Potassium Fertilizers*, 1966; C. J. Pratt and R. Noyes, *Nitrogen Fertilizer Processes*, 1965; A. V. Slack, *Chemistry and Technology of Fertilizers*, 1966; A. V. Slack, *Fertilizer Developments and Trends*, 1968.

EFFECT ON SURFACE AND UNDERGROUND WATER

Interest in public regulation of farm fertilizer use to prevent degradation of surface and underground water quality has been high because of the recent demands of some environmentalists and the hearings held by the Illinois Pollution Control Board in 1971. The board did not impose regulations but referred the question of nitrogen restriction to the Illinois Institute for Environmental Quality for study of feasibility and cost to the farmer and consumer. If restriction proves to be unfeasible, then the current Public Health Standard of 10 parts per million of nitrate nitrogen (45 ppm NO_3^-) in potable water is to be investigated by the institute to determine if the standard should be relaxed for Illinois.

Interest in controlling overuse and misuse of fertilizers will probably decline, but there will be an era of fact finding, better assessment of the various sources of mineral pollution of water, and accelerated educational activity by public agencies and fertilizer vendors on improved methods of using conventional fertilizers. There will also be more thorough consideration of various alternatives to maintain the food supply with a minimum of pollution.

Problems. Commercial (inorganic) fertilizers contain nitrogen (N) and phosphorus (P) that may contribute to eutrophication or excess fertility of surface waters. Leaching of nitrate into either surface or underground waters may contribute to high nitrate concentrations that in the past have been implicated in methemoglobinemia or cyanosis of babies and various disorders in cattle and sheep that drink the water. The sources of nutrients in surface waters are multiple, including sewage and industrial effluents, animal and wildlife wastes, precipitation, and runoff and sediment from all kinds of landscapes, fertilized or not. Thus the actual input of nutrients from fertilizers to water is difficult to assess because of the diffuse nature of agricultural drainage in contrast to that from a sewage plant or other point source. Also, a controversy exists among limnologists and others about the relative importance of N, P, and organic wastes in producing nuisance algal blooms. The problems are different in various sections of the country and in different kinds of water bodies. A fast-flowing river is much less likely to become eutrophic than a large lake with a low flow-through rate. *See* EUTROPHICATION, CULTURAL.

Fertilizer use on farms has increased enormously in the last 30 years. Nitrogen use in the United States (expressed as elemental N) increased from about 335,000 tons in 1940 to over 8×10^6 tons in 1971. Phosphorus use on farms has increased less rapidly than N use, but the increase in industrial use (for detergents) has been much greater. The phenomenal increase in the use of these nutrients on farms has been part of the circumstantial evidence used by environmentalists in calling for controls.

Direct evidence indicating that fertilizers can pollute is scanty and is usually obtained under conditions of known overuse or misuse in relation to recommended practices.

Phosphate. Many studies have shown that fertilizer phosphate does not percolate in soil water or wash off in solution in significant amounts because of the high adsorption capacity of the mineral fraction of the soil. D. R. Carter and associates in Idaho showed that irrigation water applied to a 200,000-acre irrigated tract near Twin Falls contained 66 ppb of P but that the drainage contained only 12 ppb. The soils retained P equivalent to 70% of that in the irrigation water, and all P in the fertilizers too. Practically all evidence indicates that most P reaching surface waters is carried in, or adsorbed on, sediment eroded from soils or washes out of dead vegetation. C. H. Wadleigh calculated that about 3×10^9 tons of sediment reach waterways annually from all kinds of land, and that this sediment contains about 2×10^6 tons of P. This amount of P is equal to all of the P added to farm soils annually. D. B. Timmons and associates in Minnesota have shown that phosphate leached from dead vegetation by rain and snowmelt can be a significant source of water enrichment if soil erosion and runoff are excessive. Old-fashioned soil conservation methods, including terracing, maintenance of vegetation, contour and minimum tillage, and incorporation of P fertilizers into the soil, appear to be the best practices for control of runoff of P. Hence the U.S. Environmental Protection Agency has endorsed the erosion control practices advocated by the Soil Conservation Service and other Federal and state agencies as the best control procedures rather than restrictions on fertilizer use.

Nitrogen. F. G. Viets and R. H. Hageman concluded that the literature on nitrate accumulation in soil, water, and plants provides little evidence of upward trends in nitrate content except in groundwaters in some closed basins and coastal areas of California and a few other isolated areas in the world. Nitrate, like P, has multiple sources and the specific effects of nitrogen fertilizer use are difficult to isolate from geologic accumulations and other activities of man. They concluded that an N crisis was not imminent. They suggested that in areas where upward trends in nitrate were detected in surface water or groundwater, the sources be identified on a watershed or basin basis and controls of the various inputs be applied where they would be most effective. H. L. Nightingale published convincing graphs of upward trends of nitrate in groundwater under the urban and surrounding agricultural land of the Fresno-Clovis area in California. S. R. Aldrich stated that the rivers of east-central Illinois show an upward trend in nitrate, but other rivers do not. Large fluctuations in nitrate occur from year to year that cannot be explained by fertilizer use. Nitrate in the Mississippi River appears to have stabilized since 1960, although fertilizer use in the watershed increased almost eightfold between 1958 and 1970.

A novel approach was used by D. H. Kohl and associates in studying the input of N from fertilizer into the Sangamon River in Illinois. The technique makes use of the fact that the heavy nitrogen isotope (N^{15}) is not a perfect tracer of N. Evidence indicates that soils become slightly enriched in the heavy isotope, whereas fertilizer N has an isotopic ratio close to that of N_2 in air. They concluded that a minimum of 55% of the nitrate entering the Sangamon came from fertilizer N. This conclusion is being challenged by soil scientists thoroughly familiar with isotopic techniques on the basis that the small virgin sample of soil incubated to determine the isotopic ratio of the nitrate produced may not accurately represent the 900-mi^2 watershed and that the fertilizer sample assayed may not represent the fertilizer N used on the drainage area. Nevertheless, the new analytical approach is interesting. The planned production by the Atomic Energy Commission's ICONS program of large quantities of pure N^{14} at a low price may facilitate much new research using tracer N in fertilizers. The heavy isotope N^{15} has been too expensive except for laboratory and greenhouse experiments and an occasional small field experiment.

Some scientists, such as Aldrich, doubt whether the important question is what is the specific contribution of a fertilizer application, but rather, what is the leakage of nitrate from a soil that produces a nearly maximum crop yield. Nitrate can originate from fertilizers, animal manures, and decomposition of soil humus. The high nitrate in well waters in the 1940s in the upper Midwest may

have been associated with the long-term loss of organic matter from the formerly highly fertile grassland soils when they were put under cultivation. Many agronomists contend that the balance of nature was upset when the forests were cut down and the grasslands were broken, and not when the practice was begun to use N fertilizers to restore the productivity of depleted soils.

Excessive fertilizer N can contribute nitrate to groundwater. For example, studies by F. T. Bingham and associates in southern California have shown that nitrate leakage from highly fertilized citrus groves can be excessive. In a 3-year study of a 1000-acre watershed planted mostly to citrus, they found that the nitrogen in the subsurface drainage was equal to almost 45% of the fertilizer N applied. Growers of citrus in southern California formerly used over 300 lb of N per acre annually, and leached nitrate may have contributed to the increase in the nitrate in the groundwater of the upper Santa Ana River basin. However, better control of N application rates in the last 5 years through leaf analysis has reduced N application rates by half and improved the fruit quality without sacrifice of yield. Soil and plant tissue testing for determining the N fertilizer rates needed is being adapted on a wider scale nationwide. Aldrich states that nitrate loss to water is an inevitable consequence of food production. The nitrate loss per unit of production may be least when the best technology is used.

Alternatives to fertilizer. Fertilizers containing N and P at about present rates of application are absolutely essential if the United States is to maintain its present supply of food, meat, and natural fiber for domestic consumption and export. Recycling of all urban wastes in the way that agricultural wastes are recycled now would supply only a fraction of the nutrients needed to maintain soil fertility. Meeting food requirements by expanding acreage onto marginal lands that should be left in forest or grass would only result in more erosion, sedimentation, and nutrient pollution. These are substantially conclusions of S. A. Aldrich of Illinois Pollution Control Board and Frank G. Viets, Jr. *See* WATER CONSERVATION; WATER POLLUTION.

[FRANK G. VIETS, JR.]

Bibliography: S. R. Aldrich, *BioScience*, 22: 90–95, 1972; S. R. Aldrich, *In the Matter of Plant Nutrients*, R 71-15, Illinois Pollution Control Board, 1972; F. T. Bingham, S. Davis, and E. Shade, *Soil Sci.*, 112:410–418, 1971; D. L. Carter, J. A. Bondurant, and C. W. Robbins, *Soil Sci. Soc. Amer. Proc.*, 35:331–335, 1971; D. H. Kohl, G. B. Shearer, and B. Commoner, *Science*, 174:1331, 1971; H. L. Nightingale, *Ground Water*, 8:22–28, 1970; D. B. Timmons, R. F. Holt, and J. J. Latterell, *Water Resour. Res.*, 6:1367–1375, 1970; F. G. Viets, Jr., *BioScience*, 21:460–467, 1971; F. G. Viets, Jr., and R. H. Hageman, *USDA Handbook 413*, 1971.

Fisheries conservation

The term fishery is used in this article to mean the place for taking fish or other aquatic life, particularly in sea waters. The term has various legal connotations depending upon whether the right to take fish is founded on ownership of the underlying soil and therefore exclusive (several fishery), whether the right to fish in public waters is enjoyed in common with others (common fishery), or whether the right is an exclusive privilege, derived from royal or public grant, and independent of the ownership of the underlying soil (free fishery).

A fishery may be operated for pleasure (sport fishery) or for profit (commercial fishery). Fisheries include finny fish; mollusks, such as clams, mussels, and oysters; shellfish, including lobsters, shrimp, and crayfish; aquatic plants; sponges; coral; sea cucumbers; amphibians, chiefly frogs; reptiles such as turtles, alligators, and crocodiles; and mammals, including whales, seals, and walruses. *See* MARINE RESOURCES.

Distribution and types. Fisheries are important for the production of food and of raw materials for industry, and for recreation. They occur in all the many kinds of waters that together compose nearly three-fourths of the Earth's surface. The two principal types are marine (salt-water) fisheries, which annually yield 25,000,000–30,000,000 metric tons of products; and fresh-water (inland, continental), producing annually a recorded catch of 3,000,000–4,000,000 metric tons plus a sporting take of great but unknown magnitude. Both types have potentials far greater than present realizations.

Marine fisheries are mostly commercial and are located predominantly in the Northern Hemisphere. The most important fishery centers and the principal products they yield are the northeastern Atlantic (flatfishes, including halibut and flounder, cod, haddock, coalfish, herring, and shrimp); the northwestern Pacific (salmon, flatfish, herring, crab, shrimp, lobster, squid, and octopus); the northwestern Atlantic (flatfish, cod, haddock, herring, lobster, and crab); and the Indo-Pacific (herring, bonito, mackerel, and shrimp).

The major fresh-water fishing areas and their products are Asia (ayu, salmon, milkfish, carp); Soviet Union (salmon, whitefish); Africa (many kinds of fish); and central and northern North America (trout, whitefish, bass, perch and their relatives). Fresh-water fisheries in the more highly civilized parts of the world are used primarily for sport. However, they are important largely for food in primitive regions, including parts of Africa and South America, and in densely populated lands, especially those on a rice economy, such as India, China, Japan, and the smaller nations of Southeast Asia.

Problems. The major problem in fishery conservation is how to control both man and the aquatic community to ensure the production and the harvest of aquatic crops for the present and for the future when the demand will probably be greater than now. Lack of knowledge or care by fishermen may make their fishing efforts inefficient, their treatment of captured fish wasteful, and their methods of capture destructive to stocks needed as "seed" for future harvests. Handlers, marketers, and consumers may cause waste by careless preservation (refrigeration, salting, and canning) and by inefficient preparation for the table. Lack of information may also be responsible for inadequate harvest of many underexploited or unexploited segments of the resource. Consumers need to be educated away from preferential buying habits that accelerate the demand and price for certain species but lead to discard of other aquatic organisms with equally sound food values.

The destruction of the aquatic habitat by man accentuates fishery problems, especially in inland waters. Deforestation and destructive agriculture, resulting in soil erosion and excessive warming of waters, have changed the fish-producing characteristics of many streams, usually for the worse. Sewage and industrial wastes reduce the quality of water suited for desired aquatic life. Organic pollutants such as domestic sewage may remove life-important oxygen from the water (waters generally are undesirable for aquatic life if the dissolved oxygen in them falls below 3 ppm). Pollutants may also be directly poisonous to fish and other water organisms (for example, most chemicals that are poisonous to man, such as cyanides, are also toxic to desirable aquatic life). Habitats may be destroyed by smothering them with silt (washings from mineral refineries such as coal and iron). *See* FOREST MANAGEMENT AND ORGANIZATION; SOIL CONSERVATION; WATER POLLUTION.

Man has likewise created problems for aquatic life in continental waters by changing phases of the water cycles. Deforestation and agricultural land drainage have lowered the ground water table and have lessened stream flows during dry periods. The construction of dams for water storage, power, navigation, industry, flood control, domestic water supply, and irrigation has interfered with the movements (usually for spawning) of native migratory fish. Outstanding problems created by dams are those affecting the migrations of the Columbia River salmon in the western United States and the Atlantic salmon throughout its range in both northeastern North America and Europe. To date there has been little success in developing devices (fish ladders) that enable fish to progress over high dams to upstream spawning areas. Irrigation channels may lead fish to doom by leaving them stranded when the channels are drained, and fish may be destroyed by being jammed against intake screens of power developments. Fluctuations in water level may destroy fish nests by making the water over them too shallow or too deep.

Problems of fisheries conservation also arise from fishing. The greatest problem is that of managing each fishery so as to provide sustained yields of desired species. The solutions of these problems lie in research, public education, legislation and law enforcement, and continuing reevaluation of management procedures. Fishing itself may be destructive; the gear may injure the young while capturing harvestable adults. Fishing may exceed the capacity of a species to maintain itself through reproduction and growth. Although the species may not be exterminated by such overfishing, the fishery may become unattractive economically and for recreation. If it "collapses" entirely, this may bring considerable hardship to the fishermen. Similarly, underfishing is wasteful. In small inland waters, underfishing may destroy the quality of a fishery by leading to overpopulation and thus dense stands of undersized fish, which are unattractive to fishermen. Selective fishing for preferred species, often for predatory ones, may cause coarse, unwanted kinds to usurp the fish-producing capacity of a body of water. *See* POPULATION DYNAMICS.

Some problems of fisheries conservation arise naturally. Included are diseases, natural pollution such as "red tide" of protozoa in marine water, parasites, predators, and the gradual evolution of lakes toward ponds and dry land and of streams toward sluggish, base-level waterways.

Management. Laws to regulate fishing are the chief instruments of fishery management. This is particularly true in the commercial fisheries. Commonly, laws control who may fish, as well as where, when, and how they may fish. These laws regulate the kind, size, and amount of fishes that may be taken during a prescribed period. Legal measures such as antipollution legislation also protect the habitat of aquatic organisms. Because of changing conditions, the efficacy of laws should be continually reviewed.

The artificial propagation of fishes and other aquatic organisms, reputedly practiced in China for several centuries B.C., is a means of increasing the quantities of preferred species. In Southeast Asia aquiculture combined with rice culture is an important source of fish for human food. Preferred food fishes in Europe (carp, trout) and North America (trout) are also produced for sale at fish farms or hatcheries. In addition, bait fish are propagated (and sold) extensively in North America. Aquatic farm production methods are also applied to oysters (as food or for pearls), frogs, turtles, ornamental fishes (such as goldfish), and water plants (ornamental, as the lotus, or edible, as the water chestnut).

Some artifically propagated fishes are used in the management of sport and commercial fisheries. Many kinds have been established successfully in waters far removed from their native ranges or in newly created water areas. Still others have been stocked successfully to bolster numbers available to sport fishermen. However, countless others have been placed in waters not suited to them and they soon disappeared, or they have been planted where adequate spawning stocks of the species were already present, in which cases they were wasted or they aggravated an already bad situation of overpopulation. In other situations, the artificially stocked fish have increased so rapidly that they have destroyed or seriously menaced desirable species already present.

Aquatic life in continental lakes, ponds, and streams requires a stable supply of water, with chemical and physical factors varying according to species requirements. Consequently, much effort is spent in learning the best conditions for life of preferred organisms and in managing habitat to provide the prescribed conditions. Often the environment can be improved. Starting in the headwaters of a stream system, land-use practices may be adjusted with the water world in mind, for example, by retarding surface runoff and encouraging percolation of water into the ground. In the water area itself, chemical, physical, and biological changes can be made to enhance stability, productivity, and yield to fishermen. Pollution and erosion may be controlled and improvement made in food, feeding, and shelter conditions. Species composition, competition, predation, and fishing pressure can be regulated to some extent. *See* LAND-USE PLANNING.

Ponds have been built for many centuries to provide fish and fishing in areas having few natural surface water bodies. In the middle years of the 20th century, there was a surge of artificial pond or lake construction in the United States and hundreds of thousands of small impoundments resulted from the farm pond program, especially in states not bordering the Great Lakes.

Fishery management measures also include educational programs on fish conservation. They likewise encompass basic and developmental research in government and university laboratories. Moreover, these measures include the training of professional fishery scientists to develop and apply the most effective fishery methods.

In general, governments regulate inland, coastal, and boundary-water fisheries. Fishermen are commonly required to buy licenses to fish. In the United States, regulation is primarily at the state level, with the state retaining ownership of the fishes in public waters. In many international waters, international agreements and treaties are used. Often these are in the form of international commissions such as the North Pacific Fishery Commission, Northwest Atlantic Fishery Commission, and the International Great Lakes Fishery Commission.

There are numerous governmental aids to the conservation of fisheries at the local, national, and international levels. Educational programs constitute one of these. Another aid is in the form of exploration for new stocks and the development of more effective gear and methods for handling, storing, manufacturing, shipping, and marketing the products. There is also direct Federal aid to states for research and the development of fishing. This is particularly true in the United States, where certain taxes are earmarked for the purpose. The Food and Agriculture Organization of the United Nations, as well as certain individual nations, extends international aid and mutual assistance in conservation. *See* WILDLIFE CONSERVATION.

[KARL F. LAGLER]

Food chain

The scheme of feeding relationships which unites the member species of a biological community. The idea was introduced to modern ecology by C. Elton to describe the linear series of species, usually involving plants (autotrophs), herbivores, and one or two successive sets of predators (a predator chain), or alternatively, the series of parasites and hyperparasites exploiting a host (a parasite chain). Saprophytic chains, exploiting dead tissues, are now known to be most important. The successive categories in such a chain are widely known as trophic levels.

The number of stages in a given chain does not usually exceed five and the number of organisms (or better, their biomass) diminishes rapidly. Although the species involved in a given food chain vary in space and time, similar sets of relationships recur in different habitats. For instance, the Arctic fox feeds on guillemot eggs in winter and seal carrion left by polar bears in summer; the spotted hyena has similar habits in Africa involving ostrich eggs and zebra killed by lions. Such professions were termed niches by Elton. *See* BIOMASS; ECOLOGICAL PYRAMIDS.

Food web. A neat classification by trophic levels often fails because a species may occupy more than one level during its life cycle or in response to changes in availability of food. For instance, hover flies change from decomposers to herbivores, whereas many Diptera, when adult, suck both

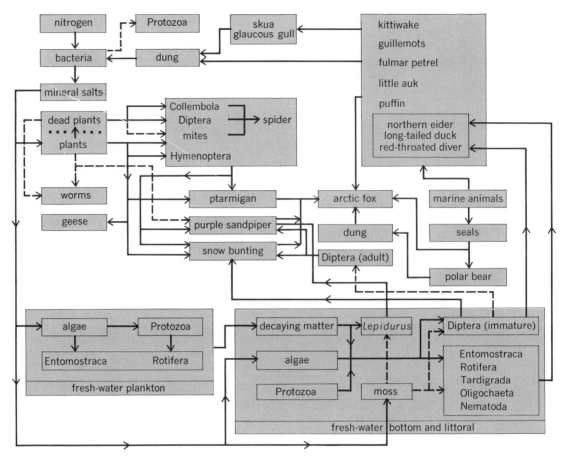

Food web among the animals on Bear Island, a barren spot in the Arctic Zone, south of Spitsbergen (the broken lines represent probable food relations that have not yet been proved). To read the diagram, start at the marine animals and follow the arrows. (*From C. S. Elton, Animal Ecology, Wiley, 1927*)

plant and animal juices. Therefore, the system of feeding relationships often resembles a web rather than a chain (see illustration).

Quantitative studies. In view of the complexity of food chains, attempts have been made to measure the size of the feeding links between species in order to select the most important. The only universal currency for such measurements is the caloric content of the food. A number of quantitative food-chain studies of this type permit comparison between different levels in a chain and also the chains in different communities. However, it must be stressed that certain species have an importance in the community which is out of proportion to their energy intake; for instance, the removal of pollinating insects may greatly affect the productivity of orchards and meadows; the excretion of external metabolites and antibiotics in minute amounts may change the balance among microorganisms. *See* ECOLOGY.   [AMYAN MACFADYEN]

Bibliography: C. S. Elton, *Animal Ecology*, 1927; P. Farb, *Ecology*, 1963; E. P. Odum, *Ecology*, 1963; E. P. Odum, *Fundamentals of Ecology*, 2d ed., 1959.

Forest and forestry

Forestry involves the management of forest lands for wood, forage, water, wildlife, and recreation. A forest is much more than an assemblage of trees. It is rather a community—technically an ecosystem—consisting of plants and animals and their environment.

Trees are the dominant form of vegetation, but shrubs, herbs, mosses, fungi, insects, reptiles, birds, mammals, soil, water, and air are essential parts of the community, each of which

Fig. 1. Loading Douglas fir log onto logging truck, Olympic National Forest, Wash. (*U.S. Forest Service*)

reacts on all the others. Therefore, the first prerequisite for forestry is knowledge of the characteristics of the living members of the community, of the environment in which they live, and of the interrelations among organisms and environments. *See* FOREST ECOLOGY; FOREST SOIL.

Management. The many goods and services afforded by the forest are considered in light of their relative values and a decision made as to which should be favored in the management of the forest under consideration. Wood for direct use or for manufacture into innumerable materials is the major product of forests and has been the chief concern of forest management techniques (Fig. 1). However, there is growing recognition of the value of other tree products, such as paper, naval stores, maple syrup, quinine, and rubber. Forage both for domestic livestock and for wildlife is often important. In addition to these tangible products, numerous services are rendered by the forest.

Forests increase the water-holding capacity of the underlying soil and facilitate the entrance of water into it. They thus reduce the surface runoff of water and tend to prevent erosion and to regulate the flow of springs and streams. By decreasing wind movement and reducing evaporation from the surrounding area, forests temper the local climate and increase production of crops in their lee. Forests also furnish some of the most attractive sites for hunting, fishing, picnicking, camping, and other recreational activities. *See* CONSERVATION OF RESOURCES.

Management may be centered on any one or a combination of these various goods and services. The usual criterion of profit used in comparing the desirability of different courses of action is difficult to apply here, even when costs and returns can be expressed in dollars and cents, as they may be with timber products. It becomes much more difficult when returns cannot be so expressed and when they accrue to someone other than the owner, as is commonly the case with watershed protection and with recreation. When mathematics fails, the forest manager must rely to a considerable degree upon his own judgment.

Following adoption of a basic policy for the handling of a specific forest property, the next step is to place it under actual management. It is appropriate to mention some of the major considerations that control forest policies and practices.

Multiple use. The variety of goods and services obtainable from the forest gives special importance to the principle of multiple use. In essence, this principle requires the use of different parts of a forest for the purpose or combination of purposes for which each is best suited. Timber may be grown for a single product or, preferably, for the integrated use of several products, such as sawlogs, veneer logs, pulpwood, piling, posts, and fuelwood. Often there are areas in the forest in which management for the simultaneous production of timber, forage, wildlife, water, and recreation is both feasible and desirable. Multiple use, however, has definite limitations. Some uses are incompatible under certain conditions; maximum production of all kinds of goods and services on the same area is impossible.

Modern forestry, although chiefly concerned with timber management, usually includes some

Fig. 2. Forest opening providing forage for livestock, Challis National Forest, Idaho. (*U.S. Forest Service*)

Fig. 3. Mechanized planting to speed up reforestation, DeSoto National Forest, Miss. (*U.S. Forest Service*)

Fig. 4. Timber stand improvement by light thinning in dense areas in order to forestall mortality and concentrate growth on better trees. McCleary Experimental Forest, Wash. (*U.S. Forest Service*)

aspects of range, wildlife, watershed, and recreation management. These activities may be conducted in the forest itself or in forest openings which for administrative reasons must be handled as part of the forest property (Fig. 2). Recreation and watershed protection in particular are steadily receiving increasing emphasis. Because of the close relation between the production and the utilization of wood, the forester is also concerned with properties of wood and the processes by which it is manufactured into thousands of products. Although wood technology is not strictly a part of forestry, the two are so closely related that

Fig. 5. Fire detection tower in the Olympic National Forest, Wash. (*U.S. Forest Service*)

Fig. 6. Airplane spraying insecticide, a modern technique to control insect epidemics in the forest, Boise National Forest, Idaho. (*U.S. Forest Service*)

they are commonly handled together in education and research.

Sustained yield. The basic principle of sustained yield involves the continuous production of forest goods and services. In timber management, not only must reforestation, either by natural or artificial means, follow cutting, but the forest property as a whole must be organized so that approximately the same amount of wood may be harvested year after year (Fig. 3). Growth and drain must balance at a sufficiently high level to meet the current demand. The same principle that provides for continuous wood production without reduction in quantity or impairment in quality also applies to other products such as wildlife, water, and recreation. As population increases and standards of living rise, pressure on the forests pushes to higher and higher levels the sustained yield which will be adequate to meet mounting needs. The higher the level and the larger the number of products and services involved, the more difficult the task of management becomes (Fig. 4). *See* FOREST MANAGEMENT AND ORGANIZATION; FOREST PLANTING AND SEEDING.

Government participation. Forests play so vital a role in so many ways in the well-being of a nation that governmental action to assure their effective management has long been recognized as desirable virtually throughout the world. That action may take the form of public ownership, public regulation, or public cooperation.

In the United States, governmental activity has been chiefly in the fields of ownership and cooperation, with little attempt at regulation. Since 1891 the Federal government has reserved large areas of land in the public domain, and since 1911 it has purchased much smaller but still considerable areas. As of Jan. 1, 1963, the Federal government owned 113,000,000 acres of commercial forest land. The states (notably Pennsylvania, Michigan, Minnestoa, and Washington) owned an additional 21,000,000 acres of commercial forest land, and counties and other local governmental units, 8,000,000 acres. While the areas in these different forms of ownership are subject to continual change, there is nevertheless little doubt that the policy of public ownership on a substantial scale is a permanent one.

Cooperation with private owners is extended by both state and Federal governments in a wide variety of ways. Most prominent is protection of forests from fire (Fig. 5), an activity handled largely by the states, with assistance of grants-in-aid from the Federal government. Other cooperative activities include participation in protection of forests from insects and diseases, sale of planting stock at cost of production, and educational and service assistance to private owners in the production, harvesting, and marketing of forest crops (Fig. 6). Some states have special legislation relating to the taxation of forest lands which is intended to encourage their improved management. A few states exercise some control over the management of private lands by establishing minimum standards of forest practice.

As a managerial activity dealing with many products and services of major value, forestry makes full use of the biological and physical sciences, of engineering, and of the social sciences. It requires technical skill of a high order in the practical application of scientific knowledge with due regard to economic and social considerations. See separate articles on important forest trees listed under their common names.

[SAMUEL T. DANA]

Bibliography: S. W. Allen and G. W. Sharpe, *An Introduction to American Forestry*, 3d ed., 1960; F. S. Baker, *Principles of Silviculture*, 1950; H. Clepper and A. B. Meyer, *American Forestry: Six Decades of Growth*, 1960; K. P. Davis, *Forest Fire: Control and Use*, 1959; R. C. Hawley and D. M. Smith, *The Practice of Silviculture*, 1954; H. L. Shirley, *Forestry and Its Career Opportunities*, 1952; H. L. Shirley and P. F. Graves, *Forest Ownership for Pleasure and Profit*, 1967; A. E. Wackerman, *Harvesting Timber Crops*, 1949; *Trees*, USDA Yearb. Agr., 1949.

Forest ecology

The science that deals with the relationship of forest trees to their environment, to one another, and to other plants and animals in the forest. The term silvics is synonymous, but is occasionally also used to include forest tree genetics and forest tree physiology as well as forest ecology. All these areas include the fundamental biological knowledge upon which rests the manipulation of the forest for man's own purposes (silviculture).

Forest ecology is commonly subdivided into forest autecology, the study of trees as individuals in relation to their environment, and forest synecology, the study of the forest as a community.

The forest environment or site consists of the physical environment surrounding the aerial portions of the tree (climatic factors) and that surrounding the subterranean portion (edaphic or soil factors). Site is subject to continual change. Atmospheric conditions within the forest fluctuate diurnally, seasonally, annually, and over a period of years; and they are changed by the developing forest itself. The forest soil owes its original nature to the parent geological material and to the weathering processes that formed the soil. It, too, is affected by the climate, the vegetation and animal life it supports, and the passage of time. External influences, particularly fire, grazing and browsing animals, and man, affect markedly the nature of the forest site and its capacity to support tree growth.

Site evaluation. Forest site quality, the capacity of an area to produce forests, is best measured by the recorded growth of trees under forest conditions. Permanent sample plots, 0.1–0.5 hectare (0.246–1.23 acres) in size, are measured every 5 to 10 years for many decades to provide the most accurate measure of site for the conditions they sample. Such plots, however, are relatively few. The height of trees grown free from overhead shade to a specified age (usually 50 or 100 years) provides the best general index of site. In the boreal forest, particularly in northern Europe, the presence of certain "plant indicators" has shown an excellent correlation with site quality.

In situations where forest trees and associated plants cannot be used as an index to site quality, reliance must be placed on soil properties. In a given climatic zone, one or more soil properties may be found to exhibit strong correlation with overall site quality. Among the properties that have been successfully used as indices in given situations are the parent geologic material, direction and degree of slope, relative topographic position, depth of soil, size of the soil particles, water-holding capacity of the soil, and depth to soil water

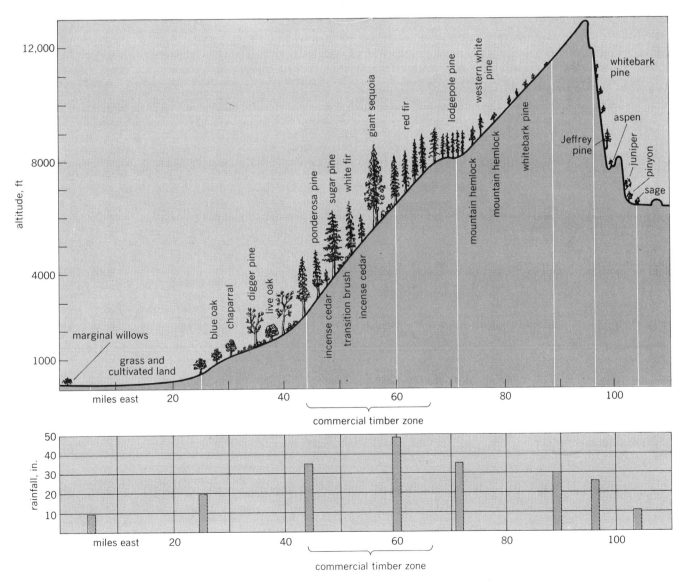

Fig. 1.   Profile of the central Sierra Nevada showing altitudinal limits of the principal forest types. (*USDA*)

Fig. 2.   Much urban and agriculture water flows from forested lands, particularly from those at higher elevations. (*U.S. Forest Service*)

table. Among climatic factors, the amount and distribution of annual precipitation have often been related to site. *See* FOREST SOIL.

Influence of site upon life history. From seed to old age, forest trees are greatly influenced by the site factor. Vigorous trees growing under favorable site conditions produce large numbers of seeds. However, very few seeds, usually far less than 1%, actually grow into trees. Many seeds are destroyed by insects, fungi, rodents, and adverse weather. Only those seeds that come to rest under suitable conditions of moisture and temperature will germinate. Most of the resulting seedlings will be killed by soil fungi ("damping-off" species), solar heating at the ground line, animals, and drought. The most favorable sites for successful forest regeneration usually have moist but well-drained and exposed mineral soil with sufficient dead material on the surface to provide some shade. Regeneration will

usually fail where living plants already fully occupy the surface or the surface root zone. *See* FOREST PLANTING AND SEEDING.

As the seedlings develop, site plays an important part in determining the growth rate of the different species and the outcome of the competition that reduces the typical initial reproduction stand from several thousand seedlings to a few hundred mature trees per hectare (2.46 acres). Some trees, including most of the pines and oaks, will grow well under a wide variety of site conditions, while others, such as black walnut, sycamore, and white ash, can thrive only under specific site conditions. Consequently, the former groups are found on many sites, whereas the latter are less widely distributed. *See* ECOLOGICAL INTERACTIONS.

Different groupings of forest tree species (forest types or associations) thus tend to develop on different sites. In the northeastern United States, for example, a spruce-fir type is characteristic of higher elevations and of cold, wet bogs. Pine types predominate on drier sites, whether sandy or rocky, unless eliminated by logging or fire. Hardwoods occupy the moister lowland soils, the oaks frequently predominating on the somewhat warmer and drier situations, and the birches, beech, and maples on the somewhat cooler and moister. In the western United States, spruces and firs are again found at the higher elevations, with many pines and woodland oaks being found in successively lower belts (Fig. 1).

Influence of forest on site. As the forest becomes established and develops, the site itself is greatly changed. Within the forest, daytime temperatures are lower and nighttime temperatures higher than in the open, thus resulting in more uniform conditions. Air temperature in the forest during the growing season often averages somewhat cooler than in the open. An abrupt forest front may occasionally cause a slight increase (about 1%) in local precipitation by causing moisture-laden air to rise and become cooled. A far more important effect of trees on precipitation is observed in regions where fog or heavy mist occurs. Tiny droplets of atmospheric moisture are deposited on the foliage and on the branches and, when the accumulation is sufficient, water drops to the ground, as "fog drip." Less of the rainfall and snowfall reaches the ground, however, as the leafy crowns of the trees can intercept up to several centimeters of gently falling rain or snow.

On the forest floor, the accumulating layer of leaves, twigs, and other litter attracts a characteristic grouping of plants and animals that live on such decaying organic matter and on each other. The organic matter is gradually incorporated into the top layer of mineral soil, creating a soil horizon rich in nutrients and humus. Water, leaching downward through the soil, tends to deplete the topsoil of soluble materials, depositing these in a lower soil horizon. Some of these materials are taken up by deeply penetrating tree roots, utilized in the development of leaves, and later returned to the surface soil in the annual leaf fall.

Under spruces and firs in cool moist climates, the forest soil is characterized by a whitish leached layer (podzol) virtually devoid of nutrients and located under the layer of unincorporated humus.

Fig. 3.   An abandoned farm which is now reverting to a condition characterized by a mixed hardwood-pine forest. (*Farm Security Administration*)

Under most forest species in the warmer parts of the temperate zone, this underlying zone of leaching is less pronounced. Under the tropical rain forest, the soil is frequently leached of all but iron and aluminum minerals as a result of the wet and warm climate. Here, soil fertility is maintained largely by the rotation of soluble nutrients through the tree roots to the tree crowns and back to the soil as litter. After land clearing for agriculture, many tropical rain-forest soils become infertile in a few months as a result of accelerated leaching and the destruction of the tree root activity that formerly brought the nutrients back to the surface.

Forest hydrology. The influences of the forest on the site result in a characteristic pattern of streamflow differing from that on unforested land in that peak streamflow is lower, and the flow after storms is prolonged. Since much urban and agricultural water is produced by forested lands, particularly those in mountains and at other higher elevations, the management of forests for water production is attaining great importance (Fig. 2).

Forests generally use more water than comparable grassland or open land because tree crowns are effective evaporating systems, and because tree roots usually penetrate the soil deeper than those of smaller plants and can thus reach a greater amount of soil moisture. Such transpiration (water-vapor loss from tree crowns) frequently will utilize from one-quarter to one-third the annual precipitation. In contrast, grass cover under similar conditions may utilize as little as half as much, primarily because of the shallower depth to which most grass roots penetrate.

On the other hand, forest soils are generally much less compact than comparable grassland or open soils, and can quickly absorb and retain far greater quantities of water. Forest soils can ordinarily absorb all available water unless the soil surface has been compacted by grazing animals or man's activities. In contrast, surface runoff under nonforest conditions may often result in severe soil erosion and in flash floods downstream.

The present tendency in forest management for water production purposes, therefore, is to avoid clear-cutting because of the accelerated erosion and flooding that may result, and instead to thin out the forest so as to maintain favorable forest soil conditions and reduce the number of trees using water. Also, attempts are being made to manage water use by trees whose roots reach the soil water table.

Forest succession. Since the developing forest changes the forest site, it follows that the changed site may itself be more favorable to a new group of tree species other than those currently occupying it. As a result, successive forest associations occupy the site in the absence of fire, logging, windstorm, or other outside disturbance. Thus, in the northeastern United States, pioneer associations characterized by aspen, cherries, birches, and pines occupy open forest sites. These in time will be replaced by intermediate associations containing oaks, maples, ash, and other species until, in several hundred years, the climax association is attained. This latter is the forest type that is capable of maintaining itself without major change, in the absence of disturbance. Species such as hemlock, sugar maple, and beech have the capacity of reproducing under their own cover and thus qualify as members of the climax type. *See* CLIMAX COMMUNITY; SUCCESSION, ECOLOGICAL.

In much of the world, the influence of fire, grazing animals, and man's activities is so great that the climax is seldom reached. In much of the southeastern United States, for instance, pine types predominate and constitute the most important commercial forests. Yet these owe their existence to frequent burning and logging and will be replaced by a mixed hardwood climax on most sites unless fire or similar disturbances are introduced from time to time (Fig. 3). Similarly, in the western United States, the extensive pure stands of Douglas fir owe their existence to periodic clear-felling by wind, fire, or man; they would be replaced by less valuable, mixed coniferous types if all disturbances were eliminated. In present-day silviculture, the professional forester utilizes logging, fire, chemicals, and mechanical treatment to maintain the forest on a given site at the stage of forest succession most valuable to man. *See* ECOLOGY; FOREST AND FORESTRY.

[STEPHEN H. SPURR]

Bibliography: *See* FOREST AND FORESTRY.

Forest management and organization

Since forests are renewable natural resources, the central purpose of forest management is to organize the forest for continued production of its many goods and services. This is particularly exemplified in management for wood production, from which, for many reasons, a sustained yield of forest products is desired. *See* FOREST RESOURCES.

The essential requirements of a regulated forest, that is, one that can maintain a sustained yield of harvestable products, is that tree age and size classes be maintained in balanced proportion so that an approximately equal annual or periodic yield of products of desired size and quality may be harvested. There must be a progression of tree age of size classes from small to large so that merchantable trees are regularly available for cutting in approximately equal volume.

From a management standpoint, there are only two general kinds of forests: those composed of even-aged and those composed of uneven-aged groups or stands of trees. Each is considered separately.

Even-aged stands. An even-aged forest stand is one in which the individual trees composing it originated at about the same time and occupy an area large enough to grow under essentially full-light conditions. Such a stand consequently has a beginning and end in time. A stand is established either naturally or by planting or seeding. Thinnings or other intermediate cuttings may be made during its life. The stand is finally harvested, a new one is established, and the growth process is repeated. A typical lifeline of an even-aged stand managed on a rotation or total life of 30 years is shown in Fig. 1. *See* FOREST PLANTING AND SEEDING.

Total harvest. The total harvest obtained from the stand over its span of life, shown in Fig. 1 as the cumulative bar at the right, is the sum of intermediate cuts made plus those from the final harvest or regeneration cuttings. This is the total growth of the stand that is taken in harvestable

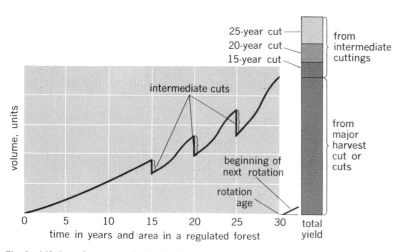

Fig. 1. Lifeline of even-aged stand and regulated forest of even-aged stands.

timber. The number of intermediate cuttings made and their kind and intensity are a function of biology, economics, the quality of the land, and the tree species. The objective is to keep the stand growing vigorously and to capture as much of the total growth potential of the land as possible in usable wood products.

Age-class organization. The organization of a forest area, made up of many such even-aged stands, for continuous production is indicated in Fig. 1 by the vertical broken lines. The spacing between the lines denotes age-class groups of individual stands as well as points in the age of a single stand. The total forest unit could be of any size. In a forest fully regulated for continuous production, there would be an approximately equal area of stands classed by age groups such as 0–5, 5–10, and up to 25–30. The uniform spacing of the vertical broken lines in the diagram indicates equal areas in each age group. In practice, the spacing would not be equal if it were necessary to adjust areas for differences in site quality between age classes. The basic aim is to have age groups of equal productivity, not necessarily of equal area.

Rotation and annual yield. Each year, consequently, an approximately equal area of timber at the rotation age of 30 years is available for final harvest cutting. The area, on the average, would be total area divided by rotation age. The final harvest volume cut would be this area multiplied by the average volume per unit of area at this age. Similarly, there would be an approximately equal area available for intermediate cuts through thinnings, taken at 15, 20, and 25 years in this instance. For thinning at each of these ages, there would again be total area divided by rotation units of area.

The decision regarding the length of the forest rotation is of key importance because the whole forest structure is organized around it; the average rotation of a forest unit cannot be quickly changed. The rotation length is determined by an integration of the biological growth capacity of the forest for different tree species and production costs with the value of forest products produced and the particular purposes of management by the owner. Similarly, the number, timing, and kind of intermediate cuts that may be taken, methods of stand regeneration, desired species, and the general intensity of management practiced are determined by a complex of biological and managerial objectives. Forests are managed not only for timber but for watershed, recreation, wildlife, and other uses. The particular "mix" can vary considerably but this variation does not invalidate the basic principles of even-aged forest management for continuous timber production.

Uneven-aged stands. An uneven-aged stand is one in which trees originated at different times so that the stand includes trees of varying ages. The age-class distribution is seldom perfect, nor need it be. The essential need is that there be a good distribution of trees from small to harvestable size that are capable of making normal growth. Specific age is not particularly important. The distinguishing feature of an uneven-aged stand, large enough to be of practical importance as a management unit, is that it has no beginning in point of time. Nor has it an end, because if it did the next stand would necessarily be even-aged.

Stock structure maintenance. The lifeline of an uneven-aged stand is shown by the solid line in Fig. 2. As shown, the stand has a certain volume of reserved growing stock. It grows for a certain number of years, termed a cutting cycle, a harvest cut is again made, and the process indefinitely repeated. If the reserve growing stock, measured by tree ages and diameters composing it, is considered to be the correct size, the volume cut is equal to the net growth made during the cutting cycle. If the amount of growing stock is to be decreased or increased for economic and silvicultural reasons, the cut is accordingly more or less than current growth.

The trees composing the residual growing stock should be well distributed by species, sizes, and ages so that a continuing crop of harvestable trees keeps coming along. There should be more small trees than large to allow for mortality and to give opportunity to select the best trees for future growth. In contrast with even-aged management, cuts are uniformly made at regular cyclic intervals only and there is no major single harvest cut nor a guiding rotation. Intermediate and harvest cuts are merged in the single cyclic cut in even-aged management.

Cutting cycle. The organization of a large forest

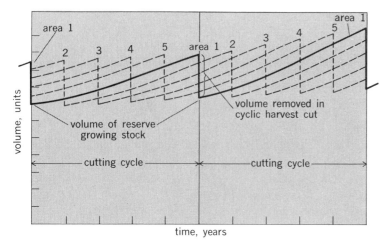

Fig. 2. Lifeline of uneven-aged stand and regulated forest of uneven-aged stands.

unit for continuous production is diagrammatically shown in Fig. 2. A 5-year cutting cycle is assumed, which means that the various stands constituting the forest unit are grouped into five areas of approximately equal productivity, as shown by number in the diagram.

Each year, a harvest cutting is made covering one of these five areas, equal in volume to 5 years growth, plus or minus whatever changes are considered appropriate to maintain the average amount of reserve growing stock. In Fig. 2 a small increase in the amount of reserve growing stock over the two cycles shown is indicated. The entire forest management unit is cut over every 5 years.

Determining the length of the cutting cycle is a decision of critical importance, comparable to the decision concerning rotation length in even-aged management, as the organization of the forest is built around the cutting cycle and quick changes are not possible. As with rotation length, the answer is derived from an integration of biological and financial factors with the objectives of management. Establishing and maintaining a desirable level of reserve growing stock and control of its composition by tree sizes and species of desired quality are other important problems.

Application in practice. In practice, forest management units are usually far from being fully regulated and especially when unmanaged areas are put under management. Continual adjustment in age- and size-class distribution, species composition, and growing stock structure is necessary in working toward a better-organized forest in consonance with changing economic conditions and management objectives. Basically, the purpose is to attain the degree of harvest continuity that is desired and considered attainable. A well-managed forest has a high degree of flexibility. Except to salvage mortality, trees do not have to be harvested at a particular time as with crop agriculture. An overall objective is to keep forest stands well stocked and growing. More can be cut at one time than at another without upsetting the balance of a well-managed forest, provided the overcuts and undercuts average out reasonably.

Even-aged management is applied, worldwide, in commercial timber production considerably more than is uneven-aged management. The reasons are that species control in regeneration, stand and area treatments, and harvest cuts can be more cheaply and effectively applied uniformly over fairly large areas, as is possible in even-aged management. In uneven-aged management, cutting is much more dispersed over the total management unit, and uniform areas treatment is less practicable. *See* FOREST AND FORESTRY.

[KENNETH P. DAVIS]

Bibliography: K. P. Davis, *Forest Management, Regulation and Valuation*, 2d ed., 1966.

Forest planting and seeding

The establishment of a new forest by seeding or, more generally, by planting of nursery-grown trees on forest land that fails to restock naturally with adequate numbers of the right species of trees (reforestation), or by such planting on nonforested land (afforestation).

Since artificial reforestation began about 1368 in Europe, it has become an integral part of forest practice on all continents on which trees grow. By 1966 almost 30,000,000 acres had been planted in the United States. The leading states were Georgia (2,400,000 acres) and Florida (2,300,000 acres). Other states with more than 1,000,000 acres of forest plantings were Alabama, Louisiana, Michigan, Mississippi, New York, Oregon, South Carolina, and Wisconsin. Probably 40,000,000 acres still require planting or seeding in the United States.

The principal purpose of forest planting is production of wood for such products as lumber, structural timber, pulpwood, posts, and poles. About 1,300,000 acres of the total acreage planted to 1966 in the United States was for windbreaks and shelterbelts to protect homes, fields, and feed lots from severe winds. Lesser but increasing amounts of tree planting have been to protect watersheds, provide food and cover for wildlife, and improve the scenic values of roadsides, parks, and other recreational areas (Fig. 1). Tree planting also is being tested for noise abatement along freeways and to give early indication of air pollution in industrial areas. *See* FOREST ECOLOGY; FOREST RESOURCES; SOIL CONSERVATION; WATER CONSERVATION.

Seed-producing areas. Until the 1950s most tree seed for forestry purposes was collected from wild stands or ordinary plantations. Since 1951 there has been an increasing development of seed-production areas (Fig. 2a) (high-quality stands or plantations in which the best trees are left for seed production and all poorer trees removed) and seed orchards (Fig. 2b) (established with seedlings or grafted trees from selected parents). As of 1966, there were 10,000 acres of seed-production areas and almost 4000 acres of seed orchards in the United States, largely in the South. To facilitate control

Fig. 1. A plantation of white pine established in 1896, one of the oldest in Michigan, photographed in 1952. It has provided wood products and is also providing recreational values. (*U.S. Forest Service*)

Fig. 2. Seed-producing areas. (*a*) Stand of loblolly pine converted to a seed-production area. Selected trees are given wide growing space so they will yield heavy crops of cones. (*b*) Seed orchard established with grafts of superior loblolly pine. Only superior progeny will be continued. Eventually such an orchard will produce certified tree seed. (*North Carolina State University*)

of these improved seeds, by 1967 some 15 states had laws requiring labeling of tree seeds, and 8 states had seed certification standards.

Seed handling. Reforestation by planting or direct seeding requires an adequate supply of good seed of species and origins well adapted to the planting site.

Collection. To assure high germinability and keeping qualities, seeds should be collected when they are ripe and before they have suffered deterioration on the tree or on the ground. Often the best time to collect is when the first seeds begin to fall naturally, but large-scale operations must begin sooner than that to avoid substantial losses of good seed. The best time for seed gathering varies for each species from season to season and place to place. Seeds of most species are best collected in the fall, but those of the aspens, cottonwoods, most elms, red maple, silver maple, poplars, and willows are collected in the spring. Seeds of cherries, choke cherries, Douglas fir, mulberries, Siberian pea tree, and plums can be gathered in the summer, and those of some ashes, yellow birch, box elder, Osage orange, black spruce, Norway spruce, sycamore, and walnuts can also be collected in the winter.

Forest tree seeds commonly are gathered from standing trees. Sometimes, felled trees provide a cheap source. In the Pacific Northwest cones often are gathered from squirrel hoards.

Seeds of many tree species must be extracted from the fruits, such as cones, berries, or pods, and cleaned to facilitate seed storage and handling.

Storage. Commonly, forest tree seeds must be stored for a few months up to several years. Seeds of many trees can be kept reasonably viable for 5 – 10 years if stored at subfreezing temperatures.

Seeds of most legumes can be stored dry at room temperature for several years.

Pretreatment. Of some 400 species of woody plants studied, 33% have seeds that germinate readily, but 7% have seeds with impermeable coats, 43% have seeds with internal dormancy, and 17% have more than one kind of seed dormancy. All these require special pretreatment to induce reasonably prompt germination.

Seed-coat impermeability often can be overcome by treatment with acids or abrasives. Internal dormancy usually requires cold, moist treatment or fall sowing to promote germination. Seeds with more than one kind of dormancy often require either a combination of seed-coat softening and cold, moist treatment, or sowing soon after collection in the later summer or early fall.

Planting conditions and requirements. Except where plant competition is slight and moisture conditions good, special ground preparation usually is needed. On rough, hilly, or mountainous land with considerable grass, herbaceous, or brushy cover, a spot about 18 in.² is scalped off with a heavy hoe or mattock just before planting each tree. Where machinery can be operated efficiently, heavy brush plows, disks, bulldozers, root rakes, or shearing blades are used to clear off the brush and expose mineral soil in lanes or over the entire area on which to plant the trees (Fig. 3). On sod an ordinary one-bottom plow or a forestry plow may be used to open parallel furrows 6 to 10 ft apart in which trees are planted at 6- to 10-ft intervals; or narrow strips are treated with aminotriazole or similar herbicides that may permit planting without furrowing. On rolling land, furrows are placed on the contour to conserve moisture and reduce soil washing.

Chemical sprays of 2, 4-D or 2, 4, 5-T, alone or in

combination, applied by ground equipment or by aircraft are often used to eliminate unwanted shrub or tree growth. Similarly, spraying simazine will control grass and herbs. Controlled burns, sometimes in combination chemical sprays applied afterward, are used to prepare brushy or recently logged areas. *See* HERBICIDE.

Tree size for planting. Trees usually are grown 1–4 years in nurseries so that they have 4- to 12-in. tops and 8- to 10-in. roots before they are field-planted. Some are transplanted once in the nursery before the final planting.

Tree planting season. Most tree planting is done at the beginning of the growing season when the trees are still dormant and can be moved with little damage and when the soil is moist. In most regions such conditions occur in spring, but in the southern United States and other areas of long growing season, planting often is best done in early winter.

Species planting regions. Tree species commonly planted are as follows: northern and northeastern United States—red, eastern white, and jack pines, and white, black, and Norway spruces; southern United States—shortleaf, loblolly, slash, and longleaf pines; western United States—Douglas fir, ponderosa, western white, and sugar pines, and Engelmann and Sitka spruces; Europe—Scotch and black pines, Norway spruce, larches, Douglas fir, and beech; Asia—native or well-adapted species of pine, spruce, fir, cypress, cedar, larch, eucalypts, oak, ash, and other hardwoods; Australia—Monterey and southern pines; South America—Monterey and southern pines, eucalypts, and hybrid populars; and South Africa—Mexican pines and acacias.

Tree planting by machine. Level to moderately rolling land may be planted by special machines largely developed since 1930. Most of them are one-row or two-row units that are pulled by a crawler tractor (Fig. 4) or, on fairly smooth terrain, by a wheel tractor. Two men and a machine can plant 10,000 or more trees per day. A large part of the tree planting in the United States is now done by machine.

Tree planting by hand. On very small areas or where the land is steep, rocky, or stumpy, trees are planted with a mattock, planting hoe, or planting bar in exposed mineral soil. One man can plant about 200–500 trees per day.

Tree planting in pots or bands. In drier regions where planting failures are frequent with bare-rooted stock, trees are sometimes grown in pots of clay, cement, or fiber or bands of asphalt paper, and transported to the planting sites in the container. The plant with its block of soil held intact by the roots is removed from the permanent container and planted on the site. If asphalt bands or fiber pots are used, container and all are planted because water can reach the roots and the containers deteriorate naturally.

Tree "bullets." In Canada and northern United States tree seeds have been sown in bottomless, slit, plastic tubes about 1 in. in diameter and 6 in. long. The tube with the germinated seedling is set out in the field with a special tool. Planting can be done throughout the growing season.

Seeding. Often the nut species such as oak, walnut, and beech are sown directly in the field by dropping the seeds or by drilling them in furrows, prepared terraces, scalped spots, or recently

Fig. 3. A poor aspen and shrub area being prepared for planting by an Athens-type disk, which is pulled by a crawler tractor. (*U.S. Forest Service*)

Fig. 4. Tractor-drawn tree-planting machines reforesting sandy old field in Michigan. (*U.S. Forest Service*)

Fig. 5. Jack pine seedlings established by direct seeding on an area that had been burned in northeastern Minnesota. (*U.S. Forest Service*)

burned-over areas. Pine, spruce, and Douglas fir seeds also have been direct-seeded successfully (Fig. 5), but since the results usually are sufficiently uncertain, planting is more economical in most cases. Of the total area reforested on the National Forests through 1966, about 9% was direct-seeded. More than two-thirds of this area is in the South and in the Pacific Northwest, where milder climatic conditions, the use of seed treated to protect from birds and rodents, and large-scale operations with aircraft and helicopters have given reasonably good success at moderate cost.

Plantation care. Tools or chemicals are used to remove overtopping weeds, brush, or inferior trees that frequently retard plantation development. Plantations also may require sprays to protect them against insects or diseases. Some diseases can be controlled by removal of the alternate host plants. Plantations need protection from forest fires. Injury from hares, rabbits, rodents, and larger mammals may be reduced by use of rodenticides, repellents, or fences. *See* FOREST AND FORESTRY; FOREST ECOLOGY; FOREST MANAGEMENT AND ORGANIZATION.   [PAUL O. RUDOLF]

Bibliography: S. W. Allen and G. W. Sharpe, *An Introduction to American Forestry*, 1962; D. M. Smith, *The Practice of Silviculture*, 1962; J. W. Toumey and C. F. Korstian, *Seeding and Planting in the Practice of Forestry*, 1942; U.S. Forest Service, *Wood-Plant Seed Manual*, 1948; P. C. Wakeley, *Planting the Southern Pines*, 1954.

Forest resources

Forest resources consist of two separate but closely related parts: the forest land and the trees (timber) on that land. In the United States forests cover one-third of the total land area of the 50 states, in total, some 759,000,000 acres. The fact that one out of every three acres of the United States is tree-covered makes this land and its condition a matter of importance to every citizen. Recognizing this importance, Congress has charged the U.S. Forest Service with the responsibility of making periodic appraisals of the national timber situation. Most of the following data are taken from the appraisal published in 1965.

Some 509,000,000 acres, or about two-thirds, of the nation's total forest area is classified as commercial forest land, that is, land suitable and available for growing commercial crops of wood. The remaining 250,000,000 acres is noncommercial forest consisting either of land with soils so poor or rocky that it is incapable of growing a commercially useful timber stand or of forest land being held for recreation or other nontimber purposes. Table 1 gives details of the distribution of forest land in the four regions of the nation.

Most of the noncommercial forests are in public ownership, including approximately 16,000,000 productive forest acres legally withdrawn for uses such as national parks, state parks, and national forest wilderness areas. Of the remaining unproductive forest, about 112,000,000 acres are in Alaska (part of the Pacific Coast region). Most discussions of forest resources, however, concentrate upon the commercial areas, and the following discussion primarily concerns this commercial forest land.

The South alone has some 39% of the total commercial area, while 34% is in the North and the remaining 27% in the West. Within this general pattern are very large differences: Maine, for example, has fully 87% of its surface covered with commercial forests, while North Dakota and Nevada are at the opposite extreme with less than 1% of their area similarly utilized.

Changes in total forest area tend to be relatively small, even though substantial shifts in land use may take place in particular states. Between 1953 and 1963 total commercial forest acreage increased by about 1 1/2%, that is, almost 7,600,000 acres (as compared with the net gain of 24,000,000 acres for 1943–1953). Most of this addition (6,700,000 acres) occurred in the South, while the West underwent a small reduction. The rapid changes in United States agriculture have led to abandonment of many marginal farms—a major source of "new" forest land today, and more than enough to offset the forest areas lost to highways, new farm establishment, pipelines, rights-of-way, and other enterprises.

Forest types. There are literally hundreds of tree species used for commercial purposes in the United States. The most general distinction made is that between softwoods (the conifers, or cone-bearing trees, such as pine, fir, and spruce) and hardwoods (the broad-leafed trees, such as maple, birch, and aspen). Viewed nationally, the forests of the United States are very evenly divided between these two major types: softwoods, occupying 47%, and hardwoods, occupying 53% of the commercial forest area. The division is very uneven, however, between the East and the West: Hardwoods dominate the East by a ratio of more than 2:1, while in the West there are more than 11 acres of softwoods for every hardwood acre. Nearly half the softwood acreage of the East is composed of loblolly and shortleaf pine, and another quarter by longleaf pine types. Together these southern pines make up nearly 40% of the South's commercial forest land. In the northern Lake states and New England the most important softwoods are the spruce-fir type and the white-red-jack pine type, but together these account for only about 20% of the North's commercial forest area.

Of the eastern hardwood forests, nearly half are oak-hickory types, containing a large number of species but characterized by the presence of one or more species of oak or hickory. Other important eastern types are the maple-beech-birch (found throughout the New England, Middle Atlantic, and Lake states regions), the very valuable swampland and bottomland hardwoods of the South, and the aspen-birch types which cover so much of the cut-over land of the North.

Douglas fir and ponderosa pine dominate the western forests, each making up almost 30% of the region's total commercial forest area. These two types are the principal sources of the nation's lumber and plywood. Nearly all the commercial forest land in coastal Alaska is composed of the hemlock-sitka spruce type. Hardwoods, mostly in Oregon and Washington, occupy only about 8% of the West's commercial forest land.

Table 1. Distribution of forest land in the United States

Type of forest land	Area, thousands of acres				
	Total	North	South	Rocky Mountains	Pacific Coast
Commercial forest	508,845	171,789	201,069	65,623	70,364
Noncommercial forest					
Unproductive	234,012	2,589	17,956	70,499	142,968
Productive-reserved	16,008	4,062	1,279	7,200	3,467
Total	250,020	6,651	19,235	77,699	146,435
Total forest land	758,865	178,440	220,304	143,322	216,799

Growth. Growth on these areas has been more than enough to match the harvest over the past decade, but the fact that growth and drain are in approximate balance provides no assurance that all is well. Much of the growth is by low-quality hardwoods in the East, yet fully 70% of United States demand for raw materials is met from the softwood production of the West and South. Potentially the most productive forest land is that of the West Coast states, where it is estimated that 24,000,000 acres are capable of growing more than 120 ft³ annually and another 15,000,000 could exceed 85 ft³ per acre per year. The next most productive area is the South, with only 16,000,000 acres of the highest quality forest (more than 120 ft³/acre annually) but more than 60,000,000 acres of good quality land (85–120 ft³). The North is considerably less well off as far as forest growth rates are concerned.

Stocking. After the potential productivity of the soil itself, the most important factors in growth of timber are probably stocking (the number of trees per acre) and the age of the trees. Gradually, as fire protection and forest management have been extended to larger and larger areas and as the practice of forestry has won more widespread acceptance by industry and by the general public, the stocking of forest lands has been increased and is still increasing. There is still a long way to go, however, before commercial forest areas can be classified as well stocked. Fully one-fifth of all forest land was estimated to be less than 40% stocked in 1962, while on 35,000,000 acres less than 10% of the land was tree-covered.

The commercial forests contain a truly vast amount of sound wood—almost 700,000,000,000 ft³ at the end of 1962. Only 10% of this timber consisted of trees of poor or diseased condition or otherwise useless for harvesting, or of dead trees that might still be utilized. And almost two-thirds of the total was in trees sufficiently large to yield at least one sawlog and called, therefore, sawtimber (Fig. 1). More than 80% of this sawtimber inventory

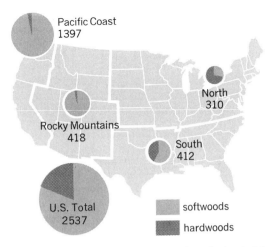

Fig. 1. Map indicating distribution of sawtimber in billions of board feet for all species as of Jan. 1, 1963. (*U.S. Department of Agriculture*)

was in conifers, with the remainder in hardwoods, a fact that highlights an important facet of timber distribution in the United States: The western states have only one-fourth of the nation's commercial forest land, but on this land 57% of the growing stock and 72% of the sawtimber are found. Further emphasizing the importance of the West to timber supplies is the fact that almost one-fourth of all United States sawtimber is Douglas fir. Oaks are the principal hardwood sawtimber species.

Timber age and cut. The other factor affecting timber growth, that of age, must be considered before forest growth can be understood. The very heavy stands of virgin timber in the West are growing relatively slowly. Only as these old stands are replaced with young vigorous trees will growth of the western forests begin to balance the cut they support. At the time these stands were inventoried, growth of western softwood sawtimber averaged only 61% of the cut, with most of the imbal-

Table 2. Summary of growth and cut situation on commercial forest area of the United States (by regions and species group)

Section and species group	Commercial forest area		Growing stock			Sawtimber		
	Millions of acres	Percent	Growth*	Cut*	Growth/cut ratio	Growth†	Cut†	Growth/cut ratio
North								
Softwoods	34.1	6.7	1,044	560	1.86	2,799	1,881	1.49
Hardwoods	137.7	27.1	3,792	1,136	3.34	9,676	4,245	2.28
South								
Softwoods	80.6	15.8	4,363	2,492	1.75	16,948	8,406	2.01
Hardwoods	120.5	23.7	3,107	1,744	1.78	8,382	6,968	1.20
Rocky Mountains								
Softwoods	59.7	11.7	866	646	1.34	3,462	3,822	0.91
Hardwoods	5.9	1.2	65	5	13.35	108	18	6.15
Pacific Coast								
Softwoods	65.2	12.8	2,755	3,492	0.79	12,657	22,638	0.56
Hardwoods	5.1	1.0	275	73	3.79	821	423	1.94
United States totals								
Softwoods	239.6	47.0	9,027	7,191	1.25	35,866	36,748	0.98
Hardwoods	269.2	53.0	7,238	2,957	2.45	18,988	11,653	1.63
Totals	508.8	100.0	16,265	10,148	1.60	54,853	48,401	1.13

*In millions of cubic feet.   †In millions of board feet.

ance in the highly important Douglas fir and ponderosa pine types. With southern pines, on the other hand, total growth was 1.7 times the cut, and in the important sawtimber size classes the margin was two to one. The hardwood growth/cut ratio appeared to be even better: the ratio of hardwood growth to cut was 2:4, and in sawtimber was 1:6. Data on growth and cut are summarized in Table 2.

Even with reasonably accurate data of growth and cut, there is room for misinterpretation of the forest situation. Total growth exceeds cut in most regions, but much of that growth is occurring on trees of low quality. Only 11% of the saw-log-sized hardwoods in the East, for example, were classed as no. 1 factory lumber logs by Forest Survey crews, with another 18% falling in the no. 2 category. Moreover, a substantial fraction of the growth is being made by trees too widely scattered to be commercially harvestable or located in areas economically inaccessible under current conditions. It is one thing—and a very good thing—to have the national timber cut balanced by forest growth; it is quite another to have that cut balanced by quality growth of species that are demanded by forest industries and located so that the timber can be logged profitably.

Ownership. The future condition of forest land in the United States is dependent to a very large extent on the decisions of the people who own these areas. The key factors in understanding the forest situation, therefore, are the forest-land ownership pattern and the attitude these owners have toward forest management and hence toward the future of the forest lands they hold.

About 72% of the commercial forest is in private ownership (367,000,000 acres). Of the remaining 28% held by various public owners, some 19% is in national forests, about 3% in other Federal ownership, 4% in state holdings, and 2% in county and municipal control. Among private owners one can distinguish three major classes: industrial, farm, and a very heterogeneous group labeled miscellaneous. Table 3 gives the details as well as the distribution between the four major regions of the United States.

Private owners. About 13% of the privately owned commercial forest area is owned by industries, with the remainder almost equally divided between farmers (30%) and the miscellaneous private owners (29%).

Today some of the most productive forest land of the United States is in private industrial holdings. Pulp and paper companies lead the forest-based industries with nearly 35,000,000 acres of forest land, much of it concentrated in the southern states. More and more, however, as forest industries become integrated, the distinction between pulp and paper, lumber, and plywood companies grows more difficult to make, and certainly is much less meaningful than in former years.

The 29% in the "miscellaneous private" category consists of a tremendous variety of individuals and groups, ranging from housewives to mining companies. Relatively few such owners are holding this land for commercial timber production. Some, such as railroad companies or oil corporations, may well be interested in producing timber while holding the subsurface mineral

Table 3. Area of commercial forest land by owner group and region*

Owners	Regions of the United States			
	North	South	Pacific Coast	Rocky Mountains
Public owners				
Federal				
National forest	10,265	10,476	32,665	43,398
Bureau of Land Management	81	27	3,242	2,076
Bureau of Indian Affairs	1,198	251	2,196	2,816
Other	964	3,308	182	31
Total	12,508	14,062	38,285	48,321
State	12,751	2,164	3,589	2,340
County and municipal	6,748	656	361	83
Private owners				
Industry				
Pulp and paper	10,797	21,614	2,611	0
Lumber	2,996	12,551	8,031	2,535
Other	523	3,257	1,713	0
Total	14,316	37,422	12,355	2,535
Farmer	55,503	78,897	7,848	8,769
Miscellaneous	69,963	67,868	7,926	3,575
Total	171,789	201,069	70,364	65,623

*From U.S. Forest Service, *Timber Trends in the United States,* Forest Resource Rep. no. 17, USDA, February, 1965.

rights, but most "miscellaneous private" owners are interested in other, nontimber, objectives.

Farmers own what is potentially the best forest land in the country, but it is also the poorest managed and the poorest stocked of all ownership categories. Most farm forests have been cut over several times, and because farmers are primarily interested in the production of other kinds of crops, they are seldom concerned with the condition of their woodlands. The relatively poor condition of farmer-owned woodlands has been a source of much disappointment, discussion, and considerable action on the part of public conservation agencies. Whether this is of crucial significance to the forest reserves of the nation is a matter of dispute. In past years some professional foresters have argued that these farm woodlands should somehow be made to contribute their full share to the timber supply of the nation. Others, of a more economic persuasion, have argued that as long as the farmer has opportunities for investing time and money in ways that will yield a greater return, he should not be expected to worry about timber production. Farm owners are often unfamiliar with forest practices, usually lack the capital required for long-term investments, and in many cases are simply uninterested in growing trees.

Public owners. Public agencies of several kinds hold large forest areas, the most important being the national forests. These contain 96,800,000 acres of commercial forest land and are managed and administered by the U.S. Forest Service, a bureau of the Department of Agriculture. The Bureau of Land Management oversees some 5,400,000 acres, and the Bureau of Indian Affairs another 6,500,000 acres. Various other agencies, especially those of the Armed Services, administer the remaining 4,500,000 acres under Federal supervision. State ownerships total 20,800,000 acres, while counties and municipalities control another 7,800,000 acres. Most of the public holdings are managed under multiple-use principles. As is true on most forest land, wood has

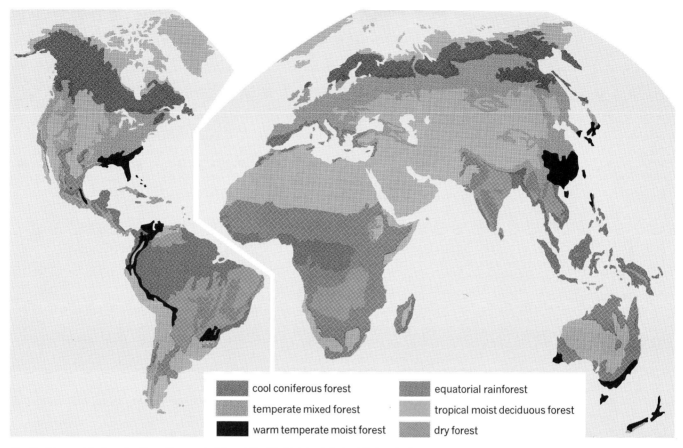

Fig. 2. The world's forests. (*From Economic Atlas of the World, 3d ed., Oxford University Press, 1965*)

cool coniferous forest

temperate mixed forest

warm temperate moist forest

equatorial rainforest

tropical moist deciduous forest

dry forest

always been the principal product of the national forests, but in recent years increasing attention has been given to recreation, wildlife, forage, and water. *See* CONSERVATION OF RESOURCES.

Public holdings now include over 60% of the softwood growing stock but only 16% of the hardwoods. Fully two-thirds of all softwood sawtimber is publicly owned, mostly on national forests. This high concentration of sawtimber makes many of the wood-based industries of the nation highly dependent upon government-owned raw material supplies.

World forest resources. In terms of forest land, the United States is far better off than many nations but not as rich as others. South America, for example, has more than 50% of its land surface forested, which may be contrasted with Great Britain with only a little over 7% of its surface so utilized. Actually, about one-third of the world's land surface is forested—the same average percentage as that observed for the United States as a whole.

In many countries, and especially in tropical areas, forests have not been well explored, nor have their boundaries been defined. In such circumstances only approximate areas and general conditions can be given. Moreover, since forests reflect differences in soil, climate, situation, and past land use, merely listing the forest types of the world would take several pages. Yet there is value in distinguishing very generalized forest types by broad locational patterns (Fig. 2).

Coniferous and temperate mixed forests. In Europe, the Soviet Union, North America, and Japan, the forests are predominantly coniferous, a fact that has been of great significance in shaping the pattern and nature of wood use in the industrialized part of the world. Closely associated with the industrialized nations one also finds the temperate mixed forests, which normally contain a high proportion of conifers along with a few broad-leafed species.

Most of the temperate mixed forests are in use, for these heavily populated areas have well-developed transportation systems. Growth rates in both the coniferous and temperate mixed forests of the North Temperate Zone are similar to those already given for the United States, which is an excellent example of the general region.

Tropical rainforests. These are made up exclusively of broad-leafed species and include the bulk of the volume of the world's broad-leafed woods. These forests are concentrated in and around the Amazon basin in South America, in western and west-central Africa, and in southeastern Asia. Generally characterized by sparse population and only slight industrial development, these forests have been little used. The fact that they typically contain many species within a small area (as many as 100 species per acre) has also served to limit their use, even though tropical rainforests include some of the most valuable of all woods, such as mahogany, cedar, and greenheart from South America; okoume, obeche, lima, and African mahogany from Africa; and rosewood and teak from Asia.

Rate of growth in the tropics can be very high, and from the more than 2,000,000,000 acres of these forests man may someday learn to obtain a major part of his needed wood fiber. Today, however, so little is known about many of the species and so few areas are under any form of systematic management, that little reliance can be placed on these vast areas to satisfy the future wood needs of mankind.

Savanna. Most of the other forests in the tropics and subtropics are dry, open woodlands, or savannas. These forests contain low volumes per acre, mostly in small sizes, and only a few of the species are of commercial value. Much of Africa (excluding western and west-central Africa) is of this low-yielding type.

Management. In recent years intensive afforestation has received considerable attention. By using measures such as careful soil preparation prior to planting, application of fertilizer, and even irrigation to adapt the environment to high-yielding species, returns from forests can be tremendously increased. For example, yields 400–500 ft³/acre are common with eucalyptus in both South America and Africa or with poplars in southern Europe. Only slightly smaller yields are obtained from the fast-growing pines under similarly intensive management. The high-yield potential of man-made forests makes them much more important than their area might indicate. According to plans already reported, most nations intend them to make an even greater contribution in the future. *See* FOREST AND FORESTRY: FOREST MANAGEMENT AND ORGANIZATION.

[G. R. GREGORY]

Bibliography: Food and Agriculture Organization of the United Nations, *Wood: World Trends and Prospects*, Freedom from Hunger Basic Study no. 16, Rome, 1967; U.S. Forest Service, *Timber Trends in the United States*, Forest Resource Rep. no. 17, USDA, February, 1965.

Forest soil

The natural medium for growth of roots of trees and associated forest vegetation. This relationship with forest vegetation gives rise to characteristics that distinguish forest soils from soils formed under other vegetation systems. The most obvious feature is the humus layer, or horizon, which is peculiar to the microenvironment imposed by the forest (Fig. 1). The humus horizon affects germination of seeds, influences soil moisture distribution, serves as a reservoir of nutrient elements, influences the susceptibility of forest stands to fire, and represents the energy source for most soil-inhabiting organisms.

The process of humification starts when residues of forest plants and animals fall to the soil surface and are gradually decomposed. The course and extent of decomposition, which determines humus characteristics, depends chiefly on the microclimate, tree species, associated flora and fauna, and the nature of the underlying mineral material. Organic matter may accumulate on the soil surface with little or no mixing with mineral

Fig. 1. Soil under an uncut forest is friable and porous, and it is deeply penetrated by roots and infiltrated with organic matter. (*USDA*)

material. This form is called mor, raw humus, or ectohumus. In the mull form, most of the organic matter is incorporated into the mineral soil, and only coarse debris, such as twigs and petioles, remains on the surface (Fig. 2). Transitional types combining characteristics of both mor and mull also occur. Mor humus is usually associated with a cool, moist microclimate and predominantly coniferous or ericaceous vegetation growing on acidic parent material. In these situations, the animal life generally responsible for mixing organic matter with mineral soil is lacking and the population of microflora tends to be dominated by fungi.

Leaf litter associated with mull soils is usually

Fig. 2. Forest cover returns large quantities of organic matter to the soil each year. The depression in the center shows the depth of litter. (*USDA*)

richer in calcium and other nutrient elements and hence supports a much larger variety of small animals and microorganisms than is found in mor layers. Metabolic activity of these organisms is stimulated by somewhat higher surface soil temperatures so that litter decomposition proceeds more rapidly. Earthworms, millipedes, and similar animals mechanically break down litter and incorporate it into mineral soil. Simultaneously, the litter is inoculated with a great host of fungi and bacteria which gradually converts the organic matter into relatively stable complexes of lignin and protein. These processes darken the upper portion of the mineral soil.

Humus relations. Factors that govern humus formation are extremely complex, and their relative importance is poorly understood. A more definite relationship exists between humus characteristics and the horizons that occur deeper in the soil profile. Many of the biochemical reactions responsible for profile development originate in the humus layer, and by-products of these reactions are transmitted downward by percolating water. Mor humus is often associated with pronounced weathering of primary minerals and leaching of the breakdown products, particularly iron, to positions lower in the profile, whereas in mull soils the profile transitions may be less distinct. Because of the delicate balances that function in organic matter decomposition, the humus layer responds rather quickly to changes in the microenvironment that may be caused by modification of the forest stand by wind damage, fire, or silvicultural operations. By contrast, deeper soil horizons are relatively stable and usually will show little change unless the soils are mechanically disturbed, as when trees are uprooted.

Root relations. The humus layer is the focal point of biological activity in forest soils. In addition to litter-decomposing animals and microbes, it usually contains roots of forest vegetation. Portions of the root systems of many trees and other woody plants exhibit modifications called mycorrhizae (Fig. 3), which result from invasion of the roots by certain fungi. The fungus is apparently seeking soluble carbohydrates transported from the leaves to the younger portions of the roots, and the root thus altered is evidently more efficient in absorbing water and nutrients to be used by the tree.

Because the supply of moisture and nutrients is generally favorable, most of the absorbing roots of trees and other forest vegetation are located in and immediately below the humus layer. This arrangement allows maximum utilization of nutrient elements released by decomposition of organic material, and although the root network and its associated microorganisms are extremely efficient in nutrient recovery, some leaching loss does occur. Such loss is balanced by rhizospheric weathering of rocks and minerals by roots occupying deeper soil layers. Forest soil depth is determined by root penetration which may vary from a few inches to more than 50 ft. Trees usually utilize a much larger volume of soil to maintain growth than do agricultural crops.

Classification. Forest soils may be classified in various ways. The classification of the Soil Survey Division of the Bureau of Plant Industry, Soils, and

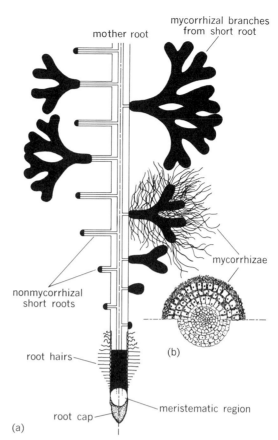

Fig. 3. Development of mycorrhizae on a pine root. Solid areas indicate absorbing surfaces. (a) The main axis, a mother root; (b) cross section, representing a mycorrhizal root. (*From P. J. Kramer, Plant and Soil Water Relationships, McGraw-Hill, 1949*)

Agricultural Engineering is sometimes used, but since this system is based on soil features important to agriculture, it usually must be modified for forestry purposes. Forest composition, minor vegetation, parent material, topography, depth to groundwater, total soil depth, and growth and yield of forest stands all have been used, individually and in various combinations in classifying forest soils. Classifications based on forest productivity have been found to be fairly successful. In such systems, tree growth measured by height, diameter, or volume is related to soil properties, texture, structure, aeration, water infiltration, moisture-retention capacity, depth, type and distribution of organic-matter nutrient supply, and other characteristics known to influence growth of the root systems. These relationships are then used to predict potential productivity of understocked or deforested lands with similar soil characteristics. Because of the large areas involved, aerial photographs are frequently used as a mapping base on which forest cover types and soil boundaries are delineated.

Hydrologic cycle relations. In addition to their role in producing cellulose, timber, and other forest benefits, forest soils perform an important regulatory function in the hydrologic cycle. Well-managed forested watersheds are often characterized by soils with a high water-infiltration capacity. Such soils are usually well protected from the destructive energy of falling raindrops that cause surface sealing and promote overland flow. Excessive overland flow results in soil erosion and reduction in water quality. Where falling rainwater or meltwater from snow has maximum opportunity to penetrate into the soil, normal drainage channels such as streams and rivers are recharged gradually through underground flow and a steady supply of high-quality water is assured. *See* HYDROLOGY; SOIL.   [GARTH K. VOIGT]

Fresh-water ecosystem

An ecosystem is the functional unit of ecology, consisting of living organisms and the nonliving environment interacting upon each other to produce an exchange of energy and materials between the living and nonliving components. A lake or pond in its entirety is such an ecosystem. A stream is also an ecosystem, although less well differentiated. Limnology is the comprehensive study of all the components of inland aquatic ecosystems and their interrelationships. *See* ECOLOGY; LIMNOLOGY.

The energy of fresh-water ecosystems is derived mainly from photosynthesis accomplished by the algae suspended in the water and by the higher plants and algae growing on or in the bottom. A variable proportion of the total energy available is derived from allochthonous organic matter, such as leaves and pollen, produced by terrestrial communities. The materials cycled within an ecosystem are derived ultimately from the weathering of rocks and the leaching of soil in the watershed, and to a lesser extent from the air.

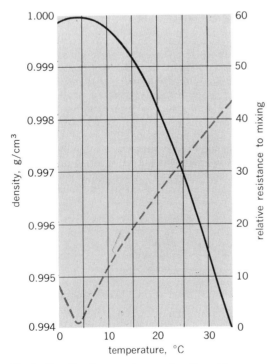

Fig. 1. Density of pure water as a function of temperature (solid line). Maximum density is at 3.94°C. The dashed line shows the relative resistance to mixing per degree difference in temperature in a water column 1 m long. Resistance to mixing between 4 and 5° is taken as unity. (*From R. C. Weast, ed., Handbook of Chemistry and Physics, 48th ed., Chemical Rubber Publications, 1967*)

Fresh-water habitats are conveniently divided into a lenitic or basin series, such as lakes, reservoirs, ponds, and bogs, and a lotic or channel series, such as rivers, streams, brooks, springs, and groundwater. The lotic series is distinguished by a continual flow of water in one direction. The limnology of the lenitic series of habitats is better understood because of the greater amount of research done on it.

LENITIC ECOSYSTEMS

The biota and biological processes of lakes and ponds operate within a framework of physical factors imposed by the external climatic environment, of which the most important are temperature stratification, light stratification, and waves and currents.

Temperature stratification. Heating of natural waters occurs mainly by the direct and rapid absorption of infrared radiation of wavelength > 0.76 microns (μ). Because 1 m of distilled water absorbs 90% of this radiation, the direct heating of water is important only in a relatively thin surface layer. Heat at greater depths has been transported there largely by wind-generated currents and turbulence. All heat in a lake in excess of 4°C is commonly referred to as wind-distributed heat.

Two water masses of different temperatures offer a resistance to mixing by the wind which is proportional to the difference in density between them. Because of the nature of the temperature-density relationship of water, the resistance to mixing is greater at high than at low temperatures (Fig. 1). As a consequence, in most lakes of sufficient depth, 10 m or greater, the wind is unable to circulate the entire lake in summer. Therefore, the lake becomes stratified (Fig. 2) into an upper, warm, freely circulating zone, the epilimnion; a lower, cold, relatively noncirculating zone, the hypolimnion; and an intermediate zone of rapid temperature change, the metalimnion, or thermocline in North American usage.

The metalimnion effectively isolates the hypolimnion from surface processes, and as a result biological and biochemical processes occurring in deep water are cumulative during the period of stratification. This is one of the most important interactions of the lenitic ecosystem.

Depth of the metalimnion varies with the size and shape of the lake, the degree of protection from wind action, and the wind regime during the warming phase of the annual cycle.

In temperate lakes there is typically a spring overturn, during which the lake is actively circulating from top to bottom; a period of summer stratification as described above; a period of fall overturn after the lake, by cooling, has again become homothermal; and, especially under ice cover, a period of winter stratification (Fig. 3). Winter stratification is often spoken of as an inverse stratification, because the warmest water is at the bottom rather than the top. A lake not covered with ice may continue to circulate throughout the winter. *See* THERMOCLINE.

Lake classifications. Based on these temperature relationships, an early system of lake classification was devised by F. A. Forel and later modified by G. C. Whipple. All lakes were divided into three classes, polar, temperate, and tropical. This classification depended on the annual range

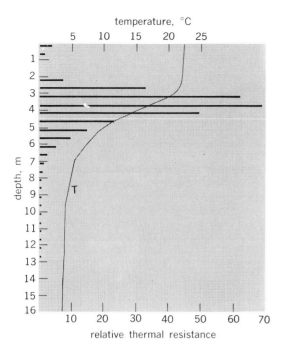

Fig. 2. A summer temperature profile (continuous curve) and relative thermal resistance to mixing (bars) for Little Round Lake, Ontario. The great thermal resistance to mixing in the metalimnion indicates a great stability. (*From J. R. Vallentyne, Principles of modern limnology, Amer. Sci., 45(3):218–244, 1957*)

of surface temperature, and each class was further subdivided into three orders based on the annual range in bottom temperatures. The first, second, and third orders are, respectively, the deep, moderate, and shallow lakes. One of the chief objections to the terminology employed in this system is that some of the best examples of tropical lakes are in Scotland and British Columbia. In these lakes, the surface temperature is always above 4°C.

A more realistic system proposed by G. E. Hutchinson and H. Löffler (Fig. 4) bases the classification on the number of periods of circulation a year, and whether the circulation occurs after warming or cooling of the lake. Dimictic lakes circulate twice a year, warm monomictic circulate once a year after a period of cooling, and cold monomictic lakes circulate once a year after a period of warming. These correspond, respectively, to the temperate, tropical, and polar lakes of the Forel-Whipple system. In addition there are the permanently frozen amictic lakes, low-altitude tropical oligomictic lakes with irregular circulation, and high-altitude tropical polymictic lakes with continuous circulation.

During turnover, most lakes circulate completely from top to bottom. Such lakes are called holomictic. In a relatively small number of lakes the water nearest the bottom has a sufficiently greater density from dissolved or possibly suspended substances so that it does not become involved in the

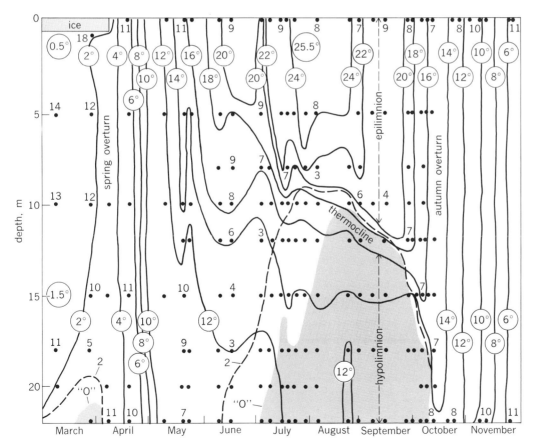

Fig. 3. Seasonal changes in water temperature and dissolved oxygen during 1906 in Lake Mendota, Wis. The figures that have no circles around them are parts per million (ppm) of oxygen. The dashed lines represent 2 ppm. The tinted area represents less than 0.2 ppm. (*After C. H. Mortimer, from G. C. Sellery, E. A. Birge, A Memoir, University of Wisconsin Press, 1956*)

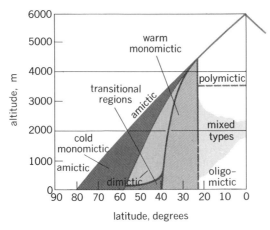

Fig. 4. The Hutchinson-Löffler system of thermal lake types. The two equatorial types occupy the unshaded areas which are labeled oligomictic and polymictic; they are separated by a region of mixed types, mainly variants of the warm monomictic type. (*From G. E. Hutchinson and H. Löffler, The thermal classification of lakes, Proc. Nat. Acad. Sci., 42(2):84–86, 1956*)

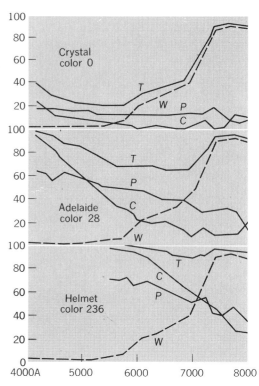

Fig. 5. Graph indicating the percentile absorption of light by water from three lakes in northeastern Wisconsin, ranging from visually uncolored (Crystal) to very dark brown (Helmet). Colors are listed as ppm of platinum units. *T*, total absorption; *P*, absorption due to suspended particulate matter; *C*, absorption due to dissolved color; *W*, absorption due to pure water. (*After H. James and E. Birge, from G. E. Hutchinson, A Treatise on Limnology, vol. 1, Wiley, 1957*)

circulation. Such lakes are called meromictic. In the noncirculating layer, or monimolimnion, of meromictic lakes the same processes occur as in the hypolimnion of a holomictic lake, but they are generally more intense and extreme because of their permanence. Sometimes the monimolimnion has a higher temperature in summer than does the overlying holomictic zone, mixolimnion, an anomalous temperature condition known as dichothermy.

Light stratification. Light is important as the source of metabolic energy in the ecosystem. The energy incorporated into organic substances by green plants through the process of photosynthesis supports the activities of all the heterotrophic organisms in the ecosystem.

Light intensity is maximal at the surface, and declines exponentially as the depth increases. Since the amount of photosynthesis accomplished is related to the intensity of light, there is a depth at which the production of organic matter by photosynthesis is equal to its utilization by respiration. This is the compensation level. The zone above is the trophogenic zone, characterized by an excess of production over consumption of organic matter. The region below is the tropholytic zone, characterized by a dominance of energy-consuming processes. In most lakes the compensation level lies in the epilimnion, which because of its continuous circulation during summer is generally regarded as the trophogenic zone.

Light passing through water also undergoes a change in its spectral composition. In any water that is sufficiently deep, the light tends to become monochromatic with increasing depth.

The factors responsible for the exponential absorption of light and its changing spectral composition are the water itself, dissolved yellow-brown color, and suspended particles or seston (Figs. 5 and 6). The effect of any dissolved color or turbidity is to reduce the thickness of the trophogenic zone and to shift maximum transmission from the blue end of the spectrum in very pure water toward the red end of the spectrum in turbid or col-

ored water. Dissolved color is generally more important than turbidity in producing these changes.

Waves and currents. Tides are negligible in even the largest inland lakes, being at most a few centimeters in the Great Lakes and in Lake Baikal. Wind, rather than gravity, is the important force generating waves and currents in inland waters. Such water movements erode and transport material alongshore, provide the turbulence necessary to keep microorganisms suspended, circulate the epilimnion or the entire lake, and provide for some turbulent transport across the metalimnial barrier.

The simplest waves in deep water are waves of oscillation, in which there is vertical rotation of particles without any net forward transport. The orbit of rotation decreases by one-half for each increase in depth below the surface amounting to one-ninth of the wavelength. In most inland lakes the water is essentially still at a depth of one-half the wavelength.

When a wave of oscillation enters water shallower than the orbit of rotation, frictional resistance with the bottom produces an unstable configuration, and the crest of the wave now plunges shoreward as a wave of translation. It is these onshore waves that erode headlands and transport sand and other material alongshore to form spits and bars, and build subaqueous terraces and benches.

With a steady wind from one direction the

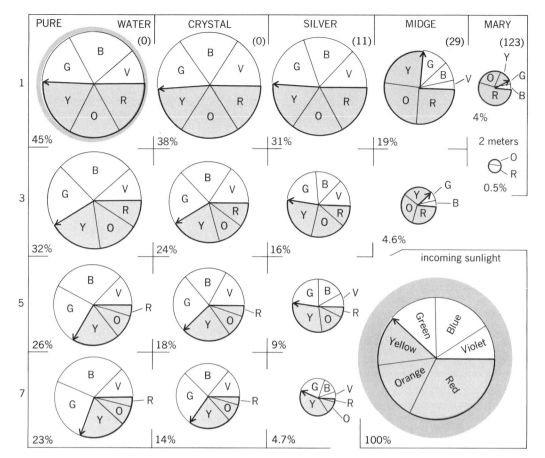

Fig. 6. Changes in color composition of light and intensity relative to the surface in a hypothetical lake of pure water and in four lakes of different dissolved color (in parentheses) in northeastern Wisconsin. Total intensities, relative to the incoming radiation, are represented by the areas of the circles and by the percentage values at their lower left. The tinted areas of the "incoming sunlight" and of "pure water" represent invisible radiation, chiefly infrared. Changing position of the clock hand in each horizontal series demonstrates selective absorption of shorter wavelengths by dissolved color in water. The numerals at extreme left show depth in meters. *(After C. H. Mortimer, from G. C. Sellery, E. A. Birge, A Memoir, University of Wisconsin Press, 1956)*

gravity-stable horizontal stratification of the lakes becomes displaced. The lake surface tilts upward slightly in a downwind direction, and the epilimnion thickens. At the same time the metalimnion tilts upward in an upwind direction, and the hypolimnion thickens here (Fig. 7). When the generating force ceases, the now unstable density surfaces begin to rock or oscillate as standing waves. The rocking motion of the lake surface, called a seiche, dampens out quickly. The rocking movement of the thermocline, known as an internal seiche, is of greater amplitude and duration and continues long after the surface may have ceased rocking (Fig. 8).

Both surface and internal seiches can be uninodal, binodal, multinodal, or even rotational in some of the larger lakes. Such internal waves are important in affecting the vertical and horizontal distribution of organisms and in generating currents and turbulence in the otherwise relatively quiet hypolimnion. In extreme instances the hypolimnion can even reach the surface of a lake on the upwind side.

BIOTA

Within this physical framework the biota and its chemical environment react upon each other. Although all organisms in an aquatic ecosystem are mutually interdependent regardless of where they occur, they are conveniently grouped according to life habit. Microorganisms living free in the water and generally independent of the bottom are known collectively as plankton. Animals living on the bottom comprise the benthos, and the larger plants rooted in the bottom the phytobenthos. Fish and a few other large animals that swim actively in the water constitute the nekton. Sessile algae and fungi and their associated groups of microscopic animals constitute the periphyton or Aufwuchs. Organisms directly dependent on the surface film comprise the neuston. The algae and microanimals living in the interstices of sand are known as the psammon. *See* BIOLOGICAL PRODUCTIVITY.

Plankton. This group constitutes the most characteristic assemblage in basin ecosystems and in the oceans and is the most important in terms of overall production. Typically the plankton consists of algae, protozoa, rotifers, copepods, cladocera, and the phantom midge larva. Unlike those in the oceans, almost none of the larger invertebrates or fishes have temporary planktonic stages. Because the organisms in this assemblage have at best only limited powers of locomotion, they are kept suspended mainly by the turbulence of the water.

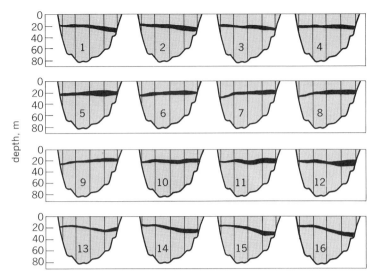

Fig. 7. Successive hourly positions of metalimnion (solid area, bounded above and below by 11 and 9° isotherms, respectively), Aug. 9, 1911, in Loch Earn, Scotland. Standing wave is an internal seiche. (*After E. Wedderburn, from C. H. Mortimer, The resonant response of stratified lakes to wind, Schweiz. Z. Hydrol., 15:94, 1953*)

Adaptations aiding this passive flotation are either a reduction in weight relative to the water, or an increase in relative surface area. Small size is an adaptation, and in many situations the bulk of the plankton is too small to be retained by the finest silk bolting cloth. This minute plankton is referred to as nannoplankton, in contrast to the larger net plankton.

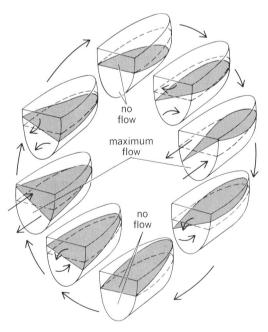

Fig. 8. Three-dimensional model showing effects of the Earth's rotation on a unimodal internal seiche in a large lake, resulting in a rotational seiche. Model fits approximately the condition observed in Loch Ness, Scotland. The metalimnion is tinted. The arrows show the direction and the relative magnitude of currents in the epilimnion and hypolimnion during various phases of the seiche. (*After C. H. Mortimer, from G. E. Hutchinson, A Treatise on Limnology, vol. 1, Wiley, 1957*)

Rate of sinking varies inversely with viscosity, which in turn varies inversely with temperature. Hence, the same organisms will sink faster in warm water than in cold. Among the planktonic cladocerans there is a summer increase in relative surface area among genetically identical generations, a phenomenon known as cyclomorphosis. The capacity for "helmet" formation is triggered by a critical temperature during development, and the helmet is maintained largely in response to turbulence in the water. Some rotifers and dinoflagellates also show increased surface area in summer.

Although a list of plankton organisms in a particular ecosystem often numbers several hundred species or more, chiefly algae, the number of species present at any one time is generally small. Thus, at any instant the majority of small to medium-sized lakes have only 1–3 species of copepods, 2–4 cladocerans, and 3–7 rotifers. In each of these groups of animals the dominant species comprises about three-fourths of the total population (Fig. 9).

Variability. A chief characteristic of plankton is its variability both from season to season and from year to year (Fig. 10). Species succeed one another often in bewildering array. Succession seems to be governed by the ever-changing combination of temperature and light, nutrient depletion, such as silica for diatoms, and by metabolites, long-chain fatty acids, carbohydrates, antibiotics, vitamins, and other substances given off to the environment that can be self-limiting to the producing species, and antagonistic or synergistic to another species. Parasitic fungi also are claimed to have a significant effect.

A peak development of a single species is known as a pulse, and when a pulse is great enough to be apparent to the unaided eye it is called a bloom. All seasons of the year exhibit pulses of at least some species, although these are often too small to affect materially the total population. Considering plankton as a whole there is generally a peak population or bloom in spring, mainly diatoms, a reduction during the summer, a second although lesser peak in fall roughly coinciding with the autumnal overturn, followed by a decline during the winter to the lowest population of the year (Fig. 11).

Vertical distribution. The vertical distribution of plankton within a lake is the result of many physical, chemical, and biotic factors. Because of the dominating influence of light, however, most species tend to have their maximum population density within the uppermost 10 m. Some of the larger zooplankters migrate toward the surface at night, and back into deeper water toward morning. Such diel vertical migration is particularly well developed in the deep-scattering layer of the oceans.

Plankton rain. In spite of turbulence there is a continual settling of senescent and dead plankton to the bottom of the lake. This plankton rain supplies energy to the deep-water benthos. It also serves as the chief cause of the extensive chemical differences between surface and bottom water that develop during summer or temporary stratification and under the ice in winter. The surface water tends to become depleted of nutrients, and the bottom water becomes enriched in them. Moreover, the utilization of plankton rain by bacteria and animals results in a reduction of the oxygen supply in deep water. The extent of oxygen reduction is also influenced by the amount of allochthonous organic matter that gets into deep water.

Plankton reproduction. Cladocerans and rotifers reproduce by parthenogenesis during most of the summer. This permits the rapid utilization of developing plankton pulses. During the deterioration of environmental conditions, such as the approach of winter, most of the cladocerans and many of the rotifers produce resistant bisexual or resting eggs that enable the species to endure during the unfavorable conditions. Resting eggs also facilitate the distribution of species by wind and birds. As a result, many fresh-water plankters, including species of algae, are virtually cosmopolitan in distribution. Copepods, on the other hand, reproduce almost exclusively by fertilized eggs. Some species form resistant eggs, but more commonly they survive unfavorable conditions by the encystment of immature stages. One species in Douglas Lake, Mich., produced more than 1,000,000 cysts per square meter of bottom in the hypolimnion. One species of *Daphnia* in the Arctic produces resting eggs by parthenogenesis.

Zooplankton. The majority of the zooplankton are herbivorous, and feed on bacteria, small algae, and organic detritus which they strain from the water by a variety of means. Their grazing activity can effectively reduce the standing crops of phytoplankton. A few cladocerans and rotifers, the phantom midge larva, and adult cyclopoid copepods are predacious. When abundant, the zooplankton can markedly reduce the oxygen content of restricted strata.

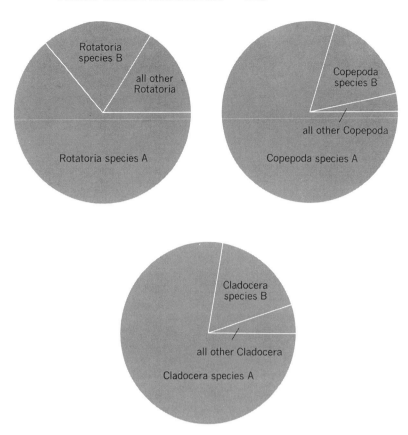

Fig. 9. Instantaneous percentage species composition of three dominant groups of zooplankters in a small- to medium-sized lake. (*From R. W. Pennak, Species composition of limnetic zooplankton communities, Limnol. Oceanogr., 2(3):222–232, 1957*)

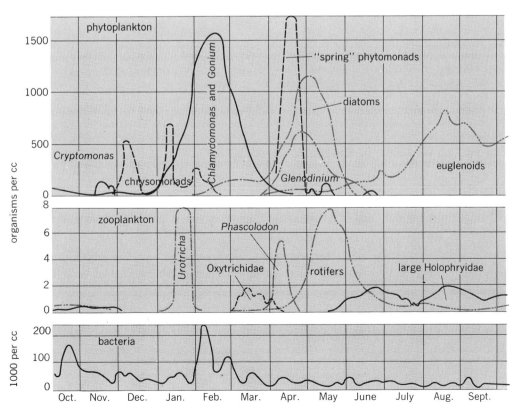

Fig. 10. Graphs illustrating seasonal succession of dominant organisms in a small permanent pond in Pennsylvania. (*From S. S. Bamforth, Ecological studies on the planktonic Protozoa of a small artificial pond, Limnol. Oceanogr., 3(4):398–412, 1958*)

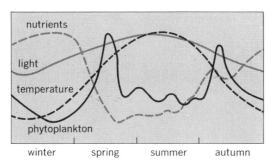

nutrients

light

temperature

phytoplankton

winter   spring   summer   autumn

Fig. 11. Typical standing-crop curve for phytoplankton in response to temperature, light, and nutrients. Decline of the spring bloom is at least in part a response to nutrient depletion, and establishment of the autumn bloom is at least in part a response to greater nutrient availability. (*From E. P. Odum and H. T. Odum, Fundamentals of Ecology, 2d ed., Saunders, 1959*)

Phytobenthos. The rooted aquatic plants form the other major photosynthetic element in fresh water. The relative importance of the phytobenthos and phytoplankton in overall production varies according to basin morphometry. As a lake becomes shallower with increasing ecological age, the phytobenthos assumes a progressively more important role. Because of its longer life cycle it ties up nutrients for longer periods than does the phytoplankton, and it contributes more organic matter to the lake bottom and causes a faster rate of filling. *See* PHYTOPLANKTON.

The phytobenthos typically occurs as three concentric zones around a lake. At the shoreline and in shallow water is the zone of emergent plants, followed by the zone of floating leaf plants, in turn followed by the zone of submersed plants. At still greater depths sometimes occur meadows of the moss *Fontinalis*, which is able to thrive in weaker light than that which the higher plants require. In many situations, particularly in hard-water lakes, the coarse, branched algae *Chara* and *Nitella* form dense beds.

The phytobenthos and the thermal stratification are used to define the zonation of the bottom. From the shoreline to the lower limit of rooted aquatics is the littoral zone, the lower limit of which corresponds roughly to the compensation level. Between the littoral zone and the top of the metalimnion is the sublittoral zone, and at still greater depths is the profundal zone.

Periphyton. One of the chief functions of the phytobenthos in the overall economy of an ecosystem is to increase the amount of colonizable substrate in the trophogenic zone. The substrate in the littoral zone is often thickly overgrown with a mat of sessile algae and fungi and a large number of associated and dependent animals such as protozoa, *Hydra*, microcrustacea, rotifers, oligochaetes, insect larvae, and snails. This community of organisms, complete with producing and consuming elements, is almost the same as an ecosystem within an ecosystem. It helps contribute to the high productivity and the diversity of microhabitats of littoral regions.

Benthos. The great differences between the littoral and profundal benthos are controlled in part by the environmental conditions and in part by the

life cycles of the individual species. In the littoral zone with its coarser sediments, wave action, a continuous supply of oxygen, warm temperatures, plus light and plants, diversity of species is the chief characteristic. Insects and mollusks are dominant, many of them confined to the littoral. All other groups of fresh-water animals also occur here. In the profundal zone, on the other hand, soft sediments, virtual absence of light, low temperatures, and potentially severe chemical conditions, especially reduction in oxygen content, severely restrict the species composition. Midge larvae, oligochaetes, and fingernail clams are the most characteristic of the larger animals. In the ooze-film assemblage there may be a great diversity of microscopic animals.

The profundal benthos is largely dependent for its livelihood on the plankton rain from above. Most of the organisms feed either directly on this detritus or on the bacteria that are decomposing it. Largely confined by steep chemical gradients to the uppermost few centimeters of the sediment, these organisms can thoroughly work through the sediments in much the same way that earthworms do through the soil. The relatively few species are sometimes present in tremendous numbers of individuals, and the total population density often exceeds that of the littoral zone.

The general absence of sessile animals in the littoral zone is one of the chief biotic differences between the oceans and inland waters. Seasonal redistribution of organisms commonly occurs. Insect larvae and nymphs move to shallow water for pupation and emergence, and mollusks move shoreward to breed. Emergence of insects in spring and summer helps to produce a minimum littoral population at this time. The maximum occurs in the winter.

Psammon. In the interstices of sand grains along lake shores is a highly specialized community consisting chiefly of pennale diatoms, bacteria, protozoa, rotifers, copepods, and primitive copepodlike crustaceans of the order Mystacocarida. Even though the sand is shifting under the influence of waves and currents, the psammon organisms are quite distinct from those in the overlying water. The shallow penetration of light into the sand enables the diatoms and other algae to make the community largely self-sustaining.

Nekton. In temperate waters fish dominate the nekton. Species requiring cold water and high levels of oxygen such as trout, coregonids, and burbot can survive in the hypolimnion of many of the larger lakes. Species tolerant of or requiring higher temperatures occur in the surface waters. Among these, certain species live offshore and others occur in the littoral region, either among weed beds or among rocks of shores exposed to strong wave action. Nearly all species, however, regardless of where the adults occur, spawn in shallow water. Here the young find an abundant food supply, and cover to help protect them from predators. In tropical lakes decapod shrimps of the families Atyidae and Palaemonidae can be important components of the nekton.

Higher vertebrates. Frogs, salamanders, turtles, snakes, crocodilians, various aquatic birds such as loons, herons, kingfishers, mergansers, and ospreys, and various mammals such as muskrats, beavers, racoons, minks, and otters are peripheral members of the aquatic biota. Many of them are essentially terrestrial animals that derive only a part of their sustenance from the water, and because they feed mainly on fish, their total demand on the energy resources of the ecosystem is relatively minor. For these reasons they are often omitted when aquatic ecosystems are considered.

Neuston. A minor life-habit assemblage is that associated with the surface film. Here belong the various insects that run, skate, or swim on the surface of the water, and some truly aquatic organisms such as bacteria, fungi, and certain algae that are suspended from the underside of the film. Quite a number of aquatic animals such as snails, *Hydra*, aquatic bugs, and beetles regularly or adventitiously occur on the underside of the film for brief periods, and some such as anopheline mosquito larvae feed on the microorganisms attached there.

Trophic relationships. The phytoplankton, phytobenthos, and algae of the periphyton and psammon are the producers. Their raw materials are light energy, carbon dioxide, and soluble inorganic sources of nitrogen, phosphorus, and sulfur. In the process of photosynthesis they manufacture organic compounds and give off oxygen to the environment.

The animals are the consumers. They utilize the organic compounds elaborated by the plants and modify them for their own metabolic needs. Through their activities they are continually converting chemical energy to heat energy which is irretrievably lost to the environment. Because energy can enter the system only at the producer level, there is a continuous and steady decrease in supply of bound energy with each successive level of utilization. In their metabolism animals use oxygen and give off carbon dioxide.

Bacteria are the reducers or transformers. They ultimately mineralize the metabolic wastes and dead organisms to liberate the nutrients in a form which may be used again by the producers. Thus, in an ecosystem there is a continuous recycling of materials. Aerobic bacteria utilize free oxygen and give off carbon dioxide. Anaerobic bacteria utilize chemical oxygen and give off carbon dioxide, methane, and other anaerobic gases.

CHEMISTRY

Changes in the chemistry of the water reveal much about the metabolism of the ecosystem and about general limnological relationships.

Oxygen. Dissolved oxygen, more than any other single substance, has advanced knowledge of limnology by permitting diagnosis of what is happening in an ecosystem.

The epilimnion, as a result of circulation and photosynthesis, is generally saturated with oxygen. In the hypolimnion, however, decomposition of the plankton rain and other organic matter results in a utilization of oxygen, which cannot be replenished until the next overturn. The extent of this depletion depends on the amount of organic matter being furnished to the hypolimnion, that is, the rate of production in the epilimnion, the volume of oxygen available for utilization, which is a morphometric factor roughly equivalent to the volume of the hypolimnion, and the temperature of the

hypolimnion, a climate dependent factor, as influencing the rate of metabolism.

Classification based on productivity. Based on the rate of production in the epilimnion, lakes have been classified as oligotrophic or unproductive and eutrophic or productive, with a continuous spectrum between (Fig. 12). Off to one side are the brown-water, or dystrophic, lakes. A. Thienemann defined an oligotrophic lake as one that has more oxygen in the hypolimnion than in the epilimnion, and a eutrophic lake as the reverse. This, however, equates the level of production with the hypolimnetic oxygen supply and ignores the other two controlling factors. B. Aberg and W. Rodhe have resolved this situation by proposing descriptive terms for oxygen distribution without reference to level of production. An orthograde oxygen curve exhibits little or no decline in the hypolimnion. A clinograde curve shows a marked decrease, having a general shape similar to that of a temperature curve. Heterograde curves are special types, having either a maximum (plus heterograde) or a minimum (minus heterograde) oxygen content in the metalimnion. Orthograde and clinograde oxygen distributions correspond, respectively, to oligotrophy and eutrophy as the terms are still commonly used.

Change in productivity. Whatever the causes, most lakes, with time, experience a decline in the oxygen content of the hypolimnion. This is commonly the result of increasing production in the epilimnion and of the gradual accumulation of sediments, which brings about a reduction in the volume of the hypolimnion. The rate of change can be accelerated by man through pollutional and agricultural enrichment, resulting in a higher rate of production in the epilimnion.

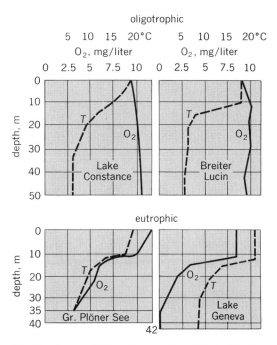

Fig. 12. Oxygen (O_2) and temperature (T) stratification in one Wisconsin and three European lakes. The two upper oxygen curves are orthograde, and the two lower ones clinograde. (*From F. Ruttner, Fundamentals of Limnology, University of Toronto Press, 1953*)

As the oxygen content declines, various species of animals are eliminated from the profundal benthos. A lake with adequate oxygen during the summer may have as many as 200 species of profundal benthos exclusive of protozoa, whereas with reduction in oxygen content the number of species is reduced, until in the most severe cases there are no higher animals at all except perhaps the facultatively anaerobic phantom midge larva. Thienemann first demonstrated this relationship for midges in the lakes of northern Germany and pointed out that lakes with only slightly reduced oxygen content have *Tanytarsus*, those with greatly reduced oxygen *Chironomous*, and those with intermediate levels *Stictochironomus* or some other genus. *Tanytarsus* has been used as an indicator of oligotrophy and *Chironomus* of eutrophy.

With respect to fish, lakes with adequate oxygen in deep water can be called two-story lakes, meaning that there is an assemblage of cold-water fishes in deep water distinct from the warm-water fishes in the epilimnion. Reduction in oxygen ultimately eliminates the cold-water fishes to produce a one-story lake. In North America the cisco has been able to persist in some lakes with a plus-heterograde oxygen distribution but with no oxygen in the hypolimnion.

A critical point in lake ontogeny is the disappearance of oxygen from water in contact with the bottom. Aerobic processes are replaced by anaerobic. Methane and hydrogen sulfide are formed in quantities and get into the water. Lowering of the oxidation-reduction potential results in the conversion of previously insoluble iron and manganese to a soluble condition, allowing them to diffuse into the water. Any excess iron can create a trap for phosphate under these conditions and prevent it from returning to the epilimnion at the next overturn. Methane bubbling through the hypolimnion helps reduce the oxygen content and induces turbulence. This is known as methane circulation.

Carbon dioxide. Carbon dioxide occurs in water as the dissolved gas, as carbonic acid, and as the carbonates and bicarbonates of calcium and magnesium and sometimes iron. All of these substances are in chemical equilibrium with one another. This carbon dioxide complex is the single most important factor that controls pH level in inland waters.

Unlike the oceans, inland waters do not have a constant percentage composition. However, in most waters the world over the dominant ions occur in the same order of abundance: $Ca^{++} > Mg^{++} > Na^+ > K^+$, and $HCO_3^- > SO_4^{--} > Cl^-$. As a result, the pH of most waters varies between 6.5 and 8.5. Higher pH occurs regularly in waters of arid regions and temporarily in hard-water lakes during vigorous photosynthesis. Lower pH is produced by humolimnic acids in dystrophic lakes and by inorganic acids, chiefly sulfuric acid (H_2SO_4) in special situations.

During photosynthesis carbon dioxide (CO_2) is withdrawn by the plants, causing the CO_2 equilibrium to shift with the production of the relatively insoluble calcium carbonate, which settles to the bottom. In littoral regions the precipitate tends to accumulate as marl. In the profundal region much or all of the carbonate can be reconverted to soluble bicarbonate by the excess CO_2 of the hypolimnetic metabolism.

A soft-water lake is one which has only small amounts of calcium and magnesium. Being dependent largely on the atmosphere for an additional supply of CO_2, such lakes are said to have a chronic CO_2 deficiency. Hard-water lakes have a CO_2 reserve in the calcium and magnesium bicarbonates, thus permitting a greater amount of production to be accomplished. Even they, however, can have all their reserve CO_2 utilized, resulting in an acute CO_2 deficiency. In such lakes some of the calcium carbonate can continue to be present in nonsettling colloidal form. This causes the methyl orange titration to give erroneous results as to the CO_2 reserve.

Other elements. Phosphorus and nitrogen are the two elements most commonly in short supply. Of these, phosphorus is generally more critical because of the nature of its supply. Nitrogen is fixed in the atmosphere during electrical storms, and in the aquatic ecosystem by certain blue-green algae and bacteria. Although these two elements and other nutrients a priori must exert control on production and succession in the ecosystem, the specific effects have usually been difficult to demonstrate in the field.

Production and productivity. The organisms present at any one time constitute the standing crop, which is the result of production but is not indicative of the rate of production. The quantity of plankton beneath a unit area of lake surface is approximately the same for oligotrophic and eutrophic lakes, although the quantity per unit volume of the trophogenic zone is greater in eutrophic lakes. Here the plankton is so dense that it limits its own production because it has reduced the transmission of light by its density.

Measurement of productivity. Production in planktonic diatoms has been measured by collecting and counting the cells that settle into jars sus-

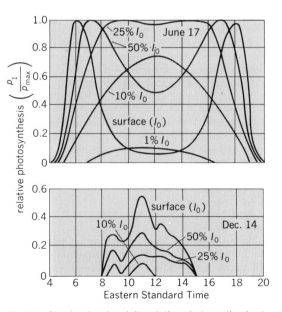

Fig. 13. Graphs showing daily relative photosynthesis at various depths in the water corresponding to stated percentages of surface light intensity (I_0) for typical days in summertime and in wintertime at Newport, R.I. (*From J. H. Ryther, The measurement of primary production, Limnol. Oceanogr., 1:72–84, 1956*)

pended in the water. In several lakes greatly different in size a reproduction rate of about 10% per day was necessary to maintain the standing crop at a particular level. Thus, a low population density may be the result of a rapid rate of sinking rather than a low rate of production.

Rates of production of phytoplankton are commonly estimated by measuring chemical changes developing between clear and opaque bottles containing portions of the same phytoplankton population. In the clear bottle photosynthesis and respiration occur, in the opaque bottle only respiration. The difference between the bottles represents gross production, and the difference between the light bottle and the initial value represents net production. The latter represents the organic matter and energy available for utilization by the heterotrophic components of the ecosystem. Changes measured in the bottles are oxygen, pH, electrical conductivity, and radiocarbon (C^{14}) uptake. The latter apparently measures net production directly, although a correction must be made for fixation of carbon in the dark.

Photosynthetic capacity. Since photosynthesis is dependent on light intensity, on bright days the trophogenic zone is thicker than on dull days, and it is thicker in summer than in winter. Because of the photoreduction of chlorophyll by bright light, maximum photosynthesis may be some distance below the surface of the water. From the photosynthetic capacity of the phytoplankton community present, it is possible to approximate the amount of photosynthesis accomplished over a 24-hr period by integrating the various subsurface light-intensity curves. In epilimnetic communities there is also a relatively constant relationship between chlorophyll content and photosynthesis. Because of this direct dependence of photosynthesis on light intensity it is not surprising that the curve for production accomplished closely follows the curve for total illumination during the year (Fig. 13). Maximum production occurs in summer, when the standing crop is low relative to spring and fall (Fig. 14). The biomass is kept at low levels by the grazing activities of herbivores in the plankton.

Consumer level. At the consumer level, production is more difficult to study because of the increased number of pathways into which the energy can be channeled. Under these conditions the concept of production is somewhat meaningless without reference to a particular product. In general, production is greater, although the standing crop is less, when the product organism is being utilized by some higher-level consumer. In two ponds studied, fish reduced the standing crop of their food benthos but increased its production (Fig. 15). Total production was about 17 times the standing crop, which compares favorably with production coefficients determined for diatoms.

Much of the organic matter fixed in the epilimnion is consumed and mineralized there to enable the nutrients to be reutilized immediately. Only a relatively small proportion of the total primary production, about one-fourth in north German lakes, gets into the hypolimnion as plankton rain. However, the chemical changes induced in the hypolimnion can be used as an index of the relative levels of production in lakes (Fig. 16). If the changes are related to unit surface area of the hypolimnion, through which the plankton sinks,

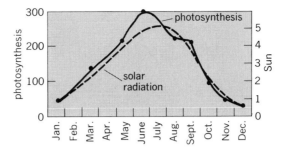

Fig. 14. Annual course of photosynthesis in relation to solar radiation under completely natural conditions in western Lake Erie. Note that maximum photosynthesis occurs in summer when the standing crop is not maximum (see Fig. 11). (*From J. Verduin, Primary production in lakes, Limnol. Oceanogr., 1:85–91, 1956*)

the results are largely independent of the relative size of the epilimnion. The two quantities most commonly measured are the rate of generation of the hypolimnetic oxygen deficit and the rate of generation of the hypolimnetic carbon dioxide accumulation. In strongly eutrophic lakes the latter measure is better, since CO_2 continues to be produced, although at a lesser rate, under anaerobic conditions after the oxygen may have long since disappeared.

LOTIC ECOSYSTEMS

A river system consists of a treelike arrangement of small channels, joining into progressively large channels, until finally all the water flows in a single large channel, corresponding to the trunk of the tree. The distributaries in deltas and alluvial

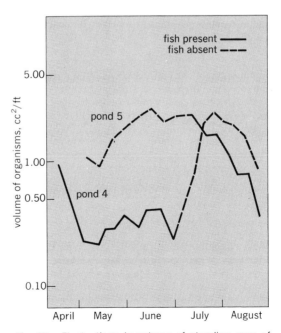

Fig. 15. Fluctuations in volume of standing crop of those benthic organisms eaten by fish in two ponds in Michigan. The action of the fish in each case was to reduce the size of the standing crop but to increase its rate of production. (*From D. W. Hayne and R. C. Ball, Benthic productivity as influenced by fish production, Limnol. Oceanogr., 1:162–175, 1956*)

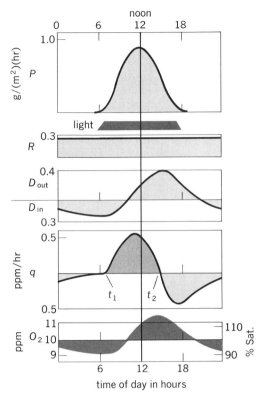

Fig. 16. Series of graphs showing component processes in the oxygen metabolism of a section of a hypothetical stream during the course of a cloudless day. The rates for production *P*, respiration *R*, and diffusion *D* are combined in *q*. The area delimited by t_1 and t_2 multiplied by the volume of flow in m³/hr and divided by the area in square miles of the stream section studied gives gross primary production per day. The method has been successfully used in streams, coral reefs, and other marine littoral areas for measuring primary production. (*From H. T. Odum, Primary production measurements in eleven Florida springs and a marine turtle-grass community, Limnol. Oceanogr., 1(2)85–97, 1956*)

fans are morphologically analogous to a reduced root system. In limestone regions, particularly, various headwater streams can be in direct contact with an underground system of caverns and channels.

Water source. Water in a stream is derived from surface drainage during precipitation and the melting of snow and ice, and from groundwater. Surface drainage contributes the bulk of inorganic turbidity to a stream, groundwater the bulk of the dissolved solids. During low water flow the turbidity is generally low and the amount of dissolved solids high. During periods of precipitation both the volume of flow and turbidity increase, and dissolved solids decline. After peak flow is reached and the current velocity begins to subside, turbidity again declines. The response of a stream to a unit storm is known as a unit hydrograph. The ascending limb of the hydrograph is always much steeper than the descending, because of the lingering effect of increased groundwater reserves.

The headwater portion of a river is dual in nature, consisting of a stream of water and all its contained substances, and a stream of heavy materials constituting the bed load, which moves along the bottom intermittently or continuously in the nature of a giant file. The molar action of this material continually erodes the stream channel deeper, and when severe, such as during a flood, can virtually eliminate the biota from large areas. Production of organisms is directly related to stability of the bottom. By eroding, streams constantly tend to cut down their channels to base level.

Gradients. Except for light, gradients in lotic ecosystems are longitudinal rather than vertical. Most commonly the slope profile of a stream is concave upward, with the steepest slopes in the headwaters. This creates gradients of current velocity, coarseness of bottom, and other conditions. Because the headwaters are generally at considerably greater elevations and in more direct contact with the groundwater system, there tend to be gradients in temperature, such as a warming downstream in summer, and in magnitude of annual temperature variation, which is least upstream.

A stream is a continuum only in a physical sense. The various longitudinal gradients interacting bring about a longitudinal succession of species and of communities. Generally the number of species in a taxonomic group increases in a downstream direction. This is known to be true of fishes in the Rhine and Colorado rivers, and of protozoa and insects in Pennsylvania. The replacement of one species of turbellarian by another is response to a changing temperature regime was first found in western Europe, and has since been found in Japan and Brazil. Undoubtedly all groups of organisms react to these gradients.

In western Europe streams are commonly divided into four biological zones on the basis of the dominant fish species. Uppermost is a trout zone, with steep gradient, rapid flow, close alternation of pools and riffles, and low temperatures. Below this is the grayling zone, with reduced current velocity, greater annual fluctuations in temperature, and a more open valley with the stream beginning to meander in a small flood plain. The barbel zone is in some respects transitional between the two cold-water zones above and the warm-water bream zone below, dominated by cyprinids, percids, pike, and other warm-water fishes. The several zones are associated with various stages in the physiographic cycle, from the youthful V-shaped valleys and steep gradients in the headwaters to mature, relatively flat valleys and extensive flood plains downstream. By the erosional activity of a stream there is a gradual headward migration of stream zones or habitats and their associated biotas.

Biota. The various biotal components of a stream are the following:

Periphyton. In headwater streams the dominant producing element is the sessile algae on the bottom, which is constantly being broken loose by the current and drifted downstream. The microscopic material (plankton) in the overlying water is sometimes described as a "pale image" of the periphyton community.

Potamoplankton. In more sluggish sections of a stream a true plankton can develop. Most of the organisms present are the same as those occurring in lakes and ponds. Their source is backwaters, sloughs, and flood-plain pools. Plankton leaving a lake tends to be rapidly removed from the stream by entrapment in the water immediately in contact with the bottom or other substrate. The plankton of the Rhine, however, is dominated for long distances by organisms derived from the Swiss lakes via the Aare River.

Benthos. The benthos of still-water sections is similar to that of lakes. In currents the two distinctive communities are the stone fauna and the moss fauna. Organisms in the stone fauna tend to be greatly flattened for greater contact with the substrate. Many are directly attached to the substrate, or they live in webs and cases that are attached. In temperate latitudes the net-winged midges and in the tropics fishes and tadpoles have suckers for attachment. Those organisms when wrested loose and drifted by the current are known as syrton. They consitute another example of the downstream transport of organic matter, and they are important in colonizing new or depopulated areas. The tendency to be swept downstream is compensated for in part by a positive rheotaxis, which directs the animals into the current and makes them move upstream against it. Animals of the moss fauna are small and well provided with grappling structures. The moss serves primarily as a substrate, with relatively few organisms feeding on it directly.

Phytobenthos. Rooted aquatics occur abundantly in areas of reduced current. Their stems and roots further reduce the current velocity and accumulate fine sediment, which changes the nature of the stream bottom and helps delay the downstream transport of nutrients.

Nekton. Fishes that require low temperature and high oxygen tend to occur in the headwaters, and the more eurytopic fishes occur in the lowland sections. Fishes in swift currents tend to be almost round in cross section, whereas many of those in quiet water are found to be strongly compressed from side to side.

Many, perhaps most, stream fishes have a restricted home range in which they spend their entire lives. If displaced downstream, for instance during a flood, they are able to find their way back. In the long-eared sunfish this is accomplished by the sense of smell more than by sight.

Continental fish faunas are dominated by cypriniform fishes, consisting of the carplike fishes such as the characinoids, gymnotoids, and cyprinoids, and the catfishlike fishes. Approximately 80% of the fishes of Southeast Asia are of this order, 50% of the Great Lakes fishes, and more than 80% in the Amazons. Peripheral fish faunas have smaller percentages of cypriniforms.

Production relationships. R. Butcher attempted to estimate productivity by the number of algal cells settling on glass slides submerged for 28 days. Oligotrophic streams yielded fewer than 2000 cells/mm², whereas eutrophic streams yielded 2500–10,000 cells/mm². Regions with moderate organic pollution yielded higher numbers, up to 100,000/mm². J. Yount has found for sessile diatoms that as the population density increases, the number of species represented decreases.

In the upper sections of streams, riffles are generally more productive than pools. Current produces a eutrophication effect, and animals living in a current have a higher rate of metabolism than those in still water. In mayflies the area of gill surface decreases as the current velocity increases.

Estimates of production have been made from the 24-hr cycles of dissolved oxygen and carbon dioxide with suitable corrections for diffusion. Attempts have also been made to approximate primary production from the quantity of chlorophyll per unit area of bottom. In the outflow of a spring, production and the pathways of energy through the ecosystem have been measured by a modification of the light bottle, dark bottle principle. Because the substrate is so important in primary production in streams, the upstream shallow portions may be considered a trophogenic zone which exports its surplus production, such as plankton and syrton, to a downstream, deeper tropholytic zone. Production of detritus-feeding benthonts such as midges, mayflies, oligochaetes, and small clams in the lowermost portions of large rivers is at times tremendous.

Pollutional relationships. Quantities of organic matter entering a stream, from domestic sewage, a paper mill, or a canning factory, can markedly change the balance of biological processes. Bacteria attack the organic matter, resulting in a utilization of oxygen and a release of plant nutrients. The extent of oxygen utilization depends on the relative dilution of the organic matter by the river water. Engineers refer to the oxygen sag curve and the process of reaeration downstream. In severe cases the oxygen can be completely utilized, resulting in a septic zone or polysaprobic zone. This is a zone of bacteria, with anaerobic processes predominating. Farther downstream conditions in the stream begin to recover, forming a zone of partial recovery, the mesosaprobic zone, and still farther downstream is the zone of complete recovery or oligosaprobic zone. In this self-purification process there is a longitudinal succession of bacteria, protozoa, algae, and insects. There is also a definite and predictable longitudinal succession of chemical conditions.

Other types of pollution such as toxic wastes and inert materials may affect the biota of a stream without changing the oxygen content. A stream is considered healthy if a great variety of microhabitats in it permit establishment of a diversity of species and trophic relationships in balance with one another. Pollution of any kind eliminates various of these microhabitats and their adapted species and often results in the increased abundance of some of those remaining. The severity of pollution has been measured by comparing the number of species in several taxonomic groups with the average numbers of species present at nonpolluted sites in the same watershed. Fishes, insects, and crustaceans are the most sensitive to pollution. Blue-green algae, bdelloid rotifers, oligochaetes, leeches, and pulmonate snails are least sensitive.

Since the early work of R. Kolkwitz and M. Marsson, attempts have been made to find indicator species for different severities or kinds of pollution. Most species, however, are sufficiently adaptable in their requirements that they can occur in more than one zone. The community of organisms present is more important than the individual species in indicating community metabolism. *See* ECOSYSTEM; ENVIRONMENT; MARINE ECOSYSTEM.   [DAVID G. FREY]

Bibliography: W. T. Edmondson (ed.), *Freshwater Biology*, 1959; G. E. Hutchinson, *A Treatise on*

Limnology, vols. 1 and 2, 1957, 1967; H. B. N. Hynes, *The Biology of Polluted Waters*, 1960; H. B. N. Hynes, *Stream Limnology*, 1969; T. T. Macan, *Freshwater Ecology*, 1963; K. Read, *Ecology of Inland Waters Estuaries*, 1961; F. Ruttner, *Fundamentals of Limnology*, 1953; J. R. Vallentyne, Principles of modern limnology, *Amer. Sci.*, 45(3):218–244, 1957; P. S. Welch, *Limnology*, 2d ed., 1952.

Front

A sloping surface of discontinuity in the troposphere, separating air masses of different density or temperature. The passage of a front at a fixed location is marked by sudden changes in temperature and wind and also by rapid variations in other weather elements, such as moisture and sky condition. *See* AIR MASS.

Although the front is ideally regarded as a discontinuity in temperature, in practice the temperature change from warm to cold air masses occurs over a zone of finite width, called a transition or frontal zone. The three-dimensional structure of the frontal zone is illustrated in Fig. 1. In typical cases the zone is about 3000 ft (1 km) in depth and 100 mi (100–200 km) in width, with a slope of approximately 1/100. The cold air lies beneath the warm in the form of a shallow wedge. Temperature contrasts are generally strongest at or near the Earth's surface. In the middle and upper troposphere, frontal structure tends to be diffuse, though sharp, narrow fronts of limited extent are common in the vicinity of strong jet streams. *See* JET STREAM.

The surface separating the frontal zone from the adjacent warm air mass is referred to as the frontal surface, and it is the line of intersection of this surface with a second surface, usually horizontal or vertical, that strictly speaking constitutes the front. According to this more precise definition, the front represents a discontinuity in temperature gradient rather than in temperature itself. The boundary on the cold air side is often ill-defined, especially near the Earth's surface, and for this reason is not represented in routine analysis of weather maps. In typical cases about one-third of the temperature difference between the Equator and the pole is contained within the narrow frontal zone, the remainder being distributed within the

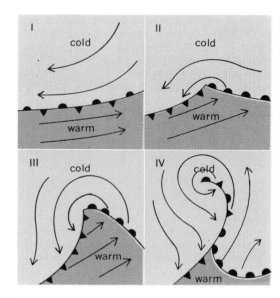

Fig. 2.   The life cycle of the wave cyclone, surface projection. Arrows denote airflow. Patterns depicted are for the Northern Hemisphere; while their mirror images apply in Southern Hemisphere.

warm and cold air masses on either side. *See* WEATHER MAP.

The wind gradient, or shear, like the temperature gradient, is large within the frontal zone and discontinuous at the boundaries. An upper-level jet stream normally is situated above the zone, the strong winds of the jet inclining downward along or near the warm boundary.

Fronts are formed when converging wind currents bring air of different origin into juxtaposition. A completely satisfactory theoretical understanding of the frontogenetical process has yet to be achieved. For a discussion of this process see the section on frontogenesis below.

Frontal waves. Many extratropical cyclones begin as wavelike perturbations of a preexisting frontal surface. Such cyclones are referred to as wave cyclones. The life cycle of the wave cyclone is illustrated in Fig. 2. In stage I, prior to the development, the front is gently curved and more or less stationary. In stage II the front undergoes a wavelike deformation, the cold air advancing to the left of the wave crest and the warm air to the right. Simultaneously a center of low pressure and of counterclockwise wind circulation appears at the crest. The portion of the front which marks the leading edge of the cold air is called the cold front. The term warm front is applied to the forward boundary of the warm air. During stage III the wave grows in amplitude and the warm sector narrows. In the final stage the cold front overtakes and merges with the warm front, forming an occluded front. The center of low pressure and of cyclonic rotation is found at the tip of the occluded front, well removed from the warm air source. At this stage the cyclone begins to fill and weaken. *See* STORM.

Cases have been documented in which the occluded structure depicted in panel IV (Fig. 2) forms in a different manner than described above.

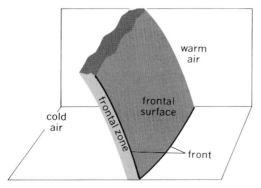

Fig. 1.   Schematic diagram of the frontal zone, angle with Earth's surface much exaggerated.

In such cases, sometimes referred to as pseudo-occlusions, the low-pressure center is observed to retreat into the cold air and frontogenesis takes place along a line joining the low center and the tip of the warm sector. Since the classical occlusion process, in which the cold front overtakes and merges with the warm front, has never been adequately verified, it is possible that most occlusions form by the alternative process.

A front moves approximately with the speed of the wind component normal to it. The strength of this component varies with season, location, and individual situation but generally lies in the range of 0–50 mph; 25 mph is a typical frontal speed.

Cloud and precipitation types and patterns bear characteristic relationships to fronts, as depicted in Fig. 3. These relationships are determined mainly by the vertical air motions in the vicinity of the frontal surfaces. Since the motions are not unique but vary somewhat from case to case and, in a given case, with the stage of development, the features of the diagram are subject to considerable variation. In general, though, the motions consist of an upgliding of the warm air above the warm frontal surface, a more restricted and pronounced upthrusting of the warm air by the cold front, and an extensive subsidence of the cold air to the rear of the cold front. *See* CLOUD; CLOUD PHYSICS.

Fast-moving cold fronts are characterized by narrow cloud and precipitation systems. When potentially unstable air is present in the warm sector, the main weather activity often breaks out ahead of the cold front in prefrontal squall lines.

Polar front. A front separating air of tropical origin from air of more northerly or polar origin is referred to as a polar front. Frequently only a fraction of the temperature contrast between tropical and polar regions is concentrated within the polar frontal zone, and a second or secondary front appears at higher latitudes. In certain locations such a front is termed an arctic front.

In winter the major or polar frontal zones extend from the northern Philippines across the Pacific Ocean to the coast of Washington, from the southeastern United States across the Atlantic Ocean to southern England, and from the northern Mediterranean eastward into Asia. An arctic frontal zone is located along the mountain barriers of western Canada and Alaska. In summer the average positions of the polar frontal zones are farther north, the Pacific zone extending from Japan to Washington and the Atlantic zone from New Jersey to the British Isles. In addition to a northward-displaced polar front over Asia, an arctic front lies along the northern shore and continues eastward into Alaska.   [RICHARD J. REED]

Frontogenesis. The formation of a front or a frontal zone requires an increase in the temperature gradient and the development of a wind shift. The frontogenesis mechanism operates even when the front is in a quasi-steady state; otherwise the turbulent mixing of heat and momentum would rapidly destroy the front.

The transport of temperature by the horizontal wind field can initiate the frontogenesis process as is shown for two cases in Fig. 4. The two wind fields shown would, in the absence of other effects, transport the isotherms in such a way that they

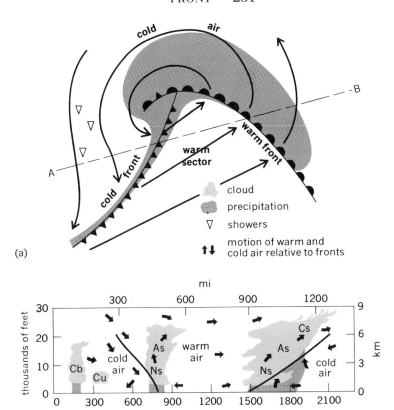

(a)

Fig. 3. Relation of cloud types and precipitation to fronts. (a) Surface weather map. (b) Vertical cross section (along A-B in diagram a). Cloud types: Cs, cirrostratus; As, altostratus; Ns, nimbostratus; Cu, cumulus; and Cb, cumulonimbus.

would become concentrated along the A-B lines in both cases. Since the temperature gradient is inversely proportional to the spacing of the isotherms, it is clear that frontogenesis would occur along the A-B lines. Vertical air motions will modify this frontogenesis process, but these modifications will be small near the ground where the vertical motion is small. Thus, near the ground the temperature gradients will continue to increase in the frontal zones as long as the horizontal wind field does not change. As the temperature gradient increases, a circulation will develop in the vertical plane through C-D in each case. The thermal wind relation is valid for the component of the horizontal wind which blows parallel to the frontal zone. The thermal wind, which is the change in the geostrophic wind over a specific vertical distance, is directed along the isotherms, and its magnitude is proportional to the temperature gradient. As the frontogenesis process increases the temperature gradient, the thermal wind must also increase. But if the thermal wind increases, the change in the actual wind over a height interval must also increase (since the thermal wind closely approximates the actual wind change with height). This corresponding change in the actual wind component along the front is accomplished through the action of the Coriolis force. A small wind component perpendicular to the front is required if the Coriolis force is to act in this manner. This leads to the circulation in the vertical plane that is shown in Fig. 5.

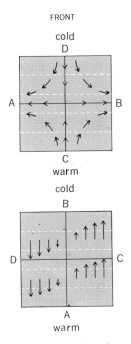

Fig. 4. Two horizontal wind fields which can cause frontogenesis. Broken lines represent isotherms and arrows show wind directions and speeds.

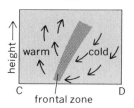

FRONT

height →

warm   cold

C   D

frontal zone

Fig. 5. The circulation in the vertical plane through C-D for both cases of Fig. 4.

This circulation plays an important role in the frontogenesis process and in determining the frontal structure. The rising motion in the warm air and the sinking motion in the cold air are consistent with observed cloud and precipitation patterns. The circulation helps give the front its characteristic vertical tilt which leaves the relatively cooler air beneath the front. Near the ground the circulation causes a horizontal convergence of mass. This speeds up the frontogenesis process by increasing the rate at which the isotherms move together. The convergence field also carries the momentum lines together in the frontal zone in such a way that a wind shear develops across the front. This wind shear gives rise to the wind shift which is observed with a frontal passage. Eventually the front reaches a quasi-steady state in which the turbulent mixing balances the frontogenesis processes. Other frontogenesis effects are important in some cases, but the mechanism presented above appears to be the predominant one.

[ROGER T. WILLIAMS]

Bibliography: H. R. Byers, *General Meteorology*, 3d ed., 1959; C. L. Godske et al., *Dynamic Meteorology and Weather Forecasting*, Carnegie Inst. Wash. Publ. no. 605, 1957; P. H. Stone, Frontogenesis by horizontal wind deformation fields, *J. Atmos. Sci.*, 23:455–465, 1966; R. T. Williams, Atmospheric frontogenesis: A numerical experiment, *J. Atmos. Sci.*, 24:627–641, 1967.

Frost

A covering of ice in one of several forms produced by the freezing of supercooled water droplets on objects colder than 32°F. The partial or complete killing of vegetation, by freezing or by temperatures somewhat above freezing for certain sensitive plants, also is called frost. Air temperatures below 32°F sometimes are reported as "degrees of frost"; thus 10°F is 22 degrees of frost.

Frost forms in exactly the same manner as dew except that the individual droplets that condense in the air a fraction of an inch from a subfreezing object are themselves supercooled, that is, colder than 32°F. When the droplets touch the cold object, they freeze immediately into individual crystals. When additional droplets freeze as soon as the previous ones are frozen, and hence are still close to the melting point because all the heat of fusion has not been dissipated, amorphous frost or rime results.

At more rapid rates of condensation, the drops form a film of supercooled water before freezing, and glaze or glazed frost ("window ice" on house windows, "clear ice" on aircraft) generally follows. Glaze formation on plants, buildings and other structures, and especially on wires sometimes is called an ice storm, or a silver frost storm, or thaw.

At slower deposition rates, such that each crystal cools well below the melting point before the next joins it, true crystalline or hoar frosts form. These include fernlike assemblages on snow surfaces, called surface hoar; similar feathery plumes in cold buildings, caves, and crevasses, called depth hoar; and the common window frost or ice flowers on house windows.

So-called killing frosts occur on clear autumn nights, when radiative cooling of ground, air, and vegetation causes plant fluids to freeze. At such times, the air temperature measured in a shelter 5–7 ft above ground usually is at least 5°F below freezing, but such standard level temperatures are poor indicators of frost severity. Air temperature varies greatly with height in the first few feet above the ground, and also with topography and vegetation around the shelter.

When wind is absent, the air layer immediately above the ground, rather than the ground itself, loses the most heat by radiation. Then the lowest temperature is 2–6 in. above, and about 1°F colder than, a bare ground surface. Above this near-ground minimum, an inversion of temperature develops for several to many feet thick. Plants radiate their heat faster than the air or ground, and may be colder than this air minimum temperature.

Frost damage can be prevented or reduced by heating the lowest air layers, or by mixing the cold surface air with the warmer air in the inversion above the tops of plants or trees.

Valley bottoms are much colder on clear nights than slopes, and may have frost when slopes are frost-free. In some notable frost pockets or hollows the air temperature may be 40°F colder than at nearby stations on higher ground, because of cold air drainage and the greater radiative cooling of level areas than of slopes. *See* CLOUD PHYSICS; DEW POINT; TEMPERATURE INVERSION; WEATHER MODIFICATION.

[ARNOLD COURT]

Fungistat and fungicide

Synthetic or biosynthetic compounds used to control fungal infections in man and plants.

Chemotherapy in man. The results from use of antibiotics and other drugs in the treatment of fungus infections of man have been discouraging; however, actinomycosis responds to large doses of penicillin, nocardiosis to a combination of penicillin and sulfadiazine, and paracoccidioidomycosis to sulfadiazine or sulfamerazine.

Sporotrichosis is the only mycotic disease for which a specific chemotherapeutic agent exists—potassium iodide. This compound has been used in the treatment of pulmonary candidiasis (moniliasis) with some success; it also has been used in conjunction with autogenous vaccines in pulmonary blastomycosis with less than satisfactory results.

Actidione, an antibiotic isolated from *Streptomyces griseus*, is selective for fungi not pathogenic to man. Although in vitro studies have indicated that it is active against *Cryptococcus neoformans*, a yeastlike organism causing serious infections of the central nervous system, this drug is too toxic for clinical use. Undecylenic acid, sodium propionate, and ointments containing sulfur and salicylic acid have been used in the treatment of ringworm of the smooth skin; sodium propionate also has been used with success in instances of vulvovaginitis due to species of *Candida*. Fungus infections of the hair and nails, as well as the more resistant infections of the skin caused by *Trichophyton rubrum*, have failed to respond to these drugs.

The development of three new antifungal agents has given new hope in the treatment of mycotic diseases of man. Nystatin, produced by *Streptomyces nounsei*, is being used in the treatment of candidiasis of the skin, mucous membranes, intestinal tract, and lungs. It has little effect upon the course of disseminated candidiasis or infections caused by other fungi.

Amphotericin B, also isolated from a species of *Streptomyces*, promises to be valuable in the treatment of histoplasmosis, blastomycosis, and coccidioidomycosis. Although the clinical response of patients with cryptococcal meningitis treated with amphotericin B has been disappointing, it still remains the only chemotherapeutic agent available for this disease.

Griseolfulvin is an antifungal agent, produced by *Penicillium griseofulvum*, with a marked activity against fungi causing ringworm. Although it only became available commercially in 1959, limited experience with griseofulvin indicates that ringworm of the hair or nails, as well as the resistant skin lesions due to *T. rubrum*, may undoubtedly be treated successfully with this drug. Few data are available with regard to the rate of relapse following the use of griseofulvin or the failure of lesions to respond to the drug. Despite the lack of such information, griseofulvin is the first antifungal agent that promises to bring relief to the many individuals who have ringworm.

[LEANOR D. HALEY]

Agricultural fungicides. Chemical compounds are used to control plant diseases caused by fungi. American farmers spend more than $100,000,000 annually to buy and apply fungicides; nevertheless, plant diseases destroy more than $3,500,000,000 worth of crops each year. Fungicides now used include both inorganic and organic compounds. Agricultural fungicides must have certain properties and conform to very strict regulations, and of thousands of compounds tested, few have reached the farmer's fields. Special equipment is needed to apply fungicides. *See* PLANT DISEASE.

Inorganic fungicides. Inorganic fungicides, such as bordeaux mixture and sulfur, are still used in the greatest amounts. Bordeaux mixture is made by mixing a solution of copper sulfate with a suspension of lime (calcium hydroxide). In 1965 about 47,000,000 lb of copper sulfate, worth about $18,000,000, was sold to make bordeaux mixture. In the same year 200,000,000 lb of processed sulfur was sold for use as a fungicide; this quantity of sulfur was valued at about $60,000,000.

Organic fungicides. Organic fungicides have become increasingly important since 1934. Some of the most useful are derivatives of dithiocarbamic acid. Examples are ferbam and ziram, the iron and zinc salts respectively of dimethyldithiocarbamic acid, and nabam, zineb, and maneb, the sodium, zinc, and manganese salts respectively of ethylenebis(dithiocarbamic acid). Another related fungicide of importance is thiram or bis-(dimethylthiocarbamoyl)sulfide. Other representative fungicides widely used are captan, *N*-(tri-chloromethylthio)-4-cyclohexene-1,2-dicarboximide; glyodin, 2-heptadecyl-2-imidazoline acetate; chloranil, tetrachloro-*p*-benzoquinone; dichlone, 2,3-dichloro-1,4-naphthoquinone; and dodine, *n*-dodecylguanidine acetate. An organic compound commonly used to control powdery mildews is 4,6-dinitro-2-(1-methylheptyl)phenylcrotonate. Cycloheximide, an antibiotic, is used for cherry leaf spot.

Requirements and regulations. These stipulations must be met by manufacturers before they sell fungicides. Agricultural fungicides must be registered with the U.S. Department of Agriculture. To get acceptance, manufacturers must prove that their products control stated diseases, without plant injury, when properly applied. In addition, the Miller law, administered by the Food and Drug Administration, will not allow materials to leave poisonous residues on edible crops. These requirements needed to protect the public have increased the cost of developing new agricultural fungicides. Labels on agricultural fungicides must state the uses for which the product is intended and the conditions of use.

Formulation. The manner in which these compounds are applied, that is, as wettable powders, dusts, or emulsions, is often essential to the success of agricultural fungicides. Raw fungicides must be pulverized to uniform particles of the most effective size, mixed with wetting agents, or dissolved in solvents. These carriers or diluents must not degrade the fungicides or injure the plants.

Foliage fungicides. This type of fungicide is applied to aboveground parts of plants, usually to prevent disease rather than cure it. Because they are intended to form a protective coating on the plant surface that kills fungus spores before infection occurs, foliage fungicides must adhere to foliage despite weathering. Fungicides also must be sufficiently stable chemically to resist degradation by water, oxygen, carbon dioxide, and sunlight. Sometimes, as in the case of zineb, specific chemical changes by weathering are necessary to produce highly fungicidal derivatives. Protective fungicides must be insoluble in water in order to remain on foliage. Certain foliage fungicides, however, are water-soluble. These materials destroy the fungus in disease spots after infection. Fungicides of this type are called eradicant or contact fungicides. An example is dodine, *n*-dodecylguanidine acetate.

Seed and soil treatments. Seeds and seedlings are protected against fungi in the soil by treating the seeds and the soil with fungicides. Seed-treating materials must be safe for seeds and must resist degradation by soil and soil microorganisms. Some soil fungicides are safe to use on living plants. An example is pentachloronitrobenzene, which can be drenched around seedlings of cruciferous crops and lettuce to protect them against root-rotting fungi. Other soil fungicides, such as formaldehyde, chloropicrin, and methyl isothiocyanate, are injurious to seeds and living plants. These compounds are useful because they are volatile. Used before planting, they have a chance to kill soil fungi and then escape from the soil.

Chemotherapeutants. Chemotherapeutants are compounds that permeate the plant to protect new growth or to eliminate infections that have already occurred. At present there is a great deal of interest in finding plant chemotherapeutants. The most promising of these materials are DuPont Benlate, 1-(butylcarbamoyl)-2-benzimidazole carbamic acid methyl ester, and Uniroyal Vitavax, 2,3-dihydro-5-carboxanilido-6-methyl-1,4-oxathiin.

Application methods. Dusters and sprayers are used to apply foliage fungicides. Conventional sprayers apply 300–500 gal/acre at pressures up to 600 psi. This equipment ensures the uniform, adequate coverage necessary for control. Recent developments in spray equipment are the mist blower and the low-pressure, low-volume sprayer. The mist blower uses an air blast to spray droplets onto foliage. Mist blowers have been successful for applying fungicides to trees but are less satisfactory for applying fungicides to row crops. The low-pressure, low-volume sprayers are lightweight machines that apply about 80 gal of concentrated spray liquid per acre at a pressure of about 100 psi. These have been successfully used to protect tomatoes and potatoes against diseases caused by fungi. The most recent refinement of this method is ultralow-volume (ULV) spraying. With ULV spraying, which employs special spinning cage micronizers, growers can protect certain crops by applying as little as 0.5–2 gal of spray liquid per acre.

[JAMES G. HORSFALL; SAUL RICH]

Bibliography: J. G. Horsfall, Principles of Fungicidal Action, 1956; Society of Chemical Industry, Fungicides in Agriculture and Horticulture, 1961.

Geomorphology

Geomorphology, once defined as "the scientific study of scenery," is being taken with greater seriousness these days. And with good reason: describing and classifying landforms, and delving into landform origin and development—particularly the rate and manner of landform change with time—are all activities that can help concerned citizens preserve, control, and intelligently exploit their environment. Through geomorphology, investigators can predict the effectiveness—or countereffectiveness—of flood and erosion controls, and the effects of mining, water diversions, irrigation, and a host of other projects on the landscape.

During the early part of this century, geomorphologists were concerned with developing and refining a model of the erosional cycle—a unified set of rules for predicting the evolution of landscapes under the influence of erosion. Their model, however, was too simple. Their approach could not succeed until a great deal more was learned about the way a particular landform (a hillslope, alluvial fan, floodplain, or river channel, for example) responds to the erosional agents acting on it and, of course, about the mechanics and rates of processes of erosion and deposition.

THE APPROACH SINCE 1945

Therefore, since about 1945, geomorphologists have been asking questions concerning the details of landscapes and the factors that control these details. What, for example, determines the width, depth, and shape of a river channel? How does the channel respond to changes in the quantity of water moving through it and to different types of flood events? Answers to such questions not only provide the information necessary for progress in geomorphology, but also are of immediate practical value in attempts to evaluate and to anticipate the influence of man on his environment. The answers will aid in predicting the effects of weather modification, urbanization, changing agricultural patterns, and erosion and flood control structures and programs on rivers and in drainage basins.

Today, research in geomorphology depends heavily on data collected by climatologists and hydrologists. This is because geomorphologists must understand how average climatic and hydrologic conditions affect landforms. In addition, they must also have data on individual events to understand what effect major storms and floods have on the landscape and its components. One could assume that the greater the stress (intensity of precipitation or velocity of floodwaters), the greater the strain on or response of a geomorphic system. However, the problem is not so simple. The complex landscape, with its valley sides, drainage divides, floodplains, and stream channels, is strongly influenced by vegetation type and density, by the erosional character of soils, and by man's modification of these variables. Therefore, the response of a landscape to natural or man-induced changes cannot usually be predicted in detail. See CLIMATOLOGY; HYDROLOGY.

MAGNITUDE AND FREQUENCY OF EVENTS

The relative significance of the large infrequent event and the small frequent event is one of the crucial questions confronting the science of geomorphology. From a purely practical viewpoint, the river control engineer asks whether he can, through control of only the largest floods, produce a smaller and relatively stable channel. The geomorphologist asks whether a change in storm characteristics alone, with no change in average precipitation, can cause dramatic changes in landscape characteristics and in the progress of the erosion cycle. To answer such questions, investigators must know more about the influence of the magnitude and frequency of events acting on the Earth's surface.

Importance of an average event. While attempting to design stable irrigation canals in India, Pakistan, and Egypt in the late 19th and early 20th centuries, British engineers developed statistical relations between canal dimensions and the quantity and velocity of water flowing in these artificial channels. The engineers discovered that channel width and depth and flow velocity were significantly related to the average discharge of water passing through a channel. They found, moreover, that these relations also applied to river systems (Fig. 1). Although the specific relations change from one river to another, channel dimensions, gradient, and meander wavelengths and amplitude are significantly related to the average quantity of water that moves through a river channel. These correlations have been used to predict changes in river channel dimensions, when discharge changes are expected due to water diversion or storage.

Importance of a major event. However, it has also been demonstrated that the character of a river system can be greatly altered by a change in only the flood discharges, without any change in average flow. For example, the South Platte River in eastern Colorado and western Nebraska has changed dramatically during the last 80 years as a result of reduction of flood peaks (Fig. 2). The average flow has remained almost constant, but large floods no longer occur because the flow is regulated upstream by dams. As a result, the channel has narrowed significantly. However, large floods have increased the width of the Cimarron River in southwestern Kansas from an average of 50 ft (15 m) in 1890 to 1200 ft (360 m) in 1942. Hence, either a change in average flow or a change in flood flows, or both, can significantly alter the morphology of a river channel.

Analytical studies. M. G. Wolman of Johns Hopkins University and J. P. Miller of Harvard University in 1960 evaluated the influence of erosional events of different magnitudes and frequencies of occurrence. They studied the amount of work (erosion and sediment transport) performed by each event and the ability of each to produce major landform changes. In effect, Wolman and Miller asked themselves the following question: Is a rare but large event more effective than numerous small events? "It is widely believed," they state, "that the infrequent events of immense magnitude are most effective in the progressive denudation of the earth's surface. Although this belief might seem to be supported by observations of some individual events such as large floods, tsunamis and dust storms, the catastrophic event is not necessarily the critical factor responsible for the development of landforms. Available evidence indicates that the evaluation of the effectiveness of a specific mechanism and of the relevant importance of different geomorphic processes in molding specific forms involves the frequency of occurrence as well as the magnitude of the individual events."

Sediment transport. For sediment transport, Wolman and Miller found that most of the work of moving sediment from drainage basins is done by relatively frequent flows of moderate magnitude. By "frequent" they mean events that reoccur at least once each year or two and generally several or more times per year.

Movement curves. This finding is illustrated by curves representing rate of sediment movement in relation to the magnitude of an event (applied stress) and the frequency of occurrence of the events. Curve A in Fig. 3 shows, for example, quantity of sediment moved within a stream channel (rate of movement) in relation to magnitude of flood (applied stress). This curve shows that large events move much more sediment. But curve B, for the frequency of events, shows that there are very few extremely-low-magnitude events and few large-magnitude events—as one might suspect, events of intermediate magnitude are most common. Curve C, obtained by multiplying the rate of sediment movement for a given magnitude of event by the number of such events, shows in a general way the total quantity of sediment moved over a period of years in relation to applied stress or flood magnitude.

The obvious conclusion is that, over a period of years, maximum work, or the maximum quantity of sediment moved from a drainage basin, is accomplished by storms and floods of intermediate magnitude and frequency. The few very-low-magnitude events do not move much sediment, and although the very-high-magnitude events transport considerable quantities of sediment from the system, they occur too infrequently to be responsible

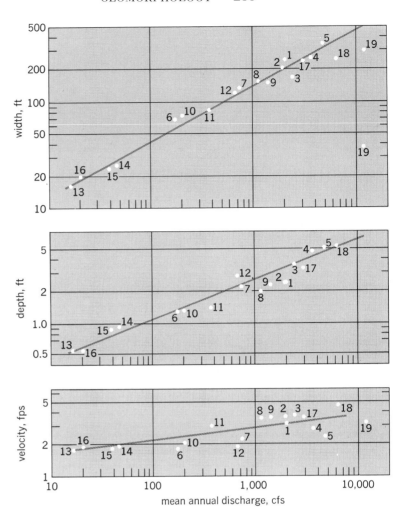

Fig. 1. Relation between width, depth, and velocity and mean annual water discharge. Data are for Bighorn River (points 1–5), Wind River (6–9), Yellowstone River (17–19), and general tributaries in Wyoming and Montana—Greybull River (10, 11), Popo Agie River (12), North Fork Owl Creek (13), Owl Creek (14), Medicine Lodge Creek (15), and Gooseberry Creek (16). When a stream has more than one data point, higher numbers are farther downstream. (*From L. B. Leopold and Thomas Maddock, Jr., The Hydraulic Geometry of Stream Channels and Some Physiographic Implications, U. S. Geol. Surv. Prof. Pap. no. 252, 1953*)

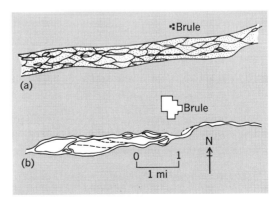

Fig. 2. Sketches of South Platte River at Brule, Nebr. (*a*) Based on surveys made in 1897, showing typical braided character of channel. (*b*) Based on aerial photographs taken in 1959. (*From S. A. Schumm, Geomorphic implications of climatic changes, in R. J. Chorley, ed., Water, Earth and Man, Methuen, 1969*)

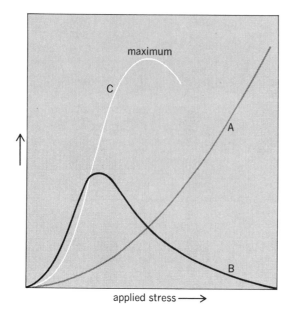

3. Generalized relations between rate or quantity of sediment movement and magnitude of storm or flood (applied stress). Curve A indicates the rate of movement, curve B the frequency of occurence, and curve C the product of rate and frequency. *(From M. G. Wolman and J. P. Miller, Magnitude and frequency of forces in geomorphic processes, J. Geol., 68: 54–74, 1960)*

for a significant effect over a long period of time. Therefore, it is the relatively frequent events of moderate size that do the most work in natural drainage systems.

Other sediment movement studies. According to data collected by the U.S. Agricultural Research Service on sediment movement from 72 small watersheds in 17 states, large storms that occur only once every 2 years, or at longer intervals, produce from 3 to 46% of the total annual suspended sediment yield, whereas storms that occur more frequently than once each year produce from 34 to 92% of total annual suspended sediment yield. On the average, therefore, the large storms are responsible for only a small part of total sediment

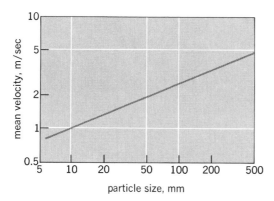

Fig. 4. Mean velocity of flow required to initiate movement of spherical rocks. Flow depth is 2 m. *(From C. R. Neill, Mean-velocity criterion for scour of coarse uniform bed material, Proceedings of the 12th Congress of the International Association for Hydraulic Research, vol. 13, pp. 6.1–6.9, 1967)*

movement. Moreover, measurements of wind transport of sand at the Kharga Oasis in Egypt show that sand movement is primarily affected by winds having lower velocity than a sandstorm but greater velocity than winds that occur a large part of the time. Such studies support the conclusion that the large storm or flood is less significant geomorphicly than one might expect.

MAJOR HYDROLOGIC EVENTS

Nevertheless, one cannot ignore the fact that large infrequent events, although they may not perform the most work, can cause significant and sometimes catastrophic alteration of stream channels and hillslopes.

Effects of Hurricane Agnes. As stated earlier, major floods destroyed the floodplain of the Cimarron River, and an absence of flooding narrowed the South Platte River (Fig. 2). However, the effects of Hurricane Agnes in 1972 on slopes and channels in the eastern United States were not so consistent. During this storm about 28 in. (70 cm) of rain fell within 8 hr in central Virginia. Within this region, many slopes became unstable and landslides were common, and valley sides and headwater channels were altered significantly. These occurrences certainly seem to confirm that infrequent events can be most effective in modifying landforms.

Light-damage areas. But elsewhere in the eastern United States, the effects of Hurricane Agnes were not impressive. In a forested drainage basin in Maryland, although 12 in. (30 cm) of rain fell during one night, only modest channel widening occurred in response to a very high discharge that was an extremely rare occurrence. Agnes caused a high discharge in the Conestoga River of southwestern Pennsylvania, but erosion and deposition on the extensive floodplain were minor. Similarly, major floods in the past on the Connecticut River have also produced relatively little alteration of the channel and floodplain.

THEORY OF GEOMORPHIC THRESHOLDS

Evidently, some data indicate that a major event can be of major importance in modifying the landscape, while other data say that such an event has only minor effect. The conflicting evidence can be explained by the theory of the geomorphic threshold.

Threshold concept. Earth materials, like structural materials, have "strength." They resist stress until it rises above some critical value, at which point there is failure, or erosion. An example is movement of sediment in the bed of a stream. There is a critical velocity of flow below which no sediment movement occurs and above which sediment movement begins (Fig. 4). The precise value of this threshold velocity depends on the size of the sediment particles. This concept of erosion threshold can also be applied to the landscape. Briefly, some landscapes or components of a landscape have apparently evolved to a condition of geomorphic instability. These landforms have a low threshold of failure. They will be significantly modified by a large infrequent event whereas others, depending on the stage of their development, will be unaffected. Therefore, there can be, even within the same region, different responses to the same

Fig. 5. Rio Puerco, an infamous arroyo that has contributed much sediment to the Rio Grande in New Mexico.

conditions of stress.

Thresholds involving slope. When a landscape is subdivided into its components, the result is an assortment of slopes of different lengths, inclinations, and shapes. A landform, then, is an assemblage of slopes, and therefore the business of the geomorphologist is the study of slopes.

When some components in a landscape fail by erosion whereas others do not, it is clear that erosion thresholds have been exceeded locally. When

Fig. 6. Abandoned cabin above the modern channel of East Fork Douglas Creek, Colo. This structure was built during the latter part of the 19th century, when the creek flowed at about the level of the cabin. Channel incision has made the step from the front door a very high one. Figure 7 shows the approximate location of the cabin.

the thresholds involve slope, they are called geomorphic thresholds.

Although the selective failure of slopes in Virginia during Hurricane Agnes could have been related to slope thresholds, the effects of soil character and vegetative density complicate the situation to a great extent. Ready identification of geomorphic thresholds can be made where the influence of vegetation and soil is modest—for example, in semiarid and arid regions. For this reason, the southwestern part of the United States has become a geomorphic laboratory in which the roles of the various agents of landscape change can be discerned more clearly.

Arroyos. In the southwestern United States, a period of major gullying began in the 1880s when the apparently stable alluvial deposits in many western valleys were cut by large channels or arroyos (Figs. 5 and 6). Controversial reasons for the sudden onset of gullying have been advanced. It has been attributed both to overgrazing and to climatic fluctuations. But neither of these explanations was completely satisfying because, even though the arroyos developed widely throughout the West, not all of the alluvial valleys became trenched.

Nongeomorphic explanations. It seems that preconceived notions prevented geologists and geomorphologists from considering that landform or slope instability is inherent in the erosion cycle. Geomorphologists have long been aware of the major climatic changes of the past, and it has always been convenient to explain discontinuities in the erosional evolution of an area as a result of climatic change. Moreover, since the Earth's surface is unstable, tectonic explanations can also be put forward for local landform deviations. And since many of the most recent variations in the landscape can be attributed to man, to his animals, or to his poor land-management practices, land managers quickly seized this explanation for recent erosional and depositional events.

Incipient instability. Obviously, climate change, uplift, and man have influenced the landscape, but both archeological and geomorphic investigations support the concept that—at least in the drylands of the world—epicycles of erosion and deposition are an integral part of the landform development cycle. It can be shown that sediment is stored in semiarid valleys, causing the slope of these valley floors to increase progressively with time. Such a situation cannot continue indefinitely, and there must be some critical slope at which a major flood will cause erosion of the sedimentary deposits. Indeed, this sequence has been documented in Wyoming and New Mexico, where the valleys of small drainage basins can be found in all stages of stability and instability. The inescapable conclusion is that, with time, some parts of a landscape evolve to a condition of incipient instability. A large storm or flood will then cause an adjustment by slope failure, floodplain destruction, or gullying, but not all the landscape will be involved in these dramatic changes. This hypothesis agrees both with the observations made following Hurricane Agnes and with the information on the distribution of arroyos in the western United States.

EVIDENCE FOR GEOMORPHIC THRESHOLDS

Recent field and experimental work carried out by the staff and graduate students of Colorado State University (with the support of the National

Fig. 7. *ERTS-1* satellite image of the Piceance Creek Basin of western Colorado. Parallel alignment of tributaries to Piceance and Yellow creeks reflects fracture patterns in the rock. Continuous patterns of the valleys are visible, but the satellite was too far away for discontinuous gullies to be visible.

Science Foundation, the Army Research Office, and the Colorado Agricultural Experiment Station) has explained, in terms of geomorphic thresholds, the distribution of the discontinuous gullies in the oil shale region of western Colorado. This work has also explained the variation of the channel pattern of the Mississippi River. In addition, the results pertain to the paradoxical effects of large storm and flood events.

Discontinuous gullies. Field studies in valleys of Wyoming, Colorado, New Mexico, and Arizona revealed that discontinuous gullies (short trenches in valley floors) can be related to the slope of the valley-floor surface. Gully erosion in these valleys tends to begin on steeper reaches of the valley floor. Given this fact, the Colorado State researchers hypothesized that, for a given region of uniform geology, land use, and climate, a critical threshold valley slope exists above which the valley floor is unstable. In order to test this hypothesis, measurements were made in the Piceance Creek Basin of western Colorado (Fig. 7). (The area is underlain by oil shale, and the potential environmental problems that could result from development of this resource gave added impetus to the study.) Within the area, valleys were selected in which discontinuous gullies were present. The drainage area and slope of each valley were measured. Since no records for runoff or flood events exist, the valley drainage area was selected as a variable reflecting runoff and flood discharge.

Slope versus area. When valley slope is plotted against drainage area for all the gullied valleys, the relationship is inverse (Fig. 8), with gentler slopes characteristic of large drainage areas. As a basis for comparison, similar measurements were made for valleys in which there were no gullies, and these data are also plotted in the figure. For drainage areas greater than about 5 mi^2 (15×10^6 m), there is a clear line of demarcation—a threshold line—separating gullied (unstable) valleys from ungullied (stable) valleys. In other words, for a given drainage area it is possible to define a valley slope above which the valley floor is unstable.

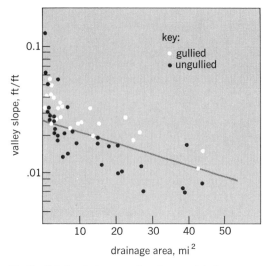

Fig. 8. Relation between valley slope and drainage area for valleys of the Piceance Creek area in western Colorado. (*From P. C. Patton, unpublished M.S. thesis*)

In only two instances do stable valleys plot above the threshold line, and these are probably incipiently unstable. A major flood will cause erosion and trenching of the now stable alluvium in these two valleys. Interestingly, the threshold relationship does not pertain to drainage basins smaller than 4 mi² (10×10^6 m). In these small valleys variations in vegetative cover—perhaps related to the aspect (direction of orientation of the basin) of the drainage system, or variations in the properties of the alluvium, or geologic differences—prevent investigators from recognizing a critical threshold slope.

Land management. Using Fig. 8, one can define the slope above which trenching will take place. This ability has obvious implications for land management because, if the valley slope at which valleys are incipiently unstable can be identified, corrective measures can be taken to artificially stabilize these critical reaches of the basin. *See* LAND-USE PLANNING.

Man's influence. Figure 8 can be used to predict

the influence of man in the Piceance Creek Basin. As man uses the basin more, vegetational cover will decrease and ruts or trails will develop. As a result, the threshold slope line will be lowered, and eventually stable valleys will become unstable. In addition, waste water from mining or irrigation will increase discharge, which has the same effect as an increase in drainage area, shifting a given point in Fig. 8 to the right. If the shift is far enough, the threshold line will be crossed and the stable valley floor will become unstable.

It seems possible that future work will demonstrate that the type of relationship shown in Fig. 8 can be established for alluvial deposits elsewhere. For example, in deserts the trenching of alluvial fans, which is common, is usually explained by renewed uplift of the mountains or by climatic fluctuations. However, the concept of a geomorphic threshold is also applicable to this situation. That is, as the fan grows, it steepens until it exceeds a threshold slope. In each case it should be possible to define a critical threshold slope above which trenching of the sedimentary deposit will take place.

This concept is illustrated by Fig. 9, where the increasing instability of an alluvial fill is represented by a line indicating change of valley or fan slope with time. Superimposed on the ascending line of increasing slope are vertical lines showing the variations of valley-floor stability caused by flood events of various magnitudes. The effect of even large events is minor until the stability of the deposit has been so reduced by steepening that, during one major storm, erosion begins (time A). This storm is only the most apparent cause of failure,

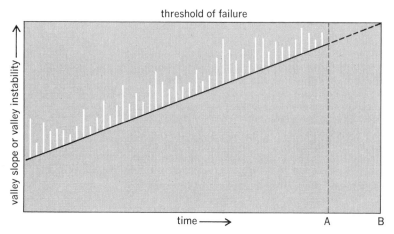

Fig. 9. Hypothetical relation between valley-floor gradient and valley-floor instability with time. Failure occurs at time A, as the apparent direct result of a major storm or flood, but would occur at time B eventually.

Fig. 10. Experimental study of a river meandering in a large flume. (*Photograph by D. Edgar*)

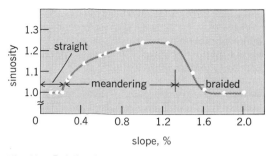

Fig. 11.   Relation between valley slope and sinuosity (ratio of channel length to length of flume or length of valley). The absolute value of the slope at which changes occur will also be influenced by discharge, which here was maintained at 0.15 cfs. (*From S. A. Schumm and H. R. Khan, Experimental study of channel patterns, Geol. Soc. Amer. Bull., 83:1755–1770, 1972*)

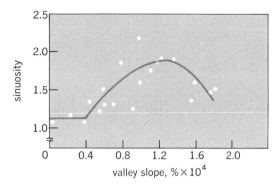

Fig. 12.   Relation between valley slope and sinuosity for the Mississippi River between Cairo, Ill., and Head of Passes, La. (*From S. A. Schumm et al., Variability of river patterns, Nature, 237:75–76, 1972*)

which would have occurred at time B in any case. These studies of alluvial deposits in drylands give further evidence that large infrequent storms can be significant, but only when a geomorphic threshold has been exceeded.

River patterns. Experimental studies of river meandering have been performed by both hydraulic engineers and geologists over many years. A similar study performed at the Engineering Research Center, Colorado State University, investigated the influence of slope and sediment loads on channel patterns (Fig. 10).

Water discharge experiments. It was found, during a series of experiments in which the water discharge was held constant, that if a straight channel was cut in slightly sloping alluvial material, the channel would remain straight. However, at steeper slopes the channel meandered. As the slope of the alluvial surface (the valley slope) steepened, the velocity of flow increased and shear forces acting on the bed and bank of the channel were increased. At some critical value of shear, bank erosion and shifting of sediments on the channel floor produced a sinuous course. The conversion from straight to sinuous channel occurred at a threshold slope (Fig. 11). As slope increased beyond this threshold, meandering increased until at another higher threshold the sinuous channel was converted into a straight braided channel, like the braided South Platte River in Fig. 2.

Abrupt pattern change. The experiments revealed that stream patterns do not change continuously from straight through meandering to braided, but instead, changed abruptly at two threshold slopes. The values of the threshold slopes probably depend on the sediment load and the hydraulic character of the flow; nevertheless, the relationship can be used to explain the variability of a stream pattern.

Mississippi patterns. If, as a stream flows along a valley floor, the slope of the floor varies, the stream channel should reflect these variations by a change in its pattern and sinuosity along its course. The Colorado State researchers had an opportunity to check this conjecture with the aid of data on the Mississippi River furnished by the Potamology Section, Vicksburg District, U.S. Army Corps of Engineers. The data show that variations in channel pattern of the Mississippi River

conform to the Colorado State experimental data, and they can be related to changes of the slope of the valley floor (Fig. 12). If the valley slope exceeds a geomorphic threshold, the river will respond dramatically by a change of channel pattern.

Misguided interference. When the valley slope is near a threshold, major flood events may significantly alter the stream pattern. Similarly, an attempt by humans to straighten a sinuous stream by cutting off some meanders will steepen the slope of the channel. This could convert the channel from a relatively stable meandering pattern to a highly unstable braided pattern. This kind of misguided interference can cause a drastic and undesirable alteration of channel character. *See* RIVER ENGINEERING.

COMPLEX RESPONSE OF GEOMORPHIC SYSTEMS

Throughout the world, geologists and archeologists have studied the most recent erosional and depositional history of valleys. They have been able to identify the sequence in which alluvial deposits were emplaced and then eroded. With

Fig. 13.   Drainage Evolution Research Facility at Colorado State University. Storms are simulated by a sprinkler system so that the erosional development of a small 30 × 50 ft drainage system can be documented through time.

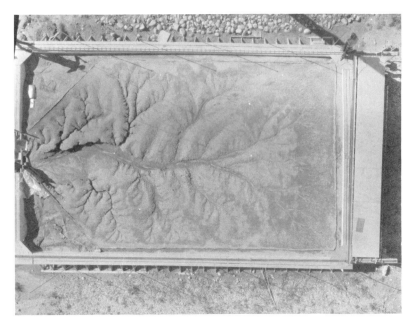

Fig. 14. Aerial view of the drainage pattern developed in the Drainage Evolution Research Facility. The shadow is from the crane boom used to suspend the camera above the surface of the box. The cross-section changes illustrated in Fig. 15 were measured about 5 ft (1.5 m) from the mouth of the drainage system.

this knowledge, and with knowledge of worldwide climate changes during the Quaternary, it should be possible to establish an alluvial chronology. That is, a particular alluvial layer should be identifiable on a regional or even a continental basis.

Alluvial chronologies. There is no question that major climatic changes have affected erosional and depositional events. But when the alluvial chronologies of the last 10,000 to 15,000 years are examined, it is not convincing that each event was in response to one simultaneous climatic change. In fact, in the southwestern United States during the last 10,000 years, the number, magnitude, and duration of erosional and depositional events not only varied from valley to valley but also varied within the same valley. Erosional and depositional events were not in phase. Perhaps this is attributable to different valleys reaching the geomorphic threshold at different times.

Rejuvenation. There is, however, another explanation for the differences from valley to valley: the response of a drainage system to rejuvenation (renewal of erosion activity). Climate change, uplift, lowering of the water base level, or a low-frequency, high-magnitude event can cause incision of a channel. This incision converts the floodplain to a terrace, which is the geomorphic evidence of the erosional episode. However, a drainage system is composed of channels, hillsides, divides, floodplains, and terraces. The response of this complex system to change may also be complex.

Simulation experiments. Such a response was unexpectedly demonstrated during an experimental study in the Drainage Evolution Research Facility at Colorado State University. The facility employs a 30×50 ft (9×15 m) box, containing 400 yd³ (304 m³) of sediment (Fig. 13). A sprinkler system provides simulated precipitation, making it possible to study the erosional evolution of a small

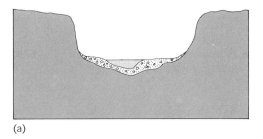

(a)

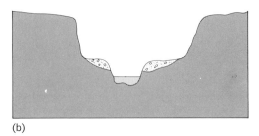

(b)

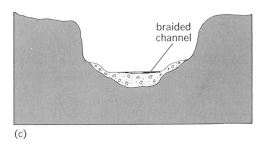

braided channel

(c)

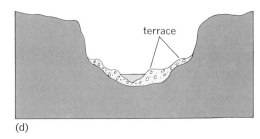

terrace

(d)

Fig. 15. Diagrammatic cross sections of an experimental channel 5 ft upstream from the outlet of the drainage system. (a) Valley and the alluvium deposited during previous run, before base-level lowering. The low width-depth channel flows on alluvium. (b) After base level has been lowered 4 in., channel incises into alluvium and bedrock floor of valley to form a terrace. Following incision, bank erosion widens channel and partially de-

stroys terrace. (c) An inset alluvial fill is deposited, as the sediment discharge from upstream increases. The high width-depth-ratio channel is braided and unstable. (d) A second terrace is formed, as the channel incises slightly and assumes a low width-depth ratio in response to reduced sediment load. In nature, channel migration will eventually destroy part of the lower terrace, and a floodplain will form at a lower level.

drainage system through time (Fig. 14).

Base-level lowering. During experimentation (Fig. 15*a*), the drainage system was rejuvenated by a slight change of base level, that is, the outlet was lowered 4 in. (10 cm). As anticipated, the base level lowering caused incision of the main channel and the development of a terrace (Fig. 15*b*). Incision occurred first at the mouth of the system, then progressed upstream, successively rejuvenating tributaries and scouring the alluvium previously deposited in the valley. As erosion progressed upstream, the main channel became a conveyor of upstream sediment in increasing quantities, and the inevitable result was that aggradation occurred in this newly cut channel and its gradient steepened. Hence, channel incision and drainage network rejuvenation were followed in a relatively short time by deposition. When deposition began, the main channel became braided, and the valley was widened by lateral erosion (Fig. 15*c*). However, as the tributaries eventually became adjusted to the new base level, sediment loads decreased, and renewed channel erosion occurred, forming a low alluvial terrace (Fig. 15*d*). This low surface formed into a better-defined channel of lower width-depth ratio as a result of the decreased sediment loads. The low surface was not a floodplain, however—it did not become flooded at maximum discharge.

Negative feedback. It appears that an event causing an erosional response within a drainage basin automatically creates a negative feedback (high sediment production) which results in deposition. This is eventually followed by incision of the alluvium, as sediment loads in the stream decrease. Within a complex natural system one event can trigger a complex reaction (morphologic or stratigraphic or both) as the components of the system respond progressively to change. This principle provides an explanation of the complexities of the alluvial chronologies.

EVALUATION OF THE MAGNITUDE-FREQUENCY CONCEPT

It appears that storm or flood events of moderate frequency and magnitude do, in fact, most of the geomorphic work. Nevertheless, large infrequent events, although of less significance in the long-term erosional evolution of a landscape, can precipitate large and even spectacular landform modification. However, the changes induced by a major event may be mostly the result of natural landform instability. Geomorphic thresholds and the complex response of geomorphic systems permit high-magnitude events to play a major role in landscape evolution.

Without doubt, these new concepts can help to predict landform response to both natural and man-induced changes. The new ability of geomorphologists to quantitatively define thresholds of instability, and thus permit recognition of potentially unstable landforms, provides a sound basis for preventive erosion control.

[STANLEY A. SCHUMM]

Geothermal power

Generation of electrical power from the stored heat in the Earth is now seen possible in two forms, natural and drilled (or artificial). In the natural form, existing emissions of steam or hot water, such as fumaroles, geysers, or even quiescent underground hot reservoirs, are harnessed to generate high-pressure steam, which in turn drives a conventional turbine generator. In the case of drilled geothermal power, the deep heat in the Earth's crust is used directly by drilling a deep hole, lining it with a pipe, sending cold water down, and bringing back hot water or steam through a smaller pipe in the center.

Natural geothermal power generation is generally limited to volcanic regions of the Earth, which rarely coincide with regions of large power demand. The drilled geothermal power concept, on the other hand, proposes to harness the almost unlimited body of high-temperature heat that is everywhere no farther than 10–15 mi (16–24 km) beneath the surface and therefore equally accessible to every center of power demand.

Natural geothermal power. Additions have been made in the principal existing installations in California, Mexico, Italy, and New Zealand, with further expansions planned, as shown in Table 1. Of greater significance are the active plans to develop the many potential sites listed, plus the widespread exploration now going on for new geothermal sites.

Although the planned capacity in each case listed is only a small percentage of the total national demand, the potential capacity is many times greater. In the United States, for instance, the 300-MW installation at The Geysers in California represents 0.1% of the nation's total, but geologists estimate its ultimate capacity at 5000 MW. The Salton Sea area has recently been measured to contain 6×10^9 acre-feet (7.4×10^{12} m³) of water under high pressure at temperatures above 500°F, a re-

Table 1. Natural geothermal power developments

Site	First installation	Total capacity, MW	
		Installed	Planned
Ecuador			*
Ethiopia (Danakil)			*
Hungary			*
Italy (Larderello)	1904	400	~600
Iceland	1969	5	*
Indonesia			*
Japan (Honshu)	1966	20	120
Kenya			*
Kinshasa (Kiabukwa)		~1	*
Mexico			
Cerro Prieto	1970	75	400
Hidalgo	1959	~5	*
New Zealand			
Wairekai	1959	350	700
Kawerau	1955	50	*
St. Lucia			*
Salvador		~1	*
United States			
The Geysers, Calif.	1960	300	600
Inyo County, Calif.			*
Salton Sea, Calif.			*
Hawaii			*
Eureka County, Nev.			*
Alaska			*
Soviet Union			*
Kamchatka	1967	5	*
50 other basins			*

*Sites with large known potential, still in planning stage.

Table 2. Comparison of approximate data for three natural geothermal installations

	The Geysers, Calif.	Larderello, Italy	Wairekai, New Zealand
Installed capacity, MW	300	390	340
Well depth, ft	800–5000	1000–3000	574–3200
Steam pressure, psia			
With no flow	175	400	235
At full power	115	80	195
Steam temperature, °F	348	266–446	315–395
Steam condition	SH	SH	Wet
Condenser pressure, in. Hg	4	4–30	2
Steam flow, M lb/hr	~6000	~9000	~6000
Noncondensable gases, %	0.7	5	0.5

source sufficient to produce 20,000 MW for southern California. In fact, known total geothermal resources in the West are now estimated to embrace 1.35×10^6 acres (5500 km²), and it is entirely possible that natural geothermal power could ultimately supply 25% of the western demand.

The power-generating equipment employed to harness these natural emissions is now fairly standard. Deep wells are drilled, and live steam is piped directly to a steam turbine. The optimum depth of the well is determined by the quantity of steam available and its superheat at a given pressure. In the most efficient cases, the steam leaves the turbine at very low pressure by virtue of direct condensation with cooling water, thus greatly increasing turbine power. Where a local source of cooling water is not available, as at The Geysers, the condensate is recirculated through cooling towers. The small percentage of noncondensable gases (CO_2, CH_4, and H_2S) in the geothermal steam is removed by steam jet ejectors. These gases, plus other corrosive impurities in the steam, require that piping and turbine blading be of stainless steel. Nevertheless the capital cost per kilowatt is low, and the absence of any fuel cost makes geothermal energy the most inexpensive source of electricity.

Approximate data for the three principal installations in the world today are shown in Table 2.

Drilled geothermal power. Drilled geothermal power is now emerging as an equally attractive concept. It permits the harnessing of the Earth's heat at any site without the need of any natural steam emission, thus avoiding the problems of corrosion and geographical location. However, this method requires the development of ultradeep drilling and casing of large holes, about 24 in. (60 cm) in diameter, to a depth of 50,000 ft (15,000 m) or more, at a cost of $10,000,000 in order to prove economical. This goal is considered quite possible in the light of recent achievements in oil well technology and of new techniques in seismology.

Drilled geothermal power is based on the temperature gradient of about 1.5°F per 100 ft (2.75°C per 100 m) in the Earth's crust, so that, at a depth of 50,000 ft (15,000 m), rock temperatures of about 800°F (420°C) can be expected. A 24-in. (60-cm) cased hole drilled to this depth would have an effective heat-transfer surface in excess of 300,000 ft² (about 30,000 m²). Water sent down such a well would drain heat from the surrounding rock as it descended, with the temperature and pressure rising to 800°F (420°C) and 16,500 psi (114×10^6 N/m²) at the bottom. With an insulated inner pipe in the well, this water would then return to the top at 700°F (370°C), and the density difference of the two columns of water would be sufficient to drive the loop without pumps. By keeping the emerging water at about 4000 psi (28×10^6 N/m²), its heat could be transferred in heat exchangers to generate superheated steam at lower pressure for conventional turbine generators. The spent water, still at about 400°F (200°C) and high pressure, could generate further power in a hydraulic turbine and de-

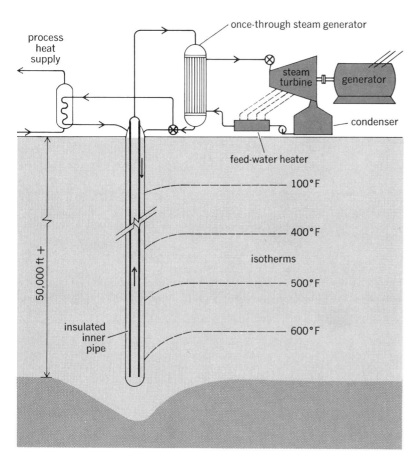

process heat supply

once-through steam generator

steam turbine

generator

condenser

feed-water heater

100°F

400°F

isotherms

500°F

600°F

50,000 ft +

insulated inner pipe

Fig. 1.   Drilled geothermal power-generating system.

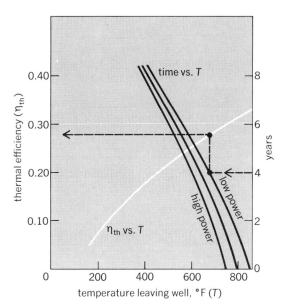

Fig. 2. Typical performance curves (given well depth and diameter) for drilled geothermal power system.

liver substantial heat for municipal central heating and other process industries before being returned to the wellhead. The entire process is shown schematically in Fig. 1.

A 24-in.-diameter (60 cm) well 50,000 ft (15,000 m) deep should generate about 20,000 kW efficiently for many years with practically no attention and with no surface contamination or air pollution. In addition it would furnish some 100×10^6 Btu/hr (30,000 kW) of process heat between 100 and 400°F (40 and 200°C). At greater depths and diameters, both efficiency and capacity increase. A typical operating cycle is shown in Fig. 2. The well is to be thought of, in economic terms, as "mining" heat stored in the Earth, analogous to mining coal or oil, with the one important difference that the stored heat will replenish itself from an essentially inexhaustible reservoir deeper in the Earth.

The economic problem, evident in Fig. 2, is to determine the optimum rate of removal of the heat, followed by the optimum dormant period while the heat returns. A high rate of removal gives a short-term high return on the capital investment, with a rapid fall in efficiency as the well cools down. A low rate of removal gives a longer useful life (between replenishment periods) but at a small return. If the heat is removed at such a low rate that the depression of the Earth isotherms away from the well reach steady state, the power generation would appear to be uneconomically low. Hence cyclical operation would appear to be optimum.

The objective in optimizing the design of such a plant is to determine the flow rate, and therefore the heat removal rate, for a given well depth and diameter, at which long-term power generation would be cheapest. The average properties of the Earth's crust now available for such analysis are a density of 170 lb (specific gravity of 2.73), a specific heat of 0.17 B/lb-°F (0.17 cal/g-°C), and a thermal diffusivity of 0.05 ft²/hr (0.013 cm²/sec).

The concept is sufficiently sound and the economic and environmental advantages are so significant that serious research and development are now being considered by both government and power companies. [ROBERT L. WHITELAW]

Bibliography: A Bible, in *U.S. Sen. Rep.*, no. 1160, Sept. 4, 1970; J. R. McNitt, *Mining Information Service*, California Division of Mines, March 1960; R. L. Moxey, *Compressed Air*, September-October 1966; F. D. Stacey, *Physics of the Earth*, 1969.

Grassland

In the broad sense, this term refers to the many kinds of herbaceous (not woody) terrestrial vegetation in which the dominant species are grasses and grasslike plants (graminoids). These include seeded stands of introduced grasses (mostly perennial) that are used in all vegetation zones for production of forage and pasture for domesticated livestock. When a forest is removed and prevented from regenerating, seminatural grasslands develop in its place. Natural grasslands occupy vast areas of the Earth's surface in regions unfavorable for the development of forests or scrub.

Approximately 30–40% of the world's land surface is exposed to a climate in which natural grassland occurred before civilization. The most extensive of these areas in the temperate zone are located in north-central Asia, eastern Europe, North and South America, and South Africa. Subtropical grasslands and savannas are most extensive in Africa, South America, and Australia.

There is considerable difficulty in judging the boundary between desert and grassland, particularly where the grasses may have been depleted by grazing. Thus the arid areas surrounding the Mediterranean and extending through Asia Minor to India, which are now desertlike, may once have been grasslands. Part of this area is suited to economical production of cereals (annual grasses) and other crops, and these contribute a major proportion of feed and food supplies. This area would revert to natural grassland, through plant succession, if retired from tillage. A large portion of the natural grassland has never been tilled and is used as a grazing resource for domesticated livestock and wild game. *See* SUCCESSION, ECOLOGICAL.

Tree-inhibiting factors. The general treelessness of natural grasslands is most frequently attributed to aridity. Forest trees cannot survive the periods of drought characteristic of grassland regions. However, drought-enduring trees and shrubs found in desert areas, although usually absent from grasslands because seedling establishment is difficult in competition with grasses, have invaded grasslands when depleted by grazing. Other factors that have been cited to account for the absence of trees in various grassland areas are impermeable soil, excessive salinity, and poor drainage.

In subtropical parts of South America, Africa, and northern Australia, savannas are formed when specialized trees occur singly throughout extensive areas of grassland. In semiarid to dry subhumid parts of the temperate zone, for example, eastern Argentina, southern Ukraine, northern Kazakhstan, and central North America, treeless areas frequently occur for hundreds of miles interrupted only by narrow bands of forest in river valleys. *See* SAVANNA.

Natural grasslands have developed in areas where fires are characteristic, since herbaceous species are much better adapted than trees and shrubs to withstand the effects of fire. This is because the perennating buds are located near the soil surface where they are less exposed and because perennial stems are absent. Some geographers have suggested that grasslands are not natural because they may owe their existence to a disturbance factor (fire); others view fire as a natural factor, as inseparable from the grassland environment as aridity.

Herbaceous communities occurring above the tree line in mountains and in the Arctic are sometimes dominated by graminoids and can be considered natural grasslands. These communities exist because of the failure of trees to invade excessively cold or windy areas. *See* PLANT FORMATIONS, CLIMAX.

Terminology. Various descriptive terms have been applied to grassland areas according to geographic distribution: the "llanos" of Venezuela, the "pampas" of Argentina, the "campos" of Brazil, the "steppes" of northern Asia, and the "prairies" and "plains" of North America. These designations have local applications but are of little value in classification.

Climate. The climate of natural grassland is intermediate between that of forests and deserts. The low precipitation and high temperatures in grassland results in a limited supply of soil moisture for plant growth. The long, dry periods associated with these conditions result in a high rate of evaporation. These water stress factors on plants are characteristic of the grassland environment.

The transition from forest to grassland is associated with a decrease in annual precipitation, decreased dependability of plants on precipitation, and a pronounced dry season. Grassland gives way to desert where precipitation is so unreliable that perennial grasses can no longer maintain a continuous cover. The higher temperature and longer growing season associated with the transition from temperate to subtropical grasslands (savannas) requires a higher level of precipitation to maintain a similar vigor of growth. Accordingly, the annual rainfall needed for the transition from forest to grassland in the temperate zone (western Canada and north-central Asia) is as low as 15 in.; in Africa savannas do not give way to tropical forest until the annual rainfall exceeds 60 in.

This wide range in grassland climates results in considerable differences in the vigor of growth that is attained. Thus shortgrasses dominate in areas adjacent to deserts where precipitation ranges from 5 to 15 in. per annum; mid- and tallgrasses abound in semiarid to subhumid conditions where the climate is sufficiently favorable to produce tilled crops; and high grasses (as tall as 15 ft) occur in subtropical regions where growing conditions are very favorable during the wet season.

Structure. The structure of a grassland community is dependent upon the plant species that occur. There are often several layers. For example, a midgrass community may have an understory of shortgrasses and forbs (nongrasslike herbaceous plants) and an overstory of forbs. A lower layer may even occur below the shortgrass layer and be composed of various dwarf species, including mosses, lichens, and club mosses. Occasionally dwarf shrubs occupy grassland, and these may compose another layer.

Climax grassland is usually considered a closed community composed almost entirely of perennial species. The annual habit is not common because in the grassland situation seedlings have difficulty in becoming established. Usually the grasses develop sufficiently so that their foliage covers the soil surface, although their bases may only occupy from 10–25% of it.

In most grasslands the graminoids comprise 75% or more of the plant growth. The graminoids are thus noted to be more aggressive than forbs, and various explanations have been given for this. Speculation has centered around the resistance of graminoids to grazing and trampling, a situation that seems to be associated with the presence of meristimatic (growth) zones in each joint of the stem and in each leaf that permits grazed leaves to continue growth and bent-over stems to straighten.

Root layering is characteristic of a grassland community. There is a relationship between the height of growth and the depth of rooting: The taller grasses have the deepest roots. In addition, the depth of rooting is related to the moisture in the soil since roots penetrate only to the depth of moisture penetration. The roots of plants in sandy soil tend to grow deeper than those in clayey soil, presumably because of the deeper penetration of moisture in the former. This relationship bears on the prominence of tallgrasses in sand-dune areas occurring within a climate capable of supporting only mid- or shortgrasses on silty to clayey soils (for example, the sandhills of Nebraska). Some forbs have roots that do not branch until they reach a depth below that of the grasses. These deep-rooted species apparently do not compete vigorously with the grasses and have remarkable abilities to endure drought. Weedy species are those unpalatable (to livestock) ones that compete with the grasses by producing roots and shoots in the same layers as grasses.

Composition. While there are a large number of grass species in the grasslands of the world, usually very few dominate in any one location; commonly 50–100 species of flowering plants occur in the same grassland community, but only a few of these are considered dominants. The complexity (in terms of numbers of plant species) of grasslands tends to increase with increasing favorableness of habitat and to decrease with increasing aridity. Among the genera of grasses that dominate over large areas are spear grasses (*Stipa* sp.), bluestems (*Andropogon* sp.), grama grasses (*Bouteloua* sp.), wheat grasses (*Agropyron* sp.), fescues (*Festuca* sp.), switch grasses (*Panicum* sp.), and themeda (*Themeda* sp.). Most commonly the composites (Compositae), constitute the next family of importance after the grasses (Gramineae) and sedges (Cyperaceae), with legumes (Leguminosae) coming next. South African grasslands are known for the preponderance of geraniums in the forb component.

The species that occupy a natural grassland can be grouped according to the season during which each develops. Thus there are species that complete their growth cycle and form seeds before taller growing species reach maximum activity. The changes which occur in grassland with season, as a result of the activity of different species at different times, has been referred to as aspect.

Exploitation and management. The effect of grazing, particularly overgrazing, on the composition and structure of grasslands has received considerable attention because of reduction in the grazing capacity that results from overuse of many natural grasslands. Grazing generally has an effect on native grassland similar to decreasing moisture supply. There is a tendency for the most palatable and most productive species to decrease in abundance (decreasers) and for the less desirable species to increase in quantity (increasers).

When grasslands become depleted through continued overuse, the cover is no longer "closed" and "invaders" become very common. This has been the explanation given in respect to the change in cover of the Californian Valley grasslands from a perennial cover of spear grasses to an annual cover of weedy grasses, including wild oats (*Avena* sp.) and bromes (*Bramus* sp).

Range management is concerned with the management of natural grasslands for the sustained production of domestic animals, an interest closely related to that of wildlife management, particularly in respect to combined use of the same lands for domesticated and game animals. There are difficulties in assessing the annual production of natural grasslands because of aspect. The various species produce shoots at different times, and the early shoots of any one grass plant often die before the last ones are produced. Therefore, the harvest of shoots at any one time is only part of the production since growth began. *See* RANGELAND CONSERVATION; WILDLIFE CONSERVATION.

Productivity. Natural grassland is a conservative type of plant community capable of withstanding wide fluctuations in environment and possibly better adjusted to surviving under adverse conditions than to taking full advantage of favorable conditions. There is a tendency toward tying up considerable amounts of nutrients in the form of dead plant materials (litter) both above and below ground. Disturbance of the community increases the rate of breakdown of these materials and increases the production of shoots. It is at least partly due to this feature that newly broken grasslands are highly productive of seeded crops. Since such areas have only been exploited for tilled agriculture from about the middle of the 19th century, it is not possible yet to determine how long this exploitation can be continued on a sustained basis.

The soils that develop under natural grasslands are chernozemic, that is, they have a deep, rich humus horizon. Such soils are not leached to a marked degree and maintain the basic ions in circulation within the zone of activity of plants. These soils are very fertile and can be cultivated without preconditioning. *See* SOIL.

Ecology. While the above discussion has emphasized the plant (producer) component of the grassland, it is logical to include both the animals (consumers) and microorganisms (microconsumers or decomposers) in a consideration of such natural systems. The organismal components must likewise be considered together with the associated soil and aerial environment in what is called the ecosystem. *See* ECOSYSTEM; PLANT COMMUNITY.   [ROBERT COUPLAND]

Bibliography: A. W. Sampson, *Range Management*, 1952; H. B. Sprague (ed.), *Grasslands*, AAAS Publ. no. 53, 1959; L. A. Stoddart and A. D. Smith, *Range Management*, 1955; J. E. Weaver, *North American Prairie*, 1954; J. E. Weaver and F. W. Albertson, *Grasslands of the Great Plains*, 1956.

Greenhouse effect, terrestrial

The Earth's atmosphere acts as the glass walls and roof of a greenhouse in trapping heat from the Sun. Like the greenhouse, it is largely transparent to solar radiation, but it strongly absorbs the longer-wavelength radiation from the ground. Much of this long-wave radiation is reemitted downward to the ground, with the paradoxical result that the Earth's surface receives more radiation than it would if the atmosphere were not between it and the Sun.

The absorption of long-wave (infrared) radiation is effected by small amounts of water vapor, carbon dioxide, and ozone in the air and by clouds. Clouds actually absorb about one-fifth of the solar radiation striking them, but unless they are extremely thin, they are almost completely opaque to infrared radiation. The appearance even of cirrus clouds after a period of clear sky at night is enough to cause the surface air temperature to increase rapidly by several degrees because of radiation from the cloud.

The greenhouse effect is most marked at night, and usually keeps the diurnal temperature range below 20°F. Over dry regions such as New Mexico and Arizona, however, where the water-vapor content of the air is low, the atmosphere is more transparent to infrared radiation, and cool nights may follow very hot days.   [LEWIS D. KAPLAN]

Groundwater

The water in the zone in which the rocks and soil are saturated, the top of which is the water table. The zone of saturation is the source of water for wells, which provide about one-fifth of the water supplies of the United States. It is also the source of the water that issues as springs and seeps, and maintains the dry-weather flow of perennial streams. The saturated zone is a great natural reservoir which absorbs and stores precipitation during wet periods and pays it out slowly during dry periods; it is, therefore, a natural regulating mechanism which tempers the severity of floods and droughts. The amount of ground water stored in the rocks in the United States is estimated to be several times as great as that stored in all lakes and reservoirs, including the Great Lakes.

Subterranean water. Water beneath the land surface may be subdivided into two parts, water in the zone of aeration above the water table and water in the zone of saturation below the water table. Water in the zone of aeration, also called vadose water, is divided into soil water or rhizic water, intermediate vadose water or argic water, and water of the capillary fringe or anastatic water.

Water in the capillary fringe is connected with the zone of saturation and is held above it by capillary forces. The lower part of the fringe may be saturated but is not a part of the zone of saturation because the water is under less than atmospheric pressure and will not flow into a well, although the walls of the well are moist. When the well reaches the zone of saturation, water will begin to enter it and will stand at the level of the water table.

Rock formations capable of yielding significant volumes of water are called aquifers. Some wells are artesian; that is, the water rises above the top of the water-bearing bed. This is a special case and will be discussed later. Other wells encounter water above the saturated zone and lose their water if extended through the impermeable bed upon which the water rests. Such bodies of water are said to be perched. This also is a special case.

Subterranean water occurs in a geologic environment, and therefore a knowledge of geology is essential to an understanding of the occurrence of water. For this reason, the study of groundwater is sometimes called hydrogeology or geohydrology. The hydrologist is particularly concerned with the number and size of the openings in rocks and soils and the manner in which they are interconnected. Variations in these openings are almost infinitely diverse. Openings are practically absent in some of the igneous rocks. They are numerous but microscopic in clay. They are large and interconnected in many sands and gravels. There are huge caverns and tubes in many limestones and lavas. Their distribution and types are as diverse as the geology itself, so that general statements about them applicable to one area may be incorrect for another.

Openings in rocks. These are of two broad types. Primary are those which existed when the rock was formed, and secondary, those which resulted from the action of physical or chemical forces after the rock was formed. Primary openings are found in sedimentary rocks such as sand and clay and certain kinds of limestone composed of triturated shells. Certain openings in lava are formed at the stage when the lava is partly liquid and partly solid and also are considered primary. Most rocks containing primary openings are relatively young geologically. Those which contain primary openings large enough to carry useful amounts of water are represented, for example, by the seaward-dipping strata of the Atlantic and Gulf coastal plains, including the coquina limestone of Florida, the intermontane valleys of the western United States, the glacial deposits of the United States, and the lava rocks of the Pacific Northwest.

Secondary openings are common in older rocks. Sand and gravel that have been cemented by chemical action, limestone indurated by compression or recrystallization, schist, gneiss, slate, granite, rhyolite, basalt and other igneous rocks, and shale generally contain few primary openings; but they all may contain fractures that will carry water. Limestone in particular is subject to solution which, beginning along small cracks, may develop channels ranging from openings a fraction of an inch across to enormous caverns capable of carrying large amounts of water.

Porosity. The property of rocks for containing voids, or interstices, is termed porosity. It is expressed quantitatively as the percentage of the total volume of rock that is occupied by openings. It ranges from as high as 80% in newly deposited silt and clay down to a fraction of 1% in the most compact rocks.

Permeability. This is the characteristic capability of rock or soil to transmit water. The porosity of a rock or soil has no direct relation to the permeability or water-yielding capacity. This capacity is closely related to the size and the degree to which the pores or openings are interconnected. If the pores are small, the rock will transmit water very slowly; if they are large and interconnected, they will transmit water readily. The standard coefficient of permeability (the Meinzer unit) used in the hydrologic work of the U.S. Geological Survey is defined as the rate of flow of water at 60°F, in gallons per day through a cross section 1 ft², under a hydraulic gradient of 100% (1 ft of head loss per foot of water travel). Under field conditions, the adjustment to standard temperature is commonly ignored, and permeability is expressed as a field coefficient at the prevailing water temperature.

Transmissibility. Another coefficient that is commonly used, transmissibility, expresses the rate at which water moves through a saturated body of rock. It is expressed as the rate of flow of water at the prevailing temperature, in gallons per day through a vertical strip of aquifer 1 ft wide, extending the full saturated height of the aquifer under a hydraulic gradient of 100%. This coefficient is especially useful for expressing the total yield of an aquifer.

Controlling forces. Water moves through permeable rocks under the influence of gravity from places of higher head to places of lower head; that is, from areas of intake or recharge to areas of discharge, such as wells or springs. Water moving through rocks is acted upon also by friction and by molecular forces. The molecular forces are the attraction of rock surfaces for the molecules of water (adhesion) and the attraction of water molecules for one another (cohesion). When wetted, each rock surface is able to retain a thin film of water despite the effect of gravity. In very fine-grained rocks, such as clay and fine silt, the interstices may be so small that molecular attraction extends from one side of a pore to the opposite side. Molecular force then becomes dominant, and water moves through the rock only very slowly under the gradients typical of natural conditions.

The amount of water that drains from a saturated rock under the influence of gravity, expressed as a percentage of the total volume of the rock, is called the specific yield. The percentage of water retained in the rock is called the specific retention. Specific yield is often called effective porosity because it represents the pore space that will surrender water to wells and so is effective in supplying water for human use. The term porosity is poorly defined and its use should be discontinued. A part of the water stored in the rocks is held by molecular forces and may have only a small share in supplying springs or wells. This latter portion is of special interest to the agriculturalist because it sustains plant life. A soil that is highly permeable permits water to pass through it easily and little is retained for the nourishment of plant life, whereas a soil that is relatively impermeable retains much of its water until it is extracted by plants or by evaporation.

Sources. It has been firmly established that the chief means of replenishment of groundwater is downward percolation of surface water, either direct infiltration of rainwater or snowmelt or infiltration from bodies of surface water which themselves are supplied by rain or snowmelt. Evidence on the replenishment of groundwater is furnished by analysis of data on the downward movement of precipitation through the soil and subsoil, the rise and fall of groundwater levels and spring discharge in response to precipitation and seepage losses from streams, and the slope of the hydraulic gradient from known areas of intake to areas of discharge. Some groundwater may originate by chemical and physical processes that take place deep within the Earth. Such water is called juvenile water or primary water to indicate that it is reaching the Earth's surface for the first time. The available evidence indicates that such water is always highly mineralized. Some water is stored in deep-lying sedimentary rocks and is a relic of the ancient seas in which these rocks were deposited. It is called connate water. The total quantity of water from juvenile and connate sources that enters the hydrologic cycle is insignificant when compared with the quantities of water derived from precipitation (meteoric water). It is balanced to some extent by withdrawal of water from the hydrologic cycle by such processes as deposition of minerals that include water in their crystalline structure. *See* HYDROLOGY.

Infiltration. Replenishment of water in the zone of saturation involves three steps: (1) infiltration of water from the surface into the rock or soil that lies directly beneath the surface, (2) downward movement through the zone of aeration of the part of the water not retained by molecular forces, and (3) entrance of this part of the water into the zone of saturation, where it becomes groundwater and moves, chiefly laterally, toward a point of discharge. Infiltration is produced by the joint action of molecular attraction and gravity. The rate of infiltration then becomes chiefly a function of the permeability of the soil. This permeability is almost infinitely diverse. Under conditions of unsaturated flow such as dominate in the zone of aeration, it varies with moisture content as well as with pore size. It varies also with the geology. For example, in the Badlands, S.Dak., where the soil and rocks are of low permeability, the infiltration capacity of the soil is low and is reached quickly after rainfall or snowmelt begins. Hence, there is not much infiltration, and any excess of precipitation or snowmelt over infiltration runs off over the surface and enters the streams. If the excess is large, serious floods and erosion result. On the other hand, the soils of the Sand Hills, Nebr., and the glacial outwash deposits of Long Island, N.Y., are so permeable that they absorb the water of the most violent storms and permit little or no direct runoff.

The permeability of the rock materials beneath the soil zone also is important. Since the soil is commonly formed by weathering of the underlying rock, the permeability of the rocks is generally comparable to that of the soil.

Water-table and artesian conditions. Water that moves downward through the soil and subsoil in excess of capillary requirements continues to move downward until it reaches a zone whose permeability is so low that the rate of further downward movement is less than the rate of replenishment from above. A zone of saturation then forms, its thickness depending on the opportunity for lateral escape of water in relation to replenishment. The top of this zone is the water table (see illustration). Under these circumstances, the water is said to be unconfined, or under water-table conditions. However, since much of the crust of the Earth has a more or less well-defined layered structure in which zones of high and low permeability alternate, situations are common in which groundwater moving laterally in a permeable rock passes between layers of relatively low permeability. Although the permeable layer contains unconfined groundwater in the area where there is no impermeable layer (confining bed) above, the part of the layer, or aquifer, that passes beneath the confining bed contains water that is pressing upward against the confining bed, and if a well is

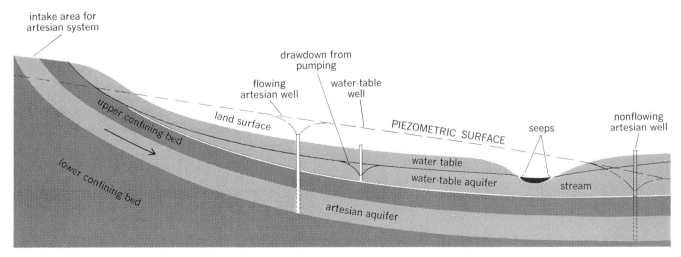

Water-table and artesian conditions.

drilled in this area the water in it will rise. It tends to rise to the level of the water in the unconfined area, but fails to reach that level by the amount of pressure head lost by friction as the water moves from the unconfined area to the well. Confined water is also called artesian water, and wells tapping it are called artesian whether or not their head is sufficient for them to flow at the land surface.

Chemical qualities. Water is said to be the universal solvent. When it condenses and falls as rain or snow, it absorbs small amounts of mineral and organic substances from the air. After falling, it continues to dissolve some of the soil and rocks through which it passes. Thus, no groundwater is chemically pure. Its commonest mineral constituents are the bicarbonates, chlorides, and sulfates of calcium, magnesium, sodium, and potassium, in ionized or dissociated form. Silica also is an important constituent. Common also are small, but significant, concentrations of iron, and manganese, fluoride, and nitrate. The concentration of the dissolved minerals varies widely with the kind of soil and rocks through which the water has passed. Ordinarily, water that contains more than 1000 parts per million (ppm) of dissolved solids is considered unfit for human consumption, and water that contains more than 2000 ppm, unfit for stock. However, both human beings and animals may become accustomed to greater concentrations.

[A. NELSON SAYRE/RAY K. LINSLEY]

Bibliography: S. N. Davis and R. J. M. DeWiest, *Hydrogeology*, 1966; R. J. Kazmann, *Modern Hydrology*, 1965; D. K. Todd, *Groundwater Hydrology*, 1959; R. C. Ward, *Principles of Hydrology*, 1964.

Halogens, atmospheric geochemistry of

The halogens are common trace constituents of the natural atmosphere and are also being added artificially as air pollutants. As objects for study of chemical reactions in the atmosphere, the halogen elements chlorine, bromine, and iodine have special interest because of the ease of their oxidation-reduction reactions under natural conditions and their existence both in the gas phase and as particles. Some phenomena which may be better understood by current research on the atmospheric geochemistry of the halogens include the nucleation of raindrops by sea salt particles, migration of fission-product iodine-131, biological uptake of iodine as an essential trace element, behavior of silver iodide as a cloud-seeding agent, and air pollution by lead bromide and lead chloride particles from automotive ethyl fluid combustion.

Halogens are added naturally to the atmosphere from the sea surface by water droplets cast upward during the bursting of air bubbles. The evidence indicates that bromide and chloride are transferred in the sea-water weight ratio Br/Cl = 0.0034 without fractionation. Iodine, which is present in the sea in the proportion I/Br = 0.001, is strongly enriched in the atmosphere probably either by inorganic volatilization or because of concentration of iodine-rich organic material at the sea-air interface. Airborne droplets and salt crystals smaller than about 10 μ in radius are carried aloft by atmospheric eddies and become aerosols whose atmospheric residence time is limited mainly by the scavenging action of precipitation. The halogen composition of aerosol particles is variable and is evidence of delicate interplay between gaseous and particulate forms of these elements in the atmosphere. *See* ATMOSPHERIC CHEMISTRY; SEA WATER.

Experimental methods. From an analytical standpoint trace constituents in the atmosphere demand sensitive methods for their detection. Neutron activation analysis has the required sensitivity for many elements, and the halogen group chlorine, bromine, and iodine has been under intensive investigation by using this method. With neutron fluxes available in many nuclear reactors, a few cubic meters of air or a few milliliters of rainwater may be analyzed for all three elements. An aqueous solution is irradiated with neutrons for a few minutes, the halogens are separated radiochemically from each other by using solvent extraction, and induced radioactivity is measured by a beta proportional counter. The amount of neutron-induced radioactivity is a sensitive indicator of the amount of chlorine, bromine, and iodine originally present in the sample.

Halogens are found in the atmosphere as particles, as vapor, and as dissolved components of rain and snow. Aerosol particles of radius greater than 0.1 μ may be collected by using a cascade impactor, which sorts particles into size groups, or by using impactor devices of other design suitable for mounting on aircraft. Alternatively, particles are efficiently extracted from the air by membrane filters. Gaseous halogens have been sampled with an aqueous bubbler device, indicating that substantial amounts exist in this form.

Natural atmosphere. Because halogens of the natural atmosphere are derived mainly from the sea, the unpolluted marine atmosphere of Hawaii has been selected for intensive sampling. Precipitation collected over land on the mountainous island of Hawaii contains 1–10 parts per million of chloride near sea level, but the mean value decreases with altitude and is tenfold lower at 7000 ft. Bromine and iodine decrease with altitude somewhat more slowly so that the ratio Br/Cl actually increases from the sea-water value at sea

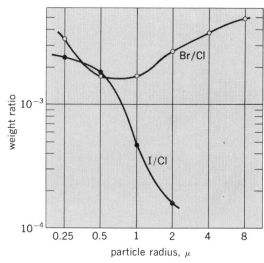

Variation of halogen composition with particle radius in aerosols over the open ocean near Hawaii.

level to fourfold greater at 7000 ft, and the ratio I/Br is roughly constant at 0.4 at all altitudes in this range. Since raindrops form around and scavenge particles of sea salt from the air, the composition of the rain should parallel that of aerosol particles themselves. Direct sampling of aerosols in the same location shows the parallel.

In samples of fresh aerosols over the open ocean, I/Cl is found to increase with decreasing droplet radius from 4 to 0.25 μ, as shown in the figure. This is approximately linearly with increasing surface-to-volume ratio. Both adsorption of inorganic vapors of iodine onto the surfaces of droplets and the concentration of iodine-rich organic films onto the surfaces may be occurring. Generally, over the sea Br/Cl tends to be less than the sea-water value, especially near 1-μ particle radius. It is likely that bromide, which is more easily volatilized through photochemical oxidation of after reaction with atmospheric oxidants, is selectively lost from the particles to the gas phase. Separate measurement of gaseous halogens indicates that the major part of both Br and I exist in this form.

Studies of the unpolluted atmosphere far from open sea water indicate that iodine and bromine, and to a lesser extent chlorine, travel far through the atmosphere before removal. In the winter atmosphere at Point Barrow, Alaska, particulate iodine averages about 1 ng/m³ (nanogram/m³) of air and bromine averages about five times greater—values similar to those observed over the open sea in Hawaii—but particulate chlorine is usually at least tenfold less in Alaska than in Hawaii. An intricate chain of events for the transfer of halogens between the particulate and vapor forms underlies this behavior.

Polluted atmosphere. Because the atmosphere above cities and adjacent rural areas contains halogens derived both naturally and from the combustion of fuels and other sources, the fundamental relationships between natural and pollution halogens are difficult to resolve by sampling in these areas alone. Fairbanks, Alaska, during winter is a model locality for studying the pollution component without interference from sea salt because of its remoteness from the sea and low wind speeds, and because the persistent low atmospheric inversion levels during winter lead to a small dilution volume and high concentrations of air pollutants. Samples taken here have shown a selective loss of bromine to the gas phase from lead chloride-bromide pollution aerosols—a deficiency which parallels that in natural marine aerosols. Lead concentrations of several micrograms per cubic meter of air in pollution aerosols are frequently observed in cities, and the significance of similar amounts of pollution bromine and its chemical reactivity in the atmosphere is now under investigation. *See* ATMOSPHERIC POLLUTION.

[JOHN W. WINCHESTER]

Bibliography: R. A. Duce, J. W. Winchester, and T. W. Van Nahl, Iodine, bromine, and chlorine in the Hawaiian marine atmosphere, *J. Geophys. Res.*, 70:1775, 1965; R. A. Duce, J. W. Winchester, and T. W. Van Nahl, Iodine, bromine, and chlorine in winter aerosols and snow from Barrow, Alaska, *Tellus*, 18:238, 1966; R. A. Duce, A. H. Woodcock, and J. L. Moyers, Variation of ion ratios with size among particles in tropical oceanic air, *Tellus*, 19:369, 1967; J. W. Winchester et al., Lead and halogens in pollution aerosols and snow from Fairbanks, Alaska, *Atmos. Environ.*, 1:105, 1967; J. W. Winchester and R. A. Duce, The global distribution of iodine, bromine and chlorine in marine aerosols, *Die Naturwiss.*, 54:110, 1967.

Health physics

The science that deals with problems of protection from the hazards of ionizing radiation or prevention of damage from exposure to this radiation while making it possible for man to make full use of this great source of energy. Health physics is a border field of physics, biology, chemistry, mathematics, medicine, engineering, and industrial hygiene. It is concerned with radiation protection problems involving research, engineering, education, and applied activities. It involves research on the effects of ionizing radiation on matter with a goal of developing a coherent theory of radiation damage. It deals with methods of measuring and assessing radiation dose, devices for reducing or preventing radiation exposure, the effects of ionizing radiation on man and his environment, radioactive waste disposal, and the establishment of maximum permissible exposure levels.

Health physics began in 1942 along with the nuclear energy and reactor programs at the University of Chicago. It is estimated that by 1967 there were more than 6000 practicing health physicists throughout the world. In 1956 the Health Physics Society was organized in the United States and now has more than 3000 members. It is one of 15 organizations affiliated with the International Radiation Protection Association, which was organized and held its first international congress in Rome, 1966. There is now a total membership of more than 5000 from more than 50 countries. *See* RADIATION INJURY (BIOLOGY).

[KARL Z. MORGAN]

Bibliography: J. D. Abbott et al., *Protection Against Radiation*, 1961; J. S. Handloser, *Health Physics Instrumentation*, 1959; Second International Conference on Peaceful Uses of Atomic Energy, Geneva, 1958, *Progress in Nuclear Energy*, ser. 12: *Health Physics*, vol. 1, 1959.

Heat balance, terrestrial atmospheric

The disposition, distribution, and transformation of the heat received by the Earth-atmosphere system from the Sun. An area of 1 cm² at the Earth's mean distance from the Sun and perpendicular to the Sun's rays would receive about 2 calories (cal) of solar radiation per minute if there were no intervening atmosphere. This quantity is called the solar constant. Since the area of a sphere is four times that of the circle it presents to parallel radiation, the outside of the Earth's atmosphere receives an average of $\frac{1}{2}$ cal/(cm²)(min) or $\frac{1}{2}$ langley/min. One-third of this radiation is reflected and scattered back to space, mostly by clouds. The remaining two-thirds is absorbed by the Earth, clouds, and atmosphere and acts to raise their temperature and to evaporate water from the oceans and clouds.

The Earth and atmosphere can lose heat to space only by radiation. Since the temperature changes on the Earth over long periods of time are small compared to the changes that would have

+25   −100   +9   space   +66

+19
absorbed
by atmos-
phere and
clouds

$\bigcirc\bigcirc$ −52 net loss by
atmosphere

+23 +10

| +17 | +24 | +6 | −119 | +105 | −23 | −10 |
| diffuse from clouds | direct solar | down-scattered | longwave radiation | latent heat | sensible heat |

Schematic representation of heat balance of Earth and atmosphere. Solar radiation reflected from Earth's surface is not shown separately but is included in solar radiation returned to space. (*After H. G. Houghton*)

occurred had a significant fraction of the solar radiation been retained as a net heat gain, the outward radiation must be equal to the two-thirds of the solar radiation that is not reflected or scattered directly to space.

H. G. Houghton's estimate of the disposition of the $\frac{1}{2}$ langley/min of solar radiation, taken as 100 units, is shown by the arrows on the lefthand side of the illustration. Of the 100 units, about 25 are reflected and scattered by clouds and about 9 units by the Earth's surface and atmosphere. About 19 units are absorbed by clouds and atmosphere, and the remainder are absorbed at the surface of the Earth. The middle part of the illustration shows the fluxes of long-wave terrestrial radiations at the top and bottom of the atmosphere. The atmosphere loses about 52 more units to Earth and space than it gains from the Earth. The Earth also transfers heat to the atmosphere by evaporation of water from the surface followed by condensation in the atmosphere (release of latent heat) and by convection from the heated ground (sensible heat). *See* GREENHOUSE EFFECT, TERRESTRIAL.

There are, of course, marked geographical deviations from this average picture of the heat balance. In particular, the tropical regions of the Earth receive considerably more solar radiation than is lost by terrestrial radiation to space, and the polar regions, considerably less. To maintain balance, heat must be transferred from low to high latitudes by the wind systems and, to some extent, by the ocean circulations. *See* ATMOSPHERIC GENERAL CIRCULATION.   [LEWIS D. KAPLAN]

Bibliography: A. H. Gordon, *Elements of Dynamic Meteorology*, 1962; H. G. Houghton, On the annual heat balance of the Northern Hemisphere, *J. Meteorol.*, 11(1):1–9, 1954.

Heating, comfort

The maintenance of the temperature in a closed volume, such as a home, office, or factory, at a comfortable level during periods of low outside temperature. Two principal factors determine the

amount of heat required to maintain a comfortable inside temperature: the difference between inside and outside temperatures and the ease with which heat can flow out through the enclosure.

Heating load. The first step in planning a heating system is to estimate the heating requirements. This involves calculating heat loss from the space, which in turn depends upon the difference between outside and inside space temperatures and upon the heat transfer coefficients of the surrounding structural members.

Outside and inside design temperatures are first selected. Ideally, a heating system should maintain the desired inside temperature under the most severe weather conditions. Economically, however, the lowest outside temperature on record for a locality is seldom used. The design temperature selected depends upon the heat capacity of the structure, amount of insulation, wind exposure, proportion of heat loss due to infiltration or ventilation, nature and time of occupancy or use of the space, difference between daily maximum and minimum temperatures, and other factors. Usually the outside design temperature used is the median of extreme temperatures.

The selected inside design temperature depends upon the use and occupancy of the space. Generally it is between 66 and 75°F.

The total heat loss from a space consists of losses through windows and doors, walls or partitions, ceiling or roof, and floor, plus air leakage or ventilation. All items but the last are calculated from $H_t = UA (t_i - y_o)$, where heat loss H_t is in British thermal units per hour, U is overall coefficient of heat transmission from inside to outside air in Btu/(hr)(ft²)(°F), A is inside surface area in square feet, t_i is inside design temperature, and t_o is outside design temperature.

Values for U can be calculated from heat transfer coefficients of air films and heat conductivities for building materials or obtained directly for various materials and types of construction from heating guides and handbooks.

The heating engineer should work with the architect and building engineer on the economics of the completed structure. Consideration should be given to the use of double glass or storm sash in areas where outside design temperature is 10°F or

Table 1. Effectiveness of double glass and insulation*

| Heat-loss members | Area, ft² | Heat loss, Btu/hr | |
		With single-glass weather-stripped windows and doors	With double-glass windows, storm doors, and 2-in. wall insulation
Windows and doors	439	39,600	15,800
Walls	1,952	32,800	14,100
Ceiling	900	5,800	5,800
Infiltration		20,800	20,800
Total heat loss		99,000	56,500
Duct loss in basement and walls (20% of total loss)		19,800	11,300
Total required furnace output		118,800	67,800

*Data are for two-story house with basement in St. Louis, Mo. Walls are frame with brick veneer and 25/32-in. insulation plus gypsum lath and plaster. Attic floor has 3-in. fibrous insulation or its equivalent. Infiltration of outside air is taken as a 1-hr air change in the 14,400 ft³ of heating space. Outside design temperature is −5°F; inside temperature is selected as 75°F.

lower. Heat loss through windows and doors can be more than halved and comfort considerably improved with double glazing. Insulation in exposed walls, ceilings, and around the edges of the ground slab can usually reduce local heat loss by 50–75%. Table 1 compares two typical dwellings. The 43% reduction in heat loss of the insulated house produces a worthwhile decrease in the cost of the heating plant and its operation. Building the house tight reduces the normally large heat loss due to infiltration of outside air.

Insulation and vapor barrier. Good insulating material has air cells or several reflective surfaces. A good vapor barrier should be used with or in addition to insulation, or serious trouble may result. Outdoor air or any air at subfreezing temperatures is comparatively dry, and the colder it is the drier it can be. Air inside a space in which moisture has been added from cooking, washing, drying, or humidifying has a much higher vapor pressure than cold outdoor air. Therefore, moisture in vapor form passes from the high vapor pressure space to the lower pressure space and will readily pass through most building materials. When this moisture reaches a subfreezing temperature in the structure, it may condense and freeze. When the structure is later warmed, this moisture will thaw and soak the building material, which may be harmful. For example, in a house that has 4 in. or more of mineral wool insulation in the attic floor, moisture can penetrate up through the second floor ceiling and freeze in the attic when the temperature there is below freezing. When a warm day comes, the ice will melt and can ruin the second floor ceiling. Ventilating the attic helps because the dry outdoor air readily absorbs the moisture before it condenses on the surfaces. Installing a vapor barrier in insulated outside walls is recommended, preferably on the room side of the insulation. Good vapor barriers include asphalt-impregnated paper, metal foil, and some plastic-coated papers. The joints should be sealed to be most effective.

Infiltration. In Table 1, the loss due to infiltration is large. It is the most difficult item to estimate accurately and depends upon how well the house is built. If a masonry or brick-veneer house is not well calked or if the windows are not tightly fitted and weather-stripped, this loss can be quite large. Sometimes, infiltration is estimated more accurately by measuring the length of crack around windows and doors.

Illustrative quantities of air leakage for various types of window construction are shown in Table 2. The figures given are in cubic feet of air per foot of crack per hour.

Design. Before a heating system can be designed, it is necessary to estimate the heating load for each room so that the proper amount of radiation or the proper size of supply air outlets can be selected and the connecting pipe or duct work designed.

Heat is released into the space by electric lights and equipment, by machines, and by people. Credit to these in reducing the size of the heating system can be given only to the extent that the equipment is in use continuously or if forced ventilation, which may be a big heat load factor, is not used when these items are not giving off heat, as in a factory. When these internal heat gain items are

Table 2. Infiltration loss with 15-mph outside wind

Building item	Infiltration, $ft^3/(ft)(hr)$
Double-hung unlocked wood sash windows of average tightness, non-weather-stripped including wood frame leakage	39
Same window, weather-stripped	24
Same window poorly fitted, non-weather-stripped	111
Same window poorly fitted, weather-stripped	34
Double-hung metal windows unlocked, non-weather-stripped	74
Same window, weather-stripped	32
Residential metal casement, 1/64-in. crack	33
Residential metal casement, 1/32-in. crack	52

large, it may be advisable to estimate the heat requirements at different times during a design day under different load conditions to maintain inside temperatures at the desired level.

Cost of operation. Design and selection of a heating system should include operating costs. The quantity of fuel required for an average heating season may be calculated from

$$F = \frac{Q \times 24 \times DD}{(t_i - t_o) \times \text{Eff} \times H}$$

where F = annual fuel quantity, same units as H
  Q = total heat loss, Btu/hr
  t_i = inside design temperature, °F
  t_o = outside design temperature, °F
  Eff = efficiency of total heating system (not just the furnace) as a decimal
  H = heating value of fuel
  DD = degree-days for the locality for 65°F base, which is the sum of 65 minus each day's mean temperature for all the days of the year. *See* DEGREE-DAY.

If a gas furnace is used for the insulated house of Table 1, the annual fuel consumption would be

$$F = \frac{56,500 \times 24 \times 4699}{[75 - (-5)] \times 0.80 \times 1050} = 94,800 \text{ ft}^3$$

For a 5°F, 6- to 8-hr night setback, this consumption would be reduced by about 5%.

[GAYLE B. PRIESTER]

Bibliography: American Society of Heating and Air Conditioning Engineers, *Handbook of Fundamentals*, 1967; C. Strock (ed.), *Handbook of Air Conditioning, Heating, and Ventilating*, 1966.

Heavy metals, pathology of

Man has evolved in an environment containing metals and has developed protective means of defense against natural concentrations. Indeed, some metals have become essential for life. However, industry is pouring metals, particularly heavy metals, into man's environment at an unprecedented and constantly increasing rate. As a result, technological society is exposing the population to some metals in unnaturally high concentrations, in unusual physical or chemical forms, and through unusual portals of entry.

This trend is certainly likely to continue. It is vital, therefore, that the effects of metals on living

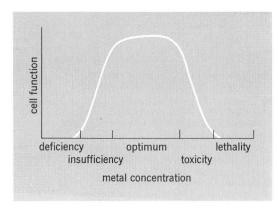

Fig. 1. Effects of metals on cells. Metals have an optimum range of concentrations within which a cell is healthy. The actual values and range vary from metal to metal.

cells be understood, as a first step toward controlling the problems. This article summarizes the present understanding of the effects and selectively describes the pathology of some environmentally important heavy-metal poisonings that illustrate the diversity of pathologic lesions.

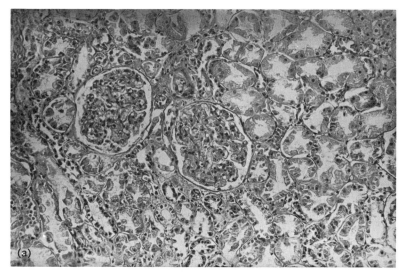

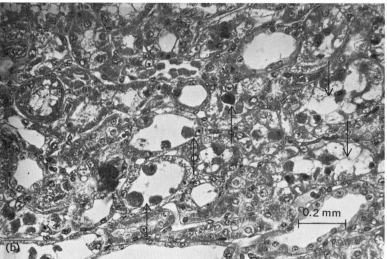

Fig. 2. Kidney cells. (a) Normal kidney cells. (b) Methyl mercury–poisoned cells. Arrows indicate dead cells.

NATURE OF METAL TOXICITY

Heavy metals, the principal subject of this article, are arbitrarily defined as those metals having a density at least five times greater than that of water. Although metals have many physical properties in common, their chemical reactivity is quite diverse, and their toxic effect on biological systems is even more diverse. *See* TOXICOLOGY.

A metal can be regarded as toxic if it injures the growth or metabolism of cells when it is present above a given concentration. Almost all metals are toxic at high concentrations and some are severe poisons even at very low concentrations. Copper, for example, is a micronutrient, a necessary constituent of all organisms, but if the copper intake is increased above the proper level, it becomes highly toxic. Like copper, each metal has an optimum range of concentration, in excess of which the element is toxic (Fig. 1). When the optimum range for a particular metal is narrow, the risk of toxicity increases; thus even a minor environmental increase can be serious.

Metals exert their toxicity on cells by interfering in any of several ways with cell metabolism. The most important of these affects enzyme systems. The more strongly electronegative metals (such as copper, mercury, and silver) bind with amino, imino, and sulfhydryl groups of enzymes, thus blocking enzyme activity. Another mechanism of action of some heavy metals (gold, cadmium, copper, mercury, lead) is their combination with the cell membranes, altering the permeability of the membranes. Others displace elements that are important structurally or electrochemically to cells which then can no longer perform their biologic functions.

Conditions for toxicity. The toxicity of a metal depends on its route of administration and the chemical compound with which it is bound. For example, mercury is highly toxic when injected intravenously or inhaled as a vapor, but is less toxic when taken by mouth. Cadmium, chromium, and lead are also highly toxic when they are injected intravenously, whereas a large number of metals (gold, cobalt, manganese, tin, thallium, nickel, zinc, and others) require greater intravenous doses to be toxic. When taken by mouth, gold and iron are only slightly toxic; however, copper, mercury, lead, and vanadium are much more toxic. Tantalum, on the other hand, is nontoxic no matter how it is administered.

The combining of a metal with an organic compound may either increase or decrease its toxic effects on cells. For example, the combination of mercury with a methyl organic radical makes the element more toxic, whereas the combination of the cupric ion with an organic radical, such as salicylaidoxine, makes the metal less toxic. Generally the combination of a metal with a sulfur to form a sulfide results in a less toxic compound than the corresponding hydroxide or oxide, because the sulfide is less soluble in body fluids than the oxide.

Many other heavy metals besides copper are essential to life, even though they occur only in trace amounts in the body tissues. They are taken up by the living cell in the form of cations and are admitted into an organism's internal environment in carefully regulated amounts, under normal circumstances, thus avoiding the effects of toxic levels. Toxicity results (1) when an excessive concentration is presented to an organism over a prolonged period of time, (2) when the metal is presented in an unusual biochemical form, or (3) when the metal is presented to an organism by way of an unusual route of intake.

Specificity. Metals have a remarkably specific effect; seldom can an excess of one essential metal prevent the damage caused by the deficiency of another. In fact, such an excess often increases the injurious effect of the deficiency. The heavy metals that appear to be essential for living cells are copper, iron, manganese, molybdenum, cobalt, and zinc. Some lighter metals, such as calcium, magnesium, sodium, and potassium, are also essential. There is some evidence that other metals, such as aluminum, cadmium, chromium, and vanadium, are essential. Many other metals found in the human body tissues have no apparent function.

Cellular response. On a molecular level of organization, metals produce diverse responses in cellular behavior. If the toxic action of a metal on a cell is interference with an essential part of cell metabolism, the cell will, of course, die. Figure 2, for example, shows two microscopic views of kidney cells. In Fig. 2a two glomeruli are surrounded by cross sections of normal kidney tubules, and Fig. 2b shows kidney tubule cell degeneration and necrosis (death) 72 hr after exposure to methyl mercury. The degenerating cells are swollen and vacuolated. The necrotic cells (indicated by arrows) are small, round, and dense and have small nuclei. Cell death provokes, in turn, an inflammatory response by the body and an attempt to repair the damage. Indeed, some metals (beryllium, for instance) provoke an excessive reparative process consisting of extensive proliferation of connective-tissue cells (histiocytes) that produce nodules of inflammatory tissue (granulomata) and extensive scarification. The microscopic views of lung tissue in Fig. 3 illustrate this effect. Figure 3a shows normal lung cells, with characteristically thin alveolar septa surrounding the air sacs. Figure 3b shows cells after exposure to beryllium. Nodular proliferations of cells and thickened septa (arrows) are evident.

Lesser degrees of injury to cells may alter their structure and function and may be associated with degenerative changes such as accumulation of fat or may limit the cells' principal biologic activity, which in turn may alter the performance of the organism as a whole. For example, degeneration of kidney tubule cells by mercury degrades the excretory function of the kidneys and leads to severe fluid and electrolytic imbalance in an individual, and possibly to death.

Carcinogens and teratogens. Less well understood but perhaps of equal significance are the carcinogenic properties of some metals. Nickel, cobalt, and cadmium have clearly been shown to produce cancers of the muscle cell type (rhabdomyosarcoma) when injected subcutaneously. However, many other metals similarly injected do not produce aberrant new cell growth (neoplasms). Nickel compounds also produce cancers of the respiratory tract. Lead has been shown to produce kidney cancers in experimental animals. The carcinogenic potential of many other metals is suspected but not proven.

Birth defects (teratogenesis) as a consequence of excessive metal intake by pregnant women do occur, but their precise nature and frequency are poorly documented. Chick embryos, when exposed to low doses of thallium or chromium, subsequently hatch with a high incidence of defective long-bone formation. Lead and cobalt have been shown to produce defective brain development in chick embryos. Many diverse metals produce growth inhibition in developing chick embryos. Manganese is a powerful mutation-producing agent for the bacterium *Escherichia coli* and the T4 bacteriophage; it is known to alter the nuclear enzyme DNA polymerase in lower forms of life, but whether it similarly affects mammals is not known. Mercury produces brain damage and mental retardation in children born to mothers exposed to excessive amounts of that metal. Lead, too, has teratogenic effects; specific congenital skeletal malformations have been induced in hamster embryos by treating the pregnant mother with various salts of lead. *See* MUTATION.

Chromosome damage. In addition to producing injury to the developing embryo, some metals produce damage to chromosomes. Both mercury and lead compounds have been shown to produce

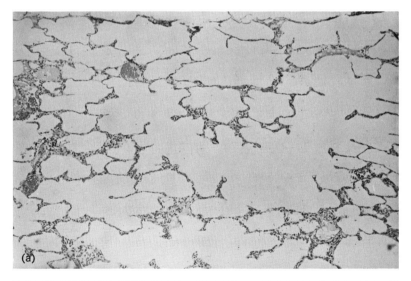

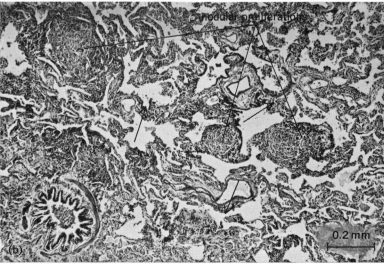

Fig. 3. Lung cells. (a) Normal cells. (b) Cells exposed to beryllium. Nodular proliferations of cells are indicated. Thickened septa are shown by arrows.

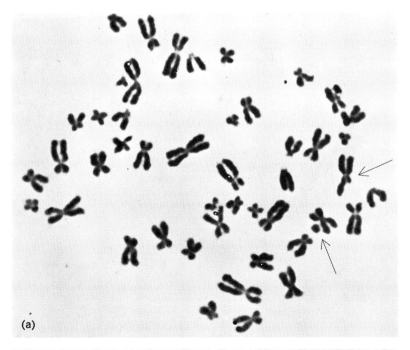

(a)

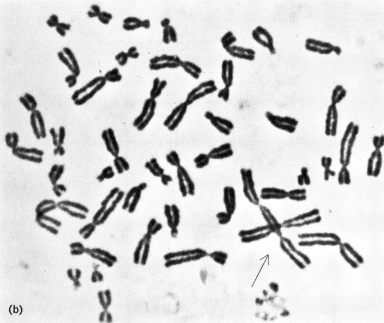

(b)

Fig. 4. Chromosomes damaged by mercury poisoning. (a) Arrows show a broken chromosome and an extra fragment. (b) Arrow shows chromatid fragments without centromeres. (*From S. Skerfving, Chromosome breakage in humans exposed to methyl mercury through fish contamination, Arch. Env. Health, 21:133–139, 1970*)

breaks in the chromosomes of somatic cells. Figure 4 shows chromosome damage from mercury. The lymphocytes have a broken chromosome and extra fragment (Fig. 4a), and three sister-chromatid fragments that lack centromeres (Fig. 4b). Steffan Skerfving, of the National Institute of Public Health, Stockholm, Sweden, cultured the lymphocytes from the blood of nine patients who were known to have excessively high tissue levels of methyl mercury from eating contaminated fish. He found a statistically significant rank correlation with chromosome breaks and mercury concentration. The full significance of these findings remains to be determined.

Many questions remain to be answered regarding the effects of metals on chromosomes. Many agents produce chromosome breaks in the test tube, but do they have the same effect in living cells? Are chromosome breaks produced in germ cells as well as somatic cells? Does the genetic damage lead to cell death, carcinogenesis, or teratogenesis? Whereas the effects of the deficiency of many trace elements on the developing embryo have been extensively studied, little is known about the teratogenic effects of an excess of metals. These questions will undoubtedly be answered as research proceeds.

CADMIUM DISEASE

Cadmium occurs in trace quantities—less than 1 part per million (ppm)—throughout the Earth's crust. There are no known deposits rich enough to justify separate mining of the metal, but it is frequently found in zinc and lead ores. Industrially cadmium has important and diverse uses in electroplating and in alloys, solders, batteries, paints (heat-resistant pigments), and metal bearings. It is also used as a neutron absorber in nuclear reactors and as an insecticide for fruit trees.

Cadmium also occurs naturally in trace quantities in many plant and animal tissues (Table 1). It therefore forms part of our normal diet, although the exact amount normally ingested is not known. Cadmium is very slowly absorbed from the gastrointestinal tract and is eliminated in both urine and feces. In healthy human adults the cadmium levels are low (Table 2), except for the kidneys, which are uniquely high in cadmium. Much lower concentrations occur in the tissues of the human fetus and newborn. With increased age, the tissue cadmium level progressively increases.

Ingestion of 15 ppm of cadmium in food, or about 15 times the normal amount, can produce mild symptoms of poisoning. Food poisonings from

Table 1. Metals in our environment in parts per million

Metal	Rock	Coal	Sea water	Plants (dry wt.)	Animals (dry wt.)
Cadmium	0.2	0.25	0.0001	0.1–6.4	0.15–3.0
Chromium	100.0	60.0	0.00005	0.8–4.0	0.02–1.3
Cobalt	25.0	15.0	0.00027	0.2–5.0	0.3–4.0
Lead	12.5	5.0	0.00003	1.8–50.0	0.3–35.0
Mercury	0.08	—	0.00003	0.02–0.03	0.05–1.0
Nickel	75.0	35.0	0.0045	1.5–36.0	0.4–26.0
Silver	0.07	0.1	0.0003	0.07–0.25	0.006–5.0
Thallium	0.45	0.05–10.0	0.00001	1.0–80.0	0.2–160.0
Vanadium	135.0	40.0	0.002	0.13–5.0	0.14–2.3
Gold	0.004	0.125	0.00001	0–0.012	0.007–0.03

Table 2. Distribution of some metals in mammalian tissues in parts per million dry weight

Metal	Skin	Lung	Liver	Kidney	Brain
Cadmium	1.0	0.08	6.7	130.0	3.0
Chromium	0.29	0.62	0.026	0.05	0.12
Cobalt	0.03	0.06	0.23	0.05	0.005
Lead	0.78	2.3	4.8	4.5	0.24
Mercury	–	0.03	0.022	0.25	–
Nickel	0.8	0.2	0.2	0.2	0.3
Silver	0.022	0.005	0.03	0.005	0.04
Thallium	0.2	0.3	0.4	0.4	0.5
Vanadium	0.02	0.05	0.04	0.05	0.3
Gold	0.2	0.3	0.0001	0.5	0.5

contamination of food and drink by cadmium-plated containers have been recorded frequently. For example. several outbreaks have occurred as the result of the action of the acids in fruit juice acting on cadmium-plated ice trays. Cadmium is soluble in weak acids such as acetic acid (vinegar), citric acid, and organic acids commonly found in food. The principal human risk is from industrial exposure, namely, inhalation of cadmium fumes. Cadmium is primarily a respiratory poison and has higher lethal potential than most other metals with a mortality rate of 15% in poisoning cases.

Recent occurrence. A remarkable case occurred in 1969 when a 28-year-old Japanese girl. Takako Nakamura. who had been working as a lathe operator in a zinc fabricating factory, hurled herself before a speeding train. Her death was listed as suicide and nothing more was thought of it for a time. Recently her body was exhumed and an autopsy performed. After correlating the levels of cadmium within the tissues with the anatomic lesions observed. the pathologists reported that she was a victim of cadmium poisoning.

Symptoms. Mild cadmium intoxication causes smarting of the eyes, dryness and irritation of the throat. and tightness in the chest and headache. With increased exposure, the respiratory distress increases in severity with uncontrollable cough and gastrointestinal pain, nausea, retching, vomiting, and diarrhea. Takako, the suicide, recorded in her diary the early symptoms: "The doctors cannot diagnose my disease; I am afraid it is cadmium poisoning. It is running through my whole body. pain eats away at me; I feel I want to tear out my stomach; tear out all my insides and cast them away."

Tissue reaction to the inhalation of cadmium fumes results in the initial deposition of cadmium in the lungs. From there the metal is widely distributed throughout the body and accumulates in the liver, kidneys, and spleen. The high concentrations may remain in the tissues for many years after the cessation of exposure. Excretion in the urine is extremely low.

Cadmium acting on the alveoli of the lungs produces a severe proliferation of alveolar septal cells which, in some cases, may completely fill the air spaces. Chronic exposure to cadmium produces emphysema, an irreversible lesion of the lungs characterized by an enlargement of the air sacs and accompanied by destruction of the walls of many of the sacs. This results in the conversion of many small air sacs into fewer large ones, with the inevitable decrease in surface area for oxygen and carbon dioxide transfer.

The kidneys are also injured in chronic cadmium poisoning. Diffuse extensive scar tissue develops between the tubules and subsequently destroys many glomeruli and tubules. Serum protein is then lost in the urine.

When administered to rats in dose levels comparable to that ingested by humans, cadmium produces high blood pressure, enlargement (hypertrophy) of the muscle of the heart (left ventricle). and hard patches (sclerotic plaques) in the very small arteries (arterioles) of the internal organs. (Communities in the United States with the highest environmental cadmium content also have the highest incidence of arteriosclerosis.) Another consequence of chronic cadmium exposure is an increased brittleness of bone, to an extent that the mere act of coughing may produce fractures of the ribs. Cadmium nitrate produces gene mutation in plants, but its effects. if any. on mammalian genes remain unknown.

MERCURY POISONING (MERCURIALISM)

The biological effects of excessive intake of mercury illustrate some important similarities and differences relative to cadmium. Mercury is unique in that it is the only metal that is a liquid at ordinary temperatures. Like cadmium. mercury is a relatively rare element. ranking near the bottom of the list of elements found in abundance on Earth.

Food grown under natural conditions contains traces of mercury. Five different studies. using a variety of methods from 1934 to the present. have shown that dietary meats contain on the average less than 0.01 ppm mercury. Fish generally have twice and vegetables have half as much (the commonly accepted limit for mercury in foods is 0.5 ppm). Some algae contain more than 100 times more mercury than the sea water in which they grow. Fish eating the algae concentrate the mercury, and predators that eat the fish in turn concentrate the mercury further.

Increasing prevalence of element. Man has evolved in an environment containing mercury. and throughout his history he has ingested plants, animals, and fish containing mercury. Through his evolution he has developed a tolerance, or possibly a need, for mercury as a trace constituent of his body. How has our urban technology affected this ecosystem?

First of all, mercury is being added to the North American environment at a rapid rate. Of greatest significance is the unmeasured increase of mercury to our atmosphere from the burning of fossil fuels. Coal and petroleum contain significant amounts of mercury; the mercury content of 36 American coals ranged from 0.10 to 33.0 ppm. In addition, organic mercurials are widely used as a fungicide and pesticide in agriculture. Other mercury compounds have been successfully used as a dressing for seed grains to prevent mildew, as a conditioner for lumber to prevent fungal discoloration, and in laundries to prevent garment mildew.

Not only is there more mercury in the environment, but evidence suggests that more of it is in a form that is most toxic to man. Liquid (metallic) mercury itself is not ordinarily toxic, since the body does not absorb mercury in this form from the digestive tract. However, many intoxications occur when metallic mercury is vaporized and in-

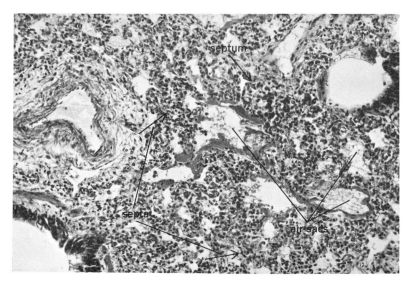

Fig. 5.   Cells from lung of a child poisoned by mercury vapor. Arrows indicate swollen septa. The air sacs are filled with fluid.

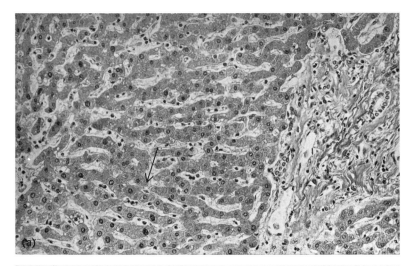

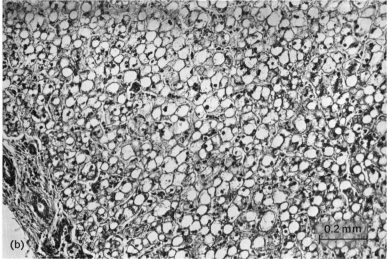

Fig. 6.   Liver cells. (a) Normal cells. (b) Mercury-poisoned cells.

haled. Figure 5 shows the effect of mercury vapor on lung tissue (it can be compared to the normal lung tissue in Fig. 3a). The tissue is from the lung of a 1-year-old child who died 6 days after exposure to the mercury vapor. The septa are swollen and thickened, and the air sacs contain fluid. Aside from such acute injuries to lung tissue, chronic low-level exposure to mercury vapor may produce symptoms or death because of its destruction of the nervous system.

Organic mercurials, too, which are readily assimilated, are becoming more prevalent. It has been proven that elemental, or inorganic, mercury, when released into the hydrosphere, is converted to an organic mercurial by the linkage of methyl or other carbon chains to the mercury by marine life. Repeated predation by animals in the food chain concentrates organic mercurials in their tissues, ultimately reaching toxic levels for man and animal.

Case histories. The increasing prevalence of organic mercurials has led to some serious episodes of intoxication of man. A case of direct ingestion of organic mercury occurred in Alamogordo, N. Mex., when several members of a family ate pork from a pig that had been inappropriately fed seed-wheat previously treated with a methyl mercury compound. Three children became seriously ill (but none died), whereas two other children who had not eaten pork were unaffected. The mother was pregnant at the time of onset of the symptoms in the three older children but she was symptom-free. The baby appeared normal at birth; however, soon thereafter this child developed convulsions and the characteristic symptom complex of chronic mercurialism, from which he seems to be gradually recovering.

Indirect chronic mercurial poisoning with a much higher morbidity and mortality rate occurred in Minamata and Niigata, Japan. The poisoning, called Minamata disease, was caused by discharge into Minamata Bay from an acetaldehyde and vinyl factory. The effluent contained large amounts of inorganic and organic mercury. There were 121 human cases of mercury poisoning recorded, with 46 deaths. About one-third of the afflicted were infants and children, some of whom had acquired mercury poisoning through the placenta prior to birth. The disease occurred mainly among fishermen and fish-eating families. Fish-eating animals such as cats, dogs, pigs, and seabirds were often affected, but herbivores such as rabbits, horses, and cows were not.

Symptoms. Chronic mercurialism presents five main symptoms: numbness, staggering gait, constriction of visual field ("tunnel vision"), garbled speech (dysarthria), and tremor. In addition, the children with Minamata disease exhibited emotional disturbances, and infants with congenital intoxication also suffered from impaired chewing or swallowing. Convulsions and mental retardation often occurred in the childhood cases. These symptoms of chronic poisoning are almost totally traceable to lesions in the central nervous system.

Autopsies of those who died soon after exposure revealed degenerative lesions in the liver, heart muscle fibers, and tubules of the kidneys. These changes were similar to those seen with inorganic mercury poisoning. Normal human liver cells (Fig. 6a) are of uniform size and take up stain uniformly. Mercury-poisoned liver cells (Fig. 6b) accumulate fat, assume various sizes, and stain unevenly. This type of fatty degeneration of cells can be seen throughout the liver and is characteristic of other metal poisonings in addition to mercury.

Acute focal inflammation without ulceration was often seen in the stomach or duodenum. The autopsies also revealed the presence of a generalized edematous swelling of the brain and a decrease in blood cell production by the bone marrow.

Another important early lesion involved the blood vessels, especially those of the central nervous system. Blood vessels became cuffed with white blood cells, and in some there was degeneration of the vessel wall and capillary proliferation.

The lesions of chronic mercurialism are principally in the brain. In the autopsies, destruction of the nerve cells of the gray matter of the cerebral cortex was found in several regions, especially the visual cortex. Extensive necrosis of nerve cells in the visual area and an increase in the nonneuronal supporting cells had taken place. These lesions principally account for the "tunnel vision" of methyl mercury poisoning. Similar changes were seen in the cerebellum. An astounding destruction of the granular cell layer was found, as shown in Fig. 7 (at arrows). This layer contains nerve cells that receive stimuli from the muscles, tendons, and joints throughout the body and transmit them to the Purkinje cell processes in the molecular layer, where the many stimuli are correlated and integrated. The cerebellum is the area of the brain where position sense, balance, and fine muscle coordination are controlled. In most cerebellar pathologic processes the sensitive cells are the large Purkinje cells; mercury is the only known metallic toxin from an external source that results in the preferential destruction of the granular cell layer.

LEAD POISONING (PLUMBISM)

The toxicity of lead has been known since antiquity. Lead and lead-containing products are extremely important industrially and are of great health significance. Exposure of man may result from inhalation of fumes and dust in the smelting of lead, from the manufacture of insecticides, pottery, and storage batteries, or from contact with gasoline containing lead additives. Intoxication by the inhalation of lead fumes is a most serious mode of exposure. The ingestion of soluble lead compounds accidentally, or with suicidal intent, is another portal of entry of the body. Only tetraethyllead in gasolines can be absorbed through the intact skin. Traces of lead occur in the diet; small amounts are absorbed into the body when it is present in the food as a soluble salt. Lead is absorbed mainly through the small and large intestines.

Irrespective of the route of entry, lead is absorbed very slowly into the human body. Even this slow and constant chronic absorption is sufficient to produce lead poisoning because the rate of elimination of lead is even slower and a slight excess in intake may result in its accumulation in the body. Much of the lead is taken up by red blood cells and circulated throughout the body. Organic lead compounds such as tetraethyllead become distributed throughout the soft tissues, with especially high concentrations in the liver and kidneys. Over a period of time, the lead may be redistributed, becoming deposited in bones, teeth, or brain. In bones, lead is immobilized and does not contribute to the general toxic symptoms of the patient. Or-

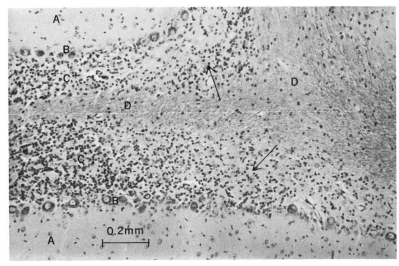

Key: A = molecular level   C = granular cell level
        B = Purkinje cells   D = white matter

Fig. 7. Cells in cerebellum. Arrows show cells destroyed by methyl mercury.

ganic lead compounds have an affinity for the central nervous system and produce lesions there.

Acute form. Acute plumbism is ordinarily seen as a result of the ingestion of inorganic soluble lead salts for suicidal, accidental, or abortion-inducing reasons. They produce a metallic taste, a dry burning sensation in the throat, cramps, retching, and persistent vomiting. The gastrointestinal tract is encrusted with the coagulated proteins of the necrotic mucosa, thereby hindering further absorption of the lead. Muscular spasms, numbness, and local palsy may appear.

Chronic form. Chronic lead poisoning is much more common. Two general patterns of symptoms relate to the gastrointestinal and nervous systems. One or the other may predominate in any particular patient; in general, the central nervous system changes, which predominate in children, are of greater significance. The abdominal symptoms in chronic lead poisoning are similar to those for acute cases, but are less severe: loss of appetite, a feeling of weakness and listlessness, headache, and muscular discomfort. Nausea and vomiting may result. Chronic excruciating abdominal pain, sometimes referred to as lead colic, may be the most distressing feature of plumbism.

An autopsy does not ordinarily reveal specific gross lesions, but a marked congestion and petechial hemorrhages are sometimes seen in the brain. Microscopic examination reveals inclusion bodies within cells of the kidney, brain, and liver. Focal necrosis of the liver is found in some cases. Lesions in the central nervous system are primarily vascular and consist of scattered hemorrhages, often in the perivascular tissue; degenerative and necrotic changes in the small vessels, surrounded by a zone of edema; and sometimes a fibrinous exudate.

THALLIUM DISEASE (THALLOTOXICOSIS)

Since thallium was discovered more than 100 years ago, it has been responsible for many therapeutic, occupational, and accidental poisonings. During the first 50 years since its discovery, thal-

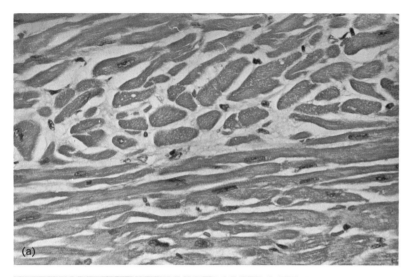

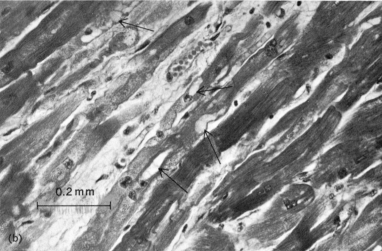

Fig. 8. Heart muscle fibers. (a) Normal fibers. (b) Fibers from a victim of beer drinker's syndrome. Arrows point to fatty vacuoles.

Shortly thereafter abdominal pain, vomiting, and diarrhea may ensue. Neurologic symptoms appear after 2 or more days of exposure, with delirium hallucinations, convulsions, and coma occurring after severe poisoning, and death may follow in about a week due to respiratory paralysis.

When smaller doses are taken, a loss of muscle coordination (ataxia) and sensations of tingling or burning of the skin (paresthesia) may be the principal symptoms. These may be followed by weakness and atrophy of the muscles. Tremor, involuntary movements (choreic athetosis), and mental aberrations, with changes in the state of consciousness, may ensue.

Alopecia is one of the best known symptoms of chronic thallium poisoning. The hair loss begins about 10 days after the ingestion of thallium, and complete loss of hair may be reached in a month. Hair in the axillary and facial regions, including the inner one-third of the eyebrows, is usually spared. The evidence suggests thallium acts directly on the hair follicles. Cardiac symptoms— rapid heartbeat, irregular pulse, and high blood pressure, with angina-like pain—have been recorded. Heart muscle fiber degeneration accompanies these symptoms. Thallium affects the sweat and sebaceous glands, and the nails are white with transverse bands and may assume unusual shapes. A blue line may develop along the gums of an individual who has ingested thallium.

The turnover of thallium in the human body is extremely slow. Under normal circumstances the tissues contain trace amounts of the metal. The biological mechanism by which thallium produces its effects on the human body is not well understood, but as with other metals, it appears to block some enzyme systems. Autopsies on fatal human cases reveal the presence of thallium in all organs and tissues, with the highest concentrations in the kidneys, intestinal mucosa, thyroid, and testes. Fat, liver, and all types of nerve tissue are uniformly low in thallium content. The autopsies reveal bleeding at many points (punctate hemorrhages) in the gastric and upper intestinal mucosa, fatty change in the liver and kidney, and small punctate hemorrhages and focal necrosis of the surface layers of the adrenal glands. Congestion of the central nervous system blood vessels is evident, and there is some degree of cerebral edema.

Neurons show varying degrees of degenerative change, especially in the locomotor fiber pathways and associated centers. In the more chronic cases, the ganglion cells of the sensory and motor horns of the spinal cord degenerate, with chromatolysis, swelling, and fatty change. Examination of peripheral nerves shows marked degeneration in nerve cells, axons, and myelin sheath. Thallium poisoning is a serious problem not only because approximately 15% of those poisoned die, but also because there is persistent neurologic damage in over half the cases.

COBALT TOXICITY

Cobalt is a metal with a very industrially important property: It is immune to attack by air or water at ordinary temperatures and imparts this property to many alloys. Under natural circumstances cobalt does not produce a toxic syndrome in man. Cobalt in most naturally occurring foodstuffs is

lium sulfate gained general use as a treatment for syphilis, gonorrhea, gout, dysentery, and tuberculosis. It was subsequently discarded as a medicine because of its unpleasant side effects, principally the temporary loss of hair (depilation). Later, dermatologists utilized this feature as a method of removing unwanted hair.

Occurrence. By 1934 more than 700 poisonings with thallium were reported in the medical literature, with more than 90% of these the result of the use of thallium as a depilatory agent. In recent years hundreds of cases of thallotoxicosis have been related to its use as a rodenticide and insecticide. For example, a group of 31 Mexican workers in California were poisoned, six fatally, after eating tortillas made from a bag of thallium-treated barley intended for rodents.

Symptoms. The symptoms of thallotoxicosis vary extensively with the dosage, age of the patient, and the acuteness of the intoxication. Inflammation involving many nerves, loss of hair, gastrointestinal cramps, and emotional changes are the principal symptoms. For large doses, the first symptoms are gastrointestinal hemorrhage, gastroenteritis, a rapid heartbeat, and headache.

mostly unabsorbed and is eliminated in the feces.

It takes an unusual set of circumstances to produce cobalt toxicity. Intravenously injected, large doses of cobalt have been observed to cause paralysis and enteritis, and sometimes death. When cobalt is injected into the bloodstream, the amount distributed throughout the tissues is very small; higher concentrations are present in the pancreas, liver, spleen, kidney, and bone. Elimination of cobalt is rapid.

Carcinogen. Researchers in England have shown that cobalt, like nickel and cadmium, produces a high incidence of cancer, namely rhabdomyosarcoma, when it is injected in powdered form into rats. A number of other metals, such as iron, copper, zinc, manganese, beryllium, and tungsten, are not carcinogenic under the same conditions. Cobalt has an inhibitory effect on cell oxidative metabolism and is toxic to connective-tissue cells grown in culture.

Beer drinker's syndrome. Another unusual kind of cobalt toxicity is the "beer drinker's syndrome." During the 1960s, some breweries in Nebraska and Quebec added cobalt to their beer to improve the stability of the foam. Subsequently, numerous fatalities occurred among people who consumed large quantities of beer. Following discontinuance of the cobalt additive, no new cases have been reported. Whether the cobalt acted independently or synergistically with other constituents of beer to enhance its toxicity is not known. The principal lesion was found in the heart; autopsies revealed enlarged, flabby hearts that were more than twice normal size. Microscopically the myocardial cells were vacuolated with numerous fat droplets. The microscopic view of heart muscle fibers from a normal heart is shown in Fig. 8*a*, and from a beer drinker's syndrome case in Fig. 8*b*. The fatty vacuoles (arrows) resulting from degenerative changes give a "moth-eaten" appearance to the fibers. Progressive degenerative changes in the myofibrils were also found in the autopsies, and complete necrosis and lysis of some heart muscle fibers were noted.

Cobalt chloride has been used as a therapeutic in the treatment of anemia, and dosages in excess of 100 mg per day for prolonged periods have been unattended by changes in the cardiac muscle. In contrast, the cobalt intake of the most avid beer drinker was calculated to be only 10–15 mg per day. This suggests that the production of the cardiac lesions was not due to cobalt alone. In addition to the myocardial lesions, the beer drinkers also frequently had gastrointestinal inflammation and extensive hemorrhagic necrosis of the liver. The production of similar lesions in experimental animals given cobalt in their water affirms the importance of cobalt in their genesis. However, the heart lesions are similar in appearance to those of beriberi, a thiamine vitamin deficiency disease, and other nutritional factors may have contributed to the beer drinker's syndrome.

Some experiments have linked cobalt intake to the development of hardening of the arteries (atherosclerosis).

NICKEL POISONING

Nickel is widely distributed throughout the Earth's crust and is a relatively plentiful element.

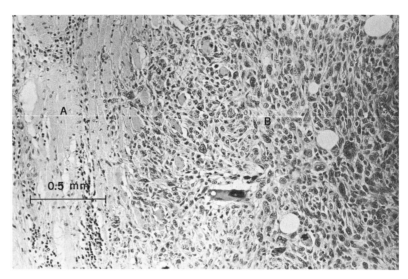

Fig. 9. Skeletal muscle fibers. Cells in region A are normal; those in region B are cancer cells.

It occurs in marine organisms, is present in the oceans, and is a common constituent of plant and animal tissues. The ordinary human diet contains 0.3–0.5 mg of nickel per day; diets rich in vegetables invariably contain much more nickel than those with foods from animal origin.

Depending on the dose, the organism involved, and the type of compound involved, nickel may be beneficial or toxic. It appears to be essential for the survival of some organisms.

Industrial hazard. The use of nickel in heavy industry has increased markedly over the last few decades, principally in the production of stainless steel and other alloys and in plating. Because of the excellent corrosion resistance of nickel and high-nickel alloys, these metals are used widely in the food processing industry. In fact foods can be contaminated with nickel during handling, processing, and cooking by utensils containing large quantities of nickel. Nickel is also used frequently

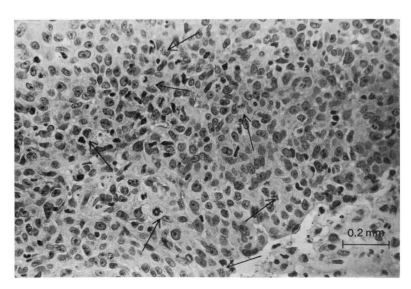

Fig. 10. Paranasal sinus carcinoma cells. Arrows indicate dividing cells.

as a catalyst; one of its most significant catalytic applications is in the hydrogenation of fat. Nickel carbonyl, $Ni(CO)_4$, is one of the most toxic nickel compounds and is a major industrial hazard. Lesser quantities of nickel are used in ceramics, as a gasoline additive, in fungicides, in storage batteries, and in pigments.

Nickel usually is not readily absorbed from the gastrointestinal tract except as nickel carbonyl. This compound has caused most of the acute toxicity of nickel. Recently it has been shown that nickel has a carcinogenic property and may be involved in hypersensitivity reactions. Figure 9 shows normal skeletal muscle fibers (region A) and rhabdomyosarcoma cells (region B) characteristic of the type produced by nickel, cobalt, or cadmium injection. Nickel dermatitis is reported with increasing frequency in industrial workers, especially nickel platers. It produces an allergic dermatitis on almost any skin area.

Nickel workers have approximately 150 times the cancer of the nasal passages and sinuses as the general population, and approximately five times the lung cancer. Figure 10 shows a paranasal sinus carcinoma. This type of neoplasm is seen with increasing frequency in industrial employees exposed to nickel carbonyl. The numerous dividing cells (arrows) suggest rapid growth. Several reports from Great Britain have established conclusively that inhaled nickel is a carcinogenic agent, and in Great Britain it is a compensable industrial disease.

Toxic action. The mechanism of toxic action of nickel is its capacity to inhibit oxidative enzyme systems. Within the body the main storage depots of nickel are the spinal cord, brain, lungs, and heart, with lesser amounts widely distributed throughout the organ systems. Acute poisoning causes headache, dizziness, nausea and vomiting, chest pain, tightness of the chest, dry cough with shortness of breath, rapid respiration, cyanosis, and extreme weaknesses. The lesions resulting from acute exposure are mainly in the lung and brain. In the lungs hemorrhage, collapse (atelectasis), and necrosis occur. Deposits of brown-black pigments are found in the lung phagocytic cells. In the brain extensive damage to blood vessels is seen. Lung cancers due to nickel exposure are usually of the squamous carcinoma variety and are indistinguishable microscopically from lung cancer of other causes.

CONCLUSION

Some metals are extremely injurious to the body tissues, whereas others are less so or are nontoxic. Exposure to them therefore produces diverse patterns of pathologic change. Some metals may produce extensive degeneration and necrosis of the cells in a particular organ, whereas others may produce cancerous change or birth defects.

Because of the increasing incidence of such effects—and the inevitability of further additions of metals to the environment—some practical goals can be set: Man should learn more about the human biological effects of excessive metal exposure through research, and should manage the type and quantity of exposure to minimize its adverse effects.

[N. KARLE MOTTET]

Herbicide

Chemicals used to kill weeds, that is, unwanted plants. Use of herbicides has developed chiefly in the 20th century, slowly at first and then rapidly and dramatically after 1945 when 2,4-D was introduced. A few milestones are: 1896–1900, introduction of selective sprays for controlling broadleaved weeds in cereal crops; 1919, discovery of the translocation of arsenic in wild morning glory by George Gray in California; 1925, introduction of sodium chlorate for killing weeds by application to the soil; 1934, importation from France to the United States of sodium dinitrocresylate as the first organic selective weed killer; 1945, advent of 2,4-D, the growth-regulator type of herbicide. Since the introduction of 2,4-D, new chemicals and new methods have come along at an accelerated rate. The latest trend has been toward the use of selective, preemergence herbicides; such chemicals are now available for use in corn, cotton, soybeans, rice, and other major crops. The manufacture and use of herbicides has become a multimillion-dollar business annually. United States production of organic herbicides is approaching 250,000,000 lb annually.

As is true of medical science and of other aspects of pest control, the control of weeds by means of herbicides has benefited practically everyone. The values of agricultural crops have been increased by billions of dollars; forests, rangelands, parks, and recreational areas have been freed of noxious plants; roadsides and industrial areas are kept clean and sightly; and millions of people have been relieved of the suffering caused by pollens and poisonous plants. Looking to the future, these benefits will be multiplied greatly as new and better herbicides are discovered. In fact, the reclamation of hundreds of millions of acres now covered by brush and undesirable trees, not only in the Americas but in Africa, Asia, and Australia, constitutes the greatest single source of land available for agricultural use.

Classification of herbicides. There are available dozens of chemicals, hundreds of formulations, and a multitude of methods for killing weeds. The following classification of herbicides by method of application has proved useful in discussing weed control problems: (1) selective foliage contact sprays; (2) selective foliage translocated sprays; (3) selective root applications; (4) nonselective foliage contact sprays; (5) nonselective foliage translocated sprays; (6) nonselective root applications.

A great majority of the available herbicides and methods for their use fall naturally into these six groups. The few exceptions will be discussed separately.

Selective herbicides are those that kill some members of a plant population with little or no injury to others. Examples are sulfuric acid, which at proper strength will kill wild mustard in barley or knotweed in young onions.

Nonselective herbicides are those that kill all vegetation to which they are applied. Examples are diquat, paraquat, or aromatic oil, which are used to keep roadsides, ditch banks, and rights-of-way open and weedfree. A rapidly expanding use for such chemicals is the destruction of vegetation before seeding in the practice of chemical plowing or nontillage. They are also used to kill annual grasses in preparation for seeding perennial grasses in pastures. Additional uses are fire prevention, elimination of highway hazards, destruction of plants that are hosts for insects and plant diseases, and killing of poisonous plants and allergen-bearing plants.

Application methods fall into two natural groups, as indicated in the above scheme of classification: application to foliage and application to soil for uptake by roots. In regions of frequent summer rains, or by using sprinkler irrigation, both these methods may be used to advantage. By spraying the foliage, leaf absorption may take place. If this is followed by rainfall or sprinkler irrigation, the chemical not absorbed by foliage may be washed into the soil and taken up by roots. Sodium chlorate may be used in this way.

Herbicide action. There are many factors that influence the action of herbicides on plants.

Differential wetting. Most leaf surfaces are waxy; many are corrugated or ridged; some are covered by small particles of wax (bloom). Such leaves differ in their retention of spray droplets. Many selective sprays such as copper salt solutions, sulfuric acid, and sodium dinitrocresylate kill weeds selectively because they wet the weeds but bounce off of the crop plants.

Orientation of leaves. Most grass leaves stand in a relatively vertical position. Many weed leaves are arranged horizontally so that they retain more spray than do grass leaves.

Location of growing points. Growing points and buds of most cereal plants are located in a crown, at or below the soil surface. Furthermore, they are wrapped within the mature bases of the older leaves. Hence they may be protected from herbicides applied as sprays. Buds of many broadleaved weeds are located at the tips of shoots and in axils of leaves. These are much more exposed to spray solutions.

Growth habits. Some perennial crops, such as alfalfa, vines, and trees, have a dormant period in winter. At that time a general contact weed killer may be safely used to get rid of weeds that later would compete with the crop for water and plant nutrients.

Application methods. By arranging spray nozzles close to the soil level, it is possible to kill low growing weeds but not an upstanding crop. In this way, general contact sprays may be used in corn, sugarcane, or sorghum. Such directed spraying is used to kill young grass in cotton with oil, to kill wild morning glory in beans with 2,4-D, and to control annual weeds in corn and sorghum with diuron in an oil emulsion.

Protoplasmic selectivity. Just as some people are immune to the effects of certain diseases while others succumb, so some weed species resist the toxic effects of herbicides whereas others are injured or killed. This results from inherent properties of the protoplasm of the respective species. One example is the use of 2,4-DB or MCPB (the butyric acid analogs of 2,4-D and MCPA) on weeds lacking β-oxidizing enzymes that are growing in crops (certain legumes) that have such enzymes. The weeds are killed because the butyric acid

compounds are broken down to 2,4-D or MCPA. Another example is the control of a wide variety of weeds in corn by simazine. Corn contains a compound that removes the chlorine from the simazine atom, rendering it nontoxic; most weeds lack this compound. A third important protoplasmic selectivity is shown by trifluralin, planavin, and a number of other herbicides that are applied through the soil. These compounds inhibit secondary root growth. Used in large, seeded crops having vigorous taproots, they kill shallow-rooted weed seedlings; the roots of the crops extend through the shallow layer of topsoil containing the herbicides, and the seedlings survive and grow to produce a crop.

Turning now to the six categories given in the classification of herbicides, it appears that each involves a group of chemicals having a common mode of action. These will be discussed in turn.

Selective contact sprays kill weeds in crops as a result of absorption into leaves where they destroy the cells. Examples are sulfuric acid, sodium dinitrocresylate, ammonium dinitro-*sec*-butyl phenylate, barban, and bromoxynil.

Selectivity is largely a matter of differential wetting. Selective oils used in carrot and other umbellifer crops, cotton, and forest tree nurseries are contact-selective materials that wet the crops as well as the weeds. Selectivity is protoplasmic.

Selective translocated spray chemicals move within the sprayed plants. They are useful as selective sprays in low-volume application and as killers of perennial weeds. Examples are 2,4-D, 2,4,5-T, MCPA, the corresponding propionic and butyric acids, amino triazole, dalapon, and picloram. Many herbicides are applied through the soil; a good number of these are selective. Most of these herbicides are applied preemergence, that is, before the emergence of a specified weed or crop; some may be applied after emergence. There has been a very great increase in the number of these herbicides available on the market since 1960. Out of 95 herbicides described in the *Herbicide Handbook of the Weed Society of America* (see table), 50 are in this class. A few examples are the benzoic acid herbicides, the chloroacetamides, thiocarbamates, phthalamic acids, and low dosages of the substituted ureas and symmetrical triazines.

Nonselective contact sprays kill all vegetation to which they are applied. Examples are sodium arsenite, sodium chlorate, penta- and metaborates in solution, DN compounds in oil, aromatic oils, PCP, acrolein, diquat, paraquat, cacodylic acid, and erbon.

Nonselective translocated sprays are used on perennial weeds primarily. Examples are acid arsenicals, chlorates, thiocyanates, sulfamates, amitrole, picloram, mono- and disodium methyl arsonates, TCA, and 2,3,6-TBA.

Materials that cause a general killing of all plants through the soil include fumigants and some of the older chemicals. Examples are carbon disulfide (CS_2), chloropicrin, methyl bromide, allyl-alcohol, EPTC, sulfamates, thiocyanates, arsenates, chlorates, borates, bromacil, dichlobenil, Tritac, fenac and high dosages of the substituted ureas, and symmetrical triazines.

In addition to the six categories mentioned, there are a few recognized methods that do not fit

Some important herbicides*

Common name and (trade name)	Chemical name	a = Soluble in b = Available as	Use
Acrolein	Acrolein	a Water b Liquid	Aquatic weed killer
Pyrazon (Pyramin)	5-Amino-4-chloro-2-phenyl-3 (2H)-pyridazinone	a Methanol b Wettable powder	Annual weed control in beet crops
Amiben	2,5-Dichloro-3-aminobenzoic acid	a Organic solvents b Liquid concentrate; granules	Selective preemergence herbicide in vegetables and soybeans
Amitrole	3-Amino-1,2,4-triazole	a Water b 50% soluble powder	Translocated spray on perennial weeds
Picloram (Tordon)	4-Amino-3,5,6-trichloro-picolinic acid	a Water b Liquid concentrate; granules	Translocated spray on perennial species
AMS (Ammate)	Ammonium sulfamate	a Water b Dry crystals	Translocated spray on woody species
Prometryne (Gesagard)	2,4-Bis (isopropanolamino)-6-methylmercapto-s-triazine	a Water (slightly) b Wettable powder; granules	Selective preemergence herbicide
(Lambast)	2,4-Bis (3-methoxypropylamino)-6-mercapto-s-triazine	a Organic solvents b Emulsifiable concentrate	Postemergence selective herbicide
Bromacil (Hyvarx)	5-Bromo-3-sec-butyl-6-methyluracil	a Organic solvents b Wettable powder; aqueous solution, granules	Soil sterilant; general at high dosage and selective at low
Metobromuron (Patoran)	3-Bromophenyl-1-methoxy-1-methylurea	a Water, organic solvents b Wettable powder	Selective preemergence herbicide
Benefin (Balan)	N-Butyl-N-ethyl-a,a,a-trifluoro-2,6-dinitro-p-tuluidine	a Organic solvents b Emulsifiable concentrate	Selective preemergence herbicide
CDED (Vegadex)	2-Chloroallyl diethyldithiocarbamate	a Organic solvents b Emulsifiable concentrate	Selective grass killer
Simazine	2-Chloro-4,6-bis (ethylamino)-s-triazine	a Insoluble b Wettable powder; granules	Selective pre- or post-emergence spray
Propazine	2-Chloro-4,6-bis (dipropyl-amino)-s-triazine	a Insoluble b Wettable powder	Selective preemergence herbicide
Barban (Carbyne)	4-Chloro-2-butynyl-N-(3-chlorophenyl) carbamate	a Organic solvents b Emulsifiable concentrate	Selective postemergence spray against wild oats
CDAA (Randox)	2-Chloro-N,N-diallylacetamide	a Organic solvents b Emulsifiable concentrate; granules	Selective grass killer
Atrazine	2-Chloro-4-(ethyl-6-isopropyl-amino)-s-triazine	a Insoluble b Wettable powder	Selective preemergence herbicide
(Ramrod)	2-Chloro-N-isopropylanilide	a Organic solvents b Wettable powder; granules	Selective preemergence herbicide
Solan	3'-Chloro-2-methyl-p-toluidide	a Organic solvents b Emulsifiable concentrate	Selective postemergence herbicide
Cloroxuron (Tenoran)	N'-4-(4-Chlorophenoxy) phenyl-N,N-dimethylurea	a Organic solvents b Wettable powder	Selective preemergence herbicide
Monuron (Telvar)	3-(p-Chlorophenyl)-1,1-dimenthylurea	a Insoluble b Wettable powder; granules	Soil sterilant, selective at low dosage, general at high
OCS-21799	2-[(4-Chloro-o-tolyl)oxy]-N-methoxyacetamide	a K salt water b Soluble powder; granules	Selective preemergence herbicide
Bromoxynil (Brominil)	3,5-Dibromo-4-hydroxybenzo-nitrile	a Organic solvents b Emulsifiable concentrate	Selective postemergence herbicide
Diallate (Avadex)	S-2,3-Dichloroallyl-N,N-diisopropylthiolcarbamate	a Organic solvents b Emulsifiable concentrate	Selective preplant soil-incorporated herbicide
Dichlobenil (Casoron)	2,6-Dichlorobenzonitrile	a Low solubility b Wettable powder; granules	General preemergence Useful in orchards and vineyards

*Herbicides listed are in *Herbicide Handbook of the Weed Society of America*, W. F. Humphrey Press, 1967.

Some important herbicides (cont.)

Common name and (trade name)	Chemical name	a = Soluble in b = Available as	Use
(Sirmate; UC 22463)	3,4-Dichlorobenzyl methylcarbamate	a Organic solvents b Emulsifiable concentrate; granules	Selective preemergence herbicide
Cypromid (Clobber)	3′,4′-Dichlorocyclopropanecar-boxanilide	a Organic solvents b Emulsifiable concentrate	Selective postemergence directional spray
Dicryl	3′,4′-Dichloro-2-methacryl-anilide	a Organic solvents b Emulsifiable concentrate	Selective postemergence directional spray
2,4-D	2,4-Dichlorophenoxyacetic acid	a Salts in water; esters in oil; acid in organic solvents b Water-soluble salts; oil-soluble esters; oil-soluble amine; emulsifiable acid	Selective translocated spray and preemergence herbicide; broad-leaved weeds
4-(2,4-DB) (Butyrac)	4-(2,4-Dichlorophenoxy) butyric acid	a Salts in water; esters in oil b Salts and esters	Selective translocated spray
Dichloroprop; 2-(2,4-DP)	2,4-Dichlorophenoxy propionic acid	a Salts in water; esters in oil; acid in organic solvents b Liquid concentrate	Selective translocated spray; oaks and other woody species
Diuron (Karmex)	3-(3-4-Dichlorophenyl)-1,1-dimethylurea	a Insoluble b Wettable powder	Soil sterilant; selective at low dosages, general at high dosage
Linuron (Lorox)	3-(3,4-Dichlorophenyl)-1-methoxy-1-methylurea	a Organic solvents b Wettable powder	Selective pre- or post-emergence herbicide
Tok E-25	2,4-Dichlorophenyl-4-nitrophenyl ether	a Organic solvents b Emulsifiable concentrate; granules	Selective preemergence herbicide
Propanil (Stam-F34; Rogue)	3′,4′-Dichloropropionanalide	a Organic solvents b Emulsifiable concentrate	Selective contact spray against watergrass in rice
Dalapon (Dowpon)	Sodium 2,2-dichloropropionate	a Water b Dry powder	Selective translocated grass killer
Diquat (Reglone)	6,7-Dihydrodipyrido(1,2-a:2′,1′-c) pyrazidiinium dibromide	a Water soluble b Aqueous solution	General contact, aquatic weed herbicide; crop desiccant
MH (Maleic Hydrazide)	1,2-Dihydro-pyridazine-3,6-dione, diethanolamine salt	a Water b Liquid concentrate	Grass killer; spray plus tillage on quack grass; growth inhibitor
Ioxynil (Certrol)	3,5-Diiodo-4-hydroxybenzonitrile	a Water soluble b Sodium salt; oil-soluble amine; octanoic ester	Postemergence broadleaf weed control in cereals
Cacodylic acid (Phytar)	Dimethylarsinic acid	a Organic solvents b Sodium salt	General contact herbi-cide and desiccant
Paraquat (Gramoxone)	1,1′-Dimethyl-4,4′-bipyridinium dichloride or bis(methyl sulfate) salts	a Water soluble b Aqueous solution	General contact herbi-cide for noncrop usage; crop desiccant
Diphenamid (Dymid; Enide)	N,N-Dimethyl-2,2-diphenyl-acetamide	a Solubility low b Wettable powder	Selective preemergence herbicide
DCPA (Dacthal)	Dimethyl-2,3,5,6-tetrachloro-terephthalate	a Solubility low b Wettable powder	Selective preemergence herbicide
(Glenbar; OCS-21944)	O,S-Dimethyltetrachlorothiotere-phthalate	a Solubility low b Emulsifiable concentrate; granules	Selective preemergence herbicide
DNBP (Premerge; Dow General)	4,6-Dinitro-o-sec-butyl phenol 4,6-Dinitro-o-sec-butyl phenol	a Water soluble b Alkanolamine salt a Oil soluble b Solution in oil	Selective preemergence herbicide General contact herbicide
Terbutol (Azak; Hercules 9573)	2,6-Di-tert-butyl-p-tolylmethylcarbamate	a Solubility low b Wettable powder; granules	Selective preemergence crabgrass herbicide for turf

Some important herbicides (cont.)

Common name and (trade name)	Chemical name	a = Soluble in b = Available as	Use
Ametryne	2-Ethylamino-4-isopropyl-amino-6-methylmercapto-s-triazine	a Organic solvents b Wettable powder; emulsifiable concentrate	Selective pre- or post-emergence or directed-spray herbicide
EPTC (Eptam)	Ethyl N,N-di-n-propyl-thiolcarbamate	a Organic solvents b Emulsifiable concentrate	Soil fumigant at low volatility
Molinate (Ordram)	S-Ethylhexahydro-1H-azepine-1-carbothioate	a Organic solvents b Emulsifiable concentrate; granules	Selective incorporated grass killer for rice
HCA	Hexachloroacetone	a Oil b Liquid	Fortifier for weed oil Contact and soil action
Norea (Herban)	3-(Hexahydro-4,7-methanoindan-5-yl)-1, l-dimethylurea	a Organic solvents b Wettable powder	Selective preemergence herbicide
(PH-40-25; THO73-H)	N-hydroxymethyl-2,6-dichloro-thiobenzamide	a Solubility low b Wettable powder; granules	Selective preemergence herbicide
CIPC (Chloro IPC)	Isopropyl N-(3-chlorophenyl)-carbamate	a Organic solvents b Emulsifiable concentrate; wettable powder; granules	Grass killer used by soil application
IPC (Chemhoe)	Isopropyl N-phenylcarbamate	a Organic solvents b Emulsifiable concentrate; wettable powder; granules	Grass killer used through the soil; temporary
Bensulide (Betasan)	N-(2-Mercaptoethyl)benzene-sulfonamide	a Organic solvents b Emulsifiable concentrate; granules	Preemergence herbicide; crabgrass and turfweed control
MAA (Ansar)	Methanearsonic acid	a Water soluble b Soluble powder; liquid concentrate	Postemergence grass killer
Prometone (Pramitol)	2-Methoxy-4,6-bis (isopropyl-amino)-s-triazine	a Organic solvents b Wettable powder; emulsifiable concentrate	Nonselective pre- and postemergence herbicide
Dicamba (BanvelD)	2-Methoxy-3,6-dichlorobenzoic acid dimethylamine salt	a Organic solvents b Liquid concentrate; granules	Selective pre-and post-emergence herbicide
Tricamba (BanvelT)	2-Methoxy-3,5,6-trichloro-benzoic acid dimethylamine salt	a Organic solvents b Liquid concentrate; granules	Selective pre-and post-emergence herbicide
(IT 3456)	Methyl-2-chloro-9-hydroxy-fluorene-(9)-carboxylate	a Organic solvents b Emulsifiable concentrate	Selective postemergence herbicide
MCPA (Methoxone; Agroxone)	2-Methyl-4-chlorophenoxyacetic acid	a Salts in water; esters in oil b Water-soluble salts; oil-soluble esters	Selective translocated spray; broad-leaved weeds
Mecoprop (MCPP)	2-Methyl-4-chlorophenoxy-propionic acid	a Salts in water; esters in oil; acid in organic solvents b Liquid concentrate	Selective translocated spray
Siduron (Tupersan)	1-(2-Methylcyclohexyl)-3-phenylurea	a Organic solvents b Wettable powder	Selective preemergence weed killer for lawns
(OCS-21693)	Methyl-2,3,5,6-tetrachloro-N-methoxy-N-methyltere-phthalamate	a Organic solvents b Emulsifiable concentrate; granules	Selective preemergence herbicide
NPA (Alanap)	N-1-Naphthylphthalamic acid	a Insoluble b Granules	Selective preemergence herbicide
Endothall	Disodium 3,6-endoxohexahydro-phthalate	a Water b Liquid concentrate	Preemergence and general contact herbicide
PCP	Pentachlorophenol	a Aromatic oil b Concentrate in oil	Fortifier for oil and oil-emulsion general contact sprays

Some important herbicides (cont.)

Common name and (trade name)	Chemical name	a = Soluble in b = Available as	Use
NaPCP	Sodium pentachlorophenate	a Water b Dry crystals	Fortifier for oil-emulsion general contact sprays
Fenuron (Dybar)	3-Phenyl-l,l-dimethylurea	a Solubility low b Granules	Herbicide for undesirable woody plants
Fenuron TCA (Urab)	3-Phenyl-l,l-dimethylurea trichloroacetate	a Water soluble b Liquid concentrate; granules	Nonselective spray or dry application
PBA (Benzac; Zobar)	Polychlorobenzoic acid dimethylamine salt	a Water soluble b Liquid concentrate	Translocated perennial weed herbicide
Allylalcohol	Propen-l-ol-3	a Water soluble b Liquid	Soil drench to control all weeds in seedbeds
Pebulate (Tillam)	S-Propyl butylethylthio-carbamate	a Organic solvents b Emulsifiable concentrate	Selective incorporated herbicide
Vernolate (Vernam)	S-Propyl dipropylthiocarbamate	a Organic solvents b Emulsifiable concentrate	Selective incorporated herbicide
Sodium arsenite	Sodium arsenite	a Water soluble b Liquid concentrate	General contact spray; debarking trees
Chlorate	Sodium chlorate	a Water b Dry crystals	General translocated spray; soil sterilant
Sesone	Sodium 2,4-dichlorophenoxy-ethyl sulfate	a Water soluble b Water-soluble powder	Selective preemergence herbicide
Borate (Monobor-chlorate)	Sodium metaborate with sodium chlorate	a Water b Dry powder	Contact spray and soil sterilant
Borax (Borascu)	Sodium tetraborate	a Solubility low b Dry powder	Mixed with chlorate, TBA, bromacil, monuron, diuron, etc.; soil sterilant
TCA	Trichloroacetic acid	a Water b Dry powder	Grass killer through soil application
Triallate (Avadex BW)	S-2,3,3-Trichloroallyl-N,N-diisopropylthiolcarbamate	a Organic solvents b Liquid concentrate	Pre- and postplant soil-incorporated herbicide
2,3,6-TBA	2,3,6-Trichlorobenzoic acid	a Salts in water; esters in oil b Liquid concentrate	Selective translocated spray, soil-applied perennial weed killer
(PH 40-21; TH 052-H)	4,5,7-Trichlorobenzthia-diazole-2,1,3	a Organic solvents b Wettable powder; granules	Selective preemergence herbicide
TCBC (Randox-T)	Trichlorobenzylchloride used with CDAA	a Solubility low b Liquid concentrate; granules	Selective preemergence herbicide
(Tritac)	2,3,6-Trichlorobenzylpropanol	a Organic solvents b Liquid concentrate	Translocated perennial weed killer
2,4,5-T	2,4,5-Trichlorophenoxyacetic acid	a Organic solvents b Water-soluble salts, oil-soluble esters; emulsifiable concentrates	Selective translocated spray for woody species
Erbon (Baron)	2-(2,4,5-Trichlorophenoxy)-ethyl-2,2-dichloropropionate	a Oil b Emulsifiable concentrate	General soil sterilant
Fenac	2,3,6-Phenylacetic acid	a Organic solvents b Liquid concentrate; granules	Selective against certain grasses and broad-leaved weeds; pre- or post-emergence
Trifluralin (Treflan)	a,a,a-Trifluoro-2,6-dinitro-N,N-dipropyl-p-toluidine	a Organic solvents b Emulsifiable concentrate; granules	Selective preemergence soil-incorporated herbicide
(Cotoran)	3-(m-Trifluoromethylphenyl)-1,1-dimethylurea	a Organic solvents b Wettable powder	Selective preemergence herbicide
2,4-DEP (Falone)	Tris-(2,4-dichlorophenoxyethyl)-phosphite	a Organic solvents b Emulsifiable concentrate; granules	Selective pre- or post-emergence spray or dry application

into the scheme. In the jar method, a toxic chemical in water solution is placed in a jar on the ground and the tops of the weed are bent over and submerged in the solution. As the foliage is injured and rendered permeable, the solution enters the stems and roots and it may pass through interconnected roots to other unsubmerged tops. This method has been used for treating wild morning glory in strawberry beds and garden areas; for eliminating poison oak in camping areas; and for eradicating camel thorn on ranges. Sodium arsenite is the cheapest chemical to use, but its poison hazard must be kept in mind.

Killing of aquatic weeds with chlorinated benzene or aromatic solvents is accomplished by pumping the chemical into irrigation water flowing in a ditch.

Properties. The table gives some of the important properties and methods of using available herbicides. Both common and chemical names are given. Solubility is expressed only in relative terms. Some chemicals termed insoluble may dissolve in the soil solution in sufficient quantities to be toxic, but they cannot be conveniently formulated or applied in water solution. Availability again refers to the forms in which the chemicals are available on the market or are submitted for testing by the manufacturers.

Many of the chemicals listed in the table are used in proprietary mixtures. These are usually marketed under brand names and, if properly formulated, may combine the properties of two or more toxicants. Examples are the borate-chlorate mixtures that combine the quick-killing action of sodium chlorate with the residual action of the borates. They have the additional virtue of safety in contrast to sodium chlorate by itself. The substituted ureas are also used in combination with borates and chlorates. Some chemicals combine two distinct toxicants in a single molecule; examples are Erbon and Urox. *See* AGRICULTURAL SCIENCE (PLANT); DEFOLIANT AND DESICCANT.

[A. S. CRAFTS]

Bibliography: L. J. Andus, *The Physiology and Biochemistry of Herbicides*, 1964; A. S. Crafts, *The Chemistry and Mode of Action of Herbicides*, 1961; A. S. Crafts and W. W. Robbins, *Weed Control*, 1962; L. J. King, *Weeds of the World: Biology and Control*, 1966; G. C. Klingman, *Weed Control as a Science*, 1961.

Highway engineering

Highway planning, location, design, and maintenance. Before the design and construction of a new highway or highway improvement can be undertaken, there must be general planning and consideration of financing. As part of general planning it is decided what the traffic needs of the area will be for a considerable period, generally 20 years, and what construction will meet those needs. To assess traffic needs, the highway engineer collects and analyzes information about the physical features of existing facilities, the volume, distribution, and character of present traffic, and the changes to be expected in these factors. He must determine the most suitable location, layout, and capacity of the new routes and structures. Frequently, a preliminary line, or location, and several alternate routes are studied. The detailed design is normally begun only when the preferred location has been chosen.

In selecting the best route careful consideration is given to the traffic requirements, terrain to be traversed, value of land needed for the right-of-way, and estimated cost of construction for the various plans. The photogrammetric method, which makes use of aerial photographs, is used extensively to indicate the character of the terrain on large projects, where it is most economical. On small projects ground-mapping methods are preferred.

Financing considerations determine whether the project can be carried out at one time or whether construction must be in stages, with each stage initiated as funds become available. In deciding the best method of financing the work, the engineer makes an analysis of whom it will benefit. Improved highways and streets benefit, in varying degrees, three groups: users, owners of adjacent property, and the general public.

Users of improved highways benefit from decreased cost of transportation, greater travel comfort, increased safety, and saving of time (Fig. 1). They also obtain recreational and educational benefits. Owners of abutting or adjacent property may benefit from better access, increased property value, more effective police and fire protection, improved parking along the street, greater safety for pedestrian traffic, and the use of the street right-of-way for the location of public utilities, such as water lines and sewers.

Evaluation of various benefits from highway construction is often difficult but is a most important phase of highway engineering. Some benefits can be measured with accuracy, but the evaluation of others is more speculative. As a result numerous methods are used to finance construction, and much engineering work may be involved in selecting the best procedure.

Right-of-way acquisition. Highway engineers must also assist in the acquisition of right-of-way needed for new highway facilities. Acquisition of the land required for construction of expressways leading into the central business areas of cities has proved extremely difficult; the public is demanding that traffic engineers work closely with city planners, architects, sociologists, and all groups interested in beautification and improvement of cities to assure that expressways extending through metropolitan areas be built only after coordinated evaluation of all major questions, including the following:

(1.) Is sufficient attention being paid to beautification of the expressway itself? (2.) Would a change in location preserve major natural beauties of the city? (3.) Could a depressed design be logically substituted for those sections where an elevated expressway is proposed? (4.) Can the general design be improved to reduce the noise created by large volumes of traffic? (5.) Are some sections of the city being isolated by the proposed location?

Because of the large land areas needed for expressways, in several communities air rights above expressways are being sold to permit apartment buildings, parking garages, and similar structures to be built over the expressway, or such areas are being put to public use (Fig. 2).

At other locations, playgrounds are being built

beneath elevated highways, as are parking garages. In Honolulu, Hawaii, construction of a new post office under an elevated section of a freeway is being studied.

Detailed design. Detailed design of a highway project includes preparation of drawings or blueprints to be used for construction. These plans show, among other information, the exact location, the dimensions of such elements as roadway width, the final profile for the road, the location and type of drainage facilities, and the quantities of work involved, including earthwork and surfacing.

Soil studies. In planning the grading operations the design engineer considers the type of material to be encountered in excavating or in cutting away the high points along the project and how the material removed can best be utilized for fill or for constructing embankments across low areas elsewhere on the project (Fig. 3). For this the engineer must analyze the gradation and physical properties of the soil, determine how the embankments can best be compacted, and calculate the volume of earthwork to be done. Electronic calculating procedures are now sometimes used for the last step. Electronic equipment has also speeded up many other highway engineering calculations.

In recent years powerful and highly mobile earth-moving machines have been developed to permit rapid and economical operations. For example, now in use in the United States is a self-propelled earthmover weighing 125 tons and capable of hauling 100 yd³ of earth. This unit is powered by a 600-hp diesel electric motor.

Surfacing. Selection of the type and thickness of roadway surfacing to be constructed is an important part of design. The type chosen depends upon the maximum loads to be accommodated, the frequency of these loads, and other factors. For some routes, traffic volume may be so low that no surfacing is economically justified and natural soil serves as the finished roadway. As traffic increases, a surfacing of sandy clay, crushed slag, crushed stone, caliche, crushed oyster shells, or a combination of these materials may be applied. If gravel is used, it usually contains sufficient clay and fine material to help stabilize the surfacing. Gravel surfaces may be further stabilized by application of calcium chloride, which also aids in controlling dust. Another surfacing is composed of portland cement and water mixed into the upper few inches of the subgrade and compacted with rollers. This procedure forms a soil-cement base that can be surfaced with bituminous materials. Roadways to carry large volumes of heavy vehicles must be carefully designed and made of considerable thickness.

Drainage structures. Much of highway engineering is devoted to the planning and construction of facilities to drain the highway or street and to carry streams across the highway right-of-way.

Removal of surface water from the road or street is known as surface drainage. It is accomplished by constructing the road so that it has a crown and by sloping the shoulders and adjacent areas so as to control the flow of water either toward existing natural drainage, such as open ditches, or into a storm drainage system of catchbasins and underground pipes. If a storm drainage system is used, as it would be with city streets, the design engineer

Fig. 1. Model of the interstate highway system in St. Paul, Minn. Elaborate expressway work near the state capitol building (upper portion of photo) gives greatly improved access to the area. The entire interstate system throughout the United States is expected to be finished by 1974. (*Minnesota Department of Highways*)

must give consideration to the total area draining onto the street, the maximum rate of runoff expected, the duration of the design storm, the amount of ponding allowable at each catchbasin, and the proposed spacing of the catchbasins along the street. From this information the desired capacity of the individual catchbasin and the size of the underground piping network are calculated.

In designing facilities to carry streams under the highway the engineer must determine the area to

Fig. 2. Footbridge over the vehicular roadway crossing the Mississippi River in Minneapolis, Minn. (*Minnesota Department of Highways*)

Fig. 3. Construction of costly bridge structures and large-scale earth-moving were required for this section of a freeway north of Oakland, Calif.

be drained, the maximum probable precipitation over the drainage basin, the highest expected runoff rate, and then, using this information, must calculate the required capacity of the drainage structure. Generally designs are made adequate to accommodate not only the largest flow ever recorded for that location but the greatest discharge that might be expected under the most adverse conditions for a given number of years.

Factors considered in calculating the expected flow through a culvert opening include the size, length, and shape of the opening, the roughness of the walls, the shape of the entrance and downstream end of the conduit, the maximum allowable height of water at the entrance, and the water level at the outlet.

There is a trend to use designs that permit drainage structures to be assembled from standard sections manufactured at a central yard. Such procedures permit better control of the work, quicker construction, and less field work. For example, with precast concrete pipe sections it is often pos-

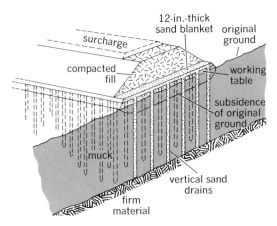

Fig. 4. Cross section of a vertical sand drain installation. Drains are 13–28 in. in diameter and are spaced 9 ft apart. Broken lines represent positions of staggered sand drains. (From L. I. Hewes and C. H. Oglesby, Highway Engineering, Wiley, 1954)

sible to avoid building small box culverts in the field. Circular culverts of large diameter are now also constructed in this way. However, culverts built of corrugated metal are specified for many projects for economy and to avoid placing small volumes of concrete at numerous locations.

Numerous small bridges are being designed to permit precast beams or girders to be placed side by side across the bridge opening to form the support for the roadway. These members are frequently of prestressed concrete. When precast members are used, the need for falsework to construct the bridge deck is eliminated, an especially beneficial move if the bridge is being built over a railroad or busy street.

Construction operations. Although much engineering and planning must be done preliminary to it, the actual construction is normally the costliest part of making highway and street improvements.

Staking out. With the award of a construction contract following the preparation of the detailed plans and specifications, engineers go onto the site and lay out the project. As part of this staking out, the limits of the earthwork are shown, location of drainage structures indicated, and profiles established.

Compaction. Heavy rollers are used to compact the soil or subgrade below the roadway in order to eliminate later settlement. Pneumatic-tired rollers and sheepsfoot rollers (steel cylinders equipped with numerous short steel teeth or feet) are often employed for this operation. Vibratory rollers have been developed and used on some projects in recent years. One type vibrates up to 3400 times/min, compacting the underlying material to an appreciable depth.

Vertical sand drains are sometimes employed to help stabilize fills or embankments constructed across wet and unstable ground (Fig. 4). Holes are drilled into the existing material and filled with sand. A horizontal layer of pervious sand or gravel is laid over the network of drains. As the fill is added over the pervious layer and the drains, the undesired water is forced upward through the columns of sand. By allowing the water to move horizontally to the edge of the embankment, the pervious layer prevents collection of water under the roadway.

To obtain quicker compaction of the unstable soil, a surcharge may be placed above the compacted fill. The surcharge material, usually intended for building the compacted fill elsewhere on the project, is removed once the unstable soil has been sufficiently compacted.

Maintenance and operation. Highway maintenance consists of the repair and upkeep of surfacing and shoulders, bridges and drainage facilities, signs, traffic control devices, guard rails, traffic striping on the pavement, retaining walls, and side slopes. Additional operations include ice control and snow removal. Because it is valuable to know why some highway designs give better performance and prove less costly to maintain than others, engineers supervising maintenance can offer valuable guidance to design engineers. Consequently, maintenance and operation are important parts of highway engineering.

[ARCHIE N. CARTER]

Bibliography: Highway Research Board, *Work-*

shop Conference on Economic Analysis, Washington, D.C., Spec. Rep. no. 56, 1960; A. R. Legault, *Highway and Airport Engineering*, 1960; P. O. Roberts and J. H. Suhrbier, *Highway Location Analysis: An Example Problem*, 1966; K. B. Woods, D. S. Berry, and W. H. Goetz (eds.), *Highway Engineering Handbook*, 1960.

Hill and mountain terrain

Land surfaces characterized by roughness and strong relief. The distinction between hills and mountains is usually one of relative size or height, but the terms are loosely and inconsistently used. Because of the prevalence of steep slopes, hill and mountain lands offer many difficulties to human occupancy. Cultivable land is scarce and patchy in occurrence, and transportation routes are often difficult to construct and maintain. The higher mountain ranges, by forcing moist air to rise in passing over them, induce large-scale condensation and precipitation. They also set up major disturbances in the broad pattern of atmospheric circulation and thus affect climates over extensive areas. Within the rough lands, large differences in elevation and in exposure to sunlight and to wind produce a complex pattern of local climatic contrasts.

Development of rough terrain. Mountain areas generally reflect disturbed structures in a portion of the Earth's crust that has been subjected to strong folding, buckling, doming, or other disordering of rock structures. Sometimes the most intense crustal deformation is known to have occurred in the distant geologic past. However the existence of elevated ranges, on which the destructive work of erosion has only begun, is a clear indication that these are areas in which crustal deformation has continued to recent times, though perhaps less violently than before. Further evidence of crustal instability lies in the fact that the cordilleran belts are currently foci of earthquakes and volcanic activity. Hill lands, with their lesser relief, indicate only lesser uplift, not a fundamentally different course of development. Some hill and low-mountain land, however, is carved by streams cutting valleys into uplifted but little-distorted rock structures.

The elevated portions of the crust may have the form of broad, warped swells, somewhat smaller domes or arched folds, upthrust broken blocks, or folded and broken masses of extreme complexity. Some definitely limited areas owe their altitude to eruptive activity that has poured out thick sheets of lava or ejected immense quantities of volcanic ash.

Although crustal uplift is necessary in order to give mountain and hill lands their elevation, most details of surface character of these lands are erosional in origin, having been carved out of the uplifted masses by streams and glaciers. The principal exceptions are fresh volcanic cones and the occasional scarps and swells that have been produced by unusually rapid deformation of the crust in geologically recent time.

Distribution pattern of rough land. Although complex, rough-land distributions are not wholly without system. Hill and mountain terrain occupies about 36% of the Earth's land area. The greater portion of that amount is concentrated in the great cordilleran belts that surround the

Pacific Ocean, the Indian Ocean, and the Mediterranean Sea. Additional rough terrain, generally low mountains and hills, occurs outside the cordilleran systems in eastern North and South America, northwestern Europe, Africa, and western Australia. Eurasia is the roughest continental land, more than half of its total area and most of its eastern portion being hilly or mountainous. Africa and Australia lack true cordilleran belts; their rough lands are scattered about in patches and interrupted bands that rarely show marked complexity of geologic structure. *See* TERRAIN AREAS, WORLDWIDE.

Predominant surface character. The features of hill and mountain lands are chiefly valleys and divides produced by sculpturing agents, especially running water and glacier ice. Local peculiarities in the form and pattern of these features reflect the arrangement and character of the rock materials within the upraised crustal mass that is being dissected.

Stream-eroded patterns. Stream-eroded features impart much of the character to most hill and mountain landscapes. In common with stream-sculptured surfaces of lesser relief, the principal differences from place to place are in the size, cross-section form, spacing, and pattern of occurrence of the stream valleys and the divides between them. These differences reflect variations in the original form and structure of the uplifted mass and in the stage to which erosion has proceeded.

In consequence of the uplift, streams have a large range of elevation through which they can cut, and as a result usually possess steep gradients early in their course of development. At this stage they are very swift, have great erosive power, and as a rule are marked by many rapids and falls. Because of the rapid downcutting, the valleys are characteristically canyonlike with steep walls and with floors no wider than the stream itself. At this same youthful stage of erosional development, divides are likely to be high and continuous, and broad ridge crests not uncommon. In hill lands such ridges often provide easier routes of travel than do the valleys, whereas in high mountain country the combination of gorgelike valleys and continuous high divides makes crossing unusually difficult. Thus in the Ozark Hills of Missouri, most of the highways and railroads follow the broad ridge crests. In the Himalayas and the central Andes, and to a lesser degree in the Rocky Mountains of Colorado, the narrow canyons are very difficult of passage, and the divides are so continuously high as to provide no easy pass routes across the ranges.

As erosional development continues, the major streams achieve gentle gradients and their valleys begin to widen. Divides become narrower and deep notches develop at valley heads, with well-defined peaks remaining between them. At this stage, which is fairly well represented by much of the Alps and by the Cascade Mountains of Washington, the principal valleys and the relatively low passes at their heads furnish routes of no great difficulty across the mountain belts. If other conditions are favorable, the wide valleys may afford significant amounts of cultivable or meadow land, as is true in the Alps. Still further erosion continues the widening of valleys and reduction of di-

Fig. 1. Head of a glaciated mountain valley ending in a cliff-walled amphitheater, called a cirque, showing a *steplike valley bottom, with lakes and waterfalls. (Photograph by Hileman, Glacier National Park)*

Fig. 2. The upper reaches of Susitna Glacier, Alaska, showing cirques and snowfields. The long tongues of *glacier ice carry bands of debris scoured from valley walls. (Photograph by Bradford Washburn)*

vides until the landscape approaches the state of an erosional plain upon which stand only small ranges and groups of mountains or hills. The mountains of New England and many of the mountains and hill groups of the Sahara and of western Africa represent late stages in the erosional sequence.

Glaciated rough terrain. Glacial features of mountain and hill lands may be produced either by the work of local glaciers in the mountain valleys or by overriding glaciers of continental size. Continental glaciation has the general effect of clearing away crags, smoothing summits and spurs, and depositing debris in the valleys. The resulting terrain is less angular than is usual for stream-eroded hills, and the characteristic glacial trademark is the numerous lakes, most of them debris-dammed, a few occupying shallow, eroded basins. Examples of rough lands overridden by ice sheets are the mountains of New England, the Adirondacks, the hills of western New York, the Laurentian Upland of eastern Canada, and the Scandinavian Peninsula.

In contrast, mountains that have been affected by local glaciers are made rougher by the glacial action. The long tongues of ice that move slowly down the valleys are excellent transporters of debris, and like their more extensive counterparts they are able to erode actively on shattered or weathered rock material. Valleys formerly occupied by glaciers are characteristically steep-walled and relatively free of projecting spurs and crags, with numerous broad cliffs, knobs, and shoulders of scoured bedrock. At their heads they generally end in cliff-walled amphitheaters called cirques (Figs. 1 and 2).

The valley bottoms commonly exhibit steplike profiles, with stretches of gentle gradient alternating with abrupt rock-faced risers. Especially in the lower parts of the glaciated sections are abundant deposits of rocky debris dropped by the ice. Sometimes these form well-defined ridges (moraines) that run lengthwise along the valley sides or swing in arcs across the valley floor. Lakes are strung along the streams like beads, most of them dammed by moraines but some occupying eroded basins. Because of rapid erosion, either that effected directly by the ice or that attendant upon the exposure of the rock surface by continual removal of the products of weathering, glaciated mountains are likely to be unusually rugged and spectacular. This is true not only of such great systems as the Himalayas, the Alps, the Alaskan Range, and the high Andes, but also of such lesser ranges as those of Labrador, the English Lake District, and the Scottish Highlands.

Most of the higher mountains of the world still bear valley glaciers, though these are not as large or as widespread as formerly. Dryness and long, warm summers limit glaciers in the United States to a few large groups on the higher peaks of the Pacific Northwest and numerous small ones in the northern Rockies.

Effects of geologic structure. Form and extent of the elevated areas, pattern of erosional valleys and divides, and, to some extent, sculptured details of slope and crest reflect geologic structure.

Some areas, such as the Ozarks, the western Appalachians, and the coast ranges of Oregon, are simply upwarped plains of homogeneous rocks that have been carved by irregularly branching streams into extensive groups of hills or mountains (Fig. 3). Others, like the Black Hills or the ranges of the Wyoming Rockies, are domes or arched folds, deeply eroded to reveal ancient granitic rocks in their cores and upturned younger stratified rocks around their edges. The Sierra Nevada of California is a massive block of the crust that has been uplifted and tilted toward the west so that it now displays a high abrupt eastern

Fig. 3. Hills in the western Appalachians, W.Va. (*Photograph by John L. Rich, Geographical Review*)

face and a long canyon-grooved western slope. The central belt of the Appalachians displays long parallel ridges and valleys that have been hewn by erosion out of a very old structure of parallel wrinkles in the crust. The upturned edges of resistant strata form the ridges; the weaker rocks between have been etched out to form the valleys. The Alps and the Himalayas are eroded from folded and broken structures of incredible complexity involving almost all varieties of rock materials. Local variations in form and pattern reflect the differences in underlying material.

Most volcanic mountains, like the Cascades of the northwestern United States or the western Andes of Peru and Bolivia, are actually erosional mountains sculptured from thick accumulations of lava and ash. In these areas of volcanic activity, however, individual eruptive vents give rise to volcanic cones that range from small cinder heaps to tremendous isolated mountains. The greater cones, such as Fuji, Ararat, Mauna Loa, or Shasta, are among the most magnificent features of the Earth's surface.   [EDWIN H. HAMMOND]

Homeostasis

In living substance, the maintenance of internal constancy and independence of the environment. The essence of living substance is that it differs chemically from the surrounding medium and yet maintains a dynamic equilibrium with its environment. Homeostasis occurs both at the cellular level and in the fluids and organ systems of multicellular organisms. Referring to homeostasis in mammals, the 19th-century French physiologist Claude Bernard coined the expression "constancy of the internal environment is the condition of life." The American physiologist Walter B. Cannon, who first used the word homeostasis, referred to the systems of checks and balances which maintain internal constancy as "the wisdom of the body." Homeostasis is usually considered as occurring in animals, although the term may apply as well to plants and microorganisms.

The cells of multicellular organisms are bathed in fluid which is kept relatively constant in its ionic composition, osmotic concentration, pH, level of sugar, and organic composition. No organism could be as dilute as fresh water and survive; terrestrial organisms must be protected against desiccation. The closest approach to conformity with the medium is in some endoparasitic worms and protozoans and in some marine invertebrates.

Patterns of adaptive responses. Two patterns of adaptive responses of animals to environmental change are recognized: (1) An animal may alter a given property with the environment; it conforms to the medium, and homeostasis consists in cellular adjustment such that metabolism continues in an altered state. For example, in cold-blooded animals the temperature conforms to that of the environment, yet activity may be high. (2) An animal may regulate its internal state and maintain internal constancy despite an altered environment. Such regulation is by a series of automatic feedback controls as environmental stress is applied, and at some environmental limits regulation fails and the animal cannot long survive. For example, warm-blooded animals maintain a constant body temperature over a range limited by both heat and cold. Both conformity and regulation are homeostatic in that they permit survival in an altering environment. A given animal may conform with respect to one physical parameter and regulate with respect to others.

Cellular level. Homeostasis at the cellular level is universal in living organisms. The first aggregates which could be called living cells must have been separated from their marine environment by a bounding layer which prevented free interchange. The cells of all plants and animals differ chemically from extracellular fluid. An important mechanism of cellular individuation resides in the cell surface, which has selective permeability and which, in many cells, provides some mechanical rigidity. The cell surface consists of a relatively inert pellicle plus a plasma membrane of protein and lipid; the surface permits entry and exit of relatively few kinds of organic molecule and varies in its permeability to inorganic ions according to activity. The surface also has some enzymes or carriers for active transport. Intracellular homeostasis consists of regulation of metabolism in its broadest meaning, that is, oxidation of foodstuff at such a level that sufficient energy will be available for necessary work and yet not at such a pace as to burn excessively.

Homeostatic mechanisms. A catalog of the mechanisms of homeostasis would be a textbook of the physiology of organ systems and of cells. Morphological changes may occur in response to environmental stress. Arctic birds and mammals tend to increase their coats of feathers or fur and insulating fat in winter; tadpoles grow larger gills when reared in low oxygen; men living at high altitudes have increased concentrations of blood hemoglobin; and bone structure varies according to mechanical stress. The sequence of homeostatic responses to a severe injury-type stress varies according to the kind of animal. In mammals a sequence of nonspecific responses may occur to any one of numerous severe stresses, the so-called general adaptation syndrome described by H. Selye. This consists, according to severity of the stress, of an initial vascular "shock" reaction and counterreactions in which, under pituitary activation, the adrenal cortex is activated, lymphocytes decrease in the blood, and other defense reactions occur.

Typical response controls. The sequence of responses to cold illustrates some of the principles of sequential reactions in complex homeostasis. First, the stimulation of cold receptors in the skin results in constriction of peripheral blood vessels and erection of hairs. If the body is then chilled, the temperature-sensing portion of the hypothalamus is stimulated, shivering begins, and metabolism increases, partly under endocrine control. The first line of defense is insulative; the second is increased heat production.

The complex reactions and activities of fear, rage, courtship, and care of young require a variety of sensory and effector structures integrated by circulatory, nervous, and endocrine systems. Epinephrine, the secretion from the adrenal medulla, has physiological action similar to impulses in the sympathetic nervous system. The digestive system has a sequence of proteolytic enzymes, overlapping in function and yet somewhat specific. The hierarchy of controls is noted, particularly for endocrines and nervous system. The anterior pituitary regulates the secretions of the adrenal cortex, thyroid, pancreas, and other glands; the posterior pituitary regulates kidney function.

Another principle is that of checks and balances. The vagus nerves slow the heart and stimulate the intestine, while the sympathetics accelerate the heart and relax the intestine. When motor nerve centers for one set of muscles are active, those for antagonistic muscles are inhibited. Various regions of the brain counterbalance other regions. Many sense organs show spontaneous activity which is modulated by stimulation, as either an increase or decrease in signals to the central nervous system. Many "conforming" animals compensate biochemically for environmental change with the result that relative internal constancy is maintained. Most aquatic poikilotherms acclimated to cold, for example, fish, have a higher oxygen consumption than individuals acclimated to warmth when both are measured at an intermediate temperature. Animal behavior is often compensatory for environmental change.

Cellular controls. In individual cells homeostasis is by different mechanisms but by similar principles to those in the whole organism. Regulation of enzyme levels is by a variety of feedback controls on protein synthesis. These include the amount of substrate (enzyme induction), the amount of products (repression or inhibition), modifying cofactors, extrinsic factors such as hormones, ionic conditions, and temperature. Similarly, the entrance of both ions and nonelectrolytes into cells is controlled by intracellular regulating processes and concentrations of critical materials. In energy liberation alternate metabolic pathways exist; the selection of a pathway is determined by intracellular needs. Also, numerous enzymes exist in one of several forms, as isozymes, and one may be produced predominantly under given conditions. The net effect of cellular homeostasis is to provide constancy of energy and structure under different extracellular conditions.

Significance. In all systems of homeostasis—hierarchies of controls, checks and balances, alternate mechanisms, compensatory reactions, and both negative and positive feedback—there are analogies to servo systems. Sensing elements are poised at critical levels and, upon deviation, control mechanisms are brought into action to restore equilibrium. Homeostatic models have been provided by computer science and cybernetics.

[C. LADD PROSSER]

Humidity

Atmospheric water-vapor content, expressed in any of several measures, especially relative humidity, absolute humidity, humidity mixing ratio, and specific humidity. Quantity of water vapor is also specified indirectly by dew point (or frost point), vapor pressure, and a combination of wet-bulb and dry-bulb (actual) temperatures. See DEW POINT.

Relative humidity is the ratio, in percent, of the moisture actually in the air to the moisture it would hold if it were saturated at the same temperature and pressure. It is a useful index of dryness or dampness for determining evaporation, or absorption of moisture.

Human comfort is dependent on relative humidity on warm days, which are oppressive if relative humidity is high but may be tolerable if it is low. At other than high temperatures, comfort is not much affected by high relative humidity.

However, very low relative humidity, which is common indoors during cold weather, can cause drying of skin or throat and adds to the discomfort of respiratory infections. The term indoor relative humidity is sometimes used to specify the relative humidity which outside air will have when heated to a given room temperature, such as 72°F, without addition of moisture. It always has a low value in cold weather and is then a better measure of the drying effect on skin than is outdoor relative humidity. This is even true outdoors because, when air is cold, skin temperature is much higher and may approximate normal room temperature.

Absolute humidity is the weight of water vapor in a unit volume of air expressed, for example, as grams per cubic meter or grains per cubic foot.

Humidity mixing ratio is the weight of water vapor mixed with unit mass of dry air, usually expressed as grams per kilogram. Specific humidity is the weight per unit mass of moist air and has nearly the same values as mixing ratio.

Dew point is the temperature at which air becomes saturated if cooled without addition of moisture or change of pressure; frost point is similar but with respect to saturation over ice. Vapor pressure is the partial pressure of water vapor in the air. Wet-bulb temperature is the lowest temperature obtainable by whirling or ventilating a thermometer whose bulb is covered with wet cloth. From readings of a psychrometer, an instrument composed of wet- and dry-bulb thermometers and a fan or other means of ventilation, values of all other measures of humidity may be determined from tables.

[J. R. FULKS]

Hydrology

The science that treats of the waters of the Earth, their occurrence, circulation, and distribution; their chemical and physical properties; and their reaction with their environment, including their relation to living things. The domain of hydrology embraces the full life history of water on the Earth.

Hydrologic cycle. The central concept of hydrology is the hydrologic cycle, a term used to describe the circulation of water from the oceans through the atmosphere to the land and back to the oceans over and under the land surface (see illustration). Water vapor moves over sea and land. It condenses into clouds and is precipitated as rain, snow, or sleet when one air mass rises to pass over another or over a mountain. Some of that which falls is evaporated while still in the air, or is intercepted by vegetation. Of the precipitation that reaches the ground surface, some evaporates quickly, some penetrates the soil, and some runs off over the land surface into streams, lakes, or ponds. Of that which penetrates the soil, some is held for a time and then returned to the atmosphere by evaporation or plant transpiration. The remainder penetrates below the soil zone to become a part of the groundwater. Study of the water in the atmosphere, although closely related to hydrology, lies more properly in the field of meteorology. The science relating to the oceans also touches closely on hydrology but is regarded as a separate science. See ATMOSPHERIC EVAPORATION; EVAPOTRANSPIRATION; METEOROLOGY; OCEANOGRAPHY.

Water is important for domestic, agricultural, and industrial uses. Thus the study of water and the means by which it may be obtained and controlled for use is of utmost importance to the welfare of mankind. Water is also a destructive agent of awesome force. Great floods periodically inundate valleys, causing death and destruction. Less dramatic are the rising water tables which, especially in irrigated areas, cause deterioration of the soil and make worthless large areas that would otherwise produce crops. Erosion of soil by flowing water and ultimate deposition of the sediment in lakes, reservoirs, stream channels, and coastal harbors is also a problem for hydrologists.

Functions. Hydrology is concerned with water after it is precipitated on the continents and before it returns to the oceans. It is concerned with measuring the amount and intensity of precipitation; quantities of water stored as snow and in glaciers, and rates of advance or retreat of glaciers; discharge of streams at various points along their courses; gains and losses of water stored in lakes and ponds; rates and quantities of infiltration into the soil and movement of soil moisture; changes of water levels in wells as an index of groundwater storage; rate of movement of water in underground reservoirs; flow of springs; dissolved mineral matter carried in water and its effects on water use; quantities of water discharged by evaporation from lakes, streams, and the soil and vegetation; and sediment transported by streams. In addition to devising methods for making these diverse measurements and storing the data in usable form, hydrology is concerned with analyzing and interpreting them to solve practical water problems. Rigorous studies of all the basic data are required to determine principles and laws involved in the occurrence, movement, and work of the water in the hydrologic cycle.

Engineering hydrology studies are basic to the design of projects for irrigation, water supply, flood control, storm drainage, and navigation. The hydrologist must estimate the probability of floods and droughts, volumes of water available for use,

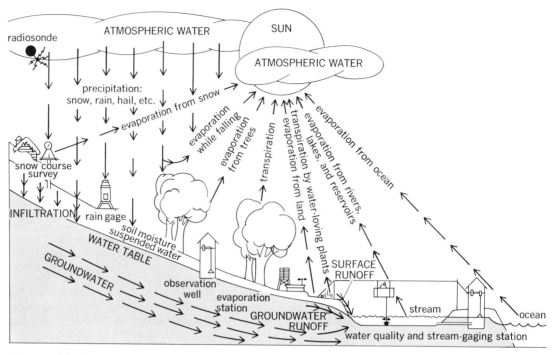

Diagram of the hydrologic cycle.

and probable maximum floods for spillway design. Often these estimates are required at locations where streamflow has not been measured. Digital computers are playing an increasing role in these studies and electrical analog computers are widely used in groundwater studies. *See* GROUNDWATER.

[ALBERT N. SAYRE/RAY K. LINSLEY]

Bibliography: V. T. Chow, *Handbook of Applied Hydrology*, 1964; S. N. Davis and R. J. M. De-Wiest, *Hydrogeology*, 1966; R. J. Kazmann, *Modern Hydrology*, 1965; R. K. Linsley, M. A. Kohler, and J. L. H. Paulhus, *Hydrology for Engineers*, 1958; D. K. Todd, *Groundwater Hydrology*, 1959; R. C. Ward, *Principles of Hydrology*, 1967.

Hydrometeorology

The study of the occurrence, movement, and changes in the state of water in the atmosphere. The term is also used in a more restricted sense, especially by hydrologists, to mean the study of the exchange of water between the atmosphere and continental surfaces. This includes the processes of precipitation and direct condensation, and of evaporation and transpiration from natural surfaces. Considerable emphasis is placed on the statistics of precipitation as a function of area and time for given locations or geographic regions.

Water occurs in the atmosphere primarily in vapor or gaseous form. The average amount of vapor present tends to decrease with increasing elevation and latitude and also varies strongly with season and type of surface. Precipitable water, the mass of vapor per unit area contained in a column of air extending from the surface of the Earth to the outer extremity of the atmosphere, varies from almost zero in continental arctic air to about 6 g/cm² in very humid, tropical air. Its average value over the Northern Hemisphere varies from around 2.0 g/cm² in January and February to

around 3.7 g/cm² in July. Its average value is around 2.8 g/cm², an amount equivalent to a column of liquid water slightly greater than 1 in. in depth. Close to 50% of this water vapor is contained in the atmosphere's first mile, and about 80% is to be found in the lowest 2 mi.

Atmospheric water cycle. Although a trivial proportion of the water of the globe is found in the atmosphere at any one instant, the rate of exchange of water between the atmosphere and the continents and oceans is high. The average water molecule remains in the atmosphere only about 10 days, but because of the extreme mobility of the atmosphere it is usually precipitated many hundreds or even thousands of miles from the place at which it entered the atmosphere.

Evaporation from the ocean surface and evaporation and transpiration from the land are the sources of water vapor for the atmosphere. Water vapor is removed from the atmosphere by condensation and subsequent precipitation in the form of rain, snow, sleet, and so on. The amount of water vapor removed by direct condensation at the Earth's surface (dew) is relatively small.

A major feature of the atmospheric water cycle is the meridional net flux of water vapor. The average precipitation exceeds evaporation in a narrow band extending approximately from 10°S to 15°N lat. To balance this, the atmosphere carries water vapor equatorward in the tropics, primarily in the quasi-steady trade winds which have a component of motion equatorward in the moist layers near the Earth's surface. Precipitation also exceeds evaporation in the temperate and polar regions of the two hemispheres, poleward of about 40° lat. In the middle and higher latitudes, therefore, the atmosphere carries vapor poleward. Here the exchange occurs through the action of cyclones and anticyclones, large-scale eddies of air with axes of spin

Table 1. Meridional flux of water vapor in the atmosphere

Latitude	Northward flux, 10^{10} g/sec
90°N	0
70°N	4
40°N	71
10°N	−61
Equator	45
10°S	71
40°S	−75
70°S	1
90°S	0

normal to the Earth's surface.

For the globe as a whole the average amount of evaporation must balance the precipitation. The subtropics are therefore regions for which evaporation substantially exceeds precipitation. The complete meridional cycle of water vapor is summarized in Table 1. This exchange is related to the characteristics of the general circulation of the atmosphere. It seems likely that a similar cycle would be observed even if the Earth were entirely covered by ocean, although details of the cycle, such as the flux across the Equator, would undoubtedly be different.

Complications in the global pattern arise from the existence of land surfaces. Over the continents the only source of water is from precipitation; therefore, the average evapotranspiration (the sum of evaporation and transpiration) cannot exceed precipitation. The flux of vapor from the oceans to the continents through the atmosphere, and its ultimate return to atmosphere or ocean by evaporation, transpiration, or runoff is known as the hydrologic cycle. Its atmospheric phase is closely related to the air mass cycle. In middle latitudes of the Northern Hemisphere, for example, precipitation occurs primarily from maritime air masses moving northward and eastward across the continents. Statistically, precipitation from these air masses substantially exceeds evapotranspiration into them. Conversely, cold and dry air masses tend to move southward and eastward from the interior of the continents out over the oceans. Evapotranspiration into these continental air masses strongly exceeds precipitation, especially during winter months. These facts, together with the extreme mobility of the atmosphere and its associated water vapor, make it likely that only a small percentage of the water evaporated or transpired from a continental surface is reprecipitated over the same continent. *See* ATMOSPHERIC GENERAL CIRCULATION; EVAPOTRANSPIRATION; HYDROLOGY.

Precipitation. Hydrometeorology is particularly concerned with the measurement and analysis of precipitation data. Since 1950 increasing attention has been paid to the use of radar in estimating precipitation. By relating the intensity of radar echo to rate of precipitation, it has been possible to obtain a vast amount of detailed information concerning the structure and areal distribution of storms.

Precipitation occurs when the air is cooled to saturation. The ascent of air towards lower pressure is the most effective process for causing rapid cooling and condensation. Precipitation may therefore be classified according to the atmospheric process which leads to the required upward motion. Accordingly, there are three basic types of precipitation: (1) Orographic precipitation occurs when a topographic barrier forces air to ascend. The presence of significant relief often leads to large variations in precipitation over relatively short distances. (2) Extratropical cyclonic precipitation is associated with the traveling regions of low pressure of the middle and high latitudes. These storms, which transport sinking cold dry air southward and rising warm moist air northward, account for a major portion of the precipitation of the middle and high latitudes. (3) Air mass precipitation results from disturbances occurring in an essentially homogeneous air mass. This is a common precipitation type over the continents in mid-latitudes during summer. It is the major mechanism for precipitation in the tropics, where disturbances may range from areas of scattered showers to intense hurricanes. In most cases there is evidence for organized lifting of air associated with areas of cyclonic (counterclockwise) vorticity. There are many questions still to be answered about the character and distribution of tropical precipitation, even though observations from the orbiting and geosynchronous meteorological satellites have already contributed substantially toward an increased understanding of these problems.

Precipitation may, of course, be in liquid or solid form. In addition to rain and snow there are other forms which often occur, such as hail, snow pellets, sleet, and drizzle. If upward motion occurs uniformly over a wide area measured in tens or hundreds of miles, the associated precipitation is usually of light or moderate intensity and may continue for a considerable period of time. Vertical velocities accompanying such stable precipitation are usually of the order of several centimeters per second. Under other types of meteorological conditions, particularly when the density of the ascending air is less than that of the environment, upward velocities may locally be very large (of the order of several meters per second) and may be accompanied by compensating downdrafts. Such convective precipitation is best illustrated by the thunderstorm. Intensity of precipitation may be extremely high, but areal extent and local duration are comparatively limited. Storms are sometimes observed in which local convective regions are embedded in a matrix of stable precipitation.

Analysis of precipitation data. Precipitation is essentially a process which occurs over an area. However, despite the experimental use of radar, most observations are taken at individual stations. Analyses of such "point" precipitation data are most often concerned with the frequency of intense storms. These data are of particular importance in evaluating local flood hazard, and may be used in such diverse fields as the design of local hydraulic structures, such as culverts or storm sewers, or the analysis of soil erosion. Intense local precipitation of short duration (up to 1 hr) is usually associated with thunderstorms. Precipitation may be extremely heavy for a short period, but tends to decrease in intensity as longer intervals are considered. Several record point accumula-

Table 2. Record observed point rainfalls*

Duration	Depth, in.	Station	Date
1 min	1.23	Unionville, Md.	July 4, 1956
8 min	4.96	Füssen, Bavaria	May 25, 1920
15 min	7.80	Plumb Point, Jamaica	May 12, 1916
42 min	12.00	Holt, Mo.	June 22, 1947
2 hr 45 min	22.00	Near D'Hanis, Texas	May 31, 1935
24 hr	73.62	Cilaos, La Reunion (Indian Ocean)	Mar. 15–16, 1952
1 month	366.14	Cherrapunji, India	July, 1861
12 months	1041.78	Cherrapunji, India	August, 1860 to July, 1861

*From J. L. H. Paulhus, Indian Ocean and Taiwan rainfalls set new records. *Mon. Weather Rev.*, 93:331–335, 1965.

tions of rainfall are shown in Table 2.

A typical hydrometeorological problem might involve estimating the likelihood of occurrence of a storm of given intensity and duration over a specified watershed to determine the required spillway capacity of a dam. Such estimates can only be obtained from a careful meteorological and statistical examination of large numbers of storms selected from climatological records. In the United States the U.S. Army Corps of Engineers, in cooperation with the Weather Bureau, has embarked on a continuing program of analysis to make such historical depth-area-duration data available to the practicing engineer.

Evaporation and transpiration. In evaluating the water balance of the atmosphere, the hydrometeorologist must also examine the processes of evaporation and transpiration from various types of natural surfaces, such as open water, snow and ice fields, and land surfaces with and without vegetation. From the point of view of the meteorologist, the problem is one of transfer in the turbulent boundary layer. It is complicated by topographic effects when the natural surface is not homogeneous. In addition the simultaneous heating or cooling of the atmosphere from below has the effect of enhancing or inhibiting the transfer process. Although the problem has been attacked from the theoretical side, empirical relationships are at present of greatest practical utility. *See* METE-OROLOGY.   [EUGENE M. RASMUSSON]

Bibliography: R. D. Fletcher, Hydrometeorology in the United States, in T. F. Malone (ed.), *Compendium of Meteorology*, 1951; E. N. Lorenz, *The Nature and Theory of the General Circulation of the Atmosphere*, 1967; H. Riehl, *Tropical Meteorology*, 1954; W. D. Sellers, *Physical Climatology*, 1965.

Hydrosphere

Approximately 74% of the Earth's surface is covered by water, in either the liquid or solid state. These waters, combined with minor contributions from groundwaters, constitute the hydrosphere:

World oceans	1.3×10^9 km³
Fresh-water lakes	1.3×10^5 km³
Saline lakes and inland seas	1.0×10^5 km³
Rivers	1.3×10^3 km³
Soil moisture and vadose water	6.7×10^4 km³
Groundwater to depth of 4000 m	8.4×10^6 km³
Icecaps and glaciers	2.9×10^7 km³

The oceans account for about 97% of the weight of the hydrosphere, while the amount of ice re-flects the Earth's climate, being higher during periods of glaciation. (Water vapor in the atmosphere amounts to 1.3×10^4 km³.) The circulation of the waters of the hydrosphere results in the weathering of the landmasses. The annual evaporation of 3.5×10^5 km³ from the world oceans and of 7.0×10^4 km³ from land areas results in an annual precipitation of 3.2×10^5 km³ on the world oceans and 1.0×10^5 km³ on land areas. The rainwater falling on the continents, partly taken up by the ground and partly by the streams, acts as an erosive agent before returning to the seas. *See* GROUNDWATER; HYDROLOGY; LAKE; RIVER; SEA WATER; TERRESTRIAL FROZEN WATER. [EDWARD D. GOLDBERG]

Bibliography: R. L. Nace, Water resources, *Environ. Sci. Technol.*, 1:550–560, 1967.

Hydrosphere, geochemistry of

The oceans of the world constitute a principal reservoir for substances in the major sedimentary cycle, which involves the processes of transport of material from the Earth's crust to the sea floor. The cycle begins with the precipitation of water, acidified by the uptake of carbon dioxide in the atmosphere, onto the continents. This results in the physical and chemical breakdown of exposed surfaces. A part of the weathered material, in dissolved or solid states, is borne by the rivers to the oceans. Evaporation at the oceanic surfaces provides atmospheric water, which precipitates in part upon the continents. This latter process completes the cycle. Table 1 gives the quantitative details by contrasting the marine areas and the respective land areas draining into the world's oceans. Clearly, per unit area, the Atlantic receives the weathering products from an integrated drainage area six times as large as that of the Pacific. The interior drainage areas are responsible for such water bodies as the Great Salt Lake, the Caspian Sea, and the Dead Sea.

Oceanic waters. The reactivities of chemical species in the oceans are reflected in the average

Table 1. Oceanic and land-drainage areas

Ocean	Area, 1000 km²	Land area drained, 1000 km²	Ratio, area drained/ ocean area
Atlantic	98,000	67,000	0.684
Indian	65,500	17,000	0.260
Antarctic	32,000	14,000	0.440
Pacific	165,000	18,000	0.110
Interior drainage		32,000	

Table 2. Residence times of elements in the oceans

Element	Residence time, years	Element	Residence time, years
Na	2.6×10^8	Mg	4.5×10^7
Ca	8.0×10^6	Li	1.2×10^7
U	6.5×10^5	K	1.0×10^7
Cu	6.5×10^4	Rb	6.1×10^6
Si	1.0×10^4	Ba	4.0×10^4
Mn	7.0×10^3	Zn	2.0×10^4
Ti	1.6×10^2	Pb	4.0×10^2
Al	1.0×10^2	Ce	3.2×10^2
Fe	1.4×10^2	Th	1.0×10^2

times spent there before precipitation to the sea floor. Those elements with short residence time in the oceans engage more readily in chemical reactions that result in the formation of solid phases than those elements with long residence times.

Residence times. The calculations of residence times are based upon a simple reservoir model of the oceans, whose chemical composition is assumed to be in steady state; that is, the amount of a given element introduced by the rivers per unit time is exactly compensated by that lost through sedimentation. An elemental residence time may be defined then by $t = A/(dA/dt)$, where A is the amount of the element in the oceans and (dA/dt) is the rate of introduction or the rate of removal of the element from the marine hydrosphere. Table 2 gives values for a representative group of elements.

The element of longest residence, sodium, has a residence time within an order of magnitude of the age of the oceans, several billion years. Similarly, the more abundant alkali and alkaline-earth metals all have residence times in the range of 10^6 to 10^8 years, resulting from the relative lack of reactivity of these elements in marine waters.

Elements which pass rapidly through the marine hydrosphere in the major sedimentary cycle, such as titanium, aluminum, and iron, not only enter the oceans in part as rapidly settling solids but also are reactants in the formation of the clay and ferromanganese minerals. For a discussion of the inorganic regulation of the composition *see* SEA WATER.

Although the relatively low values of these residence times are significant, the absolute values are probably unrealistic inasmuch as they are in conflict with an assumption used in their derivation. In treating the oceans as a simple reservoir, the mixing times of the oceans are assumed to be much less than the residence times of the elements. Yet the oceans are believed to mix in times of the order of thousands of years. Nonetheless, it is significant that one may expect to find the concentration of such elements varying from one oceanic water mass to another.

Those elements with residence times of intermediate length, periods of the order of tens and hundreds of thousands of years, are probably of nearly uniform concentration in the oceans of the world, as are the chemical species of longer marine lives. Typical members of the group are such metals as barium, lead, zinc, and nickel, elements in extremely dilute solution but actively involved in the inorganic chemistries of the seas.

Such behavior is confirmed by the observation that these elements are in states of undersaturation in oceanic waters.

Photosynthesis. The presence of the large photosynthesizing biomass in the oceans gives rise to dramatic concentration changes. The amount of photosynthesis in the oceans, calculated to be of the order of 1×10^{17} tons of carbon dioxide consumed per year, compares with estimates for land of 2×10^{16} tons/year. The depth of the photosynthetic zone can extend downward from the surface to depths of 100 meters (m) or so, depending upon the transparency of the water, season of the year, and latitude. In waters of active plant growth, carbon dioxide and oxygen (the intake and release gases of photosynthesis) are often observed in states of depletion and supersaturation, respectively.

The photosynthesizing plants require a group of dissolved chemical species, the nutrients, which are necessary for growth and multiplication. Ions

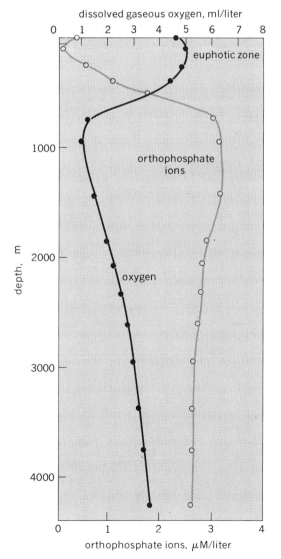

Distribution of dissolved gaseous oxygen (low values) and nutrient species (high concentrations) of orthophosphate ions at 26°22'.4N and 168°57'.5W in the Pacific Ocean. (*Data from Chinook Expedition of 1956 of the Scripps Institution of Oceanography*)

of the orthophosphoric acids, nitrate, nitrite, and ammonia, as well as monomeric silicic acid, have very low concentrations in the regions of plant productivity as compared to regions of lesser fertility. Certain other substances concerned with plant growth, such as vitamins and trace-metal ions, may have their marine concentrations governed by biological activity in surface waters, but as yet no definite relationships have been established.

The primary production of plant material furnishes the basis of nutrition for the animal domain of the oceans. The plant material which is removed from the photosynthetic zones but is not consumed by higher organisms, together with the organic debris resulting from the metabolic waste products or death of members of the marine biosphere, is oxidized principally in the oceanic waters below but adjacent to the photosynthetic zone. This results not only in low values of dissolved gaseous oxygen at such depths but also in high concentrations of the nutrient species which are released subsequent to the combustion of the organic matter (see illustration).

The dissolved organic substances, arising from life in the sea, are of the order of 0.3 mg of carbon per liter. Higher amounts of such materials are found in surface or coastal waters. *See* SEA-WATER FERTILITY.

Salt content. For many problems in the physics of the sea and in engineering, the significant parameter is the total salt content, which governs the density as a function of pressure and temperature, rather than elemental, ionic, or molecular compositions.

The total salt content is expressed by either the salinity or the chlorinity, both terms given in units of parts per thousand °/oo. The salinity is defined as the weight in grams in vacuo of solids which can be obtained from a water weight of 1 kg in vacuo under the following conditions: (1) the solids have been dried to a constant weight at 480°C; (2) the carbonates have been converted to oxides; (3) the organic matter has been oxidized; and (4) the bromine and iodine have been replaced by chlorine. A weight loss from the solid phases results from such chemistries, and hence the salinity of a given sample of sea water is somewhat less than its actual salt content.

In practice, the salt content is ascertained by the precipitation of the halogens with silver nitrate or by such physical methods as electrical conductivity, sound velocity, and refractive index, with the first of these techniques having widespread use. In the chemical titration with silver nitrate the mass of halogens contained in 1 kg of sea water, assuming the bromine and iodine are replaced by chlorine, is designated as the chlorinity. The values so obtained are dependent upon the atomic weights of both silver and chlorine. Inasmuch as changes in the salt content of water are of interest when taken over long time periods, the chlorinity has been made independent of atomic weight changes by redefining it in terms of the weight of silver precipitated, as in Eq. (1). Chlorinity deter-

$$Cl°/oo = 0.3285234 \, Ag \qquad (1)$$

minations, based either on chemical or physical methods, are related to standard sea water, a water of known salinity which is obtainable from the Hydrographic Laboratory in Copenhagen, Denmark. Chlorinity is related to the salinity, Eq. (2).

$$S°/oo = 0.030 + 1.8050 \, Cl°/oo \qquad (2)$$

Salinity of open ocean waters varies regionally between 32 and 37 °/oo. Areas where evaporation exceeds precipitation, such as enclosed basins, are characterized by higher values. Salinities of 38–39 °/oo are representative of the Mediterranean Sea, while the northern part of the Red Sea has values ranging from 40 to 41 °/oo. Coastal bays, subject to land drainage, and those waters which mix with meltwaters from cold regions, possess salinities which have all degrees of dilution comparable to those of the open ocean. *See* OCEAN-ATMOSPHERE RELATIONS.

Although sea waters exhibit marked regional and depth differences in salt content, the ratios of the major dissolved constituents to one another, listed below, are almost invariable. (Ratios are grams of given element per kilogram of seawater divided by chlorinity in parts per thousand.)

Constituents	Ratio	Constituents	Ratio
Na/Cl	0.5555	S/Cl	0.0466
Mg/Cl	0.0669	Br/Cl	0.0034
Ca/Cl	0.0213	Sr/Cl	0.00040
K/Cl	0.0206	B/Cl	0.00024

Development of the hydrosphere. Many hypotheses on the origin and evolution of the Earth's hydrosphere have been advanced over the past 50 years. Most of them can be placed in one of two categories: (1) the hypothesis of an original ocean, which proposes that the present ocean has had much the same size and composition since the beginning of the geologic record; and (2) the hypothesis of continuous accumulation, which considers that the ocean has been growing continuously, but not necessarily uniformly, since its inception.

Such considerations have certain common assumptions. First, the rock-forming species in sea water, sodium, potassium, silicon, iron, magnesium, and others, have been derived from the weathering of the Earth's surface, whereas the marine quantities of water and the anionic constituents, such as chlorine and sulfur, cannot be adequately supplied by such a mechanism. These latter substances, as gases or dissolved species, have apparently evolved from the Earth's interior by the degradation of the surface rocks. This proposal has been reached through the following arguments. The abundances of the noble gases—neon, nonradiogenic argon, and krypton—are many orders of magnitudes less on the Earth than in the universe relative to other elements. It is thus assumed that they were lost by the Earth during its formative period. Therefore, substances which existed as gases and had comparable molecular weights were similarly not retained.

The hypothesis of the permanency of the oceans through geologic time complements the hypothesis of the permanence of the continents. Since old rocks are found in the basement of the central shield areas, with greater thicknesses of younger rocks at the edges, it has been assumed that the continents have grown laterally. Consequently, the reduction in area occupied by the oceans is compensated for by an increase in average depth. However, the calculated amounts of weathered

Table 3. The chemical compositions of fresh waters and rain (values in parts per million)

Substance	Rivers*	Irish lakes†	English rain‡
Ca^{++}	15	4.0	0.1 – 2.0
Na^+	6.3	8.6	0.2 – 7.5
Mg^{++}	4.1	1.0	0.0 – 0.8
K^+	2.3	0.5	0.05 – 0.7
CO_3^{--}	58.4	8.8	0.0 – 2.7
SO_4^{--}	11.2	5.2	1.1 – 9.6
Cl^-	7.8	14.7	0.2 – 12.6
SiO_2	13.1		
NO_3^-	1		
Fe	0.67		

*Daniel A. Livingston, *Data of Geochemistry*, USGS Prof. Pap. no. 440 – G, 1963.

†E. Gorham, The chemical composition of some western Irish fresh waters, *Proc. Roy. Irish Acad.*, vol. 58B, 1957.

‡E. Gorham, On the acidity and salinity of rain, *Geochim. Cosmochim. Acta*, vol. 7, 1955.

by some geochemists to one of an initial ocean of water with the gradual accretion of anionic substances.

The hypothesis of the slow growth of the oceans gains strength with the following observations. If only 1% of the hot-spring water is juvenile, the amounts found today, if extrapolated over geologic time, give a sufficient volume to produce the present oceans. Further, the ratios of the major anionic constituents in sea water are similar to those found in plutonic gases.

Fresh waters and rain. Fresh continental waters show enormous variations in total salt content and in the relative concentrations of their various components, and such parameters in any given water body vary seasonally as well. Chemical analyses of a number of representative waters are given in Table 3.

The factors governing the composition of freshwater bodies are many, and some are poorly understood. Significant amounts of the dissolved phases materials that would form by the action of an initial ocean containing all the various anionic constituents as acids are far greater than those estimated by geologists to have been decomposed over all of geologic time. This hypothesis has been modified

Table 4. The composition of waters from chloride, sulfate, and carbonate lakes*

Substance	Dead Sea	Little Manitou Lake	Pelican Lake, Ore.
Na^+	11.14	16.8	29.25
K^+	2.42	1.0	3.58
Mg^{++}	13.62	10.9	2.62
Ca^{++}	4.37	0.48	2.27
CO_3^{--}	Trace	0.47	30.87
SO_4^{--}	0.28	48.4	22.09
Cl^-	66.37	21.8	7.97
SiO_2	Trace	0.009	1.21
Al_2O_3 } Fe_2O_3 }		0.21	0.02
Salinity	226,000	106,851	1983

*Data from G. E. Hutchinson, *A Treatise on Limonology*, vol. 1, Wiley, 1957.

appear to come from rock weathering, organic-decomposition products, atmospheric dusts, fuel combustion, air-borne particles from the marine environment, and volcanic emanations. Such materials, with the exception of the first, carried from their sources in the atmosphere, can be taken from the air to the river and lakes below by rain. Soil and rock weathering provide the major supply of ions to rivers and lakes. This can be seen from Table 3, where rainwaters have less dissolved solids than river or lake waters by an order of magnitude.

Most fresh waters can be characterized as bicarbonate waters, with HCO_3^- exceeding all other anions, although in some instances chloride or sulfate is the dominant anion.

The sodium, chlorine, and often magnesium are dominantly of marine origin. They leave the oceans as sea spray and are air-borne to the continents. These elements show dramatic decreases in concentration in surface waters going from the coastal to the interior regions. Exceptions can be found in certain waters which derive their salts mainly by the denudation of igneous areas.

Calcium, and sometimes magnesium, can originate from drainage areas, as in the case of the Wisconsin fresh waters which drain over ancient magnesian limestones. Sulfate not only has sources in the marine environment but is also produced by the combustion of sulfur-containing fuels and from the oxidation of sulfur dioxide which results from the atmospheric burning of hydrogen sulfide, a product of the decomposition of organic matter.

The average salt content of fresh waters is of the order of 120 parts per million (ppm). Lower amounts of dissolved solids (50 ppm and less) are found in waters draining igneous rock beds, while open lakes and rivers carrying high salt contents (200 ppm and above) normally result either from the leaching of salt beds or from contamination by man.

Closed basins. The chemical compositions of closed basins, water bodies in which evaporation is the mechanism for the loss of water, are illustrated in Table 4, which contains representative examples of the three classical types, the carbonate, sulfate, and chloride waters. These classes appear in sequence during the removal of water from a system with the composition of average river or lake water. The carbonate types exist up until evaporation leads to the precipitation of calcium carbonate and to liquid phases enriched in sulfate and chloride ions. Further removal of water results in the precipitation of calcium carbonate and the subsequent precipitation of gypsum, $CaSO_4 \cdot 2H_2O$. The residual waters hence contain chloride as the dominant anion.

[EDWARD D. GOLDBERG]

Bibliography: M. N. Hill (ed.), *The Sea*, vol. 2, 1963; G. E. Hutchinson, *A Treatise on Limnology*, vol. 1, 1957.

Imhoff tank

A sewage treatment tank named after its developer, Karl Imhoff. Imhoff tanks differ from septic tanks in that digestion takes place in a separate compartment from that in which settlement occurs. The tank was introduced in the United States

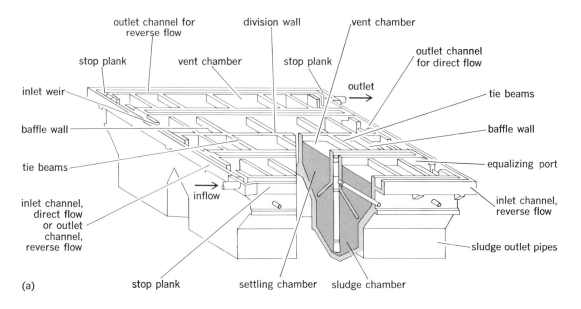

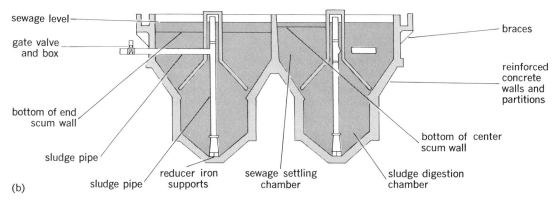

Diagram of typical large Imhoff tank for sewage treatment. (a) General arrangement. (b) Cross section. (From *H. E. Babbitt and E. R. Baumann, Sewerage and Sewage Treatment, 8th ed., Wiley, 1958*)

in 1907 and was widely used as a primary treatment process and also in preceding trickling filters. Developments in mechanized equipment have lessened its popularity, but it is still valued as a combination unit for settling sewage and digesting sludge. *See* SEPTIC TANK; SEWAGE.

The Imhoff tank is constructed with the flowing-through chamber on top and the digestion chamber on the bottom (see illustration). The upper chamber is designed according to the principles of a sedimentation unit. Sludge drops to the bottom of the tank and through a slot along its length into the lower chamber. As digestion takes place, scum is formed by rising sludge in which gas is trapped. The scum chamber, or gas vent, is a third section of the tank located above the lower chamber and beside the upper chamber. As gases escape, sludge from the scum chamber returns to the lower chamber. The slot is so constructed that particles cannot rise through it. A triangle or sidewall deflector below the slot prevents vertical rising of gas-laden sludge. Sludge in the lower chamber settles to the bottom, which is in the form of one or more steep-sloped hoppers. At intervals the sludge can be withdrawn. The overall height of the tank is 30–40 ft, and sludge can be expelled under hy-

draulic pressure of the water in the upper tank. Large tanks are built with means for reversing flow in the upper chamber, thus making it possible to distribute the settled solids more evenly over the digestion chamber.

Design. Detention period in the upper chamber is usually about $2\frac{1}{2}$ hr. The surface settling rate is usually 600 gal/(ft²)(day). The weir overflow rate is not over 10,000 gal/ft of weir per day. Velocity of flow is held below 1 ft/sec. Tanks are dimensioned with a length-width ratio of 5:1–3:1 and with depth to slot about equal to width. Multiple units are built rather than one large tank to carry the entire flow. Two flowing-through chambers can be placed above one digester unit. The digestion chamber is normally designed at 3–5 ft³ per capita of connected sewage load. When industrial wastes include large quantities of solids, additional allowance must be made. Ordinarily sludge withdrawals are scheduled twice per year. If these are to be less frequent, an increase in capacity is desirable. Some chambers have been provided with up to 6.5 ft³ per capita. The scum chamber should have a surface area 25–30% of the horizontal surface of the digestion chamber. Vents should be 24 in. wide. Top freeboard should be at least 2 ft to contain rising scum. Water under pressure must be available to combat foaming and knockdown scum.

Efficiency. The efficiency of Imhoff tanks is equivalent to that of plain sedimentation tanks. Effluents are suitable for treatment on trickling filters. The sludge is dense, and when withdrawn it may have a moisture content of 90–95%. Imhoff sludge has a characteristic tarlike odor and a black granular appearance. It dries easily and when dry is comparatively odorless. It is an excellent humus but not a fertilizer. Gas vents may occasionally give off offensive odors.   [WILLIAM T. INGRAM]

Industrial meteorology

The commercial application of weather information to the operational problems of business, industry, transportation, and agriculture in a manner intended to optimize the operation with respect to the weather factor. The weather information may consist of past weather records, contemporary weather data, predictions of anticipated weather conditions, or an understanding of physical processes which occur in the atmosphere. The operational problems are basically decisions in which weather exerts an influence.

Because meteorological data are not available to specify past, present, and future states of the atmosphere with absolute precision and knowledge of physical processes in the atmosphere is incomplete, the application of weather information to operational decisions is properly viewed as a specialized case of decision-making in the face of uncertainty. What is involved here is best illustrated by the following example developed by J. C. Thompson. A hypothetical construction company was confronted daily with the decision whether to pour concrete. If the concrete were poured and 0.15 in. or more of rain fell in the subsequent 36 hr, damage of $5000 would result. The cost of protecting the newly poured concrete from rain would be $400. To minimize the total expense of a series of such repetitive decisions (optimization of the operation), it follows from the principle of the calculated risk that protective measures should be taken only when $P > C/L$, where P is the probability of 0.15 in. or more of rain within 36 hr, C is the cost of protective measures ($400), and L is the contingent loss ($5000). For this particular case, under the actual weather occurring during a season's operations, the total expense (cost plus loss) would be $85,000 if no protective measures were taken, $72,800 if protective measures were taken every day regardless of anticipated weather, $32,600 if protective measures were taken only on days when there was a 50:50 chance of this amount of rain, and $24,400 if protective measures were taken only on days when the probability of the critical amount of rainfall exceeded 0.08, that is, in ratio of $400:$5000.

This approach can be generalized to include more complex decisions, and relations were developed by Thompson and G. W. Brier to measure the economic utility of weather information. In practice, less formal methods are frequently used and the fundamental principle described here is handled in a qualitative manner by close collaboration between the meteorologist and the user of weather information.

The substantial economic significance of weather in problems of business and industry has been emphasized by a survey conducted by the U.S. Weather Bureau in which an attempt was made to assign monetary values to savings or profits realized through applications of daily weather reports, forecasts, storm warnings, and past weather records. The total for the United States was on the order of $1,000,000,000 annually.

Some measure of the economic toll of severe or unusual weather conditions is provided by data assembled by the National Board of Fire Underwriters, which indicated that claims totaling $866,000,000 were paid as a result of 72 major hurricanes, tornadoes, windstorms, hailstorms, and rainstorms during the period 1949–1957, inclusive. These losses represent insurance losses, not total property damage, and include only instances in which claims within a single state exceeded $1,000,000.

The economic benefits that could be achieved in a single industry, the petroleum industry, by only a modest improvement in forecasts of anticipated average temperature conditions 1 month in advance have been analyzed with the result that potential reduction in tankage and inventory costs has been conservatively estimated at $100,000,000 per year. These savings would be realized by improvements in heating oil scheduling made possible through better estimates of expected space-heating requirements available from forecasts of anticipated temperature conditions.

As a result of the economic significance of the weather factor in a variety of industrial applications, there exists in meteorology a specialized activity on the part of professional meteorologists to serve the specific needs which lie outside the general public responsibilities of the U.S. Weather Bureau. This service is provided either by staff meteorologists employed by particular companies or by consultant meteorologists who work for several clients. Approximately 9% of the more than 7000 professional meteorologists in the United States are engaged in some facet of industrial meteorology. A substantial number of these are employed by commercial airlines where the need for specialized weather information is vital for safe and efficient operations. Gas and electric utility load estimates, highway and street maintenance, outdoor construction work, marine transportation, retail merchandising and advertising, flood control design, air pollution, building and plant design, atmospheric corrosion, agricultural planning and production scheduling, and air conditioning and heating design are but a few of the activities in which the industrial meteorologist has found a demand for his services. *See* AGRICULTURAL METEOROLOGY.

Annual conferences on industrial meteorology are sponsored by the Committee on Industrial Meteorology of the American Meteorological Society. That society has established a program for the certification of consulting meteorologists who meet rigorous standards of knowledge, experience, and adherence to high standards of ethical practice.

[THOMAS F. MALONE]

Bibliography: American Meteorological Society, A selective annotated bibliography on industrial meteorology, *Weatherwise*, vol. 6, no. 2, 1953;

S. Petterssen, *Introduction to Meteorology*, 3d ed., 1969; J. C. Thompson, A numerical method for forecasting rainfall in the Los Angeles area, *Mon. Weather Rev.*, 78(7):113–124, 1950; J. C. Thompson and G. W. Brier, The economic utility of weather forecasts, *Mon. Weather Rev.*, 83(11):249–254, 1955.

Insect control, biological

The term biological control was proposed in 1919 to apply to the use or role of natural enemies in insect population regulation. The natural enemies involved are termed parasites, predators, or pathogens. This remains the preferred usage, although other biological methods of insect control have been proposed or developed, such as the release of mass-produced sterile males to mate with wild females in the field, thereby greatly reducing or suppressing the pests' production of progeny. Classical biological control is an ecological phenomenon which occurs everywhere in nature without aid from, or sometimes even understanding by, man. However, man has utilized the ecological principles involved to develop the field of applied biological control of insects, and the great majority of practical applications have been achieved with insect pests. Additionally, such diverse types of pest organisms as weeds, mites, and certain mammals have been successfully controlled by use of natural enemies.

History. Man observed the action of predacious insects early in his agricultural history, and a few crude attempts to utilize predators have been carried on for centuries. However, the necessary understanding of biological and ecological principles, especially those of population dynamics, did not begin to emerge until the 19th century. The first great applied success in biological control of an insect pest occurred 2 years after the importation into California from Australia in 1888–1889 of the predatory vedalia lady beetle. This insect feeds on the cottony-cushion scale, a notorious citrus pest that was destroying orange trees at that time, and the vedalia was credited with saving the citrus industry. This successful control firmly established the field or discipline of applied biological control as it is known today. *See* POPULATION DYNAMICS.

Ecological principles. Natural enemy populations have a feedback relationship to prey populations, termed density-dependence, which results in the increasing or decreasing of one group in response to changes in the density of the other group. Such reciprocal interaction prevents indefinite increase or decrease and thus results in the achievement of a typical average density or "balance." However, this balance may be either at high or low levels, depending on the inherent abilities of the natural enemies. If the natural enemy is highly density-dependent, it can regulate the prey population density at very low levels. An effective enemy achieves regulation by rapidly responding to any increase in the prey population in two different ways: killing more prey by increased feeding or parasitism, and producing more progeny for the next generation. The net effect is a more rapid increase in prey mortality as prey population tends to increase, so that first the trend is stopped and then it becomes reversed, lowering the prey

population. This results in relaxing the pressure caused by the enemy so that ultimately the prey population is enabled to increase again, and the cycle is repeated over and over.

Applications of classical method. Since the vedalia beetle controlled the cottony-cushion scale some 80 years ago, many projects in applied biological control have been undertaken and a large number of successes achieved. There are about 275 recorded cases of applied biological control of insects in 70 countries. Of these recorded cases 83 have been so completely successful that insecticides are no longer required. In the other cases the need for chemical treatment has been more or less greatly reduced. Also, there are about 47 cases of biological control of weeds by insect natural enemies. In California alone it is estimated that the agricultural industry has been saved $195,000,000 during the past 45 years because of reduction in insect pest-caused crop losses and diminished need for chemical control. A few other outstanding successes include the biological control of Florida red scale in Israel and subsequently in Mexico, Texas, and Florida; Oriental fruit fly in Hawaii; green vegetable bug in Australia; citrus blackfly and purple scale in Mexico; dictyospermum scale and purple scale in Greece; and olive scale in California.

All the cases of applied biological control mentioned above involved foreign exploration and importation and colonization of new exotic natural enemies of the insect pest in question. In the majority of cases the pest was an invader, being native to another country, and its natural enemies had been left behind. By searching the native habitat for effective enemies and sending them to the new home of the invader pest, the biological balance was reconstituted. This method is considered to be classical biological control, and its scientific application has been responsible for most of the outstanding results obtained. There are, however, two other major phases of biological control which are categorized under the headings of conservation and augmentation.

Conservation. This phase involves manipulation of the environment to favor survival and reproduction of natural enemies already established in the habitat, whether they be indigenous or exotic. In other words, adverse environmental factors are modified or eliminated, and requisites which are lacking such as food or nesting sites may be provided. Even though potentially effective natural enemies are present in a habitat, adverse factors may so affect them as to preclude attainment of satisfactory biological control. Insecticides commonly produce such adverse effects, causing so-called upsets in balance or pest population explosions.

Augmentation. This phase concerns direct manipulation of established enemies themselves. Potentially effective enemies may be periodically decimated by environmental extremes or other adverse factors which are not subject to man's control. For example, low winter temperatures may seriously decrease certain enemy populations each year. The major means of solving such problems has involved laboratory mass culture of enemies and their periodic colonization in the field, generally after the adverse period has passed. This practice is gaining rapidly in application. *Tricho-gramma* sp., common egg parasites of lepidopterous pests, have been utilized in this manner with reportedly good results in many countries, and various microorganisms likewise have been successfully used.

Advantages over chemical control. Although research and development costs are modest for biological control projects, they are high for new insecticides. Biological control application costs are minimal and nonrecurring, except where periodic colonization is utilized, whereas insecticide costs are high annually. There are no environmental pollution problems connected with biological control, whereas insecticides cause severe problems related to toxicity to man, wildlife, birds, and fish, as well as causing adverse effects in soil and water. Biological control causes no upsets in the natural balance of organisms, but these upsets are common with chemical control; biological control is permanent, chemical control is temporary, usually one to many annual applications being necessary. Additionally, pests more and more frequently are developing resistance or immunity to insecticides, but this is not a problem with pests and their enemies. Both biological and chemical control have restrictions as far as general applicability to the control of all pest insect species is concerned, although the application of biological control remains greatly underdeveloped compared to chemical control. See ENTOMOLOGY, ECONOMIC; INSECTICIDE.   [PAUL DE BACH]

Bibliography: C. P. Clausen, *Entomophagous Insects*, 1962; P. DeBach (ed.), *Biological Control of Insect Pests and Weeds*, 1964; W. W. Kilgore and R. L. Doutt, *Pest Control: Biological, Physical and Selected Chemical Methods*, 1967; R. L. Rudd, *Pesticides and the Living Landscape*, 1964; *Scientific Aspects of Pest Control*, Nat. Acad. Sci.–Nat. Res. Counc. Publ. no. 1402, 1966; E. A. Steinhaus (ed.), *Insect Pathology*, 1963; L. A. Swan, *Beneficial Insects*, 1964.

Insecticide

A material used to kill insects and related animals by disruption of vital processes through chemical action. Chemically, insecticides may be of inorganic or organic origin. The principal source is from chemical manufacturing, although a few are derived from plants. Insecticides are classified according to type of action, as stomach poisons, contact poisons, residual poisons, systemic poisons, fumigants, repellents, or attractants. Many act in more than one way. Stomach poisons are applied to plants so that they will be ingested as insects chew the leaves. Contact poisons are applied directly to insects and are used principally to control species which obtain food by piercing leaf surfaces and withdrawing liquids. Residual insecticides are applied to surfaces so that insects touching them will pick up lethal dosages. Systemic insecticides are applied to plants or animals and are absorbed and translocated to all parts of the organisms, so that insects feeding upon them will obtain lethal doses. Fumigants are applied as gases, or in a form which will vaporize to a gas, to be inhaled by insects. Repellents prevent insects from coming in contact with their hosts. Attractants induce insects to come to specific locations in preference to normal food sources.

In the United States, about 500 species of insects are of primary economic importance, and losses caused by insects range from $4,000,000,000 to $8,000,000,000 annually.

Inorganic insecticides. Prior to 1945, large volumes of lead arsenate, calcium arsenate, paris green (copper acetoarsenite), sodium fluoride, and cryolite (sodium fluoaluminate) were used. The potency of arsenicals is a direct function of the percentage of metallic arsenic contained. Lead arsenate was first used in 1892 and proved effective as a stomach poison against many chewing insects. Calcium arsenate was widely used for the control of cotton pests. Paris green was one of the first stomach poisons and had its greatest utility against the Colorado potato beetle. The amount of available water-soluble arsenic governs the utility of arsenates on growing plants, because this fraction will cause foliage burn. Lead arsenate is safest in this respect, calcium arsenate is intermediate, and paris green is the most harmful. Care must be exercised in the application of these materials to food and feed crops because they are poisonous to man and animals as well as to insects.

Sodium fluoride has been used to control chewing lice on animals and poultry, but its principal application has been for the control of household insects, especially roaches. It cannot be used on plants because of its extreme phytotoxicity. Cryolite has found some utility in the control of the Mexican bean beetle and flea beetles on vegetable crops because of its low water solubility and lack of phytotoxicity.

Organic insecticides. These began to supplant the arsenicals when DDT [2,2-bis(p-chlorophenyl)-1,1,1-trichloroethane] became available in 1945. During World War II, the insecticidal properties of γ-benzenehexachloride (γ-1,2,3,4,5,6-hexachlorocyclohexane of γ-BHC) were discovered in England and France. The two largest-volume insecticides are DDT and γ-BHC. Certain insects cannot be controlled with either, and there are situations and crops where they cannot be used. For these reasons, other chlorinated hydrocarbon insecticides have been discovered and marketed successfully. These include TDE [2,2-bis(p-chlorophenyl)-1,1-dichloroethane], methoxychlor [2,2-bis(p-methoxyphenyl)-1,1,1-trichloroethane], Dilan [mixture of 1,1-bis(p-chlorophenyl)-2-nitropropane and 1,1-bis(p-chlorophenyl)-2-nitrobutane], chlordane (2,3,4,5,6,7,8-8-octachloro-2,3,3a,4,7,7a-hexahydro-4,7-methanoindene), heptachlor (1,4,5,6,7,8,8-heptachloro-3a,4,7,7a-tetrahydro-4,7-methanoindene), aldrin (1,2,3,4,10,10-hexachloro-1,4,4a,5,8,8a-hexahydro-1,4-*endo*, *exo*-5,8-dimethanonaphthalene), dieldrin (1,2,3,4,10,10-hexachloro-6,7-epoxy-1,4,4a,-5,6,7,8,8a-octahydro-1,4-*endo,exo*-5,8-dimethanonaphthalene), endrin (1,2,3,4,10,10-hexachloro-6,7-epoxy-1,4,4a,5,6,7,8,8a-octahydro-1,4-*endo*, *endo*-5,8-dimethanonaphthalene), toxaphene (camphene plus 67–69% chlorine), and endosulfan (6,7,8,9,10,10-hexachloro-1,5,5a,6,9,9a-hexahydro-6,9-methano-2,4,3-benzodioxathiepin-3-oxide). TDE is considerably less toxic than DDT but in general is also less effective. It has given outstanding results in the control of larvae of several moths, however. Methoxychlor is even less toxic than TDE and has found considerable use in the control of houseflies and also of the Mexican bean beetle which is not susceptible to DDT. Restrictions have been placed on its use on dairy cattle for the control of flies, and its consumption is declining. Dilan has found some use in the control of the Mexican bean beetle, as well as of some thrips and aphids. Chlordane was the first cyclopentadiene insecticide to reach commercial status. It has been the most effective chemical available for the control of roaches. In addition to lengthy residual properties, chlordane also possesses fumigant action. Related chemicals include heptachlor, aldrin, dieldrin, and endrin. They are effective against grasshoppers and are especially useful for the control of insects inhabiting soil. Registrations for uses of aldrin and dieldrin were reduced in 1966, following extensive investigations of their metabolism and persistence in animal tissues. Toxaphene is used principally for the control of the cotton boll weevil and other insect pests of cotton. Endosulfan, in addition to showing promise for the control of numerous insects, also shows promise of controlling a number of species of phytophagous (plant-feeding) mites that are not generally susceptible to chlorinated hydrocarbon insecticides.

Insect resistance. The resistance of insects to DDT was first observed in 1947 in the housefly. By the end of 1967, 91 species of insects had been proved to be resistant to DDT, 135 to cyclodienes, 54 to organophosphates, and 20 to other types of insecticides, including the carbamates. Among the chlorinated hydrocarbon insecticides, two types of resistance occur. One applies to DDT and its analogs, such as TDE, methoxychlor, and Dilan, and the other to the cyclodiene compounds, such as chlordane, heptachlor, aldrin, dieldrin, endrin, and also γ-BHC and toxaphene. Insecticides are not mutagenic, indicating that resistance preexists in a small part of the natural population even before exposure.

Nearly every country in the world, except mainland China, has reported the presence of resistant strains of the housefly. In many heavily populated urban areas of the United States, it is difficult to obtain control of roaches with chlordane. During 1957 and 1958, many growers of cotton in the southern states changed from toxaphene and γ-BHC to organic phosphorus chemicals because of the resistance of the cotton boll weevil to chlorinated hydrocarbon insecticides. By 1967, resistant strains of the cotton bollworm and the tobacco budworm had developed and proliferated to the extent that in numerous areas the use of chlorinated hydrocarbon insecticides was of doubtful value. The onion maggot has developed widely spread strains resistant to aldrin, dieldrin, and heptachlor, especially in the northeastern United States and in Ontario, Canada. Three species of corn rootworms also are resistant to these insecticides, principally in the corn-growing regions west of the Mississippi River. The development of chlorinated hydrocarbon resistance among several species of disease-transmitting mosquitoes continues to pose a threat to world health. The control of typhus in the Far East could be at stake because strains of vector lice are resistant to DDT. Evidence for insect resistance to the plant-derived insecticides pyrethrum and rotenone has only recently been established.

Organic phosphorus insecticides. The development of this type of insecticide paralleled that of the chlorinated hydrocarbons. Since 1947, more than 50,000 organic phosphorus compounds have been synthesized in academic and industrial laboratories throughout the world for evaluation as potential insecticides. Parathion [O,O-diethyl O-(p-nitrophenyl) phosphorothioate] and methyl parathion [O,O-dimethyl O-(p-nitrophenyl) phosphorothioate] are estimated to have had a world production of 70,000,000 lb during 1966.

A great diversity of activity is found among organophosphorus insecticides. Many are extremely toxic to man and other warm-blooded animals, but a few show a very low toxicity. The more important include tetraethylpyrophosphate, dicapthon [O,O-dimethyl O-(3-chloro-4-nitrophenyl) phosphorothioate], malathion [O,O-dimethyl S-(1,2-dicarbethoxyethyl) phosphorodithioate], dichloro- [O,O-dimethyl O-(2,2-dichlorovinyl) phosphate], diazinon {O,O-diethyl O-[2-isopropyl-4-methylpyrimidyl (6)] phosphorothioate}, dioxathion [2,3-p-dioxanedithiol S,S-bis (O,O-diethylphosphorodithioate)], azinphosmethyl [O,O-dimethyl S-4-oxo-1,2,3-benzotriazin-3-(4H)-yl-methylphosphorodithioate], carbophenothion [S-(p-chlorophenylthiomethyl) O,O-diethylphosphorodithioate], ethion {bis[S-(diethoxyphosphinothioyl) mercapto] methane}, EPN [O-ethyl O-(p-nitrophenyl) phenylphosphonothionate], trichlorfon [O,O-dimethyl (2,2,2,-trichloro-1-hydroxyethyl) phosphonate], dimethoate [O,O-dimethyl S-(methylcarbamoylmethyl) phosphorodithioate], and fenthion {O,O-dimethyl-O-[4-(methylthio)-m-tolyl] phosphorothioate}.

Schradan [bis(dimethylamino) phosphoric anhydride] was unique among organic insecticides in that it showed systemic properties when applied to plants. By direct contact, it has a relatively low order of activity. When sprayed on plants, it is absorbed from areas receiving treatment, is translocated throughout the entire plant, and is metabolized to yield a product highly toxic to such sucking pests as aphids and phytophagous mites. It is selective in that it affects only aphids ingesting juices from treated plants and does not kill predators which destroy aphids. With schradan, it is possible to protect the growing parts of plants without resorting to frequent spraying, because the insecticide is translocated to these growing parts, whereas with most stomach poison or residual insecticides plants outgrow the protection. Several chemicals showing systemic properties have reached commercial or near commercial status. These include demeton [O,O-diethyl O (and S)-(2-ethylthio) ethylphosphorothioates], disulfoton [O,O-diethyl S-(2-ethylthio) ethylphosphorodithioate], phorate [O,O-diethyl S-(ethylthiomethyl)-phosphorodithioate], and mevinphos [2-methoxycarbonyl-1-methyl vinyl dimethyl phosphate].

The use of organic phosphorus chemicals for the systemic control of animal parasites is another facet of interest. During 1958, semicommercial application began of coumaphos [O-(3-chloro-4-methyl-2-oxo-2-H-1-benzopyran-7-yl) O,O-diethylphosphorothioate] and ronnel [O,O-dimethyl O-(2,4,5-trichlorophenyl) phosphorothioate] for the control of grubs in cattle. Coumaphos is applied externally as a spray and is absorbed and translocated to kill the cattle grubs. Ronnel is most effective when administered internally. Dimethoate, fenthion, famphur [O-p-(dimethyl-sulfam-oyl)phenyl O,O-dimethylphosphorothioate], and menazon [S-(4,6-diamino-s-triazin-2-ylmethyl) O,O-dimethylphosphorodithioate] also show activity for this type of application.

Activity of organic phosphate insecticides results from the inhibition of the enzyme cholinesterase, which performs a vital function in the transmission of impulses in the nervous system. Inhibition of some phenyl esterases occurs also. Inhibition results from direct coupling of phosphate with the enzyme. Phosphorothionates are moderately active but become exceedingly potent upon oxidation to phosphates.

Other types of insecticides. Synthetic carbamate insecticides are attracting increased interest. These include dimetan (5,5-dimethyldihydro-resorcinol dimethylcarbamate), Pyrolan (3-methyl-1-phenyl-5-pyrazolyl dimethylcarbamate), Isolan (1-isopropyl-3-methyl-5-pyrazolyl dimethylcarbamate), pyramat [2-n-propyl-4-methylpyrimidyl-(6)-dimethylcarbamate], carbaryl (1-naphthyl-N-methylcarbamate), Bagon, (o-isopropoxyphenyl methylcarbamate), Zectran (4-dimethylamino-3,5-xylyl methylcarbamate), TRANID [5-chloro-6-oxo-2-norbornanecarbonitrile O-(methylcarbamoyl)-oxime], dimetilan [1-(dimethylcarbamoyl)-5-methyl-3-pyrazolyl dimethylcarbamate], Furadan (2,3-dihydro-2,2-dimethyl-7-benzofuranyl methylcarbamate), and Temik [2-methyl-2-(methylthio) propionaldehyde O-(methylcarbamoyl) oxime]. They are also cholinergic.

Insecticides obtained from plants include nicotine [L-1-methyl-2-(3′-pyridyl)-pyrrolidine], rotenone, the pyrethrins, sabadilla, and ryanodine, some of which are the oldest-known insecticides. Nicotine was used as a crude extract of tobacco as early as 1763. The alkaloid is obtained from the leaves and stems of *Nicotiana tabacum* and *N. rustica*. It has been used as a contact insecticide, fumigant, and stomach poison and is especially effective against aphids and other soft-bodied insects.

Rotenone is the most active of six related alkaloids found in a number of plants, including *Derris elliptica*, *D. malaccensis*, *Lonchocarpus utilis*, and *L. urucu*. *Derris* is a native of East Asia, and *Lonchocarpus* occurs in South America. The highest concentrations are found in the roots. Rotenone is active against a number of plant-feeding pests and has found its greatest utility where toxic residues are to be avoided. Rotenone is known also as derris or cubé.

The principal sources of pyrethrum are *Chrysanthemum cinerariaefolium* and *C. coccineum*. Pyrethrins, which are purified extracts prepared from flower petals, contain four chemically different active ingredients. Allethrin is a synthetic pyrethroid. The pyrethrins find their greatest use in fly sprays, household insecticides, and grain protectants because they are the safest insecticidal materials available.

Synergists. These materials have little or no insecticidal activity but increase the activity of chemicals with which they are mixed, especially that of the pyrethrins. Piperonyl butoxide {α-[2-(2-butoxyethoxy)-ethoxy]-4,5-methylenedioxy-2-propyltoluene}, sulfoxide {1,2,methylenedioxy-4-[2-octyl-(sulfinyl),propyl] benzene}, and N-(2-ethyl-hexyl)-5-norbornene-2,3-dicarboximide are commercially available. Sesamex [acetaldehyde 2-(2-ethoxyethoxy)ethyl 3,4-methylenedioxyphenyl acetal] and Tropital {piperonal bis[2-(2-butoxy-ethoxy)-ethyl]- acetal} are active but not fully developed synergists. These synergists have their greatest utility in mixtures with the pyrethrins. Some have been shown to enhance the activity of carbamate insecticides as well.

Formulation and application. Formulation of insecticides is extremely important in obtaining satisfactory control. Common formulations include dusts, water suspensions, emulsions, and solutions. Accessory agents, including dust carriers, solvents, emulsifiers, wetting and dispersing agents, stickers, deodorants or masking agents, synergists, and antioxidants, may be required to obtain a satisfactory product. Insecticidal dusts are formulated for application as powders. Toxicant concentration is usually quite low. Water suspensions are usually prepared from wettable powders, which are formulated in a manner similar to dusts except that the insecticide is incorporated at a high concentration and wetting and dispersing agents are included. Emulsifiable concentrates are usually prepared by solution of the chemical in a satisfactory solvent to which an emulsifier is added. They are diluted with water prior to application. Granular formulations are an effective means of applying insecticides to the soil to control insects which feed on the subterranean parts of plants. Proper timing of insecticide applications is important in obtaining satisfactory control. Dusts are more easily and rapidly applied than are sprays. However, results may be more erratic and much greater attention must be paid to weather conditions than is required for sprays. Coverage of plants and insects is generally less satisfactory with dusts than with sprays. It is best to make dust applications early in the day, while the plants are covered with dew, so that greater amounts of dust will adhere. If prevailing winds are too strong, a considerable proportion of dust will be lost. Spray operations will usually require the use of heavier equipment, however. Application of insecticides should be properly correlated with the occurrence of the most susceptible stage in the life cycle of the pest involved.

During the past decade, attention has focused sharply on the impact of the highly active synthetic insecticides upon the total environment—man, domestic and wild animals and fowl, soil-inhabiting microflora and microfauna, and all forms of aquatic life. Effects of these materials upon populations of beneficial insects, particularly parasites and predators of the economic species, are being critically assessed. The study of insect control by biological means has expanded. The concepts and practices of integrated pest control and pest management are expanding rapidly. Among problems associated with insect control which must receive major emphasis during the coming years are the development of strains of insects resistant to insecticides; the assessment of the significance of small, widely distributed insecticide residues in and upon the environment; the development of better and more reliable methods for forecasting insect outbreaks; and the evolvement of control programs integrating all methods—physical, physiological, chemical, biological, and cultural—for which practicality may have been demonstrated.

See ENTOMOLOGY, ECONOMIC; INSECT CONTROL, BIOLOGICAL.

[GEORGE F. LUDVIK]

Bibliography: A. B. Borkovec, *Insect Chemosterilants*, 1966; A. W. A. Brown, *Insect Control by Chemicals*, 1951; D. E. H. Frear, *Pesticide Handbook: Entoma*, 1966; D. E. H. Frear, *Pesticide Index*, 1963; M. Jacobson, *Insect Sex Attractants*, 1965; E. E. Kenega, Commercial and experimental organic insecticides, *Bull. Entomol. Soc. Amer.*, 12(2):161–217, 1966; W. W. Kilgore and R. L. Doutt, *Pest Control: Biological, Physical and Selected Chemical Methods*, 1967; H. Martin, *Insecticide and Fungicide Handbook*, 1965; C. L. Metcalf, W. P. Flint, and R. L. Metcalf, *Destructive and Useful Insects*, 1962; R. L. Metcalf, *Organic Insecticides*, 1955; R. D. O'Brien, *Insecticides: Action and Metabolism*, 1967; U.S. Department of Agriculture, *Agr. Handb. no. 331*, Agricultural Research Service and Forest Service, 1967.

Island faunas and floras

Islands generally have fewer species of animals and plants than do comparable continental areas, at least if the islands lie some distance from continents. Often there are rather few genera, but some genera have many local species that sometimes exhibit strong adaptive radiation. The proportion of endemic species that are present generally increases with the degree of isolation and with the complexity of the habitat. Gigantism, flightlessness, and other unusual characteristics are relatively frequent.

Island floras and faunas are of far greater interest than either the number of species or their economic importance might seem to justify. How the plants and animals came to be on the islands, why so many of them are found nowhere else, why so many have special or strange forms, and where their ancestors came from, are fascinating questions which present problems of great scientific importance to the biologist. The theory of organic evolution emerged from studies of island faunas by Charles Darwin and Alfred Russell Wallace, and islands are still among the most advantageous places to study evolutionary processes. They also offer uniquely suitable sites for the investigation of the nature and functioning of ecosystems. *See* ECOLOGY.

Nature and classification of islands. Islands are themselves very diverse, and their biotas (floras and faunas) are correspondingly varied. Islands are commonly separated into high and low islands; into volcanic, limestone, granitic, metamorphic, and mixed islands; and, perhaps most important from a biogeographic standpoint, into continental and oceanic islands. Naturally, most of these types are not sharply separated. On the contrary, as with wet and dry islands, there is a continuous series of intermediates between any pair or set of extremes. Usually significantly different biotas are associated with these categories. Low islands have small, impoverished biotas; high islands, richer ones. Oceanic islands are those considered never to have been part of, or connected with, any continental land mass. Their biotas are commonly poor in genera and unbalanced or disharmonic, that is, with the families—or larger groups—very unevenly represented, compared with those of continents or continental islands. The continental islands are believed to be the remnants of former continental land masses or at least to have had land connections with continents.

Oceanic islands or groups with outstanding or much-studied biotas are Rapa, the Hawaiian Islands, the Galapagos, Juan Fernandez, the Fiji Islands, the Marianas, the Carolines, the Marshalls, the Society Islands, the Azores, the Canary Islands, Tristan d'Acunha, Kerguelen, the Seychelles, and Aldabra.

Continental islands or groups are Madagascar, Japan, Formosa, the Philippines, New Guinea, the Malay Archipelago, Ceylon, New Zealand, the Antilles, the California Islands, and the British Isles.

Endemism. On almost all islands, except those immediately contiguous to other land and most low coral islands, are species and varieties that are endemic to the particular island or group, that is, found nowhere else. Even coastal islands, such as those off southern California, have a few endemics. On high islands the percentage of the total biota that is endemic usually increases with the degree of isolation from larger land masses. Thus, the endemics form over 95% of the indigenous vascular flora in the Hawaiian Islands, which are very remote from other land. On all islands there are some widespread species. They are usually seacoast species, aquatics, or marsh dwellers; and, of course, there are many introduced weeds.

The first basic type of endemics includes species that have differentiated on the islands where they are now found, usually the products of relatively recent evolution. In certain genera there may be several to many species of a given genus adapted to particular habitats—a phenomenon called adaptive radiation. Examples of this are the famous Darwin finches (Geospizidae) of the Galapagos; *Hedyotis* (Rubiaceae) in the Hawaiian Islands; and many insect genera with species adapted to different plant hosts. Other genera have differentiated into several or many species which occupy very similar habitats, apparently an evolutionary result of geographic isolation only, for example, *Cyrtandra* (Gesneriaceae) in the Hawaiian group.

The other basic type of endemics includes relicts, isolated remnants of populations that were much more widespread in the past. A good example is *Lyonothamnus floribundus*. Though it is now confined to the California Islands, fossil evidence shows that it was widespread in California during the Tertiary Period. Islands provide refuges for these species where they may not be exposed to the competition or other unfavorable circumstances that eliminated them elsewhere. For example, islands may provide a more equable climate when severe conditions develop in continental areas. Disjunct species, those found on widely separated islands, may likewise be relicts, or they may be the result of successful long-distance transport. A case in point is possibly *Charpentiera obovata* (Amaranthaceae), a shrub found only in the Hawaiian and Austral islands, separated by almost 3000 mi.

Geography of floras and faunas. The geographical distribution of insular biotas shows very definite patterns, making it possible to group species into "elements" with similar distributions and, in some cases, apparently common geographic origins. Relationships of island genera and species may often be guessed at on the basis of their taxonomic relationships with groups in other areas. Thus a preponderance of the Hawaiian vascular flora and terrestrial invertebrate fauna have their affinities to the southwest in the Indo-Malaysian region. Another element has its connections, and likely its derivation, in the Australia-New Zealand area; still another, but smaller, element finds its relationships in America; another in eastern Polynesia. The numbers of species in the predominant Indo-Pacific element are very high in the western islands of the Pacific, but fall off very markedly from island to island, eastward, except where there has been strong local evolution of species, as in Hawaii. *See* BIOGEOGRAPHY.

Dispersal. One of the central problems in the study of island biotas is that of dispersal. How did the species, or their ancestors, reach the islands where they are now found? This problem is especially fascinating on oceanic islands. To solve it, those with little faith in the effectiveness of transoceanic dispersal would assume that very few islands are truly oceanic, that at different times in the past land bridges have existed in all directions, or that ancient continents foundered or drifted away, leaving the islands and their biotas as scattered remnants. A more plausible theory is that given time enough, rare events, such as accidental transport of individuals or propagules (seeds, eggs, or spores) by wind, water, or birds, will account for the ancestors of all present indigenous biotas of oceanic islands. It is also reasonable to assume that there were many former islands, now marked only by shoals and underwater mountains, that could have served as stepping stones along which some species may have traveled to their present isolated homes. *See* POPULATION DISPERSAL.

Morphological peculiarities. A notable and little-understood feature of island biotas is the frequency of unusual morphologic features—flightlessness in birds and insects, gigantism and woody habit in usually small, herbaceous groups of plants, as well as gigantism in animals, for example, the giant tortoises of the Galapagos and Aldabra and the moas of New Zealand, and the "rosette-tree" habit in many island plant genera. Some of these features are also found in the biotas of tropical mountains. Another feature of the plants of oceanic islands is that, having evolved in the absence of large herbivores, they have weak defenses against grazing and trampling, if any. Spines and thorns are infrequent on indigenous species.

Evolution on islands. The origin of these morphological peculiarities, as well as of the more ordinary diversity found in island biotas, poses evolutionary problems that are extremely interesting and that seem amenable to solution. Islands not only provided early information on which the theory of organic evolution was founded; much study has since been devoted to insular evolution to test various concepts.

In addition to generic and specific diversity, island organisms frequently show an extraordinary degree of polymorphism within species. For example, the tree *Metrosideros collina* of Polynesia, especially Hawaii, shows a complexity of forms

bewildering for taxonomic arrangement.

At the other extreme are species with almost no genetic plasticity, usually existing in small populations. The extinct *Clermontia haleakalae* and *Hedyotis cookiana* of Hawaii were presumably of this nature, as are the Hawaiian hawk, crow, and goose.

The adaptive radiation and geographic speciation mentioned above are important evolutionary patterns notable on islands. Isolation, as well as the existence of many unoccupied niches and reduced competition during the early stages in evolution of insular biotas, is probably responsible for the evolutionary persistence of certain features that would be speedily eliminated in a more complex continental biotic situation.

The evolution of major ecosystems themselves may also be elucidated by careful and long-continued study of islands, as island ecosystems are simpler and better defined than those of larger land masses, and the development of their biotas can be better correlated with that of their physical features. Also, a great range of size and complexity is exhibited. *See* ECOSYSTEM.

Biotas of volcanic islands. New volcanic substrata are bare of all plant and animal life. But from the time the lava or ash surface is cool, it is subject to colonization by plants and animals. The biotas of volcanic islands are the products of a sequence: sporadic colonization; repeated partial destruction by new eruptions; isolation of local populations, with evolution of local races; breaking down of such isolation by mixing of populations, development of genetic diversity, and renewed isolation by dissection of volcanic domes which leads to further evolution of local forms, species, and even genera with time; new arrivals may occupy any open habitats available or may displace current tenants. New open habitats become available as long as volcanic activity continues and erosion makes still other habitats.

As volcanic islands grow older some subside, or get partly flooded by changing sea level. Parts or all may be eroded down to sea level by wave action. In the tropics coral reefs grow up around their shores and provide habitats for still other species, those characteristic of limestone shores and strands, mostly species of very wide distribution. As the high volcanic mountains gradually subside and wear away, the species dependent on high elevations, orographic rainfall, or rain shadows gradually disappear. New ones evolve, or colonize from elsewhere, that are adapted to the old worn-down topography and deeply weathered volcanic substrata. Even most of these finally disappear as subsidence gradually changes the island to an atoll. *See* SUCCESSION, ECOLOGICAL.

Biotas of coral atolls and limestone islands. Scattered through most tropical seas are flat, often ring-shaped, islands made up of the limestone skeletons of marine plants and animals and resulting from the processes described above. Most rise just above sea level. In this relatively uniform environment, varying principally in rainfall, biotas are very impoverished and are largely made up of strand species. Numbers of species are lowest on the drier atolls. As an example of the variation, the native vascular flora of the Marshall Islands ranges from 9 species per atoll in the driest northern ones, to 75 or 100 in the very wet southern atolls. Endemics are very few. With even a slight elevation, as on Aldabra Atoll, the number of species increases very sharply. Strand species are still prominent, but species requiring moderate habitats increase in numbers. There are some endemics. High limestone islands have rich, strongly endemic biotas.

Interesting and important as insular floras and faunas are, they are disappearing with distressing rapidity. The introduction of large herbivorous animals, rats, and aggressive exotic plants capable of quickly invading disturbed situations, as well as the complete destruction of whole habitats by human activities, has resulted in the reduction or total disappearance of many species. The growth of human populations and development, disturbance, and destruction even in remote islands are accelerating.

More and more species are becoming rare and threatened with extinction as man achieves greater capacity to change his environment. Many fascinating features of island floras and faunas will not long be available for study and enjoyment unless adequate measures for protecting substantial areas of natural habitats on islands are taken at once. Conservation is an immediate necessity if the study of island biotas is to have a future. *See* PLANT GEOGRAPHY.

[F. R. FOSBERG]

Bibliography: R. I. Bowman, *The Galapagos*, 1966; S. Carlquist, *Island Life*, 1965; F. R. Fosberg, *Man's Place in the Island Ecosystem*, 1963; J. L. Gressitt, *Insects of Micronesia: Introduction*, 1954; J. L. Gressitt (ed.), *Pacific Basin Biogeography*, 1963; D. Lack, *Darwin's Finches*, 1947; R. N. Philbrick, *Biology of the California Islands*, 1967; M. H. Sachet and F. R. Fosberg, *Island Bibliographies*, 1955; E. C. Zimmerman, *Insects of Hawaii*, 1948.

Jet stream

A relatively narrow, fast-moving wind current flanked by more slowly moving currents. Jet streams are observed principally in the zone of prevailing westerlies above the lower troposphere and in most cases reach maximum intensity, with regard both to speed and to concentration, near the tropopause. At a given time, the position and intensity of the jet stream may significantly influence aircraft operations because of the great speed of the wind at the jet core and the rapid spatial variation of wind speed in its vicinity. Lying in the zone of maximum temperature contrast between cold air masses to the north and warm air masses to the south, the position of the jet stream on a given day usually coincides in part with the regions of greatest storminess in the lower troposphere, though portions of the jet stream occur over regions which are entirely devoid of cloud.

Characteristics. The specific characteristics of the jet stream depend upon whether the reference is to a single instantaneous flow pattern or to an averaged circulation pattern, such as one averaged with respect to time, or averaged with respect both to time and to longitude.

If the winter circulation pattern on the Northern Hemisphere is averaged with respect to both time and longitude, a westerly jet stream is found at an elevation of about 13 km near latitude (lat) 25°. The speed of the averaged wind at the jet core is about 148 km/hr (80 knots). In summer this jet is displaced poleward to a position near lat 42°. It is found at an elevation of about 12 km with a maximum speed of about 56 km/hr (30 knots). In both seasons a speed equal to one-half the peak value is found approximately 15° of latitude south, 20° of latitude north, and 5–10 km above and below the location of the jet core itself.

If the winter circulation is averaged only with respect to time, it is found that both the intensity and the latitude of the westerly jet stream vary from one sector of the Northern Hemisphere to another. The most intense portion, with a maximum speed of about 185 km/hr (100 knots), lies over the extreme western portion of the North Pacific Ocean at about lat 22°. Lesser maxima of about 157 km/hr (85 knots) are found at lat 35° over the east coast of North America, and at lat 21° over the eastern Sahara and over the Arabian Sea. In summer, maxima are found at lat 46° over the Great Lakes region, at lat 40° over the western Mediterranean Sea, and at lat 35° over the central North Pacific Ocean. Peak speeds in these regions range between 74 and 83 km/hr (40–45 knots). The degree of concentration of these jet streams, as measured by the distance from the core to the position at which the speed is one-half the core speed, is only slightly greater than the degree of concentration of the jet stream averaged with respect to time and longitude. At both seasons and at all longitudes the elevation of these jet streams varies between 11 and 14 km.

Variations. On individual days there is a considerable latitudinal variability of the jet stream, particularly in the western North American and western European sectors. It is principally for this reason that the time-averaged jet stream is not well defined in these regions. There is also a great day-to-day variability in the intensity of the jet stream throughout the hemisphere. On a given winter day, speeds in the jet core may exceed 370 km/hr (200 knots) for a distance of several hundred miles along the direction of the wind. Lateral wind shears in the direction normal to the jet stream frequently attain values as high as 100 knots/300 nautical miles (185 km/hr/556 km) to the right of the direction of the jet stream current and as high as 100 knots/100 nautical miles (185 km/hr/185 km) to the left. Vertical shears below and above the jet core are often as large as 20 knots/1000 ft (37 km/305 m). Daily jet streams are predominantly westerly, but northerly, southerly, and even easterly jet streams may occur in middle or high latitudes when ridges and troughs in the normal westerly current are particularly pronounced or when unusually intense cyclones and anticyclones occur at upper levels.

Insufficiency of data on the Southern Hemisphere precludes a detailed description of the jet stream, but it appears that the major characteristics resemble quite closely those of the jet stream on the Northern Hemisphere. The day-to-day variability of the jet stream, however, appears to be less on the Southern Hemisphere.

It appears that an intense jet stream occurs at high latitudes on both hemispheres in the winter stratosphere at elevations above 20 km. The data available, however, are insufficient to permit the precise location or detailed description of this phenomenon. *See* AIR MASS; ATMOSPHERE.

[FREDERICK SANDERS]

Lake

An inland body of water, small to moderately large in size, with its surface exposed to the atmosphere. Most lakes fill depressions below the zone of saturation in the surrounding soil and rock materials. Generically speaking, all bodies of water of this type are lakes, although small lakes usually are called ponds, tarns (in mountains), and less frequently pools or meres. The great majority of lakes have a surface area of less than 100 square miles (mi²). More than 30 well-known lakes, however, exceed 1500 mi² in extent, and the largest freshwater body, Lake Superior, North America, covers 31,180 mi² (see table).

Most lakes are relatively shallow features of the Earth's surface. Even some of the largest lakes have maximum depths of less than 100 ft (Winnipeg, Canada; Balkash, Soviet Union; Albert, Uganda). A few, however, have maximum depths which approach those of some seas. Lake Baikal in the Soviet Union is about a mile deep at its deepest point, and Lake Tanganyika, Africa, is about 0.9 mi.

Because of their shallowness, lakes in general may be considered evanescent features of the Earth's surface, with a relatively short life in geological time. Every lake basin forms a bed onto which the sediment carried by inflowing streams is deposited. As the sediment accumulates, the storage capacity of the basin is reduced, vegetation encroaches upon the shallow margins, and eventually the lake may disappear. Most lakes also have surface outlets. Except at elevations very near sea level, a stream which flows from such an outlet gradually cuts through the barrier forming the lake basin. As the level of the outlet is lowered, the capacity of the basin is also reduced and the disappearance of the lake assured.

Dimensions of some major lakes

Lake	Area, mi²	Volume (approx), 1000 acre-ft	Shore-line, mi	Depth Av, ft	Depth Max, ft
Caspian Sea	169,300	71,300	3,730	675	3,080
Superior	31,180	9,700	1,860	475	1,000
Victoria	26,200	2,180	2,130		
Aral Sea	26,233*	775			
Huron	23,010	3,720	1,680		
Michigan	22,400	4,660			870
Baikal	13,300*	18,700		2,300	5,000
Tanganyika	12,700	8,100			4,700
Great Bear	11,490*		1,300		
Great Slave	11,170*		1,365		
Nyasa	11,000	6,800		900	2,310
Erie	9,940	436			
Winnipeg	9,390*		1,180		
Ontario	7,540	1,390			
Balkash	7,115				
Ladoga	7,000	745			
Chad	6,500*				
Maracaibo	4,000*				
Eyre	3,700*				
Onega	3,764	264			
Rudolf	3,475*				
Nicaragua	3,089	87			
Athabaska	3,085				
Titicaca	3,200	575			
Reindeer	2,445				

*Area fluctuates.

Variations in water character. Lakes differ as to the salt content of the water and as to whether they are intermittent or permanent. Most lakes are composed of fresh water, but some are more salty than the oceans. Generally speaking, a number of water bodies which are called seas are actually salt lakes; examples are the Dead, Caspian, and Aral seas. All salt lakes are found under desert or semiarid climates, where the rate of evaporation is high enough to prevent an outflow and therefore a discharge of salts into the sea. Many lesser arid-region lakes are intermittent, sometimes existing only for a short period after heavy rains and disappearing under intense evaporation. These lakes are called playas in North America, shotts in North Africa, and other names elsewhere. In such regions the surface area and volume of permanent lakes may differ enormously from wet to dry season.

The water of the more permanent salt lakes differs greatly in the degree of salinity and the type of salts dissolved. Compared to typical ocean water (approximately 35 parts per thousand, °/oo) some salt lakes are very salty. Great Salt Lake water has a dissolved solids content about four times that of sea water (150°/oo) and the Dead Sea about seven times (246°/oo). Some of the larger salt lakes have a much lower dissolved solids content, as the Aral Sea (11°/oo) and the Caspian Sea (6°/oo). The composition of salts depends in part on the geological character of the drainage area discharging into the lake, in part on the age of the lake, and in part on the excess of evaporation over inflow. As saturation is approached, the salts common in surface waters are precipitated in such an order that magnesium chloride and calcium chloride remain in solution after other salts have precipitated.

Lakes with fresh waters also differ greatly in the composition of their waters. Because of the balance between inflow and outflow, fresh lake water composition tends to assume the composite dissolved solids characteristics of the waters of the inflowing streams — with the lake's age having very little influence. Lakes with a sluggish inflow, particularly where inflowing waters have much contact with marginal vegetation, tend to have waters with high organic content. This may be observed in small lakes or ponds in a region where drainage moves through a topography of glacial moraines. Lakes formed within drainage areas having a crystalline, metamorphic, or volcanic country rock tend to have low dissolved solids content. Thus Lake Superior, with its major drainage from the Laurentian Shield, has a dissolved solids content of 0.05°/oo. The water of Grimsel Lake in the high Alps, Switzerland, has a dissolved solids content of only 0.0085°/oo. Lakes within limestone or dolomitic drainage areas have a pronounced calcium carbonate and magnesium carbonate content. As in all surface water, dissolved gases, notably oxygen, also are present in lake waters. Under a few special situations, as crater lakes in volcanic areas, sulfur or other gases may be present in lake water, influencing color, taste, and chemical reaction of the water. *See* FRESH-WATER ECOSYSTEM; HYDROSPHERE, GEOCHEMISTRY OF; LAKE, MEROMICTIC; SURFACE WATER.

Basin and regional factors. Most lakes are natural, and a large proportion of them lie in depressions of glacial origin. Thus alpine locations and regions with ground moraine or glacially eroded exposures are the sites of many of the world's lakes. The lakes of Switzerland, Minnesota, and Finland are illustrations of these types.

Lakes may be formed in depressions of differing glacial origin: (1) terrain eroded by continental glaciers, with the surface differentially deepened by ice abrasion of rocks of varying hardness and resistance; (2) valleys eroded differentially by valley glaciers; (3) cirques (glacially eroded valley heads in mountains); (4) lateral moraine barriers; (5) frontal moraine barriers; (6) valley glacier barriers; (7) irregularities in the deposition of glacial drift or ground moraine.

Lakes are particularly important surface features in the peneplaned ancient rocks of the Laurentian Shield and on the Fennoscandian Shield of northern Europe. The lake region of northern North America, which centers on the Laurentian Shield, probably has one-fifth to one-quarter of its surface in lake. Many streams in these areas are interrupted over more than half their total length by lakes. Lakes on the shields as a rule are island-studded and have extremely irregular outlines. The most permanent lakes lie on the shields themselves. Many such lakes on recently ice-scoured shields have fresh hard-rock rims with high resistance to erosion at their outlets. Because of generally low stream gradients and little sediment carried, the abrading and depositing stream actions are slow. As a result, the life of all but the small lakes in these areas probably will be measured in terms of a whole geologic period.

Some lakes in glacially formed depressions, as well as in other basins, may be considered barrier lakes. These glacial lakes are formed on glacial drift behind lateral or frontal moraine barriers. In addition, depressions of sufficient depth to contain a lake may be formed by (1) sediment deposited by streams (alluvium), and also stream-borne vegetative debris, such as tree dams on the distributaries and braided river courses of the lower Mississippi Valley; (2) landslides in mountainous areas; (3) sand dunes; (4) storm beaches and current-borne sediments along the shores of large bodies of water; (5) lava flows; (6) artificial or man-made barriers.

A large percentage of lakes is found in either the glacially formed or barrier-formed depressions. However, a few other types of depressions contain lakes: (1) craters of inactive volcanoes, or calderas (Crater Lake in Oregon is a famous example); (2) depressions of tectonic or structural origin (Great Rift Valley of Africa includes Lakes Albert, Tanganyika, and Nyasa); (3) solution cavities in limestone country rock; (4) shallow depressions cause a dotting of lakes in many parts of the tundra of high latitudes.

Conservation and economic aspects. Lakes created behind manmade barriers are becoming common features and serve multiple purposes. Examples are Lake Mead behind Hoover Dam on the Colorado River, Lake Roosevelt behind Grand Coulee Dam on the Columbia, Kentucky Lake and other lakes of the Tennessee Valley, and Lake Tsimlyanskaya on the Don.

Both natural and manmade lakes are economically significant for their storage of water, regula-

tion of stream flow, adaptability to navigation, and recreational attractiveness. A few salt lakes are significant sources of minerals. Recreational utility, long important in the alpine region of Europe and in Japan, is now a major economic attribute of many American lakes. Economic value is generally increased by location near substantial human settlement. Most of the world's lakes, however, are located in regions where they have only minor economic significance at present. *See* EUTROPHICATION, CULTURAL.   [EDWARD A. ACKERMAN]

Bibliography: R. Gresswell (ed.), *Standard Encyclopedia of the World's Rivers and Lakes*, 1966; International Association for Great Lakes Research, *Proceedings of the 10th Conference on Great Lakes Research*, 1967; P. H. Kuenen, *Realms of Water*, 1955; R. H. Lowe-McConnell (ed.), *Man Made Lakes*, 1965.

Lake, meromictic

A lake whose water is permanently stratified and therefore does not circulate completely throughout the basin at any time during the year. In the temperate zone this permanent stratification occurs because of a vertical, chemically produced density gradient. There are no periods of overturn, or complete mixing, since seasonal fluctuations in the thermal gradient are overridden by the stability of the chemical gradient.

The upper stratum of water in a meromictic lake is mixed by the wind and is called the mixolimnion. The bottom, denser stratum, which does not mix with the water above, is referred to as the monimolimnion. The transition layer between these chemically stratified strata is called the chemocline (Fig. 1).

In general, meromictic lakes in North America are restricted to (1) sheltered basins that are proportionally very small in relation to depth and that often contain colored water, (2) basins in arid regions, and (3) isolated basins in fiords. *See* LAKE.

Prior to 1960 only 11 meromictic lakes had been reported for North America. However, during 1960–1968 that number was more than doubled. There are seven known meromictic lakes in the state of Washington, six in Wisconsin, four in New York, three in Alaska, two in Michigan, and one each in Florida, Nevada, British Columbia, Labrador, and on Baffin Island. Research activity on meromictic lakes has been focused on studies of biogeochemistry, deepwater circulation, and heat flow through the bottom sediments.

Chemical studies. Very few detailed chemical studies have been done on meromictic lakes, particularly on a seasonal basis. The most typical

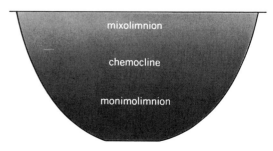

Fig. 1. Cross-sectional diagram of a meromictic lake.

chemical characteristic of meromictic lakes is the absence of dissolved oxygen in the monimolimnion. Large quantities of hydrogen sulfide and ammonia may be associated with this anaerobic condition in deep water. J. Kjensmo proposed that the accumulation of ferrous bicarbonate in the deepest layers of some lakes may have initiated meromixis under certain conditions.

Meromictic lakes are exciting model systems for many important biogeochemical studies. The isolation of the monimolimnetic water makes these studies quite interesting and important. E. S. Deevey, N. Nakai, and M. Stuiver studied the biological fractionation of sulfur and carbon isotopes in the monimolimnion of Fayetteville Green Lake, N.Y. They found the fractionation factor for sulfur to be the highest ever observed. T. Takahashi, W. S. Broecher, Y. Li, and D. L. Thurber have undertaken detailed, comprehensive geochemical and hydrological studies of meromictic lakes in New York.

Sediments from meromictic lakes are among the best for studies of lake history, since there is little decomposition of biogenic materials. G. J. Brunskill and S. D. Ludlam have obtained information on seasonal carbonate chemistry, sedimentation rates, and sediment composition and structure in Fayetteville Green Lake. Also, deuterium has been used by several workers in an attempt to unravel the history of meromictic lake water.

Physical studies. Radioactive tracers have been used by G. E. Likens and A. D. Hasler to show that the monimolimnetic water in a small meromictic lake (Stewart's Dark Lake) in Wisconsin is not stagnant but undergoes significant horizontal movement (Fig. 2). While the maximum radial spread was about 16–18 m/day, vertical movements were restricted to negligible amounts because of the strong density gradient. W. T. Edmondson reported a similar pattern of movement in the deep water of a larger meromictic lake (Soap Lake) in Washington. V. W. Driggers determined from these studies that the average horizontal eddy diffusion coefficient is 3.2 cm²/sec in the monimolimnion of Soap Lake and 17 cm²/sec in Stewart's Dark Lake.

Because vertical mixing is restricted in a meromictic lake, heat from solar radiation may be trapped in the monimolimnion and can thereby produce an anomalous temperature profile. G. C. Anderson reported monimolimnitic water temperatures as high as 50.5°C at a depth of 2 m in shallow Hot Lake, Wash.

N. M. Johnson and Likens pointed out that some meromictic lakes are very convenient for studies of geothermal heat flow because deepwater temperatures may be nearly constant. Studies of terrestrial heat flow have been made in the sediments of Stewart's Dark Lake, Wis., by Johnson and Likens and in Fayetteville Green Lake by W. H. Diment. Steady-state thermal conditions were found in the sediments of Stewart's Dark Lake, and the total heat flow was calculated to be 2.1×10^{-6} cal/cm² sec. However, about one-half of this flux was attributed to the temperature contrasts between the rim and the central portion of the lake's basin. R. McGaw found that the thermal conductivity of the surface sediments in the center of Stewart's Dark Lake was 1.10×10^{-3} cal cm/cm²

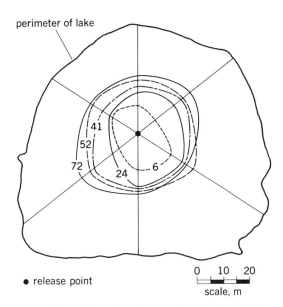

perimeter of lake

• release point

0 10 20
scale, m

Fig. 2. The outlines show the maximum horizontal displacement of a radiotracer (sodium-24) following its release at the 8-m depth in Stewart's Dark Lake, Wis. The numbers indicate the hours elapsed after release. (*After A. D. Hasler and G. E. Likens*)

sec °C, a value substantially lower than that for pure water at the same temperature but consistent with measurements on colloidal gels.

Biological studies. Relatively few kinds of organisms can survive in the rigorous, chemically reduced environment of the monimolimnion. However, anaerobic bacteria and larvae of the phantom midge (*Chaoborus* sp.) are common members of this specialized community. Using an echo sounder, K. Malueg has been able to observe vertical migrations of *Chaoborus* sp. larvae in meromictic lakes. The sound waves are reflected by small gas bladders on the dorsal surface of the larvae. The migration pattern is similar to that shown by the deep-scattering layers in the sea. The larvae come into the surface waters at night when the light intensity is low and sink into deeper waters during the daylight hours. Hasler and Likens found that biologically significant quantities of a radioisotope (iodine-131) could be transported from the deep and relatively inaccessible part of a meromictic lake to the surface and thence to the adjacent terrestrial environment by these organisms. The radiotracer appeared in flying adult *Chaoborus* sp. along the shoreline of the lake within 20 days after it had been released within the region of the chemocline. *See* FRESH-WATER ECOSYSTEM; LIMNOLOGY.

[GENE E. LIKENS]

Bibliography: E. S. Deevey, N. Nakai, and M. Stuiver, Fractionation of sulfur and carbon isotopes in a meromictic lake, *Science,* 139:407–408, 1963; W. T. Edmonson and G. C. Anderson, Some features of saline lakes in central Washington, *Limnol. Oceanogr.,* suppl. to vol. 10, pp. R87–R96, 1965; D. G. Frey (ed.), *Limnology in North America,* 1963; G. E. Hutchinson, *A Treatise on Limnology,* vol. 1, 1957; D. Jackson (ed.), *Some Aspects of Meromixis,* 1967; N. M. Johnson and G. E. Likens, Steady-state thermal gradient in the sediments of a meromictic lake, *J. Geophys. Res.*, 72:3049–3052, 1967; J. Kjensmo, Iron as the primary factor rendering lakes meromictic, and related problems, *Mitt. Int. Ver. Limnol.*, 14:83–93, 1968; G. E. Likens and P. L. Johnson, A chemically stratified lake in Alaska, *Science*, 153:875–877, 1966; F. Ruttner, *Fundamentals of Limnology*, 3d ed., 1963; K. Stewart, K. W. Malueg, and P. E. Sager, Comparative winter studies on dimictic and meromictic lakes, *Verh. Int. Ver. Limnol.*, 16:47–57, 1966.

Land drainage (agriculture)

The removal of water from land to improve the soil as a medium for plant growth and a surface for crop management operations. Water in excess of that needed by the plants may inhibit growth or the production of the economically important portion of the plant. High water content also lubricates the soil particles and frequently leads to unstable conditions unsuitable to machine and other crop operations. Drainage needs, or the amount of excess water, therefore, varies depending upon the soil, the demands of the crop, and the stability needs of the management practices. If the crops are water-tolerant and only light equipment is needed to manage the crop, the water excess may be small, but for an identical location where either the crop is not tolerant or the management practices place heavy loads on the soil, the water excess may be great.

Excess water creates problems in agricultural production over vast areas. Estimates of the acreage in need of drainage in the United States vary widely. G. D. Schwab stated that 22% of the total cropland, or 94,000,000 acres, has a dominant drainage problem, and Q. C. Ayres indicated that reclamation by drainage would be a benefit on about 216,000,000, or about 24% of all potential agricultural land in the United States. Ayres also indicated that about 33,000,000 acres have already been drained in the humid regions, and about 17,000,000 acres at least partly drained in the irrigated lands of the arid western states. Similar drainage problems exist in other countries.

Water source and disposal. The excess water may be due to rainfall overflow from streams, swamps, or other bodies of water; seepage or runoff from higher areas; or irrigation. The source of water should be identified before a solution is proposed. In general, the solution must fit the soil, the topography, the source of water, the crops being grown, and the management scheme used, including machinery, and must be economically feasible. Obviously, no one ideal solution exists, but a range of solutions may be proposed which vary in advantage with the individual situation.

Before discussing drainage systems, water disposal must be considered. All drainage systems must have an outlet for disposal of the water collected (Fig. 1); the outlet places restrictions on the type of system that may be used. A good outlet is low enough to permit water removal from the lowest area needing drainage; is stable, neither eroding nor filling rapidly; and is capable of accepting all design flows. Such an outlet is not always easily found; frequently deficiencies in the outlet must be corrected before a drainage system may be designed. In general, drainage outlets may be either natural or artificial, and the water may flow naturally by gravity or be moved by pumps; again, many combinations are possible.

Methods. There are two basic methods of draining land, and these may be combined to form a third. The first method, surface drainage, attempts to remove excess water before it enters the soil; the second, subsurface drainage, attempts to remove it after it is within the soil. When both are used, the system is called combined drainage. In practice, surface drainage is difficult to isolate from subsurface flow or vice versa, but separation is useful for a discussion of principles.

Surface drainage. This is usually accomplished by using shallow (less than 2 ft deep) open ditches to collect the surface water; the land surface, either between or along the ditches, is either graded or smoothed, or both, to promote movement of water into the ditches (Fig. 2). The ditches are constructed so that they slope toward a collector ditch or the outlet, and water flows naturally down the slope. Surface systems are usually less costly than subsurface or combined systems, and because the ditches are shallow, an adequate outlet is easier to find. Heavy soils that are slow to absorb rainfall, soils that are shallow over impermeable layers, or drainage problems due to surface flow are ideally suited to surface drainage.

Subsurface drainage. This is usually accomplished by burying conduits within the soil. They are buried so that they slope toward a collector or the outlet, and flow is the result of gravity. Outlets for subsurface systems must be lower in elevation than outlets for surface systems and thus are more difficult to locate. Conduits must be buried at least 2 ft below the soil surface to prevent damage by machines traveling over the surface. They must also be buried deep enough to promote water movement toward the drains at a rapid enough rate to prevent damage to crops and stabilize the soil for machinery operation. The water movement within the soil toward a drainage conduit is primarily by gravity; thus, unless the ability of the soil to conduct water is restricted as depth increases, a deeper drain will provide more rapid drainage over a wider area than will a shallower drain. Soils which are deep and permit rapid movement of water into and through the soil can be said to be ideally suited to subsurface drainage.

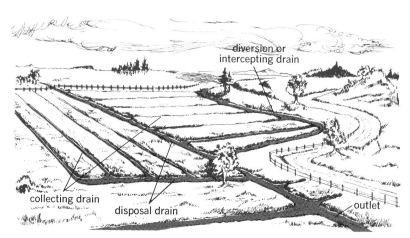

Fig. 1.   Parts of surface-drainage system. (*From Engineering Handbook for Work Unit Staffs, USDA Soil Conservation Service, 1964*)

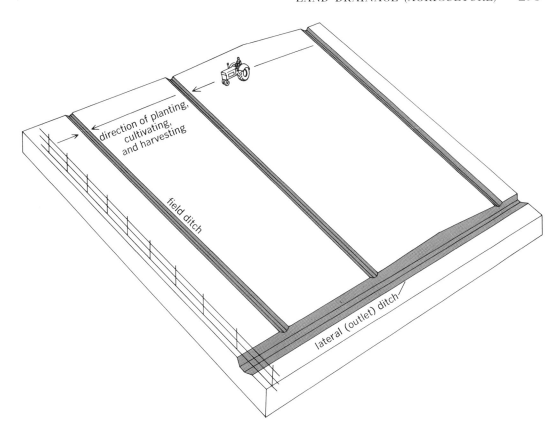

typical cross section of ground surface that has some general slope
in one direction and is covered with many small depressions and pockets

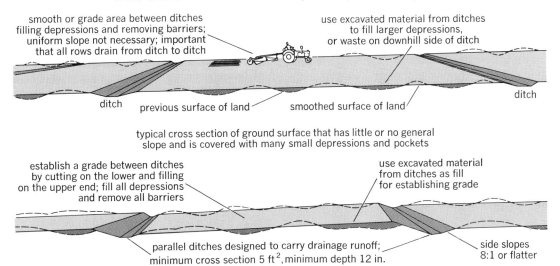

typical cross section of ground surface that has little or no general
slope and is covered with many small depressions and pockets

Fig. 2. Shallow-ditch system for surface drainage. Field ditches should be about parallel but not necessarily equidistant; and the outlet should be about 1 ft deeper than field ditches. It is necessary to clean ditches after each farming operation.

Practical applications. Ideal conditions suited exclusively to either surface or subsurface drainage are rare. Most systems operate, by either design or nature, as combined drainage systems. Ditches, even shallow ones, promote some subsurface drainage, and water in excess of that which may infiltrate the soil frequently flows over the surface to some outlet.

Drainage-system patterns. Patterns for drainage systems are of two general types. Where the drainage problem is general over an area, drainage channels may be provided at regular intervals (Fig. 3a and b); where the problem exists in isolated areas within a larger area, channels may be provided at random to include only those areas which need drainage (Fig. 3c). Random systems are usually less expensive to install, and a high proportion of drainage problems first appear as isolatable areas within a larger block. As time passes, however, the second-wettest areas become

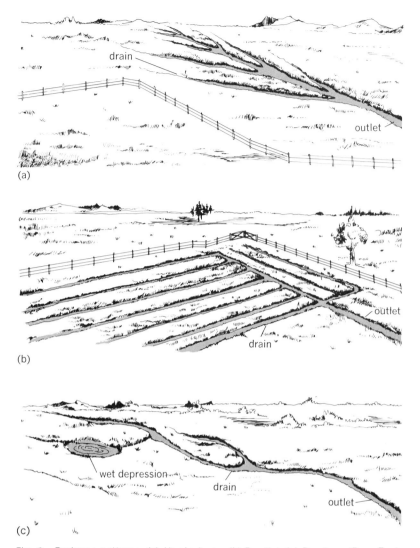

Fig. 3. Drainage patterns. (a) Herringbone. (b) Parallel. (c) Random. (*From Engineering Handbook for Work Unit Staffs, USDA Soil Conservation Service, 1964*)

the limiting factor, and random systems are extended until they look like regularly spaced systems. If expansion at a future date is considered in the initial design, the first cost is increased, but the final system is adequate. Conversely, if expansion is not considered, the initial cost may be much lower, but much of the system may have to be replaced or avoided in order to improve it in the future.

The quantity of water which must be removed by a drainage system is variable and based primarily upon experience. Most drainage system designs permit temporary flooding and require that the system dispose of a quantity of water expressed in inches of depth over the area to be drained in a period of 24 hr (called drainage coefficient). Increased protection is provided in design by increasing the drainage coefficient.

Systems construction. Most drainage systems are constructed with power equipment. Scrapers, graders, bulldozers, draglines, backhoes, plows, and special trenching machines are used to dig open ditches, and the land surface may be shaped and smoothed with some of the same equipment.

There are also special machines, land levelers, for smoothing the land. Surface drainage channels are designed with gentle side slopes to permit machines and equipment to cross easily. The earth removed from the channels is deposited where it will not interfere with drainage. Soil from the channels and high spots in the field is used to fill holes or depressions, or it may be used to raise the level of the ground surface to create increased slope into the channel (Fig. 2).

Subsurface drains are of two types. The most common type consists of buried pipes. Special pipes, made of ceramic materials, concrete, plastic, bituminous impregnated paper, or zinc-coated steel, are constructed for subsurface drainage. Openings are provided into the pipes by holes and slots cut through the pipe walls or by space left between pipe sections and, in rare cases, by permeable wall materials. A machine digs a trench to the required depth with the bottom sloping toward the outlet. The pipes are then placed into the trench and the earth returned over the pipe. In particular soils, special materials may be placed immediately over or around the pipes to prevent soil particles from entering the opening or to promote more rapid drainage. Gravel, sand, fiber glass sheets, and organic materials such as corncobs, hay, grass, or sawdust are used for this purpose. These materials are more open than the surrounding soil but present smaller openings than those present in the pipes. They also contact a larger area of soil than the openings in the pipe and thus promote better drainage.

In some soils, subsurface channels may be provided by pulling a solid object through the soil at the proper slope and depth. This type of drain is called a mole drain and the object used is called a mole. The mole is usually a steel cylinder formed to a wedge-shaped point on one end and attached to a chain or metal plate. The mole is placed in the outlet and then pulled through the soil. The mole is shaped so that it will stay in the ground at a fixed depth; therefore, the land surface must have the desired slope. If the soil has sufficient clay for a binder and enough silt, sand or stones for stability, the channel will stay open for several years and forms a cheap method of subdrainage.

Systems maintenance. All drainage systems require care in design, construction, and maintenance. Erosion or silting may occur in open ditches. Vegetation such as grass may be used along or within the ditch to stabilize the soil against either erosion or siltation, but the grass must be mowed, fertilized, and occasionally replaced. Outlets from subsurface drains may erode, removing support from around the outlets and permitting the pipe to break away, causing further erosion. Animals must be prevented from entering the pipes, and occasionally pipe sections break or collapse and require replacement. Surface entrances into buried conduits may become plugged and need cleaning; roots of perennial plants, for instance, may enter and clog a pipe.

A surface drainage system should last at least 10 years without major improvement if yearly maintenance is provided. Subsurface drainage systems should last 50 years or more if the pipe material is durable. Adequate agricultural drainage, however, is not static. As crops, management, and machines change, different demands are placed upon drainage systems and changes must be made. The soils, crops, water, and technology involved in drainage are dynamic, and the assistance of specialists is needed in devising well-designed drainage systems. The investment in drainage systems is frequently as great or greater than the original price of agricultural land. Such designs should, therefore, receive careful attention.

[RICHARD D. BLACK]

Bibliography: J. N. Luthin, *Drainage Engineering*, 1966; J. N. Luthin, *Drainage of Agricultural Lands*, 1957; G. O. Schwab et al., *Soil and Water Conservation Engineering*, 1966.

Land-use classes

Categories into which land areas can be grouped according to present use of suitability or potential suitability for specified use or according to limitations which restrict their use. The term is most commonly applied to land uses for productive purposes, such as agriculture or forestry, but may be applied for any use including engineering, architecture, urban development, wildlife, and recreation.

A complete description and assessment of land involves climate, land form, surface details, rock type, soil, vegetation, subsurface characteristics, hydrological features, and geographically associated factors, such as availability of water for irrigation purposes, location, and accessibility. Components of the land complex vary considerably from place to place, individually and in combination. The many attributes of land interact and are not equally important for all use purposes in all situations. This makes both the assessment of land potential and the specification of land requirements for a particular purpose difficult and often highly subjective. There is a trend toward, and a need for, more quantitative precision in methods. Approaches to land-use classification are based on present land use, land component surveys, landscape analysis, mathematical procedures, or a combination of these.

Present land use. The World Land Use Survey sponsored by the International Geographical Union aimed to encourage countries to map their own lands in terms of a uniform series of nine broad categories of use: settlements and associated nonagricultural land, horticulture, tree and other perennial crops, crop lands, improved permanent pasture, unimproved grazing land, woodland, swamps and marshes, and unproductive lands. Many countries which have reported on land-use surveys have introduced numerous subdivisions appropriate to the local situation.

More detailed classifications in terms of present land use required for statistical, administrative, or management purposes may be based on a variety of criteria, such as areas of individual crops or forest types, yields, disease and pest occurrence, climatic hazards, pasture types, animal grazing capacity, land-management systems, or purely economic factors (for example, input-output ratios and land values). If used in conjunction with potential land-use classifications, these can be used for comparative purposes or, in the area studied, to indicate where further productivity can be achieved or present land use modified.

Land-component surveys. The common basis for most methods is subdividing the land surface into unit areas which, at the scale of working, are essentially homogeneous in relation to use possibilities and follow with an assessment of use potential by comparison with known situations or responses. The characteristics used for subdivision are the inherent features of land, such as geology, land form, soil, climate, and vegetation. Although classifications on the basis of individual factors such as vegetation or land form have their use, single components of land are inadequate to determine land usefulness precisely. Most attention has been given to classification based on soil surveys, but associated features such as climate and topography are usually taken into account as well.

A widely used system for grouping soil taxonomic or mapping units into land-use classes is that of the USDA handbook *Land-Capability Classification*. The USDA system aims to assess suitability of soils for adapted or native plants and the kind of management the soils require to maintain continued productivity. Assessments are made largely on the kind and degree of hazards or limitations to productivity. This system provides for eight capability classes with a number of subclasses and units identified by these limitations.

The first four classes include groups of soils suitable for cultivation, but from class I to class IV the choice of suitable plants becomes more restricted or the need for more careful management or conservation practices increases. Classes V to VII soils are generally restricted to pasture, range, woodland, and wildlife use. Class VIII soils are restricted to nonagricultural purposes.

As moderately high levels of land management and of inputs and a favorable ratio of inputs to outputs are assumed, the method as a whole can be applied with precision only in areas where knowledge of land use is well advanced. In less developed areas, classification must be limited to the broader categories. In undeveloped countries with low economic ceilings, judgments need to be modified according to local standards.

The same general principles are applied, with appropriately selected criteria, for grouping soils into land classes with different degrees of suitability for a variety of purposes, such as woodland establishment, recreation, wildlife, and engineering. *See* SOIL, ZONALITY OF.

A method adopted by the Canada Land Inventory for forestry purposes illustrates a variant of this approach. It adopts a division of the land surface into homogeneous units determined by physical characteristics, followed by a rating of these units into seven capability classes according to environmental factors which influence their inherent ability to grow commercial timber. These factors are the subsoil, soil, surface, local and regional climate, and the tree species. A feature of this sytem is that a productivity rating is set in quantitative terms for each class expressed as volume of merchantable timber per acre per year. Regional inventories include a reference to the indicator tree species present for each class.

A system of land classification for the specific

purpose of establishing the extent and degree of suitability of lands for sustained profitable production under irrigation has been developed by the U.S. Bureau of Reclamation. Suitability of land is measured in terms of payment capacity and involves consideration of potential productive capacity, costs of production, and costs of land development. These are assessed from soil characteristics and topographic and drainage factors. Six basic classes of land are recognized. Classes I to III are all arable lands, but they decrease in payment capacity and become more restricted in usefulness in that order. Class IV includes lands with special uses. Class V lands are at least temporarily nonirrigable, and Class VI lands are unsuitable for irrigation use.

Landscape analysis. An alternative to the subdivision of land according to single components is the subdivision of the landscape itself into natural units, each characterized by a combination of geologic, land-form, soil, and vegetation features. This approach, referred to as the land-system approach, has been developed in a number of countries but especially in Australia. It is particularly well suited to the use of aerial photographs and has special value in the reconnaissance survey of little-known lands. The approach is based on the concept that there are discernible natural patterns of landscape covering areas with common and distinctive histories of landscape genesis. The boundaries of land systems coincide with major changes in geology or geomorphology. The pattern of a land system is formed by a number of associated and recurring land units. Each occurrence of the same land unit within a specific land system represents a similar end product of land-surface evolution from common parent material by common processes through the same series of past and present climates over the same period of time. Thus, in addition to being described similarly in terms of observable slopes and surfaces, vegetation cover, and soils, they are assumed to have a similar array of natural and potential habitats for land use. This record of the inherent features of the landscape can be interpreted in terms of land-use classes in the light of the technical knowledge available at any subsequent time. Surveys are made by concurrent and integrated studies of all the observable land features by a team of specialists in the fields of geomorphology, soils, and plant ecology. Conclusions to be drawn about immediate land use must be derived by analogy with known areas or from basic principles. The method has particular value in less developed regions. It is also being applied to special purposes such as forestry and engineering. *See* FOREST MANAGEMENT AND ORGANIZATION; GEOMORPHOLOGY.

Mathematical approaches. Land-use classification dependent upon descriptive data of land characteristics and analog processes of assessment suffer from inherent inadequacies of descriptive processes and involve a good deal of intuition and subjectivity. For this reason, effort has been made to introduce more quantitative approaches to the assessment of potentials. Foresters have developed methods of site evaluation based on the measurement of environmental factors and productivity at different sites. Key parameters are identified from multiple regressions.

This information can then be applied to classification of areas in terms of potential production.

With the advent of modern computers capable of handling and storing masses of data, there is a rapidly growing trend toward quantifying land characteristics and land-use responses and using mathematical models which relate the numerous land parameters to a variety of use responses in agricultural, forestry, and engineering fields.

The automatic scanning equipment for aerial photography and the remote sensing and automatic recording devices used with Earth satellites are opening up completely new approaches to land description and subdivision and hence to land classification. *See* LAND-USE PLANNING; LIFE ZONES.

[C. S. CHRISTIAN]

Bibliography: C. S. Christian and G. A. Stewart, Methodology of integrated surveys, *Aerial Surveys and Integrated Studies: Proceedings of the Toulouse Conference 1964*, UNESCO, 1968; *Field Manuals for Land Capability Classifications*, Canada Land Inventory (ARDA), Department of Forestry and Rural Development, various years; G. A. Hills, *The Ecological Basis for Land-Use Planning*, Ont. Dep. Lands Forests Res. Rep. no. 46, December, 1961; *Irrigated Land Use*, Bureau of Reclamation Manual, U.S. Department of the Interior, vol. 5, pt. 2, 1951; A. A. Klingebiel and P. H. Montgomery, *Land-Capability Classification*, Soil Conservation Service, USDA Handb. no. 210, 1961; D. S. Lacate, *Forest Land Classification for the University of British Columbia Research Forest*, Can. Dep. Forest. Publ. no. 1107, 1965; R. J. McCormack, *Land Capability for Forestry: Outline and Guidelines for Mapping*, Canada Land Inventory (ARDA), Department of Forestry and Rural Development, 1967; D. L. Stamp, *Land Use Statistics of the Countries of Europe*, World Land Use Surv. Occas. Pap. no. 3, 1965; D. L. Stamp, *Our Developing World*, 1960; G. A. Stewart (ed.), *Land Evaluation*, in press.

Land-use planning

Humanity has gained unprecedented physical and technical resources to regulate use and misuse of lands and resources. Planning for the distant as well as immediate consequences of man's actions is essential for long-run economy, and such investigations can warn of irreversible and irreparable deterioration in the quality of environment and life. *See* ECOLOGY; ECOLOGY, APPLIED; ECOSYSTEM; VEGETATION MANAGEMENT.

The objective of land-use planning on an individual ownership may exceed maximizing the net income of the owner. The objective of land-use planning by a public agency is even more complex and is intended to maximize long-range community benefits. Three characteristics of social and economic history nurtured contemporary land-use problems.

First, strong competitive forces, promoted by a philosophy of free private enterprise, urged rapid uncoordinated exploitation of land and its resources. This exploitation was characterized by extensive use instead of intensive methods known to modern technology.

Second, early land exploitation demonstrated

real but primitive understanding of natural land capability. The plow followed the ax in many areas where agriculture cannot do as well as forests; the plow broke the plains in many places where crops are not as compatible with the climate as more drought-resistant forage; and the plow has turned the sod of fields whose soil and slope cannot support cultivation without rapid erosion and soil depletion.

Third, early city development, in the absence of a land-use policy and plan, resulted in an indiscriminate mixture of residential, industrial, and commercial land uses. Residential areas often suffered deflated values, and commercial and industrial enterprises failed to achieve possible economies in transportation, power, and waste disposal. As deterioration of the central city set in, residences, stores, service establishments, and industries favored the suburbs for new location. This situation suggested needs for public decisions and controls on major aspects of land use in the central city and in the suburbs of every metropolitan center.

Rural land planning emphasizes the development of land-use patterns which reflect the physical and biological limits beyond which long-run depletion of the land resource will result. Increasingly, private land operators are learning that production according to land capability is good business. Since the great depression of the 1930s, both Federal and state governments have cooperatively developed plans for improved productivity of land and related resources of various large natural regions of the nation. The Great Lakes Cut-Over Region and the Northern Great Plains were areas early subject to such regional studies. The Tennessee Valley Authority (1933) marked a further emphasis upon regional resource planning. TVA and Appalachia development programs have had as their primary planning objective the discovery of economic opportunities and social amenities.

The importance of regional and major drainage basin resource development is indicated by rapid expansion since World War II. In addition, hundreds of local organizations such as watershed associations, various special districts, and intergovernmental planning committees have sprung up in response to land-use problems. These smaller regions may be dominated by a vigorous urban center. Here rural-type resource development problems intermingle with the problems of space allocation and design of the spreading urban area. Land-use planning during the later years of the 20th century is challenged to meet the needs of a planning area, composed of admixtures of rural and urban land-use problems, and recreation fringes.

Planning and the future. Trends in land-use planning suggest four developments in concept and practice. First, integrating the space-use considerations of urban-exurban planning and the resource-use considerations of wild-land, rural planning will demand increasing attention. More interchange and adaptation should take place between the space design orientation of the city planner, architect, and landscape designer and the resource capability orientation of the resource planner. The overlapping metropolitan areas must be planned to meet the basic regional needs for water, waste disposal, transportation, and open spaces for light, air, and recreation. More regard to the natural capability of each environment is needed to sustain an optimum level of regional economic opportunity and social values. Community requirements and the environmental capability should find reconciliation in a new type of integrated regional design.

Second, standards and criteria to guide land-use allocations must be developed by the students and practitioners of land-use planning and become incorporated in public land policies. *See* FOREST AND FORESTRY; RANGELAND CONSERVATION; SOIL CONSERVATION; WATER CONSERVATION; WILDLIFE CONSERVATION.

Third, if land planning is to be made an instrument of democracy, improvements are required in the methods by which social choices of the majority can find expression in the planning process, without ignoring individualism and diverse goals of a pluralistic society. Approving of general criteria for land-use allocation or of specific plans, should be made by people of the community through the operation of the political process. New communication channels can be more articulate in serving the political process and in responding to it. Experimentation should be applied to assure representation of the community and provide protection against misuse from self-seeking interests.

Fourth, to assure that plans will be carried out, closer relationships must be established between the planning authority and the agencies which exercise powers of implementation. A city planning commission may be established somewhat apart from the government authority available to implement the plan. Regional planning bodies, whether oriented primarily to large metropolitan regions or to large resource development regions, have a planning area and a scope of functions for which no single governmental body can serve as the implementing authority. Both of these situations make it difficult to mesh the "planning gears" with the "administrative gears." Political and administrative sciences are challenged to discover institutional forms and procedures through which the early and final planning process and the public power of implementation can interact responsibly.

By the 21st century, science and technology can wrest from the earth practically anything that is left over from prodigal use in the 20th. Critical problems in land-use planning of the future will arise from the difficulty of determining what is sought. These problems are not scientific and technological but they are human ones. They involve reconciling different human desires and organizing for cooperative decisions and programs of action. The challenge is to mold the natural sciences and technologies into an environmental framework with the applied social sciences to produce a land-use scheme to serve man best.

Zoning of land. Industrial societies often have zoned their lands (if at all) according to current economic values, forgetting long-range costs until struck by catastrophe. Residential areas and factories are built on floodplains, maybe unwittingly, because development is easier and cheaper on level lands than on hills. If costs of flood damage or of flood control are considered, major investments might well be allocated to uplands.

Farms, especially on rich soil, in large blocks suitable for mechanized treatment are frequently removed from best natural uses by urban sprawl. Many need legal protection from encroachment and fragmentation, and taxation policies which unfairly penalize open land or continued farming.

Developed parks and protected reserves, demonstrating natural plant and animal communities typical of diverse sites and regions, can include some areas too wet, dry, or shallow-soiled for other purposes. For educational and scientific as well as recreational purposes they also deserve high priority in the face of competing uses, according to each area's quality and location. National parks and monuments preserve unique scientific and scenic treasures, but even these are becoming degraded through lack of internal park zoning or control. Trampling of plants, compaction of soil, other abuse by pack animals or vehicles, wind or water erosion, noise and other disturbance, and unwise attraction of animals by feeding increasingly upset Nature's balances in the very places that attract most public use.

American legislation protecting wilderness areas in national forests and other areas as well as parks is a landmark of policy recognizing different levels and kinds of use, and sometimes calls for hard choices of clear priority. Since the first hearings under this legislation (for the Great Smoky Mountains National Park) the record of widespread public preference has commonly been for more inclusion of areas, and more exclusion of highways and resorts than were favored by local developers, and by officials who feel committed to the latter. To deflect commercialization within parks has demanded long-range plans in a larger regional context, commonly favoring wider private service to visitors while firmly limiting massive encroachment and destruction in the heartland.

At village, city, county, and state levels as well, there is a need for planning and implementing a balance of the most suitable areas for (1) direct urban and suburban uses, (2) use and, where pertinent, renewal of natural resources, and (3) acquisition or at least options allowing future control of landscapes and water areas. Nature conservancies (private in the United States, part of the Natural Resources Research Council in the United Kingdom) have been alert in finding and acting on opportunities.

Awareness and promptness. Urban blight and depressing obsolescence of huge, monotonous suburban developments are more symptoms — very expensive ones — of complex problems of a technologically oriented society. Great cities, stimulated to haphazard growth by industry in the 19th century, now may put on a front of greenery along modern roadways and developments; but this often masks failure to adapt the mix of land uses to the possibilities and real limitations of resources. *See* LAND-USE CLASSES.

Population explosions may be slower in industrial countries or regions than elsewhere; still their pressures and mobility commit space for industry, highways, airports, shopping centers, asphalt parking "deserts," and housing developments too fast for collecting and weighing relevant factors.

Democracies assert the right of each person, from the open country to the concrete city, to learn the choices that affect his life, to use expert counsel on the repercussions of the choices, and to exert his responsibility for influencing decisions. Yet even where mechanisms exist to implement these rights, irrevocable choices are often committed legally, or turned into reality, before their full consequences are either explored or identified. Even where conscientious citizens warn legislators or administrators of dangers or preferred alternatives, lobbying often speaks fastest and loudest from the sources having financial interest in a particular scheme.

Pollution of air, water, and landscapes is a problem which recently gained a public spotlight, without yet having wise enforcement against obvious abuses, or recognition of subtle ones. Disposal of residues in ways that will not upset the environment is a growing social problem. Effective management requires knowing the path and potential effects of each waste product until it becomes neutralized. So much public attention was rightly drawn to increasing radioactive waste in a nuclear-powered economy and to the analysis of "maximum credible" accidents (and of some incredible contingencies as well) that the far-reaching effects have been subjected to study as never before. *See* RADIOECOLOGY.

Modern man cannot escape costs of containment of by-products, of monitoring the inevitable releases to maintain low and tolerable levels, and of providing for emergency action in case of accidents. Such costs have been charged to governments and to increasing numbers of other producers of nuclear energy and materials. To control sulfur dioxide (from coal), exhaust effluents (from internal combustion engines), petroleum, and other chemical wastes, it is necessary to decide on passing along similar costs to consumers or taxpayers as a payment for maintaining or improving environmental quality. Land-use planning in site selection of major facilities, such as electric generators (which all produce waste heat, regardless of radioactive or combustion effluents), is essential for limiting these costs and maintaining compatibility with social values.   [JERRY S. OLSON]

Bibliography: S. Chase, *Rich Land, Poor Land*, 1936; Environmental Pollution Panel, President's Science Advisory Committee, *Restoring the Quality of Our Environment*, 1965; R. Lord, *The Care of the Earth: A History of Husbandry*, 1962; V. Obenhaus, L. Walford, and J. Olson, Technology and man's relation to his natural environment, in C. P. Hall (compiler), *Human Values and Advancing Technology*, 1967.

Life zones

Large portions of the Earth's land area which have generally uniform climate and soil and, consequently, a biota showing a high degree of uniformity in species composition and adaptations to environment. Related terms are vegetational formation and biome.

Merriam's zones. Life zones were proposed by A. Humboldt, A. P. DeCandolle and others who emphasized plants. Around 1900 C. Hart Merriam, then chief of the U.S. Biological Survey, related life zones, as observed in the field, with broad climatic belts across the North American continent designed mainly to order the habitats of

Characteristics of Merriam's life zones

Zone name	Example	Vegetation	Typical and important plants	Typical and important animals	Typical and important crops
Arctic-alpine zone	Northern Alaska, Baffin Island	Tundra	Dwarf willow, lichens, heathers	Arctic fox, muskox, ptarmigan	None
Hudsonian zone	Labrador, southern Alaska	Taiga, coniferous forest	Spruce, lichens	Moose, woodland caribou, mountain goat	None
Canadian zone	Northern Maine, northern Michigan	Coniferous forest	Spruce, fir, aspen, red and jack pine	Lynx, porcupine, Canada jay	Blueberries
Western division					
Humid transition zone	Northern California coast	Mixed coniferous forest	Redwood, sugar pine, maples	Blacktail deer, Townsend chipmunk, Oregon ruffed grouse	Wheat, oats, apples, pears, Irish potatoes
Arid transition zone	North Dakota	Conifer, woodland sagebrush	Douglas fir, lodgepole, yellow pine, sage	Mule deer, whitetail, jackrabbit, Columbia ground squirrel	Wheat, oats, corn
Upper Sonoran zone	Nebraska, southern Idaho	Piñon, savanna, prairie	Junipers, piñons, grama grass, bluestem	Prairie dog, blacktail jackrabbit, sage hen	Wheat, corn, alfalfa, sweet potatoes
Lower Sonoran zone	Southern Arizona	Desert	Cactus, agave, creosote bush, mesquite	Desert fox, four-toed kangaroo rats, roadrunner	Dates, figs, almonds
Eastern division					
Alleghenian zone	New England	Mixed conifer and hardwoods	Hemlock, white pine, paper birch	New England cottontail rabbit, wood thrush, bobwhite	Wheat, oats, corn, apples, Irish potatoes
Carolinian zone	Delaware, Indiana	Deciduous forest	Oaks, hickory, tulip tree, redbud	Opossum, fox, squirrel, cardinal	Corn, grapes, cherries, tobacco, sweet potatoes
Austroriparian zone	Carolina piedmont, Mississippi	Long-needle conifer forest	Loblolly, slash pine, live oak	Rice rat, woodrat, mocking bird	Tobacco, cotton, peaches, corn
Tropical zone	Southern Florida	Broadleaf evergreen forest	Palms, mangrove	Armadillo, alligator, roseate spoonbill	Citrus fruit, avocado, banana

America's important animal groups. The first-order differences between the zones, as reflected by their characteristic plants and animals, were related to temperature: moisture and other variables were considered secondary.

Each life zone correlated reasonably well with major crop regions and to some extent with general vegetation types (see table). Although later studies led to the development of other, more realistic or detailed systems, Merriam's work provided an important initial stimulus to bioclimatologic work in North America. *See* VEGETATION AND ECO-SYSTEM MAPPING.

Work on San Francisco Mountain in Arizona impressed Merriam with the importance of temperature as a cause of biotic zonation in mountains. Isotherms based on sums of effective temperatures correlated with observed distributions of certain animals and plants led to Merriam's first law, that animals and plants are restricted in northward distribution by the sum of the positive temperatures (above 43°C) during the season of growth and reproduction. The mean temperature for the six hottest weeks of the summer formed the basis for the second law, that plants and animals are restricted in southward distribution by the mean temperature of a brief period covering the hottest part of the year. Merriam's system emphasizes the similarity in biota between arctic and alpine areas and between boreal and montane regions. It was already recognized that latitudinal climatic zones have parallels in altitudinal belts on mountain slopes and that there is some biotic similarity between such areas of similar temperature regime.

In northern North America Merriam's life zones, the Arctic-Alpine, the Hudsonian, and the Canadian, are entirely transcontinental (Fig. 1 and table). Because of climatic and faunistic differences, the eastern and western parts of most life zones in the United States (the Transition, upper Austral, and lower Austral) had to be recognized separately. The western zones had to be further subdivided into humid coastal subzones and arid inland ones.

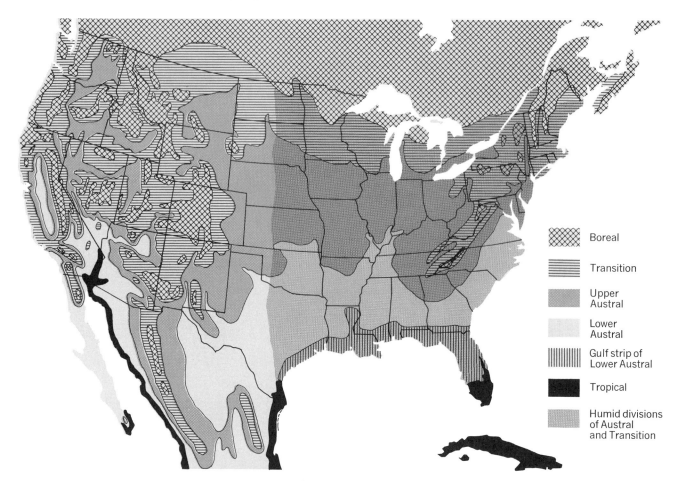

Fig. 1. Life zones of the United States. (*C. H. Merriam, 1898*)

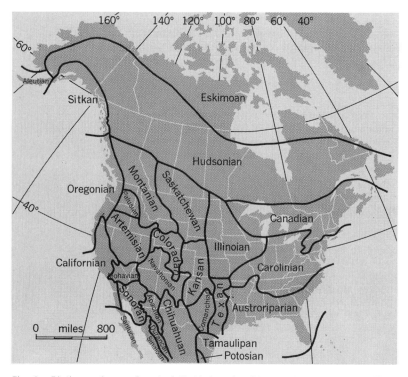

Fig. 2. Biotic provinces of part of North America (Veracruzian not shown). (*From L. R. Dice, The Biotic Provinces of North America, University of Michigan Press, 1943*)

The Tropical life zone includes the extreme southern edge of the United States, the Mexican lowlands, and Central America.

Although once widely accepted, Merriam's life zones are little used today because they include too much biotic variation and oversimplify the situation. However, much of the terminology he proposed persists, especially in North American zoogeographic literature.

Dice's zones. Another approach to life zones in North America is the biotic province concept of L. R. Dice. Each biotic province covers a large and continuous geographic area and is characterized by the occurrence of at least one important ecological association which is distinct from those of adjacent provinces. Each biotic province is subdivided into biotic districts which are also continuous, but smaller, areas distinguished by ecological differences of lesser extent or importance than those delimiting provinces. Life belts, or vertical subdivisions, also occur within biotic provinces. These are not necessarily continuous but often recur on widely separated mountains within a province where ecological conditions are appropriate.

Boundaries between biotic provinces were largely subjective, supposedly drawn where the dominant associations of the provinces covered approximately equal areas. In practice, however, too few association data including both plants and animals

tory mechanism resolves differences in sound intensity.

Decibel. In practical work dealing with sound, the strength of sounds is usually expressed in decibel (dB) units of physical intensity or sound pressure level. Decibels are relative units expressing a ratio between a given sound intensity and a reference sound. A bel is the logarithm to the base 10 of the ratio of a sound of given intensity to the reference intensity; a decibel is one-tenth of a bel. One convenient reference sound which is widely used as a standard for a decibel scale is the sound pressure of a tone of 1000 Hz (0.0002 dynes/cm^2) which is just barely heard by the normal ear. This value is sometimes called zero loudness. A sound 10 times as intense in energy has a value of 10 dB; one 100 times as intense, 20 dB. The intensity levels of some ordinary sounds are as follows: a whisper, 10 dB; a quiet office, 30 dB; street noises, 60–70 dB; a plane, 110 dB. Very loud thunder, at 120 dB, would be 1,000,000,000,000 times as intense as a barely perceptible sound of 1000 Hz (Fig. 1).

Phon. Decibels are useful units of physical intensity, but do not specify the psychological loudness of a sound. A scale of loudness level, based on intensity level, uses a unit called the phon. The loudness level of a tone is the intensity level of the 1000-Hz tone to which it sounds equal in loudness. Thus a 1000-Hz tone at 40 dB has a loudness of 40 phons. Other frequencies which are judged equal in loudness also have a loudness level of 40 phons, although their intensity level may be considerably higher than 40 dB. Figure 2 shows the relationship between decibels and phons in terms of several equal loudness contours. The lowest curve represents the absolute threshold, a loudness level of 0 phon. It should be noted that equal loudness contours at progressively higher loudness levels become more and more flattened out. In other words, the higher the intensity, the more closely does loudness level correspond with intensity level throughout the range of frequencies.

Sones. Phons are based on psychological judgments of equal loudness, but do not constitute a true loudness scale. That is, a sound of 80 phons is not necessarily twice as loud as a sound of 40 phons. To obtain a loudness scale based on equal psychological intervals, observers were asked to

adjust sounds to be "twice as loud" or "half as loud" as reference sounds, or to be midway between two reference sounds. Such a scale has been established for units called sones. One sone is defined as the loudness of a 1000-Hz tone at 40 dB (Fig. 2). If both loudness and intensity are represented logarithmically, loudness increases very rapidly at low intensities, and less and less rapidly at higher intensities.

The loudness of complex sounds varies with the distribution of their components along the frequency range. If the components are sufficiently separated, the loudness of the total complex sound is equal to the sum of the loudnesses of the components presented separately. If the components are not widely separated, the total loudness is less than the sum of the component loudnesses. This is apparently due to the fact that regions of cochlear or neural activity set up by the components overlap when they are close together.

Noise pollution. The most significant phenomenon of sound loudness is related to excessive noise in the human environment. Excessive noise has been called noise pollution. Exact standards of noise pollution have not been defined, but working standards based on extensive studies since the 1930s can be established which encompass both noise level and time duration specifications. Generally, standards of noise pollution must be defined in human factor terms or tolerance levels. These levels are (1) deafness-producing or intolerance level; (2) dangerous or deafness-promoting; (3) disturbing or deafness-related; and (4) excessive or harmful exposure. Sounds of 100 dB, that is, 20 dB above the noise of a busy street, can be defined as deafness-producing under any circumstances. Sounds of 85–100 dB may be classed as deafness-producing if exposure lasts for more than a few minutes each day and as deafness-promoting if exposure is momentary and periodic. Sounds of 65–85 dB may be classed as of moderate noise level; they are disturbing or deafness-related if persistent in daily work, and excessive or constitute harmful exposure if periodic in day-to-day activity. Workers should be protected from all levels of noise pollution. Guides to these levels can be obtained in terms of common sounds indicated in the intensity scale in Fig. 1. *See* NOISE CONTROL.

A certain amount of noise in the environment is essential to the feeling of well-being. The physiological basis of background sound is that the auditory system utilizes such noise to keep the visual system maximally sensitive. Although the ear system does not habituate to sound, noise detection is closely related to maintaining alertness and neural activation.

[KARL U. SMITH]

Bibliography: G. A. Briggs and J. Moir, *Audio and Acoustics*, 1964; I. J. Hirsh, *The Measurement of Hearing*, 1952.

Malnutrition

A state in which there is a deficiency in one or more of the essential metabolites necessary for maintenance and growth of the intact organism. The deficiency usually arises from a dietary lack, most commonly of protein or vitamins, although deficiencies arising from a lack of essential fatty acids, minerals, and carbohydrates are occasional-

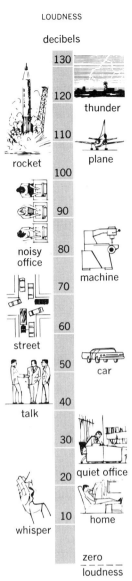

Fig. 1. Decibel scale of common sounds.

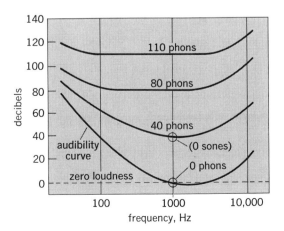

Fig. 2. The definitive relationship between loudness level, in phons, and intensity level, in decibels.

ly seen. The severity and consequences of the malnutritional state will depend not only on the type of deficiency but also on the length of time the deficiency exists. Most malnutritional states, however severe, are easily correctable if accurately diagnosed and treated at an early stage.

Deficiencies may also be the result of increased degradation of essential metabolites or presumably because of antimetabolites which interfere or compete with essential compounds in the cell. There apparently exists a hierarchy in the degree of importance of various cells in the body. Those that appear to be more essential include the cells controlling the activities of other cells, such as brain cells; those doing constant unexpendable work, as heart cells; certain cells in constant division, such as intestinal cells; and those producing large amounts of necessary extracellular proteins, as the liver and pancreas. These cells may be more susceptible to nutritional deficiencies than supportive cells exemplified by connective tissue cells.

Cell requirements. Most deficiencies resulting in a severe malnutritional state are complex multiple deficiencies and do not result from the loss of a single essential compound. There are seven major constituents of the cell which, in the proper concentrations and proper balance, are necessary in order for the cell to maintain itself. Many of these constituents have a common precursor and their pathways are closely interrelated. It is well known that the labeled carbons of radioactive glucose may eventually appear both in the carbohydrates of the cell and in the proteins, nucleic acids, and fats. The essential constituents of the cell include water, mineral ions, proteins, nucleic acids, carbohydrates, lipids, and porphyrins. Although these substances are usually found in complexes of large molecules within the cell such as the glycolipoproteins, large complex macromolecules are not transported across the cell membrane. Each individual cell is considered to be the site of synthesis of the macromolecular complexes, usually from small precursors which include amino acids, purines, pyrimidines, fatty acids, and glucose.

Essential metabolite deficiencies. When considering a deficiency of the above seven essential cellular constituents, a malnutritional state resulting from protein lack would appear to be not only the most common but the most serious. Proteins make up the large portion of the solid constituents of the cell and serve not only a major structural function but also an important metabolic function, particularly when functioning as enzymes and hormones.

Proteins. Of the 20 amino acids commonly found in man, 8 have been found to be indispensable. Although all 20 are required in the synthesis of most proteins, 12 of these may be manufactured by the cell itself from small carbon fragments, or other amino acids. However, if even 1 of the 8 essential amino acids is missing from the diet, the individual is unable to synthesize any protein and growth ceases. Death does not occur immediately because many proteins are in a constant state of degradation as well as synthesis. Some unessential amino acids become available from protein breakdown for resynthesis of the more essential proteins. The essential amino acids in man are leucine, isoleucine, lysine, phenylalanine, tryptophan, threonine, methionine, and valine. Most malnutritional states do not involve an absolute deficiency of the essential amino acids but rather a relative deficiency when evaluated with the growth and maintenance requirements of the organism or cell.

Fats. With regard to the fats, only four fatty acids appear to be essential, again in the sense that only four cannot be synthesized by the cell: arachidonic, γ-linolenic, linolenic, and linoleic. Fats apparently serve a structural function, especially on membranes and interfaces, and usually are combined with proteins as lipoprotein. They also serve as an excellent secondary source of energy when carbohydrates are deficient or unavailable. Malnutritional states involving only fat deficiency are not well recognized. *See* LIPID METABOLISM.

Carbohydrates. The carbohydrates have a somewhat lesser role as structural components within the cell, although their importance as a structural component in the form of extracellular polysaccharide ground substances is well known. Carbohydrates, glucose especially, are considered to be the cell's chief source of energy after degradation, oxidation, and the resultant production of high-energy adenosinetriphosphate (ATP) molecules.

Nucleic acids and porphyrins. Nucleic acids are essential to the chromosomal structure and apparently to protein synthesis, but they may usually be synthesized from other smaller substrates. There is no evidence that malnutritional states are due to lack of nucleic acids. Some of the vitamins, however, are essential for nucleic acid synthesis and a vitamin deficiency will decrease new nucleic acid formation. Porphyrins are likewise easily synthesized in the body. *See* NUCLEIC ACID; PORPHYRIN.

Mineral ions. The mineral ions play a very important role in cellular metabolism. The cations sodium and potassium as well as the anions carbonate, phosphate, sulfate, and chloride are responsible for regulation of the water content of the cell, and a few of them, especially potassium and magnesium, are important as activating ions for many enzymatic reactions. Malnutrition may result in an anemia due to a deficiency in iron, an ion of crucial importance in the synthesis of hemoglobin. Although sodium deficiency, hyponatremia, and potassium deficiency, hypokalemia, are seen in disease states, they are not commonly associated with malnutritional states alone. Iodine deficiency, without other evidence of malnutrition, may result in a colloid goiter of the thyroid gland. Deficiency of the other essential trace elements such as zinc, manganese, cobalt, and copper are apparently present only in severe starvation states.

Generalized malnutrition. The effect of generalized malnutrition, less than 1600 cal per day for a man weighing 70 kg, is usually obvious, resulting in extreme malaise, weakness, lack of growth, anemia, and, in severe cases, edema. The organs shrink in weight as a result of the cells shrinking in size; the fat cells in particular show characteristic changes of shrinkage and the appearance of a clear vacuolar space. Glycogen deposits in the liver and muscle disappear, and the protein structure appears reduced. Generalized malnutrition also results in a lack of resistance to any insult including drastic changes in temperature and infectious agents such as bacteria and viruses. Many of the factors concerned with this resistance are intangible. However, an important factor in resistance is the presence of antibodies. These are specialized proteins formed in response to and combined with antigenic foreign agents such as viruses. Antibody formation and thus resistance to infection is markedly low in malnutritional states.

Kwashiorkor disease. A specific disease associated with malnutrition and more specifically with a lack of dietary protein is called kwashiorkor disease. It is found most commonly in parts of Africa and Asia. The patients, usually children, show severe liver disturbances. There is a reduction of the protein and nucleic acid content of the liver cells, reduction of the serum albumin, which is synthesized in the liver, and increased fat in the liver cells. Analysis of the contents of the small bowel shows markedly decreased intestinal enzyme levels. Other organs commonly affected are the muscle and pancreas, which show markedly altered structure and decreased function.

[DONALD W. KING]

Bibliography: J. S. Fruton and S. Simmonds, *General Biochemistry*, 2d ed., 1958; R. H. S. Thompson and E. J. King (eds.), *Biochemical Disorders in Human Disease*, 2d ed., 1964.

Mangrove swamp

A swamp forest of low to tall trees and some shrubs, commonly associated with some salt marsh herbs. Swamp forests occur along the borders of many tropical shores where wave action is not intense and mud and peat are deposited (Fig. 1a). Most plants of this community are halophytes that are well adapted to salt water and fluctuations of tide level. Some have stilt or prop roots to help hold them on the shifting sediments and others have erect root structures (pneumatophores) that crop out above the surface (Fig. 1b and c). Several species have well-developed vivipary of their seeds, the hypocotyl developing while the fruit is held on the tree. These seedlings are usually so shaped and weighted that they float long distances in the sea and thus extensive migration is ensured.

The mangrove community often develops as a distinct halosere. In such areas it is zoned from open water landward in a series of different species (Fig. 2). The landward zone species develop on sediments and peats which were initially deposited in the seaward zone. These changes due to deposition in the swamp often extend the coast outward and form incipient islands in shallow, quiet waters. The swamps when dense also afford some protection against erosion resulting from violent storms. Thus, mangrove swamps have a significant geologic role. *See* ECOLOGY; PLANT FORMATIONS, CLIMAX; SUCCESSION, ECOLOGICAL.

The composition of mangrove swamps is strikingly different in two regions. Only four (Western) species predominate in the Americas and along the west coast of Africa, but over 10 (Eastern) species are frequent in swamps of eastern Africa, Asia, and the western Pacific region. The genera belong to a number of families. They are examples of convergent adaptations to their saline habitat.

[JOHN H. DAVIS]

The Eastern mangrove swamp forms a monotonous forest between low- and high-tide marks. The vegetation is extensive, luxuriant, and tall on muddy beaches, lagoons, deltas, and estuaries; but it is

Fig. 1. Mangrove swamps, Great Barrier Reef. (a) Detail of mangroves on north side of Howick Island. (b) Interior of mangrove swamp, Newton Island. (c) Mangroves at northwestern corner of King Island. (*From J. A. Steers, Salt marshes, Endeavour, 18(70):75–82, 1959*)

narrow, dwarf, and sparse along sandy and rocky shores and old coral reefs. The exact limits of such a forest continually change through silting, colonizing, and erosion on the seaward side, and through rise in level, improved drainage, and reduced inundation in the landward direction.

The plant community, up to 40 m in height, is composed of a few viviparous and gregarious species forming an unbroken canopy with no distinct understoreys. Epiphytes, parasites, and undergrowth are either scarce or absent. Other characteristics include special root formation, such as branched stilt roots (*Rhizophora* sp.), pneumatophores (*Avicennia* and *Sonneratia* sp.), and knee-bend-like roots (*Bruguiera* and *Lumnitzera* sp.);

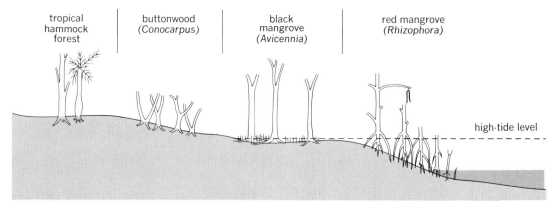

Fig. 2. Zonation of mangroves from open water landward, with their associated vegetation, in Florida. (After *J. H. Davis, The Ecology and Geologic Role of Mangroves in Florida, Carnegie Inst. Wash. Publ. no. 517, 1940*)

Fig. 3. A pioneer species (*Avicennia*) in formation of mangrove forest, colonizing mainly the exposed side, with spreading root system and asparaguslike pneumatophores. (*Courtesy of Forest Department, Malaya*)

Fig. 4. A mud chimney made by *Thalassina anomala*, the characteristic mangrove shrimp.

and xerophytic features, such as thick coriaceous foliage with closely packed mesophyll, sunken stomata, and aqueous tissue. All these characteristics indicate convergent evolution in response to peculiar conditions of nutrition, assimilation, and survival in anaerobic and shifting substratum.

The development and composition of a man-grove forest depend largely on soil type, salinity, duration and frequency of inundation, accretion of silt, strength of tides, exposure or shelter of the site, and unplanned exploitation. These factors interact in a complex manner; accordingly, the distribution, zonation, and succession of mangrove species are highly variable even within narrow geographical limits. Mangroves are most luxuriant and complex on the coasts of Malaysia and the surrounding islands. From here the diversity of the Eastern species decreases toward the Red Sea and southern Japan, where only a single species may be found. Of the four characteristic families, Rhizophoraceae, which forms extensive forests, is by far the most important. Other characteristic sub-mangrove tree-forming species belong to the genera *Lumnitzera, Acanthus, Hibiscus,* and *Scyphiphora*.

The Eastern vegetation is taller and richer in diversity than its Western counterpart. All the Western genera are found in the Eastern vegetation, but their species composition, zonation, and succession are different. The pioneers are species of *Avicennia* and *Sonneratia,* the former colonizing mainly the firmer, exposed seaward side (Fig. 3) and the latter largely the soft, rich mud along sheltered river mouths. *Ceriops decandra* plays a similar role on the sheltered east coast estuaries of the Malay Peninsula. Depending on soil type and tidal height, these pioneers are normally succeeded by either pure or mixed forests of various species of *Bruguiera* and *Rhizophora,* such as *B. cylindrica* and *B. sexangula* on firmer clay beyond the reach of ordinary tides behind *Avicennia; B. parviflora* in pure crop or mixed with *R. apiculata* on wetter muds flooded by normal high tides; *B. gymnorhiza,* often mixed with submangrove species *Acrostichum aureum, Acrostichum speciosum* (ferns), and *Oncospermum filamentosa;* and *Nipah fruticans* (palms) farther inland on drier ground with less saline soil. *R. mucronata* is more tolerant of sandy, firmer bottoms than *R. apiculata* and forms tangled thickets on banks of tidal creeks and in estuaries.

Economically mangroves are a great source of timber, poles, pilings, and fuel. The bark is used in tanning and batik industries. Some species have either food or medicinal value. For better productivity, exploitation and management must be planned carefully. *See* FOREST MANAGEMENT AND ORGANIZATION.

Animals in the Eastern mangroves are abundant and are zoned both horizontally and vertically. On the mud floor brightly colored species of *Uca* and amphibious mudskipper gobies belonging to *Boliophthalmus* and *Periophthalmus* genera can be seen. There are a variety of snails on different levels of the mangrove trees, and bivalves and worms hidden away under the mud are also very common and exclusive. *Thalassina anomala*, with its characteristic mud chimneys, is found in the drier areas (Fig. 4). Other less common but interesting forms include snakes, monitor lizards, and long-tailed crab-eating monkeys. Many birds, including the Brahminy kite, green pigeons, terns, waders, sea eagles, and kingfishers, can be seen visiting the mangroves either regularly or seasonally, but only a few species are exclusively confined there, nesting commonly in tall *Sonneratia* trees. The nuisance of leeches, which are absent from mangrove forests, is more than compensated for by the myriads of small mosquitoes. *See* ECOSYSTEM; VEGETATION AND ECOSYSTEM MAPPING.

[F. C. VOHRA]

Bibliography: F. C. Craighead, Land, mangroves, and hurricanes, *Fairchild Trop. Gard. Bull.*, vol. 19, no. 4, 1964; N. Hotchkiss, *Marsh Wealth*, 1964; D. W. Scholl, Recent sedimentary record in mangrove swamps and rise in sea level over the southwest coast of Florida, pts. 1 and 2, *Mar. Geol.*, 1:344–366, and 2:343–364, 1964.

Manure

Any plant and animal residue that may contain excreta of animals. Manures, classified into animal manure, plant manure, and compost, are added to the soil in various stages of decomposition. In the soil they undergo further degradation through the action of soil microorganisms, raising the soil fertility level and improving soil texture by increasing humus content.

Animal manure includes plant residues like straw, used as litter, as well as solid and liquid excreta. Microorganisms attack the nitrogenous components and the more readily fermentable carbohydrates during storage of the manure. Intense microbial respiration, favored by a loose texture and adequate moisture in the manure, causes a rise in temperature so that thermophilic forms of bacteria, actinomycetes, and fungi grow.

Compost consists of plant and animal residues allowed to rot before being applied to soil. Because they have a lower ratio of nitrogenous to carbohydrate material than animal manure, composts are usually reinforced with inorganic nitrogen and available phosphorus to facilitate microbial action. With moisture and aeration, decomposition proceeds accompanied by a rise in temperature.

Green manure is plant material in the form of a growing crop plowed into the soil. Because the object of manuring is to increase the supply of nitrogen, leguminous crops are grown and then plowed into the soil. Green plants are higher in soluble carbohydrates, nitrogen, and minerals than plant residues used in manures and composts; as a result decomposition sets in more rapidly. Green manures, being low in cellulose and lignin, have little effect on the humus content of soil.

[ALLAN G. LOCHHEAD]

Marine ecosystem

The ocean with its shores and estuaries is the largest conceptual unit in marine ecology. Within it, ecological systems of various sizes are recognized, for example, the particular seaweed ecosystems, tidepool ecosystems of the seashore, and estuarine-bay ecosystems. Thus small natural communities of organisms with their immediate environment, the unit marine ecosystems in the narrow sense, exist within and together compose larger systems (Figs. 1 and 2). The size limits, or boundaries, of a particular ecosystem that distinguish it from others may depend upon physical barriers to community dispersal, such as a submarine mountain; environmental factors that restrict the area of the biotic community, such as the salinity factor; the productivity cycles within the community; or the reproductive and dispersal potentials of the community. This conceptual unit

Fig. 1. A community of ribbed mussels attached to a decayed log in the intertidal zone.

Fig. 2. A marsh grass–mussel community, representing a small marine ecosystem of the intertidal zone.

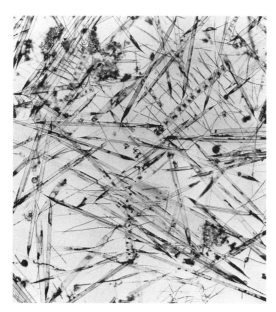

Fig. 3. Plankton diatoms. Their light spiny shells are suitable for floating on the surface of the water. (*Photograph by P. Conger, Smithsonian Institution*)

of ecological science, whether large or small, possesses individuality, a degree of stability and permanence, characteristic functional cycles, and readily recognizable components, either living or nonliving or both. *See* COMMUNITY; ECOSYSTEM.

The nonliving, or abiotic, materials in a marine ecosystem cycle comprise not only a variety of water-soluble inorganic nutrient salts, such as the phosphates, nitrates, and sulfates of calcium, potassium, and sodium and the dissolved gases oxygen and carbon dioxide, but also organic compounds, such as the various amino acids, vitamins, and growth substances. Most of the solid material dissolved in the sea originated from the weathering of the crust of the Earth.

MARINE ENVIRONMENT

The marine environment is a subject of study by oceanographers who investigate the physical, chemical, and biological properties of ocean waters, ocean currents, and ocean basins. *See* ENVIRONMENT; OCEANOGRAPHY.

Chemical factors. The chemical properties of ocean waters constitute an important aspect of the marine environment. Such chemical factors as the dissolved solids and gases in the ocean waters have been investigated. *See* SEA WATER.

Dissolved solids. The total quantity of dissolved solids in the waters of the world's oceans or in the marine ecosystem as a whole approximates 5×10^{16} metric tons, enough to form a layer 153 in. thick over the land area of the Earth. Typical ocean water contains about 34.9 grams (g) per liter of dissolved materials, in which there are 19.3 g of chlorine, 10.7 g of sodium, 2.69 g of sulfate, and 1.31 g of magnesium. Some of the most important nutrient elements are present in very small amounts, for example, the phosphates needed for the growth of algae. Phosphorus exists in ocean waters as phosphate ions. It is an essential component of living things, and in ocean water the amount present may limit the production of plants.

The inorganic form occurs in amounts varying from zero concentration to over 0.10 mg per liter. Phytoplankters, principally microscopic algae, may absorb inorganic phosphorus and reduce the amount remaining in the water to a minimum. Some is permanently lost to recycling by being bound in nonsoluble forms and deposited on the sea bottom. By their death and decay, phosphorus is returned to the aquatic environment. Certain of the 40 or more elements in sea water which, like phosphorus, exist in extremely small amounts are concentrated to important degrees. For example, macroscopic algae concentrate potassium and iodine in relatively large amounts. Of special importance is silicon, utilized by diatoms and other silica-secreting organisms.

Dissolved gases. The essential gas, carbon dioxide, is readily dissolved in sea water, but in equilibrium with the air it would contain only about 0.5 milliliters (ml) of free CO_2 per liter at 0°C. Actually, sea water contains about 47 ml of CO_2 per liter because appreciable amounts are present in the form of carbonate and bicarbonate ions. This gives great importance to the marine ecosystem as the world's reservoir of CO_2 although only about 1% is in the free form. Since, in relation to the atmosphere, its CO_2 content is 50 times greater, oceans regulate the concentration in the atmosphere. The CO_2 content of sea water originates not only from the atmosphere but also from biochemical processes in the sea—respiration and decay—and in the soil of the ocean bottom and coastal shores. The free carbon dioxide in water exists in simple solution and in the form of carbonic acid, H_2CO_3. Combined CO_2 is in the bicarbonate ions (HCO_3^-) and the carbonate ions (CO_3^{--}). These and other ions of weak acids constitute the buffer system of sea water that stabilizes its hydrogen-ion concentration, or pH, at about 8–8.4 at the surface and at 7.4–7.9 at the deeper levels. The high buffer capacity of sea water is a dominant environmental characteristic. Chemical and biochemical interactions and actions of biological agents which would otherwise produce pH changes that could seriously modify living conditions are largely negated by this buffer system. The ocean environment is thus chemically stable. Marine plants, the producers, may utilize primarily either free CO_2, or some free CO_2 and some combined CO_2. In any case, lack of CO_2 is not a limiting factor to growth of ecosystems in the sea. High concentrations, however, are a limiting factor to many marine animals such as fish.

The dissolved oxygen content in the sea-water environment varies from 0 to over 8.5 ml per liter. The oxygen in water comes from the atmosphere by diffusion and from aquatic plants by their photosynthetic action. Low temperatures and low salinities of water favor high solubilities of oxygen gas. There are low-oxygen strata in ocean waters, but in general the supply is adequate for animals. The considerable oxygen content of the deeper water layers of the ocean reached these layers when they were in the photosynthetic surface strata.

Physical factors. The physical aspects of marine ecological systems, temperature and salinity, are especially significant because of their degree of constancy away from land. The mean annual temperatures in different latitudes on the Earth remain unchanged. Seasonal variations in the ocean are small compared to those on land. Also, the entire temperature range in ocean waters is within the tolerance limits of numerous plants and animals. In polar and tropical seas, temperatures do not vary more than about 5°C during the year. Temperate seas commonly vary about 10–15°C. The extent of seasonal variations decreases in the deeper strata. In temperate and tropical regions a permanent thermal gradient, the thermocline, in which temperature decreases rapidly with depth, lies between the surface mixed layer and the deep unmixed strata. Low temperatures, around 3°C, in the deep and bottom waters of the ocean exist because the waters of greatest density, formed in high latitudes, sink to the bottom or to levels of similar density, then spread out and move toward warmer latitudes, to form a pattern of oceanic circulation. The salinity of the world's oceans varies only a few parts per thousand and averages around 35 ‰. Major latitudinal differences in density result from temperature differences. As expected, there is a vast system of oceanic circulation involving all depths and modified by large water masses contributed by adjoining seas. Thus, this offshore environment of the unit marine ecosystem is characterized by relative constancy of salinity over exceedingly large areas. The massive and relatively constant properties of this environment are reflected in the distribution, abundance, and morphological traits of its biota.

Biota. The organisms that comprise the biota of the marine ecosystem are characterized by a lack of diversity among the plants, in contrast to the animals. This generalization applies to both microscopic and macroscopic organisms. With few exceptions, all types of animals inhabiting land and fresh-water environments occur in the marine ecosystem. Six major animal groups, including ctenophores, starfishes, and certain worms, are restricted to the marine environment. Other major taxonomic units are predominantly marine. The kinds of plants and animals in the marine environment fall into three major groups: organisms of the plankton, nekton, and benthos.

Plankton organisms. These organisms are small, mostly microscopic, and have little or no power of locomotion, being distributed by water movements. There are two main types, the phytoplankton and zooplankton. The former includes all of the floating plants, such as the small algae, fungi, and sargassum weeds. Of these, the most important in the economy of the sea are algae—diatoms and dinoflagellates. They are the major producers in marine plankton (Fig. 3). Diatoms are microscopic, unicellular plants, some of which form chains. They possess characteristic shells composed of translucent silica, and have a great variety of form and sculpture. The shell structure consists of two nearly equal valves, one of which fits over the other and hence may be compared to a box with a telescoping lid. The valves are joined by connecting bands. The protoplasm within the shell is exposed by a slit or by small pores to permit metabolic interchanges with the environment. During reproductive division of the protoplasm of the diatom, one of the two protoplasmic daughter cells retains the larger epivalve, or lid valve, the other the hypovalve, or box valve. The daughter cells then lay down the needed complementary

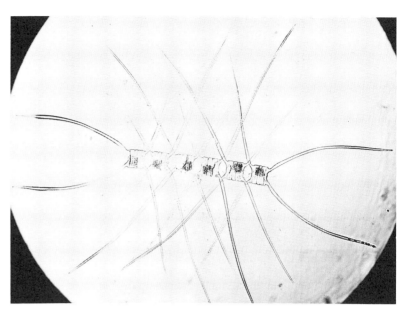

Fig. 4. *Chaetoceros atlanticus* Cleve, a branched-type oceanic diatom. (*Photograph by P. Conger, Smithsonian Institution*)

valves. This simple binary form of division is the most common one. It permits rapid production of vast populations in favorable environments of an ecosystem. During successive binary divisions, the size attainable by individuals is progressively reduced until a minimum limit is reached when, usually, the protoplasmic content of the shells escapes from the parted valves. It is enclosed in a flexible pectin membrane and is called an auxospore. These specialized spores grow in size and finally form the characteristic two valves. Diatoms possess one or more chromatophores, ranging in color from yellow to brown.

Diatoms occur as fossil, siliceous shell deposits, called diatomaceous earth, and as living producers in practically all habitats of the broad marine ecosystem. They are found floating in water, attached to the bottom, on larger plants, on animals, and, as spores, enclosed in Arctic ice. Free-floating diatoms possess structural adaptations that permit adjustments in depth. The bladder-type diatom is relatively larger, and in some forms such as *Planktoniella*, the shape is disklike, so that a zigzag course is followed in sinking. The hair type, such as *Rhizosolenia*, is a long and slender diatom, which sinks slowly when the long axis is horizontal to the pull of gravity, but more rapidly when oriented vertically. The ribbon-type cells, such as *Fragillaria*, are broad and flat, and are attached to form chains. The most abundant diatoms in the offshore oceanic waters are of the branched type, such as *Chaetoceros*. They possess numerous spiny projections that resist sinking (Fig. 4).

Dinoflagellates possess whiplike flagella that provide a slight degree of locomotion. Like diatoms, they possess structural modifications that indicate adaptation to environmental conditions. Some, such as *Dinophysis*, possess winglike structures that favor suspension; some have cellulose plates; others are naked cells. Many kinds are luminescent.

Phytoplankton organisms are much more abundant in nutrient-rich coastal waters than in offshore oceanic waters. They are the primary pro-

Fig. 5. A benthic community of brittle stars and isopods at a depth of 1200 m in the San Diego Trough, off California, 32°54'N and 117°36'W. (*Photograph by G. A. Shumway, U.S. Naval Electronics Laboratory*)

ducers upon which large and small marine animals feed.

Zooplankton organisms are the floating or weakly swimming animals, which include the eggs and larval stages as well as adult forms. The principal kinds include numerous Protozoa, such as Foraminiferida and Radiolaria; great numbers of small Crustacea, such as ostracods and copepods, with their various larval stages; various jellyfishes; numerous worms; a few mollusks; and also the eggs and early developmental stages of most of the nonplanktonic organisms in the sea. Plankton organisms are grouped on a basis of size. The smallest organisms range from about 5 to 60 μ, the next largest range up to about 1 mm in size, the next up to 1 cm, and the largest plankton group over 1 cm.

Nekton organisms. These are the actively swimming animals of the marine ecosystem. They comprise adult stages of such familiar forms as crabs, squids, fish, and whales. Some of these undergo long horizontal migrations over hundreds of miles; some migrate periodically to great depths; and a few live mostly in deep waters, for which habitat they possess marked adaptations.

Benthos organisms. These inhabit the bottom, and range from high-tide level on shore to the deep-sea bottom. There are relatively few kinds and numbers of animals on the deep-sea floor. They are mainly mud dwellers, possessing characteristic structures permitting life in a quiet, dark, muddy environment where food is scarce. Certain isopods, sponges, hydroids, brittle stars, sea urchins, and shrimp are typical animals of the deep-sea biota (Fig. 5). Inshore bottom communities at depths less than 50 m are rich in plants and animals. Although mosses and ferns are entirely absent in the sea, there are approximately 30 species of marine flowering plants. Of these, eelgrass (*Zostera*) is a characteristic shore plant. It is a perennial flowering plant, not actually a grass, that is worldwide in distribution, and is most abundant on soft bottoms in protected, coastal areas. Large brown algae, such as *Fucus* and *Ascophyllum*, are widely distributed conspicuous plants of exposed rock surfaces in the intertidal zone. *Ulva* and *Enteromorpha* are typical green algae of mud flats in quiet waters. Red algae, such as dulse (*Rhodymenia*), are commonly found below the intertidal zone. These are important food plants for the bottom-living animals of the coasts.

Conspicuous members of the benthic animal communities of the sea coast are barnacles, snails, mussels, clams, oysters, sea anemones, sea urchins, sea cucumbers, and starfish. Shore corals are largely restricted to warmer seas. The bottom animals of the marine ecosystem have structural modifications facilitating adhesion, burrowing, feeding, and protection.

MAJOR DIVISIONS

The marine ecosystem can be divided into two large areas, the pelagic and benthic divisions. Each of these consists of various zones.

Pelagic division. This division embodies all the waters of the oceans and their adjacent salt-water bodies. It is divisible into the neritic zone, extending offshore to the edge of the continental shelf to a depth of over 200 m, and the oceanic zone, embracing the remaining offshore waters. The neritic zone is rich in plant nutrients, especially phosphates and nitrates, which originate from coastal

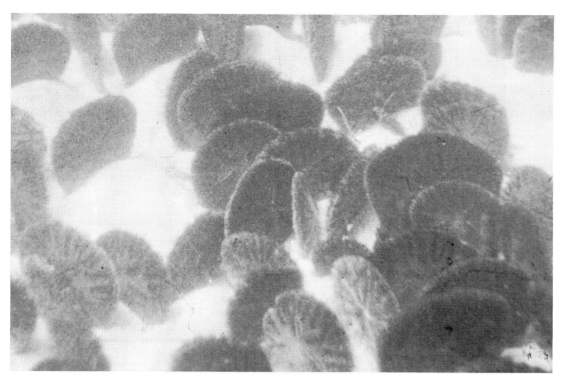

Fig. 6. A littoral community with abundant sand dollars. This photograph was taken at a depth of 4 m in Mission Bay Channel, San Diego, Calif. (*Photograph by R. F. Dill, U.S. Naval Electronics Laboratory*)

tributary waters and from bottom deposits carried upward to surface water layers by upwelling, diffusion, turbulence, or convection. The water is more variable in density and in chemical content than oceanic waters. It is also far more productive of plankton, fish, and shellfish. Many of the inshore, coastal forms are adapted to withstanding brackish waters of coastal tributaries. The oceanic zone has a well-populated upper, lighted, 200-m stratum, and deeper, relatively dark, and sparsely populated layers, characterized by great pressures, animals modified for life in darkness and under great pressures, and very few bottom animals. Since there is less plankton and suspended organic material in the water, light pentrates further than in the neritic zone. Also the water is usually very transparent. Its salt content is uniformly high and not subject to major variations in time and space. In the upper photosynthetic zone, or zone of productivity, plant nutrients are much less concentrated, and the cycle for replenishment is much longer than in coastal waters that receive nutrients and organic detritus from the land.

Benthic division. This division of the marine environment embraces the entire ocean floor, both coastal bottom and deep-sea bottom, properly termed the littoral and deep-sea systems, respectively. The littoral system consists of the eulittoral zone and the sublittoral zone.

Eulittoral zone. This extends from high-tide level to about 60 m, the approximate depth below which attached plants do not grow abundantly. This is probably the richest zone of the marine ecosystem in respect to numbers and kinds of organisms, as well as in variety of ecological types and habitat modifications. The upper intertidal portion of the zone extends between high- and low-water marks, a vertical distance that varies on different continental shores from over 12 m to a few centimeters. Changes in such environmental factors as light, temperature, salinity, and time of exposure vary tremendously within short vertical distances across the zone. These variations are reflected in the shapes, movements, tolerances, and life histories of the characteristic animals and plants. Numerous small, partly independent ecosystems thrive in this area because the substratum includes a variety of rock exposures, gravel, sand, and mud types admixed in all degrees (Fig. 6). The lower, permanently submerged portion of the eulittoral zone is characterized by abundant sessile plants, such as conspicuous rock weeds (*Fucus*), bladder wrack (*Ascophyllum*), green sea lettuce (*Ulva*), and kelp (*Laminaria* and *Nereocystis*), many of which are common to both portions of the zone. Certain of those algae, such as the giant kelp, form productive algal forests that extend down below the typical eulittoral zone into the sublittoral zone. Here, they utilize the dimly lighted, lower, and least productive portion of the littoral system.

In tropical littoral waters, the coral reef communities abound with characteristic types and forms of both plants and animals. Visible algal growth is sparse, animals often greatly predominate over plants, and depth range is to about 60 m; hence the communities occur in both zones of the littoral system. The littoral system terminates at depths varying in different latitudes between about 200 and 400 m, depending upon light and temperature factors that modify the distribution of benthic animals and plants.

Deep-sea system. This portion of the benthic division is subdivided into an upper part, the archibenthic zone or, more meaningfully, the continental deep-sea zone, extending from the edge of the continental shelf (200–400 m) to depths of about 800–1100 m, and the abyssal-benthic zone that embraces the remainder of the benthic deep-sea sytem (Fig. 5). These zones have little or no light, relatively constant conditions of salinity and temperature, and steadily decreasing numbers and kinds of organisms. In the abyssal regions of great depth, perpetual darkness, and extremely low temperature (5 to −1°C), the extreme in monotony of environment prevails. However, bacteria and at least some animals exist at about all known depths. The remains of plants and particulate organic detritus continuously rain down to settle over the bottom as a light coating, utilized by the bottom dwellers for food. The pelagic food supply from above decreases with the increase in offshore distance for two reasons: The offshore oceanic waters contain less particulate matter than inshore waters, and the longer period required to descend the deeper strata results in greater disintegration while sinking.

Estuaries. These are coastal adjuncts of the marine ecosystem. They embrace bodies of water which, by virtue of their position, are directly subject to the combined action of river and tidal currents. Compared with offshore ocean waters, they lack constancy, possessing characteristic horizontal and vertical gradients in physical, chemical, and biological properties. These gradients are subject to pronounced changes in space and time. Estuarine waters are characterized by rapidity of response to changing external conditions. Their temperature, salinity, and turbidity conditions are distinctly not uniform. They are usually rich in plant nutrients of land origin, which they transport to coastal waters, thus fertilizing these waters. Estuarine waters contain an environmentally selected biota containing representative fresh- and salt-water forms. *See* ESTUARINE OCEANOGRAPHY.

Marshes. These transitional land-water areas, covered part of the time at least by estuarine or coastal waters, comprise parts of the peripheral area of the marine ecosystem. Mangroves are characteristic woody marshes in tropical tidal waters of flat, muddy shores. More generally, coastal marshes are dominated by grasses and sedges. Marshes have characteristic biota for a certain latitude. In temperate regions, they are inhabited by characteristic birds such as bitterns and rails, and by crustaceans such as sand hoppers and fiddler crabs. Typical marsh animals and plants have tolerances to fresh water and salt water that differ from those of related forms living in fresh-water or salt-water habitats. *See* MANGROVE SWAMP.

Sediments. Inanimate, particulate materials of organic and inorganic origin that have settled on the bottom in aquatic environments comprise the sediments. Vast quantities of particulates, eroded from land surfaces by natural waters, are carried to the sea by rivers and estuarine waters. They settle to the bottom and provide food for benthic animals. Sediments may be carried to the surface, or photosynthetic stratum, either by upwelling of coastal waters or by ocean currents, and thus enter a biochemical cycle of the marine ecosystem. Marine sediments are made up of microscopic fragments of weathered rock, partly decomposed plant and animal remains, skeletal remains of organisms, inorganic precipitates from sea water, terrestrial particulates, and particulate material of volcanic origin.

[CURTIS L. NEWCOMBE]

Bibliography: J. Fraser, *Nature Adrift: The Story of Marine Plankton,* 1962; C. P. Idyll, *Abyss: The Deep Sea and the Creatures That Live in It,* 1964; H. B. Moore, *Marine Ecology,* 1958; C. L. Newcombe, Mussels, *Turtox News,* vol. 25, no. 1, 1947; E. P. Odum, *Fundamentals of Ecology,* 1953; H. U. Sverdrup, M. W. Johnson, and R. H. Fleming, *The Oceans: Their Physics, Chemistry, and General Biology,* 1942.

Marine resources

The oceans cover 71% of the Earth's surface, to an average depth of 3795 m, with a total volume of $1.37 \times 10^9 \, km^3$. Their living and nonliving contents constitute the basis of several extractive industries. Many of the sea's resources, however, cannot be used profitably at the present stage of knowledge, but a moderate and reasonably certain advance in technology would make them valuable. The development of these latent resources is an important frontier of modern science. Both extractive and nonextractive resources of the oceans are discussed here.

Extractive resources. These include (1) nonrenewable resources, or resources for which the rate of renewal is so slow as to be negligible, such as petroleum and natural gas under the sea floor, mineral deposits on the ocean bottom, dissolved minerals in the water, and the water itself; and (2) renewable resources, such as the living resources of the sea.

Petroleum and natural gas. The continental shelves (the land submerged under less than 600 ft of water) under the margins of the seas extend over about 11,800,000 mi² and include some 30,000,000 mi³ of possible oil-bearing sediments. By comparison with the petroleum content of such sediments on land, it is estimated that they contain about 400,000,000,000 barrels of recoverable crude oil, plus large amounts of natural gas.

Extensive geophysical and geological prospecting has located some of these deposits in the Gulf of Mexico, off the coast of California, in the Persian Gulf, and elsewhere. Successful drilling to recover them has been accomplished in depths of water up to 700 ft and at distances up to 100 mi from shore. Rapidly developing new techniques are extending the water depths and distances from shore in which drilling can be economically conducted. In 1968 about 16% of the free-world petroleum was produced from offshore wells. *See* OIL AND GAS, OFFSHORE.

Minerals on sea floor and in water. The floor of the deep sea is known to contain low-grade deposits of cobalt, nickel, and copper (0.1–0.7% by weight of the metals) associated with deposits of iron and manganese. None of these are now utilized, but fairly large deposits of manganese nodules (see illustration), discovered in 1957 on the tops of some Pacific seamounts, offer commercial possibilities.

Sea water itself contains a large variety of elements as dissolved salts. In 1,000,000 lb of sea water there are, for example, 18,980 lb of chlorine, 10,561 lb of sodium, 1272 lb of magnesium, 380 lb of potassium, and 65 lb of bromine. Extraction of sea salt by evaporation is an ancient industry and is now highly developed both for the recovery of sodium chloride and for the production of sodium sulfate, potassium chloride, magnesium chloride, and magnesium oxychloride cements. Commercial extraction of bromine from sea water was initiated by the Ethyl Corp. in 1924, for the manufacture of the gasoline additive ethylene dibromide. The Dow Chemical Co. initiated, in 1941 at Freeport, Tex., the production of magnesium metal from sea water, employing a combination of chemical and electrolytical processes.

With sufficient cheap power and with depletion of other sources of minerals, the production of some minerals from sea water will become increasingly feasible. *See* MINING, UNDERSEA.

Living resources. The living resources of the sea support the largest, by far, of the extractive marine industries. The world's sea fisheries, producing protein products for human consumption and for animal foods and other purposes, yielded a catch in 1967 of 52,000,000 metric tons of fishes and marine invertebrates. Of this, 2,410,000 metric tons were landed in the United States. In addition, the annual catch of whales, mostly from the Antarctic, yields more than 500,000 metric tons of usable products.

About 60% of the fish catch is made in temperate waters of the Northern Hemisphere, mostly within a few hundred miles of the land, despite the fact that the southern oceans constitute 57% of the world's sea area. The disproportionately large yield from the Northern Hemisphere is related to three factors: (1) Human populations are heavily concentrated there; (2) the major fishing nations are the industrialized maritime nations, which are located there; and (3) except for some tuna, salmon, and herring fisheries, the major sea fisheries are located in the relatively shallow areas along the continents, and the extent of these shallow areas is much greater in the Northern than in the Southern Hemisphere.

That the sea fisheries can yield greatly increased harvests is quite certain. There are vast areas of the sea, especially in the Southern Hemisphere, which are scarcely fished at all, and even in presently utilized areas there are a large number of known fish stocks which are not being harvested to their productive capacity. An example of a newly developed resource is the fishery for anchovy off the coasts of Peru and Chile that in 1967 produced over 10,000,000 tons, being the largest single-species fishery in the world. The sea-fish harvest has increased rapidly, from 17,000,000 metric tons in 1948 to 52,000,000 in 1967, but the protein deficit in many parts of the world will allow an even more rapid increase as technological advances in fish catching, processing, and distribution make possible the economic exploitation of the unused resources.

The plant life of the open sea consists of microscopic plants, phytoplankton, not amenable to direct harvesting. The larger seaweeds (algae), which are of commercial importance, occur only along the shallow edges of the sea. The *Phaeophyta*, which include the giant kelps, are the basis of important industries in the United Kingdom, Japan, and the United States. They contain a colloidal chemical substance, algin, similar to cellulose, which is of wide application in food and pharmaceutical products and in rubber and textile manufacturing. From certain genera of the *Rhodophyta*, or red algae, are produced agar and carrageenin. The world seaweed harvest amounted in 1956 to 370,000 metric tons and is capable of a large increase.

Despite the fact that the sea fishes are not being utilized anywhere near the limit of their biological potential, some scientists have become fascinated by the possibility of gathering organic material from the sea at a level lower in the food chain, such as zooplankton, which feed on microscopic plants and are, in turn, fed upon by the fishes and other higher organisms. It is, however, overlooked that, although the zooplankton production in the sea is large relative to that of the fishes, the standing crops of either represent a very small volume in a very large volume of water. The schooling habit of most commercial fishes makes it possible to catch them in economically feasible quantities. For the more dispersed zooplankton, which constitutes only a few parts in a million parts of water, the problem of profitably straining out the organisms from the water is formidable. At some future time, when man's protein foods needs are more pressing than now and when new techniques may be developed, plankton harvesting may become commercially possible. In the foreseeable future, however, the food harvest must continue to depend mainly on the fishes.

Water. Fresh water, which is a critical resource in arid regions, may be recovered from the sea by distillation, ion-exchange, and other processes. Research and development studies of these techniques are receiving much attention, but a sizable cost reduction is yet required to make such water economically usable for agriculture on a large scale, although it is already feasible for some domestic and industrial uses.

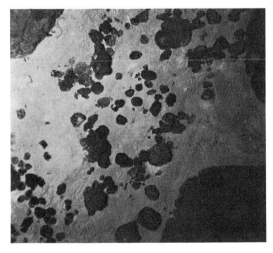

Calcareous ooze with large nodules of manganese in the southeastern Pacific, latitude 42°50′ S, longitude 125°32′ W. Depth 4560 m. (*Photograph by C. Shipek, U.S. Navy Electronics Laboratory, San Diego, Calif.*)

Nonextractive resources. Other aspects of marine resources include disposal of waste products and use of the sea as a source of energy, as a medium of transportation, and for recreational purposes.

Waste disposal. Disposal of domestic sewage and industrial wastes is conveniently accomplished near coastal population centers by running them into the adjacent sea. There, the large volume and rapid mixing of the waters dilute the wastes, and the bacteria in the sea break down the organic constituents. It is necessary, however, to consider in some detail the local effects of tidal and wind currents, density stratification, rates and volumes of mixing, the character of the bottom sediments, and the rate of disappearance of human bacteria as a basis for planning such disposal without incurring a pollution hazard.

In the coming era of large nuclear fission power plants, it may be necessary, particularly for countries with densely populated land areas and long sea coasts, to dispose of some fission waste products in the sea. Safe ocean disposal of radioactive wastes involves the selection of sites where rapid and profound dilution will occur, or where sufficient decay will take place before the radioactive waters and their contained organisms come into contact with human beings. Some low-level radioactive wastes are already being safely disposed of in coastal waters, but it is certainly not safe to introduce large quantities of high-level wastes there. Deep-ocean disposal may be possible. However, much more must be known about the deep-ocean circulation, and the transfer of elements between the deep sea and the surface waters (where man uses the sea) by physical and biological processes, before it can be stated with certainty where and under what circumstances specified quantities of radioactive waste products can safely be introduced.

Energy. Of the several forms of energy in the sea that are capable of being used to produce power, the most apparent is the ebb and flow of the tide. Several attempts have been made to harness it, but the only major tidal power plant in operation is near the mouth of the Rance River in France.

Transportation. Long-distance transportation of large cargoes by sea is the indispensable basis of international commerce. Major opportunities for more efficient use of this resource lie at the boundary between sea and land, since major problems and costs of ocean transportation are involved with getting cargoes on and off the ships. Increased knowledge of the effects of waves, currents, and tides on dredged channels and structures can provide the basis of improved harbor design and development. The possibility of creating large artificial harbors by using nuclear explosives may make possible the creation of good harbors on sea coasts where none exist.

Recreation. The recreational aspects of the sea are of importance to coastal populations in the temperate and subtropical regions, not only in providing healthful sports and satisfaction of man's curiosity and desire for beauty, but as the basis of large tourist and service industries. Important technical problems arise in rectifying conflicts between this use of coastal waters and other uses, such as commercial fishing, waste disposal, and oil drilling.   [MILNER B. SCHAEFER]

Bibliography: J. F. Brahtz (ed.), *Ocean Engineering*, 1968; H. W. Menard, *Marine Geology of the Pacific*, 1964; J. L. Mero, *Mineral Resources of the Sea*, 1964; J. E. G. Raymont, *Plankton and Productivity in the Oceans*, 1963; R. Revelle et al., *The Effects of Atomic Radiation on Oceanography and Fisheries*, Nat. Acad. Sci.–Nat. Res. Counc. Publ. no. 551, 1957; M. B. Schaefer and R. R. Revelle, *Natural Resources*, 1959; D. K. Tressler and J. M. Lemon, *Marine Products of Commerce*, 2d ed., 1951.

Meteorology

The science concerned with the atmosphere and its phenomena. Meteorology is primarily observational; its data are generally "given." The meteorologist observes the atmosphere—its temperature, density, winds, clouds, precipitation, and other characteristics—and aims to account for its observed structure and evolution (weather, in part) in terms of external influence and the basic laws of physics.

Empirical relations between observed variables, as those between the patterns of wind and weather, are developed to pose more effectively the problems to be investigated and explained and to provide essential material for the application of the science. Weather forecasting serves as an example of such application because theory still remains insufficiently developed to provide more certain applications. Little controlled experiment has been made on the atmosphere, but more is probable. *See* WEATHER FORECASTING AND PREDICTION; WEATHER MODIFICATION.

This background article has a threefold organization. The first portion presents a summary of the general physics of the air; this has been the principal approach and basis for meteorological science and its applications, such as to weather phenomena (the condition of the atmosphere at any time and place) and climate (a composite generalization of weather conditions throughout the year). A second portion, synoptic meteorology, discusses the character of the atmosphere on the basis of simultaneous observations over large areas. Concurrently, and at an accelerating pace, dynamic principles (thermodynamic and hydrodynamic) are being applied to meteorological investigations. This study of naturally produced motions in the atmosphere is forming much of the scientific basis for modern weather forecasting and physical climatology. Hence, the third portion of this article deals with dynamical meteorology. *See* CLIMATOLOGY.

GENERAL PHYSICS OF THE AIR

The components of dry air, excluding ozone, and their relative volumes (mol fractions of gases) up to a height of at least 50 km are given in Table 1. Some of the rarer gases, such as CO_2, are continually entering and leaving the atmosphere through the Earth's surface, and the fractions quoted are thus mean values. The effective molecular weight of the mixture is 28.966 and its equation of state, to 1 in 10^4, is $p = R\rho T$, where p is pressure, ρ density, T absolute temperature, and R, the specific gas constant, is 2.8704×10^6 erg/(g)(°K). *See* ATMOSPHERE.

Above 50 km, O_2 becomes progressively dissociated to O, which is probably dominant above about 150 km. N_2 probably dissociates at apprecia-

Table 1. Components of dry air

Gas	Symbol	% vol
Nitrogen	N_2	78.09
Oxygen	O_2	20.95
Argon	Ar	0.93
Carbon dioxide	CO_2	0.03
Neon	Ne	1.8×10^{-3}
Helium	He	5.2×10^{-4}
Krypton	Kr	1×10^{-4}
Hydrogen	H_2	5×10^{-5}
Xenon	Xe	8×10^{-6}
Nitrous oxide	N_2O	3.5×10^{-5}
Radon	Rn	6×10^{-16}
Methane	CH_4	1.5×10^{-4}

bly higher levels. There is no good evidence for diffusive gravitational separation of the lighter from the heavier gases below about 100 km—nor indeed is this likely. Above 100 km, however, such separation is probable and in the levels of escape (exosphere), at several hundreds of kilometers, helium and hydrogen may be dominant with proportions varying during the solar 11-year cycle.

Water vapor and ozone are highly variable additional components of air. The fraction of the former commonly decreases rapidly with height, from about 10^{-2} at sea level to 10^{-6} or 10^{-7} at about 16 km, with dissociation at much higher levels. Ozone, the product of photochemical action in sunlight at high levels, has a maximum fraction ($\sim 10^{-8}$) at about 25 km. The importance of these two constituents is far greater than their fractions might suggest because (1) both are radiatively active, water vapor and ozone in the infrared of terrestrial emission, ozone in the near ultraviolet of solar emission; and (2) water vapor condenses to the liquid or solid, giving cloud, which reflects upward a major part of solar radiation incident upon it, and in condensing releases a large latent heat to the air. Carbon dioxide, CO_2, is the only other constituent with strong infrared activity.

Thermal structure. It is convenient to divide the atmosphere into a number of layers on the basis of its thermal structure. The first layer (Fig. 1) is the troposphere, in which temperature on average decreases with height 6°C/km (the "lapse rate") everywhere except near the winter pole. The troposphere is about 16 km deep in the tropics and about 10 km deep, with substantial variations, in higher latitudes. At any one level in the layer, the mean temperature decreases from Equator to pole, by about 40°C in winter and 25°C in summer.

The second layer is the stratosphere, in which the temperature varies at first very little with height and then increases to near the surface temperature at about 50 km. In this layer temperature increases poleward, except quite near the winter pole. The boundary between troposphere and stratosphere is termed the tropopause, which may drop abruptly in altitude at about 30° lat and again in higher latitudes.

From 50 km to 80 km the temperature again decreases with height to an absolute minimum of about −80°C. This layer has been called the mesosphere and its lower and upper boundary the stratopause and mesopause, respectively.

Above 80 km the temperature again increases with height, at a rather uncertain rate. The layer is strongly ionized by solar ultraviolet and other radiations and is variously termed the ionosphere or thermosphere.

Radiation and thermal structure. During the passage of solar radiation downward through the atmosphere, the following, in broad outline, takes place. The shorter ultraviolet waves are absorbed by the thermosphere, more above than below, so that temperature may be expected to fall along the path of the beam. Entering the mesosphere, O_3 (ozone) is encountered, and new absorption of energy takes place in the near ultraviolet. The increase of ozone concentration along the path outweighs the depletion of the beam energy by O_3 at upper levels so that the temperature increases along this portion of the path. Beneath the stratopause, however, depletion in the wavelengths concerned has become large and the O_3 concentration itself ultimately falls so that temperature decreases along the path. There is little absorption in the troposphere—the slight near-infrared absorption by water vapor is practically uniform along the path because of the increasing concentration of vapor—but there is substantial backscatter by air molecules and by clouds. The Earth's surface, except snow, absorbs the incident solar radiation

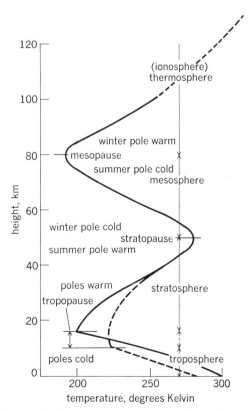

Fig. 1. Vertical structure and nomenclature of atmospheric layers in relation to temperature. Values of temperatures shown are approximate; the form of the temperature variation with height determines the nomenclature. The two curves shown for troposphere and stratosphere refer to lower (full line) and higher (dashed line) latitude conditions. The sense of the meridional temperature gradient at various levels is shown by entries "poles cold" and "poles warm." A change of meridional gradient appears to coincide with a change of vertical gradient (pause) only at the tropopause.

strongly, and this leads to the heating of the atmosphere from below by various kinds of convection and to a lapse of temperature in the troposphere.

Because the atmosphere and underlying surface maintain their temperature over the years, the terrestrial emission of radiation to space by water vapor, carbon dioxide, and ozone in the far infrared, must nearly balance the absorption of solar energy. The balance is achieved globally but not locally, the winds providing the necessary transport of heat from the regions of net absorption (lower latitudes) to those of net emission (higher latitudes). While the broad form of the temperature profile of Fig. 1 is determined by the pattern of solar absorption, actual temperatures are also due to terrestrial emission and wind transport.

Pressure and wind. The pressure at any point in the atmosphere is the weight of a column of air of unit cross section above that point. Therefore pressure decreases with height. At any level it decreases more rapidly where the air is cold and dense than where it is warm and light. The pressure at sea level is generally a little over 1 kg/cm², varying by a few percent in space and time because of the inflow of air over some regions and its outflow over others. The horizontal pressure pattern necessarily changes with height when, characteristically, the temperature varies horizontally.

In an unheated, nonrotating atmosphere, air would flow horizontally from high pressure to low to remove the pressure difference. Other forces arise in the actual atmosphere to provide a flow which is more nearly along, than perpendicular to, the isobars (lines on a map connecting points of equal barometric pressure). Therefore pressure patterns persist for days, with gradual modification, and are translated with speeds often comparable with those of the winds blowing in them. The effect of the Earth's rotation on the wind is nearly always dominant, except very near the Equator. If it precisely balances the pressure force, the wind blows with speed V along the isobars (for an observer with his back to the wind, low pressure is to the left in the Northern Hemisphere). The speed is given by the equation below,

$$V = \frac{(\partial p/\partial n)}{2\omega\rho \sin\phi}$$

where $\partial p/\partial n$ is the horizontal pressure gradient, ω the angular velocity of the Earth, ρ the air density, and ϕ the latitude. This wind V is called the geostrophic wind and is a good approximation to the actual except near the surface, where friction always intervenes. Since, as seen above, the pressure gradient generally varies with height, so do both V and the actual wind. The change of V with height is proportional to the horizontal temperature gradient, low temperature being to the left of the vector change of V in the Northern Hemisphere. See AIR PRESSURE; WIND.

General atmospheric circulation. The mean wind in nearly all parts of the atmosphere is predominantly along the latitude circles. The pattern of this zonal motion in latitude and height is shown in Fig. 2 and is the basis of the general circulation of the atmosphere. Surface easterlies (trade winds) are found in the tropics, surface westerlies in middle latitudes, and surface easterlies again near the poles. But everywhere in the troposphere, except near the Equator, winds increase in strength from the west with height; in the tropics and polar regions easterlies decrease before giving place aloft to increasing westerlies. Above the tropopause the westerlies decrease with height except over the winter polar cap where, at around 40 km, there occurs a winter westerly jet stream. This pattern of winds is consistent with the meridional gradient of temperature in the troposphere and its reversal (except over the winter polar cap) in the stratosphere, since the winds are quasi-geostrophic. The absolute maximum of zonal wind in Fig. 2 is found in the upper troposphere at about 30° lat and is the core of the subtropical jet stream. See JET STREAM.

The general circulation is maintained against frictional dissipation by a "boiler-condenser" arrangement provided by the heat absorbed, directly or indirectly, from the Sun and that emitted to space by the Earth. The wind thereby generated transports heat from source to sink to maintain thermal balance, disturbances on the mean motion being important in the process. See ATMOSPHERIC GENERAL CIRCULATION.

Water vapor, cloud, and precipitation. The motion of the air is rarely quite horizontal; it is commonly moving upward or downward at a few centimeters per second over large areas, while locally in thunderclouds the vertical velocity may be several meters per second. Air which rises is cooled by expansion into lower pressure and that which descends is correspondingly compressed and warmed. Since air is always more or less moist, a sufficient rise will cause the temperature to drop to the saturation point and cloud will then form on nuclei present in the air. Large-scale ascent leads to extensive sheets of cloud called stratus: cirrostratus at high, ice-forming levels, altostratus at medium levels, nimbostratus at lower levels. Local, strong ascent produces heap clouds (cumulus or cumulonimbus). If the cloud base is sufficiently warm or the cloud depth sufficiently great, or both, the condensed water or ice in the cloud falls out as rain, snow, or other precipitation. Certain lenticular clouds are due to local ascent and descent brought about by hills, and these clouds do not move with the wind. Cloud formation is practically confined to the troposphere because this alone contains sufficient water vapor and sufficiently sustained vertical motion. See CLOUD; CLOUD PHYSICS.

Atmospheric disturbances. The mean motion, described under the general circulation of the atmosphere, is disturbed by many patterns of motion of varying scale. A chart of the flow well above the surface over the Northern Hemisphere will generally show, on any day, large meanderings of the air poleward and equatorward superposed on a general westerly flow. These are the long waves of the westerlies, with a few thousand kilometers between the turning points. They are generally associated with a jet stream in the upper troposphere. Another class of large-scale disturbances, seasonal in nature and most apparent in the lower troposphere, is designated the monsoons; these are most developed in the south Asian cyclone of summer and Siberian anticyclone of winter.

Proceeding downward in scale are the traveling depressions (cyclones) and anticyclones, ridges

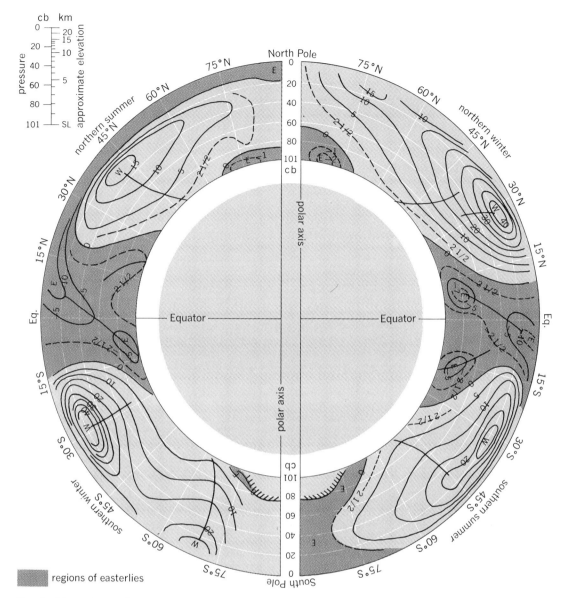

Fig. 2. The pattern of mean zonal (east-west) wind speed averaged over all longitudes as a function of latitude, height, and season (*after Y. Mintz*). Height, greatly enlarged relative to Earth radius, is shown on a linear pressure (∝ mass) scale with geometrical equivalent given at upper left. The mean zonal wind (westerly positive, easterly negative) has the same value along any one line and is shown in meters per second (1 m/sec ≅ 2 knots) on the line. Note subtropical westerly jet in high troposphere at about 30° lat (or more in northern summer). This jet should not be confused with the polar front jet of higher latitudes; the latter is migratory and so does not appear in the mean.

and troughs of extratropical latitudes, 1000–2000 km in horizontal extent, and then the somewhat shorter waves in the easterlies (trades). Next in size is the tropical cyclone (~100 km)—the hurricane or typhoon, according to location—which is mainly confined to the oceans and eastern seaboards of continents. Still smaller (~1 km) is the tornado, mainly confined to land and associated with cumulonimbus cloud, and the waterspout associated with similar cloud over the sea. The smallest "revolving storm" is the dust devil (~10 m), which occurs immediately above a hot, dry surface in a very light general wind. *See* FRONT; STORM; TORNADO.

In addition to the above well-defined patterns of

flow there are randomly distributed fluctuations of flow over several orders of magnitude of scale from about 1 cm upward, particularly evident in the bottom kilometer of the atmosphere and in and around cumulus. They are related to friction, convection, and wind shear and are important agents of vertical transfer of heat, matter (water vapor, dust, ozone), and momentum.

Air masses and fronts. The horizontal variations of air temperature referred to in previous paragraphs are commonly concentrated into narrow zones, or even discontinuities, with low gradients in intervening areas. These narrow zones or discontinuities are called fronts, and the air in a region of small temperature gradient is called an air mass. If a front moves so that a warm air mass replaces a cold air mass, it is called a warm front, and conversely a cold front. Fronts may also be stationary. They slope upward at a slope ratio of about 1 in 100, with the cold air as a wedge beneath the warm air.

Most extratropical depressions first appear as dents or waves on a frontal surface, warm air rising slowly over an extended area in the forward part of the wave. This results in stratiform cloud, and a stronger updraft immediately ahead of the rear part of the wave yields cumulonimbus.

The tropopause is higher and its temperature lower above a warm air mass than above a neighboring cold air mass, and a break, or offset, appears at the frontal boundary with a jet stream in the warm air mass near the break in the tropopause. *See* AIR MASS.

Optical phenomena. Scattering, refraction, and reflection of light by the air or by particulate matter (dust, cloud particles, or rain) in the air give rise to a variety of optical phenomena, constituting the field of meteorological optics.

Electrical properties. The atmosphere and the underlying Earth are like a leaky electrical capacitor. Positive charge is separated vertically from negative charge in thunderclouds and some other areas of disturbed weather; the net result is that the Earth's surface is left in fine-weather areas with an average negative charge σ of 2.7×10^{-4} esu/cm² to which corresponds a vertical field $F_0 (= 4\pi\sigma)$ at the surface of 100 volts/m. The other "plate" of the capacitor is the highly conducting upper atmosphere, and the leak arises from the small conductivity of the air between the plates, produced by ionization by radioactive matter in the soil and air, and by cosmic radiation; the conductivity increases with height. The resistance R of a 1-cm² column of atmosphere is about 10^{21} ohms and an air-Earth current $i = V_\infty/R$ (V_∞ being the potential of the upper conducting layer) of about 2×10^{-16} amp/cm² flows as a discharge current. V_∞ is thus about 2×10^5 volts above Earth. The increase of conductivity with height implies a proportionate decrease of the field with height, and this in turn implies a small positive ionic space charge in the air.

The air-Earth current would discharge the capacitor in about 1/2 hr if V_∞ (or σ) were not maintained by the charge separation in thunderclouds. This separation is more than adequate, the excess providing lightning flashes within the cloud—a shorting of the generator. [P. A. SHEPPARD]

SYNOPTIC METEOROLOGY

This branch of meteorology comprises the knowledge of atmospheric phenomena connected with the weather, applied mainly in weather forecasting, and based on data acquired by the synoptic method. This method involves the study of weather processes through representations of atmospheric states determined by synchronous observations at a network of stations, most of which are at least 10 km apart. By international agreement, the data taken from the Earth's surface and aloft at certain international hours of observation are inserted on weather maps, upper-air maps, vertical cross sections, time sections, and sounding diagrams with international symbols

according to fixed rules. These crude representations are then analyzed and critically evaluated in accordance with the knowledge of existing structure models in the (lower) atmosphere, in order to ascertain the best approximation to a three-dimensional image of the true atmospheric state at the hour of observation. From one such representation, or a series of them, future atmospheric states are then derived with the aid of empirical knowledge of their behavior and by application of the theoretical results of dynamic meteorology.

General atmospheric circulation. Figure 3 shows some of the main constituents of the average state in the bottom layer of the atmosphere as to flow and pressure patterns, main air masses and fronts. The illustration also indicates how these large-scale mechanisms form part of a general atmospheric circulation with trade winds, monsoons, high-reaching middle-latitude westerlies, and shallow polar easterlies (see the vertical cross section of Fig. 2).

The planetary high-pressure belt at 30°N and S is split into subtropical high-pressure cells mainly as a result of the joint dynamic and thermodynamic effect of the great continents and mountain ridges of the Earth, especially the Cordilleran highlands of South and North America. These cells again determine the formation and average position of the different polar fronts and air masses at the Earth's surface.

Other general structures in the atmospheric circulation of importance to weather are the quasi-horizontal tropopause layers at different heights within different air masses and the tropical fronts, maintained within the Zones of Intertropical Convergence (ITC). The former generally mark the top of any considerable convection or upglide motion and of clouds in the atmosphere, but represent no store of potential energy. The latter form at the meeting of two opposite trade wind systems in the doldrums and have functions partly similar to those of the polar fronts, although they are much weaker and less distinct.

Air masses and front↔jet systems. Air masses may be classified in two distinctly different ways, thermodynamically and geographically.

Thermodynamic classification. This classification is based on their recent path and life history, distinguishing mainly the two opposite cases: the air being warmer or colder than the Earth's surface, resulting in warm mass and cold mass. The warm mass, usually flowing poleward, is much warmer than the seasonal normal of the region, at least aloft. With the cooling from below, it gradually acquires a stable stratification, which damps turbulence and vertical mixing. Thus, the wind is relatively steady, the visibility low, and advection fog or stratus clouds often form within it, at least at sea, sometimes even yielding drizzle. This air mass is most typically found on the poleward side of the warm subtropical highs (Figs. 3 and 8), where the air is subsiding. These highs may get displaced poleward (Fig. 8) and will then bring periods of steady and rather dry weather—very warm in summer on land—to middle and higher latitudes. *See* WIND.

The cold mass, mostly flowing equatorward, is by definition colder than normal, at least aloft. Due to heating from below it rapidly acquires an unsta-

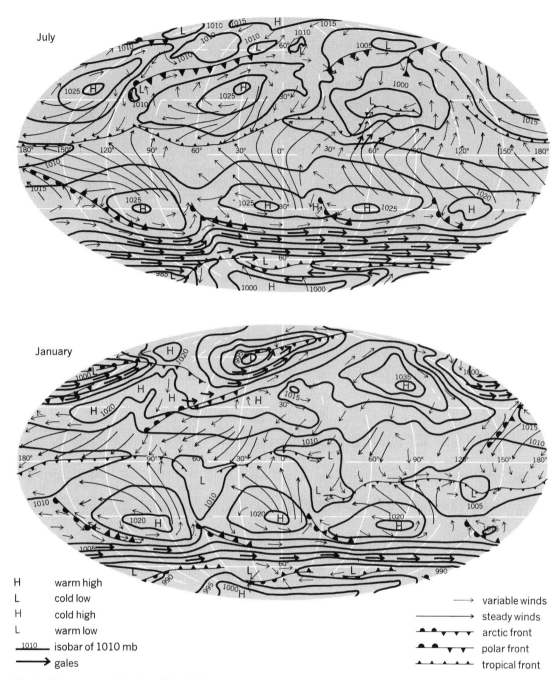

July

January

H   warm high
L   cold low
H   cold high
L   warm low
<u>1010</u>   isobar of 1010 mb
⟶   gales

⟶   variable winds
⟶   steady winds
⟶   arctic front
⟶   polar front
⟶   tropical front

Fig. 3. Air masses and fronts as links in the general atmospheric circulation.

ble stratification which favors turbulence and vertical mixing or convection. Thus, the wind is gusty, visibility is good, and usually where the air is moist enough convective clouds form: cumulus → cumulonimbus with showers → thunderstorms and even hail. This air mass is most typically found within (upper) cold lows (Fig. 10), where, if there is sufficient moisture, the instability, general convergence, and lifting tendency of the air may favor the formation even of nonfrontal rain areas, which are discussed later. Therefore, such a low usually brings a period of wet and stormy weather to the equatorward part of middle latitudes, and in winter to the subtropics. However, at night over land, even this air may be stabilized so ef-

ficiently that radiation fog occurs within it.

Geographic classification. The second classification is based on the geographical origin or position of the air and the values of characteristic properties (such as temperature and specific humidity).

Tropical air occupies all the space between the polar-front↔jet systems of both hemispheres; aloft it may reach far into the polar region. At the midtroposphere this air is 10–20°C warmer and much more humid than the polar air at the same level and latitude. At low levels, the tropical air emerges from the quasi-stationary subtropical highs, reaching middle latitudes from the southwest, particularly within the warm sectors of migrating cyclones (Fig. 7), as a mild or warm, moist

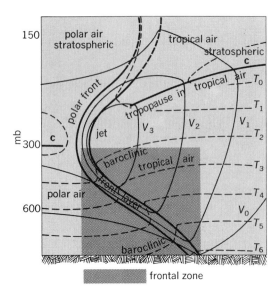

Fig. 4.   Cross section of a polar-front ↔ jet system.

and hazy air current (the stable warm mass). Aloft, it appears in high latitudes within the warm highs (Fig. 8). Below, the tropical air also flows equatorward and westward within the northern and southern trade wind systems. When approaching the tropical front (Fig. 3) it appears, at sea, as an unstable cold mass with intense convection and shower activity on both sides of the front: the equatorial rains. Over land, for example, in North Africa, only the southwestern monsoon (that is, the southeastern trade wind that has invaded the other hemisphere), undercutting the very hot and dry northeast trade wind, is moist enough to produce any convective clouds and rain.

Polar air is found on the polar side of the polar fronts, as shown in Fig. 3. Below, it emerges in winter from continental subpolar highs of 40–60°N, and in summer from the polar basin. The polar masses, and still more so the arctic and antarctic air masses, are mainly characterized by very

low temperatures and low specific humidities aloft.

Arctic air is produced over the ice- and snow-covered parts of the arctic region during the colder seasons, when the Earth's surface is a marked cold source. In the North American sector its seat, for dynamic as well as thermal reasons, lies on the average near Baffin Island, being separated from the polar air south of it by the American arctic front (Fig. 3).

Fronts and frontogenesis. Because of the Earth's rotation a surface of density discontinuity, a front, seeks a tilted position of dynamic equilibrium. A front is defined as a dynamically important, tilting layer of transition between two air masses of markedly different origin, temperature, density, and motion. Colder air lies as a wedge below the warmer; therefore, the front will coincide with a trough or bend in the (moving) isobaric system, at whose passage the wind will veer, and as a rule with falling pressure ahead and stationary or rising pressure behind. Frictionally produced convergence, or static and inertial instability, or both, preferably within the warmer air, favors the ascent of air along the front. As a result, a vast cloud mass, a cloud system, may form, with an area of continuous precipitation near the front (Fig. 6).

The main fronts reach into the stratosphere (Fig. 4) and have a horizontal extension of several thousand kilometers. They originate as links in the atmospheric circulation, within frontogenetic zones, between two stationary anticyclones (highs) and two cyclones (lows) when the axis of stretching, or confluence, of this deformation field (Fig. 5) has some extension west to east. This pattern is in its turn determined by the large-scale orography of the Earth. Because of the north-south temperature contrast, air masses of different temperature, specific humidity, and density are then brought together in a front zone along this axis. On the warm side of the polar fronts, which together more or less encircle the hemisphere, the thermal wind, as discussed in dynamical meteorology below, corresponding to the mass distribution of the fronts, implies a narrow band of intense flow near the tropopause level, a jet stream. The thermal wind is most pronounced at middle and higher latitudes, where the front is marked, steep, and usually reaches into the stratosphere (Fig. 5). There, a main front and its jet together represent a zone of maximum potential and kinetic energy. Below, in lower latitudes, the frontal tilt is often small, and there is an outflow of polar air into the tropics.

A front may appear as warm or cold. At a warm front, the warmer air gains ground and slides evenly upward above the retreating cold-air wedge (tilt or slope 1:200 to 1:100), producing a wedge-shaped upglide cloud system (Fig. 6a). At the approach of a marked warm front, therefore, a typical cloud sequence invades the sky: cirrostratus → altostratus → nimbostratus, the last yielding continuous and prolonged precipitation ahead of the front line.

At a cold front the rather steep cold-air wedge (with a slope of 1:100 to 1:50) pushes forward under the warm air and forces it upward according to one of the two flow patterns shown in Fig. 6b and c. At the approach of a cold front of the more common type (Fig. 6b) there is, therefore, another typical cloud sequence: altocumulus, partly lentic-

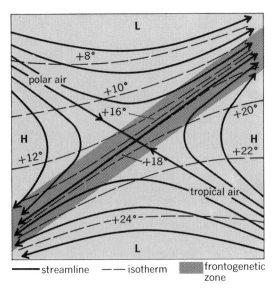

streamline   — — isotherm   frontogenetic zone

Fig. 5.   Frontogenesis in a field of deformation.

ular, rapidly thickening into nimbostratus. The precipitation is usually more intense but of shorter duration than with the warm front. *See* CLOUD; CLOUD PHYSICS.

Waves and vortices (the weather). The actual atmospheric states affecting the weather may be regarded as composed of the general circulation and its disturbances. Together they form a multitude of weather mechanisms, some of which have already been described, in which the water-vapor cycle (evaporation → transport and lifting → condensation → precipitation) evidently has a fundamental role. The atmospheric disturbances consist of a spectrum of waves of different wavelengths λ (Table 2), corresponding circulations (vortices), or both. Only the larger of these are studied synoptically. By their size the circulations may be classified as planetary (or geographical), secondary, or tertiary:

1. The long waves forming in the front↔jet zone, that is, the region of maximum energy, may represent a steady state (see section on dynamical meteorology) when their wavelength corresponds to about four circumpolar waves at low, three at medium, and possibly two at high latitudes. The planetary waves have $\lambda \sim 10,000$ km. Thereby they partly determine the shape of the general circulation (Fig. 3). The shorter long waves ($3000 < \lambda < 8000$ km) propagate slowly eastward and, because of inertial instability, mostly develop as shown in Figs. 8 and 10 through the stages wave → tongue → cutoff vortex. Since the air aloft flows rapidly through these quasi-stationary long-wave patterns, isotherms will approach coincidence with the streamlines and isobars. Thus, equatorward tongues and cutoffs will be cold and will tend to coincide with lows (at least aloft); the poleward ones will tend to be warm and coincide with highs. The former will contain a polar-air hourglass (at least in higher latitudes) or dome, possibly with an arctic-air dome inside it (Fig. 9). Correspondingly, the poleward tongues and cutoff vortices will contain tropical air and a much higher tropopause.

2. The secondary waves are the short waves ($1000 < \lambda < 3000$ km). The short waves in a front↔jet system, when unstable, will form secondary circulations, developing according to the scheme of Fig. 7, that is, initial front wave → young cyclone → initial occlusion → backbent occlusion. Important secondary circulations also form outside the front ↔ jet systems: the easterly wave, the tropical hurricane, and the convective system, occurring primarily in lower latitudes, the last two without an obvious preceding wave stage.

3. The tertiary circulations, barely observable by the ordinary synoptic network, require a mesoscale network ($d < 10$ km), time sections, or both, for a detailed study and forecast. Weather mechanisms of this size are land and sea breezes, mountain and valley winds, katabatic and other local winds (foehn, chinook, bora), local showers, tornadoes, orographic cloud and precipitation systems, and lee waves. Moreover, numerous different orographic factors and the daily period of radiation affect most meteorological elements and weather mechanisms. Their detailed study is basic for both climatology and weather forecasting.

Life cycle of extratropical cyclones. An extratropical cyclogenesis starts as a wavelike front

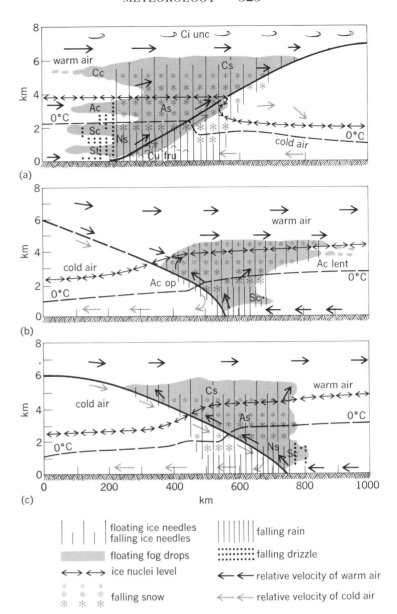

Fig. 6. Schematic cross sections of three kinds of fronts. (*a*) Warm front. (*b*) Fast-running cold front. (*c*) Slow-moving cold front. Ac, altocumulus; Ac lent, altocumulus lenticularis; Ac op, altocumulus opacus; As, altostratus; Ci unc, cirrus uncinus; Cc, cirrocumulus; Cs, cirrostratus; Cu fru, cumulus fractus; Ns, nimbostratus; St, stratus; and Sc, stratocumulus.

Table 2. Atmospheric waves

Name	Type*	Wavelength, km
Ultrasound	C	$<2 \times 10^{-5}$
Ordinary sound (tones)	C	2×10^{-5} to 10^{-2}
Explosion waves	C	10^{-2} to 10^{-1}
Helmholtz waves (Sc und, Ac und)	G	10^{-1} to 1
Short lee-waves (Ac lent, Cc lent)	G	1 to 2×10
Long lee-waves (nacreous clouds, precip.)	G(I)	2×10 to 10^2
Short jet-waves (frontal)	GI(V)	5×10^2 to 5×10^3
Long jet-waves	V	5×10^3 to 10^4
Tidal waves	(G)	$\leqq 2 \times 10^4$

*C, compression, longitudinal; G, gravitational; I, inertial; and V, vorticity-gradient; the last three are transversal.

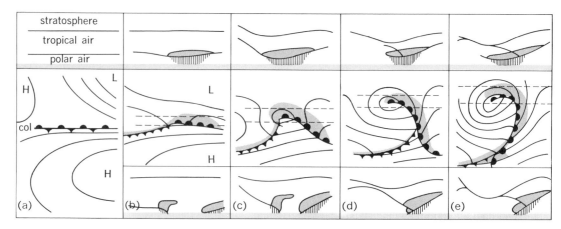

Fig. 7. Life history of cyclones from wave to vortex. The vertical sections (top and bottom) show clouds and precipitation along the two dashed lines in the map bulge. sequence. (*From C. L. Godske et al., Dynamic Meteorology and Weather Forecasting, Waverly Press, 1957*)

bulge. Ahead of the wave the front becomes a warm front, behind it a cold front. These two sections then flank a warm tongue (Fig. 7b) moving along the front. At first there is only a shallow low around the tip of this warm sector, but even the passage of such an initial frontal wave may cause

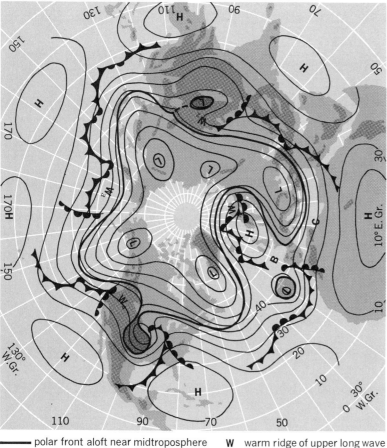

— polar front aloft near midtroposphere
⌢ polar front at Earth's surface
— pressure line aloft
H L high and low aloft
▧ ground > 2000 m above sea level

W warm ridge of upper long wave
C cold upper trough
B blocking
◀L cutoff polar-air dome

Fig. 8. Schematic circumpolar upper pressure pattern in relation to the polar fronts at the Earth's surface. (*After E. Palmén*)

sudden, severe, and unexpected weather changes. If the front wave has an appropriate size ($\lambda \sim$ 1500 km, amplitude $\geqq$ 200 km), it will usually be unstable, narrow gradually, and at last overshoot as does a breaking sea wave. The cold front overtakes the warm front at the ground, and the warm-sector air is lifted and spreads aloft. As a result, the common center of gravity of the system sinks, and potential energy is transformed into the kinetic energy of an increasing cyclonic circulation. As long as this front occluding continues, the circulation increases, and the ensuing low deepens, whereas when the warm sector has disappeared from the interior of the main cyclone, the latter will weaken, and the low fill. These features serve as good, physically comprehensible, prognostic rules. The occluded front is called an occlusion. The occluding process gives the clue to the life history of cyclones and anticyclones (Fig. 7). At its last phase the occluded front often lags behind and bends (Fig. 7d and e), a "false warm sector" forming between the direct and the backbent occlusion. Several cyclones (and lows), together constituting a cyclone series, may form on one and the same front, moving with the upper, steering current, the first one being in the most advanced stage of development (Fig. 8).

Important front↔jet systems in the Northern Hemisphere are the North Pacific polar front and the North Atlantic polar front (Fig. 3). The former extends in winter on the average from near the Philippines north of Hawaii to the northwest coast of the United States; the cyclone series forming in it then brings wet and unsteady weather to the northwestern part of North America. The latter front, extending in winter on an average from near the Bermudas to England, plays the analogous role for almost all Europe, and during certain periods for eastern United States. In summer both these systems are much weaker and lie on an average farther north. The North American arctic front and the European arctic front (Fig. 3) are fundamental for the weather in their vicinity. The former accounts for many severe weather vagaries, such as the blizzards and glaze storms of northeastern United States, and the most severe cold waves and their killing frosts farther south; it is thus of special importance to North American weather forecasting. The production of real arctic air as defined here seems to cease in summer, or at least in July. On the other hand, the corresponding air mass in the Southern Hemisphere, antarctic air, and the antarctic fronts, exist throughout the year, the antarctic ice plateau being an enormous cold source even in summer. Evidently most migrating extratropical precipitation and storm areas originate at front⟷jet systems, separating air masses of radically different motion and weather type. Therefore, these systems are of utmost importance to weather forecasting, being the real atmospheric zones of danger and main sources of the salient aperiodic weather changes of synoptic size outside the tropics.

Large-scale tropical weather systems. In the tropics (and in the subtropics in summer) nonfrontal convective and other weather phenomena may grow to hundreds of kilometers in width; consequently, they can be studied and forecast individually by synoptic methods. Both on land and at sea, short waves ($\lambda \sim 500$ km) form in the frontless trade winds outside the doldrums. These easterly waves propagate slowly westward, showing a characteristic intensification of the shower activity in their eastern part.

Over tropical seas in late summer and early fall (Northern Hemisphere, August–October; Southern Hemisphere, February–April), conditions exist that favor the formation of tropical hurricanes. Tropical hurricanes are cyclonic vortices (smaller but much more intense than the extratropical ones) of 100–400-km width; they have wind velocities often exceeding 50 m/sec (100 knots), 100–500 mm total rainfall, and very low central pressure. They move on the whole poleward, often recurving around a subtropic high. Necessary conditions for their formation seem to be (1) air (and sea-surface) temperature above, say, 27°C, implying enough lability energy to drive such large-scale convections, (2) a preexisting cyclonic motion and frictional inflow (either at the tropical front or in an easterly wave), needed to order, and possibly trigger, the convections, (3) a divergence mechanism aloft to dispose of the air that converges and rises in their interior, possibly also triggering their formation, (4) sufficient Coriolis force (that is, the hurricanes cannot form too near the Equator), and (5) no disturbing land surface within the area of formation. Whether these conditions are sufficient is not certain. The widespread destructive power of hurricanes (shore flooding, wind pressure, downpour) makes their study a major task of tropical synoptic meteorology and weather forecasting. Tracking those already formed can be done with radar, reconnaissance flights, and satellites, but discovering their imminent formation is still an unsolved problem.

Over land, conditions 1 and 2 never lead to hurricane formation but may instead — within the tropics and also in warmer seasons farther poleward — favor the formation of a migrating convective system, with a forerunning pseudo cold front (in the United States also called a squall line) at its outer edge; these systems have roughly the same extent and precipitative power as a tropical hurricane. Outside the tropics they mainly form in the warmer seasons in cyclonic warm sectors, especially in the midwestern United States, where they provide the main water supply. Because of the flood hazards, soil erosion, and other aspects, and the sudden violent squalls, or even tornadoes, sometimes attending the pseudo cold front, these systems, therefore, form another major problem of weather forecasting. Within the tropics, migrating convective systems, often in the easterly waves, will cause more variation of weather from day to day than is generally recognized, whereas the small-scale showers usually have a daily period; together they constitute the equatorial rains.

Apart from the polar-front jet (meandering between 30 and 70° lat), there is a subtropical jet, and a corresponding front, at about 30° lat near the tropopause level (~15 km altitude). As a rule, it makes only small meridional excursions. Therefore, it shows up (instead of the polar-front jet) as the main wind maximum aloft (at 30° lat) in the average meridional cross section of Fig. 2. But for the same reason, and because its front is confined to the tropopause, its influence on daily weather, at least in the tropics, seems slow and diffuse.

Circumpolar aerology. Since 1940, the technical facilities of meteorology have undergone an explosive development, partly due to the great exigencies during World War II. The vast gaps formerly in the meteorological network over the oceans are partly bridged by stationary weather ships, or ocean weather stations, with complete air-sounding equipment. This huge technical improvement has been followed by equally outstanding scientific achievements in the understanding of the dynamics and thermodynamics of the free atmosphere. The weather mechanisms and weather processes described above, which before 1940 had been mainly observed and studied within the lower half of the troposphere, can now be treated in their entirety and fitted into their general relationships with the rest of the atmosphere. On the whole, an intimate interaction takes place between the disturbances seen on the ordinary synoptic map (surface map) and the upper large-scale patterns shown by Fig. 8 and described above. Figure 9 shows this interdependence three dimensionally and in detail for disturbances of front⟷jet systems (including the arctic front). Particularly, it shows the eight weather regions that are conditioned by such systems, these naturally being of fundamental importance to synoptic meteorology and weather forecasting.

In winter, the polar-front jet lies on an average at 25–45° lat, in summer at 40–60° lat, at 7–11 km alt; it is only a few hundred kilometers broad, with wind speed surpassing 100 m/sec (200 knots). It displays two fundamentally different patterns: the zonal or high index type, where the jet has waves of only small amplitudes; and the meridional or low index type, where its meridional excursions are huge and often are cut off. Figure 10 gives an example of the isothermal, isobaric, and flow patterns at the midtroposphere, represented by the isotherms and isohypses ("contour lines") of the 500-mb surface. It shows three long waves w_1 w_2 w_3, one deep trough T, and four cutoff lows $L_1 L_2 L_3 L_4$. Where the isotherms diverge markedly from the parallelism with the streamlines and "contours," there will be a considerable horizontal advection of colder and denser air or of warmer air (Fig. 10), and ensuing advective changes of pressure and wind at most levels. These observations gave another powerful tool for quantitatively forecasting the thermal changes of flow patterns. Cal-

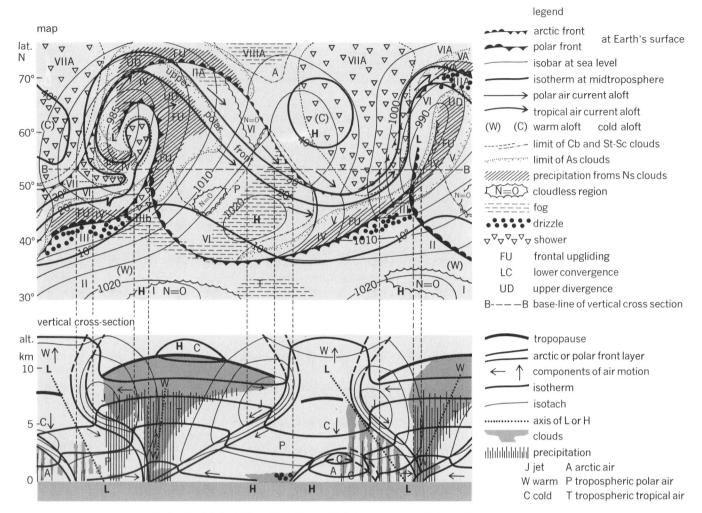

Fig. 9. Model of middle-latitude front↔jet disturbances and their weather regions (fall and spring).

culating the vertical motion by the divergence and vorticity equations has proved even better for this and similar purposes (as discussed below in dynamical meteorology).

Since the planetary waves represent a rather stable steady state, they offer no real forecast problem. The shorter long waves will move east, conserving most of their absolute vorticity, and may therefore be forecast numerically, using a very simplified general model of the atmosphere, the barotropic model, where production and annihilation of kinetic energy is excluded. The propagation and development of the short front ↔ jet waves, that is, the weather mechanisms, will to a certain extent be steered by this long-wave pattern. However, in these mechanisms frontal potential energy is transformed into cyclonic kinetic energy, and the short jet waves will often react upon the large-scale pattern. Such processes, which may lead to intense frontal cyclogenesis and the formation of large-scale cold tongues and cutoffs, confront numerical forecasting with considerable difficulties. However, the fact that the upper steering flow patterns are so simple and move much more slowly than the lower ones has proved a great help to short-range forecasting and has made an extended forecasting (2–5 days) pos-

sible on a synoptic-physical basis. Attempts are being made, with some success, to use baroclinic models of the atmosphere as well, and to incorporate the effect of heat sources and large-scale orography into the numerical short-range and extended forecasting. A main obstacle to real long-range forecasting (7–30 days) is the fact that the causes for a definitive change from the zonal to the meridional upper flow pattern are not yet known.

Aided by radar, continuous detailed mapping and tracking of rain areas is gradually being introduced. Analogously, photographic and television cloud surveys from satellites are used in studying the large-scale development of weather regions.

[TOR BERGERON]

DYNAMICAL METEOROLOGY

This branch of meteorology is the science of naturally produced motions in the atmosphere and of the related distributions in space of pressure, density, temperature, and humidity. Based on thermodynamic and hydrodynamic theories, it forms the main scientific basis of weather forecasting and climatology. *See* CLIMATOLOGY; WEATHER FORECASTING AND PREDICTION.

Thermodynamic properties of air. Atmospheric air contains water vapor and sometimes suspended

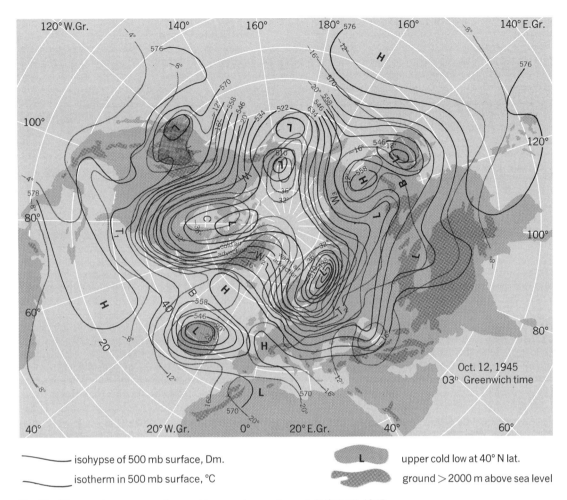

isohypse of 500 mb surface, Dm.

isotherm in 500 mb surface, °C

L   upper cold low at 40° N lat.

ground > 2000 m above sea level

Fig. 10. Circumpolar pressure, flow and temperature pattern aloft, Oct. 12, 1945.

water droplets or ice particles or both. In dynamical meteorology it suffices to use a simplified cloud physics; thus it may be assumed that the water vapor pressure cannot exceed the saturation pressure over a plane water surface and that the vapor pressure equals this saturation pressure whenever the air contains water (cloud) droplets. The distinction between water and ice is ignored most of the time.

Nonsaturated air. Nonsaturated air contains no condensation products and behaves nearly as a single ideal gas. Its pressure p, density ρ, and absolute temperature T satisfy very closely gas equation (1). Its enthalpy per unit mass h is

$$\frac{p}{\rho} = RT \qquad (1)$$

given by Eq. (2), and its entropy per unit mass may

$$h = c_p T + \text{constant} \qquad (2)$$

be written $s = c_p \ln \theta$. The potential temperature θ is given by Eq. (3), where $p_0 = 1000$ mb. In these

$$\theta = T \left(\frac{p_0}{p}\right)^{R/c_p} \qquad (3)$$

expressions the gas constant per unit mass R and the specific heat at constant pressure c_p both depend slightly on the specific humidity or water-

vapor mass concentration m. It is often sufficiently accurate to ignore this dependence and to use the values which hold for dry air; these are $R = 287 \ m^2 \ \sec^{-2} \mathrm{C}^{\circ -1}$ and $c_p = 1004 \ m^2 \ \sec^{-2} \mathrm{C}^{\circ -1}$.

Saturated air. A sample of nonsaturated air becomes saturated if its pressure is reduced or heat is removed or both while its water-vapor concentration m is kept constant. If this process continues, cloud droplets form by condensation at a rate just sufficient to maintain the vapor pressure at the saturation value. If the process is reversed, evaporation takes place. During such phase transitions the release or binding of latent heat causes temperature changes which are important for many motion processes. This property of saturated air is expressed in the formula for its enthalpy per unit mass, Eq. (4). Here L is the latent heat, and

$$h = c_p T + L m_s(p, T) + \text{constant} \qquad (4)$$

m_s is the mass concentration of saturated water vapor, which is a known function of p and T. On the other hand, the liquid (or solid) phase contributes very little to the total mass or volume of the air, so that the gas equation (1) holds with good approximation also for saturated air. Note, however, that the approximate equations (1) and (4) are not fully consistent with the second law of thermodynamics.

Thermodynamic state. The thermodynamic state

of an air sample, whether saturated or not, is defined by three state variables, for example, pressure p, density ρ, and mass concentration m of the water component (in all phases). It is often convenient to use T instead of ρ. Comparison between m and the saturation vapor concentration $m_s(p,T)$ shows whether the sample is saturated, and if so, the mass of the condensed phase.

There are many motion phenomena in which the released latent heat is unimportant. When dealing with such phenomena, the air may be considered dry and its thermodynamical state as defined by the two variables p and ρ (or T) only.

Instantaneous state of atmosphere. The instantaneous state of the atmosphere may be characterized by the distributions in space of the thermodynamic variables p, ρ, and possibly m and three components of the air velocity $\mathbf{v}$ relative to the solid Earth. These basic quantities are regarded as functions of three space coordinates and time, and as such they characterize a sequence of states or a motion process in the atmosphere.

Pressure and density are free to vary independently in space, and the density is usually variable in isobaric surfaces of constant pressure; the density field is then said to be baroclinic. Only in special cases is the density field barotropic, that is, constant in each isobaric surface.

A basic difficulty in dynamical meteorology results from the complexity of the field of motion in the real atmosphere. Superimposed upon the large-scale motion systems revealed by weather maps is a fine structure of motions of all scales down to small eddies of millimeter size. Such motion fields cannot be dealt with in all details; it is necessary to deal instead with smoothed motion fields, where details smaller than a certain scale have been left out. As a consequence of the nonlinearity of the basic equations, these details still exert a certain influence upon the larger-scale motions. This influence can be taken into account only in a statistical sense, in the form of so-called eddy terms which express the transport of momentum, enthalpy, and water substance by small-scale eddies.

Equations. The dependent variables p, ρ, m, and $\mathbf{v}$ satisfy Eqs. (5)–(8), which express, respectively, the conservation of total mass (continuity equation), Newton's second law (equation of motion), the thermodynamic energy equation (or first law), and the conservation of mass of the water component.

$$\frac{\partial \rho}{\partial t} + \mathrm{div}\,(\rho \mathbf{v}) = 0 \qquad (5)$$

$$\rho \frac{D\mathbf{v}}{Dt} = -\mathbf{k} \times f\rho\mathbf{v} - \rho g\mathbf{k} - \mathrm{grad}\, p - \frac{\partial \mathbf{F}_M}{\partial z} \qquad (6)$$

$$\rho \frac{Dh}{Dt} - \frac{Dp}{Dt} = -\frac{\partial}{\partial z}(F_{\mathrm{rad}} + F_h) + \Delta \qquad (7)$$

$$\rho \frac{Dm}{Dt} = -\frac{\partial}{\partial z}(F_{\mathrm{precip}} + F_m) \qquad (8)$$

Here Eqs. (9) and (10) apply, and the notation

$$f = 2\Omega \sin \Phi \qquad (9)$$

$$\frac{D}{Dt} = \frac{\partial}{\partial t} + \mathbf{v} \cdot \mathrm{grad} \qquad (10)$$

is as follows: t denotes time; D/Dt individual time derivative or rate of change as experienced by a moving air particle; $\mathbf{k}$ a vertical unit vector; f Coriolis parameter; Φ latitude; $\Omega = 0.7292 \cdot 10^{-4}$ sec^{-1} angular velocity of the Earth's rotation; g acceleration of gravity; $\mathbf{F}_M$ vertical eddy-flux density of momentum; F_{rad} vertical radiative-heat-flux density; F_h vertical eddy-flux density of enthalpy; F_{precip} vertical flux density of water substance by precipitation; F_m vertical eddy-flux density of water substance; and Δ viscous dissipation. To these equations one must add boundary conditions, expressing the kinematic constraint at the Earth's surface, and the various fluxes into the atmosphere from below and above.

In Eqs. (6)–(8) horizontal flux densities (that is, fluxes through vertical surfaces) have been neglected. Moreover, in Eq. (6) the expression for the Coriolis force has been simplified by including only the effect of the vertical component of the Earth's rotation ($\Omega \sin \Phi$). These simplifications are common in dynamical meteorology.

In order that Eqs. (5)–(8) with boundary and initial conditions constitute a closed system of equations, it is necessary that $\mathbf{F}_M$, F_{rad}, F_h, Δ, F_{precip}, and F_m be expressible in terms of the dependent variables. The radiative flux F_{rad} depends upon the distribution of temperature, water vapor, cloud, and the properties of the underlying surface; it can be calculated quite accurately when these quantities are known, although the calculation is time-consuming. A major difficulty is the determination of eddy-flux densities $\mathbf{F}_M$, F_h, and F_m. It is usually assumed that a major part of these fluxes is contributed by turbulence in the atmospheric boundary layer, and this part of the fluxes can be estimated from turbulence theory. Thus the last term of Eq. (6) represents the force of turbulent friction. The calculation of F_{precip} presents another difficulty, since it depends strongly on small-scale eddy motions and microprocesses within the clouds.

Note that Eqs. (5)–(8) are nonlinear, since the operator Eq. (10) produces product terms when applied to the dependent variables. Nonlinearity may also enter in the relations between eddy fluxes and dependent variables, and in the relation between condensation heat and dependent variables.

Equations (5)–(8) can be treated either analytically or numerically, and one can distinguish between analytical and numerical dynamic meteorology.

Analytical dynamical meteorology. It is not possible to find analytical solutions of Eqs. (5)–(8) which represent motion processes of the composite kind found in the real atmosphere. Instead, one deals with various kinds of idealized motion systems, for which simplifications of the equations can be justified on the basis of scale analysis and simplified geometry.

Motion without heat sources and friction. A common simplification consists in ignoring all heat sources, including condensation heat, so that Eq. (7) may be integrated to give Eq. (11) for isen-

$$\theta = \text{constant in time for each air particle} \qquad (11)$$

tropic motion. From Eq. (3) there is then for each particle a relation between p and ρ or T; such a fluid is said to be piezotropic. Another simplification consists in ignoring the force of eddy friction $(-\partial \mathbf{F}_M/\partial z)$ in Eq. (6). The bulk of the work in analytical dynamical meteorology has been done under the assumption that heat sources and friction are absent.

Equilibrium. In the absence of friction and heat sources and disregarding the slight centripetal acceleration which exists because an air current is bound to follow the Earth's curvature, Eqs. (5)–(8) permit a particularly simple solution which represents a steady, straight, horizontal current with parallel streamlines. With vanishing particle accelerations, Eq. (6) expresses balance of forces. In the vertical direction there is balance between gravity and vertical pressure force. This hydrostatic equilibrium is expressed by Eq. (12),

$$-g\rho - \frac{\partial p}{\partial z} = 0 \qquad (12)$$

where z represents height. In the horizontal direction the Coriolis force must balance the horizontal pressure force $\mathrm{grad}_h p$. Thus this geostrophic equilibrium can be expressed by Eq. (13).

$$-\mathbf{k} \times f\rho\mathbf{v} - \mathrm{grad}_h p = 0 \qquad (13)$$

Ignoring the Earth's curvature, one may place a cartesian coordinate system with the z axis pointing upward and the x axis along the horizontal velocity u. Then Eq. (13) gives Eq. (14). Thus the

$$u = -\frac{1}{f\rho}\frac{\partial p}{\partial y} \qquad (14)$$

geostrophic wind velocity is directed along the isobars (lines of constant pressure) in level surfaces, with pressure increasing to the right (or left) in the Northern (or Southern) Hemisphere (Fig. 11). At the Equator $f = 0$, and the formula breaks down. The dependent variables u, p, and ρ may depend upon y and z, but not on x if the current is steady. From Eqs. (12) and (14) one may derive Eq. (15), showing that the rate of change of u

$$\frac{\partial u}{\partial z} = \frac{g}{f}\frac{1}{\rho}\left(\frac{\partial \rho}{\partial y}\right)_p = -\frac{g}{f}\frac{1}{T}\left(\frac{\partial T}{\partial y}\right)_p \qquad (15)$$

with height is proportional to the baroclinicity (the subscript p indicates derivative in a surface of constant pressure). Thus in a barotropic current the velocity is constant with height; in particular, a state of persistent rest is necessarily barotropic.

Atmospheric boundary layer. Next to the Earth's surface is a turbulent boundary layer approximately 1 km thick, inside which the force of turbulent friction $(-\partial \mathbf{F}_M/\partial Z)$ is strong enough to disturb significantly the geostrophic equilibrium. The vertical eddy-flux density of momentum $\mathbf{F}_M$ is given by Eq. (16), where the coefficient of eddy

$$\mathbf{F}_M = -K_M \partial \mathbf{v}_h/\partial z \qquad (16)$$

viscosity K_M depends upon the intensity of the turbulence. In the lowest $20-30$ m, K_M increases rapidly with height, and the wind velocity is typically a logarithmic function of height; this wind profile is modified if the eddy flux of heat F_h is numerically large.

Overlaying this shallow layer is a much deeper layer, in which K_M changes less with height. Assuming balance between the Coriolis force, the horizontal pressure force, and turbulent friction, one obtains for constant K_M the Ekman spiral, which represents the variation of wind velocity

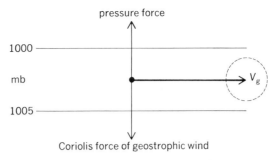

Fig. 11. The geostrophic balance. (*From S. Petterssen, Introduction to Meteorology, 3d ed., McGraw-Hill, 1969*)

with height (Fig. 12). As z increases, the wind approaches the geostrophic value.

Stability. A steady state is termed stable if any small disturbance remains small indefinitely and unstable if a significant deviation from the steady state can develop from an initial disturbance, however small. In this definition it is usually understood that the motion is isentropic, as in Eq. (11).

For a state which deviates only slightly from a known steady state, the basic nonlinear equations (5)–(8) turn into a set of linear perturbation equations in the small perturbation quantities which define the deviations of the variables ($\mathbf{v}$, p, ρ) from their steady-state values. These linear equations can be treated analytically; they have solutions representing various kinds of wave motions, whose amplitudes may either remain small or grow indefinitely. Existence or nonexistence of growing wave solutions is assumed to correspond to instability or stability of the steady state, respectively. Several types of stability may be identified.

Static stability. When in an atmosphere in equilibrium a nonsaturated air particle is displaced vertically with unchanged potential temperature ($D\theta = 0$), its temperature changes at a rate $DT/Dz = -0.01\,°\mathrm{C}\,m^{-1}$ (the dry-adiabatic lapse rate). Therefore, if the distribution of temperature with height in the atmosphere is such that at all levels $\partial\theta/\partial z > 0$ or $\partial T/\partial z > -0.01\,°\mathrm{C}\,m^{-1}$, then a particle displaced upward (or downward) becomes colder (or warmer) than its environment, and the sum of the buoyancy and the weight acting on the particle represents a net restoring force. The state is then said to be statically stable. A rigorous proof of the stability can be given by showing that the sum

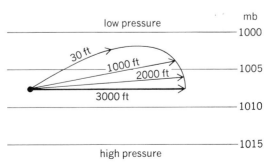

Fig. 12. Ekman spiral. Length and direction of arrows represent wind velocity for different elevations. (*From S. Petterssen, Introduction to Meteorology, 3d ed., McGraw-Hill, 1969*)

of potential and internal energy of a bounded part of the atmosphere is a minimum in the state of rest if $\partial\theta/\partial z > 0$.

If $\partial\theta/\partial z < 0$ (or $\partial T/\partial z < -0.01\,°C\,m^{-1}$) at some levels, then the state is statically unstable. Saturated air is less stable than nonsaturated air with the same temperature distribution, because the temperature of a saturated particle which is displaced vertically changes at a rate which is slower than the dry-adiabatic lapse rate, the rate depending upon p and t.

Static instability is frequently caused by heating the air from the underlying surface and produces convection currents, whose horizontal scale is of the same order as the depth of the unstable layer. Above a certain condensation level, condensation takes place in the rising currents which then become visible as cumuliform clouds.

Inertial stability. In a steady current satisfying Eqs. (12) and (14), inertial stability results from excessive Coriolis forces acting on a chain of particles parallel to the current which have been displaced along the isentropic surface normal to the current. The criterion of stability, valid in both hemispheres, is $f[f - (\partial u/\partial y)_\theta] > 0$, where the subscript θ indicates that the derivative is taken in an isentropic surface ($\theta = $ constant) and f is reckoned positive in the Northern, negative in the Southern Hemisphere.

Shearing instability. If the vertical shear has a sufficiently pronounced maximum at some level, shearing instability results. The layer of maximum shear breaks up into a system of equally spaced vortices with horizontal axes normal to the flow. Condensation may occur in the upper part of the vortices and form what is known as billow clouds.

Wave motions. In a statically and inertially stable atmosphere without shearing instability there is a spectrum of possible wave motions, ranging from high-frequency acoustic waves through gravity waves at intermediate frequencies to inertia-gravity waves at frequencies comparable with the Coriolis parameter f. Typical gravity waves have orbital periods which are short compared with a pendulum day (12 hr/sin Φ) and can therefore be studied without taking the Earth's rotation into account. Their frequency cannot exceed the Väisälä-Brunt frequency $[(g/\theta)\partial\theta/\partial z]^{1/2}$, which in the troposphere mostly corresponds to a minimum orbital period of about 8 min.

Organized stationary gravity wave motions occur in the atmosphere when a stably stratified air current blows over hilly terrain. Their properties can be explained quite well from linear theory. The terrain corrugations act to draw kinetic energy from the current and transform it into wave energy, and the process may be looked upon as a radiation of wave energy upward and horizontally downstream. Depending upon the wavelength and the distributions of temperature and wind with height, the waves may either be transmitted to very high levels, absorbed, or trapped at low levels. Condensation may occur in the wave crests making them visible as stationary clouds, even on satellite photographs; their spacing is mostly 10–50 km. Such waves cause a drag (wave resistance) on the mountains and a corresponding braking force on the airstream.

Organized long gravity-inertia waves occur as forced tidal waves, the predominant forcing effect being the diurnal heating and cooling. To some extent gravity-inertia waves are also generated, as disorganized noise, by nonlinear interaction with other motion types.

Kinematics of motions in atmosphere. Apart from sound waves, which may be treated as a separate phenomenon, all motions in the atmosphere have the character of circulations along closed streamlines. The distribution of horizontal velocity $\mathbf{v}_h$ in a horizontal surface may be characterized by the fields of two scalar quantities, namely, the relative vertical vorticity $\zeta = \mathbf{k} \cdot \mathrm{curl}\,\mathbf{v}$ and the horizontal divergence $\delta = \mathrm{div}_h\mathbf{v}_h$. The former represents twice the angular velocity of an air particle around a vertical axis relative to the solid Earth (reckoned positive when counterclockwise). Since the solid Earth rotates around the local vertical with the angular velocity $f/2$, it follows that the absolute vorticity of the air particle relative to a nonrotating frame is $f + \zeta$. The horizontal divergence δ represents the relative rate of expansion of an infinitesimal horizontal area moving with the air.

The field of $\mathbf{v}_h$ in a horizontal surface may be broken up into one field $\mathbf{v}_\zeta$ which carries all of the vorticity but no divergence, and another field $\mathbf{v}_\delta$ which carries all of the divergence but no vorticity. This is expressed by Eqs. (17).

$$\mathbf{v}_h = \mathbf{v}_\zeta + \mathbf{v}_\delta \qquad \zeta = \mathbf{k} \cdot \mathrm{curl}\,\mathbf{v}_\zeta \qquad \delta = \mathrm{div}_h\,\mathbf{v}_\delta \quad (17)$$

The continuity equation may be written approximately as Eq. (18), where ρ^* represents a standard

$$\rho^*\delta + \frac{\partial(\rho^*w)}{\partial z} = 0 \qquad (18)$$

density distribution, depending upon z only. It follows that the horizontal field $\mathbf{v}_\delta$ together with the vertical velocity w forms a system of circulations in the vertical, whereas $\mathbf{v}_\zeta$ represents a horizontal circulatory motion. These two parts of the motion field behave very differently; for an individual air particle, δ may change quickly as a result of divergent pressure forces, whereas ζ changes more slowly. For this reason the motions in the atmosphere may be classified into two categories: predominantly vertical circulations, in which $\mathbf{v}_\delta$ is the dominating component, and $\mathbf{v}_\zeta$ is of secondary importance; and predominantly horizontal circulations, in which $\mathbf{v}_\zeta$ is the dominating component, and $\mathbf{v}_\delta$ is secondary. All types of gravity-inertia waves and small-scale convection currents belong to the first category; in the atmosphere these motions generally have small amplitudes and carry little energy, and they are mostly too weak or of too small scale to show up on weather maps. The large-scale motion systems revealed by weather maps, which contain the bulk of the energy and are the carriers of weather systems, are all predominantly horizontal circulations and thus belong to the second category above.

Quasi-static approximation. For motion systems whose vertical scale is small compared to the horizontal scale, the vertical motion is weak and its momentum can be neglected. Disregarding also vertical eddy friction, the hydrostatic equation, Eq. (12), holds everywhere and replaces the vertical component of Eq. (6). In the atmosphere, where the vertical scale of motion systems is limited by the high static stability of the stratosphere, this quasi-static approximation applies to motion systems of horizontal scale larger than about 200 km.

Large-scale quasi-horizontal circulations. Equations (5)–(8) apply equally well to motion systems of all types. In the study of the important large-scale quasi-horizontal circulations, it is convenient to derive equations which apply specifically to such motions, whereas predominantly vertical circulations have been eliminated at the outset. This is achieved by a filtering approximation: The horizontal momentum due to the velocity component v_δ in Eqs. (17) is neglected; that is, the horizontal acceleration is approximated by Dv_ζ/Dt. It follows that the horizontal divergence of the forces on the right-hand side of Eq. (6) must balance. This gives a relation between v_ζ and the pressure field, which for small Rossby numbers U/fL (U is characteristic velocity and L is horizontal scale) reduces to the geostrophic relationship, Eq. (13). In extratropical latitudes the Rossby number of the large-scale quasi-horizontal circulations is of the order 10^{-1}, and these motions are therefore also quasi-geostrophic, within 10–20% error.

The evolution of the velocity field is approximately determined by the vorticity equation, which is obtained from Eq. (6) by taking the vertical vorticity. In a somewhat simplified form, Eq. (19),

$$D(f+\zeta)/Dt=-(f+\zeta)\delta \qquad (19)$$

it expresses the familiar mechanical principle that the absolute rotation of an air particle around a vertical axis speeds up when the particle contracts horizontally.

In a current moving up a mountain slope, air columns must shrink vertically and hence, by Eq. (18), diverge horizontally ($\delta > 0$); from Eq. (19) it follows that the absolute vorticity ($f+\zeta$) must decrease numerically. The opposite process takes place when the motion is downslope. This explains the observed predominance of anticyclonic vorticity ($\zeta < 0$ in the Northern Hemisphere) over large mountain ranges which are crossed by air currents.

Over level country there is no such net vertical stretching or shrinking through the whole atmospheric column; therefore, in the mean for the column there is a tendency toward conservation of absolute vorticity. This tendency shows up at the middle level of the atmosphere about 5 km above sea level, where in a crude approximation Eq. (20)

$$f+\zeta = \text{constant in time for each particle} \qquad (20)$$

holds (the barotropic model). Since f increases northward, Eq. (20) requires ζ to decrease for particles moving northward and increase for particles moving southward. As a consequence, large-scale motion systems have a tendency to move westward relative to the air.

The barotropic model has been used to study the properties of a broad zonal current, such as the westerlies in middle latitudes. It has been found that barotropic instability (a kind of shearing instability) occurs if the absolute vorticity of the undisturbed current has a pronounced maximum at some latitude; in that case large-scale vortices may grow spontaneously, feeding upon the kinetic energy of the zonal current. If the zonal current does not possess a vorticity maximum, the current is barotropically stable. In the case of a uniform current, perturbations propagate westward relative to the air as permanent Rossby waves. Such waves may be studied approximately in a cartesian coordinate system, where f increases with y (northward) at a rate $\beta = df/dy$. A wave of wavelength L, superimposed upon a uniform westerly current u, will be propagated at a speed c, satisfying the Rossby formula, Eq. (21).

$$c = u - \frac{\beta}{4\pi^2} L^2 \qquad (21)$$

In spherical coordinates one obtains instead the more accurate Rossby-Haurwitz formula, Eq. (22),

$$\gamma_n = \omega - \frac{2(\omega+\Omega)}{n(n+1)} \qquad (22)$$

which gives the angular velocity of propagation around the Earth's axis (γ_n) for a horizontal motion system where the stream function is a spherical harmonic of the order n (with any longitudinal wave number), assuming that the atmosphere has a solid rotation ω relative to the solid Earth. Some of the large-scale motion systems of the atmosphere are of this kind, and there is a satisfactory agreement between their observed propagation and Eq. (22).

The barotropic model describes only a very special mode of motion; for instance, it cannot describe conversion of potential and internal energy into kinetic energy or vice versa. A more general use of the filtered equations has revealed other types of large-scale motion systems. In the case of a zonal current which is sufficiently baroclinic, the linearized perturbation equations have solutions representing growing disturbances, which are similar to extratropical cyclones with respect to size, growth rate, and structure. Such a zonal current is said to possess baroclinic instability, and it is characteristic that the growing disturbances feed on the potential and internal energy of the zonal current.

Extratropical cyclones are observed to form in the zone of middle-latitude westerlies, characterized by pronounced baroclinicity and also by a strong maximum of cyclonic wind shear. Baroclinic and barotropic instability may both operate to cause the sudden growth of cyclones that is often observed.

Numerical dynamic meteorology. High-speed electronic computers have made it possible to find approximate numerical solutions to equation systems which are untractable by analytical methods. This has opened up a new branch of dynamical meteorology. It has become possible to drop many of the simplifying assumptions which were necessary for the analytical treatment; in particular, it is no longer necessary to linearize the equations.

Weather prediction. Starting from an initial state of the atmosphere which is known from observations, later states may be calculated from Eqs. (5)–(8) by numerical integration in time. To achieve this, the fields are represented by their values in a grid of points in space, and the differential equations are approximated by finite difference equations; thus one obtains a numerical model of the atmosphere. The total amount of computation for a given integration period depends on the approximation used in Eqs. (5)–(8), on the spacing of grid points in horizontal and vertical direction, and on the area over which the computation is extended. The quasi-static approximation is always used; the filtering approximation is used in some cases, but the equations may also be applied in nonfiltered, "primitive" form.

For use in practical weather forecasting, it is necessary that the computation proceed considerably faster into the future than real weather. This strongly restricts the size of the numerical model and makes considerable simplifications necessary, depending upon the available computer capacity.

The simplest possible model is the barotropic model, which is obtained when only one grid point is used along each vertical, at the 500-mb level. The set of governing equations then reduces to Eq. (20) or a slight modification thereof, and this equation may be expressed in terms of a stream function as the single dependent variable; all other physical processes are ignored. Although simplified to the extreme, the barotropic model can give predictions of practical value up to about 3 days ahead.

By increasing the number of grid points along the vertical, one obtains baroclinic models of increasing fidelity, requiring increasing amounts of computer time. A simplification which is frequently made in such models designed for weather prediction consists in neglecting friction and heat sources (also latent heat) and omitting humidity as a variable. The error due to this simplification grows relatively slowly with the prediction period, so that predictions of value may be obtained up to about 4 days ahead. Although humidity has been omitted, such models can still be used to predict precipitation, because this is known to occur wherever there is ascending motion.

Influences in the atmosphere propagate at the speed of sound and reach one from the remotest parts of the globe in less than 20 hr. Strictly speaking, the integration should therefore be extended to the entire atmosphere, even in a prediction for only 1 day ahead. This is prohibitive in practice, partly because the amount of computation would be too large, and partly because data which define the initial state of the entire atmosphere are not yet available. Fortunately, the bulk of the influences do not propagate at sound speed, and integrations of value can be carried out for a limited part of the world.

Integrations in time carried out with baroclinic numerical models of the atmosphere have confirmed and extended many of the results of analytical dynamic meteorology, such as the growth of cyclonelike disturbances in a baroclinically unstable zonal current and the formation of the associated frontal systems.

General circulation of atmosphere. The motions in the atmosphere are caused by solar heat. The combined process of absorption and emission of radiation causes a distribution of heat and cold sources which continually disturbs the equilibrium and maintains the motion. During this process the volume and pressure of individual air particles change with time, so that expansion takes place on the average at higher pressure than contraction. As a result, heat is converted into mechanical energy, just as in a man-made heat engine. The bulk of the mechanical energy thus released is used to overcome friction within the atmosphere itself; only a very small fraction is spent to do work on the ocean surfaces, thus maintaining ocean currents and waves.

The composite motion of the entire atmosphere is termed the general circulation. It has become possible to simulate the general circulation by numerical models. Some requirements which a general-circulation model must satisfy are: (1) It must cover the entire globe (or possibly one hemisphere, assuming symmetry at the Equator); (2) have enough vertical resolution so that all energy conversions can be represented; (3) contain radiative heat sources; and (4) contain turbulent eddy fluxes of momentum, enthalpy, and water vapor.

A numerical model for study of the general circulation must necessarily be exceedingly complicated. However, it is not intended for weather prediction, and it is therefore not necessary that the numerical integration proceed faster than real time; and it is not necessary to start the integration from a real state determined by global observations. By performing long-term numerical integrations with general-circulation models, starting from a constructed initial state, it has been possible to reproduce the main statistical characteristics of the real atmosphere, including the distribution of climate. *See* ATMOSPHERIC GENERAL CIRCULATION.

Special motion systems. Another important branch of numerical dynamic meteorology is the numerical study of various mesoscale and small-scale motion systems, such as hurricanes, fronts, squall lines, cumulus clouds, and land and sea breezes.   [ARNT ELIASSEN]

Bibliography: H. Byers, *General Meteorology*, 3d ed., 1959; A. Eliassen and E. Kleinschmidt, *Dynamic Meteorology*, in S. Flügge (ed.), *Handbuch der Physik*, vol. 48, 1957; C. L. Godske et al., *Dynamic Meteorology and Weather Forecasting*, 1957; R. M. Goody, *Atmospheric Radiation*, 1964; G. J. Haltiner and F. L. Martin, *Dynamical and Physical Meteorology*, 1957; S. L. Hess, *Introduction to Theoretical Meteorology*, 1959; E. N. Lorenz, *The Nature and Theory of the General Circulation of the Atmosphere*, 1967; T. F. Malone (ed.), *Compendium of Meteorology*, 1951; B. J. Mason, *The Physics of Clouds*, 1957; P. D. Thompson, *Numerical Weather Analysis and Prediction*, 1961.

Micrometeorology

A branch of atmospheric dynamics and thermodynamics which deals primarily with the interaction between atmosphere and ground; the interchange of masses, momentum, and energy at the earth-air interface; in short, with the lower boundary conditions of atmospheric processes. Thus, there are strong interconnections with those sciences that deal with the media underlying the atmosphere. Micrometeorology, along with other branches of meteorology, is also concerned with investigating and predicting the transport and dispersion of pollution from such sources as smokestacks and automotive exhausts in the lower atmosphere. Micrometeorology typically deals with the transport (or flux) of air properties in the vertical direction (conduction and convection) and the vertical variation of these fluxes. *See* ATMOSPHERIC POLLUTION; OCEAN-ATMOSPHERE RELATIONS; SEA WATER; SOIL.

Scale of focus. Micrometeorology differs from general meteorology (in particular, meso- and macrometeorology) in scale and with respect to the features of the atmosphere studied. All of the various branches of meteorology deal with the temporal-spatial variations of the weather elements (such as air density, temperature, components of momentum, and mixing ratios). Meso- and macrometeorological studies are typically based on standard synoptic observations (from a nationally and internationally organized network with about 20 to 200 km distance between stations), and of primary interest are large-scale air currents which serve to transport "conservative" atmospheric properties (such as heat, admixtures, and absolute vorticity) over relatively large horizontal trajectories. *See* METEOROLOGY.

Micrometeorology is concerned with the great variety of atmospheric processes which are incompletely covered by the synoptic network. Examples of such small-scale processes are mountain and valley circulations, sea breezes, and airflow past mountains and islands. The prefix micro is justified by the fact that the detailed study of the temporal-spatial variations of air properties in the lower atmosphere requires close spacing of instruments of relatively small size and low inertia (lag time).

Instrumental needs. The development and use of nonstandard mast-supported thermometers, anemometers, and other sensing instruments for the measurement of mean vertical profiles and gradients of temperature, wind speed, and other air properties is a basic requirement of micrometeorological research. Further miniaturization of sensing instruments is necessary for the study of short-time fluctuations of meteorological variables. A significant range of such fluctuations occurs with frequencies between several cycles per second to several cycles per minute.

Another basic requirement is the development and use of instruments for the direct measurement of boundary fluxes (such as evaporation, surface stress, and energy transfer, including net radiation). Micrometeorological studies are frequently concerned with the diurnal cycle of heating and cooling, the hydrologic cycle, and the energy dissipation in large-scale air currents due to surface stress (friction).

Micrometeorology is sometimes referred to as fair-weather meteorology. This appellation can be justified by the fact that of the total solar energy intercepted in space by the Earth about 70% reaches the ground under clear skies, with only about 10% scattered back to space (the remainder being absorbed in the air), while under an average cloud deck the corresponding percentages are about 30% reaching the ground, with nearly 60% being scattered back to space. Solar energy arriving at the bottom of the atmosphere is partly reflected ("albedo" radiation, normally about 15% of incoming radiation but varying between extremes of about 5–90%, for very dark-colored ground to very bright, fresh snow cover) and partly absorbed. A primary task of micrometeorology is to analyze and predict the distribution and transformation of the absorbed solar irradiation by establishing a complete energy budget of the air-submedium interface. The respective energy balance equation takes into consideration the vertical fluxes of terrestrial (or long-wave) radiation, of sensible heat conducted into the submedium as well air, and of latent energy (most important, phase transformations of H_2O such as evaporation, sublimation, and melting, but also photosynthesis wherever significant). The partitioning of energy and the subsequent changes in surface temperature, in direct response to variation in intensity of the solar forcing function, depend strongly on the strength of the existing overall air motion in the region. Because of prevailing nonuniformity in the physical structure of natural ground and its various (vegetative or nonvegetative) covers, and the relatively strong energy flux density of solar radiation, the atmosphere within a few meters of the ground is a surprisingly complex and variable structure. Strong fluctuations and gradients of temperature, density, and moisture affect the propagation of sound and light, creating irregular scattering and refraction phenomena which account for optical mirages, shimmer, and "boiling," as well as "ducting" of radio and radar beams.

Significant applications. Micrometeorological research is important in that it supplies detailed information about the physical processes in the region of the atmosphere where life is most abundant. It closes the gaps of information from synoptic networks. It produces results useful for applications in various fields such as climatology, oceanography, soil physics, agriculture, biology, chemical warfare, and air pollution. Among the branches of atmospheric physics, micrometeorology is the one whose subjects are most amenable to fairly complete experimental description, and to testing of the theoretical models.

The characteristic scale is small enough that it is both feasible and economical to attempt control of natural processes in the lower atmosphere, for example, by artificial changes of albedo or other surface characteristics such as aerodynamic roughness; by thermal admittance of the submedium (through mulching of soil or other means); by windbreaks; by the use of sunshades and smoke screens; by heat supply and artificial stirring of air; and by irrigation. *See* AGRICULTURAL METEOROLOGY; CLIMATOLOGY; INDUSTRIAL METEOROLOGY; MICROMETEOROLOGY, CROP; WEATHER MODIFICATION.

Turbulence research and applications. The mechanism of the vertical flux is essentially one of eddy mixing or turbulent exchange of air properties. It is convenient to distinguish two methods of approach: One deals with the mean vertical gradients caused by turbulence, while the other is based on statistical treatment of turbulent fluctuations recorded by low-inertia (fast-response) instruments. The connecting link between the two approaches is presented by the Reynolds expression of the mean flux of an air property as the average covariance of the fluctuations of vertical wind speed and the property considered.

A powerful tool for research is the spectrum

analysis of fluctuations by which the distribution of the variance of a meteorological element over the various frequencies is measured. According to J. Van der Hoven the power spectrum (in the range 0.001–1000 cycles per hour) of horizontal wind speed recorded at the upper levels of the meteorological tower at Brookhaven, Long Island, N.Y., shows two major peaks of energy, one at about 1 cycle per 4 days, another at about 1 cycle per minute. A broad minimum of energy at frequencies of about 1–10 cycles per hour seems to exist under varying terrain and synoptic conditions. This spectral gap appears to separate objectively the macrometeorological from the micrometeorological scale in the atmosphere.

Owing to the relatively high Reynolds number of atmospheric flow, all air motions are turbulent, which means that individual particles do not describe straight and parallel trajectories (as in truly laminar flow) even though the originating force (pressure gradient force) may be constant and uniform and the ground smooth and flat. Turbulence is sometimes visible in the shapes assumed by smoke, and felt in the variable pressure and cooling power of the wind. The intensity of atmospheric turbulence (as measured by the ratio standard deviation of a wind component divided by mean wind speed) is normally between 0.2 and 0.4, which is significantly larger than that of wind tunnel turbulence.

Low-level turbulence derives its energy basically from large-scale air motions and depends on the surface roughness. This mechanical turbulence is intensified by surface heating and damped by nocturnal cooling. Individual gusts can be ascribed to a disturbance, or eddy, of a certain extent. There is a wide range of eddy sizes (eddy spectrum). Near the ground, eddy sizes increase in proportion to distance from the ground, then decrease with further increase in height.

Atmospheric turbulence affects the flight of airplanes and ballistic missiles (rough air). It is especially troublesome when intensified by surface heating (thermal turbulence) and additional buoyancy forces generated by latent heat release in updrafts that ascend above the condensation level (cumulus convection). There are many features of turbulence research that are of common interest to micrometeorology and aeronautics.

[HEINZ H. LETTAU]

Bibliography: R. Geiger, *The Climate near the Ground*, 4th ed., 1965; H. Lettau and B. Davidson (eds.), *Exploring the Atmosphere's First Mile*, 2 vols., 1957; J. L. Lumley and H. A. Panofsky, *The Structure of Atmospheric Turbulence*, 1964; R. E. Munn, *Descriptive Micrometeorology*, 1966; O. G. Sutton, *The Challenge of Atmosphere*, 1961; O. G. Sutton, *Micrometeorology*, 1953; J. Van der Hoven, Power spectrum of horizontal wind speed in the frequency range from 0.0007 to 900 cycles per hour, *J. Meteorol.*, 14:160, 1957.

Micrometeorology, crop

Crop micrometeorology deals with the interaction of crops and their immediate physical environment. Especially, it seeks to measure and explain the overall performance of crops in net photosynthesis (photosynthesis minus respiration) and water use (transpiration plus evaporation from the soil). Because the intricate array of leaves, stems, and fruits modifies the local environment of each crop and because the processes involving energy transfers and conversions are interrelated, these studies are necessarily complex and difficult. As a pure science, crop micrometeorology is closely related to plant anatomy and physiology, meteorology, and hydrology. A practical goal is to provide improved designs for individual plants for plant breeders and improved planting patterns for farmers. *See* AGRICULTURAL METEOROLOGY; AGRICULTURAL SCIENCE (PLANT); MICROMETEOROLOGY.

Unifying concepts. Conservation laws for energy and matter are central to crop micrometeorology. Energy fluxes involved are radiation, consisting of visible radiation, near-infrared solar radiation, and terrestrial radiation (corresponding roughly to wavelengths of 0.4–0.7, 0.7–3, and 3–60 μ); convection in the air; molecular heat conduction in and near the plant parts and in the soil; and the latent heat carried by water vapor. The material substances of chief interest are water and carbon dioxide. These move by molecular diffusion near the leaves and by convection in the airflow. During the daytime generally, and sometimes at night, airflow among and above the plants is strongly turbulent. The frictional and thermal effects of the plants contribute to the pattern of air movement and influence the efficiency of turbulent transfer. Transfer of momentum from air to leaves is partly analogous to the transfers of sensible heat, water vapor, and carbon dioxide.

Both field studies and mathematical models have dealt mostly with tall, close-growing crops, such as maize and wheat, which can be treated statistically as composed of horizontal layers. Vertical streams of solar (visible and near-infrared) radiation are partly absorbed, partly reflected, and partly transmitted by each layer. For the far-infrared the plants act also as good emitters of radiation. Transfers of momentum, heat, water vapor, and carbon dioxide can be considered as composed of two parts: a general vertical transfer and a leaf-to-air transfer. At each height in the vegetation layer, intensive parameters such as temperature vary markedly with time and location. As a bare minimum, two mean or representative temperatures are needed for each height: an average air temperature between the plants and a representative temperature at the surfaces of leaves and stems. When some leaves are in direct sunlight and some are shaded, then, first, the sampling problem for measurements becomes large and, second, the adequacy of a single mean temperature value becomes dubious.

Photosynthesis. Advances in the submicroscopic chemistry and physics of photosynthesis and respiration have contributed hardly anything to a quantitative understanding of how a community of plants acts as a system to produce organic matter and useful yield. For all practical purposes, a knowledge of the anatomy and gross physiology of plants and plant parts is more important. How do respiration, transpiration, and photosynthesis of an individual plant leaf depend upon the local conditions of light, humidity, temperature, and carbon dioxide concentration? How do they depend upon the condition of the leaf itself, and how is this dependence to be characterized and measured?

The synthesis of organic matter by crops growing in the field is, of course, not understood merely from knowing the behavior of isolated plants or parts of plants under various standard conditions. There remains the problem of how the complex array of plants themselves modifies the local conditions of light, wind, temperature, humidity, and carbon dioxide. Theoretical studies and field observations have already begun to give a general picture that makes sense. Whereas many species of plants give maximum photosynthesis by individual leaves at less than full sunlight (field-grown maize is an exception), a growing crop in good condition has a higher rate of photosynthesis with a higher rate of illumination. Extremely dense plantings give higher rates of net photosynthesis than do moderately dense ones, though seed yield may be reduced. This implies that the most-shaded lower leaves do not become a liability to the plants on this account; browned leaves which appear near the ground generally are a consequence of either a shortage of water or a shortage of mineral nutrients.

Transpiration. Besides the use of a rather small fraction (2–6%) of the visible solar radiation in photosynthesis, heat conditions are important since they affect plant temperatures and transpiration rates. Heat capacities of leaves are small, so that their temperatures very nearly correspond to thermal equilibrium with their surroundings. The general concepts of heat transfer and micrometeorology appear to be adequate for this study. Field research has been advanced by several improvements in instrumentation and by the ability to collect and analyze great masses of data. Among the instruments are infrared radiometers for measuring surface temperatures and the diffusion porometer for measuring diffusive resistance of leaf surfaces, which depends on stomatal opening.

Plant parameters. From a micrometeorological viewpoint, rates of net photosynthesis and evapotranspiration appear as terms in micrometeorological equations. But in these equations the plants have physiological as well as physical characteristics. Aside from growth processes, the physiological actions of interest are plant movements, as they affect light interception and leaf-to-air transfers; stomatal opening and closing, as they affect gains and losses of carbon dioxide and water vapor; and photosynthesis and respiration characteristics, especially as functions of light interception, carbon dioxide concentration at the leaf surfaces, diffusive resistance of the leaf surfaces, and temperature and water status of the leaves. To understand these roles of the plants, functional relations, not merely static physical parameters, are needed.

The strictly physical properties of plants and plant communities are coefficients of absorption, transmission, and reflection of the three kinds of radiation; geometrical properties as they affect radiation; and geometrical and mechanical properties affecting airflow and transfers to the air. These are being more closely defined and more accurately measured for the major field crops. Crop micrometeorology aims to show how these physical and physiological properties of the plants are related to crop production and water requirements under the conditions of growing in the field.

See ECOLOGY, PHYSIOLOGICAL (PLANT); EVAPOTRANSPIRATION.

[WINTON COVEY]

Bibliography: K. W. Brown and W. Covey, Energy-budget evaluation of the micrometeorological transfer processes within a cornfield, *Agr. Meteorol.*, 3:73–96, 1966; I. R. Cowan. Mass, heat and momentum exchange between stands of plants and their atmospheric environment, *Quart. J. Roy. Meteorol. Soc.*, 94:523–544, 1968; E. Lemon, Aerodynamic studies of CO_2 exchange, in A. San Pietro, F. A. Greer, and T. J. Army (eds.), *Harvesting the Sun*, 1967; E. Lemon, Micrometeorology and the physiology of plants in their natural environment, in F. C. Steward (ed.), *Plant Physiology*, vol. 4, 1965; R. E. Munn, *Descriptive Micrometeorology*, 1966.

Mineral resources conservation

The application of public policies which increase the future supply of mineral commodities. These policies can be applied in the form of production restrictions, when competition could lead to wasteful exploitation of mineral deposits, or in the form of rules that require the producer to be efficient in the recovery and processing of minerals of economic value. It should be noted that a large number of public policies, such as taxation and import controls, also influence the conservation of mineral resources, although these policies are not necessarily applied to conserve minerals. The development of new mining, processing, and exploration techniques also has an effect on the conservation of mineral resources.

Reserves and resources. Mineral reserves are the known stock of economically exploitable mineral deposits. Reserves are quantities given in tons, cubic feet, or barrels. Within a mineral deposit, reserves are usually classified as follows: Proved reserves are definitely known to exist; probable reserves are known to exist with a fair degree of certainty; and possible reserves are the unexplored parts of the mineral deposit. Possible reserves, in contrast to proved or probable reserves, cannot be measured with any degree of certainty and are usually referred to as small or large instead of by a given numerical value. When the reserves of a mine, or a number of mines, are referred to, only proved and probable reserves are meant.

Mineral resources comprise the three classes of reserves mentioned above plus potential reserves. Potential reserves include known occurrences of minerals which are not minable because of grade, distance from consumption centers, depth of occurrence, or difficulty of metallurgical treatment. Potential reserves are expected to become profitable as economic conditions change or as technology advances. For example, shales occurring in Colorado, Utah, and Wyoming are known to contain vast quantities of kerogen, which can be converted to oil. These oil shales are not competitive with other sources of oil, but it is agreed that someday they will be, when the problems of extraction have been solved. Mineral resources conservation is thus concerned with the efficient use of known proved and probable reserves and with the addition of possible and potential reserves to the known stock.

Classification of minerals. Minerals may be conveniently classified into conventional fuels, nuclear fuels, nonfuel metallics, and other nonfuel minerals (Table 1). Depending on their use, minerals can be regarded as exhaustible fund resources or revolving fund resources.

Table 1. General classification of selected minerals and some of their uses

Minerals	Uses
Conventional fuels	
Bituminous and anthracite coal	Direct fuel, thermal electricity, gas, chemicals
Lignite	Electricity, gas, chemicals
Petroleum	Gasoline, other fuels, chemicals, plastics
Natural gas	Fuel
Nuclear fuels	
Uranium	Nuclear bombs, power, tinting glass
Thorium	Nuclear bombs, power, gas mantles
Metallic minerals (nonfuel)	
Ferrous metals	
Iron	Steel
Ferroalloys	
Manganese	Alloy steels, disinfectants
Cobalt	Alloys, catalysts, radiographic, therapeutic
Columbium (niobium)	Stainless steels, nuclear reactors
Chromium	Metallurgical, refractory, and chemical
Molybdenum	Alloy steels
Nickel	In over 3000 alloys
Tungsten	Alloys and chemicals
Vanadium	Alloys
Nonferrous metals	
Copper	Electrical products, alloys
Lead	Batteries, gasoline, paints, alloys
Tin	Tinplate, solder, chemicals
Zinc	Galvanizing, die casting, chemicals
Light metals	
Aluminum	Aircraft, rockets, building materials, liquid fuel
Magnesium	Structural, refractories
Titanium	Pigments, aircraft, alloys
Zirconium	Refractories, ceramics, metals, chemicals
Other metals	
Beryllium	Copper alloys, refractories, atomic energy field
Gold	Monetary, jewelry, dentistry
Radium	Medical, industrial radiography
Nonmetallic minerals	
Asbestos	Insulation, textiles
Borides	Glass, ceramics, gasoline, solid propellants
Corundum	Abrasives
Feldspar	Ceramic flux, artificial teeth
Fluorspar	Flux, acid, refrigerants, propellants
Phosphates	Fertilizer, chemicals
Salt	Chemicals, glass, metallurgical
Sulfur	Fertilizer, acid, iron and steel industries

Petroleum is an example of an exhaustible fund resource, for the supply may be exhausted through use. Rainwater is a typical revolving fund resource which is renewed naturally despite its use by man. Minerals available for reuse are called revolving fund resources. The same minerals may be classed as exhaustible fund resources in certain uses, for example, lead in gasoline, and as revolving fund resources in other uses, for example, lead in car batteries. Problems are particularly acute where exhaustible fund resources are in great demand but in short supply. In such cases attempts are made to discover and use lower-priced substitutes that are in greater supply.

Supply of minerals. Although the total supply of mineral deposits at any one time can be regarded as fixed, very little is known about the likelihood of new discoveries of particular kinds of deposits. Only a relatively small part of the Earth's crust has been thoroughly explored, and it is hoped that future developments in exploration techniques will increase the ability to detect new supplies of minerals. Also, the uncertainty about developments in mining and processing techniques is a factor which makes an estimation of the possible size of the world's resources a rather hazardous undertaking. Because of these and other uncertainties, such as future demand and prices, there is no sound way by which a mineral conservationist can decide whether any of the known reserves should be set aside for future generations. Mineral conservation is therefore oriented toward the efficient use of known reserves and the development of policies and techniques which will favor the increase of these reserves. Special attention is given to those mineral commodities of which present-day reserves are small when they are compared to annual consumption.

Reserve index. The ratios of reserves to annual production and reserves to annual consumption are called the reserve-production and reserve-consumption indexes. These indexes can be calculated on a worldwide, national, or regional basis. For example, in 1964 the United States had a reserve-consumption index of 120 years for iron, if total demand were to be satisfied by domestic ores. If it is taken into account that in that year about one-third of the demand was satisfied by imports, then the reserve-production index can be calculated as being about 180 years. The 1964 world reserve index for iron was 380 years. Table 2 gives the world reserve index and the United States reserve-production indexes for several important mineral commodities.

It is obvious that the mineral conservationist in the United States is concerned mainly with mineral commodities of which the United States reserve indexes are small. This will be especially the case when the commodity has to be imported from distant sources which are politically insecure. An example is tungsten, which is predominantly mined in Asia and which is of military importance. A situation such as this is a strong inducement for the United States government to store a large supply in the strategic stockpile. Industry tries to diminish its reliance on distant sources by finding substitutes, thus enhancing the conservation of limited domestic reserves. The aluminum reserve-production index for the United States is also low, but large reserves are known in relatively secure and nearby areas, thus decreasing the need for stockpiling. It should be noted here that stockpiling in itself is not meant to enhance conservation. However, the sale of mineral commodities from the stockpile in periods when industry cannot meet demand does increase price stability, and this increase has a positive effect on conservation, as will be discussed later in this article.

Government conservation policies. There are a number of government policies which are of great importance for the conservation of mineral resources. First of all, a government can intervene in the production process if it considers the way in which production takes place wasteful. An excellent example of such intervention is the production regulation of oil and gas which is enforced by many of the oil-producing states of the United States. Many states have passed laws to prevent oil wells from being placed too close to one another and pumped too fast, as the result is a considerable loss of reservoir energy or gas drive. Restrictions on the gas-oil ratio permitted have led to the return of excess gas to the reservoir. Such regulations, as well as technological advances such as gas and water injection to maintain pressure, have aided recovery of greater proportions of the oil in reservoirs.

Another government policy which has great influence, especially in the oil and gas industry, is the depletion allowance. This is based on the consideration that producers of mineral commodities deplete their sources of revenue in the course of production, and that these sources have to be replaced by costly exploration. The United States and many other governments allow producers to deduct a certain percentage of their gross revenues from the part of their incomes which is subject to the Federal income tax, thus decreasing the effective tax base. This allowance increases the funds such companies have available for exploration and attracts newcomers to the industry, which in turn increases the known reserves of mineral commodities. Opponents of the depletion allowance argue that it leads to overinvestment and lower prices of mineral commodities, and that it results in wasteful utilization of these resources. If this is the case, the depletion allowance is not desirable from the viewpoint of the conservationist.

Table 2. World and United States reserve – production indexes in 1964*

Mineral	World	United States
Oil†	33	11
Uranium	24‡	12
Manganese	146	6§
Molybdenum	100	200§
Copper	40	26
Lead	18	4
Zinc	20	20
Tungsten	40	12
Aluminum	160	4
Chromium	532	6§
Gold	20	14

*Data from P. T. Flawn, *Mineral Resources*, Rand-McNally, 1966.
†Proved reserves only.
‡Excluding communist countries.
§Reserve-consumption indexes.

The decision of mining and oil companies to explore in a certain area is strongly influenced by the legal climate which prevails in the area under consideration. A country may increase its reserves of minerals considerably by providing an attractive mineral code. Governments themselves also can undertake mineral surveys, make topographical and geological maps, and construct transport facilities to remote areas. All these factors contribute to making investments in exploration and exploitation profitable. Some governments impose import restrictions in order to enhance production and exploration within their own borders. On the one hand this increases known domestic reserves; on the other hand it leads to a more rapid depletion. However, these import restrictions are usually not based on conservational arguments only. Import restrictions on oil, as applied in the United States, were considered necessary for security reasons. Because it is self-sufficient in oil, disturbances in other oil-producing countries cannot deprive the United States of its main source of energy.

Stabilization. One of the main problems facing producers and consumers of some mineral commodities is the volatility of the price level. Examples are copper, silver, and platinum. Obviously, if the price of a metal fluctuates, producers will have problems in making long-term decisions such as opening up new mines. Consumers, if possible, use other commodities. For example, copper and aluminum are used for similar purposes in many industries. Of the two aluminum has a much more stable price level than copper. Aluminum has replaced copper in many uses, probably to a considerable degree as a result of the difference in price stability. To alleviate this problem companies or countries can form cartels to ensure a stable price. Examples are the tin agreements and copper cartels which have been established over the years. The problem with these cartels or consortia has generally been that they work well when prices are moving upward, but tend to disintegrate when prices go down, because individual producers try to sell their material before prices decline further.

Conservation through use of scrap. The production of scrap by means of recovery of used metals is of considerable importance for the conservation of mineral resources. In the United States about 50% of steel furnace feed, 40% of domestic lead consumption, and 25% of domestic copper production are derived from scrap. It has been estimated that the copper scrap reserve of the United States totals 35,000,000 tons in cartridge cases, pipe, wire, autoradiators, bearings, and so on.

Technological change. New developments have had considerable influence in increasing the mineral supply. In 1900 it took about 8 times as much coal to generate a kilowatt-hour of electricity as it did in the 1960s. Recovery of oil from oil pools doubled in the period 1920–1970. Technological advances since the early 1940s have resulted in the development of new ways of drilling blast holes in the troublesome, low-grade taconities (hard, iron-containing rock). Progress in flotation and magnetic beneficiation processes has helped to provide growing amounts of taconite and jasper concentrates from the Lake Superior region. The United States has thus added appreciably to its supplies of usable iron ore. Technology also helps in the development and use of renewable resource substitutes for scarce minerals and in the substitution of minerals which are more abundant. In addition, technology helps in the reconcentration of used minerals, thus permitting waste reduction and a wider application of the principle of recycling. Greater amounts of minerals can be recoverd from waste products and effluents, either directly or as by-products. These and many other developments have so far been successful in satisfying the increase in demand.

In the period since 1900 the world has consumed more minerals than in all preceding times. In 1962 the United States consumed nearly 8 times as much copper and 10 times as much steel as in 1900. It is impossible to tell exactly how large the growth in consumption will be in the next 60 years or so. Considerable increases in demand are very likely to continue, especially when less-developed countries are able to achieve a higher standard of living. Considerable technological improvements will be necessary in order to meet this demand.

Demand and supply problems. The search for solutions to problems which arise from imbalances between mineral supply and demand is of particular concern to the conservationist. He invariably uses the ecological approach, thus giving due cognizance to the interrelationships and interdependences existing between various minerals and between minerals and other resources. Solutions for the problems will differ under conditions of peace, cold war, and hot war. Other factors influencing the formulation and character of solutions include the economic level and the rate and direction of economic growth of the economies concerned.

The cost of utilizing minerals needs further study. This includes the need to reduce further the destruction of land by open-pit mining; the prevention of pollution of rivers and air so as to reduce the harmful effects of such pollution on health, scenery, recreation facilities; and the reuse of water downstream.

Some people feel that technology will provide all the solutions to problems which spring from increasing demands for nonrenewable resources of limited supply. Others are much less optimistic. *See* Conservation of resources; Land-use planning; Water conservation.

[ERNST H. VON METZSCH]

Bibliography: P. T. Flawn, *Mineral Resources*, 1966; J. F. McDivitt, *Minerals and Men*, 1965; E. H. Robie (ed.), *Economics of the Mineral Industries*, American Institute of Mining, Metallurgical, and Petroleum Engineers, 1964; A. Scott, *Natural Resources: The Economics of Conservation*, 1955; U.S. Bureau of Mines, *Mineral Facts and Problems*, USBM Bull. no. 630, 1965.

Mining, open-pit

The extraction of ores of metals and minerals by surface excavations. This method of mining is applicable for deposits which have large tonnage reserves, which can be exploited at a high rate of production and which do not have a ratio of overburden (waste material that must be moved) to ore that would make the operation uneconomic. Where these criteria are met, large scale earth moving equipment can be used to give low unit

Fig. 1. Roads and rounds that have developed open cuts on Cerro Bolivar in Venezuela. (*U.S. Steel Corp.*)

mining costs. For a discussion of other methods of surface mining *see* Mining, placer; Mining, strip.

Most open-pit mines are developed in the form of an inverted cone with the base of the cone on surface. Exceptions are open-pit mines developed in hills or mountains (Fig. 1). The walls of all open pits are terraced with benches to permit shovels to excavate the rock and to provide haul roads for trucks to transport the rock out of the pit. The final depth of the pits ranges from less than 100 ft to nearly 3000 ft and is dependent on the depth and value of the ore and the cost of mining. The cost of mining usually increases with the depth of the pit because of the increased distance the ore must be hauled to the surface and the increased amount of overburden, or waste material, that must be removed to expose the ore. Numerous benches or terraces in the pit permit a number of areas to be worked at one time and are necessary to give a slope to the sides of the pit. Using multiple benches also permits a balanced operation in which much of the overburden is removed from upper benches at the same time ore is being mined from lower benches. It is undesirable to remove all of the overburden from an ore deposit before mining the ore because the initial cost of developing the ore is then extremely high.

The principal operations in mining are (1) drilling, (2) blasting, (3) loading, and (4) hauling. These operations are usually required for both ore and overburden. Occasionally, the ore or the overburden is soft enough that drilling and blasting is not required.

Rock drilling. Drilling and blasting are interrelated operations. The primary purpose of drilling is to provide an opening in the rock for the placing of explosives. If the ore or overburden is soft enough to be easily excavated without blasting by explosives, it is not necessary for it to be drilled. The cuttings from the hole are often used to fill the hole after the explosive charge has been placed at the bottom. Blast holes drilled in ore serve a secondary function in that the cuttings can be sampled and assayed to determine the mineral content of the ore. Sampling of drill holes is frequently done for ores not readily identified by visual means.

The basic methods of drilling rock are (1) rotary, (2) percussion, (3) jet piercing, and (4) churn drilling. The drill holes range in size from $1\frac{1}{2}$ in. to 15 in. in diameter and are drilled to varying depths and spacings as required for the particular type of rock being mined. The drill hole depth ranges from 20 to 50 ft depending on the bench height used in the mine. The drill hole spacing (distance between holes) is governed by the depth of the holes and the hardness of the rock, and generally ranges from 12 to 20 ft. For short holes or hard rock, the hole spacing must be close; and for long holes and soft rock, the spacing may be increased.

Rotary drilling. Drilling is accomplished by rotating a bit under pressure. Compressed air is forced down a small hole in the drill shank or steel, and is allowed to escape through small holes in the bit for cooling the bit and blowing the rock cuttings out of the hole. Rotary drill holes range from a minimum diameter of 4 in. to a maximum diameter of 15 in. and are drilled by machines mounted on trucks or a crawler frame for mobility (Figs. 2 and 3). The weight of these machines ranges from 30,000 to 200,000 lb and, in general, the heavier machines are required for the larger-diameter holes. The rotary method is the most common type of drilling used in open-pit mines because it is the

Fig. 3. Crawler-mounted drilling machine. Horizontal toe holes are drilled by percussion. (*Kennecott Copper Corp.*)

cheapest method of drilling; however, it is restricted to soft and medium-hard rock.

Percussion drilling. Percussion drilling is accomplished with a star-shaped bit which is rotated while being struck with an air hammer operated by high-pressure air. Air is also forced down the drill steel to cool the bit and to blow the cuttings out of the hole. Small amounts of water are frequently added with the air to reduce the dust. The hole diameters range from $1\frac{1}{2}$ to 9 in. depending upon the type of drill. The drilling machine used for the smaller-diameter holes ($1\frac{1}{2}$–$4\frac{1}{2}$ in.) is small and usually mounted on a lightweight crawler or rubber-tired frame weighing a few thousand pounds. The air hammer and rotating device are mounted on a boom on the machine and the hammer blows are transferred by the drill steel to the drill bit. For the larger-diameter and deeper holes, the air hammer is attached directly to the bit and is lowered down the hole because the impact of the hammer blows is dissipated in the larger and longer columns of drill steel. Percussion drilling is most applicable to brittle rock in the medium-hard to hard range. The depth of the smaller-diameter holes is limited to about 25 or 30 ft, but the larger holes can be drilled to 40 or 50 ft without significant loss of efficiency.

Jet piercing. The jet piercing method of drilling was developed to drill the very hard iron ores (taconite). In this method, a hole is drilled by applying a high-temperature flame produced by fuel oil and oxygen to the rock. The holes range in size from 6 to 18 in., tend to be irregular in diameter, and require careful control of the flame to prevent overenlargement. The drilling machines are integrated units which control the fuel oil and oxygen mixture and lower the jet piercing bit down the hole. The jet piercing method is limited to very hard rock for which other types of drilling are more costly.

Churn drilling. Churn drilling was one of the first methods of drilling used in open-pit mining and was common until the 1950s; however, it has been almost entirely supplanted by rotary drilling. Churn drilling is accomplished by repeatedly raising a heavy bit weighing several tons and dropping it to shatter the rock in the bottom of the hole. Water is used in the hole to keep the cuttings in a muddy suspension; and after several feet of hole is

Fig. 2. Rotary drill used for drilling blast holes in large open-pit mine. (*Kennecott Copper Corp.*)

drilled, it is necessary to remove the mud with a bailing device. Churn drilling is relatively slow and not economically competitive with other methods of drilling.

Blasting. The type and quantity of explosive are governed by the resistance of the rock to breaking. The primary blasting agents are dynamite and ammonium nitrate, which are detonated by either electric caps or a fuselike detonator called primacord.

Dynamite is available in varying strengths for use with varying rock conditions. It is commonly used either in cartridge form or as a pulverized, free-flowing material packed in bags. It can be used for almost any blasting application, including the detonation of ammonium nitrate, which cannot be detonated by the conventional blasting cap.

Commercial, or fertilizer-grade, ammonium nitrate has become a popular blasting medium because of its low cost. It is also safer to handle, store, and transport than most explosives. Granular or prill-size ammonium nitrate is commonly packed in paper, textile, or polyethylene bags. The carbon necessary for the proper detonation of ammonium nitrate is usually provided by the addition of fuel oil. Granular or prilled ammonium nitrate is highly soluble and becomes insensitive to detonation if placed in water, and therefore its use is restricted to water-free holes. The primary advantages of ammonium nitrate over dynamite are low cost (less than one-third that of dynamite), safety, and ease of handling. Ammonium nitrate is not, however, quite as powerful as dynamite, and a higher ratio of powder to rock is required with ammonium nitrate than with dynamite.

During the early 1960s ammonium nitrate slurries were developed by mixing calculated amounts of ammonium nitrate, water, and other ingredients such as TNT and aluminum. These slurries have several advantages when compared to dry ammonium nitrate: They are more powerful because of higher densities and added ingredients, and they can be used in wet holes. Further, the slurries are generally safer and less costly than dynamite and are being used extensively in mines which are wet

or have rock difficult to blast with the less powerful ammonium nitrate. Both ammonium nitrate and slurries are detonated by a small charge of dynamite.

Mechanical loading. Ore- and waste-loading equipment in common use includes power shovels for medium to large pits and tractor-type front-end loaders for the smaller pits. The loading unit must be selected to fit the transportation system, but because the rock will usually be broken to the largest size that can be handled by the crushing plant, the size and weight of the broken material will also have a significant influence on the type of loading machine. Of equal importance in determining the type of loading equipment is the required production or loading rate, the available working room, and the required operational mobility of the loading equipment.

Power shovels. Shovels in open-pit mines range from small machines equipped with 2-yd³ buckets to large machines with 25-yd³ buckets. A 6-yd³ shovel will, under average conditions, load about 6000 tons per shift, while a 12-yd³ shovel will load about 12,000 tons per shift. However, the nature of the material loaded will have a significant effect on productivity. If the material is soft or finely broken, the shovel productivity will be high; but if the material is hard or poorly broken with a high percentage of large boulders, the shovel production will be adversely affected. Power may be derived from diesel or gasoline engines or diesel-electric or electric motors. The use of diesel or gasoline engines for power shovels is usually limited to shovels of up to 4-yd³ capacity, while diesel-electric drives are not common in shovels of more than 6-yd³ capacity. Electric drives are used in shovels ranging from 4-yd³ to 25-yd³ capacity and are the most widely used power sources in the open-pit mining operations (Figs. 4 and 5).

Draglines. A dragline is similar to a power shovel but uses a much longer boom. A bucket is suspended by a steel cable over a sheave at the end of the boom. The bucket is cast out toward the end of the boom and is pulled back by a hoist to gather a load of material which is deposited in an ore hau-

Fig. 4.   An 8-yd³ shovel loading a 75-ton haulage truck. (*Kennecott Copper Corp.*)

Fig. 5.   Typical open-pit loading. Electric shovel loads ore into railroad cars. (*Kennecott Copper Corp.*)

lage unit or on a waste pile. Draglines range in size from machines with buckets of a few yards capacity to machines with buckets of 150-yd³ capacity. Draglines are extensively used in the phosphate fields of Florida and North Carolina. The overburden and the ore (called matrix in phosphate operations) are quite soft and no blasting is required. The phosphate ore is deposited in large slurrying pits where it is mixed with water and transported hydraulically to the concentrating plant.

Front-end loaders. Front-end loaders are tractors, both rubber-tired and track-type, equipped with a bucket for excavating and loading material (Fig. 6). The buckets are usually operated hydraulically and range in capacity from 1 to 15 yd³. Front-end loaders are usually powered by diesel engines and are much more mobile and less costly than power shovels of equal capacity. On the other hand, they are not generally as durable as a power shovel nor can they efficiently excavate hard or poorly broken material. A front-end loader has about one-half the productivity of a power shovel of equal bucket capacity; that is, a 10-yd³ front-end loader will load about the same tonnage per shift as a 5-yd³ shovel. However, the loader is gaining in popularity where mobility is desirable and where the digging is relatively easy.

Mechanical haulage. The common modes of transporting ore and waste from open-pit mines are trucks, railroads, inclined skip hoists, and belt conveyors. The application of these methods or combination of these methods depends upon the size and depth of the pit, the production rate, the length of haul to the crusher or dumping place, the maximum size of the material, and the type of loading equipment used.

Truck haulage. Truck haulage is the most common means of transporting ore and waste from open-pit mines because trucks provide a more versatile haulage system at a lower cost than rail, skip, or conveyor systems in most pits. Further, trucks are frequently used in conjunction with rail and conveyor systems where the haulage distance is greater than 2 or 3 mi. In this case trucks haul from the pit to a permanent loading point on the surface from which the material is transferred to one of the other systems. Trucks range in size from 20-ton to 200-ton capacity and are powered by diesel engines or a combination of diesel-electric units. Diesel drives are used almost exclusively in trucks up to 75-ton capacity. In the larger trucks, the diesel-electric drive, in which a diesel engine drives a generator to provide power for electric motors mounted in the hubs of the wheels, is most common. The diesel engines in the conventional drives range from 175 to 1000 hp, while the engines in the diesel-electric units range from 700 to 2000 hp. Most haulage trucks can ascend road grades of 8–12% fully loaded and are equipped with various braking devices, including dynamic electrical braking, to permit safe descent on equally steep roads. Tire cost is a major item in the operating cost of large trucks, and the roads must be well designed and maintained to enable the trucks to operate efficiently at high speed.

The size of truck used is primarily dependent upon the size of the loading equipment, but the required production rate and the length of haul are also factors of consideration. Usually 20–40-ton

Fig. 6. A 3-yd³ front-end loader loading a haulage truck in a small open-pit mine. (*Eaton Yale and Towne, Inc.*)

capacity trucks are used with 2–4-yd³ shovels, while the 70–100-ton trucks are used with 8–10-yd³ shovels.

In the early 1960s major advances were made in truck design which put the truck haulage system in a favorable competitive position compared to the other methods of transportation. Trucks are used almost exclusively in small and medium-size pits and are being used in conjunction with rail haulage in the larger pits. Trucks have the advantage of versatility, mobility, and low cost when used on short-haul distances.

Rail haulage. When mine rock must be transported more than 1½ or 2 mi, rail haulage is generally employed. Since rail haulage requires a larger capital outlay for equipment than other systems, only a large ore reserve justifies the investment. As a rough rule of thumb, the reserve should be large enough to support for 25 years a production rate of 30,000 tons of ore per day and an equivalent or greater tonnage of stripping. Adverse grades should be limited to a maximum of 3% on the main lines and 4% for short distances on switchbacks. Good track maintenance requires the use of auxiliary equipment such as mechanical tie tampers and track shifters. The latter are required for relocating track on the pit benches as mining

Fig. 7. Large open-pit copper mine showing rail haulage and trolley-electric locomotive. (*Kennecott Copper Corp.*)

Fig. 8. Dumping copper-mine waste by trolley electric haulage on side-dump cars. (*Kennecott Copper Corp.*)

progresses and on the waste dumps as the disposed material builds up adjacent to the track. Ground movement in the pit resulting from disturbance of the Earth's crust or settling of the waste dumps makes track maintenance a large part of mining cost.

Locomotives in use range from 50 to 125 tons in weight, with the largest sizes coming into increased use for steeper grades and larger loads. Most mines operate either all-electric or diesel-electric models. The use of all-electric locomotives creates the problem of electrical distribution in the pit and on the waste dumps, and requires the installation of trolley lines adjacent to all tracks (Figs. 7 and 8).

Mine cars range in capacity from 50 to 100 tons of ore, and to 40 yd³ of waste. Ore is transported in various types of cars: solid-bottom, side-dump, or bottom-dump. The solid-bottom car is cheapest to maintain, but requires emptying by a rotary dumper. Waste is mostly handled by the side-dump car. Truck haulage has replaced much of the rail haulage in recent years because of advances in truck design and reduction in truck haulage costs.

Inclined skip hoist. Skip hoists consist of two counterbalanced cars of 20–40-ton capacity which operate on steeply inclined rails placed on a pit slope or wall. The skips are hoisted by a cable from the hoist house at the top of the pit. The skip system was developed to eliminate the need to haul ore and waste on long spiraling roads or tracks from the bottom of the pit. Either truck or rail haulage can be used to haul the rock from the shovel to the skip. Skips were very popular in the 1950s, but major advances in truck design and efficiency made skips all but obsolete by 1960.

Conveyors. Rubber-belt conveyors may be used to transport crushed material from the pit at slope angles up to 20°. Conveyors are especially useful for transporting large tonnages over rugged terrain and out of pits where ground conditions preclude building of good haulage roads. Improved belt design is permitting greater loading, higher speeds, and the substitution of single-flight for multiple-flight installations. The chief disadvantage of this transport system is that, to protect the belt from damage by large lumps, waste as well as ore must be crushed in the pit before it is loaded on the belt.

Waste-disposal problems. To keep costs at a minimum, the dump site must be located as near

the pit as possible. However, care must be taken to prevent location of waste dumps above possible future ore reserves. In the case of copper mines, where the waste contains quantities of the metal which can be recovered by leaching, the ground on which such waste is deposited must be impervious to leach water. Where the creation of dumps is necessary, problems of possible stream pollution and the effect on farms and on real-estate and land values must all be considered.

Slope stability and bench patterns. In open-cut and pit mining, the material ranges from unconsolidated surface debris to competent rock. The slope angle, that is, the angle at which the benches progress from bottom to top, is limited by the strength and characteristics of the material. Faults, joints, bedding planes, and especially groundwater behind the slopes are known to decrease the effective strength of the material and contribute to slides. In practice, slope angles vary from 22 to 60° and under normal conditions are about 45°. Steeper slopes have a greater tendency to fail but may be economically desirable because they lower the quantity of waste material that must be removed to provide access to the ore.

Recently developed technology permits an engineering approach to the design of slopes in keeping with measured rock and water conditions. Quantitative estimates of the factor of safety of a given design are now possible. Precise instrumentation has been developed to detect the boundaries of moving rock masses and the rate of their movement. Instrumentation to give warning of impending failure is being used in some pits.

Communication. Efficiency of mining operations, especially loading and hauling, is being improved by the use of communication equipment. Two-way, high-frequency radiophones are proving useful for communicating with haulage and repair crews and shovel operators.

[DONALD P. BELLUM]

Mining, placer

Placer mining is the working of deposits of sand, gravel, and other alluvium and eluvium containing concentrations of metals or minerals of economic importance. For many years gold has been the most important product obtained, although considerable platinum, cassiterite (tin mineral), phosphate, monazite, columbite, ilmenite, zircon, diamond, sapphire, and other gems have been produced. Other valuables recovered include native bismuth, native copper, native silver, cinnabar, and other heavy weather-resistant metals or minerals.

In addition to onshore placer mining, offshore mining for gold, diamonds, iron ore, lime sand, and oyster shells is being done. Future possibilities exist for mining of manganese nodules containing manganese, iron, copper, nickel, and cobalt and phosphorite nodules which overlie large areas of deep ocean floors. Other apparent concentrations of metals and minerals exist in vast tonnages of clays, muds, and oozes on the ocean floors.

Mining claims. In the United States placer mining claims are initially obtained by complying with mining and leasing laws of the Federal or state government or both. Concessions for mining ground in foreign countries can be obtained by following appropriate procedures of the country concerned. Specific information pertaining to acquiring mining claims in the United States can be obtained through offices of the U.S. Bureau of Land Management and the state land office having jurisdiction over the area in which the mining ground is located.

Prospecting, sampling, and valuation. A uniform grid pattern is best for sampling placers. This generally consists of equally spaced sample points in lines approximately perpendicular to the longest direction of the deposit. The distance between lines may be five times the distance between holes. If the apparent deposit seems equidistant in all directions, a grid of equally spaced samples is often used. Prior to actual laying out a prospecting grid, geophysical prospecting techniques may be used to determine approximate depth to bedrock and to outline probable old stream drainage systems. This information will allow more judicious planning of a sample point pattern. Samples are usually obtained by one or several of the following methods: shaft or caisson sinking, churn, and other drilling, and dozer or backhoe opencut trenches. Drilling in areas covered by water is accomplished by placing equipment on a barge or, in the case of the Northland, on ice. The information obtained allows the value of the deposit to be calculated and gives descriptions of the physical characteristics of the material in the deposit. The latter information is valuable for designing the mining method. Plotting unit values, obtained from sampling, on a map will indicate whether a concentration of mineral exists and the location of a pay streak. Calculations will indicate the total value and yardage in the pay streak. A decision can then be made as to whether or not the deposit can be mined on an economic basis.

Recovery methods. A form of sluice box is the type of unit most often used to separate and recover the valuable metals or minerals. Such a unit may vary in width from 12 to 60 in., 30 to 60 in. being common, and in lengths from 40 to several hundred feet. The sluice box is placed on a grade of about $1\frac{1}{2}$ in. vertical to 12 in. horizontal, but adjusted for the particular conditions. Recovery units used in the sluice include the following.

Riffles. Rocks 3–6-in., wood blocks, pole riffles, and Hungarian riffles (Figs. 1 and 2) form pockets in which valuable heavy particles will settle and be retained.

Amalgamation. Mercury is sprinkled at the head end of the sluice box to combine with and hold gold

in various types of riffles and amalgam traps. Copper plates, mercury-coated and protected from coarse gravel by suitable screen, aid in collecting gold under some conditions.

Undercurrents. These are auxiliary sluices, parallel or at right angles to the main sluice. A variety of grizzly plates, screens, and mattings are used to separate various sizes of material and to catch small particles of concentrates that may be lost when using only a conventional sluice box.

Jigs. This is essentially a box with a screen top upon which rests a 3–6-in. bed, usually of steel shot, through which pulsating water currents act. Feed to the jig comes onto the bed in a water stream and the valuable heavy particles pass through the pulsating bed and are recovered as a concentrate. Lighter waterborne material passes over the bed and into other recovery units or emerges as a finished tailing.

Other. In some plants, units such as cyclones, spirals, and grease plates are used to effect recovery. More sophisticated equipment may be used to further concentrate and separate products. Details of design and application of various recovery units appear in technical literature.

Water supply. A large volume of water is essential to nearly all types of placer mining. Water may be brought into the mining area by a ditch, in which case the water is often under sufficient head to give ample gravity pressure for the mining operation. In other cases, water is pumped from its source. Where the water supply is limited, the water is collected after initial use in a settling pond and then pumped back to the mining operation for reuse. Water rights are obtained in accordance with Federal and state regulations in the United States.

Mining methods. A number of mining methods exist, but all have elements of preparing the ground, excavating the pay material, separating and recovering the valuable products, and stacking the tailings. A general classification of present placer mining methods is small-scale hand methods, hydraulicking, mechanical methods, dredging, and drift mining.

Small-scale hand methods. Perhaps most widely known because of gold rush notoriety are methods using pans, rockers, long toms, and other equipment that are responsible for a very small percent of all placer production. Other small-scale methods include shoveling into boxes, ground sluicing, booming, and the use of dip boxes, puddling boxes, dry washers, surf washers, and small-scale mechanical placer machines.

Pan and batea. Panning currently is mostly used for prospecting and recovering valuable material from concentrates. The pan is a circular metal dish that varies in diameter from 6 to 18 in., 16 in. being quite common. Many such pans are 2 or 3 in. deep and have 30–40° sloping sides. The pan with the mineral-bearing gravel, immersed in water, is shaken to cause the heavy material to settle toward the bottom of the pan, while the light surface material is washed away by swirling and overflowing water. These actions are repeated until only the heavy concentrates remain.

In some countries a conical-shaped wood unit called a batea (12–30-in. diamter with about 150° apex angle) is used to recover valuable metals from

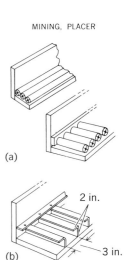
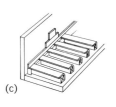

Fig. 2. Types of riffles in partial plan views. (*a*) Pole riffles. (*b*) Hungarian riffle. (*c*) Oroville Hungarian riffle. (*d*) All-steel sluice. (*From G. J. Young, Elements of Mining, 4th ed., McGraw-Hill, 1946*)

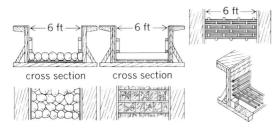

Fig. 1. Cross-sectional diagrams illustrating construction of sluices. (*From G. J. Young, Elements of Mining, 4th ed., McGraw-Hill, 1946*)

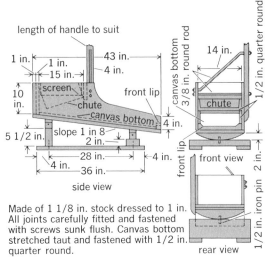

length of handle to suit

1 in.
1 in.
15 in.
43 in.
4 in.
screen
chute
10 in.
front lip
canvas bottom
slope 1 in 8
5 1/2 in.
2 in.
28 in.
4 in.
4 in.
36 in.

side view

canvas bottom
3/8 in. round rod
1/2 in. quarter round
14 in.
chute
4 in.
front lip
front view

1/2 in. iron pin   2 in.
1/2 in. quarter round
rear view

Made of 1 1/8 in. stock dressed to 1 in. All joints carefully fitted and fastened with screws sunk flush. Canvas bottom stretched taut and fastened with 1/2 in. quarter round.

Fig. 3. Diagram of plan details of a rocker. (*Modified from a drawing in Eng. Mining J., in R. S. Lewis, Elements of Mining, 2d ed., Wiley, 1941*)

river channels and bars.

Rockers. Rockers are used to sample placer deposits or to mine high-grade areas (Nome, Alaska, beaches during the gold rush) when installation of larger equipment is not justified (Fig. 3). At present, various types of engine-operated mechanical panning machines and mechanical sluices are used in concentrating samples.

Long tom. A long tom is essentially a small sluice box with various combinations of riffles, matting, and expanded metal screens and, in the case of gold, amalgamating plates.

Hydraulicking. Hydraulic mining utilizes water under pressure, forced through nozzles, to break and transport the placer gravel to the recovery plant (often sluice box), where it is washed. The valuable material is separated and retained in the sluice, and the tailings pass through and are stacked by water under pressure. Hydraulicking is a low-cost method of mining if a cheap, plentiful supply of water is available and streams are not objectionably polluted.

Mechanical equipment. In this category, equipment such as bulldozers, draglines, front-end loaders, pumps, pipelines, hydraulic elevators, and recovery units are used in a number of different combinations depending on the physical characteristics of placer deposits. A typical example is to use bulldozers to prepare the ground and push the placer material to the head end of a sluice box, from which the gravel is washed through the sluice box. The tailings are stacked with a dragline, and a pump is used to return the water for reuse.

Dredging. Large flat-lying areas are best suited for dredging because huge volumes of gravel can be handled fairly cheaply. Prior to dredging a deposit, the ground must be prepared. This preparation may consist of removing trees and other vegetation and overburden such as the frozen loess (muck) found in many parts of Alaska by use of equipment and water under pressure (stripping). Barren gravel may also be removed. If the placer gravel is frozen, the deposit may be thawed by the cold-water method, as has been done in northern Canada (Yukon Territory) and Alaska.

Bucket-ladder dredges (California type). For onshore placer mining, this mass-handling method now largely supersedes others, such as the chain-bucket dredge and one-bucket dredge for mining placer gold and other heavy placer material. Some large modern dredges can dig to depths of 150 ft and handle 10,000 yd^3 or more/24 hr.

Modern dredges are for the most part steel hulls of the compartment or pontoon types. On the hull is mounted the necessary machinery to cause an endless bucketline to revolve and dig the placer gravels as the dredge swings from side to side in a pond (Fig. 4). At the end of each swing, the buckets on the ladder are dropped and the cycle repeated. When the bucketline reaches the bottom of the pay streak, the ladder with buckets is raised, the dredge is moved ahead or into a parallel cut, and the cycle is repeated. The buckets discharge their load into a hopper, where it is washed with water into a long inclined revolving cylindrical screen (trommel) having holes that commonly vary in diameter from $\frac{1}{4}$ in. to as much as $\frac{3}{4}$ in. In this revolving screen, a stationary manifold supplies water (common pressure 60 psi) to nozzles spaced equally in a line the length of the screen. Water from these nozzles washes the sand and gravel and causes the fines and valuable minerals and metals to work through the holes in the revolving trommel and fall on the tables (series of parallel sluices), which may consist of various recovery units, such as mercury traps, jigs, riffles, matting, expanded metal screens, and undercurrents. The placer concentrates are held in the recovery units until cleanup time (the end of a day to 4 weeks, but commonly 2 weeks). The accompanying sand passes over the tables and into the dredge pond. The coarse material that will not pass through the holes in the screen is discharged at the lower end of the trommel screen onto an endless conveyor belt (stacker) which causes the coarse tailings to be deposited in back of the dredge. Overflow muddy pond water passes from the mining area through a drain (Fig. 5).

In the case of gold mining, during cleanup time, the gold amalgam is further concentrated by paddling in the sluices, then picked up, cleaned by panning, retorted to separate the gold from the mercury, and the remaining gold is cast into bricks for appropriate sale. The mercury is reused.

Dragline dredges. A washing and recovery plant, mounted on pontoons or crawler treads, is fed directly or by a conveyor from a shovel or dragline unit (Fig. 6). Flow of material and recovery of prod-

Fig. 4. Yuba bucket-ladder dredge in operation near Marysville, Calif. (*Yuba Manufacturing Co.*)

Fig. 5. Aerial oblique view up-valley over the site of a placer dredging, in which successive cuts, tailing patterns, and dredge pond are shown. The dredge appears in the left foreground. (*Pacific Aerial Survey, Inc.*)

ucts are essentially the same as for a bucketline dredge. Such dragline dredge installation is mostly used on deposits too small to justify the installation of a bucket dredge.

Offshore suction dredges. Exploration for offshore mineral deposits in the unconsolidated material of the ocean's floors is receiving a considerable amount of attention with the ultimate hope of increasing the efficiency of mining for gold, tin, diamonds, heavy mineral sands, iron sands, lime shells, and sand and gravel (also excavation for channel excavation). In addition to bucket-ladder dredges, grab dredges (clamshell buckets), hydraulic (suction) dredges, and airlift dredges are being used. In hydraulic dredges water under pressure is released near the intake and, helped by a suction created with pumps, returns to the surface, carrying with it sand and gravel from the ocean floor. In airlift dredges, air replaces water as described for the hydraulic dredge. Flow of material and recovery of products are essentially the same as for a bucketline dredge.

Small-scale venturi suction dredges have been used by skin divers to recover gold from river bottoms.

In the future, development of more sophisticated equipment for offshore mining of minerals can be anticipated.

Drift mining. Because of relatively high costs, drift mining of placer deposits is not used to a large extent. It consists of sinking a shaft to bedrock, driving drifts up- and downstream in the valley deposits for 250–300 ft, and then extending cross-

Fig. 6. Dragline dredge and pond. (*Bucyrus-Erie Co.*)

cuts the width of the pay streak. In thawed ground, timbering is necessary, whereas in frozen ground a minimum of timber is required and steam thawing plants are used to thaw the gravels before mining is done. Usually, the thawed gravel is hoisted to the surface and then washed to recover the valuable material. Research utilizing modern technique and equipment should make this method more economically competitive. *See* MINING, UNDERGROUND.

Water pollution. Recent proposed regulations on both state and Federal levels tend to place more restrictions on discharge of water carrying sediments into stream drainages. Eventually, treatment of some type may be necessary to remove such sediments, and proposed new operations will no doubt consider this factor.

Tailing disposal. Increased restriction on tailings in placer mining is being proposed in various areas. Such regulations may require smoothing of tailing piles and in some cases resoiling. In Alaska, in many cases, dredge tailings have added to the value of the ground by removing permafrost and offering a solid foundation for road and building construction in place of the original swampland surface.

Power. Electric power for operation of equipment may be generated by several methods, such as utilizing coal, oil, gas, or water (hydro). Units may be permanently located or may be diesel power units aboard dredges or movable shore plants for dredges.

Developments in technology. In past years a number of improvements have been made to increase the efficiency of placer mining. These may be summarized as follows.

1. Development of the diesel engine allowed diesel-powered pumps, tractors, draglines, dredges, and the like to be manufactured and gave much more compactness, economy, and flexibility to placer mining methods. This in turn resulted in greater yardage at lower unit costs.

2. The adaptation and use of jigs in placer recovery plants have increased the recovery of valuable minerals and metals with less work and more efficiency of compact recovery units.

3. The use of the sluice plate in mechanical mining methods has been an important improvement in recent years. This method does away with the conventional nozzle setup in front of the sluice box and the accompanying labor expense. Gravel is pushed directly into the head of the sluice into a stream of free-flowing water. This results in a fast system of coordinated operation, especially for mining placer gravels that contain a minimum of clay and cemented gravels.

4. Development of more efficient transportation carriers has allowed relatively inaccessible areas to become accessible. These include large airplanes to haul heavy equipment and fuel and land machines for crossing swampy and difficult terrain.

5. Use of rippers on tractors to aid in the mining process is becoming successful.

6. Electronic devices and television cameras for control units further increase efficiency of operation.

7. Modern lightweight pipe with the "snap-on" type of couplings and continuously improving design of equipment such as the automatic moving nozzle (Intelligiant) tend to increase the efficiency of mining.

[EARL H. BEISTLINE]

Bibliography: American Institute of Mining, Metallurgical and Petroleum Engineers, *Surface Mining*, 1968; Institution of Mining and Metallurgy, London, *Opencast Mining, Quarrying and Alluvial Mining*, 1965; C. F. Jackson and J. B. Knaebel, *Small-Scale Placer-Mining Methods*, USBM Inform. Circ. no. 6611, 1932; R. S. Lewis and G. B. Clark, *Elements of Mining*, 3d ed., 1964; R. Peele and J. A. Church (eds.), *Mining Engineers' Handbook*, 2 vols., 3d ed., 1941; E. B. Wilson, *Hydraulic and Placer Mining*, 3d ed., 1918.

Mining, strip

A surface method of mining by removing the material overlying the bed and loading the uncovered mineral, usually coal. It is safer than underground mining because neither the workers nor the equipment is subjected to such hazards as roof falls and explosions caused by gas or dust ignitions. Coal near the outcrop or at shallow depth can be stripped not only more cheaply but more completely than by deep mining, and the need for leaving pillars of coal to support the mine roof is eliminated. The roof over coal at shallow depth is weak and difficult to support in underground workings by conventional methods, yet this same weakness, of cover and of coal seam, makes stripping less difficult.

Stripping techniques. Power shovels, draglines, bulldozers, and other types of earth-moving equipment slice a cut through the overburden down to the coal. The cut ranges from 40 to 150 ft wide, depending on the type and size of equipment used. The stripped overburden (spoil) is stacked in a long ridge (spoil bank) parallel with the cut and as far as possible from undisturbed overburden (high wall). The slope of a spoil bank is approximately 1.4:1 and that of a high wall under average conditions is 0.3:1. The uncovered coal (berm) is then fragmented, loaded, and transported from the pit.

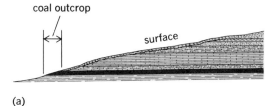

(a)

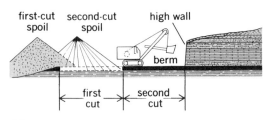

(b)

Fig. 1. Representative cross-section profile diagrams of contour strip mining of coal. (*a*) Section before stripping. (*b*) Section after second cut.

Spoil from each succeeding cut is stacked overlapping and parallel with the previous ridge and also fills the space left by the coal removed (Fig. 1).

Techniques of stripping methods are similar, but the size of equipment used depends on whether the mine is in prairie or hill country. In prairie areas the thickness of overburden is nearly uniform, the coal bed is extensive, and equipment can be used for years at one mine without dismantling and moving to another location. Large-capacity shovels costing $1,000,000 or more and requiring many months to erect on the site are used at prairie mines. A unit of this type is the 60-yd³ rig shown in Fig. 2.

In 1958 the world's largest power shovel was the 70-yd³ unit at Peabody Coal Co.'s River King mine near Freeburg, Ill. By 1968 the 180-yd³ shovel at Southwestern Illinois Coal Corp.'s Captain mine near Percy, Ill., was the largest.

Most coal underlying hills is mined by underground methods, but where the working approaches the outcrop and the overburden is thin, the roof becomes difficult and expensive to support. The coal between the actual or potential underground workings and the outcrop is then more suitable for stripping. Usually, only two or three cuts 40–50 ft wide can be made on the contour of the coal bed, after which the shovel has to be moved to another site. Thus, in contour stripping, mobility of shovels up to 5-yd³ capacity is more important than those of larger capacity.

Large draglines are used instead of shovels to strip pitching beds of anthracite to depths surpassing 400 ft; however, this use of large draglines could more properly be classed as open pit. *See* MINING, OPEN-PIT.

Removal of unconsolidated overburden by hydraulic monitoring is a technique used especially in Alaska. Water under a high-pressure head is directed through a nozzle against the overburden to wash it into deep valleys where swift streams carry it away.

Although the character of the overburden determines the thickness of overburden that can be stripped, the maximum for shovels up to 5-yd³ capacity is about 50 ft and for the largest equipment about 110 ft. To reach these goals it frequently is necessary to use a dragline, carryall, or bulldozer on the high wall or a dragline in tandem with the shovel on the berm to strip the upper few feet of overburden.

Digging equipment known as the wheel can be used ahead of a large power shovel to remove the upper 20–40 ft of unconsolidated soil, clay, or weathered strata. This spoil is discharged onto a belt conveyor, then onto a stacker, and finally deposited several hundred feet from the high wall. Overburden thus removed improves the shovel productivity rate materially. Also, coal reserves can be mined that would have been too deep for the power shovel alone to handle.

Rocks overlying coal beds present some diversity of conditions for removal. Materials generally comprise shale, sandstone, and limestone with shale predominating. Proper fragmentation before stripping may be necessary to produce sizes that are smaller than the shovel dipper. Probably more research has been done on overburden drilling and blasting than on any other phase of stripping. The

Fig. 2. Large electric shovel of the type used in prairie regions, high wall (right), and spoil bank (left) at Hanna Coal Co.'s Georgetown mine, eastern Ohio. (*U.S. Bureau of Mines and Marion Power Shovel Co.*)

diameter, depth, and spacing of drill holes, the type of drill (whether vertical or horizontal), and the amount and type of explosive for each blast hole are the variables that must be determined for optimum production. Truck-mounted rotary drills have replaced churn drills for drilling vertical holes, and for horizontal drilling, auger drills are used. Package explosives, Airdox, Cardox, and commercial ammonium nitrate mixed with diesel fuel are used for blasting. The ammonium nitrate – diesel fuel mixture is one of the cheaper explosives and has gained favor rapidly. Equipment for mixing this explosive and automatically injecting it through a plastic tube into horizontal drill holes is being tested. If perfected, it will mechanize the only manual operation remaining in the stripping cycle.

Before coal is loaded, spoil remaining on the berm is removed by bulldozer, grader and rotary brooms, or at small mines by hand brooms. Small-diameter vertical holes are augered into the bed and blasted with small charges of explosive to crack the coal. A ripper pulled by a bulldozer is effective in replacing coal drilling and blasting. Broken coal is loaded by ½–5-yd³ capacity shovels into trucks ot 5–80-ton capacity and transported from the stripping.

Land reclamation. The disfiguration of the Earth's surface and scars left by high walls and spoil piles of early stripping activity have been well publicized. Over the years, the states in which strip mining was done have enacted legislation to compel the restoration of strip-mined land to some usable form. Responsible operators have been involved in such reclamation work to preserve good public relations even before being required to do so by law. Water was impounded for recreation lakes; spoil banks were partially graded and trees planted; and some areas were graded and planted in grasses for livestock grazing.

Most states require a bond from the operator as assurance that the stripped land will be reclaimed as specified by law. If the restoration is not completed satisfactorily, the bond is forfeited and the state uses that money to do the work.

Some state strip mine reclamation laws are more demanding than others, and in 1967 the U.S. Department of the Interior published guides for restoring stripped areas. Legal requirements for reclamation have become so detailed in some states that strip mine operators usually integrate the work as part of the mining cycle rather than waiting until the stripping is completed. By so doing, some stripping equipment is operated continuously rather than intermittently, as was done in the past. Furthermore, prompt restoration lessens silting of waterways and can eliminate the formation of acid water that would otherwise contaminate them.

[JAMES J. DOWD]

Bibliography: *Mining Guidebook and Buying Directory*, annual; R. Peele, *Mining Engineers' Handbook*, 3d ed., 1948; A. L. Toenges et al., *Some Aspects of Strip Mining of Bituminous Coal in Central and South Central States*, U.S. Bur. Mines, Inform. Circ. no. 6959, 1937; U.S. Department of the Interior, *Surface Mining and Our Environment*, 1967.

Mining, underground

An underground mine is a system of underground workings for the removal of ore from its place of occurrence to the surface, and involves the deployment of men and services.

There are several basic physical elements in an underground mining system. The passageways (openings) in a mine are called drifts if they are parallel to the geological structure, and cross-cuts if they cut across it. They range in size about 60–200 ft² in cross section, depending on their functions. The workings on a level (horizontal plane) are joined with those on another level by passageways of similiar cross section, called raises if they are driven upward and winzes if driven downward.

The passageways give access to, and provide transportation routes from, the stopes, which are the excavations where the ore is mined. The stopes are between levels. There may be rooms on the level, such as pump rooms, service shops, and lunchrooms. This article discusses exploration of a mine site, methods of removing ore material, and design of underground openings for ore removal and mine facilities.

EXPLORATION

A mine is designed and the mine openings specified after the exploration phase of a mine's history. Exploration in this context is not to be confused with prospecting. Exploration is the process of finding the characteristics of the mineralized rocks and the environmental rocks that make up the mine site. These attributes are absolute and unchanging, but they can only be predicted from sampling so there is always the risk that the predictions may be wrong. Some of the attributes of a mineral deposit and their limits of variation that are found by sampling and measurement are: shape, tabular or curvilinear; attitude, flat or vertical; dimensions, thick or thin, uniform or variable, long or short, or shallow or deep; physical character, hard or soft, strong or weak, laminar, jointed, or massive; mineralization, massive, globular, or disseminated, intense or sparse, or chemically stable or unstable; and surface and overlying formations, expendable or not expendable.

During the exploration there is a feedback from predictions of the revenue and expense that would result from operation. The end result of the exploration phase is a forecast of the grade (amount of valuable mineral per ton) and tonnage that can be mined at a specified rate. The ore grade acceptable could be different at another mining rate.

Parts of the risk involved are (1) a change in the mineralization or the environmental rocks, or both, as mining progresses, (2) drastic changes in the exchange value of the production relative to wages, equipment, and supplies, (3) the availability of new or better equipment, and (4) a change in the governmental attitude, such as in taxation. Any of these will affect the grade and the tonnage for the mine site.

MINE OPERATION

A mine is designed, developed, and worked in blocks of levels and stopes. The size of the blocks may be determined by the amount of ore that has been sufficiently explored, by geological boundaries, or by the need for effective supervision. The design must meet ventilation requirements and the openings must be maintained as long as they are needed. When mining in a block is completed it may, if expedient, be cut off from the ventilating system and the workings may be allowed to collapse. This can be done only if the failure will not disrupt other operations, for example, by causing rock bursts.

There are two basic plans of attack. The choice of plan depends on whether or not the surface, or the rocks overlying the ore, may be disturbed, and on stress redistribution problems.

Longwall. The principle of longwall mining is to advance in line all the stopes and pillars being mined in a block. No remnants are left either to support the back (overlying rocks) or to constitute stress concentrators. The line of attack may advance toward the shaft or other entrance or retreat from it. In the latter case, or if there are blocks beyond the current mining to be mined later, passageways must be maintained. Longwall mining is the method generally used to mine flat-lying deposits such as coal. Rooms with uniform dimensions separated by pillars with uniform dimensions are mined. The pillars support the backs. If the preservation of the backs or the surface is not a factor, the pillars may be systematically mined (robbed) in a longwall retreat or advance.

The result of mining with rooms and pillars is a cellular pattern. Most mining methods available to the mine designer involve the creation of a cellular pattern made up of stopes and pillars. Generally the ore from the stopes is won with less expense than is involved in recovering the pillars. Often the pillars are not mined because they are worth more as pillars than they would be as ore. They are stronger than any material that could economically be used to replace them, and they fit better.

Fill. There are few situations in which it is possible to recover a worthwhile amount of pillar ore unless the adjacent stopes or rooms have been filled. No filling will support the overlying rocks as well as the ore or pillars can, but if there is some settlement onto the fill, it will be limited because the rock is dilated and occupies more volume than the solid rock does. The swell will finally support the back.

The fill may be any incombustible available material—waste rock, sand or gravel, or mill tailings. The tailings from the mill are treated to meet the mine specifications for settling and percolation rates by removing some of the fine sizes. They are then transferred from the mixing plant to the stopes as a slurry by a pipeline and distributed in the stopes through hoses. Fine-grained natural sand may be placed that way as well. The water must be taken out by decantation or by percolation, or both. Some operators add portland cement to form a weak concrete. A mixture of smelter slag and sulfide-bearing mill tailings has been used to form a weak rock by chemical action.

When fill is placed in a stope alongside a pillar that is to be recovered, a partition, usually of light timber, is installed between the fill and the pillar. Some operators use an enriched mixture of cement at the interface to form a concrete.

MINE DESIGN

Fundamentally, mining is materials handling, and a mine is designed accordingly. Three functions are involved: breaking the ore from the face, delivering it to the surface, and delivering supplies to where they are needed. Ancillary functions are getting men and services, air, power, and water to where they are needed. Power may be electrical or compressed air.

Ore breaking and transporting together constitute 30–55% of the total cost of mining. If the ore has to be broken by explosives, the drilling and blasting cost is 10–20% of the total mine cost and the transportation portion is 20–35%.

Primary breaking cost varies inversely, and the transportation cost varies directly, with the size of the broken ore. There is an optimum size for the product of blasting. Each stage of the transportation phase has a limiting size that can be accommodated, so expense saved in the breaking phase may be exceeded by that of secondary breaking between stages in the transportation system.

A mine is designed around the method of primary breaking that is chosen, and the choice is governed by the forecasts from exploration. Stopes may be open or filled if the ore has to be broken by explosives. Caving may be used if conditions are favorable, and the ore will be broken by natural forces that make up the stress field in the ore and in the environmental rocks. Mining methods will be described under major headings, but since each ore body is unique and operators are ingenious, there are variations and hybrid methods.

Open stopes. There is a further qualification to this type of stope—with or without delayed filling. Using open stopes without delayed filling may be expedient, but if there are other extensive workings, the open stopes may redistribute the ground stresses in a manner which will interfere with subsequent work.

The mining method is chosen according to the thickness and inclination of the deposit. The breaking point between thick and thin in a tabular deposit is about 16 ft. That is about the maximum length of a stull (round timber) that can be handled conveniently in an open stope to give casual support where it is thought that loose rock might develop, or to provide a working platform for the miners. The breaking point between steep and flat is about 40° from horizontal, the limit at which rock will move by gravity on a rough surface.

The stope may be worked either overhand or underhand. If the ore will move by gravity to the drawpoints, the overhand attack is by advancing a breast (face) parallel to the level and so breaking out a slice. Because a platform to work off must be constructed with stulls and lagging or plank for each drilling site, the method is limited to thin deposits. The underhand attack is started from the top of the stope, usually from the access raise, and a block is broken out by drilling and blasting down holes. The bench must be cleaned off after each blast before drilling is resumed. This method has an advantage in gold mines, where coarse gold may lodge on the footwall and have to be swept out. Once cleaned, an area will not have to be cleaned again.

If the deposit is steep enough to deliver the broken ore to the drawpoints by gravity and thick enough to prevent the broken ore from arching over the opening, it probably should not be mined off platforms in an open stope.

Shrinkage stoping. An open stope requires successive platforms for the miners to work off. If the broken ore is drawn off just enough to give working room for the miners and they can work off the broken ore, the stope is called a shrinkage stope. The length of the stope is established by two raises, one at each end, which serve as manways and service entrances for air and water lines. By cribbing up with timber, they are maintained through the broken ore. The draw is from one-third to one-half of the break. The rest of the ore is retained in the stope until the stope is complete, and then the stope may be drawn empty through chutes which

Fig. 1. Sublevel stoping.

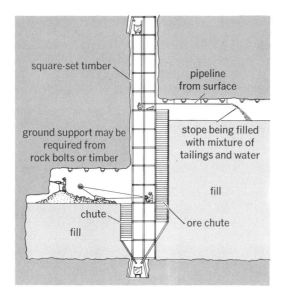

square-set timber

pipeline from surface

ground support may be required from rock bolts or timber

stope being filled with mixture of tailings and water

fill

chute

fill

ore chute

Fig. 2. Hydraulic-fill stope.

(a)

(b)

Fig. 3. A cut-and-fill stope. (a) Ready for hydraulically placed sand fill, except for the burlap lining. (b) Placing the floor over the fill.

are installed during stope preparation.

The broken ore, which moves when drawn, has little ground-support capability. Shrinkage stoping may be used only up to the limit where the back is not self-supporting, or the walls will slab off and give unacceptable dilution. The limit may be extended by using casual timber support, or rock bolts for the back, and by rock-bolting potential slabs to the walls. Once a series of shrinkage stopes has been established and some of the stopes completed, the storage available provides flexibility in production beyond that of any other method.

If the ore body is too wide for the span of the back to be self-supporting or to be cheaply supported when mined parallel to the long dimension of the ore, it may be worked with transverse shrinkage stopes and intervening pillars. This involves delayed filling to permit mining the pillars.

Sublevel stoping. A deposit that is wide enough (about 40 ft) and has walls and ore sufficiently strong to permit shrinkage stoping may be worked by sublevel stopes (Fig. 1).

A vertical face is maintained. At vertical intervals, spaced to accommodate the drilling method to be used for breaking, sublevels are driven in the long direction of the stope from the entry raise. Two procedures are available after the initial slot has been made, which is generally done by widening a raise to the width of the stope. Holes may be drilled radially from the sublevels to give an acceptable distribution of explosive behind the face or slice to be broken off, or a cut may be slashed out across the face from each sublevel and drilled with vertical holes, quarry fashion, to give a better explosive distribution. The choice depends in part on the equipment available and in part on the control wanted at the sides. When the stope is wide, or for better control of the sides (walls), the radial drilling may be done from each of two sublevels at the same level which have been driven on either side of the stope.

The stope may be longitudinal with respect to the long dimension of the ore body, or it may be transverse and separated from the adjacent stope by a pillar. It follows that it will be a delayed-fill

stope if the pillar is to be recovered. Even delayed-fill stopes are not satisfactory if the pillar material is sufficiently valuable to require complete and clean recovery.

Filled stopes. When the wall rock or ore is not sufficiently strong to permit the use of one of the open-stope methods, or if clean pillar recovery is important, methods have been devised to mine with only a small area of unsupported rock or ore exposed.

Cut-and-fill stoping. The breaking phase of this method is not different from that which has been described for the overhand shrinkage stope, but the broken ore is removed after each breast (slice) is completed and a layer of fill is placed. The cycle is: breaking, removing the broken ore, picking up the floor, raising the fill level, and replacing the floor (Figs. 2 and 3). Before the new fill is placed, the manways and the ore chute are raised with cribbing. The cribbing is covered with burlap if hydraulic fill is used. The manway at one end of the stope is built with two compartments. One compartment is used as the ore chute, called a mill hole in filled stopes.

The back must be strong enough to be self-supporting over the span, but the method is sufficiently lower in cost than the alternative so that the back may be supported with casual timber or rock bolts, which are broken out with the new breast.

Square-set stopes. When the backs, and perhaps the walls, are not strong enough to permit cut-and-fill stoping, temporary support may be provided by carefully framed and placed timbers called sets. The sets are filled when no longer needed, at a rate that depends on the rate at which they take weight and might collapse (Fig. 4).

A set is made up of posts 8 or 9 ft long, and caps and girts about 6 ft long, cut to exact lengths and framed to give a good fit at the corners. The timber is usually about 8 in. square, though round timber may be used.

A set is installed in an opening just large enough to accommodate it, and then blocked against the surrounding unstable rock or ore. Little blasting is needed because of the characteristics of the rock that make square-setting the best choice. The problem is to hold back the broken rock until the set can be placed. This may be accomplished by extending boom timbers out over the caps.

Some of the ore may be moved manually. Generally, however, by retaining open sets in the fill and fitting them with inclined slides, provision is made for gathering the ore for scraping to the mill hole.

The sets may be installed either overhand or underhand, depending on the problem, or, if the choice is not critical, on the skill of the miners. The overhand technique is more common but in some camps the miners work better underhand.

Square-setting is the usual method for removing pillars if clean, complete extraction is needed. It is a flexible method and can be used to recover ore in offsets from the main deposit.

Caving. When the surface is expendable and other characteristics are favorable, one of the caving methods may be used.

Top slicing. In some respects this is like square-setting but it is less expensive after it is underway. It is used when the surface is expendable and the ore is too weak to stay in place over a useful span.

The mining block is developed by driving a two-compartment raise through the ore to serve as an access manway and a mill hole, and by driving a longitudinal drift from the raise, at the top of the ore, to the extremity of the ore or the end of the proposed stope.

The initial unit of mining is a timbered crosscut driven each way from the drift to the edge of the block to be mined. Subsequent units are crosscuts driven adjacent to each preceding crosscut to take out a slice of ore. As the face of the slice is retreated toward the mill hole, the timbers in abandoned workings (several sets back) are permitted to fail, or forced to fail by blasting. The routine is continued until the slice is completed to the raise. The overlying formations collapse onto the broken timbers. As the routine is continued by taking successive slices, the broken timbers form a feltlike mat that has some tensile strength, and little timber support is needed in the crosscuts. Several slices

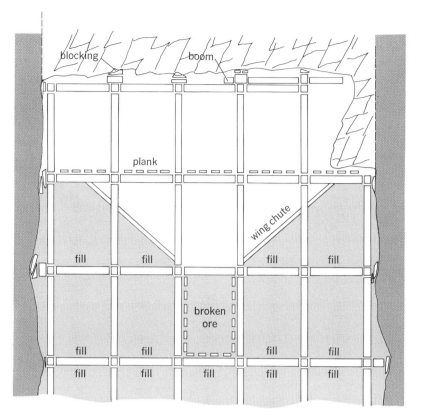

Fig. 4. Square-set stoping.

are mined concurrently, step fashion. The overlying caving formations must follow the mat. No caverns can be left which could collapse and create an air blast.

Sublevel caving. If the ore is sufficiently strong, and after a timber mat has been developed, one or more slices may be omitted. The cantilever shelf

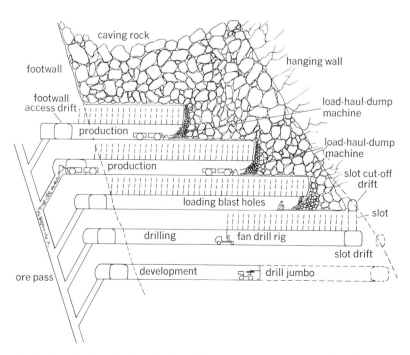

Fig. 5. An adaption of sublevel caving. Cutaway view shows progress of caving.

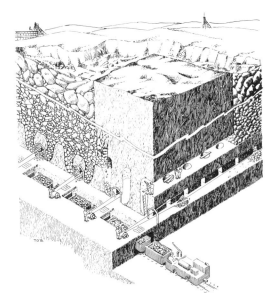

Fig. 6. Block-caving.

formed when the next slice is taken will collapse under the load of caved material and its own weight. The broken ore is moved to the mill hole as in top slicing. Several slices are advanced simultaneously as in top slicing.

An adaption has been used in which the slices are taken out as sublevels in open stoping are taken, and the over lying formations are caved against the face. The sublevel slices are advanced in steps as in top slicing (Fig. 5). There is no mat and some ore is lost into the cave material. However, it is low-grade and the overall low cost of mining makes up for the loss.

Block-caving. If the ore texture (blockiness) and strength are suitable, and if the primitive stress field is favorable, an entire block 150–250 ft on a side and several hundred feet high may be induced to cave after it is undercut. The broken ore is drawn off through bell-shaped drawpoints (Fig. 6).

The drawing cycle is critical. It must keep the undersurface of the block unstable and continuing to fail. The lateral dimensions of the block are controlled by weakening the perimeter with raises and lateral workings, or even short shrinkage stopes. No large cavities are permitted to develop. In the final phase, when caving has reached the overlying formation, care must be taken to avoid drawing it with the ore. There is no primary breaking expense but the cost of secondary breaking for transportation may be high.

Ground control. Mine openings must be kept open as long as they are needed. Mining engineers recognize that rock is not necessarily solid or inert. The study of the behavior of rocks when subjected to force is called rock mechanics. It is a comparatively new field, although knowledge of the phenomena under study has been utilized for years without formal analyses of what was going on.

The observations that rocks around a mine opening do not always behave in a manner that would be predictable by classical mechanics imply that there are other than gravitational forces involved, and that there is strong lateral component of strain energy. The source and reservoir of the

strain energy have been less obscure since geologists have measured the rate of spreading of the North Atlantic Ocean floor and associated continental land masses (average 6 cm. per year since Carboniferous times). The resultant force vector from the combination of gravitational force and tectonic force is referred to as primitive stress in this description of underground mining. The rock mass is in equilibrium until a mine opening is made; that is, it is in equilibrium for a relatively short time involved in the mine operation. The mine opening accepts no force and the force is diverted to around the opening.

Ground support. The ideal support is a pillar of appropriate size, but the use of pillars is not always feasible.

Timber. Traditionally timber has been the usual support for the perimeters of mine openings. It is usually supplied as stulls or as lagging, depending on the slenderness ratio, diameter to length, and to some extent on the use. If it is slender, it is lagging. If a log (stull) is placed vertically, it is called a post. If it is placed nearly horizontally, it is usually called a stull, whether it is acting as a beam or as a column. Both posts and stulls are installed with lower ends in hitches in the rock and upper ends loosely fitted to the back or wall, depending on the location. The final fit is achieved by driving wooden wedges between the end of the timber and the rock. The hitch may be chiseled into the rock, but in hard rock a natural recess is generally used.

When a lateral working requires timbered support, the stulls are usually framed to give a neat fit at the corners, and flatted on two sides to save space in the working. Sawn square timber is often used. The unit is two posts and a cap (stull), usually with a sill on the floor. Whether or not the sill is used depends on the expected loading. The posts and the cap are wedged tightly to the walls and the back at the corners. Lagging or plank is laid over the caps to provide overhead protection and placed behind the posts if a loose wall is expected. Raises and winzes are similarly protected unless they are to be used for hoisting and more precise timbering is needed. Steel is frequently used in the same manner as timber.

Concrete. Openings are frequently lined with concrete if permanence or added strength is needed, if the ventilation friction factor must be reduced, or if the operator does not trust timber because of the fire hazard. Generally the opening is made round or ovaloid so the concrete will be in compression. Forms and poured concrete are commonly used, generally with reinforcing bars. Circumferential steel reinforcement is not effective if the concrete is loaded in circumferential compression.

Concrete may be blown onto the rock surface with a cement gun (guniting). Sand and small-sized aggregate are mixed dry and blown through a hose to the face to be coated. Water is added as the mixture passes the nozzle. The low-moisture mixture hits the face and a portion of the aggregate falls out. A tight bond is formed at the concrete-rock interface. It is thought that the peining action of the aggregate helps to make the bond and to produce a dense concrete. In treacherous rock quite large rooms, such as underground hoist rooms 30 ft or more across, have been successfully secured in this way. The angle of impingement for

Fig. 8.   Testing the tension on a rock bolt by using a simple instrument to measure torque.

Fig. 7.   The reinforcement of a mine opening with concrete. (a) A drift reinforced with gunite. (b) The same drift beyond the gunite.

the application is critical. The thickness of the coating is not more than a few inches (say 3 or 4) over the depressions in the rock surface and thinner over the bumps (Fig. 7).

Rock bolts. The systematic use of rock bolts for rock reinforcement has increased rapidly. These are steel bolts about 3/4 in. in diameter and generally 3–5 ft long, anchored at the bottoms of holes drilled at a right angle to the rock surface, and tensioned by a nut over a small plate at the rock surface. The anchorage is generally a split shell forced against the wall of the bore hole by a wedge as tension is applied at the bolt end. Some suppliers offer a method to anchor the bolt in an epoxy resin, and some others supply a bolt that is to be embedded in concrete or cement for its entire length.

There is no consensus as to the reason that rock bolts are effective, but it is agreed that they should be installed as soon as possible after an opening is made, and should be under high tension (Fig. 8). The mechanism offered most frequently for the effectiveness of bolts in a bedded formation is that a compound beam is built up by binding several laminar beds together to act as a single thick beam. For massive rocks that have no bedding it is commonly accepted that the bolts must extend into the compression arch that is postulated to be formed above the opening, and for that reason long bolts are often specified.

Actually, the abutments of the arch are restrained from moving outward, and the tendency is to move inward, especially if there is a high primitive stress lateral component. In narrow openings there is compression close to the skin of the opening and the function of the bolt in tension is to reduce the tendency for failure in oblique shear by preventing the thickening of the rock. At some width, as an opening is enlarged from narrow to wide, the compressive primitive stress is neutralized and the back goes into tension. Rock has little tensile strength because of discontinuities. If the rock is blocky, the blocks may be held together by bolts and form a flat, or nearly flat, voussoir arch.

Transportation. Gravity is used wherever it can be effective in the movement of ore toward the surface. Ore from open stopes that are steep enough for gravity flow is loaded through chutes directly into the level transportation units. When a stope is wider than about 25 ft, bell-shaped openings (drawpoints) are driven into the floor of the stope. If more than one row of them is needed, these drawpoints are driven on about 25-ft centers on a regular pattern so one crosscut can serve the outlets of several drawpoints. When the ore is loaded into the haulage equipment, it is taken to an ore pass and moved by gravity either to a loading pocket at the shaft or to a crushing plant, and thence to the shaftpocket. An inclined ore pass will give considerable lateral movement. A mine will also have a system of waste passes.

A drawpoint may discharge through a chute into a haulage vehicle or into a short branch off the haulage line and be loaded into the main-line vehicle mechanically. If the stope is wide, the drawpoints may discharge onto the floor of a scraper drift at a higher elevation than the back of the haulage level. The ore is then delivered to the main-line vehicle by scraping (Fig. 6).

When a lode is too flat to permit the use of gravity, ore is moved to a central gathering point by a scraper, either in one stage or two. If the lode is flat enough to permit the use of wheeled or crawler-tracked vehicles, the broken ore may be loaded into a gathering vehicle and taken either to the ore pass or to the main-line transportation unit. An alternative is to use a load-haul-dump vehicle. Smaller versions of this type of machine are being introduced into large stopes. They are displacing the scraper, which in turn had displaced the small railcar and the wheelbarrow (Fig. 5).

Equipment is designed to do a specific job and its value in use beyond that job decreases rapidly. The primary gathering equipment is designed for a short haul.

Entry from surface. When the topography of the area has low relief, the entry will be by a shaft or a ramp. Sometimes both means are employed, the ramp being used for moving heavy, large equipment within the mine. If the relief is high, an adit (tunnel) may be used.

A shaft is usually located in the footwall far enough from the mine workings to avoid ground movement. It is designed for specific functions which determine the area (cross section) and shape, if the shape is not modified to accommodate ground stresses and sinking problems. It may be vertical or inclined, though the vertical shaft is the more common. Functionally either kind should be rectangular to accommodate the equipment used in it.

Many vertical shafts are circular or elliptical, but a rectangular framework is fitted in them to guide the shaft vehicles. There is an exception, not common in North America, when rope (steel-cable) guides are hung in the shaft to guide the vehicles. On the other hand, a round or elliptical shaft is better for ventilation because it may be smooth-lined and offers less air resistance.

A shaft is designed after its functions have been decided and the rock conditions have been forecast. It may be multipurpose and the cross section (plan) must include space for each of the functions, as well as a ladderway for an emergency exit. A shaft may be specialized, that is, designed exclusively for ore hoisting, for services, or for ventilation. A mine must have two shafts to provide alternate routes to the surface in case one shaft is out of commission in an emergency.   [A. V. CORLETT]

Bibliography: R. S. Lewis and G. B. Clark, *Elements of Mining*, 3d ed., 1964; R. Peele (ed.), *Mining Engineers' Handbook*, vol. 1, 3d ed., 1941.

Mining, undersea

The process of recovering mineral wealth from sea water and from deposits on and under the sea floor. Unknown except to technical specialists before 1960, undersea mining is receiving increasing attention. Frequent references to marine mineral resources and marine minerals legislation by national and international policy makers, increasing activity in marine minerals exploration, and the launching of major new seagoing mining dredges for South Africa and Southeast Asia all indicate the beginnings of a viable and expanding industry.

There are sound reasons for this sudden emphasis on a previously little-known source of minerals. While the world's demand for mineral commodities is increasing at an alarming rate, most of the developed countries have been thoroughly explored for surface outcroppings of mineral deposits. The mining industry has been required to advance its capabilities for the exploration and exploitation of low-grade and unconventional sources of ore. Corresponding advances in oceanology have highlighted the importance of the ocean as a source of minerals and indicated that the technology required for their exploitation is in some cases already available.

There is a definite realization that the venture into the oceans will require large investments, and the trend toward the consortium approach is very noticeable, not only in exploration and mining activities but in research. Undersea mining has become an important diversification for most major oil and aerospace companies, and a few mining companies appear to be taking an aggressive approach. Exploration and mining consortia have been formed by such companies as Rio Tinto Zinc, Simons Laboratory, and Charter Consolidated (all United Kingdom), Bethlehem Steel and Ocean Science and Engineering Corp. (both United States), N. V. Billiton Maatschappij (Netherlands), C. Itoh (Japan), Broken Hill Proprietary Co. (Australia), and Ocean Mining A. G. (Switzerland). A six-company Commercial Oceanology Study Group has been formed in Great Britain to investigate, among other things, the commercial exploitation of undersea minerals. In the United States, the Department of the Interior through the Bureau of Mines has taken a lead in encouraging industry by implementing cooperative programs of research and development at their Marine Minerals Technology Center at Tiburon, Calif.

While mineral resources to the value of trillions of dollars do exist in and under the oceans, their exploitation is not simple. Many environmental problems must be overcome and many technical advances must be made before the majority of these deposits can be mined in competition with existing land resources.

The marine environment may logically be divided into four significant areas: the waters, the deep sea floor, the continental shelf and slope, and the seacoast. Of these, the waters are the most significant, both for their mineral content and for their unique properties as a mineral overburden. Not only do they cover the ocean floor with a fluid medium quite different from the earth or atmosphere and requiring entirely different concepts of ground survey and exploration, but the constant and often violent movement of the surface waters combined with unusual water depths present formidable deterrents to the use of conventional seagoing techniques in marine mining operations.

The mineral resources of the marine environment are of three basic types: the dissolved minerals of the ocean waters; the unconsolidated mineral deposits of marine beaches, continental shelf, and deep-sea floor; and the consolidated deposits contained within the bedrock underlying the seas. These are described in Table 1, which shows also the subclasses of surficial and in-place deposits,

Table 1. Marine mineral resources

| Dissolved | Unconsolidated | | Consolidated | |
	Surficial	In place	Surficial	In place

Metals and
salts of:
  Magnesium
  Sodium
  Calcium
  Bromine
  Potassium
  Sulfur
  Strontium
  Boron
  Uranium
  And 30 other
    elements
 	
Fresh water
	Shallow beach or
	  offshore placers
	    Heavy mineral sands
	    Iron sands
	    Silica sands
	    Lime sands
	    Sand and gravel
	
	Authigenic deposits
	    Manganese nodules
	      (Co, Ni, Cu, Mn)
	    Phosphorite nodules
	    Phosphorite sands
	    Glauconite sands
	
	Deep ocean floor
	  deposits
	    Red clays
	    Calcareous ooze
	    Siliceous ooze
	    Metalliferous ooze
		Buried beach and
		  river placers
		    Diamonds
		    Gold
		    Platinum
		    Tin
		
		Heavy minerals
		    Magnetite
		    Ilmenite
		    Rutile
		    Zircon
		    Leucoxene
		    Monazite
		    Chromite
		    Scheelite
		    Wolframite
			Exposed stratified
			  deposits
			    Coal
			    Iron ore
			    Limestone
			
			Authigenic coatings
			    Manganese oxide
			    Associated Co, Ni, Cu
			    Phosphorite
				Disseminated
				  massive, vein,
				  or tabular deposits
				    Coal
				    Iron
				    Tin
				    Gold
				    Sulfur
				    Metallic sulfides
				    Metallic salts

characteristics which have a very great influence on the economics of exploration and mining. *See* SEA WATER.

As with land deposits, the initial stages preceding the production of a marketable commodity include discovery, characterization of the deposit to assess its value and exploitability, and mining, including beneficiation of the material to a salable product.

Exploration. Initial requirements of an exploration program on the continental shelves are a thorough study of the known geology of the shelves and adjacent coastal areas and the extrapolation of known metallogenic provinces into the offshore areas. The projection of these provinces, which are characterized by relatively abundant mineralization, generally of one predominant type, has been practiced with some success in the localization of certain mineral commodities, overlain by thick sediments. As a first step, the application of this technique to the continental shelf, overlain by water, is of considerable guidance in localizing more intensive operations. Areas thus delineated are considered to be potentially mineral-bearing and subject to prospecting by geophysical and other methods.

A study of the oceanographic environment may indicate areas favorable to the deposition of authigenic deposits in deep and shallow water. Some deposits may be discovered by chance in the process of other marine activities.

Field exploration prior to or following discovery will involve three major categories of work: ship operation, survey, and sampling.

Ship operation. Conventional seagoing vessels are used for exploration activities with equipment mounted on board to suit the particular type of operation. The use of submersibles will no doubt eventually augment existing techniques but they are not yet advanced sufficiently for normal usage.

One of the most important factors in the location of undersea minerals is accurate navigation. Ore bodies must be relocated after being found and must be accurately delineated and defined. The accuracy of survey required depends upon the phase of operation. Initially, errors of 1000 ft or more may be tolerated.

However, once an ore body is believed to exist in a given area, maximum errors of less than 100 ft are desirable. These maximum tolerated errors may be further reduced to a few feet in detailed ore body delineation and extraction.

There are a variety of types of electronic navigation systems available for use with accuracies from 3000 ft down to approximately 3 ft. Loran, Lorac, and Decca are permanently installed in various locations throughout the world. Small portable systems are available for local use that provide high accuracy within 30–50-mi ranges. For deep-ocean survey, navigational satellites have completely revolutionized the capabilities for positioning with high accuracy in any part of the world's oceans.

During sampling and mining operations, the vessel must be held steady over a selected spot on the ocean floor. Two procedures that have been fairly well developed for this purpose are multiple anchoring and dynamic positioning.

A three-point anchoring system is of value for a coring vessel working close to the surf. A series of cores may be obtained along the line of operations by winching in the forward anchors and releasing the stern anchor. Good positive control over the vessel can be obtained with this system, and if conditions warrant, a four-point anchoring system may be used. Increased holding power can be obtained by multiple anchoring at each point.

Dynamic positioning is useful in deeper water, where anchoring may not be practical. The ship is kept in position by use of auxiliary outboard propeller drive units or transverse thrusters. These can be placed both fore and aft to provide excellent maneuverability. Sonar transponders are held submerged at a depth of minimum disturbance, or the system may be tied to shore stations. The auxiliary power units are then controlled manually or by computer to keep the ranges at a constant value.

Survey. The primary aids to exploration for mineral deposits at sea are depth recorders, subbottom profilers, magnetometers, bottom sampling devices, and subbottom sampling systems. Their use is dependent upon the characteristics of the ore being sought.

For the initial topographic survey of the sea floor, and as an aid to navigation, in inshore waters, the depth recorder is indispensable. It is usually carried as standard ship equipment, but precision recorders having a high accuracy are most useful in survey work.

In the search for marine placer deposits of heavy minerals, the subbottom profiler is probably the most useful of all the exploration aids. It is one of several systems utilizing the reflective characteristics of acoustic or shock waves.

Continuous seismic profilers are a development of standard geophysical seismic systems for reflection surveys, used in the oil industry. The normal energy source is explosive, and penetration may be as much as several miles.

Subbottom profilers use a variety of energy sources including electric sparks, compressed air, gas explosions, acoustic transducers, and electromechanical (boomer) transducers. The return signals as recorded show a recognizable section of the subbottom. Shallow layers of sediment, configurations in the bedrock, faults, and other features are clearly displayed and require little interpretation. The maximum theoretical penetration is dependent on the time interval between pulses and the wave velocity in the subbottom. A pulse interval of 1/2 sec and an average velocity of 8000 ft/sec will allow a penetration of 2000 ft, the reflected wave being recorded before the next transmitted pulse.

Penetration and resolution are widely variable features on most models of wave velocity profiling systems. In general, high frequencies give high resolution with low penetration, while low frequencies give low resolution with high penetration. The general range of frequencies is at the low end of the scale and varies from 150 to 300 Hz, and the general range of pulse energy is 100–25,000 joules for nonexplosive energy sources. The choice of system will depend very much on the requirements of the survey, but for the location of shallow placer deposits on the continental shelf the smaller low-powered models have been used with considerable success.

With the advent of the flux gate, proton precession, and the rubidium vapor magnetometer, all measuring the Earth's total magnetic field to a high degree of accuracy, this technique has become much more useful in the field of mineral exploration.

Anomalies indicative of mineralization such as magnetic bodies, concentrations of magnetic sands, and certain structural features can be detected. Although all three types are adaptable to undersea survey work, the precession magnetometer is more sensitive and more easily handled than the flux gate, and the rubidium vapor type has an extreme degree of sensitivity which enhances its usefulness when used as a gradiometer on the sea surface or submerged.

Once an ore body is indicated by geological, geophysical, or other means, the next step is to sample it in area and in depth.

Sampling. Mineral deposit sampling involves two stages. First, exploratory or qualitative sampling locate mineral values and allow preliminary judgment to be made. For marine deposits this will involve such simple devices as snappers, drop corers, drag dredges, and divers. Accuracy of positioning is not critical at this stage, but of course is dependent on the type of deposit being sampled. Second, the deposit must be characterized in sufficient detail to determine the production technology requirements and to estimate the profitability of its exploitation. This quantitative sampling requires much more sophisticated equipment than does the qualitative type, and for marine work few systems in existence can be considered reliable and accurate. However, in particular cases, systems can be put together using available hardware which will satisfy the need to the accuracy required. Specifically, qualitative sampling of any mineral deposit offshore can be carried out with existing equipment. Quantitative sampling of most alluvial deposits of heavy minerals (specific gravity, less than 8) can be carried out at shallow depths (less than 350 ft overall) using existing equipment but cannot be carried out with reliability for the higher-specific-gravity minerals such as gold (specific gravity 19). Quantitative sampling of any consolidated mineral deposit offshore can be carried out within limits.

Any system that will give quantitative samples can be used for qualitative sampling, but in many cases heavy expenses could be avoided by using the simpler equipment.

To obviate the effects of the sea surface environment, the trend is toward the development of fully submerged systems, but it should be noted that the deficiencies in sampling of the heavy placer minerals are not due to the marine environment. Even on land the accuracy of placer deposit evaluation is not high and the controlling factors not well understood. There is a prime need for intensive research in this area.

Evaluation of surficial nodule deposits has been attempted over small areas using dredges and photographic interpretation. Accuracy of this method has not been confirmed but it can reasonably be assumed to offer a semiquantitative solution.

Exploitation. Despite the intense interest in undersea mining, new activities have been limited mostly to conceptual studies and exploration. The volume of production has shown little change, and publicity has tended to overemphasize some of the smaller, if more newsworthy, operations. All production comes from nearshore sources, namely, sea water, beach and nearshore placers, and nearshore consolidated deposits.

Minerals dissolved in sea water. Commercial separation techniques for the recovery of minerals dissolved in sea water are limited to chemical precipitation and filtration for magnesium and bromine salts and solar evaporation for common salts and fresh-water production on a limited scale. Other processes developed in the laboratory on pilot plant scale include electrolysis, electrodialy-

Table 2. Production from dissolved mineral deposits offshore

Mineral	Location	Number of operations	Annual production	Year	Value (millions of dollars)
Sodium, NaCl	Worldwide	90+	10,000,000 tons	1963	57.5
Magnesium, Mg, MgO, Mg(OH)$_2$	United States, United Kingdom, Germany, Soviet Union	6+	300,000 tons	1963	64.7
Bromine, Br	Worldwide	7	75,000 tons ±	1964	22.6
Fresh water	Middle East, Atlantic region, United States	41	15,850,000,000 gal	1964	20.6
Total		145+			165.4

sis, adsorption, ion exchange, chelation, oxidation, chlorination, and solvent extraction. The intensive interest in the extraction of fresh water from the sea has permitted much additional research on the recovery of minerals, but successful commercial operations will require continued development of the combination of processes involved for each specific mineral.

As shown in Table 2, three minerals or mineral suites are extracted commercially from sea water: sodium, magnesium, and bromine. Of these, salt evaporites are the most important, almost $60,-000,000 being produced annually from known operations. Japan's total production of salt products comes from the sea. Magnesium extracted from sea water accounts for 75% of domestic production of this commodity in the United States, and fresh water compares with bromine in total production value.

Unconsolidated deposits. Unconsolidated deposits include all the placer minerals, surficial and in place, as well as the authigenic deposits found at moderate to great depths.

The offshore mining of these deposits has become the most widely publicized facet of marine mining, largely because of the sudden awareness of the potential of manganese nodules as a mineral source and because of the important and exciting developments in the exploitation of offshore dia-

monds in South-West Africa. Despite the fact that there are presently no operations for nodules and only one operation actively mining diamonds, unconsolidated deposits have for some years presented a major source of exploitable minerals offshore.

So far the methods of recovery which have been used or proposed have been conventional, namely, by dredging using draglines, clamshells, bucket dredges, hydraulic dredges, or airlifts. All these methods (Fig. 1) have been used in mining to maximum depths of 200 ft, and hydraulic dredges for digging to 300 ft are being built. Extension to depths much greater than this does not appear to present any insurmountable technical difficulties.

More than 70 dredging operations were active in 1968, exploiting such diverse products as diamonds, gold, heavy mineral sands, iron sands, tin sands, lime sand, and sand and gravel (Table 3). The most important of all of these commodities is the least exotic: 60% of world production from marine unconsolidated deposits, or about $100,-000,000 annually, is involved in dredging and mining operations for sand and gravel. Other major contributors to world production are the operations for heavy mineral sands (ilmenite, rutile, and zircon), mostly in Australia, and the tin operations in Thailand and Indonesia, which account for more than 10% of the world's tin. Marine diamond operations currently account for less than 3% of the

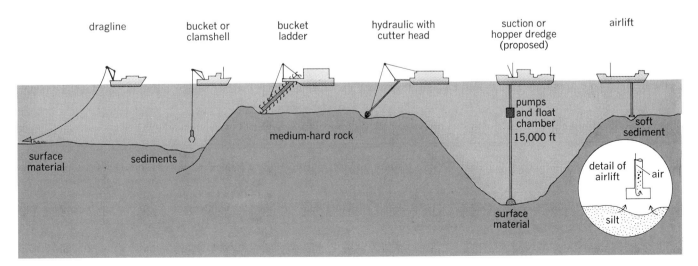

Fig. 1. Methods of dredging used in the exploitation of unconsolidated mineral deposits offshore.

Table 3. Production from unconsolidated mineral deposits offshore

Mineral	Location	Number of operations	Annual production	Year	Value (millions of dollars)
Diamonds	South-West Africa	1	221,500 yd³	1964–1965	8.9
Gold	Alaska	1	—	1966	—
Heavy mineral sands	North America, Europe, Southeast Asia, Australia	15	1,307,000 tons	1965	13.1
Iron sands	Japan	3	36,000 tons	1962	3.6
Tin sands	Southeast Asia, United Kingdom	4	10,000 tons concentrates	1965	24.2
Lime shells	United States, Iceland	9	20,000,000 yd³	1965	30.0
Sand and gravel	Britain, United States	38	100,000,000 yd³	1966	100.0
Total		71			179.8

total mineral values annually produced from offshore and beach operations.

Economics of these operations are dictated by many conditions. Table 4 shows a comparative range of throughput and cost for typical dredging operations both onshore and offshore. The spectacular range of costs offshore is indicative of the effects of different environmental conditions. In general, it may be said that offshore operations in 1967 were more costly than similar operations onshore.

The operations of Marine Diamond Corp. are of considerable interest. The first pilot dredging commenced in 1961 with a converted tug, the *Emerson K*, using an 8-in. airlift. The operation expanded until in 1963, with the fleet consisting of 3 mining vessels, all using an air- or jet-assisted suction lift, 11 support craft, and 2 aircraft. Production totaled over $1,700,000 of stones during that year from an estimated 322,000 yd³ of gravel. At that time, the estimation of mining cost was $2.33 per yd³, showing a profit of nearly $3 per yd. Subsequent unexpected problems, including severe storms, operating difficulties, and loss of one of the mining units led to a reduction of profits and transfer of company control. In the year ending June 30, 1965, the company reported an operating loss of $2.02 per yd³ in the treatment of 220,000 yd³ of gravel, valued at $28 per yd.

Production operations of Marine Diamond Corp. have fluctuated considerably to date (1968). *Diamantkus*, a vessel designed to produce 7000 yd/hr, was withdrawn as uneconomic after only 30 months of service. Only two mining units, *Barge III* and *Colpontoon*, were in operation in 1968, both converted pipe-laying barges using combination airlifts and suction dredging equipment. A third and larger unit, the *Pomona*, a multiple-head suction dredge was commissioned in March, 1967, but was damaged by storms on the first trial run. The characteristics of the bedrock, with its many potholes and extremely irregular surface, add to the difficulty in recovering the maximum amount of diamonds.

Liberal offshore mining laws introduced in 1962 in the state of Alaska have resulted in an upsurge of exploration activity for gold, particularly in the Nome area. As of 1968 only one mining operation was attempted, using a 20-in. hydrojet dredge, in submerged gravels about 60 mi east of Nome. No production was reported.

Over 70% of the world's heavy mineral sand production is from beach sand operations in Australia, Ceylon, and India. Only two oceangoing dredges are used. The majority being pontoon-mounted hydraulic dredges, or draglines, with separate washing plants.

The Yawata Iron and Steel Co. in Japan used a 10.5-yd³ barge-mounted grab dredge and a hydraulic cutter dredge to mine iron sand from the floor of Ariake Bay in water depths of 50 ft. These operations were suspended finally in 1966, the reason being given that the reserves had not been accurately surveyed and the cost of mining had not been competitive with Yawata's alternate sources of supply.

An interesting comparison between a clamshell dredge from Aokam Tin and a bucket-ladder dredge from Tongka Harbor, working offshore on the same deposit in Thailand was made. The clamshell was set up as an experimental unit using an oil tanker hull. It was designed for a digging depth of 215 ft, and mobility and seaworthiness were prime factors in its favor. However, in practice, it was never called upon to dig below 140 ft, its mobility was superfluous, and the ship hull proved very unsatisfactory in terms of usable space. Although it was able to operate in sea states which prevented the operation of the neighboring bucket dredge, its mining recovery factor was low and its operating costs much higher than anticipated. It was withdrawn from service after only 9 years, in favor of the bucket dredge.

Another major operation is run by Indonesian

Table 4. Range of throughput and costs for typical dredging operations (1967)*

	Onshore	Offshore
Yd³ per month	100,000–500,000	2,500–350,000
Dollars per yd³	0.08–0.25	0.15–74.5

*Values within these ranges are dependent on system characteristics.

State Mines off the islands of Bangka and Belitung. The operations are as far as 3 mi from shore in waters which are normally calm. They do have storms, however, which necessitate delays in the operation and the taking of precautions unnecessary onshore. The operations employ nine dredges, one of which, the *Bangka I*, constructed in 1965, is the world's largest mining bucket-line dredge. With a maximum digging depth of 135 ft it is designed to dig and treat 420,000 yd³ per month of 600 hr and is expected to produce 1000 tons of tin metal per year, depending on the richness of the ground. The dredge is working up to 15 mi offshore.

Lime shells are mined as a raw material for portland cement. Two United States operations for oyster shells in San Francisco Bay and Louisiana employ barge-mounted hydraulic cutter dredges of 16- and 18-in. diameter in 30–50 ft of water. The Iceland Government Cement Works in Akranes uses a 150-ft ship to dredge sea shells from 130-ft of water, with a 24-in. hydraulic drag dredge.

In the United Kingdom, several hopper dredges were converted for mining undersea reserves of sand and gravel. Drag suction dredges up to 38-in. in diameter are most commonly used with the seagoing hopper hulls. Similar deposits are mined in the United States, and the same type of dredge is employed for the removal of sand for harbor construction or for beach replenishment. Some sand operations use beach-mounted drag lines for removal of material from the surf zone or beyond.

Consolidated deposits. The third and last area of offshore mineral resources has an equally long history. As Table 5 shows, the production from in-place mineral deposits under the sea is quite substantial, particularly in coal deposits. Undersea coal accounts for almost 30% of the total coal production in Japan and just less than 10% in Great Britain.

Extra costs have been due mainly to exploration, with mining and development being usually conventional. In the development of the Grand Isle sulfur mine off Louisiana, some $8,000,000 of the $30,000,000 expended was estimated to be due to its offshore location. There is no doubt that costs will be greater, generally, but on the other hand, in the initial years of offshore mining as a major industry, the prospects of finding accessible, high-grade deposits will be greater than they are at present on land.

Some of the mining methods in use in 1968 are

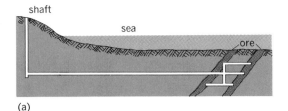

(a)

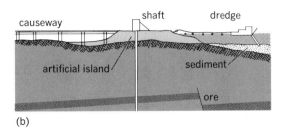

(b)

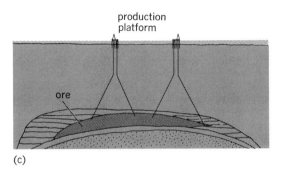

(c)

Fig. 2. Methods of mining for exploitation of consolidated mineral deposits offshore. (*a*) Shaft sunk on land, access by tunnel. (*b*) Shaft sunk at sea on artificial island. (*c*) Offshore drilling and in-place mining.

illustrated in Fig. 2. For most of the bedded deposits which extend from shore workings a shaft is sunk on land with access under the sea by tunnel. Massive and vein deposits are also worked in this manner. Normal mining methods are used, but precautions must be taken with regard to overhead cover. Near land and in shallow water a shaft is sunk at sea on an artificial island. The islands are constructed by dredging from the sea bed or by transporting fill over causeways. Sinking through the island is accompanied by normal precautions for loose, waterlogged ground, and development and mining are thereafter conventional. The same

Table 5. Production from consolidated mineral deposits offshore

Mineral	Location	Number of operations	Annual production	Year	Value (millions of dollars)
Iron ore	Finland, New-foundland*	2	1,700,000 tons	1965	17.0
Coal	Nova Scotia, Taiwan, United Kingdom, Japan, Turkey	57	33,500,000 tons	1965	335.0
Sulfur	United States	1	600,000 tons†	1965	15.0
Total		60			367

*Closed June, 1966.   †Estimated.

Table 6. Summary of production from mineral deposits offshore

Type	Minerals	Number of operations	Annual value (millions of dollars)
Dissolved minerals	Sodium, magnesium, calcium, bromine	145	165.4
Unconsolidated minerals	Diamonds, gold, heavy mineral sands, iron sands, tin sands, lime shells, sand and gravel	71	179.8
Consolidated minerals	Iron ore, coal, sulfur	60	367.0
Total		276	712.2

method is also used in oil drilling. Offshore drilling and in-place mining are used only in the mining of sulfur, but this method has considerable possibilities for the mining of other minerals for which leaching is applicable. Petroleum drilling techniques are used throughout, employing stationary platforms constructed on piles driven into the sea floor or floating drill rigs. *See* MINING, UNDERGROUND; OIL AND GAS, OFFSHORE.

In summary, Table 6 shows offshore mining production, valued at $700,000,000 annually. Though only a fraction of world mineral production, estimated at $700 billion, the results of the extensive exploration activity taking place off the shores of all five continents may alter this considerably in the future.

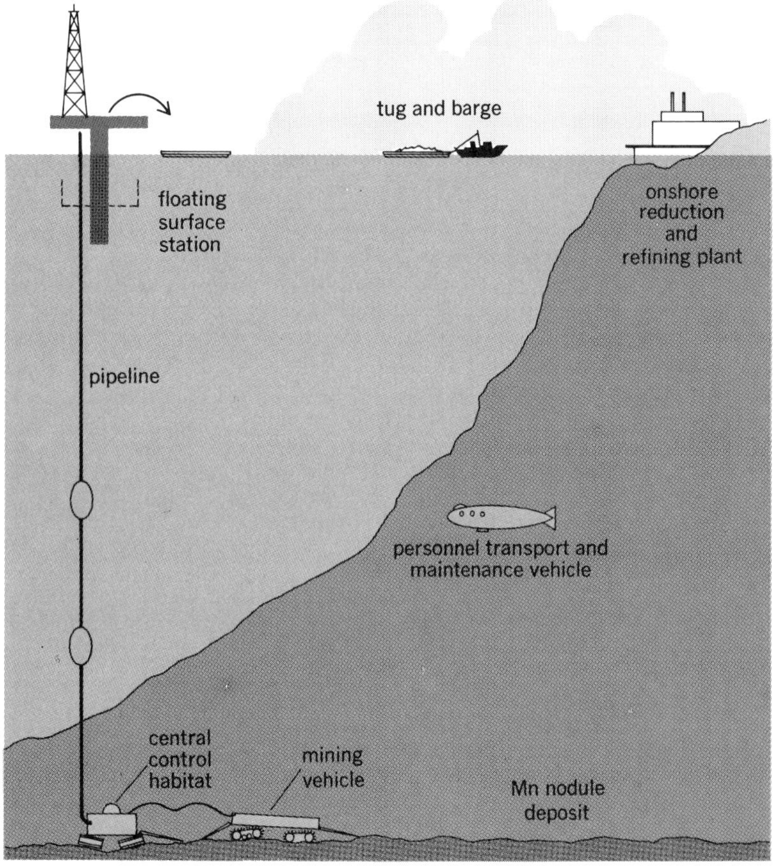

The future. Despite the technical problems which still have to be overcome, the future of the undersea mining industry is without doubt as potent as it is fascinating.

Deposits of hot metalliferous brines and oozes enriched with gold, silver, lead, and copper have been located over a 38.5-mi² area in the middle of the Red Sea at depths of 6000–7000 ft.

Major problems of dissolved mineral extraction must be solved before their exploitation, and significant advances must be made in the handling of these corrosive media at such depths and distances from shore.

The mining of unconsolidated deposits will call for the development of bottom-sited equipment to perform the massive earth moving operations that are carried out by conventional dredges today. The remarkable deposits of Co-Ni-Cu-Mn nodules covering the deep ocean floors will require new concepts in materials handling, and while some attempts may be made to mine them from the sea surface, it is almost certain that future operations will include some form of manned equipment operating on the sea floor (Fig. 3).

Consolidated deposits may call for a variety of new mining methods which will be dependent on the type, grade, and chemistry of the deposit, its distance from land, and the depth of water. Some of these methods are illustrated in Fig. 4. The possibility of direct sea floor access at remote sites through shafts drilled in the sea floor is already being given consideration under the U.S. Navy's Rocksite program and will be directly applicable to some undersea mining operations. In relatively shallow water, shafts could be sunk by rotary drilling with caissons. In deeper water the drilling equipment could be placed on the sea floor and the shaft collared on completion. The laying of large-diameter undersea pipelines has been accomplished over distances of 25 mi and has been planned for greater distances. Subestuarine road tunnels have been built using prefabricated sections. The sinking of shafts in the sea floor from the extremities of such tunnels should be technically feasible under certain conditions.

Submarine ore bodies of massive dimensions and shallow cover could be broken by means of nuclear charges placed in drill holes. The resulting broken rock could then be removed by dredging. Shattering by nuclear blast and solution mining is

Fig. 3. Proposed system for mining deep-sea nodule deposits using manned bottom vehicles. A nodule density of 2 lb/ft² will require daily coverage of 115 acres for a production of 5000 tons/day. (*U.S. Bureau of Mines*)

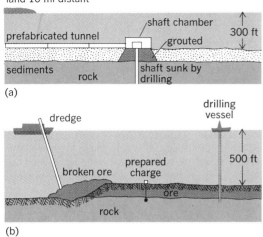

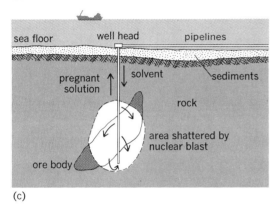

Fig. 4. Possible future methods of mining consolidated mineral deposits under the sea. (a) Shaft sinking by rotary drilling from tunnels laid on the sea floor. (b) Breaking by nuclear blasting and dredging. (c) Shattering by nuclear blast and solution mining.

a method applicable in any depth of water. This method calls for the contained detonation of a nuclear explosive in the ore body, followed by chemical leaching of the valuable mineral. Similar techniques are under study for land deposits.

There are many other government activities which will have a direct bearing on the advancement of undersea mining technology but possibly none as much as the proposed "International Decade of Ocean Exploration." The discovery of new deposits will bring with it new incentives to overcome the multitude of problems encountered in mining undersea.   [MICHAEL J. CRUICKSHANK]

Bibliography: C. F. Austin, In the rock: A logical approach for undersea mining of resources, Eng. Mining J., August, 1967; G. Baker, Detrital Heavy Minerals in Natural Accumulates, Australian Institute of Mining and Metallurgy, 1962; H. W. Bigelow, Electronic positioning systems, Undersea Technol., April, 1964; H. R. Cooper, Practical Dredging, 1958; M. J. Cruickshank, Mining Offshore Alluvials: Symposium on Opencast Mining, London Institute of Mining and Metallurgy, 1964; M. J. Cruickshank, C. M. Romanowitz, and M. P. Overall, Offshore mining: Present and future: Eng. Mining J., January, 1968; W. E. Hibbard, Jr., The government's program for encouraging the development of a marine mining industry, Transactions of the Second Annual Marine Technology Society Conference, 1966; J. L. Mero, The Mineral Resources of the Sea, 1965; E. P. Pfleider (ed.), Surface Mining, 1968; C. M. Romanowitz, The dredge of tomorrow, Eng. Mining J., April, 1962; J. A. Tallmadge et al., Minerals from sea salt, Ind. Eng. Chem., 56:44, 1964; R. D. Terry (ed.), Ocean Engineering, vol. 6: Mineral Exploitation, 1965; U.S. Army, Corps of Engineers, The Hopper Dredge: Its History, Development and Operation, 1954; B. Webb, Technology of sea diamond mining, Marine Technology Society–American Society of Limnology and Oceanography, Ocean Science and Ocean Engineering Conference, vol. 1, pp. 8–23, 1965; C. G. Welling and M. J. Cruickshank, Review of available hardware needed for undersea mining, Transactions of the Second Annual Marine Technology Society Conference, 1966; Proceedings of World Dredging Conference, 1967.

Mutation

An abrupt change in the genotype of an organism, not resulting from genetic recombination. Three main types of mutational change may be delineated: In the first type the genetic material is altered quantitatively, that is, the addition or removal of whole chromosomes, parts of chromosomes, or whole chromosome sets, and at the other extreme the addition or deletion of single base pairs from the nucleic acid of the gene. The second type are qualitative alterations of the genetic material, such as the substitution of one base pair for another. In the third type the existing genetic material may be rearranged without altering its quantity or quality. Thus, mutational changes range from those visible microscopically to those which are beyond the limits of visibility. The latter are inferred from their effects on the function of the organism.

Mutations may arise in the genetic material of any type of cell. If a mutation occurs in a somatic cell, it will be transmitted to the daughter cells unless it produces a dominant cell–lethal effect. Likewise, the daughter cells in turn will pass along the mutation to their daughters. If these cells give rise only to somatic tissue, the mutation will be eliminated with the death of the individual. However, if the mutation occurs in a cell lineage which gives rise to gametes, or if the mutation takes place in a gamete, then it can pass from one generation to the next. Therefore, germinal mutations form the basis of the inherited differences between individuals.

If there were no mutations there would be no genetic differences among organisms and no evolution of life. In fact, mutations have often been called the building blocks of evolution. For this reason, the mutation process is fundamental not only to genetics, but to the continuation of life itself in a world that is changing.

CHROMOSOMAL ABERRATIONS

Some chromosome mutations involve changes in the quantity of genetic material, while others involve only the rearrangement of genetic material.
Polyploidy and aneuploidy. The presence of three or more full sets of chromosomes is referred

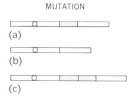

Fig. 1. Chromosome aberrations which involve segments. (a) Normal. (b) Deficiency. (c) Duplication.

to as polyploidy; that is, each chromosome homolog is represented at least three times.

The term aneuploidy describes all situations in which whole chromosomes are lost or gained. This may occur if there is segregational failure at either meiosis or mitosis. The loss of a single chromosome from a diploid complement gives rise to monosomy; that is, one chromosome is represented only once. The presence of an extra chromosome homologous with one of the chromosome pairs of a genome constitutes trisomy. Such changes are often deleterious, particularly in animals. Plants are more tolerant of chromosome loss or gain, and this has been exploited in genetic analysis.

Deficiences and duplications. The loss or repetition of a chromosomal segment may be represented diagrammatically, as shown in Fig. 1. Mutational events of this type may occur in a variety of ways. One of these is by way of the breakage-fusion-bridge cycle: During a cell division one divided chromosome suffers a break near its tip, and the sticky ends of the daughter chromatids fuse. When the centromere divides and the halves begin to move toward opposite poles, a chromosome bridge is formed, and breakage may occur again along this strand (Fig. 2). Since new broken ends are produced, this sequence of events can be repeated. The extent to which such lesions are lethal depends upon their size but, in general, duplications are tolerated more easily than deletions are.

Rearrangements. There are two main types of rearrangement: translocations and inversions. Translocations are usually reciprocal: nonhomologous chromosomes become broken, switching their broken ends in the process of rejoining (Fig. 3a).

As a result of this exchange, a block of genes is transferred to a new position in the genome. With the exception of the "position effect" described below, this does not alter the phenotype of the organism, but it does alter the linkage relation-

ships of the translocated genes, and it causes inviability of a proportion of the progeny of individuals which carry two chromosomes reciprocally translated and two structurally normal chromosomes. These individuals are called translocation heterozygotes.

Inversions occur when two breaks take place in a chromosome, and the chromosomal fragment between the breaks is rotated through 180° before rejoining (Fig. 3b).

Again, the phenotype of the organism may not be affected, but inversions often interfere with crossing-over and segregation during meiosis. A proportion of the gametes of an inversion heterozygote are thus often inviable.

Position effects. These are gametes in the production of which an odd number of crossing overs has occurred between the inverted and the noninverted segment. In some cases a change in the relative position of genes may affect the phenotype. When this happens, there is said to be a "position effect." An example of position effect is found in the fruit fly (*Drosophila melanogaster*). As a result of an inversion, the nonmutant eye-color gene is moved from its normal position in a euchromatic region at the tip of the X chromosome to the heterochromatic region near the centromere. Although it can be shown that no change has taken place in the gene itself, its function is changed. The fly has mosaic eyes with patches of red and white tissue. The mechanism of this effect is not understood.

Gene mutations. Since gene alterations cannot be observed directly, their presence must be inferred from the effects they have on the function of the organism, and from their behavior in genetic crosses. Further information can be obtained by studying their response to specific mutagenic agents.

Quantitative changes. Genetic tests suggest that deletions can mean the loss of one or a few genes or only part of a gene. When a few genes are lost the results are often lethal in haploid organisms, presumably because genes which specify some essential function have been removed. The extreme example of a quantitative change is the single base pair addition or deletion, which has the effect of determining the production of a polypeptide chain composed of a completely altered sequence of amino acids beyond the point corresponding to the position of the addition or deletion.

Qualitative changes. There are theoretically two possibilities for the substitution of one base pair for another at a particular site in the sequence of a deoxyribonucleic acid (DNA) molecule: transition, in which a pyrimidine is replaced by another pyrimidine or a purine by a purine; and transversion, in which a purine is replaced by a pyrimidine and vice versa. Transitions apparently arise as errors in the normal basepairing behavior in the replicating DNA. They can be induced by treating cells (of microorganisms) with analogs of the normal DNA bases. The origin of transversions is not well understood.

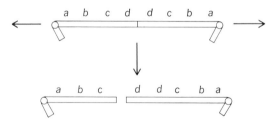

Fig. 2. Breakage-fusion-bridge cycle showing formation of mutant chromosomes following fusion of daughter chromatids prior to cell division.

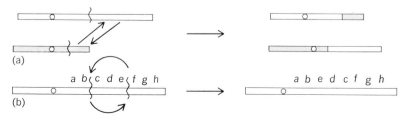

Fig. 3. Chromosome rearrangement mutations. (a) Translocation. (b) Inversion.

TECHNIQUES FOR STUDYING MUTATION

While chromosome mutations may be studied directly by cytological means in suitable material, their effects, as well as the effects of gene mutation, are studied by genetic tests.

Cytological tests. In spite of the fact that spontaneous mutational changes are very infrequent, it is usually possible to observe chromosomal aberrations following mutagenic treatment of a suitable material. A few organisms which have proved good cytological material are listed below.

The spiderwort (*Tradescantia*) is a higher plant which has been used extensively in studies of the induction of chromosomal aberrations, because the chromosomes are large and there are only six homologous pairs. Treatment is made by immersing the inflorescence in a solution of the mutagenic agent and then, after a predetermined period, fixing the material. The nuclei to be studied are haploid, since chromosome mutations are scored at the metaphase of the microspore division following meiosis.

The broad bean (*Vicia faba*) is also very favorable material for this type of study, because the cells of the root tip have very large, easily visible chromosomes. The roots are treated by immersion in a mutagenic solution.

The fruit fly (*D. melanogaster*) is particularly useful, since it allows the researcher to combine sophisticated genetical techniques with cytological study. The salivary glands of this organism contain massive polytene chromosomes which undergo precise pairing, making inversions and other aberrations easily observed. It has been possible in some cases to correlate genetical data with cytological observation.

Mutation studies in man include such indirect methods as sex ratio studies and pedigree analysis. Sex-linked recessive and dominant lethals affect the sex ratio. For example, if a man is exposed to radiation, any sex-linked dominant lethals induced in his X-bearing sperm will cause the death of his female offspring. Thus on the average more male than female offspring will be produced. Conversely, if a woman is exposed, dominant lethals will be distributed equally between her sons and daughters, but recessive lethals will be uncovered in her sons only, resulting on the average in more daughters than sons.

Aside from sex-ratio studies, mutation studies in man are mainly limited to techniques of pedigree analysis, or to indirect techniques involving analysis of gene frequencies and population structure. Direct tests of mutation may become possible when suitable genetic techniques are developed for tissue culture systems.

Genetic tests. These techniques deal principally with changes not detectable cytologically. Any cell bearing genetic material can mutate. In multicellular organism, however, for a mutation to be detected and investigated, it is essential that the change be induced in the germ line so that it can be perpetuated and studied in progeny. Somatic mutations are lost on the death of the organism.

Sex-linked recessive lethal test. Recessive lethal mutations may be induced in the X chromosomes of *Drosophila*. Heterozygous females bearing such a lethal gene do not die, but all male offspring receiving the lethal will die since the gene is not masked. To obtain heterozygous females, males are treated with mutagenic agents and mated to untreated females. Each daughter from this cross bears one treated-X chromosome and its untreated homolog. These females are mated individually to untreated males, and the progeny of some of these matings will have only one-half the normal proportion of males. Cultures which display this unusual sex ratio originated from a female which bore a sex-linked recessive lethal. Many sex-linked recessive lethals are point mutations but other changes, notably small deficiencies, also occur. Because so many genes are involved in this test, it is a particularly useful one for testing potential mutagens.

Specific locus test. In diploid organisms induced mutations are often recessive and can only be detected by the use of special technique. One such technique involves the use of a strain which carries several known recessive mutants in the heterozygous condition. Germ cells from a nonmutant strain are treated to induce recessive mutations and are then used in crosses to the test strain. Any induced recessive mutations which are allelic with those of the test strain will be expressed in the progeny. This technique has been used effectively in the mouse, as well as in *Drosophila* and maize.

Dominant lethal mutations. When the germ cells of an organism are exposed to a mutagen, fewer progeny may be produced. It has been shown that a substantial proportion of the dominant lethals which cause reduced progeny arise from chromosome breaks which have not healed. Death during the early developmental stages may result from these events, as well as other nonchromosomal anomalies.

Studies with microorganisms. Since cytological evidence cannot be obtained from these organisms, a mutation must be recognized by an alteration in function of the organism. The advantages of using microorganisms in mutation experiments lie in the large numbers of genetically uniform individuals which are readily available, and in the selective methods which can be used to screen mutants from nonmutants. Furthermore, many microorganisms have an almost entirely haploid life cycle, which simplifies the detection of mutants.

The more frequently used selective techniques are the following.

Mutations to drug resistance. These can be selected readily by plating treated or untreated organisms on a medium containing levels of the drug which kill all but the resistant mutants. A variation of this technique for bacteria is to screen for resistance to a particular bacteriophage.

Reverse mutations. An auxotrophic mutant is one which requires the supply of a specific growth factor, say an amino acid, for its growth. It is sometimes possible to induce mutations in these mutants which restore the non-requiring, prototrophic phenotype. Treated cells from the auxotroph are plated on a medium which lacks the specific requirement, and only mutants which have no need of the growth factor will grow. Although these mutants are referred to as "revertants," it is necessary to test them genetically to be sure that they are not the result of further mutations elsewhere in the genome which compensate for the original mutation without reversing it; such mutants are called "suppressor mutations." Microorganisms which have been used extensively in mutation research are the fungi *Neurospora*, yeasts, and

Aspergillus and the bacteria *Escherichia* and *Salmonella*.

ORIGIN OF MUTATIONS

Mutations can be induced by various physical and chemical agents or occur spontaneously without any artificial treatment with known mutagenic agents.

Induced mutations. In the absence of mutagenic treatments, mutations are very rare. In 1927 H. J. Muller discovered that x-rays significantly increased the frequency of mutation in *Drosophila*. Subsequently, other forms of ionizing radiation, for example, gamma rays, beta particles, fast and thermal neutrons, and alpha particles, were also found to be effective. By employing the specific locus mutation test to analyze the effects of x-rays, it has been estimated that the average mutation frequencies per locus in spermatogonia of *Drosophila* and in mice are 1.5×10^{-8} and $25 \times 10^{-8}/$roentgen, respectively.

Induced point-mutation frequencies increase proportionately to dose: Extrapolation of the curve of mutation frequency versus dose back to zero dose gives zero mutations. There is no reason to believe that there is any dose of ionizing radiation which is too low to produce a mutagenic effect proportional to its magnitude. However, in mouse spermatogonia and oocytes there is evidence to suggest that doses administered at extremely low rates are less effective than those administered at higher rates. This may result from the operation of cellular repair mechanisms.

The relationship between radiation dose and chromosome aberration frequencies has been determined in the case of *Tradescantia*; individual chromosome breaks, such as point mutations, increase in direct proportion to dose. However, the increase in chromosomal aberrations is approximately in proportion to the dose squared. This is to be expected since each aberration requires two independent events (breaks) to take place before it can be formed. An important factor which determines the frequency of these aberrations is the intensity of the given dose of radiation. When a dose is given over an extended period, fewer aberrations are found than if the same dose is given in a short time. This is due primarily to the tendency for breaks to rejoin in the original way. Thus, if two breaks are induced, but at very different times, the first may rejoin before it can interact with the second to give an aberration. The same effect can be seen if an intense dose is fractionated, and the fractions separated by several hours. Again, fewer aberrations are obtained than when an unfractionated dose of the same intensity is given.

Radiations of different ionization density have been found to differ somewhat in their efficiencies of the induction of recessive, sex-linked, point mutations. Neutrons which produce dense ionizations are less efficient than gamma- or x-rays. This effect is explained by the target theory: When an ionizing particle traverses matter, its energy is expended in the ejection of electrons from atoms in its path with the resultant production of ion pairs. From the linear relationship between dose and point mutation frequency, the assumption is made that within a certain sensitive volume there is a probability of 1 that an ionization will produce a

mutation. It is possible to calculate the size of this volume, which has been considered to represent the size of the mutating gene. If i is the chance that an ionization of an atom in the sensitive volume will occur and there are a atoms in this volume, the chance of a mutation m is ai. From the physical properties of the atoms, the value of i can be estimated and m can be determined experimentally, so m/i gives the number of atoms in the sensitive volume. From density considerations this can be converted into a sensitive volume or gene size. The validity of the target theory can be questioned on the grounds that it is not only ionizations which produce mutations, excitations are also effective. Further, it is by no means certain that all the mutations produced can be detected. However, the concept can help to account for the lower efficiencies of densely ionizing radiations. If a genetic change is produced by i ionizations, any extra ionizations over and above i are wasted.

Ultraviolet light. Radiations from 2000 to 3000 A are called ultraviolet. The wavelength most employed experimentally is 2537 A which corresponds to the peak absorption of nucleic acids. Ultraviolet radiations produce excitations in the material by which they are absorbed. One of the first indications of the genetic role of DNA came from experiments which showed a close correspondence between the action spectrum for mutation induction and the absorption spectrum of nucleic acid. It is known that an important ultraviolet effect on DNA is the production of pyrimidine dimers, formed by covalent bonding of two adjacent pyrimidine bases on the DNA chain. These compounds interfere with DNA replication, and are known to cause mutations.

Modification of radiation damage. It is realized that at least some part of the mutagenic effect of radiation is indirect. Evidence for this is obtained from experiments in which ancillary, nonmutagenic treatments are found to influence the yield of mutations from a given dose of radiation. For example, the concentration of oxygen is an important factor in determining the yield of x-ray-induced mutants, since there is a direct relationship between the oxygen tension and the amount of damage. Infrared light and temperature also seem to be important, possibly by altering the sensitivity of the chromosomes to breakage. Oxidative metabolism is necessary for chromosome rejoining to take place. Repair of ultraviolet damage is brought about by an enzyme system that works only in the visible light range, and restores pyrimidine monomers from dimers formed as ultraviolet photoproducts. In bacteria it has been shown that certain enzymes occur which cut out the damaged portions of one of the DNA strands, while other enzymes then patch up the gaps using the undamaged DNA strand as a template.

Finally, sensitivity to radiation is strongly dependent on both the type of cell used and the stage in the cell cycle at which treatment is applied.

Chemical mutagens. Reports concerning mutagenic effects of chemicals were first made in the 1930s, but it was not until 10 years later that these claims were substantiated. Since then many chemical substances have been tried and have been found to be effective mutagens (Table 1).

In spite of the fact that the chemical nature of

Table 1. Some known chemical mutagens and type of effect

Mutagen	Effective in producing	
	Mutations	Chromosomal aberrations
Urethane	+	+
Diepoxide	?	+
Maleic hydrazide	?	+
Diepoxybutane	+	+
Triazine	+	+
Formaldehyde	+	+
Mustard gas	+	+
Nitrogen mustard	+	+
Hydrogen peroxide and organic peroxides	+	+
Caffeine	+	+

the genetic material is known, it is still not always apparent how some of the substances act. The main difficulty is that the major product of the reaction of a chemical with DNA may not be the mutagen. Mutations are so rare, even after induction, that a minor chemical product could be the important one in causing the mutation. Chemicals can sometimes act specifically; base analogs, for instance, cause the substitution of one base pair for another, while acridines seem to act by causing base pairs to be added to, or deleted from, the gene. Specific behavior of this type has added a new dimension to the characterization of mutational events in microorganisms. If a mutation was originally the result of a base pair substitution, it should only revert with a similar change. The same is true for mutations which arise from base additions or deletions.

The early hope with chemical mutagens was that they would lead to the specific induction of the mutation of particular genes. This seems unlikely in light of present knowledge of DNA structure and composition.

Spontaneous mutation. Until the discovery of x-rays as mutagens, all the mutants studied were spontaneous in origin; that is, they were obtained without the deliberate application of any mutagen. In Table 2 some data are presented which show the order of rates for spontaneous mutation in different organisms.

Some mutational changes appear to occur with a

Table 2. Spontaneous mutation frequencies

Organism	Mutation	Incidence, 10^6 genes
Man	Huntington's chorea	1
	Retinoblastoma	4
	Muscular-dystrophy	8
	Hemophilia (combined)	32
Maize	R (anthocyanin production)	490
	I (colorless aleurone)	110
	Pr (purple aleurone)	11
	Sh (shrunken endosperm)	1
Neurospora	Adenine reversion (allele no. 38701)	0.06
	Inositol reversion (allele no. 37401)	0.01
Bacteriophage T4	r101 (rapid lysis) reversion	0.3
	r51 (reversion)	11

much higher frequency than others. In general, forward mutations, that is, from wild types to mutant phenotype, have higher mutation frequencies than mutations in the opposite direction. This is probably because a nonmutant gene presents many more possibilities for mutational damage than a mutant, which can only revert in a very limited number of ways.

Besides point mutations, spontaneous events also include chromosomal aberrations. However these events are extremely rare, and accurate measurements of these rates are difficult to obtain.

Several hypotheses have been put forward to account for spontaneous mutation. Natural ionizing radiation, once thought to be important, has been shown to be insufficient alone to account for the spontaneous rates observed. Another suggestion, temperature shocks, also appears unable to produce sufficient spontaneous mutations to explain the frequencies. It seems possible that naturally occurring mutagens (of which many are known) may cause some of the spontaneous changes, when they accumulate as the organism ages. Evidence for this view comes from work with seeds sown after various lengths of storage time. As the seed age increases, the probability that it will give rise to a seedling which shows chromosomal aberrations also increases. It has also been suggested that mutations could occur naturally as a result of mistakes during the replication of the DNA. Such events would occur if spontaneous tautomeric changes took place in a base, causing it to pair with the wrong partner base, or if the DNA-replicating enzymes themselves made mistakes in the incorporation of DNA bases into new DNA chains. There is some evidence to suggest that a large number of the spontaneous mutations which occur in microorganisms are base pair substitutions that might be expected from the mispairing of bases.

MUTATIONS AND EVOLUTION

Mutations provide all the primary genetic differences which exist among different organisms in nature. The phenotypic variations which result from the alterations in genotype are actively selected or eliminated by the prevailing environment. As the environmental conditions change, the types of phenotype favored will also change. Therefore, if the species is to survive, sufficient variability must be available and must be supplied rapidly enough to meet this challenge of the environment.

As has been shown already, spontaneous mutation is a very rare process. Some genes show a mutation rate of 1 in 10,000 cells, while for other genes the occurrence is only 1 in 100,000,000 cells. For organisms which multiply rapidly, these rates are not too low to provide the variability necessary for survival. Bacteria, for example, can multiply at a rate which allows 1 cell to give rise to enormous numbers in a few hours. For higher organisms with long generation times the range of variability required is supplied by storage of genetic variation in the form of heterozygotes in diploid cells and release of it by means of recombination. In this way maximum use is made of the variability "pool" kept repleted by the mutations occurring at low rates.

Since selective processes have been acting for a very long time, it is not surprising that existing organisms are well adapted to their environments. New mutations are generally harmful since they upset the balance between the organism and its environment; only rarely is a new mutation beneficial.

[BRIAN J. KILBEY]

Bibliography: C. Auerbach, *Mutation: An Introduction to Research on Mutagenesis*, 1962; C. Auerbach, *Science*, 158 (3805):1141–1147, 1967; C. Yanofsky, *Sci. Amer.*, 216:80, 1967.

Nitrogen cycle

The continuous cyclic exchange between combined nitrogen in the soil and molecular nitrogen in the atmosphere. It includes all the transformations concerned in the mineralization of nitrogenous organic substances and in the loss or gain of nitrogen by the soil.

Soil nitrogen occurs naturally in organic and inorganic forms as a result of plant, animal, and microbial growth. Nitrogen is stored in soil primarily in organic combinations not utilizable by higher plants, but made available as ammonia through the activities of soil microorganisms. The ammonia may be used by both higher plants and microorganisms either directly or after oxidation to nitrate-nitrogen. Both ammonia and nitrate may be lost from soil by leaching or through microbial action. Soil gains nitrogen chiefly through the addition of fertilizers and through microbial fixation of atmospheric nitrogen. The nitrogen cycle comprises the process of ammonification, nitrification, denitrification, and nitrogen fixation (Fig. 1).

Some authorities further subdivide ammonification into proteolysis (protein degradation to amino acids) and ammonification, and nitrification into nitritation (formation of nitrite from ammonia) and nitratation (formation of nitrate from nitrite).

Ammonification. Ammonification refers to the release of nitrogen as ammonia from organic compounds in plant, animal, and microbial residues. This is accomplished chiefly under aerobic conditions through the participation of bacteria, fungi, actinomycetes, and other microscopic forms of life. The first step in the process involves the hydrolytic cleavage of proteins, nucleic acids, and related compounds to amino acids and other simple nitrogenous substances. These are then broken down to ammonia. Uric acid and urea, the excretory products of animals, are rapidly mineralized. The ammonia-nitrogen liberated through microbial action is in excess of the requirements of these organisms for growth. Consequently when a substance that is high in nitrogen, such as protein, is added to a soil, considerable ammonia is liberated, whereas a substance relatively low in nitrogen, such as straw, yields comparatively little, if any, ammonia. Furthermore, if an excessive amount of nonnitrogenous carbonaceous material, for example, carbohydrate, is added to a soil, much available nitrogen, such as nitrate or ammonia, will be used by the rapidly developing microbial population, thus decreasing the available supply for higher plants. Not all organic nitrogen is ammonified, however; a certain portion is retained in slowly decomposable complexes and becomes an integral part of the residual soil organic matter or humus.

Nitrification. Nitrification is the bacterial oxidation of ammonia to nitrate, the chief source of readily available nitrogen for higher plants. It consists of two steps: first, ammonia is oxidized to nitrite by organisms of the genera *Nitrosomonas* and *Nitrosococcus*, and second, the resulting nitrite is oxidized to nitrate by *Nitrobacter*. These highly specialized, autotrophic bacteria obtain their energy from these oxidations and their carbon from the carbon dioxide of the atmosphere. Generally the process of nitrite oxidation is faster than that of nitrite production, so that the level of nitrite in soil is too low to induce toxic effects. The few types of microbes known to be involved in nitrification require a much more restricted set of conditions for optimal activity than do the many types engaged in ammonification. A well-aerated, fertile, neutral to slightly alkaline soil provides optimum conditions for nitrification. The nitrate so formed is utilized by plants and microorganisms, or may be lost from the soil by leaching. Under anaerobic conditions, nitrate may be reduced by the soil microflora.

Denitrification. In denitrification nitrate-nitrogen is reduced to nitrite, nitrous oxide, ammonia, and principally molecular nitrogen. Under conditions of low oxygen tension a variety of soil microorganisms utilize nitrate as a source of oxygen and reduce it to forms which may be lost by leaching or which may escape as gas into the atmosphere. The absence of oxygen, as in waterlogged soil, and the presence of an abundant supply of soluble organic matter provide favorable conditions for this process. Normal agricultural soil is well-aerated, not too moist, and contains moderate amounts of or-

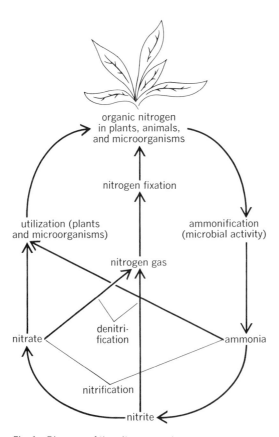

Fig. 1. Diagram of the nitrogen cycle.

Fig. 2. Effect of inoculation with two strains (cultures 175 and 378) of the nitrogen-fixing *Rhizobium leguminosarum* on the development of peas.

ganic matter or nitrate. Here, denitrification is of little economic importance.

Nitrogen fixation. Molecular atmospheric nitrogen is returned to the soil primarily by man through chemical fixation, and by soil microorganisms through biological fixation. Biological nitrogen fixation is accomplished by symbiotic and nonsymbiotic microorganisms. The symbiotic organisms are bacteria living in the modules of leguminous plants and belong to the genus *Rhizobium*. The nonsymbiotic organisms are free-living bacteria which function either aerobically, as *Azotobacter*, or anaerobically, as *Clostridium*. Certain blue-green algae such as *Nostoc*, *Anabaena*, and *Calothrix* species can fix atmospheric nitrogen. Several other groups of bacteria, including photosynthetic types, and fungi possess this characteristic to a limited degree, as has been demonstrated with isotopic nitrogen. The most important of the nitrogen-fixing bacteria are those that produce nodules on the roots of legumes such as peas, beans, alfalfa, and clover. In this mutualistic association they may add well over 100 lb of atmospheric nitrogen to an acre of soil annually; soils are usually inoculated with active preparations of these organisms wherever legumes are grown. Figure 2 shows the effect of inoculating cultures of symbiotic nitrogen-fixing bacteria on the development of peas. The nonsymbiotic forms, such as *Azotobacter* and *Clostridium*, fix much less nitrogen, with a fair average in fertile soils being about 10 lb/acre annually. Under certain conditions in the tropics, blue-green algae contribute significant amounts to the soil. *See* SOIL MICROORGANISMS; SOIL MINERALS, MICROBIAL UTILIZATION OF.

[HARRY KATZNELSON/R. E. KALLIO]

Noise, acoustic

Unwanted sound. This definition of acoustic noise, while purely general, implies that some criterion must exist before the sound can be termed unwanted. Whether a sound is noise, insofar as humans are concerned, is a subjective matter.

Criteria. Considerable effort has been expended to analyze unwanted sounds in an effort to specify objective criteria for the subjective human reactions to noise. These criteria have included annoyance, interference with speech, damage to hearing, and reduction in efficiency of work performance.

Noise is usually thought of in terms of its effect on humans; an equally important aspect, however, is its effect on the fatigue or malfunction of physical structures and equipment. In these instances criteria can in theory be established on a completely objective basis. The "unwanted" aspect in the definition of noise applies here to the fact that it is generally considered undesirable to have a structure such as an aircraft experience fatigue, or that it is undesirable to have electronic equipment guiding the aircraft fail because of malfunctions brought on by intense sound waves.

A third major criterion for describing sound as noise arises in conjunction with the perception or detection of a wanted sound in the presence of other sounds which tend to mask it. Thus in sonar the reflected sound from an object being detected is a signal which is wanted, whereas all other sound detected by the system is termed noise.

Physical specifications. The generality of the preceding definitions gives no clue to the physical specifications of sound waves called noise. The sound can be composed of definite pure-tone, or sine-wave components, or it can be a completely random phenomenon made up of an infinite number of components, each having purely random amplitude and phase characteristics. Automobile exhaust noise, for example, contains pure-tone components, related to the engine rotational speed, whereas an air jet hiss is a random noise.

The physical specification of such noises is given by their radiated intensity, frequency, and spatial distribution.

Random noises are usually described in terms of statistical values rather than in terms of the discrete variable which is used for single-frequency sounds.

The statistical description of random sound waves parallels that used for electrical noise. The magnitude of the noise is usually specified in terms of its radiated intensity in a 1-Hz frequency band, also known as the intensity spectrum level. If the random noise has a relatively uniform distribution of intensity as a function of frequency, it is often described as intensity in a frequency band more than 1 Hz wide. This may be done in terms of a constant bandwidth, such as 5, 50, or 500 Hz, or in terms of a constant percentage bandwidth, such as 1/10, 1/3, or 1 octave of frequency. The most common usage in industrial noise control is specification of intensity in octave frequency bands.

For most noise measurements the frequency range of practical interest can be covered by eight-octave frequency bands, the lowest band having a center frequency of 63 Hz, and the highest band a center frequency of 8000 Hz. The overall intensity of a random noise is the sum of the intensities in all the frequency bands by which it is specified.

It is often useful to convert the physical specification of acoustic noise to a psychophysical measure such as loudness or perceived noisiness. These measures are computed from the sound-pressure level values in 1/3- or 1-octave frequency bands.

White noise. Random noise having the same intensity, in a 1-Hz band, at every frequency in the range of interest is called white noise. Although white noise is a fairly common type of electrical noise, it is rarely encountered in acoustic noise. Most random acoustic noises tend to have a definite nonuniform distribution of intensity as a function of frequency.

Ambient noise. The residual noise present at any location of interest is called ambient noise. It is the sum of all noises present. Thus ambient noise in an office could be the result of ventilating systems, distant conversations, office machines, and so forth. Background noise is a term often used to describe the ambient noise when a particular source of sound being studied is not in operation.

[WILLIAM J. GALLOWAY]

Nuclear power

Power (or energy) derived from the fission (splitting) of the nuclei of heavy elements such as uranium, or the fusion of light elements such as deuterium or tritium. The amount of energy released per atom in fission and fusion reactions exceeds the amount for combustion reactions by factors of several millions. The fission of 1 lb of nuclear fuel, for example, liberates an amount of energy equivalent to that produced in the combustion of about 3,000,000 lb of coal. The fission and fusion reactions also differ from normal combustion reactions in that they can generate much higher temperatures and take place in much shorter times, they require no oxidants, and they release ionizing radiations and generate radioactive by-products.

Considerations. The unique aspects of the fission process make nuclear power particularly attractive for specialized applications, such as submarine propulsion, space power sources and unattended remote power stations.

The main advantage of nuclear power, however, is that nuclear fuels are a cheaper, more abundant source of energy than conventional fuels. Although estimated resources of low-cost uranium ($5–10/lb) make up an energy source no greater than that of coal, higher-cost sources of nuclear fuel are relatively inexhaustible. Certain granite rocks, for example, contain about 30–60 ppm of uranium or thorium or both. Even at this low concentration, the nuclear fuel in each ton of rock would yield an energy equivalent to that of 30 to 60 tons of coal, depending on the efficiency of recovery and fuel utilization. Entire mountains of such granites exist, making up an energy source sufficient for several thousands of years. When these are consumed, even the Earth's crust, containing on the order of 10 ppm of nuclear fuel, can be considered a potential source of energy. Problems associated with the conversion of such low-grade sources into cheap electricity remain to be solved; however, there is no inherent reason why these problems cannot be solved.

Similarly, deuterium can be recovered from sea water and, when controlled fusion is a reality, energy can be released and utilized in a thermonuclear reactor. Thus both rocks and the ocean represent a limitless source of nuclear energy.

Unlike the combustion of fossil fuels, the extraction of energy from nuclear fuels involves a large number of complicated chemical and metallurgical operations and must be carried out before the nuclear fuel can be used in a nuclear reactor. The reactor must be designed to do many things, such as control the reaction rate, remove the heat efficiently, utilize excess neutrons for new fuel production, and prevent the generated radiation from escaping. Because of these varied functions, the designer of the reactor must simultaneously consider a variety of nuclear, engineering, safety, and economic factors. Many of these factors can only be established by experiment and through the construction and operation of a large number of prototype reactors.

The development of a controlled thermonuclear reaction, necessary for the utilization of fusion fuels, is more difficult technically than the development of fission power. It involves containing an ionized gas (plasma) at temperatures above 100,000,000°F in a magnetic field for a time long enough for a self-sustaining reaction to take place. Considerable advances in current technology will be required merely to establish the technical feasibility of fusion power.

Power generation. The heat generated in nuclear fuel elements and subsequently transferred to a coolant can be recovered by using the coolant to produce steam (indirect cycle) or as the working fluid for driving a turbine (direct cycle). These cycles are depicted schematically in Fig. 1. The direct cycle is shown by solid lines and the indirect cycle by dotted lines. The direct cycle is usually associated with the use of boiling water as the coolant; the indirect cycle is most applicable to gas-cooled, water-cooled, and liquid-metal-cooled reactors. The direct cycle has the advantage of a higher thermal efficiency for a given coolant pressure; however, in such a cycle, power generation equipment must be shielded to protect against possible radioactive products in the steam. In the case of the indirect cycle, the steam generator tubes act as a barrier against such radioactivity. Both the indirect and direct cycles are used in large-scale nuclear power plants and because of

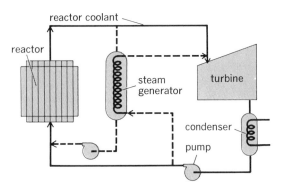

Fig. 1. Power cycles for nuclear plants.

their relative advantages and disadvantages are economically competitive with one another.

For the most efficient utilization of reactor-produced heat, coolant outlet temperatures should be as high as possible. Water-cooled reactors, therefore, are operated at as high a pressure as is feasible. Gas-cooled reactors are also operated at high pressure to reduce pumping power. Liquid-metal and organic coolants, because of their low vapor pressures, are limited only by the permissible operating temperature of fuel materials or structural metals in the system. In general, water-cooled reactors produce steam in the range of 500–550°F/700–1050 psi compared to 1050°F/2000 psi achieved in modern coal-fired plants. Although nonaqueous coolants can produce high-quality steam comparable to that of coal-fired plants, the production of cheap electricity from nuclear power does not depend solely on thermal efficiency. Optimum steam temperatures and pressures, therefore, vary with each type of reactor and in some cases are considerably lower than in conventional plants.

Central-station electrical plants. Reactors sold in the United States for the production of electricity in large thermal steam plants use pressurized or boiling light water as the coolant and moderator. As of Mar. 31, 1969, 91 reactors of this type, making up a total installed capacity of about 66,000,000 kilowatts of electricity (kwE), were in operation or were being built or on order by United States utilities. It is anticipated, moreover, that within the next decade an additional 120,000,000 kwE of light-water reactors will be ordered, bringing the total investment in this type of central station electrical plant to about $25,000,000,000 by 1980.

The favorable economic position of light-water reactors relative to that of coal-fired boilers, however, depends on the continued availability of low-cost uranium ore, that is, < $10/lb U_3O_8. Present indications are that United States resources of uranium at such costs are limited. Thus the development of nuclear reactors with improved nuclear-fuel-utilization characteristics is necessary so that nuclear fission can become a long-term economical source of electricity. Reactors in this category are (1) high-temperature graphite-moderated reactors cooled with helium; (2) heavy-water-moderated reactors cooled with D_2O, boiling H_2O, or organics; (3) graphite-moderate molten-salt thermal breeders cooled with fuel-bearing lithium-beryllium fluoride; and (4) unmoderated fast breeders cooled with sodium, helium, or steam.

Pressurized water reactor (PWR). The PWR is cooled and moderated by highly purified (that is, demineralized) ordinary water under a pressure of about 2200 psi. The core of the reactor consists of bundles of closely packed rods of slightly enriched (~ 3% U^{235}) uranium dioxide clad with a zirconium alloy called Zircaloy. These fuel assemblies are contained inside a thick steel pressure vessel, through which the pressurized water is pumped. The water, heated during its passage through the core, transfers its heat by means of an external heat exchanger to generate saturated steam at about 600°F and 800 psi. The steam drives a turbine generator set producing electricity at a net station efficiency of about 33%. The reactor and the primary coolant system are shielded and enclosed within a pressure-tight containment building as protection against the possibility of escape of highly radioactive materials from the reactor.

Boiling water reactor (BWR). The BWR is similar in many respects to the PWR; however, in this concept the water coolant is allowed to boil as it passes upward through the core, exiting as a mixture of water plus about 14 wt % steam. The steam is separated and dried within the reactor vessel and sent directly to a turbine generator. This approach eliminates the costly PWR steam generator; however, compensating cost factors make the PWR and BWR economically competitive. Because of the different method of heat removal, the BWR core optimizes at a lower pressure, lower enrichment, higher fuel inventory, larger fuel rod diameter, and lower uranium burnup than the PWR.

Advanced gas-cooled reactors (AGR). Two reactors of this type are under construction at Dungeness, England. The AGR has steel-clad uranium dioxide fuel elements and is cooled by carbon dioxide gas. It is designed to produce steam under conventional conditions (2315 psi, 1050°F, and 1050°F reheat). The reactor core and primary cooling system, including the steam generators, are housed within a prestressed concrete reactor vessel which serves to shield and contain the radioactive fission products.

High-temperature graphite-moderated reactor (HTGR). The HTGR was developed by the British and represents the most advanced in the line of gas-cooled graphite reactors in the United States and elsewhere. Unlike earlier gas-cooled reactors, the helium-cooled HTGR uses nonmetallic fuel assemblies, permitting higher gas temperatures than possible with metal-clad fuels. Outlet coolant temperatures above 1500°F are possible, resulting in the production of 1050°F, 3500-psi steam, and net station efficiencies as high as 43%. Various methods of incorporating the nuclear fuel in a graphite matrix have been investigated; apparently the most promising method contains the fuel as an oxide or carbide in tiny spherical particles coated with several layers of impervious carbon or other ceramic material. Particle diameters, including coatings, are of the order of 0.01 to 0.02 in. These fuel particles are bonded with low-density carbon into fuel sticks, which are inserted in holes in the graphite core structure. The helium coolant passes through adjacent holes to remove the heat generated. Generally, the HTGR concept utilizes the thorium-U^{233} fuel cycle to best advantage; however, slightly enriched uranium or plutonium plus thorium may also be employed.

Heavy-water-moderated reactors (HWR). These reactors can be of the pressure-vessel type or pressure-tube type. The pressure-vessel type is similar to the light-water reactors in that the same fluid (D_2O) serves both as coolant and moderator. Also, the coolant can be pressurized or boiling, as in the case of light-water reactors.

In the pressure-tube approach, on the other hand, the D_2O moderator is contained within a low-pressure calandria vessel pierced with numerous pressurized channels containing the fuel elements and coolant stream. This approach permits the use of any one of a number of coolants, such as pressurized or boiling D_2O, boiling H_2O, CO_2, or organic terphenyls. Moreover, the moderator can be kept at a relatively low temperature for better neutron economy. The use of D_2O as the moderator also allows one to use a wide variety of fuel compositions, ranging from natural or slightly enriched uranium to fully enriched uranium plus thorium. Corresponding fuel elements vary from bundles of uranium dioxide rods to concentric cylinders of thorium-uranium metal. Many design studies have been carried out to find the best coolant-fuel combination; however, no single system has a clearly demonstrated superiority.

Molten-salt reactor (MSR). In the MSR the fissile and fertile material in the form of fluoride salts is dissolved in the coolant, which is a molten mixture of lithium fluoride and beryllium fluoride. This fuel-coolant fluid is pumped through channels in a graphite core and then through a heat exchanger for transfer of heat to a non-fuel-bearing molten salt. The highly radioactive fuel salt is thus isolated from the steam-generating system. Both the primary fuel salt and secondary salt operate at low pressures and high temperatures, resulting in an overall plant efficiency of about 44%. The use of a fluid fuel not only eliminates the need for fuel fabrication but also permits fuel reprocessing (that is, fission product removal) to be carried out continuously within the reactor itself. Other attractive features of the MSR concept are high neutron economy, low fissile inventory, advantageous safety features, and low overall power costs. Although the MSR appears to be most economical when operated as a thorium-U^{233} breeder, it can also be fueled with enriched U^{235} or plutonium without serious economic penalty.

Fast breeder reactors (FBR). These reactors consist of closely packed bundles of small-diameter metal tubes containing plutonium and uranium as oxide or carbide and cooled with steam, helium, or liquid metallic sodium. Most of the development effort on fast breeders is being devoted to liquid metal coolants because of their better heat-transfer properties. As in the case of the MSR, heat generated in the primary sodium system is transferred to a secondary coolant loop to minimize and contain the highly radioactive portion of the plant. The low-pressure high-temperature secondary coolant generates steam at conditions comparable to those in a modern fossil-fueled plant and achieves an overall station efficiency of about 44%.

The core of the FBR is surrounded by U^{238} blanket, which captures leakage neutrons to produce excess plutonium. It is estimated that the annual net production of plutonium may amount to 10% or more of the total fissile inventory in the reactor. The sale of this excess plutonium would result in low power costs and would also provide the means for a nuclear power industry to sustain itself on bred fissile material only.

Economics of central-station power. Some of the factors influencing power costs which must be considered in the design of reactors for central-station power production are thermal efficiency; neutron economy, that is, the efficient use of excess neutrons to produce by-product fissile material; and the amount of power than can be extracted per unit of core volume and per unit of investment in fuel and other costly nuclear materials. These factors vary in importance with the type of reactor, cost of nuclear fuel, cost of fuel process steps, and annual charges on fuel investment. Thus, the relative importance of any one factor depends on the economic environment of each individual situation.

The cost of producing electricity in a nuclear power station is made up of the sum of annual charges for taxes, depreciation, and return on capital investment in plant and nuclear fuel; fuel cycle charges, including fuel fabrication, uranium consumption, spent fuel shipping, and fuel reprocessing; and operating labor, maintenance, and insurance costs. Costs of a central-station nuclear plant

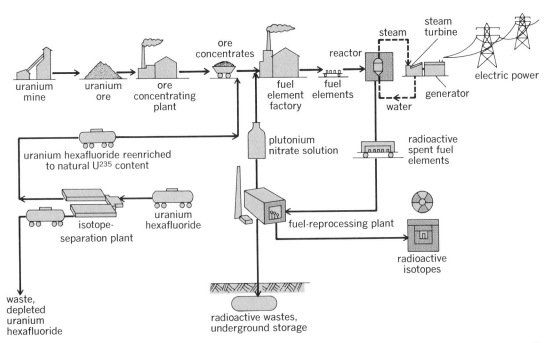

Fig. 2.   Nuclear plant operations. (*From M. Benedict and T. Pigford, Nuclear Chemical Engineering, McGraw-Hill, 1957*)

depend mainly on its type and size. Regarding the latter point, light-water nuclear plants with outputs less than 300,000-kwE capacity are not being offered for sale by United States nuclear equipment suppliers because such plants are not likely to be economically competitive with those using conventional fuels. Small nuclear plants, however, are under development in countries outside the United States and may become competitive at a future date.

Unit capital costs of the large light-water nuclear plants sold in the United States during 1967 varied widely as the result of the (1) size of individual steam-generating units, (2) number of units on a site, (3) site conditions, (4) contingency allocation, and (5) pricing policy of the vendor. Representative costs for first units on a site for various plant capacities were 400,000 kwE (that is, 400 MwE), $170/kwE; 600 MwE, $152/kwE; 800 MwE, $141/kwE; and 1000 MwE, $134/kwE. Nuclear plant prices in 1968 were about 25% higher than these figures, reflecting an increase in labor and other costs. Corresponding costs for coal-fired and oil-fired plants during the same period were $15–30/kwE lower than nuclear plant costs. To compensate for this difference, nuclear fuel costs had to be about 0.3–0.6 mill/kwhr cheaper than coal or oil in cases where nuclear plants were competitive.

Operations associated with two different nuclear fuel cycles are shown schematically in Fig. 2. In the first, as indicated in the upper left-hand corner of the figure, the operation starts with mined natural uranium, and fuel elements are manufactured directly for use in a reactor. Such a reactor must use a graphite or heavy-water moderator rather than light-water. In the case of light-water reactors, the process starts with natural uranium in the form of uranium hexafluoride, which is enriched in an isotope separation plant before fuel elements are manufactured. The cost of such enriched fuel elements is an important part of the total fuel cycle costs. An alternative method of enriching natural uranium is also shown in Fig. 2, namely, using the fissile plutonium recovered from irradiated fuel elements.

It is evident from Fig. 2 that the nuclear fuel cycle consists of a sequence of steps, starting with natural uranium, enriching it, fabricating fuel elements, burning the fuel in the reactor, and finally recovering the unconsumed fissile material from spent fuel elements. Thus nuclear fuel costs depend on the sum of the costs of all these operations and on the number of kilowatt-hours of electricity produced from a fuel element before it must be replaced with a fresh one. Fuel costs in a nuclear plant, therefore, are a function of the cost of fissionable material in fresh fuel minus its value in spent fuel, the cost of fabricating fuel elements and recovering unburned fuels, and miscellaneous costs for storage, shipping of new and spent fuel elements, and the unrecoverable fuel losses during processing. To these costs must be added fuel inventory charges and other fixed charges associated with fuel cycle operations. Before 1970, uranium was leased from the Atomic Energy Commission (AEC) for $4\frac{3}{4}$% per year because the 1954 Atomic Energy Act did not permit private ownership of nuclear fuel. Since 1970, reactor operators have been required to purchase, rather than lease, nu-

Estimated power costs in typical light-water-moderated nuclear plants in the United States, 1967 pricing conditions

Cost	Size of plant, MwE			
	400	600	800	1000
Capital cost, $/kwE	170	152	141	134
Power cost, mills/kwhr				
Annual fixed charges*	3.33	2.97	2.76	2.62
Fuel cycle costs	1.56	1.54	1.52	1.50
Operation, maintenance, and insurance	0.39	0.36	0.33	0.30
Total	5.28	4.87	4.61	4.42
Competitive coal cost				
¢/million Btu	24.3	24.1	23.9	23.7
$/short ton	6.32	6.27	6.21	6.16

*At 13.7% and 7000-hr operation per year.

clear fuel. Under these conditions annual fixed charges on fuel and special nuclear materials amount to 10–12% of the total investment in uranium, D_2O, fuel fabrication, and other costs associated with the fuel cycle. Although these so-called "working capital" costs are part of the total plant investment costs, because of their close relation they are usually listed with fuel cycle costs.

Nuclear power costs. Estimated nuclear power costs in typical large-scale water-cooled and water-moderated nuclear plants are summarized in the table as a basis for indicating the competitiveness of nuclear and coal-fired plants. Estimated power costs in nuclear plants in the size range of 400–1000 MwE under 1967 pricing conditions ranged from 5.28 to 4.42 mills/kwhr. As indicated in the table, competitive coal prices at about $0.24/1,000,000 Btu varied only slightly with plant size. Increases in nuclear plant costs which occurred in 1968 would probably add not more than $0.35/ton to the competitive coal price. It might be expected that during the 1970s nuclear plant costs will stabilize as the result of competition and standardization, in which case more than one-half of all new electric capacity ordered will be nuclear.

Power plants for propulsion. The design of reactors for propulsion of ships, planes, and rockets involves a consideration of performance (measured in terms of power output per unit of weight or volume) as well as costs. Because the nuclear power unit requires no fuel other than that initially in the reactor, the attractiveness of nuclear propulsion depends on the size and cost of the reactor relative to the size, cost, and amount of fuel that must be carried by the conventional system. Thus, nuclear propulsion appears to have an advantage when the size of the power unit is the dominant factor, or when the combination of high speeds and long ranges makes conventional fueling too costly.

Miscellaneous reactor applications. The heat, radiations, and radioactive by-products of the fission process can be utilized in a number of ways in addition to generating electricity or power. Exhaust steam from a nuclear power plant, for example, can be used to evaporate sea water to produce fresh water. Such dual-purpose plants are being considered seriously throughout the world. The ionizing radiations emitted during fission can also be used in special cases as a means of producing chemicals. Reactor-produced radioactive isotopes and fission products, moreover, can be used in many ways, such as for small remote power sources, for sterilizing foods, for producing biodegradable detergents, or for polymerizing certain chemicals. It is estimated that about $100,000,000 worth of radioisotopes will be used annually for these purposes in the early 1970s.

[JAMES A. LANE]

Bibliography: C. F. Bonilla (ed.), *Nuclear Engineering*, 1957; H. Etherington (ed.), *Nuclear Engineering Handbook*, 1958; Federal Power Commission, *National Power Survey*, 1964; S. Glasstone and A. Sesonske, *Nuclear Reactor Engineering*, 1963; D. B. Hoisington, *Nucleonics Fundamentals*, 1959; J. A. Lane, Economics of nuclear power, *Annu. Rev. Nucl. Sci.*, vol. 16, 1966; S. McLain and J. Martens, *Reactor Handbook*, vol. 4: *Engineering*, 1964; R. L. Murray, *Introduction to Nuclear Engineering*, 1954; R. Stephenson, *Introduction to Nuclear Engineering*, 2d ed., 1958; USAEC, *Civilian Power Reports: Liquid Metal-Cooled FBR* (WASH 1098), *MSR and Thorium Reactors* (WASH 1097), *Organic and BLW-Cooled HWR* (WASH 1087), *PWR, BWR* (WASH 1082), *Steam and Helium-Cooled FBR* (WASH 1090), 1968; USAEC, *Supplement to 1962 Civilian Nuclear Power Report*, 1967.

Nutrition

The science of nourishment, including the study of the nutrients that each organism must obtain from its environment in order to maintain life and reproduce. Although each kind of organism has its distinctive needs which can be studied separately, a far-reaching biochemical unity in nature has been discovered which gives vastly more coherence to the whole subject. Many nutrients, such as amino acids, minerals, and vitamins, needed by higher organisms may also be needed by the lowest forms of life — single-celled bacteria and protozoa. The recognition of this fact has made possible highly important developments in biochemistry.

Mammals need for their nutrition (aside from water and oxygen) a highly complex mixture of chemical substances, including amino acids; carbohydrates; certain lipids; a great variety of minerals, including several which are required only in minute amounts, commonly referred to as trace elements; and vitamins, organic substances of diverse structure which are treated as a group only because as nutrients they are required in relatively small amounts.

Most nutrients were recognized in the 19th century, but the vitamins and some trace minerals did not become known as fundamental cogs in the machinery of all living things until the early 20th century. The discovery of vitamins, and some of the trace elements, originally came about through the recognition of deficiency diseases, such as beriberi, scurvy, pellagra, and rickets, which arise because of specific nutritional lacks and can be cured or prevented by supplying the needed nutrients.

Different species of mammals have distinctive nutritional needs. Guinea pigs, monkeys, and human beings, for example, require an exogenous supply of ascorbic acid (vitamin C) to maintain life and health, whereas many experimental animals, including rats, do not. It is significant, however, that ascorbic acid is an essential part of the metabolic machinery of animals that do not need an exogenous supply. Rat tissues, for example, are relatively rich in ascorbic acid; unlike guinea pigs these animals are genetically endowed with biochemical mechanisms for producing ascorbic acid from carbohydrate.

One of the bases for current and continued interest in nutrition is the fact that individuals who have differing genetic backgrounds have differing nutritional needs, considered quantitatively; for this reason various human ills may arise because the individuals concerned do not get all of the nutrients in amounts compatible with their own distinctive requirements.

Every cell and tissue in the entire body requires continued adequate nutrition in order to perform its functions adequately. Since a multitude of functions, involving the production of specific chemical substances (hormones, for example) and the regulation of numerous processes, are performed by cells and tissues, it is clear that improper nutrition may produce or contribute to almost every conceivable type of illness. *See* DISEASE.

[ROGER J. WILLIAMS]

Bibliography: G. H. Beaton and E. W. McHenry, *Nutrition: A Comprehensive Treatise*, 3 vols., 1964–1966; R. L. Pike and M. L. Brown, *Nutrition: An Integrated Approach*, 1967; R. J. Williams, *Nutrition in a Nutshell*, 1962.

Ocean-atmosphere relations

This field of investigation is concerned with the boundary zone between sea and air and the dynamic relationships between oceanographic and meteorologic studies. When it is considered that nearly one-half of the heat energy that the atmosphere receives for maintaining its circulation is derived from the condensation of water vapor originating primarily from oceanic evaporation, it becomes evident that an understanding of processes occurring at the air-sea boundary is fundamental to an understanding of atmospheric behavior. The oceanic energy supply to the atmosphere is highly regionalized because of the character of ocean currents, which in turn implies that the atmospheric circulation itself (and resulting weather) is greatly influenced by the oceanic circulation. Conversely, the oceanic circulation represents a state of equilibrium in which the effects of the frictional stresses of the wind on the sea surface are balanced by changes in the distribution of density of oceanic waters. These compensating density changes are in turn related to time and space variations in radiation, heat conduction, evaporation, and precipitation. It is therefore equally manifest that an understanding of the oceanic circulation (and the resulting distribution of properties within the ocean) requires a thorough knowledge of atmospheric processes at the air-sea boundary.

The conclusion is that neither ocean nor atmosphere should be treated independently, but that they should be considered together as a single dynamical-thermodynamical system. However, the interaction of ocean and atmosphere is so complicated that it is not yet possible to completely separate cause and effect. Therefore separate discussions are presented for each of the major classes of atmospheric influences upon the ocean, as well as the oceanic influences upon the atmosphere.

Effects of wind on ocean surface. The frictional stresses of the wind on the surface of the sea produce ocean waves (and storm surges) and ocean currents. The former are transitory phenomena and are not discussed in the present article because they are of little direct meteorological interest, even though waves at sea, coastal breakers, and storm tides are of considerable maritime, as well as oceanographic, importance. *See* STORM SURGE.

Wind-induced ocean currents, on the other hand, are of large-scale significance from both the oceanographic and meteorological points of view. The wind exerts a twofold effect upon the surface layers of the ocean. In the first instance the stress of the wind leads to the formation of a shallow surface wind drift. The resulting transport of surface water by the wind drift leads in the second instance to pressure variations with depth and a changed distribution of mass (density) throughout the ocean. In the final analysis it is the resulting fields of density which account for the major current systems of the oceans. The total transport due to the wind drift is directed to the right of the wind (in the Northern Hemisphere), but the final density (slope) current which results from the sloping sea surface tends to flow in the direction of the prevailing wind, except where coastal configuration prevents the realization of such flow. Nevertheless, it should again be emphasized that the wind effect is not the sole meteorological factor that serves to determine the distribution of mass or the slope of isobaric surfaces in the oceans; heating and cooling, freezing and thawing, and evaporation and precipitation all exert their influences.

Major ocean current systems. For the reasons just outlined, the major ocean currents of the world conform closely with the prevailing anticyclonic wind circulatons of the oceans. With the exception of the northern Indian Ocean, warm currents flow poleward to about 40° lat in the western portion of all oceans, with easterly flow in the higher latitudes, equatorward drift (relatively cold) in the eastern portions of the oceans, and westerly flowing currents near the Equator. The low temperature of the waters in the eastern portions of the oceans is due partly to the high-latitude origin of the currents and partly to coastal upwelling of cold subsurface waters. Of all the currents the poleward-flowing warm currents of the western portions of the oceans are the best developed and the most important. Examples are the Gulf Stream of the North Atlantic and the Kuroshio of the North Pacific, each of which transports a tremendous volume of warm tropical water into higher latitudes.

Hydrologic and energy relations. For the Earth as a whole and for the entire year, the amounts of energy received and lost through radiation are in balance for all practical purposes. However, this is not true for any given portion of the Earth's surface, particularly during any given fraction of the annual solar cycle. The ratio of insolation to outgoing radiation decreases from the Equator toward the poles. Between the Equator and the 35th parallel, the Earth receives more energy through radiation than it loses; the reverse is true poleward from about the 35th parallel.

Because observations indicate that the lower latitudes are not becoming progressively warmer and the higher latitudes colder, it must be assumed that considerable heat is transported from lower to higher latitudes by both atmosphere and ocean. According to H. U. Sverdrup (see bibliography, G. P. Kuiper, 1954), the meridional transport of energy in the Northern Hemisphere reaches a maximum a little north of latitude 35°N. At latitude 30°N, Sverdrup computes the total energy transport across the latitude circle to be 6.5×10^{16} cal/min of which 1.9×10^{16} cal/min (or 29%) is accomplished by ocean currents, principally by the Gulf Stream and Kuroshio. The remaining energy transport is accomplished by the atmosphere.

The largest fraction of radiant energy absorbed by the oceans is utilized in evaporating sea water. A much smaller fraction, about 10%, is utilized in more direct heating of the atmosphere in contact with the sea surface. The latent energy of vaporization subsequently becomes available to the atmosphere as either sensible heat or gravitational potential energy when condensation takes place, often in a region far removed from the area where the evaporation occurred. The precipitation resulting from the condensation of atmospheric water vapor is then returned to the ocean, either directly as rainfall or snowfall, or indirectly as runoff and discharge from land areas. The hydrologic and energy cycle is thereby completed. *See* HYDROLOGY; HYDROSPHERE, GEOCHEMISTRY OF.

The maximum evaporation and heat exchange between sea and atmosphere take place where cold air flows over warm-water surfaces. The ideal locations for maximum moisture or energy transfer are therefore those areas where cold continental air flows out over warm poleward-moving ocean currents. Such ideal conditions exist during winter off the eastern coasts of the continents and over warm currents such as the Gulf Stream and Kuroshio. Radiant energy that was absorbed and stored by the oceans at lower latitudes is given off to the atmosphere by this process at places and during seasons of marked deficiency in radiative energy. Any change in the oceanic transport by ocean currents must be reflected in corresponding changes in the rates of evaporation and must finally have significant effects on atmospheric circulation.

Relations with sea-water salinity. In the absence of horizontal flow, the surface salinity of any portion of the ocean is mainly determined by three processes: decrease of salinity by precipitation, increase of salinity by evaporation, and change of salinity by vertical mixing. Salinities thus tend to be high in regions where evaporation exceeds precipitation and low where precipitation exceeds evaporation. However, horizontal transport of surface waters by wind-induced ocean currents serves to displace the areas of maximum or minimum salinity in the direction of surface flow away from the areas of maximum differences (positive or negative) between evaporation and precipitation. The conclusion, of course, is that the distributions of surface salinities (as well as other properties) in the ocean are determined almost completely by atmospheric circumstances.

[WOODROW C. JACOBS]

Bibliography: W. C. Jacobs, The energy exchange between sea and atmosphere and some of its consequences, *Bull. Scripps Inst. Oceanogr. Univ. Calif.*, 6:27–122, 1951; G. P. Kuiper (ed.), *The Solar System*, vol. 2, 1954; J. S. Malkus, *Large-scale Interactions*, in M. N. Hill (ed.), *The Sea*, vol. 1, 1962; G. Newman and W. J. Pierson, *Principles of Physical Oceanography*, 1966; H. U. Roll, *Physics of the Marine Atmosphere*, 1965; H. U. Sverdrup, *Oceanography for Meteorologists*, 1942.

Oceanography

The scientific study and exploration of the oceans and seas in all their aspects, including the sediments and rocks beneath the seas; the interaction of sea and atmosphere; the body of sea water in motion and subject to internal and external forces; the living content of the seas and sea floors and the behavior of these organisms; the chemical composition of the water; the physics of the sea and sea floor; the origin of ocean basins and ancient seas; and the formation and interaction of beaches, shores, and estuaries. Hence oceanography, sometimes called the science of the seas, consists of the marine aspects of several disciplines and branches of science: geology, meteorology, biology, chemistry, physics, geophysics, geochemistry, fluid mechanics, and in its more theoretical aspects, applied mathematics. Oceanography is also an environmental science which describes and attempts to explain all processes in the ocean, and the interrelation of the ocean with the solid and gaseous phases of the Earth and with the universe. *See* EARTH SCIENCES.

Because of the fluid nature of its contents, which permits vertical and horizontal motion and mixing, and because all the waters of the world oceans are in various degrees of communication, it is necessary to study the oceans as a unit. Further unification results from the technological necessity of studying the ocean from ships. Many phases of oceanic research can be carried out in a laboratory, but to study and understand the ocean as a whole, scientists must go out to sea with vessels adapted or built especially for that purpose. Furthermore, data must be obtained from the deepest part of the ocean and, if possible, scientists must go down to the greatest depths to observe and experiment. Another unifying influence is the fact that many oceanic problems are so complex that their geological, biological, and physical aspects must be studied by a team of scientists. Because of the unity of processes operating in the ocean, and because some writers have separated marine biology from oceanography (implying the term oceanography to embrace primarily physical oceanography, bottom relief, and sediments), the term "oceanology" is sometimes used as embracing all the science divisions of the marine hydrosphere. As used in this article, the term oceanography applies to the whole of sea science. *See* HYDROSPHERE.

Development. The early ocean voyages by Frobisher, Davis, Hudson, Baffin, Bering, Cook, Ross, Parry, Franklin, Amundsen, and Nordenskiold were undertaken primarily for geographical exploration and in search of new navigable routes. Information gathered about the ocean, its currents, sea ice, and other physical and biological phenomena was more or less incidental. Later, the polar expeditions of the Scoresbys, Parry, Markham, Greeley, Nansen, Peary, Scott, and Shackleton were also voyages of geographical discovery, although scientific observations about the sea and its inhabitants were made by some of them. William Scoresby took soundings and observed that discolored water containing living organisms (now known to be diatoms) was related to whale movements. Ross made dredge hauls of bottom-living animals. Nansen contributed to the improvement of plankton nets and suggested the existence of internal waves.

More closely related to the beginning of oceanography as comprehensive study of the seas are the 19th-century activities of naturalists Ehrenberg, Humboldt, Hooker, and Örstedt, all of whom contributed to the eventual recognition of plankton life in the sea and its role in the formation of bottom deposits. Charles Darwin's observations on coral reefs and Müller's invention of the plankton net belong to this phase of developing interest in marine science, in which men began to investigate ocean phenomena as biologists, chemists, and physicists rather than as oceanographers. In this group should also be included such physicists and mathematicians as Kepler, Vossius, Fournier, Varenius, and Laplace, who provided the background for the development of modern theories and investigations of ocean currents and air circulation.

Toward the middle of the 19th century a few scientists began to study the oceans as a whole, rather than as an incidental part of an established discipline. Forbes, as a result of his work at sea, first developed a scheme for vertical and horizontal distribution of life in the sea. On the physical side Matthew Fontaine Maury, developing and extending Franklin's earlier work, made comprehensive computations of wind and current data and set up the machinery for international cooperation. His book *Physical Geography of the Sea* has been regarded as the first text in oceanography.

Forbes and Maury were followed by a distinguished group of men whose interest in oceanography led them to make the first truly oceanographic expeditions. Most famous of these was the three-year around-the-world voyage of HMS *Challenger*, which followed earlier explorations of the *Lightning* and *Porcupine*. Instrumental in organizing these was Wyville Thompson, later joined by John Murray. Later in succession were the Norwegian Johan Hjort and the *Michael Sars* North Atlantic exploration; Louis Agassiz; and Albert Honoré Charles, Prince of Monaco, in a series of privately owned yachts named *Hirondelle I* and *II* and *Princess Alice I* and *II*. Other important contributions were made by Michael and G. O. Sars, Björn Helland-Hansen, Carl Chum, Victor Hansen, Otto Petterson, Gustav Ekman, and the vessels *Valdivia*, the Danish *Dana*, the British *Discovery*, the German *National* and *Meteor*, and the Dutch *Ingold*, *Snellius*, and *Siboga*, the French *Travailleur* and *Talisman*, the Austrian *Pola*, and the North American *Blake*, *Bache*, and *Albatross*. Among the North American pioneers were Alexander Agassiz, L. F. de Pourtales, and J. D. Dana. Pioneers in modern oceanographic work are M. Kunelsen, Sven Ekman, A. S. Sverdrup, A. Defant, Georg Wüst, Gerhard Schott, and Henry Bigelow.

Modern oceanography relies less upon single

explorations than upon the continuous operation of single vessels belonging to permanent institutions, such as *Atlantis II* of the Woods Hole Oceanographic Institution, *Argo* of Scripps Institution of Oceanography, *Vema* of the Lamont Geological Observatory, the French *Calypso*, and the large Soviet vessels *Vitiaz* and *Mikhail Lomonosov*. Single explorations continue to be made, as exemplified by the Swedish *Albatross* and Danish *Galatea*.

The reduction of data and study of collections from earlier expeditions were carried out generally in research institutions, museums, and universities not solely or primarily engaged in oceanography. The first marine laboratories were interested principally in fishery problems or were designed as biological stations to accommodate visiting investigators. Many of the former have extended their activities to cover chemical and physical oceanography during their growth and development. The latter, often active as extensions of university biological departments, are exemplified by the Naples Zoological Station and the Marine Biological Laboratory of Woods Hole. Visitors to such stations contribute greatly to the development of biology, generally in such fields as embryology and physiology.

The number of institutions devoted to organized oceanographic investigations with permanent scientific staffs has gradually grown. At first the requirements of fishery research provided the stimulus in countries adjacent to the North Sea, but in later years laboratories in other countries wholly or mainly devoted to oceanography have grown considerably in number. A few may be mentioned here. In England, among other important institutions, are the National Institute of Oceanography, the Marine Biological Laboratory at Plymouth, and the Fisheries Laboratory at Lowestoft. In the United States are the Woods Hole Oceanographic Institution in Massachusetts, the Scripps Institution of Oceanography in California, the Lamont-Doherty Geological Observatory in New York, the University of Miami Marine Laboratory in Florida, the Texas A. & M. College Department of Oceanography, and the Oceanography Laboratories of the University of Washington at Seattle, the University of Hawaii, and Oregon State University. In Germany oceanographic laboratories are located at Kiel and at Hamburg. In Denmark the Danish Biological Station is at Copenhagen. Other European laboratories include those at Bergen, Norway; Göteborg and Stockholm, Sweden; Helsinki, Finland; and Trieste. Laboratories are located at Tokyo, Japan; Namaimio and Halifax, Canada; and Hawaii. This list is not inclusive and necessarily leaves out a considerable number of important institutions.

Surveys. Oceanographic surveys require careful planning because of high cost. Provision must be made for the proper type of vessel, equipment, and laboratory facilities, adapted to the nature and duration of the survey.

Research ships. Ships of all types and sizes have been gathering information about the oceans since earliest times. Vessels of less than 300 tons displacement seldom range farther than several hundred miles from land, whereas ships larger than 300 tons displacement may work in the open ocean for several months at a time. Research ships

Fig. 1.   Deep-sea drilling vessel *Glomar Challenger*, equipped with satellite navigation equipment and capable of holding position to within 100 ft for several days, using bottom-mounted sonic beacons, tunnel thrusters, and computers.

of all sizes must be seaworthy and must provide good platforms from which to work (Fig. 1). More specifically, a ship must have comfortable quarters, adequate laboratory and deck space for preliminary analyses, plus storage space for equipment, explosives, samples, and scientific data. Machinery, usually in the form of winches and booms, is necessary for handling the complex and often heavy scientific equipment needed to probe the ocean depths (Fig. 2).

A number of the larger oceanographic vessels are equipped with general-purpose digital computers, including tapes, disk files, and process-inter-

Fig. 2.   Trawl winch used on research vessel *Vema*. (*Lamont-Doherty Geological Observatory*)

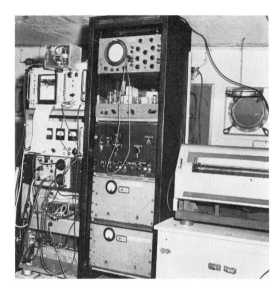

Fig. 3. Special laboratory aboard research vessel. (*Lamont-Doherty Geological Observatory*)

rupt equipment. The result is that data can be reduced on board for experimental work, and all the operations taking place on the vessel can be centralized, including the satellite navigation equipment.

Standard oceanographic equipment includes collecting bottles (Nansen bottles) for obtaining water samples and thermometers (both reversing thermometers and bathythermographs) for measuring temperatures at all depths. In addition, there are various devices for obtaining samples of ocean bottom sediments and biological specimens. These include heavy coring tubes which punch cylindrical sediment sections out of the bottom, dredges which scrape rock samples from submerged mountains and platforms, plankton nets for collecting very small planktonic organisms, and trawls for collecting large free-swimming organisms at all oceanic depths. Echo sounders provide accurate profiles of the ocean floor.

Specialized equipment for oceanic exploration includes seismographs for measuring the Earth's crustal thickness, magnetometers for measuring terrestrial magnetism, gravimeters for measuring variations in the force of gravity, hydrophotometers for measuring the distribution of light in the sea, heat probes (earth thermometers) for measuring the flow of heat from the Earth's interior, deep-sea cameras to photograph the sea bottom, bioluminescence counters for measuring the amount of luminescent light emitted by organisms, salinometers for measuring directly the salinity of sea water, and current meters to clock the speed of ocean currents.

Positioning of a ship is very important for accurate plotting of data and detailed charting of the oceans. Celestial navigation is in wide use now as in the past. Navigational aides such as electronic positioning equipment (loran, shoran, and radar) are increasing the accuracy of positioning to within several tens of yards of the ship's true position. Navigational and radio communications equipment normally is situated near the captain's bridge, but often is duplicated in the scientific lab-

oratories in order that complete communication between the ship's operators, scientists, and other participating ships can be carried on at all times.

Probably the single most important advance in deep-sea oceanography has been the introduction of satellite navigation. Combined with a computer, satellite navigation permits fixes to within several hundred yards while the ship is moving and to within several hundred feet when the ship is located with respect to bottom-mounted beacons, for example, in deep-sea drilling.

Ship's laboratories. Laboratories must be adaptable for a large number of operations. In general they are of two categories, namely, wet and special laboratories. Wet laboratories are provided with an open-drain deck so that surplus ample water can be drained out on deck. Such a laboratory is located near the winches used for running out and retrieving a long string of water-sample bottles (hydrocasts). Adjoining the wet laboratory are special laboratories equipped with benches for measuring chemical properties of the recovered water and for examination of biological and geological samples. Electronics laboratories are either part of, or adjacent to, the special laboratory, depending upon the size of the ship. Here, numerous recording devices, amplifiers, and computers are set up for a variety of purposes, such as measurements of underwater sound, measurement of the Earth's magnetic and gravity fields, and seismic measurements of the Earth's crustal thickness (Fig. 3).

Marine technology. The development of nuclear power plants permits extended voyages without the necessity of refueling. Nuclear power plants, used in conjunction with inertial guidance in the submarines *Nautilus* and *Skate*, made possible the first extended journey under the Arctic ice pack.

Fig. 4. Remote-controlled underwater television camera mounted in self-propelled vehicle, which can make visual surveys to depths of 1000 ft. Self-buoyant unit moves about or hovers at desired depth in currents or tides of several knots. (*Vare Industries, Roselle, N.J.*)

Uncharted regions of the oceans are within the reach of exploration.

Direct visual observations of the ocean depths are fast becoming a reality, both by "manned" submersibles (bathyscaph) and by television cameras. Deep-sea cameras have been developed to the point whereby motion pictures of even the deepest parts of the sea bottom can be taken (Fig. 4). However, such observations yield no information about the subsurface material. Major crustal features are determined by seismic measurement. The shallower features of the subbottom structure and deposits were not really observed until the advent of the subbottom acoustic probe. This device is a very-high-energy echo sounder capable of penetrating below the sediment-water interface and yielding a continuous profile of the subbottom strata.

Information as to the physical and chemical makeup of the underlying material, however, is dependent upon penetration and actual recovery. Commonly used coring devices rarely penetrate more than 10–20 m (on occasion to about 33 m) below the surface. A new "incremental" coring device has been developed for taking successive 2-m sediment cores to depths of possibly 100 m or more.

The most impressive achievement in this area, however, has been in deep-sea drilling. The *Glomar Challenger*, operated by the Global Marine Corp., was designed primarily for the purpose of taking cores throughout the full column of sediment in mid-ocean. It is capable of handling 24,000 ft of drill string in mid-ocean and drilling through more than 2500 ft of sediment, taking 30-ft cores in the process. By mid-1969 the vessel had successfully operated at more than 40 sites in both the Atlantic and Pacific. This operation is sponsored by the National Science Foundation, with scientific guidance supplied by JOIDES (Joint Oceanographic Institutions Deep Earth Sampling), which include the University of Washington; Institute for Marine Sciences, University of Miami; Woods Hole Oceanographic Institution; Lamont-Doherty Geological Observatory of Columbia University; and Scripps Institution of Oceanography. The last-mentioned is the operating institution. Besides incorporating the necessary innovations in drilling and coring, the *Glomar Challenger* is an example of the many important advances in oceanography, such as satellite navigation and positioning.

Oceanographic stations. Work on station consists of sampling and measuring as many marine properties as possible within the limitations of an expedition. Water, sea-bottom, and biological samples are successively collected on the long cables extended to the ocean floor. In some surveys one cable lowering may include samplers for all these items, but this is not the usual procedure. Stations are systematically located at predetermined points along the ship's path. At hydrographic stations, observations of water temperature, salinity, oxygen, and phosphate content are determined upon sample recovery. Seismic stations generally are carried out by two ships, one running a fixed course and dropping explosives while the second remains stationary and records the returning subbottom reflected or refracted sound waves. Biological stations may consist of vertical net hauls

or horizontal net tows depressed to sweep the ocean at a fixed depth. Geological stations are usually coring or bottom-dredging operations. Wherever possible the recovered data are given a preliminary reduction aboard ship so that interesting discoveries are not bypassed before sufficient information is obtained. Detailed analyses aboard ship are seldom possible because of the limitations of time, space, and laboratory equipment. Instead, the carefully processed, labeled, and stored material is preserved for intensive study ashore.

Home laboratory. This phase of the work may entail many months of careful examination and detailed analyses. Batteries of sophisticated scientific instruments are often necessary: data computers for reduction of physical oceanography information; spectrographic apparatus consisting of emission units; infrared, ultraviolet, x-ray, and mass spectrometers for chemistry; aquaria, pressure chambers, and chemostats for the biologist; electron microscopes and high-powered optical microscopes for examination of inorganic and organic constituents; radioisotope counters; and numerous standard physical, chemical, geological, and biological instruments. The great variety of measurable major and minor properties is reduced to statistical parameters which may then be integrated, correlated, and charted to increase the knowledge of sea properties and show their relationships. The essentially descriptive properties lead to an understanding of the principles which control the origin, form, and distribution of the observed phenomena. The present knowledge of the oceans is still fragmentary, but the increasing store of information already is being applied to a rapidly expanding number of man's everyday problems.

Research problems. Since the middle of the 19th century, man has learned more and more about that 71% of the Earth which is covered with sea water. The rate of increase of knowledge is being accelerated by improved tools and methods and increased interest of scientists and engineers. Some of the present and future research problems that are attracting scientists are mentioned below.

One of the oldest and still unsolved problems is the motion of ocean waters, involving surface currents, deep-sea currents, vertical and horizontal turbulent motion, and general circulation. New methods such as distribution of radioactive substances, deep-sea current meters and neutrally buoyant floats, high-precision determination of salt and gas content, and hot-wire anemometers for turbulence studies and current measurements have increased present knowledge considerably. The surface movement of water is wind-produced, and a general theory of the motion has been worked out. The deep-sea currents, which are known to be caused in part by variations in the thermohaline circulation, are still an open problem. Superimposed on these movements is turbulent motion, which ranges over the whole spectrum from large ocean eddies transporting millions of cubic meters of water per second to the tiniest vibrations of water particles. Very little is known about turbulence.

The mixing of water masses and the formation of new water masses cannot yet be completely and adequately described, as many of the thermodynamic parameters are not precisely known. Laboratory experiments and measurements of thermal expansion, saline contraction, and specific heat at constant pressure must be carried out. A further problem is the composition of sea water and the extent to which the ratio among the components is constant. In connection with these problems it has been urged that a library of water samples be established. Improved techniques of measuring sound velocity, electrical conductivity, refractive index, and density must be developed to enable scientists to follow many processes in the ocean. It is therefore necessary to study the small variations of these parameters in the sea. *See* SEA WATER.

The tides in the oceans are rather well known at the surface but are almost completely unknown in the deep sea; also, the influence of land boundaries on deep-sea tides is not yet understood. Further research also must be devoted to the interesting phenomenon of internal waves.

The study of ocean waves is one of the most advanced topics in oceanography, but the energy exchange between atmosphere and sea surface by friction must be studied further. Another problem is that of the heat exchange between ocean and atmosphere, an important link in the heat mechanism which determines the weather and the oceanic circulation. *See* OCEAN-ATMOSPHERE RELATIONS.

The climate of the past, in particular that of the last 1,000,000 years, is best studied in the ocean. Isotopic methods in paleoclimatologic research allow the determination of temperature variations in the ocean with a high degree of accuracy. The rapid growth of geochemistry and the increased sampling of deep-sea sediments through improved techniques have solved some of the problems of deep-sea sedimentation. At the same time a number of new ones have been created, such as: Why is the sediment carpet only about 300 m thick? What is the mechanism of sediment transport? What is the history of sea water? Of the ocean basin? What is the cause of the ice ages?

The results of the Deep Sea Drilling Project have confirmed the expectations of its most optimistic supporters. It has strikingly confirmed the sea-floor-spreading hypothesis of the development of the ocean basins and the newer concepts of plate tectonics. A core to the Moho (about 3 mi deep) may help answer many of the questions about the structure of the Earth's crust. The problem of the mechanism of formation of the ridges and island chains may be near solution.

The age determination of sediments by radioactivity methods, which was thought impossible 30 years ago, is now used on deep-sea sediments older than 10,000,000 years. Very little is known about the formation of minerals on the sea floor, the diffusion and adsorption of elements in and on sediments, and the reaction at slow rates in sediments. Certainly microbiological processes on the sea floor are an important factor, as they seem to produce chemical energy in sediments. *See* HYDROSPHERE, GEOCHEMISTRY OF.

In marine biology the systematics and ecology remain the major aspects. It is still the science of the "naturalist." The interest of marine biology is many-sided and not grouped around a few central problems. Ocean life offers to the general biologist the best opportunities to study such complex problems as the structure of communities and the flux of energy through these communities. The zonation of animals on the shore and in the open ocean is not yet fully understood; the cause of patchiness in the distribution must be found. On the other hand, the distribution of species by currents and eddies must be studied, and large-scale experiments on behavior must be carried out. Observation at sea has been largely neglected, and therefore the equilibrium between sea observation and laboratory experiment must be restored. The great advances in genetics, biochemistry, physiology, and microbiology also will advance the study of life in the sea. *See* MARINE ECOSYSTEM.

Applications of ocean research. Directly and indirectly the ocean is of great importance to man. It is valuable as a reservoir of natural resources, an outlet for waste disposal, and a means of transport and communication. The ocean is also important as a harmful agent causing biological, chemical, and mechanical destruction of life and property. In addition to the peaceful exploration of the oceans, there are many military applications of surface and submarine phenomena. In all of these aspects oceanography provides basic information for engineers who seek to increase its benefits and to avoid its harmful effects. *See* MARINE RESOURCES.

Food resources. The food resources of the ocean are potentially greater than those of the land since its larger area receives a proportionately larger amount of solar radiation, the source of living energy. Nevertheless, this potential is only in part realized. Oceanographic studies provide information which can help to increase fishing yields through improved exploratory fishing, economical harvesting methods, fisheries forecasts, processing techniques at sea, and aquaculture.

Fishes are dependent in their distribution upon food organisms and plankton, vertical and horizontal currents which bring nutrients to the plankton, bottom conditions, and physical and chemical characteristics of the water. A knowledge of the relation of food fishes to these environmental conditions and of the distribution of these conditions in the oceans is vital to successful extension of fishing areas. Satisfactory measurements of the basic organic productivity of the sea may become essential in the selection of regions for extended fishery exploration. *See* SEA-WATER FERTILITY.

The catching of fishes may be facilitated and new and more efficient methods devised through a knowledge of the reaction of fishes to stimuli and of their habits in general. This knowledge may result in better design of nets and in the use of electrical, sonic, photic, and chemical traps or baits (Fig. 5). The harvest also may be increased by using improved methods of locating schools of fish by sonic or other means.

The biology of fishes, their food preferences, their predators, their relation to oceanographic conditions, and the fluctuations in these conditions seasonally and from year to year are important factors in forecasting fluctuations in the fisheries. This information will aid in preventing the economic waste of alternating glut and scarcity. Similar information is essential to good management of the fisheries and to sound regulation by conservation agencies.

Other anticipated advances which require further oceanographic study include (1) the improvement of fishing by transplanting the young of existing stocks or by introducing new stocks; (2) farming or cultivation of sea fishes (although this does not seem feasible at present, scientific research has improved the cultivation of oysters and mussels in France and Japan); and (3) the use of planktonic vegetation as a source of food or animal nutrition. *See* FISHERIES CONSERVATION.

Mineral resources. Although most of the valuable chemical elements in sea water are in very great dilution, the great volume of water of the oceans (about 300,000,000 mi³) provides a limitless and readily accessible reservoir, if such dilute concentrations can be economically extracted. Magnesium is produced largely from sea water, and bromide also has been extracted commercially. High concentrations of manganese are found in manganese nodules, which are very common on certain areas of the sea floor.

Other elements occur in too great dilution to be extracted by present methods. Possibly a better understanding of the ability of certain marine plants and animals to accumulate and concentrate elements from sea water in their tissues may lead to new methods of recovering these elements. *See* MINING, UNDERSEA.

Energy and water source. Sea water contains deuterium and would be a limitless source of this element in the event of successful nuclear fusion developments. Further oceanographic knowledge and advances in engineering may lead to the increased utilization of tidal energy sources, or of the heat energy available from temperature differences in the ocean. The development of new methods for removing salt from sea water offers promise that the sea may become a practical source for potable water.

Disposal outlet. Because of its large volume, the sea is frequently used for the disposal of chemical wastes, sewage, and garbage. A knowledge of local currents and tides, as well as of the bottom fauna, is essential to avoid pollution of beaches or commercial fishing grounds. Radioactive waste disposal in offshore deeps poses problems of the rate of movement of deep waters and the transfer of radioactive materials through migration and food chains of marine organisms. A problem of rapidly growing concern is oil pollution, which is caused in part by the rapidly expanding exploitation of the continental-shelf-oil resources. The Santa Barbara incident in 1969 was one example. There must be considerable development in marine engineering and better ecological understanding if these oil resources are to be fully utilized. A second major cause of oil pollution is the breakup of giant oil tankers, such as the *Torrey Canyon*. Prevention of such accidents requires improved vessels and more stringent navigation controls.

Traffic and communication. The sea still remains an important highway; thus the knowledge and forecasting of waves, currents, tides, and weather in relation to navigation are of great practical importance. New developments include the continuous rerouting of ships at sea in order that they may follow the most economic paths in the face of changing weather conditions. A knowledge of submarine topography, geologic processes, and temp-

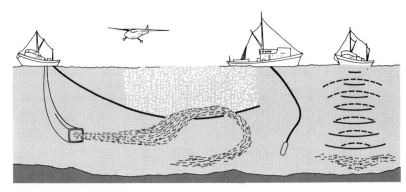

Fig. 5.   Sketch showing new ways of fishing. Fish are spotted by aircraft and by vessels using sonar. Electric and air-bubble fences trap the fish. (*Adapted from A. Spilhaus, Turn to the Sea, NAS-NRC, 1959*)

erature conditions is important for the satisfactory location, operation, and repair of submarine cables. The use of Sofar in air-sea rescue operations is based upon the principles of submarine acoustics.

Defense requirements. Defense aspects of marine research involve not only the navigation of surface vessels but also undersea craft with special navigational problems related to submarine topography, echo sounding, and the distribution of temperature, density, and other properties. Research in submarine acoustics has improved communication between, and detection of, undersea craft. In spite of these advances natural conditions, such as warm water pockets and subsurface magnetic irregularities, can conceal submarines from conventional means of detection. Investigation of these conditions is essential to any defense against missile-carrying submarines.

Property and life. Damage to docks and ships by marine borers and fouling organisms is controlled by methods that utilize a knowledge of the biology, behavior, and physiology of the destructive organisms, and of the oceanographic conditions which control their distribution. Loss of life caused by the attacks of sharks and other fishes may be reduced through an understanding of their behavior and the development of repellants and other protective devices. The chemical characteristics of sea water pose special problems of corrosion of metals. Beach erosion, wave damage to harbor and offshore structures, the effects of tsunamis and internal waves, and storms cause loss of property and life. Much of this damage may be minimized by the application of oceanographic knowledge to forecasting methods and warning systems. *See* STORM SURGE.

Indirect benefits of oceanography arise from the application of marine meteorology to weather prediction, not only over the sea areas but also over the land. The study of marine geology and marine ecology aid in the understanding of the character of oil-bearing sedimentary rocks that are found on land.

[WILLIAM A. NIERENBERG]

Bibliography: H. Barnes, *Oceanography and Marine Biology*, 1959; H. B. Bigelow, *Oceanography: Its Scope, Problems, and Economic Importance*, 1931; G. Dietrich and K. Kalle, *General*

Oceanography, 1963; W. A. Herdman, *Founders of Oceanography and Their Work: An Introduction to the Science of the Sea*, 1923; M. N. Hill (ed.), *The Sea*, 3 vols., 1962, 1963; F. G. W. Smith and H. Chapin, *The Sun, the Sea, and Tomorrow*, 1954; A. F. Spilhaus, *Turn to the Sea*, 1959; H. U. Sverdrup, M. W. Johnson, and R. H. Fleming, *The Oceans*, 1942.

Oil and gas, offshore

Oil and gas prospecting and exploitation on the continental shelves and slopes. Since Mobil Oil Co. drilled what is considered the first offshore well off the coast of Louisiana in 1945, exploration for petroleum and natural gas on the more than 8,000,000 mi² of the world's continental shelves and slopes, lying between the shore and 1000-ft water depth, has expanded rapidly to include exploration or drilling or both off the coasts of more than 75 nations.

More than 119 national and private petroleum companies have joined in the worldwide offshore search for oil and gas with a fleet of 213 mobile and 60 fixed platform drilling units. This offshore armada was increased by 16 new mobile rigs in 1969 at costs ranging between 9,000,000 and $12,000,000 each. Many authorities predict an offshore rig total of 400 units by the late 1980s, which will expand rig investment to more than $3,000,000,000.

More than half of the offshore rigs in operation are working in United States water, predominantly in the Gulf of Mexico. In mid-1969 worldwide offshore rig distribution encompassed all the continents except Antarctica.

Paralleling the development of drilling vessels able to withstand the rigors of operation in the open sea have been remarkable technological developments in equipment and methods. More than 125 companies in the United States devote a large share of their efforts to the development and manufacture of material and devices in support of offshore oil and gas.

For many years petroleum companies stopped at the water's edge or sought and developed oil and gas accumulations only in the shallow seas bordering onshore producing areas. These activities were usually confined to water depths in which drilling and producing operations could be conducted from platforms, piers, or causeways built upon pilings driven into the sea floor. Major accumulations along the Gulf Coast of the United States, in Lake Maracaibo of Venezuela, in the Persian Gulf, and in the Baku fields of the Caspian Sea were developed from such fixed structures.

Exploration deeper under the sea did not begin in earnest until the world's burgeoning appetite for energy sources, coupled with a lessening return from land drilling, provided the incentives for the huge investments required for drilling in the open sea.

Geology and the sea. There is a sound geologic basis for the petroleum industry turning to the continental shelves and slopes in search of needed reserves. Favorable sediments and structures exist beneath the present seas of the world, in geologic settings that have proven highly productive onshore. In fact, the subsea geologic similarity — or in some cases superiority — to geologic conditions on land has been a vital factor in the rapid expansion of the free world's investment in offshore exploration and production. More than $12,000,000,000 had been invested in the free world offshore effort by 1969, with an additional $5,000,000,000 slated by 1972. More than $1,000,000 per day is being spent on production expansion along the United States Gulf Coast alone.

Subsea geologic basins, having sediments considered favorable for petroleum deposits, total approximately 6,000,000 mi², or about 57% of the world's total continental shelf. This 6,000,000-mi² area is equivalent to one-third of the 18,000,000 mi² of geologic basins on land, and in 1968 it was producing about 4,000,000 bbl per day or about 15% of the free world's total production.

Free world proven oil reserves are estimated to total 375×10^9 bbl, of which 71×10^9 or 19%, are located offshore. Estimates of future offshore oil reserves remaining to be found if an economic incentive is present go as high as 700×10^9 bbl.

Furthermore, this production rate has been achieved despite the fact that only 166,000 mi², or about 2%, of the world's continental shelves out to 1000-ft water depths has been tested by the drill. Estimates indicate that only a small percentage of the seismic work necessary to evaluate the continental shelves has been accomplished, and that even by pursuing a work rate nearly double the average for the 1960s, it would take another 127 years to complete the seismic survey of all those areas judged to be prospective. Surveying of the world's total favorable continental shelf area will take an estimated 68,400 seismic crew months, with 8300 months actually accomplished.

Virtually all of the world's continental shelf has received some geological study, with active seismic or drilling exploration of some sort either planned or in effect. Offshore exploration planned or underway includes action in the North Sea; in the English Channel off the coast of France; in the Red Sea bordering Egypt; in the seas to the north, west, and south of Australia; off Sumatra; along the Gulf Coast and east, west, and northwest coast of the continental United States; in the Cook Inlet and off the west coast of Alaska; in the Persian Gulf; off Mexico; off the east and west coasts of Central America and South America; and off Nigeria and North Africa. In addition, offshore exploration has been undertaken in the Caspian and Baltic seas off the coasts of East Germany, Poland, Latvia, and Lithuania; the China seas; Gulf of Thailand; Irish Sea; Arctic Ocean; and Arctic islands.

Successes at sea. The price of success in the offshore drilling is staggering in the amount of capital required, the risks involved, and the time required to achieve a break-even point.

Offshore Louisiana, long the center of much of the world's offshore drilling, has yielded a large number of oil and gas fields. And though cumulative production by mid-1967 had reached nearly 5×10^9 bbl of oil or gas equivalent, the industry has yet to recover some $3,500,000,000 to break even on their investments in offshore Louisiana.

Major gas strikes have been made in the North Sea, and oil and gas strikes have been made in the Bass Straits off Australia. These have the potential for changing historical energy sources and the

economies of the countries involved. Discoveries
made since 1964 in the Cook Inlet of Alaska are
making a major impact on the West Coast market
of the United States; although they involve be-
tween 1.5×10^9 to 2×10^9 bbl of oil, the extremely
high exploration and development costs will cause
the break-even point to be somewhere near 10
years. The most active offshore areas outside the
United States are the North Sea, Australia, Nigeria, and the Persian Gulf.

The Okan field of Nigeria reached 45,000 barrels
per day (bpd) production in 1966. Total production
from Nigeria, where fields other than Okan are in
swamps at the edge, reached 325,000 bpd in 1966.

The Persian Gulf area, where the world's largest
fields on land have been found, has also produced
the world's largest offshore field, Safaniya, in Sau-
dia Arabia. This field is capable of producing more
than 1,000,000 bpd.

Mobile drilling platforms. The underwater
search has been made possible only by vast improvements in offshore technology. Drillers first
took to the sea with land rigs mounted on barges
towed to location and anchored or with fixed platforms accompanied by a tender ship (Fig. 1). A
wide variety of rig platforms has since evolved,
some designed to cope with specific hazards of the
sea and others for more general work. All new
types stress characteristics of mobility and the
capability for work in even deeper water.

The world's mobile platform fleet can be divided
into four main groupings: self-elevating platforms,
submersibles, semisubmersibles, and floating drill
ships.

The most widely used mobile platform is the self-
elevating, or jack-up, unit (Fig. 2). It is towed to
location, where the legs are lowered to the sea
floor, and the platform is jacked up above wave
height. These self-contained platforms are especially suited to wildcat and delineation drilling.
They are best in firmer sea bottoms with a depth
limit out to 300 ft of water.

The submersible platforms have been developed
from earlier submersible barges which were used
in shallow inlet drilling along the United States
Gulf Coast. The platforms are towed to location
and then submerged to the sea bottom. They are
very stable and can operate in areas with soft sea
floors. Difficulty in towing is a disadvantage, but
this is partially offset by the rapidity with which
they can be raised or lowered, once on location.

Semisubmersibles (Fig. 3) are a version of submersibles. They can work as bottom-supported
units or in deep water as floaters. Their key virtue
is the wide range of water depths in which they can
operate, plus the fact that, when working as
floaters, their primary buoyancy lies below the action of the waves, thus providing great stability.
The "semis" are the most recent of the rig-type
platforms.

Floating drill ships (Fig. 4) are capable of drilling
in depths from 60 to 1000 ft or more. They are built
as self-propelled ships or with a ship configuration
that requires towing. Several twin-hulled versions
have been constructed to give a stable catamaran
design. Floating drill ships use anchoring or ingenious dynamic positioning systems to stabilize their
position, the latter being necessary in deeper waters. Floaters cannot be used in waters much shal-

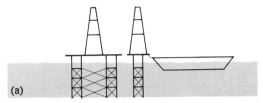

Fig. 1. Offshore fixed drilling platform. (a) Underwater
design (*World Petroleum*). (b) Rig on a drilling site
(*Marathon Oil Co., Findlay, Ohio*).

lower than 70 ft because of the special equipment
required for drilling from the vessel subject to vertical movement from waves and tidal changes, as
well as minor horizontal shifts due to stretch and
play in anchor lines. Exploration in deeper waters
necessitates building more semisubmersible and
floating drill ships. Only 18 rigs in the Gulf of Mexi-

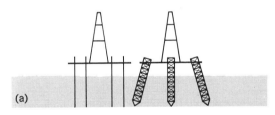

Fig. 2.   Offshore self-elevating drilling platform.
(a) Underwater design (*World Petroleum*). (b) Self-elevating drilling platform (*Marathon Oil Co., Findlay, Ohio*).

Fig. 3. Offshore semisubmersible drilling platform. (a) Underwater design (*World Petroleum*). (b) Santa Fe Marine's Blue Water no. 3 drilling rig on a drilling site (*Marathon Oil Co., Findlay, Ohio*).

co are capable of drilling in more than 200 ft of water.

Production and well completion technology. The move of exploration into the open hostile sea has required not only the development of drilling vessels but a host of auxiliary equipment and techniques. A whole new industrial complex has developed to serve the offshore industry.

Of particular interest is the development of diving techniques and submersible equipment to aid

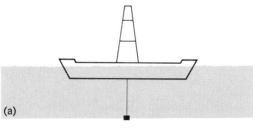

Fig. 4. Floating drill ship. Such ships can drill in depths from 60 to 1000 ft or more. (a) Underwater design (*World Petroleum*). (b) Floating drill ship on a drilling site (*Marathon Oil Co., Findlay, Ohio*).

in exploration and the completion of wells. Present (1969) completion platform technology is limited to water depths of slightly more than 400 ft. Economics will soon force sea-bottom completions which require men or robots or both to make the necessary pipe and well connections. Such work is necessary even in the water depths now being developed.

A robot device has been developed which operates from the surface and uses sonar and television for viewing; it can excavate ditches for pipelines and make simple pipe connections and well hookups in water depths to 2500 ft. Limitations of robot devices are such that the more complex needs of well completions and service require the actions of men. To fill this need, diving specialists have been used to sandbag platform bases, recover conductor pipe, survey and remove wreckage, and make pipeline connections and well hookups. Diving depths have been increased to the point where useful work has been performed at depths in excess of 600 ft. Pressure chambers to take divers to the bottom and to return them to the surface to be decompressed are operational. One has been operating routinely in 425 ft of water for Esso Exploration, Norway. This deep diving is made possible by the development of saturation diving, which uses a mixture of oxygen, helium, nitrogen, and argon. This technique has allowed divers to remain below 200 ft for 6 days while doing salvage work on a platform. It has also allowed prolonged submergence at 600 ft in preparation for actual work on wells at this depth.

Miniature submarines have taken their place in exploration and completion work allowing the viewing of conditions, the gathering of samples, and simple mechanical tasks. Their depth range is for all practical purposes unlimited.

Technical groups are experimenting with the design of drilling and production units that would be totally enclosed and be set on the sea bottom. Living in and working from these units, personnel would be able to carry out all the necessary oil field operations. In effect, such units would resemble a miniature city on the sea floor, from which a man would need to return to the surface only when his tour of work was completed. *See* OIL AND GAS FIELD EXPLOITATION.

Concomitant with the progress of the petroleum industry in its venture into the open sea has been a vast increase in the knowledge of the sea and its contained wealth. Mining of the sea floor using some of petroleum's technology has started in several areas of the world, and actual farming or ranching of the life in the sea is being planned. *See* MINING, UNDERSEA; OCEANOGRAPHY.

Hazards at sea. As the petroleum industry has pushed farther into the hostile environment of the sea, it has sustained a series of disasters, reflected by the doubling of offshore insurance rates in April, 1966. Between 1955 and 1968, 23 offshore units were destroyed by blowout and 6 by hurricane and breakup and collapse at sea. The United States Gulf Coast, where a large percentage of the world's offshore drilling has taken place, was severely hit in 1964 and 1965 by hurricanes, which claimed over $7,500,000 in tow and service vessels, $21,000,000 in fixed platforms, and $28,000,000 in mobile platforms.

These figures do not include expenses sustained from loss of wells, removal of wrecked equipment, and loss of production. Such liabilities have raised insurance rates on a $5,000,000 platform to as much as $500,000 or more per year, depending on platform type and location. Much current design work is aimed at engineering better safety features for the benefit of both the crews and structures.

Despite the hazards and monumental cost involved in extracting oil and gas from beneath the sea, the world's population explosion and its ever increasing demand for petroleum energy will force the search for new reserves into even deeper waters and more remote corners of the world. In truth, the search is only just beginning.

[G. R. SCHOONMAKER]

Bibliography: J. L. Fisher and N. Potter, *World Prospects for Natural Resources*, 1964; F. Gardner, Offshore report, *Oil Gas J.*, vol. 65, no. 28, 1967; Ocean engineering, *Petrol. Eng.*, vol. 39, no. 11, 1967; Offshore environment complicates drilling, *Petrol. Eng.*, vol. 40, no. 2, 1968; Rig fleet grows, *Oil Gas J.*, vol. 65, no. 28, 1967; L. C. Rogers, Offshore drillers gear for deep water, *Oil Gas J.*, vol. 65, no. 28, 1967; Unmanned subsea device, *World Oil*, vol. 165, no. 7, 1968; Worldwide offshore activity roundup, *World Oil*, vol. 165, no. 5, 1965.

Oil and gas field exploitation

In the petroleum industry, a field is an area underlain without substantial interruption by one or more reservoirs of commercially valuable oil or gas, or both. A single reservoir (or group of reservoirs which cannot be separately produced) is a pool. Several pools separated from one another by barren, impermeable rock may be superimposed one above another within the same field. Pools have variable areal extent. Any sufficiently deep well located within the field should produce from one or more pools. However, each well cannot produce from every pool, because different pools have different areal limits.

DEVELOPMENT

Development of a field includes the location, drilling, completion, and equipment of wells necessary to produce the commercially recoverable oil and gas in the field.

Related oil field conditions. Petroleum is a generic term which, in its broadest meaning, includes all naturally occurring hydrocarbons, whether gaseous, liquid, or solid. By variation of the temperature or pressure, or both, of any hydrocarbon, it becomes gaseous, liquid, or solid. Temperatures in producing horizons vary from approximately 60° to more than 300°F, depending chiefly upon the depth of the horizon. A rough approximation is that temperature in the reservoir sand, or pay, equals 60°F, plus 0.017°F/ft of depth below surface. Pressure on the hydrocarbons varies from atmospheric to more than 11,000 psi. Normal pressure is considered as 0.465 psi/ft of depth. Temperatures and pressure vary widely from these average figures. Hydrocarbons, because of wide variations in pressure and temperature and because of mutual solubility in one another, do not necessarily exist underground in the same phases in which they appear at the surface.

Petroleum occurs underground in porous rocks of wide variety. The pore spaces range from microscopic size to rare holes 1 in. or more in diameter. The containing rock is commonly called the sand or the pay, regardless of whether the pay is actually sandstone, limestone, dolomite, unconsolidated sand, or fracture openings in relatively impermeable rock.

Development of field. After discovery of a field containing oil or gas, or both, in commercial quantities, the field must be explored to determine its vertical and horizontal limits and the mechanisms under which the field will produce. Development and exploitation of the field proceed simultaneously. Usually the original development program is repeatedly modified by geologic knowledge acquired during the early stages of development and exploitation of the field.

Ideally, tests should be drilled to the lowest possible producing horizon in order to determine the number of pools existing in the field. Testing and geologic analysis of the first wells sometimes indicates the producing mechanisms, and thus the best development program. Very early in the history of the field, step-out wells will be drilled to determine the areal extent of the pool or pools. Step-out wells give further information regarding the volumes of oil and gas available, the producing mechanisms, and the desirable spacing of wells.

The operator of an oil and gas field endeavors to select a development program which will produce the largest volume of oil and gas at a profit. The program adopted is always a compromise between conflicting objectives. The operator desires (1) to drill the fewest wells which will efficiently produce the recoverable oil and gas; (2) to drill, complete, and equip the wells at the lowest possible cost; (3) to complete production in the shortest practical time to reduce both capital and operating charges; (4) to operate the wells at the lowest possible cost; and (5) to recover the largest possible volume of oil and gas.

Selecting the number of wells. Oil pools are produced by four mechanisms: dissolved gas expansion, gas-cap drive, water drive, and gravity drainage. Commonly, two or more mechanisms operate in a single pool. The type of producing mechanism in each pool influences the decision as to the number of wells to be drilled. Theoretically, a single, perfectly located well in a water-drive pool is capable of producing all of the commercially recoverable oil and gas from that pool. Practically, more than one well is necessary if a pool of more than 80 acres is to be depleted in a reasonable time. If a pool produces under either gas expansion or gas-cap drive, oil production from the pool will be independent of the number of wells up to a spacing of at least 80 acres per well (1866 ft between wells). Gas wells often are spaced a mile or more apart. The operator accordingly selects the widest spacing permitted by field conditions and legal requirements, as discussed later.

Major components of cost. Costs of drilling, completing, and equipping the wells influence development plans. Having determined the number and depths of producing horizons and the producing mechanisms in each horizon, the operator must decide whether he will drill a well at each location to each horizon or whether a single well can produce from two or more horizons at the same location. Clearly, the cost of drilling the field can be sharply reduced if a well can drain two, three, or more horizons. The cost of drilling a well will be higher if several horizons are simultaneously produced, because the dual or triple completion of a well usually requires larger casing. Further, completion and operating costs are higher. However, the increased cost of drilling a well of larger diameter and completing the well in two or more horizons is 20–40% less than the cost of drilling and completing two wells to produce separately from two horizons.

In some cases, the operator may reduce the number of wells by drilling a well to the lowest producible horizon and taking production from that level until the horizon there is commercially exhausted. The well is then plugged back to produce from a higher horizon. Selection of the plan for producing the various horizons obviously affects the cost of drilling and completing individual wells, as well as the number of wells which the operator will drill. If two wells are drilled at approximately the same location, they are referred to as twins, three wells at the same location are triplets, and so on.

Costs and duration of production. The operator wishes to produce as rapidly as possible because the net income from sale of hydrocarbons is obviously reduced as the life of the well is extended. The successful operator must recover from his productive wells the costs of drilling and operating those wells, and in addition he must recover all costs involved in geological and geophysical exploration, leasing, scouting, and drilling of dry holes, and occasionally other operations. If profits from production are not sufficient to recover all exploration and production costs and yield a profit in excess of the rate of interest which the operator could secure from a different type of investment, he is discouraged from further exploration.

Most wells cannot operate at full capacity because unlimited production results in physical waste and sharp reduction in ultimate recovery. In many areas, conservation restrictions are enforced to make certain that the operator does not produce in excess of the maximum efficient rate. For example, if an oil well produces at its highest possible rate, a zone promptly develops around the well where production is occurring under gas-expansion drive, the most inefficient producing mechanism. Slower production may permit the petroleum to be produced under gas-cap drive or water drive, in which case ultimate production of oil will be two to four times as great as it would be under gas-expansion drive. Accordingly, the most rapid rate of production generally is not the most efficient rate.

Similarly, the initial exploration of the field may indicate that one or more gas-condensate pools exist, and recycling of gas may be necessary to secure maximim recovery of both condensate and of gas. The decision to recycle will affect the number of wells, the locations of the wells, and the completion methods adopted in the development program.

Further, as soon as the operator determines that secondary oil-recovery methods are desired and expects to inject water, gas, steam, or, rarely, air to provide additional energy to flush or displace oil from the pay, the number and location of wells may be modified to permit the most effective secondary recovery procedures.

Legal and practical restrictions. The preceding discussion has assumed control of an entire field under single ownership by a single operator. In the United States, a single operator rarely controls a large field, and this field is almost never under a single lease. Usually, the field is covered by separate leases owned and operated by different producers. The development program must then be modified in consideration of the lease boundaries and the practices of the other operators who are in the field.

Oil and gas know no lease boundaries. They move freely underground from areas of high pressure toward lower-pressure situations. The operator of a lease is obligated to locate his wells in such a way as to prevent drainage of his lease by wells on adjoining leases, even though he may own the adjoining leases. In the absence of conservation restrictions, an operator must produce petroleum from his wells as rapidly as it is produced from wells on adjoining leases. Slow production on one lease results in migration of oil and gas to nearby leases which are more rapidly produced.

The operator's development program must provide for offset wells located as close to the boundary of his lease as are wells on adjoining leases. Further, the operator must equip his wells to produce as rapidly as the offset produces and must produce from the same horizons which are being produced in offset wells. The lessor who sold the lease to the operator is entitled to his share of the recoverable petroleum underlying his land. Negligence by the operator in permitting drainage of a lease makes the operator liable to suit for damages or cancellation of the lease.

A development program acceptable to all operators in the field permits simultaneous development of leases, prevents drainage, and results in maximum ultimate production from the field. Difficulties may arise in agreement upon the best development program for a field. Most states have enacted statutes and have appointed regulatory bodies under which judicial determination can be made of the permissible spacing of the wells, the rates of production, and the application of secondary recovery methods.

Drilling unit. Commonly, small leases or portions of two or more leases are combined to form a drilling unit in whose center a well will be drilled. Unitization may be voluntary, by agreement between the operator or operators and the interested royalty owners, with provision for sharing production from the well between the parties in proportion to their acreage interests. In many states the regulatory body has authority to require unitization of drilling units, which eliminates unnecessary offset wells and protects the interests of a landowner whose acreage holding may be too small to justify the drilling of a single well on his property alone.

Pool unitization. When recycling or some types of secondary recovery are planned, further unitization is adopted. Since oil and gas move freely across lease boundaries, it would be wasteful for an operator to repressure, recycle, or water-drive a lease if the adjoining leases were not similarly operated. Usually an entire pool must be unitized for efficient recycling of secondary recovery operations. Pool unitization may be accomplished by agreement between operators and royalty owners. In many cases, difference of opinion or ignorance on the part of some parties prevents voluntary pool unitization. Many states authorize the regulatory body to unitize a pool compulsorily on application by a specified percentage of interests of operators and royalty owners. Such compulsory unitization is planned to provide each operator and each royalty owner his fair share of the petroleum products produced from the field regardless of the location of the well or wells through which these products actually reach the surface.

EXPLOITATION—GENERAL CONSIDERATIONS

Oil and gas production necessarily are intimately related, since approximately one-third of the gross gas production in the United States is produced from wells that are classified as oil wells. However, the naturally occurring hydrocarbons of petroleum are not only liquid and gaseous but may even be found in a solid state, such as asphaltite and some asphalts.

Where gas is produced without oil, the production problems are simplified because the product flows naturally throughout the life of the well and does not have to be lifted to the surface. However, there are sometimes problems of water accumulations in gas wells, and it is necessary to pump the water from the wells to maintain maximum, or economical, gas production. The line of demarcation between oil wells and gas wells is not definitely established since oil wells may have gas-oil ratios ranging from a few cubic feet per barrel to many thousand cubic feet of gas per barrel of oil. Most gas wells produce quantities of condensable vapors, such as propane and butane, that may be liquefied and marketed for fuel, and the more stable liquids produced with gas can be utilized as natural gasoline.

Factors of method selection. The method selected for recovering oil from a producing formation depends on many factors, including well depth, well-casing size, oil viscosity, density, water production, gas-oil ratio, porosity and permeability of the producing formation, formation pressure, water content of producing formation, and whether the force driving the oil into the well from the formation is primarily gas pressure, water pressure, or a combination of the two. Other factors, such as paraffin content and difficulty expected from paraffin deposits, sand production, and corrosivity of the well fluids, also have a decided influence on the most economical method of production.

Special techniques utilized to increase productivity of oil and gas wells include acidizing, hydraulic fracturing of the formation, the setting of screens, and gravel packing or sand packing to increase permeability around the well bore.

Aspects of production rate. Productive rates per well may vary from a few barrels per day to several thousand barrels per day, and it may be necessary to produce a large percentage of water along with the oil.

Field and reservoir conditions. In some cases reservoir conditions are such that some of the wells flow naturally throughout the entire economical life of the oil field. However, in the great majority of cases it is necessary to resort to artificial lifting methods at some time during the life of the field, and often it is necessary to apply artificial lifting means immediately after the well is drilled.

Market and regulatory factors. In some oil-producing states of the United States there are state bodies authorized to regulate oil production from the various oil fields. The allowable production per well is based on various factors, including the market for the particular type of oil available, but very often the allowable production is based on an engineering study of the reservoir to determine the optimum rate of production.

Crude oil production in the United States in 1967 as reported by the *Oil and Gas Journal* averaged 8,807,000 barrels per day (bpd). This represents an increase of 6.2% over 1966 production. Imports amounted to 1,111,000 bpd in 1967. World production in 1967 was estimated at 35,128,100 bpd. Total net natural gas production during 1967 was estimated at 18.25×10^{12} ft^3.

Useful terminology. A few definitions of terms used in petroleum production technology are listed below to assist in an understanding of some of the problems involved.

Porosity. The percentage porosity is defined as the percentage volume of voids per unit total volume. This, of course, represents the total possible volume available for accumulation of fluids in a formation, but only a fraction of this volume may be effective for practical purposes because of possible discontinuities between the individual pores. The smallest pores generally contain water held by capillary forces.

Permeability. Permeability is a measure of the resistance to flow through a porous medium under the influence of a pressure gradient. The unit of permeability commonly employed in petroleum production technology is the darcy. A porous structure has a permeability of 1 darcy if, for a fluid of 1 centipoise (cp) viscosity, the volume flow is 1 cm^3/(sec)(cm^2) under a pressure gradient of 1 atm/cm.

Productivity index. The productivity index is a measure of the capacity of the reservoir to deliver oil to the well bore through the productive formation and any other obstacles that may exist around the well bore. In petroleum production technology, the productivity index is defined as production in barrels per day per pound drop in bottom-hole pressure. For example, if a well is closed in at the casinghead, the bottom-hole pressure will equal the formation pressure when equilibrium conditions are established. However, if fluid is removed from the well, either by flowing or pumping, the bottom-hole pressure will drop as a result of the resistance to flow of fluid into the well from the formation to replace the fluid removed from the well. If the closed-in bottom-hole pressure should be 1000 psi, for example, and if this pressure should drop to 900 psi when producing at a rate of 100 bbl/day (a drop of 100 psi), the well in question would have a productivity index of one.

Barrel. The standard barrel used in the petroleum industry is 42 U.S. gal.

API gravity. The American Petroleum Institute (API) scale that is in common use for indicating

Degrees API corresponding to specific gravities of crude oil at 60°/60°F

Specific gravity, in tenths	Specific gravity, in hundredths									
	.00	.01	.02	.03	.04	.05	.06	.07	.08	.09
0.60	104.33	100.47	96.73	93.10	89.59	86.19	82.89	79.69	76.59	73.57
0.70	70.64	67.80	65.03	62.34	59.72	57.17	54.68	52.27	49.91	47.61
0.80	45.38	43.19	44.06	38.98	36.95	34.97	33.03	31.14	29.30	27.49
0.90	25.72	23.99	22.30	20.65	19.03	17.45	15.90	14.38	12.89	11.43
1.00	10.00									

specific gravity, or a rough indication of quality of crude petroleum oils, differs slightly from the Baume scale commonly used for other liquids lighter than water. The table shows the relationship between degrees API and specific gravity referred to water at 60°F for specific gravities ranging from 0.60 to 1.0.

Viscosity range. Viscosity of crude oils currently produced varies from approximately 1 cp to values above 1000 cp at temperatures existing at the bottom of the well. In some areas it is necessary to supply heat artificially down the wells or circulate lighter oils to mix with the produced fluid for maintenance of a relatively low viscosity throughout the temperature range to which the product is subjected.

In addition to wells that are classified as gas wells or oil wells, the term gas-condensate well has come into general use to designate a well that produces large volumes of gas with appreciable quantities of light, volatile hydrocarbon fluids. Some of these fluids are liquid at atmospheric pressure and temperature; others, such as propane and butane, are readily condensed under relatively low pressures in gas separators for use as liquid petroleum gas (LPG) fuels or for other uses. The liquid components of the production from gas-condensate wells generally arrive at the surface in the form of small droplets entrained in the high-velocity gas stream and are separated from the gas in a high-pressure gas separator.

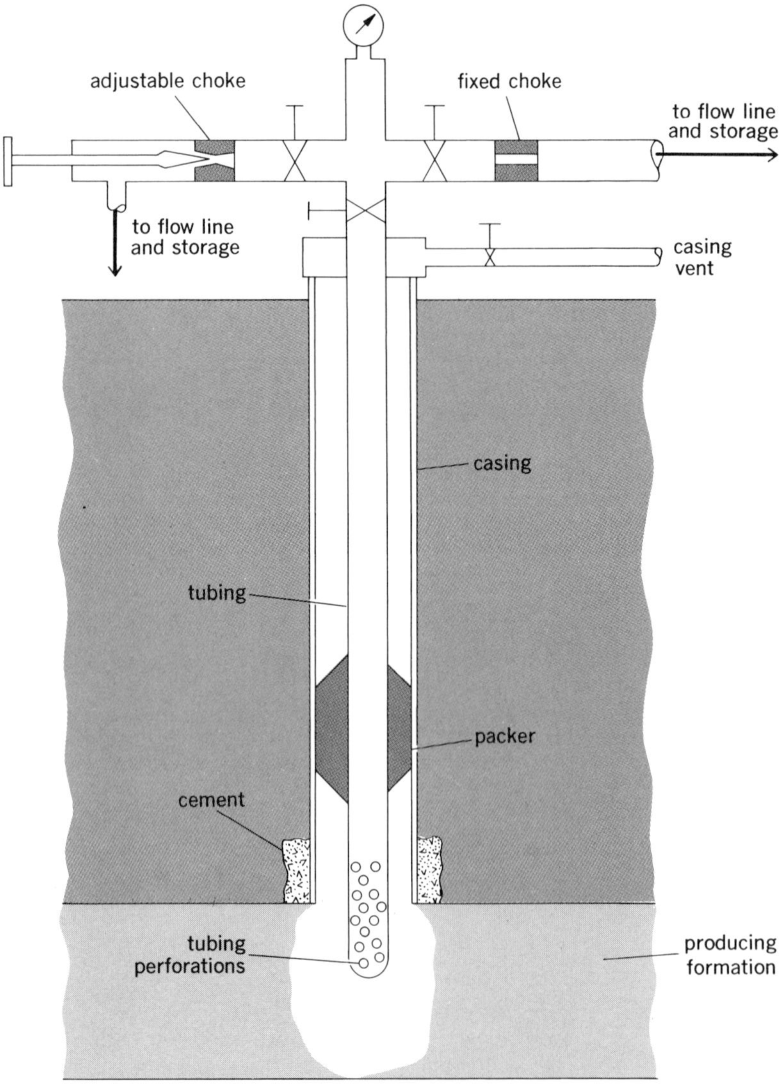

PRODUCTION METHODS IN PRODUCING WELLS

The common methods of producing oil wells are (1) natural flow; (2) pumping with sucker rods to actuate a pump located in the well fluid; (3) gas lift; (4) hydraulic subsurface pumps, in which the subsurface unit consists of a reciprocating hydraulic motor connected to a reciprocating pump and driven by circulating oil or other power fluid down the well to actuate the hydraulic engine; (5) electrically driven centrifugal well pumps; and (6) swabbing.

Numerous other methods, including jet pumps and sonic pumps, have been tried and are used to slight extent. The sonic pump is a development in which the tubing is vibrated longitudinally by a mechanism at the surface and acts as a high-speed pump with an extremely short stroke.

The total number of producing oil wells in the United States on July 1, 1967, was reported by the *Oil and Gas Journal* as 698,958, while the total number of producing wells in the entire free world was 762,737.

A total of 33,558 wells were drilled in the United States during 1967. Of this number 15,320 were productive oil wells, 3619 were classified as gas wells, 13,223 were nonproductive (dry holes), and 1396 were service wells. Service wells are utilized for various purposes, such as water injection for water flooding operations, salt-water disposal, and gas recycling.

A brief discussion of production methods, in the approximate order of their relative importance and popularity, follows.

Natural flow. Natural flow is the most economical method of production and generally is utilized as long as the desired production rate can be maintained by this method. It utilizes the formation

Fig. 1. Schematic view of well equipped for producing by natural flow.

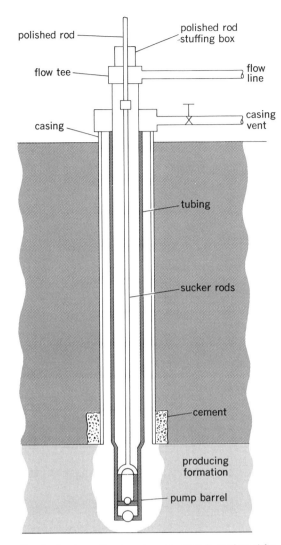

Fig. 2. A schematic view of a well which is equipped for pumping with sucker rods.

polished rod
polished rod stuffing box
flow tee
flow line
casing
casing vent
tubing
sucker rods
cement
producing formation
pump barrel

not essential but is often used to reduce the free gas volume in the casing.

Flow rates for United States wells seldom exceed a few hundred barrels per day because of enforced or voluntary restrictions to regulate production rates and to obtain most efficient and economical ultimate recovery. However, in some countries, especially in the Middle East, it is not uncommon for natural flow rates to exceed 10,000 bpd/well.

Lifting. Most wells are not self-flowing. The common types of lifting are outlined here.

Pumping with sucker rods. Approximately 90% of the wells made to produce by some artificial lift method in the United States are equipped with sucker-rod–type pumps. In these the pump is installed at the lower end of the tubing string and is actuated by a string of sucker rods extending from the surface to the subsurface pump. The sucker rods are attached to a polished rod at the surface. The polished rod extends through a stuffing box and is attached to the pumping unit, which produces the necessary reciprocating motion to actuate the sucker rods and the subsurface pump. Figure 2 shows a simplified schematic section through a pumping well. The two common variations are mechanical and hydraulic long-stroke pumping.

1. Mechanical pumping. The great majority of pumping units are of the mechanical type, consisting of a suitable reduction gear, and crank and pitman arrangement to drive a walking beam to produce the necessary reciprocating motion. A counterbalance is provided to equalize the load on the upstroke and downstroke. Mechanical pumping units of this type vary in load-carrying capacity from about 2000 to about 43,000 lb, and the torque rating of the low-speed gear which drives the crank ranges from 6400 in.-lb in the smallest API standard unit to about 1,500,000 in.-lb for the largest units now in use. Stroke length varies from about 18 to 192 in. Usual operating speeds are from about 6 to 20 strokes/min. However, both lower and higher rates of speed are sometimes used. Figure 3 shows a modern pumping unit in operation.

Production rates with sucker-rod–type pumps

energy, which may consist of gas in solution in the oil in the formation; free gas under pressure acting against the liquid and gas-liquid phase to force it toward the well bore; water pressure acting against the oil; or a combination of these three energy sources. In some areas the casinghead pressure may be of the order of 10,000 psi, so it is necessary to provide fittings adequate to withstand such pressures. Adjustable throttle values, or chokes, are utilized to regulate the flow rate to a desired and safe value. With such a high-pressure drop across a throttle valve the life of the valve is likely to be very short. Several such valves are arranged in parallel in the tubing head "Christmas tree" with positive shutoff valves between the chokes and the tubing head so that the wearing parts of the throttle valve, or the entire valve, can be replaced while flow continues through another similar valve.

An additional safeguard that is often used in connection with high-pressure flowing wells is a bottom-hole choke or a bottom-hole flow control valve that limits the rate of flow to a reasonable value, or stops it completely, in case of failure of surface controls. Figure 1 shows a schematic outline of a simple flowing well hookup. The packer is

Fig. 3. Pumping unit with adjustable rotary counterbalance. (*Oil Well Supply Division, U.S. Steel Corp.*)

vary from a fraction of 1 bpd in some areas, with part-time pumping, to approximately 3000 bpd for the largest installations in relatively shallow wells.

2. Hydraulic long-stroke pumping. For this the units consist of a hydraulic lifting cylinder mounted directly over the well head and are designed to produce stroke lengths of as much as 30 ft. Such long-stroke hydraulic units are usually equipped with a pneumatic counterbalance arrangement which equalizes the power requirement on the upstroke and downstroke.

Hydraulic pumping units also are made without any provision for counterbalance. However, these units are generally limited to relatively small wells, and they are relatively inefficient.

Gas lift. Gas lift in its simplest form consists of initiating or stimulating well flow by injecting gas at some point below the fluid level in the well. With large-volume gas-lift operations the well may be produced through either the casing or the tubing. In the former case, gas is conducted through the tubing to the point of injection; in the latter, gas may be conducted to the point of injection through the casing or through an auxiliary string of tubing. When gas is injected into the oil column, the weight of the column above the point of injection is reduced as a result of the space occupied by the relatively low density of the gas. This lightening of the fluid column is sufficient to permit the formation pressure to initiate flow up the tubing to the surface. Gas injection is often utilized to increase the flow from wells that will flow naturally but will not produce the desired amount by natural flow.

There are many factors determining the advisability of adopting gas lift as a means of production. One of the more important factors is the availability of an adequate supply of gas at suitable pressure and reasonable cost. In a majority of cases gas lift cannot be used economically to produce a reservoir to depletion because the well may be relatively productive with a low back pressure maintained on the formation but will produce very little, if anything, with the back pressure required for gas-lift operation. Therefore, it generally is necessary to resort to some mechanical means of pumping before the well is abandoned, and it may be more economical to adopt the mechanical means initially than to install the gas-lift system while conditions are favorable and later replace it.

This discussion of gas lift has dealt primarily with the simple injection of gas, which may be continuous or intermittent. There are numerous modifications of gas-lift installations, including various designs for flow valves which may be installed in the tubing string to open and admit gas to the tubing from the casing at a predetermined pressure differential between the tubing and casing. When the valve opens, gas is injected into the tubing to initiate and maintain flow until the tubing pressure drops to a predetermined value; and the valve closes before the input gas-oil ratio becomes excessive. This represents an intermittent-flow—type valve. Other types are designed to maintain continuous flow, proper pressure differential, and proper gas injection rate for efficient operation. In some cases several such flow valves are spaced up the tubing string to permit flow to be initiated from various levels as required.

Other modifications of gas lift involve the utiliza-tion of displacement chambers. These are installed on the lower end of the well tubing where oil may accumulate, and the oil is displaced up the tubing with gas injection controlled by automatic or mechanical valves.

Hydraulic subsurface pumps. The hydraulic subsurface pump has come into fairly prominent use. The subsurface pump is operated by means of a hydraulic reciprocating motor attached to the pump and installed in the well as a single unit. The hydraulic motor is driven by a supply of hydraulic fluid under pressure that is circulated down a string of tubing and through the motor. Generally the hydraulic fluid consists of crude oil which is discharged into the return line and returns to the surface along with the produced crude oil.

Hydraulically operated subsurface pumps are also arranged for separating the hydraulic power fluid from the produced well fluid. This arrangement is especially desirable where the fluid being produced is corrosive or is contaminated with considerable quantities of sand or other solids that are difficult to separate to condition the fluid for use as satisfactory power oil. This method also permits the use of water or other nonflammable liquids as hydraulic power fluid to minimize the fire hazard in case of a failure of the hydraulic power line at the surface.

Centrifugal well pumps. Electrically driven centrifugal pumps have been used to some extent, especially in large-volume wells of shallow or moderate depths. Both the pump and the motor are restricted in diameter to run down the well casing, leaving sufficient clearance for the flow of fluid around the pump housing. With the restricted diameter of the impellers the discharge head necessary for pumping a relatively deep well can be obtained only by using a large number of stages and operating at a relatively high speed. The usual rotating speed for such units is 3600 rpm, and it is not uncommon for such units to have 50 or more pump stages. The direct-connected electric motor must be provided with a suitable seal to prevent well fluid from entering the motor housing, and electrical leads must be run down the well casing to supply power to the motor.

Swabs. Swabs have been used for lifting oil almost since the beginning of the petroleum industry. They usually consist of a steel tubular body equipped with a check valve which permits oil to flow through the tube as it is lowered down the well with a wire line. The exterior of the steel body is generally fitted with flexible cup-type soft packing that will fall freely but will expand and form a seal with the tubing when pulled upward with a head of fluid above the swab. Swabs are run into the well on a wire line to a point considerably below the fluid level and then lifted back to the surface to deliver the volume of oil above the swab. They are often used for determining the productivity of a well that will not flow naturally and for assisting in cleaning paraffin from well tubing. In some cases swabs are used to stimulate wells to flow by lifting, from the upper portion of the tubing, the relatively dead oil from which most of the gas has separated.

Bailers. Bailers are used to remove fluids from wells and for cleaning out solid material. They are run into the wells on wire lines as in swabbing, but differ from swabs in that they generally are run only in the casing when there is no tubing in the well. The capacity of the bailer itself represents the volume of fluid lifted each time since the bailer does not form a seal with the casing. The bailer is simply a tubular vessel with a check valve in the bottom. This check valve generally is arranged so that it is forced open when the bailer touches bottom in order to assist in picking up solid material for cleaning out a well.

Jet pumps. A jet pump for use in oil wells operates on exactly the same principle as a water-well jet pump. Advantage is taken of the Bernoulli effect to reduce pressure by means of a high-velocity fluid jet. Thus oil is entrained from the well with this high-velocity jet in a venturi tube to accelerate the fluid and assist in lifting it to the surface, along with any assistance from the formation pressure. The application of jet pumps to oil wells has been insignificant.

Sonic pumps. Sonic pumps essentially consist of a string of tubing equipped with a check valve at each joint and mechanical means on the surface to vibrate the tubing string longitudinally. This creates a harmonic condition that will result in several hundred strokes per minute, with the strokes being a small fraction of 1 in. in length. Some of these pumps are in use in relatively shallow wells, but it appears that their field of application is very limited.

Lease tanks and gas separators. Figure 4 shows a typical lease tank battery consisting of four 1000-bbl tanks and two gas separators. Such equipment is used for handling production from wells produced by natural flow, gas lift, or pumping. In some pumping wells the gas content may be too low to justify the cost of separators for saving the gas.

Natural gasoline production. An important phase of oil and gas production in many areas is the production of natural gasoline from gas taken from the casinghead of oil wells or separated from the oil and conducted to the natural gasoline plant. The plant consists of facilities for compressing and extracting the liquid components from the gas. The natural gasoline generally is collected by cooling and condensing the vapors after compression or by absorbing in organic liquids having high boiling points from which the volatile liquids are distilled. Many natural gasoline plants utilize a combination of condensing and absorbing techniques. Figure 5 shows an overall view of a natural gasoline plant operating in western Texas.

PRODUCTION PROBLEMS AND INSTRUMENTS

To maintain production, various problems must be overcome. Numerous instruments have been developed to monitor production and to control production problems.

Corrosion. In many areas the corrosion of production equipment is a major factor in the cost of petroleum production. The following comments on the oil field corrosion problem are taken largely from *Corrosion of Oil- and Gas-Well Equipment* and reproduced by permission of NACE-API.

For practical consideration, corrosion in oil and gas-well production can be classified into four main types.

1. Sweet corrosion occurs as a result of the presence of carbon dioxide and fatty acids. Oxy-

Fig. 4. Lease tank battery with four tanks and two gas separators. *(Gulf Oil Corp.)*

gen and hydrogen sulfide are not present. This type of corrosion occurs in both gas-condensate and oil wells. It is most frequently encountered in the United States in southern Louisiana and Texas, and other scattered areas. At least 20% of all sweet oil production and 45% of condensate production are considered corrosive.

2. Sour corrosion is designated as corrosion in oil and gas wells producing even trace quantities of hydrogen sulfide. These wells may also contain oxygen, carbon dioxide, or organic acids. Sour corrosion occurs in the United States primarily throughout Arbuckle production in Kansas and in the Permian basin of western Texas and New Mexico. About 12% of all sour production is considered corrosive.

3. Oxygen corrosion occurs wherever equipment is exposed to atmospheric oxygen. It occurs most frequently in offshore installations, brine-handling and injection systems, and in shallow producing wells where air is allowed to enter the casing.

4. Electrochemical corrosion is designated as that which occurs when corrosion currents can be readily measured or when corrosion can be mitigated by the application of current, as in soil corrosion.

Corrosion inhibitors are used extensively in both

Fig. 5. Modern natural gasoline plant in western Texas. *(Gulf Oil Corp.)*

oil and gas wells to reduce corrosion damage to subsurface equipment. Most of the inhibitors used in the oil field are of the so-called polar organic type. All of the major inhibitor suppliers can furnish effective inhibitors for the prevention of sweet corrosion as encountered in most fields. These can be purchased in oil-soluble, water-dispersible, or water-soluble form.

Paraffin deposits. In many crude-oil–producing areas paraffin deposits in tubing and flow lines and on sucker rods are a source of considerable trouble and expense. Such deposits build up until the tubing or flow line is partially or completely plugged. It is necessary to remove these deposits to maintain production rates. A variety of methods are used to remove paraffin from the tubing, including the application of heated oil through tubular sucker rods to mix with and transfer heat to the oil being produced and raise the temperature to a point at which the deposited paraffin will be dissolved or melted. Paraffin solvents may also be applied in this manner without the necessity of applying heat.

Mechanical means often are used in which a scraping tool is run on a wire line and paraffin is scraped from the tubing wall as the tool is pulled back to the surface. Mechanical scrapers that attach to sucker rods also are in use. Various types of automatic scrapers have been used in connection with flowing wells. These consist of a form of piston that will drop freely to the bottom when flow is stopped but will rise back to the surface when flow is resumed. Electrical heating methods have been used rather extensively in some areas. The tubing is insulated from the casing and from the flow line, and electric current is transmitted through the tubing for the time necessary to heat the tubing sufficiently to cause the paraffin deposits to melt or go into solution in the oil in the tubing. Plastic coatings have been utilized inside tubing and flow lines to minimize or prevent paraffin deposits. Paraffin does not deposit readily on certain plastic coatings.

A common method for removing paraffin from flow lines is to disconnect the line at the well head and at the tank battery and force live steam through the line to melt the paraffin deposits and flow them out. Various designs of flow-line scrapers have also been used rather extensively and fairly successfully. Paraffin deposits in flow lines are minimized by insulating the lines or by burying

Fig. 6.　Two pumping wells with tank battery. (*Oil Well Supply Division, U.S. Steel Corp.*)

the lines to maintain a higher average temperature.

Emulsions. A large percentage of oil wells produce various quantities of salt water along with the oil, and numerous wells are being pumped in which the salt-water production is 90% or more of the total fluid lifted. Turbulence resulting from production methods results in the formation of emulsions of water in oil or oil in water; the commoner type is oil in water. Emulsions are treated with a variety of demulsifying chemicals, with the application of heat, and with a combination of these two treatments. Another method for breaking emulsions is the electrostatic or electrical precipitator type of emulsion treatment. In this method the emulsion to be broken is circulated between electrodes subjected to a high potential difference. The resulting concentrated electric field tends to rupture the oil-water interface and thus breaks the emulsion and permits the water to settle out. Figure 6 shows two pumping wells with a tank battery in the background. This tank battery is equipped with a wash tank, or gun barrel, and a gas-fired heater for emulsion treating and water separation before the oil is admitted to the lease tanks.

Gas conservation. If the quantity of gas produced with crude oil is appreciably greater than that which can be efficiently utilized or marketed, it is necessary to provide facilities for returning the excess gas to the producing formation. Formerly, large quantities of excess gas were disposed of by burning or simply by venting to the atmosphere. This practice is now unlawful. Returning excess gas to the formation not only conserves the gas for future use but also results in greater ultimate recovery of oil from the formation.

Salt-water disposal. The large volumes of salt water produced with the oil in some areas present serious disposal problems. The salt water is generally pumped back to the formation through wells drilled for this purpose. Such salt-water disposal wells are located in areas where the formation already contains water. Thus this practice helps to maintain the formation pressure as well as the productivity of the producing wells.

Offshore production. Offshore wells present additional production problems since the wells must be serviced from barges or boats. Wells of reasonable depth on land locations are seldom equipped with derricks for servicing because it is more economical to set up a portable mast for pulling and installing rods, tubing, and other equipment. However, the use of portable masts is not practical on offshore locations, and a derrick is generally left standing over such wells throughout their productive life to facilitate servicing. There are a considerable number of offshore wells along the Gulf Coast and the Pacific Coast of the United States, but by far the greatest number of offshore wells in a particular region is in Lake Maracaibo in Venezuela. Figure 7 shows a considerable number of derricks in Lake Maracaibo with pumping wells in the foreground. These wells are pumped by electric power through cables laid on the lake bottom to conduct electricity from power-generating stations onshore. An overwater tank battery is visible at the extreme right. All offshore installations, such as tank batteries, pump stations, and the derricks and pumping equipment, are supported on

Fig. 7. Numerous offshore wells located in Lake Maracaibo, Venezuela. (*Creole Petroleum Corp.*)

pilings in water up to 100 ft or more in depth. There are approximately 2300 oil derricks in Lake Maracaibo. A growing number of semipermanent platform rigs and even bottom storage facilities are being used in Gulf of Mexico waters at depths of more than 100 ft.

Instruments. The commoner and more important instruments required in petroleum production operations are included in the following discussion.

1. Gas meters, which are generally of the orifice type, are designed to record the differential pressure across the orifice, and the static pressure.

2. Recording subsurface pressure gages small enough to run down 2-in. ID (inside diameter) tubing are used extensively for measuring pressure gradients down the tubing of flowing wells, recording pressure buildup when the well is closed in, and measuring equilibrium bottom-hole pressures.

3. Subsurface samplers designed to sample well fluids at various levels in the tubing are used to determine physical properties, such as viscosity, gas content, free gas, and dissolved gas at various levels. These instruments may also include a recording thermometer or a maximum reading thermometer, depending upon the information required.

4. Oil meters of various types are utilized to meter crude oil flowing to or from storage.

5. Dynamometers are used to measure polished-rod loads. These instruments are sometimes known as well weighers since they are used to record the polished-rod load throughout a pumping cycle of a sucker-rod–type pump. They are used to determine maximum load on polished rods as well as load variations, to permit accurate counterbalancing of pumping wells, and to assure that pumping units or sucker-rod strings are not seriously overloaded.

6. Liquid-level gages and controllers are used. They are similar to those used in other industries, but with special designs for closed lease tanks.

A wide variety of scientific instruments find application in petroleum production problems. The above outline gives an indication of a few specialized instruments used in this branch of the industry, and there are many more. Special instruments developed by service companies are valued for a wide variety of purposes and include calipers to detect and measure corrosion pits inside tubing and casing and magnetic instruments to detect microscopic cracks in sucker rods.

[ROY L. CHENAULT]

Bibliography: American Petroleum Institute, *History of Petroleum Engineering*, 1961; E. L. DeGolyer (ed.), *Elements of the Petroleum Industry*, 1940; T. C. Frick (ed.), *Petroleum Production Handbook*, vol. 1: *Mathematics and Production Equipment*, 1962; T. C. Frick (ed.), *Petroleum Production Handbook*, vol. 2: *Reservoir Engineering*, 1962; V. B. Guthrie, *Petroleum Products Handbook*, 1960; W. F. Lovejoy and P. T. Homan, *Economic Aspects of Oil Conservation Regulation*, 1967; W. F. Lovejoy and P. T. Homan, *Methods of Estimating Reserves of Crude Oil, Natural Gas, and Natural Gas Liquids*, 1965; B. W. Murphy (ed.), *Conservation of Oil and Gas: A Legal History*, 1949; M. Muskat, *Physical Principles of Oil Production*, 1949; S. J. Pirson, *Oil Reservoir Engineering*, 2d ed., 1958; L. C. Uren, *Petroleum Production Engineering: Oil Field Development*, 4th ed., 1956; L. C. Uren, *Petroleum Production Engineering: Oil Field Exploitation*, 3d ed., 1953.

Oil mining

The surface or subsurface excavation of petroleum-bearing sediments for subsequent removal of the oil by washing, flotation, or retorting treatments. Oil mining also includes recovery of oil by drainage from reservoir beds to mine shafts or other openings driven into the oil rock, or by drainage from the reservoir rock into mine openings driven outside the oil sand but connected with it by bore holes or mine wells.

Surface mining consists of strip or open-pit mining. It has been used primarily for the removal of oil shale or bituminous sands lying at or near the surface. Strip mining of shale is practiced in Sweden, Manchuria, and South Africa. Strip mining of bituminous sand is conducted in Canada.

Subsurface mining is used for the removal of oil sediments, oil shale, and Gilsonite. It is practiced in several European countries and in the United States. Some authorities consider this the best method to recover oil when oil sediments are involved, because virtually all of the oil is recovered.

European experience. Subsurface oil mining was used in the Pechelbronn oil field in Alsace, France, as early as 1735. This early mining involved the sinking of shafts to the reservoir rock, only 100–200 ft below the surface, and the excava-

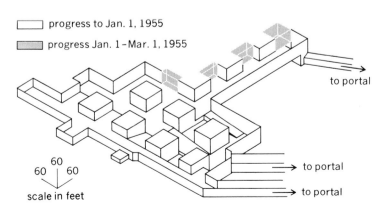

Fig. 1. Isometric drawing of U.S. Bureau of Mines experimental mine for oil shale at Rifle, Colo. Developed by 39-ft top or advanced heading, a 34-ft following bench had just been initiated when a roof fall made it necessary to conclude operations. (*U.S. Bur. Mines Rep. no. 5237, 1956*)

Fig. 2. Mine locomotive and cars removing shale from U.S. Bureau of Mines shale mine at Rifle, Colo. At left, in middle ground, is the Colorado River, nearly 3000 ft below. (*After R. Fleming, U.S. Bureau of Mines*)

Fig. 3. Close-up view of Union Oil Co. of California shale-oil retort near Grand Valley, Colo. Right part of structure is portion of the system for removing oil vapors that would otherwise escape in the gas stream.

Fig. 4. Krupp wheel excavator removing bituminous sand near Mildred Lake, Athabaska region, Alberta, Canada. (*Cities Service Co.*)

tion of the oil sand in short drifts driven from the shafts. These oil sands were hoisted to the surface and washed with boiling water to release the oil. The drifts were extended as far as natural ventilation permitted. When these limits were reached, the pillars were removed and the openings filled with waste.

This type of mining continued at Pechelbronn until 1866, when it was found that oil could be recovered from deeper, more prolific sands by letting it drain in place through mine openings, without removing the sands to the surface for treatment.

Subsurface mining of oil shale also goes back to the mid-19th century in Scotland and France. It is not so widely practiced now because of its high cost as compared with that of usual oil production, particularly in the prolific fields of the Middle East.

United States oil shale mining. The U.S. Bureau of Mines carried out an experimental mining and processing program at Rifle, Colo., between 1944 and 1955 in an effort to find economically feasible methods of producing oil shale.

One of the more important phases of this experimental program was a large-scale mine dug into what is known as the Mahagony Ledge, a rich oil shale stratum that is flat and strong, making it favorable for mining. This stratum lies under an average of about 1000 ft of overburden and is 70–90 ft thick.

The Bureau of Mines adopted the room-and-pillar system of mining, advancing into the 70-ft ledge face in two benches. The mine roof was supported by 60-ft pillars staggered at 60-ft intervals and supplemented by iron roof bolts 6 ft long (Fig. 1).

Multiple rotary drills mounted on trucks made holes in which dynamite was placed to shatter the shale; the shale was then removed from the mine by electric locomotive and cars (Fig. 2).

The experimental mining program ended in February, 1955, when a roof fall occurred. Despite this occurrence, however, the Bureau is convinced that the room-and-pillar method used in coal, salt, and limestone mines is feasible for shale oil mining in Colorado.

Oil shale does not contain oil, as such. Draining methods, therefore, are not applicable. It does, however, contain an organic substance known as kerogen. This substance decomposes and gives off a heavy, oily vapor when it is heated above 700°F in retorts. When condensed, this vapor becomes a viscous black liquid called shale oil, which resembles ordinary crude but has several significant differences.

Colorado's Mahagony Ledge yields an average of about 30 gal of oil per ton. This means that large amounts of oil shale must be mined, transported, retorted, and discarded for production of commercial quantities of oil. Various types of retort also have been tested in Colorado, but none is in commercial use (Fig. 3). *See* OIL SHALE.

Gilsonite. Gilsonite is a trade name, registered by the American Gilsonite Co., for a solid hydrocarbon found in the Uinta Basin of eastern Utah and western Colorado. The American Gilsonite Co. uses a subsurface wet-mining technique to extract about 700 tons of Gilsonite daily from its mine at Bonanza, Utah.

Conventional mining methods were found unsuitable for mass output of Gilsonite because it is friable and produces fine dust when so mined. This dust can be highly explosive. In the system now being used, tunnels are driven from the main shaft by means of water jetted through a 1/4-in. nozzle under pressure of 2000 psi. The stream of water penetrates tiny fissures and the ore falls to the bottom of the drift. The drifts are cut on a rising grade of about 2.5°. The ore is washed down to the main shaft where it is screened. Particles of sizes smaller than 3/4 in. are pumped to the surface in a water stream; larger pieces are hoisted in buckets. A long rotary drill with carbide-tipped teeth is used to remove ore that cannot be broken with water jets.

Gilsonite is moved through a pipeline in slurry form to a refinery, where it is dried and melted and then heated to about 450°F. The melted oil is fed to a coker and other processing units to make gasoline and other petroleum products.

North American tar sands. Strip and open-pit mining techniques have been established for the bituminous sands in the Athabaska region of Alberta, Canada, where the largest deposits of these sands are found. Bituminous sand, also know as asphalt rock, oil sand, and tar sand, is impregnated with heavily viscous oil that can be extracted by various methods, including hot- and cold-water washing, solvents, centrifuges and low-temperature distillation.

Once the overburden, which varies in thickness from a few feet to as much as 1800 ft, has been removed (where not too thick), the sand can be stripped in some cases directly by a power shovel or walking dragline, in other cases, after a small amount of blasting. It seems unlikely that known underground mining methods could ever be used to recover more than 90% of the bituminous sand that lies under the heavier overburden, although it is known that some of the oil in these sands flows freely, leaving a possibility for some form of recovery by drainage.

Cities Service Research and Development Co. used a Krupp wheel excavator during test operations in the Mildred Lake area of the Athabaska region, beginning in 1958 (Fig. 4). The excavator was designed to remove the oil sand from the bank in the quarry and carry it to a mobile extraction unit. The digging is accomplished by six 1.8-ft³ buckets mounted on a wheel. The loosened oil sand is dropped on the conveyor belts, which transfer the oil sand to the vibrating grizzly on the extraction unit. The excavator is capable of digging on a mining face 20 ft high by 30 ft wide by making three swinging cuts, each 6.6 ft high. The maximum cutting depth is 11.5 in. The mining capacity is 40 tons/hr. The generator and diesel-engine drive, plus the necessary switchgear, starter batteries, fuel tank, and connecting lines, are mounted on a four-wheel trailer. All operating parts are enclosed in a sheet-metal housing for protection against the weather, an important element in northerly locations. The trailer also is designed for easy movement across the mining face. *See* OIL SAND.

Mine draining of reservoirs. Drainage of oil in place, although less efficient than actual mining, is cheaper and can be carried out with much less risk. This method has been practiced for years in

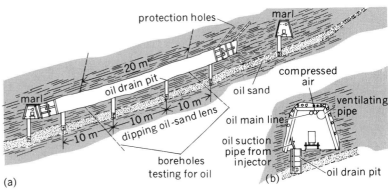

Fig. 5. Sketches illustrating (a) the method of conducting oil mining operations in the Pechelbronn oil field in Alsace, France, and (b) a detail section of the heading. (*After G. S. Rice, U.S. Bureau of Mines*)

the Pechelbronn field and the Wietze field, near Hanover, Germany. A similar venture was successfully used in Ventura County, Calif., in 1866, as well as in other states since then.

In the Pechelbronn field, shafts were sunk to the level of the oil reservoir and drifts driven into the oil rock. Large drainage surfaces were available from wells drilled from the surface in previous years and later abandoned. Unrecovered oil drained down these abandoned wells into the mine workings, where it was gathered in underground storage sumps and then pumped to the surface. All the tunnels driven from the shafts directly through the oil rocks afforded drainage surface too. This oil accumulated in a trench cut in the floor of each tunnel. Oil flowed by gravity down these trenches to the central storage sump. This method of recovery was so efficient that Pechelbronn became one of the largest mining projects in the world, with more than 200 mi of tunnels in three separate mines. In addition, hundreds of miles of 1.5–2 in. bore holes have been drilled for exploratory purposes and to check for high-pressure gas or oil pockets.

In another mining method that became more prevalent after 1925 at Pechelbronn, tunnels were driven through the impermeable rock cap above the oil sand, and at 33-ft intervals, 20–30 in. square pits were dug through the tunnel floor to the oil sands, usually 6–8 ft below. The oil drains into these pits, which are timber lined, and is lifted periodically by a pneumatic system into a pipeline extending to surface tanks (Fig. 5).

One variation of this method, known as the Ranney oil-mining system, involves driving mine galleries in impermeable strata above or below the oil sands. These galleries are driven from shafts communicating with the surface. Holes are drilled at short intervals along these galleries, and gas or oil is withdrawn through pipes sealed into the bore holes. These mine wells communicate with a system of pipe drains in the tunnels, leading to separate tanks from which the oil is pumped to the surface, or the gas permitted to flow to the surface. There are several other methods which are similar to the Ranney system, in that they also provide for oil drainage through holes drilled from mine tunnels outside the oil beds (Fig. 6).

Another method proposed for mining oil from

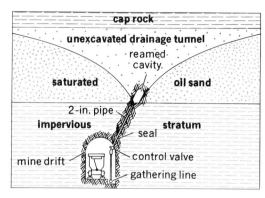

Fig. 6. Diagrammatic illustration of unexcavated drainage tunnels in connection with the Ranney system of oil mining. (*After L. C. Uren, Petroleum Production Engineering: Oil Field Exploitation, 3d ed., McGraw-Hill, 1953*)

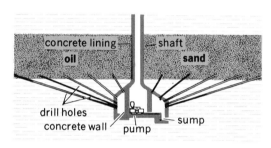

Fig. 7. Sketch of Wright system for draining oil sands by a series of bore holes drilled from a mine shaft. (*After L. C. Uren, Petroleum Production Engineering: Oil Field Exploitation, 3d ed., McGraw-Hill, 1953*)

partially drained sands involves drilling a shaft through the productive strata, followed by long, slanting holes drilled radially in all directions from the shaft bottom into the oil sands. These bore holes can be as long as 1/2 mi, and 2–5 in. in diameter. This method was used in shallow oil sands in southeastern Ohio and in Pennsylvania (Fig. 7).

Selection of method. Use of the various mining methods depends largely on conditions found. They are not usable where high gas pressures may be present. Water can likewise be a major problem. In some cases oil sands are poorly consolidated, so that they flow under pressure, making mine support difficult, if not impossible; in addition, these sands tend to mix with the oil.

Where subsurface conditions are good, that is, where gas pressure is not high, water is not excessive, and sand conditions are favorable, mining can be used, provided costs can be kept at a competitive level.

[ADE L. PONIKVAR]

Bibliography: F. L. Hartley and C. S. Brinegar, Oil shale and bituminous sand, *Energy Resources Conference*, 1956; J. B. Jones, Jr. (ed.), *Hydrocarbons from Oil Shale, Oil Sands and Coal*, 1964; *Synthetic Liquid Fuels: Annual Report of the Secretary of the Interior for 1955*, pt. 11: *Oil From Shale*, U.S. Bur. Mines Rep. no. 5237, 1956; L. C. Uren, *Petroleum Production Engineering: Oil Field Exploitation*, 3d ed., 1953.

Oil sand

A bituminous sand, asphalt rock, or tar sand; a loose sand or sandstone impregnated with very viscous oil. Among the best-known oil sand deposits are those near Vernal, Utah, and those adjacent to the Athabaska River surrounding Fort McMurray in Alberta, Canada.

The Athabaska tar sand is the largest deposit of this type known, having been estimated to contain from 100,000,000,000 to 300,000,000,000 or possibly even 500,000,000,000 bbl of oil. Of this quantity only about 5,000,000,000 bbl of oil can be recovered by open-pit mining methods. The origin of the Athabaska oil is unknown; some suggestions have been made that it is a naturally stripped petroleum reservoir (fossil oil field), but other work has indicated that it might be of nonmarine origin and not directly related to petroleum.

A number of attempts have been made to recover the oil from oil sands by solvent extraction, hot- or cold-water washing, retorting, or centrifugal separation techniques. Once isolated, the oil can be refined by conventional processes. *See* OIL MINING.

The total reserve of oil in major oil sand formations of the United States has been estimated to be 2,000,000,000–3,000,000,000 bbl.

[IRVING A. BREGER]

Oil shale

A fine-grained, usually dark-colored sedimentary rock containing complex organic matter which, on heating, decomposes to yield oil. In its strictest sense oil shale is a commercial term used to designate those shales that yield commercial quantities of oil (over 10–15 gal/ton).

Origin and composition. Oil shales may be either marine or lacustrine in origin. Each shale represents the slow accumulation of inorganic sediment together with the organic debris contributed by aquatic flora and fauna. A major contribution to the organic constituents consists of pollen and plant fragments carried into the sedimentary basin by wind or streams. Following deposition, this organic mass undergoes biochemical degradation under reducing conditions to yield a complex organic mixture. Where spores have contributed the major portion of the organic detritus, the shale takes on the properties of a cannel coal. Humic substances derived from land plants may contribute substantially to the organic constituents of certain shales.

Most oil shales contain only small percentages (about 5%) of organic matter extractable under the usual laboratory conditions. The major part of the organic matter is present in the form of high-molecular-weight, insoluble complex material.

Quartz and silicate minerals, and especially such clay minerals as hydromica, kaolinite, and montomorillonite, account for most of the inorganic constituents of an oil shale. These minerals are either supplied to the sediment by streams or may be formed from other minerals during the diagenesis of the sediment to produce the shale. Pyrite or marcasite, along with minor amounts of feldspars, is usually present. Feldspars may also be formed in position in a shale. The massive Green River oil shale of Colorado, Utah, and Wyoming is in reality largely composed of dolomite and calcite in addition to clays and other minerals commonly associated with shales.

In 1947 nahcolite, $NaHCO_3$, and dawsonite, $NaAl(OH)_2CO_3$, along with smaller amounts of related minerals, were discovered in the Green River shale of the Piceance Creek basin of northwest Colorado. Evaluation of the occurrence since then has demonstrated that the reserves of dawsonite contain about 42,000,000 tons of alumina per square mile, or about 1.5 times as much alumina as is represented by all the bauxite reserves of the United States. Current indications are that this shale may be of value not only as a source of oil, but also as a major source of alumina and soda ash. The economic value of the shale is, therefore, greatly enhanced.

Many oil shales are thinly laminated (Fig. 1), and parts of the Green River shale are known as paper shale; other shales exhibit a subconchoidal or conchoidal fracture.

The reducing conditions necessary for the preservation of organic matter in a shale-forming environment are conducive to the precipitation of available trace elements and the formation of certain minerals. Pyrite, FeS_2, occurs in large amounts in most shales. The Kupferschiefer of Mansfeld, Germany, contains an unusually high content of copper, and the Swedish alum shale has been exploited for its uranium content (about 0.04%). This shale also contains in its upper units small lenticular masses of organic matter, called kolm, that contain up to 0.5% uranium. The Chattanooga shale, containing about 0.006% uranium, has been extensively studied as a potential low-grade source of this element. Other shales may contain minor percentages of vanadium, selenium, zinc, silver, or other elements.

Current economic factors require a shale to yield approximately 15 gal/ton of oil before it can be profitably exploited. During the World War II emergency, the Swedish shale, yeilding about 12 gal/ton, was retorted as a source of oil. The Eocene Green River shale averages 15 gal/ton, but selected segments of the formation, such as the Mahogony Ledge, average about 30 gal/ton. Only this latter part of the Green River formation is considered to be of present economic significance. Most other shales of the United States yield an average of only 10–12 gal/ton of oil, making them commercially unattractive. Other shales, with indicated oil yields, are shown below:

Source	Oil yield gal/ton (approx)
France	10–17
Germany	12
Sweden	12–15
Manchuria	15
Canada	22
Scotland (torbanite)	22
Soviet Union	35–50
England	40
Estonia (kukersite)	48–86
Australia (torbanite)	80–180
South Africa (torbanite)	106

As indicated, those shales with high yields of oil

Fig. 1. A sample of the thinly laminated type of oil shale. (*Union Oil Co. of California*)

have been given special names indicating them to be somewhat extraordinary in appearance, origin, and composition.

The following shales are also well-known and have been studied in detail:

Name	Age	Occurrence
Um-Barek	Cretaceous	Israel
Frontier	Cretaceous	Wyoming
Kimmeridge	Jurassic	England
Ermelo	Carboniferous	South Africa
St. Hillaire	Carboniferous	France
Autun	Carboniferous	France
Lothian	Carboniferous	Scotland
Albert Mines	Mississippian	New Brunswick, Canada

Many additional shales, ranging in geologic age from the Paleozoic to the Tertiary, occur in the Soviet Union, Poland, Brazil, Australia, and other countries.

The composition of crude shale oil depends upon the shale from which it is produced and the method of pyrolysis. The oil, which contains paraffinic, olefinic, naphthenic, and aromatic constituents, may have a pour point ranging from 90°F (semisolid) to below 5°F. There is usually a preponderance of straight-chain hydrocarbons (approximately 50%) among the identifiable compounds in the distillate; sulfur compounds are mainly thiophenic. Conversion of most of the distillate to motor fuel can be carried out by thermal or catalytic cracking, hydrogenation, or other processes common in the petroleum industry.

Shale oil is a source not only of fuel but also of a paraffin-type wax, tar acids, and tar bases. Tar acids consist of phenol, cresols, xylenols, and carboxylic acids; tar bases contain homologs of pyridine and quinoline.

Oil-forming shales. Oil shales occur in many parts of the world, usually as relatively thin (several hundred feet) widespread formations. Some of the marine carbonaceous shales, such as the Miocene Monterey shale of California, are thought to have been source beds in which petroleum was formed; in others, such as the Chattanooga shale of the southeastern United States or the Swedish alum shale, there is little or no associated crude oil and no indication that these shales were source beds. Little is known about the differences between shales that are, or are not, likely source beds, but differences may be related to a great extent to types of organic matter present at the time of deposition.

The organic constituents of an oil shale can be derived from a single source or from several sources. This variability is reflected in the fact that the Chattanooga shale, containing 25% organic matter, yields only 10–12 gal/ton of oil on pyrolysis, whereas the Mahagony Ledge of the Green River formation, with only 15–20% organic matter yields up to 60 and an average of 30 gal/ton. In the former, the organic matter is, as has been noted, derived from terrestrial sources and behaves like coal, yielding relatively little oil on pyrolysis; in the latter, the organic matter is derived from an aquatic source (lacustrine) and, having a higher hydrogen content, produces much oil on distillation. The oil yield of a shale on pyrolysis is, therefore, dependent on the nature of the kerogen as well as its percentage.

Most geochemists are now of the opinion that petroleum hydrocarbons are formed from the lipids derived from aquatic organisms. Lack of a significant contribution of organic matter from such a source may explain the absence of crude oil in the Chattanooga shale, where the kerogen is related to terrestrial organic debris and is similar to subbituminous coal. There is evidence that some crude oil has formed in certain parts of the Green River shale, but most of the aquatic organic debris of this formation was of such chemical composition that it preferentially led to the formation of kerogen or such other substances as gilsonite, wurtzilite, and elaterite.

Recovery of oil. In the past, oil was generally recovered from shale by techniques in which the ground rock was fed into a retort, the distillate was condensed, and the gases were vented. Most retorts operated on a batch basis and employed combustion of part of the organic matter of the shale to provide the heat necessary for distillation. The usual retort had a daily capacity of 40–50 tons of shale. Continuous-type retorts were also used, in which raw shale was fed into one end of a heated tunnel and expended shale withdrawn from the other end. Economic recovery of oil from shale was also dependent to a great extent upon the development of satisfactory mining methods for the shale and upon means for the disposal of large quantities of spent shale.

To obviate mining and disposal problems, the Swedish Shale Oil Company in 1944 applied the Ljundstrom in-place method for the distillation of oil from the Swedish alum shale. Electrical heating elements were inserted into a hexagonal pattern of drill holes sunk vertically through the shale bed, and the shale was gradually heated to 400°C over a period of 5 months. The approximate yield per square meter of the field was gasoline, 515 liters; kerosine, 160 liters; heating oil, 350 liters; liquefiable gas, 80 liters; sulfur-free gas, 650 liters; sulfur, 350 kg; and ammonia, 8 kg. This process was discontinued shortly after the end of World War II when sufficient crude oil again became available in Sweden.

In 1953 and 1954 a subsidiary of the Sinclair Oil Company again attempted the in-place retorting of shale. Operating on the Green River shale near Grand Valley, Colo., some oil was produced, but the results of the experiment were inconclusive. In 1964 the same company announced that it was about to conduct new underground retorting experiments at a site 28 mi west-wouthwest of Meeker, Colo. The process called for drilling a well into the Green River shale, casing the hole nearly to its bottom, fracturing the shale hydraulically, and using a propane-air mixture to burn part of the shale underground. It was hoped that the heat and pressure so developed would permit much of the oil produced to flow to the well through the fractures. The rate and pattern of heat flow were considered the governing factors in the volume of shale that could be retorted about a well. This underground retorting procedure is expected to be cheaper than conventional mining and aboveground retorting.

The Estonian oil shale, known as kukersite, has been the basis of a major industry. The deposit, which crops out along an east-west trend parallel to the Gulf of Finland, is mined and retorted, and the gas that is produced has been piped to Leningrad and Tallinn. The Oil Shale Institute at Kohtla Jarve, founded about 1950, is said by the Soviets to be the largest research institution of its kind in the world. An underground mine under construction at Kohtla Jarve will cover an area of 50 mi² and is scheduled to produce 33,000 metric tons of shale per day.

Research in the United States on the production of oil from shale has been nearly exclusively conducted on the Green River shale of Colorado, Utah, and Wyoming. As early as 1945 the U.S. Department of the Interior constructed a demonstration plant near Rifle, Colo., that was operated by the Bureau of Mines until 1956. During these years much progress was made in the development of mining techniques and in the design of retorts. The Union Oil Company, operating a pilot plant (Fig. 2) in Grand Valley, Colo., subsequently developed a capacity for retorting 1200 tons of crushed shale daily to produce 750 barrels per day (bpd) of oil.

A large-scale incursion in oil shale operations came about in 1964 with the formation of the Colony Development Company. This company was organized to explore development of 7500 acres of Green River shale near Rifle, Colo., and 1300 acres of land in Utah. Interests were later acquired for the mineral rights to 8400 acres of land in Uintah County, Utah. The Oil Shale Corporation, part owner of the Colony Development Company, owns the patented Swedish TOSCO process for recovery of oil from crushed shale. It is clear from the widespread industrial interest that by 1964 a large-scale oil shale industry based on the Green River shale was considered feasible.

By mid-1965 Colony Development had in operation a multimillion-dollar prototype shale oil production facility about 17 mi from Grand Valley, Colo., and indicated plans to begin construction of a 50,000–100,000 bpd refinery by mid-1968. Conventional room-and-pillar mining techniques are being used, in which the shale is hauled out for crushing by special rock-moving equipment. To reduce mining costs, a Le Tourneau diesel electric hauler-loader was designed that is capable of transporting 20 tons of shale at 20 mph.

After crushing the shale to less than 0.5 in., it is retorted by the TOSCO process, in which preheated alumina balls are introduced into a horizontal rotating kiln along with the shale. Heat from the balls raises the temperature of the shale to 900°F to pyrolyze the kerogen. The kerogen is decomposed in the absence of air in contrast to the Bureau of Mines and Union Oil processes, where pyrolysis was supported by burning gas. The efficiency of the TOSCO process is dependent upon the recovery of heat from spent shale and on the heating of the alumina balls.

In the spring of 1964 the Bureau of Mines demonstration plant in Rifle, Colo., closed since 1956, was leased to the Colorado School of Mines Research Foundation by the Department of the Interior for a 5-year period. The Colorado School of Mines Research Foundation in turn contracted for operation of the Anvil Points Oil Shale Research Center by a partnership of several oil companies in what was to be a 3-year, $5,000,000 program for research leading to the production of oil. The first 18-month period was to be devoted to a study of crushing and retorting procedures, and the second 18-month period was to be spent studying mining techniques and methods for scaling up the retorts.

Oil produced from the shale by both the Bureau of Mines and Union Oil Company processes was waxy and unsaturated, and contained oxygen, nitrogen, and sulfur compounds. Union Oil eventually intended to catalytically hydrogenate and crack the distillate or otherwise to improve the quality of the product, simultaneously extracting values from petroleum coke, sulfur, and ammonia by-products. The TOSCO process operated by Colony Development similarly depends upon catalytic hydrogenation to remove nitrogen and sulfur, as well as to saturate the product and produce a crude oil that can be easily piped.

An alternate technique has been patented (U.S. Patent 3,106,521) by M. C. Huntington, president of Pyrochem Corporation, in which hydrogenation precedes pyrolysis of the shale resulting in what amounts to a one-step process. The Pyrochem process uses 20- to 60-mesh shale, the shale and a stream of hydrogen are separately preheated, and the hydrogenated shale and hydrogen are mixed in a pyrolysis zone at 300 psi and 900°F. The pyrolysis gas is then immediately passed over a cobalt-molybdenum catalyst to produce a colorless, low-viscosity, saturated liquid free of nitrogen and sulfur compounds. This product, moveover, distills predominantly in the gasoline range. The hydrogen in this process is not only used for hydrogenation, but also serves as a thermal diffuser and as a heat recovery medium. In 1967 the Battelle Memorial Institute was engaged in a $2,000,000 program for development of a high-capacity gravity-feed kiln for use in this process.

Besides these techniques for extracting oil from shale, U.S. Patent 3,342,275 has been assigned to the Standard Oil Comany of Indiana for the use of a nuclear device to retort shale in place.

Reserves and development. The reserves of shale oil have been conservatively estimated to be 960,000,000,000 bbl for the Green River formation alone; the rich part of the formation, the Mahagony Ledge, has been estimated to contain 90,000,000,000 bbl. The enormity of this reserve is fully grasped when it is realized that the entire world has produced only 90,000,000,000 bbl of crude oil between 1859 and 1957. Because of the tremen-

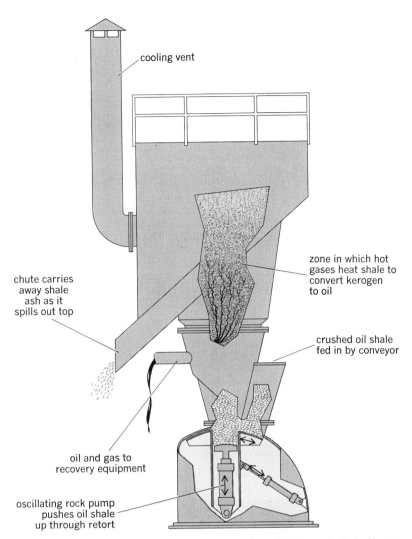

cooling vent

zone in which hot gases heat shale to convert kerogen to oil

crushed oil shale fed in by conveyor

chute carries away shale ash as it spills out top

oil and gas to recovery equipment

oscillating rock pump pushes oil shale up through retort

Fig. 2. Pilot plant model of oil shale retort. Shale is fed at bottom and retorted in center section. (*Union Oil Co. of California*)

dous value of the reserves of the Green River formation, serious efforts have been made on the one hand to open the affected land for private development and, on the other hand, to reserve it for federally administered development to assure conservation practices and maximum benefit to the country. Most of the Green River shale, including 65% of Colony Development's holdings, is under Federal title.

In 1964 Secretary of the Interior Stewart Udall rejected 200 oil shale claims covering approximately 40,000 acres of land in Colorado. Each of these claims had been canceled by the Department of the Interior prior to 1935, and oil shale deposits of Colorado, Utah, and Wyoming had been withdrawn from leasing in 1930. The decision of 1964 to reject the 200 claims was stated to be in the interest of an orderly and expeditious program of development.

Foreseeing enhanced interest in the development of the oil shale reserves of the Colorado Plateau, the Department of the Interior invited opinions from industry in 1963 as to the preferred manner in which to undertake development of the area. An Interior–Citizens Advisory Committee was organized, and in 1965 urged commercial development. There was serious discord, however, as to the approach to such development with some committee members wanting all land opened to private development, whereas others recommended concentration on research and development and a ban on leasing until more technological progress was indicated. Secretary of the Interior Udall suggested that the public interest should be protected, land speculation prevented, a policy on depletion allowances determined, and confusion over mining claims resolved. The Internal Revenue Service has since set a 15% depletion allowance based on the value of crushed shale.

With various representatives and senators from the affected states pressing for clarification of the status of land claims and for rapid development of the shale by private industry, Secretary Udall in 1967 laid down a series of rules related to the development of the deposit. Features of research leases concerning royalty payments and dedication to the public of results of research by lessees are controversial.   [IRVING A. BREGER]

Bibliography: I. A. Breger (ed.), *Organic Geochemistry*, 1963; Colorado School of Mines, Fourth Symposium on Oil Shale, *Quarterly of the Colorado School of Mines*, vol. 62, 1967; F. L. Hartley and C. S. Brinegar, Oil shale and bituminous sand, *Sci. Mon.*, 84:275–289, 1957; J. B. Jones, Jr. (ed.), *Hydrocarbons from Oil Shale, Oil Sands and Coal*, 1964; G. Sell (ed.), *Oil Shale and Cannel Coal*, vol. 2, 1951.

Permafrost

Perennially frozen ground, occurring wherever the temperature remains below 0°C for several years, whether the ground is actually consolidated by ice or not and regardless of the nature of the rock and soil particles of which the earth is composed. Perhaps 25% of the total land area of the Earth contains permafrost; it is continuous in the polar regions and becomes discontinuous and sporadic toward the Equator. During glacial times permafrost extended hundreds of miles south of its present limits in the Northern Hemisphere.

Permafrost is thickest in that part of the continuous zone that has not been glaciated. The maximum reported thickness, about 1600 m, is in northern Yakutskaya, U.S.S.R. Average maximum thicknesses are 300–500 m in northern Alaska and Canada and 400–600 m in northern Siberia. In Alaska and Canada the general range of thickness in the discontinuous zone is 50–150 m and in the sporadic zone less than 30 m. Discontinuous permafrost in Siberia is generally 200–300 m.

Temperature of permafrost at the depth of no annual change, about 10–30 m, crudely approximates mean annual air temperature. It is below −5°C in the continuous zone, between −1 and −5°C in the discontinuous zone, and above −1°C in the sporadic zone. Temperature gradients vary horizontally and vertically from place to place and from time to time. Deep temperature profiles record past climatic changes and geologic events from several thousand years ago.

Ice is one of the most important components of permafrost, being especially important where it exceeds pore space. Physical properties of permafrost vary widely from those of ice to those of normal rock types and soil. The cold reserve, that is,

the number of calories required to bring the material to the melting point and melt the contained ice, is determined largely by moisture content. Ice occurs as individual crystals ranging in size from less than 0.1 mm to at least 70 cm in diameter. Aggregates of ice crystals are common in dikes, layers, irregular masses, and ice wedges. These forms are derived in many ways in part when permafrost forms. Ice wedges grow later and characterize fine-grained sediments in continuous permafrost, joining to outline polygons. Microscopic study of thin sections of ice wedges reveals complex structures that change seasonally.

Permafrost develops today where the net heat balance of the surface of the Earth is negative for several years. Much permafrost was formed thousands of years ago but remains in equilibrium with present climates. Permafrost eliminates most groundwater movement, preserves organic remains, restricts or inhibits plant growth, and aids frost action. It is one of the most important factors in engineering and transportation in the polar regions.

[ROBERT F. BLACK]

Bibliography: J. B. Bird, *The Physiography of Arctic Canada*, 1967; R. F. Black, Permafrost: A review, *Bull. Geol. Soc. Amer.*, 65:839–856, 1954; R. J. E. Brown, Comparison of permafrost conditions in Canada and the USSR, *Polar Rec.*, 13: 741–751, 1967; *Proceedings of the Permafrost International Conference*, Nat. Acad. Sci.–Nat. Res. Counc. Publ. no. 1287, 1966.

Pesticide, persistence of

An expanding use of pesticidal chemicals in agriculture, public health, forestry, warfare, and home gardens continues to be a source of controversy. For a wide variety of reasons many attempts have been made, particularly in North America, to restrict the use of such chemicals. Considerable scientific evidence has recently been accumulated that documents the deleterious effects on the environment of several of the chlorinated hydrocarbons. On this basis, a number of environmental scientists have attempted to eliminate the use of these compounds, especially DDT. Industry has responded with the considerable resources at its disposal, and the bureaucratic machinery of government has frequently been caught in the middle. Industry has accused the anti-DDT forces of being antitechnology, and environmental scientists have replied that the technology of 1945 is antiquated and that a new technology must be developed that will be compatible with the continued existence of life on Earth. The DDT controversy might therefore be considered a prologue to a much larger controversy of the future that will decide how increasing numbers of people, with decreasing resources, will achieve a technology that will be compatible with both human values and the long-term survival of man.

The term pesticide may be defined as any substance used to kill, or to inhibit the growth and reproduction of, the members of a species considered in a local context as pests. About 800 compounds are now used as pesticides and, with increasing technological sophistication, the list can be expected to grow. Until World War II pesticides consisted of inorganic materials containing sulfur, lead, arsenic, or copper, all of which have biocidal properties, or of organic materials extracted from plants such as pyrethrum, nicotine, and rotenone. During World War II a technological revolution began with the introduction of the synthetic organic biocides. DDT was widely used during World War II to control the vectors of diseases which until that time had been serious health hazards over much of the world. Since then, DDT has been extensively used with spectacular success to reduce the incidence of malaria in the tropics. In the public mind DDT has become almost synonymous with pesticide, yet it is important to point out that DDT is only one of an increasing number of organic biocides. They vary widely in toxicity, in specificity, in persistence, and in the production of undesirable derivatives. The factors that determine how, when, and where a particular pesticide should be used, or whether it should be used at all, are therefore necessarily different for each pesticide. During the recent controversy over pesticide usage, the industry spokesmen have consistently attempted to associate those who believe it is time to ban DDT with those who campaign against the use of all pesticides. The use of chemicals to achieve a measure of control over the environment, however, has become an integral part of technology, and there are no valid arguments to support the proposition that the use of all such chemicals should now cease.

Some take the view that pests include insects, fungi, weeds, and rodents. Many insects, however, particularly the predatory species that prey upon the insects that damage crops, are clearly not pests. Fungi are essential in the processes that convert dead plant and animal material to the primary substances that can be recycled through living systems. In many ecosystems rodents are also an important link in the recycling process. On the other hand, an increasing number of species traditionally considered desirable are becoming "pests" in local contexts. Thus chemicals are used to destroy otherwise useful vegetation along roadsides or in pine plantations, or vegetation that might provide food and shelter to undesirable human populations. The control of any component of the global environment through the use of biocidal chemicals is therefore a problem much more complex than is initially suggested by the connotation of the word pest. The highly toxic chemicals that kill many different species are clearly undesirable in situations in which control over only one or two is sought.

Like any other technological product, pesticides, once used, become waste products and some become pollutants. It can no longer be assumed that pollutants will disappear in the vastness of the sea, since the global environment has suddenly become small. The total amount of waste materials produced by man is now approximately as great as the amount of organic material cycled yearly through nature. Waste products that are readily degraded to elementary materials that can be recycled clearly do not threaten the stability of the environment. In traditional terms they might be compared with the biblical ashes that return to ashes and the dust that returns to dust.

Biodegradable compounds. Of the approximately 800 biocidal compounds used as pesticides, many obey this fundamental ecological law and are degraded to elementary materials that can be recy-

cled. These compounds include the organophosphate insecticides and the carbamates, since no evidence now available suggests that any derivative of these compounds becomes a persistent pollutant. Malathion, diazinon, and parathion are examples of organophosphorous insecticides. Parathion is particularly toxic to many forms of life and has caused the deaths of many persons applying it to crops or who have been accidentally exposed. Other pesticides in this group, however, have been extensively used without causing human fatalities. They are readily metabolized in the mammalian body, and in the environment are degraded to elementary materials. Unwise use of these compounds may result in local fish kills or other loss of wildlife, but with careful use such losses can be minimized. Because of the chemical instability of these biocides they do not travel to areas for removed from the sites of application. Environmental damage, if any, is therefore local. The organophosphorous insecticides inhibit the enzyme cholinesterase, which normally breaks down acetylcholine following the transmission of the nerve impulse across the synapse.

The carbamates are another group of biocides that also inhibit the enzyme cholinesterase. Carbaryl, or sevin, is an example. These biocides are even less persistent than the organophosphates and pose fewer health hazards to man. Indiscriminate use, however, may cause local environmental problems. Carbaryl, for example, is particularly toxic to bees.

Other pest-control measures. Sophisticated methods of pest control are continually being developed. One technique involves the raising of a large number of insect pests in captivity, sterilizing them with radioactivity, and then releasing them into the environment, where they mate with the wild forms. Very few offspring are subsequently produced. Highly specific synthetic insect hormones are being developed. In an increasing number of pest situations, a natural predator of an insect has been introduced, or conditions are maintained that favor the propagation of the predator. The numbers of the potential pest species are thereby maintained below a critical threshold.

Chlorinated hydrocarbons. DDT belongs to a class of biocides that are chlorinated hydrocarbons. The chlorine-carbon bond is rare in nature and is comparatively stable. Few bacteria and fungi are equipped to break it. These biocides are therefore relatively persistent, much more so than are the carbamates and the organophosphates. The chlorinated hydrocarbons are also much less soluble in water but are very soluble in nonpolar media such as lipids. They are therefore readily concentrated by organisms from the environment, and accumulate in fatty tissues. Species higher in the food chains will subsequently accumulate the chlorinated hydrocarbons present in their food. Mobility is another property which makes some of the chlorinated hydrocarbons hazardous to the global environment. DDT and dieldrin become vapors and can be transported anywhere in the world to accumulate in distant ecosystems. Since the chlorinated hydrocarbon biocides are toxic to a wide spectrum of animal life, they have the capacity to cause harm to many nontarget species. *See* POLYCHLORINATED BIPHENYLS.

Several of the chlorinated hydrocarbons, however, do not appear to possess the combined properties of intensive use, persistence, and mobility that create the potential for environmental degradation. Lindane is more readily degraded than the others, and chlordane and toxaphene do not appear to possess the mobility of DDT and dieldrin, since they have not so far been detected in marine organisms away from contaminated estuaries.

Of all the pesticides now in use, only five have aroused the concern of environmental scientists who are attempting to consider the long-term effects upon the global environment. Aldrin, dieldrin, endrin, and heptachlor are extremely toxic, persistent, and mobile, and have the capacity to inflict harm upon nontarget species.

The fifth compound, DDT, has yet another property that renders it especially dangerous to the environment. The insecticidal compound, p,p'-DDT, is relatively persistent, but much of it is eventually degraded to another DDT compound, p,p'-DDE. This compound, DDE, is more persistent than the original DDT and appears to be the most abundant of the synthetic pollutants in the global environment. Most of the concern about DDT in the environment is actually about DDE.

Many marine biologists still find it difficult to accept the findings that DDE is now more abundant in marine birds and fish than in many terrestrial organisms. It is not the only pollutant to be so widely dispersed. Polychlorinated biphenyls, another group of chlorinated hydrocarbons that are industrial pollutants, have become similarly dispersed throughout the global ecosystem. The sea is no longer the vast entity that it was traditionally conceived to be. World production figures of many chemicals, including DDT, are now a significant fraction of the amount of organic material that is recycled each year through the world's biomass.

DDE was originally considered to be harmless because of its relative nontoxicity to organisms. Evidence is now rapidly accumulating that DDE has physiological effects that may not kill the organism but which are nevertheless deleterious and may impair the long-term survival of the species. Experiments have shown that DDE may cause several species of birds to lay eggs with abnormal amounts of calcium carbonate. Many species of raptorial and fish-eating birds have experienced low rates of reproductive success over the past 20 years, and in most cases one of the symptoms has been abnormally thin eggshells. In the most spectacular cases, such as the brown pelicans and other sea birds of the West Coast, the shells are so thin that they break during incubation. Increasing amounts of evidence link this phenomenon to DDE.

Physiological effects of DDE that may be associated with thin eggshells and other abnormalities include the inhibition of membrane enzymes that are adenosinetriphosphatases involved with ion transport, the inhibition of the enzyme carbonic anhydrase, and the induction of nonspecific liver enzymes that degrade a wide variety of lipid-soluble compounds, including steroids. All these effects of DDE have been discovered only recently, and more are likely to be discovered in the future.

In 1969 fish caught in the Pacific Ocean were seized and condemned by the U.S. Food and Drug Administration because they had been found to contain high concentrations of the DDT compounds. A number of monitoring programs for pesticides in the environment had previously been initiated in North America, but none had been designed to find the accumulation site of a pesticide pollutant or to find the rate at which it accumulates there. It is now obvious that DDE has been accumulating in the sea, but since there are no data on how fast it is accumulating, it is difficult to know what to expect in the future.

At the present time the environmental science community is convinced that DDE is contributing to the extinction of a number of species of birds. Only 4 years ago the idea that DDE might cause thin eggshells would have been dismissed by most environmental scientists as somewhat preposterous. Moreover, there is no information on what additional effects DDE or another pollutant might have, once a critical threshold level is reached.

The spokesmen for the DDT industry have maintained that DDT has no effects upon man at dosages that might be encountered. No deleterious effects have so far been found in man. A preliminary screening test indicated, however, that p,p'-DDT or a metabolic derivative might be carcinogenic. Almost no work has been done with DDE, the DDT compound that is present in food and in the environment. For reasons that are not clear to an environmental scientist, almost all of the work with DDT has been done with the insecticidal compound, p,p'-DDT, which is applied to fields. Conclusions based upon work with this compound are of doubtful validity if they are meant to include DDE also. Conclusions about the effects of DDE upon man will require many analyses for DDE and other pollutants in human tissues that can be correlated with causes of death, similar to the work that has showed the correlation between cigarette smoking and lung cancer.

Future course of action. The confiscation of the marine fish is an instance of environmental food becoming unsuitable for man because of increasing pollution. The continued release of persistent pollutants into the environment will mean that more and more food will become unavailable.

In considering the expanding populations, the increasing amounts of waste materials produced per capita, and both the known and unknown effects of persistent pollutants, the environmental scientist is forced to assume a very conservative position. Any of the persistent pollutants that are released in amounts comparable to those of the DDT compounds has the capacity to produce an irreversible effect upon the global environment. The chances, therefore, simply are not worth taking if other compounds that will do the same job are available. To preserve the global environment as a fit habitat not only for pelicans and bald eagles but for man himself, a commitment must now be made to a technology consistent with the biblical and ecological law that ashes must return to ashes. Although laws have been passed to ban the use of DDT, the threat of a major insect infestation to forests or crops may bring about a relaxation of the law.

[ROBERT W. RISEBROUGH]

Bibliography: J. W. Gillett (ed.), *The Biological*

Impact of Pesticides in the Environment, Oregon State University, August, 1969; J. J. Hickey (ed.), *Peregrine Falcon Populations: Their Biology and Decline*, 1969; M. W. Miller and G. G. Berg (eds.), *Chemical Fallout: Current Research on Persistent Pesticides*, 1969; M. Sobelman, *DDT: Selected Statements from State of Washington DDT Hearings and Other Related Papers*, 1970.

Phytoplankton

Microscopic plants forming communities which live most, if not all, of their lives suspended in the water unattached to a substrate. The communities are found in marine, brackish, and fresh-water environments.

Representation and occurrence. These organisms belong to the most primitive plant groups: various divisions of algae, bacteria, and fungi. Of these, only the algae possess chlorophylls and other pigments together with biochemical systems which enable them to carry out carbon assimilation in the presence of light, carbon dioxide, water, nutrients, and trace organic substances, such as vitamins. Because of the ability of many planktonic algae to synthesize organic matter in excess of their own respiratory requirements, the phytoplankton are called primary producers. In many aqueous environments they form the basis of the food chain or web and are frequently called the "grass of the sea." *See* ECOLOGY; FOOD CHAIN.

Other phytoplankton, including bacteria, yeasts, molds, and even some algae, cannot photosynthesize, and obtain the energy for their life processes by oxidizing organic materials originally produced by the photosynthetic plants. Sometimes occurring as planktonic forms, bacteria and fungi are then extremely important in oxidizing organic matter and in returning organically bound plant nutrients, such as nitrogen and phosphorus compounds, back into the water in inorganic form. However, their identification and enumeration require substantially different techniques from those employed for the algae. As a result, the fungi and bacteria are generally studied separately from the other planktonic plants.

Among these other planktonic plants, the following groups are often represented: the blue-green algae (Cyanophyta), the green algae (Chlorophyta), the yellow-green algae (Chrysophyta), the diatoms (Bacillariophyta), the dinoflagellates (Pyrrophyta), and the silicoflagellates (a group of uncertain affinities often placed in the Crysophyta). Members of one or more of these groups are often present in natural waters, but it is only when their abundance becomes extremely great that their presence is apparent to the unaided eye.

Dense concentrations of these organisms, generally called blooms, will often cause natural waters to become discolored and may be accompanied by unpleasant tastes or odors. A planktonic blue-green alga, *Tricodesmium (Skujaella) erythraeum*, often reaches sufficient abundance in tropical seas such as the Coral, Tasman, or Red seas to turn the color of the water to red. Blooms of other blue-green algae occur in fresh water and often give it an unpleasant taste. Under extreme conditions the algae may rise to the surface, coalesce, and form dense mats. The death and subsequent decay of

such material may give rise to obnoxious odors and may produce toxic organic products or hydrogen sulfide. Green algae may bloom and cause similar conditions in fresh water. Their virtual absence in the marine environment precludes their blooming in the sea. But coccolithophorids, belonging to the Chrysophyta, have been reported in such numbers in Oslo Fjord, Norway, as to give the water a milk-like appearance. In both marine and fresh waters, diatom blooms may cause water to turn a brownish color. Although dinoflagellates, members of the Pyrrophyta, do not normally discolor fresh water, they are the organisms chiefly responsible for red water, or red tides in brackish and ocean waters. These blooms frequently cause spectacular displays of bioluminescence which are mistakenly called "phosphorescence of the sea."

Ecology. Some dinoflagellates contain toxins. *Gymnodinium breve*, a species producing a toxin lethal to fish, caused the massive fish kills off the western coast of Florida in 1946 and 1947. Paralytic shellfish poisoning in man results from eating shellfish which have been accumulating a neurotoxin contained in the dinoflagellates *Gonyaulax catenella* or *G. tamarensis*.

Not all fish kills associated with "red water" are the result of toxin produced by dinoflagellates. In enclosed bays and some coastal areas fish kills following red water are the result of a marked decrease of dissolved oxygen in the water, causing the fish to suffocate. The decrease in oxygen after such blooms is the result of the consumption of oxygen by microorganisms which consume dying and dead dinoflagellates.

Collection and identification. The methods used to study phytoplankton vary with the purpose of the investigation. Until the early 1900s interest was mainly centered on the kinds of phytoplankton that inhabited different bodies of water and how the composition changed with the seasons. For such studies the phytoplankton were collected by slowly towing through the water conical nets of very fine mesh. By 1911 it was apparent that an appreciable fraction of the phytoplankton escaped the finest mesh nets (with mesh apertures as small as 0.064 mm), and therefore that nets did not yield quantitative samples. New methods soon appeared, several of which are still in use, and additional techniques are being developed.

Two methods commonly used are the filter and the sedimentation methods. In the former, a known volume of a sample of water containing phytoplankton is filtered through a very fine filter (pore size usually less than 0.001 mm), the material retained on the filter is often dehydrated with an alcohol dilution series, the filter is made transparent with a clearing agent which has the same refractive index as the filter, and finally the filter is mounted on a microscope slide with a mounting medium such as Canada balsam. The phytoplankton are then counted and identified on the surface of the transparent filter with a compound microscope.

The sedimentation method (often called the Utermöhl method after its inventor) uses an inverted compound microscope by which the phytoplankton are examined after they have settled onto the bottom of a volumetric cylinder. Both of these methods normally use "fixed" water samples.

Unfortunately, fixatives in common use are not perfect and an unknown fraction of the delicate algae are destroyed or distorted to such an extent as to be unrecognizable. As unsatisfactory as these methods may appear, a considerable amount of information is available on the depth distribution size, and seasonal and geographical occurrence of the more robust phytoplankton.

Quantitative analyses. Because studies by enumeration are time-consuming and require trained specialists, the amount of phytoplankton is often estimated indirectly by pigment determinations. By the use of suitable conversion factors the pigment concentration (usually that of chlorophyll *a*) can be converted to a measure of the carbon in the phytoplankton. Photosynthetic rates can be estimated by measuring carbon-14 uptake or oxygen production, but the role of photorespiration is obscure. Chlorophylls and photosynthetic rates are useful as a reference level in assessing the metabolic activity of the phytoplankton, and also provide numerical information related, in some unknown degree, to the amount and turnover rate of organic material available to organisms which feed upon the phytoplankton. When knowledge has increased sufficiently, such estimates may be useful in predicting the amount of protein available for harvesting by mankind in the form of fish, animal plankton, or phytoplankton.

Phylogenetic significance. Phytoplankton remains are also used by scientists to help unravel the past history of the Earth and ocean basins. Diatoms, coccolithophorids, silicoflagellates, and to a lesser extent the dinoflagellates, by virtue of their resistant skeletal materials or in other cases their spores, have been incorporated into the fossil record. These remains provide important keys to the manner in which these organisms evolved and possibly also to the environmental conditions which obtained during recent geological eras. *See* FRESH-WATER ECOSYSTEM; MARINE ECOSYSTEM; SEA-WATER FERTILITY.

[ROBERT W. HOLMES]

Bibliography: J. E. Raymont, *Plankton and Productivity in the Oceans*, 1963; G. A. Rounsefell and W. R. Nelson, *Red-tide Research Summarized to 1964 Including an Annotated Bibliography*, U.S. Fish Wildlife Serv. Spec. Sci. Rep. Fish. 535, 1966; R. S. Wimpenny, *The Plankton of the Sea*, 1966.

Plant, minerals essential to

Beginning in the middle of the 19th century, botanists sought to determine the chemical elements required for the growth of plants. Elements such as nitrogen, phosphorus, and potassium, required in relatively large amounts, were among the first shown to be essential. The essentiality of certain elements, such as copper and molybdenum, was not established until chemical compounds of greater purity were produced. Earlier, many of the elements were present in sufficient amounts as impurities in a nutrient solution to preclude their detection as essential elements when they were "omitted" from the solution. Elements which are required in small amounts, such as copper, molybdenum, manganese, zinc, boron, iron, and chlorine, are sometimes called trace elements. They have also been called minor elements, but this term erroneously implies that these elements play only a minor role in plant nutrition. The term micronutrients is generally favored, since it implies that small amounts are required and that they perform a nutritive role.

Most plant scientists agree that the following elements are essential for higher plants: carbon, hydrogen, oxygen, phosphorus, potassium, nitrogen, sulfur, calcium, magnesium, iron, boron, manganese, zinc, copper, chlorine, and molybdenum. The last seven are micronutrients. Nitrogen, phosphorus, and potassium are the three most important and are the components of a 5-10-5 fertilizer (5% nitrogen, 10% phosphorus, calculated as phosphorus pentoxide, and 5% potassium, calculated as potassium oxide). Certain algae require vanadium, sodium, and cobalt. Additional elements may possibly be added to the present list of essential elements for plants.

Criteria of essentiality. In order to be considered essential, an element must meet the following criteria: (1) Absence of the element results in abnormal growth, injury, or death of the plant; (2) the plant is unable to complete its life cycle without the element; (3) the element is required for plants in general; and (4) no other element can serve as a complete substitute. Most scientists prefer to add a fifth criterion, namely, that the element have a specific and direct role in the nutrition of the plant. This last criterion is difficult to determine since known, direct roles for potassium and calcium, for example, are not as yet agreed upon; their essentiality, however, is unquestioned. Without exception, when the essential elements were experimentally removed from the external environment, drastically reduced growth ensued. The direct, specific roles of the elements were then pursued, most of which have been elucidated.

Roles of essential elements. The following paragraphs discuss the roles in plants of carbon, hydrogen, oxygen, phosphorus, nitrogen, sulfur, potassium, calcium, magnesium, iron, manganese, copper, zinc, molybdenum, boron, and chlorine.

Carbon. Carbon may constitute as much as 44% of the dry weight of a typical plant such as corn. It is a component of all organic compounds, such as sugars, starch, proteins, and fats, in plants (and animals). One of its main roles, is as a constituent of a vast array of compounds which in turn are synthesized from sugar. Carbon may indeed be said to be involved in all the roles played by all the carbon-containing compounds. This would include such compounds as the plant hormones, which regulate plant growth, flowering, and reproduction.

Hydrogen. This element is also a constituent of all organic compounds, and what applies to carbon similarly applies to hydrogen. Hydrogen constitutes about 6% of the dry weight of a plant.

Oxygen. This element is a constituent of all carbohydrates, fats, and proteins and, in fact, of most organic compounds. Some organic compounds, such as carotene, which gives rise to vitamin A, are composed only of carbon and hydrogen. Since oxygen is a constituent of most organic compounds, it is involved in whatever functions those compounds perform. Oxygen constitutes about 43% of the dry weight of a plant.

Carbon, hydrogen, and oxygen are constituents of the bicarbonate ion (HCO_3^-), which is believed to be one of the chief anions (along with hydroxyl,

OH⁻) exchanged by the plant for anions absorbed from the soil. Because of this role, these three elements are involved in the process of salt absorption by roots.

In the process of oxidation (aerobic respiration) of foods by the cell, oxygen is the final acceptor of hydrogen. When a food, such as sugar, is respired by the removal of hydrogen and carbon dioxide, the latter and water appear as end products. Water is formed in the terminal step in these oxidation reactions when oxygen and hydrogen unite.

Phosphorus. Certain special proteins in all cells, the nucleoproteins, contain phosphorus. Nucleoproteins are in the nucleus of the cell and hence in the chromosomes which carry the hereditary units, the genes. Since phosphorus is a constituent of the nucleoproteins, cell division is dependent on phosphorus. The transmission of hereditary characteristics also depends on this element.

Cellular membranes are believed to consist, in part, of special phosphorus-containing fats or lipids called phospholipids. These, along with hydrated protein molecules, are very likely the chief components of cellular membranes. Differentially permeable membranes regulate the entry and exit of materials from the cells; phosphorus therefore plays a role in the permeability of cells to various substances and in the retention of substances by cells.

Phosphorus is a constituent of the special compounds diphosphopyridine nucleotide (DPN) and triphosphopyridine nucleotide (TPN). DPN and TPN are commonly called NAD (nicotinamide, adenine dinucleotide) and NADP (nicotinamide adenine dinucleotide phosphate), respectively. These compounds are involved in the transfer of hydrogen in aerobic respiration, and life itself depends on this important energy-liberating process.

The early stages in the combustion or utilization of sugar by cells involve the addition of phosphorus to the sugar molecule. Only after phosphorus is added to both ends of the sugar molecule is it cleaved and prepared for further transformations which release the chemical energy stored in the sugar molecule.

Also, phosphorus is a constituent of adenosinediphosphate (ADP) and adenosinetriphosphate (ATP). In ADP one of the two phosphate bonds is a high-energy bond, and in ATP two of the three phosphate bonds are high-energy bonds. This unique concentration of energy in these special phosphorus bonds is one of the ways in which potential energy is stored within the cell. The bond is important not only because it represents a form of stored energy, but also when it is broken the released energy can accomplish "work." Many synthetic reactions, such as the syntheses of sucrose and starch from glucose, require energy which the high-energy phosphate bond delivers to the reactions.

The reduced forms of diphosphorpyridine nucleotide (DPNH₂) and triphosphopyridine nucleotide (TPNH₂) constitute the other major form of energy storage in cells. This chemically stored energy is also available for work—driving certain chemical reactions that would otherwise proceed at imperceptibly low rates.

Nitrogen. Along with carbon, hydrogen, and oxygen, nitrogen is a constituent of all amino acids: the building blocks of proteins. Protoplasm usually has a high percentage of water, but the substance portion is primarily proteinaceous. Thus nitrogen and certain other elements such as carbon, hydrogen, and oxygen are a part of the living substance, protoplasm.

All enzymes thus far isolated are protein in nature. Therefore nitrogen is a constituent of these remarkable organic catalysts, which accomplish at room temperature, or below, chemical reactions that man can perform only with high temperature, pressure, or other special conditions.

As a constituent of chlorophyll (four nitrogen atoms in each molecule), nitrogen is required in photosynthesis, the food-manufacturing process which only plants can accomplish.

Sulfur. Certain amino acids, such as cystine, cysteine, and methionine, contain sulfur and are often components of plant proteins. Sulfur is also a constituent of the tripeptide, glutathione, which may function as a hydrogen carrier in the respiration of plants and animals. Biotin, thiamine, and coenzyme A are examples of still other sulfur-containing compounds in plants.

Potassium. Although potassium was one of the first elements shown to be essential and often accounts for 1% or more of the dry weight, there is no known potassium-containing compound in plants. Despite numerous researches, it still is not known why plants require potassium in such seemingly large amounts. Although it functions as a cofactor for certain enzyme systems, the need for such high concentrations is not known. Virtually all the potassium in plants appears to be water-soluble, emphasizing further that potassium is not a constituent of any compound, and certainly not of the larger, relatively insoluble and immobile compounds.

Calcium. Although calcium may typically be present in plants to the extent of 0.2% of the dry matter, it also is not clear why plants require so much calcium. Many workers consider the cementing substance between cells, the middle lamella, to be composed of calcium pectate. No other calcium-containing compounds of biological significance have been reported for plants, and yet they contain far more calcium than would be required for its postulated role in the middle lamella. Excesses of oxalic and other organic acids may appear in the cell as crystalline calcium salts of low solubilities. These salts, however, are considered waste products and serve no vital function. Their removal from solution by calcium may prevent a toxicity that would otherwise result from such acids.

Magnesium. Magnesium is a constituent of chlorophyll molecule, each molecule containing one atom of magnesium in the center. Although other metallic ions may be made to replace the magnesium, chlorophyll functions in photosynthesis only when it contains a magnesium atom. Despite much speculation, it has not yet been determined why only magnesium is effective in chlorophyll and in photosynthesis.

In addition to the unique role which magnesium plays in chlorophyll and photosynthesis, it is also required for the action of a host of enzymes. It may have two roles in protein synthesis: as an activator of some enzymes involved in the synthesis of nucleic acids, or as an important binding agent in microsomal particles where protein synthesis occurs. It is also apparently required for an enzyme concerned with oil formation in plants, since oil droplets are not formed in the alga *Vaucheria* in the absence of magnesium. The seeds of plants, which contain large amounts of oil, are consistently high in magnesium.

Iron. Iron is required for the formation of chlorophyll but is not a constituent of the molecule. In plant leaves about 80% of the iron is associated with chloroplasts, the chlorophyll-containing plastids. Iron is a constituent of cytochrome *f*, which may have a role in photosynthesis.

Iron is a constituent of the enzymes cytochrome oxidase, peroxidase, and catalase. Cytochrome oxidase is involved in respiration, and catalase catalyzes the breakdown of any hydrogen peroxide that forms in cells as a result of certain metabolic reactions. Why plants require so much iron is not known, since a smaller quantity would appear sufficient for its known roles. Iron is also a component of ferredoxin.

Manganese. There are no known compounds in plants of which manganese is a constituent. There is considerable evidence that the element may be a cofactor or an activator in certain enzyme systems. For example, manganese may be involved in nitrate reduction and hence in nitrogen metabolism. Manganese is known to be needed for photosynthesis, particularly by algae growing in an inorganic medium.

Copper. Copper is a constituent of laccase, ascorbic acid oxidase, plastocyanin, and tyrosinase (polyphenoloxidase). The last enzyme is believed to be involved in most plants with the terminal step in aerobic respiration, the transfer of hydrogen to oxygen to form water. This action thus links copper with energy release in plant cells.

Zinc. Although the enzyme has not been isolated, it has been shown that the enzyme which synthesizes the amino acid tryptophan requires zinc. Tryptophan in turn is the precursor from which the

plant hormone indoleacetic acid is made. Zinc then is directly necessary for the formation of tryptophan and indirectly necessary for the production of indoleacetic acid.

Two zinc-containing enzymes have been isolated from plants, namely, carbonic anhydrase and alcohol dehydrogenase.

Molybdenum. Molybdenum is required in the external solution to the extent of 1 or 2 parts per billion. In the dry matter of the plant it may be present only to the extent of about 10 parts per billion. It has been calculated that the number of molybdenum atoms required per cell of *Scenedesmus obliquus* and *Azotobacter* is 3000 and 10,000, respectively.

Molybdenum is the metal component of xanthine oxidase and of the enzyme nitrate reductase, which effects reduction of nitrate nitrogen to the reduced form of nitrogen that is incorporated into amino acids and then into proteins. Molybdenum-deficient tomato plants may accumulate nitrate to the extent of 12% of the dry weight of the plant. If such plants are given a few parts per billion of molybdenum in the external medium, the nitrate content will drop to around 1% within 2 days. *Aspergillus niger*, *Scenedesmus obliquus*, and *Chlorella pyrenoidosa* also require molybdenum for the reduction of nitrate nitrogen. Fixation of atmospheric nitrogen by one of the free-living, nitrogen-fixing bacteria, *Azotobacter*, and by a blue-green alga, *Anabaena cylindrica*, requires molybdenum. The element is therefore intimately associated with nitrogen metabolism and synthesis of protein and hence, synthesis of protoplasm. *See* NITROGEN CYCLE.

Certain species of plants appear to require molybdenum for one or more unidentified roles other than nitrate reduction or nitrogen fixation. For example, cauliflower plants grown on urea and ammonium, as reduced nitrogen sources, nevertheless develop characteristic molybdenum-deficiency symptoms known as whip tail when molybdenum is withheld.

Boron. The essentiality of this element was established around 1910. Approximately 1/2 part per million (ppm) in the external solution suffices for growth of most plants. Garden and sugarbeets, as well as alfalfa, have a somewhat higher boron requirement, 5–10 ppm being optimal.

There are no known compounds in plants of which boron is a constituent, and no enzyme system has been shown to require boron. In most plants boron is immobile, suggesting that it is combined with large, immobile molecules, and plants therefore have to receive boron continually throughout the life cycle.

Numerous functions have been proposed for boron, including roles in carbohydrate and protein metabolism. One theory states that boron is required for the translocation of sugar from the leaves (where sugar is made) to the flowers, fruits, and the growing points of stems and roots. In the absence of boron, stem and root tips die, and flowering and fruiting are drastically reduced or altogether curtailed. A certain degree of deficiency of boron, for example, that which results in almost complete failure to set seeds in alfalfa, may not materially reduce the size of the plants. This well-established phenomenon signifies its unique role

Fig. 1. Young tobacco seedling showing potassium-deficiency symptoms consisting of interveinal chlorosis and marginal and apical scorch of older leaves. (*From G. Hambidge, ed., Hunger Signs in Crops: A Symposium, American Society of Agronomy and National Fertilizer Association, 2d ed., 1950*)

in flowering and fruiting. Successful germination of pollen grains and the production of the pollen tubes require boron.

Boron-deficient plants lose the normal response to gravity, indicating that boron is involved in the production, movement, or action of the natural plant hormones that cause the stem of a horizontally placed plant to turn up and the roots to turn down.

Chlorine. In the tomato plant, chlorine deficiency results in wilting of the leaf tips and chlorosis (yellowing), bronzing, and necrosis (death) of the leaves. If chlorine is added early enough, as little as 3 ppm banishes the symptoms and normal growth proceeds. Tomato plants show deficiency when they contain about 200 ppm of chlorine (dry-weight basis), whereas they show molybdenum deficiency when the concentration is about 0.1 ppm. Therefore, the tomato plant requires several thousand times more chlorine than molybdenum.

It should be made clear that plants cannot tolerate more than a few parts per million of chlorine in the molecular, gaseous state. Ordinarily plants absorb chlorine in the ionic form as chloride. Most plants tolerate 500 ppm or more of chloride without much effect upon growth, and certain halophytes (salt plants) can grow vigorously in high concentrations of chloride slats.

Other elements. Vanadium is required for the growth of *Scenedesmus obliquus*, and it plays a role in photosynthesis in *Chlorella*. There is no evidence of its essentiality for plants other than the green algae. Silicon is required for diatoms.

Sodium is an essential element for certain blue-green algae but is not required for green algae or higher plants.

Colbalt is required only for certain blue-green algae and the nitrogen-fixing bacteria in nodules of leguminous plants.

Deficiency symptoms. A deficiency of any one of the 16 essential elements results in stunted growth and reduced yield.

Deficiency symptoms are best identified by specially trained persons, since a deficiency of a particular element has different symptoms on different plants, for example, corn and beans. Furthermore, the application of nutrients to correct deficiencies, particularly of boron, copper, manganese, zinc, and molybdenum, requires specialists in plant nutrition.

The elements which are most likely to have a limiting effect on growth are nitrogen, phosphorus, and potassium: these are present in a typical, commercially available fertilizer, such as 5-10-5. Some generalizations can be made about the deficiency symptoms of these three main elements. When nitrogen moves out of the older, hence lower, leaves of a plant, the deficiency is generally characterized by yellowing of these leaves. Phosphorus deficiency is often characterized by a purpling of the stem, leaf, or veins on the underside of the leaves.

In corn, phosphorus deficiency causes purpling of the stem and sometimes of the leaf blades. Potash (potassium) deficiency results in burn or scorch of the margins of the leaves, particularly the older, lower leaves (Fig. 1). Recognition of the deficiency symptoms of these three elements can

Fig. 2. Boron deficiency in branch of a grape plant, showing interveinal chlorosis of terminal leaves and necrotic terminal growing point. (*From J. A. Cook et al., Light fruit set and leaf injury from boron deficiency in vineyards readily corrected when identified, Calif. Agr., 15(3):3–4, 1961*)

be corrected by the application of readily available commercial fertilizer.

Chemical tests (tissue tests) can often be made of key plant tissues to determine whether a particular element is lacking. These tissue tests can detect a near-deficiency state before the symptoms become manifest. In general, these tests should be made by persons trained to conduct and interpret them.

The best approach for the average homeowner or farmer, however, is to have the soil tested if there is any question as to its productive capacity. Commercial laboratories and state agricultural experiment stations provide this service. A soil test can predict in advance of planting what nutrients are lacking. By the time deficiencies appear, plant growth and yield are usually irretrievably retarded. If they are used early enough, tissue tests can detect an incipient deficiency in time for correction.

In addition to the widespread need for nitrogen, phosphorus, and potassium, it is often necessary to add other elements. The following elements have been found to be deficient in one or more areas of the United States: boron (Fig. 2), magnesium, copper, manganese, zinc, iron, calcium, sulfur, and molybdenum. A deficiency of chlorine has not been observed under field conditions. In one soil or another, a deficiency of every essential element except chlorine has been found. Considering the number of years that some soils have been under cultivation and the amounts of essential elements which have been removed by crops, it is not surprising that agricultural soils are becoming deficient in certain essential elements. When plants are unusually low in calcium or phosphorus, for example, people and particularly grazing animals may develop certain deficiency diseases. Thus there is a very intimate relationship between plant and animal nutrition that is receiving considerable attention.   [HUGH G. GAUCH]

Bibliography: J. Bonner and J. E. Varner (eds.), Plant Biochemistry, 1965; R. M. Devlin, Plant Physiology, 1966; C. A. Lamp, O. G. Bentley, and J. M. Beattie (eds.), Trace Elements, 1958; W. D. McElroy, Cell Physiology and Biochemistry, 2d ed., 1964; H. B. Sprague (ed.), Hunger Signs in Crops: A Symposium, American Society of Agronomy and National Fertilizer Association, 3d ed., 1964; F. C. Steward (ed.), Plant physiology, 1963.

Plant community

An association of plants. Plants of various species are found growing together as vegetation, and certain combinations of species are found repeated in homogeneous areas of similar ecology, or biotopes, so often that generalizations can be made concerning these combinations. A plant community, then, has a certain species composition. A list of the plants occurring in a stand can be made by species names and by life forms. A list of all species is desirable: usually only vascular plants, bryophytes, and lichens can be recognized in the field. It is often necessary to take herbarium specimens, and such vouchers will document the study permanently. Ordinary taxonomic nomenclature is generally used, but a constant effort to improve this and to split the species into biotypes of more uniform relationships to environments must be made. The species list is limited, because within a given community the rate of increase of species number with increasing area is inversely proportional to the area investigated.

Characteristic species. Some kinds of plants are characteristic of a particular species combination; they are found only in one kind of combination wherever they occur, or regionally, or perhaps locally. Other plants are always found in a particular plant community. Still other plants occur in several kinds of communities; some are almost ubiquitous. Advantage is taken of such facts to classify the plants found into characteristic species which are exclusive to a given kind of vegetation or always found in it, differential species which occur in only one of two related communities, and accompanying species which show little or no preferences. The value of a given plant community as an indicator of habitat is determined largely by its characteristic and differential species, and it is by these species that plant communities are recognized.

Properties. It is possible to arrange the lists of species and associated ecological habitat data made for various stands of vegetation in other ways than into types of plant communities. They can be arranged to describe gradients, series, continua, or functions in correspondence with various habitat factors. Properties of the vegetation other than species composition can be used to help characterize plant communities. Total yields per unit area, such as tons of forage/hectare or cubic meters of wood/hectare, life forms, dispersal or pollination spectra, and total contents of certain chemical elements, can be used as properties of plant communities.

Plants and animals which are associated also form a community, a biocoenose. Usually the plant community forms the fixed substratum for the animals, which may be mobile.

Structure. Plant communities have a structure varying in complexity from a many-layered forest to a unistratal polar or a hot desert cryptogam community. Moss, low and tall herb, low and tall shrub, and several tree layers may be present, and there may be epiphytic societies on the tree trunks and branches. In complex communities the aspect changes throughout the year as various groups of plants go through the stages of their life cycles at different times. Given the species list for a stand of vegetation, it is possible to predict much about the structure of the plant community from knowledge of the species concerned. However, one reason for studying plant communities in addition to individual plants is that in various communities individual species behave differently. Thus fireweed (Epilobium angustifolium) occurs in many forest communities of the Northern Hemisphere. It is usually sterile, but it flowers and proliferates abundantly when the forest is destroyed by fire or cutting, producing a new habitat and opportunity for a new plant community. See SUCCESSION, ECOLOGICAL.

Dynamics. The functioning of plant communities is analogous to the physiological processes taking place in the individual plants of which the community is composed, but significant interactions between plants modify, for example, the water regime of forest floor plants and the carbon dioxide made available for photosynthesis by the green plants. The physiological tolerances of individual plants to features of their habitats are thus modified to ecological tolerances by competition with other plants in the community.

The relations of plant communities to environment are systematized under various factors of the environment. Thus, plants react to such features of the environment as regional climates, soil parent materials, topographical features as these condition local climates, ground water, wind, snow deposition, fire, and the biota available to the biotope concerned. Man is a most important part of this biota, both in uncivilized and in civilized states. Plant communities in different geographical regions differ perhaps first of all because the floras available in the different regions differ, even though ecologically the regions may be quite similar. The combined and interacting effects of all these groups of factors produce at a given time a particular ecosystem or combination of plant community in its environment in which the vegetation and environmental properties stand in functional relationship to each other. Thus, on a continental scale the change in regionally representative plant communities—from the shortgrass high plains of Colorado east through the tallgrass of Kansas to the deciduous forest of Ohio—can be interpreted as a reaction of decreasing moisture along a given annual isotherm, such as 11°C. In the mountain ranges of the western United States, transitions from shadscale (*Atriplex confertifolia*) desert to sagebrush (*Artemisia tridentata*) semi-desert to oakbrush (*Quercus gambellii*) chaparral to spruce-fir (*Picea engelmanni—Abies lasiocarpa*) coniferous forest to alpine herbaceous vegetation are related to altitudinal changes in these continental climates. Precipitation increases from 100 to 1000 mm whereas temperatures drop from 10 to 0°C as annual means. *See* ECOSYSTEM.

Other factors. These regional changes in vegetation can be found when only climate changes; the other factors of the environment are fixed at some particular values. If they are fixed at another set of values, the sequence will be quite different. A modification of the relief factor in the case of the midwestern United States, a shift from the well-drained uplands to river floodplains, will result in riverine forest communities of various types all along the isotherm. If temperature is drastically lowered, as in the Arctic, even 100 mm of precipitation will result in bog vegetation. The sequence of plant communities which corresponds to a change in one factor of the ecosystem is a function of those other factors of the ecosystem which have been constant.

Climax. Static situations are described above. However, vegetation is dynamic; it evolves. Given a fixed set of the environmental factors operating on a bare area, this area will change in the types of plant communities it supports, at a constantly decreasing rate, until a steady state is attained. These equilibrium stages are climax plant communities. At such an equilibrium, which seems to be reached in a few hundred years depending on the ecosystem, the effects of climate in determining the kind of vegetation often become paramount. Although the effects of climate may become paramount in many ecosystems, in others with extremes of one of the other factors, the effect of this latter factor may persist indefinitely. Thus, very coarse-grained, sandy-soil parent material may continue to support a plant community quite different from that on the surrounding hard land, as in the Sand Hills of Nebraska with their tallgrass vegetation surrounded by climax mid- and shortgrass.

In addition to the short-term genesis mentioned above there are changes of the environmental factors themselves which result in historical changes in plant communities. Invasion of plants new to the flora, as the chestnut blight into the hardwood forest of eastern North America which almost totally killed one of the former leading dominants in this forest, or the postglacial climatic changes which have been so well documented by pollen analyses of bog sections, are examples. If one of the factors determining a plant community changes, it is an axiom of plant ecology that the community will change. *See* CLIMAX COMMUNITY.

Distribution. The distribution of plant communities over the face of the Earth has been studied more from a physiognomic than from a floristic viewpoint. Repetitions of physiognomically similar types of vegetation do occur in widely separated parts of the earth with similar climates. The evergreen sclerophyll chaparral, of winter-wet, summer-dry, mild climates in the Mediterranean region of Europe, and in Australia, South Africa, California, and Chile is an example of floristically completely diverse regions having at least superficial similarities in the appearance of plant communities because of their similar structure. *See* PLANT FORMATIONS, CLIMAX.

Classification. Finally, plant communities may be classified. The most widespread system is that developed by J. Braun-Blanquet and used extensively in Europe. Floristically similar stands of vegetation with some characteristic species in common are abstracted into associations denoted by the terminus -etum. Associations are combined into alliances (-ion), these into orders (-etalia), and these into classes (-etea). Classes in general coincide with broad, physiognomically defined kinds of vegetation or formations. The next higher unit is a floristic one recognizing such differences as those between the Mediterranean flora and that of central and northern Europe. Obviously, if two regions have different floras, they must also have different plant communities. *See* ANIMAL COMMUNITY; COMMUNITY.

[JACK MAJOR]

Bibliography: H. P. Handson and E. D. Churchill, *The Plant Community*, 1961.

Plant disease

A great obstacle to the successful production of cultivated plants, plant disease is also sometimes destructive in natural forests and grasslands. Despite large expenditures for control measures, diseases annually destroy close to 10% of the crop plants in the United States, before and after harvest, resulting in a financial loss of more than $3,000,000,000.

Diseases may destroy plant parts outright by rotting, or may cause stunting or other malformations. Most diseases are caused by parasitic microscopic organisms such as bacteria, fungi, algae, and nematodes or roundworms, although a few are caused by parasitic higher plants, such as dodder and mistletoe. Many are caused by viruses, and some are caused by poor soil conditions, unfavorable weather, or harmful gases in the air.

The living organisms and viruses which cause disease are called pathogens. Most pathogens can multiply extremely rapidly, the bacteria by simple division, the fungi by producing spores which behave as seeds but are much smaller and simpler in structure. Bacteria are about 0.0005 in. long, and fairly large fungus spores about 0.001 in. Virus particles are not visible with ordinary microscopes; they can multiply a millionfold in a short time. Roundworms reproduce by means of eggs.

Most pathogens can be disseminated quickly and widely by wind, water, insects, man, and other animals. They infect plants through wounds or pores (stomata) or by penetrating plant surfaces. Each kind of pathogen can attack only certain kinds of plants or plant parts. Once inside the plant, living pathogens obtain their nourishment from it in various ways, destroying plant tissues or weakening the plant by robbing it of its food substances. The rapidity of growth and reproduction

Fig. 1. Common bacterial blight of bean. (*From J. C. Walker, Plant Pathology, 3d ed., McGraw-Hill, 1969*)

of pathogens and of disease development varies with the kind of pathogen and host and with soil and weather conditions. Some pathogens thrive best in hot weather, others in cool weather. Extensive and destructive epidemics develop when all conditions favor the most rapid development of the pathogen.

Good cultural practices, chemical disinfestation of planting materials, spraying or dusting with appropriate chemicals to protect against airborne infection, and the use of resistant varieties are the principal control measures.

Discussed in the following sections are the economic importance, nature, and causes of plant diseases; the characteristics, growth, and reproduction of pathogens; the infection stage and development of diseases; the dissemination of pathogens; and the diseases to which plants are subject in storage. Discussion of other aspects can be found under separate titles or under the names of plants infected.

Economic importance. All plants and their parts are subject to diseases which may be caused at various stages of their life cycles not only by microorganisms, but also by higher plants, injurious salts in the soil, and harmful gases in the air. Diseases may rot the seed, kill plants, or make them poor and unsightly; they may cause root rots, stem cankers and rots, leaf spots and blights, blossom blights, and fruit scabs, molds, spots, and rots. In transit and storage they cause rots of fleshy fruits and vegetables; mold sickness of wheat, rice, corn, and other grains; and discoloration or rotting of wood and wood products.

When weather favors their development, some diseases become epidemic and ruin vast acreages of economically important plants. The historic potato famine in Ireland in the 1840s, resulting in the death of 1,000,000 people, was due to epidemics of potato late blight. Chestnut blight ruined the chestnut forests of the United States. Stem rust destroyed about 300,000,000 bu of wheat in the United States and Canada in 1916. In the United States it destroyed 60% of the spring wheat in 1935, and 75% of the macaroni wheat and 25% of the spring bread wheat in both 1953 and 1954. Stem rust, only one of more than 3000 kinds of plant rusts, has been similarly destructive in other wheat-growing areas of the world, and it continually menaces wheat, oats, barley, rye, and many grasses. The *Helminthosporium* disease of rice was the principal cause of a famine in which a million or more people died in India in 1943.

Fig. 2. Southern bacterial wilt of tomato. Plant shows leaf epinasty and wilt. (*After Kelman, from J. C. Walker, Plant Pathology, 3d ed., McGraw-Hill, 1969*)

During the decade 1951–1960, plant diseases of all crops in the United States caused an estimated annual loss of about $3,250,000,000. This includes losses in both yield and quality. The corn crop alone was reduced by an estimated 413,051,000 bu or 12% of the potential crop. To the figure $3,250,000,000 must be added an estimated annual loss of $325,000,000 by air pollutants. Cost of controlling diseases is estimated at $115,800,000 annually for a grand total of $3,690,800,000.

Plant diseases are a dangerous threat to man's future subsistence. Much of the world is now underfed, and acute food shortages often occur in many areas. The situation tends to become worse as population increases by many millions each year. Plant diseases, old and new, are a critical limiting factor in food production. The degree to which they can be controlled will help determine whether the world can feed its rapidly growing population.   [E. C. STAKMAN]

Nature of diseases. In the broad sense, disease in plants may be considered as any physiological abnormality which produces pathological symptoms, reduces the economic or esthetic value of plant products, or kills the plant or any of its parts. Damage, caused by wind or lightning or predation of insects or other animals, is not usually called disease, although such injury to living plants may result in a physiological disturbance which is truly disease. Decay of storage organs, such as tubers and roots, is disease because such plant parts are living; decay of lumber is disease only by extension of the definition, although the processes may be similar.

Disease in plants is usually evidenced by abnormalities in appearance, called symptoms, or by the presence of a pathogen in or on the plant. Some diseases, however, have no obvious symptoms; potato virus X, for example, reduces the yield of potatoes without apparent changes in the appearance of the plants.

The symptoms of plant diseases may be death (necrosis) of all or any part of the plant, loss of turgor (wilt), overgrowths (hypertrophy and hyperplasia), stunting (hypoplasia), or various other changes in the structure and composition of the plant. Necrosis may affect any part of the plant at any stage of growth. A rapid death of foliage is often called blight (Fig. 1), whereas localized necrosis results in leaf spots and fruit spots. Necrosis of

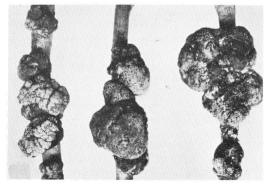

Fig. 3. Crown gall of apple. (*A. J. Riker, from J. C. Walker, Plant Pathology, 3d ed., McGraw-Hill, 1969*)

stems or bark results in cankers. Wilting may be slow or rapid, and it is usually more pronounced in dry than in moist soil. Necrosis eventually follows persistent wilting (Fig. 2). Overgrowths composed primarily of undifferentiated cells are called galls (Fig. 3), the term tumor being less commonly used to designate these structures. A bunch of small, abnormal shoots is often referred to as a witches'-broom. Underdevelopment or stunting may affect the entire plant or only certain of its parts.

Chlorosis (lack of chlorophyll in varying degree) is the most common nonstructural evidence of disease. For example, in leaves it may occur in stripes or in irregular spots (mosaic). Various degrees of curling and crinkling of the foliage often accompany chlorosis. Sometimes there is also other abnormal coloration, such as shades of red and brown.

A number of diseases may cause similar symptoms. These may be characteristic enough to permit diagnosis, but often it is necessary to identify the causal organism for exact diagnosis.

Causative agents. Usually two or more causes operate simultaneously to produce plant disease. For example, if a parasite is involved, the weather will influence the growth of the parasite as well as the plant's susceptibility to the parasite. The following subsections describe the influence on plant diseases of animals, plants, and viruses; soil conditions; weather; agricultural practices; industrial by-products; and plant metabolism.

Animals, plants, and viruses. Nematodes and insects are the animals that most commonly cause plant disease (Fig. 4). Although herbivorous animals, including many insects, bite off and swallow plant parts, the parts removed are not diseased and the animals are predators, not pathogens. However, the loss of the parts eaten may cause the rest of the plant to become diseased. Conversely, some insects are true pathogens because they remain on or in the plant and cause disease symptoms typically associated with the insects involved. Such symptoms may include yellowing, leaf curl, and overgrowths. Many nematodes are true parasites, hence pathogens, since they cause rots, overgrowths, and other plant abnormalities.

Certain algae, fungi, and bacteria are plant pathogens that cause disease. Most plant diseases are due to fungi; less than 200 are known to be caused by bacteria, and even fewer are caused by algae and parasitic seed plants, such as dodder and mistletoe (Fig. 5).

Many plant diseases are caused by viruses, which are neither plants nor animals but are similar to living things in many ways and may properly be called pathogens.

Soil conditions. Deficiencies of mineral nutrients in the soil are a frequent cause of plant disease (Fig. 6). Often the deficiency can be identified by characteristic plant symptoms. For example, yellowing of the leaf tip and midrib of corn indicates nitrogen deficiency; yellowing of the margins, potassium deficiency. However, the symptoms may vary somewhat in different plant species. In addition, deficiency diseases may be difficult to diagnose, since they sometimes resemble those caused by viruses.

Besides nitrogen, potash, and phosphorus, which plants need in relatively large amounts, smaller quantities of sulfur, calcium, and magne-

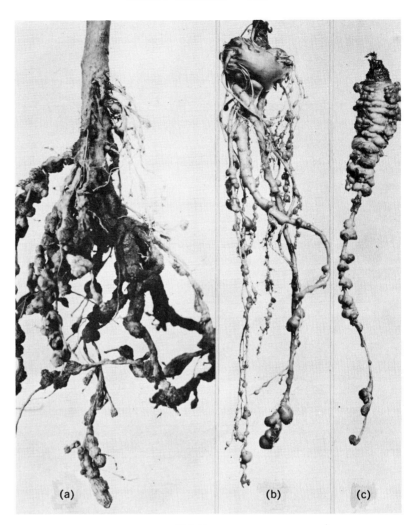

Fig. 4. Nematode galls incited by *Meloidogyne* sp. (*a*) On tomato. (*b, c*) On parsnip. (*After Cox and Jeffers, from J. C. Walker, Plant Pathology, 3d ed., McGraw-Hill, 1969*)

sium are required. Boron, iron, copper, manganese, molybdenum, zinc, and other minerals are used in such minute amounts that they are called trace elements. However, if one of the latter is missing, a typical disease may result, such as dry

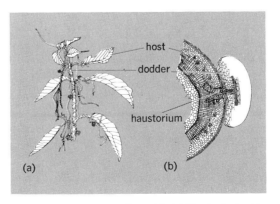

Fig. 5. Dodder (genus *Cuscuta*). (*a*) Plant attached to host. (*b*) Section through host showing haustorium of dodder extending into host. (*From F. W. Emerson, Basic Botany, 2d ed., Blakiston, 1954*)

Fig. 6.   Mineral deficiencies. (a) Potassium-deficiency disease of cabbage. (b) Iron-deficiency disease of cabbage. (c) Magnesium-deficiency disease of bean. (*From J. C. Walker, Plant Pathology, 3d ed., McGraw-Hill, 1969*)

Fig. 7.   Boron-deficiency disease of garden beet. (a) Internal necrosis of tissue occurs in the secondary cambial rings. (b) Leaves become stunted, and dormant buds of the crown are stimulated but form small distorted leaves. Internal necrosis near the exterior of the root leads to collapse of the outer tissue to form cankers. (*From J. C. Walker, Plant Pathology, 3d ed., McGraw-Hill, 1969*)

rot of rutabagas, which is due to boron deficiency (Fig. 7). See PLANT, MINERALS ESSENTIAL TO.

Frequently deficiencies of minerals cannot be determined by soil analysis alone, because the minerals may be present in chemical combinations that plants cannot use. For example, iron is often unavailable on high-lime soils, even if it is present in the soil in appreciable quantities.

Besides lime, excess amounts of many other chemicals may be present in the soil and cause plant disease. Excess of soluble salts causes "alkali injury" and aggravates drought damage; too much nitrogen may stimulate abnormal growth; while an excess of boron may cause necrosis and stunting. Unfavorable chemical balance in the soil may also result in excess acidity (low pH) or alkalinity (high pH), either of which may inhibit normal plant growth.

The soil is the principal source of water, which all plants need in varying amounts, depending upon the species. Too little available water slows growth and, below certain limits, results in wilting. Plants can recover from limited (transient) wilting, but if it is prolonged the affected parts die.

Conversely, too much water in the soil results in oxygen deficiency, which causes suffocation of the tissues of roots and other underground parts of most plants. Water-inhabiting plants, such as rice, are exceptions. Excess water in the soil favors certain kinds of fungus and bacterial diseases, and these are often confused with the purely physiologic effects of too much water.

High soil temperature during the growing season may also cause disease, for instance, internal necrosis of potato. Lack of oxygen in storage may result in blackheart of potato tubers, especially at temperatures above 40°F (Fig. 8).

The structure of soil (particle size and organic content) determines its water-holding capacity and hence affects both the conditions mentioned above and the ease with which plant roots penetrate the soil. See SOIL.

Weather conditions. Wind, lightning (Fig. 9), and hail may injure plants and cause true diseases such as those resulting from unfavorable temperatures (Fig. 10). Temperature effects range from poor development of plants grown in climates too cold or too warm to actual frost or heat damage. For example, tomatoes grow poorly and drop their blossoms in cool weather; direct sunlight on the fruit may kill the tissue, causing sunscald; and the foliage is severely damaged by even light frosts that would not harm cabbage.

Although high temperatures may literally cook plant tissue, with such results as "heat canker" of young flax and sunscald of tomato fruit, the commonest effect is to increase water loss by transpiration, resulting in drought damage. Wind has the same effect, the degree depending upon its velocity and relative humidity.

In most green plants deficiency of light causes weak, spindly growth and chlorosis, although some species can endure much shade. House plants are frequently affected in this manner, but excess shading by buildings or other plants will produce the same effect outdoors.

Agricultural practices. Mismanagement of soil, including untimely applications of irrigation water and fertilizer, can cause plant disease, but other agricultural practices are frequently injurious. The more common of these injuries result from the improper use of chemicals such as fungicides, insecticides, and herbicides. See FUNGISTAT AND FUNGICIDE; HERBICIDE; INSECTICIDE.

Nearly all fungicides are injurious to plants as well as to fungi, although the damage to the plants is usually much less than the potential injury from the diseases controlled by the fungicides. Examples of effects are increased transpiration caused by Bordeaux mixture on tomatoes, russeting of fruits caused by lime sulfur on apples, and yield reductions without visible symptoms caused by other chemicals. Conversely, the fungicide may contain a nutrient such as zinc which, if deficient in the soil, may result in much better growth of the plant.

Chemicals used for seed treatment are frequently toxic, especially to some species of plants. For example, plants of the cabbage family are stunted by copper-containing seed treatment materials. Vegetative organs, such as potato tubers, are very susceptible to chemical injury, and strong poisons, such as mercuric chloride, often do more harm than good. Materials applied to the soil to control fungi, bacteria, and nematodes may injure plants grown in the soil too soon after treatment.

Some crop plants are very sensitive to herbicides, being affected by very minute amounts of such things as 2,4-D (2,4-dichlorophenoxyacetic acid). Tomatoes may be affected from sources far removed. Symptoms of 2,4-D are sometimes confused with those of virus diseases.

Industrial by-products. The fumes from ore smelters frequently cause widespread symptoms of plant disease, including stunting, yellowing, and necrosis. Where atmospheric inversion layers prevent their escape, even traffic and domestic fumes may be toxic. *See* ATMOSPHERIC POLLUTION.

Plant metabolism products. Brown areas on stored apples (scald) may be caused by ethylene gas produced by the apples (Fig. 11). This gas occurs in small quantities in many healthy plant tissues but is produced in greater amounts by diseased and aging cells. Ethylene gas may also cause yellowing in plants, and it accelerates ripening in certain fruits such as banana.

PLANT DISEASE PATHOGENS

Most pathogens are grouped primarily on the basis of their structure; but bacteria, being morphologically simple, are classified to a considerable extent by physiological characters. Viruses represent a special problem, and such considerations as means of transmission and host symptoms are used in classifying and naming them.

Fungi, bacteria, and a few seed plants are heterotrophs; that is, they lack chlorophyll and consequently are dependent, directly or indirectly, upon green plants (autotrophs) for carbohydrates. Animals and some fungi and bacteria are also dependent upon other organisms for nutrients such as amino acids and vitamins. Viruses (Fig. 12) multiply or are replicated by synthesizing virus nucleic acids and proteins from amino acids and other compounds present in the host cell.

Plant pathogens usually penetrate into the host plant and grow within or between the cells (Fig. 13). Viruses are usually intracellular, and some are confined to the phloem, whereas plant pathogenic bacteria are usually intercellular or occur in the xylem. Fungi are composed of microscopic tubes called hyphae, by means of which plant pathogenic species penetrate into or between the host cells (Fig. 14). The powdery mildew fungi grow principally outside of the plant but send special absorptive organs (haustoria) into the host cells (Fig. 15). Some intercellular species of fungi also produce haustoria. Pathogenic seed plants, such as mistletoe and dodder, usually penetrate the host by means of rootlike absorptive organs. Pathogenic insects and nematodes may be wholly within the

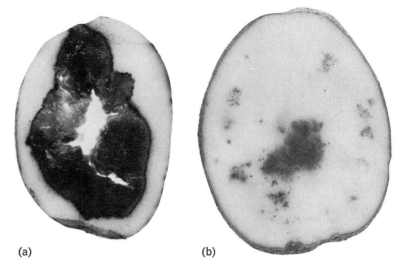

(a)   (b)

Fig. 8. Potato disease. (a) Blackheart. (b) Internal necrosis, due to high soil temperature. (*From J. C. Walker, Plant Pathology, 3d ed., McGraw-Hill, 1969*)

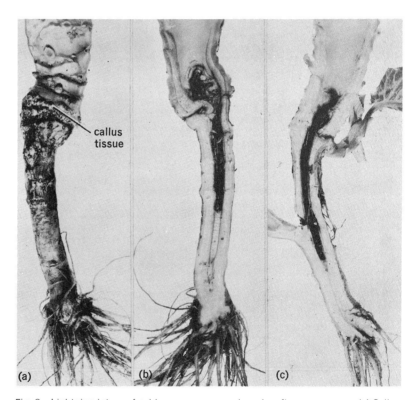

Fig. 9. Lightning injury of cabbage, seen several weeks after occurrence. (a) Callus tissue on stem at ground level where charge entered plant. (b) Interior of plant. Paths whereby the charge passed through the cortex and the vascular ring are evident. The pith was killed, and as the tissue collapsed, adventitious roots formed in the cavity. (c) Dormant buds stimulated to growth at leaf axes just below entry point of the charge. (*From J. C. Walker, Plant Pathology, 3d ed., McGraw-Hill, 1969*)

plant, or they may remain superficial and penetrate the host with specialized mouthparts.

Most plant pathogens are parasites. Some, such as the rusts and powdery mildews, are obligate parasites, that is, can grow only on a living host plant. Viruses are also in this category, although they are not typical organisms. Fungi and bacteria

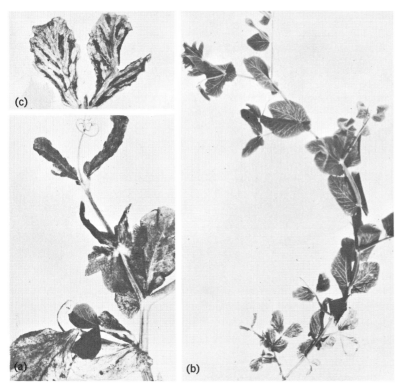

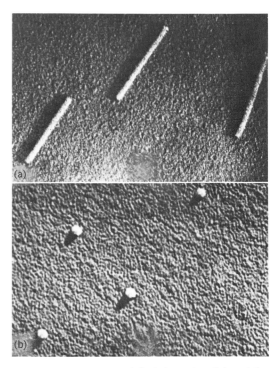

Fig. 10.   Freezing injury of pea, several weeks after injury. (*a*) Enlargement of the injured growing point in *b*; in the youngest leaf the stipules and the first pair of leaflets have assumed abnormal shapes, and the second pair of leaflets did not form. (*b*) Following killing of the growing point at the left, a lower dormant bud grew out to form the main stem. (*c*) Necrotic bands in a pair of leaflets which were developing at the time of injury. (*From J. C. Walker, Plant Pathology, 3d ed., McGraw-Hill, 1969*)

Fig. 12.   Plant viruses. (*a*) Rod-shaped particles of the tobacco mosaic virus. (*b*) Polyhedral particles of the squash mosaic virus. Electron micrographs of preparations made by the freeze-drying technique. Magnification approximately 100,000×. (*P. Kaesberg, from J. C. Walker, Plant Pathology, 2d ed., McGraw-Hill, 1957*)

Fig. 11.   Apple scald. (*USDA*)

that can use only nonliving food sources are called saprophytes.

Most fungi and all plant pathogenic bacteria can grow on nonliving organic matter as well as parasitically on living matter; these are called facultative saprophytes. Some organisms live primarily as saprophytes but also have the ability to parasitize weakened plants and are therefore called facultative parasites. Many plant pathogens have both a parasitic (or pathogenic) and a saprophytic phase of development.

Symbiotic relations of organisms. Parasitism is the one of a series of associations characterized by

intimate physical union of taxonomically dissimilar organisms. Such relationships are known as symbiosis, and may be neutral, beneficial, or harmful to the symbionts. An association such as that of legumes and nodule bacteria, beneficial to both partners, is called mutualistic symbiosis. Parasitism is antagonistic symbiosis.

There are different degrees of parasitism. In the early stages, the association between rust fungi and their hosts may appear to be almost neutral, harming the plants little. Other fungi, such as those rotting fruit, can become established only in dead tissue, producing enzymes or toxins that kill adjacent living cells which they then inhabit. Some biologists say that such organisms are saprophytes, not parasites, because they never colonize living host tissue. But the term parasitism is generally used to refer to a relationship with the host plant as a whole, because the degree of intimate relationship is often difficult to determine.

Ecologic relations of organisms. Associations of organisms in the same environment without physical union are called ecologic and are often very important in plant disease. As in symbiosis, the effects may be beneficial, neutral, or harmful. Metabiosis occurs when one organism uses a substance for food and produces a by-product that enables another to grow. If the benefits are reciprocal, the relationship is called synergism, as when the fungus *Mucor ramannianus* produces pyrimidine and *Rhodotorula rubra*, a nonsporulating yeast, makes thiazole. These chemicals are components of thiamine, which both organisms need but which neither can produce alone. If deleter-

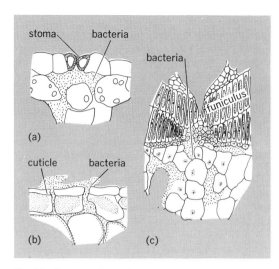

Fig. 13. Common bacterial blight of bean. (*a*) Invasion through stomata. (*b*) Invasion through rift in the cuticle of the cotyledon. (*c*) Invasion of the seed through the tissue of the funiculus. (*After Zaumeyer, from J. C. Walker, Plant Pathology, 3d ed., McGraw-Hill, 1969*)

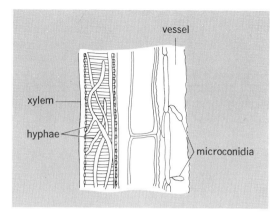

Fig. 14. The cabbage-yellows organism in trachae of the cabbage plant. Note formation of microconidia in the vessel. (*After Gilman, from J. C. Walker, Plant Pathology, 3d ed., McGraw-Hill, 1969*)

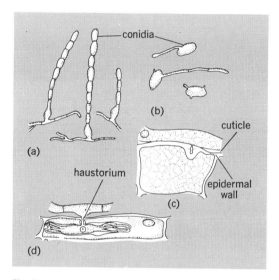

Fig. 15. *Erysiphe graminis.* (*a*) Conidiphores and conidia (spores); (*b*) germinating conidia (*after Reed*). (*c*) Penetration of cuticle and epidermal wall; (*d*) haustorium. (*after G. Smith*). (*From J. C. Walker, Plant Pathology, 3d ed., McGraw-Hill, 1969*)

ious substances (antibiotics) are produced, the relationship is called antibiosis. Usually the term antibiosis refers only to the deleterious effects of one microorganism on another, but similar relationships exist between plants of all kinds. *See* ECOLOGICAL INTERACTIONS.

All of these relationships may be important to the survival of certain plant pathogens, especially some of those which live in the soil part of the time. Metabiotic and synergistic relationships may help them to survive; antagonistic relationships hinder survival. One of the goals of the plant pathologist is to encourage antibiosis that eliminates certain soil-inhabiting pathogens.

Ecologic associations may exist between two or more pathogens inhabiting the same host plant as a common environment. When fire-blight bacteria parasitize apple twigs and permit the entrance of canker and wood-rotting fungi, the relationship between the bacteria and the fungi is metabiotic. The molds *Oospora citri aurantii* and *Penicillium digitatum* can rot fruit more rapidly together than either can alone; this is synergism. Antagonism seems to exist between races of the potato late blight fungus, and one will replace the other when they parasitize a potato plant together.

Even the relationship of host and pathogen may be ecologic at first. For example, *Rhizoctonia solani* in the soil causes visible injury to the roots of soybean before touching them. Accordingly, the fungus is at first toxic to soybean; later it becomes parasitic and pathogenic.   [CARL J. EIDE]

Growth and reproduction. Many plant pathogens, especially among the bacteria, fungi, and viruses, multiply with amazing rapidity under favorable conditions. Viruses, although not generally considered living organisms, may increase a millionfold a few days after introduction into the right place in the right kind of living plant, when temperature and other environmental conditions are favorable to the virus.

Food requirements of bacteria and fungi. Although lack of chlorophyll prevents these organisms from using solar energy to synthesize basic carbohydrates from carbon dioxide and water as green plants do, their basic nutrient requirements are essentially the same as those of higher plants. They require carbon, hydrogen, oxygen, nitrogen, phosphorus, and sulfur as structural elements. In addition, they need the metallic elements potassium, magnesium, iron, zinc, copper, calcium, gallium, manganese, molybdenum, vanadium, and scandium. Potassium and magnesium, needed in relatively large amounts, are designated macroelements; the others, some of which are needed in minute amounts, are often designated microelements. Vitamins are also needed by some species for growth and reproduction.

For experimental purposes, pure cultures of facultative saprophytes are grown in the laboratory on sterilized synthetic media containing sugars or some other source of carbon, salts of the other necessary elements, and essential vitamins and

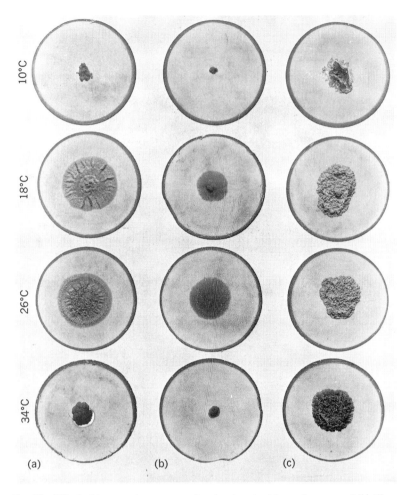

Fig. 16. Effect of temperature on growth rate of mutant lines of corn smut (*Ustilago maydis*). Note difference in ability of lines to grow at extremes. (*a*) Line W. Va. A8-3-1 is intermediate; (*b*) line W. Va. A8-5-4 scarcely grows at extremes; and (*c*) line W. Va. A8-5-2-1 grows fairly well at 10°C. (*Univ. Minn. Agr. Exp. Sta. Tech. Bull. no. 65*)

Fig. 17. Spore-producing branches of *Penicillium* similar to the one from which the drug penicillin is obtained. Chains of spores are produced on the ends of branches. (*Univ. Minn. Agr. Exp. Sta.*)

tions are solidified with gelatin or agar. Nutrient requirements for growth and reproduction are best determined by varying the composition of synthetic media. Studies on the effects of temperature, light, and other environmental factors are facilitated when organisms can be grown on culture media. Although all pathogenic organisms have some requirements in common, they differ greatly in special requirements, both on artificial media and on host plants. By growing pathogens artificially, much is learned about them which enables the development of better control measures.

Host selectivity of bacteria and fungi. Among the approximately 150 species of pathogenic bacteria and the many thousands of fungi, there are wide differences with respect to the kinds of plants and plant parts on which they can grow. Flax rust (*Melampsora lini*) grows only on wild and cultivated flax, asparagus rust (*Puccinia asparagi*) principally on asparagus; *Xanthomonas campestris*, the bacterium which causes black rot of cabbage, cauliflower, and related plants, parasitizes members of the mustard family only. On the other hand, the fungus *Rhizoctonia solani* causes root rot of potatoes, alfalfa, clover, and hundreds of other species in many different plant families; the bacterium *Agrobacterium tumefaciens* causes crown gall on grape, raspberry, chrysanthemum, and numerous other plants; the bacterium *Erwinia carotovora* causes soft rot of almost all kinds of fleshy vegetables.

Some pathogens attack only a few plant parts or tissues, others attack many. Some attack roots only, others attack stems, others cause leaf spots, and still others attack fruits. Some attack young tissues, others attack old ones. There are diseases of youth and of age, of herbaceous plants, and of woody plants. Some pathogens parasitize all plant parts of susceptible hosts at all stages of development. To understand and control the numerous diseases of thousands of kinds of plants, it is necessary to learn the conditions under which each pathogen thrives.

Environmental factors. The rate and kind of growth and reproduction of pathogens are affected by nutrition, moisture, temperature (Fig. 16), light, the acidity or alkalinity of the medium, the relative amounts of oxygen and carbon dioxide, and by other microorganisms with which they must compete. Most pathogens require free moisture for germination and infection, although some powdery mildews can germinate in dry air. Soil moisture sometimes is a determining factor in growth and reproduction. Some pathogens that live in the soil thrive best at high moisture content, some at low. Temperature determines the geographical and seasonal occurrence of many diseases, since the cardinal temperatures—the minimum, optimum, and maximum—differ for different pathogens. The peach leaf curl fungus, the potato late blight fungus, and yellow rust of wheat develop best at a relatively low temperature; the peach brown rot fungus, the potato wilt and brown rot bacterium, and stem rust of wheat develop best at a relatively high temperature. Light has less influence than temperature on the growth of pathogens in nature, but it strongly affects reproduction of some fungi. Some soil organisms, such as the potato scab bacterium, like an alkaline (high pH) soil; some,

amino acids for those organisms which cannot synthesize their own. Natural plant products, such as potato broth, steamed cornmeal, or oatmeal, often are used as nutrient bases. Liquid media are used for some purposes; for others, the nutrient solu-

such as the cabbage clubroot fungus, like an acid soil (low pH).

Reproduction of bacteria and fungi. A single bacterium divides into 2, the 2 into 4, and so on. As division may occur every 20 to 30 min, a single bacterium could produce a progeny of 300,-000,000,000 within 24 hr. The rate, however, varies with the kind of bacterium, its nutrition, temperature, and other environmental conditions.

Most fungi, however, reproduce both asexually and sexually. In many fungi asexual reproduction results in rapid multiplication (Fig. 17), whereas sexual reproduction results in the production of spores that can survive unfavorable conditions. In general, fungi continue to grow and produce asexual spores while the environment is favorable and nutrients are easily available, but they tend to produce sexual spores when growth is checked. Thus an asexually produced urediospore (summer spore) of wheat stem rust (*Puccinia graminis* var. *tritici*) can cause infection, the resulting mycelium grows for a time and then forms a pustule containing 50 to 400,000 new urediospores. The time required is only about a week at 75°F, but it increases to a month at 50°F and even longer as temperature decreases. Each new spore can cause a new infection, and this process continues at a rate that varies greatly with temperature, moisture, and light, until the wheat starts to ripen or growth is otherwise checked. Then the winter spores (teliospores) are produced; these differ from urediospores in appearance and cannot normally germinate until they have been exposed to winter weather. The apple scab fungus (*Venturia inaequalis*) may produce many successive crops of asexual spores (conidia) on the fruit and leaves during the growing season. But it does not produce sexual spores until the following spring, on infected leaves that have fallen to the ground the previous autumn. Some fungi, such as the ergot fungus (*Claviceps purpurea*), produce sclerotia, bodies made up of densely interwoven hyphae, which may survive winters or other unfavorable conditions and then produce fruiting structures under appropriate conditions in the spring.

Special stimuli are sometimes necessary to initiate the formation of fruit bodies (Fig. 18); some fungi require the stimulus of light for fructification, although they grow well in darkness; some require special temperature; others require certain nutrients or vitamins.

How, where, when, and the rate at which fungi grow and reproduce depend on their inheritance and their environment. The inheritance determines the limits within which the behavior of each kind of fungus can vary, and the environment determines its behavior under particular combinations of conditions.

[CARL J. EIDE; E. C. STAKMAN]

INFECTION AND DEVELOPMENT OF DISEASE

Infection of plants by a pathogen terminates a series of events that begins with inoculation, which is the contact of a susceptible part of a plant by the inoculum. Inoculum is any infectious part of the pathogen, such as spores, bacterial cells, and virus particles. Typically, inoculation is followed by entrance into the host, and infection follows entrance. A plant is infected when the pathogen

Fig. 18. Fruit bodies of a tree-inhabiting mushroom, *Schizophyllum*. Basidiospores are produced on the sides of the gills. In *Schizophyllum* gills are split lengthwise; in dry weather they curl up to conceal and protect surface on which spores are borne. (*Univ. Minn. Agr. Exp. Sta.*)

starts taking nourishment from it. However, some pathologists consider penetration part of the infection process.

The time between inoculation and infection is the incubation period. Because it is often difficult to tell when infection occurs, the incubation period is usually counted as the time between inoculation and the appearance of the first symptoms of infection.

The probability that infection will follow inoculation depends upon the susceptibility of the host, the virulence and amount of inoculum in contact with the host, and the duration of favorable environmental conditions. A single unit of inoculum (propagule) of many pathogens can infect a plant; this is probably true of all pathogens when conditions are ideal. Actually only part of the propagules on a plant cause infection, and hence the amount of inoculum on a plant helps to determine how many infections result. This fact has given rise to the concept of inoculum potential, which has been defined as the product of the number of propagules per unit area of the host plant and their physiological vigor. Because of the usually haphazard dissemination of inoculum of plant pathogens, only a small part of that produced actually reaches a susceptible plant. Consequently, most plant pathogens survive as species and are destructive partly because they produce fantastically large amounts of inoculum.

The inoculum. Inoculum of viruses and bacteria consists of the individual virus particles or bacterial cells, respectively; the inoculum of fungi may be spores, pieces of hyphae, or specialized structures, such as sclerotia. Pathogenic plants such as dodder produce true seed, and nematodes produce eggs, both of which function as inoculum.

Bacteria and viruses produce billions of cells or virus particles in infected plants, and each new unit theoretically can infect another plant. Fungi produce spores on the surface of hyphal growth or in a variety of specialized structures which may be large, as the giant puffball, or almost invisible to the unaided eye (Fig. 19). Some of the spore-producing structures function over a considerable period of time and, like bacteria, produce prodigious amounts of inoculum.

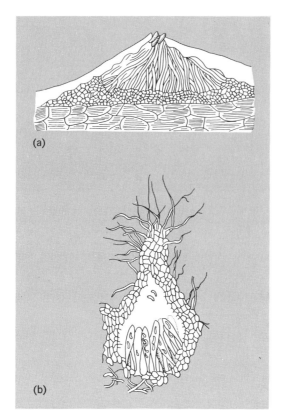

Fig. 19. *Glomerella cingulata*. (*a*) Acervulus on apple fruit. (*b*) Perithecium. (*From J. C. Walker, Plant Pathology, 3d ed., McGraw-Hill, 1969*)

PLANT DISEASE

(a)

conidium

(b)

germ tube

appressorium

Fig. 20. *Glomerella cingulata*. (*a*) Conidium (spore). (*b*) Conidium that has become septate during germination; appressorium at tip of germ tube. (*From J. C. Walker, Plant Pathology, 3d ed., McGraw-Hill, 1969*)

Bacteria and viruses are somewhat restricted as pathogens by having no special means of liberating themselves from the host, although the bacteria may ooze out in sticky droplets. For dissemination or transmission these pathogens depend chiefly upon plant contact, insects, or man, although bacteria may be spattered short distances by rain. Some fungi produce spores in sticky masses, like bacterial ooze, and are disseminated in much the same ways as bacteria. Other fungi have ways to liberate or forcibly eject spores into the air, where they can be carried by the wind. This gives fungus pathogens the potential of much farther and faster spread than the bacteria or viruses, although their arrival on a susceptible plant is much more a matter of chance than if insects carry the inoculum, because insects often seek similar plants for food. (Dissemination is considered in greater detail in a later section of this article.)

Dormant inoculum is one of the most important, but not the only, means by which plant pathogens survive during periods when parasitic life is impossible. If the pathogen is within a perennial host, it is usually quiescent during the rest period of the host. Sclerotia and even the vegetative hyphae of some fungi may survive periods of drought and cold independently of the host. Other pathogens require the protection of the dead host plant, not so much against cold and drought as against antagonistic organisms. This is especially true of plant pathogenic bacteria, few of which survive long if separated from host tissue. Some viruses can live only minutes apart from the living host; others,

such as tobacco mosaic virus, remain infective for years in dried leaves.

At the beginning of the growing season, the first inoculum of a pathogen is called primary inoculum; that which is produced later on infected plants, secondary inoculum. The primary inoculum of fungi may be resting spores or the surviving hyphae or sclerotia; often the hyphae or sclerotia produce spores which function as primary inoculum.

Many fungi produce two or more kinds of spores. Those formed late in the growing season (resting spores) usually will not germinate until after a period of dormancy and will survive more cold and drought than spores produced during the growing season. Some, like the spores of the cabbage clubroot fungus and the chlamydospores of the onion smut fungus, stay dormant for several years, thus assuring the species of survival if susceptible hosts are not grown on the land for several seasons. Such diseases are difficult to control by crop rotation.

"Repeating" spores typically are morphologically distinct from the resting spores and are produced in great numbers on diseased plants during the growing season. They usually germinate rapidly whenever environmental conditions are favorable. Before germination, repeating spores can survive for periods ranging from several hours to several weeks, depending upon the species. This determines largely how far and under what conditions a pathogen will spread during the growing season.

Spore germination. Germination, as applied to spores or seeds, means the resumption of vegetative growth leading to the development of a new individual. In fungi this usually means the production of a hypha, called a germ tube (Fig. 20). Cell division of bacteria and the hatching of nematode and insect eggs are comparable processes, so far as their function as pathogens is concerned.

Germination occurs if the spore is not dormant and if environmental conditions are favorable. This usually requires a certain temperature range and liquid water, although a few species of fungi (powdery mildews) germinate in humid air. Certain species also require the presence of food substances, special stimulants associated with the host, absence of inhibitors that may be produced by the pathogen or associated organisms, or certain degrees of acidity. Such requirements limit germination but may be a benefit to the species. For example, the necessity for a host stimulant prevents wastage of spores in the absence of the host.

When a nondormant spore is placed under favorable conditions, germination may follow in 45 min or only after several days, depending upon the species, age of the spores, and variations in the environment. Since conditions change rapidly, germination is a critical time for a fungus, because if it does not penetrate the host quickly the germ tube may be killed, especially by dryness. It is at this stage that fungi are most easily killed by fungicides.

Establishment in the host. For bacteria and viruses, entering a host is a passive process. Bacteria accidentally get into injuries or are put there by insects or other agencies; they may also be drawn by water into stomata, hydathodes, or nectaries. Viruses often are placed in the host by insects, but many can be transmitted when the sap from infected plants comes in contact with minute wounds in healthy plants.

Spores of fungi may also be carried into plants by various agencies, but many species have active means of penetration, the method usually being characteristic of the species. In some, germ tubes enter stomata by producing a flat structure (appressorium) over the stoma from which a hypha grows through the opening (Fig. 21). Others ignore the stomata: instead the appressorium adheres to the cuticle of the plant and forces a slender infection peg directly through the protective layer (Fig. 22). This apparently is accomplished entirely by pressure, as no enzyme action has been demonstrated.

Animal pathogens, such as nematodes, have special mouthparts that pierce the plant, and the nematode may remain external or it may actually enter the plant.

Even after penetration, pathogens may fail to invade the plant because of the presence of mechanical barriers, lack of proper nutrients, or the presence of inhibiting toxic substances. These factors depend not only on specific interactions between host and pathogen but also upon the environment. Successful establishment of the pathogen may mean killing the host cells and living upon the dead tissue, with or without the actual penetration of living cells.   [CARL J. EIDE]

Development within the host. After a pathogen has become established in a susceptible host, the rate of disease development under favorable conditions follows a sigmoid (S-shaped) curve with three major aspects: (1) the lag phase or incubation period, when infection is not evident externally; (2) the exponential phase, when the pathogen spreads rapidly in host tissues and symptoms and signs of disease appear; and (3) the senescence phase, when limiting mechanisms of either host or pathogen restrict further extension.

Disease development varies with genetic susceptibility of the host, genetic aggressiveness of the pathogen, and with many environmental factors that influence the host, the pathogen, and the interactions between the two. Environmental factors influence the growth rates and the metabolism of the host and the pathogen; and the interrelations between these activities determine the pattern of disease development. Furthermore, the effects of past environmental conditions on the host may affect disease development, a condition known as predisposition when host susceptibility is increased. The combined effects of these factors on growth and development of healthy crop plants in nature are poorly understood, and the problem becomes increasingly complex when the plants become diseased.

Climate often determines the adaptability of plant species to geographic areas and may also determine the geographic distribution of their diseases. For each disease there are minimum, optimum, and maximum values for each critical environmental factor. The mean measurements of weather, however, are often less important than the exact combinations of weather at critical times. Those environmental factors that deviate

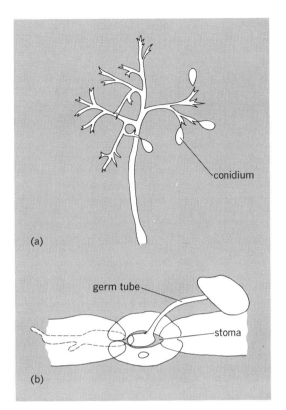

Fig. 21. *Peronospora destructor.* (*a*) Conidiophore bearing conidia. (*b*) Conidium germinating by a germ tube, the latter penetrating a stoma. (*From J. C. Walker, Plant Pathology, 3d ed., McGraw-Hill, 1969*)

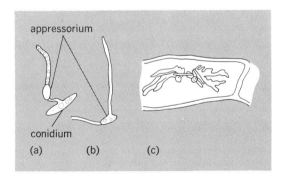

Fig. 22. *Colletotrichum circinans.* (*a*) Conidium which has germinated and formed an appressorium, which in turn has germinated. (*b*) A germinating appressorium. (*c*) Conidia and appressoria on surface of host and subcuticular mycelium developing after penetration. (*From J. C. Walker, Plant Pathology, 3d ed., McGraw-Hill, 1969*)

most from the optimum limit the development of a disease.

Effect of temperature. Temperature has a major effect on disease development and determines the seasonal and regional incidence of most diseases. For example, a succession of diseases attacks certain creeping bent grasses on golf greens in northern United States: snow mold occurs beneath melting snow in winter; red thread during the moderate temperatures of spring and fall; dollar spot during the warm temperatures of early and late summer; and brown patch during the hot weather of midsummer. The fungi causing these diseases may attack the leaves, grow into the crowns of the plants, and may kill the plants entirely in patches of characteristic size for each disease. These diseases produce similar effects, but temperature determines when each disease is most destructive.

The length of the incubation period of disease is governed by prevailing temperatures. Many diseases, such as rusts, mildews, and leaf spots, cause only small lesions on aboveground plant parts. The damage to the plant depends on the number of lesions, which in turn depends on the number of disease cycles. The elapsed time from infection to spore production—the length of the incubation period—determines the frequency of disease cycles. Thus temperature often determines whether pathogens can produce enough disease cycles for development of an epidemic. Temperature likewise may influence the symptom expression. Thus symptoms of many virus diseases disappear or are masked at high temperatures. Temperature also determines whether certain wheat varieties are susceptible or resistant to certain parasitic races of stem rust. The effect of temperature on disease development may be principally on the pathogen, or it may be on the host. When the cardinal temperatures are the same for growth of the pathogen in culture and for development of the disease, the effect is principally on the pathogen. However, when the optimum temperature for growth of the pathogen in culture differs from that for maximum disease development, the temperature probably predisposes the host plant by weakening it.

Effect of light. Light affects disease development principally by its effect on photosynthesis and the assimilative processes of the host. Obligate parasites, such as rusts and powdery mildews, generally develop best when assimilation is maximal, although the severity of the disease lesion caused by some pathogens on some hosts may be decreased by high light intensities. Low light often weakens plants and thus predisposes them to diseases caused by facultative saprophytes.

Effect of moisture. Moisture is a major factor in germination and entrance of pathogens into the host. The moisture requirements of the established pathogen are supplied by the host, since the osmotic value (water absorption capacity) of the hyphae of the pathogen is always greater than that of the parasitized host cells. Transpiration (water vapor loss) from diseased aboveground plant parts is greater than that of healthy parts. The water economy of the host is disrupted in wilt diseases by the effects of the pathogen on the translocation of water in the xylem and the osmotic permeability of foliage parenchyma, and in root diseases by the destruction of the tissues for water absorption and conduction. The rate of symptom development and death of the plant tissue in wilts and root rots is accelerated by excessively low atmospheric humidities and low soil moisture availability.

Relation of soil. Soil reaction, as regards hydrogen ion concentration, affects the development of many diseases in the soil. Potato scab is less severe in acid soils (below pH 5.2) while cabbage clubroot is not so severe in less acid soils (above pH 5.7). However, the extent to which the soil reaction affects the infectivity of these pathogens and the subsequent development of the diseases has not been determined. As the hydrogen ion concentration of the plant cell is relatively constant despite differences in the range of soil reaction, soil pH probably affects disease development indirectly by its effects on the availability to the host or pathogen of mineral nutritional elements in the soil.

Soil oxygen and carbon dioxide concentrations affect the development of root diseases. The effects on infectivity of the pathogen, predisposition of the host, and disease development have not been distinguished, although the development of the host is more adversely affected by high carbon dioxide and low oxygen tensions in the soil than is the growth of many fungal pathogens.

Effects of nutrients. The effects of nutrients are largely indirect since plants and their pathogens require the same essential mineral elements. However, the available amount of each mineral element and the balance between them affect the structure and physiology of the host and thus may be either favorable or unfavorable to the development of different pathogens. The principal mineral elements in fertilizers (nitrogen, phosphorus, potassium, and calcium) have the most pronounced effects. Diseases caused by obligate parasites such as rusts, powdery mildews, and many viruses develop best in "normal" plants having optimal mineral nutrition; while subnormal plant development due to inadequate or unbalanced mineral nutrition favors the development of many diseases, such as root rots, that are caused by facultative saprophytes. Some vascular pathogens are affected directly by the concentration of nitrogen compounds in the conductive tissues of the host.

[J. B. ROWELL]

Plant disease epidemics. When a disease spreads rapidly in a crop or other plant population, it constitutes an epidemic. Strictly speaking, the increase does not have to be spectacularly rapid to be an epidemic; the essential feature is that the pathogen, and hence the disease, is increasing in a population of host plants.

The population of plants may be small, as a single field of potatoes, or large as the total of all wheat fields from Texas to Canada. Thus epidemics may be local or regional in extent.

The time required for an epidemic to reach a destructive climax and subside may be a few days or weeks or it may go on for years. For example, an epidemic of stem rust (*Puccinia graminis* var. *tritici*) occupied roughly 3 weeks from its beginning until 100% of the plants were infected (Fig. 23). On the other hand, chestnut blight (*Endothia parasitica*) was introduced into the United States in 1904 and continued to spread through the chestnut forests for about 40 years, after which nearly all the chestnut trees were dead.

Typically the progress of an epidemic follows the same course as the growth of a population of any organism. The pathogen (for example, a fungus such as *Phytophthora infestans*, which causes late blight of potatoes) infects the host plants; in a few days the infected spots produce a crop of spores (propagules) which in turn infect new host tissue. As long as there is fresh host tissue to infect and the weather is favorable, this cycle will be repeated every 5 to 7 days. During the period of optimum development the population of the pathogen, and hence the severity of the disease, increases logarithmically, and it is possible to express the rate of increase mathematically. *See* POPULATION DYNAMICS.

Some pathogens, such as *Phytopthora infestans* and *Puccinia graminis*, may increase very rapidly, causing true population explosions. These two fungi may, on the average, double in numbers every 2 to 7 days during the logarithmic phase of development (Fig. 23). Disease of perennial plants may increase much more slowly, but still do so logarithmically.

The rate of logarithmic increase of a pathogen is dependent upon the existing population of individuals at any given time, that is, the numbers which are producing reproductive units or propagules. Other factors that affect the rate of increase can be put into three categories: (1) the number of host plants in a given area; (2) the susceptibility of the host plants; and (3) the inanimate environment.

Crops grown in fields of only one kind of plant provide ideal conditions for epidemics, because propagules of the pathogen (for example, fungus spores) can encounter more susceptible plants. Plants growing wild are usually (but not always) mixed with other species, and propagules of a pathogen, disseminated at random, have less chance to encounter susceptible plants. By growing pure stands of crop plants, disease problems are aggravated.

The pathogen itself affects the population of host plants. As an epidemic develops, plants or areas of tissue in plants become diseased and are no longer available for infection. Thus fewer of the propagules that are produced cause new infections, and the rate of increase of the pathogen and the disease slows down (Fig. 23).

Similarly, differences in host resistance influence reproduction of the pathogen. If a variety is immune, of course there is no disease, but such immunity is usually effective against only certain races of the pathogen. Varieties that are not immune may have different degrees of resistance that reduce penetration by the pathogen, increase its incubation time, and reduce the number of propagules produced. This slows down its rate of reproduction.

Environmental factors that affect individual organisms, such as temperature, light, moisture, and soil in which the plant is growing, naturally also affect the increase in the population of the pathogen and hence the rate at which the disease increases. Adverse environment tends to limit both the number of propagules that infect and the rate of development of the pathogen within the host, thus reducing the production of more propagules.

Fungicides applied to host plants reduce epidemics by killing some of the propagules of the pathogen before they can infect. Most fungicides are not completely effective, their essential effect is to reduce reproduction of the pathogen.

Epidemics, then, are characterized by the increase of successive generations of a pathogen in a population of host plants. Host and environmental

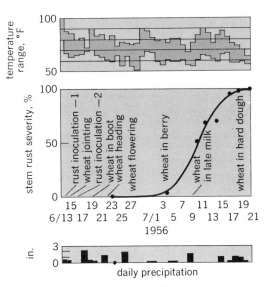

Fig. 23. Development of an epidemic of stem rust on Marquis wheat. (*From J. B. Rowell, Oil inoculation of wheat with spores of Puccinia graminis var. tritici, Phytopathology, 47(11):689–690, 1957*)

factors affect the rate of increase, and anything that can be done to manipulate such factors may reduce the increase of the pathogen to relatively harmless proportions.

DISSEMINATION OF PLANT PATHOGENS

Insect and nematode plant pathogens can within limits move about and find plants on which to live; fungi, bacteria, viruses, and seed plants have no such means of locomotion and are dependent upon other agencies for dissemination. Most fungi produce spores that are adapted for transport by the wind, but bacteria and viruses are usually carried from place to place by insects, although bacteria are often spread by water. Occasionally other animals carry plant pathogens, and man, because of his ability to travel fast and far, often becomes the unwitting means of dissemination of destructive plant pathogens.

Dissemination, which means to disperse or spread abroad, is often closely associated with transmission, which refers to the inoculation of a new host plant, as well as to the transport of inoculum. This is often accomplished by insects. However, a pathogen such as *Xanthomonas phaseoli*, which causes common bean blight, usually penetrates the seed of the plant which it has parasitized. When such a seed is planted, the new plant becomes infected. Such a pathogen is said to be seed-transmitted. Therefore this bacterium can be disseminated long distances by transporting infected seed and can also be disseminated from plant to plant in the field by splashing rain or insects.   [CARL J. EIDE]

Dissemination by wind. The plant spore is a form in which many pathogens of plants and animals spread between and within populations. Dispersal by air operates both locally and over great distances. Many viruses and bacteria that attack animals and man can be dispersed in exhaled droplets. Those that infect plants are mostly spread by vectors which range from fungi to nematodes, but the most common are insects, such as aphids or leafhoppers, which are also easily carried by wind. A few viruses can be spread within pollen and spores.

Local dissemination. Fungi are unique in the variety of physical processes they utilize to traverse the boundary layer of relatively still or smooth-flowing air adjacent to the substrates on which they grow. Powdery accumulations of detached spores, such as formed by many rust and smut fungi (or pollen), are removed by the shearing stress of moving air and, at least in the light winds usual among crops, in approximately logarithmically increasing numbers with linear increases in wind speed. Many ascomycetes and basidiomycetes have developed delicately adapted hydrostatic pressure mechanisms for releasing spores. Conidia were long assumed to be released without specialized discharge mechanisms, but some are now known to be released by the rupture of weak points in cell walls forced apart by internal pressures, by hygroscopic twisting, or (as D. S. Meredith showed) by drying mechanisms that produce great tensions in specially strengthened structures that are released when the cohesive force of the cell sap is abruptly overcome, with the formation of a gas phase within the cell. The kinetic energy of falling water drops is also used by some fungi with cup-shaped fructifications, but more often by the collision of raindrops with spore-bearing surfaces. Large transient increases of spore concentration can result from "rain tap and puff" the dual effects of the vibration of inpact and the fast radial displacement of air caused by the spreading of raindrops falling on dry surfaces. Once the surfaces are wetted, these processes are replaced by the dispersal of spores in splash droplets. Most splashed spores are in large droplets and are soon deposited, but those released in small droplets that quickly evaporate must become dry, airborne spores.

Weather and airflow factors. As so many fungi form spores on surfaces and need particular weather to liberate them, it is not surprising that the concentration of airborne spores changes dramatically with changing weather. Changes are abrupt, for example, when rain falls or cyclical in time with circadian rhythms. Spores of some pathogenic fungi are liberated only into dry air, others only into damp air, and so some are most abundant by day and others by night. Such properties have important effects on dispersal and establishment. Spores released during rain, although probably soon deposited, are more likely to find congenial conditions for germination than are spores liberated in hot dry weather; but if these can survive, they may be carried much greater distances.

Concentrations of airborne spores are determined both by the rate they are liberated and the rate they are diluted by diffusion when they reach turbulent air. The diffusion processes that meteorologists have described for aerosols show that velocities in eddies often greatly exceed the terminal velocity of spores (ranging from about 0.02 to 5.0 cm/sec for fungus spores and even up to 20 cm/sec for large pollen grains). Deposition by sedimentation is certainly important in sheltered places, but elsewhere it may be a less frequent method than impaction from moving air or washout by rain.

Range of dispersal. It is often necessary to measure the proportions of spores deposited at increasing distances from sources, propose safe isolation distances, explain the development of epidemics, and help forecast diseases. Typically, the intensity of deposition on the ground decreases fast within the first tens of meters from sources and even faster near sources close to the ground than around higher ones. P. H. Gregory drew attention to the paradox that, despite these steep deposition gradients, enough spores must escape local deposition to travel far and to establish disease hundreds or perhaps thousands of miles from their sources. Local deposition is much greater in the calmer air at night than on hot windy days with both thermal and frictional turbulence. As spores are carried higher, their distribution is increasingly determined by vertical temperature lapse rates and atmospheric pressure systems, although sedimentation and diffusion in small eddies continue. Aircraft sampling over the North Sea downwind of the British Isles has revealed the remnants of spore clouds produced over Britain on previous days and nights. Understanding these transport processes better helps to explain disease outbreaks and the dispersal tracks suggested by J. C. Zadoks.

Application of aerobiological information. Advances in aerobiology have given much information on the formation and transport of biological aerosols, but its relevance to disease attacks is often uncertain because too little attention has been paid to the viability of the spores being dispersed. This situation will not be remedied until there are better methods of growing the spores of obligate parasites and of fungi too featureless to be identified except in culture and better ways of relating contemporary spore concentration in the air to deposition on crops.   [J. M. HIRST]

Dissemination by water. Nematodes and certain fungi and bacteria which inhabit the soil may be carried from place to place by surface water. This becomes important when a pathogen such as *Plasmodiophora brassicae*, which causes clubroot of cabbage, is introduced into a field. Spores from a small center of infection can be carried all over the field. Dissemination from field to field in drainage or irrigation ditches occurs, but such dissemination for long distances is much less frequent than by wind or insects.

Splashing rain, especially if accompanied by wind, is often responsible for local spread of pathogens, especially bacteria and those fungi which produce spores in sticky masses.

Dissemination by insects. Bacteria and fungi which produce sticky spores are often disseminated by insects that sometimes are attracted to them by sweet- or putrid-smelling substances. The spores or bacteria stick to the mouthparts or other parts of the insect body and are carried from plant to plant in an apparently incidental manner. Actually this is a more effective way of dissemination than by either wind or water, because through evolutionary adaptation many insects and pathogens have developed an affinity for the same kinds of plants. Consequently a greater proportion of insect-carried spores arrive on plants they can infect than if they were scattered by the wind.

In other instances the relationship between insect and fungus is much more highly specialized. Dutch elm disease is caused by *Ceratocystis ulmi*, the spores of which are introduced into the vascular system of the tree by bark beetles (*Scolytus multistriatus* or *Hylurgopinus rufipes*) when they feed on the small twigs of healthy elm. After feeding, the beetles lay their eggs on weak or dying trees. The larvae feed beneath the bark and then pupate; when the new adults emerge, they are contaminated with the spores of the fungus which are in the pupal chambers. If the beetles did not feed on the young twigs, the fungus probably would not cause such a destructive disease, although it might still be a relatively harmless bark fungus as are similar species found only in weak trees. This is a case of transmission being more important than dissemination, although the beetle is the agent of both.

Many plant viruses are completely dependent upon insects for transmission from one plant to another. In the process the virus is also disseminated. Although a few viruses are transmitted by chewing insects, by far the most important vectors are the sucking insects, including aphids, leafhoppers, thrips, whiteflies and mealy bugs.

The diversity of relationships between viruses, plants, and insects is very great. Some viruses are transmitted by only one species of insect, some by several, and a few by many. Some insects, for example, the peach aphid (*Myzus persicae*), transmit many viruses; some only one. In some instances the insect can transmit the virus immediately after feeding upon a virus-infected plant; in others hours or days elapse before it can do so. Some viruses persist in the insect only a few hours after it has fed on a diseased plant, others persist for the rest of the life of the insect, and in some the virus is transmitted from adult to offspring through the eggs for an indefinite number of generations. In such instances the virus increases in the insect as it does in the plant, and the insect can be considered an alternate host of the virus, not merely a carrier.

The epidemiology of insect-transmitted viruses depends upon the ease and frequency with which an insect transmits the virus, which varies a great deal, and upon the life habits of the insect. If it is active, like leafhoppers, the virus may be spread far and rapidly; if it is sedentary, like mealybugs, spread may be very slow.

Dissemination by other animals. Besides insects, mites and nematodes transmit and disseminate a number of plant viruses. Among higher animals few are important as agents of dissemination of plant pathogens, though occasionally they may be very important. *Endothia parasitica*, the fungus which causes chestnut canker, produces sticky spores which adhere to the feet of birds. These birds carry the spores over considerable distances; in fact, they made it impossible to eradicate the fungus after it had been accidentally introduced into the eastern United States from the Orient.

Dissemination by man. Man is an important disseminator of plant pathogens if they lack other means of transport. Some pathogens are not adapted for widespread dissemination by wind, water, insects, or other natural means. Others have natural means for widespread dissemination, but still may be unable to surmount natural barriers such as oceans, high mountain ranges, or deserts.

Plant pathogens, as other flora and fauna, were usually confined to certain areas of the Earth before man began to alter the terrestrial environment. Frequently in his commercial, agricultural, or recreational activities, man has carried pathogens to new localities where they have caused tremendous damage, even though they were relatively harmless in their native habitat. For example, the chestnut blight fungus (*Endothia parasitica*) is a relatively mild parasite on the chestnut (*Castanea molissima*) in the Orient. When the fungus was brought to the United States about 1904 on imported chestnut trees, it spread to the native American chestnut (*Castanea dentata*), which is much more susceptible. The American chestnut forests were completely destroyed in about 40 years. Such lack of resistance in species of plants which have not evolved in the presence of the pathogen is very common, and it is the principal reason for the great danger involved in carrying pathogens to areas where they have not been before.

Dissemination by man may be hazardous even where natural barriers do not exist. Soil-borne pathogens such as *Plasmodiophora brassicae* and *Fusarium oxysporum* f. *conglutinans*, both of which attack cabbage, have been widely distributed on young cabbage seedlings which were grown in infested soil and sold for transplanting elsewhere. Neither fungus is disseminated by wind or insects, and water ordinarily carries them only to other parts of the field where they have already been introduced on infected transplants.

Plant pathogens disseminated by man are most frequently carried on infected plants. Some pathogens, such as certain cereal smuts (*Ustilago* sp.), bacterial blight (*Xanthomonas phaseoli*), and mosaic virus of bean, are transmitted and disseminated through seeds. However, this is less common and less hazardous than transmission in vegetative propagative parts such as potato seed tubers, flower bulbs, and nursery stock. Plant viruses in particular are more likely to be in the vegetative parts than in the seed.

Pathogens in either seed or planting stock have a good chance to become established in a new area because they have already infected the plant and need only to continue development when the plant starts to grow again. Similar plants and a favorable environment are likely to be present to permit further spread. Spores or bacteria disseminated independently are much less likely to survive or to find a host or favorable conditions for growth. The same may be said of pathogens in seeds, fruits, or vegetables to be used for food or in plant parts, such as lumber, to be used for manufacture.

Control of dissemination by man is attempted by quarantine.   [CARL J. EIDE]

PLANT DISEASES IN STORAGE

Tubers, fruits, and fresh vegetables are subject to spoilage by a variety of pathogenic and nonpathogenic agents during storage and transit, and often this hazard remains acute up to the time of consumption. Seeds such as those of wheat, corn, barley, soybeans, and flax, which often are stored in bulk for months or years, also are subject to deteri-

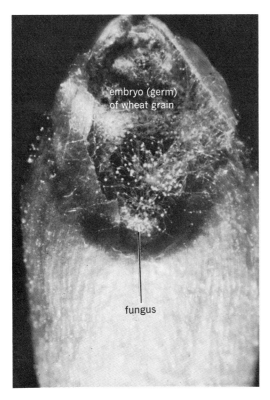

Fig. 24. Damaged grain of wheat, from commercial storage bin, with *Aspergillus* (fungus) growing from germ (embryo). Approximately 250,000 bu in this bin were affected, with a loss of more than $240,000.

oration. At times, the losses in transit and storage equal those that occur while the plants are growing. In general, storage diseases are divided into those caused by nonpathogenic factors and those caused by living organisms or pathogens.

Nonpathogenic storage diseases. Fruits and vegetables in storage suffer from a number of serious nonpathogenic or physiological diseases. Typically, these show up as discolored spots or areas on the surface of or within the affected parts, sometimes accompanied by collapse of the tissues, leaving pits on the surface or hollows within. These diseases are caused mainly by an excess of gases, such as certain esters or carbon dioxide, given off by the fruits or vegetables themselves or by chemicals introduced into the storage rooms. These diseases can be controlled by maintaining proper storage conditions, including temperature, humidity, and aeration. Fruits, vegetables, and seeds harbor abundant microflora, and damage beginning from nonpathogenic causes may be increased greatly by subsequent invasion of the tissues by bacteria and fungi able to cause rapid decay.

Pathogenic storage diseases. Common fungi, such as *Botrytis*, *Penicillium*, *Rhizopus*, and *Sclerotinia*, invade and rot many fruits and vegetables. Losses up to 25% of a shipment between harvest and consumption are common in fruits such as oranges, apples, peaches, pears, and plums and in vegetables such as potatoes, sweet potatoes, tomatoes, and peppers. Bacteria, or a combination of fungi and bacteria, often rot stored potatoes and root vegetables. These diseases may be controlled

by harvesting only sound, disease-free products, careful handling to prevent bruising, the use of clean containers, maintenance of low (about 40°F) temperatures in transit and storage, and at times the use of fungicides.

Grains stored in bulk are subject to invasion by a number of fungi, principally those in the genus *Aspergillus* (Fig. 24), which have the ability to grow at moisture contents in equilibrium with relative humidities above 70%. These reduce germinability of seeds, which is important in those to be used for malting or planting, and may reduce the quality of the grains or seeds for processing. Some storage fungi produce potent toxins; the resulting deterioration may not be detected until most of the damage has been done. Research is gradually making available to men in the grain trade the facts and principles that will enable them to reduce such losses. *See* PLANT DISEASE CONTROL.

[CLYDE M. CHRISTENSEN]

Bibliography: F. G. Bawden, *Plant Viruses and Virus Diseases*, 4th ed., 1964; W. J. Dowson, *Plant Diseases due to Bacteria*, 2d ed., 1957; P. H. Gregory, *The Microbiology of the Atmosphere*, 1961; W. R. Jenkins and D. P. Taylor, *Plant Nematology*, 1967; E. C. Stakman and J. G. Harrar, *Principles of Plant Pathology*, 1957; J. E. van der Plank, *Plant Diseases: Epidemics and Control*, 1963; J. C. Walker, *Plant Pathology*, 3d ed., 1969; R. K. S. Wood, *Physiological Plant Pathology*, 1967.

Plant disease control

Plant diseases may be controlled by a variety of methods which can be classified as cultural practices, chemicals, resistant plant varieties, eradication of pathogens, and quarantines. Often a combination of methods is most effective.

Cultural practices. Cultural practices help to control disease by eliminating or reducing the effectiveness of the pathogen or by altering the susceptibility of the host plant.

If the primary inoculum is seed-borne, its prevalence may be reduced by getting seed from an area where the disease is absent or is controlled by other means. For example, bean growers in the humid parts of the United States get seed from arid western states to avoid seed-borne *Xanthomonas phaseoli*, which causes bacterial blight. Potato growers buy certified seed which is almost free of viruses that are kept under control through isolation and rogueing, that is destroying infected plants.

Some pathogens, for example, the bacterium *X. phaseoli*, live in infected plant debris so long as the debris remains undecayed. Accordingly, bean growers not only buy western seed but also use a 2–3 year rotation plan, which allows the debris to decay, thereby eliminating the bacteria. Sometimes it is practical to remove or burn crop residue, thus reducing the amount of the pathogen that is returned to the soil.

Bacteria such as *X. phaseoli* that survive only so long as the crop residue is undecayed probably are killed by antagonistic microorganisms in the soil. Attempts have been made to reduce or eliminate some of the more persistent soil-borne plant pathogens by introducing antagonistic species into the soil or by increasing those already present. Most of these efforts have not been successful.

Certain plant pathogens, for example, viruses, infect biennial or perennial weeds which supply primary inoculum. Eradication of the weeds is often an effective means of controlling the disease in crop plants. *See* PLANT VIRUS.

The elimination of primary inoculum usually is not complete by any of the above methods and is less effective as a means of control if the pathogen increases rapidly during the growing season from a few initial infections. Sometimes the rate of spread can be reduced. For example, the spread of viruses transmitted by insects can be delayed by controlling the insect. *See* INSECT CONTROL, BIOLOGICAL.

The inherent resistance of plants to disease is usually altered by environmental factors such as temperature, moisture, and available nutrients. Moisture sometimes can be controlled, as in irrigated areas, and soil nutrients can be regulated to some degree by fertilizer practice. For example, apple trees are more susceptible to fire blight, caused by *Erwinia amylovora*, if they are growing rapidly; growth can be reduced by withholding fertilizers, especially nitrogen. *See* PLANT, MINERALS ESSENTIAL TO.

Since some pathogens enter plants through wounds, avoiding injury often reduces infection. This applies particularly to fleshy vegetables, such as potato tubers, and to fruits which are subject to storage diseases caused by bacteria and fungi. Even seeds, for example, soybean, can be cracked if harvested when too dry and become infected by soil organisms when planted.

[CARL J. EIDE]

Chemical control. Infectious plant diseases are caused by four major classes of agents: fungi (molds), nematodes (minute worms), bacteria, and viruses. These are arranged in the order of importance in chemical control, bacteria and viruses being the most difficult to control by chemicals. Hence the major tonnage of chemicals are fungicides and nematicides. *See* FUNGISTAT AND FUNGICIDE.

The last published list of antidisease chemicals showed 267 trade materials, many of which are duplicate trade names for the same chemicals.

The chemical control of plant disease differs sharply from the chemical control of human disease. Chemicals must be used to protect plants rather than to cure them, by killing the parasites while they are still in the environment and before they invade the plant.

External protection. Plant diseases attack all parts of the plant: seeds, roots, stems, leaves, and fruits. They attack plants in the field, packing shed, trains and trucks in transit, market, and home. This means that a variety of compounds and methods must be used to control parasitic forms.

Many fungi and a few nematodes attack seeds. If the organism is inside the seed, an organic mercury compound is sometimes effective. If the organism is merely on the surface of the seed, an organic sulfur compound, called a dithiocarbamate, may be used. Seeds are commonly treated by tumbling them with the dry compound in a drum or by mixing them in slurries and then drying.

Nematodes and fungi which attack plant roots must be flushed out of the soil with fumigant gases that can seep into all the crevices and spaces in the soil. A common fumigant is a mixture of methylisothiocyanate and 1,3-dichloropropene. Soil fumigants are commonly inserted several inches deep in the soil through holes in the lower end of special thin "chisels" on a machine pulled with a tractor. Soil and seed may be treated together by spilling the chemical over the seed and adjacent soil as the seed is planted. Nematicides and fungicides for use in the soil are being developed rapidly.

The biggest tonnage of chemicals for plant disease control goes onto foliage and developing fruit for protection from fungi. Foliage fungicides are known as protectants; that is, they must be applied to the plant before an invading fungus arrives. Since the waiting period may be long, serious limitations are imposed on a compound. To work effectively, it must resist erosion by rain and dew, and it must withstand the heat of the summer sun.

Few chemicals have proved to be effective. The first was elemental sulfur, used since about 1803 to control powdery mildew diseases. Although it sublimes somewhat in the heat, it survives sufficiently for effective use. The next was Bordeaux mixture, a copper-containing substance that saved the French wine industry in the 1880s. It is both exceedingly resistant to rain erosion and does not sublime. It was over 60 years before another really good type of chemical was discovered, namely, the dithiocarbamates—organic sulfur compounds which are now used worldwide. Other fungicides were developed in rapid succession, including the imidazolines, dinitrocaprylphenyl crotonate, trichloromethylthioimides, and guanidines.

Foliage and fruit fungicides are generally applied as sprays delivered by powerful machines that move through the crops. In areas where fields are large, farmers use airplanes as flying sprayers. Compounds are seldom effective if applied as dry dusts to foliage and fruit because the dusts do not stick well and consequently do not resist rain action (Fig. 1).

Fruits and vegetables are subject to disease and decay after picking. Much of this postharvest loss is not yet subject to chemical control, but oranges and other citrus fruits are widely treated in the packing shed with such compounds as diphenyl and sodium *o*-phenylphenate.

Chemotherapy. The newest frontier of the science of plant disease control is chemotherapy, the control of disease by internal treatment rather than by external protection (Fig. 2).

Although extensive researches have gone into

Fig. 1. Chemical control of potato late blight, Toluca Valley, Mexico. (*Rockefeller Foundation*)

Fig. 2. Control of downy mildew of lima beans with a streptomycin sulfate compound, Agristrep. (a) Untreated, 100% infection. (b) Sprayed with 100 ppm of streptomycin, excellent control. (*W. J. Zaumeyer, USDA*)

chemotherapy of plant disease, the biological obstacles to success are great. Effective compounds have been discovered, but to move them inside the plant to the right place at the right time and to keep them there for a long enough time is extremely difficult. Sap in the sap stream passes a given point only once; therefore if the compound cannot cure the infection in one pass, it fails. An even more severe limitation is that plants have nothing even faintly similar to white blood cells. In animals the white cells clean up the few straggler germs left by the drug. Unless the chemical treatment kills every pathogen in a plant, however, the infection flares again and eventually kills the plant. *See* PESTICIDE.   [JAMES G. HORSFALL]

Disease resistance. Plants are surrounded by a vast number of microorganisms that inhabit the air and soil. A few invade plants and injure them by disrupting their normal growth and development (disease). Such microorganisms are called pathogens. It is remarkable that plants are not injured by most microorganisms around them. Many microorganisms actually are essential for the growth and development of plants. The ability of plants to grow in an environment of microorganisms is evidence of how well they resist the harmful effects caused by pathogens (disease resistance). Most pathogens cause disease to only a certain plant, for example, pathogens of corn do not cause disease on wheat or geraniums. In addition, there are certain varieties of plants that are resistant and others that are susceptible, for example, a pathogen may attack a certain kind of wheat but not another.

What makes a plant resistant to disease? Each plant cell carries the genetic information that can be translated into chemical reactions and structures which determine whether a plant is resistant or susceptible. The mechanisms for disease resistance in plants are highly efficient since susceptibility is the rare exception in nature. Resistance may be due to the inability of the microorganism to penetrate the plant, inability to develop in the plant because the microorganism lacks nutrients or encounters compounds that inhibit its growth or reproduction, or the ability of the plant to resist microbial toxins.

Barriers to penetration. To cause disease, mi-

croorganisms must first establish physiological contact with the plant. Plants are covered by a coating of waxlike material called the cuticle, which serves as a nonspecific physical barrier. Plant storage organs also have outer protective tissues commonly referred to as peel. In addition to serving as a physical barrier, the outer coverings of many plants also contain chemical compounds that are toxic to many microorgansims. The peel of the Irish potato tuber contains chlorogenic acid, caffeic acid, α-solanine, and α-chaconine, whereas the outer scales of red onions contain protocatechuic acid and catechol. Natural openings (stomata and lenticels) exist in the barrier layer, and also this outer barrier is often broken by injury. Some pathogens invade plants through the natural openings and others only through wounds, but a few can grow through the barriers by mechanical pressure or by producing enzymes which dissolve parts of the protective barrier. The thickness of cuticle or peel is important in the resistance of certain plants to disease. In general, however, the contribution of cuticle or peel to resistance is thought to be small, except with microorganisms that invade plants only through wounds.

Nutrient factors. The absence of a factor necessary for microbial growth may be part of the disease-resistance mechanism of a plant. Certain bacteria can attack plants only if nutrients are available on the plant surface. Invasion of plum tree bark by *Rhodosticta quercina*, a fungus requiring myoinositol, is related to the myoinositol content of the bark. Susceptible varieties contain more than 10 times the amount of myoinostol found in resistant varieties. Except for microorganisms that can grow only in living plants, the presence or lack of nutrient factors is not considered a major mechanism for disease resistance. Most microorganisms grow on plant extracts or dead plant tissue from healthy susceptible or resistant plants.

Production of inhibitors. Once the microorganism has established physiological contact with the plant, it can influence the chemical reactions (metabolism) in the plant. Often this interaction of plant and microorganism results in the production of compounds (phytoalexins) around sites of penetration which inhibit the growth and reproduction of the microorganism. Phytoalexins include chlorogenic and caffeic acids, ipomeamarone, pisatin, phaseollin, 6-methoxy-mellein, orchinol, gossypol, oxidation products of phloretin, and an inhibitor not yet characterized from soybeans. Ian Cruickshank and coworkers established a relationship between the resistance of pea pods to fungi and the production of the phytoalexin, pisatin. Fungi unable to cause disease in pea pods were found to stimulate production of pisatin in amounts which markedly inhibited their growth, whereas pathogens of peas stimulated production of less pisatin. The amount of phytoalexin produced by the plant, therefore, is not the sole factor in disease resistance. The sensitivity of the microorganism to the amount of phytoalexin produced is also important. This suggests that at least two distinct biochemical mechanisms are involved in determining disease resistance—one controlling biosynthesis of phytoalexin and the other controlling the sensitivity of the microorganism to the inhibitor.

It is not surprising therefore to encounter situations where more of phytoalexin is produced by a plant after inoculation with a pathogen than with a nonpathogen. Some plants produce more than one phytoalexin after infection. Chlorogenic acid and 6-methoxy-mellein are produced by carrot and reach toxic levels around sites of microbial penetration within 24 hr after inoculation with microorganisms that do not cause disease on carrot. Microscopic examination of the inoculated tissue indicates that the inhibition of microbial growth in the carrot coincides with the production of 6-methoxy-mellein and chlorogenic acid at levels toxic to the microorganisms. It has also been demonstrated that the foliage of resistant but not susceptible varieties of soybeans produce a phytoalexin when inoculated with the pathogen *Phytophthora sojae*. Nonpathogens induce the production of the phytoalexin regardless of the susceptibility of the soybean variety of *P. sojae*. The appearance of phloretin in apple leaves following injury or inoculation with the pathogen *Venturia inaequalis* illustrates not the synthesis, but the liberation of a phenol from its nontoxic glycoside.

Phloridzin, the glucoside of phloretin, is found in leaves of apple varieties susceptible or resistant to attack by *V. inaequalis*. There is no correlation between the phloridzin content of leaves and resistance to the pathogen. When the pathogen penetrates the leaf of a highly resistant variety, the plant cells around the point of penetration immediately collapse, phloridzin is hydrolyzed by the enzyme β-glycosidase to yield phloretin and glucose, and phloretin is oxidized by phenol oxidases to yield highly fungitoxic compounds. In susceptible varieties the fungus penetrates the leaves and makes extensive growth beneath the cuticle for 10–14 days without causing collapse of plant cells. Thus phloridzin is not hydrolyzed and the pathogen is not inhibited. After 10–14 days the fungus sporulates on the leaves of susceptible varieties, and the affected tissue collapses. As with the resistance of soybeans to *P. sojae*, the potential for resistance appears to be present in all plants, and resistance may be determined by the ability of the microorganism to trigger synthesis or liberation of an inhibitor in the host. A similar series of reactions involving arbutin and hydroquinone may be important in determining the resistance of pear to the bacterium *Erwinia amylovora* that causes the disease fire blight.

Some compounds that accumulate around sites of infection or injury, for example, chlorogenic and caffeic acids, are widely distributed throughout the plant kingdom. The synthesis of others appears limited to a narrow host range, for example, 6-methoxy-mellein in carrot, phaseollin in the green bean, pisatin in the garden pea, and ipomeamarone in sweet potato. Apparently the plant has the potential for synthesis of the compound, and the microorganism used for inoculation determines the quantitative response of the plant. Where two or more compounds are produced in response to infection, the microorganism controls the relative concentration of each.

Resistance to toxins. The ability of plants to resist factors arising from plant-microbial interaction, which lead to tissue disintegration or impaired metabolic activity, may also be part of a disease-resistance mechanism. This resistance mechanism may include the presence of resistant structural components and metabolic pathways, mechanisms for detoxication, and the presence of alternate pathways for metabolism in the plant.

The resistance of immature apple fruit to many fungi that cause rots has been related to the resistance of pectic compounds in the cell walls of the green fruit to enzymes produced by the fungi. In mature fruit the microbial enzymes dissolve the cell walls, but the cell walls of immature fruit are not destroyed. During the growing season the amount of water-insoluble, resistant cell-wall material decreases as the apple matures. A major drop in the polyvalent cation content of cell-wall material occurs at about the time the fruit becomes susceptible. Conversely, the potassium content of the cell-wall material is higher in susceptibility than resistant fruit, with a major increase occurring with the onset of susceptibility. It appears that a pectin-protein polyvalent cation complex making up the cell-wall material of resistant fruits is responsible for the resistance of the fruits. The cell walls become susceptible as the polyvalent cations are lost.

Further evidence for the role of polyvalent cations is provided in the resistance of old bean seedlings to the fungus *Rhizoctonia*. Calcium ions accumulate in and immediately around developing lesions caused by the fungus. Barium, calcium, and magnesium ions inhibit tissue maceration by enzymes produced by the fungus, whereas potassium and sodium ions do not significantly influence the process. Bean tissue with lesions is more difficult to macerate with enzymes produced by the fungus than comparable healthy tissue. The cementing material (middle lamella) between plant cells around lesions is more difficult to dissolve than the middle lamella of cells more distant from lesions. The accumulation of polyvalent cations in advance of the fungus makes pectic substances of the tissue resistant to breakdown by the fungus.

In addition to producing enzymes which destroy plant cells, microorganisms have also been reported to produce toxic compounds that injure the plant without dissolving the plant tissue. The toxins that are produced are very specific for the plant in which the microorganism can cause disease. Varieties of the plant that are resistant to the microorganism are also resistant to the action of the toxin. The mechanism by which the plants are able to resist the toxin is not known; however, resistant plants may be able to change the toxin into nontoxic forms, or the toxin may be unable to penetrate into vital parts of the resistant cell. Microorganisms known to produce host-specific toxins include *Alternaria kikuchiana*, *Helminthosporium victoriae*, *Periconia circinata*, and *Helminthosporium carbonum*.　　　　　[JOSEPH KUC]

Breeding and testing for resistance. The use of disease-resistant varieties that prevent or limit infection is one of the most widely used methods of disease control. Much of man's food and fiber supply depends upon the growth of disease-resistant crops.

In dealing with disease resistance, modern scientists must recognize not only the genetic system of the host plant but also the genetic system of the pathogen. Infectious diseases, in the final analysis, are the end result of gene-controlled chemical processes that are modified to some extent by the environment.

Use. The use of disease-resistant varieties with satisfactory yield and quality characteristics is the best and most economic means of disease control. It is the only feasible means of control for many virus and bacterial diseases, most soil-borne diseases, and many foliage diseases of extensively grown crops. Over 75% of the diseases of field crops and over 50% of the diseases of vegetable crops are controlled by means of host resistance.

The meat supply for human consumption depends on corn and sorghum hybrids resistant to blights, smuts, and other diseases, and the vegetable-processing industry depends on varieties of corn, peas, cucumber, beans, and tomatoes resistant to wilts, viruses, and other diseases. The list can be extended manyfold. Most of these crops are annual plants with which rapid breeding programs are possible. Resistance is less frequently used in crops with a long life-span or in high-value crops possessing special qualities or ornamental characteristics.

Disease tolerance. Tolerance is the ability of a variety to endure disease attack without suffering the same reduction in yield or quality as another variety. Although disease tolerance is a valuable attribute of a plant, tolerant varieties have the disadvantage of allowing pathogen reproduction to take place, thus exposing nearby nontolerant varieties to infection.

Disease escape. Some plant varieties, normally susceptible to disease, escape being inoculated and therefore are less damaged by disease than are other varieties grown in the same area at the same time. For example, early varieties of plants may mature before the pathogen reaches the area in which they are grown. Plants with an upright growth habit may be less frequently infected than plants with a prostrate type of growth. A plant may be resistant or unattractive to an insect that is a carrier for a virus and thereby may escape infection.

Genetics of resistance. The expression of resistance is an inherited character in the plant and can take several forms. The simplest mode of inheritance is a single gene. Gene action may be completely dominant, incompletely dominant, or recessive. When studied in detail, genes for resistance are frequently found to exist in a large number of separate functional forms or alleles. Resistance in a variety may also be due to two or more genes acting individually or through some form of gene interaction. When two or more genes must be present simultaneously in a variety for resistance to occur, gene action is spoken of as complementary. Modifier genes may individually show no effect on disease reaction, but together with another gene for resistance may either enhance or reduce the expression of resistance.

Disease reaction sometimes is under the control of a large number of gene loci. In these situations disease reaction in segregating populations usually grades continuously between, and sometimes beyond, the limits of the susceptible parent reactions. Gene action is largely additive, but sometimes dominant and epistatic effects are seen.

Stability of resistance. Resistance can fail to function because of certain environmental conditions or because of genetic changes in the pathogen which may enable it to overcome the resistance of the host.

Cabbage varieties with multiple-gene resistance to yellows become susceptible at soil temperatures above 24°C, whereas varieties with single-gene resistance do not. Certain cereal varieties are resistant to rust in cool summers but may be susceptible when air temperatures are high. Resistance to root-invading fungi may break down when roots are injured by nematode feeding.

The most common failure of resistance, however, originates from genetic changes in the pathogen. Pathogens are living organisms with systems for the storage and release of genetic variation. Most rust, smut, and powdery mildew fungi occur in the form of races that differ from each other in ability to infect varieties of their host plants. New races that have the ability to attack a currently grown variety can appear in nature. Resistance to the new race must then be found and the breeding program repeated. Experiments indicate that this phenomenon is true for certain forms of resistance but not for others. Thus certain forms of generalized plant resistance give protection even against pathogens that are made up of many races. In other organisms, specialized races with reference to pathogenicity are not known. This is true for many organisms that cause root and culm rots, vascular wilts, and numerous foliage diseases. Resistance to these organisms seems to be lastingly effective.

Testing for disease resistance. In breeding for disease resistance or in selecting among established varieties, a reliable method for determining differences in disease reaction is necessary. Sometimes natural infection must be used; more commonly, however, carefully designed, artifical inoculation procedures are employed in field, greenhouse, or laboratory tests.

The inoculation method must be reliable so that escapes do not occur. Conditions for disease development should not be so severe, however, that different gradations of reaction to disease do not appear. It is desirable that a large number of plants be inoculated and that the method does not interfere with the breeding and selection scheme for the crop.

The inoculum should be a pure culture of the pathogen with an optimal degree of virulence. In many instances pathogenic races must be carefully selected so that plants with the desired type of resistance can be identified.

Environmental control, particularly of temperature and moisture, is needed during the inoculation and testing period. Comparable results from test to test can then be achieved, and different types of resistance can be distinguished.

Sources of resistance. Resistant germ plasm for use in breeding programs has been found in native and exotic varieties and even in wild species. When adapted native varieties are available, breeding objectives are more rapidly achieved. Resistant selections may need only to be identified and put into production. Exotic varieties have been valuable sources of resistance. For example, cottons from India and Africa have provided bacterial blight resistance; wheats from Australia and Kenya have furnished valuable genes for rust resistance; barleys from Ethiopia have provided yellow-dwarf-virus resistance; and sugarcanes from Java and India have provided the resistance to mosaic needed in American varieties of these crops. Noncultivated plant species frequently have more resistance to disease than cultivated species. Sterility, lack of chromosome pairing, and presence of undesirable characters are limitations in the use of wild plants in breeding programs. Nevertheless, wild relatives of tobacco, tomato, potato, sugarcane, wheat, oats, and other crops have contributed valuable genes for disease resistance to cultivated species.

Employment of resistance. Superior genes for disease resistance are usually exploited rapidly by plant breeders, and numerous varieties with similar resistances are put into production. With the recognition of different types of disease resistance, more careful attention is now given to the type of resistance used in the breeding program. Greater attention is directed to the less complete but more generalized types of resistance. Diversity of resistance and combinations of resistant types are also regarded as important. This has been necessary because of the repeated failure of varieties with only specific types of resistance to maintain this resistance in nature.

To maximize the effectiveness of genes for specific resistance, multiline varieties and hybrids are being developed. Multiline varieties are composed of a mechanical mixture of backcross-derived lines; each line contains different genes for resistance, but all are similar in maturity, appearance, and other respects. The components of the mixture can be varied from year to year, depending upon the pathogen races present. In addition, since some plants in the mixture are resistant each year, disease development on susceptible plants is delayed, enabling the plants to mature with light damage.

Breeding methods. Methods of breeding for disease resistance do not differ greatly from those employed for other characters. Selection, varietal hybridization and selection, and backcrossing are the methods most commonly employed. Simple selection procedures have resulted in the isolation of resistant varieties from heterogeneous crops. Where agricultural technology is more advanced and pure-line varieties of crops are grown, genetic variation must be achieved. Simple or complex crosses are made between sources of disease resistance and varieties possessing other desirable qualities. This is followed by careful selection and testing during the segregating generations so that superior disease-resistant varieties can be identified. As breeding programs have become more advanced, the only improvement needed in a variety may be additional disease resistance. In these situations the backcross breeding method is commonly employed. The method consists of repeatedly crossing each generation of resistant plants with the susceptible variety. After several generations, this is followed by selfing and selection. The breeding procedure allows for the recovery of nearly all the characteristics of the original variety but with resistance added.

Breeding for disease resistance has been a process of challenge and response. As each new disease has threatened the destruction of a crop, resistant varieties have been developed and put into production. Although breeding for resistance has been highly effective, it is expected that further gains in the efficiency and effectiveness of the method will be made. To achieve this, modern research is aimed at the identification of superior forms of disease resistance and a greater understanding of the genetical and biochemical nature of resistance. *See* BREEDING (PLANT).

[ARTHUR L. HOOKER]

Eradication campaigns. These are designed either to eliminate recently introduced pathogens completely or to protect economic plants by destroying alternate, or weed, hosts. Success in eliminating pathogens depends on early detection of the pathogen and on the efficiency of eradication measures.

Attempts made in the United States to eradicate chestnut blight and the Dutch elm disease were unsuccessful. The citrus canker disease, however, was eliminated from Florida by burning infected trees. Flag smut of wheat, which was introduced locally into Mexico, was also successfully burned out. Similarly, persistent eradication of infected plants has helped restrict many diseases.

Certain rusts can be controlled wholly or partly by eradicating alternate hosts. For example, the destruction of red cedars near apple orchards protects apples against the *Gymnosporangium* rust, because this rust cannot maintain itself on either host alone. To help control stem rust of wheat and other small grains, the growing of barberries, *Berberis* sp., has been prohibited by law in some countries. Denmark began a successful campaign against barberries in 1904, and in the United States about 500,000,000 barberries have been destroyed since 1918, with substantial reduction of the stem-rust menace (Figs. 3 and 4). Likewise in the United States white pines and other susceptible species are partly protected from blister rust by eradicating nearby currants and gooseberries, *Ribes* sp.

Like legal public health measures for human beings and for domestic animals, those for plants are essential in keeping many diseases in check.

[E. C. STAKMAN]

Fig. 3. Hormone-type chemical sprays used for killing rust-susceptible barberry plants. This process is much faster and less costly than common salt. (*USDA*)

Fig. 4. More than 513,500,000 rust-susceptible barberry bushes were destroyed on 153,000 rural and urban properties in 19 barberry-eradication states. (*USDA*)

Quarantines. Plant disease quarantines are legal measures taken by Federal or state governments to prevent the introduction of foreign plant diseases or pests into an area. Quarantines are based on the philosophy that government has the right and obligation to protect its agricultural resources and industry from the destructive effects of exotic plant diseases and pests.

The transportation of plants and plant parts was long a matter of private concern, with the result that many plant pathogens became widely distributed by international travelers or through unrestricted trade channels. The dangers of this situation became dramatically apparent following the accidental importation of the chestnut-blight fungus into the United States from Asia between 1900 and 1905 and the ultimate destruction of the American chestnut forests.

As a consequence of this and other bitter lessons, the United States government in 1912 passed the national Plant Quarantine Act. Today essentially all nations have enacted protective quarantine regulations. Quarantine laws authorize Federal or state officials to intercept and inspect shipments of plant materials and to release, fumigate, or confiscate the shipment in accordance with legal provisions. Quarantine inspectors are stationed at ports of entry, border stations, and at receiving and distributing points for freight and mail.

Value of quarantines. The value of quarantines has long been disputed. Antagonists claim that man is unable to prevent the movement of microscopic pathogens, that many quarantines are scientifically unsound, and that on occasions quarantines have been used as economic sanctions in restraint of free trade and have caused unnecessary economic losses. Supporters insist that even though not 100% effective, quarantines do prevent the introduction of many pests and diseases and retard the movement of others, giving scientists time to combat them before they become well established; that quarantines annually save the agricultural industry millions of dollars; and that these economic gains are many times greater than any possible business losses resulting from the application of quarantine measures.

Improvements in the practice of quarantining may provide assurance that all quarantines will be established on sound biological bases for maximum effectiveness, that they can be lifted with equal facility when it becomes clear that they are no longer necessary, and that, insofar as possible, new quarantine laws would be preceded by international consultation in an attempt to obtain mutual agreement and to ensure minimum disruption of the international exchange of commodities.

International disease protection. Joint efforts are made by nations to protect their agricultural resources and industries without impairment of exchange of commodities. Ideally, available knowledge on plant pests and pathogens is utilized to devise methods to limit their geographic spread and to prevent the outbreak of epidemics. Changes in cropping patterns, trade agreements, and the distribution of pests and pathogens necessitate a continuing program consisting of (1) annual plant disease surveys by the several nations with free exchange of results, (2) rigorous practice of local sanitation and plant protection, (3) prompt distribution of resistant varieties of crop plants, (4) exchange of information in improved control measures, and (5) international consultation with respect to the establishment and enforcement of quarantines.

International plant protection can be successful only when regulatory activities are fortified by scientists investigating the etiology of plant diseases, life cycles of pathogens, host-parasite relationships, and chemical and other control measures. Exchange visits by scientific personnel further strengthen understanding and lead to logical and amicable agreements. Among international organizations active in plant protection are the Food and Agriculture Organization of the United Nations, and the International Commission on Plant Disease Losses.

[J. GEORGE HARRAR]

Bibliography: E. Evans, *Plant Diseases and their Chemical Control*, 1968; R. N. Goodman, Z. Kiraly, and M. Zaitlin, *The Biochemistry and Physiology of Infectious Plant Disease*, 1967; J. B. Harborne (ed.), *Biochemistry of Phenolic Compounds*, 1964; A. L. Hooker, The genetics and expression of resistance in plants to rusts of the genus *Puccinia*, *Annu. Rev. Phytopathol.*, 5:163–182, 1967; T. Johnson et al., The world situation of the cereal rusts, *Annu. Rev. Phytopathol.*, 5:183–200, 1967; J. Kuć, Resistance of plants to infectious agents, *Annu. Rev. Microbiol.*, 20:337–370, 1966; E. E. Leppik, Relation of centers of origin of cultivated plants to sources of disease resistance, *USDA Agr. Res. Serv. Plant Introd., Invest. Pap.*, no. 13, 1968; C. J. Mirocha and I. Uritani (eds.), *The Dynamic Role of Molecular Constituents in Plant-Parasite Interaction*, 1967; K. S. Quisenberry and L. P. Reitz (eds.), *Wheat and Wheat Improvement*, 1967; E. C. Stakman and J. G. Harrar, *Principles of Plant Pathology*, 1957; J. C. Walker, *Plant Pathology*, 3d ed., 1969; W. Williams, *Genetical Principles and Plant Breeding*, 1964; R. K. S. Wood, *Physiological Plant Pathology*, 1967.

Plant formations, climax

All the Earth's surface is occupied to some extent by vegetation except for polar and alpine regions of permanent snow and ice and local physiographically active areas such as sand dune complexes (Sahara desert, in part). Despite the hundreds of thousands of species of plants and the thousands of communities (associations) in which they occur, there is a limited number of major world formation types. Each formation occurs over wide areas, most of them on all continents, and each has a recognizable structure because of the characteristic life forms (tree form, grass form) of the dominant plants. Yet the floristic composition of a formation may be so different in its different areas of occurrence that no species is common to the major areas. Each formation occupies a distinctive type of general climate, although variable local climates occur within its region.

Plant geographers differ in their classifications of formations, particularly the subdivisions of the world formations as they are intensively analyzed in local regions. However, the several formations fall into three major groups of terrestrial vegetation: forests and woodlands, grasslands, and deserts. These formations are listed in the classification shown. *See* PLANT GEOGRAPHY.

Forest formations (including woodlands)
   Tropical rainforest (including subtropical types)
   Temperate rainforest (including mountain types)
   Tropical deciduous (monsoon) forest
   Summer-green deciduous forest
   Needle-leaved forest (circumboreal type especially)
   Evergreen hardwood (Mediterranean) forest
   Savanna (transitional to grassland)
   Thorn forest (transitional to desert)
Grassland formations
   Tall (prairie) grassland
   Short (steppe) grassland
Desert formations
   Dry desert (including semidesert)
   Cold desert (tundra, including alpine tundra)

Some minor types of stable vegetation are omitted from this general classification, such as the heather-dominated moors of oceanic northwest Europe, which are characterized by *Erica* and *Calluna*, and the mangrove thickets with species of *Avicennia* and *Rhizophora* of tropical ocean shores. Likewise the vegetation of inland waters and of the sea is not considered, and all animals are arbitrarily ignored, although characteristic ones occur with each formation. The major formations have typical climax, dynamically stable communities. However, within each formation area most of the land usually is occupied by vegetation of disturbed terrain, by communities of plants under edaphic control (conditions of the substratum, terrain, or associated microclimate that are extreme for the region), or by successional communities that have not had time to develop to climax status. The formation, when thought of in a regional sense, includes all of the great variety of such communities that are subordinate to the typical climax. Climax plant formations not only occupy vast areas, but they have been in existence during millions of years and have migrated as their controlling climates have shifted on the face of the Earth. *See* VEGETATION ZONES, WORLD.

Tropical rainforest. This formation occurs extensively in broad lowland areas in the Amazon basin of South America, in the Congo of West Africa, in Malaya, and in Indonesia, with lesser areas

Fig. 1. Tropical rainforest of upland type in the lower Amazon region, Belém, Pará, Brazil. Tall trees with crowns and plank-shaped buttresses are typical. Such forests may have 100 tree species on a hectare.

elsewhere that never have frost or a period of inadequate precipitation. The tropical rainforest (Fig. 1) is very rich in woody species (100 or more tree species per hectare) and is usually without dominance by one or a few species, which is in contrast with the monsoon forest and the temperate rainforest, in which dominance may occur. It is structured in several layers, the uppermost being a discontinuous layer of occasional emergent trees that extend far above the general canopy of the forest. Internally there usually are one or two layers of lower trees and shrubs. The ground cover often consists of reproduction of woody plants, although openings in the generally dark forest permit a luxuriant growth of ferns and other herbs. The tropical rainforest is famous for its climbing vines, such as *Bauhinia* and *Carludovica*, which vary from thin cords to massive stemmed lianas for the abundance of vascular epiphytes (many genera of Bromeliaceae, Orchidaceae, and Pteridophyta); and for cauliflory, or the bearing of flowers and fruits on old stems and even on the trunks of trees, as in *Theobroma* and *Diospyros*. Trees are commonly tall, with emergent ones very tall (250 ft or more in *Ceiba* and *Bertholletia*), and usually have columnar trunks which are unbranched up to the relatively small crowns. Jungle is a term more appropriate for the forest edge, such as areas along streams where the light has full effect, or where second growth has occurred. The interior of old, high, upland rainforest is comparatively open and easy to move through. Many species of trees have planklike buttresses rising along the trunk from the main roots, as seen in *Mora*, *Ceiba*, and *Terminalia*; and others, such as *Tovomita*, *Clusia*, *Symphonia*, and *Ficus*, stand on stilt roots. Many lianas and some trees (*Ficus*) start as epiphytes and later make contact with the ground. Tree leaves are medium to large, although many species, such as the Leguminosae, have compound leaves with medium to small leaflets. Leaves are typically evergreen, with entire margins and attenuated tips called drip points, as in *Catostemma*, and are somewhat coriaceous in upper layers and thinner in lower ones. Although the forest is luxurious in appearance, tree growth is slow because of the comparatively infertile, strongly leached soil and the intense competition for light and nutrients.

Fig. 2. Summergreen deciduous forest. (a) Dominated by broad-leaved hardwoods, including *Betula, Aesculus, Fagus, Acer, Quercus, Carya*, and *Ilex*. Southern state-line region, Great Smoky Mountains National Park, Ten- nessee–North Carolina. (b) Interior view with *Quercus alba*, numerous herbs, and flowering shrub *Azalea nudiflora*; Blue Ridge Mountains in Virginia.

In contrast, after clearings are abandoned, secondary growth of trees such as *Cecropia* is rapid, and many species have large leaves (Musaceae, Zingiberaceae, Marantaceae). The tropical rainforest is the home of aboriginal shifting cultivation, as cut and burned patches of the forest soon lose their fertility. It produces many valuable and beautiful hardwoods, such as ebony, mahogany, and rosewood; it is the region of industrial rubber and cacao plantations.

Temperate rainforest. The trees in this formation are medium or tall in height and have medium or small, hard, evergreen leaves, although some associated trees have scalelike or needle-shaped leaves. Tree ferns and bamboos are abundant in certain areas of the temperate rainforest, and moss and liverwort growth may cover the tree branches and the ground. The formation occurs in many moist warm regions, but does not occupy extensive areas because it is impinged upon from all sides and seems to be in delicate balance. With increasing heat and moisture, it grades into subtropical and tropical rainforest; with increased dryness or periodicity of precipitation, it grades into tropical deciduous forest; and with coolness, into summergreen and needle-leaved forest types. It occurs on all continents and is variously composed floristically, but has some widely occurring "binding" genera, such as *Podocarpus, Weinmannia, Drimys, Araucaria, Persea, Dacrydium, Laurelia, Cedrela*, and *Nothofagus* in the Southern Hemisphere. Some examples of the temperate rainforest are the Kauri pine (*Agathis*) and southern beech (*Nothofagus*) forests of New Zealand, the *Nothofagus* forest in Chile, the sweetgum (*Liquidambar*) forests of Mexico, the coastal redwood (*Sequoia*) forests of California, the Parana pine (*Araucaria*) forest of Brazil, and the laurel (*Laurus*) forest of the Canary Islands. Other types occur on tropical and subtropical mountains where the climate is temperate, and are called cloud forests when there is a high frequency of fogginess. *See* RAINFOREST.

Tropical deciduous (monsoon) forest. This is an intermediate type characteristically making transition to rainforest and savanna woodland. It is usually referred to as the monsoon forest because its leaves are borne during the rainy season and many of the trees are deciduous during the dry season. The leaves are small to medium and some evergreen elements are present. It usually is a tall stratified forest with many life forms present. Lianas and vascular epiphytes occur but are not as abundant as in the tropical rainforest, which lacks a dry season. With a more pronounced and usually drier season, this formation grades into savanna woods which may be closed or in more extreme forms may be open and grassy. Other transition may be with thorn woodland in Africa and South America. Typical monsoon forests occur in southern Asia, where teak (*Tectona grandis*) is important, and there are related forests in Africa and South America. When monsoon forests occur with a less pronounced dry period and with more evergreen elements, the semideciduous forest types are sometimes lumped with rainforest.

Summergreen deciduous forest. This formation occurs in cool to cold temperate regions of the Northern Hemisphere where there is a long winter and a summer of several months without a pronounced dry period (Fig. 2a). The rhythm of the forest is set by the contrast between the leafless, profusely branched trees ending in a high eventopped crown of fine twigs and the heavy summer canopy formed by broad, membranaceous, mostly simple leaves. This formation is comparatively rich in tree, shrub, and herbaceous species, and is typically structured in several layers by naturally tall and short trees, shrubs, and herbs (Fig. 2b). Further rhythm is produced by a pronounced spring flora that comes quickly into bloom from subterranean bulbs, rhizomes, and corms. Many of the plants of this spring flora are species of *Anemone, Erythronium, Viola, Geranium, Dicentra*, and *Claytonia* which complete their life cycles before the full shade of the forest canopy has developed in early summer. Taller summer and autumnal layers of herbs replace the spring flora and greatly change the interior appearance of the forest except where the shade is most dense. Climbing vines, species of *Vitis* and *Psedera*, are infrequent and epiphytes are usually mosses, liverworts, and lichens. This forest formation is well represented in the eastern United States, in Europe and contiguous Asia, except for high latitudes, in the Mediterranean region, and in southeastern Asia. Some minor extensions into the warm temperate and subtropical zones occur at higher altitudes in Mexico. It makes transition with grassland toward continental interiors, with conifer forests northward and at higher altitudes, and with various types of vegetation southward in warm temperate regions. It is absent from the Southern Hemisphere in regions with a pronounced dry season, and is poorly represented in western North America. In the United States the northern hardwood forest is mostly characterized by maple (*Acer*), beech (*Fagus*), birch (*Betula*), and basswood (*Tilia*), with lesser amounts of other genera such as ash (*Fraxinus*), oak (*Quercus*), and hickory (*Carya*). Toward the drier continental interior the forest is dominated by oaks and hickories, finally forming a transition with the tall-grass prairies. The southern aspect of the forest has yellow poplar (*Liriodendron*), gum (*Nyssa*), walnut (*Juglans*), holly (*Ilex*), buckeye (*Aesculus*), sweetgum (*Liquidambar*), and magnolia (*Magnolia*) as well as species of genera of the other regions of the formation. Chestnut (*Castanea*) was a prominent member of the eastern phase of the formation before it was eliminated by a blight. This forest produces many fine and commercially important hardwoods.

Needle-leaved forest. This is best represented by the circumboreal forest dominated by spruce (*Picea*) and fir (*Abies*) (Fig. 3a). It occurs south of the tundra and north of the summer-green deciduous forest or grassland (Fig. 3b). The taiga is the northern part of the formation and lacks a closed tree canopy. The tundra may be interpenetrated on higher and drier land. The southern part is penetrated by broad-leaved deciduous trees in the valleys and on warmer slopes or by grassland openings. Upper elevations of mountain masses to the south of the Canadian spruce-fir formation in the Northern Hemisphere have belts of needle-leaved forest with spruce and fir, or a variety of pines (*Pinus*) and other conifer genera, such as *Juniperus*, *Cupressus*, *Cedrus*, *Linocedrus*, *Larix*, *Tsuga*, and *Thuja*. Some representatives of this formation (*Pinus*) occur on mountains in the tropical zone and in the Southern Hemisphere. Some conifers belong to formations not dominated by needle-leaved species, such as the hemlock (*Tsuga canadensis*), or are dominant in successional stages ear-

lier than the climax, such as the junipers (*Juniperus*) and pines (*Pinus*). The name of the forest comes from the needlelike or scalelike, hard, evergreen leaves. Mature trees, according to latitude or altitude and the nature of the species, may be comparatively low or tall, such as the Sitka spruce (*Picea sitchensis*) and redwood (*Sequoia sempervirens*). Typically the canopy is closed and the forest is dark, although the central trunk is usually unforked. Structure of the forest is simple because there are few layers in the vegetation and the dominant tree species are few in number, but the ground cover, although scanty in dry types, often is a dense mat of mosses and lichens with associated ferns, broad-leaved herbs, and low, often evergreen shrubs (*Vaccinium*, *Ledum*, *Kalmia*). The formation is typically moist, but conifer forests range into semiarid regions, which have species of *Pinus* and *Juniperus*, and into swamps, where *Taxodium* is common. In the tropical formation the climate is moist, without a dry season, and cool or cold. Winters are long and snow cover may be deep. Soils are acid and podzolized, comparatively shallow, and not fertile. The boreal forest occupies vast areas in Asia, Europe, and America (principally in Canada), which are also the zones of abundant acid bogs and peat formation. Other types of needle-leaved forests occur in different climates and on different soils, such as the juniper-piñon woodlands of the southern Rocky Mountains. This forest formation produces the world's most important commercial softwood lumber for building and pulp for paper manufacture.

Evergreen hardwood forest. Frequently, this is called the Mediterranean forest after the area where it was first studied, although it also occurs in the southwestern United States, Chile, southwestern Africa, and southwestern Australia. It occurs in regions where there is a Mediterranean-type climate with cool to cold moist winters and long dry summers. The result is that conditions are favorable for active growth in the spring, with many herbs rising from bulbs, and again in the autumn when the dry summer is over. In the Mediterranean region the forest dominants are mostly oaks (*Quercus ilex* and *W. suber*), but centuries of abuse have reduced most of the region to scrubby vegetation called maquis or garigue. In California the formation is represented by the oak-madrone type (*Quercus-Arbutus*), with fire and other disturb-

Fig. 3.   Needle-leaved (conifer) forest. (*a*) Dominated by *Picea glauca*, *P. mariana*, *P. rubens*, and *Abies balsamea*, as seen in Laurentides National Park, Quebec. (*b*) Dom-

inated by *Picea engelmani* and *Abies lasiocarpa*, as seen in central Colorado, in the Silverton region. Alpine tundra region above timberline.

Fig. 4. Savanna. (a) *Juniperus-Pinus* savanna, as found in southwestern Colorado, with open-spaced trees and xerophytic grasses. (b) Tropical savanna near Pirapora, Minas Gerais, Brazil. The low, contorted, largely ever-green, broad-leaved trees and a good cover of semi-xerophytic grasses are typical.

ances producing a variety of types of chaparral (*Arctostaphylos, Rhamnus*); in Chile by Rosaceae (*Quillaja*) and Lauraceae (*Persea, Cryptocarya*); in Australia by gums (*Eucalyptus*) and many associ-ated species; and in the Cape region of South Afri-ca by genera of Ericaceae, Bruniaceae, Rutaceae, Rhamnaceae, and Thymelaceae. This forest for-mation is highly subject to fire and to degradation by grazing animals, such as sheep and goats, and by erosion, so that it is seldom seen in mature form. In the Mediterranean region it is the home of olive cultivation.

Savanna. This is a transitional formation be-tween forest and grassland in which trees may occur generally but are widely spaced in islands of woodland or more or less as gallery woodland along drainage channels. It can occur in temper-ate regions as a rather narrow ecotone, with patches of grassland within the forest region and islands of trees within the grassland (Fig. 4a), as in Minnesota to Illinois. The formation may be exten-sive in tropical and subtropical regions, as in the big-game country of eastern Africa (*Acacia* savan-na) and in the central plateau of Brazil, the Campo Cerrado. The savanna of Brazil (Fig. 4b) sharply abuts the rainforest in many places and varies from a closed or nearly closed woodland of short, contorted, thick-barked trees through more open structure with increasing density of grasses (Andropogoneae) and broadleaved herbs until there are no more trees present. The soil is gener-ally infertile and lateritic, and sometimes there is a massive subterranean layer of stonelike laterite close to the surface. Because of the strong season-al distribution of rain, the water table is fluctuating and sometimes very deep, and vegetational activ-ity is limited to the rainy season. Common use of the savanna is for extensive animal husbandry, because the forage value and carrying capacity of the vegetation are low, and for subsistence agricul-ture along drainage channels. Fire is frequent, and some students of vegetation believe that all savan-na has been produced from woodland by centur-ies or even millenniums of human interference, especially set fires. *See* SAVANNA.

Thorn forest. This formation is dominated by small trees and shrubs, many of which are armed with thorns and spines. Leaves are absent, succu-lent, or deciduous during the long dry periods,

which may also be cool. Succulent plants of many forms usually accompany the thorny bushes and trees (Cactaceae in North America, Euphorbi-aceae in Africa). There is an ephemeral layer of herbs that appears after rains, and sometimes a thin scattering of dry grasses. The type is mostly tropical and subtropical and is represented by the caatinga of northeastern Brazil and the South Afri-can thornbush. This type is intermediate between desert and steppe.

Tall grassland. In the United States this is called the prairie. It is a continental formation lying next to broad-leaved summer-green forest (oak-hickory type) on the more humid side and short-grass plains (steppe) on the drier side. Rain-fall (20–40 in.) is strongly seasonal with a dry summer and a cool or cold winter, often with snow cover. True dominants of the prairie include such grasses as *Andropogon scoparius, Sporobolus as-per* and *S. heterolepis, Stipa spartea* and species of *Koeleria, Agropyron, Muhlenbergia,* and *Panicum,* and the sedge genus *Carex.* Numerous associated broad-leaved herbs include many species of the families Leguminosae, Rosaceae, Scrophulari-aceae, and Umbelliferae. Tall trees are sometimes present along the streams and some shrubs occur within the grassland, such as species of *Salix* and

Fig. 5. Protected grassland in the Black Hills area east of Colorado Springs, Colo., showing tall grasses and mid-grasses in a generally short-grass (steppe) region.

Symphoriocarpus. Most vegetative activity stops after the effects of the spring snowmelt and early summer rains are over. This is now a rich agricultural region concentrating on corn, small grains, and livestock. The Hungarian puszta and moist grasslands of Eurasia belong to this subformation.

Short grassland. This is generally called the steppe. In the United States it is the short-grass plains (Fig. 5). It makes contact with the tall-grass prairies on the moist side and with savanna, thorn-forest, and desert on the dry side. It is a continental formation with less precipitation (10–20 in.) than the prairie. The rainfall is less reliable and annual droughts are a hazard to agriculture. Summers may be long and dry, and winters long and cold at higher latitudes and cool to warm at lower ones. Winds are strong and wind erosion occurs where the vegetation is disturbed. The formation is typically dominated by short to medium-high grasses. Short-grass dominants in the United States include *Bouteloua hirsuta* and *B. gracilis, Buchloë dactyloides,* and species of *Muhlenbergia* and *Carex.* Mid-grasses include *Agropyron smithi, Hilaria jamesi, Sporobolus cryptandrus,* and species of *Stipa, Oryzopsis, Festuca,* and *Elymus.* Among the broad-leaved herbs, Compositae are especially abundant (*Aster, Solidago, Grindelia, Senecio, Artemisia*), as are Leguminosae (*Oxytropis, Psoralea, Petalostemon, Lupinus*). This formation was the main home of the American bison and now is largely devoted to the cattle industry and hazardous wheat raising in the less arid parts. It is the home of the man-induced dust bowls. In temperate regions the steppe plays a role corresponding to that of savanna in tropical regions. In addition to the North American grasslands, extensive areas occur in Argentina and Uruguay, in Africa on both sides of the Sahara desert, and in Asia mostly between latitudes 30 and 50°N from the Middle East to Mongolia and Manchuria. *See* GRASSLAND.

Dry desert. The dry, or true, desert is both a climatic condition and the vegetation of dry regions. Precipitation is low, from essentially nothing to a variable upper limit of about 10 in., and is erratic. Atmospheric humidity is low in the day, but dew is common at night. Temperature varies from warm to hot for all months or may be cold in the winter at high latitudes and altitudes. There is a strong daily change in temperature from hot bright days to clear cool nights. In the southwestern United States the Chihuahua Desert extends from southern New Mexico and western Texas southward into Mexico, occupying the land between mountain ranges. Organ-pipe and columnar tree cacti, treelike yuccas, and species of *Agave, Dasylirion,* and *Hechtia* form a conspicuous part of the vegetation, along with ocotillo (*Fouquieria*), mesquite (*Prosopis*), and creosote bush (*Larrea*). The low rainfall occurs largely in the summer. The Sonoran Desert occupies the lowlands around the Gulf of California in Sonora and Baja California and the adjacent United States. It is characterized by the giant saguaro cactus (*Carnegiea gigantea*), many species of prickly-pear cactus (*Opuntia*), and the shrubby *Larrea* and *Franseria* (Fig. 6). Ephemeral annuals are abundant after late winter rains. Temperatures in the Mojave Desert of California may exceed 100°F for long periods of time, and precipitation around the Gulf is about 2 in. annual-

Fig. 6. Sonoran Desert, in Arizona. Saguaro (*Carnegiea gigantea*) is mainly on south-facing slopes.

ly. Northward of these true severe desert areas, in the Great Basin region, the semideserts are less dry, but colder, and the northern part has severe winters. Species of sagebrush, especially *Artemisia tridentata* in the north and *A. spinescens* in the south, commonly are dominant, with shadscale (*Atriplex*) as an associate. Many parts of this desert are salt encrusted, with halophytes such as *Kochia, Sarcobatus, Allenrolfea,* and *Salicornia* being important. Desert vegetation is typically open, with shrubs, cacti, and bushy herbs widely spaced, but often there is root competition for the scanty water supplies, and some species are deep-rooted. The long, severe dry periods are met variously by desert and semidesert plants. Some are annual, avoiding drought; others lose their leaves or are leafless all the time; and many store water in succulent organs. In addition to the American deserts described above, there is the coastal Atacama Desert in Pacific South America, the Kalahari Desert of southwestern Africa, the great Sahara desert of northern Africa, which crosses the continent between 15 and 30°N latitude, the great Arabian deserts of the Middle East, and the deserts of Asia which extend from Kara-Kum eastward to the Gobi. Much of central Australia is occupied by the Simpson, Victoria, Gibson, and other deserts. Desert mountain ranges are covered at appropriate elevations by other plant formations, often widely disjunct, culminating even in boreal-type needle-leaved forest and alpine tundra. Oases, where the water table is high or aquifers can be tapped by wells, may support luxuriant gardens of melons, vegetables, and date palms. With modern irrigation, desert soils often prove to be exceedingly fertile. Careful management is necessary to prevent the soil from becoming useless because of salt accumulation. Deserts sustain some browsing and semideserts can be useful country for cattle, horses, and camels, but overgrazing is easy and this can result in the spread of weedy shrubs, such as the spread of sagebrush in the Great Basin Desert and of mesquite around the Chihuahua Desert. *See* DESERT VEGETATION.

Tundra. This formation comprises the vegetation of the treeless zone of high latitudes where winters are long and cold, with temperatures falling to −60°F or lower. Summers are short, with no month averaging above 50°F. Frost may occur in midsummer, although daytime maxima may reach 80°F. Precipitation usually is low, snow cover thin,

Fig. 7. Alpine vegetation (tundra), Medicine Bow Peak (12,005 ft altitude), Snowy Range, Wyo. Dwarfed, matted trees at timberline are mostly the spruce *Picea engelmanni* and the fir *Abies lasiocarpa*.

and winds strong, although fogs may be frequent and cloudiness high. There are, according to the latitude, long winter nights when the Sun rises a few hours low on the horizon or not at all, and summer days when it never sets. The ground is perennially frozen to great depths (permafrost) below a variable surface layer subject to summer thaw. Although mountains, hills, and stream valleys occur, much of the Arctic tundra landscape is of low relief. However, the microrelief is extremely important in controlling the patterns of occurrence of the plants and the communities which they form in the tundra formation. Because all conditions of life are marginal, small changes in relief, gradient, and substratum produce striking changes in vegetational pattern. Part of this is the result of abrupt changes of dominant species with slight changes in temperature and moisture. The controlling forces in geomorphological processes, and consequently in vegetational development and patterning, are largely cryopedologic or caused by freezing-thawing disruptions that result in polygons, ice wedges, surface ridges, and various kinds of pools and ponds. Although the general cast of the vegetation is low and monotonous, there is intricate complexity and regional variation in community composition and patterning despite a relatively impoverished flora. Frequent and often dominant plants are sedges, grasses, and grasslike herbs, together with lichens, mosses, and liverworts. In places there is an abundance of low, creeping shrubs or densely compact ones, and many herbs have cushionlike forms. A few inches or feet of elevation produce dry situations with distinctive vegetation, but there are many moist or wet sites, and in places many ponds and lakes. Tundra typically makes transition with coniferous forest by an ecotonal zone sometimes called the taiga. Alpine tundra (Fig. 7) occurs above the altitudinal tree line on mountain masses all over the world, even near the Equator. Its lower limit is higher in lower latitudes. It is distinguished from Arctic tundra by greater precipitation and by the different light conditions which are the result of the latitude. Some Arctic species extend far southward, even in isolated alpine tundra. *See* TUNDRA.   [STANLEY A. CAIN]

Bibliography: S. A. Cain, *Foundations of Plant Geography*, 1944; F. E. Clements and V. E. Shelford, *Bio-ecology*, 1939; P. Dansereau, *Biogeography*, 1957; H. A. Gleason and A. Cronquist, *The Natural Geography of Plants*, 1964; W. B. McDougall, *Plant Ecology*, 4th ed., 1949; H. J. Oosting, *The Study of Plant Communities*, 2d ed., 1956; P. W. Richards, *The Tropical Rain Forest*, 1952.

Plant geography

That major subdivison of botany concerned with all aspects of the spatial distribution of plants. This science is also known as phytogeography, phytochorology, or geographical botany. By tradition, it also involves some aspects of the distribution of plants in time, or historical plant geography, paleoecology, and paleobotany. In its historical development, plant geography has been intimately connected with the rise of evolution, ecology, and genetics. It is not yet consistently segregated from ecology. From a second and equally logical viewpoint, plant geography is a major subdivision of the science of geography, although by custom few geographers deal mainly with plants. The word geobotany, undesirable etymologically, is a confusing term because of numerous and contradictory usages.

The function of plant geography is to record the observed, empirical facts of plant distribution and also to understand and interpret these facts. Where possible, the study includes the prediction and control of distributional phenomena, especially as these relate to plant pests and to the introduction and spread of desirable species and vegetation types. Such practical aspects are pertinent to the fields of forestry, agriculture, range and pasture management, wildlife habitat management, horticulture, and soil and water conservation. Plant geography is, essentially, not an experimental science, and in general does not involve laboratory procedures and technological equipment.

Flora and vegetation. There are two major subdivisions of plant geography paralleling one logical breakdown of botany itself. Floristic plant geography embraces the spatial distribution of the flora while vegetational plant geography is the spatial distribution of the vegetation. A clear understanding of the two terms is essential.

Flora is a scientific term, with no common usage. The flora of an area or period of time is the totality of all species within that geographical unit, independent of their relative abundances and their relationships to one another. The technical term population, in this connection, refers collectively to all the individuals of any one species within a locality. *See* SPECIES POPULATION.

Vegetation is a term of popular origin and refers to the mass of plant life that forms the natural or seminatural landscape. The vegetation of a region is the tapestry, or carpet, of plant life, developed by differential and varying combinations and growths of the numerous elements of the local flora. Technically, it is an organized and integrated whole, at a higher level of integration than the separate species, composed of those species and their populations. Sometimes vegetation is very weakly integrated, as the pioneer plants of an abandoned field. Sometimes it is highly integrated, as in the tropical rain forest. Vegetation possesses emergent properties not necessarily found in the species themselves, and is referred to by nonbiologists as a system, or organization, a type of "organism"

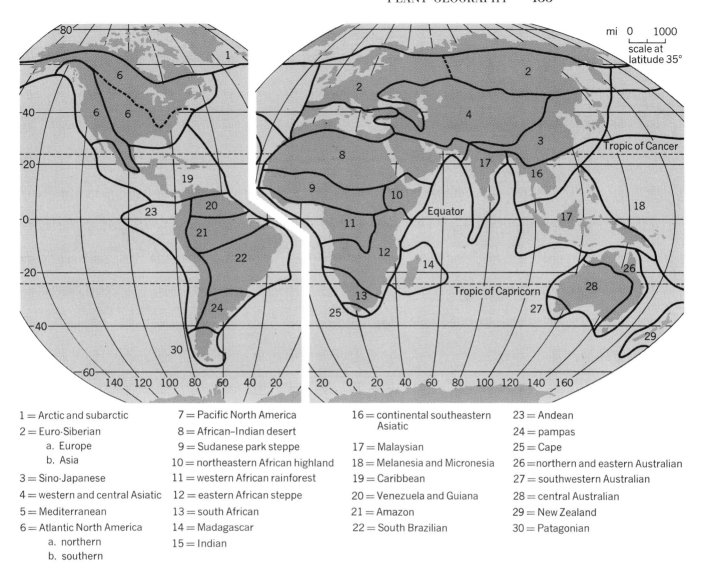

Fig. 1. Floristic regions of world. (*Modified from R. Good, Geography of Flowering Plants, 3d ed., Longmans, 1964*)

1 = Arctic and subarctic

2 = Euro-Siberian
   a. Europe
   b. Asia

3 = Sino-Japanese

4 = western and central Asiatic

5 = Mediterranean

6 = Atlantic North America
   a. northern
   b. southern

7 = Pacific North America

8 = African–Indian desert

9 = Sudanese park steppe

10 = northeastern African highland

11 = western African rainforest

12 = eastern African steppe

13 = south African

14 = Madagascar

15 = Indian

16 = continental southeastern Asiatic

17 = Malaysian

18 = Melanesia and Micronesia

19 = Caribbean

20 = Venezuela and Guiana

21 = Amazon

22 = South Brazilian

23 = Andean

24 = pampas

25 = Cape

26 = northern and eastern Australian

27 = southwestern Australian

28 = central Australian

29 = New Zealand

30 = Patagonian

(a different and more inclusive term than the organism of biologists). *See* VEGETATION.

Floristic plant geography. The basic components of any flora are the kinds of plants composing it, commonly referred to as species. The species can be grouped into various kinds of floral elements which are not mutually exclusive. For example, a genetic element has a common evolutionary origin; a migration element has a common route of entry into the territory; a historical element is distinct in terms of some past event; and an ecological element is related to an environmental preference. Aliens, escapes and very widespread species are given special treatment. An endemic species is restricted to an area, usually small and of some special interest. *See* POPULATION DISPERSAL.

The idea of area is fundamental to the science and is itself the subject of a specialized section called areography. An area is the entire region of distribution or occurrence of any species, element, or even an entire flora. The local distribution within the area as a whole, as that of a swamp shrub, is

the "topography" of that area. Areas are of interest in regard to their general size and shape, the nature of the margins, whether they are continuous or disjunct, and in their relationships to other areas. Groups of areas are unicentric or polycentric when they segregate into one or several geographically distinct territories. Areas of closely related plants that are mutually exclusive are said to be vicarious. A relict area is one surviving from an earlier and more extensive occurrence.

On the basis of areas and their floristic relationships the Earth's surface is divided into floristic regions (Fig. 1), each with a distinctive flora.

The understanding and interpretation of floras and of their distribution have been predominantly in terms of their history and ecology. Historical factors, in addition to the evolution of the species themselves, include consideration of theories of continental drift, land bridges, and orographic and climatic changes in geologic time that have affected migrations and perpetuation of floras. Ecological factors, more amenable to observation, and thus, unfortunately, to post hoc reasoning, in-

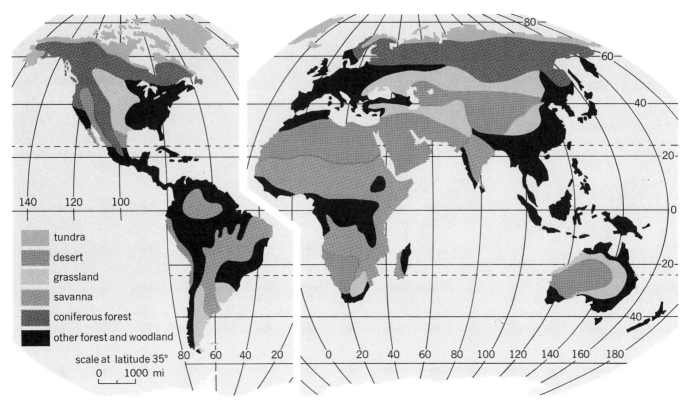

Fig. 2. Map of the world, showing the distribution of physiognomic vegetation types. (*After Brockmann-Je-* *rosch, modified from R. Good, Geography of Flowering Plants, 3d ed., Longmans, 1964*)

clude the contemporary roles played by precipitation, humidity, water levels, temperature, wind, soil, animals and man.

Vegetational plant geography. The basic components of the vegetation of any landscape are the plant communities. The science of plant communities is known as plant ecology, in the American sense, also as plant sociology, vegetation science, and phytocoenology. Many definitions of the plant community have been attempted, but none has gained universal acceptance. In part, this problem is inherent in the nature of the community itself, which is a natural phenomenon composed of elements, the species, which themselves usually maintain a high degree of independence. Thus, the community is often only a relative continuity in nature, bounded by a relative discontinuity, as judged by competent botanists. *See* PLANT COMMUNITY.

Vegetational plant geography has emphasized the mapping of so-called vegetation regions, and the interpretation of these in terms of environmental, or ecological, influences. *See* VEGETATION AND ECOSYSTEM MAPPING.

There are many aspects of a mosaic of plant communities which could serve to identify a geographic unit of vegetation, but those which have been predominant in the literature had their origins in folk knowledge. The physiognomic distinctions between grassland, forest, and desert, with such variants as woodland (open forest), savanna (scattered trees in grassland), and scrubland (dominantly shrubs) are most often emphasized. Within forest, the chief breakdown has been into coniferous evergreen forest, broadleaved decid-

uous forest, and broadleaved evergreen forests, mostly tropical. Furthermore, the attempt is made to map original "virgin" vegetation as opposed to cover types obviously due to the influence of man (Fig. 2). There is increasing dissatisfaction with this approach, but no accepted alternatives have arisen. Dissatisfaction arises from improved understanding of so-called virgin vegetation, frequently found to be influenced by ancient human populations. Furthermore, the segregation of coniferous from deciduous types is found to separate vegetations closely related in all other aspects, such as yellow birch and hemlock in North America, and to unite types otherwise unrelated, like the pine stands which are found from the tropics to the tundra edge. In addition, disturbance of grassland may allow the invasion of apparently self-perpetuating woody vegetation, or vice versa, in a manner that makes a physiognomic classification less fundamental. Unlike floristic botany, where evolution provides a single unifying principle for classification, the nature of vegetation in its geographical distribution is such that many types of regions and many types of classifications may have equally valid significance in rationalizing the natural phenomena. *See* VEGETATION ZONES, WORLD.

The interpretation of the distribution of vegetation has been overwhelmingly in terms of the existing average environment. Catastrophic factors, such as fires, hurricanes, droughts, and other abnormal weather, are receiving increasing attention. There has been relatively little emphasis on differences due to the genetic nature of the species. For example, bristle cone pine trees have a life-span of over 4000 years, and Australian eucalypts were absent from, but by nature amenable to, the environmentally similar but treeless California chaparral region. From one viewpoint, it is the varying genetic demands of different species upon their environments which permit their segregation into communities. The fact that arboreta and botanical gardens are so successful in growing many species outside their normal ranges is being recognized as a refutation of the more extreme environmentalist views.

The uniformitarian environmentalist interpretation of vegetation regions is the most completely documented. Climate is considered of primary importance. Numerous empirical formulas combining various features of temperature and moisture have been derived so as to correlate with the distribution of physiognomic vegetation types. Soil is recognized as secondary in importance. In addition, biotic factors, including both man and other animals, have limiting effects. Although analysis of the normal environment is essential to the full understanding of the distribution of vegetation types, it is not likely that, except for trigger factors, direct and simple cause-and-effect relationships will be found between vegetation types and those elements of the total environment which man isolates and studies. *See* ECOLOGY, HUMAN; ECOLOGY, PHYSIOLOGICAL (PLANT); TERRESTRIAL ECOSYSTEM.

[FRANK E. EGLER]

Bibliography: S. A. Cain, *Foundations of Plant Geography*, 1944; A. Engler, *Das Pflanzenreich*, 1900; A. Engler, *Die Vegetation der Erde*, 1896; H. A. Gleason and A. Cronquist, *The Natural Geography of Plants*, 1964; R. Good, *Geography of Flowering Plants*, 3d ed., 1964; N. V. Polunin, *Introduction to Plant Geography*, 1960; E. V. Wulff, *An Introduction to Historical Plant Geography*, 1943.

Plant societies

Assemblages of plants which constitute structural parts of plant communities. They may be components in spatial arrangement, such as layers, life-form groups, or seasonally or locally prominent populations of plants. There is no agreement as to the precise usage of the term beyond the generally accepted notion that it should be used for structural vegetation elements of a rank below or within the plant community as a whole. A hierarchy of progressively larger units in vegetation structure with an example is as follows:

Individual plant → clump of similar individuals →
(rose)   (clone of rose bushes)

society   → community →
(shrub society) (oak-hickory)
(including roses) (forest community)
(or association)

biome   → biochore
(broadleaf winter) (forest biochore)
(deciduous) (all forest formations)
(forest biome)
(or world plant)
(formation with)
(associated animals)

Societies can be defined on the basis of structure, dominance, season, and life form.

Structural societies. These societies are groups of plants within a community which attain approximately the same height and which bear their foliage at about the same level above ground. They are also known as layer societies or unions. Such societies may form a more or less continuous layer throughout the area occupied by the community. In a forest, for example, there is the canopy society of tall trees, low-tree society, shrub society, herb society, and ground-layer society. These may be refined further if necessary. Some authors have implied or stated that there is a certain cohesion among the component species of a layer society, giving it the status of a community within a community. However, since there is also often a strong interdependence between members of different layer societies, as in the influence of the canopy upon the density of lower vegetation in a forest, such societies should not be regarded as independent entities. A useful set of symbols for the recording of spatial arrangement of vegetation has been proposed by P. Dansereau.

Dominance societies. These societies were defined by J. Weaver and F. Clements as aspect societies or series of stands clearly belonging to a certain community. In addition to the characteristic dominants, these societies possess certain subdominants or codominants of another life form or aspect than the dominant elements of the community. The concept is useful especially in grasslands, marshes, heaths, and other vegetation types in which dominance is an important feature. For instance, prairie communities may be defined on the basis of grass species. Within these communities, certain conspicuous, broadleaved herbs form local aspect societies. The extent and development of such societies have been used as indicators of the recent history of grassland stands, especially with regard to climatic fluctuations.

Seasonal societies. Weaver and Clements also applied the term society more precisely to groups of plants which determine the seasonal aspects of plant communities. Examples of such seasonal societies are the carpet of trillium, bloodroot, dogtooth violet, and spring beauty in deciduous forests in the spring; the asters and goldenrods of abandoned pastures in late summer; and the masses of short-lived annuals which suddenly develop after spring rains in the deserts of the western United States. Such societies are structural units by virtue of the uniform timing of phenologic response of the plant species involved.

Within a community there may be a progression of seasonal societies, such as prevernal → vernal → estival → serotinal → autumnal → hiemal. Each society makes its own demand upon the resources of the habitat. Therefore, a full understanding of a community requires observation of all its seasonal aspects.

Life-form societies. Plants in a community which have their permanent vegetative axes and buds at the same level in or above the soil constitute life-form societies, or synusiae. Members of such societies are therefore subject to similar growth conditions and frequently develop according to a similar pattern. The illustration shows how a plant community may be analyzed structurally, using life forms as a criterion to distinguish societies, such as the following: (1) society of clumped,

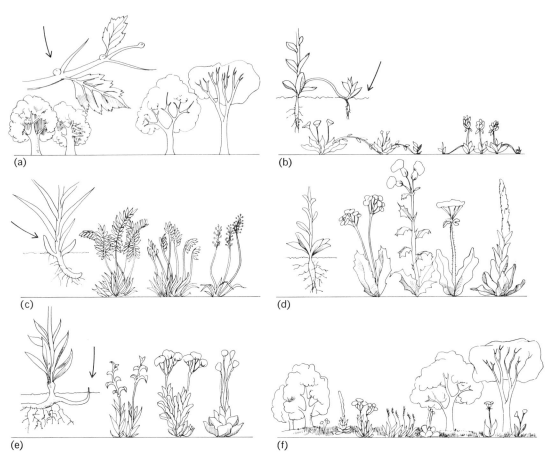

Analysis of a hawthorn-crabapple community in terms of life forms (with seasonal correlations). With each society a diagram on the left shows overwintering structure (arrow pointing to bud). (a) Society of low deciduous trees. (b) Society of sod-forming graminoids. (c) Society of rosette-forming hemicryptophytes. (d) Society of stoloniferous chamaephytes. (e) Society of rosette-biennials. (f) Structure of the entire community.

low, deciduous trees (hawthorn); (2) society of sod-forming graminoids flowering in midsummer (bluegrass, redtop, and so forth); (3) society of short-rhizomatous hemicryptophytes with winter rosettes and autumnal flowering period (goldenrod and aster); (4) stoloniferous chamaephytes with winter rosettes, flowering in early summer (pussytoes, cinquefoil, and strawberry); and (5) society of biennial rosette plants (wild carrot, thistle, and mullen). *See* PLANTS, LIFE FORMS OF.

Such divisions into societies are useful because they demonstrate the arrangement in space of the aerial and underground parts of the species in the community as well as the timing of their vegetative and reproductive cycles.

From the enumeration of criteria it is evident that all types of societies, regardless of the criteria used to define them, have certain common features; and that there is a lack of agreement regarding the precise meaning and definition of the term "society." Pending a definitive, internationally acceptable vocabulary for ecology, the word society should be used only with a qualifying adjective, such as seasonal society or structural society. Where another term with a more specific meaning is available, such as union for structural society or synusia for life-form society, the word should be avoided.

The society concept remains useful to ecologists in the analysis of vegetation as a general term for structural elements. *See* PLANT COMMUNITY.

[KORNELIUS LEMS/ARTHUR W. COOPER]

Bibliography: S. A. Cain and G. M. de Oliveria Castro, *Manual of Vegetation Analysis*, 1959; J. R. Carpenter, *An Ecological Glossary*, 1956; R. Daubenmire, *Plant Communities*, 1968; H. C. Hanson and E. D. Churchill, *The Plant Community*, 1961; J. E. Weaver and F. E. Clements, *Plant Ecology*, 2d ed., 1938.

Plants, life forms of

A term for the vegetative (morphological) form of the plant body. A related term is growth form but a theoretical distinction is often made: life form is thought by some to represent a basic genetic adaptation to environment, whereas growth form carries with it no connotation of adaptation and is a more general term applicable to structural differences.

Life-form systems are based on differences in gross morphological features, and the categories bear no necessary relationship to reproductive structures, which form the basis for taxonomic classification. Features used in establishing life-form classes include: deciduous versus evergreen leaves, broad versus needle leaves, size of leaves, degree of protection afforded the perennating tissue, succulence, and duration of life cycle (annual, biennial, or perennial). Thus the garden bean (family Leguminosae) and tomato (Solanaceae) belong to the same life form because each is an annual, finishing its entire life cycle in 1 year, while black locust (Leguminosae) and black walnut (Juglandaceae) are perennial trees with compound, deciduous leaves.

Climate and adaptation factors. There is a clear correlation between life forms and climates. For example, broad-leaved evergreen trees clearly dominate in the hot humid tropics, whereas broad-leaved deciduous trees prevail in temperate climates with cold winters and warm summers, and succulent cacti dominate American deserts. Although cacti are virtually absent from African deserts, members of the family Euphorbiaceae have evolved similar succulent life forms. Such adaptations are genetic, having arisen by natural selection. However, since there are no life forms confined only to a specific climate and since it is virtually impossible to prove that a given morphological feature represents an adaptation with survival value, some investigators are content to use life forms only as descriptive tools to portray the form of vegetation in different climates.

Raunkiaer system. Many life-form systems have been developed. Early systems which incorporated many different morphological features were difficult to use because of this inherent complexity. The most successful and widely used system is that of C. Raunkiaer, proposed in 1905; it succeeded where others failed because it was homogeneous and used only a few obvious morphological features representing important adaptations.

Reasoning that it was the perennating buds (the tips of shoots which renew growth after a dormant season, either of cold or drought) which permit a plant to survive in a specific climate, Raunkiaer's classes were based on the degree of protection afforded the bud and the position of the bud relative to the soil surface (see illustration). They applied to autotrophic, vascular, self-supporting plants. Raunkiaer's classificatory system is:

Phanerophytes: bud-bearing shoots in the air, predominantly woody trees and shrubs; subclasses based on height and on presence or absence of bud scales

Chamaephytes: buds within 25 cm of the surface, mostly prostrate or creeping shrubs

Hemicryptophytes: buds at the soil surface, protected by scales, snow, and litter

Cryptophytes: buds underneath the soil surface or under water

Therophytes: annuals, the seed representing the only perennating tissue

Subclasses were established in several categories, and Raunkiaer later incorporated leaf-size classes into the system.

By determining the life forms of a sample of 1000 species from the world's floras, Raunkiaer showed a correlation between the percentage of species in each life-form class present in an area and the climate of the area. The results (see table) were expressed as a normal spectrum, and floras of other areas were then compared to this. Raunkiaer concluded that there were four main phytoclimates: phanerophyte-dominated flora of the hot humid

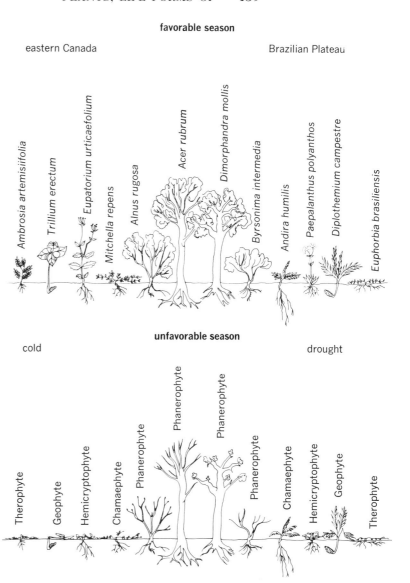

Life forms of plants according to C. Raunkiaer. (*From P. Dansereau, Biogeography: An Ecological Perspective, Ronald Press, 1957*)

tropics, hemicryptophyte-dominated flora in moist to humid temperate areas, therophyte-dominated flora in arid areas, and a chamaephyte-dominated flora of high latitudes and altitudes.

Subsequent studies modified Raunkiaer's views. (1) Phanerophytes dominate, to the virtual exclusion of other life forms, in true tropical rainforest floras, whereas other life forms become proportionately more important in tropical climates with a dry season, as in parts of India. (2) Therophytes are most abundant in arid climates and are prominent in temperate areas with an extended dry season, such as regions with Mediterranean climate (for example, Crete). (3) Other temperate floras have a predominance of hemicryptophytes with the percentage of phanerophytes decreasing from summer-green deciduous forest to grassland. (4) Arctic and alpine tundra are characterized by a flora which is often more than three-quarters chamaephytes and hemicryptophytes, the percentage of chamaephytes increasing with latitude and altitude.

Examples of life-form spectra for floras of different climates

Climate and vegetation	Area	Life form*				
		Ph	Ch	H	Cr	Th
Normal spectrum	World	46	9	26	6	13
Tropical rainforest	Queensland, Australia	96	2	0	2	0
	Brazil	95	1	3	1	0
Subtropical rainforest (monsoon)	India	63	17	2	5	10
Hot desert	Central Sahara	9	13	15	5	56
Mediterranean	Crete	9	13	27	12	38
Steppe (grassland)	Colorado, United States	0	19	58	8	15
Cool temperate (deciduous forest)	Connecticut, United States	15	2	49	22	12
Arctic tundra	Spitsbergen, Norway	1	22	60	15	2

*Ph = phanerophyte; Ch = chamaephyte; H = hemicryptophyte; Cr = geophyte (cryptophyte); and Th = therophyte.

Most life forms are present in every climate, suggesting that life form makes a limited contribution to adaptability. Determination of the life-form composition of a flora is not as meaningful as determination of the quantitative importance of a life form in vegetation within a climatic area. However, differences in evolutionary and land-use history may give rise to floras with quite different spectra, even though there is climatic similarity. Despite these problems, the Raunkiaer system remains widely used for vegetation description and for suggesting correlations between life forms, microclimate, and forest site index. *See* PLANT GEOGRAPHY.

Mapping systems. There has been interest in developing systems which describe important morphologic features of plants and which permit mapping and diagramming vegetation. Descriptive systems incorporate essential structural features of plants, such as stem architecture and height; deciduousness; leaf texture, shape, and size; and mechanisms for dispersal. These systems are important in mapping vegetation because structural features generally provide the best criteria for recognition of major vegetation units. *See* VEGETATION AND ECOSYSTEM MAPPING; VEGETATION ZONES, ALTITUDINAL; VEGETATION ZONES, WORLD.   [ARTHUR W. COOPER]

Bibliography: S. A. Cain, Life forms and phytoclimate, *Bot. Rev.*, 16(1):1–32, 1950; S. A. Cain and G. M. de Oliveria Castro, *Manual of Vegetation Analysis*, 1959; P. Dansereau, *A Universal System for Recording Vegetation*, 1958; R. Daubenmire, *Plant Communities*, 1968; A. Kuchler, *Vegetation Mapping*, 1967; C. Raunkiaer, *The Life Forms of Plants and Statistical Geography*, 1934.

Polychlorinated biphenyls

Polychlorinated biphenyls (PCBs) is a generic term covering a family of partially or wholly chlorinated isomers of biphenyl. The commerical mixtures generally contain 40–60% chlorine with as many as 50 different detectable isomers present (Fig. 1). The PCB mixture is a colorless, viscous fluid, relatively insoluble in water, that can withstand very high temperatures without degradation. However, they can be destroyed in a special industrial incinerator at 2700°F. PCBs do not conduct electricity and the more highly chlorinated isomers are not readily degraded in the environment.

PCBs have become an important subject in national and international discussions concerned with man's pollution of his environment. Attention was drawn to PCBs by widespread reports of their presence in fish, poultry, humans, and packaging materials, even from remote areas of the world. PCBs have been in use since the 1930s in most industrial nations.

Production and uses. The major uses of PCBs are a result of the properties described above and can be grouped in three major categories: open uses, partially closed systems, and closed systems.

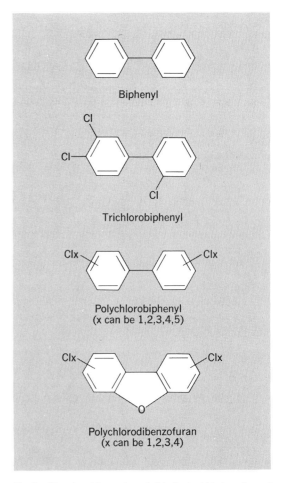

Fig. 1. Structural formulas of chlorinated biphenyls and related compounds.

Examples of uses in the first category are in paints, inks, plastics, and paper coatings. The PCBs in all these products are in direct contact with the environment and can be leached out by water or vaporize into the atmosphere. The so-called carbonless carbon paper contains PCBs in the encapsulated ink and has been claimed to be responsible for the PCBs found extensively in recycled paper. They have been used as plasticizers in polyvinyl chloride (PVC) and chlorinated rubbers. In an unusual industrial action the sole United States manufacturer of PCBs voluntarily released its production figures and stopped sales in 1971 to all users in this open system category, which amounted to about 30% of total production (Fig. 2). Swedish and Japanese producers followed this precedent, but manufacturers in Europe and the Soviet Union have not. It is generally believed that United States production is about one-half of world production. The United States has produced about 400,000 tons in the last 15 years.

Uses of PCBs regarded as partially closed systems include the working fluid in heat exchangers and hydraulic systems. These systems are subject to leakage of the PCB fluid either during use or after being discarded. PCB-contaminated poultry feed has been traced to a leaking heat exchanger used in the manufacturing process.

The electrical industry is the single major consumer of PCBs, mainly in transformers and capacitors. The fluid is generally sealed or welded into the unit so that loss is small. Of the 125,000 PCB-filled transformers put into service in the United States since 1932, over 99% are still in operation. Transformers and capacitors account for about 63% of all PCB uses.

Distribution in environment. It is not known exactly how PCBs are released into the environment or in what quantities. However, sewage outfalls, industrial and municipal disposal, leaking from dumps, and burning of refuse are certainly important sources. PCBs are found universally in the sewage of major cities, although the atmosphere is the major pathway of global transport.

Recently 50,000 turkeys in Minnesota, 88,000 chickens in North Carolina, and many thousands of eggs from various localities were destroyed after they were found to contain very high concentrations of PCBs. The Swedish government has declared the cod liver products from the Baltic Sea contain several parts per million of PCB and are unfit for human consumption. Analyses of water samples from 30 major tributaries to the Great Lakes indicate widespread contamination, with 71% of all samples having detectable concentrations (greater than 10 parts per trillion). PCBs have been found in all organisms analyzed from the North and South Atlantic, even in animals living under 11,000 ft of water. The U.S. Environmental Protection Agency has reported that one-third of the human tissue sampled in the United States contains more than 1 part per million (ppm) of PCBs. According to Robert Risebrough of the University of California's Bodega Marine Laboratory, PCB concentrations in cormorants and ospreys may be higher than in any other wildlife in North America, reaching levels of 300–1000 ppm. Birds from the South Atlantic and Central Pacific generally have lower residues, while the wildlife of the

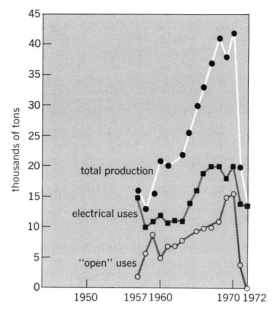

Fig. 2. United States production of polychlorinated biphenyls.

Antarctic appear to have the lowest.

Once in the environment, PCBs appear to persist for a very long time. Evidence for this can be seen in the fact that in most areas of the continents and throughout the Atlantic Ocean more PCB than DDT is found in the animals, even though three times more DDT is produced each year and all of it is put directly into the environment. Based on available data, it seems safe to assume that PCBs are present in varying concentrations in every species of wildlife on Earth.

Toxicity. Although PCBs have been in use for more than 30 years, studies of their toxic properties have begun only recently and very little is known at this time. The U.S. Food and Drug Administration has placed a limit of 5 ppm of PCBs in food products. The problem is complicated by the presence, in most PCB mixtures, of very toxic impurities believed to be polychlorodibenzofurans. Laboratory tests in the Netherlands have shown that fertile chicken eggs injected with chlorinated dibenzofurans produced seriously deformed chicks. Liver damage is a common effect of PCBs, while the occurrence of edema, skin lesions, and reproductive failure depends on the species. Hatchability of eggs is noticeably decreased by exposure.

In 1968 more than 1000 Japanese ate rice oil seriously contaminated with PCBs from a faulty heat exchanger during manufacture. The affected persons developed darkened skins, eye discharge, severe acne, and other symptoms. This came to be known as yusho, oil disease. The PCB was able to penetrate the placental barrier, since several infants born after the incident had yusho symptoms. In many cases of yusho, the symptoms were still present 3 years later. It is still a matter of controversy whether the subsequent deaths among the patients can be attributed to acute PCB poisoning, chlorinated dibenzofurans, or neither.

The history of the development and use of PCBs, followed by the discovery of its widespread occurrence in the environment, is very similar to the DDT story but with one exception: PCBs were seldom deliberately released into the environment. *See* ECOLOGY; OCEANOGRAPHY.

[GEORGE HARVEY]

Bibliography: H. Hays and R. W. Risebrough, *Nat. Hist.*, November 1971; A. V. Holden, *Nature*, 228:783, 1970; G. D. Veith and G. F. Lee, *Water Res.*, 4:265, 1970; V. Zitko and P. M. K. Choi, *PCB and Other Industrial Halogenated Hydrocarbons in the Environment*, Fisheries Research Board of Canada Rep. no. 272, 1971.

Population dispersal

The process by which groups of living organisms expand the space or range within which they live. Because of their reproductive capacity, all populations have a natural tendency to expand. As increased area supports more individuals, dispersal and reproduction are intimately correlated.

Distinction should be made between dispersal and seasonal migration. Birds, butterflies, salmon, and others migrate regularly without necessarily expanding their geographic range, since they usually return to their original areas or die out.

Dispersal phases. Dispersal consists of several phases: (1) the production of units, that is, of individuals or parts of individuals (disseminules) fit or adapted for dispersal; (2) the transportation of individuals or disseminules to the new habitat; and (3) ecesis, the process of becoming established through germination, rooting, or physiological and psychological adjustment.

Dispersal units. Certain disseminules (propagules or diaspores) may represent various stages of the life cycle of the individual. Many free-living animals do not produce special dispersal structures but rely upon the ability of the entire organism to move about (vagility). Organisms attached to a substratum, such as most plants and certain animals, produce disseminules adapted to certain agents of dispersal. In order to be effective, a disseminule must have the ability to develop into one or more complete individuals. The structures listed in Table 1 are examples of disseminules. Sperm cells, unfertilized eggs, and pollen grains, although capable of migration, are not true disseminules because they cannot give rise to new individuals.

It is possible to analyze plant communities on the basis of morphological features of the disseminules. By assigning species to dispersal types it is possible to construct dispersal spectra comparable to life form spectra in purpose and usefulness.

Transportation. Individuals or disseminules are transported in five general ways: self-dispersal (autochory), water dispersal (hydrochory), wind dispersal (anemochory), animal dispersal (zoochory), and dispersal by man (anthropochory).

In active self-dispersal, or autochory, the organism spreads in the course of its normal activities. The flight of starlings resulting in their gradual spread through the United States and the motility of bacteria resulting in gradual spread through the nutrient media are examples. Certain plants possess mechanisms of self-dispersal, such as the auxochores and ballochores listed in Table 2. In passive dispersal, one or more agents carry the dispersal unit to a new location. These agents, or

Table 1. Examples of dispersal stages in life cycle of plants and animals

Environment	Organism	Disseminule	Dispersal by
Sea	Kelp	Zoospore	Currents
	Coral	Planula	Currents
	Sea worm	Trochophore	Currents
	Clam	Trochophore	Currents
	Barnacle	Adult	Driftwood, ships
	Crab	Zoea	Currents
	Sea urchin	Pluteus	Currents
	Fish	Adult	Autochory
	Lamprey	Adult	Fish
Terrestrial	Mushroom	Spore	Wind
	Fern	Spore	Wind
	Pine	Seed	Wind
	Blueberry	Fruit	Birds
	Tumbleweed	Entire plant	Wind
	Insect	Adult	Autochory, wind
	Spider	Young animal	Wind
	Reptiles, birds, mammals	Adult	Autochory
Parasitic	Bacteria	Entire cell	Water, food, air
	Intestinal ameba	Cyst	Water, food, man
	Malaria parasite	Gamete, sporozoite	Mosquito
	Tapeworm	Egg	Pig
	Blood fluke	Egg, cercaria	Water, snail

vectors, are water currents, wind, animals and any of man's vehicles, such as trains, ships, and airplanes.

Water dispersal, or hydrochory, is prevalent in all marine and other aquatic populations. Plankton usually contains larval forms of bottom-dwellers (Fig. 1). Terrestrial forms associated with shore habitats are commonly dispersed by water. Buoyancy and resistance to salt water are a prerequisite for ocean dispersal. The first invaders of new islands such as Surtsey are often of this type. Transoceanic similarities in floras and faunas have been explained partly by ocean currents.

Wind dispersal, or anemochory, has various effects. It moves rolling disseminules in open deserts and grasslands (cyclochores, Fig. 2a); it deflects falling winged disseminules (pterochores, Fig. 2b–d); and it carries lightweight spores and disseminules with plumes for great distances (sporochores and pogonochores, Fig. 2e and h). Insects, spiders, and other light animals have been found many miles in the air, together with poplar seeds and other disseminules (Fig. 2f and g). Thus they may be carried hundreds of miles.

Table 2. Plant dispersal types based upon morphological adaptations

Name	Definition	Example
Auxochores	Deposited by parent plant	Walking fern
Cyclochores	Spherical framework	Tumbleweed
Pterochores	Disseminules winged	Maple
Pogonochores	Disseminules plumed	Milkweed
Desmochores	Disseminules sticky or barbed	Cocklebur
Sarcochores	Disseminules fleshy	Cherry
Sporochores	Disseminules minute, light	Fern
Sclerochores	Disseminules without apparent adaptations	Violet
Barochores	Disseminules heavy	Oak
Ballochores	Shot away by parent plant	Touch-me-not

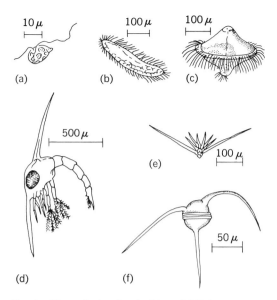

Fig. 1. Disseminules in plankton. (*a*) Kelp zoospore. (*b*) Coral planula. (*c*) Worm trochophore. (*d*) Crab zoea. (*e*) Brittle star pluteus. (*f*) Ceratium tripos.

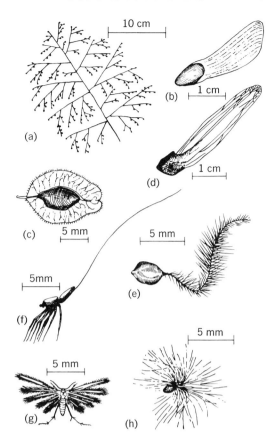

Fig. 2. Disseminules dispersed by wind. (*a*) Panic grass. (*b*) Pine seed. (*c*) Elm samara. (*d*) Tulip tree carpel. (*e*) Clematis carpel. (*f*) Spider. (*g*) Moth. (*h*) Cottonwood.

Animal dispersal, or zoochory, is divided into epizoochory (barbed or sticky disseminules, desmochores, Fig. 3*a–c*) and endozoochory (disseminules eaten and egested by animals). Disseminules adapted to endozoochory are those, such as arillate seeds (Fig. 3*d*), common in the tropics, and fruits with a fleshy mesocarp (Fig. 3*f* and *g*). Survival in the digestive tract of animals is a prerequisite. Bright fruit colors are frequent.

Dispersal by man, or anthropochory, involves purposely dispersed organisms such as domesticated animals and plants and those accidentally transported such as weeds along railroads, beetles in grain shipments, birds, rats, barnacles, and starfish on and in ships.

Ecesis. Success in population dispersal depends upon three factors: fitness of the new habitat, fitness of the migrating individuals, and the chance juxtaposition of these two which, in the long run, depends on the number of individuals invading the new habitat. The probability for a new habitat to be favorable is greatest close to the parent population. Spores blown over great distances have less chance of landing in spots suited for germination than have seeds falling close to the parent plant. In wide-range dispersal larger numbers of disseminules are usually necessary than at close range, to insure ecesis.

The fitness of the individuals depends partly upon their genetic makeup. Offspring of organisms that reproduce without sexual union (apomictic), such as aphids, dandelions, and similar organisms, are likely to succeed only in identical habitats. Offspring from self- or cross-fertilizing parents may succeed in a variety of situations. However, some hybrids which are sexually sterile are known to perpetuate themselves through apomixis. These are usually very successful locally.

Barriers to dispersal. A barrier is any discontinuity in the habitat greater than the maximum distance traveled by organisms in their normal dispersal. Oceans separating terrestrial habitats, continents separating marine habitats, mountain ranges intercepting wind dispersal, and deserts interrupting the continuity of forested land are all effective major barriers. Through the intervention of man these barriers are broken down in many cases. Since the development of frequent world travel thousands of species have become established on new continents as a result of anthropochory. *See* POPULATION DISPERSION.

[KORNELIUS LEMS/HERBERT G. BAKER]

Bibliography: P. Dansereau and K. Lems, The

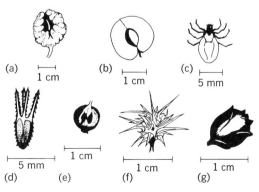

Fig. 3. Disseminules transported by animals. (*a*) Arillate legume seed. (*b*) Cherry. (*c*) Tick. (*d*) Beggar's tick fruit. (*e*) Currant berry. (*f*) Sandbur spikelet. (*g*) Juniper cone.

grading of dispersal types in plant communities and their ecological significance, *Contrib. Inst. Bot. Univ. Montréal*, vol. 71, 1957; P. A. Fryxell, Mode of reproduction of higher plants, *Bot. Rev.*, 23:135–233, 1957; P. A. Glick, *The Distribution of Insects, Spiders and Mites in the Air*, USDA Tech. Bull. no. 673, 1939; R. Hesse, W. C. Allee, and K. P. Schmidt, *Ecological Animal Geography*, 1937; H. N. Ridley, *The Dispersal of Plants Throughout the World*, 1931; E. J. Salisbury, *The Reproductive Capacity of Plants*, 1942.

Population dispersion

The spatial distribution at any particular moment of the individuals of a species of plant or animal. Under natural conditions organisms are distributed either by active movements, or migrations, or by passive transport by wind, water, or other organisms. The act or process of dissemination is usually termed dispersal, while the resulting pattern of distribution is best referred to as dispersion. Dispersion is a basic characteristic of populations, controlling various features of their structure and organization. It determines population density, that is, the number of individuals per unit of area, or volume, and its reciprocal relationship, mean area, or the average area per individual. It also determines the frequency, or chance of encountering one or more individuals of the population in a particular sample unit of area, or volume. The ecologist therefore studies not only the fluctuations in numbers of individuals in a population but also the changes in their distribution in space. *See* POPULATION DISPERSAL.

Principal types of dispersion. The dispersion pattern of individuals in a population may conform to any one of several broad types, such as random, uniform, or contagious (clumped). Any pattern is relative to the space being examined; a population may appear clumped when a large area is considered, but may prove to be distributed at random with respect to a much smaller area.

Random or haphazard. This implies that the individuals have been distributed by chance. In such a distribution, the probability of finding an individual at any point in the area is the same for all points (Fig. 1a). Hence a truly random pattern will develop only if each individual has had an equal and independent opportunity to establish itself at any given point. In a randomly dispersed population, the relationship between frequency and density can be expressed by Eq. (1), where F is percen-

$$F = 100(1 - e^{-D}) \qquad (1)$$

tage frequency, D is density, and e is the base of natural or Napierian logarithms. Thus when a series of randomly selected samples is taken from a population whose individuals are dispersed at random, the numbers of samples containing 0, 1, 2, 3, . . . , n individuals conform to the well-known Poisson distribution described by notation (2).

$$e^{-D}, De^{-D}, \frac{D^2}{2!} e^{-D}, \frac{D^3}{3!} e^{-D}, \ldots, \frac{D^n}{n!} e^{-D} \quad (2)$$

Randomly dispersed populations have the further characteristic that their density, on a plane surface, is related to the distance between individuals within the population, as shown in Eq. (3), where $\bar{r}$ is the mean distance between an individ-

ual and its nearest neighbor. These mathematical properties of random distributions provide the

$$D = \frac{1}{4\bar{r}^2} \qquad (3)$$

principal basis for a quantitative study of population dispersion. Examples of approximately random dispersions can be found in the patterns of settlement by free-floating marine larvae and of colonization of bare ground by airborne disseminules of plants. Nevertheless, true randomness appears to be relatively rare in nature, and the majority of populations depart from it either in the direction of uniform spacing of individuals or more often in the direction of aggregation.

Uniform. This type of distribution implies a regularity of distance between and among the individuals of a population (Fig. 1b). Perfect uniformity exists when the distance from one individual to its nearest neighbor is the same for all individuals. This is achieved, on a plane surface, only when the individuals are arranged in a hexagonal pattern. Patterns approaching uniformity are most obvious in the dispersion of orchard trees and in other artificial plantings, but the tendency to a regular distribution is also found in nature, as for example in the relatively even spacing of trees in forest canopies, the arrangement of shrubs in deserts, and the distribution of territorial animals.

Contagious or clumped. The most frequent type of distribution encountered is contagious or clumped (Fig. 1c), indicating the existence of aggregations or groups in the population. Clusters and clones of plants, and families, flocks, and herds of animals are common phenomena. The degree of aggregation may range from loosely connected groups of two or three individuals to a large compact swarm composed of all the members of the local population. Furthermore, the formation of groups introduces a higher order of complexity in the dispersion pattern, since the several aggregations may themselves be distributed at random, evenly, or in clumps. An adequate description of dispersion, therefore, must include not only the determination of the type of distribution, but also an assessment of the extent of aggregation if the latter is present.

Analysis of dispersion. If the type or degree of dispersion is not sufficiently evident upon inspection, it can frequently be ascertained by use of sampling techniques. These are often based on counts of individuals in sample plots or quadrats. Departure from randomness can usually be demonstrated by taking a series of quadrats and testing the numbers of individuals found therein for their conformity to the calculated Poisson distribution which has been described above. The observed values can be compared with the calculated ones by a chi-square test for goodness of fit, and lack of agreement is an indication of nonrandom distribution. If the numbers of quadrats containing zero or few individuals, and of those with many individuals, are greater than expected, the population is clumped; if these values are less than expected, a tendency towards uniformity is indicated. Another measure of departure from randomness is provided by the variance:mean ratio, which is 1.00 in the case of the Poisson (random) distribution. If the ratio of variance to mean is less than 1.00, a

POPULATION DISPERSION

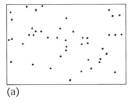

(a)

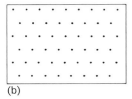

(b)

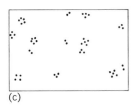

(c)

Fig. 1. Basic patterns of the dispersion of individuals in a population. (a) Random. (b) Uniform. (c) Clumped, but groups random. (*E. P. Odum, Fundamentals of Ecology, Saunders, 1953*)

regular dispersion is indicated; if the ratio is greater than 1.00, the dispersion is clumped.

In the case of obviously aggregated populations, quadrat data have been tested for their conformity to a number of other dispersion models, such as Neyman's contagious, Thomas' double Poisson, and the negative binomial distributions. However, the results of all procedures based on counts of individuals in quadrats depend upon the size of the quadrat employed. Many nonrandom distributions will seem to be random if sampled with very small or very large quadrats, but will appear clumped if quadrats of medium size are used. Therefore the employment of more than one size of quadrat is recommended.

The fact that plot size may influence the results of quadrat analysis has led to the development of a number of techniques based on plotless sampling. These commonly involve measurement of the distance between a randomly selected individual and its nearest neighbor, or between a randomly selected point and the closest individual. At least four different procedures have been used (Fig. 2). The closest-individual method (Fig. 2a) measures the distance from each sampling point to the nearest individual. The nearest-neighbor method (Fig. 2b) measures the distance from each individual to its nearest neighbor. The random-pairs method (Fig. 2c) establishes a base line from each sampling point to the nearest individual, and erects a 90° exclusion angle to either side of this line. The distance from the nearest individual lying outside the exclusion angle to the individual used in the base line is then measured. The point-centered quarter method (Fig. 2d) measures the distance from each sampling point to the nearest individual in each quadrat.

In each of these four methods of plotless sampling, a series of measurements is taken which can be used as a basis for evaluating the pattern of dispersion. In the case of the closest-individual and the nearest-neighbor methods, a population whose members are distributed at random will yield a mean distance value that can be calculated by use of the density-distance equation, Eq. (3). In an aggregated distribution, the mean observed distance will be less than the one calculated on the assumption of randomness; in a uniform distribution it will be greater. Thus the ratio $\bar{r}_A/\bar{r}_E$, where $\bar{r}_A$ is the actual mean distance obtained from the measured population and $\bar{r}_E$ is the mean distance expected under random conditions, affords a measure of the degree of deviation from randomness.

Additional information about the spatial relations in a population can be secured by extending these procedures to measurement of the distance to the second and successive nearest neighbors, or by increasing the number of sectors about any chosen sampling point. However, since all of these methods assume that the individuals are small enough to be treated mathematically as points, they become less accurate when the individuals cover considerable space.

Factors affecting dispersion. The principal factors that determine patterns of population dispersion include (1) the action of environmental agencies of transport, (2) the distribution of soil types and other physical features of the habitat, (3) the influence of temporal changes in weather and cli-

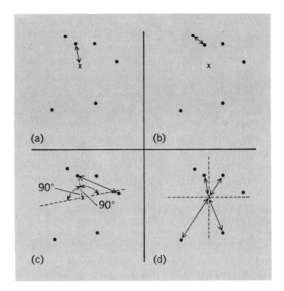

Fig. 2. Distances measured in four methods of plotless sampling. (a) Closest individual. (b) Nearest neighbor. (c) Random pairs, with 180° exclusion angle. (d) Point-centered quarter. The cross represents the sampling point in each case. (P. Greig-Smith, Quantitative Plant Ecology, Butterworths, 1957)

mate, (4) the behavior pattern of the population in regard to reproductive processes and dispersal of the young, (5) the intensity of intra- and interspecific competition, and (6) the various social and antisocial forces that may develop among the members of the population. Although in certain cases the dispersion pattern may be due to the overriding effects of one factor, in general populations are subject to the collective and simultaneous action of numerous distributional forces and the dispersion pattern reflects their combined influence. When many small factors act together on the population, a more or less random distribution is to be expected, whereas the domination of a few major factors tends to produce departure from randomness.

Environmental agencies of transport. The transporting action of air masses, currents of water, and many kinds of animals produces both random and nonrandom types of dispersion. Airborne seeds, spores, and minute animals are often scattered in apparently haphazard fashion, but aggregation may result if the wind holds steadily from one direction. Wave action is frequently the cause of large concentrations of seeds and organisms along the drift line of lake shores. The habits of fruit-eating birds give rise to the clusters of seedling junipers and cherries found beneath such perching sites as trees and fencerows, as well as to the occurrence of isolated individuals far from the original source. Among plants, it seems to be a general principle that aggregation is inversely related to the capacity of the species for seed dispersal.

Physical features of the habitat. Responses of the individuals of the population to variations in the habitat also tend to give rise to local concentrations. Environments are rarely uniform throughout, some portions generally being more suitable for life than others, with the result that population density tends to be correlated directly with the favorability of the habitat. Oriented reactions, either positive or negative, to light intensities, moisture gradients, or to sources of food or shelter, often bring numbers of individuals into a restricted area. In these cases, aggregation results from a species-characteristic response to the environment and need not involve any social reactions to other members of the population. *See* ENVIRONMENT.

Influence of temporal changes. In most species of animal, daily and seasonal changes in weather evoke movements which modify existing patterns of dispersion. Many of these are associated with the disbanding of groups as well as with their formation. Certain birds, bats, and even butterflies, for example, form roosting assemblages at one time of day and disperse at another. Some species tend to be uniformly dispersed during the summer, but flock together in winter. Hence temporal variation in the habitat may often be as effective in determining distribution patterns as spatial variation.

Behavior patterns in reproduction. Factors related to reproductive habits likewise influence the dispersion patterns of both plant and animal populations. Many plants reproduce vegetatively, new individuals arising from parent rootstocks and producing distinct clusters; others spread by means of rhizomes and runners and may thereby achieve a somewhat more random distribution. Among animals, congregations for mating purposes are common, as in frogs and toads and the breeding swarms of many insects. In contrast, the breeding territories of various fishes and birds exhibit a comparatively regular dispersion.

Intensity of competition. Competition for light, water, food, and other resources of the environment tends to produce uniform patterns of distribution. The rather regular spacing of trees in many forests is commonly attributed largely to competition for sunlight, and that of desert plants for soil moisture. Thus a uniform dispersion helps to reduce the intensity of competition, while aggregation increases it.

Social factors. Among many animals the most powerful forces determining the dispersion pattern are social ones. The social habit leads to the formation of groups or societies. Plant ecologists use the term society for various types of minor communities composed of several to many species, but when the word is applied to animals it is best confined to aggregations of individuals of the same species which cooperate in their life activities. Animal societies or social groups range in size from a pair to large bands, herds, or colonies. They can be classified functionally as mating societies (which in turn are monogamous or polygamous, depending on the habits of the species), family societies (one or both parents with their young), feeding societies (such as various flocks of birds or schools of fishes), and as migratory societies, defense societies, and other types. Sociality confers many advantages, including greater efficiency in securing food, conservation of body heat during cold weather, more thorough conditioning of the environment to increase its habitability, increased facilitation of mating, improved detection of, and defense against, predators, decreased mortality of the young and a greater life expectancy, and the possibility of division of labor and specialization of activities. Disadvantages include increased competition, more rapid depletion of resources, greater attraction of enemies, and more rapid spread of parasites and disease. Despite these disadvantages, the development and persistence of social groups in a wide variety of animal species is ample evidence of its overall survival value. Some of the advantages of the society are also shared by aggregations that have no social basis.

Optimal population density. The degree of aggregation which promotes optimum population growth and survival, however, varies according to the species and the circumstances. Groups or organisms often flourish best if neither too few nor too many individuals are present; they have an optimal population density at some intermediate level. The condition of too few individuals, known as undercrowding, may prevent sufficient breeding contacts for a normal rate of reproduction. On the other hand, overcrowding, or too high a density, may result in severe competition and excessive interaction that will reduce fecundity and lower the growth rate of individuals. The concept of an intermediate optimal population density is sometimes known as Allee's principle. *See* POPULATION DYNAMICS.

[FRANCIS C. EVANS]

Bibliography: W. C. Allee, *Animal Aggregations: A Study of General Sociology*, 1931; P. Greig-Smith, *Quantitative Plant Ecology*, 1957; K. A. Kershaw, *Quantitative and Dynamic Ecology*, 1965.

Population dynamics

The aggregate of processes that determine the size and composition of any population. In this context a population is considered to consist of organisms of a single species. It is characterized by definite time rate of birth and death and often by a definite composition with respect to the ratio between the sexes and between the numbers of individuals belonging to different age classes. An aggregation of individuals brought together fortuitously may or may not constitute a population in this sense. *See* SPECIES POPULATION.

Population size and density. Population size is normally measured in terms of numbers of individuals, while productivity is often expressed as the number of new individuals produced per unit time. There are exceptions such as stands of timber or populations of commercially valuable fish. In these instances, productivity and population size are appropriately measured in terms of mass or volume rather than numbers. The study of dynamics in such populations merely requires consideration of individual growth rates in addition to numbers and ages, and the discussion here will be limited to populations measured by enumeration.

In practice, it is difficult to define the limits of a population, and enumeration normally measures some type of population density. Crude density is the number of individuals per unit of selected space or volume. Examples of this are the number of deer in a county or other areal unit, or the mean number of fish per acre of water surface or per cubic meter of water. This measure suffers from the fact that the units of area and volume are heterogeneous and the population does not utilize all of the available space. Ecological or economic density refers to the mean number of individuals per unit of space actually utilized. Sometimes it is easiest to enumerate a population under conditions of maximum density, as when fur seals or colonial birds are on their breeding grounds, when deer are in winter yards, or when snakes are congregated in dens. Often, the only practicable measures of natural populations involve relative density and are designed to show whether a population is increasing or decreasing without determining its actual size. Thus, the number of birds seen per man-hour of walking and the number of squirrels treed per dog-hour have been used to compare population densities in different years and places. Formidable statistical and practical problems are involved in a mensuration of natural populations.

Population growth. A population can gain in numbers only by birth and immigration, and it can decrease only by death and emigration. In considerations of theoretical population dynamics, migratory movements are customarily ignored and attention is concentrated on the birth and death processes. Individuals of every species have the potentiality for producing more offspring than are required to replace the parents. Without this potential the species could not meet emergencies and would necessarily become extinct. Some organisms reproduce once per lifetime, others many times. Some produce tremendous numbers of gametes, others few. Also, the age at which reproductive maturity occurs varies tremendously —from a few minutes in bacteria to more than a century in the giant *Sequoia* tree. These life history features determine the potential growth rate of the population or the biotic potential of the species.

If these life history features remained the same in successive generations, the population would ultimately grow in accordance with the equation $N_t = Ae^{rt}$, for which the growth rate at time t is $dN/dt = rN_t$. In this formula N_t is the population size at time t, A is a constant, e is the base of the natural logarithmic system, and r is a measure of population growth. The value of r, which is determined by natural history features, is commonly referred to as the intrinsic rate of natural increase, and it has been proposed to define biotic potential as the normal maximum value of r for a given population.

This exponential form of potential population growth implies exceedingly rapid expansion. A single individual of an annual plant in which each individual produces only 2 viable seeds would leave 1,000,000 descendants in 20 years if all seeds survived, and, in fact, every species is theoretically capable of overflowing the Earth. In actuality, shifts in natality (birth rates) and mortality inhibit unlimited exponential growth.

Population control or regulation. Since potential population growth is exponential, while real population size is limited, much interest and controversy has centered about the form of actual population growth. Commonly, when the size of a growing population is plotted against time, the result is a broad S-shaped (sigmoid) form of growth curve which often appears to be symmetrical about a central point of inflection, where the rate of growth shifts from increase to decrease (Fig. 1).

Attempts to give a generalized mathematical formulation of population growth have led to consideration of equations of the form $dN/dt = rNf(N)$. This equation differs from that describing exponential growth in that the factor $f(N)$ serves as a governor or damping factor which takes the value zero when N is very large, indicating that there is some finite upper limit to the size of a population that is capable of further growth.

Various proposals have been offered concerning the form of this governing factor. If $f(N)$ is assumed to be merely a decreasing function of time, the result is one form of the discredited doctrine of racial senescence, the supposition that populations age and die as do individual organisms. Numerous workers have supposed that $f(N)$ is a random variable, sometimes positive and sometimes negative, so that populations normally fluctuate about a steady-state, or equilibrium, size. Modern probability theory shows this position to be untenable in so simple a form. By this doctrine, random extinction would be the eventual fate of all populations, but contrary to experience, small populations of obscure forms often should be observed to grow to tremendous size, at least approaching the conditions of overflowing the Earth.

In a somewhat more realistic approach, the governing factor is regarded as a decreasing function of population density so that the population inhibits its own growth beyond a certain size. It is obvious, however, that maximum population size must also be affected by the quality of the habitat. Populations of the same species are commonly more dense in some regions than in others owing to differences in the availability of essential resources such as food and nesting sites. Therefore, it is usual to speak of a carrying capacity for any given habitat and to define this as the maximum steady-state population of a given species that can be supported there.

Carrying capacity. No general agreement has been reached on a precise definition of carrying capacity, but it seems essential to recognize that this hypothetical upper limit may sometimes be exceeded temporarily. For example, populations as dense as 17 individuals per square yard have been observed in house mice, and the hordes of locusts, chinch bugs, lemmings, and other animals

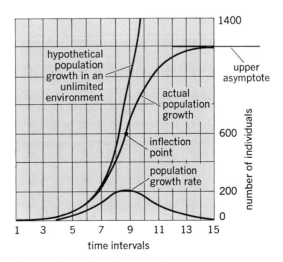

Fig. 1. Population dynamics. Three types of growth curves which use the same data.

that occur in outbreak years far exceed the capacity of the occupied land to provide food and shelter.

Environmental resistance. In practice, students of natural populations often think of population density not in absolute terms but as a density relative to some standard, such as the maximum that could be supported. Thus, the factor $f(N)$ governing population growth is most realistically regarded as a function of the unfulfilled possibilities for growth. The term environmental resistance has been employed to express this same concept that the resistance to further growth increases as the environment approaches saturation.

It is known from observations in the field and from laboratory experiments that crowding does promote increased death rates and inhibit reproduction. Pathogenic organisms can spread rapidly through populations where there is close contact between individuals, metabolic wastes may accumulate to toxic levels, malnutrition or other deficiency conditions may weaken the individuals, and in very crowded populations there may be interference with feeding and mating. Also, various symptoms of physiological stress apparently result from crowding. There are, therefore, sound reasons for considering increased population density to operate in limiting and finally inhibiting population growth; that is to say, the growing population exhibits negative feedback. Theoretically also, any cause of mortality or sterility that is always independent of population size or density, in the mathematical sense, could not prevent growing populations from occasionally overflowing the Earth.

Density factors. Numerous students have tried to classify environmental influences into density-independent factors which are theoretically incapable of regulating population size and density-dependent factors which can exert a governing effect. Much controversy has surrounded these concepts and the varying definitions given to the various classes of factors. Thus, it has been claimed that a density-dependent factor must be what others have called density-responsive; that is, the intensity of the factor must be altered by changing population density. Since population density cannot ordinarily be considered to alter weather or climate, it has been contended that meteorological conditions are density-independent and cannot regulate population size. Others have recognized that weather does sometimes operate in a density-dependent manner, or they have maintained that the true density-dependent regulating factor in such cases is competition for shelter between the individuals which forces the losers to be exposed to unfavorable weather conditions. Another prominent school of thought is utterly opposed to this general approach and maintains that populations seldom attain a level where density effects are important but are normally held at lower levels by environmental inadequacies and mortality factors such as extremes of weather.

Underpopulation effects. At the other end of the scale of size there is also a great deal of evidence in many species for underpopulation effects on population growth. In sparse populations, females may have difficulty finding mates. A small population lacks the adaptability to changed conditions that is provided by a large stock of genetic variability, and this defect may be aggravated by inbreed-ing. Also, populations often condition their surroundings and modify the impact of environmental factors. The forest literally protects the trees from wind damage, excessive insolation, and evaporation. The advantages of maintaining the population above some minimum level are obvious in gregarious animals such as bird flocks and herds of ungulates, and even more so in the social insects where ants, bees, and termites, for example, exercise considerable control over the climate inside the colonial structure. In addition, many cases are known where, often for obscure reasons, populations seem unable to resist extinction if the numbers fall below some minimum level.

From these observations it follows that any generalized concept of the growth governor $f(N)$ must provide for this factor to be small in very small populations, to rise to a maximum at some optimum population size, and to decline to zero before the population overflows the Earth. In other words, the growing population may exhibit positive feedback up to a certain size range and negative feedback at higher levels. To date, however, very little use has been made of highly generalized governing factors.

Actual population growth. The logistic function has been the equation most employed for representing actual population growth. This equation in its differential form is

$$\frac{dN}{dt} = rN\left(1 - \frac{K}{N}\right)$$

where K represents the upper asymptote or maximum size attainable by the population in question. Here the intensity of the governing factor decreases linearly with population size so that no account is taken of underpopulation effects. The integrated logistic curve, however, is sigmoid-shaped and symmetrical about its central point of inflection. It often gives a very good representation of the course of population growth. It occupies a prominent, though controversial, position in modern theories of population dynamics.

Optimum yield. An important consequence of the sigmoid form of population growth is the fact that populations of intermediate size are capable of more rapid growth and greater productivity than either very large or very small populations. If growth were strictly logistic, the most rapid growth would occur at a population size of $K/2$ and the growth rate at this point would be $rK/4$.

When man begins to exploit a large population, as in commercial fishing, the effects of his catch will be to reduce population size. If exploitation is not too intense, the smaller population will lie on a steeper portion of the sigmoid growth curve and will therefore be more productive than the larger population. The population is said to compensate for the increased mortality. In theory, productivity will increase with rate of exploitation to the point where population size reaches the inflection in the growth curve. Hence the maximum possible sustained harvest would be obtained by reducing the population to the inflection point and harvesting at a rate just sufficient to maintain this size. There are many practical difficulties in all actual attempts to determine the optimum rate of harvest and the general problem has become widely known as the optimum-yield problem. It is noteworthy that if the population is "overfished" so that its size passes below the inflection point, productivity will decline with each further decrease in size. Then each increase in the effort to harvest a crop will have the effect of reducing the long-term yield. There is reason to believe that many commercial fisheries are reducing their total catch by fishing too intensively.

The same principles apply to attempts to control noxious species. Rodent populations, for example, compensate for mortality and it is possible to harvest a large annual crop of rats without actually reducing the population. Programs of killing are often discontinued before the population passes below the inflection point where control would become progressively easier. Consequently, the most effective way of dealing with noxious forms is often to reduce the carrying capacity of the environment; programs for improving garbage disposal and for ratproofing buildings will often be much more effective than programs of killing.

Fluctuations. Populations of many species fluctuate in size from year to year, and a very large literature exists on this subject. Plagues of rodents and locusts are recorded in the Old Testament and similarly ancient sources, and the migrations of the lemmings and the eruptions, outbreaks, or gradations of various insect populations have often attracted popular attention. There is a tendency for eruptions to be most conspicuous in high latitudes and other regions where the biota is composed of relatively few species of plants and animals. Outbreak years typically follow periods of buildup during which the population nearly realizes its potential of exponential growth. Eventually, population size exceeds the capacity of the environment to sustain it and the population "crashes," often dropping abruptly to a very low level. *See* POPULATION DISPERSAL.

Cycles. The most discussed of the fluctuating populations have been those of certain gallinaceous birds, rodents, rabbits, and fur-bearing mammals of northern regions. The records of the Hudson's Bay Company, for example, provide figures for a long series of annual catches, and many students of populations have considered that the rhythms, or cycles, in such records indicate a regular periodicity in the rise and fall of population size. Although cycles of various length have been postulated, most competent opinion has considered that there are two predominant cycle lengths: a short cycle of approximately 3 or 4 years and a longer one often referred to as the 10-year cycle.

Numerous explanations have been advanced. One of the most popular has been the belief that the populations follow some extraterrestrial rhythm, especially the "sunspot cycle." Such hypotheses suffer from numerous observations indicating that populations in different regions may be out of phase with each other. Others have based explanations on population dynamics, claiming, in effect, that the cyclic species are deficient in feedback mechanisms so that exponential growth is not inhibited until disastrously high densities are attained. Still others have attributed the cycles to interactions between two species: herbivores and their food plants, or predators and their prey. The predator is visualized as growing until it exhausts

its food supply and then undergoing violent decline until the prey population has time to recover. These hypotheses are not entirely satisfying because it is difficult to see why many species with diverse life histories should adhere to two basic cycle lengths. The Canadian lynx and the chinch bug, for example, are both considered to exhibit 10-year cycles.

Other factors. It has also been noted that random variables such as the sizes of the numbers turning up on a roulette wheel will, when plotted on graph paper, give an appearance of regularity and show a series of peaks occurring at a mean interval of three or four numbers. It has been postulated that the great variety of haphazard factors affecting population size constitute a causal system comparable in complexity to that governing the roulette wheel and that the appearance of peak population years may therefore be considered to be governed by a random variable.

Whatever their causes, population fluctuations are often of great practical importance. Much remains to be learned about possibilities for predicting peak years in crop damage and in the harvest of food, game, and fur-bearing animals, or for minimizing the expectation of financial loss resulting from these fluctuations.

Age structure. Not only the size but also the composition of a population is governed by the age schedules of natality and mortality. If the life history features and the death rates for individuals of each age remain constant, a population will eventually attain a stable age distribution such that the individuals of any particular age constitute a fixed proportion of the total. In human populations, about one-half of the individuals typically fall into the age range of 15–50 years, but the ratio of older individuals to very young differs greatly from one nation to another. Consequently, diseases of old age attain greatest importance in populations where life expectancy is greatest. Problems relating to age structure are also of great significance in nonhuman populations. It is apparent that a predator can benefit man by selectively killing superannuated game animals, thus increasing productivity, or that such a predator can work against man's efforts to control noxious species even while killing many of the undesirable forms.

In populations exploited by man, age structure is affected by the intensity of exploitation. It is obvious that imposing a new mortality factor on the population will reduce life expectancy and so reduce the average age of the population members. In some animals, including man, rats, and some insects, older individuals cease to be reproductive, and the lowering of average age by moderate exploitation leads to a direct increase in birth rate. Also, by making additional resources available to promote the survival of young, a younger population results. With very intense exploitation an opposite result occurs: Reproduction is impaired by depletion of the reproductive age classes, and the remnant members become aged, producing an unproductive population of increased average age.

Demography. Demography is commonly and most concisely defined as the statistical study of populations. Demographers are usually students of human populations and are concerned not only with such fundamental and general phenomena as

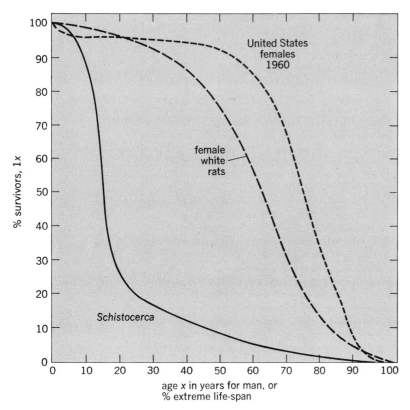

Fig. 2.   Some forms of survivorship curves. (*After L. C. Cole from M. C. Sheps and J. C. Ridley, eds., Public Health and Population Change, Pittsburgh Press, 1965*)

different species and for different environmental conditions. The survivorship curves vary from a convex to a very concave shape as viewed from above (Fig. 2). The convex shape indicates slight mortality until advanced age and the concave shape indicates heavy early mortaility followed by very gradual attrition of the few individuals that actually attain maturity. There is evidence that advances in medicine and sanitation have changed the survivorship curve for man from concave to convex.

By making certain assumptions about the temporal stability of mortality factors and the number of annual births it is possible to use the life table to derive "life expectancy" figures representing the average number of years of life remaining for an individual of a given age. This technique is fundamental for life insurance practice.

Given the form of the life table and a corresponding table of age-specific birth rates, it is possible to project the future of any population and to calculate the results of changing birth or death rates. *See* ECOLOGY.   [LA MONT C. COLE]

Bibliography: L. I. Dublin, A. J. Lotka, and M. Spiegelman, *Length of Life: A Study of the Life Table*, 1949; F. Osborn (ed.), *Our Crowded Planet*, 1962; Population studies: Animal ecology and demography, *Cold Spring Harbor Symposium on Quantitative Biology*, vol. 22, 1957; M. C. Sheps and J. C. Ridley (eds.), *Public Health and Population Change*, 1965.

Prairie ecosystem

Prairies (grasslands) are inhabited by large numbers of animals throughout the year. Many of these animal groups have been able to survive in prairies by developing protective mechanisms against the rigorous climate; they developed specific grazing habits, and complement each other in ecosystem functioning. Some species migrate to escape the winter; others move below ground and thus serve a vital role in reducing plants to a decomposable form for nutrient release, and also in churning and modifying the soil to allow new plants to grow.

Animal groups. Broadly speaking, prairie animals may be classified as herbivores, predators, and scavengers; however, the classification may depend upon the season, that is, the availability of food. Herbivores are numerous; grasses, well adapted for survival under constant grazing, and grazing animals have been inseparable components of prairie ecosystems since long before the advent of man. Groups of large herbivores found in the prairies include those listed in the table. Before pioneer settlements, some 20,000,000 to 40,-000,000 buffalo grazed in the mixed prairie portion of the Great Plains of North America, as did other herbivores, including 4,000,000 to 8,000,000 antelope, large herds of elk, and deer. The buffalo exist today only in relatively small herds on scattered game preserves; there are no prairies of sufficient size to allow large herds free migrational movement. The buffalo probably grazed heavily on one site and then moved on, baring the ground and allowing many forbs and weeds to invade, which are the plants preferred by pronghorn antelope, deer, and elk. Thus a dynamic interaction of these herbivores was generated.

Jackrabbits, cottontails, ground squirrels, and

natality, mortality, migratory movements, and age and sex structure, but also with such factors as population composition by social status, ethnic groups, and income level; rates of marriage, divorce, and illegitimacy; the size of the labor force; and such esoteric questions as the motivation for a particular family size and survival rates in various occupations. However, many ecologists insist on recognizing a field of general demography embracing the study of the fundamental phenomena that are common to all populations of living things.

The basic technique for the statistical analysis of any population is the construction of a life table. The fundamental features of a life table is a column of numbers showing, for a "cohort" consisting of a large number of individuals born simultaneously, how many would be alive on each subsequent birthday until the cohort became extinct. This column is designated "survivorship" and is represented by the symbol l_x as the proportion surviving to the xth birthday.

For a long-lived species such as man it is impracticable actually to follow a cohort throughout its history and, in any case, mortality factors would probably change so drastically over the course of a century that a life table constructed by this straightforward technique would be worthless by the time the data for its construction became available. Consequently, actual tables are constructed indirectly from observed "age-specific death rates," which are the number of individuals dying at a particular age divided by the total number entering that age class.

Many life tables have been constructed for

Distribution of large, herbivorous prairie animals

Families and examples	Area
Equidae	
Wild Horse	Mongolia, Chinese Turkestan
Onager (Asiatic wild ass)	Mongolia, Tibet, Syria
Wild ass	E. Africa to Somaliland
Zebras	E. Africa, Abyssinia, Somaliland
Camelidae	
Camels	Central Asia
Huanaco and vicuña	South America (S. Ecuador to Tierra del Fuego)
Cervidae	
Deer	N.W. Africa, the Americas
Caribou and reindeer	Eurasia, North America, Greenland
Red deer and elk	Eurasia, North America
Giraffidae	
Giraffe	Africa below the Sahara
Antilocapridae	
Pronghorn antelope	Canada to Mexico
Bovidae	
Bighorn sheep	Eurasia, Siberia, North America
Musk ox	Alaska to Greenland
Wild cattle	N. Europe, Asia to Indo-Australian Archipelago
Bison	North America, central Asia to Siberia

mice are important small prairie herbivores. One of the burrowing rodents characteristic of the pre-settlement plains and mixed prairies is the black-tailed prairie dog.

Numerous small passerine birds subsist largely on seeds, while other birds mix seeds and insects in their diet. The insects have important influences on prairies; yet their exact position in the food webs, their total effect on the plants, and their influences on each other are not well known. Grasshoppers are among the more important prairie insects, occasionally becoming very abundant and causing drastic consequences, such as the locust plagues of Africa and grasshopper plagues of North America in which many acres of vegetation have been destroyed. Although many grasshopper species are highly specific in their diets and are primarily herbivores, they sometimes act as predators and scavengers.

Most of the reptiles and amphibians of the prairie ecosystems are predators.

Historical development. Fossil records show that grasses occurred in the Tertiary. However, true grasses arose perhaps during the Quaternary; prairie plant forerunners were spreading throughout the world at least by late Cretaceous times.

The grazing animals and their food plants developed in parallel. The horse of 50,000,000 years ago, no larger than a small dog, originated in subtropical lowlands. Prairies, responding to cooler, drier climates, spread throughout the world during the Miocene; the primitive horse then developed hooves and became adapted to life on open prairies. During subsequent cooling periods there was a retreat and advance of great ice caps over the world. Intercontinental migration of certain grazing animals occurred. Strangely, the ancestor of the horse, which for millions of years had lived on the North American continent, disappeared and horses did not occur again until the Spanish conquistadors reintroduced them.

Feast or famine. Animals living on the prairies must adapt to the periodic feasts and famines; to survive they follow one of three strategies.

One group of animals, including many land mollusks and insects, feed intensively when lush herbage is available, store up food reserves as fat, and succeed through suspended physiological activities (brought on either by summer drought or winter cold.

Some prairie animals fatten during a period of plenty and then sleep during the winter in burrows or other protected niches. Such animals include most of the rodents, for example, North American prairie dogs and Eurasian marmots.

Another group of animals undertakes seasonal migrations to succeed in the prairies. Large grazing animals and many birds are important migratory species. Migrations may cover hundreds of miles or may be local from adjacent but contrasting vegetation types; some birds fly across the Equator.

Specificity in diets. Many prairie animals are specific in their dietary habits and some may starve when their specific food requirement is no longer available. Competition among prairie herbivores is important, for they subsist primarily on nutrients obtained from separate parts of the same native plants, including grasses, forbs, shrubs, and trees. Some animals prefer the leaves of certain plants, many have a preference for the seeds, and others gnaw and chew upon the roots, rhizomes, and other underground parts.

Population dynamics. Prairie animals have adapted their reproductive patterns to meet expected availability of lush, green, nutritious herbage. Thus, from early spring through late summer the young grassland animals abound. Some mice reproduce as early as in March in the Northern Hemisphere; many rodents produce several litters in one growing season. The young of most large herbivores are born somewhat later in the year; antelope kids and buffalo calves, for example, are born in late May and early June. While calves and lambs of domestic herds may be born yearlong, special feed and shelter must be provided by man for those born in the winter.

Insects are sporadic in their population densities, being partly keyed to sequences of wet and dry years or summer and winter extremes. The number of insects that overwinter in the prairies depends largely upon the severity of the winter and the food supply and cover that was available to them the previous summer.

Rodents, known for dramatic fluctuations in population densities, tend to undergo cyclic variations in density and at times reach fantastic numbers.

Changes in number of any group of herbivores soon induce changes in the number of predators and parasites and changes in the vegetation. The prairie operates as an intricate feedback system, with each component affecting the others, always in an attempt to bring population densities back to equilibrium, at least temporarily. *See* POPULATION DYNAMICS.

Man's impact on grassland animals. Most of the wild animals are influenced by man's large grazing animals in the lower-rainfall regions. Some of these animals, particularly small herbivores, benefit from livestock grazing, as noted by the fact that they are more abundant in heavily grazed areas; for example, there is a tendency for jackrabbits to be more abundant on heavily grazed areas than on the open range. In higher-rainfall areas the prairie dog perhaps develops best on places grazed by buffalo, or by man's domestic animals, since taller grasses, more sensitive to grazing, are thinned out and shorter grasses remain and spread, providing the prairie dog with better opportunity to view his surroundings. The prairie dog depends on visual acuity since he does not have the speed or agility to evade predators without escaping into burrows. The full impact of domestic animals on the wild animals is yet unknown. Man has been grazing on prairie intensively for only 100–200 years in many areas. The native vegetation is resistant to change and the full impact of abusive grazing in many regions has not yet been seen. *See* ECOSYSTEM; GRASSLAND.

[GEORGE VAN DYNE]

Bibliography: D. L. Allen, *The Life of Prairies and Plains*, 1967; P. J. Darlington, *Zoogeography: The Geographical Distribution of Animals*, 1957; National Geographic Society, *Wild Animals of North America*, 1960; J. E. Weaver and F. W. Albertson, *Grasslands of the Great Plains*, 1956.

Predator-prey relationships

The natural history of predator-prey relationships (food habits and the like) in aquatic communities is moderately well known, except for environments such as the deep sea that are difficult to sample or observe. Recent attention has centered on the quantitative implications of these relationships: (1) the effect of predation on the population size of each species of prey, (2) the strategies that predatory populations have evolved through natural selection to ensure their own survival, and (3) the broader effects of predation on the species composition of the whole community. *See* FOOD CHAIN.

Species population size. In the simple case of two species, the predator (as distinct from a parasite or scavenger) affects the dynamics of the prey population by killing its members. The extent to which predation tends to stabilize the population size of a particular prey species depends on whether the intensity of predation—the fraction of the prey population removed by predation per unit of time—increases as the size of the prey population increases and on how much time lag is involved. The intensity of predation may increase through a disproportionate increase in abundance of the predators due to immigration or reproduction or both, and through a disproportionate increase in the number of prey eaten by each predator. Such an increase in the number of prey eaten occurs when the predator forms a "search image" and concentrates its feeding activities on the prey species in question (usually through learning on the part of the predator) or when increase in the body size of predators with indeterminate growth enables them to remove disproportionately more (or larger and hence more fecund) prey.

Carnivores that are not predators in the strictest sense may indirectly control the population dynamics of the organisms on which they feed. Young flatfish "prey" on buried bivalves by biting off the exposed siphons when bivalve density exceeds a threshold level. The bivalves are able to regenerate these respiratory-feeding tubes but do so at the expense of producing more gonadal tissue and hence produce fewer offspring.

Strategies for survival. The evolutionarily correct strategy for the predatory population is to feed so as to maintain the prey population at a size and age structure that will maximize the predator's sustainable intake of energy. Ideally, the pattern of predation should maximize the growth rate and food-chain efficiency (for example, for carnivorous predators, the ratio between the yield to the predator and the production of the prey species' food) of the prey population and minimize the metabolic cost to the predator of searching for, pursuing (if necessary), capturing, and digesting prey and the risk it experiences from its own predators. The intertidal predatory snail *Thais* attacks a limited number of barnacles per day but preferentially attacks large ones, in effect maximizing its ingestion per attack and removing slow-growing individuals from the prey population. However, the metabolic cost per attack probably also increases with increasing size of barnacle, and large barnacles are more fecund than small ones are. When prey species become infertile in later life, predation concentrated upon postreproductive individuals (which often have a low growth rate) also tends to maintain a high rate of replenishment of prey. For example, most of the mortality of pelagic salps (*Thalia democratica*) caused by the predacious copepod *Sapphirina* occurs with postreproductives, so that the very rapid rate of population growth of the salps is reduced rather slightly even by relatively high concentrations of this predator. Predation upon the surplus, territoryless individuals of a species of animal prey that requires possession of a territory for mating and successful rearing of young has the same effect. Removal of injured individuals is a more obvious example; carnivores preying on herds or schools often single out for attack individuals conspicuously different in appearance or behavior from their fellows.

Metabolic costs of predation are sometimes minimized, relative to the benefits, by varying the degree of selectivity for particular species or sizes of prey. As the abundance of potential prey increases relative to the searching range of the predator, so that a relatively large amount of energy is involved in pursuit (if any) and capture, some kinds of predators tend to be specialists, having a restricted search image and limited, but presumably efficient, feeding behavior. Predators whose prey is widely dispersed, so that most of the overall cost of predation is in searching, tend to be generalists. An individual predator may also change foraging behavior depending on the relative abundances of different prey; anchovies switch from a generalized filtration of small prey items, which is relatively ineffective at natural concentrations, to a directed, biting attack on individual, larger prey when such prey become sufficiently abundant.

Community species composition. Some predators markedly affect the composition of communities because those species on which they prey directly compete for resources with other species, or are predators themselves. Predation increases diversity (the number of species in a community and the distribution of individuals among species) when the species that would dominate the community is preyed upon preferentially because of its palatability, visibility, chemical attractiveness, or size. For example, moderate grazing by domestic sheep on rich pasture increases the diversity of the vegetation since the competitively superior plants are highly palatable; sea urchins have a similar effect on the seaweeds in tide pools; and grazing by limpets on young plants prevents the establishment of competitively superior macroscopic algae in the intertidal zone.

When planktivorous fish remove the larger and more visible zooplankters in a lake, smaller species which are competitively inferior may thrive, and the resulting change in composition of the zooplankton sometimes affects the composition of the phytoplankton as well, since larger zooplankters feed preferentially on larger phytoplankton cells. Mussels, which outcompete other sessile, intertidal species for space, are preyed upon preferentially by starfish, and space is thus made available to other species in the lower intertidal, where starfish are active. Predation by the coral-eating starfish *Acanthaster* apparently increases the diversity of coral in areas where the starfish is abundant but not so dense as to decimate all coral, although the total biomass of living coral is reduced by even moderate densities of *Acanthaster*.

There are also situations in which predation leads to a decrease in diversity. If the dominant species is relatively unpalatable, predators may increase its dominance by selectively removing more palatable species, as do sheep grazing on poor pasture, in which case diversity of vegetation may thus be reduced. Even in rich pasture, heavy overgrazing may lead to the dominance of a few toxic, spiny, or otherwise unpalatable species. *See* COMMUNITY.   [MICHAEL M. MULLIN]

Bibliography: T. W. Schoener, *Annu. Rev. Ecol. System.*, 2:369–404, 1971.

Radiation biochemistry

The study of the response of the constituents of living matter to radiation, a specific injurious agent. Biochemistry, in the ordinary sense, deals with the chemistry of the building stones of living tissues and organisms, and with the balance and integrated metabolic reactions in which these take part. This article deals with the effect of ionizing radiations, radiations that ionize matter through which they pass.

The chance of exposure to radiation has increased in this atomic age. Not only is radiation more widely used in medicine for diagnostic and therapeutic purposes, but the applications of radiation in industry have increased. For example, the hazards connected with the development of atomic energy, and the possible use of atomic weapons in war make research in the effects of radiation important.

The capability of penetrating to every part of the interior of cells, without being obstructed by membranes or defensive barriers, puts ionizing radiations in a unique position as compared with other noxious agents, which are limited in penetration, or selectively active on special cell constituents.

The cells of living matter consist of a cell membrane surrounding protoplasmic protein, in which are embedded the nucleus and various small granular bodies, such as mitochondria and microsomes. The whole cell structure is permeated with water. The content of the cell is inhomogeneous, highly organized, and equipped with a series of enzymes which make complicated metabolic reactions possible.

The indirect and the direct modes of action of radiation have been proposed as mechanisms of radiation effects. These modes of action are discussed in the following paragraphs.

Indirect action. The water content of cells is about four times greater than their dry weight. The effect of radiation on the water has consequences for the solid matter contained in it.

When water is irradiated, it is split into the primary radiation products, hydroxyl (OH) radicals, hydrated electrons, and hydrogen (H) atoms, which are highly reactive, uncharged chemical entities. Oxidation and reduction reactions are brought about when the primary radiation products collide with solutes, the substances dissolved in water. In the absence of solutes, however, they quickly recombine to form water. The irradiation thus acts indirectly on solutes via the water. Consequences of this indirect action are the dilution effect, the protection effect, and the oxygen effect.

Dilution effect. This effect causes a dilute solution to appear more sensitive to radiation than a concentrated one, because a given dose of radiation results in the formation of a given number of radicals which will react with a corresponding number of solute molecules. Therefore, the proportion of solute molecules chemically changed will be large in dilute and small in concentrated solutions.

Protection effect. This comes into play when there are two or more solutes in solution which are capable of reacting with radicals. All these solutes will compete for the existing radicals and, therefore, fewer radicals will be available for each species of solute than would be the case if only these solutes were present in the solution. In other words, there is a mutual diminution of the radiation effect and each solute appears protected by the others against radiation. This protection effect will vary in accordance with the absolute amounts of solutes and with their specific capability of reacting with radicals, that is, with their capability of

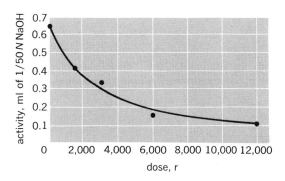

Fig. 1. Exponential relation between degree of inactivation and x-ray dose for carboxypeptidase solutions. (*W. M. Dale, 1940; reproduced by permission of the editors of Biochemical Journal*)

acting as acceptors for radicals.

The protection effect operates also in a one-solute solution if the irradiation products formed from this solute can still react with radicals. Before radiation starts, the solution of an enzyme (catalyst of living matter) contains the dissolved active enzyme molecules only. Each fraction of radiation dose delivered inactivates some enzyme molecules which, though no longer active, can still react with radicals. The inactive molecules then represent a second solute which competes with the still-active molecules for radicals. The result is that equal but subsequent increments of radiation become less and less effective, so that an activity-dose curve takes on an exponential shape, as is illustrated in Fig. 1.

Oxygen effect. The primary OH radicals, hydrated electrons, and H atoms give rise to other radicals if oxygen is present in solution, namely to HO_2 or O_2^- radicals and to the stable product hydrogen peroxide (H_2O_2). These, as oxidative agents, can react with solutes and thereby enhance the radiation effect when compared with oxygen-free solutions, hence the term oxygen effect. Hydrogen peroxide may also lead to the formation of organic peroxides which can be responsible for an after-effect, that is, a continuation of decomposition of solutes after irradiation has ceased. The interference of hydrogen peroxide complicates the clarification of reaction mechanisms. It has been found that nitric oxide increases sensitivity to radiation under conditions of anoxia to an extent similar to that observed with oxygen. The explanation which is put forward for this effect is that both nitric oxide and oxygen have an equal affinity for carbon radicals.

The target theory as originally proposed did not leave room for modification of radiation effects by chemical means. It was an all-or-none effect. Yet it has often been found that the presence of oxygen or of nitric oxide during irradiation increased the radiation effect not only in solutions but also in dry matter. It has now been proposed that the theory be modified by assuming that immediately after the passage of an ionizing particle the target molecule is left in a highly reactive state, facilitating chemical reaction. The reaction will depend on the chemical environment, the physical state, or both. Thus, one can visualize the possibility of modifying the effect of the primary dissipation of energy so as to restore the target molecule (a healing effect) or to cause its irreversible injury, depending on the reaction with the modifying agent.

The promising method of microwave spectroscopy for detecting and measuring concentrations of free radicals has established that radiation causes the formation of radicals of various lifetimes in biological matter. The subsequent interaction between these and the presence of gases like oxygen and nitric oxide, which are themselves radicals, offers an explanation of the oxygen effect and the nitric oxide effect that is alternative or supplementary to the explanation based on interaction with radical-reaction products.

Direct action. Since the dissipation of radiation energy is not confined to the solvent, direct ionization with subsequent chemical change will occur in solute molecules themselves. The frequency of such an event (single hit) increases in proportion to the concentration of the solution, and reaches its maximum when one hit is scored per molecule of a dry substance.

Some investigators believe that the nucleus of a cell contains a vital and sensitive structure in which the primary event, that is, an ionization, has to occur, or that an ionizing particle has to pass near or through it in order to cause a chemical alteration that results in the biological effect which is subsequently observed. This is the target-hit theory.

An important practical application of the direct-hit theory is the determination of the molecular weight—sometimes even the shape of large, biologically active molecules irradiated in the dry state. The underlying assumption is that the destruction of the biological function, for example, of enzymic activity, or of the ability of a virus to infect, is caused by a primary ionization produced by a fast charged particle passing through the molecule. It is possible to calculate the radiation dose which will produce, on the average, one ionization per molecule. From the number of ionizations occurring per unit volume the target size can be assessed. This method has been successful in a number of cases.

The protection effect which was previously the prerogative of the indirect action has been found to occur also in solid matter when relatively small amounts of additional substances are incorporated in the solid. Some form of intra- and intermolecular transfer of energy is assumed to occur whereby the energy is channeled preferentially to the added substances.

Direct plus indirect action. The share taken by the direct and indirect modes of action is illustrated in Fig. 2, which shows the inactivation of an aqueous solution of an enzyme (carboxypeptidase from pancreas) at various concentrations.

Since water is always in excess in living tissues, except in bone structure and fatty tissue, the opportunity of reactions with radicals is always preponderant. A distinction between the direct and indirect modes of action becomes increasingly trivial as the source of active radicals approaches the molecule or structure upon which they act. As a result, the two modes of action merge in the immediate vicinity of the target, constituting a direct hit.

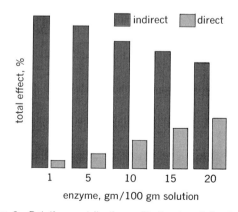

Fig. 2. Relative contributions of indirect and direct actions to total effect of x-rays on carboxypeptidase in solution. (*W. M. Dale, 1947; reproduced by permission of the editors of British Journal of Radiology*)

Measurement of radiation sensitivity. The problem of a more detailed analysis of the radiation products from substances of biochemical importance has been solved in only a few instances. However, the protection effect has made it possible to assess overall radiation sensitivity over a wide range of substances, extending from small molecules to the large molecules of proteins. From the degree of inactivation of an enzyme solution of known concentration in the absence and in the presence of a protective substance, a value of the reactivity with radicals of the substance in question can be derived. In this way it has been found that specificity of radiation effect can be detected in small molecules. Thus, it has been shown that sulfur in organic molecules causes a high degree of radiation sensitivity. In general, specific effects are not discernible in large molecules, because of the great number of their reactive groups, but it can be said that the protective power of large molecules is approximately proportional to their molecular weight.

Effect of radiation. The effect of radiation upon biological material is discussed in the following paragraphs.

Biological reduction-oxidation systems. Most reactions of solutes with radicals consist of oxidations and reductions, and it has been proposed that the OH radicals, hydrated electrons, and H atoms in irradiated water constitute a redox system, having an equivalent redox potential (ERP). This system reacts with a range of redox systems as solutes in such a way that for redox potentials greater than -0.52 volt, oxidation occurs, and for potentials less than -1.1 volts, only reduction takes place. This depends on such things as the type of radiation, pH, and presence or absence of oxygen. In the range between these values both oxidation and reduction are possible. The redox potentials of most redox systems in cells are in the oxidation range but will display different resistance to oxidative changes, according to their total concentration as well as to the respective ratio of their reduced state to their oxidized state. Biological redox systems are normal constituents in cells and form the link in many metabolic steps. There is opportunity, therefore, for interference with the normal metabolism. Some instances of biochemical redox systems are cysteine to cystine, sulfhydryl to disulfide (SH—SS), glutathione, prosthetic groups of enzymes, such as flavoprotein, coenzymes I and II, and ascorbic acid.

Redox systems may also occur in which the reversibility of the reaction is impaired. If the oxidized or the reduced state of the reacting compound suffers secondary changes, such as the formation of a polymer, or if the reduced form is capable of forming a molecular compound with the oxidized form, no proper equilibrium will be established.

Proteins. One of the most important constituents of living matter, proteins consist of long chains of amino acids occurring in characteristic proportions and in specific sequence, linked together by the peptide linkage CONH. Side chains protruding from the main chain can form cross linkages between neighboring chains. In solution, they form finely dispersed colloidal systems. The variety of existing proteins is very great, but all proteins have one reaction in common, namely, the property of denaturation. This is usually an irreversible change which occurs when proteins combine with certain chemicals or when heat or radiation is applied.

The denaturation manifests itself as coagulation with subsequent insolubility. It has been found that ionizing radiations lower the resistance of proteins to thermal denaturation. After irradiation, protein solutions contain different denatured protein derivatives and show a marked decrease in energy content, signifying deep-seated structural changes. It has been established that the oxidative attack by OH radicals is directed toward the peptide linkage. This results in the formation of high-molecular-weight carbonyl ($C=O$) compounds and of keto acids, and the release of ammonia. The radiation-sensitive SH group of cysteine, an amino acid occurring in egg albumin, can be oxidized to a disulfide (S-S), which can form a cross-linkage between neighboring chains. Oxidation can occur even further, beyond the S-S stage, without denaturation or marked instability of the protein. It has further been observed that after hydrolysis of irradiated serum albumin several amino acids are partly destroyed. Examination of irradiated protein solutions by spectrophotometry also reveals changes. Bovine serum albumin, serum globulin, and egg albumin show an increase in optical density which apparently is due to the action of radiation on their tyrosine component; if, however, the proteins contain more tryptophan than tyrosine, a decrease in density is observed.

Not only is protein, as such, changed by radiation but its building stones, the amino acids, are affected.

Amino acids. The principal effect is the loss of ammonia from, or deamination of, the amino acids. The extent of deamination varies with experimental conditions, but an important point is that it also varies with the chemical configuration of the amino acid itself. If the amino group is in the alpha (α) position, attached to the carbon atom next to the carboxyl group (COOH), as in α-alanine, $CH_3 \cdot CH(NH_2) \cdot COOH$, the loss of ammonia is nearly twice as great as for β-alanine, $CH_2(NH_2) \cdot CH_2 \cdot COOH$. In this compound, the amino group is attached to a second carbon atom, that is, in the β position. This is an example of the specificity of radiation effects. More ammonia is split from histidine, where the glyoxaline part may contribute to the yield.

In experiments in which radiation products other than ammonia were examined, it was found that alanine irradiated in a vacuum yielded acetaldehyde, pyruvic acid, propionic acid, ethylamine, and carbon dioxide. Products from glycine included glyoxylic acid, formaldehyde, acetic acid, formic acid, and carbon dioxide. It is, however, claimed that the appearance of the various products depends on whether radiation doses applied have been moderate or massive. This claim is made because some of the radiation products are not due to initial reaction, but rather such products are formed by further oxidation or decarboxylation of glyoxylic acid.

Enzymes. These are proteins which differ from other proteins in their ability to act as catalysts. Enzymes speed up specific chemical reactions.

They either have special active groups in their makeup, enabling them to combine with their specific substrates on which they act, or they are more or less firmly linked to a nonprotein partner, or prosthetic group, which, in cooperation with the protein part, functions as a highly specific catalytic system.

As far as their protein nature is concerned, enzymes will undergo the same general changes as described for other proteins, with consequent loss of activity. There are some enzymes, the S-H enzymes, in which that group is essential for enzymic activity. This S-H group is particularly sensitive to radiation and may undergo changes before deeper-seated modification of the protein has taken place. If the inactivation of the S-H enzyme has not gone far, it can be restored to its original activity by the addition of the tripeptide glutathione, which contains S-H.

An example of an enzyme containing a prosthetic group is D-amino acid oxidase, which specifically oxidizes D-amino acids only. This enzyme can be split into flavin adenine dinucleotide, a nonprotein, and a specific protein. Neither part, on its own, has enzymic activity, but when combined they constitute the complete active enzyme. Each part can be chemically changed by radiation and, when rejoined, shows a lower activity than after irradiation of the complete enzyme.

In a comparative study of the effect of the densely ionizing alpha radiation versus x-radiation on the enzyme carboxypeptidase in solution, it was found that alpha radiation was only one-twentieth as effective as x-radiation. The effect appeared to be entirely due to the δ-rays, which branch off the α-ray track as spurs and which have an ion density similar to that of x-rays, not to the primary ionization column of the α-ray track. Quite generally, the lower efficiency of alpha radiation on substances in aqueous solution is in contrast to its higher efficiency on biological systems, such as the breakage of chromosomes in cells.

Nucleic acid and nucleoproteins. These are the important chemical building stones of the genetic material contained in the chromosomes of cell nuclei. The nucleoproteins are saltlike unions of a nucleic acid with basic proteins, such as protamine or a histone. Two types of nucleic acids exist, ribonucleic acid (RNA) and deoxyribonucleic acid (DNA), both of which are built up from nucleotides containing a nitrogenous base, a pentose sugar, and phosphoric acid. The base forms an ester with the phosphoric acid. Both nucleic acids contain the bases adenine, guanine, and cytosine but RNA has uracil and the sugar D-ribose, whereas DNA contains thymine and the sugar D-deoxyribose. RNA is formed predominately in the cytoplasm and DNA in the nucleus. Since the molecular weight is of the order of $6-8 \times 10^6$, each molecule must contain a great number of nucleotide units. Nucleic acids are structured as two complementary, helical, nucleotide threads, joined together by hydrogen bonds between the basic guanine and cytosine and between adenine and thymine. The nucleic acids are similar in organization to protein, with its amino acid units and its cross-linkages. Irradiation breaks down the hydrogen bonds between the DNA threads, and denaturation by heat is facilitated after irradiation. The instability of nucleic acids is evident from the short heating (15 min in boiling water) required for a marked decrease in viscosity of DNA accompanied by decrease in molecular weight. A similar decrease of viscosity in dilute DNA solution is effected by relatively small doses of radiation.

Chemical changes require large doses of radiation. Among the reactions observed are deamination, liberation of free purine, decrease in optical density, increase in amino nitrogen, breakage of the pyrimidine ring, oxidation of the sugar moiety (ribose portion), and liberation of some inorganic phosphate. The conditioning effect of radiation is shown by the fact that acid hydrolysis subsequent to irradiation liberates free phosphate more quickly than would have occurred without irradiation. In the irradiation of the breakdown products of nucleic acids, such as nucleotides, nucleosides, and the purine and pyrimidine bases, the radiation effects resemble those from nucleic acids.

[WALTER M. DALE/M. EBERT]

Bibliography: M. Ebert and A. Howard (eds.), *Current Topics in Radiation Research*, vol. 1, 1964; M. Errera and A. Forssberg (eds.), *Mechanisms in Radiobiology*, vols. 1–2, 1961; A. Hollaender (ed.), *Radiation Biology*, vol. 1, 1954; A. M. Kuzin, *Radiation Biochemistry*, 1964.

Radiation injury (biology)

Radiobiology deals with the reaction of living cells and organisms to ionizing radiation. Only that radiation, primary or scattered, which is absorbed is effective. The absorbed dose of radiation is measured in rads. For perspective, the usual amount of radiation delivered in the course of a chest x-ray is now about 27 to 30 millirads (mrad) at the skin surface.

The millirad is one-thousandth of a rad, the physical unit of absorbed radiation and hence produces 0.1 erg/gram of tissue. Natural whole-body background radiation in the United States is estimated at about 100 mrad per year. This article will discuss the deleterious effect of ionizing radiation, especially on man and animals.

As x-rays and radium became more important in medicine (today nearly one-third of the crucial decisions in practice are based on x-ray evidence), their potential hazards to both physician and patient became apparent, and in 1928 an International Commission on Radiation Protection was set up, with a United States counterpart formed thereafter—the National Council on Radiation Protection and Measurement. In 1959 the Federal Radiation Council was established to serve as a policy-forming group to advise the President on Standards to be used by all Federal agencies. Individual states are still responsible for regulation of radiation health for their citizens.

Members of the United States population receive radiation from natural background averaging about 102 millirem (mrem) per year, from medical exposures a genetically significant dose of about 73 mrem per year, from global fallout up to 4 mrem per year, and from nuclear power generation 0.003 mrem per year. The millirem is one-thousandth of the rem, a biologic unit of radiation absorbed by man modified from the physical unit by the relative biologic effectiveness or quality factor.

Mechanism of action. When the molecules of living matter absorb electromagnetic and particulate radiation, they are ionized or excited or both. The excited molecules respond very rapidly, within microseconds, by ionization, by emission of light, by formation of triplet states, or by vibrational and thermal losses. Some molecules form new chemically reactive states, often free radicals. The free radicals, which can be measured by electro-paramagnetic resonance techniques, may interact with each other, forming new stable compounds, but they may also react with stable molecules. Thus, there is a series of direct effects on cellular components as well as a variety of indirect effects due to action of the free radicals.

While the speed of the initial reaction is very fast, the ultimate effect of that reaction may not be apparent until after a number of cell divisions have occurred, perhaps with a latent period of years.

Types of radiation. Certain types of radiation (such as alpha particles and neutrons) have a high transfer of energy for unit length of path. Also, beta particles (electrons) near the end of their path cause high-density ionization. In these there tends to be a linear relationship between dose and effect, whereas more penetrating radiations with a low linear energy transfer, such as high-energy x- or gamma rays, give a curvilinear relationship with effect, rising as the dose increases. However, in planning protection against radiation, it has been conventional to assume a direct linear relationship between dose and effect so that if there is error it would be on the side of safety.

Most exposure to radiation whether for therapy, for diagnosis, or by accident is from external sources. Through ingestion, inhalation, or injection of either natural radioactive substances such as radium and thorium or man-made radioisotopes such as ^{131}I, ^{137}Cs, and ^{239}Pu, these may gain access to the body of man or animals. Figure 1 illustrates the various types of radiation to which man, in this instance an astronaut, may be subjected.

The effects may be general or localized, depending on the degree of concentration and the chemical affinities of the radioisotope. Thus, the radium which the luminous dial workers absorbed in the 1920s, having chemical properties similar to those of lead, became deposited in the bones and gradually damaged the bone marrow or injured the bone-forming cells so that sarcomas of the bone resulted.

Because of the peristence of longer-lived radioactive isotopes, there has been concern that they might be concentrated by links in the food chain and so prove dangerous to man and other animals. D. S. Woodhead studied a fish population in the Irish Sea adjacent to the waste discharge from the Windscale Nuclear Reactor. The dose rate from the radioisotopes absorbed by plaice could have reached as high as 5 mrad per hour. This radiation has not produced discernible changes either in individual fish or in their total population and is not expected to do so.

^{90}Sr and ^{137}Cs are some of the isotopes that have been most carefully studied, partly because of their long (30-year) half-life, and they are widely dispersed. ^{90}Sr in particular is metabolized much like calcium. It can be safely assumed that if the present population permissible dose limit of 0.17

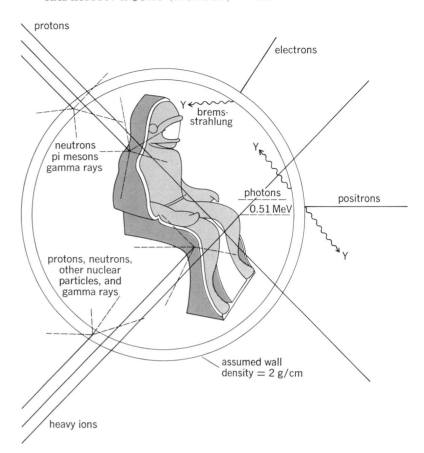

Fig. 1. A diagram of the possible ionizing radiations to which astronauts might be exposed. The sources of these radiations might be from cosmic radiation and solar flares. *(From J. F. Parker, Jr., and V. R. West, eds., Bioastronautics Data Book, 2d ed., 9: 431, NASA, 1973)*

rem per year is maintained, there will be no detectable impact on any plant, animal, or human population so exposed. *See* RADIATION BIOCHEMISTRY.

Rate of radiation. If not too seriously damaged, cells and tissues can repair to a considerable degree so that the rate at which radiation is delivered becomes very important. With an acute dose (within hours) damage may be severe, but if the rate is low, partial repair occurs and much less injury results. The dose of whole-body radiation lethal in a 30-day period to half of the population exposed (expressed as $LD_{50/30}$) is between 400 and 450 rads. $LD_{50/30}$ is the dose that will kill 50% of an exposed population within 30 days. The permissible occupational dose is 5 rem per year. (The rem is the unit of dose when modified by the quality factor.)

Tissue sensitivity and resistance. Living matter is much more susceptible to radiation damage in the wet than in the dry state, and since most cells contain a high proportion of water, they tend to react readily. Those types of cells that tend to proliferate rapidly, such as leukocytes and intestinal epithelium, are quite sensitive to radiation injury, whereas cells that normally do not proliferate, such as those of voluntary or cardiac muscle or neurons, are relatively resistant.

Animal life in general is much more sensitive than plant life, doses of hundreds of rads killing most animals, while many plants or bacteria can survive doses of thousands, and some viruses millions of rads.

Increased tissue saturation with oxygen tends to increase radiosensitivity. Sensitivity may be slightly reduced by some compounds, usually containing sulfhydryl groups.

The increased resistance that some treated cancers show to a second dose of radiation is probably more the result of changes in the local supporting tissue and vascular supply than to any acquired immunity in the cells themselves. There are a few experiments in which lightly irradiated populations of animals (receiving about 50 rads) have remained healthy and outlived the controls. In general, however, it is assumed that all radiation is injurious.

Invertebrate animals are more resistant than vertebrate. For example, an adult cockroach can withstand a dose of over 100,000 rads. Mammals are rather more radiosensitive than other vertebrates. Man is more sensitive than the rat or mouse, extensively used for experimental studies, but less sensitive than the dog or goat. There is no evidence of acquired immunity to radiation.

Cell injury. If a cell damaged by radiation is examined promptly, both nucleus and cytoplasm will show injury such as vacuolization, swelling even to the point of membrane damage and breaks in chromosomal structure. Damage to the nucleus tends to be transmitted and sometimes is enhanced with successive cell divisions as the imperfect DNA replicates. The damage done, if key enzyme systems are interfered with, may be many times as great as would be expected from the amount of energy absorbed.

There is a broad similarity to the damage in all types of cells irradiated. Cell division is retarded or stopped. Following this retardation, cell division may be resumed but may become abnormal. If the cell is of germinal rather than somatic type and is fertilized, the resulting organism may die or may evidence mutational change due to death of key cells during the developmental process.

Genetic injury. Radiation causes genetic damage by the production of gene mutations and chromosomal aberrations. At very low dose rates of radiation, a given dose is followed by fewer mutations than if the same dose had been given within a short period of time. In general, it is assumed that any mutation is deleterious, even though these mutations provide the variability that permits natural selection to occur. Dominant gene mutations show effect in the first generation. Currently available information suggests that about two such mutations might appear for each rad of low-dose radiation per million descendants. Recessive mutations which require that the damaged gene be present in duplicate may not become apparent until at least the second generation, and sometimes they may appear after many generations.

Chromosomal damage may be due to changes in number of chromosomes, breakage of chromosomes, or translocations of them. A large proportion of congenital defects, physical and mental as well as spontaneous miscarriages, are caused by chromosomal abnormalities. Experimental data from mice indicate that translocations (exchange of portions between chromosomes) may be responsible for a number of abnormalities in immediate offspring. Many cases will die before birth. Gene mutations may persist through many more generations than chromosomal aberrations and so affect a larger number of individuals. *See* MUTATION.

Radiation does not induce new or specific mutations, but increases the frequency of such mutations as already exist in the population.

In general, geneticists assume that a doubling of the present genetic burden of mutations in man would not be an intolerable burden for society. One may estimate about 3% of live-born children may have some form of genetically determined disease to provide a reference figure for the probable spontaneous genetic burden. Estimates of the amount of radiation required for this doubling dose range from about 20 to 300 rem. The studies on the children of the survivors of Hiroshima and Nagasaki indicate that the doubling dose is probably between 75 and 150 rem. However, since no clear-cut evidence of increased genetic abnormalities has been found, this estimate is essentially speculative.

In man there is no clear evidence of a genetic deleterious effect of radiation short of occurrence of partial or complete sterility, which is well established. However, the evidence as to genetic effects in experimental animals is so overwhelming that investigators must regard the results of the animal experiments as applicable to man.

Somatic injury. Damage to somatic cells may appear promptly or remain latent for long periods. The greater the dose of radiation, the more promptly the effect appears. Thus, with doses of hundreds of rads to much of the body, destruction of white blood cells or intestinal lining epithelium may occur within hours with prompt appearance of symptoms, whereas with a dose of 100 rads symptoms may be delayed for days. In general, radiation damage appears promptly in rapid-growing tissues but very slowly in slow-growing tissues. Thus, in almost static tissue such as the lens of the eye, radiation-induced cataract might not develop for decades.

The larger the volume of tissue irradiated, the greater the damage for a given dose. For example, if the whole body is irradiated, the $LD_{50/30}$ is about 400 rads, while if a volume of only a few cubic centimeters is irradiated, as in the treatment of a skin cancer, a dose of ten times that can be tolerated without serious damage. In general, tissue which is not directly irradiated is not damaged, although there is slight evidence that an abscopal effect may be possible.

Radiation tends not only to damage cells but through this damage to change intracellular materials that the cells control. Thus, collagen is increased in amount and physically changed so that it is less fibrillar, elastic tissue is damaged, and bone is rendered brittle and dense due to damage to osteoblasts and osteocytes as well as osteoclasts.

Since the immune response which enables man to ward off infection hinges to a considerable degree on lymphocytes and since the lymphocytes are easily damaged or killed by radiation, this may impair immunity. Fortunately, the immune system has a considerable factor of safety so that acute doses of less than 100 rads do not appear to affect it seriously. With acute whole-body doses of 200 rads or over, loss of immunity is considerable and may permit severe or fatal infections by organisms usually with no or little virulence.

The endothelial cells lining blood vessels and other vascular components are very sensitive to radiation, and vascular damage with partial occlusion or sometimes dilatation of the smaller vessels leads to impaired circulation of heavily irradiated tissue. Thus, a region that has been irradiated heavily tends to heal poorly if injured and to be susceptible to infection. Much of the late pathologic change seen following irradiation is due to this impaired vascularity.

During intrauterine life heavy radiation, hundreds of thousands of rem, may produce death of the embryo and abortion. Susceptibility to injury is greatest in the first trimester of pregnancy, largely because the embryo is then very rapidly growing and its key structures are being formed. Most of the injuries are due to destruction of groups of cells essential for organ or limb development. Somewhat smaller doses may impair growth, prevent full development of the brain, cause mental retardation, and produce other teratogenic effects. There is equivocal evidence that doses as low as 2 to 3 rem may cause in some cases subsequent development of leukemia and other cancers in childhood or adolescence.

Because of the relatively rapid cellular proliferation in infants and children, it is not surprising that they are more susceptible to radiation injury than adults. This susceptibility is recognized by the widely accepted regulation that persons under 18 years should not be occupationally exposed to radiation.

Heavy local radiation of the growing child (usually therapeutic for control of tumors) may lead to retardation of growth of the developing part or to the later appearance of local skin damage, atrophy, or cancer. The epiphyseal plates of bone are particularly sensitive so that irradiated bones may remain short.

Acute radiation effects. Acute radiation of the whole body or a considerable portion leads to radiation sickness. Usually a dose of over 100 rads is required to show this effect. There tend to be three distinct phases: a short prodromal period with symptoms such as nausea, fatigue, or fever, then a latent period with relative freedom from symptoms of varying length (hours to days). This is followed by more serious illness which may be manifested by hematopoietic effects: severe loss of white blood cells with susceptibility to infection, hemorrhagic manifestations due to loss of blood platelets, and anemia due to impaired replacement of red blood cells. At times, damage to the lining of the intestine may be severe. This is characterized by nausea, diarrhea, and fatal infection with the usual intestinal bacteria. The response of the skin to radiation is initially reddening, which in severe cases changes to blistering and ulceration. If the region is hairy, epilation usually occurs. These effects may appear slowly over weeks, but if the dose is high, over 1000 rads, and the time of delivery short, they may appear within hours. With very heavy doses, 5000 rads or more, prompt damage will be done to the brain, with almost immediate disorientation or coma, followed shortly by death.

Late radiation effects. Since many of the physicians who were pioneers in the use of x-rays or radium did not realize the dangers the repeated additive effects of small increments of radiation would

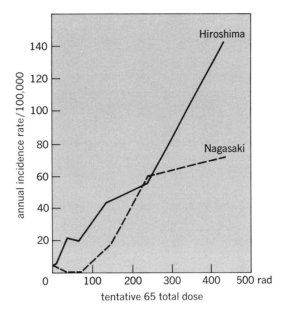

Fig. 2. The annual incidence rate of leukemia in relation to the dose of radiation received by survivors at Hiroshima and Nagasaki. The tentative 65 total dose is the tentative dose calculated by the Oak Ridge National Laboratory for the exposed populations of Nagasaki and Hiroshima based on the locality of the patient, the type of shielding if any, and checked against experimental data from a simulated source tested in 1965. (From Advisory Committee on the Biological Effects of Ionizing Radiations, The Effects on Populations of Exposure to Low Levels of Ionizing Radiation, National Academy of Sciences–National Research Council, 1972)

have, many did not protect themselves adequately. As a result, the excessive exposures led to both local and general effects—the local, chiefly radiation damage to the skin of the hands sometimes progressing to cancer; the general, a life shortening, an increase in leukemia, and an increase in some other types of cancer.

The chief somatic risk other than immediate is the delayed causation of cancer. The latent period may be many years in duration. Thus, in the Hiroshima-Nagasaki survivors the incidence of leukemia rose above that in the general population, reaching its peak about 1955. The incidence rate tended to decrease during the 1960s and has returned nearly to the spontaneous rate. The probable doses received ranged between 50 and 500 rads (Fig. 2).

In the radium-watch-dial workers and others who were exposed to radiation, small amounts of the radium deposited in the bones eventually so altered the bone cells that sarcoma resulted. In some instances the sinuses of the skull, largely surrounded by bone, became the site of cancer also. The cancers appeared among those who had received the heavier doses of radiation. In Fig. 3 this is shown, together with an approximate relationship between dose and effect. It will be noted that there appears to be a "practical threshold" below which bone cancer has not developed.

In addition to leukemia and cancer of the bone, other cancers have been proved to develop following heavy doses of radiation, and on the basis of the linear hypothesis are assumed to have developed at lower doses. Cancers of the thyroid,

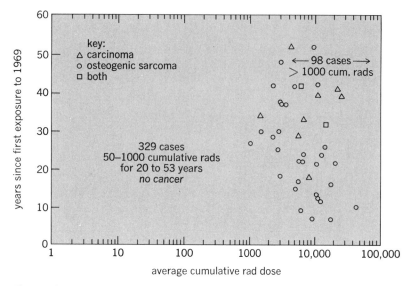

Fig. 3. The incidence of bone cancer and sinus cancer plotted against the average cumulative dose in rads for dial workers and others with internally absorbed radium. (*Modified from R. D. Evans et al., Radiogenic tumors in the radium and mesothorium cases studied at M.I.T., in C. W. Mays et al., eds., Delayed Effects of Bone-seeking Radionuclides, pp. 157–194, University of Utah Press, 1969*)

lung, and breast have developed in excess in a number of irradiated populations. The high incidence of lung cancer in uranium miners has been recognized for years. The miners worked in an atmosphere with a high content of radon and its daughter products. The disease appears to be more prevalent with longer exposure and at higher levels. Interestingly enough, in the uranium miners in the United States the incidence of cancer appears to be far higher among those who were cigarette smokers than those who were not. There is also evidence that cancer of the breast may be induced by the heavy local radiation as was experienced by some women who had been repeatedly fluoroscoped during treatment of tuberculosis years before.

While there are many uncertainties, enough data are in hand to suggest that cancer is the most serious consequence to be weighed in calculating permissible dose limits. It can be assumed that for heavily exposed populations such as the Japanese atomic bomb survivors, and some patients treated in the past with radiation for noncancerous diseases, the radiation may produce an excess of cancer deaths corresponding from 50 to 165 per million persons per rem of exposure during the first quarter century after that exposure. But estimates such as this may be made only for specific dose ranges and with the further assumption that a linear relationship exists between dose and effect.

One of the effects of radiation, probably through the retardation of cell division, is increased rate of maturation of cells. Radiation of large portions or all of the body may also lead to generalized aging effects, ranging from such things as graying of the hair and induction of cataracts to actual shortening of lifespan. There is also a similarity between the ability of radiation to induce tumors and the increasingly high incidence of tumors seen with senescence in both animals and man. Radiation not only increases the numbers of cancers in the ex-

posed population but causes them to appear much earlier in life than is usually the case.

[SHIELDS WARREN]

Bibliography: Advisory Committee on the Biological Effects of Ionizing Radiations, *The Effects on Populations of Exposure to Low Levels of Ionizing Radiation*, National Academy of Sciences–National Research Council, 1972; N. M. Bleehen (ed.), Biological basis of radiotherapy, *British Medical Bulletin*, vol. 29, January 1973; A. W. Oughterson and S. Warren, *Medical Effects of the Atomic Bomb in Japan*, Division VIII – vol. 8, National Nuclear Energy Series, Manhattan Project Technical Section, 1956; United Nations Scientific Committee on the Effects of Atomic Radiation, *Ionizing Radiation: Levels and Effects*, Official Records of the General Assembly, Suppl. no. 25 (A/8725), 1972; S. Warren, *The Pathology of Ionizing Radiation*, Monograph, Carl V. Weller Lecture, 1961; D. S. Woodhead, *Health Physics*, vol. 25, August 1973.

Radioactive fallout

The radioactive material which results from a nuclear explosion. In particular the term applies to the debris which is deposited on the ground, but common usage has extended its coverage to include airborne material as well.

Nuclear explosion. A nuclear explosion results when the fission of uranium-235 or plutonium-239 proceeds in a rapid and relatively uncontrolled way, as opposed to the controlled fission in a reactor. The energy release of an atomic (fission) bomb is usually expressed in terms of thousands of tons of TNT equivalent, and such explosions may have yields into the range of hundreds of kilotons. Still larger explosions can be produced by using the fission device as a trigger for a fusion reaction to produce a thermonuclear explosion. The yield of such thermonuclear devices can range up to hundreds of thousands of kilotons (hundreds of megatons).

The radioactivity from the explosion is produced by fission products and by activation products. The basic reaction in atomic fission is the splitting of an atom of the fissionable material (uranium or plutonium) into two lighter elements (fission products). These lighter elements are all unstable and emit beta and gamma radiation until they reach a stable state. During the fission reaction a number of neutrons are released, and they can interact with the surrounding materials of the device or of the environment to produce radioactive activation products. The fusion process does not produce fission products but does release neutrons which can add to the activation.

The fissioning of a single atom of uranium or plutonium yields almost 200 million electron volts (Mev) instantaneously, and just over 50 g of fissionable material are required to produce a yield of 1 kiloton. The tremendous energy release produces the explosive shock and temperatures ranging upward from 1,000,000°C. The device itself and the material immediately surrounding it are vaporized into a fireball which then rises in the atmosphere. The altitude at which the fireball cools sufficiently to stabilize depends very much on the yield of the explosion. In general, an atomic explosion fireball stabilizes as a cloud high in the troposphere, while a thermonuclear fireball tends to break through into the stratosphere and stabilize at altitudes above 10 mi.

As the fireball cools, the vaporized materials condense to fine particles. If the explosion has taken place high above the Earth, the only material present is that of the device itself, and the fine particulates are carried by the winds and distributed over a wide area before they descend to Earth. Bursts at or near the surface carry large amounts of inert material up with the fireball. A large fraction of the radioactive debris condenses out onto the large inert particles of soil and other material, and many small radioactive particulates attach themselves to the larger inert particles. The larger particles settle rapidly by gravity, and large percentages of the radioactivity may be deposited in a few hours near the site of the explosion. In contrast, the radioactivity from a thermonuclear explosion at high altitude takes months or years to reach the surface.

The first nuclear explosion took place in New Mexico in 1945 and was rapidly followed by the two bombings at Hiroshima and Nagasaki, Japan. Through 1969 five nations have tested a large number of nuclear and thermonuclear weapons. The United States has tested at its sites in Nevada and the Pacific; the United Kingdom has tested in Australia and Christmas Island. The Soviet Union has several areas, including Novaya Zemlya in the Arctic. The testing was heaviest in 1954, 1958, and 1961 and was stopped following negotiation of a test-ban treaty. France and China did not sign the treaty, however, and have carried out a number of tests since, France in the Sahara and the South Pacific and China at an inland test site.

Radioactive products. The total fission yield for all tests has been about 200 megatons, the total explosive yield being over 500 megatons. The production of long-lived fission products has been about 20 megacuries of strontium-90 and 30 megacuries of cesium-137. Figure 1 shows the time distribution for strontium-90 fallout in New York City, which may be considered typical of the Northern Hemisphere mid-latitudes. Figure 2 shows the latitude distribution as of 1966. Testing through 1969 does not modify this picture.

Local fallout. Local fallout is the deposition of large radioactive particles near the site of the explosion. It may present a hazard near the test site or in the case of nuclear warfare. The smaller particulates, which are spread more widely, constitute the worldwide fallout. The radioactivity comes from the isotopes (radionuclides) of perhaps 60 different elements formed in fission, plus a few others formed by activation and any unfissioned uranium or plutonium. Each of the radioactive isotopes produced in the fission process goes through three or four successive radioactive decays before becoming a stable nuclide. The radioactive half-lives of these various fission products range from fractions of a second to about 100 years. The overall beta and gamma radioactivity decays according to the minus 1.2 power of the activity at any time. Thus decay is very rapid at first and then slows down as the short-lived isotopes disappear.

The hazard from local fallout is largely due to the external gamma radiation which a person

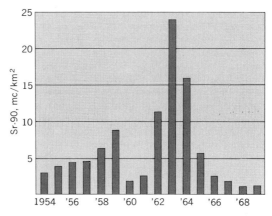

Fig. 1. Time distribution for strontium-90 fallout in New York City. (*Data from Health and Safety Laboratory, U.S. Atomic Energy Commission*)

would receive if this material were deposited near him; fallout shelters are designed to shield survivors of an explosion from this radiation. Actually, the effects of an explosion are so great that a shelter which would resist the shock would shield the occupants from fallout radiation.

In so-called clean nuclear weapons the ratio of explosive force to the amount of fission products produced is high. On this basis, small fission weapons are considered "dirty," although the absolute amount of fission products is much less than with a large clean bomb. It is also possible to increase the radioactivity produced, and thus the fallout, by surrounding the device with a material which is readily activated. Cobalt has been frequently mentioned in this connection.

Tropospheric and stratospheric fallout. As previously mentioned, the large particulates from a nuclear explosion are deposited fairly close to the site. The smaller particles are carried by the winds in the troposphere or stratosphere, generally in the direction of the prevailing westerlies. The debris in the troposphere circles the Earth in about 2 weeks, while material injected into the strato-

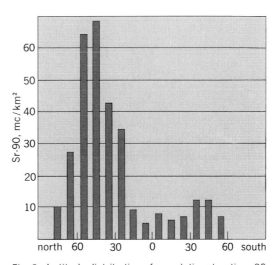

Fig. 2. Latitude distribution of cumulative strontium-90 deposit as of 1966. (*Data from Health and Safety Laboratory, U.S. Atomic Energy Commission*)

sphere may travel much faster. The horizontal distribution in the north-south direction is much slower, and it may take several months for radioactivity to cover the Northern Hemisphere, for example, following a test in the tropics. Transfer between the Northern and Southern hemispheres is slight in the troposphere but can take place in the stratosphere. Radioactive debris in the troposphere is brought down to the surface of the Earth mainly in precipitation, with perhaps 10–15% of the total amount brought down by dry deposition. Radioactive material remains about 30 days in the troposphere. The mechanism of transfer from the stratosphere to the troposphere is not completely understood, but it is obvious from measurements that the favored region for deposition is the mid-latitudes (30–50°) of the Northern and Southern hemispheres. Since most early testing took place in the Northern Hemisphere, the deposition there was about three times as great as in the Southern Hemisphere. French tests in the Southern Hemisphere have reduced this ratio slightly.

Hazards. Once the fallout has actually been deposited, its fate depends largely on the chemical nature of the radionuclides involved. Since the level of gamma radiation from worldwide fallout is negligible, attention has been paid to those nuclides which might possibly present some hazard when they enter the biosphere. Plants, for example, may be contaminated directly by foliar deposition or indirectly by deposition on the soil that is followed by root uptake. Animals may be contaminated by eating the plants, and man by eating plant or animal foods. The inhalation of radioactive material is not considered to be a significant hazard as compared to ingestion.

Based on the various considerations mentioned, the three radionuclides of greatest interest are strontium-90, cesium-137, and iodine-131. Strontium-90 is a beta emitter which follows calcium metabolically and tends to deposit in the bone. The radiation emitted could result in bone cancer or leukemia if the levels were to become sufficiently high. Cesium-137 is distributed throughout soft tissues and emits both beta and gamma radiation. It is considered to be a possible genetic hazard but not as dangerous to the individual as strontium-90. Iodine-131 is relatively short-lived and probably is a hazard only as it appears in tropospheric fallout. It is readily absorbed by cows on pasture and is transferred rapidly to their milk. The iodine in milk in turn is concentrated in the human thyroid and presents a possible high radiation dose, particularly to children, who consume larger amounts of milk and who have smaller thyroids than adults.

Sampling. Radioactive fallout is monitored by many countries. The usual systems involve networks that collect samples of airborne dust, deposition, and milk and other foods. The levels of iodine-131 or cesium-137 can be checked in living individuals by external gamma counting. Strontium-90 on the other hand can be measured only in autopsy specimens of human bone. These national data are also combined and evaluated by the Scientific Committee on the Effects of Atomic Radiation of the United Nations. Their reports are issued at intervals and offer broad reviews and summaries of the available information on levels of fallout and on possible hazards.

The studies of radioactive fallout have also included sampling in the stratosphere by balloons and high-flying aircraft, as well as taking aircraft samples in the troposphere. Very elaborate monitoring systems have been set up, and these have provided considerable scientific information in addition to their original monitoring purposes. The major benefits have been the understanding of stratospheric transfer processes as significant to meteorology and information on the uptake and metabolism of various elements by plants, animals, and man. These studies are usually reported in the specialty journals for meteorology, agriculture, and biology. *See* RADIATION INJURY (BIOLOGY).

[JOHN H. HARLEY]

Bibliography: *Reports of the United Nations Committee on the Effects of Atomic Radiation*, 1958, 1962, 1964, 1966, 1969; U.S. Public Health Service, *Radiological Health Data and Reports*, monthly.

Radioactive materials, decontamination of

The removal of radioactive contamination which is deposited on surfaces or may have spread throughout a work area. Personnel decontamination is also included. The presence of radioactive contamination is a potential health hazard, and in addition, it may interfere with the normal functioning of plant processes, particularly in those plants using radiation detection instruments for control purposes. Thus, the detection and removal of radioactive contaminants from unwanted locations to locations where they do not create a health hazard or interfere with production are the basic purposes of decontamination.

There are four ways in which radioactive contaminants adhere to surfaces, and these limit the decontamination procedures which are applicable. The contaminant may be (1) held more or less loosely by such physical forces as electrostatic or surface tension, (2) absorbed in porous materials, (3) adsorbed on or by the surface in the form of ions, atoms, or molecules, or (4) mechanically bonded to surfaces through oil, grease, tars, paint, and so on.

Methods. Decontamination methods follow two broad avenues of attack, mechanical and chemical. Commonly used mechanical methods are vacuum cleaning, sand blasting, blasting with other abrasives, flame cleaning, scraping, and surface removal (for example, removal of concrete floors with an air hammer). The principal chemical methods of decontamination are water washing, steam cleaning, and scrubbing with detergents, acids, caustics, and solvents.

Another important method of handling contamination is to store the contaminated object, or temporarily abandon the contaminated space. This can be done when the use of the material or space is not necessary for a period of time and the half-life of the contaminant is relatively short. For example, tools contaminated with short-lived fission products may be stored, or a building contaminated with such material may be sealed off and barred from use, until the natural radioactive decay has reduced the contamination to an acceptable level.

Other methods involve covering the contamination by some means, such as painting, and disposing of part or all of the contaminated equipment or facility. Considerations which determine the methods used for decontamination or removal of contamination include (1) the hazards involved in the decontamination procedure, (2) the cost of removal of the contamination, and (3) the permanency of removal of the contamination (for example, painting over a surface contaminated with a long-lived radioactive material only postpones ultimate disposal considerations).

Personnel. Personnel decontamination methods differ from those used for materials primarily because of the possibilities of injury to the person being decontaminated. Procedures used for normal personal cleanliness usually will remove radioactive contaminants from the skin, and the method used will depend upon its form and associated dirt (grease, oil, soil, and so on). Soap and water (sequestrants and detergents) normally remove more than 99% of the contaminants. If it is necessary to remove the remainder, chemical methods which remove the outer layers of skin upon which the contamination has been deposited can be used. These chemicals — citric acid, potassium permanganate, and sodium bisulfite are examples — should be used with caution and preferably under medical supervision, because of the increased risk of injury to the skin surface. The use of coarse cleansing powders should be avoided for skin decontamination, because they may lead to scratches and abraded skin which can permit the radioactive material to enter the body. Similarly, the use of organic solvents should be avoided for skin decontamination because of the probability of penetration through the pores of the skin. It is very difficult to remove radioactive material once it is fixed inside the body, and the ensuing hazard depends very little on the method of entry into the body, that is, through wounds, through pores of the skin, by injection, or by inhalation. When certain of the more dangerous radioactive materials, such as radium or plutonium, have been taken into the body, various chemical treatments have been attempted to increase the body elimination, but the results of these treatments are not very encouraging. In the case of plutonium and certain other heavy metals, the most effective treatment for removal from the body is the administration of chelating agents, such as calcium ethylenediaminetetraacetate (CaEDTA) or a sodium citrate solution of zirconyl chloride. In any case, the safest and most reliable procedure for preventing internal exposure from radioactive material is not the application of health physics procedures to prevent entry of radioactive material into the body.

Air and water. Air contaminants frequently are eliminated by dispersion into the atmosphere. Certain meteorological conditions, such as prevailing wind velocities, wind direction, and inversion layers, seriously limit the total amount of radioactive material that may be released safely to the environment. Consequently, decontamination of the airstream by filters, cyclone separators, scrubbing with caustic solutions, and entrapment on charcoal beds is often resorted to. The choice of method used is guided by such things as the volume of airflow, the cost of heating and air conditioning, the hazards associated with the airborne radioactive material, and the isolation of the operation from other populated areas.

Water decontamination processes can use one or both of the two opposing philosophies of maximum dilution or maximum concentration (and subsequent removal) of the contaminant. Water concentration methods involve the use of water purification processes, that is, ion exchange, chemical precipitation, flocculation, filtration, and biological retention.

Certain phases of radioactive decontamination procedures are potentially hazardous to personnel. Health physics decontamination practices include the use of protective clothing, respiratory devices, localized shielding, isolation or restriction of an area, provisions for the proper disposal of the attendant wastes, and application of the recommended rules and procedures for limiting the internal and external doses of ionizing radiation. *See* RADIATION INJURY (BIOLOGY); RADIOACTIVE WASTE DISPOSAL.   [KARL Z. MORGAN]

Bibliography: *Control and Removal of Radioactive Contamination in Laboratories*, Natl. Bur. Std. Handb. no. 48, 1951; International Brotherhood of Electrical Workers Staff, *Radiation Hazards and Control*, 1965; S. Kinsman, *Radiological Health Handbook*, U.S. Department of Health, Education and Welfare, PB-121784, 1957.

Radioactive waste disposal

The handling and disposal of radioactive wastes are problems present in some degree in all nuclear energy operations. Wastes in liquid, solid, or gaseous form are produced in the mining of ore, production of reactor fuel materials, reactor operation, processing of irradiated reactor fuels, and a great variety of related operations. Wastes also result from the use of radioactive materials, for example, in research laboratories, industrial operations, and medical research and treatment. The problems of waste disposal undoubtedly will increase as the nuclear energy program is further extended and diversified and as a large and widespread nuclear power industry is developed.

In the handling and disposal of radioactive wastes, the principal problem is the prevention of radiation damage to man and his environment by controlling the dispersion of radioactive materials. Damage to man may result from irradiation by external sources or by the intake (by ingestion, by inhalation, or through the skin) of radioactive materials, their passage through the respiratory and gastrointestinal tract, and their partial incorporation into the body. Radioactive waste contaminants in air, water, food, and other elements of the human environment must be kept below the maximum permissible concentrations for the particular radionuclide or mixture of radionuclides present in the wastes. Liquid or solid waste products containing significant quantities of the more dangerous radioactive materials require ultimate disposal in isolated and permanent containment media where they never again can find their way into man's environment. The more dangerous radioactive materials are those that may be readily incorporated into the body and that have relatively long half-lives, ranging from a few years to several thousand years. Both the short-lived radionuclides and those with extremely long half-lives are less hazardous to man. This is because short-lived radionuclides disappear rapidly by natural radioactive decay, while radionuclides with extremely long half-lives have such low specific activity, that is, so few microcuries per gram, that the probability of dangerous quantities entering the body is very low. Some of the more dangerous radioisotopes include fission products such as strontium-90 and cesium-137, transuranic elements such as plutonium-238 and americium-241, and naturally occurring radionuclides such as radium-226 and actinium-227. *See* RADIATION INJURY (BIOLOGY).

The highly radioactive liquid wastes associated with chemical reprocessing of reactor fuels constitute the major waste-disposal problem. However, the liquid, solid, and gaseous wastes of low or intermediate levels of activity from hospitals, industrial laboratories, research reactors, and so on must be controlled also and their dispersion limited as necessary to prevent health hazards.

The basic methods for disposal of radioactive wastes are (1) dilution and dispersion, and (2) concentration and permanent containment. When only small amounts of radioactive materials are involved and the local situation is favorable, wastes may be diluted and dispersed in water or air without danger. In cases in which dilution is not feasible, wastes must be concentrated and stored in a safe manner; and much research and development work has been devoted to methods for concentrating and storing gaseous, liquid, and solid radioactive wastes.

Liquid wastes. High-level liquid wastes, which result from experimental or operational processing of irradiated reactor fuels, are relatively small in volume but high in specific activity. The radioactivity of fuel-processing solutions may range from several hundred to thousands of curies per gallon, depending upon the processes employed. These wastes, which may be highly acid or alkaline, present extremely difficult problems in shielding, handling, and ultimate disposal.

High-level liquid wastes have been stored in underground tanks of steel and concrete with special preventive measures taken against corrosion and deterioration of the tanks or leakage of the wastes. Tank storage is not considered to be permanent disposal but must be used until feasible methods of waste treatment and ultimate disposal are developed. Up to Jan. 1, 1965, in the United States over 90,000,000 gal of high-level liquid wastes were in storage in approximately 200 tanks. The cost of waste storage in tanks has ranged from about $0.75 to $2 per gallon.

Liquid wastes of intermediate levels of activity from various chemical processes or relatively large experimental projects are of greater volume than high-level wastes. They may contain as much as 1/10 curie or more of radioactivity per gallon and are often high in dissolved chemical content. Such wastes must be shielded to prevent external radiation and are not suitable for release to the general environment without extremely effective treatment for removal of the radioactive components. When the radioactive components are removed, they must be concentrated and stored as in the case of high-level waste. In some locations intermediate-level liquid wastes have been disposed of by dispersion in shallow soil formations in which most of the radioactive elements are absorbed and retained for long periods by the soil materials.

Shallow-ground disposal is a subject of extensive study to determine the capacity of various soils in controlling the radioactivity of different waste solutions, particularly the more hazardous radionuclides, which must not be dispersed to groundwaters and surface waters. A waste-disposal system of this type has been used at the Oak Ridge National Laboratory since 1952. Through 1958 about 9,000,000 gal of liquid wastes containing about 115,000 curies of activity had been discharged into the seepage-pit system. The major operation of this kind has been at Hanford, Wash., where an unusual situation exists because of the soil formation and the isolated location, and where large quantities of liquid and radioactive materials are disposed of in shallow soil formations.

Low-level liquid wastes are present in large volumes of waste water from laboratory areas, decontamination operations, water used in basins to shield operators during work on radioactive materials, and other slightly contaminated liquids. In favorable situations, where there are large volumes of surface water in isolated areas, such wastes are diluted and dispersed untreated or following partial decontamination by waste-treatment processes. At the Oak Ridge National Laboratory, for example, the contaminated waste-water volume has been about 700,000 gal/day, of which roughly half is discharged without treatment. The remainder is decontaminated by the lime-soda water-softening process.

Solid wastes. Solid radioactive wastes include such materials as machine turnings, nonusable contaminated equipment, and contaminated trash. The activity may vary from a few times the background level to levels requiring shielding or remote handling. In general, the disposal has been by land burial in selected areas where the soil has a capacity for retaining the radionuclides and where the danger of excessive contamination of groundwater is minimal. The potential hazards and the care required in selecting, operating, and monitoring solid-waste burial grounds depend, of course, on the particular radionuclides present as well as on the levels of activity. To a limited extent, solid wastes and concentrated low-level liquid wastes have been packaged and disposed of by dumping at selected places in the ocean.

Gaseous wastes. Gaseous radioactive wastes originate from such diverse sources as air-cooled reactors, chemical processing plants, laboratory hoods, and fissionable-material fabrication facilities. The levels of radioactivity vary with the type of operation, and the pollutants may be either gaseous or in the form of particles. Deep-bed sand and fiber filters have been developed for the removal of particulate contaminants. Equipment for absorbing iodine and other reactive gases is available, and inert gases can be removed by adsorption on charcoal or silica gel. Disposal is usually by discharge into the atmosphere through tall stacks which provide dilution. Continuous air monitors are used to determine the suitability of gaseous wastes for discharge under the particular conditions and to check the levels of air contamination that result after dispersion in the atmosphere. Projections by the Atomic Energy Commission on civilian nuclear power economy in the United States and the free world indicate that by the year 2000 there will be 3.3×10^6 Mw (thermal) nuclear power. Health physicists at Oak Ridge National Laboratory estimate there will have accumulated in the Earth's atmosphere 3200 megacuries of Kr^{85} and 96 megacuries of H^3. The resulting dose from H^3 will be negligible, but it is estimated that the dose from Kr^{85} will be about 2 milli-roentgen-equivalent-man (mrem) per year at the Earth's surface if no measures are taken for its removal at nuclear-fuel processing plants. Methods are now under development to remove Kr^{85} from this waste gas.

Future wastes from industry. From predictions such as those above it may be estimated that by the year 2000 high-level liquid wastes will be produced at the rate of 250,000–2,500,000 gal/day. By that time the total accumulated volume of high-level wastes will be of the order of $2.5-1.5 \times 10^{10}$ gal, and the total accumulated radioactivity will be more than 5×10^{11} curies. Because a considerable part of this accumulated activity will be due to strontium-90 and other long-life radionuclides, methods for ultimate disposal of these wastes must provide containment and control for at least several hundred years. In a nuclear power industry large volumes of low- and intermediate-level wastes also will be produced. Although the accumulated activity at any time will be orders of magnitude less than in the high-level wastes, safe and economical disposal of these less-concentrated wastes must be achieved.

The safe control of wastes from a nuclear power industry will involve a complex scheme of waste treatment, handling, and disposal. Treatment may serve to remove the more hazardous radionuclides or to prepare the waste for disposal. Temporary storage and transportation of the wastes, followed by one or more methods of ultimate disposal, may be necessary. Prospective methods for ultimate disposal as investigated by health physicists at Oak Ridge National Laboratory include (1) the fixation of the hazardous high-level waste in a stable solid medium and subsequent permanent storage or burial of the stable solid in selected locations such as abandoned salt mines, and (2) the mixing of intermediate- to high-level waste with cement and clays to form a slurry which is injected into horizontal shale formations at hundreds of feet below the Earth's surface. Hydraulic shale-fracturing techniques have been demonstrated to be feasible, safe methods of depositing this slurry as a hard, relatively insoluble, thin grout sheet that is expected to be permanently isolated from man's environment. *See* NUCLEAR POWER.

[KARL Z. MORGAN]

Bibliography: H. Etherington (ed.), *Nuclear Engineering Handbook*, 1958; A. Frye, *The Hazards of Atomic Wastes*, 1962; R. F. Lumb (ed.), *Management of Nuclear Materials*, 1960; C. A. Mawson, *Management of Radioactive Wastes*, 1965; K. Saddington and W. L. Templeton, *Disposal of Radioactive Waste*, 1959; R. Stephenson, *Introduction to Nuclear Engineering*, 2d ed., 1958.

Radioecology

The interdisciplinary study of organisms, radionuclides, ionizing radiation, and the environment. Radioecological findings are expressed in interactions at the population or community level of organization. Ecologists commonly distinguish between the use of radionuclides in tracer studies and in effects studies. Much overlap occurs, however, and it is often necessary to show that a given tracer radioactivity level has no effect on the process being studied. In practice, therefore, the distinction in tracer versus effects studies has little utility beyond that of a convenient arrangement for discussion.

The terminology used in radioecology includes the type of radiation (alpha, beta, or gamma), the amount of radionuclide (usually measured in curies), and the dose of ionizing radiation (measured in rads).

Alpha radiation has little penetrating ability and is not often involved in radioecological studies. Beta radiation consists of low-mass, low-energy particles with medium penetrating ability. However, because beta radiation is rapidly attenuated by the air or protoplasm, its use in effects studies is limited to internal dosages or short-range work. Beta emitters such as C^{14} and P^{32} are useful radioecological tracers. Because of the penetrating ability of gamma radiation, radionuclides such as Cs^{13} and Zn^{65} have found great utility both as tracers and as sources of ionizing in effects studies.

Radioactive materials are measured in curies, a curie equaling 2.22×10^{12} radioactive disintegrations per minute (dpm). Various decimal fractions of the curie are commonly employed: A millicurie (mc) is 2.22×10^9 dpm; and a microcurie (μc) is 2.22×10^6 dpm. The dose of radiation received by an organism is a function of the size of the source (in curies), the distance to the source, the energy level of the radiation and its penetrability, and the physical characteristics of the organism. The standard of measure of radiation is the rad, the absorbed dose of 100 ergs of energy per gram of protoplasm. In terms of effects on living organisms, the rad is similar to an older unit of measurement, the roentgen.

RADIOECOLOGICAL TRACERS

Tracer studies provide information about basic structure and process. Animals can be located and monitored by the use of radioactive tags. Radionuclide tracers are also used to determine food uptake rates, metabolic rates, and cycling of mineral nutrients, and to identify food chains and ecosystems.

Marking studies. A major problem of ecological field studies is that of locating a given organism. However, small animals carrying radioactive tags can be easily located with a suitable radiation detector. Usually a radioactive pin or wire is introduced into the body, although sometimes a small capsule containing the radioisotope, or the radioisotope alone, may be ingested by or injected into the animal. The investigator carries a detector mounted on a long pole and sweeps the area traversed until a tagged individual is located. Such techniques used in conjunction with field enclosures have permitted the daily monitoring of the distribution of small rodent populations. Automatic monitoring of position and activity patterns has been accomplished by using a circular enclosure and mounting the detector on a rotating boom. A continuous graphic record of the count rate permits the investigator to pinpoint the position of the animal each time the boom rotates one full turn. Radionuclide tags have also proved useful in ascertaining the seasonal or annual movements of animals that are inaccessible—hibernating toads and 17-year cicadas, for example. Thus, radionuclide tracers provide information obtainable in no other way.

Uptake and elimination of tracers. Two important ecological processes difficult to measure in the field are ingestion rates of food and the metabolic rate of animals. Radionuclides used as tracers provide information about both of these processes. When an organism ingests food containing radionuclides, a certain portion, the "body burden," is absorbed into the body and then is excreted at a rate dependent on a number of variables, including the type of radionuclide and the species of animal.

The rate of excretion can be measured by periodic capture of the animals and measurement of decrease in the radioactive body burden. From this excretion rate, the amount of food ingested in nature can be estimated provided the following are also known: (1) the concentration of the radioisotope in the food plant, and (2) the percentage of absorbtion of the isotope in the gut of the organism. The former can be measured directly; the latter can be measured by suitable controlled experiments for it is less likely than is the excretion rate to vary between the laboratory and the field. In the procedure for measurement, an insect is introduced into an environment containing tagged food. As soon as a steady state has been reached (radionuclide elimination and absorption occurring at the same rate), the insect is removed and given untagged food, and the excretion rate is measured. The loss rate multiplied by the body burden must equal the intake under the steady-state conditions.

This method of estimating ingestion in the field is applicable only in areas in which plants have a measurable level of some radioactive material. Field environments tagged with radionuclides and available for ecological work are becoming more common.

The rate of excretion of radionuclides is also affected by temperature and activity factors. This led to the idea that the rate of excretion of certain radionuclides might provide an index to the general metabolic rate of the animal and that this could be estimated by periodic capture and counting of tagged organisms. Preliminary studies on the relationship between metabolic rate and loss rates of radionuclides have been encouraging (Fig. 1). Definitive experiments on the effects of factors other than metabolic rate have not yet (1969) been completed.

Cycling and fallout. Direct monitoring of the uptake, storage, and recycling of mineral nutrients is an active field in radioecology. The injection of calcium-45 in various species of deciduous and evergreen trees, for example, has permitted a determination of the long-term retention time of this mineral nutrient, its turnover in each species, and the various pathways of loss, leaching, and other functions. Large differences have been found between species with respect to retention of essential nutrients. Nutrients also differ greatly in the rate of loss by leaching. Rain removes an appreciable amount of calcium from green leaves, although phosphorus is lost very slowly in this manner.

By utilizing the radionuclides resulting from nuclear explosions as their own tracers, studies of their concentration through food chains have been made, such as the passage of iodine-131 through the simple food chain—grass, cow, milk, man. Such investigations have had immense practical value in predicting sites of concentrations of nuclear fallout. The high concentration of cesium-137 in the people of northern latitudes (for example, Alaska) who depend on the lichen-reindeer or the lichen-caribou food chains is a case in point. Recognition of the general principle of "biological concentration" of potential harmful materials (including pesticides) can be an important first step toward preventing serious contamination of man's environment.

Food chain studies. Isotopes can be experimentally introduced, and the concentrations within each species measured, to identify functional food chains and natural ecosystems as well as the trophic or feeding positions of organisms composing these food chains. Such techniques, originally developed in the studies of stream ecosystems, have now been applied to simple terrestrial ecosystems. If an isotope (phosphorus-32 has been used with great success) is introduced into the plants or primary producers of a system, the radionuclide will be transferred most rapidly to the plant feeders (primary consumers), somewhat less rapidly to predators (secondary consumers), and so on. Thus the time that is necessary after tagging for the peak body burden or the steady-state situation to be reached will indicate the probable trophic position or feeding level of the group. In Fig. 2, uptake curves for plant, herbivore, predator, and detritus feeders are shown. Note the differences in both rate of uptake and relative time to the peak con-

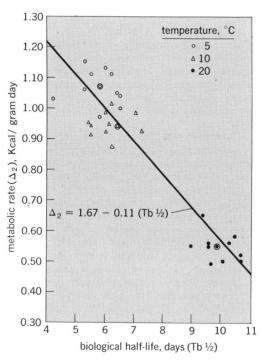

Fig. 1. Relationship of the loss rate of Zn^{65} and the metabolic rate at three temperatures for feral house mice (*Mus musculus*). (*From R. Pulliam, G. Barrett, and E. P. Odum, Bioelimination of tracer Zn^{65} in relation to metabolic rates in mice, in Proceedings of the 2d National Symposium on Radioecology, U. S. Atomic Energy Comm., 1969*)

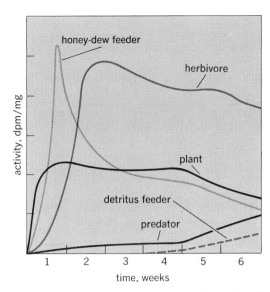

Fig. 2. Phosphorus-32 uptake curves exhibited by old-field organisms differing in trophic level, food source, or both. Phosphorus-32 levels in plants were relatively constant throughout 6-week period. (*From R. G. Wiegert, E. P. Odum, and J. H. Schnell, Ecology, vol. 48, 1967*)

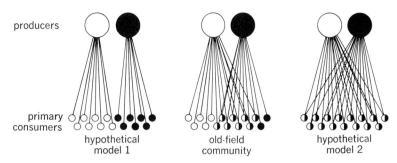

Fig. 3. Comparison of trophic relationships between two dominant species of producers and 15 species of primary consumers in a South Carolina old field with the theoretical maximum and minimum interaction models. (*From R. G. Wiegert and E. P. Odum, Radionuclide tracer measurement of food web diversity in nature, in Proceedings of the 2d National Symposium on Radioecology, U.S. Atomic Energy Commission, 1969*)

Fig. 4. The gamma source used in irradiating a forest ecosystem at Brookhaven National Laboratory. The source can be raised or lowered into a lead-shielded container through operation of a winch in a building a safe distance away. (*From G. M. Woodwell, Science, vol. 138, 1962*)

centration. *See* FOOD CHAIN; FRESH-WATER ECOSYSTEM; TERRESTRIAL ECOSYSTEM.

A modification of this technique is to tag only a single species of plant or animal in the system and thus determine which feeding interactions or food chains are actually functional in the given natural community. The result may be surprising. Figure 3, for example, shows the actual number of plant-animal interactions which occurred between the two dominant species of plant and the most abundant species of animal in a first-year weed community. These food chain interactions are compared with the theoretical maximum and minimum numbers.

EFFECTS STUDIES

Investigations dealing with the reactions to the stress imposed by ionizing radiation are referred to as effects studies. Factors which must be considered are the duration of the radiation and the responses of specific populations and communities.

Duration of radiation. From an ecological standpoint, the effects of ionizing radiation on a population or an ecosystem are strongly influenced not only by the dose received, but by the time taken to deliver the dose. If the total dose of radiation is given in a few minutes or hours, the exposure is said to be acute. The same dose delivered over a period of a few days may be short-term, over weeks or months long-term, and if a system is continuously subjected to radiation, exposure is chronic.

Population response. The effects of radiation delivered to a population of organisms can be assessed in many ways; often the simple question of survival is most important. Thus a criterion might be the average length of survival in nature, as compared to a control group or population that received no radiation. Animals that have been raised under stress (and this includes wild individuals that have been captured), can, under certain conditions, be more resistant to acute doses of ra-

diation than are animals of the same species that have been raised in the laboratory. Naturally the response of populations that are exposed to different levels of chronic ionizing radiation may range from no effect to complete extinction. Within tolerable dose rates, that is, doses that do not produce extinction, responses to the radiation stress often take the form of changes in the population birth and death rates; individual metabolic or growth rates seldom change. In other words, the effect of chronic radiation often shows up first as an ecological effect at the population level, leaving the individual outwardly unaffected.

Community response. Drastic community effects such as death, desolation and destruction occur only after the most intense radiation. Interesting community effects studies are those which involve subtle changes in the structure or function or both of the ecological community or ecosystem. The response of even the simplest community to the stress of ionizing radiation is of course the net result of a series of direct effects on the component populations, plus some indirect changes which

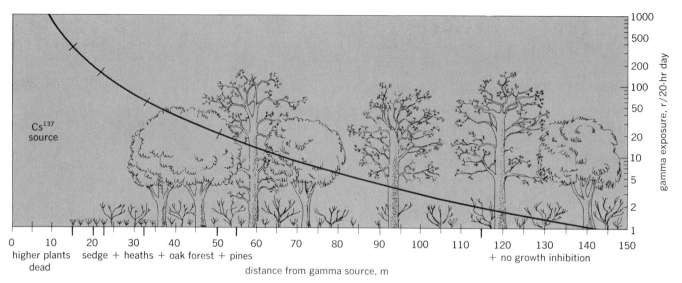

Fig. 5. Radiation damage to oak-pine forest after 6-month exposure. (*From G. M. Woodwell, Science, vol. 156, 1967*)

may be brought about because of the interaction of the radiation-affected populations in the ecosystem.

Natural communities have many homeostatic mechanisms; that is, they tend to persist unchanged if given a reasonably stable physical environment. This stable or climax state is the result of a process of development or replacement of species associations known as ecological succesion. Violent perturbations of the physical environment, fire for example, generally cause the community to return to some earlier preclimax state. If the change is temporary, succession proceeds to restore the steady-state system. If the change is permanent, annual fire for example, a successional process occurs until some steady state is established which is stable under the new environment. *See* PLANT FORMATIONS, CLIMAX; SUCCESSION, ECOLOGICAL.

In the above sense ionizing radiation is an environmental stress. If the radiation is applied to a community over a period of days or weeks and then terminated, the response of the community is characterized by successional setback and subsequent recovery. Recovery rate depends upon the initial effect of the radiation. In the case of forests, for example, the recovery time depends on whether the trees are entirely killed or whether only the aboveground parts are killed so that the roots can sprout to quickly restore the forest community.

Radioecological effects studies at the community level usually employ some form of point radiation source which can be put out in the community, and can be opened or closed from a safe distance (Fig. 4).

If the radiation is of short duration (acute), subsequent study involves zones of radiation damage beginning at the center where the most drastic effects are observed, and continuing out to some peripheral limit at which only barely perceptible ecological effects are noted, and these only months or years after the event. Such experimental studies enable investigators to predict what would happen to a particular community type under given condi-

tions of radiation. Because the radiation level is attenuated by distance, such studies are, in effect, several simultaneous experiments varying only in dose rate of radiation.

If the radiation stress is permanent (chronic), succession is permanently delayed and a new steady-state develops; the new community type is, as in the case of an acute dose, dependent on the radiation gradient with the most radiation-resistant species developing closest to the source (Fig. 5). Although much of the early work on the irradiation of natural communities was conducted in such permanent radiation fields under chronic conditions, portable irradiators have now been developed. Some of these are capable of delivering dose rates of several thousand rads per day within a few meters of the source. This has made possible the short-term (1–3 months) irradiation of natural ecosystems ranging from old fields to tropical rainforests. Studies of the response and recovery of these systems following the removal of the radiation stress not only are providing valuable data regarding basic ecological processes but are also helping to answer questions concerning the effects of irradiation on the natural environment. *See* ECOSYSTEM.

[RICHARD G. WIEGERT]

Bibliography: F. P. Hungate (ed.), *Radiation and Terrestrial Ecosystems*, 1966; D. J. Nelson and F. C. Evans (eds.), *Symposium on Radioecology*, USAEC TID-4500, 1969; G. A. Sacher (ed.), *Radiation Effects on Natural Populations*, 1966; V. Schultz and A. W. Klement (eds.), *Radioecology*, 1963.

Rainforest

A term loosely used in plant geography for forests of broad-leaved (dicotyledonous), mainly evergreen, trees found in continually moist climates in the tropics, subtropics, and some parts of the temperate zones. Sometimes the term is unjustifiably extended to include other very wet forests such as the "Olympic Rain Forest" of Washington State in which the dominant trees are conifers.

The tropical rainforest, which includes the vast Amazonian forest, large areas in West and Central Africa, the Malay Peninsula and the neighboring islands, is the home of an enormous number of plant and animal species. The trees, sometimes arranged in strata, are of various heights up to 150 or even 200 ft. As many as 200 different species of trees 1 ft or more in girth may be found in areas of 5 acres. Tropical rainforests, unlike temperate forests of beech or oak, are usually mixed in composition, no one species forming a large proportion of the whole stand; but in some parts of the tropics there are rainforests dominated by a single species, for example, the *Gilbertiodendron dewevrei* forest, which stretches over large areas in the eastern part of the Congo.

Certain structural features are characteristic of tropical rainforests, such as the thin flangelike buttresses of the larger trees, and flowers and fruit produced, as in the cacao (*Theobroma cacao*), on the trunk (cauliflorous) instead of on the branches. Orchids and other epiphytes and woody vines (lianes) are common. In rainforest which has not been culled for timber or recently disturbed, the undergrowth is generally thin and visibility is about 20 yd or more on the ground (see illustration). *See* CACAO.

Climatic factors. In typical tropical rainforest climates there are no winter and no dry seasons; consequently plant activities are possible year-round. Nevertheless, many plants are periodic in their growth, flowering, and leaf production, but the intervals may be irregular, or longer or shorter than a year. Rainforest trees rarely show annual growth rings. The flowering and leafing of different species, different individuals of the same species, or even different branches of the same tree are often not well synchronized.

In tropical regions with a marked dry season, the rainforest is replaced by semideciduous and deciduous (monsoon) forests. In these a considerable proportion of the trees becomes wholly or partly bare of leaves for some part of the year. The transition from the humid rainforest to these more or less deciduous types is usually gradual, and opinions differ as to where the dividing line should be drawn.

Productivity. Rainforest timbers are mostly very hard and are sometimes so dense that they will not float in water. Many, such as the African mahoganies (Meliaceae), provide high-quality furniture woods for which there is an increasing demand in Europe and the United States. Exploitation of the timber is often difficult because the economically valuable trees are scattered among large numbers of less useful species.

The organic productivity of tropical rainforest, though considerable, is less than might be expected from the favorable environment, perhaps because the soils tend to be infertile. The luxuriant appearance of the vegetation is deceptive, most of the available plant nutrients circulating rapidly between the plants and the superficial layers of the soil; there is thus a heavy loss of fertility when the forest is felled and the trees burned.

Primary and secondary communities. Rainforests which have never been cleared (virgin forests), or which have been undisturbed long enough to become indistinguishable from such forests, are called primary. At the present time these are being replaced by secondary communities of various kinds at an ever-increasing rate. Young secondary rainforests are less tall but denser and more tangled than primary rainforests; they are mainly composed of fast-growing, short-lived, soft-wooded trees with wind- or animal-dispersed fruits or seeds which enable them to colonize rapidly the natural or man-made clearings. *See* SUCCESSION, ECOLOGICAL.

Agricultural practices. Considerable areas of tropical rainforest have been felled to grow commercial plantation crops such as rubber and cacao, but much greater areas are cleared yearly by native cultivators to grow manioc, rice, corn, and other subsistence crops under shifting systems of cultivation. After a few harvests have been gathered, a new area is cleared, preferably in primary forest, and secondary forest is allowed to grow on the abandoned land. When the population pressure is heavy, there is a tendency to clear this forest again after too short an interval; the soil then deteriorates, especially if the vegetation is grazed and frequently burned. Under such conditions

Tropical rainforest of Brunei (Borneo). Buttressing and cauliflory are characteristic. (P. S. Ashton, Ecological Studies on the Mixed Dipterocarp Forests of Brunei State, Oxford Forest. Mem. no. 25, Clarendon Press, Oxford, 1964)

grasses such as alang-alang or cogon (*Imperata cylindrica*) invade the secondary forest, and "derived savanna," closely resembling the true savannas of less humid climates, replaces the forest. *See* SAVANNA.

Distribution. The tropical rainforest occupies lowland areas where the annual rainfall is not less than about 80 in. and there are not more than 3 or 4 consecutive months with less than about 4 in. At higher elevations it gives way to montane rainforest, and with increasing latitude it gradually merges into subtropical and temperate rainforests. These other types of rainforest are less species-rich and usually shorter than the tropical rainforest; features such as buttressing and cauliflory, which are so characteristic of the latter, are absent or much less well developed. *See* PLANTS, LIFE FORMS OF; VEGETATION ZONES, ALTITUDINAL; VEGETATION ZONES, WORLD. [PAUL W. RICHARDS]

Bibliography: P. H. Allen, *The Rain Forests of Golfo Dulce*, 1956; P. S. Ashton, *Ecological Studies in the Mixed Dipterocarp Forests of Brunei State*, Oxford Forest. Mem. no. 25, 1964; P. W. Richards, *The Tropical Rain Forest*, 3d ed., 1966; E. Aubert de la Rüe, F. Bourlière, and J.-P. Harroy, *The Tropics*, 1957; W. D. Francis, *Australian Rain Forest Trees*, 1951; J. P. Schulz, *Ecological Studies on Rain Forest in Northern Suriname*, 1960.

Rangeland conservation

The major purpose of rangeland conservation in the United States is to secure maximum forage production for each site consistent with ecological stability of the vegetation and of the soil. Also, where applicable, rangeland conservation is concerned with the preservation of watersheds, with timber production, and with recreation. Range conservation is one of the youngest fields of natural resource management. *See* ECOLOGY; SOIL.

Grazing land economics. The public range is an integral part of many livestock operations in the West, where approximately 10,000,000 head of livestock receive about one-third of their annual forage requirements from public lands. Consequently, the public range is an important element in the national production of meat, wool, and leather. Great but unestimated value is added to the nation's economy by land management on the range which improves water yield and reduces erosion and downstream siltation.

Types of grazing lands. Artificial pastures and the tall-grass prairies (now largely in corn and wheat) are not considered rangeland. The midgrass and short-grass plains west of the prairies and east of the Rocky Mountains, the arid and semiarid grasslands of the Southwest and the Great Basin, the grasslands of intermountain valleys, the brush and open woodlands of the mountains, and montane and subalpine meadows are all parts of the western range of the United States. Depending on the local climate, some of it provides only winter grazing whereas other parts, especially at higher elevations, provide from a few weeks to a few months of summer grazing. *See* GRASSLAND.

Overgrazing, corrective legislation. Federally owned land suitable for grazing, under the administration of the Bureau of Land Management of the Department of Interior, forms nearly one-tenth of the United States. Previous to the Taylor Grazing Act (June 28, 1934), western grazing lands of about 170,000,000 acres were open to free access and use without public control and management. Now 156,000,000 acres are organized into grazing districts 3,000,000–9,000,000 acres in extent that are administered to conserve and regulate the public grazing land and to help stabilize the livestock industry. The public range is used by about 20,000 private stockmen under a system of permits and a code that seeks to guarantee proper use of the range and to return to the government fair compensation for use.

A long history of cutthroat competitive grazing and of conflict between cattle and sheep operators and between them and agricultural settlers had resulted in widespread deterioration of the range. Productivity, which was naturally low in the more arid regions (averaging about 70 lb of air-dry forage per acre on the public range in contrast to about 2800 lb per acre for average hay land in the United States), was reduced by overgrazing to extremely low carrying capacity in many places. Erosion was accelerated and water loss by excessive runoff was increased on unprotected soils. Palatable and nutritious species, especially perennial grasses, were reduced in amount, and weeds and other undesirable species were increased. This is well illustrated in Texas, New Mexico, Arizona, and southwestern California by mesquite, which now occupies about 50,000,000 acres (twice what it did about 1900) and which cuts grass forage to one-third or less of full production when the mesquite bushes increase to 100 or more per acre. In the West, overgrazing also caused the spread of sagebrush, which now covers about 96,000,000 acres.

The Bureau of Land Management program since the passage of the Taylor Grazing Act has many accomplishments to its credit: increase of range forage, more livestock products, greater protection of public lands, more usable water, less erosion and downstream sedimentation, more flood control, more wildlife, and more recreational opportunities; yet the Bureau estimated that in the quarter century following the act not more than 10% of the needed range improvement work was done, largely because of inadequate Federal appropriations. The Bureau's program includes range inventories and management plans, improved fire protection, watershed treatment works (regrassing, water conservation and erosion control dams, and water spreading structures), pest and rodent control, and range management improvements (stock water developments, range fencing, corrals, and livestock and truck trails). In spite of this program, about 50% of Federal rangelands are still in a state of severe to critical erosion, 32% are suffering moderate erosion, and only 18% are in a condition of unaccelerated or no erosion.

Range inventory. The problem is to determine what the vegetation, soil, and climate will permit with regard to forage production as measured by the land's animal-carrying capacity. The range inventory includes a quantitative evaluation of the vegetation, its palatability, the nutritional value of each species, the status of the vegetation in the plant successional process, and the time and degree of permissible use.

The soil is evaluated to determine whether it is normal, eroded, or compacted and whether its productive potential is increasing or decreasing. It is helpful to know what the forage-producing capacity of the soil was originally, what it is at present, and what its possibilities are. The water infiltration rate and the moisture retention capacity of the soil are also measured, and hence its watershed management condition and its erosion potential are determined.

Existing water facilities and possibilities for water development are evaluated to determine whether artificial vegetation rehabilitation is feasible. Accessibility by roads and trails, fire history, and an analysis of fire potential are also determined. In addition, the presence or absence of poisonous plants, predators, and destructive rodents is noted.

Range management plan. The range management plan, developed from the inventory, sets forth the season of year that the range should be grazed, the duration (usually in days) of the grazing period, and the kinds and numbers of heavy livestock (cattle and horses) or lighter animals (sheep and goats) or both to be permitted. The range plan also states the need for and nature of fire protection, the need for additional trails and roads or the abandonment of old ones, the location and types of fences required to direct the control of trailing and herding, the water development needed, any special treatment of soil advisable for eroded or compacted areas, the steps required for control of poisonous plants, and the need for reduction of predatory animals and rodents. The range management plan also details the reseeding needed—the location, type of species, planting methods to be used, and use of the vegetation after planting—and considers whether rehabilitation may be expected to take place without artificial seeding under a plan for temporary retirement from use and suitable protection.

Significant advances. Quantitative surveys of range vegetation, now a fairly well-developed technique, probably will be improved through use of aerial photos, greater field mobility of survey technicians, and better statistical techniques.

Since about 1935 much has been learned about the ecology and physiology of range plants. This has resulted in the classification of such plants with respect to their physiological condition and the ecological trends indicated—so-called condition and trend classification. *See* ECOLOGY, APPLIED.

Through quantitative study of animal ingestion of various plant species, a significant advance has been made in understanding the relation of floristic composition to actual animal-carrying capacity of rangeland. One method contributing to this knowledge has been the use of the esophageal fistula and fecal collecting bags. Associated with this is an improved and intensive before-and-after range analysis. The actual nutritional value of the forage species and the usable portions of the plants under various methods of management are determined by laboratory study. This indicates to what extent supplemental feeding of livestock is necessary. *See* POPULATION DYNAMICS.

It is now accepted that the management of deer, elk, and antelope herds is also part of range man-agement. Much has been learned concerning the problem of competition of domestic livestock with big-game animals, thereby contributing to improved management of rangelands for both kinds of animals.

There is growing knowledge of ways to improve range plant composition by using large machinery and selective sprays, often applied by airplane, for destruction of undesirable brush and weeds. Still to be studied are the adaptability of plants to various locations, methods of planting, physiological adaptiveness of species, time of planting, and the time and type of use after planting. The use of airplanes or helicopters for seeding has not yet solved all range planting problems.

Problems, present and future. There is still a question as to whether any significantly large, new range areas can be opened up by an improved or different type of management. The "discovery" for range use of ex-timber lands in the coastal plain of the Southeast is an example.

Many problems of multiple use of rangeland are yet to be resolved. In the western summer range there probably will always be the problem of correlating grazing and timber production. Demands for recreation and big-game hunting areas are becoming more important. With the increasing population of the West and the more intensive use of all its lands, the demand for water is growing rapidly, making efficient management of rangelands imperative for water conservation. *See* FISHERIES CONSERVATION; FOREST MANAGEMENT AND ORGANIZATION; LAND-USE PLANNING; MINERAL RESOURCES CONSERVATION; SOIL CONSERVATION; WATER CONSERVATION; WILDLIFE CONSERVATION.

[LEWIS TURNER]

Bibliography: *See* CONSERVATION OF RESOURCES.

River

A water stream of natural origin which flows across the surface of a continent or island. A river is part of a river system which drains a topographically related section of land surface known as a river basin (Fig. 1). The system begins in the precipitation which falls on a rock-, soil-, or vegetation-covered surface and immediately becomes surface runoff, or eventually appears as snow and ice meltwater or underground drainage. Such a system may be divided into headwater streams, tributary streams, and the main stem. The headwaters are in springs, marshes, lakes, or small upper streams, generally in the highest relative elevation in a basin. A river ends in a mouth, where it may discharge into a major lake, a dry basin of interior drainage (playa), an inland sea, or the ocean.

Terminology. Like many words which have long been in general use, the term river is somewhat elastic in meaning. In English usage the main stem of a stream system is nearly always designated as a river, but so are all important tributaries and even many secondary tributaries. A tributary may also be known as a fork, branch, or creek, and may have the same volume of flow as other streams called rivers. Smaller headwater streams are usually creeks or brooks.

Rivers flow in channels or watercourses and develop many distinctive valley features by erosion and deposition.

RIVER

boundaries of secondary drainage basins

boundary of main drainage basin

Fig. 1. Maplike diagram of a drainage basin. Note that such basins are composed of a system of secondary basins.

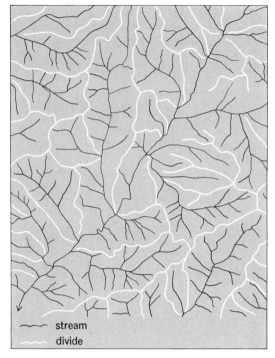

Fig. 2. Cartographic diagram illustrating stream, divide, and basin patterns in a dendritic drainage system.

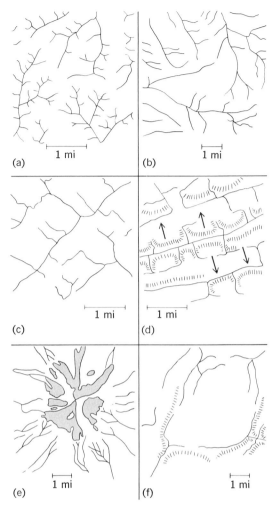

Fig. 3. Stream patterns. (*a*) Dendritic drainage in horizontal rocks, West Virginia. (*b*) Dendritic drainage in crystalline rocks, Rocky Mountains. (*c*) Rectangular drainage in jointed crystalline rocks, Adirondacks. (*d*) Trellis drainage in folded rocks, Pennsylvania. (*e*) Radial drainage on a volcano, Mount Hood, Ore. (*f*) Annular drainage on dome, Turkey Mountain, N.Mex. (*After A. K. Lobeck, Geomorphology, McGraw-Hill, 1939*)

Rivers may be described by the pattern of the system of which they are part and by their length, velocity, volume of discharge, and the nature of water flowing within them. Most rivers are part of a dendritic drainage pattern (Fig. 2), but some, responding to the underlying geologic structure, are in radial, annular, rectangular, or trellised (lattice-like) pattern. In some limestone regions a karst (enclosed depression) drainage may be found, with associated underground rivers. A few rivers, such as the Nile in its lower reaches, are exotic and flow for considerable distances without receiving drainage of any consequence from tributaries. Such river reaches always occur in arid regions.

Regime and flow patterns. The regime is directly dependent on the climate of the region or regions involved. It also is influenced by the size of the drainage basin funneling upon the stream; the direction of flow; the conditions of vegetative cover; and the nature of the surface geology (Fig. 3), topography, and soil conditions in the basin. Few if any streams have completely stable conditions of flow; the rule is variation from day to day, season to season, and year to year. Study of these variations and their causes is an important part of the science of hydrology. *See* SURFACE WATER.

In arid regions, intermittent streams are common. The flow of an intermittent stream may fluctuate markedly from nothing to flood stage within a matter of minutes if a storm of sufficient extent and intensity covers part or all of its drainage area. Normally dry channels of these streams are called arroyos or wadis.

Under more humid climatic conditions, the channels of streams of sufficient volume to be called rivers are only occasionally dry. Fluctuations of flow nonetheless are found everywhere.

For example, natural flow near the mouth of the Tennessee has varied between 4500 and 500,000 ft^3/sec. In middle latitudes the season of low flow is generally summer, when evaporation and transpiration within the basin are greatest. High water may come during autumn, winter, or spring, depending on temperature conditions and the time of heaviest precipitation. Storage of large volumes of water, such as snow over frozen ground, characteristically causes early spring floods in the Great Plains region of the United States when a rapid thaw takes place.

Within high latitude areas of the Northern Hemisphere, high water inevitably occurs in spring on the northward flowing rivers because melting progresses from headwater to mouth, and the flow of water released upstream is barred by ice dams remaining downstream. The rivers of Siberia are notable examples of this condition, with broad flooding over lowland plains.

In low latitude areas, on the other hand, high

Discharge, basin area, and length of some of the world's major rivers

River	Average discharge, ft³/sec	Basin area, mi²	Length, mi
Amazon	4,000,000	2,772,000	3900
La Plata-Paraná	2,800,000	1,198,000	2450
Congo	1,400,000	1,425,000	2900
Yangtze	770,000	750,000	3100
Brahmaputra	700,000	361,000	1680
Ganges	660,000	450,000	1640
Mississippi-Missouri	620,000	1,243,000	3892
Yenisei	615,000	1,000,000	3550
Orinoco	600,000	570,000	1600
Lena	547,000	1,169,000	2860
St. Lawrence	500,000	565,000	2150
Ob	441,000	1,000,000	2800
Mekong	390,000	350,000	2600
Volga	350,000	592,000	2325
Amur	338,000	787,000	2900
Mackenzie	280,000	682,000	2525
Columbia	256,000	258,200	1214
Zambesi	250,000	513,000	2200
Danube	218,000	347,000	1725
Niger	215,000	584,000	2600
Indus	196,000	372,000	1700
Yukon	180,000	330,000	2100
Huang	116,000	400,000	2700
Nile	100,000	1,293,000	4053
São Francisco	100,000	252,000	1811
Euphrates	30,000	430,000	1700

water is directly related to seasonal maxima of rainfall, but high altitude conditions may complicate the regime in most zones. Where there is a pronounced dry season, as on the Indian peninsula, high water occurs soon after the onset of the rainy season, when the moisture requirements of hitherto dormant vegetation are still low. In all parts of the world, altitudinal conditions may influence the regime of a stream in another manner. Where headwaters are in extensive high mountain areas with heavy winter accumulations of snow and ice, high water occurs at the season of heaviest melt, early summer or midsummer. The Columbia, Ganges, Indus, and Rhine rivers show this influence in their regimes.

Surface materials and the nature of vegetative cover also influence the regime. The more continuous the forest or grass cover, in general the more stable the volume of discharge. Soil conditions which favor easy infiltration also promote more equitable flow, as on the sands of the Atlantic and Gulf Coastal Plain of the United States.

Water qualities. Every river is an agent of erosion, as well as an agent of drainage. Many mineral materials other than water consequently are constantly in motion where a river flows. These materials are transported by water in solution, in suspension, and as bed load. For discussion of stream erosion, transport, deposition, and associated landforms see STREAM TRANSPORT AND DEPOSITION.

The high capacity of water as a solvent imparts many different qualities to river water as a solution. The great majority of rivers are fresh water, but a few are saline (relatively high salt content). All rivers, however, contain perceptible amounts of mineral material in water solution. In most cases this is calcium, the most common cause of "hard" water, but any of the elements soluble in water may be found, such as magnesium, potassium, sodium, silicon, nitrogen, and the elements which combine with them to form salts. The content of salts in solution is highest in the rivers of regions under desert or semiarid climates, but calcareous materials derived from limestone may yield hard water in humid regions.

Like most other bodies of water on the Earth's surface, a river also is a medium for the support of life, from bacteria and simple forms of plant life to fish, and amphibian, mammal, and bird wildlife. This is related not only to the capacity of water to carry nutrient minerals in solution but also to dissolved gases, particularly oxygen.

Management. The characteristics of rivers have made them important to human society. No other natural feature, except the soil, has been more closely tied to the past progress of civilization for the majority of human beings. Means of counteracting the vagaries of flow have been an important part of civil engineering for centuries. This has been true in part because of the attractiveness of floodplains to agricultural occupance, and the consequent need to avoid natural flooding. It has also followed from man's need for water storage in order to live through drought seasons. In modern times the problem of river management or river control has become much more difficult because of the rapid increase of population, its concentration in dense settlements, the extent to which manufacturing and other economic functions have encroached on floodplains, the vastly increased disposal of wastes in rivers, and the larger number of purposes that rivers must serve simultaneously. The general objects of river management are the conservation of natural flow for release at the times needed by man, the confinement of flood flow to the channel and planned areas of floodwater storage, and the maintenance of water quality at a level which will yield optimum benefit through multiple use. The techniques of river management are well understood; their practice is still very incomplete, in part because the economics of river development is not well known. Domestic river development is an important responsibility of the U. S. Army Corps of Engineers. It also is the central responsibility of the Tennessee Valley Authority and is an important objective of the Bureau of Reclamation of the Department of the Interior. *See* RIVER ENGINEERING.

Of the major rivers in the world (see table) none is yet controlled or managed in the manner which modern engineering, administrative, and biological techniques would permit. The closest approach to such management is made on some medium-sized streams, the Tennessee, the Rhine, and the Rhône, for example. Some other rivers, such as the San Joaquin in California, have been fully developed for a single purpose, irrigation. Commencing in the 1930s, the greatest river-regulation works of all history were undertaken. The United States, the Soviet Union, and (since 1946) France have been foremost in supporting work of this kind. Among the notable achievements have been the series of great dams on the Columbia, Missouri, and Colorado and the regulation of the Tennessee in the United States; the Volga-Don Canal, the lower Volga dams, and the very large dams on the Angara and Yenisei in the Soviet Union; and the Rhône regulation in France.

The greatest and potentially most productive works remain for the future. These include plans for important work on the three largest streams of all the Amazon, the La Plata-Paraná, and the Congo. These basins contain storage and power-generation sites of several times the capacity of the largest hitherto developed. Of the eight rivers having basins of 1,000,000 mi² or more in extent, only the Mississippi and Nile have more than minor control works. Still other great streams offering major possibilities for physical development are the Yenisei, Yangtze, Huang, Amur, Mekong, Chao Phraya, Tigris-Euphrates, Niger, Zambesi, Orinoco, Sâo Francisco, Danube, Mackenzie, and Yukon. The extent and timing of such development will depend upon economic need, availability of investment funds, and political cooperation. The need is patent for development of the Yangtze, Huang, Nile, Niger, Tigris-Euphrates, Danube, Sâo Francisco, and lesser streams in densely settled, underdeveloped areas. It is therefore probable that the latter half of the 20th century will be a period of extending control of these streams, as political conditions permit.

[EDWARD A. ACKERMAN; DONALD J. PATTON]

Bibliography: S. Leliavsky, *An Introduction to Fluvial Hydraulics*, 1966; L. B. Leopold, M. G. Wolman, and J. P. Miller, *Fluvial Processes in Geomorphology*, 1964; M. Morisawa, *Streams: Their Dynamics and Morphology*, 1968; R. J. Russell, *River Plains and Sea Coasts*, 1967; United States President's Water Resources Policy Commission, *Ten Rivers in America's Future*, vol. 2, 1950.

River engineering

A branch of transportation engineering consisting of the physical measures which are taken to improve the river and its banks.

Most centers of civilization developed in the valleys of the world's rivers. The people depended on alluvial plains for their agricultural economy and upon streams for domestic water and transportation. Subsequently, this reliance upon waterways has been expanded to include water for industrial as well as domestic consumption, to provide economical waterpower, and to utilize the river for waste disposal. With this expansion, use of the streams for transportation has continued, despite extensive developments of other transportation facilities. Today the improvement of rivers for inland navigation is actively prosecuted in all parts of the civilized world. Inland waterway traffic within the United States increased from 118,057,000,000 ton-miles in 1940 to 281,400,000,000 ton-miles in 1967.

The measures that are taken to improve the river and its banks may include contraction of the river channel to improve navigation depths; bank stabilization to minimize erosion which would destroy farm land, cities, and bridges; creation of slackwater pools by means of locks and dams; or combinations of these means. It may also include improvement of the channels to assist them in carrying flood flows and regulation of the rivers' flows by upstream reservoir storage. In approaching the problems of river engineering, consideration must be given to the characteristics of the stream: the slope, meandering, sediment load, flow variations, and other factors (Figs. 1 and 2).

River characteristics. A stream is said to be in regimen if the major dimensions of the channel remain relatively constant and if it is neither aggrading (raising of the bed) nor degrading (lowering of the bed). The channel need not be fixed in position; however, many streams in regimen are constantly shifting their channels by eroding the banks at one location while building them at another. *See* RIVER.

Most natural streams are in regimen, with channel dimensions that are more or less characteristic of that stream. This implies a balance between the energy forces of the flow, the forces required to erode the bed and banks, the sediment load, and possibly other factors. There is no universally accepted theory relating these factors but, in general, a stream in erodible alluvium will be wide and shallow if the banks are readily erodible or narrow and deep if the banks are erosion resistant.

Channels may be generally classified as straight, meandering (following an alignment consisting principally of pronounced bends), or braided (a number of interconnected channels presenting the appearance of a braid). These forms are influenced by many factors, including the stream discharge, the nature of the soils, and the sediment load; however, they may be best correlated with the slope of the valley in which they are located. Straight channels are found in valleys of either flat or steep slope; meandering channels occupy valleys of intermediate slope; and braided channels of streams in regimen occur on steep slopes. Braided channels of streams not in regimen may occur on either flat, intermediate, or steep slopes.

Technical knowledge is inadequate to explain fully the relationship between stream form and valley slope, but it is necessary in river engineering to recognize it. For example, many attempts to improve meandering channels by excavating straight channels have failed because the stream immediately began to erode its banks to reassert the meanders, meanwhile dumping excessive quantities of sediment into the channel downstream. In like manner, attempts to impose bends or curves on a channel in steep slopes must be considered with due caution.

River channel improvements. These may consist of revetting the banks to prevent erosion and shifting of the channel, realignment of the channel to provide smoother bends and a more regular alignment, contraction of an existent channel (particularly the contraction of a braided low-water reach to provide a single effective low-water channel), the provision of slack-water pools by the construction of low dams and ship locks, or the complete excavation of a new channel. Problems involving revetment, realignment, or contraction occur most frequently in meandering streams in erodible alluvium.

Contraction works. Used primarily to confine the low and moderate flows of a wide, shallow, or braided stream to a single effective channel, contraction works are required predominantly in meandering streams in erodible alluvium. It is important that the rectified channel be planned with due regard to maintenance of regimen. Alignment of the channel should be generally similar to that of the existent stream, following a series of smooth bends rather than straight lines and maintaining essentially the same channel slope.

Structures used in contraction consist of revetment in the concave portions of bends and of guide structures. The latter are normally of pile dike (Fig. 3) or other permeable fence-type construction designed to utilize the sedimentary and erosive characteristics of the stream in the initial shaping of the channel. *See* STREAM TRANSPORT AND DEPOSITION.

Where the position of the rectified channel deviates materially from that of the original channel, the concave banks of bends may be excavated and revetted in the dry, a pilot channel excavated, and the stream encouraged to scour the channel to the desired dimensions. In other cases the guide structures are constructed in stages, contracting the channel and causing the opposing bank to erode progressively to the desired location. The permeable guide structures serve to turn the current as desired yet permit deposition of sediments to build up the abandoned area behind them.

Locks and dams. In some streams, navigable depths are secured by relatively low-head dams, which create a series of slack-water pools. Locks, consisting of gated chambers, are provided to pass boats and barges around the dams (Fig. 4). A vessel is brought into the lock chamber from below the dam, the gates are closed, and the lock chamber is filled with water drawn from the upper reservoir. When the chamber has been filled to the level of the upper pool, the upper gates are opened and the vessel passes through into the upper pool. The reverse process is followed in going from the upper pool to the lower pool.

The lift of the lock may vary from a few feet to over 100 ft. The locks may be supplemented by gates through the dam. These gates may be lowered to the stream bed to permit free naviga-

Fig. 1. Aerial view showing head of navigation project and uncontrolled and stabilized sections of Missouri River near Sioux City, Iowa. (*U.S. Army Corps of Engineers*)

Fig. 2. Uncontrolled river.

Fig. 3. Pile dike contraction works.

Fig. 4. Lock and dam of Minneiska, Minn.

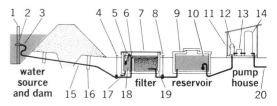

water source and dam   filter   reservoir   pump house

1 = hedge post or pipe
2 = screen suspended from post 3 ft under water
3 = flexible pipe
4 = hand valve
5 = aspirator (alum feeder)
6 = float valve
7 = hinged wood cover
8 = hand valve
9 = reinforced concrete top
10 = foot valve and strainer

11 = insulated pump house
12 = automatic pump
13 = automatic chlorinator
14 = pressure tank
15 = 2 in. iron pipe or plastic pipe muting
16 = concrete cutoff collar
17 = drain when needed
18 = coagulation sedimentation chamber
19 = washed river sand screened through ⅛ in. sieve
20 = purified water to house (below frost line)

Fig. 1. Farm-pond water-treatment system.

tion during periods of adequate flow. *See* DAM.

Canals. Canals are constructed to provide connections between waterways or to bypass critical river reaches. They may range from essentially open waterways, such as the Suez Canal, to complex systems of excavated waterways, dams, and high lift locks such as the Panama Canal; or they may be included in a system with locks, dams, and open-river navigation as in the St. Lawrence River. They may also include channels excavated through low-slope braided streams or swamp areas, as in the Illinois River.

[WENDELL E. JOHNSON; DONALD C. BONDURANT]

Bibliography: T. Blench, *Regime Behavior of Canals and Rivers,* 1957; Committee on Channel Stabilization, Corps of Engineers, U.S. Army, *State of Knowledge of Channel Stabilization in Major Alluvial Rivers,* Tech. Rep. no. 7, October, 1969; M. M. Hufschmidt and M. B. Fiering, *Simulation Techniques for Design of Water Resource Systems,* 1966; J. V. Krutilla and O. Eckstein, *Multiple Purpose River Development,* 1958; A. Maass et al., *Design of Water Resources Systems,* 1962; R. S. Rowe, *Bibliography of Rivers and Harbors and Related Fields in Hydraulic Engineering,* 1953; U.S. Office of Civil Engineers, *Seminars on River Basin Planning,* Fort Belvoir, Va., May 27–31, 1963.

Rural sanitation

Those procedures, employed in areas outside incorporated cities and not governed by city ordinances, that act on the human environment for the purpose of maintaining or improving public health. The purpose of these procedures is the furtherance of community cleanliness and orderliness for esthetic as well as health values.

Water. Purification of water supplies since 1900 has helped to prolong human life more than any other single public-health measure. Organisms which produce such diseases as typhoid, dysentery, and cholera may survive for a long time in polluted water, and prevention of contamination of water supplies is imperative to keep down the spread of such diseases. Watertight covers for wells are important means of preventing surface contamination to the water supply in rural areas.

Purification of a surface water supply, such as that from a lake or a farm pond, is accomplished by means of sedimentation, filtering, and chlorination. Sedimentation can be effected in a storage chamber by the addition of aluminum sulfate, which flocculates the finer particles of soil and other undesirable matter in suspension in the water. Filtering through fine sand removes the flocculated particles. Finally, the water is purified by means of a chlorine solution at the rate of $\frac{1}{2}-1$ part of chlorine to 1,000,000 parts of water. A system of treatment for farm-pond water is diagrammed in Fig. 1.

The colon bacillus is the usual indicator of pollution of water supplies by human waste. Chlorine kills such organisms, and therefore it is widely used for purification of water supplies.

Sewage disposal. The problem of safe disposal of sewage becomes more complex as population increases. The old practice of piping sewage to the nearest body of water has proved to be dangerous. Sanitary engineering techniques are now being used in rural areas, as well as in cities, for the disposal of household and human wastes. Where sewage-plant facilities are not available, the most satisfactory method of sewage disposal is by

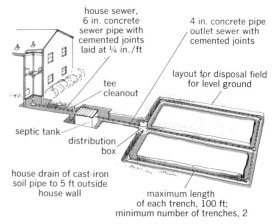

house sewer, 6 in. concrete sewer pipe with cemented joints laid at ¼ in./ft

4 in. concrete pipe outlet sewer with cemented joints

tee
cleanout

layout for disposal field for level ground

septic tank

distribution box

house drain of cast-iron soil pipe to 5 ft outside house wall

maximum length of each trench, 100 ft; minimum number of trenches, 2

Fig. 2. Typical family-size sewage-disposal system.

means of the septic tank system (Fig. 2).

The septic tank system makes use of a watertight tank for receiving all sewage. Bacterial action takes place in the septic tank and most of the sewage solids decompose, are given off as gases, or go out into the drainage lines as liquid. The gas and liquid are then released from the top 2 ft of soil without odor or sanitary problems. The solids that do not decompose settle to the bottom of the tank, where they can be easily removed and disposed of safely. Such a sewage-disposal system has made it possible for all farm homes and rural communities to have modern bathroom equipment and sanitary methods of sewage disposal. *See* SEPTIC TANK; SEWAGE DISPOSAL; WATER TREATMENT.

[HAROLD E. STOVER]

Saltmarsh

A maritime habitat characterized by special plant communities. They occur primarily in the temperate regions of the world, but typical saltmarsh communities can be formed in association with mangrove swamps in the tropics and subtropics. In inland areas where saline springs emerge or where there are salt lakes, typical saltmarsh communities can be found, dominated sometimes by the same species that occur on maritime saltmarsh. The extensive areas of inland salt desert, while exhibiting some features of similarity with saltmarsh, are nevertheless best regarded as a separate entity. Excess sodium chloride is the predominant environmental feature of salt marsh (maritime or inland). In the case of salt deserts sodium chloride is only one of the alkali salts that may occur in excess.

Occurrence of different types. Maritime saltmarsh can be found on stable, emerging, or sinking coastlines. On emerging or sinking coasts the actual extent of saltmarsh depends upon an adequate degree of wave protection and also upon the rate of change of coast level in relation to the rate of silt deposition. Mud or sand flats are raised by silt deposition to a level at which the characteristic phanerogamic plants can colonize. Saltmarshes therefore are common features of estuaries and protected bays, provided the seabed is shallow and does not shelve steeply. They also occur behind spits, barrier beaches, and offshore sand, shell, and shingle islands.

On emerging coastlines the true saltmarsh zone tends to be narrow, though older saltmarsh areas, recognizable by remnant saltmarsh species, have subsequently been invaded and dominated by the local terrestrial species. On sinking coastlines the extent of saltmarsh depends essentially on the rate of accretion from silt deposition in relation to rate of sinking. The greater the former in relation to the latter, the more extensive the saltmarshes are likely to be. Because of the dependence of saltmarshes upon accretion, they are likely to be best developed in association with eroding soft rock coastlines and estuaries of rivers that bring down abundant silt from soft rock upland. In some cases, where the amount of silt may not be very great, for example, southwestern Ireland and the New England coast of the United States, the roots and plant remains combine with the silt to form a peaty soil.

Physiography. Saltmarsh can form on mud, muddy sand, or sandy mud, but not on pure sand because mobile sand does not provide sufficiently

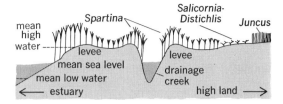

Fig. 1. Diagrammatic cross section of a marine marsh on the southern Atlantic coast of the United States; vertical exaggeration is about 10 to 1.

stable substrate, and it develops into sand dunes in the presence of plants. Typical physiographic features associated with saltmarsh are the creeks, which serve as drainage channels, and pools, which are known as pans. The type of creek system depends upon the initial substrate, whether muddy or sandy, local fresh-water drainage channels, and the type of primary colonist, for example, annual species of samphire (*Salicornia*) or clumps of ricegrass (*Spartina*) (Fig. 1). Various types of pan have been recognized, including primary, secondary, and creek (cutoff ends of creeks).

Plant zonation (succession). The colonizing plants are subject to considerable tidal inundation, but with the advent of plants the rate of accretion is hastened and the land level rises. The physical factors of the environment change, there are fewer submergences, and new species invade. A characteristic feature of saltmarsh vegetation, therefore, is the zonation of plant communities associated with changes in the environment. The zonation or succession for maritime saltmarsh is essentially dynamic since the saltmarsh is continuously, albeit slowly, building up toward the sea. *See* SUCCESSION, ECOLOGICAL.

When maritime saltmarsh develops between two lateral ridges with only a narrow creek entry, the full succession may be passed through in a few hundred years, even on a subsiding coastline. The zonation of inland saltmarshes, a static zonation, is related to decreasing salinity in proportion to the distance from the source of salt.

The early stages on any saltmarsh are sufficiently rapid to be observed in 25–50 years. The final stages of the succession depend upon whether the saltmarsh borders sand dune, meadow, or fresh-water inflow. In the last case the succession continues into brackish communities and

Fig. 2. View of *Spartina* marsh on the coast of Georgia.

finally into fresh-water swamp.

Plant species and life forms. The plants that grow on salt marshes must tolerate the excess sodium chloride and are termed halophytes. These plants possess features associated with the halophytic environment, for example, development of succulence, waxy cuticle, and salt-excreting glands; overall there is a tendency for the vegetation to exhibit a drab grayness. Members of the Chenopodiaceae are common (*Salicornia, Suaeda,* and *Allenrolfia*) and also the Plumbaginaceae (*Limonium* or *Statice*). Among the grasses *Spartina* and *Puccinellia* are important genera in different parts of the world (Fig. 2). Other genera that are widely represented in this habitat include *Plantago, Triglochin, Cotula, Scirpus,* and *Juncus*. See PLANTS, LIFE FORMS OF.

Inasmuch as there are characteristic phanerogams, so also there are characteristic algae associated with saltmarsh vegetation. Particular communities of green and blue-green algae are common, and in the Atlantic – North Sea area there may be extensive communities of free-living brown fucoids. Characteristic red algal communities are dominated by species of *Bostrychia* and *Catenella*.

As the number of species capable of growing under the specialized conditions is limited, the type of succession tends to be similar for major areas. The following groups have been proposed for saltmarshes: Arctic, European (with four subdivisions), Mediterranean, Atlantic North American (with three subdivisions), South American; Pacific North American, Japanese, and Australasian (with two subdivisions), each with a characteristic succession.

Environment. The principal feature of the environment is the excess sodium chloride, with the resulting effect of the sodium ion upon the soil colloids and the chloride ion upon plant metabolism. The frequency of tidal inundations is very important, with their effect upon salinity, as is the water table and soil aeration. At lower marsh levels the existence of an aerated layer seems essential for growth of the plants. A lowering of the soil salinity at some season also appears necessary for successful seed germination.

Productivity. From the scanty data available it appears that wild saltmarsh is highly productive biologically and compares favorably with fresh-water reed swamp. Less than 10% of the net production appears to be consumed by grazing herbivores, most of it going into the detritus path of energy flow. For this reason wild saltmarsh dominated by grasses forms excellent grazing for stock except in spring-tide periods. In most parts of the world high saltmarsh is eventually enclosed by seawalls and converted to valuable agricultural land. See VEGETATION ZONES, WORLD.

[VALENTINE J. CHAPMAN]

Bibliography: V. J. Chapman, *Coastal Vegetation*, 1964; V. J. Chapman, *Salt Marshes and Salt Deserts of the World*, 1960.

Sanitary engineering

A specialty field generally developed in civil engineering but not limited to that branch. The National Research Council defines the sanitary engineer as "a graduate of a full 4-year, or longer, course leading to a Bachelor's, or higher, degree at an educational institution of recognized standing with major study in engineering, who has fitted himself by suitable specialized training, study, and experience (1) to conceive, design, appraise, direct and manage engineering works and projects developed, as a whole or in part, for the protection and promotion of the public health, particularly as it relates to the improvement of man's environment, and (2) to investigate and correct engineering works and other projects that are capable of injury to the public health by being or becoming faulty in conception, design, direction, or management."

Sanitary engineering practice includes surveys, reports, designs, reviews, management, operation and investigation of works or programs for (1) water supply, treatment and distribution; (2) sewage collection, treatment and disposal; (3) control of pollution in surface and underground waters; (4) collection, treatment, and disposal of refuse; (5) sanitary handling of milk and food; (6) housing and institutional sanitation; (7) rodent and insect control; (8) recreational place sanitation; (9) control of atmospheric pollution and air quality in both the general air of communities and in industrial work spaces; (10) control of radiation hazards exposure; and (11) other environmental factors which have an effect on the health, comfort, safety, and well-being of people.

Sanitary engineers engage in research in engineering sciences and such related sciences as chemistry, physics, and microbiology and apply these in development of works for protection of man and control of his environment. See AIR-POLLUTION CONTROL; RADIATION INJURY (BIOLOGY); SEWAGE; SEWAGE DISPOSAL; SEWAGE TREATMENT; WATER POLLUTION; WATER SUPPLY ENGINEERING.   [WILLIAM T. INGRAM]

Bibliography: H. E. Babbitt and E. R. Baumann, *Sewerage and Sewage Treatment*, 8th ed., 1958; H. Blatz (ed.), *Radiation Hygiene Handbook*, 1959; J. C. Collins (ed.), *Radioactive Wastes, Their Treatment and Disposal*, 1960; V. M. Ehlers and E. W. Steel, *Municipal and Rural Sanitation*, 5th ed., 1958; G. M. Fair, J. C. Geyer, and D. A. Okun, *Water and Wastewater Engineering,* vol. 2: *Water Purification and Wastewater Treatment and Disposal,* 1968; G. M. Fair, J. C. Geyer, and D. A. Okun, *Water Supply and Wastewater Removal,* vol. 1, 1958; *Glossary Water and Wastewater Control Engineering,* Joint Editorial Board, APHA, ASCE, AWWA, WPCF, 1969; W. C. L. Hemeon, *Plant and Process Ventilation,* 1955; R. K. Linsley, Jr., and J. B. Franzini, *Elements of Hydraulic Engineering,* 1955; F. S. Merritt (ed.), *Standard Handbook for Civil Engineers,* sect. 22, Sanitary Engineering, 1968; *Municipal Refuse Disposal,* Commission on Refuse Disposal, APWA Research Foundation Project 104, 2d ed., 1966; F. A. Patty et al. (eds.), *Industrial Hygiene and Toxicology,* vol. 2, 2d ed., 1963; *Refuse Collection Practice,* APWA Commission on Solid Wastes, 3d ed., 1966; P. A. Sartwell, *Maxcy-Rosenau Preventive Medicine and Public Health,* 9th ed., 1965; E. W. Steel, *Water Supply and Sewerage,* 4th ed., 1960; A. C. Stern (ed.), *Air Pollution,* vol. 1, 2d ed., 1968; H. H. Uhlig, *Corrosion and Corrosion Control,* 1963.

Savanna

The term savanna was originally used to describe a tropical grassland with more or less scattered dense tree areas. This vegetation type is very abundant in tropical and subtropical areas, primarily because of climatic factors. The modern definition of savanna includes a variety of physiognomically or environmentally similar vegetation types in tropical and extratropical regions. The physiognomically savannalike extratropical vegetation types (forest tundra, forest steppe, and everglades) differ greatly in environment and species composition.

In the widest sense savanna includes a range of vegetation zones from tropical savannas with vegetation types such as the savanna woodlands ("campo cerrado", Fig. 1) to tropical grassland and thornbush. In the extratropical regions it includes the "temperate" and "cold savanna" vegetation types known under such names as taiga, forest tundra, or glades. For further details on the synonyms and terminology *see* VEGETATION ZONES, WORLD.

During the growing season the typical tropical savanna displays a short-to-tall, green-to-silvery shiny cover of bunch grasses, with either single trees or groups of trees widely scattered. This is followed by a rest period of several months during which, because of severe drought, the vegetation appears quite different, with the brown-gray dead grasses bent over and the trees either without leaves or with stiff or wilted gray-green foliage. The heat and drought during this season of the year exert a high selective pressure upon the floral and faunal composition of the savanna.

Floral and faunal composition. The physiognomic similarity of the tropical savannas is underlined by the similarity among certain floristic components. All savannas contain members of the grass family (Gramineae) in the herbaceous layer. Most savannas of the world also have one or more members of the tree family (Leguminosae), particularly of the genus *Acacia*. Also included among the trees are the families Bombacaceae, Bignoniaceae, and Dilleniaceae, and the genera *Prosopis* and *Eucalyptus*: these are abundant when they occur. Palms are also frequently found. One of the most outstanding savanna trees is *Adansonia digitata* (Bombacaceae), which achieves one of the biggest trunk diameters known for all trees (Fig. 2). The grass species, although mostly from the genera *Panicum*, *Paspalum*, and *Andropogon*, include numerous other genera such as *Aristida*, *Eragrostis*, *Schmidtia*, *Trachypogon*, *Axonopus*, *Triodia*, and *Plectrachne*, all of regional importance.

The fauna of the savannas is among the most interesting in the world. Savannas shelter herds of mammals such as the genera *Antelopus*, *Gazellus*, and *Giraffus*, and the African savannas are especially famous for their enormous species diversity, including various members of the Felidae (for example, the lion).

Numerous species of birds are indigenous to the savannas. Among these is the biggest bird, the ostrich (*Struthio camelus*), found in Africa. Many birds from extratropical regions migrate into the tropical savannas when the unfavorable season occurs.

Fig. 1. Savanna with gallery forest north of Guiaba, Campo Cerrado, Brazil.

Fig. 2. *Adansonia digitata*, a tree with large trunk, in the savanna in Senegal, Africa. (*Courtesy of H. Lieth*)

Among the lower animals, the ants and termites are most abundant. Termite colonies erect large, conical nests above the ground, which are so prominent in some savannas that they partly dominate the view of the landscape, especially during the dry season.

Environmental conditions. The climate of the tropical savannas is marked by high temperatures with more or less seasonal fluctuations. Temperatures rarely fall below 0°C. The most characteristic climatic feature, however, is the seasonal rainfall, which usually comes during the 3–5 months of the astronomic summertime. Nearly all savannas are in regions with average annual temperatures from 15 to 25°C and an annual rainfall of 32 in.

The soil under tropical savannas shows a diversity similar to that known from other semiarid regions. Black soils, mostly "chernozem," are common in the moister regions. Hardpans and occasional surface salinities are also found, and lateritic conglomerations occur along the rivers. The soils vary in mineral nutrient level, depending on geologic age, climate, and parent material. Deficiencies of minerals, specifically trace elements, are reported from many grassland areas in the continents of Africa, South America, and Australia. *See* SOIL.

Certain savanna areas suffer from severe erosion and no soil can be accumulated. Plant growth in these areas is scarce, often depending on cracks and crevices in the ground material to support tree roots. The herbaceous cover opens up under these conditions, with many stones appearing on the land surface.

In the majority of the tropical savannas water is the main limiting environmental factor: total amount and seasonality of precipitation are unfavorable for tree growth. Additional stresses to forest vegetation are caused by frequent fires, normal activities of animals, excess of salt, or nutrient deficiencies. Wherever a river flows, most of these factors change in favor of tree growth. This explains the existence of extensive gallery forests along the rivers. The gallery forest is missing only where severe local floods after storms cause soil erosion and the formation of canyons.

Geographic distribution. Tropical savannas exist between the areas of the tropical forests and deserts; this is most apparent in Africa and Australia. In the New World tropics, different conditions exist because of the circumstances created by the continental relief, and the savannas are situated between tropical forests and mountain ranges. In Madagascar and India there is a combination of both conditions.

The transitional (ecotonal) position of the savannas between forest and grassland or semidesert is the basis for the differences in opinion among authors about the size and geographical distribution of savannas. Most authors include, however, the savanna in East Africa and the belt south of the Sahara, the bush veld in South Africa, the Llanos in northern South America, and some types of scrub vegetation in Australia. Some areas in southern Madagascar, on several tropical islands (in leeward position), in Central America, in southern North America, and in India are usually included. The two latter regions, however, are subtropical. Still other areas included in the savanna concept are the Campo Cerrado and parts of the Chaco in South America; portions of the Miombo in southern Africa; wide portions of northern Africa; south of the Sahara; and wide portions of Madagascar, the Indian peninsula, and Australia. Because of the variations in the savanna concept, it is difficult to give a correct estimate of the total surface area

covered by savanna vegetation. The Food and Agriculture Organization (FAO) considers that about one-third of the total land surface is covered by predominantly grassland vegetation. Of this area, one-third can be assumed to be tropical grassland, most of which can be called savanna.

Agricultural practices. Most of the original savanna areas throughout the world are currently farmed. Ranch farming is the predominant type, with sheep being raised in the drier areas and cattle in the moister regions. In the hotter regions zebus are raised, along with several hybrids of zebus and European cattle, which are the preferred stock for this climate. The yield in meat per unit area is low, and even under extensive management it seems to be lower than the meat production of the natural animal herds of the savanna, including antelopes, giraffes, and zebras. Ranch farming does not change the character of the vegetation very much if it is well managed. The adjacent dry woodlands very often resume a physiognomy similar to the natural savannas if good farm management is applied.

Agricultural crops are of many varieties in the savanna areas, where with careful protection and management any crop can be cultivated, provided that enough water is available. Drought-tolerant crops are usually preferred among the perennials.

The majority of human settlements are small in the savanna regions of the world: the hot temperatures during part of or the entire year, together with problems of water supply, limit interest in larger settlements. Settlements are usually found along the rivers, close to the coast, or in the higher elevations. Most of the land is managed by small tribal villages or large plantation or ranch owners, with separate groups of tenants, sharecroppers or employees, or single families.

Extratropical types. The main structural character of tropical savannas is the scattered trees standing within a close cover of herbaceous vegetation. This structural character is also found in several extratropical vegetation types, but these differ greatly in the forces that limit a close tree cover, including drought (areas intermediate between steppe or prairie and forest); excess water or water combined with soil that has a shortage of oxygen and nutrients (peatbogs, marshes, or glades); short vegetation periods because of extended cold temperatures below the freezing point; excessive, long snow covers; and low light intensity (forest tundra, taiga, and cold savanna).

The forest tundra and the forest steppe are usually considered ecotonal units between their adjacent vegetation types which are, respectively, boreal coniferous forest and tundra, or deciduous forest and steppe. The ecology and species populations of these areas are so different from that of tropical savannas that it is hardly desirable to combine extratropical and tropical open woodlands under the heading savanna.

An intermediate condition is exhibited by the savannas and everglades of the southeastern United States. The intermittent soaked or dry conditions of a peaty soil, the tropically hot summers and mild winters (with frost periods, however), and the generally low nutrient level of the soil give these areas the characteristics of the cold savanna-like vegetation types and the tropical and subtropical types.

The economic potential of the three extratropical savannalike areas also varies greatly. The majority of the forest tundra is beyond agricultural exploitation, while the potential for farming in the forest steppe is better than in the steppe. The conditions in the southeastern United States vary. Farming is potentially possible in most regions, but it must be determined by economic considerations whether a given area should be developed.

Some savanna regions in many parts of the world are the last survival territories for many plant and animal species. This implies the need for conserving some savanna pieces, both tropical and extratropical. *See* FOREST AND FORESTRY; GRASSLAND; TAIGA; TUNDRA.   [HELMUT LIETH]

Bibliography: P. M. Dansereau, *Biogeography*, 1957; H. Walter, *Vegetation der Erde*, vol. 1, 1962; A. W. Kuechler, Natural vegetation of the world, Goode's World Atlas, 11th ed., 1960.

Scattering layer

A layer of organisms in the sea which causes sound to scatter and returns echoes. Recordings by sonic devices of echoes from sound scatterers

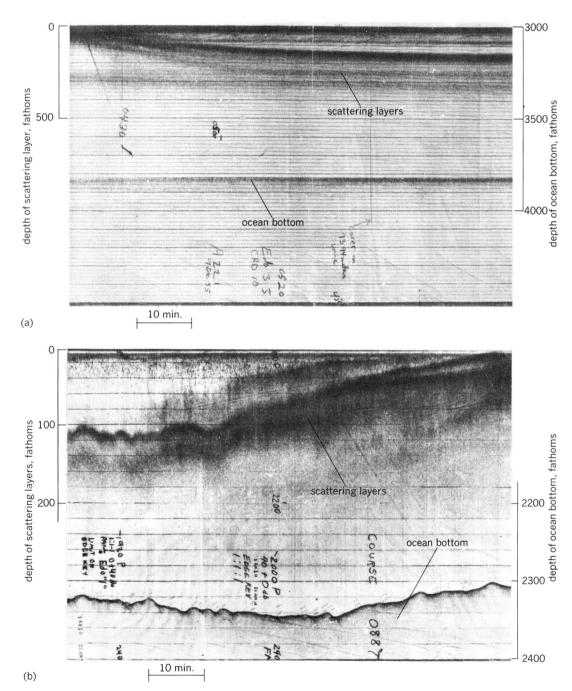

Fig. 1.   Scattering layers recorded by a 12-kHz echo sounder. (*a*) Sunrise descent of deep scattering layers in eastern Pacific off northern Chile. A layer, which appears to have remained at depth throughout the night, is shown near 300 fathoms. (*b*) Sunset ascent of deep scattering layers in western North Atlantic near 40°30'N, 50°W. (*From R. H. Backus and J. B. Hersey, Sound scattering by marine organisms, in M. N. Hill, ed., The Sea, vol. 1, Interscience, 1962*)

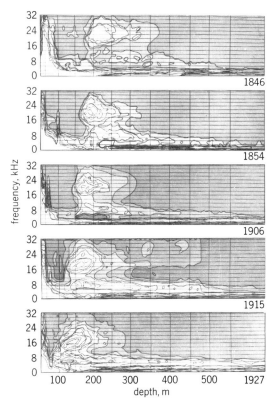

Fig. 2. Sequence of sunset observations showing scattering as a function of depth and frequency. Contours of equal sound level are 2 dB apart, with lightest areas denoting highest levels. Time of day of each observation is indicated by the number at the lower right-hand corner of each record. (*From R. H. Backus and J. B. Hersey, Sound scattering in marine organisms, in M. R. Hill, ed., The Sea, vol. 1, Interscience, 1962*)

indicate that the scattering organisms are arranged in approximately horizontal layers in the water, usually well above the bottom. The layers are found in both shallow and deep water.

Shallow water. In the shallow water of the continental shelves (less than 200 m deep), scattering layers and echoes from individuals or compact groups are very irregularly distributed and are probably made up of a variety of sea animals and possibly plants. While some fishes, notably herring, and some zooplankton have been identified by fishing, many others have not been identified.

Deep scattering layers. In deep water (greater than 200 m) one or more well-defined layers generally are present. Though commonly variable in detail, they are found to be very widely distributed. They are readily detected by echo-sounding equipment operating in the frequency range 3–60 kHz, the sound spectrum of each layer generally having maximum scattering at a somewhat different frequency than others found at the same place. Commonly, but not universally, the deep-water layers migrate vertically in apparent response to changes in natural illumination. The most pronounced migration follows a diurnal cycle, the layers rising at night, sometimes to the surface, and descending to greater depths during the day (Fig. 1). The common range of daytime depths is 200–800 m. The migration is modified by moonlight and has been

observed to be modified during the day by heavy local cloud cover, for example, a squall. Occurence of the layers in deep water was first demonstrated by C. Eyring, R. Christiansen, and R. Riatt.

Deep scattering organisms. All animals and plants, as well as nonliving detritus, contrast acoustically with sea water and, hence, any may be responsible for observed scattering in a particular instance. Many animals and plants have as part of their natural structure a gas-filled flotation organ which scatters sound many times more strongly than would be inferred from the sound energy they intercept. These are the strongest scatterers of their size. M. W. Johnson pointed out in 1946 that the layers which migrate diurnally must be animals that are capable of swimming to change their depth, rather than plant life or some physical boundary such as an abrupt temperature change in the water. In 1953 V. C. Anderson demonstrated that some of the deep-water scatterers have a much smaller acoustical impedance than sea water. This fact fits the suggestion by N. B. Marshall in 1951 that the scatterers may be small fishes with gas-filled swim bladders, many of which are known to be geographically distributed much as the layers are. In 1954 J. B. Hersey and R. H. Backus found that the principal layers in several localities migrate in frequency of peak response while migrating in depth, thus indicating that the majority of scatterers fit Marshall's suggestion (Fig. 2). In deep water, corroboration of Marshall's suggestion has come from several independent combined acoustical and visual observations made during dives of the deep submersibles *Trieste, Alvin, Soucoupe, Deep Star,* and others. Nearly all these dives were made within a few hundred miles of the east or west coasts of the continental United States. The fishes observed most commonly to form scattering layers are the lantern fishes, or myctophids, planktonic fishes that are a few inches long and possess a small swim bladder. The siphonophores (jellyfish) also have gas-filled floats and have been correlated with scattering layers by observation from submersibles. It is not clear whether fishes with swim bladders and siphonophores make up the principal constituents of the deep-water scattering layers in deep-ocean areas, but it is nearly certain that they are not exclusively responsible for all the widely observed scattering.

[JOHN B. HERSEY]

Bibliography: R. H. Backus and J. B. Hersey, Sound scattering by marine organisms, in M. N. Hill (ed.), *The Sea*, vol. 1, 1962; N. B. Marshall, Bathypelagic fishes as sound scatterers in the ocean, *J. Mar. Res.*, 10:1–17, 1951.

Sea, oil pollution of

The problem of oil spillage came to the public's attention following the grounding of the tanker *Torrey Canyon* in March 1967 at the southwest coast of England near the entrance to the English Channel. Subsequent major oil spills such as the Santa Barbara channel California oil spill in January 1969 have further raised the level of concern until today the terminology "oil spill" has become a household word. Few problems have had greater impact on the petroleum industry than those associated with oil spills. This industry in the United States is faced with the problem of supplying the

Oil consumption in North America

Area	Consumption, 1000 bbl/day*		
	1970	1975	1980
North America	15,870	20,800	25,200
Canada	1,500	1,800	2,200
United States	14,370	19,000	23,000

*A barrel equals 42 gallons (U.S.).
SOURCE: *International Petroleum Encyclopedia*. Petroleum Publishing Co., 1972.

ever-increasing demand for oil and petroleum products to customers who are demanding that the oil be supplied without a risk of oil spills. The magnitude of this problem can be appreciated by reference to the table, which shows the demand for oil in North America. In 1971 the oil consumption in North America was at a level of 15,870,000 bbl/day, and this demand is expected to grow to a level of 25,200,000 bbl/day in 1980. In other parts of the world the problem is equally as great. In 1971 the oil shipped by tanker from the Middle East primarily to Europe and Japan reached a level of 14,944,000 bbl/day.

To counter the threat of environmental damage as a result of oil spills, extensive research is being performed in the United States by private industry as well as by the Environmental Protection Agency and the Coast Guard. This research is primarily directed at developing methods to combat oil spills which minimize the damage to the environment. Treating the spilled oil with dispersants was the primary method used to fight oil spills at the time of the *Torrey Canyon*. Dispersants cause oil to spread farther and disperse in a manner similar to the way soap removes oil from one's hands, allowing the oil to be emulsified and washed away with the water. The dispersants used during the *Torrey Canyon* cleanup effort were not developed specifically for use in waters containing marine life and contained aromatic solvents which are toxic. Since that time specific dispersants less toxic to marine life and biota have been developed. Today it is generally accepted that the most extensive damage to marine life resulting from the *Torrey Canyon* incident was caused by the excessive use of dispersants in the coastal zone. In fact, the areas of the shore where dispersants were not used, but which were heavily polluted with oil alone, showed very minimum damage according to J. E. Smith, director of the Plymouth Laboratory, who has studied the biological effects of the *Torrey Canyon* oil spill. At present in the United States regulations severely limit the use of dispersants, and research efforts place emphasis on containment and recovery of oil by mechanical means.

Effects of oil pollution. When oil is spilled on water, it spreads rapidly over the surface. The forces which cause the oil to spread include the force of gravity, which results in the lighter oil seeking constant level by spreading horizontally on the heavier water. A second force is the surface-tension force, which acts at the edge of the oil slick as shown in Fig. 1. It is the surface-tension force which can result in the oil spreading to a thickness approaching a monomolecular layer. This limiting thickness is almost never achieved in large oil spills, however, because the oil interfacial surface tensions change, and the net surface tension becomes negative. The interfacial surface-tension forces change as a result of the natural processes that affect oil. One of the most important natural processes is the evaporation of the oil. Evaporation occurs rapidly, the rate depending upon the nature of the oil, the rate of thinning of the slick, wave intensity, strength of the wind, temperature, and so forth. Crude oil is a mixture of a very large number of components, each with its own properties. The most volatile components evaporate first, but with all crude oils there will undoubtedly be a residue left which is virtually involatile. In addition to evaporation, some of the oil goes into solution with the water, some is oxidized, and some is utilized by microorganisms. The most important of these processes, and the one receiving the most extensive research, is the process of microbial degradation. Many microorganisms present in seas, fresh-water lakes, and rivers have a great capacity to utilize hydrocarbons. The hydrocarbons are used as an energy source and are incorporated into new cell mass. Seeding oil slicks with special bacterial cultures has been suggested to accelerate the rate of microbial decay. However, the rate of microbial degradation of oil is limited not only by the quantity of organisms but by the availability of the oxygen and nutrients needed to support the metabolic process. Acceleration of the natural process by adding nutrients such as phosphorus or nitrogen compounds, particularly in open seas where the nutrients are not naturally available, is presently being considered. Unfortunately the rate of bacterial degradation (accelerated or natural) of floating oil is slow and is therefore not effective if the oil is threatening a coastline.

As with most types of problems, the short-term deleterious effects associated with an oil spill are better understood than the long-term effects. Marine birds, especially diving birds, appear to be the most vulnerable of the living resources to the effects of oil spillage. Harm to birds from contact with oil is reported to be a result of breakdown of the natural insulating oils and waxes which shield the birds from water, as well as due to plumage damage and ingestion of oil. Efforts to cleanse or rehabilitate birds have been generally unsuccessful because of the excessive stress that the bird

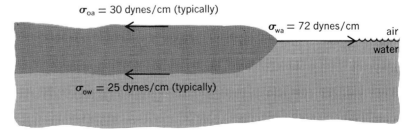

net surface tension force:
$$\sigma = \sigma_{wa} - \sigma_{oa} - \sigma_{ow}$$
$$\sigma = 17 \text{ dynes/cm}$$

Fig. 1.   Cross section of oil-water-air interface of a spreading oil slick showing the relevant surface-tension forces; 1 dyne is equivalent to 10^{-5} N.

experiences. If treatment is prolonged for any reason, most if not all of the birds will die. Shellfish are another segment of marine life directly affected by oil spillage in the coastal zone. Many shellfish have a relatively high tolerance to oil, but their flesh can become tainted for a period subsequent to heavy pollution. Shellfish are particularly vulnerable to most chemical dispersants. Fish are not generally affected by an oil spill because of their mobility which allows them to avoid heavily contaminated areas. The effects of oil on the marine food chain which consists of plants, bacteria, and small organisms is not well understood because of its complexity and because of the wide fluctuations that occur naturally and are independent of the effects of oil. In contrast to the ecological damage, the damage caused by an oil spill which is associated with recreational beach areas, coastline areas used for water sports, and areas where personal property such as docks or boats are located is well understood. There is a large expense associated with the loss of the use of these areas for even a few days.

Cleanup procedures. Experience in attempting to clean up an oil spill has shown that no perfect method exists for all situations. Cleanup methods must be evaluated and chosen on a case-by-case basis. Spills can be more easily dealt with if they are confined to a small area on the water surface. At present, however, confinement devices have not been developed which are successful in all situations. The methods being studied and which are being used to dispose of oil floating on the surface of the sea include mechanical removal of the float-

ing oil, the use of absorbents to facilitate the removal of the oil, sinking the oil, dispersion of the oil, and burning the oil.

Mechanical removal of oil. The primary ingredient of many mechanical oil spill cleanup systems is the use of mechanical booms or barriers. Containment of the oil spill at its source is the most important single action which can be taken when the oil spill is first detected. The use of booms has only had limited success to date because at moderate currents as low as 1 knot many booms have failed by allowing oil to pass below the boom. Figure 2 shows the types of failure which booms can experience in the presence of currents. At low currents, approximately 1.5 knots or less, oil will pile down against the barrier until the boom reaches its capacity. Additional oil will cause the boom to fail as shown in the figure. At higher currents the situation is less manageable since the boom will fail before a large quantity of oil has been collected within the barrier. For the low-current situation booms will perform satisfactorily if the oil is continually skimmed from the region in front of the barrier. Booms also are satisfactory for directing or sweeping oil, provided the angle between the boom and the current or drift direction of the oil is small so that oil does not accumulate along the boom. In addition, the relative velocity of the water at right angles to the boom must be less than the critical velocity for boom failure.

The performance of skimmers depends upon the thickness of the oil. When the film thickness is below ¼ in., many techniques require pumping large amounts of water and very small amounts of oil. However, once the oil-and-water mixture is removed from the water surface, the separation of the oil from the water is easily accomplished by gravity when the oil and water are allowed to settle in a tank. Skimmers now available generally fall into one of two categories. The first, mechanical surface skimming, removes the top layers of the water and oil from the surface. These devices suffer particularly in wave action, where they gulp large amounts of water unless some provision is provided to allow the weir or suction port to follow the water surface. A second type of skimmer operates on the principle of selective wetting of a surface by oil rather than by water. Rotating metal disks or conveyor belts dip into the water surface through the oil slick. When the moving surface is drawn from the water, a surface layer of oil is removed.

Absorbents. Absorbents are used to facilitate the cleaning up of oil spills. When they are applied to the slick, they absorb the oil and prevent it from spreading, and when the absorbent material is removed from the water, the oil is removed. A class of absorbents which is commonly used consists of natural materials such as peat moss, straw, sawdust, pine bark, talc, and perlite. A second class of absorbents is derived from synthetics or plastics such as high-molecular-weight polyethylene and polystyrene, polypropylene, and polyurethane. Of all synthetic absorbents, polyurethane is generally accepted to be the most promising. The natural absorbents are generally less expensive and are attractive whenever there is a chance of losing the absorbent material. The natural absorbents are either inert or biodegrade more quick-

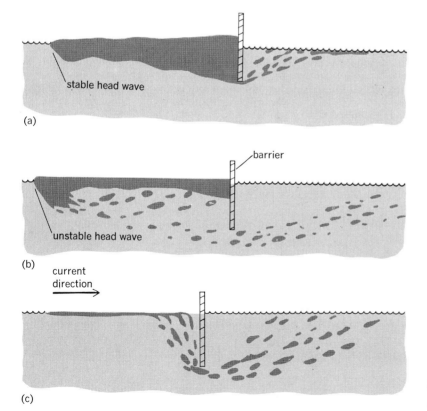

Fig. 2. Types of mechanical boom failure. (a) Low current. (b) Moderate current. (c) High current.

ly than the synthetic materials. Alternatively the synthetic material has a greater buoyancy and a higher affinity for oil. One of the problems with absorbents is distributing them in large enough quantities on the slick. Unless the absorbents can be applied and recovered from the shore, special equipment must be available. One of the most recent concepts including absorbents involves recycling the absorbent material by wringing the oil from the absorbent and returning the absorbent material to the slick. Synthetic absorbents are most suitable for this application, and a feasibility study of such a system for operation in offshore conditions is presently being conducted by the Environmental Protection Agency.

Sinking the oil. A sinking agent which consisted of 3000 tons of calcium carbonate with about 1% of sodium stearate was applied to an oil slick which originated from the *Torrey Canyon* and reportedly resulted in the sinking of about 20,000 tons of oil. The oil was sunk in the Bay of Biscay off the coast of France in 60 to 70 fathoms of water. The sinking of the oil prevented the French coast from being contaminated, and after a period of 14 months no sign of the oil was found. Other materials such as specially treated sand, fly ash, and similar synthetic material have also been used to sink oil. Opinion is still divided as to the possible environmental effects of treating the oil with a dense material and sinking it. Opposition to sinking centers around the fact that sinking the oil reduces the contact surfaces between the oil and air and between the oil and water by preventing natural diffusion of the oil. Hardening of the oil subsequent to sinking would lead to a more persistent and concentrated pollution of the sea bed compared with a lower level of more dispersed pollution on the surface. In any case, utilization of this technique would be most advantageous in deeper waters outside the heavy fishing zones and where there will be a minimum of adverse effects to productive biological life in the coastal zones.

Dispersants. A dispersant is a substance which, when applied to an oil slick, causes the oil to spread farther and disperse. A dispersant contains a surfactant, a solvent, and a stabilizer. The solvent usually comprises the bulk of the dispersant and enables the surface-active agent or surfactant to mix with, and penetrate into, the oil slick and thus form an emulsion. The stabilizer fixes the emulsion and prevents it from coalescing once it is formed. The process of dispersion of the slick is shown in Fig. 3. Dispersion is similar to sinking in that it simply displaces the oil from the water surface rather than removing it from the water altogether. Dispersion has one advantage over sinking in that it increases the slick surface area and allows a rapid increase in the rate of microbial decomposition. Dispersants are not useful in coastal regions because the process of dispersion leads to an increased extent of contamination. In addition, the most effective dispersants use solvents which are toxic to marine life. To reduce this toxic effect, reduced effectiveness must be accepted. Primarily due to the question of toxicity, the use of dispersants in the open sea appears doubtful, although their use there has potential pending additional study and field data.

Burning. Burning oil slicks on the open sea has

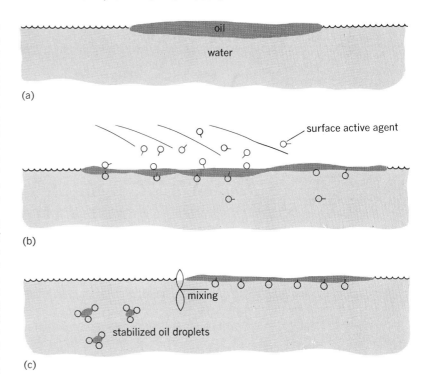

Fig. 3. Mechanism of oil slick dispersion. (*a*) Initial slick. (*b*) Application of chemical dispersant. (*c*) Mixing forms droplets which become stable emusion.

generally met with little success because the more volatile light ends evaporate quickly from the oil slick. Also, the water generally can remove heat faster than it can be created to support the combustion. Various attempts have been made to treat oil slicks to facilitate burning. The most promising of these techniques involves spreading a wicking material over the oil slick which acts to physically separate the flame from the water. Wicking agents also aid in confining the fire to a particular location, but air pollution must be expected when oil is burned in this fashion.

Prevention of oil spills. Although oil spills can probably not be eliminated entirely, steps are being taken to reduce the probability that they will occur. The United States Environmental Protection Agency and others have sponsored studies that apply reliability engineering principles to the problem and that will result in recommending procedures to be adopted to reduce the oil-spill threat. Steps are also being taken to quickly discover oil spills and to develop methods to continually monitor the water where spills are likely so that, in the event of a spill, it can be discovered quickly and proper action can be taken. The United States Coast Guard is sponsoring considerable research toward the development of remote sensing techniques. These studies have shown that, using radar and passive microwave techniques, large areas can be surveyed on a 24-hr basis even in adverse weather conditions. For the purpose of cleaning up oil spills, 67 private cooperatives are now in operation throughout the United States. These cooperatives will become more numerous in the future, judging from the fact that of the 67 in existence, 25 became operational in 1971. Twenty-two new cooperatives are in the planning stage at present. Cooperatives operate in coordination with the Coast Guard, the Army Corps of Engineers, and the Environmental Protection Agency, but in every case the motivating force for their creation has come from private industry. Assessing the ability of these cooperatives, or of any group for that matter, to clean up a large spill is difficult, but it appears that near the shore mechanical techniques can be used effectively to remove oil from the water. In the offshore situation, oil recovery is more complicated and will depend upon the given situation and other environmental conditions encountered. Most current research efforts are directed toward developing systems which will operate effectively in offshore conditions. *See* MARINE ECOSYSTEM.

[ROBERT A. COCHRAN]

Bibliography: D. S. Moulder and A. Varley (eds.), *A Bibliography on Marine and Estuarine Oil Pollution*, Laboratory of the Marine Biological Association of the United Kingdom, 1971; *Oil Containment Systems*, Edison Water Quality Laboratory, 1970; *Proceedings of the Joint Conference on Prevention and Control of Oil Spills*, Washington, D.C., June 15–17, 1971; D. Straughan (ed.), *Biological and Oceanographical Survey of the Santa Barbara Channel Oil Spill 1969–1970*, vol. 1: *Biology and Bacteriology*, 1971.

Sea ice

Ice formed by the freezing of sea water is referred to as sea ice. Ice in the sea includes sea ice, river ice, and land ice. Land ice is principally icebergs which are prominent in some areas, such as the Ross Sea and Baffin Bay. River ice is carried into the sea during spring breakup and is important only near river mouths. The greatest part, probably 99% of ice in the sea, is sea ice. *See* TERRESTRIAL FROZEN WATER.

Properties. The freezing point temperature and the temperature of maximum density of sea water vary with salinity (Fig. 1). When freezing occurs, small flat plates of pure ice freeze out of solution to form a network which entraps brine in layers of cells. As the temperature decreases more water freezes out of the brine cells, further concentrating the remaining brine so that the freezing point of the brine equals the temperature of the surrounding pure ice structure. At temperatures of −23°C

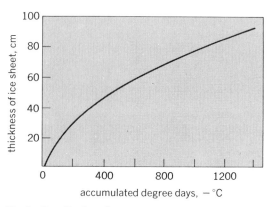

Fig. 2. Growth of undisturbed ice sheet.

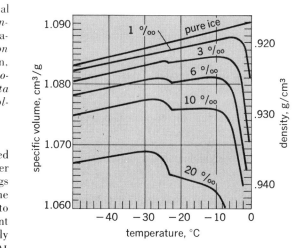

Fig. 3. Specific volume of sea ice for varying salinity and temperature, computed on basis of chemical model. (*By D. L. Anderson, based on data in Arctic Sea Ice, NAS-NRC Publ. no. 598, 1958*)

Fig. 4. Pancake ice interspersed with blocks of young ice. (*U.S. Naval Oceanographic Office*)

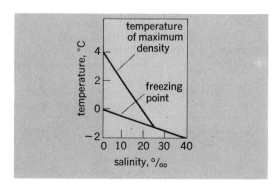

Fig. 1. Change of freezing point and temperature of maximum density with varying salinity of sea water.

Fig. 5.   Honeycombed structure of an overturned rotten block. (*U.S. Naval Oceanographic Office*)

(−9.4°F) or lower some salt in the brine crystallizes out.

The brine cells migrate and change size with changes in temperature and pressure. The general downward migration of brine cells through the ice sheet leads to freshening of the top layers to near zero salinity by late summer. During winter the top surface temperature closely follows the air temperature, whereas the temperature of the underside remains at freezing point, corresponding to the salinity of water in contact. Heat flux up through the ice permits freezing at the underside. In summer freezing can also take place under sea ice in regions where complete melting does not occur. Surface melt water (temperature 0°C) runs down through cracks in the ice to spread out underneath and contact the still cold ice masses and underlying colder sea water. Soft slush ice forms with large cells of entrapped sea water which then solidifies the following winter.

The salinity of recently formed sea ice depends on rate of freezing; thus sea ice formed at −10°C (14°F) has a salinity from 4 to 6 parts per thousand ($^\circ/_{oo}$), whereas that formed at −40°C may have a salinity from 10 to 15$^\circ/_{oo}$. Sea ice is a poor conductor of heat and the rate of ice formation drops appreciably after 4−6 in. are formed. An undisturbed sheet grows in relation to accumulated degree-days of frost. Figure 2 shows an empirical relation between ice thickness and the sum of the mean diurnal negative air temperature (degrees Celsius). The thermal conductivity varies greatly with air bubble content, perhaps between 1.5 and 5.0 ×

Fig. 6.   Hummocky floes that have weathered. (*U.S. Naval Oceanographic Office*)

Fig. 7. Unweathered pressure ridges, formed by rafting of floes. (*U.S. Naval Oceanographic Office*)

10^{-3} cal/(cm)(sec)(°C).

The specific gravity of sea ice varies between 0.85 and 0.95, depending on the amount of entrapped air bubbles. The specific heat varies greatly because changing temperature involves freezing or melting of ice. Near 0°C, amounts that freeze or melt at slight change of temperature are large and "specific heat" is anomalous. At low temperatures the value approaches that of pure ice; thus, specific heat for 4°/oo saline ice is 4.6 cal/(g) (°C) at −2°C and 0.6 at −14°C; for 8°/oo saline ice, 8.8 at −2°C and 0.6 at −14°C.

Sea ice of high salinity may expand when cooled because further freezing out occurs with attendant increase of specific volume, for example, ice of salinity 8°/oo at −2°C expands at a rate of about 93×10^{-4} cm³/g per degree Celsius decrease in temperature, at −14°C expands 0.1×10^{-4}, but at −20°C contracts 0.4×10^{-4} per degree Celsius decrease. Change of specific volume with temperature and salinity is illustrated in Fig. 3.

Sea ice is viscoelastic. Its brine content, which is very sensitive to temperature and to air bubble content, causes the elasticity to vary widely. Young's modulus measured by dynamic methods varies from 5.5×10^{10} dynes/cm² during autumn freezing to 7.3×10^{10} in winter to 3×10^{10} at spring breakup. Static tests give much smaller values, as low as 0.2×10^{10}. The flexural strength varies between 0.5 and 17.3 kg/cm² over salinity range of 7−16°/oo and temperatures −2 to −19°C.

Types and characteristics. The sea ice in any locality is commonly a mixture of recently formed ice, old ice which has survived one or more summers, and possibly old ridges of ice that formed against a coast and contain beach material. The various descriptive forms are shown in Figs. 4−7. Except in sheltered bays, sea ice is continually in motion because of wind and current. It is constantly breaking up: floes are driven together, rafting or piling one on another to form pressure ridges which may reach a thickness of 90 ft in the open sea, or if pressed against a beach may reach a thickness of 120 ft or more. Under wind stress there are always leads, or lanes of open water, which soon close, while others open somewhere else.   [WALDO LYON]

Bibliography: T. Armstrong, B. Roberts, and C. Swithinbank, *Illustrated Glossary of Snow and Ice*, 1966; W. D. Kingery (ed.), *Ice and Snow*, 1963; *Proceedings of the Conference on Arctic Sea Ice*, NAS−NRC Publ. no. 598, 1958; H. Oura (ed.), *Physics of Snow and Ice*, 1967.

Sea state

The description of the ocean surface or state of the sea surface with regard to wave action. Wind waves in the sea are of two types: Those still growing under the force of the wind are called sea; those no longer under the influence of the wind that produced them are called swell. Differences between the two types are important in forecasting ocean wave conditions. Properties of sea and swell and their influence upon sea state are described in this article.

Sea. Those waves which are still growing under the force of the wind have irregular, chaotic, and unpredictable forms (Fig. 1a). The unconnected wave crests are only two to three times as long as the distance between crests and commonly appear

Table 1. Sea height code*

Code	Height, ft	Description of sea surface
0	0	Calm, with mirror-smooth surface
1	0−1	Smooth, with small wavelets or ripples with appearance of scales but without crests
2	1−3	Slight, with short pronounced waves or small rollers; crests have glassy appearance
3	3−5	Moderate, with waves or large rollers; scattered whitecaps on wave crests
4	5−8	Rough, with waves with frequent whitecaps; chance of some spray
5	8−12	Very rough, with waves tending to heap up; continuous whitecapping; foam from whitecaps occasionally blown along by wind
6	12−20	High, with waves showing visible increase in height, with extensive whitecaps from which foam is blown in dense streaks
7	20−40	Very high, with waves heaping up with long frothy crests that are breaking continuously; amount of foam being blown from the crests causes sea surface to take on white appearance and may affect visibility
8	40+	Mountainous, with waves so high that ships close by are lost from view in the wave troughs for a time; wind carries off crests of all waves, and sea is entirely covered with dense streaks of foam; air so filled with foam and spray as to affect visibility seriously
9		Confused, with waves crossing each other from many and unpredictable directions, developing complicated interference pattern that is difficult to describe; applicable to conditions 5−8

*Modified from *Instruction Manual for Oceanographic Observations*, H.O. Publ. no. 607, 2d ed., U.S. Navy Hydrographic Office, 1955.

Table 2. Swell-condition code*

Code	Description	Height, ft	Length, ft
0	No swell	0	0
	Low swell	1–6	
1	Short or average		0–600
2	Long		600+
	Moderate swell	6–12	
3	Short		0–300
4	Average		300–600
5	Long		600+
	High swell	12+	
6	Short		0–300
7	Average		300–600
8	Long		600+
9	Confused		

*Instruction Manual for Oceanographic Observations, H.O. Publ. no. 607, 2d ed., U.S. Navy Hydrographic Office, 1955.

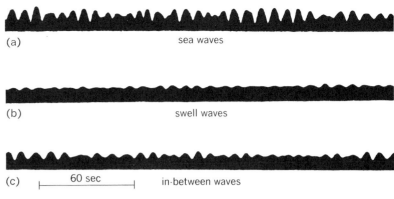

(a) sea waves

(b) swell waves

(c) |⊢ 60 sec ⊣| in-between waves

Fig. 1. Records of surface waves. (a) Sea, (b) swell, and (c) in-between waves. (Adapted from W. J. Pierson, Jr., et al., Observing and Forecasting Ocean Waves, H.O. Publ. no. 603, U.S. Navy Hydrographic Office, 1955)

to be traveling in different directions, varying as much as 20° from the dominant direction. As the waves grow, they form regular series of connected troughs and crests with wave lengths commonly ranging from 12 to 35 times the wave heights. Wave heights only rarely exceed 55 ft. The appearance of the sea surface is termed state of the sea (Table 1).

The height of a sea is dependent on the strength of the wind, the duration of time the wind has blown, and the fetch (distance of sea surface over which the wind has blown).

Swell. As sea waves move out of the generating area into a region of weaker winds, a calm, or opposing winds, their height decreases as they advance, their crests become rounded, and their surface is smoothed (Fig. 1b). These waves are more regular and more predictable than sea waves and, in a series, tend to show the same form or the same trend in characteristics. Wave lengths generally range from 35 to 200 times wave heights.

The presence of swell indicates that recently there may have been a strong wind, or even a severe storm, hundreds or thousands of miles away. Along the coast of southern California long-period waves are believed to have traveled distances greater than 5000 mi from generating areas in the South Pacific. Swell can usually be felt by the roll of a ship, and, under certain conditions, extremely long and high swells in a glassy sea may cause a ship to take solid water over its bow regularly.

A descriptive classification of swell waves is given in Table 2. When swell is obscured by sea waves, or when the components are so poorly defined that it is impossible to separate them, it is reported as confused.

In-between state. Often both sea waves and swell waves, or two or more systems of swell, are present in the same area (Fig. 1c). When waves of one system are superimposed upon those of another, crests may coincide with crests and accentuate wave height, or troughs may coincide with crests and cancel each other to produce flat zones (Fig. 2). This phenomenon is known as wave interference, and the wave forms produced are extremely irregular. When wave systems cross each other at a considerable angle, the apparently unrelated peaks and hollows are known as a cross sea.

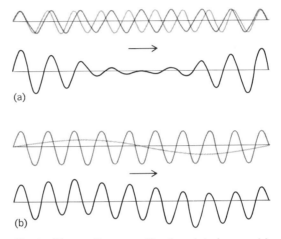

(a)

(b)

Fig. 2. Wave patterns resulting from interference. (a) Interference of waves of equal height and nearly equal length, forming wave groups. (b) Interference between short wind waves and long swell. (From Techniques for Forecasting Wind Waves and Swell, H.O. Publ. no. 604, U.S. Navy Hydrographic Office, 1951)

Breaking waves. The action of strong winds (greater than 12 knots) sometimes causes waves in deeper water to steepen too rapidly. As the height-length ratio becomes too large, the water at the crest moves faster than the crest itself and topples forward to form whitecaps.

Breakers. As waves travel over a gradually shoaling bottom, the motion of the water is restricted and the wave train is telescoped together. The wave length decreases, and the height first decreases slightly until the water depth is about one-sixth the deep-water wave length and then rapidly increases until the crest curves over and plunges to the water surface below. Swell coming into a beach usually increases in height before breaking, but wind waves are often so steep that there is little if any increase in height before breaking. For this reason, swell that is obscured by wind waves in deeper water often defines the period of the breakers.

Surf. The zone of breakers, or surf, includes the region of white water between the outermost breaker and the waterline on the beach. If the sea is rough, it may be impossible to differentiate between the surf inshore and the whitecaps in deep water just beyond.

[NEIL A. BENFER]

Bibliography: W. Bascom, *Waves and Beaches: The Dynamics of the Ocean Surface*, 1964; H. J. McLellan, *Elements of Physical Oceanography*, 1965; G. Neumann and W. J. Pierson, Jr., *Principles of Physical Oceanography*, 1966.

Sea water

Water is most often found in nature as sea water ($\cong 98\%$). The rest is found as ice, water vapor, and fresh water. Sea water is an aqueous solution of salts of a rather constant composition of elements. Its presence determines the climate and makes life possible on the Earth. The boundaries of sea water are the boundaries of the oceans, the mediterranean seas, and their embayments. The physical, chemical, biological, and geological events in the hydroplane within these boundaries are the studies which are grouped together and called oceanography. The basic properties of sea water, the distribution of these properties, the interchange of properties between sea and atmosphere or land, the transmission of energy within the sea, and the geochemical laws governing the composition of sea water and sediments are the fundamentals of oceanography.

The discussion of sea water which follows is divided into six sections: (1) physical properties of sea water; (2) interchange of properties between sea and atmosphere; (3) transmission of energy within the sea; (4) composition of sea water; (5) distribution of properties; and (6) sampling and measuring techniques. For further treatment of related aspects of physical character, composition, and constituents *see* HYDROSPHERE, GEOCHEMISTRY OF; MARINE RESOURCES; SEA-WATER FERTILITY.

PHYSICAL PROPERTIES OF SEA WATER

Sea water is basically a concentrated electrolyte solution containing many dissolved salts. The ratio of water molecules to salt molecules is about 100 to 1. Since nearly all the salt exists as electrically conducting ions, this means that the ratio of water molecules to ions is about 50 to 1; consequently, the ions are on the average no farther than about 10^{-7} cm from each other, a distance equivalent to the diameter of about five water molecules. Because pure water has a relatively open structure with tetrahedral coordination, the water molecules by virtue of their electric dipole moment (arising from the separation of the positive and negative charges) can be readily oriented or polarized by an electric field. The polarizability manifests itself in the high dielectric constant of pure water. Since electrostatic attraction between ions is inversely proportional to the dielectric constant, a high dielectric constant facilitates ionization of electrolytes because of the reduced forces between ions of opposite sign.

Salinity effects. In the neighborhood of ions, extremely high electric fields exist (around 100,000 volts/cm) and water molecules near them become aligned; water molecules that remain in the vicinity of the ions for a long time constitute a hydration shell, and the ions are said to be solvated. The alignment of water molecules produces local dielectric saturation (that is, no further alignment is possible) around the ions, thereby lowering the dielectric constant of the solution below that of pure water.

Details of ion-ion and ion-solvent interactions and their effects on the physical property of solutions are treated in the theory of electrolyte solutions.

As a consequence of the salts in sea water, its physical properties differ from those of pure water, the difference being closely proportional to the concentration of the salts or the salinity. Salinity measurements (which can be conveniently made using electrical conductivity apparatus), along with pressure and temperature data, are used to differentiate water masses. In studying the movement of water masses in the oceans and their

Table 1. Some physical properties of sea water (salinity, 35°/∞) at sea level

Property	Temperature, °C			
	0	10	20	30
Specific volume, cm³/g	0.9726	0.9738	0.9757	0.9784
Isothermal compressibility $\times 10^6$, bars^{-1}	46.7	44.3	42.7	41.7
Thermal coefficient of volume expansion $\times 10^5$, °C^{-1}	5.4	16.6	25.8	33.4
Sound speed, m/sec	1449.4	1490.1	1521.7	1545.7
Electric conductivity $\times 10^3$, (ohm-cm)$^{-1}$	29.04	38.10	47.92	58.35
Molecular viscosity, centipoise	1.89	1.39	1.09	0.87
Specific heat, cal/g°C	0.953	0.954	0.955	0.956
Optical index of refraction $(n - 1.333,338) \times 10^6$, $\lambda = 0.5876\,\mu$	6966	6657	6463	6337
Osmotic pressure, bars	23.4	24.3	25.1	26.0
Molecular thermal conductivity coefficient $\times 10^3$, cal/cm sec °C	1.27	1.31	1.35	1.38

small- and large-scale circulation patterns, including geostrophic flow, properties such as density, compressibility, thermal expansion coefficients, and specific heats need to be known as functions of temperature, pressure, and salinity.

The large value of the osmotic pressure of sea water is of great significance to biology and desalination by reverse osmosis; for example, at a salinity of 35 ‰ (parts per thousand) the osmotic pressure relative to pure water is around 25 atm. Related to osmotic pressure and very important to the formation of ice is the reversal of the freezing point and temperature of maximum density of sea water compared to pure water: The freezing point temperature is lowered to −1.9°C, and the temperature of maximum density is decreased from just below 4°C for pure water to about −3.5°C for 35 ‰ salinity sea water. Some other properties which show significant changes between sea water (salinity 35 ‰) and pure water at atmospheric pressure are shown in Tables 1 and 2. Although the world oceans show a wide range in temperature and salinity, 75% by volume occurs within a range of 0 to 6°C and 34 to 35 ‰ salinity.

Pressure effects. Since the greatest ocean depths exceed 10,000 m and more than 54% of the oceans' area is at pressure above 400 bars, it is necessary to consider the effect of pressure, as well as temperature, on the physical properties of sea water. The pressure corresponding to the maximum ocean depth is about 1100 bars. Table 3 shows the percent change in some properties at 500 and 1000 bars, corresponding to depths of about 5000 and 10,000 m at two temperatures, 0 and 20°C. The unusual pressure dependence of viscosity is a consequence of the open structure of water which is altered by pressure, temperature, and solutes.

Sound absorption. Because electromagnetic radiation can propagate in the ocean for only limited distances, sound waves are the principal means of communication in this medium. The equation for attenuation of the intensity of a plane wave (without geometrical spreading losses) is given by Eq. (1) where I_0 is the initial intensity, and I is the

$$I = I_0 e^{-2\alpha x} \qquad (1)$$

intensity at a distance of x km. Sound absorption in sea water, shown in Table 2, is considerably greater than in fresh water, about 30 times greater between frequencies of 10 and 100 kHz. This arises from a pressure-dependent chemical reaction involving magnesium sulfate with a relaxation frequency around 100 kHz. Another relaxation frequency around 1 kHz has been found in the ocean; the origin of this phenomenon has not been determined. [F. H. FISHER]

Bibliography: A. Bradshaw and K. E. Schleicher, The effect of pressure on the electrical conductance of sea water, *Deep-Sea Res.*, 12:151–162, 1965; L. A. Bromley et al., Heat capacities of sea water solutions at salinities of 1 to 12% and temperatures of 2° to 80°, *J. Chem. Eng. Data*, 12:202–206, 1967; A. Defant, *Physical Oceanography*, vol. 1, 1961; F. H. Fisher, Ion pairing of magnesium sulfate in sea water: Determined by ultrasonic absorption, *Science*, 157:823, 1967; H. W. Menard and S. M. Smith, Hypsometry of ocean basin provinces, *J. Geophys. Res.*, 71:4305–4325, 1966; R. B.

Table 2. Sound attenuation coefficient α in sea water*†

	Sound frequency, Hz			
	100	1000	10,000	100,000
Attenuation coefficient α, km⁻¹	0.00023	0.0069	0.15	6.3

*Depth ~ 1200 m, temperature ~ 4°C, and salinity = 35 ‰.
†After W. H. Thorp, Deep ocean attenuation in the sub- and low-kilocycle-per-second region, *J. Acoust. Soc. Amer.*, 38:648–654, 1965.

Montgomery, Water characteristics of Atlantic Ocean and of world ocean, *Deep-Sea Res.*, 5:134–148, 1958; G. Neumann and W. J. Pierson, Jr., *Principles of Physical Oceanography*, 1966; W. S. Reeburgh, Measurements of electrical conductivity of sea water, *J. Mar. Res.*, 23:187–199, 1965; M. Schulkin and H. W. Marsh, Sound absorption in sea water, *J. Acoust. Soc. Amer.*, 34:864–865, 1962; E. M. Stanley and R. C. Batten, *Viscosity of Sea Water at High Pressures and Moderate Temperatures*, Nav. Ship Res. Dev. Center Rep. no. 2827, 1968; W. D. Wilson, Speed of sound in sea water as a function of temperature, pressure and salinity, *J. Acoust. Soc. Amer.*, 32:641–644, 1960; W. D. Wilson and D. S. Bradley, Specific volume of sea water as a function of temperature, pressure and salinity, *Deep-Sea Res.*, 15:355–363, 1968.

INTERCHANGE BETWEEN SEA AND ATMOSPHERE

The sea and the atmosphere are fluids in contact with one another, but in different energy states — the liquid and the gaseous. The free surface boundary between them inhibits, but by no means totally prevents, exchange of mass and energy between the two. Almost all interchanges across this boundary occur most effectively when turbulent conditions prevail: a roughened sea surface, large differences in properties between the water and the air, or an unstable air column that facilitates the transport of air volumes from sea surface to high in the atmosphere.

Heat and water vapor. Both heat and water (vapor) tend to migrate across the boundary in the direction from sea to air. Heat is exchanged by three processes: radiation, conduction, and evaporation. The largest net exchange is through evaporation, the process of transferring water from sea to air by vaporization of the water.

Evaporation depends on the difference between the partial pressure of water vapor in the air and

Table 3. Percent change of sea water properties at elevated pressures

Property	500 bars pressure		1000 bars pressure	
	0°C	20°C	0°C	20°C
Electrical conductivity	6.76	3.88	11.19	6.49
Specific volume	−2.16	−1.97	−4.00	−2.28
Compressibility	−13.0	−12.0	−24.0	−22.0
Thermal expansion coefficient	224.0	19.0	386.0	34.3
Sound speed	5.85	5.66	12.02	11.08
Sound absorption coefficient		−47.0		−65.0
Molecular viscosity	−3.8	−0.3	−4.7	0.5

the vapor pressure of sea water. Vapor pressure increases with temperature, and partial pressure increases with both temperature and humidity; therefore, the difference will be greatest when the sea (always saturated) is warm and the air is cool and dry. In winter, off east coasts of continents, this condition is most ideally met, and very large quantities of water are absorbed by the air. On the average, 100 g water per square centimeter of ocean surface are evaporated per year.

Since it takes nearly 600 cal to evaporate 1 g water, the heat lost to each square centimeter of the sea surface averages 150 cal/day. This heat is stored in the atmospheric volume but is not actually transferred to the air parcels until condensation takes place (releasing the latent heat of vaporization) perhaps 1000 mi away and 1 week later.

Radiation of heat from the water surface to the atmosphere and back again are both large—of the order of 800 cal/(cm²)(day) according to E. R. Anderson (1953). However, the net flux is out from the sea; it averages about 100 cal/day.

Conduction usually plays a much smaller role than either of the above; it may transfer heat in either direction, but usually it contributes a small net transfer from sea to air. *See* HEAT BALANCE, TERRESTRIAL ATMOSPHERIC; OCEAN-ATMOSPHERE RELATIONS.

Momentum. Momentum can be exchanged between these two fluids by a process related to evaporation, that is, migration of molecules of air or water across the boundary, carrying their momentum with them. However, in natural conditions the more effective mechanism is the collision of "parcels" of the fluids, as distinct from motions of individual molecules. Also, momentum is usually transferred from air to sea, rather than vice versa. Winds whip up waves; these irregular shapes are more easily attacked by wind action than is the flat sea surface, and both waves and currents are initiated and maintained by the push and stress of the wind on the water surface.

[JUNE G. PATTULLO]

Tritium variations. The radioactive hydrogen isotope tritium occurred with an abundance of about 0.4×10^{-12} molecule of HTO per 10^6 molecules of H_2O in oceanic surface waters in the prenuclear era, representing steady-state production of tritium by cosmic rays in the atmosphere. In 1968 the concentration of this molecule was about 10 times higher in surface waters because of nuclear weapon testing. The concentration in deep waters, below the thermocline, has always been too low to be measurable because of the short tritium half-life (12.5 years) compared to oceanic mixing times of the order of hundreds to thousands of years. Studies of tritium variations in present-day ocean waters have been made by Arnold Bainbridge, W. F. Libby, and others. Calculations of mixing rates of surface and deep waters and of atmospheric residence times for tritium can be made from these data. The models used in the various studies of this subject have neglected the molecular exchange of tritium between atmosphere and sea, which is important relative to the simple input by precipitation. Calculations, based on the stable isotope effects in molecular exchange, show that the molecular-exchange input of tritium into the

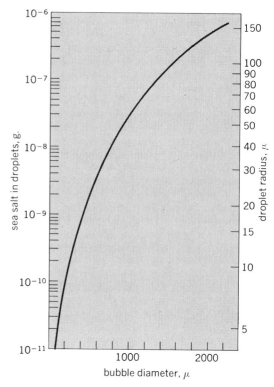

Fig. 2. Approximate relationships between the sizes of the bursting bubbles and the sizes and salt contents of the ejected droplets.

ocean is about 2.5 times greater than the input by precipitation in both the prenuclear and postnuclear epochs. It is therefore necessary to reevaluate the tritium calculations, taking this finding into account.   [HARMON CRAIG]

Projection of droplets. Water, salts, organic materials, and a net electric charge are transferred to the air through the ejection of droplets by bubbles bursting at the sea surface. The exchange of these properties between the sea and the atmosphere is of importance in meteorology and geochemistry. Upon evaporation of the water, the droplet residues are carried great distances by winds. These particles become nuclei for cloud-drop and raindrop formations and probably represent a large part of the cyclic salts of geochemistry.

Air bubbles are forced into surface waters of the sea by wave action, impinging raindrops, melting snowflakes, and other means. The larger bubbles rise to the surface, burst, and eject droplets. Many of the smaller bubbles dissolve before reaching the surface. The photomicrograph (Fig. 1) shows stages in the collapse of a bubble, and the jet and droplet formations which result. The graph (Fig. 2) shows the approximate relationships between the sizes of the bubbles, the sizes of the ejected droplets, and the weight of sea salt in these droplets.

The amounts of water which become airborne as droplets near the sea surface are not known. The best estimates which can now be made (from the limited information about the number and size of bubbles in the sea) range from about 2 to 10 g/(m²)(day) during fresh winds.

The average amounts of sea salt which become

SEA WATER

Fig. 1. Collapse of air bubble and formation of jet and droplets. (a) High-speed motion pictures of stages in the process. (b) Oblique view of jet and droplets from bubble 1.0 mm in diameter.

Table 4. Airborne sea salts in relation to wind force

Beaufort wind force	Concentrations,* µg/m³	Flux,* mg/(m²)(day)	Total,† mg/m²
2–3	2.7	0.42	6.0
4–5	9.9	3.9	11.6
6–7	21.3	24.0	21.8

*At about 500 m.
†Integrated through lowest 2000 m.

airborne at considerable altitudes are shown in Table 4. The total range of observed amounts in individual samples is from about 4×10^{-13}g/ml in a wind of Beaufort force 1 to 10^{-9}g/ml in a wind of force 12. *See* WIND.

Parts of marine organisms are seen in droplets ejected from plankton-rich water. The droplets also become coated with organic monolayers when they arise through contaminated surfaces. During moderate winds in oceanic trade wind areas, organic materials can equal 20 to 30 percent of the airborne sea salt.   [ALFRED H. WOODCOCK]

Bibliography: A. E. Bainbridge, Tritium in the North Pacific surface water, *J. Geophys. Res.*, 68: 3785, 1963; B. B. Benson, *Some Thoughts on Gases Dissolved in the Oceans*, Univ. Rhode Island Occas. Publ. no. 3, 1965; D. C. Blanchard, The electrification of the atmosphere by particles from bubbles in the sea, *Progr. Oceanogr.*, 1:71–202, 1963; H. Craig and L. I. Gordon, Deuterium and oxygen-18 variations in the ocean and marine atmosphere, in E. Tongiorgi (ed.), *Stable Isotopes in Oceanographic Studies and Paleotemperatures*, 1965; H. Craig, L. I. Gordon, and Y. Horibe, Isotopic exchange effects in the evaporation of water, *J. Geophys. Res.*, 68:5079, 1963; H. Craig, R. F. Weiss, and W. B. Clarke, Dissolved gases in the Equatorial and South Pacific Ocean, *J. Geophys. Res.*, 72:6165–6181, 1967; W. Dansgaard, Stable isotopes in precipitation, *Tellus*, 16:436, 1964; G. Ewing and E. D. McAlister, On the thermal boundary layer of the ocean, *Science*, 131(3410): 1374–1376, 1960; I. Friedman et al., The variation of the deuterium content of natural waters in the hydrologic cycle, *Rev. Geophys.*, 2:177, 1964; C. E. Junge, *Air Chemistry and Radioactivity*, 1963; R. Revelle and H. E. Suess, Gases, in M. N. Hill (ed.), *The Sea*, 1962; A. H. Woodcock, Salt nuclei in marine air as a function of altitude and wind force, *J. Meteorol.*, 10:362–371, 1953.

TRANSMISSION OF ENERGY

Electromagnetic and acoustic energy from various natural sources permeates the sea, supplying it with heat, supporting its ecology, and providing for sensory perception by its inhabitants; artificial sources afford man the means for underwater communication and detection.

Light. The primary source of energy which heats the ocean and supports its ecology is light from the Sun. On a clear day as much as 1 kw of radiant power from the Sun and sky may impinge on each square meter of sea surface. Of this power, 4–8% is reflected and the remainder is absorbed within the water as heat or as chemical potential energy

due to photosynthesis. The peak of the irradiation is close to the wavelength of greatest transparency for clear sea water, 480 mµ, but nearly half of the radiant power is infrared radiation which water absorbs so strongly that virtually none penetrates more than 1 m beneath the surface. As much as one-fifth of the incident power may be ultraviolet (below 400 mµ), and this radiation may penetrate a few tens of meters if little or no "yellow substance" (humic acids and other materials associated with organic decomposition) is present. Only a narrow spectral band of blue-green light, representing less than 10% of the total irradiation, penetrates deeply into the sea. This radiation has been detected by multiplier-phototube photometers at depths of more than 600 m. Visibility, important to predators in the feeding grounds of the sea, is possible chiefly because of this blue-green light.

Irradiance. Irradiance on a flat surface oriented in any manner decreases exponentially with depth, as illustrated by Fig. 3, which depicts experimental values of irradiance on an upward-facing surface. Irradiance on any other surface could be represented by a curve parallel (within 5%) to the one shown; irradiance on downward-facing surfaces is approximately one-fiftieth of the irradiance on upward-facing surfaces at all depths.

Absorption. Light, to be useful for heating or for photosynthesis, must be absorbed. The quantity of radiant power absorbed per unit of volume depends upon the amount of power present and the magnitude of the absorption coefficient; to a useful (5%) approximation power absorbed per unit of volume at any depth can be calculated by multiplying the irradiance at that depth, as in Fig. 3, by the slope of the curve expressed in natural log-units per unit of depth, that is, the attenuation coefficient K. Thus in Fig. 3, at a depth of 64 m where

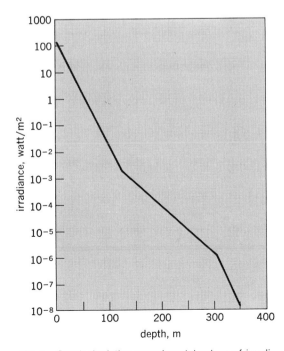

Fig. 3. Graph depicting experimental values of irradiance on an upward-facing surface.

the irradiance is 0.5 watt/m² and is decreasing with depth at the rate of 0.08 natural log-units/m, approximately $0.5 \times 0.08 = 0.04$ watt of radiant power is absorbed by every 1 m³ of sea water.

Visibility. Visibility under water is accomplished by image-forming light (rays) which must pass from the object to the observer without being scattered. The transmission of water for image-forming light is less than for diffused light, since scattering in any direction constitutes a loss of image-forming light, whereas only scattering in rearward directions is a loss for diffused light. Image-forming light is exponentially attenuated with distance, but the attenuation coefficient α averages 2.7 times greater than the attenuation coefficient for irradiance K, defined above. Apparent contrast of an underwater object having deep water as a background is exponentially attenuated with distance, the effective attenuation coefficient being $\alpha + K \cos \theta$, where θ is the inclination angle of the path of sight, and $\cos \theta = 1$ when the observer looks straight down. See discussion of water color and transparency in section on sampling and measuring techniques.   [SEIBERT Q. DUNTLEY]

Compensation intensity and depth. As daylight penetrating into the sea diminishes, the photosynthesis of plants is reduced but respiration remains approximately the same. The light value at which the rates of photosynthesis and respiration are equal is the compensation intensity. The depth at which the compensation intensity is found is the compensation depth. Both of the foregoing have also been termed compensation point, but since ambiguity may occur, it is best to avoid this term. The compensation intensity varies according to the species, the physiological condition of the plants, and other factors, particularly temperatures. Lowered temperature depresses respiration more than photosynthesis. The compensation depth depends upon the intensity of the incident radiation, the transparency of the water, and the

period considered, since illumination varies with time. Compensation intensities of 10–200 foot-candles (ftc) have been measured for phytoplankton and of 17–45 ftc for filamentous algae. Compensation depths for 24-hr periods for phytoplankton range from less than 1 m in turbid water to more than 30 m in coastal areas and to 80 m or more in the clearest tropical waters, and for attached plants, to 50 m along the coast and to 160 m in especially clear water, as in the Mediterranean. Generally the compensation depth is found where daylight is reduced to about 1% of its value at the surface for phytoplankton or about 0.3% for bottom plants. The compensation depth is of particular significance since it marks the lower limit of the photic zone within which green plants can carry on primary production necessary as an energy source for the whole marine ecosystem.

[GEORGE L. CLARKE]

Electromagnetic fields. In sea water, as in any conductor, the electromagnetic behavior is determined by the magnetic permeability μ and the electrical conductivity σ. From these one may find the skin depth δ and the characteristic impedance η given by Eqs. (2) and (3).

$$\delta = (\pi f \mu \sigma)^{-1/2} \qquad (2)$$

$$\eta = (2\pi f \mu / \sigma)^{1/2} \qquad (3)$$

Both δ and η relate to electromagnetic waves of frequency f, for which the wavelength is $2\pi\delta$, the absorption over a path length x reduces the amplitude in the ratio $e^{-x/\delta}$, and the ratio of electric to magnetic field amplitude in a plane wave is η, the former leading in phase by 45°. *See* ELECTROMAGNETIC RADIATION; SKIN EFFECT (ELECTRICITY); WAVE EQUATION.

For sea water, magnetic permeability μ is nearly the same as for free space, and electrical conductivity σ is given to about 1% by Eq. (4), where t is

$$\sigma = [4.00 + a(t - 12)][1 + .0269(S - 35)] \qquad (4)$$

temperature in °C, S is salinity in parts per thousand by weight, and a is .10 for $t > 12$ or $a = .092$ for $t < 12$ (Fig. 4). Using $\sigma = 4.0$ mho/m as a typical value, one obtains Eqs. (5) and (6). These formulas

$$\delta = 250 f^{-1/2} \text{ meter} \qquad (5)$$

$$\eta = .0014 f^{1/2} \text{ ohm} \qquad (6)$$

are expected to hold for all frequencies below about 900 MHz.

Absorption limits the penetration of a field, either inward from a boundary or outward from an electric or magnetic source, to a small multiple of the skin depth δ. A submerged horizontal dipole source near the surface will, however, have a more extensive field in the air and a shallow layer of water. At .01/Hz, δ is 2.5 km; this is a rough upper frequency limit for field fluctuations (such as those of the geomagnetic field) which can penetrate the entire thickness of the ocean layer. At 10 kHz, δ is 2.5 m. Fields of this and higher frequencies, existing, for example, in the natural atmospheric noise, can penetrate only a thin surface layer. For radio signals, the sea acts as an excellent ground plane, involving lower losses than transmission over land.

[PHILIP RUDNICK]

Bibliography: V. M. Albers (ed.), *Underwater*

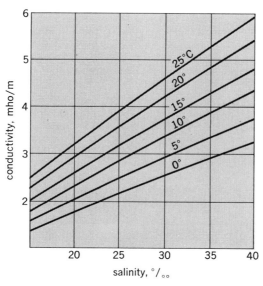

Fig. 4. Electrical conductivity of sea water. (*From B. D. Thomas, T. G. Thompson, and C. L. Utterback, The electrical conductivity of sea water, J. Cons. Perma. Int. Explor. Mer. 9:28–35, 1934*)

Acoustics, 1963; G. L. Clarke, *Elements of Ecology*, 1954; R. W. Holmes, Solar radiation, submarine daylight, and photosynthesis in *Treatise on Marine Ecology and Paleoecology*, Geol. Soc. Amer. Mem. no. 67, vol. 1, 1957; R. H. Lien, Radiation from a horizontal dipole in a semi-infinite dissipative medium, *J. Appl. Phys.* 24(1):1–4, 1953; J. R. Wait, The radiation fields of a horizontal dipole in a semi-infinite dissipative medium, *J. Appl. Phys.*, 24(7): 958–959, 1953.

COMPOSITION OF SEA WATER

The concentrations of the various components of sea water are regulated by numerous chemical, physical, and biochemical reactions.

Inorganic regulation of composition. The present-day compositions of sea waters (Table 5) are controlled both by the makeup of the ultimate source materials and by the large number of reactions, of chemical and physical natures, occurring in the oceans. This section considers the nonbiological regulatory mechanisms, most conveniently defined as those reactions occurring in a sterile ocean. For a discussion of the weathered and weathering substances that give rise to the waters of the world *see* HYDROSPHERE, GEOCHEMISTRY OF.

Interactions between the ions results in the formation of ion pairs, charged and uncharged species, which influence both the chemical and physical properties of sea water. For example, the combination of magnesium and sulfate to form the uncharged ion pair accounts for the marked absorption of sound in sea water. A model accounting for such interactions has been developed for the principal dissolved ions in sea water (Table 6).

pH and oxidation potential. The pH of surface sea waters varies between 7.8 and 8.3, with lower values occurring at depths. The pH normally goes through a minimum with increasing depth in the ocean, and this depth dependence shows a marked resemblance to the profiles of oxygen.

The oxidation potentials of sea-water systems are determined by oxygen concentration in aerobic waters and by hydrogen sulfide concentration in anoxic waters. For 25°C, 1 atm pressure, and a pH of 8, the oxidation potential is about 0.75 volt for waters containing dissolved oxygen in a state of saturation. In anoxic waters the oxidation potentials can vary from about −0.2 to −0.3 volt.

Solubility. Only calcium, among the major cations of sea water, is present in a state of saturation, and such a situation generally occurs only in surface waters. Here its concentration is governed by the solubility of calcium carbonate. Barium concentrations in deep waters can be limited by the precipitation of barium sulfate. The noble gases and dissolved gaseous nitrogen have their marine concentrations determined by the temperature at which their water mass was in contact with the atmosphere and are in states of saturation or very nearly so.

Authigenic mineral formation. The formation and alteration of minerals on the sea floor apparently are responsible for controlling the concentrations of the major cations Na, K, Mg, and Ca. Such clay minerals as illite, chlorite, and montmorillonite are presumably synthesized from these dissolved species and the river-transported weathered solids (aluminosilicates, such as kaolinite) by

Table 5. Chemical abundances in the marine hydrosphere

Element	Concentration, mg/liter	Element	Concentration, mg/liter
H	108,000	Ag	0.0003
He	0.000007	Cd	0.00011
Li	0.17	In	0.000004
Be	0.0000006	Sn	0.0008
B	4.6	Sb	0.0003
C	28	Te	–
N	15	I	0.06
O	857,000	Xe	0.00005
F	1.2	Cs	0.0003
Ne	0.0001	Ba	0.03
Na	10,500	La	1.2×10^{-5}
Mg	1350	Ce	5.2×10^{-6}
Al	0.01	Pr	2.6×10^{-6}
Si	3.0	Nd	9.2×10^{-6}
P	0.07	Pm	–
S	885	Sm	1.7×10^{-6}
Cl	19,000	Eu	4.6×10^{-7}
A	0.45	Gd	2.4×10^{-6}
K	380	Tb	–
Ca	400	Dy	2.9×10^{-6}
Sc	<0.00004	Ho	8.8×10^{-7}
Ti	0.001	Er	2.4×10^{-6}
V	0.002	Tm	5.2×10^{-7}
Cr	0.00005	Yb	2.0×10^{-6}
Mn	0.002	Lu	4.8×10^{-7}
Fe	0.01	Hf	<0.000008
Co	0.0004	Ta	<0.000003
Ni	0.007	W	0.0001
Cu	0.003	Re	0.0000084
Zn	0.01	Os	–
Ga	0.00003	Ir	–
Ge	0.00006	Pt	–
As	0.003	Au	0.00001
Se	0.00009	Hg	0.0002
Br	65	Tl	<0.00001
Kr	0.0002	Pb	0.00003
Rb	0.12	Bi	0.00002
Sr	8.0	Po	–
Y	0.00001	At	–
Zr	0.00002	Rn	0.6×10^{-15}
Nb	0.00001	Fr	–
Mo	0.01	Ra	1.0×10^{-10}
Tc	–	Ac	–
Ru	0.0000007	Th	0.000001
Rh	–	Pa	2.0×10^{-9}
Pd	–	U	0.003

Table 6. Distribution of major cations as ion pairs with sulfate, carbonate, and bicarbonate ions in sea water of chlorinity 19°/oo and pH 8.1*

Ion	Free ion, %	Sulfate ion pair, %	Bicarbonate ion pair, %	Carbonate ion pair, %
Ca^{++}	91	8	1	0.2
Mg^{++}	87	11	1	0.3
Na^+	99	1.2	0.01	–
K^+	99	1	–	–

Ion	Free ion, %	Ca ion pair, %	Mg ion pair, %	Na ion pair, %	K ion pair, %
SO_4^{--}	54	3	21.5	21	0.5
HCO_3^-	69	4	19	8	–
CO_3^{--}	9	7	67	17	–

*From R. M. Garrels and M. E. Thompson, A chemical model for sea water at 25°C and one atmosphere total pressure, *Amer. J. Sci.*, 260:57–66, 1962.

reactions of the type shown in Eq. (7). The SiO_2

$$\text{Aluminosilicates} + SiO_2 + HCO_3^- + \text{Cations}$$
$$= \text{Cation aluminosilicate} + CO_2 + H_2O \quad (7)$$

is introduced in part as diatom frustules. Such a process implies control of the carbon dioxide pressure of the atmosphere.

Table 7. Observed concentrations of some trace metals in sea water*

Ion	Observed sea-water concentration, moles/liter	Limiting compound	Calculated limiting concentration, moles/liter
Mn^{++}	4×10^{-8}	$MnCO_3$	10^{-3}
Ni^{++}	4×10^{-8}	$Ni(OH)_2$	10^{-3}
Co^{++}	7×10^{-9}	$CoCO_3$	3×10^{-7}
Zn^{++}	2×10^{-7}	$ZnCO_3$	2×10^{-4}
Cu^{++}	5×10^{-8}	$Cu(OH)_2$	10^{-6}

*Calculated on the basis of their most insoluble compound.

The formation of ferromanganese minerals, sea-floor precipitates of iron and manganese oxides, may govern the concentrations of a suite of trace metals, including zinc, manganese, copper, nickel, and cobalt. These elements are in highly undersaturated states in sea water (Table 7) but are highly enriched in these marine ores, the so-called manganese nodules, which range in size from millimeters to about 1 m in the form of coatings and as components of the unconsolidated sediments.

Cation and anion exchange. Cation-exchange reactions between positively charged species in sea water and such minerals as the marine clays and zeolites appear to regulate, at least in part, the amounts of sodium, potassium, and magnesium, as well as other members of the alkali and alkaline-earth metals which are not major participants in mineral formations. High charge and large radius influence favorably the uptake on cation-exchange minerals. It appears, for example, that while 65% of the sodium weathered from the continental rocks resides in the oceans, only 2.5, 0.15, and 0.025% of the total amounts of potassium, rubidium, and cesium ions (ions increasingly larger than sodium) have remained there. Further, magnesium and potassium are depleted in the ocean relative to sodium on the basis of data obtained from igneous rock.

The curious fact that magnesium remains in solution to a much higher degree than potassium is not yet resolved but may be explained by the ability of such ubiquitous clay minerals as the illites to fix potassium into nonexchangeable or difficultly exchangeable sites.

Similarly, anion-exchange processes may regulate the composition of some of the negatively charged ions in the oceans. For example, the chlorine-bromine ratio in sea water of 300 is displaced to values around 50 in sediments. Such a result may well arise from the replacement of chlorine by bromine in clays; however, the meager amounts of work in this field preclude any unqualified statements.

Physical processes. Superimposed upon these chemical processes are changes in the chemical makeup of sea water by the melting of ice, evaporation, mixing with runoff waters from the continents, and upwelling of deeper waters. The net effects of the first three processes are changes in the absolute concentrations of all of the elements but with no major changes in the relative amounts of the dissolved species.

Changes with time. Changes in the composition of sea water through geologic time reflect not only differences in the extent and types of weathering processes on the Earth's surface but also the relative intensities of the biological and inorganic reactions. The most influential parameter controlling the inorganic processes appears to be the sea-water temperature.

Changes in the abyssal temperatures of the oceans from their present values of near 0 to 2.2, 7.0, and 10.4°C in the upper, middle, and lower Tertiary, respectively, have been postulated from studies on the oxygen isotopic composition of the tests of benthic foraminifera. Such temperature increases would of necessity be accompanied by similar ones in the surface and intermediate waters. One obvious effect from the recent cooling of the oceans is an increase in either or both of the calcium and carbonate ions, since the solubility product of calcium carbonate has a negative temperature coefficient. Similarly, the saturated amounts of gases that can dissolve in sea water in equilibrium with the atmosphere increase with decreasing water temperatures.

[EDWARD D. GOLDBERG]

Biological regulation of composition. In the open sea all the organic matter is produced by the photosynthesis and growth of unicellular planktonic forms. During this growth all the elements essential for living matter are obtained from the sea water. Some elements are present in great excess, such as the carbon of CO_2, the potassium, and the sulfur (as sulfate). Other elements—for example, phosphorus, nitrogen, and silicon—are present in small enough quantities so that plant growth removes virtually all of the supply from the water. During photosynthesis, as these elements are being removed from the water, oxygen is released.

The biochemical cycle. The organic matter formed by photosynthesis and growth of the unicellular plants may be largely eaten by the zooplankton, and these in turn form the food for larger organisms. At each step of the food chain a large proportion of the eaten material is digested and excreted, and this, along with dead organisms, is decomposed by bacterial action. The decomposition process removes oxygen from the water and returns to the water those elements previously absorbed by the phytoplankton.

The distribution of oxygen and essential nutrient elements in the sea is modified by the spatial separation of these biological processes. Photosynthesis is limited to the surface layers of the ocean, generally no more than 100 m or so of depth, but the decomposition of organic material may take place at any depth. Reflecting this separation of processes, the concentration of nutrient elements in the surface is low, rises to maximum values at intermediate depths (300–800 m), and decreases slightly to fairly constant values which extend nearly to the bottom. Frequently, a slight increase in the concentration of essential elements is observed near the bottom. The oxygen distribution is the opposite of the one just described, with high values at the surface, a minimum value at mid-depth, and intermediate values in the deep water. The oxygen-minimum–nutrient-maximum level in the ocean is the result of two processes working simultaneously. In part it is formed by the decomposition of organic matter sinking from the surface, and in part it results from the fact that this water was originally at the surface in high latitudes, where it contained organic matter and subsequently cooled, sank, and spread out over the oceans at the appropriate density levels.

Because of the nearly constant composition of marine organisms, the elements required in the formation of organic material vary in a correlated way. Analyses of marine organisms indicate that in their protoplasm the elements carbon, nitrogen, and phosphorus are present in the ratios of 100:15:1 by atoms. In the production of organic matter these elements are removed from the water in these ratios, and during the decomposition of organic matter they are returned to the water in the same ratios. However, since the decomposition of organic material is not an instantaneous process which releases all elements simultaneously, it is not unusual to find different ratios of concentration of these elements in the sea. Particularly in coastal waters and in confined seas the ratio of concentration of nitrogen to phosphorus, for example, may differ widely from the 15:1 ratio of composition within the organisms.

The biochemical circulation. Unlike the major elements in sea water, the concentrations of these nutrients are widely different in different oceans of the world. Pacific Ocean water contains nearly twice the concentration of nitrogen and phosphorus found at the same depth in the North Atlantic, and intermediate concentrations are found in the South Atlantic and Antarctic oceans. The lowest concentrations for any extensive body of water are found in the Mediterranean, where they are only about one-third of those in the North Atlantic.

These variations can be attributed to the ways in which the water circulates in these oceans and to the effect of the biological processes on the distribution of elements. The Mediterranean, for example, receives surface water, already low in nutrients, from the North Atlantic and loses water from a greater depth through the Straits of Gibraltar. While the water is in the Mediterranean, the surface layers are further impoverished by growth of phytoplankton, and the organic material formed sinks to the bottom water and is lost in the deeper outflow. A similar process explains the low nutrient concentrations in the North Atlantic, which receives surface water from the South Atlantic and loses an equivalent volume of water from greater depths. *See* SEA-WATER FERTILITY.

[BOSTWICK H. KETCHUM]

Buffer mechanism. The constituents of sea water include a number of cations, all of which are weak acids, and a smaller number of anions, some of which are strong bases. Thus sea water is always somewhat on the alkaline side of neutrality, ranging in pH roughly between the limits of 7.5 and 8.3.

In chemical oceanography the term alkalinity is used to denote not the concentration of hydroxyl ions, as might be expected, but the concentration of strong bases. Alkalinity can be defined as the number of equivalents of strong acid required to convert stoichiometrically all the strong bases to weak acids.

The addition or subtraction of weak acids therefore does not affect the alkalinity of sea water, although through the operation of the buffer mechanism it changes the pH.

The principal weak acid in sea water is carbonic, resulting from the hydrolysis of dissolved carbon dioxide. Boric acid is also present in significant amounts. Salts of these two weak acids are the strong bases which make up the alkalinity. Total combined boron, whether as acid or borate, is in virtually constant proportion to chlorinity, the ratio of boron to chlorinity being about 0.00024, which is equivalent to a specific ratio of boric acid to chlorinity of about 0.022 (concentrations in millimoles per liter).

The total alkalinity is more variable in the ocean. Near the surface it can be increased by addition of dissolved carbonates in river discharge or decreased by the precipitation of lime in the formation of coral and shells. In deeper layers it can be increased by the solution of calcareous debris sinking from the surface. F. Koczy (1956) gives a thorough discussion of the variation of specific alkalinity in the oceans. The average ratio of alkalinity to chlorinity is about 0.120 for surface sea water, increasing somewhat with depth (concentrations in milliequivalents per liter).

Even more variable than the alkalinity is the total dissolved CO_2. At the surface, CO_2 moves between the sea and the atmosphere, and in the euphotic zone CO_2 is removed by photosynthesis to be incorporated into organic matter. Below the euphotic zone, CO_2 is regenerated through the biological oxidation of organic matter. Total CO_2 thus typically may vary from 2.0 or 2.1 mM/liter at the surface to 2.8 or more at the depth of the oxygen minimum.

Some of this CO_2 is in physical solution, and some is undissociated carbonic acid; but most of it is in ionic form, mainly bicarbonate ion with some carbonate. The proportions of the various forms are governed by the dissociation constants of carbonic and boric acids, which vary with temperature and pressure, and by the activity coefficients of the various ions concerned, which vary with temperature, pressure, and salinity. In practice, the activity coefficients and the dissociation constants are combined into apparent dissociation constants, which are tabulated by H. W. Harvey (1957) as functions of temperature and salinity at 1 atm pressure.

[JOHN LYMAN]

Bibliography: R. M. Garrels and M. E. Thompson, A chemical model for sea water at 25°C and one atmosphere total pressure, *Amer. J. Sci.*, 260: 47–66, 1962; H. W. Harvey, *Chemistry and Fertility of Sea Waters*, 1957; M. N. Hill (ed.), *The Sea*, vol. 2, 1963; F. Koczy, The specific alkalinity, *Deep-Sea Res.*, 3:279–288, 1956; J. Lyman and R. B. Abel, Chemical aspects of physical oceanography, *J. Chem. Educ.*, 35:113–115, 1958; F. T. MacKenzie and R. M. Garrels, Chemical mass balance between rivers and oceans, *Amer. J. Sci.*, 264:507–525, 1966; A. C. Redfield, The biological control of chemical factors in the environment, *Amer. Sci.*, 46(3):205–221, 1958.

DISTRIBUTION OF PROPERTIES

The distribution of physical and chemical properties in the ocean is principally the result of the following: (1) radiation (of heat); (2) exchange with the land (of heat, water, and solids such as salts) and with the atmosphere (of water, salt, heat, and dissolved gases); (3) organic processes (photosynthesis, respiration, and decay); and (4) mixing

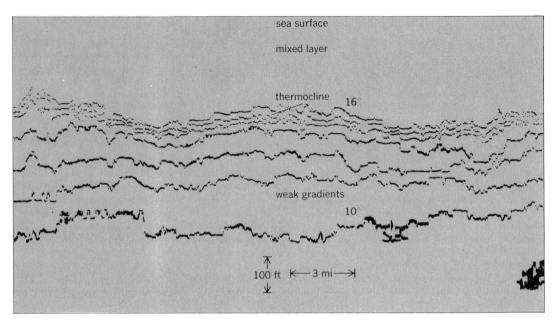

Fig. 5.   Thermal structure most commonly found in the sea.

and stirring processes. These processes are largely responsible for the formation of particular water types and ocean water masses.

Detailed thermal structure. A detailed knowledge of the sea's thermal structure and its relation to oceanic dynamic processes has been accumulated. The details of two-dimensional structure measurement of the sea have been provided by the temperature structure profiler. Although thermal changes occur during the towing of the profiler, the spatial plot is realistic since heat transfer at the surface is negligible, and advective changes proceed more slowly than the 6-knot movement of the ship.

In Fig. 5 the vertical scale represents depth from the surface to 800 ft. The horizontal scale may be interpreted as either time or distance, be-

cause the towing ship has a constant speed of 6 knots. Since the vertical scale is about 100 times the horizontal, the isotherm lines appear to be steeper than they are. Each isotherm is a whole-degree Celsius from 16 to 10°. This example reflects the common, mixed layer, without isotherms, found from the surface to a depth, in this case, of 260 ft. Below these are the closely spaced isotherms that compose the sharp thermocline.

Isotherm oscillations. The profiler made possible a determination of detailed horizontal changes in isotherms. An important large-scale feature revealed is the vertical, wavelike undulation in the main thermocline, with wave heights of 50–100 ft and crests about every 12 mi. Smaller oscillations are conspicuous on all isotherms. These are the ever-present internal waves, which occur on the average thermocline at the rate of 1–4 per mile. Their height varies inversely with the strength of the thermocline and usually amounts to 3–20 ft.

The two-dimensional operation discloses the complexity of the thermal structure in the sea. From the data acquired, it is possible to derive quantitative information from the slope of isothermal surfaces. By scaling the depth of isotherms every 304 ft, it is possible to obtain the angle that the isotherm makes with the horizontal (Table 8).

The isotherms in the thermocline are unusually flat, with a median slope of 17 min. By use of a power spectrum, it was found that the wavelengths vary widely. There are frequently several peaks in the power spectrum corresponding to wavelengths between 0.25 and 1.0 mi. Repeated peaks occur at 0.3 and 0.7 mi. The greatest power is in the long-period waves. Other analyses from data collected in a single place reveal (Fig. 6) that vertical changes in the isotherms have definite cycles corresponding to the short-period internal waves (5–8 min), seiches (14 and 20 min), tidal phenomena (6, 8, 12, and 24 hr), meteorological (3–5 days), lunar cycles (14 and 28 days), and seasonal cycles (1 year).

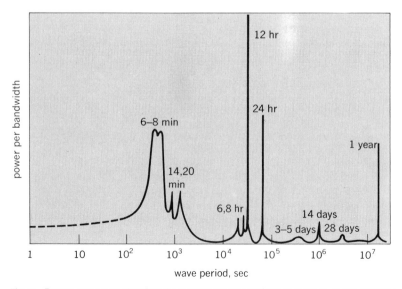

Fig. 6.   Power spectrum showing the most common oscillation periods in the internal thermal structure of the ocean.

It is believed that the internal waves in the open sea move in different and changing directions. Analyses of tows in different directions have provided information on the direction of propagation of dominant waves. Up to 200 mi off the coast of southern California the dominant direction is shoreward.

An attempt has been made to measure a three-dimensional structure. This is done by towing the profiler in circles and box patterns. Preliminary data indicate that the ocean-temperature structure possesses numerous small thermal domes or humps moving in different directions.

Circulation and boundary influences. The detailed vertical thermal sections reveal the dynamic processes going on in the sea. Domes or ridges in the thermal structure, caused by eddies or oppositely flowing adjacent currents, have been observed to be 150 ft in height within a distance of 6 mi. A similar change in the depth and character of a thermocline has been found across major current systems. Areas of upwelling are apparent where the thermocline tilts upward toward the surface and generally becomes weaker.

The structure is also influenced by the boundaries of water masses and land features. A strong horizontal thermocline may serve as a vertical boundary. If one water layer moves in a different direction from another on the opposite side of the thermocline, a large-scale turbulence, with temperature inversion, can occur (Fig. 7).

Vertical water mass boundaries, or fronts, are detected easily by taking detailed two-dimensional profiles. The isotherms extend to the surface to form a horizontal thermal gradient with turbulence along the thermal boundary.

Islands, points of land, and shoals also influence thermal structure and internal waves. The progressive nature of internal waves is slowed down over shoals, where they have a shorter wavelength. This happens when the waves approach shore, where they not only become more closely spaced but also refract and move shoreward with long crests. Another effect on the thermal structure is created by the tide, which causes a large mass of water to move in and out across the continental shelf off California. The colder, heavier water mass coming shoreward along the bottom is affected by the bottom friction to form a wall-like front (Fig. 8). This is essentially an internal tidal bore, which occurs at flooding spring tides. The front of this bore contains turbulence and sometimes weak thermal inversions. It is also evident that the strength of the thermocline changes with tide. When internal

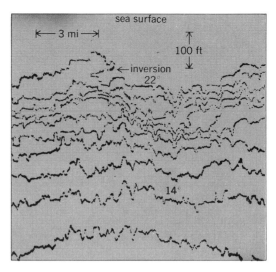

Fig. 7. Extensive turbulence with temperature inversion.

waves approach the surface boundary, their crests become flat and their troughs sharp. If they converge with the sea floor, the troughs grow flat, and the crests become steep and sharp.

[E. C. LA FOND]

Stirring and mixing processes. Stirring and mixing are processes of prime importance in determining the distribution of properties and in the formation of water masses in the ocean. Stirring refers here to motions which increase the average magnitude of the gradient of a property throughout a specified region. Mixing is employed in a narrow sense to denote the molecular diffusion or conduction processes by which gradients are decreased. Stirring and subsequent mixing can decrease the gradients in a region much more quickly than molecular processes acting alone. Combined stirring and mixing are often called turbulent diffusion.

Stirring motions involve shearing or more precisely, deformations of the water. Stirring scales range from those of the permanent currents down nearly to molecular scale. Smaller-scale motions, often highly complex, are the most effective in stirring.

With very important exceptions noted below, the preferred direction of stirring is along surfaces

Table 8. Slope of an isotherm*

Slope, minutes from horizontal	Observations, %
0–10	34
10–20	20
20–30	11
30–40	7
40–50	6
50–60	5
>60	17

*Based on 60,000 depth observations.

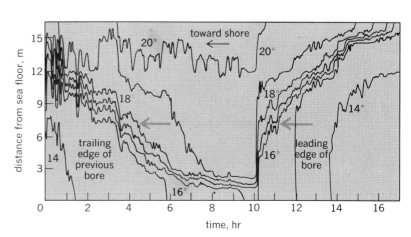

Fig. 8. Thermal structure changes caused by an internal tidal bore.

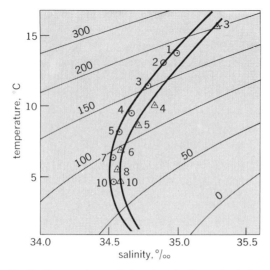

Fig. 9. Temperature-salinity values for Carnegie station 40 (circles) and Dana station 3756 (triangles); see Fig. 11 for locations. Depths of observations in hectometers (100 m). Dashed lines represent definition of Pacific equatorial water. Solid lines represent specific volume as thermosteric anomaly in centiliters (.01 liter) per ton.

the stabilizing forces to produce and maintain a shallow, homogeneous top layer over much of the ocean. Stirring induced at the surface penetrates across the potential density surfaces just below the homogeneous layer but is damped out as it extends deeper into the thermocline or halocline layer. Thus, at some depth in these layers, the more usual stirring along potential density surfaces becomes dominant. In great depths and in high latitudes, the dominance is less strong since the vertical gradient of potential density is small. Stirring across potential density surfaces may also be brought about by tidal currents, but these are strong only in the shallow, coastal parts of the ocean.

Because of the variety and complexity of stirring motions, no general method of treating them quantitatively has proved really satisfactory. It is often expedient to assume that stirring and mixing follow rules analogous to those for molecular diffusion. Sometimes analysis of the motions in detail is possible, although difficulties in both theory and observation are great. Statistical treatment of the detailed motion seems to provide the most realistic approach.

[JOHN D. COCHRANE]

Ocean water masses. Ocean water masses are extensive bodies of subsurface ocean water characterized by a relatively constant relationship between temperature and salinity or some other conservative dissolved constituent. The concept was developed to permit identification and tracing of such water bodies. The assumption is made that the characteristic properties of the water mass were acquired in a region of origin, usually at the surface, and were subsequently modified by lateral and vertical mixing. The observed characteristics in place thus depend both on the original properties and on the degree of modification en route to the region where observed.

A water mass is usually defined by means of a characteristic diagram, on which temperature or some other thermodynamic variable is plotted against an expression for the amount of one component of the mixture. A point on such a diagram defines a water type (representing conditions in the region of origin), and the line between points observes, at least approximately, the property of mixtures; that is, on the line connecting the two points the proportion lying between the point representing one water type and the point for any mixture equals the proportion of the second water type in the mixture. The resulting curve for a vertical water column has been called a characteristic curve, because for a given water mass its shape is invariant, regardless of depth. The existence of such a curve implies continuous renewal of water types, since otherwise mixing would lead to homogeneous water, represented by a point on the diagram.

Temperature-salinity relationships. In oceanography the characteristic diagram of temperature against salinity is usually used in studies of water masses, and the resulting temperature-salinity curve (the *T-S* curve) is used to define a water mass (Fig. 9).

In drawing such a curve, data from the upper 100 m are usually omitted because of seasonal variation and local modification in the surface

of constant potential density. (Potential density is the density of water when brought adiabatically to atmospheric pressure.) The preference is due to the fact that potential density almost always increases with depth. Under this condition, vertical displacement of a water parcel from a potential density surface is resisted by a stabilizing force proportional to the vertical gradient of potential density.

At the sea surface, powerful mechanisms exist which induce stirring across potential density surfaces. These are the wind stresses and the thermal convections resulting from cooling and evaporation. Near the sea surface, they greatly outweigh

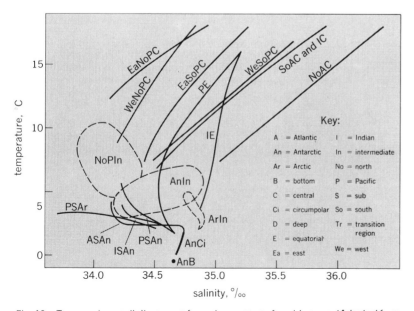

Fig. 10. Temperature-salinity curves for water masses of world ocean. (*Adapted from H. U. Sverdrup, R. H. Fleming, and M. W. Johnson, The Oceans, Prentice-Hall, 1942*)

layer, so that strictly speaking, a water mass as defined extends only to within 100 m of the sea surface. Although ideally a water mass is defined by a single *T-S* curve, because of random errors in field measurements and perhaps fine structure in the water mass itself, in practice an envelope of values provides a more useful definition.

On the *T-S* diagram any property which is a function only of temperature and salinity can be represented by the appropriate family of isopleths (such as values at constant pressure of density expressed as σ_t, or thermosteric anomaly, sound speed, and saturation concentration of dissolved gases). Therefore, the *T-S* diagram with isopleths can be used to determine values of such temperature-salinity dependent functions. Since the ocean is inherently stable (that is, density increases monotonically with depth), examination of the slope of a *T-S* curve (on which depth is indicated) relative to the isopleths of density permits an estimate of the vertical distribution of stability. The diagram is often useful for the detection of faulty observations and as a guide to interpolation on neighboring stations. When a uniform series of data is available, it can also be used for the quantitative representation of the frequency distribution of water characteristics.

Water-mass types. The most important and best-established water masses (characterized by *T-S* curves in Fig. 10; distribution shown in Fig. 11) occur in the upper 1000 m of the ocean. These are of three general types: (1) polar water, present south of 40°S in all oceans, and north of 40°N in the Pacific; (2) central water, occurring at mid-latitudes over most of the world ocean; and (3) equatorial water, present in the equatorial zones of the Pacific and Indian oceans.

Polar waters, including the Subarctic, Subantarctic, and Antarctic Circumpolar water masses, are formed at the surface in high latitudes and thus are cold and have relatively low salinity. Subantarctic water is bounded in the south by the well-defined Antarctic Convergence, south of which circumpolar water is found; Subarctic water has no clearcut northern boundary.

The central water masses appear to sink in the regions of the subtropical convergences (35–40°S and N), where during certain seasons of the year horizontal *T-S* relations at the surface are similar to the vertical distributions characteristic of the various water masses. The great differences in their properties are attributed to differences in the amounts of evaporation and precipitation, heating and cooling, atmospheric and oceanic circulation, and the distributions of land and sea in the source regions.

The widespread and well-defined equatorial waters (Fig. 9) separate the central water masses of the Indian and Pacific oceans. These equatorial water masses are apparently formed by subsurface mixing at low latitudes, although the place and manner of their formation is not well known.

Intermediate waters underlie the central water masses in all oceans. Antarctic Intermediate Water sinks as a water type along the Antarctic Convergence; the water mass then formed by subsequent mixing is characterized by a salinity minimum. Arctic Intermediate Water, of little importance in the Atlantic, is widespread in the Pa-

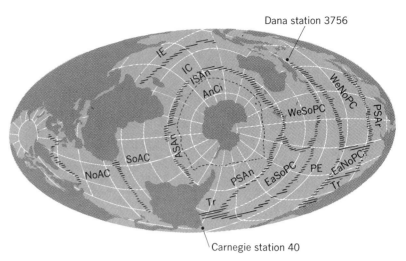

Fig. 11. Distribution of representative water masses of upper 1000 m (symbols as in Fig. 10). Dashed line around Antarctica represents Antarctic Convergence. (*Adapted from H. U. Sverdrup, R. H. Fleming, and M. W. Johnson, The Oceans, Prentice-Hall, 1942*)

cific and is apparently formed northeast of Japan. Other important intermediate water masses are formed in the Atlantic and Indian oceans by addition of Mediterranean and Red Sea water, respectively. Deep and bottom waters of the world ocean are formed in high latitudes of the North Atlantic, South Atlantic (Weddell Sea), and Indian oceans.

[WARREN S. WOOSTER]

Isotopic variations. Studies of the isotopic composition of ocean water have characterized the major deep-water masses and investigated their origin and mixing, have investigated isotopic exchange with dissolved ions, and have been concerned with the geological history of the sea. Other studies have been concerned with the exchange of moisture between the atmosphere and the sea, and the nature of the boundary layer at the air-sea interface.

Variations in the ratios of the stable isotopes are uniformly reported in terms of the ratio in an arbitrarily defined standard mean ocean water (SMOW), whose isotopic composition is numerically defined relative to a water standard distributed by the International Atomic Energy Agency in Vienna, Austria. Absolute concentrations of the isotopes are much more difficult to measure than variations in ratio, but the absolute concentrations of SMOW are believed to correspond to a D/H ratio of 1/6328, or 158 parts per million (ppm) of D, and an O^{18}/O^{16} ratio of 1993×10^{-6}, or 1989 ppm of O^{18}. The isotopic variations are reported as δ values relative to SMOW, defined from the relation $R/R_{\text{SMOW}} = 1 + \delta$, where R is the ratio D/H or O^{18}/O^{16}. The delta values are always given in parts per thousand (°/oo), the same units in which salinity is recorded by oceanographers.

The isotopic variations in surface ocean waters reflect the net effects resulting from the overall balance between precipitation, evaporation, molecular exchange, and mixing with other surface and deep waters. In equatorial latitudes and in high latitudes, precipitation exceeds evaporation, while the reverse is true in the latitudes of the trade winds. The isotopic relationships are best

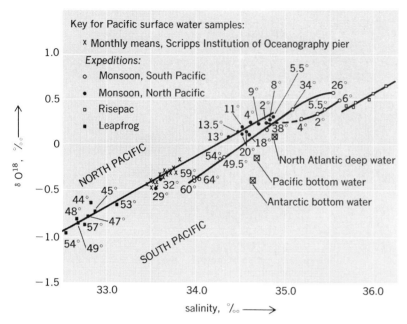

Fig. 12. Diagram illustrating the relationship between the O¹⁸ content and salinity in the surface and deep waters of the Pacific, Antarctic, and North Atlantic oceans.

seen by plotting the δ values versus salinity, as shown in Fig. 12.

The δ-S relationships are seen to have different slopes in the North and South Pacific surface waters because of the varying intensities of the different processes operating. At higher latitudes the relationships are approximately linear in regions where precipitation exceeds evaporation, but in low latitudes the relationships become very complicated in passing through the trade-wind regions to the equatorial zone of high precipitation. These variations can be related quantitatively to the net evaporation, precipitation, and mixing rates in local regions if mixing models are used in which the number of equations does not exceed the variables, and preliminary models along these lines have been made.

Deuterium-O¹⁸ relationships. In equatorial and temperate latitudes, surface ocean waters show simple linear relationships between D and O¹⁸ of the form $\delta D = M\delta O^{18}$, with zero intercept and a slope M which decreases with increasing ratio of evaporation to precipitation in a region. Characteristic values of the slope M are 7.5 in the North Pacific, 6.5 in the North Atlantic, and 6.0 in the Red Sea.

Freezing effects. The freezing of sea water concentrates salt in the liquid so that water of high salinity tends to form and sink. This process is especially important in the formation of Antarctic bottom water on, for example, the Weddell Sea Shelf in the Atlantic portion of Antarctica. The isotopic separation factors in the freezing process are so small (1.0203 for D and 1.0027 for O¹⁸) that very little change in isotopic composition is produced. Ocean waters generated by such a process will thus be expected to lie along an almost horizontal line, parallel to the abscissa in a plot, such as Fig. 12, of isotopic composition against salinity. This has been observed in measurements on both Arctic and Antarctic waters.

For example, in Fig. 12 the Antarctic Bottom Water plotted in the diagram is related to the points shown for samples of South Pacific surface waters in latitudes 59–64°S. Monsoon, Risepac, and Leapfrog refer to Scripps Institution of Oceanography (SIO) expeditions. Risepac samples were along an east-west track in the South Pacific between the Society Islands and South America. The Antarctic Bottom Water samples represented were actually taken from the Weddell Sea area. The surface waters in comparable latitudes in the Atlantic plot at the same points shown for the South Pacific surface waters at 59–64°S, because all these waters are restricted to the same isotopic composition and salinity by the rapid circumpolar circulation around Antarctica.

In the Weddell Sea region, however, the points for surface and intermediate waters connecting the high-latitude surface waters and the Antarctic Bottom Water point are parallel to the abscissa in Fig. 12, indicating that the bottom water is directly generated by the freezing process. This is completely in accord with the classical picture worked out by oceanographers. There is no evidence, either from classical or isotopic data, that this process operates to a significant extent anywhere else in the oceans, although it may also operate on the Ross Sea Shelf, where no detailed winter studies have yet been made.

Deep-water relationships. The oxygen-18 values for the core waters in the major deep-ocean-water masses are shown in Fig. 12. Indian Ocean Deep Water is indistinguishable from Pacific Bottom Water, and the Pacific Deep Water has this composition over the entire Pacific Basin, except in very high southern latitudes where mixing with Antarctic Bottom Water is observed. The precision of the isotopic data shown is about ±0.02‰. The isotopic data are sufficiently precise that mixing between North Atlantic Deep Water and underlying, northerly flowing Antarctic Bottom Water can be seen directly on such a diagram.

The North Atlantic Deep Water point plots directly on the line relating the salinity and δ values of surface waters in the North Atlantic (not shown). Hence, the convective origin of this water from sinking of cooled surface water (discovered by F. Nansen) is directly verified by the isotopic data. However, Pacific and Indian Ocean deep and bottom waters nowhere plot on the lines relating the surface waters in the δ-S diagram. This is seen directly for the Pacific waters in Fig. 12. Thus, the deep water in these oceans cannot be significantly replenished by convective mixing with surface waters in the Pacific and Indian Ocean basins; they must originate by the mixing of other deep waters. This is in agreement with the classical oceanographic picture based on density and stability considerations.

The data in Fig. 12 further indicate that the Pacific and Indian deep waters do not lie on a line connecting the North Atlantic Deep Water and the Antarctic Bottom Water, the only major sources which can provide deep water for the Pacific and Indian oceans. Pacific and Indian deep waters are almost 50–50 mixtures of North Atlantic Deep Water and Antarctic Bottom Water, but there must be a so-called "third component" in small amounts, which almost certainly represents inter-

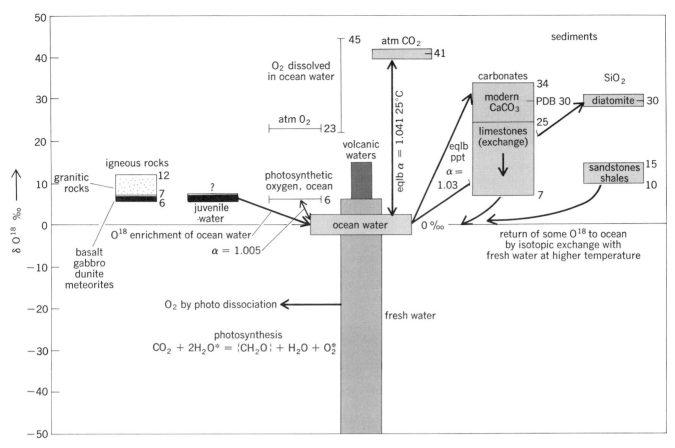

Fig. 13. The oxygen isotope geochemical cycle. The isotopic variations are the per mil deviations from standard mean ocean water (SMOW) with approximate isotopic abundances $O^{16} = 1$, $O^{17} = 1/2500$, and $O^{18} = 1/500$.

Single-stage isotopic fractionation factors (α) between oxygen in water and in the other substances are noted by connecting arrows.

mediate water. It is not yet possible to decide whether the third component is intermediate water added in the Atlantic, where the major mixing of the two main sources takes place, or whether it represents Pacific and Indian Ocean intermediate water added locally in these latter oceanic basins; the former interpretation seems more probable and is supported by some measurements, but detailed studies are necessary. In any case, the major characteristics of the deep-ocean mixing processes seem to be understood.

Changes in isotopic composition. During the Pleistocene Epoch, the isotopic composition of the sea has oscillated about its present composition because of the periodic formation and disappearance of continental ice sheets greatly depleted in D and O^{18}. These variations are estimated to be of the order of $1°/_{oo}$ for O^{18} and $7°/_{oo}$ for D. On the geological time scale of hundreds of millions of years the changes may have been even greater. It is known that H and D atoms can escape from the Earth's gravitational field, and it is believed that H atoms will diffuse upward and escape at a greater rate than D. These atoms are formed by photolysis of water vapor in the atmosphere, and it thus seems likely that the D/H ratio of the ocean may have been increasing with time because of preferential loss of H atoms. Hydrogen found in igneous rocks generally averages about $80°/_{oo}$ lower in D

than ocean water and is, therefore, at least consistent with the idea that this may be juvenile hydrogen which is of the same composition as the early oceans.

The geochemical cycle of O^{18} is much more complicated because of the variety of oxidation-reduction reactions in oxygen geochemistry. A schematic outline of the isotopic oxygen cycle is shown in Fig. 13. According to present knowledge of oxygen isotope fractionation effects, ocean water of $0°/_{oo}$ would be in isotopic exchange equilibrium with igneous rocks at about 250°C; at oceanic temperatures it is far removed from equilibrium. The ocean water observed today is therefore not a direct sample of juvenile water which has escaped from the interior of the Earth at high temperatures—such water would be about $9°/_{oo}$ on the SMOW scale. It is possible, however, that there is a continual input of juvenile water into the ocean, so that the ocean is changing to, or has reached, a steady state in which the net oxygen removed from the ocean has the composition of the incoming juvenile water.

Calcium carbonate and silica remove oxygen of about $30°/_{oo}$ from the ocean, but some of this O^{18} is cycled back into the sea by isotopic exchange with fresh water on the contentnts. The average oxygen removed in this way is probably about $20°/_{oo}$. It is difficult to see how other processes can account for significantly more back-cycling of O^{18}, and it seems likely that the ocean may have undergone a progressive depletion of O^{18} through time, especially since Cretaceous times when large amounts of limestone began to be deposited in the deep sea by organisms. Studies are currently being made of oxygen isotope ratios in ancient carbonates, phosphates, and sulfates to obtain definitive evidence on this problem. When more is known about the actual variations and secular changes in isotopic compositions of the ocean, this information can in turn be applied to questions of the origin and growth of the ocean itself.

[HARMON CRAIG]

Bibliography: J. L. Cairns, Asymmetry of internal tidal waves in shallow coastal water, *J. Geophys. Res.*, 72:3563–3565, 1967; J. D. Cochrane, The frequency distribution of water characteristics in the Pacific Ocean, *Deep-Sea Res.*, 5:111– 127, 1958; H. Craig, Abyssal carbon and radiocarbon in the Pacific, *J. Geophys. Res.*, 74:5491–5506, 1969; H. Craig, Isotopic composition and origin of the Red Sea and Salton Sea geothermal brines, *Science*, 154:1544, 1966; H. Craig and L. I. Gordon, Deuterium and oxygen 18 variations in the ocean and marine atmosphere, in E. Tongiorgi (ed.), *Stable Isotopes in Oceanographic Studies and Paleotemperatures*, 1965; H. Craig and B. Hom, Deuterium–oxygen 18–chlorinity relationships in the formation of sea ice (abstract), *Trans. Amer. Geophys. Union*, 49:216, 1968; C. Eckart, An analysis of the stirring and mixing processes in incompressible fluids, *J. Mar. Res.*, 7:265, 1948; S. Epstein and T. Mayeda, Variation of O^{18} content of waters from natural sources, *Geochim. Cosmochim. Acta*, 4:213, 1953; I. Friedman et al., Variation of deuterium content of natural waters in the hydrologic cycle, *Rev. Geophys.*, 2:177, 1964; Y. Horibe and N. Ogura, Deuterium content as a parameter of water mass in the ocean, *J. Geophys. Res.*, 73:1239, 1968; E. C. LaFond, in *Encyclopedia of Oceanography*, 1966; E. C. LaFond and K. G. La-Fond, Internal thermal structure in the ocean, *J. Hydronautics*, 1:48–53, 1967; E. C. LaFond and K. G. LaFond, in *The New Thrust Seaward*; Transactions of the 3d Annual Marine Technology Society Conference, 1967; R. B. Montgomery, Water characteristics of Atlantic Ocean and of world ocean, *Deep-Sea Res.*, 5:134–148, 1958; D. Rochford, Total phosphorus as a means of identifying East Australian water masses, *Deep-Sea Res.*, 5:89–110, 1958.

SAMPLING AND MEASURING TECHNIQUES

Observations of conditions in the sea generally must be made in situ and the information transmitted back to the observer, or the result must be recorded in situ and retained for reading when withdrawn from the sea. Consequently, the marine scientist is faced with peculiar problems of technique and the use of special equipment to obtain much of his information about sea water. Some of the more commonly used methods and devices for obtaining observations relative to the physical properties of sea water are described here.

Temperature-measuring devices. Temperature measurements in the surface layer of the ocean, to depths of 900 ft (275 m), are usually made with the bathythermograph, or BT, a nonelectric device which gives a continuous record of water temperature and depth as it is lowered and raised. The instrument can be operated at frequent intervals while underway and therefore provides a rapid means of obtaining a detailed picture of temperature distribution within the surface layer.

Reversing thermometers. The most reliable and widely used temperature-measuring device is the deep-sea reversing thermometer. This mercury-in-glass thermometer records the temperature at the time it is inverted. It is reliable to about 0.01°C with proper corrections. These thermometers are usually in a pressure-proof glass tube. In this manner the thermometer is protected from the effect of pressure, and the true water temperature is given. These same thermometers are available with a mercury bulb which can be exposed to sea pressure. The compression of the bulb on these unprotected thermometers causes them to read about 1°C high for each 100 m depth. When protected and unprotected reversing thermometers are used in pairs, they give both temperature and depth (thermometric depth). Reversing thermometers usually are attached to a reversing water bottle which collects a water sample when it is inverted.

Electrical thermometers. Since 1950 a rapidly increasing number of electric temperature recorders have been developed around thermistor beads encased in glass. A typical thermistor will change resistance 4% or about 100 ohms/°C and will have a thermal time constant of a few tenths of 1 sec. Electric temperature recorders are usually made to plot temperature against time. They are often used to study microstructure and may have a sensitivity of 0.01–0.001°C. When measuring elements are to be lowered from a ship, the recorder is usually made to plot temperature against depth.

In 1958 a system was developed utilizing electrical thermometers attached to a special cable which could be towed behind a vessel at 500 ft and at speeds of 10 knots. The thermometers are attached to the cable at 25-ft intervals. A facsimile type of recorder draws continuous isotherms on a depth-distance plot. The depth of each degree or tenth-degree isotherm is plotted every 2 sec.

Radiation thermometers have been built and flown from low-flying aircraft. These measure changes in temperature of the water surface to about 0.1°C. Radiotelemetering buoys permit water and air temperatures to be observed on unattended buoys and transmitted to land.

Water samplers. For many chemical and gas analyses it is essential to obtain samples of 100– 1000 ml of sea water. These usually are obtained by a series of Nansen bottles attached on $\frac{3}{16}$-in. hydrographic cable. The bottles are designed to flush continuously when lowered in an upright position. A weight messenger is then sent down the cable. As it strikes the tripping device of the Nansen bottle, the bottle is inverted and the lids are closed (Fig. 14). At the same time another messenger is released to trip the next lower bottle and so on. Usually two reversing thermometers are attached to each bottle. The thermometer and water bottle give accurate results, but the method is very time-consuming. To obtain a synoptic picture of temperature and salinity, or density, a number of

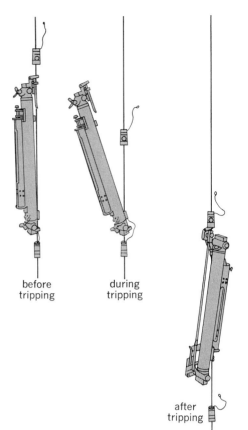

before
tripping

during
tripping

after
tripping

Fig. 14. Nansen bottle in three positions. (*From Instruction Manual for Oceanographic Observations, H.O. Publ, no. 607, 2d ed., U.S. Navy Hydrographic Office, 1955*)

closely spaced observations must be taken in a relatively short period of time.

In an effort to eliminate contamination from a metallic case, samplers such as the Van Dorn have been made from plastic tubing. Large rubber stoppers on each end are pulled shut by rubber bands. Simple open-tube-type samplers of $\frac{1}{10}$ – 10-liter capacity can easily be made.

Measurements such as carbon-14 require samples as large as 200 – 400 liters. Flushing of such large samplers requires either a large open-ended hose construction or a barrel with two flapper-type ports and water scoops to aid ventilation. After the sample is brought to the surface, some large samplers are retrieved on deck while full. Others are emptied while still in the water by means of a hose.

Continuous electrical temperature recorders that obtain temperatures with a single probe will almost certainly be used in the future. This will increase the need for a single collecting device that will take many water samples when used at the end of the cable.　　　　[ALLYN C. VINE]

Serial observations. These are measurements of temperature, salinity, and other properties at a series of depths at some location in the ocean (an oceanographic station), by which the distribution in space and time of these properties (and others computed from them such as density and geopotential) may be described.

Bottles in series. A number of water samplers with thermometers attached are usually lowered

on the same cast. As many as 26 samplers have been lowered at once. The number depends on the number of levels to be sampled, the strength of the wire, and the extent of possible damage to the equipment which may result from the roll of the ship or from dragging the bottom.

After the first (deepest) bottle is attached, the wire is paid out and the next bottle and its messenger are attached. When all bottles have been lowered, it is necessary to wait for the thermometers to approach equilibrium (about 10 min) before releasing a messenger to trip the cast. If the wire is nearly vertical, the messenger falls about 200 m/min. At high wire angles (60° and more have occurred under high wind conditions or strong current shear) the messenger will fall more slowly and may stick. The angle can sometimes be reduced by maneuvering the ship.

After allowing time for the final messenger on the cable to trip the deepest bottle, the wire is pulled in and the bottles removed. Water samples are drawn into laboratory bottles and the thermometers are read.

Thermometric depths. When protected and unprotected thermometers are reversed at the same depth, the unprotected will give a higher reading because of the pressure on its bulb. The difference in the two readings depends upon the pressure at reversal, and since this is proportional to depth the "thermometric depth" can be computed. With information from both protected and unprotected thermometers at several of the levels, the shape of the wire can be estimated, and the depths of the other samplers computed. Unprotected thermometers are ordinarily used at depths greater than 200 m, since the depth of the upper bottles can be computed from the wire angle and length. Depth computations are estimated to be accurate to ±5 m in the upper 1000 m and to about 0.5% of wire length below that.

Standard depths. In 1936 the International Association of Physical Oceanography proposed certain standard levels at which observations should be made or values interpolated in reporting. They are (in meters) 0, 10, 20, 30, 50, 75, 100, 150, 200, (250), 300, 400, 500, 600, (700), 800, 1000, 1200, 1500, 2000, 2500, and 3000 and by 1000-m intervals at greater depths (depths in parentheses being optional). These values are recommended as a convenient standard of comparison, not as being sufficient for measuring the ocean everywhere. Where the precise level of maxima or minima in the various properties is to be determined, the standard depths may not be adequate and more depths must be sampled.

[JOSEPH L. REID]

Analysis of water samples. The development of analytical methods to measure the kind and quantities of dissolved and suspended substances in sea water has paralleled advances in analytical chemistry. In addition to the usual considerations of accuracy, precision, speed, and cost that control the choice of analytical methods in most applications, methods for sea-water analyses are further restricted by the necessities of performing some analyses on shipboard immediately after the samples are obtained and of storing other samples for analyses that can be performed only in a shore-based laboratory. For example, analyses for biologically active substances, especially those present in trace quantities, are performed on shipboard, whereas analyses for which the highest precision and accuracy are demanded, frequently those which require precision weighing, are performed in shore-based laboratories.

In-place techniques. During the 1960s considerable effort was expended in the development of in-place analytical techniques. All of the new techniques depend upon the generation of an electrical signal in which either a voltage or a frequency is made to be proportional to the concentration of the constituent or the intensity of the property being measured. All such methods produce a continuous record in which the property being measured is displayed either as a function of time or of depth of water.

Three kinds of in-place instruments have been used successfully. The first two utilize a transducer lowered on a cable beneath the ship. One version (the older) requires an electrical cable from transducer to ship with recording of the signal on shipboard; the second has the recorder and transducer on the cable end. With the recorder on shipboard there can be continuous monitoring of the signal, and the raising and lowering can be changed to rerecord or emphasize a feature. However, recording on shipboard requires the use of at least one electrical conductor insulated from the sea, a system which requires much maintenance. Placing the recording equipment on the end of the cable with the transducer eliminates the need for long lengths of insulated electrical conductors but introduces the possibility of damaging the entire instrument by flooding with sea water should a seal on the pressure case develop a leak under pressure. A third, the newest type of instrument, is a free-fall device in which all components are in a single container which is dropped into the water with no attachment to the ship. The device, which by itself has positive buoyancy, enters the water with jettisonable weights attached and is carried either to the bottom, where the weights are released, or to some predetermined depth, where a pressure-activated release drops the weights. Once back on the surface the device can be located with the aid of radar reflectors, flashing lights, and radio beacons. The primary advantage of free-fall devices, other than eliminating winches with long lengths of cable, is the possibility of a rapid survey of a region. Many devices can be set out in a predetermined pattern and later retrieved without the time-consuming "on station" intervals required by the older hydrographic casts with cable and sampling bottles.

In-place devices having transducers for electrical conductivity, temperature, and pressure (depth) are commercially available. Many of the so-called single-ion electrodes being developed and marketed have potential application with in-place devices. Included are electrodes for measuring dissolved oxygen, halides, fluoride, calcium, magnesium, and sulfide.

Examination of Table 5 shows that the dissolved substances in sea water fall into two main groups: the major constituents, which include sodium, potassium, calcium, magnesium, chloride, sulfate, bromide, and the sum of carbonate, bicarbonate, and carbonic acid; and the minor constituents, which include all of the other elements in the table. The major constituents show the unique property of being present in very nearly contant ratio in open sea water. For this reason, direct analyses for these substances are rarely performed today, except when estimates of small variations in concentrations that may be produced by biochemical and geochemical processes are sought. When analyses are made for these constituents (except for the CO_2 system constituents) samples are returned to a shore-based laboratory. Storage containers for such samples must be carefully chosen, because changes in concentration of some constituents may result from the interaction of sea water with the glass in many common, soft glass containers.

The most complete study on record of the concentrations of major sea-water constituents still is that conducted by W. Dittmar on 77 samples taken during the round-the-world cruise of the HMS *Challenger* in 1873–1876. Dittmar's analyses were made by what are now considered classical gravimetric and weight titration procedures: Chloride was analyzed by the Volhard method; calcium, by precipitation of the oxalate followed by ignition to, and weighing of, the oxide; magnesium, by precipitation of magnesium ammonium phosphate followed by ignition to, and weighing of, pyrophosphate; sulfate, by precipitation and weighing of barium sulfate; potassium, by precipitation of potassium chloroplatinate followed by weighing of metallic platinum after reduction with hydrogen; and sodium, indirectly, as the difference between the measured sum of all cations as sulfates and the sum of the magnesium, calcium, and potassium calculated as sulfates. Despite empirical corrections and analyses by difference in Dittmar's study, more recent analyses have produced only small changes in the mean values of the concentrations of the major constituents.

Activation analysis. Several modern analytical techniques are ideally suited for analyses of many of the minor constituents in sea water. Activation analysis is the production of radioactive isotopes of elements present in a sample by controlled exposure of the sample to the neutron flux in a nuclear reactor followed by measurement of the decay properties, such as half-life and rate of decay of the artificial activities produced. This method has been used to provide estimates of the concentrations of some 18 trace elements and minor constituents in both sea and fresh waters.

Two sample preparation procedures have been used in activation analysis studies. One uses scavenging or carrier techniques to concentrate minor elements by precipitating a naturally occurring major constituent. For example, trace quantities of iron can be carried by precipitation of magnesium hydroxide. Carrier techniques provide a means of concentrating minor elements from large volumes of sea water and in some cases allow selective removal of one or more elements.

The second method of sample preparation involves the freeze-drying of a sample and irradiation of all of the solids produced. This method has the advantage of low probability of sample contamination; however, the massive quantities of sodium and halide activities produced require "cooling" for many days to allow the short-lived activities of the major elements to die out before postirradiation chemistry can be conducted. Another distinct advantage of neutron-activation analysis of whole (freeze-dried) samples is that several elements can be determined in one sample and with one irradiation.

Gas chromatography. Gas-chromatographic methods have been adapted to the analyses of dissolved atmospheric gases and to a rapidly growing list of dissolved organic substances. As with the adaption of many other analytical systems to sea-water analysis, sample preparation for gas-chromatographic analyses requires special attention. Not only does water interfere in many gas-chromatographic procedures but, in addition, the concentration of many gaseous and organic constituents is below the useful range of gas-chromatographic instruments. Sample preparation then always requires separation of water from the test substances and usually a concentrating step.

The development of gas-chromatographic methods for the analysis of dissolved oxygen, nitrogen, and argon gives a quick and convenient means of separating indicators of biological and physical processes from one another. This is possible because all three gases are atmospheric components, and the sea surface, under most circumstances, can be considered to be in equilibrium with the atmosphere. The concentration of argon in sea water, because it is not involved in any naturally occurring biological or chemical processes, provides a record of the physical conditions which existed at the sea surface when a sample of water now at some subsurface depth was last at the surface. Present-day argon concentration in subsurface samples then allows one to estimate what the oxygen and nitrogen concentrations would have been in the same samples had there been no changes of biological origin. The concentration of oxygen in subsurface waters, on the other hand, responds rapidly to increases by photosynthesis (in the upper 75–100 m) and to decreases by plant and animal respiration and bacteriological decomposition at all depths. Nitrogen concentration may be either increased or decreased during nitrification or denitrification processes, which are much slower than oxygen-controlling processes.

A single 5-mm sample is sufficient for gas-chromatographic analyses for oxygen, nitrogen, and argon. The three test gases are removed from sea water in an attachment to the chromatograph, which acts as a scrubber through which the carrier gas, usually helium, bubbles through the sample and scrubs the other gases from solution. A special sampling bottle, which can be interspersed among the usual Nansen-type samples on a cable and can be attached directly to the gas chromatograph, reduces sampling errors and eliminates the possibility of contamination by contact with the atmosphere.

Studies, using gas-chromatographic analyses, have shown some unique features concerning the distribution of carbon monoxide between the atmosphere and the sea.

Mass spectrometry. The high sensitivity of the mass spectrometer has been utilized in a method for the analyses of helium, neon, krypton, xenon, and argon in the sea. A single 5.5-ml sample is sufficient for these analyses. The method uses a unique sampling procedure, in which a piece of thin-walled metal tubing is crimped off at both ends at the sampling location. The spacing between the two crimping tools establishes the volume size.

The sample is released in a vacuum chamber within the mass spectrometer system by simply puncturing the thin-walled tubing.

Anodic stripping. Anodic-stripping techniques, which consist of modifications of the older polarographic method of analysis, provide measures of the extent to which some of the biochemically reactive elements — zinc, copper, cobalt, lead, and iron, for example — are complexed with both inorganic and organic ligands. Anodic-stripping methods have useful sensitivities down to $10^{-12}\,M$ for some elements and have shown complex formation to be a common occurrence and, therefore, a significant factor in biochemical and geochemical processes.   [DAYTON E. CARRITT]

Bibliography: H. Barnes, *Apparatus and Methods of Oceanography*, 1959; H. Barnes, *Oceanography and Marine Biology*, 1959; D. E. Carritt, Analytical chemistry in oceanography, *J. Chem. Educ.*, 35:119–122, 1958; J. B. Hersey, Electronics in oceanography, *Advan. Electron. Electron Phys.*, 9:239–295, 1957; E. O. Hulburt, Optics of distilled and natural water, *J. Opt. Soc. Amer.*, 35(11):698–705, 1945; Instrument Society of America, *Marine Sciences Instrumentation*, vol. 1, 1962; J. J. Myers, C. H. Holm, and R. F. McAllister (eds.), *Handbook of Ocean and Underwater Engineering*, 1969; H. U. Sverdrup, M. W. Johnson, and R. H. Fleming, *The Oceans*, 1942.

Sea-water fertility

The fertility of a given area of the sea may be defined in terms of the production of living organic matter by the organisms which it contains. The primary productive process in the sea, as on land, is the photosynthetic reduction of carbon dioxide by plants, and the rate of this process for unit area or volume, expressed as mass of carbon fixed in unit time, is a measure of fertility. It may be measured directly in several ways, or it may be estimated from chemical changes in the water. The quantity of living organisms present at any time in a given place (standing stock or standing crop), although it may be related to fertility, is not a measure of it, being the result of a balance between production and removal. Coastal and oceanic waters differ considerably in the types of organisms which they support; the chief difference is in the bottom fauna, which is usually abundant in shallow water and sparse in the deep oceans. Along some shores there may be dense growths of kelp which may have exceptionally high rates of carbon fixation. *See* BIOLOGICAL PRODUCTIVITY.

Fertility of the oceans. This depends on the production of phytoplankton in the upper layer, called the euphotic zone, which receives ample sunlight for the photosynthetic processes of plants. The plants are mostly diatoms and flagellates which reproduce by division, on average probably less than once per day. If not carried downward by water movement, they tend to remain for some time in the euphotic zone since they often have structures which offer resistance to motion through the water. They sink slowly but are usually eaten by the small zooplankton before reaching the bottom. It has been estimated that about 80% of these plants are so utilized; the remainder, comprising those uneaten, or eaten, partly digested, and excreted by zooplankton, descend to nourish the benthos.

For growth the phytoplankton need radiant energy, carbon dioxide and water, nitrogen, phosphorus, and a range of trace elements of which sea water usually contains enough, with possible occasional exceptions of iron and manganese. Diatoms also need silicon, which is present in the water as monomeric silicic acid, generally in sufficient quantity. There is also a requirement, varying in nature and amount for different species, for accessory organic nutrient factors, but this is not yet fully understood. There is always enough carbon dioxide and water, so that plant growth is normally limited by the supply of radiant energy, nitrogen, and phosphorus.

About one-half of the energy in the solar spectrum can be used for photosynthesis, since only the visible spectrum of 3800–7200 A is used. Carbon fixation (gross production) is proportional to light intensity up to about 20 g-cal/cm^2 per day (saturation intensity); above this amount photosynthetic efficiency decreases. It is inhibited at intensities about one-third that of full sunlight, so that carbon fixation at the surface may be less than at intermediate depths. Net production is positive when the radiation intensity is sufficient for carbon gained by photosynthesis to exceed that lost by respiration. This compensation intensity has been found to be somewhat less than 100 g-cal/cm^2 per day, and the maximum depth at which it is attained (compensation depth) varies from a few meters to more than 100 m, depending on the transparency of the water. Compensation depth corresponds very roughly to the depth at which the surface intensity in the middle of the day has been attenuated to about 1%. At low intensities it has been found that to convert 1 mole of carbon dioxide to carbohydrate, the algae require about 10 quanta of radiant energy; this corresponds, when various corrections are made, to a maximum energy-conversion efficiency of about 17%. Making further allowance for decreased efficiency at higher intensities and correcting for loss by respiration, it is possible to calculate the maximum possible yields of organic material at various levels of surface irradiation. These yields range from zero at the compensation intensity of 100 g-cal/cm^2 per day to some 25 g/m^2 (day weight) at 600 g-cal/cm^2 per day, which may be attained in spells of fine summer weather. Over longer periods intensities in most parts of the world average between 200 and 400 g-cal/cm^2 per day, and the calculated production comes out at some 10–20 g/m^2 organic matter daily. Such values are in fact approached by the highest values recorded for single days in shallow water. Open-ocean values in spring are usually around 2–5 g/m^2 per day, and annual averages are 0.4–1.0 g/m^2 per day.

When irradiation is adequate for any length of time, the limitation on growth is set by the availability of nitrogen and phosphorus. The amount of these elements depends on the locality. Deep water in the oceans contains relatively high concentrations of nitrate and phosphate ions, maintained

by decomposition of descending dead organisms. However, these nutrients are hindered from reaching the surface by the stable vertical density gradient in the ocean. Surface concentrations of nitrogen and phosphorus normally range up to about 200 μg N/liter and 30 μg P/liter in temperate latitudes in winter. In spring, however, active plant growth can reduce these concentrations to undetectable levels, which may remain fairly constant throughout summer. New growth is then dependent upon nutrients regenerated from dead plants and from excretory matter from the animals. A similar state of affairs generally exists throughout the year in the tropics, and relatively low apparent rates of carbon fixation are normal. Nevertheless, the total yearly production is comparable with that of other regions since radiation intensities are adequate throughout the whole year and also because high transparency of the water allows light to penetrate deeply.

In regions such as the equatorial divergences and the west coasts of Africa or South America, where physical processes cause nutrient-rich deep water to be brought to the surface, high instantaneous productivities may occur, resulting in large standing stocks of all organisms in the food chain.

The chemical composition of marine phytoplankton, unlike that of land plants, shows high protein (40–50%) and high fat (20–27%) contents and is resembled by that of the animals which succeed it in the food chain. This food chain has more links than most terrestrial ones and is in fact a fairly complex web. Some estimates of production of each species in the food chain have been made in areas, such as the North Sea, where there are important fishing grounds, and it has been possible to make reasonably reliable estimates of fish populations. To account for these estimated rates of production of the various links in the chains leading to pelagic and to demersal fish, it is necessary to postulate transfer efficiencies well over 10%, a suggestion which is supported by experimental evidence. Indeed, feeding experiments with herbivorous zooplankton have shown efficiencies greater than 50%. Compared with those on land, marine food chains appear to be efficient. It is perhaps remarkable that when the annual fishery yield of the North Sea is compared with the estimated primary production, it is found that nearly 0.5% of primary production is available for human use. The productivity of all the oceans has been put at 1.6–15.5 × 10^{10} tons carbon/year, compared with 1.9 × 10^{10} tons carbon/year for the land areas. The table gives estimates of the annual production of living material in two nearshore environments. The estimates are based on many assumptions. *See* FOOD CHAIN.   [FRANCIS A. J. ARMSTRONG]

Productivity and its measurement. The production of organic matter in the sea, as on land, is accomplished by the photosynthetic activity of autotrophic plants. In coastal waters, where sunlight penetrates to the bottom, both rooted plants and benthic algae contribute to this process. In the open sea, organic production is limited to unicellular algae, the phytoplankton, which live suspended in the upper layers of all ocean waters.

Productivity of benthic plant communities may be determined by periodic harvest and measurement of their growth over discrete time intervals. Such direct methods are impossible in the study of the short-lived plankton because of unmeasurable losses from natural death, predation, sedimentation, and advection.

A more satisfactory approach to both benthic and planktonic plant production is through measurement of chemical changes of the water accompanying photosynthesis and growth. These may be followed in the natural environment for periods ranging from 1 day to several weeks, or in the laboratory by exposing representative samples of the plant population to natural conditions for periods not exceeding 1–2 days.

Photosynthetic activity is indicated by the changes in the water of nitrogen and phosphorus salts, oxygen, and carbon dioxide, and by the degree of acidity (pH). Calculations based on natural-environment changes of these indicators must allow for gas exchanges across the water surface and the effects of vertical mixing between surface and deep waters. In such calculations the horizontal advection is generally neglected, and complete chemical recycling between sampling periods cannot be accounted for. For the last reason, the method tends to give conservative estimates of productivity.

Experimental laboratory studies include measurement of oxygen production, carbon dioxide assimilation, and pH change. Both natural-environment and laboratory studies of changes of these properties represent the net effect of photosynthesis and respiration (by both plants and animals), and hence measure net production. Respiration may be measured separately in the laboratory in dark-bottle experiments. This measurement, when added to the net change observed in transparent bottles, gives a measure of real photosynthesis or gross production. Oxygen-bottle experiments lack the necessary sensitivity for use in the open sea, where plankton are sparse; such experiments have been largely replaced by the extremely sensitive method of measuring CO_2 uptake using C^{14} as a tracer. $C^{14}O_2$ uptake appears to be equivalent to net production (photosynthesis minus respiration) by the plant community.

A third method for estimating productivity is based on the premise that photosynthesis is a function of two independent variables, the chlorophyll content of the plants and the light intensity which they receive. Production may be calculated from simultaneous measurements of these factors in the ocean and from their experimentally derived relationship to photosynthesis.

The few existing measurements of dense benthic plant communities indicate that they may produce as much as 20 g organic matter/(m²) (day), an amount equivalent to the best agricultural yields on land. Plankton production seldom if ever attains this level, though values half as great are not uncommon. The productivity of shallow, inshore

waters is generally higher than that of the open sea, but the seasonal range of most marine areas includes two orders of magnitude. The mean annual rate of production in the oceans as a whole is a matter of some controversy, but probably lies between 100 and 300 g organic matter/m² sea surface, which represents an efficiency of utilization of 0.1–0.2% of incident, visible solar energy.

[JOHN H. RYTHER]

Geographic variations in productivity. Strictly speaking, variations in productivity imply variations in the rate of entry of carbon into the organic cycle, or gross photosynthesis. The extent to which this takes place in the sea is determined by the amount of photosynthesizing plant tissue present, the temperature, and the available light energy. The net productivity is the rate of plant growth. This is of greater value as a measurement because it eliminates from the determination the respiratory and excretory losses of the plants and specifies the production of food for the planktonic animals. Variations in net productivity depend on the physiological and oceanographical factors affecting algal growth in the sea, to which the availability of nutrient salts is of prime importance.

The limited penetration of daylight into the sea restricts plant growth to the upper euphotic layer where the nutrients can be assimilated. The plants sink to deeper layers where the nutrients are released by decomposition of plant material by microorganisms and returned to the sea. Acting against this downward transport of nutrients is eddy diffusion (produced by turbulence), which brings nutrients up to the surface from the richer deeper waters. This process is facilitated where the water is homogeneous and it is suppressed by stratification. Currents perform the major transport of nutrients, and where surface divergences occur, upwelling currents bring deeper water to the surface (see illustration). In the illustration the northern Indian Ocean and the China Sea appear under southwestern monsoon conditions; these currents are reversed, with a shift in regions of divergence, in the northeastern monsoon.

In temperate and higher latitudes there is a pronounced seasonal cycle, with suppressed production in winter because of excessive turbulence and lack of light, a rapid burst of growth in spring, and limitation of this growth by lack of nutrients when the waters stabilize in summer. Frequently in autumn there is a subsidiary flowering before growth is again limited by lack of light.

In tropical and subtropical regions a more permanent stratification limits nutrient supply to the euphotic zone, and there appears to be a low net production rate. Exceptions are found in regions of divergence, principally on the western coasts of the continents and to a lesser extent in the open ocean in the equatorial region, where upwelling of rich deeper waters permits a high productivity, often throughout the year.

The question of relative production in high and low latitudes is still undecided, for the high productivity of the polar seas is of short seasonal duration; the tropics may in fact equal it on an annual basis, although running at a lower instantaneous rate. As yet, measurements of production rates are inconclusive on this point.

Some estimate of the production of higher animals can be obtained from commercial fishing and

Annual production of living organisms, g/m²

Organism	English Channel	North Sea
Phytoplankton	730–910	1000
Zooplankton	275	160
Pelagic fish	2.9	6
Demersal fish	1.9	1.7
Benthos	55	50

whaling statistics. The correlation with net production rate appears to be fairly good; however, it is modified by feeding and breeding requirements of the animals in question and by the fact that many higher animals of no commercial interest may be produced in some areas.   [RONALD I. CURRIE]

Size of populations and fluctuations. In temperate waters fish tend to spawn at the same place at a fixed season. The larvae drift in a current from spawning ground to nursery ground, and the adults migrate from the feeding ground to the spawning ground. Adolescents leave the nursery to join the adult stock on the feeding ground. The migration circuit is based on the track of the larval drift, and the population is isolated by its unique time of spawning. Because larvae suffer intense mortality, numbers in the population are perhaps regulated naturally during the period of larval drift.

Methods of measuring the size of marine populations are of three basic kinds. The first is by a census based on samples which together constitute a known fraction either of the whole population, as in the case of sessile species such as shellfish, or of a particular age range, as in the case of fish which have pelagic eggs whose total abundance can be measured by fine-meshed nets hauled vertically through the water column. The second is by marking or tagging, in which a known number of marked individuals are mixed into the population and the ratio of unmarked to marked individuals is subsequently measured from samples. The third is applicable to commercially exploited populations where the total annual catch is known: the mortality rate caused by exploitation is measured, based on the age composition of the populations, thus establishing what fraction of the population the catch is. The last two methods are most generally used.

The largest measured populations are of pelagic fish, particularly of the herring family and related species, Clupeidae. One of these is the Atlantic herring (*Clupea harengus*) which contains on the order of 1,000,000,000 mature individuals and ranges over hundreds of miles of the northeast Atlantic.

Populations of bottom-living fish tend to fluctuate slowly over long time periods of 50–75 years. However, those of pelagic fish, like the Atlantic herring, may fluctuate dramatically. If fish spawn at a fixed season, they are vulnerable to climatic change. During long periods there are shifts in wind strength and direction, with consequent delays or advancements in the timing of the production cycle. If the cycle becomes progressively delayed, the fish larvae, hatched at a fixed season, become progressively short of food, and the populations decline with time. Climatic change affects the population during the period of larval drift when isolation is maintained and when numbers are normally regulated.   [D. H. CUSHING]

Biological species and water masses. Biogeographical regions in the ocean are related to the distribution of water masses. Their physical individuality and ecological individuality are derived from partly closed patterns of circulation and from amounts of incident solar radiation characteristic of latitudinal belts. Each region may be described in terms of its temperature-salinity property and of the biological species which are adapted to all or part of the relatively homogeneous physical-chemical environment.

Cosmopolitan species. The discrete distributions of many species are circumscribed by the regions of oceanic convergence bounding principal water masses. Other distributions are limited to current systems. Cosmopolitan species are distributed across several of the temperature-salinity water masses or oceans: their wider specific tolerances

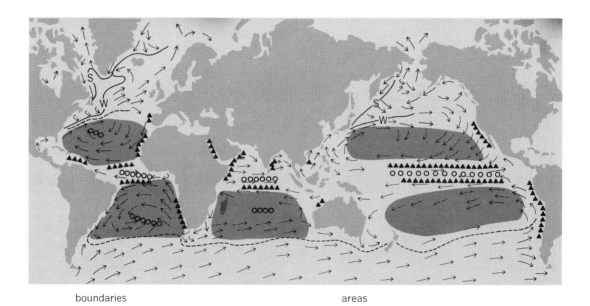

boundaries
— S — polar waters, summer
— W — polar waters, winter
- - - - - - - temperate waters

areas
▲▲▲▲▲▲ divergence and upwelling
ooooooo convergence and sinking
█ subtropical and tropical areas of low productivity

A world map (Mercator projection) which shows the ocean surface currents, the boundaries between areas of high and low productivity, and the principal regions of divergence and sinking.

reflect adaptations to broadly defined water types. No pelagic distribution is fully understood in terms of the ecology of the species.

A habitat is integrated and maintained by a current system: oceanic gyral, eddy, or current, with associated countercurrents. This precludes species extinction that could occur if a stock were swept downstream into an alien environment. The positions of distribution boundaries may vary locally with seasonal or short-term changes in temperature, available food, transparency of the water, or direction and intensity of currents.

Phytoplankton species are distributed according to temperature tolerances in thermal water masses, but micronutrients (for example, vitamin B_{12}) are essential for growth in certain species. The cells of phytoplankton reproduce asexually and sometimes persist in unfavorable regions as resistant resting spores. New populations may develop in prompt response to local change in temperature or in nutrient content of the water. Such species are less useful in tracing source of water than are longer-lived, sexually reproducing zooplankton species.

Indicator organisms. The indicator organism concept recognizes a distinction between typical and atypical distributions of a species. The origin of atypical water is indicated by the presumed affinity of the transported organisms with their established centers of distribution.

Zooplankton groups best understood with respect to their oceanic geography are crustaceans such as copepods and euphausiids, chaetognaths (arrow worms), polychaetous annelids, pteropod mollusks, pelagic tunicates, foraminiferids, and radiolarians. Of these the euphausiids are the strongest diurnal vertical migrants (200–700 m). The vertical dimension of euphausiid habitat agrees with the thickness of temperature-salinity water masses, and many species distributions correspond with the positions of the masses. In the Pacific different species, some of which are endemic to their specific waters, occupy the subarctic mass (such as *Thysanoessa longipes*), the transition zone, a mixed mass lying between subarctic and central water in midocean and between subarctic and equatorial water in the California Current (for example, *Nematoscelis difficilis*), the barren North Pacific central (such as *Euphausia hemigibba*) and South Pacific central masses (for example, *E. gibba*), the Pacific equatorial mass (such as *E. diomediae*), a southern transition zone analogous to that of the Northern Hemisphere (represented by *Nematoscelis megalops*), and a circumglobal subantarctic belt south of the subantarctic convergence (such as *E. lucens*).

Epipelagic fishes and other strongly swimming vertebrates are believed to be distributed according to temperature tolerances of the species and availability of food. However, distributions of certain bathypelagic fishes (such as *Chauliodus*) have been related to water mass. *See* SEA WATER.

[EDWARD BRINTON]

Bibliography: R. J. H. Beverton, Long term dynamics of certain North Sea fish populations, in E. D. Lecren and M. W. Holdgate (eds.), *The Exploitation of Natural Animal Populations*, 1962; R. J. H. Beverton and S. J. Holt, On the dynamics of exploited fish populations, *Fish Invest.* (London), vol. 19, 1957; D. H. Cushing, Biological and hydrographic changes in British seas during the last thirty years, *Biol. Rev.*, 41:221–258, 1966; A. W. Ebeling, *Melamphaidae, I: Systematics and Zoogeography of the Species in the Bathypelagic Fish Genus Melamphaes*, Dana Rep. no. 58, 1962; H. W. Harvey, *The Chemistry and Fertility of Sea Waters*, 2d ed., 1957; M. N. Hill (ed.), *The Sea*, vol. 2: *The Composition of Sea Water, Biological Species, Water-Masses and Currents*, 1963; F. R. H. Jones, *The Migration of Fish*, 1968; J. E. G. Raymont, *Plankton and Productivity in the Oceans*, 1963; J. P. Riley and G. Skirrow, *Chemical Oceanography*, vol. 1, 1965.

Sensible temperature

The temperature at which air with some standard humidity, motion, and radiation would provide the same sensation of human comfort as existing atmospheric conditions. Of the many sensible temperature formulas thus far proposed, none is completely satisfactory or generally accepted. Most are intended for warm, moist conditions; a few, like "wind chill," are for cold weather. Some are purely empirical, modifying the actual temperature according to the humidity; others are theoretical, and express estimated heat loss rather than an equivalent temperature. *See* HUMIDITY.

Heat is produced constantly by the human body at a rate depending on muscular activity. For body heat balance to be maintained, this heat must be dissipated by conduction to cooler air, by evaporation of perspiration into unsaturated air, and by radiative exchange with surroundings. Air motion (wind) affects the rate of conductive and evaporative cooling of skin, but not of lungs; radiative losses occur only from bare skin or clothing, and depend on its temperature and that of surroundings, as well as sunshine intensity.

As air temperatures approach body temperature, conductive heat loss decreases and evaporative loss increases in importance. Hence, at warmer temperature, humidity is the second most important atmospheric property controlling heat loss and hence comfort, and the various sensible temperature formulas incorporate some humidity measure. Most used is C. P. Yaglou's effective temperature, represented by lines of equal comfort on a chart of dry-bulb versus wet-bulb temperature; it relates existing comfort to that in motionless, saturated air.

Effective temperature is approximated by E. Thom's temperature-humidity index (THI), called discomfort index until certain southern United States cities objected, given in °F by Eq. (1), where

$$THI = 15 + 0.4t \qquad (1)$$

t is the sum of the dry-bulb and wet-bulb temperatures. The THI is routinely tabulated at major Weather Bureau stations, and accumulated into "cooling degree days." Similar is humiture, first applied by O. F. Heavener in 1937 to the average of temperature in °F and relative humidity, and redefined by V. E. Lally and B. F. Watson in 1960, as Eq. (2), where t is Fahrenheit temperature and e is

$$Humiture = t + e - 10 \qquad (2)$$

vapor pressure in millibars. Many other "comfort temperatures" and "sultriness indices" have been proposed.

Under cold conditions, atmospheric moisture is

negligible and wind becomes important in heat removal. P. A. Siple's wind chill, given by Eq. (3),

$$\text{Wind chill} = (10.45 + 10\sqrt{v} - v)(33 - t) \quad (3)$$

does not estimate sensible temperature as such, but heat loss in kcal/m² hr, for wind speed v in m/sec and air temperature t in °C. It is used extensively by the U.S. Army, government agencies, construction contractors, and others in cold areas. *See* ENVIRONMENTAL PROTECTION.

[ARNOLD COURT]

Bibliography: S. Licht (ed.), *Medical Climatology*, 1964; S. W. Tromp (ed.), *Medical Biometeorology*, 1963.

Septic tank

A single-story settling tank in which settled sludge is in immediate contact with sewage flowing through the tank while solids are being decomposed by anaerobic bacterial action. Such tanks have limited use in municipal treatment, but are the primary resource for the treatment of sewage from individual residences. There are probably well over 4,000,000 septic tanks in use in home

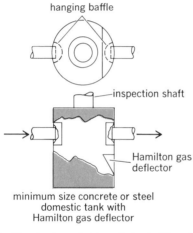

Fig. 1. Circular household septic tank. (*From H. E. Babbitt and E. R. Baumann, Sewerage and Sewage Treatment, 8th ed., Wiley, 1958*)

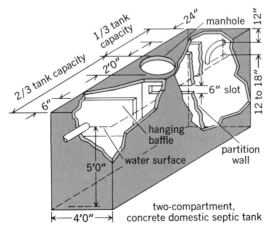

Fig. 2. Two-compartment rectangular household septic tank. (*From H. E. Babbitt and E. R. Baumann, Sewerage and Sewage Treatment, 8th ed., Wiley, 1958*)

disposal systems in the United States. Septic tanks are also used by isolated schools and institutions and for sanitary sewage treatment at small industrial plants.

Home disposal units. Septic tanks have a capacity of approximately 1 day's flow. Since sludge is collected in the same unit, additional capacity is provided for sludge. One formula for sludge storage that has been used is $Q = 17 + 7.5y$, where Q is the volume of sludge and scum in gallons per capita per year, and y is the number of years of service without cleaning. About one-half of a 500-gal tank is occupied by sludge in 5 years in an ordinary household installation. The majority of states require a minimum capacity of 500 gal in a single tank. Some states require a second compartment of 300-gal capacity. Single- and double-compartment tanks are shown in Figs. 1 and 2. Such units are buried in the ground and are not serviced until the system gives trouble because of clogging or overflow. Commercial scavenger companies are available in most areas. A tank truck equipped with pumps is brought to the premises, and the tank content is pumped out and taken to a sewer manhole or a treatment plant for disposal. In rural areas the sludge may be buried in an isolated place.

Municipal and institutional units. These are designed to hold 12–24 hours' flow, with additional sludge capacity provided. Provision is made for sludge withdrawal about once a year. Desirable features of design are (1) watertight and corrosion-resistant material (concrete and well-protected metal have been used); (2) a vented tank; (3) manhole openings in the roof of the tank to permit inspection; (4) baffles at the inlet and the outlet to a depth below the probable scum line, usually 18–24 in. below the water surface; (5) sludge draw-off lines—although seldom used, they should be designed so that they can be rodded or unplugged by some positive mechanism; (6) hoppers or sloped bottoms so that digested sludge can be withdrawn as required; and (7) provision for safe handling of septic tank effluent by disposal underground or by chlorination before discharge to a stream, or both.

Tank efficiency. Septic tank effluent is dangerous and odorous. It will contain pathogenic bacteria and sewage solids. Particles of sludge and scum are trapped in the flow and will cause nuisance at the point of discharge unless properly handled. Efficiency in removal of solids is less than that for plain sedimentation. While 60% suspended solids removal is used theoretically, it is seldom obtained in practice. Improvement is noted when tanks are built with two compartments. Shallow tanks give somewhat better results than very deep tanks. *See* SEWAGE TREATMENT.

[WILLIAM T. INGRAM]

Bibliography: U.S. Public Health Service, *Manual of Septic Tank Practice*, 1960; U.S. Robert A. Taft Sanitary Engineering Center, *Studies on Household Sewage Disposal Systems*, U.S. Public Health Service, pts. 1–3, 1949, 1950, and 1954.

Sewage

A combination of (1) liquid wastes conducted away from residences, institutions, and business buildings and (2) the liquid wastes from industrial establishments with (3) such surface, ground, and storm

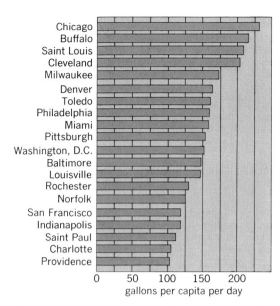

Fig. 1. Estimated water use in 20 major cities of the United States. (*Research Division, New York University College of Engineering*)

water as may find its way or be admitted into the sewers. Category 1 is known as sanitary or domestic sewage; category 2 is usually referred to as industrial waste; category 3 is known as storm sewage.

Relation to water consumption. Sewage is the waste water reaching the sewer after use; hence it is related in quantity and in flow fluctuation to water use. The quantity of sewage is generally less than the water consumption since some portion of water used for fire fighting, lawn irrigation, street washing, industrial processing, and leakage does not reach the sewer. These losses are compensated for partly by the addition of water from private wells, groundwater infiltration, and illegal connections from roof drains. Water consumption increases with size of community served and many other community characteristics. Characteristics of each city must be studied and analyzed for

specific information. As a general average estimate, communities with population under 1000 use about 60 gal per capita per day (gcd), while communities of 100,000 use about 140 gcd. In a study of large cities of the United States, the median consumption was 154 gcd and the median population was 658,000 (Fig. 1). An accepted unit flow for domestic sewage as shown in the table is 100 gcd.

Infiltration of groundwater should be held to a minimum. It may be expected to be equal to or less than 30,000 gal/day/mile of sewer including house connections. Much depends on the quality of sewer construction. Water may enter through poorly made joints and, in quantity, through poorly constructed, leaky manholes and illegal and abandoned sewers. Sewers in wet ground with a high water table will have more infiltration. Sewers under pressure may have infiltration or leakage to the surrounding ground. The danger of groundwater pollution from leaky sewers should be avoided.

Fluctuations in sewage flow are related to water use characteristics but tend to dampen out since there is a time lag from the time of use to appearance in the sewer mains and trunks (Fig. 2). Hourly, daily, and seasonal fluctuations affect design of sewers, pumping stations, and treatment plants.

Daily and seasonal variations depend largely on community characteristics. Weekend flows may be lower than weekday. Industrial operations of seasonal nature influence the seasonal average. The seasonal average and annual average are about equal in May and June. The seasonal average is about 124% in late summer and may drop to about 87% at the end of winter. Peak flows may reach 200% of average at the treatment plant and may be more than 300% of average in the laterals. Laterals are designed for 400 gcd and mains and trunks for 250 gcd.

Design periods. These are dependent on the proposed sewer construction. Lateral sewers may be designed for ultimate flow of the area to be sewered. Mains may be designed for periods of 10–40

Rates of sewage flow from various sources*

Character of district	Gal per capita per day	Gal per acre per day	Source of sewage	Gal per capita per day
Domestic			Trailer courts	50†
Average	100		Motels	53†
High-cost dwellings	150	7,500	State prisons	
Medium-cost dwellings	100	8,000	Maximum	280‡
Low-cost dwellings	80	16,000	Average	176
Commercial			Minimum	104
Hotels, stores, and			Mental hospitals	
office buildings		60,000	Maximum	216‡
Markets, warehouses,			Average	123
wholesale districts		15,000	Minimum	38
Industrial			Grade school	4.4§
Light industry		14,000	High school	3.9§

*From H. E. Babbitt and E. R. Baumann, *Sewerage and Sewage Treatment*, 8th ed., Wiley, 1958.

†From report of State Sanitary Engineers, *Public Works*, p. 108, March, 1957.

‡From J. C. Frederick, *Public Works*, p. 112, April, 1957.

§Average of 4.4 gal per day per pupil between 7:30 A.M. and 5:30 P.M. The average for the high school is spread over more hours per day. From C. H. Coberly, *Public Works*, p. 143, May, 1957.

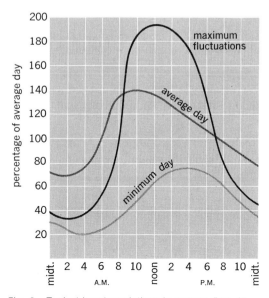

Fig. 2. Typical hourly variations in sewage flow. (*From H. E. Babbitt and E. R. Baumann, Sewerage and Sewage Treatment, 8th ed., Wiley, 1958*)

years. Trunk sewers may be planned for long periods with provision made in design for parallel or separate routings of trunks of smaller size to be constructed as the need arises. Economics, available funds, and engineering judgment affect selection of the design periods. Appurtenances may have a different life, since replacement of mechanical equipment will be necessary. A span of 20–25 years is often selected and a timetable of additions during that period is then scheduled in the overall improvement plan.

Storm sewage. Storm sewage is liquid flowing into sewers during or following a period of rainfall and resulting therefrom. An estimate of the quantity of storm sewage is necessary in sewage design.

Estimating quantity of storm sewage requires a knowledge of intensity and duration of storms, distance water travels before reaching sewer, permeability and slope characteristics of the surface over which water flows to sewer inlet, and shape and amount of area to be drained to inlet. These general considerations are included in the equation $Q = CAIR$ expressing the runoff from a watershed having no retention or storage of water. Q is expressed in cubic feet per second (cfs), A is area, I is the relative imperviousness of the surface expressed as a decimal, and R is the rate of rainfall in inches per hour. C is a coefficient permitting the expression of the factors in convenient units; in the above units it may be taken as 1 so that the equation becomes $Q = AIR$.

Time of concentration is a combination of the theoretical time required for a drop to run from the most distant point to the inlet and time from sewer inlet to point of concentration. The inlet time may range from 3 min for a steep slope on an impervious area to 20 min on a city block. The time of flow is assumed to be the velocity in the full flowing sewer divided by the length of sewer from inlet to point of concentration. Flood crest and storage time while the sewer is filling are usually neglected, the effect being that of assuming a larger rate of flow and therefore providing a safety factor in design.

Values of I, the runoff coefficient, range from 0.01 in wooded areas to 0.95 on roofed surfaces. A common value used in residential areas with considerable land in lawn, garden, and shrubbery is 0.30–0.40. In built-up areas, values of 0.70–0.90 may be used.

Rainfall intensity values are selected on the basis of frequency and duration of storms. In some sewer design it is necessary to select a value for the expected occurrence of maximum runoff. This is done by using one of the several formulas which will allow a prediction of R for 5, 10, or 15 years. The element of calculated risk is combined with engineering judgment in deciding which R to choose. For lesser structures in residential areas a 5-year frequency may be used with reasonable safety. Where failure would endanger property, the 10-, 15-, or 25-year frequency of occurrence provides a more conservative design basis; 50-year frequency may be selected where flooding could cause lasting damage and disrupt facilities. In such instances cost-benefit studies may be made to guide the selection of a suitable frequency. *See* Hydrology.

Pumping sewage. Not all sewage will flow by gravity without unnecessary expense in circuitous

routing or deep excavation: therefore, pumping stations may be advantageous. Pumping stations may be required in the basements of large buildings. Pumping stations are provided with two or more pumps of sufficient capacity so that with one unit out of service the remaining unit or units will pump the maximum flow. Motive power is required from at least two sources, usually electric motors and auxiliary fuel-fed motor drive. Care must be taken to have motive power above flood level and protected from the elements. Screening is usually required ahead of pumping stations, unless the pumps themselves are self-cleaning. Many states require that pumping units be installed in a dry well and that sewage be confined to a separate wet well. Buildings above ground should fit the surroundings. Small pumping stations are often one unit and made fully automatic so that minimum attendance is required. Safety measures must be considered. Centrifugal pumps are used almost exclusively in larger stations. Air ejector units may be installed in smaller stations.

Examination of sewage. Sewage is actually water with a small amount of impurity in it. Examination of sewage is required to know the effects of these impurities. Various tests are used to aid in determining the characteristics, composition, and condition of sewage. These include physical examination, solids determination, tests for determining the oxygen requirement of organic matter, chemical and bacteriological tests, and examination under the microscope.

Physical tests for turbidity, odor, color, and temperature are made. Normal fresh sewage is gray and somewhat opaque, has little odor, and has a temperature slightly higher than the water supply. Decomposition of organic matter darkens the sewage, and odors are characteristic of stale or septic sewage.

Tests for residue or solid matter provide an indication of the types of solids, the strength of the sewage, and the physical state of the solids. Total solids determinations measure both suspended and dissolved solids. A sample of the sewage is filtered. The suspended solids can be determined by drying the material recovered on the filter. The dissolved solids can be determined by evaporation of the filtered portion. Heating the solids residue until organic matter gasifies separates volatile solids from fixed solids or inorganic ash. Loss on ignition represents the volatile or organic fraction and is a good measure of sewage strength.

Measurement of the part of the suspended solids heavy enough to settle is made in an Imhoff cone. The settleable-solids test is useful in determining sludge-producing characteristics of sewage.

Tests for organic matter are made principally to determine the oxygen requirement of sewage. These tests include the biochemical oxygen demand test (BOD), the chemical oxygen demand test, the oxygen consumed test, and the relative stability test. Organisms in sewage require oxygen for growth, and the BOD measures the amount of dissolved oxygen required for decomposition of organic solids for a measured time at a constant temperature. The standard measurement is made for 5 days at 20°C and is a good measure of sewage strength. Since the BOD measurement includes both biological and chemical oxygen requirement, another test, the chemical oxygen demand, is sometimes used to measure the chemical oxygen requirement. Sewage is heated in the presence of an oxidizing agent such as potassium dichromate. The oxygen requirement is that of chemical digestion since all organisms have been killed. This test is increasing in use. The oxygen-consumed test uses potassium dichromate as the oxidizing agent. The result offers some index of the readily oxidizable carbonaceous material. The relative stability test indicates when the oxygen present in plant effluent or polluted water is exhausted. The data express as a percentage the approximate amount of oxygen available in water in relation to the amount required for complete stability. The test is a color test using methylene blue. Reducing agents, precipitation of color, concentration of dye, amount of dissolved oxygen in the sample, and other factors affect the reliability of this test, and it is considered generally as a rough or screening test of the condition of plant effluent. Tests for nitrogen include those for free ammonia, albuminoid ammonia, organic nitrogen, nitrites, and nitrates. The latter are indications of oxidation change and stabilization and are used in checking condition of plant effluent.

Bacteriological tests are made primarily to determine the presence of organisms of the coliform group. The organisms exist in the intestines of warm-blooded animals and are used as an index of the presence of fecal material. The coliform test is made on chlorinated effluents to determine the efficiency of chlorination. Occasionally other bacteriological determinations are made in special studies to determine the presence of organisms of the *Salmonella* group or dysentery group in polluted water and sewage.

Microscopic tests are not normally made on raw sewage. They are used as part of plant operator control in treatment processes. Examinations for the presence of algae, protozoa, bacteria, fungi, rotifers, and worms are made when necessary.

[WILLIAM T. INGRAM]

Bibliography: APHA-AWWA-WPCA Joint Committee, *Standard Methods for the Examination of Water, Sewage, and Industrial Wastes*, 12th ed., 1965; G. M. Fair, J. C. Geyer, and D. A. Okun, *Water Supply and Waste-water Removal*, vol. 1, 1958; W. T. Ingram, *Water Fluoridation Practices in Major Cities of the United States*, pt. 1, Research Division, New York University College of Engineering, 1958; F. A. Kristal and F. A. Annett, *Pumps*, 2d ed., 1953; H. F. Seidel and J. L. Cleasby, Statistical analysis of water works data for 1960, *J. AWWA*, 58 (12): 1507, 1966.

Sewage disposal

The discharge of waste waters into surface-water or groundwater courses, which constitute the natural drainage of an area. Most waste waters contain offensive and potentially dangerous substances, which can cause pollution and contamination of the receiving water bodies. Contamination is defined as the impairment of water quality to a degree that creates a hazard to public health. Pollution refers to the adverse effects on water quality that interfere with its proper and beneficial use.

In the past, the dilution afforded by the receiving water body was usually great enough to render waste substances innocuous. Since the turn of the century, however, the dilution of many rivers has

been inadequate to absorb the greater waste discharges caused by the increase in population and expansion of industry.

The principal sources of pollution are domestic sewage and industrial wastes. The former includes the used water from dwellings, commercial establishments, and street washings. Industrial wastes constitute acids, chemicals, oils, and animal and vegetable matter carried by cleaning or used process waters from factories and plants. For a discussion of sources of wastes *see* SEWAGE.

Regulation of water pollution. This is primarily a responsibility of the state, in cooperation with the Federal and local governments. The health departments of many states are given statutory power and responsibility for the control of water pollution, and they have established specific water quality standards. There are two basic types of standards—stream standards, dealing with the quality of the receiving water, and effluent standards, referring to strength of wastes discharged. Both types are based on the capacity of the receiving waters to absorb waste substances and on the beneficial uses made of the water.

The self-purification capacity is determined by the available dilution, the biophysical environment of the stream, and the strength and characteristics of the wastes. Beneficial uses include drinking, bathing, recreation, fish culture, irrigation, industrial uses, and disposal of wastes without creation of pollution.

Adjustment of these conflicting interests and equitable distribution of water resources is complex from the technical, economic, and political viewpoints. These considerations have led to the establishment of interstate commissions, which provide a means of coordinated control of the larger rivers.

Water-quality criteria deal with the physical, chemical, and biological parameters of pollution. The most common standards are concerned with physical appearance, odor production, dissolved-oxygen concentration, pathogenic contamination, and potentially toxic or harmful chemicals. The allowable quantity and concentration of these characteristics and substances vary with the water usage.

Absence of odor and unsightliness, and the presence of some dissolved oxygen are common minimum standards. Preliminary or primary treatment of waste waters is usually required for the maintenance of these standards. Highest-quality waters require clarity, oxygen saturation, low bacteriological counts, and absence of harmful substances. In these cases, intermediate or complete treatment may be required. *See* SEWAGE TREATMENT; WATER ANALYSIS; WATER POLLUTION.

Stream pollution. Biological, or bacteriological, pollution is indicated by the presence of the coliform group of organisms. While nonpathogenic itself, this group is a measure of the potential presence of contaminating organisms. Because of temperature, food supply, and predators, the environment provided by natural bodies of water is not favorable to the growth of pathogenic and coliform organisms. Physical factors, such as flocculation and sedimentation, also help remove bacteria. Any combination of these factors provides the basis for the biological self-purification capacity of natural water bodies.

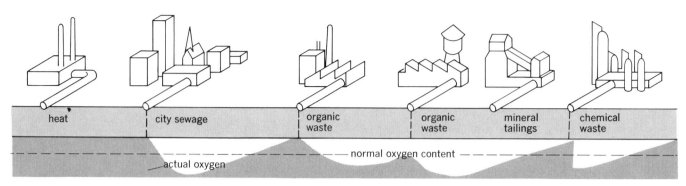

Fig. 1. Variation of oxygen content of polluted stream.

When subjected to a disinfectant such as chlorine, bacterial die-away is usually defined by Chick's law, which states that the number of organisms destroyed per unit of time is proportional to the number of organisms remaining. This law cannot be directly applied in natural streams because of the variety of factors affecting the removal and death rates in this environment. The die-away is rapid in shallow, turbulent streams of low dilution, and slow in deep, sluggish streams with a high dilution factor. In both cases, higher temperatures increase the rate of removal.

The concentration of many physical characteristics and chemical substances may be calculated directly if the relative volumes of the waste stream and river flow are known. Chlorides and mineral solids fall into this category. Some substances in waste discharges are chemically or biologically unstable, and their rates of decrease can be predicted or measured directly. Sulfites, nitrites, some phenolic compounds, and organic matter are examples of this type of waste.

These simple relationships, however, do not apply to the concentration of dissolved oxygen. This factor depends not only on the relative dilutions, but upon rate of oxidation of the organic material and rate of reaeration of the stream.

Nonpolluted natural waters are usually saturated with dissolved oxygen. They may even be supersaturated because of the oxygen released by green water plants under the influence of sunlight. When an organic waste is discharged into a stream, the dissolved oxygen is utilized by the bacteria in their metabolic processes to oxidize the organic matter. The oxygen is replaced by reaeration through the water surface exposed to the atmosphere. This replenishment permits the bacteria to continue the oxidative process in an aerobic environment. In this state, reasonably clean appearance, freedom from odors, and normal animal and plant life are maintained.

An increase in the concentration of organic matter stimulates the growth of bacteria and increases the rates of oxidation and oxygen utilization. If the concentration of the organic pollutant is so great that the bacteria use oxygen more rapidly than it can be replaced, only anaerobic bacteria can survive and the stabilization of organic matter is accomplished in the absence of oxygen. Under these conditions, the water becomes unsightly and malodorous, and the normal flora and fauna are destroyed. Furthermore, anaerobic decomposition

proceeds at a slower rate than aerobic. For maintenance of satisfactory conditions, minimal dissolved oxygen concentrations in receiving streams are of primary importance.

Figure 1 shows the effect of municipal sewage and industrial wastes on the oxygen content of a stream. Cooling water, used in some industrial processes, is characterized by high temperatures, which reduce the capacity of water to hold oxygen in solution. Thermal pollution, however, is significant only when large quantities are concentrated in relatively small flows. Municipal sewage requires oxygen for its stabilization by bacteria. Oxygen is utilized more rapidly than it is replaced by reaeration, resulting in the death of the normal aquatic life. Further downstream, as the oxygen demands are satisfied, reaeration replenishes the oxygen supply.

Any organic industrial waste produces a similar pattern in the concentration of dissolved oxygen. Certain chemical wastes have high oxygen demands which may be exerted quickly, producing a sudden drop in the dissolved oxygen content. Other chemical wastes may be toxic and destroy the biological activity in the stream. Strong acids and alkalies make the water corrosive, and dyes, oils, and floating solids render the stream unsightly. Suspended solids, such as mineral tailings, may settle to the bed of the stream, smother purifying microorganisms, and destroy breeding places. Although these latter factors may not deplete the oxygen, the polutional effects may still be serious.

Deoxygenation. Polluted waters are deprived of oxygen by the exertion of the biochemical oxygen demand (BOD), which is defined as the quantity of oxygen required by the bacteria to oxidize the organic matter. The rate of this reaction is assumed to be proportional to the concentration of the remaining organic matter, measured in terms of oxygen. This reaction may be expressed as Eq. (1),

$$\frac{dL}{dt} = -K_1 L \tag{1}$$

which integrates to give Eq. (2) or Eq. (3), in which

$$L_t = L_0 e^{-K_1 t} \tag{2}$$

$$y = L_0 (1 - e^{-K_1 t}) \tag{3}$$

L_t is BOD remaining at any time t, L_0 is ultimate BOD, y is BOD exerted at end of t, and K_1 is coefficient defining the reaction velocity. The coefficient is a function of temperature given by

Eq. (4), in which T is temperature in degrees Celsius,

$$K_T = K_{20} \cdot 1.047^{T-20} \qquad (4)$$

K_T is value of the coefficient at T, and K_{20} is value of the coefficient at 20°C.

The BOD of a waste is determined by a standard laboratory procedure and is reported in terms of the 5-day value at 20°C. From a set of BOD values determined for any time sequence, the reaction velocity constant K_1 may be calculated. Knowledge of this coefficient permits determination of the ultimate BOD from the 5-day value in accordance with the above equations. For municipal sewages and many industrial wastes the value of K_1 at 20°C is between 0.15 and 0.75 per day. A common value for sewage is 0.4 per day.

The coefficient determined from laboratory BOD data may be significantly different from that calculated for stream BOD data. The determination of the stream rate may be made from a reexpression of Eqs. (1)–(4) in the form of Eq. (5),

$$K_r = \frac{1}{t} \log \frac{L_A}{L_B} \qquad (5)$$

where L_A is the BOD measured at an upstream station, L_B is the BOD at a station downstream from A, and t is the time of flow between the two stations. Values of K_r range from 0.10 to 3.0 per day. The difference between the laboratory rate K_1 and the stream rate K_r is due to the turbulence of the stream flow, biological growths on the stream bed, insufficient nutrients, and inadequate bacteria in the river water. These factors influence the rate of oxidation in the stream as well as the removal of organic matter. Such processes as flocculation, sedimentation, and scour of the organic material in the river affect the removal rate but do not necessarily influence the rate of oxidation and the associated dissolved oxygen concentration. Field surveys are usually required to determine the pollution assimilation capacity of a stream.

When a significant portion of the waste is in the suspended state, settling of the solids in a slow-moving stream is probable. The organic fraction of the sludge deposits decomposes anaerobically, except for the thin surface layer which is subjected to aerobic decomposition due to the dissolved oxygen in the overlying waters. In warm weather, when the anaerobic decomposition proceeds at a more rapid rate, gaseous end products, usually carbon dioxide and methane, rise through the supernatant waters. The evolution of the gas bubbles may raise sludge particles to the water surface. Although this phenomenon may occur while the water contains some dissolved oxygen, the more intense action during the summer usually results in depletion of dissolved oxygen.

Reoxygenation. Water may absorb oxygen from the atmosphere when the oxygen in solution falls below saturation. Dissolved oxygen for receiving waters is also derived from two other sources: that in the receiving water and the waste flow at the point of discharge, and that given off by green plants. The latter source is restricted to daylight hours and the warmer seasons of the year and, therefore, is not usually used in any engineering analysis of stream capacity.

Unpolluted water maintains in solution the maximum quantity of dissolved oxygen. The saturation value is a function of temperature and the concentration of dissolved substances, such as chlorides. When oxygen is removed from solution, the deficiency is made up by the atmospheric oxygen, which is absorbed at the water surface and passes into solution. The rate at which oxygen is absorbed, or the rate of reaeration, is proportional to the degree of undersaturation and may be expressed as in Eq. (6), in which D is dissolved oxy-

$$\frac{dD}{dt} = -K_2 D \qquad (6)$$

gen deficit, t is time, and K_2 is reaeration coefficient.

The reaeration coefficient depends upon the ratio of the volume to the surface area and the intensity of fluid turbulence. An approximate value of the coefficient may be obtained from Eq. (7), in

$$K_2 = \frac{D_L U^{1/2}}{H^{3/2}} \qquad (7)$$

which D_L is coefficient of molecular diffusion of oxygen in water, U is average velocity of the river flow, and H is average depth of the river section.

The effect of temperature on this coefficient is identical with its effect on the deoxygenation coefficient. A common range of K_2 is from 0.20 to 5.0 per day. Many waste constituents, such as surface-active substances, interfere with the molecular diffusion of oxygen and reduce the value of the reaeration rate from that of pure water. Winds, waves, rapids, and tidal mixing are factors which create circulation and surface renewal and enhance reaeration.

Oxygen balance. The oxygen balance in a stream is determined by the concentration of organic matter and its rate of oxidation, and by the dissolved oxygen concentration and the rate of reaeration. The simultaneous action of deoxygenation and reaeration produces a pattern in the dissolved oxygen concentration known as the dissolved oxygen sag. The differential equation describing the combined action of deoxygenation and reaeration is given in Eq. (8), which states that the rate of

$$\frac{dD}{dt} = K_1 L - K_2 D \qquad (8)$$

change in the dissolved oxygen deficit D is the result of two independent rates. The first is that of oxygen utilization in the oxidation of organic matter. This reaction increases the dissolved oxygen deficit at a rate that is proportional to the concentration or organic matter L. The second rate is that of reaeration, which replenishes the oxygen utilized by the first reaction and decreases the deficit. Integration of this equation yields Eq. (9), where L_0

$$D_t = \frac{K_1 L_0}{K_2 - K_r} (e^{-K_r t} - e^{-K_2 t}) + D_0 e^{-K_2 t} \qquad (9)$$

and D_0 are the initial biochemical oxygen demand and the initial dissolved oxygen deficit, respectively, and D_t is the deficit at time t. The proportionality constants K_1 and K_2 represent the coefficients of deaeration and reaeration, respectively, and K_r the coefficient of BOD removal in the stream.

Figure 2 shows a typical dissolved oxygen sag curve resulting from a pollution of amount L_0 at $t = 0$. The sag curve is shown to result from the deoxy-

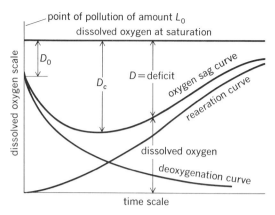

Fig. 2. Dissolved-oxygen sag curve and its components.

genation curve and the reaeration curve. A point of particular significance on the sag curve is that of minimum dissolved oxygen concentration, or maximum deficit. At this location, the rate of change of the deficit is zero, which results in the numerical equality of the opposing rates of deoxygenation and reoxygenation. The balance at this critical point may be written as Eq. (10), where the BOD at

$$K_2 D_c = K_1 L = K_1 L_0 e^{-K_r t_c} \qquad (10)$$

the critical point has been replaced by its equivalent at zero time (the location of the waste discharge). The value of the time t_c may be calculated from Eq. (11).

$$t_c = \frac{1}{K_2 - K_r} \log \frac{K_2}{K_r} \left[1 - \frac{D_0 (K_2 - K_r)}{K_1 L_0} \right] \qquad (11)$$

Allowable pollutional load. The pollutional load L_0 that a stream may absorb is a function of the dissolved oxygen deficit D_c, the coefficients $K_1, K_r,$ and K_2, and the initial deficit D_0. The dissolved oxygen deficit is usually established by water pollution standards of the health agency, and the initial deficit is determined by upstream pollution. The engineering problem is usually associated with the assignment of representative values of the coefficients K_1, K_r, and K_2 for a given flow and temperature condition.

Seasonal temperatures influence the saturation of oxygen and the rates of deaeration and reaeration. Variation in stream flow with the seasons affects the dilution factor. The most critical conditions occur during the summer when the stream runoff is low and the temperatures are high.

Pollution in lakes and estuaries. In lakes self-purification is slower than in streams because of the low rates of dispersion of the waste waters. There is no turbulence characteristic as in flowing rivers, and mixing depends primarily on winds, waves, and currents. Waste-water outfalls are designed to take advantage of the dispersion induced by these factors and to prevent the development of concentrated sewage fields.

In estuaries, the dispersion of waste waters is complicated by the tides, which carry various portions of the pollutant back and forth over many cycles, and by the difference of density in fresh water, waste water, and salt water. The equation defining the oxygen balance must be modified to allow for the greater time that an average particle of pollution is detained within the estuary; the

flushing mechanism of such bodies is therefore of primary concern. *See* ESTUARINE OCEANOGRAPHY.

Each estuary presents problems of density currents, configuration, and exchange that distinguish it from others. Field measurements of salinities, currents, and cross sections, in addition to the measurement of physical, chemical, and biological characteristics, are necessary to evaluate the pollution capacity of these watercourses. Dilution and dispersion in ocean waters is complicated by many of the same factors as in estuaries. The death rates of the coliform bacteria are greater in sea water. The outfalls must be designed and located to promote effective dispersion and to prevent the accumulation of sewage fields.

Oxidation ponds and land disposal. The forces of natural purification are utilized in shallow ponds, called oxidation ponds. Successful operation of these basins usually requires relatively high temperatures and sunshine. Carbon dioxide is released by means of the bacterial decomposition of the organic matter. Algae growth develops, consuming the carbon dioxide, ammonia, and other waste products and releasing oxygen under proper climatic conditions.

Oxidation ponds are efficient and relatively economical.

Instead of relying on the algae as a primary source of oxygen, mechanical aeration of the pond contents may be employed. Lagoons aerated in this manner are not as susceptible to climatic conditions as the oxidation ponds.

Land disposal of sewage is occasionally practiced by surface or flood irrigation. The former is the discharge of sewage upon the ground, from which it evaporates and through which it percolates. However, a significant portion remains which must be collected in surface drainage channels. Although this method is not particularly efficient for domestic sewage, a modification of it, spray irrigation, has been successfully employed in the treatment of a few industrial wastes. In flood irrigation, all the sewage is permitted to seep through the ground and is usually collected in underdrains. This method takes advantage of the mechanical filtration and biological purification afforded by the soil. Unless the sewage is treated before irrigation, odors and clogging usually occur and possible contamination of ground or surface water can result. *See* SANITARY ENGINEERING.

[DONALD J. O'CONNOR]

Bibliography: E. P. Anderson, *Audel's Domestic Water Supply and Sewage Disposal Guide,* 1960; G. M. Fair and J. C. Geyer, *Water Supply and Waste-Water Disposal,* 1954; H. W. Streeter and E. B. Phelps, *A Study of the Pollution and Natural Purification of the Ohio River,* Public Health Bull. no. 146, 1925; Texas Water and Sewage Works Association, *Manual for Sewage Plant Operators,* 3d ed., 1964; U.S. Public Health Service, *Oxygen Relationships in Stream,* Tech. Rep. no. W58−2, 1958.

Sewage treatment

Any process to which sewage is subjected in order to remove or alter its objectionable constituents and thus render it less offensive or dangerous. These processes may be classified as preliminary, primary, secondary, or complete, depending on the degree of treatment accomplished. Preliminary

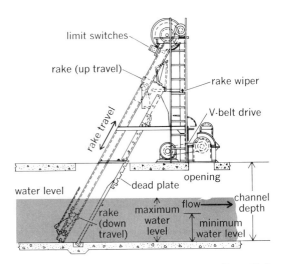

Fig. 1. Mechanically cleaned bar screen. (*From H. E. Babbitt and E. R. Baumann, Sewerage and Sewage Treatment, 8th ed., Wiley, 1958*)

treatment may be the conditioning of industrial waste prior to discharge to remove or to neutralize substances injurious to sewers and treatment processes, or it may be unit operations which prepare the water for major treatment. Primary treatment is the first and sometimes the only treatment of sewage. It is the removal of floating solids and coarse and fine suspended solids. Secondary treatment utilizes biological methods of treatment, that is, oxidation processes following primary treatment by sedimentation. Complete treatment removes a high percentage of suspended, colloidal, and organic matter.

Septic tanks and Imhoff tanks are considered secondary treatment methods because sedimentation is combined with biological digestion of the sludge. *See* IMHOFF TANK; SEPTIC TANK.

Coarse solids removal. This is accomplished by means of racks, screens, grit chambers, and skimming tanks. Racks are fixed screens composed of parallel bars placed in the waterway to catch debris. The bars are usually spaced 1 in. or more apart. Screens are devices with openings usually of uniform size 1 in. or less placed in the line of flow. Screens may be fixed or movable and vary in construction as bar screens, band screens, or cage screens. Such screens are hand cleaned or mechanically cleaned (Fig. 1). Grit chambers remove inorganic solids but may also trap heavier particles of organic nature such as seeds (Fig. 2). Grit chambers are designed so that the flow in the chamber is at 1 ft/sec or more. At less than that velocity, organic material also settles. Removal of grit is done either by hand or mechanically. Devices are added to mechanically cleaned units which wash most of the organic material out of the grit. Skimming chambers are devices for removing floating solids and grease. Air has been used to coagulate greases which then float and are skimmed off mechanically or by hand.

Fine solids removal. This is accomplished by screens with very small openings 1/16 or 1/32 in. wide, by sedimentation, or by both.

Fine screens are set in the line of flow and operated mechanically. Band screens, drum screens, plate screens, and vibratory screens are in use and the finer particles of floating solids are removed as well as coarse solids passing a rack. In some treatment plants screenings are passed through a grinder and returned to the flow so that they will settle out in the sedimentation tank. Another device, the comminutor, barminutor, or griductor, has high-speed rotating edges working in the flow of sewage (Fig. 3). These blades cut, chop, and shred the solids, which then pass on to the sedimentation unit.

Sedimentation. Sedimentation has one objective, the removal of settleable solids. Some floating materials are also removed by skimming devices, called clarifiers, built into sedimentation units. The basins are either circular or rectangular. In the circular unit sewage flows in at the center and out over weirs along the circumference (Fig. 4). In the rectangular tanks sewage flows into one end and out the other (Fig. 5).

The efficiency of a settling basin is dependent on a number of factors other than particle size, specific gravity, and settling velocity. Concentration of suspended matter, temperature, retention period, depth and shape of basin, baffling, total length of flow, wind, and biological effects all have an effect on solids removal. Density currents and short-circuiting may negate theoretical detention computations. Improper baffling may have the effect of reducing the effective surface area and creating dead or nonflow areas within the tank. In general a settling tank of good design with surface settling rates of 600 gal/(ft²) (day) and a 2-hr detention period will remove 50–60% of the suspended solids and at the same time remove 30–35% of the biochemical oxygen demand (Fig. 6).

The settling velocity of a particle is a function of specific gravity of the particle, specific gravity of water, viscosity of liquid, and particle diameter. Settling rates of particles larger than 0.1 mm are determined empirically. Sizes less than 0.1 mm settle in accordance with Stokes' law. Theoretically, if the forward motion of the water is less than

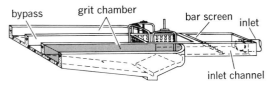

Fig. 2. Grit chamber. (*From Engineering Extension Department of Iowa State College, Bull. no. 58, 1953*)

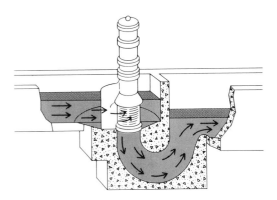

Fig. 3. A comminutor in place. (*Chicago Pump Co.*)

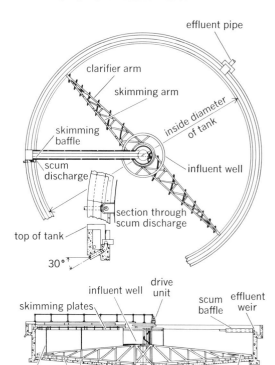

Fig. 4. Typical circular clarifier, a skimming device. (*From H. E. Babbitt and E. R. Baumann, Sewerage and Sewage Treatment, 8th ed., Wiley, 1958*)

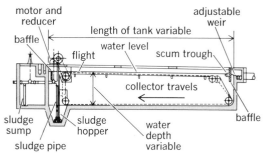

Fig. 5. Longitudinal section of typical rectangular clarifier. (*From E. W. Steel, Water Supply and Sewerage. 4th ed., McGraw-Hill, 1960*)

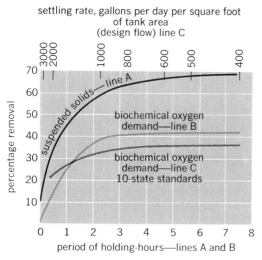

Fig. 6. Probable performance of sedimentation basins. (*From H. E. Babbitt and E. R. Baumann, Sewerage and Sewage Treatment, 8th ed., Wiley, 1958*)

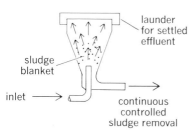

Fig. 7. Diagram of a vertical-flow sedimentation tank. (*From H. E. Babbitt and E. R. Baumann, Sewerage and Sewage Treatment, 8th ed., Wiley, 1958*)

the vertical settling rate of the particle, a particle at the surface will settle some distance below the surface in a given time interval. After that time interval the surface layer of water could be removed and it would contain no solids. The term surface settling rate is introduced as a practical measure of the rate of flow through the basin, if the rate of flow is equal to the surface area times the settling velocity of the smallest particle to be removed. Hence the selection of an overflow or surface settling rate expressed as gallons per day per square foot of surface area establishes a relationship between flow and area.

Flocculent suspensions have little or no settling velocity. These may occur in raw sewage but occur more frequently in secondary settling of effluents from activated sludge units. Such suspensions

may be removed by passing the inflowing water upward through a blanket of the material (Fig. 7). Theoretically there is a mechanical sweeping action in which smaller particles are attached to larger particles which then have sufficient weight to settle. Another type of treatment for such material is provided by an inner chamber equipped with baffles which rotate and stir the liquor and aid the formation of larger and heavier floc (Fig. 8). The same purpose is also achieved by agitation with air. Some of the settled sludge is raised by airlift and mixed in with the material, thus forming a mixture with improved settling characteristics.

Sedimentation basin design. Practical considerations and engineering judgment must be applied in designing sedimentation basins. Depth is usually held at 10-ft sidewall depth or less. The surface area requirement is usually 600 gal/(ft²) (day) for primary treatment alone and 800–1000 for all other tanks. The detention period is normally 2 hr. These three parameters of design must be adjusted since each is dependent on the other for a given design flow (average daily flow at a plant). When mechanical sludge-removal equipment is used, the tank dimensions are usually sized to a conventional equipment specification. Rectangular tanks are built-in units with common walls between units and unit width up to 25 ft. The length-width ratio, frequently determined by economical design dimension, should not be greater than 5:1. The mini-

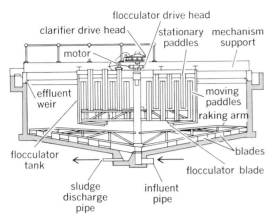

Fig. 8. The Dorr clariflocculator. (*From E. W. Steel, Water Supply and Sewerage, 4th ed., McGraw-Hill, 1960*)

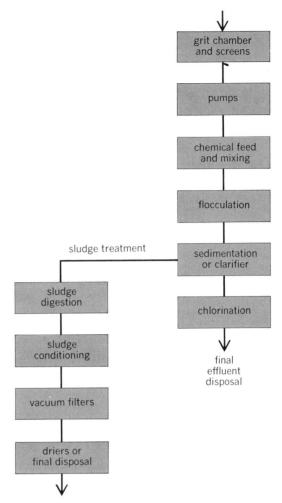

Fig. 9. Flow-through diagram of a chemical treatment plant. (*From H. E. Babbitt and E. R. Baumann, Sewerage and Sewage Treatment, 8th ed., Wiley 1958*)

mum length should be 10 ft. Final sizing may be fitted to convenient equipment dimensions.

Sludge removal on a regular schedule is mandatory in separate sedimentation tanks. If sludge is not removed, gasification occurs and large blocks of sludge begin to appear on the surface. These must then be removed by scum-removal mechanism or broken up so that they will settle. In circular tanks radial blades move the sludge to a center sludge hopper. In rectangular tanks the hopper is located at the inlet end and blades on a traveling chain move sludge in reverse of sewage flow. The heavier solids settle at the inlet and have a short travel path. These same blades may rise to the surface and move scum with the sewage flow to the outlet end where it is held by a baffle and removed by some form of scum-removal device. Sludge-removal mechanisms are often operated intermittently by time-clock relay mechanisms.

Appurtenances in the form of skimmers, scrapers, and other mechanical devices are many. Manufacturers have variants to offer, and competition is keen. Manufacturers' literature should be studied carefully and specifications should be carefully written to procure equipment meeting the requirements of engineering design.

Detention periods are theoretical. The actual flowthrough time is influenced by the inlet and outlet construction. On circular tanks inlets are submerged. Water rises inside a baffle extending downward to still the currents. Rectangular tank inlets may be submerged or, more commonly, sewage is brought to a trough which has a weir extending the width of the tank. The flow then moves forward with less short-circuiting. The outlet device on circular tanks is nearly always a circumferential weir adjusted to level after installation. The weir may be sharp-edged and level or provided with a sawtoothlike series of V-notches. On rectangular tanks, in order to provide enough weir length, a device known as a launder is used. A launder is a series of fingerlike shallow conduits set to water level and receiving flow from both sides of the conduit. Each of the fingers is connected to a common exit trough. The normal weir loading should not exceed 10,000 gal/linear ft of weir per day in small plants, or 15,000 in units handling more than 1,000,000 gal per day (1.0 mgd).

Chemical precipitation. Many attempts have been made to utilize chemical coagulants in the flocculation of sewage. The process, if used, is similar to that used in water treatment. The cost of chemicals and the somewhat intermediate treatment obtained with chemicals have kept this process out of general use. Its principal use today is in the preparation of sludge for filtration. Various steps in chemical precipitation are shown in Fig. 9. Alum, ferric sulfate, ferric chloride, and lime are used to form an insoluble precipitate which adsorbs colloidal and suspended solids. The entire floc settles and is removed as sludge. Sixty-four patents for proprietary chemical treatment processes were granted in the United States from 1873 to 1935. The Guggenheim process employs ferric chloride and aeration. The Scott-Darcy process employs ferric chloride made by treating scrap iron with chlorine solution. *See* WATER TREATMENT.

Oxidation processes. These are secondary treatment processes, although a few activated-sludge plants have been built without primary sedimentation. Oxidation process methods are (1) filtration by intermittent sand filters, contact filters, and trickling filters; (2) aeration by the activated-sludge process or by contact aerators; and (3) oxidation ponds. There are three basic oxidation methods, all depending on biological growth. Each provides a method of bringing organic matter in suspension or solution in sewage into immediate contact with a population of microorganisms living under aerobic conditions. The processes are called filtration, activated sludge, and contact aeration.

Filtration. Intermittent sand filters are sand beds provided with underdrains. Sewage is dosed intermittently by siphon or by pump, at rates from 20,000 gal per acre per day (gad) to a maximum of 125,000 gad when operated as a secondary treatment process. Rates may go to 500,000 gad (or 0.5 mgad) when operated as a tertiary process. Beds are usually 2 1/2–3 ft deep and are constructed with 6–12 in. of gravel at the bottom. The sand is sized to a uniformity coefficient of 5.0 or less (3.5 preferred), with effective size of 0.2–0.5 mm. The uniformity coefficient is the ratio between the sieve size that will pass 60% and the effective size. The effective size is the sieve size in millimeters that permits 10% of the sand by weight to pass. A mat of solids is formed in the surface layer of sand and must be removed periodically. The dry surface mat can be scraped clean, but periodically the top 6 in. or so of mat must be removed and replaced. Plants with sand filters operate at better than 95% removal of biochemical oxygen demand (BOD).

Trickling filters are beds of media, usually rock, over which settled sewage is sprayed. Microorganisms form a slime layer on the media surface and the water passes down over the surface in a thin film. Nutrients from the sewage are adsorbed in the slime layer and absorbed as food by organisms. Filters are ventilated through the underdrainage system or by other means, and thus oxygen, sewage, and organisms are brought together. Plants with trickling filtration have been operated at 90–95% efficiency of BOD removal.

Filter media include various materials such as stone, crushed rock, ceramic shapes, slag, and plastics. Preferred media are stone and crushed rock which do not fragment, flour, or soften on exposure to sewage. Rock sizes range from 1 to 6

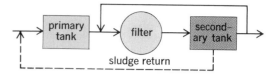

Fig. 11. Flow diagram of a single-stage high-rate trickling filter plant. (*From ASCE-FSIWA Joint Committee, Sewage Treatment Plant Design, 1959*)

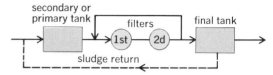

Fig. 12. Flow diagram of a two-stage high-rate trickling filter plant. (*From ASCE-FSIWA Joint Committee, Sewage Treatment Plant Design, 1959*)

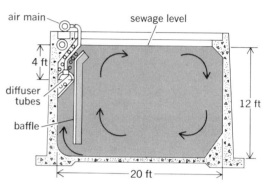

Fig. 13. Cross section of a spiral-flow activated-sludge tank with cylindrical diffusers. (*From E. W. Steel, Water Supply and Sewerage, 4th ed., McGraw-Hill, 1960*)

in.: however, current practice employs sizes between 2- and 4-in. nominal diameter. Plastic corrugated sheets have been employed on very deep filters. Pretreatment of sewage is normally required. When the waste contains a concentration of dissolved solids, as with milk waste, without any great concentration of settleable solids, the waste may be applied directly to the filter. Some advantage is gained by preaeration so that the waste applied to the filter has some dissolved oxygen.

Filters are classified as standard or low-rate filters, high-rate filters, and controlled filters. The filter introduced in the United States early in the 20th century was a bed of stone 6–8 ft deep with a distribution system of fixed nozzles. This type of filter is called a standard or low-rate filter. The allowable organic loading is about one-third that of a high-rate filter having 3- to 6-ft depth introduced during 1930–1940 and developed with many variations of recirculation and application of sewage since that time. In 1956 controlled filtration on sectionalized units composing a deep filter was introduced. The loading rate with no recirculation on such filters is 10–12 times that of low-rate filters.

Low-rate filters are dosed at a rate of 1–4 mgad by siphon through nozzles so spaced that water reaches every part of the filter surface during a dosing cycle. The application of water by this method is intermittent. The rotary distributor (Fig. 10) may also be operated by siphon. This type of

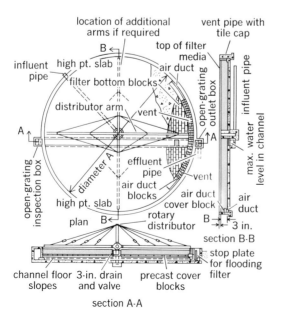

Fig. 10. Circular trickling filter. (*Link-Belt Co.*)

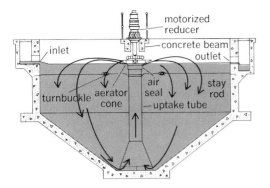

Fig. 14. Simplex aerator. (*From E. W. Steel, Water Supply and Sewerage, 4th ed., McGraw-Hill, 1960*)

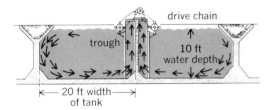

Fig. 15. Cross section of a Link-Belt mechanical aerator. (*From E. W. Steel, Water Supply and Sewerage. 4th ed., McGraw-Hill, 1960*)

distributor has two or four radial arms supported on a center pedestal. Hydraulic force of water passing through the nozzles fixed to the arm causes the arm to rotate. The distributor may be operated in continuous rotation by feeding from a weir box. In either case the filter is sprayed as the arm passes over a given section and the dosing is intermittent with a short time interval between doses. With the fixed-nozzle method the interval may be 5 min, but with the rotary distributor the dosing interval may be no more than 15 sec.

High-rate filters depend on recirculation. The hydraulic loading rate is about 20 mgad with a range of 9–44 mgd. Rotary distributors are used. Pumps pick up settled effluent and return it. Filters are often set up as primary and secondary filters with recirculation of water to each. Several alternative flow arrangements are demonstrated in Figs. 11 and 12. Recirculation ratios range from 1:1 to about 5:1. Final sedimentation is required for both low- and high-rate filters as filter slime and organic debris are washed free.

Aeration. Aeration is accomplished in tanks in which compressed air is diffused in liquid by various devices: filter plates, filter tubes, ejectors, and jets; or in which air is mixed with liquid by mechanical agitation. The high degree of treatment possible with conventional activated sludge, 95–98% BOD removal, has made it a popular method of treatment. Sewage organisms seeded in sludge which has passed through treatment are returned to incoming sewage and mixed thoroughly with the liquor. In this way the biota, oxygen supply, and sewage are brought together. Contact aeration utilizes air diffusion to keep a biota suspension thoroughly mixed; however, the biota are also maintained in active growth on plates of impervious material such as cement-asbestos suspended in the mixed liquor of the aeration tank. Slime growth forms on the plates, and liquid passing by the plates furnishes the plate biota with a source of nutrients.

Activated-sludge process, the conventional process, requires an aeration period of 4–8 hr. Much of the oxidation takes place in the first 3 hr of detention. Aeration tanks are usually long, narrow, rectangular tanks with porous plates or diffusers along the length to keep the liquor well agitated throughout (Fig. 13). Widths are 15–30 ft and depths about 15 ft. Length-width ratio is about 5:1.

Air requirements are 0.2–1.5 ft³ air/gal of sewage treated. It is necessary to maintain dissolved oxygen (DO) levels at 2 ppm or higher.

Mechanical aeration is done in square or rectangular aeration tanks, depending on the mechanism. In the Simplex method liquor is drawn by impeller up a draft tube and expelled over the tank surface (Fig. 14). In the Link-Belt unit, brushes introduce a spiral motion with considerable agitation. The period of aeration may be up to 8 hr with this method (Fig. 15). Modifications of the aeration process include modified aeration, step aeration, tapered aeration, stage aeration, biosorption, bioactivation, dual aeration, and others.

Recirculation of sludge is one of the essentials of the process. About 25–35% of the sludge settled in the final sedimentation tank is returned to the aeration tank (Fig. 16). Concentration of solids in mixed liquor may be about 3000 mg/liter in diffused air units and a little less in mechanical aeration units. The ratio of sludge volume settled to suspended solids is known as the Mohlmann index:

$$\text{Mohlmann index} = \frac{\text{volume of sludge settled in 30 min, \%}}{\text{suspended solids, \%}}$$

A good settling sludge has an index below 100. Sludge age, another important factor, is the average time that a particle of suspended solids re-

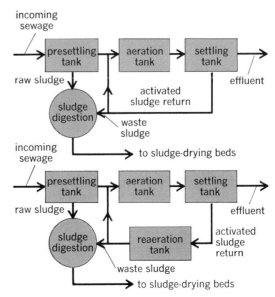

Fig. 16. Flow diagrams of typical activated-sludge plants. (*From E. W. Steel, Water Supply and Sewerage, 4th ed., McGraw-Hill, 1960*)

mains under aeration and is the ratio of the dry weight of sludge in the tank in pounds to the suspended solids load in pounds per day.

Contact aerators provide an aeration period of 5 hr or more. Aeration is usually preceded by preaeration of the raw sewage before primary settling. The preaeration lasts 1 hr. Loadings are based on two factors: pounds per day per 1000 ft² of contact surface (6.0 or less), and pounds per day per 1000 ft² per hour of aeration (1.2 or less). Air supply of 1.5 ft³/gal of flow is required. The process has an overall plant efficiency of about 90% BOD removal.

Chlorination. Chlorination of treated sewage has one major purpose: to reduce the coliform group of organisms. Sufficient chlorine to satisfy demand and provide a residual of 2.0 mg/liter should be added. The following magnitude of dosage is possible: primary effluent, 20 mg/liter; trickling filter plant effluent, 15 mg/liter; activated sludge plant effluent, 8 mg/liter; sand filter effluent, 6 mg/liter. The contact period should be at least 15 min at peak hourly flow.

Oxidation ponds. These are ponds 2–4 ft in depth designed to allow the growth of algae under suitable conditions in sewage media. Oxygen is absorbed from the air, but the conversion of CO_2 to O_2 by *Chlorella pyrenoidosa* and other algae provides an additional source of oxygen of great value. Oxidation ponds should be preceded by primary treatment. A loading figure of 50 lb BOD/acre is recommended. BOD removal efficiency may range from 40 to 70%.

[WILLIAM T. INGRAM]

Bibliography: ASCE-FSIWA Joint Committee, *Sewage Treatment Plant Design*, Geyer, 1959; H. E. Babbitt and E. R. Baumann, *Sewerage and Sewage Treatment*, 8th ed., 1958; G. M. Fair and J. C. Geyer, *Water Supply and Waste-Water Disposal*, 1954.

Smog, photochemical

The word "smog" can be defined in several ways. It was coined near the beginning of the century to refer to a combination of coal smoke and fog, the particles of the smoke often serving as nuclei on which the water vapor condensed. Later, the term was used to refer to any dirty urban atmosphere. A unique type of dirty atmosphere was recognized in the late 1940s and early 1950s and came to be known as photochemical smog. It can be defined as that type of air pollution which owes many of its properties to the products of photochemical reactions involving the vapor of various organic substances, especially hydrocarbons, oxides of nitrogen, and atmospheric oxygen. Occasionally, fog may also be involved in the formation of such smog.

Photochemical smog is particularly apparent in cities where little coal is burned, there is little industrial pollution, and there are large concentrations of automobiles. Undoubtedly, photochemical smog contributed to the pollution in many cities long before it was recognized as a special type of pollution.

History. Photochemical smog was first recognized in Los Angeles, and most of the early research was undertaken there. Ozone in high concentrations (0.2 parts per million by volume, ppmv, is typical) was first detected in Los Angeles and other western United States cities by A. W. Bartel and J. W. Temple. A. J. Haagen-Smit, C. E. Bradley, and M. M. Fox demonstrated that the ozone is formed photochemically from a mixture of oxides of nitrogen, hydrocarbon vapor, and air. At about this time F. W. Went found that the products of chemical reactions of ozone with unsaturated hydrocarbons such as those in automobile exhaust gases produced the same type of damage to plants as did Los Angeles smog, and a group at Stanford Research Institute demonstrated that automobiles were the principal contributors to the reactive organic vapor such as hydrocarbons and much of the oxides of nitrogen in the Los Angeles atmosphere. R. D. Cadle and H. S. Johnston pointed out that the reaction that initiated the ozone formation was almost certainly the photochemical decomposition of nitrogen dioxide to form nitric oxide and atomic oxygen followed by reaction of the latter with hydrocarbons and with molecular oxygen of the air. While a large amount of very competent, valuable research into the causes, effects, and cures of photochemical smog has been undertaken since the early 1950s, it has all been based on that early research.

Meteorology. The intensity of photochemical smog at any time and place, like that of any type of air pollution, is markedly influenced by meteorological conditions and the geographical features of the region. If a city is in a bowl defined by mountains or hills, and an atmospheric inversion clamps a lid over the area, smog of any kind is apt to be intense. Even so, the smog intensity in a particular part of a city subject to such conditions may be very different from that in another part of the same city, due to a number of factors such as local air motions. *See* TEMPERATURE INVERSION.

Effects. Usually the most obvious and immediately annoying aspect of urban pollution is the murkiness of the air which it produces. This visibility decrease is mainly caused by the suspended particles. The median particle size in photochemical smog is so small that the particles preferentially scatter blue light, and depending upon the nature and angle of illumination, the air may appear blue or brown. The brown appearance is accentuated by the presence of nitrogen dioxide (NO_2), which is itself brown in color and results from various combustion operations, including driving automobiles.

Another especially annoying feature of photochemical smog is the eye irritation and lacrimation it produces. This irritation seems to result from the combined effects of several gases, especially for-

Typical concentrations of some pollutants in photochemical smog parts per hundred million by volume (pphmv)

Pollutant	Concentration
CO	3000
O_3	25
Hydrocarbons (paraffins)	100
Hydrocarbons (olefins)	25
Acetylene	25
Aldehydes	60
SO_2	20
Oxides of nitrogens	20
NH_3	2

maldehyde, acrolein, peroxyacetyl nitrate (PAN), and possibly peroxybenzoyl nitrate.

As mentioned above, photochemical smog produces an unusual type of plant damage (phytotoxic effect). It has been variously described as silvering, bronzing, or glazing, and usually first appears on the underside of the leaves. This type of damage has resulted in serious economical loss, the extent of damage varying markedly with the type of plant. For example, spinach is especially susceptible. The damage is, of course, not limited to agricultural crops. Forests in the vicinity of Los Angeles have been severely damaged.

The phytotoxicants almost certainly are formed by the photochemical reactions mentioned above, but the substances responsible for the plant damage have not been identified. Ozone, PAN, and various organic peroxides that may be produced by smog reactions all damage plants, but the symptoms are different from those described above.

Although the main acute physiological effect of photochemical smog on humans—eye irritation—has been studied extensively, little is known about possible chronic effects, and the little that is known is mainly circumstantial evidence obtained by statistical comparisons of people living in rural and urban environments. Such studies appear to implicate smog of both types in causing heart disease as well as various types of cancer. Furthermore, various carcinogens, such as 3,4-benzpyrene, have been isolated from Los Angeles smog. Skin cancers have been produced by applying extracts of gasoline engine and diesel engine exhaust gases to the skin of mice. Although certainly not proved, photochemical smog may be considerably less carcinogenic than that of the coal-burning variety; the concentrations of 3,4-benzpyrene in western United States cities where the smog is primarily photochemical are lower than in the highly industrialized, coal-burning, eastern areas.

Chemical reactions. The primary and secondary photochemical reactions leading to unpleasant products in photochemical smog were mentioned earlier. Although these reactions and their products render the smog much more unpleasant, the atmosphere would be very unpleasant even in the absence of these reactions because of the accumulations of automobile exhaust and other anthropogenic emissions to the atmosphere. In simplified form, these are reactions (1) through (11).

$$NO_2 + h\nu \rightarrow NO + O \qquad (1)$$
$$O + O_2 + M \rightarrow O_3 + M \qquad (2)$$
$$O_3 + NO \rightarrow NO_2 + O_2 \qquad (3)$$
$$O + Olefins \rightarrow R + R^1O$$
$$or \quad R—R^1 \qquad (4)$$
$$\underset{O}{\diagdown}$$
$$O_3 + Olefins \rightarrow Products \qquad (5)$$
$$R + O_2 \rightarrow RO_2 \qquad (6)$$
$$RO_2 + O_2 \rightarrow RO + O_3 \qquad (7)$$
$$RO_2 + NO \rightarrow RO + NO_2 \qquad (8)$$
$$O_2(^1\Delta) + RH \rightarrow ROOH \qquad (9)$$
$$RO_2 + SO_2 \rightarrow RO + SO_3 \qquad (10)$$
$$SO_3 + H_2O \rightarrow H_2SO_4 \qquad (11)$$

Several of these reactions are highly speculative, such as (7), (8), and (10). Reaction (9) has often been suggested but is probably too slow to be important. Reactions (1), (2), and (3) lead to a pseudo-steady-state concentration of ozone that is much

smaller than that occurring in photochemical smog. Two general processes have been proposed to explain the large accumulation of ozone. One (the more likely) is that some reaction, such as (8), removes NO so rapidly that reaction (3) is less effective in destroying ozone. The other is that some process in addition to (2), such as (7), produces ozone.

The chemical reactions produce large quantities of particles. A large percentage of these is produced by reactions involving free radicals such as R, RO, and RO_2. These radicals undergo complicated sequences of reactions producing high-molecular-weight substances of low vapor pressure that condense as particles from the air. Sulfuric acid droplets are also present in smog, and in the photochemical variety they are largely produced by rapid oxidation of sulfur dioxide (SO_2) emitted by the burning of sulfur-containing fuels followed by reaction (11). This oxidation seems to be intimately associated with the other smog reactions. The SO_2 may be oxidized to SO_3 by reaction (10) or a similar reaction involving the hydroperoxy radical (HO_2).

The free-radical reactions produce a large variety of organic compounds, such as PAN, organic acids, and aldehydes, in addition to the high-molecular-weight products. There is mounting evidence that chain reactions involving hydroxyl radicals (OH) may be important, the chains being initiated by reactions such as (12).

$$O + RCHO \rightarrow OH + RCO \qquad (12)$$

Obviously the chemistry is very complex, and details will remain unresolved for many years. The chemical reactions occurring in smog are not limited to cities, but are taking place in the atmosphere all over the globe. The trace atmospheric constituents are both natural and man-made—hydrocarbons, for example, being emitted in vast amounts by many species of plants.

Control. The alleviation of the unpleasant effects of photochemical smog, like that of all smog, remains primarily control at the source. Many suggestions, some highly ingenious, have been made for removing smog, once it has formed. The usual difficulty is the tremendous weight of air that must be removed or treated. Since automobile exhaust is the major precursor of photochemical smog, methods for greatly decreasing the rates of emission of harmful constituents in such exhaust have received much attention. Three such constituents are especially undesirable: organic vapors, particularly hydrocarbons; oxides of nitrogen; and carbon monoxide (CO). While little is known about the possible participation of CO in smog reactions, it is highly toxic. Control is achieved by engine modification, by afterburners, and by combinations of these.

Of course, automobiles are not the only contributors to photochemical smog. Almost all combustion produces oxides of nitrogen. Various industrial operations, especially refineries, emit organic vapors; evaporation of gasoline occurs at service stations; and backyard incinerators can contribute appreciably.

In principle, all pollution of the air could be stopped by eliminating all pollutant emissions, but this is not economically or technologically feasible. Much research on photochemical smog is designed to provide the basis for sound decisions concerning where to place the emphasis for control. The results of such research have already been very effective. Attempts are being made to make the results of the chemical, meteorological, and economic studies even more useful by combining them into mathematical models which can be used to predict the amount of pollution at any given place depending upon the input into the models. Such models can be used to predict the effectiveness of various control measures. *See* AIR-POLLUTION CONTROL; ATMOSPHERIC POLLUTION.

[RICHARD D. CADLE]

Bibliography: R. D. Cadle, *J. Coll. Interface Chem.*, 39:25–31, 1972; R. D. Cadle and E. R. Allen, *Science*, 167:243–249, 1970; P. A. Leighton, *Photochemistry of Air Polution*, 1961; C. S Thesday (ed.), *Chemical Reactions in Urban Atmospheres*, 1971.

Soil

Freely divided rock-derived material containing an admixture of organic matter and capable of supporting vegetation. Soils are independent natural bodies, each with a unique morphology resulting from a particular combination of climate, living plants and animals, parent rock materials, relief, the groundwaters, and age. Soils support plants, occupy large portions of the Earth's surface, and have shape, area, breadth, width, and depth. Soil, as used here, differs in meaning from the term as used by engineers, where the meaning is unconsolidated rock material. *See* SOIL CHEMISTRY.

This article is divided into four parts: origin and classification of soils, physical properties of soil, soil management, and soil erosion.

ORIGIN AND CLASSIFICATION OF SOILS

Soil covers most of the land surface as a continuum. Each soil grades into the rock material below and into other soils at its margins, where changes occur in relief, groundwater, vegetation, kinds of rock, or other factors which influence the development of soils. Soils have horizons, or layers, more or less parallel to the surface and differing from those above and below in one or more properties, such as color, texture, structure, consistency, porosity, and reaction (Fig. 1). The horizons may be thick or thin. They may be prominent, or so weak that they can be detected only in the laboratory. The succession of horizons is called the soil profile. In general, the boundary of soils with the underlying rock or rock material occurs at depths ranging from 1 to 6 ft, though the extremes lie outside of this range.

Origin of soils. Soil formation proceeds in stages, but these stages may grade indistinctly from one into another. The first stage is the accumulation of unconsolidated rock fragments, the parent material. Parent material may be accumulated by deposition of rock fragments moved by glaciers, wind, gravity, or water, or it may accumulate more or less in place from physical and chemical weathering of hard rocks.

The second stage is the formation of horizons. This stage may follow or go on simultaneously with the accumulation of parent material. Soil horizons are a result of dominance of one or more processes over others, producing a layer which differs from the layers above and below.

Major processes. The major processes in soils which promote horizon differentiation are gains, losses, transfers, and transformations of organic matter, soluble salts, carbonates, silicate clay minerals, sesquioxides, and silica. Gains consist normally of additions of organic matter, and of oxygen and water through oxidation and hydration, but in some sites slow continuous additions of new mineral materials take place at the surface or soluble materials are deposited from groundwater. Losses are chiefly of materials dissolved or suspended in water percolating through the profile or running off the surface. Transfers of both mineral and organic materials are common in soils. Water moving through the soil picks up materials in solution or suspension. These materials may be deposited in another horizon if the water is withdrawn by plant roots or evaporation, or if the materials are precipitated as a result of differences in pH (degree of acidity), salt concentration, or other conditions in deeper horizons.

Other processes tend to offset those that promote horizon differentiation. Mixing of the soil occurs as the result of burrowing by rodents and earthworms, overturning of trees, churning of the soil by frost, or shrinking and swelling. On steep slopes the soil may creep or slide downhill with attendant mixing. Plants may withdraw calcium or other ions from deep horizons and return them to the surface in the leaf litter.

Saturation of a horizon with water for long periods makes the iron oxides soluble by reduction from ferric to ferrous forms. The soluble iron can move by diffusion to form hard concretions or

Fig. 1. Photograph of a soil profile showing horizons. The dark crescent-shaped spots at the soil surface are the result of plowing. The dark horizon lying 9–18 in. below the surface is the principal horizon of accumulation of organic matter that has been washed down from the surface. The thin wavy lines were formed in the same manner.

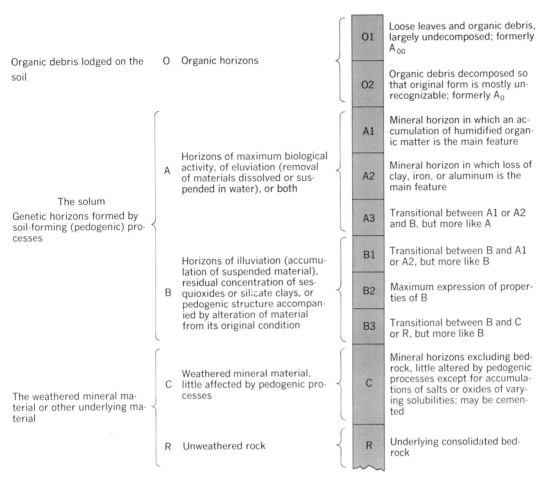

Organic debris lodged on the soil   O   Organic horizons

O1   Loose leaves and organic debris, largely undecomposed; formerly A_{00}

O2   Organic debris decomposed so that original form is mostly unrecognizable; formerly A_0

The solum
Genetic horizons formed by soil-forming (pedogenic) processes

A   Horizons of maximum biological activity, of eluviation (removal of materials dissolved or suspended in water), or both

A1   Mineral horizon in which an accumulation of humidified organic matter is the main feature

A2   Mineral horizon in which loss of clay, iron, or aluminum is the main feature

A3   Transitional between A1 or A2 and B, but more like A

B   Horizons of illuviation (accumulation of suspended material), residual concentration of sesquioxides or silicate clays, or pedogenic structure accompanied by alteration of material from its original condition

B1   Transitional between B and A1 or A2, but more like B

B2   Maximum expression of properties of B

B3   Transitional between B and C or R, but more like B

The weathered mineral material or other underlying material

C   Weathered mineral material, little affected by pedogenic processes

C   Mineral horizons excluding bedrock, little altered by pedogenic processes except for accumulations of salts or oxides of varying solubilities; may be cemented

R   Unweathered rock

R   Underlying consolidated bedrock

Fig. 2. A hypothetical soil profile having all principal horizons. Other symbols are used to indicate features subordinate to those indicated by capital letters and numbers. The more important of these are as follows: ca, as in Cca, accumulations of carbonates; cs, accumulations of calcium sulfate; cn, concretions; g, strong gleying (reduction of iron in presence of groundwater); h, illuvial humus; ir, illuvial iron; m, strong cementation; p, plowing; sa, accumulations of very soluble salts; si, cementation by silica; t, illuvial clay; x, fragipan (a compact zone which is impenetrable by roots).

splotches of red or brown in a gray matrix. Or if the iron remains, the soil will have shades of blue or green. This process is called gleying, and can be superimposed on any of the others.

The kinds of horizons present and the degree of their differentiation, both in composition and structure, depend on the relative strengths of the processes. In turn, these relative strengths are determined by the way man uses the soil as well as by the natural factors of climate, plants and animals, relief and groundwater, and the period of time during which the processes have been operating.

Composition. In the drier climates where precipitation is appreciably less than the potential for evaporation and transpiration, horizons of soluble salts, including calcium carbonate and gypsum, are normally found at the average depth of water penetration.

In humid climates some materials normally considered insoluble may be gradually removed from the soil or at least from the surface horizons. A part of the removal may be in suspension. The movement of silicate clay minerals would be an example. The movement of iron oxides is accelerated by the formation of chelates with the soil organic matter. Silica is removed in appreciable amounts in solution or suspension, though quartz sand is relatively unaffected. In warm humid climates free iron and aluminum oxides and silicate clays accumulate in soils, apparently because of low solubility relative to other minerals.

In cool humid climates solution losses are evident in such minerals as feldspars. Free sesquioxides tend to be removed from the surface horizons and to accumulate in a lower horizon, but mixing by animals and falling trees may counterbalance the downward movement.

Structure. Concurrently with the other processes, distinctive structures are formed in the different horizons. In the surface horizons, where there is a maximum of biotic activity, small animals, roots, and frost action keep mixing the soil material. Aggregates of varying sizes are formed and bound by organic matter, microorganisms, and colloidal material. The aggregates in the immediate surface tend to be loosely packed with many large pores among them. Below this horizon of high biotic activity, the structure is formed chiefly by volume changes due to wetting, drying, freezing, thawing, or shaking of the soil by roots of trees swaying with the wind. Consequently, the sides of any one aggregate, or ped, conform in shape to the sides of adjacent peds.

Water moving through the soil usually follows root channels, wormholes, and ped surfaces. Accordingly, materials that are deposited in a horizon commonly coat the peds. In the horizons that have received clay from an overlying horizon, the peds usually have a coating or varnish of clay making the exterior unlike the interior in appearance. Peds formed by moisture or temperature changes normally have the shapes of plates, prisms, or blocks.

Horizons. Pedologists have developed sets of symbols to identify the various kinds of horizons commonly found in soils. The nomenclature originated in Russia, where the letters A, B, and C were applied to the main horizons of the black soils of the steppes. A designated the dark surface horizon of maximum organic matter accumulation, C the unaltered parent material, and B the intermediate horizon. The usage of the letters A, B, and C spread to western Europe, where the intermediate or B horizon was a horizon of accumulation of free sesquioxides or silicate clays or both. Thus the idea developed that a B horizon is a horizon of accumulation. Some, however, define a B horizon by position between A and C. Subdivisions of the major horizons have been shown by either numbers or letters, for example, Bt or B2. No internationally accepted set of horizon symbols has been developed. In the United States the designations shown in Fig. 2 have been widely used since about 1935, with minor modifications made in 1962. Lowercase letters were added to numbers in B horizons to indicate the nature of the material that had accumulated. Generally, "h" is used to indicate translocated humus, "t" for translocated clay, and "ir" for translocated iron oxides. Thus, B2t indicates the main horizon of clay accumulation.

Classification. Systems of soil classification are influenced by concepts prevalent at the time a system is developed. Since ancient times, soil has been considered as the natural medium for plant growth. Under this concept, the earliest classifications were based on relative suitability for different crops, such as rice soils, wheat soils, and vineyard soils.

Early American agriculturists thought of soil chiefly as disintegrated rock, and the first comprehensive American classification was based primarily on the nature of the underlying rock.

In the latter part of the 19th century, some Russian students noted relations between the steppe and black soils and the forest and gray soils. They developed the concept of soils as independent natural bodies formed by the influence of environmental factors operating on parent materials over time. The early Russian classifications grouped soils at the highest level, according to the degree to which they reflected the climate and vegetation. They had classes of Normal, Abnormal, and Transitional soils, which later became known as Zonal, Intrazonal, and Azonal. Within the Normal or Zonal soils, the Russians distinguished climatic and vegetative zones in which the soils had distinctive colors and other properties in common. These formed classes that were called soil types. Because some soils with similar colors had very different properties that were associated with differences in the vegetation, the nature of the vegetation was some-times considered in addition to the color to form the soil type name, for example, Gray Forest soil and Gray Desert soil. The Russian concepts of soil types were accepted in other countries as quickly as they became known. In the United States, however, the name soil type had been used for some decades to indicate differences in soil texture, chiefly texture of the surface horizons; so the Russian soil type was called a Great Soil Group. *See* SOIL, ZONALITY OF.

Many systems of classification have been attempted but none has been found markedly superior; most systems have been modifications of those used in Russia. Two bases for classification have been tried. One basis has been the presumed genesis of the soil; climate and native vegetation were given major emphasis. The other basis has been the observable or measurable properties of the soil. To a considerable extent, of course, these are used in the genetic system to define the great soil groups. The morphologic systems, however, have not used soil genesis as such, but have attempted to use properties that are acquired through soil development.

The principal problem in the morphologic systems has been the selection of the properties to be used. Grouping by color, tried in the earliest systems, produces soil groups of unlike genesis.

The Soil Survey staff of the U.S. Department of Agriculture and the land-grant colleges adopted a new classification scheme in 1965. Although the new system has been widely tested, only time can tell how much more useful it will be than earlier systems. As knowledge of soil genesis increases, modifications of classification systems will continue to be necessary.

The system differs from earlier systems in that it may be applied to either cultivated or virgin soils. Previous systems have been based on virgin profiles, and cultivated soils were classified on the presumed characteristics or genesis of the virgin soils. The new system has six categories, based on both physical and chemical properties. These categories are the order, suborder, great group, subgroup, family, and series, in decreasing rank.

The nomenclature. The names of the taxa or classes in each category are derived from the classic languages in such a manner that the name itself indicates the place of the taxa in the system and usually indicates something of the differentiating properties. The names of the highest category, the order, end in the suffix "sol," preceded by formative elements that suggest the nature of the order. Thus, Aridisol is the name of an order of soils that is characterized by being dry (Latin *aridus*, dry, plus *sol*, soil). A formative element is taken from each order name as the final syllable in the names of all taxa of suborders, great groups, and subgroups in the order. This is the syllable beginning with the vowel that precedes the connecting vowel with "sol." Thus, for Aridisols, the names of the taxa of lower classes end with the syllable "id," as in Argid and Orthid.

Suborder names have two syllables, the first suggesting something of the nature of the suborder and the last identifying the order. The formative element "arg" in Argid (Latin argillus, clay) suggests the horizon of accumulation of clay that defines the suborder.

Great group names have one or more syllables to

suggest the nature of the horizons and have the suborder name as an ending. Thus great group names have three or more syllables but can be distinguished from order names because they do not end in "sol." Among the Argids, great groups are Natrargids (Latin *natrium*, sodium) for soils that have high contents of sodium, and Durargids (Latin *durus*, hard) for Argids with a hardpan cemented by silica and called a duripan.

Subgroup names are binomial. The great group name is preceded by an adjective such as "typic," which suggests the type or central concept of the great group, or the name of another great group, suborder, or order converted to an adjective to suggest that the soils are transitional between the two taxa.

Family names consist of several adjectives that describe the texture (sandy, silty, clayey, and so on), the mineralogy (siliceous, carbonatic, and so on), the temperature regime of the soil (thermic, mesic, frigid, and so on), and occasional other properties that are relevant to the use of the soil.

Series names are abstract names, taken from towns or places near where the soil was first identified. Cecil, Tama, and Walla Walla are names of soil series.

Order. In the highest category 10 orders are recognized. These are distinguished chiefly by differences in kinds and amounts of organic matter in the surface horizons, kinds of B horizons resulting from the dominance of various specific processes, evidences of churning through shrinking and swelling, base saturation, and lengths of periods during which the soil is without available moisture. The properties selected to distinguish the orders are reflections of the degree of horizon development and the kinds of horizons present.

The orders, the formative elements in the names, and the general nature of the included soils are given in Table 1.

Suborder. This category narrows the ranges in soil moisture and temperature regimes, kinds of horizons, and composition, according to which of these is most important. Moisture or temperature or soil properties associated with them are used to define suborders of Alfisols, Mollisols, Oxisols, Ultisols, and Vertisols. Kinds of horizons are used for Aridisols, compositions for Histosols and Spodosols, and combinations for Entisols and Inceptisols.

Great group. The taxa (classes) in this category group soils that have the same kinds of horizons in the same sequence and have similar moisture and temperature regimes. Exceptions to horizon sequences are made for horizons so near the surface that they are apt to be mixed by plowing or lost rapidly by erosion if plowed.

Subgroup. The great groups are subdivided into subgroups that show the central properties of the great group, intergrade subgroups that show properties of more than one great group, and other subgroups for soils with atypical properties that are not characteristic of any great group.

Family. The families are defined largely on the basis of physical and mineralogic properties of importance to plant growth.

Series. The soil series is a group of soils having horizons similar in differentiating characteristics and arrangement in the soil profile, except for texture of the surface portion, and developed in a particular type of parent material.

Table 1. Soil orders

Order	Formative element in name	General nature
Alfisols	alf	Soils with gray to brown surface horizons, medium to high base supply, with horizons of clay accumulation; usually moist, but may be dry during summer
Aridisols	id	Soils with pedogenic horizons, low in organic matter, and usually dry
Entisols	ent	Soils without pedogenic horizons
Histosols	ist	Organic soils (peats and mucks)
Inceptisols	ept	Soils that are usually moist, with pedogenic horizons of alteration of parent materials but not of illuviation
Mollisols	oll	Soils with nearly black, organic-rich surface horizons and high base supply
Oxisols	ox	Soils with residual accumulations of inactive clays, free oxides, kaolin, and quartz; mostly tropical
Spodosols	od	Soils with accumulations of amorphous materials in subsurface horizons
Ultisols	ult	Soils that are usually moist, with horizons of clay accumulation and a low supply of bases
Vertisols	ert	Soils with high content of swelling clays and wide deep cracks during some season

Type. This category of earlier systems of classification has been dropped but is mentioned here because it was used for almost 70 years and many references to it are found in the literature about soils. The soil types within a series differed primarily in the texture of the plow layer or equivalent horizons in unplowed soils. Cecil clay and Cecil fine sandy loam were types within the Cecil series. The texture of the plow layer is still indicated in published soil surveys if it is relevant to the use of the soil, but it is now considered as one kind of soil phase. Soil surveys are discussed in the next section of this article.

Classifications of soils have been developed in several countries based on other differentia. The principal classifications have been those of the Soviet Union, Germany, France, Canada, Australia, and New Zealand, and the United States. Other countries have modified one or the other of these to fit their own conditions. Soil classifications have usually been developed to fit the needs of a government that is concerned with the use of its soils. In this respect soil classification has differed from classifications of other natural objects, such as plants and animals, and there is no international agreement on the subject.

Many practical classifications have been developed on the basis of interpretations of the usefulness of soils for specific purposes. An example is

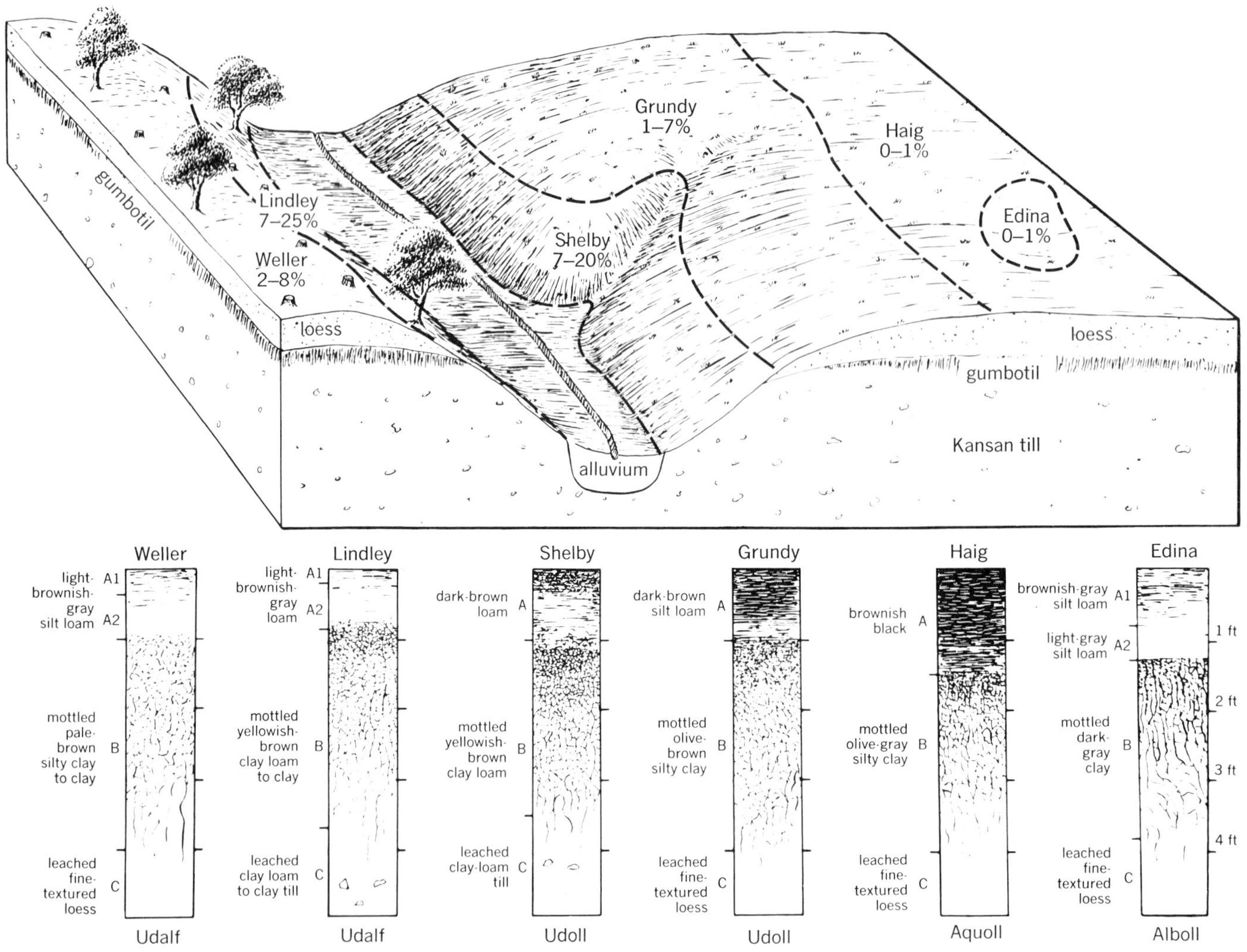

Fig. 3. Sketch showing the relation of the soil pattern to relief, parent material, and native vegetation on a farm in south-central Iowa. The soil slope gradient is expressed as a percentage. (*Modified from R. W. Simonson, F. F. Riechen, and G. D. Smith, Understanding Iowa Soils, Brown, 1952*)

the capability classification, which groups soils according to the number of safe alternative uses, risks of damage, and kinds of problems that are encountered under use.

Surveys. Soil surveys include those researches necessary (1) to determine the important characteristics of soils, (2) to classify them into defined series and other units, (3) to establish and map the boundaries between kinds of soil, and (4) to correlate and predict adaptability of soils to various crops, grasses, and trees; behavior and productivity of soils under different management systems; and yields of adapted crops on soils under defined sets of management practices. Although the primary purpose of soil surveys has been to aid in agricultural interpretations, many other purposes have become important, ranging from suburban planning, rural zoning, and highway location, to tax assessment and location of pipelines and radio transmitters. This has happened because the soil properties important to the growth of plants are also important to its engineering uses.

Soil surveys were first used in the United States in 1898. Over the years the scale of soil maps has

been increased from 1/2 or 1 in. to the mile, to 3 or 4 in. to the mile for mapping humid farming regions, and up to 8 in. to the mile for maps in irrigated areas. After the advent of aerial photography, planimetric maps were largely discontinued in favor of aerial photographic mosaics. The United States system has been used, with modifications, in many other countries.

Two kinds of soil maps are made. The common map is a detailed soil map, on which soil boundaries are plotted from direct observations throughout the surveyed area. Reconnaissance soil maps are made by plotting soil boundaries from observations made at intervals. The maps show soil and other differences that are of significance for present or foreseeable uses.

The units shown on soil maps usually are phases of soil series. The phase is not a category of the classification system. It may be a subdivision of any class of the system according to some feature that is of significance for use and management of the soil, but not in relation to the natural landscape. The presence of loose boulders on the surface of the soil makes little difference in the growth of a forest, but is highly significant if the soil is to be plowed. Phases are most commonly based on slope, erosion, presence of stone or rock, or differences in the rock material below the soil itself. If a legend identifies a phase of a soil series, the soils so designated on a soil map are presumed to lie within the defined range of that phase in the major part of the area involved. Thus, the inclusion of lesser areas of soils having other characteristics is tolerated in the mapping if their presence does not appreciably affect the use of the soil. If there are other soils that do affect the use, inclusions up to 15% of the area are tolerated without being indicated in the name of the soil.

If the pattern of occurrence of two or more series is so intricate that it is impossible to show them separately, a soil complex is mapped, and the legend includes the word "complex," or the names of the series are connected by a hyphen and followed by a textural class name. Thus the phrase Fayette-Dubuque silt loam indicates that the two series occur in one area and that each represents more than 15% of the total area.

In places the significance of the difference between series is so slight that the expense of separating them is unwarranted. In such a case the names of the series are connected by a conjunction, for example, Fayette and Downs silt loam. In this kind of mapping unit, the soils may or may not be associated geographically.

It is possible to make accurate soil maps only because the nature of the soil changes with alterations in climatic and biotic factors, in relief, and in groundwaters, all acting on parent materials over long periods of time. Boundaries between kinds of soil are made where such changes become apparent. On a given farm the kinds of soil usually form a repeating pattern related to the relief (Fig. 3).

Because concepts of soil have changed over the years, maps made 30–50 years ago may use the same soil type names as maps made in recent years, but with different meanings. The older maps must therefore be interpreted with caution.

[GUY D. SMITH]

PHYSICAL PROPERTIES OF SOIL

The physical properties of soil are important in agriculture because of their influence on plant growth and on the management requirements of the land. They influence plant growth from seeding to maturity by regulating the supply of air, water, and heat. The absorption of essential nutrients by plant roots is dependent upon an available supply of oxygen, water, and heat. Thus, physical properties indirectly regulate the nutrition of plants and their response to liming and fertilization. The more favorable the supply of air, water, and heat in each soil layer or horizon, along with the absence of mechanical impedance to root growth, the greater is the potential rooting system zone for plants.

Physical properties of the soil also determine the kind, amount, and ease of tillage, the runoff and erosion potential, and the type of plants which can or should be grown on a given soil.

Many people use the word tilth in referring to the physical condition of the soil. Tilth has been defined as the physical condition of the soil in its relation to plant growth. The physical condition of the soil is controlled by, or is the result of, whatev-

er set of physical properties the soil has at any given time.

Soil physics is that branch of soil science which is concerned with the study of the physical properties of the soil. These physical properties include texture, particle density, structure, bulk density, porosity, water, air, temperature, consistency, compactibility, and color. Just as important as the amount of water, air, and heat in the soil at any one time is the soil's conductivity for these constituents. All of these properties are interrelated.

The four major components of the soil are inorganic particles, organic matter, water, and air. The proportions of these components vary greatly from place to place in a field, from one layer or horizon to another, and in different parts of the world. The amount of air, water, and heat in the soil changes from day to day and from season to season.

Soil texture. About one-half of the total volume of mineral soils consists of solid matter, of which 80–99% is inorganic and 1–20% is organic material. The inorganic fraction consists of rock and mineral particles of many sizes and shapes. They are classified into five major size groups called separates. The two largest separates are stone and gravel. Stone particles are greater than 76 mm (3 in.) and gravel particles are 2–76 mm along their greatest diameter. Sand particles are 0.05–2.00 mm in diameter. Sand particles may be graded by size as very fine, fine, medium, coarse, and very coarse. Silt has particles 0.002–0.005 mm in diameter. Clay, the smallest of the soil particles, has a diameter of less than 0.002 mm.

After separating the coarser separates by sieving, the amount of silt and clay is determined by methods that depend upon the rate of settling or sedimentation (based on Stoke's law) of these two separates from a water suspension in which they have been well dispersed with the aid of a dispers-

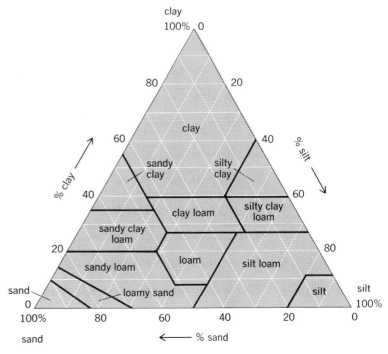

Fig. 4.   Triangle showing percentage of sand, silt, and clay in each textural class.

ing agent. The stone, gravel, and sand separates of a soil can be seen with the naked eye. Clay can be examined only with an electron microscope.

Determination of the particle-size distribution in a soil is called a particle-size or mechanical analysis. The texture of a soil is determined by its content of sand, silt, and clay. The percentages of sand, silt, and clay in the 12 textural classes are shown in Fig. 4. With this texture triangle one can determine the textural class of a soil from its percentages of sand, silt, and clay. The textural class is combined with the series name of a soil to give the soil type, such as Sassafras sandy loam, Miami silt loam, or Houston clay loam.

The stone, gravel, and larger sand particles usually act as separate particles. They may be rounded, angular, or platelike in shape. They are composed of rock fragments and of primary minerals such as quartz. Soils with large amounts of stone, gravel, and coarse sand have low plant-nutrient and water-holding capacities, and permit rapid air, water, and heat movement through them. A high content of fine sand may increase water-holding capacity, but it also often increases a soil's susceptibility to wind and water erosion. Sand imparts a grittiness to the feel of a soil.

The clay fraction controls most of the important properties of a soil. In soils of the cold and temperate regions, clay is composed chiefly of secondary crystalline alumina silicates. These consist of the kaolinite, illite, and montmorillonite groups of clay minerals. Hydrated oxides of iron and aluminum are the main components of the clay in the more highly weathered soils typical of many parts of the tropics.

Because of their extremely small size, clay particles have a very large specific surface which is responsible for the great adsorptive capacity of clay soils for water, gases, ions, and organic molecules. Clays are well known for their plasticity and stickiness when wet. They also expand or swell with wetting and contract or shrink upon drying. Movement of air and water through clay soils is often very slow because of the small size of the pores between the clay particles. Clay particles are platelike in shape.

Silt particles exhibit some of the properties of sand and clay. They are usually angular in shape with quartz being the dominant mineral. The available water-holding capacity of soils often is proportional to their silt content. Many silt particles have a coating of clay particles. Without this clay coating silt has a floury or a talcum-powder feel when dry and loose. Soils with large amounts of silt and clay have very poor air and moisture relations and are very difficult to manage. They are often very erodible. The loam soils generally have the most desirable texture for crop growth and ease of management.

It is seldom feasible to try to change the texture of a soil in the field. However, sand often is mixed with clay soils to change their texture to a sandy loam for special uses, as in greenhouses. The texture of surface soils may change as a result of removal of the smaller particles by wind and water erosion or by eluviation (movement within the soil).

Organic matter. The organic matter in the soil is made up of the partially decomposed remains of plant and animal tissues as well as the bodies of living soil microorganisms and plant roots. Humus is the more or less stable fraction of the organic matter or its decomposition products remaining in the soil. Many good and some bad effects accompany the decomposition of organic matter. During decomposition of organic matter by the soil microorganisms, gluelike soil-aggregate bonding substances are produced. With knowledge of the great importance of these natural soil-conditioning materials, the chemical industry has produced a number of synthetic soil conditioners.

Much of the soil organic matter has colloidal properties. It has two to three times the absorptive capacity for water, gases, ions, and other colloids as the same amount of clay. Its superior water- and nutrient-holding capacity makes it an ideal substitute for clay in improving droughty, infertile sandy soils, and its good tilth-promoting qualities make it the universally recognized ameliorator of tight, sticky, or hard and lumpy clay soils.

Density of particles. The inorganic soil particles may consist of many kinds of minerals with a wide range in particle density. The average particle density for most mineral soils varies between 2.60 and 2.75 g/cm³. The average density of humus particles ranges between 1.2 and 1.4 g/cm³. For general calculations the average particle density of soil is taken to be 2.65 g/cm³. The pycnometer is used to determine soil particle density. The plowed layer weighs about 2,000,000 lb/acre.

Soil structure. Soil structure refers to the arrangement of soil particles into aggregates or peds of different sizes and shapes. Pure sands have single grain structure. Because of the adhesive and cohesive properties of clay and organic matter, the inorganic and organic particles have been combined to form the following types of structure as found in the A and B horizons of most soils: platy, prismatic, columnar, blocky, nuciform, granular, and crumb (Fig. 5).

These types of structure have been developed from the bonding together of the individual particles (accretion) or the breakdown of large massive mixtures of gravel, sand, silt, clay, and organic matter (disintegration). The formation or genesis of a given type of structure and the stability of the aggregate produced seem to be associated with (1) the contraction and expansion resulting from hydration and desiccation of the clay-organic matter upon wetting and drying, as well as freezing and thawing; (2) the physical activity of roots and soil animals; (3) the influence of humus and decomposing organic matter and of the slimes and mycelia of the microorganisms that provide bonding substances with which aggregates are held together; and (4) the effect of absorbed cations which bring about flocculation or dispersion of the colloidal matter.

The prism, columnar, block, and sometimes the platelike types of structure are found mostly in subsoils. Granules and crumbs are found in largest numbers in surface soils (Fig. 6). Compacted layers in the soil often have a platy structure.

The size, shape, arrangement, and particularly the amount of overlap of soil aggregates and individual sand, gravel, and stone particles are extremely important because they largely determine the size, shape, arrangement, and continuity of pores in the soil.

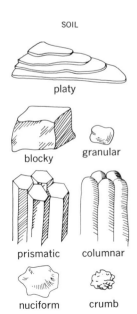

Fig. 5. Types of soil structure.

There are a number of ways of attempting to characterize the structure of a soil. The first and most direct is by visual examination of an undisturbed section of soil. Much can be learned about the size, shape, and arrangement of the soil particles, and the pore space, by close inspection of each horizon with the naked eye or with a magnifying glass. Micromorphology is the microscopic observation and photography of the structure of soil.

A second method is to measure how much of the soil has been aggregated into granules or crumbs with diameters above a given dimension, 0.25 mm being the most common. In well-granulated soils 70–80% of the total mass may be aggregated into granules or crumbs greater than 0.25 mm, as determined by wet-sieving a sample of soil through a 60-mesh screen. Aggregation values of 40–50% are more commonly found in soils under ordinary management. Sandy soils or clay soils having poor structure may have only 10–20% aggregation. Except in very sandy soils, such a low amount of aggregation usually forecasts a physical condition very unsatisfactory for plant growth.

Measurement of the permeability of the soil to water and air provides another means of evaluating its structure.

Bulk density. Bulk density is the mass (weight) of a unit volume of dry soil usually expressed in grams per cubic centimeter. It is determined by the density of the particles and by their arrangement.

The soil structure is the major factor in accounting for changes in the bulk density of a soil from time to time or from layer to layer in the profile. Soils with many particles closely packed together have high bulk densities and correspondingly low total pore space. Bulk density is a measure of the amount of compaction in soils. Traffic by farm machinery in intensively cultivated soils, trampling of cattle in heavily grazed pastures, and foot traffic on lawns and recreational areas result in severe compaction as reflected in bulk densities of 1.7–2.0 g/cm³. Bulk densities of uncompacted, porous soils are about 1.2–1.3 g/cm³. In undisturbed forest or grassland soils densities may be 0.9–1.0 g/cm³. High amounts of organic matter will lower the bulk density.

Pore space. The voids or openings between the particles of the soil are spoken of collectively as the pore space. It makes up roughly one-half the volume of the soil. In very loose, fluffy soils with low bulk density it may occupy 60–65% of the total volume. In very compact soil layers it may be reduced to 35–40%. Pore space is calculated by Eq. (1).

$$\frac{\%\text{ total}}{\text{pore space}} = 100 - \left(\frac{\text{bulk density}}{\text{particle density}} \times 100\right) \quad (1)$$

Pore space in a soil is occupied by air and water in reciprocally varying amounts. Very dry soils have most of their pore spaces filled with air. The opposite is true for very wet soils.

There is considerable variation in the size, shape, and arrangement of pores in the soil. The effective size of a pore can be estimated by the amount of force required to withdraw water from the pore. These suction values, expressed in centimeters of water, can be translated into equivalent

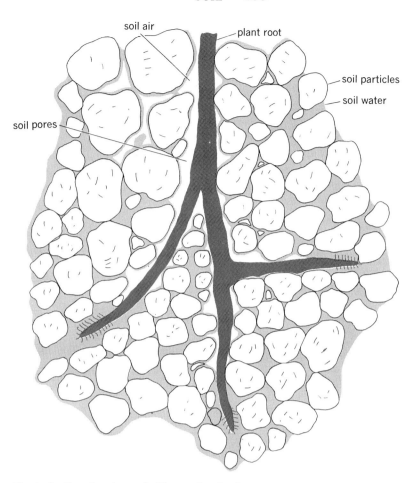

Fig. 6. Portion of surface soil with granular structure.

pore diameters using the capillary rise equation, Eq. (2), where r = radius of pore in centimeters,

$$r = \frac{2T}{hdg} \quad (2)$$

T = surface tension of water, d = density of water, g = acceleration of gravity, and h = suction force in centimeters of water.

The ideal soil should have the proper assortment of large, medium, and small pores. A sufficient number of large or macro pores (with diameters greater than 0.06 mm), connected with each other, are needed for the rapid intake and distribution of water in the soil and for the disposal of excess water by drainage into the substratum or into artificial drains. When without water they serve as air ducts. Cracks, old root channels, and animal burrows may serve as large pores. Soils with insufficient functional macro porosity lose a great deal of rainfall and irrigation water as runoff. They drain slowly and often remain poorly aerated after wetting. One of the first effects of compaction is the reduction of the size and number of the larger pore spaces in the soil.

The primary purpose of the small pores (less than 0.01 mm in diameter) is to hold water. It is through medium-sized pores (0.06–0.01 mm in diameter) that much of the capillary movement of water takes place. Loose, droughty, coarse, sandy soils have too few small pores. Many tight clay soils could well afford to have a greater number of larger pores.

Soil water. The movement and retention of water in the soil is related to the size, shape, continuity, and arrangement of the pores, their moisture content, and the amount of surface area of the soil particles. Movement and retention of water may be characterized by the energy relationships or forces which control these two phenomena.

Water retention. Some water is held in the soil pores by the force of adhesion (the attraction of solid surfaces for water molecules) and by the force of cohesion (the attraction of water molecules for each other). Water held by these two forces keeps the smaller pores full of water and also maintains relatively thick films on the walls of many larger pores. Not until the pores in one layer of soil are filled with all the water they can hold does water move into the layer below.

Water is found in the soil in both the liquid and vapor state. The soil air in all the pores, except those in the surface inch or so of very hot, dry soils is saturated with water vapor.

The liquid water may be characterized by the suction force or tension with which it is held in the soil by adhesion or cohesion. These suction or tension values may be expressed as (1) height in centimeters of a unit water column whose weight just equals the force under consideration; (2) pF, the logarithm of the centimeter height of this column; (3) atmospheres or bars; or (4) pounds per square inch (psi). For example, 1000 cm water tension= pF_3 = 1 atm = 14.7 psi. The moisture content of the soil is determined by drying the soil at 105°C until it reaches constant weight and then dividing the weight of water lost by the weight of oven-dry soil. This value times 100 equals the percentage of water in the soil on a dry-weight basis. The percentage of moisture on the volume basis for a given depth of soil is calculated by Eq. (3). Tensiometers,

$$\begin{aligned} \% \text{ soil moisture (dry weight basis)} &\div 100 \\ \times \text{ bulk density} \times \text{depth soil (in.)} \\ = \text{in. of water per in. soil depth} \end{aligned} \quad (3)$$

electrical resistance blocks, and neutron gages are used to measure the moisture content of the soil in place. Thus changes in moisture content in the soil can be followed within the effective range of each instrument.

There are several soil-moisture "equilibrium" points. Water remaining at oven dryness is held at tensions above 10,000 atm. The hygroscopic coefficient is a rough measurement of the water held by air-dry soil at a tension of about 31 atm. The wilting point, or wilting percentage, represents that moisture content or moisture tension (15 atm) at which plant roots cannot absorb water rapidly enough to offset losses by transpiration, causing the plant to wilt, first temporarily and then permanently. Certain plants of desert and dry farming regions are able to stay alive and even grow on water held at tensions up to 25–30 atm by the soil. The field capacity represents the water remaining in a soil layer 2 or 3 days after having been saturated by rain or irrigation when the rate of downward movement of water held at low suction forces (0–1 atm) has decreased to a progressively slower rate of water removal. A definite tension value cannot be assigned to this equilibrium point, although the

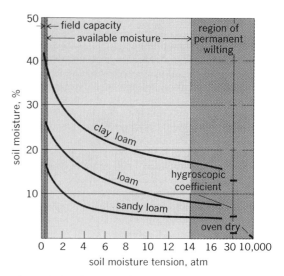

Fig. 7. Soil moisture tension curves.

water held at a tension of 0.33 atm often is used to estimate the upper limit of a soil's available water-holding capacity. The maximum retentive capacity is the moisture content of a soil when all of its pores are filled or saturated with water, and under zero tension.

The moisture in a soil which is available for plant use is usually assumed to be that held between field capacity and the permanent wilting percentage. This is called the available soil moisture. Sandy loams hold 1–1½, loams 1½–2, and clay loams 1¾–2½ in. of available water per foot of soil. The retentivity of soils of different textures for moisture at different tensions is shown in Fig. 7. These are called soil moisture tension curves. Much of the water in sandy soils is held at low tensions. The opposite is true for clay soils.

Water held at tensions less than 1–2 atm is the most easily available for root absorption and plant growth. An adequate supply of this water should be maintained in the root zone by rainfall or irrigation, especially during periods of critical water need by plants.

Water movement. Water moves in the soil as a gas and as a liquid. Vapor transfer takes place by diffusion in response to a vapor pressure gradient. Vapor movement is through air-filled pores from a moist to a dry layer and from a warm to a cool layer. Drying of a wet soil surface on a hot, dry, windy day and the condensation of water droplets on the undersurface of a plastic mulch on a cool, summer morning are the result of vapor transfer.

Liquid movement may be expressed by the equation $V = Ki$. V is the volume of water crossing the unit area perpendicular to the flow in the unit time. The proportionality factor K is the hydraulic conductivity or the permeability of the soil to water. It is controlled by the size, shape, arrangement, and moisture content of the soil pores. The value i is the water-moving force. It has two force components, the force of gravity and a suction or tension gradient force. The force or pull of gravity is of constant magnitude and always acts in a downward direction. The drainage or removal of most of the water from large pores is by this force, and the drainage is sometimes referred to as gravitational water. The suction or tension gradient force may vary both in magnitude and direction. The flow or capillary conduction of water in unsaturated soil is due to the gradient or difference in suction between two points in the soil. The flow is in the direction of the increase in suction. This accounts for the movement of water toward roots which have depleted the supply of water held at low tension in the soil at the soil-root interface, for the upward capillary conduction of water from an underlying saturated layer (water table), and for the slow downward movement of water after a rain or irrigation. Rate and amount of water movement by capillary conduction to the root system is not usually sufficient to meet the demands of the plant for water, except when a sufficient amount of water held at suctions less than 1 atm is present around the root. Since capillary conductivity decreases rapidly as the soil becomes drier, the water needs of plants are satisfied also by an extension of their root systems into fresh supplies of water held at low tension in hitherto untapped or recently refilled soil pores. It is very important, therefore, that soil structure be such as to permit the rapid extension of the root system through the whole soil mass.

Soil air and aeration. Soil air differs from the atmosphere above the soil in that it usually contains 5–100 times as much carbon dioxide (0.15–0.65%) and slightly less oxygen, and is saturated with water vapor. In deep, poorly drained soil layers or in heavily manured soils the CO_2 content may reach 10% and the O_2 content decrease to 1%. In water-logged soils, anaerobic conditions may result in methane and hydrogen sulfide production.

Aeration refers to the movement of gases in and out of each soil layer or horizon. The movement of gas within the soil as well as to and from the atmosphere is by diffusion in a direction determined by its own partial pressure. The rate of diffusion of each gas in and out of the soil depends on differences in concentration of each gas in the soil and in the atmosphere, and on the ability of the soil to transmit the gases. Diffusion or aeration is proportional to the volume of air-filled pores in any soil layer.

Temperature. The temperature of field soils shows rather definite changes at different depths, at different times during the day and night, and at different seasons of the year. These changes are determined by the amount of radiant energy that reaches the soil surface and by the thermal properties of the soil. Only that part of the heat energy which is absorbed causes changes in soil temperature. Heat produced by intense microbial decomposition of fresh organic matter mixed with the soil will also increase soil temperature. Dark-colored soils capture a much higher proportion of the radiant energy than do light-colored soils. The insulating effects of vegetative cover and surface mulches keeps the soil cooler than bare, fallow soil. Wet soils warm more slowly than dry soils because the heat capacity of water is five times that of the mineral soil particles. The energy absorbed by the soil surface is disposed of in one or more of the following ways: radiation to the atmosphere, heating of the air above the soil by convection, increasing the temperature of the surface soil, or conduction to the deeper soil layers.

Consistency and compactibility. As the moisture content of a soil changes from air dryness to saturation, its consistency varies from a state of hardness or brittleness, to loose, soft, or friable, to tough or plastic, to sticky or viscous. The reaction of the soil to physical manipulation, as in tillage, is primarily an expression of the properties of cohesion, adhesion, and plasticity. These properties are largely determined by the structure, organic matter content, kind and amount of clay, nature of adsorbed bases, and moisture content, which regulates the thickness of the water films around the soil particles. Tillage should be done only after the soil attains a soft, friable condition—when it breaks apart or can be worked into granules 1–5 mm in diameter. This is a very desirable range of particle size for good seed and root bed conditions.

Each soil has a critical moisture range, often near field capacity, at which pressure by foot or machinery traffic results in maximum compaction. Bulk density, permeability, porosity, and penetrometer measurements are used to indicate the degree of compaction as found in traffic pans in the surface soil, the plow sole, or natural hardpans. Soil compaction is a very serious problem because it reduces the permeability of the soil to air and water, and increases the resistance of the soil to root penetration. Hard, dry surface crusts may also prevent seedling emergence.

Color. Soil color may be influenced by, and indicates the kind of, parent material, chemical composition, organic matter content, drainage, aeration, or oxidation. A blotched or mottled yellow, gray, and blue subsoil indicates poor drainage, aeration, or oxidation. A clear red, yellow, or brown color indicates good drainage. The color of some soils is inherited from the parent material. Organic matter gives a brown to black color to that horizon where it is concentrated. Color of soils is determined by comparison with standard colors of known hue, value, and chroma in the Munsell soil color charts.   [RUSSELL B. ALDERFER]

SOIL MANAGEMENT

Soil management may be defined as the preparation, manipulation, and treatment of soils for the production of crops, grasses, and trees. Good soil management involves practices which will maintain a high level of production on a sustained basis. Ideally, these practices should provide the crop with an adequate supply of air, water, and nutrients; maintain or improve the fertility of the soil for subsequent crops; and prevent the development of conditions which might be injurious to plants.

Several systems of land-use classification have been developed which help a farmer to know the kinds of soils he has on his farm and their suitability for various types of farming. One of these systems, developed by the U.S. Soil Conservation Service, involves land-use capability ratings.

Land capability survey maps. Land capability survey maps are worked out in conjunction with the soil survey and serve as a guide to the suitability of land for cultivation, grazing, forestry, wildlife, watersheds, or recreation, with primary consideration given to erosion control. There are eight capability classes which describe the characteristics of the land and the difficulty or risk involved in using it for one kind of crop production or another. These eight classes are sometimes distinguished on land capability survey maps by roman numerals as well as by standard colors. Four of the eight classes include land that is suited for regular cultivation with varying degrees of erosion control measures and management practices required; three classes of land are not suited for cultivation but require permanent vegetation and impose severe limitations on land use; and one class includes lands suited only for wildlife, recreation, or watershed purposes. For a description of the land capability classes and the management practices recommended for each class *see* SOIL CONSERVATION.

Cropping system. A cropping system refers to the kind and sequence of crops grown on a given area of soil over a period of time. It may be a regular rotation of different crops, in which the sequence of crops follows a definite order, or it may consist of a single crop grown year after year in the same location. Other cropping systems include different crops but have no definite or planned sequence.

Cropping systems that involve the systematic rotation of different crops generally include hay and pasture crops, small grains, and cultivated row crops. Legumes, such as alfalfa, clover, and vetch, are usually grown alone or mixed with grasses in the hay and pasture sequence in the rotation because they supply nitrogen and contribute to good soil tilth. The beneficial effect of legumes and grasses on tilth may be attributed to the fact that (1) the soil is not tilled while these crops are being grown, and (2) the organic matter returned to the soil by the extensive root systems and in the plowed-under top growth is particularly suited to the development of a stable, porous soil structure.

Small grains function somewhat like legumes and grasses in giving protection against soil erosion, but they add no nitrogen and remove moderate quantities of plant food from the soil. Since small grains do not provide maximum economic return from the high nitrogen residues left in the soil by legumes, and are likely to lodge owing to the stimulation of growth from these residues, they are not planted in the rotation following legumes. Small grains are generally planted either at the end of a rotation following row crops or as a companion crop for legumes.

Row crops, such as corn, potatoes, cotton, and sugarbeets, are an excellent choice to follow legumes because they utilize the nitrogen supplied by the legumes and bring good cash returns. Since row crops in early stages of growth provide little protection against erosion and require considerable cultivation which breaks down soil structure, it is not considered desirable to plant them continuously.

A cropping system that involves growing the same crop year after year generally depletes the soil and results in lower crop yields. This is particularly true if the crop is cultivated frequently and returns little crop residue to the soil. Weeds, diseases, and insects also become more of a management problem when the cropping system does not involve rotation. Thus the farmer who intends to grow one crop year after year becomes completely dependent on disease-resistant varieties of plants, chemical insecticides and fungicides, soil fumigation, and other methods of controlling diseases, insects, and pests. Through the appropriate use of improved varieties, pesticides, and adequate amounts of fertilizers, farmers have succeeded in maintaining a high level of production on land repeatedly planted with the same crop. While such results are causing farmers to take another look at cropping systems that do not involve rotation, they are well aware that more intensive practices and costly supplements are required to maintain production.

Organic matter and tilth. The value of adding organic matter to the soil in the form of animal manures, green manures, and crop residues for producing favorable soil tilth has been known since ancient times. Research has provided information that helps to explain the mechanisms for this effect.

Experiments reveal that during the decomposition of organic matter in the soil, microorganisms synthesize a variety of gumlike substances, at least partly polysaccharide in nature which, when dried with the soil, bind the soil particles together into a porous, water-stable structure. While these binding substances are produced in relatively large quantities during stages of rapid organic-matter decomposition, they may in turn be decomposed by other organisms. Thus, to maintain a continuous supply of binding substances, organic matter must be added to the soil frequently.

In addition to the beneficial action of microbially synthesized binding substances, roots and fungal mycelia also contribute to the development of favorable soil tilth by molding the smaller gum-cemented granules into still larger aggregates. The aggregates adhering together form large pores that permit the rapid movement of air and water and form small pores that store water. Both conditions are essential features of good tilth.

Unfortunately not all the effects of organic matter on tilth are desirable. The growth of organisms in a fine-textured soil may interfere with the downward movement of water whenever the soil pores become clogged with microbial bodies. This condition is of particular significance where water is ponded to recharge underground water supplies or to leach excessive amounts of soluble salts from the soil.

Even the characteristic property of organic matter of promoting aggregate formation is not always desirable. Some surface mulches during decomposition induce the formation of a layer of small surface aggregates which are more susceptible to wind erosion than the fine soil particles initially present. In such cases, aggregation intensifies the hazard of severe wind erosion.

In spite of these negative effects of organic matter on the physical properties of soil, the incorporation of organic matter with soil is the most suitable and practical way of developing and maintaining good tilth.

Conditioners and stabilizers. Soil conditioners and stabilizers include a wide variety of natural and synthetic compounds that, upon incorporation with the soil, improve its physical properties. The term soil amendment also is applied to these compounds but is a more general term since it includes any material, exclusive of fertilizers, that is worked into the soil to make it more productive, regardless of whether it benefits the physical, chemical, or microbiological properties of the soil.

Soluble salts of calcium, such as calcium chloride and gypsum, or acid and acid formers, including sulfur, sulfuric acid, iron sulfate, and aluminum sulfate, have been used as conditioners to improve the physical properties of soils that were made unfavorable by excessive quantities of sodium ions adsorbed on the soil colloids.

The tilth of dense clay soils, which are slow to take water and have a marked tendency to become cloddy, may be improved by the addition of gypsum and by-product lime from sugarbeet processing factories. Limestone has improved the physical condition of acid soils, apparently by stimulating the activity of microorganisms to synthesize substances that bind soil particles into aggregates.

The discovery that soil microorganisms synthesize substances that improve soil structure stimulated the search for synthetic compounds that would be more effective than the natural products. While a wide variety of compounds have improved soil structure temporarily, three water-soluble, polymeric electrolytes of high molecular weight which are very resistant to microbial decomposition have been developed commercially for use in ameliorating poor soil structure. These are modified hydrolyzed polyacrilonitrile (HPAN), modified vinyl acetate maleic acid (VAMA), and a copolymer of isobutylene and maleic acid (IBMA). High cost relative to yield increase has limited these materials to experimental use.

Mixed with the soil in amounts ranging from 0.02 to 0.2% of soil by weight, these compounds are readily adsorbed by moist soils and tend to stabilize or fix the existing structure. They are therefore synthetic binding agents and should be added only to soils that have previously been worked into a desirable physical condition. These materials are not equally effective on all soils, and if improperly used can stabilize a poor physical condition.

Fertility. Soil fertility may be defined as that quality of a soil which enables it to provide nutrient elements and compounds in adequate amounts and in proper balance for the growth of specified plants, when other growth factors such as light, moisture, temperature, and the physical condition of the soil are favorable.

Testing. Even though relatively fertile and of good physical condition, a soil may be lacking in one or more of 16 elements presently known to be essential to plant growth, or it may be strongly acidic, alkaline, or salty, and thus unsuitable for plant growth. Fortunately, soil tests are available that indicate the existence of possible deficiencies or excesses in the soil. In most instances, these tests involve the use of various reagents for extracting from the soil the total or proportionate amount of the nutrient or compound in question. The amount of material extracted is then compared with values that have been correlated previously with crop response on the same or similar soil. No single test is reliable for all crops on all soils. See PLANT, MINERALS ESSENTIAL TO.

Control of pH. The availability of soil nutrients for plants is influenced greatly by the reaction of the soil. Soils may be classified as acid, neutral, or alkaline in reaction. The method commonly used in measuring and expressing degrees of acidity or alkalinity is in terms of pH. The pH value of soil may range from less than 4 to more than 8; the lower the value, the more acid the soil. Under most conditions lime is applied to acid soils to maintain their pH between 6.5 and 7.0. Under special conditions it may be desirable to maintain pH values either higher or lower than these. In any case the desirability of applying lime should be determined by the pH of the soil and the requirements of the plants to be grown.

It is occasionally necessary to make soils more acid. Materials commonly used to decrease pH are sulfur, sulfuric acid, iron sulfate, and aluminum sulfate.

Control of salinity. Restricted drainage caused by either slow permeability or a high water table is the principal factor in the formation of saline soils. Such soils may be improved by establishing artificial drainage, if a high water table exists, and by subsequent leaching with irrigation water to remove excess soluble salts.

Soils can be leached by applying water to the surface and allowing it to pass downward through the root zone. Leaching is most efficient when it is possible to pond water over the entire surface.

The amount of water required to leach saline soils depends on the initial salinity level of the soil and the final salinity level desired. When water is ponded over the soil about 50% of the salt in the root zone can be removed by leaching with 6 in. of water for each foot of root zone; about 80% can be removed with 1 ft of water per foot of soil to be leached; and 90% can be removed with 2 ft of water per foot of soil to be leached.

Because all irrigation waters contain dissolved salts, nonsaline soils may become saline unless water is applied in addition to that required to replenish losses by plant transpiration and evaporation, to leach out the salt that has accumulated during previous irrigations and through the addition of fertilizer.

Regulating nutrient supplies. The nutrients supplied to crops can be regulated by modifying the availability of nutrients already present in the soil. This can be accomplished by changing soil reaction, turning under green manure crops, including legumes which add nitrogen, and adding fertilizers. See FERTILIZER.

By changing soil reaction through the addition of lime, acidulating agents such as sulfur, or residually acid fertilizers such as ammonium sulfate, solubility and availability of compounds of phosphorus, iron, manganese, copper, zinc, boron, and molybdenum can be increased or decreased. Phosphorus compounds are generally more available in the slightly acid to neutral pH range, whereas compounds of iron, manganese, zinc, and copper become more available as the acidity of the soil increases. The activity of microorganisms responsible for the transformation of nitrogen, sulfur, and phosphorus compounds into forms available for plants also is influenced by soil reaction. A reaction which is too acid or too alkaline retards the activities of these organisms.

The decay of turned-under green manures and plant residues produces carbon dioxide, rendering

Fig. 8.   Sheet erosion showing how soil has been brought from entire cultivated hillside. A large soil deposit has collected on flat area at bottom of slope. (*USDA*)

Fig. 9.   Rill erosion showing how water has followed the old corn rows. (*USDA*)

Fig. 10.   Large gully could be repaired by plowing in and seeding to grass. (*USDA*)

soluble the nutrients from soil particles, and the nutrients which were absorbed from the soil during the growth of these crops are also made available. Although the turning under of a green manure affects the availability of nutrients, it does not add to the total nutrient supply unless the green manure is a legume which fixes atmospheric nitrogen.

The system of farming determines to a considerable extent the manner in which fertilizers are used to regulate nutrient supplies. Each system of farming depends upon the crop, soil, climate, kinds and rates of fertilizers applied, and available equipment; and for each system of farming there are many ways of applying fertilizers.

Common methods of applying fertilizers include broadcasting, banding, deep placement, and foliar applications. Broadcasting fertilizer on the soil is usually less desirable than localized placement of the fertilizer in relation to the seed or plant. Banding fertilizers to the side of the rows in furrow bottoms or beds and drilling fertilizer with the seed give the best response from limited quantities of fertilizer. Deep placement of fertilizers is effective in arid regions where soils dry out to a considerable depth or where deep-rooted crops are grown. Foliar applications of fertilizers, particularly those containing micronutrients, circumvent soil interactions. Such interactions within the soil may render the applied fertilizer unavailable to the crop. Foliar applications also make it possible for the farmer to supply his crops directly with a number of essential plant nutrients at critical stages of growth.

[DANIEL G. ALDRICH, JR.]

SOIL EROSION

Soil erosion is that physical process by which soil material is weathered away and carried downgrade by water or moved about by wind. Two categories of erosion are recognized. The first, called geologic erosion, is a natural process that takes place independent of man's activities. This kind of erosion is always active, wearing away the surface features of the Earth. The second kind, referred to as accelerated erosion, occurs when man disturbs the surface of the Earth or quickens the pace of erosion in any way. It produces conditions that are abnormal and poses a problem for the future food supply of the world. To combat erosion successfully, it is important that man recognize the erosion processes and have a knowledge of the factors which affect erosion.

Types. Erosion by running water is usually recognized in one of three forms: sheet erosion, rill erosion, and gully erosion.

Sheet erosion. The removal of a thin layer of soil, more or less uniformly, from the entire surface of an area is known as sheet erosion. It usually occurs on plowed fields that have been recently prepared for seeding, but may also take place after the crop is seeded. Generally only the finer soil particles are removed. Although the depth of soil lost is not great, the loss of relatively rich topsoil from an entire field may be serious (Fig. 8). If continued for a period of years, the entire surface layer of soil may be removed. In many parts of the world only the surface layer is suited for cultivation.

Rill erosion. During heavy rains runoff water is

concentrated in small streamlets or rivulets. As the volume or velocity of the water increases, it cuts narrow trenches called rills. Erosion of this type can remove large quantities of soil and reduce the soil fertility rapidly (Fig. 9). This type of erosion is particularly detrimental because all traces of the rills are removed after the land is tilled. The losses which occurred are often forgotten and adequate conservation measures are not taken to prevent further loss of soil.

Gully erosion. This type of erosion occurs where the concentrated runoff is sufficiently large to cut deep trenches, or where continued cutting in the same groove deepens the incision. Gullying often develops where there is a water overfall. The stream bed is cut back at the overfall and the gully lengthens headward or upslope. Once started, gullying may proceed rapidly, particularly in soils that do not possess much binding material. Gully erosion requires intensive control measures (Fig. 10), such as terracing or the use of diversion ditches, check dams, sod-strip checks, and shrub checks.

Affective factors. The rate and extent of soil erosion depend upon such interrelated factors as type of soil, steepness of slope, climatic characteristics, and land use.

Type of soil. Soil types vary greatly in physical and chemical composition. The amounts of sand, silt, and clay constituents, colloidal material, and organic matter all have a bearing upon the ease with which particles or aggregates can be detached from the body of the soil. Such detachment is caused chiefly by the beating action of raindrops. The particles are then transported downgrade by moving water. Sandy or gravelly soils often have little colloidal material to bind particles together, and hence these materials are easily detached. However, because of their size, sand particles are more difficult to move than fine particles. For this reason sand particles are moved chiefly by rapidly flowing water on steep slopes or by streams at flood stage. Finer particles, such as silt, clay, and organic matter, can be carried by water moving at a slower rate. On gently sloping fields there is a tendency for more of the fine particles to be carried away, leaving the heavier sand particles behind. However, if rainfall is intense and the volume of runoff great, sand may be moved even on gentle slopes.

Slope. The relation of slope to the amount of erosion on different classes of soil is illustrated in Fig. 11. The amount of total runoff from rainfall increases only slightly with increase in the slope of the land above 1–2%, but the speed of the flowing water, or rate of runoff, may increase greatly. Since the capacity of moving water to transport soil particles increases in geometric ratio to the rate of flow, the amount of erosion increases greatly with increase in the slope of the land (Fig. 12).

Climate. In cold climates the frozen soil is not subject to erosion for several months of the year. However, if such areas receive heavy snow, serious erosion may take place when the snow melts. This is particularly true if the snow melts as the ground gradually thaws. As the water moves over the thin unfrozen layer of soil, it transports much of this soil material downgrade.

In warm climates soils are susceptible to erosion

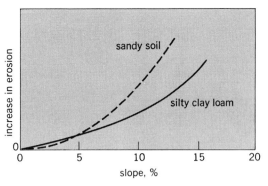

Fig. 11. Generalized diagram illustrating greater loss of fine-grained soil (silty clay loam) on gentle slopes (0–5%) and greater loss of sandy soil on steeper slopes.

any time there are heavy rains. Such soils are particularly vulnerable to erosion if rains fall in winter and there is little vegetative cover.

The amount of rainfall is an important factor in determining the erosion that occurs in a given region. However, the character of the rainfall is usually a much more important factor than the total amount in determining the seriousness of erosion. A rain falling at the rate of 2 in./hr may cause three to five times as much erosion as a rainfall of 1 in./hr. Regions where most of the precipitation comes in the form of mist or gentle rain may undergo little erosion, even though the total rainfall may be high and other conditions conducive to erosion.

In some areas of dry climates strong winds cause soil movement and serious loss of soil. Wind erosion is more common on sandy soils, but it is by no means confined to them. Heavier soils, which have a fluffy physical condition produced by freezing and thawing or drying, may be moved in great quantities by the wind.

Land use. The type of crops and the system of management influence the amount and type of erosion. Bare soils, clean uncultivated soils, or land in intertilled row crops permit the greatest amount of erosion. Crops that give complete ground cover throughout the year, such as grass or forests, are most effective in controlling erosion. Small-grain crops, or those that provide a fairly dense cover for only part of the year, are intermediate in their effect on erosion. Table 2 gives results of some of the earliest experiments in the

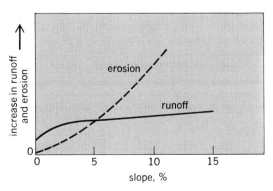

Fig. 12. Effect of slope on total amount of runoff and on rate of runoff and soil erosion.

Table 2. Relative runoff and erosion from soil under different land uses, with mean rainfall of 35.87 in.*

Land use and treatment	Runoff, %	Tons soil per acre eroded annually	Years required to lose surface 7 in. of soil
Plowed 8 in. deep, no crop; fallowed to keep weeds down	28.4	35.7	28
Plowed 8 in., corn annually	27.4	17.8	56
Plowed 8 in., wheat annually	25.2	6.7	150
Rotation; corn, wheat, red clover	14.1	2.3	437
Bluegrass sod	11.6	0.3	3547

*Missouri Res. Bull. no. 63, 1923.

United States on differences in land use and the effect on runoff and erosion. These results show that cultivated land, especially without a crop or protective cover of vegetation, is particularly vulnerable to erosion. In addition, excessive erosion usually occurs where cultivated crops like corn, cotton, and tobacco are grown on hilly or sloping land that is subjected to increased rates of runoff. In some areas where row crops have been grown continually the soil has been removed to the depth of the plow layer within a lifetime.

Pastures in humid areas usually have a tough continuous sod that prevents or greatly reduces sheet or surface erosion. Natural range cover, if in good condition, is usually effective in controlling erosion, but in areas of limited rainfall, where bunch grasses form most of the cover on range land, occasional heavy rain may cause severe erosion of the bare soil exposed between the bunches of grass. Forest lands, with their overhead canopies of trees and surface layers of decaying organic matter, have much greater water intake and much less surface runoff and erosion.

Wind erosion. In the western half of the United States and in many other parts of the world, great quantities of soil are moved by the wind. This is particularly so in arid and semiarid areas. Sandy soils are more subject to wind erosion than silt loam or clay loam soils. The latter, however, are easily eroded when climatic conditions cause the soil to break into small aggregates, ranging from 0.4 to 0.8 mm in diameter. The coarse particles usually are moved relatively short distances, but the fine dust particles may be carried by strong winds for hundreds or even thousands of miles.

In some areas the coarse, or sand, particles are moved by the wind and deposited over extensive areas as dunes. The dunes move forward in the same direction as the prevailing winds, the particles being moved from the windward side of the dune to the lee side. If dunes become covered with grass or other vegetation, they cease to move. The sandhill region of Nebraska is a good example of such an area. *See* DUNE.

Control. The following are a few fundamental principles which will help control erosion and greatly reduce the damage done by soil erosion.

1. Keep land covered with a growing crop or grass as much time as possible. Cover increases intake of water and reduces runoff. The extent of erosion control will be roughly in proportion to the effective cover.

2. When there is no growing crop, retain a cover of stubble or crop residue on the land between crops and until the next crop is well started. This can be done by using a system known as stubble-mulch farming. It ultilizes the idea of preparing a seedbed for a new crop without burying the residue from the previous crop. Tillage tools that work beneath the surface and pulverize the soil without necessarily inverting it or burying the residue are used instead of moldboard plows. This system is best adapted to regions of low rainfall or warm climates.

3. Avoid letting water concentrate and run directly downhill. By doing this the soil is protected against water at its maximum cutting power. Construct terraces with gentle grades to carry the runoff water around the hill at slow speeds. These diversions should empty onto grassed waterways or on meadowland to prevent creation of gullies.

4. Plant crops and till the soil along the contours.

5. Control wind erosion by keeping land covered with sod or planted crops as much of the time as possible. Maintain crop residue on the land between crops and while the next crop is getting started.

6. If wind erosion begins on a bare field or where a crop is just getting started, the soil drifting may be stopped temporarily by cultivation. An implement with shovels that will throw up clods or chunks of soil to give a rough surface is usually effective. Often, only strips through the field need be so treated to stop erosion on the whole area.

7. Moving dunes may require artificial cover or mechanical obstructions on the windward side, followed by vegetative plantings, depending on climatic conditions. Along shorelines, beach grasses followed by woody plants and forests may be required.

For a discussion of the physical, economic, and social effects of soil erosion *see* SOIL CONSERVATION.   [FRANK L. DULEY]

Bibliography: L. D. Baver, *Soil Physics,* 3d ed., 1956; C. E. Black (ed.), *Methods of Soil Analysis,* Amer. Soc. Agron. Monogr. no. 9, 1964; H. Jenny, *Factors of Soil Formation,* 1941; H. Kohnke, *Soil Physics,* 1968; C. E. Millar, L. M. Turk, and H. D. Foth, *Fundamentals of Soil Science,* 3d ed., 1958; G. W. Robinson, *Soils,* 3d ed., 1951; B. T. Shaw (ed.), *Soil Physical Conditions and Plant Growth,* vol. 2, Amer. Soc. Agron. Monogr., 1952; L. M. Thompson, *Soils and Soil Fertility,* 2d ed., 1957; USDA, *Soil: The Yearbook of Agriculture,* 1957; USDA, *Soil Classification, A Comprehensive System: 7th Approximation,* 1960; USDA, *Soil Survey Manual,* Handb. no. 18, 1951; USDA, *Soils and Men: The Yearbook of Agriculture,* 1938; USDA, *Supplement to Soil Classification System: 7th Approximation,* 1967; E. L. Worthen and S. R. Aldrich, *Farm Soils: Their Fertilization and Management,* 5th ed., 1956.

Soil, zonality of

Many soils that are geographically associated on plains have common properties that are the result of formation in similar climates with similar vegetation. Because climate determines the natural vegetation to a large extent and because climate changes gradually with distance on plains, there are vast zones of uplands on which most soils have many common properties. This was first observed in Russia toward the end of the 19th century by V. V. Dokuchaev, the father of modern soil science. He also observed that on floodplains and steep slopes and in wet places the soils commonly lacked some or most of the properties of the upland soils. In mountainous areas, climate and vegetation tend to vary with altitude, and here the Russian students observed that many soils at the same altitude had many common properties. This they called vertical zonality in contrast with the lateral zonality of the soils of plains.

Zonal classification of soils. These observations led N. M. Sivirtsev to propose about 1900 that major kinds of soil could be classified as Zonal if their properties reflected the influence of climate and vegetation, as Azonal if they lacked well-defined horizons, and as Intrazonal if their properties resulted from some local factor such as a shallow groundwater or unusual parent material. This concept was not accepted for long in Russia. It was adopted in the United States in 1938 as a basis for classifying soil but was dropped in 1965. This was because the Zonal soils as a class could not be defined in terms of their properties and because they had no common properties that were not shared by some Intrazonal and Azonal soils. It was also learned that many of the properties that had been thought to reflect climate were actually the result of differences in age of the soils and of past climates that differed greatly from those of the present.

Zonality of soil distribution is important to students of geography in understanding differences in farming, grazing, and forestry practices in different parts of the world. To a very large extent, zonality is reflected but is not used directly in the soil classification currently being used in the United States. The Entisols include most soils formerly called Azonal. Most of the soils formerly called Intrazonal are included in the orders of Vertisols, Inceptisols, and Histosols and in the aquic suborders such as Aquolls and Aqualfs. Zonal soils are mainly included in the other suborders in this classification. For a discussion of this classification *see* SOIL.

The soil orders and suborders have been defined largely in terms of the common properties that result from soil formation in similar climates with similar vegetation. Because these properties are important to the native vegetation, they have continuing importance to farming, ranching, and forestry. Also, because the properties are common to most of the soils of a given area, it is possible to make small-scale maps that show the distribution of soil orders and suborders with high accuracy.

Zonal properties of soils. A few examples of zonal properties of soils and their relation to soil use follow. The Mollisols, formerly called Chernozemic soils, are rich in plant nutrients. Their natural ability to supply plant nutrients is the highest of any group of soils, but lack of moisture often limits plant growth. Among the Mollisols, the Udolls are associated with a humid climate and are used largely for corn (maize) and soybean production. Borolls have a cool climate and are used for spring wheat, flax, and other early maturing crops. Ustolls have a dry, warm climate and are used largely for winter wheat and sorghum without irrigation. Yields are erratic on these soils. They are moder-

ately high in moist years, but crop failures are common in dry years. The drier Ustolls are used largely for grazing. Xerolls have a rainless summer, and crops must mature on moisture stored in the cool seasons. Xerolls are used largely for wheat and produce consistent yields.

The Alfisols, formerly a part of the Podzolic soil group, are lower in plant nutrients than Mollisols, particularly nitrogen and calcium, but supported a permanent agriculture before the development of fertilizers. With the use of modern fertilizers, yields of crops are comparable to those obtained on Mollisols. The Udalfs are largely in intensive cultivation and produce high yields of a wide variety of crops. Boralfs, like Borolls, have short growing seasons but have humid climates. They are used largely for small grains or forestry. Ustalfs are warm and dry for long periods. In the United States they are used for grazing, small grains, and irrigated crops. On other continents they are mostly intensively cultivated during the rainy season. Population density on Ustalfs in Africa is very high except in the areas of the tsetse fly. Xeralfs are used largely for wheat production or grazing because of their dry summers.

Ultisols, formerly called Latosolic soils, are warm, intensely leached, and very low in supplies of plant nutrients. Before the use of fertilizers, Ultisols could be farmed for only a few years after clearing and then had to revert to forest for a much longer period to permit the trees to concentrate plant nutrients at the surface in the leaf litter. With the use of fertilizers, Udults produce high yields of cotton, tobacco, maize, and forage. Ustults are dry for long periods but have good moisture supplies during a rainy season, typically during monsoon rains. Forests are deciduous, and cultivation is mostly shifting unless fertilizers are available.

Aridisols, formerly called Desertic soils, are high in some plant nutrients, particularly calcium and potassium, but are too dry to cultivate without irrigation. They are used for grazing to some extent, but large areas are idle. Under irrigation some Aridisols are highly productive, but large areas are unsuited to irrigation or lack sources of water.   [GUY D. SMITH]

Bibliography: M. Baldwin, C. E. Kellogg, and J. Thorp, Soil classification, in USDA, *Soils and Men: The Yearbook of Agriculture,* 1938; B. T. Bunting, *The Geography of Soil,* 1965; H. C. Byers et al., Formation of soil, in USDA, *Soils and Men: The Yearbook of Agriculture,* 1938; C. E. Kellogg, Soil and society, in USDA, *Soils and Men: The Yearbook of Agriculture,* 1938.

Soil balance, microbial

The equilibrium between the diverse types of microorganisms in soil. Although the qualitative and quantitative composition of the soil microflora and microfauna fluctuates with temperature, moisture, and treatment (such as fertilization, cultivation, and cropping) of the soil, a balance exists which is characteristic of a given soil. The balance is determined chiefly by the available supply of nutrients required by groups of microorganisms of different nutritional needs. The numbers and types of the microorganisms also depend on the available nutrient supply. Associative and antagonistic effects exerted by certain organisms on others are factors in establishing the balance. The equilibrium is not easily upset. Natural soil resists balance change when organisms are introduced.

Associative action. The process whereby one type of microorganism produces a substance required by another is widespread in soil. This action may be extended through successive groups to give a chain effect. Thus ammonia formed through decomposition of proteins by proteolytic microorganisms is used by nitrite-forming bacteria. Nitrite is required by nitrate-forming bacteria. Many cellulose-decomposing organisms utilize nitrate in hydrolyzing cellulose, and form glucose and organic acids which may be used by still other forms. Many soil bacteria synthesize vitamins needed by other organisms. Syntrophism is a form of associative action in which two organisms are mutually dependent, each producing a factor needed by the other.

Antagonisms between soil microbes. These are a factor in maintaining the equilibrium and are manifested in different ways. Many protozoa depend upon bacteria for food and ingest certain species in preference to others. Antagonism may rest on a competition for nutrients. It may also depend upon the production of substances inhibitory to other organisms, particularly antibiotics. Though synthesized only in small amounts, the antibiotics exert their effects in the microenvironments in which soil microbes are active. The advantage possessed by such microorganisms does not lead to their predominance, since capacity for antibiotic production is but one factor in the competition with other microbes.

[ALLAN G. LOCHHEAD]

Soil chemistry

The composition and chemical properties of plant-supporting earth materials. This article discusses the minerals in some common soil types and soil chemical processes, such as cation exchange, anion exchange, and metal complexing. Examples are given of plant dependence on soil nutrients.

Table 1. Average content of less abundant constituents in surface soil

Constituent	Podzol	Gray-brown podzolic	Red-yellow podzolic	Prairie	Latosol
MnO	0.03	0.07	0.51	0.12	0.66
TiO$_2$	0.61	0.79	1.02	0.71	4.3
CaO	0.95	0.59	0.22	0.88	0.66
MgO	0.49	0.40	0.16	0.68	0.84
K$_2$O	1.75	1.65	0.88	1.97	0.42
Na$_2$O	0.53	1.39	0.16	1.03	0.08
P$_2$O$_5$	0.12	0.15	0.17	0.22	0.43
SO$_3$	0.11	0.17	0.07	0.17	0.24

For a description of the origin, classification, and management of soil *see* SOIL: SOIL CONSERVATION.

Elemental composition. The elemental compositions of soils vary over wide ranges, and only a few generalizations can be made. Soils containing more than 20% organic matter are termed organic; those with less, mineral. Organic soils such as peats and mucks may contain as much as 95% carbonaceous material. Carbon, oxygen, and hydrogen are the most abundant elements in soil organic matter. Nitrogen, phosphorus, and sulfur are important constituents. Carbon-nitrogen ratios vary between 10 and 20, the lower value being characteristic of upland soils found in the temperate regions. Soil organic matter contains about one-tenth as much phosphorus or sulfur as it does nitrogen.

At least 95% of the mass of most upland soils is mineral. The oxides of hydrogen, silicon, aluminum, and iron are the most abundant components of mineral soils. Except for calcareous soils and some latosols (lateritic soils), the oxides of silicon, SiO$_2$, aluminum, Al$_2$O$_3$, and iron, Fe$_2$O$_3$, make up at least 90% of the elemental composition of the mineral portion. Silicon, aluminum, and iron are present as primary clay and oxide minerals.

The chemical composition in a given soil varies with depth. Because of accumulations of clay and oxide minerals in deeper layers, Al$_2$O$_3$ and Fe$_2$O$_3$ content usually is higher in B than in A horizons, while that of SiO$_2$ is lower. Organic matter content decreases rapidly with depth.

Figure 1 shows mean values and standard deviations of the SiO$_2$, Al$_2$O$_3$, and Fe$_2$O$_3$ content of the A and B horizons of certain zonal soils. Average surface soil content of compounds present in smaller amounts is shown in Table 1. Cations are present as exchangeable ions or in mineral lattices. Ranges of trace element contents of soils are shown in Fig. 2. *See* SOIL, ZONALITY OF.

Minerals. The minerals in soils are derived from parent rock and soil materials. These materials are mixtures of minerals which are broken down into separate minerals by physical and chemical weathering processes.

Primary minerals in sand and silt. Most of the common rock-forming minerals in soils occur in sand and silt fractions. As a result of weathering, as well as differences in the parent rock, mineral proportions vary tremendously from soil to soil. Quartz, accumulated at the expense of less resistant pyroxenes, amphiboles, micas, and feldspars, predominates in the sand and silt of soils in humid regions (Table 2). In soils of subhumid and arid regions, contents of nonquartz minerals in sand-

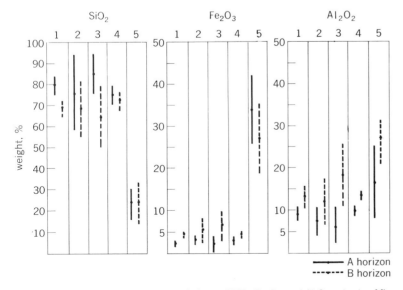

Fig. 1.   Mean values and standard deviations of SiO$_2$, Fe$_2$O$_3$, and Al$_2$O contents of five great soil groups. Key: 1, podzols (means of 7 soils); 2, gray-brown podzols (23 soils); 3, red-yellow podzols (35 soils); 4, prairie (11 soils); 5, latosols (7 soils).

Table 2. Percentage of primary minerals in topsoil

Topsoil	Quartz	Feldspars	Micas
Coarse sand	92.7 ± 6.5	3.2 ± 5.5	1.6 ± 2.5
Fine sand	86.4 ± 11.0	4.5 ± 3.6	1.4 ± 2.2
Silt	56.4 ± 15.8	8.0 ± 6.8	8.9 ± 15.0

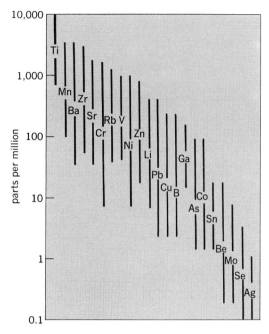

Fig. 2.   Range of contents of some trace elements commonly found in mineral soils.

size fractions are greater, averaging 20% and 37%, respectively.

Minerals in clay fractions. Clay and sesquioxide minerals make up the bulk of the clay-size fraction of soils, though others may be present. The proportions of various mineral species vary from soil to soil and within a particular profile. In the following list mineral compositions of clay fractions are given in the order of their increasing resistance to alteration. Minerals near the top of the list generally are found in clays of slightly weathered soils and those near the bottom of the list in soils which have been exposed to drastic weathering.

Mineral	*Soils*
Feldspars	Glacial soils of southern Canada
Muscovite-illite	Subhumid and arid soils; glacial soils of temperate regions
Vermiculite – interstratified clay	Subhumid and arid soils (from weathering of micas); red-yellow podzolic soils
Montmorillonite-beidellite	Volcanic ash, limestone, and basic rock soils; soils weathered from micas in temperate regions
Kaolinite-halloysite	Red-yellow podzolic soils, some latosols
Gibbsite-allophane	Red-yellow podzols, latosols
Hematite-goethite	Red-yellow podzolic soils, 5–30%; latosols, 30–75%
Anatase	Up to 35% in some latosols

In zonal soils of humid-cool to subhumid-temperate areas, illite is. the predominant mineral. Mixtures of kaolin-vermiculite-interstratified minerals occur in humid-temperate regions. Kaolins with admixtures of smaller amounts of gibbsite and hematite predominate in warm-humid regions. Under tropical conditions, the proportions of sesquioxide minerals are high and in extreme cases only these minerals remain. Arid, basic conditions favor the formation of montmorillonoids. These also occur where soil development is retarded, as by imperfect drainage.

Cation exchange. Many small soil particles, both mineral and organic, possess net negative charges. The charges are balanced by cations which exist in more or less diffuse swarms near the surfaces of the particles. The balancing ions are called exchangeable cations and are in kinetic equilibrium with the soil solution. Their quantity, usually expressed as milliequivalents (meq) per 100 g of dry soil, is the cation exchange capacity (CEC).

Lattice substitution in clays. In clay minerals, as in micas, there is extensive proxying of ions in octahedral and tetrahedral lattice positions, with a resulting unbalance of charge. The net charges of clays are negative. They are large for micas and vermiculite (135–270 meq/100 g), intermediate for montmorillonoids (80–135 meq/100 g), and quite small for kaolins (2–10 meq/100 g). In clay mineral micas and in vermiculite, much of the lattice charge is balanced by cations which occur between fixed lattice layers. Interlayer potassium ions in micas are not displaced under ordinary conditions, and are not counted with the exchangeable cations, though they are bound by the same kinds of forces. The cation exchange capacities of mica clays vary between 20 and 40 meq/100 g.

The common interlayer ions of vermiculite are calcium and magnesium, though vermiculitelike minerals with interlayer aluminum exist in many soils. Rates of ion exchange are sluggish for such minerals, particularly when ammonium or potassium is one of the exchanging ions, and the cation exchange capacity is not well defined.

Cations balancing the isomorphous substitution charges of montmorillonoids are almost completely exchangeable, and there are excellent correlations between the exchange capacities of these minerals and their chemical compositions.

Isomorphous substitution also appears to contribute to the cation exchange capacity of kaolins, though the charge here is small.

Lattice termination in clays. Lattice terminations can result in further development of negative charge, particularly when the pH is above 6. Protons may ionize from SiOH groups around the edges of clay particles, and edge aluminum ions may adsorb OH^- or may shift in coordination number from 6 to 4 as pH is raised. Maximum de-

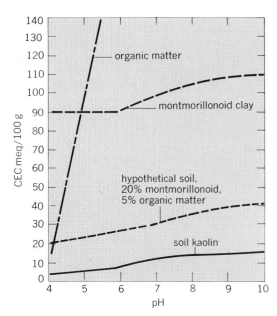

Fig. 3. Idealized cation-exchange capacities (CEC) of clays and organic matter.

velopment of edge charge occurs at around pH 10.

The magnitude of the negative charge which can develop on lattice terminals depends on edge area. In meq/100 g, it is about 20 for montmorillonite, 10 for illite, and 2–10 for kaolins, depending on both the particle size and the crystal form.

Allophanes have cation exchange capacities (at pH 7) of 60–120 meq/100 g. Probably this results largely from coordination shifts involving aluminum.

Organic matter. The organic matter in soils contains carboxyl, phenol, enol, and imide groups. Ionization of protons from these results in negatively charged particles, the charge increasing with the extent of ionization. The apparent ionization constant for carboxyl, COOH, groups of soil organic matter is about 5. Other acid groups ionize appreciably only when pH is above 7.

The carboxyl content of soil organic matter varies between about 150 and 275 meq/100 g. Organic matter can bind as much as 400 meq of cations per 100 g at high pH.

Cation-exchange capacities of soils. The cation-exchange capacities of soils usually are measured by treatment with a salt solution, such as neutral normal ammonium acetate, to achieve saturation with one species of cation. This is an arbitrary procedure, since the quantity of cation adsorbed after such treatment is not a unique value characteristic of the soil alone, but depends as well on the con-

centration, the ion composition, and the pH of the saturating solution. Cations of a saturating solution displace positive ions which are bonded electrostatically to soil particles and also displace protons which ionize from weakly acidic groups on clays and organic colloids. The proportion of protons which is replaced depends particularly upon the pH of the saturating solution, the proportion becoming larger as the pH is raised.

Cation-exchange capacities of soils can be generalized by referring to information on mineralogical makeup and organic matter percentages. For given mineral types CEC varies directly with amount of clay and organic matter (Fig. 3); kaolins contribute little, montmorillonoids a great deal, and micas are intermediate.

Exchangeable cations. Exchangeable cations are those balancing the negative charges on soil particles. Experimentally, they are the cations displaced upon leaching with a salt solution. These can be divided into two groups, the exchangeable metal cations (Ca, Mg, K, Na, Al are most abundant) and exchangeable hydrogen. The former are bonded largely through electrostatic forces, the latter almost entirely to weakly acidic spots.

To a first approximation, exchangeable metal cations can be regarded as neutralizing isomorphous substitution charges and those weakly acidic groups which are ionized under the prevailing conditions.

Exchangeable hydrogen presents more difficulty. As determined experimentally in most procedures, it is the sum of hydrolyzable metal cations displaced, particularly Al, and of ionization of weakly acidic groups caused by contact with the displacing solution. This is incorrect, and exchange acidity can be resolved into its two major components by appropriate measurements.

Soils which are less weathered because of youth, low rainfall, and temperate or cold climate have as exchangeable cations largely calcium and magnesium. Highly weathered soils, unless derived from basic parent material, have large proportions of their permanent charges (20–95% under native conditions) countered by exchangeable aluminum. Some soils of dry areas contain large amounts of exchangeable sodium.

Exchangeable cation populations of some representative soils are given in Table 3. The common exchangeable cations differ in their affinity for ion exchange spots on soil clays and organic colloids.

For permanent charges:

$$Al > Ca \gtreqless Mg > K > Na \gtreqless H$$

For weak-acid charges of clays:

$$H > Ca \gtreqless Mg > K > Na$$

For carboxyl groups of organic matter:

$$H \gg Ca \gtreqless Mg \gg K > Na$$

Ion-exchange equilibria can be described fairly accurately by means of mass-action expressions.

Anion exchange. Ion exchange refers to the association of ions with solid surfaces in such a way that they can be replaced stoichiometrically by other ions of the same charge. Anion exchange in soils has a different connotation. Anion exchange does occur in many soils, but is often com-

Table 3. Exchangeable cations of some representative soils (meq/100 g)

Soil	CEC_7	Ca	Mg	K	Na	H
Chernozem	55	46	6	1	2	0
Prairie	22	13	7	1.5	1	0
Alkali	21	12	2	0.3	6.7	0
Podzol	17.7	2.0	4.2	0.1	0	11.4
Red-yellow podzolic	8	2.0	1.0	0.1	0	4.9

charges of soil particles increase as the pH is lowered, and in the case of some latosols, positive charge may exceed negative charge when pH is below 4. All clay minerals, however, as well as temperate region soils, possess net negative charges at every value of pH.

The development of positive charges on soil minerals may result in the exchange adsorption of anions. Halide, nitrate, and sulfate ions, as well as phosphate, fluoride, and carboxylic acid anions, are sorbed and are mutually replaceable. Capacity for anion sorption by this mechanism varies from near zero at neutrality for all minerals to as much as 1 meq/100 g for kaolins, and 20 meq/100 g for kaolin-iron and aluminum oxide combinations in latosols. Montmorillonoids and other clays with large lattice charges do not sorb small anions via this mechanism unless salt contents are very high. Halides and similar anions are negatively adsorbed by such clays under most circumstances.

Certain anions, such as phosphate and fluoride, coordinate strongly with ferric iron and aluminum and are sorbed by soils through another mechanism. Phosphate ions, for example, can react with clay and oxide minerals in soils to form basic iron and aluminum phosphates similar to strengite, variscite, and palmerite. Phosphate is displaced from such minerals by hydroxyl, fluoride, or other ions which coordinate strongly with ferric iron and aluminum. Such reactions are viewed as involving solution-precipitation rather than anion exchange.

Soils containing kaolin clays and free iron and aluminum oxides can fix large amounts of phosphate through decomposition-precipitation reactions. It appears that aluminum phosphate complexes are more prevalent in soils than are ferric compounds.

Acid soils. Soils with pH less than 7 are termed acid. Acid pH usually results from the presence of exchangeable hydrogen and aluminum ions, the former bonding to weakly acidic exchange spots, the latter to permanent lattice charges. Clays with electrostatically bonded exchangeable hydrogen plicated by the occurrence of other reactions which also result in the lowering of solution concentrations of anions.

Clay and oxide minerals in soils can develop positive charges on certain exposed surfaces. This can happen through proton acceptance, hydroxyl ionization, or perhaps other events. The positive are unstable, and decompose to yield silicic acid and sufficient Al or Mg to counter the exchange spots occupied initially by H ions.

The pH of acid soils varies between 3 and 7, the reaction depending on the ion saturation and the soluble electrolyte content. In the absence of excessive amounts of soluble salts or acids, a pH of less than 4 is rare.

Soil acidity can be discussed in terms of buffer curves relating soil pH to the amount of a basic metal cation such as Ca neutralizing exchange spots. Idealized examples are shown in Fig. 4.

Montmorillonoids, vermiculites, and illites are similar in their neutralization behavior. Permanent lattice charge accounts for the bulk of their exchange capacity and Al for their acidity. Buffer curves for kaolins have two sections, one indicating displacement of exchangeable Al by Ca, the other reflecting the development of weak acid

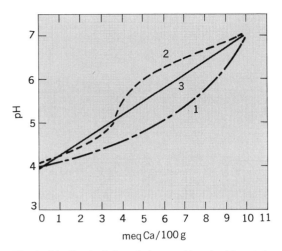

Fig. 4. Idealized relations between pH and calcium saturation of soils. Key: 1, soil containing approximately 10% montmorillonoid clay, $CEC_7 = 10$ meq/100 g; 2, soil containing approximately 60% kaolin clay $CEC_7 = 10$ meq/100 g; 3, soil containing approximately 5% organic matter, $CEC_7 = 10$ meq/100 g.

charge. Below pH 7, organic colloids behave as weak acids with dissociation constants of around 10^{-5}.

Percentage base saturation, reflecting the proportion of soil CEC countered by basic cations, has no general significance, since CEC at an arbitrary pH such as 7 consists of permanent and weak-acid components, whose proportions vary from soil to soil. In the absence of organic colloids and with Ca and Mg as exchangeable ions, pH 6 corresponds closely to 100% base saturation of permanent charge.

Soil acidity often is characterized through the measurement of exchangeable hydrogen. More meaningful determinations are for exchangeable Al and for weak-acid charge at a given pH.

Many soils of humid regions are too acid for optimum plant growth. Application of calcium and magnesium carbonates is a remedial measure.

Calcareous soils. Soils containing accumulations of calcium and magnesium carbonate are referred to as calcareous. Carbonates may occur throughout the profile or may be concentrated in certain horizons, their distributions reflecting the nature of the parent material and the weathering regime. Percentages of carbonates vary from less than 1 to 70.

Calcareous soils are saturated largely with calcium and magnesium. In contrast, with acid soils, pH and concentrations of ions in the soil solution are controlled not by ion-exchange equilibria, but by the $CaCO_3$-CO_2-H_2O system.

Equilibria in this system can be described by the equation

$$pH - \tfrac{1}{2}pCa = 4.85 - \tfrac{1}{2}\log pCO_2$$

Here pCO_2 is the partial pressure of CO_2 in the atmosphere in contact with a $CaCO_3$ water system, and pCa is the negative logarithm of the molar activity of calcium in the soil solution. The manner in which pH and solution calcium concentration vary with CO_2 partial pressure is shown in Fig. 5.

Since soil-air CO_2 content varies from 0.003%

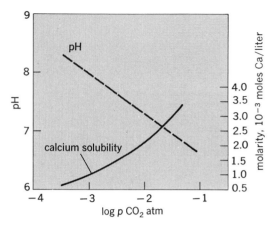

Fig. 5. Calcium solubility and pH of calcareous soil related to CO_2 partial pressure.

(the content in the atmosphere) to nearly 20%, calcium solubility and the reaction of calcareous soils can vary widely.

Salted soils. Soils of dry areas often contain sufficiently large amounts of soluble salts, exchangeable sodium, or both to have peculiar chemical and physical properties. Such soils are referred to as salted.

The U.S. Salinity Laboratory classification of salted soils is based on the electrical conductivity of a saturation extract rather than on salt percentage as such. Soils yielding saturation extracts with specific conductivities greater than 4 millimhos are saline. The concentration of a solution of this conductivity varies with the nature of the salt, averaging 45 meq/liter for the electrolytes commonly found in salted soils. The lower-limiting salt percentages of saline soils vary from around 0.1 for coarse-textured soils to 0.25 for heavier ones.

The second classification criterion for salted soils is their percentage sodium saturation. If this is above 15, the soil is alkali.

Salted soils can be placed into four groups (Table 4). The salts in salted soils are largely chlorides and sulfates of calcium, magnesium, and sodium. The proportions of the cations in solution are controlled by ion-exchange equilibria, as follows (ESP is exchangeable sodium percentage; SAR is sodium adsorption ratio):

$$ESP = -0.0126 + 0.01475\ SAR$$

$$SAR = \frac{1}{2} \frac{\text{concentration sodium}}{[\text{concentration (calcium + magnesium)}]}$$

$$ESP = \frac{\text{exchangeable sodium}}{\text{exchangeable (calcium + magnesium)}}$$

Table 4. Four categories of salted soils

Category	Exchangeable cations saturation extract conductivity, mmho	Exchangeable sodium percentage	pH	Sodium adsorption ratio
Saline	>4	<15	7.5–8.5	10–14
Saline-alkali	>4	>15	7.5–8.5	20–70
Nonsaline-alkali	<4	>15	8.5–10	14–40
Degraded alkali	<4	>15	6.5–7.5	

Saline-alkali soils are changed into alkali soils when salts are leached away. Similarly, normal or saline soils can be converted to alkali soils through use of irrigation water containing excessive amounts of sodium.

Reclamation of salted soils involves the removal of excess salts by leaching and of excess sodium through ion exchange with calcium from gypsum or from soil carbonates. The latter are dissolved by the addition of acidifying materials such as sulfuric acid, aluminum sulfate, or elemental sulfur.

[NATHANIEL T. COLEMAN]

Metal complexing. Many of the metallic elements required by plants and animals in small amounts form coordination compounds with electron donors in soils, giving rise to metal-organic complexes. Copper deficiencies in organic soils and the large amount of copper required to overcome these deficiencies bear witness to the importance of these kinds of reactions in soil systems. Considerable speculation and some lines of evidence have also been directed toward the possibility of soluble metal-organic complexes existing in soil systems. Substantial progress has been made in identifying reactive groups of organic matter available for complexing trace metals, and measurements are being made of the presence and degree of metal-organic complexing in the soil solution. The latter may help to explain some field observations that were difficult to interpret in terms of what had been previously understood about the chemistry of these elements in soils.

Plant residues of various sorts have been known for some time to mobilize a number of metals in soils. Most workers now regard soluble complex formation as being the principal explanation of this effect.

Work in Egypt indicates that copper in soil extracts is present in part as stable or as highly inert complexes. The fraction of the copper complexed increased from 20 to about 99% of the total copper as manure was added to the soil in increasing rates. The presence of the stable complexes was detected by passing a water extract of the soil through cation-exchange columns. Copper not adsorbed by the column was interpreted as being in a noncationic form, presumably as a chelate. Labile complexes would disassociate when exposed to such treatment and would not be measured.

The complexes measured by the researchers in Egypt may be too inert to be available to plants. Certainly any more labile complexes that may be present in soils are of interest to plant nutrition, but a complex that disassociates readily is more difficult to detect. Some means must be provided for assessing the mass-action effect of the unassociated ion in the system. When the amount of unassociated ion is subtracted from the total concentration of the element in solution, the remainder is equal to the fraction of the element associated with complexing agents or as ion pairs. (For convenience, ion-pair formation has been grouped with complexing.)

Unassociated ion assessment. Several means are available for assessing the mass-action effect of unassociated ions. The most common is the specific ion electrode. Electrodes are not generally available for trace elements, however, and their sensitivity is somewhat restricted for ions at very low activities in solution. The technique used in soils is to introduce either a small amount of a resin or a competing complexing agent that forms complexes that can be separated from the rest of the system by extraction with an immiscible solvent. By adding only very small quantities of complexing agent or resin, the fraction of the metal combining with them can be restricted to a point where the original equilibrium of the system is not significantly disturbed. In this way the presence of the unassociated ion is assessed, and calculation of the complexed ion is made possible.

This approach has been used to evaluate the fraction of zinc, copper, manganese and, to a lesser extent, cobalt that is complexed in the solutions of several soils. Of these elements, cobalt does not appear to be complexed appreciably, but from 50 to 75% of the zinc, 90% of the manganese, and as much as 99% of the copper in soil solutions is present as an organic complex. As might be expected, more of the element is complexed in the surface soils than in subsoils.

Copper and zinc complexes. The importance of soluble metal complexes in soils cannot yet be judged, but there is some indication that the phenomenon may clarify the hitherto unexplained behavior of at least copper. The distribution of copper deficiencies, for example, cannot be explained solely in terms of adsorption by mineral or organic surfaces (reactions thought to make copper unavailable). Copper, like zinc, is bound increasingly strongly as the pH of the soil increases, which would lead to reduced availability at high pH values. Zinc deficiencies, as one might anticipate from this, occur mostly in calcareous soils but, contrary to expectations, copper deficiencies are found mostly in acid organic soils. It is proposed that the complexing of copper, which appears to be more effective under alkaline than acid conditions, compensates for other reactions to keep the element relatively available in calcareous soils. The complexing of zinc, while also relatively high in calcareous soils, is not great enough to compensate for the reactions taking it out of solution. In the same manner, complexing of copper in the soil solution may contribute to its increased uptake after liming—a phenomenon that has been observed by some.

Organic material effect. The organic material in soils that may contribute to complexing of micronutrients varies enormously and is frequently very complex. A number of simple compounds, such as aliphatic acids and amino acids, have been found in soil solutions. More complicated molecules have also been identified, including the very stable vitamin B_{12} that chelates cobalt. But probably the most important sources of complexing in soils are the so-called humic and fulvic acids and possibly the polyphenols. The structures of these compounds have not been determined as such. Rather, they are characterized in terms of the linkages in their central chains and their functional groups. Particular attention has been paid to the presence of carboxyl, hydroxyl, and carbonyl groups, but small amounts of other groups, particularly esterified phosphate, although present in smaller amounts, may prove to be as important in the complexing of heavy metals.   [J. F. HODGSON]

Nutrients. Plants tolerate larger amounts of molybdenum than do animals and, except for legumes, do not need cobalt for growth. Ruminant

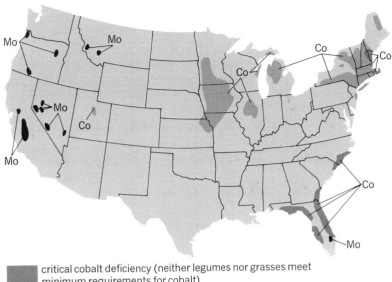

critical cobalt deficiency (neither legumes nor grasses meet minimum requirements for cobalt)

sporadic cobalt deficiency (legumes but not grasses meet minimum requirements for cobalt)

molybdenum toxicity (legumes have 10 ppm or more of molybdenum)

Fig. 6. Areas of cobalt deficiency and molybdenum toxicity.

animals require a minimum of about 0.07 ppm (parts per million) of cobalt in feed and, under grazing conditions, cannot tolerate much more molybdenum than 10–20 ppm. Areas where common forage plants have too little cobalt for grazing animals occur principally in the eastern United States, and areas of plants with too much molybdenum are in the West (Fig. 6). Soil and geologic materials determine the distribution of cobalt deficiency and molybdenum toxicity in grazing animals.

Molybdenum. Molybdenum toxicity is a problem among ruminant animals principally in parts of California, Nevada, Oregon, and Montana. The problem areas are largely wet, narrow floodplains and alluvial fans of small streams. The extent of the problem areas tends to be exaggerated because the problem areas are interspersed locally with broad areas of productive soils.

Size of streams and the rock areas they drain determine how much molybdenum is present in alluvium. Most areas of molybdenum toxicity in the western United States are on granitic alluvium that is not mixed with materials from other streams. Small areas also occur on alluvium derived from some shales in northwestern Oregon. Because all alluvium from granite and shales does not give rise to molybdenum-toxic areas, there must be a source of the molybdenum in the higher-lying areas that the streams drain. Broad floodplains of large rivers have materials from many streams, and the large amounts of molybdenum that any stream may contribute are diluted. Thus, molybdenum toxicity is not a problem on broad floodplains.

The molybdenum in alluvium is readily released to plants if soils are wet. If soils are well drained, the release of molybdenum to plants is slow; and plants do not accumulate large amounts of molybdenum, even though the soil may contain large

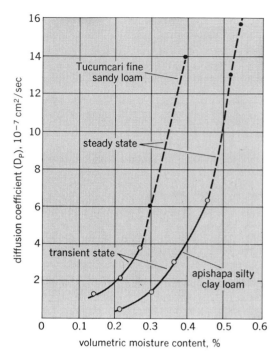

Fig. 7. Diffusion coefficient for phosphorus in soil as a function of soil moisture content.

amounts. Molybdenum is also more available to plants in alkaline than in acid soils. Thus, the incidence of molybdenum toxicity is greatest in wet, neutral-to-alkaline soil areas; but molybdenum toxicity can also occur in wet, acid soil areas if the soils have enough molybdenum. The release of molybdenum from these acid soils may not be as rapid as from alkaline soils, but plants may have from 10 to 20 ppm or more of molybdenum.

Cobalt. Cobalt deficiency occurs in the eastern United States. Unlike molybdenum-toxic areas, cobalt-deficient areas cover broad glacial-drift plains in New England and the lower coastal plain from North Carolina to Florida.

Geology also plays an important role in the distribution of areas where deficiencies of cobalt occur. In New England the glacier that overrode the White Mountains left drift deposits on broad plains to the southeast that contribute very little cobalt to soils. The small amounts of cobalt contributed by granites of the White Mountains are today traceable through the low cobalt content in soils on floodplains of the Saco and Merrimac rivers, both of which originate in the White Mountains. Cobalt deficiency in this part of New England was a problem for the earliest settlers who tried to raise cattle.

Only very small amounts of cobalt were contributed by the sandy coastal-plain deposits of the southeastern United States. Here the cobalt deficiency probably resulted from weathering and loss of cobalt as the sandy materials which were weathered in the Piedmont and mountains were carried to the sea to form coastal-plain deposits.

The development of spodosols (podzols) in granitic deposits of New England and humaquods (humus groundwater podzols) in sandy coastal-plain deposits has caused further loss of whatever

cobalt these sandy materials had. Cobalt is leached from these soils much as are iron and organic matter. The humaquods especially have very small amounts of reactive forms of cobalt left because of intense leaching. The cobalt leached from humaquods contrasts strikingly with an apparent biopedogenic recycling of soil cobalt from the subsoil to the ground surface in ultisols (red-yellow podzolic soils) of the southeastern United States. Cobalt deficiency is not a problem in ruminant animals on farms with ultisols, because plants have adequate amounts of cobalt to meet their nutritional requirements.　　　　[JOE KUBOTA]

Diffusion of nutrients. Plant roots obtain nutrients from the soil by processes of diffusion, mass flow of water, and root extension. The relative importance of each process depends on the nutrient content in the plant and the soil, and on the length and size of roots.

Diffusion becomes important when nutrient demands by the plant cause a lowering of the concentration in solution at the root surface. S. A. Barber indicated that such conditions often occur for potassium, nitrogen, phosphorus, and occasionally for sulfur but rarely for calcium and magnesium. For example, the amount of phosphorus arriving at the root surface depends on the diffusion coefficient (D_P), the concentration gradient between the root surface and the soil solution, the root radius, and a capacity factor. This capacity factor depends on how much phosphorus the solid phase will supply to the solution for a unit change in the solution concentration of phosphorus.

D_p *in soil.* Diffusion coefficients in porous media such as soil are always smaller than in solutions. The ions diffuse through the soil water, which can occupy generally 10–40% of the cross-sectional area, and they follow a more tortuous path through the soil pores. In 1965, S. R. Olsen, W. D. Kemper, and J. van Schaik measured the D_P for phosphorus in two types of soil as a function of soil moisture content, as illustrated in Fig. 7. An important development in this measurement was to evaluate

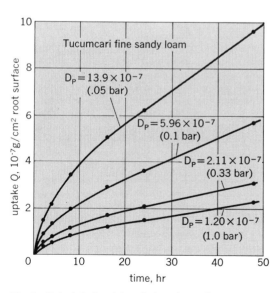

Fig. 8. Calculated uptake of phosphorus by corn roots as related to diffusion coefficient of phosphorus and soil moisture tension.

the role of the solid-phase soil in replenishing the phosphorus concentration in solution. For comparison the diffusion coefficient for phosphorus in solution is 5.0×10^{-6} cm²/sec.

The importance of D_P in absorption of phosphorus by corn roots is shown in Fig. 8. The plant required an eightfold increase in time to absorb the same amount of phosphorus as D_P varied from 13.9×10^{-7} to 1.20×10^{-7} cm²/sec because of a change in moisture content. The larger D_P represents a moisture content commonly found a few hours after a rainfall or irrigation, while the lower D_P typifies a soil where the plant has used most of the available soil water.

Concentration gradient. With corn (*Zea mays*), the phosphorus concentration at the root surface was reduced to 0.65 of its initial value after 24 hr of uptake. Barber, J. M. Walker, and E. H. Vesey illustrated by radioautographs how the roots reduce the phosphorus concentration around them during uptake. Plants may differ in their ability to reduce the concentration at their root surfaces. Thus, the capacity of the soil to supply phosphorus cannot be defined accurately without taking into account the action of the roots.

Root radius. Calculations show that small root hairs (0.001-cm radius) can absorb phosphorus four times faster than large roots (0.05-cm radius) for equal root surface areas. This difference is important especially for plants growing on soil with small amounts of available phosphorus.

Capacity factor. Clay soils have a larger capacity factor than sandy soils, as well as a greater D_P. When the concentration of phosphorus in solution is the same, the roots will absorb more phosphorus from the clay soils. These observations provide a basis from which better soil tests may be developed to estimate the need for phosphorus fertilizer in soil.   [STERLING R. OLSEN]

Bibliography: F. E. Bear (ed.), *Chemistry of the Soil*, 2d ed., 1964; G. T. Felbeck, Jr., Structural chemistry of soil humic substances, *Adv. Agron.*, 17:327–368, 1965; H. R. Geering, J. F. Hodgson, and C. Sdano, Micronutrient cation complexes in soil solution, IV: The chemical state of manganese in soil solution, *Soil Sci. Soc. Amer. Proc.*, 33:81–85, 1969; J. F. Hodgson, W. L. Lindsay, and J. F. Trierweiler, Micronutrient cation complexes in soil solution, II: Complexing of zinc and copper in displaced solution from calcareous soils, *Soil Sci. Soc. Amer. Proc.*, 30:723–726, 1966; J. Kubota, Distribution of cobalt deficiency in grazing animals in relation to soils and forage plants of the United States, *Soil Sci.*, 106:122–130, 1968; J. Kubota et al., The relationship of soils to molybdenum toxicity in cattle in Nevada, *Soil Sci. Soc. Amer. Proc.*, 25:227–232, 1961; J. Kubota et al., The relationship of soils to molybdenum toxicity in grazing animals in Oregon, *Soil Sci. Soc. Amer. Proc.*, 31:667–671, 1967; S. R. Olsen and W. D. Kemper, Movement of nutrients to plant roots, *Adv. Agron.*, 20:91–151, 1968; S. R. Olsen and F. S. Watanabe, Effective volume of soil around plant roots determined from phosphorus diffusion, *Soil Sci. Soc. Amer. Proc.*, 30:598–602, 1966; S. K. Tobia and A. S. Hanna, Effect of copper sulfate added to irrigation water on copper status of Egyptian soils, II: Free and complexed copper, *Soil Sci.*, 92:123–126, 1961.

Soil conservation

The practice of arresting and minimizing artificially accelerated soil deterioration. Its importance has grown because cultivation of soils for agricultural production, deforestation and forest cutting, grazing of natural range, and other disturbances of the natural cover and position of the soil have increased greatly in the last 100 years in response to the growth in world population and man's technical capacity. Accelerated soil deterioration has been the consequence.

Erosion extent and intensity. Accelerated erosion has been known throughout history wherever men have tilled or grazed slopes or semiarid soils. There are many evidences of the physical effects of accelerated erosion in the eastern and central parts of the Mediterranean basin, in Mesopotamia, in China, and elsewhere. Wherever the balance of nature is a delicate one, as on steep slopes in regions of intense rainstorms, or in semiarid regions of high rainfall variability, grazing and cultivation eventually have had to contend with serious or disabling erosion. Irrigation works of the Tigris and Euphrates valleys are thought to have suffered from the sedimentation caused by quickened erosion on the rangelands of upstream areas in ancient times. The hill sections of Palestine, Syria, southern Italy, and Greece experienced serious soil losses from grazing and other land use mismanagement many centuries ago. Accelerated water erosion on the hills of southern China and wind erosion in northwestern China also date far back into history. Exactly what effects these soil movements may have had on history has been a debated question, but their impact may have been serious on some cultures, such as those of the Syrian and Palestinian areas, and debilitating on others, as in the case of classical Rome and the China of several centuries ago.

The exact extent of accelerated soil erosion in the world today is not known, particularly as far as the rate of soil movement is concerned. However, it may be safely said that nearly every semiarid area with cultivation or long-continued grazing, every hill land with moderate to dense settlement in humid temperate and subtropical climates, and all cultivated or grazed hill lands in the Mediterranean climate areas suffer to some degree from such erosion. Thus recognized problems of erosion are found in such culturally diverse areas as southern China, the Indian plateau, south Australia, the South African native reserves, the Soviet Union, Spain, the southeastern and midwestern United States, and Central America.

Within the United States the most critical areas have been the hill lands of the Piedmont and the interior Southeast, the Great Plains, the Palouse area hills of the Pacific Northwest, southern California hills, and slope lands of the Midwest. The high-intensity rainstorms of the Southeast, and the cyclical droughts of the Plains have predisposed the two larger areas to erosion. The light-textured A horizon formed under the Plains grass cover was particularly susceptible to wind removal, while the high clay content of many southeastern soils predisposed them to water movement. These natural susceptibilities were repeatedly brought into play by agricultural systems which stressed corn and

Fig. 1.   Erosion of sandstone caused by strong wind and occasional hard rain in an arid region. (*USDA*)

cotton in the Southeast, corn in the Midwest, and intensive grazing and small grains on the Plains, Palouse, and California. The open soil surface left in the traditional cotton, corn, and tobacco cultivation of the Southeast furnished almost ideal conditions for water erosion, and at the same time caused heavy nutrient depletion of soils thus cropped. The open fields of seasons between crops have also been susceptible to soil depletion. Open fields have been especially disastrous to maintenance of soil cover during the droughts of the Plains. Soil mismanagement thus has been a common practice in parts of the United States where the stability of soil cover hangs in delicate balance.

Types of soil deterioration. Soil may deteriorate either by physical movement of soil particles from a given site or by depletion of the water-soluble elements in the soil which contribute to the nourishment of crop plants, grasses, trees, and other economically usable vegetation. The physical movement generally is referred to as erosion. Wind, water, glacial ice, animals, and man's

Fig. 2.   Improper land use. Corn rows planted up and down the slope rather than on the contour. Note better growth of plants in bottom (deeper) soil in foreground as compared to stunted growth of plants on slope. (*USDA*)

tools in use may be agents of erosion. For purposes of soil conservation, the two most important agents of erosion are wind and water, especially as their effects are intensified by the disturbance of natural cover or soil position. Water erosion always implies the movement of soil downgrade from its original site. Eroded sediments may be deposited relatively close to their original location, or they may be moved all the way to a final resting place on the ocean floor. Wind erosion, on the other hand, may move sediments in any direction, depositing them quite without regard to surface configuration. Both processes, along with erosion by glacial ice, are part of the normal physiographic (or geologic) processes which are continuously acting upon the surface of the Earth. The action of both wind and water is vividly illustrated in the scenery of arid regions (Fig. 1). Soil conservation is not so much concerned with these normal processes as with the new force given to them by man's land use practices. *See* LAND-USE PLANNING.

Depletion of soil nutrients obviously is a part of soil erosion. However, such depletion may take place in the absence of any noticeable amount of erosion. The disappearance of naturally stored nitrogen, potash, phosphate, and some trace elements from the soil also affects the usability of the soil for man's purposes. The natural fertility of virgin soils always is depleted over time as cultivation continues, but the rate of depletion is highly dependent on management practices. *See* SOIL.

Accelerated erosion may be induced by any land use practice which denudes the soil surfaces of vegetative cover (Fig. 2). If the soil is to be moved by water, it must be on a slope. The cultivation of a corn or a cotton field is a clear example of such a practice. Corn and cotton are row crops; cultivation of any row crop on a slope without soil-conserving practices is an invitation to accelerated erosion. Cultivation of other crops, like the small grains, also may induce accelerated erosion, especially where fields are kept bare between crops to store moisture. Forest cutting, overgrazing, grading for highway use, urban land use, or preparation for other large-scale engineering works also may speed the natural erosion of soil (Fig. 3).

Where and when the soil surface is denuded, the movement of soil particles may proceed through splash erosion, sheet erosion, rill erosion, gullying, and wind movement (Fig. 4). Splash erosion is the minute displacement of surface particles caused by the impact of falling rain. Sheet erosion is the gradual downslope migration of surface particles, partly with the aid of splash, but not in any defined rill or channel. Rills are tiny channels formed where small amounts of water concentrate in flow. Gullies are V- or U-shaped channels of varying depths and sizes. A gully is formed where water concentrates in a rivulet or larger stream during periods of storm. It may be linear or dendritic (branched) in pattern, and with the right slope and soil conditions may reach depths of 50 ft or more. Gullying is the most serious form of water erosion because of the sharp physical change it causes in the contour of the land, and because of its nearly complete removal of the soil cover in all horizons. On the edges of the more permanent stream channels, bank erosion is another form of soil movement.

Causes of soil mismanagement. One of the chief causes of erosion-inducing agricultural practices in the United States has been ignorance of their consequences. The cultivation methods of the settlers of western European stock who set the pattern of land use in this country came from a physical environment which was far less susceptible to erosion than North America, because of the mild nature of rainstorms and the prevailing soil textures in Europe. Corn, cotton, and tobacco, moreover, were crops unfamiliar to European agriculture. In eastern North America the combination of European cultivation methods and American intertilled crops resulted in generations of soil mismanagement. In later years the plains environment, with its alternation of drought and plentiful moisture, was also an unfamiliar one to settlers from western Europe.

Conservational methods of land use were slow to develop, and mismanagement was tolerated because of the abundance of land in the 18th and 19th centuries. One of the cheapest methods of obtaining soil nutrients for crops was to move on to another farm or to another region. Until the 20th century, land in the United States was cheap, and for a period it could be obtained by merely giving assurance of settlement and cultivation. With low capital investments, many farmers had little stimulus to look upon their land as a vehicle for permanent production. Following the Civil War, tenant cultivators and sharecroppers presented another type of situation in the Southeast where stimulus toward conservational soil management was lacking. Management of millions of acres of Southeastern farmland was left in the hands of men who had no security in their occupancy, who often were illiterate, and whose terms of tenancy and meager

(a)

(b)

Fig. 4.   Two most serious types of erosion. (a) Sheet erosion as a result of downhill straight-row cultivation. Note onions washed completely out of ground. (b) Gully erosion destroying rich farmland and threatening highway. (*USDA*)

Fig. 3.   Rill erosion on highway fill. The slopes have been seeded (horizontal lines) with annual lespedeza to bind and stabilize soil. (*USDA*)

training forced them to concentrate on corn, cotton, and tobacco as crops.

On the Plains and in other susceptible western areas, small grain monoculture, particularly of wheat, encouraged the exposure of the uncovered soil surface so much of the time that water and wind inevitably took their toll (Fig. 5). On rangelands, the high percentage of public range (for whose management little individual responsibility could be felt), lack of knowledge as to the precipitation cycle and range capacity, and the urge to maximize profits every year contributed to a slower, but equally sure denudation of cover.

Finally, the United States has experienced extensive erosion in mountain areas because of forest mismanagement. Clearcutting of steep slopes, forest burning for grazing purposes, inadequate fire protection, and shifting cultivation of forest lands have allowed vast quantities of soil to wash out of the slope sites where they could have pro-

Fig. 5.   Wind erosion. Accumulation of topsoil blown from bare field on right. (*USDA*)

duced timber and other forest values indefinitely. In the United States the central and southern Appalachian area and the southern part of California have suffered severely in this respect, but all hill or mountain forest areas, except the Pacific Northwest, have had such losses. *See* FOREST MANAGEMENT AND ORGANIZATION; FOREST RESOURCES.

Economic and social consequences. Where the geographical incidence of soil erosion has been extensive, the damages have been of the deepest social consequence. Advanced stages of erosion may remove all soil and therefore all capacity for production. More frequently it removes the most productive layers of the soil—those having the highest capacity for retention of moisture, the highest soil nutrient content, and the most ready response to artificial fertilization. Where gullying or dune formation takes place, erosion may make cultivation physically difficult or impossible. Thus, depending on extent, accelerated erosion may affect productivity over a wide area. At its worst, it may cause the total disappearance of productivity, as on the now bare limestone slopes of many Mediterranean mountains. At the other extreme may be the slight depression of crop yields which may follow the progress of sheet erosion over short periods. In the case of forest soil losses, except where the entire soil cover disappears, the effects may not be felt for decades, corresponding to the growth cycle of given tree species. Agriculturally, however, losses are apt to be felt within a matter of a few years.

Moderate to slight erosion cannot be regarded as having serious social consequences, except over many decades. As an income drepressant, however, it does prevent a community from reaching full productive potentiality. More severe erosion has led to very damaging social dislocation. For those who choose to remain in an eroding area or who do not have the capacity to move, or for whom migration may be politically impossible, the course of events is fateful. Declining income leads to less means to cope with farming problems, to poor nutrition and poor health, and finally to family exist-

ence at the subsistence level. Communities made up of a high proportion of such families do not have the capacity to support public services, even elementary education. Unless the cycle is broken by outside financial and technical assistance or by the discovery of other resources, the end is a subsistence community whose numbers decline as the capacity of the land is further reduced under the impact of subsistence cultivation. This has been illustrated in the hill and mountain lands of southeastern United States, in Italy, Greece, Palestine, China, and elsewhere for many millions of peasant people. Illiteracy, short life-spans, nutritional and other disease prevalence, poor communications, and isolation from the rest of the world have been the marks of such communities. Where they are politically related to weak national governments, idefinite stagnation and decline may be forecast. Where they are part of a vigorous political system, their rehabilitation can be accomplished only through extensive investment contributed by the nation at large. In the absence of rehabilitation, these communities may constitute a continued financial drain on the nation for social services such as education, public health, roads, and other public needs.

Effects on other resources. Accelerated erosion may have consequences which reach far beyond the lands on which the erosion takes place and the community associated with them. During periods of heavy wind erosion, for example, the dust fall may be of economic importance over a wide area beyond that from which the soil cover has been removed. The most pervasive and widespread effects, however, are those associated with water erosion. Removal of upstream cover changes the regimen of streams below the eroding area. Low flows are likely to be lower and their period longer where upper watersheds are denuded than where normal vegetative cover exists. Whereas flood crests are not necessarily higher in eroding areas, damages may be heightened in the valleys below eroding watersheds because of the increased deposition of sediment of different sizes, the rapid lifting of channels above floodplains, and

Artificial storage becomes necessary to derive the services from water which are economically possible and needed. But even the possibilities of storage eventually may disappear when erosion of upper watersheds continues. Reservoirs may be filled with the moving sediment and lose their capacity to reduce flood crests, store flood waters, and augment low flows. For this reason plans for permanent water regulation in a given river basin must always include watershed treatment where eroding lands are in evidence. *See* WATER CONSERVATION.

Conservation measures and technology. Measures of soil management designed to reduce the effects of accelerated erosion have been known in both the Western world and in the Far East since long before the time of Christ. The value of forests for watershed protection was known in China at least 10 centuries ago. The most important of the ancient measures on agricultural lands was terrace construction, although actual physical restoration of soil to original sites also has been practiced. Terrace construction in the Mediterranean countries, in China, Japan, and the Philippines represents the most impressive remaking of the face of the Earth before the days of modern earth-moving equipment (Fig. 6). Certain land management practices which were soil conserving have been a part of western European agriculture for centuries, principally those centering on livestock husbandry and crop rotation. Conservational management of the soil was known in colonial Virginia and by Thomas Jefferson and others during the early years of the United States. However, it is principally since 1920 that the technique of soil conservation has been developed for many types of environment in terms of an integrated approach. The measures include farm, range, and forest management practices, and the building of engineered structures on land and in stream channels.

Farm, range, and forest. A first and most important step in conservational management is the determination of land capability—the type of land the choking of irrigation canals.

A long chain of other effects also ensues. Because of the extremes of low water in denuded areas during dry seasons, water transportation is made difficult or impossible without regulation, fish and wildlife support is endangered or disappears, the capacity of streams to carry sewage and other wastes safely may be seriously reduced, recreational values are destroyed, and run-of-the-river hydroelectric generation reaches a very low level. use and economic production to which a plot is suited by slope, soil type, drainage, precipitation, wind exposure, and other natural attributes. The objective of such determination is to achieve permanent productive use as nearly as possible. The United States Soil Conservation Service has developed one of the more easily understood and widely employed classifications for such determination (Fig. 7). In it eight classes of land are recognized within United States territory. Four classes represent land suited to cultivation, from the class 1 flat or nearly flat land suited to unrestricted cultivation, to the steeper or eroded class 4 lands which can be cultivated only infrequently. Three additional classes are grazing or forestry land, with varying degrees of restriction on use. The eighth class is suited only to watershed, recreation, or

wildlife support. The aim in the United States has been to map all lands from field study of their capabilities, and to adjust land use to the indicated capability as it becomes economically possible for the farm, range, or forest operator to put conservational use into force.

Once the capability of land has been determined, specific measures of management come into play. For class 1 land few special practices are necessary. After the natural soil nutrient minerals begin to decline under cultivation, the addition of organic or inorganic fertilizers becomes necessary. The return of organic wastes, such as manure, to the soil is also required to maintain favorable texture and optimum moisture-holding capacity. Beyond these measures little need be added to the normal operation of cultivation.

On class 2, 3, and 4 lands, artificial fertilization will be required, but special measures of conservational management must be added. The physical conservation ideal is the maintenance of such land under cover for as much of the time as possible. This can be done where pasture and forage crops are suited to the farm economy. However, continuous cover often is neither economically desirable nor possible. Consequently, a variety of devices has been invented to minimize the erosional results from tillage and small grain or row crop growth. Tillage itself has become an increasingly important conservational measure since it can affect the relative degree of moisture infiltration and soil grain aggregation, and therefore erosion. Where wind erosion is the danger, straw mulches or row or basin listing may be employed and alternating strips of grass and open-field crops planted. Fields in danger of water erosion are plowed on the contour (not up- and downslope), and if lister cultivation is also employed, water-storage capacity of the furrows will be increased. Strip-cropping, in which alternate strips of different crops are planted on the contour, may also be employed. Crop rotations that provide for strips of closely planted legumes and perennial grasses alternating with

Fig. 6.   The Ifugao rice terraces, Philippines. (*Philippine Embassy, Washington, D.C.*)

grains such as wheat and barley and with inter-tilled strips are particularly effective in reducing soil erosion. Fields may also be terraced, and the terraces strip-cropped. The bench terrace, which interrupts the slope of the land by a series of essentially horizontal slices cut into the slope, is now comparatively little used in the United States in contrast to the broad-base terrace, which imposes comparatively little impediment to cultivation. A broad-base terrace consists of a broad, shallow surface channel, flanked on the downward slope by a low, sloping embankment. If the terrace is constructed with a slight gradient (channel-type or graded terrace), it serves to reduce erosion by conducting excess water off the slope in a controlled manner. If it is constructed on the contour (level or ridge-type terrace), its primary purpose is to conserve moisture. The embankment on the downslope side of a broad-base terrace may be constructed from soil taken from the upper side only (sometimes referred to as a Nichols terrace) or from both sides (Mangum terrace).

Design of conservational cultivation also includes provision for grass-covered waterways to collect drainage from terraces and carry it into stream courses without erosion. Where suitable conditions of slope and soil permeability are found, shallow retention structures may also be constructed to promote water infiltration. These are of special value where insufficient soil moisture is a problem at times. Additional moisture always encourages more vigorous cover growth.

The measures just described may be considered preventive. There are also measures of rehabilitation where fields already have suffered from erosion and offer possibilities of restoration. Grading with mechanical equipment and the construction of small check dams across former gullies are examples.

For the remaining four classes of land, whose principal uses depend on the continuous maintenance of cover, management is more important than physical conditioning. In some cases, however, water retention structures, check dams, and other physical devices for retarding erosion may be applied on forest lands and rangelands. In the United States such structures are not often found in forest lands, although they have been commonly employed in Japanese forests. In forestry the conservational management objective is one of maximum production of wood and other services while maintaining continuous soil cover. The same is true for grass and other forage plants on managed grazing lands. For rangelands, adjustment of use is particularly difficult because grazing must be tolerated only to the extent that the range plants still retain sufficient vitality to withstand a period of drought which may arrive at any time.

A last set of erosion-control measures is directed toward minimizing stream bank erosion which may be large over the length of a long stream. This may be done through revetments, retaining walls, and jetties, which slow down current undercutting banks and hold sand and silt in which soil-binding

Fig. 7. Land capability classes. Suitable for cultivation: 1, requires good soil management practices only; 2, moderate conservation practices necessary; 3, intensive conservation practices necessary; 4, perennial vegetation—infrequent cultivation. Unsuitable for cultivation (pasture, hay, woodland, and wildlife): 5, no restrictions in use; 6, moderate restrictions in use; 7, severe restrictions in use; 8, best suited for wildlife and recreation. (*USDA*)

Fig. 8. Stream bank erosion control. Construction of a new conservation pool which will help reduce flooding, retard downstream erosion, and store water. (*USDA*)

willows, kudzu, and other vegetation may become established. Sediment detention reservoirs also reduce the erosive power of the current, and catchment basins or flood-control storage helps reduce high flows (Fig. 8).

Conservation agencies and programs. Whereas excellent soil-conserving soil management was maintained for generations by some farmers and farm groups, as in Lancaster County, Pa., a major amount of the soil-conservation activities in the United States is derived from Federal government assistance. The Soil Conservation Service of the USDA has been a focal agency in spreading knowledge of soil conservation in farmland and rangeland management and aiding in its application. In practice, the local administration of a soil-conserving program is within a Soil Conservation District, which usually is coincident with a county, and is organized under state law. The district is the liaison unit between the farmer and public assistance agencies at the state and Federal levels. It is managed by a board or committee, generally composed of five members, and usually elected by farmers within the district. Other local public bodies which may have soil-conservation objectives include conservancy districts, wind-erosion districts, drainage or irrigation districts, Agricultural Stabilization and Conservation Service County Committees, and Farmers' Home Administration County Committees. In addition there are private groups with conservational interests, such as the farmers' cooperatives and national farm organizations like the Farm Bureau Federation.

The local districts may be aided technically and financially in their program. Much of the financial aid stems from Federal sources, and theoretically it is on a matching fund basis. In actual practice, however, a major part of the expenditures for special soil-conserving programs is from Federal funds. Technical aid is provided throughout the nation by the Soil Conservation Service, and also by the U.S. Forest Service for its special fields of forestry and grazing-land management. Technical aid also has been provided by the Agricultural Extension Services and the Land Grant Colleges of the several states. The Tennessee Valley Authori-

ty has maintained a program of its own design, with the cooperation of the colleges and the Extension Services. The Soil and Moisture Conservation Operations Office of the Indian Service, U.S. Department of the Interior, likewise has conducted a program limited to specific Indian lands.

Financial assistance for soil-conservation measures has been provided by the Federal government through the Soil Conservation Service, the Agricultural Stabilization and Conservation Service, the TVA, and the Farmers' Home Administration. Assistance has been particularly in the form of loans from the FHA, in low-cost fertilizer from the TVA, and as direct cash outlay from other agencies. Over the years, the program of the Agricultural Stabilization and Conservation Service has been the largest single source of financial aid for these purposes.

In addition to technical and financial aid, the farmers or other land operators of the United States are given valuable indirect assistance through the many research programs, basic and applied, which treat the fields related to soil conservation. The work of the Agricultural Research Service, of the Soil Conservation Service, of the Tennessee Valley Authority, and of the Land Grant Colleges has been especially helpful. Through these works new soil-conserving plants, new fertilizers, improved means of physical control, and new methods of management have been developed. Through them soil conservation has not only become important but also an increasingly efficient public activity in the United States. *See* FOREST AND FORESTRY; RANGELAND CONSERVATION.

[EDWARD A. ACKERMAN; DONALD J. PATTON]

Bibliography: I. Burton and R. W. Kates, *Readings in Resource Management and Conservation*, 1965; R. B. Held, *Soil Conservation in Perspective*, 1965; R. M. Highsmith, J. G. Jensen, and R. D. Rudd, *Conservation in the United States*, 1962; G. O. Schaub et al., *Soil and Water Conservation Engineering*, 2d ed., 1966.

Soil mechanics

The application of the laws of solid and fluid mechanics to soils and similar granular materials as a basis for design, construction, and maintenance of stable foundations and earth structures. Soil mechanics differs from other applications of mechanics to engineering, such as the design of concrete and steel structures, in that soil and similar materials have a much wider range of mechanical and other physical properties than concrete and steel. In addition, soil materials are usually present in layers and strata that vary widely in composition and physical character.

Soil mechanics as an applied science has a relatively small system of legitimate theory and a large and important methodology concerned with soil exploring, sampling, and testing. It also makes use of information from other fields, such as geology and soil science.

Soil mechanics will always be an important part of soil engineering. However, the tools of soil engineering are no longer predominantly the principles of solid and fluid mechanics but to an ever-increasing extent those of other subdivisions of physics, such as thermodynamics, electricity and magnetism, acoustics, optics, and chemical, atomic, and nuclear physics. This is true also in the present development of soil exploration and testing and in endeavors to improve soil properties.

Soil, in the engineering sense, comprises all accumulations of solid particles in the earth mantle that are loose enough to be moved with spade or shovel. Soils range from deep-lying geologic deposits to agricultural surface soils. Soil mechanics is also applicable to similar granular assemblies of artificial origin, such as fills of mineral waste materials. *See* SOIL.

SOIL TYPES AND COMPOSITION

Soils vary widely in composition and physical properties. At one extreme are the inert, granular, cohesionless sands and gravels; at the other are clay soils of great water affinity and well-developed cohesion in the moist and dry state. According to the ease or difficulty of working them, rather than their density, soils are called light when predominantly granular and noncohesive and heavy when predominantly clayey and cohesive.

Mineral origins of soils. The mineral composition of gravels, stones, and boulders is essentially that of the parent rock from which they have been derived, predominantly by mechanical weathering action.

Sands are largely quartzitic and siliceous in humid climates but may be any kind of mineral in dry climates or under special circumstances. The white sands of New Mexico consist of gypsum particles; coral and shell beach sands may have more than 90% of calcium carbonate particles; the black sand of Yellowstone Park, Wyo., and some of the blue and purple beaches of the Pacific consist of obsidian particles.

The silt particles resemble quite closely the minerals in the parent rock, with feldspars, micas, and quartz usually well represented. In certain tropical soils, however, the silt and even larger particles may have been formed by stable agglomeration of smaller-sized chemical rock decomposition products.

Clay particles are submicroscopic, plate-shaped crystalline minerals that have been divided into three main groups: the kaolinite, the hydrous mica or illite, and the montmorillonite group. The clays possess great affinity for water as a result of the large amount of surface per volume of particle and as a function of the number and kind of adsorbed cations that neutralize unbalanced electrical charges in the clay mineral structures.

The boulder, gravel, and sand fractions are called coarse-grained or granular; they may be considered as the bones, and the combined silt, clay, and water fractions as the meat of a soil. Depending on their size, composition, or granulometry, soils may have a continuous granular skeleton, with the pores between the sand and gravel particles either empty or filled to various degrees with the silt-clay-water phase; they may possess a matrix of the latter in which the granular materials are discontinuously dispersed, or they may consist entirely of silt components, clay components, or both. The physical and mechanical properties of soils with sand and gravel skeletons are markedly different from those without. In practice, the limiting size between granular and nongranular (silt-

clay) fractions is the opening of a 200-mesh sieve (74μ).

Types by deposition. Engineering soils include unconsolidated sediments transported to their present place by glaciers, water, and air and soils formed at their present site by climatic and biologic forces from solid igneous, sedimentary, or metamorphic rock or from loose sediments transported by the above-named agents. The different agents have different carrying capacities and also affect the properties of their loads in different ways. It is, therefore, important to recognize and name such soils in accordance with the means of their transportation.

Glacial soils have been transported by glaciers whose action may be likened to giant bulldozers that push all sorts of materials ahead, with droppings on the side and grinding underneath. Spring melting of the glaciers stops their forward movement and permits the settling out of finely ground material. Thus, glacial soils may vary in size composition from boulders to varved clays. Typical for these soils is a disordered landscape, often with inhibited drainage and development of bogs and peat.

Aeolian soils range from sand dunes to loess deposits whose particles are predominantly of silt size. The valley loesses (Missouri, Mississippi, Rhine, and others) are formed from glacier-ground particles stirred up and dissipated by wind from the dry bottoms of glacier-draining rivers during the winter when the glacier supplies no water.

Fluvial soils are river deposits of relatively uniform particle size within the range from gravel to clay. The size itself depends upon the speed of water flow at the specific location.

Lacustrine soils are sediments formed on the bottom of lakes, and marine soils are sediments in ocean and other salt-water bodies. The salt content of the latter often flocculates the fine sediments and gives them a special type of structure. Because the carrying capacity of the respective agents varies with their speed, thus with weather and season, sedimentary deposits are usually stratified, that is, built up in layers of similarly sized particles.

Soils developed in place from either solid- or loose-rock parent material by the action of climate, plant growth, and animal life are usually shallow in temperate and cold regions. They are of importance mainly in the case of shallow foundations, as in highway and airport engineering. In tropical regions, however, they may extend to considerable depth and acquire importance for deep foundations. The science of pedology is concerned with their formation and characteristics.

Soil structure. Natural soil systems are characterized not only by the sizes and types of their component particles but also by the arrangement of these particles in relation to each other. For granular, noncohesive soils the porosity, or its supplement, the portion of the total volume filled by the solid particles, often suffices as an indicator of structure or packing. In natural clay and silt soils, typical secondary structures are formed whose disturbance can greatly alter their mechanical properties. The structure may be caused by and be typical of flocculation, as in marine clays, or may have been developed by wetting and drying

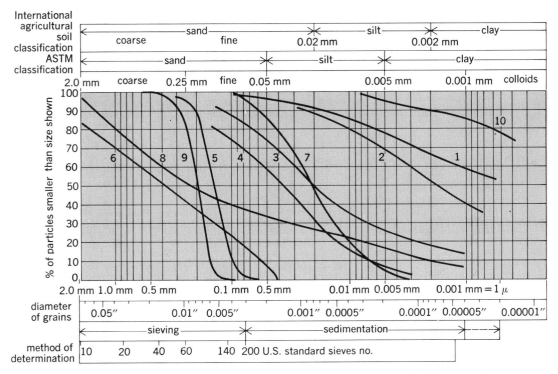

Fig. 1. Grain-size accumulation curves, plotted on semilogarithmic paper. Coarser fractions (left) measured by sieving; finer fractions (right) determined by sedimentation. Two of several systems of classification by grain size are shown (top). Curves 1 and 2, clay soils of the Nile Delta; 3 and 4, silts from the Nile Delta; 5, Port Said beach sand; 6, sand artificially graded for maximum density; 7, Vicksburg loess; 8, New Mexico adobe brick; 9, Daytona Beach sand; 10, Wyoming bentonite. (*From G. P. Tschebotarioff, Soil Mechanics, Foundations and Earth Structures, McGraw-Hill, 1951*)

and freezing and thawing cycles. Since soils may be employed in engineering in the undisturbed, partly disturbed, or greatly disturbed condition, the existence of soil structure must be taken into account.

For each specific purpose, the soil must be tested in a condition as close as possible to the one in which it is to be used.

Particle size and weight. The grain-size distribution of soils lacking or having negligible amounts of particles smaller than 74 μ is determined by dry sieving, and that of materials with all particles smaller than 74 μ by sedimentation methods. The latter are based on Stokes' law for the rate of fall in a viscous medium of lesser density, Eq. (1). Here ΔG = difference in specific gravity

$$v = \Delta G g d^2/1800n \quad \text{cm/sec} \qquad (1a)$$

$$d = \sqrt{1800\, nv/\Delta Gg} \qquad (1b)$$

between particle and viscous medium, g = acceleration of gravity, d = diameter of particle in millimeters, n = viscosity of medium in poises, and v = rate of fall in centimeters per second. Practical methods utilize the decrease of suspended particles with time at a definite distance from the surface of a soil-in-water suspension. This decrease is usually calculated from hydrometer measurements of the density of the suspension. For soils having both coarse and fine constituents, sieving (both dry and wet) is combined with sedimentation analysis.

At least four different sieve sizes are used, with openings of 4760, 2000, 420, and 74 μ, respectively. For specific purposes, other sieves are used to advantage. The results of mechanical analysis are best presented in the form of a grain-size accumulation curve as in Fig. 1. The naming of soils in accordance with their texture is shown in Fig. 2.

Determination of the specific gravity of soil particles is necessary for the sedimentation analysis and also for calculating (from the weight per unit volume and the moisture content) the actual volume relationships in the soil, which are important for its mechanical-strength properties. The ratio of the volume of water- and air-filled voids to the volume of solids in the sample is called the voids ratio e. Because of the variability of water content of soils, their unit weight is usually given in pounds per cubic foot of dry soil.

Engineering soil classifications. Soils have been classified for engineering purposes by several public agencies and individuals. The most widely used system at the present time is the so-called unified classification of the U.S. Bureau of Reclamation. While the different systems vary in details and nomenclature, the physical classification principles are essentially the same. The first division is made between soils that contain a coarse granular skeleton of gravel, sand, or both and those without such a skeleton, the silt-clay soils. The coarse granular materials are then subdivided into gravel and sand soils, which are further differentiated by the degree of water affinity of silt-clay material found in the intergranular spaces. The silt-clay soils which may or may not contain separate oc-

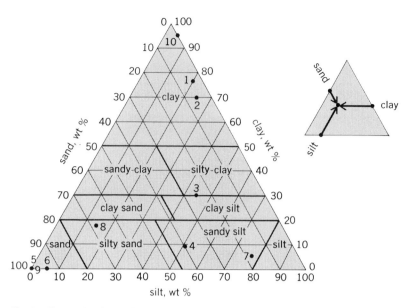

Fig. 2. Textural soil classification chart. The nomenclature of soil is determined by the percentages of sand, clay, and silt that are contained. The 10 numbered soils refer to those in Fig. 1. (*From G. P. Tschebotarioff, Soil Mechanics, Foundations and Earth Structures, McGraw-Hill, 1951*)

cluded coarse particles are subdivided into groups that differ in water affinity, plasticity, and swelling and shrinkage characteristics. A separate class of boulders and cobbles may be added to the coarse granular category and one of fibrous organic soil (peat) of great compressibility to the fine-grained category. The physical distinctions underlying engineering soil classifications entrain definite differences in other important physical properties, such as mechanical strength, elasticity, compressibility, shear and flow behavior, and permeability. The classification of a soil thus may serve to indicate its suitability for various engineering uses.

Soil-water relationships. Relationships between soil and water are of primary importance in soil mechanics. Their understanding and utilization presupposes a thorough knowledge of the properties of the water substance and of the physical and physicochemical characteristics of the surfaces of the soil minerals. Water is a peculiar substance. According to its molecular weight, it should be a gas at room temperature, but it is a liquid. Even as a liquid, it possesses structural properties commonly associated with solids and provable by means of x-rays. Its peculiarities, which are the consequences of the geometric-electric structure of the H_2O molecule, assert themselves also in its interactions with any type of surface.

Hygroscopic water. Dry soil minerals adsorb water from the atmosphere; the actual amounts depend on the physicochemical character of the surfaces and increase with rising relative humidity. The amount adsorbed is called hygroscopic water; it is usually measured at room temperature and expressed in percent of the dry soil weight. Very small amounts of water may be in solid solution within the surface of the minerals. Additional increments build up water films whose consistency may range from solid near the particle, through plastic, to that of normal water at a certain dis-

tance from the particle surface. A plastic water film 10^{-6} cm thick may not even equalize the roughness of a sand or gravel particle, but if it is around a plate-shaped clay mineral 10^{-5} cm thick, the water film represents more than 20% of the total clay-water volume. Therefore, the smaller the soil minerals, the more important are their water affinity and its consequences, such as swelling and shrinkage.

Gravitational and capillary water. In addition to this physicochemically restrained water, there may be free water in the soil pores, which is called gravitational water if it moves freely under the force of gravity, or capillary water if it is controlled by the forces of capillarity. The height at which capillary water can exist above the groundwater level depends on the effective pore radius. The and of breaking existing bonds does not exceed that of forming new bonds between atoms or molecules. Even in plastic bodies, very rapid deformation produces brittle fracture. In solid crystals, these requirements are best fulfilled if their building units are arranged with the highest degree of symmetry, as is the case with many metals. In macroparticle systems, such as soils, the requirements for plasticity are essentially the same. Large masses of relatively uniform sands and gravels may deform plastically at a slow rate and without change in voids ratio.

Small soil masses, and especially laboratory samples, show plasticity only if they are moist and cohesive. If so, the reforming of bonds broken during deformation of the mass and the reestablishment of the original symmetry of the bonding elements pertain essentially to the molecules in the water films around the particles. These water films play the same role in soils (increasing the distances between gliding planes and decreasing cohesion) as elevated temperature does in crystalline solids. Because plasticity increases with increasing total area of gliding planes per unit volume, the greatest plasticity is possessed by moist systems of pure clays of smallest particle size and greatest force of interparticle attraction. Admixture of coarser-grained material to clay interferes with the normal development of gliding planes and "shortens" the soil to an extent that depends on the amount, type, and size distribution of the coarser components.

Consistency limits. The term consistency is preferentially employed for mechanical resistance properties in the twilight zone between true elastic solid and simple liquid behavior. Since several physical phenomena are usually involved in consistency, this property is normally defined by the use of specific apparatus and standardized procedure. Typical are the slump test for fresh concrete, the penetration test for asphalt, and the consistency limits tests for soils. The last indicate what water content (percent of dry weight of soil) will bring soils to states of analogous consistency. The liquid limit is the water content at the transition between the plastic and the liquid state, the plastic limit is the water content at the solid-plastic transition, and the shrinkage limit is the moisture content that will just fill the soil pores in the solid state if that state is reached by the drying out of a soil paste. The difference between the liquid and plastic limits is the plasticity index. It indicates the moisture range in which a soil shows plastic behavior. These consistency indices are valuable tools if properly understood, but they do not as such make it possible to predict under what circumstances large soil systems will show rupture, plastic flow, or creep. This depends on complex factors of granulometry; amount, type, and water content of soil fines; and stress conditions. Analogous factors govern the behavior of all building materials.

SEEPAGE AND FROST

Water movement and frost action in soil must be considered in projects such as dam building and construction on seasonally and permanently frozen soils.

Seepage. Water movement in soils is normally caused by hydraulic pressure or tension gradients. This flow is ordinarily viscous or laminar rather than turbulent. Darcy's law is fundamental: $V = Akit$, where V = volume of water flowing in time t through soil with a cross section A and where pressure gradient $i = \Delta p/\Delta l$ (pressure drop per unit length of flow) and k = coefficient of permeability. The permeability coefficient varies from 100 cm/sec for clean gravel to 10^{-9} cm/sec for heavy clay.

Since water possesses an electric structure that interacts with the electrically charged soil minerals, it also moves in soil capillaries upon application of electric potentials. In this case, k of the Darcy equation is replaced by k_e, the electroosmotic transmission coefficient, and i by the electric potential gradient. Since the electric soil-mineral surface interaction structure is temperature-susceptible, water also moves in dense clay soils upon application of a thermal gradient with coefficient of thermoosmosis k_{th}. In unsaturated soils of high porosity, thermal gradients also cause water movement in the vapor phase.

Measurement of permeability. The coefficient of permeability may be determined either in the labowater affinity and the capillarity of clay and silt soils cause the entrance of water in either the liquid or vapor phase and also affect the ease with which water moves through a soil, or the difficulty of its removal.

SHEAR AND PLASTICITY

Soil under stress will deform by rupture, plastic flow, or creep. The mechanism will depend on factors such as stress conditions, water content, and soil cohesiveness.

Shear in soils. Soils are composed of many separate particles of great range in size and shape. The particles may or may not be held together by water films or by a clay-water cement. Analogous systems are encountered in the molecular world. Systems composed of a single kind of molecule or atom have, at constant pressure, definite temperatures of transition from the solid to the liquid state. This transition generally involves an expansion, that is, an increase in the interparticle spacing, which represents the only real difference between the solid and the liquid phases at the melting point.

Introduction of molecules of different size and character into a pure substance lowers the melting point; also, the mixture will soften over a range of temperature instead of melting sharply at a single temperature. In multicomponent materials, such as asphalts and pitches, a wide softening and liquefaction range replaces a definite melting point. The same phenomenon occurs in macroparticle

systems, such as soils. Densely packed gravels and sands are macromeritic (large-particle) solids. If submitted to shear stresses, they must expand in the shear zone to voids ratios characteristic of the molten state, in order that shear may take place without breaking of the individual particles. At voids ratios above the critical (melting range) ratio, gravel and sand soils can be "liquefied" by vibration that reduces interparticle friction or, if sheared, they may collapse to the critical voids ratio.

In granular noncohesive soils, the shear resistance obeys the equation $S = N \tan \varphi$, in which S = shear resistance in psi, N = effective normal pressure on the shear plane in psi, and $\tan \varphi$ = coefficient of friction, or tangent of angle of friction related to angle of repose of a pile of the granular material. If the soil possesses a granular skeleton bonded together by moist or dry clay, the shear resistance can be approximated by $S = C + N \tan \varphi$, in which C denotes the cohesion of the system in psi. The numerical value of C depends on the amount, type, and water content of the clay binder, its physicochemical interaction with the coarse particles, and the history of the system. In soils without granular skeleton, S approaches C.

Plasticity. The property of plasticity is possessed by many crystalline solids within a certain temperature range below and adjoining the melting point.

Conditions for plasticity. Plasticity denotes the ability of a body to deform permanently without rupture under applied stresses. This presupposes that during deformation (1) no marked expansion takes place in the shearing zones which would alter the cohesive forces; (2) the particulate components (molecules, atoms, micro- and macroparticles) are in similar geometric arrangement after as before deformation; and (3) the rate of deformation ratory on disturbed or undisturbed samples by means of the constant- or falling-head permeameters, or in the field by pumping or injection tests. In the constant-head permeameter, used for mate-

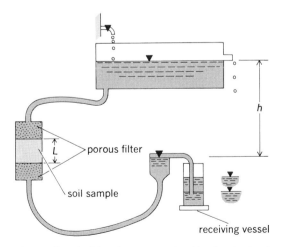

Fig. 3. Constant-head permeameter. Head water level is kept constant. Water percolates through soil sample of thickness L. Porous filters hold soil in place. The tail water level is kept constant by overflow. Volume of discharge is measured in receiving vessel. (*From G. P. Tschebotarioff, Soil Mechanics, Foundations and Earth Structures, McGraw-Hill, 1951*)

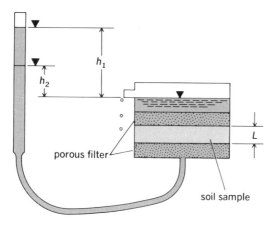

Fig. 4. Falling-head permeameter. Water level h_1 in thin glass tube decreases to h_2 as water percolates through soil sample of thickness L, restrained between porous filters. (*From G. P. Tschebotarioff, Soil Mechanics, Foundations and Earth Structures, McGraw-Hill, 1951*)

rials of high permeability, water maintained at a constant level flows through a soil sample. Permeability is computed from the rate of flow. In the falling-head permeameter, used for materials of low permeability, the permeating water is supplied by means of a standpipe with cross-sectional area a in which the hydraulic head falls from h_1 to h_2 during the time t while the water volume $a(h_1 - h_2)$ flows through a soil specimen of thickness L (Figs. 3 and 4). Then k is given by Eq. (2).

$$k = 2.3 \frac{La}{At} \log \frac{h_1}{h_2} \qquad (2)$$

In calculating k from pumping and injection tests in the field, spherical symmetry is assumed in the pressure distribution around the well point or the injection nozzle. Moving water transmits mo-

mentum to the contacting soil particles. If in non-cohesive soils the flow is upward and the hydrostatic uplift is sufficient to sustain the weight of the individual soil particles, a quick condition results. The minimum hydraulic gradient i_{cr} to cause quick condition can be estimated from Eq. (3), where $G =$ specific gravity and $e =$ the voids ratio.

$$i_{cr} = \frac{G-1}{1-e} \qquad (3)$$

Flow nets. The study of seepage through natural and man-made soil structures is greatly facilitated by the use of flow nets. These are two nests of curves, one representing the flow lines and the other the equipotential lines. In accordance with the laws of laminar flow, flow lines, which follow the path of the water, may not intersect each other; and the equipotential lines, each connecting points of equal hydraulic head, must cross the flow lines at right angles. While an infinite number of flow lines exists, only a few are drawn and in such a manner that the quantities of flow between adjacent lines are equal. Then the equipotential lines are drawn so that they intersect the flow lines at right angles and form areas that resemble squares as closely as possible (Fig. 5).

A conventional flow net has the following properties: (1) All paths, each of which lies between two adjacent flow lines, carry the same seepage quantities; (2) potential differences are the same between all adjacent equipotential lines; (3) flow lines intersect equipotential lines at right angles; (4) all figures formed by adjacent pairs of lines resemble squares as closely as possible if the soil is homogeneous; and (5) at any point of the net, the spacing of equipotential lines is inversely proportional to the hydraulic gradient and the spacing of flow lines inversely proportional to the seepage velocity. For the drawing of flow nets, the boundary conditions must be known. Two-dimensional presentation is of course applicable only if the profile remains the same in the third dimension. From a satisfactory flow net, the seepage can be calculated using Eq. (4), where $Q =$ quantity of

$$Q = \frac{n_f}{n_d} k h_i \qquad (4)$$

the seepage in cubic feet per second per linear foot of the length of the structure, $n_f =$ number of flow paths, $n_d =$ number of spaces between equipotential lines in the net, $k =$ coefficient of permeability in feet per second, and $h_i =$ head loss in feet between surfaces of the head and tail waters. Methods are available to correct for differences in permeability in different directions. Seepage in complicated structures is studied on models employing, instead of water, electric or other energy forms whose transmission obeys the same mathematical laws expressed in the Laplace equations for conjugate functions.

Frost action in soils. Frost action is of great engineering importance. The forceful expansion of confined water when it freezes can destroy porous building materials and loosen soil. Of greater importance is the accumulation of water in the form of ice lenses, with resulting frost heave in winter and morass formation during thawing.

A freezing front penetrating into moist soil starts

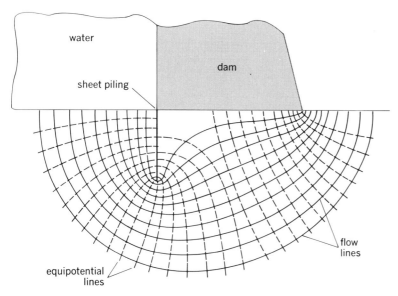

Fig. 5. Flow net indicated under the cutoff wall of a dam. (*From D. P. Krynine, Soil Mechanics, 2d ed., McGraw-Hill, 1947*)

only a limited number of crystallization centers. Water moves to these from the surrounding soil, especially from lower depths. As it freezes, it gives off about 80 cal/g of water. If the heat released by the freezing water just balances the heat lost by conduction to the earth surface, the freezing front becomes stationary and forms thick ice lenses, especially if the groundwater level or other water reservoir is close by. Without such close supply of free water, the frost may advance until a new moist layer is encountered. Here nuclei and ice formation begin anew.

Daily and other short-period temperature variations make the larger lenses grow at the expense of the smaller. In large areas of Siberia and also in Canada and Alaska, permanently frozen soil (permafrost) is encountered at depths that depend on the climate and the thermal conductivity of the surface soil. This frozen ground prevents the drainage of water from spring thawing. Structures founded on permanently frozen soil must be separated from it by insulating materials. Some of the permafrost is not in equilibrium with the present climate and if once melted would not be reformed. *See* PERMAFROST.

EXPLORATION, SAMPLING, AND TESTING

The extent to which field testing is necessary depends on the type, magnitude, and importance of the job. Before a sampling and testing program is started, all readily available information should be utilized. Such information may be in the form of construction experience records in the same general location; air photos on which recognizable erosion and vegetation features indicate the types of surface and subsurface soil to be encountered, as on a proposed highway route; geological maps that show the profiles of solid and loose rock and serve as reminders of troubles usually associated with certain rock types, as sinkholes are with limestone; and pedologic maps, which are especially useful in highway soil work. Valuable information is often obtained from old maps which show soil and drainage patterns that have been covered up in urban or industrial developments.

Site investigations. These are made either to find out whether the soil in its natural-site condition will support the planned structure or to establish the qualities of soils as construction materials for dams, embankments, subgrades, and other uses. For the first case, the natural-site condition and the samples taken for laboratory testing must be disturbed as little as possible. For the second case, disturbed samples are useful, but they should not be permitted to dry out before being tested.

For exploration of relatively large sites or extended strips, such as airports and highways, electrical and mechanical energy-transmission phenomena may be employed. The electric method furnishes information on the electric resistivity of the soil at various depths. Previous determination of relationships between soil type, water content, salinity, and electric resistivity permits the plotting of a soil profile and the detection of strata of specific materials.

Mechanical energy is employed in seismic and vibration tests. These tests utilize the reflection and refraction of earth waves at interfaces of different strata and the variation of their speeds of propagation with the character of the conducting medium. The seismic velocity of compression waves varies for different soils within a range of 500–8000 ft/sec and for solid rock from 6000 to more than 25,000 ft/sec. The velocity in water is about 4700 ft/sec. In refraction shooting, one usually explodes a blasting cap or a small charge of dynamite and records the first signal picked up by the seismographs located at three different distances. The methods employing excited vibration utilize either the difference in rate of propagation in different soils or the characteristic frequency of the soil-vibrator system.

Sampling. Soil sampling may be divided into shallow and deep and disturbed and undisturbed. Shallow disturbed samples of cohesive soils are usually obtained with a soil auger. Slightly disturbed samples of both cohesive and noncohesive soils may be obtained by carefully pushing thin-walled metal cylinders into the ground. The cylinder is subsequently capped on both ends to prevent moisture loss.

Deep samples can be taken by digging trenches and pits so that the actual profile is exposed. Also, machines are available for making open holes into the ground 10–36 in. wide and well over 10 ft deep. The most common method for taking deep samples is by driving a casing into the ground. The casing is at least 2½ in. wide. A record is kept of the hammer weight, height of drop, and the number of blows required to drive through each linear foot. While the boring proceeds, the casing is cleaned out with water conducted to the bottom of the casing by a pipe of about 1-in. diameter. Inspection of washings can give an idea of the character of the soil layer reached. If an undisturbed sample is to be taken at a certain depth, then the wash water is shut off above that depth and the wash pipe replaced by or changed into a pushing rod, by which a thin-shelled sampler (Shelby tube) is pushed into the stratum to be sampled. The tubes with the samples are then extracted, sealed on both ends, and sent to the laboratory for testing. A number of modifications of procedure and of type of sampler exists, but the essence of the method remains the same.

For important construction, the casing is driven to bedrock and through it for 5–10 ft in order to make sure that a boulder is not mistaken for a rock stratum. Because of the possibility of sinkholes, core drilling in limestone and dolomite should be as close to the actual construction site as possible.

Field tests. Static or dynamic tests at the site may be made for mechanical properties or other physical characteristics, such as density, moisture content, permeability, and thermal resistivity. The in-place density and moisture content may be measured by digging a clean hole, measuring its volume, and weighing and determining the water content of the earth removed. Densities and moisture contents of surface soils and of deeper layers reached by boreholes may also be determined by nondestructive methods, employing γ-rays for the former and neutrons for the latter. Thermal resistivity is usually determined with the thermal needle.

Common mechanical-resistance tests include sounding, that is, pushing a steel rod or $\frac{3}{4}$–1 in.-

wide pipe into the soil and noticing the resistance to driving as a function of depth; penetration tests, where the resistance to penetration by a cone of standardized form is measured at various levels reached by a borehole; and vane shear tests, where a vane is pushed into a soil layer with subsequent application of horizontal shear forces. These tests possess only indicative value.

Load or bearing-power tests are performed in open pits at foundation level. Round or square plates of practically undeformable material are employed for transmission of the applied load. The pressure bulb, or zone, created under the loaded plate bears a definite geometrical relationship to the shape and dimensions of the plate. Thus, the loading of small plates will not detect layers of low bearing power and large compressibility that may be reached by the pressure bulb of the actual foundations. Within limitations, load tests are valuable tools, especially for shallow foundations. To determine the effect of surface loading by a foundation on deeper soil layers, the stress acting on them is calculated from theory, and undisturbed samples taken from the location and depth concerned are submitted in the laboratory to the calculated stresses.

Stresses in soil masses may be due to their own weight or to outside forces. In either case, the geometry of the system is an important factor in the resulting stress distribution. The other decisive factor is the physical state of the soil mass. This physical state may range from an elastic solid to a macromeritic liquid like quicksand. Over the entire range, plasticity may be observable to various degrees. Every type of internal structure and response to stresses of different intensity, known from other construction materials, may be encountered in undisturbed and disturbed soils. This emphasizes the importance of obtaining representative soil samples and testing them carefully in the laboratory. Even then the results must be used with engineering judgment, since large masses of a material behave differently from small ones.

THEORETICAL STRESSES

Calculation of stresses would become too cumbersome if all pertinent soil factors were carried along. Theories are therefore based on idealized earth masses, which for calculation of stress distribution are homogeneous, isotropic, elastic bodies and for nonelastic deformation or stability problems are idealized masses endowed with friction, cohesion, or both. The methods of stress analysis are the same as in statics and strength of materials.

Pressure bulb. Stress distribution in a perfectly elastic continuum under a vertical point load at the horizontal boundary was first derived by J. Boussinesq. If the point of application of a force P serves as the origin of a coordinate system with vertical coordinates z and horizontal coordinates x and y, and $x^2 + y^2 = r^2$, then vertical pressure is given by Eq. (5). By connecting points of equal vertical

$$\sigma_z = \frac{3P}{2\pi} z^3 (x^2 + y^2 + z^2)^{-5/2} \qquad (5)$$

stress, a pressure bulb is obtained. Stress distribution in the elastic medium under a loaded area, rather than a point, is obtained by resolving the distributed force into a large number of point forces and then employing the principle of superposition where stresses from different points overlap. Such solutions are available only for simple forms, such as line, strip rectangle, square, and circle. However, the form of most foundations can be approximated by summation, subtraction, or both, of shapes for which solutions are available, and the stresses calculated, using the principle of superposition. A pressure bulb under a loaded area of simple shape is shown in Fig. 6.

Accuracy of analysis. Despite the drastic simplifications in theoretical treatment, loading tests have shown that, except in the immediate vicinity of the loaded areas, the calculated data agree quite well with the experimental results.

Direct loading tests have shown differences in load acceptance by different soils. Contrary to theoretical assumptions, movement of soil material occurs close to the plate where stresses are intensive. Coarse-grained soils in which shear resistance is directly proportional to normal stress resist movement most in the zone under the center of the plate and have lower resistance and more pronounced outward movement as the plate edges are approached. As a result, the center cone accepts most of the applied load, and the stress intensity under the center of the plate has been found in the case of sand to be two or three times the average stress under the plate. With plastic clay, where the shear resistance is pressure-independent, material moves from the central position outward until it encounters the resistance of the earth outside the plate edge. This equalizes the pressure under the plate and produces a maximum under the rim (Fig. 7).

CONSOLIDATION AND SETTLEMENT

The consequences of soil loading are affected by the presence of pore water. Normal pore water may be under simple hydraulic pressure γz if located at a distance z below the groundwater level and

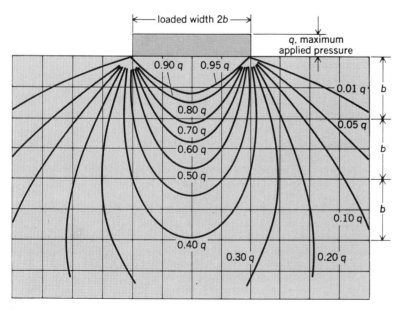

Fig. 6. Stress distribution for loaded areas of simple shape. Distribution pattern of vertical pressures through soil from vertical load on surface is called pressure bulb, from shape of equal pressure curves. (*From Road Research Laboratory of Department of Scientific and Industrial Research, London, Soil Mechanics for Road Engineers, 1954*)

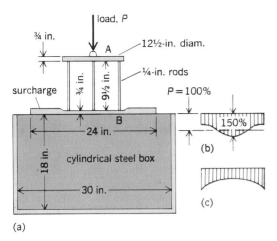

(a)

Fig. 7. Stress acceptance by sand and clay. O. Faber's experiments in 1933, in diagram a, showed that under a load applied through a circular plate B, stresses in sand would follow the pattern shown in b, reaching a maximum at the center of the plate and that stresses in clay would follow the pattern shown in c, reaching a maximum around the edges. Soil stresses were determined from contraction of the rods between the two disks (A and B) when load P was applied. (*From D. P. Krynine, Soil Mechanics, 2d ed., McGraw-Hill, 1947*)

having a density γ. It may be stressless, if located at the groundwater level, or under tension, if in capillaries above the latter. Sudden decrease in pore volume, occasioned by sudden loading, may result in a pressure increment which is called excess hydrostatic or neutral pressure. Presence of hydrostatic water decreases the weight of the solid soil particles by the weight of an equal volume of water. This must be considered in calculating the total effective stress due to the weight of overburden; it affects the friction part of shear resistance, which is proportional to the normal pressure on the shear plane. Excess hydrostatic pressure as well as hydraulic uplift must be subtracted from the total interparticle pressure to obtain the effective pressure which governs frictional resistance. Excess hydrostatic pressure is dissipated by expulsion of water, at a rate determined by the hydraulic gradient and the permeability, with consolidation of the system until the total stress is carried by the skeleton of solid particles.

Consolidation theory. Consolidation has been studied theoretically only for systems of simple geometry, such as a soft-clay stratum bounded by one or two pervious sand or gravel strata, under the assumptions that the soil is saturated, soil particles and water are incompressible, Darcy's law is valid and the coefficient of permeability remains constant during consolidation, the time lag of consolidation is due entirely to low permeability, the soil is laterally confined, the total and effective normal stresses are the same for any point on a horizontal plane and water flows out of the voids only in a vertical direction, and the change in effective pressure results in a corresponding change in voids ratio.

Loading of sand within its elastic range produces an immediate elastic deformation supplemented by some consolidation due to grain adjustment. Loading of a saturated clay produces an immediate, though small, elastic deformation and

a slow deformation that can be called elastoid since, though water is expelled, no shear takes place in the system.

Settlement analysis. For actual structures, settlement analysis involves the following: (1) detection of a soil stratum of high compressibility by borings; (2) determination of pressure-voids ratio relationships and time factor in the laboratory on undisturbed soil samples from borings; (3) calculation of existing pressures due to natural overburden and of pressure increase by added foundation loads; (4) initial condition of consolidation; (5) theoretical curve of time factor versus consolidation; (6) predicted time-settlement curve assuming instantaneous loading and curve corrected for construction period; and (7) comparison with settlement experience of other buildings in location and follow-up by actual settlement observations on structure.

If plastic under the applied stresses, soil will flow sideways, resulting in secondary consolidation that may, like creep in metals, continue at a constant rate or may decrease with time. The rigidity of the foundation influences the stress acceptance of the contracting soil and thereby the consolidation behavior. Proper design of a foundation minimizes total settlement and eliminates as much objectionable differential settlement as possible.

STABILITY

Stability of an earth structure is primarily resistance to shear failure. Such resistance involves geometrical and soil-physical factors. Wide and deep foundations are more stable than narrow and shallow ones.

Bearing power. Several building codes contain so-called safe bearing values, which may have ranges of $25-40$ tons/ft^2 for rock to 1 ton/ft^2 for soft clay. The minimum load causing failure is the ultimate bearing power which, divided by an arbitrary factor of safety, is called safe bearing power. More meaningful values are obtained from loading tests, shear tests, and unconfined and triaxial compression tests. Safe construction must be designed for the weakest possible condition of the soil on site. There exist a number of semiempirical laws with respect to foundation stability against shear failure.

Pile foundations are used for carrying building loads through soft compressible layers to layers strong enough to carry them. They may be end-bearing if resting on solid rock or floating if driven into sand or clay; if floating, the load is transmitted by skin and point resistance. Groups of floating piles develop, by superposition, pressure bulbs characteristic of the geometry of the group rather than of the individual pile.

Landslides. These may be due to slow soil flow, such as creep, to fast liquid flow, to sinking into caves or mines, and to sliding. Often they occur along water-lubricated boundaries of inclined rock strata. Soil mechanics is particularly concerned with ensuring cuts, embankments, earth dams, and other earth structures against failure by sliding.

Stability analysis employs the concept of a sliding wedge, one surface of which is the surface of maximum shear stress in the soil. This surface is found from the geometry of the system. Next, the shear resistance on this surface is determined as a

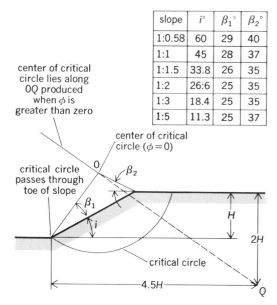

slope	$i°$	$\beta_1°$	$\beta_2°$
1:0.58	60	29	40
1:1	45	28	37
1:1.5	33.8	26	35
1:2	26:6	25	35
1:3	18.4	25	35
1:5	11.3	25	37

Fig. 8. Stability analysis of embankments. Fellenius' construction for location of critical slip circle. Tan ϕ is coefficient of internal friction. (*From Road Research Laboratory of the Department of Scientific and Industrial Research, London, Soil Mechanics for Road Engineers, 1954*)

function of soil character and normal stresses. Specific calculation methods for plane and curved sliding surfaces have been developed. Since soil cohesion remains constant, while friction increases with pressure, slopes of cohesive soils possess critical heights, while slopes of purely frictional materials may be stable up to any height. This demonstrates the importance of proper choice of materials and densification to greatest possible shear resistance of soil used for embankment and earth dam construction (Fig. 8).

Retaining walls. Stability analysis of retaining walls, that is, determination of the magnitude, distribution, and direction of soil pressure on such walls, is also based on the concept of a sliding wedge. One differentiates between neutral pressure (if no movement at all has taken place), active pressure (in the case of sufficient movement to mobilize the shear resistance of the sliding surface), and passive pressure (if the wall has pushed the soil back sufficiently to mobilize shear in a wedge which for geometric reasons involves a larger sliding surface than the active pressure). Special methodology for calculating earth pressures with varying simplifying assumptions is available. Further progress appears to depend more on a better understanding of soil-physical properties than on employing more complicated mathematics.

Conduits. Conduits are rigid or flexible pipes for the conduction of liquids, especially water, and are placed in ditches with backfilling or under embankments. Present knowledge and design formulas are based mainly on experimentation, and the closer the actual conditions resemble those of the experiment, the more reliable are the formulas.

LUNAR AND PLANETARY SOILS

Lunar and planetary soils, the same as their terrestrial counterparts, are products of environmental forces acting on the original rock substrate plus acquired meteoritic material and the substance of incident corpuscular radiation. The latter is especially important in the case of the Moon, which because of its low gravity ($\frac{1}{6}$ of the terrestrial) has an extremely thin atmosphere (10^{-14} torr) with very little shielding power against electromagnetic and corpuscular radiation. The protons (H^+) of the solar wind impinging on the lunar surface can react chemically with the oxides of the lunar rock and form hydroxides or even H_2O, which subsequently evaporates into space. This may account for the vesicular character of lunar rock and soil particles and for their dark color. Because of the practical absence of atmosphere on the Moon, dry soil particles have no adsorbed air films, which on Earth satisfy their attractive force fields. As a result, the "raw" lunar soil particles have a tendency to cohere and behave somewhat like moist soils, even in the absence of water. Modern developments in remote sensing permit the collection of compositional and environmental data by surveying spacecraft to a degree that is sufficient for prediction of the most significant features of planetary soils on the basis of present knowledge of the physics and chemistry of matter.

[H. F. WINTERKORN]

Bibliography: American Society for Testing and Materials, *Procedures for Testing Soils*, 1958; D. P. Krynine, *Soil Mechanics*, 2d ed., 1947; National Research Council, *Water and Its Movement in Soils*, Highway Res. Board Spec. Rep. no. 40, 1958; F. F. Scott, *Principles of Soil Mechanics*, 1963; K. Terzaghi and R. B. Peck, *Soil Mechanics in Engineering Practice*, 1948; G. P. Tschebotarioff, *Soil Mechanics, Foundations and Earth Structures*, 1951; H. F. Winterkorn, Principles and practice of soil stabilization, *Colloid Chem.*, 6: 459–492, 1946.

Soil microorganisms

Microorganisms in the soil include protozoa, fungi, slime molds, green algae, diatoms, blue-green algae, and bacteria. The bacteria are a heterogeneous group and include the procaryotic mycelial forms called actinomycetes as well as the simple unicellular forms called eubacteria.

Bacteria, actinomycetes, and fungi are the groups most active in decomposing organic residues and in rendering inorganic nutrients soluble. The final result of this activity is the liberation of such elements as carbon, nitrogen, phosphorus, potassium, and sulfur in forms available to plants.

Eubacteria. The eubacteria exceed all other soil microorganisms in numbers and in the variety of their activities. Numbers may surpass 100,000,000/g of soil by plate count or 1,000,000,000/g by microscopic count. The bacteria vary in size, shape, growth requirements, energy utilization, and function. Morphologically they are divided into straight or irregular rods, of both spore-forming and non-spore-forming types, thin flexible rods, cocci, vibrios, and spirilla. Short rods and cocci are most frequent, but many of the cocci are the coccoid stage of pleomorphic (varying in shape and size) rods or spores of actinomycetes.

Members of most taxonomic groups, with the exception of certain animal and human parasites, occur in soil. Some taxonomic groups are characteristic of soil alone. Although the identity of many bacteria engaged in specific processes, such as nitrification and nitrogen fixation, is known, a large proportion of the indigenous (autochthonous) organisms have not been classified. Of these, many have not been grown in any culture medium, a requisite for systematic study. Consequently, taxonomic knowledge of the autochthonous microflora is imperfect.

On the basis of their nutrition, soil bacteria are divided into autotrophs and heterotrophs. Autotrophs are able to use carbon dioxide as the sole source of carbon for their body tissues; heterotrophs must obtain carbon from organic foods.

Autotrophic bacteria. These comprise two groups (photosynthetic and chemosynthetic) according to their source of energy. The purple and the green sulfur bacteria are photosynthetic because of the presence of bacteriochlorophyll or chlorobium chlorophyll pigments. Like chlorophyll-containing algae and higher plants, they obtain energy from sunlight.

Other autotrophs are chemosynthetic, deriving energy from various oxidation reactions. Their requirements for food and energy are met by inorganic sources. These autotrophs carry out the process of nitrification, in which two stages are distinguished, the oxidation of ammonia to nitrite and that of nitrite to nitrate. Autotrophic sulfur bacteria derive energy from the oxidation of elemental sulfur, sulfides, sulfites, thiosulfates, and thiocyanates to sulfuric acid, which reacts with soil bases to form sulfates.

Heterotrophic bacteria. Heterotrophic bacteria, which constitute the great majority of soil bacteria, derive both food and energy from the decomposition of organic substances. They embrace a wide variety of morphological and taxonomic types including spore-formers or zymogenous forms, which are bacteria that develop in soil in response to the addition of certain substances like organic matter, or certain processes like aeration. Also included are the far more numerous non-spore-formers which make up the great majority of the autochthonous microflora. The most abundant forms are short rods and pleomorphic rods.

The majority of the heterotrophs require combined nitrogen to build cell substance. The nitrogen-fixing bacteria utilize elemental nitrogen of the air. These bacteria include symbiotic organisms, such as species of the genus *Rhizobium* that live in symbiosis with leguminous plants, and nonsymbiotic organisms, such as species of the aerobic genera *Azotobacter*, *Beijerinckia*, and *Azotomonas*, and the anaerobic genus *Clostridium*. The indigenous heterotrophic bacteria may be classified according to their nutritional requirements. Though all require a source of energy, such as a simple sugar, the additional needs of some are satisfied by inorganic salts. Other soil bacteria require amino acids or more complex food sources, and some more exacting bacteria require factors present in soil extract. The proportion of each of the nutritional groups is fairly constant in soil of definite type. However, increased fertility, such as that resulting from fertilizer treatment, is reflected in an increase in the proportion of bacteria with complex requirements at the expense of those with simple nutritional needs. Although there is no precise correlation between nutritional requirements and morphological type or taxonomic grouping, *Pseu-*

domonas species are more abundant among the nutritional group with simpler needs; the pleomorphic types, particularly *Arthrobacter* species, are relatively more numerous in the group with the most complex requirements.

As much as 25% of the indigenous soil bacteria capable of being isolated may require one or more vitamins for growth. The vitamins most essential are, in order of frequency, thiamine, biotin, and vitamin B_{12}. A smaller percentage require other B vitamins as well as the terregens factor, a substance found only in soil that promotes bacterial growth. *See* NITROGEN CYCLE; SOIL SULFUR, MICROBIAL CYCLE OF.

Actinomycetes. Next to the bacteria in numbers, the actinomycetes range from hundreds of thousands to several millions per gram of soil. They are more abundant in dry and warm soils than in wet and cold soils. With increasing depth of soil, their numbers are reduced proportionately less than those of bacteria. Actinomycetes are particularly abundant in grassland soil. The three genera of Actinomycetales occurring most commonly in soil are *Nocardia*, *Streptomyces*, and *Micromonospora*.

Thermophilic forms, represented by the genus *Thermoactinomyces*, are active in rotting manure and also may be present, though inactive, in normal soil. *Streptomyces* species are the dominant types. Although they are largely saprophytes, a few species, such as those associated with potato scab, are parasitic.

Less is known of the function of the actinomycetes in soil than of the bacteria. They are heterotrophic and are nutritionally an adaptable group, less demanding in growth requirements than many bacteria. They take part in the decomposition of a wide range of carbon and nitrogen compounds, including the more resistant celluloses and lignins, and are important in humus formation. Actinomycetes are responsible for the earthy or musty odor characteristic of soil rich in humus.

As much as 60% of the actinomycetes isolated from soil by plating methods may show antagonism toward bacteria or fungi in artificial culture. The importance of this antibiotic-producing capacity under normal soil conditions is not known. However, although antibiotics can rarely be detected in soil and then only under abnormal conditions, they may be important in microenvironments where intense microbial activity takes place.

Fungi. Fungi are present in numbers ranging from several thousand to several hundred thousand per gram of soil. They occur extensively in the mycelial state, as well as in the form of spores. Since plate colonies may develop from fragments of mycelium or from spores, plate counts give only an approximation of the abundance of fungi in soil. As with bacteria, some fungi do not grow on plates; consequently plating methods give minimum counts. Most fungi require humid, aerobic conditions for growth and spore formation. They are most common near the surface of soil and are more abundant in lighter, well-aerated soils than in heavier soils. Because the optimum pH range for fungi is 4.5–5.5, they are more prevalent in acid soils which are less favorable to bacteria and actinomycetes.

Ecologically, two broad groups of soil fungi may be recognized, the soil-inhabiting and the root-inhabiting types. The soil-inhabiting fungi are able to survive indefinitely as saprophytes and have a general distribution in soil. They include not only obligate saprophytes but also some unspecialized parasites which are able to infect plant roots, but whose parasitism is only incidental to their saprophytic existence. Root-inhabiting fungi are specialized parasites that invade living root tissues. Their distribution in soil is localized and depends upon the presence of the host plant. Their activity diminishes following death of the plant and they persist in soil only as resting spores or sclerotia. Mycorrhizal fungi are included among the root-inhabiting fungi.

Soil fungi are heterotrophic and have a wide variety of food requirements. All obtain their carbon entirely from simple carbohydrates, alcohols, or organic acids. But although some fungi can utilize inorganic nitrogen, others require more complex forms or vitamins, chiefly thiamine and biotin.

Soil fungi do not comprise as many physiological groups as do the bacteria, but as a group they are more versatile in their ability to decompose a great variety of organic compounds. The saprophytes, the true soil inhabitants, may be divided into groups depending upon the nature of the substrate favoring their development. Two such groups are the sugar fungi and the cellulose-decomposing fungi. Other groups attack some of the most resistant substances, such as lignins, vegetable gums, and waxes. When plants die, or when fresh plant material is added to soil, the growth of fungi is greatly stimulated. Those able to attack the more soluble constituents, such as sugars and other simpler carbon compounds, develop rapidly. Chief among such forms are the Phycomycetes. As the special substrate is exhausted, other types flare up and attack progressively more resistant components of organic residues. Cellulose and hemicelluloses are decomposed by a variety of fungi including species of *Penicillium, Aspergillus, Sporotrichum,* and *Fusarium*. Fungi are the predominant lignin-decomposing organisms. Various simple fungi can attack the lignin of straw and leafy plant material, although the higher Basidiomycetes are most active in decomposing lignin-rich residues.

Although polysaccharide-forming bacteria play a part, fungi are chiefly responsible for improving the physical structure of soil by exerting a binding effect on loose particles, thus forming water-stable aggregates. This binding effect is caused by the growth of mycelia which form fine networks that entangle the smaller particles. The soil-binding effect is favored by addition of fresh organic material whose decomposition products provide cementing substances.

Yeasts. A group of simple fungi, yeasts occur in soil only to a limited extent in the surface layers. In field soils their numbers are small, some samples being devoid of yeasts. They are found most frequently in the soils of orchards, vineyards, and apiaries where special conditions, particularly the presence of sugars, favor growth of yeasts which invade the soil. Soil is not a favorable medium for the growth of yeasts, and they do not play a significant part in soil processes.

Algae. These are widely distributed in soils, developing most abundantly in moist, fertile soils well supplied with nitrates and available phosphates. They contain chlorophyll and in the soil surface layers, where they are chiefly confined, function as green plants converting carbon dioxide and inorganic nitrogen into cell substance by means of energy derived from sunlight. Smaller numbers occur at lower depths where, in the absence of sunlight, they exist heterotrophically. The soil algae comprise the green algae (Chlorophyceae), the blue-green algae (Myxophyceae), and the diatoms (Bacillariaceae). In acid soils green algae predominate, whereas in neutral or alkaline soils the other groups are more prominent. Numbers of algae in soil vary widely, ranging from a few hundred to several hundred thousand per gram of soil.

As autotrophs, algae are of importance in adding to the organic matter of soils. They play a fundamental ecological role on barren and eroded lands by colonizing such areas and synthesizing protoplasm from inorganic substances. Several blue-green algae are able to fix atmospheric nitrogen and are of agricultural significance, particularly in rice culture. Under the water-logged conditions needed for this crop, blue-green algae develop abundantly and may increase the nitrogen supply by as much as 20 lb/acre.

Protozoa. Occurring in all arable soils, protozoa are largely confined to the surface layers although in drier, sandy soils they may penetrate more deeply. Numbers usually range from a few hundred in dry soil to several hundred thousand per gram in moist soils rich in organic matter. Most soil protozoa are flagellates and amebas; ciliates are less frequent although they are often found in wet soils and swamps. Protozoa are active in soil only when living in a water film. The majority are able to form cysts, and in this inactive state they can withstand desiccation.

Although a few flagellates, such as *Euglena,* have chlorophyll and are autotrophic and others can live saprophytically by absorbing nutrients from solution, the majority of soil protozoa feed by ingesting solid particles, mainly bacteria. Not all bacteria are suitable as food. Amebas have decided preferences for certain bacterial species and will not ingest others, particularly pigmented bacteria. The formation of cysts by protozoa is favored by some bacterial types, not by others. Excystment of some amebas requires the presence of bacteria; others are independent of bacteria. Though protozoa are a factor in maintaining the microbial equilibrium in soil through their selective action on bacteria, their effect is limited and is not considered detrimental to the activities of the micropopulation as a whole.

Myxomycetes. Myxomycetes, or slime fungi, form a minor group of soil microorganisms intermediate in character between the flagellated protozoa and the fungi. They possess a motile, flagellated stage in their life cycle, and later form large aggregates of cells or coalesce into jellylike masses of naked protoplasm. These eventually form spores which give rise to flagellated forms. Like the protozoa, the myxomycetes feed on bacteria.

Viruses and phages. Although these ultramicroscopic organisms exist in soil, little is known of the part they play in soil processes. Viruses that attack plants and animals can in some cases be transmitted from the soil. Phages, which are active against bacteria and actinomycetes, limit susceptible microorganisms and thus affect the microbial balance. Those that attack the various species of symbiotic nitrogen-fixing bacteria may prevent effective inoculation of legumes, particularly in soils in which the same crop has been repeatedly grown. The deleterious effect of phage on the nitrogen-fixing bacteria was formerly ascribed to direct lytic action (dissolution of the bacterial cell). However, this is now considered to result from the development of phage-resistant mutants which are less effective in fixing nitrogen than are the parent strains.

[ALLAN G. LOCHHEAD]

Soil minerals, microbial utilization of

Microorganisms utilize soil minerals for their own growth, and also help make the essential nutrients available to higher plants. A soil is fertile mainly because it can supply growing plants with nutrients (calcium, magnesium, potassium, and phosphate) in forms which can be readily taken up by plants. The conversion of the nutrients in soil minerals into forms available to plants is of course partly carried out by the plant roots themselves, but experiments have begun to show that microorganisms play a part in this conversion.

Soil formation. New soil is perpetually being formed all over the world by the breaking up of rocks into fine particles and the dissolving out of minerals from them. This process, the so-called weathering of rocks, was at one time supposed to be caused entirely by physical and chemical agents—heat, frost, water, and the oxygen of the air. It is now known that exposed rocks are subject also to microbial attack. When the island of Krakatau was visited 3 years after the great volcanic explosion which blew off its top, microscopic blue-green algae were found growing on the bare rocks, the first living things to appear there. These blue-green algae are the most self-supporting of all forms of life: they are photosynthetic, obtaining carbon from the carbon dioxide in the air, and also nitrogen-fixing, obtaining nitrogen from the atmosphere. In 1946, Soviet microbiologists found that the blue-green algae on freshly exposed rocks are soon accompanied by bacteria, among which nitrogen fixers and autotrophic nitrifiers are prominent. The nitrogen fixers add to the nitrogen supply, and the nitrifiers, because they fix carbon dioxide, add to the carbon supply in the film of growth on the rock. The two groups pave the way for other bacteria and fungi. The film of organic growth, which is continually dissolving mineral nutrients from the rock below it, increases until it can support the growth first of lichens, then of mosses, and finally of higher plants.

Buried rocks are also broken down to form new soil; if they are not buried too deeply, there is probably a succession of microbial growth on them. This has not been proved, because plant roots, at depths which they can reach, also break up rocks. Rock minerals are dissolved by carbon dioxide and by organic acids; both of these are given off by plants and by microorganisms. Since it is impossible to distinguish between them, in most cases it must be assumed that both plant roots and microorganisms are taking part in soil formation.

Utilization of minerals in formed soils. In soil which is already formed, microorganisms may release nutrients from the soil minerals and make them available to plants. In soil where a particular nutrient is scarce, they may render it unavailable by assimilating the nutrient as fast as they release it from the minerals: in this case they may be said to be competing successfully with the growing plants.

The mechanism of release of nutrients consists in dissolving out the nutrient from the soil mineral and converting it into a soluble salt or (in the case of heavy metals) into a complex with a chelating compound. Carbon dioxide is probably the most abundant dissolving agent, as all microorganisms and plant roots produce it in respiration. Organic acids such as lactic acid are also important dissolving agents which are secreted into the soil by plant roots and microorganisms, particularly fungi. Some fungi and bacteria from podzols (relatively infertile soils) are evidently adapted to dissolve minerals in a poor soil because in cultures of them more acid is formed in poor than in rich nutrient media. The gum, or slime, produced by some bacteria can take up phosphate, sulfate, and possibly potassium from culture media: if the same mechanism operates in soil, it should supply the bacteria with more of these nutrients.

Inorganic acids are produced by some autotrophic soil bacteria. For example, the nitrifiers *Nitrosomonas* and *Nitrobacter* produce nitric acid as the end result of their combined activity, and the sulfur-oxidizing *Thiobacillus* species produce sulfuric acid. These acids quickly combine with metals in the soil minerals to form nitrates and sulfates. The autotrophic bacteria are useful to plants not only because they supply nitrogen or sulfur in an assimilable form but also because they make soluble salts of most of the metals essential for plant growth, such as potassium and magnesium. Hydrogen sulfide, produced during the bacterial decomposition of proteins, may render tertiary phosphates soluble. Even if it causes the precipitation of iron and other metals, the sulfides are subject to bacterial oxidation, with the formation of soluble sulfates.

Some of the bacteria that reduce ferric hydroxide convert the iron into a chelated form held in an organic complex, but the nature of these organic complex-forming substances is quite unknown.

Phosphates. The most definite evidence for the beneficial effect of soil microorganisms on plant growth has been obtained in work with phosphates in the Netherlands, the Soviet Union, and Australia since 1950. Most of the phosphate in soil is combined with calcium or iron in insoluble compounds; many soil bacteria and fungi, when they are first isolated from soil, can dissolve these insoluble phosphates, though they lose this ability when kept in artificial culture. These phosphate dissolvers are commoner in the rhizosphere (the zone immediately surrounding plant roots) than elsewhere, and it has been found that plants grow better and take up more phosphate from insoluble calcium phosphate in soil with rhizosphere organisms present than in sterile soil. Cultures of phosphate-dissolving microorganisms are added to compost heaps in the Soviet Union and are said to improve the quality of the compost.

The mycorhiza fungi on the roots of pine trees are supposed to supply the trees with phosphate, but this is unlikely since trees without mycorhiza, such as oaks, also increase soluble phosphate in the soil around their roots. It is more probable that phosphate is brought up from the subsoil by the tree roots themselves.

Microorganisms may deprive plants of phosphate under certain conditions. All microorganisms need phosphate, and some soil species, the nitrogen-fixing *Azotobacter* sp. for instance, need quite large amounts. Fungi have a particularly high phosphate requirement; if much organic matter with a low phosphate content is added to soil, the fungi that develop on it may fix so much phosphate in organic compounds that a temporary phosphate starvation is induced in plants growing there.

Potassium, calcium, and magnesium. Very little is definitely known about the effect of microorganisms on the supply of these three elements to plants. It may reasonably be assumed that all three are turned into soluble salts through the action of the sulfur oxidizers and the nitrifying bacteria. Potassium is liberated in the breakdown of complex silicates by the so-called silicate-decomposing bacteria. It is quite probable that the breakdown of silicates is not a specific enzymatic process carried out by a special group of bacteria, but is caused rather by acids produced by many different soil bacteria and fungi.

Iron and manganese. Bacteria which reduce oxides of iron and manganese to soluble ferrous and manganous salts are very common in soil. The most efficient of them appear to be so because they make either organic acids or chelating substances. It is probable, but by no means proved, that these bacteria improve the supply of iron and manganese to plants. They also induce movement of both metals in the soil profile.

Microbial oxidation of iron and manganese compounds to insoluble oxides and hydroxides also takes place in soil. It has been claimed that the autotrophic iron bacteria, which are common in water and in bogs, also occur in drier soils. They may make the iron concretions and hardpans in tropical lateritic soils, but there is no experimental evidence for this claim.

It is therefore probable, though not yet proved, that soil microorganisms can increase the supply of essential nutrients to plants. A great deal more experimental work will be necessary to decide how much microbes contribute to plant nutrition by modifying soil minerals. *See* NITROGEN CYCLE; SOIL MICROORGANISMS.

[JANE MEIKLEJOHN]

Soil phosphorus, microbial cycle of

This microbial cycle is essentially a phosphate cycle. Plants assimilate phosphorus as phosphate, the $H_2PO_4^-$ ion, and build it into organic compounds such as phytin, nucleic acids, and phospholipids. The microbial breakdown of dead plant tissues liberates the phosphate again, so that there is an alternation in the soil between organic and inorganic phosphate (see illustration). Micro-

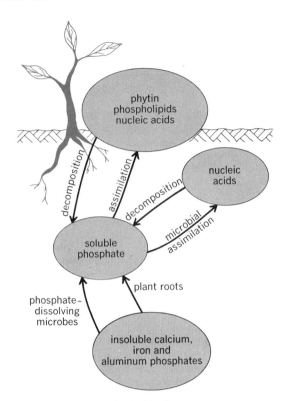

The microbial cycle of soil phosphorus.

organisms themselves assimilate phosphate, and consequently much of the organic phosphate in soil is contained in microbial cells. Plants may be deprived of phosphate by the addition of phosphate-deficient organic wastes to soil because the fungi which develop on the waste assimilate all the available phosphate.

There is also an alternation in soil between soluble and insoluble inorganic phosphates. Microorganisms can increase the phosphate supply to plants by dissolution of insoluble, tertiary phosphates through acid production; such microbes are particularly active in the rhizosphere. Others may precipitate soluble primary or secondary phosphates as tertiary salts as a result of the production of alkali. *See* SOIL MINERALS, MICROBIAL UTILIZATION OF.

[JANE MEIKLEJOHN]

Soil sulfur, microbial cycle of

Plants and microorganisms assimilate sulfur from soil sulfates and convert it into organic sulfur compounds. Sulfates are formed in soil by oxidation of sulfides (see illustration), which are derived from (1) the parent rock from which the soil is formed, (2) the breakdown of organic sulfur compounds, and (3) the reduction of sulfates by anaerobic bacteria of the genus *Desulfovibrio*. Corrosion of water pipes and other buried iron structures is caused by species of *Desulfovibrio*. The oxidation of sulfides to sulfur in soil is probably not a microbial process but a purely chemical one. The sulfur is oxidized microbially to sulfates by the autotrophic *Thiobacilli*. *See* SOIL MINERALS, MICROBIAL UTILIZATION OF.

[JANE MEIKLEJOHN]

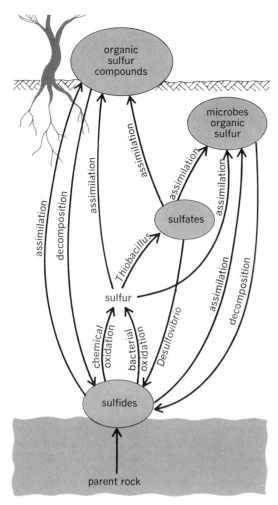

The microbial cycle of soil sulfur.

Solar energy

Solar energy is the broad-band electromagnetic energy radiated by the Sun into space. The majority of this energy is radiated in the visible portion of the spectrum, peaking at roughly 5000 A (green light; 10 A = 1 nm). The Earth intercepts roughly one half-billionth of this total radiated energy, or 173,000 × 10[12] W.

This intercepted solar energy is the primary source of power for the global ecosystem, driving the great hydrological and atmospheric cycles and providing the energy for photosynthesis which sustains life in its myriad forms. Thirty percent of incoming solar radiation is directly reflected into space, with the remaining 70% absorbed by the atmosphere, oceans, and land; is transformed in processes of evaporation and precipitation in the hydrological cycle and in driving oceanic and atmospheric currents; and is captured through photosynthesis. Through radiation of infrared energy into space, the Earth remains in thermal equilibrium with the radiation from the Sun (Fig. 1).

Conversion into useful form. Solar energy can be converted to many forms of energy directly useful to society through a broad menu of existing and emerging solar conversion technologies (Fig. 2). In principle, the entire present energy budget of the

United States (67 × 10[15] Btu/year; 1 Btu = 1055.06 J) could be provided by converting solar energy with 20% conversion efficiency to other forms of energy using an area of 50,000 mi[2] in the southwestern United States. This would be equivalent to adding roughly 10% to the existing land now dedicated to the country's only major solar conversion industry — agriculture. In reality, it seems unlikely that solar production of various forms of energy could reach such production levels in less than a century. However, a number of economic, social, environmental, and technical trends could result in major industries developing to produce significant quantities of energy through solar conversion before the end of this century.

The forms of useful energy which can be produced from solar conversion include low-temperature heat (for water heating and space conditioning), high-temperature heat (for specialized solar furnace work), electricity, mechanical energy, and gaseous and liquid fuels, in addition to the traditional agricultural products. A variety of systems which combine various processes of energy conversion, storage, and transport have been built or are technically feasible to facilitate such conversion.

Advantages and disadvantages. As an energy source, solar energy has both advantages and disadvantages. It is a high-grade and clean form of energy, available at predictable rates, widely distributed, and not subject to import quotas, resource depletion, or international politics. On the other hand, the ground-level power density is low (averaging 10–30 W/ft[2] over 24 hr; 1 ft[2] = 0.093 m[2]), and is subject to the diurnal cycle, seasonal variations, and patterns of weather and climate. Any systems designed to capture and convert solar energy are themselves subject to weather problems and, because large areas of collection are required (20 mi[2] of collectors would be required for a 1000-MWe power plant, compared with a fraction of a square mile for a conventional fossil-fuel or nuclear plant; 1 mi[2] = 2.59 km[2]), the systems may be very expensive to build. The challenge for science and technology for an alternative — fusion power — is the demonstration of technical feasibility, and the challenge for fission reactors is to make the systems and the storage of their by-products safe. The challenge to technology for solar energy

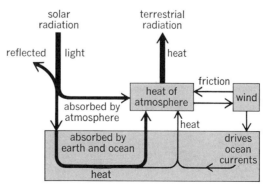

Fig. 1. Flow of solar energy through the biosphere. *(From M. King Hubbert, The Energy Resources of the Earth, Scientific American, September 1971)*

key:

----- electricity
───── thermal or chemical energy
++++++ mechanical energy

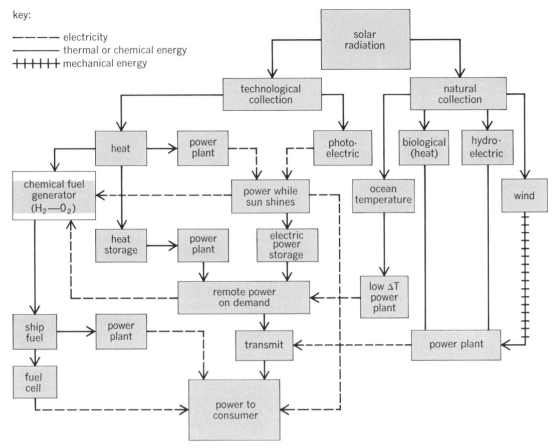

Fig. 2. Diagram showing alternative processes for the conversion of solar energy to electricity. (*From Solar Energy as a National Energy Resource, NASA/NSF Solar Energy Study, University of Maryland, 1973*)

conversion technologies is to make them *competitive*.

Energy collection systems. The solar energy systems that have been built or are technically feasible can be divided into natural and technological collection systems. Natural systems involve the initial conversion of solar energy through a variety of natural processes, followed by additional conversion by technological means. Traditional forms of technologies designed to take advantage of natural collection include windmills and wind turbines, waterwheels, and hydroelectric power plants.

Annual hydroelectric power generation is roughly 45,000 MWe in the United States and has reached its ecologically sensible limits, although potential generation is estimated at 160,000 MWe, roughly half the present generation capacity from all sources. Estimates for the global potential for hydroelectric generation equal the total global rate of consumption of fossil fuels today; however, the problems of silting and larger ecological problems may inhibit such enormous hydro power development.

Wind power is one of the oldest sources of mechanical energy. By 1915 some 100 MW of electrical energy were being produced in Denmark. Recent estimates indicate that wind-generated electricity could be competitive with alternative sources. Such systems would include offshore facilities and provide hydrogen to fuel-cell generat-

ing stations or for other purposes. Installations off New England alone could, according to recent estimates, provide as much as 300×10^9 kWhe annually (equal to 10% of present United States electrical production). Estimates of the maximum possible annual electricity production by the end of the century, using wind turbines of New England, in the Great Lakes region, through the Great Plains, offshore along the Texas Gulf Coast, and along the Aleutian Chain, indicate 1.5×10^{12} kWhe annually may be possible.

Technological systems directly intercept and convert solar energy. Such systems include solar stills, water heaters, space-heating and air-conditioning systems, solar-powered engines, solar/thermal electric power generation systems, photovoltaic or solar-cell converters which convert light directly into electricity, and photosynthesis and other biological processes which convert solar energy to liquid and gaseous forms of chemical energy.

All such systems require a "collector"—a device which intercepts sunlight. "Focusing" collectors are required where high temperatures or high power densities are required, and can focus only direct sunlight. Flat-plate, or "nonfocusing," collectors are used when concentration of sunlight is not required, and can use both direct and diffuse solar radiation. Another kind of nonfocusing collector is a pond in which various photosynthetic and related processes could convert incident sunlight into methane, hydrogen, and other gaseous fuels suitable for further conversion into electricity, heat, or mechanical energy — through combustion systems or fuel cells.

Biological conversion of sunlight to various energy forms has been proposed many times. Oil, coal, and gas are the fossil remnants of biological material representing solar energy that has been stored over billions of years. Plants and trees convert sunlight and other resources to a form which could be used for energy conversion through combustion. However, recent proposals have emphasized the potential use of various types of algaes to produce methane, hydrogen, and other gaseous fuels at solar conversion efficiencies as high as 10%.

Recent estimates suggest that the use of a variety of bioconversion technologies could result in synthetic fuel production which would provide 10% of the liquid fuel requirements of the United States by the year 2020, assuming increasing use of such fuels. Such solar-generated fuels may be competitive with alternatives well before the end of the century.

Thermal energy for buildings. Domestic water heating, space heating, and air conditioning require thermal energy in the range 140–190°F (60–87°C). Flat-plate solar collectors are well suited to supplying heat at these temperatures. These collectors consist basically of a black absorbing surface with a layer or two of glass on the front side and insulation on the back. Air or liquids flowing over the surface or through channels attached to or integrally part of the absorbing surface carry heat away. Simple collectors like these, connected to a hot-water storage tank, have provided hundreds of thousands of residences in Israel, Florida, and Australia with domestic hot water.

In some applications, the use of a gas or electric booster to supplement solar heating during overcast or cloudy periods permits an uninterrupted supply of hot water. These heaters are manufactured in Israel, Australia, Japan, the Soviet Union, and, on a smaller scale, in the United States. The aggregate world business is perhaps $2 million per year. Such systems, designed to interface with the complex requirements of the United States housing industry, could compete with electric water heating today if the comparisons were made on the basis of annual operating (including capitalization) expenses, and may compete in some applications, such as apartment buildings, with natural gas in the next decade. No systems which are of commercial interest in the United States construction industry have yet been developed.

Widespread commercial application of such systems could eventually make a significant contribution to low-grade thermal energy needs now being supplied by high-grade forms of energy (electricity and natural gas). In some regions, such as southern California, domestic water heating accounts for 6% of primary energy use and is the largest single end use of primary energy in new apartment buildings.

Solar space heating has been demonstrated in perhaps 20 experimental buildings around the world, using a variety of solar collection, storage, and control technologies. Estimates of the costs and performance of such systems if produced commercially in large quantities indicate that, on

an annual basis, solar heating could compete with all-electric heating in most parts of the United States today.

Solar space cooling is in an even more experimental stage, although experiments with solar-heat-actuated absorption air conditioners have been carried out, and techniques for driving a compression air conditioner from a solar-driven organic Rankine-cycle turbine are under study.

The development, introduction, and commercial diffusion of solar/thermal technologies for water heating and space conditioning will require a concerted effort on the part of government, utilities, builders, equipment suppliers and code officials, and labor. Experience within the housing industry has demonstrated that the appearance of technical feasibility and economic competitiveness is no guarantee that a new technology will be accepted or widely used, due to the unique economic and decision-making environment in which this industry operates. If, however, a successful commercialization program was carried out nationally, solar-energy applications to buildings could save several billion dollars annually through fuel savings toward the end of the century.

Power generation. Solar electric power generation is technically feasible today, and has been for almost a century; many solar-driven pumps and heat engines were built and operated in the United States and elsewhere, with a flourish of activity between 1870 and 1914. Some experts have estimated that the use of focusing collectors to provide high-temperature thermal energy to run an engine to drive, in turn, a generator system would provide electricity at a cost of 5–10 cents/kWhe. Conventionally produced electricity costs approximately 2 cents/kWhe today and may cost 5–6 cents/kWhe before the end of the century. Solar electric power systems using modern technology and mass-produced components may well supply electrical energy at similar costs.

Recent proposals have been made to build a prototype model of a large-scale solar power system using contemporary technologies. Such systems would work by concentrating sunlight onto thermal collectors to generate high-temperature steam in

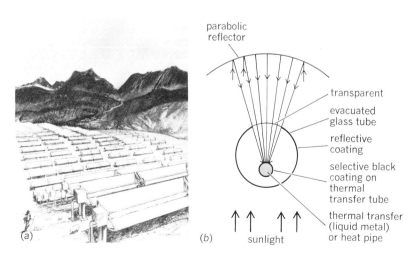

Fig. 3. Concept of a solar/thermal electric power plant. (a) Trough-shaped collectors for harvesting sunlight. (b) Diagram of the trough-shaped collector. (From Minneapolis-Honeywell, Inc.)

order to run a conventional steam turbine electric power generator. Determination of the costs of electric power produced in this manner await detailed engineering design and feasibility studies and prototype fabrication which will be made under NSF and NASA sponsorship this decade. Preliminary studies indicate that such systems may be competitive with other sources of commercial electricity toward the end of the next decade (Fig. 3).

Solar cells. Solar cells (small rectangular wafers of treated silicon) convert sunlight directly into electricity at efficiencies up to 15%. Virtually hand-made and tailored to each individual satellite and spacecraft application, these cells are no more representative of the kinds of direct conversion systems which might be mass-produced for the consumer market than the multimillion-dollar moon buggy is of a production automobile. Technologies for making complex multilayered materials by the square mile have been developed to an extraordinary level by the photography industry. Similar techniques have promise for the development of a thin-film version of the spacecraft solar cell, but might require a decade of solid-state research and development before commercial prototypes would be available.

Such a material on the roof of a home in the southwestern United States, operating at 15% efficiency, could provide up to 100 kWhe/day — enough to run all the lights and appliances in the modern home and provide sufficient energy to power a small electric commuter vehicle over 50 mi/day. Since these converters would absorb over 90% of the incident solar radiation, the heat remaining after 10–20% of this sunlight was converted to electricity could be used for solar water heating and space conditioning. Such systems would produce electricity with no water, air, thermal, or radioactive pollution.

The development of a commercial analog of the silicon solar cell, producing electricity at costs comparable with costs of alternatives over the next few decades, could — if combined with commercial electrical and chemical energy storage systems, and integrated into living systems designed to optimize the use of energy and nonenergy resources — contribute to a long-term global shift in patterns of development of the built environment toward ecologically stable systems.

Regardless of the method of conversion of solar energy into electricity, the diurnal and seasonal variations in sunlight are poorly matched to patterns of use of electricity. It has long been suggested that solar-produced electricity be used to produce hydrogen and oxygen through electrolysis. These, in turn, can be stored, shipped economically through pipelines, and combined in a variety of ways, including combustion in steam-turbine generators, fuel cells, industrial processes, and space-conditioning systems to produce work, electricity, heat, and distilled water. The emerging interest in a "hydrogen economy" may lead to technologies which can effectively match the variable output of a solar electric power generator to diverse patterns of energy demand.

Microwave systems. The most complex technical option for the use of solar energy has been proposed by Peter Glaser of the Arthur D. Little Company. His proposal is for the development of a synchronous satellite system which would convert incident sunlight in space to microwave energy, beam the microwaves to Earth, and then collect and convert the microwave energy back to electrical energy on the ground. This would represent the most formidable application of technology for solar energy yet considered. It would require the development of a space shuttle system capable of putting payloads into orbit at a cost below \$100/lb (1 lb = 454 g), and the development of a lightweight solar array at costs of \$30/ft² — far below those associated with the current spacecraft technology.

The present conception is to have giant solar panels using lightweight mirror concentrators and thin silicon solar converters to provide 10,000 MWe on Earth. Such a system would weigh over 25,000 tons (1 short ton = 0.907 metric ton) and would require perhaps a thousand space shuttle round trips. At a projected cost of \$20 billion for the complete commercial system (\$2000/kW), the projected cost would be less than 5 cents/kWhe. Whether or not such a system could be built for such costs can be decided only by a major research effort running parallel with the development of a sophisticated shuttle technology.

Outlook in developed nations. Solar technologies have been of little economic interest in the developed nations in the past due to the low cost of abundant alternative forms of energy, including natural gas, oil, coal, and electricity. With prices of energy in all forms certain to rise steeply over the next few decades, and growing concern over the need for environmental quality and energy resource conservation, a competitive market for solar technologies is likely to be competitive with alternatives during the last few decades of the century and beyond.

These solar technologies constitute an important set of potential options to provide significant quantities of energy in an environmentally compatible way. Increasing the diversity of major energy production technologies is an important element in stabilizing and increasing the adaptability of societies, since such diversity constitutes an insurance policy against future uncertainties in the eventual economic, environmental, social, and technical characteristics of any widely used energy technology. The commercial development of widespread use of a broad menu of solar technologies could significantly increase the diversity of energy options available over the next three decades. These technologies are inherently low in traditional kinds of environmental impact, including air pollution, water pollution, and land destruction associated with fossil-fuel combustion, oil spills, and strip mining. However, the widespread use of any technology will result in some impact on society, often unsuspected in form or magnitude. Solar technologies are no exception. Widespread use of solar water-heating and space-conditioning technologies will require new legal precedents in three-dimensional zoning and establishment of urban sun rights, to protect the owners of solar equipment against shadowing by construction of neighboring buildings. Optimization of the form of buildings to maximize solar energy conversion could compromise such structures in serious esthetic and functional ways, unless considerable design care is taken. Land used for solar/thermal electrical conversion must be picked with extreme sensitivity to such issues as desert ecology, esthetics, impact on local cultures, and water rights. Solar energy conversion represents one of the few sets of energy technologies whose environmental impact will be evaluated before and during the actual commercial development and deployment of such systems.

Commercial development, implementation, and diffusion of such options in the developed nations will require bold and innovative programs. Planned diversity is not yet viewed as an important insurance policy by most developed nations, and most technological planning is characterized by very short time horizons. Unfortunately, the commercial development and evaluation of many major new solar energy technologies will require one to two decades, with several additional decades required before sufficient diffusion of such technologies can occur to provide some assistance in resolving energy/environment problems. In addition, many of these technologies have uncertain economic characteristics, and traditional industry cannot afford to invest too heavily in options which are both uncertain in terms of potential competitiveness and long-term in payoff. Finally, there is little experience with major, long-term development programs which are sensitive from the beginning to social, environmental, and other issues. Long-term commitments, involving consortia of government agencies, industry, utilities, universities, and entrepreneurial individuals and organizations, appear necessary to stimulate the research, development, implementation, and diffusion of such options, even in countries where the economic and technical resources to accomplish such programs exist.

Outlook in Third World. The potential role of advanced solar energy technologies for the underdeveloped countries may be more significant than has generally been discussed. Substantial improvements in the nutrition, health, housing, and education of the two-thirds of the world's population living in underdeveloped regions can be achieved only by economic development in these regions, coupled with reductions in the high rates of population growth that have recently prevailed there. Worldwide development in the pattern established by the rich nations, however, implies local environmental burdens that the developing nations should wish to avoid and a total global burden that may prove unsustainable. The resolution of this dilemma may lie in technologies and lifestyles that bypass the environmental and social pitfalls which have plagued established industrial processes and patterns of human development. The bypass or overleap process, if it is possible at all, will require substantial contributions of money and technological expertise from today's industrial nations.

Energy technology goes to the core of the development/environment dilemma: energy is an indispensable ingredient of prosperity, a major contributor to environmental disruption, and an important determinant of patterns of living. The prosperity gap between rich and poor nations corresponds closely to an energy gap: the developing regions, with about two-thirds of the world's population, account for only 15% of the world's energy consumption. Prospects for narrowing the energy gap are clouded by the uneven geographical distribution of fossil fuels (these being especially deficient in Latin America and Africa), by the high

economic costs of technology to extract, convert, and usefully employ energy, and by the environmental and social liabilities of the various energy sources. Hydro power, with enormous potential in Latin America and Africa, may flood fertile land, drown revenue-bearing scenic attractions, increase evaporative losses of water, displace indigenous populations, impair soil fertility downstream, and facilitate the spread of parasitic diseases (such as schistosomiasis). Nuclear power is economically attractive only in plant sizes too large to suit most developing countries, and it bears, among other threats, the potential for proliferation of nuclear weapons capability.

Solar energy technology offers possible solutions to many of the foregoing problems. Historically, most advocates of solar energy for the developing regions have confined attention to low-technology, very-small-scale applications, such as cookers, solar stills, and food drying. Convergence of several technical and social trends now makes it apparent, however, that sophisticated and innovative uses of solar energy can play a much larger role in ecologically sensible development. These trends include *(a)* renewed interest in direct use of solar energy for space heating, air conditioning, and water heating in industrial nations, owing to rising prices of conventional fuels and electricity and to environmental concerns; *(b)* technical progress in selective optical coatings, solid-state engineering, heat and electrical storage, and heat transfer technology, opening the possibility of solar-thermal-electric and solar photovoltaic power plants at interesting efficiencies and costs and with great flexibility as to size; *(c)* recognition in industrial nations that energy-efficient design of homes, buildings, transportation systems, industrial processes—indeed, patterns of living—can greatly reduce energy requirements per unit of economic good; and *(d)* growing awareness that the achievement of a decent standard of living in developing regions will under any circumstances require ambitious and imaginative transfers of capital and technical knowledge from the rich countries to the poor ones.

In a world with limited resources and an expanding demand for thermal power, it seems evident that science should make every effort to develop the means to harness one of the cleanest, most abundant, and most basic forms of energy known—sunlight. To ignore this opportunity in the face of a growing energy-environment dilemma would constitute a major act of ecological irresponsibility.　　　　　　[JEROME M. WEINGART]

Bibliography: P. Donovan et al., *Solar Energy as a National Energy Resource*, NASA/NSF Solar Energy Panel, 1973; International Solar Energy Society, *Proceedings of the 1973 International Solar Energy Society Conference: The Sun in the Service of Mankind*, Paris, 1973; International Solar Energy Society, *Solar Energy: The Journal of Solar Energy Science and Technology*, quarterly; A. M. Zarem and D. D. Erway (eds.), *Introduction to the Utilization of Solar Energy*, 1963.

Solid-waste disposal

The systems approach to the disposal of solid waste. Until the relatively recent past, Americans have not seemed greatly concerned with the problems of environmental pollution, and least of all with the pollution resulting from inadequate solid-waste management. The most convenient disposal method—usually an open dump—would suffice. The land was large, the population pressure was small, and natural resources were seemingly without limit. However, the 20th century, and particularly the period since World War II, has witnessed dramatic changes in the United States—most importantly its transformation from a predominately rural population of approximately 76×10^6 in 1900 to a population now over 200×10^6. Of particular significance is the fact that over 70% of this population was counted as urban in the 1970 census. As striking as the increase in the population and its shift from rural to urban character has been the growth in the productivity of American agriculture and industry.

TYPES AND VOLUME OF WASTES

All these population changes have had profound effects upon the types and volume of solid wastes generated in the United States and encourage their accumulation in urban areas.

Municipal wastes. The rise of certain industries and marketing techniques, such as the packaging industry and the wide acceptance of packaged and canned foods (primarily for convenience), has contributed to a very great increase in per capita generation of solid wastes. For example, in the United States in 1920 on a per capita basis, an average of 1.25 kg of wastes were collected per day; today this figure has grown to 2.4 kg, and by 1980 it is estimated that this will rise to 3.6 kg. In a typical year Americans will discard over 27.2×10^6 metric tons of paper, 3.6×10^6 metric tons of plastics, 48×10^9 cans, and 26×10^9 bottles. The numerous disposable products on the market and an economy which frequently makes it cheaper to replace worn items than to repair them, contribute to the high per capita waste production.

In addition to increasing waste volumes, these technological changes have also had an important impact upon the characteristics of refuse, and in some cases the point at which the refuse is generated. For example, the trend toward consumption of processed foods (canned and frozen) is largely responsible for a relatively low percentage of putrescibles and high percentage of paper, plastics, and metals in the waste stream. Fruit and vegetable trimmings now accumulate in enormous quantities at canneries and food-processing plants, whereas they formerly accumulated in the kitchens of individual dwellings.

Although there is some geographical variation in characteristics of municipal solid wastes, social, political, and economic factors cause the major variation in physical and chemical composition. However, with standardization of data, reporting methods, and so forth, it has been possible to develop data meaningful from a national standpoint. In gross terms, the following list is descriptive of the physical composition of United States municipal solid waste, by weight: 50% paper; 10% metal; 10% glass; 20% food wastes; 3% yard waste (grass clippings, tree trimmings); 1% wood; 1% plastic; 1% cloth and rubber; and 4% inert material.

Agricultural wastes. The great rise in productivity of American agriculture is largely due to newer and more efficient methods of stock raising and cultivation of the land. These same methods, however, have tended to aggravate the problem of solid-waste disposal in rural areas. Herds of cattle and other animals, once left to graze over large open meadows, are now often confined to feedlots, where they fatten more rapidly for the market. However, on feedlots they produce enormous and concentrated quantities of manures that cannot be readily and safely assimilated into the soil, as under conventional open grazing practices (Fig. 1). Animal manures from concentrated livestock production are associated with such undesirable effects as fish-kills, eutrophication of lakes, off-flavors in surface waters, excessive nitrate contamination of aquifers, odors, dusts, and the wholesale production of flies and other noxious insects. Some agricultural wastes, such as those from concentrated poultry and egg production enterprises, may contain elements which, if allowed to leach into ground or surface waters, may constitute a localized threat to public health. Examples of these are the arsenic, manganese, and zinc substances normally found in poultry manure. The problem of agricultural-waste disposal is compounded by the fact that the superior convenience and nutrient value of chemical fertilizers have resulted in a lessened demand for animal manures to be used as soil fertilizers. *See* WASTES, AGRICULTURAL.

Mineral and fossil-fuel wastes. During the past 30 years, more than 18.1×10^9 metric tons of mineral solid wastes have been generated in the United States by the mineral and fossil fuel mining, milling, and processing industries. Because of transportation costs, solid-waste problems in the minerals industries are largely local in nature. Slag heaps, culm piles, and mill tailings generally accumulate in proximity to the extraction or processing operation. Prior to 1965 an estimated 18,000 km², or approximately 2×10^6 hectares, were covered

Fig. 1. Changes in agricultural practices have served to increase production but have also aggravated agricultural waste problems. For example, raising livestock on feedlots, rather than in open pasture, has resulted in large accumulations of manure which cannot be readily assimilated by the soil. (*U.S. Environmental Protection Agency*)

Fig. 2. Open dumps provide food and harborage for rats, flies, and other pests. Nearly half of all open dumps contribute to water pollution, and three-fourths to local air pollution. (*U.S. Environmental Protection Agency*)

with unsightly mineral and solid fossil fuel mining and processing wastes or otherwise devastated as a result of mineral and fuel production.

Although some 80 mineral industries generate quantities of solid waste, 8 major industries are responsible for 80% of the total. These are copper, which contributes the largest waste tonnage, followed by iron and steel, bituminous coal, phosphate rock, lead, zinc, alumina, and anthracite industries.

Total solid wastes. In summary, the total solid-waste load generated from municipal and industrial sources in the United States amounts to more than 326×10^6 metric tons annually; this figure includes 227×10^6 metric tons of household, commercial, and municipal wastes and 99×10^6 metric tons resulting from industrial activities. The annual total of agricultural wastes, including animal manures and crop wastes, is estimated to be over 1.8×10^9 metric tons. The present annual rate of mineral solid-waste generation is an estimated 10^9 metric tons, with an anticipated rise to 1.8×10^9 metric tons by 1980. Even this projection for mineral wastes may prove low if ocean and oil shale mining become large-scale commercial enterpris-

es. Altogether, over 3.2×10^9 metric tons of solid wastes are generated in the United States every year.

PUBLIC HEALTH, NATURAL RESOURCES, AND ECONOMY

Nature has demonstrated its capacity to disperse, degrade, absorb, or otherwise dispose of unwanted residues in the natural sinks of the atmosphere, inland waterways, oceans, and the soil. But the major concern are those residues that may poison, damage, or otherwise affect one or more species in the biosphere, with resultant changes in the ecological balance.

Public health. The relationship between public health and improper disposal of solid wastes has long been recognized. Rats, flies, and other disease vectors breed in open dumps and in residential areas or other places where food and harborage are available (Fig. 2). An extensive search of the medical literature has indicated an association between solid wastes and 22 human diseases. Perhaps the most obvious and direct relationship between solid wastes and human health can be observed for that segment of the population which experiences occupational exposure—namely, refuse collectors and processing/disposal plant operators. A study performed jointly by the American Public Works Association and the National Safety Council revealed the startling fact that the work injury (frequency) rate among solid-waste employees is nearly nine times greater than the average for all United States industries.

Implications for public health and other problems associated with water and air pollution have been linked to mismanagement of solid wastes. Leachate from open refuse dumps and poorly engineered landfills has contaminated surface and groundwaters. Contamination of water from mineral tailings may be especially hazardous if the leachate contains such toxic elements as copper, arsenic, and radium. Open burning of solid wastes or incineration in inadequate facilities frequently results in gross air pollution. Many residues resulting from mismanagement of solid wastes are not readily eliminated or degraded. Some are hazardous to human health; others adversely affect desirable plants and animals. Although attempts to set standards for air and water quality have been made and are continuing, in many cases harmful levels of many individual contaminants and of their combined or synergistic effects are not yet known. *See* AIR POLLUTION.

Natural resources. The Earth's mineral and other resources are not unlimited. Man extracts metals from ores and transforms other materials from a natural state to man-made products, but when these products have fulfilled their usefulness and are classified as solid wastes, the valuable and nonrenewable materials are often lost. Iron, for example, is a nonrenewable resource concentrated in ores over periods involving millions of years of geological processes. This element is extracted, processed, and widely dispersed over the Earth and serves many useful purposes, from cans to automobiles. When discarded, these objects rust

away and that amount of iron may never be available again for use by future generations. Shortages in resources for some of the less common elements have been created already by many indiscriminate waste disposal practices.

Economic aspects. Implicit in public attitudes toward waste materials is the problem of economics. National concern must transcend both the concept of what the public can "afford" to pay and the question of why the expenditure of about 4.5×10^9 each year for solid-waste collection and disposal has not staved off the present mounting problems with solid wastes. The truth is that the national standard of living depends to a significant degree upon processes leading to waste generation. It is reasonable to presume that a departure from traditional wastefulness of resources might reduce the volume of wastes to be managed. Thus there is a need for finding ways in which wastes themselves may be salvaged, reworked, and recycled back as a part of resources that are being processed.

A system for managing solid wastes must be economically as well as technologically feasible. In waste management, as in many other fields, a particular facility or item of equipment may be technically suitable but prohibitively expensive. Many communities are unable or unwilling to pay the price necessary to take advantage of the best presently available solid-wastes disposal systems or devices.

Location and quality of operation of solid-wastes disposal facilities may have an important impact upon the economics of an area, affecting land-use patterns and the value of surrounding property. Although incinerators, sanitary landfills, and compost plants generally do not enhance the value of surrounding property, there are isolated examples in southern California and other areas in which expensive residential neighborhoods are found

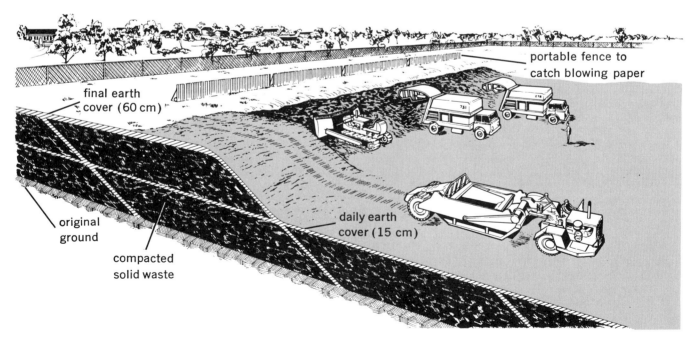

Fig. 3. Area method of sanitary landfill. The bulldozer spreads and compacts solid wastes. The scraper (foreground) is used to haul the cover material at the end of the day's operations. A portable fence catches any blowing debris. This is used with any landfill method. (*U.S. Environmental Protection Agency*)

daily earth cover (15 cm)

earth cover obtained
by excavation
in trench

original
ground

compacted
solid waste

Fig. 4. Trench method of sanitary landfill. The waste collection truck deposits its load into the trench, where the bulldozer spreads and compacts it. At the end of the day the dragline excavates soil from the future trench; this soil is used as the daily cover material. Trenches can also be excavated with a front-end loader, bulldozer, or scraper. (*U.S. Environmental Protection Agency*)

adjacent to sanitary landfill sites, planned for ultimate conversion to parks, golf links, and other recreational uses. Three-quarters of the 4.5×10^9 yearly cost of United States solid-waste disposal is attributed to the collection and transport of solid wastes. Yet this expenditure has purchased only hit-or-miss collection systems, with little satisfaction for the majority of American communities. The one-quarter of expenditures currently allocated for processing and disposal of wastes has, despite obvious public objections, resulted in the use of the open dump as the predominant disposal method.

MANAGEMENT TECHNIQUES

Solid-waste management differs in important respects from air- and water-pollution control. The difference derives from the fact that there are two pollutant transport systems: natural (air and water) and artificial (vehicular transport). In general, air and flowing water carry pollutants across political boundaries in response to natural laws that are not subject to legislative repeal. In contrast, solid wastes must be left where they are generated or transported by mechanical means.

The bulk of solid wastes are deposited on land, and disposal tends to be a local problem. Thus the principal solutions to solid-waste management lie in providing operational systems that employ physical procedures rather than in regulation. Such handling, along with reclamation and reuse as the solid-waste management goal, offers the ultimate solution.

Although open dumping—with all of its attendant problems—is the predominant means of solid-waste disposal in the United States, several acceptable alternatives to open dumping and open burning are presently available.

Sanitary landfill. The sanitary landfill is defined by the American Society of Civil Engineers as a method of disposing of refuse on land without creating nuisances or hazards to public health or safe-

ty by utilizing the principles of engineering to confine the refuse to the smallest practical area, to reduce it to the smallest practical volume, and to cover it with a layer of earth at the conclusion of each day's operation or at such more frequent intervals as may be necessary.

Such a landfill is a well-controlled and truly sanitary method of disposal of solid wastes upon land. It consists of four basic operations: (1) The solid wastes are deposited in a controlled manner in a prepared portion of the site. (2) The solid wastes are spread and compacted in thin layers. (3) The solid wastes are covered daily or more frequently, if necessary, with a layer of earth. (4) The cover material is compacted daily.

Two general methods of landfilling have evolved: the area method and the trench method (Fig. 3 and 4). Some schools of thought also include a third, the slope, or ramp, method (Fig. 5). In some operations, a slope or ramp is used in combination with the area or trench methods.

In an area sanitary landfill the solid wastes are placed on the land; a bulldozer or similar equipment spreads and compacts the wastes; then the wastes are covered with a layer of earth; and finally the earth cover is compacted. The area method is best suited for flat areas or gently sloping land, and is also used in quarries, ravines, valleys, or where other suitable land depressions exist. Normally the earth cover material is hauled in or obtained from adjacent areas.

In a trench sanitary landfill a trench is cut in the ground and the solid wastes are placed in it. The solid wastes are then spread in thin layers, compacted, and covered with earth excavated from the trench. The trench method is best suited for flat land where the water table is not near the ground surface. Normally the material excavated from the trench can be used for cover with a minimum of hauling. A disadvantage is that more than one piece of equipment may be necessary.

In the ramp or slope method (a variation of the

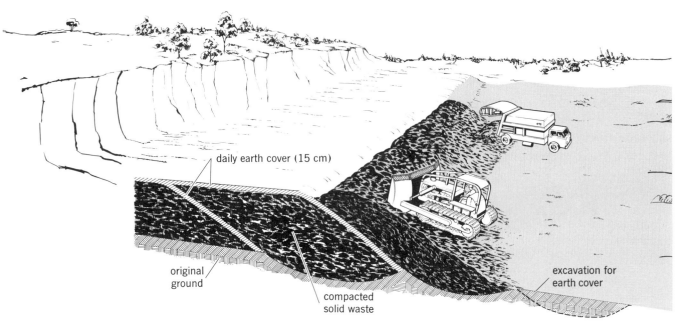

daily earth cover (15 cm)

original
ground

compacted
solid waste

excavation for
earth cover

Fig. 5. Ramp variation of sanitary landfill. Solid wastes are spread and compacted on a slope. The daily cell may be covered with earth scraped from the base of the ramp. This variation is often used with either the area or trench method. (*U.S. Environmental Protection Agency*)

area and trench landfills) the solid wastes are dumped on the side of an existing slope. After spreading the material in thin layers on the slope, the bulldozing equipment compacts it. The cover material, usually obtained just ahead of the working face, is spread on the ramp and compacted. As a method of landfilling, this variation is generally suited to all areas. The advantage of utilizing only one piece of equipment to perform all operations makes the ramp or slope method particularly applicable to smaller operations. The slope or ramp is commonly used with either area or trench sanitary landfill.

Completed landfills have been used for recreational purposes, for example, in parks, playgrounds, or golf courses. Parking and storage areas or botanical gardens are other final uses. Because of settling and gas problems, construction of buildings on completed landfills generally has been avoided; in several locations, however, one-story rambling-type buildings and airport runways for light aircraft have been constructed directly on sanitary landfills. In such cases it is important for the designer to avoid concentrated foundation loading, which can result in uneven settlement and cracking of the structure.

Incineration. Incineration is a controlled combustion process for burning solid, liquid, or gaseous combustible wastes to gases and to a residue containing little or no combustible material. In this regard, incineration is a disposal process because incinerated materials are converted to water and gases that are released to the atmosphere. The end products of municipal incineration, however, must be disposed of. These end products include the particulate matter carried by the gas stream, incinerator residue, siftings, and process water. Incinerator residue consists of noncombustible materials, such as metal and glass, as well as com-

bustible materials not completely consumed in the burning process.

The advantages of incineration are numerous, especially where land within economic haul distance is unavailable for disposal of solid waste by the sanitary landfill method. A well-designed and carefully operated incinerator may be centrally located and has been found acceptable in industrial areas so that haul time and distance can be shortened (Fig. 6). The solid waste is reduced in weight and volume, and the residue produced can be nuisance-free and satisfactorily used as fill material. In a properly designed incinerator, the operation can be adjusted to handle solid waste of varying quantity and character.

An incinerator requires a large capital investment, and operating costs are higher than for sanitary landfill. Skilled labor is required to operate, maintain, and repair the facility. Thus capital and operational costs must be compared with the costs of alternate disposal methods, and full consideration must be given to the effects of the methods on the community and its neighbors.

Oversized or bulky burnable wastes (logs, tree stumps, mattresses, large furniture, tires, large signs, demolition lumber, and so on) usually are not processed in a municipal incinerator since they are either too large to charge, burn too slowly, or contain frame steel of dimension and shape that could foul grate operation or the residue removal systems. A few incinerators include grinding or shredding equipment for reducing incinerable bulky items to sizes suitable for charging. In recent years, special incinerators have been designed and constructed to handle portions of bulky, combustible solid wastes without pretreatment. Unless these materials can be incinerated, their bulk and abundance will add greatly to the amount of land necessary for final disposal. Other

Fig. 6. Incineration of solid wastes in a modern plant can effectively reduce the volume of wastes to be disposed. However, this process is comparatively expensive and is therefore used primarily by large cities. (*U.S. Environmental Protection Agency*)

discarded large items, such as washing machines, refrigerators, water heater tanks, stoves, and large auto parts, cannot be handled by incineration and they too add considerable volume to a fill.

Stringent air-pollution restrictions will require the installation and maintenance of air-pollution control devices on municipal incinerators. The best means of preventing air pollution is to provide efficient combustion; that is, adequate grate surface and combustion air, proper temperature, and constant operation. Since even the well-designed, well-operated incinerators may occasionally produce particulates or odor because of uncontrollable variations in the refuse, most incinerators will include fly ash (airborne particles) removal systems. The larger particles are removed by means of screens or baffles, and slow the velocity of the gases so that large airborne particles fall out. To reduce the particulate load even further, water sprays or water-wetted impingement baffles may be installed. Cyclones (devices which remove particles through centifugal force) may also be used for removal of fine particulates. Electrostatic precipitators are used in some incinerator plants.

Composting. Composting is a method of handling and processing solid-waste material which produces a sanitary, nuisance-free humuslike substance, which may be used as a soil conditioner (Fig. 7). Technically composting is a biological degradation process, employing microorganisms to decompose the raw organic materials under controlled conditions of ventilation, temperature, and moisture.

Although many different methods of composting have been developed, there are certain common operations in all:

1. Sorting and separating. This operation is carried out manually or mechanically or both to remove bulky wastes, metals, glass, cloth, cardboard, and other materials that will not compost or that have salvage value.

2. Reduction of particule size. Reduction is achieved by grinding, shredding, rasping, or tumbling, sometimes accompanied by preliminary screening to develop a near homogeneous mass with uniform particle size, which will degrade more rapidly.

3. Composting (or biological decomposition) in the presence of air to prevent anaerobic conditions. Composting is accomplished in either enclosed chambers containing mechanical means of turning the ground mass and introducing air, or in open rows (windrows) which are turned at intervals to maintain aerobic conditions.

4. Screening and further grinding of the composted material to prepare it for use as a soil conditioner. The larger undesired particles are either rejected or returned to the process.

A basic problem area in composting is development of a market for the product. Compost is not a fertilizer but a soil conditioner which has been marketed primarily to some segment of agriculture. If sold in small bagged lots to the nursery industry or to the homeowner, municipal compost must compete with animal manures, peat moss, and other similar products. Furthermore the qual-

Fig. 7. Although composting is a minor method of solid-wastes disposal, it has the advantage of permitting use of organic wastes as a soil conditioner. Here a mechanical device is turning a compost heap to aerate it and thereby promote the process of biological degradation of the organic wastes. (*U.S. Environmental Protection Agency*)

ity of the compost has to be more stringently controlled if it is to compete with these organic garden products.

As a method for treatment of solid wastes, composting is not widely used in the United States, and only a very small percentage of the total waste load is subjected to this process.

RECYCLING AND REUSE OF SOLID WASTES

A traditional view of the solid-waste problem focuses on difficulties inherent in collection and disposal operations. However, solid wastes must also be regarded as a problem in the proper management of resources. This difference in point of view could well hold revolutionary implications for those involved in solid-waste management. Increasingly, solid wastes are being viewed as a "resource out of place," to be recovered and reused whenever possible.

If the meaning of this concept is widely applied, it will result not only in conserving natural resources but in reducing to a minimum the amount of waste material that must be ultimately disposed. In the long run we may have no choice but to accept and apply this principle. As waste tonnages accumulate and even greater demands are placed upon the Earth's resources, every indication is that we will run short of both disposal sites and certain mineral and forest products.

The primary barrier to reusing valuable elements in municipal wastes lies in the expense and difficulty of hand separation. Two possibilities exist for overcoming this barrier: separation and segregation of waste materials at the source, and development and improvement of mechanical separation techniques. The first solution would require cooperation on the part of the entire community to place household wastes into separate containers reserved for paper, glass, tin cans, aluminum cans,

putrescibles, or other categories. While this method is simple and can be made to work, the suggestion sometimes meets with opposition from disposal service patrons. This approach also increases collection costs, due to the necessity for keeping various classes of wastes separate.

The alternate approach—efficient mechanical separation—is presently the object of considerable research and development effort.

Whether or not an increased fraction of the solid-waste load will be salvaged and reused depends upon the existence of secondary materials markets. Improved technology in the area of separation will have no practical effect unless salvaged materials can be sold and utilized. It is likely that in many areas of the country a potential, but as yet unexploited, market already exists for certain materials.

In some cases it may be necessary to develop hitherto untried uses for solid-waste materials and to stimulate new markets for them. For example, it is possible by a combination of various chemical, physical, and biochemical processes to convert cellulosic wastes, such as paper or bagasse (sugarcane waste), into protein. This source of protein could be a valuable supplement for animal feeds or perhaps even foods for human consumption.

Experimental use of crushed glass as a road paving material is another example of current efforts being made to develop new markets for a common waste material.

While important technological and economic barriers serve to limit the amounts of solid waste that are now recycled and reused, it seems certain that these limits can be greatly extended. However, new technology and imaginative ideas in developing markets for waste materials are needed.

FUTURE

The Federal government, under the Solid Waste Disposal Act of 1965 (P.L. 89-272) as amended, has provided impetus for better solid-waste disposal planning and has sponsored an extensive research and development program to improve technology in all aspects of solid-waste management.

Fluidized-bed incinerator. One of the most promising research and development projects in the solid-waste field is the CPU-400 now being developed under contract to the U.S. Environmental Protection Agency (EPA) by the Combustion Power Co. of Palo Alto, Calif. Basically the CPU-400 is a fluidized-bed incinerator that burns solid waste at high pressure to produce hot gases to power a turbine, which in turn drives an electrical generator. Municipal solid wastes constitute a better fuel than generally imagined—having a heating value of 2268 Btu/kg, or approximately one-half that of a good grade coal. As designed, the CPU-400 should produce approximately 15,000 kW of electric power while burning 363 metric tons of municipal refuse daily. The generator unit should supply 10% of the electric power requirements of the community providing the refuse, thus offsetting part of the cost of waste disposal. Under the concept envisaged with the CPU-400, refuse haul distances could be greatly reduced by locating the units at strategic points in the urban area. For example, five such units would be required to process all of the refuse from the city of San Francisco, while 40 would be required for New York City. In addition to power generation, up to three-fourths of the heat in the gas-turbine cycle would be available for such auxiliary functions as steam production, drying sewage sludge, or saline water conversion. Developmental work on the CPU-400 also contemplates use of the vacuum produced by the gas turbine. This vacuum should be useful in conveying refuse to the incinerator from collection points in the city through pipes buried in the streets.

Classification of wastes. In an effort to automate the separation of mixed wastes into homogeneous batches for salvage and reuse, EPA has contracted with the Stanford Research Institute, Menlo Park, Calif., for development of an "air classification" process. Waste materials are separated by a high-velocity airstream passing upward through a vertical zigzag column; particles in a waste mixture are separated as a function of density, size, and aerodynamic properties (Fig. 8).

Paper fiber recovery. One of the most advanced systems for recovery of municipal wastes is being demonstrated by the city of Franklin, Ohio, with EPA grant support. The basic technology applied in the system was developed for use in producing paper from pulp. Mixed refuse is fed by conveyor belt to a large tank of water with rotating blades at the bottom. Large and heavy materials are removed from the bottom of the tank and passed under an electromagnet, which separates the ferrous metals. The water slurry which contains the smaller and lighter materials is passed through a battery of screens and centrifuges, which extract cellulose fiber—for use in making paper—and a separate mixture of glass, aluminum, and other nonferrous metals.

An additional step will involve extraction of glass from this gritty mixture, with separation into various colors by an optical sorting device. The remaining mixture has a relatively high percentage of aluminum, which has potential for reclamation by the aluminum industry.

Conclusion. It is unlikely that a research "breakthrough," affording some universally applicable solution to the solid-waste problem, will ever be achieved. The varying nature and quantities of wastes generated in different locations and the economic differences prevailing in different parts of the country will require varying approaches to solution. Ideally, solid-waste systems should allow for maximum salvage and reuse of waste materials, with hygienic and pollution-free collection and disposal for the remaining fraction of the waste load.   [RICHARD D. VAUGHAN]

Sonic boom

Strong pressure waves (shock waves) generated by aircraft in supersonic flight and heard along the ground as explosivelike sounds called booms or bangs. Contrary to popular belief, a sonic boom does not occur at only one location at the instant the aircraft breaks the sound barrier. Instead, aircraft flying faster than the speed of sound (approximately 1100 ft/sec) generate shock waves that are dragged with the aircraft as long as it is flying supersonic. Where this trailing shock wave intercepts the ground (Fig. 1), it is heard as an impulsive type of sound. In some situations, depending on the shock-wave pattern, an observer may hear two distinct booms or a double boom rather than hear the single sound.

Some factors affecting the sonic boom signal heard on the ground, such as speed, altitude, and route of the aircraft, can be reasonably well controlled. Others, such as meteorological conditions, topography, and ground-level air turbulence, cannot be modified by man. Consequently, the extent to which sonic booms can be predicted and controlled for known flight profiles and flying conditions is limited by those environmental factors beyond man's control. During acceleration, sonic booms two to three times greater in magnitude (superbooms) than those occurring during cruise can be created as the vehicle accelerates to supersonic speeds.

Typical pressure-time functions or signatures for sonic booms generated by large and small aircraft in supersonic flight are shown in Fig. 1. Because of the N-like shape of the pressure signatures, they are frequently referred to as "N-waves." Aircraft A and B are identical, on the same course flying at the same speed, but differing in altitude. Aircraft B and C differ only in size as represented. The durations of the sonic boom pressure waves Δt are directly related to the length of the aircraft. The range of durations of sonic booms for current aircraft and for the proposed

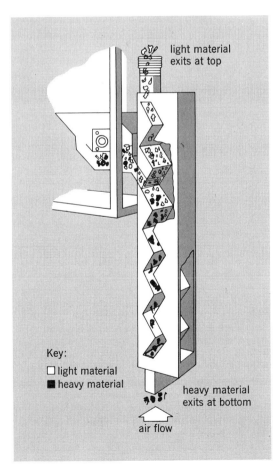

Fig. 8.   Schematic diagram of a device for classifying solid wastes in a high-velocity airstream. Light materials, such as paper and metal foil, are carried up through the zigzag column and are thus separated from falling heavy materials. (*Stanford Research Institute*)

light material exits at top

Key:
☐ light material
■ heavy material

heavy material exits at bottom

air flow

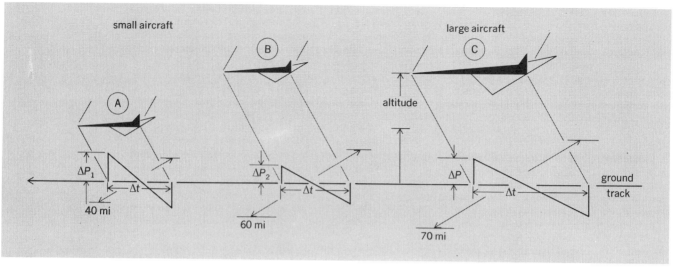

Fig. 1. Duration, intensity, and lateral spread of sonic booms as a function of aircraft size and altitude. (*From* *H. E. von Gierke and C. W. Nixon, Sonic boom, McGraw-Hill Yearbook of Science and Technology of 1967*)

supersonic transports (SST) for commercial operation varies from about 0.05 sec for a fighter aircraft to about 0.5 sec for a large commercial supersonic transport. Duration varies only slightly with altitude (compare aircraft A and B in Fig. 1), being shorter and more directly related to aircraft length for lower altitudes where shock waves have less opportunity to disperse. The significant difference between Δts of the two different-sized aircraft is easily seen.

The loudness of the sonic boom perceived by an observer is a function of, among other things, the initial rise time of the primary shock-wave signature. Signals with a steep or fast rise time have been expected and shown in actual test to sound louder than signals with the same peak overpressure and duration and slower rise times. Sonic booms with fast rise times have much more energy in the frequency bands where the ear is more sensitive and, therefore, seem to give louder acoustic signals than the others. A sonic boom with a slower rise time is a more effective stimulus inside buildings. The shaking and rattling of the building and its contents due to the shock wave add to the overall acoustic stimulus caused by the boom and it may be perceived as more intense or more objectionable than an equal outdoor boom with a fast rise time. *See* LOUDNESS.

The intensity of the sonic boom, that is, the magnitude of its pressure peak ΔP, and the lateral distance from the ground track at which it will be heard are dependent upon the size (lift and drag) of the aircraft and its altitude. Increasing the altitude of the aircraft reduces the overpressure on the ground; however, at the same time it increases the lateral spread or width of the area which is being exposed to the sonic boom. This altitude effect is shown by the two identical aircraft A and B, with the boom at the lower altitude being higher in intensity (ΔP_1 compared to ΔP_2) and with a narrower sonic boom path (40 mi as compared to 60 mi).

Sonic boom is measured in pounds per square foot (psf) of overpressure or pressure above the normal atmospheric pressure. The intensity of normal sonic booms typically experienced these days in communities from aircraft above 30,000–40,000 ft is seldom above 2.0 psf and rarely as high as 5.0 psf. Maximum sonic booms of 120–144 psf produced by aircraft flying at 50–100 ft are about 100 times greater than the usual community sonic boom exposures.

Aircraft altitude is the primary contributor to the magnitude of the sonic boom and is one of the factors most accessible to control measures. Current U.S. Air Force aircraft are restricted from supersonic flight over inhabited areas below an altitude of 30,000 ft except for special missions or in the event of emergency.

Maximum overpressures during transoceanic flight of SST-type aircraft are assumed to become 2.5 psf during acceleration and 1.7 psf during cruise and descent phases. Minimum altitudes of 50,000–60,000 ft during cruising will produce booms of approximately 1.5 psf; however, it has not been determined, and is unlikely, that this level would be acceptable to the general population for frequent operations so that flights over land could be permitted.

It appears that there is no possibility to significantly reduce or eliminate the sonic boom from current and next-generation aircraft which fly at supersonic speeds. Longer booms from the larger aircraft may be psychoacoustically no less acceptable assuming a constant magnitude ΔP. No major breakthrough in the reduction of sonic booms appears on the horizon. The most promising approach to minimizing boom exposure is that of regulating flights through operational control. Commercial supersonic transportation over land will require regulation in a manner that will neither annoy or disturb the general population nor financially penalize the SST to an unbearable extent. Judicious scheduling of supersonic flights, care in acceleration and maneuver, cruise at high altitudes, and avoidance of population centers are major factors of practical significance. It is possible that the extent of operational control and flight regulation necessary to insure acceptable levels of sonic boom on the ground may, in fact, be intolerable for the use of commercial SSTs over land.

Human responses. The complex problem of describing man's reactions to sonic booms is represented in Fig. 2. The occurrence of a sonic boom can affect man's environment in many ways. The two major situations experienced by residents in a community are the outdoor and the indoor sonic boom exposures. The outdoor exposure is nominally a clean signal or N-wave with fast initial rise time. The indoor sonic boom exposure is quite different because it is affected by the transmission properties of the building, that is, the cavity resonance and absorption properties of the room and mechanical excitation of the main building modes. Indoor booms are lower in magnitude, longer in duration, have slower rise times, and visibly vibrate objects within the building which in turn generate noises. Indoor sonic boom exposures, with the added visual, vibratory, and acoustic cues, are generally less acceptable than those out of doors.

The auditory component of the booms which results in loudness sensations is the primary stimulus affecting the individual. When people are not used to sonic boom exposure, they express concern for this sudden loud sound which can visibly shake and rattle buildings and their contents, and they have worried about its possible harmful effects. Close analysis of considerable experience from human and animal exposures to blast waves of varying duration (which are similar to sonic boom waves but much more intense) as well as to rather intense sonic booms leaves little doubt that present-day sonic booms and those expected in the future do no direct physiological harm to humans. A widespread margin of safety exists between what appears to be a loud boom in a community and corresponding pressures required to do any possible damage to man. This is demonstrated in Table 1. Data on mechanical response phenomena leading to pathological damage such as eardrum rupture and lung damage are taken from blast literature. These pressures are 5 times higher for eardrum rupture and 15 times higher for lung damage than the highest booms ever observed in special

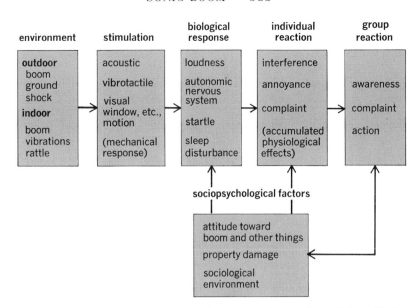

Fig. 2. Man's individual and group reactions to sonic booms. (*From H. E. von Gierke, Effects of sonic boom on people: Review and outlook, J. Acoust. Soc. Amer., vol. 39, no. 5, pt. 2, 1966*)

test flights from aircraft at minimum altitudes and many hundred times higher than the booms observed in communities from typical supersonic flying.

If the auditory system, which is extremely sensitive to changes in pressure, is not injured or harmed by sonic booms, the rest of the human organism may be expected and has been proven safe for such exposures. Specific auditory observations made during overflight programs, essentially all negative findings, are given in Table 2. As indicated, no temporary or permanent hearing loss has been observed following boom exposures as intense as 120 psf and as frequent as 30 booms per day for 30 consecutive days at levels above 5 psf. The probability of a measurable temporary hearing

Table 1. Estimates and observations of effects of sonic boom exposures of various peak overpressures*

Peak overpressure		Predicted or measured effects	
lb/ft²	dynes/cm²		
0–1	0–478		No damage to ground structures; no significant public reaction day or night
1.0–1.5	478–717	Sonic booms from normal operational altitudes: typical community exposures (seldom above 2 lb/ft²)	Very rare minor damage to ground structures; probable public reaction
1.5–2.0	717–957		Rare minor damage to ground structures; significant public reaction particularly at night
2.0–5.0	957–2393		Incipient damage to structures
20–144	957×10^3– 6.8×10^4	Measured sonic booms from aircraft flying at supersonic speeds at minimum altitude: experienced by humans without injury	
720	3.44×10^5	Estimated threshold for eardrum rupture (maximum overpressure)	
2160	1.033×10^6	Estimated threshold for lung damage (maximum overpressure)	

*From H. E. von Gierke, Effects of sonic booms on people: Review and outlook, *J. Acoust. Soc. Amer.*, vol. 39, no. 5, pt. 2, 1966.

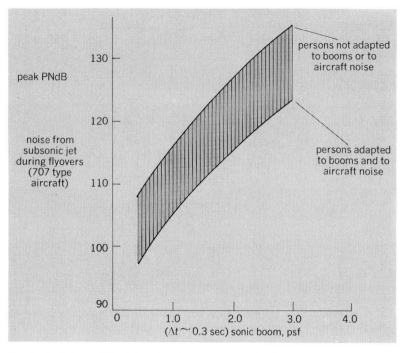

Fig. 3. Sonic boom (Δt ≅ 0.3 sec) peak overpressures (psf) judged equal in noisiness or acceptability (PNdB) to noise from subsonic jet aircraft (Boeing 707) during flyovers at altitudes of approximately 400 ft. (*Adapted from K. Kryter, P. J. Johnson, and J. R. Young, Psychological Experimentation on Noise from Subsonic Aircraft and Sonic Booms at Edwards Air Force Base, Contract no. AF 49(638)-1758, Final Report, NSBEO-4-67, Stanford Research Institute, 1967*)

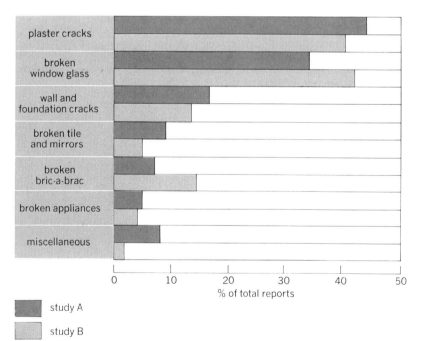

☐ study A

☐ study B

Fig. 4. Types of damage to property reported during two different community overflight programs. (*From C. W. Nixon and P. Borsky, Effects of sonic boom on people: St. Louis, Missouri, 1961-1962, J. Acoust. Soc. Amer., vol. 39, no. 5, pt. 2, 1966*)

Table 2. Observed and predicted auditory responses to sonic booms*

Nature of auditory response	Sonic boom experience or prediction	
Rupture of the tympanic membrane	None expected below 720 psf; none observed up to 144 psf	
Aural pain	None observed up to 144 psf	
Short temporary fullness, tinnitus	Reported above 95 psf	
Hearing loss: permanent	None expected from frequency and intensity of boom occurrence	
Hearing loss: temporary	None measured: 1. 3–4 hr after exposure up to 120 psf 2. Immediately after boom up to 30 psf	
Stapedectomy	No ill effects reported	After booms up to 3.5 psf
Hearing aids	No ill effects reported	

*C. W. Nixon, *Proceedings of Noise as a Public Health Hazard*, ASHA Rep. no. 4, February, 1969.

demonstrate that pressure variations thousands of times greater than those experienced in communities are necessary to even approach the threshold of damage to the human ear. The margin of safety is obvious.

Startle is another psychophysiological response resulting from sudden, loud sonic boom exposures. The possibility of completely eliminating the startle response to occasional sonic booms is not promising. The consistency of this reaction from one person to another suggests that it is an inborn reaction that is modified very little by learning and experience. Although it is hoped that such reactions may be greatly minimized in situations where sonic booms occur with scheduled regularity, the extent to which such adaptive behavior would occur has not been determined. When an adequate startle stimulus occurs a startle reaction may be expected. It is noted that these same startle reactions also occur in response to sharp, loud noises resulting from slamming doors, thunder, dropping of heavy objects, or even automotive backfires.

It is now clear that man need not be concerned about direct physiological harm from sonic booms. However, the fact that repeated exposure to sonic booms, particularly at night, would constitute another increment to the overall environmental stress of modern living, such as neighborhood and traffic noise and the five o'clock rush hour, must be recognized.

Individual reaction to sonic booms, which in turn is the basis for the reaction of larger groups, has been studied in depth during community overflight studies in St. Louis, Mo., in 1961–1962; in Oklahoma City, Okla., in 1964; and at Edwards Air Force Base, Calif., in 1966–1967. During the program in Oklahoma City a total of 1253 sonic booms were generated over the area at a rate of eight per day during a 6-month period. The intensity of the booms was scheduled for 1.5 psf during most of the study and 2.0 psf during the latter stage. Almost 3000 adults were personally interviewed three times during the 6-month period to determine their reactions to the sonic booms.

Substantial numbers of residents reported interferences with ordinary living activities, such as house rattles and vibrations, startle, and interruption of sleep and rest, and annoyance with these

loss from exposures to booms at 2–5 psf is nearly zero. It has been estimated that it would take 75 booms within 3 min repeated daily for many years to produce a modest permanent hearing loss. Consequently, both theoretical and empirical data

interruptions. Although the majority felt they could learn to live with the numbers and kinds of booms experienced, at the end of the study 27% felt they could not accept the booms.

Annoyance is a key factor in human acceptance of sonic booms. Individuals with feelings of annoyance are more prone to complain and to take some action against the sonic booms than are others. The belief that sonic booms have damaged personal property and homes (see response of structures below) contributes strongly to feelings of annoyance. Individuals who ordinarily would not complain about booms do so freely when they believe their personal property is damaged. Clearly a supersonic transport, to be acceptable, must generate sonic booms at a level below the threshold of damage to personal property and homes.

Answers to many questions in the physiological, psychoacoustic, and behavioral areas can be obtained in the laboratory. A scale has been developed which expresses the noisiness or acceptability of sound exposures and this scale forms the basis of criteria and the guidelines presently in use for estimating community reaction to aircraft noise. The level of the noisiness or the perceived noise level (PNL) may be calculated from physical measures of the sound and expressed in units called PNdB. Appropriate psychoacoustical studies have demonstrated and experience has confirmed the ranges of acceptability which correspond to the various values along the PNdB scale. Sonic boom exposures have been empirically equated with this well-established noise acceptability scale (PNdB) as demonstrated in Fig. 3. With this procedure a sonic boom of 2.0 psf is judged equivalent in noisiness or acceptability to an aircraft noise of about 120 PNdB. The perceived noise level concept appears to be a more appropriate means of representing acceptability of sonic booms than does the physical unit of pounds per square foot, which has been customarily employed. However, even though some preliminary answers to practical questions are available and more will be obtained in a few years by ongoing research, an understanding of the overall long-term human response to stimuli of this type and potential health hazards will require long and continuing efforts and may be realized only when SSTs fly regularly, if they ever do so, over land.

Response of structures. Supersonic flights over populated areas must generate sonic booms at levels that will not damage structures and personal property. Sonic booms above this level will not be accepted by the general public. Window glass and other minor damage to isolated structures may appear as a result of sonic boom exposures and such damage can be an important factor in community response. The types of damage to property most often reported (Fig. 4) relate to secondary or decorative elements and consist of cracks in brittle surface treatments, such as plaster, tile, window glass, and masonry. Such damage is noted to be superficial in nature, is restricted to non-load-bearing members, and thus does not affect the strength of the primary structure. Further, it is judged that the superficial damage usually reported is, in large measure, associated with stress concentrations in the structure.

Stress concentrations in buildings may be due to such factors as curing of green lumber, dehydra-

tion of cementious materials, settling of foundations, and poor workmanship. Such factors exist in varying degrees in all structures and can contribute to failures when a triggering load is applied. The overpressure of a sonic boom has this triggering action capability as do vehicle traffic, thunder, windstorms, heavy falling objects, and even many routine household operations. Well-constructed buildings in good repair would not be expected to experience serious damage. Superficial damage would not be expected either, except in situations where critical load concentrations existed.

An overview of the sonic boom situation today, based upon the pool of available knowledge, strongly suggests that SST aircraft as presently planned would create sonic booms that would not be acceptable to the general public. Therefore, their use over land is considered unlikely. Contemporary factors such as mobility in the community, technological advances, political climate, national and international affairs, as well as many other things, motivate beliefs, attitudes, and opinions that influence community response at any one time period. See NOISE, ACOUSTIC.

[HENNING E. VON GIERKE; CHARLES W. NIXON]

Bibliography: H. H. Hubbard, Sonic booms, *Phys. Today*, February, 1968; C. W. Nixon, Human response to sonic boom, *J. Aerosp. Med.*, 36:399, 1965; *Sonic Boom Experiments at Edwards Air Force Base*, NSBEO, AD no. 655310, July 27, 1967; Sonic boom symposium, *J. Acoust. Soc. Amer.*, vol. 39, no. 5, pt. 2 1966; H. A. Wilson, Sonic boom, *Sci. Amer.*, vol. 206, January, 1962.

Species population

A group of similar organisms residing in a defined space at a certain time. Although species have geographic ranges, their individuals typically are not scattered over the entire area but occur in groups, the species populations. These follow more or less discontinuous spatial patterns. The size of such groups and the number of populations into which a species may be divided vary. In the almost extinct whooping crane (*Grus americana*) the entire species forms a single population. For a common bird like the English sparrow (*Passer domesticus*) many groupings may be defined whose size depends upon convenience for some particular study. All the sparrows of a single farm or of Great Britain or even America may constitute a species population. Often a population is not completely separable from neighboring groups.

One can describe populations as static units at some instant of time, but they can be explained only in developmental terms. The component living individuals, which are born, respond to their environment, and ultimately die, confer on the population certain statistical attributes. These attributes provide the basis for the group concept. Birth, death, immigration, and emigration rates determine density, age distribution, and sex ratio and are related to dispersion and genetic constitution of the population. See POPULATION DYNAMICS.

Population density. This is a familiar concept in human populations, where it is customarily expressed as the number of people per unit area. For other organisms different scales are more suitable; bacteria are measured in numbers per cubic centimeter. Often the biomass, the amount of living material, is more instructive than are numbers alone. Numbers of fish in a lake are complemented by information about their size. Among many populations, whose individuals move about, density must be estimated by elaborate sampling techniques. The development of better methods of estimation is itself a significant part of population research. *See* BIOMASS.

Dispersion. Dispersion modifies the interpretation of density. Organisms within a population are often not randomly distributed but occur in clusters. Such clumping may be historically caused—heavy seeds of a plant fall close to the parent. Local differences in habitat may be responsible—in dry weather earthworms concentrate in moist spots. Animals are often social—conspicuous examples occur in the family groups of some warm-blooded vertebrates, but less apparent yet significant sociality is a widespread phenomenon. In contrast to clumping, which is common, uniform spacing is rare. Except in artificial examples such as a corn field, carefully planted with stalks equidistant, spacing is caused by intolerance between individuals, as in the territories established by nesting birds. *See* POPULATION DISPERSION.

Age distribution and sex ratio. In some organisms, as in annual plants, all members of the population are equally old. More commonly many age groups exist simultaneously whose populational effects are not the same. Age distribution affects subsequent population growth through birth and death rates and is in turn affected by these. A constant proportion of organisms of any age group is attained by any population with constant age-specific birth, death, emigration, and immigration rates. Age distributions often fluctuate as a result of variations in these rates. In certain species similar changes in sex ratio are common. Characteristically a growing population has a high proportion of young individuals, with age structure shifting more toward older organisms when population growth rate declines or even becomes negative. Age distribution in some, but far from all, populations can be determined as a result of some discontinuous growth process. Annual rings of trees, fish scales, and certain bones yield information as to the age of individuals.

Birth, death, immigration, and emigration rates. Crude rates state the number of individuals born, dying, and moving in or out per unit of population for some time interval. Although they describe what has happened to the group, these rates have little predictive power and reveal almost nothing about the age structure of the remainder. Age-specific rates which treat births, deaths, and movements as functions of age are the major immediate determinants for the population. Knowledge of the causes for their change is therefore essential for predicting population growth. Variation of these rates within and between species is enormous. The only generalization possible is that high birth rates normally imply high death rates and vice versa. Many components of environment influence the rates. One major factor may be the population itself, which is part of its own environment.

Acclimation and adaptation. Individuals in a population adjust to spatial and temporal variations in environment by numerous morphological, physiological, and behavioral mechanisms that may collectively be termed plasticity. Plastic responses permit populations to succeed in more areas and to persist longer than if these responses did not exist. Genetic variability within a population provides an additional mechanism for adjusting to environment. Where there is systematic change of the environment over an area, populations often exhibit gradual changes in some morphological or physiological characteristics, that is, clines. These are based on cumulative changes in gene frequencies. Where environmental variation is more abrupt, genetically distinct ecotypes may result from the conjunction of selective and isolating forces.

Maintained variation within a single, randomly breeding population (for example, balanced polymorphism), may enable a group of organisms simultaneously to be adjusted to several alternate adapted states. Different breeding systems, by modifying the extent of inbreeding, and chromosomal inversions regulate genetic recombination and thus prevent dissolution of favorable genetic combinations. The extent to which reversible genetic changes accompany fluctuations in numbers and in environment is not well enough established to assess its general significance. In an evolutionary sense the genetic constitution of a species population may perhaps be visualized as analogous to a multiple strand in a complex, irregularly anastomosing braid that represents the evolutionary history of the species. *See* MUTATION.

Dynamics. Knowledge of the properties of populations results in an understanding of their dynamics. During their existence, populations require a constant input of energy and materials. The energy is in part transformed into metabolic heat and in part is stored as new protoplasm and body food reserves. Some remains in dead bodies. Usually little of the energy used by animals, and none of that used by plants, becomes available twice to the same population. The materials required, however, often recirculate through the population any number of times. A population newly released into an adequate environment grows in characteristic fashion and modifies its environment. Given a continuing source of energy and raw materials and a sufficiently stable environment, it may reach a more or less steady state. Otherwise a population peak and decline ensue. This may have cyclic characteristics. Particular causes for growth, stability, decline, and extinction must be sought in complex interactions between the physical and biological environment and birth, death, immigration, and emigration rates modified by genetic change. Species populations often are systems with considerable powers of self-regulation. As a result, they may persist as recognizable entities for indefinite periods.   [PETER W. FRANK]

Bibliography: W. C. Allee et al., *Principles of Animal Ecology*, 1949; H. G. Andrewartha and L. C. Birch, *The Distribution and Abundance of Animals*, 1954; T. O. Browning, *Animal Populations*, 1963; P. R. Ehrlick and R. W. Holm, *The Process of Evolution*, 1963; L. B. Slobodkin, *Growth and Regulation of Animal Populations*, 1961.

Spring, water

A place where a concentrated flow of groundwater reaches the land surface or discharges into a body of surface water. Springs issue where the water table intersects the land surface. Likewise, where the piezometric surface of an artesian aquifer stands above the land surface, water may issue as a spring if a suitable conduit such as a fault or a solution channel is available through which the groundwater can reach the land surface. *See* GROUNDWATER.

Springs, including some less concentrated flows called seeps, are the chief source of the dry-weather flow of most streams. They range from mere zones of seepage along river and pond banks to single or multiple orifices that discharge hundreds of millions of gallons of water per day. Springs may be classified also by such characteristics as mineral content of the water, temperature, geologic structure, and periodicity of flow.

The largest single spring in the United States is probably Silver Springs in Florida, which has an average flow of more than 700,000,000 gal/day. The water issues from limestone into a pool and then flows to the sea in a sizable river. Giant Springs in Montana issues from sandstone and has a discharge of nearly 400,000,000 gal/day. Big Spring in Missouri discharges nearly 300,000,000 gal/day from a limestone aquifer. Comal Springs in Texas discharges a similar amount from a limestone aquifer, the water being brought to the surface by a large fault which provides a conduit. Groups of springs such as Malad Springs in Idaho discharge more than 700,000,000 gal/day, and the Thousand Springs in Idaho discharge about 550,000,000 gal/day from lavas of the Snake River Plain.

[ALBERT N. SAYRE/RAY K. LINSLEY]

Bibliography: S. N. Davis and R. J. M. DeWiest, *Hydrogeology*, 1966; D. K. Todd, *Ground Water Hydrology*, 1959.

Squall line

A line of thunderstorms, near whose advancing edge squalls occur along an extensive front. The thundery region, 20–50 km wide and a few hundred to 2000 km long, moves at a typical speed of 15 m/sec (30 knots) for 6–12 hr or more and sweeps a broad area. In the United States, severe squall lines are most common in spring and early summer when northward incursions of maritime tropical air east of the Rockies interact with polar front cyclones. Ranking next to hurricanes in casualties and damage caused, squall lines also supply most of the beneficial rainfall in some regions.

A squall line may appear as a continuous wall of cloud, with forerunning sheets of dense cirrus, but severe weather is concentrated in swaths traversed by the numerous active thunderstorms. Their passage is marked by strong gusty winds, usually veering at onset, rapid temperature drop, heavy rain, thunder and lightning, and often hail and tornadoes. Turbulent convective clouds, 10–15 km high, present a severe hazard to aircraft, but may be circumnavigated with use of radar.

Formation requires an unstable air mass rich in water vapor in the lowest 1–3 km, such that air rising from this layer, with release of heat of condensation, will become appreciably warmer than the surroundings at upper levels. Broad-scale flow patterns vary; Fig. 1 typifies the most intense outbreaks. In low levels, warm moist air is carried northward from a source such as the Gulf of Mexico. This process, often combined with high-level cooling on approach of a cold upper trough, can

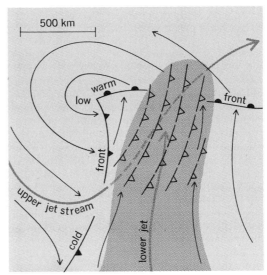

Key: ▨ moist air in warm sector of a cyclone

Fig. 1. Successive locations of squall line moving eastward through unstable northern portion of tongue of moist air in warm sector of a cyclone. Solid arrows show general flow in low levels; broken arrows, axes of strongest wind at about 1 km aboveground and at 10–12 km.

rapidly generate an unstable air mass. *See* THUNDERSTORM.

The instability of this air mass can be released by a variety of mechanisms. In the region downstream from an upper-level trough, especially near the jet stream, there is broad-scale gentle ascent which, acting over a period of hours, may suffice; in other cases frontal lifting may set off the convection. Surface heating by insolation is an important contributory mechanism; there is a marked preference for formation in midafternoon although some squall lines form at night. By combined thermodynamical and mechanical processes, they often persist while sweeping through a tongue of unstable air, as shown in Fig. 1. Squall lines forming in midafternoon over the Plains States often arrive over the midwestern United States at night. *See* STORM.

Figure 2 shows, in a vertical section, the simplified circulation normal to the squall line. Slanting of the drafts is a result of vertical wind

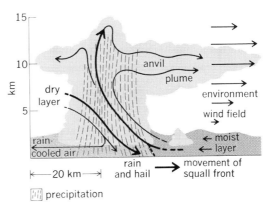

▨ precipitation

Fig. 2. Section through squall-line-type thunderstorm.

shear in the storm environment. Partially conserving its horizontal momentum, rising air lags the foot of the updraft on the advancing side. In the downdraft, air entering from middle levels has high forward momentum and undercuts the low-level moist layer, continuously regenerating the updraft. Buoyancy due to release of condensation heat drives the updraft, in whose core vertical speeds of 30–60 m/sec are common near tropopause level. Rain falling from the updraft partially evaporates into the downdraft branch, which enters the storm from middle levels where the air is dry, and the evaporatively chilled air sinks, to nourish an expanding layer of dense air in the lower 1–2 km that accounts for the region of higher pressure found beneath and behind squall lines. In a single squall-line thunderstorm about 20 km in diameter, 5–10 kilotons/sec of water vapor may be condensed, half being reevaporated within the storm and the remainder reaching the ground as rain or hail.   [CHESTER W. NEWTON]

Bibliography: L. J. Battan, *The Thunderstorm*, 1964; C. W. Newton, Severe convective storms, in *Advances in Geophysics*, vol. 12, 1967.

Storm

An atmospheric disturbance involving perturbations of the prevailing pressure and wind fields on scales ranging from tornadoes (1 km across) to extratropical cyclones (2–3000 km across); also, the associated weather (rain storm, blizzard, and the like). Storms influence man's activities in such matters as agriculture, transportation, building construction, water impoundment and flood control, and the generation, transmission, and consumption of electric energy.

The form assumed by a storm depends on the nature of its environment, especially the large-scale flow patterns and the horizontal and vertical variation of temperature; thus the storms most characteristic of a given region vary according to latitude, physiographic features, and season. This article is mainly concerned with extratropical cyclones and anticyclones, the chief disturbances over roughly half the Earth's surface. Their circulations control the embedded smaller-scale storms.

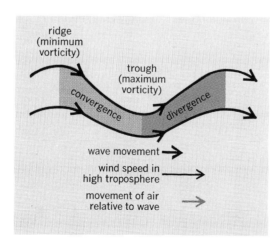

Fig. 1. Convergence and divergence in a wave pattern at a level in the upper troposphere or lower stratosphere. The lengths of arrows indicate the speed and relative motion of wind and wave.

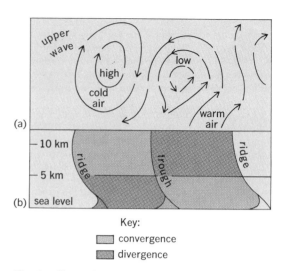

Fig. 2. Air motion associated with a wave pattern. (a) Circulation pattern in low levels in relation to upper wave. (b) West-east vertical section showing simplified regions of convergence and simplified regions of divergence. (*Adapted from J. Bjerknes*)

Large-scale disturbances of the tropics differ fundamentally from those of extratropical latitudes. *See* SQUALL LINE; THUNDERSTORM; TORNADO.

Extratropical cyclones mainly occur poleward of 30° latitude, with peak frequencies in latitudes 55–65°. Those of appreciable intensity form on or near fronts between warm and cold air masses. They tend to evolve in a regular manner, from small, wavelike perturbations (as seen on a sea-level weather map) to deep waves to occluded cyclones. *See* FRONT; METEOROLOGY.

Dynamical processes. The atmosphere is characterized by regions of horizontal convergence and divergence in which there is a net horizontal inflow or outflow of air in a given layer. Regions of appreciable convergence in the lower troposphere are always overlaid by regions of divergence in the upper troposphere. As a requirement of mass conservation and the relative incompressibility of the air, low-level convergence is associated with rising motions in the middle troposphere, and low-level divergence is associated with descending motions (subsidence).

Fields of marked divergence are associated with the wave patterns seen on an upper-level chart (for example, at the 300-millibar level), in the manner shown in Fig. 1. The flow curvature indicates a maximum of vorticity (or cyclonic rotation about a vertical axis) in the troughs and minimum vorticity in the ridges. An air parcel moving from trough to ridge would undergo a decrease of vorticity, which by the principle of conservation of angular momentum implies horizontal divergence. In the upper troposphere, where the wind exceeds the speed of movement of the wave pattern, the divergence field is as shown in Fig. 1. To sustain a cyclone, a relative arrangement of upper and lower flow patterns as shown in Fig. 2 is ideal. This places upper divergence over the region of low-level convergence that occupies the central part and forward side of the cyclone, and upper convergence over the central and forward parts of the anticyclone where there is lower-level divergence.

The upper- and lower-level systems, although broadly linked, move relative to one another. Cyclogenesis commonly occurs when an upper-level trough advances relative to a slow-moving surface front (stages 1 and 2 in Fig. 3). The region of divergence in advance of the trough becomes superposed over the front, inducing a cyclone that develops (stage 3 in Fig. 3) in proportion to the strength of the upper divergence. In a pattern such as Fig. 1, the divergence is strongest if the waves have short lengths and large amplitudes and if the upper-tropospheric winds are strong. For the latter reason, cyclones form mainly in close proximity to the jet stream, that is, in strongly baroclinic regions where there is a large increase of wind with height. *See* JET STREAM.

Frontal storms and weather. Weather patterns in cyclones are highly variable, depending on moisture content and thermodynamic stability of air masses drawn into their circulations. Warm and occluded fronts, east of and extending into the cyclone center, are regions of gradual upgliding motions, with widespread cloud and precipitation but usually no pronounced concentration of stormy

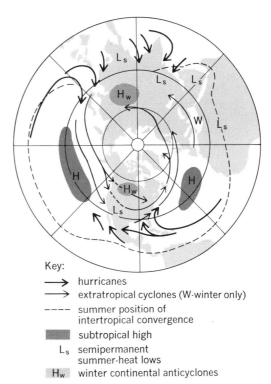

Key:

⟶ hurricanes

⟶ extratropical cyclones (W-winter only)

---- summer position of intertropical convergence

▨ subtropical high

L_s semipermanent summer-heat lows

H_w winter continental anticyclones

Fig. 4. Principal tracks of extratropical cyclones and hurricanes with significantly associated features in the Northern Hemisphere.

conditions. Extensive cloudiness also is often present in the warm sector.

Passage of the cold front is marked by a sudden wind shift, often with the onset of gusty conditions, with a pronounced tendency for clearing because of general subsidence behind the front. Showers may be present in the cold air if it is moist and unstable because of heating from the surface. Thunderstorms, with accompanying squalls and heavy rain, are often set off by sudden lifting of warm, moist air at or near the cold front, and these frequently move eastward into the warm sector.

Middle-latitude highs or anticyclones. Extratropical cyclones alternate with high-pressure systems or anticyclones, whose circulation is generally opposite to that of the cyclone. The circulations of highs are not so intense as in well-developed cyclones, and winds are weak near their centers. In low levels the air spirals outward from a high; descent in upper levels results in warming and drying aloft.

Anticyclones fall into two main categories, the warm "subtropical" and the cold "polar" highs. The large and deep subtropical highs, centered over the oceans in latitudes 25–40° and separating the easterly trade winds from the westerlies of middle latitudes, are highly persistent.

Cold anticyclones, forming in the source regions of polar and arctic air masses, decrease in intensity with height. Such highs may remain over the region of formation for long periods, with spurts of cold air and minor highs splitting off the main mass, behind each cyclone passing by to the south. Following passage of an intense cyclone in middle latitudes, the main body of the polar high may move southward in a major cold outbreak.

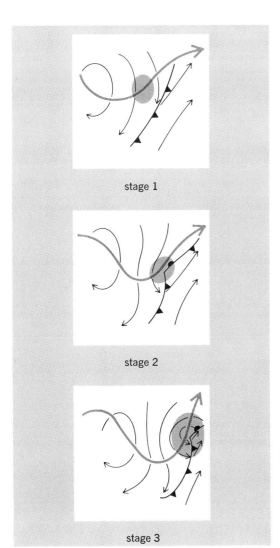

stage 1

stage 2

stage 3

Key: ▨ region of strongest upper-level divergence

Fig. 3. Upper-level trough advancing relative to surface front and initiating cyclone. (*After S. Petterssen*)

Blizzards are characterized by cold temperatures and blowing snow picked up from the ground by high winds. Blizzards are normally found in the region of a strong pressure gradient between a well-developed arctic high and an intense cyclone. True blizzards are common only in the central plains of North America and Siberia and in Antarctica.

Principal cyclone tracks. Principal tracks for all cyclones of the Northern Hemisphere are shown in Fig. 4. In middle latitudes cyclones form most frequently off the continental east coasts and east of the Rocky Mountains.

Movements of cyclones, both extratropical and tropical, are governed by the large-scale hemispheric wave patterns in the upper troposphere. The character of these waves is reflected in part by large circulation systems such as the subtropical highs, and greatest anomalies from the principal cyclone tracks occur when these highs are displaced from the mean positions shown in Fig. 4. Warm highs occasionally extend into high latitudes, blocking the eastward progression suggested by the average tracks and causing cyclones to move from north or south around the warm highs.

Over the Mediterranean, cyclones form frequently in winter but rarely in summer, when this area is occupied by an extension of the Atlantic subtropical high. Both the subtropical highs and the cyclone tracks in middle latitudes shift northward during the warmer months: on west coasts cyclones are infrequent or absent in summer south of latitudes 40–45°.

Role in terrestrial energy balance. Air moving poleward on the east side of a cyclone is warmer than air moving equatorward on the west side. Also, the poleward-moving air is usually richest in water vapor. Thus both sensible- and latent-heat transfer by disturbances contribute to balancing the net radiative loss in higher latitudes and the net radiative gain in the tropics. Air that rises in a disturbance is generally warmer than the air that sinks (Fig. 2), and latent heat is released in the ascending branches by condensation and precipitation. Hence the disturbance also transfers heat upward, as is required to balance the net radiative loss in the upper part of the atmosphere. *See* ATMOSPHERIC GENERAL CIRCULATION.

[CHESTER W. NEWTON]

Bibliography: E. Palmén and C. W. Newton, *Atmospheric Circulation Systems*, 1969; S. Pettersen, *Introduction to Meteorology*, 3d ed., 1969; H. Riehl, *Introduction to the Atmosphere*, 1965.

Storm detection

The methods and techniques used in detecting the formation of severe storms. These methods include procedures for locating, tracking, and forecasting a storm's movement. To assist in storm detection, the meteorologist has adapted a number of special tools, such as radar and satellites, to supplement the usual surface and upper-air charts and analyses, routine meteorological observations, and reports of visual observations from cooperative storm-warning networks. *See* STORM; WEATHER FORECASTING AND PREDICTION; WEATHER MAP.

Radar. The development of radar as a meteorological instrument for the continuous represen-

tation of precipitation has resulted in a major advance in the detection of areas of frontal precipitation, hurricanes and typhoons, squall lines, thunderstorms, and in some cases tornadoes. Although the 100–200-mi observing radius of one weather radar station is not adequate to cover the entire extent of extratropical cyclones and attendant frontal systems, it is possible, where radar coverage is fairly complete, to combine several observations so that the precipitation areas associated with an entire system can be mapped and followed. Areas of precipitation, indicating vertical motion, can be watched to see how they change as a result of the dynamic processes at work, the latter being deduced from synoptic analyses. Thus critical changes in the shape, the rate of development and decay, and the movement of such areas can be noted as they occur.

Radar serves one of its most valuable purposes in locating, tracking, and defining hurricanes which approach or enter continental coastal areas, and in detecting their more dangerous parts. The familiar spiral-band patterns consisting of lines of intense convective activity or squalls, comparable to severe continental squall lines, are generally well defined, and the time of their arrival at a given point prior to the main storm center can be estimated with fair accuracy. The spiral structure of these bands often defines the center of the storm to within 5–10 mi (Fig. 1). The extent of the rain area and at least a qualitative estimate of precipitation amount, based on echo intensity, and the expected path and rate of movement of the storm can be derived from radar observations.

Radar is most commonly used for the detection and tracking of thunderstorms, squall lines, and

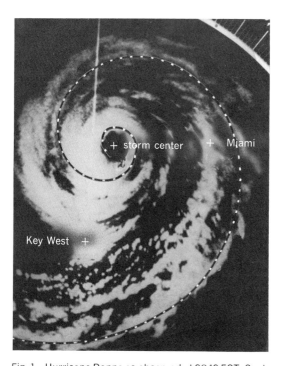

Fig. 1. Hurricane Donna as observed at 0840 EST, Sept. 10, 1960, on the WSR-57, 10-cm radar set at Key West, Fla. A spiral overlay of cross over angle α = 15° has been fitted to the precipitation bands in order to indicate the location of the storm center. (*ESSA photograph*)

quite frequently tornadoes. This is a routine operation with modern equipment which detects precipitation rates as small as 0.01 in./hr to ranges of 125 mi. Since the very small cloud particles are not significantly detected by radar, sharply formed echoes or groups of echoes imply precipitation of shower intensity. The meteorologist carefully tracks more intense echoes: echoes showing rapid lateral growth, hail-indicating protuberances, or other unusual characteristics: and echoes of storms which are known by visual observations to be producing severe weather at the surface. The position of such echoes is plotted at frequent intervals.

The intensity of local storms can be estimated from the echo intensity, which increases with increasing precipitation rate and is particularly great with hail. Advanced radar systems include contour mapping capability, which shows echo intensity at a glance. Systems which provide a display like that shown in Fig. 2 are being obtained for installation at all major U.S. Weather Bureau radar installations. The vertical development of echoes is also a useful storm parameter, and is given by use of a range-height indicator (RHI) or by scanning the radar antenna in azimuth at an increasing elevation angle until the echo disappears.

Since observations indicate a high correlation between tornadoes and intense thunderstorms, the radar meteorologist views the most intense, rapidly growing echoes with greatest suspicion when conditions are otherwise favorable for tornado formation. Particularly intense and enduring tornadoes have also been identified in a large proportion of instances by telltale protrusions in the form of a hook or a figure 6 extending from the right-rear quadrant of an echo (Fig. 3). Such observations, however, are usually made within about 50 mi of the radar location and require favorable antenna elevation, receiver gain, and view of the storm. Sometimes similar echo configurations are not associated with tornadoes. Nevertheless, whenever unusual echoes are observed, immediate efforts are made to obtain local visual observations. If the echo is found to be associated with a tornado, it is tracked continuously and communities in its path are warned.

A proposed use of radar for tornado and turbulence detection involves the Doppler frequency shift associated with the motion of the radar targets. Significantly, a rapid relative motion of reflecting particles simultaneously detected by the radar receiver is associated with a correspondingly rapid fluctuation of the signal intensity, which can be recorded and analyzed by means of equipment attached to radars similar to those in general operational use today. However, the highly localized nature of tornadoes, the large range of velocities characteristic of them, and low values of reflectivity sometimes associated with the region of strongest winds pose difficult problems for the retrieval of information suitably specialized: applications of Doppler radar techniques to studies of severe storm identification and analysis are presently the subject of intensive research (Fig. 4). *See* THUNDERSTORM; TORNADO.

Sferics. The electromagnetic radiations produced by lightning discharges, or sferics, most commonly experienced as static in AM radio re-

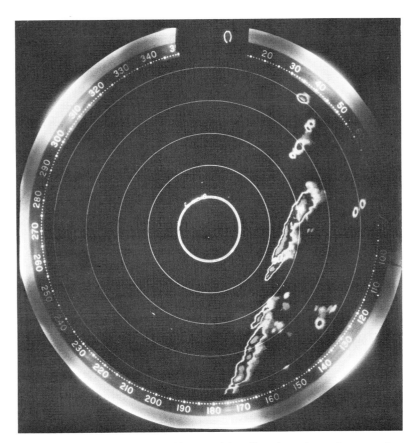

Fig. 2. Plan-position indicator with storm intensities shown by contour mapping. Bright cores show where hail is likely. (*ESSA photograph*)

ception, provide an effective means for detecting thunderstorms. Simple "lightning counters" have been constructed on this principle and, while nondirectional and of short reception range, they serve as fairly reliable indicators of local thunderstorm activity. During the early 1960s an experimental network of sferics detectors was operated in the central part of the United States to pinpoint, map, and follow areas of sferics sources or thunderstorm activity. However, the problems involved in determining storm locations by a triangulation technique, involving identification of signals simultaneously generated and received at stations several hundred miles apart, were not sufficiently overcome to warrant retention of the system for operational use. The development of a sferics detection system remains an attractive goal, and advances in electronics techniques and signal-processing capabilities in the 1960s have stimulated renewed research directed toward this end.

Infrasonics. Experiments by the National Bureau of Standards have indicated that tornadoes can be associated with sound or pressure waves of very low frequency—about 1/10 to 1/50 hertz (Hz), far below an auditory threshold of 15 Hz—and that these sounds are propagated through the atmosphere to great distances. It is believed that these waves are essentially trapped between the ground and a layer of high temperature (fast sound speed) at a height of about 50 km and propagated with relatively little loss in intensity or waveform. Networks of ultrasensitive microbarographs equipped

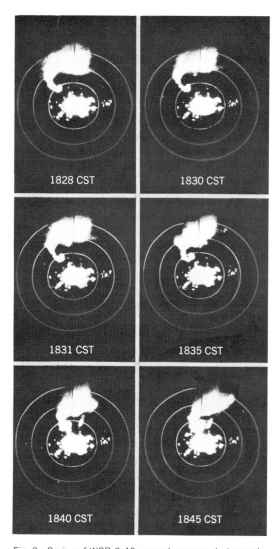

Fig. 3. Series of WSR-3, 10-cm radarscope photographs taken at Topeka, Kan., May 19, 1960, showing the development of a pronounced hook-shaped echo extending from the main thunderstorm cell, associated with a tornado at Meriden, Kan. Range marks indicate 10-mi intervals. (*ESSA photograph*)

Fig. 4. Radar systems for research at the National Severe Storms Laboratory, Norman, Okla. The feasibility of Doppler radar techniques is being examined with the 30-ft antenna in the foreground. (*ESSA photograph*)

with special noise-reducing line microphones and appropriate frequency pass bands are used to detect the signals, and by comparison of the times of arrival, the azimuth of the source of the wavefront can be determined. Determination of additional azimuths by other, properly spaced, similar arrays of instruments would enable the source to be fixed, as in the case of sferics. As with sferics, however, acoustical methods are considered less promising for the near future than techniques involving radar and satellites.

Microseisms. Another form of very-low-frequency waves, called microseisms, have periods of 4–7 sec. These waves are propagated along the surface of the Earth rather than in the atmosphere. Microseisms have been more fully explored by seismologists. The oscillations can be applied to the detection of storms at sea, from which these oscillations originate. The exact mechanism of their production is not agreed upon, but their specific source in intense cyclonic storms, hurricanes, and typhoons suggests a pumping action of the vortex on the surface of the sea. Multiple tripartite networks of seismograph stations are needed to determine azimuths and fixes on a storm. Such networks have been used to detect storms well over 1000 mi away.

Aircraft and satellites. High-level observational platforms facilitate detecting and tracking of storms through observations of the large-scale cloud patterns. Such pictures give the meteorologist an invaluable perspective, not only of the form and extent of a particular weather system, but also of the environment in which the system is embedded.

Aircraft. When synoptic reports from conventional sensors or satellites indicate the formation of a hurricane beyond the range of land-based radar, direct verification and subsequent observation may be provided by aircraft. The "hurricane hunters" of the U.S. Navy and special U.S. Weather Bureau aircraft fly over the area and penetrate the storm after it develops, taking observations of the basic meteorological elements and recording visually, photographically, and by radar the cloud and precipitation patterns of the entire storm. The location of the storm center, or eye, the movement of the center, and the storm size, intensity, and rate of growth are continuously observed and the data are recorded and relayed to forecast centers. If the hurricane is sufficiently far from land areas, experiments to modify the hurricane by cloud seeding may also be attempted. *See* WEATHER MODIFICATION.

The eye of a hurricane has also been tracked by means of radio transmissions from a free-floating, constant-level balloon, or hurricane beacon, released from an aircraft in the storm center at an altitude of about 15,000 ft.

Satellites. Tremendous strides in the development of meteorological satellites during the 1960s have led to timely detection of all types of large storms over all portions of the Earth (Fig. 5). Satellites in polar orbit and equipped with infrared sensors have provided sufficient observations both by day and night to provide global maps of cloud patterns, and cloud top temperatures. Comparison of such patterns with information obtained by other means has provided effective techniques for locating hurricane centers and estimating their inten-

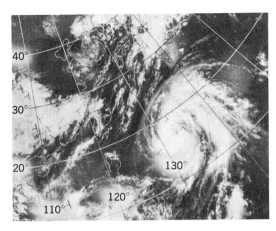

Fig. 5. *ESSA-7* montage of several pictures taken Oct. 16–17, 1968. Typhoon Gloria at right; outlines of China and Japan at left. (*ESSA photograph*)

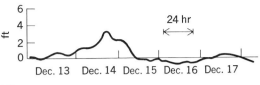

(a)

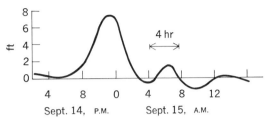

(b)

Surge hydrographs. (*a*) Winter storm of 1954 at Sandy Hook, N.J. (*b*) Hurricane of 1944 at Newport, R.I.

sity. Advanced technology satellites (ATS) in geosychronous (Earth-stationary) orbits approximately 22,000 mi above the Equator transmit photographs of practically the entire facing hemisphere two times or more each hour. Interpretation of photographs at data centers yields quick and accurate assessment of large-scale storm activity on a global basis.

Meanwhile, the data of various other satellites are being received by rather simple receiving equipment and reproduced in picture format for local use on facsimile equipment at some 200 receiving stations around the world. The automatic picture transmission (APT) technique is a means for reproducing the satellite-eye view of the Earth beneath for use by the meteorological service of any country or organization which has acquired the relatively inexpensive equipment needed. The recording of daily photographic data over a region 200 mi in radius surrounding a typical receiving site gives an immense amount of valuable information while reducing the need for aircraft reconnaissance of hurricanes, for example. Some potential also exists for effective monitoring by satellites of smaller-scale disturbances, such as severe thunderstorms, although the view of thunderstorm tops from satellite heights is not as indicative of heavy rain as the view in radar photographs.

[EDWIN KESSLER]

Bibliography: American Meteorological Society, *Proceedings of the 5th Conference on Severe Storms*, 1967; D. Atlas (ed.), *Severe Local Storms*, Amer. Meteorol. Soc. Monogr., vol. 5, no. 27, 1964; L. J. Battan, *Radar Meteorology*, 1959; G. E. Dunn and B. I. Miller, *Atlantic Hurricanes*, 1960; J. S. Marshall (ed.), *Proceedings of the 13th Conference on Radar Meteorology*, American Meteorological Society, 1968.

Storm surge

A transient, localized disturbance at sea level, resulting from the action of a tropical cyclone, an extratropical cyclone, or a squall over the sea. Storm surges, or storm tides, are not to be confused with tsunamis, or tidal waves, which result from seismic or molar disturbances of the Earth. In the Northern Hemisphere those coastal regions which are particularly vulnerable to storm surges

include the periphery of the Gulf of Mexico, the Atlantic Coast of the United States, the Gulf of Bengal, Japan and other islands of the western Pacific which lie in the typhoon belt, and the coastal regions of the North Sea. The surges occurring in the North Sea originate from the actions of large-scale extratropical storms, particularly winter storms. On the eastern coast of the United States, hurricane-induced surges, as well as surges originating from intense winter storms, occur. In the Great Lakes and the Gulf of Mexico, surges resulting from squalls are known to occur; however, hurricane-induced surges pose a more serious threat to the low-lying coastal areas of the Gulf.

The time history of the surge at a given location at shore is represented by the surge hydrograph. This is a time sequence of the difference between the measured tide and the predicted periodic tide (see illustration). Maximum surge elevations of 15 ft above predicted tide are not uncommon. In the case of hurricane-induced surges, the peak water level seems to depend primarily upon the atmospheric pressure at the hurricane center. However, the horizontal scale, the direction and speed of propagation of the hurricane, and the coastal geometry and bottom topography are important influencing factors in the storm surge behavior. When a hurricane crosses the coast from the sea, the greatest surge alongshore usually occurs to the right of the hurricane path.

A storm surge is essentially a forced inertiogravitational wave of great wave length. This implies that the duration or speed of the storm determines the dynamic augmentation of the water level at shore above that which would occur if the storm were stationary. Also, the inertial character of surges can explain quasi-periodic resurgences that often follow the primary forced surge.

[ROBERT O. REID]

Bibliography: J. C. Freeman, Jr., L. Baer, and G. H. Jung, The bathystrophic storm tide, *J. Mar. Res.*, 16(1):12–22, 1957; D. L. Harris, The hurricane surge, *Proceedings of the 6th Conference on Coastal Engineering*, Council on Wave Research, pp. 96–114, 1958; U.S. Department of Commerce, *Characteristics of Hurricane Storm Surge*, 1963: P. Welander, *Numerical Prediction of Storm Surges*, in H. E. Landsberg and J. Van Mieghem (eds.), *Advances in Geophysics*, vol. 8, 1961.

Stream transport and deposition

The sediment debris load of streams is a natural corollary to the degradation of the landscape by weathering and erosion. Eroded material reaches stream channels through rills and minor tributaries, being carried by the transporting power of running water and by mass movement, that is, by slippage, slides, or creep. The size represented may vary from clay to boulders. At any place in the stream system the material furnished from places upstream either is carried away or, if there is insufficient transporting ability, is accumulated as a depositional feature. The accumulation of deposited debris tends toward increased ease of movement, and this tends eventually to bring into balance the transporting ability of the stream and the debris load to be transported.

Stream loads. Because streams form and adjust their own channels, the debris load to be carried and the ability to carry load tend to reach and maintain a quasi-equilibrium. A reach of stream (part of the course) which attains this equilibrium is considered graded.

Much has been written concerning the concept of the graded stream. At one time absence of waterfalls or other discontinuities of longitudinal profile was considered necessary and, in fact, evidence for the condition of grade. Because much remains to be learned about the mechanics of debris transportation, the criteria for the graded condition may be expected to be extended and revised. In the present state of knowledge, however, it appears acceptable to think of reaches or segments of channel being graded, even when separated by reaches not so adjusted. A graded stream is one in which, over a period of years, slope and channel characteristics are delicately adjusted to provide, with available discharge, the shear forces required for the transportation of the load supplied from the drainage basin.

Two terms which have been useful to geologists and engineers dealing with rivers are competence and capacity. Competence was used by G. K. Gilbert to mean the ability to move debris, and its measure is the maximum size of material which can be barely be moved. Capacity of a stream is the total load which it can carry under given conditions and is measured as weight of debris moved per unit of time. The usefulness of these terms has lessened with demand for increasingly quantitative description of stream action. Sampling equipment now in general use measures only the suspended portion of the debris and not that moving close to the streambed. Thus, except in special situations, the carrying capacity of a stream cannot be precisely measured, and available theory allows only an approximation of total load by computation.

The maximum size of debris which can be carried varies, depending on subtle variations of several factors. Thus competence, a highly useful concept, cannot be determined with satisfaction either in the field or by computation. The concepts implied by these terms will gain even greater value and importance as both theory and field measurement techniques improve. The following review of the present status of theory of debris transport will perhaps indicate how the usefulness of these concepts depends greatly on ability to determine quantitative values for them.

Debris transport theory. Debris transport is inextricably associated with the hydromechanics of flow in open channels. It is now known that the introduction of sediment grains into a fluid alters in an important manner many of the hydraulic relationships which applied to a fixed bed. For example, in a movable-bed channel, boundary roughness is not merely the rugosity of the nonmoving bed and banks. Once the particles begin to move, the shear-resisting flow is altered. Particles can assume many different configurations, among which are dunes or ripples or a plane, and these bed forms depend on the transportation process. Thus the resisting shear at the buundary depends on the debris transport itself.

When shear applied by water to a grain bed of uniform-size particles becomes sufficient to move a layer of grains, successive layers do not progressively peel off indefinitely. After some layers are put in motion, an equilibrium is reached. Transport then continues without further degradation. R. A. Bagnold showed by theory and experiment that the grains in transport add a new force normal to the bed which holds the particles exposed at the

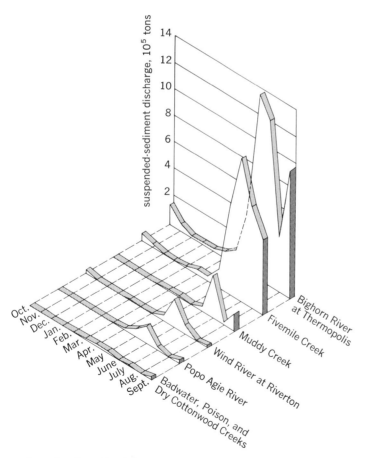

Comparison of sediment loads for several streams in the Wind River drainage basin, in Wyoming, 1950 water year. (*From B. R. Colby, C. H. Hembree, and F. H. Rainwater, Sedimentation and Chemical Quality of Surface Waters in the Wind River Basin, Wyoming, USGS Water-Supply Pap. no. 1373, 1956*)

bed against the stress of the overlying fluid-grain mixture. This force, the dispersive stress between sheared grains, makes a fundamental difference in the stress structure between fixed and movable-bed channels.

Of all the theories put forward to elucidate the physics of sediment transport, the most objectively derived from first principles rather than from empirical data is that of Bagnold. This theory is based on the general idea that work involved in sediment movement comes from the energy expended by the flowing water. The equations are predicated on the idea that only a portion of the total power spent by the water is used for debris transport, that is, that the available power times an efficiency factor equals that part utilized for carrying clastic load.

The formulation most widely used for computing sediment transport is one derived by H. A. Einstein which involves both theory and several empirically derived coefficients. His computational procedure has been simplified by B. R. Colby and C. H. Hembree for application to field problems.

Field measurement. Nearly without exception, the field measurement of sediment in transport is capable of sampling only the suspended part of the load. That which is of size larger than sand and that transported in the 2 in. nearest the bed are not measured. At a few experimental sites it has been possible to measure both the suspended load at a normal river cross section and the total load. In streams carrying material in the sand-size range and smaller, but no gravel, the total load may be measured with the usual suspended-load sampler by creating turbulence sufficient to throw all debris in transit into a condition of suspension for a brief time. Such a procedure was used on the Niobrara River near Cody, Neb., by Colby and Hembree, who reported that for that location the suspended load concentration averaged 51% of the concentration represented by the total sediment load.

In western rivers flowing through alluvial valleys, suspended-load concentrations tend to occur in the range of 100 to 5000 ppm (parts per million). The concentration increases geometrically, not linearly, with increased water discharge. Such concentrations can result in large amounts of sediment in a day, a month, or a year. An example is shown in the figure. A river of moderate size, Bighorn River at Thermopolis, Wyo., averages 3.6 in. of runoff from the drainage basin of 8080 mi.^{2} This

amounted to 1,360,000 acre-feet annually during a 41-year period. In the figure it can be seen that in 1950 about 1,400,000 tons of sediment passed Thermopolis in the month of June alone.

In the humid East, sediment concentrations are considerably smaller, but runoff is larger than in the semiarid West. Some average values of suspended sediment contributed from different areas are shown in the table.   [LUNA B. LEOPOLD]

Bibliography: R. A. Bagnold, *An Approach to the Sediment Transport Problem from General Physics*, USGS Prof. Pap. no. 422-I, 1966; C. B. Brown, *Rates of Sediment Production in Southwestern United States*, U.S. Soil Conserv. Serv. Tech. Pap. no. 58, 1945; G. M. Brune, *Rates of Sediment Production in Midwestern United States*, U.S. Soil Conserv. Serv. Tech. Pap. no. 65, 1948; B. R. Colby and C. H. Hembree, *Computations of Total Sediment Discharge Niobrara River Near Cody, Nebraska*, USGS Water-Supply Pap. no. 1357, 1955; B. R. Colby, C. H. Hembree, and F. H. Rainwater, *Sedimentation and Chemical Quality of Surface Waters in the Wind River Basin, Wyoming*, USGS Water-Supply Pap. no. 1373, 1956; H. A. Einstein, *The Bed-Load Function for Sediment Transportation in Open Channel Flows*, USDA Tech. Bull. no. 1026, 1950; G. K. Gilbert, *Transportation of Debris by Running Water*, USGS Prof. Pap. no. 86, 1914; R. F. Hadley and S. A. Schumm, *Hydrology of the Upper Cheyenne River Basin*, USGS Water-Supply Pap. no. 1531-B, 1961.

Succession, ecological

A gradual process brought about by the change in the number of individuals of each species of a community and by the establishment of new species populations which may gradually replace the original inhabitants. Succession depends upon several major factors: (1) the floristic and faunistic resources of the general region, that is, what organisms are in a position to invade a site; (2) the rate of change of the habitat and its receptivity to these potential invaders; and (3) chance factors which may influence these interactions. *See* PLANT COMMUNITY; POPULATION DISPERSAL.

Terminology. The term sere is used to describe all of the temporary communities which occur during a successional sequence on a given site. Eventually succession ends, changes in species composition of the community cease or fluctuate within some bounds, and the result is a climax community.

Successional sequences may be classified on the basis of starting habitat. Thus, hydroseres are communities in which pioneer plants invade open water, eventually forming some kind of soil such as peat or muck; xeroseres are communities on dry, sterile ground, such as rock, sand, or clay.

If an open habitat has never been occupied before, the way is opened for primary succession. If the habitat has been disturbed, leaving vestiges of soil, seeds, and other organic debris from previous occupancy, a secondary succession will begin.

Stages of succession. As an area is invaded and taken over by successive populations of plants and animals, the physiognomy of the community changes. Isolated pioneer plants eventually give way to a consolidation of the plant cover. With continued exploitation of the habitat, new species take over which have life forms able to utilize

---

Rates of suspended sediment production

Region	Sediment yield, tons/mi^2-year
Streams in western U.S.*	1200–1400
Cheyenne Basin, Wyo.†	3900
Streams in midwestern U.S.‡	240–850

*From C. B. Brown, *Rates of Sediment Production in Southwestern United States*, U.S. Soil Conserv. Serv. Tech. Pap. no. 58, 1945.

†From R. F. Hadley and S. A. Schumm, *Hydrology of the Upper Cheyenne River Basin*, USGS Water-Supply Pap. no. 1531-B, 1961.

‡From G. M. Brune, *Rates of Sediment Production in Midwestern United States*, U.S. Soil Conserv. Serv. Tech. Pap. no. 65, 1948.

more fully the resources of the habitat than previous life forms until finally the climax community is established. Such a community can perpetuate itself indefinitely, at least in theory, because each species reproduces and maintains a stable number of individuals. *See* PLANTS, LIFE FORMS OF.

It was traditional to divide succession into pioneer, consolidation, subclimax, and climax stages on the basis of physiognomy. However, such discrete stepwise changes are not always shown by close study of the populations, because there is a continuously changing array of species. Therefore, the subdivision of a successional continuum must depend on local conditions and purposes of study. *See* ECOLOGY, APPLIED; VEGETATION MANAGEMENT.

The following characteristics of ecological succession have been listed by R. Whittaher: Its nature is continuous; comparable habitats are not always settled at the same rate or by the same group of species; and certain stages may be prolonged, telescoped into one another, or omitted entirely in different habitats or in different parts of the same habitat.

Methods of study. Any unexploited or undisturbed habitat to which organisms have access is suitable for succession studies. These habitats include microhabitats, such as dead pine cones with changing fungal populations, dead animal carcasses with changing insect and microorganism populations, and hay infusions (Fig. 1) and other types of laboratory microcosms, and the more extensive habitats such as ponds, sections of rivers, mud flats, sand dunes, abandoned fields, lava flows, or other terrain types. *See* ECOSYSTEM.

To study the population changes occurring during natural succession, permanent sample plots may be used which have been established in the area being studied and which are observed year after year with the changes recorded. However, succession in such plots tends to slow down after a few years as more and more perennial plants become established. It is then necessary to deduce the course of succession from a comparison with similar habitats, which were invaded at earlier times and hence are further advanced in succession. This procedure introduces a number of possi-

ble errors. Environmental conditions of the compared plots may be slightly different; also, the availability of disseminules is never exactly the same in different places and at different times. Because of these difficulties, knowledge of succession is partly extrapolated from observed short-term changes and partly synthesized from isolated examples.

Studies of succession have been carried out in the laboratory in carefully controlled, small habitats known as aquarium ecosystems. These offer advantages in terms of size, types and numbers of flora and fauna, environmental control, and the time element, which is greatly telescoped. Such systems, however, are not similar to natural ones in that they are closed; that is, there is no means for constant addition of new species and there is no natural means for removing accumulated organic substances. Despite this, on the basis of such studies many useful hypotheses are being framed, and these can be tested against natural systems.

Examples of succession. The general patterns which occur during succession are illustrated by several examples.

Primary succession is exemplified by the development of forests following deglaciation in southeastern Alaska. After the ice has retreated, mosses, fireweed, and herbs pioneer on the raw, calcareous glacial till. Willows then begin to form a prostrate mat which eventually becomes an erect, dense scrub. Alder dominates the next stage, eventually forming pure thickets; these are succeeded by trees, with sitka spruce appearing first and western hemlock later. On well-drained sites the hemlocks are the climax, but wet sites may revert to more or less open muskeg.

In the southeastern United States, secondary succession has been widespread where cultivated fields were abandoned (Fig. 2). During the first summer after cultivation horseweed and crabgrass dominate the fields. In the second year, dominance abruptly changes to aster. Beginning with the third year there is a shift toward a grass called broomsedge (*Andropogon virginicus*). By the fifth year this plant may be replaced by aster or it may persist as brambles and shrubs invade. Young pines may invade fields at any stage, but they typically become large enough to shade and litter the ground only after the field reaches the broomsedge stage. The pines mature rapidly, shading out the broomsedge so that by the tenth to twelfth year what was once a field has become a young pine forest. Pines typically do not reproduce in their own shade and litter unless suitable fires and site conditions occur; but young hardwoods (oaks, hickories, and sweet gum) do become established under the pine canopy. As the pines mature and die, the young hardwoods replace them. Thus, within a period of 150–200 years (or less if pines are cut without regenerating well), an oak-hickory forest can be reestablished on the abandoned field. *See* PLANT FORMATIONS, CLIMAX.

Causes and nature of succession. Although there are variations in successions among different geographic areas and different habitats, certain general characteristics can be observed.

Many successional changes can be explained by changes in the physical environment occurring in conjunction with the changes in species composi-

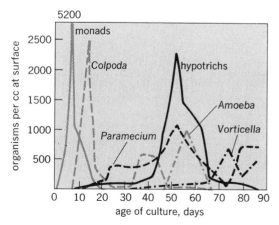

Fig. 1. Succession of protozoan populations in hay infusion. (*After Woodruff, 1912, from E. P. Odum, Fundamentals of Ecology, 2d ed., Saunders, 1959*)

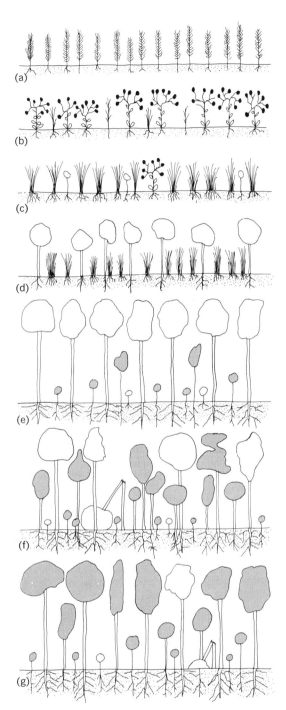

Fig. 2. Plant succession on abandoned field in North Carolina Piedmont. (a) First year dominated by horseweed. (b) Second year dominated by aster with dead horseweed. (c) Third to fifth year dominated by broomsedge and young pine. (d) Young pines overtopping broomsedge. (e) Pure stand of pine with hardwood seedlings after 25–30 years. (f) Old-growth pine being replaced by vigorous young hardwoods. (g) Climax hardwoods of oak and hickory.

tion. Some of these changes may be induced by the activities of the organisms themselves and are referred to as autogenic; that is, as a community exploits the resources of its habitat, physical and chemical changes are brought about. For example, in the sere following deglaciation, the alder stage is accompanied by a rapid increase in the amount of available soil nitrogen, due to the activity of nitrogen-fixing organisms living in nodules on the alder roots. Trees enter the community when there has been a significant increase in nitrogen in the surface soil and woody debris from pioneer trees. The resources needed by one population are altered, and conditions become favorable for the invasion of others; communities pave the way for one another in ideal autogenic succession, but its time phasing may depend on rates of aging, death, and regeneration in each "wave."

In some cases a factor not directly controlled by the organisms in the community acts to modify the environment. Such forces are called allogenic and include continued deposition of sand along riverbanks or silt on lake bottoms. Autogenic and allogenic forces operate together in most successions. See BIOSPHERE, GEOCHEMISTRY OF; FOOD CHAIN.

In some cases successional changes seem to occur without habitat change and appear to be explained by characteristics of the life cycles of the organisms involved. In the old fields of the southeastern United States, horseweed dominates first-year fields because its seed require no cold treatment and because they ripen early enough to germinate in the autumn of the year the field is abandoned. Aster seed, although not requiring cold treatment either, ripen too late to germinate in the autumn. Important trends may depend on chemical inhibition or allelopathic relations.

Mosaic patterns. Succession proceeds at different rates in different parts of the habitat. New invaders, at first randomly occupying small spots, soon become centers for the establishment of clusters of some new species, or for inhibition of others. Thus a mosaic of vegetation pattern comes about which may long persist into a climax. Other patterns may develop as individual trees die and as gap phase replacement occurs in sunny and shaded sides or openings.

Zonation. Succession frequently progresses from the edge to the center of an open habitat, creating belts of different species. The belts in the center of such areas are still in the early stages of succession, while the edge has progressed to a later stage. Thus, communities which will succeed one another in time become laid out in spatial arrangements, making possible the study of successional stages across the different zones (Fig. 3). See VEGETATION.

Zonation, however, is not always evidence of succession, nor do the communities found in successive zones necessarily reflect the true course of succession. On a riverbank with a zonation of willows, cottonwoods, sycamores, ashes, and maples, each species occupies a slightly different site, with different degrees of flooding, drainage, and soil deposition. Hence they are not always successive stages in a sere but merely adjacent stands of trees in different habitats.

Succession-retrogression cycles. The progress of succession is often interrupted by natural or manmade disturbances which open up closed communities, clear much of the habitat, or kill off certain key species of the community. If this disturbance does not recur, succession is simply set back (retrogression) and starts again from an ear-

Fig. 3. Zonation of vegetation in the succession on open sand flat with jack pine as climax.

lier stage. Several examples of regular alternation of succession and retrogression in a cyclic pattern have been discovered. Such is the case in raised bogs, dunes, heaths, and certain arctic areas where the permafrost level fluctuates. However, if the disturbance continues or recurs at regular or irregular intervals, the community will adjust permanently and will stabilize in a kind of disturbance-controlled climax or mosaic pattern of alternating communities.

Convergence toward the climax. A key observation about the process of succession is that, within a given climatic region, the initial stages of various seres may differ greatly in environment and species composition; but as succession proceeds, later stages bear a progressively greater resemblance to one another, first in their physiognomy and later in species composition. This convergence is largely dependent upon the development of the soil, and research on succession must be firmly based upon a study of soil in different stages of genesis.

Succession and soil development. In most primary successions the initial substrate is not differentiated into soil horizons. In the course of succession, plants take up certain minerals and deposit organic matter on the surface, thus starting the process of soil development. Products of decay may infiltrate the soil to various depths, where they are deposited and can be reutilized. If the initial substrate is rocky, plant roots and rhizoids may break it up while their stems and leaves trap dust and sand particles. Gradually a layered soil is formed. *See* SOIL.

The climax. Ecological succession eventually leads to a community, the climax, with greater stability and permanence than any of the successional stages because such a community results from the ability of the species which compose it to reproduce and exclude potential dominants under the conditions which prevail in the community.

Climate and related factors. The basic characteristics of the climax are determined by climate. In humid climates some type of forest will develop with its basic properties determined by interactions of temperature and moisture. In more arid climates savanna, grassland, or even desert may be climax. Within a given climatic region the termination of succession may be related to various specific combinations of environmental conditions,

such as repeated fires, drainage patterns, and topographic factors. Thus a mosaic of several communities of similar appearance, but with different species composition, make up a regional complex of climaxes. *See* CLIMAX COMMUNITY.

It is known from the analysis of pollen grains found in lake deposits that for many thousands of years the vegetation of the Northern Hemisphere has been in continuous change as a result of increasing mildness of climate. Similar, but less notable, changes of climate have occurred during the past 100 years. These have led some ecologists to conclude that the necessary climatic stability for attainment of the climax does not exist. Such a position would lead to viewing the climax as a community in which the rate of change has slowed down to the point where it is essentially nondetectable with current methods.

Monoclimax concept. Early in the 20th century F. Clements proposed that, within a given climatic area, all seres would end in a single kind of climax community. Thus the regional climax would be identical for all seres within the area. This concept, known as the monoclimax concept, was based primarily on the observation that successions in different habitats tended to converge on a similar sort of climax community. Clements felt that over a long period of time the earlier differences in site conditions, such as soil texture and chemical composition of the parent rock, would be overcome and what were originally very different sites would then be sufficiently similar to support an identical regional climax community. Close examination of the theory showed that, although there were trends of the sort Clements proposed, certain environments never could become sufficiently similar to support the same climax community. For example, there are certain unalterable properties peculiar to each soil. A soil deficient in calcium will not improve no matter how many plants grow on it; in fact, calcium reserves as carbonate may disappear by solution. Although the texture of a soil may be slightly modified by additional organic matter, some basic properties are not changed and others may become less favorable. For these reasons convergence toward a uniform habitat with a uniform climax is never complete. Thus the monoclimax concept fell into disfavor.

Polyclimax concept. An alternative theory, the polyclimax concept, recognized that a variety of different sites might exist in an area and that each of these could potentially support a climax community. The polyclimax theory took note of four factors: (1) the effect of local relief, creating a mosaic rather than a homogeneous community cover; (2) topographic differences responsible for different topographic climaxes on various exposures in mountains; (3) soil inhibitions creating edaphic climaxes; and (4) climatic changes leaving relic communities as climaxes of the past, persisting in isolated refuges. Thus only a portion of the landscape may actually support the regional climax as viewed by Clements. Despite this, within a climatic area there generally is a complex of species which tend to be present frequently enough on similar sites so that the concept of a regional climax has a degree of validity. However, it should not be expected to develop on all sites within the area. Such regional climaxes are the units often portrayed on maps of community types of large sections of continents (Fig. 4).

Trends in succession. E. P. Odum has outlined a series of trends which summarizes important functional and compositional changes during succession. Species composition changes rapidly during the early stages of a sere and more slowly as the climax is approached. The total weight of living and nonliving organic matter increases through early succession but it may decrease later. Feeding relationships, generally simple at first, become more complex in some climax and many mixed stages. Early in some stages of succession a greater amount of organic matter tends to be produced by organisms than is used by them as food; consequently, organic matter accumulates. As succession proceeds, however, the number or activity (or both) of heterotrophic forms increases, causing an increase in the total amount of community respiration. Thus community production exceeds community respiration in some pioneer or transitional stages, and the two processes are theoretically balanced (averaging over space and tine) in a climax community. *See* BIOMASS; ECOLOGY, PHYSIOLOGICAL (PLANT).

Importance of succession. An understanding of succession is important for effective management of communities by man.

Man's activities generally have the effect of stopping succession at an early stage. Fire was used by most primitive people to keep forests suitable for hunting by clearing away the underbrush or to clear land for agriculture. Since almost all major crop plants are annuals of the pioneer stage, regular plowing is needed to hold down succession. Other methods of checking succession include mowing and grazing, which favor sod-forming grasses and kill most tree seedlings, and selective weed control with chemicals. *See* LAND-USE PLANNING.

Succession can be either retarded or accelerated by manipulation of drainage. Such manipulation may be used to establish a cranberry bog or a stand of productive timber. Many timber trees occur naturally as successional species rather than in climaxes. Thus the forest must be managed so as to permit these species to grow free from the competition of the climax species which will eventually replace them. Fire, herbicides, and cutting practices are all useful in regulating the growth and composition of forest stands so that early successional stages are maintained. *See* CONSERVATION OF RESOURCES; ECOLOGY; FOREST MANAGEMENT AND ORGANIZATION; VEGETATION MANAGEMENT; VEGETATION ZONES, WORLD.

[ARTHUR W. COOPER]

Bibliography: W. C. Allee et al., *Principles of Animal Ecology*, 1949; B. W. Allred and E. S. Clements (eds), *Dynamics of Vegetation*, 1949; E. L. Braun, *Deciduous Forests of Eastern North America*, 1950; F. E. Clements, *Plant Succession and Indicators*, 1928; R. Daubenmire, *Plant Communities*, 1968; K. A. Kershaw, *Quantitative and Dynamic Plant Ecology*, 1964; E. P. Odum, *Fundamentals of Ecology*, 2d ed., 1959; R. L. Smith, *Ecology and Field Biology*, 1966; J. E. Weaver and F. E. Clements, *Plant Ecology*, 1929.

Surface water

A term commonly used to designate the water flowing in stream channels. The term is sometimes used in a broader sense as opposed to "subsurface water." In this sense, surface water includes water in lakes, marshes, glaciers, and reservoirs as well as that flowing in streams. In the broadest sense, surface water is all the water on the surface of the Earth and thus includes the water of the oceans. Subsurface water includes water in the root zone of the soil and groundwater flowing or stored in the rock mantle of the Earth. Subsurface water differs from surface water in the mechanics of its movement as well as in its location.

Surface and subsurface water are two stages of the movement of the Earth's water through the hydrologic cycle. The world's ocean and atmospheric moisture are two other main stages of the grand water cycle of the Earth. At any time there is a certain quantity of water in the various stages of this cycle. For example, the 11.5×10^9 acre-ft of water in the atmosphere is much greater than the 0.25×10^9 acre-ft of water in the stream channels. *See* HYDROLOGY.

Considerably more may be outlined concerning water's rate of movement through the several stages of the hydrologic cycle. In some stages, glaciers for example, water is locked up for long periods of time; but water in the atmosphere or in the streams is transient. A numerical value for the time of transit of water is its detention period in years—specifically, the ratio between the bulk or volume of water in a given stage of the hydrologic cycle and its mean rate of flow through that stage.

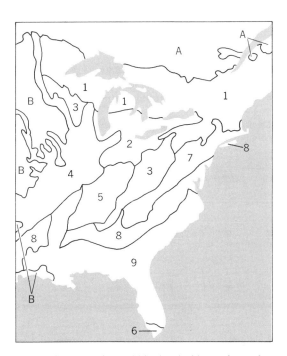

Fig. 4. Climax regions within the deciduous forest formation: 1, hemlock, northern hardwoods region; 2, beech-maple region; 3, maple-basswood region; 4, oak-hickory region; 5, western mesophytic region; 6, mixed mesophytic region; 7, oak-chestnut region; 8, oak-pine region; 9, southeastern evergreen forest region; A, boreal needleleaf forest formation; B, the grassland formation. (*Redrawn after E. L. Braun, 1950*)

For example, the Earth's supply of water as a whole amounts to about $1,100,000 \times 10^9$ acre-ft. The mean rate of flow through the hydrologic cycle is of the order of 320×10^9 acre-ft per year. Thus the detention period is about 3000 years; that is, on the average, each particle of the Earth's supply of water partakes in movement through that cycle once in 3000 years. This is an average—some particles may move more than once, some part way, and some not at all in this period of time.

Table 1 gives estimates of the amounts of water in various parts of the hydrologic cycle and their detention periods.

It may be noted that surface water on the continents is but a small part of the world's water and that the bulk of that is in fresh-water lakes. However, the detention period is also short. This means that the surface-water part, and especially the water in the streams, is rapidly discharged and replenished. That is why surface water, as well as the shallower groundwater, is called a renewable resource. Water that has a detention period of more than a generation is not renewed within sufficient time to be so considered. *See* GROUNDWATER; RIVER; TERRESTRIAL FROZEN WATER.

Source of water in streams. Precipitation that reaches the Earth is subdivided by processes of evaporation and infiltration into various routes of subsequent travel. Evaporation from wet land surfaces and from vegetation returns some of the water to the atmosphere immediately. Precipitation that falls at rates less than the local rate of infiltration enters the soil. Some of the infiltrated water is retained in the soil, sustaining plant life, and some reaches the groundwater.

Because of the slope of the land surface, the precipitation that exceeds the capacity of the soil to absorb water flows overland in the direction of the steepest slope and concentrates in rills and minor channels. During storms most of the water in surface streams is derived from that portion of the precipitation which fails to infiltrate the soil. In some forested areas of high relief and probably in some other areas, stormflow in stream channels is composed, in large part, of water which was infiltrated into the surface soil but which moved rapidly through the surficial mantle of litter and humus to the channels.

When streams are low, on the other hand, the bulk of the water in channels is the contribution of groundwater derived from precipitation that infiltrated during storms. The flow of surface streams during rainless periods represents the gradual draining of water stored temporarily in the ground. Dry-weather streamflow is the overflow of a groundwater reservoir.

The distinction between surface and subsurface water, though useful, should not obscure the fact that water on the surface and water underground is physically connected through pores, cracks, and joints in rock and soil material. In many areas, particularly in humid regions, surface water in stream channels is the visible part of a reservoir, which is partly underground; the water surface of a river is the visible extension of the surface of the groundwater.

Disposition of precipitation. Streamflow represents only a small percentage, on the average, of the water that falls as precipitation. The flow in streams under natural conditions is called runoff. The ratio of average annual runoff to average annual precipitation in the United States ranges from 20 to 40% in humid parts of the United States and from 2 to 4% in semiarid areas. On the average, the annual budget of water over the United States is roughly as follows:

Average precipitation	30 in.
Runoff by rivers to sea	9 in.
Evapotranspiration from plants and soil	21 in.
Transport of atmospheric moisture from oceans to continental area	9 in.

The 30 in. of water contributed by precipitation must be balanced by a return of water to the atmosphere. There are 21 in. returned to the atmosphere by evapotranspiration from the continental area and 9 in. flow to the oceans. Thus, to balance the atmospheric budget, the 9 in. of water that is transported as vapor from the oceans to the continents must be included.

Average runoff. There are great geographic variations within this average balance as can be visualized by a map showing annual runoff in the United States. The total runoff is greatest in areas of highest precipitation and lowest losses. In the mountains of the Northwest, where precipitation is as much as 150 in. annually, the runoff in surface streams is more than 50 in. annually over a considerable part of the high mountain country.

Much of the semiarid parts of New Mexico, Arizona, and parts of California and Nevada yield an annual runoff of only a few tenths of an inch from a precipitation of 4–8 in. In the mountainous parts of the same areas, precipitation reaches 25–30 in. and, locally, the runoff may average as much as 10 in. annually from small areas. Much of the United States west of the 100th meridian has an average annual runoff of less than 3 in.

Discharge in nearly all streams has a marked annual cycle. Spring and early summer are usually periods of high flow resulting from snowmelt and rain during a period of relatively low water loss by

Table 1. Distribution of the world's supply of water

Location	Volume of water, 10^9 acre-ft	Percentage of total	Detention period, years
World's oceans	1,060,000	97.39	5,000
Surface water on the continents			
Glaciers and polar ice caps	20,000	1.83	2,000
Fresh-water lakes	100	0.0093	100
Saline lakes and inland seas	68	0.0063	50
Average in stream channels	0.25	0.00002	0.05
Total surface water	20,200		700 av
Subsurface water on the continents			
Root zone of the soil	10	0.00094	0.25
Groundwater above 2500 ft	3,700	0.339	5
Groundwater below 2500 ft	4,600	0.425	100
Total subsurface water	8,300		
Atmospheric water	115	0.0011	0.03
Total world water (rounded)	1,088,000	100	3,000

Table 2. Representative values of extreme peak flows

River	Date	Drainage area, mi^2	Peak discharge cfs	Peak discharge cfs/mi^2
Big Branch near Waynesville, N.C.	Aug. 30, 1940	0.4	4,500	11,000
Big Creek near Waynesville, N.C.	Aug. 30, 1940	1.32	12,900	9,800
Laurel Creek above White Pine, W.Va.	August, 1943	2.42	7,400	3,060
Cameron Creek near Tehachapi, Calif.	Sept. 30, 1932	3.59	13,500	3,760
Unnamed Creek near York, Nebr.	July 9, 1950	6.93	23,000	3,320
Meyers Canyon near Mitchell, Ore.	July 13, 1956	12.7	54,500	4,290
Alazan Creek, below Martinez Creek, Tex.	Sept. 9, 1921	17.1	25,900	1,510
Salem Creek below Woodstown, N.J.	Sept. 1, 1940	17.5	26,100	1,490
Morgan Creek near Chapel Hill, N.C.	Aug. 4, 1924	29.1	30,000	1,030
Pine Tree Canyon, 12 mi north of Mohave, Calif.	Aug. 12, 1931	35.0	59,500	1,700
Elkhorn Creek, Keystone, W.Va.	June, 1901	44	60,000	1,360
Little Nemaha River at Syracuse, Nebr.	May 9, 1950	218	225,000	1,030
Guadalupe River near Ingram, Tex.	July 1, 1932	336	206,000	613
W. Nueces River near Brackettville, Tex.	June 14, 1935	402	580,000	1,440
W. Nueces River near Cline, Tex.	June 14, 1935	880	536,000	609
Eel River at Scotia, Calif.	Dec. 22, 1955	3,113	541,000	174
Devils River near Del Rio, Tex.	Sept. 1, 1932	4,060	597,000	147
Neosho River near Parsons, Kans.	July 14, 1951	4,817	410,000	85.1
Little River at Cameron, Tex.	Sept. 10, 1921	7,000	647,000	92.4
Ohio River at Sewickley, Pa.	Mar. 18, 1936	19,500	574,000	29.4
Susquehanna River at Marietta, Pa.	Mar. 19, 1936	25,900	787,000	30.4
Ohio River at Evansville, Ky.	Jan. 29, 1937	107,000	1,410,000	13.2
Ohio River at Metropolis, Ill.	Feb. 1, 1937	203,000	1,850,000	9.1
Columbia River at The Dalles, Ore.	June 6, 1894	237,000	1,240,000	5.2

transpiration. Late summer is often the period of lowest flow owing to the infrequency of precipitation and to the maximum use of water by leafy vegetation. The distribution of runoff throughout the year is not the same for the whole country, however, because it varies regionally depending on the seasonal distribution of precipitation and depending on the importance of snowmelt as a source of runoff.

Individual streams and rivers in the United States have been measured over varying periods of time at about 12,000 sites. Daily discharge at most of them is published by the U.S. Geological Survey (USGS) in the series of Water-Supply Papers entitled *Surface Water Supply of the United States.* The data are tabulated for each measuring station and are grouped in volumes by river basins.

If a stream goes dry occasionally or does not have enough flow to satisfy a desired use, storage reservoirs can be built to conserve high flows for release during low-flow periods. The design of such reservoirs requires a knowledge of how low a flow is likely to be experienced, how long it may last, and how frequently it can be expected to recur. Streamflow records can be analyzed to answer these questions.

Extremes of runoff. The amount of streamflow available during periods of extremely low flow can be shown by flow-duration curves and other types of low-flow frequency analysis. Flow-duration curves, which express the per cent of time the flow of a particular stream has been equal to or greater than any given quantity, have long been used in water-power studies, and curves showing the frequency and severity of annual lows are now being used in water-supply studies and stream-sanitation studies. The preparation of such curves is facilitated by the use of electronic computers. Many such

computations have been made and published, but there is no single source of systematic publication of such material. Summaries of published data on low flows for some rivers in the United States may be obtained from the USGS.

Much more information is organized and published on floods than on low flows. Representative values of peak discharges showing the magnitude of extreme flows experienced in drainage basins of various sizes are given in Table 2.

Data similar to those included in Table 2 are published in a systematic manner under the headings of "extremes" in the tabulated flow data for individual gaging stations in the Water-Supply Papers of the USGS. Flood expectancy, even for ungaged places, may be estimated from curves the USGS is publishing in a series of papers dealing with the frequency and magnitude of floods by individual states or areas.

Relation of runoff to drainage area. Average water yield or annual runoff in a physically homogeneous area increases in direct proportion to the size of the drainage basin, but this is not true of flood potentiality. Small drainage basins produce larger peak flows per unit of drainage area than do large basins. The relation between magnitude of flood peak and the contributing drainage area may be expressed by Eq. (1), where Q is discharge

$$Q = aA^c \qquad (1)$$

in cubic feet per second (cfs), A is drainage area in square miles (mi^2), a and c are coefficients. If Q represents the average annual water yield, then the value of c is approximately unity for most basins in humid areas. If Q represents peak flood discharge of a given frequency or recurrence interval, such as a 10- or 50-year average recurrence interval, then past observations indicate that the

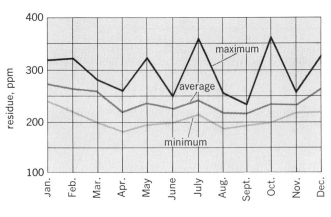

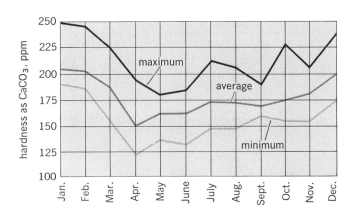

Mississippi River at Minneapolis, Minn.

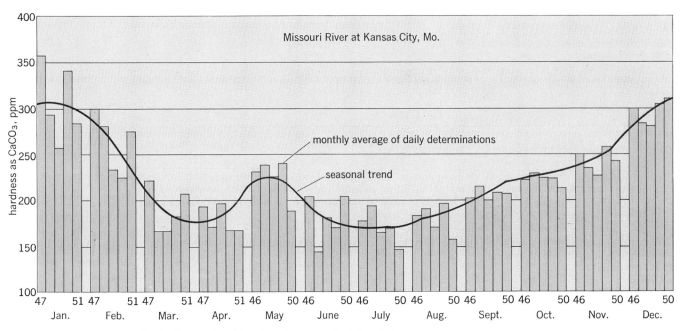

Fig. 1.   Seasonal variation in quality characteristics of streams.

value of c is between 0.7 and 0.8 for a large variety of drainage basins.

Some reasons for these differences in exponents are as follows. Each square mile of a homogeneous drainage basin contributes an equal quantity of water over the course of many years. Therefore, average annual runoff from a drainage basin is the summation of the contribution of each unit area.

Peak rates of runoff during floods, on the other hand, may be viewed roughly as the contributions from the parts of the drainage basin at varying distances away. The different distances mean that the peak flood contributions reach the several points downstream at different times, and the farther downstream one goes, the greater the floodwave is spread out. The flattening action is increased by the temporary storage of water in the stream channels and in the bordering floodplains. As it moves downstream, the flood crest is increased by the contributions of tributaries, but because these contributions are not synchronized, the peak contributions are not simply additive. Owing to the delays due to channel storage, flood-peak dis-

charges increase with drainage area, but to a power less than unity; peak flows per unit of drainage area decrease with drainage area.

Quality of water. The usefulness of available water is often limited by its quality. Good quality often is considerably more important than unlimited quantity, particularly in industry.

It is characteristic of river waters to vary in chemical and physical quality almost continuously. The chemical quality of the water in lakes, particularly large ones, remains relatively constant throughout the year. Differences between streams are caused by several factors. These include (1) the nature of soils and rocks over and through which the water flows, (2) the length of time the water is in contact with various rock types, (3) the water quality of tributary streams, (4) proportion of flow due to groundwater discharge, (5) flood and drought conditions, and, of course, (6) man-made pollution. Figure 1 illustrates seasonal variations in amounts of dissolved substance in two large streams.

Figure 2 illustrates variation in dissolved solids

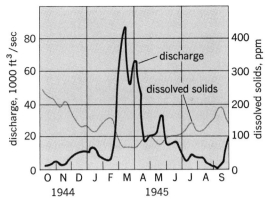

Fig. 2. Discharge and dissolved solids in Allegheny River at Kitanning, Pa.

mittent streams of arid regions. The sediment concentration of the Colorado River at Grand Canyon, Ariz., averaged about 0.6% (6000 ppm) for a period of nearly 30 years.

Data on the chemical and sediment loads of streams in the United States are published systematically in the Water-Supply Papers of the USGS. *See* STREAM TRANSPORT AND DEPOSITION.

Characteristics of river channels. Rivers and streams form channels which are the routes of transport for water and debris-load delivered to them by the basin during the slow process of landscape degradation. Channels have certain characteristics that are amazingly universal regardless of the location of the river basin. The basic mechanics which lead to these common characteristics are only imperfectly understood.

Water only partly fills the channel during periods of low flow. Generally, on more than half the days in a year, only the lowest one-fifth of the channel depth is filled with water. The channel flows bankful about once a year on the average. Though this varies somewhat from reach to reach and from one river basin to another, the generalization that the channel is constructed by the river so that it overflows once every year or every 2 years is one of the most interesting and potentially useful items of information about natural channels. Because a flood is, by definition, a flow which exceeds channel capacity, the above generalization emphasizes the fact that flooding is a natural characteristic of rivers.

Natural channels tend to be roughly trapezoidal rather than elliptical or semicircular. The width of the water surface along the channel at any given frequency of flow (high flow or low flow) generally increases as the square root of the discharge, as

with streamflow. In general, streams flowing into the Atlantic Ocean, eastern Gulf of Mexico, and the northern Pacific Ocean are of good to excellent quality for general purposes. Dissolved matter and hardness are usually below 100 parts per million (ppm) (milligrams of dissolved substance in a liter of water) and often below 50 ppm. This applies to streams in their natural state and does not take into account the effects of pollution.

Midcontinent and southwestern streams generally have high concentrations of dissolved matter, some excessively so, and must be treated extensively for a large variety of general uses. For example, the water of the Missouri River at Kansas City has about 220–500 ppm of dissolved matter and hardness of about 190–300 ppm during a typical year. A generalized picture of the variations in surface-water quality throughout the United States is given in Fig. 3.

The temperature of streams varies continuously throughout the year, ranging from a minimum of freezing (about 32°F) in winter in northern latitudes to a maximum of about 90°F in summer in southern latitudes. The monthly average water temperature generally follows rather closely the monthly average air temperatures, except in areas where the flow is made up largely of melting snow or ice, or of groundwater. The water temperature of streams fed by snowmelt is lower than the air temperature for some distances downstream from the snow fields. The water temperature of streams whose flow comes from groundwater tends to be more uniform and in summer months is usually colder than the average air temperature.

Sediment also affects surface-water quality. Nearly all streams are turbid during flood periods, some carrying tremendous quantities of sediment which must be removed before the water is suitable for industrial and most other uses. In eastern United States the amount of suspended matter in typical streams seldom exceeds 0.3% (3000 ppm) and generally averages a few hundred parts per million. In many midcontinent and western streams sediment concentrations are much greater, a maximum of 10% not being uncommon in some streams; frequently the maximum is considerably higher.

The sediment-carrying characteristics of western streams range from relatively clear-flowing mountain streams to near mud flows in the inter-

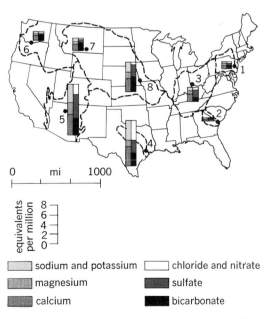

Fig. 3. Chemical composition of water in several basins. Data are for (1) Delaware River at Trenton, N.J., 1951; (2) Savannah River near Clyo, Ga., 1940; (3) Ohio River at Cincinnati, Ohio, 1951; (4) Brazos River at Richmond, Tex., 1951; (5) Colorado River near Grand Canyon, Ariz., 1951; (6) Columbia River near Rufus, Ore., 1951; (7) Yellowstone River at Billings, Mont., 1951; and (8) Missouri River at Nebraska City, Nebr., 1951.

discharge increases downstream with the addition of tributaries. The shape of the channel cross section is asymmetric at bends, the deepest part being near the concave bank.

The depth increases downstream but not as rapidly as the width. The width-depth ratio increases downstream as about the 0.1 power of the discharge as given by expression (2), where w and d

$$\frac{w}{d} \propto Q^{0.1} \qquad (2)$$

are respectively mean width and mean depth, and Q is discharge of a given frequency.

Channel slope or gradient decreases downstream generally following an exponential or logarithmic law. Also, size of debris composing the bed tends to diminish downstream. Despite the decreasing slope in the downstream direction, mean water velocity increases slightly along the length of a river or maintains a roughly constant value.

[LUNA B. LEOPOLD]

Bibliography: V. T. Chow (ed.), *Advances in Hydroscience*, vols. 1 and 2, 1964 and 1965; V. T. Chow (ed.), *Handbook of Hydrology*, 1964; W. G. Hoyt and W. B. Langbein, *Floods*, 1955; W. B. Langbein, *Annual Runoff in the United States*, USGS Circ. no. 52, 1949; L. B. Leopold and W. B. Langbein, *A Primer on Water*, USGS, 1959; L. B. Leopold and T. Maddock, Jr., *The Hydraulic Geometry of Stream Channels and Some Physiographic Implications*, USGS Prof. Pap. no. 252, 1953; O. E. Meinzer (ed.), *Hydrology*, 1949; R. L. Nace, *Water Management, Agriculture, and Ground-Water Supplies*, Annu. Meet. AAAS, 1958.

Swamp, marsh, and bog

Wet flatlands, where mesophytic vegetation is areally more important than open water, are commonly developed in filled lakes, glacial pits and potholes (see illustration), or poorly drained coastal plains or floodplains. Swamp is a term usually applied to a wet land where trees and shrubs are an important part of the vegetative association, and bog implies lack of solid foundation. Some bogs consist of a thick zone of vegetation floating on water.

Unique plant associations characterize wet lands in various climates and exhibit marked zonation characteristics around the edge in response to different thicknesses of the saturated zone above the firm base of soil material. Coastal marshes covered with vegetation adapted to saline water are common on all continents. Presumably many of these had their origin in recent inundation due to post-Pleistocene rise in sea level.

The total area covered by these physiographic features is not accurately known, but particularly in glaciated regions many hundreds of square miles are covered by marsh. *See* MANGROVE SWAMP.   [LUNA B. LEOPOLD]

Systems ecology

The combined approaches of systems analysis and the ecology of whole ecosystems and subsystems, when viewed as interdependent and functionally interacting components. Among the organisms and organs in any subdivision of vegetation or animal communities or microorganisms, there are many relations from one part to another and from each to the local environment. Physiological functions and other relations can sometimes be expressed in terms of mathematical functions. These need not be quantitative functions (one-to-one correspondence for real numbers) but may express logical alternatives (all-or-nothing) in the direction of flow of material or energy; they may express also alternative pathways of development or direction of change in the organic or inorganic parts of the system. *See* ECOLOGY; ECOLOGY, PHYSIOLOGICAL (PLANT); ECOSYSTEM.

SYSTEMS ANALYSIS AND OPERATIONS RESEARCH IN BIOLOGY

Operations research is generally oriented toward maximizing some benefit from a system whose operation and immediate objective, or several kinds of long-term goals, are already fairly well understood. The constraints of nature, economics, and perhaps other social values must be formulated explicitly, usually quantitatively. Systems analysis may enter sooner to formulate the goals and the most relevant aspects of the system for disciplining the research. To reveal the operation's values and constraints may be harder than reasoning from them. Problems relating to systems of landscapes, water bodies, and people are still very challenging. Special cases are being worked out, but the most reliable general approaches are still not established. Teams including a number of ecologists, resource specialists, and mathematical consultants are tackling such problems with several relatively new methods. *See* ECOLOGY, APPLIED.

Motivations and style have varied notably. An intellectual goal is probing the analogies of mathematical formalism and ecological structure. This has been fascinating but also frustrating because the most relevant information for tests "in nature" seem to be lacking. Recognition of the weakness of information as presently organized has guided the search for missing data, well-completed case studies, and new research. Estimations and deci-

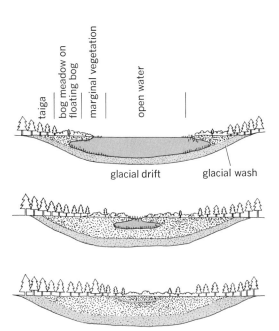

Cross-sectional diagram representing the progressive filling by vegetation of a pit lake in recently glaciated terrain. (*After C. A. Davis*)

sions about hypotheses and action have put high priority on harnessing computer power for massive data sets. Analysis of available but hitherto incompletely processed data is helpful in an orientation stage. Yet "postmortem" treatment of results that were sought for one purpose usually cannot make up for differences between current needs and the need that was originally structured for other goals. Hence investigators are coming to realize the need for very early partnership of systems analysts and experimenters in the environmental research of the future. This is important for individual projects and even more so for research and development programs with many projects. *See* WILDLIFE CONSERVATION.

Watt, Holling, and others further illustrate the background of these problems. Their own early studies emphasize promising directions of approach to submodels that ask more specific questions, emphasizing certain processes or target populations that are known (or revealed in the preliminary analysis to be particularly important in the context of a large resource problem).

Networks: qualitative structure. A simple network model of an ecosystem's structure consists of closed or open connections (links) between the network's nodes. The latter represent organisms or environmental variables, or both. Some kinds of damage for natural change for an ecosystem embody the "turnoff" relation of a switching circuit. Disconnection is formalized by a change from 1 (meaning connected) to 0 in the simple algebra of George Boole. There is also an isomorphism or analogy between Boolean algebra and circuits and sets (1 = intersection; 0 = nonintersecting). Verbal logic (1 = true statement; 0 = false statement) should aid in the understanding of a simple description of the ecosystem, the basic ecological concepts themselves, and the digital operation of computer tools which are being increasingly used in ecology.

With the preceding notions, the ecosystem concept can be defined in several ways. Such a system may be called the sum or union of specified sets of organisms, of habitat features, and of certain signals (photon input, for example) crossing the arbitrary boundaries which describe the horizontal and vertical limits of our system. Instead of stressing concrete objects, the same ecosystem may be redefined as a union of variables representing states and transformations of this local ecosystem.

Formally, the state of an ecosystem can be specified by a point in n-dimensional space, or by the equivalent idea of an n-component vector in this space. Abstractly, a vector is an ordered set of variables with specific components that are examined in some detail under separate articles. *See* FRESH-WATER ECOSYSTEM; MARINE ECOSYSTEM; TERRESTRIAL ECOSYSTEM.

Much of systems ecology then proceeds to consider how such a vector actually changes: during the formation and subsequent development of a new ecosystem; during the phase of stationary state or dynamic equilibrium, if it can be attained; and during changes in any such equilibrium that are brought about by natural disruption or by human management of resources.

Networks: quantitative change. The foregoing qualitative structure of a web of life and environment as a network can sometimes be carried rather

directly to a quantitative expression of change, and rates of change. For the most pertinent environmental properties or living components, different variables describe the state at any one time. The state vector mentioned above moves through n-dimensional space to describe how these variables change as a function of time. To each point along the time axis (the independent variable) there corresponds one point along the axis of each dependent variable. The simultaneous changes of several dependent variables imply various kinds of functions or other mutual relations between them, for example, the successional changes of an ecosystem which were described either as fairly abrupt or gradual transformation from one condition or state (named community type) to another as functions of time—either discrete or continuous changes. *See* SUCCESSION, ECOLOGICAL.

It is clearly important to distinguish systems with steady-state from systems which have not even approximately attained a balance of income and loss for their most important parts. Equally important differences may exist between exceedingly slow changes, which are hardly distinguishable from a constant state, and changes which are more or less rapid. The rate of change is as fundamental for the concept of ecological succession and development as is the question of whether system change exists or not (that is, that rates for all parts either do or do not equal zero).

An average rate of change per increment or unit of time can be expressed for both discontinuous and continuous processes. In dealing with continuous change in time t or space s, the step size of either independent variable ($\Delta t, \Delta s$ or $\Delta x, \Delta y, \Delta z$ in a three-dimensional coordinate system) is diminished toward a limit of zero so that an average rate of change of a variable v approaches the instantaneous rate or derivative of the original function. This is represented by v, v', dv/dt, or $\partial v/\partial t$. The original function v is the integral of its own rate of change. The classic models of population ecology are related to these elementary concepts of calculus, and to very simple differential equations whose positive terms are related to birth (or immigration) b, and negative terms to death or loss. Readers having no prior instruction on derivatives and integrals, or having only hazy recall of the notation, should be able to follow many extensions of this basically simple balance or imbalance of rates if they take time to review the background material. Other sections well illustrate the key ideas of simulating behavior of real systems with a vividness that should aid the grasp of the mathematical background. Extension of simulation models to some corresponding ecological topics can be made clear not only for populations of individual species but also for multiple species–environment systems, without restriction to linear systems of equations. *See* POPULATION DYNAMICS.

Chance processes in ecosystems. The deterministic models of calculus and differential equations initially neglect the aspect of chance variation which is realized to be very important in the processes of nature. Concepts of probability and random variables are important extensions of concepts of sets and functions which have helped to overcome this limitation. Probability models describe the invasion, survival, and dispersal of pioneer species which create and change a new ecosystem in a hitherto barren area. For example, the Poisson process is one of several stochastic processes which describes events occurring independently of one another in time (or space) with uniform probability of a happening over the whole domain. The Poisson distribution of the number of happenings, and other distributions (binomial, normal, exponential, contagious), illustrate the consequence of independence and uniformity, or lack thereof, when multiple events are considered. For example, the arrival and survival of propagules and organisms may be independent and uniform at first, but interdependent (or nonuniform, or both) later on.

MAINTENANCE OF ECOSYSTEMS

Once an ecosystem has been created, how is it maintained? It is obviously not a static system, but nevertheless tends (in many cases) to approximate a dynamic equilibrium or to fluctuate around a stationary state. Incomes and losses (birth and death of populations, intake and expenditure of energy, uptake and return of nutrients in exchange with the environment) tend to become equal, by definition, if an ecosystem reaches a stage of approximate balance in nature. A reexamination of trends of such incomes and losses will clarify loose concepts of stability in nature. For a mathematical formulation of the much discussed ecological concept of climax, one asks: What kinds of balances must all be maintained as necessary and sufficient conditions for some approximation to zero rate of change? More broadly, one also must consider processes that maintain the metabolism of an ecosystem, even if it is not quite in a steady state, that is, if there is a net rate of change which is equal to income rate − loss rate. Thus some topics formerly introduced here were already mentioned. *See* CLIMAX COMMUNITY.

Matrix formulation of state and change of state. An additional dimension generalizing the vector concept provides formal coherence to many topics in a shorthand that is conveniently available for matrices. The previous description of the ecosystem as a vector was a one-dimensional list or array of variables describing the system's components. A simple table or two-dimensional array briefly reviews some of the kinds of quantities describing each component, and is called a matrix. Row and column definitions are basic to the manipulation of matrices. Learning them is well repaid, not only by simple notation but by power of programming data in arrays in a digital computer.

With such tools matrices and statistics can be used to extend the concepts of process and also of pattern of populations in space and time. Aside from the small-scale (random, contagious, regular) patterns within a seemingly homogeneous ecosystem, there are larger-scale patterns of the landscape or of water bodies which maintain themselves because of the dependence of organisms, soils, and microclimates on certain independent variables of the landscape. Regression methods, most conveniently summarized in matrix form, provide the analysis of observational data to estimate functions and relations of the local ecosystem to the conditioning factors in the larger ecosystem which surrounds it.

Energy flow and cycles of nutrients. Within a local unit or module within this large-scale pattern of many local ecosystems distributed over the lands and waters of the Earth, the following prerequisites are needed to maintain them, in either a steady or not-so-steady state. Since these systems constantly dissipate energy (as long as they are alive), they require replenishment of this energy from outside, and a redistribution of food throughout the food-web network. Rates of this redistribution can be represented by either a deterministic or a stochastic matrix, according to the ecologist's stage of model building. The matrix for nutrient redistribution differs from that for energy flow because of the recycling of nutrients and contaminants within the ecosystem. The relations between energy flow and the second law of thermodynamics are expressed by the continuing loss of low-grade heat or dispersed energy into an "absorbing barrier." *See* ECOLOGY.

Matrix formulation of ecosystem change or stability. The previous generalization from one variable at a time to several variables can also be formulated to cover the redistribution of land or water area (or volume in three dimensions) as a community complex undergoes successional change. In this case, the vector variable in Eq. (1)

$$\mathbf{v} = (v_1, v_2, \ldots, v_n) \qquad (1)$$

may be taken as representing the area covered by each type of ecosystem. The same notations could be used for the preceding questions for representing the amount of a given nutrient or energy on any one landscape unit or type.

Redistribution is controlled for area, material, or energy in the following way. (Several related formulations can be applied to any one of these cases.) In essence, all reduce to an operator matrix which is simply a table giving the rates of transfer from the condition represented by row j to that represented by column k, as shown in notation (2).

$$\mathbf{P} = \begin{bmatrix} p_{11} & p_{12} & p_{13} & p_{14} \\ p_{21} & p_{22} & p_{23} & p_{24} \\ p_{31} & p_{32} & p_{33} & p_{34} \\ p_{41} & p_{42} & p_{43} & p_{44} \end{bmatrix} \qquad (2)$$

Thus p_{12} is a number (perhaps a function of time) that represents the rate of transfer from condition, or part, 1 to 2; p_{13} represents transfer from 1 to 3; p_{14} represents transfer from 1 to 4 if the state of the system is represented by a horizontal row, as in Eq. (1). (If the state is represented by a column vector, the rules of matrix operation simply require reversing the left-right convention of subscript notations, which is equivalent to transposing the matrix along its main diagonal from upper left to lower right.)

The main diagonal terms here represent the amount which remains in 1 at the end of the time increment, which must of course be specified to avoid confusion. Thus p_{11} represents the variable initially in v, which remains in v_1 after a step; p_{22} similarly for v_2, and so on. With these conventions, the rules of matrix multiplications represent the state of the system from the combination of all incomes and losses between time t and $t + \Delta t$, which shall here be called $t + 1$, assuming unit steps of time as in Eq. (3). This was discussed as if the

$$[v_1\, v_2\, v_3\, v_4]_t\, P = [v_1\, v_2\, v_3\, v_4]_{t+1} \qquad (3)$$

process were exactly determined, but many of the

same properties would carry over in stochastic processes. If the state depends only on the previous state and the matrix, then a Markov process is implied, but this is not a necessary or even a realistic restriction for many ecological situations, any more than for technological situations.

Similarly the hypothesis of constant probability (homogeneous stochastic process) or constant coefficients may be relaxed in real applications, even though most mathematical solutions in closed form are developed for those restrictions.

[JERRY S. OLSON]

Bibliography: M. S. Bartlett, *Stochastic Population Models in Ecology and Epidemiology*, 1960; T. W. Kerlin, *An Introduction to the Methods of Systems Analysis*, Oak Ridge Nat. Lab. Rep. ORNL-NEUT-3028-1, 1969; A. J. Lotka, *Elements of Mathematical Biology*, 1956; R. V. O'Neill and J. Hett, *A Preliminary Bibliography on Mathematical Modeling in Ecology*, Oak Ridge Nat. Lab. Int. Biol. Prog. Rep. ORNL-IBP-70, 1970; E. C. Pielou, *An Introduction to Mathematical Ecology*, 1969; D. E. Reichle (ed.), *1970 Ecological Studies, 1: Analysis of Temperate Forest Ecosystems*, 1970; K. E. F. Watt (ed.), *Ecology and Resource Management*, 1968; R. H. Whittaker, *Communities and Ecosystems*, 1970; G. M. Woodwell and H. H. Smith (eds.), *Diversity and Stability in Ecological Systems*, Brookhaven Symp. Biol. no. 22, BNL 50175 (C-56), 1969.

Taiga

A zone of forest vegetation encircling the Northern Hemisphere between the arctic-subarctic tundras in the north and the steppes, hardwood forests, and prairies in the south. The chief characteristic of the taiga is the prevalence of forests dominated by conifers. The taiga varies considerably in tree species from one major geographical region to another, and within regions there are distinct latitudinal subzones. The dominant trees are particular species of spruce, pine, fir, and larch. Other conifers, such as hemlock, white cedar, and juniper, occur locally, and the broad-leaved deciduous trees, birch and poplar, are common associates in the southern taiga regions. Taiga is a Siberian word, equivalent to "boreal forest." *See* FOREST AND FORESTRY; TUNDRA.

Climate. The northern and southern boundaries of the taiga are determined by climatic factors, of which temperature is most important. However, aridity controls the forest-steppe boundary in central Canada and western Siberia. In North America there is a broad coincidence between the northern and southern limits of the taiga and the mean summer and winter positions of the arctic air mass. In the taiga the average temperature in the warmest month, July, is greater than 50°F, distinguishing it from the forest-tundra and tundra to the north; however, less than four of the summer months have averages above 50°F, in contrast to the summers of the deciduous forest further south, which are longer and warmer. Taiga winters are long, snowy, and cold — the coldest month has an average temperature below 32°F. Permafrost occurs in the northern taiga. It is important to note that climate is as significant as vegetation in defining taiga. Thus, many of the world's conifer forests, such as those of the American Pacific Northwest, are excluded from the taiga by their high precipitation and mild winters.

Subzones. The taiga can be divided into three subzones (see illustration) in almost all of the regions which it occupies; these divisions are recognized mainly by the particular structure of the forests rather than by changes in tree composition. These subdivisions are the northern taiga, the middle taiga, and the southern taiga.

Northern taiga. This subzone is characterized on moderately drained uplands by open-canopy forests, dominated in Alaska, Canada, and Europe by spruce and in Siberia by spruce, larch, and pine. The well-spaced trees and low ground vegetation, usually rich lichen carpets and low heathy shrubs, yield a beautiful parkland landscape; this is exemplified best in North America by the taiga of Labrador and Northern Quebec. This subzone is seldom reached by roads and railways, in part because the trees seldom exceed 30 ft in height and have limited commercial value. These forests are important as winter range of Barren Ground caribou, but in many parts of the drier interior of North America their area has been decreased by

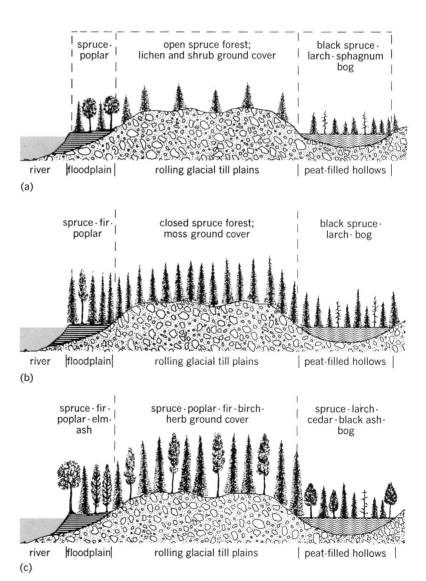

A schematic profile of the three main subzones of the North American taiga, showing the main forest assemblages on three of the more important landform types. (*a*) Northern taiga. (*b*) Middle taiga. (*c*) Southern taiga.

fire, started both by lightning and man.

Middle taiga. This subzone is a broad belt of closed-canopy evergreen forests on uplands. The dark, somber continuity is broken only where fires, common in the drier interiors of the continents, have given temporary advantage to the rapid colonizers, pine, paper birch, and aspen poplar. The deeply shaded interior of mature white and black spruce forests in the middle taiga of Alaska and Canada permits the growth of few herbs and shrubs; the ground is mantled by a dense carpet of mosses. Here, as elsewhere in the taiga, depressions are filled by peat bogs, dominated in North America by black spruce and in Eurasia by pine. Everywhere there is a thick carpet of sphagnum moss associated with such heath shrubs as bog cranberry, Labrador tea, and leatherleaf. Alluvial sites bear a well-grown forest yielding merchantable timber, with fir, white spruce, and black poplar as the chief trees.

Southern taiga. This subzone is characterized on moderately drained soils throughout the Northern Hemisphere by well-grown trees (mature specimens up to 95 ft) of spruce, fir, pine, birch, and poplar. These trees are represented by different species in North America, Europe, and Siberia. Of the three taiga zones, this has been exploited and disturbed to the greatest extent by man, and relatively few extensive, mature, and virgin stands remain. In northwestern Europe this subzone has been subject to intensive silviculture for several decades and yields forests rich in timber and pulpwood. In Alaska and Canada the forests have a much shorter history of forest management, but they yield rich resources for the forest industries. *See* FOREST MANAGEMENT AND ORGANIZATION.

Fauna. In addition to caribou, the taiga forms the core area for the natural ranges of black bear, moose, wolverine, marten, timber wolf, fox, mink, otter, muskrat, and beaver. The southern fringes of the taiga are used for recreation. *See* VEGETATION ZONES, WORLD. [J. C. RITCHIE]

Bibliography: A. Bryson, *Geographical Bulletin*, vol. 8, 1966; *Good's School Atlas*, 1950; *New Oxford Atlas*, 1951; W. Scotter, *Effects of Forest Fires on the Winter Range of Barren-ground Caribou in Northern Saskatchewan*, Wildlife Manage. Bull. no. 18, Canadian Wildlife Service, 1965.

Temperature inversion

The increase of air temperature with height; an atmospheric layer in which the upper portion is warmer than the lower. Such an increase is opposite, or inverse, to the usual decrease of temperature with height, or lapse rate, in the troposphere of about 6.5°C/km or 3.3°F/1000 ft, and somewhat less on mountain slopes. However, above the tropopause, temperature increases with height throughout the stratosphere, decreases in the mesosphere, and increases again in the thermosphere. Thus inversion conditions prevail throughout much of the atmosphere much or all of the time, and are not unusual or abnormal. *See* ATMOSPHERE.

Inversions are created by radiative cooling of a lower layer, by subsidence heating of an upper layer, or by advection of warm air over cooler air or of cool air under warmer air. Outgoing radiation, especially at night, cools the Earth's surface, which in turn cools the lowermost air layers, creating a nocturnal surface inversion a few centimeters to several hundred meters thick. Over polar snowfields, inversions may be a kilometer or more thick, with differences of 30°C or more. Solar warming of a dust layer can create an inversion below it, and radiative cooling of a dust layer or cloud top can create an inversion above it. Sinking air warms at the dry adiabatic lapse of 10°C/km, and can create a layer warmer than that below the subsiding air. Air blown from over cool water onto warmer land or from snow-covered land onto warmer water can cause a pronounced inversion that persists as long as the flow continues. Warm air advected above a colder layer, especially one trapped in a valley, may create an intense and persistent inversion.

Inversions effectively suppress vertical air movement, so that smokes and other atmospheric contaminants cannot rise out of the lower layer of air. California smog is trapped under an extensive subsidence inversion; surface radiation inversions, intensified by warm air advection aloft, can create serious pollution problems in valleys throughout the world; radiation and subsidence inversions, when horizontal air motion is sluggish, create widespread pollution potential, especially in autumn over North America and Europe. *See* ATMOSPHERIC POLLUTION. [ARNOLD COURT]

Terrain areas, worldwide

Subdivisions of the continental surfaces distinguished from one another on the basis of the form, roughness, and surface composition of the land. These areas of distinctive landforms are the product of various combinations and sequences of events involving both deformation of the Earth's crust and surficial erosion and deposition by water, ice, gravity, and wind. The pattern of landform differences is strongly reflected in the arrangement of such other features of the natural environment as climate, soils, and vegetation. These regional associations must be carefully reckoned with by man in his planning of activities as diverse as agriculture, transportation, city development, and military operations.

The illustration distinguishes among eight classes of terrain, on the basis of steepness of slopes, local relief (the maximum local difference in elevation), cross-sectional form of valleys and divides, and nature of the surface material. Approximate definitions of terms used and percentage figures indicating the fraction of the world's land area occupied by each class are as follows: (1) flat plains: nearly level land, slight relief, 4%; (2) rolling and irregular plains: mostly gently sloping, low relief, 30%; (3) tablelands: upland plains broken at intervals by deep valleys or escarpments, moderate to high relief, 5%; (4) plains with hills or mountains: plains surmounted at intervals by hills or mountains of limited extent, 15%; (5) hills: mostly moderate to steeply sloping land of low to moderate relief, 8%; (6) low mountains: mostly steeply sloping, high relief, 14%; (7) high mountains: mostly steeply sloping, very high relief, 13%; and (8) ice caps: surface material, glacier ice, 11%.

The continents differ considerably. Australia, the smoothest continent, has only one-fifth of its area occupied by hill and mountain terrain as

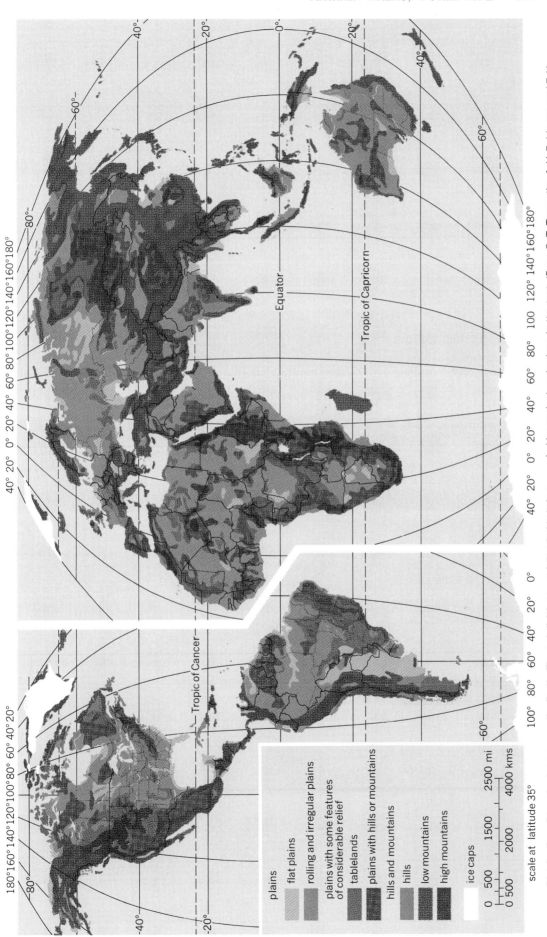

plains

flat plains

rolling and irregular plains

plains with some features
of considerable relief

tablelands

plains with hills or mountains

hills and mountains

hills

low mountains

high mountains

ice caps

scale at latitude 35°

0   500   1500   2500 mi

0 500   2000   4000 kms

Distribution of terrain classes over the Earth. The subdivisions are intended to bring out only the most striking contrasts among land surfaces. Percentage of area occupied by each class is given in the text. *(From G. T. Trewartha, A. H. Robinson, and E. H. Hammond, Elements of Geography, 5th ed., McGraw-Hill, 1967)*

against one-third of North America and more than one-half of Eurasia. Antarctica is largely ice covered; the only other great ice cap is on Greenland.

North America, South America, and Eurasia are alike in that most of their major mountain systems are linked together in extensive cordilleran belts. These form a broken ring about the Pacific basin, with an additional arm extending westward across southern Asia and Europe. The principal plains of Eurasia and the Americas lie on the Atlantic and Arctic sides of the cordilleras, but are in part separated from the Atlantic by lesser areas of rough terrain.

Most of Africa and Australia, together with the eastern uplands of South America and the peninsulas of Arabia and India, show great similarity to one another. They lack true cordilleran belts, and are composed largely of upland plains and tablelands, locally surmounted by groups of hills and mountains, and in many places descending to the sea in rough, dissected escarpments.

[EDWIN H. HAMMOND]

Terrain sensing, remote

Remote sensing of terrain may be thought of as the gathering and recording of information about terrain surfaces without actual contact with the object or area being investigated. This article discusses a limited aspect of remote sensing—the use of photography, radar, and infrared sensing in airplanes and artificial satellites.

Man has always used remote sensing in a primitive form. In ancient times he climbed a tree, or stood on a hill and looked and listened, or sniffed for odors borne on the wind. In fact, taste and touch are the only ways man can be aware of his environment without remote sensing. Although man is able to sense his environment without the use of instruments, he is not able to record the sensed information without the aid of instruments which are referred to as remote sensors. These instruments are a modern refinement of the art of reconnaissance, an early example being the first aerial photograph, taken in 1858 from a balloon floating over Paris.

The eye is sensitive only to visible light, a very small portion of the electromagnetic spectrum (Fig. 1). Cameras, operating like the eye, can sense and record in a slightly larger portion of the spectrum. For gathering of invisible data, instruments operating in other regions of the spectrum are employed. Remote sensors include devices that are sensitive to force fields, such as gravity gradient systems, and devices that record the reflection or emission of electromagnetic energy. Both passive electromagnetic sensors (those that rely on natural sources of illumination, such as the Sun) and active ones (those that utilize an artificial source of illumination) are considered to be remote sensors. Several remote sensing instruments and their applications are listed in the table.

Terrain characteristics. Each type of surface material (for example, soils, rocks, and vegetation) absorbs and reflects solar energy in a characteristic manner depending upon its atomic and molecular structure (Fig. 1c). In addition, a certain amount of internal energy is emitted which is partially independent of the solar flux. The absorbed, reflected, and emitted energy can be detected by remote sensing instruments in terms of characteristic spectral signatures and images. These signatures can usually be correlated with known rock, soil, crop, and other terrain features. Chemical

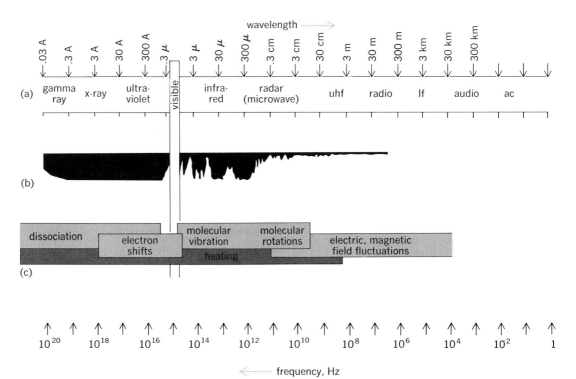

Fig. 1. Characteristics of the electromagnetic energy spectrum which are of significance in remote sensing. (a) Regions of electromagnetic spectrum. (b) Atmospheric transmission. The dark-colored areas are regions through which electromagnetic energy is not transmitted. (c) Phenomena detected.

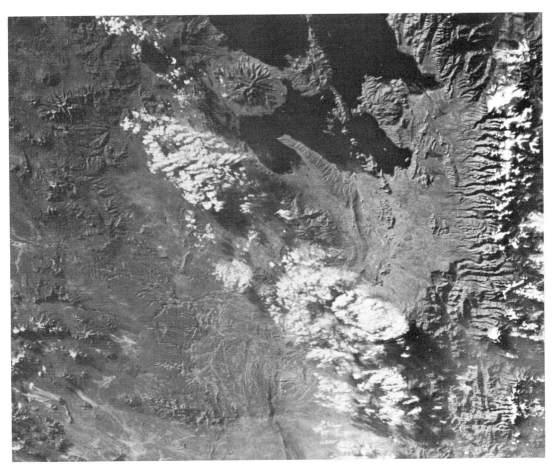

Fig. 2. Terrain features of Lake Titicaca area, Peru-Bolivia, detected and recorded by hand-held camera on board *Gemini 5* spacecraft. North is at the top of the picture. Three physiographic regions of the Andes can be seen—the Cordillera Occidental, the Altiplano, and the Cordillera Oriental. Note volcanic mountain structure on west side of Lake Titicaca and geologic folds (south and southeast portion of the picture). (*NASA*)

composition, surface irregularity, degree of consolidation, and moisture content are among the parameters that are known to affect the records obtained by electromagnetic remote sensing devices. Selection of the specific parts of the electromagnetic spectrum to be utilized in terrain sensing is governed largely by the photon energy, frequency, and atmospheric transmission characteristics of the spectrum (Fig. 1).

Types of remote sensing. Inasmuch as remote sensing is a composite term which includes many types of sensing, its meaning can be best understood by describing several of the types. Remote sensing is generally conducted by means of remote sensors installed in aircraft and satellites, and much of the following discussion refers to sensing from such platforms.

Photography. Photography is probably the most useful remote sensor system because it has the greatest number of known applications, it has been developed to a high degree, and a great number of people are experienced in analyzing the imagery obtained. Much of the experience gained over the years from photographs of the terrain taken from aircraft is being drawn upon for use in space.

Results of photographic experiments carried out in Gemini and unmanned spacecraft have vividly shown the applicability of these systems in space. These results indicate that such sensors may pro-

vide valuable data for delineating and identifying various terrain features, such as those shown in Fig. 2. Photographic techniques using a multispectral approach are also being used. Multispectral photography can be defined as the isolation of the reflected energy from a surface in a number of given wavelength bands and the recording of each spectral band separately on film. This technique allows the scientist to select the significant bandwidths in which a given area of terrain displays maximum tonal contrast and, hence, increases the effective spectral resolution of the system over conventional black-and-white or color systems.

Because of its spectral selectivity capabilities, the multispectral approach provides a means of collecting a great amount of specific information. In addition, it has less sensitivity to temperature, humidity, and reproduction variables than does conventional color photography and retains the high resolution associated with broadband black-and-white mapping film.

It is possible to reconstruct a conventional color photograph or the equivalent color-distortion photographs (such as Aerial Ectachrome Infrared) from the set of multispectral images. Work is progressing on the application of color-enhancing techniques to this type of photography, and laboratory studies on the sensitivity of the human eye to detection of color are being conducted.

Some areas of application for remote sensor instruments

Technique	Agriculture and forestry	Geology and planetology	Hydrology	Oceanography	Geography
Visual photography	Soils, plants, vigor, disease	Surface structure, surface features	Drainage patterns	Sea state, erosion, turbidity, hydrography	Cartography, land use, transportation, terrain and vegetation characteristics, thematic mapping
Multispectral photography			Soil moisture	Sea color, productivity	
Infrared imagery and spectroscopy	Terrain composition, plant condition	Thermal anomaly, minerals	Areas of cooling	Ocean currents, sea ice	Energy currents and land use
Radar: ranging imagery and scatterometry	Soil characteristics	Surface roughness, tectonics	Soil moisture, runoff slopes	Sea state, ice flow and ice, tsunami warning	Land ice, cartography, geodesy
Radio-frequency reflectivity		Subsurface layering, minerals	Soil moisture	Ice thickness, sea state	Land ice thickness, vegetation
Passive microwave radiometry and imagery	Thermal state of terrain	Subsurface layering	Snow, ice	—	Snow and ice
Absorption spectroscopy (remote geochemical sensing)	—	Mineral deposits, trace metals, oil	—	Surface flora	—

Radar. The ground clutter that radar engineers have, by tradition, sought to suppress has taken on a new significance as a scientific tool for terrain investigations. It has been demonstrated that radar return amplitude is affected by the composition of the illuminated area, its moisture content, vegetation extent and type, surface roughness, and even temperature in certain circumstances.

Radar returns are recorded in various forms to aid in their analysis. The forms having primary geoscience interest can be classed in one of the following categories: (1) scattering coefficients, which are a powerful tool in studying the nature of radar return and which have been correlated with different terrain types and have been directly applied to oceanic surface studies; (2) altimetry data;

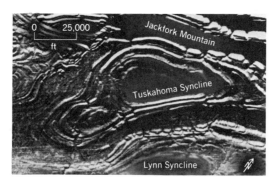

Fig. 3. Geologic structures in the Ouachita Mountains of Oklahoma mapped by imaging radar. The aircraft was flying NW-SE above the top of this area. The dark areas are radar shadow since the energy transmitted at low angle does not pick up terrain information on the southwest side of the ridges. (*NASA*)

(3) penetration measurements, which have been utilized for mineral exploration, for example, detection of faults through moisture associated with them; and (4) radar images.

Of these techniques, imagery generally presents the optimum geoscience information content of radar return. This is partly because well-developed photographic interpretation techniques are applicable to radar image analysis. These images are especially valuable for delineating various structural phenomena (Fig. 3).

The image record of the terrain return is affected by the frequency, angle of incidence, and polarization of the radar signal. For example, if the terrain being imaged is covered by vegetation, a K-band (35 GHz) signal records the vegetation, whereas a P-band (0.4 GHz) signal is likely to penetrate the vegetation and thus record a combination of vegetation and soil surface. In general, each frequency band represents a potential source of unique data. The angle of incidence of the incident wave affects the image because of radar shadowing on the backside of protruding objects. The angle of incidence can also show differences because of changes in orientation of the many facets on a surface which affect the return strength. Although information is lost in the shadow region, the extent of the shadow indicates the height of the object and hence has been useful in emphasizing linear features such as faults and lineations reflecting joint systems.

Infrared. Infrared is electromagnetic radiation having wavelengths of 0.7 to about 800 μ. All material things continuously emit infrared radiation as long as they have a temperature above absolute zero in the Kelvin scale. This radiation involves molecular vibrations as modified by crystal lattice motions of the material. The total amount and the wavelength distribution of the infrared radiation

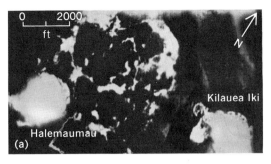

Fig. 4. Infrared imagery of Kilauea Volcano, Hawaii, permits detection of onset of volcanic activity. (a) Infrared image. Tonal variation shows distribution of radiant heat; the brighter the tone, the warmer the surface. (b) Aerial photograph. Large vent at left and small vent at right are within large crater (caldera). AT-11 mapping camera. (U.S. Geological Survey)

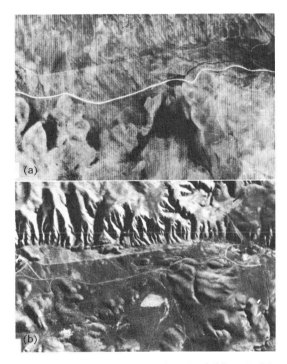

Fig. 5. Infrared imagery can indicate landslide material. (a) A presunrise infrared image showing lobate light (warm) patches that are lobelike masses of solid and rocks which have moved as landslides in a direction from bottom toward top of frame. Dark areas are undrained or poorly drained zones between landslide lobes. (b) Aerial photograph showing topographic expression of landslide area (bottom) and hills forming fault scarp (top). (U.S. Geological Survey and NASA)

are dependent on two factors: the temperature of the material and its radiating efficiency. This efficiency factor is called emissivity.

The imagery provided by an infrared scanning system gives much information that is not available from ordinary photography. The brightness with which an object appears on an infrared image depends on its radiant temperature. The hotter the object, the brighter it will appear on a positive image (Fig. 4). Radiant temperature is largely dependent upon the chemistry, grain size, surface roughness, and thermal properties of the material.

Not all infrared radiation received by a sensor is emitted. During the hours of sunlight the received power, out to the wavelength of approximately 3.5 μ, is predominantly reflected radiation, that is, thermal infrared.

Examination of infrared images may yield extensive information concerning various surface qualities. At night only the emitted components are present; therefore, in the $1.0-3.5-\mu$ region, completely different effects are imaged in presunrise images as compared to photographic pictures taken in daylight. The significance of this information is just beginning to be studied, and much additional research remains to be done (Fig. 5).

[PETER C. BADGLEY]

Bibliography: P. C. Badgley et al., The application of remote sensing instruments in earth resource surveys, Geophysics, 32:583–601, 1967; P. C. Badgley et al., Earth resources, in A Survey of Space Applications, NASA Spec. Publ. no. 142, 1967; R. N. Colwell et al., Basic matter and energy relationships involved in remote reconnaissance, Photogram Eng., 29:761–799, 1965; M. Holter et al., Fundamentals of Infrared Technology, 1962; R. K. Moore, Radar as a Sensor, Center Res. Eng. Sci. Rep. no. 61-7, 1966; D. C. Parker and M. F. Wolfe, Remote sensing, Int. Sci. Technol., 43:20–31, 1965; Proceedings of the Fourth Symposium on Remote Sensing of Environment, 1966.

Terrestrial ecosystem

A term that distinguishes the complex of ecosystems of the land surfaces of the Earth from freshwater and marine ecosystems. It encompasses ecosystems that exist on the continents and islands of the world and comprehends a series of dynamic open interaction systems that include living forms (animals, plants, and microorganisms) and their nonliving environment (soils, geological formations, and atmospheric constituents) and the activities, interrelations, chemical reactions, physical changes, and all other phenomena of each. Energy that enters these systems, chiefly in the form of sunlight, circulates through the systems and powers the life processes of the organisms, influences the rate and nature of chemical reactions and physical changes, and is partially accumulated in the bodies of organisms and in other chemical and physical states. See ECOSYSTEM; FRESH-WATER ECOSYSTEM; MARINE ECOSYSTEM.

Comparison of ecosystems. Terrestrial ecosystems differ from aquatic ecosystems in several important respects. The most obvious difference in the abiotic components of the systems is the basic physical contrast between the media. Terrestrial organisms are surrounded by air, a mixture comprised of gaseous elements and compounds. Water vapor is present in the atmosphere but forms only a small portion of the total volume of the air.

Liquid water, the medium of the aquatic environment, is much more dense, less fluid, and less transparent, and has a much greater thermal stability than air. The aquatic environment is marked by a superabundance of moisture, a lack of temperature extremes, slow changes in temperature, the dispersion of most of the essential mineral elements in the engulfing medium (usually in low concentration), a relatively short supply of oxygen and a high content of carbon dioxide, and a sharp decrease in illumination with slight increase in depth. On the other hand, in the terrestrial environment water is often in critically short supply, extremes and rapid changes of temperature are common, mineral elements are limited in occurrence to the substrate (generally in relatively high concentrations), and light is intense, at least at upper levels of the vegetation.

Adaptations to the habitat. The contrast in density, composition, and physical properties of the media in terrestrial and aquatic habitats is accompanied by and is responsible for, contributory to, or correlated with, contrasts in the geomorphologic development of the habitat, climatological features of the environment, pedological development, and morphologic and physiologic characteristics of the biota. Many organisms that live in the water have not developed, or have lost, devices that afford protection against water loss; indeed, many are able to absorb moisture and nutrients through the entire surface of their bodies. Many terrestrial organisms have evolved impervious cuticular coatings, surficial cuticular coverings, behaviorial adaptations, and physiologic adaptations that reduce water loss or make exceptionally efficient use of available water. Land plants, particularly the ferns and seed plants, have developed various water- and nutrient-collecting structures like rhizoids and roots. Correlated with these is the development of a complex translocation system through which water and dissolved minerals move from root to leaves and food circulates through the plant. Water plants generally have poorly developed roots and translocation systems.

Structural adaptations. Particularly in animals, the body shape of terrestrial species generally presents a small surface area per unit of volume, whereas marine animals, especially the mobile forms, are often elongate or flattened and hence present a large amount of surface per unit of volume. To some extent, body shape in motile forms is correlated with the density of the media and represents a streamlining adaptation that facilitates the movement of aquatic animals. Another structural adaptation to the contrasting density of the media is the development of rigid or semirigid supporting tissues in terrestrial plants and the virtual lack of such development in aquatic plants, which are buoyed up by the water. Virtually all woody plants are terrestrial or semiaquatic. Indeed, in both plants and animals, there is a tremendous taxonomic (evolutionary) division between aquatic and terrestrial forms. Seed plants, ferns, mosses, liverworts, fungi, and lichens are most characteristic and abundantly represented on land, while water is the chief domain of the algae. Thus the diversity in gross structure and life

form, as well as in species, is much greater on land. A tremendous variety of forests, grasslands, savannas, scrubs, succulent deserts, tundras, and other forms of vegetation characterize the terrestrial environment. The range of physiognomy of aquatic vegetation is much less and stratification or layering is absent or weakly developed in aquatic communities. *See* PLANT FORMATIONS, CLIMAX.

Habitat distribution of organisms. Warm-blooded animals and snakes and lizards typically are land dwellers. Insects are abundant and ecologically important on land, less so in fresh water, and virtually are absent from salt-water habitats. Fish, bivalves, and various lower animals are limited to or at least most abundant in aquatic environments. The bulk of aquatic vegetation, particularly where the water is deep, is floating rather than rooted, while that of land is almost entirely rooted and stationary. In contrast, many aquatic animals are immobile and depend on water currents to deliver food to their vicinity; most terrestrial animals are mobile. Land animals are generally capable of much more rapid movement in their dispersed medium than are comparable aquatic animals, which occupy a very dense medium.

Extent of terrestrial ecosystem. Although the aerial environment is well lighted from the upper limits of the atmosphere to the ground, the thickness of the terrestrial ecosystem is limited by two sets of circumstances. First the vertical development of land vegetation is restricted by the characteristics and limitations of supporting tissues, by the periodic stresses and breakage produced by strong winds, by the limitations of water-raising systems, and various other factors. Secondly, the soil-air interface is a tremendously important physical boundary. Below this interface there is no light, oxygen becomes a limiting factor, and water may saturate the soil at a depth of a few inches or feet or it may be in critically short supply. Thus the tallest living tree is a redwood 364 ft high, and the roots of only a few species of plants extend to a depth of 100 ft. Discounting airborne spores, insects, and similar material accidentally carried into the upper atmosphere, the maximum thickness of the terrestrial ecosystem in a given locality is less than 400 ft. Usually, it does not exceed 150 ft in thickness in forested areas and is much thinner in areas that support only grassland or tundra. On the other hand, the inhabited zone of a marine environment may be greater than 30,000 ft thick, although food-producing plants are limited to the surface layers due to limits of light penetration.

Temperature. The great seasonal temperature variations and rapid temperature changes in most terrestrial regions force periods of dormancy in plants and poikilothermic animals whose body temperatures approximate the environment. Homeothermic or warm-blooded animals, on the other hand, can remain active even during very cold weather if sufficient food is available. Aquatic plants and animals are not exposed to rapid thermal changes and the amplitude of the seasonal thermal cycle is much less in the water than in most terrestrial habitats. Not only is temperature regulation a greater problem in terrestrial habitats, but the functioning of the entire terrestrial ecosystem is controlled to a much greater extent by temperature changes. The fluidity of the atmosphere, the cyclonic circulation of huge air masses, and the variations in solar energy result in considerable diurnal thermal variations as well as seasonal and secular variations. During the growing season, or nondormant periods, these thermal variations impose corresponding variations in the rate of productivity of the green plants (through their effect on the rate of photosynthesis), on the rate of herbivore and carnivore harvesting activity (through their effect on the physiology of the organisms, especially as they impose dormancy due to low temperatures or estivation due to high temperatures), and on the energy expenditures of homeothermic animals (through their effect on the temperature gradient between the air and the bodies of the animals and the respiratory activity required to maintain body temperature). *See* HOMEOSTASIS.

Seasonal effects. During the cold or dry dormant seasons, the primary production of the terrestrial ecosystem is drastically reduced or even halted. Most herbivorous animals are inactive. The activities of herbivores that remain operative and the activities of saprophytes and scavengers that continue to feed during the unfavorable seasons result in a net decrease in the standing crop, for they feed on materials elaborated during the preceding favorable season or seasons. Most poikilothermic carnivores are forced to remain dormant or die during cold seasons, but homeothermic carnivores may remain active. In the latter instance, the carnivores' food preferences may change during the cold season, or their methods of hunting may be changed to compensate for the cold-season habits of their prey. Migrations of populations are related closely to seasonal changes. Migration of some birds, for example, has been found to be triggered by changes in daylength. But avian populations that are dependent on insects migrate to and from an area at times that correspond closely with the vernal increase and autumnal decrease in insect populations. Other species that are dependent on plant foods migrate at times when the daylight period becomes too short to allow the consumption of sufficient plant food to carry the animals through the longer dark period. Many grazing and browsing mammals migrate vertically in mountainous areas or horizontally in level areas to obtain food and to benefit from more favorable climatic conditions.

Unfavorable seasons also affect aquatic communities, especially fresh-water communities, but the seasonal changes generally are less intense, the activity of primary producers generally does not ebb as low as in terrestrial communities, and the limiting factors are most often a lack of oxygen and nutrients rather than extremely low temperatures or the unavailability of water. Animal activity in aquatic habitats is also reduced during unfavorable seasons, but the level of minimum activity in the aquatic environment of deeper fresh-water bodies and in the oceans is considerably higher than that in the terrestrial environment.

Catastrophic agents. Fire, windthrow, and many other catastrophic agents are peculiar to the terrestrial environment and play important roles in the functioning of the ecosystem. Furthermore, the terrestrial environment is the habitat of man and is more subject to human interference than the aquatic environment, particularly the marine segment of it. Human influence is so great that man has substituted artificially maintained ecosystems for natural ecosystems. The artificial ecosystems are represented on land by farmed areas, managed forests, urbanized areas, and similar developments. *See* ECOLOGY, APPLIED.

Energy. Terrestrial ecosystems differ from aquatic ecosystems in another significant aspect. They are dependent on stored nutrients that may have been residual for hundreds of thousands of years. They are marked by a continual net loss of nutrients and other physical components through erosion and leaching, while there is a concomitant net gain to the aquatic ecosystems. Except in certain restricted areas, the use of aquatic food by terrestrial animals, the phenomenon of salt spray, the filling of lakes and bogs, and the harvest of marine and fresh-water organisms by human activities represent a small return of materials to the terrestrial environment. Much larger returns are made by crustal upheavals that expose sections of the ocean floor and once again make available to terrestrial organisms the materials accumulated on the ocean bottom. The reverse also takes place, and former land areas may be depressed beneath the surface of the oceans. These crustal movements result in a long-term recirculation of mineral elements as well as a periodic renewal of erosion cycles. The terrestrial and aquatic ecosystems are also connected in a great many other direct ways, so that it is most logical to consider the entire Earth and its envelope of gases to be the ultimate ecosystem. However, this world ecosystem is an open system and is dependent upon an external supply of energy, chiefly from the Sun, and is influenced by a variety of forces of external origin.

Trophic levels. If the weight (biomass) of the living components of an ecosystem at a given moment is determined, separated into weights for the various trophic levels (green plants, herbivores, predators, scavengers, and saprophytes), and figured graphically, a pyramidal form of graph results. In the terrestrial ecosystem, the green plants are relatively long-lived; some live for several weeks, others for years, a few for centuries. Because the green vegetation is the primary producer level and thus limits the bulk of organisms that can feed upon it and because the green vegetation is long-lived, its biomass is greater than that of any other trophic level or of all other trophic levels combined. In aquatic ecosystems, however, this is not necessarily the case. Where short-lived plankton forms are the predominant green vegetation and longer-lived fish and bottom organisms are the herbivores and carnivores, the biomasses of the higher trophic levels often exceed those of the primary producers. Of course, when the annual production of the various trophic levels in either ecosystem is calculated, the production of the green vegetation is greatest.

Productivity. From the meager quantitative data available on the primary productivity of various ecosystems, it appears that no generalization can be made concerning the comparative productivity of terrestrial and aquatic communities. Further, it is apparent that productivity in either environment is not a regional characteristic, but is dependent

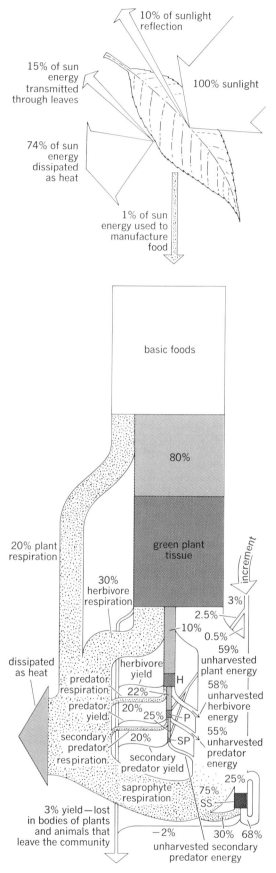

Pathways and magnitude of energy transfers in a beech-maple forest in the north-central United States. (*From J. S. McCormick, The Living Forest, Harper, 1959*)

wholly upon the ecosystem involved. According to summaries made by E. Odum (1953), an Ohio cornfield was found to produce 862 metric tons of organic carbon per square kilometer per year, the trees in a New York apple orchard produce 526, European forests produce an average of 225, cultivated land on the average produces 160, a dry grassland produces 48, and a desert produces only 6 metric tons. H. Odum and coworkers (1958) present data that indicate that a montane rainforest in Puerto Rico produces approximately 1100 metric tons, and data presented by A. Krogh (1934) indicate that a Danish beech forest may produce 1500–2000 metric tons.

For comparison, data cited by E. Odum (1953) show that several lakes in the north-central United States average 111–480 metric tons per square kilometer per year and that several marine communities average 60–1000 metric tons. Significantly higher annual primary production has been reported by H. Odum (1957) for a fresh-water spring in Florida, 6390 metric tons, and by E. Odum and H. Odum (1955) for a coral atoll in the Pacific Ocean, 8328 metric tons.

Trophic structure. The trophic structure of terrestrial and aquatic ecosystems is strikingly similar (see table). There is more variability within the series of analyzed terrestrial communities than there is between those communities and the aquatic types. In addition, the relative efficiency of the vegetation, in terms of the proportion of incident solar radiation converted to stored form in organic carbon compounds, is virtually identical for the terrestrial and fresh-water communities listed, although the efficiency of the tropical atoll community was considerably greater.

In all the ecosystems, a relatively low proportion of available solar energy is utilized directly by the biota.

Energy transfer. The energy transfers that occur in a beech-maple forest community in the central United States have been estimated by J. S. McCormick (1959) and are an example of terrestrial systems in general (see illustration). The energy utilized each year by the organisms in an acre of this forest is approximately equivalent to the electricity required to supply an average New York City household with power for nearly half a century. Virtually all the energy that enters the system directly is in the form of sunlight, but only about 1% of the available solar energy is actually transformed by green plants into the chemical energy of food. Approximately 10% of the solar energy is reflected from the plant surfaces, 15% is passed through the leaves, and 74% is dissipated as heat.

A portion of the energy stored in basic foods or in photosynthetic products manufactured by green plants is utilized by the plants in respiration and the remainder is stored in the form of plant tissue. Part of the energy of plant tissues harvested and utilized by herbivores such as insects, rodents, and deer is dissipated as heat by respiration, and part is stored in the body tissues of the herbivores. The energy contained in the tissues of herbivores is utilized by predators and the tissues of predators are, in turn, utilized by secondary predators. At each step, some energy is lost to the community through respiration and part is unharvested or unassimilat-

Trophic structure of terrestrial and aquatic communities on basis of biomass or energy content of standing crop

Category	Terrestrial communities			Fresh-water communities			Marine communities	
	Blue-grass meadow, Michigan	Beech forest, Denmark	Montane rainforest, Puerto Rico	Cedar Bog Lake, Minnesota	Webster's Lake, Wisconsin	Silver Springs, Florida	Coral reef, Eniwetok Atoll	Eel-grass, North Sea
Primary producers, %	78	93	98.4	89.3	86.8	94.3	83.1	79.7
Herbivores, %	16	7	1.6	9.0	10.2	4.3	15.6	19.9
Carnivores, %	6			1.6	3.4	1.5	1.3	0.3
Efficiency of primary producers*	1.2	1.0				1.2	5.8†	
Basis of figures	Energy	Biomass	Biomass	Energy	Biomass	Biomass	Biomass	Biomass

*Ratio of energy transformation to incident solar energy.

†Computed on basis of light reaching community rather than light reaching water surface.

ed. The unharvested and unassimilated energy accumulates chiefly on the forest floor in the form of dead leaves, twigs, flowers, fruits, fallen trunks, dead bodies, feces, and liquid wastes and is utilized by scavengers and saprophytes. Ultimately, all the energy that enters the forest community is dissipated as heat by respiratory processes, is lost in the bodies of plants and animals that leave or are taken from the forest, or is lost in the form of heat evolved by forest fires. In natural communities, these losses are balanced, or at least offset, by materials that enter the community from other sources. *See* ECOLOGY. [JACK S. MC CORMICK]

Bibliography: A. Krogh, Conditions of life in the ocean, *Ecol. Monogr.*, 4(3):421–429, 1934; J. McCormick, *The Living Forest*, 1959; E. P. Odum, *Fundamentals of Ecology*, 2d ed., 1959; H. T. Odum, Trophic structure and productivity of Silver Springs, *Ecol. Monogr.*, 27:55–112, 1957; H. T. Odum, W. Abbott, and R. Selander, Studies on the productivity of the lower montane rainforest of Puerto Rico (Abstract), *Bull. Ecol. Soc. Amer.*, 39: 85, 1958; H. T. Odum and E. P. Odum, Trophic structure and productivity of a windward coral reef community on Eniwetok Atoll, *Ecol. Monogr.*, 25(3):291–320, 1955.

Terrestrial frozen water

Seasonally or perennially frozen waters of the Earth, exclusive of the atmosphere. Water in the frozen or solid state is the hexagonally crystallized, birefringent mineral known as ice. Terrestrial ice occurs in the form of temporary seasonal accretions during the cold months and in the perennial ice cover represented by glaciers, land-fast ice (ice shelves), and subsurface ground ice in permafrost regions.

Glaciers, past and present. Under present climatic conditions the semipermanent terrestrial ice cover is essentially glacial. Glaciers cover approximately 10% of the world's land area (14,972,138 km², estimated by R. F. Flint, 1957). Of this ice, 96% lies in Greenland (1,726,400 km²) and Antarctica (12,650,000 km²), leaving only 4% of the world's glaciers in mountains and subpolar regions. This is to be compared with approximately 32% of the world's land covered by glaciers during their maximum extension in the Pleistocene Ice Age. A close estimate of the total volume of terrestrial ice is not possible because the thickness of the Antarctic ice sheet is not yet adequately known. Flint, in 1957, tentatively estimated the volume of glacier ice existing today as about 24,-000,000 km³, "equivalent to a layer of water having the area of the present oceans and approximately 59 m thick." No estimate has been attempted of the volume of subsurface terrestrial frozen water existing in permafrost regions.

Properties of terrestrial ice. Ice is one of the more abundant minerals on the surface of the Earth. It is usually observed as colorless and transparent, but in large, dense masses it shows a vivid light blue color. The other physical properties of terrestrial frozen water vary considerably under changed conditions of internal temperature, load, crystalline orientation, and mass density. For example, ice has an indentation hardness which varies with the mass temperature and a scratch microhardness which differs with the orientation of the crystal plane. On the microhardness or Mohs scale at 0°C it has a hardness of about 2 and at −44°C a hardness of about 4. Coincident with this property is its variable plasticity or ability to deform under stress. The deformation rate is considerably reduced under colder temperatures. At a temperature of −1°C, the deformation (flow) rate of polycrystalline ice may be expected to be slowed to approximately one-fifth of that at 0°C. For example, ice chilled to −10°C has been observed by J. W. Glen to have 25 times as much ability to resist deformation as ice at 0°C.

Special hydrothermal relationships. No significant influence on the deformation rate appears in ice through changes in hydrostatic pressure, the relationship being as negligible as in liquids;

Table 1. Temperature and pressure relationships for forms of ice and related liquid water*

Forms	Temperature, °C	Pressure, atm
Water; ice I, III	−22.0	2,047
Ice I, II, III	−34.7	2,100
Water; ice III, V	−17.0	3,417
Ice II, III, V	−24.3	3,397
Water; ice V, VI	+0.16	6,175
Water; ice VI, VII	+81.6	21,700

*Based on data reported in 1940 by N. E. Dorsey.

Table 2. Thermal constants of various forms of frozen water compared with liquid water and other substances*

Water forms and other substances	Conductivity, cal/(°C)(cm)(sec)	Specific heat, cal/(°C)(g)	Density, g/cm³	Thermal diffusivity, cm²/sec	Relative diffusivity to ice (approx. ratio)
Frozen water					
New snow	0.0003	0.5	0.20	0.0030	0.27
Old snow	0.0006	0.5	0.30	0.0040	0.36
Average firn	0.0019	0.5	0.55	0.0070	0.64
Firn ice	0.0038	0.5	0.75	0.0100	0.91
Ice	0.0050	0.5	0.92	0.0110	1
Comparative materials					
Water (0°C)	0.0014	1.0	1.00	0.0014	0.13
Rubber	0.0005	0.40	0.92	0.0014	0.13
Steel (mild)	0.1100	0.12	7.85	0.12	11
Aluminum	0.4800	0.21	2.70	0.86	78
Copper	0.9300	0.09	8.94	1.14	104

*Based on data from U.S. Army Corps of Engineers, *Review of the Properties of Snow and Ice*, 1951, and other sources as reported in M. M. Miller, *Glaciothermal Studies on the Taku Glacier, Alaska*, 1954.

but the melting point decreases as the pressure increases. This amounts to a reduction in temperature of 0.0075°C for each atmosphere of increase in pressure. As a result of this property, the melting temperature of an ice mass at any depth is fundamentally conditioned by the weight of the overlying mass. The term used to refer to this is pressure melting temperature. It is because of this characteristic that ice skating is possible. At the edge of a skate blade sufficient pressure is exerted to cause formation of a thin film of water, which serves as a lubricant. Similarly, when a wire is pressed into a block of ice, or even when pieces of ice are pressed together, melting occurs at the contact surface. When the pressure is released, the water refreezes, uniting the ice into a continuous mass. This is the process of regelation.

Classification of ice forms. Although there is only one ordinary form or phase of terrestrial frozen water, ice I, five other stable phases of ice exist in addition to one unstable phase. These, however, are only the product of great pressure and are not found in normal conditions outside of the experimental laboratory. The stable, extraordinary forms are known as ice II, III, V, VI, and VII; IV is the unstable category. Each of these reverts to normal terrestrial ice, water, or both when the pressure is released. Table 1 shows pressure and temperature parameters (triple points) under which these forms of ice and related liquid water exist.

Massive ice constants. Because massive ice is an aggregate of individual ice crystals, it may also be considered a monocrystalline rock. In the form of snow or firn (granular summer equivalent), it has been genetically likened to a sedimentary rock. (Similarly, pond, river, sea, or refrigerator ice may be likened to an igneous rock, and glacier ice to a metamorphic rock.) Any of these forms when dry are poor conductors of electricity. Cold dry ice, for example, has a resistivity of 10^9 ohm-cm for direct current and low-frequency alternating current and 3×10^6 ohm-cm at 60 kHz. The dielectric constant for low frequencies at 0°C is 74.6, and for high frequencies and very low temperatures it drops to 3.0. Ice is a poor conductor of heat.

A list of the thermal constants of various forms of frozen terrestrial water under normal conditions is in Table 2 (as compiled from various sources for the Juneau Icefield Research Program by M. M. Miller, 1954). The general conductivities for snow and average firn noted in this table are based on data from the U.S. Army Corps of Engineers (*Snow, Ice and Permafrost Research Establishment*, 1951). The density of firn ice is chosen arbitrarily between given values for average firn and ice. For comparison, the thermal properties listed in the International Critical Tables for rubber, steel, aluminum, and copper are also noted. The diffusivity figures in Table 2 are rounded off for convenient reference. *See* GROUNDWATER; HYDROLOGY; SURFACE WATER.

[MAYNARD M. MILLER]

Bibliography: N. E. Dorsey, *Properties of Ordinary Water-Substance*, ACS Monogr. no. 81, reprint, 1954; R. F. Flint, *Glacial and Pleistocene Geology*, 1957; J. W. Glen, The creep of polycrystalline ice, *Proc. Roy. Soc. London Ser. A*, 228: 528–529, 1955; M. M. Miller, *Glaciothermal Studies on the Taku Glacier, Alaska*, Ass. Int. Hydrol. Publ. no. 39, 1954; E. R. Pounder, *Physics of Ice*, 1965; *Review of the Properties of Snow and Ice*, Snow, Ice, and Permafrost Research Establishment, U.S. Army Corps of Engineers Rep. no. 4, 1951.

Thermal spring

A spring with water temperature substantially above the average temperature of springs in the region in which it occurs. The average temperature of springs is ordinarily within a few degrees of the mean annual temperature of the atmosphere. Thus waters of thermal springs range in temperature from as low as 60°F, in an area where normal groundwater has a temperature of 40–50°F, to well above the boiling point.

The two main considerations in the origin of thermal springs are the source of the water and the source of the heat. The water may be ordinary groundwater that percolates slowly downward, is heated by the Earth's normal thermal gradient (the temperature of the Earth normally increases about 1°F for each 50–100 ft of depth), and then returns to the surface without losing all the added heat. The water of thermal springs may be in part juvenile, a product of the crystallization or recrystallization of rock at depth. Since juvenile water is virtually certain to become mixed with connate or meteoric water on its way to the surface, there are no thermal springs whose water can be demonstrated to be wholly juvenile.

Investigations of Warm Springs, Ga., and of other thermal waters in the eastern United States indicate that the water entered the aquifer by normal recharge from precipitation, percolated deep into the Earth by reason of the geologic structure, and there received its heat before returning to the surface. On the other hand, the springs in Yellowstone Park, Wyo., Steamboat Springs, Nev., and many other localities in the western United States may derive part of their water and much of their heat from bodies of superheated rocks, perhaps in the last stages of cooling from the molten state. Many of the springs in the western United States discharge water that is near the boiling point.

Where spring water is above the boiling point it has been tapped to provide steam for power production. Such power installations are found in New Zealand, Italy, and California. Hot springs have also been used to provide heat for homes and swimming pools. *See* GROUNDWATER; SPRING, WATER.

[ALBERT N. SAYRE/RAY K. LINSLEY]

Thermocline

A layer of sea water in which the temperature decrease with depth is greater than that of the overlying and underlying water. Such layers are semipermanent features of the oceanic temperature structure, and their depth and thickness show marked variation with season, latitude and longitude, and local environmental conditions. Since the three-dimensional temperature structure has a great effect on many oceanic properties, such as the transmission of sound, the study of the nature and behavior of the thermocline is of extreme importance to many oceanographic interests, both economic and military. In general, two major types of thermocline may be identified: the permanent thermocline and the seasonal thermocline. In addition to these types, shallow thermoclines or similar stable layers often occur, owing to diurnal heating of the surface waters.

Permanent thermocline. This feature is so named because its character is virtually unchanged seasonally. In Arctic and Antarctic regions, the water is cold from top to bottom. As this dense water flows south and north, respectively, it sinks beneath warmer water which moves outward from the Equator. This gives rise to the temperature discontinuity known as the permanent thermocline. The cold water flowing slowly through the deep ocean basins exhibits conservative properties throughout all the oceans; however, on top of this dense layer lie a number of shallow layers whose character varies from ocean to ocean. The top of the permanent thermocline is quite shallow at the Equator, reaches maximum depth at mid-latitudes, and becomes shallow again at about 50° latitude. The thermocline disappears between 55 to 60°N or S. In general, as the permanent thermocline deepens, it becomes thicker and the temperature gradient within it decreases. Figure 1 indicates schematically variations with latitude in the char-

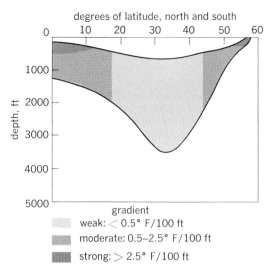

Fig. 1. The permanent thermocline, based on averages for depth, thickness, and gradient within thermocline.

acteristics of the permanent thermocline. *See* SEA WATER.

Seasonal thermocline. This feature is a summer phenomenon found at shallower water depth than the permanent thermocline in all the world's oceans except those perennially ice-infested. As air temperatures rise above ocean temperatures in the spring season and the sea surface receives more heat than it loses by radiation and convection, the surface water begins to warm so that a negative temperature gradient develops in the first few feet (Fig. 2a). (Numbers 1–8 show sequence in development and disappearance of thermocline; profiles show temperature structure.) The surface waters are then mixed by transfer of energy from the wind. Although this mixing serves to lower the surface temperature, the net effect is a downward

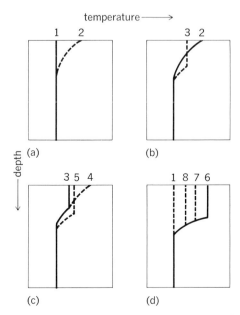

Fig. 2. Formation and breakup of seasonal thermocline. (a) Formation, 1st stage. (b) Formation, 2d stage. (c) Formation, 3d stage. (d) Breakup.

transport of heat and formation of an isothermal layer whose temperature is warmer than the underlying water (Fig. 2b). A strong temperature gradient, or seasonal thermocline, is thus formed between the isothermal surface layer and water beneath. This process repeats itself (Fig. 2c) until the gradient in the seasonal thermocline becomes so strong that summer winds cannot impart sufficient energy to drive the isothermal layer deeper. From July through September such a surface layer of mixed water underlain by a strong negative temperature gradient is found in most of the ocean. As air temperatures fall in autumn, the water loses heat to the atmosphere by convective and radiative processes, and the surface layer is cooled to the temperature of the water below. The seasonal thermocline breaks up (Fig. 2d), to form again the following spring. Seasonal thermoclines may be affected locally by vertical wind mixing, currents, and heat exchange across the interface between ocean and atmosphere. Further distortions may occur because the density discontinuity associated with thermoclines provides a favorable environment for internal waves. Practically all physical processes occurring in the sea have an effect on thermocline characteristics.

[JOHN J. SCHULE, JR.]

Thunderstorm

A convective storm accompanied by lightning and thunder, rain or rarely snow showers, and often hail. Gusty squall winds are observed near the onset of precipitation. The characteristic cloud is the cumulonimbus, a towering turbulent cloud often having an anvil-shaped top. Sheets or fragments of heavy cirrus, altocumulus, and low stratus or stratocumulus are often associated.

In many regions, particularly the tropics, thunderstorms furnish much or most of the total annual rainfall. Cloudbursts, squall winds, hail, and lightning associated with thunderstorms do many millions of dollars damage annually. For discussions of electric phenomena associated with thunderstorms *see* ATMOSPHERIC ELECTRICITY.

Although much heavier, thunderstorm rain is localized compared with the frontal rain in a cyclone. Thunderstorms may be chaotically scattered over a wide area or may appear as large clusters, often in line array. An isolated thunderstorm is perhaps 3–5 mi across; clusters 20–50 mi across are common, while lines of clusters may be several hundred miles in length. Bases of thunderstorm clouds range from 1000 to 3000 ft in moist air masses, up to 10,000–15,000 ft in dry air. Tops commonly reach 30,000–50,000 ft, but are higher in the tropics and lower in high latitudes. *See* SQUALL LINE.

A thunderstorm consists of one or more cells, identified by distinct patterns of vertical motion, and it goes through characteristic stages of development. In the cumulus or growing stage, a thunderstorm consists of updrafts, or vigorously rising columns of air; in the mature stage (see illustration) both updrafts and pronounced downdrafts are found; and in the dissipating stage gentle downdrafts characterize the whole cloud or cell. Vertical speeds in the drafts have median values near 8 m/sec, but may be several times stronger.

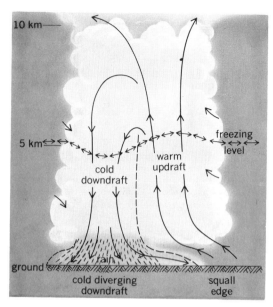

Diagram showing simplified circulation in a vertical section through a mature thunderstorm. (*After G. A. Suckstorff; H. R. Byers and R. R. Braham, Jr.*)

If unsaturated air is lifted without addition or loss of heat, it cools by expansion at the dry adiabatic lapse rate of 9.8°C/km. If saturated, it cools at the lesser moist adiabatic lapse rate because released latent heat of condensation partly counteracts the expansional cooling. If the existing decrease of temperature with height outside the cloud is greater than the moist adiabatic rate, a rising saturated air parcel becomes warmer and less dense than its surroundings, and thermodynamic instability exists because the parcel would then be accelerated upward because of its relative buoyancy.

For such instability to be realized, the air rising from lower levels must have high moisture content. Initiation of convection can result from surface heating, which causes local bubbles or columns of air to rise to saturation level, or from mechanical lifting by a front or over rising ground.

Moisture for cloud and rain formation is condensed in the updrafts, because rising air as it cools can hold less water in vapor form. Once rain begins, downdrafts form in the heavy rain area. These are colder than the cloud environment, and sink by gravity force. Coldness of the downdraft is due to partial evaporation of raindrops, causing loss of heat by the air. Such cooling is intensified by entrainment, or drawing in through the sides of the cloud, of dry air which can be cooled by evaporation to a lower temperature than the air which originated in the initial updraft.

On approaching the Earth's surface, the downdrafts are forced to spread out laterally. Gusty squall winds, often of considerable force, are found near the edge of the cold air diverging outward from the rain area. Formation of new updraft cells results from lifting of the neighboring unstable air by the cold air spreading out from a mature thunderstorm. By this process a cluster of thunderstorm cells may be continuously regenerated even though some cells dissipate.

Because the ground and lower layers of the air are warmest in midafternoon, instability is most pronounced then, and thunderstorm incidence shows a decided peak at that time. Thunderstorms are most common in the tropics and over middle latitude land areas in summer, but they occur occasionally even at high latitudes. In the United States, most frequent occurrences are in the mountainous regions of the Southwest and over the Florida peninsula. Most thunderstorm activity is found in regions of general low-level convergence and weak upward motions, such as on the advancing sides of cyclones and in the intertropical convergence. *See* STORM.

[CHESTER W. NEWTON]

Bibliography: L. J. Battan, *The Thunderstorm*, 1964; T. F. Malone (ed.), *Compendium of Meteorology*, 1951; U.S. Weather Bureau, *The Thunderstorm*, 1949.

Tornado

An intense rotary storm of small diameter, the most violent of weather phenomena. Tornadoes always extend downward from the base of a convective-type cloud, generally in the vicinity of a severe thunderstorm.

Appearance ranges from a broad funnel with smallest diameter at the ground, to a narrow rope-like vortex which may not reach the ground or may intermittently lift and dip. An ill-defined cloud of dust or debris often surrounds the true tornado cloud near the ground (see illustration). In surface layers, air spirals inward toward the vortex, generally rotating in a counterclockwise sense, rising rapidly in the funnel. From structural damage and other evidence, probable wind speeds up to 300 mph or more have been calculated.

The visible funnel consists of cloud droplets condensed because of expansional cooling resulting from markedly lower (probably by 100–200 millibars) pressure in the vortex than in the surroundings. Height of the visible funnel depends upon the cloud base and may be 1000–10,000 ft; however, the tornado vortex probably extends a considerable distance upward within the accompanying cloud.

Width of the path of destruction varies from a few yards to a mile, averaging 700 ft. Length of the path ranges from very short up to 300 mi, averaging 4 or 5 mi. Movement is generally from southwest, but may be in any direction. Speed of movement averages 35 mph but is highly variable.

Damage in the millions of dollars, with loss of many lives, takes place occasionally when tornadoes strike heavily populated areas. Structural damage to buildings results partly from explosion when the atmospheric pressure outside is suddenly reduced and partly from force of the extremely strong winds. Damage from explosion may be reduced by venting or by prior opening of windows to allow rapid equalization of pressure inside and outside the building.

Tornadoes occur on all continents, but are common only in the United States and Australia. They have been observed in all states, being most frequent in Iowa, Kansas, Missouri, Illinois, and Oklahoma. Although tornadoes occur in all months, greatest seasonal frequency is in late spring and early summer. Most frequent occur-

Tornado, at Fargo, N.Dak. (*Fargo Forum photograph by C. Gebert, Grand Prize Winner, 11th Annual Graflex Contest*)

rence is in the Southern states in early spring; the locus of greatest activity shifts northward into the Central states in summer. Occurrences are noted at all times of day, but there is a strong peak in incidence during midafternoon.

Requisite conditions for tornado formation are pronounced thermodynamic instability combined with sufficient amounts of water vapor to produce thunderstorms, along with the presence of strong winds in the upper troposphere. These are the conditions that favor development of a squall line; at times 10–30 tornadoes occur as parasites to such a disturbance. Tornado probability increases with the thermal instability, and it is greatest with thunderstorms that have an intense radar reflectivity (large precipitation content) and high tops penetrating into the stratosphere. However, some tornadoes are not connected to the thunderstorm itself, but to vigorously developing cumulus clouds adjacent to the thunderstorm. *See* SQUALL LINE.

Both laboratory simulations and observations of natural vortexes suggest that rising motions (up to 150 mph observed) take place mainly in the sheath of the funnel, and that there is likely to be descending motion in or near its core. The intense winds are accounted for by inward-moving rings of air increasing in rotary motion under conservation of angular momentum, and speeds are consistent with movement toward the inner region of very low pressure, allowing for some frictional loss. The cause of the low pressure is inadequately understood.

Tornadoes in the United States are mostly found on the south sides of the parent thunderstorms. Heavy rain and hail often follow (but sometimes precede) passage of a tornado. The heaviest rain is likely to fall a few miles north of the tornado track; sometimes no rain falls along the track itself. Widespread thundersqualls are often observed outside the actual tornado path.

Accurate location of tornadoes by use of radar and radio direction-finding devices (frequent electrical discharges are characteristic) sometimes enables useful short-period prediction of likely future movements.   [CHESTER W. NEWTON]

Bibliography: S. D. Flora, *Tornadoes of the United States*, rev. ed., 1954; T. F. Malone (ed.), *Compendium of Meteorology*, 1951; S. Petterssen, *Weather Analysis and Forecasting*, 2d ed., 1956.

Toxicology

The study of poisons and their effects, mechanisms of action, and methods of treatment. The term poison is difficult to define because any material, if used in large enough amounts or administered in certain ways, will produce harmful effects on some structure or function of the body. Other factors which must be considered include an individual's general state of health, age, sex, and whether previous exposure to an agent has increased his tolerance to it or has caused a cumulative effect which renders him more susceptible to an added dose.

For all practical purposes, poisons are substances which cause tissue damage or malfunction of a potentially serious degree when given to an average individual in small amounts. This amount is usually considered to be less than 50 g. In addition, poisons must exert their effects through chemical or physicochemical mechanisms. The above definition eliminates hypersensitivity reactions which do not occur in the average person, and also excludes agents that cause purely physical damage.

Categories of poison. Different poisons produce their damage in various ways. Four broad categories may be enumerated.

1. Some agents act so rapidly that no perceptible direct effects are seen; death occurs swiftly as a result of anoxia, usually following circulatory collapse. Carbon monoxide and cyanide are asphyxiant gases of this category. The alcohols, ethers, and other hydrocarbons which are central nervous system depressants are also included. In each case, a lesser amount of the same substance may not be rapidly fatal, and then direct chemical effects may be seen in susceptible tissues.

2. Many agents produce damage at point of contact or entry into the body. Such locally destructive poisons include such corrosives as acids or alkalies, and also irritant gases, volatile oils, and aconite.

3. Systemic poisons exert their principal effects after they are absorbed in the body. They may also cause some local irritation at the point of entry to the body and therefore produce compound injury. The damage is usually directed toward specific organs or systems such as the liver, kidney, nervous system, or blood-forming tissues. Agents of this group include arsenic, which in acute cases causes fatal blood vessel injury, and other heavy metals that inhibit enzymic activity at cellular levels. Lead, mercury salts, thallium, cadmium, phosphorus, manganese, and chromium each cause fairly specific injury to certain tissues.

4. Certain systemic poisons exert their effects on some part of the body after absorption, but no damage is seen at the point of entry. These include the blood-destroying (hemolytic) poisons, such as nitrobenzene and arsine, and the central nervous system depressants, such as hydrogen sulfide and the war gases of the nerve-gas type.

Another point of view in classification of poisons considers the source of each noxious substance, even though chemical characteristics and effects may be quite different in each group. The categories of insecticides, industrial and occupational compounds, drugs, vegetable poisons, and animal venoms illustrate the major groups; such a classification is useful in both preventive and diagnostic medicine.

Other descriptive terms used in reference to poisons reflect the mechanism of action or the principal tissues involved. Examples include corrosives, irritants, asphyxiants, enzyme inhibitors, neurotoxins, and hemotoxins.

Statistical data. Statistical studies of poisonings vary a great deal in their conclusions, chiefly because of differences in geographic location, occupational and social status of those involved, and the availability of good laboratory facilities for toxicological analysis.

About 55% of deaths from poison are accidental, 45% are suicidal, and fewer than 1% are homocidal. Children are most often involved in accidental injury following ingestion or exposure to various substances. Of the deaths among children, 55% follow aspirin ingestion, 25% result from damage produced by other drugs, and the remainder include the more common cleaning agents, cosmetics, kerosine, lye, paint products, pesticides, and animal venoms.

Industrial and agricultural workers are frequently exposed to poisons through accidents or by improper use of safety measures. Chlorinated hydrocarbons, nitro and amino aromatics, benzene, carbon monoxide, chromium, mercuric salts, methyl alcohol, hydrogen sulfide, arsenic, DDT, and other insecticides account for a large proportion of cases and, frequently, fatalities.

Carbon monoxide and overdoses of barbiturates are probably the most commonly used chemicals in suicides. Many other materials are used, despite the agonizing results known to be produced by some, such as lye.

Ethyl alcohol, carbon monoxide, acids, alkalies, cyanide compounds, mercury preparations, arsenicals, sedatives, and benzene are probably the most commonly encountered agents in poisonings, if one excludes the special categories of insect and snake bites, food poisoning, and certain plants, such as inedible mushrooms, that contain harmful materials.

History. The history of poisons is as old as man. Even today some of the most primitive peoples of the world possess substances whose development is shrouded in mysticism and superstition. Often the ingredients have remained impervious to modern analysis; occasionally one of these poisons is purified, standardized, and introduced into accepted medical use. For example, curare originated as arrow poison for the South American Indians of the Amazon and Orinoco regions. Although a synthetic compound now has been developed, the action is similar, but instead of using amounts that cause fatal muscle paralysis, the physician uses lesser quantities to relax the muscular system.

Other medicinals which have been handed to civilization from witch doctors and ancient priesthoods include strychnine, opium, caffeine, strophanthin, cocaine, atropine, digitalis, and ergotamine. Still more compounds, usually prepared from certain plants, have been used from time to time to remove an unwanted person, or more commonly, have been used by large numbers of a population as partial poisons to produce certain effects. The most notable of these are opium, hashish, caffeine, nicotine, marijuana, heroin, betel, and fly agaric. The deadly nightshade, henbane, mandrake root, and the thorn apple are examples of plants which have been used and misused because of their "magical" properties, which permit the mind to experience fantastic states or to be plunged into a narcolepsy, impervious to pain, fatigue, or sorrow.

Modern man has added innumerable chemicals capable of potential injury, even though most have been isolated or synthesized for beneficial reasons. Modern chemical and pharmaceutical processes have developed hundreds of new products which are used daily. Many are poisons whose usefulness outweighs the occasional instances of sickness or death they cause.

Treatment for poison. Treatment depends on the nature of the poison, the amount, and the route of administration. First-aid measures are directed toward prevention of further absorption by the body, removal of poisons when possible, neutralization by general or specific antidotes, and support of failing body functions. For inhaled poisons,

fresh air, oxygen, artificial respiration, and maintenance of body temperature may be employed as needed until trained aid is available. For ingested poisons, a specific antidote is given, if known, or the universal antidote, if the agent is unknown. The universal antidote consists of two parts powdered charcoal, one part tannic acid, and one part magnesium oxide. A household substitute can be made from two pieces of burned toast, one part of strong tea, and one part of milk of magnesia. Other materials may be given to dilute the poison and to slow absorption, such as milk, beaten eggs, or a suspension of flour, starch, or eggs in water. Vomiting should be induced in conscious patients unless a corrosive has been ingested, in which case regurgitation may cause further damage.

For skin and eye contaminations, water is used to flush off the poison. No chemical antidotes should be employed by the untrained.

Injected poisons from snake bites or drug overdosage should be immobilized as much as possible at the site of entry. The patient is kept still, a tourniquet is placed above the site, an ice pack is applied to reduce spread, and incision with suction may be used. Polyvalent venom antiserum should be available in areas with a high incidence of snake bite.

In the United States, the establishment of poison control centers has increased rapidly in the past few years. These are set up by qualified personnel and are usually located in or near a cooperating hospital. Their purpose is to be available for disseminating information on the toxicity and ingredients of both general and trade name poisons, and on treatment of poisoning. First-aid information is given to individuals who telephone, together with information on availability of medical treatment. Physicians may call to obtain specific information for diagnosis and treatment. Most of these centers also have treatment facilities available for local cases. In addition, information on epidemiology, treatment, diagnostics, and mechanisms of action of new poisons, as well as the results of research studies, is passed on to other centers and interested persons through a national clearing house.

[EDWARD G. STUART/N. KARLE MOTTET]

Tree diseases

Forest and shade tree diseases are discussed separately here, although the same causal agents can be involved with both groups of trees. In forests, generally the concern is with the effect of diseases on stands of trees rather than with individuals, and although the value of individual trees in forests can be high, usually this is much less than the value of shade trees. Diseases that result in disfiguration of trees, such as leaf spots, are important on shade trees but not on forest trees and, in contrast, shade trees can serve very well with substantial amounts of heart rot, which in forest trees is of major consequence. Thus the emphases on control are different, and the corrective measures themselves are often quite distinct.

Diseases of forest trees. From seed to maturity forest trees are subject to a succession of diseases. Losses due to pests and fire amount to 92% of the net saw timber growth, and diseases account for 45% of these losses. Young succulent seedlings, especially conifers, are killed by soil-inhabiting

Fig. 1. Oak wilt. (a) Oaks killed by the oak wilt fungus. (b) Mycelial mat that the oak wilt fungus produces beneath the bark of a wilted tree.

fungi (damping off); the root systems of older seedlings may be attacked by complexes of fungi and nematodes, including in particular fungi in the genera *Cylindrocladium*, *Sclerotium*, and *Fusarium*. Chemical treatment of soil with biocides, such as formulations containing methyl bromide, and fungicides, as well as cultural practices unfavorable to the development of pathogens, will help control seedling diseases. *See* FOREST ECOLOGY; FOREST PLANTING AND SEEDING; FUNGISTAT AND FUNGICIDE.

Leaf diseases. In forests, leaf diseases usually cause negligible losses, but brown spot needle blight, caused by *Scirrhia acicola*, can cause excessive defoliation in nurseries and plantations and prevent longleaf pine from starting height growth. Fungicides in the nurseries and prescribed burning in plantations have been used successfully for control. There are other important leaf diseases, such as Dothistroma needle blight, which have caused losses in plantations of Austrian pine (*Pinus nigra*) and ponderosa pine (*P. ponderosa*) in the central and northwestern United States. Copper fungicides applied at the right time control some of these diseases.

Wilt diseases. Diseases such as oak wilt (Fig. 1) are systemic (infecting entire plant), killing susceptible species in a matter of weeks or months by plugging the tree's vascular system. The fungus *Ceratocystis fagacearum* is disseminated to nearby healthy trees through root grafts and to longer distances by unrelated insects, including the Nitidulidae. The mycelial mats of the fungus are covered with spores that can be disseminated by insects. Control is possible by eradicating infected trees, especially the recently wilted red oaks on which spores are produced, and by disrupting root grafts with the chemical called Vapam, or VPM.

Canker diseases. These result from localized killing of the cambium and range from chestnut blight, which since its introduction in 1904 has practically annihilated the American chestnut (Fig. 2), to less important cankers caused by species of *Nectria* and *Eutypella* which seldom result in death of the tree. Large living chestnut trees are rarely seen in northern states, and they are vanishing rapidly in the Smoky Mountains of the southern Appalachians. Satisfactory resistant varieties of chestnut are being developed, but this fine tree species may never regain its prominent place in eastern forests. Control measures for other canker diseases usually consist of removing infected trees while thinning stands.

Rust diseases. Rusts, such as white pine blister rust and southern fusiform rust, which result in death of trees are very important diseases (Fig. 3), while other rusts which involve only the leaves are not usually serious problems. Many of the important rust fungi occur on two unrelated hosts, such as the currant for white pine blister rust and the oak for fusiform rust. Eradication of the alternate host, resistant varieties, pruning, and selection of planting sites less favorable for the fungus are the major control measures.

Heart rots. All tree species, even woods resistant to decay such as redwood, are subject to one or more fungi which can decay the center or heartwood portion of the main stem (Fig. 4). Some of these fungi enter through wounds and branch stubs and decay the heartwood; others enter roots and decay these roots and then the lower portion of the main stem. Most heart-rotting fungi invade trees approaching maturity, and losses by these fungi can be avoided by shortening rotation and by avoiding wounds. A few decay fungi such as *Fomes annosus*, *Polyporus tomentosus*, and *Poria weirii* can invade young, vigorous trees and kill them. Thining plantations increases the incidence of *F. annosus*, which can quickly invade fresh stumps and move from there through the roots to surrounding healthy trees. Losses caused by this fungus can be reduced by less thinning and by treating fresh stumps with chemicals (borax, urea).

Miscellaneous diseases. Dwarf mistletoe is a small, partially parasitic plant that causes its host to produce witches'-brooms (dense clusters of branches) and eventually kills ponderosa pine, other western conifers, and black spruce (Fig. 5). Control consists of eradicating infected trees.

In the 1930s birch dieback, little-leaf disease of shortleaf pine, and pole blight of western white pine appeared, and as yet the causes are not

Fig. 2. Chestnut blight in American chestnut tree (*USDA*). Inset is mycelial fan of chestnut blight caused by *Endothia parasitica*, advancing through bark of American chestnut. The tip of the fan on the left is surrounded by cortical tissue. The contents of the cortical cells back from the tip of the fan are discolored to a yellowish brown, as indicated by the darkened cells (*after W. C. Bramble, from J. S. Boyce, Forest Pathology, 2d ed., McGraw-Hill, 1948*).

Fig. 3. White pine blister rust (*Cronartium ribicola*). (*Photograph by Robert Campbell*)

Fig. 4. Shelf fungus (*Fomes applanatus*) on a dead aspen tree. This wood-rotting fungus enters through roots and wounds and decays the heartwood and sapwood of both hardwoods and softwoods.

Fig. 5. Pistillate shoots of ponderosa pine mistletoe (*Arceuthobium vaginatum forma cryptopodum*) on branch of sapling. Note unripe fruits. (*U.S. Forest Service*)

completely known. Birch dieback may be due in part to increased soil temperature which causes excessive root mortality. Little-leaf results from a combination of poor soil drainage and the root parasite *Phytophthora cinnamomi*. Pole blight is due in part to soil moisture deficiencies. These are examples of other similar diseases, which are all due in part to unfavorable weather and which may or may not involve parasites.

Diseases of shade trees. Shade trees are often grown in abnormal habitats, and because appearance is more important than the wood produced, shade tree problems are different from those in the forest. The desire for new and different trees has resulted in the introduction of exotic species, many of which are unsuited to the new climate. Even native species are often moved from their normal habitats and placed where they are predisposed to secondary pathogens.

Fungus diseases. A tree species such as Colora-

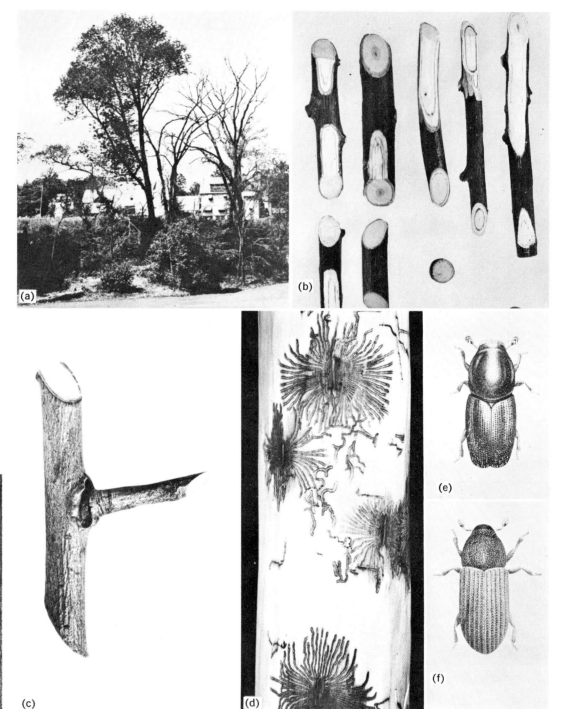

TREE DISEASES

Fig. 7. Elm infected with bacterial wetwood. The bacterial ooze is coming from a small wound.

Fig. 6. Dutch elm disease. (a) Group of trees affected by the disease. (b) Discoloration in the sapwood of infected trees. (c) Feeding scar in small elm crotch made by an adult of the smaller European elm bark beetle (*Scolytus multistriatus*). (d) Brood galleries made by female beetles and larvae. (e) European elm bark beetle, the most important carrier of the Dutch elm disease. (f) Native elm bark beetle. (a,c,d, USDA; b,e,f, Michigan State University)

do blue spruce in its natural environment is seldom attacked by the spruce canker fungus (*Cytospora kunzei*), yet this fungus can kill this tree species when it is planted in other parts of the world. The effects of this secondary parasite can be minimized by selecting protected sites with better soils and avoiding dry, southwest-facing slopes. As a general rule, many diseases which can be attributed to an unfavorable environment as well as those caused by weakly parasitic fungi can be avoided by selecting native tree species from a local seed source.

Dutch elm disease, introduced to the United States in 1930 or earlier, is an important shade tree disease because of the value of the elm for lining streets (Fig. 6). The Dutch elm disease fungus (*Ceratocystis ulmi*) is introduced to the vascular system of healthy trees by the smaller European elm bark beetle (*Scolytus multistriatus*) and the American species (*Hylurgopinus rufipes*). Resistant varieties of elm and possibly systemic chemicals to prevent bark beetle feeding may be developed, but until that time sanitation (destruction of dead and dying elm wood) and a protective insecticide are the best means of control.

Virus diseases. The elms are also attacked by a virus, and the disease is known as phloem necrosis. This is the best known of the virus diseases of trees. Infected trees wilt and die, the inner bark turning brown and emitting a wintergreen odor. The virus is transmitted by the elm leafhopper (*Scaphoides luteolus*).

Bacterial diseases. One of the few known bacterial diseases of trees is wetwood of elm, caused by *Erwinia nimipressuralis* (Fig. 7). The wetwood condition occurs in many other tree species and presumably is caused by bacteria. Satisfactory control measures have not been developed.

Leaf spot and heart rot. Diseases of minor importance in the forest, such as leaf spots, may be objectionable on shade trees. Leaf spots occur in years when the weather is favorable for the fungi involved, usually coinciding with a wet spring season. Such diseases can be controlled by application of the right fungicides, usually in early spring. Heart rot, on the other hand, is a major problem in the forest but is of minor importance in shade trees.

Many trees may remain alive for decades with extensive heart rot. Expensive cavity work is of questionable value since it does not stop the decay process and may not add much strength to the tree. *See* PLANT DISEASE.

[DAVID W. FRENCH]

Tundra

An area supporting some vegetation beyond the northern limit of trees, between the upper limit of trees and the lower limit of perennial snow on mountains, and on the fringes of the Antarctic continent and its neighboring islands. The term is of Lapp or Russian origin, signifying treeless plains of northern regions. Biologists, and particularly plant ecologists, sometimes use the term tundra in the sense of the vegetation of the tundra landscape. Tundra has distinctive characteristics as a kind of landscape and as a biotic community, but these are expressed with great differences according to the geographic region.

Patterns. Characteristically tundra has gentle topographic relief, and the cover consists of perennial plants a few centimeters to a meter or a little more in height. The general appearance during the growing season is that of a grassy sward in the wetter areas, a matted spongy turf on mesic sites, and a thin or sparsely tufted lawn or lichen heath on dry sites. In winter, snow mantles most of the surface with drifts shaped by topography and surface objects including plants; vegetation patterns are largely determined by protecting drifts and local areas exposed to drying and scouring effects of winter winds. By far, most tundra occurs below the mean annual temperature is below the freezing point of water, and perennial frost (permafrost) accumulates in the ground below the depth of annual thaw and to depths at least as great as 500 m. A substratum of permafrost, preventing downward percolation of water, and the slow decay of

water-retaining humus at the soil surface serve to make the tundra surface moister during the thaw season than the precipitation on the area would suggest. Retention of water in the surface soils causes them to be subject to various disturbances during freezing and thawing, as occurs at the beginning and end of, and even during, the growing season. Where the annual thaw reaches depths of less than about 50 cm, the soils undergo "swelling," frost heaving, frost cracking, and other processes that result in hummocks, polygonal ridges or cracks, or "soil flows" that slowly creep down slopes. As the soils are under this perennial disturbance, plant communities are unremittingly disrupted and kept actively recolonizing the same area. Thus topography, snow cover, soils, and vegetation interact to produce patterns of intricate complexity when viewed at close range.

Plant species, life-forms, and adaptations. The plants of tundra vegetation are almost exclusively perennial. A large proportion have their perennating buds less than 20 cm above the soil (chamaephytes in the Raunkiaer life-form system), especially among the abundant mosses and lichens. Another large group has the perennating organs at the surface of the soil (hemicryptophytes in the Raunkiaer system). Vegetative reproduction is common—by rhizomes (many of the sedges), stolons (certain grasses and the cloudberry, *Rubus chamaemorus*), or bulbils near the inflorescence (*Polygonum viviparum, Poa vivipara, Saxifraga hirculis*); thus clone formation is common in plant populations. Apomixis, the short-circuiting of the sexual reproduction process, is found frequently among flowering plants of tundra. Seed is set regularly by agamospermy, for example, in many dandelions (*Taraxacum* sp.), hawkweeds (*Hieracium* sp.), and grasses (*Calamogrostis* sp., *Poa* sp., *Festuca* sp.). The high incidence of apomixis in tundra flowering plants is coincident with high frequency of polyploidy, or multiple sets of chromosomes, in some circumstances a mechanical cause of failure of the union of gametes by the regular sexual process. Asexual reproduction and polyploidy tend to cause minor variations in plant species populations to become fixed to a greater extent than in populations at lower latitudes, and evolution tends to operate more at infraspecific levels without achieving major divergences. Adaptations are more commonly in response to physical factors of the stressful cold environment rather than to biotic factors, such as pollinators or dispersal agents, of the kinds that exert such control in the congenial warm, moist climates.

Soil conditions. Tundra soils are azonal, without distinct horizons, or weakly zonal. Soils on all but very dry and windswept sites tend to accumulate vegetable humus because low temperatures and waterlogging of soils inhibit processes of decay normally carried out by bacteria, fungi, and minute animals. Where permafrost or other impervious layers are several meters or more beneath the surface in soils with some fine-grained materials, leaching produces an Arctic Brown Soil in which there is moderately good drainage and cycling of mineral nutrients. In the greater part of tundra regions not mantled by coarse, rocky "fjell-field" materials (Fig. 1), the soils are more of the nature of half-bog or bog soils. These are characterized by heavy accumulations of raw or weakly decayed

Fig. 1.   Fjell-field tundra of the high Arctic. Sedges, mosses, and lichens form a thin and discontinuous sod. Late-persisting snowbanks are withdrawing from surfaces that are lighter in color because they lack many of the common plants, including dark-colored species of lichens. (*Photograph by W. S. Benninghoff, U.S. Geological Survey*)

humus at the surface overlying a waterlogged or perennially frozen mineral horizon that is in a strongly reduced state from lack of aeration. Such boggy tundra soils are notoriously unproductive from the standpoint of cultivated plants, but they are moderately productive from the standpoint of shallowly rooted native plants. In Finland, forest plantations are being made increasingly productive on such soils by means of nutrient feeding to aerial parts. *See* SOIL, ZONALITY OF.

Productivity. By reason of its occurrence where the growing season is short and where cloudiness and periods of freezing temperatures can reduce growth during the most favorable season, tundra vegetation has low annual production. Net radia-

tion received at the Earth's surface is less than 20 kg-cal/cm²/year for all Arctic and Antarctic tundra regions. Assuming a 2-month growing season and 2% efficiency for accumulation of green plant biomass, 1 cm² could accumulate biomass equivalent to 66.6 g-cal/year. This best value for tundra is not quite one-half the world average for wheat production and about one-eighth of high-yield wheat production. The tundra ecosystem as a whole runs on a lower energy budget than ecosystems in lower latitudes; in addition, with decomposer and reducer organisms working at lower efficiency in cold, wet soils, litter and humus accumulate, further modifying the site in unfavorable ways. Grazing is one of the promising management techniques (Fig. 2) because of its assistance in speeding up the recycling of nutrients and reducing accumulation of raw humus. *See* BIOMASS; ECOSYSTEM; VEGETATION MANAGEMENT.

Fauna. The Arctic tundras support a considerable variety of animal life. The vertebrate herbivores consist primarily of microtine mammals (notably lemmings), hares, the grouselike ptarmigan, and caribou (or the smaller but similar reindeer of Eurasia). Microtine and hare populations undergo cyclic and wide fluctuations of numbers; these fluctuations affect the dependent populations of predators, the foxes, weasels, hawks, jaegers, and eagles. Alpine tundras generally have fewer kinds of vertebrate animals in a given area because of greater discontinuity of the habitats. Arctic and Alpine tundras have distinctive migrant bird faunas during the nesting season. Tundras of the Aleutian Islands and other oceanic islands are similar to Alpine tundras with respect to individuality of their vertebrate faunas, but the islands support more moorlike matted vegetation over peaty soils under the wetter oceanic climate. Tundras of the Antarctic continent have no vertebrate fauna strictly associated with it. Penguins and other sea birds establish breeding grounds locally on ice-free as well as fringing ice-covered areas. The only connection those birds have with the tundra ecosystem is the contribution of nutrients from the sea through their droppings. All tundras, including even those of the Antarctic, support a considerable variety of invertebrate animals, notably nematode worms, mites, and collembola on and in the soils, but some other insects as well. Soil surfaces and mosses of moist or wet tundras in the Arctic often teem with nematodes and collembola. Collembola, mites, and spiders have been found above 20,000 ft in the Himalayas along with certain molds, all dependent upon organic debris imported by winds from richer communities at lower altitudes. *See* TAIGA; VEGETATION ZONES, WORLD.   [WILLIAM S. BENNINGHOFF]

Bibliography: Arctic Institute of North America, *Arctic Bibliography*, 13 vols., 1953–1967; G. A. Doumani (ed.), *Antarctic Bibliography*, 2 vols., 1965, 1966; M. J. Dunbar, *Ecological Development in Polar Regions*, 1968; H. P. Hansen (ed.), *Arctic Biology*, 18th Annual Biology Colloquium, Oregon State College, 1957; N. Polunin, *Introduction to Plant Geography and Some Related Sciences*, 1960; J. C. F. Tedrow (ed.), *Antarctic Soils and Soil Forming Processes*, American Geophysical Union, Antarctic Research Series, vol. 8, 1966; H. E. Wright, Jr., and W. H. Osburn, *Arctic and Alpine Environments*, 1968.

Fig. 2.   Alpine tundra in French Alps. Altitudinal limit of trees occurs in valley behind building in middle distance. Although similar in vegetation structure to tundra of polar regions, Alpine tundras of lower latitudes are usually richer in vascular plant species than tundras of polar regions, and structure and composition of the vegetation have been modified by pasturing. (*Photograph by W. S. Benninghoff*)

Utility industry, electrical

Successful startup of a record 44,000,000 kW of new generating capability during 1973 enabled electrical utilities in the United States to carry a record summer peak of 350,400,000 kW with a reserve margin of slightly over 20%, and to build a little surplus toward 1974 requirements. Of this total new generating capability, about 9,800,000 kW was in nuclear-fueled units, bringing the total nuclear capability to about 25,200,000 kW.

Nuclear power. Under the conditions that prevail in the electrical utility industry today, the increase of 9,800,000 kW in nuclear generation is of vital importance. In the first place, a nuclear power station has an attractive appearance and emits no combustion gases to pollute the atmosphere. In addition, every kilowatt of nuclear generation in service saves about 10 barrels (1 barrel petroleum = 0.16 m³) per year of low-sulfur fuel oil that otherwise would have been needed for power station fuel. Finally, because nuclear fuel costs about one-fourth as much per kilowatt-hour as low-sulfur fuel oil, the use of nuclear generating plants helps to hold down the rising cost of electricity. *See* NUCLEAR POWER.

But the building of nuclear power stations has not been easy. Because the basic technology of harnessing nuclear fission was developed originally as a wartime bomb project, many costly precautions are prescribed to safeguard civilian applications. These safeguards have proliferated to the point where upward of 40 separate reviews and regulatory approvals are required before a nuclear power unit can be built and put into commercial operation. Hence there are many points where delays can occur.

As a matter of fact, more than half of the nuclear generation planned for completion during 1972 was not licensed in time to carry load until 1973. And continuing delays of this nature, in recent years, forced electrical utilities to build nearly

10,000,000 kW of alternative generation—mostly driven by combustion (gas) turbines—to supply their customers until the nuclear units could be put on line. This was costly in two ways. In the first place, the duplication of effort added well over $1,000,000,000 to utility construction expenditures. Then the less efficient use of premium fuels in these combustion turbines boosted operating costs by as much as $100,000,000 per year.

As 1973 drew to a close, however, the prospects for building and licensing new nuclear power stations essentially on schedule seemed to be improving. This situation stems not so much from less vigorous voicing of objections at the multitudinous reviews, as from the electrical utilities allowing more time for reviews in their construction schedules. If the units already under construction or planned for service before 1980 are completed on schedule, the electrical utility industry will have nearly 120,000,000 kW of nuclear stations on line generating 720,000,000,000 kWhr per year and saving 1,200,000,000 bbl per year of imported oil. This savings could ease the rise in cost of electrical service by as much as 10%. Figure 1 indicates projected generating capacity of nuclear steam and other systems.

Conventional steam units. Other generating capability added to electrical utility systems during 1973 included 22,896,000 kW in 46 conventional steam units. Primary fuel is coal for 26 of these and oil for 14. Only 6 were designed primarily for natural gas, which, during the several years required to build large generating units, became increasingly difficult to procure on long-term contracts.

Reflecting this growing shortage of natural gas, stations planned for completion after 1974 are swinging heavily to coal or oil, even though the coal must be hauled as much as 1000 mi from low-sulfur fields and coal-burning stations cost as much as $100/kW more to build than stations designed solely to burn gas. This differential stems largely from the cost of coal- and ash-handling systems, the need for pollution-control devices, and the additional size of coal-burning boilers.

Combustion turbines. Combustion turbines, so called because most of them today burn oil instead of gas, contributed about 5,500,000 kW to the 1973 generation additions, largely as replacements for several nuclear units that were delayed. The rate at which they will be added in future years probably will slacken as the industry becomes more successful in meeting its construction schedules.

As this situation develops, combustion turbines will be reassigned to the peaking service for which they are best suited, or upgraded to more efficient cycling service by adding heat recovery boilers in their exhaust ducts and generating steam there to drive steam turbines. Such a system, known as a combined cycle, permits extensive preassembly in factories and speedy erection on site. It also affords extremely good overall efficiency and discharges less than half as much heat to the cooling system as a conventional steam system.

Thermal pollution. This point about heat discharge is vitally important to the electrical utility industry because of mounting governmental restrictions on the discharge of warm water, dubbed

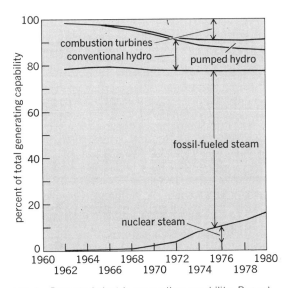

Fig. 1. Projected electric generating capability. Recent completions of nuclear combustion turbine and pumped-storage generation move the industry toward more economical mix. (*Courtesy of Electrical World*)

Fig. 2. Artist's drawing of the sulfur removal system of the Commonwealth Edison plant near Lemont, Ill. Apparatus in open framing (subsequently enclosed) and low building and tanks at right were required to retrofit stack-gas scrubber to one generating unit. (*Courtesy of Commonwealth Edison Company*)

July 1, 1977, and electrical utilities are striving strenuously, although not too hopefully, to comply.

Air pollution. Methods under test for removing pollutants from combustion gases differ somewhat, but they generally involve passing the gases through a chemical spray, most commonly a water solution of pulverized limestone, in a scrubber to wash most of the sulfur dioxide and dust into holding tanks. The washed combustion gases are then blown out to the atmosphere via the usual tall stack. The process does not end there, however, as it is still necessary to dispose of the pollutants captured in the wash liquid without causing pollution in a different form, and a variety of methods for doing this are still under test. An architect's drawing of one such installation is shown in Fig. 2.

An additional problem in adding antipollution devices to existing generating stations is the power required to operate them. Some cooling towers require huge fans to drive air through the falling water, and stack-gas scrubbers have numerous motor-driven pumps and, sometimes, additional fans to compensate for gas velocity loss. On the average, these devices absorb 5 to 8% of the main generator's output. The power system must be enlarged by that amount, nearly a normal year's growth, to maintain its capability of meeting its customers' demands.

Thus the cost of electric service is increased by the use of additional fuel, as well as by charges on the investment in antipollution and cooling devices and in additional generating plant. All of these have been boosted by continuing inflation and the scramble for nonpolluting fuels. As a result, the average rate for residential service in 1973 was 2.37¢ per kWhr, up 13.4% over the low of 2.09¢ set in 1969. Projections indicate a continuing rise to about 2.65¢ per kWhr during the 1980s, increased by an unpredictable amount by further inflation.

Hydroelectric plants. Conventional hydroelectric additions during 1973 totaled a mere 1,104,000 kW, largely in the form of additional generating units in existing stations. This reflects the fact that few suitable sites remain for development, and the best of these, High Mountain Sheep on the Snake River, is snarled in controversy.

Pumped-storage hydro, on the other hand, gained some 4,561,000 kW, mostly at two sites—Ludington, Mich., and Northfield Mountain, Mass. As indicated in the table, the sum of these two projects almost equals the total pumped-storage capability in service prior to 1973. Generation of this type, however, is not a prime energy source. Instead, it functions much like a storage battery to absorb surplus electricity from other types of generation during light-load periods, generally late at night, and to feed it back to the electrical system during heavy-load periods.

Electricity usage. The resumption of industrial activity to a near-capacity level during 1973 caused a 10.4% increase in the industrial consumption of electricity, stemming in part from the exceedingly rapid installation of air- or water-pollution-control devices. A survey of 2100 large customers by the Edison Electric Institute revealed that 7.3% of the electricity they bought in 1972 was for such uses. About 5% of the industrial consumption went to processing fuel for nuclear stations, one of the

thermal pollution for propaganda purposes, into natural lakes, rivers, or oceans. A recent proposal would require the addition of evaporative cooling towers on all nuclear units and on base-loaded fossil-fueled units. This proposal, if it becomes mandatory in its present form, could force the accelerated construction of enormous cooling towers for 400 to 500 generating units at a cost approaching $2,000,000,000. Such a system sprays the warm water into a massive flow of air, which removes heat from the water partly by contact and partly by evaporation. The cooled water is returned to the generating units, and the part that evaporates is dissipated to the atmosphere, often in the form of a vapor plume that can be seen for miles.

This cooling tower program would be in addition to retrofitting antipollution devices to coal- and oil-burning stations at a cost of some $10,000,000,000. This cost is by no means firm, because extensive tests on some 20 trial installations have yet to demonstrate that any system is sufficiently developed for dependable commercial operation. Rulings in effect at this time require compliance with extremely rigid restrictions no later than

cleanest sources of electricity. Industrial consumption will continue to climb as industries tool up to meet new demands for their products.

Residential customers used 32.3% of the electrical utility industry's output in 1973, an increase of 8.5% over their consumption the previous year. About a third of this growth stemmed from a 3% increase in number of customers and two-thirds from increased usage per average customer. A substantial portion of this increased usage is attributable to the use of electricity for heating 7,250,000 dwelling units, a gain of nearly 1,000,000 during 1973 that reflects a growing preference for this convenient and pollution-free form of heating, and concern about the future availability and cost of gas.

Commercial consumption of electricity shared the increase in 1973, gaining 8% and accounting for 22.8% of the total electricity output. Much of this increase probably stemmed from the completion of enormous shopping centers, with superbly lighted stores and parking areas, in all regions of the nation.

All three classifications of electricity usages seem assured of continuing growth in the years ahead, although the rates of growth will decline gradually as the country's population approaches its ultimate level sometime after the year 2000. Consequently, the electrical utility industry must continue adding generating capability at a rate of 40,000,000 to 50,000,000 kW each year. It must do this in the face of growing environmental restraints, dislocations of fuel supplies, and difficulty in raising the money to pay for this generating capability.

Transmission circuit construction. An unusually heavy construction program during 1973 added 13,000 mi of transmission circuits to connect the additional generating capability to load areas and to strengthen interconnections among load areas. Nearly 19% of this mileage was in overhead circuits for operation at 345,000 V or higher, and hence had carrying capabilities upward of 600,000 kW per circuit. Additions at this extra-high-voltage level were built in all regions of the country, but about 65% of the mileage was in the North Central and Pacific regions.

Lower-voltage transmission lines in the 115,000 to 230,000-V range accounted for 80% of the total mileage but only about one-third of the carrying capacity. These additions also were in all regions, but nearly 50% of the total mileage was in the Southeast and South Central regions.

Only 1.2% of the transmission mileage added during 1973 was underground, practically all of it in the metropolitan areas of the Middle Atlantic, North Central, and South Atlantic regions. In addition, 66% of the mileage was for operation in the 115,000- to 161,000-V range, where its carrying capacity would average only about 60,000 kW per circuit.

Stepping up to these transmission voltages at generating stations and down again at load centers required the construction of substations containing nearly 150,000,000 kVA of transformer capacity. About a third of this capacity was the main power transformers at generating stations, another third was used in major substations to interconnect transmission circuits at different voltage levels, and most of the remainder was installed in new or enlarged distribution substations serving local load areas.

Housing construction continued at a 2,170,000-dwelling-unit level in 1973, slightly down from the 2,379,000 units that were started during 1972. Consequently, distribution circuits had to be extended to serve about 2,000,000 new residential and 200,000 new commercial customers. In addition, the average usage per residential customer rose 5.8%, to 8133 kWhr, boosting total energy sales in the residential category to 555,050,000,000 kWhr. During the same period, commercial sales rose 8%, to 390,808,000,000 kWhr. Accordingly, distribution substation capability had to be expanded by nearly 50,000,000 kVA, and primary circuits extended 93,000 mi, of which about 25% was put underground.

Cost of system additions. Expenditures for system additions during 1973 approached $20,600,000,000, some 23% higher than in 1972. Of these expenditures, some 57% went for generating stations, 14% for transmission, 24% for distribution, and the remaining 5% for such items as control centers, load-control systems, operating headquarters, and a wide variety of construction and transportation devices to improve the productivity of electrical utility operation. *Electrical World's* 24th Annual Electrical Industry Forecast indicates that electrical utility expenditures for plant additions will remain at about the 1973 level, with increases due to continuing inflation, each year through 1975, then resume the traditional upward trend. This high level of expenditures is severely taxing the industry's ability to raise the required funds and burdening it with extremely high interest charges that must be reflected in future rates for electrical service.

[LEONARD M. OLMSTED]

Bibliography: The 1970 National Power Survey, Federal Power Commission, pts. 1 and 4, 1972; 1973 Annual Statistical Report, *Elec. World*,

United States electric power industry statistics for 1973

Parameter	Amount	Increase over 1972, %
Generating capability, kW ($\times 10^3$)		
Total	443,524	11.0
Conventional hydro	53,551	2.1
Pumped-storage hydro	8,680	111.0
Fossil-fueled steam	317,922	7.8
Nuclear steam	25,195	64.6
Combustion turbine and		
internal combustion	38,276	17.0
Energy production, kWhr ($\times 10^6$)	1,905,406	8.6
Energy sales, kWhr ($\times 10^6$)		
Total	1,719,906	9.0
Residential	555,050	8.5
Commercial	390,808	8.0
Industrial	706,071	10.4
Miscellaneous	67,977	4.6
Revenue, total ($\times 10^6$)	$31,250	11.9
Capital expenditures, total ($\times 10^6$)	$20,582	23.6
Customers ($\times 10^3$)		
Residential	69,300	3.0
Total	78,350	2.9
Residential usage, kWhr (average)	8,133	5.8
Residential bill, ¢/kWhr (average)	2.37	3.4

From *Elec. World*, Sept. 15, 1973, and extrapolations from Edison Electric Institute monthly data.

179(6):35–66, March 15, 1973; *Statistical Year-book of the Electric Utility Industry*, Edison Electric Institute, 1973; 24th Annual Electrical Industry Forecast, *Elec. World*, 180(6):39–56, Sept. 15, 1973.

Vegetation

The total mass of plant life that occupies a given area. The plant cover in any landscape consists of a matrix of individuals usually belonging to many different species. The different species (irrespective of their abundance) are collectively referred to as the flora. Thus the flora of Walden Pond (Massachusetts), of Cheboygan (Michigan), of Manitoba (Canada), or of Switzerland is merely a list of all the different kinds of plants that have been found in a certain pond, township, province, or country. The flora of wheat fields, city streets, woodlots, and the like may also be inventoried. About 350,000 species of plants have been described. The tropics contain by far the largest numbers: the Lower Amazon Basin, for instance, may have as many as 42,000 higher plants, many arctic areas have less than 100, some high mountains have 10 or so, a large tract of desert in Mauritania is reported to have 4, and vast Antarctica has only 2. In this respect, some areas are considered to be floristically rich and others poor.

Flora provides the raw materials or building blocks of vegetation. The pieces of this mosaic are the communities, or plant societies, which can be recognized and described in terms of their composition (flora), their appearance or physiognomy (structure), and their ecology or site requirements. To this list may be added some consideration of origin and migration and of relation to man's influence.

Origin of flora. In any one region, plant species (and whole masses of vegetation, for that matter) are likely to have come from the four points of the compass, and quite often the stamp of past migrations is quite visible on the land. Thus, at high altitudes in the southern Appalachians, spruces and firs bear witness to a colder period in the past, as do the hemlocks that have taken refuge on the cool slopes at lower altitudes. Similarly, in Spain and Portugal, evergreen cherries and rhododendrons in moist ravines are relicts of a wetter climatic period. In all parts of the world, local floras can be analyzed to estimate, for instance, the importance of species of prairie origin in a now forested area, of species of desert origin in an alpine area, of species of montane origin in the plain, or of species of tropical origin in a temperate area.

Structure. The organization in space of the mass (or structure) of vegetation confers a particular appearance or physiognomy upon the countryside: The height, branching, and spread or coverage of the plants, the size, texture, and density of their foliage, the ratio of woody to herbaceous to mossy tissues, and the emergence of definite layers contrast rather strongly from place to place. Also, the times of development and withering impose an alternation of physical conditions that is sometimes extreme. For instance, the deciduous forests lose their whole canopy of leaves for a long period, allowing the light and warmth of the Sun to reach the soil. At that time a thick layer of herbs frequently develops. Conversely, the evergreenness of the boreal forest of Canada retards the wasting of

snow on the forest floor in the spring. *See* PLANT COMMUNITY; PLANTS, LIFE FORMS OF.

In any one landscape, vegetation masses vary in structure: A hillside may be forested at the top and have a savannalike shrubby field lower down and a grassy swamp at the base: emerging sandbanks or sharp rock outcrops may be almost completely barren. Flooded or very wet places harbor a vegetation very different from that of the driest areas. In fact, it is often possible to assign a particular community (or group of communities) to each site: Dunes, rocky outcrops, riverbanks, and poorly drained flats all support characteristic assemblages of plants. Some species indicate the presence of lime (the Canada violet), poor drainage (the cardinal flower), the passage of fire (the fireweed), the prevalence of acidity (the pitcher plant), or the periodic recurrence of floods (the skunk cabbage). These are called indicator species. It is in their physiology that these plants differ and thereby draw attention to contrasting ecosystems, since the condition of each habitat imposes its particular stresses. *See* ECOSYSTEM.

Not only are some kinds exclusive to wet or dry soils, or to acid or salt soils, but also the overall development of the plant mass (structure) is different: Only herbs and low shrubs appear in a marsh; grasses and other herbs occur in a field; and a multilayered distribution of trees, shrubs, herbs, and mosses characterizes a forest.

Ecology. Even a summary investigation of site conditions reveals that slope, exposure, drainage, chemical composition, texture and structure of soil, and relative acidity are major factors that most visibly influence the local distribution of vegetation and induce the emergence of different ecosystems. Thus, if sampled in many stands, dune communities, floodplain forests, bog heaths, and ravine scrubs are likely to have the same composition and the same physiognomy or at least to fluctuate within a given pattern in any one region.

These landforms are not static, and as the dunes become stabilized, as the marshes fill in, as the slopes erode, and as the riverbanks are silted, the vegetation itself changes. This process, known as succession, makes for a more or less gradual replacement of one plant community by another, until a climax is reached. Such relative stability (or dynamic equilibrium) features, as a rule, the best-developed vegetation on the most completely stratified soil of the well-drained uplands. *See* CLIMAX COMMUNITY; PLANT FORMATIONS, CLIMAX; SUCCESSION, ECOLOGICAL.

Energy cycles. Especially where green plants are involved, vegetation is a transformer of environmental resources: Water, oxygen, and carbon dioxide in the air and water, and various organic and inorganic compounds in the soil are taken up by plants able to utilize solar energy and are transformed into living tissue, and substances are often accumulated as reserves (starch, sugar, and cellulose), eventually to be returned to the air or soil. In the fulfillment of this process the potential of different kinds of vegetation contrasts greatly: Algal or lichen crusts on rocks have a very slow uptake, whereas a tall forest or a deeply rooted grassland effects a tremendous turnover.

This energy relationship, which expresses the productivity of the plant cover, is of course related to environmental controls, mostly to climate and soil, but also to historical factors, such as dissemination, migration, and physiographic change, as well as to human and animal action.

Management. Man's interference (lumbering, plowing, damming, burning, and other activities) creates new habitats and new resources for plant and animal life. A landscape modified by man and later abandoned tends to revert to its primeval conditions, but frequently cannot because of the permanent damage or change to the soil or to the other factors of the site. *See* VEGETATION MANAGEMENT; VEGETATION ZONES, ALTITUDINAL; VEGETATION ZONES, WORLD.   [PIERRE DANSEREAU]

Bibliography: W. D. Billings, *Plants and the Ecosystem*, 1964; S. A. Cain, *Foundations of Plant Geography*, 1944; P. Dansereau, *Biogeography: An Ecological Perspective*, 1957; F. E. Egler, Vegetation as an object of study, *Phil. Sci.*, 9(3):245–260, 1942; R. Good, *The Geography of the Flowering Plants*, 3d ed., 1964; H. J. Oosting, *The Study of Plant Communities*, 2d ed., 1956.

Vegetation and ecosystem mapping

The drawing of maps which locate different kinds of plant cover in a geographic area is a rapidly developing art and science. It is relevant to man's wise use of many of the Earth's limited resources. Practical ecological relations between dominating plants and other parts of the ecosystem complex suggest that ecosystems be represented in many mapping activities. *See* ECOLOGY; ECOSYSTEM; TERRESTRIAL ECOSYSTEM.

BASIC CONSIDERATIONS

Relations between plants, animals, and the resources in a local ecosystem make the vegetation significant as the dominating biomass and food source. Primary producers also influence many factors in the larger environment as they impinge upon man, on other consumers (or visiting animals), and on decomposing organisms and residues in the system. Thus in fundamental study of ecosystem relations and in many applications of ecological sciences, vegetation maps are useful. *See* CONSERVATION OF RESOURCES; ECOLOGY, APPLIED; ENVIRONMENT; FOREST AND FORESTRY; LAND-USE PLANNING; WILDLIFE CONSERVATION.

Probably since ancient times, and certainly since the mapmaking art developed in the Renaissance, the location of vegetation has been drawn in relation to the natural and cultural patterns. Küchler, however, emphasizes the recent traditions in America and Europe for clarifying basic considerations of the variables or classes to be mapped, and of the techniques and applications. He illustrates many ways in which important terrain features have become combined with terminology that applies more strictly to vegetation in its structural, floristic, or ecological aspects. Some other geographers and many botanists have felt obliged to purge such hybrid terminology for the sake of legitimate definitions by one or more of these aspects. Experienced mappers and committees, such as that of UNESCO, have often compromised (see the table).

Yet many scientists, resource managers, and especially laymen revert to words and concepts which carry a message about climatic, terrain, or other variables of the ecosystem because these are functionally related to vegetation itself. Be-

cause of such functional correlations, certain vegetation maps well delineate the main contrasts between ecosystem types (or gradients) as distributed over the world or over regions. Depending greatly on the scale of mapping, there is a challenge to reexamine the question of how well the ecosystem can be used as the focus for future mapping efforts. *See* PLANT GEOGRAPHY; VEGETATION MANAGEMENT.

TECHNICAL REQUIREMENTS AND METHODS

It may seem premature to anticipate this challenge of ecosystem mapping before the principles of mapping the vegetation itself have been clarified and practiced more widely around the world. However, a variety of new techniques of gathering and processing data make it timely to consider this problem.

Base maps and scales. Topographic maps have been improved in showing details of slope and aspect (direction) of a mapping unit, in addition to its altitude and exposure. Simple, pale contour lines provide this information without distracting from the markings superimposed to represent vegetation or ecosystems. Topographic shading makes relief and drainage stand out more sharply, but may interfere with the biological patterns to be shown.

Maps on a scale of 1:24,000 from the U.S. Geological Survey may be a bit too small for working purposes with plant associations and diverse cultural features. However, they are widely available and can be enlarged photostatically for field work —perhaps to the size of aerial photographs. U.S. Army Map Service sheets at 1:250,000 are available in consistent style over wide areas (including many not mapped recently or not at all on large scale). If not too outdated by land clearing, green forest overlay printing on both of the foregoing map types offers a first-order distinction between forest-woodland and other kinds of vegetation: subdivision can then proceed to any appropriate degree for the purpose of the mapping. Such maps at least serve well to show relations of altitudinal zonation, and may further suggest relations to geologic substrate (and soil type associations). *See* GEOMORPHOLOGY; VEGETATION ZONES, ALTITUDINAL.

The mapping of wide areas at the larger scales, even with benefit of remote sensing techniques mentioned below, may be of quite varying accuracy for regions with different economic and landscape conditions. A common denominator of small-scale mapping at a widely available and acceptable scale convention of 1:1,000,000 has been advocated. The UNESCO legend of types as given (with abbreviations) in the table was nominally oriented toward this scale, although some of the subdivisions advocated by Ellenberg and others would become most useful on larger-scale maps.

For large countries, wall maps with good color printing can show many details by using patterns and symbols at smaller scales. For the United States, Küchler's 1:3,168,000 map of 116 kinds of "Potential Natural Vegetation" has considerable detail related to patterns of rivers (shown) and mountains (not shown, but readily recognized by distribution of montane conifer or meadow types). For one-sixth of the Earth's continental area in the Soviet Union, an eight-sheet wall map at

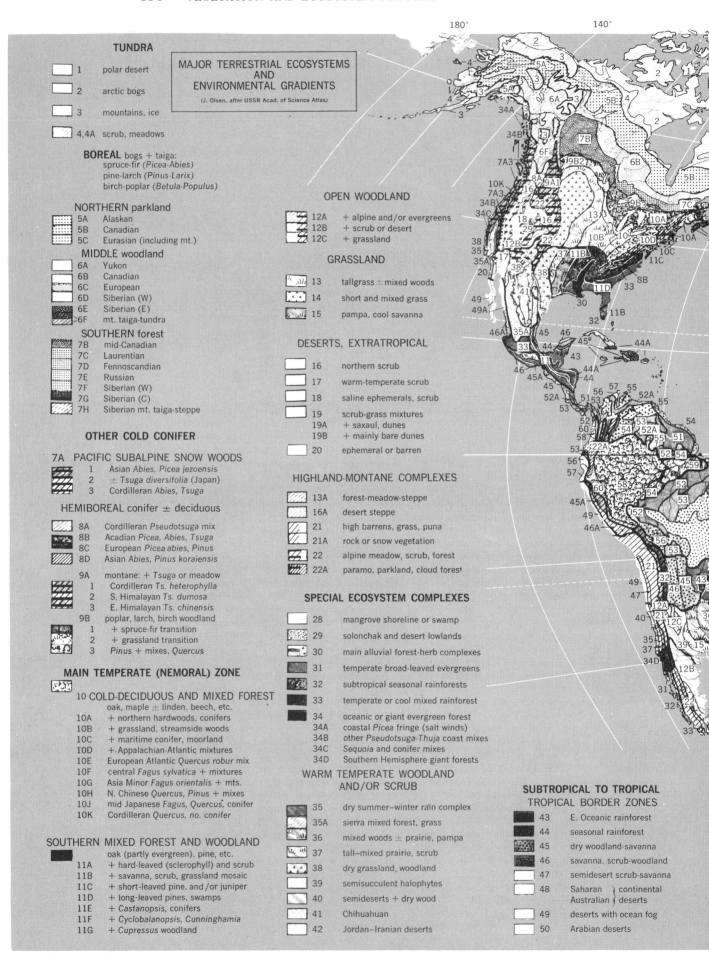

TUNDRA

1   polar desert
2   arctic bogs
3   mountains, ice
4,4A   scrub, meadows

BOREAL bogs + taiga:
spruce-fir (*Picea-Abies*)
pine-larch (*Pinus-Larix*)
birch-poplar (*Betula-Populus*)

NORTHERN parkland
5A   Alaskan
5B   Canadian
5C   Eurasian (including mt.)

MIDDLE woodland
6A   Yukon
6B   Canadian
6C   European
6D   Siberian (W)
6E   Siberian (E)
6F   mt. taiga-tundra

SOUTHERN forest
7B   mid-Canadian
7C   Laurentian
7D   Fennoscandian
7E   Russian
7F   Siberian (W)
7G   Siberian (C)
7H   Siberian mt. taiga-steppe

OTHER COLD CONIFER

7A PACIFIC SUBALPINE SNOW WOODS
1   Asian *Abies*, *Picea jezoensis*
2   ± *Tsuga diversifolia* (Japan)
3   Cordilleran *Abies*, *Tsuga*

HEMIBOREAL conifer ± deciduous
8A   Cordilleran *Pseudotsuga* mix
8B   Acadian *Picea*, *Abies*, *Tsuga*
8C   European *Picea abies*, *Pinus*
8D   Asian *Abies*, *Pinus koraiensis*

9A   montane: + *Tsuga* or meadow
1   Cordilleran *Ts. heterophylla*
2   S. Himalayan *Ts. dumosa*
3   E. Himalayan *Ts. chinensis*
9B   poplar, larch, birch woodland
1   + spruce-fir transition
2   + grassland transition
3   *Pinus* + mixes, *Quercus*

MAIN TEMPERATE (NEMORAL) ZONE

10 COLD-DECIDUOUS AND MIXED FOREST
oak, maple ± linden, beech, etc.
10A   + northern hardwoods, conifers
10B   + grassland, streamside woods
10C   + maritime conifer, moorland
10D   + Appalachian-Atlantic mixtures
10E   European Atlantic *Quercus robur* mix
10F   central *Fagus sylvatica* + mixtures
10G   Asia Minor *Fagus orientalis* + mts.
10H   N. Chinese *Quercus*, *Pinus* + mixes
10J   mid Japanese *Fagus*, *Quercus*, conifer
10K   Cordilleran *Quercus*, no. conifer

SOUTHERN MIXED FOREST AND WOODLAND
oak (partly evergreen), pine, etc.
11A   + hard-leaved (sclerophyll) and scrub
11B   + savanna, scrub, grassland mosaic
11C   + short-leaved pine, and/or juniper
11D   + long-leaved pines, swamps
11E   + *Castanopsis*, conifers
11F   + *Cyclobalanopsis*, *Cunninghamia*
11G   + *Cupressus* woodland

MAJOR TERRESTRIAL ECOSYSTEMS AND ENVIRONMENTAL GRADIENTS
(J. Olsen, after USSR Acad. of Science Atlas)

OPEN WOODLAND
12A   + alpine and/or evergreens
12B   + scrub or desert
12C   + grassland

GRASSLAND
13   tallgrass ± mixed woods
14   short and mixed grass
15   pampa, cool savanna

DESERTS, EXTRATROPICAL
16   northern scrub
17   warm-temperate scrub
18   saline ephemerals, scrub
19   scrub-grass mixtures
19A   + saxaul, dunes
19B   + mainly bare dunes
20   ephemeral or barren

HIGHLAND-MONTANE COMPLEXES
13A   forest-meadow-steppe
16A   desert steppe
21   high barrens, grass, puna
21A   rock or snow vegetation
22   alpine meadow, scrub, forest
22A   paramo, parkland, cloud forest

SPECIAL ECOSYSTEM COMPLEXES
28   mangrove shoreline or swamp
29   solonchak and desert lowlands
30   main alluvial forest-herb complexes
31   temperate broad-leaved evergreens
32   subtropical seasonal rainforests
33   temperate or cool mixed rainforest
34   oceanic or giant evergreen forest
34A   coastal *Picea* fringe (salt winds)
34B   other *Pseudotsuga-Thuja* coast mixes
34C   *Sequoia* and conifer mixes
34D   Southern Hemisphere giant forests

WARM TEMPERATE WOODLAND AND/OR SCRUB
35   dry summer—winter rain complex
35A   sierra mixed forest, grass
36   mixed woods ± prairie, pampa
37   tall—mixed prairie, scrub
38   dry grassland, woodland
39   semisucculent halophytes
40   semideserts + dry wood
41   Chihuahuan
42   Jordan—Iranian deserts

SUBTROPICAL TO TROPICAL
TROPICAL BORDER ZONES
43   E. Oceanic rainforest
44   seasonal rainforest
45   dry woodland-savanna
46   savanna, scrub-woodland
47   semidesert scrub-savanna
48   Saharan } continental
      Australian { deserts
49   deserts with ocean fog
50   Arabian deserts

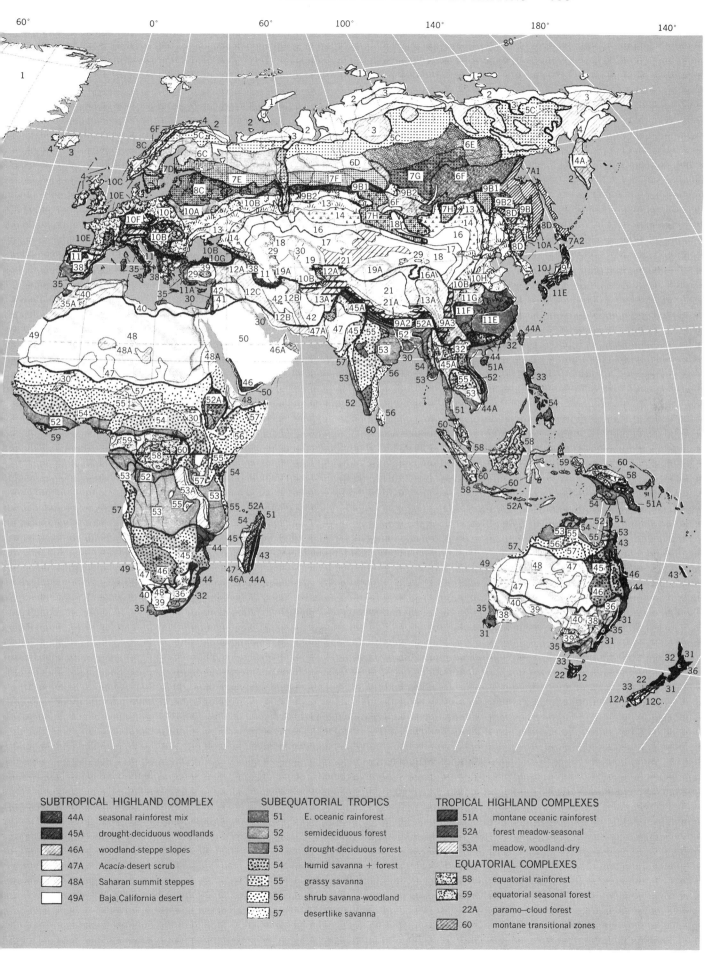

SUBTROPICAL HIGHLAND COMPLEX

■	44A	seasonal rainforest mix
■	45A	drought-deciduous woodlands
▨	46A	woodland-steppe slopes
▫	47A	Acacia-desert scrub
▫	48A	Saharan summit steppes
□	49A	Baja California desert

SUBEQUATORIAL TROPICS

▨	51	E. oceanic rainforest
▨	52	semideciduous forest
▨	53	drought-deciduous forest
▨	54	humid savanna + forest
▨	55	grassy savanna
▨	56	shrub savanna-woodland
▫	57	desertlike savanna

TROPICAL HIGHLAND COMPLEXES

■	51A	montane oceanic rainforest
■	52A	forest meadow-seasonal
▨	53A	meadow, woodland-dry

EQUATORIAL COMPLEXES

▨	58	equatorial rainforest
▨	59	equatorial seasonal forest
	22A	paramo—cloud forest
▨	60	montane transitional zones

Major vegetation and ecosystem groups

This legend of the UNESCO Committee on Classification and Mapping (mainly for scales 1:1,000,000 and smaller; larger scale is required for many of the finer subdivisions, however) is only slightly changed from that published provisionally by Ellenberg and Mueller-Dombois (1965–1966). A less abbreviated adaptation is available in the "Geographic Index" of Reichle (1970), which is keyed to a map of regions which are coded according to their estimated pre-iron-age carbon mass of vegetation (above and below ground). The IBP Index Code for major groups of biomes was proposed by the International Biological Program section on Terrestrial Productivity (PT) for indexing projects, sites, and ecosystems.

	I. Formation Class
	A. FORMATION SUBCLASS
IBP	1. *Formation Group*
Index	a. Formation
Code	1) Subformation

I. Closed Forests (> 5 m tall, crowns touching in wind—except that immature, cutover and grazed forest types may be shorter or more open without being called scrub or woodland, respectively)

A. MAINLY EVERGREEN FOREST: Canopy never without foliage, although individual trees may shed leaves

Fr	1. *Tropical Rainforest* (= ombrophilous)
	Little or no bud protection, nor cold or drought resistance; "drip-tip" leaves common
Frd	2. *Tropical and Subtropical Evergreen Seasonal Forest:*
	Some bud protection and noticeable dry-season shedding
	a–d. Lowland, Submontane, Montane, dry "Subalpine"
Fdr	3. *Tropical and Subtropical Semi-Deciduous Forest:* Upper canopy mostly drought-deciduous; evergreen trees in canopy layers or understory
Fr	4. *Subtropical Seasonal Rainforest*
Fr	5. *Mangrove Forest*
Fr	6. *Temperate and Subpolar Rainforest*
Fd	7. *Temperate Evergreen Seasonal Broad-leaved Forest*
Fs	8. *Winter-rain Hard-Broad-leaved* (Sclerophyll) *Evergreen Forest*
Fn	9. *Coniferous Evergreen Forests*
	a. Giant evergreen conifers (> 50 m tall)
	b. Conifers rounded or flattened
	c. Conifers mostly conical
	d. Conifers cylindro-conical, with short branches

B. MAINLY DECIDUOUS FORESTS

Fd	1. *Drought-deciduous* (Monsoon) *Forest* (tropical, subtropical)
	2. *Cold-deciduous Forest with Evergreens:*
Fc	a. With Evergreen broad-leaved trees and climbers
Fc	b. With Hard-broad-leaved evergreen shrubs
Fcn	c. With Evergreen needle-leaved trees (cool)
Fcn	d. With Evergreen needle-leaved trees (warm)
Fcn	e. With Conifers and/or broad-leaved evergreens
Fc	3. *Cold-deciduous (Summergreen) Forests:*
	Evergreens (if any) mostly shrubs, or scattered
	a. Temperate lowland and submontane ("nemoral")
	b. Montane, boreal and humid-site
	c. Subalpine or subpolar (< 20 m; commonly gnarled)
	d. Alluvial, flooded:
	1) Occasionally or never
	2) Regularly
	e. Swamp or bog forest

C. DRY FORESTS (commonly grading to open woodlands)

Fd	1. *Hard-leaved Forests:*
	Some with swollen underground bases (xylopods)
Fd	2. *Thorn Forests*
	a, b. Mixed deciduous-evergreen, deciduous
Ds	3. *Mainly Succulent Forests* (trees and/or shrubs)

II. Open Woodlands (< 5 m tall, crowns projecting over 30% of surface; may be grassy, grading to savanna)

A. MAINLY EVERGREEN WOODS

Fs	1. *Evergreen broad-leaved woodlands*
Fn	2. *Evergreen needle- or scale-leaved woodlands*
	a. Conifers rounded, flattened or irregular (e.g. pine)
	1) With hard-broad-leaved understory
	2) Without hard-broad-leaved understory
	b. Conifers mostly conical or dense (e.g. juniper)
	c. Conifers cylindro-conical, or sheared

B. MAINLY DECIDUOUS WOODLANDS

Fd	1. *Drought-deciduous Woodlands*
Fcn	2. *Cold-deciduous Woodland with Evergreens* (see IB2)
Fc	3. *Cold-deciduous Woodland* (summergreen)
D	C. Dry Woodlands (divided as for IC, but sparser)

Major vegetation and ecosystem groups (cont.)

	III. Scrub (mainly 0.5–5 m; thicket or shrubland, with grass)
	A. MAINLY EVERGREEN SCRUB
	1. *Evergreen Broad-leaved Shrubland or Thickets*
	2. *Evergreen needle-leaved and microphyll scrub*
	B. MAINLY DECIDUOUS SCRUB
Fd	1. *Drought-deciduous scrub with evergreens*
Fd	2. *Drought-deciduous scrub without evergreens*
Fc	3. *Cold-deciduous scrub*
	C. DESERT SHRUBLANDS
Dx	1. *Mainly evergreen subdesert*
Dx	2. *Deciduous subdesert*
	a. Without succulents
Ds	b. With succulents

	IV. Dwarf Scrub and Related Ecosystems
	A. MAINLY EVERGREEN DWARF SCRUB
Th	1. *Evergreen dwarf scrub thickets* (heath)
T	2. *Evergreen dwarf shrubland*
Tg	3. *Evergreen dwarf scrub – herb mixture*
T	B. MAINLY DECIDUOUS DWARF SCRUB
Dx	C. SEMIDESERT DWARF SHRUB
Tt	D. TUNDRAS
	1. *Moss tundra*
	2. *Lichen tundra*
Tp	E. MOSSY BOGS
	1. *Raised Bogs* (oceanic, montane, subcontinental)
	2. *Nonraised Bogs*
	a. Blanket bog (oceanic, submontane, montane)
	b. String bogs ("aapa")
	c. Lake bogs

	V. Terrestrial Herbaceous Ecosystems
Gs	A. SAVANNAS AND RELATED GRASSLANDS
	(tropical, subtropical)
Gp	B. STEPPES AND RELATED GRASSLANDS ("prairie")
	1. *Tall-grass Prairie* (steppe)
	2. *Mid-grass (and mixed) Prairie* (steppe)
	3. *Short-grass Prairie* (steppe)
	4. *Forb-rich Meadow Steppes*
	C. MEADOWS, PASTURES AND RELATED GRASSLANDS
Gf	1. *Hay Meadow and/or Grazing Pasture* (below treeline)
Ga	2. *Meadow or Pasture* (above treeline)
Gw	D. SEDGE SWAMPS AND FLUSHES
	1. *Sedge Peat Swamps*
	2. *Flushes* (with seepage water ± lime)
Gr	3. *Reed and Tall graminoid Swamp*
Gh	E. HERBACEOUS AND SEMIWOODY SALT SWAMPS AND SHORES
	1. *Succulent or Nonsucculent Salt Marsh*
	2. *Salt Meadows* (Marine and Inland)
	F. FORB ECOSYSTEMS
G	1. *Mainly Perennial Forb Complexes*
De	2. *Mainly Ephemeral Forb Complexes*
De	3. *Episodical Forb Complexes*
Ge	G. CROP COMPLEXES, VILLAGES, CITIES

	VI. Deserts and other Sparsely Vegetated Land
D	A. SCARCELY VEGETATED ROCKS AND SCREES
Db	B. SCARCELY VEGETATED SAND
	1. *Presently (or Recently) Active Dunes*
	2. *Bare Migrating Dune*
D	C. DESERT BARRENS

1:4,000,000 shows even finer detail for the actual vegetation with 109 colored mapping units—many subdivided by letters and supplemented by symbols for significant species. Still smaller scales were required (varying with continent and type of map) in the Soviet Union's "Physical-Geographic Atlas of the World," but with remarkably little loss of detail, as compared with atlases in countries not using the Cyrillic alphabet.

Aerial photographs. Aerial photos have been used since the 1920s to provide direct images of forests and other visibly distinct vegetation cover, with or without extensive ground checking to provide species recognition and other calibration data. Canada, with vast underpopulated terrain (like the Soviet Union's), has found aerial photography especially useful. Large areas are stereoscopically covered from high altitudes for surveys, road and harvest plans, and fire control operations. Also, very-large-scale photographs of statistically selected plots have permitted measurement of shadow and tree height and crown form to provide quantitative measures of tree volume (and biomass, by implication).

Soil surveys are routinely made on aerial photographs because of the convenience of landmarks, and the need for precise measurement of areas for which conservation payments are made to individual farmers. Even when land is freshly plowed, there are limits to what can be inferred from the air or from observation of surface soil horizons. Based on local experience, the soil surveyor, conservation agent, forester, or landowner may examine pits or augur borings to observe the hidden lower soil horizons, but necessarily the interpolations between a few points of detailed observation are made indirectly. Sometimes soil hue and darkness as related to moisture conditions are helpful, but weather may blur the value of these aids.

The experienced mapper relies on the "lay of the land" (perhaps perceptibly slight differences in concavity or convexity) and relations to quality and vigor of vegetation to help fill in the incomplete observations of substrate conditions. The practical man shares with the research scientist a real need for "working hypotheses" about where a boundary or transition should be looked for, and then he guides his next observations (augur holes, field vegetation plots) to test his ideas. The balance of intensive and extensive work, and of "calibration" checks and routine extension of the mapped area, depends on the economics of available time and competing duties. An understandable administrative need for controlling these constraints tends to favor statistical control and an objective measure of the amount of unexplained variance within the somewhat arbitrarily grouped mapping units. Yet for the understanding of vegetation, soil, or the other terrain variables (which correlate with the whole ecosystem's condition or which help to explain that condition), nonroutine detectivelike insights are needed, as well as the orderly use of what has been guessed, observed, and confirmed. *See* SOIL CONSERVATION.

Multiband remote sensing. Methods using various combinations of the electromagnetic spectrum of signals from the Earth to aircraft or to satellites may expand the scope of mapping in ways which can hardly be estimated in detail. Color photography, especially in the brief periods of spring and autumn color change, has already made striking improvements in recognition of species—and hence of the special conditions of site or history. Infrared photography indicates that reflection is much greater for broad-leaved trees than for conifers. There has been hope of detecting differences between reflectivity of diseased (and hence often drying or dying) trees before the eye or regular film can discriminate. Diagnoses based on infrared photography have been promising for the relatively uniform conditions of field crops, but more experience will be needed before pest or disease infestations can be detected reliably in wild vegetation and on rough terrain. *See* TERRAIN SENSING, REMOTE.

Nonphotographic electronic devices for scanning longer-waved infrared (thermal) radiation offer more opportunities than have been publicized widely. Such techniques may be used for detecting landscape patterns empirically, and perhaps for explaining some of these in terms of the microclimatic energy balance. Interference by moisture absorption in certain parts of the infrared spectrum poses limitations on the less expensive instruments (for example, pistollike bolometers for pointing out "hot spots" of fire or warm microclimate). *See* ENVIRONMENT.

Scanners which can separate many bands of radiation offer untapped versatility. To take advantage of the right "signature" (or wavelength combination) for the cover condition of interest, investigators can use complex optical, electronic, and computer processing, but this makes the operation expensive. Even the best engineering methods, applied with a narrow empirical viewpoint, leave much to be desired in carrying experience from a few early test areas into the wide range of natural conditions which demand insight about why certain clues are consistently more valuable than others. The patterns of signals, interpreted subjectively and by filtering procedures (for example, procedures related to tree and shadow sizes), are even less fully studied than wavelength signatures. However, these patterns may prove more valuable in the long run in the hands of interpreters and computer programmers with field experience and knowledge of the fundamentals of physical-biological relationships.

Early progress with radar scanning of landscapes has been promising, suggesting, for example, geologic and landform patterns not recognized from the ground or from the usual map and photo analysis. Some of these patterns may affect soils and plants. It is not yet clear what direct information about vegetation can be obtained with K-band radar. *See* RADAR.

WORLD ECOSYSTEM MAPS AND ZONES

Local detail must be generalized even further than for continents when the scale is reduced for an overview of the Earth as an ecosystem. Some problems must be treated eventually for the whole Earth system, for example, exchange of elements such as carbon and radiocarbon by the atmosphere. *See* BIOSPHERE, GEOCHEMISTRY OF.

Inside the front cover of Reichle's *Analysis of Temperate Forest Ecosystems*, Lieth's map shows predicted annual rates of carbon exchange over the continents and oceans, but without including direct exchange of respiration by green plants; omission of some other exchanges (for example, by animal consumption of vegetation) may make the estimates low, but probably will not change the striking geographic pattern shown in colors.

J. Olson's map of the living inventory of carbon (inside the back cover of Reichle's book) is modified only in various details and scale conversions from a map of vegetation mass by Bazilevich, prepared with Rodin. Their data clearly cannot be based on interpolation of data from ecosystems in all world zones, because discussion by Bazilevich and Rodin shows very uneven sources of productivity and biomass data. Rather, their approach was to take 30 broadly generalized ecosystem types (34 for Olson), with boundaries based primarily on vegetation mapping, and to assign for the whole of each mapping unit the general range of biomass (or, in other 1969 maps, annual productivity and approximate annual exchanges of chemical elements). Improvements in these estimates as well as carbon exchange and biomass are expected from field studies of the International Biological Program by 1974. Details of patterns of ecosystem rates and parameters, on the scale of the world zones and altitudinal zones and patterns discussed below, and on the more detailed scale showing the accelerating changes due to man, will then remain a major challenge for generations of ecologists.

A framework for this ongoing, long-term research on ecosystems is provided by the map in this article. It is intermediate in detail between those mentioned in the last two paragraphs and the refined continental maps of the "Physical-Geographic Atlas of the World." Its working base is Plate 75 (Landscape Types) of the Atlas, with a legend which has been simplified in the process of translation. Other details have been changed to aid future matching of the UNESCO legend of ecosystem types in the table in the manner suggested by the "Geographic Index of World Ecosystems" given in the back of Reichle's book.

Broadly, the world's environmental gradient pattern is simple, starting with brief summer energy input into the polar and tundra complexes (see 1 to 4 on the map). Zones 5 to 7 correspond approximately to the northern, middle, and southern taiga. Mountain conifer forests are called "oroboreal" by some authors but in many respects differ from the most typical frozen-soil spruce-fir, pine-larch, and birch-*Populus* forests of boreal lowlands, which are here subdivided by letters in regions where different species and local conditions dominate.

[JERRY OLSON]

Bibliography: A. W. Küchler, *Potential Natural Vegetation of the Conterminous United States* (with manual), Amer. Geogr. Soc. Special Publication no. 26, 1964; A. W. Küchler, *Vegetation Mapping*, New York, 1967; D. E. Reichle (ed.), *Ecological Studies 1:Analysis of Temperate Forest Ecosystems*, Berlin-Heidelberg-New York, 1970; D. Riley and A. Young, *World Vegetation*, London, 1966.

Vegetation management

The art and practice of manipulating vegetation, such as timber, forage, crops, or wildlife, so as to produce a desired part or aspect of that material in higher quantity or quality. The term vegetation describes the complex of plant communities in the botanical landscape. It is a technical term not to be confused with the use of the word in reference to vegetable life and plants in general.

The field of vegetation management lies intermediate between the broader and more inclusive field of natural resource management (which involves both renewable and nonrenewable natural resources) and specialties such as forest management and range management. Inasmuch as certain types of vegetation are correlated with certain land types, such as shortgrass plains vegetation with the plains as a geographic unit, vegetation management also grades imperceptibly into fields collec-

tively known as land management. *See* CONSERVATION OF RESOURCES; FOREST MANAGEMENT AND ORGANIZATION; LAND-USE PLANNING.

In the total landscape, aside from urban and cropland areas, vegetation is the most easily manipulatable element. Although fauna, flora, climate, and soil are part of the natural environment and can be altered, it is generally difficult, uncertain, and costly to do so. The vegetation, however, can often be changed relatively cheaply and effectively.

MAN AND MANAGEMENT

Vegetation management involves an interrelationship between vegetation and man in which man becomes an external factor of the environment, acting upon, and interrelating with, the vegetation in ways which are presumably purposeful and beneficial.

Mismanagement. The history of the human race is commingled with the use and abuse of forests and grasslands. Prehistoric man, dependent upon edible roots, fruits, and nuts, scavenged for food until the area was completely barren. The biblical cedars of Lebanon and the ginkgo trees that have survived only in Chinese monastery gardens are symbolic of the all but total destruction of forest lands by early civilizations. The areas around the Mediterranean, with their distinctive semiarid winter-rain climate, have witnessed an almost total destruction of forest cover, which resulted in replacement by various kinds of scrubs and grasslands, the loss of soil blankets, and alterations of the water regimes. In turn, these changes have had an impact on the survival of civilizations; in these cases, mismanagement reached the point of essential irreversibility.

Overexploitation of vegetation resources was common in North America until the start of the 20th century. This was characterized by ranges that were overgrazed to quickly produce the highest numbers of stock. These practices permanently destroyed desirable species of plants which were replaced by species of less desirable quality and of reduced quantity, not adequate for feeding the stock. Changes in rodent populations, soil deterioration and erosion, and other alterations make recovery extremely slow. Comparable situations occur in forest management, in which highly profitable short-term methods of lumber exploitation lead to increasingly lower productivity of the forest lands.

Effective practices. Although increased knowledge has brought an improvement in contemporary vegetation management, these practices have not been in operation long enough to prove themselves capable of maintaining continued high levels of resource productivity without deterioration of the total ecosystem. Furthermore, techniques are being developed that, while they are shortcuts to the objectives, may in turn produce destructive side effects.

The widespread use of herbicides and insecticides, for example, has led to spectacular short-term gains, but it took more than a decade for extremely undesirable alterations to the ecosystem to become evident. *See* HERBICIDE; INSECTICIDE.

Costs, benefits, values, judgments, and hazards of vegetation management must be evaluated not in the light of short-term profits to the individual or to a minor social unit, but in relation to long-term gains to society itself, qualitatively as well as quantitatively.

MANAGEMENT OF VEGETATION CHANGE

Artificial and anthropogenic changes in vegetation are intimately related to reasonable "natural change," that is, change which occurs essentially independently of man's influence. Such a change may be autogenic, induced directly by the vegetation itself, or heterogenic, induced directly by some other factor such as drought, flood, disease, or fire. An autogenic change is exemplified by the chemical reactions in the soil which result from leaf fall and decay.

Changes may be very rapid, as on old abandoned agricultural lands where a forest may develop from bare soil within a century; or the changes may be very slow, involving thousands of years, as when new volcanoes arise from the ocean or when granite islands are left by a receding sea. Any one stage of plant development may vary greatly in duration, as 1 year for a particular annual weed type, to 4000 years for redwood trees.

The autogenic changes which occur on abandoned agricultural lands typify the principles of vegetation management. On such old fields the influence of man had predominated up to the time of abandonment, but his influence may continue after this. For example, high nutrient levels in the soil due to previous fertilization may persist.

Old-field vegetation change is often described in terms of external appearances, such as the physiognomy of the vegetation. The original crop is succeeded by annual weeds, then in turn by grasses, heavy forbs (broad-leaved herbs), and shrubs; finally trees are predominant. Such a description gives no indication of the time when certain species first appear on the land, and when they finally disappear. Since the entire goal of vegetation management is the production of particular plant species, it is of great importance to know when each species does, or can, enter or leave the community.

Theory of floristic relays. Physiognomic change is brought about by a succession of plants invading in relays, each relay being a stage of what has been called plant succession. As traditionally described, this process begins with hardy pioneer species which invade an exposed or plowed area; these plants are adapted to survive in a harsh environment; and their survival causes the environment to be altered. They bind and change the soil and thus create new conditions to which they are less well adapted but which are suitable for new species with other adaptations. The new organisms slowly replace the pioneers and then dominate the community until their activities alter the environment, allowing replacement by better-adapted forms. The cycle continues until the most highly adapted forms succeed and dominate. One important effect of this process is to produce, at each succeeding step, a community that tempers the environment more successfully than its predecessors did. *See* SUCCESSION, ECOLOGICAL.

Theory of initial floristic composition. Physiognomic change is interpreted in terms of a floristic composition determined largely or entirely at the time the land was abandoned. Physiognomic change is thus related not to newly invading plants, but to plants present from the start that develop into prominence only at some later time. The predominant plants eventually terminate their short-lived existence, or are overtopped, or fail in competition with still other plants which also had been there from the start. Initial floristic composition probably never occurs as the sole determining factor in vegetation change, but is a component in the total complex of change.

Actual change. In at least the temperate summer-rain regions of the world there is a predominance of species that entered initially into the bare land, including most of the shrubs and also some of the trees. Shrubs rarely invade dense grassland as seedlings, though underground root invasion may be significant at the peripheries of the grassland. Solid shrub land is rarely invaded by trees (if trees occur with the shrubs, they are likely to be as old as, or older than, the shrubs). The forest is composed partly of trees that invaded original bare soil and partly of trees that were able to invade grassland, such as elm, ash, and pine. *See* GRASSLAND.

On the basis of natural vegetation change, vegetation management involves altering the time of invasion, the time of disappearance, and the degree of predominance of the different species in the normal pattern of vegetation change. So-called weed species are to be eliminated, while desirable species, those of economic value that produce a resource, are to be encouraged.

MANAGEMENT METHODS

Desirable alterations in the processes of natural vegetation change are accomplished by a wide variety of techniques and methods which directly and indirectly affect the phenomenon. Among the methods that directly affect vegetation are fire, ax, sickle, and plow.

Fire. Fire has probably been used to manage, and mismanage, vegetation as long as it has been used by man to cook his food. The effects of anthropic fires are anastomosed with those of natural fires, which are caused, for example, by lightning and by spontaneous combustion. When man entered North America, he found a continent of vegetation that was conditioned by fire. The development of the wise use of fire in forest and range practice has been a slow and painful process, but it is now a recognized tool in the management of certain forest types, particularly the pines of the southeastern United States.

Tools. When primitive man first used the ax, he was selective, choosing only those kinds of trees which bore the fruits he wished to eat, the best hearth wood, or the best timber for building. This selectivity quickly changed the species composition of the forest, sometimes encouraging the advance of species that bore little if any resemblance to those of the original condition. For example, parts of the northern fringe of forest in the Canadian Arctic were thought to be composed naturally of deciduous hardwoods. Eventually it was realized that these populations had completely removed the more desirable conifers.

The ax was soon supplemented by the saw, and, as technology developed, massive machines were produced that moved over the land, snipping trees, stripping the branches, and neatly stacking the poles in a carrier.

The sickle is to grassland what the ax is to woodland. Though originally used to harvest the heads from grains, this selective act turned to overexploitation which tended to destroy the very species that was desirable. As technology led to the development of machines, cutting tended to be indiscriminate and woody species were also cut and controlled. A pattern emerged in which there was a sharp segregation between mowed grasslands and unmowed brushy lands; this is found in all parts of the world today. Mowing machines, steadily increasing in size and efficiency, now cut swaths 50 ft wide and stems up to an inch or more in diameter.

Although the plow is essentially an agricultural tool, its use is important in the history of much land that has been abandoned and now bears seminatural vegetation. The plow not only turns over and homogenizes the upper layers of the soil, which previously might have been distinctly stratified, but also chops up the rhizomes and roots of certain species of plants, tending to make them far more abundant than they otherwise would be. The temporarily bare soil, relatively free from competition among plants, can be ideal for the establishment of certain species, especially trees, which would otherwise not become established. The effects of plowing may thus be discernible in managed vegetation for decades, even centuries.

Chemicals. Since World War II a direct method of affecting vegetation has been the use of chemical pesticides and herbicides. Chemicals have been used, however, since the first caveman noticed that salt water killed vegetation. The effectiveness and safety of chemicals are factors based on their application, that is, whether this practice is indiscriminate and nonselective, as by airplane, or whether it is a discriminate, selective local application.

A classic case is the indiscriminate aerial application of herbicides to the demilitarized zone of Vietnam, which affects whatever surrounding areas the chemicals may be blown to by winds. The indiscriminate aerial application of broad-spectrum persistent insecticides such as DDT for the control of one species of insect has vast and ramifying effects throughout the total ecosystem. Since all pollinating insects are affected, as well as animals which feed on these insects, the indirect effects on vegetation are felt for years beyond the time of application. *See* ECOSYSTEM; PLANT DISEASE CONTROL.

Indirect methods. The major indirect methods of managing vegetation may be classified as those affecting water, soil, and air.

Water. Among the more obvious methods of managing water is the technology of dam building. Beavers have long dominated the events of many upstream tributaries throughout the world, even to the extent of being recognized as a force shaping the local physiography. The effects of beavers become insignificant when compared to those of dam engineers, who are turning many rivers into a series of lakes with enormous consequences to the entire landscape.

Controlled changes in lake levels can be unde-sirable, as in the wide bare shores during seasonal drawdowns, but such changes can also be desirable, as in certain wildlife-management practices in which temporary drawdowns control excessive growth of undesirable aquatic plants. However, the lowering of water tables due to excessive "mining" of water for human populations, such as for irrigation, is having subtle effects on natural and seminatural vegetation types, and there are problems of polluted water. Overfertilization by phosphates and nitrates from sewage disposal systems and septic tanks leads to population explosions of undesirable aquatic plants. Balances are further upset by the addition of herbicides to control these plants, while the basic nutrient problem which causes the imbalance is unattended. Pollution by persistent insecticides, washed from the land, carried in the atmosphere, or transported by organisms which accumulate them, is becoming ever more hazardous. If, for example, such poisons affected the diatom populations of the ocean, the oxygen balance would be upset, in turn affecting the oxygen content of the atmosphere, with effects on all animal life including man. *See* WATER CONSERVATION.

Soil. Fertilization is one of the most common methods of managing soil, and thus of indirectly managing the vegetation. Fertilization is generally thought of as only an agricultural practice, but it has potentialities in the management of forest lands and rangelands. Another approach in soil management, the analog of agricultural contour plowing, is the terracing of rangelands and watershed lands. The terraces greatly reduce superficial runoff of precipitation and increase infiltration into the soil, thus leading to marked changes in the nature of the local vegetation. One of the problems in soil management is polluted soil, that is, soil that because of previous agricultural practices has accumulated significant amounts of arsenic, copper, and lead and thus affects the composition and structure of vegetation. For example, copper smelters have not only destroyed considerable vegetation in their vicinity, but the land has continued bare and in a state of erosion for several decades. In addition, the accumulation of persistent pesticides in the soil has definite but still largely unknown effects on soil fauna and flora and thus on the associated grassland and woodland vegetation.

Air. Temperature and precipitation control is a primary consideration among the approaches to managing the air. Temperature control is possible on a local scale in agricultural practices, such as by the use of smudge pots in orange groves: By dispersing particles in the lower atmosphere, thermal radiation from the soil is lowered and critical freezing temperatures are ameliorated. Precipitation control, linked with the concept of climate control, is accomplished largely by seeding silver iodide at cloud-forming levels or by electric discharges. *See* CLIMATE, MAN'S INFLUENCE ON.

Pollution and smog of urban air are known not only to limit the kinds of plants that can satisfactorily be grown in cities but also to eliminate completely such plants as lichens from the vegetation. However, the chronic effects of minor pollution in the landscape are not easily open to scientific study. In aerial spraying with insecticides, for example, large percentages of the chemical do not fall on the target area but are wafted away to points unknown. This drift can account for damage to crops at distances of 30 mi or more. *See* AIR-POLLUTION CONTROL.

PRODUCTS OF MANAGEMENT

Various methods applied to managing vegetation can be correlated with normal alterations to lead to desirable or undesirable "products." The undesirable products are called weeds, and their control promptly triggers still further management practices. It is convenient to distinguish two groups of desirable products: specific products such as wood derived from vegetation, and complex products inherent in the landscape itself, such as watershed vegetation.

Specific products. Among the specific products of vegetation are wood, forage, and wildlife. Emphasis on these three products has given rise to the three major fields of forest management, range management, and wildlife management. *See* RANGELAND CONSERVATION.

Wood. In forestry, timber production maintains precedence, while pulp and paper manufacture is secondary. The art and practice of forest management is often known as silviculture, while the science behind silviculture is known as silvics. *See* FOREST RESOURCES.

Forage. Range management is concerned with the natural and seminatural grasslands, scrub lands, and woodlands that contain a ground layer of herbaceous vegetation. The desirable forage is viewed primarily in terms of cattle and sheep, but also goats, camels, asses, and many other domesticated grazing mammals, each of which plays a role by selecting the plants it prefers and eliminating others. Long-term continued high productivity of rangeland is the goal of range management. Excessive grazing increases the rodent populations, which take their toll of the remaining forage but in turn attract eagles, predators who control the numbers of rodents and thus help to maintain the balance of this ecosystem.

Wildlife. Although problems of population numbers, predators, disease, and nutrition all play important roles in wildlife management, it is recognized that food and cover—the habitat—rank first, and that they are largely attributes of the vegetation. *See* WILDLIFE CONSERVATION.

Complex products. Three important vegetation-controlled landscapes that must be established for purposive vegetation management are watershed management, right-of-way and roadside vegetation management, and naturalistic landscaping.

Watershed management. The purpose of watershed management is to produce the most water for human consumption by controlling the vegetation and thus regulating the flow of water to prevent damaging floods. Watershed management attempts to control vegetation either by directing streams to flow into reservoirs or by incrementing the water table from available surface water. Some forms of exploitive forestry in which deforestation leads to floods, soil erosion, and change of climate are unwise. On the other hand, rational management practices can convert forest to stable seminatural grassland. *See* WATER SUPPLY ENGINEERING.

Right-of-way and roadside management. Right-of-way and roadside management is a facet of veg-etation management. In this domain are included the seminatural unmown sides of highway and railroad systems, the land under electric power and telephone lines, and the land above pipelines. In 1953 approximately 50,000,000 acres, more than the combined acreage of all six New England states, fell into this category. Since the power lines alone of two states, Michigan and North Dakota, have been estimated at over 1,000,000 acres each, it is reasonable to assume that this land now totals over 70,000,000 acres in the 50 states. Vegetation management on these lands is largely a problem in disposing of unwanted trees by rootkilling and in developing a stable low cover of herbs or shrubs that is also highest in scenic, recreational, wildlife, and conservation values. *See* LAND-USE CLASSES; LAND-USE PLANNING.

Landscape management. Landscape management, a field closely akin to landscape gardening, involves naturalistic landscaping. The goal is to produce a terrain of varied plant communities that forms a scenic mosaic of vegetation of high esthetic values. The ideas are applicable to small country estates, to the margins of parkways, and to other semiwild lands where such economically valuable products as timber or forage are not paramount. This field of management has emerged primarily since World War II.

Another concern is the management of complex vegetation-dominated landscapes known (and seemingly misnamed) as "natural areas" or "non-managed lands." A natural area is a tract kept as free as possible from human interference because of its scientific, educational, and cultural values. Problems that arise in the management of these complex ecosystems are: (1) how to control high populations of a browsing animal when the tract is not large enough to support the large predators; (2) whether an infestation of insects may result in the death of overmature trees; (3) how to control extreme droughts; and (4) how to regulate excessive rains that result in floods. Nature operates in cycles, the peaks of which may be catastrophic from the short-term view. Furthermore, there may be infiltrations of polluted groundwater or of passing winds contaminated by urban smog or agricultural pesticides. In such cases no local control or remedy is possible.

Summary. The field of vegetation management is a vast and ramifying system composed of complexes of plant communities in many interrelationships with the environment, including atmosphere, soil, animals, fire, and chemicals. These aspects of the environment are all in purposeful relationships with man and are assumed to be beneficial at least for short periods, but may prove detrimental in the long term. Vegetation management is thus part of a still larger ecological whole, involving man in his total environment. *See* ECOLOGY; VEGETATION.

[FRANK E. EGLER]

Vegetation zones, altitudinal

The extensive, even transcontinental, bands of physiognomically similar vegetation that range from tropical rainforest to arctic tundra; the more or less distinct belts of vegetation that occur, one above another, on high mountains; and the concentric or linear bands of distinctive vegetation around ponds, along rivers and coasts, and in similar sites. Such zones are correlated with vertical and horizontal gradients of environmental conditions. In most situations, these graded conditions are related directly or indirectly to solar radiation (heat and light) and moisture. See ENVIRONMENT.

Latitudinal variations in solar radiation. The two sensible components of solar radiation are heat and light; the Sun is the principal source of the heat and light energy impinging upon the Earth. If the Earth were a plane oriented perpendicularly to the axis of the Sun's rays, its entire surface would be exposed equally to radiation. Because the Earth is spherical, however, only a small section, known as the tropics, is exposed regularly to more or less vertical radiation. As one approaches the Earth's poles, the angle of incidence of the Sun's rays decreases and their energy is diluted by being spread over a greater area. In addition, the Earth is enveloped by layers of gases collectively known as the atmosphere. These gases and the materials suspended in them reflect and absorb solar radiation. The path of solar rays through the atmosphere is shortest in the tropics and lengthens toward the poles as the angle of incidence decreases. Thus, from the tropics poleward the intensity of solar radiation generally decreases. Because moist air absorbs and reflects more of radiation than dry air, however, the pattern of decrease is not uniform. Through their effects on atmospheric humidity and their action as heat exchangers, large bodies of water, including the oceans and major lakes, modify the climatic patterns of the Earth. The pattern is further modified because the Earth's rotational axis is not perpendicular to the plane of the Earth's orbit around the Sun. In effect, this inclination causes the directly vertical noontime rays of the Sun to migrate annually from the Tropic of Cancer to the Tropic of Capricorn and back again and produces the familiar seasonal changes in day length and temperatures in extratropical regions. These factors and others combine to form an intricate pattern of climatic regions over the Earth's surface. These regions generally are oriented in parallel bands, from the tropics around the Equator to the arctic and antarctic regions around the poles, but there are many irregularities occurring in the pattern. See CLIMATOLOGY.

Altitudinal variations in solar radiation. Within any major region there are variations in solar radiation associated with elevation and, outside the tropics, with aspect. With an increase in elevation above sea level, air becomes less dense and is not able to hold as much moisture per unit volume. The lessening of density results in a decrease of barometric pressure of about 1 in./1000 ft. With a greater dispersion of gas molecules and a lower concentration of water vapor, the heat-absorbing power of the air is diminished. In addition, as altitude increases, the distance through the atmosphere that the Sun's rays must travel to reach the Earth diminishes. Therefore, more light and more heat energy reach a surface at a high elevation than reach a similar surface at a low elevation. It has been estimated that about 20% of solar radiation is reflected or absorbed by the atmosphere before it reaches sea level, whereas only 11% is lost before it reaches a 14,000-ft-high peak. The difference in light intensity is not of sufficient importance to photosynthesis to be a critical ecol-

ogical factor, but the heating of the soil may produce very high surface temperatures during the day. The heat is lost rapidly by reradiation at night, and because the air is not able to absorb appreciable quantities of heat, nocturnal temperatures are relatively low. Measurements on Pike's Peak, Colo., for example, occasionally indicate surface soil temperatures near 140°F at a time when the air temperature 5 ft above the soil is about 70°F. Nocturnal reradiation also prevents the heat from being conducted into the deeper soil layers. On Pike's Peak the soil at a depth of 10 in. may be 85°F cooler than the surface soil.

Altitudinal variations in temperature. Air masses that encounter mountains are forced upward and over the barrier. The rising air expands and is cooled adiabatically at a rate that averages about 1°F decrease in temperature per 330 ft rise in altitude in summer and 1°F per 400 ft rise in winter. The thermic rate varies considerably because of moisture, wind, slope exposure, and other factors.

The reduction of temperature as altitude increases also brings about later frosts in the spring and earlier frosts in the autumn and thus shortens the frost-free or growing season. In Arizona, for example, a station at an elevation slightly above sea level has an average frost-free season of 322 days, a station at an elevation of 2660 ft has an average frost-free season of 246 days, and another station at an elevation of 9000 ft, Mount Lemmon, has a frost-free season that averages only 122 days. However, there is a considerable variation in the length of the frost-free season from place to place in a given vegetation type. Also, temperature alone may not suffice to determine the growing season, because many low plants may be covered by snow for several days or even weeks after the air temperature becomes favorable. Thus, it appears that the absolute length of the season of favorable temperatures is not closely correlated with altitudinal changes in vegetation.

Mountainous precipitation and evaporation. A portion of the water vapor contained in rising air masses condenses as the air mass is cooled and may form clouds or fall as rain or snow. Thus, mountains are often referred to as islands of greater precipitation. In the mountains, the annual precipitation generally increases in amount with an increase in altitude, at least up to a certain elevation. Mountains subject to moist winds from a prevailing direction may have heavy precipitation on windward slopes and a rain shadow or dry belt on the lee side. If the mountains are high enough, there is generally a maximum of precipitation on the intermediate slopes and a gradual decrease in average precipitation from that point to the mountain crests. Furthermore, the lower temperatures at high altitudes result in a lessening of evaporation and transpiration, so that precipitation may be more effective at higher elevations. The rate of diminution of the evaporation rate caused by cooling is partly offset by the fact that the evaporating power of the air increases with altitude as atmospheric pressure decreases. Decreases of evaporation losses that result from cooling have been demonstrated. However, these losses are also offset by lower soil temperatures, which reduce water uptake, and by increases in wind velocity.

The decreased atmospheric pressure at higher

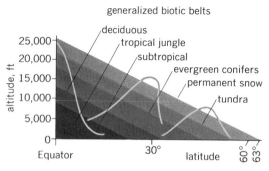

generalized biotic belts

Fig. 1. Diagram which illustrates the variations of altitudinal positions of some vegetational zones with latitudinal changes. (*From A. M. Woodbury, Principles of General Ecology, McGraw-Hill, 1954*)

altitudes also is accompanied by reductions in the partial pressures of oxygen and carbon dioxide. Thus, the oxygen concentration in the soil atmosphere decreases with altitude and may become a limiting factor for some species. Likewise, the lower partial pressure of carbon dioxide may be limiting to photosynthesis in many species.

Mountain environments and vegetation. From the preceding summary of altitudinal gradation of climatic phenomena, it can be seen that mountain environments are complex and highly variable. The conditions are never the same on two mountain masses, because many factors, such as slope gradient, exposure, direction of prevailing winds, nearness to oceans, geologic structure, and local topography, lead to individuality. In addition, plant life may respond to different limiting factors at various elevations. For example, studies in eastern Washington and northern Idaho have indicated that ecotones between vegetation types are related to critical deficiencies of heat at upper elevations, and to critical water shortages at lower elevations.

Generally, mountains in tropical, subtropical, and temperate areas are covered by forests, re-

gardless of the nature of the surrounding vegetation. The forests, often of several types, such as tropical rainforest, summergreen deciduous forest, and needle-leaved evergreen forest, are found at different elevations, and will extend upslope to a point at which they yield to meadows composed of grasses, perennial herbs, and low shrubs. If the mountain is high enough, there may be an alpine tundra and an area of perennial snow without vegetation (Fig. 1). Even in widely separated areas with different floras, similar environmental conditions tend to support similar vegetation types. *See* LIFE ZONES.

The variety of altitudinal zones thus increases with the height of the mountain. However, it also increases with the nearness of the mountain to the Equator (Fig. 1), because as one travels farther from the Equator, he experiences climatic and vegetational changes parallel to those experienced when one climbs a mountain. Thus, fewer low-elevation forms of vegetation are available in higher latitudes. Ultimately, in arctic regions, only tundra vegetation may occur on the mountains, and in the antarctic there are many snow-covered ranges with no vegetation.

Altitudinal zonation. The altitudinal zonation of vegetation in the northern intermountain region of Idaho and eastern Washington can be taken as an example of the arrangement of zones in a temperate mountainous area (Fig. 2). In this area, the lower elevations are dominated by sagebrush (*Artemisia tridentata*). At higher elevations, several grassland types may occur, and above these may be a poorly developed woodland of juniper (*Juniperus scopulorum*), with or without limber pine (*Pinus flexilis*). The ponderosa pine zone is the lowest of the well-formed conifer forests. It is often composed entirely of ponderosa pine (*P. ponderosa*), with an undergrowth of prairie grasses and forbs or shrubs. Douglas fir (*Pseudotsuga taxifolia*) occupies the zone above the ponderosa pine, but is often found in mixed stands with the pine or with species from the arborvitae-hemlock zone

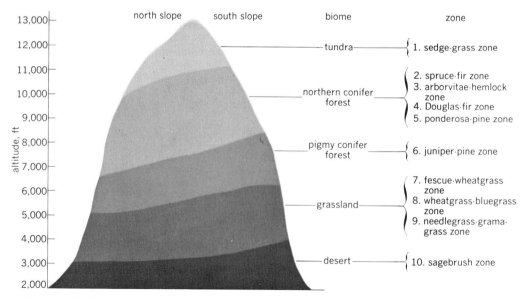

Fig. 2. The altitudinal zonation of vegetation in the northern intermountain region of Idaho and eastern Washington can be used as an example of the arrangement of zones in a temperate mountainous area. (*Based on data by R. F. Daubenmire, from E. Odum, Fundamentals of Ecology, Saunders, 2d ed., 1959*)

that occurs above. The highest coniferous forest zone is the spruce-fir zone in which the Engelmann spruce (*Picea engelmanni*), alpine fir (*Abies lasiocarpa*), and mountain hemlock (*Tsuga heterophylla*) occur in various mixtures. The sedge-grass zone, or tundra, occupies all of the areas above the limits of forest growth.

Topographic effects. Most mountainous areas are composed of a variety of geological formations, are interrupted by deep canyons, ravines, or passes, and have slopes with various exposures. In such areas, vegetation zones, therefore, generally are very irregular and may have projections or outliers that extend hundreds of feet in elevation above or below the main body. Usually, a given zone is higher on a warm, Equator-facing slope than on a cooler, pole-facing slope. Measurements in several areas have shown that temperatures on an Equator-facing slope are usually a few degrees warmer than those on a pole-facing slope as a result of the greater intensity of insolation on the former.

In narrow valleys, ravines, or canyons, inversions of temperature may occur frequently as masses of cold, more dense airflow down the valley walls and along the valley bottom. These cold air masses often accumulate to some depth in places where their paths are obstructed by constrictions in the valley or by other barriers. Thus, the minimum temperatures increase upslope from the valley bottom to a point at which the inversion ends, sometimes as high as 300 m above the valley floor; then they decrease in accord with the normal thermic gradient. As a response to the lower temperatures and greater moisture in such valleys, they are often occupied by pennantlike projections of vegetation zones that occur at much higher elevations on the mountain slopes. Thus, inversions of vegetation zones can be seen in many narrow valleys. In the northern intermountain region, for example, the bottom of a cool canyon may be occupied by a spruce-fir forest, the lower slope may be covered with a Douglas fir forest that intergrades with ponderosa pine; and the upper slopes and adjacent mountain sides may be covered with ponderosa pine. Thus, the ponderosa pine zone is higher in altitude in such localities than is the spruce-fir zone. *See* CLIMAX COMMUNITY; PLANT COMMUNITY; TERRESTRIAL ECOSYSTEM; VEGETATION ZONES, WORLD.

[JACK MC CORMICK]

Bibliography: R. F. Daubenmire, The life-zone problem in the northern intermountain region, *Northwest Sci.*, 20(2):28–38, 1946; R. F. Daubenmire, Vegetational zonation in the Rocky Mountains, *Bot. Rev.*, 9(6):325–393, 1943; F. Shreve, *The Vegetation of a Desert Mountain Range as Conditioned by Climatic Factors*, Carnegie Inst. Wash. Publ. no. 217, 1915.

Vegetation zones, world

Plant communities assembled into regional patterns by the region's physiography, geological parent material, and history. The physiography determines the drainage of the region; the geological parent material provides a definite physicochemical composition for the plant community; and the history of the region reflects the major climatic shifts and the more or less disturbing interference of man.

Fig. 1. Pineapple air plant, an epiphyte. (*Photograph by D. N. Sprunt, from National Audubon Society*)

The vegetation of the world forms a mosaic that can be viewed at different orders of magnitude. A contemplation of the concentric belts that circle a pond reveals that, from the deep to the shallow water and on to permanently flooded ground, different species of plants, and therefore different communities, dominate: The floating pads of water lilies and pondweeds are strewn across the open water and rooted in the ooze below; the rushes form an emergent belt of stiff culms rising to an even height; and, nearer the shore, arrowheads expand their broad halberd-shaped leaves. Similar zonations can be seen in many places where the water level varies or where some other feature of the substratum undergoes a change induced by soil or light. These individual and closely knit plant communities respond in a characteristic way to site factors, to the quality and quantity of resources and conditions necessary to their life: light, heat, water, oxygen, carbon dioxide, nutrients, penetrability, and relative acidity of the substratum. *See* PLANT COMMUNITY; VEGETATION.

On the broader geographical scale, where vegetation can actually be mapped, it is impossible to show such features as the rings of vegetation that surround a pond or even to account for the many variations of height and density of forest on a mountain or for the interwoven scheme of marsh and upland in a river basin. Some generalization must therefore be made.

As far as land is concerned, it is generally agreed that the prevailing upland vegetation is the most characteristic of the region as a whole; it very often covers the greatest percentage of the total surface. It is also argued that the plant cover of well-drained land holds the key to the whole regional dynamics, in that the gradual improvement of drainage in both the wet and dry sites is conducive to a replacement of marshy (hygrophytic) or desertlike (xerophytic) types by intermediate (mesophytic) types, which are better balanced and more stable. This concept is no longer adhered to in such a simplified form: The theoretical convergence of all vegetation changes toward a single climax type does not operate with uniform efficiency throughout. For instance, in regions of low or irregular rainfall, the arrests in succession are more pronounced, and consequently the effect of soil becomes as great as that of climate in determining the plant cover. *See* PLANT FORMATIONS, CLIMAX; SUCCESSION, ECOLOGICAL.

There remains some validity, however, in retaining the prevalent upland vegetation as an indicator of total regional conditions, primarily climate. It is in this connection that one of the following 20 vegetation classes or formation classes can be recognized as a characteristic of each region. It may be added that they differ very much in their features of productivity: The huge mass of plant tissue which the tropical rainforest elaborates daily from soil and air materials is in strong contrast to the long dormancy of the desert and the tundra and their explosive but rather limited growth. *See* CLIMAX COMMUNITY; ECOLOGY, PHYSIOLOGICAL (PLANT).

Important as each of these climax types may be to its region, their assignment to an area hardly provides a true geographical description. Even when the prevailing climatic control, as well as the predominant soil-forming mechanism, has been detected, and the principal historical influences have been traced, other forms of vegetation, especially in the xerophytic and hygrophytic situations, remain to be considered. The prevalence of dunes, cliffs, and permanent floods may also be such as to allow overwhelmingly more space to the xerophytic or hygrophytic than to the mesophytic. Finally, so much of the surface of the Earth has been modified by man that the vegetation of the manmade and man-maintained landscape, as well as the main crops that serve the regional economy, must be brought into focus.

Therefore, the following 20 vegetation classes are to be considered as areal formations. In the summary outlines hereafter, the following information will be given for each class: a description of the vegetation that is typical of the class on a regional basis; the kind of climate that prevails and the obvious soil-forming process; the principal kinds of vegetation to be expected in drier or wetter habitats; and some actual geographical examples of each one. It is certainly of major interest to point also to the peculiar play of climate-soil-vegetation as it operates in each situation. This permits an evaluation of quantity and diversity of environmental resources, of their availability, of the relative adequacy of the tapping method of the plants, and of the limitations of plants in resource utilization. Thus, formation of peat, susceptibility to fire, succulence, and other responses resulting from the reaction of vegetation as a whole to environmental forces are very unevenly, although predictably, distributed.

Much of the prevailing vocabulary used to describe vegetation is misleading, for two reasons. First, it all too frequently evokes a characteristic of the terrain that is assumed to induce a particular plant formation (such as plains, tundra, barrens, or shallows) rather than a feature of the plants themselves (such as forest, meadow, or palm brake). Thus rainforest is a type prevalent in areas of constantly high rainfall. It would be better to refer to it as "a tall, dense, broad-leaved evergreen forest," a descriptive expression which evokes qualities of the vegetation. The second reason is that many expressions are given a different meaning in different areas. Such is the case for bush, woodland, jungle, savanna, and steppe. Nothing short of a new approach (involving a new vocabulary) would do away with such equivocations. Synonyms are usually partial, since they have often been applied to a regional phase and not to the formation class as a whole.

Tropical rainforest. A class also known as selva, hylaea, forêt dense, forêt ombrophile, forêt tropicale humide, Regengehölze, and pluviilignosa. It consists of tall, close-growing trees, often buttressed at the base, their columnar trunks more or less unbranched in the lower two-thirds, and forming a spreading and frequently flat crown. Some of the outstanding ones protrude above the canopy. The number of species in any one area is so great that often no more than one individual of each can be seen from any given observation point. Several families are represented, among them the laurel, fig, brazil nut, locust, and myrtle families. However, they resemble one another in that their leaves are large and evergreen and that they often support a mass of epiphytic plants that rest upon their boughs but are not parasitic (Fig. 1). These are ferns and orchids and, in America, bromeliads. Lianas are also common. They are twining rather than climbing, their wood often weak and needing the support of trees and shrubs. They can therefore attain great heights in the forest only if they are carried by a growing tree. The dark floor of the forest supports rather sparse plant life, and thus it is easy to travel through. The tropical rainforest prevails over large tracts of the wet and warm lowlands near the Equator in South America (the Lower Amazon Valley), Asia (parts of India and Indonesia), and Africa (Congo Basin). *See* RAINFOREST.

Such vegetation occurs only in areas of high temperature and high rainfall where both are also evenly distributed and constant: that is, no shocks are experienced by the plants and the seasons are rather poorly defined. The soils are lateritic (latosolic): they accumulate no humus at the surface, the silicates are washed down, and the iron compounds remain near the surface, lending a red color.

Coastal areas harbor mangrove, a forest of stilted trees rooted in the fine silt of tidal swamps. The flooded and riparian forests, for example, the Amazonian igapó, contrast with the uplands by the thickness of their undergrowth and the abundance of palms. Coconut groves are common on strands, where they alternate with screw pines, casuarinas, and other drought- and wind-resisting plants. But where the rainforest itself has been disturbed, the low growth becomes very dense, with tangles of lianas, bristling bamboo scrub, thorny palms, and thickly branching shrubs. This secondary condition is known as jungle. The soils of these regions are quite vulnerable once they have been stripped of their native cover. Therefore, clearing agriculture, with shifting plots of small dimensions, is not a fundamentally bad practice. Subsistence on upland rice, squashes, and corn is possible. However, large tracts of land are now cultivated to coffee, bananas, sugarcane, pineapple, rubber trees, and other crops, or they are grazed. It is a great problem to maintain the fertility of these lands at a productive level. Cassava thrives on very poor sites. Rice must be grown under periodic artificial flooding, although the less productive upland varieties are also in use.

Temperate rainforest. Terms used to describe this class are laurisilva, laurel forest, cloud forest, notohylaea, moss forest, and sometimes subtropical forest. It differs from the tropical rainforest in that it has comparatively few species and therefore large populations of one kind. The trees are somewhat shorter, with smaller, evergreen leaves (especially the conifers, podocarps, and araucarioids, which are frequently present). Large tree ferns are usually abundant, in fact more so than in the tropical rainforest. Palms are sometimes present, as are lianas and epiphytes (with a prevalence of bryophytes over vascular plants). The herb layer of the forest floor is frequently well developed. Sometimes a sheath of mosses (Fig. 2) invests the trunk of trees at a certain level (moss line).

Areas of high and evenly distributed rainfall also having relatively little change in temperature and no frost (much of New Zealand, Madeira, southern Japan, and the Paraná Plateau of Brazil) are favorable to this kind of vegetation. It is probably more accurate to class the western hemlock–red cedar forest of Pacific Northwest America here than in the needle-leaved evergreen forest. Tropical mountains commonly grow stands of temperate rainforest (then called cloud forest) at the cloud level (Mexico, West Indies, Congo, and East Indies). This vegetation prevailed over very large portions of the Earth during the Tertiary, and some uninterruptedly equable areas (such as Madeira and the Canaries or the coastal strip of central California) retain vestiges of ancient communities (laurel or redwood forests). Some podocarps or araucarioids, several laurels, Myrsinaceae, the southern beeches, and trees of the *Weinmannta*, *Clethra*, and *Drimys* genera are typical of temperate rainforests.

Pedogenesis (soil formation) oscillates between laterization and podzolization. A blocking of drainage often results not only in marsh formation but in bog formation. Dune, scree, cliff vegetation, and practically all the very abundant second growth are evergreen. Forest yield is excellent in quantity and quality, although exotic species often do better than native ones. Much of the land is grazed or given to mixed farming. The permanent moisture is extremely favorable to tropical and subtropical garden crops as well as to temperate ones. Thus in Madeira oranges and apples, bananas and grapes, and sugarcane and wheat are grown side by side.

Tropical deciduous forest. Also known as monsoon forest, subtropical forest, semideciduous for-

Fig. 2. Mossy maple in rainforest, Olympic National Park, Wash. (*Photograph by J. O. Sumner, from National Audubon Society*)

est, forêt tropophile, forêt mésophile, and Monsundwald, this type consists of trees, sometimes almost as tall as those of the rainforest, which shed their leaves periodically. This condition makes for rather a conspicuous development of one or more lower layers. Thus the teak forest of Burma shows good stands of bamboo underbrush that retain their foliage all year. Lianas and epiphytes are locally abundant. *See* FOREST AND FORESTRY.

The shedding of leaves is a response to the long dry season which usually alternates with a very wet one, as in monsoon Asia. It may result, in very warm areas, from a shorter dry season (West Indies). However, the light period is often involved also, and many trees lose their leaves even in exceptionally wet years. The soil processes tend toward laterization. West Africa and Malaysia probably have the greatest areas of this vegetation, which is also present more sparsely in South and Central America, often in contact with the boundary of tropical rainforest.

The galleria forests that grow along the streams are quite different, inasmuch as moisture is available to them constantly. Jungle, a thicker, more obstructed vegetation, is common as second growth. However, where lateritic processes have been extreme or where soil degradation is otherwise very advanced, savanna is likely to predominate. These are areas of tropical agriculture (with the exception of the higher-moisture-demanding crops such as pineapple and cacao) where the yam, sweet potato, and sugarcane are cultivated and where pasturing is a common practice.

Summergreen deciduous forest. Other terms used to describe this class are temperate deciduous forest, forêt décidue, forêt à feuilles caduques, and Sommergrün Laubholzwald. This forest is tall and close-growing, the trees unbranched in their lower half and shading a rather scattered underbrush of small trees and shrubs. The herbs are exceedingly abundant in the spring, before the leaves fully develop in the canopy, but many of them disappear completely from above the soil surface to be replaced in the summer by well-distanced tufts of taller ones, many of which are short-day plants.

This kind of forest (dominated by oaks, maples, chestnuts, beeches, ashes, cherries, hickories, walnuts, and lindens) is practically restricted to the Northern Hemisphere (eastern North America, western Europe, and eastern Asia) under a continental climate of cold winters and very warm summers. The warm summers partly inhibit podzolization, so that the zonal soil is brown podzolic or gray-brown podzolic. With the possible exception of a small belt of alder forest in Argentina, this formation is absent from the Southern Hemisphere.

After lumbering, in this type of forest pines very often take over in a spectacular way, spreading from a former position on ridges, sandy, or other dry sites. The periodically flooded lowlands that accumulate a rich muck are also normally forested (Fig. 3), although the tree species (poplars, willows, ashes, and elms) are not the same as on the upland. Poorly drained areas, on the other hand, harbor extensive swamps and marshes, both fresh-

and salt-water. Completely blocked drainage retains peat, which has often accumulated to a great depth in previous (often colder) times. *See* FOREST MANAGEMENT AND ORGANIZATION.

Since some of the most densely populated parts of North America, Europe, and Asia occur in this area, few vestiges of primeval forest remain; besides the domesticated groves, there are many second-growth stands of timber (mostly pine, oak, and birch). The nonurbanized or unindustrialized land is usually devoted to cereals, mixed farming, and horticulture.

Needle-leaved evergreen forest. This class, also termed coniferous forest, Nadelwald, softwood forest, and resinous forest, consists of stiff, upright trees (conifers), usually conical at the top, with rather short lateral branches, growing quite close together and casting deep shade on very sparse shrubs, patches of herbs, and a sometimes continuous moss layer on the ground. The leaves are small, hard, and evergreen. Specific composition is most homogeneous in spite of the large circumboreal range. The tree flora is richest around the Pacific (*Pseudotsuga*, *Keteleeria*, and *Libocedrus*), poorer in the continental masses (*Picea* and *Abies*), and poorest on some of the European mountain ranges. The subordinate plants are often ericads, orchids, and sedges.

The podzolized soil is acid as a combined result of a cold, humid climate and the slow decomposition of the rather stable resins. Where blocked drainage causes anaerobic decomposition, peat accumulates indefinitely, so that bogs are an outstanding feature of rough, glaciated terrain such as the Canadian and the Scandinavian Shields. The Canadian muskeg shows all stages of development, from reed or sedge meadows to ericaceous scrub on *Sphagnum*-moss mats to closed larch and spruce bog forest. Rocky outcrops are commonly covered at first by lichen crusts and later by small grasses and trailing shrubs.

The needle-leaved evergreen forest is extremely flammable, and natural causes, as well as human carelessness, induce large-scale destruction in the western United States, northern Canada, Siberia, and northwestern Europe. This is often followed by a rapid invasion of deciduous-leaved trees (poplars and birches) that last hardly more than a generation.

The production of lumber for construction or pulpwood, hydraulic power, fishing, and hunting dominate. However, subsistence agriculture is scattered throughout, and some of the broader alluvial lands have higher fertility and harbor a prosperous dairy industry. *See* FOREST RESOURCES.

Evergreen hardwoods. This class, also termed sclerophyll forest, Hartlaubgehölze, and durilignosa, occurs as low forest or woodland characterized by hard- and rather small-leaved trees with a thick bark and sometimes an open crown. The undergrowth consists largely of evergreen broad-leaved shrubs, with some lianas (often spiny) and a spring herb layer. In the Northern Hemisphere, live oaks predominate, whereas proteaceous types do in the Southern Hemisphere.

The yearly growth cycle has the two peaks (spring and autumn) that distinguish the Mediterranean regime (whether in Spain, coastal California and Chile, South Africa, or southwestern Australia): cool, rainy winters, and warm (but not

Fig. 3. American elms in the floodplain of northern summergreen deciduous forest region. (*Photograph by H. Mayer, from National Audubon Society*)

tropical) dry summers. The tendency in soil formation oscillates between laterization and podzolization.

Although bogs are relictual, marshes are extremely well developed (Pontine, Camargue, and Doñana) and are in contact with extensive saltmarshes (Fig. 4).

A great deal of the vegetation in these areas has been modified by combinations of lumbering, fire, grazing, and agriculture, and the secondary types are mostly maquis or chaparral, a rather dense scrub, often spiny and interspersed with aromatic herbs. Garigue or matorral are even lower and more open and show ultimate degradation. Some areas are also occupied by ericaceous scrub (heath) or by pine or oak barrens (landes).

Mediterranean lands, where they are not too badly eroded by grazing goats and sheep, grow olives, citrus fruits, vines, and also cereals, especially wheat and corn.

Tropical woodland. This class has also been referred to as savanna-woodland, parkland, forêt claire, parc, and Savannenwald. It differs from the forest not so much in the size of the trees as in the wider spacing that allows more growth in the subordinate layers. Thus, the *Eucalyptus* woodlands in Australia, the broad-leaved-evergreen woodlands of western Africa, and a wavering belt in Brazil between the Amazonian rainforest and the campo cerrado follow the same pattern of tree-shrub-herb distribution. The lower strata are usually sparse but almost always characterized by evergreen shrubs and seasonal graminoids. *See* SAVANNA.

The climate is warm and moist enough for vigorous tree growth and yet too dry to allow closure of the canopy. These conditions prevail between tropical-wet and tropical-dry zones. The climatic cycle involved, however, may be one of short-term shift between these extremes or of relatively stabilized alternation. Pedogenesis reflects these opposing tendencies, with the drier sites tending to calcification (or even salinization), and the moister to laterization.

In ravines and other places where a more constant water supply is available, true forest is likely to develop. On the contrary, on ridges, dunes, and cliffs, the plant cover is reduced to low scrub. These areas are vulnerable to fire, which, combined with grazing, creates extensive erosion hazards.

Temperate woodland. Woodland, parkland, forêt-parc, and Savannenwald are other names used to describe this vegetation class. Similar to the tropical woodland in spacing, height, and stratification, it can be either deciduous or evergreen, broad-leaved or needle-leaved. The California yellow pine zone is fairly typical: It stands between the sagebrush scrub and the needle-leaved evergreen forest. The parkland of Saskatchewan offers a very different physiognomy with its medallions of aspen trees in a matrix of grassland, a sort of "emulsion" of forest and prairie.

The climate shows a moist and a dry season and, as with the tropical woodland, there may be short-term fluctuations. Soil formation here is also unstable and highly susceptible to conflicting chernozem-podzol, glei-podzol, or saline-podzol influences.

In washes and flats, grassy or shrubby forma-

Fig. 4. Narrow section of a coastal marsh showing vegetation zonation ranging from the sea through tidal flats to the upland sites. (*Photograph by W. Jahoda, from National Audubon Society*)

tions tend to prevail, whereas in ravines and on river edges a curtain of trees appears, generally poplars and willows.

Tropical savanna. This is defined as a vegetation of widely scattered or bunched small trees growing out of lower vegetation, and it undergoes many variations. It is also known as sabana, Savannen, campo cerrado, savane, and savane arborée. The trees may be deciduous or evergreen, and the low growth, shrubby or grassy. The seasonal rhythm is very pronounced and induces a pulsation of vegetative activity which strongly marks the aspect of the landscape. Thus the grasses in the Brazilian campo cerrado and the Kenyan acacia savanna rise, green and dense, and eventually fade to a straw mat. Savanna plants are frequently very deeply rooted: Some shrubs are able to tap the enormous reserve at the phreatic (underground water) level itself.

The tropical savanna regime rests upon a pronounced dry season, often two. The fluctuation of water availability is due to decreased and even interrupted precipitation, usually accompanied by the highest temperatures. An extreme swing of laterization is likely to develop hardpan.

Savanna regions generally exhibit a forest on stream edges (galleria) and harbor rather sparse scrub on dry topography. In Brazil, as in many parts of Africa, relict forests are still included within savanna territory. At all events, the exact boundary of climatic savanna is not always reliably traced, as this formation is extremely subject to recurrent fires. Some of the woody elements are especially resistant, whereas the extensive mats of seasonally dry graminoids are extremely flammable.

Pasturing by large herds (elephants, giraffes, and antelopes in Africa) also has the effect of reducing local coverage of plants over the years. The tropical savanna lends itself well to many extensive crops (millet, sorghum, cassava, coffee, and several types of fruit trees).

Temperate and cold savanna. Forest-tundra, lichen woodland, boreal woodland, and taiga are terms used to describe the temperate and cold savanna. This regional formation, very extensively represented in North America and in Eurasia at high latitudes, consists of scattered or clumped trees (very often conifers and mostly needle-leaved evergreens) and a shrub layer of varying coverage. Mosses and, even more abundantly, lichens often form an almost continuous carpet.

The very peculiar high-altitude vegetation known as páramo in South America (and duplicated in Africa) is also a cold savanna, which is characterized by rosette trees and cushion plants.

In most areas where cold savanna prevails, especially at high latitudes, the extreme continentality (possibly the world record of temperature ranges) effectively reduces total growth and precludes agglomeration above the shrub level. The balance of gleization and podzolization shows many oscillations.

Congested depressions harbor muskeg; there is relatively little marsh formation; and the drier ridges, especially the rocky ones, harbor scrub or tundra.

Large herds of reindeer and caribou which feed upon the lichens and the buds of woody plants exert a strong modifying influence. Fire is less important than in the needle-leaved forest. Although some urban centers have developed around the mines of northern Canada and Siberia, agricultural exploitation is virtually impossible. *See* TAIGA.

Thornbush. Thorn scrub, savane armée, savane épineuse, Dorngeholz, Dorngestrauch, and dornveld are other names for this class. The dominants are principally of two kinds: tall succulents and profusely branching smooth-barked deciduous hardwoods which vary a great deal in density, from an inextricable mesquite bush in the Caribbean to an open spurge thicket in Central Africa. The northeastern Brazilian caatinga consists of thorny tall shrubs and cacti in its upper layer and mostly of annuals among the herbs. The annuals tend to be quite ephemeral. The South African dornveld is rather similar, except that spurges replace the cacti.

The climate is that of warm desert, except for a rather short but intense rainy season that causes much lateral transport of soil and strongly interferes with the pedogenic processes.

In certain favored positions, palm savannas or grasslands appear, whereas the soils that can hold no moisture have desert vegetation.

The hard soils are not favorable to cultivation, although certain fiber plants (sisal in Mexico and caroá in Brazil) do very well. They are extensively if unproductively pastured.

Tropical scrub. This class is also known as bush, brush, thicket, mallee, and fourré. It is composed of low woody plants (shrubs), sometimes growing quite close together, but more often separated by large patches of bare ground. Clumps of herbs are scattered throughout. This formation class is not very frequent, for it tends to be replaced by tropical savanna or tropical woodland. The Ghanaian evergreen coastal thicket and the Australian mallee are good examples of a rather narrowly confined zone, a wedge between better-developed formations. The pedological units are

thus formed as a result of conflicting tendencies.

Grassy marshes and meadows are common in these areas; islands of forest and savanna also occur. The natural thickness is often reduced, especially for pasturing, but the woody plants always tend to reinvade.

These areas, like their temperate analogs, are most often rangelands.

Temperate and cold scrub. Synonymous terms for this class are heath, bosque, and fourré. The density of this formation varies a good deal, as does its periodicity. The shrubs may be evergreen (Tierra del Fuego beeches and Irish heathers) or deciduous (subarctic willows and alders). An undergrowth of ferns and other large-leaved herbs is quite frequent, especially at the subalpine level.

A considerable amount of moisture is necessary, whether in the soil (snowmelt), in the atmosphere (mist), or as a result of great seasonal downpour. Wind shearing or very cold winters prevent tree growth, although occasionally widespread soil conditions or historic factors are responsible (Azorean laurel–juniper scrub).

Closed depressions allow bog formation which, in the superhumid regions, tends to creep upland.

The warmer scrub areas (sagebrush zone of Wyoming and Azorean laurel heath) are most often converted to pasture and are excellent grass producers. This is also true of moist subalpine regions (mugo pine and *Rhododendron* scrub in the Pyrenees).

Tundra. This formation grows very low, and consists of trailing or matted shrubs (willows, heaths, and sometimes conifers), some of them evergreen, compact, and small-leaved; of cushion plants (with foreshortened stems and deep taproots as in the alpine campion); of tufted herbs; and of a great abundance of mosses and lichens. The geographical extent of many arctic-alpine species is almost coextensive with the greatest development of tundra at the high latitudes of the Northern Hemisphere, descending to the 60th parallel and lower in some places, and reaching all the major mountain systems. In many mountain systems, however, grassy formations (meadows) are more prevalent. Alpine areas of the Southern Hemisphere (New Zealand, Chile, and Falklands), having few floristic analogies with the boreal, nevertheless harbor a similarly structured tundra.

A short growing season and extremely severe winters are the determining climatic factors, whether at high latitudes or altitudes. The higher altitudes, however, usually have higher summer temperatures and do not always exhibit a soil with a permanently frozen layer (permafrost) which inhibits deep rooting; but the seasonal freezing is often drastic enough to cause a churning of soil particles which results in polygonal structures at the surface. The main soil-forming process is gleization, the development of a bluish-gray layer of soil at the bottom, with organic accumulation at the surface.

The apparent uniformity of tundra regions, which is due to the very restricted flora, is broken by the aridity of certain fell-fields (or felsenmeer) where crusts of lichens clothe the evenly spaced stones or boulders, by small mossy bogs and meadows, and by dunes and cliffs. An occasional ravine harbors willow scrub. The fauna show adaptive features of several kinds: The migrant birds form a very rich and abundant group, whereas migrants of intra-arctic amplitude or local residents (barren-ground caribou, ptarmigan, arctic hare, and fox) modify their fur, plumage, and feeding habits with the season.

There is virtually no direct utilization of vegetation by man, and no cultivation even where radio, aviation, military, or mining settlements have developed.

Prairie. Tall-grass prairie, Wiesen, and high tussockland are terms that describe the vegetation of this class. This formation is dominated by tall grasses and contains almost no woody elements, at least none that overtop the grasses. Frequently, the seasonal turnover can be very striking: Grasses leaf and flower in the spring and early summer, whereas the forbs (broad-leaved herbs, most of them composites) remain in leaf until the late summer, when the days become shorter, and then flower. This is a phenomenon of middle and high latitudes, as in Iowa and the Hungarian puszta. The Argentine pampa also belongs to this formation, although it contains some shrubs, as does part of the New Zealand tussockland.

Typically continental temperature extremes, seasonal rains, and cold winters combine to induce a development of deep black soil (chernozem) overlying an accumulation of calcium in depth. River edges and bluffs are often wooded, since they have more available water during the drought and because wind action is somewhat reduced. The immediate river edge has a floodplain regime with willows, poplars, and elms (in the Northern Hemisphere), but the humid bluffs may harbor extensions of upland forest from the adjacent climatic zone. On the other hand, marshy ground is occupied by "wet prairie," that is, by graminoid formations. Stripped areas are easily windblown and quickly turn to steppe and even desert or dune. Fire is so frequent that it appears to be a normal agent in the natural cycle.

Large bands of grazing animals (bison in North America and Europe) were part of the original balance. In some areas (northeastern Oklahoma) they have been replaced by domesticated cattle. But for the most part prairie country is considered the best cereal land in the world and has been used for growing wheat, barley, and maize. *See* PRAIRIE ECOSYSTEM.

Steppe. The steppe, or short-grass prairie, differs from the prairie in that the grasses are of lower stature and bunched or sparsely distributed and also in that low shrubs are somewhat conspicuous. Much bare ground remains exposed which may be used by annuals after the rains. Low-growing succulents are not unusual. *See* GRASSLAND.

The mid-latitude continental regime which prevails, for instance, over the extensive flat plains of the Soviet Union and of central North America, is favorable to the development of steppe. Drought, which usually follows winter rains, is greater than in the prairie and does not allow the accumulation of a deep black horizon at the surface of the soil, but rather a characteristic chestnut layer. Excessively drained sites lend themselves to alkaline accumulations.

Salt lakes and saltmarshes are fairly common with their concentric belts of saltwort, shadscale, and other halophytes. In protected areas, small groves of deciduous woodland (aspen groves in Saskatchewan) or patches of savanna prevail, and in badly eroded areas an open scrub virtually without grasses is prevalent.

The steppe is pastured and cultivated (cereals and cotton) and with proper exploitation is quite fertile. However, where abusive practices have prevailed, eolian erosion has advanced very far, and sandstorms have resulted in "badlands" on which rehabilitation is very slow.

Meadow. This class, also known as pelouse and Wiesen, is a low grassland, dense and continuous, variously interspersed with forbs (and even with mosses, which locally prevail) but very few if any shrubs.

It is characteristic of the formation class only in areas of high moisture, its optimum being in some alpine regions, where relatively high day temperatures in summer offset the rigor of winter. Some parts of the Arctic (for example, the Mackenzie Lowlands eastward to Hudson Bay) also have meadow rather than tundra. The rather dense "short-grass prairie" of Saskatchewan probably should be classed here also. The growing season is necessarily rather short (or else quite cold, as in the subantarctic islands and parts of Patagonia), and the yearly increment is not very abundant. The soil consists of a well-developed humus layer resting on gravel or on some other parent material which is usually not strongly modified.

On the rockier sites (fell-fields) low woody vegetation becomes more important and tundra is rather likely to take over. In some of the sheltered ravines either a megaphorbia (a prairie dominated by forbs) or a scrub will develop.

Natural alpine meadows are used for pasturing, although they are much less productive than secondary meadows at the subalpine level.

Warm desert. This is a vegetation zone, also named désert, Wüst, and Trockenwüst, in which little vegetation of any kind grows, although succulent and small broad-leaved shrubs are characteristic. In a sense, tiny annuals (sometimes called "belly plants") are even more typical in that they crop up in large numbers shortly after one of the infrequent and usually inconstant rains. Less than 10% of the space is covered by plants, and they economize water in a variety of ways: succulence (cacti, Fig. 5), reduction of leaf surface (creosote

Fig. 5. Prickly pear cactus (*Opuntia engelmannii*) in the Sonoran Desert. (*Photograph by H. L. Parent, from National Audubon Society*)

bush), and great extension of root system (smoke tree), among others. Accordingly, deserts are of many kinds: the dunes and rocky wastes of the Sahara that harbor hard grasses (*Stipa*) and gnarled evergreen shrubs, the hard caliche that bears creosote bush in the Sonoran region of the United States and Mexico, the coarse pebbly surfaces of the Karroo with its small succulent plants (*Lithops*, *Faucaria*, and *Fenestraria*), and the saltbush areas of Australia, Algeria, and the southwestern part of the United States. *See* DESERT VEGETATION.

The climate is one of extremely low rainfall (possibly no rain at all for years) and very high temperatures (in fact, much higher than in the equatorial zone). Yearly amplitudes of temperature are considerable. Pedogenesis leads to salinization, and even to the formation of alkali crusts on the soil surface.

The rockier sites are usually taken over by succulent "gardens." The stream beds (known as draws, arroyos, and oueds) usually have a characteristic vegetation, closer or taller than those of the rocky sites, consisting of shrubs or even trees very different in kind from those of other habitats. Whereas the salt lakes are bordered with a belt of halophytic vegetation (such as *Salicornia* and *Atriplex*), the fresh-water bodies have palm copses (*Washingtonia* in California and *Phoenix* in Africa) and grass.

When irrigated, desert soils are frequently quite rich and will grow big crops of dates, grapefruit, oranges, and even fodder such as alfalfa.

Cold desert. The cold desert is also known as rock desert and frigorideserta. Extreme cold combined with adverse edaphic conditions such as a very coarse rocky substratum allow neither tundra nor meadow to develop. Instead, tufts of herbs (often graminoids) occupy shallow crevices, open sandy areas, or loose schistose screes. Mosses and lichens are not very conspicuous.

This condition is frequently found on nunataks, on emergent mountain peaks, and at very high latitudes (northern Ellesmere Island).

Crust vegetation. Although it is to be seen literally everywhere, crust vegetation does not prevail over large enough sections of the Earth's surface sufficiently often to be considered characteristic of a regional formation-class. It consists of algae, lichens, mosses, or liverworts and has a thickness of only a few centimeters and a rather variable coverage, ranging from the full carpeting by mosses of some Antarctic rocky islands to the sparse dotting by lichens of some Saharan rocky ranges.

The climate must be one of very sparse precipitation and of extremely high or extremely low temperature. The ice margin and nunataks of Greenland, part of northern Ellesmere Island, all of ice-free Antarctica, and some of the snow-free high altitudes of various mountain ranges have practically no other vegetation, with the exception of a few herbs and shrubs in protected sites. The contrastingly richer fauna (for example, migrant birds) occasionally provides good soil fertilizer and permits small patches of lush growth.

Such regions are uninhabited, and it cannot be said that their soil or vegetation are of any utility to man. *See* VEGETATION ZONES, ALTITUDINAL.

[PIERRE DANSEREAU]

Bibliography: P. Dansereau, *A Universal System For Recording Vegetation*, 1958; P. Dansereau, *Biogeography: An Ecological Perspective*, 1957; P. Dansereau, *Vegetation of the Continents*, in preparation; S. R. Eyre, *Vegetation and Soils: A World Picture*, 1963; R. W. J. Keay et al., *Vegetation Map of Africa South of the Tropic of Cancer*, 1959; A. W. Küchler, National vegetation, in E. B. Espenshade, Jr. (ed.), *Goode's World Atlas*, 11th ed., 1960; D. L. Linton, Vegetation, in C. G. Lewis (ed.), *Oxford Atlas*, 1951; A. Strahler, *Introduction to Physical Geography*, 1965.

Wastes, agricultural

The food and fiber industry, including production and processing, produced over 2×10^9 tons (1.81×10^9 metric tons) of waste in 1970, or about 58% of all the solid wastes produced in the country. These wastes include the obvious manure and organic residues from farms and forests, as well as various solid materials and water discharged from processing or manufacturing plants that use any form of an agricultural product as a raw material. Some of these wastes are utilized or recycled, but most of them require disposal. Many of the present disposal techniques simply move the waste to another place rather than solve the problem. Therefore the only true solution to the waste problems of agriculture is to consider the material as a national resource to be recovered by recycling and utilization.

Solid wastes. The following sections discuss the recycling of various types of solid waste.

Animal wastes. Disposal of animal wastes poses a major problem to the agricultural industry. Tremendous amounts of manure are produced (Table 1), but it is the concentration of this manure in small areas that constitutes the real problem. Agricultural science and technology met the challenge to increase animal production by developing procedures to raise large numbers of animals in confinement, as in cattle feedlots. Unfortunately the technology for managing animal waste has not developed as rapidly. Where usual disposal techniques are being curtailed by social pressure or by possible danger to the environment, the animal producer is forced to operate under a narrowing profit margin and higher overhead costs. Recycling and utilization may partly solve these problems.

Many systems for injecting manure into the subsoil have been investigated. Such systems reduce the insect and odor problems, but the operations are not economical and probably will be used only in areas where the social pressures are extreme. At present, land spreading must be considered as a least expensive rather than economical or profit-making method of waste handling.

Many methods have been investigated for the potential utilization of animal wastes, for example, selling as fertilizer, composting, animal feeding, and energy (methane) production. The primary purpose of composting is to eliminate putrescible organic matter while conserving much of the original plant nutrients. Manures may be composted alone but frequently are combined with high-carbon, low-nitrogen wastes such as sawdust, corn cobs, paper, and municipal refuse. The resulting material is suitable for use as a soil conditioner and organic fertilizer. Full-scale operations have

Table 1. Numbers of livestock and their total waste production in the United States in 1970

Livestock	Total population, millions	Solid wastes, millions of metric tons/year	Liquid wastes, millions of metric tons/year	Total wastes, millions of metric tons/year
Cattle	107	1015.0	391.0	1406.0
Hogs	57	62.5	36.5	99.0
Sheep	21	9.6	5.8	14.4
Horses	3	17.5	4.4	21.9
Chickens	2950	54.0	–	54.0
Turkeys	106	21.7	–	21.7
Ducks	11	1.6	–	1.6
Total	–	1181.9	–	1618.6

been technically successful with poultry, beef, and dairy manure in a unique regional situation, but in general the market for this product is very small. Without a market, essentially all the dry matter remains for further disposal. Manure, either as a fertilizer or soil conditioner, is difficult to sell profitably, regardless of preparation procedures, because of the low cost of commercial fertilizers. *See* MANURE.

Dried and dehydrated manure is sometimes sold as a soil conditioner, organic fertilizer, or animal feed supplement, but the cost of drying and dehydration is generally greater than the return realized.

Use of animal wastes in the feed of animals has recently received much publicity as a potential method for utilizing agricultural wastes. When nutritional principles are followed, the technique has produced good results, especially if the waste of a single-stomached animal is added to the rations of ruminants or if the waste of the ruminant is chemically treated before use. However, a Food and Drug Administration regulation (1970) prohibits the use of animal wastes as feed supplements because of the possible transmission of drugs, feed additives, and pesticides to another animal and to some agricultural products, such as milk and eggs. On-going research should clarify the situation.

Much of the nutritional value of animal wastes might be captured by growing housefly larvae and insects as a source of protein for animal feed supplement. The method has been tested and shows considerable promise. Algae can be commercially grown on a manure substrate and can also be used as a feed supplement. The remaining growth media could be used as a soil conditioner or fertilizer.

Manure can be used directly as a fuel or as the substrate to produce methane anaerobically. Manure must be collected and dried to be used as a fuel, a costly operation. Therefore it is doubtful that any appreciable quantity will ever be utilized in this manner.

Regardless of the disposal procedure or recovery process, the soil will be the ultimate disposal site of most of the bulk of animal wastes. Therefore the challenge is to develop techniques of recycling to incorporate animal wastes in land management programs without damaging the environment or causing a nuisance to the human population. This will most likely be accomplished by convincing the farmer that manure improves the physical condition of a soil as well as supplies plant nutrients.

Crop and orchard residues. The tonnage of plant residues left on the farms exceeds by far the tonnage of the crops taken to market. These wastes consist of straw, stubble, leaves, hulls, vines, tree limbs, and similar trash. Most of the residue is burned to eliminate troublesome plant diseases, pests, and weeds. Unfortunately this pollutes the

Table 2. Wastes produced by selected agricultural industries in the United States during 1970

Type of processing industry	Total solid waste discharged, millions of metric tons/year	Solid waste salvaged, millions of metric tons/year	Solid waste requiring disposal, millions metric tons/year	Comment
Meat	17.5	17.5	0.003	Values include all livestock and broilers
Vegetables and fruit	12.7	8.7	4.0	Organic and inorganic wastes in disposed fraction
Dairy	0.2	*	*	As fat-free solids
Forestry	36.0	*	*	Logging residue from lumber mill
	55.0	*	*	As sawdust and edging
	41.5	*	*	From pulp and paper mill

*No estimates are made because the amount of salvaged and disposed by-product depends upon the location and size of the plant or operation.

air with smoke and volatile organic compounds. A small part of these materials is being used for mulch and as a soil builder when the mulch is plowed under, ensilage, bedding for animals, and as bulk material in the manufacture of corrugated cartons, insulated board, and specialty paper. Although the handling of crop residues is not a major problem in terms of amount, these wastes can be a focal point of major infestation by harboring insects and plant diseases. Obviously, new and better methods are needed to handle these wastes without polluting the environment.

Food processing wastes. The meat processing industry produces few wastes, other than water, that require disposal (Table 2). This industry has illustrated that the utilization of wastes can be profitable. In recent years over 35×10^9 lb (15.89×10^9 kg) of meat and similar amounts of other animal parts, or wastes, have been processed annually. These wastes, including those collected during water disposal procedures, are the raw material for the manufacture of soap, leather goods, glue, gelatin, and animal feeds. Glands and organs are processed to produce hormones, vitamins, enzymes, liver products, bile acids, and sterols. Other animal parts are the source of certain fatty acids, oils, grease, and glycerine. Bones can be processed to produce proteins and fats, as well as bone meal. Finally fuels and solvents can be made from the waste products of the above industries. The waste waters leaving the processing plants still contain small amounts of solids, and disposal procedures are necessary. The utilization of these suspended materials must be preceded by a more efficient use of water which will result in concentrating the wastes and improving the economics of recovery.

Although the fruit and vegetable processing industry does not produce large amounts of solid wastes relative to other segments of agriculture, these wastes are difficult to handle because of their varied nature. Peels, skins, pulp, seeds, and fibers are suspended in billions of gallons of water which may be saline, alkaline, or acidic and may contain a wide variety of soluble organic compounds. Of the 12.7×10^6 tons (11.52×10^6 metric tons) of solid waste produced, only about 4×10^6 tons require disposal—but at a cost of $\$25 \times 10^9$. Most of the remaining 8.7×10^6 tons are utilized as animal feed and consist principally of the by-products of the processing of citrus fruits, potatoes, and corn. These by-products are usually given away, resulting in a cost-free disposal procedure which is very valuable to the industry. Increasing the efficiency of the in-plant use of water is a necessity before any new salvage operations will be economical.

About 118×10^9 lb (53.4×10^9 kg) of milk were converted in 1970 to fluid milk, cheese, butter, condensed and powdered milk, and ice cream. The amount of solid wastes coming from this industry is small compared with the amounts of other agricultural wastes, but they require costly handling. Milk solids have very high pollution potentials. Therefore these solids must be removed from millions of gallons of water daily before this water can be released to streams or used for irrigation. The collected milk solids do have value as animal feed, feed supplements, raw material for the production of some chemicals such as alcohol, and as a basic ingredient in growth medium for microorganisms in the production of pharmaceutical chemicals. Lowering the cost of production and increased utilization are, as before, related to a more efficient use of water.

Forest waste products. A large part of a tree harvested for pulp paper or lumber becomes waste. It has been estimated that 19% of the tree is left in the forest. Approximately 16% of the log is wasted as sawdust and 34% as slabs and edgings during milling. Finally pulp and paper mills discharge effluents containing about 50% of the log. Wastes from the wood industry have great pollution potential and are usually destroyed by burning because other disposal procedures are too costly. Burning of forest debris is deemed to be necessary in the control of forest diseases and insects, but it results in air pollution.

There have been many processes developed by wood science and technology laboratories to recover a large variety of by-products, but these processes have been only sparingly applied, primarily because of economic limitations. Wood is composed primarily of lignin and cellulose. Lignin can be processed to obtain many valuable chemicals, such as artificial vanilla, or used directly as a binder, a dispersant, an emulsion stabilizer, and a sequestrant. Lignin can also be utilized as the raw material for plastic production. Cellulose is readily processed to simple sugars, alcohol, fodder, yeasts, and chemicals such as furfural. Utilization methods that may be economically feasible in the future include the use of wastes for the production of specific chemicals, through fermentation, and in new building products such as board, paper, and blocks.

Recycling of paper does not seem to have a promising future. Now less than 20% is recycled as compared to about 30% 20 years ago. If the cost of removing the adulterations of the paper, such as ink, plastic, and metal clips, is decreased, waste paper recycling probably will be increased, thereby decreasing the overall wastes produced by the wood industry.

Liquid wastes. The agricultural industry uses water in almost every operation of production, processing, and manufacturing. Irrigation and food processing use the largest amounts. These waters must be considered agricultural wastes in the same sense as are solid wastes and must either be recovered or disposed of. In the past the axiom "control pollution by dilution" was readily accepted. Consequently, large volumes of water were used to dilute the agricultural wastes before they were discharged into lagoons or other surface waters, sprinkled on a field, or discharged into a municipal sewage system. The great increase of soluble and solid wastes to be transported, the lack of usable water in some areas, and social pressures are causing a reevaluation of water use for this purpose. Processing plant procedures are being changed to increase recycling of water in the plant by segregating highly contaminated water and using the clean discharged water to irrigate crops. The highly contaminated water is handled by conventional disposal procedures before it is utilized or recycled.

Irrigation of agricultural crops uses vast quantities of water. In the past this was not carefully controlled and streams were contaminated by runoff and drainage waters containing sediment, soluble salts, pesticides, and so forth. However, important changes are taking place, such as better controls in the ditches and fields and reuse of runoff irrigation waters. Irrigation water is reused by collecting the normal runoff water in shallow ponds and then pumping it to other fields. Less than 10% of the liquid water is lost by this method. Also, most of the soluble and solid materials usually lost in runoff are kept on the farm, thereby decreasing the possible spread of a polluting agent, although salt buildup may be a problem in some areas.

Agricultural waters can also be used to recharge groundwater. Uncontaminated water can be discharged directly into the groundwater. Contaminated water can be processed, or in some cases the soil can be used as a filter. See SOLID-WASTE MANAGEMENT; WATER POLLUTION.

[WALTER R. HEALD]

Bibliography: R. B. Enghahl, *Public Health Service Publ. No. 1856*, 1969; E. P. Taigamides, *Proceedings of the 24th Purdue Industrial Waste Conference*, pp. 542–549, 1969; C. H. Wadleigh, *USDA Misc. Publ. No. 1065*, 1968; N. H. Wooding, Jr., *Spec. Circ. No. 113*, Pennsylvania State University Extension Service, 1970.

Water analysis

The application of bacteriologic tests to determine the sanitary quality of potable and recreational waters. The objective is detection of animal and human pollution rather than detection of pathogenic bacteria since it may be presumed that water thus polluted is unsafe.

Quantitative testing. The standard plate-count method is used. Measured portions of water are plated with nutrient agar, and colonies that develop are counted. Preferably, two sets of plates, incubated at 20 and 35°C respectively, are used per sample: using a single set, incubation is at 35°C, a temperature favorable for bacteria of human origin. Counts greater than 1000 bacteria per milliliter usually suggest pollution, as does a 35°C count higher than that at 20°C. Quantitative testing is of greatest value in showing deviations from normal conditions in a routinely tested water source.

Qualitative testing. This detects particular bacteria or groups of bacteria whose presence indicates fecal pollution. In the United States the coliform group is almost exclusively employed as the indicator: *Escherichia coli* per se is used in many other areas; and the enterococci have also been suggested as indicators. Coliforms are defined as aerobic and facultative anaerobic, gram-negative, non-spore-forming bacilli which ferment lactose with gas formation, and include *E. coli* and other less common fecal organisms as well as certain nonfecal bacteria. The coliform group is a less selective indicator than *E. coli* alone, but errors introduced are on the safe side. Highly selective media or environments, such as MacConkey broth containing bile salts or the Eijkman test using 45.5°C for incubation, are used for *E. coli* detection, and the American Public Health Association Standard Methods include procedures for distinguishing between fecal and nonfecal coliforms based on a 45.5°C incubation temperature. Coliform detection may require the following series of three tests.

Presumptive test. Lactose or lauryl tryptose broth in Durham fermentation tubes (Fig. 1) is inoculated with the sample and incubated at 35°C. The presence of gas in the inverted vials of the fermentation tubes within 48 hr constitutes a positive result. Absence of gas is interpreted as a coliform-free sample. Gas is presumptive but not conclusive evidence of coliforms since other bacteria, mainly gram-positive forms, may also produce gas. Consequently further testing is required.

Confirmed test. Either liquid or solid media are used. All confirming media are selective, inhibiting bacteria which produce false positive presumptives while permitting coliform growth. Solid media are also differential and make possible recognition of certain coliforms from their colonial morphology. Brilliant green bile broth, the liquid

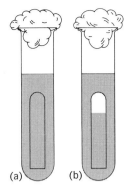

Fig. 1. Durham fermentation tube. (*a*) Before incubation. (*b*) After incubation.

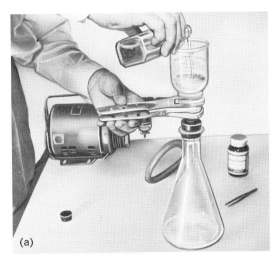

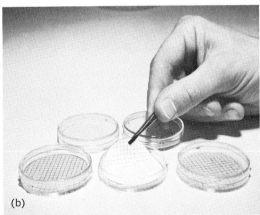

Fig. 2. Membrane filter technique. (*a*) Water sample filtered through membrane filter. (*b*) Membrane filter placed on nutrient pad. (*c*) Typical coliform colonies on membrane filter. (*Millipore Filter Corp.*)

confirming medium, is inoculated from positive presumptive tubes (that is, those tubes which have gas), and the broth is incubated at 35°C. Gas production within 48 hr is interpreted as indicative of the presence of coliforms. Eosin-methylene blue or Endo agar, the solid confirming media, are streaked from positive presumptive tubes and incubated at 35°C for 24 hr. A positive result is the presence of recognizable colonies of coliforms on the plates.

Completed test. Since not all coliforms give recognizable colonies on solid confirming media, a completed test is necessary when atypical colonies are found. A transfer is made from an atypical colony to both a nutrient agar slant and a lactose fermentation tube, which are then incubated at 35°C. If colonies of gram-negative nonsporulating rods are found on the slant and gas is produced in lactose broth, coliforms are proven to be present.

Most probable number (MPN). Evaluation of water quality is based on the numbers of coliforms present. To determine numbers, replicate portions of a series of decimal dilutions of the sample are inoculated into presumptive tubes. The number of portions and the range of dilutions used vary with the character of the water examined. A statistical analysis of the distribution of positive and negative tubes allows an estimate of the MPN of coliforms in the sample. MPN tables have been prepared that cover the most frequent combinations of sample portions used.

Membrane filters. This technique (Fig. 2) for coliform detection has been approved as an alternative standard method by the U.S. Public Health Service. Water is filtered through a cellulosic membrane which collects bacteria on its surface. The membrane is placed on a pad saturated with selective medium: the diffusion of nutrients supports development of recognizable coliform colonies on the membrane surface. The technique has the advantages of speed, economy of materials and space, and gives a direct rather than statistical estimate of coliform population.

Sanitary standards. Routine bacteriologic testing cannot provide an absolute evaluation of the quality of a water. Consequently, quality standards based on MPN determinations are somewhat arbitrary. For potable waters, most communities in the United States follow the U.S. Public Health Service Drinking Water Standards of 1946. These specify the minimum number of samples to be examined each month and the maximum permissible percentage of positive portions, but not specific permissible MPNs. Roughly speaking, to meet these standards, a water must show an MPN of less than 1 per 100 ml for most samples analyzed. Swimming pool and recreational water standards are generally less exacting.

[SYDNEY C. RITTENBERG]

Bibliography: American Public Health Association, *Standard Methods for the Examination of Water and Wastewater*, 12th ed., 1965.

Water conservation

The protection, development, and efficient management of water resources for beneficial purposes. Water occupies more than 71% of the Earth's surface. Its physical and chemical properties make it essential to life and modern civilization. Water is combined with carbon dioxide by green plants in the synthesis of carbohydrates, from which all other foods are formed. It is a highly efficient medium for dissolving and transporting nutrients through the soil and throughout the bodies of plants and animals. It can also carry deadly organisms and toxic wastes, including radioactivity. Water is an indispensable raw material for a multitude of domestic and industrial purposes.

UNDERGROUND WATER

Water occurs both underground and on the surface. Usable groundwater in the United States is estimated to be 47,500,000,000 acre-feet. Annual runoff from the land averages 1,299,000,000 acre-feet. The volume of groundwater greatly exceeds that of all fresh-water lakes and reservoirs combined. It occurs in several geologic formations (aquifers) and at various depths. Groundwater under pressure is known as artesian water, and it may become available either by natural or artificial flowing wells. Groundwater, if abundant, may maintain streams and springs during extended dry periods. It originates from precipitation of various ages as determined by measurements of the decay of tritium, a radioisotope of hydrogen found in groundwater. The water table is the upper level of saturated groundwater accumulation. It may appear at the surface of the Earth in marshes, swamps, lakes, or streams, or hundreds of feet down. Seeps or springs occur where the contour of the ground intercepts the water table. In seeps the water oozes out, whereas springs have distinct flow. Water tables fluctuate according to the source and extent of recharge areas, the amount and distribution of rainfall, and the rate of extraction. The yield of aquifers depends on the porosity of their materials. The yield represents that portion of water which drains out by gravity and becomes available by pumping. Shallow groundwater (down to 50 ft) is trapped by dug or driven wells, but deep sources require drilled wells. The volume of shallow wells may vary greatly in accordance with fluctuations in rainfall and degree of withdrawal. *See* GROUNDWATER.

SURFACE WATER

Streams supply most of the water needs of the United States. Lakes, ponds, swamps, and marshes, like reservoirs, represent stored streamflow. The natural lakes in the United States are calculated to contain 13,000,000,000 acre-feet. Swamps and other wet lands along river deltas, around the borders of interior lakes, and in coastal regions add millions more to the surface supplies. The oceans and salty or brackish sounds, bays, bayous, or estuaries represent almost unlimited potential fresh-water sources. Brackish waters are being used increasingly by industry for cooling and flushing. Reservoirs, dammed lakes, farm ponds, and other small impoundments have a combined usable storage of 300,000,000 acre-feet. The smaller ones furnish water for livestock, irrigation, fire protection, flash-flood protection, fish and waterfowl, and recreation. However, most artificial storage is in reservoirs of over 5000 acre-feet. Lake Mead, located in Arizona and Nevada and formed by Hoover Dam, is the largest (227 mi^2) of the 1300 reservoirs, and it contains 10% of the total stored-water capacity, or over 31,000,000 acre-feet. These structures regulate streamflow to provide more dependable supplies during dry periods when natural runoff is low and demands are high. They store excess waters in wet periods, thus mitigating damaging floods.

Water use. Water withdrawals for all purposes totaled 1,740,000,000,000 gallons per day (1740 bgd) in 1955 in the United States, but actual consumption was only 94 bgd. Exclusive of reservoir evaporation, daily use was 71 bgd, 444 gal per capita or 6% of the nation's runoff. Not counted is an additional 20 bgd transpired by western phreatophytes, plants growing along stream banks, canals, and reservoirs. Because reservoirs evaporate 23 bgd, interest has developed in underground storage and in coating water surfaces with monomolecular films of oil-derived detergents, such as hexadecanol, to reduce evaporation. *See* ECOLOGY.

The 115,000,000 people in the United States served by public water supplies in 1955 withdrew 11.3 bgd, and rural dwellers and their livestock used 2.5 bgd. Crop irrigation practiced in all 48 states withdrew 110 bgd, the heaviest use being in the 17 drier Western states. Only 40% of irrigation water is returned to streams or aquifers, whereas 90% of other public water supplies is returned. However, pollution prevents reuse of part of the return flow of public water supplies.

Withdrawals for waterpower account for 1500 bgd out of the total in the United States, but this use consumes practically none of the water. Industry takes nearly 116 bgd, mostly for condenser cooling and for boilers, sanitary services, and cooling machinery in power plants, but industry consumes only 2% of the water utilized. Air conditioning averages only 1.1 bgd but may constitute a severe drain on available supplies during hot dry spells in local areas.

Fish, wildlife, and recreation water requirements are nonconsumptive. Clean natural waters and the aquatic environment they create constitute major attractions in undeveloped wilderness areas, national parks, and even in more highly developed agricultural, forest, or suburban localities. However, artificial impoundments, unless properly located, designed, and operated, can destroy or depreciate priceless natural environment.

Water pollution. Streams have traditionally served for waste disposal. Towns and cities, industries, and mines provide thousands of pollution sources. Pollution dilution requires large amounts of water. Treatment at the source is safer and less wasteful than flushing untreated or poorly treated wastes downstream. However, sufficient flows must be released to permit the streams to dilute, assimilate, and carry away the treated effluents. *See* WATER POLLUTION.

Hydrologic cycle. This term refers to the continuous circulation of the Earth's moisture. Because of this characteristic, water is considered a renewable natural resource, but underground water is "mined" when it is pumped out faster than its natural renewal and thus may be like a fund resource. The oceans furnish most of the moisture for evaporation and precipitation. Part of the precipitation evaporates, part is returned directly to the oceans, part runs off quickly into watercourses and lakes, part enters the soil or other porous material where it is retained, and the balance enters deep aquifers where it is stored or flows along impermeable underground layers into streams or springs. Solar

radiation provides the energy for the hydrologic cycle. Unequal heating of the Earth's surface creates air currents of varying temperatures and pressures. Warm air masses carry moisture from the oceans. When these air masses are cooled, their capacity to retain moisture is reduced and precipitation results. The source of the air masses and the pressure and temperature gradients aloft determine the form, intensity, and duration of precipitation. Precipitation may occur as cyclonic or low-pressure storms (mostly a winter phenomenon responsible for widespread rains), as thunderstorms of high intensity and limited area, or as mountain storms wherein warm air dumps its moisture when lifted and cooled in crossing high land barriers (Fig. 1).

Precipitation is characteristically irregular. Generally, the more humid an area and the nearer the ocean, the more evenly distributed is the rainfall. Whatever the annual average, rainfall in arid regions tends to vary widely from year to year and to fall in a few heavy downpours. Large floods and active erosion result from heavy and prolonged rainfall or rapid melt of large volumes of snow. Flash floods often follow local intense thunderstorms.

Runoff depends on depth, porosity, and compactness of the soil and the underlying material, steepness and configuration of the surface, and character and density of the vegetation. Plant crowns, ground cover, litter, and humus dissipate the force of rainfall, thus reducing its power to compact and dislodge mineral soil particles and seal the surface pores. The quantity of water entering the soil during a given time depends on the rate of rainfall in relation to the size and distribution of the pores in each soil horizon and the thickness of

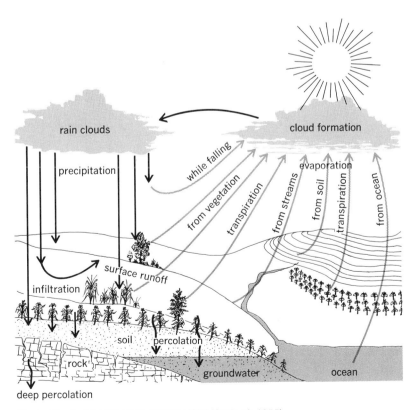

Fig. 1. The hydrologic cycle. (*Water, USDA Yearbook, 1955*)

each horizon. Surface runoff follows when rainfall exceeds the rate at which water is absorbed and transmitted downward. Soil water is retained by adhesion against the pull of gravity in the capillary pores of the soil until their capacity is filled. Such water may evaporate or is available for plant use including photosynthesis and transpiration. Excess moisture in the large pores drains slowly into watercourses or wet-weather springs, or enters rock crevices, limestone sinks, and shales, sands, or other permeable materials. *See* HYDROLOGY; SOIL.

Land management vitally influences the distribution and character of runoff. Inadequate vegetation or surface organic matter; compaction of farm, ranch, or forest soils by heavy vehicles; frequent crop-harvesting operations; repeated burning; or excessive trampling by livestock, deer, or elk all expose the soil to the destructive energy of rainfall or rapid snowmelt. On such lands little water enters the soil, soil particles are dislodged and quickly washed into watercourses, and gullies may form. *See* LAND-USE PLANNING; SOIL CONSERVATION.

Water management problems. These involve economic, social, and intangible values. Efforts to plan and develop river systems for multiple purposes often generate conflicts among different water uses, for example, irrigation versus navigation on the Missouri River, or hydropower versus salmon or trout fisheries, wildlife, national park, wilderness, or historic resources on such rivers as the Columbia, Colorado, and Potomac. Other conflicts stem from actual or threatened dumping of municipal, agricultural (silt), industrial, or acid mine wastes into streams, as occurs upstream from Philadelphia, Pa.; Washington, D.C.; Cumberland, Md.; and other cities, or from operations of power dams, irrigation projects, or other uses which restrain or divert flows to the extent of destroying fish habitats or impairing recreational or wildlife resources. Another source of conflict is the mining of groundwaters in areas of critically low groundwater supplies.

Water management technology. This involves the application of biological and engineering principles to attain desired goals. Biological methods for growing upland vegetation having low moisture requirements are being studied by the U.S. Forest Service and other agencies. Mechanical methods include the practice of water spreading, which is utilized to desilt floodwaters and to promote the percolation of water into the soil for crop use and groundwater recharge. Water that would be a strong pollutant of streams and lakes may be spread on the land. Seabrook Farms in southern New Jersey releases 50 ft/acre of food-processing wash water annually onto an 80-acre oak woodland on sandy soil. About 103,000,000 acres of wet lands have been drained in 40 states. Drainage has failed, however, where the suitability of soils for such practice was not adequately determined, or where erosion from adjacent slopes of improperly farmed land silted up the drainage structures. In some instances drainage has drastically reduced waterfowl habitat or aggravated downstream flood damages. *See* FOREST AND FORESTRY.

The avoidance of water waste takes several forms. Recycling has permitted huge savings of water, especially in petroleum plants, chemical

factories, and steel mills. In some cases reductions have amounted to 96%. Artificial groundwater recharge is successfully practiced in Long Island, N.Y., where over 311 injection wells conduct water used in air conditioning to underground storage areas for cooling and reuse. A National Association of Manufacturers survey reported that 45% of 3343 industrial establishments applied some kind of water purification treatment. Flows in pipes can be reduced, warmer water can be used, or several grades can be applied by means of separate pipelines. Metering of water stimulates more economical use and encourages repair of leaky connections.

Water rights. In the United States early rights to water followed the riparian doctrine, which grants the property owner reasonable use of surface waters flowing past his land unimpaired by upstream landowners. The drier West, however, has favored the appropriation doctrine, which advocates the prior right of the person who first applied the water for beneficial purposes, whether or not his land adjoins the stream. Rights to groundwater are generally governed by the same doctrines. Both doctrines are undergoing intensive study.

State laws generally are designed to protect riparian owners against pollution. States administer the regulatory provisions of their pollution-control laws, develop water quality standards and waste-treatment requirements, and supervise construction and maintenance standards of public service water systems. Some states can also regulate groundwater use to prevent serious overdrafts. Artesian wells may have to be capped, permits may be required for drilling new wells, or reasonable use may have to be demonstrated.

Federal responsibilities consist largely of financial support or other stimulation of state and local water management. Federal legislation permits court action on suits involving interstate streams where states fail to take corrective action following persistent failure of a community or industry to comply with minimum waste-treatment requirements. Federal legislation generally requires that benefits of water development projects equal or exceed the costs. It specifies that certain costs be allocated among local beneficiaries but that most of the expense be assumed by the Federal government. In 1955, however, the Presidential Advisory Committee on Water Resources Policy recommended that cost sharing be based on benefits received, and that power, industrial, and municipal water-supply beneficiaries pay full cost. These phases of water resource development present difficult and complex questions, because many imponderables enter into the estimates of probable monetary and social benefits from given projects as well as into the cost allocation aspect.

Watershed control. This approach to planning, development, and management rests on the established interdependence of water, land, and people. Conditions on the land are often directly reflected in the behavior of streamflow and in the accumulation of groundwater. The integrated approach on smaller watersheds is illustrated by projects under the Watershed Protection and Flood Prevention Act of 1954 (Public Law 566) as amended by P. L. 1018 in 1956. This act originally applied to floods on the smaller tributaries whose watersheds largely are agricultural, but more recently the applica-

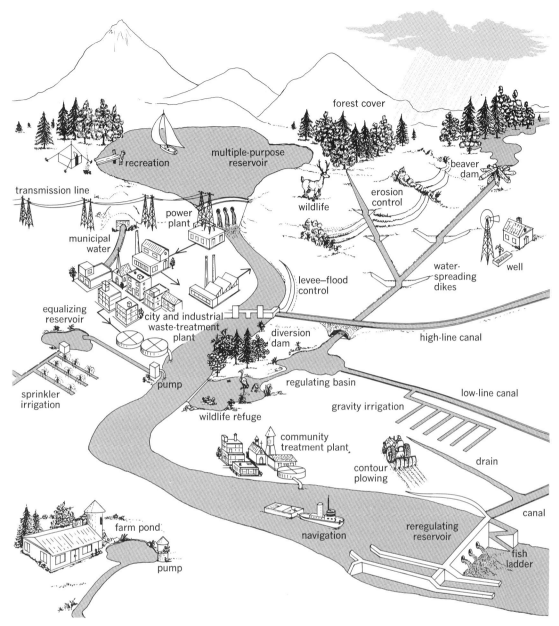

Fig. 2. A multiple-purpose river basin development, encompassing a wide variety of integrated land and water developments, services, and research. (*A Water Policy for the American People, vol. 1, Report of President's Water Resources Policy Commission, 1950*)

tion of the act has been broadened to include mixed farm and residential areas. Damages from such frequent floods equal half the national total from all floods. Coordination of structures and land-use practices is sought to prevent erosion, promote infiltration, and retard high flows. The Soil Conservation and Forest Services of the U.S. Department of Agriculture administer the program. The Soil Conservation Service also cooperates with other Federal and state agencies and operates primarily through the more than 2000 soil conservation districts.

River basins may be large and complex watersheds. For example, the Tennessee River Basin comprises 40,000 mi² in contrast to the 390-mi² upper limit specified in Public Law 1018. Basin projects may involve systems of multipurpose storage reservoirs, intensive programs of watershed

protection, and improvement and management of farm, forest, range, and urban lands. They may call for scientific research, industrial development, health and educational programs, and financial arrangements to stimulate local initiative. The most complete development to date is the Tennessee River Basin, where well-planned cooperative activities have encompassed a wide variety of integrated land and water developments, services, and research (Fig. 2).

Water conservation organizations. Organizations for meeting water problems take various forms. Local or intrastate drainage, irrigation, water-supply, or flood-control activities may be handled by special districts, soil conservation districts, or multipurpose state conservancy districts with powers to levy assessments. Interstate compacts have served limited functions on a regional level. To date Congress has not given serious consideration to proposals for establishing a special Federal agency with powers to review and coordinate the recommendations and activities of development services such as the Corps of Engineers, Bureau of Reclamation, Fish and Wildlife Service, Soil Conservation Service, and Forest Service and to resolve conflicts among agencies and citizen groups. Some national and regional civic groups, such as the League of Women Voters, are studying alternative approaches to the administration of river basin programs.

International agreements. Cooperation in the control, allocation, and utilization of international waters is authorized by treaties with Canada and Mexico. Permanent commissions have been established to deal with specific streams such as the Rio Grande and Colorado, or with boundary waters generally, as provided in the treaty with Canada. The United Nations, through its Technical Assistance Program and through regional commissions, is promoting cooperative studies and developments among underdeveloped nations having common boundaries.   [BERNARD FRANK]

COASTAL WATERS

Management of coastal waters and water basins is needed to protect them from pollution and to provide maximum beneficial use. The salt content of coastal waters limits their direct use to man, but the coastal zone is rich in resources, yielding 75% of the United States seafood harvest of 4,500,000,000 lb per year and most of the sport-fishing catch of 1,000,000,000 lb. Coastal waters also provide recreation for tens of millions of people who sport-fish, swim, sail, water-ski, skin-dive, and study nature. Because the health of coastal waters is in jeopardy from pollution and other alterations, conservation programs are underway along all coastlines of the United States.

Origin and description. Coastal waters are a mixture of fresh water draining off the land and salty water from the ocean. They occupy the transition zone between land and sea, including the wetlands fringe of marshes and tideflats, the estuarine complex of rivers, lagoons, and bays, and the inner band of ocean waters around the continent. Coastal waters are identified by salinity, which is the amount of sodium chloride and other dissolved salts present. Salt content is measured in parts of salt per thousand parts of water, abbreviated ppt. Oceanic waters have a salinity of 35 to 36 ppt, coastal waters vary from 32 ppt (about 90% sea water) along the open coast to 0.3 ppt (about 1% sea water) in estuarine headwaters.

In terms of geological time the coast of North America is but a temporary waterline at the edge of the continent. When the Indians first came to the continent via Siberia and Alaska 12,000 years ago, the Atlantic coastline was 50 mi farther into the ocean. The Earth was colder, much more of the water was locked up in glaciers, and sea level was 300 ft lower than today. If the Earth continues to warm or the glacial ice caps to melt, the present coastal zone will disappear as the sea moves hundreds of miles inland.

The coastal zone of the United States varies greatly in form. The Georgia coast is fringed with boundless expanses of marshland. The Washington coast is abrupt and rocky, indented by deep cold sounds. The Texas coast is flat, bordered by long narrow lagoons behind a ribbon of thin sandbars called barrier beaches. The coasts are so ramified with lagoons, bays, and islands of the estuarine complex that the total United States shoreline is 88,600 mi, while the general outline of the shore is only 12,400 mi. The richest of coastal waters are the estuaries, the protected inner waters. They provide sanctuary for multitudes of the young of oceanic fishes that feed there and escape from predatory fish in ocean waters outside. They nurture prodigious amounts of clams, oysters, crabs, waterfowl, and fish.

Coastal resources. In the development of the country, fish resources have been the economic mainstay of hundreds of coastal communities. Since the mid-1950s the commercial fishing industry has declined; the fishing industry is now able to supply only 25% of the 20,000,000,000 lb of seafood consumed annually in the United States. For example, Maryland oyster production, once over 72,000,000 lb per year, is now down to less than 8,000,000 lb. The decline is mostly from increasing scarcity of fish and shellfish. Many seashore communites have found renewed prosperity in salt-water sport fishing. Increased leisure time and mobility have raised the number of salt-water anglers to 10,000,000. They support tackle shops, charter-boat fleets, marinas, motels, and restaurants. Now even sport fishermen are seeing depletion of many favorite species.

Depletion of fish and shellfish is caused by pollution and other activities of man that affect the delicate balance of coastal life. Abundance of species in the coastal zone is governed by many natural forces, including oxygen content, salinity, temperature, chemistry, circulation, turbidity, and landform. Conservation action is necessary to enhance survival conditions of living coastal resources.

Sewage. Pollution by sewage is the foremost problem of today in most coastal communities. Raw sewage contains pathogenic bacteria that cause human diseases such as typhoid, typhus, and hepatitis. These and various gastroenteric diseases may be contracted from eating raw shellfish that live in sewage-polluted waters. Health authorities have closed more than a million acres of the best shellfish beds and put segments of the shellfish industry abruptly out of business, with economic losses running to tens of millions of dollars per year. Sewage-polluted water is also unfit for uses such as swimming or water-skiing.

Some sewage flows directly into coastal waters without treatment, 500,000,000 gal per day from New York City alone. Some enters from seepage or overflow of septic tanks. But most sewage is processed to some extent. Purification of sewage is expensive but absolutely necessary for healthy coastal waters. The cost increases as the degree of treatment is intensified. Most treatments remove only a fraction of dissolved fertilizing minerals such as nitrates and phosphates. These nutrients from sewage plants overfertilize coastal waters where the effluent is discharged.

Excess nutrients cause overgrowth of algae and other organisms, reducing useful productivity. This process is called eutrophication. The murky water of polluted estuaries admits little sunlight, plant photosynthesis below the surface is at a low level, and the amount of dissolved oxygen is reduced. Dead and dying plants either rise to the surface to decay, often on the shore, or fall to the bottom to decay and mix with the silt in a thick mass of putrefying matter where anaerobic bacteria produce hydrogen sulfide gas. In heavily polluted waters marine bacteria thrive. As they scavenge organic wastes entering coastal waters, the bacteria use up oxygen in their respiration and produce carbon dioxide. Some bacteria cause outbreaks of disease in salt-water fish. Jellyfish and other nuisance organisms prosper on the sewage-enriched waters. These conditions reduce the desirable aquatic life—fish, shellfish, and waterfowl. Sewage plants of the future will be required to eliminate dissolved fertilizing minerals as part of the routine treatment.

Contamination by toxic materials. Insecticides reach coastal waters via runoff from the land, often causing heavy fish kills. Any amount above one-tenth part of insecticide to 1,000,000 parts of water can be lethal to some fish for most of the following: DDT, parathion, malathion, endrin, dieldrin, toxaphene, lindane, and heptachlor. Contamination of fish eggs by DDT is fatal to a high proportion of young. Insecticides function mainly as paralytic nerve poisons with resulting lack of coordination, erratic behavior, loss of equilibrium, muscle spasms, convulsions, and finally suffocation. Recent Federal and state legislation has all but eliminated DDT, the worst of the pesticides, from future use in the United States. There will be more stringent regulation of use of other dangerous pesticides in the future.

Chemical pollutants such as metals, acids, and gases result from industrial activities. Paper and pulp mills discharge wastes that are dangerous to aquatic life because the wastes have a high oxygen demand and deplete oxygen. Other factories discharge lead, copper, zinc, nickel, mercury, cadmium, and cobalt, which are toxic to coastal life in concentrations as low as 0.5 part per million (ppm). Cyanide, sulfide, ammonia, chlorine, and fluorine are also poisonous. To prevent chemical pollution, factories are now being required to remove contaminants from their wastes before discharging them into coastal waters or into local sewage systems.

Oil pollution arises from various sources. Most cases of fish poisoning are from accidental spillage from tankers, storage depots, or wells. However, slow but constant leakage from refineries ruins waterways and is difficult to remedy. Oysters seem unable to breed in the vicinity of refineries. Careless handling at plants also results in water pollution by poisonous by-products, such as cresols and phenols that are toxic in amounts of 5–10 ppm. In past years tankers used to pump oil into the water while cleaning their tanks, but this is now being corrected by stronger Federal laws. Oil, gasoline, kerosine, and tar kill fish and reduce oxygen exchange from air to water. In less than lethal amounts, petroleum products taint the flesh of fishes, making them inedible.

Thermal pollution. This is caused by the discharge of hot water from power plants or factories and recently from desalination plants. Power plants are the main source of heated discharges. They are placed on streams, lakes, or bays to secure a ready source of coolant. A large power installation may pump in 1,000,000 gal per minute and discharge it at a temperature 20° above that of the water. Although temperatures of coastal waters range from summer highs of 95°F in southern lagoons to winter lows of 29°F in northern estuaries, each area has a typical pattern of seasonal temperature to which life there has adapted. In a shallow bay with restricted tidal flow the rise in temperature can cause gross alterations to the natural ecology. The facts on this new threat are being searched out while many companies must wait for approval to build plants. Recent Federal standards prohibit heating of coastal waters by more than 4°F.

Physical alterations. Many factors other than pollution have diminished living resources of the coastal zone. For example, dredging in estuarine waters to fill wetlands for house lots, parking lots, or industrial sites destroys the marshes that provide sanctuary for waterfowl and for the young of estuarine fishes. As the bay bottom is torn up, the loosened sediments shift about with the current and settle in thick masses on the bottom, suffocating animals and plants. In this way, the marshes are eliminated and the adjoining bays are degraded as aquatic life zones. The northeast Atlantic states have lost 45,000 acres of coastal wetlands in only 10 years; Connecticut has lost half of its original coastal marshlands. San Francisco Bay has been nearly half obliterated by filling. Dredging to remove sand and gravel has the same disruptive effects as dredging for landfill or other purposes, whether the sand and gravel are sold for profit or used to replenish beach sand eroded away by storms. The dredging of boat channels adds to the siltation problem.

Dredging and filling to build highways through wetland and estuarine areas are disruptive. A poorly planned highway serves as a long dam, cutting off marshes and transforming them into lakes or semistagnant fresh-water impoundments. Cut off from the sea, such areas often become highly polluted and fill up with silt from land runoff. Conservation interests are recommending open passage for tidal waters under all coastal roadways.

In addition to DDT and other mosquito toxins threatening coastal life, there are damaging physical control activities as well. An example is the impounding of marshes with dikes or levees to prevent mosquito breeding. This tends to make fresh-water lagoons of the marshes and to eliminate them as places which can support marine and brackish water life. A better method now recommended for salt-marsh mosquitoes is ditching the marshes with tide-level canals which provide passage for small fish to enter and eat the mosquito larvae before hatching.

Conservation agencies and legislation. The principal Federal agencies active in coastal water conservation are the Fish and Wildlife Service, the Federal Water Pollution Control Administration, and the Public Health Service. Coastal states exercise jurisdiction through their counterpart conservation and health departments. Private organizations with strong interests in coastal water conservation are the Izaak Walton League (Washington, D.C.), the Audubon Society (New York), and the American Littoral Society (Sandy Hook, N.J.). These organizations stress the need for citizen action in coastal water conservation.

Federal legislation has provided the major impetus for cleaning up coastal waters. The most effective legislation is the Water Quality Act of 1965, an amendment to Public Law 84-660 of 1956, the basic antipollution law. Under this act each state has been required to submit for approval specific water quality standards that it would enforce. To implement these standards, some states have voted bond issues to assist community water clean-up programs in amounts up to a billion dollars.

The Fish and Wildlife Coordination Act of 1956 requires the U.S. Army Corps of Engineers to consult with the Fish and Wildlife Service before allowing dredging, landfill, or other development in the coastal zone. Developers are required to modify their projects to minimize damage to coastal water; often they are prevented from any activity. The Federal government and most states now have funds to purchase lands for conservation purposes, including marshlands and other shore property vital to conservation of coastal waters.

[JOHN R. CLARK]

Bibliography: J. Clark, *Fish and Man: Conflict in the Atlantic Estuaries*, American Littoral Society, Highlands, N.J., 1967; National Technical Advisory Committee, *Water Quality Criteria*, Federal Water Pollution Control Administration, 1969.

Water pollution

Any change in natural waters which may impair their further use, caused by the introduction of organic or inorganic substances, or a change in temperature of the water. The growth of population and the concomitant expansion in industrial and agricultural activities have rapidly increased the importance of the field of water-pollution control. In the attack on environmental pollution, higher standards for water cleanliness are being adopted by state and Federal governments, as well as by interstate organizations.

Historical developments. Ancient man joined into groups for protection. Later, he formed communities on watercourses or the seashore. The waterway provided a convenient means of transportation, and fresh waters provided a water supply. The watercourses then became receivers of his waste water along with contaminants. As industries developed, they added their discharges to those of the community. When the concentration of added substances became dangerous to man or so degraded the water that it was unfit for further use, water-pollution control began. With development of wide areas, pollution of surface water supplies became more critical because waste water of an upstream community became part of the water supply of the downstream community.

Serious epidemics of waterborne diseases such as cholera, dysentery, and typhoid fever were caused by underground seepage from privy vaults into town wells. Such direct bacterial infections through water systems can be traced back to the late 18th century, even though the germ or bacterium as the cause of disease was not proved for nearly another century. The well-documented epidemic of the Broad Street Pump in London during 1854 resulted from direct leakage from privies into the hand-pumped well which provided the neighborhood water supply. There were 616 deaths from cholera among the users of the well within 40 days.

Eventually, abandoning wells in such populated locations and providing piped water to buildings improved public health. Further, sewers for drainage of waste water were constructed, but then infections between communities rather than between the residents of a single community became apparent. Modern public health protection is provided by highly refined and well-controlled plants both for the purification of the community water supply and treatment of the waste water.

Relation to water supply. Water-pollution control is closely allied with the water supplies of communities and industries because both generally share the same water resources. There is great similarity in the pipe systems that bring water to each home or business property, and the systems of sewers or drains that subsequently collect the waste water and conduct it to a treatment facility. Treatment should prepare the flow for return to the environment so that the receiving watercourse will be suitable for beneficial uses such as general recreation, and safe for subsequent use by downstream communities or industries.

The volume of waste water, the used water that must be disposed of or treated, is a factor to be considered. Depending on the amount of water used for irrigation, the amount lost in pipe leakage, and the extent of water metering, the volume of waste water may be 70–130% of the water drawn from the supply. In United States cities, waste-water quantities are usually 75–200 gal per capita daily. The higher figure applies to large cities with old systems, limited metering, and comparatively cheap water; the lower figure to smaller communities with little leakage and good metering. Probably the average in the United States for areas served by sewers is 125–150 gal of waste water per person per day. Of course, industrial consumption in larger cities increases per capita quantities.

Related scientific disciplines. The field of water-pollution control encompasses a part of the broader field of sanitary or environmental engineering. It includes some aspects of chemistry, hydrology, biology, and bacteriology, in addition to public administration and management. These scientific disciplines evaluate problems and give the civil and sanitary engineer basic data for the designing of structures to solve the problems. The solutions usually require the collection of domestic and industrial waste waters and treatment before discharge into receiving waters. *See* HYDROLOGY; SANITARY ENGINEERING.

Self-purification of natural waters. Any natural watercourse contains dissolved gases normally found in air in equilibrium with the atmosphere. In this way fish and other aquatic life obtain oxygen for their respiration. The amount of oxygen which the water holds at saturation depends on temperature and follows the law of decreased solubility of gases with a temperature increase. Because water temperature is high during the summer, oxygen dissolved in the water is then at a low point for the year.

Degradable or oxidizable substances in waste waters deplete oxygen through the action of bacteria and related organisms which feed on organic waste materials, using available dissolved oxygen for their respiration. If this activity proceeds at a rate fast enough to depress seriously the oxygen level, the natural fauna of a stream is affected; if the oxygen is entirely used up, a condition of oxygen exhaustion occurs which suffocates aerobic

Table 1. General nature of industrial waste waters

Industry	Processes or waste	Effect
Brewery and distillery	Malt and fermented liquors	Organic load
Chemical	General	Stable organics, phenols, inks
Dairy	Milk processing, bottling, butter and cheese making	Acid
Dyeing	Spent dye, sizings, bleach	Color, acid or alkaline
Food processing	Canning and freezing	Organic load
Laundry	Washing	Alkaline
Leather tanning	Leather cleaning and tanning	Organic load, acid and alkaline
Meat packing	Slaughter, preparation	Organic load
Paper	Pulp and paper manufacturing	Organic load, waste wood fibers
Steel	Pickling, plating, and so on	Acid
Textile manufacture	Wool scouring, dyeing	Organic load, alkaline

organisms in the stream. Under such conditions the stream is said to be septic and is likely to become offensive to the sight and smell.

Domestic waste waters. Domestic waste waters result from the use of water in dwellings of all types, and include both water after use and the various waste materials added: body wastes, kitchen wastes, household cleaning agents, and laundry soaps and detergents. The solid content of such waste water is numerically low and amounts to less than 1 lb per 1000 lb of domestic waste water. Still, the character of these waste materials is such that they cause significant degradation of receiving waters, and they may be a major factor in spreading waterborne diseases, notably typhoid and dysentery.

Characteristics of domestic waste water vary from one community to another and in the same community at different times. Physically, community waste water usually has the grayish colloidal appearance of dishwater, with floating trash apparent. Chemically, it contains the numerous and complex nitrogen compounds in body wastes, as well as soaps and detergents and the chemicals normally present in the water supply. Biologically, bacteria and other microscopic life abound. Waste waters from industrial activities may affect all of these characteristics materially.

Industrial waste waters. In contrast to the general uniformity of substances found in domestic waste waters, industrial waste waters show increasing variation as the complexity of industrial processes rises. Table 1 lists major industrial categories along with the undesirable characteristics of their waste waters.

Because biological treatment processes are ordinarily employed in water-pollution control plants,

Table 2. Dilution ratios for waterways

Type	Stream flow, ft^3/sec/1000 population
Sluggish streams	7–10
Average streams	4–7
Swift turbulent streams	2–4

large quantities of industrial waste waters can interfere with the processes as well as the total load of a treatment plant. The organic matter present in many industrial effluents often equals or exceeds the amount from a community. Accommodations for such an increase in the load of a plant should be provided for in its design.

Discharge directly to watercourses. The industrial revolution in England and Germany and the subsequent similar development in the United States increased problems of water-pollution control enormously. The establishment of industries caused great migrations to the cities, the immediate result being a great increase in wastes from both population and industrial activity. For some years discharges were made directly to watercourses, the natural assimilative power of the receiving water being used to a level consistent with the required cleanliness of the watercourse. Early dilution ratios required for this method are shown in Table 2. Because of the more rapid absorption of oxygen from the air by a turbulent stream, it has a high rate of reaeration and a low dilution ratio; the converse is true of slow-flowing streams.

Development of treatment methods. With the passage of time, the waste loads imposed on streams exceeded the ability of the receiving water to assimilate them. The first attempts at wastewater treatment were made by artificially providing means for the purification of waste waters as observed in nature. These forces included sedimentation and exposure to sunlight and atmospheric oxygen, either by agitated contact or by filling the interstices of large stone beds intermittently as a means of oxidation. However, practice soon outstripped theory because bacteriology was only then being born and there were many unknowns about the processes.

In later years testing stations were set up by municipalities and states for experimental work. Notable among these were the Chicago testing station and one established at Lawrence by the state of Massachusetts, a pioneer in the public health movement. From the results of these direct investigations, practices evolved which were gradually explained through the mechanisms of chemistry and biology in the 20th century.

Thermal pollution. An increasing amount of attention has been given to thermal pollution, the raising of the temperature of a waterway by heat discharged from the cooling system or effluent wastes of an industrial installation. This rise in temperature may sufficiently upset the ecological balance of the waterway to pose a threat to the native life-forms. This problem has been especially noted in the vicinity of nuclear power plants. Thermal pollution may be combated by allowing the waste water to cool before it empties into the waterway. This is often accomplished in large cooling towers.

Current status. The modern water-pollution engineer or chemist has a wealth of published information, both theoretical and practical, to assist him. While research necessarily will continue, he can draw on established practices for the solution to almost any problem. A challenging problem has been the handling of radioactive wastes. Reduction in volume, containment, and storage constitute the principal attack on this problem. Because of the fundamental characteristics of radioactive

Table 3. Federal aid for treatment plant construction

Fiscal year	Congressional authorization	Congressional appropriation
1967	$150,000,000	$150,000,000
1968	450,000,000	203,000,000
1969	700,000,000	214,000,000
1970	1,000,000,000	300,000,000
1971	1,250,000,000	–

wastes, the development of other methods seems unlikely.

Federal aid. Because of public demands and the actions of state legislatures and the Congress of the United States, there has been a surge of interest in, and a demand for, firm solutions to water-pollution problems. Although the Federal government granted aid for construction of municipal treatment plants as an employment relief measure in the 1930s, no comprehensive Federal legislation was enacted until 1948. This was supplemented by a major change in 1956, when the United States government again offered grants to municipalities to assist in the construction of water-pollution control facilities. These grants were further extended to small communities for the construction of both water and sewer systems.

In 1966 President Lyndon B. Johnson moved the Water Pollution Control Administration from the U.S. Public Health Service to the Department of the Interior. This brought its activity under cabinet-level surveillance. Unfortunately, Federal appropriations were not made in accordance with authorizations. Federal annual authorization and appropriation figures for 1967–1971 are given in Table 3. State and Federal regulations are increasing constantly in severity. This tendency is expected to continue until the problem of water pollution is brought under complete control. Even then, water quality will be monitored to make certain that actual control is achieved on a day-to-day or even an hour-to-hour basis. The increase in activity in water-pollution control is apparent, and the 50,000–60,000 workers in the field in the United States are expected to double in the 1970s. *See* SEPTIC TANK; SEWAGE; SEWAGE DISPOSAL; SEWAGE TREATMENT. [RALPH E. FUHRMAN]

Bibliography: H. E. Babbitt and J. J. Baumann, *Sewerage and Sewage Treatment*, 1958; E. B. Besselievre, *The Treatment of Industrial Waste*, 2d ed., 1969; G. M. Fair et al., *Water and Wastewater Engineering*, vol. 2, 1968; J. Snow, *Mode of Communication of Cholera*, 1855; Water Pollution Control Federation, *Careers in Water Pollution Control*.

Water supply engineering

A branch of civil engineering concerned with the development of sources of supply, transmission, distribution, and treatment of water. The term is used most frequently in regard to municipal water works, but applies also to water systems for industry, irrigation, and other purposes.

SOURCES OF WATER SUPPLY

Underground waters, rivers, lakes, and reservoirs, the primary sources of fresh water, are re-plenished by rainfall. Some of this water flows to the sea through surface and underground channels, some is taken up by vegetation, and some is lost by evaporation.

Groundwater. Water obtained from subsurface sources, such as sands and gravels and porous or fractured rocks, is called groundwater. Groundwater flows toward points of discharge in river valleys and, in some areas, along the seacoast. The flow takes place in water-bearing strata known as aquifers. The velocity may be a few feet to several miles per year, depending upon the permeability of the aquifer and the hydraulic gradient or slope. A steep gradient or slope indicates relatively high pressure, or head, forcing the water through the aquifer. When the gradient is flat, the pressure forcing the water is small. When the velocity is extremely low, the water is likely to be highly mineralized; if there is no movement, the water is rarely fit for use. *See* GROUNDWATER.

Permeability is a measure of the ease with which water flows through an aquifer. Coarse sands and gravels, and limestone with large solution passages, have high permeability. Fine sand, clay, silt, and dense rocks (unless badly fractured) have low permeability.

Water table. In an unconfined stratum the water table is the top or surface of the groundwater. It may be within a few inches of the ground surface or hundreds of feet below. Normally it follows the topography. Aquifers confined between impervious strata may carry water under pressure. If a well is sunk into such an aquifer and the pressure is sufficient, water may be forced to the surface, resulting in an artesian well. The water table elevation and artesian pressure may vary substantially with the seasons, depending upon the amount of rainfall recharging the aquifer and the amount of water taken from the aquifer. If pumpage exceeds recharge for an extended period, the aquifer is depleted and the water supply lost.

Salt-water intrusion. Normally the groundwater flow is toward the sea. This normal flow may be reversed, however, by overpumping and lowering of the water table or artesian pressure in an aquifer. Salt water flowing into the fresh-water aquifer being pumped is called salt-water intrusion.

Springs. Springs occur at the base of sloping ground or in depressions where the surface elevation is below the water table, or below the hydraulic gradient in an artesian aquifer from which the water can escape. Artesian springs are fed through cracks in the overburden or through other natural channels extending from the confined aquifer under pressure to the surface. *See* SPRING, WATER.

Wells. Wells are vertical openings, excavated or drilled, from the ground surface to a water-bearing stratum or aquifer. Pumping a well lowers the water level in it, which in turn forces water to flow from the aquifer. Thick, permeable aquifers may yield several million gallons daily with a drawdown (lowering) of only a few feet. Thin aquifers, or impermeable aquifers, may require several times as much drawdown for the same yield, and frequently yield only small supplies.

Dug wells, several feet in diameter, are frequently used to reach shallow aquifers, particularly for small domestic and farm supplies. They furnish small quantities of water, even if the soils penetrated are relatively impervious. Large-capacity dug wells or caisson wells, in coarse sand and gravel, are used frequently for municipal supplies. Drilled wells are sometimes several thousand feet deep.

The portion of a well above the aquifer is lined with concrete, stone, or steel casing, except where the well is through rock that stands without support. The portion of the well in the aquifer is built with open-joint masonry or screens to admit the water into the well. Metal screens, made of perforated sheets or of wire wound around supporting ribs, are used most frequently. The screens are galvanized iron, bronze, or stainless steel, depending upon the corrosiveness of the water and the expected life of the well.

The distance between wells must be sufficient to avoid harmful interference when the wells are pumped. In general, economical well spacing varies directly with the quantity of water to be pumped, and inversely with the permeability and thickness of the aquifer. It may range from a few feet to a mile or more.

Infiltration galleries are shafts or passages extending horizontally through an aquifer to intercept the groundwater. They are equivalent to a row of closely spaced wells and are most successful in thin aquifers along the shore of rivers, at depths of less than 75 ft. The galleries are built in open cuts or by tunneling, usually with perforated or porous liners to screen out the aquifer material and to support the overburden.

Ranney wells consist of a center caisson with horizontal, perforated pipes extending radially into the aquifer. They are particularly applicable to the development of thin aquifers at shallow depths.

Specially designed pumps, of small diameter to fit inside well casings, are used in all well installations, except in flowing artesian wells or where the water level in the well is high enough for direct suction left by a pump on the surface (about 15 ft). Well pumps are set some distance below the water level, so that they are submerged even after the drawdown is established. Well-pump settings of 100 ft are common, and they may exceed 300 ft where the groundwater level is low. Multiple-stage centrifugal pumps are used most generally. They are driven by motors at the surface through vertical shafts, or by waterproof motors attached directly below the pumps. Wells are sometimes pumped by air lift, that is, by injecting compressed air through a pipe to the bottom of the well.

Surface water. Natural sources, such as rivers and lakes, and impounding reservoirs are sources of surface water. See DAM; SURFACE WATER.

Water is withdrawn from rivers, lakes, and reservoirs through intakes. The simplest intakes are pipes extending from the shore into deep water, with or without a simple crib and screen over the outer end. Intakes for large municipal supplies may consist of large conduits or tunnels extending to elaborate cribs of wood or masonry containing screens, gates, and operating mechanisms. Intakes in reservoirs are frequently built as integral parts of the dam and may have multiple ports at several levels to permit selection of the best water. The location of intakes in rivers and lakes must take into consideration water quality, depth of water, likelihood of freezing, and possible interference with navigation. Reservoir intakes are usually de-

signed for gravity flow through the dam or its abutments. In lakes and rivers, the water flows by gravity through the intake to a pumping station on the shore.

TRANSMISSION AND DISTRIBUTION

The water from the source must be transmitted to the community or area to be served and distributed to the individual customers.

Transmission mains. The major supply conduits, or feeders, from the source to the distribution system are called mains or aqueducts.

Canals. The oldest and simplest type of aqueducts, especially for transmitting large quantities of water, are canals. Canals are used where they can be built economically to follow the hydraulic gradient or slope of the flowing water. If the soil is suitable, the canals are excavated with sloping sides and are not lined. Otherwise, concrete or asphalt linings are used. Gravity canals are carried across streams or other low places by wooden or steel flumes, or under the streams by pressure pipes known as inverted siphons.

Tunnels. Used to transmit water through ridges or hills, tunnels may follow the hydraulic grade line and flow by gravity or may be built below the

Pipelines. Pipelines are a common type of transmission main, especially for moderate supplies not requiring large aqueducts or canals. Pipes are of cast iron, steel, reinforced concrete, cement-asbestos, or wood. Pipeline material is determined by cost, durability, ease of installation and maintenance, and resistance to corrosion. The pipeline must be large enough to deliver the required grade line to operate under considerable pressure. Rock tunnels may be lined to prevent the overburden from collapsing, to prevent leakage, or to reduce friction losses by providing a smooth interior. amount of water and strong enough to withstand the maximum gravity or pumping pressure. Pipelines are usually buried in the ground for protection and coolness.

Distribution system. Included in the distribution system are the network of smaller mains branching off from the transmission mains, the house services and meters, the fire hydrants, and the distribution storage reservoirs. The network is composed of transmission or feeder mains, usually 12 in. or more in diameter, and lateral mains along each street, or in some cities along alleys between the streets. The mains are installed in grids so that lateral mains can be fed from both ends where possible. Mains fed from one direction only are called dead ends; they are less reliable and do not furnish as much water for fire protection as do mains within the grid. Valves at intersections of mains permit a leaking or damaged section of pipe to be shut off with minimum interruption of water service to adjacent areas.

House services. The small pipes, usually of iron, copper, or plastic material, extending from the water main in the street to the customer's meter at the curb line or in the cellar are called house services. In most cities each service is metered, and the customer's bill is based on the water actually used.

Fire hydrants. Fire hydrants have a vertical barrel extending to the depth of the water main, a quick-opening valve with operating nut at the top, and connections threaded to receive fire hose.

Hydrants must be reliable, and they must drain upon closing to prevent freezing.

Distribution reservoirs. These are used to supplement the source of supply and transmission system during peak demands, and to provide water during a temporary failure of the supply system. In small waterworks the reservoirs usually equal at least one day's water consumption; in larger systems the reservoirs are relatively smaller but adequate to meet fire-fighting demands. Ground storage reservoirs, elevated tanks, and standpipes are used for distribution reservoirs.

Ground storage reservoirs, if on high ground, can feed the distribution system by gravity, but otherwise it is necessary to pump water from the reservoir into the distribution system. Circular steel tanks and basins built of earth embankments, concrete, or rock masonry are used. Earth reservoirs are usually lined to prevent leakage and entrance of dirty water. The reservoirs should be covered to protect the water from dust, rubbish, and bird droppings.

Elevated storage reservoirs are tanks on towers, or high cylindrical standpipes resting on the ground. Storage reservoirs are built high enough so that the reservoir will maintain adequate pressure in the distribution system at all times.

Elevated tanks are usually of steel plate, mounted on steel towers. Wood is sometimes used for industrial and temporary installations. Standpipes are made of steel plate, strong enough to withstand the pressure of the column of water. The required capacity of a standpipe is greater than that of an elevated tank because only the upper portion of a standpipe is sufficienty elevated for normal use.

Distribution-system design. To assure the proper location and size of feeder mains and laterals to meet normal and peak water demands, a distribution system must be expertly designed. As the water flows from the source of supply or distribution reservoir across a city, the water pressure is lowered by the friction in the pipes. The pressures required for adequate service depend upon the height of buildings, need for fire protection, and other factors, but 40–60 psi is the minimum for good service. Higher pressures for fire fighting are obtained by booster pumps on fire engines which take water from fire hydrants. In small towns adequate hydrant flows are the controlling factor in determining water-main size; in larger communities the peak demands for air conditioning and lawns sprinkling during the summer months control the size of main needed. The capacity of a distribution system is usually determined by opening fire hydrants and measuring simultaneously the discharge and the pressure drop in the system. The performance of the system when delivering more or less water than during the test can be computed from the pressure drops recorded.

An important factor in the economical operation of municipal water supplies is the quantity of water lost from distribution because of leaky joints, cracked water mains, and services abandoned but not properly shut off. Unaccounted-for water, including unavoidable slippage of customers' meters, may range from 10% in extremely well-managed systems to 30–40% in poor systems. The quantities flowing in feeder mains, the friction losses, and the amount of leakage are frequently measured by means of pitometer surveys. A pitometer is a portable meter that can be inserted in a water main under pressure to measure the velocity of flow, and thus the quantity of flow.

Pumping stations. Pumps are required wherever the source of supply is not high enough to provide gravity flow and adequate pressure in the distribution system. The pumps may be high or low head depending upon the topography and pressures required. Booster pumps are installed on pipelines to increase the pressure and discharge, and adjacent to ground storage tanks for pumping water into distribution systems. Pumping stations usually include two or more pumps, each of sufficient capacity to meet demands when one unit is down for repairs or maintenance. The station must also include piping and valves arranged so that a break can be isolated quickly without cutting the whole station out of service.

Centrifugal pumps have displaced steam-driven reciprocating pumps in modern practice, although many of the old units continue to give good service. The centrifugal pumps are driven by electric motors, steam turbines, or diesel engines, with gasoline engines frequently used for standby service. The centrifugal pumps used most commonly are designed so that the quantity of water delivered decreases as the pumping head or lift increases. Both horizontal and vertical centrifugal pumps are available in a wide capacity range. In the horizontal type, the pump shaft is horizontal with the driving motor or engine at one end of the pump. Vertical pumps are driven by a vertical-shaft motor directly above the pump or are driven by a horizontal engine through a right-angle gear head.

Automatic control of pumping stations is provided to adjust pump operations to variations in water demand. The controls start and stop pumps of different capacity as required. In the event of mishap or failure of a unit, alarms are sounded. The controls are activated by the water level in a reservoir or tank, by the pressure in a water main, or by the rate of flow through a meter. Remote control of pumps is frequently employed, with the signals transmitted over telephone wires. *See* WATER POLLUTION; WATER TREATMENT.

[RICHARD L. HAZEN]

Bibliography: R. W. Abbett, *American Civil Engineering Practice*, 3 vols., 1956; H. E. Babbitt, J. J. Doland, and J. H. Cleasby, *Water Supply Engineering*, 6th ed., 1962; C. V. Davis, *Handbook of Applied Hydraulics*, 2d ed., 1952; G. M. Fair and H. L. Geyer, *Water Supply and Waste-Water Disposal*, 1954; W. A. Hardenbergh and E. B. Rodie, *Water Supply and Waste Disposal*, 1961; R. K. Lensley and J. B. Franzini, *Water Resources Engineering*, 1964; L. S. Nielsen, *Standard Plumbing Engineering Design*, 1963; E. W. Steel, *Water Supply and Sewerage*, 4th ed., 1960.

Water table

The upper surface of the zone of saturation in permeable rocks not confined by impermeable rocks. It may also be defined as the surface underground at which the water is at atmospheric pressure. Saturated rock may extend a little above this level, but the water in it is held up above the water table by capillarity and is under less than atmospheric pressure; therefore, it is the lower part of the capillary fringe and is not free to flow into a well by gravity. Below the water table, water is free to move under the influence of gravity. The position of the water table is shown by the level at which water stands in wells penetrating an unconfined water-bearing formation.

Where a well penetrates only impermeable material, there is no water table and the well is dry. But if the well passes through impermeable rock into water-bearing material whose hydrostatic head is higher than the level of the bottom of the impermeable rock, water will rise approximately to the level it would have assumed if the whole column of rock penetrated had been permeable. This is called artesian water, and the surface to which it rises is called the piezometric surface.

The water table is not a level surface but has irregularities that are commonly related to, though less pronounced than, those of the land surface. Also, it is not stationary but fluctuates with the seasons and from year to year. It generally declines during the summer months, when vegetation uses most of the water that falls as precipitation, and rises during the late winter and spring, when the demands of vegetation are low. The water table usually reaches its lowest point after the end of the growing season and its highest point just before the beginning of the growing season. Superimposed on the annual fluctuations are fluctuations of longer period which are controlled by climatic variations. The water table is also affected by withdrawals, as by pumping from wells. *See* GROUNDWATER.   [ALBERT N. SAYRE/RAY K. LINSLEY]

Bibliography: S. N. Davis and R. J. M. DeWiest, *Hydrogeology*, 1966.

Water treatment

Physical and chemical processes for making water suitable for human consumption and other purposes. Drinking water must be bacteriologically safe and comparatively free of turbidity, color, and taste-producing substances. Excessive hardness and high concentration of dissolved solids are also undesirable, particularly for boiler feed and industrial purposes. The more important treatment processes are sedimentation, coagulation, filtration, disinfection, softening, and aeration.

Plain sedimentation. Silt, clay, and other fine material settle to the bottom if the water is allowed to stand or flow quietly at low velocities. Sedimentation occurs naturally in reservoirs and is accomplished in treatment plants by basins or settling tanks. The detention time in a settling basin may range from an hour to several days. The water may flow horizontally through the basin, with solids settling to the bottom, or may flow vertically upward at a low velocity so that the particles will settle through the rising water. Settling basins are most effective if shallow, and rarely exceed 10–20 ft in depth. Plain sedimentation will not remove extremely fine or colloidal material within a reasonable time, and the process is used principally as a preliminary to other treatment methods.

Coagulation. Fine particles and colloidal material are combined into masses by coagulation. These masses, called floc, are large enough to settle in basins and to be caught on the surface of filters. Waters high in organic material and iron may coagulate naturally with gentle mixing. The term is usually applied to chemical coagulation, in which iron or aluminum salts are added to the water to form insoluble hydroxide floc. The floc is a feathery, highly absorbent substance to which color-producing colloids, bacteria, fine particles, and other substances become attached and are thus removed from the water.

The coagulant dose is a function of the physical and chemical character of the raw water, the adequacy of settling basins and filters, and the degree of purification required. Moderately turbid water coagulates more easily than perfectly clear water, but extremely turbid water requires more coagulant. Coagulation is more effective at higher temperatures. Lime, soda ash, or caustic soda may be required in addition to the coagulant to provide sufficient alkalinity for the formation of floc, and regulation of the pH (hydrogen-ion concentration) is usually desirable for best results. Powdered limestone, clay, bentonite, or silica are sometimes added as coagulant aids to strengthen and weight the floc. Coagulation and sedimentation are the most important parts of modern water purification.

Filtration. Suspended solids, colloidal material, bacteria, and other organisms are filtered out by passing the water through a bed of sand or pulverized coal, or through a matrix of fibrous material supported on a perforated core. Filtration of turbid or highly colored water usually follows sedimentation or coagulation and sedimentation. Soluble materials such as salts and metals in ionic form are not removed by filtration.

Slow sand filters. Known also as English filters, these consist of beds of sand 20–48 in. deep, through which the water is passed at fairly low rates—2,500,000 to 10,000,000 gal per acre. The size of beds ranges from a fraction of an acre in small plants to several acres in large plants. An underdrain system of graded gravel and perforated pipes transmits the filtered water from the filters to the point of discharge. The sand is usually fine, ranging from 0.2 to 0.5 mm in diameter. The top of the filter clogs with use, and a thin layer of dirty sand is scraped from the filter periodically to maintain capacity.

Slow sand filters operate satisfactorily with reasonably clear waters but clog rapidly with turbid waters. The filters are covered in cold climates to prevent the formation of ice and to facilitate operation in the winter. In milder climates they are often open. Slow sand filters have a high bacteriological efficiency, but few have been built since the development of water disinfection, because of the large area required, the high construction cost, and the labor needed to clean the filters and to handle the filter sand.

Rapid sand filters. These operate at rates of 125,000,000 to 250,000,000 gal per acre per day; or 25 to 50 times the slow-sand-filter rates. The high rate of operation is made possible by the coagulation and sedimentation ahead of filtration to remove the heaviest part of the load, the use of fairly coarse sand, and facilities for backwashing the filter to keep the bed clean. The filter beds are small, generally ranging from 150 ft^2 in small plants to 1500 ft^2 in the largest filter plants. The filters consist of a layer of sand or, occasionally, crushed anthracite coal 18–24 in. deep, resting on graded layers of gravel above an underdrain system. The sand is coarse, 0.4–1.0 mm in diameter, depending upon the raw water quality and pretreatment, but the grain size must be fairly uniform to assure proper backwashing. The underdrain system serves both to collect the filtered water and to distribute the wash water under the filters when they are being washed. Several types of underdrains are used, including perforated pipes, perforated false bottoms of concrete, and tile and porous plates.

Filters are backwashed at rates 5–10 times the filtering rate. The wash water passes upward through the sand and out of the filters by way of wash-water gutters and drains. Washing agitates the sand bed and releases the dirt to flow out of the filter with the wash water. The quantity of water used for washing ranges from 1 to 10% of the total output, depending upon the turbidity of the water applied to the filters and the efficiency of the filter design.

Municipal and large-capacity filters for industry usually are built in concrete boxes or in open tanks of wood and steel. The flow through the sand may be caused by gravity, or the water may be forced through the sand under pressure by pumping. Pressure filters can be operated at higher rates than gravity filters, because of the greater head available to force the water through the sand. However, excessive pressure causes turbidity, and bacteria may appear in the discharge water. For this reason, and because pressure filters are difficult to inspect and keep in good order, open gravity filters are favored for public water supplies.

Diatomaceous earth filters. Swimming-pool installations and small water supplies frequently use this type of filter. The filters consist of a medium or septum supporting a layer of diatomaceous earth through which the water is passed. A filter layer is built up by the addition of diatomaceous earth to the water. When the pressure loss becomes excessive, filters must be backwashed and a fresh layer of diatomaceous earth applied. Filter rates of $2\frac{1}{2}$–6 gal per minute per square foot are attained.

Disinfection. There are several methods of treatment of water to kill living organisms, particularly pathogenic bacteria; the application of chlorine or chlorine compounds is the most common. Less frequently used methods include the use of ultraviolet light, ozone, or silver ions. Boiling is the favorite household emergency measure.

Chlorination is simple and inexpensive and is practiced almost universally in public water supplies. It is often the sole treatment of clear, uncontaminated waters. In most water treatment plants it supplements coagulation and filtration. Chlorination is used also as protection against contamination of water in distribution mains and reservoirs after purification.

Chlorine gas is most economical and easiest to apply in large systems. For small works, calcium hypochlorite or sodium hypochlorite is frequently used. Regardless of which form is used, the dose varies with the water quality and degree of contamination. Clear, uncontaminated water can be disinfected with small doses, usually less than one part per million; contaminated water may require several times as much. The amount of chlorine taken up by organic matter and minerals in water is known as the chlorine demand. For proper disinfection the dose must exceed the demand so that free chlorine remains in the water.

Chlorination alone is not reliable for the treatment of contaminated or turbid water. A sudden increase in the chlorine demand may absorb the full dose and provide no residual chlorine for disinfection, and it cannot be assumed that the chlorine will penetrate particles of organic matter. Chlorine is applied before filtration, after filtration, and sometimes at both times.

Chlorine sometimes causes objectionable tastes or odors in water. This may be due to excessive chlorine doses, but more frequently it is caused by a combination of chlorine and organic matter, such as algae, in the water. Some algae, relatively unobjectionable in the natural state, produce unbearable tastes after chlorination. In other cases, strong chlorine doses oxidize the organic matter completely and produce odor-free water. Excessive chlorine may be removed by dechlorination with sulfur dioxide. Also, ammonia is often added for taste control to reduce the concentration of free chlorine. Activated carbon is also effective in the reduction of both natural and chlorine tastes and odors.

Water softening. The "hardness" of water is due to the presence of calcium and magnesium salts. These salts make washing difficult, waste soap, and cause unpleasant scums and stains in households and laundries. They are especially harmful in boiler feedwater because of their tendency to form scales.

Municipal water softening is common where the natural water has a hardness in excess of 150 parts per million. Two methods are used: (1) The water is treated with lime and soda ash to precipitate the calcium and magnesium as carbonate and hydroxide, after which the water is filtered; (2) the water is passed through a porous cation exchanger which has the ability of substituting sodium ions in the exchange medium for calcium and magnesium in the water. The exchange medium may be a natural sand known as zeolite, or may be manufactured from organic resins. The exchange medium must be recharged periodically by backwashing with brine.

For high-pressure steam boilers or some other industrial processes, almost complete deionization of water is needed, and treatment includes both cation and anion exchangers. Lime-soda plants are similar to water purification plants, with coagulation, settling, and filtration. Zeolite or cation-exchange plants are usually built of steel tanks with appurtenances for backwashing the media with salt brine. If the water to be softened is turbid, filtration ahead of zeolite softening may be required.

Aeration. Aeration is a process of exposing water to air by dividing the water into small drops, by forcing air through the water, or by a combination of both. The first method uses jets, fountains, waterfalls, and riffles; in the second, compressed air is admitted to the bottom of a tank through perforated pipes or porous plates; in the third, drops of water are met by a stream of air produced by a fan.

Aeration is used to add oxygen to water and to remove carbon dioxide, hydrogen sulfide, and taste-producing gases or vapors. Aeration is also used in iron-removal plants to oxidize the iron ahead of sedimentation or filtration. *See* WATER

POLLUTION; WATER SUPPLY ENGINEERING.

[RICHARD HAZEN]

Bibliography: R. W. Abbett, *American Civil Engineering Practice*, 3 vols., 1956; C. V. Davis (ed.), *Handbook of Applied Hydraulics*, 3d ed., 1969; G. M. Fair and J. C. Geyer, *Water Supply and Waste Water Disposal*, 1954; G. V. James, *Water Treatment*, 1965; L. G. Rich, *Unit Processes of Sanitary Engineering*, 1963.

Weather forecasting and prediction

Procedures for extrapolation of the future character of weather on the basis of present and past conditions. Accurate weather prediction requires a knowledge of the past state of the atmosphere, an understanding of the physical laws governing atmospheric behavior, and the availability of necessary technological aids for the rapid dissemination of meteorological information and the preparation of the forecast. The historical development of methods for forecasting the weather can be traced to innovations in these three areas. For introductory discussions of atmospheric science *see* ATMOSPHERE; METEOROLOGY.

This article is divided into five sections. The first part emphasizes the current status of the whole field of weather forecasting and prediction, and concentrates on short-range prediction, up to 48 hr in advance. The second portion deals with long-range prediction. The third section considers statistical forecasting procedures. A fourth section summarizes the bases for the developing techniques of numerical prediction. The last portion briefly characterizes the weather offices and centers through which are funneled the data of observation for analysis and processing into forecast and prediction.

DEVELOPMENT

Information on the state of the atmosphere has been greatly expanded since 1939, when the demands of aircraft in World War II led to the installation of radio-sounding stations and the development of weather reconnaissance systems. These sources have since been supplemented by radar aids, rockets, and satellites. When these data are recorded on charts, the meteorologist has a three-dimensional picture of atmospheric structure. A series of such charts at 6-, 12-, or 24-hr intervals shows the development of weather systems in terms of the changes in wind, pressure, temperature, humidity, cloudiness, precipitation, and visibility. These maps permit an analysis of the dominant long-wave patterns (discussed under long-range forecasting) with linear dimensions of $10^3 - 10^4$ mi, the migratory cyclones and anticyclones with dimensions of $10^2 - 10^3$ mi, and the weather conditions averaged over an area on the order of 10^4 mi^2. The resolution of weather detail is less exact over oceanic regions, in the polar areas, and in most tropical areas. Small-scale weather systems such as land-sea breezes, mountain-valley winds, and convective showers cannot be depicted on the conventional weather map and can be studied only by means of a dense network of observing stations. Radar information gives precipitation detail within the 10^4-mi^2 area.

During the 1960s satellites provided the most significant developments in data acquisition. Besides cloud pictures, satellite technology will yield much needed information on upper-air winds and temperatures on a global scale.

Data to forecasting. The step from data to forecast is achieved by a variety of methods which may be classified as follows.

Semiempirical techniques. The forecaster first synthesizes the raw information into dynamically meaningful models of the atmosphere (as best exemplified by the polar-front model of J. Bjerknes and H. Solberg). Then, by a combination of methods including the extrapolation of past trends, expected changes based upon qualitative physical reasoning, and recollection of the behavior of similar situations in the past, he arrives at an estimate of the position of the prominent features on tomorrow's weather charts. These features include the location and intensity of cyclones, anticyclones, fronts, upper-level pressure ridges and troughs, and the like. Certain formulas can be applied to this phase of the forecast but, because of simplifying assumptions involved in their use, they yield questionable results. The details of weather such as types of clouds, rainfall, and temperature are deduced from this prediction by again referring to models and by considering such effects as advection, radiation, topographical influences, and expected stability changes.

The success of these techniques is limited by their semiempirical character and by the ability of an individual to handle the tremendous mass of significant information. Data are now so voluminous that this method of prediction utilizes only a fraction of the available information, and it becomes somewhat a matter of personal choice as to the selection process.

Numerical methods. The advent of an expanded network of weather observations and of high-speed computers has greatly stimulated meteorological research so that forecasting is passing from the pre-1945 qualitative phase to a quantitative era in which predictions are based upon computation, guided by physical principles. These methods are discussed in the sections on the numerical prediction of weather and the statistical prediction of weather.

The large-scale aspects of the atmospheric flow pattern are most accurately predicted by numerical methods. In addition, numerical methods are superseding the qualitative techniques of the past in the prediction of cloudiness, precipitation, wind, and temperature.

Evaluating forecasting. To analyze the accuracy of weather forecasts, it is necessary to recognize the statistical nature of the element being predicted. Precipitation, even when it covers wide areas, is rarely uniform in intensity. Small convective showers, which develop and dissipate rapidly, are often embedded in the general rain or snow and cause significant variations over distances of the order of 10 mi and over time periods of less than 1 hr. The shower or thunderstorm cells which are associated with air-mass showers or with cold fronts exhibit a maximum variability in time and space. This turbulent behavior presents a major forecasting problem. Consequently, the prediction must give wide range to the estimates of precipitation intensity in order to cover the probable variability over an area on the order of 10^4 mi^2. Only in those areas where the local variations can be attributed to orographical influences is it possible to present a more definitive estimate.

In contrast to the variability of precipitation, the turbulent fluctuations of wind and temperature are so rapid that they are not of general interest to the forecaster or the public. The specialized problem of predicting atmospheric pollution is an exception. Local variations in temperature and wind within the 1-hr, 10-mi scale can be attributed to such well-understood influences as the nature of the underlying surface, proximity to a water area, and a valley or mountain effect. Hence, detailed forecasts of these elements can be made with considerable reliability.

The final consideration with all types of forecast is the length of the time step. The larger the scale of the atmospheric system, the more persistent the phenomenon. For example, an individual summer shower has a life-span of approximately 1 hr; a cyclone is ordinarily identifiable for at least 3 days, and a particular long-wave pattern may persist for weeks. The major problem in weather prediction is the forecasting of a new development; it may be as difficult to pinpoint a summer shower a few hours in advance as to predict, a week in advance, the broad-scale features of the weather associated with a long-wave pattern. For similar reasons, the accuracy with which the weather may be predicted in detail decreases rapidly with the time elapsed since the observations were made.

Probable future developments. A major handicap in the past has been the paucity of dynamically significant information on the atmosphere over the ocean areas and over the less-populated regions of the globe. The developments in satellite technology offer real promise for filling these data gaps. With the globe's atmosphere well charted and with developments in numerical prediction it is anticipated that quantitative forecasts, via the computer, will replace completely the older qualitative techniques. The major improvement can be expected in the longer-range aspects of weather prediction. The isolated summer afternoon rain shower, which plagues many a picnic, will probably remain an unsolved forecast problem for many years to come.   [JAMES M. AUSTIN]

LONG-RANGE FORECASTING

Long-range weather forecasts are of two types. Medium- or extended-range forecasts cover periods of from 48 hr to a week in advance. Forecasts for longer periods generally extend over periods of a month, a season, or possibly two or more seasons in advance. Although meteorologists have been working on long-range prediction problems for more than 100 years, the degree of accuracy is small for all predictions exceeding a week in advance. This seems particularly true when the predictions are examined under rigid statistical controls, such as climatological probability. There is little or no evidence to indicate any sustained success for forecasts embracing periods of more than a season in advance. The reason for such limited ability lies in the utter complexity of the atmosphere's behavior—the vicissitudes of a compressible fluid responding to changing external stimuli, such as the Sun, and changing characteristics of the Earth's surface, both in space and time.

Medium or extended ranges. Scientific methods of extended-range prediction take for granted that the further out in time the forecast is pro-

jected, the more general must be the nature of the prediction. Short-range forecasts for 48 hr or less in advance specify the detail of weather in space and in time. Medium-range forecasts for a week in advance cover average conditions and trends within the week in intervals of a couple of days. Forecasts for a month or more, however, can indicate only the broad-scale (for example, areas of several hundred thousand square miles) features of average or prevailing weather. Such broad-scale aspects are usually expressed in terms of departures from seasonal norms for elements such as temperature and precipitation.

Medium-range forecast methods are apt to use one or a combination of dynamic, statistical, and synoptic techniques. Dynamic methods capitalize upon the best physical knowledge of meteorological phenomena. Statistical methods employ empirically derived equations as substitutes for physical knowledge. In the synoptic technique, various hemispheric wind and weather charts are surveyed and interpreted by an experienced meteorologist. An important part of modern dynamic methods is the principle that the vertical component of absolute vorticity remains fairly constant as air columns of the middle troposphere move from one area to another (discussed below under numerical weather prediction). When instantaneous wind and pressure charts for midtropospheric levels are averaged in time or space, certain small-scale perturbations, including short waves or vortexes in the horizontal, are suppressed. What remains are smooth, long, or planetary waves which in effect constitute a special class of motions; they are not only of larger scale (often being composed of a family of cyclones or anticyclones), but they also evolve more slowly than the individual wind charts from which they are constructed. The planetary waves which these time-averaged charts reveal are responsible for variations in the position and intensity of the well-known sea-level centers of action (like the Bermuda high, one of the subtropical oceanic highs) which largely determine prevailing weather abnormalities. For purposes of extended forecasting, the averaging process is performed on past (observed) data as well as on numerically predicted charts to 4 days in advance. Various methods of comparison of such time-averaged charts enable the synoptician to assay the continuity and trend of large-scale systems and to extrapolate them into the future for some reasonable period. Dynamic methods of prediction may also be used with some success on time-averaged charts, but the physical reasons for this are not clear.

Procedures for extended forecasting vary around the world largely in accordance with facilities and availability of scientific manpower. Many countries do not have available the high-speed computing equipment necessary to prepare the dynamic component of the forecast, nor have they even the statistical components. In these cases extended forecasts are either not prepared at all or are made by educated synoptic guesswork.

After predictions of average planetary wind flows at upper levels have been made, it is possible to infer the accompanying types of weather in different areas as well as to estimate the general regions for breeding and movement of storms and air masses. Here again the statistics of the motions and weather are more predictable than is the day-to-day detail. In fact, the translation of average wind circulation into average weather is amenable to statistical stratification procedures and is fairly objective. These methods employ as input numerically predicted charts for the 700-millibar (mb) level and also surface temperature predictions to 48 hr made by field forecasters throughout the United States. From these data multiple-regression formulas make possible high-speed computation of temperature forecasts for the weather forecast centers in the United States. A similar procedure gives precipitation estimates. Other weather elements are inferred from the predicted charts for the period.

The boundary between the domain of short- and extended-range forecasting techniques has shifted: Dynamic predictions performed by computer now form the basis for 72-hr prognoses. It is possible that before 1980 dynamic predictions will form a new base for daily forecasts up to a week in advance and that time-averaging methods will no longer be necessary in that range.

Longer-period forecasting. For periods more than a week in advance, methods of forecasting rely more upon statistics and less upon physical reasoning. Concentrated efforts to explore the physics of long-range weather phenomena began about 1955 as a result of increased availability of computing facilities and hemisphere-wide data coverage, particularly from the upper air. However, attempts are being made to prepare forecasts a month ahead by employing dynamic principles in conjunction with statistical and synoptic techniques. For these purposes, another class of mean motions is defined by construction of mean maps for 30-day periods. Although real understanding of these methods is remote, experiments indicate that such objective, machine-produced prognoses are helpful and contain a large part of the accuracy of 30-day forecasts. These prognoses take rough account of the net effect of changes in insolation associated with the change of season, the tendency of certain branches of the general circulation to persist, and the compatibility of the positions of certain large-scale features like the Bermuda high and Icelandic low. Once the circulation pattern for the Northern Hemisphere is prognosticated, the average temperature and precipitation anomalies are computed with the help of elaborate statistical specification equations. The numerical results are then adjusted subjectively by attempting to consider factors not in the equations, like snow cover and wet or dry soil.

Another less expensive and less time-consuming method of longer-range prediction involves the use of statistical analogs. The historical files of weather maps for past periods (mean maps may also be used) are searched, and a wind and weather pattern is sought which is as similar as possible to the one which has been operative, on the assumption that what transpired in the earlier case will repeat. For best results the analogy should be good for large areas of the hemisphere, should hold for upper levels as well as for sea level, and should stand up for a sequence of periods preceding the forecast. The logic of this method is appealing; similar patterns under the same stimuli (such as season of the year) tend to repeat. However, the relatively short span of time for which

meteorological records have been kept makes it difficult to find good analogs.

For periods beyond a month, statistical techniques seem to be the only ones sufficiently accurate; even here, there is some question as to whether the samples of data (length of record) are adequate to assure the stability of discovered relationships. However, experiments in seasonal forecasting indicate some reliability in predicting departures from normal of temperature for the contiguous United States.

Another avenue of approach to the long-range forecast problem which shows promise involves large-scale interactions between the ocean and atmosphere. J. Bjerknes has produced work which indicates that seasonally variable ocean temperatures near the Equator affect the rainfall there, and that the resulting variable release of heat of condensation controls the Hadley circulation, which in turn determines the strength of the subtropical anticyclones and the prevailing westerlies. Complementary work by J. Namias suggests that the longitudinal positioning and intensities of the long waves of the planetary circulation are also determined by air-sea interactions. Although these ideas are not developed to the point of utilizing them objectively in long-range forecasting, attempts are being made to consider them subjectively. *See* ATMOSPHERIC GENERAL CIRCULATION.

Finally, there is hope that numerical forecasts iterated day by day for weeks in advance may yield economically valuable statistics on the weather of the forthcoming month or season. A truly adequate global network of observations is necessary before this will be possible.

[JEROME NAMIAS]

STATISTICAL WEATHER FORECASTING

Statistical weather forecasting is the prediction of weather by rules based upon the statistics of weather behavior. A prediction may state the expected value of a specific weather element, such as a wind speed, or the probability of occurrence of a specific weather event, such as a thunderstorm. In the former case the prediction is understood to contain an error whose probable value may or may not be stated. The choice of the form of prediction may depend upon the intended audience: for example, the statement that there are 2 chances in 10 that tonight's temperature will fall below 32°F might aid a fruit grower more than the statement that tonight's expected minimum temperature will be 36°F.

Basic premises. Statistical forecasting is based upon the premise that the future worldwide state of the atmosphere is determined, at least approximately, by the present state, together with the intervening influences of the Sun and the underlying ocean and land, according to immutable physical laws. In theory, forecasting is equivalent to solving the equations representing these laws, but the equations are rather intractable, and because there are vast gaps between observing stations, the present weather is only partially known. It is sometimes more feasible to ascertain how future weather must evolve from present weather by studying how the observed portions of the atmosphere have previously behaved.

Prediction rules established from such study often relate future weather to present and past weather, instead of present weather alone, since past knowledge may partially compensate for incomplete present knowledge. A rule is commonly expressed as a mathematical formula. Sometimes the same information is more conveniently presented as a graph or a table.

A rule established for one location does not generally apply at another location. For example, a table established for San Francisco, showing the probability of occurrence of nighttime fog following various combinations of midafternoon temperature and relative humidity, would not be valid for predicting fog in New York. A new rule is usually needed for each new weather prediction in any particular area.

Statistical procedures. A general kind of procedure is used to establish a formula. The meteorologist chooses a set of weather elements for "predictors" and selects, commonly from past records, a set of data consisting of corresponding observed values of the predictors and the predictand. As the next step, he chooses a mathematical form with a limited number of degrees of freedom, ordinarily appearing as undetermined constants, and restricts the formula to this form. He specifies a process, ordinarily the minimization of the sum of squares of the prediction errors, by which the chosen data shall determine the constants. Evaluating the constants is then an objective and usually routine mathematical task.

The meteorologist may modify this procedure and classify combinations of values of the chosen predictors into categories. He may then construct a table by choosing, for each category, the average observed value of the predictand as the expected value or, alternatively, by choosing the observed frequency of occurrence as the probability.

The preparation of a forecast once the rule is established is objective and usually simple. The forecaster evaluates the formula after introducing the appropriate numerical values of the predictors, or reads the forecast from the appropriate location in the graph or table.

The data selected for establishing a formula constitute a finite sample of the total history of the weather. The formula is likely to succeed, when applied to future weather, only if the number of degrees of freedom is small compared to the number of values of each predictor in the sample, since virtually any finite set of numbers will fit a sufficiently complicated formula.

Ideally the sample should be made very large. When this is not feasible, because of the excessive labor involved or the absence of extensive past records, the degrees of freedom must be restricted. This is accomplished by limiting the number of predictors or restricting the formula to a more highly specialized form. Meteorological experience or physical reasoning should be used as a guide, since a blind choice of predictors, or of a mathematical form, is unlikely to yield a successful formula.

Statistical linear regression. The simplest mathematical form, and the one whose theory is most highly developed, is the linear formula. When many predictors have been chosen, the number may be reduced either by factor analysis, which selects a few linear combinations of the predictors in order of their ability to represent all the predictors, or by a procedure which selects a few predic-

tors in order of their independent contribution to the prediction. Widespread investigation of linear formulas has followed the advent of high-speed electronic computing machines.

Appropriate nonlinear formulas are theoretically superior to linear formulas, but since they usually involve many degrees of freedom, they are more difficult to discover.

Statistical methods are highly suitable for predicting special local phenomena such as the occurrence of fog. For preparing prognostic weather maps one or two days in advance, statistical formulas are useful, but are frequently inferior to conventional subjective forecasts. For forecasting several days in advance, linear statistical formulas show a slight positive utility and compare favorably with other methods.

[EDWARD N. LORENZ]

NUMERICAL WEATHER PREDICTION

Numerical weather prediction is the prediction of weather phenomena by the numerical solution of the equations governing the motion and changes of condition of the atmosphere. More generally, the term applies to any numerical solution or analysis of the atmospheric equations of motion.

The laws of motion of the atmosphere may be expressed as a set of partial differential equations relating the instantaneous rates of change of the meteorological variables to their instantaneous distribution in space. These are developed in dynamic meteorology. A prediction for a finite time interval is obtained by summing the succession of infinitesimal time changes of the meteorological variables, each of which is determined by their distribution at the preceding instant of time. Although this process of integration may be carried out in principle, the nonlinearity of the equations and the complexity and multiplicity of the data make it impossible in practice. Instead, one must resort to finite-difference approximation techniques in which successive changes in the variables are calculated for small, but finite, time intervals at a finite grid of points spanning part or all of the atmosphere. Even so, the amount of computation is vast, and numerical weather prediction remained only a dream until the advent of the modern high-speed electronic computing machine. These machines are capable of performing the millions of arithmetic operations involved with a minimum of human labor and in an economically feasible time span. Numerical methods are gradually replacing the earlier, more subjective methods of weather prediction in many United States government weather services. This is particularly true in the preparation of prognoses for large areas. The detailed prediction of local weather phenomena has not yet benefited greatly from the use of numerico-dynamic methods, as indicated above in the general section on weather forecasting.

Short-range numerical prediction. By the nature of numerical weather prediction, its accuracy depends on (1) an understanding of the laws of atmospheric behavior, (2) the ability to measure the instantaneous state of the atmosphere, and (3) the accuracy with which the solutions of the continuous equations of motion are approximated by finite-difference means. The greatest success has been achieved in predicting the motion of the large-scale (>1000 mi) pressure systems in the atmosphere for relatively short periods of time (1–3 days). For such space and time scales, the poorly understood energy sources and frictional dissipative forces may be largely ignored, and rather coarse space grids may be used.

The large-scale motions are characterized by their properties of being quasi-static, quasi-geostrophic, and horizontally quasi-nondivergent, as dicussed in another article. *See* METEOROLOGY.

These properties may be used to simplify the equations of motion by filtering out the motions which have little meteorological importance, such as sound and gravity waves. The resulting equations then become, in some cases, more amenable to numerical treatment.

A simple illustration of the methods employed for numerical weather prediction is given by the following example. Consider a homogeneous, incompressible, frictionless fluid moving over a rotating, gravitating plane in such a manner that the horizontal velocity does not vary with height. For quasi-static flow the equations of motion are Eqs. (1), and the equation of mass conservation is Eq.

$$\frac{\partial u}{\partial t} + u\frac{\partial u}{\partial x} + v\frac{\partial u}{\partial y} = -g\frac{\partial h}{\partial x} + 2\omega v$$

$$\frac{\partial v}{\partial t} + u\frac{\partial v}{\partial x} + v\frac{\partial v}{\partial y} = -g\frac{\partial h}{\partial y} - 2\omega u \tag{1}$$

(2), where u and v are the velocity components in

$$\frac{\partial h}{\partial t} + u\frac{\partial h}{\partial x} + v\frac{\partial h}{\partial y} = -h\left(\frac{\partial u}{\partial x} + \frac{\partial v}{\partial y}\right) \tag{2}$$

the directions of the horizontal rectangular coordinates x and y, t is the time, g is the acceleration of gravity, ω is the angular speed of rotation, and h is the height of the free surface of the fluid. Let the variables u, v, and h be defined at the points $x = i\Delta x$, $y = j\Delta x$ ($i = 0, 1, 2, \ldots, I$; $j = 0, 1, 2, \ldots, J$) and at the times $t = k\Delta t$ ($k = 0, 1, 2, \ldots, K$), and denote quantities at these points and times by the subscripts i, j, and k. Derivatives such as $\partial u/\partial t$ and $\partial u/\partial x$ may be approximated by the central difference quotients given by Eqs. (3). In this way

$$\frac{\Delta_k u_{i,j}}{2\Delta t} \equiv \frac{u_{i,j,k+1} - u_{i,j,k-1}}{2\Delta t}$$

$$\frac{\Delta_i u_{j,k}}{2\Delta x} \equiv \frac{u_{i+1,j,k} - u_{i-1,j,k}}{2\Delta x} \tag{3}$$

Eqs. (4), the finite-difference analogs of the contin-

$$u_{i,j,k+1} = u_{i,j,k-1} - \frac{\Delta t}{\Delta x}(u_{i,j,k}\,\Delta_i u_{j,k} + v_{i,j,k}\,\Delta_j u_{i,k}$$
$$+ g\,\Delta_i h_{j,k}) + 4\omega v_{i,j,k}\,\Delta t$$

$$v_{i,j,k+1} = v_{i,j,k-1} - \frac{\Delta t}{\Delta x}(u_{i,j,k}\,\Delta_i v_{j,k} + v_{i,j,k}\,\Delta_j v_{i,k}$$
$$+ g\,\Delta_j h_{i,k}) - 4\omega u_{i,j,k}\,\Delta t \tag{4}$$

$$h_{i,j,k+1} = h_{i,j,k-1} - \frac{\Delta t}{\Delta x}[u_{i,j,k}\,\Delta_i h_{j,k} + v_{i,j,k}\,\Delta_j h_{i,k}$$
$$+ h_{i,j,k}\,(\Delta_i u_{j,k} + \Delta_j v_{i,k})]$$

uous equations, are obtained. Equations (4) give u, v, and h at the time $(k+1)\Delta t$ in terms of u, v, and h at the times $k\Delta t$ and $(k-1)\Delta t$. It is then possible to calculate u, v, and h at any time by iterative application of the above equations.

It may be shown, however, that the solution of

the finite-difference equations will not converge to the solution of the continuous equations unless the criterion $\Delta s/\Delta t > c\sqrt{2}$ is satisfied, where c is the maximum value of the speed of long gravity waves $\sqrt{gh}$. Under circumstances comparable to those in the atmosphere, Δt is found to be so small that a 24-hr prediction requires some 200 time steps and approximately 10,000,000 multiplications for an area the size of the Earth's surface. The computing time on a machine with a multiplication speed of 100 μsec, an addition speed of 10 μsec, and a memory access time of 10 μsec would be about 30 min. The magnitude of the computational task may be comprehended from the fact that the more accurate atmospheric models now envisaged will require some 100–1000 times this amount of computation.

A saving of time is accomplished by utilizing the quasi-nondivergent property of the large-scale atmospheric motions. If, in the above example, the horizontal divergence $\partial u/\partial x + \partial v/\partial y$ is set equal to zero, the motion is found to be completely described by the equation for the conservation of the vertical component of absolute vorticity, as developed in another article. *See* METEOROLOGY.

The solution of this equation may be obtained in far fewer time steps since gravity wave motions are filtered out by this constraint and the velocity c in the Courant-Friedrichs-Lewy criterion becomes merely the maximum particle velocity instead of the much greater gravity wave speed.

Cloud and precipitation prediction. If, to the standard dynamic variables u, v, w, p, and ρ, a sixth variable, the density of water vapor, is added, it becomes possible to predict clouds and precipitation as well as the air motion. When a parcel of air containing a fixed quantity of water vapor ascends, it expands adiabatically and cools until it becomes saturated. Continued ascent produces clouds and precipitation.

To incorporate these effects into a numerical prediction schema one adds Eq. (5), which governs

$$\frac{Dr}{Dt} \equiv \frac{\partial r}{\partial t} + u\frac{\partial r}{\partial x} + v\frac{\partial r}{\partial y} + w\frac{\partial r}{\partial z} = S \tag{5}$$

the rate of change of specific humidity r. Here S represents a source or sink of moisture. Then it is necessary also to include as a heat source in the thermodynamic energy equation a term which represents the time rate of release of the latent heat of condensation of water vapor. The most successful predictions made by this method are obtained in regions of strong rising motion, whether induced by forced orographic ascent or by horizontal convergence in well-developed depressions. The physics and mechanics of the convective cloud-formation process make the prediction of convective cloud and showery precipitation more difficult.

Extended-range numerical prediction. The extension of numerical predictions to long time intervals requires a more accurate knowledge than now exists of the energy transfer and turbulent dissipative processes within the atmosphere and at the air-earth boundary, as well as greatly augmented computing-machine speeds and capacities. However, predictions of mean conditions over large areas may well become possible before such developments have taken place, for it is now possible to incorporate into the prediction equations estimates of the energy sources and sinks—esti-

Fig. 1. Weather data are received on teletypewriters recording on tape, which is then fed into a component for converting the information into punch-card records for further processing. (*ESSA*)

mates which may be inaccurate in detail but correct in the mean. Several mathematical experiments involving such simplified energy sources have yielded predictions of mean circulations that strongly resemble those of the atmosphere.

Numerical calculation of climate. The above-mentioned experiments lead to a hope that it will be possible to explain the principal features of the Earth's climate, that is, the average state of the weather, well before it becomes possible to predict the daily fluctuations of weather for extended peri-

Fig. 2. Automatic plotters at the National Meteorological Center prepare charts from computer-processed weather data from around the world. (*ESSA*)

ods. Should these hopes be realized it would then become possible to undertake a rational analysis of paleoclimatic variation and changes induced by artificial means. If the existing climate could be understood from a knowledge of the existing energy sources, atmospheric constituents, and Earth surface characteristics, it might also be possible to predict the effects on the climate of natural or artificial modifications in one or more of these elements.   [JULE G. CHARNEY]

CENTERS AND OFFICES OF FORECASTING

Weather forecasts for all parts of the United States are prepared by the Weather Bureau of the Environmental Science Services Administration (ESSA). Various phases of the forecast work are performed, in general, at two working levels: (1) the National Meteorological Center and (2) the weather forecast centers. However, in special areas and circumstances a limited amount of forecast work is performed at local weather service offices.

National Meteorological Center. The National Meteorological Center, located in Suitland, Md., near Washington, D.C., collects weather observations, prepares weather charts, and issues forecasts on the future state of the atmosphere on a hemispheric scale. To the extent that available data permit, analyses and forecasts are also prepared for the Southern Hemisphere. Weather observations taken throughout the Northern Hemisphere are sent to the Center at frequent intervals by landline and radioteletypewriter circuits and other rapid communication systems (Fig. 1). Reports of weather elements measured near the surface of the Earth throughout the Northern Hemisphere are collected four times a day. Every 24 hr the Center receives 25,500 surface reports, of which 22,300 are from land stations and about 3200 from ships at sea. Reports of upper-air conditions are received daily from 900 pilot balloon observations of wind direction and speed: 980 radiosonde observations of upper-air pressure, temperature, and humidity: and 500 rawinsonde (radar wind radiosonde) observations of pressure, temperature, humidity, and winds. The Center also receives some 900 reports from commercial aircraft in flight, 400 reports from scheduled military weather reconnaissance aircraft, and a varying number of indirect soundings of upper-air temperature made by the experimental Nimbus-Sirs satellite system. Hundreds of photographs are received each day from the ESSA Environmental Satellite System showing the cloud cover for the entire Earth, except that part shrouded in polar night.

The surface reports and to a limited extent the upper-air reports are plotted on hemispheric charts for analysis (Fig. 2). Most of the upper-air data go from the communication system directly into the automatic data-processing system, a part of the computer complex. The computer prepares an analysis of the "initial state" of the atmosphere as of data observation time and then quickly prepares forecasts of the state at 12-hr intervals out to 48 hr and at 24-hr intervals out to 120 hr. Some of these forecasts in the form of prognostic charts go directly from the computer to a complex device called a facsimile group-converter, and thence into the national facsimile communication systems without any manual processing. Still others are subjected to examination by expert forecasters and analysts before they are distributed.

In addition to the daily forecast charts, the National Meteorological Center prepares 5-day and 30-day forecasts and outlooks. At present (1970) the 5-day forecasts take the form of forecast charts of airflow in the Northern Hemisphere and frontal patterns at sea level (one for each of the 5 days); they also include charts of the expected departures from normal of precipitation and temperature. The 30-day outlooks are issued as charts showing the expected mean monthly airflow over the Northern Hemisphere and the expected 30-day departures from normal of precipitation and temperature.

Forecast centers. The forecast centers make use of the facsimile charts received from the National Meteorological Center in their preparation of forecasts for release to the general public. There were 30 such centers in early 1970, but the system plan calls for increasing this number to 50 as a means of decentralizing the final forecast preparation. These centers also receive additional data in the form of hourly observations taken from the North American continent and reports at 6-hr intervals from ships at sea.

Area refinements and warnings. Information received on the facsimile charts is supplemented by plotting and analysis of such additional data as may be required to introduce refinements in the shorter-range periods of the forecasts for the respective districts. Radar weather information is also utilized. A 24-hr vigil of all meteorological phenomena which might endanger life and property within a region is maintained: high winds, thunderstorms, blizzards, heavy snows, cold waves, killing frosts, fog, and heavy rain are among the weather phenomena for which warnings are issued by the forecast center to local offices.

Forecasts. In addition to issuing warnings, the forecast center makes regular forecasts at 6-hr intervals for issue to radio and the press. These forecasts contain information regarding expected cloudiness, precipitation, temperature, wind, humidity, and other weather factors of interest to the public. The center prepares 5-day forecasts on Mondays, Wednesdays, and Fridays for relatively broad areas using guidance charts received from the National Meteorological Center.

Aviation and other special services. The basic weather service to aviation consists of warnings and forecasts prepared at forecast centers. A special unit of aviation meteorologists issues several different types of forecasts at regular intervals. These include aviation terminal forecasts, area forecasts, upper-wind forecasts, and advisories of conditions considered hazardous to aircraft in flight. Through teletypewriter communications, these aviation weather forecasts are transmitted to local Weather Bureau offices; Federal Aviation Agency traffic control and communication stations; Air Force, Navy, Army, and other government agencies operating aircraft; and to many private businesses directly concerned with aviation (Fig. 3).

Certain of the forecast centers are assigned special functions such as issuing hurricane warnings, severe local storm warnings, international aviation forecasting, ocean area forecasts for merchant shipping, and river and flood forecasting. The hurricane warning centers spearheaded by the Na-

Fig. 3. Meteorologist at Washington National Airport studies latest analyzed weather chart transmitted by National Meteorological Center via facsimile network to several hundred stations. (*ESSA*)

tional Hurricane Center located in Miami, Fla., issue hurricane warnings and advisories giving the storm's position, intensity, direction and rate of movement, and a statement of the effects to be experienced, such as the onset of high tides or destructive winds with their direction and probable duration (Fig 4). The Severe Local Storm Forecast Center (SELS) at Kansas City, Mo., specializes in forecasting areas (about 100 by 200 mi) likely to experience tornadoes, hailstorms, or severe thunderstorms. The international aviation centers forecast weather and upper-air conditions for international flights originating in the United States. The marine centers prepare forecasts of surface weather and sea conditions over the high seas. Separate and distinct from the weather forecasting centers, the Weather Bureau also maintains river forecast centers. A river forecast center is respon-

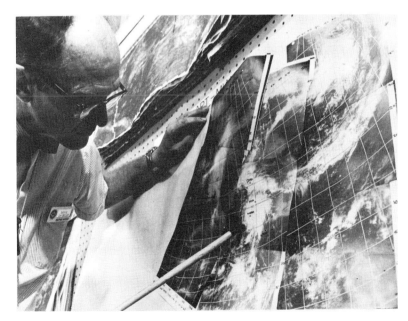

Fig. 4. A meteorologist at the ESSA National Hurricane Center, Miami, Fla., rechecks the position of a major storm on a satellite photograph before issuing a public advisory. (*ESSA*)

sible for maintaining a continuous river watch and issuing forecasts of river and flood conditions. Advance warnings minimize the damaging effects of flooding by providing time for the evacuation of people and movable property.

Weather service offices. Local Weather Bureau offices are considered to be primarily service offices and in the Weather Bureau system are referred to as weather service offices. These offices provide reports, advisories, warnings, forecasts, and other weather information services to the public in their respective communities and surrounding areas. Most of the forecast information is adapted from forecasts and warnings received from the weather forecast centers. The weather service offices do, however, have the primary responsibility for tornado warnings and for warnings of other types of hazardous weather out to a range of 3 hr. Also, certain offices located in areas of intensive cultivation of high-value, frost-sensitive crops have the responsibility for issuing overnight (12-hr) minimum temperature forecasts for specific points keyed to the local frost protection problems.

Local forecasts and reports are distributed by press, radio and television, public teletypewriter circuits, and automatic telephone forecast repeaters (in some large cities); and they are also available by telephone or personal visit at the Weather Bureau office. During periods of severe weather, local Weather Bureau offices release emergency weather warnings to all radio and television stations in the area affected for frequent broadcast to the public. In addition, the Red Cross, Civil Defense, Coast Guard, state and local police, highway commissions, and other agencies are kept fully informed and assist in distributing warning messages so that every action may be taken to safeguard life and property.

[EDWARD M. VERNON]

Bibliography: J. Bjerknes and H. Solberg, Life cycle of cyclones and polar front theory of atmospheric circulation, *Geofys. Publ.,* no. 3, 1922; T. F. Malone (ed.), *Compendium of Meteorology,* 1951; J. Namias, Long range weather forecasting: History, current status, and outlook, *Bull. Amer. Meteorol. Soc.,* 49(5), 1968; H. A. Panofsky and G. W. Brier, *Some Applications of Statistics to Meteorology,* 1958; S. Petterssen, *Weather Analysis and Forecasting,* 2 vols., 2d ed., 1956; P. D. Thompson, *Numerical Weather Analysis and Prediction,* 1961.

Weather map

A chart portraying the state of the atmospheric circulation and weather at a particular time over a wide area. It is derived from a careful analysis of simultaneous weather observations made at many observing points in the area. Such a chart gives the weather forecaster an integrated picture of the location, structure, and, when several successive charts are available, the motion and development of the various weather systems. From this study he may construct a prognostic chart, which portrays various weather features for selected times in the future.

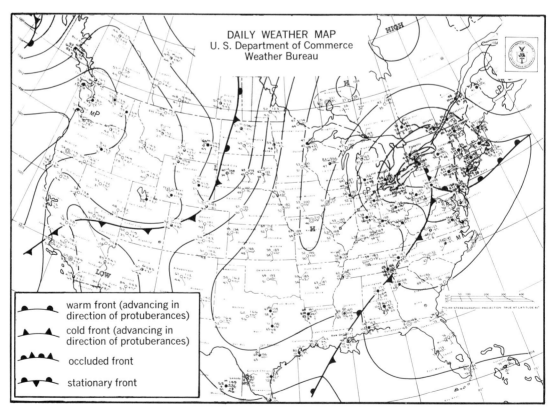

Fig. 1.   A surface weather map at 0600 Greenwich Civil Time. Sea-level isobars (thin lines) are drawn for every 4 millibars and labeled in whole millibars; fronts, or transition zones separating air masses, are indicated by heavy lines; *mT* indicates tropical maritime air; *mP* indicates polar maritime air. Areas where precipitation was falling at 0600 are shaded. Previous 6-hourly positions of low-pressure center in eastern Great Lakes are indicated by crosses connected by arrows.

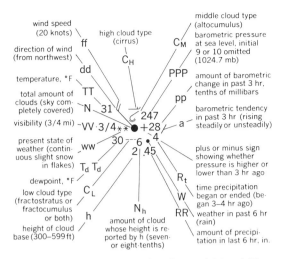

Fig. 2. Abbreviated code and station model for plotting weather elements at an observation station on the Earth's surface. Letters are symbols for the meteorological elements. In coded reports only numbers and cloud and weather plotting symbols are used. The use of symbols and abbreviations increases the density of information which can be put on the map.

Many kinds of weather maps are used, depending on the weather elements of immediate interest and their elevation above the ground. At a typical large weather analysis central, as many as 35 different charts are constructed for a given time. Among the more common of these is the surface map, which portrays the weather at the Earth's surface (Fig. 1). All the mapped weather elements except pressure are those directly observed at the weather station (Fig. 2). Except over the oceans the so-called sea-level pressure is a fictitious quantity obtained by reducing the surface pressure to sea level by a special formula so that continuous isolines (called isobars) may be drawn through regions having the same pressure value. These isobars are important in portraying the winds, the physical structure of weather systems, and the location of fronts and air masses. See AIR MASS; FRONT.

The surface map, at least over ocean regions, represents a section through the atmosphere along an approximately horizontal surface. This procedure may also be used in constructing upper-level maps, or charts showing the distribution of weather at fixed elevations above sea level. It is

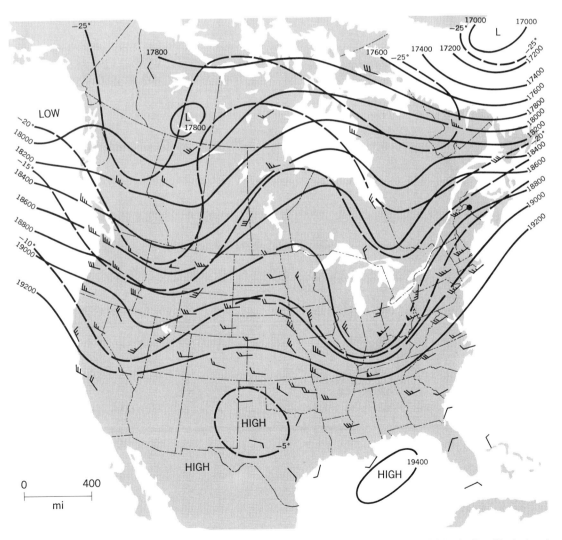

Fig. 3. Representative 500-millibar constant-pressure chart. Isolines of topography of constant-pressure surface show height in feet above sea level. Isotherms are in °C at 5° intervals. Arrows with barbs fly with wind and show wind speed in knots. Each long barb = 10 knots; triangular barb = 50 knots. (U.S. Weather Bureau)

more common, however, to portray the weather at high elevations on constant-pressure surfaces (Fig. 3). The elements plotted at each upper-air station usually include the height of the constant-pressure surface, temperature, and some measure of humidity, usually the dew point. Lines of constant elevation (called isohypses or contours) then are drawn to portray the topography of the selected pressure surface. The contours bear a relation to the winds similar to that of the isobars of a constant-level chart. Thus the winds tend to blow along the contours with low heights to the left when facing downstream. Their speed is roughly inversely proportional to the sine of the latitude and the distance between contours when these are drawn at constant-height intervals. Usually isolines of temperature, or isotherms, are also drawn, as in Fig. 3.

Other important two-dimensional or one-dimensional atmospheric sections cannot be termed weather maps, but rather meteorological charts and diagrams. These include vertical cross sections through the atmosphere, and thermodynamic diagrams similar to those used in studies of heat engines.

Weather map analysis. This branch of synoptic meteorology had its beginning at about the time the telegraph was invented, when for the first time weather observations covering large areas could be sent rapidly to a central location. Its steady development was greatly accelerated following World War I, when the techniques of air-mass analysis were developed by the Scandinavian school headed by V. Bjerknes. Upper-air charts were not commonly constructed until the 1930s, when sufficient high-level data became available.

The preparation of a weather chart at a large analysis central can be described as follows. First the encoded data at the surface and upper levels are transmitted to the central from collection centers by means of teletypes. If a map covering the Northern Hemisphere is to be prepared, data from approximately 850 surface and 400 upper-air reporting stations must be processed. This mass of data is subject to errors of observation, encoding, and transmission. Furthermore, large areas, particularly in the tropic and arctic latitudes, contain no observation stations and, hence, no data at all. The detection and correction of the errors, and the interpolation of weather in the intervals between reporting stations demand the greatest skill on the part of the chartmen who plot the data and the analysts who must interpret the data in terms of consistent physical structures of the atmosphere. After the data are corrected and plotted, the analyst then locates and draws various features such as fronts, air masses, and isobars. He must be guided by known physical principles regarding the horizontal, vertical, and temporal continuity of the atmosphere; that is, his analysis must be internally consistent. When finished, the chart is ready for the forecaster.

Data from orbiting weather satellites in the form of video cloud pictures are used on a daily basis in weather map analysis, particularly in areas of sparse surface-based observations.

Beginning in May 1969, the vertical atmospheric temperature structure over a large part of the entire Earth has been transmitted from satellites (such as *Nimbus 3* and *4* in 1969–1970) carrying satellite infrared spectrometer (SIRS) instruments. These data are used in automatic weather map analyses to supplement conventional surface-based systems. In 1970 this was done routinely over the Northern Hemisphere and experimentally over the Southern.

Automatic weather map analysis. With the introduction of high-speed electronic computers into meteorology in 1951, it was realized that the slow manual methods of data collection and processing, described above, could be greatly speeded up so as to keep up with advances in automatic weather forecasting and with the large volume of data being received from all parts of the world. For these reasons experiments in automatic analysis were begun in a number of countries around 1953, and in 1957 the Joint Numerical Weather Prediction unit at Suitland, Md. (now the National Meteorological Center), began the routine production of machine-analyzed weather maps. The objectives and final products are essentially the same as described above under weather map analysis. After the analysis has been completed by the main computer, it is stored on magnetic tapes and then displayed visually in a number of optional forms, such as machine print-outs, mechanically drawn maps using curve plotters, and photographs of pictures projected on video tubes. *See* METEOROLOGY; WEATHER FORECASTING AND PREDICTION.

[PHILIP F. CLAPP]

Weather modification

The changing of natural weather phenomena by man. Weather is the product of the interaction of atmospheric processes on many scales, reaching from the planetary circulation to the microphysical processes in the evolution of cloud droplets and ice crystals. So far, only on the microscale of condensation and freezing nuclei has man begun to exert

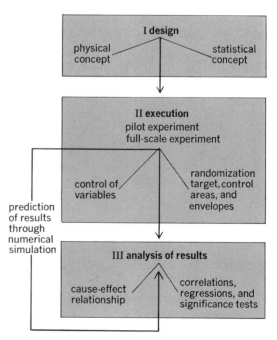

Fig. 1. Scheme for approach to weather modification experiments. Experiments must be in agreement with physical and statistical principles of experimentation.

Fig. 2. Cloud dissipation. (a) Three lines in stratocumulus cloud layer 15 min after seeding. (b) Opening in stratocumulus layer 70 min after seeding. (U.S. Army ECOM, Fort Monmouth, N.J.)

modifying influences on weather. For example, the artificial modification of rain at New York City may suggest a seeding activity just a few miles upwind of the city, but the artificial modification of a 1000-mi-diameter cyclone in the Gulf of Alaska is impossible at present.

The numerous physical and meteorological processes which are involved in the experimental approach to weather modification make it complex and difficult. In many cases it is virtually impossible to design a classical physical experiment for determining a cause-effect relationship, and it is necessary to adopt a statistically designed experiment. Here one need not know all physical processes, feedback mechanisms, and interactions in order to derive the influence of the one artificially modified parameter, but one is required to conduct a great number of identical experiments. This calls for experimental periods which have to be counted in decades. In the meantime the environmental conditions of the experiment may change, so the applicability of the statistical approach is limited. In view of these difficulties, an approach emerges which attempts to combine both experimental principles, the classical and the statistical (Fig. 1). The prediction of the outcome of the experiment by means of numerical simulation provides a powerful support for the physical experiments. Fast computers with large memories help assess the influence of major factors which can be entered as parameters into a theoretical cloud model.

Developments in experimental meteorology have caused an air of optimism in the approach to weather modification. In addition, radar, pulse Doppler radar, aircraft with sophisticated instru-

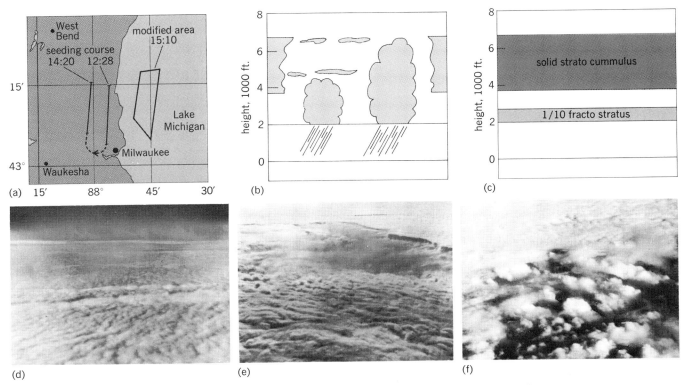

Fig. 3. Development of heavy snow showers after a seeding experiment for cloud dissipation over Milwaukee, Wis., on Nov. 24, 1953. (a) Location of experiment. (b) Vertical section before seeding and (c) 63 min after seeding. (d) Cloud appearance 15 min after seeding, (e) 31 min after seeding, and (f) 63 min after seeding. (U.S. Army ECOM, Fort Monmouth, N.J.)

Fig. 4. "Explosion" of cumulus cloud following release of heat of fusion caused by seeding with silver iodide. (a) Time of seeding. (b) Views at 9 min, (c) at 19 min, and (d) at 38 min after the seeding. (*Courtesy of J. Simpson, ESSA*)

mentation serving as observing platforms, and satellites are providing unprecedented observation and analysis facilities.

Cloud seeding. Development of pyrotechnic mixtures for the dispersal of silver iodide and other seeding reagents has greatly increased the capability of proper airborne delivery of the most widely used seeding agents. There is increasing interest in using condensation nuclei as seeding material.

Fog dispersal. In the late 1940s I. Langmuir, V. J. Schaefer, and B. Vonnegut discovered that supercooled clouds can be dissipated by seeding with dry ice or silver iodide. The method is now applied operationally for the dispersal of supercooled fogs from airports all over the world. Systems are being developed for seeding with liquid propane in France (Orly Field) and for seeding with dry ice, silver iodide, and lead iodide in the United States and the Soviet Union.

Dispersal of warm fog (temperature warmer than frost point) is an unsolved problem. This fog, unlike supercooled fog, is in a stable state which prevents the exploitation of an energy sink. Heat, seeding with hygroscopic particles, mixing of the low-level air layers by vertically blowing propellers, heat from jet engines lined up along runways, and seeding with polyelectrolytes (a strong hydration agent) are some of the approaches, but none is entirely satisfactory.

Cloud modification. Some striking results of artificial weather modification have been obtained in cloud modification. Figure 2 shows an area of about 100 km² in which supercooled stratocumulus clouds were dissipated by seeding each of three parallel 16-km-long lines 5 km apart with 100 lb of dry ice. In experiments over populated regions the weather has been modified conspicuously on the mesoscale of meteorological events: Sunshine has been made for several thousand people on a dull overcast winter day.

A contrasting case is illustrated in Fig. 3. Here, after seeding for dissipation, a miniature squall line developed over Lake Michigan consisting of heavy snow showers which lasted for about 1 hr.

Figure 4 shows another spectacular result of cloud seeding. A cumulus cloud reaching to the −8°C level was overseeded with silver iodide. Overseeding caused glaciation and hence heat of fusion was released by the glaciating cloud water. This heat increased the buoyancy of the cloud, and the cloud "exploded," growing in height and also in width. This experiment has a clear physical concept which tested a cause-effect relationship; however, randomization with seeded and control clouds contributed materially to sharpening the experimental result. Of 22 test clouds 14 were seeded and 8 were not; only 1 of the clouds not seeded grew comparatively. Of the 14 seeded clouds, 4 exploded and 10 increased in height with no increase in width.

These experiments indicate that it is possible to affect materially the life history of clouds through seeding. *See* CLOUD PHYSICS.

Rainmaking. Water is one of the most abundant but also one of the most wanted substances on Earth. The interest in artificial rainmaking is therefore understandable. A panel of the U.S. National Academy of Sciences expressed restrained optimism by stating: "There is increasing but still somewhat ambiguous statistical evidence that precipitation from some types of cloud and storm systems can be moderately increased or redistributed by seeding techniques." However, the panel also recommended the early establishment of several carefully designed, randomized, seeding experiments, planned in such a way as to permit assessment of the seedability of a variety of storm types. These recommendations reflect that much corroborative evidence is still missing.

Indeed, the evidence is often controversial, particularly where the experiments can only be based on statistical design. In all such experiments it is tacitly assumed that the seeding agent gets to the right location in the cloud system and acts as desired. Seeding with ground-based generators has uncertainties because of the unknown diffusion properties of the seeding agent in the atmospheric boundary layer. Seeding from aircraft has yielded positive results in Israel, inconclusive results in Missouri, and negative results in Arizona.

The most noteworthy progress in this area comes from increased theoretical understanding of the precipitation process, particularly for convective clouds. Well-designed seeding experiments on 33 pairs of cumulus clouds in Australia indicated in the statistical analysis that seeding at cloud base with 20 g of silver iodide produced significant rain increases, while seeding with 0.2 g was ineffective. A seedability index can be derived from these data: lifetime of cloud in hours times depth of cloud in kilometers. Figure 5 illustrates that, for an index below 3, clouds do not rain naturally or after seeding. However, the cases which statistically determined the success through seeding were all connected with high seedability indexes.

Downwind effects. Sporadic statistical analyses indicate that downwind effects occur as far away as 300 mi. Observations have been reported from the United States and Australia, with precipitation increases in most cases. In view of the rate of decay of silver iodide in daylight, the mechanism of such effects is dubious, and a strong case can be made that, downwind from any location, positive anomalies can be found simply because of the natural variability of the rate of rainfall.

Hail suppression. Scientists of the Soviet Union report the development of an operational hail suppression project which reduces hail damage by 80–90%. It involves the detection with radar of the hail-spawning cloud regions and the delivery into them of seeding material by means of grenades. In the United States a physical concept is being developed which differs somewhat from the Soviet concept, and delivery of the seeding agent from aircraft is foreseen rather than from the ground. A project in France conducts surface seeding on a large scale, discharging tons of silver iodide into the air during the hail season. The analysis of the nonrandomized experiments indicates success. Other projects are being conducted in Germany (Bavaria), Italy, and Argentina. While the theory of hail formation suggests that hail may be suppressed by a comparatively small seeding effort, there is great difficulty in designing a field experiment which is in agreement with the scheme presented in Fig. 1. The sporadic nature of hailstorms and their large natural variability practically exclude effective randomization of the experiments. For the analysis phase it is therefore necessary to have available a store of excellent historical data on the experimental area, such as exists in the Canadian Hail Research Project in Alberta.

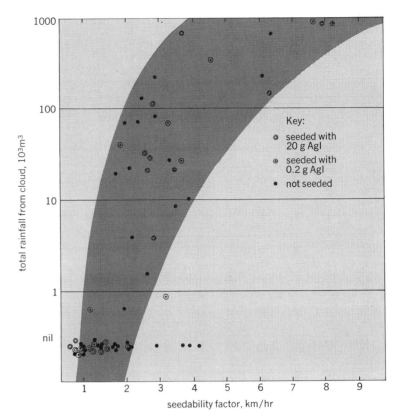

Fig. 5. Seedability factor for rainmaking from convective clouds. Seedability factor is product of cloud depth in kilometers and cloud lifetime in hours.

Lightning suppression. Two concepts have been tested in the United States. The first, developed by the Department of Agriculture, makes use of overseeding of thunderstorms with silver iodide. While physical relationships in the suppression mechanism are not fully developed, it appears that the method decreases ground strokes and increases intracloud strokes. Another approach has been developed by the U.S. Army jointly with the Environmental Science Services Administration. Discharge of the charge centers of a storm by corona discharges initiated through the introduction of metallic needles, so-called chaff. It can be shown that for thunderstorm fields, $10^6 - 10^7$ chaff particles (weighing 5 to 50 lb) can discharge several amperes, a result in agreement with the magnitude of the thunderstorm-charging current.

Modification of severe storms. Modification of severe storms such as tornadoes or hurricanes is in its infancy, essentially because of incomplete understanding of the dynamic structure of these storms. The Environmental Science Services Administration and the U. S. Navy have developed under Project Stormfury a concept of hurricane modification based upon overseeding of the wall clouds. Release of the heat of fusion is believed to alter the pressure gradient in such a way as to diminish the destructive winds near the storm center. As wind damage is proportional to the square of the wind velocity, a relatively small reduction of the velocity may mean a large reduction of damage.

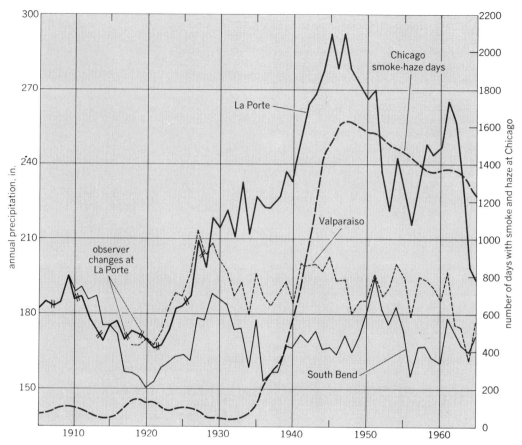

Fig. 6. Rainfall anomaly for La Porte, Ind., compared with increase of smoke and haze at Chicago. (*From S. A. Changnon, Jr., The La Porte Weather Anomaly: Fact or Fiction, Amer. Meteorol. Soc. Bull. no. 49, 1968*)

Seeding experiments were conducted on hurricanes Esther in 1961 and Beulah in 1963. A 10% reduction of wind speed was observed: however, the reduction is well within the natural wind variability of the storm.

Inadvertent weather modification. Modern civilization affects weather in many ways. The great industrial centers of the world are plagued with haze and smog due to effluents from factories and combustion of hydrocarbons. Haze and smog affect the heat radiation budget of the atmosphere as well as precipitation processes. Aerosols discharged by metal foundries act as freezing nuclei, and strong evidence now appears for such inadvertent weather modification. In La Porte, Ind., the rate of precipitation has increased considerably since the construction of foundries at Gary, upwind of La Porte (Fig. 6). While the combustion of hydrocarbons causes formation of smog in large urban areas, a more subtle climate modification takes place on a global scale. The combustion process not only liberates aerosols but also carbon dioxide gas, an important constituent of the air that contributes to the greenhouse effect of the atmosphere. The gas is transparent for the Sun's visible radiation but traps the Earth's thermal radiation; thus the more carbon dioxide released, the warmer the climate becomes. Computer calculations project an average temperature increase of 1–2°C within two to three generations from this effect. Though this change may appear small, it would have profound influences on weather and climate. Glaciers and ice caps would melt, and coastal cities might become inundated. It is possible that other changes in the atmosphere will counteract the effect of carbon dioxide. *See* ATMOSPHERIC POLLUTION.

Effect on ecology and society. While influences on ecological systems appear to be often exaggerated in view of the great natural variability of weather, socioeconomic considerations are important. Fog dispersal methods at airports enable many passengers to travel, effective hail suppression would save millions of dollars, and rainmaking would be the least expensive answer to the water dilemma of modern civilization. But as interests of individuals differ, rainmaking may be a blessing to one and blight for another; socioeconomic considerations are therefore important at a time when weather modification is still in its infancy.

[HELMUT K. WEICKMANN]

Wildlife conservation

The science and art of making decisions and taking action to manipulate wild animal populations and their environments, and to influence men to achieve specific benefits from the wildlife resource. Although still called wildlife conservation by some people, wildlife management, now the preferred term, has more than its earlier connotations of preservation. Wildlife management is concerned not only with increasing populations of rare or endangered species, such as the whooping crane (Fig. 1), but also with stabilizing certain pop-

Fig. 1. The whooping crane, a rare North American migratory species, is in danger of extinction. There were only 55 individuals in existence in 1969. (*Bureau of Sport Fisheries and Wildlife*)

ulations, such as the American bison, increasing or stabilizing desirable species taken for sport recreation (game), and decreasing some populations of birds and harmful mammals.

The problems of management are complex and involve conflicts between groups with opposing interests. For example, orchard owners want fewer bark-eating rabbits, while hunters want more rabbits of all kinds; fox hunters want increased populations, while other people wish to reduce foxes (Fig. 2) as vectors of rabies; hunters who want only native game animals differ with hunters who would stock or import foreign species into any area; hunters and farmers are at odds on the proper number of raccoons to have in forests near cornfields; foresters who normally manage their forests to supply deer to hunters do not want deer in their nurseries; city dwellers want birds in parks but not on their buildings.

Efforts to resolve or to balance these opposing interests require the best possible individual and group decisions. These decisions, to be satisfying for many people, require a great deal of information, including information about wildlife values. *See* CONSERVATION OF RESOURCES.

WILDLIFE VALUES

Wildlife has many values. It contributes to the natural "character" of land; that is, land of whatever type does not seem really wild if one knows animals are absent. Animals contribute to a sense of completeness or wholeness of the landscape. Wildlife has great esthetic values—geese flying in

Fig. 2. The fox is a valuable fur animal. It eats animals that damage crops and fruit trees, but it also eats rabbits and is a vector of rabies in some areas.

spring, bighorn sheep silhouetted on the horizon, even the chipmunk on a mossy log—all produce real, though hard-to-measure, values. Wild creatures have a great capacity for inspiring creative reactions to their beauty as evidenced in works of art, music, and poetry. Their past values in motivating explorers and pioneers to open the

Fig. 3. Ducks and geese require proper nesting, rearing, and feeding areas throughout their migration flyways. In addition, they need protection under wisely set hunting regulations. (*After J. Von Huizen, U.S. Fish and Wildlife Service*)

way westward and hasten the development of America are well known.

A more easily measured value of wildlife is its contribution to personal, community, and national economy. Furs are an obvious monetary resource, but the market fluctuates with women's and men's clothing styles. An estimated $4,046,440 is spent each year in the United States on hunting and sport fishing. Hunters and fishermen, 23.2% of all people 12 years old or older in the United States, each spend an average of $123.06, according to the 1965 survey by the U.S. Fish and Wildlife Service. The money spent by them is a minimum measure of the recreational values attached to the wildlife resource, since such values go beyond the actual taking of game animals, birds, and fish, but also include the time and thought spent anticipating and reliving the outdoor experience. Wildlife also has the potential of providing insights for investigators into behavior and other scientific phenomena not similarly available from studying other species. Finally, there is the contribution of animals to the proper functioning of natural systems. Their function has been considered a major part of the balance of nature, an idea that suggests the interdependence of plants on animals, animals on plants, and both to each other and their environment. *See* ECOSYSTEM.

Fig. 4. Beaver fur was the "gold" that once lured trappers to first explore America. The beaver is still a valuable animal, not only for fur but for its beneficial influence on water supplies. However, it is a forest and agricultural pest in some areas.

Management task. To achieve a desired abundance of a particular species of wildlife, modern wildlife managers work intensively with the environment in which the animals live. Wildlife populations naturally respond very sensitively to their habitats. When water levels are down, ducks produce fewer young (Fig. 3); when dams of beaver (Fig. 4) are torn out and the rich land within beaver ponds is planted to crops, beaver disappear; where the woody food of deer and elk is not abundant or after the supply of forage has been eaten, populations of these animals are reduced or they have fewer young.

Land has a carrying capacity. Fish ponds can support only a given number of fish. A particular pond may carry 500 fish, each weighing 1/2 lb, or 250 fish weighing 1 lb, based on the size of the pond, food in the water, temperature, and other factors. Only by fertilization, planting, or other cultural practices can carrying capacity be increased. Mammal populations are similar; each area has a limited ability to support animals. Usually where the soil is poorest, animals will be fewest. Where the soil is rich, food and cover are abundant and if other factors are not restrictive, then there will be more animals, but the particular kinds may vary with population cycles and other factors discussed below. *See* POPULATION DYNAMICS.

Wildlife managers are concerned with the natural ability of populations to reproduce. The reproduction potential is set by genetics but strongly influenced by the quality of food. When food supplies are low or the quality of food is reduced (as a result of soil erosion or leaching of nutrients from soil), reproduction is reduced. When food quantity or quality is low, animals have fewer young per year. Another factor called stress influences populations in several complex ways. Much research on stress is now in progress, but evidence supports the conclusion that, as populations become more crowded, reproduction is reduced, more food is required, and diseases have greater effects. Thus there are several interacting natural limits on wildlife population growth imposed both by the environment and by the population itself. Overcoming these limits is the job of the wildlife manager interested in increasing populations. Working with these factors, learning more about them, and trying to influence the role of each factor in the context of other values are the job of the wildlife manager responsible for decreasing damage to wildlife. *See* ECOLOGY, APPLIED.

History. Past efforts at species preservation were too few and too late to save animals such as the passenger pigeon, heath hen, Labrador duck, Carolina parakeet, several species of grizzly bears, the eastern bison, great sea mink, plains wolf, Badlands bighorn sheep, and several species of fish. The problem of preventing extinction is quite real today, and threatened species include the whooping crane, black-footed ferret, ivory-billed woodpecker, Florida Everglade kite, California condor, Atwater's prairie chicken, and Hawaiian goose.

Threats to populations of wildlife usually result from a very complex set of factors. The factor usually at the top of the list is destruction of wildlife homes and habitat (Fig. 5). Interest in habitat has appeared late in the history of efforts to stabilize or increase wildlife or game populations. As species decreased, protective laws were passed first, then stocking was done from elaborate game farms, and then predators were controlled. Following these efforts, protected areas or refuges were set aside, and then finally habitat was managed as an ecosystem. Establishing areas of complete wildlife protection, stocking, and broad-scale predator control have proved over and over to be insufficient. Habitat management has been a more efficient technique.

Now the trend is away from an approach of any single technique for influencing a population. The newer approach, a systems approach, is that of selecting a group of techniques specifically designed for benefiting one target species for each area and for achieving the specific objectives of the landowner. The system involves (1) intensive study and research by experts to isolate problems, (2) definition of goals, (3) education of the public (and specialists) by many media, (4) programs to retain and improve the habitat, (5) goal-oriented game harvest regulations or population control programs, (6) manipulation of sex and age ratios, as well as density of animals, to achieve desired population change, (7) actions by agents to deter or apprehend poachers or game-law violators, (8) control of disease and parasites, and of select individual predators and the causes of accidents, and (9) well-researched transplanting of wildlife from areas where it has bred successfully to areas with high potential for population buildups. *See* SYSTEMS ECOLOGY.

Wildlife management or conservation began in the late 1800s when sensitive men saw that many wildlife populations could not take care of themselves. The tedious, costly, and often dangerous experiences of early conservationists is well told by James B. Trefethen in *Crusade for Wildlife.* The picture of success is confusing, for what worked in one place failed in others; what was an acceptable practice for one species was harmful to another. Only after a long period of trial and error with game species did the need for scientific wildlife research become obvious.

The ring-necked pheasant is an example of an introduced bird that succeeded brilliantly over much of the northern part of the United States. Hundreds of other species have been tried under almost every conceivable program of introduction. Where the habitat is good for pheasants they will reproduce abundantly and fill every available space. Where the conditions of the environment are not right, where there are one or more limiting factors (such as inadequate natural calcium in the soil or critical moisture-temperature relations in the nest), there will be no natural populations of birds. Continued stocking is wasteful. There are yet no naturalized ring-necked pheasant populations in the southeastern states although several million dollars have been spent in work to achieve such populations. Extra birds stocked onto areas where birds occur naturally are also wasted unless immediately harvested by hunters; stocking "before the gun" is practiced in many states. The cost to the public of each pheasant taken by a hunter in some areas exceeds $35. The cost of producing some bobwhite quail has exceeded $150 per bird taken by the hunter (Fig. 6). Nature quickly cuts populations to its carrying capacity.

Sanctuaries, or areas of complete protection for upland game, are generally considered useless

Fig. 5. Destruction of wildlife habitat by bulldozers, fire, highways, herbicides, land drainage, and changing land use contributes to reduction in wildlife numbers and changes in the species present in an area.

Fig. 6. Game farm with pens for quail. (*Virginia Commission of Game and Inland Fisheries*)

Fig. 7. Migratory wildlife need protection as well as year-around habitat if they are to retain their present abundance or to increase.

since animals do not "spill out" of such areas as once thought. On the other hand, protected breeding areas and special areas are considered essential for migratory waterfowl (Fig. 7). Some refuges concentrate populations, stop their normal migration, and increase hunter harvests along the route. Canada geese in the Atlantic and Mississippi flyways are an example.

Predator control may be needed on select individual animals, but is generally found undesirable and even very hazardous when practiced indiscriminately by individual or public impulse. Often there are no net benefits from control. Extra costs may be incurred as rodents or rabbits and their damage increase without the presence of the animals that once exercised some restriction on their populations.

Winter feeding attracts wildlife to some areas for increased viewing and appreciation by man, but generally does little good for most game popu-

Fig. 8. For man to have the wildlife populations he desires, habitats in which all of the food and other year-around needs of animals are met are essential. (USDA)

lations in the long run. Artificial winter feeding of elk or deer is frequently harmful because it may allow more animals to survive the winter than the environment can support. The range is damaged by the extra animals; plants lose their vigor; erosion occurs; poaching, predation, and disease are increased; and mass die-offs occur in subsequent years. Also, winter feeding of high-quality food may even kill starving animals.

Today's needs. The public and the professional wildlife staffs of state and Federal conservation agencies can influence wildlife populations by the following major practices: (1) laws to protect endangered species, (2) regulations to allow a recreational harvest of game animals and manipulation of other populations that put each population in balance with its food supply, (3) protection of wildlife areas, both large land holdings of the government and small areas, such as woodlots and fencerows, needed by certain species of birds and animals (Fig. 8), (4) research into the needs of and responses of animals to each other and their environment, (5) investment in select areas for maximum wildlife production, maintaining moderate production cheaply under existing conditions, and (6) education of the public: increasing their appreciation of wildlife (and thus the benefits which they derive from it) and also their awareness of how actions (hunting, voting, destroying food and cover, changing the environment, and contributing to research and law enforcement) influence wildlife and the larger complex of man's own environment. See LAND-USE PLANNING.

Government programs. Every state now has employees with university degrees in wildlife management. The public can gain assistance, advice, or referral to appropriate experts or agents from such people. In addition, the Federal government has professional wildlife managers in the Soil Conservation Service, the U.S. Forest Service, the U.S. Park Service, the U.S. Atomic Energy Commission, and the U.S. Fish and Wildlife Service. The last, an agency of the Department of the Interior, has extensive activities in the areas of management and enforcement, research, education, wildlife damage and depredation control, allocation of sportsmen's taxes for fish and wildlife restoration, and operation of the U.S. National Wildlife Refuge System. The Fish and Wildlife Service works with the states and even other countries to develop practices for the wisest use of the wildlife resource. The Migratory Bird Treaty Act brings Canada, Mexico, the United States, and other countries together in work to secure the protection of all migratory birds, particularly the protection of waterfowl.

Many private groups work alone or with these agencies to aid in the best use of the wildlife resource. The *Conservation Directory* (National Wildlife Federation) lists such groups as the Wildlife Society, the Wildlife Management Institute, Ducks Unlimited, and the International Association of Fish and Game Commissioners. These organizations, as well as individuals throughout the United States, attach different values to wildlife and desire different benefits. Some have short-range goals; others, long-range. Some are well educated and have abundant information about wildlife populations; others have little information but have strong emotional (or commercial) attachments to certain animals or inflexible ideas about them. See ENVIRONMENT; FISHERIES CONSERVATION; FOREST AND FORESTRY.

The problem of wildlife management in a democracy is that of resolving conflicts between groups, such as those arising from the above differences. Recently the difficulty has been formulated as an optimizing problem and thus is subject to the methods of wildland operations research.

WILDLAND OPERATIONS RESEARCH

There has emerged within the past 25 years an art called operations research (OR) based on the science of analyzing large and complex systems of all types and designing means to achieve specified goals in some optimal fashion. Wildland operations research is the application of the concepts and methods of OR to wildland problems. The term wildlands usually refers to all of the resources (space, water, soil, air, minerals, forests, and wildlife) of nonagricultural lands, such as forests, rangelands, marshes, deserts, and undeveloped parks. The differences within and between these categories may be great or indistinct (for example, differences between highly managed tree farms, farm woodlots, and wilderness forests). Every system can be considered a subsystem of a larger system. See TERRESTRIAL ECOSYSTEM.

At the root of OR is the concept of an optimum system. A system is almost anything that can be logically analyzed in the categories of inputs, processes, outputs, and feedback (Fig. 9). It is possible to conceive of machines, decisions, organs of the body, or even the entire body or populations of animals in the four components of systems. OR is concerned with analyzing, designing, or manipulating systems in order to understand them, to be able to predict their function, or to be able to manipulate them to achieve goals according to certain criteria.

Wildland operations research (WOR) deals with natural resource systems such as the ecosystem, with populations of fish or other animals, with forests and wood-dependent industries, with rangelands and marshlands, with migratory-bird flyways, and with pest species and their effects. See ECOLOGY.

WOR is a natural, evolutionary response of science to the recognition that certain problems have no simple solutions, perhaps no solutions at all, and that choices between alternatives, even

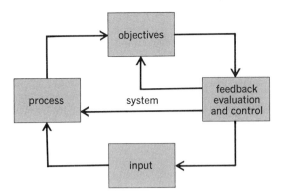

Fig. 9. Wildland systems, as well as others, can be described with the components shown here. Arrows represent the flow of thought and action.

wrong ones, may at times be necessary. The problems of natural resources are often of this type. A solution to a timber harvest problem may have profound effects that cause "vibrations" throughout the forest system. A typical large problem arises when roads must be built to get the logs from a remote timber tract to market; the local economics may be such that profit cannot be made if a high-quality road is built; if the timber is not harvested, it will be "wasted" and local mills will suffer; low-grade roads cause silt; silt fills dams, reducing their net long-term effectiveness; silt reduces stream suitability for fisheries; sport fishing influences the local economy and thus impinges on the cost-benefit decisions of the road; logging roads increase hunter access to areas as well as providing some recreational sites; roads decrease the wild character of land; logging usually increases deer food and deer populations; logging reduces gray squirrel and wild turkey populations. In the long run logging may increase air and water pollution from mills; increased numbers of people have increased needs for all types of wood and wood products.

The problem is large and complex with many decisions throughout, all requiring inputs of both information and value judgments. The problems become increasingly more complex and difficult, and each decision more risk-laden. The latter is true since each decision that is made today by wildland managers influences resources that are becoming increasingly limited; each decision influences more people; and more money or values now ride on each choice. WOR is thus an interdisciplinary scientific way of arriving more objectively at optimal solutions to large problems concerning the use of natural resources.

The ability of the computer to handle thousands of calculations rapidly, to make repetitive calculations, to modify its own instructions, and to help make decisions have made it the major tool for coping with the strategy and tactics of analysis and decision making. Among the useful methods are modeling, linear programming, dynamic programming, game theory, simulation, network analysis, and decision theory.

Modeling. To work with complex realities one's ideas must be reduced to a manageable form. Some ideas, once written, can be satisfactorily dealt with as a "word model". Some ideas cannot

be communicated adequately by words, and thus mathematical formulations are needed. These mathematical models, often equations or a series of inequalities, are ways of communicating the size and relations of certain variables. For example, it can be said that a population of game animals is the result of the interaction of the proportion of females in the previous population, the life expectancy of that population, and the average number of young produced by females in the population. Mathematically this statement can be expressed as

$$P = S \cdot e_1 \cdot m$$

where P is the population, S is the proportion of females, e_1 is the life expectancy of the first age class, and m is the birth rate. A graphical model (Fig. 10) can be developed from such expressions to provide the game manager with a tool to help him understand and make decisions for modifying the factors that influence the size and changes in his populations.

Usually mathematical models are developed to represent all known or major features of the reality of a situation. These models are needed by computers for making calculations or generating alternatives for a vast array of decisions made in wildland systems. Models can be based on theoretical relations or can be generalizations of quantities of data. Much research in the wildland sciences is devoted to collecting better data in order to build models or to explain the differences between predictions and real-world events. One exciting aspect of developing mathematical abstractions of reality is that during the process unexpected new results or insights are frequently gained. Model building still requires compromises; natural resource sytems have so many variables that only a part of them can be dealt with at any time. Some models are now accurate enough

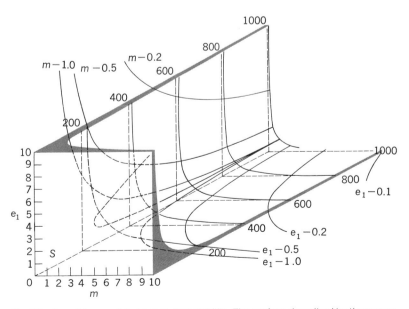

Fig. 10. A working model of population stability. The surface described by the curves represents population stability. When values of P fall on the surface, the population is stable; when they fall below the surface, a decreasing population is indicated; and when values are above the surface and not within the volume, an increasing population is indicated.

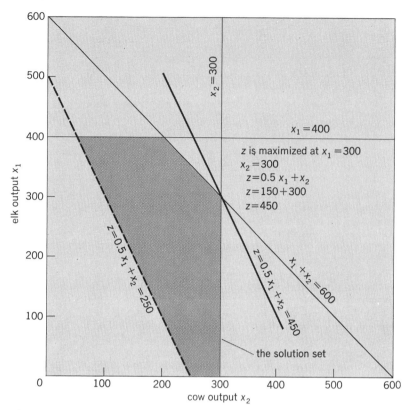

Fig. 11. A graphical solution to an elementary linear-programming problem. Within the colored area lie all possible solutions; the point $x_1 = 300$; $x_2 = 300$ is optimal and thus the optimum mix is 150 elk and 300 cows.

ical LP solution (Fig. 11) to a problem of how to achieve an optimum mix of the number of elk and cattle on a range to maximize the animal units produced. Both elk and cattle can be produced on a given unit of land. Elk yield 0.5 animal unit; cows yield 1 animal unit. There is only range enough (because of different feeding habits) to support 400 elk or 300 cows. The capacity of the land is such that not more than a total of 600 animals can be supported.

George Van Dyne, a noted practitioner of WOR, has combined linear programming with multiple regression equations. The equations he uses relate yield of forest or rangeland vegetation to soil and topographic variables, and these are used to determine optimum yields.

Efforts are now underway to perform LP for most of the decisions on several select western U.S. Forest Service Ranger Districts. The current limitations are that objective functions (goals) of complex wildland systems are difficult to specify. For an industry, the objective functions may be to maximize profit or minimize costs. The objective function for most public wildlands is to maximize the long-term total benefits from the wildlands. The benefits are so interrelated, abundant, and variable among people that the task is very difficult. Present results encourage continuance of the task. The long-term aspects of the problem are best handled by dynamic programming.

Dynamic programming. LP requires that the relations with which it deals have no exponents. Often this requirement cannot be met; over a period of time, changes are rarely constant but often curvilinear, fluctuating, or cyclic. Dynamic programming provides a technique for handling such changes. Dynamic programming carries equations to a point in time; new calculations are made based on shifting rules or criteria that are part of the program; it moves to the next point in time and makes continuing calculations until the optimal solutions are found. The technique even allows for random developments to occur and a sequence of decisions to be made.

As with LP, large dynamic-programming problems require a computer but the prospective user need not write the complex program of instructions for the computer unless he is reformulating the nature of the problem. A practitioner can concentrate on getting the problem into proper form, preparing the data and checking that there is not some peculiarity of his problem that makes any surprising solutions unrealistic.

Game theory. The possibilities for employing game theory to game populations are intriguing and have been scarcely touched. Game theory is the mathematical formulation of conflicts of interest. The types of wildland conflicts already outlined may depend on questions like: who are the players, what are the probabilities of various moves or plays of the game, what are the risks and uncertainties involved with each decision or play, what are appropriate strategies or plans of play, and what are the game payoffs or outcomes? In so-called war games, the quest is the formulation and selection of strategies to win most often in the long run with complete or incomplete information. Industries have similarly used game theory to analyze markets, to make investment decisions, and to win against their competitors. In games against

to allow predictions in which users can be very confident. Certain risks are involved in decisions based on models, but the risks are usually not as great as when no models are made clear in words, symbols, and pictures.

Linear programming. Linear programming (LP) is a technique for arriving at an optimum solution to certain problems—maximizing a desired outcome subject to constraints which limit that outcome. If these constraints can be described as linear equations or inequalities, simple problems can be handled graphically but most problems require a computer to search within the "space" described by all of the equations to find an optimal solution.

For example, given a forest with existing distributions of ages and types of trees by acreage, how can a specific timber harvest schedule be satisfied and how can the resulting forest types be distributed in an acceptable fashion within a predefined time and at the least possible cost? The implications of such a decision and its impact on the habitat of forest game animals, such as elk, deer, and wild turkey, are obvious. Wildlife management problems might be formulated: Given a region with different deer populations, each with different sex and age ratios, and each consuming food at different rates, what combination of regulations of hunting-season opening dates, season length, legal weapons, sex limits, and number of deer taken will maximize the kill while stabilizing the residual population at a level compatible with available forage?

The following simple example presents a graph-

nature, in decisions of what types of crops and how much to plant in the face of drought or insect attacks, and in flood and forest-fire loss reduction game theory deals with situations of various sorts of conflict. The players have limited control over the variables of the situation; in conflicting human interests each player may wish to maximize his selfish payoff, but all may realize that a better outcome is to allow the game to continue and not press for termination.

Games have been constructed for university wildlife management students in which the students make many decisions about the setting of hunting seasons or the manipulation of populations. They play against a computer simulation of nature, and in so doing, learn whether they win or lose without ever harming the resource. Game theory has been used to decide on the proper allocation of marshes to certain types of waterfowl-food planting, to determine suggested moves for law-enforcement agents against poachers, and for allocating equipment and personnel for standby for forest-fire fighting.

Decision theory is grossly defined and includes decision making based on sampling, statistics, and probabilities; the construction and analysis of decision trees or branching networks of related alternatives; and network analysis employing the critical path method (CPM) or project evaluation and review technique (PERT). Decision theory overlaps game theory. A decision tree is presented (Fig. 12) showing the method. A tree is a model, and when employed experimentally, it is a simulator. CPM and PERT are techniques used by management to plan and schedule events. They are used to schedule the expected progress and possible delays in programs that must meet deadlines. Wildland planting practices on wildlife refuges would be an example.

Simulation. Simulation is the activity of building a model and then performing experiments on it by asking questions such as "what if I change this factor?" Computers are often used and elaborate systems can be studied without disrupting ongoing physical operations and without hazards or great costs that may be associated with experimentation or change of the real world. Simulation allows explanation and prediction. It is the dynamic use of models but it is not an optimizing technique. It may display thousands of alternatives, the selection among which will be dependent upon some outside criteria or strategy of choice.

Ecosystems, forests, bogs, rangelands, grasslands, industries, physiological systems, populations, and even socioeconomic systems have now been satisfactorily simulated. Systems exist for simulating the elk and big-game forage production in national forests of the Pacific Northwest. Physical and computer-based models of major waterways now exist. Management practices for forests are now simulated.

When objectives are poorly defined, when variables are still being selected and studied, or when decisions are needed on problems for which data are scarce, simulation may provide the technique for evaluation through rapid computer generation of all of the alternatives so that they may be considered.

Simulation is one of the more powerful tools of WOR. Associated with this power is the surprise

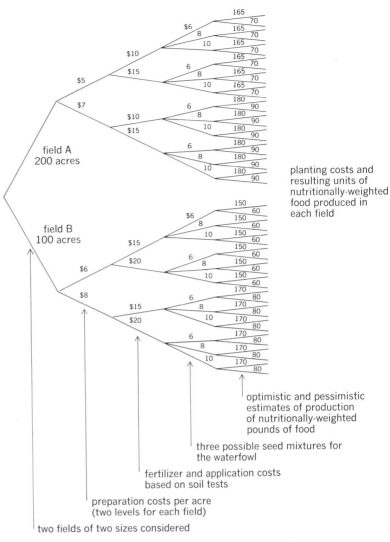

planting costs and resulting units of nutritionally-weighted food produced in each field

optimistic and pessimistic estimates of production of nutritionally-weighted pounds of food

three possible seed mixtures for the waterfowl

fertilizer and application costs based on soil tests

preparation costs per acre (two levels for each field)

two fields of two sizes considered

Fig. 12. A decision tree by which relative values can be attached to alternate strategies or methods for producing waterfowl foods.

discovery (serendipity) that is associated with both the building of models and examining of results.

Only a few universities now offer strong programs of study linking operations research and systems analysis with ecology and wildlife management. Both the analytical abilities of the wildlands scientist and manager and his manipulative and decision-making activities in nature need to be combined in predicting the consequences of various actions taken with man's environment. As an aid to making this combination, WOR will undoubtedly be crucial for designing and maintaining an environment fit for man's material and intangible needs.

[ROBERT H. GILES, JR.]

PRESERVATION OF WILDERNESS

Advocates of the preservation of wilderness have been active in the United States since the early 1900s, with Theodore Roosevelt's administration bestowing an official blessing on the concept of natural area conservation. The acknowledged leader of the contemporary wilderness movement was the forester Robert Marshall, who

A list of the locations of wilderness areas, their acreage, and date of inclusion in the Wilderness System

State areas	Acreage	Year of origin	State areas	Acreage	Year of origin
Arizona			*Nevada*		
Chiricahua (FS)	18,000	1964	Jarbidge (FS)	64,827	1964
Galiuro (FS)	52,717	1964	*New Hampshire*		
Mazatzal (FS)	205,346	1964	Great Gulf (FS)	5,552	1964
Sierra Ancha (FS)	20,850	1964	*New Jersey*		
Superstition (FS)	124,140	1964	Great Swamp National		
California			Wildlife Refuge (BSFW)	3,750	1968
Caribou (FS)	19,080	1964	*New Mexico*		
Cucamonga (FS)	9,022	1964	Gila (FS)	433,916	1964
Desolation (FS)	41,343	1969	Pecos (FS)	167,416	1964
Dome Land (FS)	62,561	1964	San Pedro Parks (FS)	41,132	1964
Hoover (FS)	42,800	1964	Wheeler Peak (FS)	6,029	1964
John Muir (FS)	504,263	1964	White Mountain (FS)	31,283	1964
Marble Mountain (FS)	214,543	1964	*Oregon*		
Minarets (FS)	109,559	1964	Diamond Peak (FS)	35,440	1964
Mokelumne (FS)	50,400	1964	Eagle Cap (FS)	221,355	1964
San Gabriel (FS)	36,137	1968	Gearhart Mountain (FS)	18,709	1964
San Gorgonio (FS)	34,718	1964	Kalmiopsis (FS)	76,900	1964
San Jacinto (FS)	21,955	1964	Mount Hood (FS)	14,160	1964
San Rafael (FS)	142,918	1968	Mount Jefferson (FS)	99,600	1968
South Warner (FS)	69,547	1964	Mount Washington (FS)	46,655	1964
Thousand Lakes (FS)	16,335	1964	Mountain Lakes (FS)	23,071	1964
Ventana (FS)	54,857	1969	Strawberry Mountain (FS)	33,653	1964
Yolla-Bolly Middle Eel	111,091	1964	Three Sisters (FS)	196,708	1964
Colorado			*Washington*		
La Garita (FS)	48,486	1964	Glacier Peak (FS)	464,741	1964
Maroon Bells – Snowmass	71,329	1964	Goat Rocks (FS)	82,680	1964
(FS)			Mount Adams (FS)	42,411	1964
Mount Zirkel (FS)	72,472	1964	Pasayten (FS)	518,000	1968
Rawah (FS)	26,674	1964	*Wyoming*		
West Elk (FS)	61,412	1964	Bridger (FS)	383,300	1964
Idaho and Montana			North Absaroka (FS)	351,104	1964
Selway-Bitterroot (FS)	1,243,669	1964	South Absaroka (FS)	483,678	1964
Minnesota			Teton (FS)	563,500	1964
Boundary Waters Canoe			Total acreage in National FS	10,174,056	
Area (FS)	1,029,257	1964	Wilderness Preservation System BSFW	3,750	
Montana				10,177,806	
Anaconda-Pintlar (FS)	159,086	1964			
Bob Marshall (FS)	950,000	1964	Total acreage reviewed by	2,789,456	
Cabinet Mountains (FS)	94,272	1964	agencies through Oct., 1969		
Gates of the Mountains (FS)	28,562	1964	Total approximate acreage	52,616,598	
			under or pending review		

wrote in 1930: "A thorough study should forthwith be undertaken to determine the probable wilderness needs of the country. Of course no precise reckoning could be attempted, but a radical calculation would be feasible. It ought to be radical for three reasons: because it is easy to convert a natural area to industrial or motor usage, impossible to do the reverse; because the population which covets wilderness recreation is rapidly enlarging; and because the higher standard of living which may be anticipated should give millions the economic power to gratify what is today merely pathetic yearning. Once the estimate is formulated, immediate steps should be taken to establish enough tracts to insure everyone who hungers for it a generous opportunity of enjoying wilderness isolation."

A wilderness system formed on the nucleus of administratively designated national forest lands, most of them high mountain country, was the dream of Marshall and his friends who created the Wilderness Society in 1935, a national organization with the goal of wilderness preservation. Since then, it has also become deeply involved with the broad complex of environmental problems brought on by the demands of technological progress and a pyramiding population.

Shaping the programs. To carry on its purpose, the Wilderness Society conducts an educational program concerning the value of wilderness in its relationship to the problems of society, the overall objective being to increase knowledge and appreciation of wilderness in the context of man's ecological perspective, as well as his spiritual and cultural needs. Wilderness, as the Society sees it, is a valuable natural resource that belongs to all, and its preservation for educational, scientific, and recreational use is part of a balanced conservation program essential to the survival of man's culture.

The founders of the Wilderness Society, knowing that the impact of civilization had already changed the face of the United States, hoped that many of the remaining Federally owned wild lands would be protected through administrative practices such as those already set up by the U.S. Forest Service to designate wild, scenic, and primitive areas and the National Park Service with its mandate from Congress to preserve and protect natural phenomena such as the Grand Canyon, Yellowstone, and Yosemite.

The work of Olaus Murie in wildlife biology helped to round out the concept of a national wilderness preservation system through consideration of refuge areas as ecosystems in which wildlife conservation purposes operated for the benefit of a whole community of species, rather than for the stocking or farming of one "target" game or fish species.

Through books, articles, and lectures, Sigurd F. Olson's interpretation of wilderness increased public awareness and appreciation of the natural scene. When he said "Wilderness to the people of America is a spiritual necessity, an antidote to the pressures of modern life and a means of regaining serenity and equilibrium," he spoke for millions, and helped lay the groundwork for congressional action.

The three decades following the creation of the Wilderness Society were punctuated by fierce struggles to withstand the encroachment of developments on Echo Park and Dinosaur National Monument (Utah), the wilderness lake country of the Quetico-Superior, the New York State Forest Preserve, and many others, demonstrating the growing concern of the public in protecting and setting aside wilderness regions.

Government regulation. Wilderness preservation was adopted as a national policy with the signing on Sept. 3, 1964, of the Wilderness Law after 8 years of public discussion and debate in Congress and across the land. In the process of its passage the attitudes of many Americans became firm on the subject of conservation in general, and groundwork was laid for the present nationwide interest in environmental quality and the close scrutiny of development activities and practices which would damage the country's diminishing resource of wild land, free-flowing water, and clean air.

Such tracts of roadless land as are already in public ownership are to be protected as wilderness upon their inclusion in the National Wilderness Preservation System, established by the act of 1964. The law defines wilderness as follows:

"A wilderness in contrast with those areas where man and his own works dominate the landscape, is hereby recognized as an area where the earth and its community of life are untrammeled by man, where man himself is a visitor who does not remain. An area of wilderness is further defined to mean . . . an area of undeveloped Federal land retaining its primeval character and influence, without permanent improvements or human habitation, which is protected and managed so as to preserve its natural conditions and which (1) generally appears to have been affected primarily by the forces of nature with the imprint of man's work substantially unnoticeable; (2) has outstanding opportunities for solitude or a primitive and unconfined type of recreation; (3) has at least five thousand acres of land or is of sufficient size as to make practicable its preservation and use in an unimpaired condition; and (4) may also contain ecological, geological, or other features of scientific, educational, scenic, or historical value."

The Wilderness Law established an initial 54 units of national forest land as wilderness areas, provided for review of roadless areas and islands within the remaining national forest primitive areas, the national parks and monuments, and the national wildlife refuges and ranges. These reviews were to be undertaken on a 10-year schedule by the agency in charge of administering the area (Forest Service of the Department of Agriculture for forests; National Park Service of the Department of the Interior for parks and monuments; Bureau of Sport Fisheries and Wildlife of the Interior Department's Fish and Wildlife Service for refuges and ranges). Other Federal lands, including wilderness within the extensive public domain under the Bureau of Land Management, are not mentioned in the law, although they could be included in the System through direct enactment by Congress at any time.

Public hearings are provided for in the Wilderness Law. These are to be held in the states where the candidate wilderness areas are located and are to be conducted by the administering agency. At these hearings local citizens and groups, as well as government officials, may comment on or suggest improvements in the agency's proposal for wilderness boundaries. A 10-year deadline, to expire Sept. 3, 1974, was established, during which time the agencies' review of roadless areas was to be completed and their proposals transmitted to the President and thence to Congress.

During the first 3-year period after passage of the Wilderness Law, the "wilderness agencies" framed their regulations for the management of the wilderness areas which would be under their jurisdictions. These were published in the Federal Register of Feb. 17 and May 31, 1966. In 1968 the first additions to the Wilderness System were made through enactments relating to five new wilderness units and an addition to one existing area (Glacier Peak); two more additions had been made by Oct. 15, 1969. Toward the end of 1969, the schedule of agency reviews and hearings had been completed on 45 wilderness areas in 18 states, while 130 more, totaling more than 40,000,000 acres, remained to be dealt with (see table).

An important feature of the Wilderness Law is its avoidance of any contradiction of the basic purposes for which the Federal lands had originally been set aside. The national forest primitive areas had been administratively protected from development and mechanized equipment since 1929. The national parks, with their many scenic treasures, had been dedicated to be preserved for the benefit and enjoyment of future generations. The wildlife conservation practices in refuges and ranges were seen as requiring in many cases an unspoiled natural environment, large enough to meet the need for a complete community of associated species. Thus the Wilderness System was set up to strengthen the authority of the wilderness administrator in resisting pressure for declassifying or downgrading the protection afforded such areas.

The all-important result of the long effort to give wilderness official recognition was public acceptance of the idea, so well expressed in the oft-quoted statement of Wallace Stegner: "Something will have gone out of us as a people if we ever let the remaining wilderness to be destroyed; if we permit the last virgin forests to be turned into comic books or plastic cigarette cases; if we drive the few remaining species into zoos or extinction; if we pollute the last clear air and dirty the last clean stream and push our paved roads through the last of the silences." See CONSERVATION OF RE-SOURCES.   [SIGURD F. OLSON]

Bibliography: D. L. Allen, *Our Wildlife Legacy*, 1962; C. W. Churchman, R. L. Ackoff, and E. L. Arnoff, *Introduction to Operations Research*, 1957; R. F. Dasmann, *Wildlife Biology*, 1964; R. H. Giles (ed.), *Wildlife Management Techniques*, Wildlife Society, 1969; J. B. Trefethen, *Crusade for Wildlife*, 1961; G. M. Van Dyne, Application and integration of multiple linear regression and linear programming in renewable resource analysis, *J. Range Manage.* 19(6):356–362, 1966; K. E. F. Watt, *Systems Analysis in Ecology*, 1966; K. E. F. Watt, *Ecology and Resource Management*.

Wind

The motion of air relative to the Earth's surface. The term usually refers to horizontal air motion, as distinguished from vertical motion, and to air motion averaged over a chosen period of 1–3 min. Micrometeorological circulations (air motion over periods of the order of a few seconds) and others small enough in extent to be obscured by this averaging are thereby eliminated. The choice of the 1- to 3-min interval has proven suitable for the study of (1) the hour-to-hour and day-to-day changes in the atmospheric circulation pattern; and (2) the larger-scale aspects of the atmospheric general circulation.

The direct effects of wind near the surface of the Earth are manifested by soil erosion, the character of vegetation, damage to structures, and the production of waves on water surfaces. At higher levels wind directly affects aircraft, missile and rocket operations, and dispersion of industrial pollutants, radioactive products of nuclear explosions, dust, volcanic debris, and other material. Directly or indirectly, wind is responsible for the production and transport of clouds and precipitation and for the transport of cold and warm air masses from one region to another. *See* ATMOSPHERIC GENERAL CIRCULATION.

Cyclonic and anticyclonic circulation. Each is a portion of the pattern of airflow within which the streamlines (which indicate the pattern of wind direction at any instant) are curved so as to indicate rotation of air about some central point of the cyclone or anticyclone. The rotation is considered cyclonic if it is in the same sense as the rotation of the surface of the Earth about the local vertical, and is considered anticyclonic if in the opposite sense. Thus, in a cyclonic circulation, the streamlines indicate counterclockwise (clockwise for anticyclonic) rotation of air about a central point on the Northern Hemisphere or clockwise (counterclockwise for anticyclonic) rotation about a point on the Southern Hemisphere. When the streamlines close completely about the central point, the pattern is denoted respectively a cyclone or an anticyclone. Since the gradient wind represents a good approximation to the actual wind, the center of a cyclone tends strongly to be a point of minimum atmospheric pressure on a horizontal surface. Thus the terms cyclone, low-pressure area, or "low" are often used to denote essentially the same phenomenon. In accord with the requirements of the gradient wind relationship, the center of an anticyclone tends to coincide with a point of maximum pressure on a horizontal surface, and the terms anticyclone, high-pres-

sure area, or "high" are often used interchangeably.

Cyclones and anticyclones are numerous in the lower troposphere at all latitudes. At higher levels the occurrence of cyclones and anticyclones tends to be restricted to subpolar and subtropical latitudes, respectively. In middle latitudes the flow aloft is mainly westerly, but the streamlines exhibit wavelike oscillations connecting adjacent regions of anticyclonic circulation (ridges) and of cyclonic circulation (troughs).

Although the atmosphere is never in a completely undisturbed state, it is customary to refer to cyclonic and anticyclonic circulations specifically as atmospheric disturbances. Cyclones, anticyclones, ridges, and troughs are intimately associated with the production and transport of clouds and precipitation, and hence convey a connotation of disturbed meteorological conditions.

A more rigorous definition of circulation is often employed, in which the circulation over an arbitrary area bounded by the closed curve S is given by Eq. (1), where the integration is taken complete-

$$C = \oint v_t \, dS \tag{1}$$

ly around the boundary of the area. Here v refers to the wind at a point on the boundary, the subscript t denotes the component of this wind parallel to the boundary, and dS is a line element of the boundary. The component v_t is considered positive or negative according to whether it represents cyclonic or anticyclonic circulation along the boundary S. In this context, the circulation may be positive (cyclonic) or negative (anticyclonic) even when the streamlines within the area are straight, since the distribution of wind speed affects the value of C. *See* ATMOSPHERE; CLOUD.

Convergent or divergent patterns. These are said to occur in areas in which the (horizontal) wind flow and distribution of air density is such as to produce a net accumulation or depletion, respectively, of mass of air. Rigorously, the mean horizontal mass divergence over an arbitrary area A bounded by the closed curve S is given by Eq. (2),

$$D = \frac{1}{A} \oint \rho v_n \, dS \tag{2}$$

where the integration is taken completely around the boundary of the area. Here ρ is the density of air, v refers to the wind at a point on the boundary, the subscript n denotes the component of this wind perpendicular to the boundary, and dS is an element of the boundary. The component v_n is taken positive when it is directed outward across the boundary and negative when it is directed inward. Convergence is thus synonymous with negative divergence. If spatial variations of density are neglected, the analogous concept of velocity divergence and convergence applies.

The horizontal mass divergence or convergence is intimately related to the vertical component of motion. For example, since local temporal rates of change of air density are relatively small, there must be a net vertical export of mass from a volume in which horizontal mass convergence is taking place. Only thus can the total mass of air within the volume remain approximately constant. In particular, if the lower surface of this volume coincides with a level ground surface, upward motion must occur across the upper surface of this volume. Similarly, there must be downward motion immediately above such a region of horizontal mass divergence.

The horizontal mass divergence or convergence is closely related to the circulation. In a convergent wind pattern the circulation of the air tends to become more cyclonic: in a divergent wind pattern the circulation of the air tends to become more anticyclonic.

Regions which lie in the path of an approaching cyclone are characterized by a convergent wind pattern in the lower troposphere and by upward vertical motion throughout most of the troposphere. Since the upward motion tends to produce condensation of water vapor in the rising air current, abundant cloudiness and precipitation typically occur in this region. Conversely, the area in advance of an anticyclone is characterized by a divergent wind pattern in the lower troposphere and by downward vertical motion throughout most of the troposphere. In such a region, clouds and precipitation tend to be scarce or entirely lacking.

A convergent surface wind field is typical of fronts. As the warm and cold currents impinge at the front, the warm air tends to rise over the cold air, producing the typical frontal band of cloudiness and precipitation. *See* FRONT.

Zonal surface winds. Such patterns result from a longitudinal averaging of the surface circulation. This averaging typically reveals a zone of weak variable winds near the Equator (the doldrums) flanked by northeasterly trade winds in the Northern Hemisphere and southeasterly trade winds in the Southern Hemisphere, extending poleward in each instance to about latitude 30°. The doldrum belt, particularly at places and times at which it is so narrow that the trade winds from the two hemispheres impinge upon it quite sharply, is designated the intertropical convergence zone, or ITC. The resulting convergent wind field is associated with abundant cloudiness and locally heavy rainfall. A westerly average of zonal surface winds prevails poleward of the trade wind belts and dominates the middle latitudes of both hemispheres. The westerlies are separated from the trade winds by the subtropical high-pressure belt, which occurs between latitudes 30 and 35° (the horse latitudes), and are bounded on the poleward side in each hemisphere between latitudes 55 and 60° by the subpolar trough of low pressure. Numerous cyclones and anticyclones progress eastward in the zone of prevailing westerlies, producing the abrupt day-to-day changes of wind, temperature, and weather which typify these regions. Poleward of the subpolar low-pressure troughs, polar easterlies are observed.

The position and intensity of the zonal surface wind systems vary systematically from season to season and irregularly from week to week. In general the systems are most intense and are displaced toward the Equator in a given hemisphere during winter. In this season the subtropical easterlies and prevailing westerlies attain mean speeds of about 15 knots, while the polar easterlies are somewhat weaker. In summer the systems are displaced toward the pole by 5 to 10° of latitude and weaken to about one-half their winter strength.

When the pattern of wind circulation is averaged with respect to time instead of longitude,

striking differences between the Northern and Southern Hemispheres are found. On the Southern Hemisphere, variations from longitude to longitude are relatively small and the averaged pattern is described quite well in terms of the zonal surface wind belts. On the Northern Hemisphere there are large differences from longitude to longitude. In winter, for example, the subpolar trough is mainly manifested in two prominent low centers, the Icelandic low and the Aleutian low. The subtropical ridge line is drawn northward in effect over the continents and is seen as a powerful and extensive high-pressure area over Asia and as a relatively weak area of high pressure over North America. In summer the Aleutian and Icelandic lows are weak or entirely absent, while extensive areas of low pressure over the southern portions of Asia and western North America interrupt the subtropical high-pressure belt. *See* CLIMATOLOGY.

Upper air circulation. Longitudinal averaging indicates a predominance of westerly winds. These westerlies typically increase with elevation and culminate in the average jet stream, which is found in lower middle latitudes near the tropopause at elevations between 35,000 and 40,000 ft. The subtropical ridge line aloft is found equatorward of its surface counterpart and easterlies occur at upper levels over the equatorward portions of the trade wind belts. In high latitudes, weak westerlies aloft are found over the surface polar easterlies. Seasonal and irregular fluctuations of the circulation aloft are similar to those which characterize the surface winds. *See* JET STREAM.

Minor terrestrial winds. In this category are circulations of relatively small scale, attributable indirectly to the character of the Earth's surface. One example, the land and sea breeze, is a circulation driven by pronounced heating or cooling of a given area in comparison with little heating or cooling in a horizontally adjacent area. During the day, air rises over the strongly heated land and is replaced by a horizontal breeze from the relatively cool sea. At night, air sinks over the cool land and spreads out over the now relatively warm sea.

Another example is formed by the mountain and valley winds. These result from cooling and heating, respectively, of the mountain slopes relative to the horizontally adjacent free air above the valley floor. During the day, air flows up from the valley along the strongly heated mountain slopes, but at night, air flows down the relatively cold mountain slopes toward the valley bottom. A similar type of descending current of cooled air is often observed along the sloping surface of a glacier. This nighttime air drainage, under proper topographical circumstances, can lead to the accumulation of a pool of extremely cold air in nearby valley bottoms.

Local winds. These commonly represent modifications by local topography of a circulation of large scale. They are often capricious and violent in nature and are sometimes characterized by extremely low relative humidity. Examples are the mistral which blows down the Rhone Valley in the south of France, the bora which blows down the gorges leading to the coast of the Adriatic Sea, the foehn winds which blow down the Alpine valleys, the williwaws which are characteristic of the fiords of the Alaskan coast and the Aleutian Islands, and the chinook which is observed on the eastern slopes of the Rocky Mountains. Local names are also given in some instances to currents of somewhat larger scale which are less directly related to topography. Examples of this type of wind are the norther, which represents the rapid flow of cold air from Canada down the plains east of the Rockies and along the east coast of Mexico into Central America; the nor'easter of New England, which is part of the wind circulation about intense cyclones centered offshore along the Middle Atlantic coastal states; and the sirocco, a southerly wind current from the Sahara which is common on the coast of North Africa and sometimes crosses the Mediterranean Sea.

[FREDERICK SANDERS]

Zoogeographic region

A major unit of the Earth's surface characterized by faunal homogeneity. This definition is more of a concept than a reality. Because different classes of animals arose and dispersed at different times during Earth history and because animals vary greatly in their vagilities and tolerances to environmental conditions, no two major groups of animals display complete coincidence in their geographic limits. As a result, delimitation of the several regions is difficult and such patterns as have been defined are not universally applicable to the entire animal kingdom.

Fortunately, insofar as existing animals are concerned, there appears to be a rough average geographic pattern. This pattern is best mirrored in the distribution of birds and mammals. Considerable coincidence in the geographic patterns of these two groups has possibly resulted from the fact that they arose and dispersed relatively recently in geologic time, and, as a result, their extant distributions have been controlled by major geographic barriers of the not too distant past. These patterns reflect to a lesser degree the distributions of other major groups. It appears, therefore, that several parts of the Earth's surface have, through isolation, served as major centers of evolution, and in the final analysis, it is these centers that are recognized as zoogeographic regions. Broad zones of transition have developed between the major zoogeographic regions owing either to the ability of some animals to transcend major barriers or to the disappearance of the barriers. The occurrence of such transitional areas has led to most of the difficulties in plotting zoogeographic boundaries.

Classification. Although earlier attempts had been made to partition the Earth into zoogeographic regions, it was in 1858 that P. Sclater, on the basis of bird distribution, presented the first practical regional classification. This was extended by A. Wallace in 1876 and has since undergone some further modification, but the pattern described by Sclater has remained basically sound. The table correlates the classical Sclater-Wallace system with K. Schmidt's 1954 review of the problem.

Although the Palearctic and Nearctic regions share many faunal elements, particularly insofar as eastern Asia and eastern North America are concerned, they do display important faunal differences. The tropical elements of the two regions, especially, show considerable divergence. That of Palearctica has been derived from the Ethiopian and Oriental regions, whereas the Nearctic tropical element has stemmed, for the most part, from Neotropica. Thus, although the two northern regions are frequently combined as Holarctica in order to express their many resemblances, they are more generally treated individually.

Palearctic region. This region includes all Eurasia north of the Tropic of Cancer, excepting parts of the Arabian Peninsula and northwestern India. Both the East (China) and the West (Europe) originally supported deciduous forest. To the North, the boreal forest and the tundra are circumpolar. The vast interior of Asia is largely grasslands with considerable desert on the high plateaus.

The Palearctic fauna is difficult to characterize. As a generalization, it is composed of three elements. One is worldwide or almost so, the second is more or less circumpolar, and the third has been derived from the Old World tropics. At the family and subfamily levels there are few exclusives. Furthermore, the extensive longitudinal spread of the region that leaves the temperate forests of East and West disjunct, because of the subhumid lands of the interior, produces a number of east-west faunal differences.

Fishes. The fish fauna is dominated by the widely distributed cyprinids or minnows. With North America, the Palearctic shares the percids or perches, paddlefish, and suckers. The last two are restricted to the Asian area of the Palearctic. The cobitids (loaches) and a few catfishes are shared with the Old World tropics.

Amphibians. Among the amphibians the hynobiid salamanders are endemic to eastern Asia. Newts are Holarctic in their distribution. The plethodontid salamander of Europe is a single representative of a large, New World group, whereas, in contrast, the hellbender of China is related to that of eastern North America. The tailless amphibians are largely the widely distributed bufonids or toads, and the hylid and ranid frogs.

Reptiles. The reptilian fauna is poor. The emydid turtles and the veranid and agamid lizards have all been derived from the South. Among the snakes, the majority belong to the harmless and almost worldwide colubrids. True vipers in the West have been derived from the South, and the pit vipers of the East are shared with the New World.

Birds. The Palearctic avifauna reflects the general faunal picture outlined above. A more or less worldwide element is represented by the hawks, woodpeckers, swallows, and finches. Circumpolar groups include the grouse, waxwings, and creepers. Tropical elements, largely migratory, include

Zoogeographic regions

Schmidt (1954)			Classical system regions
Realms	Regions	Subregions	
Arctogaean	Holarctic	Arctic	Nearctic-Palearctic
		Nearctic	Nearctic
		Caribbean	
		Palearctic	Palearctic
	Paleotropical	Oriental	Oriental
		Ethiopian	Ethiopian
		Malagasy	
Neogaean	Neotropical	Neotropical	Neotropical
Notogaean	Australian	Australian	
		Papuan	Australian
	Oceanian	New Zealandian	
		Oceanic	
		Antarctic	Unclassified

the Old World flycatchers, larks, and starlings. Such Oriental derivatives as pheasants, cuckoos, shrikes, and white-eyes occur only in eastern Asia. Only a single family, the hedge sparrows, is endemic to the region.

Mammals. The mammalian fauna resembles the avifauna in its geographic relationships. Worldwide (except for Australia) groups include rabbits, cricetid mice and rats, squirrels, and cats. Circumpolar types include the beaver, jumping mice, and the pikas. The murids (Old World mice and rats), the dormice (also African), and the panda (restricted to eastern Asia) are essentially Old World groups.

Nearctica. Nearctica includes all of North America north of the edge of the Mexican Plateau. The vegetation pattern is similar to that of Palearctica. Across the North, the boreal forest and the tundra are both transcontinental. The entire East and the coastal fringe of the West support temperate forest cover. The central region, although bisected by the forested Rocky Mountains, is largely grasslands and deserts.

In general, the faunal picture is also similar to that of Palearctica. There is considerable east-west diversity and a mixture of worldwide, tropical, mostly from South America, and holarctic groups. Endemism is, however, better marked.

Fishes. The fish fauna is richest east of the Mississippi River. Cyprinids are abundant. Holarctic or at least Nearctic and Asian groups include the suckers, paddlefishes, and perches. A few South American elements, notably characins and cichlids, extend up to the Rio Grande River. Exclusive groups are represented by the bowfin, mooneyes, trout, and basses.

Amphibians. Among the amphibians, the salamanders, especially the plethodontids, are particularly characteristic. Most tailless amphibians, like those of Palearctica, are divided among the widely distributed bufonids, the hylid tree frogs, and the ranids. A rather interesting, primitive frog, *Ascaphus*, is known only from the high mountains of the western coast. It belongs to a group that is otherwise restricted to a small island off the coast of New Zealand.

Reptiles. The reptile fauna includes such almost worldwide groups as the gekkonid and scincid lizards, and the colubrid snakes. It shares with Eurasia the emydid turtles, the anguid lizards, and the pit vipers. More restricted in their distributions (mainly in South America) are the coral snakes, the teiid lizards, the iguanid lizards (also on Fiji and in Madagascar), and the peculiar, poisonous gila monster.

Birds. For the most part, the avifauna comprises worldwide and Holarctic groups and a number of migratory neotropical elements. Ducks, pigeons, hawks, kingfishers, swallows, and finches are all widely distributed. Holarctic groups include grouse, creepers, and waxwings. Strictly New World elements are represented by the New World flycatchers, vireos, orioles, and humming-birds. The turkey is almost endemic, extending as far south as northern Central America.

Mammals. Of the mammals, the cats, bovids, rabbits, cricetid mice, and squirrels are wide-ranging. Mammals more typical of Holarctica include the jumping mice, microtine mice, and the beaver. Two exclusives occur in the region, the pronghorn of the western deserts and the mountain beaver of the Northwest.

Ethiopian region. In the modern parlance, this region refers to Africa south of about the mid-Sahara. Some zoogeographers also include the Arabian Peninsula, which with Mediterranean Africa is considered by others to be a transition province. The region is entirely tropical, except for the subtropics of the extreme south, although temperatures are moderated by the continent's upland nature. Tropical forests fringe the Gulf of Guinea and extend inland through the Congo Basin and occur also in the eastern mountains. Deserts include the Sahara, the Kalahari, and the Somaliland coast. Most of the remainder of the continent supports scrub forest and grasslands.

Faunally, the Ethiopian region is most closely related to the Oriental region. Nevertheless, it shares a number of groups with Holarctica, whereas South American relationships are evident in its fish and turtle faunas.

Fishes. Among the fishes, the widely distributed cyprinids are abundant. The lungfish has South American affinities, as do a wealth of characins and the cichlids which have undergone extensive differentiation in the East African lakes. The primitive birchirs as well as several groups of primitive bony fishes are exclusives.

Amphibians. The amphibian fauna is not particularly remarkable. It includes the widely distributed caecilians, bufonids, and ranids. It shares the primitive pipids with South America. The true tree frogs are lacking, but their niche is filled by the rhacaphorids, which are also Oriental; some authorities consider the African representatives of this group to be a separate family restricted to the continent. A family of frogs related to the worldwide narrow-mouth frogs is exclusively African.

Reptiles. Although it includes such widely distributed groups as the skinks among the lizards and many harmless colubrid snakes, the reptile fauna is essentially pantropical. Notable examples of these are the side-necked turtles, crocodiles, gekkos, and worm snakes. Oriental affinities are evident in the chameleons, an egg-eating snake, the cobras, and the true vipers.

Birds. The avifauna, although large, is represented by few endemics above the generic level. The bulk of the avifauna is composed of widely distributed groups such as the owls and hawks, kingfishers, swallows, and true finches. The region has much in common with both Europe (Old World flycatchers, starlings, and orioles) and the Orient (broadbills, hornbills, and honey guides). Exclusives include the ostrich, the secretary bird, certain genera of guinea fowl, widow birds, and tick birds.

Mammals. The spectacular big game, such as the elephant, rhinoceros, hippopotamus, a variety of antelopes, horses (zebras), cats, and giraffes, hardly serve to bring out the true nature of the mammalian fauna. The region supports such widely distributed groups as mustelids, rabbits, and many rodents. Oriental affinities are indicated by rhinoceroses, Old World monkeys, great apes, scaly anteaters, and bamboo rats. Africa also possesses a number of endemic or near-endemic mammals of which the golden mole, elephant shrews, hyraxes, and the giraffe are best known.

The fauna of Madagascar and the islands of the Indian Ocean presents a number of interesting problems that are not likely to be settled in the immediate future. Madagascar supports a fauna that has much in common with Africa—side-necked turtles, rhacophorid frogs, chameleons, and many birds and bats. In contrast, many mainland groups are absent, such as primary fresh-water fishes, pythons and cobras, the ostrich and tickbirds, hystricomorph rodents, Old World monkeys, and great apes. The island possesses a number of endemics including a group of narrow-mouthed frogs, some rollers and vangas among the birds, and a variety of lemurs. A few Oriental relationships are also apparent.

The Seychelles and Mascarene Island groups of the Indian Ocean support small but interesting faunas. The flightless birds of the Mascarenes, especially the now extinct dodo, are particularly noteworthy.

Oriental region. This region encompasses tropical Asia from the Iranian Peninsula eastward through the East Indies to and including Borneo and the Philippines. Its exact boundaries, however, are difficult to define because of broad areas of transition between it and adjacent regions. Aside from the Thar Desert and isolated areas of semidesert, especially along the eastern coast of India, the region originally supported forest growth.

The region appears to have served both as a center of dispersal for cold-blooded vertebrates and as a crossroads through which various other groups have passed. Although its fauna bears much in common with the Ethiopian region, it shows both Palearctic and Australian affinities.

Fishes. Although ancient groups of fishes that are so characteristic of tropical regions are absent, its fish fauna is, nevertheless, large and diversified. Cypriniform fishes, though widely distributed, are especially characteristic of the region, and one group, the loaches, is almost exclusively Oriental. Several families of catfishes are endemic to it, and still others of this same group are shared with Africa. The climbing perch make up another distinctive group and further emphasize African affinities.

Amphibians. The amphibian fauna is large. Caecilians are present, although the salamanders barely enter the region. Among the tailless amphibians are the widely distributed bufonids, ranids, and narrow-mouthed frogs. It has many rhacophorid frogs which further attest to African faunal affinities.

Reptiles. The reptilian fauna is largely shared with other regions. Widely distributed groups include the skinks and harmless colubrid snakes, whereas such pantropical groups as the gekkos, worm snakes, and pythons are all well represented. It shares chameleons, egg-eating snakes, and the cobras with Africa, while the agamid lizards and true vipers are widely distributed through the Old World. The pit vipers are shared with the New World. The Oriental region also supports several small endemic families of snakes, and the surrounding seas are well populated with sea snakes.

Birds. The avifauna, like that of Africa with which it shares many groups, is poor in endemics above the generic level. Widely distributed groups such as woodpeckers, pigeons, and jays make up the bulk of the fauna, and pheasants are particularly characteristic of the region. A few Australian groups such as the frogmouths and wood swallows enter the Orient from the southeast. The region boasts only a single exclusive family, the fairy bluebirds.

Mammals. The mammal fauna, although represented by such wide-ranging groups as weasels, rabbits, and squirrels, includes a number of animals shared with Africa as well as a good representation of endemics. African affinities are evident in the Old World monkeys, the scaly anteaters, the rhinoceros, the elephant, and the fruit bats. Among the endemics may be mentioned the flying lemurs, tree shrews, and tarsiers.

The island archipelago to the east and southeast, which includes Sumatra, Java, and the Philippines, supports a vertebrate fauna that is very definitely Oriental in character. As is generally true of archipelagoes, the geographic patterns of various animal groups form a complex mosaic that is accompanied by depauperization, which is particularly evident in the northernmost islands of the Philippines.

Neotropical region. This area includes Mexico south of the Mexican Plateau, the West Indies, Central America, and South America. The first three are frequently treated as a transition zone between Nearctica and Neotropica.

Although about two-thirds of the region lies within the tropics, its plateau and montane character is responsible for considerable areas with non-tropical temperatures. Generally speaking, the Amazon Basin and the eastern coasts support high rainforests. Scrub forests and grasslands clothe much of the interior of South America, whereas desert occupies much of the continent's western coast and Patagonia. The Andes and the Central American mountain systems provide a variety of vegetation belts.

Owing to a history of isolation through much of the Cenozoic Era, the South American fauna is composed of two very distinct elements. One is of considerable age with some pantropical groups. The other is a younger element that has invaded the region relatively recently from North America.

Fishes. The fish fauna, although represented by only a few major groups, is, nevertheless, very rich and with many endemics at the family level. Representatives of more ancient types include the pantropical osteoglossids and a lungfish which is shared with Africa. Other African affinities are evidenced by the characins, cichlids, and nandids. The widely distributed cyprinids are absent, and a number of other northern types barely enter Central America. Among the more notable endemics are the gymnotid (electric) eels and many families of catfishes.

Amphibians. The amphibian fauna includes a wealth of the widely distributed hylids and leptodactylids (these are especially abundant in Australia) and such pantropical or near-pantropical groups as the caecilians and brachycephalid frogs. With Africa it shares the pipid frogs and with the Orient the narrow-mouthed frogs. The wide-ranging ranids and salamanders are poorly represented. The region supports only a single exclusive frog family, the primitive *Rhinophrynus*, which is restricted to southern Mexico.

Reptiles. The reptile fauna includes skinks and harmless colubrid snakes among the near-cosmopolitan groups. Pantropical representatives include the side-necked turtles, some related to those of Africa and others to those of Australia, gekkos, worm snakes, and the coral snake family. It shares the pit vipers with Nearctica and the Orient. Other groups confined to the New World include the many iguanid lizards, also on Madagascar and one on Fiji, and the teiids. Endemic reptiles include several families of turtles that are restricted to northern Central America, the caimans, and the boas.

Birds. The rich avifauna has led to the designation of South America as the bird continent. Although many widely distributed groups such as herons, ducks, hawks, parrots, trogons, and thrushes are well represented, about 50% of the Neotropical families are endemic. About one-third of the bird species belong to the exclusive furnaroid group, antbirds, ovenbirds, and woodhewers, whereas the wood warblers, hummingbirds, and flycatchers, although shared to a greater or lesser extent with North America, make up most of the remaining endemic species. Other exclusives include the flightless rheas, tinamous, toucans, and cotingas.

Mammals. The mammal fauna further emphasizes South America's history of isolation. The list of endemics is long and includes several groups of marsupials, sloths, New World monkeys, and most of the hystricomorph families. With the Old World it shares the camels (llama) and tapir, and with North America the armadillos, peccaries, and pocket gophers. More wide-ranging groups include weasels, cats, squirrels, cricetid mice and rats, and some bats.

Central America and lowland Mexico is an area of faunal overlap between the Nearctic and Neotropical elements. The region is a pathway of considerable environmental diversity, and it has been utilized by northern groups dispersing southward and by southern groups dispersing northward, both to varying degrees. As a result a complex zoogeographic mosaic, still largely unstudied, obtains throughout the region.

The West Indies support a depauperate fauna of both Nearctic and Neotropical affinities. The representation of the various groups through the archipelago appears to be in direct proportion to their ability to cross water barriers. Although many schemes have been presented to explain the populating of the islands, most evidence indicates that most groups were transported across water barriers from Central America.

Australian region. Included in this region are continental Australia, New Guinea, Tasmania, and lesser islands through the Solomon group. The boundary between it and the Oriental region has long been debated. It was originally placed by Wallace (Wallace's Line) well to the West between Bali and Lombok, between the Celebes and Borneo, and south of the Philippines. The easternmost boundary now generally accepted lies just west of Aru and New Guinea. Modern zoogeographers recognize that the fauna of the archipelago between the two extremes is transitional. Schmidt

refers to it as the Celebesian transition subregion, the Wallacea of many authors.

Continental Australia possesses both tropical and subtropical climates. Probably no less than 75% of its area supports desert or grassland cover. Forests and forested grasslands are restricted to the northern, eastern, and southwestern fringes of the continent. Eucalyptus forest, almost endemic to Australia, predominates in the southeast. New Guinea has a cover of considerable rainforest and wet mountain forest.

As a result of its long history of isolation, the region has a fauna that includes some very ancient endemic groups. These have survived here and in South America and have undergone considerable adaptive radiation. In addition, Australia also possesses a more recent element of Oriental affinities.

Fishes. The fish fauna is poor. Aside from an osteoglossid, an ancient group shared with South America, Africa, and southeastern Asia, and an exclusive family of lungfish, only distantly related to that of South America and Africa, the fishes of Australia are salt-tolerant and of wide distributions.

Amphibians. Of the amphibians, only the frogs are represented and these by only four families. In Australia proper the leptodactylids and hylids, both almost cosmopolitan, make up about 90% of the amphibian fauna. New Guinea, in contrast, supports a wealth of the narrow-mouthed frogs.

Reptiles. Reptiles are well represented throughout the region. Pantropical groups include crocodiles, side-necked turtles, and gekkos, whereas skinks and colubrid snakes are even more wideranging. General Old World affinities are expressed by the agamid and veranid lizards and the pythons. Among the poisonous snakes the region possesses only the elapids. A single reptilian family, the pygopod lizards, is endemic.

Birds. The avifauna, although lacking many widely distributed families such as pheasants, woodpeckers, and finches, is represented by such near-cosmopolitan groups as the pigeons, kingfishers, and parrots. Endemics include the spectacular emu and cassowary (both flightless), the bird of paradise (especially characteristic of New Guinea), and the strictly Australian scrubbirds and lyrebirds.

Mammals. The mammalian fauna of the region is spectacular in the adaptive radiation displayed by the marsupials. Although not exclusive, they constitute the bulk of the mammal fauna. These and the monotremes testify to the long-continued isolation of Australia proper. Bats and the murid mice and rats are the only other mammalian groups native to the region. The dingo (wild dog) and the rabbit are introductions.

To the east of Australia and New Guinea, the faunas of the islands of the Pacific (Oceanic region of some zoogeographers) suffer gradual impoverishment. New Zealand to the southeast lacks strictly fresh-water fishes, turtles, snakes, and mammals, aside from bats and a few introduced species such as the deer. The islands have served as a refuge for such ancient groups as the lizardlike tuatara *Sphenodon*, a tailed frog which is shared with northwestern North America, and the moas and kiwis.

Toward the northeast strictly fresh-water fishes

do not extend beyond New Guinea, terrestrial mammals and frogs reach only to the Solomons, snakes extend to Fiji, and lizards fall just short of Samoa. The Hawaiian Islands, aside from introduced forms, include among their vertebrate fauna only a single bat and a variety of endemic birds of New World affinities. *See* BIOGEOGRAPHY; BIOTIC ISOLATION; ISLAND FAUNAS AND FLORAS; PLANT GEOGRAPHY.   [L. C. STUART]

Bibliography: P. J. Darlington, *Zoogeography*, 1957; K. P. Schmidt, Faunal realms, regions, and provinces, *Quart. Rev. Biol.*, 29(4):322–331, 1954; P. L. Sclater, On the general geographical distribution of the members of the class Aves, *J. Proc. Linnaean Soc. (Zool.)*, 2:130–145, 1858; A. R. Wallace, *The Geographical Distribution of Animals*, 2 vols., 1876.

Zoogeography

The science that attempts to describe and explain the distribution of animals through space and time. To accomplish this, zoogeographers inventory and analyze two sets of factors, those intrinsic to the organism (genetic), and those extrinsic to the organism (environmental). Thus the data of zoogeography are drawn from such varied fields as animal morphology, physiology and systematics, botany, paleontology, physical geography, and stratigraphy. Zoogeography, therefore, is frequently viewed as a borderline science. Nevertheless, it attains distinction as a special field of study through synthesis and integration.

Subject area. The limits of zoogeography are difficult to define. Technically, any study of animal distribution is by fiat zoogeography. During the 20th century, however, investigations into a number of special phases of animal distribution have given rise to fields of study which have attained the status of sciences in their own rights. Among these may be mentioned especially ecology and limnology. The dividing line between zoogeography and ecology, for example, is frequently merely one of scale. Zoogeography operates at the global and continental levels and deals essentially with higher systematic categories (genus and above). In contrast, ecology is concerned with more local areas and trivial systematic categories (species, subspecies, or single populations thereof). *See* ECOLOGY; LIMNOLOGY.

Objectives. The aims of zoogeography vary depending upon the interests of the investigator. One individual may seek to partition the Earth into zoogeographic regions; another may essay an explanation of a geographic pattern displayed by some particular group of animals; and a third may attempt to show how some geographic pattern is sustained in terms of faunal adaptations. Zoogeography, then, may have many aims, each characterized by a particular approach. These approaches have been examined by W. Allee and K. Schmidt, and viewed collectively they express the essential aims of the subject. Though not in complete accord with Allee and Schmidt, the following discussion serves to summarize the nature of the ends that modern zoogeographers hope to attain through the employment of the several approaches to the subject.

Systematic approach. The systematic approach takes as its point of departure a systematic group, generally some unit between genus and class. The problem is to delimit such a group, to analyze its makeup, and to describe its distribution. The assembling of these data is the chore of the systematist, and museum collections provide him with his essential materials. The resultant data may be interpreted in several ways. The investigator may undertake an explanation of the distribution of the group in terms of Earth history and the evolution of the group as revealed by the paleontological record and the extant genetic relationships of its components and near relatives (historical interpretation). Or, in contrast, he may analyze the distribution in terms of the morphological or physiological qualities or both that have permitted the group to radiate adaptively into the several environments that may be circumscribed by its range (ecological interpretation). D. Amadon's monograph of the Hawaiian honeycreepers is a superb example of a systematic study interpreted along both historical and ecological lines.

It was the systematic approach that engendered one of the most violent quarrels ever witnessed by zoogeography. Zoogeographers early recognized that certain more specialized groups of animals, like placental mammals, though occurring on the southern continents, show continuity in their distributions through boreal regions. In contrast, many more primitive groups such as the lungfishes and side-necked turtles occur only on the southern continents. To explain discontinuities in the distributions of these more primitive groups, some zoogeographers were led to postulate ancient, transoceanic land bridges connecting the southern continents. These postulates were contrary to geological facts and were not acceptable to more conservative zoogeographers. The problem was finally resolved by W. D. Matthew, who showed that the vertebrates, at least, had evolved in the north, dispersed southward, and, in the case of many primitive groups, had been exterminated in the north. This hypothesis, borne out by the paleontological record, is now generally accepted.

Regional approach. The regional approach seeks to partition the Earth into regions of various scales on the basis of animal resemblances. In general there are two types of zoogeographic regions, faunal and ecological. A faunal region is characterized by an assemblage of animals whose ranges show a high degree of accordance, in which endemism is generally considerable, and from which certain groups of animals are frequently absent. An ecological region is characterized by an assemblage of animals displaying homologous or analogous adaptive responses to a particular set of environmental conditions.

Throughout most of the last half of the 19th century students of zoogeography devoted most of their energies to the partitioning of the globe into faunal regions (see illustration). At the global level the initial practical divisioning of the Earth followed P. L. Sclater's system. Quibbling over the comparative ranks of regions, values of indicator groups, and the positions of boundaries followed. These investigations have been summarized by Schmidt. At the continental level North America in particular was subjected to the scrutiny of zoogeographers seeking a basic pattern of animal distribution. Early attempts between 1870 and

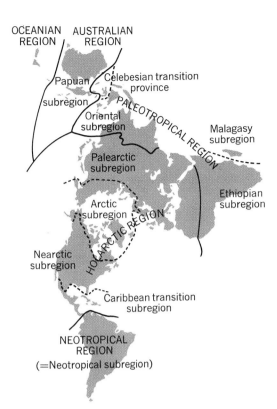

OCEANIAN REGION

AUSTRALIAN REGION

Papuan subregion

Celebesian transition province

PALEOTROPICAL REGION

Oriental subregion

Malagasy subregion

Palearctic subregion

Arctic subregion

HOLARCTIC REGION

Ethiopian subregion

Nearctic subregion

Caribbean transition subregion

NEOTROPICAL REGION

(=Neotropical subregion)

Zoogeographic regions of the world. (From K. P. Schmidt, Faunal realms, regions, and provinces, Quart. Rev. Biol., 29(4):322–331, 1954)

1890 culminated in C. H. Merriam's life zone hypothesis. Though this scheme served as a useful statement of animal distribution during the first quarter of the 20th century, it has since been largely abandoned in favor of L. R. Dice's biotic province concept. *See* LIFE ZONES; ZOOGEOGRAPHIC REGION.

Efforts to divide the globe into ecological regions have been based more upon the distribution of environment types than upon the distribution of animal assemblages. Thus the present-day ecological regions have been defined largely in terms of vegetation which seems to summate the effects of such environmental features as climate, soils, and topography. A climax vegetation type (forest, grassland, desert) with its associated fauna constitutes a biome or biome type, which is the zoographer's term for an ecological region. The fauna within any biome, through homologous or analogous structures and physiological processes, is adapted to the prevailing environmental features. In the desert biome type, for example, these features include scarcity of water, desiccating winds, extreme diurnal temperature ranges, and lack of cover. Regardless of the locales of deserts, the faunas of these regions are adapted in similar ways to cope with the environmental conditions encountered therein. The fact that phylogenetically the faunas of the Great Australian Desert and of the Colorado Desert have nothing in common is beside the point. Animal life in the two regions responds in similar ways to the desert environment. *See* BIOME; PLANT FORMATIONS, CLIMAX; VEGETATION ZONES, WORLD.

Thus, whereas a faunal region is a continuous

areal unit, an ecological region lacks areal continuity and may occur in widely separated parts of the Earth. Furthermore, a single faunal region may encompass several ecological regions. The Neotropical realm, for example, includes such diverse ecological regions as the Amazonian rainforest, the Guiana grasslands, and the Patagonian Desert.

Explanations as to how ecological regions are sustained as entities depend largely on the environmental morphologist and environmental physiologist. Data on ecological regions have been summarized by R. Hesse and expanded by Allee and Schmidt.

Faunal approach. The faunal approach essays an analysis of animal distributions in terms of faunal groups which have homologous histories in contrast to homologous ranges, that is, to the regional approach. This type of study developed more recently. The first presentation of this sort was the now classic analysis of the reptile and amphibian faunas of the Americas by E. Dunn. Development of this approach had to await the accumulation of data on specific groups by the systematist. As more and more groups of animals were analyzed systematically and historically, pattern types began to take form. It was discovered, for example, that the reptile and amphibian fauna of Central America, nominally a part of the Neotropical realm or faunal region, is historically not homogeneous but, rather, is composed of at least three distinct faunal elements. These have moved into Central America, and sometimes through it at different times. Thus the mere reference of the Central American fauna to the Neotropical realm or a region thereof indicates very little. A faunal analysis provides an understanding of the composition of that fauna and how it developed.

Because analyses of this sort present data which are of value to other sciences, especially as supporting evidence to the paleogeographer, the faunal approach has proved most useful.

It should be understood that in modern zoogeographical studies the several approaches are rarely utilized independently. Most frequently the investigator weaves the several approaches together in an effort to determine the nature of a zoogeographic pattern, how it developed, and how it is sustained. [L. C. STUART]

Bibliography: W. C. Allee and K. P. Schmidt, *Ecological Animal Geography*, 1951; D. Amadon, The Hawaiian honeycreeper (Aves, Drepaniidae), *Bull. Amer. Museum Natur. Hist.*, vol. 95, no. 4, 1950; P. J. Darlington, *Zoogeography*, 1957; L. R. Dice, *The Biotic Provinces of North America*, 1943; E. R. Dunn, The herpetological fauna of the Americas, *Copeia*, 3:106–119, 1931; H. Gadow, *The Wanderings of Animals*, 1913; P. E. James and C. F. Jones (eds.), *American Geography: Inventory and Prospect*, 1954; W. D. Matthew, Climate and evolution, *Ann. N.Y. Acad. Sci.*, 24:1–323, 1915; C. H. Merriam, Laws of temperature control of the geographic distribution of terrestrial animals and plants, *Nat. Geogr. Mag.*, 6:229–238, 1894; A. E. Ortmann, The geographical distribution of freshwater decapods and its bearing upon ancient geography, *Proc. Amer. Phil. Soc.*, 41:267–400, 1902; P. L. Sclater, On the general geographical distribution of the members of the class Aves, *Proc. Linnaean Soc. London*, 130–145, 1858.

Zooplankton

The zooplankton is the assemblage of animals in an aquatic environment that is composed of the forms which occur suspended in the water and have such limited powers of locomotion that they drift passively with the water movements. Except in certain weak swimming larval stages, the fish and the few other animals that actively control their own movements in the water are excluded from the zooplankton and are referred to collectively as nekton. Many animal species occur only as part of the zooplankton and spend their entire life cycle in this assemblage. Many other species occur in the zooplankton during part of their life cycle but spend the rest of their time as part of a different assemblage, usually the bottom community.

The zooplankton is an important component in most aquatic environments and is present in a variety of freshwater and marine habitats, from small ponds to large lakes, from the upper reaches of brackish estuaries to the deep oceans. In flowing-water environments, such as streams and rivers, the zooplankton forms are much less important because they can not maintain themselves and are continuously swept downstream. Often, however, a river will contain appreciable numbers of zooplankton organisms because there are bodies of standing water along its course that contribute a steady supply of such forms in the outflow waters.

Composition. The zooplankton of the oceans is composed of a great variety of forms. Almost every animal phylum is represented by at least a few species. Single-celled animals of many types (Protozoa), jellyfish and their relatives (Coelenterata), comb jellies (Ctenophora), polychaete worms (Annelida), snails with reduced or absent shells (Mollusca), transparent arrow worms (Chaetognatha), primitive members of man's own phylum (Chordata), and especially crustaceans of a great variety of species (Arthropoda) are the most commonly encountered forms in the open-ocean zooplankton. In fact, two of these groups, the arrow worms and the comb jellies, are found only as part of this assemblage.

Especially in nearshore areas the eggs and immature stages of bottom-dwelling animals form an important element in the marine zooplankton. Many bottom animals produce eggs or young which join the zooplankton for part of their existence and then settle to the bottom, where they complete their development. The early stages of starfish, oysters, snails, clams, barnacles, crabs, lobsters, worms of various kinds, and many other forms are found in the zooplankton. The existence of a planktonic stage for these animals aids their dispersion and explains how these forms can quickly develop on newly available habitats such as dock pilings and buoys that may be many miles from the nearest adults.

In contrast to the oceanic assemblage, the freshwater zooplankton is much more limited in variety. This is because many of the groups that are prominent in the seas are unable to survive in fresh water. Further, freshwater bottom-dwelling forms seldom possess planktonic immature stages. The limited variety of the freshwater assemblage is shown by the fact that three groups provide the great majority of the species. Two of these groups, the protozoans and the crustaceans, are also major components of the marine assemblage. The third group, the wheel animals or rotifers (Rotifera), are also present in the seas but usually only as a minor component.

One noteworthy observation concerning the zooplankton is the almost complete absence of representatives from that most abundant and diversified group, the insects. Insects are, with very minor exceptions, entirely lacking from the marine zooplankton. In fresh water, although the larvae of the phantom midge (*Chaoborus*) are of some importance in ponds and small lakes, the insects as a group are also of very minor significance.

Occurrence. Although the zooplankton is well represented in most aquatic habitats, the exact species present vary greatly in relation to the specific environmental conditions. Physical-chemical factors, such as temperature, light intensity, amount of dissolved solids (salinity), concentrations of heavy metals, and pH, play important roles in determining what species will occur in an environment. Biotic factors, such as availability of food, prevalence of predators, and abundance of competitors, also play vital roles in controlling species composition.

A physical factor of special importance in determining species composition is depth. Many species are present only near the surface, while other forms, often rather bizarre in appearance, are limited to a particular stratum of the subsurface waters. In both lakes and the oceans the depth range within which most of the individuals of a population occur may vary somewhat during the course of a day due to vertical migrations. These movements are controlled by fluctuations in light intensity as well as other factors and often result, in the individuals' ascending at night and descending during the daylight hours.

Collection and identification. With the exception of certain of the jellyfish and related forms, the members of the zooplankton are almost all microscopic. Their small size in conjunction with their dispersion in great volumes of water concealed the existence of this assemblage until the early 1800s. The development that revealed the vast community of planktonic animals and plants was the introduction of small tow nets for catching and concentrating the individuals. Such nets were made of fine silk gauze and were towed from boats in such a way that the organisms were strained from the water and retained in a small jar attached to the end of the net.

Such tow nets are still the most common equipment for collecting zooplankton, although newer materials such as nylon have largely replaced silk in the gauze. The organisms collected in a net haul are usually identified and counted. The meaning of these counts is often difficult to interpret, however, for it is usually somewhat unclear how much water was strained during the haul and thus what concentration of the animals in their environment. Also, many of the smaller individuals may have escaped through the meshes of the net, while larger forms may have seen or felt its approach and avoided capture altogether. A further problem with the use of nets is that the process of capture often severely damages some of the delicate forms such as the jellyfish, comb jellies, and primitive chordates. Newer techniques of collection are being developed to improve scientists' ability to study the zooplankton populations. Pumping large volumes of water aboard a ship and then concentrating the animals, or determining the amount of zooplankton at different depths by measuring the strength of the acoustical signals reflected from the different layers of the water column, are examples of more modern methods.

After capture the zooplankton animals are usually preserved for later examination under a microscope. Formalin and ethyl alcohol are the most commonly used preservatives. The identification to species of the preserved animals in a zooplankton sample is usually an arduous undertaking. Many tens of thousands of zooplankton species occur in the sea, a sizable number of them as yet undescribed. Even when the adults are known, the young stages are often undescribed. For general ecological studies it is usually just not possible to carry the identifications to the species level, and the animals are segregated into more general groupings. For special studies or where specific identifications are absolutely necessary, experts with many years of experience are usually called upon.

Feeding relationships. The zooplankton organisms play a vital role in the ecology of aquatic environments, especially large bodies of water such as lakes and the oceans. Members of the zooplankton are the major herbivores in these ecosystems. They filter out or otherwise capture the algae and other plants in the plankton and consume these cells as their food. The herbivorous individuals are preyed upon, in turn, by larger zooplankters (often themselves taken by larger predators), small fish, certain of the bottom-dwelling forms, and even some of the largest aquatic organisms such as basking sharks and whalebone or baleen whales. Thus, the zooplankton functions as an intermediate link in the food chain, linking the plants that produce organic matter with the larger forms such as fish. *See* FOOD CHAIN.

The herbivorous zooplankters tend to be concentrated close to the surface in large bodies of water. In the deeper waters there is not enough light for plants to grow. Thus, the herbivorous forms cannot find food there. This does not mean, however, that the deep waters of large lakes and the oceans are devoid of zooplankton. On the contrary, there are many marine forms and a few freshwater species that occur only at these depths. Dead and dying individuals from the plankton of the surface layers tend to sink into the depths. These organic particles furnish a food source for many of the deepwater forms. Other species are carnivorous and feed on the scavengers or each other.

Reproduction. As might be expected with an assemblage that contains a great variety of forms, the modes of reproduction vary greatly among the zooplankton groups. One general characteristic, however, is the production of large numbers of young. In general, the young zooplankters receive very little parental care. Further, the characteristics of the zooplankter's environment leave the young with no place to hide from predators. Thus, a large number must be produced to ensure that a few survive to maturity. Many zooplankton forms of the surface layers are also aided in survival by being nearly transparent and thereby difficult for the predators to detect.

The zooplankton forms, at least in fresh water and the upper layers of the temperate and polar seas, tend to have restricted periods of reproduction. During these periods one or more generations are produced, whereas reproduction stops or is at a very low level for the rest of the year. The periods of reproduction seem to be at least partially related to seasonal fluctuations in temperature and availability of food. Such seasonal fluctuations of environmental factors are much less prominent in the deep oceans and tropical seas, and there reproduction often seems to be more evenly spaced throughout the year.

[ANDREW ROBERTSON]

Bibliography: A. C. Hardy, *The Open Sea: The World of Plankton*, 1956; G. E. Hutchinson, *A Treatise on Limnology*, vol. 2: *Introduction to Lake Biology and the Limnoplankton*, 1967; W. D. Russell-Hunter, *Aquatic Productivity*, 1970.

McGRAW-HILL ENCYCLOPEDIA OF ENVIRONMENTAL SCIENCE

List of Contributors

List of contributors

A

Ackerman, Dr. Edward A. *Carnegie Institution of Washington.* LAKE. EARTH, RESOURCE PATTERNS OF; RIVER; SOIL CONSERVATION — all in part.

Alderfer, Dr. Russell B. *Professor of Soils, Soils and Crops Department, College of Agriculture and Environmental Science, Rutgers University.* SOIL — in part.

Aldrich, Dr. Daniel G., Jr. *Department of Soil and Plant Nutrition, University of California, Davis.* SOIL — in part.

Armstrong, Francis A. J. *Freshwater Institute, Winnipeg.* SEA-WATER FERTILITY — in part.

Austin, Prof. James M. *Professor of Meteorology, Massachusetts Institute of Technology.* WEATHER FORECASTING AND PREDICTION — in part.

B

Badgley, Dr. Peter C. *Director, Earth Sciences Division, Office of Naval Research, Department of the Navy.* TERRAIN SENSING, REMOTE.

Baker, Dr. Herbert G. *Department of Botany, University of California, Berkeley.* Validator of POPULATION DISPERSAL.

Barker, Dr. Joseph W. *Chairman of the Board (retired), Research Corp., New York.* ENGINEERING.

Baumeister, Theodore. *Consulting Engineer; Stevens Professor Emeritus of Mechanical Engineering, Columbia University.* ENGINEERING, SOCIAL IMPLICATIONS OF.

Beistline, Dr. Earl H. *Provost, University of Alaska.* MINING, PLACER.

Bellum, Donald P. *Pit Superintendent, Nevada Mines Division, Kennecott Copper Corp., McGill, NV.* MINING, OPEN PIT.

Benfer, Neil A. *BOMAP Scientific Editor, Barbados Oceanographic and Meteorological Analysis Project.* SEA STATE.

Benninghoff, Dr. William S. *Department of Botany, University of Michigan.* AEROBIOLOGY; TUNDRA.

Bergeron, Prof. Tor. *Institute of Meteorology, Uppsala.* METEOROLOGY — in part.

Billings, Dr. William D. *Department of Botany, Duke University.* DESERT VEGETATION; ECOLOGY, PHYSIOLOGICAL (PLANT).

Black, Prof. Richard D. *Department of Agricultural Engineering, New York State College of Agriculture, Cornell University.* LAND DRAINAGE (AGRICULTURE).

Black, Dr. Robert F. *Department of Geology, University of Connecticut.* PERMAFROST.

Blair, Dr. W. Frank. *Professor of Zoology, University of Texas.* COMMUNITY CLASSIFICATION.

Blanc, Milton L. *Research Climatologist, Office of Climatology, U.S. Weather Bureau, Tempe, AZ.* AGRICULTURAL METEOROLOGY — in part.

Bondurant, Donald C. *U.S. Army Engineer Division, Missouri River, Omaha.* RIVER ENGINEERING — in part.

Breger, Dr. Irving A. *U.S. Geological Survey.* OIL SAND; OIL SHALE.

Brinton, Dr. Edward. *Scripps Institution of Oceanography, La Jolla.* SEA-WATER FERTILITY — in part.

Buettner, Dr. Konrad J. K. *Department of Atmospheric Sciences and Department of Dermatology, University of Washington, Seattle.* BIOCLIMATOLOGY.

C

Cadle, Dr. Richard D. *Department of Atmospheric Chemistry, National Center for Atmospheric Research.* SMOG, PHOTOCHEMICAL.

Cain, Dr. Stanely A. *Director, Institute for Environmental Quality, and Charles Lathrop Pack Professor, Department of Resource Planning and Conservation, University of Michigan.* CONSERVATION OF RESOURCES; PLANT FORMATIONS, CLIMAX.

Carritt, Dr. Dayton E. *American Dynamics International, Inc., Ft. Lauderdale.* SEA WATER — in part.

Carter, Archie N. *President, Carter, Krueger and Associates, Inc., Minneapolis.* HIGHWAY ENGINEERING.

Chapman, Dr. Valentine J. *University of Aukland.* SALT-MARSH.

Charney, Dr. Jule G. *Department of Meteorology, Massachusetts Institute of Technology.* WEATHER FORECASTING AND PREDICTION — in part.

Chenault, Roy L. *Chief Research Engineer (retired), Oilwell Division, U.S. Steel Corp.* OIL AND GAS FIELD EXPLOITATION.

Christensen, Dr. Clyde M. *Professor of Plant Pathology, University of Minnesota.* PLANT DISEASE — in part.

Christian, C. S. *Member of the Executive, CSIRO, Canberra City, Australia.* LAND-USE CLASSES.

Clapp, Philip F. *Weather Bureau, U.S. Department of Commerce.* WEATHER MAP.

Clark, John R. *Curator, New York Aquarium.* WATER CONSERVATION — in part.

Clarke, Dr. George L. *Professor of Biology, Harvard University, and Marine Biologist, Woods Hole Oceanographic Institution.* SEA WATER — in part.

Cochran, Robert A. *Shell Development Co., Houston.* SEA, OIL POLLUTION OF.

Cochrane, Prof. John D. *Department of Oceanography, Texas A & M University.* SEA WATER — in part.

Cole, Dr. Lamont C. *Professor of Ecology, Division of Biological Sciences, Cornell University.* POPULATION DYNAMICS.

Coleman, Prof. Nathaniel T. *Department of Soils and Plant Nutrition, University of California, Riverside.* SOIL CHEMISTRY — in part.

Cooper, Dr. Arthur W. *Department of Botany, North Carolina State University, Raleigh.* LIFE ZONES; PLANTS, LIFE FORMS OF; SUCCESSION, ECOLOGICAL. Validator of PLANT SOCIETIES.

Corlett, A. V. *Mining Engineer, Kingston, Ontario.* MINING, UNDERGROUND.

Coupland, Dr. Robert T. *Department of Plant Ecology, University of Saskatchewan.* GRASSLAND.

Court, Dr. Arnold. *Professor of Climatology, Department of Geography, California State University, Northridge.* FROST; SENSIBLE TEMPERATURE; TEMPERATURE INVERSION.

Covey, Dr. Winton. *Division of Natural Sciences, Concord College.* MICROMETEOROLOGY, CROP.

Crafts, Dr. Alden S. *Department of Botany, University of California, Davis.* DEFOLIANT AND DESICCANT; HERBICIDE.

Craig, Dr. Harmon. *Professor of Geochemistry, Scripps Institution of Oceanography, La Jolla.* SEA WATER—in part.

Cruickshank, Michael J. *Research Supervisor, Marine Minerals Technology Center, U.S. Bureau of Mines, Tiburon, CA.* MINING, UNDERSEA.

Currie, Ronald I. *Secretary, Scottish Marine Biological Association, and Director, Dunstaffnage Marine Research Laboratory, Oban, Scotland.* SEA-WATER FERTILITY—in part.

Curtis, John T. *Deceased; formerly, University of Wisconsin.* CLIMAX COMMUNITY.

Cushing, Dr. D. H. *Fisheries Laboratory, Ministry of Agriculture, Fisheries and Food, Lowenstoft, England.* SEA-WATER FERTILITY—in part.

D

Dale, Dr. Walter Max. *Deceased; formerly, Professor and Head, Department of Biochemistry, Christie Hospital, Manchester, England.* RADIATION BIOCHEMISTRY.

Dana, Dr. Samuel T. *Dean Emeritus, School of Natural Resources, University of Michigan.* FOREST AND FORESTRY.

Dansereau, Dr. Pierre. *Professor of Ecology, Institut d'Urbanisme, Université de Montréal.* VEGETATION; VEGETATION ZONES, WORLD.

Davis, Dr. David E. *Head, Department of Zoology, North Carolina State University, Raleigh.* ECOLOGICAL INTERACTIONS.

Davis, Dr. John H. *Professor Emeritus of Botany, University of Florida.* DUNE VEGETATION. MANGROVE SWAMP—in part.

Davis, Prof. Kenneth P. *School of Forestry, Yale University.* FOREST MANAGEMENT AND ORGANIZATION.

DeBach, Dr. Paul. *Professor of Biological Control, University of California, Riverside.* INSECT CONTROL, BIOLOGICAL.

Deland, Dr. Raymond J. *Department of Meteorology and Oceanography, New York University.* AIR PRESSURE.

Dice, Dr. Lee R. *Professor Emeritus of Zoology, University of Michigan.* ECOLOGY, HUMAN.

Dietz, Dr. Robert S. *Atlantic Oceanographic and Meteorological Laboratories, ESSA, U.S. Department of Commerce, Miami.* CONTINENTAL SHELF AND SLOPE.

Dowd, James J. *Mining and Preparation Section, Bureau of Mines, U.S. Department of the Interior.* MINING, STRIP.

Duley, Dr. Frank L. *Principal, Agricultural College, University of Peshawar.* SOIL—in part.

Duntley, Dr. Seibert Q. *Department of Physics, University of California, La Jolla.* SEA WATER—in part.

E

Ebert, Dr. M. *Paterson Laboratories, Christie Hospital and Holt Radium Institute, Manchester, England.* Validator of RADIATION BIOCHEMISTRY.

Eddy, Dr. Samuel. *Museum of Natural History, University of Minnesota.* LIMNOLOGY.

Egler, Dr. Frank E. *Director (retired), Aton Forest, Norfolk, CT.* PLANT GEOGRAPHY; VEGETATION MANAGEMENT.

Eide, Dr. Carl J. *Department of Plant Pathology, Institute of Agriculture, University of Minnesota.* PLANT DISEASE; PLANT DISEASE CONTROL—both in part.

Eliassen, Prof. Arnt *Institute of Geophysics, University of Oslo.* METEOROLOGY—in part.

Evans, Prof. Francis C. *Professor of Zoology, University of Michigan.* POPULATION DISPERSION. Validator of ANIMAL COMMUNITY.

F

Fisher, Dr. Frederick H. *Scripps Institution of Oceanography, University of California, San Diego.* SEA WATER—in part.

Fosberg, Dr. F. R. *Smithsonian Institution.* ISLAND FAUNAS AND FLORAS.

Frank, Prof. Bernard. *Deceased; formerly, Professor of Watershed Management, Colorado State University.* WATER CONSERVATION—in part.

Frank, Dr. Peter W. *Department of Biology, University of Oregon.* SPECIES POPULATION.

French, Dr. David W. *Department of Plant Pathology, University of Minnesota.* TREE DISEASES.

Frey, Prof. David G. *Department of Zoology, Indiana University.* FRESH-WATER ECOSYSTEM.

Fuhrman, Dr. Ralph E. *Assistant Director, National Water Commission, Arlington.* WATER POLLUTION.

Fulks, J. R. *U.S. Weather Bureau, Chicago.* DEW POINT; HUMIDITY.

G

Galloway, Dr. William J. *Bolt, Beranek and Newman, Inc., Canoga Park, CA.* NOISE, ACOUSTIC.

Gardner, Prof. Victor R. *Professor of Pomology (retired), Rutgers University, and Consultant, Plant Growth Regulator Development Program, Proctor and Gamble Co.* AGRICULTURAL SCIENCE (PLANT).

Gauch, Dr. Hugh G. *Department of Botany, College of Agriculture, University of Maryland.* PLANT, MINERALS ESSENTIAL TO.

Giles, Prof. Robert H., Jr. *Division of Forestry and Wildlife Science, College of Agriculture, Virginia Polytechnic Institute.* WILDLIFE CONSERVATION—in part.

Goldberg, Dr. Edward D. *Scripps Institution of Oceanography, La Jolla.* HYDROSPHERE; HYDROSPHERE, GEOCHEMISTRY OF; SEA WATER—in part.

Gregory, Dr. G. R. *Department of Forestry, School of Natural Resources, University of Michigan.* FOREST RESOURCES.

H

Hader, Rodney N. *Secretary, American Chemical Society, Washington, DC.* AGRICULTURAL CHEMISTRY.

Haley, Dr. Leanor D. *Chief, Mycology Training Unit, Department of Health, Education and Welfare, National Communicable Disease Center, Atlanta.* FUNGISTAT AND FUNGICIDE—in part.

Hammond, Prof. Edwin H. *Professor and Head, Department of Geography, University of Tennessee.* HILL AND MOUNTAIN TERRAIN; TERRAIN AREAS, WORLDWIDE.

Harley, Dr. John H. *Director, Health and Safety Laboratory, U.S. Atomic Energy Commission.* RADIOACTIVE FALLOUT.

Harrar, J. George. *Rockefeller Foundation, New York.* PLANT DISEASE CONTROL—in part.

Harvey, Dr. George R. *Woods Hole Oceanographic Institution.* POLYCHLORINATED BIPHENYLS.

Hasler, Dr. Arthur D. *Laboratory of Limnology, University of Wisconsin.* EUTROPHICATION, CULTURAL.

Hazel, Dr. L. N. *Iowa State University.* BREEDING (ANIMAL).

Hazen, Richard. *Hazen and Sawyer, Consulting Engineers, New York.* WATER SUPPLY ENGINEERING; WATER TREATMENT.

Heald, Dr. Walter R. *Northeast Watershed Research Cen-*

ter, U.S. Department of Agriculture, University Park, PA. WASTES, AGRICULTURAL.

Hersey, Dr. John B. *Office of Naval Research, U.S. Department of the Navy.* SCATTERING LAYER.

Hewson, Dr. E. Wendell. *Chairman, Department of Atmospheric Sciences, Oregon State University.* ATMOSPHERIC POLLUTION.

Hirst, Dr. J. M. *Rothamsted Experimental Station, Harpenden, England.* PLANT DISEASE — in part.

Holmes, Dr. Robert W. *Department of Biological Sciences, University of California, Santa Barbara.* PHYTOPLANKTON.

Hooker, Dr. Arthur L. *Department of Plant Pathology, University of Illinois.* PLANT DISEASE CONTROL — in part.

Horsfall, Dr. James G. *Director, Connecticut Agricultural Experiment Station, New Haven.* FUNGISTAT AND FUNGICIDE; PLANT DISEASE CONTROL — both in part.

I

Ingram, William T. *Consulting Engineer, Whitestone, NY.* AIR-POLLUTION CONTROL; IMHOFF TANK; SANITARY ENGINEERING; SEPTIC TANK; SEWAGE; SEWAGE TREATMENT.

Irving, Dr. Laurence. *Institute of Arctic Biology, University of Alaska.* ARCTIC BIOLOGY.

J

Jacobs, Dr. Woodrow C. *Director, Environmental Data Service, Environmental Science Services Administration, Silver Spring, MD.* OCEAN-ATMOSPHERE RELATIONS. AGRICULTURAL METEOROLOGY — in part.

Johnson, Dr. Ronald R. *Department of Animal Sciences and Industry, Oklahoma State University.* AGRICULTURAL SCIENCE (ANIMAL).

Johnson, Wendell E. *Consulting Engineer, McLean, VA.* RIVER ENGINEERING — in part.

Junge, Dr. C. E. *Max-Planck-Institut für Chemie (Otto-Hahn-Institut), Mainz.* ATMOSPHERIC CHEMISTRY.

K

Kallio, Dr. R. E. *School of Life Sciences, University of Illinois.* Validator of NITROGEN CYCLE.

Kaplan, Prof. Lewis D. *Professor of Meteorology, University of Chicago.* GREENHOUSE EFFECT, TERRESTRIAL; HEAT BALANCE, TERRESTRIAL ATMOSPHERIC.

Katznelson, Dr. Harry. *Deceased; formerly, Director, Research Branch, Microbiology Research Institute, Canada.* NITROGEN CYCLE.

Kendeigh, Dr. S. Charles. *Department of Zoology, University of Illinois.* BIOME; COMMUNITY.

Kessler, Dr. Edwin. *Director, National Severe Storms Laboratory, Research Laboratories, Environmental Science Services Administration, Norman, OK.* STORM DETECTION.

Ketchum, Dr. Bostwick H. *Woods Hole Oceanographic Institution.* SEA WATER — in part.

Kilbey, Dr. Brian J. *Department of Genetics, University of Edinburgh.* MUTATION.

King, Dr. Donald W. *School of Medicine, Yale University.* MALNUTRITION.

Kubota, Dr. Joe. *Research Soil Scientist, Soil Survey Investigations, Plant, Soil and Nutrition Laboratory, U.S. Department of Agriculture, Ithaca, NY.* SOIL CHEMISTRY — in part.

Kuc, Dr. Joseph. *Department of Biochemistry, Purdue University.* PLANT DISEASE CONTROL — in part.

L

LaFond, Dr. Eugene C. *Senior Scientist and Consultant for Oceanography, Naval Undersea Center, San Diego.* SEA WATER — in part.

Lagler, Prof. Karl F. *School of Natural Resources, University of Michigan.* FISHERIES CONSERVATION.

LANE, JAMES A. *Oak Ridge National Laboratory.* NUCLEAR POWER.

Lapple, Charles E. *Senior Scientist, Chemical Engineering Department, Stanford Research Institute, Menlo Park.* DUST AND MIST COLLECTION.

Lems, Dr. Kornelius. *Deceased; formerly, Associate Professor and Chairman, Department of Biological Sciences, Goucher College.* PLANT SOCIETIES; POPULATION DISPERSAL.

Leopold, A. Starker. *School of Forestry and Conservation, University of California, Berkeley.* BIOGEOGRAPHY.

Leopold, Dr. Luna B. *Department of Geology and Geophysics, University of California, Berkeley.* STREAM TRANSPORT AND DEPOSITION; SURFACE WATER; SWAMP, MARSH, AND BOG.

Lettau, Prof. Heinz H. *Department of Meteorology, University of Wisconsin.* MICROMETEOROLOGY.

Lieth, Dr. Helmut. *Department of Botany, University of North Carolina, Chapel Hill.* SAVANNA.

Likens, Prof. Gene E. *Section of Ecology and Systematics, Division of Biological Sciences, Cornell University.* LAKE, MEROMICTIC.

Linsley, Prof. Ray K. *Professor of Civil Engineering, Stanford University, and President, Hydrocomp International.* ATMOSPHERIC EVAPORATION; EVAPOTRANSPIRATION. Validator of AQUIFER; ARTESIAN SYSTEMS; GROUNDWATER; HYDROLOGY; SPRING, WATER; THERMAL SPRING; WATER TABLE.

Lochhead, Dr. Allan G. *Canada Department of Agriculture, Microbiology Research Institute, Ottawa.* MANURE; SOIL BALANCE, MICROBIAL; SOIL MICROORGANISMS.

Lorenz, Dr. Edward N. *Department of Meteorology, Massachusetts Institute of Technology.* WEATHER FORECASTING AND PREDICTION — in part.

Ludlam, Prof. Frank H. *Department of Meteorology, Imperial College, London.* CLOUD.

Ludvik, Dr. George F. *Insecticide Application Research, Agricultural Research and Development Department, Monsanto Co., St. Louis.* INSECTICIDE.

Lyman, Dr. John. *Department of Oceanography, North Carolina State University, Raleigh.* SEA WATER — in part.

Lyon, Dr. Waldo K. *U.S. Navy Electronics Laboratory, San Diego.* SEA ICE.

M

McCormick, Dr. Jack. *Chairman, Department of Ecology and Land Management, Academy of Natural Sciences, Philadelphia.* TERRESTRIAL ECOSYSTEM; VEGETATION ZONES, ALTITUDINAL.

Macfadyen, Dr. Amyan. *Professor, School of Biological and Environmental Studies, The New University of Ulster, Coleraine, County Londonderry, Northern Ireland.* BIOLOGICAL PRODUCTIVITY; BIOMASS; ECOLOGICAL PYRAMIDS; FOOD CHAIN.

Major, Dr. Jack. *Department of Biology, University of California, Davis.* PLANT COMMUNITY.

Malone, Dr. Thomas F. *Director of Research, University of Connecticut.* INDUSTRIAL METEOROLOGY.

Mason, Dr. Basil J. *Director General, Meteorological Office, Bracknell, England.* CLOUD PHYSICS.

Meigs, Dr. Peveril. *Consultant, Wayland, MA.* ENVIRONMENTAL PROTECTION.

Meiklejohn, Dr. Jane. *Senior Research Fellow, University*

College, Salisbury, Rhodesia. SOIL MINERALS, MICROBIAL UTILIZATION OF; SOIL PHOSPHORUS, MICROBIAL CYCLE OF; SOIL SULFUR, MICROBIAL CYCLE OF.

Miller, Prof. Maynard M. *Professor of Geology, Michigan State University, and Director, Foundation for Glacial and Environmental Research, Seattle.* TERRESTRIAL FROZEN WATER.

Mitchell, Dr. J. Murray, Jr. *Environmental Data Service, U.S. Department of Commerce, Silver Spring, MD.* CLIMATOLOGY.

Morgan, Dr. Karl Z. *Director, Health Physics Division, Oak Ridge National Laboratory.* HEALTH PHYSICS; RADIOACTIVE MATERIALS, DECONTAMINATION OF; RADIOACTIVE WASTE DISPOSAL.

Mottet, Dr. N. Karle. *Professor of Pathology and Director of Hospital Pathology, University Hospital, University of Washington, Seattle.* DISEASE; HEAVY METALS, PATHOLOGY OF. Validator of TOXICOLOGY.

Mueller, Prof. George. *Institute of Molecular Evolution, University of Miami.* BIOSPHERE, GEOCHEMISTRY OF.

Mullin, Dr. Michael M. *Department of Oceanography, University of California, La Jolla.* PREDATOR-PREY RELATIONSHIPS.

Murgatroyd, Dr. R. J. *Meteorological Office, Bracknell, England.* ATMOSPHERE.

N

Namias, Jerome. *Chief, Extended Forecast Division, Weather Bureau, Environmental Science Services Administration, Washington, DC.* DROUGHT. WEATHER FORECASTING AND PREDICTION — in part.

Newcombe, Dr. Curtis L. *Director, San Francisco Bay Marine Research Center, San Francisco State College.* MARINE ECOSYSTEM.

Newell, Dr. Norman D. *American Museum of Natural History, New York.* FAUNAL EXTINCTION.

Newton, Dr. Chester W. *National Center for Atmospheric Research, Boulder.* DUST STORM; SQUALL LINE; STORM; THUNDERSTORM; TORNADO.

Nierenberg, Prof. William A. *Director, Scripps Institution of Oceanography, University of California, San Diego.* OCEANOGRAPHY.

Nixon, Dr. Charles W. *Consultant, Kettering, OH.* SONIC BOOM — in part.

Norris, Robert M. *Department of Geology, University of California, Santa Barbara.* DUNE.

O

O'Connor, Dr. Donald J. *Department of Civil Engineering, Manhattan College.* SEWAGE DISPOSAL.

Odum, Dr. Eugene P. *Director, Institute of Ecology, University of Georgia.* ECOLOGY.

Olmsted, Leonard M. *Senior Engineering Editor, "Electrical World," New York.* ELECTRIC POWER SYSTEMS; UTILITY INDUSTRY, ELECTRICAL.

Olsen, Dr. Sterling R. *Agricultural Research Service, U.S. Department of Agriculture, Colorado State University.* SOIL CHEMISTRY — in part.

Olson, Dr. Jerry S. *Oak Ridge National Laboratory.* LAND-USE PLANNING; SYSTEMS ECOLOGY; VEGETATION AND ECOSYSTEM MAPPING. ECOLOGY, APPLIED — in part. Validator of CLIMAX COMMUNITY.

Olson, Sigurd F. *President, Wilderness Society, Ely, MN.* WILDLIFE CONSERVATION — in part.

P

Palmén, Dr. Erik. *Department of Meteorology, University of Helsinki.* ATMOSPHERIC GENERAL CIRCULATION.

Patton, Dr. Donald J. *Department of Geography, Florida State University, Tallahassee.* EARTH, RESOURCE PATTERNS OF; RIVER; SOIL CONSERVATION — all in part.

Pattullo, Dr. June G. *Department of Oceanography, Oregon State University.* SEA WATER — in part.

Pimentel, Dr. David. *Professor of Insect Ecology, Department of Entomology and Section of Ecology and Systematics, New York State College of Agriculture and Life Sciences, Cornell University.* ENTOMOLOGY, ECOLOGICAL.

Platt, Dr. Robert B. *Chairman, Department of Biology, Emory University.* ENVIRONMENT.

Pomeroy, Lawrence R. *Department of Zoology, University of Georgia, Athens.* BIOGEOCHEMICAL BALANCE.

Ponikvar, Ade L. *Formerly, "Modern Plastics," McGraw-Hill Publications Co., New York.* OIL MINING.

Priester, Gayle B. *Principal Engineer, Electric Engineering Department, Baltimore Gas and Electric Co.* HEATING, COMFORT.

Pritchard, Dr. Donald W. *Director, Chesapeake Bay Institute, Johns Hopkins University.* ESTUARINE OCEANOGRAPHY — in part.

Prosser, Dr. C. Ladd. *Department of Physiology, University of Illinois, Urbana.* HOMEOSTASIS.

Q

Quirin, Edward J. *President, Frederic R. Harris, Inc., New York.* COASTAL ENGINEERING.

R

Rasmusson, Dr. Eugene M. *Geophysical Fluid Dynamics Laboratory, Environmental Science Services Administration, Princeton.* HYDROMETEOROLOGY.

Reed, Prof. Richard J. *Department of Atmospheric Sciences, University of Washington, Seattle.* FRONT — in part.

Reid, Joseph L. *Scripps Institution of Oceanography, La Jolla.* SEA WATER — in part.

Reid, Prof. Robert O. *Department of Oceanography, Texas A & M University.* STORM SURGE.

Rich, Dr. Saul. *Senior Plant Pathologist, Connecticut Agricultural Experiment Station, New Haven.* FUNGISTAT AND FUNGICIDE — in part.

Richards, Dr. Paul W. *School of Plant Biology, University College of North Wales.* RAINFOREST.

Riley, Dr. Ralph. *Plant Breeding Institute, Cambridge, England.* BREEDING (PLANT).

Risebrough, Dr. Robert. *Department of Nutritional Sciences, University of California, Berkeley.* PESTICIDES, PERSISTENCE OF.

Ritchie, Dr. J. C. *Department of Biology, Trent University, Peterborough, Ontario.* TAIGA.

Rittenberg, Prof. Sydney C. *Department of Bacteriology, University of California, Los Angeles.* WATER ANALYSIS.

Robertson, Dr. Andrew. *NOAA, Rockville, MD.* ZOOPLANKTON.

Robinson, Dr. G. D. *Center for the Environment and Man, Hartford, CT.* CLIMATE, MAN'S IMPACT ON.

Rowell, Dr. John B. *U.S. Department of Agriculture, and Department of Plant Pathology and Botany, University of Minnesota, St. Paul.* PLANT DISEASE — in part.

Rudnick, Dr. Philip. *Scripps Institution of Oceanography, La Jolla.* SEA WATER — in part.

Rudolf, Paul O. *Principal Silviculturist (retired), North Central Forest Experimental Station, St. Paul.* FOREST PLANTING AND SEEDING.

Ryther, Dr. John H. *Woods Hole Oceanographic Institution.* SEA-WATER FERTILITY — in part.

S

Sanders, Dr. Frederick. *Department of Meteorology, Massachusetts Institute of Technology.* JET STREAM; WIND.

Sayre, Dr. Albert N. *Deceased; formerly, Consulting Groundwater Geologist, Behre Dolbear and Co.* AQUIFER; ARTESIAN SYSTEMS; GROUNDWATER; HYDROLOGY; SPRING WATER; THERMAL SPRING; WATER TABLE.

Schaefer, Dr. Milner B. *Director, Institute of Marine Resources, University of California, La Jolla.* MARINE RESOURCES.

Schoonmaker, G. R. *Vice President, Production-Exploration, Marathon Oil Co., Findlay, OH.* OIL AND GAS, OFFSHORE.

Schule, John Joseph, Jr. *Acting Director, Department of Marine Sciences, U.S. Naval Oceanographic Office.* THERMOCLINE.

Schumm, Dr. Stanley A. *Department of Geology, Colorado State University.* GEOMORPHOLOGY.

Segeler, C. George. *Director, Technical Services, Dave Sage, Inc., New York.* DEGREE-DAY.

Sharpe, Dr. C. F. Stewart. *Falls Church, VA.* AVALANCHE.

Sheppard, Prof. P. A. *Department of Meteorology, Imperial College, London.* METEOROLOGY — in part.

Shrock, Prof. Robert R. *Department of Geosciences, Texas Technical University, Lubbock.* EARTH SCIENCES.

Slack, Dr. A. V. *Chief, Applied Research Branch, Division of Chemical Development, TVA, Muscle Shoals, AL.* FERTILIZER — in part.

Smith, Dr. Guy D. *Soil Conservation Service, U.S. Department of Agriculture.* SOIL, ZONALITY OF. SOIL — in part.

Smith, Prof. Karl U. *Director, Behavioral Cybernetics Laboratory, and Professor of Psychology, University of Wisconsin.* LOUDNESS.

Spurr, Prof. Stephen H. *Vice President and Dean, Graduate School, University of Michigan.* FOREST ECOLOGY. ECOLOGY, APPLIED — in part.

Stakman, Dr. E ʕ. *Professor Emeritus, Institute of Agriculture, University of Minnesota,* PLANT DISEASE; PLANT DISEASE CONTROL — both in part.

Stevenson, Dr. Robert E. *Scientific Liaison Officer, Office of Naval Research, Scripps Institution of Oceanography, La Jolla.* ESTUARINE OCEANOGRAPHY — in part.

Stover, Prof. Harold E. *Professor and Extension Agricultural Engineer (retired), Kansas State University.* RURAL SANITATION.

Stuart, Dr. Edward G. *Deceased; formerly, West Virginia School of Medicine.* TOXICOLOGY.

Stuart, Prof. L. C. *Professor of Zoology, Panajachel, Guatemala.* ZOOGEOGRAPHIC REGION; ZOOGEOGRAPHY.

Swithinbank, Dr. Charles W. *Scott Polar Research Institute, Cambridge, England.* ANTARCTICA.

T

Tankersley, G. J. *East Ohio Gas Co., Cleveland.* COAL GASIFICATION.

Turner, Dean Lewis. *Formerly, Utah State University.* RANGELAND CONSERVATION.

U

U.S. Army Corps of Engineers and U.S. Bureau of Reclamation. DAM.

V

Vallentyne, Dr. John R. *Fisheries Research Board of Canada, Freshwater Institute, Winnipeg.* BIOSPHERE.

Van Dyne, Dr. George. *Natural Resource Ecology Laboratory, Colorado State University.* PRAIRIE ECOSYSTEM.

Vaughan, Richard D. *Solid Waste Management Office, Environmental Protection Agency, Rockville, MD.* SOLID-WASTE DISPOSAL.

Vernon, Edward M. *Weather Analysis and Prediction Division, U.S. Weather Bureau, Silver Spring, MD.* WEATHER FORECASTING AND PREDICTION — in part.

Viets, Dr. Frank G., Jr. *U.S. Department of Agriculture, Fort Collins, CO.* FERTILIZER — in part.

Vine, Allyn C. *Oceanographer, Woods Hole Oceanographic Institution.* SEA WATER — in part.

Vohra, Dr. F. C. *School of Biological Sciences, University of Malaya, Kuala Lumpur.* MANGROVE SWAMP — in part.

Voigt, Dr. Garth K. *Margaret K. Musser Professor of Forest Soils, School of Forestry, Yale University.* FOREST SOIL.

Von Gierke, Dr. Henning E. *Consultant, Yellow Springs, OH.* SONIC BOOM — in part.

Von Metzsch, Ernst H. *Harvard University.* MINERAL RESOURCES CONSERVATION.

Vonnegut, Dr. Bernard. *Department of Atmospheric Science, State University of New York, Albany.* ATMOSPHERIC ELECTRICITY.

W

Warren, Dr. Shields. *Cancer Research Institute, New England Deaconess Hospital, Boston.* RADIATION INJURY (BIOLOGY).

Weickmann, Dr. Helmut K. *Director, Atmospheric Physics and Chemistry Laboratory, Environmental Science Services Administration, Boulder.* WEATHER MODIFICATION.

Weingart, Dr. Jerome M. *Senior Research Fellow in Environmental Engineering, Environmental Quality Laboratory, California Institute of Technology, Pasadena.* SOLAR ENERGY.

White, Prof. Colin. *Department of Epidemiology and Public Health, School of Medicine, Yale University.* EPIDEMIOLOGY.

Whitelaw, Dr. Robert L. *Mechanical Engineering Department, Virginia Polytechnic Institute.* GEOTHERMAL POWER.

Whittaker, Dr. Robert H. *Ecology and Systematics, Cornell University.* BIOTIC ISOLATION; ECOSYSTEM.

Wiegert, Dr. Richard G. *Department of Zoology, University of Georgia.* RADIOECOLOGY.

Willett, Prof. Hurd C. *Department of Meteorology, Massachusetts Institute of Technology.* AIR MASS.

Williams, Dr. Roger J. *Department of Chemistry, University of Texas.* NUTRITION.

Williams, Dr. Roger T. *Department of Meteorology, Naval Postgraduate School, Monterey.* FRONT — in part.

Winchester, Prof. John W. *Department of Oceanography, Florida State University, Tallahassee.* HALOGENS, ATMOSPHERIC GEOCHEMISTRY OF.

Winterkorn, Dr. Hans F. *Department of Civil and Geological Engineering, Princeton University.* SOIL MECHANICS.

Wonders, Dr. William C. *Department of Geography, University of Alberta.* ARCTIC AND SUBARCTIC ISLANDS.

Wood, Ivan D. *Consulting Engineer, Denver.* CROPS, IRRIGATION OF.

Woodbury, Dr. Angus M. *Ecological Research, University of Utah.* ANIMAL COMMUNITY.

Woodcock, Dr. Alfred H. *Institute of Geophysics, University of Hawaii, Honolulu.* SEA WATER — in part.

Wooster, Dr. Warren S. *Department of Oceanography, University of California, La Jolla.* SEA WATER — in part.

McGRAW-HILL ENCYCLOPEDIA OF ENVIRONMENTAL SCIENCE

Index

Index

A

A horizon 528
Aberg, B. 225
Aberrations, chromosomal 361–362, 365, 458
Abies 648
Abies lasiocarpa 645
Abiotic environment 151, 175–178
  atmosphere 177
  ionizing radiation 176–177
  light 176
  substratum 177
  temperature 176
  water 177
Abiotic substances 150
Abrasion (disease condition) 132
Abronia latifolia 137
Abscess (bacterial disease) 132
Abscission (defoliant) 128
Absolute humidty 275
Absorbents (oil spill cleaning) 484
Absorption spectroscopy (remote terrain sensing) 612
Absorption spectrum (atmosphere) 78
Abyssal-benthic biome 64
Abyssal communities 165
Acacia 479
Acanthaster 453
Acanthus 308
Acclimation (species population) 586
Accreting shoreline 106
Accretion zones (beach) 105
Acer 431
Acer rubrum 137, 439
Acer saccharum 137
Acid and base buffers 497
Acid soils 545
Acids (poison) 622
Aconite (poison) 622
Acoustic noise 367–368, 581–585
Acoustics (sonic boom) 581–585
Acrolein 264
Acrostichum aureum 308
Acrostichum speciosum 308
Actidione 232
Actinomycetes 565
Actinomycosis 10, 132, 232
Activated-sludge process (sewage treatment) 522–523
Activation analysis (sea water) 506–507
Acute plumbism 259
Adansonia digitata 479
Adaptation (species population) 586
Adaptive radiation (island biota) 288–289

Adaptive responses (patterns) 274
Adélie penguin 30
Adenosinediphosphate (ADP) 403
Adenosinetriphosphate (ATP) 403
ADP *see* Adenosinediphosphate
Adsorbers, gas-solid 51
Advanced gas-cooled reactors 369
Advanced Technology Satellite: satellite photo by *ATS 3* 81
  storm detection 593
Aeolian soils 556
Aeration: contact type 523
  mechanical type 523
  sewage treatment 523
  simplex type 523
  soil 535
  water treatment 668
Aeration, zone of 247–248
Aerial photograph: terrain sensing, remote 611
  vegetation and ecosystem mapping 637–638
Aeroallergens 3, 46
Aerobiology 1–4
  aeroallergens 3, 46
  aerosols 38
  air-pollution control 19–25
  atmospheric pollution 46–52
  biological aerosols 420
  global and regional monitoring 4
  long-distance transport 1–2
  new approaches 3–4
  predicting disease outbreaks 2–3
Aerology, circumpolar 325–326
Aerosol bioclimatology 54
Aerosols: biological 420
  continental 38
  inadvertent weather modification 682
  maritime 38
  natural 38
  particle size 20
  stratospheric 38
Aesculus 431
Afforestation: forest management 216
  forest planting and seeding 209–212
Africa: landform 610
  resource pattern 144
African coffee rust 2
Agamospermy 627
Agassiz, Alexander 374
Agassiz, Louis 374
Agathis 430
Agave 131, 136, 433

Age-class organization (forest) 208
Agglomerators, ultrasonic 51
Aggregation (ecology) 145
Agricultural Adjustment Act of 1933 113
Agricultural chemistry: important chemicals 5
  scope of field 5
Agricultural drought 133
Agricultural fungicides 233
Agricultural lands (crustal variation) 143
Agricultural meteorology 5–7, 334
  diverse responses to weather 6–7
  major participating agencies 6
  microfocus of investigation 6
  principal technical literature 7
  temperature factors 6
  water and moisture problems 6
Agricultural science (animal) 7–10
  agricultural chemistry 5
  artificial insemination 8
  breeding (animal) 71–72
  chemical composition of feedstuffs 8–9
  digestibility of feeds 9
  formulation of animal feeds 9
  fungus infections 10
  livestock breeding 7–8
  livestock feeding 8
  livestock judging 9–10
  livestock pest and disease control 10
  nutritional needs of different animals 9
Agricultural science (plant) 10–14
  agricultural chemistry 5
  breeding (plant) 72–75
  fertilizers and plant nutrition 11–12
  herbicides 12–13
  insecticides, fungicides, and nematocides 12
  mechanization 10–11
  photoperiodism 14
  viruses 13
Agricultural wastes 574, 655–656
  agricultural chemistry 5
  agricultural meteorology 5–7
  agricultural science (animal) 7–10
  agricultural science (plant) 10–14
  breeding (animal) 71–72
  crops, irrigation of 114–118
  drought 133

Agricultural wastes—*cont.*
  effect on climate 80–82
  efficiency limits 154
  fertilizer 191–199
  herbicide 262–268
  industrial meteorology 283–284
  insect control, biological 284–285
  insecticide use 173
  machinery in 10–11
  manure 309
  micrometeorology, crop 334–335
  remote sensing 612
  slash and burn 81
  soil ecosystem 159
Agriculture, soil and crop practices in: agricultural fungicides 233
  cropping system 536
  grasslands 247
  land drainage (agriculture) 294–297
  land-use classes 553–555
  physical properties of soil 531–535
  plant disease 410–411
  plant disease control 422–423
  rainforest 469–470
  savanna 480
  soil erosion 550
Agrobacterium tumefaciens 414
Agropyron 246, 432
Agropyron smithi 433
Agroxone 266
Air: components 316–320
  contaminants 46
  continental 88
  general physics of 316–320
  maritime 88
  nonsaturated 327
  radioactive decontamination 463
  saturated 327
  soil 535
  thermal structure 317
  thermodynamic properties 326–328
Air bioclimatology 53–54
Air-cleaning devices 21
Air conditioning: air filter 139
  solar energy 571
Air filter (dust collector) 139
Air mass 14–17, 319–320
  classification 16–17
  cycle 277
  development of concept 15
  front 230–232
  geographic classification 321–322
  origin 15

Air mass—*cont.*
  population dispersal 445
  precipitation 277
  squall line 587
  thermodynamic classification
    320–321
  weather significance 15–16
Air-monitoring instruments
  23–24
Air motion: cloud classification
  99
  cloud physics 99–102
Air pollution 46–52, 53, 641–642,
  breeding (plant) 72–75
  electric power generation 630
  halogens 251
  smog, photochemical 524–526
Air-pollution control 19–25
  air-quality control 24
  collection of contaminants 51
  containment of radioactive
    materials 51
  dispersion of contaminants 51
  laws 51–52
  monitoring radioactive fallout
    462
  municipal incinerators 579
  prevention of pollution 50–51
  smog, photochemical 524–526
  sources of pollution 19–20
Air pressure 17–19, 318
  distribution of temperature
    88–90
  horizontal and time variations
    17–18
  relation to wind and weather
    18–19
  units 17
  variation with height 17
Air Quality Act of 1967 52
Air sampling: biological material
  4
  pollutants 23–24
Air temperature: altitudinal var-
    iations 643
  control 641
  frost 232
  temperature inversion 608
Aircraft: radar 2
  sonic boom 581–585
  storm detection 592
Alanap 266
Alang-alang 470
Albedo: global 79, 80
  planetary 85
Alcaptonuria 132
Alcohol dehydrogenase 404
Alcohols: disease 132
  poison 622
Alder (plant) 136
Aldrich, S. A. 199
Aldrich, S. R. 199
Aldrin 285, 286, 400
Aleutian Islands 31–32
Aleutian low 693
Alfalfa: defoliation 128
  root zone depths 115
Alfisols 529, 541
Algae: aerial transport 3
  Antarctica 30
  blue-green 187, 401, 566
  brown 312
  eutrophic lakes 187
  green 401, 566
  marine resources 315

Algae—*cont.*
  nitrogen-fixing 155
  phytoplankton 401–402
  plant disease 407, 409
  red 312
  soil 566
  yellow-green 401
Algebra, Boolean 605
Algin 315
Algodones dunes (Calif.) 135
Alkali Act of 1906 51
Alkalic injury of plants 410
Alkalies (poison) 622
Alkaline soil (crop productivity)
  115–116
Alkalinity (sea water) 497
Allee, W. 698
Alleghenian life zone 301
Allen, K. R. 62
Allenrolfea 433, 478
Allergen: aeroallergens 3
  indoor allergens 3
Allocation of resources (conser-
    vation measure) 111
Allowable pollutional load
    (stream) 518
Allylalcohol 263, 267
Alnus 136
Alnus rugosa 439
Aloe 315
Alopecia (thallium poisoning)
  260
Alpha particles (radiation injury,
    biology) 457
Alpha radiation (radioecology)
    465–468
Alpine fir 645
Alpine tundra 434, 628
Alternaria kikuchiana 425
Alternating-current systems 169
Altitudinal vegetation zones
    642–645
Altocumulus clouds 97, 98
Altostratus clouds 97, 98, 318
Aluminum: constituent of living
    matter 70
  soil 542
  thermal constants 618
  uses 336
Amadon, D. 698
Amalgamation (mineral recov-
    ery) 343
Amaranthaceae 288
Amazon River 473
Ambient noise 368
Ambrosia 3
Ambrosia artemisiifolia 439
Ambursen type dam 118
Amebas (soil microorganisms)
    566
Amebic dysentery 132
American beach grass 107,
    136
American boreal faunal region
    26
American Gas Association 102
American jack (animal breeding)
    8
American Petroleum Institute
    scale (specific gravity)
    385–386
Ametryne 266
Amiben 264
Amictic lakes 219
Aminifying bacteria 155

Amino acids: deficiencies 306
  effect of radiation 455
  nutrition 372
5-Amino–4-chloro–2-phenyl-
    3(2H)-pyridazinone 264
Aminotriazole (ATA) 128
3-Amino-1,2,4-triazole 264
4-Amino-3,5,6-trichloropicolinic
    acid 264
Amitrole 264
Ammate 264
Ammonia 69
  ammonium nitrate production
    193
  ammonium phosphate produc-
    tion 195–196
  in aquatic forms 69
  fertilizer 192, 193
  low-troposphere concentra-
    tion 39
  mixed fertilizers 197
  nitrogen cycle 155, 366
  production 193
  urea production 194
Ammonification (nitrogen cycle)
    366
Ammonium nitrate: blasting 340
  fertilizer 193
  production 193
Ammonium phosphate: fertilizer
    192, 195–196
  production 196
Ammonium polyphosphate (fer-
    tilizer) 196
Ammonium sulfamate 264
Ammonium sulfate (fertilizer)
    193–194
Ammonium thiocyanate 128
Ammophila 135
Ammophila arenaria 136
Ammophila breviligulata 107,
    136, 137
Amphibians: Australian region
    697
  Ethiopian region 695
  Nearctic region 695
  Neotropical region 696–697
  Oriental region 696
  Palearctic region 694
  prairie ecosystem predators
    451
Amphiprion percula 145
AMS (herbicide) 264
Amundsen, Roald 30
Amur River 473
Anabaena 367
Anabaena cylindrica 404
Anabaena spiroides 187
Analysis, isotope 76
Analysis, systems 604–607
Analysis, water 514–515,
    655–656
Anastatic water 248
Anatase 543
Anderson, E. R. 492
Anderson, V. C. 482
Andira humilis 439
Andropogon 246, 479
Andropogon scoparius 432
Andropogon virginicus 596
Andropogoneae 432
Anemia: macrocytic 132
  pernicious 132
Anemochory 442
Anemone 430

Anemone patens 67
Aneuploidy 361–362
Angina pectoris 56
Angular momentum flux (atmos-
    phere) 43, 45
Animal community 25–27,
    60–61, 108–109
  biome 63–64
  classification 109–110
  consortism 27
  criteria for communities 26
  distribution 26
  divisions 26
  ecosystem 164–167
  habitat 25–26
  operation 26–27
  organization 26
  social relations 27
Animal ecology 164–167
  aerobiology 1–4
  animal community 25–27
  biogeography 59–61
  biological productivity
    61–63
  biomass 63
  biome 63–64
  biotic isolation 71
  climax community 93–96
  community classification
    109–110
  ecological interactions
    145–148
  ecological pyramids 148
  ecology 148–158
  ecology, human 160–161
  environment 175–178
  food chain 201–202
  fresh-water ecosystem
    218–230
  island faunas and floras
    288–290
  mangrove swamp 307–309
  population dispersal 442–444
  population dispersion 444–446
  population dynamics 446–450
  prairie ecosystem 450–452
  radioecology 465–468
  species population 585–586
  succession, ecological
    595–599
  zoogeographic region 694–698
  zoogeography 698–699
  zooplankton 700–701
Animal feed composition 5, 8–9,
    653
Animal geography 60–61
Animal manure 309
Animal pathology: aerobiology
    1–4
  disease 131–134
  effect of photochemical smog
    525
  epidemiology 180–182
  heavy metals, pathology of
    253–262
  malnutrition 305–307
  radiation injury (biology)
    456–460
  toxicology 622–623
Animal wastes (recycling)
    652–653
Animals: climax community
    93–96
  dispersal of disseminules 443
  dissemination of plant

Animals—cont.
  pathogens 421
  effects of atmospheric pollution 50
  nitrogen cycle 155
  population dispersal 445
Animals, social 27, 147, 446
Anion exchange: sea water 496
  soil 544–545
Annular stream drainage 472
Anodic stripping (sea water analysis) 507
Anomalies, congenital 132
Ansar 266
Antagonism (soil microbes) 542
Antagonistic symbiosis 412
Antarctic faunal region 25, 26
Antarctic front 15
Antarctic ice 75
Antarctic subregion 694
Antarctic tundra 628
Antarctica 27–30
  biology 30
  climate 29–30
  exploration and research 30
  geology 28–29
  ice sheet 28
Antelopus 479
Anthracite coal (uses) 336
Anthrax 132
Anthropochory 443
Anthroposphere 64
Antibiosis 413
Antibiotic (fungistat and fungicide) 232–233
Antiboreal faunal region 26
Anticyclones 47, 318, 589–590, 692, 693
Anticyclonic circulation 692
Antifungal agents 232–233
Anti-Locust Control Center 2–3
Aphanizomenon flosaquae 187
Apomixis 627
Apple scab fungus 415
Apple scald 412
Applications Technology Satellite (ATS) 593
Applied climatology 83
Applied ecology 158–160, 172, 298–300
Apron (geology) 114
Aqua ammonia 194
Aqualfs 541
Aquatic community (predator-prey relationships) 452–453
Aquatic ecosystem: eutrophication, cultural 187–189
  fresh-water ecosystem 218–230
  phytoplankton 401–402
  predator-prey relationships 452–453
  seasonal changes 615
  trophic structure 150
  zooplankton 700–701
Aquatic plants (eutrophic lakes) 187
Aquatic systems 57–58, 110
Aquifer 30
  artesian systems 34–35, 587
  water supply 664
Aquolls 541
Aragonite skeletons 69
Araucaria 430

Arch dam 118, 119–120
Archipelago 31
Archipelago, Canadian Arctic 32
Arctic-alpine life zone 301
Arctic and subarctic islands 30–33
  character of major islands 31–33
  climate 31
  diversity of land surfaces 31
  vegetation and soils 31
Arctic Basin 81
Arctic biology 33–34
  Arctic people 34
  life on land 33–34
  life on seas 34
  migration of life into Arctic 34
  preservation of warmth 34
Arctic bogs 634
Arctic Brown Soil 627
Arctic faunal region 26
Arctic fox 34
Arctic front 15
Arctic sea ice 81
Arctic subregion 694, 699
Arctic tundra 434, 628
Arctogaean realm 694
Arctostaphylos 137, 432
Area sanitary landfill 576, 577
Argic water 248
Argon (low-troposphere concentration) 39
Arid regions (irrigation programs) 118
Arid transition life zone 301
Arid tropical scrub (Mexico) 59
Arid vegetation zones 60
Aridisols 529, 541
Aristida 479
Armadillo, giant 191
Armstrong, Terence 33
Arroyos 130, 238
Arsenates 263
Arsenic 262, 622
Arsenic acid 128
Arsenical poison 12
Arsenicals (insecticides) 285
Arsine (poison) 622
Artemisia 433
Artemisia pychocephala 137
Artemisia spinescens 130, 433
Artemisia tridentata 130, 407, 433, 644
Artesian aquifer 35, 587
Artesian systems 34–35
Artesian water 250, 664, 667
Arthritis (epidemiology) 181
Arthrobacter 565
Artificial insemination 8, 71–72
Artificially formed beaches and dunes 107
Asbestos 143, 336
Ascariasis 132
Ascidia 66
Ascophyllum 312, 313
Ascorbic acid: deficiency 133
  disease 132
  nutrition 272
Ascorbic acid oxidase 404
Asexually propagated crops 74–75
Ash tree 431, 648
Asparagus rust 414
Aspens 136, 137
Aspergillus 422, 566

Aspergillus niger 404
Asphalt rock (oil sand) 394
Asphyxiants (disease) 132
Assimilation, biological 61–62
Associations, ecological 109–110
Aster 433
Asterionella 188
Asthma 132
Astronauts (exposure to ionizing radiations) 457
Astronomy (genesis of biosphere) 64–65
ATA see Aminotriazole
Atlantic coast (dune vegetation) 135
Atlantic faunal region 26
Atmosphere 35–37
  abiotic factor 177
  air pressure 17–19
  atmospheric chemistry 37–39
  biogeochemical cycling 57
  biosphere 64
  carbon dioxide concentration 78
  circulation 36–37
  cloud 96–99
  cloud physics 99–102
  composition 35–36
  dew point 131
  dilution of pollutants 21–22
  distribution of water 177
  disturbances 318–319
  earth sciences 144–145
  electrical properties 320
  greenhouse effect, terrestrial 247
  halogens, atmospheric geochemistry of 250–251
  heat balance, terrestrial atmospheric 251–252
  humidity 275
  hydrometeorology 276–278
  hydrosphere, geochemistry of 278–281
  jet stream 290–291
  kinematics of motions 330
  lake ecosystem nutrient flow 58
  meridional flux of water 277
  meteorology 316–332
  ocean-atmosphere relations 372–374
  pressure and wind 318
  radiation and thermal structure 317–318
  temperature change 78
  temperature inversion 608
  theories and models of climate 76–77
  thermal structure 36–37, 317
  water vapor content 78–79
  wind 692–694
Atmosphere, natural 250–251
Atmospheric analysis (air-monitoring instruments) 23–24
Atmospheric boundary layer 329
Atmospheric chemistry 35–36, 37–39, 53–54
  aerosols 38
  gases 37–38
  precipitation 39
  radioactivity 38–39
Atmospheric circulations (drought) 133

Atmospheric dynamics (micrometeorology) 332–334
Atmospheric electricity 39–40, 320
  disturbed-weather phenomena 39–40
  fair-weather field 40
Atmospheric evaporation 40–41, 189
  estimating evaporation 40–41
  heat of vaporization 40
  salt, ice, and snow factors 41
Atmospheric gases, 78, 177
Atmospheric general circulation 41–46, 86, 88, 277, 318, 332
  angular momentum flux 43
  balance requirements 43–44
  mean circulation 41–43
  mechanism of transfer processes 44–45
  seasonal changes 45–46
  wind 692–694
Atmospheric geochemistry of halogens 250–251
Atmospheric particles 79–80
  optical properties 80
  particle load 79–80
Atmospheric pollution 46–52, 53–54, 641–642
  air-pollution control 19–25
  atmospheric particles 79–80
  carbon dioxide pollution 77–79
  controls 50–51
  costs of 50
  dispersion 46–49
  effect on climate 77–80
  effects on man and environment 49–50
  electric power generation 630
  halogens 251
  inadvertent weather modification 682
  insecticides 174
  laws 51–52
  plant disease 411
  smog, photochemical 524–526
  sources 46
  temperature inversion 608
  types 46
Atmospheric sciences 145
Atmospheric stability 329–330
Atmospheric water cycle 276–277
Atolls, coral 289–290
Atomic bomb (radioactive fallout) 460
Atomic test (radioactive fallout) 39
ATP see Adenosinetriphosphate
Atrazine 264
Atriplex 130, 433, 652
Atriplex arenaria 135
Atriplex canescens 130
Atriplex confertifolia 130, 407
Atrophy (human cells) 133
Atropine 623
ATS see Applications Technology Satellite
Atyidae 224
Aufwuchs (periphyton) 221
Auroral physics (International Geophysical Year 1957–1958) 30
Australia: landform 608–610
  resource pattern 144

Australian beefwood tree 1
Australian faunal region 25
Australian region 694, 697, 699
Australian subregion 694
Austrian pine 624
Austroriparian life zone 301
Autecology 149
Autecology, forest 204–205
Authigenic mineral (formation) 495–496
Autochory 442
Automated irrigation systems 117–118
Automobile: contribution to air pollution 46
  smog, photochemical 525
Autotrophic bacteria 565
Autotrophic organisms 150
Autotrophic plants (ocean) 508–509
Auxochores 442
Avadex 264
Avadex BW 267
Avalanche breakers 52
Avalanches 52
Avalon Peninsula 32
Avena 247
Avicennia 307, 308, 429
Axel Heiberg Island 31, 32
Axonopus 479
Azak 265
Azonal soil 541
Azotobacter 69, 367, 404, 565, 568
Azotomonas 565

B

B 9 *see N*-Dimethylamine succinamic acid
B horizon 528
Baccharis 136
Bacillariaceae 566
Bacillariophyta 401
Backcrossing (plant breeding) 73
Backfill pressures (concrete dam) 119
Background noise (acoustics) 368
Backus, R. H. 482
Bacteria: aminifying 155
  autotrophic 565
  chemosynthetic 565
  decomposers 150
  denitrifying 155
  food requirements 413–414
  green sulfur 565
  heterotrophic 565
  host selectivity 414
  nitrate 155
  nitrate-forming 542
  nitrifying 568
  nitrite 155
  nitrite-forming 542
  nitrogen cycle 155, 366–367
  nitrogen-fixing 146, 155, 565
  photosynthetic 565
  phytoplankton 401–402
  plant disease 407, 409, 411, 412
  purple sulfur 565
  reproduction 415

Bacteria—*cont.*
  shade tree diseases 627
  silicate-decomposing 568
  soil microorganisms 564–567
Bacterial blight 408, 413, 421
Bacterial disease in man 132
Bacterial genetics 363
Bacterial wilt (tomato) 408
Bacteriology: examination of sewage 515
  water analysis 655–656
Baffin Island 31, 32
Baffles (incinerators) 579
Bag filters (dust collectors) 51, 138
Bagnold, R. A. 140, 594
Bagon 287
Bailers (oil well production) 388–389
Bainbridge, Arnold 492
Balan 264
Balance: biogeochemical 56–59
  community 109
  geostrophic 329
Ballochores 442
Bamboo 430
Band screens (sewage treatment) 519
Bank erosion 550
Banks Island 31
BanvelD 266
BanvelT 266
Bar (meteorology) 17
Bar-built estuaries 182, 184
Barban 264
Barber, S. A. 548
Barbiturates (poisons) 623
Barbiturism 132
Barchan dunes 134
Barge-mounted grab dredge (undersea mining) 358
Bark beetles 421
Barnacle–gastropod–brown algae biome 64
Barnacles 312
Barochores 442
Barrel (petroleum production technology) 385
Barrel cactus 130
Barrens, desert 637
Bartel, A. W. 524
Basic slag fertilizer 196
Basin, closed 281
Basin irrigation 117
Basswood 137, 431
Batea (placer mining) 343–344
Bathyscaphe 377
Bathythermographs (oceanographic survey) 376
Bauhinia 429
Baust, John 34
Bauxite 143
Beach erosion 105–106, 107
  currents 106
  littoral processes 106
  tides 106
  waves 105–106
Beach grass 135
Beach pea 137
Beach restoration 106
Beaches, artificially formed 107
Bear, polar 34
Beech 137, 430, 431, 647, 648

Beech-maple forest (energy transfer) 616
Beer drinker's syndrome 260, 261
Bees: honey and wax production 173
  pollination role 173
Beet (brown heart) 11
Beeton, Alfred M. 188
Behavior, migratory 451, 615
Behavior, reproductive 446
Beijerinckia 565
Belts, cordilleran 610
Belts, life 302
Bench border irrigation 117
Beneficiation (conservation measure) 111
Benefin 264
Benlate 233
Bensulide 266
Benthal division (ocean) 110
Benthic division: deep-sea system 314
  marine ecosystem 313–314
Benthos: lenitic ecosystems 221, 224
  littoral 224
  lotic ecosystems 229
  marine ecosystem 312
  ocean 508–509
  profundal 224
Benzac 267
Benzothiazoyl-2-oxyacetic acid 13
3,4-Benzpyrene (photochemical smog) 525
Beriberi 132, 272
Bermuda High 133
Bernard, Claude 273
Bernstein, I. L. 3
Bertholletia 429
Bertrand, D. 64
Bertrand, G. 64
Beryllium (uses) 336
Beta particles (biological radiation injury) 457
Beta radiation (radioecology) 465–468
Betasan 266
Betel 623
Betula 136, 431
γ-BHC (insecticide) 285, 286
BI-GAS process 102, 105
Bigelow, Henry 374
Bignoniaceae 479
Biocenose 108, 406
Biochemical circulation (sea water) 497
Biochemical cycle (sea water) 496–497
Biochemical oxygen demand (BOD) 516–517
Biochemical oxygen demand test 514
Biochemistry: chemical composition of living matter 66–68
  nutrition 372
  plants, minerals essential to 402–406
  radiation biochemistry 453–456, 457
Biociation 64
Bioclimate 175
Bioclimatic sweat control 54

Bioclimatology 52–56
  aerosol bioclimatology 54
  air bioclimatology 53–54
  climatotherapy 56
  extreme climates and microclimates 55–56
  ideal bioclimate 56
  paleobioclimatology 56
  photochemical bioclimatology 53
  statistical bioclimatology 56
  thermal bioclimatology 54–55
Biocycle: fresh-water 110
  marine 110
  terrestrial 109–110
Biodegradable compounds (pesticides) 399
Bioengineering 172
Biogenic molecules, degradation of 70
Biogenic salts 155
Biogeochemical balance 56–59
  analysis of cycles 58
  oxygen and carbon dioxide 58
  toxins 58
Biogeochemical cycles 57–58, 64, 68–70, 154–157
  nitrogen cycle 366–367
  nonessential and radioactive elements 156–157
  nutrients 155
  types of cycles 155–157
Biogeochemical prospecting 67–68
Biogeochemistry 64, 66–68, 164
Biogeography 59–61
  animal community 25–27
  animal geography 60–61
  Antarctica 30
  arctic biology 33–34
  biological species and water masses 510–511
  ecosystem 164–167
  human biogeography 61
  island faunas and floras 289–290
  plant formations, climax 428–434
  plant geography 59–60
  species population 585–586
  vegetation zones, altitudinal 642–645
  vegetation zones, world 645–652
  zoogeographic region 694–698
  zoogeography 698–699
Biological aerosols 420
Biological elements 67, 179
Biological insect control 173, 284–285
Biological monitoring 4
Biological oxidation 403, 455
Biological productivity 61–63, 155, 165–166
  aquatic organisms 304
  biomass 63
  controls 154
  ecological efficiencies 152–153
  ecosystems 163
  efficiency 62–63, 154
  energy and materials 151–152
  energy degradation 152
  energy flow 151

Biological productivity—*cont.*
  energy flow diagrams and
    models 152, 153
  energy transfer 152–153
  fresh-water ecosystem
    226–227
  grassland 247
  gross production and respira-
    tion 153
  lake classification 225
  lotic ecosystem 229
  matter and energy in ecosys-
    tems 62
  measurement of production
    62
  population dynamics 448–449
  primary production in the
    world 154
  rainforest 469
  saltmarsh 478
  sea-water fertility 507–511
  seasonal changes 615
  species, individuals, and
    energy flow 153–154
  standing crop and energy flow
    153
  trophic levels 615–616
  tundra 628
Biological reduction-oxidation
  systems (effects of radiation) 455
Biological rhythms 165
Biological stations (oceanography) 377
Biology: aerobiology 1–4
  Antarctica 30
  arctic 33–34
  bioclimatology 52–56
  biogeochemical balance 56–59
  biogeography 59–61
  biosphere 64
  ecology 148–158
  health physics 251
  heating, comfort 252–253
  marine 378
  radiation 453–456
Bioluminescence (algae blooms)
  401
Bioluminescence counters
  (oceanographic survey)
  276
Biomass 61, 63, 66, 150
  carbon dioxide reservoirs 78
  ecological pyramids 148
  energy flow diagrams 152
  food chain 201–202
  measurement 63
  metabolic rate 63
  pyramids 63
  species population 585–586
  standing crop 153
  trophic levels 109, 615
Biome 109, 110
Biometeorology 1–4, 52
Biophages 150
Biosphere 26, 64, 109, 151, 164
  biogeochemical cycles
    154–157
  carbon dioxide reservoirs 78
  evolution of 65
  flow of solar energy 569
  genesis of 64–65
  lake ecosystem nutrient flow
    58

Biosphere, geochemistry of
    64–71
  biogeochemical cycles
    154–157
  chemical composition of living
    matter 66–68
  degradation of biogenic
    molecules 70
  ecological succession, cause
    of 597
  genesis of biosphere 64–65
  influence of living matter
    65–66
Biota: adaptations to the habitat
    614
  climax community 93–96
  fresh-water ecosystems
    221–225
  island faunas and floras
    288–290
  lotic ecosystems 228–229
  marine ecosystem 311–312
  marine resources 315
  savanna 479
Biotic community 63–64, 108
Biotic districts 109
Biotic environment 175, 178
Biotic isolation 71
Biotic province 109, 110,
    302–303
Biotic realm 109, 110
Biphenyls, polychlorinated
    440–442
Birch 136, 431, 607
Birch dieback 624, 625
Birds: Australian region 697
  Ethiopian region 695
  migration 615
  Nearctic region 695
  Neotropical region 697
  Oriental region 696
  Palearctic region 694–695
  passerine 451
Birth rate: population growth
    447
  species population 586
Bischof, W. 77
Bis(dimethylthiocarbamoyl)sul-
  fide 77
2,4-Bis(isopropanolamino)-
  6-methylmercapto-
  s-triazine 264
2,4-Bis(3-methoxypropyl-
  amino)-6-mercapto-s-
  triazine 264
Bison 95, 151
Bison antiquus 191
Bison occidentalis 191
Bituminous coal: Lurgi
    generator 103
  Synthane process 105
  uses 336
Bituminous sands 393, 394
Biuret (urea manufacture) 194
Bivalve-annelid biome 64
Bjerknes, J. 669
Bjerknes, V. 15, 678
Black stem rust 1
Black-tailed prairie dog 451
Blackbrush (vegetation) 130
Blackheart 411
Bladder wrack 313
Blasting (open-pit mining)
    340

Blending, bulk (mixed fertilizer)
    197
Blight diseases (plants) 408
Blister blight 2
Blizzard 590
Block, continental 114
Block-caving (mining) 352
Blocky soil structure 532
Blood-destroying poisons 622
Blooms (algae and diatoms) 401
Blowout dunes 134
Blue-green algae 187, 401, 566
Blue pike 188
Bluestems 246
BOD *see* Biological oxygen de-
    mand
Bog 604, 637
Boiling water reactor 369
Boliophthalmus 309
Bolts, rock 353
Bombacaceae 479
Bombs, atomic and fission 460
Bones (fertilizer) 196
Boojum tree 131
Boole, George 605
Boolean algebra (systems ecol-
    ogy) 605
Boom, mechanical 484
Boom, sonic 581–585
Bora 693
Boralfs 541
Borascu 267
Borates 263, 267
Borax 267
Bordeaux mixture 12, 233, 410
Border strip irrigation 116–117
Boreal forest 31, 607, 634
  coniferous 142
  Mexico 59
Boreal woodland 650
Boric acid (sea water) 497
Borides (uses) 336
Boring and drilling (mineral):
    open-pit mining 338–340
  undersea mining 359–360
Borolls 541
Boron: constituent of living mat-
    ter 70
  essential element for plants
    404–405
  micronutrient 198
  nutrient 155
Boron-deficiency disease (gar-
    den beet) 410
Bosque (desert vegetation) 650
Botany: geographical 343
  plant geography 434–438
Botrytis 422
Bottom fauna (eutrophic lakes)
    187
Botulism 132
Boundary layer, atmospheric
    329
Boussinesq, J. 562
Bouteloua 246
Bouteloua gracilis 433
Bouteloua hirsuta 433
Bowhead whale 34
Bradley, C. E. 524
Brahmaputra River 473
Braided channels 474
Bramus 247
Branchiobdellidae 145
Braun-Blanquet, J. 407

Breakers, zone of 489–490
Breaking waves 489–490
Breakwaters (shore protection)
    108
Breeders: graphite-moderate
    molten-salt thermal 369
  unmoderated fast 369
Breeding (animal) 71–72
  accuracy of selection 71
  agricultural science (animal)
    7–10
  crossbreeding 72
  genetic fingerprinting 72
  methods of selection 71
  purebred breeds 72
  selection of animals 9
Breeding (plant): agricultural
    science (plant) 10–14
  asexually propagated crops
    74–75
  cross-pollinating species
    73–74
  disease resistant plants
    425–427
  recurrent selection 74
  scientific method 72–73
  self-pollinating species 73
  special techniques 75
Breeding, pedigree 73
Breezes, land and sea 693
Bridge (highway engineering)
    270
Bridge engineers 172
Brier, G. W. 283
Brittle stars 312
Broad bean (mutation study) 363
Broecher, W. S. 293
Bromacil 263, 264
Bromeliaceae 429
Bromeliads 646
Bromes 247
Bromide (sea water) 379
Bromine: constituent of living
    matter 69
  halogens, atmospheric
    geochemistry of 250–251
  sea water 315
Brominil 264
5-Bromo-3-*sec*-butylmethyluracil
    264
3-Bromophenyl-1-methoxy-1-
    methylurea 264
Bromoxynil 264
Bronchitis, chronic 49
Bronzing (plant) 405
Brook (headwater stream) 471
Broomsedge 596
Brown algae 312
Brown coal (Lurgi gasifiers)
    103
Brown heart 11
Brown patch 417
Brown spot needle blight 624
Brown-water (fresh-water
    ecosystem) 225
Bruguiera 307, 308
Bruguiera cylindrica 308
Bruguiera gymnorhiza 308
Bruguiera parviflora 308
Bruguiera sexangula 308
Bruniaceae 432
Brunskill, G. J. 293
Brush (tropical vegetation) 650
Buchloë dactyloides 433

Bucket-ladder dredge: placer mining 344
  undersea mining 358
  Yuba 344
Buckeye 431
Buckthorns 136
Bud sage (vegetation) 130
Buffalo 450
Buffers, acid and base (sea water) 497
Bugseed 137
Buildings: heating equipment 128
  response to sonic boom 585
  thermal energy systems 571
Bulk blending (mixed fertilizer) 197
Bulk density (soil) 533
Bulk-handling machines (open-pit mining) 340–342
Bulkhead (shore protection) 106–108
Bullets, tree 211
Bumelia 136
Bureau of Biological Survey 113
Burning (oil spill cleanup) 485
Burns, chemical and thermal 132
Bursera microphylla 131
Bush (tropical vegetation) 650
Butcher, R. 229
Butterfly valves 125
Buttress type dam 118
1-(Butylcarbamoyl)-2-benzimidazole carbamic acid methyl ester 233
N-Butyl-*N*-ethyl-*a*,*a*,*a*-trifluoro-2,6-dinitro-*p*-toluidine 264
Butyrac 265
Byrd, Richard E. 30
Byrsonima intermedia 439

C

C horizon 528
Cabbage (lightning injury) 411
Cabbage palm 136
Cacao 469
Cacodylic acid 128, 265
Cactaceae 432
Cactus 129, 130, 131, 136, 433, 651
Cadle, R. D. 524
Cadmium: heavy metals, pathology of 253
  poison 622
Cadmium disease (occurrence and symptoms) 256–257
Caffeic acid 424, 425
Caffeine 365, 623
CAgM *see* Commission for Agricultural Meteorology
Caisson wells 665
Cakile 135, 137
Calamogrostis 627
Calamovilfa longifolia 137
Calandria vessel 369
Calcareous soils 542, 545–546
Calcite skeletons 69
Calcium: constituent of living matter 69
  cycle 157
  essential element for plants 403

Calcium—*cont*.
  soils 116, 537
  utilization by soil microorganisms 568
Calcium arsenate (insecticide) 285
Calcium cyanamide (fertilizer) 194
Calcium nitrate (fertilizer) 194
Calcium phosphate (fertilizer) 192
Callendar, G. S. 78
Calluna 429
Calothrix 367
Calpouzos, L. 4
Cambrian (faunal extinction) 189
Camel (mass extinction) 191
Campo Cerrado 432, 649
Campos (Brazilian grasslands) 246
Canadian Arctic Archipelago 32
Canadian life zone 301
Canadian muskeg 648
Canals (water transmission) 665
Cancer 132
  caused by radiation 459–460
  lung 49
  miners 172
  nickel workers 262
  skin 53, 525
Candida 232
Candidiasis (chemotherapy) 232
Canker diseases: forest trees 624
  plants 409
Cannon, Walter B. 273
Canyon, submarine 114
Capillary water (soil) 558
Caprolactam (ammonium sulfate production) 193
Captan 233
Carbamates (biocides) 399
Carbaryl 287, 399
Carbohydrate-phosphates 69
Carbohydrates 67
  animal feed composition 8–9
  deficiencies 306
  metabolism 306
  nutrition 372
Carbon: biological element 67
  cycle 57, 58, 66, 68–69, 78
  essential element for plants 402
  soil 542
Carbon dioxide: absorption of solar radiation 78
  component of air 317
  content of the atmosphere 58
  cycle 37
  lake chemistry 226
  low-troposphere concentration 39
  pollution 77–79
  reservoirs 78
  sea water 310, 497
  urea production 194
Carbon dioxide acceptor process 102, 104–105
Carbon disulfide 263
Carbon monoxide: air pollution 46
  low-troposphere concentration 39
  poison 622

Carbonate (deposition in the oceans) 69
Carbonic acid (sea water) 497
Carbonic anhydrase 404
Carbophenothion 286
Carbosphere 64
Carbyne 264
Carcinogen: cobalt 261
  heavy metals 255
  nickel 262
  radiation 459–460
  smog, photochemical 525
  sun 53
Carcinoma (skin) 53
Carex 432, 433
Caribbean subregion 694
Caribbean transition subregion 699
Caribou 33, 142, 628
Caries, dental 132
Carludovica 429
Carnegiea gigantea 433
Carnivores: homeothermic 615
  poikilothermic 615
  predators 452
Carnotite 143
Carolinian life zone 301
Carp 188
Carrying capacity (populations) 447–448
Carter, D. R. 198
Cartography: soil maps 530–531
  vegetation and ecosystem mapping 633–639
  weather map 676–678
Carya 136, 431
Case fatality ratio (disease) 182
Casoron 264
Castanea 431
Castanea dentata 421
Castanea mollissima 421
Casuarina 1, 646
CAT *see* Clear-air turbulence
Cat, saber-toothed 191
Catalase 404
Catchbasins (highway engineering) 269
Catchment areas 143
Cations (distribution in sea water) 495
Cation exchange: sea water 496
  soil 543–544
Catochortus venustus 137
Catostemma 429
Cauliflower (whiptail) 11
Caving (mining) 351–352
CDAA *see* 2-Chloro-*N*,*N*-diallylacetamide
CDED *see* 2-Chloroallyl diethyldithiocarbamate
Ceanothus dentatus 137
Cedars 136
Cedrela 430
Cedrus 431
Ceiba 429
Celebesian transition province 699
Cell (biology): homeostasis 274
  malnutrition 306
  radiation injury 458
  response to disease 133
  response to heavy metals 255
  water content and radiation 453–454

Cell division (effect of radiation) 458
Cell membranes (phospholipids) 403
Cell permeability (homeostasis) 274
Cells, solar 572
Cellulose-decomposing fungi 566
Central heating (comfort heating) 253
Central low (statistical bioclimatology) 56
Central-station electrical plants 369–370
Central water 501
Centrifugal pumps (water pumping stations) 666
Centrifugal well pumps (oil well production) 388
Ceramic rings (packed bed) 138
Ceratocystis fagacearum 624
Ceratocystis ulmi 421, 626
Ceratophyllum 187
Cercidium floridum 131
Cercidium microphyllum 131
Cereal smuts 421
Cereus giganteus 130
Ceriops decandra 308
Cerium (biogeochemical cycles) 156–157
Certrol 265
Cesium (biogeochemical cycles) 156–157
Cesium-137 (biological radiation injury) 462
α-Chaconine 424
Chaetoceros 311
Chaetoceros atlanticus 311
Chain, food 201–202
Chamaephyte 439, 627
Chambers (physiological ecology) 163
Change of climate 75–76
Channels: braided 474
  meandering 474
  river 474, 603–604
  spillways 123–124
  straight 474
  stream 603–604
Chaoborus 293
Chara 224
Charles, Albert Honoré 374
Charpentiera obovata 288
Charts (weather) 676–678
Chattanooga shale 395, 396
Chauliodus 511
Chelation 527
Chemhoe 266
Chemical burn 132
Chemical defoliation 128
Chemical elements 66–68
Chemical engineering 172
  agricultural chemistry 5
  air-pollution control 19–25
  defoliant and desiccant 127–128
  dust and mist collection 137–140
  fertilizer 191–199
  fungistat and fungicide 232–235
  herbicide 262–268
  insecticide 285–288

Chemical engineering—cont.
  pesticides, persistence of 398–401
  polychlorinated biphenyls 440–442
  water softening 668
Chemical mutagens 364–365
Chemical oceanography 497
Chemical oxygen demand test 514
Chemical pollutants (coastal water) 661
Chemical precipitation (sewage treatment) 521
Chemicals: plant disease control 423–424
  vegetation management 641
Chemistry: agricultural 5
  atmospheric 35–36, 37–39, 53–54
  diffusion of nutrients 548–549
  health physics 251
  soil 542–549
Chemocline (lake) 292
Chemosynthetic bacteria 565
Chemotherapy: fungistat and fungicide 232–233
  plant disease control 423–424
Chemurgy 5
Chernozemic soils 541
Cherries 136, 648
Cherry leaf spot 233
Chestnut 421, 431, 648
  blight 408, 418, 427, 625
  canker disease 624
Chickenpox 132
Chick's law 516
Chihuahuan desert 130–131
Chinook (wind) 693
Chinstrap penguin 30
Chironomus 226
Chironomus plumosus 187
Chlorates 263, 267
Chlordane 285, 286, 400
Chlorella 3, 405
Chlorella pyrenoidosa 404
Chloride (sodium soils) 116
Chlorinated benzene 268
Chlorinated hydrocarbons:
  biocides 399–400
  insecticides 286
Chlorinated rubbers 441
Chlorination: sewage treatment 524
  water treatment 668
Chlorine: constituent of living matter 69
  essential element for plants 405
  halogens, atmospheric geochemistry of 250–251
  nutrient 155
  sea water 315
  water purification 476
Chlorine gas (water treatment) 668
Chlorinity (sea water) 280
Chloro IPC 266
2-Chloroallyl diethyldithiocarbamate (CDED) 264
2-Chloro-4,6-bis(dipropylamino)-s-triazine 264
4-Chloro-2-butynyl-N-(3-chlorophenyl) carbamate 264

Chlorococcum 3
2-Chloro-N,N-diallylactamide (CDAA) 264
2-Chloro-4-(ethyl-6-isopropyl-(amino)-s-triazine 264
Chlorogenic acid 424, 425
2-Chloro-N-isopropylanilide 264
3'Chloro-2-methyl-p-toluidide 264
2-[(4-Chloro-o-tolyl)oxy]-N-methoxyacetamide (OCS-21799) 264
Chloromycetin (constituent of living matter) 69
N'-4-(4-Chlorophenoxy)phenyl-(N,N-dimethylurea) 265
3-(p-Chlorophenyl)-1,1-dimethyl-urea 12, 264
Chlorophyceae 566
Chlorophyll: constituent of living matter 69
  iron 404
  magnesium 403
Chlorophyta 401
Chloropicrin 263
Chlorosis 405, 409
Cholera 132
Christiansen, R. 482
Chromatography, gas 507
Chromium: poison 622
  uses 336
Chromosome: damage due to heavy metals 255–256
  mutation 361–366
  rearrangement mutations 362
Chromosome aberrations 361–362, 365, 458
Chronic bronchitis (air pollution) 49
Chronic diseases (prevalence ratios) 181
Chronic lead poisoning 259
Chrysanthemum cinerariaefolium 287
Chrysanthemum coccineum 287
Chrysolina quadrigemina 173
Chrysophyta 401
Chum, Carl 374
Churn drilling (open-pit mining) 339–340
CIPC see Isopropyl–N-(3-chlorophenyl) carbamate
Circuit breakers 170
Circuits: logic 605
  switching 605
  transmission 169, 170, 631
Circular clarifier 520
Circular trickling filter 522
Circulation: anticyclonic 692
  atmospheric, general 41–46, 86, 88, 133, 277, 318, 320, 332, 692–694
  cyclonic 692
  meridional 41
  monsoon 89
  tropospheric zonal 41
  upper air 693
Circulation, biochemical 497
Circulation, estuarine 182–184
Circulation flux (atmosphere) 45
Circulatory disorders 56
Circumpolar aerology 325–326

Cirques (mountains) 273
Cirriform clouds 97
Cirrocumulus clouds 97, 98
Cirrostratus clouds 97, 98, 318
Cirrus clouds 96, 97
Cisco (fish) 188
Citrus blackfly 284
Citrus canker disease 427
Citrus fruit 133
Civil engineering 171–172
  coastal engineering 105–108
  conduits 564
  dam 118–127
  highway engineering 268–271
  Imhoff tank 281–283
  land-use planning 298–300
  physical alterations of the coastal zone 661
  retaining walls 564
  river engineering 474–476
  rural sanitation 476–477
  sanitary engineering 478
  septic tank 512
  sewage 512–515
  sewage disposal 515–518
  sewage treatment 518–524
  solid-waste disposal 573–581
  thermal energy for buildings 571
  water supply engineering 664–666
  water treatment 667–669
Cladocerans (fresh-water ecosystem) 223
Cladophora 187
Claims, mining 342–343
Clams 312
Clamshell dredge (undersea mining) 358
Clarifiers, circular and rectangular 520
Clariflocculator, Dorr 521
Clarke, F. W. 64
Classes: land-use 297–298
  terrain 609
  vegetation 646–652
Classification: cloud 97–99
  community 26, 109–110
  engineering soil 557–558
  soil 528–530
  wastes 581
Claus process 103, 105
Claviceps purpurea 415
Clay 94, 531
  earth dams 121–122
  mineral composition 556
  minerals 542, 543
  oil shale 394
  properties 532
Claytonia 430
Clean Air Act of 1956 51
Clean Air Act of 1963 52
Clean nuclear weapons 461
Clear-air turbulence (CAT) 334
Clements, F. 437, 598
Clermontia haleakalae 289
Clethra 647
Cliffs (influence on vegetation) 94–95
Climate: agriculture, effect of 80–82
  atmospheric, general 41–46,
  carbon dioxide pollution 77–79
  change of 75–76
  climatology 82–93

Climate—cont.
  climax community 93, 598
  community changes 108
  desert 141, 142
  effect on population dispersion 446
  global 83–92
  grassland 246
  humid 141, 142, 527
  ice 141
  ice cap 142
  Mediterranean 141, 142, 143
  numerical calculation 674
  plant life forms 439
  polar cap 141, 142
  puna 141, 142
  rainforest 141, 142
  release of heat by man's activities 82
  sea ice, effect of 81–82
  semiarid 141, 142, 143
  soil erosion 539
  surface changes, effect of 80–82
  taiga 141, 142, 607
  terrestrial communities 165
  theories and models of 76–77
  tropical rainforest 469
  tropical savannas 480
  tundra 141, 142
  upper midlatitude 141, 142
  water-deficient 142
  wet-and-dry 141, 142
  zonality of soil 540–541
Climate, man's impact on 75–82, 641, 682
  agriculture 80–82
  atmospheric particles 79–80
  atmospheric pollution 77–80
  carbon dioxide pollution 77–79
  change of climate 75–76
  heat release 82
  mathematical expression 77
  model types 77
  sea ice 81–82
  surface changes 80–82
  theories and models of climate 76–77
Climatic climax 93, 167
Climatic regions 92–93, 141
  quantitative definitions 90
  world 91
Climatic types 142
Climatography 83
Climatology 82–93
  agricultural meteorology 5–7
  air mass 14–17
  air pressure 17–19
  Antarctica 29–30
  applied 83
  arctic and subarctic islands 30–33
  atmosphere 35–37
  atmospheric chemistry 37–39
  atmospheric electricity 39–40
  atmospheric general circulation 41
  atmospheric pollution 46–52
  bioclimatology 52–56
  climate, man's impact on 75–82
  climatic regions 92–93
  cloud 96–99
  cloud physics 99–102
  degree-day 128–129

Climatology—cont.
dew point 131
drought 133
earth, resource patterns of 140–144
effect on terrestrial ecosystems 615
environmental protection 178–180
front 230–232
frost 232
greenhouse effect, terrestrial 247
heat balance, terrestrial atmospheric 251–252
humidity 275
hydrometeorology 276–278
industrial meteorology 283–284
jet stream 290–291
meteorology 316–332
micrometeorology 332–334
origins and pattern of global climate 83–92
physical and dynamic 83
regional 82
sensible temperature 511–512
smog, photochemical 524–526
squall line 587–588
statistical 83
storm 588–590
synoptic 83
temperature inversion 608
thunderstorm 620–621
tornado 621–622
weather forecasting and prediction 669–676
weather map 676–678
weather modification 678–682
wind 692–694
Climatopathology 52
Climatophysiology 52
Climatotherapy 56
Climax: climatic 93, 167
disturbance 95–96
edaphic 94
fog 96
forest 158
prevailing 167
salt spray 96
topographic 95
Climax community 93–96, 109, 167, 407, 595, 598–599
disclimax 95–96
edaphic modifiers 94
forest succession 207
forestry 158
grassland 245–247
gross production and respiration 153
kinds of climax 93
monoclimax 93–94
polyclimax 94
topographic modifiers 94–95
Climax plant formations 108, 245–247, 428–434, 596, 646
Climax species (biome) 63–64
Climbing dunes 134
Climbing vines 60
Clinograde curve 225
Clobber (herbicide) 265
Cloroxuron 264
Close, R. C. 2
Closed basins (geochemistry) 281

Closed forests 636
Clostridium 367, 565
Cloth collector (dust collector) 138–139
Clothing (environmental protection) 178, 179
Cloud 96–99
altocumulus 98
altostratus 98
cirrocumulus 98
cirrostratus 98
cirrus 97
classification 97–99
cloud physics 99–102
convective 101
cumulonimbus 98–99, 620
cumulus 97, 102
droplet 97
fibrous 97
fronts 231
heap 97
high level 97
ice 97
layer 97
low-level 97
middle-level 97
modification 680
mother-of-pearl 37
nimbostratus 98
rudiments of formation 96–97
squall line 587
stratocumulus 98
stratus 98
supercooled 97, 99, 101, 102
thunderstorm 620–621
Cloud chamber (ice crystals) 101
Cloud classification 97–99
Cloud forest 59, 647
Cloud formation 80, 96–97, 99
Cloud physics 99–102, 316
artificial stimulation of rain 102
basic aspects of cloud physics 101
cloud formation 99
mechanisms of precipitation release 99–100
precipitation from layer-cloud system 100
production of showers 100–101
Cloud prediction 673
Cloud seeding 102, 679–681
Cloudberry (tundra vegetation) 627
Cloudbursts (thunderstorm) 620
Clovers (defoliation) 128
Clumped population dispersion 444
Clusia 429
CMU see 3-(*p*-Chlorophenyl)-1,1-dimethylurea
Coagulation (water treatment) 667
Coal 66, 69
bituminous 103, 105
brown 103
noncaking 103
steam turbine fuel 168
subbituminous 103, 105
Coal engineers 172
Coal gasification 102–105
BI-GAS process 105
carbon dioxide acceptor process 104–105

Coal gasification—cont.
HYGAS gasifier 103–104
Lurgi gasifier 102–103
new synthetic pipeline gas from coal 103
raw gas purification 105
Synthane process 105
Coal mining: strip 346–348
underground 348–354
Coalescence process (precipitation release) 99–100
Coastal dunes 134–135
Coastal engineering 105–108
beach erosion 105–106
breakwaters 108
bulkheads, seawalls, and revetments 106–108
groins 107–108
shore-protective structures 106–108
Coastal plain estuary (distribution of pollutants) 182–185
Coastal waters 660–662
conservation agencies and legislation 661–662
contamination by toxic materials 661
management 660
origin and description 660
sewage 660–661
thermal pollution 661
Cobalt: constituent of living matter 70
essential element for plants 405
nutrient 155
soil 548
uses 336
Cobalt chloride (treatment of anemia) 261
Cobalt deficiency (soil) 548
Cobalt toxicity 260–261
beer drinker's syndrome 261
carcinogen 261
Cocaine 623
Coccidiomycosis 132
Coccoloba uvifera 136
Coconut 646
Coconut palm 136
Cocos nucifera 136
Coffeeberry 137
Cogon (rainforest vegetation) 470
Coke: ammonium sulfate production 193
packed bed 138
Colby, B. R. 595
Cold (safe body exposure) 55
Cold anticyclones 589
Cold conifer ecosystem 634
Cold-deciduous forest 634, 636
Cold-deciduous scrub 637
Cold-deciduous woodland 636
Cold desert 652
Cold front 231, 323
Cold savanna 480
Coleogyne ramosissima 130
Coliform test 515
Collector, cloth (pollution control) 138–139
Collector, dust (pollution control) 138
Colletotrichum circinans 417
Colorado potato beetle 12
Columbia River 473
Columbia sheep 8

Columbian mammoth (mass extinction) 191
Columbium (uses) 336
Columnar soil structure 532
Combine (agricultural machinery) 11
Combustion (inadvertent weather modification) 682
Combustion turbines (electric power generation) 629
Comfort heating 128–129, 252–253
Commensalism 145–146
Commercial fishery 200
Commercial forest 212, 214
Comminutor 519
Commission for Agricultural Meteorology 6
Common fishery 200
Communication engineers 172
Community 108–109, 149
abyssal 165
animal community 25–27, 60–61, 63–64, 108–109, 109–110, 164–167
balance 109
biogeography 59–61
biome 63–64
biotic 63–64, 108
climax 93–96, 109, 153, 158, 167, 207, 245–247, 407, 595
concepts of structure 109
differentiation 165
disturbance 95
divisions 26
dominants 108
ecological interactions 145–148
ecosystem 164–167
forest and forestry 202–204
forest ecology 204–207
fresh-water 617
functional kingdoms 166
human 160–161
marine 617
operations 26–27
plant 59–60, 63–64, 108–110, 129–130, 164–167, 202–207, 245–247, 307–309, 406–407, 436–438, 477–478, 508–509, 595–599, 632–633, 642–645
response to radiation 467–468
saltmarsh 477–478
spatial patterning 165
species, individuals, and energy flow 153–154
structure 108
succession 109
succession, ecological 595–599
terrestrial 165, 617
Community classification 26
aquatic systems 110
terrestrial biocycle 109–110
Comparative physiology (homeostasis) 273–275
Competence (rivers) 594
Competition (ecological interactions) 146–147
effect on population dispersion 446
mass extinctions 189
Complete sewage treatment 519
Complex dunes 134
Compositae 246

Composition, animal-feed 653
Compost 309
Composting (solid-waste disposal) 579–580
Compounds, biodegradable 399
Concrete (support of mine openings) 352–353
Concrete dam 118–121
   forces acting on 119
   foundation 126
   instruments 127
   stability and allowable stresses 119–120
   temperature control 120–121
Condensation nuclei (cloud formation) 99
Conditioners, soil 536–537
Conductivity, electrical 494
Conduits 564
Congenital anomalies 132
Congo River 473
Conifer swamps 95
Coniferous forests 60, 64, 95, 215, 431, 607–608 636, 648
Conservation of resources 110–113
   conservation trends 111–112
   crops micrometeorology 334
   ecology, applied 158–160
   federal legislation 113
   fisheries 159, 199–201, 379, 660
   forest management 203
   forest resources 212–216
   human resources 112
   insect control, biological 284
   land-use planning 298–300
   mineral resources 113, 335–338,
   nature of conservation 111
   nature of resources 110–111
   rangeland 159, 247, 470–471, 642
   recreation resources 112
   soil 113, 199, 203, 535–538, 549–555, 641
   vegetation management 639–642
   water 113, 118, 160, 188, 198, 199, 473–476, 641, 656–664
   wildlife 113, 159–160, 199–201, 642, 682–692
Consolidated marine mineral resources 355, 359–360
Consortism (ecological interactions) 27
Constant, solar 251
Constant-head permeameter 559
Constant-level chart 678
Consumers (ecosystem) 150, 166
Consumption: biological 61
   water 513
Consumptive use of water (plant, water relations of) 115
Contact aeration (sewage treatment) 523, 524
Contact dermatitis 132
Contact fungicides 233
Contagious population dispersion 444
Continent: Antarctica 27–30
   cointinental shelf and slope 113–114
   Eurasian 144

Continent—cont.
   resource pattern 144
   terrain areas, worldwide 608–610
Continental aerosols 38
Continental air: hydrologic cycle 88
   mass 16
Continental block 114
Continental margin 114
Continental plateaus 114
Continental precipitation patterns 91–92
Continental rise 114
Continental shelf and slope 113–114
   oil and gas, offshore 380–383
   scattering layer 482
Continental terrace 113–114
Continentality (meteorology) 91–92
Contour ditches (irrigation) 117
Control, air pollution 19–25, 579
Controlled-environment (plant research) 163
Controlled filters (sewage filters) 522
Convection (precipitation) 91
Convective clouds (precipitation) 101
Convergence: meteorology 16–17
   precipitation 91
   wind patterns 692–693
Cool coniferous forest 215
Cooling: degree-day, 55, 128–129, 511
   solar space 571
Copepod 223, 312, 452
Copper: complexes in soils 547
   constituent of living matter 70
   essential elements for plants 404
   heavy metals, pathology of 253
   micronutrient 198
   nutrient 155
   thermal constants 618
   uses 336
Coral atolls (biotas) 289–290
Coral-eating starfish 453
Coral reef (ecosystem study) 149
Corals, shore 312
Cordilleran belts 610
Coring tubes (oceanographic survey) 376
Coriolis force 231
Corispermum 137
Corn: loss due to insects 173
   phosphorus deficiency 405
   rootworms 286
   root zone depths 115
Corrosion: electrochemical 389
   oil and gas-well production 389–390
   oxygen 389
   sour 389
   sweet 389
Corrosives (disease) 132
Corundum 336
Cosmetic standards (fruits and vegetables) 174
Cosmic abundance of elements 35–36
Cosmic rays 38

Cosmopolitan species (sea water) 510–511
Cotoran 267
Cotton (root zone depths) 115
Cotton boll weevil 286
Cotton picker (agricultural machinery) 11
Cottontails 450
Cottrell precipitator 139
Cotula 478
Counters, bioluminescence 376
CPM see Critical path method
CPU-400 (solid waste disposal) 580
Crabeater seal 30
Crassulacean acid metabolism 129
Crawler-mounted drilling machine 339
Crayfish (commensalism) 145
Creeks 471
Creeping shrubs (arctic islands) 31
Creosote bush 130, 433
Cretaceous (faunal extinction) 189
Crindelia 433
Critical path method (CPM) 689
Croaker (fish) 186
Cronese Cat (falling dune) 134
Crop, standing 153
Crop destruction (defoliant and desiccant) 128
Crop micrometeorology 334–335
Crop residues (utilization and disposal) 653–654
Crop rotation 536, 553
Cropping system 536
Crops: asexually propagated 74–75
   growth regulators 13–14
   losses due to insects 173
Crops, irrigation of 114–118
   automated systems 117
   consumptive use 115
   fertilizing operations 118
   humid and arid regions 118
   land drainage (agriculture) 294–297
   methods 116–117
   outlook 118
   soil, plant, and water relationships 115
   use of water by plants 114–115
   water contamination 654–655
   water quality 115–116
Cross-pollination 72–74
Crossbreeding 8, 72
Crown gall 408, 414
Crude death rate (epidemiology) 180
Cruickshank, Ian 424
Crumb soil structure 532
Crustacea 312
Cryolite (insecticide) 285
Cryptocarya 432
Cryptococcus neoformans 232
Crystal, ice 97, 100, 101
Crystal, snow 101
Cultural eutrophication 187–189, 198
Cultural evolution 161
Culvert (highway engineering) 270

Cumuliform clouds 97
Cumulonimbus clouds 98–99
   precipitation 318
   thunderstorm type of cloud 620
Cumulus clouds 96, 97, 318
   formation 99
   seeding 102
Curie 465
Current (lakes) 220–221
Current meters (oceanographic survey) 276
Currents, littoral 106
Currents, ocean 88, 106, 373, 377, 510
Cut-and-fill stoping 350–351
Cutting cycle (forest) 208–209
Cyanide (poison) 622
Cyanocobalamin 67
Cyanophyta 401
Cycle: air mass 277
   atmospheric water 276–277
   biochemical 496–497
   biogeochemical 57–58, 64, 68–70, 154–157, 366–367
   carbon dioxide 37
   forest cutting 208–209
   geochemical 503
   hydrologic 88, 90–91, 99–102, 177, 189, 217–218, 275–276, 373, 599–604, 657–658
   nitrogen 366–367
   sedimentary 156, 278
Cycle of materials (ecosystem) 166
Cyclochores 442
Cyclogenesis 56
Cycloheximide 233
Cycloloma 137
Cyclone 318, 692, 693
   extratropical 89, 323–325, 588, 589, 593
   principal tracks 590
   storm surge 593–594
   tropical 319, 593
   wave 230
   weather patterns 589
Cyclone incinerator 579
Cyclone separator 21, 138
Cyclonic circulation 692
Cyclonic depressions 100
Cyclonic liquid scrubber 139
Cyclonic spray scrubber 21
Cylinder valves (dams) 125
Cyperaceae 246
Cypromid 265
Cyrtandra 288
Cyrtonyx montezumae 60
Cystospora kunzei 626
Cytochrome oxidase 404
Cytological tests (mutation studies) 363
Cytoplasmic male sterility (breeding, plant) 75

D

2,4-D see 2,4-Dichlorophenoxyacetic acid
Dacrydium 430
Dacthal 265
Dakota sandstone 35
Dalapon 265
Dalea polyadenia 130
Dalton, J. 40

Dam 118–127, 475
  Ambursen type 118
  arch 118, 119–120
  buttress type 118
  concrete 118–121, 126, 127
  determination of height 126
  diamond-head type 118
  diversion 118, 125, 126
  earth dams 121–122, 126, 127
  East Canyon (Utah) 121, 122
  engineers 172
  foundation treatment 126–127
  gravity 118, 119, 123
  Green Peter (Oregon) 119, 120
  inspection 127
  instrumentation 127
  multiple-purpose 118
  navigation 118, 125
  North Fork (Va.) 123
  outlet gates 126
  penstocks 125
  Pound (Va.) 124
  power 118, 125
  reservoir outlet works 124–125
  selection of sight 125–126
  selection of type 126
  slab-and-buttress type 118
  spillways 122–124, 125
  storage 118, 125
  water management 641
Damage, radiation 251
Dana, J. D. 374
Dandelions 627
Dansereau, P. 437
Dansgaard, W. 76
Danube River 473
Daphnia (fresh-water ecosystem) 223
Darwin, Charles 288
Darwin finches 288
Dasylirion 131, 433
Davis, Charles C. 187
Davis Strait (Arctic Islands) 32
Dawsonite 395
Day neutral plant 14
2,4-DB (herbicide) 263
4-(2,4-DB) see 4-(2,4-Dichloro-phenoxy) butyric acid
DCPA see Dimethyl-2,3,5,6-tetrachloroterephthalate
DDE (DDT compound) 400
DDT see Dichlorodiphenyl-trichloroethane
Deafness (noise pollution) 305
Death 131–132
Death rate: crude (epidemiology) 180
  species population 586
Debris transport theory 594–595
DeCandolle, A. P. 300
Decibels 305, 585
Deciduous forest 109, 469, 636
  climax regions 599
  summergreen 430–431
  tropical 430
Deciduous scrub 637
Deciduous woodlands 636
Deciduous xerophytes (desert) 129
Decision theory (wildland operations research) 689
Decomposers (ecosystem) 150, 151
Decontamination of radioactive materials 462–463

Deep-basin estuary 182, 184
Deep-sea drilling 377
Deep-sea drilling vessel 375
Deep-sea fauna 314
  predator-prey relationships 452–453
  scattering layers 482
Deep-sea reversing thermometer 504
Deep-sea system (marine ecosystem) 314
Deep water (scattering layer) 482
Deevey, E. S. 293
Defant, A. 374
Deficiency, nutritional 132
Defoliant and desiccant 5, 127–128
Defoliation, chemical 128
Degradation, energy 152
Degradation of biogenic molecules 70
Degree-day 128–129
  cooling 55, 511
  heating 55
Degrees of frost 232
Demeton 286
Demography 449–450
Dendritic star-shaped ice crystal 101
Dendritic stream drainage 472
Dengue 174
Denitrification (nitrogen cycle) 366–367
Denitrifying bacteria 155
Density, population 585–586
Dental caries 132
Dental excavation 127
Deoxygenation (polluted waters) 516
Deoxyribonucleic acid (DNA): effect of radiation 456
  mutation 362
2,4-DEP see Tris-(2,4-dichlo-rophenoxyethyl)-phosphite
Deposition zones (beach) 105
De Pourtales, L. F. 374
Depressions, cyclonic 100
Depth recorder (marine mineral exploration) 356
Dermatitis: contact 132
  nickel 262
Derris elliptica 287
Derris malaccensis 287
Descriptive climatology 83
Desert 64, 142–143, 429, 637
  Chihuahuan (western North America) 130–131
  climate 141, 142
  cold 652
  dry 433
  dune 134–135
  extratropical 634
  Great Basin (western North America) 130
  human bioclimatology 55
  mean temperature 90
  Mexico 59
  Mojave (western North America) 130
  precipitation 90
  rock 652
  soils 541
  Sonoran (western North America) 130–131
  vegetation 129–131
  warm 651–652

Désert (vegetation zone) 651
Desert barrens 637
Desert erosion features (dune) 134–135
Desert Land Act of 1877 113
Desert regions (western North America) 130
Desert shrublands 637
Desert vegetation 129–131, 433, 651–652
  effects of physical environment 129–130
  life-form adaptations 129
  western North America 130–131
Desiccant, defoliant and 127–128
Desiccation, preharvest 128
Design, map 633–639
Desmochores 442
Desulfovibrio 568
Detection, storm 590–593
Developmental disease 132
Devon Island (Arctic) 31, 32
Devonian (faunal extinction) 189
Dew point 131, 275
Diabetes 132, 181
Diagrams, energy flow 152
Diallate 264
Diamond-head buttress type dam 118
Diaspores 442
Diatomaceous earth filters 668
Diatoms 401, 566
  eutrophic lakes 187
  marine ecosystem 311
Diazinon 286, 399
3,5-Dibromo-4-hydroxybenzo-nitrile 264
2,6-Di-tert-butyl-p-tolyl-methylcarbamate 265
Dicamba 266
Dice, L. R. 302, 699
Dicentra 430
Dice's life zones 302–303
Dichlobenil 263, 264
Dichlone 233
S-2,3-Dichloroallyl-N,N-diisopro-pylthiolcarbamate 264
2,5-Dichloro-3-aminobenzoic acid 264
3,4-Dichlorobenzl methylcarba-mate 265
2,6-Dichlorobenzonitrile 264
3′,4′-Dichlorocyclopropanecar-boxanilide 265
Dichlorodiphenyltrichloroethane (DDT) 12, 58, 174, 285, 286, 398–400
3′,4′-Dichloro-2-methacrylanilide 265
2,3-Dichloro-1,4-naphtho-quinone 233
2,4-Dichlorophenoxyacetic acid 12, 128, 262, 263, 265
4-(2,4-Dichlorophenoxy)butyric acid 265
2,4-Dichlorophenoxypropionic acid 265
3-(3-4-Dichlorophenyl)-1,1-dimethylurea 265
3-(3,4-Dichlorophenyl)-l-methoxy-1-methylurea 265
2,4-Dichlorophenyl-4-nitrophenyl ether 265
Dichloroprop 265

3′,4′-Dichloropropionanalide 265
Dicryl 265
Dictyospermum scale 284
Dieldrin 285, 286, 399, 400
Diepoxide (mutagen) 365
Diepoxybutane (mutagen) 365
Diesel engine (placer mining) 346
Diet: malnutrition 305–307
  prairie animals 451
Digitalis 623
2,3-Dihydro-5-carboxanilido-6-methyl-1,4-oxathiin 233
6,7-Dihyrodipyrido(1,2-a:2′,1′-c) pyrazidiinium dibromide 265
1,2-Dihydro-pyridazine-3,6-dione diethanolamine salt (MH) 265
3,5,-Diiodo-4-hydroxybenzonitrile 265
Dike, pile 475
Dilan 285, 286
Dilleniaceae 479
Dilution effect (irradiation of solution) 453
Diment, W. H. 293
Dimetan 287
Dimethoate 286, 287
N-Dimethylamine succinamic acid (B9) 13
Dimethylarsinic acid 265
1,1′-Dimethyl-4,4′-bipyridinium dichloride salt 265
N,N-Dimethyl-2,2-diphenyl-acetamide 265
Dimethyl-2,3,5,6-tetrachloro-terephthalate 265
O,S-Dimethyltetrachlorothio-terephthalate 265
Dimetilan 287
Dimictic lakes 219
Dimmick, R. L. 3
Dimorphandra mollis 439
Dingell-Johnson Act 113
4,6-Dinitro-o-sec-butyl phenol 265
Dinitrocaprylphenyl crotonate 423
4,6-Dinitro-2-(1-methylheptyl) phenylcrotonate 233
Dinoflagellates 311, 401
Dinophysis 311
Diospyros 429
Dioxathion 286
Diphenamid 265
Diphosphopyridine nucleotide (DPN) 403
Diphtheria 132
Diplothemium campestre 439
Diquat 128, 265
Dire wolf (mass extinction) 191
Direct current, high-voltage 170
Direct solar pigmentation 53
Disclimax 95–96
  fire 95
  pasture 95
  stable 95
Discomfort index 511
Discosoma 145
Disease: bacterial 132
  cadmium 256–257
  chronic 181
  cytological response 133
  endogenous 132–133
  environmental factors 133
  epidemiology 180–182

Disease—*cont.*
    etiology 132–133
    exogenous 132–133
    fungus 132
    infectious 180–182
    Kwashiorkor 307
    mental 181
    noninfectious 180–182
    plant 233, 405–428, 623–627
    protozoan 132
    recognition 131–132
    respiratory 49
    rickettsial 132
    thallium 259–260
    tree 623–627
    virus 132
    water-borne 476
Disease control, plant 422–428
Disease escape, plant 426
Disease rates (epidemic)
    181–182
Disease-resistant plants 424–426
Disease-tolerant plants 426
Disinfection (water treatment)
    668
Disodium 3,6-endoxohexa-
    hydrophthalate 266
Disorders, metabolic 133
Dispersal, population 289,
    442–444
Dispersants (oil spill cleanup)
    485
Dispersion, population 442–446,
    586
Dispersion of pollution 46–49
Disposal, sewage 476–478, 515
Disposal, solid-waste 573–581
Disposal, tailing 346
Disposal, waste 316, 342
Disseminules 442
    plankton 443
    transportation 442
    transported by animals 443
    wind dispersed 443
Dissolved gas expansion well 383
Distribution reservoirs 666
Distribution systems, electric
    170
Distribution transformers 170
Districts, biotic 109
Disturbance climax 95–96
Disturbance communities 95
Disulfoton 286
Dithiocarbamic acid 233
Dittmar, W. 506
Diuron 265
Divergence: meteorology 16–17
    wind patterns 692–693
Diversion dams 118, 125
Division, cell 458
DNA *see* Deoxyribonucleic acid
DNA polymerase 255
DNBP *see* 4,6-Dinitro-*o-sec*-
    butyl phenol
n-Dodecylguanidine acetate
    233
Dodine 233
Dokuchaev, V. V. 541
Doldrums 693
Dollar spot 417
Dolomite (carbon dioxide accep-
    tor process) 105
Domestic sewage 512–515
Domestic waste 663
Dominance plant societies 437
Dominant lethal mutations 363

Dominants (community) 108
Doppler radar (storm detection)
    592
Dormancy 614, 615
Dormant inoculum (plant
    pathogens) 416
Dorngeholz 650
Dorngestrauch 650
Dornveld 650
Dorr clariflocculator 521
Douglas fir 644
Dow General 265
Downdraft (thunderstorm) 620
Down's syndrome 132
Dowpon 265
2-(2,4-DP) *see* 2,4-Dichloro-
    phenoxypropionic acid
DPN *see* Diphosphopyridine
    nucleotide
Dragline dredges (placer mining)
    344–345
Draglines (open-pit mining)
    340–341
Drain, vertical sand 270
Drainage: highway engineering
    269–270
    subsurface 294–297
    surface 294
Drainage, land 294–297, 515–518
Drainage basin 471, 601–602
Drainage patterns 296
Drainage systems: construction
    296
    land drainage (agriculture)
        294–297
    maintenance 296–297
    patterns 295–296
    shallow-ditch 295
Dredge: barge-mounted grab 358
    bucket-ladder 344, 358
    clamshell 358
    dragline 344–345
    hydraulic cutter 358
    oceanographic survey 376
    offshore suction 345
    Yuba bucket-ladder 344
Dredging: exploitation of uncon-
    solidated mineral deposits
    357–358
    placer mining 344–345
Drift, littoral 105–107
Drift mining (placer) 345–346
Driggers, V. W. 293
Drill, rotary 339
Drill ships, floating 381–382
Drilled geothermal power
    244–245
Drilled geothermal power-
    generating system 244
Drilling:churn 339–340
    deep-sea 377
    offshore 359
    oil and gas well 383–385.
    percussion 339
    rock 338–340
    rotary 339
Drilling and boring (mineral)
    338–340
Drilling machine, crawler-
    mounted 339
Drilling platform: mobile
    381–382
    offshore fixed 381
    offshore self-elevating 381
    offshore semisubmersible
        381–382

Drilling vessel, deep-sea
    375
Drimys 430, 647
Drizzle drops 99
Droplet clouds 97
Drosophila melanogaster 363
Drought 133
    agricultural 133
    desert vegetation 129–131
    hydrologic 133
Drought-deciduous forest 636
Drought-deciduous scrub 637
Drought-deciduous woodlands
    636
Drug idiosyncrasy 132
Drug resistance (mutations) 363
Drum gates (dam control) 124,
    125
Drum screens (sewage treat-
    ment) 519
Dry air components 317
Dry arroyos 130
Dry-bulb temperatures 275
Dry desert 433
Dry forest 215, 636
Dry ice (seeding clouds) 102
Dry snow avalanches 52
Dry-weather streamflow 600
Dry woodlands 636
Dulse 312
Dune 107, 134–135
    artificially formed 107
    barchan 134
    blowout 134
    clay 135
    climbing 134
    coastal 134–135
    complex 134
    desert 135
    falling 134
    fossil 135
    geographic distribution
        134–135
    gypsum 135
    materials 135
    parabolic 134
    restoration 106
    sand 107, 130
    seif 134
    silt 135
    size 135
    subtropical 135
    transverse 134
    types 134–135
Dune, E. 699
Dune vegetation 107, 130,
    135–137
    Atlantic and Gulf coasts
        135–136
    Great Lakes coasts 137
    succession and zonation 135
Dunes, Algodones (Calif.) 135
Durham fermentation tube 655
Durilignosa 648
D'Urville, Dumont 30
Dust 140
    cloud 96–99
    particle size 20
Dust and mist collection: air fil-
    ter 139
    cloth collector 138–139
    electrostatic precipitator 139
    gravity settling chamber 138
    inertial device 138
    miscellaneous equipment 139
    packed bed 138

Dust and mist collection—*cont.*
    scrubber 139
Dust-bowl conditions 81
Dust collector (air pollution con-
    trol) 138
Dust devil 319
Dust particles, mineral 101
Dust storm 140
Dutch elm disease 421, 626
Dwarf elephant (mass extinction)
    191
Dwarf mistletoe 624
Dwarf scrub 637
Dybar 267
Dyes, hydroxyquinoline 70
Dymid 265
Dynamic programming (wildland
    operation research) 688
Dynamical meteorology 326–332
    analytical 328–331
    instantaneous state of atmos-
        phere 328
    numerical 331–332
    thermodynamic properties of
        air 326–328
Dynamics, population 108–109,
    446–451, 586, 684
Dynamite (blasting) 340
Dynamometers (petroleum pro-
    duction) 391
Dysentery, amebic 132
Dystrophic lakes 225

E

E layer 37
Eagles 628
Earth (solar radiation) 83
Earth, resource patterns of
    140–144
    employed and potential re-
        sources 143
    humid macrothermal regions
        142
    humid mesothermal regions
        142
    humid microthermal regions
        142
    marine resources 143–144
    resources of the continents
        144
    rock composition and surface
        configuration 143
    water-deficient regions
        142–143
Earth dam 121–122
    foundation 126
    foundation treatment 127
    instruments 127
Earth sciences 144–145
East Antarctica 28
East Canyon Dam (Utah) 121,
    122
East Greenland Current 33
Easterlies 41, 42, 43, 319
Easterlies, polar 89
Ecesis 443
Echinocactus
Echinocereus 130
Echo sounder: detection of scat-
    tering layer 481–482
    oceanographic survey 276
Ecocline 162

Ecological associations 109–110
Ecological efficiencies 152–153
Ecological entomology 172–174
Ecological interactions 145–148,
   164, 178
   commensalism 145–146
   community 108
   competition 146–147
   consortism 27
   influence of forest site 206
   mutualism 146
   natural-resource ecosystem
      158
   neutralism 146
   parasitism 148
   predation 147–148
   symbiosis 145
Ecological pyramids 148
   biomass pyramids 63
   food chain 201–202
Ecological races 162
Ecological succession 93–96,
   109, 110, 167, 207, 468,
   477–478, 595–599, 632, 640
Ecological systems (aerobiology)
   3–4
Ecological systems, energy in
   151–154, 166–167
   fresh-water ecosystem 218,
      220
   systems ecology 606
   terrestrial ecosystems 615,
      616–617
   vegetation 632–633
Ecology 148–158
   aerobiology 1–4
   approaches 149
   arctic biology 33–34
   biogeochemical cycles
      154–157
   biotic components 149–151
   biotic influence on environ-
      ment 151
   coral reef study 149
   ecosystem development
      157–158
   ecological interactions
      145–148
   ecological pyramids 148
   ecosystem 149–151
   energy and materials 151–
      152
   energy flow 151–154
   energy flow diagrams 152
   energy transfer 152–153
   entomology, ecological
      172–174
   food chain 201–202
   forest ecology 204–207
   fresh-water ecosystem
      217–230
   functional 149
   homeostatic mechanisms 151
   insect control, biological
      284–285
   land-use classes 297–298
   marine ecosystem 309–314
   nutrients 155
   pesticides, persistence of
      398–401
   primary production in the
      world 154
   radioecology 465–468

Ecology—cont.
   systems ecology 604–607
   types of cycles 155–157
Ecology, applied 158–160
   agriculture 159
   engineering, social implica-
      tions of 172
   environmental pollution 160
   fisheries management 160
   forestry 158–159
   land-use planning 298–300
   range management 159
   water-resource management
      160
   wildlife management 159–160
Ecology, estuarine 149
Ecology, forest 204–207,
   216–218
Ecology, fresh-water 149
Ecology, functional 149
Ecology, grassland 149
Ecology, human 53–56, 61, 145,
   151, 160–161, 164
   climate, man's impact on
      75–82
   community 160–161
   cultural evolution 161
   disclimax 95–96
   environmental protection
      178–180
   regulatory mechanisms 161
   scope 161
Ecology, marine 149
Ecology, physiological (plant)
   161–163
   desert vegetation 129–131
   field measurements 162–163
   measurement in controlled
      environments 163
   micrometeorology, crop
      334–335
   scope 161–162
   systems ecology 604–607
   trends in succession 599
Ecology, stream 149
Ecology, systems 194, 604–607,
   685
Ecology, terrestrial 149
Economic entomology 159,
   284–285
Ecosystem 65, 108, 109,
   149–151, 164–167
   adult systems 157
   animal community 25–27
   aquatic 150, 187–189,
      401–402, 452–453, 615,
      700–701
   biological productivity 61–63
   biomass 63
   biome 63–64
   biotic components 149–151
   chance processes in 605–606
   climax 167
   cold conifer 634
   community adaptation
      164–165
   community differentiation
      165
   complexity and interrelations
      164–167
   coral reef 149
   cycling and functional king-
      doms 166

Ecosystem—cont.
   development 157–158
   duration of radiation 467
   ecological 148–158
   ecological pyramids 148
   energy flow 151–154
   energy-flow diagrams 152
   environmental dependence
      and change 164
   estuarine environments
      185–186
   eutrophication 57–58
   food chains and trophic levels
      166
   forb 637
   fresh-water 110, 164, 187–189,
      218–230, 293, 304, 401,
      613–614, 700–701
   grassland 245–247
   island faunas and floras
      288–290
   lake 58
   lenitic 218–221
   lotic 227–229
   maintenance 606
   major groups 636
   mangrove swamp 307–309
   marine 60–61, 110, 164,
      309–314, 378, 401–402,
      613–614, 700–701
   metabolism 225
   natural-resource 158
   pest management technology
      174
   plant community 406–407
   plant disease 412–413
   prairie 450–452
   productivity relations 165–166
   response to radiation 467–468
   rhythms 165
   soil 159
   spatial patterning 165
   steady state 166–167
   subtropical 634
   succession 167
   terrestrial 59–61, 63–64, 150,
      164, 613–617, 634
   tropical 634
   vegetation and ecosystem
      mapping 633–639
   vegetation structure 632
   wildland operations research
      686–689
   youthful types 157
Ecotone 26
Ecotypes 162
Ectohumus 216
Ectoparasites 148
Eczema 132
Edaphic climax 94
Eddy flux (atmosphere) 45
Edmondson, W. T. 187, 293
Edwards limestone (aquifer)
   30
Eelgrass 312
Effluent standards (regulation of
   water pollution) 515
Eijkman test 655
Einstein, H. A. 595
Ekman, Gustav 374
Ekman, Sven 374
Ekman spiral 329
Elastosis, skin 53

Electric distribution systems 170
Electric energy (electric power
   systems) 167–171
Electric power generation
   167–169, 629–632
Electric power plant,
   solar/thermal 571
Electric power substation 170
Electric power systems 167–171
   distribution 170
   electric utility industry
      170–171
   generation 167–169
   nuclear reactors 369–370
   substations 170
   transmission 169–170
Electric shock 132
Electric utility industry 170–171
Electrical conductivity (sea
   water) 494
Electrical engineering 172
Electrical plants, central-station
   369–370
Electrical power engineering:
   electric power systems
      167–171
   geothermal power 243–245
Electrical thermometers
   (oceanography) 504
Electrical utility industry
   629–632
   combustion turbines 629
   conventional steam units 629
   nuclear power 629
   thermal pollution 629–630
Electricity: atmospheric 39–40,
   320
   conversion of solar energy 570
   electric power systems
      167–171
   geothermal power 243–245
   hydroelectric generation 570
   solar cells 572
   usage 630–631
   utility industry, electrical
      629–632
   wind-generated 570
Electrochemical corrosion (oil
   and gas field exploitation)
   389
Electromagnetic energy (solar
   energy) 569–573
Electromagnetic fields (sea
   water) 494
Electromagnetic sensors 610
Electromagnetic spectrum (remote sensing) 610–611
Electrons (biological radiation
   injury) 457
Electrostatic precipitator 21,
   51
   dust collection 139
   horizontal-flow 22
   incinerators 579
Electrothermal HYGAS gasifier
   104
Elements, biological 67
Elements, chemical: abun-
   dances in the marine hy-
      drosphere 495
   biogeochemical cycles 68–70
   composition of living matter
      66–68

Elements, chemical—*cont.*
  cosmic abundance of 35–36
  geochemical distribution of 37–39
  sea water 310, 314–315
  soil composition 542
Elements, radioactive 156–157
Elephant, dwarf 191
Elephant tree 131
Elevated storage reservoirs 666
Ellesmere Island (Arctic) 31, 32
Ellsworth, Lincoln 30
Elm bark beetle 626, 627
Elms 648
Elton, C. S. 148, 201
Elymus 433
Embankments (stability analysis) 564
Embedded pressure cells 127
Embolism 56
Emperor penguin 30
Emphysema (air pollution) 49
Empirical models (climate) 77
Employed resources 143
Emulsions (oil well production) 390
Encephalitis 132
Endemism (island faunas and floras) 288
Endocrine mechanisms (homeostasis) 274
Endogenous disease 132–133
Endoparasites 148
Endosulfan 285, 286
Endothall 128, 266
Endothia parasitica 418, 421
3,6-Endoxohexahydrophthalic acid 128
Endrin 285, 286, 400
Energy: abiotic environment 176–178
  ecosystem 62
  electric 167–171
  electromagnetic 569–573
  geothermal power 243–245
  marine resources 316
  nuclear 160
  solar 76–77, 83–85, 569–573
  transmission in sea water 493
Energy degradation (ecosystem) 152
Energy flow (ecosystem) 151–154, 166–167
Energy flow diagrams 152
Energy flux (atmosphere) 43–44, 45
Energy in ecological systems 151–154, 166–167, 218, 220, 606, 615, 616–617, 632–633
Energy pyramids 148
Energy sources (nuclear power) 368–372
Energy transfer (ecosystem) 152–153
Engelmann spruce 645
Engine: diesel 346
  heat 76–77
  internal combustion 169
Engineering 171–172
  air-pollution control 19–25
  chemical 137–140, 172, 232–235, 262–268, 285–288, 398–401, 668

Engineering—*cont.*
  civil 105–108, 118–127, 171–172, 268–271, 298–300, 474–476, 476–477, 478, 512–515, 515–518, 518–524, 571, 573–581, 661, 664–666, 667–669
  coastal 105–108
  electrical 172
  electrical power 167–171
  environmental protection 178–180
  health physics 251
  heating 252
  highway 268–271
  industrial meteorology 283–284
  integrating influences 172
  management 172
  mechanical power 128–129, 252–253, 569–573
  military 171
  mining 102–105, 172, 338–342, 342–346, 346–348, 348–354, 354–361
  nuclear 368–372, 460–462, 462–463, 463–465
  petroleum 380–383, 383–391, 482–486
  petroleum reservoir 383
  sanitary 172, 476–477, 478, 667–669
  social implications of 172
  soil 555–564
  traffic 268–271
  transportation 474–476
  types 171–172
  water supply 642, 664–666, 667–669
Engineering geology (open-pit mining) 342
Engineering hydrology 275
Engineering soil classifications 557–558
English filters (water treatment) 667
English sparrow 585
Enide 265
Enteromorpha 312
Entisols 529, 541
Entomology (insect pests of crops) 2
Entomology, ecological 172–174
  crop loss due to insects 173–174
  environmental impact 174
  home and health hazards 174
  honey and wax production 173
  insecticide use 173
  pest control 173
  pollination role 173
  role in cycle of matter 173
  systems approach 174
Entomology, economic: forest management 159
  insect control, biological 284–285
Environment 175–178
  abiotic 151, 176–178
  air-pollution control 19–25
  atmosphere 177
  biotic 175, 178
  biotic influence 151

Environment—*cont.*
  distribution of metals 256
  distribution of polychlorinated biphenyls 441
  dynamic concept 176
  ecosystem 164–167
  effect on population dispersion 445–446
  estuarine 185–186
  fresh-water 177
  human 180
  increasing prevalance of mercury 257–258
  indoor 3
  limiting factors 176
  mass faunal extinctions 189
  marine 177, 310–312, 354–355, 400
  military 180
  nature of 175
  substratum 177
  water 177
Environmental control (plant research) 163
Environmental pollution 160
Environmental protection 178–180
  man 179
  physical geographic environment 178–179
  research on materials 179–180
  special complexities 180
Environmental resistance (population dynamics) 448
Environmental rhythms (response of communities) 165
Environmental stresses 178–179
Enzyme: cellular homeostasis 274
  effect of radiation 455
Ephedra 1
Ephemeral forb complexes 637
Ephemerals (desert) 129, 130, 131
EPIDEM (computer model) 3
Epidemic (plant disease) 418–419
Epidemiology 180–182
  case fatality ratio 182
  disease rates 181–182
  experimental 182
  morbidity rates 181
  mortality rate 180–181
Epilimnion: chemistry 225
  lenitic ecosystem 218
  productivity 225
Epiphytes 469, 646, 647
Epithelial metaplasia 132
Episodical forb complexes 637
Epoch: Paleocene 190
  Pliocene 189
Eptam 266
EPTC *see* Ethyl-*N*,*N*-di-*n*-propylthiolcarbamate
Equatorial complexes 635
Equatorial rainforest 215
Equatorial water 501
Equus asinus 8
Equus caballus 8
Eradicant fungicides 233

Eragrostis 479
Erbon 267
Ergot fungus 415
Ergotamine 623
Erica 136, 429
Ericaceae 432
Ericads 648
Eroding shoreline 106
Erosion: bank 550
  beach 105–106, 107
  complex response of geomorphic systems 241–243
  development of landforms 235–236
  dust storm 140
  glaciated rough terrain 272–273
  gully 239, 539, 550, 551
  hydrology 275–276
  rill 538–539, 550, 551
  sheet 538, 550, 551
  soil 538–540, 549–550
  splash 550
  stream-eroded features 271–272
  stream transport and deposition 594–595
  water 550
  wind 540, 550, 552
Erosion zones (beach) 105
Erosional mountains 273
Erosional valleys 273
Erwinia amylovora 425
Erwinia carotovora 414
Erwinia nimipressuralis 627
Erysiphe graminis 413
Erythema, solar 53, 132
Erythronium 430
Escherichia coli: mutation 255
  water analysis 655
Eskimos 34, 61
Estonian oil shale 396
Estuarine ecology 149
Estuarine oceanography 182–187, 314
  bar-built estuary 184
  classification 182
  deep-basin estuary 184
  estuarine circulation 182–184
  estuarine environments 185–186
  flushing of estuaries 184–185
  physical structure and circulation 182–184
  pollution 518
  sediments 186–187
Estuarine sediments 186–187
  distribution of grain sizes 186
  manner of deposition 186–187
  organic constituents 186
Estuary: bar-built 182
  coastal-plain 182–184, 185
  deep-basin 182, 184
  estuarine oceanography 182–187
  eutrophication, cultural 187–189
  inverse 182
  neutral 182
  partially mixed 183
  pollution 518
  positive 182
  salt-wedge 182–183, 185

Estuary—*cont.*
  sound 184
  true 182
  vertically homogeneous
    183–184
Ethers (poison) 622
Ethiopian faunal region 25, 694,
  695–696
Ethiopian faunal subregion 694,
  699
2-Ethylamino-4-isopropylamino-6-
  methylmercapto-*s*-triazine
  266
Ethyl-*N,N*-di-*n*-propylthiolcar-
  bamate (EPTC) 266
S-Ethylhexahydro-1-*H*-azepine-
  1-carbothioate 266
Eubacteria (soil microorganisms)
  564–565
Eucalyptus 432, 479, 679
Euglena gracilis 67
Eulittoral zone (marine ecosys-
  tem) 313
Eupatorium urticaefolium
  439
Euphausia diomediae 511
Euphausia gibba 511
Euphausia hemigibba 511
Euphausia lucens 511
Euphausiids 511
Euphorbia brasiliensis 439
Euphorbiaceae 432
Euphorbs 129
Euphotic zone (ocean) 110,
  507–508
Euphrates River 473
Eurasia (landform) 610
Eurasian continent (resource
  pattern) 144
European boreal faunal region
  26
Eutrophic lake 225
Eutrophication 57–58
Eutrophication, cultural 187–189
  action programs and future
    studies 188
  extent of problem 187–188
  fertilizers 198
Eutypella 624
Evaporation: atmospheric
  40–41, 189
  evapotranspiration 189
  hydrologic cycle 88
  hydrometeorology 278
  mountains 643–644
  ocean 276, 278
  ocean-atmosphere relations
    372–374
  ocean surface 90–91
  sea-water 491–492
  water conservation 657–658
Evaporative cooling towers (nuclear power generation) 630
Evaporimeters 40–41
Evapotranspiration 6, 189, 277
  effects of weather and climate
    53
  micrometeorology 335
  phreatophytes 189
  plant, water relations of 115
  xerophytes and mesophytes
    189
Even-aged forest stand 207–208
Everglades (southeastern United States) 480

Evergreen broad-leaved shrubland 637
Evergreen broad-leaved woodlands 636
Evergreen dwarf scrub 637
Evergreen dwarf shrubland 637
Evergreen forest 636
Evergreen hardwoods 431–432,
  648–649
Evergreen needle-leaved wood-
  lands 636
Evergreen needle-leaved scrub
  637
Evergreen scrub 637
Evergreen subdesert 637
Evergreen woods 636
Evolution: biosphere 65
  cultural 161
  faunal extinction 189–191
  geochemical evidence 66
  island biotas 289
  mutations 365–366
  organic 586
  paleobioclimatology 56
Excavator, Krupp wheel 392,
  393
Exchangeable cations (soil) 544
Excreta (manures) 309
Exobasidium vexans 2
Exogenous disease 132–133
Experimental epidemiology 182
Experiments, rainmaking 102
Explosion, nuclear 460–462
Explosive: ammonium nitrate
  340
  dynamite 340
Extended-range forecasting
  670–671
External parasites (livestock) 10
Extinct peccary (mass extinc-
  tion) 191
Extinct tapir 191
Extinction: faunal 189–191
  mass 189
  predation 148
Extratropical cyclones 89, 588
  life cycle 323–325
  principal tracks 589
  storm surge 593
Extratropical cyclonic precipita-
  tion 277
Extratropical deserts 634
Extratropical savanna 480–481
Eyring, C. 482

F

F1 layer 37
F2 layer 37
Facultative parasites 148
Fagus 431
Fagus grandifolia 137
Fair-weather field 40
Fair-weather meteorology 333
Falling dunes 134
Falling-head permeameter 560
Falone 267
False heather 136
Familial polyposis 132
Family selection 71
Famphur 287

Farm fertilizers 198–199
Farm forests 214
Fast breeder reactors 370
Fast-running cold front 323
Fat: animal feed composition
  8–9
  deficiencies 306
Fauna: deep-sea 314, 452–453,
  482
  habitat distribution 614
  lake bottom 187
  prairie ecosystem 450–451
  savanna 479
  taiga 608
  tundra 628
  wildlife conservation 682–692
  zoogeographic region 694–698
  zoogeography 696–699
  zooplankton 700–701
Faunal extinction 189–191
  causes 190–191
  evidence from fossils 189
  mechanics 190
  patterns 189–190
  predation 190
Faunal region: animal community 25–27
  marine littoral 26
  terrestrial 25
Faunas and floras, island
  288–290
Federal conservation legislation
  113
Federal Water Pollution Control
  Act of 1948 113
Feedback mechanisms
  (biogeochemical cycles) 156
Feeding, livestock 8
Feldspar 543
  dunes 135
  oil shale 394
  uses 336
Felidae 479
Fenac 263, 267
Fenestraria 652
Fenthion 287
Fenuron 267
Fenuron TCA 267
Ferbam 233
Ferns 646
Ferrel cells 41, 43
Ferroalloys (uses) 336
Ferromanganese mineral formation 496
Fertility: sea-water 280, 497,
  507–511
  soil 537
Fertilization (soil management)
  641
Fertilizer 5, 191–199, 537–538
  bulk blending 197
  farm 198–199
  liquid 118, 197
  manure 652–653
  micronutrients 198
  mixed 197
  multinutrient 195
  nitrogen 192–195
  phosphate 195–196
  plant nutrition 11–12
  potash 196–197
  types 192
Fescues 246
Festuca 246, 433
Fetch area (beach erosion) 105

Fetus (radiation injury) 459
Fibrous clouds 97
Ficus 429
Field, gas 383–391
Field, oil 383–391
Field beans (defoliation) 128
Filariasis 132, 174
Fill (underground mining)
  349
Filled stopes 350–351
Filter: sewage treatment 522
  water treatment 667–668
Filter method (phytoplankton
  collection) 401
Filtering (water purification)
  476
Filters: bag 51, 138
  controlled 522
  diatomaceous earth 668
  English 667
  high-rate 522, 523
  low-rate 522
  rapid sand 667–668
  slow sand 667
  standard 522
  trickling 522
Filtration: sewage treatment 522
  water treatment 667
Fingerling fish 186
Fingerprinting, genetic 72
Fiord 184
Fir (taiga) 607
Fire: abiotic factor 177–178
  climax community 93
  disclimaxes 95
  disturbance of climax com-
    munities 95
  forest 640
  human bioclimatology 55–56
  hydrants 665–666
  vegetation management 640
Firn (thermal constants) 618
Fish: Australian region 697
  conservation of resources
    110–113
  Ethiopian region 695
  fisheries conservation 199–201
  marine resources 315
  Nearctic region 695
  Neotropical region 696
  Oriental region 696
  Palearctic region 694
  planktivorous 453
Fish ladders 200
Fisheries conservation 159,
  199–201, 379
  coastal waters 660
  distribution and types 200
  management 201
  problems 200–201
Fishery: commercial 200
  common 200
  free 200
  management 160, 201
  marine 200, 378
  sea 315
  sport 200
Fission, nuclear (nuclear power)
  368–372
Fission bomb (radioactive fallout) 460
Fixation, nitrogen 367
Fixed-bed reactor system (Lurgi
  gasifier) 103
Fixed cone dispersion valves 125

Fjell-field tundra 628
Flag smut 427
Flat plains 608, 609
Flatlands, wet 604
Flax rust 414
Fleas (livestock parasites) 10
Flight, supersonic 581–585
Flint, R. F. 31, 617
Floating bog 604
Floating drill ships 381–382
Floating ring gates 124
Floes, hummocky 487
Flood control 275
Flood Control Act of 1927
    113
Flood method of irrigation
    116–117
Floods (river) 601
Flora: geographic distribution
    434
    savanna 479
    vegetation 632–633
Floras and faunas, island
    288–290
Floristic composition 640
Floristic plant geography
    435–436
Floristic relays 640
Flow nets (water seepage
    through soil) 560
Flow resources (renewable
    natural resources) 111
Flowering (plant) 163
Fluid mechanics (sonic boom)
    581–585
Fluidized-bed incinerator
    580–581
Fluoride salts (molten-salt reac-
    tor) 370
Fluorine (constituent of living
    matter) 69
Fluorspar (uses) 336
Flushes (vegetation and ecosys-
    tem mapping) 637
Flushing, estuarine 184–185
Fluvial soils 556
Flux, angular momentum 43
Flux gate magnetometer 356
Fly agaric 623
Foehn 56, 693
Fog: air pollution 47–48
    climax 96
    dispersal 680
    dispersion of pollution 46–47
    disruption of climax com-
        munities 95–96
Fog drip 206
Foliage fungicides 233, 423
Folic acid (disease) 132
Fomes annosus 624
Fomes applanatus 625
Fontinalis 224
Food chain 26–27, 65, 109, 152,
    166, 201–202
    biological productivity 61–63
    biomass 63
    ecological pyramids 148
    food web 201–202
    lake 304
    marine 508
    quantitative studies 202
    radioecological tracer studies
        466–467
    role of insects 173
    zooplankton 701

Food cycles (aquatic organisms)
    304
Food engineering: food proces-
        sing wastes 654
    grading of market animals
        9–10
    hazards of nickel 262
Food poisoning 132
Food processing wastes (recy-
    cling) 654
Food production (efficiency
    limits) 154
Food pyramid 27
Food resources (ocean) 378
Food web 109, 201–202
Forage: grassland 245–247
    range management 642
Foraminiferan oozes 66
Foraminiferida 312
Forb: ecosystems 637
    grasslands 246
Forb-rich meadow steppes 637
Forecast centers, weather
    675–676
Forecasting and prediction,
        weather 320–326, 331–332,
        590–593, 669–676
    extended-range 670–671
    long-range 670–671
    medium-range 670–671
    statistical weather 671–672
Forel, F. A. 218
Forel-Whipple system of lake
    classification 219
Forest 94, 429
    autecology 204–205
    beech-maple 616
    boreal 31, 59, 607, 634
    boreal coniferous 142
    climax 158
    closed 636
    cloud 59, 647
    cold deciduous 634, 636
    commercial 212, 214
    conifer 95, 431
    coniferous 60, 64, 215,
        607–608, 648
    coniferous evergreen 636
    conservation of resources
        110–113
    cool coniferous 215
    deciduous 109, 469, 599, 636
    drought-deciduous 636
    dry 215, 636
    earth, resource patterns of 140
    evergreen 636
    evergreen hardwood 431–432
    farm 214
    fire 95
    galleria 647
    growth 213–214
    hard-leaved 636
    hardwood 95, 212
    laurel 647
    lumbering 95
    mangrove 636
    Mediterranean 431
    monsoon 430, 469, 647
    moss 647
    needle-leaved 430, 431
    needle-leaved evergreen 648
    noncommercial 212
    overgrazing 95
    pine 95
    pine-oak 59

Forest—*cont*.
    rainforest 59, 468–470
    redwood 96
    resinous 648
    sclerophyll 648
    semideciduous 647
    softwood 212, 648
    southern 634
    subtropical 647
    subtropical evergreen sea-
        sonal 636
    subtropical semi-deciduous
        636
    succession, ecological 596
    succulent 636
    summergreen 636
    summergreen deciduous
        430–431, 648
    summergreen temperate 93
    swamp 307–309
    taiga 607–608
    temperate deciduous 64, 648
    temperate evergreen seasonal
        636
    temperate evergreen broad-
        leaved 636
    temperate mixed 215
    thorn 59, 432, 636
    tropical broad-leaved 64
    tropical cloud 96
    tropical deciduous 59, 430,
        647–648
    tropical evergreen 59
    tropical evergreen seasonal
        636
    tropical moist deciduous 215
    tropical semi-deciduous 636
    types 212
    warm temperate moist 215
    winter-rain hard-broad-leaved
        evergreen 636
    world resources 215–216
Forest and forestry 202–204
    ecology 204–207
    government participation 204
    management and organization
        203, 207–209
    multiple use 203–204
    planting and seeding 209–212
    resources 212–216
    soil 216–218
    sustained yield 204
    tree diseases 623–625
Forest ecology 204–207
    hydrology 207
    influence of site upon life his-
        tory
    site evaluation 205–206
    soil 216–218
    succession 207
Forest fire (vegetation manage-
    ment) 640
Forest management and organi-
        zation 158–159, 202–204,
        207–209, 216
    even-aged stands 207–208
    hydrology 207
    importance of succession 599
    measurement of growth 205
    planting and seeding 209–212
    uneven-aged stands 208–209
Forest planting and seeding 206,
        209–212
    even-aged stands 207–208
    plantation care 212
    planting conditions and re-
        quirements 210–212

Forest planting and seeding
        —*cont*.
    seed handling 210
    seed-producing areas 209–210
    sustained yield 204
Forest products 203
Forest reserves 209–212
Forest resources 158, 203,
        212–216, 642
    growth 213–214
    management and organization
        207–209, 216
    mangroves 308–309
    ownership 214–215
    rainforest 469
    soil 216–218
    types of forests 212
    world resources 215–216
Forest rotation 208
Forest soil 205, 206–207,
        216–218
    classification 217
    humus relations 217
    hydrologic cycle relations
        217–218
    resources 212
    root relations 217
Forest stand: even-aged 207–208
    uneven-aged 208–209
Forest steppe 480
Forest stock 208
Forest succession 207
Forest synecology 205
Forest trees (diseases) 623–625
Forest tundra 480, 481, 650
Forest waste products 654
Forest zone (Atlantic and Gulf
    coasts) 136
Forestry 158–159, 202–204
    ecology 204–207
    management and organization
        207–209
    planting and seeding 209–212
    remote sensing 612
    resources 212–216
    small ownership 214
    soil 216–218
Forêt à feuilles caduques 648
Forêt claire 649
Forêt décidue 648
Forêt dense 646
Forêt mesophile 647
Forêt ombrophile 646
Forêt-parc 649
Forêt tropicale humide 646
Forêt tropophile 647
Formaldehyde (mutagen) 365
Fossil (evidence for change of
    climate) 76
Fossil dunes 135
Fossil-fuel engineers 172
Fossil-fuel wastes 574–575
Fossil fuels 172
Fossil resins 70
Foster, J. 191
Foundation treatment for dams
    126–127
Foundations, pile 563
Fouquieria 433
Fouquieria splendens 129, 131
Fourré 650
Fourwing saltbush 130
Fox, M. M. 524
Foxes 34, 628
Fracture (mechanical injury) 132

Fragilaria 188, 311
Fragilaria crotonensis 187
Frank B. Black Research
    Center, Mansfield, Ohio 170
Franseria 433
Franseria chamissonis 136
Franz Josef Land 31, 32
Fraxinus 431
Free fishery 200
Freeboard height (dam) 126
Freezing (sea water) 486, 502
Fresh water: chemical composition 281
    geochemistry 281
Fresh-water biocycle 110
Fresh-water communities
    (trophic structure) 617
Fresh-water ecology 149
Fresh-water ecosystem 110, 164,
    218–230, 613–614
    biota 221–225, 228–229
    carbon dioxide 226
    chemistry 225–227
    eutrophication, cultural
        187–189
    gradients 228
    higher vertebrates 224–225
    lake classifications 218–220
    lenitic ecosystems 218–221
    light stratification 220
    limnology 304
    lotic ecosystems 227–229
    meromictic lake 293
    nekton 224
    neuston 225
    oxygen 225–226
    periphyton 224
    phytobenthos 224
    phytoplankton 401–402
    plankton 221–223
    pollutional relationships 229
    production and productivity
        226–227
    production relationships 229
    psammon 224
    temperature stratification 218
    trophic relationship 225
    water source 228
    waves and currents 220–221
    zooplankton 700–701
Fresh-water environments 177
Fresh-water fisheries 200
Freuchen, Peter 34
Frigorideserta 652
Front 14–15, 56, 100, 230–232,
    319–320, 322–323
    cold 231
    fast-running cold 323
    frontal waves 230–231
    frontogenesis 231–232
    polar 231, 324
    slow-moving cold 323
    squall line 587–588
    warm 323
Front-end loaders (open-pit mining) 341
Frontal storms 589
Frontal waves 230–231
Frontal zone (air mass) 230
Frontogenesis 231–232, 322–323,
Frost 232
    defoliation 128
    degrees of 232
    killing 232
    point 131, 275

Frostbite 56, 132
Frozen water (thermal spring)
    618
Fruit fly (mutation study) 363
Fruiting (ecology, physiological)
    163
Fucus 312, 313
Fuel cycle, thorium-U²³³ 369
Fuel gas (coal gasification)
    102–105
Fuels 172
    fossil 172
    nuclear 172, 368–372
Fumes (particle size) 20
Functional ecology 149
Functional kingdoms 166
Fungi: cellulose-decomposing
    566
    decomposers 150
    food requirements 413–414
    host selectivity 414
    mycorrhizal 566, 568
    phosphate requirement 568
    phytoplankton 401–402
    plant disease 407, 409, 411,
        412
    reproduction 415
    root-inhabiting 566
    slime 566
    soil-inhabiting 566
    soil microorganisms 564–567
    sugar 566
    tree diseases 623–627
Fungicides 5, 12
    agricultural 233
    contact 233
    eradicant 233
    foliage 233, 423
    inorganic 233
    organic 233
    plant disease 410
    plant disease control 423
Fungistat and fungicide 232–234
    agricultural fungicides 233
    application methods 233–234
    chemotherapeutants 233
    chemotherapy in man 232–233
    foliage fungicides 233
    formulation 233
    importance of a major event
        234–235
    importance of an average
        event 234
    inorganic fungicides 233
    organic fungicides 233
    requirements and regulations
        233
    seed and soil treatments 233
    threshold concept 236–237
Fungus disease 132
Fungus infections: fungistat and
    fungicide 232–233
    livestock 10
Furadan 287
Furrow method of irrigation 116
Fusarium 566, 623
Fusarium oxysporum f. conglutinans 421
Fusion, nuclear 368–372

G

Gages: liquid-level 391
    settlement 127
Galleria forests 647

Game theory (wildland operations research) 688–689
Gamma radiation: local fallout
    461
    radiation injury (biology) 457
    radioecology 465–468
Ganges River 473
Gangrene, gas 132
Garigue 649
Gas: chlorine 668
    coal gasification 102–105
    dust and mist collection
        137–140
    fuel 102–105
    natural 168, 314, 380–383,
        383–391,
    sea water 310
    synthetic pipeline 103–105
Gas-cap drive well 383
Gas chromatography (sea water
    analysis) 507
Gas-condensate well 386
Gas conservation (oil well production) 390
Gas-cooled reactors 369
Gas field 383–391
Gas gangrene 132
Gas lift (oil well production) 388
Gas meters (petroleum production) 391
Gas separators 389
Gas-solid adsorbers 51
Gas turbines (electric energy
    generation) 169
Gaseous pollutants 46
Gaseous wastes, radioactive
    464–465
Gases, atmospheric 37–38, 78,
    178
Gasification, coal 102–105
Gasifier: HYGAS 103–104
    Lurgi 102–103
Gates: drum 124, 125
    floating ring 124
    high-pressure slide 126
    jet flow 126
    outlet 126
    ring 125
    slide 125
    spillway 124, 125
    tainter 124, 125, 126
    tractor 125, 126
    vertical-lift 124, 125
Gazellus 479
Gems 143
Gene (mutation) 361–366
General atmospheric circulation
    320
Generating stations (electricity)
    167–169
Genesis of biosphere 64–65
Genetic fingerprinting 72
Genetic injury (radiation) 458
Genetic tests (mutation studies)
    363–364
Genetics: bacterial 363
    breeding (animal) 71–72
    breeding (plant) 72–75
    human 372
    plant disease resistance 426
    population 586
    mutation 361–366
Gentoo penguin 30
Geochemical cycle (oxygen
    isotope) 503

Geochemistry 145
    biogeochemical balance 56–59
    biosphere 64–71, 154–157, 597
    distribution of elements 37–39
    halogens, atmospheric
        geochemistry of 250–251
    hydrosphere 278–281, 378
    lake, meromictic 292–294
    organic 64, 70
    soil chemistry 542–549
Geographic environment, physical 178–179
Geographical botany 434
Geography: animal 60–61
    Antarctica 27–30
    arctic and subarctic islands
        30–33
    biogeography 59–61
    geomorphology 234–243
    remote sensing 612
Geography, lunar 180
Geography, physical 27–30,
    30–33, 134–135, 140–144,
    271–273, 291–292, 398
    river 471–474
    terrain areas, worldwide
        608–610
    terrain sensing, remote
        610–613
Geography, plant 59–60,
    428–434, 434–438, 468–470,
    633–639, 642–645, 645–652
Geohydrology (groundwater)
    247–250
Geologic basins, subsea 380
Geologic slope (geomorphic
    thresholds) 237–238
Geological stations (oceanography) 377
Geology: Antarctica 28–29
    aquifer 30
    artesian systems 34–35
    dust storm 140
    ecological interactions 145
    estuarine sediments 186–187
    faunal extinction 191
    genesis of biosphere 65
    groundwater 247–250
    remote sensing 612
    selection of dam site 125
    selection of type of dam 126
    stream transport and deposition 594–595
Geology, engineering 342
Geology, marine 354–361, 378,
    380
Geology, petroleum 380, 383
Geology, surficial and historical:
    avalanche 52
    continental shelf and slope 113
    dune 134–135
    geomorphology 234–243
    hill and mountain terrain
        271–273
    lake formation 292
    oil sand 394
    oil shale 394–398
    soil 526–540
    soil, zonality of 540–541
Geomagnetism (International
    Geophysical Year
    1957–1958) 30
Geomorphic systems 241–243
Geomorphic thresholds 236–238,
    238–241

Geomorphology 234–243
  alluvial chronologies 242
  analytical studies 235
  approach since 1945 234
  arroyos 238
  classification of estuaries 182
  complex response of geomorphic systems 241–243
  discontinuous gullies 239–241
  effects of Hurricane Agnes 236
  evaluation of the magnitude-frequency concept 243
  evidence for geomorphic thresholds 238–241
  geomorphic thresholds 236–238
  incipient instability 238
  light-damage areas 236
  magnitude and frequency of events 234–236
  major hydrologic events 236
  misguided interference 241
  nongeomorphic explanations 238
  rejuvenation 242–243
  river patterns 241
  sediment movement studies 236
  thresholds involving slope 237–238
Geophysics: air mass 14–17
  air pressure 17–19
  aquifer 30
  artesian systems 34–35
  atmospheric evaporation 40–41
  ecological interactions 145
  eutrophication, cultural 187–189
  evapotranspiration 189
  general physics of the air 316–320
  geothermal power 243–245
  groundwater 247–250
  heat balance, terrestrial atmospheric 251–252
  hydrology 275–276
  jet stream 290–291
  soil mechanics 555–564
  spring, water 586–587
  squall line 587–588
  surface water 599–604
  terrestrial frozen water 617–618
  thermal spring 618–619
  thermocline 619–620
  water table 666–667
Geophyte 439
Geosciences 145
Geospizidae 288
Geostrophic balance 329
Geothermal power 243–245
  drilled geothermal power 244–245
  natural geothermal power 243–244
Geranium 430
German measles 132
Germination (field measurements) 162
Gesagard 264
Gesneriaceae 288
Giant armadillo (mass extinction) 191

Giant ground sloth (mass extinction) 191
Gibberelins 5
Gibbsite-allophate minerals 543
Gilbert, G. K. 594
Gilbertiodendron dewevrei 469
Gilsonite mining 392–393
Giraffus 479
Glacial lakes 292
Glacial soils 556
Glaciated terrain 272–273, 292, 604
Glaciers 617
  Aleutian Islands 31–32
  valley 273
Glaciology (International Geophysical Year 1957–1958) 30
Glaser, Peter 572
Glass, B. 191
Glass wool (packed bed) 138
Glen, J. W. 617
Glenbar 265
Global albedo 79, 80
Global average models (climate) 77
Global climate (origins and patterns) 83–92
Glomerella cingulata 416
Glycogen storage disease 132
Goiter 132
Gold: heavy metals, pathology of 253
  mining 344
  uses 336
Gondwanaland 28–29
Gonorrhea 132
Gonyaulax catenella 401
Gonyaulax tamarensis 401
Gossypol 424
Grade (livestock breeding) 8
Graded stream 594
Grading of market animals 9–10
Grading-up (livestock breeding) 8
Grafting (plant breeding) 73
Grain crops (virus disease) 2
Gramineae 246, 479
Graminoids (grasslands) 246
Gramoxone 265
Granite rocks (nuclear fuel) 368
Granular soil structure 532
Grape 137, 405
Graphite-moderated molten-salt thermal breeders 369
Grasses: arctic islands 31
  grama 246
  root zone depths 115
  spear 246
  switch 246
  wheat 246
Grasshoppers, prairie 451
Grassland 64, 94, 109, 143, 245–247, 429, 470, 634, 637
  climate 246
  composition 246–247
  conservation of resources 110–113
  earth, resource patterns of 140
  ecology 149, 247
  exploitation and management 247
  fire 95
  management 159
  overgrazing 95

Grassland—cont.
  prairie ecosystem 450–452
  productivity 247
  savanna 479–481
  short 433
  structure 246
  tall 432–433
  terminology 246
  tree-inhibiting factors 245–246
  trophic structure 150
Gravel 531
  earth dams 121
  packed bed 138
  properties 532
  soils 557
Gravimetry (International Geophysical Year 1957–1958) 30
Gravitational settling (atmosphere cleansing) 48
Gravitational water (soil) 558
Gravity (abiotic factor) 177–178
Gravity dam 118, 123
Gravity drainage well 383
Gravity settling chamber (dust collector) 138
Grazing lands: economics 470
  types 470
Grazing pasture 637
Greasewood 130
Great Basin Desert (southwestern United States) 130
Great Lakes: dune vegetation 137
  eutrophication 187–188
Green algae 401, 566
Green manure 309, 537
Green Peter Dam (Ore.) 119, 120
Green River shale 396
Green sea lettuce 313
Green sulfur bacteria 565
Greenhouse effect 78, 85, 247, 682
Greenland 31, 32, 33
  ice 75
  ice cap 76
  ratio of oxygen-16 to oxygen-18 in ice sheet 76
Gregory, P. H. 2, 420
Griseofulvin 233
Grit chambers (sewage treatment) 519
Groin (beach stabilization) 105, 107–108
Gross production: biological 61
  ecosystem 153
Ground sloth (mass extinction) 191
Ground squirrels 450
Ground storage reservoirs 666
Groundsel bush 136
Groundwater 247–250, 600
  aquifer 30
  artesian systems 34–35
  chemical qualities 250
  conservation 656
  controlling forces 248–249
  frost action in soils 560–561
  hydrologic cycle 88
  hydrosphere 278
  infiltration 249
  land drainage (agriculture) 294–297
  openings in rocks 248

Groundwater—cont.
  salt-water intrusion 664
  seepage and frost in soil 559
  sewer construction 513
  soil-water relationships 558–559
  sources 249
  spring, water 586–587
  subterranean water 247–248
  thermal springs 618
  water supply 664–665
  water table 666–667
  water-table and artesian conditions 249–250
Growth, plant 161–163, 402–406
Growth, population 447, 448
Growth regulator (plants) 6, 13–14
Grus americana 585
Guanidines 423
Guggenheim process (sewage treatment) 521
Guinea pig (ascorbic acid) 133
Gulf coast (dune vegetation) 135
Gulf Stream 33, 373
Gully erosion 239, 539, 550, 551
Gum tree 431, 432
Gymnodinium breve 401
Gymnosporangium rust 427
Gypsum: dunes 135
  phorphoric acid production 195
  sodium soils 116

H

Haagen-Smit, A. J. 524
Habitat 25–26, 175
Hadley cells 41, 43
Hafsten, U. 1
Hageman, R. H. 199
Hail: cloud physics 99–102
  soft 100
  suppression 681
  thunderstorm 620
  tornado 622
Hailstone 100, 101
Hailstorm 681
Halocline (estuary) 183
Halogens, atmospheric geochemistry of 250–251
  experimental methods 250
  natural atmosphere 250–251
  polluted atmosphere 251
Halophytes: desert 129
  mangrove swamp 307
  saltmarsh 478
Hand, C. 149
Hansen, Victor 374
Haphazard population dispersion 444
Haplopappus erocoides 137
Harbor (breakwaters) 108
Hard-leaved forests 636
Hardwoods 212
  evergreen 431–432, 648–649
  forest fire 95
Hares 628
Hartlaubgehöltze 648
Harvest (forest) 207–209
Harvester, peanut 11
Harvey, H. W. 497
Hashish 623
Hasler, A. D. 293

Haulage: mechanical 341–342
    rail 341–342
    truck 341
Hawks 628
Hawkweeds 627
Hay fever 3
Hay meadow 637
Hayes, J. V. 3
Hays, J. D. 191
Haze (inadvertent weather modification) 682
Hazelhoff, E. 11
HCA (herbicide) 266
Health: disease 131–134
    effects of atmospheric pollution 49–50
    epidemiology 180–182
    malnutrition 305–307
    nutrition 372
    public 575–576
Health physics 251
    radioactive materials, decontamination of 462–463
    radioactive waste disposal 463–465
Heap clouds 97, 318
Hearing, human (loudness) 304–305
Heart (beer drinker's syndrome) 260
Heart rots: forest trees 624–625
    shade trees 627
Heat (safe body exposure) 55
Heat balance (human body) 54–55
Heat balance, terrestrial atmospheric 40, 41, 43–44, 84, 85, 251–252, 318, 373, 491–492, 590
Heat engine (analogy of atmosphere) 76–77
Heat insulation 253
Heat of vaporization (atmosphere) 40
Heat probes (oceanographic survey) 276
Heat rate of building 128
Heath 650
Heating, solar space 571
Heating comfort: central 253
    cost of operation 253
    degree-day 55, 128–129
    design 253
    infiltration 253
    insulation and vapor barrier 253
    load 252–253
Heating engineer 252
Heating system: comfort 252–253
    efficiency 128
Heavener, O. F. 511
Heavy metals, pathology of 253–262
    cadmium disease 256–257
    carcinogens and teratogens 255
    cellular response 255
    chromosome damage 255–256
    cobalt toxicity 260–261
    conditions for toxicity 254–255
    lead poisoning (plumbism) 259
    mercury poisoning (mercurialism) 257

Heavy metals, pathology of
    —cont.
    nature of metal toxicity 254–256
    nickel poisoning 261–262
    specificity 255
    thallium disease (thallotoxicosis) 259–260
Heavy-water-moderated reactors 369–370
Hechtia 433
Hedgehog cactus 130
Hedyotis 288–290
Hedyotis cookiana 289
Helium (low-troposphere concentration) 39
Helland-Hansen, Bjorn 374
Helminthosporium carbonum 425
Helminthosporium disease 408
Helminthosporium victoriae 425
Helminths (cause of disease) 132
Hematite-goethite minerals 543
Hembree, C. H. 595
Hemicryptophyte 439, 627
Hemlock 137, 431, 607
Hemoglobin (sickle-cell trait) 133
Hemolytic poisons 622
Hemorrhage (disease) 132
Hepatitis, viral 132
Heptachlor 285, 286, 400
Herb-shrub-tree society 108
Herban 266
Herbicides 5, 12–13, 262–268
    action 263
    classification of 262–263
    nonselective 262–263
    plant disease 410
    preemergence 262
    properties 268
    protoplasmic selectivity 263
    selective 262
Herbivores (prairie) 450, 451
Hercules 9573 (herbicide) 265
Hereditary disease 132
Heritability of traits (animal breeding) 71
Heroin 623
Herpes 132
Herringbone drainage pattern 296
Hersey, J. B. 482
Hesse, R. 699
Heterograde curve 225
Heterosis 73
Heterotrophic organisms 150, 565
Hexachloroacetone (HCA) 128, 266
Hexagenia 188
Hexagonal ice crystals 101
Hexagonal prismatic column ice crystal 101
3-(Hexahydro-4,7-methanoindan-5-yl)-1,1-dimethylurea 266
Hibernation (prairie animals) 451
Hibiscus 308
Hickories 136, 648
Hieracium 627
High, polar 589
High-level clouds 97
High mountains 608, 609
High-pressure slide gate 126
High-rate filters 522, 523

High-rate trickling filter 522
High-temperature graphite-moderated reactors 369
High tussockland 651
High-voltage direct current 170
High-voltage transmission 170
Highland-montane complexes 634
Highs: middle-latitude 589–590
    subtropical 589
    subtropical oceanic 89
    thermal 89
Highway engineering 268–271
    construction operations 270
    detailed design 269–270
    maintenance and operation 270
    right-of-way acquisition 268–269
Hilaria jamesi 433
Hill and mountain terrain 271–273, 608–610
    development of rough terrain 271
    distribution pattern of rough land 271
    predominant surface character 271–273
Hills 94–95, 608, 609
Hirst, J. M. 2
Histoplasmosis 132
Histosols 529, 541
Hjort, Johan 374
Holarctic region 694, 699
Holdridge, L. R. 303
Holdridge's life zones 303
Holly 136, 431
Holocoenosis 175
Holomictic lakes 219
Homeostasis 131
    cellular level 274
    community 166–167
    ecology 149
    ecosystem 151, 166–167
    environmental stresses 179
    homeostatic mechanisms 274
    patterns of adaptive responses 274
    significance 274–275
Homeothermic animals 614, 615
Homo sapiens 66
Honey (production by bees) 173
Honeybees (pollination role) 173
Hookworm 132
Hopkins, David 34
Horizons, soil 177, 216–217, 526–527, 527–528
Horizontal-flow electrostatic precipitator 22
Horn flies (livestock parasites) 10
Horse (mass extinction) 191
Horseflies (livestock parasites) 10
Hot springs 619
Houghton, H. G. 252
House dust (aeroallergens) 3
Howard, L. 97
HPAN see Modified hydrolyzed polyacrilonitrile
Huang River 473
Hudsonia tomentosa 136
Hudsonian life zone 301
Human bioclimatology 53–56

Human biogeography 61
Human community 160–161
Human ecology 53–56, 61, 75–82, 95–96, 145, 151, 160–161, 164, 178–180
Human environments (environmental protection) 180
Human genetics (nutrition) 372
Human populations (age structure) 449
Human resources 111, 112
Humans (response to sonic boom) 583–585
Humboldt, A. 300
Humid climates 142, 527
Humid macrothermal climatic regions 141, 142
Humid mesothermal climatic regions 141, 142
    mean temperature 90
    precipitation 90
Humid microthermal climatic regions 142
    mean temperature 90
    precipitation 90
Humid regions (irrigation programs) 118
Humid subtropical climatic region 141, 142
Humidity: absolute 275
    dew point 131
    mixing ratio 275
    relative 275
    sensible temperature 511
    specific 275
Humification (forest soil) 216–217
Humiture 511
Hummocky floes 487
Humus 532
    formation 217
    manure in soil 309
    mor 216
    mull 216
    raw 216
Humus layer (forest soil) 216–217
Hungarian riffle 343
Huntington, M. C. 397
Hurricane 319
    aircraft detection and tracking 592
    modification 681–682
    radar detection 590
    seeding 682–692
Hurricane Agnes (effects on landform) 236
Hurricane Donna (radar picture) 590
Hutchinson, G. E. 64, 67, 219
Hutchinson-Löffler system of thermal lake types 220
Hybrid (animal) 8
Hybrid vigor (plant) 73
Hybridization (plant breeding) 73
Hydra 224, 225
Hydraulic cutter dredge (undersea mining) 358
Hydraulic-fill stope 350

Hydraulic long-stroke pumping (oil well production) 388
Hydraulic subsurface pumps (oil well production) 388
Hydraulic turbine (electric energy generation) 168–169
Hydraulicking (placer mining) 344
Hydrocarbons: air pollution 46
  chlorinated 399–400
  poison 622
  smog, photochemical 524
Hydrochory 442
Hydroelectric plants 630
Hydroelectric power generation 570
Hydrogasification: BI-GAS process 105
  carbon dioxide acceptor process 104–105
  coal gasification 103
  HYGAS process 103
Hydrogen: biological element 67
  cycle 68
  essential element for plants 402
  low-troposphere concentration 39
  soil 542
Hydrogen peroxide (mutagen) 365
Hydrogen sulfide (poison) 622
Hydrogenerators 168
Hydrogeology (groundwater) 247–250
Hydrographic stations: oceanography 377
  storm surge 593
Hydroids 312
Hydrolithosphere (terrestrial heat balance) 84
Hydrologic cycle 88, 90–91, 177, 373, 657–658
  cloud formation 96–97
  cloud physics 99–102
  evapotranspiration 189
  forest soils 217–218
  surface water 599–604
Hydrology 275–276
  atmospheric evaporation 40–41
  development of landforms 236
  drought 133
  engineering 275
  forest 207
  forest soil 217–230
  frost 232
  frost action in soils 560–561
  functions 275
  groundwater 247–250
  remote sensing 612
  sea water 490–507
  seepage and frost in soil 559
  soil water 534–535
  soil-water relationships 558–559
  spring, water 586–587
  stream transport and deposition 594–595
  surface water 599–604
  terrestrial frozen water 617–618
  water table 666–667

Hydrometeorology 276–278
  analysis of precipitation data 277–278
  atmospheric water cycle 276–277
  evaporation and transpiration 278
  precipitation 277
Hydromica (oil shale) 394
Hydrophotometers (oceanographic survey) 276
Hydroplants 168
Hydrosphere 278, 595
  biosphere 64
  development 280–281
  oceanography 374–380
Hydrosphere, geochemistry of 278–281
  age determination of sediments 378
  closed basins 281
  fresh waters and rain 281
  oceanic waters 278
Hydrosphere, marine 279
N-Hydroxymethyl-2,6-dichlorothiobenzamide 266
Hydroxyquinoline dyes 70
HYGAS process 102–104
  electrothermal version 104
  oxygen version 104
  steam-iron-hydrogen version 104
Hygiene, industrial 251
Hygroscopic water (soil) 558
Hylaea 646
Hylurgopinus rufipes 421, 627
Hyperplasia 133, 408
Hypersensitivity (disease causative agent) 132
Hyperthyroidism 133
Hypertrophy 133, 408
Hypolimnion: chemistry 225
  lenitic ecosystems 218, 220
  lentic water 110
  productivity 225
Hypoplasia 133, 408
Hyvarx 264

I

IBP see International Biological Program
Ice: atmospheric evaporation 41
  avalanche 52
  dry 102
  firn 618
  frost 232
  land 486
  massive 618
  melting temperature 618
  pancake 486
  permafrost 398
  river 486
  sea 81, 486–488
  terrestrial 617–618
  thermal constants 618
Ice age, little 76
Ice Age, Pleistocene 617
Ice caps 75, 76
Ice climate 141, 142
Ice clouds 97
Ice crystal process (precipitation release mechanism) 100

Ice crystals: cloud composition 100
  cloud formation 97
  cloud physics 101
  types 101
Ice nuclei (cloud physics) 101
Ice pressure (concrete dam) 119
Ice sheet (Antarctica) 28
Ice wedges (permafrost) 398
Icebergs 486
Iceland 31, 32
Icelandic low 693
ICONS program (Atomic Energy Commission) 199
Ideal bioclimate 56
Idria columnaris 131
Igneous rock 143
Ilex 431
Ilex opaca 136
Illness (disease) 131–134
Imagery: infrared 612–613, 638
  passive microwave 612
Imhoff, Karl 281
Imhoff tank 281–283, 519
  design 282–283
  efficiency 283
Imidazolines 423
Immunity (impairment by radiation) 458–459
Imperata cylindrica 470
Imperial mammoth (mass extinction) 191
Impingement separators 138
Inbreeding 8, 73
Incas (human biogeography) 61
Inceptisols 529, 541
Incidence rate (disease) 181
Incineration (solid waste disposal) 22, 578–579
Incinerator 24
  design 22–23
  fluidized-bed 580–581
  solid-waste disposal 578–579
Inclined skip hoist (open-pit mining) 342
Indentation hardness (ice) 617
Index, Palmer 133
Indian rice grass 130
Indicator organisms (distribution of species) 511
Indigo bush 130
Individualistic concept of community 109
Indo-Pacific faunal region 26
Indoor allergens 3
Indoor environment (aeroallergens) 3
Induced mutations 364–365
Indus River 473
Industrial engineers 172
Industrial hygiene (health physics) 251
Industrial meteorology 283–284
Industrial wastes 316, 515, 663
  distribution in estuaries 185
  radioactive 465
  sewage 512–515
  stream pollution 516

Inertial device (dust collector) 138
Inertial stability (atmosphere) 330
Infections, fungus 10
Infectious diseases (epidemiology) 180–182

Infiltration: galleries, water 665
  groundwater 249
  heat loss in buildings 253
Influenza 132
Infrared imagery: terrain sensing, remote 612–613
  vegetation and ecosystem mapping 638
Infrasonics (storm detection) 591–592
Injury, radiation (biology) 456–460
Inland Waterways Commission 113
Inoculum: dormant 416
  plant pathogens 415–416
Inorganic fungicides 233
Inorganic insecticides 285
Inoue, Eiichi 2
Insect attractant 5
Insect control, biological 173, 284–285,
  advantages over chemical control 285
  applications of classical method 284
  augmentation 284–285
  conservation 284
  ecological principles 284
  history 284
Insect repellent 5
Insect vectors 174
Insecticides 5, 12
  contamination of coastal waters 661
  formulation and application 287
  inorganic 285
  insect resistance 286
  organic 285–286
  organic phosphorus 286–287
  organophosphate 399
  plant disease 410
  poison 623
  residual 285
  synergists 287
  synthetic carbamate 287
  systemic 285
  use in agriculture 173–174
Insects: crop loss 173–174
  dissemination of plant pathogens 420–421
  entomology, ecological 172–174
  home and health hazards 174
  pest control 173
  plant disease 409
  pollination role 173
  resistance to insecticide 286
  role in cycle of matter 173
Insemination, artificial 8, 71–72
Instrumentation (dam) 127
Insulation, heat 253
Integration (conservation measure) 111
Interactions, ecological 108, 145–148, 158, 164, 178, 206, 412–413,
Intermittent sand filters 522
Internal combustion engines (electric energy generation) 169
Internal seiche 221, 222
Internal wave 499

International Biological Program (IBP) 1
Intertropical convergence zone 693
Intertropical front 15
Intoxication, chemical 132
Intrazonal soil 541
Inverse estuaries 182
Inversion: chromosome 362
  radiation 47
  subsidence 48
  temperature 46, 608
Invertebrate zoology (entomology, ecological) 172–174
Iodine: biological element 67, 69
  halogens, atmospheric geochemistry of 250–251
  radiation injury (biology) 462
Ionizing radiation: abiotic factor 176–177
  faunal extinction 191
  indirect mutations 364
  radiation biochemistry 453–456
  radiation injury (biology) 456–460
  radioecology 465–468
Ionosphere 36, 317
Ionospherics (International Geophysical Year 1957–1958) 30
Ions, mineral 306
Ioxynil 265
IPC see Isopropyl-N-(3-chlorophenyl)-carbamate
Ipomeamarone 424
Ipomoea pes-caprae 135
Iron: constituent of living matter 70
  essential element for plants 404
  heavy metals, pathology of 253
  micronutrient 198
  nutrient 155
  soil 542
  uses 336
  utilization by soil microorganisms 568
Iron cycle 70
Iron-deficiency disease (cabbage) 410
Iron sulfate (sodium soils) 116
Ironwood 131
Irradiance (sea water) 493
Irrigation 275
  basin 117
  bench border 117
  border strip 116–117
  crops 114–118, 294–297, 654–655
  flood method 116–117
  furrow method 116
  humid and arid regions 118
  sprinkler 117
Irrigation systems, automated 117–118
Irving, L. 34
Island faunas and floras 288
  biotas of coral atolls and limestone islands 289–290
  biotas of volcanic islands 289
  dispersal 289
  endemism 288
  evolution on islands 289

Island faunas and floras
  —cont.
  geography of floras and faunas 288–289
  morphological peculiarities 289
  nature and classification of islands 288
Islands: arctic and subarctic 30–33
  limestone 289–290
  volcanic 289
Isobar 17, 677, 678
Isohypses 678
Isolan 287
Isolation, biotic 71
Isopods 312
Isopropyl N-(3-chlorophenyl) carbamate (IPC) 3, 13, 266
Isopropyl N-phenylcarbamate 266
Isotherm oscillations (sea water) 498
Isotherms (sea) 499
Isotope: composition of ocean water 501–504
  in protoplasm 68
  radioactive 157
  radon 38
Isotope analysis (evidence for change of climate) 76
IT 3456 see Methyl-2-chloro-9-hydroxyfluorene-(9)-carboxylate
Iva imbricata 135

J

Jackrabbits 450
Jaegers (tundra wildlife) 628
Jellyfishes 312
Jet flow gate 126
Jet piercing (open-pit mining) 339
Jet pumps (oil well production) 389
Jet stream 41, 290–291
  characteristics 290–291
  variations 290–291
Jetty 108
Jigs: mineral recovery 343
  placer mining 346
Johnson, M. W. 482
Johnson, N. M. 293
Johnston, H. S. 524
Joint meters (concrete dams) 127
Joshua tree 130
Judging, livestock 9–10
Juglans 431
Juncus 478
Jungle 55, 647
Juniper 431, 607, 644
Juniperus 136, 431
Juniperus scopulorum 644
Juvenile water 249, 618–619

K

Kalanchoe blossfieldiana 14
Kalmia 431
Kangaroos 151
Kaolinite: atmospheric ice nuclei 101
  oil shale 394
Kaolinite-halloysite minerals 543

Karmex 265
Kauri pine 430
Keeling, C. D. 77
Kelp 313
Kemper, W. D. 548
Kerguelen faunal region 26
Keteleeria 648
Kidney (methyl mercury-poisoned cells) 253
Kidney colic 56
Killing frosts 232
Kingdoms, functional (ecosystems) 166
Kjensmo, J. 293
Klamath weed 173
Klinefelter's syndrome 132
Knee-bend-like roots 307
Kochia 433
Koczy, F. 497
Kodiak Island (Arctic) 31, 32
Koeleria 432
Koeppen, W. 92
Koeppen classification 92–93
Kohl, D. H. 199
Kolkwitz, R. 229
Kolm (shale) 395
Kraus, E. J. 128
Krog, J. 34
Krogh, A. 616
Krupp wheel excavator 392, 393
Krypton (low-troposphere concentration) 39
Küchler, A. W. 633
Kuiper, G. P. 373
Kukersite 396
Kunelsen, M. 374
Kurile Island (Arctic) 31
Kuroshio 373
Kwashiorkor 132, 307

L

Laboratory, research vessel 376
Labrador Current 32, 33
Laccase 404
Laceration 132
Lacustrine soils 556
Lake 291–292
  amictic 219
  basin and regional factors 291–292
  classifications 218–220
  conservation and economic aspects 292
  dimictic 219
  dystrophic 225
  eutrophic 325
  eutrophication, cultural 187–189
  fresh-water ecosystem 218–230
  glacial 292
  holomictic 219
  light stratification 220
  limnology 304
  oligotrophic 225
  pollution 518
  salt 291
  temperate 218
  temperature stratification 218
  water supply 665
  waves and currents 220–221
Lake ecosystem (nutrient flow) 58

Lake Erie (eutrophication) 188
Lake fly (fauna) 187
Lake herring 188
Lakes, meromictic 219, 292–294
  biological studies 293
  chemical studies 292–293
  physical studies 293
Lally, V. E. 511
Lamarck, J. B. 64
Lambast 264
Laminaria 313
Land (earth, resource patterns of) 140
Land breeze 693
Land disposal of sewage 518
Land drainage (agriculture) 294–297
  drainage-system patterns 295–296
  methods 294–295
  systems construction 296
  systems maintenance 296
  water source and disposal 294
Land drainage (sewage disposal) 515–518
Land ice 486
Land management (valley slope) 240
Land reclamation (strip mining) 347–348
Land resources 140
Land-use classes 297–298
  farm, range, and forest 553–555
  land-component surveys 297–298
  landscape analysis 298
  mathematical approaches 298
  present land use 297
Land-use planning 298–300
  awareness and promptness 300
  highway engineering 268–271
  importance of succession 599
  planning and the future 299
  range management plan 471
  soil erosion 539–540
  vegetation management 639–642
  zoning of land 299–300
Landfill, sanitary 577–578
Landforms: distribution of vegetation 632
  terrain areas, worldwide 608–610
Lands, grazing 470
Landsberg, H. E. 93
Landscape: geomorphology 234–243
  land-use classes 298
  management 642
  productive 157–158
Landslides 563–564
Lanettes 135
Langmuir, I. 680
La Plata–Paraná River 473
Lapps (human biogeography) 61
Larch (taiga) 607
Larix 431
Larrea 433
Larrea divaricata 130
Late blight 418
Late solar pigmentation (human race) 53
Laterite 94

Lateritic soils 542
Lathyrus 137
Latosolic soils 541
Lauraceae 432
Laurel 430, 647
Laurel forest 647
Laurelia 430
Laurisilva 647
Laurus 430
Laws, air pollution 51–52
Layer, scattering 481–482
Layer clouds 97, 100
Layer system (clothing) 179
Leaching (soils) 537
Lead: heavy metals, pathology
    of 253
    poison 622
    uses 336
Lead arsenate (insecticide) 285
Lead poisoning: acute form 259
    chronic form 259
Leaf disease in forests 624
Leaf spot (shade trees) 627
Lease tank 389
Leather oak 137
Ledum 431
Legumes 246, 536
Leguminosae 246, 432, 433, 479
Leiophyllum buxifolium 136
Lemmings 628
Lena River (Siberia) 473
Lenitic ecosystems 218–221
    lake classifications 218–220
    light stratification 220
    temperature stratification 218
Lentic water (biocycle) 110
Lenticular clouds 318
Leopard seal 30
Leprosy 132, 181
Lesions (pathology) 132
Lethal dose 50 (radiation) 457
Leukemia (radiation) 459–460
Li, Y. 293
Liana 469, 646, 647
Libby, W. F. 492
Libocedrus 648
Lice (livestock) 10
Lichens 142
    Antarctica 30
    arctic islands 31
    tundra 637
    woodland 650
Liebig, J. 64
Liebig's law of the minimum 57
Life, origin of (genesis of bio-
    sphere) 64–65
Life, pyramid of 166
Life belts (biotic provinces) 302
Life-form societies 437–438
Life forms of plants 129,
    437–438, 438–440, 478
Life table (populations) 450
Life zone 110, 300–303
    Dice's zones 302–303
    Holdridge's zones 303
    Merriam's zones 300–302
Lifting (oil well production)
    387–389
Light: abiotic factor 176
    effect on plant disease
    418
    transmission in sea water
    493–494
    ultraviolet 364
Light-water reactors 369

Lightning: arresters 170
    production of electric charge
    101
    sferics 591
    suppression 681
    thunderstorm 620–621
Lignite: carbon dioxide acceptor
    process 105
    Lurgi gasifiers 103
    uses 336
Likens, G. E. 293
Limber pine 644
Lime sulfur (sodium soils) 116
Limestone 143
    Edwards 30
    Ocala 30
    sodium soils 116
Limestone islands 289–290
Limnetic zone: aquatic habitats
    304
    lentic water 110
Limnology 110, 304
    factors influencing productiv-
    ity 304
    food cycles 304
    fresh-water ecosystem
    218–230
    meromictic lake 293
Limonium 478
Lindane 400
Lindens 648
Line, squall 587–588
Linear programming (wildland
    operations research) 688
Lines, transmission 169, 631
Link-Belt mechanical aerator
    523
Linocedrus 431
Linuron 265
Lipids 67, 372
Liquid fertilizer 118, 197
Liquid-level gages (petroleum
    production) 391
Liquid scrubbers 51, 139
Liquid wastes: agricultural
    654–655
    radioactive 464
Liquidambar 430, 431
Liriodendron 431
Lithium (in living matter) 69
Lithops 652
Lithosphere: biosphere 64
    lake ecosystem nutrient flow
    58
Little bluestem bunchgrass 137
Little ice age 76
Little-leaf disease 624, 625
Littoral benthos (lenitic ecosys-
    tem) 224
Littoral currents 106
Littoral drift 105, 106, 107
Littoral system (marine ecosys-
    tem) 313–314
Littoral transport 105, 106, 108
Limnetic zone: aquatic habitats
    304
    lake 224
    lentic water 110
    ocean 110
Live oak 136, 137
Liver (mercury-poisoned cells)
    258
Livestock: breeding 7–8
    feeding 8
    judging 9–10

Livestock—*cont.*
    pest and disease control 10
Ljundstrom in-place method 396
Llanos 246
Loaders, front-end 341
Local fallout 461
Locks (river engineering) 475
Locomotive (mining rail haulage)
    342
Lodgepole pine 136
Löffler, H. 219
Logic circuits (systems ecology)
    605
Lonchocarpus urucu 287
Lonchocarpus utilis 287
Long-range forecasting 670–671
Long Range Mountains (Arctic)
    32
Long tom (placer mining) 344
Longwall mining 348–349
Lorox 265
Lotic ecosystems 227–229
    biota 228–229
    gradients 228
    water source 228
Lotic water (biocycle) 110
Loudness 304–305
    decibel 305
    noise pollution 305
    phon 305
    sones 305
Low: Aleutian 693
    cyclonic depressions 100, 692
    Icelandic 693
    thermal 89
Low-level clouds 97
Low mountains 608, 609
Low-pressure area 692
Low-rate filters 522
Lower atmosphere 36
Lower Sonoran life zone 301
Ludlam, S. D. 293
Lumber (forest management)
    203
Lumbering (disturbance of
    climax communities) 95
Lumnitzera 307, 308
Lumpy jaw 10
Lunar geography 180
Lunar soils 564
Lung: beryllium-poisoned cells
    255
    mercury-poisoned cells 258
Lung cancer (air pollution) 49
Lupine 137
Lupinus 433
Lupinus chamissonis 137
Lurgi gasifier 102–103
Lycopodium alpinum 70
Lymphocytes (radiation injury)
    458
Lyonothamnus floribundus 288
Lysimeters (evapotranspiration)
    189

M

MAA *see* Methanearsonic acid
MacConkey broth 655–656
McCormick, J.S. 616
Machine-design engineers 172
Mackenzie River (Canada) 81,
    473
Macroclimatology 82

Macrocommunity 164
Macroconsumers 150, 151
Macrocytic anemia 132
Macroenvironment 164
Macronutrients 155, 192
Macrophytes (eutrophic lakes)
    187
Magnesium: constituent of living
    matter 69
    essential element for plants
    403–404
    sea water 315, 379
    soils 116
    uses 336
    utilization by soil microor-
    ganisms 568
Magnesium chlorate 128
Magnesium-deficiency disease
    (bean) 410
Magnetism, remanent 191
Magnetite spherules 38
Magnetometers: marine mineral
    exploration 356
    oceanographic survey 376
    rubidium vapor 356
Magnolia 431
Mahagony Ledge 392
Malagasy subregion 694, 699
Malaria 132, 174
Malathion 286, 399
Male sterility, cytoplasmic 75
Maleic hydrazide (MH-30) 6,
    265, 365
Mallee 650
Malnutrition 133, 305–307
    cell requirements 306
    essential metabolite deficien-
    cies 306
    generalized malnutrition
    306–307
    Kwashiorkor disease 307
Mammals: Arctic region 33–34
    Australian region 697–698
    Ethiopian region 695–696
    Nearctic region 695
    Neotropical region 697
    Oriental region 696
    Palearctic region 695
    recent mass extinction in
    North America 191
Mammoth: Columbian 191
    imperial 191
    woolly 191
Man: dispersal of disseminules
    443
    dissemination of plant
    pathogens 421
    environmental protection
    178–180
    impact on climate 75–82, 641,
    682
    impact on grassland animals
    452
Management: fishery 160, 201
    range 159, 247
    right-of-way 642
    roadside 642
    soil 535–538
    solid-waste 577–580
    timber 204
    vegetation 633, 639–642
    water 658
    water-resource 160
    watershed 642, 658–659
    wildlife 159–160

Management and organization, forest 158–159, 202–204, 207–209, 209–212, 216, 599
Management engineering 172
Manch 233
Manganese: constituent of living matter 70
essential element for plants 404
micronutrient 198
nutrient 155
poison 622
uses 336
utilization by soil microorganisms 568
Mange mites (livestock parasites) 10
Mangrove: forest 636, 646
swamp 307–309
thickets 429
Manure: animal 309
green 309, 537
plant 309
recycling 652–653
Manzanitas 137
Map: soil 530–531
topographic 633–637
weather 15, 16, 19, 676–678
Map design (vegetation and ecosystem mapping) 633–639
Maple 137, 431, 648
Mapping: vegetation and ecosystem 440, 633–639
weather map 676–678
Marantaceae 430
Marcasite (oil shale) 394
March elder 135
Mare 8
Margin, continental 114
Marijuana 623
Marine biocycle 110
Marine biology 378
Marine biomes 64
Marine communities (trophic structure) 617
Marine ecology 149
Marine ecosystem 60–61, 110, 164, 309–314, 378, 613–614
benthic division 313–314
biota 311–312
chemical factors 310
estuaries 314
estuarine oceanography 182–187
major divisions 312–314
marine environment 310–312
marshes 314
pelagic division 312–313
physical factors 310–311
phytoplankton 401–402
sediments 314
zooplankton 700–701
Marine environment 177, 310–312
estuarine oceanography 182–187
mineral resources 354–355
pesticide pollution 400
Marine fisheries 200, 378
Marine food chains 508
Marine geology: mining, undersea 354–361

Marine geology—cont.
paleoclimatologic research 378
petroleum prospecting 380
Marine hydrosphere (elements) 279
Marine littoral faunal regions 26
Marine navigation (mineral exploration) 355–356
Marine resources 140, 143–144, 314–316, 378–379
extractive 314–316
nonextractive 316
Marine sediments 66, 186, 314
Marine technology 376–377
Mariposa 137
Maritime aerosols 38
Maritime air (hydrologic cycle) 88
Maritime air mass 16
Maritime saltmarsh 477–478
Marram grass 136, 137
Marsh (marine ecosystem) 314
Marsh, bog, and swamp 604
Marsh, salt 637
Marshall, N. B. 482
Mass, air 230–232, 319–320, 320–321, 321–322, 587
Mass, ocean water 500–501
Mass extinction 189
Mass selection 71
Mass spectrometry (sea water analysis) 507
Massive ice (physical constants) 618
Mastodon (mass extinction) 191
Material testing (environmental protection) 180
Materials: effects of atmospheric pollution 50
environmental protection 179–180
Mathematical models (wildland operations research) 687–688
Mathematics: health physics 251
models of climate 77
Matorral 649
Matrix theory (systems ecology) 606
Matter (ecosystem) 62
Matthew, W. D. 698
Mauna Loa Observatory 4
Maury, Matthew Fontaine 374
Maximization (conservation measure) 111
Mayans 61
Mayfly 188
MCPA see Methyl-4-chlorophenoxyacetic acid
MCPB (herbicide) 263
MCPP see 2-Methy-4-chlorophenoxy propionic acid
Meadow 637, 651
bog 694
hay 637
salt 637
Mean annual precipitation (major land areas) 89
Meandering channels 474

Measles 132
Measurement: forest 205
noise 367–368
Mechanical aeration (sewage treatment) 523
Mechanical aerator, Link-Belt 523
Mechanical booms (oil spill cleanup) 484
Mechanical engineers 172
Mechanical haulage (open-pit mining) 341–342
Mechanical injury 132
Mechanical loading (open-pit mining) 340–342
Mechanical oil spill cleanup systems 484
Mechanical power engineering: degree-day 128–129
heating, comfort 252–253
solar energy 569–573
Mechanical pumping (oil well production) 387–388
Mechanics: fluid 581–585
soil 555–564
Mechanisms: endocrine 274
homeostatic 274
Mecoprop 266
Medical microbiology (epidemiology) 180–182
Medicinals (disease) 130
Medicine: health physics 251
heavy metals, pathology of 253–262
Mediterranean climate 141, 142, 143
Mediterranean faunal region 26
Mediterranean forest 431
Medium-range forecasting 670–671
Mekong River 473
Melampsora lini 414
Meloidogyne 409
Melosira 188
Melville Island (Arctic) 31, 32
Membrane, cell 403
Membrane filter technique (water analysis) 655–656
Menazon 287
Meningitis 132
Mental diseases (epidemiology) 181
N-(2-Mercaptoethyl)benzenesulfonamide 266
Mercury: heavy metals, pathology of 253
increasing prevalence in environment 257
Mercury poisoning 257–259
case histories 258
chromosome damage 256
symptoms 258–259
Mercury salts (poison) 622
Meres (lakes) 291
Meridional circulation 41
Meromictic lakes 219, 292–294
Merriam, C. Hart 300, 699
Merriam's life zones 300–302
Mesoclimatology 82
Mesopause 36, 37, 317
Mesophytes (evapotranspiration) 189

Mesophytic vegetation (swamp, marsh, and bog) 604
Mesosphere 36, 37, 317, 608
Mesquite 129, 131, 433
Mesquite-grassland (Mexico) 59
Metabolic disorders 133
Metabolic rate (biological productivity) 63
Metabolism: carbohydrate 306
crassulacean acid 129
homeostasis 274
plant 161–163, 411
protein 306
Metabolite (deficiencies) 306
Metal: complexing in soils 546–548
heavy metals, pathology of 253–262
mineral resources conservation 335–338
mining, open-pit 338–342
mining, placer 342–346
trace 496
Metal toxicity: conditions for 254–255
nature of 254–256
Metalimnion: lenitic ecosystems 218, 221
lentic water 110
Metallic-ore mining engineers 172
Metallic poisons 132
Metallurgy (hazards of nickel) 262
Metamorphic rock 143
Metaplasia, epithelial 132, 133
Meteorological optics 320
Meterological satellites 592–593, 678
Meteorology 145, 316–332
aerobiology 1–4
agricultural 5–7, 334
air masses and front-jet systems 320–322
air masses and fronts 319–320
analytical dynamical 328–331
atmospheric disturbances 318–319
circumpolar aerology 325–326
dust storm 140
dynamical 326–332
electrical properties 320
equations 328
fair-weather 333
fronts and frontogenesis 322–323
general atmospheric circulation 318, 320
general physics of the air 316
industrial 283–284
instantaneous state of atmosphere 328
International Geophysical Year 1957–1958 30
large-scale tropical weather systems 325
life cycle of extratropical cyclones 323–325
numerical dynamic 331–332
optical phenomena 320
plant energy and water relations 162
pressure and wind 318

Meteorology—*cont.*
radiation and thermal structure 317–318
synoptic 678
thermal structure 317
thermodynamic properties of
air 326–328
water vapor, cloud, and precipitation 318
waves and vortices (the
weather) 323
Meteorology and climatology
82–93, 316–332
agricultural meteorology 5–7
air mass 14–17
air pressure 17–19
arctic and subarctic islands
30–33
atmosphere 35–37
atmospheric chemistry 37–39
atmospheric dilution of pollutants 21–22
atmospheric electricity 39–40
atmospheric general circulation 41–46
atmospheric pollution 46–52
bioclimatology 52–56
climate, man's impact on
75–82
cloud 96–99
cloud physics 99–102
dew point 131
drought 133
environmental protection
178–180
front 230–232
frost 232
greenhouse effect, terrestrial
247
heat balance, terrestrial atmospheric 251–252
humidity 275
hydrometeorology 276–278
industrial meteorology
283–284
jet stream 290–291
micrometeorology 332–334
micrometeorology, crop
334–335
ocean-atmosphere relations
372–374
sensible temperature 511–512
smog, photochemical 524–526
squall line 587–588
storm 588–590
storm detection 590–593
temperature inversion 608
thunderstorm 620–621
tornado 621–622
weather forecasting and prediction 669–676
weather map 676–678
weather modification 678–682
wind 692–694
Meter: current 276
gas 391
joint 127
oil 391
Methane (low-troposphere concentration) 39
Methanearsonic acid (MAA) 266
Methoxone 266
2-Methoxy-4,6-bis(isopropyl-
amino)-*s*-triazine 266

Methoxychlor 285, 286
2-Methoxy-3,6-dichlorobenzoic
acid dimethylamine salt
266
6-Methoxy-mellein 425
2-Methoxy-3,5,6-trichlorobenzoic
acid dimethylamine salt
266
Methyl bromide 263
Methyl caprate 13
Methyl parathion 286
Methyl-2-chloro-9-hydroxyfluo-
rene-(9)-carboxylate
(IT 3456) 266
2-Methyl-4-chlorophenoxyacetic
acid (MCPA) 266
2-Methyl-4-chlorophenoxy-
propionic acid (MCPP) 266
1-(2-Methylcyclohexyl)-3-phenyl-
urea 266
Methyl-2,3,5,6-tetrachloro-*N*-
methoxy-*N*-methyltere-
phthalamate 266
Metobromuron 264
Metrosideros collina 289
Mevinphos 286
Mexico (vegetation zones)
59–60
MH *see* 1,2-Dihydro-pyridazine-
3,6-dione, diethanolamine
salt
Mica 143
Mice 451
Microassociation 110
Microbial cycle: soil phosphorus
568
soil sulfur 565, 568
Microbial soil balance 541–542
Microbial utilization of soil minerals 567–568
Microbiology: medical 180–182
nitrogen cycle 366–367
soil 366–367, 541–542,
564–567
soil minerals, microbial utilization of 567–568
soil phosphorus, microbial
cycle of 568
water analysis 655–656
water disinfection 668
Microclimate 175
environmental stresses
179
human bioclimatology
55–56
Microclimatology, 6, 82
Microcommunities 164
Microconsumers 150
Microcurie 465
Microenvironment 164, 175
Microhabitat 25
Micrometeorology 332–334
instrumental needs
333
scale of focus 333
significant applications 333
turbulence research and applications 333–334
Micrometeorology, crop 334–335
photosynthesis 334–335
plant parameters 335
transpiration 335
unifying concepts 334
Micromonospora 565

Micronutrients 155, 192, 198
Microorganisms: aerial transport
1–2
mutation study 363
soil 366–367, 537, 541–542,
564–567, 567–568, 568
Microphyll scrub 637
Micropogon undulatus 186
Microseisms (storm detection)
592
Microwave systems (solar
energy utilization) 572
Mid-grass prairie 637
Middle-latitude highs 589–590
Middle-level clouds 97
Middle taiga 607, 608
Middle woodland 634
Midge 187
Migration: mass extinctions
189–190
species population 586
Migratory behavior 451, 615
Migratory Bird Conservation Act
113
Migratory Bird Treaty Act 686
Military engineering 171
Military environments (environ-
mental protection) 180
Miller, J. P. 235
Miller law (fungistat and fun-
gicide) 233
Millicurie 465
Millirad 456
Millirem 456
Mineral: animal feed composition 8–9
earth, resource patterns of 140
mining, open-pit 338–342
mining, placer 342–346
mining, strip 346–348
mining, undersea 348–354
plant, minerals essential to
402–406
plant deficiency disease
405–406, 410
Mineral-dust particles (atmospheric ice nuclei) 101
Mineral ions (deficiencies) 306
Mineral nutrition of plants 11–12,
191–199, 402–406, 537–538,
547–549
Mineral resources: oceans 379
patterns 143
Mineral resources conservation
113, 335–338
classification of minerals
336
demand and supply problems
338–342
government policies 337
reserve index 336–337
reserves 335–336
stabilization 337
supply of minerals 336
technological change 337–338
use of scrap 337
Mineral wastes 574–575
Minerals: clay 394
conservation of resources
110–113
ferromanganese 496
general classification
336

Minerals—*cont.*
nutrition 372
on sea floor 314–315
silicate 394
soil minerals, microbial utilization of 567–568
soils 542–543
in water 314–315
Minerals essential to plant
191–199, 402–406, 409–410
Minerals Leasing Act of 1920
113
Miners, uranium 172
Mining: coal 346–348, 348–354
drift 345–346
earth's heat 245–247
gold 344
longwall 348–349
offshore 360
oil 391–394
petroleum 391–394
rooms and pillars 348
subsurface 391–394
surface 338–342, 346–348,
391–394
Mining, open-pit 338–342
bituminous sands 393
blasting 340
mechanical haulage 341–342
mechanical loading 340–341
rock drilling 338–340
tar sands 393
waste-disposal problems
342
Mining, placer 342–346
claims 342–343
developments in technology
346
dredging 344–345
drift mining 345–346
hydraulicking 344
mechanical equipment 344
methods 343–344
power 346
prospecting, sampling, and
valuation 343
recovery methods 343
tailing disposal 346
water pollution 346
Mining, strip 346–348
bituminous sands 393
land reclamation 347–348
stripping techniques 346–347
tar sands 393
Mining, underground 348–354
caving 351–352
entry from surface 354–361
exploration 348–349
fill 349
filled stopes 350–351
ground control 352
ground support 352–353
longwall 348–349
mine design 349–352
mine operation 348–349
open stopes 349–350
transportation 353–354
Mining, undersea 354–361
exploitation 356–361
exploration 355–356
future 360–361
oil and gas offshore 380–383
Mining claims (placer mining)
342–343

Mining engineering 172
  coal gasification 102–105
  fossil-fuel 172
  open-pit 338–342
  placer 342–346
  strip 346
  underground 348–354
  undersea 354–361
Minor terrestrial winds 693
Mismanagement, soil 551–552
Mississippi-Missouri River 473
Mississippi River (channel pattern) 241
Mist and dust collection 137–140
Mistral 693
Mitchella repens 439
Mixed fertilizer 197
Mixed forest 634
Mixing (distribution of sea water properties) 499–500
Mixolimnion: lake 292
  lenitic ecosystems 220
Mobile drilling platforms (offshore technology) 381–382
Models: climate 76–77
  empirical 77
  global average 77
  mathematical 687–688
  wildland operations research 687–688
Modification, weather 102, 678–682
Modified hydrolyzed polyacrilonitrile (HPAN) 537
Modified vinyl acetate maleic acid (VAMA) 537
Mojave Desert (southwestern United States) 130
Molinate 266
Mollisols 529, 541
Mollusks 312
Molten-salt reactor 370
Molybdenum: essential elements for plants 404
  nutrient 155
  soils 547
  toxicity of soils 547
  uses 336
Monimolimnion: lake 292
  lenitic ecosystems 220
Monitoring, biological 4
Monobar-chlorate 267
Monoclimax 93–94, 598
Monsoon 318
Monsoon circulation 89
Monsoon forest 430, 469, 647
Monsundwald 647
Montadale sheep 8
Montane rainforest 470
Monterey pine 137
Montezuma quail 60
Montmorillonite (oil shale) 394
Montmorillonite-beidellite minerals 543
Monuron 264
Moon (lunar soil) 564
Moors (northwest Europe) 429
Mor humus 216
Mora 429
Morbidity rates (epidemiology) 181
Mortality (population growth) 447

Mortality rate (epidemiology) 180–181
Mosaic patterns (succession, ecological) 597
Mosquito (disease vector) 174
Moss forest 647
Moss tundra 637
Mosses: Antarctica 30
  arctic islands 31
Mossy bogs 637
Mother-of-pearl clouds 37
Motive-power engineers 172
Motor vehicle (contribution to air pollution) 46
Mountain and hill terrain 271–273
Mountain hemlock 645
Mountain wind 693
Mountains: environments and vegetation 644
  erosional 273
  evaporation 643–644
  high 608, 609
  influence on vegetation 645
  low 608, 609
  precipitation 643–644
  sources of water 143
  volcanic 273
Mouse ear (plant disease) 11
Mucor ramannianus 412
Mud chimney 308, 309
Mudskipper gobies 309
Mugo pine 650
Muhlenbergia 432, 433
Mull humus 216
Muller, H. J. 364
Multiband remote sensing 638
Multilinear plant varieties 75
Multinutrient fertilizers 195
Multiple-arch buttress type dam 118
Multiple-dome buttress type dam 118
Multiple-purpose dam 118
Multispectral photography (terrain sensing, remote) 611, 612
Mumps 132
Municipal sewage (stream pollution) 516
Municipal wastes 185, 574
Murray, John 374
Musaceae 430
Muscatine, L. 149
Muscovite-illite minerals 543
Musk-ox 33
Muskeg, Canadian 648
Mussels 312
Mustard gas 132, 365
Mutagens, chemical 364–365
Mutation 361–366
  chromosomal aberrations 361–362
  chromosome deficiencies and duplications 362
  chromosome rearrangement 362
  cytological tests 363
  evolution 365–366
  genes 362
  genetic tests 363–364
  induced 364–365
  origin of 364–365
  polyploidy and aneuploidy 361–362

Mutation—*cont.*
  position effects 362
  qualitative changes 362
  quantitative changes 362
  radiation 458
  rearrangement of chromosomes 362
  spontaneous 365
  techniques for studying of 362–364
Mutualism 146
Mutualistic symbiosis 412
Mycology, medical (livestock disease) 10
Mycorrhizae (pine root) 217
Mycorrhizal fungi 566, 568
Mycotic disease (chemotherapy) 232
Myrica 136
Myriophyllum 187
Myrsinaceae 647
Myrtles 135
Mystacocarida 224
Myxomycetes (soil microorganisms) 566
Myxophyceac 566

N

Nabam 233
NAD *see* Nicotinamide adenine dinucleotide
Nadelwald 648
NADP *see* Nicotinamide adenine dinucleotide phosphate
Nagaoka, M. 11
Nahcolite 395
Nakat, N. 293
Nannoplankton 222
Nansen, F. 502
Nansen bottles 376, 504–505
NaPCP *see* Sodium pentachlorophenate
β-Naphthoxypropionic acid 13
N-1-Naphthylphthalamic acid 266
National Aerometric Data Bank 4
National Meteorological Center 674–675
National Waterways Commission 113
Natural aerosols 38
Natural atmosphere (halogens) 250–251
Natural flow (oil well production) 386–387
Natural gas: ammonia production 193
  continental shelves 314
  oil and gas, offshore 380–383
  oil and gas field exploitation 383–391
  production 389
  steam turbine fuel 168
  uses 336
Natural geothermal power 243–244
Natural-resource ecosystem 158
Natural resources: coastal resources 660
  conservation of resources 110–113
  earth, resource patterns of 140–144

Natural resources—*cont.*
  ecology, applied 158–160
  forest and forestry 202–204
  forest management and organization 207–209
  forest resources 212–216
  marine resources 314–316
  mineral resources conservation 335–338
  nonrenewable 111
  ocean 378–379
  renewable 111
  shortages 576
  vegetation management 639–642
  water conservation 656–662
Natural ventilation (atmosphere) 47
Navajo sandstone (fossil dunes) 135
Navigation (engineering hydrology) 275
Navigation, marine 355–356
Navigation dams 118, 125
Nearctic faunal region 25
Nearctic-Palearctic region 694
Nearctic region 694, 695
Nearctic subregion 694, 699
Necrosis: animal 133
  plant 405, 408–409
Nectria 624
Needle ice crystal 101
Needle-leaved evergreen forest 648
Needle-leaved forest 430, 431
Needle valves 125
Nekton 700
  lenitic ecosystem 224
  lotic ecosystems 229
  marine ecosystem 312
Nematocide 12, 423
Nematode galls 409
Nematodes (plant disease) 407, 409
Nematoscelis difficilis 511
Nematoscelis megalops 511
Neogaean realm 694
Neon (low-troposphere concentration) 39
Neoplasia 133
Neoplasms 255
Neotropical faunal region 25
Neotropical region 694, 696–697, 699
Neotropical subregion 694, 699
Nereocystis 313
Neritic region (ocean) 110
Nervous system (vertebrate) 274
Net plankton 222
Net primary production (biological) 61
Neurospora 363
Neuston 221, 225
Neutral estuaries 182
Neutralism (ecological interactions) 146
Neutrons (biological radiation injury) 457
New Red Sandstone (fossil dunes) 135
New Siberian Island 31, 32
New snow (thermal constants) 618
New Zealand faunal region 26

New Zealandian subregion 694
Newfoundland 31, 32, 32–33
NFE see Nitrogen-free extract
Niacin (disease) 132
Niche (animal community) 25, 175
Nickel: dermatitis 262
  heavy metals, pathology of 253
  industrial hazard 262
  poisoning 261–262
  toxic action 261–262
  uses 336
Nickel carbonyl 262
Nicotiana rustica 287
Nicotiana tabacum 287
Nicotinamide adenine dinucleotide (NAD) 403
Nicotinamide adenine dinucleotide phosphate (NADP) 403
Nicotine 287, 623
Niger River 473
Nightingale, H. L. 199
Nile River 473
Nimbostratus clouds 98, 318
Niobium (uses) 336
Nipah fruticans 308
Nitella 224
Nitrate: fertilizer 192
  nitrogen cycle 155, 366
  rock 143
Nitrate bacteria 155
Nitrate-forming bacteria 542
Nitric acid: ammonium nitrate production 193
  nitric phosphate production 196
Nitric phosphate (fertilizer) 196
Nitrification (nitrogen cycle) 366
Nitrifying bacteria 568
Nitrite (nitrogen cycle) 155, 366
Nitrite bacteria 155
Nitrite-forming bacteria 542
Nitrobacter 366
Nitrobenzene (poison) 622
Nitrogen: biological element 67
  essential element for plants 403
  examination of sewage 515
  fertilizer 192–195
  organic 194
  soil 366, 542
  soil conservation 199
Nitrogen cycle 57, 69, 155, 156, 166, 366–367
  ammonification 366
  denitrification 366–367
  nitrification 366–367
  nitrogen fixation 69, 366, 367
Nitrogen dioxide: low-troposphere concentration 39
  smog, photochemical 524
Nitrogen fixing: algae 155, 156
  bacteria 146, 155, 156, 565
Nitrogen-free extract (NFE) 9
Nitrogen gas (nitrogen cycle) 366
Nitrogen mustard (mutagen) 365
Nitrogen oxides (air pollution) 46
Nitrogen solutions (fertilizer) 194
Nitrosococcus 366
Nitrosomonas 366
Nitrous oxide (low-troposphere concentration) 39
Nocardia 565

Nocardiosis (chemotherapy) 232
Noise: ambient 368
  background 368
  white 368
Noise, acoustic 367–368
  criteria 367
  physical specifications 367–368
  sonic boom 581–585
Noise level, perceived 585
Noise measurements 367–368, 585
Noise pollution 305
Noncaking coals (Lurgi gasifiers) 103
Noncommercial forests 212
Noninfectious diseases (epidemiology) 180–182
Nonmetallic inorganic poisons 132
Nonnutritive residue analysis 9
Nonprecipitating water cloud 99
Nonraised bogs 637
Nonrenewable natural resources 111, 314
Nonsaturated air 327
Nonselective foliage contact sprays 262, 263
Nonselective foliage translocated sprays 262, 263
Nonselective herbicides 262–263
Noosphere 64
Norea 266
Nor'easter 694
Normal full-pool elevation (dam) 126
North America: biotic provinces 302
  desert regions 130–131
  dune vegetation 135–137
  landform 610
  resource pattern 144
North Fork Dam (Va.) 123
Northern Ferrel cell 43
Northern Hadley cell 43
Northern Hemisphere: arctic and subarctic islands 30–33
  change of climate 76
Northern parkland 634
Northern taiga 607–608
Nostoc 367
Nothofagus 430
Notogaean realm 694
Notohylaea 647
Novaya Zemlya 31, 32
NPA see Numbering plan area
Nuciform soil structure 532
Nuclear energy (environmental pollution) 160
Nuclear engineering: nuclear power 368–372
  radioactive fallout 460–462
  radioactive materials, decontamination of 462–463
  radioactive waste disposal 463–465
Nuclear engineers 172
Nuclear explosion (radioactive fallout) 460–462
Nuclear fission 368–372
Nuclear fuels 172, 368–372
Nuclear fusion 368–372
Nuclear plants (power cycles) 368

Nuclear power 368–372
  central-station electrical plants 369–370
  considerations 368
  costs 371
  economics of central-station power 370–371
  electric power generation 629
  marine technology 376–377
  miscellaneous reactor applications 371–372
  power generation 368–369
  power plant 172
  power plants for propulsion 371
Nuclear propulsion 371
Nuclear reactor 368
Nuclear weapons, clean 461
Nuclei, condensation 99
Nuclei, ice 101
Nucleic acids 69
  deficiencies 306
  effect of radiation 456
  photochemistry of 364, 456
Nucleoproteins: effect of radiation 456
  phosphorus 403
Nuclides, radioactive 38–39
Numbering plan area (NPA) 266
Numerical weather prediction 672–674
Nunivak Island (Arctic) 31
Nutrients 155
  animal wastes 653
  effect on plant disease 418
  fertilizer 192
  lake ecosystem 58
  malnutrition 305–307
  nutrition 372
  soil 537–538
  soil chemistry 547–548
  soil minerals, microbial utilization of 567–568
Nutrition 8–9, 372
Nutritional deficiency 132
Nyssa 431
Nystatin 232

O

Oak 136, 137, 431, 648
Oak wilt 624
Oakbrush 407
Ob River (Arctic) 81, 473
Obligate parasites 148
Ocala limestone (aquifer) 30
Ocean: carbon dioxide reservoirs 78
  chemical properties 310
  earth sciences 144–145
  energy source 379
  fertility 507–508
  geochemistry 278–281
  hydrologic cycle 88, 657–658
  hydrosphere 278
  marine ecosystem 309–314
  marine resources 314–316
  oceanography 374–380
  sea ice 486–488
  sea state 488–490
  water source 379
Ocean-atmosphere relations 44, 86–88, 90–91, 276, 372–374, 378, 491–493
  effects of wind on ocean surface 373

Ocean-atmosphere relations —cont.
  hydrologic and energy relations 373
  major ocean current systems 373
  relations with sea-water salinity 373
Ocean currents 377
  beach erosion 106
  distribution of temperature 88
  major systems 373
  sea-water fertility 510
  wind-induced 373
Ocean water mass: temperature-salinity relationships 500–501
  types 501
Ocean waves 105–106, 378, 488–489
  breaking waves 489–490
  sea state 488–490
  swell 489
Oceanian region 694, 699
Oceanic region 110
Oceanic subregion 694
Oceanic zones 177
Oceanographic stations 377
Oceanographic submersibles 377, 382
Oceanographic surveys 375–377
Oceanographic vessels 375–377
Oceanography 145, 374–380
  animal life 60–61
  applications of ocean research 378–379
  chemical 497
  continental shelf and slope 113–114
  development 374–375
  estuarine oceanography 182–187, 314, 518
  marine ecosystem 309–314
  marine resources 143–144, 314–316
  ocean-atmosphere relations 372–374
  predator-prey relationships 452–453
  remote sensing 612
  research problems 377–378
  scattering layer 481–482
  sea ice 486–488
  sea state 488–490
  sea water 490–507
  sea-water fertility 507–511
  storm surge 593–594
  surveys 375–377
  thermocline 619–620
Oceanology 374
Ocotillo 129, 131, 433
OCS-21799 see 2-[(4-Chloro-o-tolyl)oxy]-N-methoxy-acetamide
Odum, E. P. 63, 149, 599, 616
Odum, H. T. 149, 616
Office of Coal Research 102
Offshore drilling 359
Offshore fixed drilling platform 381
Offshore mining production 360
Offshore oil and gas 314, 380–383, 390–391
Offshore placer mining 342
Offshore self-elevating drilling platform 381

Offshore semisubmersible drilling platform 381–382
Offshore suction dredges (placer mining) 345
Ogden, E. C. 3
Oil and gas, offshore 314, 380–383, 390–391
  geology and the sea 380
  hazards at sea 382–383
  mobile drilling platforms 381–382
  production and well completion technology 382
  successes at sea 380–381
Oil and gas field exploitation 383–391
  aspects of production rate 385
  corrosion 389–390
  development of field 383–384
  emulsions 390
  factors of method selection 385
  gas conservation 390
  general considerations 385–389
  instruments 391
  lease tanks and gas separators 389
  legal and practical restrictions 384–385
  lifting 387–389
  natural flow 386–387
  natural gasoline production 389
  offshore production 390–391
  paraffin deposits 390
  production methods in producing wells 386–389
  production problems and instruments 389–391
  related oil field conditions 383
  salt-water disposal 390
  useful terminology 385–386
Oil and gas well drilling 383–385
Oil disease (polychlorinated biphenyls) 441
Oil field (oil and gas field exploitation) 383–391
Oil-forming shales 395–396
Oil meters (petroleum production) 391
Oil mining 391–394
  European experience 391–392
  gilsonite 392–393
  mine draining of reservoirs 393–394
  North American tar sands 393
  selection of method 394
  United States oil shale mining 392
Oil-mining system, Ranney 393, 394
Oil pollution 379
  coastal waters 661
  effects of 483–484
  sea 482–483
Oil pools 383
Oil reservoirs (mine drainage) 393–394
Oil rock (oil mining) 391
Oil sand 391, 394
Oil shale 394–398
  Estonian 396
  oil-forming shales 395–396
  organic constituents 396
  origin and composition 394–395
  recovery of oil 396–397

Oil shale—cont.
  reserves and development 397–398
  retort 397
  strip mining 391
  surface mining 391
  United States mining 392
Oil sinking (oil spill cleanup) 485
Oil slick dispersion 485
Oil spill 482–486
Oil well 383, 386–389
Old snow (thermal constants) 618
Oligocene (composition of marine mud) 70
Oligotrophic lake 225
Olive scale 284
Olneya tesota 131
Olsen, S. R. 548
Olson, J. 638
Oncospermum filamentosa 308
Onshore placer mining 342
Oospora citri aurantii 413
Oozes: foraminiferan 66
  radiolarian 66
Opdyke, N. D. 191
Open energy systems: community 166
  ecosystem 166
Open-pit mining 338–342, 393
  communication 342
  slope stability and bench patterns 342
Open stopes (underground mining) 349
Open system (homeostasis) 131
Open woodland 634, 636
Operation Plant Hopper 2
Operations research: systems ecology 604–607
  wildland problems 686
Opium 623
Optics, meteorological 320
Opuntia 130, 136, 137, 433
Opuntia engelmannii 651
Opuntia serpentina 136
Orach 135
Orchard residues (utilization and disposal) 653–654
Orchidaceae 429
Orchids 469, 646, 648
Orchinol 424
Orders, soil 529, 541
Ordram 266
Organic evolution 586
Organic fungicides 233
Organic geochemistry 64, 70
Organic insecticides 285–286
Organic matter: effect on metal complexing in soils 547
  soil 532, 536, 544
Organic nitrogen fertilizer 194
Organic phosporus insecticides 286–287
Organic soils 542
Organismic concept of community 109
Organisms: indicator 511
  nitrogen-fixing 156
  periodicity in 165, 449
Organophosphate insecticides 399
Oriental faunal region 25
Oriental region 694
Oriental subregion 694, 699
Origin of life 64–65

Orinoco River 473
Orographic precipitation 91, 277
Oroville Hungarian riffle 343
Oryzopsis 433
Oryzopsis hymenoides 130
Oscillation, waves of 220
Oscillations, isotherm 498
Oscillatoria rubescens 187
Osteomalacia 132
Ostracods 312
Ostrich 479
Outbreeding (animal breeding) 8
Outer space (terrestrial heat balance) 84
Outlet gates 126
Overflow-type spillway 119, 121, 123–124
Overgrazing: corrective legislation 470
  disruption of climax communities 95
Oxalis 129
Oxalis gigantea 129
Oxidation, biological 455
Oxidation ponds 518, 524
Oxidation potentials (sea water) 495
Oxidation processes (sewage treatment) 521–524
Oxisols 529
Oxygen: bioclimatology 53
  biological element 67
  content of polluted stream 516
  cycle 57, 58, 69
  essential element for plants 402–403
  in eutrophic lakes 188
  lake chemistry 225–226
  sea water 310
  sewage requirement 514–515
  soil 542
  stream balance 517
Oxygen corrosion (oil and gas well production) 389
Oxygen effect (irradiation of solution) 454
Oxygen HYGAS gasifier 104
Oxytropis 433
Oysters 312
Ozone 37
  bioclimatology 54
  component of air 317
  low-troposphere concentration 39
  smog, photochemical 524
  stratospheric 53, 54

P

Pacific Coast (dune vegetation) 136–137
Pacific faunal region 26
Pacific subalpine snow woods 634
Pacific temperate faunal region 26
Packed bed (dust collector) 138
Packed tower 21
Paepalanthus polyanthos 439
Palaemonidae 224
Palaearctic faunal region 25
Palearctic region 694–695
Palearctic subregion 694, 699
Paleobioclimatology 56
Paleobotany 60
Paleocene Epoch 190

Paleoclimatology (ocean studies) 378
Paleomagnetics (faunal extinction) 191
Paleontology 60
  faunal extinction 189–191
  prairie ecosystem 451
Paleotropical region 694, 699
Palm 647
Palm, coconut 136
Palmer index 133
Palo verde (desert vegetation) 130
Pampas 246
Pancake ice 486
Panic grass 135
Panicum 246, 432
Panicum amarum 135
Panicum paspalum 479
Panning (placer mining) 343
Paper, recycled 581, 654
Papuan subregion 694, 699
Parabolic dunes 134
Paracoccidioidomycosis (chemotherapy) 232
Paraffin deposits (oil well production) 390
Parallel drainage pattern 296
Parameterization 77
Páramo 650
Parana pine 430
Paraquat 128, 265
Parasites 148
  animal 132
  external 10
  facultative 148
  plant pathogens 411
Parasitism 148, 412
Parasitology, medical (livestock pest and disease control) 10
Parathion 286, 399
Parc 649
Paris green (insecticide) 285
Parkland (vegetation zone) 649
Parkland, northern 634
Partially mixed estuaries 183
Particles: air pollution 46
  atmospheric 79–80
  dust and mist collection 137–140
  mineral-dust 101
  soil 531–532
Passer domesticus 585
Passerine birds 451
Passive microwave imagery (remote terrain sensing) 612
Passive microwave radiometry (remote terrain sensing) 612
Pasture 637
  disclimax 95
  grassland 245–247
  grazing 637
Pathogens: growth and reproduction 413–415
  plant 407–422
  soil 422
  water disinfection 668
Pathological lesions 132
Pathology: animal 1–4, 131–134, 180–182, 253–262, 305–307, 456–460, 525, 622–623
  heavy metals 253–262
  plant 407–422, 422–428, 623–627
Patoran 264
Patterns, stream 472

PBA see Polychlorobenzoic acid dimethylamine salt
PCP see Pentachlorophenol
Peanut harvester 11
Peat bogs (taiga) 608
Peat moss (oil absorbent) 484
Pebulate 267
Pecan (mouse ear) 11
Peccary, extinct 191
Pedigree breeding (plant) 73
Pedigree selection (animal breeding) 71
Pedogenesis: temperate rain-forest 647
    tropical woodland 649
Pedology 556
Pelagial division (ocean) 110
Pelagic biome 64
Pelagic division (marine ecosystem) 312–313
Pelagic salps 452
Pellagra 132, 272
Pelouse 651
Penguin 30
Penicillin 5
Penicillium 422, 566
Penicillium digitatum 413
Penicillium griseofulvum 233
Penstock 125
Pentachlorophenol (PCP) 266
Perceived noise level (sonic boom) 585
Perched water 248
Percussion drilling (open-pit mining) 339
Perennial forb complexes 637
Periconia circinata 425
Periodicity in organisms 165, 449
Periophthalmus 309
Periphyton 221, 224, 228
Perlite (oil absorbent) 484
Permafrost 31, 398, 561
Permanent thermocline 619
Permeability: petroleum production technology 385
    rocks 248
    soils 559
Permeability, cell 274
Permeameter: constant-head 559
    falling-head 560
Permian (faunal extinction) 189
Pernicious anemia 132
Peronospora destructor 417
Peroxidase 404
Perpetual frost: mean temperature 90
    precipitation 90
Persea 430, 432
PERT see Project evaluation and review technique
Peruvian faunal region 26
Pesticides 5, 12, 46, 132, 173
Pesticides, persistence of:
    biodegradable compounds 399
    chlorinated hydrocarbons 399–400
    future course of action 400
    other pest-control measures 399
Petalostemon 433
Petroleum: continental shelves 314
    geology 380, 383
    oil and gas, offshore 380–383

Petroleum—cont.
    oil and gas field exploitation 383–391
    oil mining 391–394
    origin of 66, 69
    prospecting 380
    reservoir engineering 383
    secondary recovery 384
    uses 336
Petroleum engineering (sea, oil pollution of) 482–486
Petroleum engineers 172
Petterson, Otto 374
pH: control in soil 537
    sea water 495, 497
PH 40-21 see 4,5,7-Trichloro-benzthiadiazole-2,1,3
Phages (soil microorganisms) 566–567
Phanerophyte 439
Phantom midge 293
Phaseollin 424
Phenothiazine 10
Phenotypic plasticity 162
2,3,6-Phenylacetic acid 267
3-Phenyl-1,1-dimethylurea 267
3-Phenyl-1,1-dimethylurea trichloroacetate 267
Phenylketonuria 132
Pheromones 2
Phloem necrosis 627
Phloridzin 425
Phoenix 652
Phon 305
Phosgene 132
Phosphate: cycle in ecosystem 166–167
    fertilizer 192, 195–196
    rock 143
    soil conservation 198–199
    uses 336
    utilization by soil microorganisms 567–568
Phosphate rock: fertilizer 196
    nitric phosphate production 196
    phosphoric acid production 195
    superphosphate production 195
    triple superphosphate production 194
Phospholipids 69, 403
Phosphoproteins 69
Phosphoric acid: ammonium phosphate production 195–196
    fertilizers 195
    production 195
    triple superphosphate production 195
Phosphorus: cycle 57, 69
    essential element for plants 403
    fertilizer 192
    poison 622
    pollutant 57
    soil 542
Photochemical bioclimatology 53
Photochemical smog 524–526
Photochemistry of nucleic acids 364, 456
Photography: aerial 611, 637–638
    multispectral 611, 612
    terrain sensing, remote 611

Photography—cont.
    underwater 377
    visual 612
Photolysis 58
Photoperiodism in plants 14
Photosynthesis 65, 66, 69, 176
    atmospheric oxygen content 58
    ecology, physiological (plant) 161–163
    effects of weather and climate 53
    field measurements 162–163
    fresh-water ecosystems 218, 220
    fresh-water plants 226
    manganese 404
    micrometeorology, crop 334–335
    oceanic region 110
    oceans 279–280
    phytoplankton 227, 401, 402
    sea water 496
    sea-water fertility 507–511
    succulents 129
    summer-green temperate forest 93
    true xerophytes 129
Photosynthetic bacteria 565
Phreatophytes 118
    desert 129
    evapotranspiration 189
Phycomycetes 566
Phylogeny (phytoplankton studies) 402
Physical and dynamic climatology 83
Physical geographic environment (stresses) 178–179
Physical geography 471–474, 608–610, 610–613
    Antarctica 27–30
    arctic and subarctic islands 30–33
    dune 134–135
    earth, resource patterns of 140–144
    hill and mountain terrain 271–273
    lake 291–292
    permafrost 398
Physics: auroral 30
    cloud 99–102, 316
    health 251, 462–463, 463–465
    soil 531
Physiological and experimental psychology (loudness) 304–305
Physiological ecology (plant) 129–131, 161–163, 599, 604–607
Physiology: bioclimatology 52–56
    comparative 273–275
    homeostasis 273–275
    plant 161–163, 402–406
Phytar 265
Phytoalexin 424, 425
Phytobenthos: lenitic ecosystems 221, 224
    lotic ecosystems 229
Phytochorology 434
Phytogeography 434
Phytopathology: aerobiology 1–4
    plant disease 407–422
    tree diseases 623–627
Phytophthora cinnamoni 625

Phytophthora infestans 6, 418, 419
Phytophthora sojae 425
Phytoplankton 401–402
    collection and identification 401–402
    ecology 401
    eutrophic lakes 187
    marine ecosystem 311–312
    ocean 310, 507–508
    phylogenetic significance 402
    phytobenthos 224
    quantitative analyses 402
    regulation of sea water composition 496–497
    representation and occurrence 401
Phytotoxic effects (photochemical smog) 525
Phytotoxicants 525
Phytotron (physiological ecology) 163
Picea 137, 648
Picea engelmanni 645
Picea sitchensis 136, 431
Picloram 128, 264
Piercing, jet 339
Piezometers 127
Pigmentation, solar 53
Pile dike 475
Pile foundations 563
Pine (taiga) 607
Pine bark (oil absorbent) 484
Pine barrens 94
Pine forest (fire) 95
Pine-oak forest (Mexico) 59
Pineapple air plant 645
Pines 136, 137, 431
Pinus 136, 431
Pinus banksiana 137
Pinus clausa 136
Pinus contorta 136
Pinus flexilis 644
Pinus nigra 624
Pinus ponderosa 624, 644
Pinus radiata 137
Pinus resinosa 137
Pinus strobus 137
Pioneer zone (Atlantic and Gulf coasts) 135
Pipe flow (penstocks) 125
Pipelines (water transmission) 665
Pisatin 424
Pitchblende 143
Pittman-Robertson Act 113
Placer mining 342–346
Plague 132
Plains 246, 608, 609
Plan-position indicator (storm intensities) 591
Planetary albedo 85
Planetary soils 564
Planetology (remote sensing) 612
Planktivorous fish 453
Plankton: disseminules 443
    fresh-water ecosystems 221–223
    marine ecosystems 311–312
    net 222, 276
    periphyton 228
    phytoplankton 401–402, 507–508
    potamoplankton 228–229
    productivity 226–227

Plankton—*cont.*
 rain 222–223
 reproduction 223
 zooplankton 223, 508, 700–701
Planktoniella 311
Planning, land-use 268–271,
  298–300, 471, 639–642
Plant: autotrophic 508–509
 bioclimatology 52–53
 causes of disease in 132
 climax community 93–96
 day neutral 14
 effects of atmospheric pollution 50
 nitrogen cycle 155
 seed 614
 short-day 14
 soil-water relations 115
 woody 614
Plant, mineral nutrition of 11–12,
  537–538, 547–549
 fertilizer 191–199
 minerals essential to 402–406
Plant, minerals essential to
  402–406
 criteria of essentiality 402
 deficiency symptoms 405–406
 fertilizer 191–199
 plant disease 409–410
 roles of essential elements
  402–405
Plant, water relations of 6
 desert vegetation 129
 ecology, physiological (plant)
  161–163
 forest hydrology 207
 irrigation practices 114–115
 plant disease 410
 summer-green temperate
  forest 93
Plant community 59–60, 108–109,
  406–407, 436
 benthic 508–509
 biome 63–64
 characteristic species 406
 classification 109–110, 407
 climax 407
 desert 129–130
 distribution 407
 dynamics 406–407
 ecosystem 164–167
 forest and forestry 202–204
 forest ecology 204–207
 grassland 245–247
 mangrove swamp 307–309
 plant societies 437–438
 properties 406
 saltmarsh 477–478
 structure 406
 succession, ecological 595–599
 vegetation 632–633
 vegetation zones, altitudinal
  642–645
Plant disease 407–422
 aerobiology 1–4
 agricultural fungicides 233
 causative agents 409–411
 chlorine deficiency 405
 development within the host
  417–418
 dissemination by animals
  421
 dissemination by insects
  420–421
 dissemination by man 421
 dissemination by water 420

Plant disease—*cont.*
 dissemination by wind 419–420
 ecologic relations of organisms
  412
 economic importance
  408
 epidemics 418–419
 growth and reproduction
  413–415
 infection and development of
  415–419
 inoculum 415–416
 mineral deficiency 11, 405–406
 nature of 408–409
 nonpathogenic storage diseases 422
 pathogenic storage diseases
  421–422
 pathogens 411–415
 spore germination 416
 symbiotic relations of organisms 412
 tree diseases 623–627
 viruses 13
Plant disease control 422–428
 breeding and testing for resistance 425–427
 chemical control 423–424
 cultural practices 422–423
 disease resistance 424–425
 eradication campaigns 427
 quarantines 427
 aerobiology 1–4
 biogeography 59–61
 biological productivity 61–63
 biomass 63
 biome 63–64
 biotic isolation 71
 climax community 93–96
 community 108–110
 desert vegetation 129–131
 dune vegetation 135–137
 ecological interactions
  145–148
 ecology, physiological 161–163
 ecosystem 164–167
 environment 175–178
 forest ecology 204–207
 grassland 245–247
 island faunas and floras
  288–290
 life zones 300–303
 mangrove swamp 307–309
 phytoplankton 401–402
 plant community 406–407
 plant formations, climax
  428–434
 plant geography 434–438
 plant societies 437–438
 plants, life forms of 438–440
 population dispersal 442–444,
  444–446
 prairie ecosystem 450–452
 radioecology 465–468
 rainforest 468–470
 saltmarsh 477–478
 savanna 479–481
 succession, ecological 595–599
 systems ecology 604–607
 taiga 607–608
 terrestrial ecosystem 613–617
 tundra 627–628
 vegetation 632–633
 vegetation and ecosystem
  mapping 633–639

Plant disease control—*cont.*
 vegetation management
  639–642
 vegetation zones, altitudinal
  642–645
 vegetation zones, world
  645–652
Plant formations, climax 108,
  428–434, 596, 646
 dry desert 433
 evergreen hardwood forest
  431–432
 grassland 245–247
 needle-leaved forest 431
 savanna 432
 short grassland 433
 summergreen deciduous forest
  430–431
 tall grassland 432–433
 temperate rainforest 430
 thorn forest 432
 tropical deciduous (monsoon)
  forest 430
 tropical rainforest 429–430
 tundra 433–434
Plant geography 59–50, 434–438
 flora and vegetation 434–435
 floristic plant geography
  435–436
 plant formations, climax
  428–434
 rainforest 468–470
 savanna 480
 vegetation and ecosystem
  mapping 633–639
 vegetation zones, altitudinal
  642–645
 vegetation zones, world
  645–652
 vegetational plant geography
  436–437
Plant growth 11
 ecology, physiological 161–163
 plant, minerals essential to
  402–406
 regulators 5, 13–14
Plant hoppers 2
Plant manure 309
Plant metabolism: disease 411
 ecology, physiological 161–163
Plant pathogen: development
  within the host 417–418
 dissemination 419–421
 establishment in the host
  416–417
 inoculum 415–416
 spore germination 416
Plant pathology: disease 407–422
 disease control 422–428
 tree diseases 623–627
Plant physiology: ecology,
  physiological 161–163
 plant, minerals essential to
  402–406
Plant Quarantine Act of 1912 428
Plant reproduction 73–74
Plant societies 437–438
 dominance societies 437
 life-form societies 437–438
 seasonal societies 437
 structural societies 437
Plant varieties, multilinear 75
Plant virus 13, 409, 411, 627
Plantago 478
Plantation (trees) 212

Planting and seeding, forest 204,
  206, 207–208, 209–212
Plants (installations): hydroelectric 630
 nuclear 368
 power 167
Plants, life forms of 437–438,
  438–440
 adaptations of desert vegeta-
  tion 129
 climate and adaptation factors
  439
 mapping systems 440
 Raunkiaer system 439–440
 saltmarsh 478
Plasmodiophora brassicae 420,
  421
Plasticity: phenotypic 162
 soils 559–560
Plastics (oil absorbent) 484
Plastocyanin 404
Plate-count method (water
  analysis) 655
Plate screens (sewage treatment)
  519
Plate tower 21
Plateaus, continental 114
Platy soil structure 532
Plectrachne 479
Pleistocene Ice Age 617
Pliocene Epoch 189
Plumbaginaceae 478
Plumbism 259
Plutonium: fast-breeder reactors
  370
 radioactive fallout 460
Pneumatophores 307
Pneumonia 132
Poa vivipara 627
Podocarpus 430
Podzolic soil 541, 648
Podzols 31
Pogonochores 442
Poikilothermic animals 614, 615
Point mutations (spontaneous
  mutation) 365
Poisoning 132
Poisons: arsenical 12
 blood-destroying 622
 heavy metals, pathology of
  253–262
 hemolytic 622
 lead 259
 mercury 257–259
 metallic 132
 nickel 261–262
 nonmetallic inorganic 132
 systemic 622
 toxicology 622–623
 treatment 623
Polar air mass 16
Polar bear 34
Polar cap climate 141, 142
Polar desert 634
Polar easterlies 89
Polar front 15, 133, 231, 324
Polar-front-jet systems 321, 322
Polar high 589
Polar meteorology (arctic and
  subarctic islands) 30–33
Polar water 501
Pole blight 624, 625
Pole riffles 343
Policies, conservation 112–113
Poliomyelitis 132

Pollination: aerial transport of
    pollen 1–2, 3
  plant breeding 72–73
  role of insects 173
Pollinosis 3
Pollution: agricultural wastes 574
  air-pollution control 19–25
  atmospheric 46–52, 53–54,
    77–80, 174, 251, 411,
    524–526, 608, 630, 641–642,
    682
  carbon dioxide 77–79
  chemical 661
  dispersion of 46–49
  engineering, social implica-
    tions of 172
  environmental 160
  in estuaries 185
  gaseous 46
  insecticide 174
  land-use planning 300
  mineral and fossil-fuel wastes
    574–575
  mismanagement of solid
    wastes 576
  municipal wastes 574
  ocean 379
  oil 379, 482–486, 661
  particulate 46
  pesticides 399, 400
  polychlorinated biphenyls 441
  radioactive 46
  sea, oil pollution of 482–483
  stream 515–518
  thermal 172, 629–630, 661,
    663
  water 57–58, 185, 187–189,
    200, 229, 346, 515–518, 654–
    655, 655–656, 657, 660–661,
    662–664, 667–669
Pollution, noise 305
Polychlorinated biphenyls 58,
    440–442
  distribution in environment 441
  production and uses 440–441
  toxicity 441–442
Polychlorobenzoic acid di-
    methylamine salt (PBA) 267
Polyclimax 93, 94, 598–599
Polygonum viviparum 627
Polyphenoloxidase 404
Polyphosphoric acid (production
    of phosphatic fertilizers) 195
Polyploidy 75, 361–362
Polyporus tomentosus 624
Polyvinyl chloride (PVC) 441
Ponderosa pine 624, 644
Ponds 291
  fresh-water ecosystem
    218–230
  limnology 304
  oxidation 518, 524
Pools (water body) 291
Pools, oil 383
Poplars 136, 137, 607, 648
Population 109, 149
  control 447–448
  density of species 585–586
  duration of radiation in
    467
  human 449
  predatory 452
  prey 452
  response to radiation 467
  species 446, 585–586

Population dispersal 442–444
  barriers to 443
  island biotas 289
  phases 442–443
Population dispersion 444–446
  analysis of 444–445
  factors affecting 445–446
  optimal population density 446
  principal types 444
  species population 586
Population dynamics 446–450
  community 108–109
  demography 449–450
  fluctuations 449
  optimal population density 446
  optimum yield 448–449
  population control or regula-
    tion 447–448
  population growth 447
  population size and density
    446–447
  prairie ecosystem 451
  species population 586
  wildlife management 684
Population genetics (species
    population) 586
Population growth 447, 448
Populus 136
Populus balsamifera 137
Populus deltoides 137
Pore space (soil) 533–534
Poria weirii 624
Porosity: petroleum production
    technology 385
  rocks 248
Porphyrins (deficiencies) 306
Port (breakwaters) 108
Position effects (mutation) 362
Positive estuaries 182
Possible reserves (mineral) 335
Potamoplankton 228–229
Potash: fertilizer 196–197
  ore 197
  rock 143
Potassium: constituent of living
    matter 69
  essential element for plants 403
  fertilizer 192
  potash 196–197
  sea water 315
  utilization by soil microor-
    ganisms 568
Potassium chloride: fertilizer 192
  potash 197
Potassium-deficiency disease
    (cabbage) 410
Potato: blight 6
  disease 411
  famine 408
  loss due to insects 173
  root zone depths 115
Potential production (biological)
    61
Potential reserves (mineral) 335
Potential resources 143
Pound Dam (Va.) 124
Powdery mildew fungi 411
Power: geothermal 243–245
  nuclear 368–372, 376–377, 629
Power dams 118, 125
Power generation: electric
    167–169, 629–632
  hydroelectric 570
  nuclear power 368–369
  solar electric 571–572

Power generation—cont.
  thermal spring 619
Power plant: electricity 167
  fossil-fuel-fired 172
  nuclear 172
Power shovels: open-pit mining
    340
  strip mining 346–347
Power systems, electric 167–171,
    369–370
Prairie 246, 432, 470
  fire 95
  mid-grass 637
  short-grass 637, 651
  tall-grass 637, 651
Prairie dog, black-tailed 451
Prairie ecosystem 450–452
  animal groups 450–451
  feast or famine 451
  historical development 451
  man's impact on grassland
    animals 452
  population dynamics 451
  specificity in diets 451
Pramitol 266
Precipitation 39, 40, 97, 316
  air mass 277
  chemical 521
  cleaning the troposphere 39
  cloud physics 99–102
  cloud seeding 681
  continental patterns 91
  convective clouds 101
  dispersion of pollution 46–47
  disposition of 600–602
  drought 133
  evapotranspiration 189
  extratropical cyclonic 277
  fronts 231
  hydrologic cycle 88, 275–276,
    657–658
  hydrometeorology 277–278
  layer cloud system 100
  mean annual 89
  mountains 277, 643–644
  prediction 673
  production of showers 100–101
  release mechanism 99–100
  source of water in streams 600
  supercooled clouds 102
  thermal 139
  thunderstorm 620–621
  tornado 622
Precipitator: Cottrell 139
  electrostatic 21, 51, 139, 579
  horizontal-flow electrostatic 22
Predation 147–148, 190
Predator chain 201
Predator control 685
Predator-prey relationships
    452–453
  community species composi-
    tion 452–453
  prairie 451
  species population size 452
  strategies for survival 452
Prediction and forecasting,
    weather 331–332, 590–593,
    669–676
Preemergence herbicides 262
Pregnancy (radiation injury)
    459
Preharvest desiccation 128
Preliminary sewage treatment
    519

Premerge 265
Preservation (conservation
    measure) 111
Pressure: abiotic factor 177–178
  air 88–90, 318
  backfill 119
  bulb 562
  cells, embedded 127
  cells, uplift 127
  ice 119
  melting temperature of
    ice 618
  silt 119
Pressure gages, recording sub-
    surface 391
Pressurized water reactor 369
Prevailing climax 167
Prevalence ratio (disease) 181
Prey-predator relationships
    452–453
Prickly pear cactus 130, 651
Primary minerals 542–543
Primary production: biological 61
  ecosystem 154
  gross 61
  micronutrients 155
  net 61
Primary rainforest 469
Primary sewage treatment:
    coarse solids removal 519
  fine solids removal 519–521
Primary succession 596
Primary water 249
Prince of Wales Island (Arctic) 31
Prismatic soil structure 532
Probable reserves (mineral) 335
Producers (ecosystem) 150, 151,
    166
Production, food 154
Production-consumption effi-
    ciency 62–63
Productive landscape 157–158
Productivity, biological 61–63,
    151–152, 152–153, 153–154,
    155, 163, 165–166, 226–227,
    229, 247, 304, 448–449, 469,
    478, 507–511, 615–616, 628
Productivity index (petroleum
    production technology) 385
Products, forest 203
Profile, soil 526, 527
Profilers, subbottom 356
Profundal benthos (lenitic
    ecosystem) 224
Profundal zone: aquatic habitats
    304
  lentic water 110
Progeny testing 71
Programming, linear 688
Project EHV (electric power sys-
    tem) 170
Project evaluation and review
    technique (PERT) 689
Prometone 266
Prometryne 264
Propagation, vegetative 73, 74–75
Propagules 442
Propanil 265
Propazine 264
Propen-1-ol-3 267
Propulsion, nuclear 371
S-Propyl butylethylthiocar-
    bamate 267
S-Propyl dipropylthiocarbamate
    267

Prosopis 131, 433, 479
Prosopis juliflora 129
Prospecting: biogeochemical
  67–68
  placer mining 343
  underground mining 348
  undersea mining 355
Prospecting, petroleum (continental shelves and slopes)
  380
Protection, environmental
  178–180
Protection effect (irradiation of
  solution) 453–454
Protective landscape 157–158
Proteins 67, 69
  animal feed composition 8–9
  deficiencies 306
  effect of radiation 455
  Kwashiorkor disease 307
  metabolism 306
  nutrition 372
  synthesis in nitrogen cycle 155
Proton precession magneto-
  meter 356
Protoplasm (nitrogen cycle) 155
Protozoa 312, 566
Protozoan disease 132
Proved reserves (mineral) 335
Province, biotic 109, 110,
  302–303
Prunus 136
Prunus pumila 137
Psammon 221, 224
Psedera 430
Pseudomonas 565
Pseudotsuga 648
Pseudotsuga taxifolia 644
Psittacosis 132
Psoralea 433
Psychoacoustics 584–585
Psychology, physiological and
  experimental 304–305
Psychrometer 275
Ptarmigan 628
Pteridophyta 429
Pterochores 442
Public health (solid-waste dis-
  posal) 575–576
Puccinellia 478
Puccinia asparagi 414
Puccinia graminis 419
Puccinia graminis vartritici
  415
Pumped-storage hydro 630
Pumping: hydraulic long-stroke
  388
  oil well production 387–388
  sewage 514
  stations 666
Pumps: centrifugal 666
  centrifugal well 388
  hydraulic subsurface 388
  jet 389
  sonic 389
  water pumping stations 666
  well 665
Puna climate 141, 142
Purebred 7–8, 72
Purification, water 515, 667–669
Purple scale 284
Purple sulfur bacteria 565
PVC *see* Polyvinyl chloride
Pygmies 61
Pyramat 287

Pyramid: biomass 63
  ecological 63, 148, 201–202
  energy 148
Pyramid of life 166
Pyramid of numbers 148
Pyramin 264
Pyrazon 264
Pyrethrins 287
Pyrite: oil shale 394
  shale 395
Pyrochem process 397
Pyrolan 287
Pyrrophyta 401

Q

Quail, Montezuma 60
Quarantine (plant disease) 428
Quartz: dunes 135
  oil shale 394
Queen Elizabeth Islands (Arctic)
  32
Quercus 136, 431
Quercus agrifolia 137
Quercus alba 137
Quercus durata 137
Quercus gambellii 407
Quercus ilex 431
Quercus rubra 137
Quercus suber 431
Quercus velutina 137
Quercus virginiana 136
Quillaja 432

R

Rabies 132
Races, ecological 162
Racks (sewage treatment) 519
Rad 456
Radar: aircraft 2
  Doppler 592
  engineers 172
  remote terrain sensing 612
  storm detection 590–591
  vegetation and ecosystem
  mapping 638
Radial stream drainage 472
Radiation: adaptive 288, 289
  biochemistry 453–456
  effect on biological material
  455
  gamma 461
  injury (biology) 456–460
  ionizing 176–177, 191, 364,
  453–456, 456–460
  radioecology 465–468
  sickness 459
  solar 40, 46–47, 84, 247, 317
  syndrome 132
  ultraviolet 53–54
Radiation, terrestrial 36, 317–318
  greenhouse effect, terrestrial
  247
  heat balance, terrestrial atmospheric 251–252
Radiation biochemistry 453–456,
  457
  direct action 454
  direct plus indirect action 454
  effect of radiation 455–456
  indirect action 453–454
  measurement of radiation sensitivity 455

Radiation damage (health
  physics) 251
Radiation injury (biology)
  456–460
  acute radiation effects 459
  cell injury 458
  genetic injury 458
  health physics 251
  late radiation effects 459–460
  mechanism of action 457
  rate of radiation 457
  somatic injury 458–459
  tissue sensitivity and resistance 457–458
  types of radiation 457
Radiation inversion (air pollution)
  47
Radiation sensitivity (measurement) 455
Radiation thermometers
  (oceanography) 504
Radio engineers 172
Radio-frequency reflectivity (remote terrain sensing)
  612
Radioactive elements
  (biogeochemical cycles)
  156–157
Radioactive fallout 39, 460–462
  hazards 462
  local fallout 461
  nuclear explosion 460–461
  radioactive products 461
  sampling 462
  tropospheric and stratospheric
  fallout 461–462
Radioactive isotopes (tracers)
  157
Radioactive materials, decontamination of 462–463
  air and water 463
  methods 462–463
  personnel 463
Radioactive nuclides: accumulated by organisms 66
  cosmic-ray-produced 38–39
Radioactive pollutants 46
Radioactive waste 663
Radioactive waste disposal
  463–465
  future wastes from industry
  465
  gaseous wastes 464–465
  liquid wastes 464
  solid wastes 464
Radiobiology (radiation injury)
  456–460
Radium: carcinogen 459
  radiation injury (biology) 457
  uses 336
Radioactivity: atmosphere 38–39
  radioactive fallout 460–462
Radioecological tracers 465–467
  cycling and fallout 466
  uptake and elimination 466
Radioecology: community response 467–468
  cycling and fallout 466
  duration of radiation 467
  effects studies 467–468
  food chain studies 466–467
  marking studies 465–466
  nonessential and radioactive
  elements 156–157
  population response 467

Radioecology—*cont.*
  radioecological tracers
  465–467
  uptake and elimination of tracers 466
Radioisotopes (radiation injury)
  457
Radiolaria 312
Radiolarian oozes 66
Radiometry, passive microwave
  612
Radionuclide: fallout 461
  radioactive waste disposal
  463–465
  radioecology 465–468
  tags 465
Radium D 38
Radon isotopes 38
Ragweed 3
Rail haulage (open-pit mining)
  341–342
Railroad vine 135
Rain 97, 99
  artificial stimulation 102
  chemical compositions 281
  cloud physics 99–102
  formation 100
  geochemistry 281
  hydrologic cycle 275–276
  hydrometeorology 277
  layer clouds 100
  plankton 222
  squall line 588
  storm sewage 514
  thunderstorm 620–621
  tornado 622
Raincloud systems 100
Rainforest 142, 468–470
  agricultural practices 469–470
  climate 141, 142
  climatic factors 469
  distribution 470
  equatorial 215
  Mexico 59, 60
  montane 470
  primary 469
  primary and secondary communities 469
  productivity 469
  secondary 469
  subpolar 636
  subtropical 470
  subtropical seasonal 636
  temperate 430, 470, 636, 647
  tropical 90, 95, 429–430,
  469–470, 636, 646–647
Rainmaking experiments 102,
  680–681
Rain-out (atmosphere cleansing)
  48
Ramp sanitary landfill 577–578
Ramrod (herbicide) 264
Random drainage pattern 296
Random population dispersion
  444
Randox 264
Randox-T 267
Range management 159, 247
Rangeland conservation 159, 247,
  470–471, 642
  grazing land economics 470
  overgrazing, corrective legislation for 470
  problems, present and future
  471

Rangeland conservation—cont.
  range inventory 470–471
  range management plan 471
  significant advances 471
  types of grazing lands 470
Ranney oil-mining system 393, 394
Ranney wells 665
Rapid sand filters 667–668
Rat bite fever 132
Rats (commensalism) 145
Rattus 145
Raunkiaer, C. 439
Raunkiaer life form system 439–440
Raw humus 216
Rayleigh scattering (attenuation of the solar beam) 78
Raynor, G. 3
Rays, cosmic 38
Reactor: advanced gas-cooled 369
  boiling water 369
  fast breeder 370
  gas-cooled 369
  heavy-water-moderated 369–370
  high-temperature graphite-moderated 369
  light-water 369
  molten-salt 370
  nuclear 368
  pressurized water 369
  water-cooled 369
Realm, biotic 109, 110
Reaper 10
Recent Epoch (composition of marine mud) 70
Recessive lethal mutation 363
Reclamation: land 347–348
  saline water 315
Recorder, depth 356
Recording subsurface pressure gages (petroleum production) 391
Recreation resources 112, 316
Rectangular clarifier 520
Rectangular stream drainage 472
Recurrent selection (breeding, plant) 73, 74
Recycling: paper 581
  solid waste 580, 652–654
  waste materials 111
Red algae 312
Red grape 136
Red oak 137
Red thread (plant disease) 417
Red water 401
Reducers (ecosystem) 166
Reduction-oxidation systems, biological 455
Redwoods 96, 137, 430, 431
Reflectivity, radio-frequency 612
Reforestation: avalanche prevention 52
  forest planting and seeding 209–212
  sustained yield 204
Regional climatology 82
Regions: climatic 90, 91, 92–93, 141
  desert 130
  humid macrothermal 141, 142
  humid mesothermal 141, 142
  humid microthermal 142

Regions—cont.
  palearctic 694–695
  tropical 94
  water-deficient 141, 142–143
  zoogeographic 694–698, 699
Reglone 265
Regulator, growth 6
Reindeer 33, 142
Rejuvenation (drainage system) 242
Relapsing fever 132
Remanent magnetism (faunal extinction) 191
Remote sensing, multiband 638
Remote sensors 610, 612
Remote terrain sensing 610–613, 638
Renewable natural resources 111, 314
Renewal time (estuary) 184–185
Reoxygenation (polluted water) 517
Reproduction, plant: cross-pollinating species 73–74
  self-pollinating species 73
Reproductive behavior (effect on population dispersion) 416
Reptiles 451
  Australian region 697
  Ethiopian region 695
  Nearctic region 695
  Neotropical region 697
  Oriental region 696
  Palearctic region 694
Research, operations (biology) 604–607
Reserve index (minerals) 336–337
Reserves: forest 209–212
  mineral 335
Reservoirs: carbon dioxide 78
  distribution 666
  elevated storage 666
  ground storage 666
  oil 393–394
  outlet works 124–125
  water supply 665
Residual insecticides 285
Residues: crop 653–654
  orchard 653–654
Resinous forest 648
Resins, fossil 70
Resistance: drug 363
  environmental 448
Resource patterns of Earth 140–144
Resources: conservation of 110–113, 199–201, 203, 212–216, 470–471, 639–642, 656–662, 682–692
  employed 143
  flow 111
  food 378
  forest 158, 203, 207–209, 212–216, 216–218, 308–309, 469, 642
  fund 111
  human 111, 112
  land 140
  marine 140, 143–144, 314–316, 378–379
  mineral 143, 379
  natural 110–113, 140–144, 158–160, 207–209, 212–216, 314–316, 335–338, 378–379, 576, 639–642, 656–662

Resources—cont.
  nonrenewable 314
  potential 143
  recreation 112
  renewable 314
  stock 111
Respiration: atmospheric oxygen content 58
  plant 163
Respiratory disease (atmospheric pollution) 49
Restoration (conservation measure) 111
Retaining walls 564
Retort, oil shale 397
Retrieval of Aerometric Data (SAROAD) system 4
Reutilization (conservation measure) 111
Reverse mutations 363–364
Reversing thermometer: deep-sea 504
  oceanography 376, 504
Reversing water bottle 504
Revetment (shore protection) 106–108
Revolving storm 319
Rhabdomyosarcoma 255
Rhamnaceae 432
Rhamnus 432
Rhamnus californica 137
Rhizic water 247–248
Rhizobium 69, 367, 565
Rhizoctonia solani 413, 414
Rhizophora 307, 308, 429
Rhizophora apiculata 308
Rhizophora mucronata 308
Rhizopus 422
Rhizosolenia 311
Rhododendron 650
Rhodotorula rubra 412
Rhodymenia 312
Rhythms: biological 165
  environmental 165
Riatt, R. 482
Ribonucleic acid (RNA) 456
Ricegrass 477
Rickets 132, 272
Rickettsial fever 132
Ridges (meteorology) 318
Riffle 343
Right-of-way: for highway facilities 268–269
  management of vegetation 642
Rill erosion 538–539, 550, 551
Ring gate 125
Ringed penguin 30
Ringed seal 34
Ringworm 132, 233
Rise, continental 114
River 471–474
  characteristics 474–475
  coastal-plain estuaries 182
  engineering 474–476
  gradients 228
  influence on vegetation 94–95
  lotic ecosystems 227–229
  management 473–474
  patterns 241
  regime and flow patterns 472–473
  seasonal variation in water quality 602–603
  stream transport and deposition 594–595

River—cont.
  terminology 471–472
  water qualities 473
  water source 228
  water supply 665
River basin (watershed) 659
River channels 474
  characteristics 603–604
  improvements 475–476
River engineering 474–476
  river channel improvements 475–476
  river characteristics 474–475
River ice 486
RNA *see* Ribonucleic acid
Roadside management (vegetation) 642
Roadway surfacing 269
Rock: asphalt 394
  composition and structure 143
  conservation of resources 110–113
  earth dams 121
  granite 368
  igneous 143
  metamorphic 143
  oil 391
  permeability 248
  phosphate 195, 196
  porosity 248
  regional classes 141
  sedimentary 143, 394–398
  serpentine 94
  transmissibility 248
Rock bolts (underground mining) 353
Rock desert 652
Rock drilling (open-pit mining) 338–340
Rockers (placer mining) 344
Rocky Mountain spotted fever 132
Rodhe, W. 225
Rogue (herbicide) 265
Rolling and irregular plains 608, 609
Ronnel 286
Room-and-pillar mining 348, 392
Root (botany): forest soil relations 217
  mangrove root structure 307
Root-inhabiting fungi 565
Root layering (grassland) 246
Root rot 414
Rosaceae 432
Roseola 132
Ross, Sir James Clark 30
Ross seal 30
Rotary drilling (open-pit mining) 339
Rotation: crop 536, 553
  forest 208
Rotenone 287
Rothamsted Experimental Station 2
Rotifers (fresh-water ecosystem) 223
Roundworms (plant disease) 407
Row crops (cropping system) 536
Rubber (thermal constants) 618
Rubiaceae 288
Rubidium 69
Rubidium vapor magnetometer 356

Rubus chamaemorus 627
Runoff: average 600–601
   extremes 601
   relation to drainage area
      601–602
Rural sanitation 476–477
   sewage disposal 476–477
   water 476
Rust (tree disease) 624
Rust fungi 624
Rutabaga (brown heart)
   11
Rutaceae 432
Ruthenium (biogeochemical cy-
   cles) 156–157
Ryanodine 287

S

Sabadilla 287
Sabal palmetto 136
Sabana 649
Saber-toothed cat (mass extinc-
   tion) 191
Safferman, R. S. 3
Sagebrush 130, 136, 407, 433, 644
Saguaro cactus 130, 433
Sailor's skin (skin elastosis) 53
St. Lawrence Island (Arctic) 31
St. Lawrence River 473
St. Peter sandstone 35
Sakhalin Island (Arctic) 31
Salicornia 433, 477, 478, 652
Salicylism 132
Saline soil (crop productivity)
   115–116
Saline water reclamation 315
Salinity: control in soil 537
   sea water 280, 373, 490–491
Salix 137, 432
Salomonsen, Finn 34
Salsola kali 135
Salt: sea water 280
   uses 336
Salt lakes 291
   steppe 651
   warm desert 652
Salt meadows 637
Salt spray (disruption of climax
   communities) 95–96
Salt water: atmospheric evapora-
   tion 41
   oil well production 390
Salt-wedge estuary 182–183, 185
Salt wort 135
Salted soils 546
Saltmarsh 477–478, 637, 649
   environment 478
   maritime 477–478
   occurrence of different types
      477
   physiography 477
   plant species and life forms 478
   plant zonation (succession)
      477–478
   productivity 478
   steppe 651
Salts, biogenic 155
Samphire 477
Samplers, subsurface 391
Sampling, soil 561
Sand 94, 531
   bituminous 393, 394

Sand—*cont.*
   earth dams 121
   mineral composition 556
   oil 391, 394
   packed bed 138
   primary minerals 542–543
   properties 532
   soils 557
   tar 393, 394
   water filters 667–668
Sand ambrosia 136
Sand cherries 137
Sand dune 107, 130, 134–135
Sand filters: intermittent 522
   rapid 667–668
   slow 667
Sand myrtle 136
Sand pine 136
Sand plains 94
Sand reed 137
Sandstone: Navajo 135
   New Red 135
   oil sand 394
   Triassic 135
Sandstorm 140
Sanitary engineering 172, 478
   rural sanitation 476–477
   water treatment 667–669
Sanitary landfill 577–578
   area 576, 577
   ramp 577–578
   slope 577–578
   trench 577
Sanitary sewage 512–515
Sanitation, rural 476–477
Santa Gertrudis sheep 8
São Francisco River 473
Sapphirina 452
Saprophages 150
Saprophytes 150
Saprophytic chains 201
Sarcobatus 433
Sarcobatus vermiculatus 130
Sarcochores 442
SAROAD *see* Storage and Re-
   trieval of Aerometric Data
Sars, G. O. 374
Sars, Michael 374
Satellites, meteorological
   592–593, 678
Saturated air 327
Saturation, zone of 247, 666–667
Sauger 188
Savane: arborée 649
   armée 650
   épineuse 650
Savanna 94, 216, 245–246, 432,
   637
   agricultural practices 480
   cold 480, 650
   environmental conditions 480
   extratropical types 480–481
   floral and faunal composition
      479
   geographic distribution 480
   management 159
   Mexico 59
   temperate 95, 650
   tropical 64, 90, 95, 479–480,
      649–650
Savanna-woodland 649
Savannen 649
Savannenwald 649
Saw palmetto 136
Sawdust (oil absorbent) 484

Saxifraga hirculis 627
Scaphoides luteolus 627
Scarlet fever 132
Scattering, Rayleigh 78
Scattering layer 481–482
   deep scattering layers 482
   deep scattering organisms 482
   shallow water 482
Scavengers, prairie 451
Scenedesmus obliquus 404, 405
Schaefer, V. J. 680
Schlichting, H. E., Jr. 2, 3
Schmidt, K. 694, 698
Schmidtia 479
Schott, Gerhard 374
Schradan 286
Science: atmospheric 145
   earth 144–145
   sea 374–380
Scirpus 478
Scirrhia acicola 624
Sclater, P. L. 698
Sclerochores 442
Sclerophyll forest 648
Sclerotia 415, 416
Sclerotinia 422
Sclerotium 624
Scolytus multistriatus 421, 627
Scott, R. F. 30
Scott-Darcy process 521
Screens: incinerators 579
   sewage treatment 519
Screw pines 646
Scrophulariaceae 432
Scrub 634, 637
   arid tropical 59
   cold-deciduous 637
   deciduous 637
   drought-deciduous 637
   dwarf 637
   evergreen 637
   evergreen dwarf 637
   evergreen needle-leaved 637
   livestock breeding 8
   oaks 136
   temperate and cold 650
   thorn 650
   tropical 650
Scrub zone (Atlantic and Gulf
   coasts) 135–136
Scrubber: cyclonic liquid 139
   cyclonic spray 21
   dust collector 139
   liquid 51
   venturi 21
Scurvy 132, 133, 272
Scyphiphora 308
Sea 374–380, 488–489
Sea, oil pollution of 482–486
   cleanup procedures 484–485
   effects of oil pollution 483–484
   prevention of oil spills 485–486
Sea anemones 145, 312
Sea breeze 693
Sea fisheries 315
Sea floor 495–496
Sea height code 488
Sea ice 486–488
   effect on climate 81–82
   properties 486–488
   specific gravity 488
   specific volume 486
   types and characteristics 488
Sea oats 135
Sea purslane 135

Sea rocket 135, 139
Sea state 488–490
   breaking waves 489–490
   in-between state 489
   swell 489
Sea urchins 64, 312
Sea wall (shore protection)
   106–108
Sea water 490–507
   analysis of water samples
      505–507
   biological regulation of com-
      position 496–497
   buffer mechanism 497
   chemical properties 310
   compensation intensity and
      depth 494
   composition 378, 495–497
   detailed thermal structure
      498–499
   distribution of properties
      497–504
   electromagnetic fields 494
   elements 279, 314–315
   heat and water vapor 491–492
   inorganic regulation of compos-
      ition 495–496
   interchange between sea and
      atmosphere 491–493
   isotopic variation 501–504
   light 493–494
   mining, undersea 354–361
   momentum 492
   ocean water masses 500
   physical properties 490–491
   pressure effects 491
   projection of droplets 492–493
   recovery of dissolved minerals
      356–357
   salinity 373, 490–491
   salt content 280
   sampling and measuring tech-
      niques 504–507
   sea ice 486–488
   serial observations 505
   sound absorption 491
   stirring and mixing processes
      499–500
   temperature-measuring de-
      vices 504
   thermocline 619–620
   transmission of energy 493–495
   tritium variations 492
   water samplers 504–505
Sea-water fertility 280, 497,
   507–511
   biological species and masses
      510–511
   fertility of the oceans 507–508
   geographic variations in pro-
      ductivity 509–510
   productivity and its measure-
      ment 508–509
   size of populations and fluctua-
      tions 510
Seal 30, 34
Search image (predator-prey relationships) 452
Seasonal plant societies 437
Seasonal thermocline 619–620
Seasons: changes of atmospheric
   circulation 45–46
   effect on terrestrial ecosystem
      615
Seaweed (marine resources) 315

Secondary production (biological) 62
Secondary rainforest 469
Secondary recovery, petroleum 384
Secondary sewage treatment 519, 521–524
Sedge peat swamps 637
Sedges 31, 246, 648
Sediment: development of landforms 235
estuarine 186–187
marine 66, 186, 314
stream transport and deposition 594–595
surface-water quality 603
Sedimentary cycle 156, 278
Sedimentary rocks 143, 394–398
Sedimentation: sewage treatment 519–520
water treatment 476, 667
Sedimentation basin (design) 520–521
Sedimentation method (phytoplankton collection) 401–402
Sedimentation tank, vertical-flow 520
Seed handling, tree 210
Seed plants (habitat) 614
Seed protection (fungicide) 233
Seeding, cloud 102, 679, 680, 681
Seeding, forest 209–212
Seepage (soils) 559
Seiche, internal 221, 222
Seif dunes 134
Seismic stations (oceanography) 377
Seismographs (oceanographic survey) 276
Seismology: International Geophysical Year 1957–1958 30
storm detection 592
Selater-Wallace system 694
Selection: animal breeding 71
plant breeding 72
recurrent 73, 74
Selective foliage contact sprays 262, 263
Selective foliage translocated sprays 262, 263
Selective root herbicides 262
Selenography 180
Self-pollination 72, 73
Self-purification (natural water) 515, 662–663
Selva (tropical rainforest) 646
Selye, H. 274
Semiarid climate 141, 142, 143
Semideciduous forest 647
Semidesert dwarf shrub 637
Senecio 433
Sensible temperature 511–512
Sensors: electromagnetic 610
remote 610
Separator: cyclone 21, 138
gas 389
impingement 138
Septic tank 477, 512, 519
home disposal units 512
municipal and institutional units 512
tank efficiency 512
Sequoia 137, 430

Sequoia sempervirens 431
Sere (ecological succession) 595
Serenoa repens 136
Serpentine rocks 94
Serum sickness 132
Sesone 267
Sesuvium portulacastrum 135
Settlement gages 127
Settling basin 667
Settling chamber 21, 51, 138
Severnaya Zemlya (Arctic) 31, 32
Severny Island (Arctic) 31
Sevin 399
Sewage 512–515
coastal water pollution 660–661
cultural eutrophication 188
design periods 513–514
domestic 512–515
examination of 514–515
municipal 516
pumping 514
relation to water consumption 513
sanitary 512–515
septic tank 512
storm 512–515
Sewage disposal 316, 515–518
oxidation ponds and land disposal 518
pollution in lakes and estuaries 518
regulation of water pollution 515
rural sanitation 476–477
sanitary engineering 478
stream pollution 515
Sewage treatment 57, 518–524
coarse solids removal 519
fine solids removal 519–521
Imhoff tank 281–283
oxidation processes 521–524
septic tank 512
water conservation 188
Sewer: design periods 513–514
trunk 514
Sex-linked inheritance (mutation study) 363
Sex-linked recessive lethal test 363
Sferics (thunderstorm detection) 591
Shackleton, Sir Ernest 30
Shade trees (diseases) 625–627
Shadscale 130, 407, 433
Shaft: underground mining 354
vertical 354
Shale: Chattanooga 395, 396
Green River 396
oil 391, 392, 394–398
Shallow-ditch drainage system 295
Shallow thermoclines 619
Shallow water (scattering layer) 482
Shear (soil) 559
Shearing instability (atmosphere) 330
Sheepshead (fish) 188
Sheet erosion 538, 550, 551
Shelby tube 561
Shelf, continental 113–114, 482
Shelf fungus 625
Shelter (environmental protection) 179–180
Sherpas 61

Ships (oceanographic surveys) 375–377
Shock, electric 132
Shore (protective structures) 106–108
Shore corals 312
Shore processes: beach erosion 105–106
waves in lakes 220
Shoreline: accreting 106
coastal engineering 105–108
eroding 106
Short-day plant 14
Short-grass prairie 637, 651
Short grassland 433
Shotgun fertilizer application 198
Shovels, power 340
Show-ring judging 9
Shower clouds (precipitation) 100–101
Shreve, F. 130
Shrimp 312
Shrinkage stoping 349–350
Shrub: semidesert dwarf 637
xerophytic 130
Shrublands: desert 637
evergreen broad-leaved 637
evergreen dwarf 637
Great Basin Desert 130
Sickle-cell trait 133
Sickness: disease 131–134
radiation 459
serum 132
Side-roll sprinkler system 117
Siduron 266
Silicate-decomposing bacteria 568
Silicate minerals (oil shale) 394
Silicoflagellates 401
Silicon: constituent of living matter 70
soil 542
solar cells 572
Silt 531
earth dams 121–122
mineral composition 556
primary minerals 542–543
properties 532
Silt-clay soils 557
Silt pressure (concrete dam) 119
Silver (heavy metals, pathology of) 253
Silver iodide: cloud seeding 102, 680
ice-nucleating substance 101
Silvics (forest ecology) 204–207
Silviculture 158, 159
forest and forestry 202–204
forest ecology 204–207
forest management and organization 207–209
Simazine 264
Simplex aerator 523
Simulation (wildland operations research) 689
Sinking, oil 485
Siple, P. A. 512
Sirmate 265
Sirocco (wind) 694
Sitka spruce 136, 431
Sivirtsev, N. M. 541
Skaergaard 33
Skeletons: aragonite 69
calcite 69
Skerfving, Steffan 256

Skimmers (oil spill cleanup) 484
Skimming tanks (sewage treatment) 519
Skin: cancer 53, 525
elastosis 53
radioactive decontamination 463
relative humidity 54
sunburn 53
water transfer 54
Skujaella erythraeum 401
Sky light (bioclimatology) 53
Slab-and-buttress type dam 118
Slab avalanches 52
Slash-and-burn agriculture 81
Sleeve valves 125
Slicing, top 351
Slide gates 125
Slime fungi (soil microorganisms) 566
Slope: continental 113–114
geologic 237–238
sanitary landfill 577–578
valley 241
Sloth, giant ground 191
Slow-moving cold front 323
Slow sand filters 667
Sludge: septic tank 512
sewage treatment 521, 523
Sluice box (placer mining) 343
Sluice plate (placer mining) 346
Smallpox 132, 181
Smelt 188
Smog 20, 53, 54, 641
atmospheric pollution 46–52
inadvertent weather modification 682
temperature inversion 608
Smog, photochemical 524–526
chemical reactions 525
control 525–526
effects 524–525
history 524
meteorology 524
Smoke (cloud) 96–99
Smoke stack (dispersion of pollution) 48
SMOW see Standard mean ocean water
Snails 312
Snake bites 623
Snow 99
atmospheric evaporation 41
avalanche 52
cloud physics 99–102
hydrologic cycle 275–276
hydrometeorology 277
new 18
old 618
production of 101
Snowsheds (avalanche prevention) 52
Snowslides 52
Social animals 27
competition 147
dispersion patterns 446
Social implications of engineering 172
Social parasitism 148
Societies, plant 437–438
Sodium: constituent of living matter 69
essential element for plants 405
nutrient 155
soils 116

Sodium arsenite 267
Sodium chlorate 262, 263, 267
Sodium *cis*-3-chloroacrylate 128
Sodium 2,4-dichlorophenoxy-
   ethyl sulfate 267
Sodium 2,2-dichloropropionate
   265
Sodium fluoride (insecticide)
   285
Sodium metaborate 267
Sodium nitrate (fertilizer) 194
Sodium pentachlorophenate
   (NaPCP) 267
Sodium propionate 232
Sodium soils 116
Sodium trichloroacetate (TCA) 12
Sofar (oceanography) 379
Soft hail 100
Soft rot 414
Softening, water 668
Softwood forest 212, 431, 648
Soil 526–540
   abiotic factor 177
   acid 545
   aeolian 556
   affective factors 539–540
   air and aeration 535
   algae 566
   alkaline 115–116
   Antarctic region 31
   Arctic Brown 627
   atmospheric ice nuclei 101
   bearing power 563
   biosphere 64
   bulk density 533
   calcareous 542, 545–546
   Chernozemic 541
   classification 528–530, 541
   climax community 93
   color 535
   conditioners and stabilizers
      536–537
   consistency and compactibility
      535
   consolidation 562–563
   control of erosion 540
   control of salinity 537
   cropping system 536
   density of particles 532
   desertic 541
   dust storm 140
   earth, resource patterns of 140
   effect on plant disease 418
   family 529
   fertility 537
   fluvial 556
   forest 205, 206–207, 212,
      216–218
   frost, action in 560–561
   glacial 556
   gravel 557
   great group 529
   groundwater 247–250
   hydrology 275–276
   lacustrine 556
   land capability survey maps
      535–536
   land-use classes 297–298
   landslides 563–564
   lateritic 542
   latosolic 541
   lunar 564
   management 535–538
   mangrove forest 308
   maps 530–531

Soil—*cont.*
   mineral origins 556
   mismanagement 551–552
   moisture 177
   nitrogen cycle 366–367
   orders 529, 541
   organic matter 532, 536, 542
   origin and classification
      526–531
   particle size and weight 557
   pathogens 422
   physical properties 531–535
   physics 531
   plant, water relations of 115
   plant communities 94
   plant disease 409
   planetary 564
   podzolic 541, 648
   pore space 533–534
   profile 526, 527
   regulating nutrient supplies
      537
   saline 115–116
   salted 546
   sampling 561
   sand 557
   series 529
   settlement 562–563
   silt-clay 557
   sodium 116
   soil-water relationships
      558–559
   stabilizers 536–537
   structure 532–533, 556–557
   subgroup 529
   suborders 528–529, 541
   succession and soil develop-
      ment 598
   surveys 530–531
   temperature 535
   testing 537
   textural classification chart
      558
   texture 531–532
   tropical savannas 480
   tundra 31, 627–628
   type 529–530
   types by deposition 556
Soil, zonality of 540–541
   zonal classification of soils 541
   zonal properties of soils 541
Soil and crop practices in agriculture 233, 247, 294–297
   410–411, 422, 469–470, 480,
   531–535, 536, 550, 553–555
Soil balance, microbial 541–542
   antagonisms between soil mi-
      crobes 542
   associative action 542
Soil chemistry 542–549
   acid soils 545
   anion exchange 544–545
   calcareous soils 545–546
   cation exchange 543–544
   elemental composition 542
   metal complexing 546–548
   minerals 542–543
   nutrients 547–548
   salted soils 546
Soil conservation 110–113,
   549–555,
   causes of soil mismanagement
      551–552
   conservation agencies and
      programs 555

Soil conservation—*cont.*
   conservation measures and
      technology 553
   economic and social conse-
      quences 552
   effects on other resources
      552–553
   erosion extent and intensity
      549–550
   farm, range, and forest 553–555
   fertilizer phosphate 198–199
   forests 203
   nitrogen fertilizer 199
   soil management 535–538
   types of soil deterioration 550
   vegetation management 641
Soil Conservation Act of 1935 113
Soil Conservation Service 113
Soil ecosystem (agriculture) 159
Soil engineering (soil mechanics)
   555–564
Soil erosion 538–540
   affective factors 539–540
   control 540
   economic and social conse-
      quences 552
   effects on other resources
      552–553
   extent and intensity 549–550
   types 538–539
   types of soil deterioration 550
Soil formation (microbial action)
   567
Soil horizons 177, 528
   composition 527
   forest soil 216–217
   formation 526–527
   structure 527–528
Soil-inhabiting fungi 566
Soil mechanics 555–564
   accuracy of analysis 562
   bearing power 563
   conduits 564
   consolidation and settlement
      562–563
   consolidation theory 563
   engineering soil classifications
      557–558
   exploration, sampling, and
      testing 561–562
   field tests 561–562
   frost action in soils 560–561
   landslides 563–564
   lunar and planetary soils 564
   mineral origins of soils 556
   particle size and weight 557
   plasticity 559–560
   pressure bulb 562
   retaining walls 564
   sampling 561
   seepage and frost 559
   settlement analysis 563
   shear and plasticity 559–561
   shear in soils 559
   site investigations 561
   soil structure 556–557
   soil types and composition
      556–559
   soil-water relationships
      558–559
   stability 563–564
   theoretical stresses 562
   types by deposition 556
Soil microbiology: nitrogen cycle
   366–367

Soil microbiology—*cont.*
   soil balance, microbial 541–542
Soil microorganisms 537,
   564–567
   actinomycetes 565
   algae 566
   eubacteria 564–565
   fungi 565–566
   myxomycetes 566
   nitrogen cycle 366–367
   protozoa 566
   soil balance, microbial 541–542
   soil minerals, microbial utiliza-
      tion of 567–568
   soil phosphorus, microbial
      cycle of 568
   soil sulfur, microbial cycle of
      568
   viruses and phages 566–567
   yeasts 566
Soil minerals, microbial utilization of 567–568
   soil formation 567
   soil phosphorus, microbial
      cycle of 568
   soil sulfur, microbial cycle of
      568
   utilization of minerals in
      formed soils 567–568
Soil particles: density 532
   soil structures 532–533
   soil texture 531–532
Soil phosphorus, microbial cycle
   of 567–568
Soil studies (highway engineering) 269
Soil sulfur, microbial cycle of
   565, 568
Soil water 247–248, 534–535
   water movement 534–535
   water retention 534
Solan 264
α-Solanine 424
Solar cells 572
Solar constant 83, 251
Solar electric power generation
   571–572
Solar energy 76–77, 569–573
   advantages and disadvantages
      569–570
   climate 83–85
   conversion into useful form 569
   energy collection systems
      570–571
   microwave systems 572
   outlook in developed nations
      572–573
   outlook in Third World 573
   power generation 571–572
   solar cells 572
   terrestrial heat balance 85
   thermal energy for buildings
      571
Solar erythema 53
Solar pigmentation 53
Solar radiation 317
   altitudinal variations 643
   atmospheric circulation 86
   atmospheric evaporation 40
   dispersion of pollution 46–47
   greenhouse effect, terrestrial
      247
   latitudinal variations 64
   terrestrial heat balance 84
Solar space cooling 571

Solar space heating 571
Solar technology: developed nations 572–573
　Third World 573
Solar/thermal electric power plant 571
Solar tides 37
Solar ultraviolet radiation (bioclimatology) 53
Solberg, H. 669
Solid waste: agricultural wastes 652–654
　radioactive 464
　recycling and reuse 580
Solid-waste disposal 573–581
　classification of wastes 581
　composting 579–580
　economic aspects 576–577
　fluidized-bed incinerator 580–581
　future 580–581
　incineration 22, 578–579
　management techniques 577–580
　municipal wastes 574
　natural resources 576
　paper fiber recovery 581
　public health, natural resources, and economy 575–577
　recycling and reuse of solid wastes 580
　sanitary landfill 577–578
　total solid wastes 575
　types and volume of wastes 574–575
Solid Waste Disposal Act of 1965 580
Solid-waste management 577–580
Solidago 433
Solomon, A. M. 3
Solutions, nitrogen 194
Sommergrün Laubholzwald 648
Sones 305
Sonic boom 581–585
　human responses 583–585
　response of structures 585
Sonic pumps (oil well production) 389
Sonneratia 307, 308, 309
Sonoran Desert (southwestern United States) 130–131, 651
Sound: abiotic factor 177–178
　decibel scale 305
　loudness 304–305
　noise, acoustic 367–368
Sound absorption (sea water) 491
Sound estuary 184
Sounder, echo 481–482
Sour corrosion (oil and gas well production) 389
South America: landform 610
　resource pattern 144
South Australian faunal region 26
South Pole 27–30
South West African faunal region 26
Southampton Island (Arctic) 31
Southern Ferrel cell 43
Southern forest 634
Southern fusiform rust 624
Southern Hadley cell 43
Southern taiga 607, 608

Soybeans (defoliation) 128
Space cooling, solar 571
Space environments (environmental protection) 180
Space heating, solar 571
Spacecraft (solar cell) 572
Spartina 477, 478
Spear grasses 246
Speciation 165, 289
Species (community composition) 108
Species population 585–586
　acclimation and adaptation 586
　age distribution and sex ratio 586
　aquatic predatory population 452–453
　birth, death, immigration, and emigration rates 586
　density 585–586
　dispersion 586
　population dynamics 446–450, 586
Specific gravity: American Petroleum Institute (API) scale 385–386
　soil particles 557
Specific humidity 275
Specific locus test 363
Spectroscopy: absorption 612
　infrared 612
Spectrum, absorption 78
Spherules, magnetite 38
Spiderwort (mutation study) 363
Spill, oil 482–486
Spillways 122–124
　channel type 123–124
　gates 124, 125
　overflow type 119, 121, 123–124
Spiral-flow activated-sludge tank 522
Spirochetes (cause of disease) 132
Splash erosion 550
Spodosols 529
Sponges 312
Spontaneous mutation 365
Spore germination (plant pathogens) 416
Spore trap, suction-type 4
Spores (aerial transport) 1–2, 3
Sporobolus asper 432
Sporobolus cryptandrus 433
Sporobolus heterolepis 432
Sporochores 442
Sporotrichosis (chemotherapy) 232
Sporotrichum 566
Sport fishery 200
Spotted fever, Rocky Mountain 132
Spray tower 21
Springs: artesian 664
　hot 619
　thermal 618–619
　water 586–587, 664
Sprinkler irrigation 117
Sprinkler system, side-roll 117
Spruces 137
　sitka 136
　taiga 607
Spruce canker fungus 626
Squall line 587–588

Squall winds (thunderstorm) 620
Square-set stopes 351
SST see Supersonic transport
Stability, atmospheric 329–330
Stabilizers, soil 536–537
Stable disclimax 95
Stable flies (livestock parasites) 10
Stakman, E. C. 1
Stam-F34 265
Standard atmosphere 17
Standard filters 522
Standard mean ocean water (SMOW) 501
Standards: effluent 515
　stream 515
Sanding crop: biomass 153
　energy flow 153
Standpipes 666
Starfish, coral-eating 453
Startle response (sonic boom) 584
State, sea 488–490
Static stability (atmosphere) 329–330
Stations: biological 377
　generating 167–169
　geological 377
　hydrographic 377
　oceanographic 377
　seismic 377
Statistical bioclimatology 56
Statistical climatology 83
Statistical weather forecasting 671–672
Steam (electric power generation) 629
Steam-iron-hydrogen HYGAS gasifier 104
Steam turbine: electric power generation 167–168
　geothermal power 244
Steel (thermal constants) 618
Steel wool (packed bed) 138
Stegner, Wallace 691
Stem rust 408, 418
Stephanodiscus 188
Stephanodiscus astrae 187
Steppe 246, 432, 637, 651
　forb-rich meadow 637
　forest 480
　mean temperature 90
　precipitation 90
Stictochironomus 226
Stilt roots 307
Stipa 246, 433, 652
Stipa spartea 432
Stirring (distribution of sea water properties) 499–500
Stock, forest 208
Stock resources 111
Stocking (wildlife management) 685
Stone 531, 532
Stoping: cut-and-fill 350–351
　filled 350–351
　hydraulic–fill 350
　open 349–350
　shrinkage 349–350
　square-set 351
　sublevel 350
Storage and Retrieval of Aerometric Data (SAROAD) 4

Storage dams 118, 125
Storage disease, plant 421–422
Storage reservoirs: control of stream flow 601
　elevated 666
　ground 666
Storm 588–590
　atmospheric electricity 39–40
　cloud formation 96
　drainage 275
　dynamical processes 588–589
　frontal storms and weather 589
　frontal waves 230–231
　hailstones 101
　middle-latitude highs or anticyclones 589–590
　modification 681–682
　principal cyclone tracks 590
　revolving 319
　role in terrestrial energy balance 590
　sewage 512–515
　squall line 587–588
　surge 593–594
　thunderstorm 620–621
　tides 593
　tornado 621–622
Storm, dust 140
Storm detection 590–593
　aircraft and satellites 592–593
　infrasonics 591–592
　microseisms 592
　radar 590–591
　sferics 591
Straight channels 474
Stratiform clouds 97
Stratocumulus clouds 98
Stratopause 36, 37, 317
Stratosphere 36–37, 38, 317
　air temperature 608
　radioactive fallout 461–462
　terrestrial heat balance 84
Stratospheric aerosols 38
Stratospheric ozone 53, 54
Stratus clouds 96, 98, 318
Straw (oil absorbent) 484
Stream: channels 603–604
　ecology 149
　erosion patterns 271–272
　flow 600–601
　graded 228, 594
　limnology 304
　loads 594
　lotic ecosystems 227–229
　patterns 472
　pollution 515–518
　river 471–474
　seasonal variation in water quality 602–603
　stream transport and deposition 594–595
　surface water 599–604
　water source 228, 600
　water temperature 603
Stream, jet 290–291
Stream standards (regulation of water pollution) 515
Stream transport and deposition 594–595, 603
　debris transport theory 594–595
　field measurement 595
　stream loads 594–595

Streptomyces 565
Streptomyces griseus 232
Streptomyces noursei 232
Stress (environmental protection) 178–179
Stress (homeostatic mechanisms) 274
Stress (mechanics): concrete dam 119
soil 559, 562
Strip-cropping 553
Strip mining 346–348, 393
Stripping, anodic 507
Strontium: biogeochemical cycles 156–157
constituent of living matter 69
Strontium-90 39
fallout 461
radiation injury (biology) 462
Strophanthin 623
Structural engineers 172
Structural plant societies 437
Structures (response to sonic boom) 585
Struthio camelus 479
Strychnine 623
Stuiver, M. 293
Stunting (plant diseases) 408
Suaeda 478
Subarctic islands 30–33
Subbituminous coals: carbon dioxide acceptor process 105
Lurgi gasifiers 103
Subbottom profilers (marine mineral exploration) 356
Subdesert, evergreen 637
Subequatorial tropics 635
Subirrigation 117
Sublevel caving 351–352
Sublevel stoping 350
Submarine canyon 114
Submarines: *Nautilus* 376
Skate 376
Submersibles, oceanographic 377, 382
Suborders of soil 528–529, 541
Subpolar rainforest 636
Subsea geologic basins (petroleum deposits) 380
Subsidence inversion (air pollution) 48
Substation, electric power 170
Substitution (conservation measure) 111
Substratum (abiotic factor) 177
Subsurface drainage 294
Subsurface mining (oil) 391–394
Subsurface pumps, hydraulic 388
Subsurface samplers (petroleum production) 391
Subsurface water 599
Subterranean ground society 108
Subterranean water 247–248
Subtropical dunes (Florida) 135
Subtropical ecosystem 634
Subtropical evergreen seasonal forest 636
Subtropical forest 647
Subtropical high cells 89
Subtropical highland complex 635

Subtropical highs 589
Subtropical oceanic highs 89
Subtropical rainforest 470
Subtropical seasonal rainforest 636
Subtropical semi-deciduous forest 636
Succession: forest 207
primary 596
Succession, ecological 109, 110, 135, 167, 595–599, 632
causes and nature of succession 596–598
climax community 93–96, 598–599
dune vegetation 135
effect of radiation 468
examples of 596
forest succession 207
importance of 599
management of vegetation change 640
methods of study 596
saltmarsh 477–478
stages of 595–596
terminology 595
trends in 599
Succession-retrogression cycles (succession, ecological) 597–598
Succulent forests 636
Succulents (desert) 129
Sucker rods (oil well production) 387
Suction-type spore trap 4
Suess, E. 64
Sugar fungi 566
Sulfamates 263
Sulfates (soil sulfur, microbial cycle of) 568
Sulfur: essential element for plants 403
organic sulfur compounds 542, 568
phosphoric acid production 195
rock 143
sodium soils 116
uses 336
Sulfur-containing dusts (plant disease control) 12
Sulfur cycle 57, 70
Sulfur dioxide: air pollution 20, 21
effects on humans and vegetation 50
low-troposphere concentration 39
Sulfur oxides (air pollution) 46
Sulfuric acid: phosphoric acid production 195
sodium soils 116
superphosphate production 195
Summergreen deciduous forest 430–431, 648
Summergreen forest 432, 636
Summergreen temperate forests (climatic climax) 93
Sunamis 593
Sunburn 53, 132
Sunlight (biological conversion) 571
Supercooled clouds 97, 99, 101, 102

Supercooled water: cloud physics 101
frost 232
Superorganism 146
Superphosphate: fertilizer 195
mixed fertilizers 197
phosphatic fertilizers 195
Supersonic flight (sonic boom) 581–585
Supersonic transport (SST) 582
Surf 489–490
Surface-drainage system 294
Surface mining: mining, open-pit 338–342
mining, strip 346–348
oil mining 391–394
Surface water 599–604
characteristics of river channels 603–604
conservation 656–660
deposition of precipitation 600–602
quality of water 602–603
source of water in streams 600
water supply 665
Surface weather map 676, 677–678
Surfacing, roadway 269
Surge, storm 593–594
Surveys, soil 530–531
Survivorship curves 450
Svalbard (Arctic) 31, 32
Sverdrup, A. S. 374
Sverdrup, H. U. 373
Swabs (oil well production) 388
Swamp: conifer 95
mangrove 307–309
sedge 637
sedge peat 637
Swamp, marsh, and bog 604
Swamp forest (mangrove swamp) 307–309
Sweat control, bioclimatic 54
Sweating (heat balance) 54–55
Sweet corrosion (oil and gas well production) 389
Sweetgum 430, 431
Swell (sea state) 489
Swell-condition code 489
Swtich grasses 246
Switches (electric power systems) 170
Switching circuit (systems ecology) 605
Symbiosis 145
antagonistic 412
mutualistic 412
Symphonia 429
Symphoriocarpus 433
Symptoms (disease) 131
Syndrome: beer drinker's 261
Down's 132
Klinefelter's 132
radiation 132
Turner's 132
Synecology, forest 205
Synedra 188
Synergism 412
Synergists (insecticidal activity) 287
Synoptic climatology 83
Synoptic meteorology 320–326
air masses and front-jet systems 320–322
circumpolar aerology 325–326

Synoptic meteorology—*cont.*
fronts and frontogenesis 322–323
general atmospheric circulation 320
large-scale tropical weather systems 325
life cycle of extratropical cyclones 323–325
waves and vortices (the weather) 323
weather map analysis 678
Synthane process 102, 105
Synthetic carbamate insecticides 287
Synthetic pipeline gas 103–105
Syntrophism 542
Syphilis 132
Systems: alternating-current 169
aquatic 110
cropping 536
front-jet 324
heating 252–253
microwave 572
open biological 131
open energy 166
polar-front-jet 321, 322
raincloud 100
surface-drainage 294
transmission 169–170
wilderness 690
Systemic insecticides 285
Systemic poisons 622
Systems analysis (systems ecology) 604–607
Systems approach (pest management) 174
Systems ecology 149, 604–607
aerobiology 3–4
change processes in ecosystems 605–606
energy flow and cycles of nutrients 606
maintenance of ecosystems 606–607
matrix formulation of ecosystem change or stability 606–607
matrix formulation of state and change of state 606
networks, qualitative structure 605
networks, quantitative change 605
systems analysis and operations research in biology 604–606
wildlife management 685

T

2,4,5-T *see* 2,4,5-Trichlorophenoxyacetic acid
T4 bacteriophage 255
Table, water 247, 249–250, 666–667
Tablelands 608, 609
Tags, radionuclide 465
Taiga 31, 142, 431, 434, 480, 604, 607–608, 650
climate 141, 142, 607
fauna 608

Taiga—cont.
  middle 607, 608
  northern 607–608
  southern 607, 608
  subzones 607–608
Tailing disposal (placer mining) 346
Tainter gate 124, 125, 126
Takahashi, T. 293
Talc (oil absorbent) 484
Tall-grass prairie 637, 651
Tall grassland 432–433
Tank: Imhoff 281–283, 519
  lease 389
  septic 477, 512, 519
Tansley, A. G. 164
Tantalum (heavy metals, pathology of) 253
Tanytarus 226
Tapeworm 132
Tapir, extinct 191
Tar sand: mining 393
  oil sand 394
Taraxacum 627
Targhee sheep 8
Tarns 291
Taxodium 431
Taylor Grazing Act 470
2,3,6-TBA see
  2,3,6-Trichlorobenzoic acid
TCA see Sodium trichloroacetate
TCBC see Trichlorobenzylchloride
TDE (insecticide) 285, 286
TDN see Total digestive nutrients
Tea (blister blight) 2
Teak 430
Technology: engineering 171–172
  marine 376–377
  solar 572–573
Tectona grandis 430
Telegraph engineers 172
Telephone engineers 172
Television, underwater 377
Television engineers 172
Teliospores 415
Telvar see 3-(p-Chlorophenyl)-1,1-dimethylurea
Temik see [2-Methyl-2-(methylthio) propionaldehyde-O-(methylcarbamoyl) oxime]
Temperate and cold savanna 650
Temperate deciduous forests 64, 648
Temperate evergreen seasonal broad-leaved forest 636
Temperate lakes 218
Temperate mixed forest 215
Temperate rainforest 430, 470, 636, 647
Temperate vegetation zones 60
Temperate woodland 649
Temperate zone: ecosystem 634
  plant communities 94
  savannas 95
  soils 94
Temperature: abiotic factor 176
  air 232, 608, 641, 643

Temperature—cont.
  degree-day 128
  dry-bulb 275
  effect of carbon dioxide pollution 78–79
  effect on plant disease 417–418
  effect on terrestrial ecosystems 614–615
  heating, comfort 252–253
  human bioclimatology 54–56
  inversion 46, 608
  pressure melting 618
  regulation 614
  sensible 511–512
  soils 535
  wet-bulb 275
Temperature-humidity index (THI) 511
Temperature-measuring devices (oceanography) 504
Temple, J. W. 524
Tenoran 264
Teratogenesis (birth defects) 255
Terbutol 265
Terminalia 429
Terrace, continental 113–114
Terracing (agriculture) 554, 641
Terrain: environmental stresses 179
  glaciated 272–273, 292, 604
  hill and mountain 271–273, 608–610
  stream-eroded 271–272
Terrain areas, worldwide 271, 608–610
Terrain classes (distribution over the earth) 609
Terrain sensing, remote 610–613
  terrain characteristics 610–611
  types of 611
  vegetation and ecosystem mapping 638
Terrestrial atmosphere 35–37
Terrestrial atmospheric heat balance 40, 41, 43–44, 251–252, 318, 373, 590
Terrestrial biocycle 109–110
Terrestrial biomes 64
Terrestrial community: climatic expression 165
  stratification 165
  trophic structure 617
Terrestrial ecology 149
Terrestrial ecosystem 164, 613–617, 634
  biogeography 59–61
  biome 63–64
  catastrophic agents 615
  comparison of ecosystems 613–614
  energy 615
  extent of terrestrial ecosystem 614
  herbaceous 637
  seasonal effects 615
  temperature 614–615
  trophic levels 615–617
  trophic structure 150
Terrestrial faunal regions 25
Terrestrial frozen water 617–618
  glaciers, past and present 617
  massive ice constants 618

Terrestrial frozen water—cont.
  properties of terrestrial ice 617–618
Terrestrial greenhouse effect 247
Terrestrial heat balance 84, 85, 491–492
Terrestrial ice 617–618
  classification of ice forms 618
  properties 617–618
Terrestrial radiation 36, 247, 251–252, 317–318
Terrestrial vegetation: concentrations of elements 67
  stratification 108
Territoriality 147
Terry, K. 191
Tests, cytological 363
Tetraethylpyrophosphate dicapthon 286
Texture, soil 531–532
TH 052-H see 4,5,7-Trichlorobenzthiadiazole-2,1,3
Thais 452
Thalassina anomala 308, 309
Thalia democratica 452
Thallium (poison) 622
Thallotoxicosis 259–260
  occurrence 260
  symptoms 260
Themeda 246
Theobroma 429
Theobroma cacao 469
Theories and models of climate 76–77
  mathematical expression 77
  model types 77
Thermal bioclimatology 54–55
Thermal burns 132
Thermal highs (climatology) 89
Thermal lows (climatology) 89
Thermal phosphates (fertilizer) 196
Thermal pollution 172, 663
  coastal waters 661
  electric power generation 629–630
Thermal precipitation (dust collection) 139
Thermal spring 618–619
Thermal wind 231
Thermoactinomyces 565
Thermocline 619–620
  permanent 619
  seasonal 619–620
  shallow 619
Thermodynamics (micrometeorology) 332–334
Thermometer: deep-sea reversing 504
  electrical 504
  radiation 504
  reversing 376, 504
Thermometric depth (sea water) 505
Thermoregulation 274, 614
Thermosphere 36, 37, 317
Therophyte 439
THI see Temperature-humidity index
Thiabendazole 10
Thiamine (disease) 132
Thickets 637, 650

Thickets, evergreen dwarf scrub 637
Thienemann, A. 225
Thiocyanates 263
Thiram 233
Thom, E. 511
Thompson, J. C. 283
Thompson, Wyville 374
Thorium: radiation injury (biology) 457
  uses 336
Thorium-U²³³ fuel cycle 369
Thorn forest 59, 432, 636
Thorn scrub 650
Thornbush 650
Thresholds, geomorphic 236–238
Thrush (fungus disease) 132
Thuja 431
Thunderstorm 620–621
  electrical activity 39–40
  precipitation 100–101
  radar detection 591
  squall line 587–588
  tornado formation 622
Thurber, D. L. 293
Thymelaceae 432
Thysanoessa longipes 511
Ticks (livestock parasites) 10
Tidal flats 186
Tidal flow (estuarine circulation) 182–184
Tidal waves 593
Tidal zones 177
Tides: beach erosion 106
  solar 37
  storm 593
Tilia 431
Tilia americana 137
Tillage 553
Tillam 267
Tilth 531, 536
Timber: age and cut 213–214
  forest management 203
  growth 213–214
  management 204
  stocking 213
  support of mine openings 352
Timmons, D. B. 198
Tin 336
Tissue: radiation injury 458–459
  radiation sensitivity and resistance 457
Titanium (uses) 336
Tok E-25 see 2,4 Dichlorophenyl-4-nitrophenyl ether
Tomato: ascorbic acid 133
  chlorine deficiency 405
Top slicing (underground mining) 351
Topographic climaxes 95
Topographic maps (vegetation and ecosystem mapping) 633–637
Topography (climax community) 93
  environmental stresses 179
  geomorphology 234–243
  hill and mountain terrain 271–273

Topography (climax community) —cont.
  influence on vegetation 94–95
  selection of dam site 125
  selection of type of dam 126
  surface configuration 143
Tordon 264
Tornado 319, 588, 661–622
  infrasonic detection 591–592
  modification 681–682
  radar detection 591
TOSCO process 396–397
Total digestible nutrients (TDN) 9
Total harvest (forest) 207–208
Total Kjeldahl nitrogen 9
Tovomita 429
Tower: packed 21
  plate 21
  spray 21
Toxaphene 285, 286, 400
Toxicity: cobalt 260–261
  metal 254–255
  molybdenum 547
Toxicology 622–623
  categories of poison 622
  history 623
  statistical data 622–623
  treatment for poison 623
Toxins: biogeochemical cycling 58
  dinoflagellates 401
  plant resistance 425
  systemic 132
Toxoplasmosis 132
TPN *see* Triphosphopyridine nucleotide
Trace metals: in protoplasm 68
  in sea water 496
Tracers (ecosystems): bio-elimination 465–467
  flow measurement in ecological networks 157
Trachoma 132
Trachypogon 479
Tractor gates 125, 126
Trade winds 89
Tradescantia 363
Traffic engineering (highway engineering) 268–271
TRANID *see* [5-Chloro-6-oxo-2-norbornanecarbonitrile *O*-(methylcarbamoxyl)-oxime]
Transantarctic Mountains 28
Transformers, distribution 170
Transition zone (air mass) 230
Translocation (chromosome) 362
Transmissibility (rocks) 248
Transmission, high-voltage 170
Transmission circuits 169, 170, 631
Transmission engineers 170
Transmission system 169–170, 631
Transpiration 162
  evapotranspiration 189
  hydrometeorology 278
  land 276
  micrometeorology 335
Transport, littoral 105, 106, 108
Transport and deposition, stream 603
Transportation: industrial meteorology 283–284

Transportation—*cont.*
  marine resources 316
  underground mining 353–354
Transportation engineering (river engineering) 474–476
Transverse dunes 134
Trawls (oceanographic survey) 276
Treatment: sewage 281–283, 512, 518–524
  water 476–477, 663, 667–669
Tree: bullet (forest planting) 211
  diseases 623–627
  ferns 430, 647
  forest and forestry 202–204, 623–625
  forest ecology 204–207
  forest management and organization 207–209
  forest planting and seeding 209–212
  planting by hand 211
  planting by machine 211
  planting season 211
  root zone depths 115
  shade 625–627
  size for planting 211
Trefethen, James B. 685
Treflan 267
Trellis stream drainage 472
Trench sanitary landfill 577
Triallate 267
Triassic (faunal extinction) 189
Triassic sandstones (fossil dunes) 135
Triazine (mutagen) 365
Tributary (river) 471
Tributylphosphorotrithioate 128
Tricamba 266
Trichinosis 132
Trichloracetic acid (TCA) 267
S-2,3,3-Trichloroallyl-*N,N*-diisopropylthiolcarbamate 267
2,3,6-Trichlorobenzoic acid (2.3,6-TBA) 267
4,5,7-Trichlorobenzthiadiazole-2,1,3 (PH 40-21) 267
Trichlorobenzylchloride (TCBC) 267
2,3,6-Trichlorobenzylpropanol 267
N-(Tri-chloromethylthio)-4-cyclohexene-1,2-imidazoline acetate 233
Trichloromethylthioimides 423
2,4,5-Trichlorophenoxyacetic acid (2,4,5-T) 12, 128, 267
2-(2,4,5-Trichlorophenoxy)-ethyl-2,2-dichloropropionate 267
Trichogramma 285
Trichomonal vaginitis 132
Trichophyton rubrum 232
Trickling filters 522
  circular 522
  high-rate 522
Tricodesmium erythraeum 401
a,a,a-Trifluoro-2,6-dinitro-*N,N*-dipropyl-*p*-toluidine 267

3-(*m*-Trifluoromethylphenyl)-1,1-dimethylurea 267
Trifluralin 267
Triglochin 478
Trillium erectum 439
Triodia 479
Triphosphopyridine nucleotide (TPN) 403
Triple superphosphate: fertilizer 195
  production 195
Tris-(2,4-dichlorophenoxyethyl)-phosphite 267
Trisomy 21, 132
Tritac 263, 267
Tritium (sea water) 492
Trockenwüst 651
Trophic levels 65, 148, 150, 152, 166, 615–617
  community balance 109
  ecological efficiencies 153
  food chain 201–202
  fresh-water ecosystem 225
Trophogenic zone (lakes) 220, 224
Tropholytic zone (lakes) 220
Tropical air mass 16
Tropical broad-leaved forest 64
Tropical cloud forests 96
Tropical cyclone 319, 593
Tropical deciduous forest, 59, 430, 647–648
Tropical ecosystem 634
Tropical evergreen forest (Mexico) 59
Tropical evergreen seasonal forest 636
Tropical highland complexes 635
Tropical life zone 301
Tropical moist deciduous forest 215
Tropical rainforest 429–430, 469–470, 636, 646–647
  fire 95
  mean temperature 90
  precipitation 90
Tropical regions: plant communities 94
  soils 94
Tropical savanna 64, 95, 479–480, 649–650
  mean temperature 90
  precipitation 90
Tropical scrub 650
Tropical semi-deciduous forest 636
Tropical vegetation zones 60
Tropical woodland 649
Tropics: subequatorial 635
  weather systems 325
Tropital 287
Tropopause 36, 317, 320, 608
Troposphere 36, 38, 39, 317
  air temperature 608
  front 230–232
  radioactive fallout 461–462
  terrestrial heat balance 84
Tropospheric zonal circulation 41
Troughs (meteorology) 319

Truck haulage (open-pit mining) 341
True estuaries 182
True heaths 136
True xerophytes (desert) 129
Trunk sewers 514
Tryptophan (zinc) 404
Tsuga 431
Tsuga canadensis 137, 431
Tsuga heterophylla 645
Tuberculosis 132, 133, 181
Tucker, W. 191
Tundra 64, 142, 431, 433–434, 627–628, 634, 637, 645, 650–651
  Alpine region 434, 628
  Antarctic region 628
  arctic islands 31
  Arctic region 33, 434, 628
  climate 141, 142
  fauna 628
  fjell-field 628
  forest 480, 481
  Greenland 33
  lichen 637
  mean temperature 90, 637
  moss 637
  patterns 627
  plant species, life-forms, and adaptations 627
  precipitation 90
  productivity 628
  soil conditions 31, 627–628
Tungsten (uses) 336
Tunnels (water transmission) 665
Tupersan 266
Turbines: combustion 629
  gas 169
  hydraulic 168–169
  steam 167–168
Turbulence, clear-air 334
Turbulent impaction (cleansing the atmosphere) 48
Turner's syndrome 132
Turnip (brown heart) 11
Typhoid 132
Typhoon 319
Typhus 132
Tyrosinase 404

U

Uca 309
Udalfs 541
Udolls 541
Udults 541
Uffen, R. 191
Ultisols 529, 541
Ultralow-volume spraying (fungicide) 234
Ultrasonic agglomerators 51
Ultraviolet light (induced mutation) 364
Ultraviolet radiation: bactericidal action 53
  biology 53–54
  solar 53
  sunburn 53
Ulva 312, 313
Umbelliferae 432
Unalaska Island (Arctic) 31, 32

Unconsolidated marine mineral resources 355, 357–359
Undecylenic acid 232
Undercurrents (mineral recovery) 343
Underground mining 348–354
Underground water (conservation) 656
Underpopulation (effects) 448
Undersea mining 354–361, 380–383
Underwater photography 377
Underwater television 377
Undulant fever 132
Uneven-aged forest stand 208–209
Uniform population dispersion 444
Unimak Island (Arctic) 31
Uniola paniculata 135
United States (life zones) 302
United States Aerobiology Program 4
Unmoderated fast breeders 369
Updrafts (thunderstorm) 620
Uplift pressure cells 127
Upper air charts 678
Upper air circulation 693
Upper atmosphere 36
Upper midlatitude climate 141, 142
Upper Sonoran life zone 301
Urab 267
Uranium: fast-breeder reactors 370
  granite rock 368
  miners 172
  radioactive fallout 460
  rock 143
  steam turbine fuel 168
  uses 336
Uranium dioxide: advanced gas-cooled reactor 369
  pressurized water reactor 369
Urea: fertilizer 192, 194
  nitrogen cycle 69, 155
  production 194
Urediospore 415
Urethane (mutagen) 365
Uric acid 69
Urocanic acid (photochemical bioclimatology) 53
Ustalfs 541
Ustilago 421
Ustolls 541
Ustults 541
Utermöhl method 401
Utility industry, electrical 170–171, 629–632
  air pollution 630
  cost of system additions 631
  electricity usage 630–631
  hydroelectric plants 630
  transmission circuit construction 631

V

Vaccinium 431
Vadose water 247–248
Vaginitis, trichomonal 132

Valley: complex response of geomorphic systems 241–243
  erosional 273
  glaciers 273
  influence on vegetation 94–95
  slope 241
  winds 693
Valves (dams) 125
VAMA *see* Modified vinyl acetate maleic acid
Van der Hoven, J. 334
Van Schaik, J. 548
Vanadium: constituent of living matter 70
  essential element for plants 405
  heavy metals, pathology of 253
  nutrient 155
  uses 336
Vapam (VPM) 624
Vapor, water 40–41, 78–79, 96–99, 177, 275–276, 276–278, 317, 318
Vapor barriers 253
Vapor pressure: air 275
  atmospheric evaporation 40
Vaporization, heat of 40
Vector (mathematics) 605
Vectors, insect 174
Vegadex 264
Vegetation 434–435, 632
  altitudinal zonation 644
  bog 604
  classes 646–652
  conservation of resources 110–113
  crust 652
  desert 129–131, 433, 651–652
  dune 107, 130, 135–137
  ecology 632
  energy cycles 632–633
  major groups 636
  marsh 604
  mesophytic 604
  origin of flora 632
  soil, zonality of 540–541
  structure 632
  swamp 604
  terrestrial 67, 108
  zonation 597, 598
Vegetation and ecosystem mapping 633–639
  aerial photographs 637–638
  base maps and scales 633–637
  basic considerations 633
  multiband remote sensing 638
  plant life forms 440
  technical requirements and methods 633–638
  world ecosystem maps and zones 638–639
Vegetation management 633, 639–642
  actual change 640
  chemicals 641
  complex products 642
  effective practices 639–640
  fire 640
  indirect methods 641
  man and management 639–640

Vegetation management—*cont.*
  management of vegetation change 640
  mismanagement 639
  products of management 642
  specific products 642
  theory of floristic relays 640
  theory of initial floristic composition 640
  tools 640–641
Vegetation zones, altitudinal 642–645
  altitudinal variations in solar radiation 643
  altitudinal variations in temperature 643
  altitudinal zonation 644–645
  latitudinal variations in solar radiation 643
  mountain environments and vegetation 644
  mountains precipitation and evaporation 643–644
  topographic effects 645
Vegetation zones, world 436, 645–652
  Antarctic region 31
  arctic biology 33–34
  arid 60
  cold desert 652
  crust vegetation 652
  desert 129–131
  dune 135–137
  evergreen hardwoods 648–649
  grassland 245–247
  life zones 300–303
  meadow 651
  Mexico 59–60
  needle-leaved evergreen forest 648
  plant formations, climax 428–434
  prairie 651
  rainforest 468–470
  saltmarsh 477–478
  savanna 479–481
  steppe 651
  summergreen deciduous forest 648
  taiga 607–608
  temperate 60
  temperate and cold savanna 650
  temperate and cold scrub 650
  temperate rainforest 647
  temperate woodland 649
  thornbush 650
  tropical 60
  tropical deciduous forest 647
  tropical rainforest 646–647
  tropical savanna 649–650
  tropical scrub 650
  tropical woodland 649
  tundra 627–628, 650–651
  warm desert 651–652
  world's forests 215
Vegetational plant geography 436–437
Vegetative propagation 73, 74–75
Ventilation, natural 47
Venturi scrubber 21
Venturia inaequalis 415, 425
Verdansky, V. I. 64

Vermiculite-interstratified clay 543
Vernam 267
Vernolate 267
Vertebrate (fresh-water) 224–225
Vertical-flow sedimentation tank 520
Vertical-lift gate 124, 125
Vertical sand drain 270
Vertical shafts (underground mining) 354
Vertically homogeneous estuaries 183–184
Vertisols 529, 541
Vessels, oceanographic 375–377
Vest-Spitsbergen (Arctic) 31
Vibratory screens (sewage treatment) 519
Vicia faba 363
Victoria Island (Arctic) 31, 32
Viets, F. G. 199
Viets, Frank G., Jr. 199
Vinogradov, A. P. 64
Vinson Massif (Antarctic) 29
Viola 430
Viral hepatitis 132
Virus: disease 132
  plant 409, 411, 627
  soil microorganisms 566–567
Viscosity (crude oils) 386
Visibility (sea water) 494
Visual photography (remote terrain sensing) 612
Vitamin: animal feed composition 8–9
  nutrition 372
Vitamin A 8, 132
Vitamin B_{12} 67, 132
Vitamin C 133
Vitamin D 8, 53, 132
Vitamin E 8
Vitamin K 8, 132
Vitavax 233
Vitis 137
Volatile oils (poison) 622
Volcanic island (biotas) 289
Volcanic mountains 273
Volga River 473
Von Liebig, Justus 11
Vonnegut, B. 680
VPM *see* Vapam
Vulvovaginitis (chemotherapy) 232

W

Waddington, C. 191
Wadleigh, C. H. 198
Waggoner, Paul E. 3
Wallace, A. 694
Wallace, Alfred Russell 288–290
Walleye (fish) 188
Walls, retaining 564
Walnut 431, 648
Waltz Mills Cable Test Center 170
War gases (poison) 622
Warfare agents (disease) 132
Warm-blooded animals 614
Warm desert 651–652
Warm front 323
Warm temperate moist forest 215
Warm temperate woodland 634

Warts 132
Washingtonia 652
Washout (cleansing the atmosphere) 48
Waste disposal 57, 379
    open-pit mining 342
    radioactive 463–465
    sea 316
    solid waste 573–581
Wastes: agricultural 574, 652–655
    animal 652–653
    classification of 581
    domestic 663
    food processing 654
    fossil-fuel 574–575
    gaseous 464–465
    industrial 316, 512–515, 516, 663
    liquid 464, 654–655
    mineral 574–575
    municipal 185, 574
    radioactive 663–664
    solid 464, 652–654
Water: abiotic factor 177
    argic 248
    agricultural wastes 654–655
    anastatic 248
    artesian 250, 667
    biosphere, geochemistry of 67
    brown 225
    capillary 558
    central 501
    cloud composition 100
    coastal 660–662
    conservation of resources 110–113, 656–662
    crops, irrigation of 114–118
    deep 482
    desert vegetation 129
    dispersal of disseminules 442
    dissemination of plant pathogens 420
    earth, resource patterns of 140
    effect on plant disease 418
    equatorial 501
    eutrophication 57–58
    evapotranspiration 189
    fresh 281
    frost action in soils 560–561
    frozen 618
    gravitational 558
    groundwater 247–250
    hardness 668
    hydrology 275–276
    hydrometeorology 276–278
    hydrosphere 278
    hygroscopic 558
    juvenile 249, 618–619
    lake 291–292
    lake, meromictic 292–294
    perched 248
    plant and soil relations 115
    polar 501
    pollution 662–664
    population dispersal 445
    primary 249
    purification 476
    radioactive decontamination 463
    resource 315
    rhizic 247–248
    rights 658
    salt 390

Water—*cont.*
    sea 279, 280, 310, 314–315, 354–361, 356–357, 378, 486–488, 490–507, 619–620
    seepage and frost in soil 559
    softening 668
    soil 247–248, 534–535
    soil-water relationships 558–559
    spring 586–587, 664
    subsurface 599
    subterranean 247–248
    supercooled 101, 232
    supply 275
    supply engineering 664–666
    surface 599–604, 656–660, 665
    table 666–667
    terrestrial frozen water 617–618
    thermal constants 618
    thermal spring 618–619
    transmission mains 665
    treatment 667–669
    tropical savannas 480
    underground 656
    use 657
    vadose 247–248
Water analysis 655–656
    completed test 656
    confirmed test 655–656
    examination of sewage 514–515
    membrane filters 656
    most probable number (MPN) 656
    presumptive test 655
    qualitative testing 655–656
    quantitative testing 655
    regulation of water pollution 515
    sanitary standards 656
Water-bearing formation 30
Water-borne disease 476
Water bottle, reversing 504
Water cloud, nonprecipitating 99
Water conservation 113, 160, 656–662
    coastal resources 660
    coastal waters 660–662
    conservation agencies and legislation 659, 661–662
    contamination by toxic materials 661
    cultural eutrophication 188
    fertilizer problems 198
    hydrologic cycle 657–658
    international agreements 660
    irrigation 118
    nitrogen fertilizer 199
    origin and description 660
    physical alterations 661
    river management 473–474
    sewage 660–661
    surface water 656–660
    thermal pollution 661
    underground water 656
    vegetation management 641
    water management problems 658
    water management technology 658

Water conservation—*cont.*
    water pollution 657
    water rights 658
    water use 657
    watershed control 658–659
Water consumption (relation to sewage) 513
Water-cooled reactors 369
Water cycle, atmospheric 276–277
Water-deficient climates 142
Water-deficient regions 141, 142–143
Water distribution system 665–666
Water drive well 383
Water erosion 550
Water heating (thermal energy) 571
Water management: problems 658
    technology 658
Water pollution 57–58, 515, 657, 662–664
    agricultural wastes 654–655
    current status 663–664
    development of treatment methods 663
    discharge directly to watercourses 663
    distribution of pollutants in estuaries 185, 518
    domestic waste waters 663
    eutrophication, cultural 187–189
    federal aid 664
    fertilizer 198–199
    fisheries conservation 200
    historical developments 662
    industrial waste waters 663
    lake 518
    placer mining 346
    regulation 515
    related scientific disciplines 662
    relation to water supply 662–664
    self-purification of natural waters 662–663
    sewage disposal 515–518, 660–661
    stream 229
    thermal pollution 663
    water analysis 655–656
    water treatment 667–669
Water power (electric energy generation) 168–169
Water Power Act of 1920 113
Water-power engineers 172
Water purification 515, 667–669
Water relations of plant 93, 114–115, 129, 161–163, 207, 410
Water-resource management 160
Water samplers (sea water) 504–505
Water supply engineering 664–666
    distribution system 665–666
    groundwater 664–665
    pumping stations 666
    sources of water supply 664–665

Water supply engineering —*cont.*
    surface water 665
    transmission and distribution 665–666
    water treatment 667–669
    watershed management 642
Water table 247, 666–667
    artesian conditions 249–250
    water supply 664
Water treatment 663, 667–669
    aeration 668
    coagulation 667
    disinfection 668
    filtration 667–668
    plain sedimentation 667
    rural sanitation 476–477
    water softening 668
Water vapor: atmospheric evaporation 40–41, 177, 318
    bioclimatology 53
    cloud 96–99
    component of air 317
    effect of carbon dioxide pollution 78–79
    humidity 275
    hydrology 275–276
    hydrometeorology 276–278
    low-troposphere concentration 39
Watershed management 203, 642, 658–659
Watershed protection 112
Watershed Protection and Flood Prevention Act of 1954 113, 658
Watson, B. F. 511
Wave cyclone (life cycle) 230
Wave motions (atmosphere) 330
Waves: beach erosion 105–106
    breaking 489–490
    frontal 230–231
    internal 499
    lakes 220–221
    ocean 105–106, 378, 488–490
    tidal 593
Waves of oscillation (lakes) 220
Wax (production by bees) 173
Weasels 628
Weather 145
    air mass 14–17
    air pressure 17–19
    atmosphere 35–37
    atmospheric electricity 39–40
    climate, man's impact on 75–82
    climatology 82–93
    cloud 96–99
    drought 133
    effect on plant disease 410
    effects of atmospheric pollution 50
    environmental protection 178–180
    front 230–232
    greenhouse effect, terrestrial 247
    human bioclimatology 56
    hydrometeorology 276–278
    industrial meteorology 283–284
    meteorology 316–332
    modification 678–682
    squall line 587–588

Weather—cont.
   storm 588–590
   synoptic meteorology
      320–326
   thunderstorm 620–621
   tornado 621–622
   weather map 676–678
   wind 692–694
Weather forecasting and prediction 331–332, 669–676
   basic premises 671–672
   centers and offices of forecast-
      ing 674–676
   cloud and precipitation pre-
      diction 673
   data to forecasting 669
   development 669–670
   evaluating forecasting
      669–670
   extended-range numerical
      prediction 673–674
   long-range forecasting
      670–671
   longer-period forecasting 671
   medium or extended ranges
      670–671
   National Meteorological
      Center 674–675
   numerical calculation of cli-
      mate 674
   numerical weather prediction
      672–674
   probable future developments
      670
   short-range numerical predic-
      tion 672–673
   statistical linear regression
      672
   statistical procedures 672
   statistical weather forecasting
      671–672
   storm detection 590–593
   synoptic meteorology
      320–326
   weather service offices 676
Weather map 15, 16, 19,
      676–678
   automatic analysis 678
   surface 676, 677–678
Weather modification 678–682
   artificial stimulation of rain
      102, 680–681
   cloud modification 680
   cloud seeding 680
   downwind effects 681
   effect on ecology and society
      682
   fog dispersal 680
   hail suppression 681
   inadvertent weather modifica-
      tion 682
   lightning suppression 681
   modification of severe storms
      681–682
Weather service offices
      676
Weathering processes 281
Weaver, J. 437
Web, food 201–202
Weddell seals 30
Wedges, ice 398
Weeks Act of 1911 113
Weightlessness (space environ-
      ment) 180
Weinmannia 430, 647

Well: artesian 35
   caisson 665
   dissolved gas expansion 383
   gas-cap drive 383
   gas-condensate 386
   gravity drainage 383
   oil 383
   pumps 388, 665
   Ranney 665
   water-driven 383
   water supply 664–665
Went, F. W. 524
West African tropical faunal re-
      gion 26
West Antarctica 29
Westerlies 41, 43, 89, 290, 693
Wet-and-dry climate 141, 142
Wet-bulb temperature 275
Wet collector 21
Wet flatlands 604
Wet snow avalanches 52
Whale, bowhead 34
Wheat: black stem rust 1
   grasses 246
   stem rust 415
Whipple, G. C. 218
Whiptail (plant disease) 11
White cedar (taiga) 607
White noise 368
White pine blister rust 624, 625
Whitecaps 490
Whittaher, R. 596
Whittaker, R. H. 166, 167
Whooping cough 132
Whooping crane 585, 683
Wiesen 651
Wilderness, preservation of
      689–692
Wilderness Law of 1964 691
Wilderness System 690
Wildland (operations research)
      686–689
Wildlife conservation 110–113,
      159–160, 642, 689–692
   dynamic programming 688
   fisheries conservation 199–201
   game theory 688–689
   government programs 686
   government regulation
      691–692
   history 684–686
   linear programming 688
   management task 684
   modeling 687–688
   preservation of wilderness
      689–692
   shaping the programs 690–691
   simulation 689
   today's needs 686
   wildland operations research
      686–689
   wildlife values 683–686
Wildlife management 159–160
Wilkes, Charles 30
Williwaws 693
Willows 137, 648
Wilt: forest trees 624
   plant diseases 408
Wind 318, 320, 692–694
   abiotic factor 177
   air pressure 18–19
   atmospheric general circula-
      tion 41–46
   atmospheric pollution 46–52
   chill 178, 512

Wind—cont.
   convergent or divergent pat-
      terns 692
   cyclonic and anticyclonic cir-
      culation 692
   dispersal of disseminules 442
   disruption of climax com-
      munities 95–96
   dissemination of plant
      pathogens 419–420
   distribution of temperature
      88–90
   dust storm 140
   effect on ocean surface 373
   erosion 540, 550, 552
   front 230–232
   frontogenesis 231–232
   jet stream 290–291
   local 693–694
   minor terrestrial 693
   ocean currents 106
   ocean waves 106
   sea state 488–490
   squall 620
   thermal 231
   trade 89
   upper air circulation 693
   waves and currents in inland
      waters 220
   zonal surface 693
Wind tunnel (air pollution study)
      23
Winged pigweed 137
Winter-rain hard-broad-leaved
      evergreen forest 636
Witches'-brooms 624
WMO see World Meteorological
      Organization
WMO cloud classification 97
Wolf, dire 191
Wolman, M. G. 235
Wood: forest management 203
   waste product handling 654
Woodhead, D. S. 457
Woodland 64, 649
   boreal 650
   cold-deciduous 636
   deciduous 636
   drought-deciduous 636
   dry 636
   evergreen broad-leaved 636
   evergreen needle 636
   lichen 650
   middle 634
   open 634, 636
   temperate 649
   tropical 649
   warm temperate 634
Woody plants (habitat) 614
Woody vines 469
Woolly mammoth (mass extinc-
      tion) 191
World floristic regions 435
World life zones 303
World Meteorological Organiza-
      tion (WMO) 6
World vegetation zones 31,
      33–34, 59–60, 129–131,
      135–137, 215, 245–247,
      300–303, 428–434, 436,
      468–470, 477–478, 479–481,
      607–608, 627–628, 645–652
Worldwide terrain areas 271,
      608–610
Worms 312

Wrangel Island (Arctic) 31, 32
Wüst (vegetation zone) 651
Wüst, Georg 374

X

Xanthine oxidase 404
Xanthomonas campestris 414
Xanthomonas phaseoli 421, 422
Xenon (low-troposphere con-
      centration) 39
Xeralfs 541
Xeric environment 177
Xerolls 541
Xerophytes: deciduous 129
   evapotranspiration 189
   true 129
Xerophytic shrub (Mojave Des-
      ert) 130
Xeroseres 595
X-rays (biological radiation in-
      jury) 456–460

Y

Yaglou, C. P. 511
Yangtze River 473
Yaws (disease) 132
Yeasts (soil microorganisms)
      566
Yellow fever 174
Yellow-green algae 400
Yellow poplar 431
Yenisei River 81, 473
Yield, biological 61
Yonge, C. M. 149
Yuba bucket-ladder dredge 344
Yucca 131, 137, 433
Yucca brevifolia 130
Yucca gloriosa 136
Yughny Island (Arctic) 31
Yukon River 473
Yusho 441

Z

Zadoks, J. C. 420
Zambesi River 473
Zectran 287
Zinc: complexes in soils 547
   essential element for plants
      404
   micronutrient 198
   nutrient 155
   uses 336
Zineb 233
Zingiberaceae 430
Ziram 233
Zirconium (uses) 336
Zobar 267
Zonal classification of soils 541
Zonal soil 541
Zonal surface winds 693
Zonality of soil 540–541
Zonation: succession, ecological
      597
   wet lands 604
Zone: erosion 105
   forest 136
   frontal 230
   pioneer 135
   scrub 135–136
   transition 230
   vegetation 108

Zone of aeration 247–248
Zone of breakers 489–490
Zone of saturation: groundwater
  247–250
  water table 666–667
Zoochory 443
Zoogeographic region 694–698,
  699
  Australian region 697
  classification 694

Zoogeographic region—*cont.*
  Ethiopian region 695–696
  Nearctic region 695
  Neotropical region 696–697
  Oriental region 696
  palearctic region 694–695
Zoogeography 60–61, 698–699
  animal community 25–27
  arctic biology 33–34
  faunal approach 699

Zoogeography—*cont.*
  objectives 698
  regional approach 698–699
  subject area 698
  systematic approach 698
Zoology, invertebrate 172–174
Zooplankton: collection and
    identification 700
  composition 700
  feeding relationships 701

Zooplankton—*cont.*
  fresh-water ecosystem 223
  marine biological sampling 700–701
  marine ecosystem 311–312
  occurrence 700
  ocean 508
  oceanic geography 511
  reproduction 701
Zoosphere 26
Zostera 312